Student & Parent

One-Stop Inter[illegible]es

Log on to

ca.algebra2.com

- Complete Student Edition
- Links to Online Study Tools

California Online Study Tools

- Extra Examples
- Self-Check Quizzes
- Vocabulary Review
- Chapter Test Practice
- Standardized Test Practice

Online Activities

- WebQuest Projects
- USA TODAY Activities
- Career Links
- Data Updates

Graphing Calculator Keystrokes

- Calculator Keystrokes for other calculators

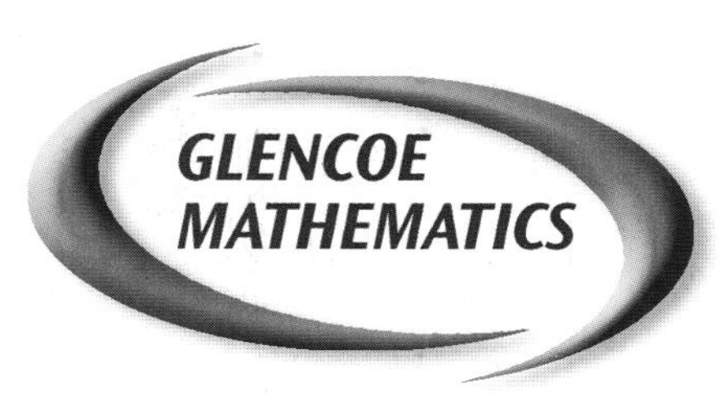

California Edition

Algebra 2

Contents

New York, New York   Columbus, Ohio   Chicago, Illinois   Peoria, Illinois   Woodland Hills, California

ISBN: 0-07-865980-9 *(California Student Edition)*

California Teacher Advisory Board

David Barker
Mathematics Department Chairperson
Los Alamitos High School
Los Alamitos, California

David J. Chamberlain
Secondary Mathematics Resource Teacher
Capistrano Unified School District
San Juan Capistrano, California

Donald O. Cowan
Mathematics Department Chairperson
Live Oak High School
Morgan Hill, California

Eric V. Johnson
Mathematics Teacher
Los Primeros Structured School
Camarillo, California

Tom Massa
Mathematics Teacher
Los Altos Middle School
Camarillo, California

Robert Newman
Mathematics Teacher
Los Altos Middle School
Camarillo, California

Patricia Taepke
Mathematics Teacher
South Hills High School
Covina-Valley Unified School District
West Covina, California

Joanne M. Wainscott
Mathematics Department Chairperson
La Jolla High School
La Jolla, California

Image Credits: **CA1** Royalty-free/CORBIS; **CA2** Getty Images; **CA5** Royalty-free/CORBIS

Agricultural Field and Mountains in California

- Go to bed early the night before the test. You will think more clearly after a good night's rest.
- Read each problem carefully, underline key words, and think about ways to solve the problem before you try to answer the question.
- Relax. Most people get nervous when taking a test. It's natural. Just do your best.
- Answer questions you are sure about first. If you do not know the answer to a question, skip it and go back to that question later.
- Become familiar with common formulas and when they should be used.
- Think positively. Some problems may seem hard to you, but you may be able to figure out what to do if you read each question carefully.
- If no figure is provided, draw one. If one is furnished, mark it up to help you solve the problem.
- When you have finished each problem, reread it to make sure your answer is reasonable.
- Make sure that the number of the question on the answer sheet matches the number of the question on which you are working in your test booklet.

California Algebra II Content Standards

🔑 = Key Standards defined by Mathematics Framework for California Public Schools

	Content Standard
🔑 1.0	Students solve equations and inequalities involving absolute value.
🔑 2.0	Students solve systems of linear equations and inequalities (in two or three variables) by substitution, with graphs, or with matrices.
🔑 3.0	Students are adept at operations on polynomials, including long division.
🔑 4.0	Students factor polynomials representing the difference of squares, perfect square trinomials, and the sum and difference of two cubes.
🔑 5.0	Students demonstrate knowledge of how real and complex numbers are related both arithmetically and graphically. In particular, they can plot complex numbers as points in the plane.
🔑 6.0	Students add, subtract, multiply, and divide complex numbers.
🔑 7.0	Students add, subtract, multiply, divide, reduce, and evaluate rational expressions with monomial and polynomial denominators and simplify complicated rational expressions, including those with negative exponents in the denominator.
🔑 8.0	Students solve and graph quadratic equations by factoring, completing the square, or using the quadratic formula. Students apply these techniques in solving word problems. They also solve quadratic equations in the complex number system.
🔑 9.0	Students demonstrate and explain the effect that changing a coefficient has on the graph of quadratic functions; that is, students can determine how the graph of a parabola changes as a, b, and c vary in the equation $y = a(x - b)^2 + c$.
🔑 10.0	Students graph quadratic functions and determine the maxima, minima, and zeros of the function.
🔑 11.0	Students prove simple laws of logarithms.
🔑 11.1	Students understand the inverse relationship between exponents and logarithms and use this relationship to solve problems involving logarithms and exponents.
🔑 11.2	Students judge the validity of an argument according to whether the properties of real numbers, exponents, and logarithms have been applied correctly at each step.
🔑 12.0	Students know the laws of fractional exponents, understand exponential functions, and use these functions in problems involving exponential growth and decay.
13.0	Students use the definition of logarithms to translate between logarithms in any base.
14.0	Students understand and use the properties of logarithms to simplify logarithmic numeric expressions and to identify their approximate values.
🔑 15.0	Students determine whether a specific algebraic statement involving rational expressions, radical expressions, or logarithmic or exponential functions is sometimes true, always true, or never true.
16.0	Students demonstrate and explain how the geometry of the graph of a conic section (e.g., asymptotes, foci, eccentricity) depends on the coefficients of the quadratic equation representing it.
17.0	Given a quadratic equation of the form $ax^2 + by^2 + cx + dy + e = 0$, students can use the method for completing the square to put the equation into standard form and can recognize whether the graph of the equation is a circle, ellipse, parabola, or hyperbola. Students can then graph the equation.

	Content Standard
🔑 18.0	Students use fundamental counting principles to compute combinations and permutations.
🔑 19.0	Students use combinations and permutations to compute probabilities.
🔑 20.0	Students know the binomial theorem and use it to expand binomial expressions that are raised to positive integer powers.
21.0	Students apply the method of mathematical induction to prove general statements about the positive integers.
22.0	Students find the general term and the sums of arithmetic series and of both finite and infinite geometric series.
🔑 23.0	Students derive the summation formulas for arithmetic series and for both finite and infinite geometric series.
24.0	Students solve problems involving functional concepts, such as composition, defining the inverse function and performing arithmetic operations on functions.
25.0	Students use properties from number systems to justify steps in combining and simplifying functions.

San Diego Skyline

How To...

Master the Content Standards

Standards Practice Countdown

Pages CA8–CA32 of this text include a section called **Standards Practice Countdown**. Each page contains 8 problems that are similar to those you might find on most standardized tests. You should plan to complete one page each week to help you master the content standards.

Plan to spend a few minutes each day working on the Standards Practice problem(s) for that day unless your teacher asks you to do otherwise. These multiple-choice questions address the California Algebra II Content Standards. If you have difficulty with any problem, you can refer to the lesson that is referenced in parentheses after the problem.

Your teacher can provide you with an answer sheet to record your work and your answers for each week. A printable worksheet is also available at ca.algebra2.com. At the end of the week, your teacher may want you to turn in the answer sheet.

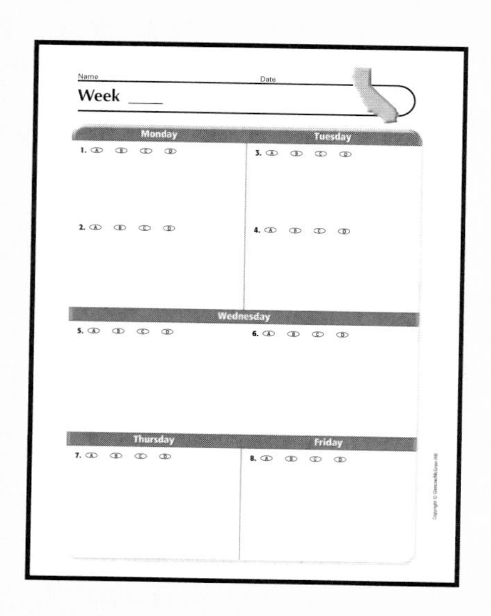

California Standards Practice Workbooks

The **CAHSEE Mathematics Standards Practice Workbook** contains examples and practice for each CAHSEE standard as well as cumulative CAHSEE Standards practice.

The **California Math Standards, Algebra 2 Practice and Sample Test Workbook** contains practice by standard and two sample tests.

By practicing the standards throughout the year, you will be better prepared for tests that assess those standards.

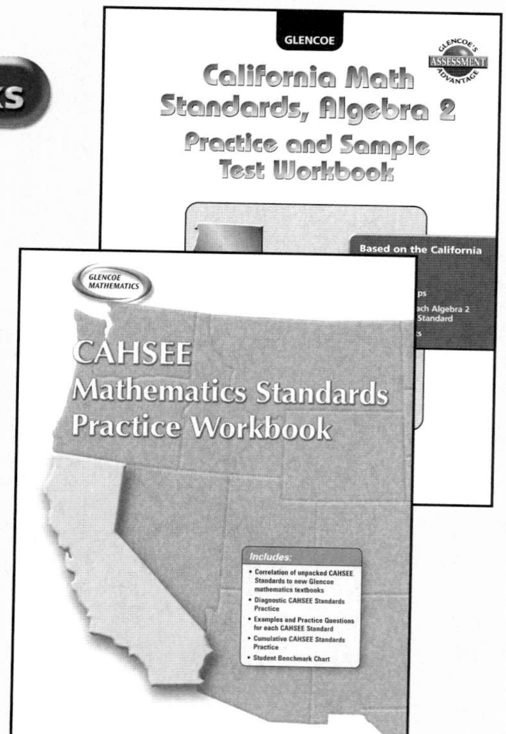

Your Textbook

Each lesson lets you know which California Algebra II Content Standard is being covered in that lesson. A complete list of the California Algebra II Content Standards can be found on pages CA4–CA5.

- The **boldface** portion of each standard indicates the specific portion of the standard that is addressed in that lesson.
- Key Standards are indicated with (Key) at the end of each standard.

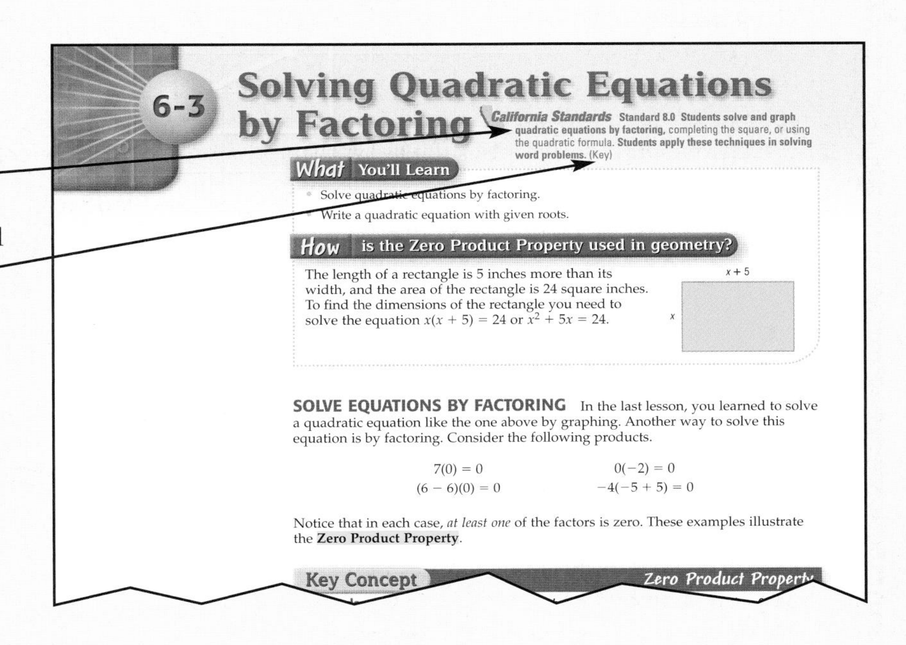
6-3 Solving Quadratic Equations by Factoring

California Standards Standard 8.0 **Students solve and graph quadratic equations by factoring,** completing the square, or using the quadratic formula. **Students apply these techniques in solving word problems.** (Key)

What You'll Learn

- Solve quadratic equations by factoring.
- Write a quadratic equation with given roots.

How is the Zero Product Property used in geometry?

The length of a rectangle is 5 inches more than its width, and the area of the rectangle is 24 square inches. To find the dimensions of the rectangle you need to solve the equation $x(x + 5) = 24$ or $x^2 + 5x = 24$.

SOLVE EQUATIONS BY FACTORING In the last lesson, you learned to solve a quadratic equation like the one above by graphing. Another way to solve this equation is by factoring. Consider the following products.

$7(0) = 0$ $\quad$ $0(-2) = 0$

$(6 - 6)(0) = 0$ $\quad$ $-4(-5 + 5) = 0$

Notice that in each case, *at least one* of the factors is zero. These examples illustrate the **Zero Product Property**.

Key Concept — Zero Product Property

Your textbook contains many opportunities for you to master the Algebra II Content Standards. Take advantage of these so you are prepared for tests that assess these standards.

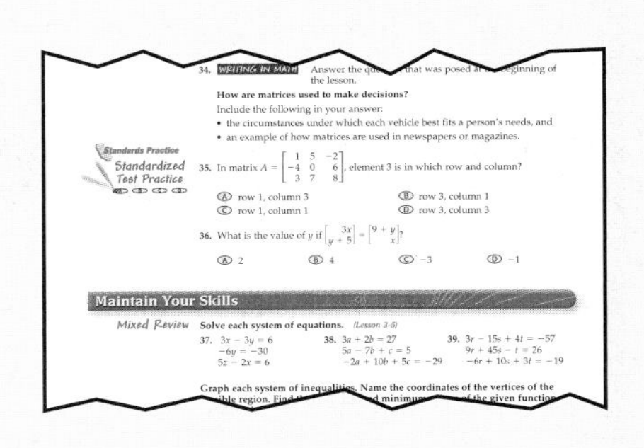
How are matrices used to make decisions?

Standards Practice — Standardized Test Practice

Maintain Your Skills

Mixed Review Solve each system of equations.

- **Each lesson** contains at least two practice problems that are similar to ones found on most standardized tests. The **Chapter Practice Test** also includes a similar practice problem.
- **Worked-out examples** in each chapter show you step-by-step solutions of problems that are similar to ones found on most standardized tests. **Test-Taking Tips** are also included.

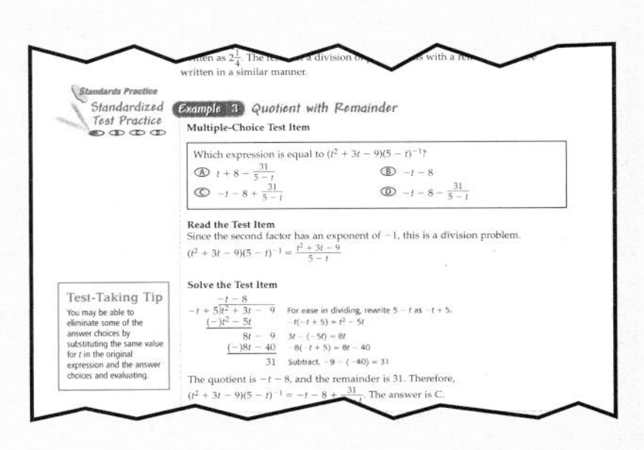
Standards Practice — Standardized Test Practice

Example 3 Quotient with Remainder

Multiple-Choice Test Item

Read the Test Item

Solve the Test Item

Test-Taking Tip

- Two pages of **Standardized Test Practice** are included at the end of each chapter. These problems may cover any of the content up to and including the chapter they follow.
- The **Preparing for Standardized Tests** section of your textbook on pages 877–892 discuss various strategies for attacking questions like those that appear on national standardized tests. Additional practice problems are also available.

Standards Practice Countdown

Week 1

Monday

1. What is the product of $(x + 3)$ and $(x - 3)$? *(Prerequisite Skill)*

Ⓐ $x^2 + 6x - 9$
Ⓑ $x^2 - 6x - 9$
Ⓒ $x^2 - 6x + 9$
Ⓓ $x^2 - 9$

2. Which answer best describes the error in the solution? *(Prerequisite Skill)*

$$x^3 + 12x^2 + 8x = 0$$
$$x(x^2 + 12x + 8) = 0$$
$$x^2 + 12x + 8 = 0$$
$$x = \frac{-12 \pm \sqrt{12^2 - 4(1)(8)}}{2(1)}$$
$$x = \frac{-12 \pm \sqrt{112}}{2}$$
$$x = \frac{-12 \pm 4\sqrt{7}}{2}$$
$$x = -6 \pm 2\sqrt{7}$$

Ⓐ error in simplifying square root
Ⓑ division by zero
Ⓒ error in use of Quadratic Formula
Ⓓ error in addition or subtraction

Tuesday

3. What is the area of the shaded region? *(Prerequisite Skill)*

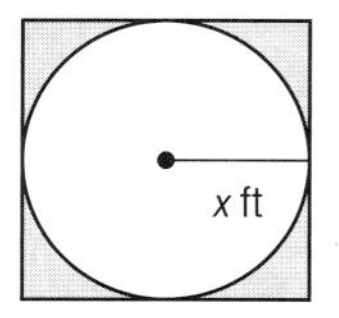

Ⓐ $(x^2 - \pi x^2)$ ft^2
Ⓑ x^2 ft^2
Ⓒ $4x^2$ ft^2
Ⓓ $x^2(4 - \pi)$ ft^2

4. Which statement in this proof is invalid? *(Prerequisite Skill)*

Given: $x = 0$

1. $x + 1 = 1$
2. $(x + 1)^2 = 1^2$
3. $x^2 + 2x + 1 = 1$
4. $x^2 + 2x = 0$
5. $x(x + 2) = 0$
6. $\frac{x(x + 2)}{x} = \frac{0}{x}$
7. $x + 2 = 0$
8. $x = -2$

Ⓐ Statement 1
Ⓑ Statement 2
Ⓒ Statement 5
Ⓓ Statement 6

Wednesday

5. What are the solutions of $|x + 6| = 9$? *(Lesson 1-4)*

Ⓐ 15 and −3
Ⓑ 9 and −9
Ⓒ 3 and −15
Ⓓ 15 and −15

6. What is the value of $\frac{(x - 3)(x + 1)(x + 6)}{(x + 1)(x - 1)}$ when $x = -1$? *(Lesson 1-1)*

Ⓐ 10
Ⓑ −10
Ⓒ −20
Ⓓ undefined

Thursday

7. What are the factors of $4x^2 - 9$? *(Prerequisite Skill)*

Ⓐ $2x - 3$ and $2x + 3$
Ⓑ $x - 3$ and $x + 3$
Ⓒ $2x - 3$ and $2x - 3$
Ⓓ $2x + 3$ and $2x + 3$

Friday

8. What are the solutions of $|x - 12| = -7$? *(Lesson 1-4)*

Ⓐ 5 and 19
Ⓑ −5 and −19
Ⓒ 5 and −5
Ⓓ There are no solutions.

Week 2

Monday

1. What is the complete factorization of $x^4 - 16$? *(Prerequisite Skill)*

(A) $(x^2 + 4)(x^2 - 4)$
(B) $(x^2 + 4)(x + 2)(x - 2)$
(C) $(x^2 - 4)(x^2 - 4)$
(D) $(x + 2)(x - 2)(x + 2)(x - 2)$

2. What is the value of $\frac{a^2 - 3bc^2}{a + b}$ when $a = 2$, $b = 3$, and $c = -1$? *(Lesson 1-1)*

(A) -1
(B) 1
(C) $\frac{11}{5}$
(D) $\frac{13}{5}$

Tuesday

3. Which number does not satisfy the inequality $|a - 3| < 9$? *(Lesson 1-6)*

(A) -6
(B) -4
(C) 0
(D) 4

4. What is the area of the shaded region? *(Prerequisite Skill)*

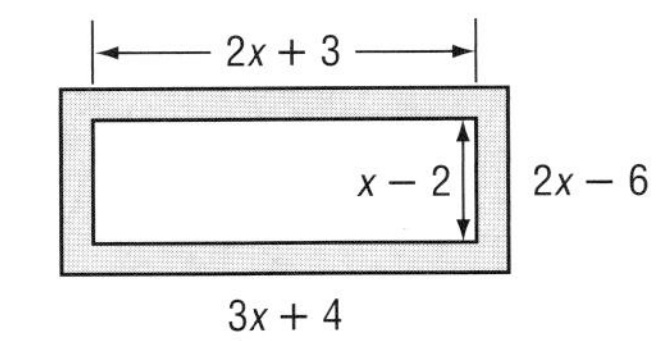

(A) $(3x + 4)(2x + 3) - (2x - 6)(x - 2)$
(B) $(3x + 4)(2x - 6)$
(C) $(3x + 4)(2x - 6) - (2x + 3)(x - 2)$
(D) $(3x + 4)^2(2x - 6)^2 - (2x + 3)^2(x - 2)^2$

Wednesday

5. What are the factors of $x^2 + 16x + 64$? *(Prerequisite Skill)*

(A) $x - 8$ and $x - 8$
(B) $x + 16$ and $x + 4$
(C) $x + 8$ and $x + 8$
(D) $x + 8$ and $x - 8$

6. Which equation is never true? (All variables represent real numbers.) *(Prerequisite Skill)*

(A) $(a + b)^2 = a^2 + b^2$
(B) $a^x = a$
(C) $x^a \cdot x^b = x^b$
(D) $x^2 - 8x + 22 = 0$

Thursday

7. How much must be added to both sides of $x^2 + 7x = 12$ to solve by completing the square? *(Prerequisite Skill)*

(A) $\frac{7}{2}$
(B) $\frac{49}{4}$
(C) $\frac{49}{2}$
(D) 49

Friday

8. What is the complete factorization of $4x^2 - 36$? *(Prerequisite Skill)*

(A) $4(x^2 - 9)$
(B) $(2x + 6)(2x - 6)$
(C) $4(x + 6)(x - 6)$
(D) $4(x + 3)(x - 3)$

Standards Practice Countdown

Week 3

Monday

1. At Amici's East Coast Pizzeria in San Francisco, the cost of 2 large pizzas and 3 small pizzas is \$64.25. The cost of 1 large and 2 small pizzas is \$37.25. What is the cost of 1 large pizza? *(Lesson 3-2)*

- (A) \$10.25
- (B) \$12.25
- (C) \$14.75
- (D) \$16.75

2. Which point belongs to the solution set of the system of inequalities? *(Lesson 3-3)*

$$\begin{cases} 3x - 2y > 4 \\ -x + y < 9 \end{cases}$$

- (A) (0, 0)
- (B) (−2, 1)
- (C) (5, 4)
- (D) (−6, 3)

Tuesday

3. Which region represents the solution of the system of inequalities? *(Lesson 3-3)*

$$\begin{cases} 2x + y \le 8 \\ x + 3y > 9 \end{cases}$$

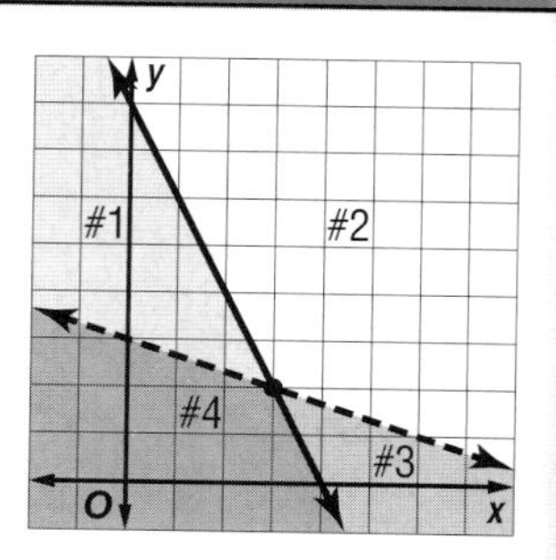

- (A) 1
- (B) 2
- (C) 3
- (D) 4

4. Which point is the solution of the system of equations? *(Lessons 3-2 and 3-3)*

$$\begin{cases} 3x + 4y = -1 \\ -x - 6y = -2 \end{cases}$$

- (A) (1, −1)
- (B) $\left(-1, \frac{1}{2}\right)$
- (C) (−4, 1)
- (D) (3, −2)

Wednesday

5. What is the value of x? *(Lesson 3-2)*

$$\begin{cases} 3x + 2y = 1 \\ -2x - 2y = -2 \end{cases}$$

- (A) −2
- (B) −1
- (C) 1
- (D) 2

6. What is the solution of the inequality $|x + 3| < 9$? *(Lesson 1-6)*

- (A) −12 and 6
- (B) all real numbers less than 9
- (C) all real numbers less than 6
- (D) all real numbers between −12 and 6

Thursday

7. Which number is not a solution to the inequality $|3x - 1| < 5$? *(Lesson 1-6)*

- (A) −1
- (B) 0
- (C) 1
- (D) 2

Friday

8. The graph of which inequality has part of its solution in the first quadrant? *(Lesson 2-7)*

- (A) $y < -2x - 1$
- (B) $y \le -4$
- (C) $y > -x + 3$
- (D) $-y > \frac{1}{2}x + 5$

Week 4

Monday

1. Use the graph to determine which ordered pair could be the solution of the system of equations. *(Lesson 3-1)*

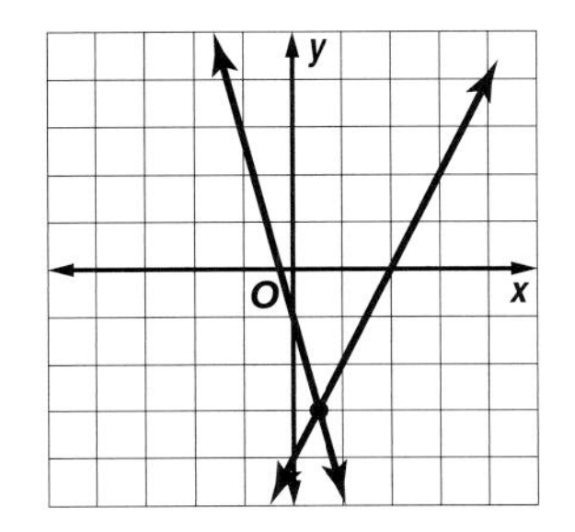

Ⓐ $\left(\frac{1}{2}, -3\right)$
Ⓑ $\left(-\frac{1}{2}, -3\right)$
Ⓒ $\left(3, \frac{1}{2}\right)$
Ⓓ $\left(-3, \frac{1}{2}\right)$

2. What is the solution of the system of equations? *(Lesson 3-5)*

$$\begin{cases} \frac{1}{2}x - y + 2z = 1 \\ 3x - 3y = -15 \\ \frac{1}{2}x + 6y - z = 2 \end{cases}$$

Ⓐ $x = 4, y = 3, z = 1$
Ⓑ $x = 6, y = 0, z = -1$
Ⓒ $x = 0, y = 5, z = 26$
Ⓓ $x = -4, y = 1, z = 2$

Tuesday

3. What is the solution of the system? *(Lesson 3-5)*

$$\begin{cases} 3x + 2y + z = 11 \\ 3x - z = 7 \\ 5x - 2y = 4 \end{cases}$$

Ⓐ $x = -2, y = -3, z = 1$
Ⓑ $x = 2, y = 3, z = -1$
Ⓒ $x = 3, y = 0, z = 2$
Ⓓ $x = 4, y = 8, z = 5$

4. The cost of terrace level seats for 2 adults and 3 children to see the Anaheim Angels at Edison Field is $41. The cost for 3 adults and 4 children is $58. What is the cost of a ticket for a child? *(Lesson 3-2)*

Ⓐ $7
Ⓑ $8
Ⓒ $9
Ⓓ $10

Wednesday

5. Which absolute value inequality corresponds to the graph shown below? *(Lesson 1-6)*

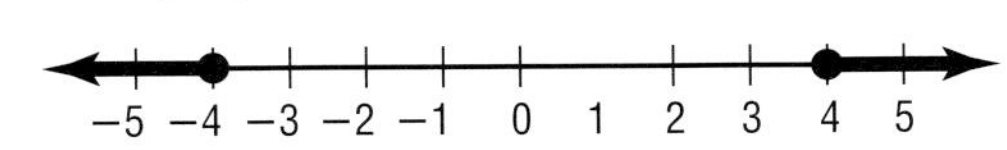

Ⓐ $|x| > 4$
Ⓑ $|x| < 4$
Ⓒ $|x| \geq 4$
Ⓓ $|x| \leq 4$

6. Which system of equations has $\left(-3, \frac{5}{3}\right)$ as a solution? *(Lesson 3-2)*

Ⓐ $\begin{cases} 2x - 3y = -11 \\ x - 3y = -2 \end{cases}$
Ⓑ $\begin{cases} x + 6y = 2 \\ x + y = -1 \end{cases}$
Ⓒ $\begin{cases} x - 3y = -8 \\ 2x + 3y = -11 \end{cases}$
Ⓓ $\begin{cases} 2x - 3y = -11 \\ x + 3y = 2 \end{cases}$

Thursday

7. What is the simplest form of $\frac{2x - 3}{x + 1} + \frac{5}{x + 1}$? *(Prerequisite Skill)*

Ⓐ 2
Ⓑ 1
Ⓒ $\frac{2x + 2}{x + 1}$
Ⓓ $\frac{10x - 15}{(x + 1)^2}$

Friday

8. What is the complete factorization of $4x^3 + 2x^2 - 6x - 3$? *(Prerequisite Skill)*

Ⓐ $2x^2(2x + 1) - 3(2x + 1)$
Ⓑ $(2x^2 - 3)(2x + 1)$
Ⓒ $(2x - 3)(2x + 3)(2x + 1)$
Ⓓ The polynomial does not factor.

Week 5

Monday

1. What should each side be multiplied by to solve for x and y? *(Lesson 4-8)*

$$\begin{bmatrix} 5 & -3 \\ 3 & 2 \end{bmatrix}\begin{bmatrix} x \\ y \end{bmatrix} = \begin{bmatrix} -2 \\ 5 \end{bmatrix}$$

(A) $\begin{bmatrix} 2 & 3 \\ -3 & 5 \end{bmatrix}$

(B) $\frac{1}{19}\begin{bmatrix} -3 & 5 \\ 2 & 3 \end{bmatrix}$

(C) $\frac{1}{19}\begin{bmatrix} 2 & 3 \\ -3 & 5 \end{bmatrix}$

(D) $\frac{1}{19}\begin{bmatrix} 5 & -3 \\ 3 & 2 \end{bmatrix}$

2. What is the solution of $\begin{bmatrix} 5 & 6 \\ -3 & -4 \end{bmatrix}\begin{bmatrix} x \\ y \end{bmatrix} = \begin{bmatrix} -2 \\ 1 \end{bmatrix}$? *(Lesson 4-8)*

(A) $\left(1, -\frac{1}{2}\right)$

(B) $(2, -2)$

(C) $(-2, 2)$

(D) $\left(-1, \frac{1}{2}\right)$

Tuesday

3. Which system of equations can be solved with the matrix equation $\begin{bmatrix} 3 & -4 \\ -1 & 2 \end{bmatrix}\begin{bmatrix} x \\ y \end{bmatrix} = \begin{bmatrix} 2 \\ 0 \end{bmatrix}$? *(Lesson 4-8)*

(A) $\begin{cases} 3x - y = 2 \\ -4x + 2y = 0 \end{cases}$

(B) $\begin{cases} 3x - 4y = 2 \\ -x + 2y = 0 \end{cases}$

(C) $\begin{cases} -3x + 4y = 2 \\ x - 2y = 0 \end{cases}$

(D) $\begin{cases} 3x + 2y = 2 \\ -x - 4y = 0 \end{cases}$

4. What is the solution of the system of equations represented by $\begin{bmatrix} 4 & -3 \\ -2 & -1 \end{bmatrix}\begin{bmatrix} x \\ y \end{bmatrix} = \begin{bmatrix} -1 \\ -2 \end{bmatrix}$? *(Lesson 4-8)*

(A) $(-1, -1)$

(B) $(1, 1)$

(C) $\left(1, \frac{1}{2}\right)$

(D) $\left(\frac{1}{2}, 1\right)$

Wednesday

5. What is the solution of $\begin{bmatrix} 2 & -1 & 3 \\ 4 & 0 & -2 \\ 1 & 1 & -1 \end{bmatrix}\begin{bmatrix} x \\ y \\ z \end{bmatrix} = \begin{bmatrix} 4 \\ -14 \\ -4 \end{bmatrix}$? *(Lesson 4-6)*

(A) $x = -2, y = 1, z = 3$

(B) $x = -3, y = 0, z = 1$

(C) $x = -3, y = -7, z = 1$

(D) $x = -2, y = 0, z = 2$

6. Which graph represents the solution of $|3x + 5| < 8$? *(Lesson 1-6)*

(A)

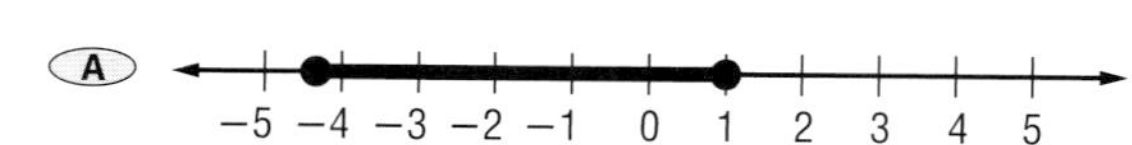

(B) −5 −4 −3 −2 −1 0 1 2 3 4 5

(C) −5 −4 −3 −2 −1 0 1 2 3 4 5

(D) −5 −4 −3 −2 −1 0 1 2 3 4 5

Thursday

7. For which inequality is $(-2, -2)$ not a solution? *(Lesson 2-7)*

(A) $|x| < 4$

(B) $|x + y| > 4$

(C) $|x - y| \leq 1$

(D) $|4y| > 6$

Friday

8. What is the <u>complete</u> factorization of $x^4 + 16$? *(Prerequisite Skill)*

(A) $(x + 2)^4$

(B) $(x^2 + 4)(x^2 - 4)$

(C) $(x^2 + 4)(x^2 + 4)$

(D) does not factor

Week 6

Monday

1. What would both sides of $\begin{bmatrix} 2 & 1 \\ 1 & 1 \end{bmatrix} \begin{bmatrix} x \\ y \end{bmatrix} = \begin{bmatrix} 2 \\ 3 \end{bmatrix}$ be multiplied by to solve for x and y? *(Lesson 4-8)*

   A $\begin{bmatrix} -2 & -1 \\ -1 & -1 \end{bmatrix}$

   B $\begin{bmatrix} 2 & 1 \\ 1 & 1 \end{bmatrix}$

   C $\begin{bmatrix} 1 & -1 \\ -1 & 2 \end{bmatrix}$

   D $\frac{1}{2}\begin{bmatrix} 1 & -1 \\ -1 & 2 \end{bmatrix}$

2. How many solutions does the system of equations have? *(Lesson 3-1)*

   $$\begin{cases} 3x + y = 5 \\ y = -3x + 9 \end{cases}$$

   A 0
   B 1
   C 2
   D infinitely many

Tuesday

3. In which quadrants do solutions of the inequality $y > |x - 2| + 1$ lie? *(Lesson 2-7)*

   A Quadrant I only
   B Quadrants I and II only
   C Quadrants III and IV only
   D Quadrants I, II, III, and IV

4. What is the solution of $\begin{bmatrix} -2 & -3 \\ 1 & 4 \end{bmatrix} \begin{bmatrix} x \\ y \end{bmatrix} = \begin{bmatrix} -1 \\ -2 \end{bmatrix}$? *(Lesson 4-8)*

   A $(-1, 1)$
   B $\left(0, -\frac{1}{2}\right)$
   C $(2, -1)$
   D $\left(4, -\frac{3}{2}\right)$

Wednesday

5. What is the area of the shaded region defined by the two squares? *(Prerequisite Skill)*

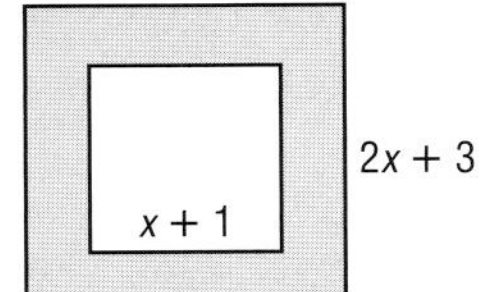

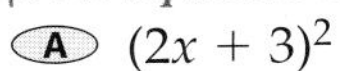

   A $(2x + 3)^2$
   B $(x + 1)^2$
   C $(x + 1)^2 - (2x + 3)^2$
   D $(3x + 4)(x + 2)$

6. What are the dimensions of the coefficient matrix for the system? *(Lesson 4-8)*

   $$\begin{cases} 2x + z = 5 \\ -y - z = -4 \\ x + 2y = 8 \end{cases}$$

   A 1×1
   B 2×2
   C 3×3
   D 4×4

Thursday

7. How many solutions does the matrix equation have? *(Lesson 4-8)*

   $$\begin{bmatrix} 2 & 3 \\ 1 & 2 \end{bmatrix} \begin{bmatrix} x \\ y \end{bmatrix} = \begin{bmatrix} 8 \\ 7 \end{bmatrix}$$

   A 0
   B 1
   C 2
   D infinitely many

Friday

8. A system of equations has no solutions if the determinant of the coefficient matrix is what number? *(Lesson 4-6)*

   A 1
   B $\frac{1}{2}$
   C 0
   D -1

Standards Practice Countdown

Week 7

Monday

1. What is the sum of $2 + 3i$ and $-6 - 5i$? *(Lesson 5-9)*
   - (A) $4 - 2i$
   - (B) $8 + 8i$
   - (C) $-4 + 2i$
   - (D) $-4 - 2i$

2. x is a real number. For which values of n is the statement true? *(Lesson 5-5)*

$$\sqrt[n]{x^n} = x$$

   - (A) no real numbers
   - (B) all natural numbers
   - (C) all even natural numbers
   - (D) all odd natural numbers

Tuesday

3. Which expression represents the area of the shaded region? *(Prerequisite Skill)*

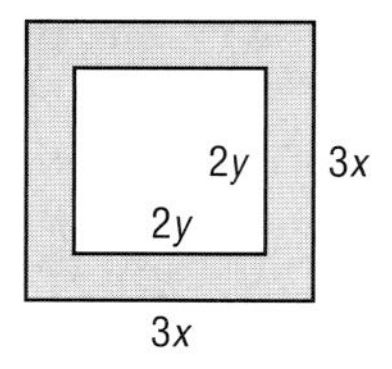

   - (A) $(3x - 2y)^2$
   - (B) $(3x + 2y)(3x - 2y)$
   - (C) $3x - 2y$
   - (D) $(3x + 2y)^2$

4. What is the coefficient of x in the quotient when $2x^3 + 3x - 4$ is divided by $x + 1$? *(Lesson 5-3)*
   - (A) -5
   - (B) -2
   - (C) 1
   - (D) 2

Wednesday

5. Which expression is equal to $x^4y^{\frac{3}{2}}$? *(Lesson 5-7)*
   - (A) $\left(x^4y^{\frac{1}{2}}\right)(xy^3)$
   - (B) $(x^2y^2)\left(x^2y^{\frac{3}{4}}\right)$
   - (C) $(xy)\left(x^3y^{\frac{3}{2}}\right)$
   - (D) $\left(x^3y^{\frac{1}{2}}\right)(xy)$

6. What is the solution of the matrix equation $\begin{bmatrix} 2 & 3 \\ -4 & 9 \end{bmatrix}\begin{bmatrix} x \\ y \end{bmatrix} = \begin{bmatrix} 2 \\ 1 \end{bmatrix}$? *(Lesson 4-8)*
   - (A) $(2, -1)$
   - (B) $\left(\frac{1}{2}, \frac{1}{3}\right)$
   - (C) $(2, 1)$
   - (D) $\left(\frac{1}{2}, -\frac{1}{3}\right)$

Thursday

7. Which is <u>not</u> a factor of $x^4 - 1$? *(Prerequisite Skill)*
   - (A) $x + 1$
   - (B) $x - 1$
   - (C) $x^3 - 1$
   - (D) $x^2 - 1$

Friday

8. What is the product of $3 - i$ and $-2 + 4i$? *(Lesson 5-9)*
   - (A) $-2 + 10i$
   - (B) $-10 + 10i$
   - (C) $-2 + 14i$
   - (D) $-10 + 14i$

Week 8

Monday

1. What is the larger of the two solutions of $|2x + 6| - 7 = 21$? *(Lesson 1-4)*

(A) -17
(B) -11
(C) 11
(D) 17

2. What is the coefficient of x in the quotient when $x^3 + 4x^2 - 3x + 8$ is divided by $x + 2$? *(Lesson 5-3)*

(A) -7
(B) 1
(C) 2
(D) 6

Tuesday

3. What is the difference between $7 - 6i$ and $4 - 2i$? *(Lesson 5-9)*

(A) $3 - 4i$
(B) $11 - 4i$
(C) $3 - 8i$
(D) $11 - 8i$

4. What is the remainder when $3x^3 - 4x + 7$ is divided by $x - 2$? *(Lesson 5-3)*

(A) -9
(B) 11
(C) 23
(D) 27

Wednesday

5. Which point is a solution of the inequality $y \geq 2|x - 3| - 2$? *(Lesson 2-7)*

(A) $(-4, 5)$
(B) $(0, 0)$
(C) $(3, 1)$
(D) $(8, 7)$

6. How many solutions does the system of equations have? *(Lesson 3-1)*

$$\begin{cases} x + y = 9 \\ 2x + 3y = 6 \end{cases}$$

(A) 0
(B) 1
(C) 2
(D) infinitely many

Thursday

7. Simplify $\frac{-3 + 2i}{4 + 3i}$. *(Lesson 5-9)*

(A) $\frac{18}{25} + \frac{17}{25}i$
(B) $-\frac{6}{25} + \frac{17}{25}i$
(C) $-\frac{6}{25} - \frac{17}{25}i$
(D) $-\frac{3}{4} + \frac{2}{3}i$

Friday

8. Which complex number is represented by point Q on the graph? *(Lesson 5-9)*

b
a
O
R
Q
P
S

(A) $-1 - 3i$
(B) $1 - 3i$
(C) $3 + i$
(D) $-3 - i$

Standards Practice Countdown

Week 9

Monday

1. The axis of symmetry of a parabola is $x = 6$. One x-intercept is at (13, 0). Where is the other x-intercept? *(Lesson 6-2)*

- (A) (20, 0)
- (B) (1, 0)
- (C) (−1, 0)
- (D) (−6, 0)

2. The graph of the parabola $-y + 6 = 4(x + 3)^2$ opens in which direction? *(Lesson 6-6)*

- (A) up
- (B) down
- (C) left
- (D) right

Tuesday

3. Use the part of the parabola shown to find the other zero of the parabola. *(Lesson 6-2)*

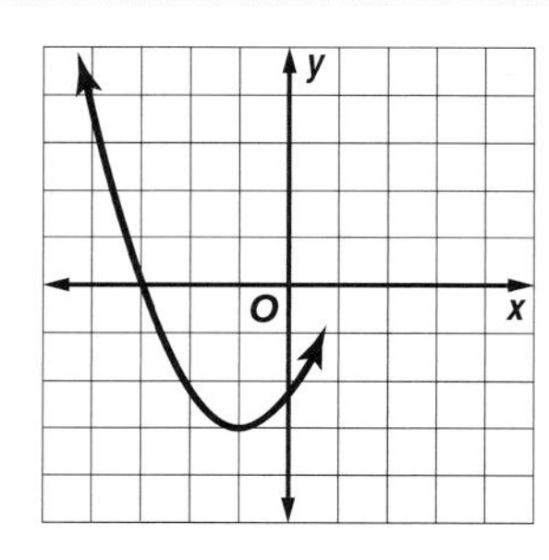

- (A) (1, 0)
- (B) (3, 0)
- (C) (0, 3)
- (D) (0, 1)

4. What is the maximum value of the function $y = -4x^2 + 4x - 9$? *(Lesson 6-1)*

- (A) −10
- (B) −9
- (C) −8
- (D) 9

Wednesday

5. What are the x-intercepts of $y + 2 = x^2 + \frac{7}{2}x$? *(Lesson 6-2)*

- (A) 4 and $-\frac{1}{2}$
- (B) 4 and $\frac{1}{2}$
- (C) −4 and $-\frac{1}{2}$
- (D) −4 and $\frac{1}{2}$

6. What is the quotient when $x^3 - 3x^2 + x - 1$ is divided by $x + 2$? *(Lesson 5-3)*

- (A) −23
- (B) $\frac{-23}{x + 2}$
- (C) $x^2 - 5x + 11$
- (D) 21

Thursday

7. In which quadrants do solutions of the inequality $y > |x| + 5$ lie? *(Lesson 2-7)*

- (A) Quadrant I only
- (B) Quadrants I and II only
- (C) Quadrants III and IV only
- (D) Quadrants I, II, III, and IV

Friday

8. How many solutions does the system of equations have? *(Lesson 3-1)*

$$\begin{cases} x = -y + 7 \\ 0 = -x - y + 7 \end{cases}$$

- (A) 0
- (B) 1
- (C) 2
- (D) infinitely many

Week 10

Monday

1. What is the equation of the axis of symmetry for the parabola $y = 2(x + 3)^2 - 9$? *(Lesson 6-6)*

   (A) $x = -3$
   (B) $x = 3$
   (C) $y = -9$
   (D) $y = 9$

2. What is the minimum value of the function $y - 3 = x^2 + 6x + 6$? *(Lesson 6-1)*

   (A) -3
   (B) 0
   (C) 3
   (D) 6

Tuesday

3. The graph of $y = 2(x + 1)^2 + 1$ is shown below. How will the graph change if $(x + 1)$ is replaced with $(x - 2)$? *(Lesson 6-6)*

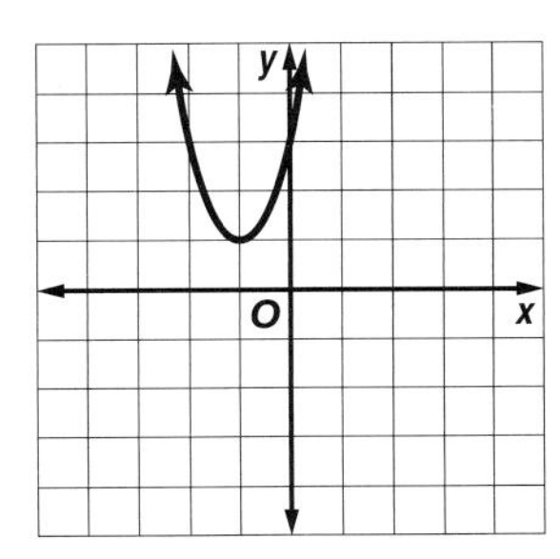

   (A) The graph will move left.
   (B) The graph will move right.
   (C) The graph will move up.
   (D) The graph will move down.

4. How many real solutions does $0 = 3(x + 4)^2 - 6$ have? *(Lesson 6-2)*

   (A) 0
   (B) 1
   (C) 2
   (D) 3

Wednesday

5. Which statement is <u>true</u>? *(Lesson 5-4)*

   (A) A difference of squares never factors.
   (B) A sum of squares never factors.
   (C) A sum of squares always factors.
   (D) A sum of squares sometimes factors.

6. What is the value of $\frac{b^2 - a}{b^2 + 3c}$ when $a = 6$, $b = -1$, and $c = 3$? *(Lesson 1-1)*

   (A) $-\frac{7}{8}$
   (B) $-\frac{1}{2}$
   (C) $\frac{1}{2}$
   (D) 2

Thursday

7. Which point is <u>not</u> a solution of the system of inequalities? *(Lesson 3-3)*

$$\begin{cases} -10x + 7y < 14 \\ 9x - 5y \geq 11 \end{cases}$$

   (A) (2, 1)
   (B) (4, 0)
   (C) (−2, 1)
   (D) (0, −4)

Friday

8. What is the maximum value of the function $y = -6(x + 4)^2 - 5$? *(Lesson 6-1)*

   (A) -6
   (B) -5
   (C) -4
   (D) 4

Week 11

Monday

1. It costs a computer chip maker \$427 to produce a microchip. The company has operating expenses of \$40,000 per month. The company sells its chips for \$728 each. What is the company's profit if it sells 10,000 chips in a month? *(Lesson 7-7)*
   - (A) \$7,280,000
   - (B) \$7,240,000
   - (C) \$3,010,000
   - (D) \$2,970,000

2. Which complex number is plotted on the coordinate plane shown? *(Lesson 5-9)*

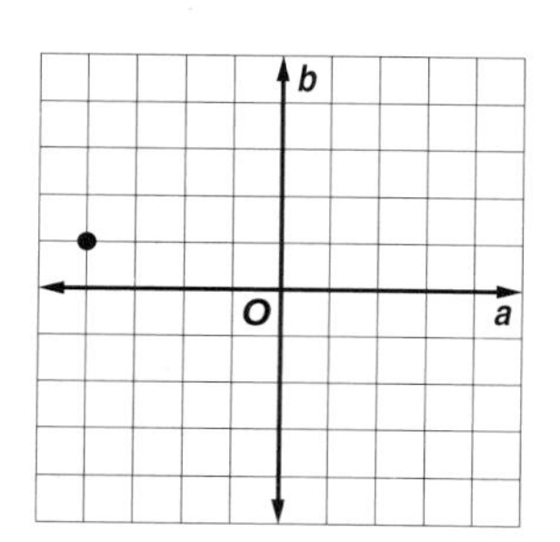

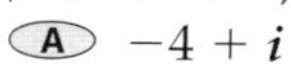

   - (A) $-4 + i$
   - (B) $1 - 4i$
   - (C) $-4 - i$
   - (D) $-4i$

Tuesday

3. Which expression represents the coefficient of x^2 in $f(x) - g(x)$ if $f(x) = ax^2 + bx + c$ and $g(x) = -3x^2 - 4$? *(Lesson 7-7)*
   - (A) $3 - a$
   - (B) $a + 3$
   - (C) $a - 3$
   - (D) $(a + 3)^2$

4. What is the remainder when $x^4 + 2x^3 - 4x^2 - x - 6$ is divided by $x + 3$? *(Lesson 7-4)*
   - (A) -12
   - (B) -6
   - (C) 0
   - (D) 6

Wednesday

5. In which quadrants do solutions of the inequality $y < -|x| + 3$ lie? *(Lesson 2-7)*
   - (A) Quadrant I only
   - (B) Quadrants I and II only
   - (C) Quadrants III and IV only
   - (D) Quadrants I, II, III, and IV

6. Which function is the inverse of $f(x) = x^3 + 1$? *(Lesson 7-8)*
   - (A) $g(x) = \sqrt[3]{x} - 1$
   - (B) $g(x) = \sqrt[3]{x - 1}$
   - (C) $g(x) = -x^3 - 1$
   - (D) $g(x) = \sqrt[3]{x^3 + 1}$

Thursday

7. What amount must be added to each side of $x^2 - 3x = \frac{3}{4}$ to solve by completing the square? *(Lesson 6-4)*
   - (A) $\frac{9}{4}$
   - (B) $\frac{3}{2}$
   - (C) $-\frac{3}{2}$
   - (D) $-\frac{9}{4}$

Friday

8. If $f(x) = x^3 + x^2 - 6x + 1$ and $g(x) = 3$, what is $[g \circ f](x)$? *(Lesson 7-7)*
   - (A) 3
   - (B) 19
   - (C) $x^3 + x^2 - 6x + 1$
   - (D) $x^3 + x^2 - 6x + 4$

Week 12

Monday

1. If $g(x)$ is the inverse function of $f(x)$, which statement is always <u>true</u>? *(Lesson 7-8)*
   - (A) $f(x) \cdot g(x) = x$
   - (B) $\frac{f(x)}{g(x)} = x$
   - (C) $[f \circ g](x) = 1$
   - (D) $[f \circ g](x) = x$

2. If $f(x) = x^2 - 11$ and $g(x) = x + 1$, what is $[f \circ g](x)$? *(Lesson 7-7)*
   - (A) $x^2 - 10$
   - (B) $x^2 + 2x - 10$
   - (C) $x^2 + 2x - 11$
   - (D) $x^2 + x - 10$

Tuesday

3. What is the product of $(5 + 3i)$ and $(5 - 3i)$? *(Lesson 5-9)*
   - (A) $25 - 9i$
   - (B) $25 + 9i$
   - (C) 34
   - (D) 16

4. A company pays 15% tax on its net profit. The profit function is $g(x) = 3x^2 + 2x - 1{,}500$. The tax function is $f(x) = 0.15x$. Which function determines the company's tax amount? *(Lesson 7-7)*
   - (A) $[f \circ g](x)$
   - (B) $[g \circ f](x)$
   - (C) $f(x) \cdot g(x)$
   - (D) $g(x) - f(x)$

Wednesday

5. Which function is the inverse of $f(x) = 3x - 4$? *(Lesson 7-8)*
   - (A) $g(x) = 3x + 4$
   - (B) $g(x) = \frac{1}{3}x + 4$
   - (C) $g(x) = \frac{x + 4}{3}$
   - (D) $g(x) = 4x + 3$

6. For which value of a does the graph shown below correspond with the inequality $|x - a| < 9$? *(Lesson 1-6)*

   - (A) −6
   - (B) −3
   - (C) 3
   - (D) 6

Thursday

7. Which statement about the graph of $y = -3(x - 1)^2 - 6$ is <u>true</u>? *(Lesson 6-1)*
   - (A) It has a maximum and 2 zeros.
   - (B) It has a maximum and no zeros.
   - (C) It has a minimum and 2 zeros.
   - (D) It has a minimum and no zeros.

Friday

8. If $f(x) = 3x^2 - 6x + 7$ and $g(x) = x^3 + 4x$, what is $f(x) - g(x)$? *(Lesson 7-7)*
   - (A) $x^3 - 3x^2 + 10x - 7$
   - (B) $2x^2 - 10x + 7$
   - (C) $-x^3 + 3x^2 - 2x + 7$
   - (D) $-x^3 + 3x^2 - 10x + 7$

Week 13

Monday

1. Which is the graph of $y < -|x| - 2$? *(Lesson 2-7)*

Ⓐ
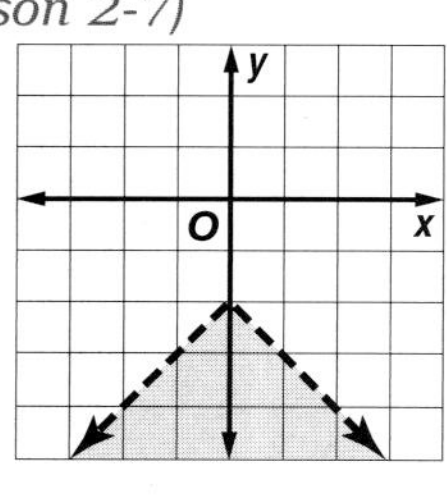

Ⓑ
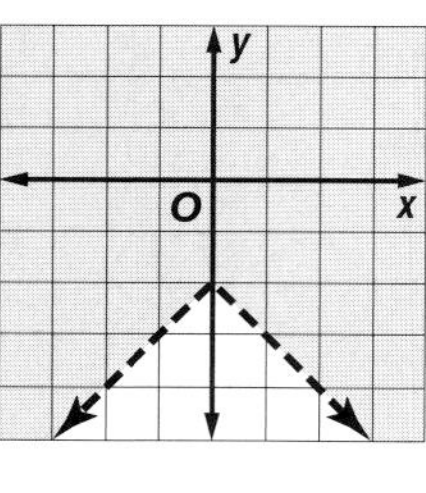

Ⓒ
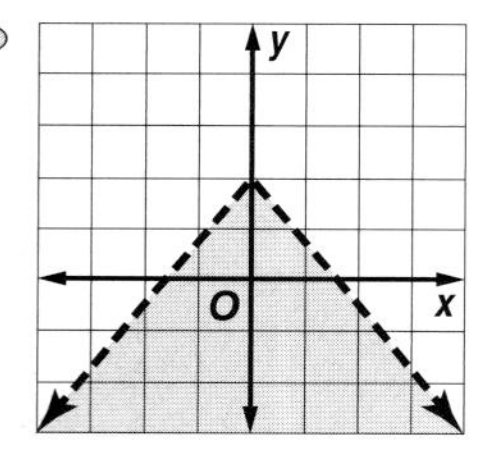

Ⓓ
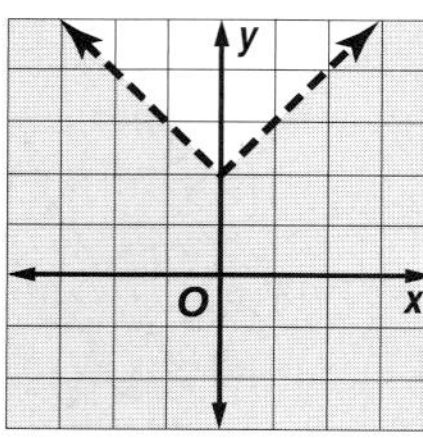

2. What is the equation of the axis of symmetry for $x = -3(y + 4)^2 + 9$? *(Lesson 8-2)*
   - Ⓐ $x = 9$
   - Ⓑ $x = -9$
   - Ⓒ $y = 4$
   - Ⓓ $y = -4$

Tuesday

3. Which function is the inverse of $f(x) = \sqrt[5]{x - 1}$? *(Lesson 7-8)*
   - Ⓐ $g(x) = (x - 1)^5$
   - Ⓑ $g(x) = x^5 + 1$
   - Ⓒ $g(x) = (x + 1)^5$
   - Ⓓ $g(x) = x^5 - 1$

4. For which value of x is the product a real number? *(Lesson 5-9)*

$$(3 + 4i)(3 + xi)$$

   - Ⓐ -4
   - Ⓑ -3
   - Ⓒ -1
   - Ⓓ 4

Wednesday

5. What are the solutions of the equation $x^2 - 2x = -3$? *(Lesson 6-5)*
   - Ⓐ $1 + \sqrt{2}, 1 - \sqrt{2}$
   - Ⓑ 1 and -1
   - Ⓒ $1 + i\sqrt{2}, 1 - i\sqrt{2}$
   - Ⓓ i and $-i$

6. What are the values of x and y? *(Lesson 4-8)*

$$\begin{bmatrix} 3 & 2 \\ -1 & -2 \end{bmatrix}\begin{bmatrix} x \\ y \end{bmatrix} = \begin{bmatrix} -5 \\ 1 \end{bmatrix}$$

   - Ⓐ $(1, -4)$
   - Ⓑ $(-1, -1)$
   - Ⓒ $(1, -1)$
   - Ⓓ $\left(-2, \frac{1}{2}\right)$

Thursday

7. What is the value of $\frac{ab - 3a^2}{2b - 5c}$ when $a = -1$, $b = 2$, and $c = 1$? *(Lesson 1-1)*
   - Ⓐ -5
   - Ⓑ -1
   - Ⓒ 1
   - Ⓓ 5

Friday

8. What is the solution of the system of equations? *(Lesson 3-2)*

$$\begin{cases} y = 5x - 4 \\ y = 3x + 8 \end{cases}$$

   - Ⓐ $(-6, -34)$
   - Ⓑ $(6, 26)$
   - Ⓒ $(7, 29)$
   - Ⓓ $(10, 46)$

Week 14

Monday

1. What is the standard form of the equation $16x^2 + 9y^2 - 32x + 36y - 92 = 0$? *(Lesson 8-6)*

Ⓐ $\dfrac{(x-1)^2}{9} - \dfrac{(y-2)^2}{16} = 1$

Ⓑ $16(x-1)^2 - 9(y+2)^2 = 1$

Ⓒ $\dfrac{(x-1)^2}{9} + \dfrac{(y-2)^2}{16} = 1$

Ⓓ $16(x-1)^2 - 9(y+2)^2 = 144$

2. What is the value of d if the equation represents a circle with radius $\sqrt{19}$? *(Lesson 8-6)*

$$x^2 + y^2 - 12x + dy = -33$$

Ⓐ 16

Ⓑ 8

Ⓒ 4

Ⓓ −16

Tuesday

3. Which answer choice shows the graph of $9x^2 + 4y^2 + 36x - 16y + 16 = 0$? *(Lesson 8-6)*

Ⓐ

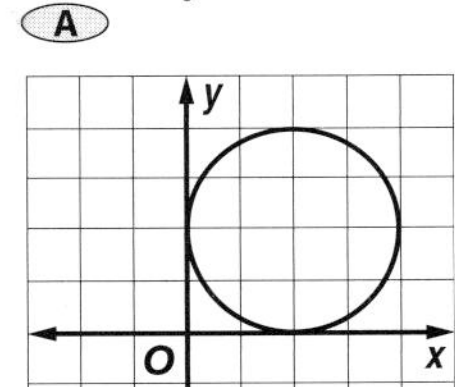

Ⓑ

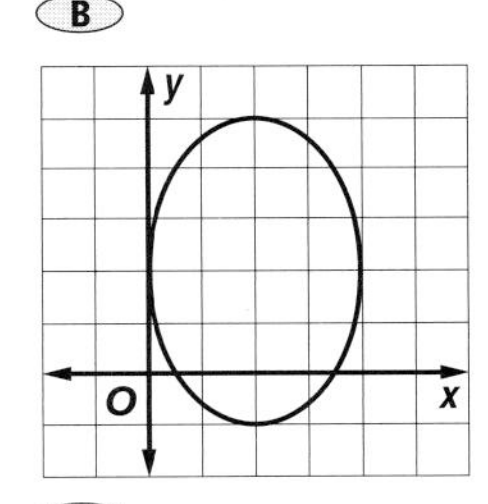

Ⓒ

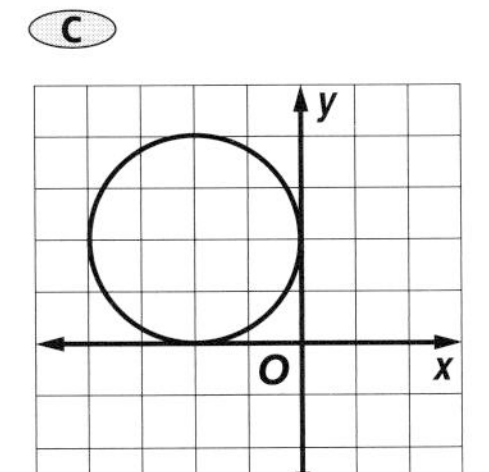

Ⓓ

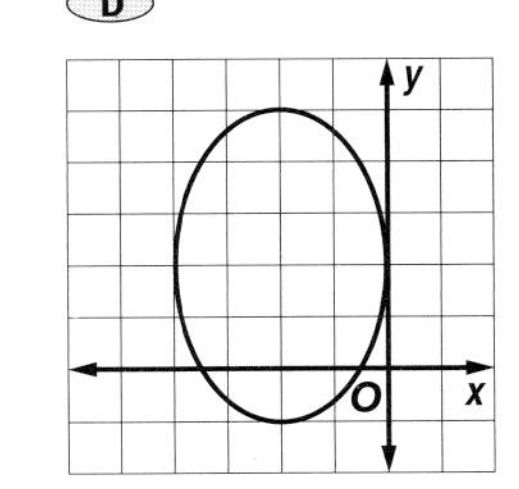

4. Which conic section is represented by the equation $4y^2 - x + 20y + 27 = 0$? *(Lesson 8-6)*

Ⓐ hyperbola

Ⓑ ellipse

Ⓒ circle

Ⓓ parabola

Wednesday

5. What is the length of the major axis for the ellipse $\dfrac{y^2}{49} + \dfrac{x^2}{16} = 1$? *(Lesson 8-4)*

Ⓐ 49

Ⓑ 16

Ⓒ 14

Ⓓ 7

6. What are the values of x and y? *(Lesson 5-9)*

$$(3 + y\boldsymbol{i}) + (x - 6\boldsymbol{i}) = -2 + \boldsymbol{i}$$

Ⓐ $x = -5, y = 7$

Ⓑ $x = -1, y = 7$

Ⓒ $x = -1, y = 5$

Ⓓ $x = -5, y = 5$

Thursday

7. What are the solutions of $x^2 - 8x + 20 = 0$? *(Lesson 6-5)*

Ⓐ 10 and −2

Ⓑ $4 + 2\boldsymbol{i}$ and $4 - 2\boldsymbol{i}$

Ⓒ $2 + 4\boldsymbol{i}$ and $2 - 4\boldsymbol{i}$

Ⓓ −10 and 2

Friday

8. Which point is not a solution of the system of inequalities? *(Lesson 3-3)*

$$\begin{cases} x + 3y > 10 \\ x - 5y \leq 9 \end{cases}$$

Ⓐ (3, 3)

Ⓑ (2, 4)

Ⓒ (−1, 4)

Ⓓ (1, 3)

Standards Practice Countdown

Week 15

Monday

1. Which way does the graph of the parabola represented by $y = 3(x - 2)^2 + 7$ open? *(Lesson 8-2)*

- (A) left
- (B) right
- (C) up
- (D) down

2. What is the equation of an asymptote of $\frac{x^2}{25} - \frac{y^2}{49} = 1$? *(Lesson 8-5)*

- (A) $y = -\frac{7}{5}x$
- (B) $y = -\frac{5}{7}x$
- (C) $x = -\frac{7}{5}y$
- (D) $y = \frac{5}{7}x$

Tuesday

3. The graph of $x^2 + y^2 + 4x - 15 = 0$ is which conic section? *(Lesson 8-6)*

- (A) parabola
- (B) circle
- (C) ellipse
- (D) hyperbola

4. What are the coordinates of one focus of the ellipse with equation $x^2 + 9y^2 - 2x + 36y + 28 = 0$ *(Lesson 8-4)*

- (A) $\left(1, 2 + 2\sqrt{2}\right)$
- (B) $\left(1, 2 - 2\sqrt{2}\right)$
- (C) $\left(1 + 2\sqrt{2}, -2\right)$
- (D) $\left(1 + 2\sqrt{2}, 2\right)$

Wednesday

5. What are the coordinates of the vertex of the parabola $y = -4\left(x + \frac{1}{2}\right)^2 + \frac{3}{2}$? *(Lesson 8-6)*

- (A) $\left(\frac{1}{2}, \frac{3}{2}\right)$
- (B) $\left(\frac{1}{2}, -\frac{3}{2}\right)$
- (C) $\left(-\frac{1}{2}, -\frac{3}{2}\right)$
- (D) $\left(-\frac{1}{2}, \frac{3}{2}\right)$

6. The graph shown is the inverse of which function? *(Lesson 7-8)*

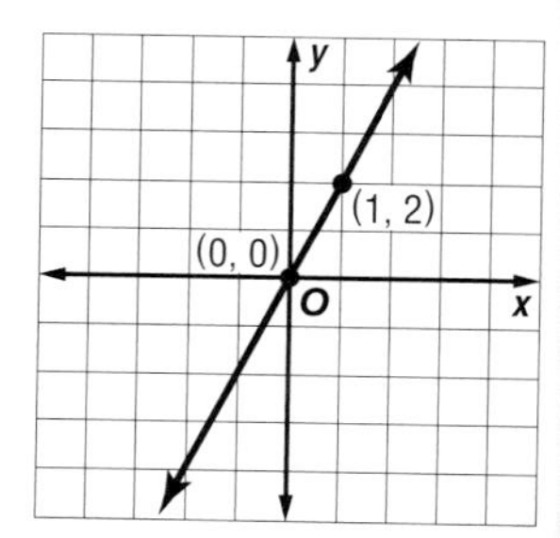

- (A) $y = -2x$
- (B) $y = -\frac{1}{2}x$
- (C) $y = \frac{1}{2}x$
- (D) $y = 2x$

Thursday

7. What is the minimum value of the quadratic function $y = 3(x + 2)^2 - 8$? *(Lesson 6-1)*

- (A) −8
- (B) −3
- (C) 3
- (D) 8

Friday

8. In which quadrant do the graphs of these equations intersect? *(Lesson 3-1)*

$$\begin{cases} y = 2x + 1 \\ y = -x - 14 \end{cases}$$

- (A) Quadrant I
- (B) Quadrant II
- (C) Quadrant III
- (D) Quadrant IV

Week 16

Monday

1. What is the simplest form of $\dfrac{1}{1+\dfrac{1}{x-1}}$? *(Lesson 9-1)*

   (A) $\dfrac{1}{\dfrac{x}{x-1}}$ (B) $\dfrac{1}{x}$

   (C) $\dfrac{x-1}{x}$ (D) $\dfrac{x}{x-1}$

2. What is the simplest form of $\dfrac{(x^2y^{-3}z^4)^2}{(x^{-2}y)^{-3}}$? *(Lesson 9-1)*

   (A) $\dfrac{x^4y^{-6}z^8}{x^6y^{-3}}$

   (B) $x^2y^3z^8$

   (C) $\dfrac{1}{x^2y^3z^8}$

   (D) $\dfrac{z^8}{x^2y^3}$

Tuesday

3. Which values of x make the expression $\dfrac{x-3}{1+\dfrac{2}{x-2}}$ undefined? *(Lesson 9-1)*

   (A) 2 and 3 (B) 0, 2, and 3

   (C) 0 and 2 (D) 2

4. What is the simplest form of $\dfrac{x^2-9}{x^2-4} \div \dfrac{x^3-27}{x^2+5x+6}$? *(Lesson 9-1)*

   (A) $\dfrac{x^2-9}{x^2-4} \cdot \dfrac{x^2+5x+6}{x^3-27}$

   (B) $\dfrac{x^2-4}{x^2-9} \div \dfrac{x^3-27}{x^2+5x+6}$

   (C) $\dfrac{(x+3)(x-3)}{(x-2)(x^2-3x+9)}$

   (D) $\dfrac{(x+3)^2}{(x-2)(x^2+3x+9)}$

Wednesday

5. What is the simplest form of $\dfrac{1}{x+1} - \dfrac{1}{x-1}$? *(Lesson 9-2)*

   (A) $\dfrac{-2}{(x+1)(x-1)}$

   (B) 0

   (C) $\dfrac{(x-1)-(x+1)}{(x+1)(x-1)}$

   (D) $\dfrac{2x}{(x+1)(x-1)}$

6. Which is the equation for the graph shown below? *(Lesson 8-6)*

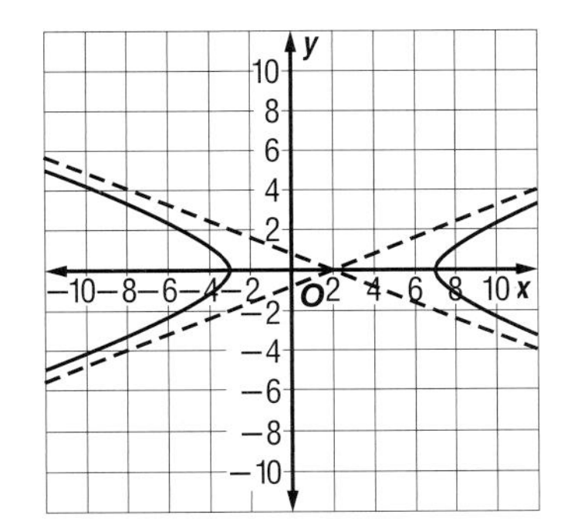

   (A) $(x-2)^2 + y^2 = 25$

   (B) $\dfrac{(x-2)^2}{25} - \dfrac{y^2}{4} = 1$

   (C) $\dfrac{(x-2)^2}{25} + \dfrac{y^2}{4} = 1$

   (D) $(y-2)^2 + x^2 = 25$

Thursday

7. Which number does <u>not</u> satisfy the inequality $|y+1| - 2 \leq 2$? *(Lesson 1-6)*

   (A) −1 (B) 2

   (C) 3 (D) 4

Friday

8. For which value of y will $(-1+3i)(1-yi)$ be a real number? *(Lesson 5-9)*

   (A) 3 (B) 1

   (C) −1 (D) −3

Standards Practice Countdown

Week 17

Monday

1. What is the simplest form of $\frac{x}{x+1} + \frac{1}{x-1}$? *(Lesson 9-2)*
   - (A) 0
   - (B) $\frac{x^2-1}{(x+1)(x-1)}$
   - (C) $\frac{x^2+1}{(x+1)(x-1)}$
   - (D) 1

2. What is the simplest form of $\frac{x^2+8x+16}{x^2+x-12} \cdot \frac{x^2-9}{x^2+7x+12}$? *(Lesson 9-1)*
   - (A) $\frac{(x+4)^2}{(x-3)^2}$
   - (B) -1
   - (C) 0
   - (D) 1

Tuesday

3. What is the least common denominator for the sum? *(Lesson 9-2)*

$$\frac{1}{x^2-9} + \frac{1}{x^2+5x+6} + \frac{1}{x^2-4}$$

   - (A) $(x+3)(x-3)(x+2)(x-2)$
   - (B) $(x+3)^2(x+2)^2(x-3)(x-2)$
   - (C) $(x^2-9)(x^2+5x+6)(x^2-4)$
   - (D) x^2-4

4. What is the simplest form of $\frac{a^2b-b^2a}{a^2-b^2}$? *(Lesson 9-1)*
   - (A) $\frac{ab(a-b)}{a^2-b^2}$
   - (B) $\frac{ab}{a+b}$
   - (C) $b-a$
   - (D) $\frac{a^2b-b^2a}{a^2-b^2}$

Wednesday

5. Which point is <u>not</u> a solution to the inequality? *(Lesson 2-7)*

$$|4x-3y| \leq 12$$

   - (A) $(3, 0)$
   - (B) $(0, 4)$
   - (C) $\left(\frac{3}{2}, -2\right)$
   - (D) $(1, -3)$

6. What is the axis of symmetry for the parabola shown? *(Lesson 6-6)*
   - (A) $y = 1$
   - (B) $y = -1$
   - (C) $x = -2$
   - (D) $x = 2$

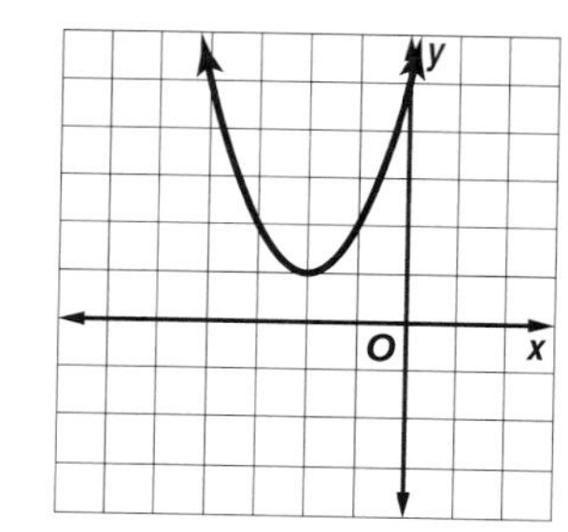

Thursday

7. What is the simplest form of $\frac{a^2b^3}{a^{-1}b^{-2}} \div \frac{a^{-2}b}{ab^{-3}}$? *(Lesson 9-1)*

   - (A) b^{-1}
   - (B) a^0b^1
   - (C) a^6b
   - (D) a^6b^9

Friday

8. What are the values of x and y? *(Lesson 5-9)*

$$(x-3i)(2+4i) = 4y + 7yi$$

   - (A) $x = -54, y = 30$
   - (B) $x = 54, y = 30$
   - (C) $x = -54, y = -30$
   - (D) $x = 54, y = -30$

Week 18

Monday

1. For $b > 1$, $\log_b 0 = x$ is true for which values of x? *(Lesson 10-2)*

(A) 0 only
(B) negative real numbers
(C) positive real numbers
(D) no real numbers

2. What is the value of x if $3^x = 4$? *(Lesson 10-3)*

(A) $\frac{4}{3}$
(B) $\frac{3}{2}$
(C) $\frac{\ln 4}{\ln 3}$
(D) $\frac{\ln 3}{\ln 4}$

Tuesday

3. Which values of x satisfy $\log_3 (x + 3) + \log_3 (x - 5) = 2$? *(Lesson 10-3)*

(A) only 6
(B) only -4
(C) 6 and -4
(D) no real numbers

4. Which quadrant contains no points in the solution of the system of inequalities? *(Lesson 3-3)*

$$\begin{cases} y > -2x + 1 \\ y > -\frac{1}{3}x + 1 \end{cases}$$

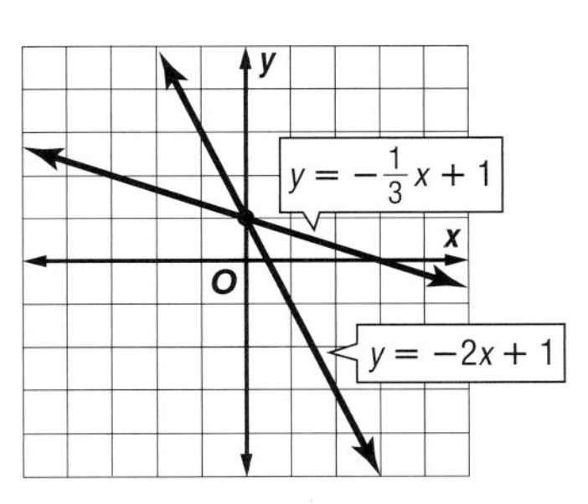

(A) Quadrant I
(B) Quadrant II
(C) Quadrant III
(D) Quadrant IV

Wednesday

5. Which choice is the best approximation for x if $\log_4 45^{10} = x$? *(Lesson 10-3)*

(A) $2 < x < 3$
(B) $20 < x < 30$
(C) $100 < x < 110$
(D) $110 < x < 120$

6. What are the values of x and y in the equation? *(Lesson 10-4)*

$$\log_6 31 = \frac{\log_{10} x}{\log_{10} y}$$

(A) $x = 6, y = 31$
(B) $x = 31, y = 25$
(C) $x = 31, y = 6$
(D) $x = 31, y = 10$

Thursday

7. What are the solutions to the quadratic equation $x^2 + 3x = -8$? *(Lesson 6-5)*

(A) $\frac{-3 \pm \sqrt{41}}{2}$
(B) $-3 \pm \sqrt{41}$
(C) $\frac{-3 \pm i\sqrt{23}}{2}$
(D) $-3 \pm i\sqrt{23}$

Friday

8. What are the coordinates of one focus of the hyperbola $\frac{y^2}{17} - \frac{x^2}{8} = 1$? *(Lesson 8-5)*

(A) (5, 0)
(B) (−5, 0)
(C) (0, 0)
(D) (0, −5)

Standards Practice Countdown

Week 19

Monday

1. It takes about 5,730 years for the Carbon-14 in an object to decay to half. If an object initially has 120 grams of Carbon-14, how much will remain after 8,000 years? *(Lesson 10-6)*
   - A about 38.0 g
   - B about 39.6 g
   - C about 45.6 g
   - D about 315.8 g

2. If $\ln 2 \approx 0.6931$, what is the approximate value of $\ln \frac{1}{2}$? *(Lesson 10-3)*
   - A -0.6931
   - B 0.1733
   - C 0.3466
   - D 1.3862

Tuesday

3. The population of California in 1990 was 29,760,021. The population in 2000 was 33,871,648. If t is the time in years after 1990, which equation represents the population of California? *(Lesson 10-6)*
   - A $y = 33{,}871{,}648e^{0.01294t}$
   - B $y = 29{,}760{,}021e^{10t}$
   - C $y = 29{,}760{,}021e^{-0.01294t}$
   - D $y = 29{,}760{,}021e^{0.01294t}$

4. A company's expenses are given by $f(x) = 12x + 1{,}000$. The company's income is given by $g(x) = 42x$. Which function represents the company's profit? (x is the number of items sold.) *(Lesson 7-7)*
   - A $h(x) = 30x - 1{,}000$
   - B $h(x) = 12x + 1{,}000$
   - C $h(x) = 42x$
   - D $h(x) = -30x + 1{,}000$

Wednesday

5. What are the values of x and y? *(Lesson 5-9)*
   $$(2x + yi)(3 - i) = -2 + 14i$$
   - A $x = 0, y = -2$
   - B $x = -1, y = 4$
   - C $x = -1, y = 14$
   - D $x = \frac{1}{2}, y = 5$

6. If $\log 3 \approx 0.4771$, what is the approximate value of $\log 81$? *(Lesson 10-3)*
   - A -1.9084
   - B 0.1193
   - C 1.9084
   - D 12.8817

Thursday

7. The graph of $y = 2(x - 1)^2 + 4$ is shown. What is true of the graph of $y = 6(x - 1)^2 + 4$ when compared to the graph shown? *(Lesson 8-2)*

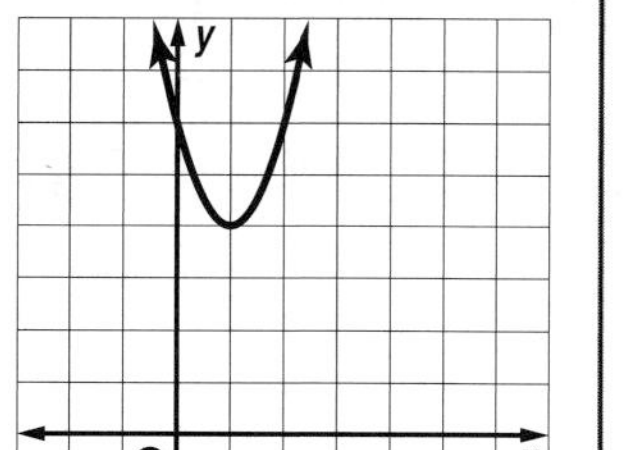

   - A It is shifted up.
   - B It is shifted down.
   - C It is wider.
   - D It is narrower.

Friday

8. Rangers release 4 beavers into a lake. If the beaver population doubles every 3 years, when will the population reach 50? *(Lesson 10-6)*
   - A after about 9 years
   - B after about 11 years
   - C after about 13 years
   - D after about 15 years

Week 20

Monday

1. What is the 18th term of the series shown below? *(Lessons 11-1 and 11-2)*

$(-8) + (-2) + 4 + \ldots$

(A) 94
(B) 102
(C) 108
(D) 110

2. What is the sum of the series $2 + 6 + 10 + \ldots + 126$? *(Lesson 11-2)*

(A) 1,800
(B) 1,922
(C) 1,984
(D) 2,048

Tuesday

3. What is the sum of the first 25 positive even integers? *(Lesson 11-2)*

(A) 300
(B) 325
(C) 600
(D) 650

4. For which values of b is the statement true? *(Lesson 10-4)*

$$\log_{11} 22 = \frac{\log_b 22}{\log_b 11}$$

(A) all real numbers
(B) all positive real numbers
(C) all positive real numbers except $b = 1$
(D) no real numbers

Wednesday

5. What is the sum of the series $1 + \left(\frac{1}{2}\right) + \left(\frac{1}{2}\right)^2 + \left(\frac{1}{2}\right)^3 + \left(\frac{1}{2}\right)^4 + \left(\frac{1}{2}\right)^5$? *(Lesson 11-4)*

(A) $\frac{15}{8}$
(B) $\frac{31}{16}$
(C) $\frac{63}{32}$
(D) $\frac{127}{64}$

6. What is the sum of the first n natural numbers? *(Lesson 11-2)*

(A) $\frac{n(n+1)}{2}$
(B) $\frac{n(n-1)}{2}$
(C) $n(n + 1)$
(D) $n(n - 1)$

Thursday

7. What is the radius of the circle with equation $x^2 + y^2 - 4x + 8y + 3 = 0$? *(Lesson 8-6)*

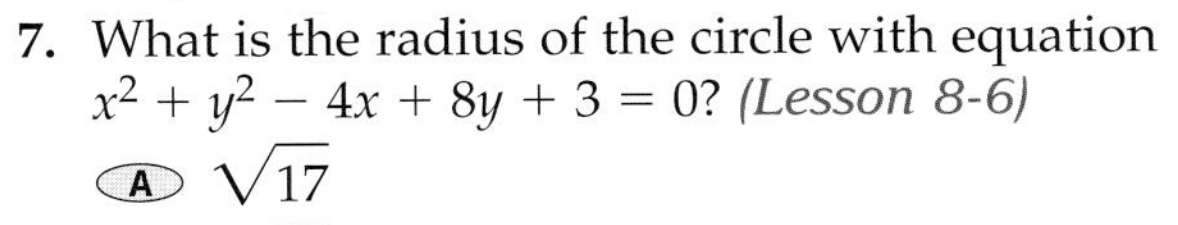

(A) $\sqrt{17}$

(B) $\sqrt{20}$
(C) 17
(D) 20

Friday

8. Which statement about the function graphed below is true? *(Lesson 7-8)*

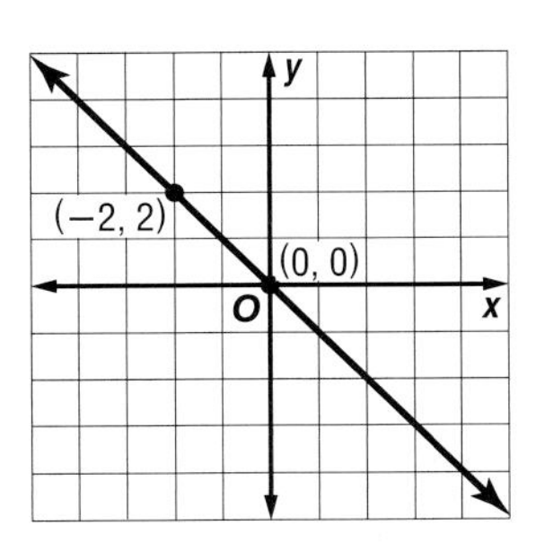

(A) The inverse function has positive slope.
(B) The function is its own inverse.
(C) The function has no inverse.
(D) The function has more than one inverse.

Standards Practice Countdown

Week 21

Monday

1. What is the fourth term in the expansion of $(2x - 3y)^6$? *(Lesson 11-7)*

A $20(8x^4)(9y^2)$
B $15(4x^2)(81y^4)$
C $15(8x^3)(27y^3)$
D $20(8x^3)(-27y^3)$

2. Which step could be used in a proof by mathematical induction to prove that a statement is true for all natural numbers? *(Lesson 11-8)*

A Assume true for $n = 1$
B Prove true for $n = 1$
C Assume false for all natural numbers to get a contradiction.
D Assume true for all natural numbers.

Tuesday

3. What is the sum of the series $1 + \left(-\frac{2}{5}\right) + \left(-\frac{2}{5}\right)^2 + \left(-\frac{2}{5}\right)^3 + \ldots$? *(Lesson 11-5)*

A $\frac{5}{7}$
B $\frac{7}{5}$
C $\frac{5}{3}$
D $\frac{3}{5}$

4. What is the 12th term of the series $1 + \left(\frac{2}{3}\right)^2 + \left(\frac{2}{3}\right)^4 + \ldots$? *(Lessons 11-3 and 11-4)*

A $\left(\frac{2}{3}\right)^9$
B $\left(\frac{2}{3}\right)^{12}$
C $\left(\frac{2}{3}\right)^{22}$
D $\left(\frac{2}{3}\right)^{24}$

Wednesday

5. Which function models population growth in which the initial population is 40 and which doubles every 10 years? Let x represent the number of years. *(Lesson 10-6)*

A $f(x) = 40e^{-0.06931x}$
B $g(x) = 40e^x$
C $h(x) = 40e^{0.06931x}$
D $k(x) = e^{40x}$

6. Which equation is true for all real numbers? *(Lessons 5-5, 5-7 and 10-3)*

A $\sqrt{x^2} = x$
B $\log x - \log y = \frac{\log x}{\log y}$
C $\sqrt[3]{x^3} = x$
D $(x^6)^{\frac{1}{2}} = x^3$

Thursday

7. What is the least common denominator for $\frac{2}{x+3} - \frac{3}{x^2-9}$? *(Lesson 9-2)*

A $x + 3$
B $(x + 3)(x - 3)$
C $(x + 3)(x^2 - 9)$
D $x^3 - 27$

Friday

8. The inverse of which function is shown below? *(Lesson 7-8)*

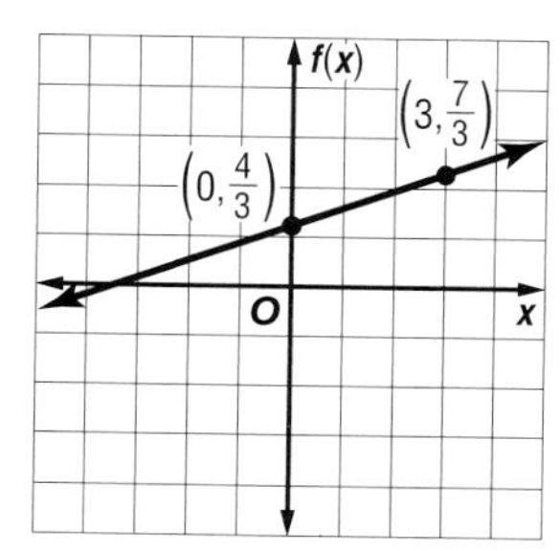

A $f(x) = 4x + 3$
B $f(x) = 3x + 4$
C $f(x) = 4x - 3$
D $f(x) = 3x - 4$

Week 22

Monday

1. When getting dressed, Miguel chooses from 2 pairs of shoes, 4 pairs of pants, and 7 shirts. How many different outfits can Miguel choose? *(Lesson 12-1)*

- (A) 13
- (B) 22
- (C) 42
- (D) 56

2. In the California Daily 3 lottery game, players choose a 3-digit number, each digit being from 0 to 9. How many different numbers can a player choose in which every digit is even? *(Lesson 12-1)*

- (A) 125
- (B) 250
- (C) 500
- (D) 1,000

Tuesday

3. In how many ways can the letters of HOLLYWOOD be arranged? *(Lesson 12-2)*

- (A) $9!$
- (B) $9! - 3! - 2!$
- (C) $(9 - 3 - 2)!$
- (D) $\frac{9!}{3!2!}$

4. How many 5-card hands dealt from a deck of 52 cards have exactly 1 club, 1 spade, 1 diamond, and 2 hearts? *(Lesson 12-2)*

- (A) $13 \cdot 12 \cdot 11 \cdot 10 \cdot 9$
- (B) $13 \cdot 13 \cdot 13 \cdot 13 \cdot 13$
- (C) $13 \cdot 13 \cdot 13 \cdot 13 \cdot 12$
- (D) $\frac{52!}{13!5!}$

Wednesday

5. What is the complete factorization of $x^6 - 64$? *(Lesson 5-4)*

- (A) $(x^3 + 8)(x^3 - 8)$
- (B) $(x + 2)(x - 2)(x^2 + 2x + 4)(x^2 - 2x + 4)$
- (C) $(x + 2)^2(x^2 + 2x + 4)^2$
- (D) $(x - 2)^2(x^2 - 2x + 4)^2$

6. How many different 3-digit combinations are there for a lock with the digits 0 through 9 on its face? *(Lesson 12-1)*

- (A) 27
- (B) 30
- (C) 729
- (D) 1,000

Thursday

7. Which is an equation of the graph shown below? *(Lesson 8-6)*

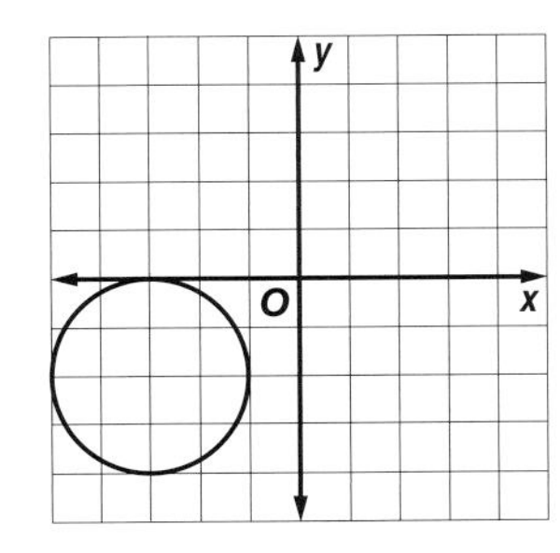

- (A) $x^2 + y^2 + 6x + 4y + 9 = 0$
- (B) $x^2 + y^2 - 6x - 4y + 9 = 0$
- (C) $x^2 - y^2 + 6x + 4y + 9 = 0$
- (D) $x^2 + 4y^2 - 6x - 16y + 9 = 0$

Friday

8. What is the sum of the first 25 positive odd integers? *(Lesson 11-2)*

- (A) 600
- (B) 625
- (C) 650
- (D) 675

Standards Practice Countdown

Week 23

Monday

1. Which inequality is graphed below? *(Lesson 2-7)*

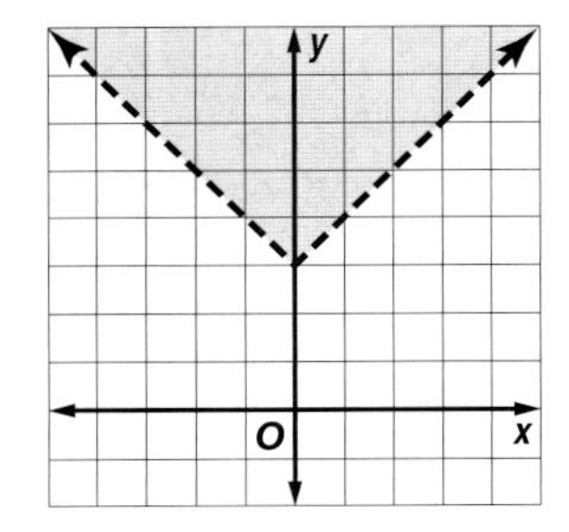

(A) $y > |x| + 3$
(B) $y < |x| + 3$
(C) $y \leq |x| + 3$
(D) $y > |x + 3|$

2. What is the 16th term of the series $2 + 1 + \frac{1}{2} + \ldots$? *(Lessons 11-3 and 11-4)*

(A) $\left(\frac{1}{2}\right)^{14}$
(B) $\left(\frac{1}{2}\right)^{15}$
(C) $\left(\frac{1}{2}\right)^{16}$
(D) $\left(\frac{1}{2}\right)^{17}$

Tuesday

3. To help identify skateboards if they are stolen, a manufacturer inscribes serial numbers on the bottom. If a serial number consists of 4 digits followed by 3 letters, how many skateboards have only even digits and vowels in the serial number? *(Lesson 12-1)*

(A) $7 \cdot 5$
(B) 7^5
(C) 5^7
(D) $4^4 \cdot 5^3$

4. To solve the system

$$\begin{cases} 3x - 5y = 12 \\ 2x + y = 21 \end{cases}$$

using elimination, which steps could be used? *(Lesson 4-6)*

(A) Add the given equations.
(B) Subtract the given equations.
(C) Multiply 2nd equation by 5 and subtract from 1st equation.
(D) Multiply 2nd equation by 5 and add to 1st equation.

Wednesday

5. How many terms are there in the expansion of $(3x + 4y)^{12}$? *(Lesson 11-7)*

(A) 11
(B) 12
(C) 13
(D) 14

6. Which point is a solution of the system of equations? *(Lesson 3-5)*

$$\begin{cases} x + z = 1 \\ 2x + 3y = -2 \\ 5y - 2z = -4 \end{cases}$$

(A) (1, 0, 0)
(B) (−1, 0, 2)
(C) (2, −2, −3)
(D) (2, 0, −1)

Thursday

7. What are the solutions of $2|x + 3| + 1 = 7$? *(Lesson 1-4)*

(A) 1 and −7
(B) 0 only
(C) −6 only
(D) 0 and −6

Friday

8. For which values of x is $\frac{x^2 + 5x + 6}{x^2 - 5x + 6}$ undefined? *(Lesson 9-1)*

(A) −2 and −3
(B) 0 and −6
(C) 2 and 3
(D) 0 and 6

Standards Practice Countdown

Week 24

Monday

1. The graph of which function is shown below? *(Lesson 6-1)*

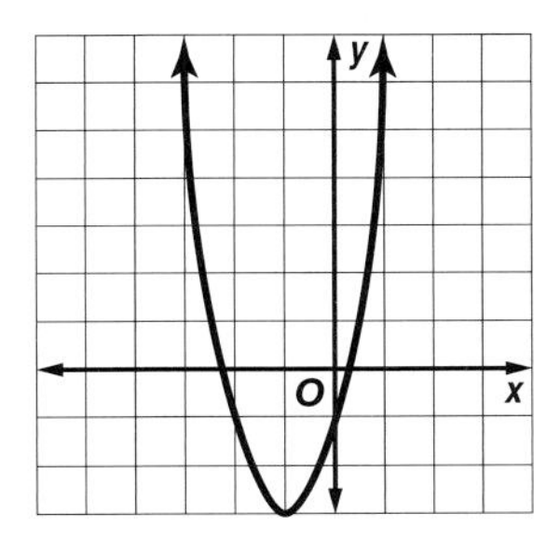

(A) $y = -2(x + 1)^2 + 3$
(B) $y = -2(x + 1)^2 - 3$
(C) $y = 2(x + 1)^2 + 3$
(D) $y = 2(x + 1)^2 - 3$

2. What are the coordinates of the center of the circle with equation $9x^2 + 9y^2 - 12x + 18y - 185 = 0$? *(Lesson 8-6)*

(A) $\left(\frac{2}{3}, 1\right)$
(B) $\left(\frac{2}{3}, -1\right)$
(C) $\left(1, -\frac{2}{3}\right)$
(D) $\left(-1, -\frac{2}{3}\right)$

Tuesday

3. To win the California Super Lotto Plus, a player must correctly choose 5 numbers from 1 to 47 and choose the correct MEGA number from 1 to 27. What is the probability of winning if a player chooses 1 group of numbers? *(Lesson 12-4)*

(A) $\frac{1}{C(47, 5)}$
(B) $C(47, 5) \cdot 27$
(C) $\frac{1}{C(47, 5)} \cdot \frac{1}{27}$
(D) $\frac{1}{C(74, 6)}$

4. What is the value of $\log 10^{40} - 37 \log 10 + \log 100^{-1}$? *(Lesson 10-3)*

(A) 1
(B) 0
(C) $\log(40 - 37 + 1)$
(D) 10

Wednesday

5. What is the product? *(Lesson 5-9)*
$(1 + 2i)(-4 + 3i)$

(A) $2 + 5i$
(B) $-10 - 5i$
(C) $-10 + 5i$
(D) $2 - 5i$

6. What is the sum of the natural numbers from 11 to 25? $\left(1 + 2 + \ldots + n = \frac{n(n + 1)}{2}.\right)$ *(Lesson 11-2)*

(A) 66
(B) 259
(C) 270
(D) 325

Thursday

7. What is the least common denominator for $\frac{1}{a^3b^2c} + \frac{2}{abc} + \frac{1}{a^2bc^3}$? *(Lesson 9-2)*

(A) abc
(B) ab^2c
(C) $a^6b^5c^5$
(D) $a^3b^2c^3$

Friday

8. A company that builds desks spends \$29 for parts and \$37 for labor for each desk that it produces. The company sells the desks for \$129. Which function represents the company's profit? *(Lesson 7-7)*

(A) $P(x) = 63$
(B) $P(x) = 63x$
(C) $P(x) = 66x$
(D) $P(x) = 129x$

Week 25

Monday

1. One type of standard California license plate uses 3 digits followed by 3 letters. The first and third letter can be any letter except I, O, or Q. How many of these plates can be made? *(Lesson 12-1)*

   A $9^3 \cdot 23^2 \cdot 26$ B $9^3 \cdot 26^3$
   C $10^3 \cdot 23^2 \cdot 26$ D $10^3 \cdot 26^3$

2. Using a standard deck of 52 cards, what is the probability of choosing a heart and then choosing a diamond if the first card is not replaced? *(Lesson 12-3)*

   A $\frac{1}{2{,}704}$ B $\frac{1}{2{,}652}$
   C $\frac{169}{2{,}704}$ D $\frac{13}{204}$

Tuesday

3. If two dice are rolled, what is the probability that the sum of the numbers facing up will be at least 9? *(Lesson 12-5)*

   A $\frac{5}{18}$ B $\frac{1}{6}$
   C $\frac{13}{18}$ D $\frac{4}{11}$

4. How many ways can 5 girls and 5 boys form a line if a girl must be first in line and boys and girls must alternate? *(Lesson 12-2)*

   A 5!
   B 10!
   C $(5!)^2$
   D $(10!)^2$

Wednesday

5. Each of the 8 multiple choice questions on a quiz has 4 answer choices. What is the probability of guessing correctly on exactly 6 of the questions? *(Lesson 12-8)*

   A About 0.4%
   B About 1%
   C About 4%
   D About 40%

6. Which whole number is closest to x if $\log_2 91 = x$? *(Lesson 10-2)*

   A 3
   B 5
   C 7
   D 9

Thursday

7. What is the coefficient of x^2y^3 in the expansion of $(3x - y)^5$? *(Lesson 11-7)*

   A −270
   B −90
   C 90
   D 270

Friday

8. The graph of $y = 3(x - 4)^2$ is shown. How does the graph of $y = 3(x - 1)^2$ compare to this graph? *(Lesson 6-6)*

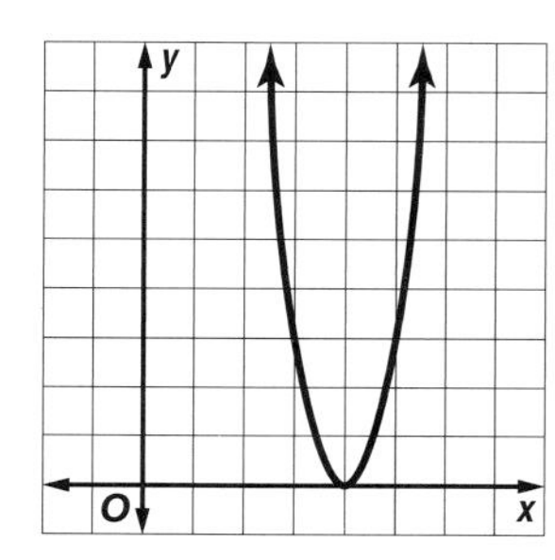

   A It is 3 units to the left.
   B It is 3 units to the right.
   C It is 3 units higher.
   D It is 3 units lower.

GLENCOE MATHEMATICS

California Edition

Algebra 2

Holliday

Cuevas

Moore-Harris

Carter

Marks

Casey

Day

Hayek

New York, New York
Columbus, Ohio
Chicago, Illinois
Peoria, Illinois
Woodland Hills, California

The McGraw-Hill Companies

Send all inquiries to:
Glencoe/McGraw-Hill
8787 Orion Place
Columbus, OH 43240

ISBN: 0-07-865980-9

2 3 4 5 6 7 8 9 10 055/071 13 12 11 10 09 08 07 06 05 04

Contents in Brief

Authors

Berchie Holliday, Ed.D.
Former Mathematics Teacher
Northwest Local
School District
Cincinnati, OH

Gilbert J. Cuevas, Ph.D.
Professor of Mathematics
Education
University of Miami
Miami, FL

Beatrice Moore-Harris
Educational Specialist
Bureau of Education
and Research
League City, TX

John A. Carter
Director of Mathematics
Adlai E. Stevenson
High School
Lincolnshire, IL

Consulting Author

Carol Malloy, Ph.D.
Associate Professor, Curriculum Instruction,
Secondary Mathematics
The University of North Carolina at Chapel Hill
Chapel Hill, NC

Authors

Daniel Marks, Ed.D.
Associate Professor of Mathematics
Auburn University at Montgomery
Montgomery, AL

Ruth M. Casey
Mathematics Teacher Department Chair
Anderson County High School
Lawrenceburg, KY

Roger Day, Ph.D.
Associate Professor of Mathematics
Illinois State University
Normal, IL

Linda M. Hayek
Mathematics Teacher
Ralston Public Schools
Omaha, NE

Contributing Authors

USA TODAY
The USA TODAY Snapshots®, created by USA TODAY®, help students make the connection between real life and mathematics.

Dinah Zike
Educational Consultant
Dinah-Might Activities, Inc.
San Antonio, TX

Content Consultants

Each of the Content Consultants reviewed every chapter and gave suggestions for improving the effectiveness of the mathematics instruction.

Mathematics Consultants

Gunnar E. Carlsson, Ph.D.
Consulting Author
Professor of Mathematics
Stanford University
Stanford, CA

Ralph L. Cohen, Ph.D.
Consulting Author
Professor of Mathematics
Stanford University
Stanford, CA

Alan G. Foster
Former Mathematics Teacher & Department Chairperson
Addison Trail High School
Addison, IL

Les Winters
Instructor
California State University Northridge
Northridge, CA

William Collins
Director, The Sisyphus Math Learning Center
East Side Union High School District
San Jose, CA

Dora Swart
Mathematics Teacher
W.F. West High School
Chehalis, WA

David S. Daniels
Former Mathematics Chair
Longmeadow High School
Longmeadow, MA

Mary C. Enderson, Ph.D.
Associate Professor of Mathematics
Middle Tennessee State University
Murfreesboro, TN

Gerald A. Haber
Consultant, Mathematics Standards and Professional Development
New York, NY

C. Vincent Pané, Ed.D.
Associate Professor of Education/ Coordinator of Secondary & Special Subjects Education
Molloy College
Rockville Centre, NY

Reading Consultant

Lynn T. Havens
Director
Project CRISS
Kalispell, MT

ELL Consultant

Idania Dorta
Mathematics Educational Specialist
Miami-Dade County Public Schools
Miami, FL

Teacher Reviewers

Each Teacher Reviewer reviewed at least two chapters of the Student Edition, giving feedback and suggestions for improving the effectiveness of the mathematics instruction.

Yvonne Adonai
Assistant Principal, Mathematics
Middle College at Medgar Evers College
Brooklyn, NY

Ann Rushing Allred
Secondary Mathematics Coordinator
Bossier Parish Schools
Bossier City, LA

Thomas J. Altonjy
Mathematics Supervisor
Montville Township Public Schools
Montville, NJ

Susan J. Barr
Department Chair/Teacher
Dublin Coffman High School
Dublin, OH

Douglas W. Becker
Math Dept Chair/Senior Math Teacher
Gaylord High School
Gaylord, MI

Dr. Edward A. Brotak
Professor, Atmospheric Sciences
UNC Asheville
Asheville, NC

Teacher Reviewers

Sonya Smith Bryant
Mathematics Teacher
Booker T. Washington High School
Shreveport, LA

Judy Buchholtz
Math Department Chair/Teacher
Dublin Scioto High School
Dublin, OH

A. G. Chase
Mathematics Teacher
Evergreen High School
Vancouver, WA

Natalie Dillinger
Mathematics Teacher
Hurricane High School
Hurricane, WV

John M. Dunford, Jr.
Chairman Mathematics
Tuba City High School
Tuba City, AZ

Diana Flick
Mathematics Teacher
Harrisonburg High School
Harrisonburg, VA

Susan Hammer
Mathematics Department Head
Gaither High School
Tampa, FL

Deborah L. Hewitt
Mathematics Teacher
Chester High School
Chester, NY

Kristen L. Karbon
Mathematics Teacher
Troy High School
Troy, MI

William Leschensky
Former Mathematics Teacher
Glenbard South High School
College of DuPage
Glen Ellyn, IL

Patricia Lund
Mathematics Teacher
Divide County High School
Crosby, ND

Wallace J. Mack
Mathematics Department Chairperson
Ben Davis High School
Indianapolis, IN

T. E. Madre
Mathematics Department Chairperson
North Mecklenburg High School
Huntersville, NC

Marilyn Martau
Mathematics Teacher (Retired)
Lakewood High School
Lakewood, OH

Ron Millard
Mathematics Department Chair
Shawnee Mission South High School
Overland Park, KS

Rebecca D. Morrisey
Assistant Principal
Leavenworth High School
Leavenworth, KS

Constance D. Mosakowsky
Mathematics Teacher
Minnie Howard School
Alexandria, VA

Anne Newcomb
Mathematics Department Chairperson
Celina High School
Celina, OH

Barbara Nunn
Secondary Mathematics Curriculum Specialist
Broward County Schools
Ft. Lauderdale, FL

Shannon Collins Pan
Department of Mathematics
Waverly High School
Waverly, NY

Aletha T. Paskett
Mathematics Teacher
Indian Hills Middle School
Sandy, UT

Holly K. Plunkett
Mathematics Teacher
University High School
Morgantown, WV

Thomas M. Pond, Jr.
Mathematics Teacher
Matoaca High School
Chesterfield County Public Schools, VA

Debra K. Prowse
Mathematics Teacher
Beloit Memorial High School
Beloit, WI

B. J. Rasberry
Teacher
John T. Hoggard High School
Wilmington, NC

Harry Rattien
A.P. Supervisor (Math)
Townsend Harris High School at QC
Flushing, NY

Becky Reed
Teacher
John F. Kennedy High School
Mt. Angel, OR

Steve Sachs
Mathematics Department Chairperson
Lawrence North High School
Indianapolis, IN

Sue W. Sams
Mathematics Teacher/Department Chair
Providence High School
Charlotte, NC

Calvin Stuhmer
Mathematics Teacher
Sutton Public Schools
Sutton, NE

Ruth Stutzman
Math Department Chair & Teacher
Jefferson Forest High School
Forest, VA

Patricia Taepke
Mathematics Teacher and BTSA Trainer
South Hills High School
West Covina, CA

Christine Waddell
Mathematics Department Chair/Teacher
Albion Middle School
Sandy, UT

Gail Watson
Mathematics Teacher
Pineville High School
Pineville, LA

Linda E. Westbrook
Mathematics Department Chair
George Jenkins High School
Lakeland, FL

Cottina Woods
Lane Technical High School
Chicago, IL

Warren Zarrell
Mathematics Department Chairman
James Monroe High School
North Hills, CA

Teacher Advisory Board

Glencoe/McGraw-Hill wishes to thank the following teachers for their feedback on Glencoe *Algebra*. They were instrumental in providing valuable input toward the development of this program.

Mary Jo Ahler
Mathematics Teacher
Davis Drive Middle School
Apex, NC

David Armstrong
Mathematics Facilitator
Huntington Beach Union High School District
Huntington Beach, CA

Berta Guillen
Mathematics Department Chairperson
Barbara Goleman Sr. High School
Miami, FL

Bonnie Johnston
Academically Gifted Program Coordinator
Valley Springs Middle School
Arden, NC

JoAnn Lopykinski
Mathematics Teacher
Lincoln Way East High School
Frankfort, IL

David Lorkiewicz
Mathematics Teacher
Lockport High School
Lockport, IL

Norma Molina
Ninth Grade Success Initiative Campus Coordinator
Holmes High School
San Antonio, TX

Sarah Morrison
Mathematics Department Chairperson
Northwest Cabarrus High School
Concord, NC

Raylene Paustian
Mathematics Curriculum Coordinator
Clovis Unified School District
Clovis, CA

Tom Reardon
Mathematics Department Chairperson
Austintown Fitch High School
Youngstown, OH

Guy Roy
Mathematics Coordinator
Plymouth Public Schools
Plymouth, MA

Jenny Weir
Mathematics Department Chairperson
Felix Verela Sr. High School
Miami, FL

UNIT 1

First-Degree Equations and Inequalities 2

Chapter 1 Solving Equations and Inequalities 4

WebQuest Internet Project

Prerequisite Skills

FOLDABLES™

Reading and Writing Mathematics

Standardized Test Practice

Lesson 1-4, page 31

Chapter 2 Linear Relations and Functions 54

Lesson 2-2, page 64

Chapter 3 Systems of Equations and Inequalities

Lesson 3-4, page 131

Chapter 4 Matrices 152

Prerequisite Skills

FOLDABLES™

Reading and Writing Mathematics

Standardized Test Practice

Lesson 4-6, page 193

Polynomial and Radical Equations and Inequalities 218

Chapter 5 Polynomials 220

WebQuest Internet Project

Lesson 5-7, page 259

Prerequisite Skills

FOLDABLES™ Study Organizer

Reading and Writing Mathematics

Standardized Test Practice

Snapshots

Chapter 6 Quadratic Functions and Inequalities 284

Lesson 6-4, page 311

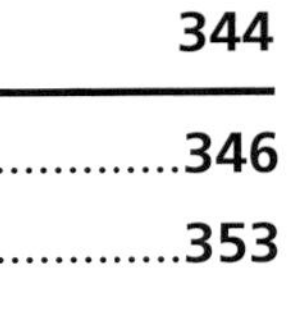

Chapter 7 Polynomial Functions 344

Lesson 7-1, page 346

UNIT 3

Advanced Functions and Relations

Chapter 8 Conic Sections

Lesson 8-4, page 435

Chapter 9 Rational Expressions and Equations 470

Lesson 9-5, page 503

Chapter 10 Exponential and Logarithmic Relations 520

Lesson 10-6, page 564

Discrete Mathematics 574

Chapter 11 Sequences and Series 576

WebQuest Internet Project

Prerequisite Skills

FOLDABLES™

Reading and Writing Mathematics

Standardized Test Practice

Lesson 11-5, page 603

Chapter 12 Probability and Statistics 630

Lesson 12-3, page 648

UNIT 5

Trigonometry 696

Chapter 13 Trigonometric Functions 698

WebQuest Internet Project

Prerequisite Skills

FOLDABLES™

Reading and Writing Mathematics

Standardized Test Practice

Lesson 13-6, page 744

Chapter 14 Trigonometric Graphs and Identities 760

Lesson 14-7, page 803

Student Handbook

Skills

Reference

Prerequisite Skills

Study Organizer 761

Reading and Writing Mathematics

Standardized Test Practice

Snapshots 797

One-Stop Internet Resources

Need extra help or information? Log on to math.glencoe.com or any of the Web addresses below to learn more.

Online Study Tools

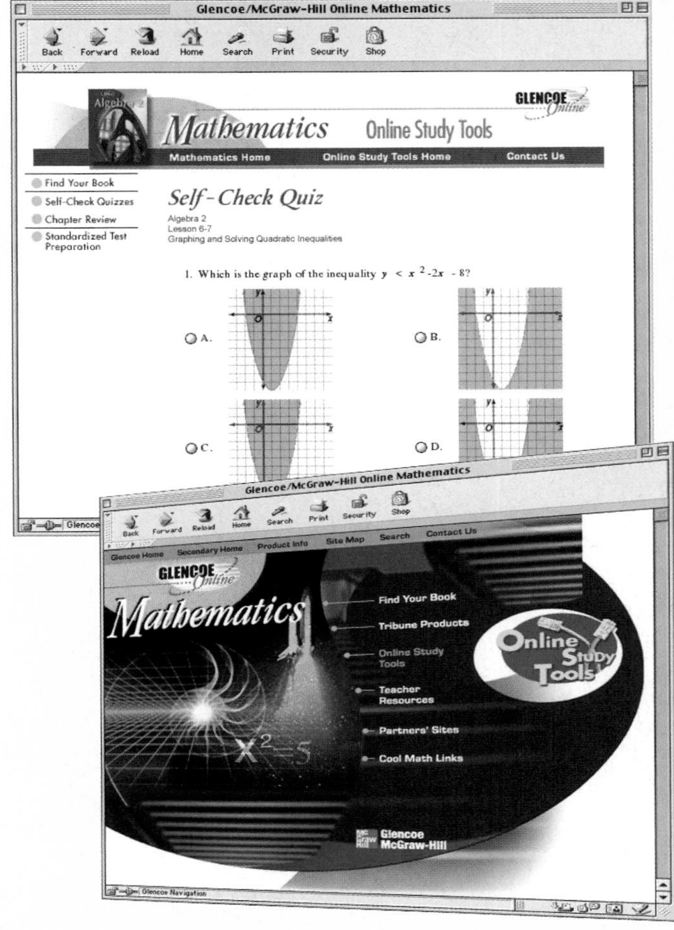

- www.algebra2.com/extra_examples shows you additional worked-out examples that mimic the ones in your book.
- www.algebra2.com/self_check_quiz provides you with a practice quiz for each lesson that grades itself.
- www.algebra2.com/vocabulary_review lets you check your understanding of the terms and definitions used in each chapter.
- www.algebra2.com/chapter_test allows you to take a self-checking test before the actual test.
- www.algebra2.com/standardized_test is another way to brush up on your standardized test-taking skills.

Research Options

- www.algebra2.com/webquest walks you step-by-step through a long-term project using the Web. One WebQuest for each unit is explored using the mathematics from that unit.
- www.algebra2.com/usa_today provides activities related to the concept of the lesson as well as up-to-date Snapshot data.
- www.algebra2.com/careers links you to additional information about interesting careers.
- www.algebra2.com/data_update links you to the most current data available for subjects such as basketball and family.

Calculator Help

- www.algebra2.com/other_calculator_keystrokes provides you with keystrokes other than the TI-83 Plus used in your textbook.

UNIT 1

First-Degree Equations and Inequalities

You can model and analyze real-world situations by using algebra. In this unit, you will solve and graph linear equations and inequalities and use matrices.

Lessons in Home Buying, Selling

Source: *USA TODAY,* November 18, 1999

"'Buying a home,' says Housing and Urban Development Secretary Andrew Cuomo, 'is the most expensive, most complicated and most intimidating financial transaction most Americans ever make.'" In this project, you will be exploring how functions and equations relate to buying a home and your income.

Log on to www.algebra2.com/webquest. Begin your WebQuest by reading the Task.

Then continue working on your WebQuest as you study Unit 1.

Lesson	1-3	2-5	3-2	4-6
Page	27	84	120	192

USA TODAY

USA TODAY Snapshots®

Household spending

The average household spent $35,535 in 1998, the most recent data available. A household averages 2.5 people. Expenditures:

Housing	$11,713
Transportation	$6,616
Food	$4,810
Insurance/pensions	$3,381
Health care	$1,903
Entertainment	$1,746
Apparel	$1,674
Other	$3,692

Source: Bureau of Labor Statistics Consumer Expenditure Survey

By Mark Pearson and Marcy E. Mullins, USA TODAY

ıapter 1 Solving Equations and Inequalities

What You'll Learn

- **Lesson 1-1** Simplify and evaluate algebraic expressions.
- **Lesson 1-2** Classify and use the properties of real numbers.
- **Lesson 1-3** Solve equations.
- **Lesson 1-4** Solve absolute value equations.
- **Lessons 1-5 and 1-6** Solve and graph inequalities.

Key Vocabulary

- order of operations (p. 6)
- algebraic expression (p. 7)
- Distributive Property (p. 12)
- equation (p. 20)
- absolute value (p. 28)

Why It's Important

Algebra allows you to write expressions, equations, and inequalities that hold true for most or all values of variables. Because of this, algebra is an important tool for describing relationships among quantities in the real world. For example, the angle at which you view fireworks and the time it takes you to hear the sound are related to the width of the fireworks burst. A change in one of the quantities will cause one or both of the other quantities to change.

In Lesson 1-1, you will use the formula that relates these quantities.

Getting Started

Prerequisite Skills To be successful in this chapter, you'll need to master these skills and be able to apply them in problem-solving situations. Review these skills before beginning Chapter 1.

For Lessons 1-1 through 1-3 — Operations with Rational Numbers

Simplify.

1. $20 - 0.16$

2. $12.2 + (-8.45)$

3. $-3.01 - 14.5$

4. $-1.8 + 17$

5. $\frac{1}{4} - \frac{2}{3}$

6. $\frac{3}{5} + (-6)$

7. $-7\frac{1}{2} + 5\frac{1}{3}$

8. $-11\frac{5}{8} - \left(-4\frac{3}{7}\right)$

9. $(0.15)(3.2)$

10. $2 \div (-0.4)$

11. $(-1.21) \div (-1.1)$

12. $(-9)(0.036)$

13. $-4 \div \frac{3}{2}$

14. $\left(\frac{5}{4}\right)\left(-\frac{3}{10}\right)$

15. $\left(-2\frac{3}{4}\right)\left(-3\frac{1}{5}\right)$

16. $7\frac{1}{8} \div (-2)$

For Lesson 1-1 — Powers

Evaluate each power.

17. 2^3

18. 5^3

19. $(-7)^2$

20. $(-1)^3$

21. $(-0.8)^2$

22. $-(1.2)^2$

23. $\left(\frac{2}{3}\right)^2$

24. $\left(-\frac{4}{11}\right)^2$

For Lesson 1-5 — Compare Real Numbers

Identify each statement as *true* or *false*.

25. $-5 < -7$

26. $6 > -8$

27. $-2 \geq -2$

28. $-3 \geq -3.01$

29. $-9.02 < -9.2$

30. $\frac{1}{5} < \frac{1}{8}$

31. $\frac{2}{5} \geq \frac{16}{40}$

32. $\frac{3}{4} > 0.8$

Relations and Functions Make this Foldable to help you organize your notes. Begin with one sheet of notebook paper.

Step 1 **Fold**

Fold lengthwise to the holes.

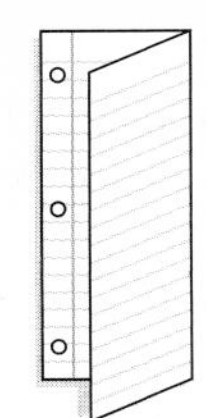

Step 2 **Open and Label**

Open and label the columns as shown.

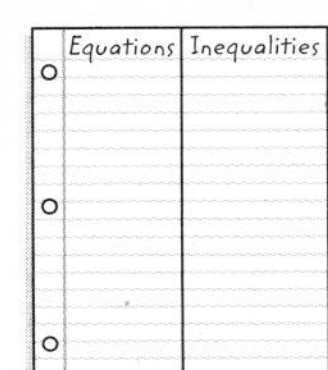

Reading and Writing As you read and study the chapter, write notes, examples, and graphs in each column.

1-1 Expressions and Formulas

California Standards Standard 7.0 **Students** add, subtract, multiply, divide, reduce, and **evaluate rational expressions with monomial and polynomial denominators** and simplify complicated rational expressions, including those with negative exponents in the denominator. (Key)

What You'll Learn

- Use the order of operations to evaluate expressions.
- Use formulas.

How are formulas used by nurses?

Vocabulary

- order of operations
- variable
- algebraic expression
- formula

Nurses setting up intravenous or IV fluids must control the flow rate F, in drops per minute. They use the formula $F = \frac{V \times d}{t}$, where V is the volume of the solution in milliliters, d is the drop factor in drops per milliliter, and t is the time in minutes.

Suppose a doctor orders 1500 milliliters of IV saline to be given over 12 hours. Using a drop factor of 15 drops per milliliter, the expression $\frac{1500 \times 15}{12 \times 60}$ gives the correct flow rate for this patient's IV.

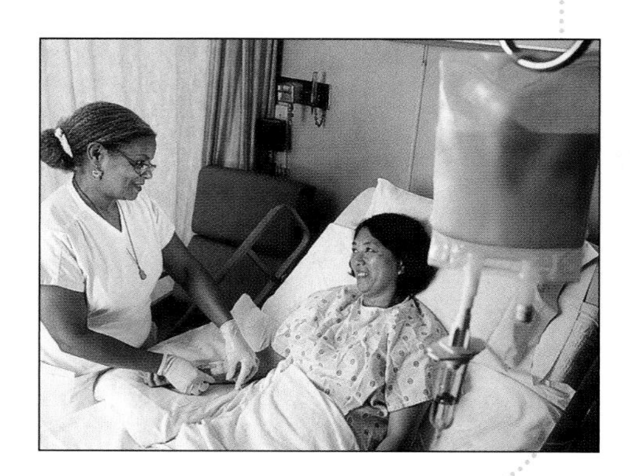

ORDER OF OPERATIONS A numerical expression such as $\frac{1500 \times 15}{12 \times 60}$ must have exactly one value. In order to find that value, you must follow the **order of operations**.

Key Concept — Order of Operations

Step 1 Evaluate expressions inside grouping symbols, such as parentheses, ( ), brackets, [ ], braces, { }, and fraction bars, as in $\frac{5+7}{2}$.

Step 2 Evaluate all powers.

Step 3 Do all multiplications and/or divisions from left to right.

Step 4 Do all additions and/or subtractions from left to right.

Grouping symbols can be used to change or clarify the order of operations. When calculating the value of an expression, begin with the innermost set of grouping symbols.

Example 1 Simplify an Expression

Find the value of $[2(10 - 4)^2 + 3] \div 5$.

$[2(10 - 4)^2 + 3] \div 5 = [2(6)^2 + 3] \div 5$ First subtract 4 from 10.

$= [2(36) + 3] \div 5$ Then square 6.

$= (72 + 3) \div 5$ Multiply 36 by 2.

$= 75 \div 5$ Add 72 and 3.

$= 15$ Finally, divide 75 by 5.

The value is 15.

Scientific calculators follow the order of operations.

Graphing Calculator Investigation

Order of Operations

Think and Discuss

1. Simplify $8 - 2 \times 4 + 5$ using a graphing calculator.
2. Describe the procedure the calculator used to get the answer.
3. Where should parentheses be inserted in $8 - 2 \times 4 + 5$ so that the expression has each of the following values?
   a. -10 **b.** 29 **c.** -5
4. Evaluate $18^2 \div (2 \times 3)$ using your calculator. Explain how the answer was calculated.
5. If you remove the parentheses in Exercise 4, would the solution remain the same? Explain.

Variables are symbols, usually letters, used to represent unknown quantities. Expressions that contain at least one variable are called **algebraic expressions**. You can evaluate an algebraic expression by replacing each variable with a number and then applying the order of operations.

Study Tip

Common Misconception
A common error in this type of problem is to subtract before multiplying.

$64 - 1.5(9.5) \neq 62.5(9.5)$

Remember to follow the order of operations.

Example 2 Evaluate an Expression

Evaluate $x^2 - y(x + y)$ if $x = 8$ and $y = 1.5$.

$x^2 - y(x + y) = 8^2 - 1.5(8 + 1.5)$	Replace x with 8 and y with 1.5.
$= 8^2 - 1.5(9.5)$	Add 8 and 1.5.
$= 64 - 1.5(9.5)$	Find 8^2.
$= 64 - 14.25$	Multiply 1.5 and 9.5.
$= 49.75$	Subtract 14.25 from 64.

The value is 49.75.

Example 3 Expression Containing a Fraction Bar

Evaluate $\frac{a^3 + 2bc}{c^2 - 5}$ if $a = 2$, $b = -4$, and $c = -3$.

The fraction bar acts as both an operation symbol, indicating division, and as a grouping symbol. Evaluate the expressions in the numerator and denominator separately before dividing.

$\frac{a^3 + 2bc}{c^2 - 5} = \frac{2^3 + 2(-4)(-3)}{(-3)^2 - 5}$	$a = 2$, $b = -4$, and $c = -3$
$= \frac{8 + (-8)(-3)}{9 - 5}$	Evaluate the numerator and the denominator separately.
$= \frac{8 + 24}{9 - 5}$	Multiply -8 by -3.
$= \frac{32}{4}$ or 8	Simplify the numerator and the denominator. Then divide.

The value is 8.

FORMULAS A **formula** is a mathematical sentence that expresses the relationship between certain quantities. If you know the value of every variable in the formula except one, you can find the value of the remaining variable.

Example 4 Use a Formula

GEOMETRY The formula for the area A of a trapezoid is $A = \frac{1}{2}h(b_1 + b_2)$, where h represents the height, and b_1 and b_2 represent the measures of the bases. Find the area of the trapezoid shown below.

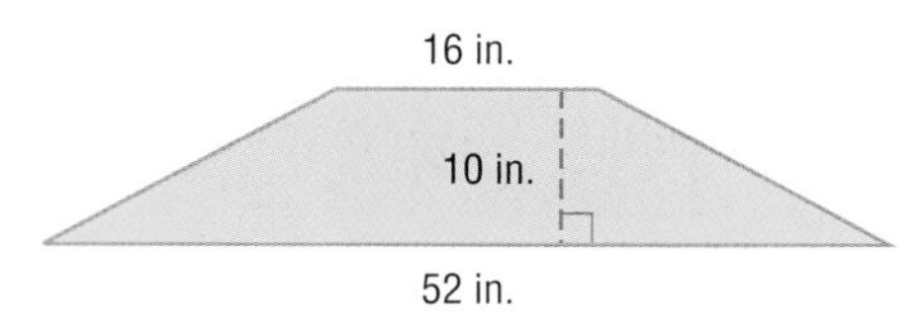

Substitute each value given into the formula. Then evaluate the expression using the order of operations.

$A = \frac{1}{2}h(b_1 + b_2)$	Area of a trapezoid
$= \frac{1}{2}(10)(16 + 52)$	Replace h with 10, b_1 with 16, and b_2 with 52.
$= \frac{1}{2}(10)(68)$	Add 16 and 52.
$= 5(68)$	Divide 10 by 2.
$= 340$	Multiply 5 by 68.

The area of the trapezoid is 340 square inches.

Check for Understanding

Concept Check

1. **Describe** how you would evaluate the expression $a + b[(c + d) \div e]$ given values for a, b, c, d, and e.
2. **OPEN ENDED** Give an example of an expression where subtraction is performed before division and the symbols ( ), [ ], or { } are not used.
3. **Determine** which expression below represents the amount of change someone would receive from a $50 bill if they purchased 2 children's tickets at $4.25 each and 3 adult tickets at $7 each at a movie theater. Explain.

   a. $50 - 2 \times 4.25 + 3 \times 7$ **b.** $50 - (2 \times 4.25 + 3 \times 7)$
   c. $(50 - 2 \times 4.25) + 3 \times 7$ **d.** $50 - (2 \times 4.25) + (3 \times 7)$

Guided Practice

Find the value of each expression.

4. $8(3 + 6)$
5. $10 - 8 \div 2$
6. $14 \cdot 2 - 5$
7. $[9 + 3(5 - 7)] \div 3$
8. $[6 - (12 - 8)^2] \div 5$
9. $\frac{17(2 + 26)}{4}$

Evaluate each expression if $x = 4$, $y = -2$, and $z = 6$.

10. $z - x + y$
11. $x + (y - 1)^3$
12. $x + [3(y + z) - y]$

Application **BANKING For Exercises 13–15, use the following information.**
Simple interest is calculated using the formula $I = prt$, where p represents the principal in dollars, r represents the annual interest rate, and t represents the time in years. Find the simple interest I given each of the following values.

13. $p = \$1800$, $r = 6\%$, $t = 4$ years

14. $p = \$5000$, $r = 3.75\%$, $t = 10$ years

15. $p = \$31{,}000$, $r = 2\frac{1}{2}\%$, $t = 18$ months

Practice and Apply

Homework Help

For Exercises	See Examples
16–37	1, 3
38–50	2, 3
51–54	4

Extra Practice See page 828.

Find the value of each expression.

16. $18 + 6 \div 3$

17. $7 - 20 \div 5$

18. $3(8 + 3) - 4$

19. $(6 + 7)2 - 1$

20. $2(6^2 - 9)$

21. $-2(3^2 + 8)$

22. $2 + 8(5) \div 2 - 3$

23. $4 + 64 \div (8 \times 4) \div 2$

24. $[38 - (8 - 3)] \div 3$

25. $10 - [5 + 9(4)]$

26. $1 - \{30 \div [7 + 3(-4)]\}$

27. $12 + \{10 \div [11 - 3(2)]\}$

28. $\frac{1}{3}(4 - 7^2)$

29. $\frac{1}{2}[9 + 5(-3)]$

30. $\frac{16(9 - 22)}{4}$

31. $\frac{45(4 + 32)}{10}$

32. $0.3(1.5 + 24) \div 0.5$

33. $1.6(0.7 + 3.3) \div 2.5$

34. $\frac{1}{5} - \frac{20(81 \div 9)}{25}$

35. $\frac{12(52 \div 2^2)}{6} - \frac{2}{3}$

36. BICYCLING The amount of pollutants saved by riding a bicycle rather than driving a car is calculated by adding the organic gases, carbon monoxide, and nitrous oxides emitted. To find the pounds of pollutants created by starting a typical car 10 times and driving it for 50 miles, find the value of the expression $\frac{(52.84 \times 10) + (5.955 \times 50)}{454}$.

37. NURSING Determine the IV flow rate for the patient described at the beginning of the lesson by finding the value of $\frac{1500 \times 15}{12 \times 60}$.

Evaluate each expression if $w = 6$, $x = 0.4$, $y = \frac{1}{2}$, and $z = -3$.

38. $w + x + z$

39. $w + 12 \div z$

40. $w(8 - y)$

41. $z(x + 1)$

42. $w - 3x + y$

43. $5x + 2z$

44. $z^4 - w$

45. $(5 - w)^2 + x$

46. $\frac{5wx}{z}$

47. $\frac{2z - 15x}{3y}$

48. $(x - y)^2 - 2wz$

49. $\frac{1}{y} + \frac{1}{w}$

50. GEOMETRY The formula for the area A of a circle with diameter d is $A = \pi\left(\frac{d}{2}\right)^2$. Write an expression to represent the area of the circle.

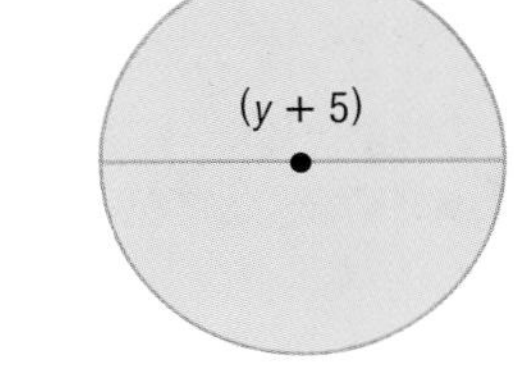

51. Find the value of ab^n if $n = 3$, $a = 2000$, and $b = -\frac{1}{5}$.

52. **MEDICINE** Suppose a patient must take a blood pressure medication that is dispensed in 125-milligram tablets. The dosage is 15 milligrams per kilogram of body weight and is given every 8 hours. If the patient weighs 25 kilograms, how many tablets would be needed for a 30-day supply? Use the formula $n = 24d \div [8(b \times 15 \div 125)]$, where n is the number of tablets, d is the number of days the supply should last, and b is the body weight of the patient in kilograms.

More About. . .

Fireworks

To estimate the width w in feet of a firework burst, use the formula $w = 20At$. In this formula, A is the estimated viewing angle of the firework display and t is the time in seconds from the instant you see the light until you hear the sound.

Source: www.efg2.com

53. **MONEY** In 1950, the average price of a car was about \$2000. This may sound inexpensive, but the average income in 1950 was much less than it is now. To compare dollar amounts over time, use the formula $V = \frac{A}{S}C$, where A is the old dollar amount, S is the starting year's Consumer Price Index (CPI), C is the converting year's CPI, and V is the current value of the old dollar amount. Buying a car for \$2000 in 1950 was like buying a car for how much money in 2000?

Year	Average CPI
1950	42.1
1960	29.6
1970	38.8
1980	82.4
1990	130.7
2000	174.0

Source: U.S. Department of Labor

Online Research **Data Update** What is the current Consumer Price Index? Visit www.algebra2.com/data_update to learn more.

54. **FIREWORKS** Suppose you are about a mile from a fireworks display. You count 5 seconds between seeing the light and hearing the sound of the firework display. You estimate the viewing angle is about 4°. Using the information at the left, estimate the width of the firework display.

55. **CRITICAL THINKING** Write expressions having values from one to ten using exactly four 4s. You may use any combination of the operation symbols +, −, ×, ÷, and/or grouping symbols, but no other numbers are allowed. An example of such an expression with a value of zero is $(4 + 4) - (4 + 4)$.

56. **WRITING IN MATH** Answer the question that was posed at the beginning of the lesson.

How are formulas used by nurses?

Include the following in your answer:
- an explanation of why a formula for the flow rate of an IV is more useful than a table of specific IV flow rates, and
- a description of the impact of using a formula, such as the one for IV flow rate, incorrectly.

57. Find the value of $1 + 3(5 - 17) \div 2 \times 6$.

Ⓐ -4 Ⓑ 109 Ⓒ -107 Ⓓ -144

58. The following are the dimensions of four rectangles. Which rectangle has the same area as the triangle at the right?

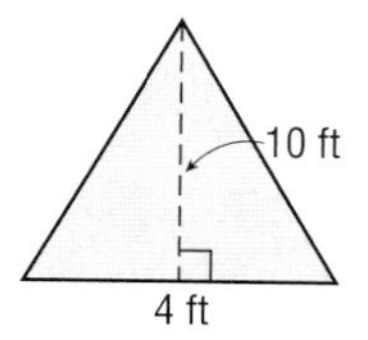

Ⓐ 1.6 ft by 25 ft Ⓑ 5 ft by 16 ft Ⓒ 3.5 ft by 4 ft Ⓓ 0.4 ft by 50 ft

Maintain Your Skills

Getting Ready for the Next Lesson **PREREQUISITE SKILL** **Evaluate each expression.**

59. $\sqrt{9}$ 60. $\sqrt{16}$ 61. $\sqrt{100}$ 62. $\sqrt{169}$

63. $-\sqrt{4}$ 64. $-\sqrt{25}$ 65. $\sqrt{\frac{4}{9}}$ 66. $\sqrt{\frac{36}{49}}$

1-2 Properties of Real Numbers

What You'll Learn

- Classify real numbers.
- Use the properties of real numbers to evaluate expressions.

How is the Distributive Property useful in calculating store savings?

Manufacturers often offer coupons to get consumers to try their products. Some grocery stores try to attract customers by doubling the value of manufacturers' coupons. You can use the Distributive Property to calculate these savings.

Vocabulary

- real numbers
- rational numbers
- irrational numbers

REAL NUMBERS All of the numbers that you use in everyday life are **real numbers**. Each real number corresponds to exactly one point on the number line, and every point on the number line represents exactly one real number.

Real numbers can be classified as either **rational** or **irrational**.

Study Tip

Reading Math
A *ratio* is the comparison of two numbers by division.

Key Concept — *Real Numbers*

Rational Numbers

- **Words** A rational number can be expressed as a ratio $\frac{m}{n}$, where m and n are integers and n is not zero. The decimal form of a rational number is either a terminating or repeating decimal.
- **Examples** $\frac{1}{6}$, 1.9, 2.575757..., -3, $\sqrt{4}$, 0

Irrational Numbers

- **Words** A real number that is not rational is irrational. The decimal form of an irrational number neither terminates nor repeats.
- **Examples** $\sqrt{5}$, π, 0.010010001...

The sets of natural numbers, $\{1, 2, 3, 4, 5, \ldots\}$, whole numbers, $\{0, 1, 2, 3, 4, \ldots\}$, and integers, $\{\ldots, -3, -2, -1, 0, 1, 2, \ldots\}$ are all subsets of the rational numbers. The whole numbers are a subset of the rational numbers because every whole number n is equal to $\frac{n}{1}$.

Real Numbers (R)

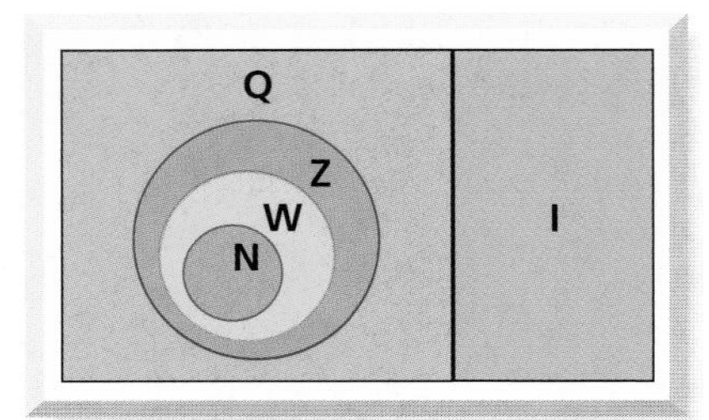

The Venn diagram shows the relationships among these sets of numbers.

R = reals | Q = rationals
I = irrationals | Z = integers
W = wholes | N = naturals

The square root of any whole number is either a whole number or it is irrational. For example, $\sqrt{36}$ is a whole number, but $\sqrt{35}$, since it lies between 5 and 6, must be irrational.

Example 1 Classify Numbers

Name the sets of numbers to which each number belongs.

a. $\sqrt{16}$

$\sqrt{16} = 4$ — naturals (N), wholes (W), integers (Z), rationals (Q), reals (R)

b. -185 — integers (Z), rationals (Q), and reals (R)

c. $\sqrt{20}$ — irrationals (I) and reals (R)

$\sqrt{20}$ lies between 4 and 5 so it is not a whole number.

d. $-\frac{7}{8}$ — rationals (Q) and reals (R)

e. $0.\overline{45}$ — rationals (Q) and reals (R)

The bar over the 45 indicates that those digits repeat forever.

Study Tip

Common Misconception
Do not assume that a number is irrational because it is expressed using the square root symbol. Find its value first.

PROPERTIES OF REAL NUMBERS The real number system is an example of a mathematical structure called a *field*. Some of the properties of a field are summarized in the table below.

Key Concepts — Real Number Properties

For any real numbers a, b, and c:

Property	Addition	Multiplication
Commutative	$a + b = b + a$	$a \cdot b = b \cdot a$
Associative	$(a + b) + c = a + (b + c)$	$(a \cdot b) \cdot c = a \cdot (b \cdot c)$
Identity	$a + 0 = a = 0 + a$	$a \cdot 1 = a = 1 \cdot a$
Inverse	$a + (-a) = 0 = (-a) + a$	If $a \neq 0$, then $a \cdot \frac{1}{a} = 1 = \frac{1}{a} \cdot a$.
Distributive	$a(b + c) = ab + ac$ and $(b + c)a = ba + ca$	

Study Tip

Reading Math
$-a$ is read the opposite of a.

Example 2 Identify Properties of Real Numbers

Name the property illustrated by each equation.

a. $(5 + 7) + 8 = 8 + (5 + 7)$

Commutative Property of Addition

The Commutative Property says that the order in which you add does not change the sum.

b. $3(4x) = (3 \cdot 4)x$

Associative Property of Multiplication

The Associative Property says that the way you group three numbers when multiplying does not change the product.

Example 3 Additive and Multiplicative Inverses

Identify the additive inverse and multiplicative inverse for each number.

a. $-1\frac{3}{4}$

Since $-1\frac{3}{4} + \left(1\frac{3}{4}\right) = 0$, the additive inverse of $-1\frac{3}{4}$ is $1\frac{3}{4}$.

Since $-1\frac{3}{4} = -\frac{7}{4}$ and $\left(-\frac{7}{4}\right)\left(-\frac{4}{7}\right) = 1$, the multiplicative inverse of $-1\frac{3}{4}$ is $-\frac{4}{7}$.

b. 1.25

Since $1.25 + (-1.25) = 0$, the additive inverse of 1.25 is -1.25.

The multiplicative inverse of 1.25 is $\frac{1}{1.25}$ or 0.8.

CHECK Notice that $1.25 \times 0.8 = 1$. ✓

You can model the Distributive Property using algebra tiles.

Algebra Activity

Distributive Property

- A 1 tile is a square that is 1 unit wide and 1 unit long. Its area is 1 square unit. An x tile is a rectangle that is 1 unit wide and x units long. Its area is x square units.

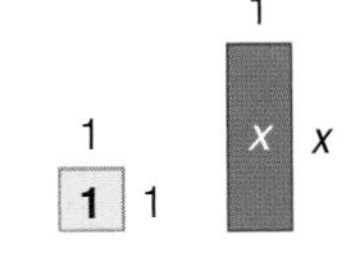

- To find the product $3(x + 1)$, model a rectangle with a width of 3 and a length of $x + 1$. Use your algebra tiles to mark off the dimensions on a product mat. Then make the rectangle with algebra tiles.
- The rectangle has 3 x tiles and 3 1 tiles. The area of the rectangle is $x + x + x + 1 + 1 + 1$ or $3x + 3$. Thus, $3(x + 1) = 3x + 3$.

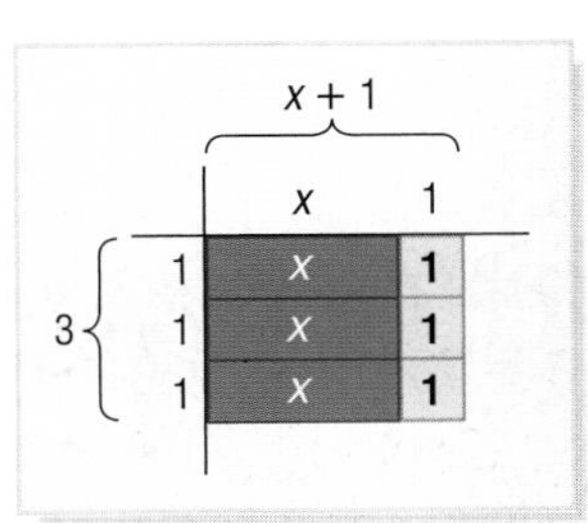

Model and Analyze

Tell whether each statement is *true* or *false*. Justify your answer with algebra tiles and a drawing.

1. $4(x + 2) = 4x + 2$

2. $3(2x + 4) = 6x + 7$

3. $2(3x + 5) = 6x + 10$

4. $(4x + 1)5 = 4x + 5$

The Distributive Property is often used in real-world applications.

Example 4 Use the Distributive Property to Solve a Problem

FOOD SERVICE A restaurant adds a 20% tip to the bills of parties of 6 or more people. Suppose a server waits on five such tables. The bill without the tip for each party is listed in the table. How much did the server make in tips during this shift?

Party 1	Party 2	Party 3	Party 4	Party 5
\$185.45	\$205.20	\$195.05	\$245.80	\$262.00

There are two ways to find the total amount of tips received.

Method 1

Multiply each dollar amount by 20% or 0.2 and then add.

$$T = 0.2(185.45) + 0.2(205.20) + 0.2(195.05) + 0.2(245.80) + 0.2(262)$$
$$= 37.09 + 41.04 + 39.01 + 49.16 + 52.40$$
$$= 218.70$$

Method 2

Add the bills of all the parties and then multiply the total by 0.2.

$$T = 0.2(185.45 + 205.20 + 195.05 + 245.80 + 262)$$
$$= 0.2(1093.50)$$
$$= 218.70$$

The server made \$218.70 during this shift.

Notice that both methods result in the same answer.

More About. . .

Food Service

Leaving a "tip" began in 18th century English coffee houses and is believed to have originally stood for "To Insure Promptness." Today, the American Automobile Association suggests leaving a 15% tip.

Source: Market Facts, Inc.

The properties of real numbers can be used to simplify algebraic expressions.

Example 5 Simplify an Expression

Simplify $2(5m + n) + 3(2m - 4n)$.

$2(5m + n) + 3(2m - 4n)$

$= 2(5m) + 2(n) + 3(2m) - 3(4n)$	Distributive Property
$= 10m + 2n + 6m - 12n$	Multiply.
$= 10m + 6m + 2n - 12n$	Commutative Property (+)
$= (10 + 6)m + (2 - 12)n$	Distributive Property
$= 16m - 10n$	Simplify.

Check for Understanding

Concept Check

1. **OPEN ENDED** Give an example of each type of number.
   a. natural **b.** whole **c.** integer
   d. rational **e.** irrational **f.** real
2. **Explain** why $\frac{\sqrt{3}}{2}$ is *not* a rational number.
3. **Disprove** the following statement by giving a counterexample. A **counterexample** is a specific case that shows that a statement is false. Explain.
   Every real number has a multiplicative inverse.

Guided Practice

Name the sets of numbers to which each number belongs.

4. -4 **5.** 45 **6.** $6.\overline{23}$

Name the property illustrated by each equation.

7. $\frac{2}{3} \cdot \frac{3}{2} = 1$ **8.** $(a + 4) + 2 = a + (4 + 2)$ **9.** $4x + 0 = 4x$

Identify the additive inverse and multiplicative inverse for each number.

10. -8 **11.** $\frac{1}{3}$ **12.** 1.5

Simplify each expression.

13. $3x + 4y - 5x$

14. $9p - 2n + 4p + 2n$

15. $3(5c + 4d) + 6(d - 2c)$

16. $\frac{1}{2}(16 - 4a) - \frac{3}{4}(12 + 20a)$

Application

BAND BOOSTERS **For Exercises 17 and 18, use the information below and in the table.**

Ashley is selling chocolate bars for $1.50 each to raise money for the band.

17. Write an expression to represent the total amount of money Ashley raised during this week.

18. Evaluate the expression from Exercise 17 by using the Distributive Property.

Ashley's Sales for One Week

Day	Bars Sold
Monday	10
Tuesday	15
Wednesday	12
Thursday	8
Friday	19
Saturday	22
Sunday	31

Practice and Apply

Homework Help

For Exercises	See Examples
19–27, 40–42, 59–62	1
28–39	2
43–48	3
63–65	4
49–58, 66–69	5

Extra Practice
See page 828.

Name the sets of numbers to which each number belongs.

19. 0 **20.** $-\frac{2}{9}$ **21.** $\sqrt{121}$ **22.** -4.55

23. $\sqrt{10}$ **24.** -31 **25.** $\frac{12}{2}$ **26.** $\frac{3\pi}{2}$

27. Name the sets of numbers to which all of the following numbers belong. Then arrange the numbers in order from least to greatest.

$2.\overline{49}, 2.4\overline{9}, 2.4, 2.49, 2.\overline{9}$

Name the property illustrated by each equation.

28. $5a + (-5a) = 0$

29. $(3 \cdot 4) \cdot 25 = 3 \cdot (4 \cdot 25)$

30. $-6xy + 0 = -6xy$

31. $[5 + (-2)] + (-4) = 5 + [-2 + (-4)]$

32. $(2 + 14) + 3 = 3 + (2 + 14)$

33. $\left(1\frac{2}{7}\right)\left(\frac{7}{9}\right) = 1$

34. $2\sqrt{3} + 5\sqrt{3} = (2 + 5)\sqrt{3}$

35. $ab = 1ab$

NUMBER THEORY **For Exercises 36–39, use the properties of real numbers to answer each question.**

36. If $m + n = m$, what is the value of n?

37. If $m + n = 0$, what is the value of n? What is n called with respect to m?

38. If $mn = 1$, what is the value of n? What is n called with respect to m?

39. If $mn = m$, what is the value of n?

More About. . .

Math History

Pythagoras (572–497 B.C.), was a Greek philosopher whose followers came to be known as the Pythagoreans. It was their knowledge of what is called the Pythagorean Theorem that led to the first discovery of irrational numbers.

Source: *A History of Mathematics*

MATH HISTORY **For Exercises 40–42, use the following information.**
The Greek mathematician Pythagoras believed that all things could be described by numbers. By "number" he meant positive integers.

40. To what set of numbers was Pythagoras referring when he spoke of "numbers?"

41. Use the formula $c = \sqrt{2s^2}$ to calculate the length of the hypotenuse c, or longest side, of this right triangle using s, the length of one leg.

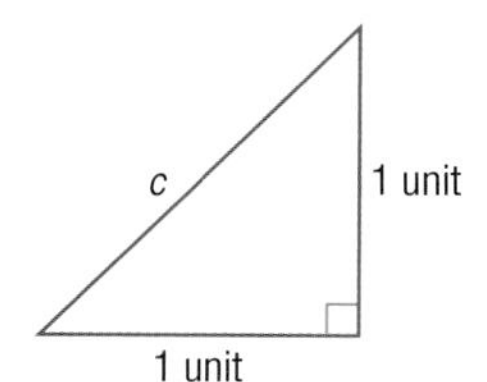

42. Explain why Pythagoras could not find a "number" to describe the value of c.

Name the additive inverse and multiplicative inverse for each number.

43. -10

44. 2.5

45. -0.125

46. $-\frac{5}{8}$

47. $\frac{4}{3}$

48. $-4\frac{3}{5}$

Simplify each expression.

49. $7a + 3b - 4a - 5b$

50. $3x + 5y + 7x - 3y$

51. $3(15x - 9y) + 5(4y - x)$

52. $2(10m - 7a) + 3(8a - 3m)$

53. $8(r + 7t) - 4(13t + 5r)$

54. $4(14c - 10d) - 6(d + 4c)$

55. $4(0.2m - 0.3n) - 6(0.7m - 0.5n)$

56. $7(0.2p + 0.3q) + 5(0.6p - q)$

57. $\frac{1}{4}(6 + 20y) - \frac{1}{2}(19 - 8y)$

58. $\frac{1}{6}(3x + 5y) + \frac{2}{3}\left(\frac{3}{5}x - 6y\right)$

Determine whether each statement is *true* or *false*. If false, give a counterexample.

59. Every whole number is an integer.

60. Every integer is a whole number.

61. Every real number is irrational.

62. Every integer is a rational number.

WORK **For Exercises 63 and 64, use the information below and in the table.**
Andrea works as a hostess in a restaurant and is paid every two weeks.

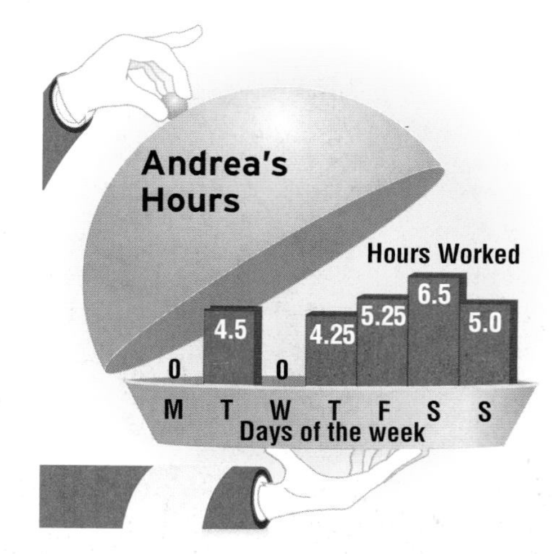

63. If Andrea earns \$6.50 an hour, illustrate the Distributive Property by writing two expressions representing Andrea's pay last week.

64. Find the mean or average number of hours Andrea worked each day, to the nearest tenth of an hour. Then use this average to predict her pay for a two-week pay period.

65. BAKING Mitena is making two types of cookies. The first recipe calls for $2\frac{1}{4}$ cups of flour, and the second calls for $1\frac{1}{8}$ cups of flour. If Mitena wants to make 3 batches of the first recipe and 2 batches of the second recipe, how many cups of flour will she need? Use the properties of real numbers to show how Mitena could compute this amount mentally. Justify each step.

BASKETBALL For Exercises 66 and 67, use the diagram of an NCAA basketball court below.

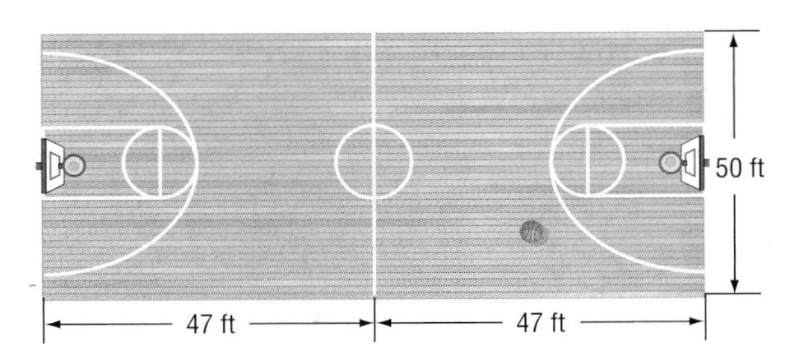

66. Illustrate the Distributive Property by writing two expressions for the area of the basketball court.

67. Evaluate the expression from Exercise 66 using the Distributive Property. What is the area of an NCAA basketball court?

SCHOOL SHOPPING For Exercises 68 and 69, use the graph at the right.

68. Illustrate the Distributive Property by writing two expressions to represent the amount that the average student spends shopping for school at specialty stores and department stores.

69. Evaluate the expression from Exercise 68 using the Distributive Property.

70. **CRITICAL THINKING** Is the Distributive Property also true for division? In other words, does $\frac{b+c}{a} = \frac{b}{a} + \frac{c}{a}$, $a \neq 0$? If so, give an example and explain why it is true. If not true, give a counterexample.

71. **WRITING IN MATH** Answer the question that was posed at the beginning of the lesson.

How is the Distributive Property useful in calculating store savings?

Include the following in your answer:

- an explanation of how the Distributive Property could be used to calculate the coupon savings listed on a grocery receipt, and
- an example of how the Distributive Property could be used to calculate the savings from a clothing store sale where all items were discounted by the same percent.

Standards Practice

Standardized Test Practice

72. If a and b are natural numbers, then which of the following must also be a natural number?

I. $a - b$ **II.** ab **III.** $\frac{a}{b}$

Ⓐ I only Ⓑ II only Ⓒ III only

Ⓓ I and II only Ⓔ II and III only

73. If $x = 1.4$, find the value of $27(x + 1.2) - 26(x + 1.2)$.

Ⓐ 1 Ⓑ -0.4 Ⓒ 2.6 Ⓓ 65

Extending the Lesson

For Exercises 74–77, use the following information.
The product of any two whole numbers is always a whole number. So, the set of whole numbers is said to be *closed* under multiplication. This is an example of the **Closure Property**. State whether each statement is *true* or *false*. If false, give a counterexample.

74. The set of integers is closed under multiplication.

75. The set of whole numbers is closed under subtraction.

76. The set of rational numbers is closed under addition.

77. The set of whole numbers is closed under division.

Maintain Your Skills

Mixed Review

Find the value of each expression. *(Lesson 1-1)*

78. $9(4 - 3)^5$

79. $5 + 9 \div 3(3) - 8$

Evaluate each expression if $a = -5$, $b = 0.25$, $c = \frac{1}{2}$, and $d = 4$. *(Lesson 1-1)*

80. $a + 2b - c$

81. $b + 3(a + d)^3$

82. GEOMETRY The formula for the surface area SA of a rectangular prism is $SA = 2\ell w + 2\ell h + 2wh$, where ℓ represents the length, w represents the width, and h represents the height. Find the surface area of the rectangular prism. *(Lesson 1-1)*

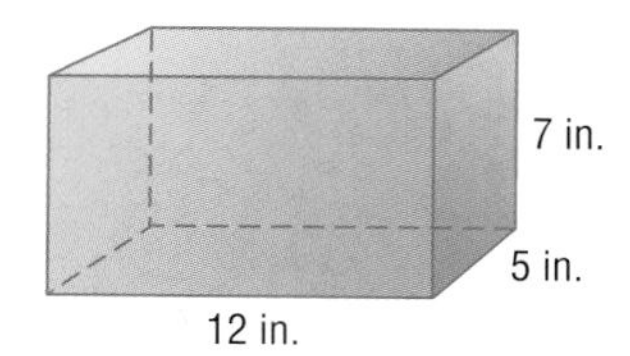

Getting Ready for the Next Lesson

PREREQUISITE SKILL **Evaluate each expression if $a = 2$, $b = -\frac{3}{4}$, and $c = 1.8$.**
(To review ***evaluating expressions****, see Lesson 1-1.)*

83. $8b - 5$

84. $\frac{2}{5}b + 1$

85. $1.5c - 7$

86. $-9(a - 6)$

Practice Quiz 1

Lessons 1-1 and 1-2

Find the value of each expression. *(Lesson 1-1)*

1. $18 - 12 \div 3$

2. $-4 + 5(7 - 2^3)$

3. $\frac{18 + 3 \times 4}{13 - 8}$

4. Evaluate $a^3 + b(9 - c)$ if $a = -2$, $b = \frac{1}{3}$, and $c = -12$. *(Lesson 1-1)*

5. ELECTRICITY Find the amount of current I (in amperes) produced if the electromotive force E is 2.5 volts, the circuit resistance R is 1.05 ohms, and the resistance r within a battery is 0.2 ohm. Use the formula $I = \frac{E}{R + r}$. *(Lesson 1-1)*

Name the sets of numbers to which each number belongs. *(Lesson 1-2)*

6. 3.5

7. $\sqrt{100}$

8. Name the property illustrated by $bc + (-bc) = 0$. *(Lesson 1-2)*

9. Name the additive inverse and multiplicative inverse of $\frac{6}{7}$. *(Lesson 1-2)*

10. Simplify $4(14x - 10y) - 6(x + 4y)$. *(Lesson 1-2)*

Algebra Activity

A Follow-Up of Lesson 1-2

Investigating Polygons and Patterns

Collect the Data

Use a ruler or geometry drawing software to draw six large polygons with 3, 4, 5, 6, 7, and 8 sides. The polygons do not need to be regular. Convex polygons, ones whose diagonals lie in the interior, will be best for this activity.

1. Copy the table below and complete the column labeled *Diagonals* by drawing the diagonals for all six polygons and record your results.

Figure Name	Sides (n)	Diagonals	Diagonals From One Vertex
triangle	3	0	0
quadrilateral	4	2	1
pentagon	5		
hexagon	6		
heptagon	7		
octagon	8		

Analyze the Data

2. Describe the pattern shown by the number of diagonals in the table above.

3. Complete the last column in the table above by recording the number of diagonals that can be drawn from one vertex of each polygon.

4. Write an expression in terms of n that relates the number of diagonals from one vertex to the number of sides for each polygon.

5. If a polygon has n sides, how many vertices does it have?

6. How many vertices does one diagonal connect?

Make a Conjecture

7. Write a formula in terms of n for the number of diagonals of a polygon of n sides. (*Hint:* Consider your answers to Exercises 2, 3, and 4.)

8. Draw a polygon with 10 sides. Test your formula for the decagon.

9. Explain how your formula relates to the number of vertices of the polygon and the number of diagonals that can be drawn from each vertex.

Extend the Activity

10. Draw 3 noncollinear dots on your paper. Determine the number of lines that are needed to connect each dot to every other dot. Continue by drawing 4 dots, 5 dots, and so on and finding the number of lines to connect them.

11. Copy and complete the table at the right.

12. Use any method to find a formula that relates the number of dots, x, to the number of lines, y.

13. Explain why the formula works.

Dots (x)	Connection Lines (y)
3	3
4	
5	
6	
7	
8	

1-3 Solving Equations

What You'll Learn

- Translate verbal expressions into algebraic expressions and equations, and vice versa.
- Solve equations using the properties of equality.

Vocabulary

- open sentence
- equation
- solution

How can you find the most effective level of intensity for your workout?

When exercising, one goal is to find the best level of intensity as a percent of your maximum heart rate. To find the intensity level, multiply 6 and P, your 10-second pulse count. Then divide by the difference of 220 and your age A.

Multiply 6 and your pulse rate	and divide by	the difference of 220 and your age.
$6 \times P$	$\div$	$(220 - A)$

VERBAL EXPRESSIONS TO ALGEBRAIC EXPRESSIONS Verbal expressions can be translated into algebraic or mathematical expressions using the language of algebra. Any letter can be used as a variable to represent a number that is not known.

Example 1 Verbal to Algebraic Expression

Write an algebraic expression to represent each verbal expression.

a. 7 less than a number $n - 7$

b. three times the square of a number $3x^2$

c. the cube of a number increased by 4 times the same number $p^3 + 4p$

d. twice the sum of a number and 5 $2(y + 5)$

A mathematical sentence containing one or more variables is called an **open sentence**. A mathematical sentence stating that two mathematical expressions are equal is called an **equation**.

Example 2 Algebraic to Verbal Sentence

Write a verbal sentence to represent each equation.

a. $10 = 12 - 2$ Ten is equal to 12 minus 2.

b. $n + (-8) = -9$ The sum of a number and -8 is -9.

c. $\frac{n}{6} = n^2$ A number divided by 6 is equal to that number squared.

Open sentences are neither true nor false until the variables have been replaced by numbers. Each replacement that results in a true sentence is called a **solution** of the open sentence.

PROPERTIES OF EQUALITY To solve equations, we can use properties of equality. Some of these *equivalence relations* are listed in the table below.

Study Tip

Properties of Equality
These properties are also known as *axioms of equality*.

$x^2 - (5x)$

$2(x-6)$

Key Concept *Properties of Equality*

Property	Symbols	Examples
Reflexive	For any real number a, $a = a$.	$-7 + n = -7 + n$
Symmetric	For all real numbers a and b, if $a = b$, then $b = a$.	If $3 = 5x - 6$, then $5x - 6 = 3$.
Transitive	For all real numbers a, b, and c, if $a = b$ and $b = c$, then $a = c$.	If $2x + 1 = 7$ and $7 = 5x - 8$, then $2x + 1 = 5x - 8$.
Substitution	If $a = b$, then a may be replaced by b and b may be replaced by a.	If $(4 + 5)m = 18$, then $9m = 18$.

Example 3 Identify Properties of Equality

Name the property illustrated by each statement.

a. If $3m = 5n$ and $5n = 10p$, then $3m = 10p$.

Transitive Property of Equality

b. If $-11a + 2 = -3a$, then $-3a = -11a + 2$.

Symmetric Property of Equality

Sometimes an equation can be solved by adding the same number to each side or by subtracting the same number from each side or by multiplying or dividing each side by the same number.

Key Concept *Properties of Equality*

Addition and Subtraction Properties of Equality

- **Symbols** For any real numbers a, b, and c, if $a = b$, then $a + c = b + c$ and $a - c = b - c$.
- **Examples** If $x - 4 = 5$, then $x - 4 + 4 = 5 + 4$.
If $n + 3 = -11$, then $n + 3 - 3 = -11 - 3$.

Multiplication and Division Properties of Equality

- **Symbols** For any real numbers a, b, and c, if $a = b$, then $a \cdot c = b \cdot c$ and, if $c \neq 0$, $\frac{a}{c} = \frac{b}{c}$.
- **Examples** If $\frac{m}{4} = 6$, then $4 \cdot \frac{m}{4} = 4 \cdot 6$. If $-3y = 6$, then $\frac{-3y}{-3} = \frac{6}{-3}$.

Example 4 Solve One-Step Equations

Solve each equation. Check your solution.

a. $a + 4.39 = 76$

$a + 4.39 = 76$	Original equation
$a + 4.39 - 4.39 = 76 - 4.39$	Subtract 4.39 from each side.
$a = 71.61$	Simplify.

The solution is 71.61.

(continued on the next page)

CHECK

$$a + 4.39 = 76 \quad \text{Original equation}$$
$$71.61 + 4.39 \stackrel{?}{=} 76 \quad \text{Substitute 71.61 for } a.$$
$$76 = 76 \checkmark \quad \text{Simplify.}$$

b. $-\frac{3}{5}d = 18$

$$-\frac{3}{5}d = 18 \quad \text{Original equation}$$
$$-\frac{5}{3}\left(-\frac{3}{5}\right)d = -\frac{5}{3}(18) \quad \text{Multiply each side by } -\frac{5}{3}\text{, the multiplicative inverse of } -\frac{3}{5}.$$
$$d = -30 \quad \text{Simplify.}$$

The solution is -30.

CHECK

$$-\frac{3}{5}d = 18 \quad \text{Original equation}$$
$$-\frac{3}{5}(-30) \stackrel{?}{=} 18 \quad \text{Substitute } -30 \text{ for } d.$$
$$18 = 18 \checkmark \quad \text{Simplify.}$$

Study Tip

Multiplication and Division Properties of Equality

Example 4*b* could also have been solved using the Division Property of Equality. Note that dividing each side of the equation by $-\frac{3}{5}$ is the same as multiplying each side by $-\frac{5}{3}$.

Sometimes you must apply more than one property to solve an equation.

Example 5 Solve a Multi-Step Equation

Solve $2(2x + 3) - 3(4x - 5) = 22$.

$2(2x + 3) - 3(4x - 5) = 22$	Original equation
$4x + 6 - 12x + 15 = 22$	Distributive and Substitution Properties
$-8x + 21 = 22$	Commutative, Distributive, and Substitution Properties
$-8x = 1$	Subtraction and Substitution Properties
$x = -\frac{1}{8}$	Division and Substitution Properties

The solution is $-\frac{1}{8}$.

You can use properties of equality to solve an equation or formula for a specified variable.

Example 6 Solve for a Variable

GEOMETRY **The surface area of a cone is $S = \pi r\ell + \pi r^2$, where S is the surface area, ℓ is the slant height of the cone, and r is the radius of the base. Solve the formula for ℓ.**

$S = \pi r\ell + \pi r^2$	Surface area formula
$S - \pi r^2 = \pi r\ell + \pi r^2 - \pi r^2$	Subtract πr^2 from each side.
$S - \pi r^2 = \pi r\ell$	Simplify.
$\frac{S - \pi r^2}{\pi r} = \frac{\pi r\ell}{\pi r}$	Divide each side by πr.
$\frac{S - \pi r^2}{\pi r} = \ell$	Simplify.

Many standardized test questions can be solved by using properties of equality.

Example 7 Apply Properties of Equality

Multiple-Choice Test Item

If $3n - 8 = \frac{9}{5}$, what is the value of $3n - 3$?

Ⓐ $\frac{34}{5}$ Ⓑ $\frac{49}{15}$ Ⓒ $-\frac{16}{5}$ Ⓓ $-\frac{27}{5}$

Test-Taking Tip

If a problem seems to require lengthy calculations, look for a shortcut. There is probably a quicker way to solve it. Try using properties of equality.

Read the Test Item

You are asked to find the value of the expression $3n - 3$. Your first thought might be to find the value of n and then evaluate the expression using this value. Notice, however, that you are *not* required to find the value of n. Instead, you can use the Addition Property of Equality on the given equation to find the value of $3n - 3$.

Solve the Test Item

$3n - 8 = \frac{9}{5}$ Original equation

$3n - 8 + 5 = \frac{9}{5} + 5$ Add 5 to each side.

$3n - 3 = \frac{34}{5}$ $\frac{9}{5} + 5 = \frac{9}{5} + \frac{25}{5}$ or $\frac{34}{5}$

The answer is A.

To solve a word problem, it is often necessary to define a variable and write an equation. Then solve by applying the properties of equality.

Example 8 Write an Equation

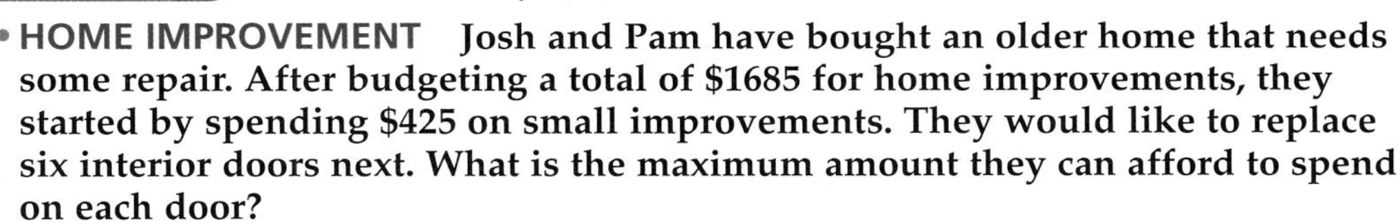

HOME IMPROVEMENT Josh and Pam have bought an older home that needs some repair. After budgeting a total of $1685 for home improvements, they started by spending $425 on small improvements. They would like to replace six interior doors next. What is the maximum amount they can afford to spend on each door?

More About. . .

Home Improvement

Previously occupied homes account for approximately 85% of all U.S. home sales. Most homeowners remodel within 18 months of purchase. The top two remodeling projects are kitchens and baths.

Source: National Association of Remodeling Industry

Explore Let c represent the cost to replace each door.

Plan Write and solve an equation to find the value of c.

The number of doors	times	the cost to replace each door	plus	previous expenses	equals	the total cost.
6	$\cdot$	c	+	425	=	1685

Solve

$6c + 425 = 1685$ Original equation

$6c + 425 - 425 = 1685 - 425$ Subtract 425 from each side.

$6c = 1260$ Simplify.

$\frac{6c}{6} = \frac{1260}{6}$ Divide each side by 6.

$c = 210$ Simplify.

They can afford to spend $210 on each door.

Examine The total cost to replace six doors at $210 each is 6(210) or $1260. Add the other expenses of $425 to that, and the total home improvement bill is 1260 + 425 or $1685. Thus, the answer is correct.

Check for Understanding

Concept Check

1. **OPEN ENDED** Write an equation whose solution is -7.
2. **Determine** whether the following statement is *sometimes, always,* or *never* true. Explain.

   Dividing each side of an equation by the same expression produces an equivalent equation.
3. **FIND THE ERROR** Crystal and Jamal are solving $C = \frac{5}{9}(F - 32)$ for F.

Crystal	Jamal
$C = \frac{5}{9}(F - 32)$	$C = \frac{5}{9}(F - 32)$
$C + 32 = \frac{5}{9}F$	$\frac{9}{5}C = F - 32$
$\frac{9}{5}(C + 32) = F$	$\frac{9}{5}C + 32 = F$

Who is correct? Explain your reasoning.

Guided Practice

Write an algebraic expression to represent each verbal expression.

4. five increased by four times a number
5. twice a number decreased by the cube of the same number

Write a verbal expression to represent each equation.

6. $9n - 3 = 6$
7. $5 + 3x^2 = 2x$

Name the property illustrated by each statement.

8. $(3x + 2) - 5 = (3x + 2) - 5$
9. If $4c = 15$, then $4c + 2 = 15 + 2$.

Solve each equation. Check your solution.

10. $y + 14 = -7$
11. $7 + 3x = 49$
12. $-4(b + 7) = -12$
13. $7q + q - 3q = -24$
14. $1.8a - 5 = -2.3$
15. $-\frac{3}{4}n + 1 = -11$

Solve each equation or formula for the specified variable.

16. $4y - 2n = 9$, for y
17. $I = prt$, for p

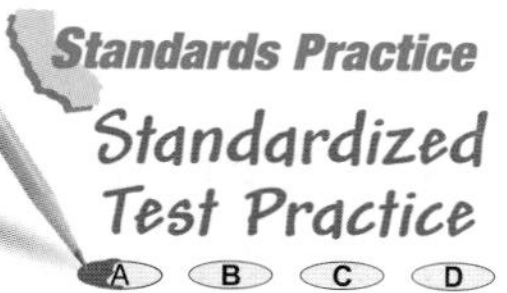

18. If $4x + 7 = 18$, what is the value of $12x + 21$?

   (A) 2.75 (B) 32 (C) 33 (D) 54

Practice and Apply

Homework Help

For Exercises	See Examples
19–28	1
29–34	2
35–40	3
41–56	4, 5
57–62	6
63–74	7

Extra Practice
See page 828.

Write an algebraic expression to represent each verbal expression.

19. the sum of 5 and three times a number
20. seven more than the product of a number and 10
21. four less than the square of a number
22. the product of the cube of a number and -6
23. five times the sum of 9 and a number
24. twice the sum of a number and 8
25. the square of the quotient of a number and 4
26. the cube of the difference of a number and 7

GEOMETRY **For Exercises 27 and 28, use the following information.**

The formula for the surface area of a cylinder with radius r and height h is π times twice the product of the radius and height plus twice the product of π and the square of the radius.

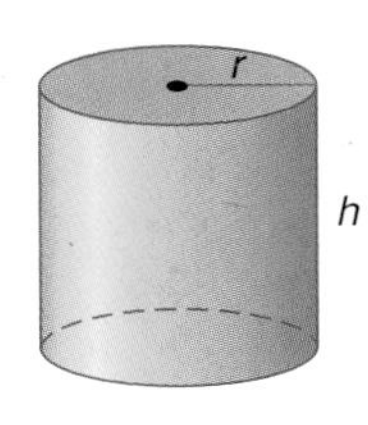

27. Translate this verbal expression of the formula into an algebraic expression.

28. Write an equivalent expression using the Distributive Property.

Write a verbal expression to represent each equation.

29. $x - 5 = 12$

30. $2n + 3 = -1$

31. $y^2 = 4y$

32. $3a^3 = a + 4$

33. $\frac{b}{4} = 2(b + 1)$

34. $7 - \frac{1}{2}x = \frac{3}{x^2}$

Name the property illustrated by each statement.

35. If $[3(-2)]z = 24$, then $-6z = 24$.

36. If $5 + b = 13$, then $b = 8$.

37. If $2x = 3d$ and $3d = -4$, then $2x = -4$.

38. If $g - t = n$, then $g = n + t$.

39. If $14 = \frac{x}{2} + 11$, then $\frac{x}{2} + 11 = 14$.

40. If $y - 2 = -8$, then $3(y - 2) = 3(-8)$.

Solve each equation. Check your solution.

41. $2p + 15 = 29$

42. $14 - 3n = -10$

43. $7a - 3a + 2a - a = 16$

44. $x + 9x - 6x + 4x = 20$

45. $\frac{1}{9} - \frac{2}{3}b = \frac{1}{18}$

46. $\frac{5}{8} + \frac{3}{4}x = \frac{1}{16}$

47. $27 = -9(y + 5)$

48. $-7(p + 8) = 21$

49. $3f - 2 = 4f + 5$

50. $3d + 7 = 6d + 5$

51. $4.3n + 1 = 7 - 1.7n$

52. $1.7x - 8 = 2.7x + 4$

53. $3(2z + 25) - 2(z - 1) = 78$

54. $4(k + 3) + 2 = 4.5(k + 1)$

55. $\frac{3}{11}a - 1 = \frac{7}{11}a + 9$

56. $\frac{2}{5}x + \frac{3}{7} = 1 - \frac{4}{7}x$

Solve each equation or formula for the specified variable.

57. $d = rt$, for r

58. $x = \frac{-b}{2a}$, for a

59. $V = \frac{1}{3}\pi r^2 h$, for h

60. $A = \frac{1}{2}h(a + b)$, for b

61. $\frac{a(b - 2)}{c - 3} = x$, for b

62. $x = \frac{y}{y + 4}$, for y

Define a variable, write an equation, and solve the problem.

63. BOWLING Jon and Morgan arrive at Sunnybrook Lanes with $16.75. Find the maximum number of games they can bowl if they each rent shoes.

SUNNYBROOK LANES

Shoe Rental: *$1.50*

Games: *$2.50 each*

For Exercises 64–70, define a variable, write an equation, and solve the problem.

64. **GEOMETRY** The perimeter of a regular octagon is 124 inches. Find the length of each side.

65. **CAR EXPENSES** Benito spent $1837 to operate his car last year. Some of these expenses are listed below. Benito's only other expense was for gasoline. If he drove 7600 miles, what was the average cost of the gasoline per mile?

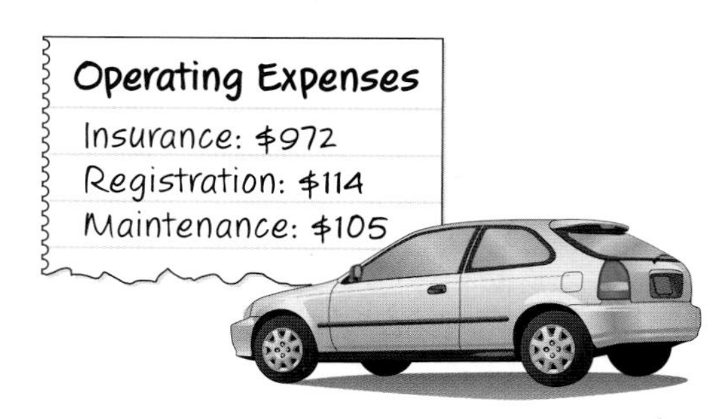

66. **SCHOOL** A school conference room can seat a maximum of 83 people. The principal and two counselors need to meet with the school's student athletes to discuss eligibility requirements. If each student must bring a parent with them, what is the maximum number of students that can attend each meeting?

67. **FAMILY** Chun-Wei's mother is 8 more than twice his age. His father is three years older than his mother is. If all three family members have lived 94 years, how old is each family member?

68. **SCHOOL TRIP** The Parent Teacher Organization has raised $1800 to help pay for a trip to an amusement park. They ask that there be one adult for every five students attending. Adult tickets cost $45 and student tickets cost $30. If the group wants to take 50 students, how much will each student need to pay so that adults agreeing to chaperone pay nothing?

69. **BUSINESS** A trucking company is hired to deliver 125 lamps for $12 each. The company agrees to pay $45 for each lamp that is broken during transport. If the trucking company needs to receive a minimum payment of $1365 for the shipment to cover their expenses, find the maximum number of lamps they can afford to break during the trip.

70. **PACKAGING** Two designs for a soup can are shown at the right. If each can holds the same amount of soup, what is the height of can A?

Career Choices

Industrial Design

Industrial designers use research on product use, marketing, materials, and production methods to create functional and appealing packaging designs.

Online Research

For information about a career as an industrial designer, visit: www.algebra2.com/careers

RAILROADS **For Exercises 71–73, use the following information.**

The First Transcontinental Railroad was built by two companies. The Central Pacific began building eastward from Sacramento, California, while the Union Pacific built westward from Omaha, Nebraska. The two lines met at Promontory, Utah, in 1869, about 6 years after construction began.

71. The Central Pacific Company laid an average of 9.6 miles of track per month. Together the two companies laid a total of 1775 miles of track. Determine the average number of miles of track laid per month by the Union Pacific Company.

72. About how many miles of track did each company lay?

73. Why do you think the Union Pacific was able to lay track so much more quickly than the Central Pacific?

74. MONEY Allison is saving money to buy a video game system. In the first week, her savings were \$8 less than $\frac{2}{5}$ the price of the system. In the second week, she saved 50 cents more than $\frac{1}{2}$ the price of the system. She was still \$37 short. Find the price of the system.

You can write and solve equations to determine the monthly payment for a home. Visit www.algebra2.com/webquest to continue work on your WebQuest project.

75. CRITICAL THINKING Write a verbal expression to represent the algebraic expression $3(x - 5) + 4x(x + 1)$.

76. WRITING IN MATH Answer the question that was posed at the beginning of the lesson.

How can you find the most effective level of intensity for your workout?

Include the following in your answer:

- an explanation of how to find the age of a person who is exercising at an 80% level of intensity I with a pulse count of 27, and
- a description of when it would be desirable to solve a formula like the one given for a specified variable.

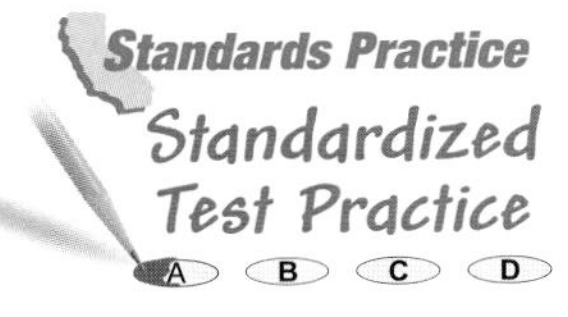

77. If $-6x + 10 = 17$, then $3x - 5 =$

Ⓐ $-\frac{7}{6}$. Ⓑ $-\frac{17}{2}$. Ⓒ 2. Ⓓ $\frac{19}{3}$. Ⓔ $\frac{5}{3}$.

78. In triangle PQR, $\overline{QS}$ and $\overline{SR}$ are angle bisectors and angle $P = 74°$. How many degrees are there in angle QSR?

Ⓐ 106 Ⓑ 121 Ⓒ 125

Ⓓ 127 Ⓔ 143

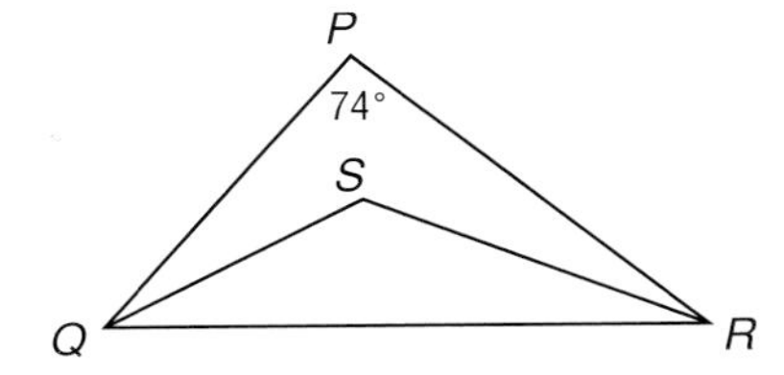

Maintain Your Skills

Mixed Review

Simplify each expression. *(Lesson 1-2)*

79. $2x + 9y + 4z - y - 8x$

80. $4(2a + 5b) - 3(4b - a)$

Evaluate each expression if $a = 3$, $b = -2$, and $c = 1.2$. *(Lesson 1-1)*

81. $a - [b(a - c)]$

82. $c^2 - ab$

83. GEOMETRY The formula for the surface area S of a regular pyramid is $S = \frac{1}{2}P\ell + B$, where P is the perimeter of the base, ℓ is the slant height, and B is the area of the base. Find the surface area of the square-based pyramid shown at the right. *(Lesson 1-1)*

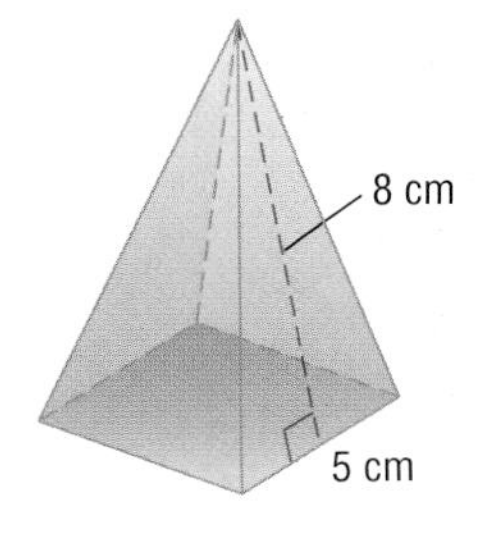

Getting Ready for the Next Lesson

PREREQUISITE SKILL Identify the additive inverse for each number or expression. *(To review **additive inverses**, see Lesson 1-2.)*

84. 5 **85.** -3 **86.** 2.5

87. $\frac{1}{4}$ **88.** $-3x$ **89.** $5 - 6y$

1-4 Solving Absolute Value Equations

California Standards Standard 1.0 Students solve equations and inequalities **involving absolute value.** (Key)

What You'll Learn

- Evaluate expressions involving absolute values.
- Solve absolute value equations.

Vocabulary

- absolute value
- empty set

How can an absolute value equation describe the magnitude of an earthquake?

Seismologists use the Richter scale to express the magnitudes of earthquakes. This scale ranges from 1 to 10, 10 being the highest. The uncertainty in the estimate of a magnitude E is about plus or minus 0.3 unit. This means that an earthquake with a magnitude estimated at 6.1 on the Richter scale might actually have a magnitude as low as 5.8 or as high as 6.4. These extremes can be described by the absolute value equation $|E - 6.1| = 0.3$.

ABSOLUTE VALUE EXPRESSIONS The **absolute value** of a number is its distance from 0 on the number line. Since distance is nonnegative, the absolute value of a number is always nonnegative. The symbol $|x|$ is used to represent the absolute value of a number x.

Key Concept — Absolute Value

- **Words** For any real number a, if a is positive or zero, the absolute value of a is a. If a is negative, the absolute value of a is the opposite of a.
- **Symbols** For any real number a, $|a| = a$ if $a \geq 0$, and $|a| = -a$ if $a < 0$.
- **Model** $|-3| = 3$ and $|3| = 3$

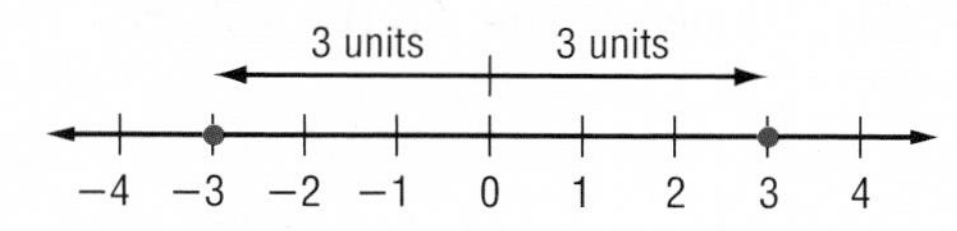

When evaluating expressions that contain absolute values, the absolute value bars act as a grouping symbol. Perform any operations inside the absolute value bars first.

Example 1 Evaluate an Expression with Absolute Value

Evaluate $1.4 + |5y - 7|$ if $y = -3$.

$1.4 + |5y - 7| = 1.4 + |5(-3) - 7|$ Replace y with -3.

$= 1.4 + |-15 - 7|$ Simplify $5(-3)$ first.

$= 1.4 + |-22|$ Subtract 7 from -15.

$= 1.4 + 22$ $|-22| = 22$

$= 23.4$ Add.

The value is 23.4.

ABSOLUTE VALUE EQUATIONS Some equations contain absolute value expressions. The definition of absolute value is used in solving these equations. For any real numbers a and b, where $b \geq 0$, if $|a| = b$, then $a = b$ or $-a = b$. This second case is often written as $a = -b$.

Example 2 Solve an Absolute Value Equation

Solve $|x - 18| = 5$. Check your solutions.

Case 1 $a = b$ **or** **Case 2** $a = -b$

Case 1:

$$x - 18 = 5$$
$$x - 18 + 18 = 5 + 18$$
$$x = 23$$

Case 2:

$$x - 18 = -5$$
$$x - 18 + 18 = -5 + 18$$
$$x = 13$$

CHECK

$$|x - 18| = 5$$
$$|23 - 18| \stackrel{?}{=} 5$$
$$|5| \stackrel{?}{=} 5$$
$$5 = 5 \quad \checkmark$$

$$|x - 18| = 5$$
$$|13 - 18| \stackrel{?}{=} 5$$
$$|-5| \stackrel{?}{=} 5$$
$$5 = 5 \quad \checkmark$$

The solutions are 23 or 13. Thus, the solution set is {13, 23}.

On the number line, we can see that each answer is 5 units away from 18.

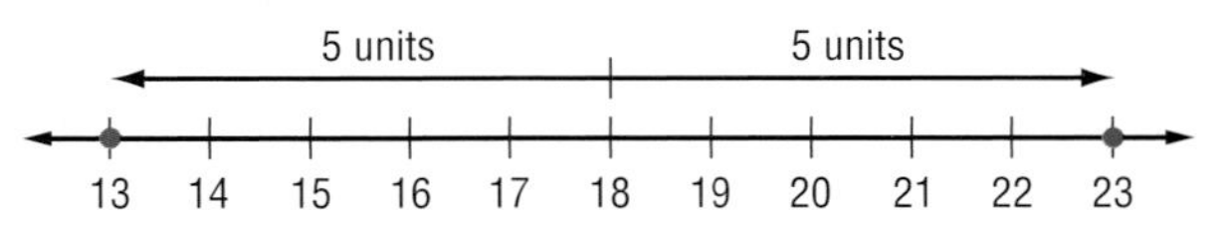

Because the absolute value of a number is always positive or zero, an equation like $|x| = -5$ is never true. Thus, it has no solution. The solution set for this type of equation is the **empty set**, symbolized by { } or $\varnothing$.

Study Tip

Common Misconception
For an equation like the one in Example 3, there is no need to consider the two cases. Remember to check your solutions in the original equation to prevent this error.

Example 3 No Solution

Solve $|5x - 6| + 9 = 0$.

$|5x - 6| + 9 = 0$ Original equation

$|5x - 6| = -9$ Subtract 9 from each side.

This sentence is *never* true. So the solution set is $\varnothing$.

It is important to check your answers when solving absolute value equations. Even if the correct procedure for solving the equation is used, the answers may not be actual solutions of the original equation.

Example 4 One Solution

Solve $|x + 6| = 3x - 2$. Check your solutions.

Case 1 $a = b$ **or** **Case 2** $a = -b$

Case 1:

$$x + 6 = 3x - 2$$
$$6 = 2x - 2$$
$$8 = 2x$$
$$4 = x$$

Case 2:

$$x + 6 = -(3x - 2)$$
$$x + 6 = -3x + 2$$
$$4x + 6 = 2$$
$$4x = -4$$
$$x = -1$$

There appear to be two solutions, 4 or -1.

(continued on the next page)

CHECK $|x + 6| = 3x - 2$ or $|x + 6| = 3x - 2$

$|4 + 6| \stackrel{?}{=} 3(4) - 2$ or $|-1 + 6| \stackrel{?}{=} 3(-1) - 2$

$|10| \stackrel{?}{=} 12 - 2$ or $|5| \stackrel{?}{=} -3 - 2$

$10 = 10$ ✓ or ~~$5 = -5$~~

Since $5 \neq -5$, the only solution is 4. Thus, the solution set is $\{4\}$.

Check for Understanding

Concept Check

1. **Explain** why if the absolute value of a number is always nonnegative, $|a|$ can equal $-a$.
2. **Write** an absolute value equation for each solution set graphed below.

a. 4 units 4 units

−4 −3 −2 −1 0 1 2 3 4

b.

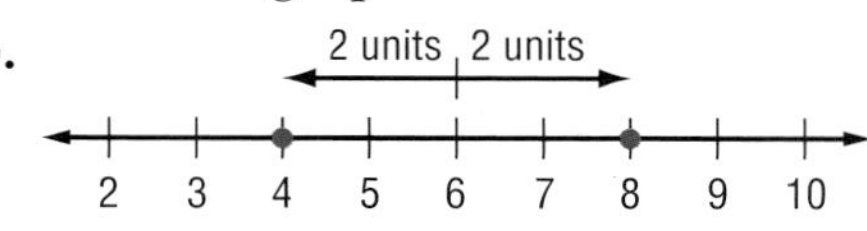

3. **Determine** whether the following statement is *sometimes, always,* or *never* true. Explain.
For all real numbers a and b, $a \neq 0$, the equation $|ax + b| = 0$ will have one solution.
4. **OPEN ENDED** Write and evaluate an expression with absolute value.

Guided Practice

Evaluate each expression if $a = -4$ and $b = 1.5$.

5. $|a + 12|$
6. $|-6b|$
7. $-|a + 21|$

Solve each equation. Check your solutions.

8. $|x + 4| = 17$
9. $|b + 15| = 3$
10. $|a - 9| = 20$
11. $|y - 2| = 34$
12. $|2w + 3| + 6 = 2$
13. $|c - 2| = 2c - 10$

Application

FOOD For Exercises 14–16, use the following information.
A meat thermometer is used to assure that a safe temperature has been reached to destroy bacteria. Most meat thermometers are accurate to within plus or minus 2°F. **Source:** U.S. Department of Agriculture

14. The ham you are baking needs to reach an internal temperature of 160°F. If the thermometer reads 160°F, write an equation to determine the least and greatest temperatures of the meat.
15. Solve the equation you wrote in Exercise 14.
16. To what temperature reading should you bake a ham to ensure that the minimum internal temperature is reached? Explain.

Practice and Apply

Evaluate each expression if $a = -5$, $b = 6$, and $c = 2.8$.

17. $|-3a|$
18. $|-4b|$
19. $|a + 5|$
20. $|2 - b|$
21. $|2b - 15|$
22. $|4a + 7|$
23. $-|18 - 5c|$
24. $-|c - a|$
25. $6 - |3c + 7|$
26. $9 - |-2b + 8|$
27. $3|a - 10| + |2a|$
28. $|a - b| - |10c - a|$

Homework Help	
For Exercises	See Examples
17–28	1
29–49	2–4

Extra Practice
See page 829.

Solve each equation. Check your solutions.

29. $|x - 25| = 17$
30. $|y + 9| = 21$
31. $|a + 12| = 33$
32. $2|b + 4| = 48$
33. $8|w - 7| = 72$
34. $|3x + 5| = 11$
35. $|2z - 3| = 0$
36. $|6c - 1| = -2$
37. $7|4x - 13| = 35$
38. $-3|2n + 5| = -9$
39. $-12|9x + 1| = 144$
40. $|5x + 9| + 6 = 1$
41. $|a - 3| - 14 = -6$
42. $3|p - 5| = 2p$
43. $3|2a + 7| = 3a + 12$
44. $|3x - 7| - 5 = -3$
45. $4|3t + 8| = 16t$
46. $|15 + m| = -2m + 3$

47. **COFFEE** Some say that to brew an excellent cup of coffee, you must have a brewing temperature of 200°F, plus or minus five degrees. Write and solve an equation describing the maximum and minimum brewing temperatures for an excellent cup of coffee.

48. **MANUFACTURING** A machine is used to fill each of several bags with 16 ounces of sugar. After the bags are filled, another machine weighs them. If the bag weighs 0.3 ounce more or less than the desired weight, the bag is rejected. Write an equation to find the heaviest and lightest bag the machine will approve.

49. **METEOROLOGY** The troposphere is the layer of atmosphere closest to Earth. The average upper boundary of the layer is about 13 kilometers above Earth's surface. This height varies with latitude and with the seasons by as much as 5 kilometers. Write and solve an equation describing the maximum and minimum heights of the upper bound of the troposphere.

More About. . .

Meteorology

The troposphere is characterized by the density of its air and an average vertical temperature change of 6°C per kilometer. All weather phenomena occur within the troposphere.

Source: NASA

CRITICAL THINKING **For Exercises 50 and 51, determine whether each statement is *sometimes*, *always*, or *never* true. Explain your reasoning.**

50. If a and b are real numbers, then $|a + b| = |a| + |b|$.

51. If a, b, and c are real numbers, then $c|a + b| = |ca + cb|$.

52. **WRITING IN MATH** Answer the question that was posed at the beginning of the lesson.

How can an absolute value equation describe the magnitude of an earthquake?

Include the following in your answer:

- a verbal and graphical explanation of how $|E - 6.1| = 0.3$ describes the possible extremes in the variation of the earthquake's magnitude, and
- an equation to describe the extremes for a different magnitude.

53. Which of the graphs below represents the solution set for $|x - 3| - 4 = 0$?

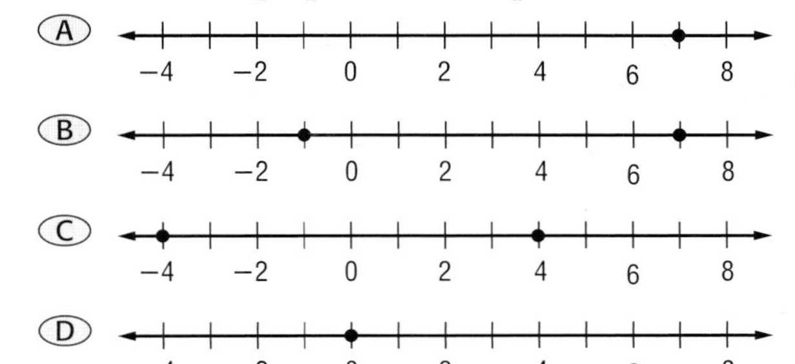

54. Find the value of $-|-9| - |4| - 3|5 - 7|$.

(A) -19 (B) -11 (C) -7 (D) 11

Extending the Lesson

For Exercises 55–58, consider the equation $|x + 1| + 2 = |x + 4|$.

55. To solve this equation, we must consider the case where $x + 4 \geq 0$ and the case where $x + 4 < 0$. Write the equations for each of these cases.

56. Notice that each equation you wrote in Exercise 55 has two cases. For each equation, write two other equations taking into consideration the case where $x + 1 \geq 0$ and the case where $x + 1 < 0$.

57. Solve each equation you wrote in Exercise 56. Then, check each solution in the original equation, $|x + 1| + 2 = |x + 4|$. What are the solution(s) to this absolute value equation?

58. **MAKE A CONJECTURE** For equations with one set of absolute value symbols, two cases must be considered. For an equation with two sets of absolute value symbols, four cases must be considered. How many cases must be considered for an equation containing three sets of absolute value symbols?

Maintain Your Skills

Mixed Review

Write an algebraic expression to represent each verbal expression. *(Lesson 1-3)*

59. twice the difference of a number and 11

60. the product of the square of a number and 5

Solve each equation. Check your solution. *(Lesson 1-3)*

61. $3x + 6 = 22$

62. $7p - 4 = 3(4 + 5p)$

63. $\frac{5}{7}y - 3 = \frac{3}{7}y + 1$

Name the property illustrated by each equation. *(Lesson 1-2)*

64. $(5 + 9) + 13 = 13 + (5 + 9)$

65. $m(4 - 3) = m \cdot 4 - m \cdot 3$

66. $\left(\frac{1}{4}\right)4 = 1$

67. $5x + 0 = 5x$

Determine whether each statement is *true* or *false*. If false, give a counterexample. *(Lesson 1-2)*

68. Every real number is a rational number.

69. Every natural number is an integer.

70. Every irrational number is a real number.

71. Every rational number is an integer.

GEOMETRY **For Exercises 72 and 73, use the following information.**

The formula for the area A of a triangle is $A = \frac{1}{2}bh$, where b is the measure of the base and h is the measure of the height. *(Lesson 1-1)*

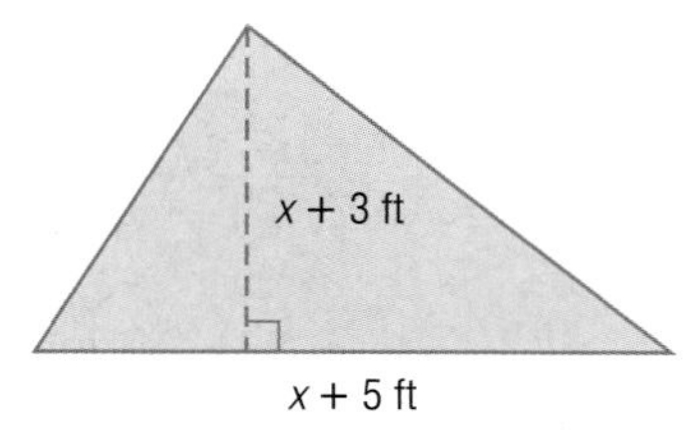

72. Write an expression to represent the area of the triangle above.

73. Evaluate the expression you wrote in Exercise 72 for $x = 23$.

Getting Ready for the Next Lesson

PREREQUISITE SKILL **Solve each equation.** *(To review **solving equations**, see page 20.)*

74. $14y - 3 = 25$

75. $4.2x + 6.4 = 40$

76. $7w + 2 = 3w - 6$

77. $2(a - 1) = 8a - 6$

78. $48 + 5y = 96 - 3y$

79. $\frac{2x + 3}{5} = \frac{3}{10}$

1-5 Solving Inequalities

What You'll Learn

- Solve inequalities.
- Solve real-world problems involving inequalities.

Vocabulary

- set-builder notation
- interval notation

How can inequalities be used to compare phone plans?

Kuni is trying to decide between two rate plans offered by a wireless phone company.

	Plan 1	Plan 2
Monthly Access Fee	$35.00	$55.00
Minutes Included	150	400
Additional Minutes	40¢	35¢

To compare these two rate plans, we can use inequalities. The monthly access fee for Plan 1 is less than the fee for Plan 2, $\$35 < \$55$. However, the additional minutes fee for Plan 1 is greater than that of Plan 2, $\$0.40 > \$0.35$.

SOLVE INEQUALITIES For any two real numbers, a and b, exactly one of the following statements is true.

$$a < b \qquad a = b \qquad a > b$$

This is known as the **Trichotomy Property** or the *property of order*.

Adding the same number to, or subtracting the same number from, each side of an inequality does not change the truth of the inequality.

Study Tip

Properties of Inequality
The properties of inequality are also known as *axioms* of inequality.

Key Concept — *Properties of Inequality*

Addition Property of Inequality

- **Words** For any real numbers, a, b, and c:
  If $a > b$, then $a + c > b + c$.
  If $a < b$, then $a + c < b + c$.
- **Example**
  $3 < 5$
  $3 + (-4) < 5 + (-4)$
  $-1 < 1$

Subtraction Property of Inequality

- **Words** For any real numbers, a, b, and c:
  If $a > b$, then $a - c > b - c$.
  If $a < b$, then $a - c < b - c$.
- **Example**
  $2 > -7$
  $2 - 8 > -7 - 8$
  $-6 > -15$

These properties are also true for $\leq$ and $\geq$.

These properties can be used to solve inequalities. The solution sets of inequalities in one variable can then be graphed on number lines. Use a circle with an arrow to the left for $<$ and an arrow to the right for $>$. Use a dot with an arrow to the left for $\leq$ and an arrow to the right for $\geq$.

Example 1 Solve an Inequality Using Addition or Subtraction

Solve $7x - 5 > 6x + 4$. Graph the solution set on a number line.

$7x - 5 > 6x + 4$	Original inequality
$7x - 5 + (-6x) > 6x + 4 + (-6x)$	Add $-6x$ to each side.
$x - 5 > 4$	Simplify.
$x - 5 + 5 > 4 + 5$	Add 5 to each side.
$x > 9$	Simplify.

Any real number greater than 9 is a solution of this inequality.

The graph of the solution set is shown at the right.

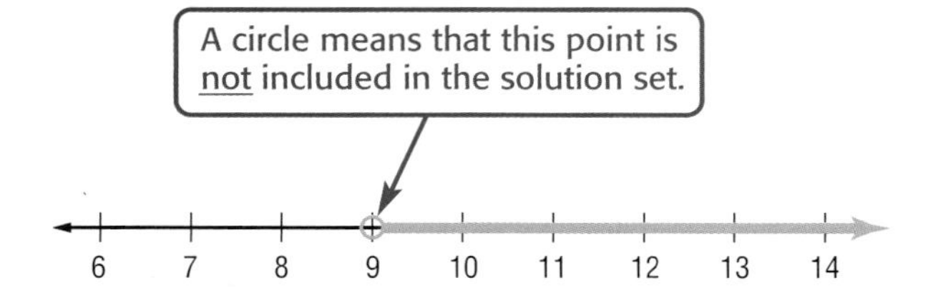

CHECK Substitute 9 for x in $7x - 5 > 6x + 4$. The two sides should be equal. Then substitute a number greater than 9. The inequality should be true.

Multiplying or dividing each side of an inequality by a positive number does not change the truth of the inequality. However, multiplying or dividing each side of an inequality by a *negative* number requires that the order of the inequality be *reversed*. For example, to reverse $\leq$, replace it with $\geq$.

Key Concept — *Properties of Inequality*

Multiplication Property of Inequality

- **Words** For any real numbers, a, b, and c, where

  c is positive: if $a > b$, then $ac > bc$. if $a < b$, then $ac < bc$.

  c is negative: if $a > b$, then $ac < bc$. if $a < b$, then $ac > bc$.

- **Examples**

  $-2 < 3$
  $4(-2) < 4(3)$
  $-8 < 12$

  $5 > -1$
  $(-3)(5) < (-3)(-1)$
  $-15 < 3$

Division Property of Inequality

- **Words** For any real numbers, a, b, and c, where

  c is positive: if $a > b$, then $\frac{a}{c} > \frac{b}{c}$. if $a < b$, then $\frac{a}{c} < \frac{b}{c}$.

  c is negative: if $a > b$, then $\frac{a}{c} < \frac{b}{c}$. if $a < b$, then $\frac{a}{c} > \frac{b}{c}$.

- **Examples**

  $-18 < -9$
  $\frac{-18}{3} < \frac{-9}{3}$
  $-6 < -3$

  $12 > 8$
  $\frac{12}{-2} < \frac{8}{-2}$
  $-6 < -4$

These properties are also true for $\leq$ and $\geq$.

Study Tip

Reading Math
$\{x \mid x > 9\}$ is read *the set of all numbers x such that x is greater than 9.*

The solution set of an inequality can be expressed by using **set-builder notation**. For example, the solution set in Example 1 can be expressed as $\{x \mid x > 9\}$.

Example 2 Solve an Inequality Using Multiplication or Division

Solve $-0.25y \geq 2$. Graph the solution set on a number line.

$-0.25y \geq 2$	Original inequality
$\frac{-0.25y}{-0.25} \leq \frac{2}{-0.25}$	Divide each side by -0.25, reversing the inequality symbol.
$y \leq -8$	Simplify.

The solution set is $\{y \mid y \leq -8\}$.

The graph of the solution set is shown below.

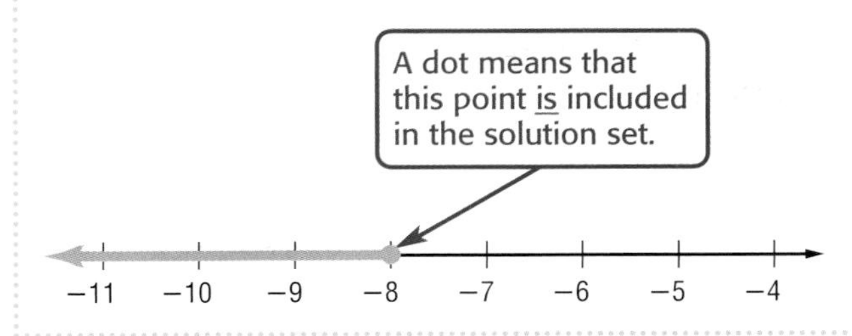

Study Tip

Reading Math
The symbol $+\infty$ is read *positive infinity*, and the symbol $-\infty$ is read *negative infinity*.

The solution set of an inequality can also be described by using **interval notation**. The infinity symbols $+\infty$ and $-\infty$ are used to indicate that a set is unbounded in the positive or negative direction, respectively. To indicate that an endpoint *is not* included in the set, a parenthesis, (or), is used.

$x < 2$ — interval notation $(-\infty, 2)$

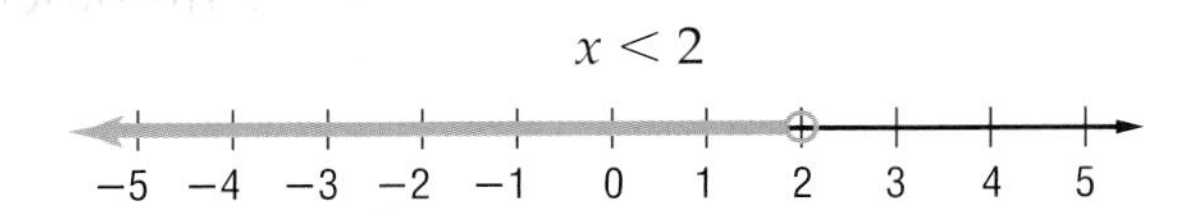

A bracket is used to indicate that the endpoint, -2, *is* included in the solution set below. Parentheses are always used with the symbols $+\infty$ and $-\infty$, because they do not include endpoints.

$x \geq -2$ — interval notation $[-2, +\infty)$

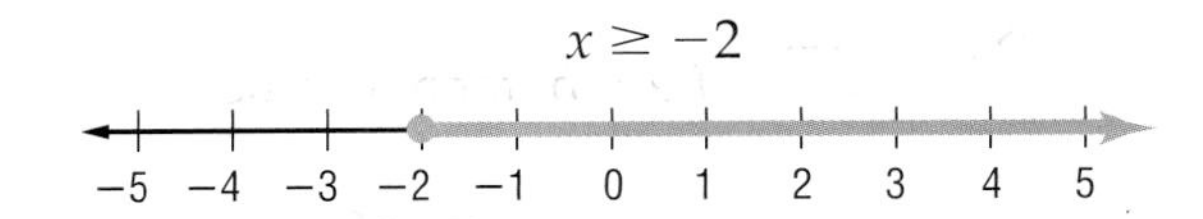

Study Tip

Solutions to Inequalities
When solving an inequality,

- if you arrive at a false statement, such as $3 > 5$, then the solution set for that inequality is the empty set, $\varnothing$.
- if you arrive at a true statement such as $3 > -1$, then the solution set for that inequality is the set of all real numbers.

Example 3 Solve a Multi-Step Inequality

Solve $-m \leq \frac{m + 4}{9}$. Graph the solution set on a number line.

$-m \leq \frac{m+4}{9}$	Original inequality
$-9m \leq m + 4$	Multiply each side by 9.
$-10m \leq 4$	Add $-m$ to each side.
$m \geq -\frac{4}{10}$	Divide each side by -10, reversing the inequality symbol.
$m \geq -\frac{2}{5}$	Simplify.

The solution set is $\left[-\frac{2}{5}, +\infty\right)$ and is graphed below.

REAL-WORLD PROBLEMS WITH INEQUALITIES Inequalities can be used to solve many verbal and real-world problems.

Example 4 Write an Inequality

DELIVERIES Craig is delivering boxes of paper to each floor of an office building. Each box weighs 64 pounds, and Craig weighs 160 pounds. If the maximum capacity of the elevator is 2000 pounds, how many boxes can Craig safely take on each elevator trip?

Explore Let b = the number of boxes Craig can safely take on each trip. A maximum capacity of 2000 pounds means that this weight must be less than or equal to 2000.

Plan The total weight of the boxes is $64b$. Craig's weight plus the total weight of the boxes must be less than or equal to 2000. Write an inequality.

Craig's weight	plus	the weight of the boxes	is less than or equal to	2000.
160	+	$64b$	$\leq$	2000

Solve

$160 + 64b \leq 2000$ Original inequality

$160 - 160 + 64b \leq 2000 - 160$ Subtract 160 from each side.

$64b \leq 1840$ Simplify.

$\frac{64b}{64} \leq \frac{1840}{64}$ Divide each side by 64.

$b \leq 28.75$ Simplify.

Examine Since he cannot take a fraction of a box, Craig can take no more than 28 boxes per trip and still meet the safety requirements of the elevator.

Study Tip

Inequality Phrases

- $<$ is less than; is fewer than
- $>$ is greater than; is more than
- $\leq$ is at most; is no more than; is less than or equal to
- $\geq$ is at least; is no less than; is greater than or equal to

You can use a graphing calculator to find the solution set for an inequality.

Graphing Calculator Investigation

Solving Inequalities

The inequality symbols in the TEST menu on the TI-83 Plus are called *relational operators*. They compare values and return 1 if the test is true or 0 if the test is false.

You can use these relational operators to find the solution set of an inequality in one variable.

Think and Discuss

1. Clear the Y= list. Enter $11x + 3 \geq 2x - 6$ as Y1. Put your calculator in DOT mode. Then, graph in the standard viewing window. Describe the graph.
2. Using the TRACE function, investigate the graph. What values of x are on the graph? What values of y are on the graph?
3. Based on your investigation, what inequality is graphed?
4. Solve $11x + 3 \geq 2x - 6$ algebraically. How does your solution compare to the inequality you wrote in Exercise 3?

Check for Understanding

Concept Check

1. **Explain** why it is not necessary to state a division property for inequalities.
2. **Write** an inequality using the > symbol whose solution set is graphed below.

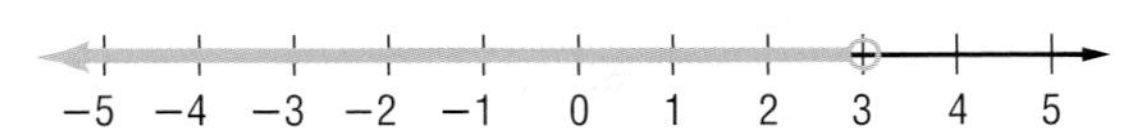

3. **OPEN ENDED** Write an inequality for which the solution set is the empty set.

Guided Practice

Solve each inequality. Describe the solution set using set-builder or interval notation. Then graph the solution set on a number line.

4. $a + 2 < 3.5$
5. $5 \geq 3x$
6. $11 - c \leq 8$
7. $4y + 7 > 31$
8. $2w + 19 < 5$
9. $-0.6p < -9$
10. $\frac{n}{12} + 15 \leq 13$
11. $\frac{5z + 2}{4} < \frac{5z}{4} + 2$

Define a variable and write an inequality for each problem. Then solve.

12. The product of 12 and a number is greater than 36.
13. Three less than twice a number is at most 5.

Application

14. **SCHOOL** The final grade for a class is calculated by taking 75% of the average test score and adding 25% of the score on the final exam. If all scores are out of 100 and a student has a 76 test average, what score does the student need to make on the final exam to have a final grade of at least 80?

Practice and Apply

Homework Help

For Exercises	See Examples
15–40	1–3
41–51	4

Extra Practice
See page 829.

Solve each inequality. Describe the solution set using set-builder or interval notation. Then, graph the solution set on a number line.

15. $n + 4 \geq -7$
16. $b - 3 \leq 15$
17. $5x < 35$
18. $\frac{d}{2} > -4$
19. $\frac{g}{-3} \geq -9$
20. $-8p \geq 24$
21. $13 - 4k \leq 27$
22. $14 > 7y - 21$
23. $-27 < 8m + 5$
24. $6b + 11 \geq 15$
25. $2(4t + 9) \leq 18$
26. $90 \geq 5(2r + 6)$
27. $14 - 8n \leq 0$
28. $-4(5w - 8) < 33$
29. $0.02x + 5.58 < 0$
30. $1.5 - 0.25c < 6$
31. $6d + 3 \geq 5d - 2$
32. $9z + 2 > 4z + 15$
33. $2(g + 4) < 3g - 2(g - 5)$
34. $3(a + 4) - 2(3a + 4) \leq 4a - 1$
35. $y < \frac{-y + 2}{9}$
36. $\frac{1 - 4p}{5} < 0.2$
37. $\frac{4x + 2}{6} < \frac{2x + 1}{3}$
38. $12\left(\frac{1}{4} - \frac{n}{3}\right) \leq -6n$

39. **PART-TIME JOB** David earns \$5.60 an hour working at Box Office Videos. Each week, 25% of his total pay is deducted for taxes. If David wants his take-home pay to be at least \$105 a week, solve the inequality $5.6x - 0.25(5.6x) \geq 105$ to determine how many hours he must work.

40. **STATE FAIR** Juan's parents gave him \$35 to spend at the State Fair. He spends \$13.25 for food. If rides at the fair cost \$1.50 each, solve the inequality $1.5n + 13.25 \leq 35$ to determine how many rides he can afford.

More About. . .

Child Care

In 1995, 55% of children ages three to five were enrolled in center-based child care programs. Parents cared for 26% of children, relatives cared for 19% of children, and non-relatives cared for 17% of children.

Source: National Center for Education Statistics

Define a variable and write an inequality for each problem. Then solve.

41. The sum of a number and 8 is more than 2.
42. The product of -4 and a number is at least 35.
43. The difference of one half of a number and 7 is greater than or equal to 5.
44. One more than the product of -3 and a number is less than 16.
45. Twice the sum of a number and 5 is no more than 3 times that same number increased by 11.
46. 9 less than a number is at most that same number divided by 2.

47. **CHILD CARE** By Ohio law, when children are napping, the number of children per child care staff member may be as many as twice the maximum listed at the right. Write and solve an inequality to determine how many staff members are required to be present in a room where 17 children are napping and the youngest child is 18 months old.

Maximum Number of Children Per Child Care Staff Member
At least one child care staff member caring for:
Every 5 infants less than 12 months old (or 2 for every 12)
Every 6 infants who are at least 12 months olds, but less than 18 months old
Every 7 toddlers who are at least 18 months old, but less than 30 months old
Every 8 toddlers who are at least 30 months old, but less than 3 years old

Source: Ohio Department of Job and Family Services

CAR SALES **For Exercises 48 and 49, use the following information.**
Mrs. Lucas earns a salary of $24,000 per year plus 1.5% commission on her sales. If the average price of a car she sells is $30,500, about how many cars must she sell to make an annual income of at least $40,000?

48. Write an inequality to describe this situation.
49. Solve the inequality and interpret the solution.

TEST GRADES **For Exercises 50 and 51, use the following information.**
Ahmik's scores on the first four of five 100-point history tests were 85, 91, 89, and 94.

50. If a grade of at least 90 is an A, write an inequality to find the score Ahmik must receive on the fifth test to have an A test average.
51. Solve the inequality and interpret the solution.

52. **CRITICAL THINKING** Which of the following properties hold for inequalities? Explain your reasoning or give a counterexample.
    a. Reflexive
    b. Symmetric
    c. Transitive

53. **WRITING IN MATH** Answer the question that was posed at the beginning of the lesson.

    How can inequalities be used to compare phone plans?

    Include the following in your answer:
    - an inequality comparing the number of minutes offered by each plan, and
    - an explanation of how Kuni might determine when Plan 1 might be cheaper than Plan 2 if she typically uses more than 150 but less than 400 minutes.

Standardized Test Practice

Standards Practice

54. If $4 - 5n \geq -1$, then n could equal all of the following EXCEPT

Ⓐ $-\frac{1}{5}$. Ⓑ $\frac{1}{5}$. Ⓒ 1. Ⓓ 2.

55. If $a < b$ and $c < 0$, which of the following are true?

I. $ac > bc$ **II.** $a + c < b + c$ **III.** $a - c > b - c$

Ⓐ I only Ⓑ II only Ⓒ III only

Ⓓ I and II only Ⓔ I, II, and III

Graphing Calculator

Use a graphing calculator to solve each inequality.

56. $-5x - 8 < 7$ **57.** $-4(6x - 3) \leq 60$ **58.** $3(x + 3) \geq 2(x + 4)$

Maintain Your Skills

Mixed Review

Solve each equation. Check your solutions. *(Lesson 1-4)*

59. $|x - 3| = 17$ **60.** $8|4x - 3| = 64$ **61.** $|x + 1| = x$

62. SHOPPING On average, by how much did the number of people who just browse, but not necessarily buy, online increase each year from 1997 to 2003? Define a variable, write an equation, and solve the problem. *(Lesson 1-3)*

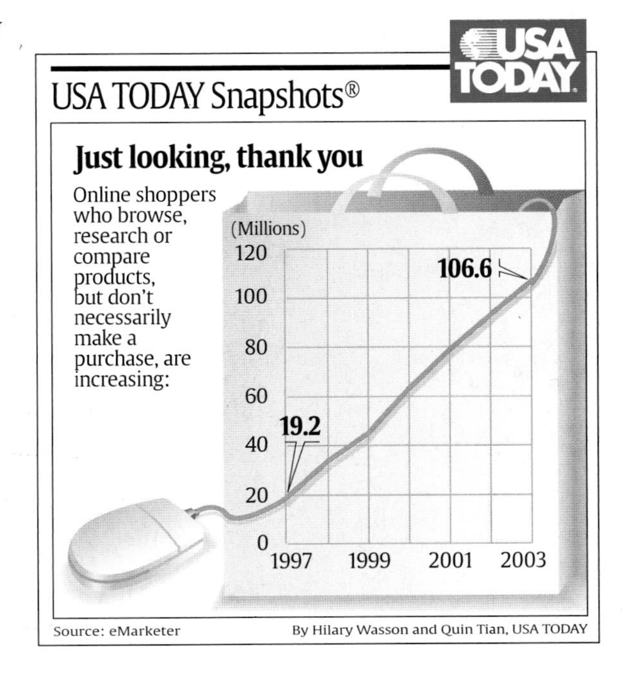

Name the sets of numbers to which each number belongs. *(Lesson 1-2)*

63. 31 **64.** $-4.\overline{2}$ **65.** $\sqrt{7}$

66. BABY-SITTING Jenny baby-sat for $5\frac{1}{2}$ hours on Friday night and 8 hours on Saturday. She charges \$4.25 per hour. Use the Distributive Property to write two equivalent expressions that represent how much money Jenny earned. *(Lesson 1-2)*

Getting Ready for the Next Lesson

PREREQUISITE SKILL Solve each equation. Check your solutions.
*(To review **solving absolute value equations**, see Lesson 1-4.)*

67. $|x| = 7$ **68.** $|x + 5| = 18$ **69.** $|5y - 8| = 12$

70. $|2x - 36| = 14$ **71.** $2|w + 6| = 10$ **72.** $|x + 4| + 3 = 17$

Practice Quiz 2

Lessons 1-3 through 1-5

1. Solve $2d + 5 = 8d + 2$. Check your solution. *(Lesson 1-3)*
2. Solve $s = \frac{1}{2}gt^2$ for g. *(Lesson 1-3)*
3. Evaluate $|x - 3y|$ if $x = -8$ and $y = 2$. *(Lesson 1-4)*
4. Solve $3|3x + 2| = 51$. Check your solutions. *(Lesson 1-4)*
5. Solve $2(m - 5) - 3(2m - 5) < 5m + 1$. Describe the solution set using set-builder or interval notation. Then graph the solution set on a number line. *(Lesson 1-5)*

1-6 Solving Compound and Absolute Value Inequalities

California Standards Standard 1.0 **Students solve** equations and **inequalities involving absolute value.** (Key)

What You'll Learn

- Solve compound inequalities.
- Solve absolute value inequalities.

Vocabulary

- compound inequality
- intersection
- union

How are compound inequalities used in medicine?

One test used to determine whether a patient is diabetic and requires insulin is a glucose tolerance test. Patients start the test in a *fasting state,* meaning they have had no food or drink except water for at least 10 but no more than 16 hours. The acceptable number of hours h for fasting can be described by the following compound inequality.

$$h \geq 10 \text{ and } h \leq 16$$

COMPOUND INEQUALITIES A **compound inequality** consists of two inequalities joined by the word *and* or the word *or*. To solve a compound inequality, you must solve each part of the inequality. The graph of a compound inequality containing *and* is the **intersection** of the solution sets of the two inequalities. Compound inequalities involving the word *and* are called *conjunctions.* Compound inequalities involving the word *or* are called *disjunctions.*

Key Concept — "And" Compound Inequalities

- **Words** A compound inequality containing the word *and* is true if and only if *both* inequalities are true.
- **Example** $x \geq -1$

  $x < 2$

  $x \geq -1$ and $x < 2$

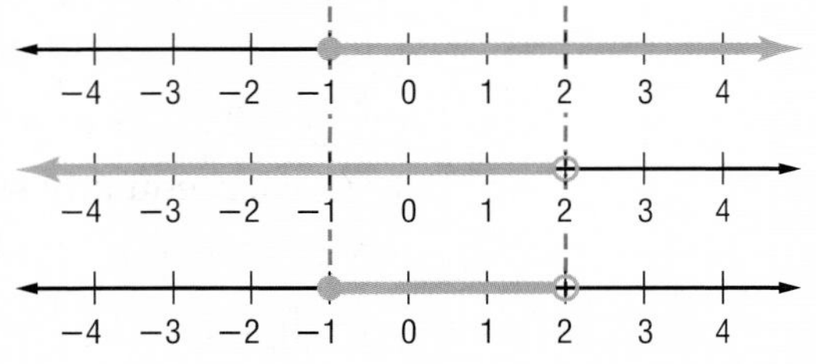

Another way of writing $x \geq -1$ and $x < 2$ is $-1 \leq x < 2$. Both forms are read *x is greater than or equal to −1 and less than 2.*

Study Tip

Interval Notation The compound inequality $-1 \leq x < 2$ can be written as $[-1, 2)$, indicating that the solution set is the set of all numbers between -1 and 2, including -1, but not including 2.

Example 1 Solve an "and" Compound Inequality

Solve $13 < 2x + 7 \leq 17$. Graph the solution set on a number line.

Method 1

Write the compound inequality using the word *and*. Then solve each inequality.

$$13 < 2x + 7 \quad \text{and} \quad 2x + 7 \leq 17$$
$$6 < 2x \qquad\qquad 2x \leq 10$$
$$3 < x \qquad\qquad x \leq 5$$
$$3 < x \leq 5$$

Method 2

Solve both parts at the same time by subtracting 7 from each part. Then divide each part by 2.

$$13 < 2x + 7 \leq 17$$
$$6 < 2x \leq 10$$
$$3 < x \leq 5$$

Graph the solution set for each inequality and find their intersection.

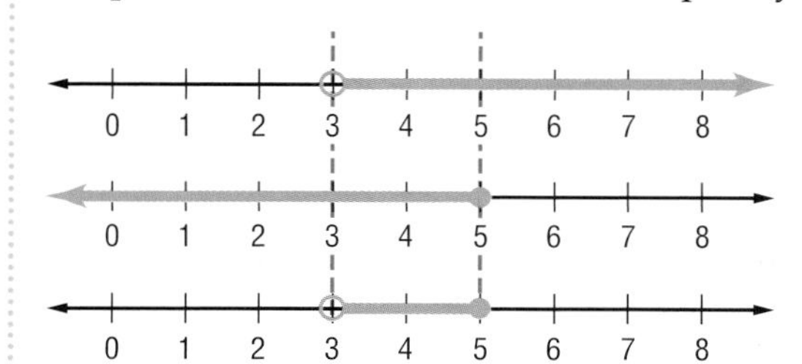

$x > 3$

$x \leq 5$

$3 < x \leq 5$

The solution set is $\{x \mid 3 < x \leq 5\}$.

The graph of a compound inequality containing *or* is the **union** of the solution sets of the two inequalities.

Key Concept — "Or" Compound Inequalities

- **Words** A compound inequality containing the word *or* is true if one or more of the inequalities is true.
- **Example**

$x \leq 1$

$x > 4$

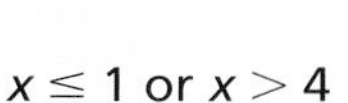

$x \leq 1$ or $x > 4$

Example 2 Solve an "or" Compound Inequality

Solve $y - 2 > -3$ or $y + 4 \leq -3$. Graph the solution set on a number line.

Solve each inequality separately.

$y - 2 > -3$ or $y + 4 \leq -3$

$y > -1$ $\qquad$ $y \leq -7$

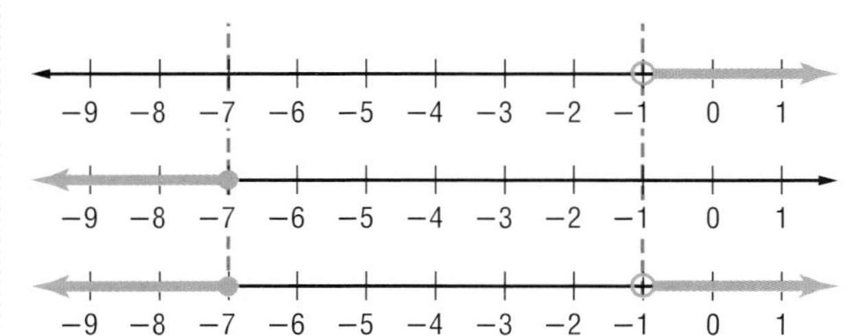

$y > -1$

$y \leq -7$

$y > -1$ or $y \leq -7$

The solution set is $\{y \mid y > -1 \text{ or } y \leq -7\}$.

Study Tip

Interval Notation

In interval notation, the symbol for the union of the two sets is $\cup$. The compound inequality $y > -1$ or $y \leq -7$ is written as $(-\infty, -7] \cup (-1, +\infty)$, indicating that all values less than and including -7 are part of the solution set. In addition, all values greater than -1, not including -1, are part of the solution set.

ABSOLUTE VALUE INEQUALITIES In Lesson 1-4, you learned that the absolute value of a number is its distance from 0 on the number line. You can use this definition to solve inequalities involving absolute value.

Example 3 Solve an Absolute Value Inequality (<)

Solve $|a| < 4$. Graph the solution set on a number line.

You can interpret $|a| < 4$ to mean that the distance between a and 0 on a number line is less than 4 units. To make $|a| < 4$ true, you must substitute numbers for a that are fewer than 4 units from 0.

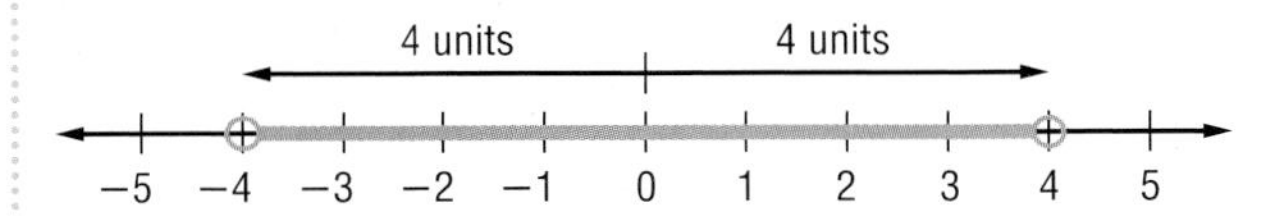

Notice that the graph of $|a| < 4$ is the same as the graph of $a > -4$ and $a < 4$.

All of the numbers between -4 and 4 are less than 4 units from 0. The solution set is $\{a \mid -4 < a < 4\}$.

Study Tip

Absolute Value Inequalities

Because the absolute value of a number is never negative,

- the solution of an inequality like $|a| < -4$ is the empty set.
- the solution of an inequality like $|a| > -4$ is the set of all real numbers.

Example 4 Solve an Absolute Value Inequality (>)

Solve $|a| > 4$. Graph the solution set on a number line.

You can interpret $|a| > 4$ to mean that the distance between a and 0 is greater than 4 units. To make $|a| > 4$ true, you must substitute values for a that are greater than 4 units from 0.

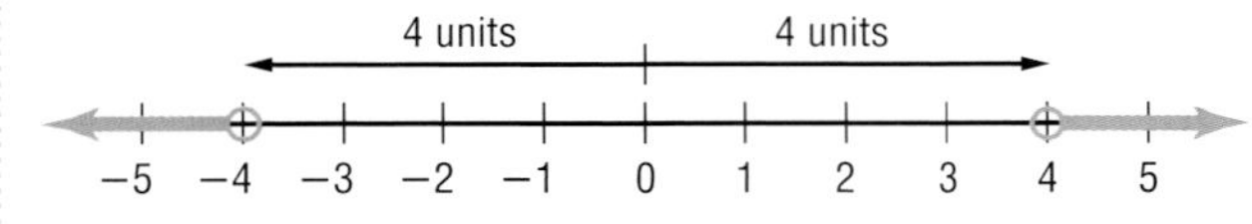

Notice that the graph of $|a| > 4$ is the same as the graph of $a > 4$ or $a < -4$.

All of the numbers *not* between -4 and 4 are greater than 4 units from 0. The solution set is $\{a \mid a > 4 \text{ or } a < -4\}$.

An absolute value inequality can be solved by rewriting it as a compound inequality.

Key Concept — *Absolute Value Inequalities*

- **Symbols** For all real numbers a and b, $b > 0$, the following statements are true.
  1. If $|a| < b$ then $-b < a < b$.
  2. If $|a| > b$ then $a > b$ or $a < -b$.
- **Examples** If $|2x + 1| < 5$, then $-5 < 2x + 1 < 5$.
  If $|2x + 1| > 5$, then $2x + 1 > 5$ or $2x + 1 < -5$.

These statements are also true for $\leq$ and $\geq$, respectively.

Example 5 Solve a Multi-Step Absolute Value Inequality

Solve $|3x - 12| \geq 6$. Graph the solution set on a number line.

$|3x - 12| \geq 6$ is equivalent to $3x - 12 \geq 6$ or $3x - 12 \leq -6$. Solve each inequality.

$3x - 12 \geq 6$ or $3x - 12 \leq -6$

$3x \geq 18$ $\quad$ $3x \leq 6$

$x \geq 6$ $\quad$ $x \leq 2$

The solution set is $\{x \mid x \geq 6 \text{ or } x \leq 2\}$.

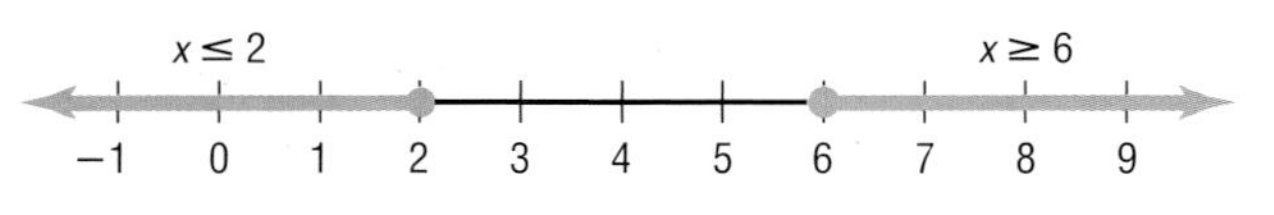

Example 6 Write an Absolute Value Inequality

More About. . .

Job Hunting

When executives in a recent survey were asked to name one quality that impressed them the most about a candidate during a job interview, 32 percent said honesty and integrity.

Source: careerexplorer.net

JOB HUNTING To prepare for a job interview, Megan researches the position's requirements and pay. She discovers that the average starting salary for the position is \$38,500, but her actual starting salary could differ from the average by as much as \$2450.

a. Write an absolute value inequality to describe this situation.

Let x = Megan's starting salary.

Her starting salary could differ from the average	by as much as	\$2450.
$\lvert 38{,}500 - x \rvert$	$\leq$	2450

b. Solve the inequality to find the range of Megan's starting salary.

Rewrite the absolute value inequality as a compound inequality. Then solve for x.

$$-2450 \leq 38{,}500 - x \leq 2450$$
$$-2450 - 38{,}500 \leq 38{,}500 - x - 38{,}500 \leq 2450 - 38{,}500$$
$$-40{,}950 \leq -x \leq -36{,}050$$
$$40{,}950 \geq x \geq 36{,}050$$

The solution set is $\{x \mid 36{,}050 \leq x \leq 40{,}950\}$. Thus, Megan's starting salary will fall between \$36,050 and \$40,950, inclusive.

Check for Understanding

Concept Check

1. **Write** a compound inequality to describe the following situation.
   Buy a present that costs at least \$5 and at most \$15.
2. **OPEN ENDED** Write a compound inequality whose graph is the empty set.
3. **FIND THE ERROR** Sabrina and Isaac are solving $\lvert 3x + 7 \rvert > 2$.

Sabrina	Isaac
$\lvert 3x + 7 \rvert > 2$	$\lvert 3x + 7 \rvert > 2$
$3x + 7 > 2$ or $3x + 7 < -2$	$-2 < 3x + 7 < 2$
$3x > -5$ $3x < -9$	$-9 < 3x < -5$
$x > -\frac{5}{3}$ $x < -3$	$-3 < x < -\frac{5}{3}$

Who is correct? Explain your reasoning.

Guided Practice

Write an absolute value inequality for each of the following. Then graph the solution set on a number line.

4. all numbers between -8 and 8
5. all numbers greater than 3 or less than -3

Write an absolute value inequality for each graph.

6. 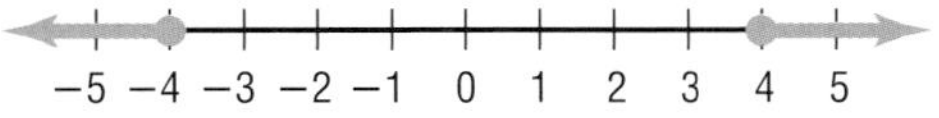

7. 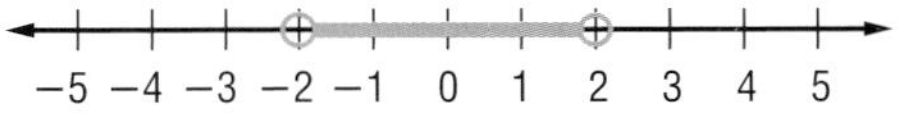

Solve each inequality. Graph the solution set on a number line.

8. $y - 3 > 1$ or $y + 2 < 1$
9. $3 < d + 5 < 8$
10. $|a| \geq 5$
11. $|g + 4| \leq 9$
12. $|4k - 8| < 20$
13. $|w| \geq -2$

Application

14. **FLOORING** Deion estimates that he will need between 55 and 60 ceramic tiles to retile his kitchen floor. If each tile costs \$6.25, write and solve a compound inequality to determine what the cost c of the tile could be.

Practice and Apply

Homework Help	
For Exercises	**See Examples**
15–26, 33–44	3–5
27–32, 51, 52	1, 2
45–50	6

Extra Practice
See page 829.

Write an absolute value inequality for each of the following. Then graph the solution set on a number line.

15. all numbers greater than or equal to 5 or less than or equal to -5
16. all numbers less than 7 and greater than -7
17. all numbers between -4 and 4
18. all numbers less than or equal to -6 or greater than or equal to 6
19. all numbers greater than 8 or less than -8
20. all number less than or equal to 1.2 and greater than or equal to -1.2

Write an absolute value inequality for each graph.

21. 22.

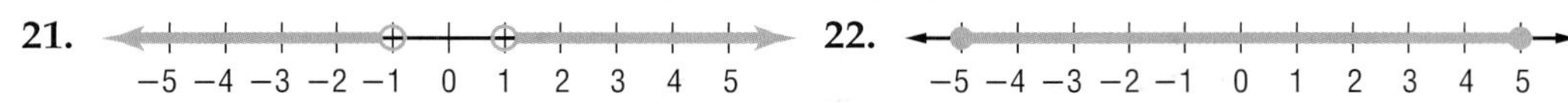

23. 24.

25. 26.

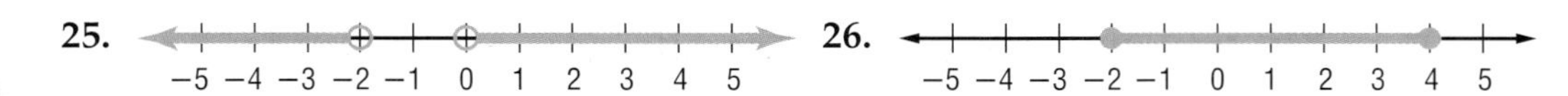

More About. . .

Betta Fish

Adult Male Size: 3 inches
Water pH: 6.8–7.4
Temperature: 75–86°F
Diet: omnivore, prefers live foods
Tank Level: top dweller
Difficulty of Care: easy to intermediate
Life Span: 2–3 years
Source: www.about.com

Solve each inequality. Graph the solution set on a number line.

27. $3p + 1 \leq 7$ or $2p - 9 \geq 7$
28. $9 < 3t + 6 < 15$
29. $-11 < -4x + 5 < 13$
30. $2c - 1 < -5$ or $3c + 2 \geq 5$
31. $-4 < 4f + 24 < 4$
32. $a + 2 > -2$ or $a - 8 < 1$
33. $|g| \leq 9$
34. $|2m| \geq 8$
35. $|3k| < 0$
36. $|-5y| < 35$
37. $|b - 4| > 6$
38. $|6r - 3| < 21$
39. $|3w + 2| \leq 5$
40. $|7x| + 4 < 0$
41. $|n| \geq n$
42. $|n| \leq n$
43. $|2n - 7| \leq 0$
44. $|n - 3| < n$

45. **BETTA FISH** A Siamese Fighting Fish, also known as a Betta fish, is one of the most recognized and colorful fish kept as a pet. Using the information at the left, write a compound inequality to describe the acceptable range of water pH levels for a male Betta.

SPEED LIMITS **For Exercises 46 and 47, use the following information.**
On some interstate highways, the maximum speed a car may drive is 65 miles per hour. A tractor-trailer may not drive more than 55 miles per hour. The minimum speed for all vehicles is 45 miles per hour.

46. Write an inequality to represent the allowable speed for a car on an interstate highway.

47. Write an inequality to represent the speed at which a tractor-trailer may travel on an interstate highway.

48. HEALTH *Hypothermia* and *hyperthermia* are similar words but have opposite meanings. Hypothermia is defined as a lowered body temperature. Hyperthermia means an extremely high body temperature. Both conditions are potentially dangerous and occur when a person's body temperature fluctuates by more than 8° from the normal body temperature of 98.6°F. Write and solve an absolute value inequality to describe body temperatures that are considered potentially dangerous.

MAIL **For Exercises 49 and 50, use the following information.**
The U.S. Postal Service defines an oversized package as one for which the length L of its longest side plus the distance D around its thickest part is more than 108 inches and less than or equal to 130 inches.

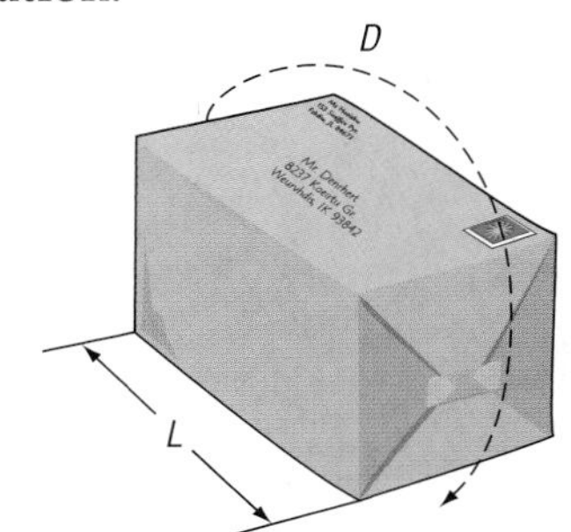

49. Write a compound inequality to describe this situation.

50. If the distance around the thickest part of a package you want to mail is 24 inches, describe the range of lengths that would classify your package as oversized.

GEOMETRY **For Exercises 51 and 52, use the following information.**
The *Triangle Inequality Theorem* states that the sum of the measures of any two sides of a triangle is greater than the measure of the third side.

51. Write three inequalities to express the relationships among the sides of $\triangle ABC$.

52. Write a compound inequality to describe the range of possible measures for side c in terms of a and b. Assume that $a > b > c$. (*Hint*: Solve each inequality you wrote in Exercise 51 for c.)

53. CRITICAL THINKING **Graph each set on a number line.**

a. $-2 < x < 4$

b. $x < -1$ or $x > 3$

c. $(-2 < x < 4)$ and $(x < -1$ or $x > 3)$ (*Hint*: This is the intersection of the graphs in part **a** and part **b**.)

d. Solve $3 < |x + 2| \leq 8$. Explain your reasoning and graph the solution set.

54. **WRITING IN MATH** Answer the question that was posed at the beginning of the lesson.

How are compound inequalities used in medicine?

Include the following in your answer:

- an explanation as to when to use *and* and when to use *or* when writing a compound inequality,
- an alternative way to write $h \geq 10$ and $h \leq 16$, and
- an example of an acceptable number of hours for this fasting state and a graph to support your answer.

Standardized Test Practice

55. SHORT RESPONSE Solve $|2x + 11| > 1$ for x.

56. If $5 < a < 7 < b < 14$, then which of the following best defines $\frac{a}{b}$?

Ⓐ $\frac{5}{7} < \frac{a}{b} < \frac{1}{2}$ Ⓑ $\frac{5}{14} < \frac{a}{b} < \frac{1}{2}$

Ⓒ $\frac{5}{7} < \frac{a}{b} < 1$ Ⓓ $\frac{5}{14} < \frac{a}{b} < 1$

Graphing Calculator

LOGIC MENU For Exercises 57–60, use the following information. You can use the operators in the LOGIC menu on the TI-83 Plus to graph compound and absolute value inequalities. To display the LOGIC menu, press [2nd] [TEST] [▶].

57. Clear the Y= list. Enter (5X + 2 > 12) and (3X − 8 < 1) as Y1. With your calculator in DOT mode and using the standard viewing window, press [GRAPH]. Make a sketch of the graph displayed.

58. Using the TRACE function, investigate the graph. Based on your investigation, what inequality is graphed?

59. Write the expression you would enter for Y1 to find the solution set of the compound inequality $5x + 2 \geq 3$ or $5x + 2 \leq -3$. Then use the graphing calculator to find the solution set.

60. A graphing calculator can also be used to solve absolute value inequalities. Write the expression you would enter for Y1 to find the solution set of the inequality $|2x - 6| > 10$. Then use the graphing calculator to find the solution set. (*Hint*: The absolute value operator is item 1 on the MATH NUM menu.)

Maintain Your Skills

Mixed Review

Solve each inequality. Describe the solution set using set builder or interval notation. Then graph the solution set on a number line. *(Lesson 1-5)*

61. $2d + 15 \geq 3$ **62.** $7x + 11 > 9x + 3$ **63.** $3n + 4(n + 3) < 5(n + 2)$

64. CONTESTS To get a chance to win a car, you must guess the number of keys in a jar to within 5 of the actual number. Those who are within this range are given a key to try in the ignition of the car. Suppose there are 587 keys in the jar. Write and solve an equation to determine the highest and lowest guesses that will give contestants a chance to win the car. *(Lesson 1-4)*

Solve each equation. Check your solutions.

65. $5|x - 3| = 65$ **66.** $|2x + 7| = 15$ **67.** $|8c + 7| = -4$

Name the property illustrated by each statement. *(Lesson 1-3)*

68. If $3x = 10$, then $3x + 7 = 10 + 7$.

69. If $-5 = 4y - 8$, then $4y - 8 = -5$.

70. If $-2x - 5 = 9$ and $9 = 6x + 1$, then $-2x - 5 = 6x + 1$.

Simplify each expression. *(Lesson 1-2)*

71. $6a - 2b - 3a + 9b$ **72.** $-2(m - 4n) - 3(5n + 6)$

Find the value of each expression. *(Lesson 1-1)*

73. $6(5 - 8) \div 9 + 4$ **74.** $(3 + 7)^2 - 16 \div 2$ **75.** $\frac{7(1 - 4)}{8 - 5}$

Chapter 1 Study Guide and Review

Vocabulary and Concept Check

- absolute value (p. 28)
- Addition Property
  - of Equality (p. 21)
  - of Inequality (p. 33)
- algebraic expression (p. 7)
- Associative Property (p. 12)
- Commutative Property (p. 12)
- compound inequality (p. 40)
- counterexample (p. 14)
- Distributive Property (p. 12)
- Division Property
  - of Equality (p. 21)
  - of Inequality (p. 34)
- empty set (p. 29)
- equation (p. 20)
- formula (p. 8)
- Identity Property (p. 12)
- intersection (p. 40)
- interval notation (p. 35)
- Inverse Property (p. 12)
- irrational numbers (p. 11)
- Multiplication Property
  - of Equality (p. 21)
  - of Inequality (p. 34)
- open sentence (p. 20)
- order of operations (p. 6)
- rational numbers (p. 11)
- real numbers (p. 11)
- Reflexive Property (p. 21)
- set-builder notation (p. 34)
- solution (p. 20)
- Substitution Property (p. 21)
- Subtraction Property
  - of Equality (p. 21)
  - of Inequality (p. 33)
- Symmetric Property (p. 21)
- Transitive Property (p. 21)
- Trichotomy Property (p. 33)
- union (p. 41)
- variable (p. 7)

Choose the term from the list above that best matches each example.

1. $y > 3$ or $y < -2$
2. $0 + (-4b) = -4b$
3. $(m - 1)(-2) = -2(m - 1)$
4. $35x + 56 = 7(5x + 8)$
5. $ab + 1 = ab + 1$
6. If $2x = 3y - 4$, $3y - 4 = 7$, then $2x = 7$.
7. $4(0.25) = 1$
8. $2p + (4 + 9r) = (2p + 4) + 9r$
9. $|5n|$
10. $6y + 5z - 2(x + y)$

Lesson-by-Lesson Review

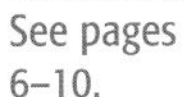

1-1 Expressions and Formulas

See pages 6–10.

Concept Summary

- Order of Operations

Step 1 Simplify the expressions inside grouping symbols, such as parentheses, ( ), brackets, [ ], braces, { }, and fraction bars.

Step 2 Evaluate all powers.

Step 3 Do all multiplications and/or divisions from left to right.

Step 4 Do all additions and/or subtractions from left to right.

Example **Evaluate $\frac{y^3}{3ab + 2}$ if $y = 4$, $a = -2$, and $b = -5$.**

$$\frac{y^3}{3ab + 2} = \frac{4^3}{3(-2)(-5) + 2} \quad y = 4,\ a = -2, \text{ and } b = -5$$

$$= \frac{64}{3(10) + 2} \quad \text{Evaluate the numerator and denominator separately.}$$

$$= \frac{64}{32} \text{ or } 2$$

Exercises **Find the value of each expression.** *See Example 1 on page 6.*

11. $10 + 16 \div 4 + 8$ **12.** $[21 - (9 - 2)] \div 2$ **13.** $\frac{14(8 - 15)}{2}$

Evaluate each expression if $a = 12$, $b = 0.5$, $c = -3$, and $d = \frac{1}{3}$.
See Examples 2 and 3 on page 7.

14. $6b - 5c$ **15.** $c^3 + ad$ **16.** $\frac{9c + ab}{c}$ **17.** $a[b^2(b + a)]$

1-2 Properties of Real Numbers

See pages 11–18.

Concept Summary

- Real numbers (R) can be classified as rational (Q) or irrational (I).
- Rational numbers can be classified as natural numbers (N), whole numbers (W), and/or integers (Z).
- Use the properties of real numbers to simplify algebraic expressions.

Example **Simplify $4(2b + 6c) + 3b - c$.**

$4(2b + 6c) + 3b - c = 4(2b) + 4(6c) + 3b - c$ Distributive Property

$= 8b + 24c + 3b - c$ Multiply.

$= 8b + 3b + 24c - c$ Commutative Property (+)

$= (8 + 3)b + (24 - 1)c$ Distributive Property

$= 11b + 23c$ Add 3 to 8 and subtract 1 from 24.

Exercises **Name the sets of numbers to which each value belongs.**
See Example 1 on page 12.

18. $-\sqrt{9}$ **19.** $1.\overline{6}$ **20.** $\frac{35}{7}$ **21.** $\sqrt{18}$

Simplify each expression. *See Example 5 on page 14.*

22. $2m + 7n - 6m - 5n$ **23.** $-5(a - 4b) + 4b$ **24.** $2(5x + 4y) - 3(x + 8y)$

1-3 Solving Equations

See pages 20–27.

Concept Summary

- Verbal expressions can be translated into algebraic expressions using the language of algebra, using variables to represent the unknown quantities.
- Use the properties of equality to solve equations.

Example **Solve $4(a + 5) - 2(a + 6) = 3$.**

$4(a + 5) - 2(a + 6) = 3$ Original equation

$4a + 20 - 2a - 12 = 3$ Distributive Property

$2a + 8 = 3$ Commutative, Distributive, and Substitution Properties

$2a = -5$ Subtraction Property (=)

$a = -2.5$ Division Property (=)

Exercises **Solve each equation. Check your solution.**
See Examples 4 and 5 on pages 21 and 22.

25. $x - 6 = -20$ **26.** $-\frac{2}{3}a = 14$ **27.** $7 + 5n = -58$
28. $3w + 14 = 7w + 2$ **29.** $5y + 4 = 2(y - 4)$ **30.** $\frac{n}{4} + \frac{n}{3} = \frac{1}{2}$

Solve each equation or formula for the specified variable. *See Example 6 on page 22.*

31. $Ax + By = C$ for x **32.** $\frac{a - 4b^2}{2c} = d$ for a **33.** $A = p + prt$ for p

1-4 Solving Absolute Value Equations

See pages 28–32.

Concept Summary

- For any real numbers a and b, where $b \geq 0$, if $|a| = b$, then $a = b$ or $a = -b$.

Example

Solve $|2x + 9| = 11$.

Case 1 $a = b$ **or** **Case 2** $a = -b$

$$2x + 9 = 11 \qquad 2x + 9 = -11$$
$$2x = 2 \qquad 2x = -20$$
$$x = 1 \qquad x = -10$$

The solution set is $\{1, -10\}$. Check these solutions in the original equation.

Exercises **Solve each equation. Check your solutions.**
See Examples 1–4 on pages 28–30.

34. $|x + 11| = 42$ **35.** $3|x + 6| = 36$ **36.** $|4x - 5| = -25$
37. $|x + 7| = 3x - 5$ **38.** $|y - 5| - 2 = 10$ **39.** $4|3x + 4| = 4x + 8$

1-5 Solving Inequalities

See pages 33–39.

Concept Summary

- Adding the same number to, or subtracting the same number from, each side of an inequality does not change the truth of the inequality.
- When you multiply or divide each side of an inequality by a negative number, the direction of the inequality symbol must be *reversed*.

Example

Solve $5 - 4a > 8$. Graph the solution set on a number line.

$5 - 4a > 8$ Original inequality
$-4a > 3$ Subtract 5 from each side.
$a < -\frac{3}{4}$ Divide each side by -4, reversing the inequality symbol.

The solution set is $\left\{a \,\middle|\, a < -\frac{3}{4}\right\}$.

The graph of the solution set is shown at the right.

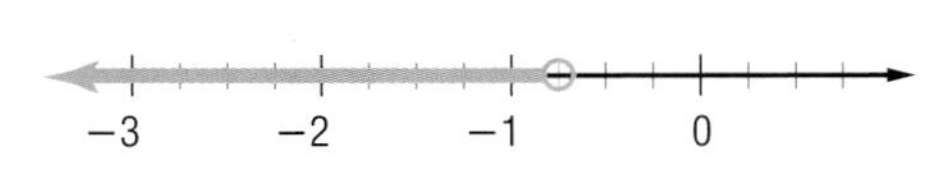

Chapter 1 **For More ...**

- Extra Practice, see pages 828–829.
- Mixed Problem Solving, see page 862.

Exercises **Solve each inequality. Describe the solution set using set builder or interval notation. Then graph the solution set on a number line.**
See Examples 1–3 on pages 34–35.

40. $-7w > 28$ **41.** $3x + 4 \geq 19$ **42.** $\frac{n}{12} + 5 \leq 7$

43. $3(6 - 5a) < 12a - 36$ **44.** $2 - 3z \geq 7(8 - 2z) + 12$ **45.** $8(2x - 1) > 11x - 17$

1-6 Solving Compound and Absolute Value Inequalities

See pages 40–46.

Concept Summary

- The graph of an *and* compound inequality is the intersection of the solution sets of the two inequalities.
- The graph of an *or* compound inequality is the union of the solution sets of the two inequalities.
- For all real numbers a and b, $b > 0$, the following statements are true.
  1. If $|a| < b$ then $-b < a < b$.
  2. If $|a| > b$ then $a > b$ or $a < -b$.

Examples **Solve each inequality. Graph the solution set on a number line.**

1 $-19 < 4d - 7 \leq 13$

$-19 < 4d - 7 \leq 13$ Original inequality

$-12 < 4d \leq 20$ Add 7 to each part.

$-3 < d \leq 5$ Divide each part by 4.

The solution set is $\{d \mid -3 < d \leq 5\}$.

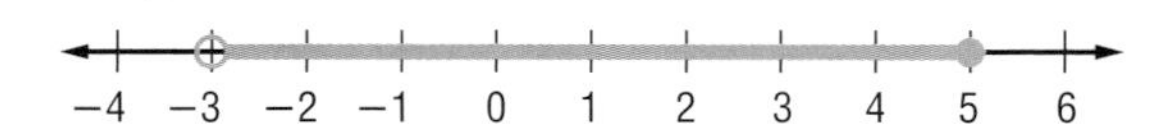

2 $|2x + 4| \geq 12$

$|2x + 4| \geq 12$ is equivalent to $2x + 4 \geq 12$ or $2x + 4 \leq -12$.

$2x + 4 \geq 12$ or $2x + 4 \leq -12$ Original inequality

$2x \geq 8$ $\quad 2x \leq -16$ Subtract 4 from each side.

$x \geq 4$ $\quad x \leq -8$ Divide each side by 2.

The solution set is $\{x \mid x \geq 4 \text{ or } x \leq -8\}$.

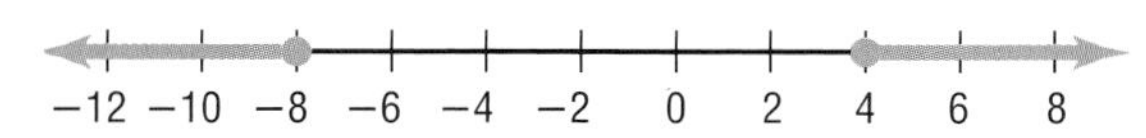

Exercises **Solve each inequality. Graph the solution set on a number line.**
See Examples 1–5 on pages 40–42.

46. $-1 < 3a + 2 < 14$ **47.** $-1 < 3(y - 2) \leq 9$ **48.** $|x| + 1 > 12$

49. $|2y - 9| \leq 27$ **50.** $|5n - 8| > -4$ **51.** $|3b + 11| > 1$

Chapter 1 Practice Test

Vocabulary and Concepts

Choose the term that best completes each sentence.

1. An algebraic *(equation, expression)* contains an equals sign.
2. *(Whole numbers, Rationals)* are a subset of the set of integers.
3. If $x + 3 = y$, then $y = x + 3$ is an example of the *(Transitive, Symmetric)* Property of Equality.

Skills and Applications

Find the value of each expression.

4. $[(3 + 6)^2 \div 3] \times 4$
5. $\frac{20 + 4 \times 3}{11 - 3}$
6. $0.5(2.3 + 25) \div 1.5$

Evaluate each expression if $a = -9$, $b = \frac{2}{3}$, $c = 8$, and $d = -6$.

7. $\frac{db + 4c}{a}$
8. $\frac{a}{b^2} + c$
9. $2b(4a + a^2)$

Name the sets of numbers to which each number belongs.

10. $\sqrt{17}$
11. 0.86
12. $\sqrt{64}$

Name the property illustrated by each equation or statement.

13. $(7 \cdot s) \cdot t = 7 \cdot (s \cdot t)$
14. If $(r + s)t = rt + st$, then $rt + st = (r + s)t$.
15. $\left(3 \cdot \frac{1}{3}\right) \cdot 7 = \left(3 \cdot \frac{1}{3}\right) \cdot 7$
16. $(6 - 2)a - 3b = 4a - 3b$
17. $(4 + x) + y = y + (4 + x)$
18. If $5(3) + 7 = 15 + 7$ and $15 + 7 = 22$, then $5(3) + 7 = 22$.

Solve each equation. Check your solution(s).

19. $5t - 3 = -2t + 10$
20. $2x - 7 - (x - 5) = 0$
21. $5m - (5 + 4m) = (3 + m) - 8$
22. $|8w + 2| + 2 = 0$
23. $12\left|\frac{1}{2}y + 3\right| = 6$
24. $2|2y - 6| + 4 = 8$

Solve each inequality. Describe the solution set using set builder or interval notation. Then graph the solution set on a number line.

25. $4 > b + 1$
26. $3q + 7 \geq 13$
27. $5(3x - 5) + x < 2(4x - 1) + 1$
28. $|5 + k| \leq 8$
29. $-12 < 7d - 5 \leq 9$
30. $|3y - 1| > 5$

For Exercises 31 and 32, define a variable, write an equation or inequality, and solve the problem.

31. **CAR RENTAL** Mrs. Denney is renting a car that gets 35 miles per gallon. The rental charge is \$19.50 a day plus 18¢ per mile. Her company will reimburse her for \$33 of this portion of her travel expenses. If Mrs. Denney rents the car for 1 day, find the maximum number of miles that will be paid for by her company.
32. **SCHOOL** To receive a B in his English class, Nick must have an average score of at least 80 on five tests. He scored 87, 89, 76, and 77 on his first four tests. What must he score on the last test to receive a B in the class?

Standards Practice

33. **STANDARDIZED TEST PRACTICE** If $\frac{a}{b} = 8$ and $ac - 5 = 11$, then $bc =$

Ⓐ 93. Ⓑ 2. Ⓒ $\frac{5}{8}$. Ⓓ cannot be determined

Chapter 1

Standardized Test Practice

Standards Practice

Part 1 Multiple Choice

Record your answers on the answer sheet provided by your teacher or on a sheet of paper.

1. In the square at the right, what is the value of x?

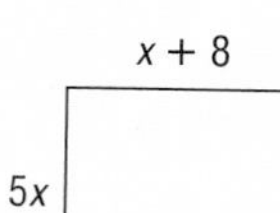

Ⓐ 1 Ⓑ 2
Ⓒ 3 Ⓓ 4

2. On a college math test, 18 students earned an A. This number is exactly 30% of the total number of students in the class. How many students are in the class?

Ⓐ 5 Ⓑ 23
Ⓒ 48 Ⓓ 60

3. A student computed the average of her 7 test scores by adding the scores together and dividing this total by the number of tests. The average was 87. On her next test, she scored a 79. What is her new test average?

Ⓐ 83 Ⓑ 84
Ⓒ 85 Ⓓ 86

4. If the perimeter of $\triangle PQR$ is 3 times the length of PQ, then $PR =$ _____.

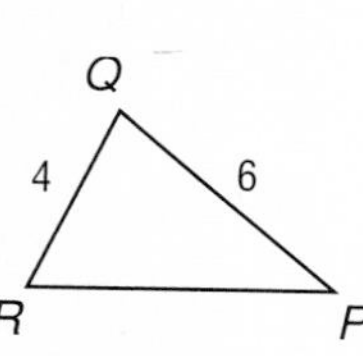

Note: Figure not drawn to scale.

Ⓐ 4 Ⓑ 6
Ⓒ 7 Ⓓ 8

5. If a different number is selected from each of the three sets shown below, what is the greatest sum these 3 numbers could have?

R = {3, 6, 7}; S = {2, 4, 7}; T = {1, 3, 7}

Ⓐ 13 Ⓑ 14
Ⓒ 17 Ⓓ 21

6. A pitcher contains a ounces of orange juice. If b ounces of juice are poured from the pitcher into each of c glasses, which expression represents the amount of juice remaining in the pitcher?

Ⓐ $\frac{a}{b} + c$ Ⓑ $ab - c$
Ⓒ $a - bc$ Ⓓ $\frac{a}{bc}$

7. The sum of three consecutive integers is 135. What is the greatest of the three integers?

Ⓐ 43 Ⓑ 44
Ⓒ 45 Ⓓ 46

8. The ratio of girls to boys in a class is 5 to 4. If there are a total of 27 students in the class, how many are girls?

Ⓐ 15 Ⓑ 12
Ⓒ 9 Ⓓ 5

9. For which of the following ordered pairs (x, y) is $x + y > 3$ and $x - y < -2$?

Ⓐ (0, 3) Ⓑ (3, 4)
Ⓒ (5, 3) Ⓓ (2, 5)

10. If the area of $\triangle ABD$ is 280, what is the area of the polygon $ABCD$?

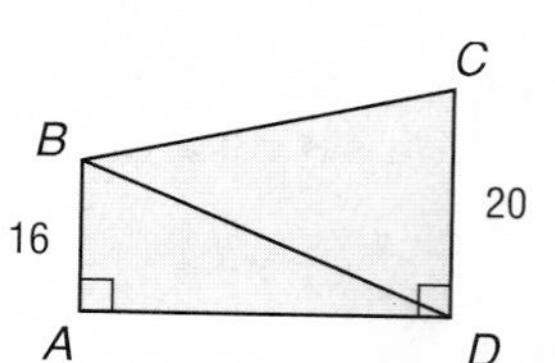

Note: Figure not drawn to scale.

Ⓐ 560 Ⓑ 630
Ⓒ 700 Ⓓ 840

Test-Taking Tip

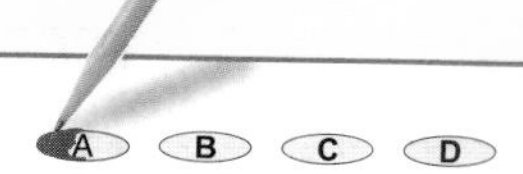

Question 9
To solve equations or inequalities, you can replace the variables in the question with the values given in each answer choice. The answer chioce that results in true statements is the correct answer choice.

Preparing for Standardized Tests
For test-taking strategies and more practice, see pages 877–892.

Part 2 Short Response/Grid In

Record your answers on the answer sheet provided by your teacher or on a sheet of paper.

11. In the triangle below, x and y are integers. If $25 < y < 30$, what is one possible value of x?

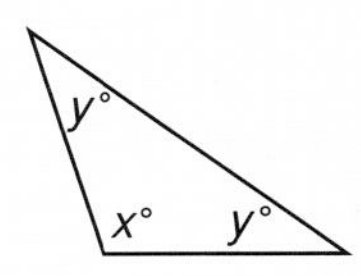

12. If n and p are each different positive integers and $n + p = 4$, what is one possible value of $3n + 4p$?

13. In the figure at the right, what is the value of x?

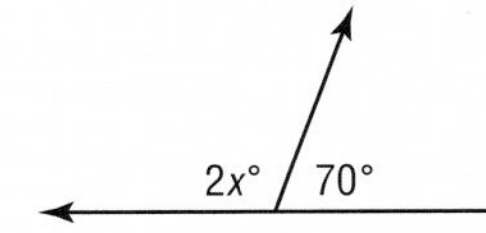

14. One half quart of lemonade concentrate is mixed with $1\frac{1}{2}$ quarts of water to make lemonade for 6 people. If you use the same proportions of concentrate and water, how many quarts of lemonade concentrate are needed to make lemonade for 21 people?

15. If 25 percent of 300 is equal to 500 percent of t, then t is equal to what number?

16. In the figure below, what is the area of the shaded square in square units?

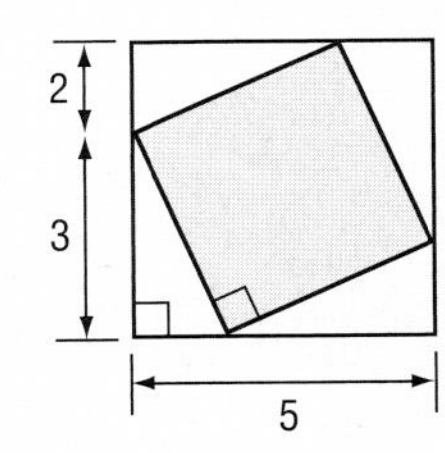

17. There are 140 students in the school band. One of these students will be selected at random to be the student representative. If the probability that a brass player is selected is $\frac{2}{5}$, how many brass players are in the band?

18. A shelf holds fewer than 50 cans. If all of the cans on this shelf were put into stacks of five cans each, no cans would remain. If the same cans were put into stacks of three cans each, one can would remain. What is the greatest number of cans that could be on the shelf?

19. The area of a trapezoid is $\frac{1}{2}h(b_1 + b_2)$, where h is the altitude, and b_1 and b_2 are the lengths of the parallel bases. If a trapezoid has an altitude of 8 inches, an area of 56 square inches, and one base 4 inches long, what is the length of its other base in inches?

Part 3 Extended Response

Record your answers on a sheet of paper. Show your work.

For Exercises 20–22, use the information below and in the table.

Amanda's hours at her summer job for one week are listed in the table below. She earns $6 per hour.

Amanda's Work Hours	
Sunday	0
Monday	6
Tuesday	4
Wednesday	0
Thursday	2
Friday	6
Saturday	8

20. Write an expression for Amanda's total weekly earnings.

21. Evaluate the expression from Exercise 20 by using the Distributive Property.

22. Michael works with Amanda and also earns $6 per hour. If Michael's earnings were $192 this week, write and solve an equation to find how many more hours Michael worked than Amanda.

Chapter 2

Linear Relations and Functions

What You'll Learn

- **Lesson 2-1** Analyze relations and functions.
- **Lessons 2-2 and 2-4** Identify, graph, and write linear equations.
- **Lesson 2-3** Find the slope of a line.
- **Lesson 2-5** Draw scatter plots and find prediction equations.
- **Lessons 2-6 and 2-7** Graph special functions, linear inequalities, and absolute value inequalities.

Key Vocabulary

- linear equation (p. 63)
- linear function (p. 63)
- slope (p. 68)
- slope-intercept form (p. 75)
- point-slope form (p. 76)

Why It's Important

Linear equations can be used to model relationships between many real-world quantities. One of the most common uses of a linear model is to make predictions.

Most hot springs are the result of groundwater passing through or near recently formed, hot, igneous rocks. Iceland, Yellowstone Park in the United States, and North Island of New Zealand are noted for their hot springs. *You will use a linear equation to find the temperature of underground rocks in Lesson 2-2.*

Getting Started

Prerequisite Skills To be successful in this chapter, you'll need to master these skills and be able to apply them in problem-solving situations. Review these skills before beginning Chapter 2.

For Lesson 2-1 — Identify Points on a Coordinate Plane

Write the ordered pair for each point.

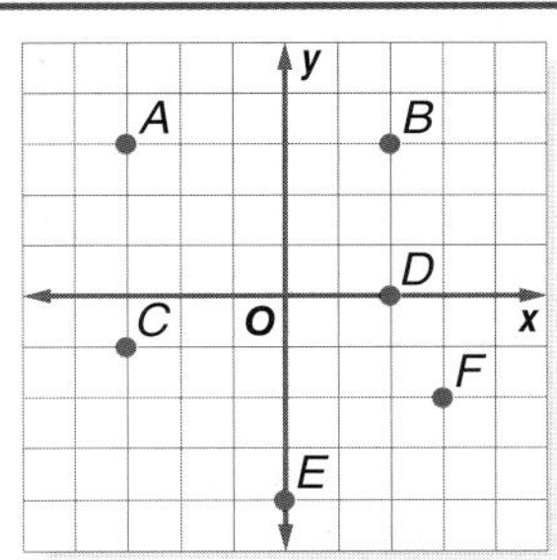

1. A **2.** B
3. C **4.** D
5. E **6.** F

For Lesson 2-1 — Evaluate Expressions

Evaluate each expression if $a = -1$, $b = 3$, $c = -2$, and $d = 0$. *(For review, see Lesson 1-1.)*

7. $c + d$ **8.** $4c - b$ **9.** $a^2 - 5a + 3$
10. $2b^2 + b + 7$ **11.** $\frac{a - b}{c - d}$ **12.** $\frac{a + c}{b + c}$

For Lesson 2-4 — Simplify Expressions

Simplify each expression. *(For review, see Lesson 1-2.)*

13. $x - (-1)$ **14.** $x - (-5)$ **15.** $2[x - (-3)]$
16. $4[x - (-2)]$ **17.** $\frac{1}{2}[x - (-4)]$ **18.** $\frac{1}{3}[x - (-6)]$

For Lessons 2-6 and 2-7 — Evaluate Expressions with Absolute Value

Evaluate each expression if $x = -3$, $y = 4$, and $z = -4.5$. *(For review, see Lesson 1-4.)*

19. $|x|$ **20.** $|y|$ **21.** $|5x|$
22. $-|2z|$ **23.** $5|y + z|$ **24.** $-3|x + y| - |x + z|$

Relations and Functions Make this Foldable to help you organize your notes. Begin with two sheets of grid paper.

Step 1 Fold

Fold in half along the width and staple along the fold.

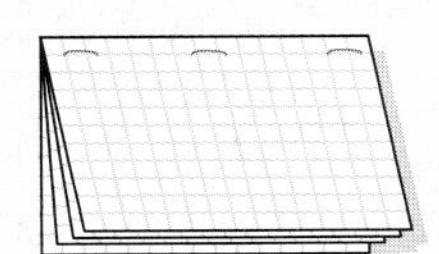

Step 2 Cut and Label

Cut the top three sheets and label as shown.

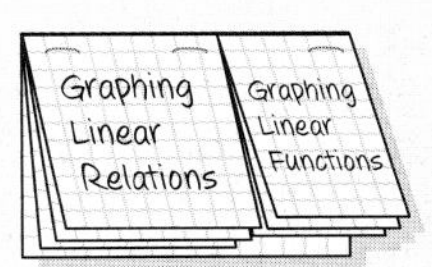

Reading and Writing As you read and study the chapter, write notes, examples, and graphs under the tabs.

2-1 Relations and Functions

What You'll Learn

- Analyze and graph relations.
- Find functional values.

Vocabulary

- ordered pair
- Cartesian coordinate plane
- quadrant
- relation
- domain
- range
- function
- mapping
- one-to-one function
- vertical line test
- independent variable
- dependent variable
- functional notation

How do relations and functions apply to biology?

The table shows the average lifetime and maximum lifetime for some animals. The data can also be represented as **ordered pairs**. The ordered pairs for the data are (12, 28), (15, 30), (8, 20), (12, 20), and (20, 50). The first number in each ordered pair is the average lifetime, and the second number is the maximum lifetime.

(12, 28)
average lifetime ↑ ↑ maximum lifetime

Animal	Average Lifetime (years)	Maximum Lifetime (years)
Cat	12	28
Cow	15	30
Deer	8	20
Dog	12	20
Horse	20	50

Source: *The World Almanac*

GRAPH RELATIONS You can graph the ordered pairs above on a *coordinate system* with two axes. Remember that each point in the coordinate plane can be named by exactly one ordered pair and that every ordered pair names exactly one point in the coordinate plane.

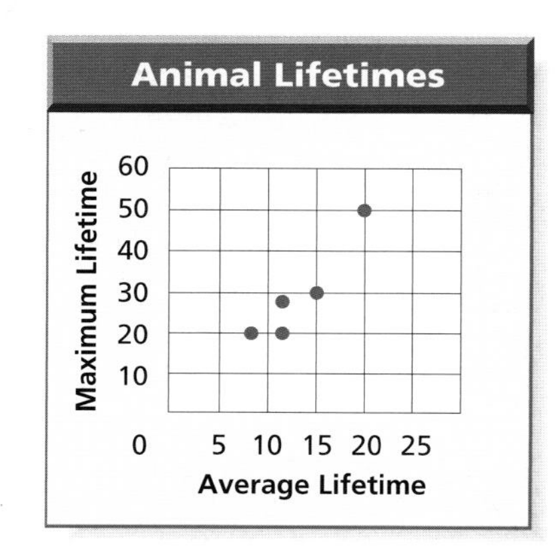

The graph of the animal lifetime data lies in only one part of the Cartesian coordinate plane—the part with all positive numbers. The **Cartesian coordinate plane** is composed of the *x-axis* (horizontal) and the *y-axis* (vertical), which meet at the *origin* (0, 0) and divide the plane into four **quadrants**. *The points on the two axes do not lie in any quadrant.*

In general, any ordered pair in the coordinate plane can be written in the form (x, y).

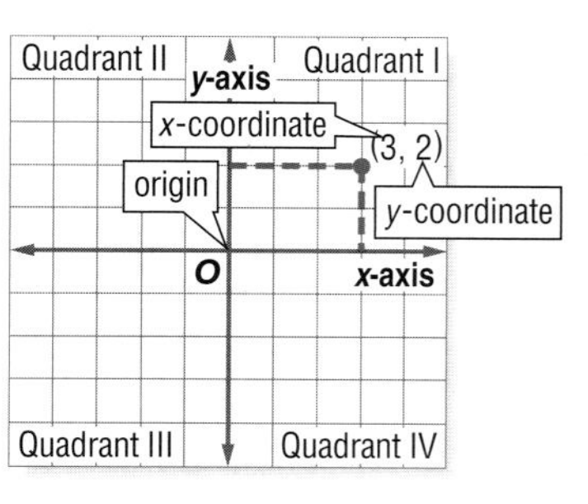

Assume that each square on a graph represents 1 unit unless otherwise labeled.

A **relation** is a set of ordered pairs, such as the one for the longevity of animals. The **domain** of a relation is the set of all first coordinates (x-coordinates) from the ordered pairs, and the **range** is the set of all second coordinates (y-coordinates) from the ordered pairs. The *graph* of a relation is the set of points in the coordinate plane corresponding to the ordered pairs in the relation.

Study Tip

Reading Math
An x-coordinate is sometimes called an *abscissa*, and a y-coordinate is sometimes called an *ordinate*.

A **function** is a special type of relation in which each element of the domain is paired with *exactly one* element of the range. A **mapping** shows how each member of the domain is paired with each member of the range.

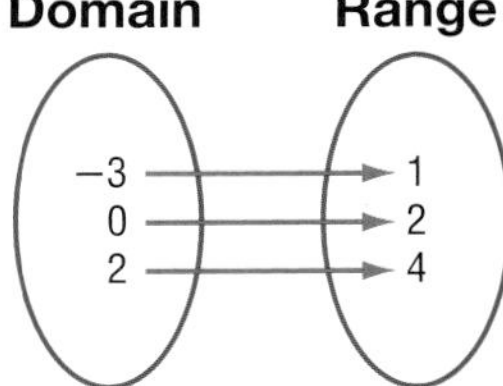

The first two relations shown below are functions. The third relation is not a function because the −3 in the domain is paired with both 0 and 6 in the range. A function like the first one below, where each element of the range is paired with exactly one element of the domain, is called a **one-to-one function**.

Concept Summary — *Functions*

{(−3, 1), (0, 2), (2, 4)}

Domain Range

−3 → 1
0 → 2
2 → 4

one-to-one function

{(−1, 5), (1, 3), (4, 5)}

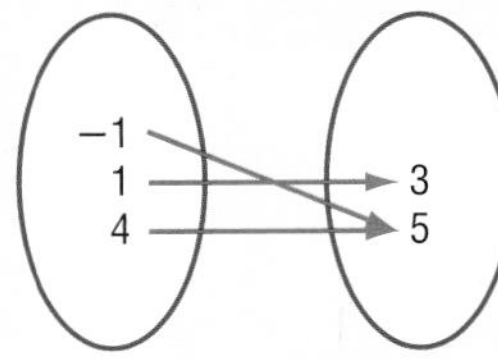

function,
not one-to-one

{(5, 6), (−3, 0), (1, 1), (−3, 6)}

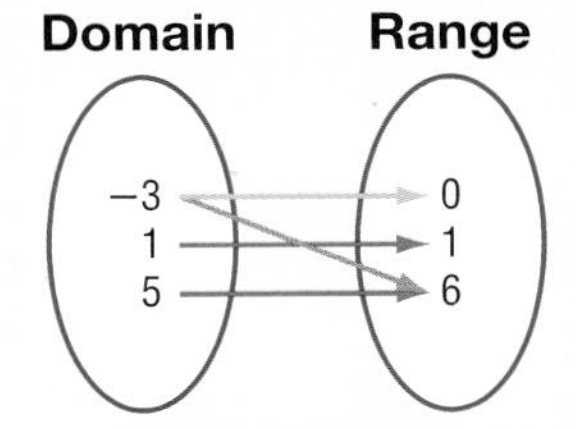

not a function

Example 1 Domain and Range

State the domain and range of the relation shown in the graph. Is the relation a function?

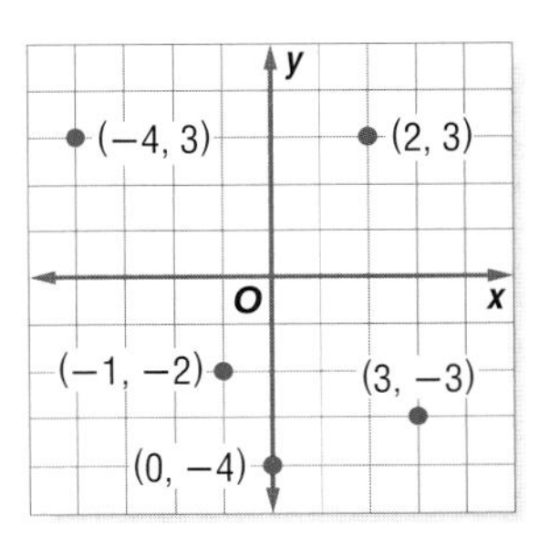

The relation is {(−4, 3), (−1, −2), (0, −4), (2, 3), (3, −3)}.
The domain is {−4, −1, 0, 2, 3}.
The range is {−4, −3, −2, 3}.

Each member of the domain is paired with exactly one member of the range, so this relation is a function.

You can use the **vertical line test** to determine whether a relation is a function.

Key Concept — *Vertical Line Test*

- **Words** If no vertical line intersects a graph in more than one point, the graph represents a function.

  If some vertical line intersects a graph in two or more points, the graph does not represent a function.

- **Models**

y O x

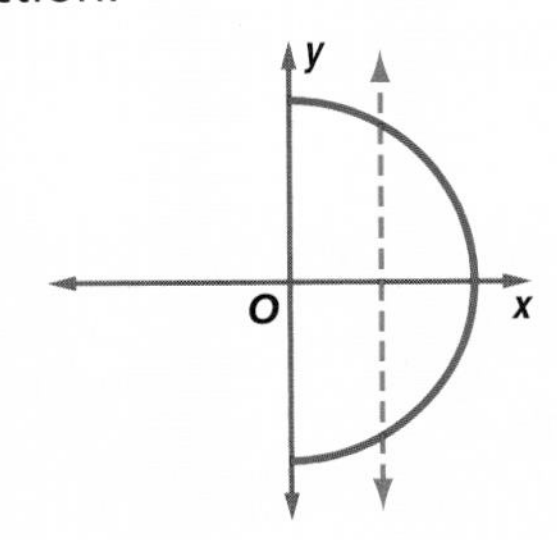

In Example 1, there is no vertical line that contains more than one of the points. Therefore, the relation is a function.

Example 2 Vertical Line Test

GEOGRAPHY The table shows the population of the state of Indiana over the last several decades. Graph this information and determine whether it represents a function.

Year	Population (millions)
1950	3.9
1960	4.7
1970	5.2
1980	5.5
1990	5.5
2000	6.1

Source: U.S. Census Bureau

Population of Indiana

Population (millions)

7 6 5 4 3 2 0

3.9 4.7 5.2 5.5 5.5 6.1

'50 '60 '70 '80 '90 '00

Year

Use the vertical line test. Notice that no vertical line can be drawn that contains more than one of the data points. Therefore, this relation is a function. *Notice also that each year is paired with only one population value.*

> **Study Tip**
>
> *Vertical Line Test*
>
> You can use a pencil to represent a vertical line. Slowly move the pencil to the right across the graph to see if it intersects the graph at more than one point.

EQUATIONS OF FUNCTIONS AND RELATIONS Relations and functions can also be represented by equations. The solutions of an equation in x and y are the set of ordered pairs (x, y) that make the equation true.

Consider the equation $y = 2x - 6$. Since x can be any real number, the domain has an infinite number of elements. To determine whether an equation represents a function, it is often simplest to look at the graph of the relation.

Example 3 Graph Is a Line

a. Graph the relation represented by $y = 2x + 1$.

Make a table of values to find ordered pairs that satisfy the equation. Choose values for x and find the corresponding values for y. Then graph the ordered pairs.

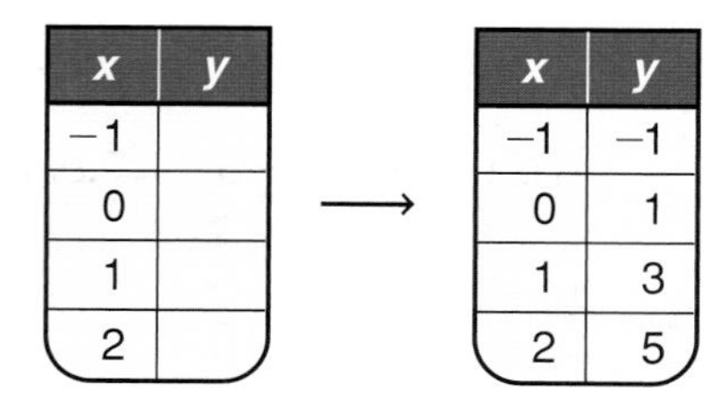

x	y
−1	
0	
1	
2	

→

x	y
−1	−1
0	1
1	3
2	5

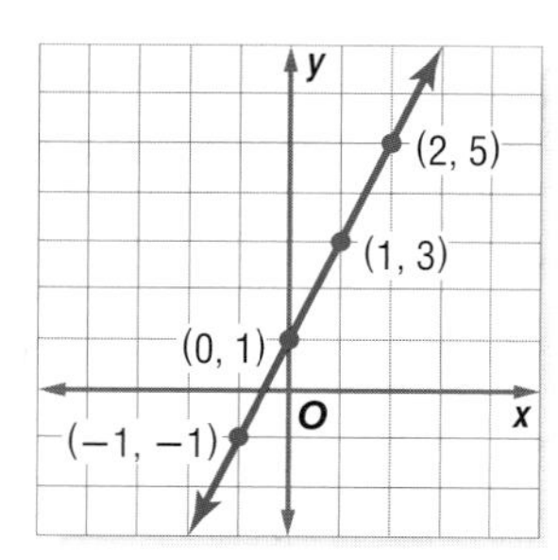

b. Find the domain and range.

Since x can be any real number, there is an infinite number of ordered pairs that can be graphed. All of them lie on the line shown. Notice that every real number is the x-coordinate of some point on the line. Also, every real number is the y-coordinate of some point on the line. So the domain and range are both all real numbers.

c. Determine whether the relation is a function.

This graph passes the vertical line test. For each x value, there is exactly one y value, so the equation $y = 2x + 1$ represents a function.

Example 4 Graph Is a Curve

a. Graph the relation represented by $x = y^2 - 2$.

Make a table. In this case, it is easier to choose y values and then find the corresponding values for x. Then sketch the graph, connecting the points with a smooth curve.

x	y
	−2
	−1
	0
	1
	2

→

x	y
2	−2
−1	−1
−2	0
−1	1
2	2

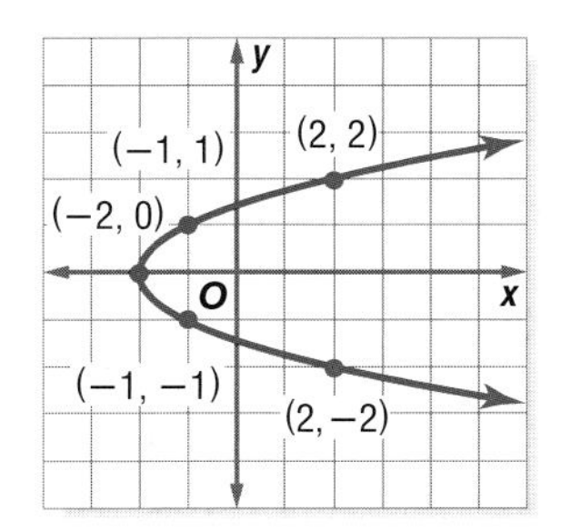

b. Find the domain and range.

Every real number is the y-coordinate of some point on the graph, so the range is all real numbers. But, only real numbers greater than or equal to −2 are x-coordinates of points on the graph. So the domain is $\{x \mid x \geq -2\}$.

c. Determine whether the relation is a function.

You can see from the table and the vertical line test that there are two y values for each x value except $x = -2$. Therefore, the equation $x = y^2 - 2$ does not represent a function.

Study Tip

Reading Math
Suppose you have a job that pays by the hour. Since your pay *depends* on the number of hours you work, you might say that your pay is a *function* of the number of hours you work.

When an equation represents a function, the variable, usually x, whose values make up the domain is called the **independent variable**. The other variable, usually y, is called the **dependent variable** because its values depend on x.

Equations that represent functions are often written in **functional notation**. The equation $y = 2x + 1$ can be written as $f(x) = 2x + 1$. The symbol $f(x)$ replaces the y and is read "f of x." The f is just the name of the function. It is not a variable that is multiplied by x. Suppose you want to find the value in the range that corresponds to the element 4 in the domain of the function. This is written as $f(4)$ and is read "f of 4." The value $f(4)$ is found by substituting 4 for each x in the equation. Therefore, $f(4) = 2(4) + 1$ or 9. *Letters other than f can be used to represent a function. For example, $g(x) = 2x + 1$.*

Example 5 Evaluate a Function

Given $f(x) = x^2 + 2$ and $g(x) = 0.5x^2 - 5x + 3.5$, find each value.

a. $f(-3)$

$f(x) = x^2 + 2$ Original function

$f(-3) = (-3)^2 + 2$ Substitute.

$= 9 + 2$ or 11 Simplify.

b. $g(2.8)$

$g(x) = 0.5x^2 - 5x + 3.5$ Original function

$g(2.8) = 0.5(2.8)^2 - 5(2.8) + 3.5$ **Estimate:** $g(3) = 0.5(3)^2 - 5(3) + 3.5$ or −7

$= 3.92 - 14 + 3.5$ Multiply.

$= -6.58$ Compare with the estimate.

c. $f(3z)$

$f(x) = x^2 + 2$ Original function

$f(3z) = (3z)^2 + 2$ Substitute.

$= 9z^2 + 2$ $(ab)^2 = a^2b^2$

Check for Understanding

Concept Check

1. **OPEN ENDED** Write a relation of four ordered pairs that is *not* a function.
2. **Copy** the graph at the right. Then draw a vertical line that shows that the graph does not represent a function.

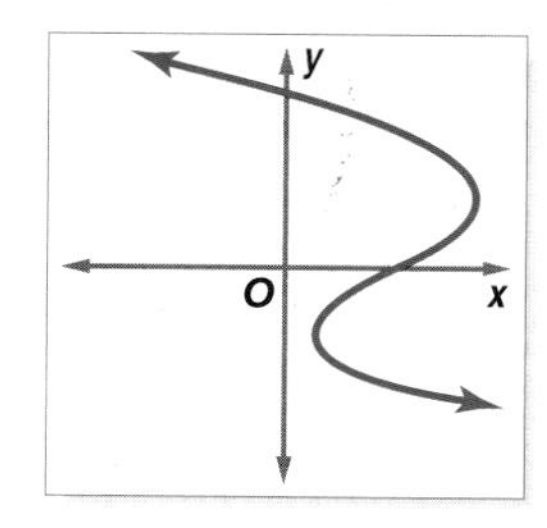

3. **FIND THE ERROR** Teisha and Molly are finding $g(2a)$ for the function $g(x) = x^2 + x - 1$.

Teisha	Molly
$g(2a) = 2(a^2 + a - 1)$	$g(2a) = (2a)^2 + 2a - 1$
$= 2a^2 + 2a - 2$	$= 4a^2 + 2a - 1$

Who is correct? Explain your reasoning.

Guided Practice **Determine whether each relation is a function. Write *yes* or *no*.**

4.

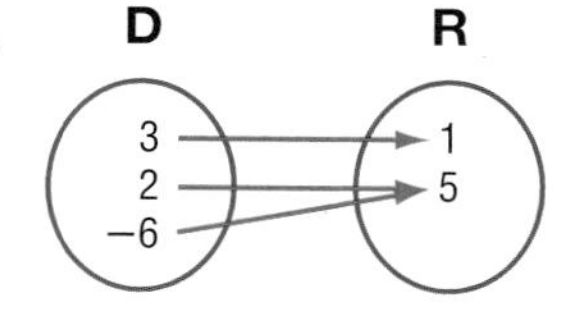

5.

x	y
5	−2
10	−2
15	−2
20	−2

6.

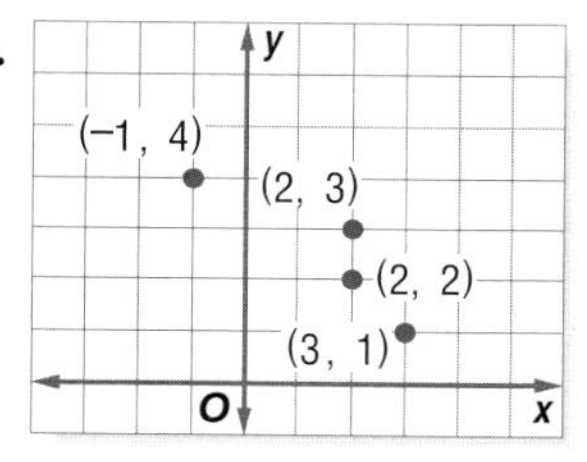

Graph each relation or equation and find the domain and range. Then determine whether the relation or equation is a function.

7. $\{(7, 8), (7, 5), (7, 2), (7, -1)\}$
8. $\{(6, 2.5), (3, 2.5), (4, 2.5)\}$
9. $y = -2x + 1$
10. $x = y^2$
11. Find $f(5)$ if $f(x) = x^2 - 3x$.
12. Find $h(-2)$ if $h(x) = x^3 + 1$.

Application **WEATHER** **For Exercises 13–16, use the table of record high temperatures (°F) for January and July.**

13. Identify the domain and range. Assume that the January temperatures are the domain.
14. Write a relation of ordered pairs for the data.
15. Graph the relation.
16. Is this relation a function? Explain.

City	Jan.	July
Los Angeles	88	97
Sacramento	70	114
San Diego	88	95
San Francisco	72	105

Source: U.S. National Oceanic and Atmospheric Administration

Practice and Apply

Homework Help

For Exercises	See Examples
17–28	1, 2
29–32	3
33, 34	4
35–45, 55	2
46–54, 56	5

Extra Practice
See page 830.

Determine whether each relation is a function. Write *yes* or *no*.

17. D: 10, 20, 30; R: 1, 2, 3

18. D: 3, 2, −1; R: 1, 3, 5, 7

19.

x	y
0.5	−3
2	0.8
0.5	8

20.

x	y
2000	$4000
2001	$4300
2002	$4000
2003	$4500

21.

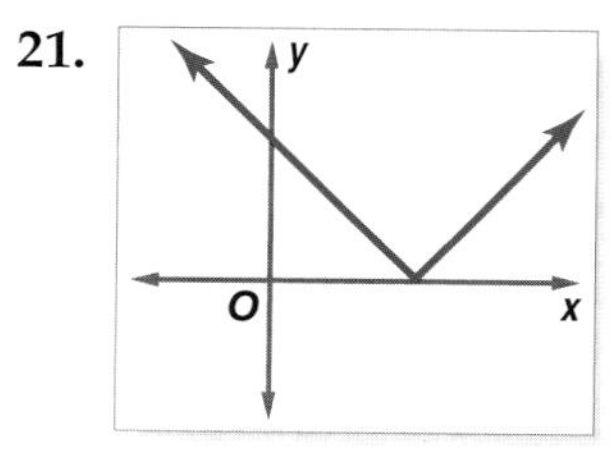

22.

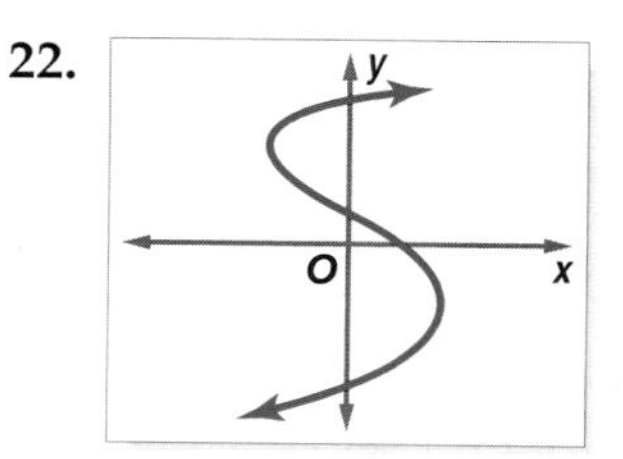

Graph each relation or equation and find the domain and range. Then determine whether the relation or equation is a function.

23. $\{(2, 1), (-3, 0), (1, 5)\}$
24. $\{(4, 5), (6, 5), (3, 5)\}$
25. $\{(-2, 5), (3, 7), (-2, 8)\}$
26. $\{(3, 4), (4, 3), (6, 5), (5, 6)\}$
27. $\{(0, -1.1), (2, -3), (1.4, 2), (-3.6, 8)\}$
28. $\{(-2.5, 1), (-1, -1), (0, 1), (-1, 1)\}$
29. $y = -5x$
30. $y = 3x$
31. $y = 3x - 4$
32. $y = 7x - 6$
33. $y = x^2$
34. $x = 2y^2 - 3$

SPORTS **For Exercises 35–37, use the table that shows the leading home run and runs batted in totals in the American League for 1996–2000.**

Year	1996	1997	1998	1999	2000
HR	52	56	56	48	47
RBI	148	147	157	165	145

Source: *The World Almanac*

35. Make a graph of the data with home runs on the horizontal axis and runs batted in on the vertical axis.
36. Identify the domain and range.
37. Does the graph represent a function? Explain your reasoning.

More About. . .

Sports

The major league record for runs batted in (RBIs) is 191 by Hack Wilson.

Source: www.baseball-almanac.com

FINANCE **For Exercises 38–41, use the table that shows a company's stock price in recent years.**

Year	Price
1997	\$39
1998	\$43
1999	\$48
2000	\$55
2001	\$61
2002	\$52

38. Write a relation to represent the data.
39. Graph the relation.
40. Identify the domain and range.
41. Is the relation a function? Explain your reasoning.

GOVERNMENT **For Exercises 42–45, use the table below that shows the number of members of the U.S. House of Representatives with 30 or more consecutive years of service in Congress from 1987 to 1999.**

Year	1987	1989	1991	1993	1995	1997	1999
Representatives	12	13	11	12	9	6	3

Source: *Congressional Directory*

42. Write a relation to represent the data.
43. Graph the relation.
44. Identify the domain and range.
45. Is the relation a function? If so, is it a one-to-one function? Explain.

Find each value if $f(x) = 3x - 5$ and $g(x) = x^2 - x$.

46. $f(-3)$
47. $g(3)$
48. $g\left(\frac{1}{3}\right)$
49. $f\left(\frac{2}{3}\right)$
50. $f(a)$
51. $g(5n)$

52. Find the value of $f(x) = -3x + 2$ when $x = 2$.

53. What is $g(4)$ if $g(x) = x^2 - 5$?

54. **HOBBIES** Chaz has a collection of 15 CDs. After he gets a part-time job, he decides to buy 3 more CDs every time he goes to the music store. The function $C(t) = 15 + 3t$ counts the number of CDs, $C(t)$, he has after t trips to the music store. How many CDs will he have after he has been to the music store 8 times?

55. **CRITICAL THINKING** If $f(3a - 1) = 12a - 7$, find $f(x)$.

56. **WRITING IN MATH** Answer the question that was posed at the beginning of the lesson.

How do relations and functions apply to biology?

Include the following in your answer:

- an explanation of how a relation can be used to represent data, and
- a sentence that includes the words *average lifetime, maximum lifetime,* and *function.*

Standards Practice

Standardized Test Practice

57. If $f(x) = 2x - 5$, then $f(0) =$

Ⓐ 0. Ⓑ −5. Ⓒ −3. Ⓓ $\frac{5}{2}$.

58. If $g(x) = x^2$, then $g(x + 1) =$

Ⓐ 1. Ⓑ $x^2 + 1$. Ⓒ $x^2 + 2x + 1$. Ⓓ $x^2 - x$.

Extending the Lesson

A function whose graph consists of disconnected points is called a *discrete function.* A function whose graph you can draw without lifting your pencil is called a *continuous function.* Determine whether each function is *discrete* or *continuous.*

59.

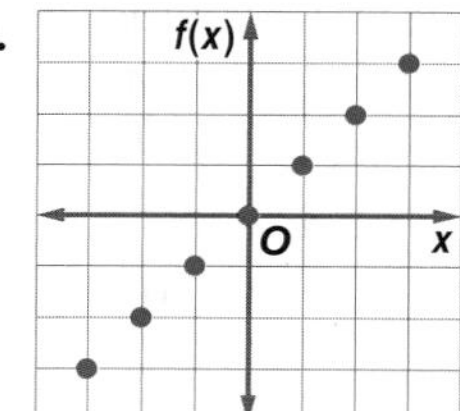

60.

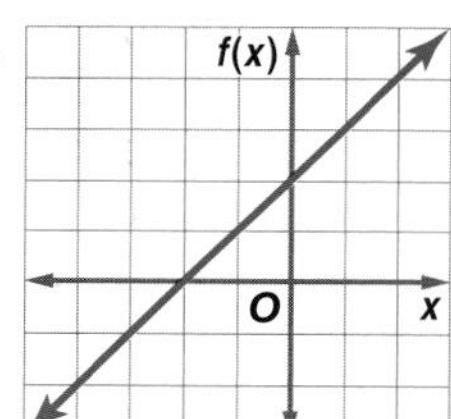

61. $\{(-3, 0), (-1, 1), (1, 3)\}$

62. $y = -x + 4$

Maintain Your Skills

Mixed Review

Solve each inequality. *(Lessons 1-5 and 1-6)*

63. $|y + 1| < 7$

64. $|5 - m| < 1$

65. $x - 5 < 0.1$

SHOPPING For Exercises 66 and 67, use the following information.
Javier had \$25.04 when he went to the mall. His friend Sally had \$32.67. Javier wanted to buy a shirt for \$27.89. *(Lesson 1-3)*

66. How much money did he have to borrow from Sally to buy the shirt?

67. How much money did that leave Sally?

Simplify each expression. *(Lessons 1-1 and 1-2)*

68. $3^2(2^2 - 1^2) + 4^2$

69. $3(5a + 6b) + 8(2a - b)$

Getting Ready for the Next Lesson

PREREQUISITE SKILL Solve each equation. Check your solution.
*(To review **solving equations**, see Lesson 1-3.)*

70. $x + 3 = 2$

71. $-4 + 2y = 0$

72. $0 = \frac{1}{2}x - 3$

73. $\frac{1}{3}x - 4 = 1$

2-2 Linear Equations

What You'll Learn

- Identify linear equations and functions.
- Write linear equations in standard form and graph them.

Vocabulary

- linear equation
- linear function
- standard form
- y-intercept
- x-intercept

How do linear equations relate to time spent studying?

Lolita has 4 hours after dinner to study and do homework. She has brought home math and chemistry. If she spends x hours on math and y hours on chemistry, a portion of the graph of the equation $x + y = 4$ can be used to relate how much time she spends on each.

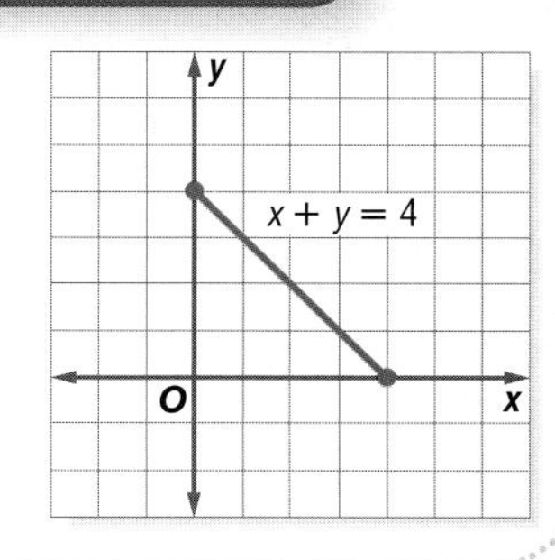

IDENTIFY LINEAR EQUATIONS AND FUNCTIONS An equation such as $x + y = 4$ is called a linear equation. A **linear equation** has no operations other than addition, subtraction, and multiplication of a variable by a constant. The variables may not be multiplied together or appear in a denominator. A linear equation does not contain variables with exponents other than 1. The graph of a linear equation is always a line.

Linear equations	Not linear equations
$5x - 3y = 7$	$7a + 4b^2 = -8$
$x = 9$	$y = \sqrt{x + 5}$
$6s = -3t - 15$	$x + xy = 1$
$y = \frac{1}{2}x$	$y = \frac{1}{x}$

A **linear function** is a function whose ordered pairs satisfy a linear equation. Any linear function can be written in the form $f(x) = mx + b$, where m and b are real numbers.

Example 1 *Identify Linear Functions*

State whether each function is a linear function. Explain.

a. $f(x) = 10 - 5x$ — This is a linear function because it can be written as $f(x) = -5x + 10$. $m = -5$, $b = 10$

b. $g(x) = x^4 - 5$ — This is not a linear function because x has an exponent other than 1.

c. $h(x, y) = 2xy$ — This is not a linear function because the two variables are multiplied together.

Example 2 Evaluate a Linear Function

MILITARY In August 2000, the Russian submarine *Kursk* sank to a depth of 350 feet in the Barents Sea. The linear function $P(d) = 62.5d + 2117$ can be used to find the pressure (lb/ft²) at a depth of d feet below the surface of the water.

a. Find the pressure at a depth of 350 feet.

$P(d) = 62.5d + 2117$ Original function

$P(350) = 62.5(350) + 2117$ Substitute.

$= 23{,}992$ Simplify.

The pressure at a depth of 350 feet is about 24,000 lb/ft^2.

b. The term 2117 in the function represents the atmospheric pressure at the surface of the water. How many times as great is the pressure at a depth of 350 feet as the pressure at the surface?

Divide the pressure 350 feet below the surface by the pressure at the surface.

$\frac{23{,}992}{2117} \approx 11.33$ Use a calculator.

The pressure at that depth is more than 11 times as great as the pressure at the surface.

More About. . .

Military

To avoid decompression sickness, it is recommended that divers ascend no faster than 30 feet per minute.

Source: www.emedicine.com

STANDARD FORM Any linear equation can be written in **standard form**, $Ax + By = C$, where A, B, and C are real numbers.

Key Concept — Standard Form of a Linear Equation

The standard form of a linear equation is $Ax + By = C$, where $A \geq 0$, A and B are not both zero.

Example 3 Standard Form

Write each equation in standard form. Identify A, B, and C.

a. $y = -2x + 3$

$y = -2x + 3$ Original equation

$2x + y = 3$ Add $2x$ to each side.

So, $A = 2$, $B = 1$, and $C = 3$.

b. $-\frac{3}{5}x = 3y - 2$

$-\frac{3}{5}x = 3y - 2$ Original equation

$-\frac{3}{5}x - 3y = -2$ Subtract $3y$ from each side.

$3x + 15y = 10$ Multiply each side by -5 so that the coefficients are integers and $A \geq 0$.

So, $A = 3$, $B = 15$, and $C = 10$.

c. $3x - 6y - 9 = 0$

$3x - 6y - 9 = 0$ Original equation

$3x - 6y = 9$ Add 9 to each side.

$x - 2y = 3$ Divide each side by 3 so that the coefficients have a GCF of 1.

So, $A = 1$, $B = -2$, and $C = 3$.

Study Tip

Vertical and Horizontal Lines
An equation of the form $x = C$ represents a vertical line, which has only an x-intercept. $y = C$ represents a horizontal line, which has only a y-intercept.

In Lesson 2-1, you graphed an equation or function by making a table of values, graphing enough ordered pairs to see a pattern, and connecting the points with a line or smooth curve. Since two points determine a line, there are quicker ways to graph a linear equation or function. One way is to find the points at which the graph intersects each axis and connect them with a line. The y-coordinate of the point at which a graph crosses the y-axis is called the **y-intercept**. Likewise, the x-coordinate of the point at which it crosses the x-axis is the **x-intercept**.

Example 4 Use Intercepts to Graph a Line

Find the x-intercept and the y-intercept of the graph of $3x - 4y + 12 = 0$. Then graph the equation.

The x-intercept is the value of x when $y = 0$.

$3x - 4y + 12 = 0$	Original equation
$3x - 4(0) + 12 = 0$	Substitute 0 for y.
$3x = -12$	Subtract 12 from each side.
$x = -4$	Divide each side by 3.

The x-intercept is -4. The graph crosses the x-axis at $(-4, 0)$.

Likewise, the y-intercept is the value of y when $x = 0$.

$3x - 4y + 12 = 0$	Original equation
$3(0) - 4y + 12 = 0$	Substitute 0 for x.
$-4y = -12$	Subtract 12 from each side.
$y = 3$	Divide each side by -4.

The y-intercept is 3. The graph crosses the y-axis at $(0, 3)$.

Use these ordered pairs to graph the equation.

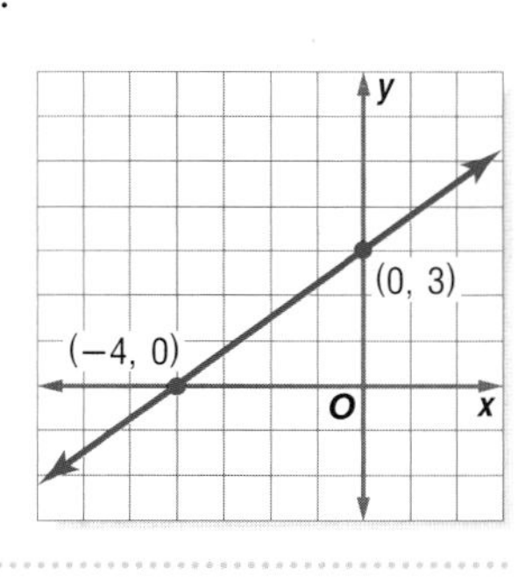

Check for Understanding

Concept Check

1. **Explain** why $f(x) = \frac{x+2}{2}$ is a linear function.
2. **Name** the x- and y-intercepts of the graph shown at the right.
3. **OPEN ENDED** Write an equation of a line with an x-intercept of 2.

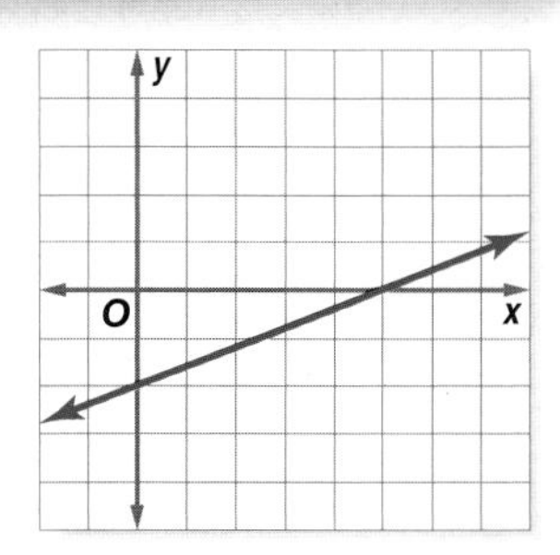

Guided Practice

State whether each equation or function is linear. Write *yes* or *no*. If no, explain your reasoning.

4. $x^2 + y^2 = 4$
5. $h(x) = 1.1 - 2x$

Write each equation in standard form. Identify A, B, and C.

6. $y = 3x - 5$
7. $4x = 10y + 6$
8. $y = \frac{2}{3}x + 1$

Find the x-intercept and the y-intercept of the graph of each equation. Then graph the equation.

9. $y = -3x - 5$
10. $x - y - 2 = 0$
11. $3x + 2y = 6$
12. $4x + 8y = 12$

Application **ECONOMICS For Exercises 13 and 14, use the following information.**
On January 1, 1999, the euro became legal tender in 11 participating countries in Europe. Based on the exchange rate on March 22, 2001, the linear function $d(x) = 0.8881x$ could be used to convert x euros to U.S. dollars.

13. On that date, what was the value in U.S. dollars of 200 euros?

14. On that date, what was the value in euros of 500 U.S. dollars?

Online Research **Data Update** How do the dollar and the euro compare today? Visit www.algebra2.com/data_update to convert among currencies.

Practice and Apply

Homework Help

For Exercises	See Examples
15–24	1
25, 26	2
27–38	3
39–52	4
53–60	2, 4

Extra Practice
See page 830.

State whether each equation or function is linear. Write *yes* or *no*. If no, explain your reasoning.

15. $x + y = 5$

16. $\frac{1}{x} + 3y = -5$

17. $x + \sqrt{y} = 4$

18. $h(x) = 2x^3 - 4x^2 + 5$

19. $g(x) = 10 + \frac{2}{x^2}$

20. $f(x) = 6x - 19$

21. $f(x) = 7x^5 + x - 1$

22. $y = \sqrt{2x - 5}$

23. Which of the equations $x + 9y = 7$, $x^2 + 5y = 0$, and $y = 3x - 1$ is not linear?

24. Which of the functions $f(x) = 2x + 4$, $g(x) = 7$, and $h(x) = x^3 - x^2 + 3x$ is not linear?

PHYSICS For Exercises 25 and 26, use the following information.
When a sound travels through water, the distance y in meters that the sound travels in x seconds is given by the equation $y = 1440x$.

25. How far does a sound travel underwater in 5 seconds?

26. In air, the equation is $y = 343x$. Does sound travel faster in air or water? Explain.

Write each equation in standard form. Identify *A*, *B*, and *C*.

27. $y = -3x + 4$

28. $y = 12x$

29. $x = 4y - 5$

30. $x = 7y + 2$

31. $5y = 10x - 25$

32. $4x = 8y - 12$

33. $\frac{1}{2}x + \frac{1}{2}y = 6$

34. $\frac{1}{3}x - \frac{1}{3}y = -2$

35. $0.5x = 3$

36. $0.25y = 10$

37. $\frac{5}{6}x + \frac{1}{15}y = \frac{3}{10}$

38. $0.25x = 0.1 + 0.2y$

Find the *x*-intercept and the *y*-intercept of the graph of each equation. Then graph the equation.

39. $5x + 3y = 15$

40. $2x - 6y = 12$

41. $3x - 4y - 10 = 0$

42. $2x + 5y - 10 = 0$

43. $y = x$

44. $y = 4x - 2$

45. $y = -2$

46. $y = 4$

47. $x = 8$

48. $x = 1$

49. $f(x) = 4x - 1$

50. $g(x) = 0.5x - 3$

CRITICAL THINKING For Exercises 51 and 52, use $x + y = 0$, $x + y = 5$, and $x + y = -5$.

51. Graph the equations on a coordinate plane. Compare and contrast the graphs.

52. Write a linear equation whose graph is between the graphs of $x + y = 0$ and $x + y = 5$.

More About. . .

Geology

Geothermal energy from hot springs is being used for electricity in California, Italy, and Iceland.

GEOLOGY **For Exercises 53–55, use the following information.**
Suppose the temperature T (°C) below Earth's surface is given by $T(d) = 35d + 20$, where d is the depth (km).

53. Find the temperature at a depth of 2 kilometers.

54. Find the depth if the temperature is 160°C.

55. Graph the linear function.

FUND-RAISING **For Exercises 56–59, use the following information.**
The Jackson Band Boosters sell beverages for $1.75 and candy for $1.50 at home games. Their goal is to have total sales of $525 for each game.

56. Write an equation that is a model for the different numbers of beverages and candy that can be sold to meet the goal.

57. Graph the equation.

58. Does this equation represent a function? Explain.

59. If they sell 100 beverages and 200 pieces of candy, will the Band Boosters meet their goal?

60. **GEOMETRY** Find the area of the shaded region in the graph. (*Hint:* The area of a trapezoid is given by $A = \frac{1}{2}h(b_1 + b_2)$.)

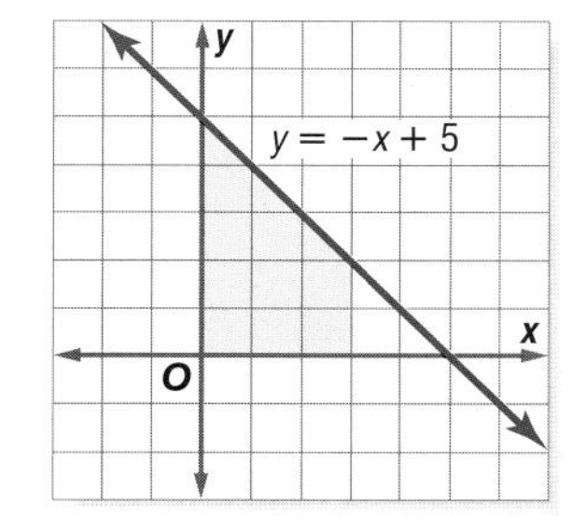

61. **WRITING IN MATH** Answer the question that was posed at the beginning of the lesson.

How do linear equations relate to time spent studying?

Include the following in your answer:
- why only the part of the graph in the first quadrant is shown, and
- an interpretation of the graph's intercepts in terms of the amount of time Lolita spends on each subject.

Standards Practice

Standardized Test Practice

62. Which function is linear?

Ⓐ $f(x) = x^2$ Ⓑ $g(x) = 2.7$ Ⓒ $g(x) = \sqrt{x - 1}$ Ⓓ $f(x) = \sqrt{9 - x^2}$

63. What is the y-intercept of the graph of $10 - x = 2y$?

Ⓐ 2 Ⓑ 5 Ⓒ 6 Ⓓ 10

Maintain Your Skills

Mixed Review

State the domain and range of each relation. Then graph the relation and determine whether it is a function. *(Lesson 2-1)*

64. $\{(-1, 5), (1, 3), (2, -4), (4, 3)\}$

65. $\{(0, 2), (1, 3), (2, -1), (1, 0)\}$

Solve each inequality. *(Lesson 1-6)*

66. $-2 < 3x + 1 < 7$

67. $|x + 4| > 2$

68. **TAX** Including a 6% sales tax, a paperback book costs $8.43. What is the price before tax? *(Lesson 1-3)*

Simplify each expression. *(Lesson 1-1)*

69. $(9s - 4) - 3(2s - 6)$

70. $[19 - (8 - 1)] \div 3$

Getting Ready for the Next Lesson

BASIC SKILL **Find the reciprocal of each number.**

71. 3

72. -4

73. $\frac{1}{2}$

74. $-\frac{2}{3}$

75. $-\frac{1}{5}$

76. $3\frac{3}{4}$

77. 2.5

78. -1.25

2-3 Slope

What You'll Learn

- Find and use the slope of a line.
- Graph parallel and perpendicular lines.

How does slope apply to the steepness of roads?

The grade of a road is a percent that measures the steepness of the road. It is found by dividing the amount the road rises by the corresponding horizontal distance.

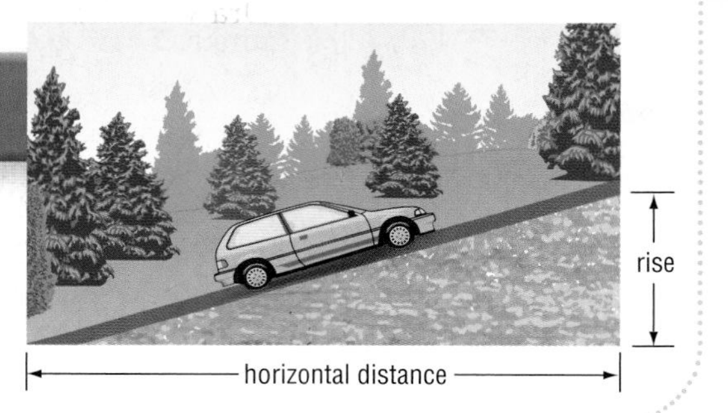

Vocabulary

- slope
- rate of change
- family of graphs
- parent graph
- oblique

SLOPE The **slope** of a line is the ratio of the change in y-coordinates to the corresponding change in x-coordinates. The slope measures how steep a line is. Suppose a line passes through points at (x_1, y_1) and (x_2, y_2).

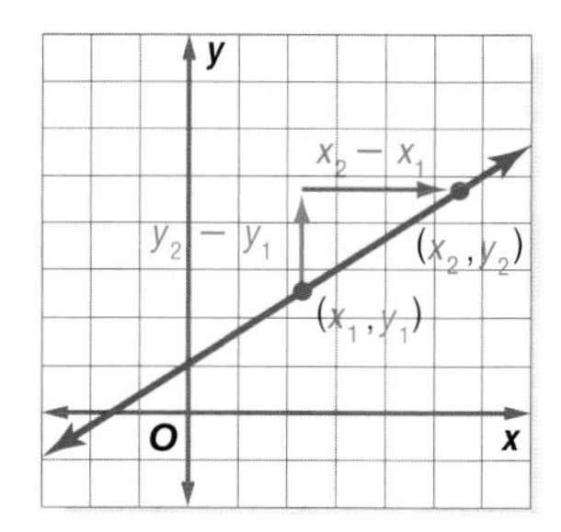

$$\text{slope} = \frac{\text{change in } y\text{-coordinates}}{\text{change in } x\text{-coordinates}}$$
$$= \frac{y_2 - y_1}{x_2 - x_1}$$

The slope of a line is the same, no matter what two points on the line are used.

Study Tip

Slope
The formula for slope is often remembered as *rise over run*, where the rise is the difference in y-coordinates and the run is the difference in x-coordinates.

Key Concept — Slope of a Line

- **Words** The slope of a line is the ratio of the change in y-coordinates to the change in x-coordinates.
- **Symbols** The slope m of the line passing through (x_1, y_1) and (x_2, y_2) is given by $m = \frac{y_2 - y_1}{x_2 - x_1}$, where $x_1 \neq x_2$.

Example 1 Find Slope

Find the slope of the line that passes through (−1, 4) and (1, −2). Then graph the line.

$m = \frac{y_2 - y_1}{x_2 - x_1}$ Slope formula

$= \frac{-2 - 4}{1 - (-1)}$ $(x_1, y_1) = (-1, 4), (x_2, y_2) = (1, -2)$

$= \frac{-6}{2}$ or -3 Simplify.

The slope of the line is -3.

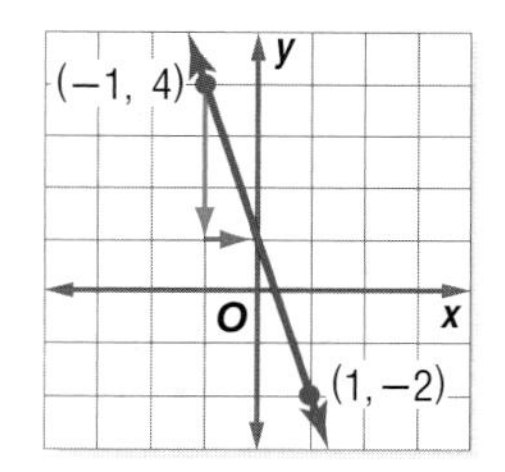

Graph the two ordered pairs and draw the line. Use the slope to check your graph by selecting any point on the line. Then go down 3 units and right 1 unit or go up 3 units and left 1 unit. This point should also be on the line.

Example 2 Use Slope to Graph a Line

Graph the line passing through (−4, −3) with a slope of $\frac{2}{3}$.

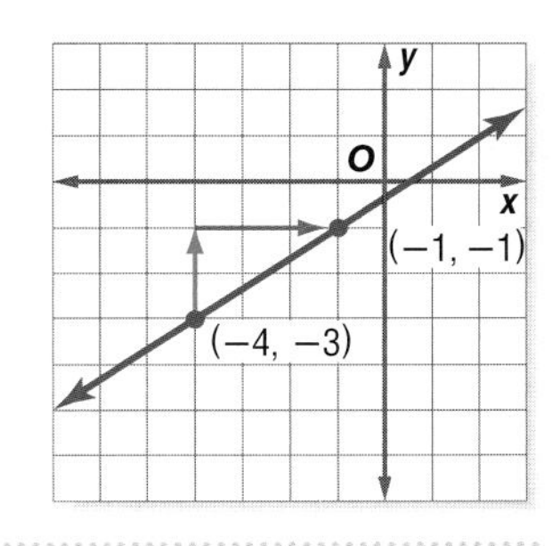

Graph the ordered pair (−4, −3). Then, according to the slope, go up 2 units and right 3 units. Plot the new point at (−1, −1). *You can also go right 3 units and then up 2 units to plot the new point.*

Draw the line containing the points.

The slope of a line tells the direction in which it rises or falls.

Concept Summary

If the line rises to the right, then the slope is *positive*.

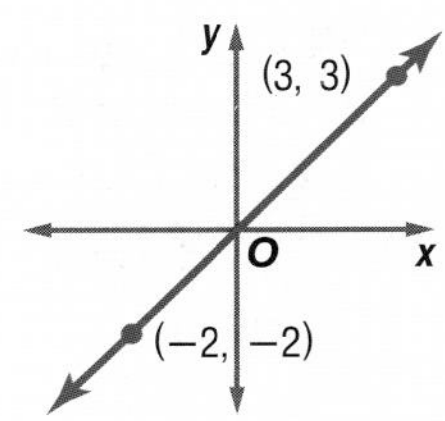

$$m = \frac{3-(-2)}{3-(-2)} = 1$$

If the line is horizontal, then the slope is *zero*.

y
(−3, 2)
(3, 2)
O
x

$$m = \frac{2-2}{3-(-3)} = 0$$

If the line falls to the right, then the slope is *negative*.

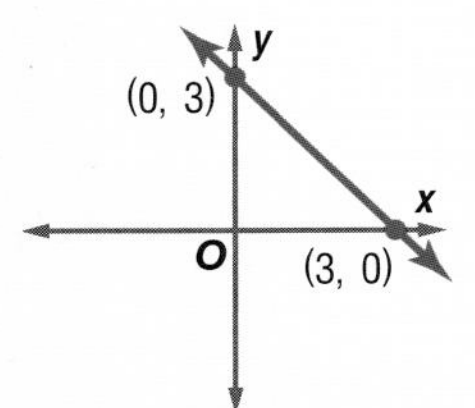

$$m = \frac{0-3}{3-0} = -1$$

If the line is vertical, then the slope is *undefined*.

y
(−2, 3)
x
O
(−2, −2)

$x_1 = x_2$, so m is undefined.

Slope is often referred to as **rate of change**. It measures how much a quantity changes, on average, relative to the change in another quantity, often time.

Example 3 Rate of Change

Log on for:
- Updated data
- More activities on rate of change

www.algebra2.com/usa_today

TRAVEL Refer to the graph at the right. Find the rate of change of the number of people taking cruises from 1985 to 2000.

$m = \frac{y_2 - y_1}{x_2 - x_1}$ Slope formula

$= \frac{6.9 - 2.2}{2000 - 1985}$ Substitute.

≈ 0.31 Simplify.

Between 1985 and 2000, the number of people taking cruises increased at an average rate of about 0.31(1,000,000) or 310,000 people per year.

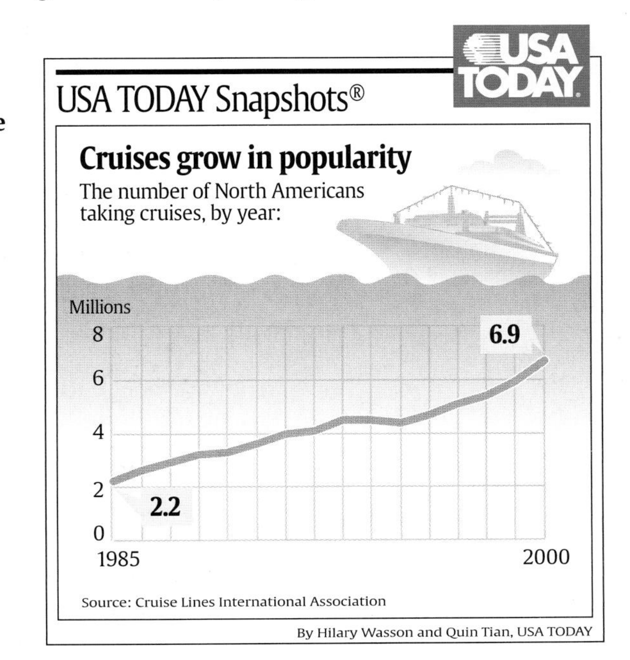

PARALLEL AND PERPENDICULAR LINES A **family of graphs** is a group of graphs that displays one or more similar characteristics. The **parent graph** is the simplest of the graphs in a family. A graphing calculator can be used to graph several graphs in a family on the same screen.

Graphing Calculator Investigation

Lines with the Same Slope

The calculator screen shows the graphs of $y = 3x$, $y = 3x + 2$, $y = 3x - 2$, and $y = 3x + 5$.

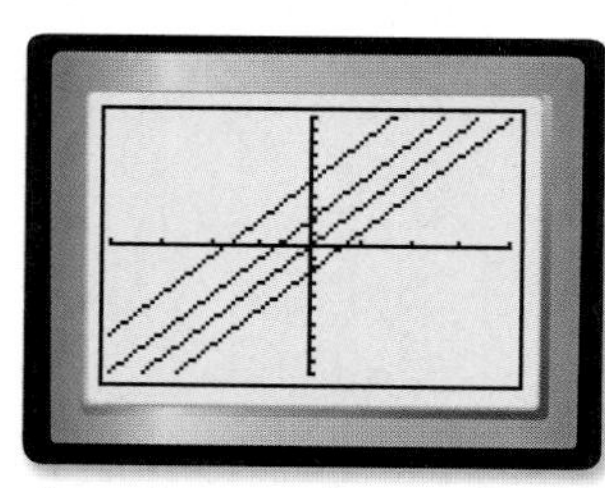

[−4, 4] scl: 1 by [−10, 10] scl: 1

Think and Discuss

1. Identify the parent function and describe the family of graphs. What is similar about the graphs? What is different about the graphs?
2. Find the slope of each line.
3. Write another function that has the same characteristics as this family of graphs. Check by graphing.

In the Investigation, you saw that lines that have the same slope are parallel. These and other similar examples suggest the following rule.

Key Concept — *Parallel Lines*

- **Words** In a plane, nonvertical lines with the same slope are parallel. All vertical lines are parallel.
- **Model**

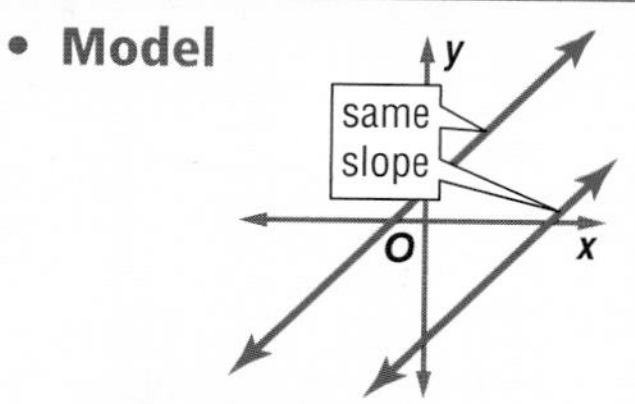

Study Tip

Horizontal Lines

All horizontal lines are parallel because they all have a slope of 0.

Example 4 *Parallel Lines*

Graph the line through (−1, 3) that is parallel to the line with equation $x + 4y = -4$.

The x-intercept is −4, and the y-intercept is −1. Use the intercepts to graph $x + 4y = -4$.

The line falls 1 unit for every 4 units it moves to the right, so the slope is $-\frac{1}{4}$.

Now use the slope and the point at (−1, 3) to graph the line parallel to the graph of $x + 4y = -4$.

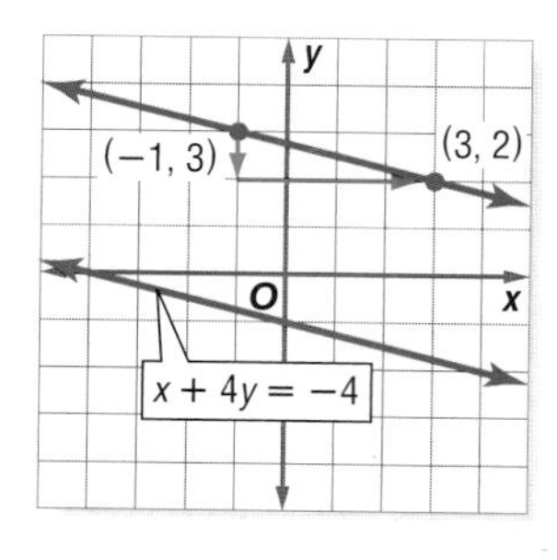

The figure at the right shows the graphs of two lines that are perpendicular. You know that parallel lines have the same slope. What is the relationship between the slopes of two perpendicular lines?

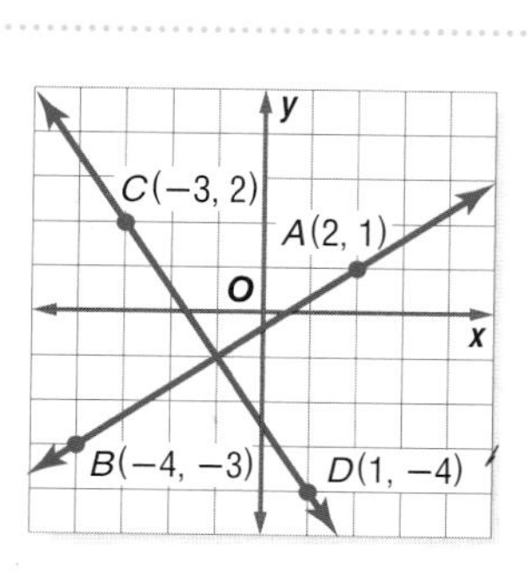

slope of line AB: $\frac{-3 - 1}{-4 - 2} = \frac{-4}{-6}$ or $\frac{2}{3}$

slope of line CD: $\frac{-4 - 2}{1 - (-3)} = \frac{-6}{4}$ or $-\frac{3}{2}$

The slopes are opposite reciprocals of each other. This relationship is true in general. When you multiply the slopes of two perpendicular lines, the product is always −1.

Study Tip

Reading Math

An *oblique* line is a line that is neither horizontal nor vertical.

Key Concept — Perpendicular Lines

- **Words** In a plane, two oblique lines are perpendicular if and only if the product of their slopes is −1.
- **Symbols** Suppose m_1 and m_2 are the slopes of two oblique lines. Then the lines are perpendicular if and only if $m_1m_2 = -1$, or $m_1 = -\frac{1}{m_2}$.
- **Model**

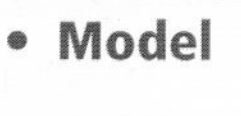

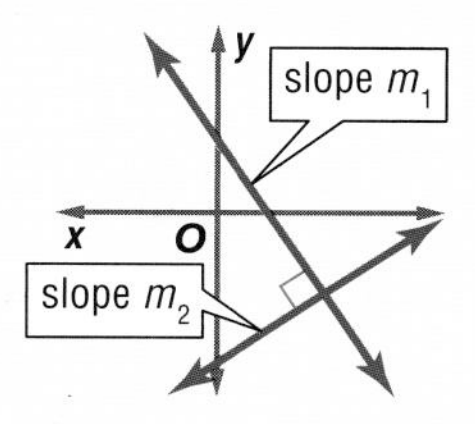

Any vertical line is perpendicular to any horizontal line.

Example 5 Perpendicular Line

Graph the line through (−3, 1) that is perpendicular to the line with equation $2x + 5y = 10$.

The x-intercept is 5, and the y-intercept is 2. Use the intercepts to graph $2x + 5y = 10$.

The line falls 2 units for every 5 units it moves to the right, so the slope is $-\frac{2}{5}$. The slope of the perpendicular line is the opposite reciprocal of $-\frac{2}{5}$, or $\frac{5}{2}$.

Start at (−3, 1) and go up 5 units and right 2 units. Use this point and (−3, 1) to graph the line.

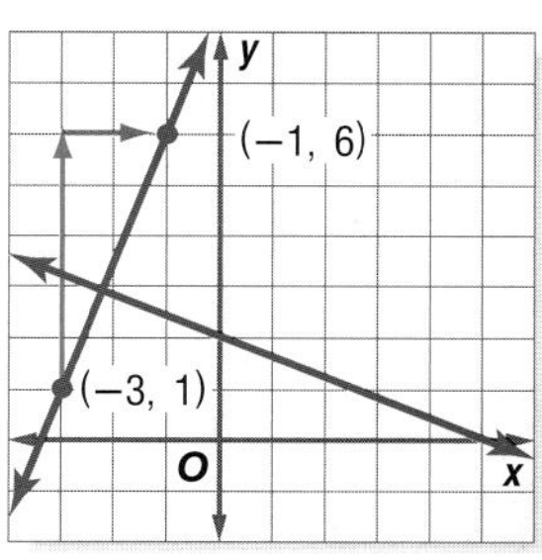

Check for Understanding

Concept Check

1. **OPEN ENDED** Write an equation of a line with slope 0.
2. **Decide** whether the statement below is *sometimes*, *always*, or *never* true. Explain.
   The slope of a line is a real number.
3. **FIND THE ERROR** Mark and Luisa are finding the slope of the line through (2, 4) and (−1, 5). Who is correct? Explain your reasoning.

Mark	Luisa
$m = \frac{5-4}{2-(-1)}$ or $\frac{1}{3}$	$m = \frac{4-5}{2-(-1)}$ or $-\frac{1}{3}$

Guided Practice

Find the slope of the line that passes through each pair of points.

4. (1, 1), (3, 1)
5. (−1, 0), (3, −2)
6. (3, 4), (1, 2)

Graph the line passing through the given point with the given slope.

7. (2, −1), −3
8. (−3, −4), $\frac{3}{2}$

Graph the line that satisfies each set of conditions.

9. passes through (0, 3), parallel to graph of $6y - 10x = 30$
10. passes through (4, −2), perpendicular to graph of $3x - 2y = 6$
11. passes through (−1, 5), perpendicular to graph of $5x - 3y - 3 = 0$

Application **WEATHER For Exercises 12–14, use the table that shows the temperatures at different times on March 23, 2002.**

Time	8:00 A.M.	10:00 A.M.	12:00 P.M.	2:00 P.M.	4:00 P.M.
Temp (°F)	36	47	55	58	60

12. What was the average rate of change of the temperature from 8:00 A.M. to 10:00 A.M.?

13. What was the average rate of change of the temperature from 12:00 P.M. to 4:00 P.M.?

14. During what 2-hour period was the average rate of change of the temperature the least?

Practice and Apply

Homework Help

For Exercises	See Examples
15–30	1
31–36	2
37–42	3
43–52	4, 5

Extra Practice
See page 830.

Find the slope of the line that passes through each pair of points.

15. $(6, 1), (8, -4)$

16. $(6, 8), (5, -5)$

17. $(-6, -5), (4, 1)$

18. $(2, -7), (4, 1)$

19. $(7, 8), (1, 8)$

20. $(-2, -3), (0, -5)$

21. $(2.5, 3), (1, -9)$

22. $(4, -1.5), (4, 4.5)$

23. $\left(\frac{1}{2}, -\frac{1}{3}\right), \left(\frac{1}{4}, \frac{2}{3}\right)$

24. $\left(\frac{1}{2}, \frac{2}{3}\right), \left(\frac{5}{6}, \frac{1}{4}\right)$

25. $(a, 2), (a, -2)$

26. $(3, b), (-5, b)$

27. Determine the value of r so that the line through $(6, r)$ and $(9, 2)$ has slope $\frac{1}{3}$.

28. Determine the value of r so that the line through $(5, r)$ and $(2, 3)$ has slope 2.

ANCIENT CULTURES Mayan Indians of Mexico and Central America built pyramids that were used as their temples. Ancient Egyptians built pyramids to use as tombs for the pharohs. Estimate the slope that a face of each pyramid makes with its base.

29.

The Pyramid of the Sun in Teotihuacán, Mexico, measures about 700 feet on each side of its square base and is about 210 feet high.

30.

The Great Pyramid in Egypt measures 756 feet on each side of its square base and was originally 481 feet high.

Graph the line passing through the given point with the given slope.

31. $(2, 6), m = \frac{2}{3}$

32. $(-3, -1), m = -\frac{1}{5}$

33. $(3, -4), m = 2$

34. $(1, 2), m = -3$

35. $(6, 2), m = 0$

36. $(-2, -3)$, undefined

ENTERTAINMENT **For Exercises 37–39, refer to the graph that shows the number of CDs and cassette tapes shipped by manufacturers to retailers in recent years.**

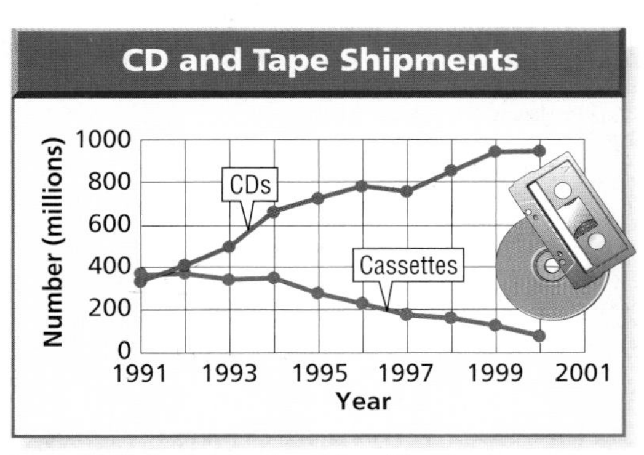

Source: Recording Industry Association of America

37. Find the average rate of change of the number of CDs shipped from 1991 to 2000.

38. Find the average rate of change of the number of cassette tapes shipped from 1991 to 2000.

39. Interpret the sign of your answer to Exercise 38.

TRAVEL **For Exercises 40–42, use the following information.**
Mr. and Mrs. Wellman are taking their daughter to college. The table shows their distance from home after various amounts of time.

Time (h)	Distance (mi)
0	0
1	55
2	110
3	165
4	165
5	225

40. Find the average rate of change of their distance from home between 1 and 3 hours after leaving home.

41. Find the average rate of change of their distance from home between 0 and 5 hours after leaving home.

42. What is another word for *rate of change* in this situation?

Graph the line that satisfies each set of conditions.

43. passes through $(-2, 2)$, parallel to a line whose slope is -1

44. passes through $(-4, 1)$, perpendicular to a line whose slope is $-\frac{3}{2}$

45. passes through $(3, 3)$, perpendicular to graph of $y = 3$

46. passes through $(2, -5)$, parallel to graph of $x = 4$

47. passes through $(2, -1)$, parallel to graph of $2x + 3y = 6$

48. passes through origin, parallel to graph of $x + y = 10$

49. perpendicular to graph of $3x - 2y = 24$, intersects that graph at its x-intercept

50. perpendicular to graph of $2x + 5y = 10$, intersects that graph at its y-intercept

51. GEOMETRY Determine whether quadrilateral $ABCD$ with vertices $A(-2, -1)$, $B(1, 1)$, $C(3, -2)$, and $D(0, -4)$ is a rectangle. Explain.

52. CRITICAL THINKING If the graph of the equation $ax + 3y = 9$ is perpendicular to the graph of the equation $3x + y = -4$, find the value of a.

53. **WRITING IN MATH** Answer the question that was posed at the beginning of the lesson.

How does slope apply to the steepness of roads?

Include the following in your answer:

- a few sentences explaining the relationship between the grade of a road and the slope of a line, and
- a graph of $y = 0.08x$, which corresponds to a grade of 8%. (A road with a grade of 6% to 8% is considered to be fairly steep. The scales on your x- and y-axes should be the same.)

Standardized Test Practice

Standards Practice

54. What is the slope of the line shown in the graph at the right?

(A) $-\frac{3}{2}$ (B) $-\frac{2}{3}$ (C) $\frac{2}{3}$ (D) $\frac{3}{2}$

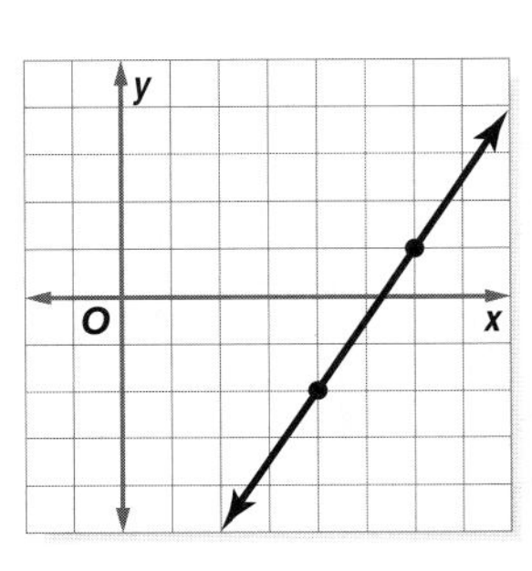

55. What is the slope of a line perpendicular to a line with slope $-\frac{1}{2}$?

(A) -2 (B) $-\frac{1}{2}$ (C) $\frac{1}{2}$ (D) 2

Graphing Calculator

FAMILY OF GRAPHS **Use a graphing calculator to investigate each family of graphs. Explain how changing the slope affects the graph of the line.**

56. $y = 2x + 3, y = 4x + 3, y = 8x + 3, y = x + 3$

57. $y = -3x + 1, y = -x + 1, y = -5x + 1, y = -7x + 1$

Maintain Your Skills

Mixed Review **Find the x-intercept and the y-intercept of the graph of each equation. Then graph the equation.** *(Lesson 2-2)*

58. $-2x + 5y = 20$
59. $4x - 3y + 8 = 0$
60. $y = 7x$

Find each value if $f(x) = 3x - 4$. *(Lesson 2-1)*

61. $f(-1)$
62. $f(3)$
63. $f\left(\frac{1}{2}\right)$
64. $f(a)$

Solve each inequality. *(Lessons 1-5 and 1-6)*

65. $5 < 2x + 7 < 13$
66. $2z + 5 \geq 1475$

67. **SCHOOL** A test has multiple-choice questions worth 4 points each and true-false questions worth 3 points each. Marco answers 14 multiple-choice questions correctly. How many true-false questions must he answer correctly to get at least 80 points total? *(Lesson 1-5)*

Simplify. *(Lessons 1-1 and 1-2)*

68. $\frac{1}{3}(15a + 9b) - \frac{1}{7}(28b - 84a)$
69. $3 + (21 \div 7) \times 8 \div 4$

Getting Ready for the Next Lesson

PREREQUISITE SKILL **Solve each equation for y.**
*(To review **solving equations**, see Lesson 1-3.)*

70. $x + y = 9$
71. $4x + y = 2$
72. $-3x - y + 7 = 0$
73. $5x - 2y - 1 = 0$
74. $3x - 5y + 4 = 0$
75. $2x + 3y - 11 = 0$

Practice Quiz 1

Lessons 2-1 through 2-3

1. State the domain and range of the relation $\{(2, 5), (-3, 2), (2, 1), (-7, 4), (0, -2)\}$. *(Lesson 2-1)*

2. Find the value of $f(15)$ if $f(x) = 100x - 5x^2$. *(Lesson 2-1)*

3. Write $y = -6x + 4$ in standard form. *(Lesson 2-2)*

4. Find the x-intercept and the y-intercept of the graph of $3x + 5y = 30$. Then graph the equation. *(Lesson 2-2)*

5. Graph the line that goes through $(4, -3)$ and is parallel to the line whose equation is $2x + 5y = 10$. *(Lesson 2-3)*

2-4 Writing Linear Equations

What You'll Learn

- Write an equation of a line given the slope and a point on the line.
- Write an equation of a line parallel or perpendicular to a given line.

Vocabulary

- slope-intercept form
- point-slope form

How do linear equations apply to business?

When a company manufactures a product, they must consider two types of cost. There is the *fixed cost*, which they must pay no matter how many of the product they produce, and there is *variable cost*, which depends on how many of the product they produce. In some cases, the total cost can be found using a linear equation such as $y = 5400 + 1.37x$.

FORMS OF EQUATIONS Consider the graph at the right. The line passes through $A(0, b)$ and $C(x, y)$. Notice that b is the y-intercept of $\overleftrightarrow{AC}$. You can use these two points to find the slope of $\overleftrightarrow{AC}$. Substitute the coordinates of points A and C into the slope formula.

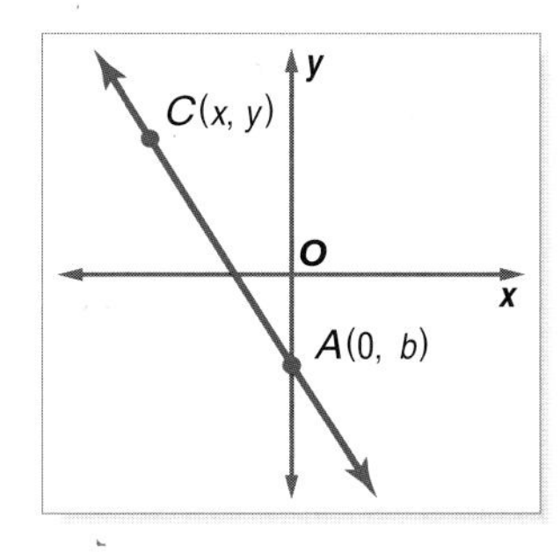

$m = \frac{y_2 - y_1}{x_2 - x_1}$ Slope formula

$m = \frac{y - b}{x - 0}$ $(x_1, y_1) = (0, b)$, $(x_2, y_2) = (x, y)$

$m = \frac{y - b}{x}$ Simplify.

Now solve the equation for y.

$mx = y - b$ Multiply each side by x.

$mx + b = y$ Add b to each side.

$y = mx + b$ Symmetric Property of Equality

When an equation is written in this form, it is in **slope-intercept form**.

Study Tip

Slope-intercept Form

The equation of a vertical line cannot be written in slope-intercept form because its slope is undefined.

Key Concept *Slope-Intercept Form of a Linear Equation*

- **Words** The slope-intercept form of the equation of a line is $y = mx + b$, where m is the slope and b is the y-intercept.
- **Symbols** $y = mx + b$ (slope: m; y-intercept: b)
- **Model**

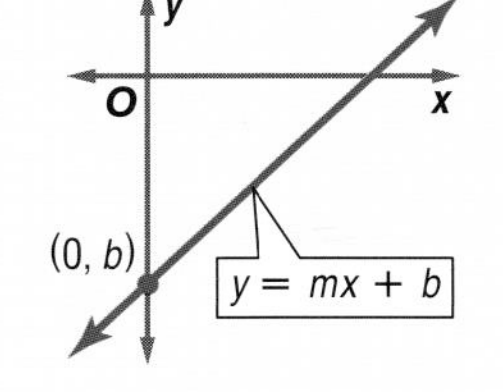

If you are given the slope and y-intercept of a line, you can find an equation of the line by substituting the values of m and b into the slope-intercept form. For example, if you know that the slope of a line is -3 and the y-intercept is 4, the equation of the line is $y = -3x + 4$, or, in standard form, $3x + y = 4$.

You can also use the slope-intercept form to find an equation of a line if you know the slope and the coordinates of any point on the line.

Example 1 Write an Equation Given Slope and a Point

Write an equation in slope-intercept form for the line that has a slope of $-\frac{3}{2}$ and passes through (−4, 1).

Substitute for m, x, and y in the slope-intercept form.

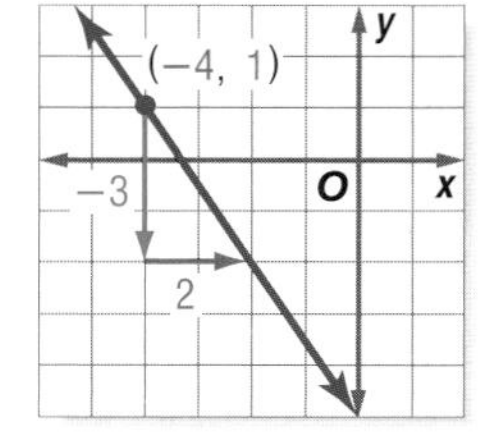

$y = mx + b$	Slope-intercept form
$1 = \left(-\frac{3}{2}\right)(-4) + b$	$(x, y) = (-4, 1)$, $m = -\frac{3}{2}$
$1 = 6 + b$	Simplify.
$-5 = b$	Subtract 6 from each side.

The y-intercept is −5. So, the equation in slope-intercept form is $y = -\frac{3}{2}x - 5$.

If you are given the coordinates of two points on a line, you can use the **point-slope form** to find an equation of the line that passes through them.

Key Concept — Point-Slope Form of a Linear Equation

- **Words** The point-slope form of the equation of a line is $y - y_1 = m(x - x_1)$, where (x_1, y_1) are the coordinates of a point on the line and m is the slope of the line.
- **Symbols** $y - y_1 = m(x - x_1)$ — m: slope; y_1, x_1: coordinates of point on line

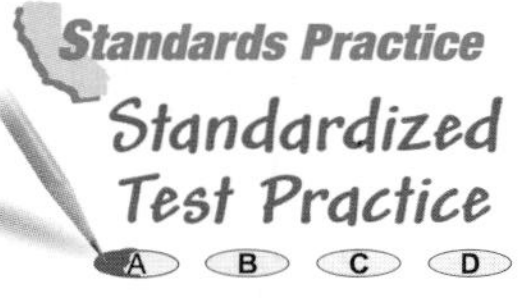

Example 2 Write an Equation Given Two Points

Multiple-Choice Test Item

What is an equation of the line through (−1, 4) and (−4, 5)?

Ⓐ $y = -\frac{1}{3}x + \frac{11}{3}$ Ⓑ $y = \frac{1}{3}x + \frac{13}{3}$ Ⓒ $y = -\frac{1}{3}x + \frac{13}{3}$ Ⓓ $y = -3x + 1$

Read the Test Item

You are given the coordinates of two points on the line. Notice that the answer choices are in slope-intercept form.

Solve the Test Item

- First, find the slope of the line.

$m = \frac{y_2 - y_1}{x_2 - x_1}$	Slope formula
$= \frac{5 - 4}{-4 - (-1)}$	$(x_1, y_1) = (-1, 4)$, $(x_2, y_2) = (-4, 5)$
$= \frac{1}{-3}$ or $-\frac{1}{3}$	Simplify.

The slope is $-\frac{1}{3}$. That eliminates choices B and D.

- Then use the point-slope form to find an equation.

$y - y_1 = m(x - x_1)$	Point-slope form
$y - 4 = -\frac{1}{3}[x - (-1)]$	$m = -\frac{1}{3}$; you can use either point for (x_1, y_1).
$y - 4 = -\frac{1}{3}x - \frac{1}{3}$	Distributive Property
$y = -\frac{1}{3}x + \frac{11}{3}$	The answer is A.

Test-Taking Tip

To check your answer, substitute each ordered pair into your answer. Each should satisfy the equation.

When changes in real-world situations occur at a linear rate, a linear equation can be used as a model for describing the situation.

Example 3 Write an Equation for a Real-World Situation

SALES As a salesperson, Eric Fu is paid a daily salary plus commission. When his sales are $1000, he makes $100. When his sales are $1400, he makes $120.

a. Write a linear equation to model this situation.

Let x be his sales and let y be the amount of money he makes. Use the points (1000, 100) and (1400, 120) to make a graph to represent the situation.

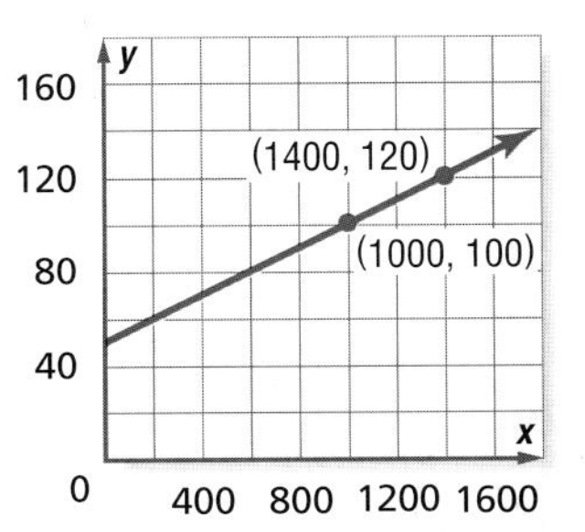

$m = \frac{y_2 - y_1}{x_2 - x_1}$ Slope formula

$= \frac{120 - 100}{1400 - 1000}$ $(x_1, y_1) = (1000, 100)$, $(x_2, y_2) = (1400, 120)$

$= 0.05$ Simplify.

Now use the slope and either of the given points with the point-slope form to write the equation.

$y - y_1 = m(x - x_1)$ Point-slope form

$y - 100 = 0.05(x - 1000)$ $m = 0.05$, $(x_1, y_1) = (1000, 100)$

$y - 100 = 0.05x - 50$ Distributive Property

$y = 0.05x + 50$ Add 100 to each side.

The slope-intercept form of the equation is $y = 0.05x + 50$.

b. What are Mr. Fu's daily salary and commission rate?

The y-intercept of the line is 50. The y-intercept represents the money Eric would make if he had no sales. In other words, $50 is his daily salary.

The slope of the line is 0.05. Since the slope is the coefficient of x, which is his sales, he makes 5% commission.

Study Tip

Alternative Method

You could also find Mr. Fu's salary in part **c** by extending the graph. Then find the y value when x is 2000.

c. How much would he make in a day if Mr. Fu's sales were $2000?

Find the value of y when $x = 2000$.

$y = 0.05x + 50$ Use the equation you found in part **a**.

$= 0.05(2000) + 50$ Replace x with 2000.

$= 100 + 50$ or 150 Simplify.

Mr. Fu would make $150 if his sales were $2000.

PARALLEL AND PERPENDICULAR LINES The slope-intercept and point-slope forms can be used to find equations of lines that are parallel or perpendicular to given lines.

Example 4 Write an Equation of a Perpendicular Line

Write an equation for the line that passes through (−4, 3) and is perpendicular to the line whose equation is $y = -4x - 1$.

The slope of the given line is −4. Since the slopes of perpendicular lines are opposite reciprocals, the slope of the perpendicular line is $\frac{1}{4}$.

(continued on the next page)

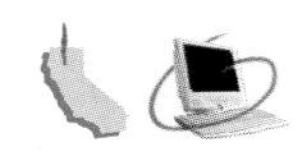

Use the point-slope form and the ordered pair $(-4, 3)$ to write the equation.

$y - y_1 = m(x - x_1)$ Point-slope form

$y - 3 = \frac{1}{4}[x - (-4)]$ $(x_1, y_1) = (-4, 3)$, $m = \frac{1}{4}$

$y - 3 = \frac{1}{4}x + 1$ Distributive Property

$y = \frac{1}{4}x + 4$ Add 3 to each side.

An equation of the line is $y = \frac{1}{4}x + 4$.

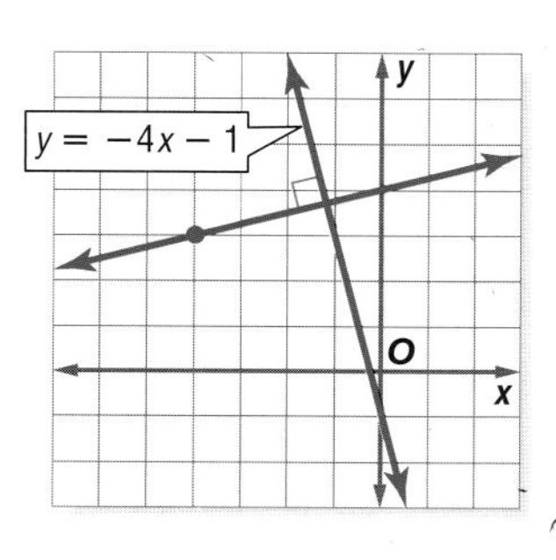

Check for Understanding

Concept Check

1. **OPEN ENDED** Write an equation of a line in slope-intercept form.
2. **Identify** the slope and y-intercept of the line with equation $y = 6x$.
3. **Explain** how to find the slope of a line parallel to the graph of $3x - 5y = 2$.

Guided Practice

State the slope and y-intercept of the graph of each equation.

4. $y = 2x - 5$
5. $3x + 2y - 10 = 0$

Write an equation in slope-intercept form for the line that satisfies each set of conditions.

6. slope 0.5, passes through $(6, 4)$
7. slope $-\frac{3}{4}$, passes through $\left(2, \frac{1}{2}\right)$
8. passes through $(6, 1)$ and $(8, -4)$
9. passes through $(-3, 5)$ and $(2, 2)$
10. passes through $(0, -2)$, perpendicular to the graph of $y = x - 2$

11. Write an equation in slope-intercept form for the graph at the right.

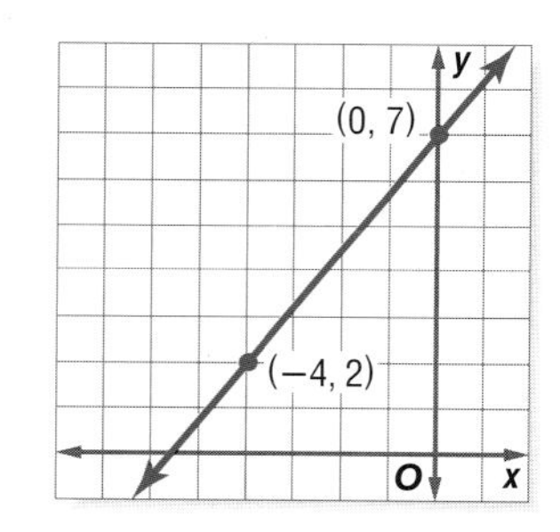

12. What is an equation of the line through $(2, -4)$ and $(-3, -1)$?

(A) $y = -\frac{3}{5}x + \frac{26}{5}$
(B) $y = -\frac{3}{5}x - \frac{14}{5}$
(C) $y = \frac{3}{5}x - \frac{26}{5}$
(D) $y = \frac{3}{5}x + \frac{14}{5}$

Practice and Apply

Homework Help

For Exercises	See Examples
13–18, 21–28	1
19, 20, 29–34, 39, 40	2, 3
35–38	4
41–52	1–3

Extra Practice
See page 831.

State the slope and y-intercept of the graph of each equation.

13. $y = -\frac{2}{3}x - 4$
14. $y = \frac{3}{4}x$
15. $2x - 4y = 10$
16. $3x + 5y - 30 = 0$
17. $x = 7$
18. $cx + y = d$

Write an equation in slope-intercept form for each graph.

19.

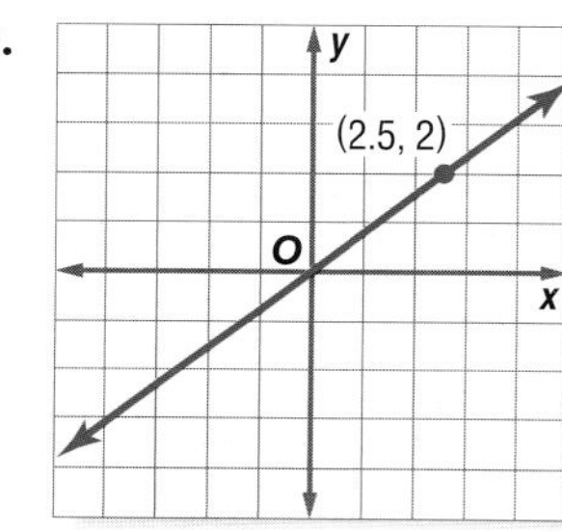

20.

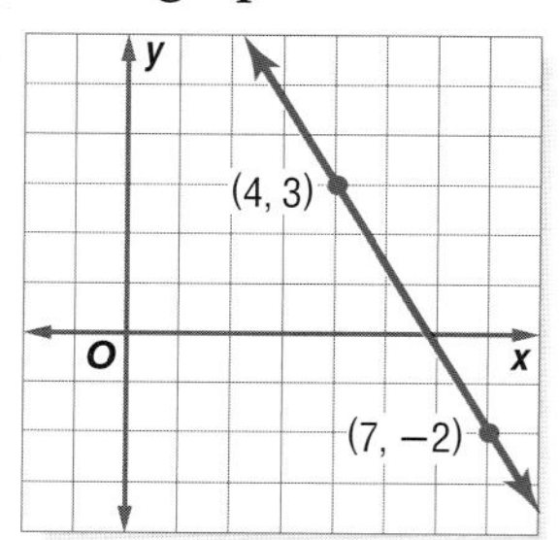

Write an equation in slope-intercept form for each graph.

21.

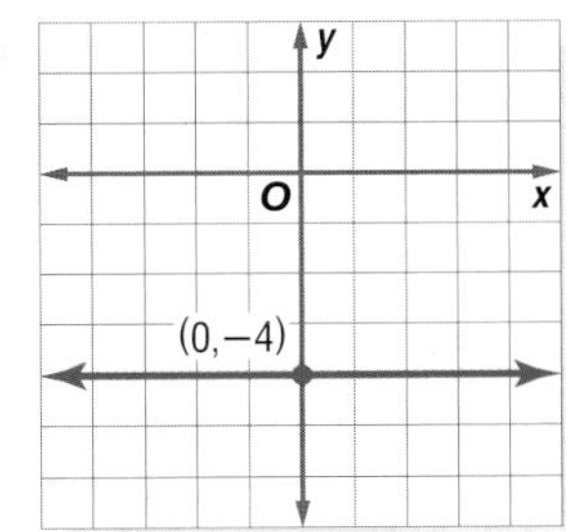

22.

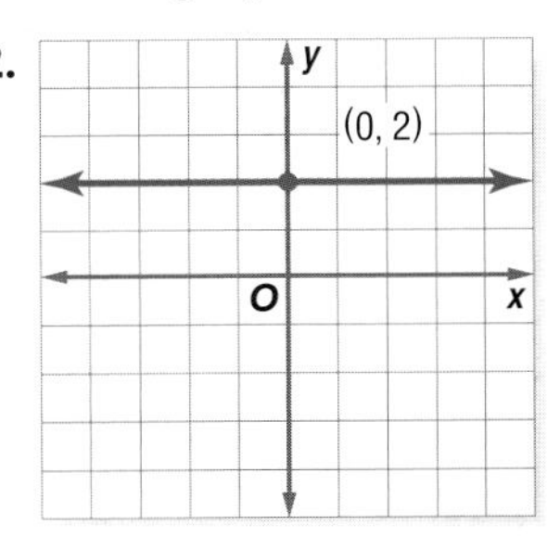

Write an equation in slope-intercept form for the line that satisfies each set of conditions.

23. slope 3, passes through $(0, -6)$

24. slope 0.25, passes through $(0, 4)$

25. slope $-\frac{1}{2}$, passes through $(1, 3)$

26. slope $\frac{3}{2}$, passes through $(-5, 1)$

27. slope -0.5, passes through $(2, -3)$

28. slope 4, passes through the origin

29. passes through $(-2, 5)$ and $(3, 1)$

30. passes through $(7, 1)$ and $(7, 8)$

31. passes through $(-4, 0)$ and $(3, 0)$

32. passes through $(-2, -3)$ and $(0, 0)$

33. x-intercept -4, y-intercept 4

34. x-intercept $\frac{1}{3}$, y-intercept $-\frac{1}{4}$

35. passes through $(4, 6)$, parallel to the graph of $y = \frac{2}{3}x + 5$

36. passes through $(2, -5)$, perpendicular to the graph of $y = \frac{1}{4}x + 7$

37. passes through $(6, -5)$, perpendicular to the line whose equation is $3x - \frac{1}{5}y = 3$

38. passes through $(-3, -1)$, parallel to the line that passes through $(3, 3)$ and $(0, 6)$

39. Write an equation in slope-intercept form of the line that passes through the points indicated in the table.

40. Write an equation in slope-intercept form of the line that passes through $(-2, 10)$, $(2, 2)$, and $(4, -2)$.

x	y
−1	−5
1	1
3	7

GEOMETRY For Exercises 41–43, use the equation $d = 180(c - 2)$ that gives the total number of degrees d in any convex polygon with c sides.

41. Write this equation in slope-intercept form.

42. Identify the slope and d-intercept.

43. Find the number of degrees in a pentagon.

44. ECOLOGY A park ranger at Blendon Woods estimates there are 6000 deer in the park. She also estimates that the population will increase by 75 deer each year thereafter. Write an equation that represents how many deer will be in the park in x years.

45. BUSINESS Refer to the signs below. At what distance do the two stores charge the same amount for a balloon arrangement?

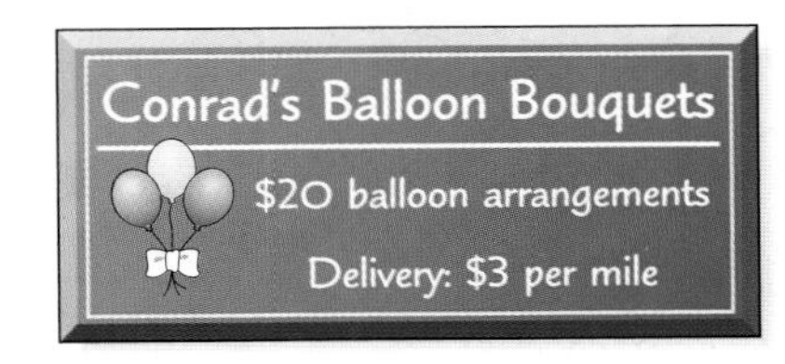

Science

Ice forms at a temperature of 0°C, which corresponds to a temperature of 32°F. A temperature of 100°C corresponds to a temperature of 212°F.

SCIENCE **For Exercises 46–48, use the information on temperatures at the left.**

46. Write and graph the linear equation that gives the number y of degrees Fahrenheit in terms of the number x of degrees Celsius.

47. What temperature corresponds to 20°C?

48. What temperature is the same on both scales?

TELEPHONES **For Exercises 49 and 50, use the following information.**
Namid is examining the calling card portion of his phone bill. A 4-minute call at the night rate cost \$2.65. A 10-minute call at the night rate cost \$4.75.

49. Write a linear equation to model this situation.

50. How much would it cost to talk for half an hour at the night rate?

51. CRITICAL THINKING Given $\triangle ABC$ with vertices $A(-6, -8)$, $B(6, 4)$, and $C(-6, 10)$, write an equation of the line containing the altitude from A. (*Hint:* The altitude from A is a segment that is perpendicular to $\overline{BC}$.)

52. **WRITING IN MATH** Answer the question that was posed at the beginning of the lesson.

How do linear equations apply to business?

Include the following in your answer:

- the *fixed cost* and the *variable cost* in the equation $y = 5400 + 1.37x$, where y is the cost for a company to produce x units of its product, and
- the cost for the company to produce 1000 units of its product.

53. Find an equation of the line through $(0, -3)$ and $(4, 1)$.

Ⓐ $y = -x + 3$ Ⓑ $y = -x - 3$ Ⓒ $y = x - 3$ Ⓓ $y = -x + 3$

54. Choose the equation of the line through $\left(\frac{1}{2}, -\frac{3}{2}\right)$ and $\left(-\frac{1}{2}, \frac{1}{2}\right)$.

Ⓐ $y = -2x - \frac{1}{2}$ Ⓑ $y = -3x$ Ⓒ $y = 2x - \frac{5}{2}$ Ⓓ $y = \frac{1}{2}x + 1$

Extending the Lesson

For Exercises 55 and 56, use the following information.

The form $\frac{x}{a} + \frac{y}{b} = 1$ is known as the **intercept form** of the equation of a line because a is the x-intercept and b is the y-intercept.

55. Write the equation $2x - y - 5 = 0$ in intercept form.

56. Identify the x- and y-intercepts of the graph of $2x - y - 5 = 0$.

Maintain Your Skills

Mixed Review

Find the slope of the line that passes through each pair of points. *(Lesson 2-3)*

57. $(7, 2), (5, 6)$ **58.** $(1, -3), (3, 3)$ **59.** $(-5, 0), (4, 0)$

60. INTERNET A Webmaster estimates that the time (seconds) required to connect to the server when n people are connecting is given by $t(n) = 0.005n + 0.3$. Estimate the time required to connect when 50 people are connecting. *(Lesson 2-2)*

Solve each inequality. *(Lessons 1-5 and 1-6)*

61. $|x - 2| \leq -99$ **62.** $-4x + 7 \leq 31$ **63.** $2(r - 4) + 5 \geq 9$

Getting Ready for the Next Lesson

PREREQUISITE SKILL **Find the median of each set of numbers.**
*(To review **finding a median**, see pages 822 and 823.)*

64. $\{3, 2, 1, 3, 4, 8, 4\}$ **65.** $\{9, 3, 7, 5, 6, 3, 7, 9\}$

66. $\{138, 235, 976, 230, 412, 466\}$ **67.** $\{2.5, 7.8, 5.5, 2.3, 6.2, 7.8\}$

2-5 Modeling Real-World Data: Using Scatter Plots

What You'll Learn

- Draw scatter plots.
- Find and use prediction equations.

How can a linear equation model the number of Calories you burn exercising?

The table shows the number of Calories burned per hour by a 140-pound person running at various speeds. A linear function can be used to model these data.

Speed (mph)	Calories
5	508
6	636
7	731
8	858

Vocabulary

- bivariate data
- scatter plot
- line of fit
- prediction equation

SCATTER PLOTS Data with two variables, such as speed and Calories, is called **bivariate data**. A set of bivariate data graphed as ordered pairs in a coordinate plane is called a **scatter plot**. A scatter plot can show whether there is a relationship between the data.

Example 1 Draw a Scatter Plot

HOUSING The table below shows the median selling price of new, privately-owned, one-family houses for some recent years. Make a scatter plot of the data.

Year	1990	1992	1994	1996	1998	2000
Price ($1000)	122.9	121.5	130.0	140.0	152.5	169.0

Source: U.S. Census Bureau and U.S. Department of Housing and Urban Development

Graph the data as ordered pairs, with the number of years since 1990 on the horizontal axis and the price on the vertical axis.

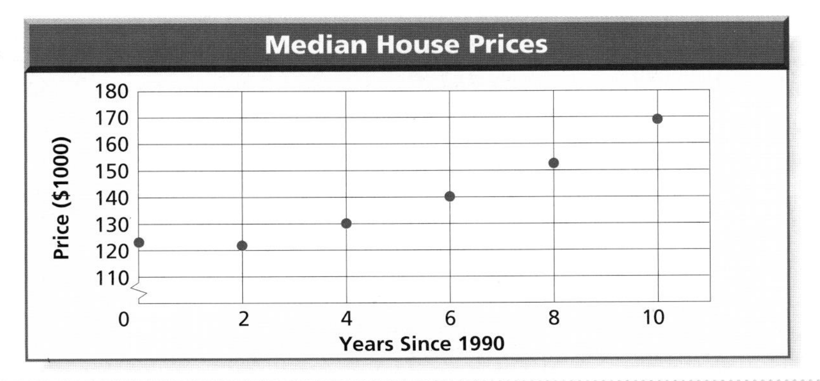

Study Tip

Choosing the Independent Variable
Letting x be the number of years since the first year in the data set sometimes simplifies the calculations involved in finding a function to model the data.

PREDICTION EQUATIONS Except for (0, 122.9), the data in Example 1 appear to lie nearly on a straight line. When you find a line that closely approximates a set of data, you are finding a **line of fit** for the data. An equation of such a line is often called a **prediction equation** because it can be used to predict one of the variables given the other variable.

To find a line of fit and a prediction equation for a set of data, select two points that appear to represent the data well. This is a matter of personal judgment, so your line and prediction equation may be different from someone else's.

Example 2 Find and Use a Prediction Equation

Study Tip

Reading Math
A data point that does not appear to belong with the rest of the set is called an **outlier**.

HOUSING **Refer to the data in Example 1.**

a. Draw a line of fit for the data. How well does the line fit the data?

Ignore the point (0, 122.9) since it would not be close to a line that represents the rest of the data points. The points (4, 130.0) and (8, 152.5) appear to represent the data well. Draw a line through these two points. Except for (0, 122.9), this line fits the data very well.

b. Find a prediction equation. What do the slope and y-intercept indicate?

Find an equation of the line through (4, 130.0) and (8, 152.5). Begin by finding the slope.

$$m = \frac{y_2 - y_1}{x_2 - x_1} \quad \text{Slope formula}$$

$$= \frac{152.5 - 130.0}{8 - 4} \quad \text{Substitute.}$$

$$\approx 5.63 \quad \text{Simplify.}$$

Study Tip

Reading Math
When you are predicting for an x value greater than any in the data set, the process is known as **extrapolation**. When you are predicting for an x value between the least and greatest in the data set, the process is known as **interpolation**.

$$y - y_1 = m(x - x_1) \quad \text{Point-slope form}$$

$$y - 130.0 = 5.63(x - 4) \quad m = 5.63, (x_1, y_1) = (4, 130.0)$$

$$y - 130.0 = 5.63x - 22.52 \quad \text{Distributive Property}$$

$$y = 5.63x + 107.48 \quad \text{Add 130.0 to each side.}$$

One prediction equation is $y = 5.63x + 107.48$. The slope indicates that the median price is increasing at a rate of about $5630 per year. The y-intercept indicates that, according to the trend of the rest of the data, the median price in 1990 should have been about $107,480.

c. Predict the median price in 2010.

The year 2010 is 20 years after 1990, so use the prediction equation to find the value of y when $x = 20$.

$$y = 5.63x + 107.48 \quad \text{Prediction equation}$$

$$= 5.63(20) + 107.48 \quad x = 20$$

$$= 220.08 \quad \text{Simplify.}$$

The model predicts that the median price in 2010 will be about $220,000.

d. How accurate is the prediction?

Except for the outlier, the line fits the data very well, so the predicted value should be fairly accurate.

Study Tip

Outliers
If your scatter plot includes points that are far from the others on the graph, check your data before deciding it is an outlier. You may have made a graphing or recording mistake.

Algebra Activity

Head versus Height

Collect the Data

- Collect data from several of your classmates. Use a tape measure to measure the circumference of each person's head and his or her height. Record the data as ordered pairs of the form (height, circumference).

Analyze the Data

1. Graph the data in a scatter plot.
2. Choose two ordered pairs and write a prediction equation.
3. Explain the meaning of the slope in the prediction equation.

Make a Conjecture

4. Predict the head circumference of a person who is 66 inches tall.
5. Predict the height of an individual whose head circumference is 18 inches.

Check for Understanding

Concept Check

1. **Choose** the scatter plot with data that could best be modeled by a linear function.

a. 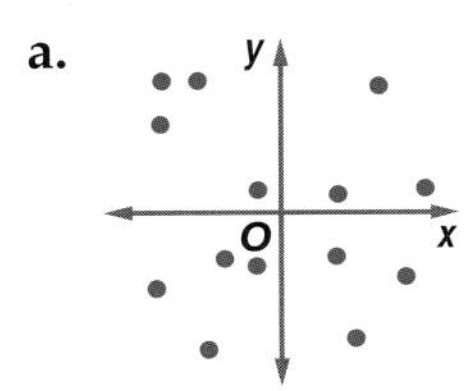

b. 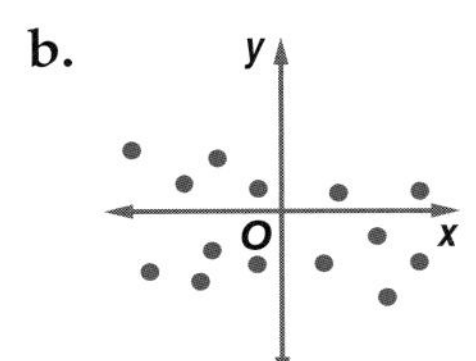

c. 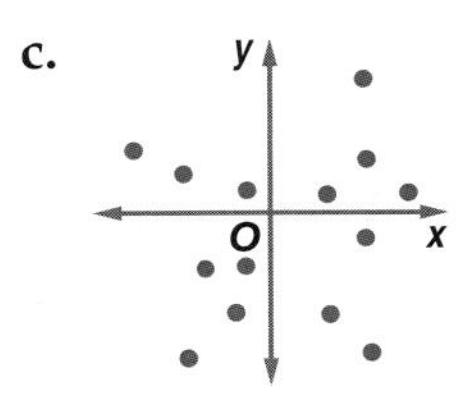

d. 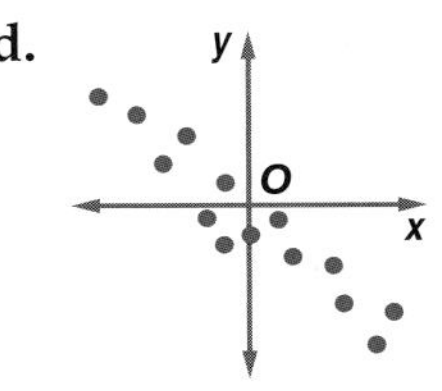

2. **Identify** the domain and range of the relation in the graph at the right. **Predict** the value of y when $x = 5$.

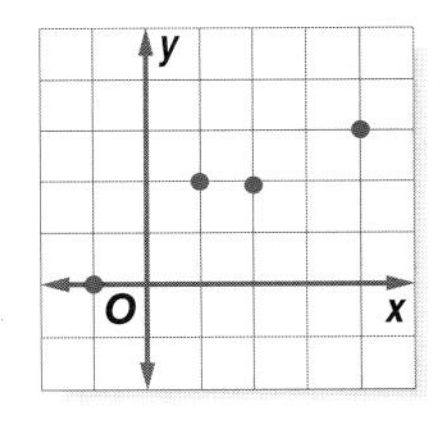

3. **OPEN ENDED** Write a different prediction equation for the data in Examples 1 and 2 on pages 81 and 82.

Guided Practice

Complete parts a–c for each set of data in Exercises 4 and 5.

a. Draw a scatter plot.

b. Use two ordered pairs to write a prediction equation.

c. Use your prediction equation to predict the missing value.

4. **SCIENCE** Whether you are climbing a mountain or flying in an airplane, the higher you go, the colder the air gets. The table shows the temperature in the atmosphere at various altitudes.

Altitude (ft)	0	1000	2000	3000	4000	5000
Temp (°C)	15.0	13.0	11.0	9.1	7.1	?

Source: NASA

5. **TELEVISION** As more channels have been added, cable television has become attractive to more viewers. The table shows the number of U.S. households with cable service in some recent years.

Year	1990	1992	1994	1996	1998	2010
Households (millions)	55	57	59	65	67	?

Source: Nielsen Media Research

Practice and Apply

Homework Help

For Exercises	See Examples
6–21	1, 2

Extra Practice
See page 831.

Complete parts a–c for each set of data in Exercises 6–9.

a. **Draw a scatter plot.**

b. **Use two ordered pairs to write a prediction equation.**

c. **Use your prediction equation to predict the missing value.**

6. **SAFETY** All states and the District of Columbia have enacted laws setting 21 as the minimum drinking age. The table shows the estimated cumulative number of lives these laws have saved by reducing traffic fatalities.

Year	1995	1996	1997	1998	1999	2010
Lives (1000s)	15.7	16.5	17.4	18.2	19.1	?

Source: National Highway Traffic Safety Administration

7. **HOCKEY** Each time a hockey player scores a goal, up to two teammates may be credited with assists. The table shows the number of goals and assists for some of the members of the Detroit Red Wings in the 2000–2001 NHL season.

Goals	31	15	32	27	16	20	8	4	12	12	?
Assists	45	56	37	30	24	18	17	5	10	7	15

Source: www.detroitredwings.com

8. **HEALTH** Bottled water has become very popular. The table shows the number of gallons of bottled water consumed per person in some recent years.

Year	1992	1993	1994	1995	1996	1997	2010
Gallons	8.2	9.4	10.7	11.6	12.5	13.1	?

Source: U.S. Department of Agriculture

A scatter plot of loan payments can help you analyze home loans. Visit www.algebra2.com/webquest to continue work on your WebQuest project.

9. **THEATER** Broadway, in New York City, is the center of American theater. The table shows the total revenue of all Broadway plays for some recent seasons.

Season	'95–'96	'96–'97	'97–'98	'98–'99	'99–'00	'09–'10
Revenue ($ millions)	436	499	558	588	603	?

Source: The League of American Theatres and Producers, Inc.

MEDICINE **For Exercises 10–12, use the graph that shows how much Americans spent on doctors' visits in some recent years.**

10. Write a prediction equation from the data for 1990, 1995, and 2000.

11. Use your equation to predict the amount for 2005.

12. Compare your prediction to the one given in the graph.

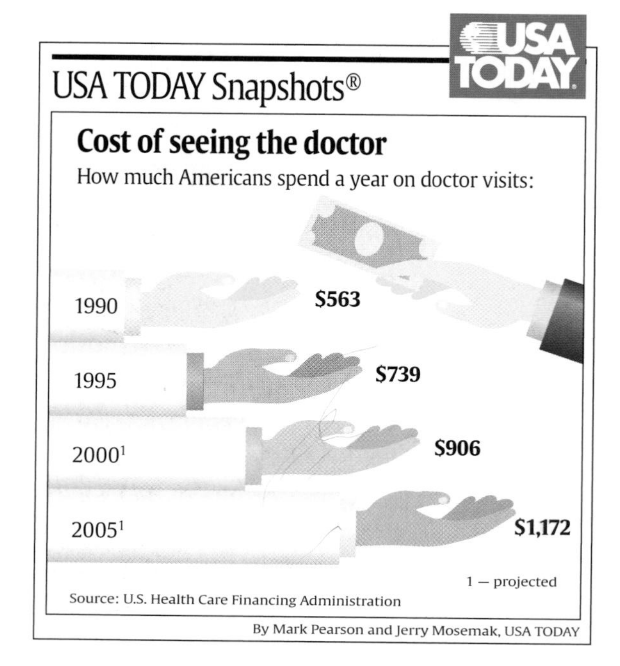

Career Choices

Finance

A financial analyst can advise people about how to invest their money and plan for retirement.

Online Research For information about a career as a financial analyst, visit: www.algebra2.com/careers

FINANCE **For Exercises 13 and 14, use the following information.**
Della has $1000 that she wants to invest in the stock market. She is considering buying stock in either Company 1 or Company 2. The values of the stocks at the ends of the last four months are shown in the tables below.

Company 1	
Month	Share Price ($)
Aug.	25.13
Sept.	22.94
Oct.	24.19
Nov.	22.56

Company 2	
Month	Share Price ($)
Aug.	31.25
Sept.	32.38
Oct.	32.06
Nov.	32.44

13. Based only on these data, which stock should Della buy? Explain.
14. Do you think investment decisions should be based on this type of reasoning? If not, what other factors should be considered?

GEOGRAPHY **For Exercises 15–18, use the table below that shows the elevation and average precipitation for selected cities.**

City	Elevation (feet)	Average Precip. (inches)	City	Elevation (feet)	Average Precip. (inches)
Rome, Italy	79	33	London, England	203	30
Algiers, Algeria	82	27	Paris, France	213	26
Istanbul, Turkey	108	27	Bucharest, Romania	298	23
Montreal, Canada	118	37	Budapest, Hungary	456	20
Stockholm, Sweden	171	21	Toronto, Canada	567	31
Berlin, Germany	190	23			

Source: World Meteorological Association

15. Draw a scatter plot with elevation as the independent variable.
16. Write a prediction equation.
17. Predict the average annual precipitation for Dublin, Ireland, which has an elevation of 279 feet.
18. Compare your prediction to the actual value of 29 inches.

CRITICAL THINKING **For Exercises 19 and 20, use the table that shows the percent of people ages 25 and over with a high school diploma over the last few decades.**

19. Use a prediction equation to predict the percent in 2010.
20. Do you think your prediction is accurate? Explain.

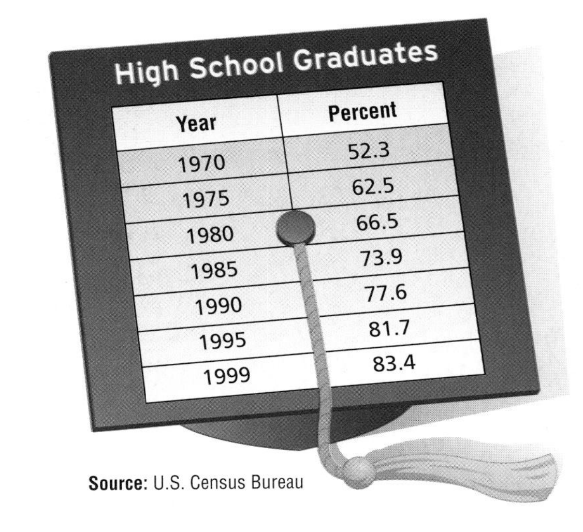

High School Graduates

Year	Percent
1970	52.3
1975	62.5
1980	66.5
1985	73.9
1990	77.6
1995	81.7
1999	83.4

Source: U.S. Census Bureau

21. **RESEARCH** Use the Internet or other resource to look up the population of your community or state in several past years. Use a prediction equation to predict the population in some future year.

22. WRITING IN MATH Answer the question that was posed at the beginning of the lesson.

How can a linear equation model the number of Calories you burn exercising?

Include the following in your answer:

- a scatter plot and a prediction equation for the data, and
- a prediction of the number of Calories burned in an hour by a 140-pound person running at 9 miles per hour, with a comparison of your predicted value with the actual value of 953.

23. Which line best fits the data in the graph at the right?

(A) $y = x$ (B) $y = -0.5x + 4$

(C) $y = -0.5x - 4$ (D) $y = 0.5 + 0.5x$

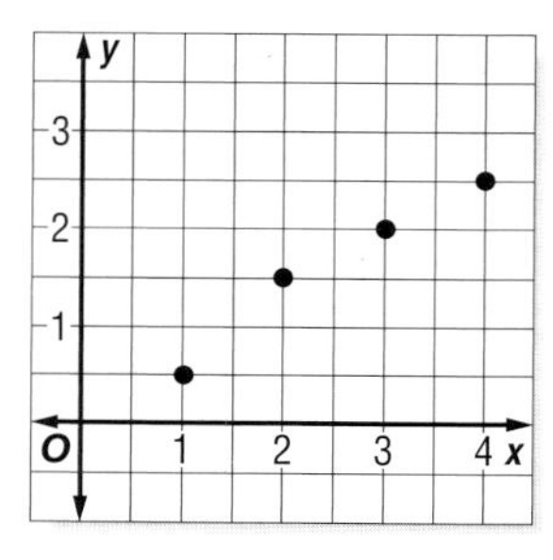

24. A prediction equation for a set of data is $y = 0.63x + 4.51$. For which x value is the predicted y value 6.4?

(A) 3 (B) 4.5

(C) 6 (D) 8.54

Extending the Lesson

For Exercises 25–30, use the following information.

A **median-fit line** is a particular type of line of fit. Follow the steps below to find the equation of the median-fit line for the data.

Federal and State Prisoners (per 100,000 U.S. citizens)								
Year	1986	1988	1990	1992	1994	1996	1998	1999
Prisoners	217	247	297	332	389	427	461	476

Source: U.S. Bureau of Justice Statistics

25. Divide the data into three approximately equal groups. There should always be the same number of points in the first and third groups. Find x_1, x_2, and x_3, the medians of the x values in Groups 1, 2, and 3, respectively. Find y_1, y_2, and y_3, the medians of the y values in Groups 1, 2, and 3, respectively.

26. Find an equation of the line through (x_1, y_1) and (x_3, y_3).

27. Find Y, the y-coordinate of the point on the line in Exercise 26 with an x-coordinate of x_2.

28. The median-fit line is parallel to the line in Exercise 26, but is one-third closer to (x_2, y_2). This means it passes through $\left(x_2, \frac{2}{3}Y + \frac{1}{3}y_2\right)$. Find this ordered pair.

29. Write an equation of the median-fit line.

30. Predict the number of prisoners per 100,000 citizens in 2005 and 2010.

Maintain Your Skills

Mixed Review

Write an equation in slope-intercept form that satisfies each set of conditions. *(Lesson 2-4)*

31. slope 4, passes through (0, 6)

32. passes through (5, −3) and (−2, 0)

Find each value if $g(x) = -\frac{4x}{3} + 7$. *(Lesson 2-1)*

33. $g(3)$

34. $g(0)$

35. $g(-2)$

36. $g(-4)$

37. Solve $|x + 4| > 3$. *(Lesson 1-6)*

Getting Ready for the Next Lesson

PREREQUISITE SKILL **Find each absolute value.** *(To review **absolute value**, see Lesson 1-4.)*

38. $|-3|$

39. $|11|$

40. $|0|$

41. $\left|-\frac{2}{3}\right|$

42. $|-1.5|$

Graphing Calculator Investigation

A Follow-Up of Lesson 2-5

Lines of Regression

You can use a TI-83 Plus graphing calculator to find a line that best fits a set of data. This line is called a **regression line** or **line of best fit**. You can also use the calculator to draw scatter plots and make predictions.

INCOME **The table shows the median income of U.S. families for the period 1970–1998.**

Year	1970	1980	1985	1990	1995	1998
Income ($)	9867	21,023	27,735	35,353	40,611	46,737

Source: U.S. Census Bureau

Find and graph a regression equation. Then predict the median income in 2010.

Step 1 *Find a regression equation.*

- Enter the years in L1 and the incomes in L2.

  KEYSTROKES: STAT ENTER 1970 ENTER

- Find the regression equation by selecting LinReg(ax+b) on the STAT CALC menu.

  KEYSTROKES: STAT ▶ 4 ENTER

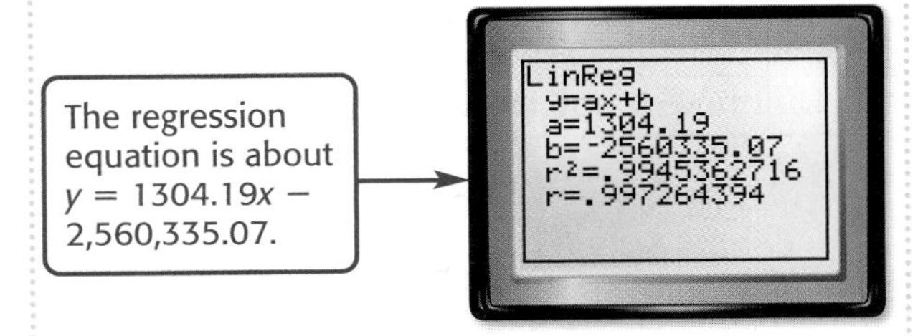

The slope indicates that family incomes were increasing at a rate of about $1300 per year.

The number r is called the **linear correlation coefficient**. The closer the value of r is to 1 or -1, the closer the data points are to the line. *If the values of r^2 and r are not displayed, use DiagnosticOn from the CATALOG menu.*

Step 2 *Graph the regression equation.*

- Use STAT PLOT to graph a scatter plot.

  KEYSTROKES: 2nd [STAT PLOT] ENTER ENTER

- Select the scatter plot, L1 as the Xlist, and L2 as the Ylist.
- Copy the equation to the Y= list and graph.

  KEYSTROKES: Y= VARS 5 ▶ ▶ 1 GRAPH

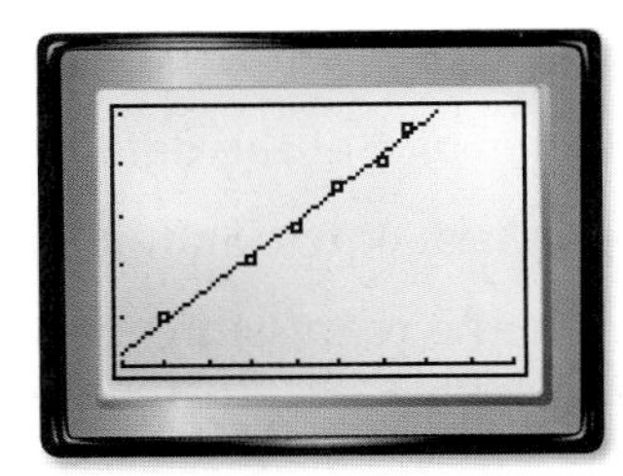

[1965, 2010] scl: 5 by [0, 50,000] scl: 10,000

Notice that the regression line does not pass through any of the data points, but comes close to all of them. The line fits the data very well.

Step 3 *Predict using the regression equation.*

- Find y when $x = 2010$. Use value on the CALC menu.

  KEYSTROKES: 2nd CALC 1 2010 ENTER

According to the regression equation, the median family income in 2010 will be about $61,087.

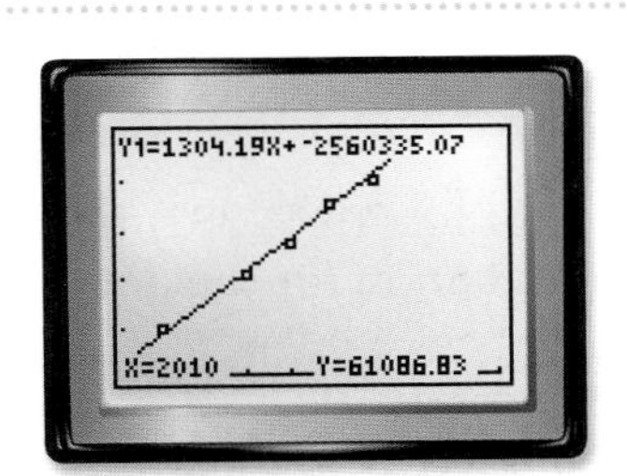

Exercises

GOVERNMENT For Exercises 1–3, use the table below that shows the population and the number of representatives in Congress for selected states.

State	CA	NY	TX	FL	NC	IN	AL
Population (millions)	29.8	18.0	17.0	12.9	6.6	5.5	4.0
Representatives	52	31	30	23	12	10	7

Source: *The World Almanac*

1. Make a scatter plot of the data.
2. Find a regression equation for the data.
3. Predict the number of representatives for Oregon, which has a population of about 2.8 million.

BASEBALL For Exercises 4–6, use the table at the right that shows the total attendance for minor league baseball in some recent years.

4. Make a scatter plot of the data.
5. Find a regression equation for the data.
6. Predict the attendance in 2010.

Year	Attendance (millions)
1985	18.4
1990	25.2
1995	33.1
2000	37.6

Source: National Association of Professional Baseball Leagues

TRANSPORTATION For Exercises 7–11, use the table below that shows the retail sales of motor vehicles in the United States for the period 1992–1999.

Motor Vehicle Sales								
Year	1992	1993	1994	1995	1996	1997	1998	1999
Vehicles (thousands)	13,118	14,199	15,413	15,118	15,456	15,498	15,963	17,414

Source: American Automobile Manufacturers Association

7. Make a scatter plot of the data.
8. Find a regression equation for the data.
9. According to the regression equation, what was the average rate of change of vehicle sales during the period?
10. Predict the sales in 2010.
11. How accurate do you think your prediction is? Explain.

RECREATION For Exercises 12–15, use the table at the right that shows the amount of money spent on skin diving and scuba equipment in some recent years.

12. Find a regression equation for the data.
13. Delete the outlier (1997, 332) from the data set. Then find a new regression equation for the data.
14. Use the new regression equation to predict the sales in 2010.
15. Compare the new correlation coefficient to the old value and state whether the regression line fits the data better.

Skin Diving and Scuba Equipment	
Year	Sales ($ millions)
1993	315
1994	322
1995	328
1996	340
1997	332
1998	345
1999	363

Source: National Sporting Goods Association

2-6 Special Functions

What You'll Learn

- Identify and graph step, constant, and identity functions.
- Identify and graph absolute value and piecewise functions.

Vocabulary

- step function
- greatest integer function
- constant function
- identity function
- absolute value function
- piecewise function

How do step functions apply to postage rates?

The cost of the postage to mail a letter is a function of the weight of the letter. But the function is not linear. It is a special function called a **step function**.

Weight not over (ounces)	Price ($)
1	0.34
2	0.55
3	0.76
4	0.97
…	…

U.S. MAIL

STEP FUNCTIONS, CONSTANT FUNCTIONS, AND THE IDENTITY FUNCTION The graph of a step function is not linear. It consists of line segments or rays. The **greatest integer function**, written $f(x) = [\![x]\!]$, is an example of a step function. The symbol $[\![x]\!]$ means *the greatest integer less than or equal to x*. For example, $[\![7.3]\!] = 7$ and $[\![-1.5]\!] = -2$ because $-1 > -1.5$. Study the table and graph below.

Study Tip

Greatest Integer Function
Notice that the domain of this step function is all real numbers and the range is all integers.

$f(x) = [\![x]\!]$	
x	$f(x)$
$-3 \le x < -2$	-3
$-2 \le x < -1$	-2
$-1 \le x < 0$	-1
$0 \le x < 1$	0
$1 \le x < 2$	1
$2 \le x < 3$	2
$3 \le x < 4$	3

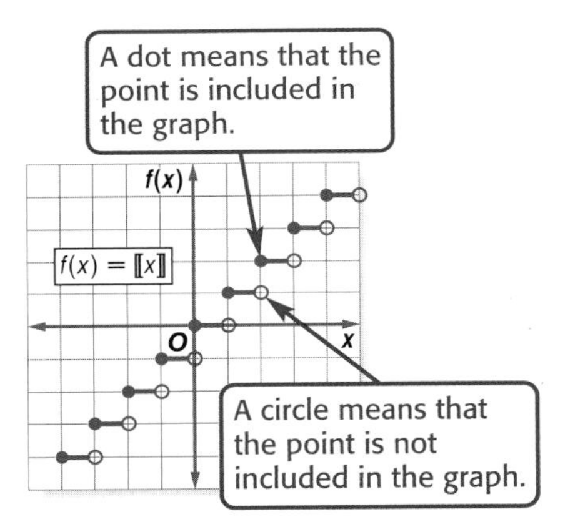

Example 1 Step Function

BUSINESS Labor costs at the Fix-It Auto Repair Shop are $60 per hour or any fraction thereof. Draw a graph that represents this situation.

Explore The total labor charge must be a multiple of $60, so the graph will be the graph of a step function.

Plan If the time spent on labor is greater than 0 hours, but less than or equal to 1 hour, then the labor cost is $60. If the time is greater than 1 hour but less than or equal to 2 hours, then the labor cost is $120, and so on.

Solve Use the pattern of times and costs to make a table, where x is the number of hours of labor and $C(x)$ is the total labor cost. Then draw the graph.

(continued on the next page)

x	$C(x)$
$0 < x \leq 1$	\$60
$1 < x \leq 2$	\$120
$2 < x \leq 3$	\$180
$3 < x \leq 4$	\$240
$4 < x \leq 5$	\$300

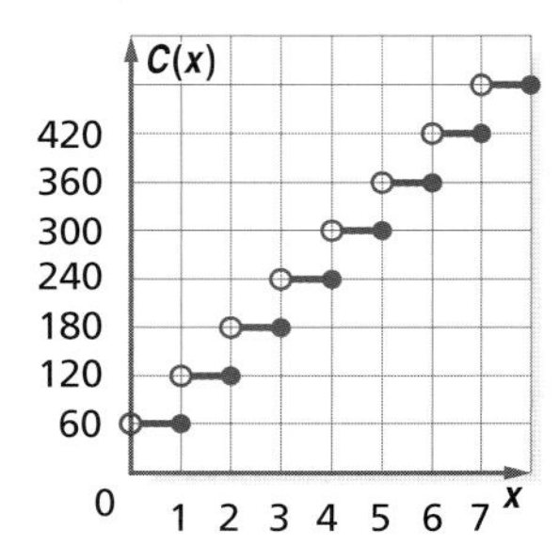

Examine Since the shop rounds any fraction of an hour up to the next whole number, each segment on the graph has a circle at the left endpoint and a dot at the right endpoint.

You learned in Lesson 2-4 that the slope-intercept form of a linear function is $y = mx + b$, or in functional notation, $f(x) = mx + b$. When $m = 0$, the value of the function is $f(x) = b$ for every x value. So, $f(x) = b$ is called a **constant function**.

The function $f(x) = 0$ is called the zero function.

Example 2 Constant Function

Graph $f(x) = 3$.

For every value of x, $f(x) = 3$. The graph is a horizontal line.

$f(x) = 3$	
x	$f(x)$
-2	3
-0.5	3
0	3
$\frac{1}{3}$	3

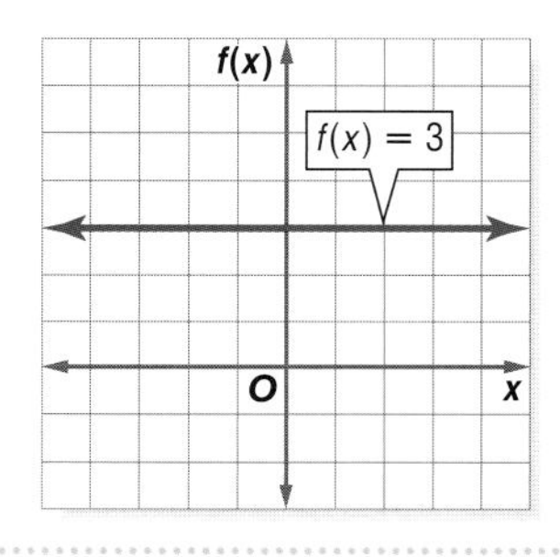

Another special case of slope-intercept form is $m = 1, b = 0$. This is the function $f(x) = x$. The graph is the line through the origin with slope 1.

Since the function does not change the input value, $f(x) = x$ is called the **identity function**.

$f(x) = x$	
x	$f(x)$
-2	-2
-0.5	-0.5
0	0
$\frac{1}{3}$	$\frac{1}{3}$

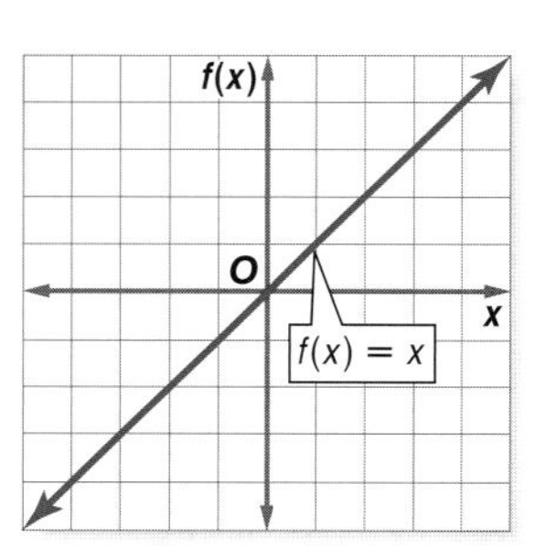

Study Tip

Absolute Value Function

Notice that the domain is all real numbers and the range is all nonnegative real numbers.

ABSOLUTE VALUE AND PIECEWISE FUNCTIONS Another special function is the **absolute value function**, $f(x) = |x|$.

$f(x) = \|x\|$	
x	$f(x)$
-3	3
-2	2
-1	1
0	0
1	1
2	2
3	3

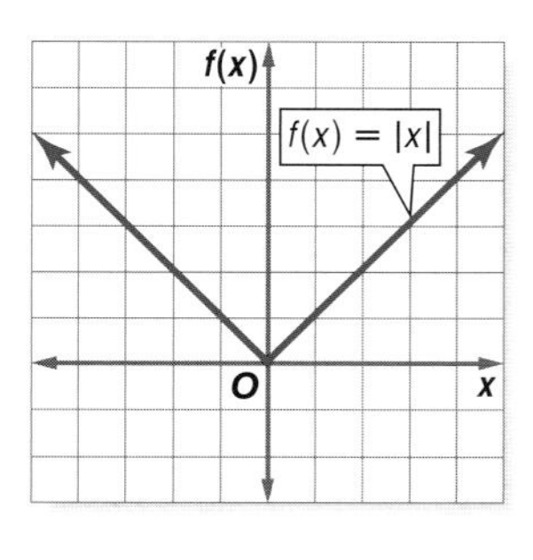

The absolute value function can be written as $f(x) = \begin{cases} -x \text{ if } x < 0 \\ x \text{ if } x \geq 0 \end{cases}$. A function that is written using two or more expressions is called a **piecewise function**.

Recall that a family of graphs is a group of graphs that displays one or more similar characteristics. The parent graph of most absolute value functions is $y = |x|$.

Study Tip

Look Back
To review **families of graphs**, see Lesson 2-3.

Example 3 Absolute Value Functions

Graph $f(x) = |x| + 1$ and $g(x) = |x| - 2$ on the same coordinate plane. Determine the similarities and differences in the two graphs.

Find several ordered pairs for each function.

x	$\|x\| + 1$
−2	3
−1	2
0	1
1	2
2	3

x	$\|x\| - 2$
−2	0
−1	−1
0	−2
1	−1
2	0

Graph the points and connect them.

- The domain of each function is all real numbers.
- The range of $f(x) = |x| + 1$ is $\{y \mid y \geq 1\}$.
  The range of $g(x) = |x| - 2$ is $\{y \mid y \geq -2\}$.
- The graphs have the same shape, but different y-intercepts.
- The graph of $g(x) = |x| - 2$ is the graph of $f(x) = |x| + 1$ translated down 3 units.

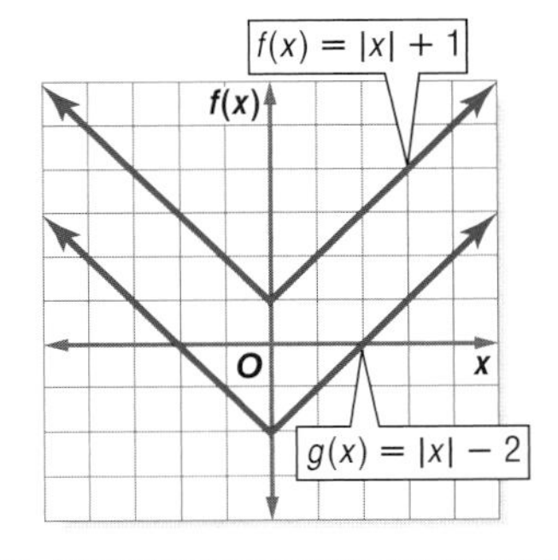

You can also use a graphing calculator to investigate families of absolute value graphs.

Graphing Calculator Investigation

Families of Absolute Value Graphs

The calculator screen shows the graphs of $y = |x|$, $y = 2|x|$, $y = 3|x|$, and $y = 5|x|$.

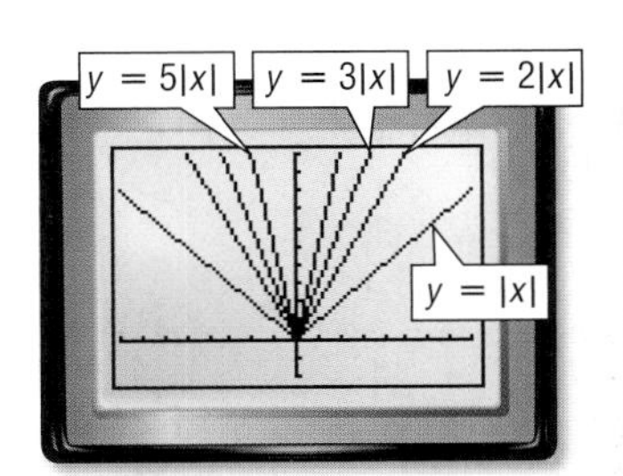

[−8, 8] scl: 1 by [−2, 10] scl: 1

Think and Discuss

1. What do these graphs have in common?
2. Describe how the graph of $y = a|x|$ changes as a increases. Assume $a > 0$.
3. Write an absolute value function whose graph is between the graphs of $y = 2|x|$ and $y = 3|x|$.
4. Graph $y = |x|$ and $y = -|x|$ on the same screen. Then graph $y = 2|x|$ and $y = -2|x|$ on the same screen. What is true in each case?
5. In general, what is true about the graph of $y = a|x|$ when $a < 0$?

To graph other piecewise functions, examine the inequalities in the definition of the function to determine how much of each piece to include.

Example 4 Piecewise Function

Graph $f(x) = \begin{cases} x - 4 \text{ if } x < 2 \\ 1 \text{ if } x \geq 2 \end{cases}$. Identify the domain and range.

Step 1 Graph the linear function $f(x) = x - 4$ for $x < 2$. Since 2 does not satisfy this inequality, stop with an open circle at $(2, -2)$.

Step 2 Graph the constant function $f(x) = 1$ for $x \geq 2$. Since 2 does satisfy this inequality, begin with a closed circle at $(2, 1)$ and draw a horizontal ray to the right.

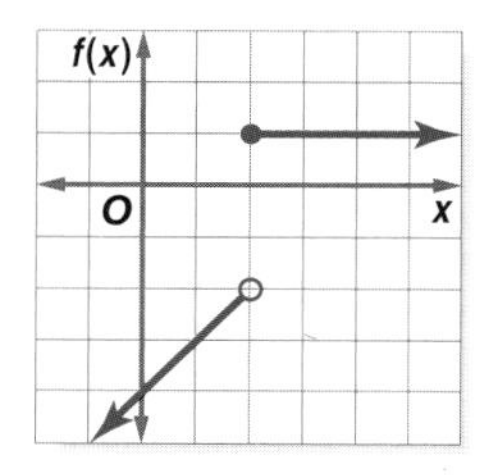

The function is defined for all values of x, so the domain is all real numbers. The values that are y-coordinates of points on the graph are 1 and all real numbers less than -2, so the range is $\{y \mid y < -2 \text{ or } y = 1\}$.

Study Tip

Graphs of Piecewise Functions

The graphs of each part of a piecewise function may or may not connect. A graph may stop at a given x value and then begin again at a different y value for the same x value.

Concept Summary — *Special Functions*

Step Function	Constant Function	Absolute Value Function	Piecewise Function
horizontal segments and rays	horizontal line	V-shape	different rays, segments, and curves

Example 5 Identify Functions

Determine whether each graph represents a step function, a constant function, an absolute value function, or a piecewise function.

a.

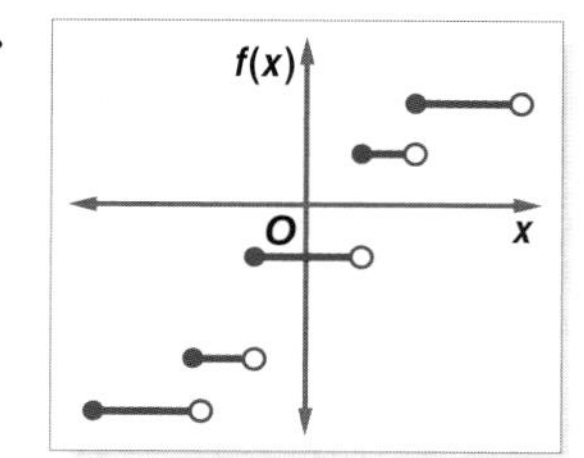

The graph has multiple horizontal segments. It represents a step function.

b.

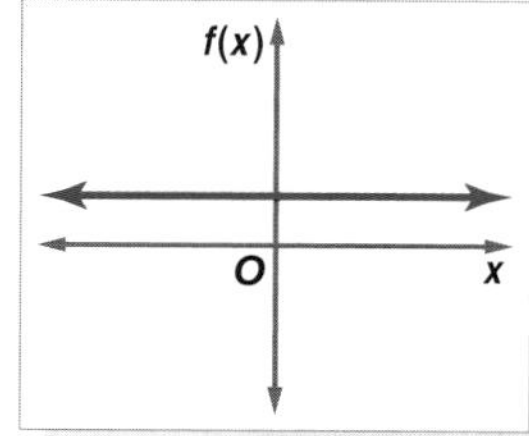

The graph is a horizontal line. It represents a constant function.

Check for Understanding

Concept Check

1. **Find a counterexample** to the statement *To find the greatest integer function of x when x is not an integer, round x to the nearest integer.*
2. **Evaluate** $g(4.3)$ if $g(x) = [\![x - 5]\!]$.
3. **OPEN ENDED** Write a function involving absolute value for which $f(-2) = 3$.

Guided Practice **Identify each function as S for step, C for constant, A for absolute value, or P for piecewise.**

4.

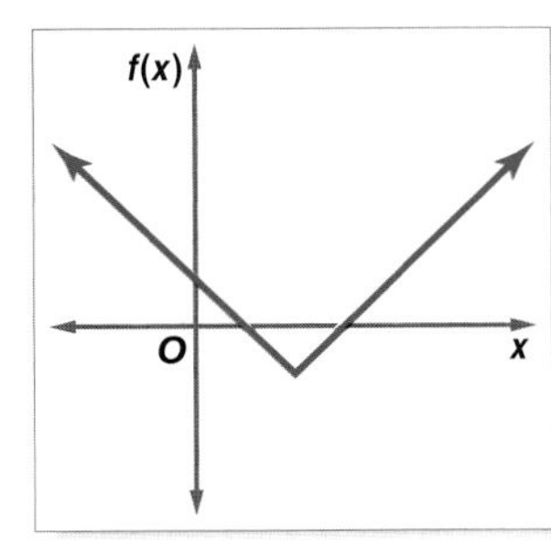

5.

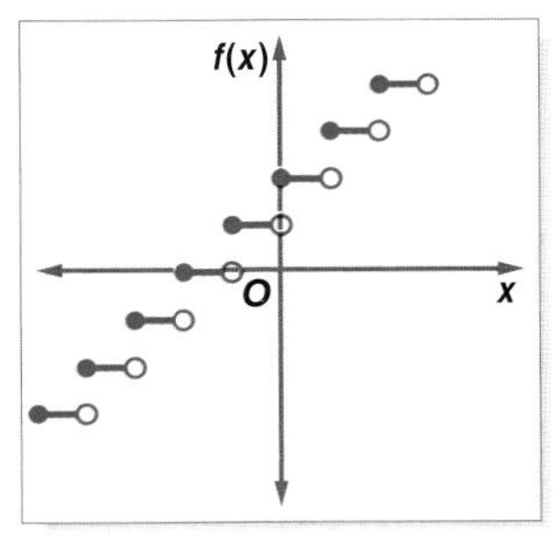

Graph each function. Identify the domain and range.

6. $f(x) = -[\![x]\!]$

7. $g(x) = [\![2x]\!]$

8. $h(x) = |x - 4|$

9. $f(x) = |3x - 2|$

10. $g(x) = \begin{cases} -1 \text{ if } x < 0 \\ -x + 2 \text{ if } x \geq 0 \end{cases}$

11. $h(x) = \begin{cases} x + 3 \text{ if } x \leq -1 \\ 2x \text{ if } x > -1 \end{cases}$

Application **PARKING** **For Exercises 12–14, use the following information.**
A downtown parking lot charges \$2 for the first hour and \$1 for each additional hour or part of an hour.

12. What type of special function models this situation?

13. Draw a graph of a function that represents this situation.

14. Use the graph to find the cost of parking there for $4\frac{1}{2}$ hours.

Practice and Apply

Homework Help

For Exercises	See Examples
15–20	5
21–29	1
30–37, 45–47, 49	3
38–41, 44, 48	2, 4
42, 43	1, 3

Extra Practice
See page 831.

Identify each function as S for step, C for constant, A for absolute value, or P for piecewise.

15.

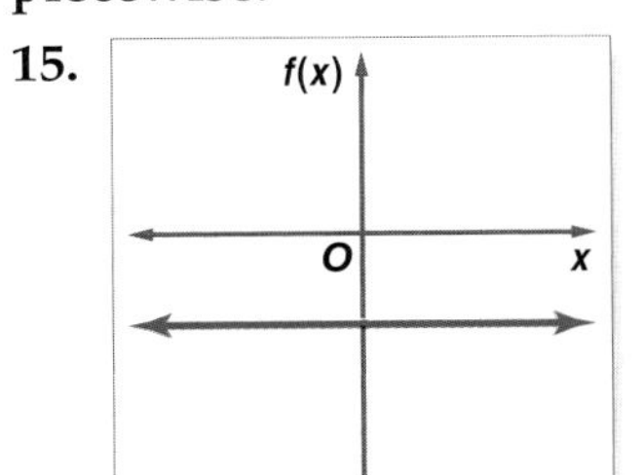

16.

17.

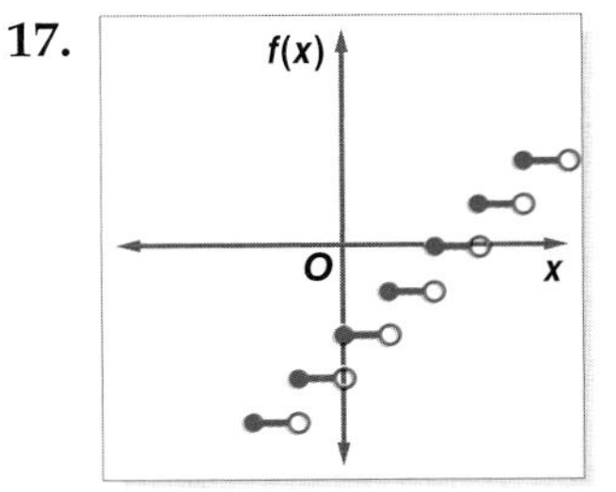

18.

19.

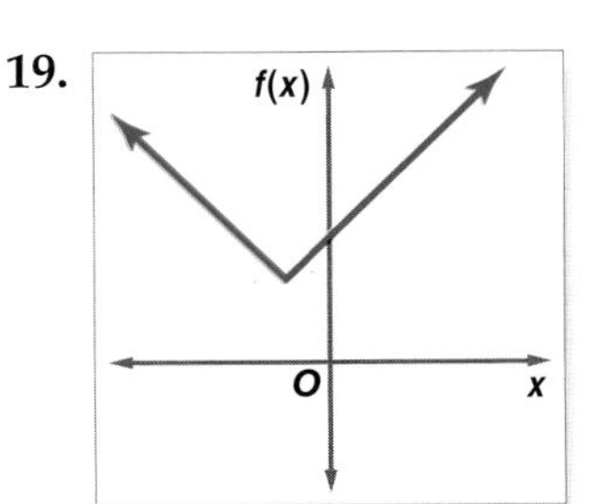

20.

21. TRANSPORTATION Bluffton High School chartered buses so the student body could attend the girls' basketball state tournament games. Each bus held a maximum of 60 students. Draw a graph of a step function that shows the relationship between the number of students x who went to the game and the number of buses y that were needed.

TELEPHONE RATES **For Exercises 22 and 23, use the following information.**
Sarah has a long-distance telephone plan where she pays 10¢ for each minute or part of a minute that she talks, regardless of the time of day.

22. Graph a step function that represents this situation.

23. Sarah made a call to her brother that lasted 9 minutes and 40 seconds. How much did the call cost?

Graph each function. Identify the domain and range.

24. $f(x) = [\![x + 3]\!]$
25. $g(x) = [\![x - 2]\!]$
26. $f(x) = 2[\![x]\!]$
27. $h(x) = -3[\![x]\!]$
28. $g(x) = [\![x]\!] + 3$
29. $f(x) = [\![x]\!] - 1$
30. $f(x) = |2x|$
31. $h(x) = |-x|$
32. $g(x) = |x| + 3$
33. $g(x) = |x| - 4$
34. $h(x) = |x + 3|$
35. $f(x) = |x + 2|$
36. $f(x) = \left|x - \frac{1}{4}\right|$
37. $f(x) = \left|x + \frac{1}{2}\right|$
38. $f(x) = \begin{cases} -x \text{ if } x \leq 3 \\ 2 \text{ if } x > 3 \end{cases}$
39. $h(x) = \begin{cases} -1 \text{ if } x < -2 \\ 1 \text{ if } x > 2 \end{cases}$
40. $f(x) = \begin{cases} x \text{ if } x < -3 \\ 2 \text{ if } -3 \leq x < 1 \\ -2x + 2 \text{ if } x \geq 1 \end{cases}$
41. $g(x) = \begin{cases} -1 \text{ if } x \leq -2 \\ x \text{ if } -2 < x < 2 \\ -x + 1 \text{ if } x \geq 2 \end{cases}$
42. $f(x) = [\![\,|x|\,]\!]$
43. $g(x) = |[\![x]\!]|$

44. Write the function shown in the graph.

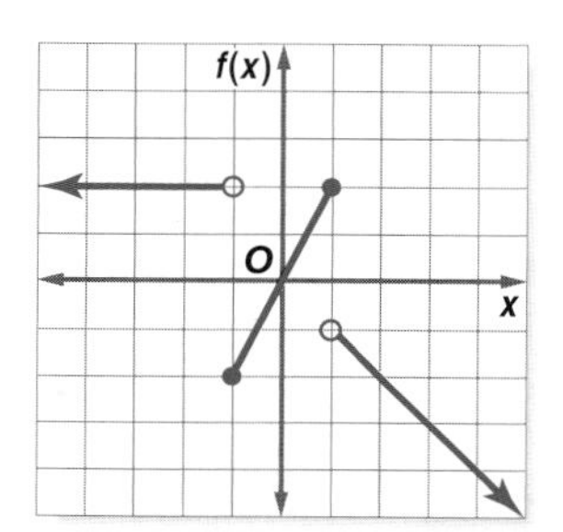

More About. . .

Nutrition

Good sources of vitamin C include citrus fruits and juices, cantaloupe, broccoli, brussels sprouts, potatoes, sweet potatoes, tomatoes, and cabbage.

Source: *The World Almanac*

NUTRITION **For Exercises 45–47, use the following information.**
The recommended dietary allowance for vitamin C is 2 micrograms per day.

45. Write an absolute value function for the difference between the number of micrograms of vitamin C you ate today x and the recommended amount.

46. What is an appropriate domain for the function?

47. Use the domain to graph the function.

48. **INSURANCE** According to the terms of Lavon's insurance plan, he must pay the first $300 of his annual medical expenses. The insurance company pays 80% of the rest of his medical expenses. Write a function for how much the insurance company pays if x represents Lavon's annual medical expenses.

49. **CRITICAL THINKING** Graph $|x| + |y| = 3$.

50. **WRITING IN MATH** Answer the question that was posed at the beginning of the lesson.

How do step functions apply to postage rates?

Include the following in your answer:

- an explanation of why a step function is the best model for this situation, while your gas mileage as a function of time as you drive to the post office cannot be modeled with a step function, and
- a graph of a function that represents the cost of a first-class letter.

51. For which function does $f\left(-\frac{1}{2}\right) \neq -1$?

Ⓐ $f(x) = 2x$ Ⓑ $f(x) = |-2x|$ Ⓒ $f(x) = [\![x]\!]$ Ⓓ $f(x) = [\![2x]\!]$

52. For which function is the range $\{y \mid y \leq 0\}$?

Ⓐ $f(x) = -x$ Ⓑ $f(x) = [\![x]\!]$ Ⓒ $f(x) = |x|$ Ⓓ $f(x) = -|x|$

Maintain Your Skills

Mixed Review

HEALTH For Exercises 53–55, use the table that shows the life expectancy for people born in various years. *(Lesson 2-5)*

Year	1950	1960	1970	1980	1990	1997
Expectancy	68.2	69.7	70.8	73.7	75.4	76.5

Source: National Center for Health Statistics

53. Draw a scatter plot in which x is the number of years since 1950.

54. Find a prediction equation.

55. Predict the life expectancy of a person born in 2010.

Write an equation in slope-intercept form that satisfies each set of conditions. *(Lesson 2-4)*

56. slope 3, passes through $(-2, 4)$

57. passes through $(0, -2)$ and $(4, 2)$

Solve each inequality. Graph the solution set. *(Lesson 1-5)*

58. $3x - 5 \geq 4$

59. $28 - 6y < 23$

Getting Ready for the Next Lesson

PREREQUISITE SKILL Determine whether (0, 0) satisfies each inequality. Write *yes* or *no*. *(To review **inequalities**, see Lesson 1-5.)*

60. $y < 2x + 3$

61. $y \geq -x + 1$

62. $y \leq \frac{3}{4}x - 5$

63. $2x + 6y + 3 > 0$

64. $y > |x|$

65. $|x| + y \leq 3$

Practice Quiz 2

Lessons 2-4 through 2-6

1. Write an equation in slope-intercept form of the line with slope $-\frac{2}{3}$ that passes through $(-2, 5)$. *(Lesson 2-4)*

BASKETBALL For Exercises 2–4, use the following information.
On August 26, 2000, the Houston Comets beat the New York Liberty to win their fourth straight WNBA championship. The table shows the heights and weights of the Comets who played in that final game. *(Lesson 2-5)*

Height (in.)	74	71	76	70	66	74	72
Weight (lb)	178	147	195	150	138	190	?

Source: WNBA

2. Draw a scatter plot.

3. Use two ordered pairs to write a prediction equation.

4. Use your prediction equation to predict the missing value.

5. Graph $f(x) = |x - 1|$. Identify the domain and range. *(Lesson 2-6)*

2-7 Graphing Inequalities

California Standards Standard 1.0 **Students solve** equations and **inequalities involving absolute value.** (Key)

What You'll Learn

- Graph linear inequalities.
- Graph absolute value inequalities.

Vocabulary

- boundary

How do inequalities apply to fantasy football?

Dana has Vikings receiver Randy Moss as a player on his online fantasy football team. Dana gets 5 points per receiving yard that Moss gets and 100 points per touchdown that Moss scores. He considers 1000 points or more to be a good game. Dana can use a linear inequality to check whether certain combinations of yardage and touchdowns, such as those in the table, result in 1000 points or more.

	Yards	TDs
Game 1	168	3
Game 2	144	2
Game 3	136	1

GRAPH LINEAR INEQUALITIES A linear inequality resembles a linear equation, but with an inequality symbol instead of an equals symbol. For example, $y \leq 2x + 1$ is a linear inequality and $y = 2x + 1$ is the related linear equation.

The graph of the inequality $y \leq 2x + 1$ is the shaded region. Every point in the shaded region satisfies the inequality. The graph of $y = 2x + 1$ is the **boundary** of the region. It is drawn as a solid line to show that points on the line satisfy the inequality. If the inequality symbol were $<$ or $>$, then points on the boundary would not satisfy the inequality, so the boundary would be drawn as a dashed line.

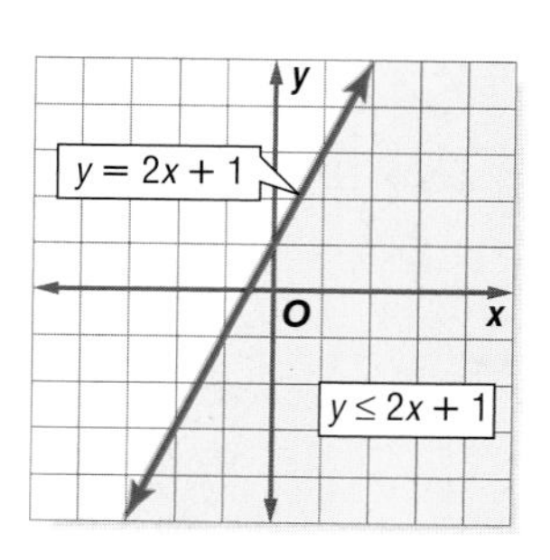

You can graph an inequality by following these steps.

Step 1 Determine whether the boundary should be solid or dashed. Graph the boundary.

Step 2 Choose a point not on the boundary and test it in the inequality.

Step 3 If a true inequality results, shade the region containing your test point. If a false inequality results, shade the other region.

Example 1 Dashed Boundary

Graph $2x + 3y > 6$.

The boundary is the graph of $2x + 3y = 6$. Since the inequality symbol is $>$, the boundary will be dashed. Use the slope-intercept form, $y = -\frac{2}{3}x + 2$.

Now test the point (0, 0). *The point (0, 0) is usually a good point to test because it results in easy calculations.*

$2x + 3y > 6$	Original inequality
$2(0) + 3(0) > 6$	$(x, y) = (0, 0)$
$0 > 6$	false

Shade the region that does *not* contain (0, 0).

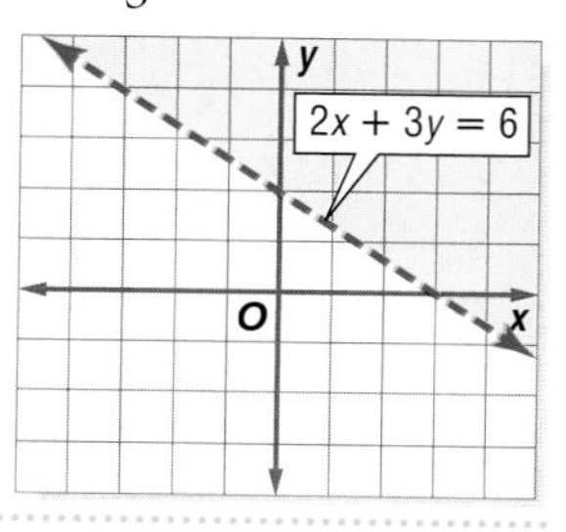

Inequalities can sometimes be used to model real-world situations.

Example 2 Solid Boundary

BUSINESS A mail-order company is hiring temporary employees to help in their packing and shipping departments during their peak season.

a. Write an inequality to describe the number of employees that can be assigned to each department if the company has 20 temporary employees available.

Let p be the number of employees assigned to packing and let s be the number assigned to shipping. Since the company can assign *at most* 20 employees total to the two departments, use a $\leq$ symbol.

The number of employees for packing	and	the number of employees for shipping	is at most	twenty.
p	$+$	s	$\leq$	20

Study Tip

Look Back
To review **translating verbal expressions to inequalities,** see Lesson 1-5.

b. Graph the inequality.

Since the inequality symbol is $\leq$, the graph of the related linear equation $p + s = 20$ is solid. This is the boundary of the inequality.

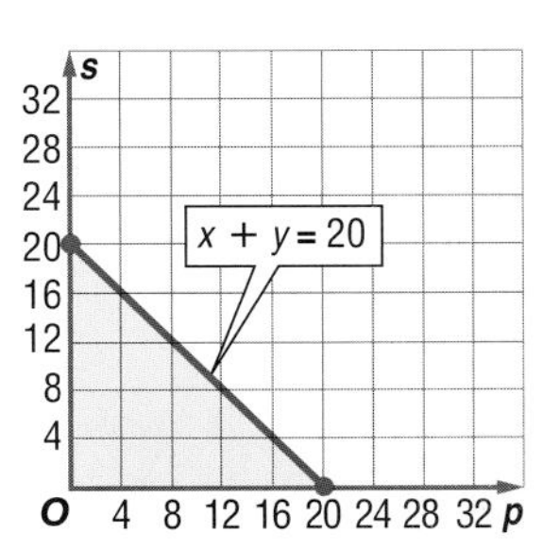

Test (0, 0).

$p + s \leq 20$	Original inequality
$0 + 0 \leq 20$	$(p, s) = (0, 0)$
$0 \leq 20$	true

Shade the region that contains (0, 0). *Since the variables cannot be negative, shade only the part in the first quadrant.*

c. Can the company assign 8 employees to packing and 10 employees to shipping?

The point (8, 10) is in the shaded region, so it satisfies the inequality. The company can assign 8 employees to packing and 10 to shipping.

GRAPH ABSOLUTE VALUE INEQUALITIES Graphing absolute value inequalities is similar to graphing linear inequalities. The inequality symbol determines whether the boundary is solid or dashed, and you can test a point to determine which region to shade.

Example 3 Absolute Value Inequality

Graph $y < |x| + 1$.

Since the inequality symbol is $<$, the graph of the related equation $y = |x| + 1$ is dashed. Graph the equation.

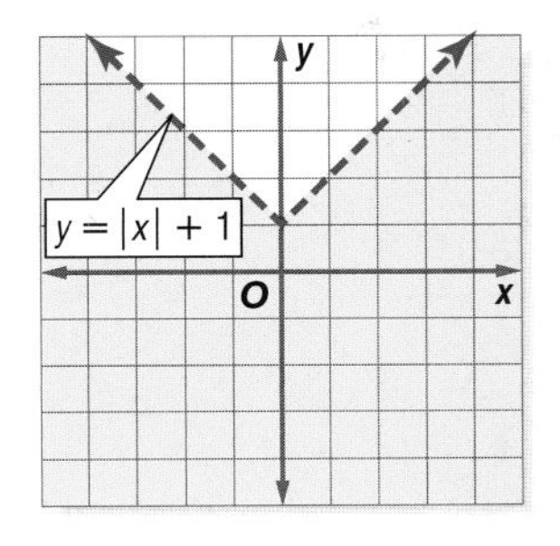

Test (0, 0).

$y < |x| + 1$	Original inequality
$0 < |0| + 1$	$(x, y) = (0, 0)$
$0 < 0 + 1$	$|0| = 0$
$0 < 1$	true

Shade the region that includes (0, 0).

Check for Understanding

Concept Check

1. **Write** an inequality for the graph at the right.

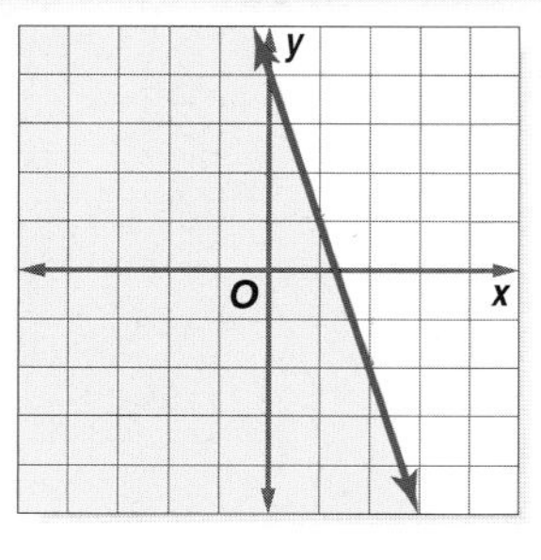

2. **Explain** how to determine which region to shade when graphing an inequality.
3. **OPEN ENDED** Write an absolute value inequality for which the boundary is solid and the solution is the region above the graph of the related equation.

Guided Practice

Graph each inequality.

4. $y < 2$
5. $y > 2x - 3$
6. $x - y \geq 0$
7. $x - 2y \leq 5$
8. $y > |2x|$
9. $y \leq 3|x| - 1$

Application

SHOPPING **For Exercises 10–12, use the following information.**
Gwen wants to buy some cassettes that cost \$10 each and some CDs that cost \$13 each. She has \$40 to spend.

10. Write an inequality to represent the situation, where c is the number of cassettes she buys and d is the number of CDs.
11. Graph the inequality.
12. Can she buy 3 cassettes and 2 CDs? Explain.

Practice and Apply

Homework Help

For Exercises	See Examples
13–24, 31, 32	1, 2
25–30, 41	3
33–40	2

Extra Practice
See page 832.

Graph each inequality.

13. $x + y > -5$
14. $3 \geq x - 3y$
15. $y > 6x - 2$
16. $x - 5 \leq y$
17. $y \geq -4x + 3$
18. $y - 2 < 3x$
19. $y \geq 1$
20. $y + 1 < 4$
21. $4x - 5y - 10 \leq 0$
22. $x - 6y + 3 > 0$
23. $y > \frac{1}{3}x + 5$
24. $y \geq \frac{1}{2}x - 5$
25. $y \leq |x|$
26. $y > |4x|$
27. $y + |x| < 3$
28. $y \geq |x - 1| - 2$
29. $|x + y| > 1$
30. $|x| \leq |y|$

31. Graph all the points on the coordinate plane to the left of the graph of $x = -2$. Write an inequality to describe these points.
32. Graph all the points on the coordinate plane below the graph of $y = 3x - 5$. Write an inequality to describe these points.

SCHOOL **For Exercises 33 and 34, use the following information.**
Rosa's professor says that the midterm exam will count for 40% of each student's grade and the final exam will count for 60%. A score of at least 90 is required for an A.

33. The inequality $0.4x + 0.6y \geq 90$ represents this situation, where x is the midterm score and y is the final exam score. Graph this inequality.
34. If Rosa scores 85 on the midterm and 95 on the final, will she get an A?

DRAMA **For Exercises 35–37, use the following information.**
Tickets for the Prestonville High School Drama Club's spring play cost \$4 for adults and \$3 for students. In order to cover expenses, at least \$2000 worth of tickets must be sold.

35. Write an inequality that describes this situation.
36. Graph the inequality.
37. If 180 adult and 465 student tickets are sold, will the club cover its expenses?

More About. . .

Finance

A *dividend* is a payment from a company to an investor. It is a way to make money on a stock without selling it.

FINANCE For Exercises 38–40, use the following information.

Carl Talbert estimates that he will need to earn at least \$9000 per year combined in dividend income from the two stocks he owns to supplement his retirement plan.

Company	Dividend per Share
Able Rentals	\$1.20
Best Bikes	\$1.80

38. Write and graph an inequality for this situation.

39. Will he make enough from 3000 shares of each company?

40. **CRITICAL THINKING** Graph $|y| < x$.

41. **WRITING IN MATH** Answer the question that was posed at the beginning of the lesson.

How do inequalities apply to fantasy football?

Include the following in your answer:

- an inequality, and an explanation of how you obtained it, to represent a good game for Randy Moss in Dana's fantasy football league,
- a graph of your inequality (remember that the number of touchdowns cannot be negative, but receiving yardage can be), and
- which of the games with statistics in the table qualify as good games.

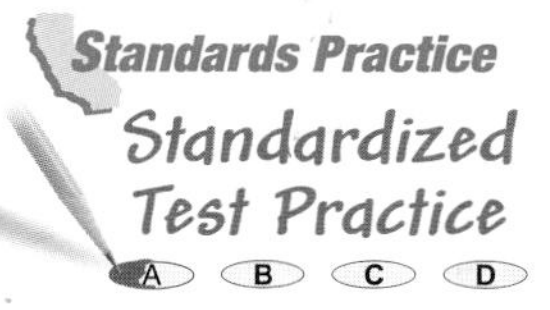

Standards Practice

Standardized Test Practice

42. Which could be the inequality for the graph?

(A) $y < 3x + 2$ (B) $y \leq 3x + 2$

(C) $y > 3x + 2$ (D) $y \geq 3x + 2$

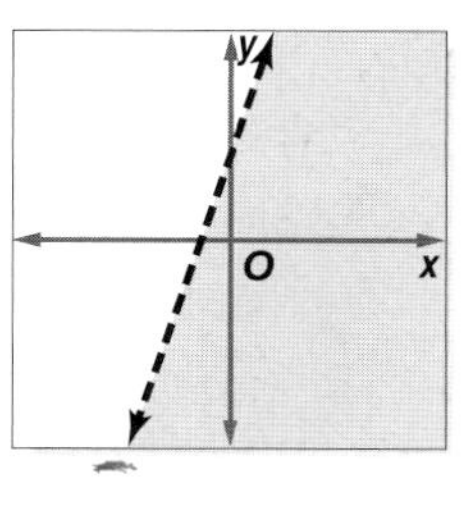

43. Which point satisfies $y > 5|x| - 3$?

(A) (2, 2) (B) (−1, 3)

(C) (3, 7) (D) (−2, 4)

Graphing Calculator

SHADE(COMMAND You can graph inequalities with a graphing calculator by using the Shade(command located in the DRAW menu. You must enter two functions.

- The first function defines the lower boundary of the region to be shaded.
- The second function defines the upper boundary of the region.
- If the inequality is "$y \leq$," use the Ymin window value as the lower boundary.
- If the inequality is "$y \geq$," use the Ymax window value as the upper boundary.

Graph each inequality.

44. $y \geq 3$ **45.** $y \leq x + 2$ **46.** $y \leq -2x - 4$ **47.** $x - 7 \leq y$

Maintain Your Skills

Mixed Review

Graph each function. Identify the domain and range. *(Lesson 2-6)*

48. $f(x) = [\![x]\!] - 4$ **49.** $g(x) = |x| - 1$ **50.** $h(x) = |x - 3|$

SALES For Exercises 51–53, use the table that shows the years of experience for eight sales representatives and their sales during a given period of time. *(Lesson 2-5)*

Years	6	5	3	1	4	3	6	2
Sales (\$)	9000	6000	4000	3000	6000	5000	8000	2000

51. Draw a scatter plot.

52. Find a prediction equation.

53. Predict the sales for a representative with 8 years of experience.

Solve each equation. Check your solution. *(Lesson 1-3)*

54. $4x - 9 = 23$ **55.** $11 - 2y = 5$ **56.** $2z - 3 = -6z + 1$

Chapter 2 Study Guide and Review

Vocabulary and Concept Check

absolute value function (p. 90)
bivariate data (p. 81)
boundary (p. 96)
Cartesian coordinate plane (p. 56)
constant function (p. 90)
dependent variable (p. 59)
domain (p. 56)
family of graphs (p. 70)
function (p. 57)
functional notation (p. 59)
greatest integer function (p. 89)
identity function (p. 90)
independent variable (p. 59)
linear equation (p. 63)
linear function (p. 63)
line of fit (p. 81)
mapping (p. 57)
one-to-one function (p. 57)
ordered pair (p. 56)
parent graph (p. 70)
piecewise function (p. 91)
point-slope form (p. 76)
prediction equation (p. 81)
quadrant (p. 56)
range (p. 56)
rate of change (p. 69)
relation (p. 56)
scatter plot (p. 81)
slope (p. 68)
slope-intercept form (p. 75)
standard form (p. 64)
step function (p. 89)
vertical line test (p. 57)
x-intercept (p. 65)
y-intercept (p. 65)

Choose the correct term to complete each sentence.

1. The (*constant, identity*) function is a linear function described by $f(x) = x$.
2. The graph of the (*absolute value, greatest integer*) function forms a V-shape.
3. The (*slope-intercept, standard*) form of the equation of a line is $Ax + By = C$, where A and B are not both zero.
4. Two lines in the same plane having the same slope are (*parallel, perpendicular*).
5. The (*domain, range*) is the set of all x-coordinates of the ordered pairs of a relation.
6. The set of all y-coordinates of the ordered pairs of a relation is the (*domain, range*).
7. The ratio of the change in y-coordinates to the corresponding change in x-coordinates is called the (*slope, y-intercept*) of a line.
8. The (*line of fit, vertical line test*) can be used to determine if a relation is a function.

Lesson-by-Lesson Review

2-1 Relations and Functions

See pages 56–62.

Concept Summary

- A relation is a set of ordered pairs. The domain is the set of all x-coordinates, and the range is the set of all y-coordinates.
- A function is a relation where each member of the domain is paired with exactly one member of the range.

Example **Graph the relation $\{(-3, 1), (0, 2), (2, 5)\}$ and find the domain and range. Then determine whether the relation is a function.**

The domain is $\{-3, 0, 2\}$, and the range is $\{1, 2, 5\}$.

Graph the ordered pairs. Since each x value is paired with exactly one y value, the relation is a function.

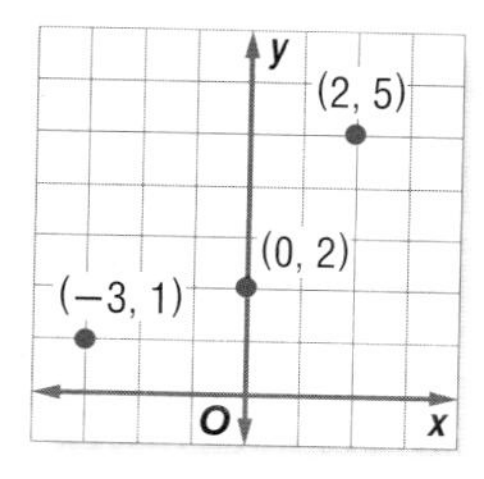

www.algebra2.com/vocabulary_review

Exercises **Graph each relation or equation and find the domain and range. Then determine whether the relation or equation is a function.**
See Examples 1 and 2 on pages 57 and 58.

9. $\{(6, 3), (2, 1), (-2, 3)\}$
10. $\{(-5, 2), (2, 4), (1, 1), (-5, -2)\}$
11. $y = 0.5x$
12. $y = 2x + 1$

Find each value if $f(x) = 5x - 9$. *See Example 5 on page 59.*

13. $f(6)$
14. $f(-2)$
15. $f(y)$
16. $f(-2v)$

2-2 Linear Equations

See pages 63–67.

Concept Summary

- A linear equation is an equation whose graph is a line. A linear function can be written in the form $f(x) = mx + b$.
- The standard form of a linear equation is $Ax + By = C$.

Example **Write $2x - 6 = y + 8$ in standard form. Identify A, B, and C.**

$2x - 6 = y + 8$ Original equation
$2x - y - 6 = 8$ Subtract y from each side.
$2x - y = 14$ Add 6 to each side.

The standard form is $2x - y = 14$. So, $A = 2$, $B = -1$, and $C = 14$.

Exercises **State whether each equation or function is linear. Write *yes* or *no*. If no, explain your reasoning.** *See Example 1 on page 63.*

17. $3x^2 - y = 6$
18. $2x + y = 11$
19. $h(x) = \sqrt{2x + 1}$

Write each equation in standard form. Identify A, B, and C. *See Example 3 on page 64.*

20. $y = 7x + 15$
21. $0.5x = -0.2y - 0.4$
22. $\frac{2}{3}x - \frac{3}{4}y = 6$

Find the x-intercept and the y-intercept of the graph of each equation. Then graph the equation. *See Example 4 on page 65.*

23. $-\frac{1}{5}y = x + 4$
24. $6x = -12y + 48$
25. $y - x = -9$

2-3 Slope

See pages 68–74.

Concept Summary

- The slope of a line is the ratio of the change in y-coordinates to the corresponding change in x-coordinates.

$$m = \frac{y_2 - y_1}{x_2 - x_1}$$

- Lines with the same slope are parallel. Lines with slopes that are opposite reciprocals are perpendicular.

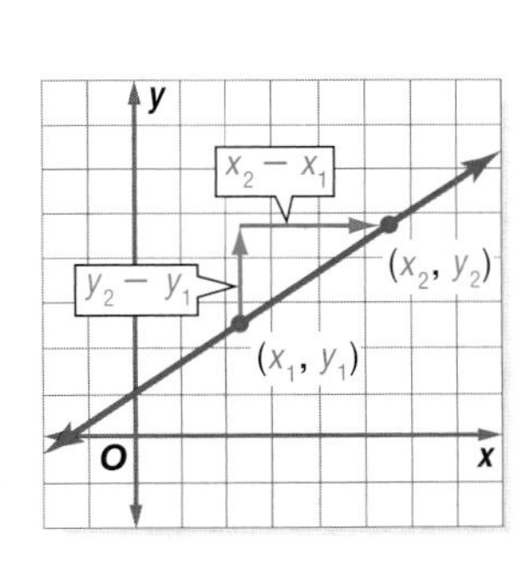

Example **Find the slope of the line that passes through (−5, 3) and (7, 9).**

$m = \frac{y_2 - y_1}{x_2 - x_1}$ Slope formula

$= \frac{9 - 3}{7 - (-5)}$ $(x_1, y_1) = (-5, 3), (x_2, y_2) = (7, 9)$

$= \frac{6}{12}$ or $\frac{1}{2}$ Simplify.

Exercises **Find the slope of the line that passes through each pair of points.**
See Example 1 on page 68.

26. (−6, −3), (6, 7) **27.** (5.5, −5.5), (11, −7) **28.** (−3, 24), (10, −41)

Graph the line passing through the given point with the given slope.
See Example 2 on page 69.

29. (0, 1), $m = 2$ **30.** (3, −2), $m = \frac{5}{2}$ **31.** (−5, 2), $m = -\frac{1}{4}$

Graph the line that satisfies each set of conditions.
See Examples 4 and 5 on pages 70 and 71.

32. passes through (2, 0), parallel to a line whose slope is 3
33. passes through (−1, −2), perpendicular to a line whose slope is $\frac{1}{2}$
34. passes through (4, 1), perpendicular to graph of $2x + 3y = 1$
35. passes through (−2, 2), parallel to graph of $-2x + y = 4$

2-4 Writing Linear Equations

See pages 75–80.

Concept Summary

- Slope-Intercept Form: $y = mx + b$
- Point-Slope Form: $y - y_1 = m(x - x_1)$

Example **Write an equation in slope-intercept form for the line through (4, 5) that is parallel to the line through (−1, −3) and (2, −1).**

First, find the slope of the given line.

$m = \frac{y_2 - y_1}{x_2 - x_1}$ Slope formula

$= \frac{-1 - (-3)}{2 - (-1)}$ $(x_1, y_1) = (-1, -3)$, $(x_2, y_2) = (2, -1)$

$= \frac{2}{3}$ Simplify.

The parallel line will also have slope $\frac{2}{3}$.

$y - y_1 = m(x - x_1)$ Point-slope form

$y - 5 = \frac{2}{3}(x - 4)$ $(x_1, y_1) = (4, 5), m = \frac{2}{3}$

$y = \frac{2}{3}x + \frac{7}{3}$ Slope-intercept form

Exercises **Write an equation in slope-intercept form for the line that satisfies each set of conditions.** *See Examples 1, 2, and 4 on pages 76–78.*

36. slope $\frac{3}{4}$, passes through (−6, 9)
37. passes through (3, −8) and (−3, 2)
38. passes through (−1, 2), parallel to the graph of $x - 3y = 14$
39. passes through (3, 2), perpendicular to the graph of $4x - 3y = 12$

2-5 Modeling Real-World Data: Using Scatter Plots

See pages 81–86.

Concept Summary

- A scatter plot is a graph of ordered pairs of data.
- A prediction equation can be used to predict one of the variables given the other variable.

Example

WEEKLY PAY **The table below shows the median weekly earnings for American workers for the period 1985–1999. Predict the median weekly earnings for 2010.**

Year	1985	1990	1995	1999	2010
Earnings ($)	343	412	479	549	?

Source: U.S. Bureau of Labor Statistics

A scatter plot suggests that any two points could be used to find a prediction equation. Use (1985, 343) and (1990, 412).

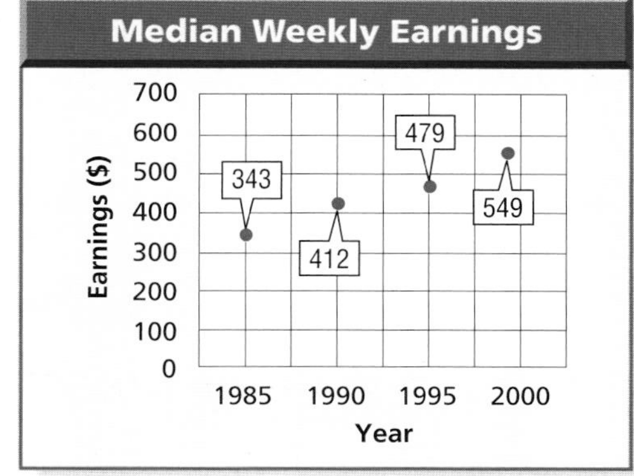

Source: U.S. Bureau of Labor Statistics

$m = \frac{y_2 - y_1}{x_2 - x_1}$ Slope formula

$= \frac{412 - 343}{1990 - 1985}$ $(x_1, y_1) = (1985, 343)$, $(x_2, y_2) = (1990, 412)$

$= \frac{69}{5}$ or 13.8 Simplify.

$y - y_1 = m(x - x_1)$ Point-slope form

$y - 343 = 13.8(x - 1985)$ Substitute.

$y = 13.8x - 27{,}050$ Add 343 to each side.

To predict the earnings for 2010, substitute 2010 for x.

$y = 13.8(2010) - 27{,}050$ $x = 2010$

$= 688$ Simplify.

The model predicts median weekly earnings of $688 in 2010.

Exercises **For Exercises 40–42, use the table that shows the number of people below the poverty level for the period 1980–1998.** *See Examples 1 and 2 on pages 81 and 82.*

40. Draw a scatter plot.
41. Use two ordered pairs to write a prediction equation.
42. Use your prediction equation to predict the number for 2010.

Year	People (millions)
1980	29.3
1985	33.1
1990	33.6
1995	36.4
1998	34.5
2010	?

Source: U.S. Census Bureau

• Extra Practice, see pages 830–832.
• Mixed Problem Solving, see page 863.

2-6 Special Functions

See pages 89–95.

Concept Summary

Greatest Integer

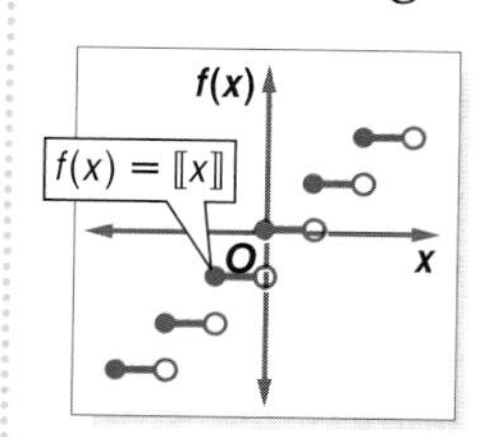

Constant

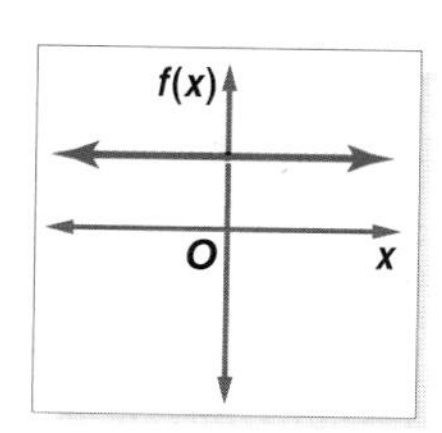

Absolute Value

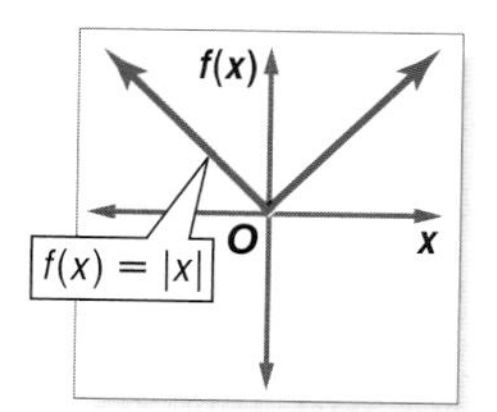

Piecewise

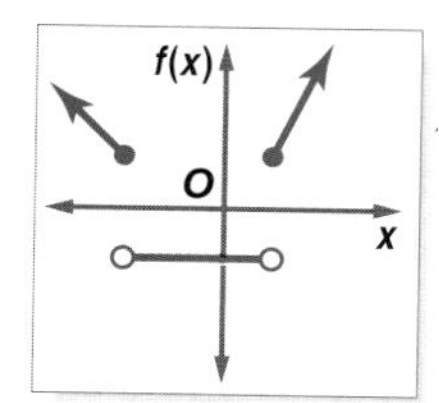

Example **Graph the function $f(x) = 3|x| - 2$. Identify the domain and range.**

The domain is all real numbers. The range is all real numbers greater than or equal to -2.

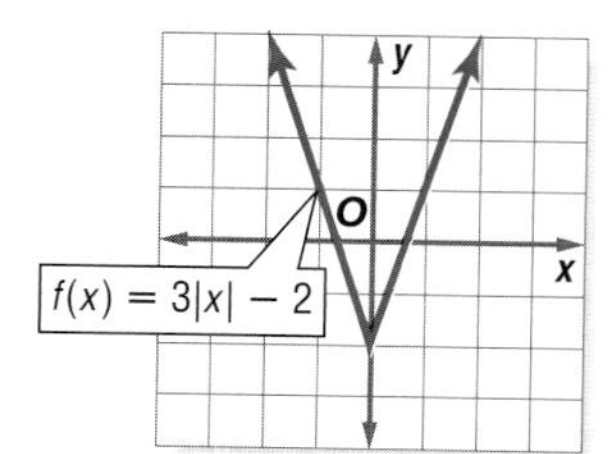

Exercises **Graph each function. Identify the domain and range.**
See Examples 1–3 on pages 89–91.

43. $f(x) = [\![x]\!] - 2$
44. $h(x) = [\![2x - 1]\!]$
45. $g(x) = |x| + 4$
46. $h(x) = |x - 1| - 7$
47. $f(x) = \begin{cases} 2 \text{ if } x < -1 \\ -x - 1 \text{ if } x \geq -1 \end{cases}$
48. $g(x) = \begin{cases} -2x - 3 \text{ if } x < 1 \\ x - 4 \text{ if } x > 1 \end{cases}$

2-7 Graphing Inequalities

See pages 96–99.

Concept Summary

You can graph an inequality by following these steps.

Step 1 Determine whether the boundary is solid or dashed. Graph the boundary.
Step 2 Choose a point not on the boundary and test it in the inequality.
Step 3 If a true inequality results, shade the region containing your test point. If a false inequality results, shade the other region.

Example **Graph $x + 4y \leq 4$.**

Since the inequality symbol is $\leq$, the graph of the boundary should be solid. Graph the equation. Test (0, 0).

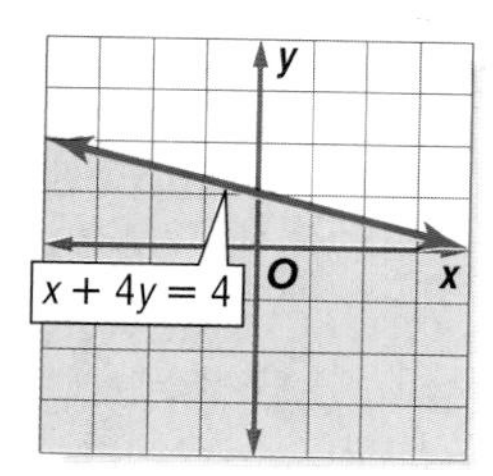

$x + 4y \leq 4$ Original inequality
$0 + 4(0) \leq 4$ $(x, y) = (0, 0)$
$0 \leq 4$ Shade the region that contains (0, 0).

Exercises **Graph each inequality.** *See Examples 1–3 on pages 96 and 97.*

49. $y \leq 3x - 5$
50. $x > y - 1$
51. $y + 0.5x < 4$
52. $2x + y \geq 3$
53. $y \geq |x| + 2$
54. $y > |x - 3|$

Chapter 2

Practice Test

Vocabulary and Concepts

Choose the correct term to complete each sentence.

1. The variable whose values make up the domain of a function is called the (*independent*, *dependent*) variable.
2. To find the (*x-intercept*, *y-intercept*) of the graph of a linear equation, let $y = 0$.
3. An equation of the form ($Ax + By = C$, $y = mx + b$) is in slope-intercept form.

Skills and Applications

Graph each relation and find the domain and range. Then determine whether the relation is a function.

4. $\{(-4, -8), (-2, 2), (0, 5), (2, 3), (4, -9)\}$
5. $y = 3x - 3$

Find each value.

6. $f(3)$ if $f(x) = 7 - x^2$
7. $f(0)$ if $f(x) = x - 3x^2$

Graph each equation or inequality.

8. $y = \frac{3}{5}x - 4$
9. $4x - y = 2$
10. $x = -4$
11. $y = 2x - 5$
12. $f(x) = 3x - 1$
13. $f(x) = [\![3x]\!] + 3$
14. $g(x) = |x + 2|$
15. $h(x) = \begin{cases} x + 2 \text{ if } x < -2 \\ 2x - 1 \text{ if } x \geq -2 \end{cases}$
16. $y \leq 10$
17. $x > 6$
18. $-2x + 5 \leq 3y$
19. $y < 4|x - 1|$

Find the slope of the line that passes through each pair of points.

20. $(8, -4), (6, 1)$
21. $(-2, 5), (4, 5)$
22. $(5, 7), (4, -6)$

Graph the line passing through the given point with the given slope.

23. $(1, -3), 2$
24. $(-2, 2), -\frac{1}{3}$
25. $(3, -2)$, undefined

Write an equation in slope-intercept form for the line that satisfies each set of conditions.

26. slope -5, y-intercept 11
27. x-intercept 9, y-intercept -4
28. passes through $(-6, 15)$, parallel to the graph of $2x + 3y = 1$
29. passes through $(5, 2)$, perpendicular to the graph of $x + 3y = 7$

RECREATION **For Exercises 30–32, use the table that shows the amount Americans spent on recreation in recent years.**

Year	1995	1996	1997	1998
Amount ($ billions)	401.6	429.6	457.8	494.7

Source: U.S. Bureau of Economic Analysis

30. Draw a scatter plot, where x represents the number of years since 1995.
31. Write a prediction equation.
32. Predict the amount that will be spent on recreation in 2010.

Standards Practice

33. **STANDARDIZED TEST PRACTICE** What is the slope of a line parallel to $y - 2 = 4(x + 1)$?

Ⓐ -4 Ⓑ $-\frac{1}{4}$ Ⓒ $\frac{1}{4}$ Ⓓ 4

Chapter 2 Standardized Test Practice

Standards Practice

Part 1 Multiple Choice

Record your answers on the answer sheet provided by your teacher or on a sheet of paper.

1. In the figure, $\angle B$ and $\angle BCD$ are right angles. $\overline{BC}$ is 9 units, $\overline{AB}$ is 12 units, and $\overline{CD}$ is 8 units. What is the area, in square units, of $\triangle ACD$?

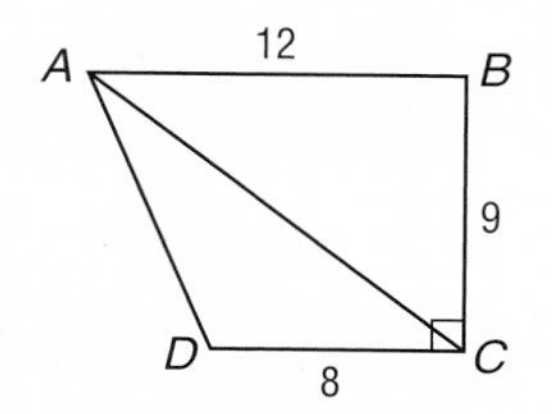

Ⓐ 36
Ⓑ 60
Ⓒ 72
Ⓓ 135

2. If $x + 3$ is an even integer, then x could be which of the following?

Ⓐ -2 Ⓑ -1
Ⓒ 0 Ⓓ 2

3. What is the slope of the line that contains the points (15, 7) and (6, 4)?

Ⓐ $\frac{1}{4}$ Ⓑ $\frac{1}{3}$
Ⓒ $\frac{3}{8}$ Ⓓ $\frac{2}{3}$

4. In 2000, Matt had a collection of 30 music CDs. Since then he has given away 2 CDs, purchased 6 new CDs, and traded 3 of his CDs to Kashan for 4 of Kashan's CDs. Since 2000, what has been the percent of increase in the number of CDs in Matt's collection?

Ⓐ $3\frac{1}{3}\%$ Ⓑ 10%
Ⓒ $14\frac{2}{7}\%$ Ⓓ $16\frac{2}{3}\%$

5. If the product of $(2 + 3)$, $(3 + 4)$, and $(4 + 5)$ is equal to three times the sum of 40 and x, then $x =$ ______.

Ⓐ 43 Ⓑ 65
Ⓒ 105 Ⓓ 195

6. If one side of a triangle is three times as long as a second side and the second side is s units long, then the length of the third side of the triangle can be

Ⓐ $3s$. Ⓑ $4s$.
Ⓒ $5s$. Ⓓ $6s$.

7. Which of the following sets of numbers has the property that the *product* of any two numbers is also a number in the set?

I the set of positive numbers
II the set of prime numbers
III the set of even integers

Ⓐ I only
Ⓑ II only
Ⓒ III only
Ⓓ I and III only

8. If $\frac{3 + x}{7 + x} = \frac{3}{7} + \frac{3}{7}$, then $x =$ ______.

Ⓐ $\frac{3}{7}$ Ⓑ 3
Ⓒ 7 Ⓓ 21

9. The average (arithmetic mean) of r, s, x, and y is 8, and the average of x and y is 4. What is the average of r and s?

Ⓐ 4 Ⓑ 6
Ⓒ 8 Ⓓ 12

Test-Taking Tip

Questions 1–9

On multiple-choice questions, try to compute the answer first. Then compare your answer to the given answer choices. If you don't find your answer among the choices, check your calculations.

Preparing for Standardized Tests
For test-taking strategies and more practice, see pages 877–892.

Part 2 Short Response/Grid In

Record your answers on the answer sheet provided by your teacher or on a sheet of paper.

10. If n is a prime integer such that $2n > 19 \geq \frac{7}{8}n$ what is one possible value of n?

11. If $\overline{AC}$ is 2 units, what is the value of t?

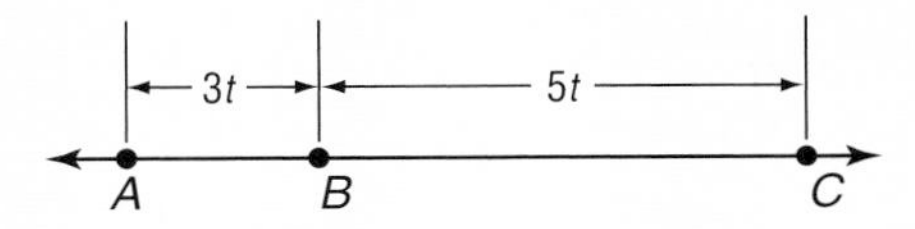

12. If $0.85x = 8.5$, what is the value of $\frac{1}{x}$?

13. In $\triangle ABC$, what is the value of $w + x + y + z$?

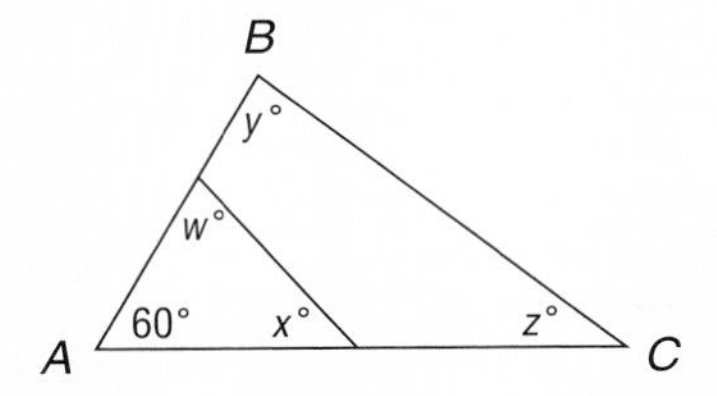

14. In an election, a total of 4000 votes were cast for three candidates, A, B, and C. Candidate C received 800 votes. If candidate B received more votes than candidate C and candidate A received more votes than candidate B, what is the least number of votes that candidate A could have received?

15. If the points $P(-2, 3)$, $Q(2, 5)$, and $R(2, 3)$ are vertices of a triangle, what is the area of the triangle?

16. How many of the first one hundred positive integers contain the digit 7?

17. A triangle has a base of length 17, and the other two sides are equal in length. If the lengths of the sides of the triangle are integers, what is the shortest possible length of a side?

Part 3 Extended Response

Record your answers on a sheet of paper. Show your work.

For Exercises 18–23, use the information below and in the table.

The amount that a certain online retailer charges for shipping an electronics purchase is determined by the weight of the package. The charges for several different weights are given in the table below.

Electronics Shipping Charges	
Weight (lb)	Shipping ($)
1	5.58
3	6.76
4	7.35
7	9.12
10	10.89
13	12.66
15	13.84

18. Write a relation to represent the data. Use weight as the independent variable and the shipping charge as the dependent variable.

19. Graph the relation on a coordinate plane.

20. Is the relation a function? Explain your reasoning.

21. Find the rate of change of the shipping charge per pound.

22. Write an equation that could be used to find the shipping charge y for a package that weighs x pounds.

23. Find the shipping charge for a package that weighs 19 pounds.

Chapter 3

Systems of Equations and Inequalities

What You'll Learn

- **Lessons 3-1, 3-2, and 3-5** Solve systems of linear equations in two or three variables.
- **Lesson 3-3** Solve systems of inequalities.
- **Lesson 3-4** Use linear programming to find maximum and minimum values of functions.

Key Vocabulary

- system of equations (p. 110)
- substitution method (p. 116)
- elimination method (p. 118)
- linear programming (p. 130)
- ordered triple (p. 136)

Why It's Important

Systems of linear equations and inequalities can be used to model real-world situations in which many conditions must be met. For example, hurricanes are classified using inequalities that involve wind speed and storm surge. Weather satellites provide images of hurricanes, which are rated on a scale of 1 to 5. *You will learn how to classify the strength of a hurricane in Lesson 3-3.*

Getting Started

Prerequisite Skills To be successful in this chapter, you'll need to master these skills and be able to apply them in problem-solving situations. Review these skills before beginning Chapter 3.

For Lesson 3-1 — Graph Linear Equations

Graph each equation. *(For review, see Lesson 2-2.)*

1. $2y = x$
2. $y = x - 4$
3. $y = 2x - 3$
4. $x + 3y = 6$
5. $2x + 3y = -12$
6. $4y - 5x = 10$

For Lesson 3-2 — Solve for a Specified Variable

Solve each equation for y. *(For review, see Lesson 1-3.)*

7. $2x + y = 0$
8. $x - y = -4$
9. $6x + 2y = 12$
10. $8 - 4y = 5x$
11. $\frac{1}{2}y + 3x = 1$
12. $\frac{1}{3}x - 2y = 8$

For Lessons 3-3 and 3-4 — Graph Inequalities

Graph each inequality. *(For review, see Lesson 2-7.)*

13. $y \geq -2$
14. $x + y \leq 0$
15. $y < 2x - 2$
16. $x + 4y < 3$
17. $2x - y \geq 6$
18. $3x - 4y < 10$

For Lesson 3-5 — Evaluate Expressions

Evaluate each expression if $x = -3$, $y = 1$, and $z = 2$. *(For review, see Lesson 1-1.)*

19. $3x + 2y - z$
20. $3y - 8z$
21. $x - 5y + 4z$
22. $2x + 9y + 4z$
23. $2x - 6y - 5z$
24. $7x - 3y + 2z$

Systems of Linear Equations and Inequalities Make this Foldable to organize your notes. Begin with one sheet of 11" × 17" paper and four sheets of grid paper.

Step 1 Fold and Cut

Fold the short sides of the 11" × 17" paper to meet in the middle. Cut each tab in half as shown.

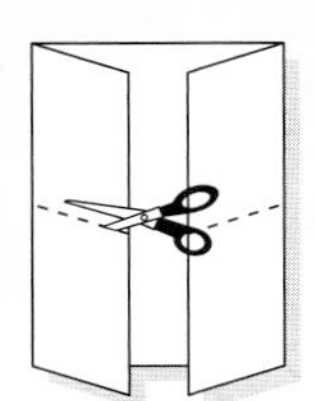

Step 2 Staple and Label

Insert 2 folded half-sheets of grid paper in each tab. Staple at edges. Label each tab as shown.

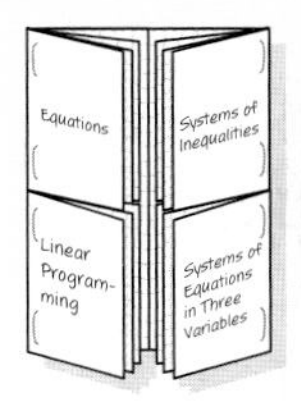

Reading and Writing As you read and study the chapter, fill the tabs with notes, diagrams, and examples for each topic.

3-1 Solving Systems of Equations by Graphing

California Standards
Standard 2.0 Students solve systems of linear equations and inequalities (**in two** or three **variables**) by substitution, **with graphs,** or with matrices. (Key)

What You'll Learn

- Solve systems of linear equations by graphing.
- Determine whether a system of linear equations is consistent and independent, consistent and dependent, or inconsistent.

Vocabulary

- system of equations
- consistent
- inconsistent
- independent
- dependent

How can a system of equations be used to predict sales?

Since 1999, the growth of in-store sales for Custom Creations by Cathy can be modeled by $y = 4.2x + 29$. The growth of her online sales can be modeled by $y = 7.5x + 9.3$. In these equations, x represents the number of years since 1999, and y represents the amount of sales in thousands of dollars.

The equations $y = 4.2x + 29$ and $y = 7.5x + 9.3$ are called a system of equations.

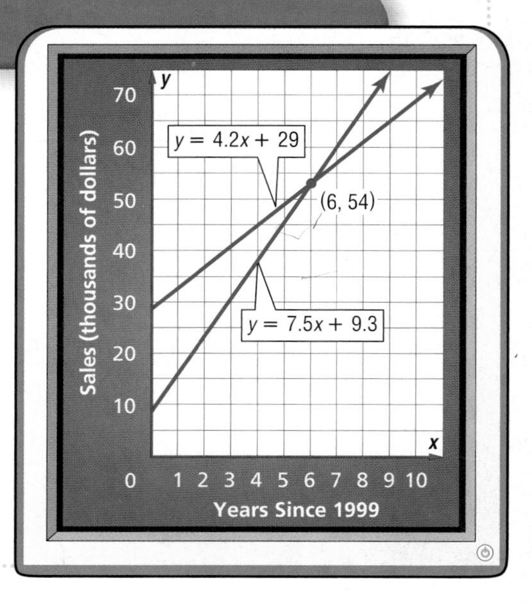

GRAPH SYSTEMS OF EQUATIONS A **system of equations** is two or more equations with the same variables. To solve a system of equations, find the ordered pair that satisfies all of the equations. One way to do this is to graph the equations on the same coordinate plane. The point of intersection represents the solution.

Example 1 Solve by Graphing

Solve the system of equations by graphing.

$$2x + y = 5$$
$$x - y = 1$$

Write each equation in slope-intercept form.

$2x + y = 5 \rightarrow y = -2x + 5$

$x - y = 1 \rightarrow y = x - 1$

The graphs appear to intersect at (2, 1).

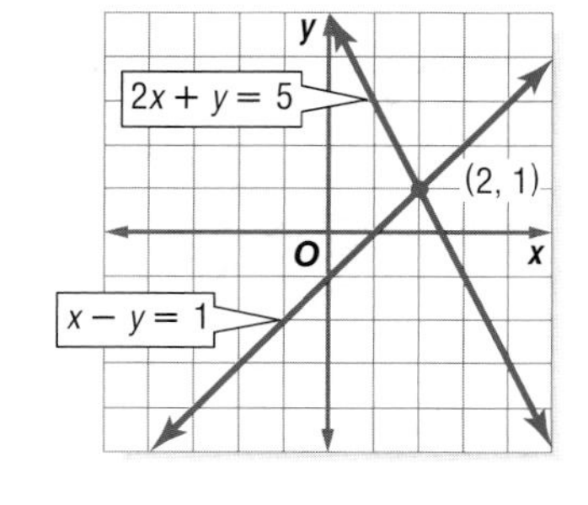

CHECK Substitute the coordinates into each equation.

$2x + y = 5$	$x - y = 1$	Original equations
$2(2) + 1 \stackrel{?}{=} 5$	$2 - 1 \stackrel{?}{=} 1$	Replace x with 2 and y with 1.
$5 = 5$ ✓	$1 = 1$ ✓	Simplify.

The solution of the system is (2, 1).

Study Tip

Checking Solutions
When using a graph to find a solution, always check the ordered pair in *both* original equations.

Systems of equations are used in businesses to determine the *break-even point*. The break-even point is the point at which the income equals the cost.

Example 2 Break-Even Point Analysis

MUSIC Travis and his band are planning to record their first CD. The initial start-up cost is $1500, and each CD will cost $4 to produce. They plan to sell their CDs for $10 each. How many CDs must the band sell before they make a profit?

Let x = the number of CDs, and let y = the number of dollars.

Cost of x CDs	is	cost per CD	plus	startup cost.
y	$=$	$4x$	$+$	1500

Income from x CDs	is	price per CD	times	number of CDs.
y	$=$	10	$\cdot$	x

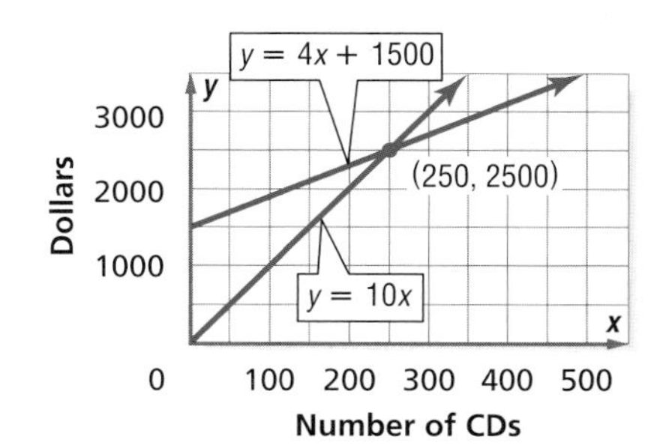

The graphs intersect at (250, 2500). This is the break-even point. If the band sells fewer than 250 CDs, they will lose money. If the band sells more than 250 CDs, they will make a profit.

More About. . .

Music

Compact discs (CDs) store music digitally. The recorded sound is converted to a series of 1s and 0s. This coded pattern can then be read by an infrared laser in a CD player.

CLASSIFY SYSTEMS OF EQUATIONS Graphs of systems of linear equations may be intersecting lines, parallel lines, or the same line. A system of equations is **consistent** if it has at least one solution and **inconsistent** if it has no solutions. A consistent system is **independent** if it has exactly one solution or **dependent** if it has an infinite number of solutions.

Example 3 Intersecting Lines

Graph the system of equations and describe it as *consistent and independent, consistent and dependent,* or *inconsistent.*

$x + \frac{1}{2}y = 5$

$3y - 2x = 6$

Write each equation in slope-intercept form.

$x + \frac{1}{2}y = 5 \rightarrow y = -2x + 10$

$3y - 2x = 6 \rightarrow y = \frac{2}{3}x + 2$

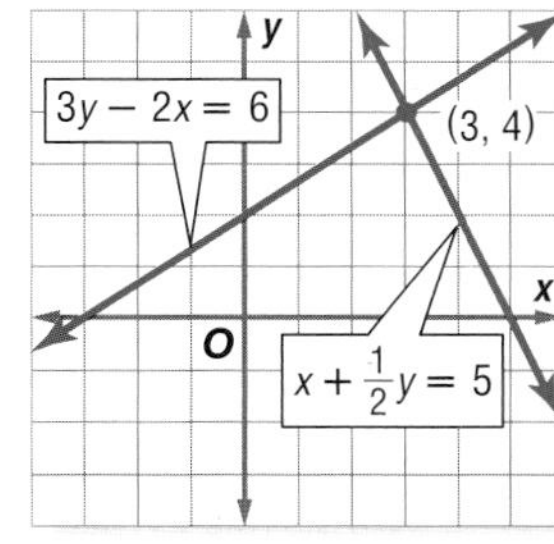

The graphs intersect at (3, 4). Since there is one solution, this system is *consistent and independent*.

Example 4 Same Line

Graph the system of equations and describe it as *consistent and independent, consistent and dependent,* or *inconsistent.*

$9x - 6y = 24$
$6x - 4y = 16$

$9x - 6y = 24 \rightarrow y = \frac{3}{2}x - 4$

$6x - 4y = 16 \rightarrow y = \frac{3}{2}x - 4$

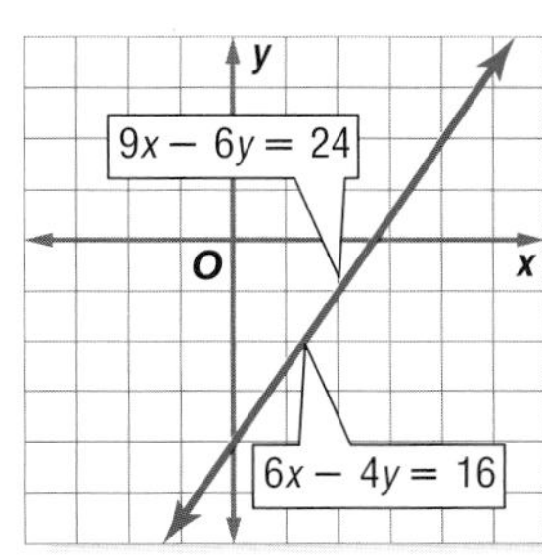

Since the equations are equivalent, their graphs are the same line. Any ordered pair representing a point on that line will satisfy both equations.

So, there are infinitely many solutions to this system. This system is *consistent and dependent*.

Example 5 Parallel Lines

Graph the system of equations and describe it as ***consistent and independent, consistent and dependent,*** **or** ***inconsistent.***

$3x + 4y = 12$
$6x + 8y = -16$

$3x + 4y = 12 \rightarrow y = -\frac{3}{4}x + 3$

$6x + 8y = -16 \rightarrow y = -\frac{3}{4}x - 2$

The lines do not intersect. Their graphs are parallel lines. So, there are no solutions that satisfy both equations. This system is *inconsistent.*

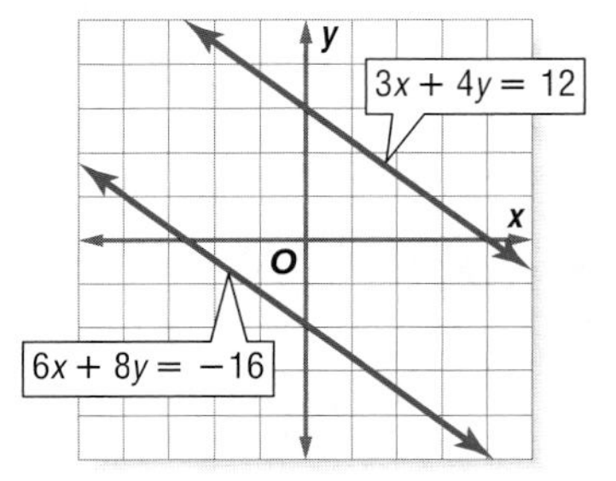

Study Tip

Parallel Lines
Notice from their equations that the lines have the same slope and different *y*-intercepts.

The relationship between the graph of a system of equations and the number of its solutions is summarized below.

Concept Summary — Systems of Equations

consistent and independent	consistent and dependent	inconsistent
intersecting lines; one solution	same line; infinitely many solutions	parallel lines; no solution

Check for Understanding

Concept Check

1. **Explain** why a system of linear equations cannot have exactly two solutions.
2. **OPEN ENDED** Give an example of a system of equations that is consistent and independent.
3. **Explain** why it is important to check a solution found by graphing in both of the original equations.

Guided Practice

Solve each system of equations by graphing.

4. $y = 2x + 9$
   $y = -x + 3$
5. $3x + 2y = 10$
   $2x + 3y = 10$
6. $4x - 2y = 22$
   $6x + 9y = -3$

Graph each system of equations and describe it as ***consistent and independent, consistent and dependent,*** **or** ***inconsistent.***

7. $y = 6 - x$
   $y = x + 4$
8. $x + 2y = 2$
   $2x + 4y = 8$
9. $x - 2y = 8$
   $\frac{1}{2}x - y = 4$

Application **PHOTOS For Exercises 10–12, use the graphic at the right.**

10. Write equations that represent the cost of developing a roll of film at each lab.
11. Under what conditions is the cost to develop a roll of film the same for either store?
12. When is it best to use The Photo Lab and when is it best to use Specialty Photos?

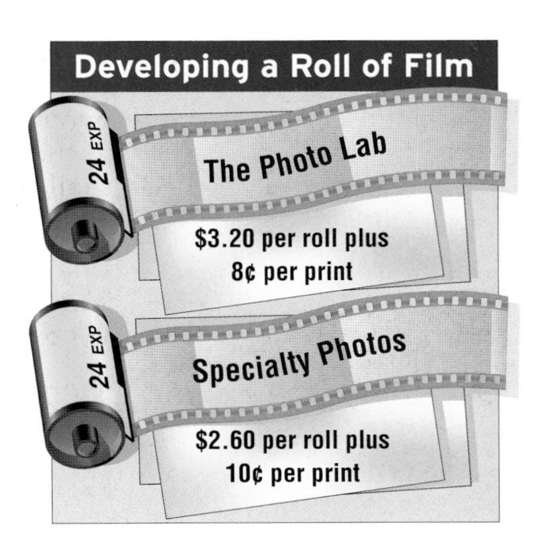

Practice and Apply

Homework Help

For Exercises	See Examples
13–24, 37, 38	1
25–36	3–5
39–47	2

Extra Practice
See page 832.

Solve each system of equations by graphing.

13. $y = 2x - 4$
 $y = -3x + 1$
14. $y = 3x - 8$
 $y = x - 8$
15. $x + 2y = 6$
 $2x + y = 9$
16. $2x + 3y = 12$
 $2x - y = 4$
17. $3x - 7y = -6$
 $x + 2y = 11$
18. $5x - 11 = 4y$
 $7x - 1 = 8y$
19. $2x + 3y = 7$
 $2x - 3y = 7$
20. $8x - 3y = -3$
 $4x - 2y = -4$
21. $\frac{1}{4}x + 2y = 5$
 $2x - y = 6$
22. $\frac{2}{3}x + y = -3$
 $y - \frac{1}{3}x = 6$
23. $\frac{1}{2}x - y = 0$
 $\frac{1}{4}x + \frac{1}{2}y = -2$
24. $\frac{4}{3}x + \frac{1}{5}y = 3$
 $\frac{2}{3}x - \frac{3}{5}y = 5$

Graph each system of equations and describe it as *consistent and independent, consistent and dependent,* or *inconsistent*.

25. $y = x + 4$
 $y = x - 4$
26. $y = x + 3$
 $y = 2x + 6$
27. $x + y = 4$
 $-4x + y = 9$
28. $3x + y = 3$
 $6x + 2y = 6$
29. $y - x = 5$
 $2y - 2x = 8$
30. $4x - 2y = 6$
 $6x - 3y = 9$
31. $2y = x$
 $8y = 2x + 1$
32. $2y = 5 - x$
 $6y = 7 - 3x$
33. $0.8x - 1.5y = -10$
 $1.2x + 2.5y = 4$
34. $1.6y = 0.4x + 1$
 $0.4y = 0.1x + 0.25$
35. $3y - x = -2$
 $y - \frac{1}{3}x = 2$
36. $2y - 4x = 3$
 $\frac{4}{3}x - y = -2$

37. **GEOMETRY** The sides of an angle are parts of two lines whose equations are $2y + 3x = -7$ and $3y - 2x = 9$. The angle's vertex is the point where the two sides meet. Find the coordinates of the vertex of the angle.

38. **GEOMETRY** The graphs of $y - 2x = 1$, $4x + y = 7$, and $2y - x = -4$ contain the sides of a triangle. Find the coordinates of the vertices of the triangle.

TRAVEL For Exercises 39–41, use the following information.
Adam and his family are planning to rent a midsize car for a one-day trip. In the Standard Rental Plan, they can rent a car for \$52 per day plus 23 cents per mile. In the Deluxe Rental Plan, they can rent a car for \$80 per day with unlimited mileage.

39. For each plan, write an equation that represents the cost of renting a car.
40. Graph the equations. Estimate the break-even point of the rental costs.
41. If Adam's family plans to drive 150 miles, which plan should they choose?

ECONOMICS **For Exercises 42–44, use the graph below that shows the supply and demand curves for a new multivitamin.**
In Economics, the point at which the supply equals the demand is the *equilibrium price.* If the supply of a product is greater than the demand, there is a surplus and prices fall. If the supply is less than the demand, there is a shortage and prices rise.

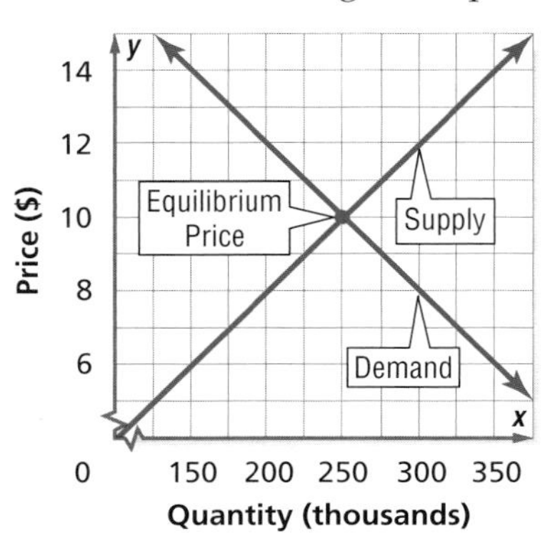

42. If the price for vitamins is $8 a bottle, what is the supply of the product and what is the demand? Will prices tend to rise or fall?

43. If the price for vitamins is $12 a bottle, what is the supply of the product and what is the demand? Will prices tend to rise or fall?

44. At what quantity will the prices stabilize? What is the equilibrium price for this product?

More About. . .

Population
In the United States there is approximately one birth every 8 seconds and one death every 14 seconds.
Source: U.S. Census Bureau

POPULATION **For Exercises 45–47, use the graphic that shows 2000 state populations.**

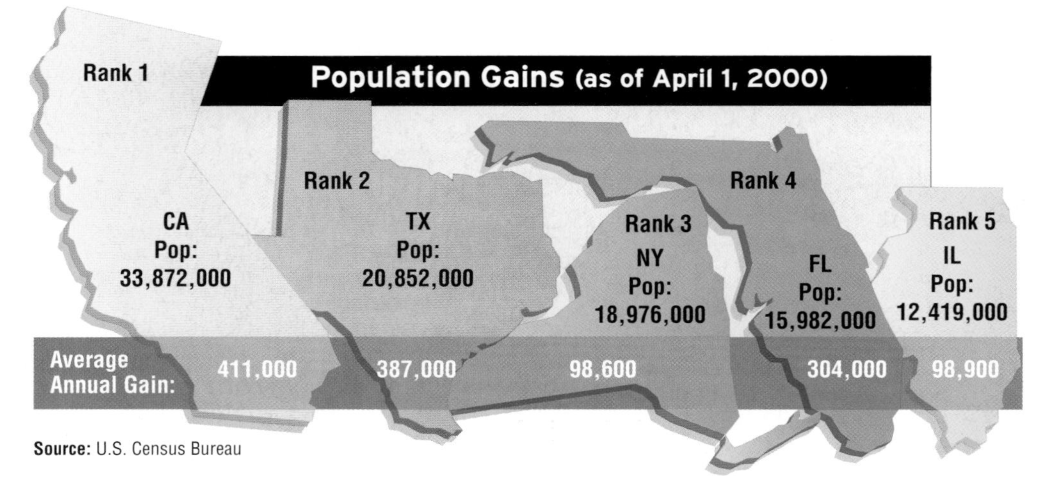

Source: U.S. Census Bureau

45. Write equations that represent the populations of Florida and New York x years after 2000. Assume that both states continue to gain the same number of residents every year. Let y equal the population in thousands.

46. Graph both equations for the years 2000 to 2020. Estimate when the populations of both states will be equal.

47. Do you think Florida will overtake New York as the third most populous state by 2010? by 2020? Explain your reasoning.

48. **CRITICAL THINKING** State the conditions for which the system below is: (a) consistent and dependent, (b) consistent and independent, (c) inconsistent.

$$ax + by = c$$
$$dx + ey = f$$

49. **WRITING IN MATH** Answer the question that was posed at the beginning of the lesson.

How can a system of equations be used to predict sales?

Include the following in your answer:
- an explanation of the real-world meaning of the solution of the system of equations in the application at the beginning of the lesson, and
- a description of what a business owner would learn if the system of equations representing the in-store and online sales is inconsistent.

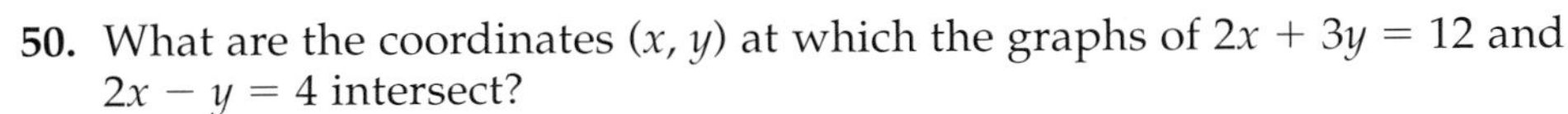

50. What are the coordinates (x, y) at which the graphs of $2x + 3y = 12$ and $2x - y = 4$ intersect?

Ⓐ $(3, 2)$ Ⓑ $(2, 3)$ Ⓒ $(1, -2)$ Ⓓ $(-3, 6)$

51. Which equation has the same graph as $4x + 8y = 12$?

Ⓐ $x + y = 3$ Ⓑ $2x + y = 3$ Ⓒ $x + 2y = 3$ Ⓓ $2x + 2y = 6$

Graphing Calculator

INTERSECT FEATURE To use a TI-83 Plus to solve a system of equations, graph both equations on the same screen. Then, select intersect, which is option 5 under the CALC menu, to find the coordinates of the point of intersection. Solve each system of equations to the nearest hundredth.

52. $y = 0.125x - 3.005$
$y = -2.58$

53. $3.6x - 2y = 4$
$-2.7x + y = 3$

54. $y = 0.18x + 2.7$
$y = -0.42x + 5.1$

55. $1.6x + 3.2y = 8$
$1.2x + 2.4y = 4$

56. $y - \frac{1}{4}x = 6$
$2y + \frac{1}{2}x = 3$

57. $\frac{1}{2}y - 5x = 8$
$\frac{1}{3}y - 8x = -7$

Maintain Your Skills

Mixed Review

Graph each inequality. *(Lesson 2-7)*

58. $y \geq 5 + 3x$

59. $2x + y > -4$

60. $2y - 1 \leq x$

Identify each function as S for step, C for constant, A for absolute value, or P for piecewise. *(Lesson 2-6)*

61.

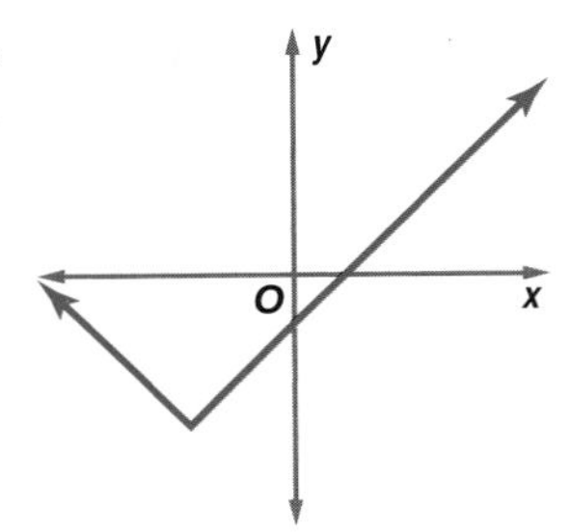

62.

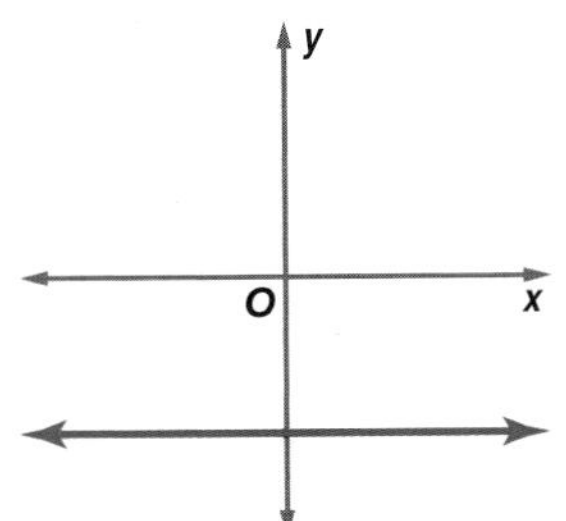

63.

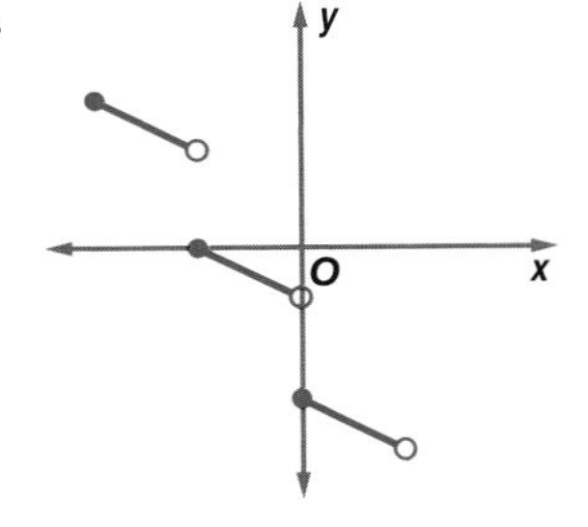

Solve each equation. Check your solutions. *(Lesson 1-4)*

64. $|x| - 5 = 8$

65. $|w + 3| = 12$

66. $|6a - 4| = -2$

67. $3|2t - 1| = 15$

68. $|4r + 3| - 7 = 10$

69. $|k + 7| = 3k - 11$

Write an algebraic expression to represent each verbal expression. *(Lesson 1-3)*

70. the sum of 8 and 2 times a number

71. six less than the square of a number

72. four times the sum of a number and 5

73. the quotient of a number and 3 increased by 1

Getting Ready for the Next Lesson

PREREQUISITE SKILL Simplify each expression.
*(To review **simplifying expressions**, see Lesson 1-2.)*

74. $(3x + 5) - (2x + 3)$

75. $(3y - 11) + (6y + 12)$

76. $(5x - y) + (-8x + 7y)$

77. $6(2x + 3y - 1)$

78. $5(4x + 2y - x + 2)$

79. $3(x + 4y) - 2(x + 4y)$

3-2 Solving Systems of Equations Algebraically

California Standards
Standard 2.0 Students solve systems of linear equations and inequalities (**in two** or three **variables) by substitution,** with graphs, or with matrices. (Key)

What You'll Learn

- Solve systems of linear equations by using substitution.
- Solve systems of linear equations by using elimination.

Vocabulary

- substitution method
- elimination method

How can systems of equations be used to make consumer decisions?

In January, Yolanda's long-distance bill was $5.50 for 25 minutes of calls. The bill was $6.54 in February, when Yolanda made 38 minutes of calls. What are the rate per minute and flat fee the company charges?

Let x equal the rate per minute, and let y equal the monthly fee.

January bill: $25x + y = 5.5$
February bill: $38x + y = 6.54$

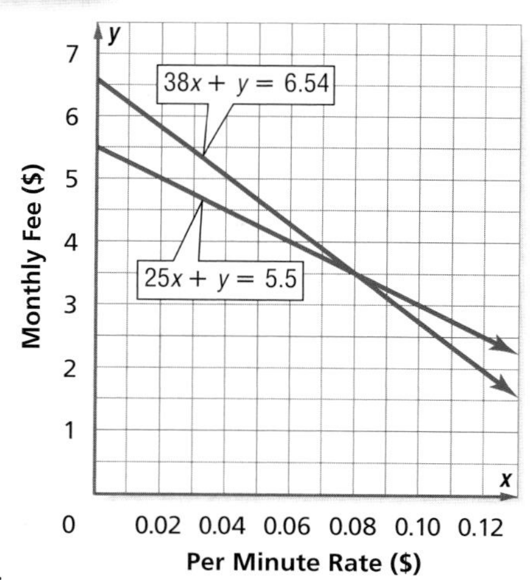

It is difficult to determine the exact coordinates of the point where the lines intersect from the graph. For systems of equations like this one, it may be easier to solve the system by using algebraic methods.

SUBSTITUTION One algebraic method is the **substitution method**. Using this method, one equation is solved for one variable in terms of the other. Then, this expression is substituted for the variable in the other equation.

Example 1 Solve by Using Substitution

Use substitution to solve the system of equations.

$$x + 2y = 8$$
$$\frac{1}{2}x - y = 18$$

Study Tip

Coefficient of 1
It is easier to solve for the variable that has a coefficient of 1.

Solve the first equation for x in terms of y.

$x + 2y = 8$ First equation
$x = 8 - 2y$ Subtract $2y$ from each side.

Substitute $8 - 2y$ for x in the second equation and solve for y.

$\frac{1}{2}x - y = 18$ Second equation
$\frac{1}{2}(8 - 2y) - y = 18$ Substitute $8 - 2y$ for x.
$4 - y - y = 18$ Distributive Property
$-2y = 14$ Subtract 4 from each side.
$y = -7$ Divide each side by -2.

Now, substitute the value for y in either original equation and solve for x.

$x + 2y = 8$ First equation
$x + 2(-7) = 8$ Replace y with -7.
$x - 14 = 8$ Simplify.
$x = 22$ The solution of the system is $(22, -7)$.

Standardized Test Practice

Example 2 Solve by Substitution

Extended Response Test Item

Matthew stopped for gasoline twice on a long car trip. The price of gasoline at the first station where he stopped was \$1.46 per gallon. At the second station, the price was \$1.55 per gallon. Matthew bought a total of 36.1 gallons of gasoline and spent \$54.29. Write and solve a system of equations to find the number of gallons of gasoline Matthew bought at each gas station. Show your work.

Read the Test Item

You are asked to find the number of gallons of gasoline that Matthew bought at each gas station.

Solve the Test Item

Step 1 Define variables and write the system of equations. Let x represent the number of gallons bought at the first station and y represent the number of gallons bought at the second station.

$x + y = 36.1$ *The total number of gallons was 36.1.*

$1.46x + 1.55y = 54.29$ *The total price was \$54.29.*

Step 2 Solve one of the equations for one of the variables in terms of the other. Since the coefficient of y is 1, it makes sense to solve the first equation for y in terms of x.

$x + y = 36.1$ *First equation*

$y = 36.1 - x$ *Subtract x from each side.*

Step 3 Substitute $36.1 - x$ for y in the second equation.

$1.46x + 1.55y = 54.29$ *Second equation*

$1.46x + 1.55(36.1 - x) = 54.29$ *Substitute 36.1 − x for y.*

$1.46x + 55.955 - 1.55x = 54.29$ *Distributive Property*

$-0.09x = -1.665$ *Simplify.*

$x = 18.5$ *Divide each side by −0.09.*

Step 4 Now replace x with 18.5 in either equation to find the value of y.

$x + y = 36.1$ *First equation*

$18.5 + y = 36.1$ *Substitute 18.5 for x.*

$y = 17.6$ *Subtract 18.5 from each side.*

Test-Taking Tip

Substituting 18.5 for x in either of the original equations would allow you to find the value of y. But the first equation is simpler and therefore easier to solve.

Step 5 Check the solution.

$x + y = 36.1$ *Original equation* $\quad 1.46x + 1.55y = 54.29$

$18.5 + 17.6 \stackrel{?}{=} 36.1$ *Substitute.* $\quad 1.46(18.5) + 1.55(17.6) \stackrel{?}{=} 54.29$

$36.1 = 36.1 \checkmark$ *Simplify.* $\quad 24.01 + 27.28 = 54.29 \checkmark$

Step 6 State the solution.

Matthew bought 18.5 gallons at the first station and 17.6 gallons at the second station.

ELIMINATION Another algebraic method is the **elimination method**. Using this method, you eliminate one of the variables by adding or subtracting the equations. When you add two true equations, the result is a new equation that is also true.

Example 3 Solve by Using Elimination

Use the elimination method to solve the system of equations.

$4a + 2b = 15$
$2a + 2b = 7$

In each equation, the coefficient of b is 2. If one equation is subtracted from the other, the variable b will be eliminated.

$$\begin{aligned} 4a + 2b &= 15 \\ (-)\ 2a + 2b &= 7 \\ \hline 2a \qquad &= 8 \quad \text{Subtract the equations.} \\ a &= 4 \quad \text{Divide each side by 2.} \end{aligned}$$

Now find b by substituting 4 for a in either original equation.

$$\begin{aligned} 2a + 2b &= 7 && \text{Second equation} \\ 2(4) + 2b &= 7 && \text{Replace } a \text{ with 4.} \\ 8 + 2b &= 7 && \text{Multiply.} \\ 2b &= -1 && \text{Subtract 8 from each side.} \\ b &= -\frac{1}{2} && \text{Divide each side by 2.} \end{aligned}$$

The solution is $\left(4, -\frac{1}{2}\right)$.

Study Tip

Common Misconception
You may find it confusing to subtract equations. It may be helpful to multiply the second equation by −1 and then add the equations.

Sometimes, adding or subtracting the two equations will not eliminate either variable. You may use multiplication to write an equivalent equation so that one of the variables has the same or opposite coefficient in both equations. When you multiply an equation by a nonzero number, the new equation has the same set of solutions.

Example 4 Multiply, Then Use Elimination

Use the elimination method to solve the system of equations.

$3x - 7y = -14$
$5x + 2y = 45$

Multiply the first equation by 2 and the second equation by 7. Then add the equations to eliminate the y variable.

$3x - 7y = -14$ Multiply by 2. $\rightarrow$ $6x - 14y = -28$

$5x + 2y = 45$ Multiply by 7. $\rightarrow$ $(+)\ 35x + 14y = 315$

$$\begin{aligned} 41x \qquad &= 287 && \text{Add the equations.} \\ x &= 7 && \text{Divide each side by 41.} \end{aligned}$$

Replace x with 7 and solve for y.

$$\begin{aligned} 3x - 7y &= -14 && \text{First equation} \\ 3(7) - 7y &= -14 && \text{Replace } x \text{ with 7.} \\ 21 - 7y &= -14 && \text{Multiply.} \\ -7y &= -35 && \text{Subtract 21 from each side.} \\ y &= 5 && \text{Divide each side by } -7. \end{aligned}$$

The solution is (7, 5).

Study Tip

Alternative Method
You could also multiply the first equation by 5 and the second equation by 3. Then subtract to eliminate the x variable.

If you add or subtract two equations in a system and the result is an equation that is never true, then the system is inconsistent and it has no solution. If the result when you add or subtract two equations in a system is an equation that is always true, then the system is dependent and it has infinitely many solutions.

Example 5 Inconsistent System

Use the elimination method to solve the system of equations.

$8x + 2y = 17$
$-4x - y = 9$

Use multiplication to eliminate x.

$8x + 2y = 17 \quad\rightarrow\quad 8x + 2y = 17$

$-4x - y = 9$ Multiply by 2. $\quad -8x - 2y = 18$

$0 = 35$ Add the equations.

Since there are no values of x and y that will make the equation $0 = 35$ true, there are no solutions for this system of equations.

Check for Understanding

Concept Check

1. **OPEN ENDED** Give an example of a system of equations that is more easily solved by substitution and a system that is more easily solved by elimination.

2. **Make a conjecture** about the solution of a system of equations if the result of subtracting one equation from the other is $0 = 0$.

3. **FIND THE ERROR** Juanita and Vincent are solving the system $2x - y = 6$ and $2x + y = 10$. Who is correct? Explain your reasoning.

Juanita

$2x - y = 6$
$(-)2x + y = 10$
$0 = -4$

The statement $0 = -4$ is never true, so there is no solution.

Vincent

$2x - y = 6$
$(+)2x + y = 10$
$4x = 16$
$x = 4$

$2x - y = 6$
$2(4) - y = 6$
$8 - y = 6$
$y = 2$

The solution is $(4,2)$.

Guided Practice

Solve each system of equations by using substitution.

4. $y = 3x - 4$
   $y = 4 + x$

5. $4c + 2d = 10$
   $c + 3d = 10$

Solve each system of equations by using elimination.

6. $2r - 3s = 11$
   $2r + 2s = 6$

7. $2p + 4q = 18$
   $3p - 6q = 3$

Solve each system of equations by using either substitution or elimination.

8. $a - b = 2$
   $-2a + 3b = 3$

9. $5m + n = 10$
   $4m + n = 4$

10. $3g - 2h = -1$
    $8h = 5 + 12g$

11. $\frac{1}{4}x + y = \frac{7}{2}$
    $x - \frac{1}{2}y = 2$

Standardized Test Practice

12. A wholesaler is offering two different packages of custom printed T-shirts and sweatshirts to groups for fund-raisers. One package contains 10 dozen T-shirts and 14 dozen sweatshirts for \$1544. The other package contains 18 dozen T-shirts and 8 dozen sweatshirts for \$1472. Write and solve a system of linear equations to find the cost, in dollars, of a dozen T-shirts.

(A) \$76

(B) \$58

(C) \$48

(D) \$36

Practice and Apply

Homework Help	
For Exercises	See Examples
13–18	1, 2
19–24	3, 4
25–49	1–5

Extra Practice
See page 832.

Solve each system of equations by using substitution.

13. $2j - 3k = 3$
$j + k = 14$

14. $2r + s = 11$
$6r - 2s = -2$

15. $5a - b = 17$
$3a + 2b = 5$

16. $-w - z = -2$
$4w + 5z = 16$

17. $6c + 3d = 12$
$2c = 8 - d$

18. $2x + 4y = 6$
$7x = 4 + 3y$

Solve each system of equations by using elimination.

19. $u + v = 7$
$2u + v = 11$

20. $m - n = -9$
$7m + 2n = 9$

21. $3p - 5q = 6$
$2p - 4q = 4$

22. $4x - 5y = 17$
$3x + 4y = 5$

23. $2c + 6d = 14$
$\frac{1}{2}c - 3d = 8$

24. $3s + 2t = -3$
$s + \frac{1}{3}t = -4$

Solve each system of equations by using either substitution or elimination.

25. $r + 4s = -8$
$3r + 2s = 6$

26. $10m - 9n = 15$
$5m - 4n = 10$

27. $3c - 7d = -3$
$2c + 6d = -34$

28. $6g - 8h = 50$
$4g + 6h = 22$

29. $2p = 7 + q$
$6p - 3q = 24$

30. $3x = -31 + 2y$
$5x + 6y = 23$

31. $3u + 5v = 6$
$2u - 4v = -7$

32. $3a - 2b = -3$
$3a + b = 3$

33. $s + 3t = 27$
$\frac{1}{2}s + 2t = 19$

34. $f = 6 - 2g$
$\frac{1}{6}f + \frac{1}{3}g = 1$

35. $0.25x + 1.75y = 1.25$
$0.5x + 2.5y = 2$

36. $0.4m + 1.8n = 8$
$1.2m + 3.4n = 16$

WebQuest

A system of equations can be used to compare home loan options. Visit www.algebra2.com/webquest to continue work on your WebQuest project.

37. Three times one number added to five times another number is 54. The second number is two less than the first. Find the numbers.

38. The average of two numbers is 7. Find the numbers if three times one of the numbers is one half the other number.

SKIING For Exercises 39 and 40, use the following information.
All 28 members in Crestview High School's Ski Club went on a one-day ski trip. Members can rent skis for \$16.00 per day or snowboards for \$19.00 per day. The club paid a total of \$478 for rental equipment.

39. Write a system of equations that represents the number of members who rented the two types of equipment.

40. How many members rented skis and how many rented snowboards?

41. **HOUSING** Campus Rentals rents 2- and 3-bedroom apartments for \$700 and \$900 per month, respectively. Last month they had six vacant apartments and reported \$4600 in lost rent. How many of each type of apartment were vacant?

42. **GEOMETRY** Find the coordinates of the vertices of the parallelogram whose sides are contained in the lines whose equations are $2x + y = -12$, $2x - y = -8$, $2x - y - 4 = 0$, and $4x + 2y = 24$.

INVENTORY For Exercises 43 and 44, use the following information.
Heung-Soo is responsible for checking a shipment of technology equipment that contains laser printers that cost \$700 each and color monitors that cost \$200 each. He counts 30 boxes on the loading dock. The invoice states that the order totals \$15,000.

43. Write a system of two equations that represents the number of each item.

44. How many laser printers and how many color monitors were delivered?

Career Choices

Teacher

Besides the time they spend in a classroom, teachers spend additional time preparing lessons, grading papers, and assessing students' progress.

Online Research
For information about a career as a teacher, visit: www.algebra2.com/careers

TEACHING For Exercises 45–47, use the following information.
Mr. Talbot is writing a test for his science classes. The test will have true/false questions worth 2 points each and multiple-choice questions worth 4 points each for a total of 100 points. He wants to have twice as many multiple-choice questions as true/false.

45. Write a system of equations that represents the number of each type of question.

46. How many true/false questions and multiple-choice questions will be on the test?

47. If most of his students can answer true/false questions within 1 minute and multiple-choice questions within $1\frac{1}{2}$ minutes, will they have enough time to finish the test in 45 minutes?

EXERCISE For Exercises 48 and 49, use the following information.
Megan exercises every morning for 40 minutes. She does a combination of step aerobics, which burns about 11 Calories per minute, and stretching, which burns about 4 Calories per minute. Her goal is to burn 335 Calories during her routine.

48. Write a system of equations that represents Megan's morning workout.

49. How long should she participate in each activity in order to burn 335 Calories?

50. **CRITICAL THINKING** Solve the system of equations.

$$\frac{1}{x} + \frac{3}{y} = \frac{3}{4}$$
$$\frac{3}{x} - \frac{2}{y} = \frac{5}{12}$$

$\left(\textit{Hint}: \text{Let } m = \frac{1}{x} \text{ and } n = \frac{1}{y}.\right)$

51. **WRITING IN MATH** Answer the question that was posed at the beginning of the lesson.

How can a system of equations be used to make consumer decisions?

Include the following in your answer:

- a solution of the system of equations in the application at the beginning of the lesson, and
- an explanation of how Yolanda can use a graph to decide whether she should change to a long-distance plan that charges \$0.10 per minute and a flat fee of \$3.00 per month.

Standardized Test Practice

Standards Practice

52. If $x = y + z$ and $x + y = 6$ and $x = 10$, then $z =$

(A) 4. (B) 8. (C) 14. (D) 16.

53. If the perimeter of the square shown at the right is 48 units, find the value of x.

(A) 3 (B) 4

(C) 6 (D) 8

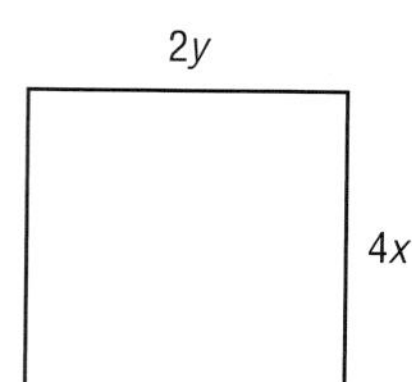

Maintain Your Skills

Mixed Review

Graph each system of equations and describe it as *consistent and independent*, *consistent and dependent*, or *inconsistent*. *(Lesson 3-1)*

54. $y = x + 2$
$y = x - 1$

55. $4y - 2x = 4$
$y - \frac{1}{2}x = 1$

56. $3x + y = 1$
$y = 2x - 4$

Graph each inequality. *(Lesson 2-7)*

57. $x + y \leq 3$

58. $5y - 4x < -20$

59. $3x + 9y \geq -15$

Write each equation in standard form. Identify *A*, *B*, and *C*. *(Lesson 2-2)*

60. $y = 7x + 4$

61. $x = y$

62. $3x = 2 - 5y$

63. $6x = 3y - 9$

64. $y = \frac{1}{2}x - 3$

65. $\frac{2}{3}y - 6 = 1 - x$

66. ELECTRICITY Use the formula $I = \frac{E}{R + r}$ to find the amount of current I (in amperes) produced if the electromotive force E is 1.5 volts, the circuit resistance R is 2.35 ohms, and the resistance r within a battery is 0.15 ohms. *(Lesson 1-1)*

Getting Ready for the Next Lesson

PREREQUISITE SKILL Determine whether the given point satisfies each inequality. *(To review **inequalities**, see Lesson 2-7.)*

67. $3x + 2y \leq 10$; $(2, -1)$

68. $4x - 2y > 6$; $(3, 3)$

69. $7x + 4y \geq -15$; $(-4, 2)$

70. $7y + 6x < 50$; $(-5, 5)$

Practice Quiz 1

Lessons 3-1 and 3-2

Solve each system of equations by graphing. *(Lesson 3-1)*

1. $y = 3x + 10$
$y = -x + 6$

2. $2x + 3y = 12$
$2x - y = 4$

Solve each system of equations by using either substitution or elimination. *(Lesson 3-2)*

3. $y = x + 5$
$x + y = 9$

4. $2x + 6y = 2$
$3x + 2y = 10$

5. AIRPORTS According to the Airports Council International, the busiest airport in the world is Atlanta's Hartsfield International Airport, and the second busiest is Chicago's O'Hare Airport. Together they handled 150.5 million passengers in the first six months of 1999. If Hartsfield handled 5.5 million more passengers than O'Hare, how many were handled by each airport? *(Lesson 3-2)*

3-3 Solving Systems of Inequalities by Graphing

What You'll Learn

- Solve systems of inequalities by graphing.
- Determine the coordinates of the vertices of a region formed by the graph of a system of inequalities.

Vocabulary

- system of inequalities

California Standards

Standard 2.0 **Students solve systems of linear** equations and **inequalities (in two** or three **variables)** by substitution, **with graphs,** or with matrices. (Key)

How can you determine whether your blood pressure is in a normal range?

During one heartbeat, blood pressure reaches a maximum pressure (systolic) and a minimum pressure (diastolic), which are measured in millimeters of mercury (mm Hg). Blood pressure is expressed as the maximum pressure over the minimum pressure—for example, 120/80. Normal blood pressure for people under 40 ranges from 100 to 140 mm Hg for the maximum and from 60 to 90 mm Hg for the minimum. This information can be represented by a system of inequalities.

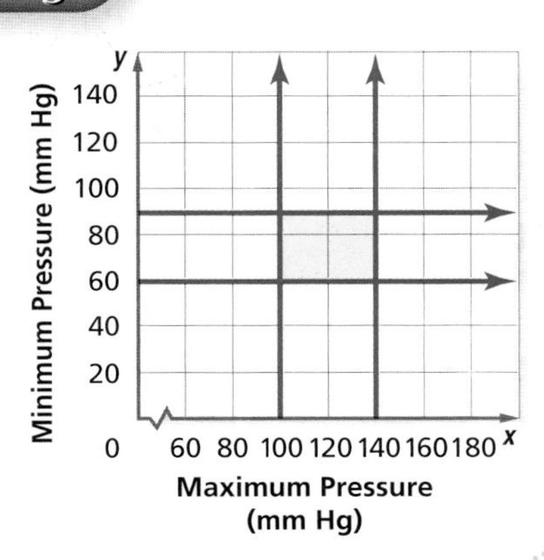

GRAPH SYSTEMS OF INEQUALITIES To solve a **system of inequalities**, we need to find the ordered pairs that satisfy all of the inequalities in the system. One way to solve a system of inequalities is to graph the inequalities on the same coordinate plane. The solution set is represented by the intersection of the graph.

Example 1 Intersecting Regions

Solve each system of inequalities by graphing.

a. $y > -2x + 4$
$y \leq x - 2$

solution of $y > -2x + 4 \rightarrow$ Regions 1 and 2
solution of $y \leq x - 2 \rightarrow$ Regions 2 and 3

The intersection of these regions is Region 2, which is the solution of the system of inequalities. Notice that the solution is a region containing an infinite number of ordered pairs.

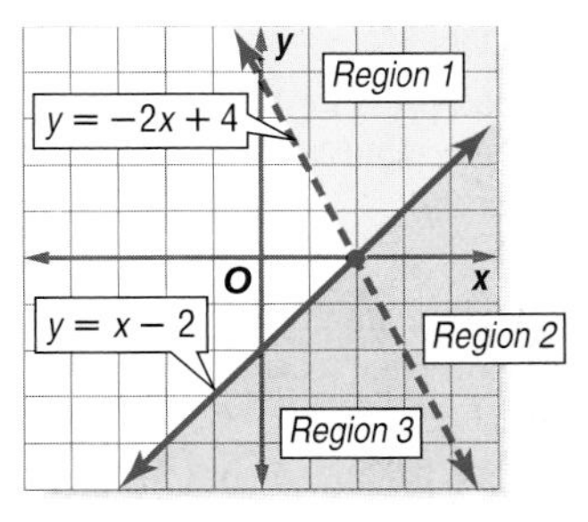

The green area represents where the yellow area of one graph overlaps the blue area of the other.

Study Tip

Look Back To review **graphing inequalities**, see Lesson 2-7.

b. $y > x + 1$
$|y| \leq 3$

The inequality $|y| \leq 3$ can be written as $y \leq 3$ and $y \geq -3$.

Graph all of the inequalities on the same coordinate plane and shade the region or regions that are common to all.

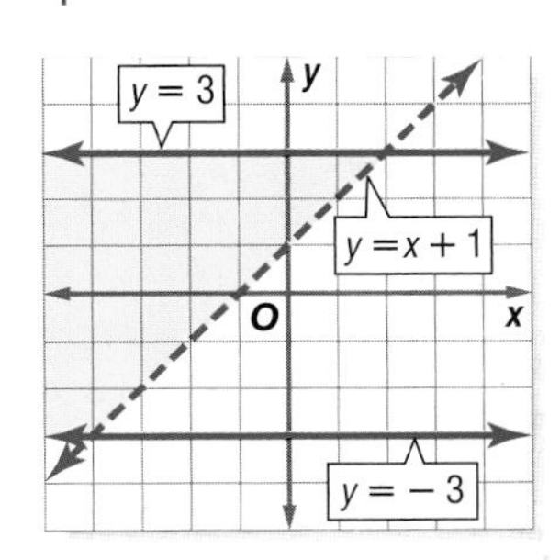

Study Tip

Reading Math
The empty set is also called the *null set.* It can be represented as ∅ or { }.

It is possible that two regions do *not* intersect. In such cases, we say the solution is the empty set ∅ and no solution exists.

Example 2 Separate Regions

Solve the system of inequalities by graphing.

$y > \frac{1}{2}x + 1$

$y < \frac{1}{2}x - 3$

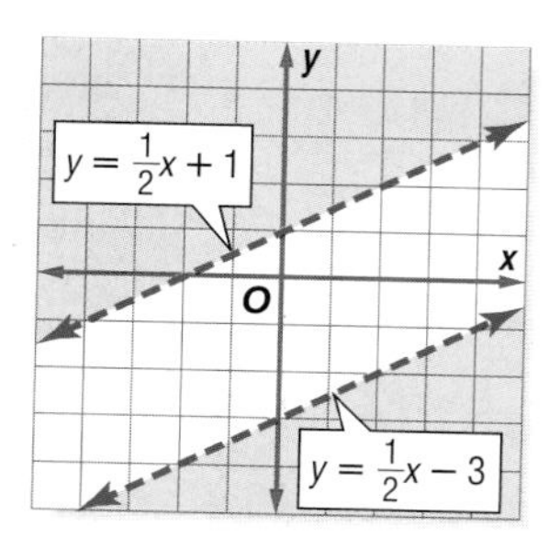

Graph both inequalities. The graphs do not overlap, so the solutions have no points in common. The solution set is ∅.

Example 3 Write and Use a System of Inequalities

SPACE EXPLORATION **In 1959, NASA wanted astronauts who were at least 5 feet 4 inches, but no more than 5 feet 11 inches tall, because of limited space inside the Mercury capsule. They were to be between 21 and 40 years of age. Write and graph a system of inequalities that represents the range of heights and ages for qualifying astronauts.**

Let h represent the height of an astronaut in inches. The acceptable heights are at least 5 feet 4 inches (or 64 inches) and no more than 5 feet 11 inches (or 71 inches). We can write two inequalities.

$64 \leq h$ and $h \leq 71$

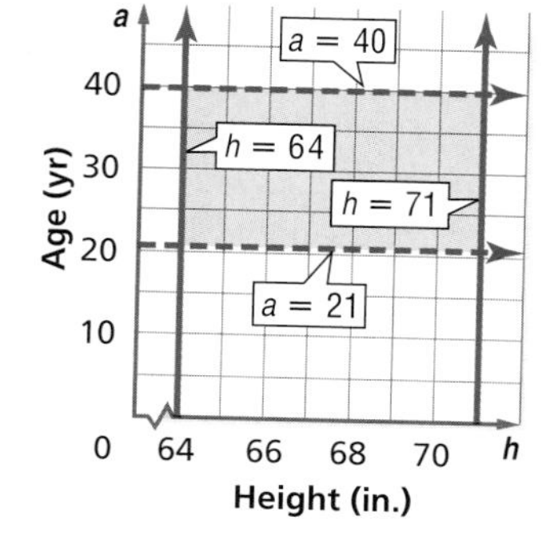

Let a represent the age of an astronaut. The acceptable ages can also be written as two inequalities.

$a > 21$ and $a < 40$

Graph all of the inequalities. Any ordered pair in the intersection of the graphs is a solution of the system.

More About...

Space Exploration

Today the basic physical qualifications for an astronaut are:

- blood pressure no greater than 140 over 90,
- distant visual acuity no greater than 20/100 uncorrected, correctable to 20/20, and
- height from 60 inches to 76 inches.

Source: NASA

FIND VERTICES OF A POLYGONAL REGION Sometimes, the graph of a system of inequalities forms a polygonal region. You can find the vertices of the region by determining the coordinates of the points at which the boundary lines intersect.

Example 4 Find Vertices

Find the coordinates of the vertices of the figure formed by $x + y \geq -1$, $x - y \leq 6$, and $12y + x \leq 32$.

Graph each inequality. The intersection of the graphs forms a triangle.

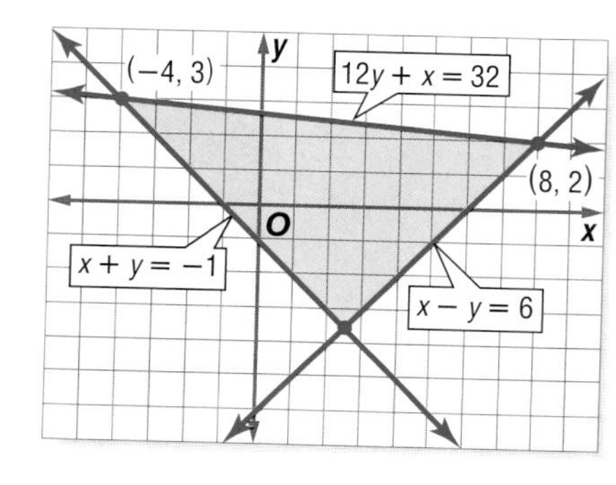

The coordinates $(-4, 3)$ and $(8, 2)$ can be determined from the graph. To find the coordinates of the third vertex, solve the system of equations $x + y = -1$ and $x - y = 6$.

Add the equations to eliminate y.

$$\begin{array}{rl} x + y = & -1 \\ (+)\ x - y = & 6 \\ \hline 2x \quad = & 5 \end{array}$$ Add the equations.

$$x = \frac{5}{2}$$ Divide each side by 2.

Now find y by substituting $\frac{5}{2}$ for x in the first equation.

$x + y = -1$ First equation

$\frac{5}{2} + y = -1$ Replace x with $\frac{5}{2}$.

$y = -\frac{7}{2}$ Subtract $\frac{5}{2}$ from each side.

The vertices of the triangle are at $(-4, 3)$, $(8, 2)$, and $\left(\frac{5}{2}, -\frac{7}{2}\right)$.

Check for Understanding

Concept Check

1. **OPEN ENDED** Write a system of inequalities that has no solution.
2. **Tell** whether the following statement is *true* or *false*. If false, give a counterexample. *A system of two linear inequalities has either no points or infinitely many points in its solution.*
3. **State** which region is the solution of the following systems of inequalities.

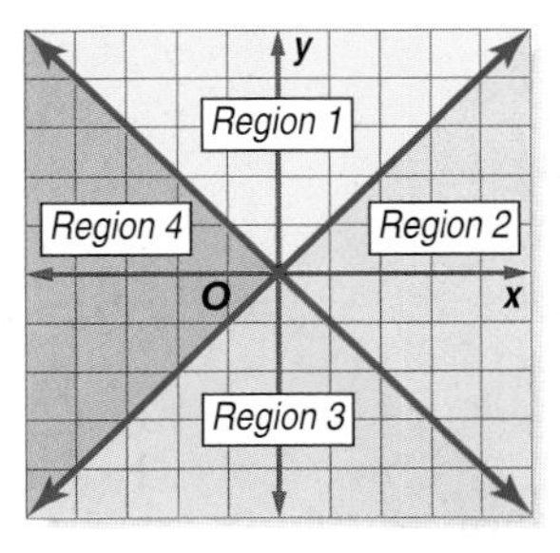

   a. $y \geq x$, $y \leq -x$
   b. $y \leq x$, $y \geq -x$
   c. $y \geq x$, $y \geq -x$
   d. $y \leq x$, $y \leq -x$

Guided Practice

Solve each system of inequalities by graphing.

4. $x \leq 4$, $y > 2$
5. $y \geq x - 2$, $y \leq -2x + 4$
6. $|x - 1| \leq 2$, $x + y > 2$
7. $x \leq 1$, $y < 2x + 1$, $x + 2y \geq -3$

Find the coordinates of the vertices of the figure formed by each system of inequalities.

8. $y \leq x$, $y \geq -3$, $3y + 5x \leq 16$
9. $y \geq x - 3$, $y \leq x + 7$, $x + y \leq 11$, $x + y \geq -1$

Application

SHOPPING For Exercises 10 and 11, use the following information.
Willis has been sent to the grocery store to purchase bagels and muffins for the members of the track team. He can spend at most \$28. A package of bagels costs \$2.50 and contains 6 bagels. A package of muffins costs \$3.50 and contains 8 muffins. He needs to buy at least 12 bagels and 24 muffins.

10. Graph the region that shows how many packages of each item he can purchase.
11. Give an example of three different purchases he can make.

Practice and Apply

Homework Help

For Exercises	See Examples
12–23	1, 2
24–31	4
32–37	3

Extra Practice
See page 833.

Solve each system of inequalities by graphing.

12. $x \geq 2$
$y > 3$

13. $x \leq -1$
$y \geq -4$

14. $y < 2 - x$
$y > x + 4$

15. $y > x - 3$
$|y| \leq 2$

16. $3x + 2y \geq 6$
$4x - y \geq 2$

17. $4x - 3y < 7$
$2y - x < -6$

18. $y < 2x - 3$
$y \leq \frac{1}{2}x + 1$

19. $3y \leq 2x - 8$
$y \geq \frac{2}{3}x - 1$

20. $|x| \leq 3$
$|y| > 1$

21. $|x + 1| \leq 3$
$x + 3y \geq 6$

22. $y \geq 2x + 1$
$y \leq 2x - 2$
$3x + y \geq 9$

23. $x - 3y > 2$
$2x - y < 4$
$2x + 4y \geq -7$

Find the coordinates of the vertices of the figure formed by each system of inequalities.

24. $y \geq 0$
$x \geq 0$
$x + 2y \leq 8$

25. $y \geq -4$
$y \leq 2x + 2$
$2x + y \leq 6$

26. $x \leq 3$
$-x + 3y \leq 12$
$4x + 3y \geq 12$

27. $x + y \leq 9$
$x - 2y \leq 12$
$y \leq 2x + 3$

28. $y \geq -3$
$x \leq 6$
$y \geq x - 2$
$2y \leq x + 5$

29. $y \geq x - 5$
$y \leq 2x + 11$
$x + 2y \leq 12$
$x + 2y \geq 2$

30. Find the area of the region defined by the system of inequalities $y + x \leq 3$, $y - x \leq 3$, and $y \geq -1$.

31. Find the area of the region defined by the system of inequalities $x \geq -3$, $y + x \leq 8$, and $y - x \geq -2$.

32. **PART-TIME JOBS** Bryan Clark makes $10 an hour cutting grass and $12 an hour for raking leaves. He cannot work more than 15 hours per week. Graph two inequalities that Bryan can use to determine how many hours he needs to work at each job if he wants to earn at least $120 per week.

Career Choices

Atmospheric Scientist

The best known use of atmospheric science is for weather forecasting. However, weather information is also studied for air-pollution control, agriculture, and transportation.

Online Research For information about a career as an atmospheric scientist, visit: www.algebra2.com/careers

HURRICANES **For Exercises 33 and 34, use the following information.**
Hurricanes are divided into categories according to their wind speed and storm surge.

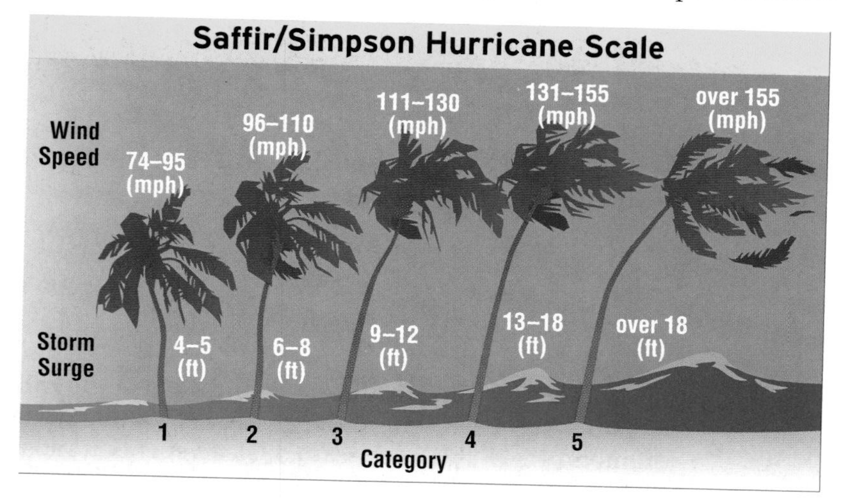

33. Write and graph the system of inequalities that represents the range of wind speeds s and storm surges h for a category 3 hurricane.

34. On September 16, 1999, Hurricane Floyd hit the United States with winds of 140 mph. Classify Hurricane Floyd, and identify the heights of its storm surges.

BAKING **For Exercises 35–37, use the recipes at the right.**
The Merry Bakers are baking pumpkin bread and Swedish soda bread for this week's specials. They have 24 cups of flour and at most 26 teaspoons of baking powder.

35. Graph the inequalities that represent how many loaves of each type of bread the bakers can make.
36. List three different combinations of breads they can make.
37. Which combination uses all of the available flour and baking soda?
38. **CRITICAL THINKING** Find the area of the region defined by $|x| + |y| \leq 5$ and $|x| + |y| \geq 2$.
39. **WRITING IN MATH** Answer the question that was posed at the beginning of the lesson.

    How can you determine whether your blood pressure is in a normal range?

    Include the following in your answer:
    - an explanation of how to use the graph, and
    - a description of the regions that indicate high blood pressure, both systolic and diastolic.

Standards Practice
Standardized Test Practice
A B C D

40. Choose the system of inequalities whose solution is represented by the graph.

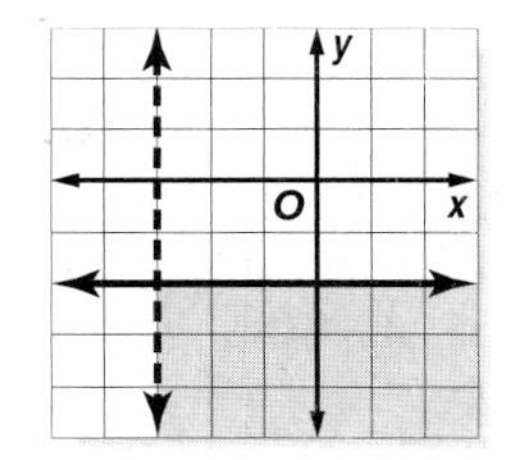

(A) $y < -2$, $x < -3$
(B) $y \leq -2$, $x > -3$
(C) $x \leq -2$, $y > -3$
(D) $x < -3$, $y < -3$

41. **EXTENDED RESPONSE** Create a system of inequalities for which the graph will be a square with its interior located in the first quadrant.

Maintain Your Skills

Mixed Review

Solve each system of equations by using either substitution or elimination. *(Lesson 3-2)*

42. $4x - y = -20$, $x + 2y = 13$
43. $3x - 4y = -2$, $5x + 2y = 40$
44. $4x + 5y = 7$, $3x - 2y = 34$

Solve each system of equations by graphing. *(Lesson 3-1)*

45. $y = 2x + 1$, $y = -\frac{1}{2}x - 4$
46. $2x + y = -3$, $6x + 3y = -9$
47. $2x - y = 6$, $-x + 8y = 12$

48. Write an equation in slope-intercept form of the line that passes through $(-4, 4)$ and $(6, 9)$. *(Lesson 2-4)*

Getting Ready for the Next Lesson

PREREQUISITE SKILL **Find each value if $f(x) = 4x + 3$ and $g(x) = 5x - 7$.**
*(To review **functions**, see Lesson 2-1.)*

49. $f(-2)$
50. $g(-1)$
51. $g(3)$
52. $f(6)$
53. $f(0.5)$
54. $g(-0.25)$

Graphing Calculator Investigation

A Follow-Up of Lesson 3-3

California Standards Standard 2.0 **Students solve systems of linear** equations and **inequalities (in two** or three **variables)** by substitution, **with graphs,** or with matrices. (Key)

Systems of Linear Inequalities

You can graph systems of linear inequalities with a TI-83 Plus calculator using the Y= menu. You can choose different graphing styles to shade above or below a line.

Example **Graph the system of inequalities in the standard viewing window.**

$y \geq -2x + 3$
$y \leq x + 5$

Step 1

- Enter $-2x + 3$ as Y1. Since y is greater than $-2x + 3$, shade above the line.
  KEYSTROKES: -2 [X,T,θ,n] [+] 3
- Use the left arrow key to move your cursor as far left as possible. Highlight the graph style icon. Press [ENTER] until the shade above icon, ◥, appears.

Step 2

- Enter $x + 5$ as Y2. Since y is less than $x + 5$, shade below the line.
  KEYSTROKES: [X,T,θ,n] [+] 5
- Use the arrow and [ENTER] keys to choose the shade below icon, ◣.

Step 3

- Display the graphs by pressing [GRAPH].
  Notice the shading pattern above the line $y = -2x + 3$ and the shading pattern below the line $y = x + 5$. The intersection of the graphs is the region where the patterns overlap. This region includes all the points that satisfy the system $y \geq -2x + 3$ and $y \leq x + 5$.

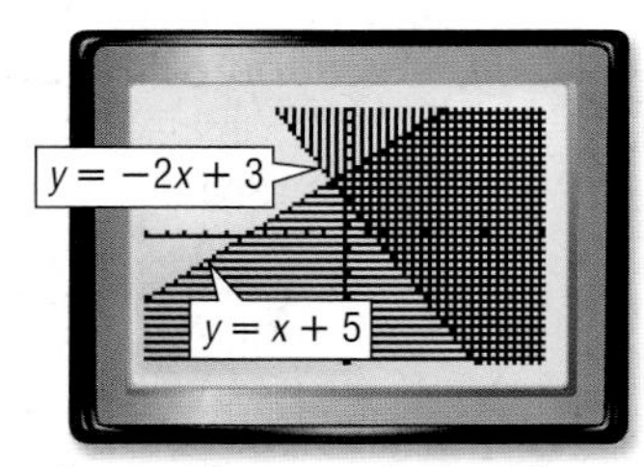

[−10, 10] scl: 1 by [−10, 10] scl: 1

Exercises

Solve each system of inequalities. Sketch each graph on a sheet of paper.

1. $y \geq 4$
$y \leq -x$

2. $y \geq -2x$
$y \leq -3$

3. $y \geq 1 - x$
$y \leq x + 5$

4. $y \geq x + 2$
$y \leq -2x - 1$

5. $3y \geq 6x - 15$
$2y \leq -x + 3$

6. $y + 3x \geq 6$
$y - 2x \leq 9$

7. $6y + 4x \geq 12$
$5y - 3x \leq -10$

8. $\frac{1}{4}y - x \geq -2$
$\frac{1}{3}y + 2x \leq 4$

www.algebra2.com/other_calculator_keystrokes

3-4 Linear Programming

What You'll Learn

- Find the maximum and minimum values of a function over a region.
- Solve real-world problems using linear programming.

Vocabulary

- constraints
- feasible region
- bounded
- vertices
- unbounded
- linear programming

How is linear programming used in scheduling work?

One of the primary tasks of the U.S. Coast Guard is to maintain the buoys that ships use to navigate. The ships that service buoys are called buoy tenders.

Suppose a certain buoy tender can carry up to 8 replacement buoys. Their crew can check and repair a buoy in 1 hour. It takes the crew $2\frac{1}{2}$ hours to replace a buoy.

MAXIMUM AND MINIMUM VALUES The buoy tender captain can use a system of inequalities to represent the limitations of time and the number of replacement buoys on the ship. If these inequalities are graphed, all of the points in the intersection are the combinations of repairs and replacements that the buoy tender can schedule.

The inequalities are called the **constraints**. The intersection of the graphs is called the **feasible region**. When the graph of a system of constraints is a polygonal region like the one graphed at the right, we say that the region is **bounded**.

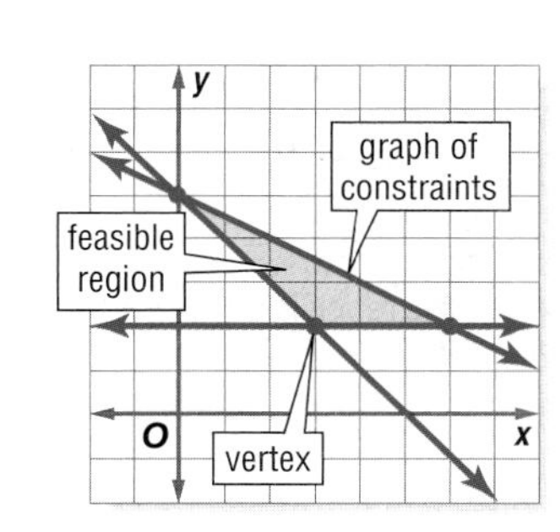

Sometimes it is necessary to find the maximum or minimum values that a linear function has for the points in a feasible region. For example, the buoy tender captain wishes to maximize the total number of buoys serviced. The maximum or minimum value of a related function *always* occurs at one of the **vertices** of the feasible region.

Example 1 Bounded Region

Study Tip

Reading Math
The notation $f(x, y)$ is used to represent a function with two variables x and y. It is read *f of x and y*.

Graph the following system of inequalities. Name the coordinates of the vertices of the feasible region. Find the maximum and minimum values of the function $f(x, y) = 3x + y$ for this region.

$x \geq 1$
$y \geq 0$
$2x + y \leq 6$

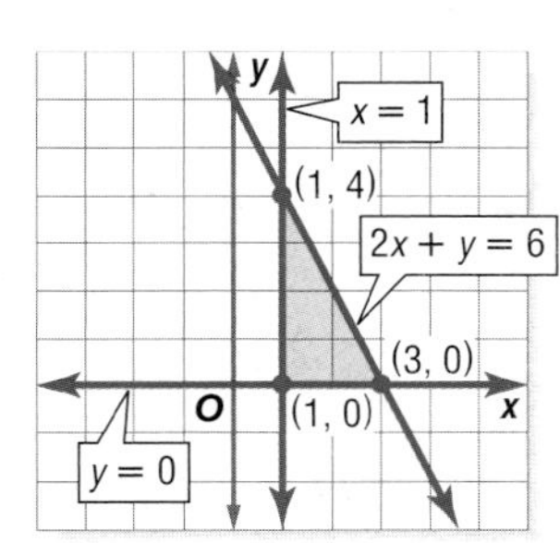

Step 1 Find the vertices of the region. Graph the inequalities.

The polygon formed is a triangle with vertices at (1, 4), (3, 0), and (1, 0).

(continued on the next page)

Step 2 Use a table to find the maximum and minimum values of $f(x, y)$. Substitute the coordinates of the vertices into the function.

(x, y)	$3x + y$	$f(x, y)$	
(1, 4)	3(1) + 4	7	
(3, 0)	3(3) + 0	9	← maximum
(1, 0)	3(1) + 0	3	← minimum

The maximum value is 9 at (3, 0). The minimum value is 3 at (1, 0).

Sometimes a system of inequalities forms a region that is open. In this case, the region is said to be **unbounded**.

Example 2 Unbounded Region

Graph the following system of inequalities. Name the coordinates of the vertices of the feasible region. Find the maximum and minimum values of the function $f(x, y) = 5x + 4y$ for this region.

$2x + y \geq 3$

$3y - x \leq 9$

$2x + y \leq 10$

Graph the system of inequalities. There are only two points of intersection, (0, 3) and (3, 4).

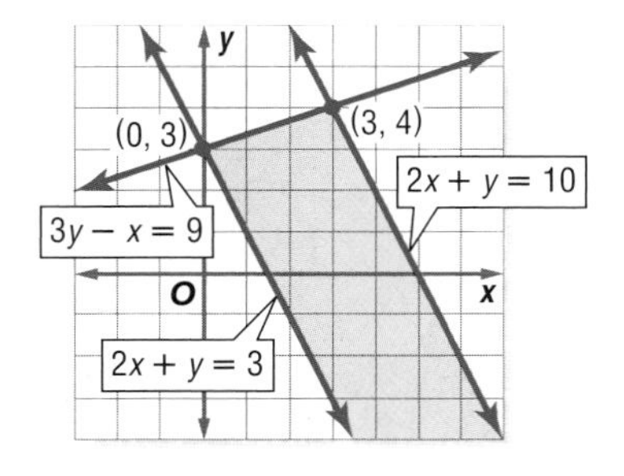

(x, y)	$5x + 4y$	$f(x, y)$
(0, 3)	5(0) + 4(3)	12
(3, 4)	5(3) + 4(4)	31

The maximum is 31 at (3, 4).

Although $f(0, 3)$ is 12, it is not the minimum value since there are other points in the solution that produce lesser values. For example, $f(3, -2) = 7$ and $f(20, -35) = -40$. It appears that because the region is unbounded, $f(x, y)$ has no minimum value.

Study Tip

Common Misconception

Always test a point contained in the feasible region when the graph is unbounded. Do not assume that there is no minimum value if the feasible region is unbounded below the line, or that there is no maximum value if the feasible region is unbounded above the line.

REAL-WORLD PROBLEMS The process of finding maximum or minimum values of a function for a region defined by inequalities is called **linear programming**. The steps used to solve a problem using linear programming are listed below.

Key Concept — *Linear Programming Procedure*

Step 1 Define the variables.

Step 2 Write a system of inequalities.

Step 3 Graph the system of inequalities.

Step 4 Find the coordinates of the vertices of the feasible region.

Step 5 Write a function to be maximized or minimized.

Step 6 Substitute the coordinates of the vertices into the function.

Step 7 Select the greatest or least result. Answer the problem.

Linear programming can be used to solve many types of real-world problems. These problems have certain restrictions placed on the variables, and some function of the variable must be maximized or minimized.

Example 3 Linear Programming

VETERINARY MEDICINE As a receptionist for a veterinarian, one of Dolores Alvarez's tasks is to schedule appointments. She allots 20 minutes for a routine office visit and 40 minutes for a surgery. The veterinarian cannot do more than 6 surgeries per day. The office has 7 hours available for appointments. If an office visit costs $55 and most surgeries cost $125, find a combination of office visits and surgeries that will maximize the income the veterinarian practice receives per day.

Step 1 Define the variables.

v = the number of office visits
s = the number of surgeries

Step 2 Write a system of inequalities.

Since the number of appointments cannot be negative, v and s must be nonnegative numbers.

$v \geq 0$ and $s \geq 0$

An office visit is 20 minutes, and a surgery is 40 minutes. There are 7 hours available for appointments.

$20v + 40s \leq 420$ 7 hours = 420 minutes

The veterinarian cannot do more than 6 surgeries per day.

$s \leq 6$

Step 3 Graph the system of inequalities.

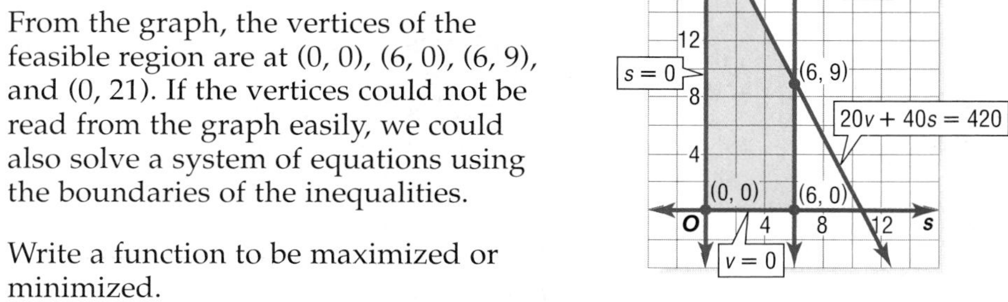

Step 4 Find the coordinates of the vertices of the feasible region.

From the graph, the vertices of the feasible region are at (0, 0), (6, 0), (6, 9), and (0, 21). If the vertices could not be read from the graph easily, we could also solve a system of equations using the boundaries of the inequalities.

Step 5 Write a function to be maximized or minimized.

The function that describes the income is $f(s, v) = 125s + 55v$. We wish to find the maximum value for this function.

Step 6 Substitute the coordinates of the vertices into the function.

(s, v)	$125s + 55v$	$f(s, v)$
(0, 0)	125(0) + 55(0)	0
(6, 0)	125(6) + 55(0)	750
(6, 9)	125(6) + 55(9)	1245
(0, 21)	125(0) + 55(21)	1155

Step 7 Select the greatest or least result. Answer the problem.

The maximum value of the function is 1245 at (6, 9). This means that the maximum income is $1245 when Dolores schedules 6 surgeries and 9 office visits.

More About...

Veterinary Medicine

Surgeries are usually performed in the morning so that the animal can recover throughout the day while there is plenty of staff to monitor its progress.

Source: www.vetmedicine.miningco.com

Check for Understanding

Concept Check

1. **Determine** whether the following statement is *always*, *sometimes*, or *never* true. *A feasible region has a minimum and a maximum value.*

2. **OPEN ENDED** Give an example of a system of inequalities that forms a bounded region.

Guided Practice

Graph each system of inequalities. Name the coordinates of the vertices of the feasible region. Find the maximum and minimum values of the given function for this region.

3. $y \geq 2$
$x \geq 1$
$x + 2y \leq 9$
$f(x, y) = 2x - 3y$

4. $x \geq -3$
$y \leq 1$
$3x + y \leq 6$
$f(x, y) = 5x - 2y$

5. $y \leq 2x + 1$
$1 \leq y \leq 3$
$x + 2y \leq 12$
$f(x, y) = 3x + y$

6. $y \geq -x + 2$
$2 \leq x \leq 7$
$y \leq \frac{1}{2}x + 5$
$f(x, y) = 8x + 3y$

7. $x + 2y \leq 6$
$2x - y \leq 7$
$x \geq -2, y \geq -3$
$f(x, y) = x - y$

8. $x - 3y \geq -7$
$5x + y \leq 13$
$x + 6y \geq -9$
$3x - 2y \geq -7$
$f(x, y) = x - y$

Application

MANUFACTURING For Exercises 9–14, use the following information.
The Future Homemakers Club is making canvas tote bags and leather tote bags for a money making project. Both types of tote bags will be lined with canvas and have leather handles. For the canvas tote bags, they need 4 yards of canvas and 1 yard of leather. For the leather tote bags, they need 3 yards of leather and 2 yards of canvas. Their advisor has purchased 56 yards of leather and 104 yards of canvas.

9. Let c represent the number of canvas tote bags and let ℓ represent the number of leather tote bags. Write a system of inequalities to represent the number of tote bags that can be produced.
10. Draw the graph showing the feasible region.
11. List the coordinates of the vertices of the feasible region.
12. If the club plans to sell the canvas bags at a profit of $20 each and the leather bags at a profit of $35 each, write a function for the total profit on the bags.
13. Determine the number of canvas and leather bags that they need to make for a maximum profit.
14. What is the maximum profit?

Practice and Apply

Homework Help

For Exercises	See Examples
15–29	1, 2
31–36, 38–42	3

Extra Practice
See page 833.

Graph each system of inequalities. Name the coordinates of the vertices of the feasible region. Find the maximum and minimum values of the given function for this region.

15. $y \geq 1$
$x \leq 6$
$y \leq 2x + 1$
$f(x, y) = x + y$

16. $y \geq -4$
$x \leq 3$
$y \leq 3x - 4$
$f(x, y) = x - y$

17. $y \geq 2$
$1 \leq x \leq 5$
$y \leq x + 3$
$f(x, y) = 3x - 2y$

18. $y \geq 1$
$2 \leq x \leq 4$
$x - 2y \geq -4$
$f(x, y) = 3y + x$

19. $y \leq x + 2$
$y \leq 11 - 2x$
$2x + y \geq -7$
$f(x, y) = 4x - 3y$

20. $y \leq x + 6$
$y + 2x \geq 6$
$2 \leq x \leq 6$
$f(x, y) = -x + 3y$

21. $x + y \leq 3$
$x + 2y \leq 4$
$x \geq 0, y \geq 0$
$f(x, y) = 3y - 4x$

22. $y \leq 7 - x$
$3x - 2y \leq 6$
$x \geq 0, y \geq 0$
$f(x, y) = 5x - 2y$

23. $y \geq x - 3$
$y \leq 6 - 2x$
$2x + y \geq -3$
$f(x, y) = 3x + 4y$

24. $x + y \geq 4$
$3x - 2y \leq 12$
$x - 4y \geq -16$
$f(x, y) = x - 2y$

25. $x + y \geq 2$
$4y \leq x + 8$
$y \geq 2x - 5$
$f(x, y) = 4x + 3y$

26. $2x + 2y \geq 4$
$2y \geq 3x - 6$
$4y \leq x + 8$
$f(x, y) = 3y + x$

27. $2x + 3y \geq 6$
$3x - 2y \geq -4$
$5x + y \geq 15$
$f(x, y) = x + 3y$

28. $x \geq 0$
$y \geq 0$
$x + 2y \leq 6$
$2y - x \leq 2$
$x + y \leq 5$
$f(x, y) = 3x - 5y$

29. $x \geq 2$
$y \geq 1$
$x - 2y \geq -4$
$x + y \leq 8$
$2x - y \leq 7$
$f(x, y) = x - 4y$

30. **CRITICAL THINKING** The vertices of a feasible region are $A(1, 2)$, $B(5, 2)$, and $C(1, 4)$. Write a function that satisfies each condition.

a. A is the maximum and B is the minimum.

b. C is the maximum and B is the minimum.

c. B is the maximum and A is the minimum.

d. A is the maximum and C is the minimum.

e. B and C are both maxima and A is the minimum.

PRODUCTION **For Exercises 31–36, use the following information.**
There are a total of 85 workers' hours available per day for production at a calculator manufacturer. There are 40 workers' hours available for encasement and quality control each day. The table below shows the number of hours needed in each department for two different types of calculators.

Calculator Production Time

Calculator Type	Production Time	Encasement and Quality Control
graphing calculator	$1\frac{1}{2}$ hours	2 hours
computer-algebra systems (CAS)	1 hour	$\frac{1}{2}$ hour

31. Let g represent the number of graphing calculators and let c represent the number of CAS calculators. Write a system of inequalities to represent the number of calculators that can be produced.

32. Draw the graph showing the feasible region.

33. List the coordinates of the vertices of the feasible region.

34. If the profit on a graphing calculator is $50 and the profit on a CAS calculator is $65, write a function for the total profit on the calculators.

35. Determine the number of each type of calculator that is needed to make a maximum profit.

36. What is the maximum profit?

37. **RESEARCH** Use the Internet or other reference to find an industry that uses linear programming. Describe the restrictions or constraints of the problem and explain how linear programming is used to help solve the problem.

FARMING **For Exercises 38–41, use the following information.**
Dean Stadler has 20 days in which to plant corn and soybeans. The corn can be planted at a rate of 250 acres per day and the soybeans at a rate of 200 acres per day. He has 4500 acres available for planting these two crops.

38. Let c represent the number of acres of corn and let s represent the number of acres of soybeans. Write a system of inequalities to represent the possible ways Mr. Stadler can plant the available acres.

39. Draw the graph showing the feasible region and list the coordinates of the vertices of the feasible region.

40. If the profit on corn is \$26 per acre and the profit on soybeans is \$30 per acre, how much of each should Mr. Stadler plant? What is the maximum profit?

41. How much of each should Mr. Stadler plant if the profit on corn is \$29 per acre and the profit on soybeans is \$24 per acre? What is the maximum profit?

42. **PACKAGING** The Cookie Factory's best selling items are chocolate chip cookies and peanut butter cookies. They want to sell both types of cookies together in combination packages. The different-sized packages will contain between 6 and 12 cookies, inclusively. At least three of each type of cookie should be in each package. The cost of making a chocolate chip cookie is 19¢, and the selling price is 44¢ each. The cost of making a peanut butter cookie is 13¢, and the selling price is 39¢. How many of each type of cookie should be in each package to maximize the profit?

43. WRITING IN MATH Answer the question that was posed at the beginning of the lesson.

How is linear programming used in scheduling work?

Include the following in your answer:

- a system of inequalities that represents the constraints that are used to schedule buoy repair and replacement,
- an explanation of the linear function that the buoy tender captain would wish to maximize, and
- a demonstration of how to solve the linear programming problem to find the maximum number of buoys the buoy tender could service in 24 hours at sea.

44. A feasible region has vertices at (0, 0), (4, 0), (5, 5), and (0, 8). Find the maximum and minimum of the function $f(x, y) = x + 3y$ over this region.

Ⓐ maximum: $f(0, 8) = 24$
minimum: $f(0, 0) = 0$

Ⓑ minimum: $f(0, 0) = 0$
maximum: $f(5, 5) = 20$

Ⓒ maximum: $f(5, 5) = 20$
minimum: $f(0, 8) = 8$

Ⓓ minimum: $f(4, 0) = 4$
maximum: $f(0, 0) = 0$

45. What is the area of square $ABCD$?

Ⓐ 25 units^2

Ⓑ $4\sqrt{29}$ units^2

Ⓒ 29 units^2

Ⓓ $25 + \sqrt{2}$ units^2

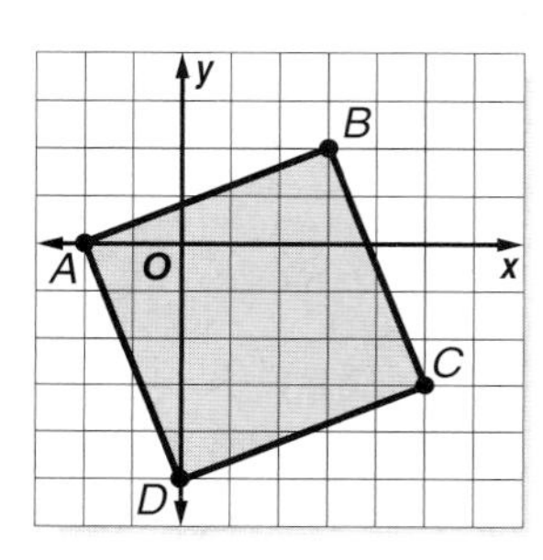

Maintain Your Skills

Mixed Review

Solve each system of inequalities by graphing. *(Lesson 3-3)*

46. $2y + x \geq 4$
$y \geq x - 4$

47. $3x - 2y \leq -6$
$y \leq \frac{3}{2}x - 1$

Solve each system of equations by using either substitution or elimination. *(Lesson 3-2)*

48. $4x + 5y = 20$
$5x + 4y = 7$

49. $6x + y = 15$
$x - 4y = -10$

50. $3x + 8y = 23$
$5x - y = 24$

SCHOOLS **For Exercises 51 and 52, use the graph at the right.** *(Lesson 1-3)*

51. Define a variable and write an equation that can be used to determine on average how much the annual per-pupil spending has increased from 1986 to 2001.

52. Solve the problem.

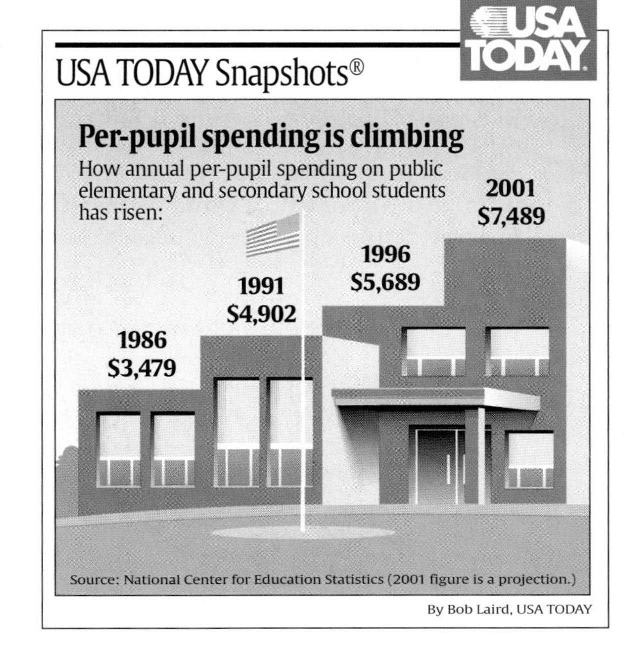

Name the property illustrated by each equation. *(Lesson 1-2)*

53. $4n + (-4n) = 0$

54. $(2 \cdot 5) \cdot 6 = 2 \cdot (5 \cdot 6)$

55. $\left(-\frac{3}{2}\right)\left(-\frac{2}{3}\right) = 1$

56. $6(x + 9) = 6x + 6(9)$

Getting Ready for the Next Lesson

PREREQUISITE SKILL **Evaluate each expression if $x = -2$, $y = 6$, and $z = 5$.**
*(To review **evaluating expressions**, see Lesson 1-1.)*

57. $x + y + z$

58. $2x - y + 3z$

59. $-x + 4y - 2z$

60. $5x + 2y - z$

61. $3x - y + 4z$

62. $-2x - 3y + 2z$

Practice Quiz 2

Lessons 3-3 and 3-4

Solve each system of inequalities by graphing. *(Lesson 3-3)*

1. $y - x > 0$
$y + x < 4$

2. $y \geq 3x - 4$
$y \leq x + 3$

3. $x + 3y \geq 15$
$4x + y \leq 16$

Graph each system of inequalities. Name the coordinates of the vertices of the feasible region. Find the maximum and minimum values of the given function for this region. *(Lesson 3-4)*

4. $x \geq 0$
$y \geq 0$
$y \leq 2x + 4$
$3x + y \leq 9$
$f(x, y) = 2x + y$

5. $x \leq 5$
$y \geq -3x$
$2y \leq x + 7$
$y \geq x - 4$
$f(x, y) = 4x - 3y$

Algebra Activity

A Preview of Lesson 3-5

Graphing Equations in Three Variables

To graph an equation in three variables, it is necessary to add a third dimension to our coordinate system. The graph of an equation of the form $Ax + By + Cz = D$, where A, B, C, and D can not all be equal to zero is a plane.

When graphing in three-dimensional space, begin with the xy-coordinate plane in a horizontal position. Then draw the z-axis as a vertical line passing through the origin. There are now three coordinate planes: the xy-plane, the xz-plane, and the yz-plane. These planes intersect at right angles and divide space into eight regions, called **octants**.

A point in space (three dimensions) has three coordinates and is represented by an **ordered triple** (x, y, z).

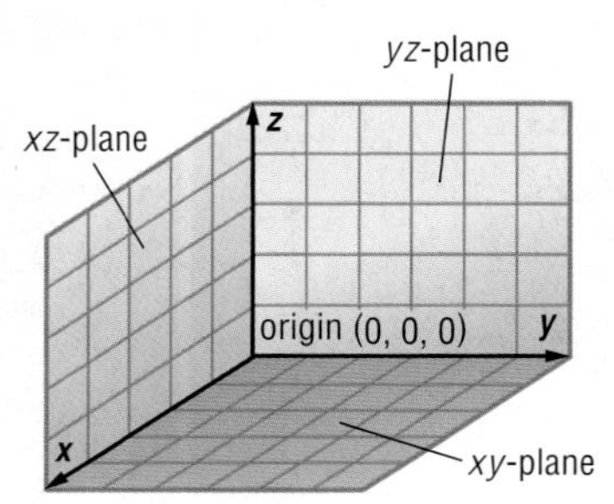

The first octant contains the points in space for which all three coordinates are positive.

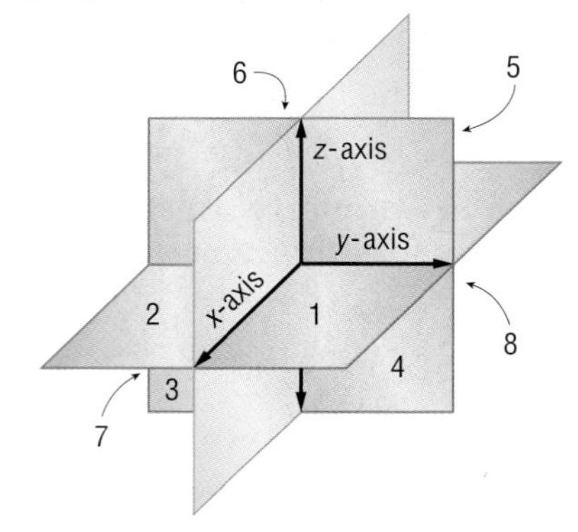

The octants are numbered as shown.

Activity 1

Use isometric dot paper to graph (3, 4, 2) on a three-dimensional coordinate system. Name the octant in which it lies.

Draw the x-, y-, and z-axes as shown.

Begin by finding the point (3, 4, 0) in the xy-plane.

The z-coordinate is 2, so move the point up two units parallel to the z-axis.

The point lies in octant 1.

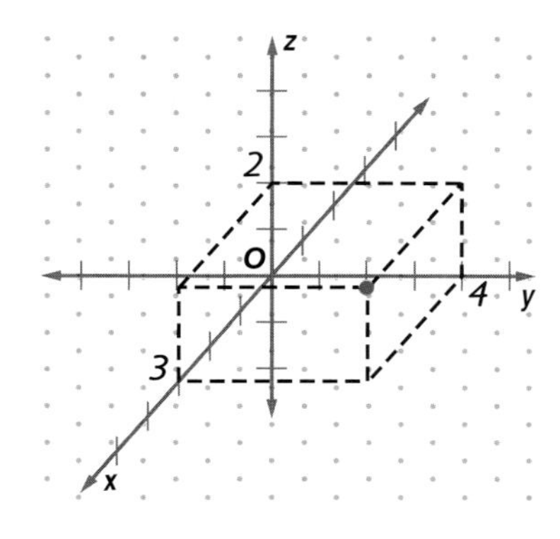

To graph a linear equation in three variables, first find the intercepts of the graph. Connect the intercepts on each axis. This forms a portion of a plane that lies in a single octant.

Activity 2

Graph $2x + 3y + 4z = 12$.

Begin by finding the x-, y-, and z-intercepts.

x-intercept	**y-intercept**	**z-intercept**
Let $y = 0$ and $z = 0$.	Let $x = 0$ and $z = 0$.	Let $x = 0$ and $y = 0$.
$2x = 12$	$3y = 12$	$4z = 12$
$x = 6$	$y = 4$	$z = 3$

To sketch the plane, graph the intercepts, which have coordinates (6, 0, 0), (0, 4, 0), and (0, 0, 3). Then connect the points. Remember this is only a portion of the plane that extends indefinitely.

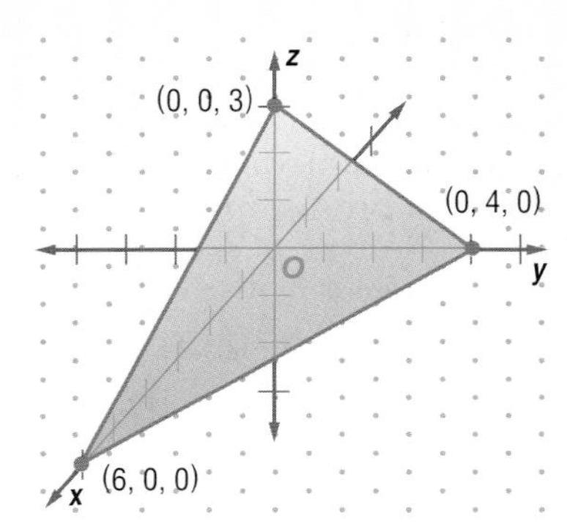

Model and Analyze

Graph each ordered triple on a three-dimensional coordinate system. Name the octant in which each point lies.

1. (5, 3, 6)
2. (−2, 4, 3)
3. (1, −5, 7)

Graph each equation. Name the coordinates for the x-, y-, and z-intercepts.

4. $3x + 6y + z = 6$
5. $2x - 5y + 4z = 20$
6. $x + 3y - 6z = 3$
7. $-3x + 5y + 10z = 15$
8. $6x + 9z = 18$
9. $4x - 6y = 24$

Write an equation of the plane given its x-, y-, and z-intercepts, respectively.

10. 8, −3, 6
11. 10, 4, −5
12. $\frac{1}{2}$, 4, −12

13. Describe the values of x, y, and z as either positive or negative for each octant.

14. Consider the graph $x = -3$ in one, two, and three dimensions.

a. Graph the equation on a number line.

b. Graph the equation on a coordinate plane.

c. Graph the equation in a three-dimensional coordinate axis.

d. Describe and compare the graphs in parts **a**, **b**, and **c**.

e. **Make a conjecture** about the graph of $x > -3$ in one, two, and three dimensions.

3-5 Solving Systems of Equations in Three Variables

What You'll Learn

- Solve systems of linear equations in three variables.
- Solve real-world problems using systems of linear equations in three variables.

California Standards
Standard 2.0
Students solve systems of linear equations and inequalities (**in** two or **three variables) by substitution,** with graphs, or with matrices. (Key)

How can you determine the number and type of medals U.S. Olympians won?

At the 2000 Summer Olympics in Sydney, Australia, the United States won 97 medals. They won 6 more gold medals than bronze and 8 fewer silver medals than bronze.

You can write and solve a system of three linear equations to determine how many of each type of medal the U.S. Olympians won. Let g represent the number of gold medals, let s represent the number of silver medals, and let b represent the number of bronze medals.

$g + s + b = 97$ The U.S. won a total of 97 medals.
$g = b + 6$ They won 6 more gold medals than bronze.
$s = b - 8$ They won 8 fewer silver medals than bronze.

SYSTEMS IN THREE VARIABLES The system of equations above has three variables. The graph of an equation in three variables, all to the first power, is a plane. The solution of a system of three equations in three variables can have one solution, infinitely many solutions, or no solution.

Key Concept — System of Equations in Three Variables

One Solution
- planes intersect in one point

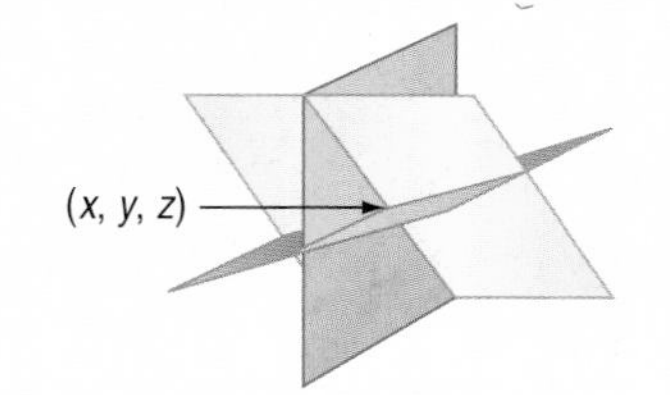

Infinite Solutions
- planes intersect in a line
- planes intersect in the same plane

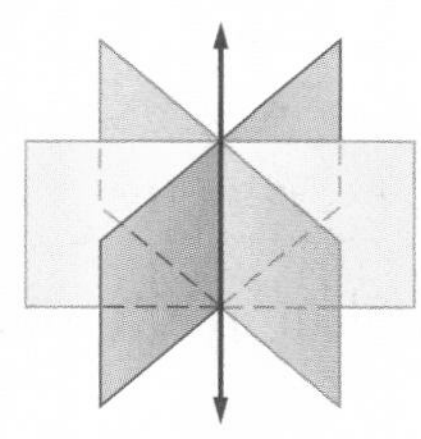

No Solution
- planes have no point in common

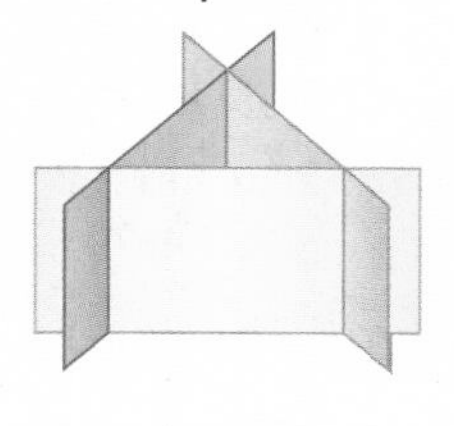

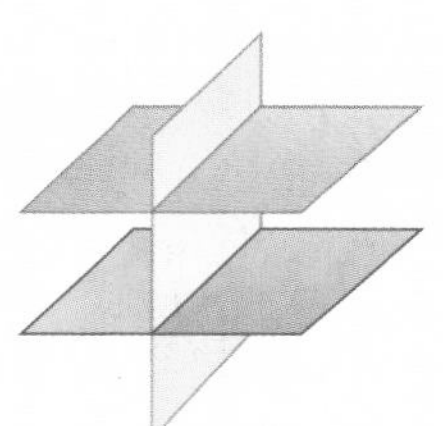

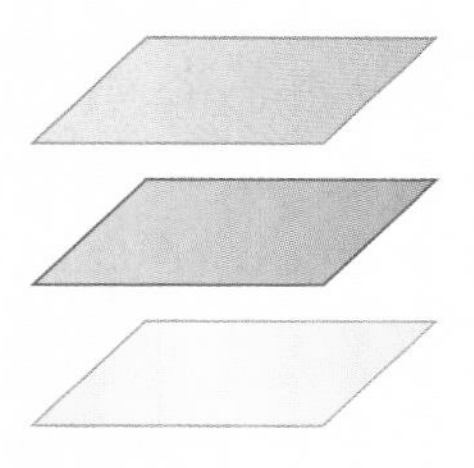

Solving systems of equations in three variables is similar to solving systems of equations in two variables. Use the strategies of substitution and elimination. The solution of a system of equations in three variables x, y, and z is called an **ordered triple** and is written as (x, y, z).

Example 1 One Solution

Solve the system of equations.

$$x + 2y + z = 10$$
$$2x - y + 3z = -5$$
$$2x - 3y - 5z = 27$$

Study Tip

Elimination
Remember that you can eliminate any of the three variables.

Step 1 Use elimination to make a system of two equations in two variables.

$x + 2y + z = 10$ Multiply by 2. $2x + 4y + 2z = 20$

$2x - y + 3z = -5$ $(-)\ 2x - y + 3z = -5$

$5y - z = 25$ Subtract to eliminate x.

$2x - y + 3z = -5$ Second equation

$(-)\ 2x - 3y - 5z = 27$ Third equation

$2y + 8z = -32$ Subtract to eliminate x.

Notice that the x terms in each equation have been eliminated. The result is two equations with the same two variables y and z.

Step 2 Solve the system of two equations.

$5y - z = 25$ Multiply by 8. $40y - 8z = 200$

$2y + 8z = -32$ $(+)\ 2y + 8z = -32$

$42y = 168$ Add to eliminate z.

$y = 4$ Divide by 42.

Substitute 4 for y in one of the equations with two variables and solve for z.

$5y - z = 25$ Equation with two variables

$5(4) - z = 25$ Replace y with 4.

$20 - z = 25$ Multiply.

$z = -5$ Simplify.

The result is $y = 4$ and $z = -5$.

Step 3 Substitute 4 for y and -5 for z in one of the original equations with three variables.

$x + 2y + z = 10$ Original equation with three variables

$x + 2(4) + (-5) = 10$ Replace y with 4 and z with -5.

$x + 8 - 5 = 10$ Multiply.

$x = 7$ Simplify.

The solution is $(7, 4, -5)$. You can check this solution in the other two original equations.

Example 2 Infinite Solutions

Solve the system of equations.

$4x - 6y + 4z = 12$
$6x - 9y + 6z = 18$
$5x - 8y + 10z = 20$

Eliminate x in the first two equations.

$$
\begin{array}{lll}
4x - 6y + 4z = 12 & \text{Multiply by 3.} & 12x - 18y + 12z = 36 \\
6x - 9y + 6z = 18 & \text{Multiply by } -2. & (+)\ -12x + 18y - 12z = -36 \quad \text{Add the equations.} \\
 & & 0 = 0
\end{array}
$$

The equation $0 = 0$ is always true. This indicates that the first two equations represent the same plane. Check to see if this plane intersects the third plane.

$$
\begin{array}{lll}
4x - 6y + 4z = 12 & \text{Multiply by 5.} & 20x - 30y + 20z = 60 \\
5x - 8y + 10z = 20 & \text{Multiply by } -2. & (+)\ -10x + 16y - 20z = -40 \quad \text{Add the equations.} \\
 & & 10x - 14y = 20 \\
 & & 5x - 7y = 10 \quad \text{Divide by the GCF, 2.}
\end{array}
$$

The planes intersect in the line. So, there are an infinite number of solutions.

Example 3 No Solution

Solve the system of equations.

$6a + 12b - 8c = 24$
$9a + 18b - 12c = 30$
$4a + 8b - 7c = 26$

Eliminate a in the first two equations.

$$
\begin{array}{lll}
6a + 12b - 8c = 24 & \text{Multiply by 3.} & 18a + 36b - 24c = 72 \\
9a + 18b - 12c = 30 & \text{Multiply by 2.} & (-)\ 18a + 36b - 24c = 60 \quad \text{Subtract the equations.} \\
 & & 0 = 12
\end{array}
$$

The equation $0 = 12$ is never true. So, there is no solution of this system.

More About. . .

Investments

A certificate of deposit (CD) is a way to invest your money with a bank. The bank generally pays higher interest rates on CDs than savings accounts. However, you must invest your money for a specific time period, and there are penalties for early withdrawal.

REAL-WORLD PROBLEMS When solving problems involving three variables, use the four-step plan to help organize the information.

Example 4 Write and Solve a System of Equations

INVESTMENTS **Andrew Chang has $15,000 that he wants to invest in certificates of deposit (CDs). For tax purposes, he wants his total interest per year to be $800. He wants to put $1000 more in a 2-year CD than in a 1-year CD and invest the rest in a 3-year CD. How much should Mr. Chang invest in each type of CD?**

Number of Years	1	2	3
Rate	3.4%	5.0%	6.0%

Explore Read the problem and define the variables.

a = the amount of money invested in a 1-year certificate
b = the amount of money in a 2-year certificate
c = the amount of money in a 3-year certificate

Plan Mr. Chang has $15,000 to invest.

$a + b + c = 15{,}000$

The interest he earns should be $800. The interest equals the rate times the amount invested.

$0.034a + 0.05b + 0.06c = 800$

There is $1000 more in the 2-year certificate than in the 1-year certificate.

$b = a + 1000$

Solve Substitute $b = a + 1000$ in each of the first two equations.

$a + (a + 1000) + c = 15{,}000$	Replace b with $(a + 1000)$.
$2a + 1000 + c = 15{,}000$	Simplify.
$2a + c = 14{,}000$	Subtract 1000 from each side.

$0.034a + 0.05(a + 1000) + 0.06c = 800$	Replace b with $(a + 1000)$.
$0.034a + 0.05a + 50 + 0.06c = 800$	Distributive Property
$0.084a + 0.06c = 750$	Simplify.

Now solve the system of two equations in two variables.

$2a + c = 14{,}000$ Multiply by 0.06. $0.12a + 0.06c = 840$

$0.084a + 0.06c = 750$ $(-)\ 0.084a + 0.06c = 750$

$$0.036a = 90$$

$$a = 2500$$

Substitute 2500 for a in one of the original equations.

$b = a + 1000$	Third equation
$= 2500 + 1000$	$a = 2500$
$= 3500$	Add.

Substitute 2500 for a and 3500 for b in one of the original equations.

$a + b + c = 15{,}000$	First equation
$2500 + 3500 + c = 15{,}000$	$a = 2500$, $b = 3500$
$6000 + c = 15{,}000$	Add.
$c = 9000$	Subtract 6000 from each side.

So, Mr. Chang should invest $2500 in a 1-year certificate, $3500 in a 2-year certificate, and $9000 in a 3-year certificate.

Examine Check to see if all the criteria are met.

The total investment is $15,000.

$2500 + 3500 + 9000 = 15{,}000$ ✓

The interest earned will be $800.

$0.034(2500) + 0.05(3500) + 0.06(9000) = 800$

$85 + 175 + 540 = 800$ ✓

There is $1000 more in the 2-year certificate than the 1-year certificate.
$3500 = 2500 + 1000$ ✓

Check for Understanding

Concept Check

1. **Explain** how you can use the methods of solving a system of two equations in two variables to solve a system of three equations in three variables.

2. **FIND THE ERROR** Melissa is solving the system of equations $r + 2s + t = 3$, $2r + 4s + 2t = 6$, and $3r + 6s + 3t = 12$.

$$r + 2s + t = 3 \rightarrow \quad 2r + 4s + 2t = 6$$
$$2r + 4s + 2t = 6 \rightarrow (-)2r + 4s + 2t = 6$$
$$0 = 0$$

The second equation is a multiple of the first, so they are the same plane. There are infinitely many solutions.

Is she correct? Explain your reasoning.

3. **OPEN ENDED** Give an example of a system of three equations in three variables that has $(-3, 5, 2)$ as a solution. Show that the ordered triple satisfies all three equations.

Guided Practice

Solve each system of equations.

4. $x + 2y = 12$
   $3y - 4z = 25$
   $x + 6y + z = 20$

5. $9a + 7b = -30$
   $8b + 5c = 11$
   $-3a + 10c = 73$

6. $r - 3s + t = 4$
   $3r - 6s + 9t = 5$
   $4r - 9s + 10t = 9$

7. $2r + 3s - 4t = 20$
   $4r - s + 5t = 13$
   $3r + 2s + 4t = 15$

8. $2x - y + z = 1$
   $x + 2y - 4z = 3$
   $4x + 3y - 7z = -8$

9. $x + y + z = 12$
   $6x - 2y - z = 16$
   $3x + 4y + 2z = 28$

Application

COOKING **For Exercises 10 and 11, use the following information.**
Jambalaya is a Cajun dish made from chicken, sausage, and rice. Simone is making a large pot of jambalaya for a party. Chicken costs \$6 per pound, sausage costs \$3 per pound, and rice costs \$1 per pound. She spends \$42 on $13\frac{1}{2}$ pounds of food. She buys twice as much rice as sausage.

10. Write a system of three equations that represents how much food Simone purchased.

11. How much chicken, sausage, and rice will she use in her dish?

Practice and Apply

Homework Help

For Exercises	See Examples
12–23	1–3
24–30	4

Extra Practice
See page 833.

Solve each system of equations.

12. $2x - y = 2$
    $3z = 21$
    $4x + z = 19$

13. $-4a = 8$
    $5a + 2c = 0$
    $7b + 3c = 22$

14. $5x + 2y = 4$
    $3x + 4y + 2z = 6$
    $7x + 3y + 4z = 29$

15. $8x - 6z = 38$
    $2x - 5y + 3z = 5$
    $x + 10y - 4z = 8$

16. $4a + 2b - 6c = 2$
    $6a + 3b - 9c = 3$
    $8a + 4b - 12c = 6$

17. $2r + s + t = 14$
    $-r - 3s + 2t = -2$
    $4r - 6s + 3t = -5$

18. $3x + y + z = 4$
$2x + 2y + 3z = 3$
$x + 3y + 2z = 5$

19. $4a - 2b + 8c = 30$
$a + 2b - 7c = -12$
$2a - b + 4c = 15$

20. $2r + s + t = 7$
$r + 2s + t = 8$
$r + s + 2t = 11$

21. $6x + 2y + 4z = 2$
$3x + 4y - 8z = -3$
$-3x - 6y + 12z = 5$

22. $r + s + t = 5$
$2r - 7s - 3t = 13$
$\frac{1}{2}r - \frac{1}{3}s + \frac{2}{3}t = -1$

23. $2a - b + 3c = -7$
$4a + 5b + c = 29$
$a - \frac{2b}{3} + \frac{c}{4} = -10$

24. The sum of three numbers is 20. The second number is 4 times the first, and the sum of the first and third is 8. Find the numbers.

25. The sum of three numbers is 12. The first number is twice the sum of the second and third. The third number is 5 less than the first. Find the numbers.

26. **TRAVEL** Jonathan and members of his Spanish Club are going to Costa Rica over spring break. Before his trip, he purchases 10 travelers checks in denominations of \$20, \$50, and \$100, totaling \$370. He has twice as many \$20 checks as \$50 checks. How many of each type of denomination of travelers checks does he have?

DINING For Exercises 27 and 28, use the following information.
Maka loves the lunch combinations at Rosita's Mexican Restaurant. Today however, she wants a different combination than the ones listed on the menu.

27. Assume that the price of a combo meal is the same price as purchasing each item separately. Find the price for an enchilada, a taco, and a burrito.

28. If Maka wants 2 burritos and 1 enchilada, how much should she plan to spend?

More About. . .

Basketball

In 2001, Katie Smith was ranked 1st in the WNBA for points per game, three-point field goals, and minutes per game. She was also ranked 5th for free-throw percentage.

Source: www.wnba.com

BASKETBALL For Exercises 29 and 30, use the following information.
In the 2000–2001 season, Minnesota's Katie Smith was ranked first in the WNBA for total points and three-point goals made. She scored 646 points making 355 shots, including 3-point field goals, 2-point field goals, and 1-point free throws. She made 27 more 2-point field goals than 3-point field goals.

29. Write a system of three equations that represents the number of goals Katie Smith made.

30. Find the number of each type of goal she made.

Online Research **Data Update** What are the current rankings for the WNBA? Visit www.algebra2.com/data_update to learn more.

31. **CRITICAL THINKING** The general form of an equation for a parabola is $y = ax^2 + bx + c$, where (x, y) is a point on the parabola. Determine the values of a, b, c for the parabola at the right. Write the general form of the equation.

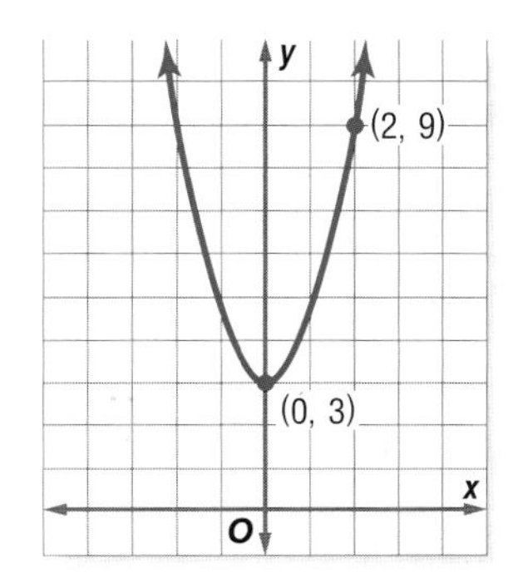

32. **WRITING IN MATH** Answer the question that was posed at the beginning of the lesson.

How can you determine the number and type of medals U.S. Olympians won?

Include the following in your answer:

- a demonstration of how to find the number of each type of medal won by the U.S. Olympians, and
- a description of another situation where you can use a system of three equations in three variables to solve a problem.

Standards Practice

Standardized Test Practice

33. If $a + b = 16$, $a - c = 4$, and $b - c = -4$, which statements are true?

I. $b + c = 12$

II. $a - b = 8$

III. $a + c = 20$

(A) I only (B) II only

(C) I and II only (D) I, II, and III

34. If $x + y = 1$, $y + z = 10$, and $x + z = 3$, what is $x + y + z$?

(A) 7 (B) 8 (C) 13 (D) 14

Maintain Your Skills

Mixed Review

35. **PAPER** Wood pulp can be converted to either notebook paper or newsprint. The Canyon Pulp and Paper Mill can produce at most 200 units of paper a day. Regular customers require at least 10 units of notebook paper and 80 units of newspaper daily. If the profit on a unit of notebook paper is \$500 and the profit on a unit of newsprint is \$350, how many units of each type of paper should the mill produce each day to maximize profits? *(Lesson 3-4)*

Solve each system of inequalities by graphing. *(Lesson 3-3)*

36. $y \leq x + 2$
$y \geq 7 - 2x$

37. $4y - 2x > 4$
$3x + y > 3$

38. $3x + y \geq 1$
$2y - x \leq -4$

STAMPS For Exercises 39 and 40, use the following information.
The table shows the price for first-class stamps since the U.S. Postal Service was created on July 1, 1971. *(Lesson 2-5)*

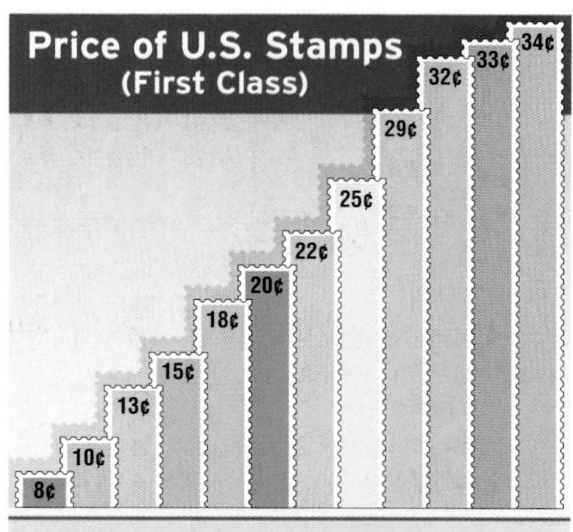

39. Write a prediction equation for this relationship.

40. Predict the price for a first-class stamp issued in the year 2010.

Simplify each expression. *(Lesson 1-2)*

41. $5x + 2y - 4x + y$

42. $(4z + 1) - (6z - 7)$

43. $(8s - 5t) + (9t + s)$

44. $4(6a + 5b) - 2(3a + 2b)$

Chapter 3

Study Guide and Review

Vocabulary and Concept Check

bounded region (p. 129)
consistent system (p. 111)
constraints (p. 129)
dependent system (p. 111)
elimination method (p. 118)
feasible region (p. 129)
inconsistent system (p. 111)
independent system (p. 111)
linear programming (p. 130)
ordered triple (p. 139)
substitution method (p. 116)
system of equations (p. 110)
system of inequalities (p. 123)
unbounded region (p. 130)
vertices (p. 129)

Choose the letter of the term that best matches each phrase.

1. the inequalities of a linear programming problem
2. a system of equations that has an infinite number of solutions
3. the region of a graph where every constraint is met
4. a method of solving equations in which one equation is solved for one variable in terms of the other variable
5. a system of equations that has at least one solution
6. a method of solving equations in which one variable is eliminated when the two equations are combined
7. the solution of a system of equations in three variables (x, y, z)
8. a method for finding the maximum or minimum value of a function
9. a system of equations that has no solution
10. a region in which no maximum value exists

a. consistent system
b. dependent system
c. constraints
d. inconsistent system
e. elimination method
f. feasible region
g. linear programming
h. ordered triple
i. substitution method
j. unbounded region

Lesson-by-Lesson Review

3-1 Solving Systems of Equations by Graphing

See pages 110–115.

Concept Summary

- The solution of a system of equations can be found by graphing the two lines and determining if they intersect and at what point they intersect.

Solve the system of equations by graphing.

$x + y = 3$
$3x - y = 1$

Graph both equations on the same coordinate plane.

The solution of the system is (1, 2).

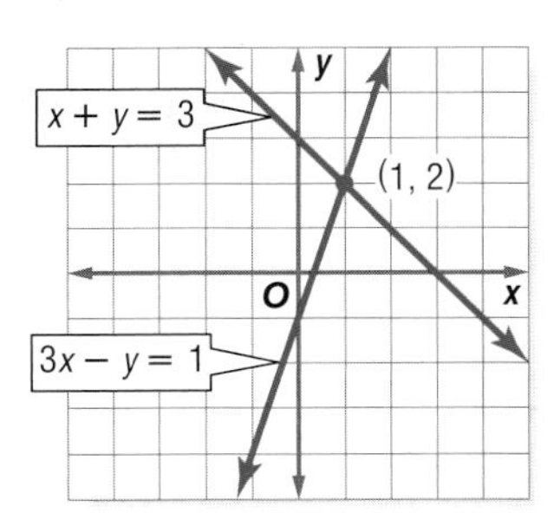

Exercises **Solve each system of equations by graphing.** *See Example 1 on page 110.*

11. $3x + 2y = 12$
 $x - 2y = 4$

12. $8x - 10y = 7$
 $4x - 5y = 7$

13. $y - 2x = 8$
 $y = \frac{1}{2}x - 4$

14. $20y + 13x = 10$
 $0.65x + y = 0.5$

3-2 Solving Systems of Equations Algebraically

See pages 116–122.

Concept Summary

- In the substitution method, one equation is solved for a variable and substituted to find the value of another variable.
- In the elimination method, one variable is eliminated by adding or subtracting the equations.

Examples

1 Use substitution to solve the system of equations.

$x = 4y + 7$
$y = -3 - x$

Substitute $-3 - x$ for y in the first equation.

$x = 4y + 7$	First equation
$x = 4(-3 - x) + 7$	Substitute $-3 - x$ for y.
$x = -12 - 4x + 7$	Distributive Property
$5x = -5$	Add $4x$ to each side.
$x = -1$	Divide each side by 5.

Now substitute the value for x in either original equation.

$y = -3 - x$	Second equation
$y = -3 - (-1)$ or -2	The solution is $(-1, -2)$.

2 Use the elimination method to solve the system of equations.

$3x - 2y = 8$
$-x + y = 9$

Multiply the second equation by 2. Then add the equations to eliminate the y variable.

$3x - 2y = 8$
$-x + y = 9$
Multiply by 2.

$$\begin{aligned} 3x - 2y &= 8 \\ (+)\ -2x + 2y &= 18 \\ x \quad &= 26 \end{aligned}$$
Add the equations.

Replace x with 26 and solve for y.

$3x - 2y = 8$	Original equation.
$3(26) - 2y = 8$	Replace x with 26.
$78 - 2y = 8$	Multiply.
$-2y = -70$	Subtract 78 from each side.
$y = 35$	The solution is (26, 35).

Exercises **Solve each system of equations by using either substitution or elimination.** *See Examples 1–4 on pages 116–119.*

15. $x + y = 5$; $2x - y = 4$

16. $2x - 3y = 9$; $4x + 2y = -22$

17. $7y - 2x = 10$; $-3y + x = -3$

18. $-2x - 6y = 0$; $3x + 11y = 4$

19. $3x - 5y = -13$; $4x + 2y = 0$

20. $x + y = 4$; $x - y = 8.5$

3-3 Solving Systems of Inequalities by Graphing

See pages 123–127.

Concept Summary

- A solution of a system of inequalities is found by graphing the inequalities and determining the intersection of the graphs.

Example **Solve the system of inequalities by graphing.**

$y \leq x + 2$

$y \geq -4 - \frac{1}{2}x$

Graph each inequality and shade the intersection.

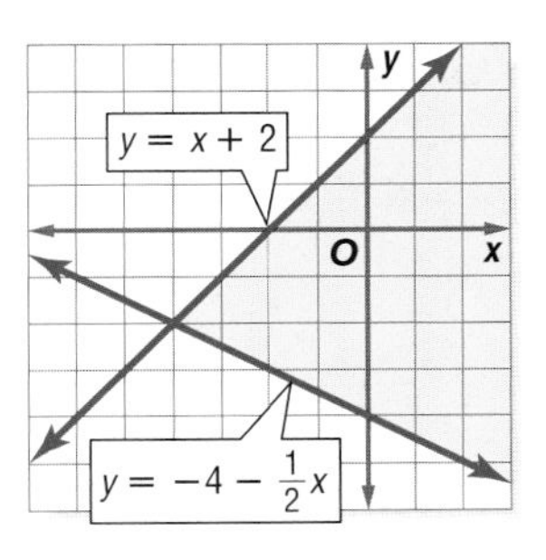

Exercises **Solve each system of inequalities by graphing.**
See Examples 1–3 on pages 123–124.

21. $y \leq 4$, $y > -3$

22. $|y| > 3$, $x \leq 1$

23. $y < x + 1$, $x > 5$

24. $y \leq x + 4$, $2y \geq x - 3$

3-4 Linear Programming

See pages 129–135.

Concept Summary

- The maximum and minimum values of a function are determined by linear programming techniques.

Example **The available parking area of a parking lot is 600 square meters. A car requires 6 square meters of space, and a bus requires 30 square meters of space. The attendant can handle no more than 60 vehicles. If a car is charged \$3 to park and a bus is charged \$8, how many of each should the attendant accept to maximize income?**

Let c = the number of cars and b = the number of buses.

$c \geq 0$, $b \geq 0$, $6c + 30b \leq 600$, and $c + b \leq 60$

Graph the inequalities. The vertices of the feasible region are (0, 0), (0, 20), (50, 10), and (60, 0).

The profit function is $f(c, b) = 3c + 8b$. The maximum value of \$230 occurs at (50, 10). So the attendant should accept 50 cars and 10 buses.

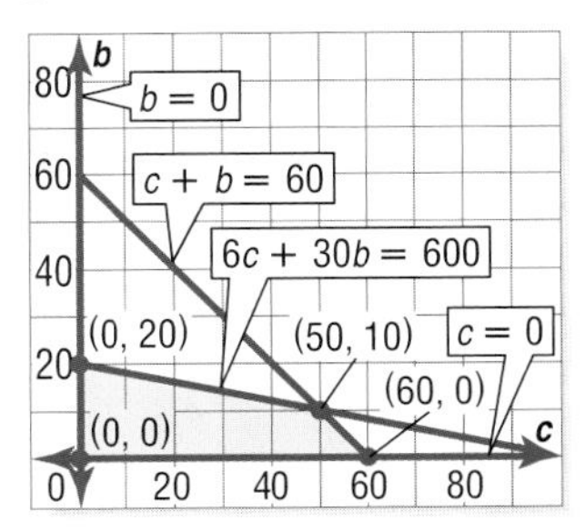

Exercise *See Example 3 on page 131.*

25. MANUFACTURING A toy manufacturer is introducing two new dolls, My First Baby and My Real Baby. In one hour, the company can produce 8 First Babies or 20 Real Babies. Because of demand, the company produces at least twice as many First Babies as Real Babies. The company spends no more than 48 hours per week making these two dolls. The profit on each First Baby is \$3.00, and the profit on each Real Baby is \$7.50. Find the number and type of dolls that should be produced to maximize profit.

Chapter 3 **For More ...**

- Extra Practice, see pages 832–833.
- Mixed Problem Solving, see page 864.

3-5 Solving Systems of Equations in Three Variables

See pages 138–144.

Summary

- A system of three equations in three variables can be solved algebraically by using the substitution method or the elimination method.

Example **Solve the system of equations.**

$$x + 3y + 2z = 1$$
$$2x + y - z = 2$$
$$x + y + z = 2$$

Step 1 Use elimination to make a system of two equations in two variables.

$x + 3y + 2z = 1$, $2x + y - z = 2$ — Multiply by 2.

$$\begin{aligned} 2x + 6y + 4z &= 2 && \text{First equation} \\ (-)\ 2x + y - z &= 2 && \text{Second equation} \\ \hline 5y + 5z &= 0 && \text{Subtract.} \end{aligned}$$

$$\begin{aligned} x + 3y + 2z &= 1 && \text{First equation} \\ (-)\ x + y + z &= 2 && \text{Third equation} \\ \hline 2y + z &= -1 && \text{Subtract to eliminate } x. \end{aligned}$$

Step 2 Solve the system of two equations.

$5y + 5z = 0$, $2y + z = -1$ — Multiply by 5.

$$\begin{aligned} 5y + 5z &= 0 \\ (-)\ 10y + 5z &= -5 \\ \hline -5y &= 5 && \text{Subtract to eliminate } z. \\ y &= -1 && \text{Divide by } -5. \end{aligned}$$

Substitute -1 for y in one of the equations with two variables and solve for z.

$$\begin{aligned} 5y + 5z &= 0 && \text{Equation with two variables} \\ 5(-1) + 5z &= 0 && \text{Replace } y \text{ with } -1. \\ 5z &= 5 && \text{Add 5 to each side.} \\ z &= 1 && \text{Divide each side by 5.} \end{aligned}$$

Step 3 Substitute -1 for y and 1 for z in one of the equations with three variables.

$$\begin{aligned} 2x + y - z &= 2 && \text{Original equation with three variables} \\ 2x + (-1) - 1 &= 2 && \text{Replace } y \text{ with } -1 \text{ and } z \text{ with } 1. \\ 2x &= 4 && \text{Add 2 to each side.} \\ x &= 2 && \text{Divide each side by 2.} \end{aligned}$$

The solution is $(2, -1, 1)$.

Exercises **Solve each system of equations.** *See Examples 2–4 on pages 140–141.*

26. $x + 4y - z = 6$
$3x + 2y + 3z = 16$
$2x - y + z = 3$

27. $2a + b - c = 5$
$a - b + 3c = 9$
$3a - 6c = 6$

28. $e + f = 4$
$2d + 4e - f = -3$
$3e = -3$

Chapter 3 Practice Test

Vocabulary and Concepts

Choose the word or term that best completes each statement or phrase.

1. Finding the maximum and minimum value of a linear function subject to constraints is called *(linear, polygonal)* programming.
2. The process of adding or subtracting equations to remove a variable and simplify solving the system of equations is called *(substitution, elimination)*.
3. If a system of three equations in three variables has one solution, the graphs of the equations intersect in a *(point, plane)*.

Skills and Applications

Solve each system of equations by graphing, substitution, or elimination.

4. $-4x + y = -5$
   $2x + y = 7$
5. $x + y = -8$
   $-3x + 2y = 9$
6. $3x + 2y = 18$
   $y = 6x - 6$
7. $-6x + 3y = 33$
   $-4x + y = 16$
8. $-7x + 6y = 42$
   $3x + 4y = 28$
9. $2y = 5x - 1$
   $x + y = -1$

Solve each system of inequalities by graphing.

10. $y \geq x - 3$
    $y \geq -x + 1$
11. $x + 2y \geq 7$
    $3x - 4y < 12$
12. $3x + y < -5$
    $2x - 4y \geq 6$

Graph each system of inequalities. Name the coordinates of the vertices of the feasible region. Find the maximum and the minimum values of the given function.

13. $5 \geq y \geq -3$
    $4x + y \leq 5$
    $-2x + y \leq 5$
    $f(x, y) = 4x - 3y$
14. $x \geq -10$
    $1 \geq y \geq -6$
    $3x + 4y \leq -8$
    $2y \geq x - 10$
    $f(x, y) = 2x + y$

MANUFACTURING **For Exercises 15 and 16, use the following information.**
A sporting goods manufacturer makes a \$5 profit on soccer balls and a \$4 profit on volleyballs. Cutting requires 2 hours to make 75 soccer balls and 3 hours to make 60 volleyballs. Sewing needs 3 hours to make 75 soccer balls and 2 hours to make 60 volleyballs. Cutting has 500 hours available, and Sewing has 450 hours available.

15. How many soccer balls and volleyballs should be made to maximize the profit?
16. What is the maximum profit the company can make from these two products?

Solve each system of equations.

17. $x + y + z = -1$
    $2x + 4y + z = 1$
    $x + 2y - 3z = -3$
18. $x + z = 7$
    $2y - z = -3$
    $-x - 3y + 2z = 11$
19. SHOPPING Carla bought 3 shirts, 4 pairs of pants, and 2 pairs of shoes for a total of \$149.79. Beth bought 5 shirts, 3 pairs of pants, and 3 pairs of shoes totaling \$183.19. Kayla bought 6 shirts, 5 pairs of pants, and a pair of shoes for \$181.14. Assume that all of the shirts were the same price, all of the pants were the same price, and all of the shoes were the same price. What was the price of each item?
20. STANDARDIZED TEST PRACTICE Find the point at which the graphs of $2x + 3y = 7$ and $3x - 4y = 2$ intersect.

www.algebra2.com/chapter_test/ca

Chapter 3 Standardized Test Practice

Standards Practice

Part 1 Multiple Choice

Record your answers on the answer sheet provided by your teacher or on a sheet of paper.

1. What is the slope of any line parallel to the graph of $6x + 5y = 9$?

(A) -6 (B) $-\frac{6}{5}$

(C) $\frac{2}{3}$ (D) 6

2. In the figure, $\triangle MOQ$ is similar to $\triangle NOP$. What is the length of $\overline{MQ}$?

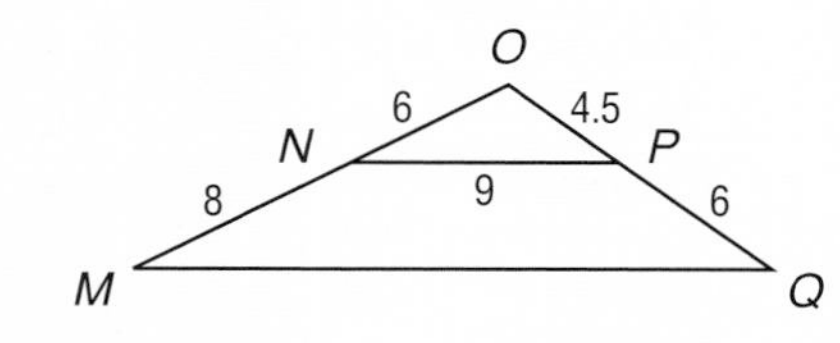

(A) 12 (B) 12.5

(C) 19 (D) 21

3. If $3x - y = -3$ and $x + 5y = 15$, what is the value of y?

(A) -3 (B) 0

(C) 1 (D) 3

4. When 3 times x is increased by 4, the result is less than 16. Which of the following is a graph of the real numbers x that satisfy this relationship?

(A) -4

(B) 4

(C) 4

(D) 4

5. What is the area of the square $ABCD$?

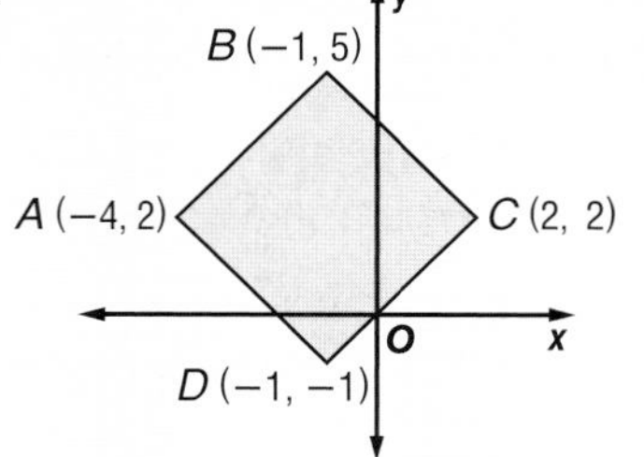

(A) 27 units^2

(B) $9\sqrt{2}$ units^2

(C) 18 units^2

(D) $12\sqrt{2}$ units^2

6. Twenty-seven white cubes of the same size are put together to form a larger cube. The larger cube is painted red. How many of the smaller cubes have exactly one red face ?

(A) 4 (B) 6

(C) 9 (D) 12

7. Find the value of $|-4| \cdot |3|$.

(A) -12 (B) -1

(C) 7 (D) 12

8. If two sides of a triangle measure 30 and 60, which of the following *cannot* be the measure of the third side?

(A) 30 (B) 31

(C) 40 (D) 60

9. Marcus tried to compute the average of his 8 test scores. He mistakenly divided the correct total S of his scores by 7. The result was 12 more than what it should have been. Which equation would determine the value of S?

(A) $8S - 12 = 7S$ (B) $\frac{S}{7} = \frac{S + 12}{18}$

(C) $\frac{S}{7} + 12 = \frac{S}{8}$ (D) $\frac{S}{7} - 12 = \frac{S}{8}$

10. If $x = -2$, then $15 - 3(x + 1) =$

(A) 6. (B) 12.

(C) 18. (D) 21.

Preparing for Standardized Tests
For test-taking strategies and more practice, see pages 877–892.

Part 2 Short Response/Grid In

Record your answers on the answer sheet provided by your teacher or on a sheet of paper.

11. Six of the 13 members of a club are boys, and the rest are girls. What is the ratio of girls to boys in the club?

12. The integer k is greater than 50 and less than 100. When k is divided by 3, the remainder is 1. When k is divided by 8, the remainder is 2. What is one possible value of k?

13. The area of the base of the rectangular box shown below is 35 square units. The area of one of the faces is 56 square units. Each of the dimensions a, b, and c is an integer greater than 1. What is the volume of the rectangular box?

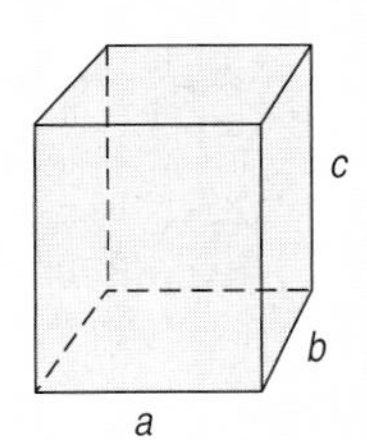

14. Four lines on a plane intersect in one point, forming 8 equal angles that are nonoverlapping. What is the measure, in degrees, of one of these angles?

15. What is the greatest of five consecutive integers if the sum of these integers equals 135?

16. If the perimeter of a rectangle is 12 times the width of the rectangle, then the length of the rectangle is how many times the width?

17. Points A, B, C, and D lie in consecutive order on a line. If $AC = \frac{4}{3}AB$ and $BD = 6BC$, then what is $\frac{AB}{CD}$?

18. The average (arithmetic mean) of the test scores of Mrs. Hilgart's first period class of x Algebra 2 students is 74. The average of the test scores of Mrs. Hilgart's fifth period class of y Algebra 2 students is 88. When the scores of both classes are combined, the average is 76. What is the value of $\frac{x}{y}$?

19. A box contains 19 cups of rice. At most, how many servings can Alicia make from this box of rice if each serving is at least $\frac{3}{4}$ cup?

Test-Taking Tip (A) (B) (C) (D)

Questions 8, 12, 16, and 17

If the question involves a geometric object but does not include a figure, draw one. A diagram can help you see relationships among the given values that will help you answer the question.

Part 3 Extended Response

Record your answers on a sheet of paper. Show your work.

For Exercises 20–22, use the information below.

Christine had one dress and three sweaters cleaned at the dry cleaner and the charge was \$19.50. The next week, she had two dresses and two sweaters cleaned for a total charge of \$23.00.

20. Let d represent the price of cleaning a dress and s represent the price of cleaning a sweater. Write a system of linear equations to represent the prices of cleaning each item.

21. Solve the system of equations using substitution or elimination. Explain your choice of method.

22. What will the charge be if Christine takes two dresses and four sweaters to be cleaned?

Chapter

4 Matrices

What You'll Learn

- **Lesson 4-1** Organize data in matrices.
- **Lessons 4-2, 4-3, and 4-5** Perform operations with matrices and determinants.
- **Lesson 4-4** Transform figures on a coordinate plane.
- **Lessons 4-6 and 4-8** Use matrices to solve systems of equations.
- **Lesson 4-7** Find the inverse of a matrix.

Key Vocabulary

- matrix (p. 154)
- determinant (p.182)
- expansion by minors (p. 183)
- Cramer's Rule (p. 189)
- matrix equation (p. 202)

Why It's Important

Data are often organized into matrices. For example, the National Federation of State High School Associations uses matrices to record student participation in sports by category for males and females. To find the total participation of both groups in each sport, you can add the two matrices. *You will learn how to add matrices in Lesson 4-2.*

Getting Started

Prerequisite Skills To be successful in this chapter, you'll need to master these skills and be able to apply them in problem-solving situations. Review these skills before beginning Chapter 4.

For Lesson 4-1 — Solve Equations

Solve each equation. *(For review, see Lesson 1-3.)*

1. $3x = 18$ **2.** $2a - 3 = -11$ **3.** $4t - 5 = 14$

4. $\frac{1}{3}y + 5 = 9$ **5.** $3k + 5 = 2k - 8$ **6.** $5m - 6 = 7m - 8$

For Lessons 4-2 and 4-7 — Additive and Multiplicative Inverses

Name the additive inverse and the multiplicative inverse for each number.
(For review, see Lesson 1-2.)

7. 3 **8.** -11 **9.** 8 **10.** -0.5

11. 1.25 **12.** $\frac{5}{9}$ **13.** $-\frac{8}{3}$ **14.** $-1\frac{1}{5}$

For Lesson 4-4 — Graph Ordered Pairs

Graph each set of ordered pairs on a coordinate plane. *(For review, see Lesson 2-1.)*

15. $\{(0, 0), (1, 3), (-2, 4)\}$ **16.** $\{(-1, 5), (2, -3), (4, 0)\}$

17. $\{(-3, -3), (-1, 2), (1, -3), (3, -6)\}$ **18.** $\{(-2, 5), (1, 3), (4, -2), (4, 7)\}$

For Lessons 4-6 and 4-8 — Solve Systems of Equations

Solve each system of equations by using either substitution or elimination.
(For review, see Lesson 3-2.)

19. $x = y + 5$; $3x + y = 19$ **20.** $3x - 2y = 1$; $4x + 2y = 20$ **21.** $5x + 3y = 25$; $4x + 7y = -3$

22. $y = x - 7$; $2x - 8y = 2$ **23.** $5x - 3y = 16$; $x - 3y = 8$ **24.** $9x + 4y = 17$; $3x - 2y = 29$

FOLDABLES™ Study Organizer

Matrices Make this Foldable to help you organize your notes. Begin with one sheet of notebook paper.

Step 1 Fold and Cut

Fold lengthwise to the holes. Cut eight tabs in the top sheet.

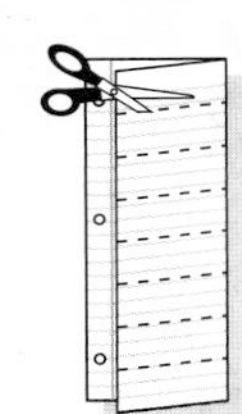

Step 2 Label

Label each tab with a lesson number and title.

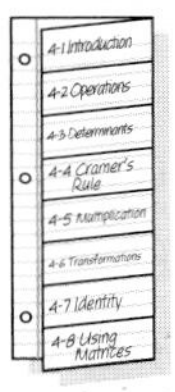

Reading and Writing As you read and study the chapter, write notes and examples under the tabs.

4-1 Introduction to Matrices

California Standards **Standard 2.0 Students solve systems of linear equations** and inequalities **(in two or three variables)** by substitution, with graphs, or **with matrices.** (Key)

What You'll Learn

- Organize data in matrices.
- Solve equations involving matrices.

Vocabulary

- matrix
- element
- dimension
- row matrix
- column matrix
- square matrix
- zero matrix
- equal matrices

How are matrices used to make decisions?

Sabrina wants to buy a sports-utility vehicle (SUV). There are many types of SUVs in many prices and styles. So, Sabrina makes a list of the qualities for different models and organizes the information in a matrix.

	Base Price	Horse-power	Towing Capacity (lb)	Cargo Space (ft³)	Fuel Economy (mpg)
Large SUV	$32,450	285	12,000	46	17
Standard SUV	$29,115	275	8700	16	17.5
Mid-Size SUV	$27,975	190	5700	34	20
Compact SUV	$18,180	127	3000	15	26.5

Source: *Car and Driver Buyer's Guide*

When the information is organized in a matrix, it is easy to compare the features of each vehicle.

Study Tip

Reading Math
The plural of *matrix* is *matrices*. Matrices are sometimes called *ordered arrays.*

ORGANIZE DATA A **matrix** is a rectangular array of variables or constants in horizontal rows and vertical columns, usually enclosed in brackets.

Example 1 Organize Data in a Matrix

Sharon wants to install cable television in her new apartment. There are two cable companies in the area whose prices are listed below. Use a matrix to organize the information. When is each company's service less expensive?

Metro Cable	
Basic Service (26 channels)	$11.95
Standard Service (53 channels)	$30.75
Premium Channels (in addition to Standard Service)	
• One Premium	$10.00
• Two Premiums	$19.00
• Three Premiums	$25.00

Cable City	
Basic Service (26 channels)	$9.95
Standard Service (53 channels)	$31.95
Premium Channels (in addition to Standard Service)	
• One Premium	$8.95
• Two Premiums	$16.95
• Three Premiums	$22.95

Organize the costs into labeled columns and rows.

	Basic	Standard	Standard Plus One Premium	Standard Plus Two Premiums	Standard Plus Three Premiums
Metro Cable	11.95	30.75	40.75	49.75	55.75
Cable City	9.95	31.95	40.90	48.90	54.90

Metro Cable has the best price for standard service and standard plus one premium channel. Cable City has the best price for the other categories.

Study Tip

Element
The elements of a matrix can be represented using double subscript notation. The element a_{ij} is the element in row i column j.

In a matrix, numbers or data are organized so that each position in the matrix has a purpose. Each value in the matrix is called an **element**. A matrix is usually named using an uppercase letter.

$$A = \begin{bmatrix} 2 & 6 & 1 \\ 7 & 1 & 5 \\ 9 & 3 & 0 \\ 12 & 15 & 26 \end{bmatrix} \text{ 4 rows, 3 columns}$$

The element 15 is in row 4, column 2.

A matrix can be described by its **dimensions**. A matrix with m rows and n columns is an $m \times n$ matrix (read "m by n"). Matrix A above is a 4×3 matrix since it has 4 rows and 3 columns.

Example 2 Dimensions of a Matrix

State the dimensions of matrix B if $B = \begin{bmatrix} 1 & -3 \\ -5 & 18 \\ 0 & -2 \end{bmatrix}$.

$$B = \begin{bmatrix} 1 & -3 \\ -5 & 18 \\ 0 & -2 \end{bmatrix} \text{ 3 rows, 2 columns}$$

Since matrix B has 3 rows and 2 columns, the dimensions of matrix B are 3×2.

Certain matrices have special names. A matrix that has only one row is called a **row matrix**, while a matrix that has only one column is called a **column matrix**. A matrix that has the same number of rows and columns is called a **square matrix**. Another special type of matrix is the **zero matrix**, in which every element is 0. The zero matrix can have any dimension.

EQUATIONS INVOLVING MATRICES Two matrices are considered **equal matrices** if they have the same dimensions and if each element of one matrix is equal to the corresponding element of the other matrix.

$\begin{bmatrix} 6 & 3 \\ 0 & 9 \\ 1 & 3 \end{bmatrix} \neq \begin{bmatrix} 6 & 0 & 1 \\ 3 & 9 & 3 \end{bmatrix}$ The matrices have different dimensions. They are not equal.

$\begin{bmatrix} 1 & 2 \\ 8 & 5 \end{bmatrix} \neq \begin{bmatrix} 1 & 8 \\ 2 & 5 \end{bmatrix}$ Corresponding elements are not equal. The matrices are not equal.

$\begin{bmatrix} 5 & 6 & 0 \\ 0 & 7 & 2 \\ 3 & 1 & 4 \end{bmatrix} = \begin{bmatrix} 5 & 6 & 0 \\ 0 & 7 & 2 \\ 3 & 1 & 4 \end{bmatrix}$ The matrices have the same dimensions and the corresponding elements are equal. The matrices are equal.

The definition of equal matrices can be used to find values when elements of equal matrices are algebraic expressions.

Example 3 Solve an Equation Involving Matrices

Solve $\begin{bmatrix} y \\ 3x \end{bmatrix} = \begin{bmatrix} 6 - 2x \\ 31 + 4y \end{bmatrix}$ for x and y.

Since the matrices are equal, the corresponding elements are equal. When you write the sentences to show this equality, two linear equations are formed.

$y = 6 - 2x$

$3x = 31 + 4y$

(continued on the next page)

This system can be solved using substitution.

$3x = 31 + 4y$	Second equation
$3x = 31 + 4(6 - 2x)$	Substitute $6 - 2x$ for y.
$3x = 31 + 24 - 8x$	Distributive Property
$11x = 55$	Add $8x$ to each side.
$x = 5$	Divide each side by 11.

To find the value for y, substitute 5 for x in either equation.

$y = 6 - 2x$	First equation
$y = 6 - 2(5)$	Substitute 5 for x.
$y = -4$	Simplify.

The solution is $(5, -4)$.

Check for Understanding

Concept Check

1. **Describe** the conditions that must be met in order for two matrices to be considered equal.
2. **OPEN ENDED** Give examples of a row matrix, a column matrix, a square matrix, and a zero matrix. State the dimensions of each matrix.
3. **Explain** what is meant by corresponding elements.

Guided Practice

State the dimensions of each matrix.

4. $\begin{bmatrix} 3 & 4 & 5 & 6 & 7 \end{bmatrix}$

5. $\begin{bmatrix} 10 & -6 & 18 & 0 \\ -7 & 5 & 2 & 4 \\ 3 & 11 & 9 & 7 \end{bmatrix}$

Solve each equation.

6. $\begin{bmatrix} x + 4 \\ 2y \end{bmatrix} = \begin{bmatrix} 9 \\ 12 \end{bmatrix}$

7. $\begin{bmatrix} 9 & 13 \end{bmatrix} = \begin{bmatrix} x + 2y & 4x + 1 \end{bmatrix}$

Application

WEATHER For Exercises 8 and 9, use the table that shows a five-day forecast indicating high (H) and low (L) temperatures.

Fri	Sat	Sun	Mon	Tue
H 88	H 88	H 90	H 86	H 85
L 54	L 54	L 56	L 53	L 52

8. Organize the temperatures in a matrix.
9. What are the dimensions of the matrix?

Practice and Apply

Homework Help

For Exercises	See Examples
10–15	2
16–25	3
26–31	1

Extra Practice
See page 834.

State the dimensions of each matrix.

10. $\begin{bmatrix} 6 & -1 & 5 \\ -2 & 3 & -4 \end{bmatrix}$

11. $\begin{bmatrix} 7 \\ 8 \\ 9 \end{bmatrix}$

12. $\begin{bmatrix} 0 & 0 & 8 \\ 6 & 2 & 4 \\ 1 & 3 & 6 \\ 5 & 9 & 2 \end{bmatrix}$

13. $\begin{bmatrix} -3 & 17 & -22 \\ 9 & 31 & 16 \\ 20 & -15 & 4 \end{bmatrix}$

14. $\begin{bmatrix} 17 & -2 & 8 & -9 & 6 \\ 5 & 11 & 20 & -1 & 4 \end{bmatrix}$

15. $\begin{bmatrix} 16 & 8 \\ 10 & 5 \\ 0 & 0 \end{bmatrix}$

Solve each equation.

16. $[2x \quad 3 \quad 3z] = [5 \quad 3y \quad 9]$

17. $[4x \quad 3y] = [12 \quad -1]$

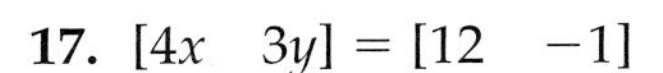

18. $\begin{bmatrix} 4x \\ 5 \end{bmatrix} = \begin{bmatrix} 15 + x \\ 2y - 1 \end{bmatrix}$

19. $\begin{bmatrix} 4x - 3 & 3y \\ 7 & 13 \end{bmatrix} = \begin{bmatrix} 9 & -15 \\ 7 & 2z + 1 \end{bmatrix}$

20. $\begin{bmatrix} x + 3y \\ 3x + y \end{bmatrix} = \begin{bmatrix} -13 \\ 1 \end{bmatrix}$

21. $\begin{bmatrix} 2x + y \\ x - 3y \end{bmatrix} = \begin{bmatrix} 5 \\ 13 \end{bmatrix}$

22. $\begin{bmatrix} 2x \\ 2x + 3y \end{bmatrix} = \begin{bmatrix} y \\ 12 \end{bmatrix}$

23. $\begin{bmatrix} 4x \\ y - 1 \end{bmatrix} = \begin{bmatrix} 11 + 3y \\ x \end{bmatrix}$

24. $\begin{bmatrix} x^2 + 1 & 5 - y \\ x + y & y - 4 \end{bmatrix} = \begin{bmatrix} 5 & x \\ 5 & 3 \end{bmatrix}$

25. $\begin{bmatrix} 3x - 5 & x + y \\ 12 & 9z \end{bmatrix} = \begin{bmatrix} 10 & 8 \\ 12 & 3x + y \end{bmatrix}$

More About. . .

Movies

Adjusting for inflation, *Cleopatra* (1963) is the most expensive movie ever made. Its $44 million budget is equivalent to $306,867,120 today.

Source: *The Guinness Book of Records*

MOVIES **For Exercises 26 and 27, use the advertisement shown at the right.**

26. Write a matrix for the prices of movie tickets for adults, children, and seniors.

27. What are the dimensions of the matrix?

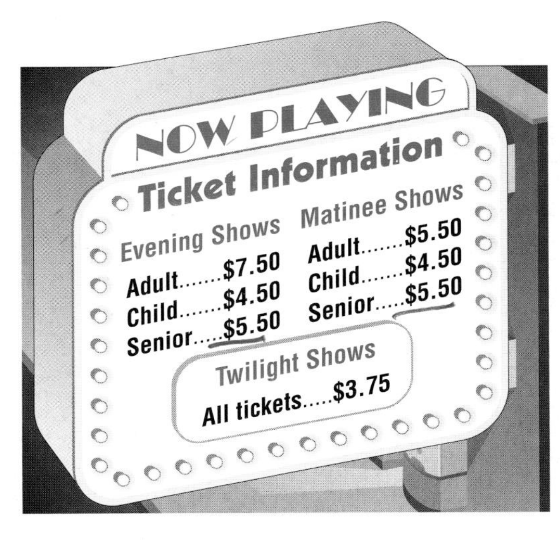

DINING OUT **For Exercises 28 and 29, use the following information.**
A newspaper rated several restaurants by cost, level of service, atmosphere, and location using a scale of ★ being low and ★★★★ being high.

Catalina Grill: cost ★★, service ★, atmosphere ★, location ★

Oyster Club: cost ★★★, service ★★, atmosphere ★, location ★★

Casa di Pasta: cost ★★★★, service ★★★, atmosphere ★★★, location ★★★

Mason's Steakhouse: cost ★★, service ★★★★, atmosphere ★★★★, location ★★★

28. Write a 4 × 4 matrix to organize this information.

29. Which restaurant would you select based on this information and why?

HOTELS **For Exercises 30 and 31, use the costs for an overnight stay at a hotel that are given below.**

Single Room: $60 weekday; $79 weekend

Double Room: $70 weekday; $89 weekend

Suite: $75 weekday; $95 weekend

30. Write a 3 × 2 matrix that represents the cost of each room.

31. Write a 2 × 3 matrix that represents the cost of each room.

CRITICAL THINKING **For Exercises 32 and 33, use the matrix at the right.**

32. Study the pattern of numbers. Complete the matrix for column 6 and row 7.

33. In which row and column will 100 occur?

$$\begin{bmatrix} 1 & 3 & 6 & 10 & 15 & \dots \\ 2 & 5 & 9 & 14 & 20 & \dots \\ 4 & 8 & 13 & 19 & 26 & \dots \\ 7 & 12 & 18 & 25 & 33 & \dots \\ 11 & 17 & 24 & 32 & 41 & \dots \\ 16 & 23 & 31 & 40 & 50 & \dots \\ \vdots & \vdots & \vdots & \vdots & \vdots & \vdots \end{bmatrix}$$

34. **WRITING IN MATH** Answer the question that was posed at the beginning of the lesson.

How are matrices used to make decisions?

Include the following in your answer:

- the circumstances under which each vehicle best fits a person's needs, and
- an example of how matrices are used in newspapers or magazines.

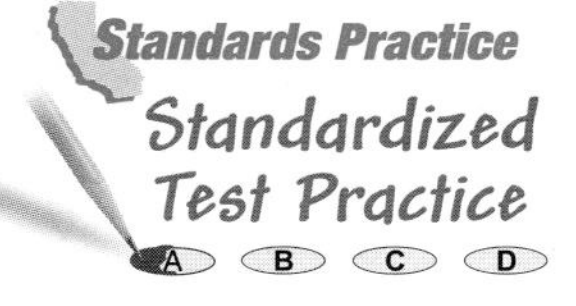

35. In matrix $A = \begin{bmatrix} 1 & 5 & -2 \\ -4 & 0 & 6 \\ 3 & 7 & 8 \end{bmatrix}$, element 3 is in which row and column?

(A) row 1, column 3
(B) row 3, column 1
(C) row 1, column 1
(D) row 3, column 3

36. What is the value of y if $\begin{bmatrix} 3x \\ y+5 \end{bmatrix} = \begin{bmatrix} 9+y \\ x \end{bmatrix}$?

(A) 2
(B) 4
(C) −3
(D) −1

Maintain Your Skills

Mixed Review

Solve each system of equations. *(Lesson 3-5)*

37. $3x - 3y = 6$
$-6y = -30$
$5z - 2x = 6$

38. $3a + 2b = 27$
$5a - 7b + c = 5$
$-2a + 10b + 5c = -29$

39. $3r - 15s + 4t = -57$
$9r + 45s - t = 26$
$-6r + 10s + 3t = -19$

Graph each system of inequalities. Name the coordinates of the vertices of the feasible region. Find the maximum and minimum values of the given function for this region. *(Lesson 3-4)*

40. $y \geq 3$
$y \leq x + 2$
$y \leq -2x + 15$
$f(x, y) = 2x + 3y$

41. $y \geq \frac{1}{3}x$
$y \geq -5x + 16$
$y \leq -x + 10$
$f(x, y) = 5x - y$

42. $y \geq \frac{1}{2}x$
$y \geq -x + 3$
$y \leq -\frac{3}{2}x + 12$
$f(x, y) = 3y - x$

BUSINESS For Exercises 43–45, use the following information.
The parking garage at Burrough's Department Store charges \$1.50 for each hour or fraction of an hour for parking. *(Lesson 2-6)*

43. Graph the function.

44. What type of function represents this situation?

45. Jada went shopping at Burrough's Department Store yesterday. She left her car in the parking garage for two hours and twenty-seven minutes. How much did Jada pay for parking?

Find each value if $f(x) = x^2 - 3x + 2$. *(Lesson 2-1)*

46. $f(3)$
47. $f(0)$
48. $f(2)$
49. $f(-3)$

Getting Ready for the Next Lesson

PREREQUISITE SKILL Find the value of each expression.
*(To review **evaluating expressions**, see Lesson 1-2.)*

50. $8 + (-5)$
51. $-2 - 8$
52. $3.5 + 2.7$
53. $6(-3)$
54. $\frac{1}{2}(34)$
55. $6(4) + 3(-9)$
56. $-5(3 - 18)$
57. $14\left(\frac{1}{4}\right) - 12\left(\frac{1}{6}\right)$

Spreadsheet Investigation

A Follow-Up of Lesson 4-1

Organizing Data

You can use a computer spreadsheet to organize and display data. Then you can use the data to create graphs or perform calculations.

Example

Enter the data on the Atlantic Coast Conference Men's Basketball scoring into a spreadsheet.

Atlantic Coast Conference 2000–2001 Men's Basketball			
Team	Free Throws	2-Point Field Goals	3-Point Field Goals
Clemson	456	549	248
Duke	697	810	407
Florida State	453	594	148
Georgia Tech	457	516	260
Maryland	622	915	205
North Carolina	532	756	189
North Carolina State	507	562	170
Virginia	556	648	204
Wake Forest	443	661	177

Source: Atlantic Coast Conference

Use Column A for the team names, Column B for the numbers of free throws, Column C for the numbers of 2-point field goals, and Column D for the numbers of 3-point field goals.

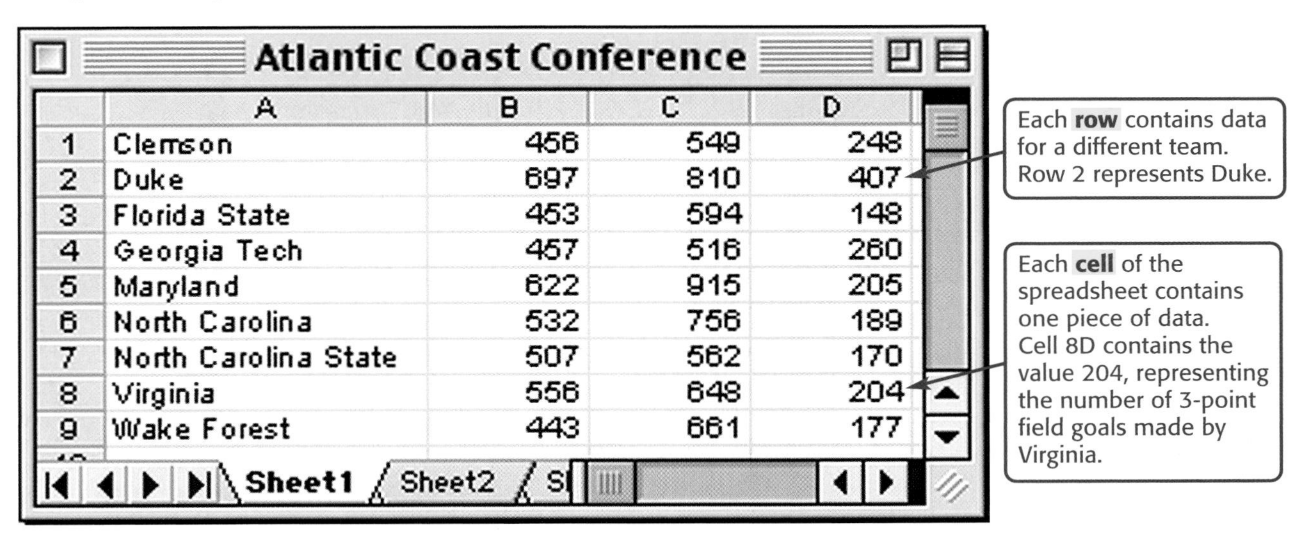

Model and Analyze

1. Enter the data about sports-utility vehicles on page 154 into a spreadsheet.
2. Compare and contrast how data are organized in a spreadsheet and how they are organized in a matrix.

4-2 Operations with Matrices

What You'll Learn

- Add and subtract matrices.
- Multiply by a matrix scalar.

How can matrices be used to calculate daily dietary needs?

Vocabulary
- scalar
- scalar multiplication

In her job as a hospital dietician, Celeste designs weekly menus for her patients and tracks various nutrients for each daily diet. The table shows the Calories, protein, and fat in a patient's meals over a three-day period.

Day	Breakfast			Lunch			Dinner		
	Calories	Protein (g)	Fat (g)	Calories	Protein (g)	Fat (g)	Calories	Protein (g)	Fat (g)
1	566	18	7	785	22	19	1257	40	26
2	482	12	17	622	23	20	987	32	45
3	530	10	11	710	26	12	1380	29	38

These data can be organized in three matrices representing breakfast, lunch, and dinner. The daily totals can then be found by adding the three matrices.

ADD AND SUBTRACT MATRICES Matrices can be added if and only if they have the same dimensions.

Key Concept — Addition of Matrices

- **Words** If A and B are two $m \times n$ matrices, then $A + B$ is an $m \times n$ matrix in which each element is the sum of the corresponding elements of A and B.
- **Symbols** $\begin{bmatrix} a & b & c \\ d & e & f \\ g & h & i \end{bmatrix} + \begin{bmatrix} j & k & l \\ m & n & o \\ p & q & r \end{bmatrix} = \begin{bmatrix} a+j & b+k & c+l \\ d+m & e+n & f+o \\ g+p & h+q & i+r \end{bmatrix}$

Example 1 Add Matrices

a. Find $A + B$ if $A = \begin{bmatrix} 4 & -6 \\ 2 & 3 \end{bmatrix}$ and $B = \begin{bmatrix} -3 & 7 \\ 5 & -9 \end{bmatrix}$.

$A + B = \begin{bmatrix} 4 & -6 \\ 2 & 3 \end{bmatrix} + \begin{bmatrix} -3 & 7 \\ 5 & -9 \end{bmatrix}$ Definition of matrix addition

$= \begin{bmatrix} 4+(-3) & -6+7 \\ 2+5 & 3+(-9) \end{bmatrix}$ Add corresponding elements.

$= \begin{bmatrix} 1 & 1 \\ 7 & -6 \end{bmatrix}$ Simplify.

b. Find $A + B$ if $A = \begin{bmatrix} 3 & -7 & 4 \\ 12 & 5 & 0 \end{bmatrix}$ and $B = \begin{bmatrix} 2 & 9 \\ 4 & -6 \end{bmatrix}$.

Since the dimensions of A are 2×3 and the dimensions of B are 2×2, you cannot add these matrices.

You can subtract matrices in a similar manner.

Key Concept — *Subtraction of Matrices*

- **Words** If A and B are two $m \times n$ matrices, then $A - B$ is an $m \times n$ matrix in which each element is the difference of the corresponding elements of A and B.
- **Symbols** $\begin{bmatrix} a & b & c \\ d & e & f \\ g & h & i \end{bmatrix} - \begin{bmatrix} j & k & l \\ m & n & o \\ p & q & r \end{bmatrix} = \begin{bmatrix} a-j & b-k & c-l \\ d-m & e-n & f-o \\ g-p & h-q & i-r \end{bmatrix}$

Example 2 Subtract Matrices

Find $A - B$ if $A = \begin{bmatrix} 9 & 2 \\ -4 & 7 \end{bmatrix}$ and $B = \begin{bmatrix} 3 & 6 \\ 8 & -2 \end{bmatrix}$.

$A - B = \begin{bmatrix} 9 & 2 \\ -4 & 7 \end{bmatrix} - \begin{bmatrix} 3 & 6 \\ 8 & -2 \end{bmatrix}$ Definition of matrix subtraction

$= \begin{bmatrix} 9-3 & 2-6 \\ -4-8 & 7-(-2) \end{bmatrix}$ Subtract corresponding elements.

$= \begin{bmatrix} 6 & -4 \\ -12 & 9 \end{bmatrix}$ Simplify.

More About. . .

Animals

The rarest animal in the world today is a giant tortoise that lives in the Galapagos Islands. "Lonesome George" is the only remaining representative of his species *(Geochelone elephantopus abingdoni)*. With virtually no hope of discovering another specimen, this species is now effectively extinct.

Source: www.ecoworld.com

Example 3 Use Matrices to Model Real-World Data

ANIMALS **The table below shows the number of endangered and threatened species in the United States and in the world. How many more endangered and threatened species are there on the world list than on the U.S. list?**

Endangered and Threatened Species				
Type of Animal	United States		World	
	Endangered	Threatened	Endangered	Threatened
Mammals	61	8	309	24
Birds	74	15	252	21
Reptiles	14	22	79	36
Amphibians	9	8	17	9
Fish	69	42	80	42

Source: Fish and Wildlife Service, U.S. Department of Interior

The data in the table can be organized in two matrices. Find the difference of the matrix that represents species in the world and the matrix that represents species in the U.S.

World **U.S.**

$\begin{bmatrix} 309 & 24 \\ 252 & 21 \\ 79 & 36 \\ 17 & 9 \\ 80 & 42 \end{bmatrix} - \begin{bmatrix} 61 & 8 \\ 74 & 15 \\ 14 & 22 \\ 9 & 8 \\ 69 & 42 \end{bmatrix} = \begin{bmatrix} 309-61 & 24-8 \\ 252-74 & 21-15 \\ 79-14 & 36-22 \\ 17-9 & 9-8 \\ 80-69 & 42-42 \end{bmatrix}$ Subtract corresponding elements.

(continued on the next page)

$$= \begin{bmatrix} 248 & 16 \\ 178 & 6 \\ 65 & 14 \\ 8 & 1 \\ 11 & 0 \end{bmatrix} \begin{matrix} \text{mammals} \\ \text{birds} \\ \text{reptiles} \\ \text{amphibians} \\ \text{fish} \end{matrix}$$

(columns: Endangered, Threatened)

The first column represents the difference in the number of endangered species on the world and U.S. lists. There are 248 mammals, 178 birds, 65 reptiles, 8 amphibians, and 11 fish species in this category.

The second column represents the difference in the number of threatened species on the world and U.S. lists. There are 16 mammals, 6 birds, 14 reptiles, 1 amphibian, and no fish in this category.

SCALAR MULTIPLICATION You can multiply any matrix by a constant called a **scalar**. This operation is called **scalar multiplication**.

Key Concept — Scalar Multiplication

- **Words** The product of a scalar k and an $m \times n$ matrix is an $m \times n$ matrix in which each element equals k times the corresponding elements of the original matrix.
- **Symbols** $k\begin{bmatrix} a & b & c \\ d & e & f \end{bmatrix} = \begin{bmatrix} ka & kb & kc \\ kd & ke & kf \end{bmatrix}$

Example 4 Multiply a Matrix by a Scalar

If $A = \begin{bmatrix} 2 & 8 & -3 \\ 5 & -9 & 2 \end{bmatrix}$, find $3A$.

$3A = 3\begin{bmatrix} 2 & 8 & -3 \\ 5 & -9 & 2 \end{bmatrix}$ Substitution.

$= \begin{bmatrix} 3(2) & 3(8) & 3(-3) \\ 3(5) & 3(-9) & 3(2) \end{bmatrix}$ Multiply each element by 3.

$= \begin{bmatrix} 6 & 24 & -9 \\ 15 & -27 & 6 \end{bmatrix}$ Simplify.

Many properties of real numbers also hold true for matrices.

Study Tip

Additive Identity
The matrix $\begin{bmatrix} 0 & 0 \\ 0 & 0 \end{bmatrix}$ is called a *zero matrix*. It is the *additive identity matrix* for any 2×2 matrix. How is this similar to the additive identity for real numbers?

Concept Summary — Properties of Matrix Operations

For any matrices A, B, and C with the same dimensions and any scalar c, the following properties are true.

Commutative Property of Addition	$A + B = B + A$
Associative Property of Addition	$(A + B) + C = A + (B + C)$
Distributive Property	$c(A + B) = cA + cB$

Example 5 Combination of Matrix Operations

If $A = \begin{bmatrix} 7 & 3 \\ -4 & -1 \end{bmatrix}$ and $B = \begin{bmatrix} 9 & 6 \\ 3 & 10 \end{bmatrix}$, find $5A - 2B$.

Perform the scalar multiplication first. Then subtract the matrices.

$5A - 2B = 5\begin{bmatrix} 7 & 3 \\ -4 & -1 \end{bmatrix} - 2\begin{bmatrix} 9 & 6 \\ 3 & 10 \end{bmatrix}$ Substitution

$= \begin{bmatrix} 5(7) & 5(3) \\ 5(-4) & 5(-1) \end{bmatrix} - \begin{bmatrix} 2(9) & 2(6) \\ 2(3) & 2(10) \end{bmatrix}$ Multiply each element in the first matrix by 5 and multiply each element in the second matrix by 2.

$= \begin{bmatrix} 35 & 15 \\ -20 & -5 \end{bmatrix} - \begin{bmatrix} 18 & 12 \\ 6 & 20 \end{bmatrix}$ Simplify.

$= \begin{bmatrix} 35 - 18 & 15 - 12 \\ -20 - 6 & -5 - 20 \end{bmatrix}$ Subtract corresponding elements.

$= \begin{bmatrix} 17 & 3 \\ -26 & -25 \end{bmatrix}$ Simplify.

Study Tip

Matrix Operations
The order of operations for matrices is similar to that of real numbers. Perform scalar multiplication before matrix addition and subtraction.

Graphing Calculator Investigation

Matrix Operations

Most graphing calculators can perform operations with matrices. On the TI-83 Plus, [2nd] [MATRX] accesses the matrix menu. Choose EDIT to define a matrix. Enter the dimensions of the matrix A using the [▶] key. Then enter each element by pressing [ENTER] after each entry. To display and use the matrix in calculations, choose the matrix under NAMES from the [MATRX] menu.

Think and Discuss

1. Enter $A = \begin{bmatrix} 3 & -2 \\ 5 & 4 \end{bmatrix}$ with a graphing calculator. Does the calculator enter elements row by row or column by column?
2. Notice that there are two numbers in the bottom left corner of the screen. What do these numbers represent?
3. Clear the screen. Find the matrix $18A$.
4. Enter $B = \begin{bmatrix} 1 & 9 & -3 \\ 8 & 6 & -5 \end{bmatrix}$. Find $A + B$. What is the result and why?

Check for Understanding

Concept Check

1. **Describe** the conditions under which matrices can be added or subtracted.
2. **OPEN ENDED** Give an example of two matrices whose sum is a zero matrix.
3. **Write** a matrix that, when added to a 3×2 matrix, increases each element in the matrix by 4.

Guided Practice

Perform the indicated matrix operations. If the matrix does not exist, write *impossible*.

4. $[5 \quad 8 \quad -4] + [12 \quad 5]$
5. $\begin{bmatrix} 3 & 7 \\ -2 & 1 \end{bmatrix} - \begin{bmatrix} 2 & -3 \\ 5 & -4 \end{bmatrix}$
6. $3\begin{bmatrix} 6 & -1 & 5 & 2 \\ 7 & 3 & -2 & 8 \end{bmatrix}$
7. $4\begin{bmatrix} 2 & 7 \\ -3 & 6 \end{bmatrix} + 5\begin{bmatrix} -6 & -4 \\ 3 & 0 \end{bmatrix}$

Use matrices A, B, and C to find the following.

$A = \begin{bmatrix} 2 & 3 \\ 5 & 6 \end{bmatrix}$ $B = \begin{bmatrix} -1 & 7 \\ 0 & -4 \end{bmatrix}$ $C = \begin{bmatrix} 9 & -4 \\ -6 & 5 \end{bmatrix}$

8. $A + B + C$ **9.** $3B - 2C$ **10.** $4A + 2B - C$

Application

SPORTS For Exercises 11–13, use the table below that shows high school participation in various sports.

Sport	Males		Females	
	Schools	Participants	Schools	Participants
Basketball	16,763	549,499	16,439	456,873
Track and Field	14,620	477,960	14,545	405,163
Baseball/Softball	14,486	455,305	12,679	340,480
Soccer	9041	321,416	7931	257,586
Swimming and Diving	5234	83,411	5450	133,235

Source: National Federation of State High School Associations

11. Write two matrices that represent these data for males and females.

12. Find the total number of students that participate in each individual sport expressed as a matrix.

13. Could you add the two matrices to find the total number of schools that offer a particular sport? Why or why not?

Practice and Apply

Homework Help

For Exercises	See Examples
14–29	1, 2, 4, 5
30–39	3

Extra Practice
See page 834.

Perform the indicated matrix operations. If the matrix does not exist, write *impossible*.

14. $\begin{bmatrix} 4 \\ 1 \\ -3 \end{bmatrix} + \begin{bmatrix} 6 \\ -5 \\ 8 \end{bmatrix}$

15. $\begin{bmatrix} -5 & 7 \\ 6 & 8 \end{bmatrix} - \begin{bmatrix} 4 & 0 & -2 \\ 9 & 0 & 1 \end{bmatrix}$

16. $\begin{bmatrix} 12 & 0 & 8 \\ 9 & 15 & -11 \end{bmatrix} - \begin{bmatrix} -3 & 0 & 4 \\ 9 & 2 & -6 \end{bmatrix}$

17. $-2\begin{bmatrix} 2 & -4 & 1 \\ -3 & 5 & 8 \\ 7 & 6 & -2 \end{bmatrix}$

18. $5[0 \quad -1 \quad 7 \quad 2] + 3[5 \quad -8 \quad 10 \quad -4]$

19. $5\begin{bmatrix} 1 \\ -1 \\ -3 \end{bmatrix} + 6\begin{bmatrix} -4 \\ 3 \\ 5 \end{bmatrix} - 2\begin{bmatrix} -3 \\ 8 \\ -4 \end{bmatrix}$

20. $\begin{bmatrix} 1.35 & 5.80 \\ 1.24 & 14.32 \\ 6.10 & 35.26 \end{bmatrix} + \begin{bmatrix} 0.45 & 3.28 \\ 1.94 & 16.72 \\ 4.31 & 21.30 \end{bmatrix}$

21. $8\begin{bmatrix} 0.25 & 0.5 \\ 0.75 & 1.5 \end{bmatrix} - 2\begin{bmatrix} 0.25 & 0.5 \\ 0.75 & 1.5 \end{bmatrix}$

22. $\frac{1}{2}\begin{bmatrix} 4 & 6 \\ 3 & 0 \end{bmatrix} - \frac{2}{3}\begin{bmatrix} 9 & 27 \\ 0 & 3 \end{bmatrix}$

23. $5\begin{bmatrix} \frac{1}{2} & 0 & 1 \\ 2 & \frac{1}{3} & -1 \end{bmatrix} + 4\begin{bmatrix} -2 & \frac{3}{4} & 1 \\ \frac{1}{6} & 0 & \frac{5}{8} \end{bmatrix}$

Use matrices A, B, C, and D to find the following.

$A = \begin{bmatrix} 5 & 7 \\ -1 & 6 \\ 3 & -9 \end{bmatrix}$ $B = \begin{bmatrix} 8 & 3 \\ 5 & 1 \\ 4 & 4 \end{bmatrix}$ $C = \begin{bmatrix} 0 & 4 \\ -2 & 5 \\ 7 & -1 \end{bmatrix}$ $D = \begin{bmatrix} 6 & 2 \\ 9 & 0 \\ -3 & 0 \end{bmatrix}$

24. $A + B$ **25.** $D - B$ **26.** $4C$

27. $6B - 2A$ **28.** $3C - 4A + B$ **29.** $C + \frac{1}{3}D$

BUSINESS **For Exercises 30–32, use the following information.**
The Cookie Cutter Bakery records each type of cookie sold at three of their branch stores. Two days of sales are shown in the spreadsheets below.

30. Write a matrix for each day's sales.

31. Find the sum of the two days' sales expressed as a matrix.

32. Find the difference in cookie sales from Friday to Saturday expressed as a matrix.

	A	B	C	D	E
1	Friday	chocolate chip	peanut butter	sugar	cut-out
2	Store 1	120	97	64	75
3	Store 2	80	59	36	60
4	Store 3	72	84	29	48

	A	B	C	D	E
1	Saturday	chocolate chip	peanut butter	sugar	cut-out
2	Store 1	112	87	56	74
3	Store 2	84	65	39	70
4	Store 3	88	98	43	60

WEATHER **For Exercises 33–35, use the table that shows the total number of deaths due to severe weather.**

Year	Lightning	Tornadoes	Floods	Hurricanes
1996	52	25	131	37
1997	42	67	118	1
1998	44	130	136	9
1999	46	94	68	19
2000	51	29	37	0

Source: National Oceanic & Atmospheric Administration

33. Find the total number of deaths due to severe weather for each year expressed as a column matrix.

34. Write a matrix that represents how many more people died as a result of lightning than hurricanes for each year.

35. What type of severe weather accounted for the most deaths each year?

More About. . .

Weather
Flash floods and floods are the number 1 weather-related killer recorded in the U.S. each year. The large majority of deaths due to flash flooding are a result of people driving through flooded areas.

Source: National Oceanic & Atmospheric Administration

Online Research **Data Update** What are the current weather statistics? Visit www.algebra2.com/data_update to learn more.

RECREATION **For Exercises 36–39, use the following price list for one-day admissions to the community pool.**

36. Write a matrix that represents the cost of admission for residents and a matrix that represents the cost of admission for nonresidents.

37. Find the matrix that represents the additional cost for nonresidents.

38. Write a matrix that represents the cost of admission before 6:00 P.M. and a matrix that represents the cost of admission after 6:00 P.M.

39. Find a matrix that represents the difference in cost if a child or adult goes to the pool after 6:00 P.M.

Daily Admission Fees

Residents

Time of day	Child	Adult
Before 6:00 P.M.	$3.00	$4.50
After 6:00 P.M.	$2.00	$3.50

Nonresidents

Time of day	Child	Adult
Before 6:00 P.M.	$4.50	$6.75
After 6:00 P.M.	$3.00	$5.25

40. **CRITICAL THINKING** Determine values for each variable if $d = 1$, $e = 4d$, $z + d = e$, $f = \frac{x}{5}$, $ay = 1.5$, $x = \frac{d}{2}$, and $y = x + \frac{x}{2}$.

$$a\begin{bmatrix} x & y & z \\ d & e & f \end{bmatrix} = \begin{bmatrix} ax & ay & az \\ ad & ae & af \end{bmatrix}$$

41. **WRITING IN MATH** Answer the question that was posed at the beginning of the lesson.

How can matrices be used to calculate daily dietary needs?

Include the following in your answer:

- three matrices that represent breakfast, lunch, and dinner over the three-day period, and
- a matrix that represents the total Calories, protein, and fat consumed each day.

Standards Practice

Standardized Test Practice

A B C D

42. Which matrix equals $\begin{bmatrix} 5 & -2 \\ -3 & 7 \end{bmatrix} - \begin{bmatrix} 3 & 4 \\ -5 & 6 \end{bmatrix}$?

(A) $\begin{bmatrix} 2 & 2 \\ -8 & 1 \end{bmatrix}$ (B) $\begin{bmatrix} 8 & -6 \\ -8 & 1 \end{bmatrix}$ (C) $\begin{bmatrix} 2 & 2 \\ 2 & 1 \end{bmatrix}$ (D) $\begin{bmatrix} 2 & -6 \\ 2 & 1 \end{bmatrix}$

43. Solve for x and y in the matrix equation $\begin{bmatrix} x \\ 7 \end{bmatrix} + \begin{bmatrix} 3y \\ -x \end{bmatrix} = \begin{bmatrix} 16 \\ 12 \end{bmatrix}$.

(A) $(-5, 7)$ (B) $(7, 5)$ (C) $(7, 3)$ (D) $(5, 7)$

Maintain Your Skills

Mixed Review

State the dimensions of each matrix. *(Lesson 4-1)*

44. $\begin{bmatrix} 1 & 0 \\ 0 & 1 \end{bmatrix}$

45. $\begin{bmatrix} 2 & 0 & 3 & 0 \end{bmatrix}$

46. $\begin{bmatrix} 5 & 1 & -6 & 2 \\ -38 & 5 & 7 & 3 \end{bmatrix}$

47. $\begin{bmatrix} 7 & -3 & 5 \\ 0 & 2 & -9 \\ 6 & 5 & 1 \end{bmatrix}$

48. $\begin{bmatrix} 8 & 6 \\ 5 & 2 \\ -4 & -1 \end{bmatrix}$

49. $\begin{bmatrix} 7 & 5 & 0 \\ -8 & 3 & 8 \\ 9 & -1 & 15 \\ 4 & 2 & 11 \end{bmatrix}$

Solve each system of equations. *(Lesson 3-5)*

50. $2a + b = 2$
$5a = 15$
$a + b + c = -1$

51. $r + s + t = 15$
$r + t = 12$
$s + t = 10$

52. $6x - 2y - 3z = -10$
$-6x + y + 9z = 3$
$8x - 3y = -16$

Solve each system by using substitution or elimination. *(Lesson 3-2)*

53. $2s + 7t = 39$
$5s - t = 5$

54. $3p + 6q = -3$
$2p - 3q = -9$

55. $a + 5b = 1$
$7a - 2b = 44$

SCRAPBOOKS For Exercises 56–58, use the following information.
Ian has $6.00, and he wants to buy paper for his scrapbook. A sheet of printed paper costs 30¢, and a sheet of solid color paper costs 15¢. *(Lesson 2-7)*

56. Write an inequality that describes this situation.

57. Graph the inequality.

58. Does Ian have enough money to buy 14 pieces of each type of paper?

Getting Ready for the Next Lesson

PREREQUISITE SKILL Name the property illustrated by each equation.
(To review the ***properties of equality****, see Lesson 1-2.)*

59. $\frac{7}{9} \cdot \frac{9}{7} = 1$

60. $7 + (w + 5) = (7 + w) + 5$

61. $3(x + 12) = 3x + 3(12)$

62. $6(9a) = 9a(6)$

4-3 Multiplying Matrices

What You'll Learn

- Multiply matrices.
- Use the properties of matrix multiplication.

How can matrices be used in sports statistics?

Professional football teams track many statistics throughout the season to help evaluate their performance. The table shows the scoring summary of the Oakland Raiders for the 2000 season. The team's record can be summarized in the record matrix R. The values for each type of score can be organized in the point values matrix P.

Oakland Raiders Regular Season Scoring	
Type	**Number**
Touchdown	58
Extra Point	56
Field Goal	23
2-Point Conversion	1
Safety	2

Source: National Football League

Record

$$R = \begin{bmatrix} 58 \\ 56 \\ 23 \\ 1 \\ 2 \end{bmatrix} \begin{matrix} \text{touchdown} \\ \text{extra point} \\ \text{field goal} \\ \text{2-point conversion} \\ \text{safety} \end{matrix}$$

Point Values

$$P = [6 \quad 1 \quad 3 \quad 2 \quad 2]$$

You can use matrix multiplication to find the total points scored.

MULTIPLY MATRICES You can multiply two matrices if and only if the number of columns in the first matrix is equal to the number of rows in the second matrix. When you multiply two matrices $A_{m \times n}$ and $B_{n \times r}$, the resulting matrix AB is an $m \times r$ matrix.

outer dimensions = dimensions of AB

$$\begin{matrix} A & \cdot & B & = & AB \\ 2 \times 3 & & 3 \times 4 & & 2 \times 4 \end{matrix}$$

inner dimensions are equal

Example 1 Dimensions of Matrix Products

Determine whether each matrix product is defined. If so, state the dimensions of the product.

a. $A_{2 \times 5}$ **and** $B_{5 \times 4}$

$$\begin{matrix} A & \cdot & B & = & AB \\ 2 \times 5 & & 5 \times 4 & & 2 \times 4 \end{matrix}$$

The inner dimensions are equal so the matrix product is defined. The dimensions of the product are 2×4.

b. $A_{1 \times 3}$ **and** $B_{4 \times 3}$

$$\begin{matrix} A & \cdot & B \\ 1 \times 3 & & 4 \times 3 \end{matrix}$$

The inner dimensions are not equal, so the matrix product is not defined.

The product of two matrices is found by multiplying columns and rows. The entry in the first row and first column of AB is found by multiplying corresponding elements in the first row of A and the first column of B and then adding.

Key Concept — Multiply Matrices

- **Words** The element a_{ij} of AB is the sum of the products of the corresponding elements in row i of A and column j of B.
- **Symbols** $\begin{bmatrix} a_1 & b_1 \\ a_2 & b_2 \end{bmatrix} \cdot \begin{bmatrix} x_1 & y_1 \\ x_2 & y_2 \end{bmatrix} = \begin{bmatrix} a_1x_1 + b_1x_2 & a_1y_1 + b_1y_2 \\ a_2x_1 + b_2x_2 & a_2y_1 + b_2y_2 \end{bmatrix}$

Example 2 Multiply Square Matrices

Find RS if $R = \begin{bmatrix} 2 & -1 \\ 3 & 4 \end{bmatrix}$ and $S = \begin{bmatrix} 3 & -9 \\ 5 & 7 \end{bmatrix}$.

$$RS = \begin{bmatrix} 2 & -1 \\ 3 & 4 \end{bmatrix} \cdot \begin{bmatrix} 3 & -9 \\ 5 & 7 \end{bmatrix}$$

Step 1 Multiply the numbers in the first row of R by the numbers in the first column of S, add the products, and put the result in the first row, first column of RS.

$$\begin{bmatrix} 2 & -1 \\ 3 & 4 \end{bmatrix} \cdot \begin{bmatrix} 3 & -9 \\ 5 & 7 \end{bmatrix} = \begin{bmatrix} 2(3) + (-1)(5) & \\ & \end{bmatrix}$$

Step 2 Multiply the numbers in the first row of R by the numbers in the second column of S, add the products, and put the result in the first row, second column of RS.

$$\begin{bmatrix} 2 & -1 \\ 3 & 4 \end{bmatrix} \cdot \begin{bmatrix} 3 & -9 \\ 5 & 7 \end{bmatrix} = \begin{bmatrix} 2(3) + (-1)(5) & 2(-9) + (-1)(7) \\ & \end{bmatrix}$$

Step 3 Multiply the numbers in the second row of R by the numbers in the first column of S, add the products, and put the result in the second row, first column of RS.

$$\begin{bmatrix} 2 & -1 \\ 3 & 4 \end{bmatrix} \cdot \begin{bmatrix} 3 & -9 \\ 5 & 7 \end{bmatrix} = \begin{bmatrix} 2(3) + (-1)(5) & 2(-9) + (-1)(7) \\ 3(3) + 4(5) & \end{bmatrix}$$

Step 4 Multiply the numbers in the second row of R by the numbers in the second column of S, add the products, and put the result in the second row, second column of RS.

$$\begin{bmatrix} 2 & -1 \\ 3 & 4 \end{bmatrix} \cdot \begin{bmatrix} 3 & -9 \\ 5 & 7 \end{bmatrix} = \begin{bmatrix} 2(3) + (-1)(5) & 2(-9) + (-1)(7) \\ 3(3) + 4(5) & 3(-9) + 4(7) \end{bmatrix}$$

Step 5 Simplify the product matrix.

$$\begin{bmatrix} 2(3) + (-1)(5) & 2(-9) + (-1)(7) \\ 3(3) + 4(5) & 3(-9) + 4(7) \end{bmatrix} = \begin{bmatrix} 1 & -25 \\ 29 & 1 \end{bmatrix}$$

So, $RS = \begin{bmatrix} 1 & -25 \\ 29 & 1 \end{bmatrix}$.

Study Tip

Multiplying Matrices

To avoid any miscalculations, find the product of the matrices in order as shown in Example 2. It may also help to cover rows or columns not being multiplied as you find elements of the product matrix.

When solving real-world problems, make sure to multiply the matrices in the order for which the product is defined.

Example 3 Multiply Matrices with Different Dimensions

TRACK AND FIELD **In a four-team track meet, 5 points were awarded for each first-place finish, 3 points for each second, and 1 point for each third. Find the total number of points for each school. Which school won the meet?**

School	First Place	Second Place	Third Place
Jefferson	8	4	5
London	6	3	7
Springfield	5	7	3
Madison	7	5	4

Explore The final scores can be found by multiplying the track results for each school by the points awarded for each first-, second-, and third-place finish.

Plan Write the results of the races and the points awarded in matrix form. Set up the matrices so that the number of rows in the points matrix equals the number of columns in the results matrix.

Results **Points**

$$R = \begin{bmatrix} 8 & 4 & 5 \\ 6 & 3 & 7 \\ 5 & 7 & 3 \\ 7 & 5 & 4 \end{bmatrix} \qquad P = \begin{bmatrix} 5 \\ 3 \\ 1 \end{bmatrix}$$

Solve Multiply the matrices.

$$RP = \begin{bmatrix} 8 & 4 & 5 \\ 6 & 3 & 7 \\ 5 & 7 & 3 \\ 7 & 5 & 4 \end{bmatrix} \cdot \begin{bmatrix} 5 \\ 3 \\ 1 \end{bmatrix}$$ Write an equation.

$$= \begin{bmatrix} 8(5) + 4(3) + 5(1) \\ 6(5) + 3(3) + 7(1) \\ 5(5) + 7(3) + 3(1) \\ 7(5) + 5(3) + 4(1) \end{bmatrix}$$ Multiply columns by rows.

$$= \begin{bmatrix} 57 \\ 46 \\ 49 \\ 54 \end{bmatrix}$$ Simplify.

The labels for the product matrix are shown below.

Total Points

Jefferson
London
Springfield
Madison

Jefferson won the track meet with a total of 57 points.

Examine R is a 4×3 matrix and P is a 3×1 matrix; so their product should be a 4×1 matrix. *Why?*

More About. . .

Track and Field

Running and hurdling contests make up the track events. Jumping and throwing contests make up the field events. More than 950,000 high school students participate in track and field competitions each year.

Source: www.encarta.msn.com

MULTIPLICATIVE PROPERTIES Recall that the same properties for real numbers also held true for matrix addition. However, some of these properties do *not* always hold true for matrix multiplication.

Example 4 Commutative Property

Find each product if $P = \begin{bmatrix} 8 & -7 \\ -2 & 4 \\ 0 & 3 \end{bmatrix}$ and $Q = \begin{bmatrix} 9 & -3 & 2 \\ 6 & -1 & -5 \end{bmatrix}$.

a. PQ

$$PQ = \begin{bmatrix} 8 & -7 \\ -2 & 4 \\ 0 & 3 \end{bmatrix} \cdot \begin{bmatrix} 9 & -3 & 2 \\ 6 & -1 & -5 \end{bmatrix}$$ Substitution

$$= \begin{bmatrix} 72 - 42 & -24 + 7 & 16 + 35 \\ -18 + 24 & 6 - 4 & -4 - 20 \\ 0 + 18 & 0 - 3 & 0 - 15 \end{bmatrix}$$ Multiply columns by rows.

$$= \begin{bmatrix} 30 & -17 & 51 \\ 6 & 2 & -24 \\ 18 & -3 & -15 \end{bmatrix}$$ Simplify.

b. QP

$$QP = \begin{bmatrix} 9 & -3 & 2 \\ 6 & -1 & -5 \end{bmatrix} \cdot \begin{bmatrix} 8 & -7 \\ -2 & 4 \\ 0 & 3 \end{bmatrix}$$ Substitution

$$= \begin{bmatrix} 72 + 6 + 0 & -63 - 12 + 6 \\ 48 + 2 + 0 & -42 - 4 - 15 \end{bmatrix}$$ Multiply columns by rows.

$$= \begin{bmatrix} 78 & -69 \\ 50 & -61 \end{bmatrix}$$ Simplify.

In Example 4, notice that $PQ \neq QP$ because $\begin{bmatrix} 30 & -17 & 51 \\ 6 & 2 & -24 \\ 18 & -3 & -15 \end{bmatrix} \neq \begin{bmatrix} 78 & -69 \\ 50 & -61 \end{bmatrix}$. This demonstrates that the Commutative Property of Multiplication does not hold for matrix multiplication. The order in which you multiply matrices is very important.

Example 5 Distributive Property

Find each product if $A = \begin{bmatrix} 3 & 2 \\ -1 & 4 \end{bmatrix}$, $B = \begin{bmatrix} -2 & 5 \\ 6 & 7 \end{bmatrix}$, and $C = \begin{bmatrix} 1 & 1 \\ -5 & 3 \end{bmatrix}$.

a. $A(B + C)$

$$A(B + C) = \begin{bmatrix} 3 & 2 \\ -1 & 4 \end{bmatrix} \cdot \left(\begin{bmatrix} -2 & 5 \\ 6 & 7 \end{bmatrix} + \begin{bmatrix} 1 & 1 \\ -5 & 3 \end{bmatrix} \right)$$ Substitution

$$= \begin{bmatrix} 3 & 2 \\ -1 & 4 \end{bmatrix} \cdot \begin{bmatrix} -1 & 6 \\ 1 & 10 \end{bmatrix}$$ Add corresponding elements.

$$= \begin{bmatrix} 3(-1) + 2(1) & 3(6) + 2(10) \\ -1(-1) + 4(1) & -1(6) + 4(10) \end{bmatrix} \text{ or } \begin{bmatrix} -1 & 38 \\ 5 & 34 \end{bmatrix}$$ Multiply columns by rows.

b. $AB + AC$

$$AB + AC = \begin{bmatrix} 3 & 2 \\ -1 & 4 \end{bmatrix} \cdot \begin{bmatrix} -2 & 5 \\ 6 & 7 \end{bmatrix} + \begin{bmatrix} 3 & 2 \\ -1 & 4 \end{bmatrix} \cdot \begin{bmatrix} 1 & 1 \\ -5 & 3 \end{bmatrix}$$ Substitution

$$= \begin{bmatrix} 3(-2) + 2(6) & 3(5) + 2(7) \\ -1(-2) + 4(6) & -1(5) + 4(7) \end{bmatrix} + \begin{bmatrix} 3(1) + 2(-5) & 3(1) + 2(3) \\ -1(1) + 4(-5) & -1(1) + 4(3) \end{bmatrix}$$

$$= \begin{bmatrix} 6 & 29 \\ 26 & 23 \end{bmatrix} + \begin{bmatrix} -7 & 9 \\ -21 & 11 \end{bmatrix}$$ Simplify.

$$= \begin{bmatrix} -1 & 38 \\ 5 & 34 \end{bmatrix}$$ Add corresponding elements.

Notice that in Example 5, $A(B + C) = AB + AC$. This and other examples suggest that the Distributive Property is true for matrix multiplication. Some properties of matrix multiplication are shown below.

Concept Summary — Properties of Matrix Multiplication

For any matrices A, B, and C for which the matrix product is defined, and any scalar c, the following properties are true.

Associative Property of Matrix Multiplication	$(AB)C = A(BC)$
Associative Property of Scalar Multiplication	$c(AB) = (cA)B = A(cB)$
Left Distributive Property	$C(A + B) = CA + CB$
Right Distributive Property	$(A + B)C = AC + BC$

To show that a property is true for all cases, you must show it is true for the general case. To show that a property is *not* true for all cases, you only need to find a counterexample.

Check for Understanding

Concept Check

1. **OPEN ENDED** Give an example of two matrices whose product is a 3×2 matrix.
2. **Determine** whether the following statement is *always*, *sometimes*, or *never* true. Explain your reasoning.
   For any matrix $A_{m \times n}$ for $m \neq n$, A^2 is defined.
3. **Explain** why, in most cases, $(A + B)C \neq CA + CB$.

Guided Practice

Determine whether each matrix product is defined. If so, state the dimensions of the product.

4. $A_{3 \times 5} \cdot B_{5 \times 2}$
5. $X_{2 \times 3} \cdot Y_{2 \times 3}$

Find each product, if possible.

6. $[3 \;\; -5] \cdot \begin{bmatrix} 3 & 5 \\ -2 & 0 \end{bmatrix}$
7. $\begin{bmatrix} 5 \\ 8 \end{bmatrix} \cdot [3 \;\; -1 \;\; 4]$
8. $\begin{bmatrix} 5 & -2 & -1 \\ 8 & 0 & 3 \end{bmatrix} \cdot \begin{bmatrix} -4 & 2 \\ 1 & 0 \end{bmatrix}$
9. $\begin{bmatrix} 4 & -1 \\ 3 & 5 \end{bmatrix} \cdot \begin{bmatrix} 7 \\ 4 \end{bmatrix}$
10. Use $A = \begin{bmatrix} 2 & -1 \\ 3 & 5 \end{bmatrix}$, $B = \begin{bmatrix} -4 & 1 \\ 8 & 0 \end{bmatrix}$, and $C = \begin{bmatrix} 3 & 2 \\ -1 & 2 \end{bmatrix}$ to determine whether $A(BC) = (AB)C$ is true for the given matrices.

Application

SPORTS For Exercises 11 and 12, use the table below that shows the number of kids registered for baseball and softball.

The Westfall Youth Baseball and Softball League charges the following registration fees: ages 7–8, \$45; ages 9–10, \$55; and ages 11–14, \$65.

Team Members		
Age	**Baseball**	**Softball**
7–8	350	280
9–10	320	165
11–14	180	120

11. Write a matrix for the registration fees and a matrix for the number of players.
12. Find the total amount of money the League received from baseball and softball registrations.

Practice and Apply

Homework Help

For Exercises	See Examples
13–18	1
19–26	2
27–30	4,5
31–41	3

Extra Practice
See page 834.

Determine whether each matrix product is defined. If so, state the dimensions of the product.

13. $A_{4 \times 3} \cdot B_{3 \times 2}$
14. $X_{2 \times 2} \cdot Y_{2 \times 2}$
15. $P_{1 \times 3} \cdot Q_{4 \times 1}$
16. $R_{1 \times 4} \cdot S_{4 \times 5}$
17. $M_{4 \times 3} \cdot N_{4 \times 3}$
18. $A_{3 \times 1} \cdot B_{1 \times 5}$

Find each product, if possible.

19. $\begin{bmatrix} 2 & -1 \end{bmatrix} \cdot \begin{bmatrix} 5 \\ 4 \end{bmatrix}$

20. $\begin{bmatrix} 3 & -2 \\ 5 & 1 \end{bmatrix} \cdot \begin{bmatrix} 4 & 1 \\ 2 & 7 \end{bmatrix}$

21. $\begin{bmatrix} 4 & -1 & 6 \\ 1 & 5 & -8 \end{bmatrix} \cdot \begin{bmatrix} 1 & 3 \\ 9 & -6 \end{bmatrix}$

22. $\begin{bmatrix} 4 & -2 & -7 \\ 6 & 3 & 5 \end{bmatrix} \cdot \begin{bmatrix} -2 \\ 5 \\ 3 \end{bmatrix}$

23. $\begin{bmatrix} 2 & -1 \\ 3 & 4 \end{bmatrix} \cdot \begin{bmatrix} 3 & -9 & -2 \\ 5 & 7 & -6 \end{bmatrix}$

24. $\begin{bmatrix} 7 & 3 \\ 0 & 2 \\ 5 & 5 \end{bmatrix} \cdot \begin{bmatrix} -2 & 1 & 4 \\ 3 & -5 & 2 \\ 4 & 3 & 1 \end{bmatrix}$

25. $\begin{bmatrix} 4 & 0 \\ -3 & 7 \\ -5 & 9 \end{bmatrix} \cdot \begin{bmatrix} 6 & 4 \\ -2 & 1 \end{bmatrix}$

26. $\begin{bmatrix} 0 & 8 \\ 3 & 1 \\ -1 & 5 \end{bmatrix} \cdot \begin{bmatrix} 3 & 1 & -2 \\ 0 & 8 & -5 \end{bmatrix}$

Use $A = \begin{bmatrix} 1 & -2 \\ 4 & 3 \end{bmatrix}$, $B = \begin{bmatrix} -5 & 2 \\ 4 & 3 \end{bmatrix}$, $C = \begin{bmatrix} 5 & 1 \\ 2 & -4 \end{bmatrix}$, and scalar $c = 3$ to determine whether the following equations are true for the given matrices.

27. $AC + BC = (A + B)C$
28. $c(AB) = A(cB)$
29. $C(A + B) = AC + BC$
30. $ABC = CBA$

PRODUCE **For Exercises 31–34, use the table and the following information.**
Carmen Fox owns three fruit farms on which he grows apples, peaches, and apricots. He sells apples for $22 a case, peaches for $25 a case, and apricots for $18 a case.

Number of Cases in Stock of Each Type of Fruit			
Farm	Apples	Peaches	Apricots
1	290	165	210
2	175	240	190
3	110	75	0

31. Write an inventory matrix for the number of cases for each type of fruit for each farm.
32. Write a cost matrix for the price per case for each type of fruit.
33. Find the total income of the three fruit farms expressed as a matrix.
34. What is the total income from all three fruit farms combined?

35. **CRITICAL THINKING** Give an example of two matrices A and B whose product is commutative so that $AB = BA$.

FUND-RAISING For Exercises 36–39, use the table and the information below.
Lawrence High School sold wrapping paper and boxed cards for their fund-raising event. The school receives \$1.00 for each roll of wrapping paper sold and \$0.50 for each box of cards sold.

Total Amounts for Each Class		
Class	Wrapping Paper	Cards
Freshmen	72	49
Sophomores	68	63
Juniors	90	56
Seniors	86	62

36. Write a matrix that represents the amounts sold for each class and a matrix that represents the amount of money the school earns for each item sold.

37. Write a matrix that shows how much each class earned.

38. Which class earned the most money?

39. What is the total amount of money the school made from the fund-raiser?

FINANCE For Exercises 40–42, use the table below that shows the purchase price and selling price of stock for three companies.
For a class project, Taini "bought" shares of stock in three companies. She bought 150 shares of a utility company, 100 shares of a computer company, and 200 shares of a food company. At the end of the project she "sold" all of her stock.

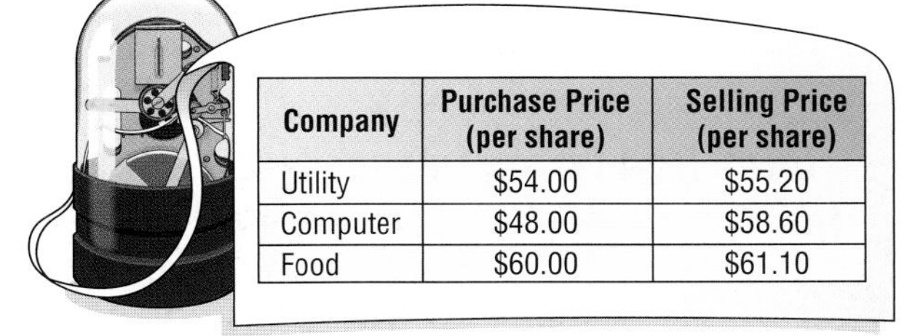

Company	Purchase Price (per share)	Selling Price (per share)
Utility	\$54.00	\$55.20
Computer	\$48.00	\$58.60
Food	\$60.00	\$61.10

40. Organize the data in two matrices and use matrix multiplication to find the total amount she spent for the stock.

41. Write two matrices and use matrix multiplication to find the total amount she received for selling the stock.

42. Use matrix operations to find how much money Taini "made" or "lost."

43. CRITICAL THINKING Find the values of a, b, c, and d to make the statement $\begin{bmatrix} 3 & 5 \\ -1 & 7 \end{bmatrix} \cdot \begin{bmatrix} a & b \\ c & d \end{bmatrix} = \begin{bmatrix} 3 & 5 \\ -1 & 7 \end{bmatrix}$ true. If the matrix $\begin{bmatrix} a & b \\ c & d \end{bmatrix}$ is multiplied by any other matrix containing two columns, what do you think the result would be?

44. WRITING IN MATH Answer the question that was posed at the beginning of the lesson.

How can matrices be used in sports statistics?

Include the following in your answer:
- a matrix that represents the total points scored in the 2000 season, and
- an example of another sport where different point values are used in scoring.

Standards Practice

Standardized Test Practice

45. If C is a 5×1 matrix and D is a 3×5 matrix, what are the dimensions of DC?

(A) 5×5 (B) 3×1 (C) 1×3 (D) DC is not defined.

46. What is the product of $[5 \quad -2 \quad 3]$ and $\begin{bmatrix} 1 & -2 \\ 0 & 3 \\ 2 & 5 \end{bmatrix}$?

(A) $[11 \quad -1]$ (B) $\begin{bmatrix} 11 \\ -1 \end{bmatrix}$ (C) $\begin{bmatrix} 5 & -10 \\ 0 & -6 \\ 6 & -15 \end{bmatrix}$ (D) undefined

Maintain Your Skills

Mixed Review

Perform the indicated matrix operations. If the matrix does not exist, write *impossible*. *(Lesson 4-2)*

47. $3\begin{bmatrix} 4 & -2 \\ -1 & 7 \end{bmatrix}$

48. $[3 \quad 5 \quad 9] + \begin{bmatrix} 5 \\ 2 \\ 6 \end{bmatrix}$

49. $2\begin{bmatrix} 6 & 3 \\ -8 & -2 \end{bmatrix} - 4\begin{bmatrix} 8 & 1 \\ 3 & -4 \end{bmatrix}$

Solve each equation. *(Lesson 4-1)*

50. $\begin{bmatrix} 3x + 2 \\ 15 \end{bmatrix} = \begin{bmatrix} 23 \\ -4y - 1 \end{bmatrix}$

51. $\begin{bmatrix} x + 3y \\ 2x - y \end{bmatrix} = \begin{bmatrix} -22 \\ 19 \end{bmatrix}$

52. $\begin{bmatrix} x + 3z \\ -2x + y - z \\ 5y - 7z \end{bmatrix} = \begin{bmatrix} -19 \\ -2 \\ 24 \end{bmatrix}$

53. CAMERA SUPPLIES Mrs. Franklin is planning a family vacation. She bought 8 rolls of film and 2 camera batteries for $23. The next day, her daughter went back and bought 6 more rolls of film and 2 batteries for her camera. This bill was $18. What is the price of a roll of film and a camera battery? *(Lesson 3-2)*

Find the x-intercept and the y-intercept of the graph of each equation. Then graph the equation. *(Lesson 2-2)*

54. $y = 3 - 2x$

55. $x - \frac{1}{2}y = 8$

56. $5x - 2y = 10$

Getting Ready for the Next Lesson

PREREQUISITE SKILL Graph each set of ordered pairs on a coordinate plane.
*(To review **graphing ordered pairs**, see Lesson 2-1.)*

57. $\{(2, 4), (-1, 3), (0, -2)\}$

58. $\{(-3, 5), (-2, -4), (3, -2)\}$

59. $\{(-1, 2), (2, 4), (3, -3), (4, -1)\}$

60. $\{(-3, 3), (1, 3), (4, 2), (-1, -5)\}$

Practice Quiz 1

Lessons 4-1 through 4-3

Solve each equation. *(Lesson 4-1)*

1. $\begin{bmatrix} 3x + 1 \\ 7y \end{bmatrix} = \begin{bmatrix} 19 \\ 21 \end{bmatrix}$

2. $\begin{bmatrix} 2x + y \\ 4x - 3y \end{bmatrix} = \begin{bmatrix} 9 \\ 23 \end{bmatrix}$

3. $\begin{bmatrix} 2 & x \\ y & 5 \end{bmatrix} = \begin{bmatrix} 2 & 1 \\ 3 & z \end{bmatrix}$

BUSINESS For Exercises 4 and 5, use the table and the following information.
The manager of The Best Bagel Shop keeps records of each type of bagel sold each day at their two stores. Two days of sales are shown below.

Day	Store	Type of Bagel			
		Sesame	Poppy	Blueberry	Plain
Monday	East	120	80	64	75
	West	65	105	77	53
Tuesday	East	112	79	56	74
	West	69	95	82	50

4. Write a matrix for each day's sales. *(Lesson 4-1)*

5. Find the sum of the two days' sales using matrix addition. *(Lesson 4-2)*

Perform the indicated matrix operations. *(Lesson 4-2)*

6. $\begin{bmatrix} 3 & 0 \\ 7 & 12 \end{bmatrix} - \begin{bmatrix} 6 & -5 \\ 4 & -1 \end{bmatrix}$

7. $\frac{2}{3}\begin{bmatrix} 9 & 0 \\ 12 & 15 \end{bmatrix} + \begin{bmatrix} -2 & 3 \\ -7 & -7 \end{bmatrix}$

8. $5\begin{bmatrix} -2 & 4 & 5 \\ 0 & -4 & 7 \end{bmatrix}$

Find each product, if possible. *(Lesson 4-3)*

9. $\begin{bmatrix} 4 & 0 & -8 \\ 7 & -2 & 10 \end{bmatrix} \cdot \begin{bmatrix} -1 & 3 \\ 6 & 0 \end{bmatrix}$

10. $\begin{bmatrix} 3 & -1 \\ 2 & 5 \end{bmatrix} \cdot \begin{bmatrix} 4 & -1 & -2 \\ -3 & 5 & 4 \end{bmatrix}$

4-4 Transformations with Matrices

What You'll Learn

- Use matrices to determine the coordinates of a translated or dilated figure.
- Use matrix multiplication to find the coordinates of a reflected or rotated figure.

Vocabulary

- vertex matrix
- transformation
- preimage
- image
- isometry
- translation
- dilation
- reflection
- rotation

How are transformations used in computer animation?

Computer animation creates the illusion of motion by using a succession of computer-generated still images. Computer animation is used to create movie special effects and to simulate images that would be impossible to show otherwise.

Complex geometric figures can be broken into simple triangles and then moved to other parts of the screen.

TRANSLATIONS AND DILATIONS Points on a coordinate plane can be represented by matrices. The ordered pair (x, y) can be represented by the column matrix $\begin{bmatrix} x \\ y \end{bmatrix}$. Likewise, polygons can be represented by placing all of the column matrices of the coordinates of the vertices into one matrix, called a **vertex matrix**.

Triangle ABC with vertices $A(3, 2)$, $B(4, -2)$, and $C(2, -1)$ can be represented by the following vertex matrix.

$$\triangle ABC = \overset{\quad A \quad B \quad C}{\begin{bmatrix} 3 & 4 & 2 \\ 2 & -2 & -1 \end{bmatrix}} \begin{matrix} \leftarrow x\text{-coordinates} \\ \leftarrow y\text{-coordinates} \end{matrix}$$

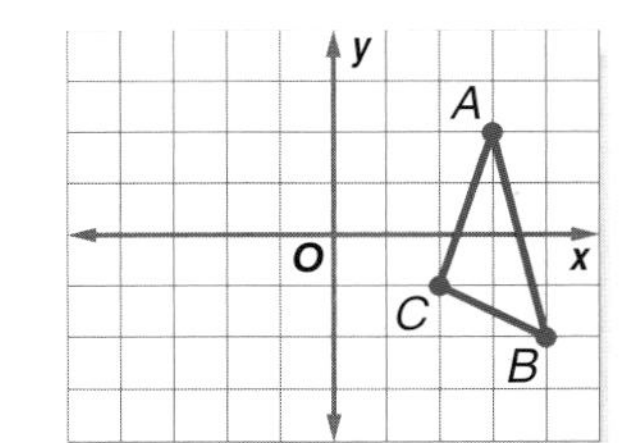

Matrices can be used to perform transformations. **Transformations** are functions that map points of a **preimage** onto its **image**. If the image and preimage are congruent figures, the transformation is an **isometry**.

One type of isometry is a translation. A **translation** occurs when a figure is moved from one location to another without changing its size, shape, or orientation. You can use matrix addition and a *translation matrix* to find the coordinates of a translated figure.

Study Tip

Reading Math
A matrix containing coordinates of a geometric figure is also called a *coordinate matrix.*

Example 1 Translate a Figure

Find the coordinates of the vertices of the image of quadrilateral $QUAD$ with $Q(2, 3)$, $U(5, 2)$, $A(4, -2)$, and $D(1, -1)$, if it is moved 4 units to the left and 2 units up. Then graph $QUAD$ and its image $Q'U'A'D'$.

Write the vertex matrix for quadrilateral $QUAD$. $\begin{bmatrix} 2 & 5 & 4 & 1 \\ 3 & 2 & -2 & -1 \end{bmatrix}$

To translate the quadrilateral 4 units to the left, add -4 to each x-coordinate. To translate the figure 2 units up, add 2 to each y-coordinate. This can be done by adding the translation matrix $\begin{bmatrix} -4 & -4 & -4 & -4 \\ 2 & 2 & 2 & 2 \end{bmatrix}$ to the vertex matrix of $QUAD$.

(continued on the next page)

Vertex Matrix of $QUAD$		Translation Matrix		Vertex Matrix of $Q'U'A'D'$

$$\begin{bmatrix} 2 & 5 & 4 & 1 \\ 3 & 2 & -2 & -1 \end{bmatrix} + \begin{bmatrix} -4 & -4 & -4 & -4 \\ 2 & 2 & 2 & 2 \end{bmatrix} = \begin{bmatrix} -2 & 1 & 0 & -3 \\ 5 & 4 & 0 & 1 \end{bmatrix}$$

The coordinates of $Q'U'A'D'$ are $Q'(-2, 5)$, $U'(1, 4)$, $A'(0, 0)$, and $D'(-3, 1)$. Graph the preimage and the image. The two quadrilaterals have the same size and shape.

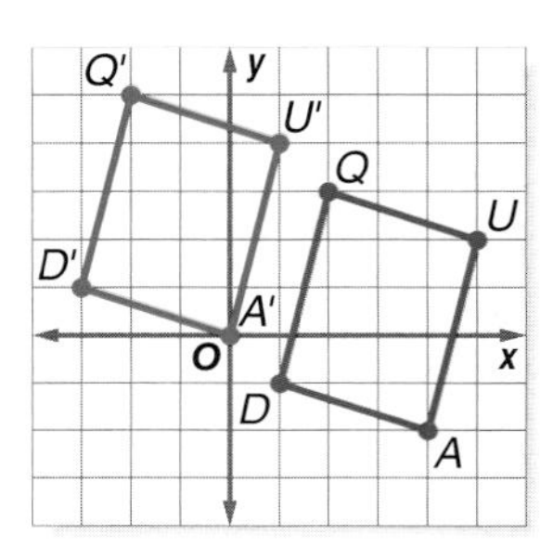

Standardized Test Practice

A B C D

Example 2 Find a Translation Matrix

Short-Response Test Item

Rectangle $A'B'C'D'$ is the result of a translation of rectangle $ABCD$. A table of the vertices of each rectangle is shown. Find the coordinates of A and D'.

Rectangle $ABCD$	Rectangle $A'B'C'D'$
A	$A'(-1, 1)$
$B(1, 5)$	$B'(4, 1)$
$C(1, -2)$	$C'(4, -6)$
$D(-4, -2)$	D'

Read the Test Item

- You are given the coordinates of the preimage and image of points B and C. Use this information to find the translation matrix. Then you can use the translation matrix to find the coordinates of A and D'.

Solve the Test Item

- Write a matrix equation. Let (a, b) represent the coordinates of A and let (c, d) represent the coordinates of D'.

$$\begin{bmatrix} a & 1 & 1 & -4 \\ b & 5 & -2 & -2 \end{bmatrix} + \begin{bmatrix} x & x & x & x \\ y & y & y & y \end{bmatrix} = \begin{bmatrix} -1 & 4 & 4 & c \\ 1 & 1 & -6 & d \end{bmatrix}$$

$$\begin{bmatrix} a + x & 1 + x & 1 + x & -4 + x \\ b + y & 5 + y & -2 + y & -2 + y \end{bmatrix} = \begin{bmatrix} -1 & 4 & 4 & c \\ 1 & 1 & -6 & d \end{bmatrix}$$

- Since these two matrices are equal, corresponding elements are equal.

Solve an equation for x.

$$1 + x = 4$$
$$x = 3$$

Solve an equation for y.

$$5 + y = 1$$
$$y = -4$$

- Use the values for x and y to find the values for $A(a, b)$ and $D'(c, d)$.

$$a + x = -1 \qquad a + 3 = -1 \qquad a = -4$$

$$b + y = 1 \qquad b + (-4) = 1 \qquad b = 5$$

$$-4 + x = c \qquad -4 + 3 = c \qquad -1 = c$$

$$-2 + y = d \qquad -2 + (-4) = d \qquad -6 = d$$

So the coordinates of A are $(-4, 5)$, and the coordinates for D' are $(-1, -6)$.

Test-Taking Tip

Sometimes you need to solve for unknown value(s) before you can solve for the value(s) requested in the question.

When a geometric figure is enlarged or reduced, the transformation is called a **dilation**. In a dilation, all linear measures of the image change in the same ratio. For example, if the length of each side of a figure doubles, then the perimeter doubles, and vice versa. You can use scalar multiplication to perform dilations.

Example 3 Dilation

$\triangle JKL$ has vertices $J(-2, -3)$, $K(-5, 4)$, and $L(3, 2)$. Dilate $\triangle JKL$ so that its perimeter is one-half the original perimeter. What are the coordinates of the vertices of $\triangle J'K'L'$?

If the perimeter of a figure is one-half the original perimeter, then the lengths of the sides of the figure will be one-half the measure of the original lengths. Multiply the vertex matrix by the scale factor of $\frac{1}{2}$.

$$\frac{1}{2}\begin{bmatrix} -2 & -5 & 3 \\ -3 & 4 & 2 \end{bmatrix} = \begin{bmatrix} -1 & -\frac{5}{2} & \frac{3}{2} \\ -\frac{3}{2} & 2 & 1 \end{bmatrix}$$

The coordinates of the vertices of $\triangle J'K'L'$ are $J'\left(-1, -\frac{3}{2}\right)$, $K'\left(-\frac{5}{2}, 2\right)$, and $L'\left(\frac{3}{2}, 1\right)$.

Graph $\triangle JKL$ and $\triangle J'K'L'$. The triangles are not congruent. The image has sides that are half the length of those of the original figure.

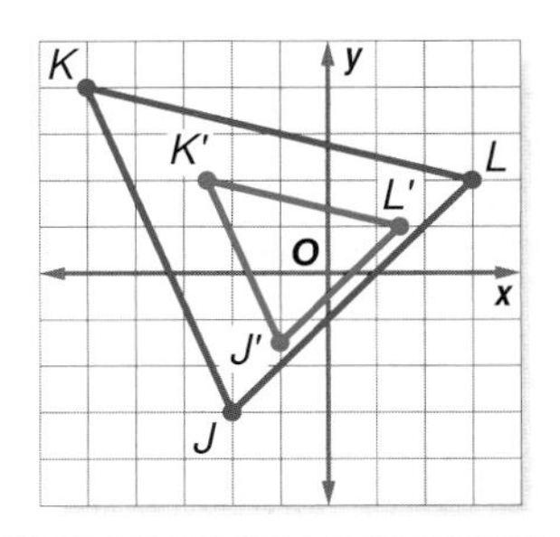

REFLECTIONS AND ROTATIONS In addition to translations, reflections and rotations are also isometries. A **reflection** occurs when every point of a figure is mapped to a corresponding image across a line of symmetry using a *reflection matrix*. The matrices used for three common reflections are shown below.

Concept Summary — *Reflection Matrices*

For a reflection over the:	x-axis	y-axis	line $y = x$
Multiply the vertex matrix on the left by:	$\begin{bmatrix} 1 & 0 \\ 0 & -1 \end{bmatrix}$	$\begin{bmatrix} -1 & 0 \\ 0 & 1 \end{bmatrix}$	$\begin{bmatrix} 0 & 1 \\ 1 & 0 \end{bmatrix}$

Example 4 Reflection

Find the coordinates of the vertices of the image of pentagon $QRSTU$ with $Q(1, 3)$, $R(3, 2)$, $S(3, -1)$, $T(1, -2)$, and $U(-1, 1)$ after a reflection across the y-axis.

Write the ordered pairs as a vertex matrix. Then multiply the vertex matrix by the reflection matrix for the y-axis.

$$\begin{bmatrix} -1 & 0 \\ 0 & 1 \end{bmatrix} \cdot \begin{bmatrix} 1 & 3 & 3 & 1 & -1 \\ 3 & 2 & -1 & -2 & 1 \end{bmatrix} = \begin{bmatrix} -1 & -3 & -3 & -1 & 1 \\ 3 & 2 & -1 & -2 & 1 \end{bmatrix}$$

The coordinates of the vertices of $Q'R'S'T'U'$ are $Q'(-1, 3)$, $R'(-3, 2)$, $S'(-3, -1)$, $T'(-1, -2)$, and $U'(1, 1)$. Notice that the preimage and image are congruent. Both figures have the same size and shape.

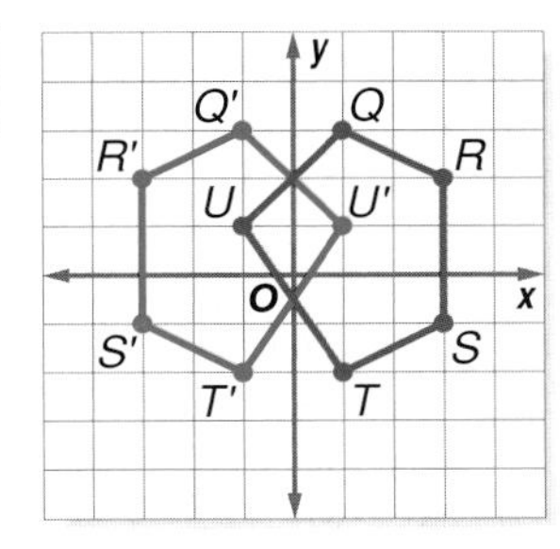

A **rotation** occurs when a figure is moved around a center point, usually the origin. To determine the vertices of a figure's image by rotation, multiply its vertex matrix by a *rotation matrix*. Commonly used rotation matrices are summarized below.

Concept Summary — Rotation Matrices

For a counterclockwise rotation about the origin of:	90°	180°	270°
Multiply the vertex matrix on the left by:	$\begin{bmatrix} 0 & -1 \\ 1 & 0 \end{bmatrix}$	$\begin{bmatrix} -1 & 0 \\ 0 & -1 \end{bmatrix}$	$\begin{bmatrix} 0 & 1 \\ -1 & 0 \end{bmatrix}$

Example 5 Rotation

Find the coordinates of the vertices of the image of $\triangle ABC$ with $A(4, 3)$, $B(2, 1)$, and $C(1, 5)$ after it is rotated 90° counterclockwise about the origin.

Write the ordered pairs in a vertex matrix. Then multiply the vertex matrix by the rotation matrix.

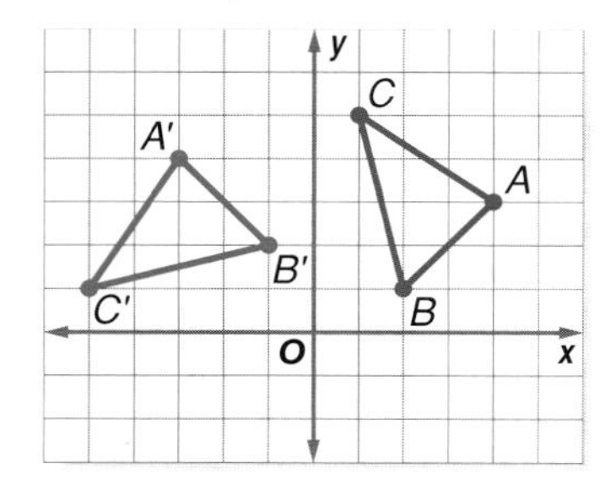

$$\begin{bmatrix} 0 & -1 \\ 1 & 0 \end{bmatrix} \cdot \begin{bmatrix} 4 & 2 & 1 \\ 3 & 1 & 5 \end{bmatrix} = \begin{bmatrix} -3 & -1 & -5 \\ 4 & 2 & 1 \end{bmatrix}$$

The coordinates of the vertices of $\triangle A'B'C'$ are $A'(-3, 4)$, $B'(-1, 2)$, and $C'(-5, 1)$. The image is congruent to the preimage.

Check for Understanding

Concept Check

1. **Compare and contrast** the size and shape of the preimage and image for each type of transformation. Tell which transformations are isometries.
2. **Write** the translation matrix for $\triangle ABC$ and its image $\triangle A'B'C'$ shown at the right.
3. **OPEN ENDED** Write a translation matrix that moves $\triangle DEF$ up and left on the coordinate plane.

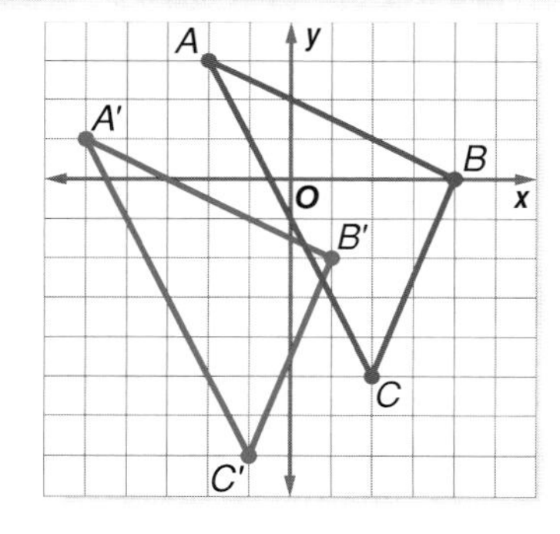

Guided Practice

Triangle ABC with vertices $A(1, 4)$, $B(2, -5)$, and $C(-6, -6)$ is translated 3 units right and 1 unit down.

4. Write the translation matrix.
5. Find the coordinates of $\triangle A'B'C'$.
6. Graph the preimage and the image.

For Exercises 7–10, use the rectangle at the right.

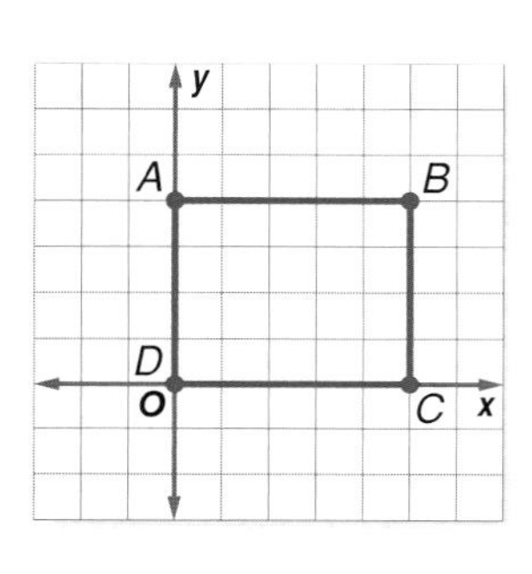

7. Write the coordinates in a vertex matrix.
8. Find the coordinates of the image after a dilation by a scale factor of 3.
9. Find the coordinates of the image after a reflection over the x-axis.
10. Find the coordinates of the image after a rotation of 180°.

11. A point is translated from B to C as shown at the right. If a point at $(-4, 3)$ is translated in the same way, what will be its new coordinates?

Ⓐ $(3, 4)$ Ⓑ $(1, 1)$

Ⓒ $(-7, 8)$ Ⓓ $(1, 6)$

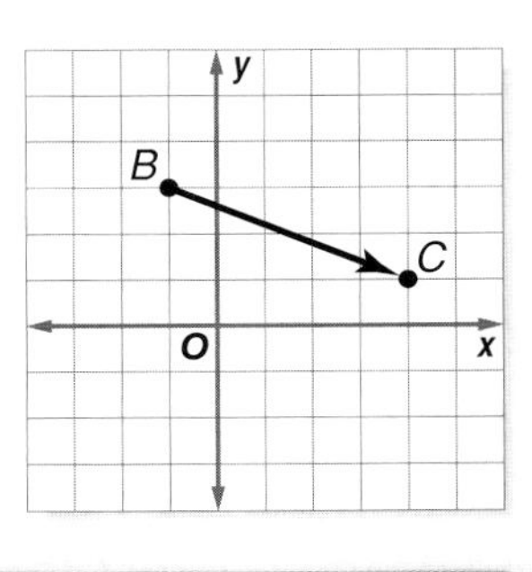

Practice and Apply

Homework Help

For Exercises	See Examples
12–14, 35, 36	1
15–17, 33, 34	3
18–20	4
21–23, 25, 37	5
24	2
26–32, 38–41	1–5

Extra Practice
See page 835.

For Exercises 12–14, use the following information.
Triangle DEF with vertices $D(1, 4)$, $E(2, -5)$, and $F(-6, -6)$ is translated 4 units left and 2 units up.

12. Write the translation matrix.

13. Find the coordinates of $\triangle D'E'F'$.

14. Graph the preimage and the image.

For Exercises 15–17, use the following information.
The vertices of $\triangle ABC$ are $A(0, 2)$, $B(1.5, -1.5)$, and $C(-2.5, 0)$. The triangle is dilated so that its perimeter is three times the original perimeter.

15. Write the coordinates for $\triangle ABC$ in a vertex matrix.

16. Find the coordinates of the image $\triangle A'B'C'$.

17. Graph $\triangle ABC$ and $\triangle A'B'C'$.

For Exercises 18–20, use the following information.
The vertices of $\triangle XYZ$ are $X(1, -1)$, $Y(2, -4)$, and $Z(7, -1)$. The triangle is reflected over the line $y = x$.

18. Write the coordinates of $\triangle XYZ$ in a vertex matrix.

19. Find the coordinates of $\triangle X'Y'Z'$.

20. Graph $\triangle XYZ$ and $\triangle X'Y'Z'$.

For Exercises 21–23, use the following information.
Parallelogram $DEFG$ with $D(2, 4)$, $E(5, 4)$, $F(4, 1)$, and $G(1, 1)$ is rotated 270° counterclockwise about the origin.

21. Write the coordinates of the parallelogram in a vertex matrix.

22. Find the coordinates of parallelogram $D'E'F'G'$.

23. Graph the preimage and the image.

24. Triangle DEF with vertices $D(-2, 2)$, $E(3, 5)$, and $F(5, -2)$ is translated so that D' is at $(1, -5)$. Find the coordinates of E' and F'.

25. A triangle is rotated 90° counterclockwise about the origin. The coordinates of the vertices are $J'(-3, -5)$, $K'(-2, 7)$, and $L'(1, 4)$. What were the coordinates of the triangle in its original position?

For Exercises 26–28, use quadrilateral *QRST* shown at the right.

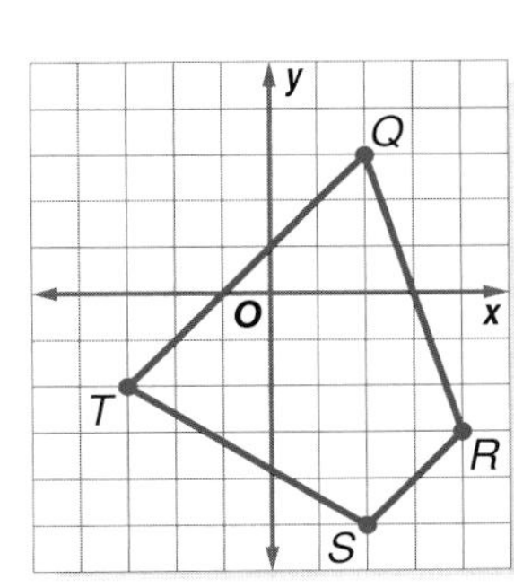

26. Write the vertex matrix. Multiply the vertex matrix by -1.

27. Graph the preimage and image.

28. What type of transformation does the graph represent?

For Exercises 29–32, use rectangle $ABCD$ with vertices $A(-4, 4)$, $B(4, 4)$, $C(4, -4)$, and $D(-4, -4)$.

29. Find the coordinates of the image in matrix form after a reflection over the x-axis followed by a reflection over the y-axis.

30. Find the coordinates of the image after a 180° rotation about the origin.

31. Find the coordinates of the image after a reflection over the line $y = x$.

32. What do you observe about these three matrices? Explain.

LANDSCAPING For Exercises 33 and 34, use the following information.
A garden design is plotted on a coordinate grid. The original plan shows a fountain with vertices at $(-2, -2)$, $(-6, -2)$, $(-8, -5)$, and $(-4, -5)$. Changes to the plan now require that the fountain's perimeter be three-fourths that of the original.

33. Determine the new coordinates for the fountain.

34. The center of the fountain was at $(-5, -3.5)$. What will be the coordinate of the center after the changes in the plan have been made?

More About. . .

Technology

Douglas Engelbart invented the "X-Y position indicator for a display system" in 1964. He nicknamed this invention "the mouse" because a tail came out the end.

Source: www.about.com

TECHNOLOGY For Exercises 35 and 36, use the following information.
As you move the mouse for your computer, a corresponding arrow is translated on the screen. Suppose the position of the cursor on the screen is given in inches with the origin at the bottom left-hand corner of the screen.

35. You want to move your cursor 3 inches to the right and 4 inches up. Write a translation matrix that can be used to move the cursor to the new position.

36. If the cursor is currently at (3.5, 2.25), what are the coordinates of the position after the translation?

37. GYMNASTICS The drawing at the right shows four positions of a man performing the giant swing in the high bar event. Suppose this drawing is placed on a coordinate grid with the hand grips at $H(0, 0)$ and the toe of the figure in the upper right corner at $T(7, 8)$. Find the coordinates of the toes of the other three figures, if each successive figure has been rotated 90° counterclockwise about the origin.

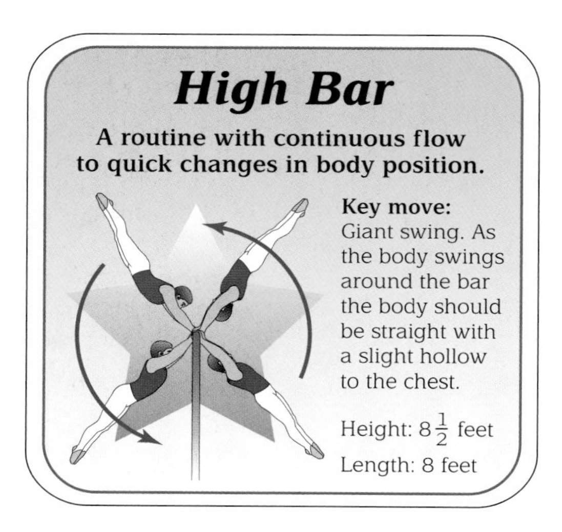

FOOTPRINTS For Exercises 38–41, use the following information.
The combination of a reflection and a translation is called a *glide reflection*. An example is a set of footprints.

38. Describe the reflection and transformation combination shown at the right.

39. Write two matrix operations that can be used to find the coordinates of point C.

40. Does it matter which operation you do first? Explain.

41. What are the coordinates of the next two footprints?

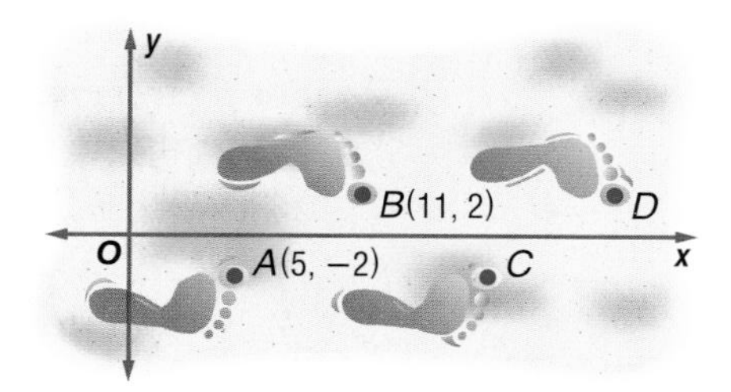

42. **CRITICAL THINKING** Do you think a matrix exists that would represent a reflection over the line $x = 3$? If so, make a conjecture and verify it.

43. **WRITING IN MATH** Answer the question that was posed at the beginning of the lesson.

How are transformations used in computer animation?

Include the following in your answer:

- an example of how a figure with 5 points (coordinates) could be written in a matrix and multiplied by a rotation matrix, and
- a description of the motion that is a result of repeated dilations with a scale factor of one-fourth.

44. Which matrix represents a reflection over the y-axis followed by a reflection over the x-axis?

(A) $\begin{bmatrix} 0 & -1 \\ -1 & 0 \end{bmatrix}$ (B) $\begin{bmatrix} -1 & 0 \\ 0 & -1 \end{bmatrix}$ (C) $\begin{bmatrix} -1 & -1 \\ -1 & -1 \end{bmatrix}$ (D) none of these

45. Triangle ABC has vertices with coordinates $A(-4, 2)$, $B(-4, -3)$, and $C(3, -2)$. After a dilation, triangle $A'B'C'$ has coordinates $A'(-12, 6)$, $B'(-12, -9)$, and $C'(9, -6)$. How many times as great is the perimeter of $A'B'C'$ as ABC?

(A) 3 (B) 6 (C) 12 (D) $\frac{1}{3}$

Maintain Your Skills

Mixed Review

Determine whether each matrix product is defined. If so, state the dimensions of the product. *(Lesson 4-3)*

46. $A_{2 \times 3} \cdot B_{3 \times 2}$ 47. $A_{4 \times 1} \cdot B_{2 \times 1}$ 48. $A_{2 \times 5} \cdot B_{5 \times 5}$

Perform the indicated matrix operations. If the matrix does not exist, write *impossible*. *(Lesson 4-2)*

49. $2\begin{bmatrix} 4 & 9 & -8 \\ 6 & -11 & -2 \\ 12 & -10 & 3 \end{bmatrix} + 3\begin{bmatrix} 1 & 2 & 3 \\ 2 & 3 & 4 \\ 3 & 4 & 5 \end{bmatrix}$

50. $4\begin{bmatrix} 3 & 4 & -7 \\ 6 & -9 & -2 \\ -3 & 1 & 3 \end{bmatrix} - \begin{bmatrix} -8 & 6 & -4 \\ -7 & 10 & 1 \\ -2 & 1 & 5 \end{bmatrix}$

Graph each relation or equation and find the domain and range. Then determine whether the relation or equation is a function. *(Lesson 2-1)*

51. (3, 5), (4, 6), (5, −4) 52. $x = -5y + 2$ 53. $x = y^2$

Write an absolute value inequality for each graph. *(Lesson 1-6)*

54. −5 −4 −3 −2 −1 0 1 2 3 4 5

55. −5.6 −4.2 −2.8 −1.4 0 1.4 2.8 4.2

56. −6 −5 −4 −3 −2 −1 0 1 2 3 4

57. −4 −3 −2 −1 0 1 2 3 4 5 6

58. **BUSINESS** Reliable Rentals rents cars for \$12.95 per day plus 15¢ per mile. Luis Romero works for a company that limits expenses for car rentals to \$90 per day. What is the maximum number of miles that Mr. Romero can drive each day? *(Lesson 1-5)*

Getting Ready for the Next Lesson

BASIC SKILL **Use cross products to solve each proportion.**

59. $\frac{x}{8} = \frac{3}{4}$ 60. $\frac{4}{20} = \frac{1}{m}$ 61. $\frac{2}{3} = \frac{a}{42}$

62. $\frac{5}{6} = \frac{k}{4}$ 63. $\frac{2}{y} = \frac{8}{9}$ 64. $\frac{x}{5} = \frac{x+1}{8}$

4-5 Determinants

What You'll Learn

- Evaluate the determinant of a 2 × 2 matrix.
- Evaluate the determinant of a 3 × 3 matrix.

Vocabulary

- determinant
- second-order determinant
- third-order determinant
- expansion by minors
- minor

How are determinants used to find areas of polygons?

The "Bermuda Triangle" is an area located off the southeastern Atlantic coast of the United States that is noted for a high incidence of unexplained losses of ships, small boats, and aircraft. You can estimate the area of this triangular region by finding the determinant of the matrix that contains the coordinates of the vertices of the triangle.

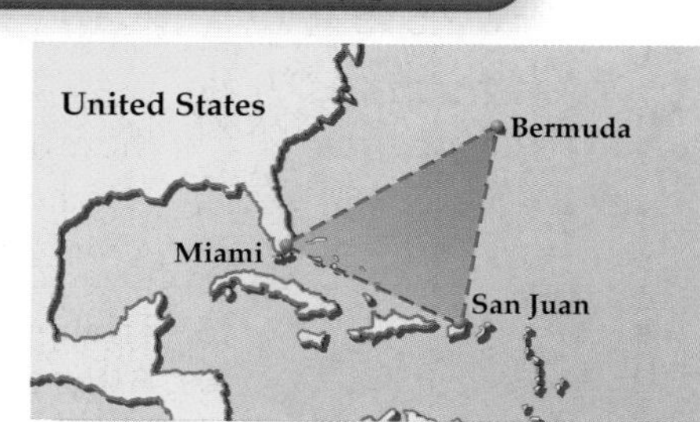

DETERMINANTS OF 2 × 2 MATRICES Every square matrix has a number associated with it called its determinant. A **determinant** is a square array of numbers or variables enclosed between two parallel lines. For example, the determinant of $\begin{bmatrix} 3 & -1 \\ 2 & 5 \end{bmatrix}$ can be represented by $\begin{vmatrix} 3 & -1 \\ 2 & 5 \end{vmatrix}$ or det $\begin{bmatrix} 3 & -1 \\ 2 & 5 \end{bmatrix}$. The determinant of a 2 × 2 matrix is called a **second-order determinant**.

Key Concept — Second-Order Determinant

- **Words** The value of a second-order determinant is found by calculating the difference of the products of the two diagonals.
- **Symbols** $\begin{vmatrix} a & b \\ c & d \end{vmatrix} = ad - bc$

Example 1 Second-Order Determinant

Find the value of each determinant.

a. $\begin{vmatrix} -2 & 5 \\ 6 & 8 \end{vmatrix}$

$\begin{vmatrix} -2 & 5 \\ 6 & 8 \end{vmatrix} = (-2)(8) - 5(6)$ Definition of determinant

$= -16 - 30$ Multiply.

$= -46$ Simplify.

b. $\begin{vmatrix} 7 & 4 \\ -3 & 2 \end{vmatrix}$

$\begin{vmatrix} 7 & 4 \\ -3 & 2 \end{vmatrix} = (7)(2) - 4(-3)$ Definition of determinant

$= 14 - (-12)$ Multiply.

$= 26$ Simplify.

Study Tip

Reading Math
The term *determinant* is often used to mean the *value* of the determinant.

DETERMINANTS OF 3 × 3 MATRICES Determinants of 3 × 3 matrices are called **third-order determinants**. One method of evaluating third-order determinants is **expansion by minors**. The **minor** of an element is the determinant formed when the row and column containing that element are deleted.

$$\begin{vmatrix} a_1 & b_1 & c_1 \\ a_2 & b_2 & c_2 \\ a_3 & b_3 & c_3 \end{vmatrix} \quad \text{The minor of } a_1 \text{ is } \begin{vmatrix} b_2 & c_2 \\ b_3 & c_3 \end{vmatrix}.$$

$$\begin{vmatrix} a_1 & b_1 & c_1 \\ a_2 & b_2 & c_2 \\ a_3 & b_3 & c_3 \end{vmatrix} \quad \text{The minor of } b_1 \text{ is } \begin{vmatrix} a_2 & c_2 \\ a_3 & c_3 \end{vmatrix}.$$

$$\begin{vmatrix} a_1 & b_1 & c_1 \\ a_2 & b_2 & c_2 \\ a_3 & b_3 & c_3 \end{vmatrix} \quad \text{The minor of } c_1 \text{ is } \begin{vmatrix} a_2 & b_2 \\ a_3 & b_3 \end{vmatrix}.$$

To use expansion by minors with third-order determinants, each member of one row is multiplied by its minor and its *position sign*, and the results are added together. The position signs alternate between positive and negative, beginning with a positive sign in the first row, first column.

$$\begin{bmatrix} + & - & + \\ - & + & - \\ + & - & + \end{bmatrix}$$

Key Concept — *Third-Order Determinant*

$$\begin{vmatrix} a & b & c \\ d & e & f \\ g & h & i \end{vmatrix} = a\begin{vmatrix} e & f \\ h & i \end{vmatrix} - b\begin{vmatrix} d & f \\ g & i \end{vmatrix} + c\begin{vmatrix} d & e \\ g & h \end{vmatrix}$$

The definition of third-order determinants shows an expansion using the elements in the first row of the determinant. However, any row can be used.

Example 2 *Expansion by Minors*

Evaluate $\begin{vmatrix} 2 & 7 & -3 \\ -1 & 5 & -4 \\ 6 & 9 & 0 \end{vmatrix}$ **using expansion by minors.**

Decide which row of elements to use for the expansion. For this example, we will use the first row.

$$\begin{vmatrix} 2 & 7 & -3 \\ -1 & 5 & -4 \\ 6 & 9 & 0 \end{vmatrix} = 2\begin{vmatrix} 5 & -4 \\ 9 & 0 \end{vmatrix} - 7\begin{vmatrix} -1 & -4 \\ 6 & 0 \end{vmatrix} + (-3)\begin{vmatrix} -1 & 5 \\ 6 & 9 \end{vmatrix} \quad \text{Expansion by minors}$$

$$= 2(0 - (-36)) - 7(0 - (-24)) - 3(-9 - 30) \quad \text{Evaluate } 2 \times 2 \text{ determinants.}$$

$$= 2(36) - 7(24) - 3(-39)$$

$$= 72 - 168 + 117 \quad \text{Multiply.}$$

$$= 21 \quad \text{Simplify.}$$

Another method for evaluating a third-order determinant is by using diagonals.

Step 1 Begin by writing the first two columns on the right side of the determinant.

$$\begin{vmatrix} a & b & c \\ d & e & f \\ g & h & i \end{vmatrix} \Rightarrow \begin{vmatrix} a & b & c \\ d & e & f \\ g & h & i \end{vmatrix} \begin{matrix} a & b \\ d & e \\ g & h \end{matrix}$$

(continued on the next page)

Step 2 Next, draw diagonals from each element of the top row of the determinant downward to the right. Find the product of the elements on each diagonal.

$$\begin{vmatrix} a & b & c \\ d & e & f \\ g & h & i \end{vmatrix} \Rightarrow \begin{vmatrix} a & b & c \\ d & e & f \\ g & h & i \end{vmatrix} \begin{matrix} a & b \\ d & e \\ g & h \end{matrix}$$

$aei \quad bfg \quad cdh$

Then, draw diagonals from the elements in the third row of the determinant upward to the right. Find the product of the elements on each diagonal.

$gec \quad hfa \quad idb$

$$\begin{vmatrix} a & b & c \\ d & e & f \\ g & h & i \end{vmatrix} \Rightarrow \begin{vmatrix} a & b & c \\ d & e & f \\ g & h & i \end{vmatrix} \begin{matrix} a & b \\ d & e \\ g & h \end{matrix}$$

Step 3 To find the value of the determinant, add the products of the first set of diagonals and then subtract the products of the second set of diagonals. The sum is $aei + bfg + cdh - gec - hfa - idb$.

Example 3 Use Diagonals

Evaluate $\begin{vmatrix} -1 & 3 & -3 \\ 4 & -2 & -1 \\ 0 & -5 & 2 \end{vmatrix}$ **using diagonals.**

Step 1 Rewrite the first two columns to the right of the determinant.

$$\begin{vmatrix} -1 & 3 & -3 \\ 4 & -2 & -1 \\ 0 & -5 & 2 \end{vmatrix} \begin{matrix} -1 & 3 \\ 4 & -2 \\ 0 & -5 \end{matrix}$$

Step 2 Find the products of the elements of the diagonals.

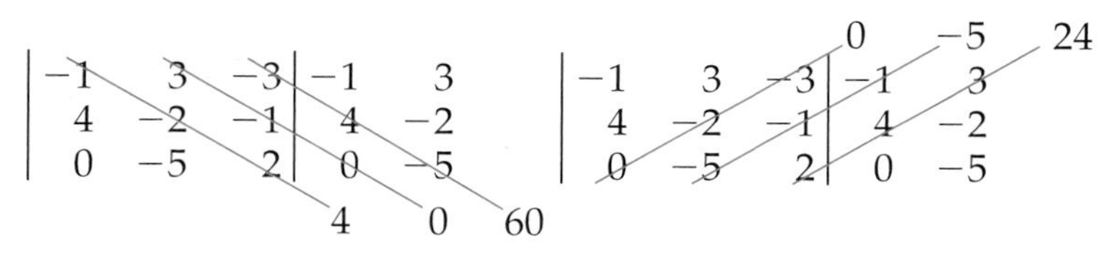

Step 3 Add the bottom products and subtract the top products.
$4 + 0 + 60 - 0 - (-5) - 24 = 45$

The value of the determinant is 45.

One very useful application of determinants is finding the areas of polygons. The formula below shows how determinants can be used to find the area of a triangle using the coordinates of the vertices.

Study Tip

Area Formula
Notice that it is necessary to use the absolute value of A to guarantee a nonnegative value for the area.

Key Concept — Area of a Triangle

The area of a triangle having vertices at (a, b), (c, d), and (e, f) is $|A|$, where

$$A = \frac{1}{2}\begin{vmatrix} a & b & 1 \\ c & d & 1 \\ e & f & 1 \end{vmatrix}.$$

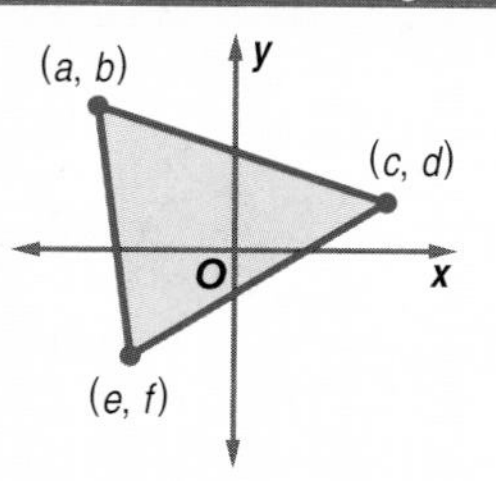

Example 4 Area of a Triangle

GEOMETRY Find the area of a triangle whose vertices are located at (−1, 6), (2, 4), and (0, 0).

Assign values to a, b, c, d, e, and f and substitute them into the Area Formula. Then evaluate.

$$A = \frac{1}{2}\begin{vmatrix} a & b & 1 \\ c & d & 1 \\ e & f & 1 \end{vmatrix} \quad \text{Area Formula}$$

$$= \frac{1}{2}\begin{vmatrix} -1 & 6 & 1 \\ 2 & 4 & 1 \\ 0 & 0 & 1 \end{vmatrix} \quad (a, b) = (-1, 6), (c, d) = (2, 4), (e, f) = (0, 0)$$

$$= \frac{1}{2}\left[-1\begin{vmatrix} 4 & 1 \\ 0 & 1 \end{vmatrix} - 6\begin{vmatrix} 2 & 1 \\ 0 & 1 \end{vmatrix} + 1\begin{vmatrix} 2 & 4 \\ 0 & 0 \end{vmatrix}\right] \quad \text{Expansion by minors}$$

$$= \frac{1}{2}[-1(4 - 0) - 6(2 - 0) + 1(0 - 0)] \quad \text{Evaluate } 2 \times 2 \text{ determinants.}$$

$$= \frac{1}{2}[-4 - 12 + 0] \quad \text{Multiply.}$$

$$= \frac{1}{2}[-16] \text{ or } -8 \quad \text{Simplify.}$$

Remember that the area of a triangle is the absolute value of A. Thus, the area is $|-8|$ or 8 square units.

Check for Understanding

Concept Check

1. **OPEN ENDED** Write a matrix whose determinant is zero.
2. **FIND THE ERROR** Khalid and Erica are finding the determinant of $\begin{bmatrix} 8 & 3 \\ -5 & 2 \end{bmatrix}$.

Khalid

$\begin{vmatrix} 8 & 3 \\ -5 & 2 \end{vmatrix} = 16 - (-15)$

$= 31$

Erica

$\begin{vmatrix} 8 & 3 \\ -5 & 2 \end{vmatrix} = 16 - 15$

$= 1$

Who is correct? Explain your reasoning.

3. **Explain** why $\begin{bmatrix} 2 & 1 & 7 \\ 3 & -5 & 0 \end{bmatrix}$ does not have a determinant.
4. **Find a counterexample** to disprove the following statement.
Two different matrices can never have the same determinant.
5. **Describe** how to find the minor of 6 in $\begin{bmatrix} 5 & 11 & 7 \\ -1 & 3 & 8 \\ 6 & 0 & -2 \end{bmatrix}$.
6. **Show** that the value of $\begin{vmatrix} -2 & 3 & 5 \\ 0 & -1 & 4 \\ 9 & 7 & 2 \end{vmatrix}$ is the same whether you use expansion by minors or diagonals.

Guided Practice

Find the value of each determinant.

7. $\begin{vmatrix} 7 & 8 \\ 3 & -2 \end{vmatrix}$
8. $\begin{vmatrix} -3 & -6 \\ 4 & 8 \end{vmatrix}$
9. $\begin{vmatrix} 0 & 8 \\ 5 & 9 \end{vmatrix}$

Evaluate each determinant using expansion by minors.

10. $\begin{vmatrix} 0 & -4 & 0 \\ 3 & -2 & 5 \\ 2 & -1 & 1 \end{vmatrix}$

11. $\begin{vmatrix} 2 & 3 & 4 \\ 6 & 5 & 7 \\ 1 & 2 & 8 \end{vmatrix}$

Evaluate each determinant using diagonals.

12. $\begin{vmatrix} 1 & 6 & 4 \\ -2 & 3 & 1 \\ 1 & 6 & 4 \end{vmatrix}$

13. $\begin{vmatrix} -1 & 4 & 0 \\ 3 & -2 & -5 \\ -3 & -1 & 2 \end{vmatrix}$

Application

14. **GEOMETRY** Find the area of the triangle shown at the right.

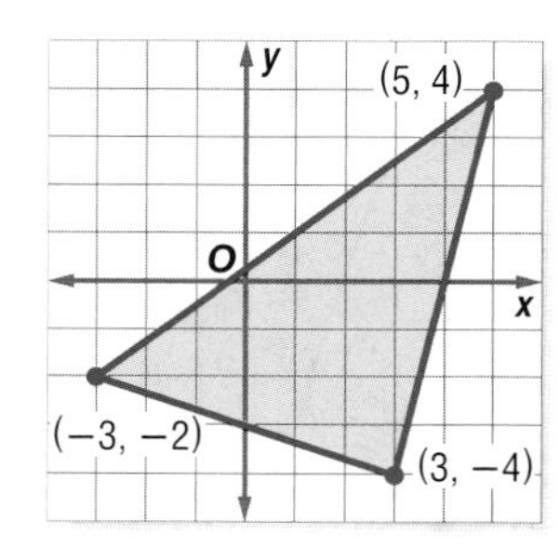

Practice and Apply

Homework Help

For Exercises	See Examples
15–26, 39	1
27–32, 40	2
33–38	3
41–44	4

Extra Practice
See page 835.

Find the value of each determinant.

15. $\begin{vmatrix} 10 & 6 \\ 5 & 5 \end{vmatrix}$

16. $\begin{vmatrix} 8 & 5 \\ 6 & 1 \end{vmatrix}$

17. $\begin{vmatrix} -7 & 3 \\ -9 & 7 \end{vmatrix}$

18. $\begin{vmatrix} -2 & 4 \\ 3 & -6 \end{vmatrix}$

19. $\begin{vmatrix} 2 & -7 \\ -5 & 3 \end{vmatrix}$

20. $\begin{vmatrix} -6 & -2 \\ 8 & 5 \end{vmatrix}$

21. $\begin{vmatrix} -9 & 0 \\ -12 & -7 \end{vmatrix}$

22. $\begin{vmatrix} 6 & 14 \\ -3 & -8 \end{vmatrix}$

23. $\begin{vmatrix} 15 & 11 \\ 23 & 19 \end{vmatrix}$

24. $\begin{vmatrix} 21 & 43 \\ 16 & 31 \end{vmatrix}$

25. $\begin{vmatrix} 7 & 5.2 \\ -4 & 1.6 \end{vmatrix}$

26. $\begin{vmatrix} -3.2 & -5.8 \\ 4.1 & 3.9 \end{vmatrix}$

Evaluate each determinant using expansion by minors.

27. $\begin{vmatrix} 3 & 1 & 2 \\ 0 & 6 & 4 \\ 2 & 5 & 1 \end{vmatrix}$

28. $\begin{vmatrix} 7 & 3 & -4 \\ -2 & 9 & 6 \\ 0 & 0 & 0 \end{vmatrix}$

29. $\begin{vmatrix} -2 & 7 & -2 \\ 4 & 5 & 2 \\ 1 & 0 & -1 \end{vmatrix}$

30. $\begin{vmatrix} -3 & 0 & 6 \\ 6 & 5 & -2 \\ 1 & 4 & 2 \end{vmatrix}$

31. $\begin{vmatrix} 1 & 5 & -4 \\ -7 & 3 & 2 \\ 6 & 3 & -1 \end{vmatrix}$

32. $\begin{vmatrix} 3 & 7 & 6 \\ -1 & 6 & 2 \\ 8 & -3 & -5 \end{vmatrix}$

Evaluate each determinant using diagonals.

33. $\begin{vmatrix} 1 & 1 & 1 \\ 3 & 9 & 5 \\ 8 & 7 & 4 \end{vmatrix}$

34. $\begin{vmatrix} 1 & 5 & 2 \\ -6 & -7 & 8 \\ 5 & 9 & -3 \end{vmatrix}$

35. $\begin{vmatrix} 8 & -9 & 0 \\ 1 & 5 & 4 \\ 6 & -2 & 3 \end{vmatrix}$

36. $\begin{vmatrix} 4 & 10 & 7 \\ 3 & 3 & 1 \\ 0 & 5 & 2 \end{vmatrix}$

37. $\begin{vmatrix} 2 & -3 & 4 \\ -2 & 1 & 5 \\ 5 & 3 & -2 \end{vmatrix}$

38. $\begin{vmatrix} 4 & -2 & 3 \\ -2 & 3 & 4 \\ 3 & 4 & 2 \end{vmatrix}$

39. Solve for x if $\det \begin{bmatrix} 2 & x \\ 5 & -3 \end{bmatrix} = 24$.

40. Solve $\det \begin{bmatrix} 4 & x & -2 \\ -x & -3 & 1 \\ -6 & 2 & 3 \end{bmatrix} = -3$ for x.

41. **GEOMETRY** Find the area of the polygon shown at the right.

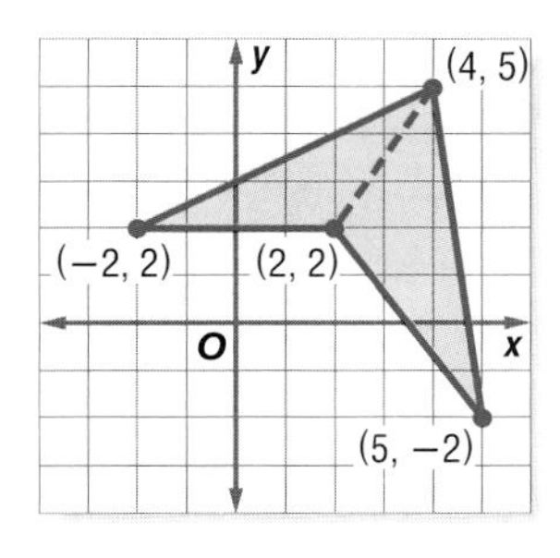

Archaeologist

Archaeologists attempt to reconstruct past ways of life by examining preserved bones, the ruins of buildings, and artifacts such as tools, pottery, and jewelry.

Source: www.encarta.msn.com

Online Research
For information about a career as an archaeologist, visit: www.algebra2.com/careers

42. **GEOMETRY** Find the value of x such that the area of a triangle whose vertices have coordinates (6, 5), (8, 2), and $(x, 11)$ is 15 square units.

43. **ARCHAEOLOGY** During an archaeological dig, a coordinate grid is laid over the site to identify the location of artifacts as they are excavated. During a dig, three corners of a rectangular building have been partially unearthed at (−1, 6), (4, 5), and (−3, −4). If each square on the grid measures one square foot, estimate the area of the floor of the building.

44. **GEOGRAPHY** Mr. Cardona is a regional sales manager for a company in Florida. Tampa, Orlando, and Ocala outline his region. If a coordinate grid in which 1 unit = 10 miles is placed over the map of Florida with Tampa at the origin, the coordinates of the three cities are (0, 0), (7, 5), and (2.5, 10). Use a determinant to estimate the area of his sales territory.

45. **CRITICAL THINKING** Find a third-order determinant in which no element is 0, but for which the determinant is 0.

46. **CRITICAL THINKING** Make a conjecture about how you could find the determinant of a 4×4 matrix using the expansion by minors method. Use a diagram in your explanation.

47. **WRITING IN MATH** Answer the question that was posed at the beginning of the lesson.

How are determinants used to find areas of polygons?

Include the following in your answer:
- an explanation of how you could use a coordinate grid to estimate the area of the Bermuda Triangle, and
- some advantages of using this method in this situation.

48. Find the value of det A if $A = \begin{bmatrix} 0 & 3 & -2 \\ -4 & 0 & 1 \\ 3 & 2 & 0 \end{bmatrix}$.

Ⓐ 0 Ⓑ 12 Ⓒ 25 Ⓓ 36

49. Find the area of triangle ABC.

Ⓐ 10 units^2
Ⓑ 12 units^2
Ⓒ 14 units^2
Ⓓ 16 units^2
Ⓔ none of these

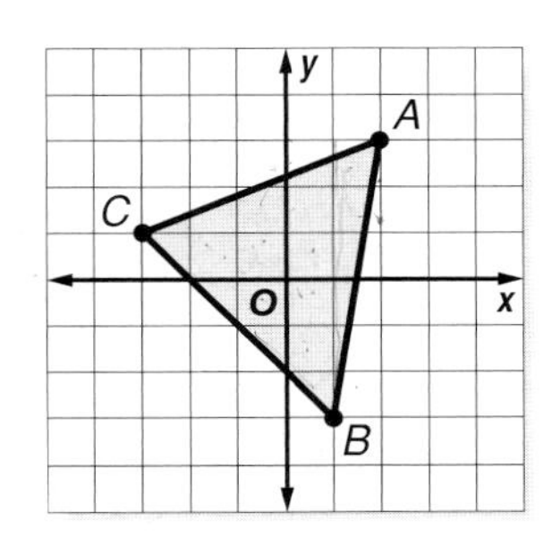

Graphing Calculator

MATRX FUNCTION You can use a TI-83 Plus to find determinants of square matrices using the MATRX functions. Enter the matrix under the NAMES menu. Then use the arrow keys to highlight the MATH menu. Choose det(, which is option 1, to calculate the determinant.

Use a graphing calculator to find the value of each determinant.

50. $\begin{vmatrix} 3 & -6.5 \\ 8 & 3.75 \end{vmatrix}$

51. $\begin{vmatrix} 1.3 & 7.2 \\ 6.1 & 5.4 \end{vmatrix}$

52. $\begin{vmatrix} 6.1 & 4.8 \\ 9.7 & 3.5 \end{vmatrix}$

53. $\begin{vmatrix} 8 & 6 & -5 \\ 10 & -7 & 3 \\ 9 & 14 & -6 \end{vmatrix}$

54. $\begin{vmatrix} 10 & 20 & 30 \\ 40 & 50 & 60 \\ 70 & 80 & 90 \end{vmatrix}$

55. $\begin{vmatrix} 10 & 12 & 4 \\ -3 & 18 & -9 \\ 16 & -2 & -1 \end{vmatrix}$

Maintain Your Skills

Mixed Review

For Exercises 56–58, use the following information.

The vertices of $\triangle ABC$ are $A(-2, 1)$, $B(1, 2)$ and $C(2, -3)$. The triangle is dilated so that its perimeter is $2\frac{1}{2}$ times the original perimeter. *(Lesson 4-4)*

56. Write the coordinates of $\triangle ABC$ in a vertex matrix.

57. Find the coordinates of $\triangle A'B'C'$.

58. Graph $\triangle ABC$ and $\triangle A'B'C'$.

Find each product, if possible. *(Lesson 4-3)*

59. $[5 \quad 2] \cdot \begin{bmatrix} -2 \\ 3 \end{bmatrix}$

60. $\begin{bmatrix} 2 & 4 \\ -2 & 3 \end{bmatrix} \cdot \begin{bmatrix} 3 & 9 \\ -1 & 2 \end{bmatrix}$

61. $\begin{bmatrix} 5 \\ 7 \end{bmatrix} \cdot \begin{bmatrix} 1 & 6 \\ -4 & 2 \end{bmatrix}$

62. $\begin{bmatrix} 7 & 4 \\ -1 & 2 \\ -3 & 5 \end{bmatrix} \cdot [3 \quad 5]$

63. $[4 \quad 2 \quad 0] \cdot \begin{bmatrix} 3 & -2 \\ 1 & 0 \\ 5 & 6 \end{bmatrix}$

64. $\begin{bmatrix} 7 & -5 & 4 \\ 6 & 1 & 3 \end{bmatrix} \cdot \begin{bmatrix} -1 & 3 \\ -2 & -8 \\ 1 & 2 \end{bmatrix}$

65. RUNNING The length of a marathon was determined by the first marathon in the 1908 Olympic Games in London, England. The race began at Windsor Castle and ended in front of the royal box at London's Olympic Stadium, which was a distance of 26 miles 385 yards. Determine how many feet the marathon covers using the formula $f(m, y) = 5280m + 3y$, where m is the number of miles and y is the number of yards. *(Lesson 3-4)*

Write an equation in slope-intercept form for the line that satisfies each set of conditions. *(Lesson 2-4)*

66. slope 1 passes through (5, 3)

67. slope $-\frac{4}{3}$ passes through (6, −8)

68. passes through (3, 7) and (−2, −3)

69. passes through (0, 5) and (10, 10)

Getting Ready for the Next Lesson

PREREQUISITE SKILL **Solve each system of equations.**
*(To review **solving systems of equations**, see Lesson 3-2.)*

70. $x + y = -3$
$3x + 4y = -12$

71. $x + y = 10$
$2x + y = 11$

72. $2x + y = 5$
$4x + y = 9$

73. $3x + 5y = 2$
$2x - y = -3$

74. $6x + 2y = 22$
$3x + 7y = 41$

75. $3x - 2y = -2$
$4x + 7y = 65$

4-6 Cramer's Rule

California Standards Standard 2.0 **Students solve systems of linear equations** and inequalities (**in two or three variables**) by substitution, with graphs, or **with matrices.** (Key)

What You'll Learn

- Solve systems of two linear equations by using Cramer's Rule.
- Solve systems of three linear equations by using Cramer's Rule.

How is Cramer's Rule used to solve systems of equations?

Two sides of a triangle are contained in lines whose equations are $1.4x + 3.8y = 3.4$ and $2.5x - 1.7y = -10.9$. To find the coordinates of the vertex of the triangle between these two sides, you must solve the system of equations. However, solving this system by using substitution or elimination would require many calculations. Another method for solving systems of equations is Cramer's Rule.

Vocabulary

- Cramer's Rule

SYSTEMS OF TWO LINEAR EQUATIONS

Cramer's Rule uses determinants to solve systems of equations. Consider the following system.

$$ax + by = e \quad \text{a, b, c, d, e, and f represent constants, *not* variables.}$$
$$cx + dy = f$$

Study Tip

Look Back
To review **solving systems of equations**, see Lesson 3-2.

Solve for x by using elimination.

$adx + bdy = de$	Multiply the first equation by d.
$(-)\ bcx + bdy = bf$	Multiply the second equation by b.
$adx - bcx = de - bf$	Subtract.
$(ad - bc)x = de - bf$	Factor.
$x = \dfrac{de - bf}{ad - bc}$	Notice that $ad - bc$ must not be zero.

Solving for y in the same way produces the following expression.

$$y = \frac{af - ce}{ad - bc}$$

So the solution of the system of equations $ax + by = e$ and $cx + dy = f$ is $\left(\frac{de - bf}{ad - bc}, \frac{af - ce}{ad - bc}\right)$.

Notice that the denominators for each expression are the same. It can be written using a determinant. The numerators can also be written as determinants.

$$ad - bc = \begin{vmatrix} a & b \\ c & d \end{vmatrix} \qquad de - bf = \begin{vmatrix} e & b \\ f & d \end{vmatrix} \qquad af - ce = \begin{vmatrix} a & e \\ c & f \end{vmatrix}$$

Key Concept — Cramer's Rule for Two Variables

The solution of the system of linear equations

$$ax + by = e$$
$$cx + dy = f$$

is (x, y), where $x = \dfrac{\begin{vmatrix} e & b \\ f & d \end{vmatrix}}{\begin{vmatrix} a & b \\ c & d \end{vmatrix}}$, $y = \dfrac{\begin{vmatrix} a & e \\ c & f \end{vmatrix}}{\begin{vmatrix} a & b \\ c & d \end{vmatrix}}$, and $\begin{vmatrix} a & b \\ c & d \end{vmatrix} \neq 0$.

Example 1 System of Two Equations

Use Cramer's Rule to solve the system of equations.

$5x + 7y = 13$
$2x - 5y = 13$

$$x = \frac{\begin{vmatrix} e & b \\ f & d \end{vmatrix}}{\begin{vmatrix} a & b \\ c & d \end{vmatrix}} \quad \text{Cramer's Rule} \quad y = \frac{\begin{vmatrix} a & e \\ c & f \end{vmatrix}}{\begin{vmatrix} a & b \\ c & d \end{vmatrix}}$$

$$= \frac{\begin{vmatrix} 13 & 7 \\ 13 & -5 \end{vmatrix}}{\begin{vmatrix} 5 & 7 \\ 2 & -5 \end{vmatrix}} \quad a = 5, b = 7, c = 2, d = -5, e = 13, \text{ and } f = 13 \quad = \frac{\begin{vmatrix} 5 & 13 \\ 2 & 13 \end{vmatrix}}{\begin{vmatrix} 5 & 7 \\ 2 & -5 \end{vmatrix}}$$

$$= \frac{13(-5) - 13(7)}{5(-5) - 2(7)} \quad \text{Evaluate each determinant.} \quad = \frac{5(13) - 2(13)}{5(-5) - 2(7)}$$

$$= \frac{-156}{-39} \text{ or } 4 \quad \text{Simplify.} \quad = \frac{39}{-39} \text{ or } -1$$

The solution is $(4, -1)$.

Cramer's Rule is especially useful when the coefficients are large or involve fractions or decimals.

Example 2 Use Cramer's Rule

More About. . .

Elections

In 1936, Franklin D. Roosevelt received a record 523 electoral college votes to Alfred M. Landon's 8 votes. This is the largest electoral college majority.

Source: *The Guinness Book of Records*

ELECTIONS In the 2000 presidential election, George W. Bush received about 8,400,000 votes in California and Texas while Al Gore received about 8,300,000 votes in those states. The graph shows the percent of the popular vote that each candidate received in those states.

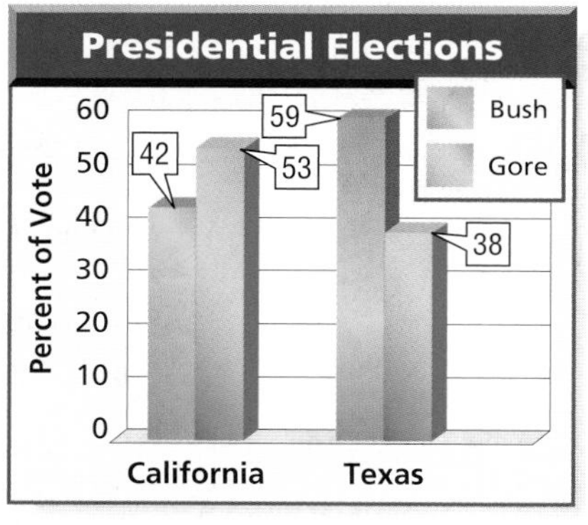

Source: States' Elections Offices

a. Write a system of equations that represents the total number of votes cast for each candidate in these two states.

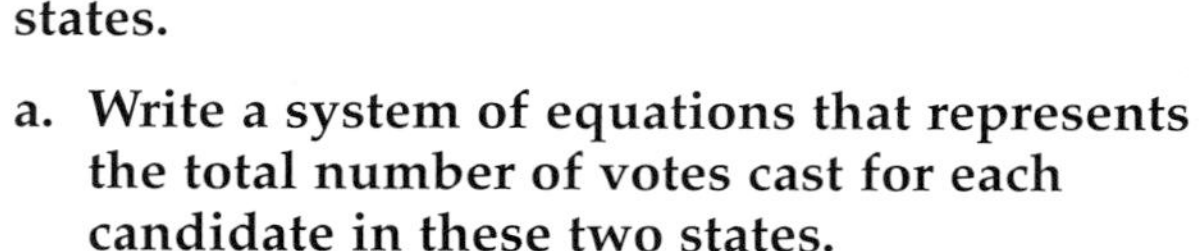

Let x represent the total number of votes in California.

Let y represent the total number of votes in Texas.

$0.42x + 0.59y = 8{,}400{,}000$ Votes for Bush

$0.53x + 0.38y = 8{,}300{,}000$ Votes for Gore

b. Find the total number of popular votes cast in California and in Texas.

$$x = \frac{\begin{vmatrix} e & b \\ f & d \end{vmatrix}}{\begin{vmatrix} a & b \\ c & d \end{vmatrix}} \quad \text{Cramer's Rule} \quad y = \frac{\begin{vmatrix} a & e \\ c & f \end{vmatrix}}{\begin{vmatrix} a & b \\ c & d \end{vmatrix}}$$

$$= \frac{\begin{vmatrix} 8{,}400{,}000 & 0.59 \\ 8{,}300{,}000 & 0.38 \end{vmatrix}}{\begin{vmatrix} 0.42 & 0.59 \\ 0.53 & 0.38 \end{vmatrix}} \qquad = \frac{\begin{vmatrix} 0.42 & 8{,}400{,}000 \\ 0.53 & 8{,}300{,}000 \end{vmatrix}}{\begin{vmatrix} 0.42 & 0.59 \\ 0.53 & 0.38 \end{vmatrix}}$$

$$= \frac{8{,}400{,}000(0.38) - 8{,}300{,}000(0.59)}{0.42(0.38) - 0.53(0.59)}$$

$$= \frac{-1{,}705{,}000}{-0.1531}$$

$$\approx 11{,}136{,}512.08$$

$$= \frac{0.42(8{,}300{,}000) - 0.53(8{,}400{,}000)}{0.42(0.38) - 0.53(0.59)}$$

$$= \frac{-966{,}000}{-0.1531}$$

$$\approx 6{,}309{,}601.57$$

The solution of the system is about (11,136,512.08, 6,309,601.57).

So, there were about 11,100,000 popular votes cast in California and about 6,300,000 popular votes cast in Texas.

SYSTEMS OF THREE LINEAR EQUATIONS

You can also use Cramer's Rule to solve a system of three equations in three variables.

Key Concept — Cramer's Rule for Three Variables

The solution of the system whose equations are

$$ax + by + cz = j$$
$$dx + ey + fz = k$$
$$gx + hy + iz = \ell$$

is (x, y, z), where $x = \frac{\begin{vmatrix} j & b & c \\ k & e & f \\ \ell & h & i \end{vmatrix}}{\begin{vmatrix} a & b & c \\ d & e & f \\ g & h & i \end{vmatrix}}$, $y = \frac{\begin{vmatrix} a & j & c \\ d & k & f \\ g & \ell & i \end{vmatrix}}{\begin{vmatrix} a & b & c \\ d & e & f \\ g & h & i \end{vmatrix}}$, $z = \frac{\begin{vmatrix} a & b & j \\ d & e & k \\ g & h & \ell \end{vmatrix}}{\begin{vmatrix} a & b & c \\ d & e & f \\ g & h & i \end{vmatrix}}$, and $\begin{vmatrix} a & b & c \\ d & e & f \\ g & h & i \end{vmatrix} \neq 0$.

Example 3 System of Three Equations

Use Cramer's Rule to solve the system of equations.

$$3x + y + z = -1$$
$$-6x + 5y + 3z = -9$$
$$9x - 2y - z = 5$$

$$x = \frac{\begin{vmatrix} j & b & c \\ k & e & f \\ \ell & h & i \end{vmatrix}}{\begin{vmatrix} a & b & c \\ d & e & f \\ g & h & i \end{vmatrix}} = \frac{\begin{vmatrix} -1 & 1 & 1 \\ -9 & 5 & 3 \\ 5 & -2 & -1 \end{vmatrix}}{\begin{vmatrix} 3 & 1 & 1 \\ -6 & 5 & 3 \\ 9 & -2 & -1 \end{vmatrix}}$$

$$y = \frac{\begin{vmatrix} a & j & c \\ d & k & f \\ g & \ell & i \end{vmatrix}}{\begin{vmatrix} a & b & c \\ d & e & f \\ g & h & i \end{vmatrix}} = \frac{\begin{vmatrix} 3 & -1 & 1 \\ -6 & -9 & 3 \\ 9 & 5 & -1 \end{vmatrix}}{\begin{vmatrix} 3 & 1 & 1 \\ -6 & 5 & 3 \\ 9 & -2 & -1 \end{vmatrix}}$$

$$z = \frac{\begin{vmatrix} a & b & j \\ d & e & k \\ g & h & \ell \end{vmatrix}}{\begin{vmatrix} a & b & c \\ d & e & f \\ g & h & i \end{vmatrix}} = \frac{\begin{vmatrix} 3 & 1 & -1 \\ -6 & 5 & -9 \\ 9 & -2 & 5 \end{vmatrix}}{\begin{vmatrix} 3 & 1 & 1 \\ -6 & 5 & 3 \\ 9 & -2 & -1 \end{vmatrix}}$$

Use a calculator to evaluate each determinant.

$x = \frac{-2}{-9}$ or $\frac{2}{9}$ $\quad$ $y = \frac{12}{-9}$ or $-\frac{4}{3}$ $\quad$ $z = \frac{3}{-9}$ or $-\frac{1}{3}$

The solution is $\left(\frac{2}{9}, -\frac{4}{3}, -\frac{1}{3}\right)$.

Check for Understanding

Concept Check

1. **Describe** the condition that must be met in order to use Cramer's Rule.
2. **OPEN ENDED** Write a system of equations that *cannot* be solved using Cramer's Rule.
3. **Write** a system of equations whose solution is $x = \frac{\begin{vmatrix} -6 & 5 \\ 30 & -2 \end{vmatrix}}{\begin{vmatrix} 3 & 5 \\ 4 & -2 \end{vmatrix}}$, $y = \frac{\begin{vmatrix} 3 & -6 \\ 4 & 30 \end{vmatrix}}{\begin{vmatrix} 3 & 5 \\ 4 & -2 \end{vmatrix}}$.

Guided Practice

Use Cramer's Rule to solve each system of equations.

4. $x - 4y = 1$
   $2x + 3y = 13$
5. $0.2a = 0.3b$
   $0.4a - 0.2b = 0.2$
6. $\frac{1}{2}r - \frac{2}{3}s = 2\frac{1}{3}$
   $\frac{3}{5}r + \frac{4}{5}s = -10$
7. $2x - y + 3z = 5$
   $3x + 2y - 5z = 4$
   $x - 4y + 11z = 3$
8. $a + 9b - 2c = 2$
   $-a - 3b + 4c = 1$
   $2a + 3b - 6c = -5$
9. $r + 4s + 3t = 10$
   $2r - 2s + t = 15$
   $r + 2s - 3t = -1$

Application

INVESTING For Exercises 10 and 11, use the following information.
Jarrod Wright has $4000 he would like to invest. He could put it in a savings account paying 6.5% interest annually, or in a certificate of deposit with an annual rate of 8%. He wants his interest for the year to be $297.50, because earning more than this would put him into a higher tax bracket.

10. Write a system of equations, in which the unknowns *s* and *d* stand for the amounts of money Jarrod should deposit in the savings account and the certificate of deposit, respectively.
11. How much should he put in a savings account, and how much should he put in the certificate of deposit?

Practice and Apply

Homework Help

For Exercises	See Examples
12–25	1
26–31	3
32–37	2

Extra Practice
See page 835.

Use Cramer's Rule to solve each system of equations.

12. $5x + 2y = 8$
    $2x - 3y = 7$
13. $2m + 7n = 4$
    $m - 2n = -20$
14. $2r - s = 1$
    $3r + 2s = 19$
15. $3a + 5b = 33$
    $5a + 7b = 51$
16. $2m - 4n = -1$
    $3n - 4m = -5$
17. $4x + 3y = 6$
    $8x - y = -9$
18. $0.5r - s = -1$
    $0.75r + 0.5s = -0.25$
19. $1.5m - 0.7n = 0.5$
    $2.2m - 0.6n = -7.4$
20. $3x - 2y = 4$
    $\frac{1}{2}x - \frac{2}{3}y = 1$
21. $2a + 3b = -16$
    $\frac{3}{4}a - \frac{7}{8}b = 10$
22. $\frac{1}{3}r + \frac{2}{5}s = 5$
    $\frac{2}{3}r - \frac{1}{2}s = -3$
23. $\frac{3}{4}x + \frac{1}{2}y = \frac{11}{12}$
    $\frac{1}{2}x - \frac{1}{4}y = \frac{1}{8}$

You can use Cramer's Rule to compare home loans. Visit www.algebra2.com/webquest to continue work on your WebQuest project.

24. **GEOMETRY** The two sides of an angle are contained in lines whose equations are $4x + y = -4$ and $2x - 3y = -9$. Find the coordinates of the vertex of the angle.
25. **GEOMETRY** Two sides of a parallelogram are contained in the lines whose equations are $2.3x + 1.2y = 2.1$ and $4.1x - 0.5y = 14.3$. Find the coordinates of a vertex of the parallelogram.

Use Cramer's Rule to solve each system of equations.

26. $x + y + z = 6$
$2x + y - 4z = -15$
$5x - 3y + z = -10$

27. $a - 2b + c = 7$
$6a + 2b - 2c = 4$
$4a + 6b + 4c = 14$

28. $r - 2s - 5t = -1$
$r + 2s - 2t = 5$
$4r + s + t = -1$

29. $3a + c = 23$
$4a + 7b - 2c = -22$
$8a - b - c = 34$

30. $4x + 2y - 3z = -32$
$-x - 3y + z = 54$
$2y + 8z = 78$

31. $2r + 25s = 40$
$10r + 12s + 6t = -2$
$36r - 25s + 50t = -10$

GAMES **For Exercises 32 and 33, use the following information.**
Marcus purchased a game card to play virtual games at the arcade. His favorite games are the race car simulator, which costs 7 points for each play, and the snowboard simulator, which costs 5 points for each play. Marcus came with enough money to buy a 50-point card, and he has time to play 8 games.

32. Write a system of equations.

33. Solve the system using Cramer's Rule to find the number of times Marcus can play race car simulator and snowboard simulator.

INTERIOR DESIGN **For Exercises 34 and 35, use the following information.**
An interior designer is preparing invoices for two of her clients. She has ordered silk dupioni and cotton damask fabric for both of them.

Client	Fabric	Yards	Total Cost
Harada	silk cotton	8 13	$604.79
Martina	silk cotton	$5\frac{1}{2}$ 14	$542.30

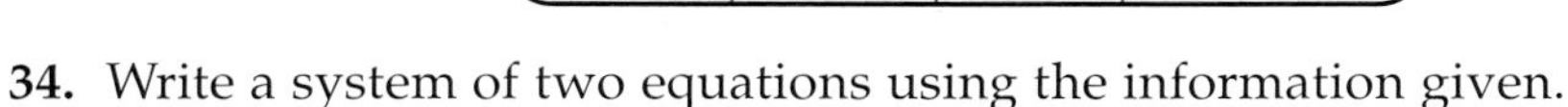

34. Write a system of two equations using the information given.

35. Find the price per yard of each fabric.

PRICING **For Exercises 36 and 37, use the following information.**
The Harvest Nut Company sells made-to-order trail mixes. Santito's favorite mix contains peanuts, raisins, and carob-coated pretzels. Peanuts sell for $3.20 per pound, raisins are $2.40 per pound, and the carob-coated pretzels are $4.00 per pound. Santito chooses to have twice as many pounds of pretzels as raisins, wants 5 pounds of mix, and can afford $16.80.

36. Write a system of three equations using the information given.

37. How many pounds of peanuts, raisins, and carob-coated pretzels can Santito buy?

38. **CRITICAL THINKING** In Cramer's Rule, if the value of the determinant is zero, what must be true of the graph of the system of equations represented by the determinant? Give examples to support your answer.

39. **WRITING IN MATH** Answer the question that was posed at the beginning of the lesson.

How is Cramer's Rule used to solve systems of equations?

Include the following in your answer.
- an explanation of how Cramer's rule uses determinants, and
- a situation where Cramer's rule would be easier to solve a system of equations than substitution or elimination and why.

Career Choices

Interior Designer

Interior designers work closely with architects and clients to develop a design that is not only aesthetic, but also functional, within budget, and meets all building codes.

Source: www.uwec.edu

Online Research
For information about a career as an interior designer, visit:
www.algebra2.com/careers

Standardized Test Practice

Standards Practice

40. Use Cramer's Rule to solve the system of equations $3x + 8y = 28$ and $5x - 7y = -55$.

(A) (3, 5) (B) (−4, 5) (C) (4, 2) (D) (−5, 4)

41. **SHORT RESPONSE** Find the measures of $\angle ABC$ and $\angle CBD$.

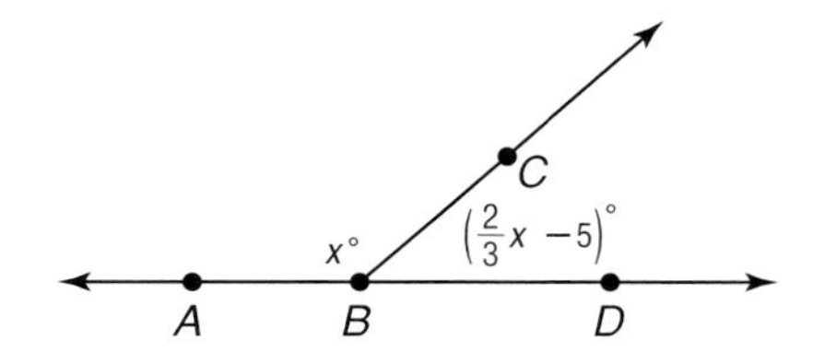

Maintain Your Skills

Mixed Review

Find the value of each determinant. *(Lesson 4-5)*

42. $\begin{vmatrix} 3 & 2 \\ -2 & 4 \end{vmatrix}$ 43. $\begin{vmatrix} 8 & 6 \\ 4 & 8 \end{vmatrix}$ 44. $\begin{vmatrix} -5 & 2 \\ 4 & 9 \end{vmatrix}$

For Exercises 45–47, use the following information.
Triangle ABC with vertices $A(0, 2)$, $B(-3, -1)$, and $C(-2, -4)$ is translated 1 unit right and 3 units up. *(Lesson 4-4)*

45. Write the translation matrix.
46. Find the coordinates of $\triangle A'B'C'$.
47. Graph the preimage and the image.

Solve each system of equations by graphing. *(Lesson 3-1)*

48. $y = 3x + 5$
$y = -2x - 5$

49. $x + y = 7$
$\frac{1}{2}x - y = -1$

50. $x - 2y = 10$
$2x - 4y = 12$

51. **BUSINESS** The Friendly Fix-It Company charges a base fee of \$35 for any in-home repair. In addition, the technician charges \$10 per hour. Write an equation for the cost c of an in-home repair of h hours. *(Lesson 1-3)*

Getting Ready for the Next Lesson

PREREQUISITE SKILL **Find each product, if possible.**
*(To review **multiplying matrices**, see Lesson 4-3.)*

52. $\begin{bmatrix} 2 & 5 \end{bmatrix} \cdot \begin{bmatrix} 3 & 1 \\ -2 & 6 \end{bmatrix}$ 53. $\begin{bmatrix} 0 & 9 \\ 5 & 7 \end{bmatrix} \cdot \begin{bmatrix} 2 & -6 \\ 8 & 1 \end{bmatrix}$ 54. $\begin{bmatrix} 5 & -4 \\ 8 & 3 \end{bmatrix} \cdot \begin{bmatrix} 5 \\ 1 \end{bmatrix}$

Practice Quiz 2

Lessons 4-4 through 4-6

For Exercises 1–3, reflect square $ABCD$ with vertices $A(1, 2)$, $B(4, -1)$, $C(1, -4)$, and $D(-2, -1)$ over the y-axis. *(Lesson 4-4)*

1. Write the coordinates in a vertex matrix.
2. Find the coordinates of $A'B'C'D'$.
3. Graph $ABCD$ and $A'B'C'D'$.

Find the value of each determinant. *(Lesson 4-5)*

4. $\begin{vmatrix} 3 & -2 \\ 5 & 4 \end{vmatrix}$ 5. $\begin{vmatrix} -8 & 3 \\ 6 & 5 \end{vmatrix}$ 6. $\begin{vmatrix} 1 & 3 & -2 \\ 7 & 0 & 4 \\ -3 & 5 & -1 \end{vmatrix}$ 7. $\begin{vmatrix} 3 & 4 & 4 \\ 2 & 1 & 5 \\ 0 & -8 & 6 \end{vmatrix}$

Use Cramer's Rule to solve each system of equations. *(Lesson 4-6)*

8. $3x - 2y = 7$
$4x - y = 6$

9. $7r + 5s = 3$
$3r - 2s = 22$

10. $3a - 5b + 2c = -5$
$4a + b + 3c = 9$
$2a - c = 1$

4-7 Identity and Inverse Matrices

What You'll Learn

- Determine whether two matrices are inverses.
- Find the inverse of a 2×2 matrix.

Vocabulary

- identity matrix
- inverse

How are inverse matrices used in cryptography?

With the rise of Internet shopping, ensuring the privacy of the user's personal information has become an important priority. Companies protect their computers by using codes. Cryptography is a method of preparing coded messages that can only be deciphered by using the "key" to the message.

The following technique is a simplified version of how cryptography works.

- First, assign a number to each letter of the alphabet.
- Convert your message into a matrix and multiply it by the coding matrix. The message is now unreadable to anyone who does not have the key to the code.
- To decode the message, the recipient of the coded message would multiply by the opposite, or inverse, of the coding matrix.

Code								
_ 0	A 1	B 2	C 3	D 4	E 5	F 6	G 7	H 8
I 9	J 10	K 11	L 12	M 13	N 14	O 15	P 16	Q 17
R 18	S 19	T 20	U 21	V 22	W 23	X 24	Y 25	Z 26

IDENTITY AND INVERSE MATRICES Recall that in real numbers, two numbers are inverses if their product is the identity, 1. Similarly, for matrices, the **identity matrix** is a square matrix that, when multiplied by another matrix, equals that same matrix. If A is any $n \times n$ matrix and I is the $n \times n$ identity matrix, then $A \cdot I = A$ and $I \cdot A = A$.

2 × 2 Identity Matrix

$$\begin{bmatrix} 1 & 0 \\ 0 & 1 \end{bmatrix}$$

3 × 3 Identity Matrix

$$\begin{bmatrix} 1 & 0 & 0 \\ 0 & 1 & 0 \\ 0 & 0 & 1 \end{bmatrix}$$

Key Concept — Identity Matrix for Multiplication

- **Words** The identity matrix for multiplication I is a square matrix with 1 for every element of the main diagonal, from upper left to lower right, and 0 in all other positions. For any square matrix A of the same dimension as I, $A \cdot I = I \cdot A = A$.
- **Symbols** If $A = \begin{bmatrix} a & b \\ c & d \end{bmatrix}$, then $I = \begin{bmatrix} 1 & 0 \\ 0 & 1 \end{bmatrix}$ such that

$$\begin{bmatrix} a & b \\ c & d \end{bmatrix} \cdot \begin{bmatrix} 1 & 0 \\ 0 & 1 \end{bmatrix} = \begin{bmatrix} 1 & 0 \\ 0 & 1 \end{bmatrix} \cdot \begin{bmatrix} a & b \\ c & d \end{bmatrix} = \begin{bmatrix} a & b \\ c & d \end{bmatrix}.$$

Two $n \times n$ matrices are **inverses** of each other if their product is the identity matrix. If matrix A has an inverse symbolized by A^{-1}, then $A \cdot A^{-1} = A^{-1} \cdot A = I$.

Example 1 Verify Inverse Matrices

Determine whether each pair of matrices are inverses.

a. $X = \begin{bmatrix} 2 & 2 \\ -1 & 4 \end{bmatrix}$ **and** $Y = \begin{bmatrix} \frac{1}{2} & \frac{1}{2} \\ -1 & \frac{1}{4} \end{bmatrix}$

Check to see if $X \cdot Y = I$.

$$X \cdot Y = \begin{bmatrix} 2 & 2 \\ -1 & 4 \end{bmatrix} \cdot \begin{bmatrix} \frac{1}{2} & \frac{1}{2} \\ -1 & \frac{1}{4} \end{bmatrix}$$ Write an equation.

$$= \begin{bmatrix} 1 - 2 & 1 + \frac{1}{2} \\ -\frac{1}{2} + (-4) & -\frac{1}{2} + 1 \end{bmatrix} \text{ or } \begin{bmatrix} -1 & 1\frac{1}{2} \\ -4\frac{1}{2} & \frac{1}{2} \end{bmatrix}$$ Matrix multiplication

Since $X \cdot Y \neq I$, they are *not* inverses.

b. $P = \begin{bmatrix} 3 & 4 \\ 1 & 2 \end{bmatrix}$ **and** $Q = \begin{bmatrix} 1 & -2 \\ -\frac{1}{2} & \frac{3}{2} \end{bmatrix}$

Find $P \cdot Q$.

$$P \cdot Q = \begin{bmatrix} 3 & 4 \\ 1 & 2 \end{bmatrix} \cdot \begin{bmatrix} 1 & -2 \\ -\frac{1}{2} & \frac{3}{2} \end{bmatrix}$$ Write an equation.

$$= \begin{bmatrix} 3 - 2 & -6 + 6 \\ 1 - 1 & -2 + 3 \end{bmatrix} \text{ or } \begin{bmatrix} 1 & 0 \\ 0 & 1 \end{bmatrix}$$ Matrix multiplication

Now find $Q \cdot P$.

$$Q \cdot P = \begin{bmatrix} 1 & -2 \\ -\frac{1}{2} & \frac{3}{2} \end{bmatrix} \cdot \begin{bmatrix} 3 & 4 \\ 1 & 2 \end{bmatrix}$$ Write an equation.

$$= \begin{bmatrix} 3 - 2 & 4 - 4 \\ -\frac{3}{2} + \frac{3}{2} & -2 + 3 \end{bmatrix} \text{ or } \begin{bmatrix} 1 & 0 \\ 0 & 1 \end{bmatrix}$$ Matrix multiplication

Since $P \cdot Q = Q \cdot P = I$, P and Q are inverses.

Study Tip

Verifying Inverses
Since multiplication of matrices is not commutative, it is necessary to check the products in both orders.

FIND INVERSE MATRICES Some matrices do not have an inverse. You can determine whether a matrix has an inverse by using the determinant.

Key Concept — Inverse of a 2 × 2 Matrix

The inverse of matrix $A = \begin{bmatrix} a & b \\ c & d \end{bmatrix}$ is $A^{-1} = \frac{1}{ad - bc}\begin{bmatrix} d & -b \\ -c & a \end{bmatrix}$, where $ad - bc \neq 0$.

Notice that $ad - bc$ is the value of det A. Therefore, if the value of the determinant of a matrix is 0, the matrix cannot have an inverse.

Example 2 Find the Inverse of a Matrix

Find the inverse of each matrix, if it exists.

a. $R = \begin{bmatrix} -4 & -3 \\ 8 & 6 \end{bmatrix}$

Find the value of the determinant.

$$\begin{vmatrix} -4 & -3 \\ 8 & 6 \end{vmatrix} = -24 - (-24) = 0$$

Since the determinant equals 0, R^{-1} does not exist.

b. $P = \begin{bmatrix} 3 & 1 \\ 5 & 2 \end{bmatrix}$

Find the value of the determinant.

$$\begin{vmatrix} 3 & 1 \\ 5 & 2 \end{vmatrix} = 6 - 5 \text{ or } 1$$

Since the determinant does not equal 0, P^{-1} exists.

$$P^{-1} = \frac{1}{ad - bc}\begin{bmatrix} d & -b \\ -c & a \end{bmatrix} \quad \text{Definition of inverse}$$

$$= \frac{1}{3(2) - 1(5)}\begin{bmatrix} 2 & -1 \\ -5 & 3 \end{bmatrix} \quad a = 3, b = 1, c = 5, d = 2$$

$$= 1\begin{bmatrix} 2 & -1 \\ -5 & 3 \end{bmatrix} \text{ or } \begin{bmatrix} 2 & -1 \\ -5 & 3 \end{bmatrix} \quad \text{Simplify.}$$

CHECK $\begin{bmatrix} 2 & -1 \\ -5 & 3 \end{bmatrix} \cdot \begin{bmatrix} 3 & 1 \\ 5 & 2 \end{bmatrix} = \begin{bmatrix} 6 - 5 & 2 - 2 \\ -15 + 15 & -5 + 6 \end{bmatrix} = \begin{bmatrix} 1 & 0 \\ 0 & 1 \end{bmatrix}$ ✓

More About. . .

Cryptography

The Enigma was a German coding machine used in World War II. Its code was considered to be unbreakable. However, the code was eventually solved by a group of Polish mathematicians.

Source: www.bletchleypark.org.uk

Matrices can be used to code messages by placing the message in a $2 \times n$ matrix.

Example 3 Use Inverses to Solve a Problem

a. CRYPTOGRAPHY Use the table at the beginning of the lesson to assign a number to each letter in the message GO_TONIGHT. Then code the message with the matrix $A = \begin{bmatrix} 2 & 1 \\ 4 & 3 \end{bmatrix}$**.**

Convert the message to numbers using the table.

G O _ T O N I G H T

7 | 15 | 0 | 20 | 15 | 14 | 9 | 7 | 8 | 20

Write the message in matrix form. Then multiply the message matrix B by the coding matrix A.

$$BA = \begin{bmatrix} 7 & 15 \\ 0 & 20 \\ 15 & 14 \\ 9 & 7 \\ 8 & 20 \end{bmatrix} \cdot \begin{bmatrix} 2 & 1 \\ 4 & 3 \end{bmatrix} \quad \text{Write an equation.}$$

$$= \begin{bmatrix} 14 + 60 & 7 + 45 \\ 0 + 80 & 0 + 60 \\ 30 + 56 & 15 + 42 \\ 18 + 28 & 9 + 21 \\ 16 + 80 & 8 + 60 \end{bmatrix} \quad \text{Matrix multiplication}$$

(continued on the next page)

Study Tip

Messages
If there is an odd number of letters to be coded, add a 0 at the end of the message.

$$= \begin{bmatrix} 74 & 52 \\ 80 & 60 \\ 86 & 57 \\ 46 & 30 \\ 96 & 68 \end{bmatrix} \quad \text{Simplify.}$$

The coded message is 74 | 52 | 80 | 60 | 86 | 57 | 46 | 30 | 96 | 68.

b. Use the inverse matrix A^{-1} to decode the message in Example 3a.

First find the inverse matrix of $A = \begin{bmatrix} 2 & 1 \\ 4 & 3 \end{bmatrix}$.

$$A^{-1} = \frac{1}{ad - bc}\begin{bmatrix} d & -b \\ -c & a \end{bmatrix} \quad \text{Definition of inverse}$$

$$= \frac{1}{2(3) - (1)(4)}\begin{bmatrix} 3 & -1 \\ -4 & 2 \end{bmatrix} \quad a = 2, b = 1, c = 4, d = 3$$

$$= \frac{1}{2}\begin{bmatrix} 3 & -1 \\ -4 & 2 \end{bmatrix} \text{ or } \begin{bmatrix} \frac{3}{2} & -\frac{1}{2} \\ -2 & 1 \end{bmatrix} \quad \text{Simplify.}$$

Next, decode the message by multiplying the coded matrix C by A^{-1}.

$$CA^{-1} = \begin{bmatrix} 74 & 52 \\ 80 & 60 \\ 86 & 57 \\ 46 & 30 \\ 96 & 68 \end{bmatrix} \cdot \begin{bmatrix} \frac{3}{2} & -\frac{1}{2} \\ -2 & 1 \end{bmatrix}$$

$$= \begin{bmatrix} 111 - 104 & -37 + 52 \\ 120 - 120 & -40 + 60 \\ 129 - 114 & -43 + 57 \\ 69 - 60 & -23 + 30 \\ 144 - 136 & -48 + 68 \end{bmatrix}$$

$$= \begin{bmatrix} 7 & 15 \\ 0 & 20 \\ 15 & 14 \\ 9 & 7 \\ 8 & 20 \end{bmatrix}$$

Use the table again to convert the numbers to letters. You can now read the message.

7 | 15 | 0 | 20 | 15 | 14 | 9 | 7 | 8 | 20

G O _ T O N I G H T

Check for Understanding

Concept Check

1. **Write** the 4 × 4 identity matrix.
2. **Explain** how to find the inverse of a 2 × 2 matrix.
3. **OPEN ENDED** Create a square matrix that does not have an inverse.

Guided Practice

Determine whether each pair of matrices are inverses.

4. $A = \begin{bmatrix} 2 & -1 \\ 1 & -3 \end{bmatrix}, B = \begin{bmatrix} \frac{1}{2} & 0 \\ 0 & -\frac{1}{3} \end{bmatrix}$

5. $X = \begin{bmatrix} 3 & 1 \\ 5 & 2 \end{bmatrix}, Y = \begin{bmatrix} 2 & -1 \\ -5 & 3 \end{bmatrix}$

Find the inverse of each matrix, if it exists.

6. $\begin{bmatrix} 8 & -5 \\ -3 & 2 \end{bmatrix}$

7. $\begin{bmatrix} 4 & -8 \\ -1 & 2 \end{bmatrix}$

8. $\begin{bmatrix} -5 & 1 \\ 7 & 4 \end{bmatrix}$

Application

9. **CRYPTOGRAPHY** Select a headline from a newspaper or the title of a magazine article and code it using your own coding matrix. Give your message and the coding matrix to a friend to decode. (*Hint*: Use a coding matrix whose determinant is 1 and that has all positive elements.)

Practice and Apply

Homework Help

For Exercises	See Examples
10–19, 32, 33	1
20–31	2
34–41	3

Extra Practice
See page 836.

Determine whether each pair of matrices are inverses.

10. $P = \begin{bmatrix} 0 & 1 \\ 1 & 1 \end{bmatrix}, Q = \begin{bmatrix} -1 & 1 \\ 1 & 0 \end{bmatrix}$

11. $R = \begin{bmatrix} 2 & 2 \\ 3 & 4 \end{bmatrix}, S = \begin{bmatrix} 2 & -1 \\ -\frac{3}{2} & 1 \end{bmatrix}$

12. $A = \begin{bmatrix} 6 & 2 \\ 5 & 2 \end{bmatrix}, B = \begin{bmatrix} 1 & 1 \\ -\frac{5}{2} & -3 \end{bmatrix}$

13. $X = \begin{bmatrix} \frac{1}{3} & -\frac{2}{3} \\ \frac{2}{3} & -\frac{1}{3} \end{bmatrix}, Y = \begin{bmatrix} 1 & 2 \\ 2 & 1 \end{bmatrix}$

14. $C = \begin{bmatrix} 1 & 5 \\ 1 & -2 \end{bmatrix}, D = \begin{bmatrix} \frac{2}{7} & \frac{5}{7} \\ \frac{1}{7} & -\frac{1}{7} \end{bmatrix}$

15. $J = \begin{bmatrix} 1 & 2 & 3 \\ 2 & 3 & 1 \\ 1 & 1 & 2 \end{bmatrix}, K = \begin{bmatrix} -\frac{5}{4} & \frac{1}{4} & \frac{7}{4} \\ \frac{3}{4} & \frac{1}{4} & -\frac{5}{4} \\ \frac{1}{4} & -\frac{1}{4} & \frac{1}{4} \end{bmatrix}$

Determine whether each statement is *true* or *false*.

16. Only square matrices have multiplicative identities.

17. Only square matrices have multiplicative inverses.

18. Some square matrices do not have multiplicative inverses.

19. Some square matrices do not have multiplicative identities.

Find the inverse of each matrix, if it exists.

20. $\begin{bmatrix} 5 & 0 \\ 0 & 1 \end{bmatrix}$

21. $\begin{bmatrix} 6 & 3 \\ 8 & 4 \end{bmatrix}$

22. $\begin{bmatrix} 1 & 2 \\ 2 & 1 \end{bmatrix}$

23. $\begin{bmatrix} 3 & 1 \\ -4 & 1 \end{bmatrix}$

24. $\begin{bmatrix} -3 & -2 \\ 6 & 4 \end{bmatrix}$

25. $\begin{bmatrix} -3 & 7 \\ 2 & -6 \end{bmatrix}$

26. $\begin{bmatrix} 4 & -3 \\ 2 & 7 \end{bmatrix}$

27. $\begin{bmatrix} -2 & 0 \\ 5 & 6 \end{bmatrix}$

28. $\begin{bmatrix} -4 & 6 \\ 6 & -9 \end{bmatrix}$

29. $\begin{bmatrix} 2 & -5 \\ 6 & 1 \end{bmatrix}$

30. $\begin{bmatrix} \frac{1}{2} & -\frac{3}{4} \\ \frac{1}{6} & \frac{1}{4} \end{bmatrix}$

31. $\begin{bmatrix} \frac{3}{10} & \frac{5}{8} \\ \frac{1}{5} & \frac{3}{4} \end{bmatrix}$

32. Compare the matrix used to reflect a figure over the x-axis to the matrix used to reflect a figure over the y-axis.

a. Are they inverses?

b. Does your answer make sense based on the geometry? Use a drawing to support your answer.

33. The matrix used to rotate a figure 270° counterclockwise about the origin is $\begin{bmatrix} 0 & 1 \\ -1 & 0 \end{bmatrix}$. Compare this matrix with the matrix used to rotate a figure 90° counterclockwise about the origin.

a. Are they inverses?

b. Does your answer make sense based on the geometry? Use a drawing to support your answer.

GEOMETRY **For Exercises 34–38, use the figure below.**

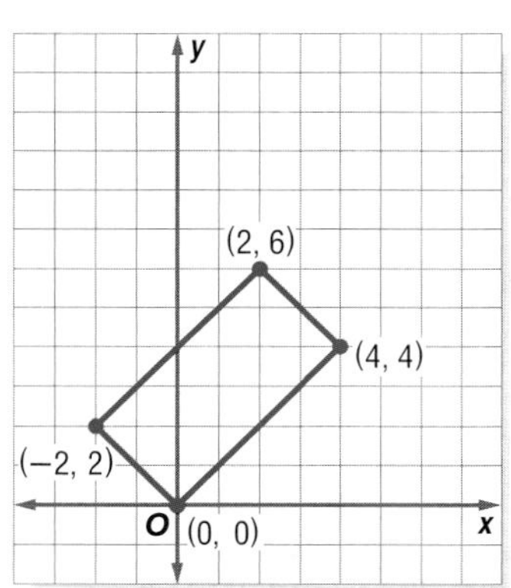

34. Write the vertex matrix A for the rectangle.

35. Use matrix multiplication to find BA if $B = \begin{bmatrix} 2 & 0 \\ 0 & 2 \end{bmatrix}$.

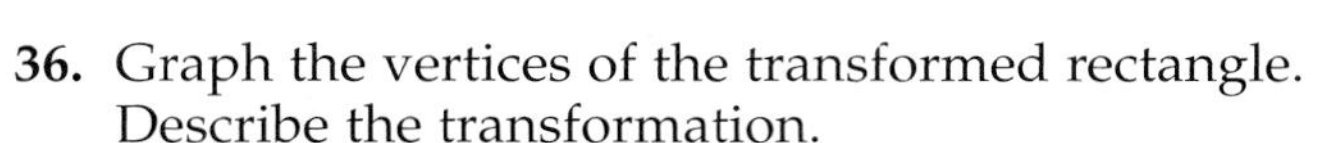

36. Graph the vertices of the transformed rectangle. Describe the transformation.

37. Make a conjecture about what transformation B^{-1} describes on a coordinate plane.

38. Test your conjecture. Find B^{-1} and multiply it by the result of BA. Make a drawing to verify your conjecture.

CRYPTOGRAPHY **For Exercises 39–41, use the alphabet table below.**
Your friend has sent you a series of messages that were coded with the coding matrix $C = \begin{bmatrix} 2 & 1 \\ 1 & 1 \end{bmatrix}$. Use the inverse of matrix C to decode each message.

Code		
A 26	**J** 17	**S** 8
B 25	**K** 16	**T** 7
C 24	**L** 15	**U** 6
D 23	**M** 14	**V** 5
E 22	**N** 13	**W** 4
F 21	**O** 12	**X** 3
G 20	**P** 11	**Y** 2
H 19	**Q** 10	**Z** 1
I 18	**R** 9	_ 0

39. 50 | 36 | 51 | 29 | 18 | 18 | 26 | 13 | 33 | 26 | 44 | 22 | 48 | 33 | 59 | 34 | 61 | 35 | 4 | 2

40. 59 | 33 | 8 | 8 | 39 | 21 | 7 | 7 | 56 | 37 | 25 | 16 | 4 | 2

41. 59 | 34 | 49 | 31 | 40 | 20 | 16 | 14 | 21 | 15 | 25 | 25 | 36 | 24 | 32 | 16

42. **RESEARCH** Use the Internet or other reference to find examples of codes used throughout history. Explain how messages were coded.

43. **CRITICAL THINKING** For which values of a, b, c, and d will $A = \begin{bmatrix} a & b \\ c & d \end{bmatrix} = A^{-1}$?

44. **WRITING IN MATH** Answer the question that was posed at the beginning of the lesson.

How are inverse matrices used in cryptography?

Include the following in your answer:

- an explanation of why the inverse matrix works in decoding a message, and
- a description of the conditions you must consider when writing a message in matrix form.

45. What is the inverse of $\begin{bmatrix} 4 & 1 \\ 10 & 2 \end{bmatrix}$?

Ⓐ $\begin{bmatrix} -1 & \frac{1}{2} \\ 5 & -2 \end{bmatrix}$ Ⓑ $\begin{bmatrix} 2 & -1 \\ -10 & 4 \end{bmatrix}$ Ⓒ $\begin{bmatrix} 1 & 5 \\ \frac{1}{2} & 2 \end{bmatrix}$ Ⓓ $\begin{bmatrix} -2 & \frac{1}{2} \\ 5 & -1 \end{bmatrix}$

46. Which matrix does *not* have an inverse?

Ⓐ $\begin{bmatrix} 5 & 3 \\ 2 & 4 \end{bmatrix}$ Ⓑ $\begin{bmatrix} 1 & 2 \\ 2 & 1 \end{bmatrix}$ Ⓒ $\begin{bmatrix} -3 & -3 \\ 6 & -6 \end{bmatrix}$ Ⓓ $\begin{bmatrix} -10 & -5 \\ 8 & 4 \end{bmatrix}$

Graphing Calculator

INVERSE FUNCTION The x^{-1} key on a TI-83 Plus is used to find the inverse of a matrix. If you get a SINGULAR MATRIX error on the screen, then the matrix has no inverse.

Use a graphing calculator to find the inverse of each matrix.

47. $\begin{bmatrix} -11 & 9 \\ 6 & -5 \end{bmatrix}$

48. $\begin{bmatrix} 12 & 4 \\ 15 & 5 \end{bmatrix}$

49. $\begin{bmatrix} 2 & -1 \\ 1 & -3 \end{bmatrix}$

50. $\begin{bmatrix} 25 & -4 \\ -35 & 6 \end{bmatrix}$

51. $\begin{bmatrix} 2 & 5 & 2 \\ 1 & 4 & 1 \\ 6 & 3 & 3 \end{bmatrix}$

52. $\begin{bmatrix} 3 & 1 & 2 \\ -2 & 0 & 4 \\ 3 & 5 & 2 \end{bmatrix}$

Maintain Your Skills

Mixed Review

Use Cramer's Rule to solve each system of equations. *(Lesson 4-6)*

53. $3x + 2y = -2$
$x - 3y = 14$

54. $2x + 5y = 35$
$7x - 4y = -28$

55. $4x - 3z = -23$
$-2x - 5y + z = -9$
$y - z = 3$

Evaluate each determinant by using diagonals or expansion by minors. *(Lesson 4-5)*

56. $\begin{vmatrix} 2 & 8 & -6 \\ 4 & 5 & 2 \\ -3 & -6 & -1 \end{vmatrix}$

57. $\begin{vmatrix} -3 & -3 & 1 \\ -9 & -2 & 3 \\ 5 & -2 & -1 \end{vmatrix}$

58. $\begin{vmatrix} 5 & -7 & 3 \\ -1 & 2 & -9 \\ 5 & -7 & 3 \end{vmatrix}$

Find the slope of the line that passes through each pair of points. *(Lesson 2-3)*

59. (2, 5), (6, 9)

60. (1, 0), (−2, 9)

61. (−5, 4), (−3, −6)

62. (−2, 2), (−5, 1)

63. (0, 3), (−2, −2)

64. (−8, 9), (0, 6)

65. **OCEANOGRAPHY** The deepest point in any ocean, the bottom of the Mariana Trench in the Pacific Ocean, is 6.8 miles below sea level. Water pressure in the ocean is represented by the function $f(x) = 1.15x$, where x is the depth in miles and $f(x)$ is the pressure in tons per square inch. Find the water pressure at the deepest point in the Mariana Trench. *(Lesson 2-1)*

Evaluate each expression. *(Lesson 1-1)*

66. $3(2^3 + 1)$

67. $7 - 5 \div 2 + 1$

68. $\frac{9 - 4 \cdot 3}{6}$

69. $[40 - (7 + 9)] \div 8$

70. $[(-2 + 8)6 + 1]8$

71. $(4 - 1)(8 + 2)^2$

Getting Ready for the Next Lesson

PREREQUISITE SKILL **Solve each equation.**
*(To review **solving multi-step equations**, see Lesson 1-3.)*

72. $3k + 8 = 5$

73. $12 = -5h + 2$

74. $7z - 4 = 5z + 8$

75. $\frac{x}{2} + 5 = 7$

76. $\frac{3 + n}{6} = -4$

77. $6 = \frac{s - 8}{-7}$

4-8 Using Matrices to Solve Systems of Equations

What You'll Learn

- Write matrix equations for systems of equations.
- Solve systems of equations using matrix equations.

How can matrices be used in population ecology?

Population ecology is the study of a species or a group of species that inhabits the same area. A biologist is studying two species of birds that compete for food and territory. He estimates that a particular region with an area of 14.25 acres (approximately 69,000 square yards) can supply 20,000 pounds of food for the birds during their nesting season.

Species A needs 140 pounds of food and has a territory of 500 square yards per nesting pair. Species B needs 120 pounds of food and has a territory of 400 square yards per nesting pair. The biologist can use this information to find the number of birds of each species that the area can support.

Vocabulary

- matrix equation

California Standards
Standard 2.0
Students solve systems of linear equations and inequalities (**in two or three variables**) by substitution, with graphs, or **with matrices.** (Key)

WRITE MATRIX EQUATIONS The situation above can be represented using a system of equations that can be solved using matrices. Consider the system of equations below. You can write this system with matrices by using the left and right sides of the equations.

$$\begin{array}{l} 5x + 7y = 11 \\ 3x + 8y = 18 \end{array} \rightarrow \begin{bmatrix} 5x + 7y \\ 3x + 8y \end{bmatrix} = \begin{bmatrix} 11 \\ 18 \end{bmatrix}$$

Write the matrix on the left as the product of the coefficients and the variables.

$$\underset{\text{coefficient matrix}}{\overset{A}{\begin{bmatrix} 5 & 7 \\ 3 & 8 \end{bmatrix}}} \cdot \underset{\text{variable matrix}}{\overset{X}{\begin{bmatrix} x \\ y \end{bmatrix}}} = \underset{\text{constant matrix}}{\overset{B}{\begin{bmatrix} 11 \\ 18 \end{bmatrix}}}$$

The system of equations is now expressed as a **matrix equation**.

Example 1 Two-Variable Matrix Equation

Write a matrix equation for the system of equations.
$5x - 6y = -47$
$3x + 2y = -17$

Determine the coefficient, variable, and constant matrices.

$$\begin{array}{l} 5x - 6y = -47 \\ 3x + 2y = -17 \end{array} \rightarrow \begin{bmatrix} 5 & -6 \\ 3 & 2 \end{bmatrix} \quad \begin{bmatrix} x \\ y \end{bmatrix} \quad \begin{bmatrix} -47 \\ -17 \end{bmatrix}$$

Write the matrix equation.

$$\overset{A}{\begin{bmatrix} 5 & -6 \\ 3 & 2 \end{bmatrix}} \cdot \overset{X}{\begin{bmatrix} x \\ y \end{bmatrix}} = \overset{B}{\begin{bmatrix} -47 \\ -17 \end{bmatrix}}$$

You can use a matrix equation to determine the weight of an atom of an element.

Example 2 Solve a Problem Using a Matrix Equation

More About. . .

Chemistry

Atomic mass units are relative units of weight because they were compared to the weight of a hydrogen atom. So a molecule of nitrogen, whose weight is 14.0 amu, weighs 14 times as much as a hydrogen atom.

Source: www.sizes.com

CHEMISTRY The molecular formula for glucose is $C_6H_{12}O_6$, which represents that a molecule of glucose has 6 carbon (C) atoms, 12 hydrogen (H) atoms, and 6 oxygen (O) atoms. One molecule of glucose weighs 180 atomic mass units (amu), and one oxygen atom weighs 16 atomic mass units. The formulas and weights for glucose and another sugar, sucrose, are listed below.

Sugar	Formula	Atomic Weight (amu)
glucose	$C_6H_{12}O_6$	180
sucrose	$C_{12}H_{22}O_{11}$	342

a. Write a system of equations that represents the weight of each atom.

Let c represent the weight of a carbon atom.
Let h represent the weight of a hydrogen atom.

Write an equation for the weight of each sugar. The subscript represents how many atoms of each element are in the molecule.

Glucose: $6c + 12h + 6(16) = 180$ Equation for glucose

$6c + 12h + 96 = 180$ Simplify.

$6c + 12h = 84$ Subtract 96 from each side.

Sucrose: $12c + 22h + 11(16) = 342$ Equation for sucrose

$12c + 22h + 176 = 342$ Simplify.

$12c + 22h = 166$ Subtract 176 from each side.

b. Write a matrix equation for the system of equations.

Determine the coefficient, variable, and constant matrices.

$$\begin{matrix} 6c + 12h = 84 \\ 12c + 22h = 166 \end{matrix} \rightarrow \begin{bmatrix} 6 & 12 \\ 12 & 22 \end{bmatrix} \quad \begin{bmatrix} c \\ h \end{bmatrix} \quad \begin{bmatrix} 84 \\ 166 \end{bmatrix}$$

Write the matrix equation.

$$\underset{A}{\begin{bmatrix} 6 & 12 \\ 12 & 22 \end{bmatrix}} \cdot \underset{X}{\begin{bmatrix} c \\ h \end{bmatrix}} = \underset{B}{\begin{bmatrix} 84 \\ 166 \end{bmatrix}}$$

You will solve this matrix equation in Exercise 11.

SOLVE SYSTEMS OF EQUATIONS You can solve a system of linear equations by solving a matrix equation. A matrix equation in the form $AX = B$, where A is a coefficient matrix, X is a variable matrix, and B is a constant matrix, can be solved in a similar manner as a linear equation of the form $ax = b$.

$ax = b$	Write the equation.	$AX = B$
$\left(\frac{1}{a}\right)ax = \left(\frac{1}{a}\right)b$	Multiply each side by the inverse of the coefficient, if it exists.	$A^{-1}AX = A^{-1}B$
$1x = \left(\frac{1}{a}\right)b$	$\left(\frac{1}{a}\right)a = 1, A^{-1}A = I$	$IX = A^{-1}B$
$x = \left(\frac{1}{a}\right)b$	$1x = x, IX = X$	$X = A^{-1}B$

Notice that the solution of the matrix equation is the product of the inverse of the coefficient matrix and the constant matrix.

Example 3 Solve a System of Equations

Use a matrix equation to solve the system of equations.
$6x + 2y = 11$
$3x - 8y = 1$

The matrix equation is $\begin{bmatrix} 6 & 2 \\ 3 & -8 \end{bmatrix} \cdot \begin{bmatrix} x \\ y \end{bmatrix} = \begin{bmatrix} 11 \\ 1 \end{bmatrix}$, when $A = \begin{bmatrix} 6 & 2 \\ 3 & -8 \end{bmatrix}$, $X = \begin{bmatrix} x \\ y \end{bmatrix}$, and $B = \begin{bmatrix} 11 \\ 1 \end{bmatrix}$.

Step 1 Find the inverse of the coefficient matrix.

$$A^{-1} = \frac{1}{-48 - 6}\begin{bmatrix} -8 & -2 \\ -3 & 6 \end{bmatrix} \text{ or } -\frac{1}{54}\begin{bmatrix} -8 & -2 \\ -3 & 6 \end{bmatrix}$$

Step 2 Multiply each side of the matrix equation by the inverse matrix.

$$-\frac{1}{54}\begin{bmatrix} -8 & -2 \\ -3 & 6 \end{bmatrix} \cdot \begin{bmatrix} 6 & 2 \\ 3 & -8 \end{bmatrix} \cdot \begin{bmatrix} x \\ y \end{bmatrix} = -\frac{1}{54}\begin{bmatrix} -8 & -2 \\ -3 & 6 \end{bmatrix} \cdot \begin{bmatrix} 11 \\ 1 \end{bmatrix}$$ Multiply each side by A^{-1}

$$\begin{bmatrix} 1 & 0 \\ 0 & 1 \end{bmatrix} \cdot \begin{bmatrix} x \\ y \end{bmatrix} = -\frac{1}{54}\begin{bmatrix} -90 \\ -27 \end{bmatrix}$$ Multiply matrices.

$$\begin{bmatrix} x \\ y \end{bmatrix} = \begin{bmatrix} \frac{5}{3} \\ \frac{1}{2} \end{bmatrix}$$ $\begin{bmatrix} 1 & 0 \\ 0 & 1 \end{bmatrix} = I$

The solution is $\left(\frac{5}{3}, \frac{1}{2}\right)$. Check this solution in the original equation.

Study Tip

Identity Matrix
The identity matrix on the left verifies that the inverse matrix has been calculated correctly.

Example 4 System of Equations with No Solution

Use a matrix equation to solve the system of equations.
$6a - 9b = -18$
$8a - 12b = 24$

The matrix equation is $\begin{bmatrix} 6 & -9 \\ 8 & -12 \end{bmatrix} \cdot \begin{bmatrix} a \\ b \end{bmatrix} = \begin{bmatrix} -18 \\ 24 \end{bmatrix}$, when $A = \begin{bmatrix} 6 & -9 \\ 8 & -12 \end{bmatrix}$, $X = \begin{bmatrix} a \\ b \end{bmatrix}$, and $B = \begin{bmatrix} -18 \\ 24 \end{bmatrix}$.

Find the inverse of the coefficient matrix.

$$A^{-1} = \frac{1}{-72 + 72}\begin{bmatrix} -12 & 9 \\ -8 & 6 \end{bmatrix}$$

The determinant of the coefficient matrix $\begin{bmatrix} 6 & -9 \\ 8 & -12 \end{bmatrix}$ is 0, so A^{-1} does not exist. There is no unique solution of this system.

Graph the system of equations. Since the lines are parallel, this system has no solution. Therefore, the system is inconsistent.

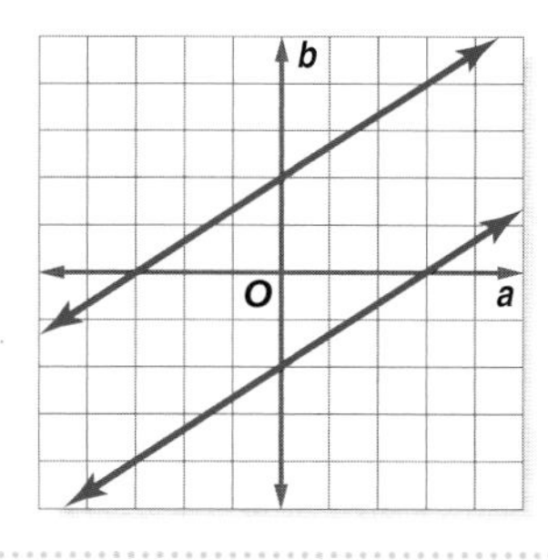

Study Tip

Look Back
To review **inconsistent systems of equations**, see Lesson 3-1.

To solve a system of equations with three variables, you can use the 3×3 identity matrix. However, finding the inverse of a 3×3 matrix may be tedious. Graphing calculators and computer programs offer fast and accurate methods for performing the necessary calculations.

Graphing Calculator Investigation

Systems of Three Equations in Three Variables

You can use a graphing calculator and a matrix equation to solve systems of equations. Consider the system of equations below.

$3x - 2y + z = 0$
$2x + 3y - z = 17$
$5x - y + 4z = -7$

Think and Discuss

1. Write a matrix equation for the system of equations.
2. Enter the coefficient matrix as matrix A and the constant matrix as matrix B in the graphing calculator. Find the product of A^{-1} and B. Recall that the x^{-1} key is used to find A^{-1}.
3. How is the result related to the solution?

Check for Understanding

Concept Check

1. **Write** the matrix equation $\begin{bmatrix} 2 & -3 \\ 1 & 4 \end{bmatrix} \cdot \begin{bmatrix} r \\ s \end{bmatrix} = \begin{bmatrix} 4 \\ -2 \end{bmatrix}$ as a system of linear equations.
2. **OPEN ENDED** Write a system of equations that does not have a unique solution.
3. **FIND THE ERROR** Tommy and Laura are solving a system of equations. They find that $A^{-1} = \begin{bmatrix} 3 & -2 \\ -7 & 5 \end{bmatrix}$, $B = \begin{bmatrix} -7 \\ -9 \end{bmatrix}$, and $X = \begin{bmatrix} x \\ y \end{bmatrix}$.

Tommy

$$\begin{bmatrix} x \\ y \end{bmatrix} = \begin{bmatrix} 3 & -2 \\ -7 & 5 \end{bmatrix} \cdot \begin{bmatrix} -7 \\ -9 \end{bmatrix}$$
$$\begin{bmatrix} x \\ y \end{bmatrix} = \begin{bmatrix} -3 \\ 4 \end{bmatrix}$$

Laura

$$\begin{bmatrix} x \\ y \end{bmatrix} = \begin{bmatrix} -7 \\ -9 \end{bmatrix} \cdot \begin{bmatrix} 3 & -2 \\ -7 & 5 \end{bmatrix}$$
$$\begin{bmatrix} x \\ y \end{bmatrix} = \begin{bmatrix} 42 \\ 31 \end{bmatrix}$$

Who is correct? Explain your reasoning.

Guided Practice

Write a matrix equation for each system of equations.

4. $x - y = -3$
   $x + 3y = 5$
5. $2g + 3h = 8$
   $-4g - 7h = -5$
6. $3a - 5b + 2c = 9$
   $4a + 7b + c = 3$
   $2a - c = 12$

Solve each matrix equation or system of equations by using inverse matrices.

7. $\begin{bmatrix} 3 & 1 \\ 4 & -2 \end{bmatrix} \cdot \begin{bmatrix} x \\ y \end{bmatrix} = \begin{bmatrix} 13 \\ 24 \end{bmatrix}$
8. $\begin{bmatrix} 8 & -1 \\ 2 & 3 \end{bmatrix} \cdot \begin{bmatrix} a \\ b \end{bmatrix} = \begin{bmatrix} 16 \\ -9 \end{bmatrix}$
9. $5x - 3y = -30$
   $8x + 5y = 1$
10. $5s + 4t = 12$
    $4s - 3t = -1.25$

Application

11. **CHEMISTRY** Refer to Example 2 on page 203. Solve the system of equations to find the weight of a carbon, hydrogen, and oxygen atom.

Practice and Apply

Homework Help

For Exercises	See Examples
12–19	1
20–31	3, 4
32–34	2

Extra Practice
See page 836.

Write a matrix equation for each system of equations.

12. $3x - y = 0$
$x + 2y = -21$

13. $4x - 7y = 2$
$3x + 5y = 9$

14. $5a - 6b = -47$
$3a + 2b = -17$

15. $3m - 7n = -43$
$6m + 5n = -10$

16. $2a + 3b - 5c = 1$
$7a + 3c = 7$
$3a - 6b + c = -5$

17. $3x - 5y + 2z = 9$
$x - 7y + 3z = 11$
$4x - 3z = -1$

18. $x - y = 8$
$-2x - 5y - 6z = -27$
$9x + 10y - z = 54$

19. $3r - 5s + 6t = 21$
$11r - 12s + 16t = 15$
$-5r + 8s - 3t = -7$

Solve each matrix equation or system of equations by using inverse matrices.

20. $\begin{bmatrix} 7 & -3 \\ 2 & 5 \end{bmatrix} \cdot \begin{bmatrix} m \\ n \end{bmatrix} = \begin{bmatrix} 41 \\ 0 \end{bmatrix}$

21. $\begin{bmatrix} 3 & 1 \\ 2 & -1 \end{bmatrix} \cdot \begin{bmatrix} a \\ b \end{bmatrix} = \begin{bmatrix} 13 \\ 2 \end{bmatrix}$

22. $\begin{bmatrix} 4 & -3 \\ 5 & 2 \end{bmatrix} \cdot \begin{bmatrix} a \\ b \end{bmatrix} = \begin{bmatrix} -17 \\ -4 \end{bmatrix}$

23. $\begin{bmatrix} 7 & 1 \\ 3 & -8 \end{bmatrix} \cdot \begin{bmatrix} x \\ y \end{bmatrix} = \begin{bmatrix} 43 \\ 10 \end{bmatrix}$

24. $\begin{bmatrix} 2 & -9 \\ 6 & 5 \end{bmatrix} \cdot \begin{bmatrix} c \\ d \end{bmatrix} = \begin{bmatrix} 28 \\ -12 \end{bmatrix}$

25. $\begin{bmatrix} 6 & 5 \\ 3 & 2 \end{bmatrix} \cdot \begin{bmatrix} a \\ b \end{bmatrix} = \begin{bmatrix} 18 \\ 7 \end{bmatrix}$

26. $6r + s = 9$
$3r = -2s$

27. $5a + 9b = -28$
$2a - b = -2$

28. $p - 2q = 1$
$p + 5q = 22$

29. $4m - 7n = -63$
$3m + 2n = 18$

30. $x + 2y = 8$
$3x + 2y = 6$

31. $4x - 3y = 5$
$2x + 9y = 6$

32. **PILOT TRAINING** Hai-Ling is training for his pilot's license. Flight instruction costs \$105 per hour, and the simulator costs \$45 per hour. The school requires students to spend 4 more hours in airplane training than in the simulator. If Hai-Ling can afford to spend \$3870 on training, how many hours can he spend training in an airplane and in a simulator?

33. **SCHOOLS** The graphic shows that student-to-teacher ratios are dropping in both public and private schools. If these rates of change remain constant, predict when the student-to-teacher ratios for private and public schools will be the same.

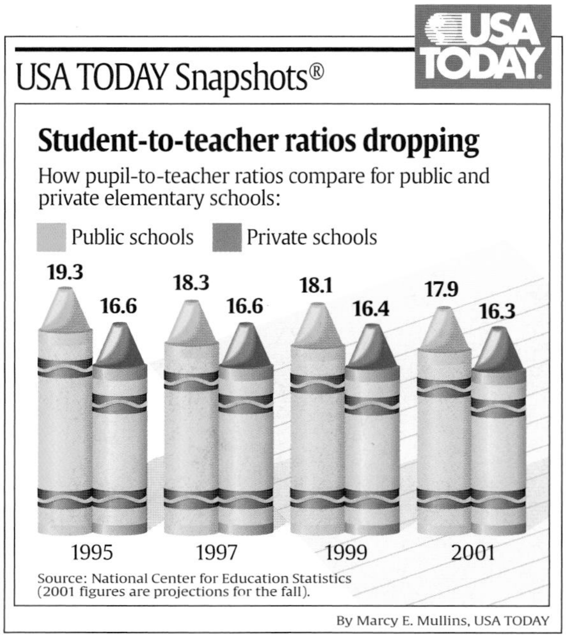

USA TODAY

USA TODAY Snapshots®

Student-to-teacher ratios dropping

How pupil-to-teacher ratios compare for public and private elementary schools:

Year	Public schools	Private schools
1995	19.3	16.6
1997	18.3	16.6
1999	18.1	16.4
2001	17.9	16.3

Source: National Center for Education Statistics (2001 figures are projections for the fall).

By Marcy E. Mullins, USA TODAY

34. **CHEMISTRY** Cara is preparing an acid solution. She needs 200 milliliters of 48% concentration solution. Cara has 60% and 40% concentration solutions in her lab. How many milliliters of 40% acid solution should be mixed with 60% acid solution to make the required amount of 48% acid solution?

35. **CRITICAL THINKING** Describe the solution set of a system of equations if the coefficient matrix does not have an inverse.

36. **WRITING IN MATH** Answer the question that was posed at the beginning of the lesson.

How can matrices be used in population ecology?

Include the following in your answer:

- a system of equations that can be used to find the number of each species the region can support, and
- a solution of the problem using matrices.

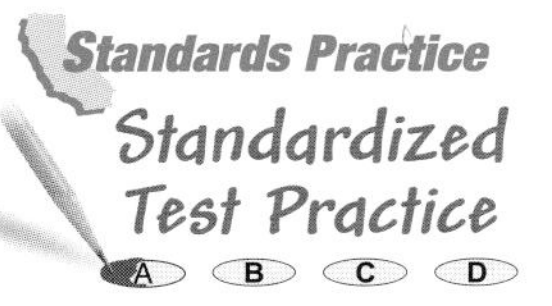

37. Solve the system of equations $6a + 8b = 5$ and $10a - 12b = 2$.

(A) $\left(\frac{3}{4}, \frac{1}{2}\right)$ (B) $\left(\frac{1}{2}, \frac{3}{4}\right)$ (C) $\left(\frac{1}{2}, -\frac{1}{2}\right)$ (D) $\left(\frac{1}{2}, \frac{1}{4}\right)$

38. **SHORT RESPONSE** The Yogurt Shoppe sells cones in three sizes: small $0.89; medium, $1.19; and large, $1.39. One day Scott sold 52 cones. He sold seven more medium cones than small cones. If he sold $58.98 in cones, how many of each size did he sell?

INVERSE MATRICES **Use a graphing calculator to solve each system of equations using inverse matrices.**

39. $2a - b + 4c = 6$
$a + 5b - 2c = -6$
$3a - 2b + 6c = 8$

40. $3x - 5y + 2z = 22$
$2x + 3y - z = -9$
$4x + 3y + 3z = 1$

41. $2q + r + s = 2$
$-q - r + 2s = 7$
$-3q + 2r + 3s = 7$

Maintain Your Skills

Mixed Review

Find the inverse of each matrix, if it exists. *(Lesson 4-7)*

42. $\begin{bmatrix} 4 & 4 \\ 2 & 3 \end{bmatrix}$

43. $\begin{bmatrix} 9 & 5 \\ 7 & 4 \end{bmatrix}$

44. $\begin{bmatrix} -3 & -6 \\ 5 & 10 \end{bmatrix}$

Use Cramer's Rule to solve each system of equations. *(Lesson 4-6)*

45. $6x + 7y = 10$
$3x - 4y = 20$

46. $6a + 7b = -10.15$
$9.2a - 6b = 69.944$

47. $\frac{x}{2} - \frac{2y}{3} = 2\frac{1}{3}$
$3x + 4y = -50$

48. **ECOLOGY** If you recycle a $3\frac{1}{2}$-foot stack of newspapers, one less 20-foot loblolly pine tree will be needed for paper. Use a prediction equation to determine how many feet of loblolly pine trees will *not* be needed for paper if you recycle a pile of newspapers 20 feet tall. *(Lesson 2-5)*

Solve each equation. Check your solutions. *(Lesson 1-4)*

49. $|x - 3| = 7$

50. $-4|d + 2| = -12$

51. $5|k - 4| = k + 8$

WebQuest Internet Project

Lessons in Home Buying, Selling

It is time to complete your project. Use the information and data you have gathered about home buying and selling to prepare a portfolio or Web page. Be sure to include your tables, graphs, and calculations. You may also wish to include additional data, information, or pictures.

www.algebra2.com/webquest

Graphing Calculator Investigation

A Follow-Up of Lesson 4-8

California Standards Standard 2.0 **Students solve systems of linear equations** and inequalities **(in two or three variables)** by substitution, with graphs, or **with matrices.** (Key)

Augmented Matrices

Using a TI-83 Plus, you can solve a system of linear equations using the MATRX function. An **augmented matrix** contains the coefficient matrix with an extra column containing the constant terms.

The reduced row echelon function of a graphing calculator reduces the augmented matrix so that the solution of the system of equations can be easily determined.

Write an augmented matrix for the following system of equations. Then solve the system by using the reduced row echelon form on the graphing calculator.

$3x + y + 3z = 2$
$2x + y + 2z = 1$
$4x + 2y + 5z = 5$

Step 1 Write the augmented matrix and enter it into a calculator.

The augmented matrix $B = \left[\begin{array}{ccc|c} 3 & 1 & 3 & 2 \\ 2 & 1 & 2 & 1 \\ 4 & 2 & 5 & 5 \end{array}\right]$.

Begin by entering the matrix.

KEYSTROKES: *Review matrices on page 163.*

Step 2 Find the reduced row echelon form (rref) using the graphing calculator.

KEYSTROKES: 2nd [MATRX] ▶ ALPHA [B] 2nd [MATRX] 2) ENTER

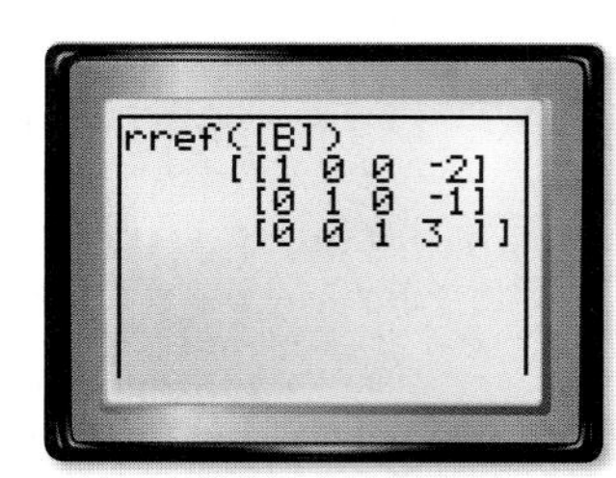

Study the reduced echelon matrix. The first three columns are the same as a 3×3 identity matrix. The first row represents $x = -2$, the second row represents $y = -1$, and the third row represents $z = 3$. The solution is $(-2, -1, 3)$.

Exercises

Write an augmented matrix for each system of equations. Then solve with a graphing calculator.

1. $x - 3y = 5$
$2x + y = 1$

2. $15x + 11y = 36$
$4x - 3y = -26$

3. $2x + y = 5$
$2x - 3y = 1$

4. $3x - y = 0$
$2x - 3y = 1$

5. $3x - 2y + z = -2$
$x - y + 3z = 5$
$-x + y + z = -1$

6. $x - y + z = 2$
$x - z = 1$
$y + 2z = 0$

www.algebra2.com/other_calculator_keystrokes

Chapter 4

Study Guide and Review

Vocabulary and Concept Check

Choose the correct term to complete each sentence.

determinant
dilation
dimensions
equal matrices
identity matrix
isometry
matrix equation
rotation
scalar multiplication
translation

1. The matrix $\begin{bmatrix} 1 & 0 & 0 \\ 0 & 1 & 0 \\ 0 & 0 & 1 \end{bmatrix}$ is a(n) _______ for multiplication.
2. When an image and a preimage are congruent, then the transformation is called a(n) _______.
3. _______ is the process of multiplying a matrix by a constant.
4. A(n) _______ is when a figure is moved around a center point.
5. The _______ of $\begin{bmatrix} -1 & 4 \\ 2 & -3 \end{bmatrix}$ is -5.
6. A(n) _______ is the product of the coefficient matrix and the variable matrix equal to the constant matrix.
7. The _______ of a matrix tell how many rows and columns are in the matrix.
8. A(n) _______ occurs when a figure is moved from one location to another on the coordinate plane.
9. The matrices $\begin{bmatrix} 3x \\ x + 2y \end{bmatrix}$ and $\begin{bmatrix} y \\ 7 \end{bmatrix}$ are _______ if $x = 1$ and $y = 3$.
10. A(n) _______ is when a geometric figure is enlarged or reduced.

Lesson-by-Lesson Review

4-1 Introduction to Matrices

See pages 154–158.

Concept Summary

- A matrix is a rectangular array of variables or constants in horizontal rows and vertical columns.
- Equal matrices have the same dimensions and corresponding elements equal.

Example **Solve $\begin{bmatrix} 2x \\ y \end{bmatrix} = \begin{bmatrix} 32 + 6y \\ 7 - x \end{bmatrix}$ for x and y.**

Since the matrices are equal, corresponding elements are equal. You can write two linear equations.

$2x = 32 + 6y$
$y = 7 - x$

(continued on the next page)

Solve the system of equations.

$2x = 32 + 6y$	First equation
$2x = 32 + 6(7 - x)$	Substitute $7 - x$ for y.
$2x = 32 + 42 - 6x$	Distributive Property
$8x = 74$	Add $6x$ to each side.
$x = 9.25$	Divide each side by 8.

The solution is (9.25, −2.25).

To find the value for y, substitute 9.25 for x in either equation.

$y = 7 - x$	Second equation
$= 7 - 9.25$	Substitute 9.25 for x.
$= -2.25$	Simplify.

Exercises **Solve each equation.** *See Example 3 on pages 155 and 156.*

11. $\begin{bmatrix} 2y - x \\ x \end{bmatrix} = \begin{bmatrix} 3 \\ 4y - 1 \end{bmatrix}$

12. $\begin{bmatrix} 7x \\ x + y \end{bmatrix} = \begin{bmatrix} 5 + 2y \\ 11 \end{bmatrix}$

13. $\begin{bmatrix} 3x + y \\ x - 3y \end{bmatrix} = \begin{bmatrix} -3 \\ -1 \end{bmatrix}$

14. $\begin{bmatrix} 2x - y \\ 6x - y \end{bmatrix} = \begin{bmatrix} 2 \\ 22 \end{bmatrix}$

4-2 Operations with Matrices

See pages 160–166.

Concept Summary

- Matrices can be added or subtracted if they have the same dimensions. Add or subtract corresponding elements.
- To multiply a matrix by a scalar k, multiply each element in the matrix by k.

Examples

1 **Find $A - B$ if $A = \begin{bmatrix} 3 & 8 \\ -5 & 2 \end{bmatrix}$ and $B = \begin{bmatrix} -4 & 6 \\ 1 & 9 \end{bmatrix}$.**

$A - B = \begin{bmatrix} 3 & 8 \\ -5 & 2 \end{bmatrix} - \begin{bmatrix} -4 & 6 \\ 1 & 9 \end{bmatrix}$ Definition of matrix subtraction

$= \begin{bmatrix} 3 - (-4) & 8 - 6 \\ -5 - 1 & 2 - 9 \end{bmatrix}$ Subtract corresponding elements.

$= \begin{bmatrix} 7 & 2 \\ -6 & -7 \end{bmatrix}$ Simplify.

2 **If $X = \begin{bmatrix} 3 & 2 & -1 \\ 4 & -6 & 0 \end{bmatrix}$, find $4X$.**

$4X = 4\begin{bmatrix} 3 & 2 & -1 \\ 4 & -6 & 0 \end{bmatrix}$

$= \begin{bmatrix} 4(3) & 4(2) & 4(-1) \\ 4(4) & 4(-6) & 4(0) \end{bmatrix}$ or $\begin{bmatrix} 12 & 8 & -4 \\ 16 & -24 & 0 \end{bmatrix}$ Multiply each element by 4.

Exercises **Perform the indicated matrix operations. If the matrix does not exist, write *impossible*.** *See Examples 1, 2, and 4 on pages 160–162.*

15. $\begin{bmatrix} -4 & 3 \\ -5 & 2 \end{bmatrix} + \begin{bmatrix} 1 & -3 \\ 3 & -8 \end{bmatrix}$

16. $[0.2 \quad 1.3 \quad -0.4] - [2 \quad 1.7 \quad 2.6]$

17. $\begin{bmatrix} 1 & -5 \\ -2 & 3 \end{bmatrix} + \frac{3}{4}\begin{bmatrix} 0 & 4 \\ -16 & 8 \end{bmatrix}$

18. $\begin{bmatrix} 1 & 0 & -3 \\ 4 & -5 & 2 \end{bmatrix} - 2\begin{bmatrix} -2 & 3 & 5 \\ -3 & -1 & 2 \end{bmatrix}$

4-3 Multiplying Matrices

See pages 167–174.

Concept Summary

- Two matrices can be multiplied if and only if the number of columns in the first matrix is equal to the number of rows in the second matrix.

Example **Find XY if $X = [6 \quad 4 \quad 1]$ and $Y = \begin{bmatrix} 2 & 5 \\ -3 & 0 \\ -1 & 3 \end{bmatrix}$.**

$XY = [6 \quad 4 \quad 1] \cdot \begin{bmatrix} 2 & 5 \\ -3 & 0 \\ -1 & 3 \end{bmatrix}$ Write an equation.

$= [6(2) + 4(-3) + 1(-1) \quad 6(5) + 4(0) + 1(3)]$ Multiply columns by rows.

$= [-1 \quad 33]$ Simplify.

Exercises **Find each product, if possible.** *See Example 2 on page 168.*

19. $[2 \quad 7] \cdot \begin{bmatrix} 5 \\ -4 \end{bmatrix}$

20. $\begin{bmatrix} 8 & -3 \\ 6 & 1 \end{bmatrix} \cdot \begin{bmatrix} 2 & -3 \\ 1 & -5 \end{bmatrix}$

21. $\begin{bmatrix} 3 & 4 \\ 1 & 0 \\ 2 & -5 \end{bmatrix} \cdot \begin{bmatrix} -2 & 4 & 5 \\ 3 & 0 & -1 \\ 1 & 0 & -1 \end{bmatrix}$

22. $\begin{bmatrix} 3 & 0 & -1 \\ 4 & -2 & 3 \end{bmatrix} \cdot \begin{bmatrix} 7 & 1 \\ 6 & -3 \\ 2 & 1 \end{bmatrix}$

4-4 Transformations with Matrices

See pages 175–181.

Concept Summary

- Use matrix addition and a translation matrix to find the coordinates of a translated figure.
- Use scalar multiplication to perform dilations.
- To reflect a figure, multiply the vertex matrix on the left by a reflection matrix.

reflection over x-axis: $\begin{bmatrix} 1 & 0 \\ 0 & -1 \end{bmatrix}$ reflection over y-axis: $\begin{bmatrix} -1 & 0 \\ 0 & 1 \end{bmatrix}$ reflection over line $y = x$: $\begin{bmatrix} 0 & 1 \\ 1 & 0 \end{bmatrix}$

- To rotate a figure counterclockwise about the origin, multiply the vertex matrix on the left by a rotation matrix.

$90°$ rotation: $\begin{bmatrix} 0 & -1 \\ 1 & 0 \end{bmatrix}$ $180°$ rotation: $\begin{bmatrix} -1 & 0 \\ 0 & -1 \end{bmatrix}$ $270°$ rotation: $\begin{bmatrix} 0 & 1 \\ -1 & 0 \end{bmatrix}$

Example **Find the coordinates of the vertices of the image of $\triangle PQR$ with $P(4, 2)$, $Q(6, 5)$, and $R(0, 5)$ after it is rotated 90° counterclockwise about the origin.**

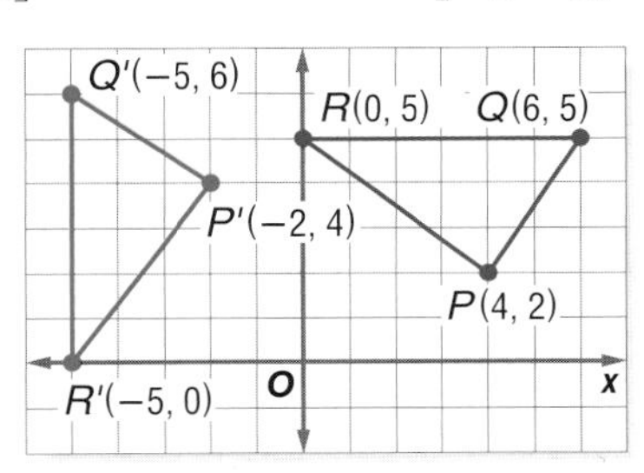

Write the ordered pairs in a vertex matrix. Then multiply the vertex matrix by the rotation matrix.

$\begin{bmatrix} 0 & -1 \\ 1 & 0 \end{bmatrix} \cdot \begin{bmatrix} 4 & 6 & 0 \\ 2 & 5 & 5 \end{bmatrix} = \begin{bmatrix} -2 & -5 & -5 \\ 4 & 6 & 0 \end{bmatrix}$

The coordinates of the vertices of $\triangle P'Q'R'$ are $P'(-2, 4)$, $Q'(-5, 6)$, and $R'(-5, 0)$.

Exercises **For Exercises 23–26, use the figure at the right.**
See Examples 1–5 on pages 175–178.

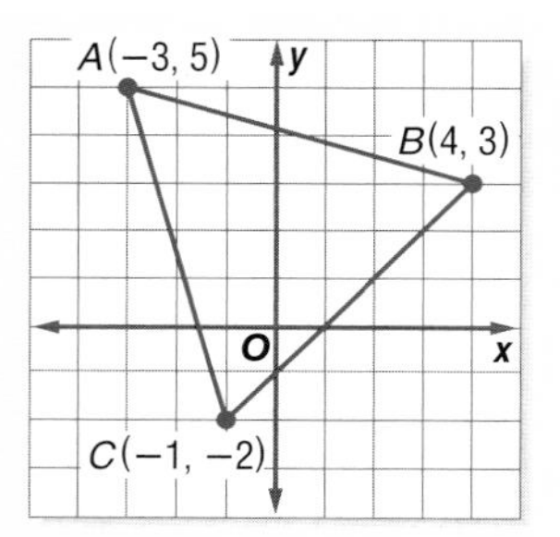

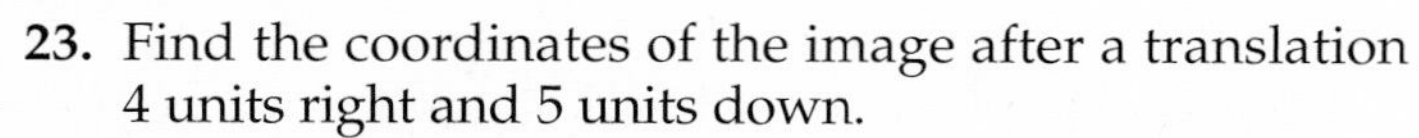

23. Find the coordinates of the image after a translation 4 units right and 5 units down.

24. Find the coordinates of the image of the figure after a dilation by a scale factor of 2.

25. Find the coordinates of the image after a reflection over the y-axis.

26. Find the coordinates of the image of the figure after a rotation of 180°.

4-5 Determinants

See pages 182–188.

Concept Summary

- Determinant of a 2 × 2 matrix: $\begin{vmatrix} a & b \\ c & d \end{vmatrix} = ad - bc$
- Determinant of a 3 × 3 matrix: $\begin{vmatrix} a & b & c \\ d & e & f \\ g & h & i \end{vmatrix} = a\begin{vmatrix} e & f \\ h & i \end{vmatrix} - b\begin{vmatrix} d & f \\ g & i \end{vmatrix} + c\begin{vmatrix} d & e \\ g & h \end{vmatrix}$
- Area of a triangle with vertices at (a, b), (c, d), and (e, f):

  $|A|$ where $A = \frac{1}{2}\begin{vmatrix} a & b & 1 \\ c & d & 1 \\ e & f & 1 \end{vmatrix}$

Examples

1 **Find the value of** $\begin{vmatrix} 3 & 6 \\ -4 & 2 \end{vmatrix}$.

$\begin{vmatrix} 3 & 6 \\ -4 & 2 \end{vmatrix} = 3(2) - (-4)(6)$ Definition of determinant

$= 6 - (-24)$ or 30 Simplify.

2 **Evaluate** $\begin{vmatrix} 3 & 1 & 5 \\ 1 & -2 & 1 \\ 0 & -1 & 2 \end{vmatrix}$ **using expansion by minors.**

$\begin{vmatrix} 3 & 1 & 5 \\ 1 & -2 & 1 \\ 0 & -1 & 2 \end{vmatrix} = 3\begin{vmatrix} -2 & 1 \\ -1 & 2 \end{vmatrix} - 1\begin{vmatrix} 1 & 1 \\ 0 & 2 \end{vmatrix} + 5\begin{vmatrix} 1 & -2 \\ 0 & -1 \end{vmatrix}$ Expansion by minors

$= 3(-4 - (-1)) - 1(2 - 0) + 5(-1 - 0)$ Evaluate 2 × 2 determinants.

$= -9 - 2 - 5$ or -16 Simplify.

Exercises **Find the value of each determinant.** *See Examples 1–3 on pages 182–184.*

27. $\begin{vmatrix} 4 & 11 \\ -7 & 8 \end{vmatrix}$

28. $\begin{vmatrix} 6 & -7 \\ 5 & 3 \end{vmatrix}$

29. $\begin{vmatrix} 12 & 8 \\ 9 & 6 \end{vmatrix}$

30. $\begin{vmatrix} 2 & -3 & 1 \\ 0 & 7 & 8 \\ 2 & 1 & 3 \end{vmatrix}$

31. $\begin{vmatrix} 7 & -4 & 5 \\ 1 & 3 & -6 \\ 5 & -1 & -2 \end{vmatrix}$

32. $\begin{vmatrix} 6 & 3 & -2 \\ -4 & 2 & 5 \\ -3 & -1 & 0 \end{vmatrix}$

4-6 Cramer's Rule

See pages 189–194.

Concept Summary

- Cramer's Rule for two variables:
  The solution of the system of equations $ax + by = e$ and $cx + dy = f$ is (x, y), where $x = \frac{\begin{vmatrix} e & b \\ f & d \end{vmatrix}}{\begin{vmatrix} a & b \\ c & d \end{vmatrix}}$, $y = \frac{\begin{vmatrix} a & e \\ c & f \end{vmatrix}}{\begin{vmatrix} a & b \\ c & d \end{vmatrix}}$, and $\begin{vmatrix} a & b \\ c & d \end{vmatrix} \neq 0$.

- Cramer's Rule for three variables:
  The solution of the system whose equations are $ax + by + cz = j$, $dx + ey + fz = k$, $gx + hy + iz = \ell$ is (x, y, z), where

$$x = \frac{\begin{vmatrix} j & b & c \\ k & e & f \\ \ell & h & i \end{vmatrix}}{\begin{vmatrix} a & b & c \\ d & e & f \\ g & h & i \end{vmatrix}},\ y = \frac{\begin{vmatrix} a & j & c \\ d & k & f \\ g & \ell & i \end{vmatrix}}{\begin{vmatrix} a & b & c \\ d & e & f \\ g & h & i \end{vmatrix}},\ z = \frac{\begin{vmatrix} a & b & j \\ d & e & k \\ g & h & \ell \end{vmatrix}}{\begin{vmatrix} a & b & c \\ d & e & f \\ g & h & i \end{vmatrix}}, \text{ and } \begin{vmatrix} a & b & c \\ d & e & f \\ g & h & i \end{vmatrix} \neq 0.$$

Example

Use Cramer's Rule to solve each system of equations $5a - 3b = 7$ and $3a + 9b = -3$.

$$a = \frac{\begin{vmatrix} 7 & -3 \\ -3 & 9 \end{vmatrix}}{\begin{vmatrix} 5 & -3 \\ 3 & 9 \end{vmatrix}} \quad \text{Cramer's Rule} \qquad b = \frac{\begin{vmatrix} 5 & 7 \\ 3 & -3 \end{vmatrix}}{\begin{vmatrix} 5 & -3 \\ 3 & 9 \end{vmatrix}}$$

$$= \frac{63 - 9}{45 + 9} \quad \text{Evaluate each determinant.} \qquad = \frac{-15 - 21}{45 + 9}$$

$$= \frac{54}{54} \text{ or } 1 \quad \text{Simplify.} \qquad = \frac{-36}{54} \text{ or } -\frac{2}{3}$$

The solution is $\left(1, -\frac{2}{3}\right)$.

Exercises **Use Cramer's Rule to solve each system of equations.**

See Examples 1 and 3 on pages 190 and 191.

33. $9a - b = 1$
$3a + 2b = 12$

34. $x + 5y = 14$
$-2x + 6y = 4$

35. $3x + 4y = -15$
$2x - 7y = 19$

36. $8a + 5b = 2$
$-6a - 4b = -1$

37. $6x - 7z = 13$
$8y + 2z = 14$
$7x + z = 6$

38. $2a - b - 3c = -20$
$4a + 2b + c = 6$
$2a + b - c = -6$

4-7 Identity and Inverse Matrices

See pages 195–201.

Concept Summary

- An identity matrix is a square matrix with ones on the diagonal and zeros in the other positions.
- Two matrices are inverses of each other if their product is the identity matrix.
- The inverse of matrix $A = \begin{bmatrix} a & b \\ c & d \end{bmatrix}$ is $A^{-1} \frac{1}{ad - bc}\begin{bmatrix} d & -b \\ -c & a \end{bmatrix}$, where $ad - bc \neq 0$.

• Extra Practice, see pages 834–836.
• Mixed Problem Solving, see page 865.

Example **Find the inverse of $S = \begin{bmatrix} 3 & -4 \\ 2 & 1 \end{bmatrix}$.**

Find the value of the determinant.

$\begin{vmatrix} 3 & -4 \\ 2 & 1 \end{vmatrix} = 3 - (-8)$ or 11

Use the formula for the inverse matrix.

$S^{-1} = \frac{1}{11}\begin{bmatrix} 1 & 4 \\ -2 & 3 \end{bmatrix}$

Exercises **Find the inverse of each matrix, if it exists.** *See Example 2 on page 197.*

39. $\begin{bmatrix} 3 & 2 \\ 4 & -2 \end{bmatrix}$ **40.** $\begin{bmatrix} 8 & 6 \\ 9 & 7 \end{bmatrix}$ **41.** $\begin{bmatrix} 2 & 4 \\ -3 & 6 \end{bmatrix}$

42. $\begin{bmatrix} 6 & -2 \\ 3 & -1 \end{bmatrix}$ **43.** $\begin{bmatrix} 0 & 2 \\ 5 & -4 \end{bmatrix}$ **44.** $\begin{bmatrix} 6 & -1 & 0 \\ 5 & 8 & -2 \end{bmatrix}$

4-8 Using Matrices to Solve Systems of Equations

See pages 202–207.

Concept Summary

- A system of equations can be written as a matrix equation in the form $A \cdot X = B$.

  $\begin{matrix} 2x + 3y = 12 \\ x - 4y = 6 \end{matrix} \rightarrow \begin{bmatrix} 2 & 3 \\ 1 & -4 \end{bmatrix} \cdot \begin{bmatrix} x \\ y \end{bmatrix} = \begin{bmatrix} 12 \\ 6 \end{bmatrix}$

- To solve a matrix equation, find the inverse of the coefficient matrix. Then multiply each side by the inverse matrix, so $X = A^{-1}B$.

Example Solve $\begin{bmatrix} 4 & 8 \\ 2 & -3 \end{bmatrix} \cdot \begin{bmatrix} x \\ y \end{bmatrix} = \begin{bmatrix} 12 \\ 13 \end{bmatrix}$.

Step 1 Find the inverse of the coefficient matrix.

$$A^{-1} = \frac{1}{-12 - 16}\begin{bmatrix} -3 & -8 \\ -2 & 4 \end{bmatrix} \text{ or } -\frac{1}{28}\begin{bmatrix} -3 & -8 \\ -2 & 4 \end{bmatrix}$$

Step 2 Multiply each side by the inverse matrix.

$$-\frac{1}{28}\begin{bmatrix} -3 & -8 \\ -2 & 4 \end{bmatrix} \cdot \begin{bmatrix} 4 & 8 \\ 2 & -3 \end{bmatrix} \cdot \begin{bmatrix} x \\ y \end{bmatrix} = -\frac{1}{28}\begin{bmatrix} -3 & -8 \\ -2 & 4 \end{bmatrix} \cdot \begin{bmatrix} 12 \\ 13 \end{bmatrix}$$

$$\begin{bmatrix} 1 & 0 \\ 0 & 1 \end{bmatrix} \cdot \begin{bmatrix} x \\ y \end{bmatrix} = -\frac{1}{28}\begin{bmatrix} -140 \\ 28 \end{bmatrix}$$

$$\begin{bmatrix} x \\ y \end{bmatrix} = \begin{bmatrix} 5 \\ -1 \end{bmatrix}$$

The solution is $(5, -1)$.

Exercises **Solve each matrix equation or system of equations by using inverse matrices.** *See Example 3 on page 204.*

45. $\begin{bmatrix} 5 & -2 \\ 1 & 3 \end{bmatrix} \cdot \begin{bmatrix} x \\ y \end{bmatrix} = \begin{bmatrix} 16 \\ 10 \end{bmatrix}$ **46.** $\begin{bmatrix} 4 & 1 \\ 3 & -2 \end{bmatrix} \cdot \begin{bmatrix} a \\ b \end{bmatrix} = \begin{bmatrix} 9 \\ 4 \end{bmatrix}$

47. $3x + 8 = -y$
$4x - 2y = -14$

48. $3x - 5y = -13$
$4x + 3y = 2$

Chapter 4 Practice Test

Vocabulary and Concepts

Choose the letter that best matches each description.

1. $\begin{vmatrix} a & b \\ c & d \end{vmatrix} = ad - bc$
2. $\begin{bmatrix} a & b \\ c & d \end{bmatrix} \cdot \begin{bmatrix} x \\ y \end{bmatrix} = \begin{bmatrix} e \\ f \end{bmatrix}$
3. $\frac{1}{ad - bc}\begin{bmatrix} d & -b \\ -c & a \end{bmatrix}$

a. inverse of $\begin{bmatrix} a & b \\ c & d \end{bmatrix}$

b. determinant of $\begin{bmatrix} a & b \\ c & d \end{bmatrix}$

c. matrix equation for $ax + by = e$ and $cx + dy = f$

Skills and Applications

Solve each equation.

4. $\begin{bmatrix} 3x + 1 \\ 2y \end{bmatrix} = \begin{bmatrix} 10 \\ 4 + y \end{bmatrix}$
5. $\begin{bmatrix} 2x & y + 1 \\ 13 & -2 \end{bmatrix} = \begin{bmatrix} -16 & -7 \\ 13 & z - 8 \end{bmatrix}$

Perform the indicated matrix operations. If the matrix does not exist, write *impossible*.

6. $\begin{bmatrix} 2 & -4 & 1 \\ 3 & 8 & -2 \end{bmatrix} - 2\begin{bmatrix} 1 & 2 & -4 \\ -2 & 3 & 7 \end{bmatrix}$
7. $\begin{bmatrix} 1 & 6 & 7 \\ 1 & -3 & -4 \end{bmatrix} \cdot \begin{bmatrix} -4 & 3 \\ -1 & -2 \\ 2 & 5 \end{bmatrix}$

Find the value of each determinant.

8. $\begin{vmatrix} -1 & 4 \\ -6 & 3 \end{vmatrix}$
9. $\begin{vmatrix} 5 & -3 & 2 \\ -6 & 1 & 3 \\ -1 & 4 & -7 \end{vmatrix}$

Find the inverse of each matrix, if it exists.

10. $\begin{bmatrix} -2 & 5 \\ 3 & 1 \end{bmatrix}$
11. $\begin{bmatrix} -6 & -3 \\ 8 & 4 \end{bmatrix}$
12. $\begin{bmatrix} 5 & -2 \\ 6 & 3 \end{bmatrix}$

Solve each matrix equation or system of equations by using inverse matrices.

13. $\begin{bmatrix} 1 & 8 \\ 2 & -6 \end{bmatrix} \cdot \begin{bmatrix} x \\ y \end{bmatrix} = \begin{bmatrix} -3 \\ -17 \end{bmatrix}$
14. $\begin{bmatrix} 5 & 7 \\ -9 & 3 \end{bmatrix} \cdot \begin{bmatrix} m \\ n \end{bmatrix} = \begin{bmatrix} 41 \\ -105 \end{bmatrix}$
15. $5a + 2b = -49$
    $2a + 9b = 5$

For Exercises 16–18, use $\triangle ABC$ whose vertices have coordinates $A(6, 3)$, $B(1, 5)$, and $C(-1, 4)$.

16. Use the determinant to find the area of $\triangle ABC$.
17. Translate $\triangle ABC$ so that the coordinates of B' are (3, 1). What are the coordinates of A' and C'?
18. Find the coordinates of the vertices of a similar triangle whose perimeter is five times that of $\triangle ABC$.
19. **RETAIL SALES** Brittany is preparing boxes of assorted chocolates. Chocolate-covered peanuts cost $7 per pound. Chocolate-covered caramels cost $6.50 per pound. The boxes of assorted candies contain five more pounds of peanut candies than caramel candies. If the total amount sold was $575, how many pounds of each candy were needed to make the boxes?

20. **STANDARDIZED TEST PRACTICE** If $\begin{bmatrix} 43 & z \\ 7x - 2 & 2x + 3 \end{bmatrix} = \begin{bmatrix} z + 3 & 2m + 5 \\ y & 37 \end{bmatrix}$, then $y =$

Ⓐ 120. Ⓑ 117. Ⓒ 22. Ⓓ not enough information

Chapter 4 Standardized Test Practice

Standards Practice

Part 1 Multiple Choice

Record your answers on the answer sheet provided by your teacher or on a sheet of paper.

1. If the average (arithmetic mean) of ten numbers is 18 and the average of six of these numbers is 12, what is the average of the other four numbers?

Ⓐ 15 Ⓑ 18

Ⓒ 27 Ⓓ 28

2. A car travels 65 miles per hour for 2 hours. A truck travels 60 miles per hour for 1.5 hours. What is the difference between the number of miles traveled by the car and the number of miles traveled by the truck?

Ⓐ 31.25 Ⓑ 40

Ⓒ 70 Ⓓ 220

3. In the figure, $a =$

Ⓐ 1.

Ⓑ 2.

Ⓒ 3.

Ⓓ 4.

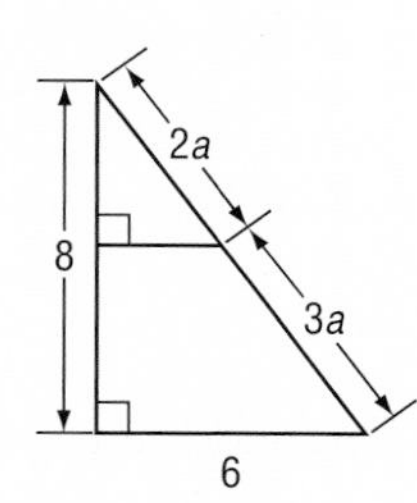

4. If the circumference of a circle is $\frac{4\pi}{3}$, then what is half of its area?

Ⓐ $\frac{2\pi}{9}$ Ⓑ $\frac{4\pi}{9}$

Ⓒ $\frac{8\pi}{9}$ Ⓓ $\frac{2\pi^2}{9}$

5. A line is represented by the equation $x = 6$. What is the slope of the line?

Ⓐ 0 Ⓑ $\frac{5}{6}$

Ⓒ 6 Ⓓ undefined

6. In the figure, $ABCD$ is a square inscribed in the circle centered at O. If $\overline{OB}$ is 10 units long, how many units long is minor arc BC?

Ⓐ $\frac{5}{2}\pi$ units

Ⓑ 5π units

Ⓒ 10π units

Ⓓ 20π units

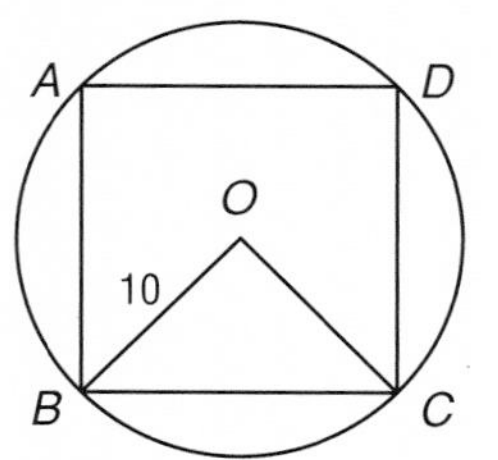

7. If $3 < x < 5 < y < 10$, then which of the following best defines $\frac{x}{y}$?

Ⓐ $\frac{3}{10} < \frac{x}{y} < 1$

Ⓑ $\frac{3}{10} < \frac{x}{y} < \frac{1}{2}$

Ⓒ $\frac{3}{5} < \frac{x}{y} < \frac{1}{2}$

Ⓓ $\frac{3}{5} < \frac{x}{y} < 1$

8. If $x + 3y = 12$ and $\frac{2}{3}x - y = 5$, then $x =$

Ⓐ 1. Ⓑ 8.

Ⓒ 9. Ⓓ 13.5.

9. At what point do the two lines with the equations $7x - 3y = 13$ and $y = 2x - 3$ intersect?

Ⓐ $(-4, -11)$ Ⓑ $(4, 11)$

Ⓒ $(4, 5)$ Ⓓ $(5, 4)$

10. If $N = \begin{bmatrix} -1 & 0 \\ 5 & -2 \end{bmatrix}$ and $M = \begin{bmatrix} -1 & 0 \\ 5 & 2 \end{bmatrix}$, find $N - M$.

Ⓐ $\begin{bmatrix} 0 & 0 \\ 0 & 0 \end{bmatrix}$ Ⓑ $\begin{bmatrix} 1 & 0 \\ 0 & 1 \end{bmatrix}$

Ⓒ $\begin{bmatrix} 0 & 0 \\ 0 & -4 \end{bmatrix}$ Ⓓ $\begin{bmatrix} -2 & 0 \\ 0 & -4 \end{bmatrix}$

Preparing for Standardized Tests
For test-taking strategies and more practice, see pages 877–892.

Part 2 Short Response/Grid In

Record your answers on the answer sheet provided by your teacher or on a sheet of paper.

11. A computer manufacturer reduced the price of its Model X computer by 3%. If the new price of the Model X computer is \$2489, then how much did the computer cost, in dollars, before its price was reduced? (Round to the nearest dollar.)

12. In square $PQRS$, $PQ = 4$, $PU = UQ$, and $PT = TS$. What is the area of the shaded region?

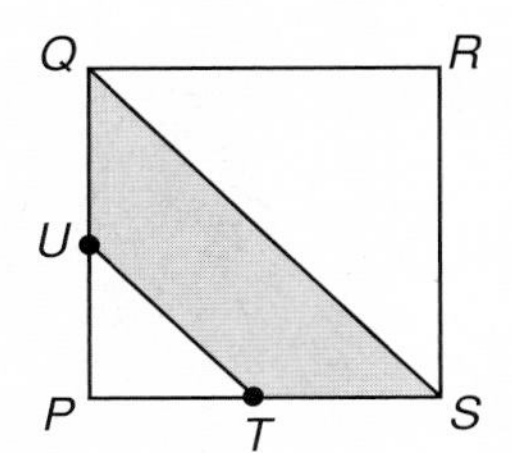

13. Write an equation of a line that passes through the origin and is parallel to the line with equation $3x - y = 5$.

14. A rectangular solid has two faces the same size and shape as Figure 1 and four faces the same size and shape as Figure 2. What is the volume of the solid in cubic units?

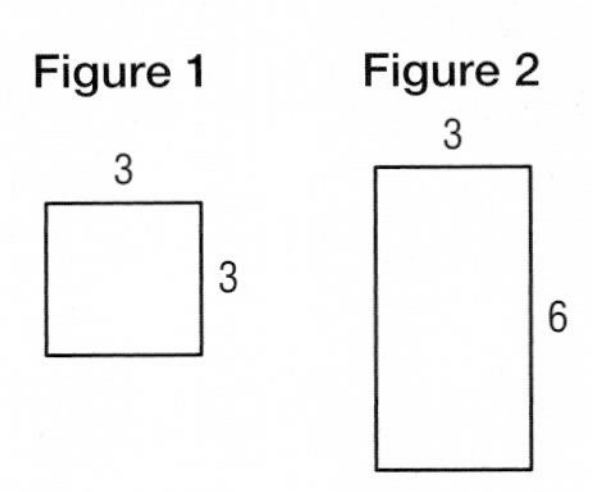

15. If the average (arithmetic mean) of three different positive integers is 60, what is the greatest possible value of one of the integers?

Test-Taking Tip

Questions 15 and 16
Watch for the phrases "greatest possible" or "least possible." Think logically about the conditions that make an expression greatest or least. Notice what types of numbers are used—positive, even, prime, integers.

16. The perimeter of a triangle is 15. The lengths of the sides are integers. If the length of one side is 6, what is the shortest possible length of another side of the triangle?

17. In this sequence below, each term after the first term is $\frac{1}{4}$ of the term preceding it. What is the sixth term of this sequence?

$$320, 80, 20, \ldots$$

18. If the sum of two numbers is 5 and their difference is 2, what is their product?

19. What positive value of k would make the lines below parallel in the coordinate plane?

$$9x + ky = 16$$
$$kx + 4y = 11$$

Part 3 Extended Response

Record your answers on a sheet of paper. Show your work.

For Exercises 20–22, use the information below.

The Colonial High School Yearbook Staff is selling yearbooks and chrome picture frames engraved with the year. The number of yearbooks and frames sold to members of each grade is shown in the table.

Sales for Each Class		
Grade	Yearbooks	Frames
9th	423	256
10th	464	278
11th	546	344
12th	575	497

20. Find the difference in the sales of yearbooks and frames made to the 10th and 11th grade classes.

21. Find the total numbers of yearbooks and frames sold.

22. A yearbook costs \$48, and a frame costs \$18. Find the total sales of books and frames sold to each class.

UNIT 2

Polynomial and Radical Equations and Inequalities

Equations that model real-world data allow you to make predictions about the future. In this unit, you will learn about nonlinear equations, including polynomial and radical equations, and inequalities.

WebQuest Internet Project

Population Explosion

The United Nations estimated that the world's population reached 6 billion in 1999. The population had doubled in about 40 years and gained 1 billion people in just 12 years. Assuming middle-range birth and death trends, world population is expected to exceed 9 billion by 2050, with most of the increase in countries that are less economically developed. In this project, you will use quadratic and polynomial mathematical models that will help you to project future populations.

Log on to www.algebra2.com/webquest. Begin your WebQuest by reading the Task.

Then continue working on your WebQuest as you study Unit 2.

Lesson	5-1	6-6	7-4
Page	227	326	369

USA TODAY Snapshots®

Tokyo leads population giants

The 10 most populous urban areas in the world:

	(Millions)
Tokyo	26.4
Mexico City	18.1
Bombay, India	18.1
Sao Paulo, Brazil	17.8
New York	16.6
Lagos, Nigeria	13.4
Los Angeles	13.1
Shanghai, China	12.9
Calcutta, India	12.9
Buenos Aires, Argentina	12.1

Source: United Nations

By Bob Laird, USA TODAY

Chapter 5

Polynomials

What You'll Learn

- **Lessons 5-1 through 5-4** Add, subtract, multiply, divide, and factor polynomials.
- **Lessons 5-5 through 5-8** Simplify and solve equations involving roots, radicals, and rational exponents.
- **Lesson 5-9** Perform operations with complex numbers.

Key Vocabulary

- scientific notation (p. 225)
- polynomial (p. 229)
- FOIL method (p. 230)
- synthetic division (p. 234)
- complex number (p. 271)

Why It's Important

Many formulas involve polynomials and/or square roots. For example, equations involving speeds or velocities of objects are often written with square roots. You can use such an equation to find the velocity of a roller coaster. *You will use an equation relating the velocity of a roller coaster and the height of a hill in Lesson 5-6.*

Getting Started

Prerequisite Skills To be successful in this chapter, you'll need to master these skills and be able to apply them in problem-solving situations. Review these skills before beginning Chapter 5.

For Lessons 5-2 and 5-9 — Rewrite Differences as Sums

Rewrite each difference as a sum.

1. $2 - 7$ **2.** $-6 - 11$ **3.** $x - y$

4. $8 - 2x$ **5.** $2xy - 6yz$ **6.** $6a^2b - 12b^2c$

For Lesson 5-2 — Distributive Property

Use the Distributive Property to rewrite each expression without parentheses.
(For review, see Lesson 1-2.)

7. $-2(4x^3 + x - 3)$ **8.** $-1(x + 2)$ **9.** $-1(x - 3)$

10. $-3(2x^4 - 5x^2 - 2)$ **11.** $-\frac{1}{2}(3a + 2)$ **12.** $-\frac{2}{3}(2 + 6z)$

For Lessons 5-5 and 5-9 — Classify Numbers

Find the value of each expression. Then name the sets of numbers to which each value belongs. *(For review, see Lesson 1-2.)*

13. $2.6 + 3.7$ **14.** $18 \div (-3)$ **15.** $2^3 + 3^2$

16. $\sqrt{4 + 1}$ **17.** $\frac{18 + 14}{8}$ **18.** $3\sqrt{4}$

Polynomials Make this Foldable to help you organize your notes. Begin with four sheets of grid paper.

Step 1 Fold and Cut

Fold in half along the width. On the first two sheets, cut along the fold at the ends. On the second two sheets, cut in the center of the fold as shown.

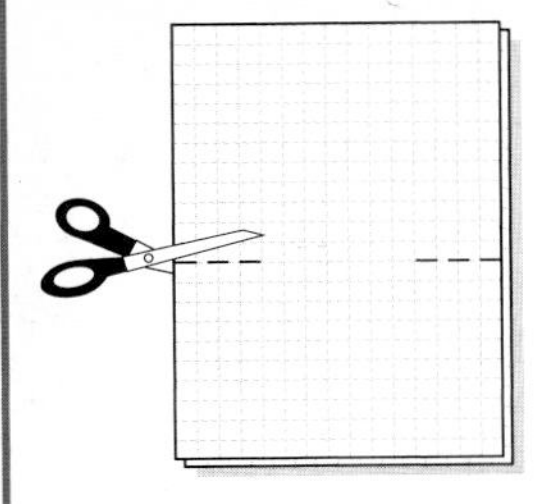

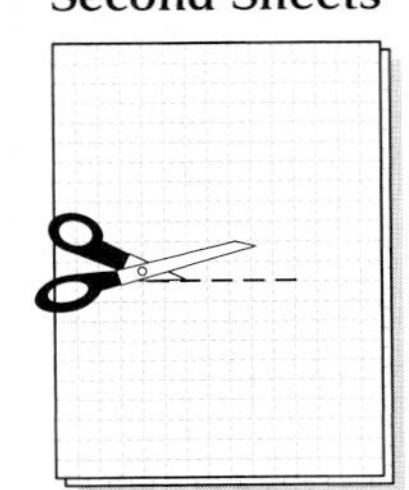

Step 2 Fold and Label

Insert first sheets through second sheets and align the folds. Label the pages with lesson numbers.

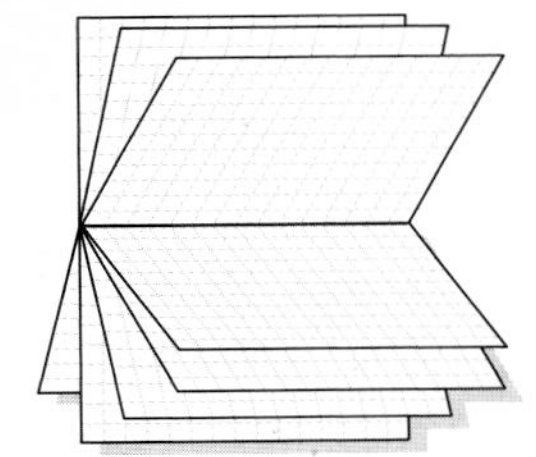

Reading and Writing As you read and study the chapter, fill the journal with notes, diagrams, and examples for polynomials.

5-1 Monomials

California Standards Standard 11.2 Students judge the validity of an argument according to whether the properties of real numbers, exponents, and logarithms have been applied correctly at each step. (Key)

What You'll Learn

- Multiply and divide monomials.
- Use expressions written in scientific notation.

Vocabulary

- monomial
- constant
- coefficient
- degree
- power
- simplify
- standard notation
- scientific notation
- dimensional analysis

Why is scientific notation useful in economics?

Economists often deal with very large numbers. For example, the table shows the U.S. public debt for several years in the last century. Such numbers, written in standard notation, are difficult to work with because they contain so many digits. Scientific notation uses powers of ten to make very large or very small numbers more manageable.

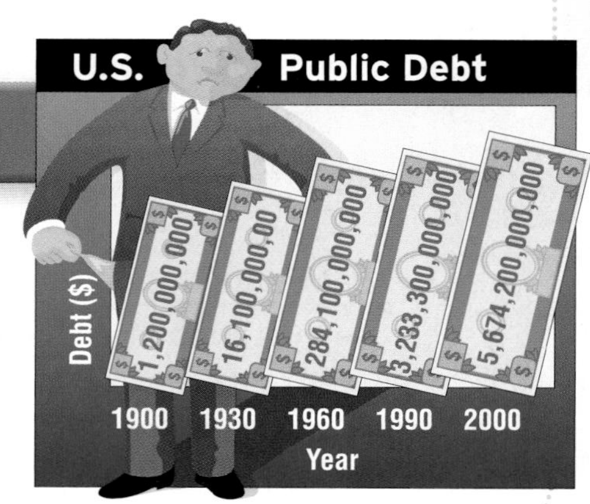

Source: U.S. Department of the Treasury

MONOMIALS A **monomial** is an expression that is a number, a variable, or the product of a number and one or more variables. Monomials cannot contain variables in denominators, variables with exponents that are negative, or variables under radicals.

Monomials	Not Monomials
$5b$, $-w$, 23, x^2, $\frac{1}{3}x^3y^4$	$\frac{1}{n^4}$, $\sqrt[3]{x}$, $x + 8$, a^{-1}

Constants are monomials that contain no variables, like 23 or -1. The numerical factor of a monomial is the **coefficient** of the variable(s). For example, the coefficient of m in $-6m$ is -6. The **degree** of a monomial is the sum of the exponents of its variables. For example, the degree of $12g^7h^4$ is $7 + 4$ or 11. The degree of a constant is 0.

A **power** is an expression of the form x^n. The word *power* is also used to refer to the exponent itself. Negative exponents are a way of expressing the multiplicative inverse of a number. For example, $\frac{1}{x^2}$ can be written as x^{-2}. Note that an expression such as x^{-2} is not a monomial. *Why?*

Key Concept — Negative Exponents

- **Words** For any real number $a \neq 0$ and any integer n, $a^{-n} = \frac{1}{a^n}$ and $\frac{1}{a^{-n}} = a^n$.
- **Examples** $2^{-3} = \frac{1}{2^3}$ and $\frac{1}{b^{-8}} = b^8$

To **simplify** an expression containing powers means to rewrite the expression without parentheses or negative exponents.

Example 1 Simplify Expressions with Multiplication

Simplify $(3x^3y^2)(-4x^2y^4)$.

$(3x^3y^2)(-4x^2y^4) = (3 \cdot x \cdot x \cdot x \cdot y \cdot y)(-4 \cdot x \cdot x \cdot y \cdot y \cdot y \cdot y)$	Definition of exponents
$= 3(-4) \cdot x \cdot x \cdot x \cdot x \cdot x \cdot y \cdot y \cdot y \cdot y \cdot y \cdot y$	Commutative Property
$= -12x^5y^6$	Definition of exponents

Example 1 suggests the following property of exponents.

Key Concept — *Product of Powers*

- **Words** For any real number a and integers m and n, $a^m \cdot a^n = a^{m+n}$.
- **Examples** $4^2 \cdot 4^9 = 4^{11}$ and $b^3 \cdot b^5 = b^8$

To multiply powers of the same variable, add the exponents. Knowing this, it seems reasonable to expect that when dividing powers, you would subtract exponents. Consider $\frac{x^9}{x^5}$.

$$\frac{x^9}{x^5} = \frac{\overset{1}{\cancel{x}} \cdot \overset{1}{\cancel{x}} \cdot \overset{1}{\cancel{x}} \cdot \overset{1}{\cancel{x}} \cdot \overset{1}{\cancel{x}} \cdot x \cdot x \cdot x \cdot x}{\underset{1}{\cancel{x}} \cdot \underset{1}{\cancel{x}} \cdot \underset{1}{\cancel{x}} \cdot \underset{1}{\cancel{x}} \cdot \underset{1}{\cancel{x}}} \quad \text{Remember that } x \neq 0.$$

$$= x \cdot x \cdot x \cdot x \quad \text{Simplify.}$$

$$= x^4 \quad \text{Definition of exponents}$$

It appears that our conjecture is true. To divide powers of the same base, you subtract exponents.

Key Concept — *Quotient of Powers*

- **Words** For any real number $a \neq 0$, and integers m and n, $\frac{a^m}{a^n} = a^{m-n}$.
- **Examples** $\frac{5^3}{5} = 5^{3-1}$ or 5^2 and $\frac{x^7}{x^3} = x^{7-3}$ or x^4

Example 2 *Simplify Expressions with Division*

Simplify $\frac{p^3}{p^8}$. Assume that $p \neq 0$.

$$\frac{p^3}{p^8} = p^{3-8} \quad \text{Subtract exponents.}$$

$$= p^{-5} \text{ or } \frac{1}{p^5} \quad \text{Remember that a simplified expression cannot contain negative exponents.}$$

CHECK

$$\frac{p^3}{p^8} = \frac{\overset{1}{\cancel{p}} \cdot \overset{1}{\cancel{p}} \cdot \overset{1}{\cancel{p}}}{\underset{1}{\cancel{p}} \cdot \underset{1}{\cancel{p}} \cdot \underset{1}{\cancel{p}} \cdot p \cdot p \cdot p \cdot p \cdot p} \quad \text{Definition of exponents}$$

$$= \frac{1}{p^5} \quad \text{Simplify.}$$

You can use the Quotient of Powers property and the definition of exponents to simplify $\frac{y^4}{y^4}$, if $y \neq 0$.

Method 1

$$\frac{y^4}{y^4} = y^{4-4} \quad \text{Quotient of Powers}$$

$$= y^0 \quad \text{Subtract.}$$

Method 2

$$\frac{y^4}{y^4} = \frac{\overset{1}{\cancel{y}} \cdot \overset{1}{\cancel{y}} \cdot \overset{1}{\cancel{y}} \cdot \overset{1}{\cancel{y}}}{\underset{1}{\cancel{y}} \cdot \underset{1}{\cancel{y}} \cdot \underset{1}{\cancel{y}} \cdot \underset{1}{\cancel{y}}} \quad \text{Definition of exponents}$$

$$= 1 \quad \text{Divide.}$$

In order to make the results of these two methods consistent, we define $y^0 = 1$, where $y \neq 0$. In other words, any nonzero number raised to the zero power is equal to 1. *Notice that 0^0 is undefined.*

The properties we have presented can be used to verify the properties of powers that are listed below.

Key Concept — *Properties of Powers*

- **Words** Suppose a and b are real numbers and m and n are integers. Then the following properties hold.
- **Examples**

Power of a Power: $(a^m)^n = a^{mn}$ — $(a^2)^3 = a^6$

Power of a Product: $(ab)^m = a^m b^m$ — $(xy)^2 = x^2y^2$

Power of a Quotient: $\left(\frac{a}{b}\right)^n = \frac{a^n}{b^n}$, $b \neq 0$ and $\left(\frac{a}{b}\right)^{-n} = \left(\frac{b}{a}\right)^n$ or $\frac{b^n}{a^n}$, $a \neq 0$, $b \neq 0$ — $\left(\frac{a}{b}\right)^3 = \frac{a^3}{b^3}$, $\left(\frac{x}{y}\right)^{-4} = \frac{y^4}{x^4}$

Example 3 Simplify Expressions with Powers

Simplify each expression.

a. $(a^3)^6$

$$(a^3)^6 = a^{3(6)} \quad \text{Power of a power}$$
$$= a^{18}$$

b. $(-2p^3s^2)^5$

$$(-2p^3s^2)^5 = (-2)^5 \cdot (p^3)^5 \cdot (s^2)^5$$
$$= -32p^{15}s^{10} \quad \text{Power of a power}$$

c. $\left(\frac{-3x}{y}\right)^4$

$$\left(\frac{-3x}{y}\right)^4 = \frac{(-3x)^4}{y^4} \quad \text{Power of a quotient}$$
$$= \frac{(-3)^4x^4}{y^4} \quad \text{Power of a product}$$
$$= \frac{81x^4}{y^4} \quad (-3)^4 = 81$$

d. $\left(\frac{a}{4}\right)^{-3}$

$$\left(\frac{a}{4}\right)^{-3} = \left(\frac{4}{a}\right)^3 \quad \text{Negative exponent}$$
$$= \frac{4^3}{a^3} \quad \text{Power of a quotient}$$
$$= \frac{64}{a^3} \quad 4^3 = 64$$

Study Tip

Simplified Expressions

A monomial expression is in simplified form when:

- there are no powers of powers,
- each base appears exactly once,
- all fractions are in simplest form, and
- there are no negative exponents.

With complicated expressions, you often have a choice of which way to start simplifying.

Example 4 Simplify Expressions Using Several Properties

Simplify $\left(\frac{-2x^{3n}}{x^{2n}y^3}\right)^4$.

Method 1

Raise the numerator and denominator to the fourth power before simplifying.

$$\left(\frac{-2x^{3n}}{x^{2n}y^3}\right)^4 = \frac{(-2x^{3n})^4}{(x^{2n}y^3)^4}$$
$$= \frac{(-2)^4(x^{3n})^4}{(x^{2n})^4(y^3)^4}$$
$$= \frac{16x^{12n}}{x^{8n}y^{12}}$$
$$= \frac{16x^{12n-8n}}{y^{12}}$$
$$= \frac{16x^{4n}}{y^{12}}$$

Method 2

Simplify the fraction before raising to the fourth power.

$$\left(\frac{-2x^{3n}}{x^{2n}y^3}\right)^4 = \left(\frac{-2x^{3n-2n}}{y^3}\right)^4$$
$$= \left(\frac{-2x^{n}}{y^3}\right)^4$$
$$= \frac{16x^{4n}}{y^{12}}$$

SCIENTIFIC NOTATION The form that you usually write numbers in is **standard notation**. A number is in **scientific notation** when it is in the form $a \times 10^n$, where $1 \leq a < 10$ and n is an integer. Scientific notation is used to express very large or very small numbers.

Study Tip

Graphing Calculators
To solve scientific notation problems on a graphing calculator, use the EE function. Enter 6.38×10^6 as 6.38 [2nd] [EE] 6.

Example 5 Express Numbers in Scientific Notation

Express each number in scientific notation.

a. 6,380,000

$6{,}380{,}000 = 6.38 \times 1{,}000{,}000$ — $1 \leq 6.38 < 10$

$= 6.38 \times 10^6$ — Write 1,000,000 as a power of 10.

b. 0.000047

$0.000047 = 4.7 \times 0.00001$ — $1 \leq 4.7 < 10$

$= 4.7 \times \frac{1}{10^5}$ — $0.00001 = \frac{1}{100{,}000}$ or $\frac{1}{10^5}$

$= 4.7 \times 10^{-5}$ — Use a negative exponent.

You can use properties of powers to multiply and divide numbers in scientific notation.

Example 6 Multiply Numbers in Scientific Notation

Evaluate. Express the result in scientific notation.

a. $(4 \times 10^5)(2 \times 10^7)$

$(4 \times 10^5)(2 \times 10^7) = (4 \cdot 2) \times (10^5 \cdot 10^7)$ — Associative and Commutative Properties

$= 8 \times 10^{12}$ — $4 \cdot 2 = 8$, $10^5 \cdot 10^7 = 10^{5+7}$ or 10^{12}

b. $(2.7 \times 10^{-2})(3 \times 10^6)$

$(2.7 \times 10^{-2})(3 \times 10^6) = (2.7 \cdot 3) \times (10^{-2} \cdot 10^6)$ — Associative and Commutative Properties

$= 8.1 \times 10^4$ — $2.7 \cdot 3 = 8.1$, $10^{-2} \cdot 10^6 = 10^{-2+6}$ or 10^4

Real-world problems often involve units of measure. Performing operations with units is known as **dimensional analysis**.

More About. . .

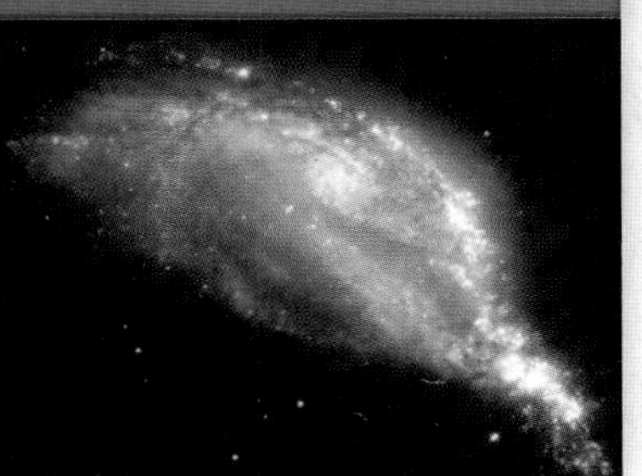

Astronomy

Light travels at a speed of about 3.00×10^8 m/s. The distance that light travels in a year is called a *light-year*.

Source: www.britannica.com

Example 7 Divide Numbers in Scientific Notation

ASTRONOMY After the Sun, the next-closest star to Earth is Alpha Centauri C, which is about 4×10^{16} meters away. How long does it take light from Alpha Centauri C to reach Earth? Use the information at the left.

Begin with the formula $d = rt$, where d is distance, r is rate, and t is time.

$t = \frac{d}{r}$ — Solve the formula for time.

$= \frac{4 \times 10^{16} \text{ m}}{3.00 \times 10^8 \text{ m/s}}$ — ← Distance from Alpha Centauri C to Earth

$= \frac{4}{3.00} \cdot \frac{10^{16}}{10^8 \text{ 1/s}}$ — **Estimate:** The result should be slightly greater than $\frac{10^{16}}{10^8}$ or 10^8.

$\approx 1.33 \times 10^8 \text{ s}$ — $\frac{4}{3.00} \approx 1.33$, $\frac{10^{16}}{10^8} = 10^{16-8}$ or 10^8

It takes about 1.33×10^8 seconds or 4.2 years for light from Alpha Centauri C to reach Earth.

Check for Understanding

Concept Check

1. **OPEN ENDED** Write an example that illustrates a property of powers. Then use multiplication or division to explain why it is true.

2. **Determine** whether $x^y \cdot x^z = x^{yz}$ is *sometimes, always,* or *never* true. Explain.

3. **FIND THE ERROR** Alejandra and Kyle both simplified $\frac{2a^2b}{(-2ab^3)^{-2}}$.

Alejandra	Kyle
$\frac{2a^2b}{(-2ab^3)^{-2}} = (2a^2b)(-2ab^3)^2$ $= (2a^2b)(-2)^2a^2(b^3)^2$ $= (2a^2b)2^2a^2b^6$ $= 8a^4b^7$	$\frac{2a^2b}{(-2ab^3)^{-2}} = \frac{2a^2b}{(-2)^{-2}a(b^3)^{-2}}$ $= \frac{2a^2b}{4ab^{-6}}$ $= \frac{2a^2bb^6}{4a}$ $= \frac{ab^7}{2}$

Who is correct? Explain your reasoning.

Guided Practice

Simplify. Assume that no variable equals 0.

4. $x^2 \cdot x^8$
5. $(2b)^4$
6. $(n^3)^3(n^{-3})^3$
7. $\frac{30y^4}{-5y^2}$
8. $\frac{-2a^3b^6}{18a^2b^2}$
9. $\frac{81p^6q^5}{(3p^2q)^2}$
10. $\left(\frac{1}{w^4z^2}\right)^3$
11. $\left(\frac{cd}{3}\right)^{-2}$
12. $\left(\frac{-6x^6}{3x^3}\right)^{-2}$

Express each number in scientific notation.

13. 421,000
14. 0.000862

Evaluate. Express the result in scientific notation.

15. $(3.42 \times 10^8)(1.1 \times 10^{-5})$
16. $\frac{8 \times 10^{-1}}{16 \times 10^{-2}}$

Application

17. **ASTRONOMY** Refer to Example 7 on page 225. The average distance from Earth to the Moon is about 3.84×10^8 meters. How long would it take a radio signal traveling at the speed of light to cover that distance?

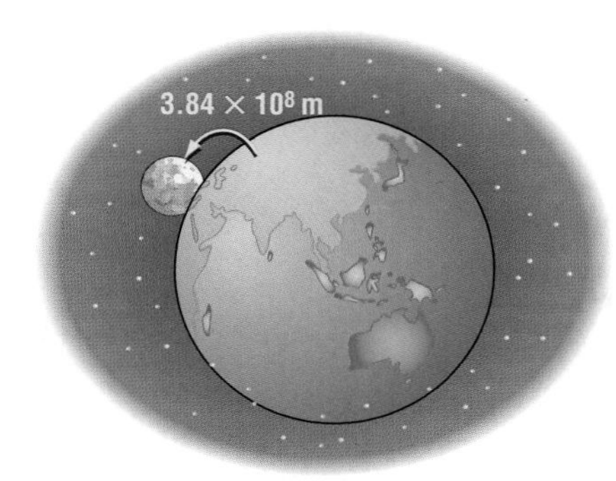

Practice and Apply

Simplify. Assume that no variable equals 0.

18. $a^2 \cdot a^6$
19. $b^{-3} \cdot b^7$
20. $(n^4)^4$
21. $(z^2)^5$
22. $(2x)^4$
23. $(-2c)^3$
24. $\frac{a^2n^6}{an^5}$
25. $\frac{-y^5z^7}{y^2z^5}$
26. $(7x^3y^{-5})(4xy^3)$
27. $(-3b^3c)(7b^2c^2)$
28. $(a^3b^3)(ab)^{-2}$
29. $(-2r^2s)^3(3rs^2)$
30. $2x^2(6y^3)(2x^2y)$
31. $3a(5a^2b)(6ab^3)$
32. $\frac{-5x^3y^3z^4}{20x^3y^7z^4}$

Homework Help

For Exercises	See Examples
18–35, 60	1–3
36–39	4
40–43	1, 2
44–49, 56, 57	5
50–55, 58, 59	6, 7

Extra Practice
See page 836.

33. $\frac{3a^5b^3c^3}{9a^3b^7c}$
34. $\frac{2c^3d(3c^2d^5)}{30c^4d^2}$
35. $\frac{-12m^4n^8(m^3n^2)}{36m^3n}$
36. $\left(\frac{8a^3b^2}{16a^2b^3}\right)^4$
37. $\left(\frac{6x^2y^4}{3x^4y^3}\right)^3$
38. $\left(\frac{x}{y^{-1}}\right)^{-2}$
39. $\left(\frac{v}{w^{-2}}\right)^{-3}$
40. $\frac{30a^{-2}b^{-6}}{60a^{-6}b^{-8}}$
41. $\frac{12x^{-3}y^{-2}z^{-8}}{30x^{-6}y^{-4}z^{-1}}$

42. If $2^{r+5} = 2^{2r-1}$, what is the value of r?

43. What value of r makes $y^{28} = y^{3r} \cdot y^7$ true?

Express each number in scientific notation.

44. 462.3
45. 43,200
46. 0.0001843
47. 0.006810
48. 502,020,000
49. 675,400,000

Evaluate. Express the result in scientific notation.

50. $(4.15 \times 10^3)(3.0 \times 10^6)$
51. $(3.01 \times 10^{-2})(2 \times 10^{-3})$
52. $\frac{6.3 \times 10^5}{1.4 \times 10^3}$
53. $\frac{9.3 \times 10^7}{1.5 \times 10^{-3}}$
54. $(6.5 \times 10^4)^2$
55. $(4.1 \times 10^{-4})^2$

56. **POPULATION** The population of Earth is about 6,080,000,000. Write this number in scientific notation.

57. **BIOLOGY** Use the diagram at the right to write the diameter of a typical flu virus in scientific notation.

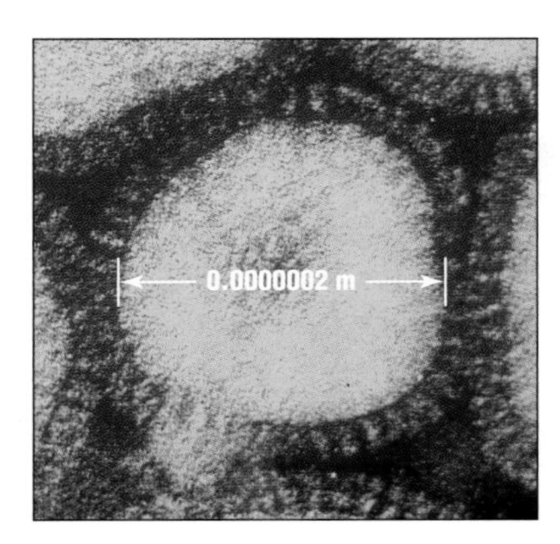

58. **CHEMISTRY** One gram of water contains about 3.34×10^{22} molecules. About how many molecules are in 500 grams of water?

59. **RESEARCH** Use the Internet or other source to find the masses of Earth and the Sun. About how many times as large as Earth is the Sun?

60. **CRITICAL THINKING** Determine which is greater, 100^{10} or 10^{100}. Explain.

WebQuest

A scatter plot of populations will help you make a model for the data. Visit www.algebra2.com/webquest to continue work on your WebQuest project.

CRITICAL THINKING **For Exercises 61 and 62, use the following proof of the Power of a Power Property.**

$$a^m a^n = \overbrace{a \cdot a \cdot \ldots \cdot a}^{m \text{ factors}} \cdot \overbrace{a \cdot a \cdot \ldots \cdot a}^{n \text{ factors}}$$
$$= \overbrace{a \cdot a \cdot \ldots \cdot a}^{m + n \text{ factors}}$$
$$= a^{m+n}$$

61. What definition or property allows you to make each step of the proof?

62. Prove the Power of a Product Property, $(ab)^m = a^m b^m$.

63. **WRITING IN MATH** Answer the question that was posed at the beginning of the lesson.

Why is scientific notation useful in economics?

Include the following in your answer:

- the 2000 national debt of $5,674,200,000,000 and the U.S. population of 281,000,000, both written in words and in scientific notation, and
- an explanation of how to find the amount of debt per person, with the result written in scientific notation and in standard notation.

Standards Practice

64. Simplify $\frac{(2x^2)^3}{12x^4}$.

Ⓐ $\frac{x}{2}$ Ⓑ $\frac{2x}{3}$ Ⓒ $\frac{1}{2x^2}$ Ⓓ $\frac{2x^2}{3}$

65. $7.3 \times 10^5 = ?$

Ⓐ 73,000 Ⓑ 730,000 Ⓒ 7,300,000 Ⓓ 73,000,000

Maintain Your Skills

Mixed Review

Solve each system of equations by using inverse matrices. *(Lesson 4-8)*

66. $2x + 3y = 8$
$x - 2y = -3$

67. $x + 4y = 9$
$3x + 2y = -3$

Find the inverse of each matrix, if it exists. *(Lesson 4-7)*

68. $\begin{bmatrix} 2 & 5 \\ -1 & -2 \end{bmatrix}$

69. $\begin{bmatrix} 4 & 3 \\ 2 & 1 \end{bmatrix}$

Evaluate each determinant. *(Lesson 4-3)*

70. $\begin{vmatrix} 3 & 0 \\ 2 & -2 \end{vmatrix}$

71. $\begin{vmatrix} 1 & 0 & -3 \\ 2 & -1 & 4 \\ -3 & 0 & 2 \end{vmatrix}$

Solve each system of equations. *(Lesson 3-5)*

72. $x + y = 5$
$x + y + z = 4$
$2x - y + 2z = -1$

73. $a + b + c = 6$
$2a - b + 3c = 16$
$a + 3b - 2c = -6$

TRANSPORTATION **For Exercises 74–76, refer to the graph at the right.** *(Lesson 2-5)*

74. Make a scatter plot of the data, where the horizontal axis is the number of years since 1970.

75. Write a prediction equation.

76. Predict the median age of vehicles on the road in 2010.

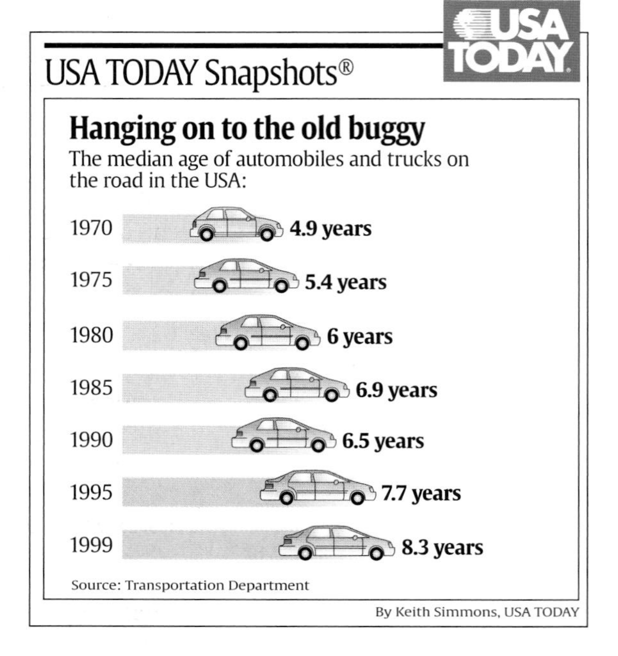

Solve each equation. *(Lesson 1-3)*

77. $2x + 11 = 25$

78. $-12 - 5x = 3$

Getting Ready for the Next Lesson

Use the Distributive Property to find each product.
(To review the ***Distributive Property****, see Lesson 1-2.)*

79. $2(x + y)$ **80.** $3(x - z)$ **81.** $4(x + 2)$

82. $-2(3x - 5)$ **83.** $-5(x - 2y)$ **84.** $-3(-y + 5)$

5-2 Polynomials

California Standards Standard 3.0 **Students are adept at operations on polynomials,** including long division. (Key)

What You'll Learn

- Add and subtract polynomials.
- Multiply polynomials.

How can polynomials be applied to financial situations?

Shenequa wants to attend Purdue University in Indiana, where the out-of-state tuition is $8820. Suppose the tuition increases at a rate of 4% per year. You can use polynomials to represent the increasing tuition costs.

Year	Tuition
1	$8820
2	$9173
3	$9540
4	$9921

Vocabulary

- polynomial
- terms
- like terms
- trinomial
- binomial
- FOIL method

ADD AND SUBTRACT POLYNOMIALS If r represents the rate of increase of tuition, then the tuition for the second year will be $8820(1 + r)$. For the third year, it will be $8820(1 + r)^2$, or $8820r^2 + 17{,}640r + 8820$ in expanded form. This expression is called a polynomial. A **polynomial** is a monomial or a sum of monomials.

The monomials that make up a polynomial are called the **terms** of the polynomial. In a polynomial such as $x^2 + 2x + x + 1$, the two monomials $2x$ and x can be combined because they are **like terms**. The result is $x^2 + 3x + 1$. The polynomial $x^2 + 3x + 1$ is a **trinomial** because it has three unlike terms. A polynomial such as $xy + z^3$ is a **binomial** because it has two unlike terms. The *degree* of a polynomial is the degree of the monomial with the greatest degree. For example, the degree of $x^2 + 3x + 1$ is 2, and the degree of $xy + z^3$ is 3.

Study Tip

Reading Math
The prefix *bi-* means *two*, and the prefix *tri-* means *three*.

Example 1 Degree of a Polynomial

Determine whether each expression is a polynomial. If it is a polynomial, state the degree of the polynomial.

a. $\frac{1}{6}x^3y^5 - 9x^4$

This expression is a polynomial because each term is a monomial.
The degree of the first term is $3 + 5$ or 8, and the degree of the second term is 4.
The degree of the polynomial is 8.

b. $x + \sqrt{x} + 5$

This expression is not a polynomial because $\sqrt{x}$ is not a monomial.

To *simplify* a polynomial means to perform the operations indicated and combine like terms.

Example 2 Subtract and Simplify

Simplify $(3x^2 - 2x + 3) - (x^2 + 4x - 2)$.

$(3x^2 - 2x + 3) - (x^2 + 4x - 2) = 3x^2 - 2x + 3 - x^2 - 4x + 2$ Distribute the -1.

$= (3x^2 - x^2) + (-2x - 4x) + (3 + 2)$ Group like terms.

$= 2x^2 - 6x + 5$ Combine like terms.

MULTIPLY POLYNOMIALS You can use the Distributive Property to multiply polynomials.

Example 3 Multiply and Simplify

Find $2x(7x^2 - 3x + 5)$.

$2x(7x^2 - 3x + 5) = 2x(7x^2) + 2x(-3x) + 2x(5)$ Distributive Property

$= 14x^3 - 6x^2 + 10x$ Multiply the monomials.

You can use algebra tiles to model the product of two binomials.

Algebra Activity

Multiplying Binomials

Use algebra tiles to find the product of $x + 5$ and $x + 2$.

- Draw a 90° angle on your paper.
- Use an x tile and a 1 tile to mark off a length equal to $x + 5$ along the top.
- Use the tiles to mark off a length equal to $x + 2$ along the side.
- Draw lines to show the grid formed.
- Fill in the lines with the appropriate tiles to show the area product. The model shows the polynomial $x^2 + 7x + 10$.

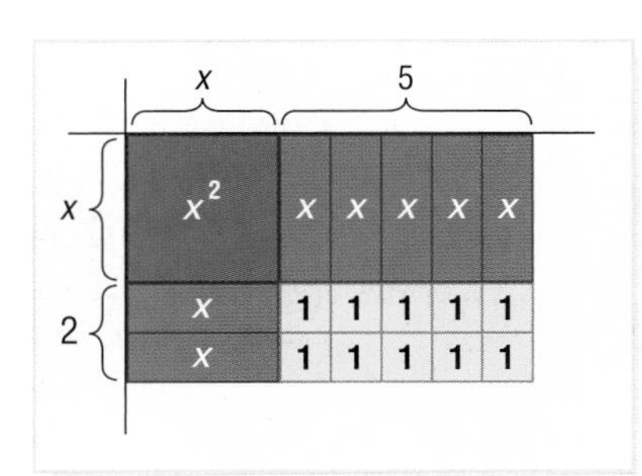

The area of the rectangle is the product of its length and width. So, $(x + 5)(x + 2) = x^2 + 7x + 10$.

The **FOIL method** uses the Distributive Property to multiply binomials.

Key Concept — *FOIL Method for Multiplying Binomials*

The product of two binomials is the sum of the products of **F** the *first* terms, **O** the *outer* terms, **I** the *inner* terms, and **L** the *last* terms.

Study Tip

Vertical Method
You may also want to use the vertical method to multiply polynomials.

$$\begin{array}{r} 3y + 2 \\ (\times)\ 5y + 4 \\ \hline 12y + 8 \\ 15y^2 + 10y \quad\ \ \\ \hline 15y^2 + 22y + 8 \end{array}$$

Example 4 Multiply Two Binomials

Find $(3y + 2)(5y + 4)$.

$(3y + 2)(5y + 4) = \underbrace{3y \cdot 5y}_{\text{First terms}} + \underbrace{3y \cdot 4}_{\text{Outer terms}} + \underbrace{2 \cdot 5y}_{\text{Inner terms}} + \underbrace{2 \cdot 4}_{\text{Last terms}}$

$= 15y^2 + 22y + 8$ Multiply monomials and add like terms.

Example 5 Multiply Polynomials

Find $(n^2 + 6n - 2)(n + 4)$.

$(n^2 + 6n - 2)(n + 4)$

$= n^2(n + 4) + 6n(n + 4) + (-2)(n + 4)$ Distributive Property

$= n^2 \cdot n + n^2 \cdot 4 + 6n \cdot n + 6n \cdot 4 + (-2) \cdot n + (-2) \cdot 4$ Distributive Property

$= n^3 + 4n^2 + 6n^2 + 24n - 2n - 8$ Multiply monomials.

$= n^3 + 10n^2 + 22n - 8$ Combine like terms.

Check for Understanding

Concept Check

1. **OPEN ENDED** Write a polynomial of degree 5 that has three terms.
2. **Identify** the degree of the polynomial $2x^3 - x^2 + 3x^4 - 7$.
3. **Model** $3x(x + 2)$ using algebra tiles.

Guided Practice

Determine whether each expression is a polynomial. If it is a polynomial, state the degree of the polynomial.

4. $2a + 5b$
5. $\frac{1}{3}x^3 - 9y$
6. $\frac{mw^2 - 3}{nz^3 + 1}$

Simplify.

7. $(2a + 3b) + (8a - 5b)$
8. $(x^2 - 4x + 3) - (4x^2 + 3x - 5)$
9. $2x(3y + 9)$
10. $2p^2q(5pq - 3p^3q^2 + 4pq^4)$
11. $(y - 10)(y + 7)$
12. $(x + 6)(x + 3)$
13. $(2z - 1)(2z + 1)$
14. $(2m - 3n)^2$

Application

15. **GEOMETRY** Find the area of the triangle.

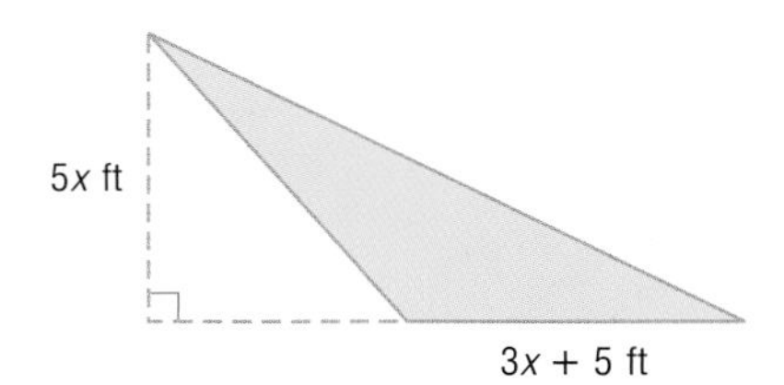

Practice and Apply

Homework Help

For Exercises	See Examples
16–21	1
22–27, 35, 36, 51	2
28–33, 47, 48	3
34	2, 3
37–46, 52, 53	4
49, 50, 54	5

Extra Practice See page 837.

Determine whether each expression is a polynomial. If it is a polynomial, state the degree of the polynomial.

16. $3z^2 - 5z + 11$
17. $x^3 - 9$
18. $\frac{6xy}{z} - \frac{3c}{d}$
19. $\sqrt{m - 5}$
20. $5x^2y^4 + x\sqrt{3}$
21. $\frac{4}{3}y^2 + \frac{5}{6}y^7$

Simplify.

22. $(3x^2 - x + 2) + (x^2 + 4x - 9)$
23. $(5y + 3y^2) + (-8y - 6y^2)$
24. $(9r^2 + 6r + 16) - (8r^2 + 7r + 10)$
25. $(7m^2 + 5m - 9) + (3m^2 - 6)$
26. $(4x^2 - 3y^2 + 5xy) - (8xy + 3y^2)$
27. $(10x^2 - 3xy + 4y^2) - (3x^2 + 5xy)$
28. $4b(cb - zd)$
29. $4a(3a^2 + b)$
30. $-5ab^2(-3a^2b + 6a^3b - 3a^4b^4)$
31. $2xy(3xy^3 - 4xy + 2y^4)$
32. $\frac{3}{4}x^2(8x + 12y - 16xy^2)$
33. $\frac{1}{2}a^3(4a - 6b + 8ab^4)$

34. **PERSONAL FINANCE** Toshiro has $850 to invest. He can invest in a savings account that has an annual interest rate of 3.7%, and he can invest in a money market account that pays about 5.5% per year. Write a polynomial to represent the amount of interest he will earn in 1 year if he invests x dollars in the savings account and the rest in the money market account.

E-SALES **For Exercises 35 and 36, use the following information.**
A small online retailer estimates that the cost, in dollars, associated with selling x units of a particular product is given by the expression $0.001x^2 + 5x + 500$. The revenue from selling x units is given by $10x$.

35. Write a polynomial to represent the profit generated by the product.
36. Find the profit from sales of 1850 units.

Simplify.

37. $(p + 6)(p - 4)$
38. $(a + 6)(a + 3)$
39. $(b + 5)(b - 5)$
40. $(6 - z)(6 + z)$
41. $(3x + 8)(2x + 6)$
42. $(4y - 6)(2y + 7)$
43. $(a^3 - \mathrm{b})(a^3 + \mathrm{b})$
44. $(m^2 - 5)(2m^2 + 3)$
45. $(x - 3y)^2$
46. $(1 + 4c)^2$
47. $d^{-3}(d^5 - 2d^3 + d^{-1})$
48. $x^{-3}y^2(yx^4 + y^{-1}x^3 + y^{-2}x^2)$
49. $(3b - c)^3$
50. $(x^2 + xy + y^2)(x - y)$

51. Simplify $(c^2 - 6cd - 2d^2) + (7c^2 - cd + 8d^2) - (-c^2 + 5cd - d^2)$.

52. Find the product of $6x - 5$ and $-3x + 2$.

More About. . .

Genetics

The possible genes of parents and offspring can be summarized in a *Punnett square*, such as the one above.

Source: *Biology: The Dynamics of Life*

53. **GENETICS** Suppose R and W represent two genes that a plant can inherit from its parents. The terms of the expansion of $(R + W)^2$ represent the possible pairings of the genes in the offspring. Write $(R + W)^2$ as a polynomial.

54. **CRITICAL THINKING** What is the degree of the product of a polynomial of degree 8 and a polynomial of degree 6? Include an example in support of your answer.

55. **WRITING IN MATH** Answer the question that was posed at the beginning of the lesson.

How can polynomials be applied to financial situations?

Include the following in your answer:

- an explanation of how a polynomial can be applied to a situation with a fixed percent rate of increase,
- two expressions in terms of r for the tuition in the fourth year, and
- an explanation of how to use one of the expressions and the 4% rate of increase to estimate Shenequa's tuition in the fourth year, and a comparison of the value you found to the value given in the table.

Standards Practice
Standardized Test Practice

A B C D

56. Which polynomial has degree 3?

Ⓐ $x^3 + x^2 - 2x^4$
Ⓑ $-2x^2 - 3x + 4$
Ⓒ $x^2 + x + 12^3$
Ⓓ $1 + x + x^3$

57. $(x + y) - (y + z) - (x + z) = ?$

Ⓐ $2x + 2y + 2z$
Ⓑ $-2z$
Ⓒ $2y$
Ⓓ $x - y - z$

Maintain Your Skills

Mixed Review

Simplify. Assume that no variable equals 0. *(Lesson 5-1)*

58. $(-4d^2)^3$
59. $5rt^2(2rt)^2$
60. $\frac{x^2yz^4}{xy^3z^2}$
61. $\left(\frac{3ab^2}{6a^2b}\right)^2$

62. Solve the system $4x - y = 0$, $2x + 3y = 14$ by using inverse matrices. *(Lesson 4-8)*

Graph each inequality. *(Lesson 2-7)*

63. $y \leq -\frac{1}{3}x + 2$
64. $x + y > -2$
65. $2x + y < 1$

Getting Ready for the Next Lesson

PREREQUISITE SKILL **Simplify. Assume that no variable equals 0.**
*(To review **properties of exponents**, see Lesson 5-1.)*

66. $\frac{x^3}{x}$
67. $\frac{4y^5}{2y^2}$
68. $\frac{x^2y^3}{xy}$
69. $\frac{9a^3b}{3ab}$

5-3 Dividing Polynomials

California Standards Standard 3.0 Students are adept at operations on polynomials, including long division. (Key)

What You'll Learn

- Divide polynomials using long division.
- Divide polynomials using synthetic division.

How can you use division of polynomials in manufacturing?

Vocabulary
- synthetic division

A machinist needed $32x^2 + x$ square inches of metal to make a square pipe $8x$ inches long. In figuring the area needed, she allowed a fixed amount of metal for overlap of the seam. If the width of the finished pipe will be x inches, how wide is the seam? You can use a quotient of polynomials to help find the answer.

Metal Needed

s, $\frac{x}{2}$, x, x, x, $\frac{x}{2}$

$8x$

s = width of seam

Finished Pipe

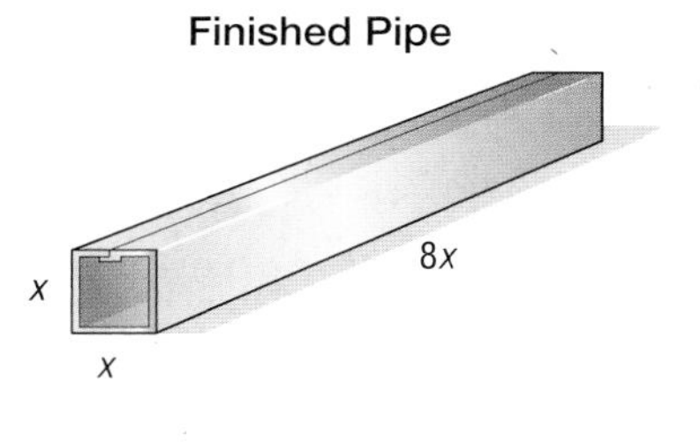

USE LONG DIVISION In Lesson 5-1, you learned to divide monomials. You can divide a polynomial by a monomial by using those same skills.

Example 1 Divide a Polynomial by a Monomial

Simplify $\dfrac{4x^3y^2 + 8xy^2 - 12x^2y^3}{4xy}$.

$$\frac{4x^3y^2 + 8xy^2 - 12x^2y^3}{4xy} = \frac{4x^3y^2}{4xy} + \frac{8xy^2}{4xy} - \frac{12x^2y^3}{4xy} \quad \text{Sum of quotients}$$

$$= \frac{4}{4} \cdot x^{3-1}y^{2-1} + \frac{8}{4} \cdot x^{1-1}y^{2-1} - \frac{12}{4} \cdot x^{2-1}y^{3-1} \quad \text{Divide.}$$

$$= x^2y + 2y - 3xy^2 \quad x^{1-1} = x^0 \text{ or } 1$$

You can use a process similar to long division to divide a polynomial by a polynomial with more than one term. The process is known as the *division algorithm*. When doing the division, remember that you can only add or subtract like terms.

Example 2 Division Algorithm

Use long division to find $(z^2 + 2z - 24) \div (z - 4)$.

$$\begin{array}{r} z \\ z-4 \overline{)z^2 + 2z - 24} \\ (-)z^2 - 4z \\ \hline 6z - 24 \end{array} \qquad \begin{array}{l} \\ \\ z(z-4) = z^2 - 4z \\ 2z - (-4z) = 6z \end{array}$$

$$\begin{array}{r} z + 6 \\ z-4 \overline{)z^2 + 2z - 24} \\ (-)z^2 - 4z \\ \hline 6z - 24 \\ (-)6z - 24 \\ \hline 0 \end{array}$$

The quotient is $z + 6$. The remainder is 0.

Just as with the division of whole numbers, the division of two polynomials may result in a quotient with a remainder. Remember that $9 \div 4 = 2 + R1$ and is often written as $2\frac{1}{4}$. The result of a division of polynomials with a remainder can be written in a similar manner.

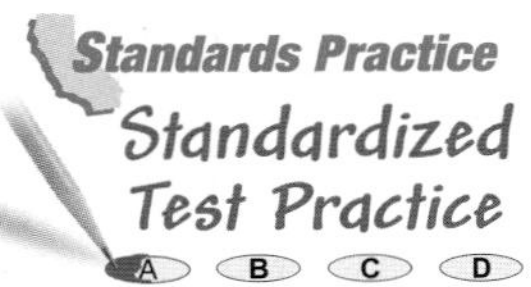

Example 3 Quotient with Remainder

Multiple-Choice Test Item

Which expression is equal to $(t^2 + 3t - 9)(5 - t)^{-1}$?

Ⓐ $t + 8 - \frac{31}{5 - t}$

Ⓑ $-t - 8$

Ⓒ $-t - 8 + \frac{31}{5 - t}$

Ⓓ $-t - 8 - \frac{31}{5 - t}$

Read the Test Item

Since the second factor has an exponent of -1, this is a division problem.

$$(t^2 + 3t - 9)(5 - t)^{-1} = \frac{t^2 + 3t - 9}{5 - t}$$

Solve the Test Item

$$\begin{array}{rl} -t - 8 & \\ -t + 5\overline{)t^2 + 3t - 9} & \text{For ease in dividing, rewrite } 5 - t \text{ as } -t + 5. \\ \underline{(-)t^2 - 5t} & -t(-t + 5) = t^2 - 5t \\ 8t - 9 & 3t - (-5t) = 8t \\ \underline{(-)8t - 40} & -8(-t + 5) = 8t - 40 \\ 31 & \text{Subtract. } -9 - (-40) = 31 \end{array}$$

The quotient is $-t - 8$, and the remainder is 31. Therefore, $(t^2 + 3t - 9)(5 - t)^{-1} = -t - 8 + \frac{31}{5 - t}$. The answer is C.

Test-Taking Tip

You may be able to eliminate some of the answer choices by substituting the same value for t in the original expression and the answer choices and evaluating.

USE SYNTHETIC DIVISION **Synthetic division** is a simpler process for dividing a polynomial by a binomial. Suppose you want to divide $5x^3 - 13x^2 + 10x - 8$ by $x - 2$ using long division. Compare the coefficients in this division with those in Example 4.

$$\begin{array}{r} 5x^2 - 3x + 4 \\ x - 2\overline{)5x^3 - 13x^2 + 10x - 8} \\ \underline{(-)5x^3 - 10x^2} \\ -3x^2 + 10x \\ \underline{(-)-3x^2 + 6x} \\ 4x - 8 \\ \underline{(-)4x - 8} \\ 0 \end{array}$$

Example 4 Synthetic Division

Use synthetic division to find $(5x^3 - 13x^2 + 10x - 8) \div (x - 2)$.

Step 1 Write the terms of the dividend so that the degrees of the terms are in descending order. Then write just the coefficients as shown at the right.

$$\begin{array}{cccc} 5x^3 & -13x^2 & +10x & -8 \\ \downarrow & \downarrow & \downarrow & \downarrow \\ 5 & -13 & 10 & -8 \end{array}$$

Step 2 Write the constant r of the divisor $x - r$ to the left. In this case, $r = 2$. Bring the first coefficient, 5, down as shown.

$$\begin{array}{c|cccc} 2 & 5 & -13 & 10 & -8 \\ \hline & 5 & & & \end{array}$$

Step 3 Multiply the first coefficient by r: $2 \cdot 5 = 10$. Write the product under the second coefficient. Then add the product and the second coefficient: $-13 + 10 = -3$.

$$\begin{array}{r|rrrr} 2 & 5 & -13 & 10 & -8 \\ & & 10 & & \\ \hline & 5 & -3 & & \end{array}$$

Step 4 Multiply the sum, -3, by r: $2(-3) = -6$. Write the product under the next coefficient and add: $10 + (-6) = 4$.

$$\begin{array}{r|rrrr} 2 & 5 & -13 & 10 & -8 \\ & & 10 & -6 & \\ \hline & 5 & -3 & 4 & \end{array}$$

Step 5 Multiply the sum, 4, by r: $2 \cdot 4 = 8$. Write the product under the next coefficient and add: $-8 + 8 = 0$. The remainder is 0.

$$\begin{array}{r|rrrr} 2 & 5 & -13 & 10 & -8 \\ & & 10 & -6 & 8 \\ \hline & 5 & -3 & 4 & \end{array}$$

The numbers along the bottom row are the coefficients of the quotient. Start with the power of x that is one less than the degree of the dividend. Thus, the quotient is $5x^2 - 3x + 4$.

To use synthetic division, the divisor must be of the form $x - r$. If the coefficient of x in a divisor is not 1, you can rewrite the division expression so that you can use synthetic division.

Example 5 Divisor with First Coefficient Other than 1

Use synthetic division to find $(8x^4 - 4x^2 + x + 4) \div (2x + 1)$.

Use division to rewrite the divisor so it has a first coefficient of 1.

$\dfrac{8x^4 - 4x^2 + x + 4}{2x + 1} = \dfrac{(8x^4 - 4x^2 + x + 4) \div 2}{(2x + 1) \div 2}$ Divide numerator and denominator by 2.

$= \dfrac{4x^4 - 2x^2 + \frac{1}{2}x + 2}{x + \frac{1}{2}}$ Simplify the numerator and denominator.

Since the numerator does not have an x^3-term, use a coefficient of 0 for x^3.

$x - r = x + \frac{1}{2}$, so $r = -\frac{1}{2}$.

$$\begin{array}{r|rrrr|r} -\frac{1}{2} & 4 & 0 & -2 & \frac{1}{2} & 2 \\ & & -2 & 1 & \frac{1}{2} & -\frac{1}{2} \\ \hline & 4 & -2 & -1 & 1 & \frac{3}{2} \end{array}$$

The result is $4x^3 - 2x^2 - x + 1 + \dfrac{\frac{3}{2}}{x + \frac{1}{2}}$. Now simplify the fraction.

$\dfrac{\frac{3}{2}}{x + \frac{1}{2}} = \frac{3}{2} \div \left(x + \frac{1}{2}\right)$ Rewrite as a division expression.

$= \frac{3}{2} \div \frac{2x + 1}{2}$ $\quad x + \frac{1}{2} = \frac{2x}{2} + \frac{1}{2} = \frac{2x + 1}{2}$

$= \frac{3}{2} \cdot \frac{2}{2x + 1}$ Multiply by the reciprocal.

$= \frac{3}{2x + 1}$ Multiply.

The solution is $4x^3 - 2x^2 - x + 1 + \frac{3}{2x + 1}$.

(continued on the next page)

CHECK Divide using long division.

$$\begin{array}{r} 4x^3 - 2x^2 - x + 1 \\ 2x + 1 \overline{)8x^4 + 0x^3 - 4x^2 + x + 4} \\ \underline{(-)8x^4 + 4x^3} \quad\quad\quad\quad\quad\quad \\ -4x^3 - 4x^2 \quad\quad\quad\quad \\ \underline{(-)-4x^3 - 2x^2} \quad\quad\quad\quad \\ -2x^2 + x \quad\quad \\ \underline{(-)-2x^2 - x} \quad\quad \\ 2x + 4 \\ \underline{(-)2x + 1} \\ 3 \end{array}$$

The result is $4x^3 - 2x^2 - x + 1 + \frac{3}{2x + 1}$. ✓

Check for Understanding

Concept Check

1. **OPEN ENDED** Write a quotient of two polynomials such that the remainder is 5.
2. **Explain** why synthetic division cannot be used to simplify $\frac{x^3 - 3x + 1}{x^2 + 1}$.
3. **FIND THE ERROR** Shelly and Jorge are dividing $x^3 - 2x^2 + x - 3$ by $x - 4$.

Shelly

4	1	-2	1	-3
		4	-24	100
	1	-6	25	-103

Jorge

4	1	-2	1	-3
		4	8	36
	1	2	9	33

Who is correct? Explain your reasoning.

Guided Practice

Simplify.

4. $\frac{6xy^2 - 3xy + 2x^2y}{xy}$
5. $(5ab^2 - 4ab + 7a^2b)(ab)^{-1}$
6. $(x^2 - 10x - 24) \div (x + 2)$
7. $(3a^4 - 6a^3 - 2a^2 + a - 6) \div (a + 1)$
8. $(z^5 - 3z^2 - 20) \div (z - 2)$
9. $(x^3 + y^3) \div (x + y)$
10. $\frac{x^3 + 13x^2 - 12x - 8}{x + 2}$
11. $(b^4 - 2b^3 + b^2 - 3b + 2)(b - 2)^{-1}$
12. $(12y^2 + 36y + 15) \div (6y + 3)$
13. $\frac{9b^2 + 9b - 10}{3b - 2}$

Standards Practice

Standardized Test Practice

14. Which expression is equal to $(x^2 - 4x + 6)(x - 3)^{-1}$?

Ⓐ $x - 1$

Ⓑ $x - 1 + \frac{3}{x - 3}$

Ⓒ $x - 1 - \frac{3}{x - 3}$

Ⓓ $-x + 1 - \frac{3}{x - 3}$

Practice and Apply

Simplify.

15. $\frac{9a^3b^2 - 18a^2b^3}{3a^2b}$
16. $\frac{5xy^2 - 6y^3 + 3x^2y^3}{xy}$
17. $(28c^3d - 42cd^2 + 56cd^3) \div (14cd)$
18. $(12mn^3 + 9m^2n^2 - 15m^2n) \div (3mn)$
19. $(2y^3z + 4y^2z^2 - 8y^4z^5)(yz)^{-1}$
20. $(a^3b^2 - a^2b + 2a)(-ab)^{-1}$

Homework Help

For Exercises	See Examples
15–20, 51	1
21–34, 49, 50, 52–54	2, 4
35–38	3, 4
39–48	2, 3, 5

Extra Practice
See page 837.

21. $(b^3 + 8b^2 - 20b) \div (b - 2)$
22. $(x^2 - 12x - 45) \div (x + 3)$
23. $(n^3 + 2n^2 - 5n + 12) \div (n + 4)$
24. $(2c^3 - 3c^2 + 3c - 4) \div (c - 2)$
25. $(x^4 - 3x^3 + x^2 - 5) \div (x + 2)$
26. $(6w^5 - 18w^2 - 120) \div (w - 2)$
27. $(x^3 - 4x^2) \div (x - 4)$
28. $(x^3 - 27) \div (x - 3)$
29. $\dfrac{y^3 + 3y^2 - 5y - 4}{y + 4}$
30. $\dfrac{m^3 + 3m^2 - 7m - 21}{m + 3}$
31. $\dfrac{a^4 - 5a^3 - 13a^2 + 10}{a + 1}$
32. $\dfrac{2m^4 - 5m^3 - 10m + 8}{m - 3}$
33. $\dfrac{x^5 - 7x^3 + x + 1}{x + 3}$
34. $\dfrac{3c^5 + 5c^4 + c + 5}{c + 2}$
35. $(g^2 + 8g + 15)(g + 3)^{-1}$
36. $(2b^3 + b^2 - 2b + 3)(b + 1)^{-1}$
37. $(t^5 - 3t^2 - 20)(t - 2)^{-1}$
38. $(y^5 + 32)(y + 2)^{-1}$
39. $(6t^3 + 5t^2 + 9) \div (2t + 3)$
40. $(2h^3 - 5h^2 + 22h) \div (2h + 3)$
41. $\dfrac{9d^3 + 5d - 8}{3d - 2}$
42. $\dfrac{4x^3 + 5x^2 - 3x - 1}{4x + 1}$
43. $\dfrac{2x^4 + 3x^3 - 2x^2 - 3x - 6}{2x + 3}$
44. $\dfrac{6x^4 + 5x^3 + x^2 - 3x + 1}{3x + 1}$
45. $\dfrac{x^3 - 3x^2 + x - 3}{x^2 + 1}$
46. $\dfrac{x^4 + x^2 - 3x + 5}{x^2 + 2}$
47. $\dfrac{x^3 + 3x^2 + 3x + 2}{x^2 + x + 1}$
48. $\dfrac{x^3 - 4x^2 + 5x - 6}{x^2 - x + 2}$

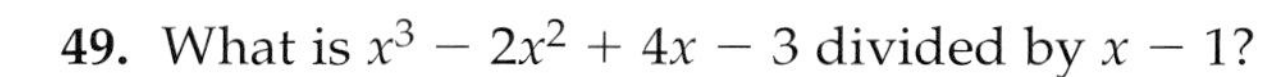

49. What is $x^3 - 2x^2 + 4x - 3$ divided by $x - 1$?
50. Divide $2y^3 + y^2 - 5y + 2$ by $y + 2$.
51. **BUSINESS** A company estimates that it costs $0.03x^2 + 4x + 1000$ dollars to produce x units of a product. Find an expression for the average cost per unit.
52. **ENTERTAINMENT** A magician gives these instructions to a volunteer.
    - Choose a number and multiply it by 3.
    - Then add the sum of your number and 8 to the product you found.
    - Now divide by the sum of your number and 2.

    What number will the volunteer always have at the end? Explain.

Career Choices

Cost Analyst

Cost analysts study and write reports about the factors involved in the cost of production.

Online Research For information about a career in cost analysis, visit: www.algebra2.com/careers

MEDICINE For Exercises 53 and 54, use the following information.
The number of students at a large high school who will catch the flu during an outbreak can be estimated by $n = \dfrac{170t^2}{t^2 + 1}$, where t is the number of weeks from the beginning of the epidemic and n is the number of ill people.

53. Perform the division indicated by $\dfrac{170t^2}{t^2 + 1}$.
54. Use the formula to estimate how many people will become ill during the first week.

PHYSICS For Exercises 55–57, suppose an object moves in a straight line so that after t seconds, it is $t^3 + t^2 + 6t$ feet from its starting point.

55. Find the distance the object travels between the times $t = 2$ and $t = x$.
56. How much time elapses between $t = 2$ and $t = x$?
57. Find a simplified expression for the average speed of the object between times $t = 2$ and $t = x$.
58. **CRITICAL THINKING** Suppose the result of dividing one polynomial by another is $r^2 - 6r + 9 - \dfrac{1}{r - 3}$. What two polynomials might have been divided?

59. **WRITING IN MATH** Answer the question that was posed at the beginning of the lesson.

How can you use division of polynomials in manufacturing?

Include the following in your answer:

- the dimensions of the piece of metal that the machinist needs,
- the formula from geometry that applies to this situation, and
- an explanation of how to use division of polynomials to find the width s of the seam.

60. An office employs x women and 3 men. What is the ratio of the total number of employees to the number of women?

Ⓐ $1 + \frac{3}{x}$ Ⓑ $\frac{x}{x+3}$ Ⓒ $\frac{3}{x}$ Ⓓ $\frac{x}{3}$

61. If $a + b = c$ and $a = b$, then all of the following are true EXCEPT

Ⓐ $a - c = b - c$. Ⓑ $a - b = 0$.

Ⓒ $2a + 2b = 2c$. Ⓓ $c - b = 2a$.

Maintain Your Skills

Mixed Review

Simplify. *(Lesson 5-2)*

62. $(2x^2 - 3x + 5) - (3x^2 + x - 9)$
63. $y^2z(y^2z^3 - yz^2 + 3)$
64. $(y + 5)(y - 3)$
65. $(a - b)^2$

66. **ASTRONOMY** Earth is an average of 1.5×10^{11} meters from the Sun. Light travels at 3×10^8 meters per second. About how long does it take sunlight to reach Earth? *(Lesson 5-1)*

Write an equation in slope-intercept form for each graph. *(Lesson 2-4)*

67.

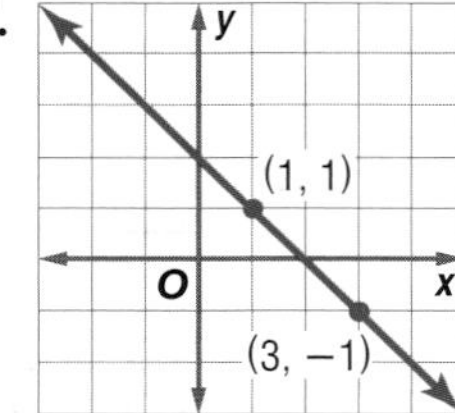

68.

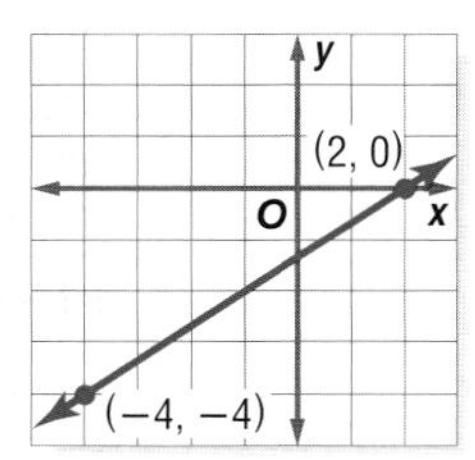

Getting Ready for the Next Lesson

BASIC SKILL **Find the greatest common factor of each set of numbers.**

69. 18, 27
70. 24, 84
71. 16, 28
72. 12, 27, 48
73. 12, 30, 54
74. 15, 30, 65

Practice Quiz 1 — Lessons 5-1 through 5-3

Express each number in scientific notation. *(Lesson 5-1)*

1. 653,000,000
2. 0.0072

Simplify. *(Lessons 5-1 and 5-2)*

3. $(-3x^2y)^3(2x)^2$
4. $\frac{a^6b^{-2}c}{a^3b^2c^4}$
5. $\left(\frac{x^2z}{xz^4}\right)^2$
6. $(9x + 2y) - (7x - 3y)$
7. $(t + 2)(3t - 4)$
8. $(n + 2)(n^2 - 3n + 1)$

Simplify. *(Lesson 5-3)*

9. $(m^3 - 4m^2 - 3m - 7) \div (m - 4)$
10. $\frac{2d^3 - d^2 - 9d + 9}{2d - 3}$

5-4 Factoring Polynomials

California Standards Standard 4.0 Students factor polynomials representing the difference of squares, perfect square trinomials, and the sum and difference of two cubes. (Key)

What You'll Learn

- Factor polynomials.
- Simplify polynomial quotients by factoring.

How does factoring apply to geometry?

Suppose the expression $4x^2 + 10x - 6$ represents the area of a rectangle. Factoring can be used to find possible dimensions of the rectangle.

? units

$A = 4x^2 + 10x - 6$ units^2 ? units

FACTOR POLYNOMIALS Whole numbers are factored using prime numbers. For example, $100 = 2 \cdot 2 \cdot 5 \cdot 5$. Many polynomials can also be factored. Their factors, however, are other polynomials. Polynomials that cannot be factored are called *prime*.

The table below summarizes the most common factoring techniques used with polynomials.

Concept Summary — *Factoring Techniques*

Number of Terms	Factoring Technique	General Case
any number	Greatest Common Factor (GCF)	$a^3b^2 + 2a^2b - 4ab^2 = ab(a^2b + 2a - 4b)$
two	Difference of Two Squares Sum of Two Cubes Difference of Two Cubes	$a^2 - b^2 = (a + b)(a - b)$ $a^3 + b^3 = (a + b)(a^2 - ab + b^2)$ $a^3 - b^3 = (a - b)(a^2 + ab + b^2)$
three	Perfect Square Trinomials	$a^2 + 2ab + b^2 = (a + b)^2$ $a^2 - 2ab + b^2 = (a - b)^2$
	General Trinomials	$acx^2 + (ad + bc)x + bd = (ax + b)(cx + d)$
four or more	Grouping	$ax + bx + ay + by = x(a + b) + y(a + b)$ $= (a + b)(x + y)$

Whenever you factor a polynomial, always look for a common factor first. Then determine whether the resulting polynomial factor can be factored again using one or more of the methods listed in the table above.

$12x^2y^5 - 6x^3y^2 + 2xy$

$2xy(6xy^4 - 3x^2y + 1)$

Example 1 GCF

Factor $6x^2y^2 - 2xy^2 + 6x^3y$.

$$6x^2y^2 - 2xy^2 + 6x^3y = (2 \cdot 3 \cdot x \cdot x \cdot y \cdot y) - (2 \cdot x \cdot y \cdot y) + (2 \cdot 3 \cdot x \cdot x \cdot x \cdot y)$$

$$= (2xy \cdot 3xy) - (2xy \cdot y) + (2xy \cdot 3x^2)$$

$$= 2xy(3xy - y + 3x^2)$$

The GCF is $2xy$. The remaining polynomial cannot be factored using the methods above.

Check this result by finding the product.

A GCF is also used in grouping to factor a polynomial of four or more terms.

Example 2 Grouping

Factor $a^3 - 4a^2 + 3a - 12$.

$a^3 - 4a^2 + 3a - 12 = (a^3 - 4a^2) + (3a - 12)$ Group to find a GCF.

$= a^2(a - 4) + 3(a - 4)$ Factor the GCF of each binomial.

$= (a - 4)(a^2 + 3)$ Distributive Property

You can use algebra tiles to model factoring a polynomial.

Algebra Activity

Factoring Trinomials

Use algebra tiles to factor $2x^2 + 7x + 3$.

Model and Analyze

- Use algebra tiles to model $2x^2 + 7x + 3$.
- To find the product that resulted in this polynomial, arrange the tiles to form a rectangle.

x^2	x^2	x
x	x	1
x	x	1
x	x	1

- Notice that the total area can be expressed as the sum of the areas of two smaller rectangles.

x^2	x^2	x	$2x^2 + x$
x	x	1	$6x + 3$
x	x	1	
x	x	1	

Use these expressions to rewrite the trinomial. Then factor.

$2x^2 + 7x + 3 = (2x^2 + x) + (6x + 3)$ total area = sum of areas of smaller rectangles

$= x(2x + 1) + 3(2x + 1)$ Factor out each GCF.

$= (2x + 1)(x + 3)$ Distributive Property

Make a Conjecture

Study the factorization of $2x^2 + 7x + 3$ above.

1. What are the coefficients of the two x terms in $(2x^2 + x) + (6x + 3)$? Find their sum and their product.
2. Compare the sum you found in Exercise 1 to the coefficient of the x term in $2x^2 + 7x + 3$.
3. Find the product of the coefficient of the x^2 term and the constant term in $2x^2 + 7x + 3$. How does it compare to the product in Exercise 1?
4. Make a conjecture about how to factor $3x^2 + 7x + 2$.

Study Tip

Algebra Tiles
When modeling a polynomial with algebra tiles, it is easiest to arrange the x^2 tiles first, then the x tiles and finally the 1 tiles to form a rectangle.

The FOIL method can help you factor a polynomial into the product of two binomials. Study the following example.

$$(ax + b)(cx + d) = \overbrace{ax \cdot cx}^{F} + \overbrace{ax \cdot d}^{O} + \overbrace{b \cdot cx}^{I} + \overbrace{b \cdot d}^{L}$$
$$= acx^2 + (ad + bc)x + bd$$

Notice that the product of the coefficient of x^2 and the constant term is $abcd$. The product of the two terms in the coefficient of x is also $abcd$.

Example 3 Two or Three Terms

Factor each polynomial.

a. $5x^2 - 13x + 6$

To find the coefficients of the x-terms, you must find two numbers whose product is $5 \cdot 6$ or 30, and whose sum is -13. The two coefficients must be -10 and -3 since $(-10)(-3) = 30$ and $-10 + (-3) = -13$.

Rewrite the expression using $-10x$ and $-3x$ in place of $-13x$ and factor by grouping.

$5x^2 - 13x + 6 = 5x^2 - 10x - 3x + 6$	Substitute $-10x - 3x$ for $-13x$.
$= (5x^2 - 10x) + (-3x + 6)$	Associative Property
$= 5x(x - 2) - 3(x - 2)$	Factor out the GCF of each group.
$= (5x - 3)(x - 2)$	Distributive Property

b. $3xy^2 - 48x$

$3xy^2 - 48x = 3x(y^2 - 16)$	Factor out the GCF.
$= 3x(y + 4)(y - 4)$	$y^2 - 16$ is the difference of two squares.

c. $c^3d^3 + 27$

$c^3d^3 = (cd)^3$ and $27 = 3^3$. Thus, this is the sum of two cubes.

$c^3d^3 + 27 = (cd + 3)[(cd)^2 - 3(cd) + 3^2]$	Sum of two cubes formula with $a = cd$ and $b = 3$
$= (cd + 3)(c^2d^2 - 3cd + 9)$	Simplify.

d. $m^6 - n^6$

This polynomial could be considered the difference of two squares or the difference of two cubes. The difference of two squares should always be done before the difference of two cubes. This will make the next step of the factorization easier.

$m^6 - n^6 = (m^3 + n^3)(m^3 - n^3)$	Difference of two squares
$= (m + n)(m^2 - mn + n^2)(m - n)(m^2 + mn + n^2)$	Sum and difference of two cubes

You can use a graphing calculator to check that the factored form of a polynomial is correct.

Graphing Calculator Investigation

Factoring Polynomials

Is the factored form of $2x^2 - 11x - 21$ equal to $(2x - 7)(x + 3)$? You can find out by graphing $y = 2x^2 - 11x - 21$ and $y = (2x - 7)(x + 3)$. If the two graphs coincide, the factored form is probably correct.

- Enter $y = 2x^2 - 11x - 21$ and $y = (2x - 7)(x + 3)$ on the Y= screen.
- Graph the functions. Since two different graphs appear, $2x^2 - 11x - 21 \neq (2x - 7)(x + 3)$.

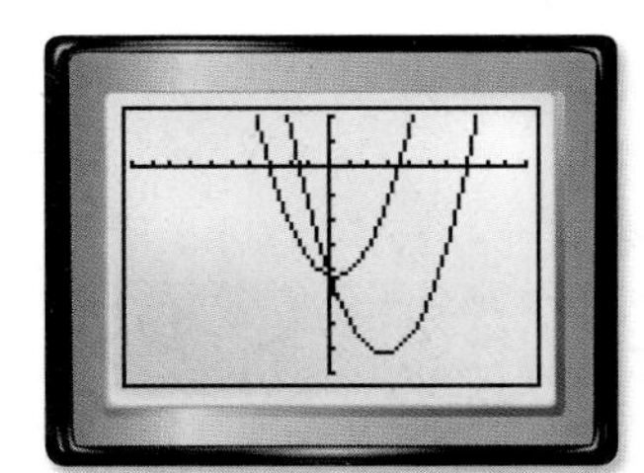

[−10, 10] scl: 1 by [−40, 10] scl: 5

Think and Discuss

1. Determine if $x^2 + 5x - 6 = (x - 3)(x - 2)$ is a true statement. If not, write the correct factorization.
2. Does this method guarantee a way to check the factored form of a polynomial? Why or why not?

SIMPLIFY QUOTIENTS In Lesson 5-3, you learned to simplify the quotient of two polynomials by using long division or synthetic division. Some quotients can be simplified using factoring.

Example 4 Quotient of Two Trinomials

Simplify $\frac{x^2 + 2x - 3}{x^2 + 7x + 12}$.

$\frac{x^2 + 2x - 3}{x^2 + 7x + 12} = \frac{\cancel{(x + 3)}^{1}(x - 1)}{(x + 4)\cancel{(x + 3)}_{1}}$ Factor the numerator and denominator.

$= \frac{x - 1}{x + 4}$ Divide. Assume $x \neq -3, -4$.

Therefore, $\frac{x^2 + 2x - 3}{x^2 + 7x + 12} = \frac{x - 1}{x + 4}$, if $x \neq -3, -4$.

Check for Understanding

Concept Check

1. **OPEN ENDED** Write an example of a perfect square trinomial.
2. **Find a counterexample** to the statement $a^2 + b^2 = (a + b)^2$.
3. **Decide** whether the statement $\frac{x - 2}{x^2 + x - 6} = \frac{1}{x + 3}$ is *sometimes*, *always*, or *never* true.

Guided Practice

Factor completely. If the polynomial is not factorable, write *prime*.

4. $-12x^2 - 6x$
5. $a^2 + 5a + ab$
6. $21 - 7y + 3x - xy$
7. $y^2 - 6y + 8$
8. $z^2 - 4z - 12$
9. $3b^2 - 48$
10. $16w^2 - 169$
11. $h^3 + 8000$

Simplify. Assume that no denominator is equal to 0.

12. $\frac{x^2 - 2x - 8}{x^2 - 5x - 14}$
13. $\frac{2y^2 + 8y}{y^2 - 16}$

Application

14. **GEOMETRY** Find the width of rectangle *ABCD* if its area is $3x^2 + 9xy + 6y^2$ square centimeters.

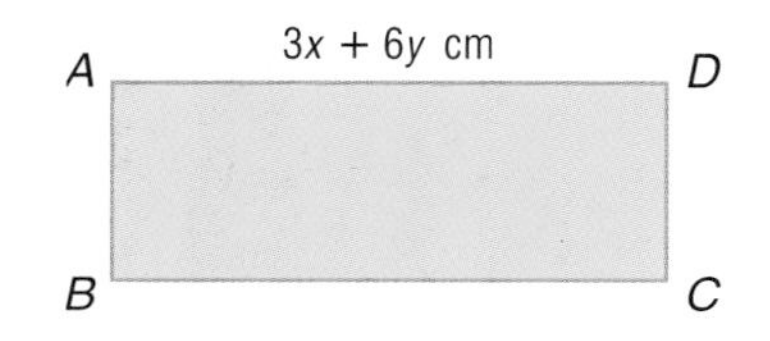

Practice and Apply

Factor completely. If the polynomial is not factorable, write *prime*.

15. $2xy^3 - 10x$
16. $6a^2b^2 + 18ab^3$
17. $12cd^3 - 8c^2d^2 + 10c^5d^3$
18. $3a^2bx + 15cx^2y + 25ad^3y$
19. $8yz - 6z - 12y + 9$
20. $3ax - 15a + x - 5$
21. $x^2 + 7x + 6$
22. $y^2 - 5y + 4$
23. $2a^2 + 3a + 1$
24. $2b^2 + 13b - 7$
25. $6c^2 + 13c + 6$
26. $12m^2 - m - 6$
27. $3n^2 + 21n - 24$
28. $3z^2 + 24z + 45$

Homework Help

For Exercises	See Examples
15–18	1
19, 20	2
21–38, 43–45, 55	3
39–42	2, 3
46–54	4

Extra Practice
See page 837.

29. $x^2 + 12x + 36$
30. $x^2 - 6x + 9$
31. $16a^2 + 25b^2$
32. $3m^2 - 3n^2$
33. $y^4 - z^2$
34. $3x^2 - 27y^2$
35. $z^3 + 125$
36. $t^3 - 8$
37. $p^4 - 1$
38. $x^4 - 81$
39. $7ac^2 + 2bc^2 - 7ad^2 - 2bd^2$
40. $8x^2 + 8xy + 8xz + 3x + 3y + 3z$
41. $5a^2x + 4aby + 3acz - 5abx - 4b^2y - 3bcz$
42. $3a^3 + 2a^2 - 5a + 9a^2b + 6ab - 15b$

43. Find the factorization of $3x^2 + x - 2$.

44. What are the factors of $2y^2 + 9y + 4$?

45. **LANDSCAPING** A boardwalk that is x feet wide is built around a rectangular pond. The combined area of the pond and the boardwalk is $4x^2 + 140x + 1200$ square feet. What are the dimensions of the pond?

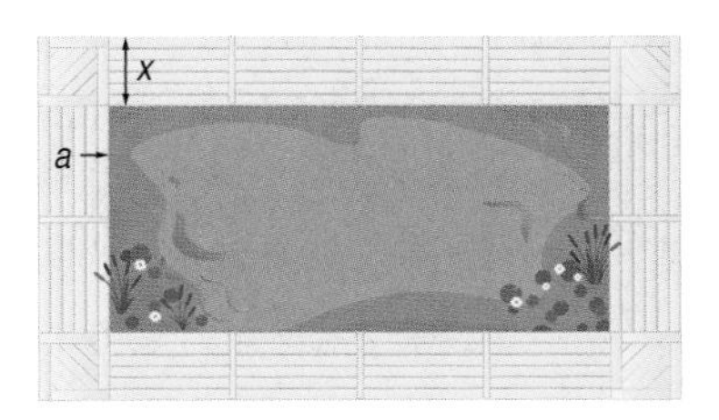

Simplify. Assume that no denominator is equal to 0.

46. $\dfrac{x^2 + 4x + 3}{x^2 - x - 12}$
47. $\dfrac{x^2 + 4x - 5}{x^2 - 7x + 6}$
48. $\dfrac{x^2 - 25}{x^2 + 3x - 10}$
49. $\dfrac{x^2 - 6x + 8}{x^3 - 8}$
50. $\dfrac{x^2}{(x^2 - x)(x - 1)^{-1}}$
51. $\dfrac{x + 1}{(x^2 + 3x + 2)(x + 2)^{-2}}$

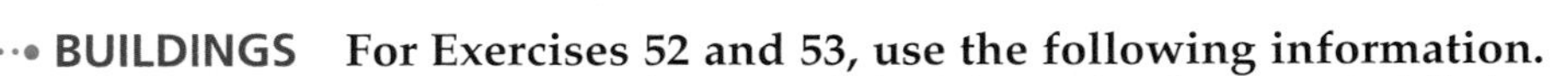

BUILDINGS **For Exercises 52 and 53, use the following information.**
When an object is dropped from a tall building, the distance it falls between 1 second after it is dropped and x seconds after it is dropped is $16x^2 - 16$ feet.

52. How much time elapses between 1 second after it is dropped and x seconds after it is dropped?

53. What is the average speed of the object during that time period?

More About. . .

Buildings

The tallest buildings in the world are the Petronas Towers in Kuala Lumpur, Malaysia. Each is 1483 feet tall.

Source: www.worldstallest.com

54. **GEOMETRY** The length of one leg of a right triangle is $x - 6$ centimeters, and the area is $\frac{1}{2}x^2 - 7x + 24$ square centimeters. What is the length of the other leg?

55. **CRITICAL THINKING** Factor $64p^{2n} + 16p^n + 1$.

56. **WRITING IN MATH** Answer the question that was posed at the beginning of the lesson.

How does factoring apply to geometry?

Include the following in your answer:

- an explanation of how to use factoring to find possible dimensions for the rectangle described at the beginning of the lesson, and
- why your dimensions are not the only ones possible, even if you assume that the dimensions are binomials with integer coefficients.

Standardized Test Practice

Standards Practice

57. Which of the following is the factorization of $2x - 15 + x^2$?

(A) $(x - 3)(x - 5)$ (B) $(x - 3)(x + 5)$

(C) $(x + 3)(x - 5)$ (D) $(x + 3)(x + 5)$

58. Which is not a factor of $x^3 - x^2 - 2x$?

(A) x (B) $x + 1$ (C) $x - 1$ (D) $x - 2$

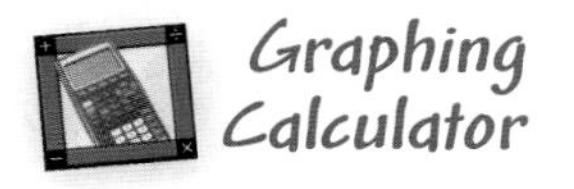

Graphing Calculator

CHECK FACTORING **Use a graphing calculator to determine if each polynomial is factored correctly. Write *yes* or *no*. If the polynomial is not factored correctly, find the correct factorization.**

59. $3x^2 + 5x + 2 \stackrel{?}{=} (3x + 2)(x + 1)$

60. $x^3 + 8 \stackrel{?}{=} (x + 2)(x^2 - x + 4)$

61. $2x^2 - 5x - 3 \stackrel{?}{=} (x - 1)(2x + 3)$

62. $3x^2 - 48 \stackrel{?}{=} 3(x + 4)(x - 4)$

Maintain Your Skills

Mixed Review

Simplify. *(Lesson 5-3)*

63. $(t^3 - 3t + 2) \div (t + 2)$

64. $(y^2 + 4y + 3)(y + 1)^{-1}$

65. $\dfrac{x^3 - 3x^2 + 2x - 6}{x - 3}$

66. $\dfrac{3x^4 + x^3 - 8x^2 + 10x - 3}{3x - 2}$

Simplify. *(Lesson 5-2)*

67. $(3x^2 - 2xy + y^2) + (x^2 + 5xy - 4y^2)$

68. $(2x + 4)(7x - 1)$

Perform the indicated operations, if possible. *(Lesson 4-5)*

69. $[3 \quad -1] \cdot \begin{bmatrix} 0 \\ 2 \end{bmatrix}$

70. $\begin{bmatrix} 1 & -4 \\ 2 & 2 \end{bmatrix} \cdot \begin{bmatrix} 0 & 3 \\ 9 & -1 \end{bmatrix}$

71. **PHOTOGRAPHY** The perimeter of a rectangular picture is 86 inches. Twice the width exceeds the length by 2 inches. What are the dimensions of the picture? *(Lesson 3-2)*

Determine whether each relation is a function. Write *yes* or *no*. *(Lesson 2-1)*

72.

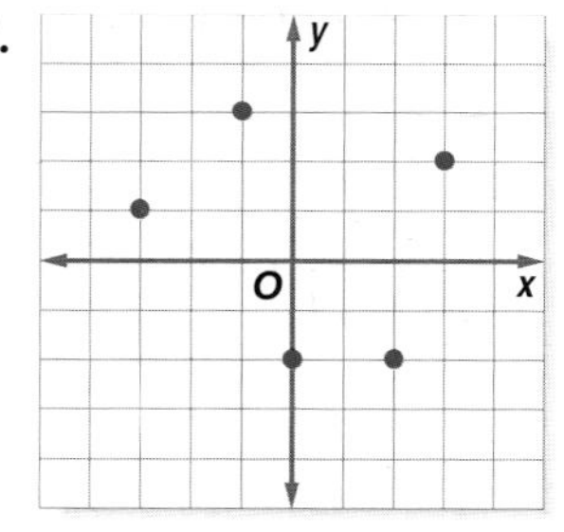

73.

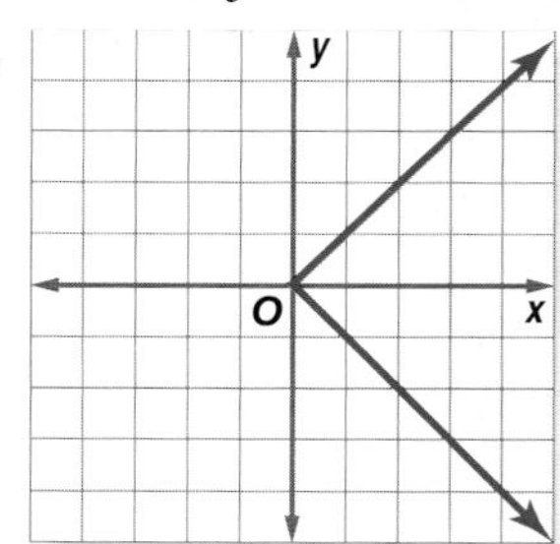

State the property illustrated by each equation. *(Lesson 1-2)*

74. $(3 + 8)5 = 3(5) + 8(5)$

75. $1 + (7 + 4) = (1 + 7) + 4$

Getting Ready for the Next Lesson

PREREQUISITE SKILL **Determine whether each number is *rational* or *irrational*.**
*(To review **rational and irrational numbers**, see Lesson 1-2.)*

76. 4.63

77. π

78. $\dfrac{16}{3}$

79. 8.333...

80. 7.323223222...

81. $9.7\overline{1}$

5-5 Roots of Real Numbers

California Standards Standard 15.0 **Students determine whether a specific algebraic statement involving** rational expressions, **radical expressions,** or logarithmic or exponential functions **is sometimes true, always true, or never true.** (Key)

What You'll Learn

- Simplify radicals.
- Use a calculator to approximate radicals.

How do square roots apply to oceanography?

The speed s in knots of a wave can be estimated using the formula $s = 1.34\sqrt{\ell}$, where ℓ is the length of the wave in feet. This is an example of an equation that contains a square root.

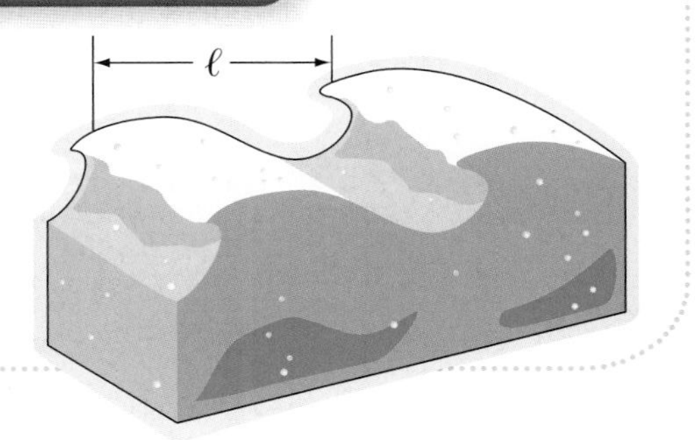

Vocabulary

- square root
- *n*th root
- principal root

SIMPLIFY RADICALS Finding the square root of a number and squaring a number are inverse operations. To find the **square root** of a number n, you must find a number whose square is n. For example, 7 is a square root of 49 since $7^2 = 49$. Since $(-7)^2 = 49$, -7 is also a square root of 49.

Key Concept — *Definition of Square Root*

- **Words** For any real numbers a and b, if $a^2 = b$, then a is a square root of b.
- **Example** Since $5^2 = 25$, 5 is a square root of 25.

Since finding the square root of a number and squaring a number are inverse operations, it makes sense that the inverse of raising a number to the nth power is finding the ***n*th root** of a number. The table below shows the relationship between raising a number to a power and taking that root of a number.

Powers	Factors	Roots
$a^3 = 125$	$5 \cdot 5 \cdot 5 = 125$	5 is a cube root of 125.
$a^4 = 81$	$3 \cdot 3 \cdot 3 \cdot 3 = 81$	3 is a fourth root of 81.
$a^5 = 32$	$2 \cdot 2 \cdot 2 \cdot 2 \cdot 2 = 32$	2 is a fifth root of 32.
$a^n = b$	$\underbrace{a \cdot a \cdot a \cdot a \cdot \ldots \cdot a}_{n \text{ factors of } a} = b$	a is an nth root of b.

This pattern suggests the following formal definition of an nth root.

Key Concept — *Definition of nth Root*

- **Words** For any real numbers a and b, and any positive integer n, if $a^n = b$, then a is an nth root of b.
- **Example** Since $2^5 = 32$, 2 is a fifth root of 32.

Study Tip

Reading Math
$\sqrt[n]{50}$ is read *the nth root of 50.*

The symbol $\sqrt[n]{\ }$ indicates an nth root.

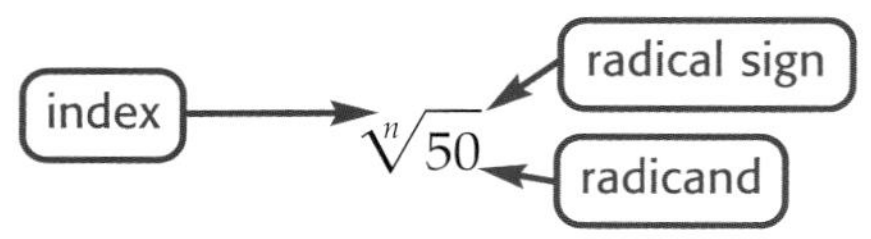

Some numbers have more than one real nth root. For example, 36 has two square roots, 6 and -6. When there is more than one real root, the nonnegative root is called the **principal root**. When no index is given, as in $\sqrt{36}$, the radical sign indicates the principal square root. The symbol $\sqrt[n]{b}$ stands for the principal nth root of b. If n is odd and b is negative, there will be no nonnegative root. In this case, the principal root is negative.

$\sqrt{16} = 4$	$\sqrt{16}$ indicates the principal square root of 16.
$-\sqrt{16} = -4$	$-\sqrt{16}$ indicates the opposite of the principal square root of 16.
$\pm\sqrt{16} = \pm 4$	$\pm\sqrt{16}$ indicates both square roots of 16. $\pm$ means positive or negative.
$\sqrt[3]{-125} = -5$	$\sqrt[3]{-125}$ indicates the principal cube root of -125.
$-\sqrt[4]{81} = -3$	$-\sqrt[4]{81}$ indicates the opposite of the principal fourth root of 81.

Concept Summary *Real nth roots of b, $\sqrt[n]{b}$, or $-\sqrt[n]{b}$*

n	$\sqrt[n]{b}$ if $b > 0$	$\sqrt[n]{b}$ if $b < 0$	$b = 0$
even	one positive root, one negative root $\pm\sqrt[4]{625} = \pm 5$	no real roots $\sqrt{-4}$ is not a real number.	one real root, 0 $\sqrt[n]{0} = 0$
odd	one positive root, no negative roots $\sqrt[3]{8} = 2$	no positive roots, one negative root $\sqrt[5]{-32} = -2$	

Example 1 Find Roots

Simplify.

a. $\pm\sqrt{25x^4}$

$$\pm\sqrt{25x^4} = \pm\sqrt{(5x^2)^2}$$
$$= \pm 5x^2$$

The square roots of $25x^4$ are $\pm 5x^2$.

b. $-\sqrt{(y^2 + 2)^8}$

$$-\sqrt{(y^2 + 2)^8} = -\sqrt{[(y^2 + 2)^4]^2}$$
$$= -(y^2 + 2)^4$$

The opposite of the principal square root of $(y^2 + 2)^8$ is $-(y^2 + 2)^4$.

c. $\sqrt[5]{32x^{15}y^{20}}$

$$\sqrt[5]{32x^{15}y^{20}} = \sqrt[5]{(2x^3y^4)^5}$$
$$= 2x^3y^4$$

The principal fifth root of $32x^{15}y^{20}$ is $2x^3y^4$.

d. $\sqrt{-9}$

$$\sqrt{-9} = \sqrt[2]{-9}$$

n is even.
b is negative.

Thus, $\sqrt{-9}$ is not a real number.

When you find the nth root of an even power and the result is an odd power, you must take the absolute value of the result to ensure that the answer is nonnegative.

$$\sqrt{(-5)^2} = |-5| \text{ or } 5 \qquad \sqrt{(-2)^6} = |(-2)^3| \text{ or } 8$$

If the result is an even power or you find the nth root of an odd power, there is no need to take the absolute value. *Why?*

Example 2 Simplify Using Absolute Value

Simplify.

a. $\sqrt[8]{x^8}$

Note that x is an eighth root of x^8. The index is even, so the principal root is nonnegative. Since x could be negative, you must take the absolute value of x to identify the principal root.

$\sqrt[8]{x^8} = |x|$

b. $\sqrt[4]{81(a+1)^{12}}$

$\sqrt[4]{81(a+1)^{12}} = \sqrt[4]{[3(a+1)^3]^4}$

Since the index 4 is even and the exponent 3 is odd, you must use the absolute value of $(a+1)^3$.

$\sqrt[4]{81(a+1)^{12}} = 3|(a+1)^3|$

APPROXIMATE RADICALS WITH A CALCULATOR Recall that real numbers that cannot be expressed as terminating or repeating decimals are *irrational numbers*. $\sqrt{2}$ and $\sqrt{3}$ are examples of irrational numbers. Decimal approximations for irrational numbers are often used in applications.

Example 3 Approximate a Square Root

PHYSICS The time T in seconds that it takes a pendulum to make a complete swing back and forth is given by the formula $T = 2\pi\sqrt{\frac{L}{g}}$, where L is the length of the pendulum in feet and g is the acceleration due to gravity, 32 feet per second squared. Find the value of T for a 3-foot-long pendulum.

Explore You are given the values of L and g and must find the value of T. Since the units on g are feet per second squared, the units on the time T should be seconds.

Plan Substitute the values for L and g into the formula. Use a calculator to evaluate.

Solve

$T = 2\pi\sqrt{\frac{L}{g}}$ Original formula

$= 2\pi\sqrt{\frac{3}{32}}$ $L = 3, g = 32$

≈ 1.92 Use a calculator.

It takes the pendulum about 1.92 seconds to make a complete swing.

Examine The closest square to $\frac{3}{32}$ is $\frac{1}{9}$, and π is approximately 3, so the answer should be close to $2(3)\sqrt{\frac{1}{9}} = 2(3)\left(\frac{1}{3}\right)$ or 2. The answer is reasonable.

Study Tip

Graphing Calculators

To find a root of index greater than 2, first type the index. Then select $\sqrt[x]{}$ from the MATH menu. Finally, enter the radicand.

Check for Understanding

Concept Check

1. **OPEN ENDED** Write a number whose principal square root and cube root are both integers.
2. **Explain** why it is not always necessary to take the absolute value of a result to indicate the principal root.
3. **Determine** whether the statement $\sqrt[4]{(-x)^4} = x$ is *sometimes*, *always*, or *never* true. Explain your reasoning.

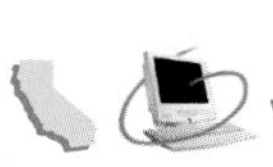

Guided Practice Use a calculator to approximate each value to three decimal places.

4. $\sqrt{77}$
5. $-\sqrt[3]{19}$
6. $\sqrt[4]{48}$

Simplify.

7. $\sqrt[3]{64}$
8. $\sqrt{(-2)^2}$
9. $\sqrt[5]{-243}$
10. $\sqrt[4]{-4096}$
11. $\sqrt[3]{x^3}$
12. $\sqrt[4]{y^4}$
13. $\sqrt{36a^2b^4}$
14. $\sqrt{(4x + 3y)^2}$

Application

15. **OPTICS** The distance D in miles from an observer to the horizon over flat land or water can be estimated using the formula $D = 1.23\sqrt{h}$, where h is the height in feet of the point of observation. How far is the horizon for a person whose eyes are 6 feet above the ground?

Practice and Apply

Homework Help

For Exercises	See Examples
16–27, 60–62	3
28–59	1, 2

Extra Practice See page 838.

Use a calculator to approximate each value to three decimal places.

16. $\sqrt{129}$
17. $-\sqrt{147}$
18. $\sqrt{0.87}$
19. $\sqrt{4.27}$
20. $\sqrt[3]{59}$
21. $\sqrt[3]{-480}$
22. $\sqrt[4]{602}$
23. $\sqrt[5]{891}$
24. $\sqrt[6]{4123}$
25. $\sqrt[7]{46{,}815}$
26. $\sqrt[6]{(723)^3}$
27. $\sqrt[4]{(3500)^2}$

Simplify.

28. $\sqrt{225}$
29. $\pm\sqrt{169}$
30. $\sqrt{-(-7)^2}$
31. $\sqrt{(-18)^2}$
32. $\sqrt[3]{-27}$
33. $\sqrt[7]{-128}$
34. $\sqrt{\frac{1}{16}}$
35. $\sqrt[3]{\frac{1}{125}}$
36. $\sqrt{0.25}$
37. $\sqrt[3]{-0.064}$
38. $\sqrt[4]{z^8}$
39. $-\sqrt[6]{x^6}$
40. $\sqrt{49m^6}$
41. $\sqrt{64a^8}$
42. $\sqrt[3]{27r^3}$
43. $\sqrt[3]{-c^6}$
44. $\sqrt{(5g)^4}$
45. $\sqrt[3]{(2z)^6}$
46. $\sqrt{25x^4y^6}$
47. $\sqrt{36x^4z^4}$
48. $\sqrt{169x^8y^4}$
49. $\sqrt{9p^{12}q^6}$
50. $\sqrt[3]{8a^3b^3}$
51. $\sqrt[3]{-27c^9d^{12}}$
52. $\sqrt{(4x - y)^2}$
53. $\sqrt[3]{(p + q)^3}$
54. $-\sqrt{x^2 + 4x + 4}$
55. $\sqrt{z^2 + 8z + 16}$
56. $\sqrt{4a^2 + 4a + 1}$
57. $\sqrt{-9x^2 - 12x - 4}$

58. Find the principal fifth root of 32.

59. What is the third root of -125?

60. **SPORTS** Refer to the drawing at the right. How far does the catcher have to throw a ball from home plate to second base?

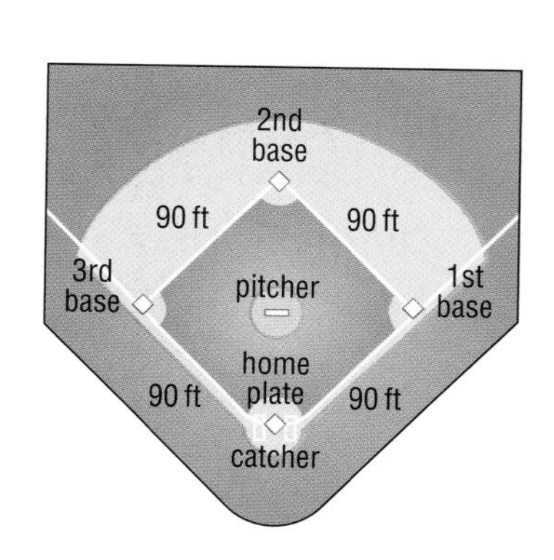

61. **FISH** The relationship between the length and mass of Pacific halibut can be approximated by the equation $L = 0.46\sqrt[3]{M}$, where L is the length in meters and M is the mass in kilograms. Use this equation to predict the length of a 25-kilogram Pacific halibut.

More About...

Space Science

The escape velocity for the Moon is about 2400 m/s. For the Sun, it is about 618,000 m/s.

Source: NASA

62. SPACE SCIENCE The velocity v required for an object to escape the gravity of a planet or other body is given by the formula $v = \sqrt{\frac{2GM}{R}}$, where M is the mass of the body, R is the radius of the body, and G is Newton's gravitational constant. Use $M = 5.98 \times 10^{24}$ kg, $R = 6.37 \times 10^{6}$ m, and $G = 6.67 \times 10^{-11}$ N · m²/kg² to find the escape velocity for Earth.

63. CRITICAL THINKING Under what conditions does $\sqrt{x^2 + y^2} = x + y$?

64. WRITING IN MATH Answer the question that was posed at the beginning of the lesson.

How do square roots apply to oceanography?

Include the following in your answer:

- the values of s for $\ell = 2, 5,$ and 10 feet, and
- an observation of what happens to the value of s as the value of ℓ increases.

Standardized Test Practice

Standards Practice

65. Which of the following is closest to $\sqrt{7.32}$?

(A) 2.6 (B) 2.7 (C) 2.8 (D) 2.9

66. In the figure, $\triangle ABC$ is an equilateral triangle with sides 9 units long. What is the length of $\overline{BD}$ in units?

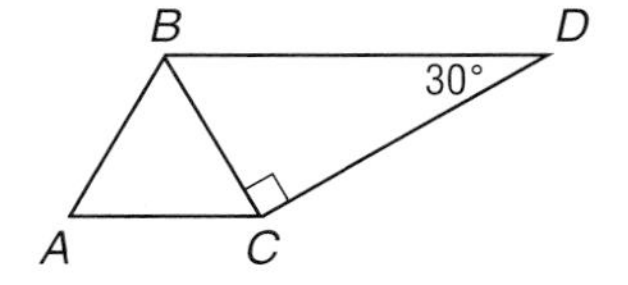

(A) 3 (B) 9

(C) $9\sqrt{2}$ (D) 18

Maintain Your Skills

Mixed Review

Factor completely. If the polynomial is not factorable, write *prime*. *(Lesson 5-4)*

67. $7xy^3 - 14x^2y^5 + 28x^3y^2$

68. $ab - 5a + 3b - 15$

69. $2x^2 + 15x + 25$

70. $c^3 - 216$

Simplify. *(Lesson 5-3)*

71. $(4x^3 - 7x^2 + 3x - 2) \div (x - 2)$

72. $\frac{x^4 + 4x^3 - 4x^2 + 5x}{x + 5}$

73. TRAVEL The matrix at the right shows the costs of airline flights between some cities. Write a matrix that shows the costs of two tickets for these flights. *(Lesson 4-2)*

	New York	LA
Atlanta	405	1160
Chicago	709	1252

Solve each system of equations by using either substitution or elimination. *(Lesson 3-2)*

74. $a + 4b = 6$
$3a + 2b = -2$

75. $10x - y = 13$
$3x - 4y = 15$

76. $3c - 7d = -1$
$2c - 6d = -6$

Getting Ready for the Next Lesson

PREREQUISITE SKILL **Find each product.**
*(To review **multiplying binomials**, see Lesson 5-2.)*

77. $(x + 3)(x + 8)$

78. $(y - 2)(y + 5)$

79. $(a + 2)(a - 9)$

80. $(a + b)(a + 2b)$

81. $(x - 3y)(x + 3y)$

82. $(2w + z)(3w - 5z)$

5-6 Radical Expressions

California Standards Standard 15.0 Students determine whether a specific algebraic statement involving rational expressions, **radical expressions,** or logarithmic or exponential functions **is sometimes true, always true, or never true.** (Key)

What You'll Learn

- Simplify radical expressions.
- Add, subtract, multiply, and divide radical expressions.

Vocabulary

- rationalizing the denominator
- like radical expressions
- conjugates

How do radical expressions apply to falling objects?

The amount of time t in seconds that it takes for an object to drop d feet is given by $t = \sqrt{\frac{2d}{g}}$, where $g = 32 \text{ ft/s}^2$ is the acceleration due to gravity. In this lesson, you will learn how to simplify radical expressions like $\sqrt{\frac{2d}{g}}$.

SIMPLIFY RADICAL EXPRESSIONS You can use the Commutative Property and the definition of square root to find an equivalent expression for a product of radicals such as $\sqrt{3} \cdot \sqrt{5}$. Begin by squaring the product.

$$(\sqrt{3} \cdot \sqrt{5})^2 = \sqrt{3} \cdot \sqrt{5} \cdot \sqrt{3} \cdot \sqrt{5}$$

$$= \sqrt{3} \cdot \sqrt{3} \cdot \sqrt{5} \cdot \sqrt{5} \quad \text{Commutative Property of Multiplication}$$

$$= 3 \cdot 5 \text{ or } 15 \quad \text{Definition of square root}$$

Since $\sqrt{3} \cdot \sqrt{5} > 0$ and $(\sqrt{3} \cdot \sqrt{5})^2 = 15$, $\sqrt{3} \cdot \sqrt{5}$ is the principal square root of 15. That is, $\sqrt{3} \cdot \sqrt{5} = \sqrt{15}$. This illustrates the following property of radicals.

Key Concept — Product Property of Radicals

For any real numbers a and b and any integer $n > 1$,

1. if n is even and a and b are both nonnegative, then $\sqrt[n]{ab} = \sqrt[n]{a} \cdot \sqrt[n]{b}$, and

2. if n is odd, then $\sqrt[n]{ab} = \sqrt[n]{a} \cdot \sqrt[n]{b}$.

Follow these steps to simplify a square root.

Step 1 Factor the radicand into as many squares as possible.

Step 2 Use the Product Property to isolate the perfect squares.

Step 3 Simplify each radical.

Example 1 Square Root of a Product

Simplify $\sqrt{16p^8q^7}$.

$$\sqrt{16p^8q^7} = \sqrt{4^2 \cdot (p^4)^2 \cdot (q^3)^2 \cdot q} \quad \text{Factor into squares where possible.}$$

$$= \sqrt{4^2} \cdot \sqrt{(p^4)^2} \cdot \sqrt{(q^3)^2} \cdot \sqrt{q} \quad \text{Product Property of Radicals}$$

$$= 4p^4|q^3|\sqrt{q} \quad \text{Simplify.}$$

However, for $\sqrt{16p^8q^7}$ to be defined, $16p^8q^7$ must be nonnegative. If that is true, q must be nonnegative, since it is raised to an odd power. Thus, the absolute value is unnecessary, and $\sqrt{16p^8q^7} = 4p^4q^3\sqrt{q}$.

Look at a radical that involves division to see if there is a quotient property for radicals that is similar to the Product Property. Consider $\frac{49}{9}$. The radicand is a perfect square, so $\sqrt{\frac{49}{9}} = \sqrt{\left(\frac{7}{3}\right)^2}$ or $\frac{7}{3}$. Notice that $\frac{7}{3} = \frac{\sqrt{49}}{\sqrt{9}}$. This suggests the following property.

Key Concept — *Quotient Property of Radicals*

- **Words** For any real numbers a and $b \neq 0$, and any integer $n > 1$, $\sqrt[n]{\frac{a}{b}} = \frac{\sqrt[n]{a}}{\sqrt[n]{b}}$, if all roots are defined.
- **Example** $\frac{\sqrt{27}}{\sqrt{3}} = \sqrt{9}$ or 3

You can use the properties of radicals to write expressions in simplified form.

Concept Summary — *Simplifying Radical Expressions*

A radical expression is in simplified form when the following conditions are met.

- The index n is as small as possible.
- The radicand contains no factors (other than 1) that are nth powers of an integer or polynomial.
- The radicand contains no fractions.
- No radicals appear in a denominator.

Study Tip

Rationalizing the Denominator
You may want to think of rationalizing the denominator as making the denominator a rational number.

To eliminate radicals from a denominator or fractions from a radicand, you can use a process called **rationalizing the denominator**. To rationalize a denominator, multiply the numerator and denominator by a quantity so that the radicand has an exact root. Study the examples below.

Example 2 Simplify Quotients

Simplify each expression.

a. $\sqrt{\frac{x^4}{y^5}}$

$\sqrt{\frac{x^4}{y^5}} = \frac{\sqrt{x^4}}{\sqrt{y^5}}$ Quotient Property

$= \frac{\sqrt{(x^2)^2}}{\sqrt{(y^2)^2 \cdot y}}$ Factor into squares.

$= \frac{\sqrt{(x^2)^2}}{\sqrt{(y^2)^2} \cdot \sqrt{y}}$ Product Property

$= \frac{x^2}{y^2\sqrt{y}}$ $\sqrt{(x^2)^2} = x^2$

$= \frac{x^2}{y^2\sqrt{y}} \cdot \frac{\sqrt{y}}{\sqrt{y}}$ Rationalize the denominator.

$= \frac{x^2\sqrt{y}}{y^3}$ $\sqrt{y} \cdot \sqrt{y} = y$

b. $\sqrt[5]{\frac{5}{4a}}$

$\sqrt[5]{\frac{5}{4a}} = \frac{\sqrt[5]{5}}{\sqrt[5]{4a}}$ Quotient Property

$= \frac{\sqrt[5]{5}}{\sqrt[5]{4a}} \cdot \frac{\sqrt[5]{8a^4}}{\sqrt[5]{8a^4}}$ Rationalize the denominator.

$= \frac{\sqrt[5]{5 \cdot 8a^4}}{\sqrt[5]{4a \cdot 8a^4}}$ Product Property

$= \frac{\sqrt[5]{40a^4}}{\sqrt[5]{32a^5}}$ Multiply.

$= \frac{\sqrt[5]{40a^4}}{2a}$ $\sqrt[5]{32a^5} = 2a$

OPERATIONS WITH RADICALS You can use the Product and Quotient Properties to multiply and divide some radicals, respectively.

Example 3 Multiply Radicals

Simplify $6\sqrt[3]{9n^2} \cdot 3\sqrt[3]{24n}$.

$6\sqrt[3]{9n^2} \cdot 3\sqrt[3]{24n} = 6 \cdot 3 \cdot \sqrt[3]{9n^2 \cdot 24n}$	Product Property of Radicals
$= 18 \cdot \sqrt[3]{2^3 \cdot 3^3 \cdot n^3}$	Factor into cubes where possible.
$= 18 \cdot \sqrt[3]{2^3} \cdot \sqrt[3]{3^3} \cdot \sqrt[3]{n^3}$	Product Property of Radicals
$= 18 \cdot 2 \cdot 3 \cdot n$ or $108n$	Multiply.

Can you add radicals in the same way that you multiply them? In other words, if $\sqrt{a} \cdot \sqrt{a} = \sqrt{a \cdot a}$, does $\sqrt{a} + \sqrt{a} = \sqrt{a + a}$?

Algebra Activity

Adding Radicals

You can use dot paper to show the sum of two like radicals, such as $\sqrt{2} + \sqrt{2}$.

Model and Analyze

Step 1 First, find a segment of length $\sqrt{2}$ units by using the Pythagorean Theorem with the dot paper.

$a^2 + b^2 = c^2$
$1^2 + 1^2 = c^2$
$2 = c^2$

Step 2 Extend the segment to twice its length to represent $\sqrt{2} + \sqrt{2}$.

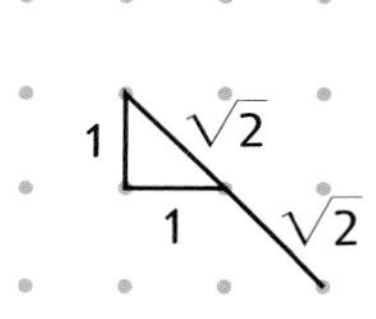

Make a Conjecture

1. Is $\sqrt{2} + \sqrt{2} = \sqrt{2 + 2}$ or 2? Justify your answer using the geometric models above.
2. Use this method to model other irrational numbers. Do these models support your conjecture?

In the activity, you discovered that you cannot add radicals in the same manner as you multiply them. You add radicals in the same manner as adding monomials. That is, you can add only the like terms or like radicals.

Two radical expressions are called **like radical expressions** if both the indices and the radicands are alike. Some examples of like and unlike radical expressions are given below.

$\sqrt{3}$ and $\sqrt[3]{3}$ are not like expressions.	Different indices
$\sqrt[4]{5x}$ and $\sqrt[4]{5}$ are not like expressions.	Different radicands
$2\sqrt[4]{3a}$ and $5\sqrt[4]{3a}$ are like expressions.	Radicands are $3a$; indices are 4.

Study Tip

Reading Math
Indices is the plural of *index*.

Example 4 Add and Subtract Radicals

Simplify $2\sqrt{12} - 3\sqrt{27} + 2\sqrt{48}$.

$2\sqrt{12} - 3\sqrt{27} + 2\sqrt{48}$	
$= 2\sqrt{2^2 \cdot 3} - 3\sqrt{3^2 \cdot 3} + 2\sqrt{2^2 \cdot 2^2 \cdot 3}$	Factor using squares.
$= 2\sqrt{2^2} \cdot \sqrt{3} - 3\sqrt{3^2} \cdot \sqrt{3} + 2\sqrt{2^2} \cdot \sqrt{2^2} \cdot \sqrt{3}$	Product Property
$= 2 \cdot 2 \cdot \sqrt{3} - 3 \cdot 3 \cdot \sqrt{3} + 2 \cdot 2 \cdot 2 \cdot \sqrt{3}$	$\sqrt{2^2} = 2$, $\sqrt{3^2} = 3$
$= 4\sqrt{3} - 9\sqrt{3} + 8\sqrt{3}$	Multiply.
$= 3\sqrt{3}$	Combine like radicals.

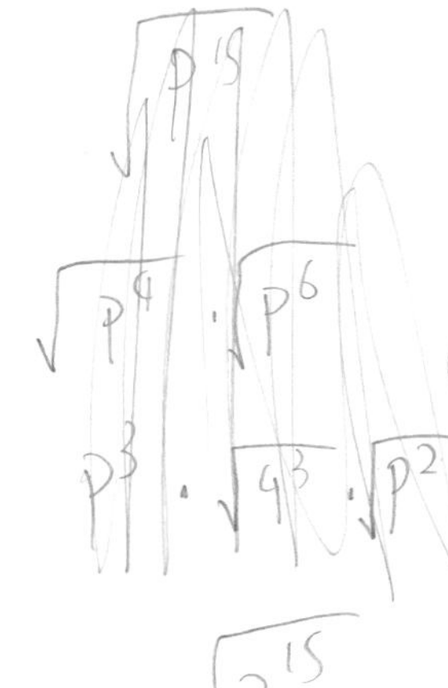

Just as you can add and subtract radicals like monomials, you can multiply radicals using the FOIL method as you do when multiplying binomials.

Example 5 Multiply Radicals

Simplify each expression.

a. $(3\sqrt{5} - 2\sqrt{3})(2 + \sqrt{3})$

$$(3\sqrt{5} - 2\sqrt{3})(2 + \sqrt{3}) = \overset{F}{3\sqrt{5} \cdot 2} + \overset{O}{3\sqrt{5} \cdot \sqrt{3}} - \overset{I}{2\sqrt{3} \cdot 2} - \overset{L}{2\sqrt{3} \cdot \sqrt{3}}$$

$= 6\sqrt{5} + 3\sqrt{5 \cdot 3} - 4\sqrt{3} - 2\sqrt{3^2}$	Product Property
$= 6\sqrt{5} + 3\sqrt{15} - 4\sqrt{3} - 6$	$2\sqrt{3^2} = 2 \cdot 3$ or 6

b. $(5\sqrt{3} - 6)(5\sqrt{3} + 6)$

$(5\sqrt{3} - 6)(5\sqrt{3} + 6) = 5\sqrt{3} \cdot 5\sqrt{3} + 5\sqrt{3} \cdot 6 - 6 \cdot 5\sqrt{3} - 6 \cdot 6$	FOIL
$= 25\sqrt{3^2} + 30\sqrt{3} - 30\sqrt{3} - 36$	Multiply.
$= 75 - 36$	$25\sqrt{3^2} = 25 \cdot 3$ or 75
$= 39$	Subtract.

Binomials like those in Example 5b, of the form $a\sqrt{b} + c\sqrt{d}$ and $a\sqrt{b} - c\sqrt{d}$ where a, b, c, and d are rational numbers, are called **conjugates** of each other. The product of conjugates is always a rational number. You can use conjugates to rationalize denominators.

Example 6 Use a Conjugate to Rationalize a Denominator

Simplify $\frac{1 - \sqrt{3}}{5 + \sqrt{3}}$.

$$\frac{1 - \sqrt{3}}{5 + \sqrt{3}} = \frac{(1 - \sqrt{3})(5 - \sqrt{3})}{(5 + \sqrt{3})(5 - \sqrt{3})}$$ Multiply by $\frac{5 - \sqrt{3}}{5 - \sqrt{3}}$ because $5 - \sqrt{3}$ is the conjugate of $5 + \sqrt{3}$.

$= \frac{1 \cdot 5 - 1 \cdot \sqrt{3} - \sqrt{3} \cdot 5 + (\sqrt{3})^2}{5^2 - (\sqrt{3})^2}$	FOIL Difference of squares
$= \frac{5 - \sqrt{3} - 5\sqrt{3} + 3}{25 - 3}$	Multiply.
$= \frac{8 - 6\sqrt{3}}{22}$	Combine like terms.
$= \frac{4 - 3\sqrt{3}}{11}$	Divide numerator and denominator by 2.

Check for Understanding

Concept Check

1. **Determine** whether the statement $\frac{1}{\sqrt[n]{a}} = \sqrt[n]{a}$ is *sometimes*, *always*, or *never* true. Explain.
2. **OPEN ENDED** **Write** a sum of three radicals that contains two like terms.
3. **Explain** why the product of two conjugates is always a rational number.

Guided Practice

Simplify.

4. $5\sqrt{63}$
5. $\sqrt[4]{16x^5y^4}$
6. $\sqrt{\frac{7}{8y}}$
7. $(-2\sqrt{15})(4\sqrt{21})$
8. $\frac{\sqrt[3]{625}}{\sqrt[3]{25}}$
9. $\sqrt{2ab^2} \cdot \sqrt{6a^3b^2}$
10. $\sqrt{3} - 2\sqrt[4]{3} + 4\sqrt{3} + 5\sqrt[4]{3}$
11. $3\sqrt[3]{128} + 5\sqrt[3]{16}$
12. $(3 - \sqrt{5})(1 + \sqrt{3})$
13. $\frac{1 + \sqrt{5}}{3 - \sqrt{5}}$

Application

14. **LAW ENFORCEMENT** A police accident investigator can use the formula $s = 2\sqrt{5\ell}$ to estimate the speed s of a car in miles per hour based on the length ℓ in feet of the skid marks it left. How fast was a car traveling that left skid marks 120 feet long?

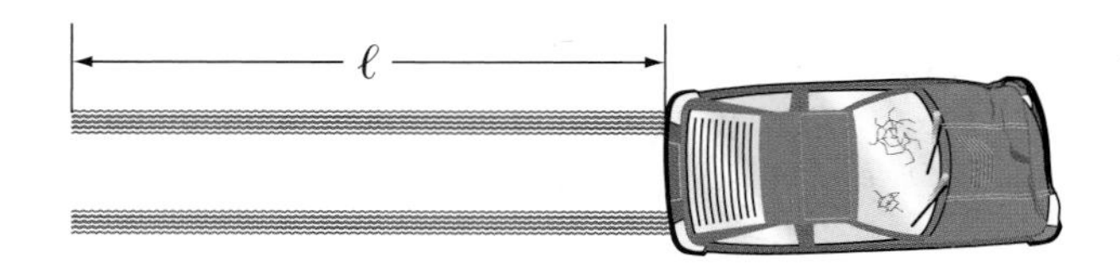

Practice and Apply

Homework Help	
For Exercises	**See Examples**
15–26	1
27–30	2
31–34	3
35–38	4
39–42	5
43–48	6

Extra Practice
See page 838.

Simplify.

15. $\sqrt{243}$
16. $\sqrt{72}$
17. $\sqrt[3]{54}$
18. $\sqrt[4]{96}$
19. $\sqrt{50x^4}$
20. $\sqrt[3]{16y^3}$
21. $\sqrt{18x^2y^3}$
22. $\sqrt{40a^3b^4}$
23. $3\sqrt[3]{56y^6z^3}$
24. $2\sqrt[3]{24m^4n^5}$
25. $\sqrt[4]{\frac{1}{81}c^5d^4}$
26. $\sqrt[5]{\frac{1}{32}w^6z^7}$
27. $\sqrt[3]{\frac{3}{4}}$
28. $\sqrt[4]{\frac{2}{3}}$
29. $\sqrt{\frac{a^4}{b^3}}$
30. $\sqrt{\frac{4r^8}{t^9}}$
31. $(3\sqrt{12})(2\sqrt{21})$
32. $(-3\sqrt{24})(5\sqrt{20})$
33. What is $\sqrt{39}$ divided by $\sqrt{26}$?
34. Divide $\sqrt{14}$ by $\sqrt{35}$.

Simplify.

35. $\sqrt{12} + \sqrt{48} - \sqrt{27}$
36. $\sqrt{98} - \sqrt{72} + \sqrt{32}$
37. $\sqrt{3} + \sqrt{72} - \sqrt{128} + \sqrt{108}$
38. $5\sqrt{20} + \sqrt{24} - \sqrt{180} + 7\sqrt{54}$
39. $(5 + \sqrt{6})(5 - \sqrt{2})$
40. $(3 + \sqrt{7})(2 + \sqrt{6})$
41. $(\sqrt{11} - \sqrt{2})^2$
42. $(\sqrt{3} - \sqrt{5})^2$
43. $\frac{7}{4 - \sqrt{3}}$
44. $\frac{\sqrt{6}}{5 + \sqrt{3}}$
45. $\frac{-2 - \sqrt{3}}{1 + \sqrt{3}}$
46. $\frac{2 + \sqrt{2}}{5 - \sqrt{2}}$
47. $\frac{x + 1}{\sqrt{x^2 - 1}}$
48. $\frac{x - 1}{\sqrt{x - 1}}$

49. **GEOMETRY** Find the perimeter and area of the rectangle.

$3 + 6\sqrt{2}$ yd

$\sqrt{8}$ yd

More About. . .

Amusement Parks

Attendance at the top 50 theme parks in North America increased to 175.1 million in 2000.

Source: *Amusement Business*

AMUSEMENT PARKS **For Exercises 50 and 51, use the following information.**
The velocity v in feet per second of a roller coaster at the bottom of a hill is related to the vertical drop h in feet and the velocity v_0 in feet per second of the coaster at the top of the hill by the formula $v_0 = \sqrt{v^2 - 64h}$.

50. Explain why $v_0 = v - 8\sqrt{h}$ is not equivalent to the given formula.

51. What velocity must a coaster have at the top of a 225-foot hill to achieve a velocity of 120 feet per second at the bottom?

Online Research **Data Update** What are the values of v and h for some of the world's highest and fastest roller coasters? Visit www.algebra2.com/data_update to learn more.

SPORTS **For Exercises 52 and 53, use the following information.**
A ball that is hit or thrown horizontally with a velocity of v meters per second will travel a distance of d meters before hitting the ground, where $d = v\sqrt{\frac{h}{4.9}}$ and h is the height in meters from which the ball is hit or thrown.

52. Use the properties of radicals to rewrite the formula.

53. How far will a ball that is hit horizontally with a velocity of 45 meters per second at a height of 0.8 meter above the ground travel before hitting the ground?

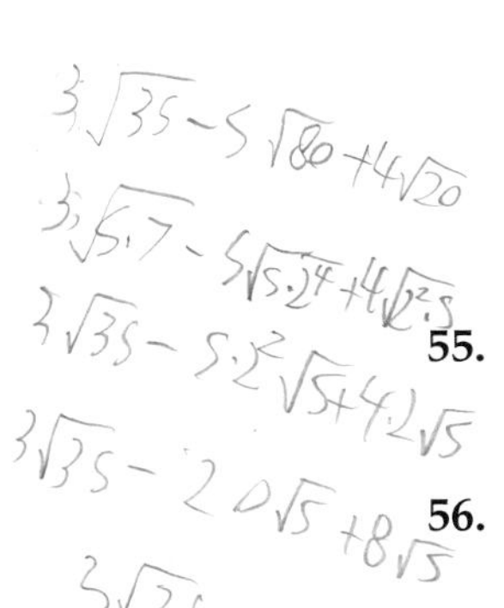

54. **AUTOMOTIVE ENGINEERING** An automotive engineer is trying to design a safer car. The maximum force a road can exert on the tires of the car being redesigned is 2000 pounds. What is the maximum velocity v in ft/s at which this car can safely round a turn of radius 320 feet? Use the formula $v = \sqrt{\frac{F_c r}{100}}$, where F_c is the force the road exerts on the car and r is the radius of the turn.

55. **CRITICAL THINKING** Under what conditions is the equation $\sqrt{x^3y^2} = xy\sqrt{x}$ true?

56. **WRITING IN MATH** Answer the question that was posed at the beginning of the lesson.

How do radical expressions apply to falling objects?

Include the following in your answer:

- an explanation of how you can use the properties in this lesson to rewrite the formula $t = \sqrt{\frac{2d}{g}}$, and
- the amount of time a 5-foot tall student has to get out of the way after a balloon is dropped from a window 25 feet above.

Standards Practice

Standardized Test Practice

57. The expression $\sqrt{180}$ is equivalent to which of the following?

Ⓐ $5\sqrt{6}$ Ⓑ $6\sqrt{5}$ Ⓒ $3\sqrt{10}$ Ⓓ $36\sqrt{5}$

58. Which of the following is *not* a length of a side of the triangle?

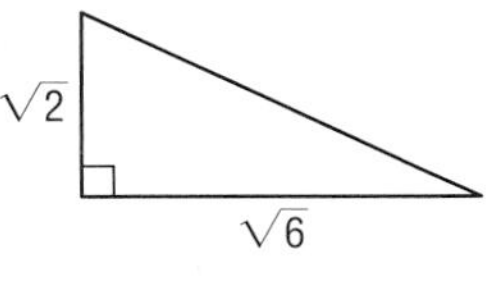

Ⓐ $\sqrt{8}$ Ⓑ $2\sqrt{2}$

Ⓒ $\sqrt{4 + 2}$ Ⓓ $\sqrt{4} + \sqrt{2}$

Maintain Your Skills

Mixed Review

Simplify. *(Lesson 5-5)*

59. $\sqrt{144z^8}$
60. $\sqrt[3]{216a^3b^9}$
61. $\sqrt{(y+2)^2}$

Simplify. Assume that no denominator is equal to 0. *(Lesson 5-4)*

62. $\frac{x^2 + 5x - 14}{x^2 - 6x + 8}$
63. $\frac{x^2 - 3x - 4}{x^2 - 16}$

Perform the indicated operations. *(Lesson 4-2)*

64. $\begin{bmatrix} 3 & -4 \\ 2 & 8 \\ 0 & 1 \end{bmatrix} + \begin{bmatrix} -5 & 0 \\ 7 & 7 \\ 3 & -6 \end{bmatrix}$
65. $\begin{bmatrix} 3 & 3 \\ 0 & -2 \end{bmatrix} - \begin{bmatrix} 2 & -1 \\ 5 & 2 \end{bmatrix}$

66. Find the maximum and minimum values of the function $f(x, y) = 2x + 3y$ for the region with vertices at $(2, 4)$, $(-1, 3)$, $(-3, -3)$, and $(2, -5)$. *(Lesson 3-4)*

67. State whether the system of equations shown at the right is *consistent and independent*, *consistent and dependent*, or *inconsistent*. *(Lesson 3-1)*

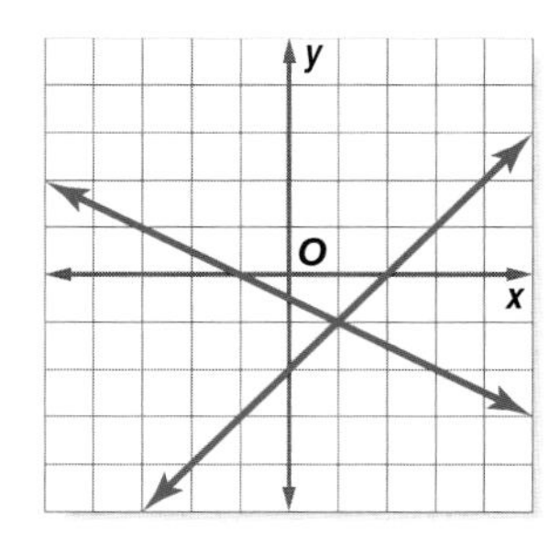

68. **BUSINESS** The amount that a mail-order company charges for shipping and handling is given by the function $c(x) = 3 + 0.15x$, where x is the weight in pounds. Find the charge for an 8-pound order. *(Lesson 2-2)*

Solve. *(Lessons 1-3, 1-4, and 1-5)*

69. $2x + 7 = -3$
70. $-5x + 6 = -4$
71. $|x - 1| = 3$
72. $|3x + 2| = 5$
73. $2x - 4 > 8$
74. $-x - 3 \leq 4$

Getting Ready for the Next Lesson

BASIC SKILL **Evaluate each expression.**

75. $2\left(\frac{1}{8}\right)$
76. $3\left(\frac{1}{6}\right)$
77. $\frac{1}{2} + \frac{1}{3}$
78. $\frac{1}{3} + \frac{3}{4}$
79. $\frac{1}{8} + \frac{5}{12}$
80. $\frac{5}{6} - \frac{1}{5}$
81. $\frac{5}{8} - \frac{1}{4}$
82. $\frac{1}{4} - \frac{2}{3}$

Practice Quiz 2

Lessons 5-4 through 5-6

Factor completely. If the polynomial is not factorable, write prime. *(Lesson 5-4)*

1. $3x^3y + x^2y^2 + x^2y$
2. $3x^2 - 2x - 2$
3. $ax^2 + 6ax + 9a$
4. $8r^3 - 64s^6$

Simplify. *(Lessons 5-5 and 5-6)*

5. $\sqrt{36x^2y^6}$
6. $\sqrt[3]{-64a^6b^9}$
7. $\sqrt{4n^2 + 12n + 9}$
8. $\sqrt{\frac{x^4}{y^3}}$
9. $(3 + \sqrt{7})(2 - \sqrt{7})$
10. $\frac{5 + \sqrt{2}}{2 + \sqrt{2}}$

5-7 Rational Exponents

California Standards **Standard 12.0 Students know the laws of fractional exponents,** understand exponential functions, and use these functions in problems involving exponential growth and decay. (Key)

What You'll Learn

- Write expressions with rational exponents in radical form, and vice versa.
- Simplify expressions in exponential or radical form.

How do rational exponents apply to astronomy?

Astronomers refer to the space around a planet where the planet's gravity is stronger than the Sun's as the *sphere of influence* of the planet. The radius r of the sphere of influence is given by the formula $r = D\left(\frac{M_P}{M_S}\right)^{\frac{2}{5}}$, where M_p is the mass of the planet, M_S is the mass of the Sun, and D is the distance between the planet and the Sun.

RATIONAL EXPONENTS AND RADICALS You know that squaring a number and taking the square root of a number are inverse operations. But how would you evaluate an expression that contains a fractional exponent such as the one above? You can investigate such an expression by assuming that fractional exponents behave as integral exponents.

$$\left(b^{\frac{1}{2}}\right)^2 = b^{\frac{1}{2}} \cdot b^{\frac{1}{2}} \quad \text{Write the square as multiplication.}$$

$$= b^{\frac{1}{2} + \frac{1}{2}} \quad \text{Add the exponents.}$$

$$= b^1 \text{ or } b \quad \text{Simplify.}$$

Thus, $b^{\frac{1}{2}}$ is a number whose square equals b. So it makes sense to define $b^{\frac{1}{2}} = \sqrt{b}$.

Key Concept $b^{\frac{1}{n}}$

- **Words** For any real number b and for any positive integer n, $b^{\frac{1}{n}} = \sqrt[n]{b}$, except when $b < 0$ and n is even.
- **Example** $8^{\frac{1}{3}} = \sqrt[3]{8}$ or 2

Example 1 Radical Form

Write each expression in radical form.

a. $a^{\frac{1}{4}}$

$a^{\frac{1}{4}} = \sqrt[4]{a}$ Definition of $b^{\frac{1}{n}}$

b. $x^{\frac{1}{5}}$

$x^{\frac{1}{5}} = \sqrt[5]{x}$ Definition of $b^{\frac{1}{n}}$

Example 2 Exponential Form

Write each radical using rational exponents.

a. $\sqrt[3]{y}$

$\sqrt[3]{y} = y^{\frac{1}{3}}$ Definition of $b^{\frac{1}{n}}$

b. $\sqrt[8]{c}$

$\sqrt[8]{c} = c^{\frac{1}{8}}$ Definition of $b^{\frac{1}{n}}$

Many expressions with fractional exponents can be evaluated using the definition of $b^{\frac{1}{n}}$ or the properties of powers.

Study Tip

Negative Base
Suppose the base of a monomial is negative such as $(-9)^2$ or $(-9)^3$. The expression is undefined if the exponent is even because there is no number that, when multiplied an even number of times, results in a negative number. However, the expression is defined for an odd exponent.

Example 3 Evaluate Expressions with Rational Exponents

Evaluate each expression.

a. $16^{-\frac{1}{4}}$

Method 1

$16^{-\frac{1}{4}} = \frac{1}{16^{\frac{1}{4}}}$ $b^{-n} = \frac{1}{b^n}$

$= \frac{1}{\sqrt[4]{16}}$ $16^{\frac{1}{4}} = \sqrt[4]{16}$

$= \frac{1}{\sqrt[4]{2^4}}$ $16 = 2^4$

$= \frac{1}{2}$ Simplify.

Method 2

$16^{-\frac{1}{4}} = (2^4)^{-\frac{1}{4}}$ $16 = 2^4$

$= 2^{4\left(-\frac{1}{4}\right)}$ Power of a Power

$= 2^{-1}$ Multiply exponents.

$= \frac{1}{2}$ $2^{-1} = \frac{1}{2^1}$

b. $243^{\frac{3}{5}}$

Method 1

$243^{\frac{3}{5}} = 243^{3\left(\frac{1}{5}\right)}$ Factor.

$= (243^3)^{\frac{1}{5}}$ Power of a Power

$= \sqrt[5]{243^3}$ $b^{\frac{1}{5}} = \sqrt[5]{b}$

$= \sqrt[5]{(3^5)^3}$ $243 = 3^5$

$= \sqrt[5]{3^5 \cdot 3^5 \cdot 3^5}$ Expand the cube.

$= 3 \cdot 3 \cdot 3$ or 27 Find the fifth root.

Method 2

$243^{\frac{3}{5}} = (3^5)^{\frac{3}{5}}$ $243 = 3^5$

$= 3^{5\left(\frac{3}{5}\right)}$ Power of a Power

$= 3^3$ Multiply exponents.

$= 27$ $3^3 = 3 \cdot 3 \cdot 3$

In Example 3b, Method 1 uses a combination of the definition of $b^{\frac{1}{n}}$ and the properties of powers. This example suggests the following general definition of rational exponents.

Key Concept — Rational Exponents

- **Words** For any nonzero real number b, and any integers m and n, with $n > 1$, $b^{\frac{m}{n}} = \sqrt[n]{b^m} = \left(\sqrt[n]{b}\right)^m$, except when $b < 0$ and n is even.
- **Example** $8^{\frac{2}{3}} = \sqrt[3]{8^2} = \left(\sqrt[3]{8}\right)^2$ or 4

In general, we define $b^{\frac{m}{n}}$ as $\left(b^{\frac{1}{n}}\right)^m$ or $(b^m)^{\frac{1}{n}}$. Now apply the definition of $b^{\frac{1}{n}}$ to $\left(b^{\frac{1}{n}}\right)^m$ and $(b^m)^{\frac{1}{n}}$.

$\left(b^{\frac{1}{n}}\right)^m = \left(\sqrt[n]{b}\right)^m$ $(b^m)^{\frac{1}{n}} = \sqrt[n]{b^m}$

Example 4 Rational Exponent with Numerator Other Than 1

WEIGHT LIFTING **The formula $M = 512 - 146{,}230B^{-\frac{8}{5}}$ can be used to estimate the maximum total mass that a weight lifter of mass B kilograms can lift in two lifts, the snatch and the clean and jerk, combined.**

a. According to the formula, what is the maximum amount that 2000 Olympic champion Xugang Zhan of China can lift if he weighs 72 kilograms?

$M = 512 - 146{,}230B^{-\frac{8}{5}}$ Original formula

$= 512 - 146{,}230(72)^{-\frac{8}{5}}$ $B = 72$

≈ 356 kg Use a calculator.

The formula predicts that he can lift at most 356 kilograms.

b. Xugang Zhan's winning total in the 2000 Olympics was 367.50 kg. Compare this to the value predicted by the formula.

The formula prediction is close to the actual weight, but slightly lower.

More About. . .

Weight Lifting

With origins in both the ancient Egyptian and Greek societies, weightlifting was among the sports on the program of the first Modern Olympic Games, in 1896, in Athens, Greece.

Source: International Weightlifting Association

SIMPLIFY EXPRESSIONS All of the properties of powers you learned in Lesson 5-1 apply to rational exponents. When simplifying expressions containing rational exponents, leave the exponent in rational form rather than writing the expression as a radical. To simplify such an expression, you must write the expression with all positive exponents. Furthermore, any exponents in the denominator of a fraction must be positive *integers*. So, it may be necessary to rationalize a denominator.

Example 5 Simplify Expressions with Rational Exponents

Simplify each expression.

a. $x^{\frac{1}{5}} \cdot x^{\frac{7}{5}}$

$x^{\frac{1}{5}} \cdot x^{\frac{7}{5}} = x^{\frac{1}{5} + \frac{7}{5}}$ Multiply powers.

$= x^{\frac{8}{5}}$ Add exponents.

b. $y^{-\frac{3}{4}}$

$y^{-\frac{3}{4}} = \dfrac{1}{y^{\frac{3}{4}}}$ $b^{-n} = \frac{1}{b^n}$

$= \dfrac{1}{y^{\frac{3}{4}}} \cdot \dfrac{y^{\frac{1}{4}}}{y^{\frac{1}{4}}}$ Why use $\frac{y^{\frac{1}{4}}}{y^{\frac{1}{4}}}$?

$= \dfrac{y^{\frac{1}{4}}}{y^{\frac{4}{4}}}$ $y^{\frac{3}{4}} \cdot y^{\frac{1}{4}} = y^{\frac{3}{4} + \frac{1}{4}}$

$= \dfrac{y^{\frac{1}{4}}}{y}$ $y^{\frac{4}{4}} = y^1$ or y

When simplifying a radical expression, always use the smallest index possible. Using rational exponents makes this process easier, but the answer should be written in radical form.

Example 6 Simplify Radical Expressions

Simplify each expression.

a. $\frac{\sqrt[8]{81}}{\sqrt[6]{3}}$

$\frac{\sqrt[8]{81}}{\sqrt[6]{3}} = \frac{81^{\frac{1}{8}}}{3^{\frac{1}{6}}}$ Rational exponents

$= \frac{(3^4)^{\frac{1}{8}}}{3^{\frac{1}{6}}}$ $81 = 3^4$

$= \frac{3^{\frac{1}{2}}}{3^{\frac{1}{6}}}$ Power of a Power

$= 3^{\frac{1}{2} - \frac{1}{6}}$ Quotient of Powers

$= 3^{\frac{1}{3}}$ or $\sqrt[3]{3}$ Simplify.

b. $\sqrt[4]{9z^2}$

$\sqrt[4]{9z^2} = (9z^2)^{\frac{1}{4}}$ Rational exponents

$= (3^2 \cdot z^2)^{\frac{1}{4}}$ $9 = 3^2$

$= 3^{2\left(\frac{1}{4}\right)} \cdot z^{2\left(\frac{1}{4}\right)}$ Power of a Power

$= 3^{\frac{1}{2}} \cdot z^{\frac{1}{2}}$ Multiply.

$= \sqrt{3} \cdot \sqrt{z}$ $3^{\frac{1}{2}} = \sqrt{3}, z^{\frac{1}{2}} = \sqrt{z}$

$= \sqrt{3z}$ Simplify.

c. $\frac{m^{\frac{1}{2}} - 1}{m^{\frac{1}{2}} + 1}$

$\frac{m^{\frac{1}{2}} - 1}{m^{\frac{1}{2}} + 1} = \frac{m^{\frac{1}{2}} - 1}{m^{\frac{1}{2}} + 1} \cdot \frac{m^{\frac{1}{2}} - 1}{m^{\frac{1}{2}} - 1}$ $m^{\frac{1}{2}} - 1$ is the conjugate of $m^{\frac{1}{2}} + 1$.

$= \frac{m - 2m^{\frac{1}{2}} + 1}{m - 1}$ Multiply.

Concept Summary — Expressions with Rational Exponents

An expression with rational exponents is simplified when all of the following conditions are met.

- It has no negative exponents.
- It has no fractional exponents in the denominator.
- It is not a complex fraction.
- The index of any remaining radical is the least number possible.

Check for Understanding

Concept Check

1. **OPEN ENDED** Determine a value of b for which $b^{\frac{1}{6}}$ is an integer.
2. **Explain** why $(-16)^{\frac{1}{2}}$ is not a real number.
3. **Explain** why $\sqrt[n]{b^m} = \left(\sqrt[n]{b}\right)^m$.

Guided Practice

Write each expression in radical form.

4. $7^{\frac{1}{3}}$
5. $x^{\frac{2}{3}}$

Write each radical using rational exponents.

6. $\sqrt[4]{26}$
7. $\sqrt[3]{6x^5y^7}$

Evaluate each expression.

8. $125^{\frac{1}{3}}$
9. $81^{-\frac{1}{4}}$
10. $27^{\frac{2}{3}}$
11. $\dfrac{54}{9^{\frac{3}{2}}}$

Simplify each expression.

12. $a^{\frac{2}{3}} \cdot a^{\frac{1}{4}}$
13. $\dfrac{x^{\frac{5}{6}}}{x^{\frac{1}{6}}}$
14. $\dfrac{1}{2z^{\frac{1}{3}}}$
15. $\dfrac{a^2}{b^{\frac{1}{3}}} \cdot \dfrac{b}{a^{\frac{1}{2}}}$
16. $(mn^2)^{-\frac{1}{3}}$
17. $z(x - 2y)^{-\frac{1}{2}}$
18. $\sqrt[6]{27x^3}$
19. $\dfrac{\sqrt[4]{27}}{\sqrt[4]{3}}$

Application

20. **ECONOMICS** When inflation causes the price of an item to increase, the new cost C and the original cost c are related by the formula $C = c(1 + r)^n$, where r is the rate of inflation per year as a decimal and n is the number of years. What would be the price of a \$4.99 item after six months of 5% inflation?

Practice and Apply

Homework Help

For Exercises	See Examples
21–24	1
25–28	2
29–40	3
41–52, 64–66	5
53–63	6

Extra Practice
See page 838.

Write each expression in radical form.

21. $6^{\frac{1}{5}}$
22. $4^{\frac{1}{3}}$
23. $c^{\frac{2}{5}}$
24. $(x^2)^{\frac{4}{3}}$

Write each radical using rational exponents.

25. $\sqrt{23}$
26. $\sqrt[3]{62}$
27. $\sqrt[4]{16z^2}$
28. $\sqrt[3]{5x^2y}$

Evaluate each expression.

29. $16^{\frac{1}{4}}$
30. $216^{\frac{1}{3}}$
31. $25^{-\frac{1}{2}}$
32. $81^{-\frac{3}{4}}$
33. $(-27)^{-\frac{2}{3}}$
34. $(-32)^{-\frac{3}{5}}$
35. $81^{-\frac{1}{2}} \cdot 81^{\frac{3}{2}}$
36. $8^{\frac{3}{2}} \cdot 8^{\frac{5}{2}}$
37. $\left(\dfrac{8}{27}\right)^{\frac{1}{3}}$
38. $\left(\dfrac{1}{243}\right)^{-\frac{3}{5}}$
39. $\dfrac{16^{\frac{1}{2}}}{9^{\frac{1}{2}}}$
40. $\dfrac{8^{\frac{1}{3}}}{64^{\frac{1}{3}}}$

Simplify each expression.

41. $y^{\frac{5}{3}} \cdot y^{\frac{7}{3}}$
42. $x^{\frac{3}{4}} \cdot x^{\frac{9}{4}}$
43. $\left(b^{\frac{1}{3}}\right)^{\frac{3}{5}}$
44. $\left(a^{-\frac{2}{3}}\right)^{-\frac{1}{6}}$
45. $w^{-\frac{4}{5}}$
46. $x^{-\frac{1}{6}}$
47. $\dfrac{t^{\frac{3}{4}}}{t^{\frac{1}{2}}}$
48. $\dfrac{r^{\frac{2}{3}}}{r^{\frac{1}{6}}}$
49. $\dfrac{a^{-\frac{1}{2}}}{6a^{\frac{1}{3}} \cdot a^{-\frac{1}{4}}}$
50. $\dfrac{2c^{\frac{1}{8}}}{c^{-\frac{1}{16}} \cdot c^{\frac{1}{4}}}$
51. $\dfrac{y^{\frac{3}{2}}}{y^{\frac{1}{2}} + 2}$
52. $\dfrac{x^{\frac{1}{2}} + 2}{x^{\frac{1}{2}} - 1}$
53. $\sqrt[4]{25}$
54. $\sqrt[6]{27}$
55. $\sqrt{17} \cdot \sqrt[3]{17^2}$
56. $\sqrt[3]{5} \cdot \sqrt{5^3}$
57. $\sqrt[8]{25x^4y^4}$
58. $\sqrt[6]{81a^4b^8}$
59. $\dfrac{xy}{\sqrt{z}}$
60. $\dfrac{ab}{\sqrt[3]{c}}$
61. $\sqrt[3]{\sqrt{8}}$
62. $\sqrt{\sqrt[3]{36}}$
63. $\dfrac{8^{\frac{1}{6}} - 9^{\frac{1}{4}}}{\sqrt{3} + \sqrt{2}}$
64. $\dfrac{x^{\frac{5}{3}} - x^{\frac{1}{3}}z^{\frac{4}{3}}}{x^{\frac{2}{3}} + z^{\frac{2}{3}}}$

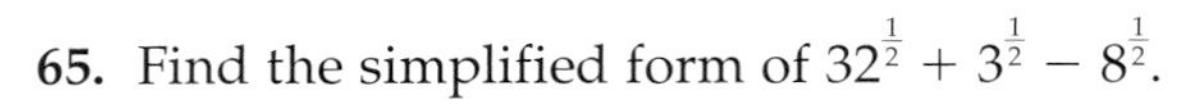

65. Find the simplified form of $32^{\frac{1}{2}} + 3^{\frac{1}{2}} - 8^{\frac{1}{2}}$.

66. What is the simplified form of $81^{\frac{1}{3}} - 24^{\frac{1}{3}} + 3^{\frac{1}{3}}$?

More About. . .

Music

The first piano was made in about 1709 by Bartolomeo Cristofori, a maker of harpsichords in Florence, Italy.

Source: www.infoplease.com

MUSIC **For Exercises 67 and 68, use the following information.**
On a piano, the frequency of the A note above middle C should be set at 440 vibrations per second. The frequency f_n of a note that is n notes above that A should be $f_n = 440 \cdot 2^{\frac{n}{12}}$.

67. At what frequency should a piano tuner set the A that is one octave, or 12 notes, above the A above middle C?

68. Middle C is nine notes below the A that has a frequency of 440 vibrations per second. What is the frequency of middle C?

69. **BIOLOGY** Suppose a culture has 100 bacteria to begin with and the number of bacteria doubles every 2 hours. Then the number N of bacteria after t hours is given by $N = 100 \cdot 2^{\frac{t}{2}}$. How many bacteria will be present after 3 and a half hours?

70. **CRITICAL THINKING** Explain how to solve $9^x = 3^{x + \frac{1}{2}}$ for x.

71. **WRITING IN MATH** Answer the question that was posed at the beginning of the lesson.

How do rational exponents apply to astronomy?

Include the following in your answer:

- an explanation of how to write the formula $r = D\left(\frac{M_P}{M_S}\right)^{\frac{2}{5}}$ in radical form and simplify it, and
- an explanation of what happens to the value of r as the value of D increases assuming that M_p and M_S are constant.

Standardized Test Practice

72. Which is the value of $4^{\frac{1}{2}} + \left(\frac{1}{2}\right)^4$?

Ⓐ 1 Ⓑ 2 Ⓒ $2\frac{1}{16}$ Ⓓ $2\frac{1}{2}$

73. If $4x + 2y = 5$ and $x - y = 1$, then what is the value of $3x + 3y$?

Ⓐ 1 Ⓑ 2 Ⓒ 4 Ⓓ 6

Maintain Your Skills

Mixed Review

Simplify. *(Lessons 5-5 and 5-6)*

74. $\sqrt{4x^3y^2}$

75. $(2\sqrt{6})(3\sqrt{12})$

76. $\sqrt{32} + \sqrt{18} - \sqrt{50}$

77. $\sqrt[4]{(-8)^4}$

78. $4\sqrt{(x - 5)^2}$

79. $\sqrt{\frac{9}{36}x^4}$

80. **BIOLOGY** Humans blink their eyes about once every 5 seconds. How many times do humans blink their eyes in two hours? *(Lesson 1-1)*

Getting Ready for the Next Lesson

PREREQUISITE SKILL **Find each power.** *(To review* ***multiplying radicals****, see Lesson 5-6.)*

81. $(\sqrt{x - 2})^2$

82. $(\sqrt[3]{2x - 3})^3$

83. $(\sqrt{x} + 1)^2$

84. $(2\sqrt{x} - 3)^2$

5-8 Radical Equations and Inequalities

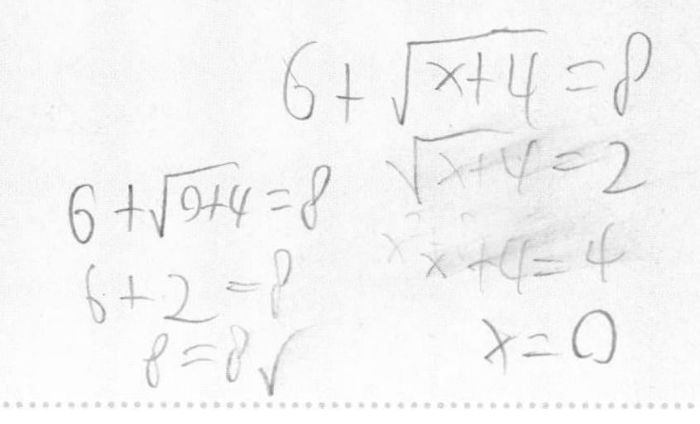

What You'll Learn

- Solve equations containing radicals.
- Solve inequalities containing radicals.

Vocabulary

- radical equation
- extraneous solution
- radical inequality

How do radical equations apply to manufacturing?

Computer chips are made from the element silicon, which is found in sand. Suppose a company that manufactures computer chips uses the formula $C = 10n^{\frac{2}{3}} + 1500$ to estimate the cost C in dollars of producing n chips. This equation can be rewritten as a radical equation.

SOLVE RADICAL EQUATIONS Equations with radicals that have variables in the radicands are called **radical equations**. To solve this type of equation, raise each side of the equation to a power equal to the index of the radical to eliminate the radical.

Example 1 Solve a Radical Equation

Solve $\sqrt{x+1} + 2 = 4$.

$\sqrt{x+1} + 2 = 4$	Original equation
$\sqrt{x+1} = 2$	Subtract 2 from each side to isolate the radical.
$(\sqrt{x+1})^2 = 2^2$	Square each side to eliminate the radical.
$x + 1 = 4$	Find the squares.
$x = 3$	Subtract 1 from each side.

CHECK

$\sqrt{x+1} + 2 = 4$	Original equation
$\sqrt{3+1} + 2 \stackrel{?}{=} 4$	Replace x with 3.
$4 = 4$ ✓	Simplify.

The solution checks. The solution is 3.

When you solve a radical equation, it is very important that you check your solution. Sometimes you will obtain a number that does not satisfy the original equation. Such a number is called an **extraneous solution**. You can use a graphing calculator to predict the number of solutions of an equation or to determine whether the solution you obtain is reasonable.

Example 2 Extraneous Solution

Solve $\sqrt{x-15} = 3 - \sqrt{x}$.

$\sqrt{x-15} = 3 - \sqrt{x}$	Original equation
$(\sqrt{x-15})^2 = (3 - \sqrt{x})^2$	Square each side.
$x - 15 = 9 - 6\sqrt{x} + x$	Find the squares.
$-24 = -6\sqrt{x}$	Isolate the radical.
$4 = \sqrt{x}$	Divide each side by -6.
$4^2 = (\sqrt{x})^2$	Square each side again.
$16 = x$	Evaluate the squares.

(continued on the next page)

CHECK $\sqrt{x - 15} = 3 - \sqrt{x}$

$$\sqrt{16 - 15} \stackrel{?}{=} 3 - \sqrt{16}$$
$$\sqrt{1} \stackrel{?}{=} 3 - 4$$
$$1 \neq -1$$

The solution does not check, so the equation has no real solution.

The graphing calculator screen shows the graphs of $y = \sqrt{x - 15}$ and $y = 3 - \sqrt{x}$. The graphs do not intersect, which confirms that there is no solution.

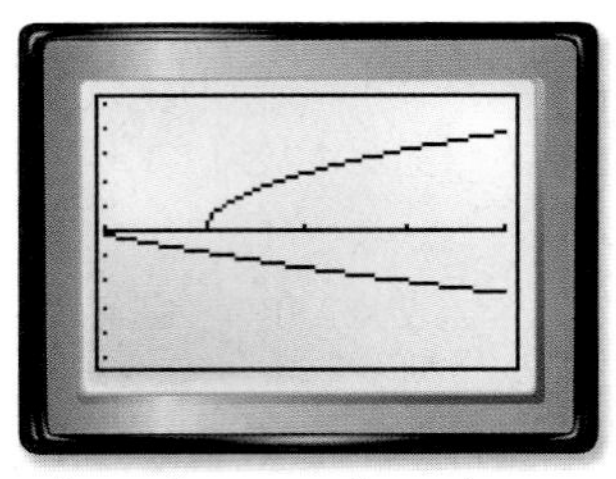

[10, 30] scl: 5 by [−5, 5] scl: 1

You can apply the same methods used in solving square root equations to solving equations with roots of any index. Remember that to undo a square root, you square the expression. To undo an nth root, you must raise the expression to the nth power.

Study Tip

Alternative Method

To solve a radical equation, you can substitute a variable for the radical expression. In Example 3, let $A = 5n - 1$.

$$3A^{\frac{1}{3}} - 2 = 0$$
$$3A^{\frac{1}{3}} = 2$$
$$A^{\frac{1}{3}} = \frac{2}{3}$$
$$A = \frac{8}{27}$$
$$5n - 1 = \frac{8}{27}$$
$$n = \frac{7}{27}$$

Example 3 Cube Root Equation

Solve $3(5n - 1)^{\frac{1}{3}} - 2 = 0$.

In order to remove the $\frac{1}{3}$ power, or cube root, you must first isolate it and then raise each side of the equation to the third power.

$3(5n - 1)^{\frac{1}{3}} - 2 = 0$	Original equation
$3(5n - 1)^{\frac{1}{3}} = 2$	Add 2 to each side.
$(5n - 1)^{\frac{1}{3}} = \frac{2}{3}$	Divide each side by 3.
$\left[(5n - 1)^{\frac{1}{3}}\right]^3 = \left(\frac{2}{3}\right)^3$	Cube each side.
$5n - 1 = \frac{8}{27}$	Evaluate the cubes.
$5n = \frac{35}{27}$	Add 1 to each side.
$n = \frac{7}{27}$	Divide each side by 5.

CHECK

$3(5n - 1)^{\frac{1}{3}} - 2 = 0$	Original equation
$3\left(5 \cdot \frac{7}{27} - 1\right)^{\frac{1}{3}} - 2 \stackrel{?}{=} 0$	Replace n with $\frac{7}{27}$.
$3\left(\frac{8}{27}\right)^{\frac{1}{3}} - 2 \stackrel{?}{=} 0$	Simplify.
$3\left(\frac{2}{3}\right) - 2 \stackrel{?}{=} 0$	The cube root of $\frac{8}{27}$ is $\frac{2}{3}$.
$0 = 0 \checkmark$	Subtract.

The solution is $\frac{7}{27}$.

SOLVE RADICAL INEQUALITIES You can use what you know about radical equations to help solve radical inequalities. A **radical inequality** is an inequality that has a variable in a radicand.

Example 4 Radical Inequality

Solve $2 + \sqrt{4x - 4} \leq 6$.

Since the radicand of a square root must be greater than or equal to zero, first solve $4x - 4 \geq 0$ to identify the values of x for which the left side of the given inequality is defined.

$$4x - 4 \geq 0$$
$$4x \geq 4$$
$$x \geq 1$$

Now solve $2 + \sqrt{4x - 4} \leq 6$.

$2 + \sqrt{4x - 4} \leq 6$ Original inequality

$\sqrt{4x - 4} \leq 4$ Isolate the radical.

$4x - 4 \leq 16$ Eliminate the radical.

$4x \leq 20$ Add 4 to each side.

$x \leq 5$ Divide each side by 4.

It appears that $1 \leq x \leq 5$. You can test some x values to confirm the solution. Let $f(x) = 2 + \sqrt{4x - 4}$. Use three test values: one less than 1, one between 1 and 5, and one greater than 5. Organize the test values in a table.

$x = 0$	$x = 2$	$x = 7$
$f(0) = 2 + \sqrt{4(0) - 4}$ $= 2 + \sqrt{-4}$ Since $\sqrt{-4}$ is not a real number, the inequality is not satisfied.	$f(2) = 2 + \sqrt{4(2) - 4}$ $= 4$ Since $4 \leq 6$, the inequality is satisfied.	$f(7) = 2 + \sqrt{4(7) - 4}$ ≈ 6.90 Since $6.90 \nleq 6$, the inequality is not satisfied.

Study Tip

Check Your Solution

You may also want to use a graphing calculator to check. Graph each side of the original inequality and examine the intersection.

The solution checks. Only values in the interval $1 \leq x \leq 5$ satisfy the inequality. You can summarize the solution with a number line.

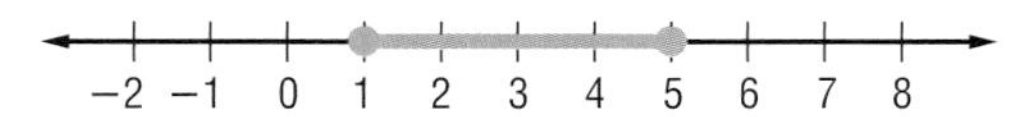

Concept Summary — Solving Radical Inequalities

To solve radical inequalities, complete the following steps.

Step 1 If the index of the root is even, identify the values of the variable for which the radicand is nonnegative.

Step 2 Solve the inequality algebraically.

Step 3 Test values to check your solution.

Check for Understanding

Concept Check

1. **Explain** why you do not have to square each side to solve $2x + 1 = \sqrt{3}$. Then solve the equation.
2. **Show** how to solve $x - 6\sqrt{x} + 9 = 0$ by factoring. Name the properties of equality that you use.
3. **OPEN ENDED** Write an equation containing two radicals for which 1 is a solution.

Guided Practice **Solve each equation or inequality.**

4. $\sqrt{4x+1}=3$
5. $4-(7-y)^{\frac{1}{2}}=0$
6. $1+\sqrt{x+2}=0$
7. $\sqrt{z-6}-3=0$
8. $\frac{1}{6}(12a)^{\frac{1}{3}}=1$
9. $\sqrt[3]{x-4}=3$
10. $\sqrt{2x+3}-4\leq 5$
11. $\sqrt{b+12}-\sqrt{b}>2$

Application

12. **GEOMETRY** The surface area S of a cone can be found by using $S=\pi r\sqrt{r^2+h^2}$, where r is the radius of the base and h is the height of the cone. Find the height of the cone.

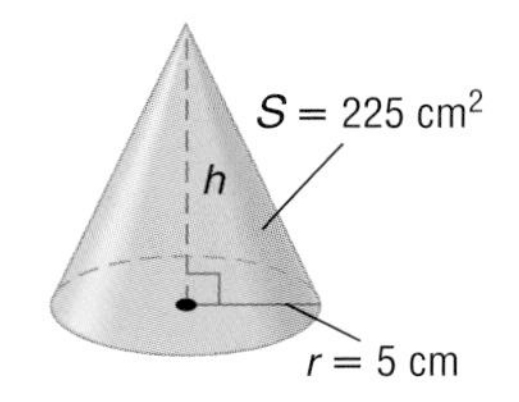

Practice and Apply

Homework Help

For Exercises	See Examples
13–24, 29–32, 37–42	1–3
25–28, 33–36	4

Extra Practice
See page 839.

Solve each equation or inequality.

13. $\sqrt{x}=4$
14. $\sqrt{y}-7=0$
15. $a^{\frac{1}{2}}+9=0$
16. $2+4z^{\frac{1}{2}}=0$
17. $\sqrt[3]{c-1}=2$
18. $\sqrt[3]{5m+2}=3$
19. $7+\sqrt{4x+8}=9$
20. $5+\sqrt{4y-5}=12$
21. $(6n-5)^{\frac{1}{3}}+3=-2$
22. $(5x+7)^{\frac{1}{5}}+3=5$
23. $\sqrt{x-5}=\sqrt{2x-4}$
24. $\sqrt{2t-7}=\sqrt{t+2}$
25. $1+\sqrt{7x-3}>3$
26. $\sqrt{3x+6}+2\leq 5$
27. $-2+\sqrt{9-5x}\geq 6$
28. $6-\sqrt{2y+1}<3$
29. $\sqrt{x-6}-\sqrt{x}=3$
30. $\sqrt{y+21}-1=\sqrt{y+12}$
31. $\sqrt{b+1}=\sqrt{b+6}-1$
32. $\sqrt{4z+1}=3+\sqrt{4z-2}$
33. $\sqrt{2}-\sqrt{x+6}\leq -\sqrt{x}$
34. $\sqrt{a+9}-\sqrt{a}>\sqrt{3}$
35. $\sqrt{b-5}-\sqrt{b+7}\leq 4$
36. $\sqrt{c+5}+\sqrt{c+10}>2.5$

37. What is the solution of $2-\sqrt{x+6}=-1$?
38. Solve $\sqrt{2x+4}-4=2$.

39. **CONSTRUCTION** The minimum depth d in inches of a beam required to support a load of s pounds is given by the formula $d=\sqrt{\frac{s\ell}{576w}}$, where ℓ is the length of the beam in feet and w is the width in feet. Find the load that can be supported by a board that is 25 feet long, 2 feet wide, and 5 inches deep.

40. **AEROSPACE ENGINEERING** The radius r of the orbit of a satellite is given by $r=\sqrt[3]{\frac{GMt^2}{4\pi^2}}$, where G is the universal gravitational constant, M is the mass of the central object, and t is the time it takes the satellite to complete one orbit. Solve this formula for t.

More About. . .

Health

A ponderal index p is a measure of a person's body based on height h in meters and mass m in kilograms. One such formula is $p = \frac{\sqrt[3]{m}}{h}$.

Source: *A Dictionary of Food and Nutrition*

41. **PHYSICS** When an object is dropped from the top of a 50-foot tall building, the object will be h feet above the ground after t seconds, where $\frac{\sqrt{50 - h}}{4} = t$. How far above the ground will the object be after 1 second?

42. **HEALTH** **Use the information about health at the left.** A 70-kilogram person who is 1.8 meters tall has a ponderal index of about 2.29. How much weight could such a person gain and still have an index of at most 2.5?

43. **CRITICAL THINKING** Explain how you know that $\sqrt{x + 2} + \sqrt{2x - 3} = -1$ has no real solution without trying to solve it.

44. **WRITING IN MATH** Answer the question that was posed at the beginning of the lesson.

How do radical equations apply to manufacturing?

Include the following in your answer:

- the equation $C = 10n^{\frac{2}{3}} + 1500$ rewritten as a radical equation, and
- a step-by-step explanation of how to determine the maximum number of chips the company could make for \$10,000.

Standardized Test Practice

Standards Practice

45. If $\sqrt{x + 5} + 1 = 4$, what is the value of x?

(A) -4 (B) 0 (C) 2 (D) 4

46. Side $\overline{AC}$ of triangle ABC contains which of the following points?

(A) (3, 4) (B) (3, 5) (C) (4, 3)

(D) (4, 5) (E) (4, 6)

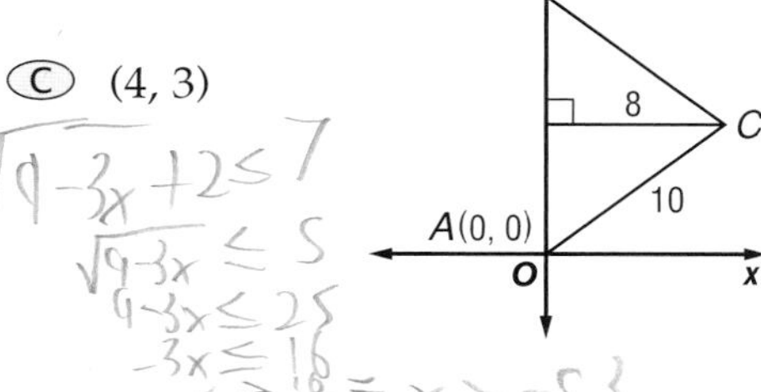

Maintain Your Skills

Mixed Review

Write each radical using rational exponents. *(Lesson 5-7)*

47. $\sqrt[7]{5^3}$

48. $\sqrt{x + 7}$

49. $\left(\sqrt[3]{x^2 + 1}\right)^2$

Simplify. *(Lesson 5-6)*

50. $\sqrt{72x^6y^3}$

51. $\frac{1}{\sqrt[3]{10}}$

52. $(5 - \sqrt{3})^2$

53. **BUSINESS** A dry cleaner ordered 7 drums of two different types of cleaning fluid. One type cost \$30 per drum, and the other type cost \$20 per drum. The total cost was \$160. How much of each type of fluid did the company order? Write a system of equations and solve by graphing. *(Lesson 3-1)*

Getting Ready for the Next Lesson

PREREQUISITE SKILL **Simplify each expression.**
*(To review **binomials**, see Lesson 5-2.)*

54. $(5 + 2x) + (-1 - x)$

55. $(-3 - 2y) + (4 + y)$

56. $(4 + x) - (2 - 3x)$

57. $(-7 - 3x) - (4 - 3x)$

58. $(1 + z)(4 + 2z)$

59. $(-3 - 4x)(1 + 2x)$

Graphing Calculator

A Follow-Up of Lesson 5-8

Solving Radical Equations and Inequalities by Graphing

You can use a TI-83 Plus to solve radical equations and inequalities. One way to do this is by rewriting the equation or inequality so that one side is 0 and then using the zero feature on the calculator.

Solve $\sqrt{x} + \sqrt{x + 2} = 3$.

Step 1 *Rewrite the equation.*

- Subtract 3 from each side of the equation to obtain $\sqrt{x} + \sqrt{x + 2} - 3 = 0$.
- Enter the function $y = \sqrt{x} + \sqrt{x + 2} - 3$ in the Y= list.

KEYSTROKES: *Review entering a function on page 128.*

Step 2 *Use a table.*

- You can use the TABLE function to locate intervals where the solution(s) lie. First, enter the starting value and the interval for the table.

KEYSTROKES: [2nd] [TBLSET] 0 [ENTER] 1 [ENTER]

Step 3 *Estimate the solution.*

- Complete the table and estimate the solution(s).

KEYSTROKES: [2nd] [TABLE]

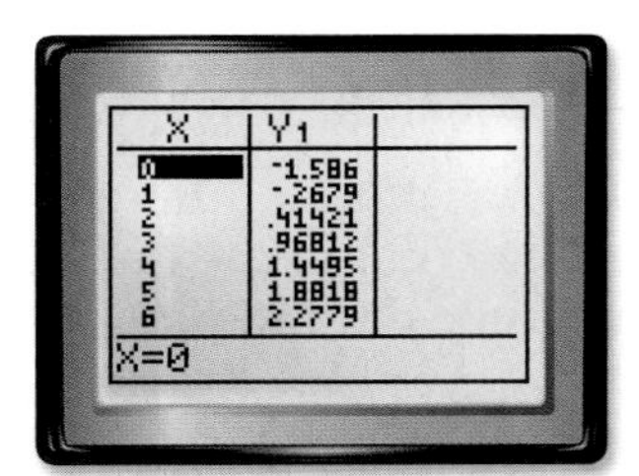

Since the function changes sign from negative to positive between $x = 1$ and $x = 2$, there is a solution between 1 and 2.

Step 4 *Use the zero feature.*

- Graph, then select zero from the CALC menu.

KEYSTROKES: [GRAPH] [2nd] [CALC] 2

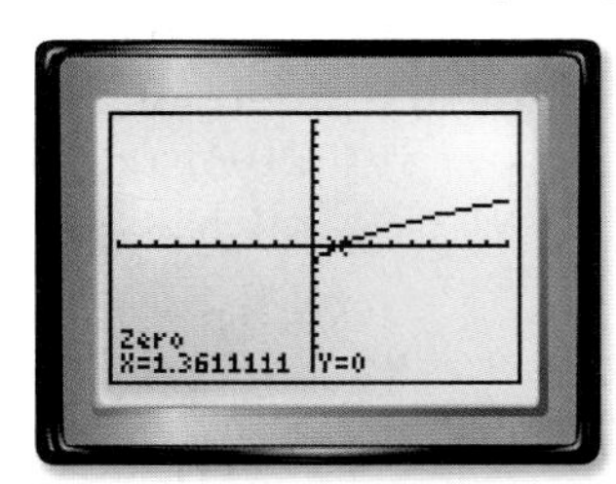

[−10, 10] scl: 1 by [−10, 10] scl: 1

Place the cursor to the left of the zero and press [ENTER] for the Left Bound. Then place the cursor to the right of the zero and press [ENTER] for the Right Bound. Press [ENTER] to solve.

The solution is about 1.36. This agrees with the estimate made by using the TABLE.

www.algebra2.com/other_calculator_keystrokes

Investigation

Instead of rewriting an equation or inequality so that one side is 0, you can also treat each side of the equation or inequality as a separate function and graph both.

Solve $2\sqrt{x} > \sqrt{x + 2} + 1$.

Step 1 *Graph each side of the inequality.*

- In the **Y=** list, enter $y_1 = 2\sqrt{x}$ and $y_2 = \sqrt{x + 2} + 1$. Then press GRAPH.

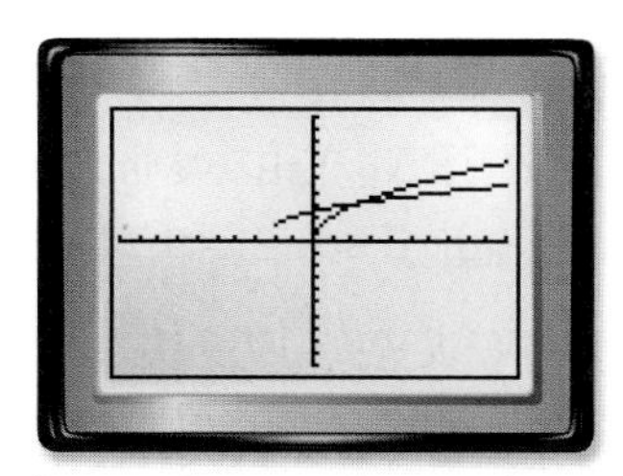

[−10, 10] scl: 1 by [−10, 10] scl: 1

Step 2 *Use the trace feature.*

- Press . You can use ▲ or ▼ to switch the cursor between the two curves.

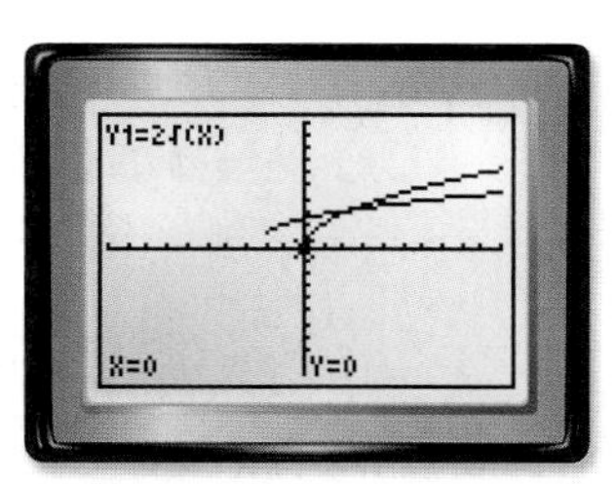

[−10, 10] scl: 1 by [−10, 10] scl: 1

The calculator screen above shows that, for points to the left of where the curves cross, $Y_1 < Y_2$ or $2\sqrt{x} < \sqrt{x + 2} + 1$. To solve the original inequality, you must find points for which $Y_1 > Y_2$. These are the points to the right of where the curves cross.

Step 3 *Use the intersect feature.*

- You can use the **INTERSECT** feature on the **CALC** menu to approximate the x-coordinate of the point at which the curves cross.

  KEYSTROKES: 2nd [CALC] 5

- Press ENTER for each of First curve?, Second curve?, and Guess?.

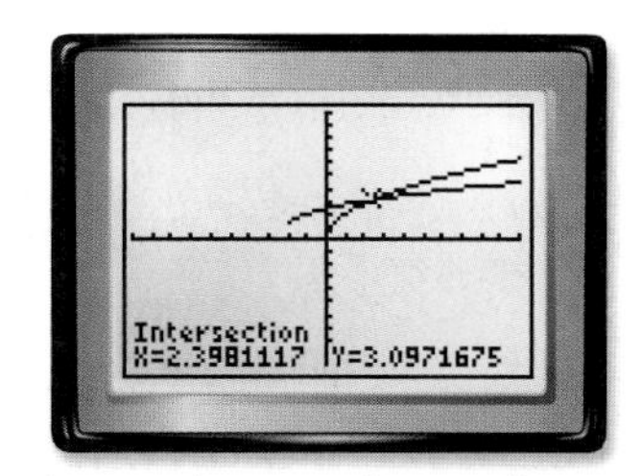

[−10, 10] scl: 1 by [−10, 10] scl: 1

The calculator screen shows that the x-coordinate of the point at which the curves cross is about 2.40. Therefore, the solution of the inequality is about $x > 2.40$. *Use the symbol $>$ instead of $\geq$ in the solution because the symbol in the original inequality is $>$.*

Exercises

Solve each equation or inequality.

1. $\sqrt{x + 4} = 3$
2. $\sqrt{3x - 5} = 1$
3. $\sqrt{x + 5} = \sqrt{3x + 4}$
4. $\sqrt{x + 3} + \sqrt{x - 2} = 4$
5. $\sqrt{3x - 7} = \sqrt{2x - 2} - 1$
6. $\sqrt{x + 8} - 1 = \sqrt{x + 2}$
7. $\sqrt{x - 3} \geq 2$
8. $\sqrt{x + 3} > 2\sqrt{x}$
9. $\sqrt{x} + \sqrt{x - 1} < 4$
10. Explain how you could apply the technique in the first example to solving an inequality.

5-9 Complex Numbers

What You'll Learn

- Add and subtract complex numbers.
- Multiply and divide complex numbers.

How do complex numbers apply to polynomial equations?

Consider the equation $2x^2 + 2 = 0$. If you solve this equation for x^2, the result is $x^2 = -1$. Since there is no real number whose square is -1, the equation has no real solutions. French mathematician René Descartes (1596–1650) proposed that a number i be defined such that $i^2 = -1$.

Vocabulary

- imaginary unit
- pure imaginary number
- complex number
- absolute value
- complex conjugates

California Standards
Standard 5.0 **Students demonstrate knowledge of how real and complex numbers are related both arithmetically and graphically. In particular, they can plot complex numbers as points in the plane.** (Key)

Standard 6.0 Students add, subtract, multiply, and divide complex numbers. (Key)

ADD AND SUBTRACT COMPLEX NUMBERS Since i is defined to have the property that $i^2 = -1$, the number i is the principal square root of -1; that is, $i = \sqrt{-1}$. i is called the **imaginary unit**. Numbers of the form $3i$, $-5i$, and $i\sqrt{2}$ are called **pure imaginary numbers**. Pure imaginary numbers are square roots of negative real numbers. For any positive real number b, $\sqrt{-b^2} = \sqrt{b^2} \cdot \sqrt{-1}$ or bi.

Example 1 Square Roots of Negative Numbers

Simplify.

a. $\sqrt{-18}$

$$\begin{aligned}\sqrt{-18} &= \sqrt{-1 \cdot 3^2 \cdot 2}\\ &= \sqrt{-1} \cdot \sqrt{3^2} \cdot \sqrt{2}\\ &= i \cdot 3 \cdot \sqrt{2} \text{ or } 3i\sqrt{2}\end{aligned}$$

b. $\sqrt{-125x^5}$

$$\begin{aligned}\sqrt{-125x^5} &= \sqrt{-1 \cdot 5^2 \cdot x^4 \cdot 5x}\\ &= \sqrt{-1} \cdot \sqrt{5^2} \cdot \sqrt{x^4} \cdot \sqrt{5x}\\ &= i \cdot 5 \cdot x^2 \cdot \sqrt{5x} \text{ or } 5ix^2\sqrt{5x}\end{aligned}$$

Study Tip

Reading Math
i is usually written before radical symbols to make it clear that it is not under the radical.

The Commutative and Associative Properties of Multiplication hold true for pure imaginary numbers.

Example 2 Multiply Pure Imaginary Numbers

Simplify.

a. $-2i \cdot 7i$

$$\begin{aligned}-2i \cdot 7i &= -14i^2\\ &= -14(-1) \quad i^2 = -1\\ &= 14\end{aligned}$$

b. $\sqrt{-10} \cdot \sqrt{-15}$

$$\begin{aligned}\sqrt{-10} \cdot \sqrt{-15} &= i\sqrt{10} \cdot i\sqrt{15}\\ &= i^2\sqrt{150}\\ &= -1 \cdot \sqrt{25} \cdot \sqrt{6}\\ &= -5\sqrt{6}\end{aligned}$$

Example 3 Simplify a Power of i

Simplify i^{45}.

$$\begin{aligned}i^{45} &= i \cdot i^{44} && \text{Multiplying powers}\\ &= i \cdot (i^2)^{22} && \text{Power of a Power}\\ &= i \cdot (-1)^{22} && i^2 = -1\\ &= i \cdot 1 \text{ or } i && (-1)^{22} = 1\end{aligned}$$

The solutions of some equations involve pure imaginary numbers.

Example 4 Equation with Imaginary Solutions

Solve $3x^2 + 48 = 0$.

$3x^2 + 48 = 0$ Original equation

$3x^2 = -48$ Subtract 48 from each side.

$x^2 = -16$ Divide each side by 3.

$x = \pm\sqrt{-16}$ Take the square root of each side.

$x = \pm 4i$ $\sqrt{-16} = \sqrt{16} \cdot \sqrt{-1}$

Study Tip

Quadratic Solutions
Quadratic equations always have complex solutions. If the discriminant is:

- negative, there are two imaginary roots,
- zero, there are two equal real roots, or
- positive, there are two unequal real roots.

Consider an expression such as $5 + 2i$. Since 5 is a real number and $2i$ is a pure imaginary number, the terms are not like terms and cannot be combined. This type of expression is called a **complex number**.

Key Concept — Complex Numbers

- **Words** A complex number is any number that can be written in the form $a + bi$, where a and b are real numbers and i is the imaginary unit. a is called the real part, and b is called the imaginary part.
- **Examples** $7 + 4i$ and $2 - 6i = 2 + (-6)i$

The Venn diagram at the right shows the complex number system.

- If $b = 0$, the complex number is a real number.
- If $b \neq 0$, the complex number is imaginary.
- If $a = 0$, the complex number is a pure imaginary number.

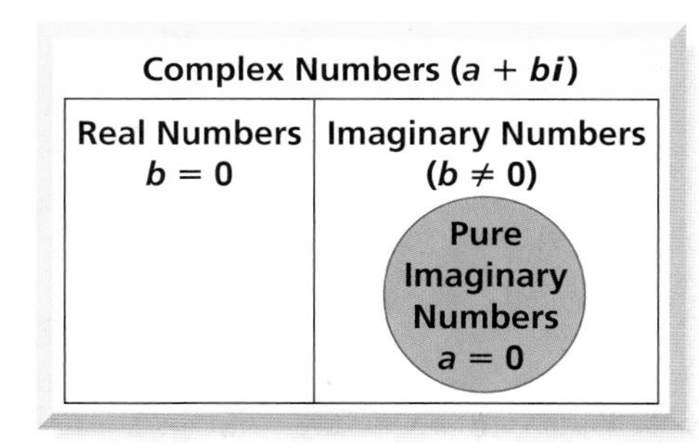

Two complex numbers are equal if and only if their real parts are equal and their imaginary parts are equal. That is, $a + bi = c + di$ if and only if $a = c$ and $b = d$.

Example 5 Equate Complex Numbers

Find the values of x and y that make the equation $2x - 3 + (y - 4)i = 3 + 2i$ true.

Set the real parts equal to each other and the imaginary parts equal to each other.

$2x - 3 = 3$ Real parts

$2x = 6$ Add 3 to each side.

$x = 3$ Divide each side by 2.

$y - 4 = 2$ Imaginary parts

$y = 6$ Add 4 to each side.

Study Tip

Reading Math
The form $a + bi$ is sometimes called the **standard form** of a complex number.

To add or subtract complex numbers, combine like terms. That is, combine the real parts and combine the imaginary parts.

Example 6 Add and Subtract Complex Numbers

Simplify.

a. $(6 - 4i) + (1 + 3i)$

$(6 - 4i) + (1 + 3i) = (6 + 1) + (-4 + 3)i$ Commutative and Associative Properties

$= 7 - i$ Simplify.

b. $(3 - 2i) - (5 - 4i)$

$(3 - 2i) - (5 - 4i) = (3 - 5) + [-2 - (-4)]i$ Commutative and Associative Properties

$= -2 + 2i$ Simplify.

You can model the addition and subtraction of complex numbers geometrically.

Algebra Activity

Adding Complex Numbers

You can model the addition of complex numbers on a coordinate plane. The horizontal axis represents the real part a of the complex number, and the vertical axis represents the imaginary part b of the complex number.

Use a coordinate plane to find $(4 + 2i) + (-2 + 3i)$.

- Create a coordinate plane and label the axes appropriately.
- Graph $4 + 2i$ by drawing a segment from the origin to (4, 2) on the coordinate plane.
- Represent the addition of $-2 + 3i$ by moving 2 units to the left and 3 units up from (4, 2).
- You end at the point (2, 5), which represents the complex number $2 + 5i$.
  So, $(4 + 2i) + (-2 + 3i) = 2 + 5i$.

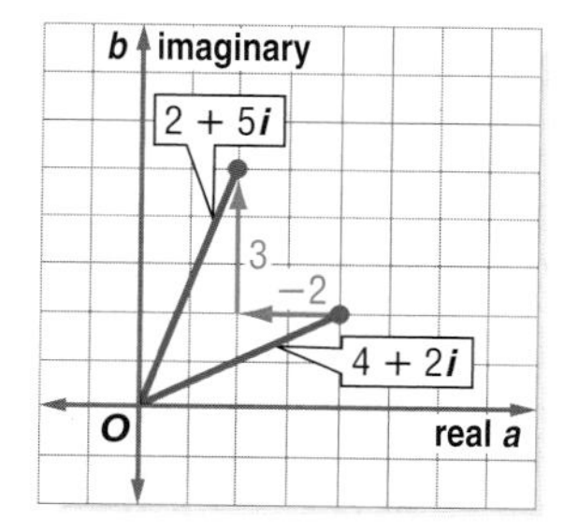

Model and Analyze

1. Model $(-3 + 2i) + (4 - i)$ on a coordinate plane.
2. Describe how you could model the difference $(-3 + 2i) - (4 - i)$ on a coordinate plane.
3. The **absolute value** of a complex number is the distance from the origin to the point representing that complex number in a coordinate plane. Refer to the graph above. Find the absolute value of $2 + 5i$.
4. Find an expression for the absolute value of $a + bi$.

MULTIPLY AND DIVIDE COMPLEX NUMBERS Complex numbers are used with electricity. In a circuit with alternating current, the voltage, current, and impedance, or hindrance to current, can be represented by complex numbers.

You can use the FOIL method to multiply complex numbers.

Example 7 Multiply Complex Numbers

Study Tip

Reading Math
Electrical engineers use j as the imaginary unit to avoid confusion with the I for current.

ELECTRICITY In an AC circuit, the voltage E, current I, and impedance Z are related by the formula $E = I \cdot Z$. Find the voltage in a circuit with current $1 + 3j$ amps and impedance $7 - 5j$ ohms.

$E = I \cdot Z$	Electricity formula
$= (1 + 3j) \cdot (7 - 5j)$	$I = 1 + 3j, Z = 7 - 5j$
$= 1(7) + 1(-5j) + (3j)7 + 3j(-5j)$	FOIL
$= 7 - 5j + 21j - 15j^2$	Multiply.
$= 7 + 16j - 15(-1)$	$j^2 = -1$
$= 22 + 16j$	Add.

The voltage is $22 + 16j$ volts.

Two complex numbers of the form $a + bi$ and $a - bi$ are called **complex conjugates**. The product of complex conjugates is always a real number. For example, $(2 + 3i)(2 - 3i) = 4 - 6i + 6i + 9$ or 13. You can use this fact to simplify the quotient of two complex numbers.

Example 8 Divide Complex Numbers

Simplify.

a. $\dfrac{3i}{2 + 4i}$

$\dfrac{3i}{2+4i} = \dfrac{3i}{2+4i} \cdot \dfrac{2-4i}{2-4i}$	$2 + 4i$ and $2 - 4i$ are conjugates.
$= \dfrac{6i - 12i^2}{4 - 16i^2}$	Multiply.
$= \dfrac{6i + 12}{20}$	$i^2 = -1$
$= \dfrac{3}{5} + \dfrac{3}{10}i$	Standard form

b. $\dfrac{5 + i}{2i}$

$\dfrac{5+i}{2i} = \dfrac{5+i}{2i} \cdot \dfrac{i}{i}$	Why multiply by $\frac{i}{i}$ intead of $\frac{-2i}{-2i}$?
$= \dfrac{5i + i^2}{2i^2}$	Multiply.
$= \dfrac{5i - 1}{-2}$	$i^2 = -1$
$= \dfrac{1}{2} - \dfrac{5}{2}i$	Standard form

Check for Understanding

Concept Check

1. **Determine** if each statement is *true* or *false*. If false, find a counterexample.
   a. Every real number is a complex number.
   b. Every imaginary number is a complex number.
2. **Decide** which of the properties of a field and the properties of equality that the set of complex numbers satisfies.
3. **OPEN ENDED** Write two complex numbers whose product is 10.

Study Tip

Look Back
Refer to Chapter 1 to review the properties of fields and the properties of equality.

Guided Practice

Simplify.

4. $\sqrt{-36}$
5. $\sqrt{-50x^2y^2}$
6. $(6i)(-2i)$
7. $5\sqrt{-24} \cdot 3\sqrt{-18}$
8. i^{29}
9. $(8 + 6i) - (2 + 3i)$
10. $(3 - 5i)(4 + 6i)$
11. $\dfrac{3 + i}{1 + 4i}$

Solve each equation.

12. $2x^2 + 18 = 0$
13. $4x^2 + 32 = 0$
14. $-5x^2 - 25 = 0$

Find the values of m and n that make each equation true.

15. $2m + (3n + 1)i = 6 - 8i$
16. $(2n - 5) + (-m - 2)i = 3 - 7i$

Application

17. **ELECTRICITY** The current in one part of a series circuit is $4 - j$ amps. The current in another part of the circuit is $6 + 4j$ amps. Add these complex numbers to find the total current in the circuit.

Practice and Apply

Homework Help

For Exercises	See Examples
18–21	1
22–25	2
26–29	3
30–33, 46, 47	6
34–37, 42, 43	7
38–41, 44, 45	8
48–55	4
56–61	5

Extra Practice
See page 839.

Simplify.

18. $\sqrt{-144}$
19. $\sqrt{-81}$
20. $\sqrt{-64x^4}$
21. $\sqrt{-100a^4b^2}$
22. $\sqrt{-13} \cdot \sqrt{-26}$
23. $\sqrt{-6} \cdot \sqrt{-24}$
24. $(-2i)(-6i)(4i)$
25. $3i(-5i)^2$
26. i^{13}
27. i^{24}
28. i^{38}
29. i^{63}
30. $(5 - 2i) + (4 + 4i)$
31. $(3 - 5i) + (3 + 5i)$
32. $(3 - 4i) - (1 - 4i)$
33. $(7 - 4i) - (3 + i)$
34. $(3 + 4i)(3 - 4i)$
35. $(1 - 4i)(2 + i)$
36. $(6 - 2i)(1 + i)$
37. $(-3 - i)(2 - 2i)$
38. $\dfrac{4i}{3 + i}$
39. $\dfrac{4}{5 + 3i}$
40. $\dfrac{10 + i}{4 - i}$
41. $\dfrac{2 - i}{3 - 4i}$
42. $(-5 + 2i)(6 - i)(4 + 3i)$
43. $(2 + i)(1 + 2i)(3 - 4i)$
44. $\dfrac{5 - i\sqrt{3}}{5 + i\sqrt{3}}$
45. $\dfrac{1 - i\sqrt{2}}{1 + i\sqrt{2}}$
46. Find the sum of $ix^2 - (2 + 3i)x + 2$ and $4x^2 + (5 + 2i)x - 4i$.
47. Simplify $[(3 + i)x^2 - ix + 4 + i] - [(-2 + 3i)x^2 + (1 - 2i)x - 3]$.

Solve each equation.

48. $5x^2 + 5 = 0$
49. $4x^2 + 64 = 0$
50. $2x^2 + 12 = 0$
51. $6x^2 + 72 = 0$
52. $-3x^2 - 9 = 0$
53. $-2x^2 - 80 = 0$
54. $\frac{2}{3}x^2 + 30 = 0$
55. $\frac{4}{5}x^2 + 1 = 0$

Find the values of m and n that make each equation true.

56. $8 + 15i = 2m + 3ni$
57. $(m + 1) + 3ni = 5 - 9i$
58. $(2m + 5) + (1 - n)i = -2 + 4i$
59. $(4 + n) + (3m - 7)i = 8 - 2i$
60. $(m + 2n) + (2m - n)i = 5 + 5i$
61. $(2m - 3n)i + (m + 4n) = 13 + 7i$
62. **ELECTRICITY** The impedance in one part of a series circuit is $3 + 4j$ ohms, and the impedance in another part of the circuit is $2 - 6j$. Add these complex numbers to find the total impedance in the circuit.

Career Choices

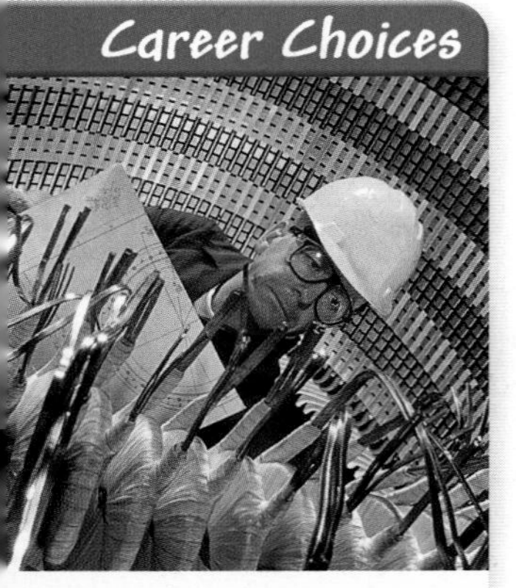

Electrical Engineering

The chips and circuits in computers are designed by electrical engineers.

Online Research
To learn more about electrical engineering, visit: www.algebra2.com/careers

ELECTRICAL ENGINEERING For Exercises 63 and 64, use the formula $E = I \cdot Z$.

63. The current in a circuit is $2 + 5j$ amps, and the impedance is $4 - j$ ohms. What is the voltage?
64. The voltage in a circuit is $14 - 8j$ volts, and the impedance is $2 - 3j$ ohms. What is the current?

65. **CRITICAL THINKING** Show that the order relation "$<$" does not make sense for the set of complex numbers. (*Hint*: Consider the two cases $i > 0$ and $i < 0$. In each case, multiply each side by i.)

66. **WRITING IN MATH** Answer the question that was posed at the beginning of the lesson.

How do complex numbers apply to polynomial equations?

Include the following in your answer:

- how the a and c must be related if the equation $ax^2 + c = 0$ has complex solutions, and
- the solutions of the equation $2x^2 + 2 = 0$.

67. If $i^2 = -1$, then what is the value of i^{71}?

(A) -1 (B) 0 (C) $-i$ (D) i

68. The area of the square is 16 square units. What is the area of the circle?

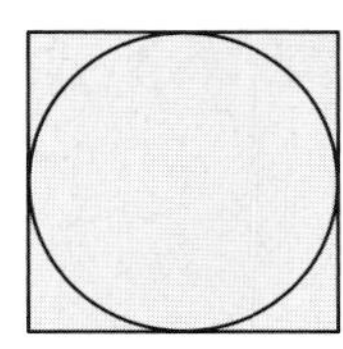

(A) 2π units^2 (B) 12 units^2

(C) 4π units^2 (D) 16π units^2

Extending the Lesson

PATTERN OF POWERS OF i

69. Find the simplified forms of i^6, i^7, i^8, i^9, i^{10}, i^{11}, i^{12}, i^{13}, and i^{14}.

70. Explain how to use the exponent to determine the simplified form of any power of i.

Maintain Your Skills

Mixed Review

Solve each equation. *(Lesson 5-8)*

71. $\sqrt{2x + 1} = 5$ 72. $\sqrt[3]{x - 3} + 1 = 3$ 73. $\sqrt{x + 5} + \sqrt{x} = 5$

Simplify each expression. *(Lesson 5-7)*

74. $x^{-\frac{1}{5}} \cdot x^{\frac{2}{3}}$ 75. $\left(y^{-\frac{1}{2}}\right)^{-\frac{2}{3}}$ 76. $a^{-\frac{3}{4}}$

For Exercises 77–80, triangle ABC is reflected over the x-axis. *(Lesson 4-6)*

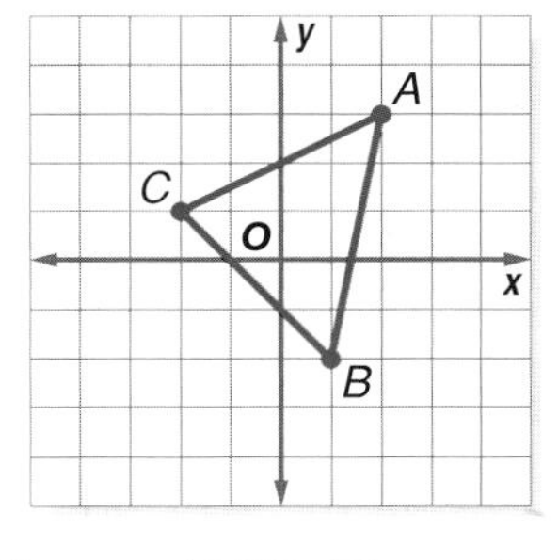

77. Write a vertex matrix for the triangle.

78. Write the reflection matrix.

79. Write the vertex matrix for $\triangle A'B'C'$.

80. Graph $\triangle A'B'C'$.

81. **FURNITURE** A new sofa, love seat, and coffee table cost \$2050. The sofa costs twice as much as the love seat. The sofa and the coffee table together cost \$1450. How much does each piece of furniture cost? *(Lesson 3-5)*

Graph each system of inequalities. *(Lesson 3-3)*

82. $y < x + 1$
$y > -2x - 2$

83. $x + y \geq 1$
$x - 2y \leq 4$

Find the slope of the line that passes through each pair of points. *(Lesson 2-3)*

84. $(-2, 1), (8, 2)$ 85. $(4, -3), (5, -3)$

Chapter 5 Study Guide and Review

Vocabulary and Concept Check

- absolute value (p. 272)
- binomial (p. 229)
- coefficient (p. 222)
- complex conjugates (p. 273)
- complex number (p. 271)
- conjugates (p. 253)
- constant (p. 222)
- degree (p. 222)
- dimensional analysis (p. 225)
- extraneous solution (p. 263)
- FOIL method (p. 230)
- imaginary unit (p. 270)
- like radical expressions (p. 252)
- like terms (p. 229)
- monomial (p. 222)
- *n*th root (p. 245)
- polynomial (p. 229)
- power (p. 222)
- principal root (p. 246)
- pure imaginary number (p. 270)
- radical equation (p. 263)
- radical inequality (p. 264)
- rationalizing the denominator (p. 251)
- scientific notation (p. 225)
- simplify (p. 222)
- square root (p. 245)
- standard notation (p. 225)
- synthetic division (p. 234)
- terms (p. 229)
- trinomial (p. 229)

Choose a word or term from the list above that best completes each statement or phrase.

1. A number is expressed in _______ when it is in the form $a \times 10^n$, where $1 \leq a < 10$ and n is an integer.
2. A shortcut method known as _______ is used to divide polynomials by binomials.
3. The _______ is used to multiply two binomials.
4. A(n) _______ is an expression that is a number, a variable, or the product of a number and one or more variables.
5. A solution of a transformed equation that is not a solution of the original equation is a(n) _______.
6. _______ are imaginary numbers of the form $a + bi$ and $a - bi$.
7. For any number a and b, if $a^2 = b$, then a is a(n) _______ of b.
8. A polynomial with three terms is known as a(n) _______.
9. When a number has more than one real root, the _______ is the nonnegative root.
10. i is called the _______.

Lesson-by-Lesson Review

5-1 Monomials

See pages 222–228.

Concept Summary

- The properties of powers for real numbers a and b and integers m and n are as follows.

$a^{-n} = \frac{1}{a^n}, a \neq 0$

$(a^m)^n = a^{mn}$

$a^m \cdot a^n = a^{m+n}$

$(ab)^m = a^m b^m$

$\frac{a^m}{a^n} = a^{m-n}, a \neq 0$

$\left(\frac{a}{b}\right)^n = \frac{a^n}{b^n}, b \neq 0$

- Use scientific notation to represent very large or very small numbers.

Examples

1 **Simplify $(3x^4y^6)(-8x^3y)$.**

$(3x^4y^6)(-8x^3y) = (3)(-8)x^{4+3}y^{6+1}$ Commutative Property and products of powers

$= -24x^7y^7$ Simplify.

www.algebra2.com/vocabulary_review

2 **Express each number in scientific notation.**

a. 31,000

$31{,}000 = 3.1 \times 10{,}000$

$= 3.1 \times 10^4$ $10{,}000 = 10^4$

b. 0.007

$0.007 = 7 \times 0.001$

$= 7 \times 10^{-3}$ $0.001 = \frac{1}{1000}$ or $\frac{1}{10^3}$

Exercises **Simplify. Assume that no variable equals 0.**
See Examples 1–4 on pages 222–224.

11. $f^{-7} \cdot f^4$ **12.** $(3x^2)^3$ **13.** $(2y)(4xy^3)$ **14.** $\left(\frac{3}{5}c^2f\right)\left(\frac{4}{3}cd\right)^2$

Evaluate. Express the result in scientific notation. *See Examples 5–7 on page 225.*

15. $(2000)(85{,}000)$ **16.** $(0.0014)^2$ **17.** $\frac{5{,}400{,}000}{6000}$

5-2 Polynomials

See pages 229–232.

Concept Summary

- Add or subtract polynomials by combining like terms.
- Multiply polynomials by using the Distributive Property.
- Multiply binomials by using the FOIL method.

Examples

1 **Simplify $(5x^2 + 4x) - (3x^2 + 6x - 7)$.**

$5x^2 + 4x - (3x^2 + 6x - 7)$
$= 5x^2 + 4x - 3x^2 - 6x + 7$
$= (5x^2 - 3x^2) + (4x - 6x) + 7$
$= 2x^2 - 2x + 7$

2 **Find $(9k + 4)(7k - 6)$.**

$(9k + 4)(7k - 6)$
$= (9k)(7k) + (9k)(-6) + (4)(7k) + (4)(-6)$
$= 63k^2 - 54k + 28k - 24$
$= 63k^2 - 26k - 24$

Exercises **Simplify.** *See Examples 2–5 on pages 229 and 230.*

18. $(4c - 5) - (c + 11) + (-6c + 17)$ **19.** $(11x^2 + 13x - 15) - (7x^2 - 9x + 19)$
20. $-6m^2(3mn + 13m - 5n)$ **21.** $x^{-8}y^{10}(x^{11}y^{-9} + x^{10}y^{-6})$
22. $(d - 5)(d + 3)$ **23.** $(2a^2 + 6)^2$ **24.** $(2b - 3c)^3$

5-3 Dividing Polynomials

See pages 233–238.

Concept Summary

- Use the division algorithm or synthetic division to divide polynomials.

Example **Use synthetic division to find $(4x^4 - x^3 - 19x^2 + 11x - 2) \div (x - 2)$.**

2	4	−1	−19	11	−2
		8	14	−10	2
	4	7	−5	1	0

→ The quotient is $4x^3 + 7x^2 - 5x + 1$.

Exercises **Simplify.** *See Examples 1–5 on pages 233–235.*

25. $(2x^4 - 6x^3 + x^2 - 3x - 3) \div (x - 3)$ **26.** $(10x^4 + 5x^3 + 4x^2 - 9) \div (x + 1)$
27. $(x^2 - 5x + 4) \div (x - 1)$ **28.** $(5x^4 + 18x^3 + 10x^2 + 3x) \div (x^2 + 3x)$

5-4 Factoring Polynomials

See pages 239–244.

Concept Summary

- You can factor polynomials using the GCF, grouping, or formulas involving squares and cubes.

Examples

1 Factor $4x^3 - 6x^2 + 10x - 15$.

$4x^3 - 6x^2 + 10x - 15 = (4x^3 - 6x^2) + (10x - 15)$ Group to find the GCF.

$= 2x^2(2x - 3) + 5(2x - 3)$ Factor the GCF of each binomial.

$= (2x^2 + 5)(2x - 3)$ Distributive Property

2 Factor $3m^2 + m - 4$.

Find two numbers whose product is $3(-4)$ or 12, and whose sum is 1. The two numbers must be 4 and -3 because $4(-3) = -12$ and $4 + (-3) = 1$.

$3m^2 + m - 4 = 3m^2 + 4m - 3m - 4$

$= (3m^2 + 4m) - (3m + 4)$

$= m(3m + 4) + (-1)(3m + 4)$

$= (3m + 4)(m - 1)$

Exercises **Factor completely. If the polynomial is not factorable, write *prime*.**
See Examples 1–3 on pages 239 and 241.

29. $200x^2 - 50$

30. $10a^3 - 20a^2 - 2a + 4$

31. $5w^3 - 20w^2 + 3w - 12$

32. $x^4 - 7x^3 + 12x^2$

33. $s^3 + 512$

34. $x^2 - 7x + 5$

5-5 Roots of Real Numbers

See pages 245–249.

Concept Summary

Real nth roots of b, $\sqrt[n]{b}$, or $-\sqrt[n]{b}$			
n	$\sqrt[n]{b}$ if $b > 0$	$\sqrt[n]{b}$ if $b < 0$	$\sqrt[n]{b}$ if $b = 0$
even	one positive root one negative root	no real roots	one real root, 0
odd	one positive root no negative roots	no positive roots one negative root	one real root, 0

Examples

1 Simplify $\sqrt{81x^6}$.

$\sqrt{81x^6} = \sqrt{(9x^3)^2}$ $81x^6 = (9x^3)^2$

$= 9|x^3|$ Use absolute value.

2 Simplify $\sqrt[7]{2187x^{14}y^{35}}$.

$\sqrt[7]{2187x^{14}y^{35}} = \sqrt[7]{(3x^2y^5)^7}$ $2187x^{14}y^{35} = (3x^2y^5)^7$

$= 3x^2y^5$ Evaluate.

Exercises **Simplify.** *See Examples 1 and 2 on pages 246 and 247.*

35. $\pm\sqrt{256}$

36. $\sqrt[3]{-216}$

37. $\sqrt{(-8)^2}$

38. $\sqrt[5]{c^5d^{15}}$

39. $\sqrt{(x^4 - 3)^2}$

40. $\sqrt[3]{(512 + x^2)^3}$

41. $\sqrt[4]{16m^8}$

42. $\sqrt{a^2 - 10a + 25}$

5-6 Radical Expressions

See pages 250–256.

Concept Summary

For any real numbers a and b and any integer $n > 1$,

- Product Property: $\sqrt[n]{ab} = \sqrt[n]{a} \cdot \sqrt[n]{b}$
- Quotient Property: $\sqrt[n]{\frac{a}{b}} = \frac{\sqrt[n]{a}}{\sqrt[n]{b}}$

Example **Simplify $6\sqrt[5]{32m^3} \cdot 5\sqrt[5]{1024m^2}$.**

$6\sqrt[5]{32m^3} \cdot 5\sqrt[5]{1024m^2} = 6 \cdot 5\sqrt[5]{(32m^3 \cdot 1024m^2)}$	Product Property of Radicals
$= 30\sqrt[5]{2^5 \cdot 4^5 \cdot m^5}$	Factor into exponents of 5 if possible.
$= 30\sqrt[5]{2^5} \cdot \sqrt[5]{4^5} \cdot \sqrt[5]{m^5}$	Product Property of Radicals
$= 30 \cdot 2 \cdot 4 \cdot m$ or $240m$	Write the fifth roots.

Exercises **Simplify.** *See Examples 1–6 on pages 250–253.*

43. $\sqrt[6]{128}$ **44.** $\sqrt{5} + \sqrt{20}$ **45.** $5\sqrt{12} - 3\sqrt{75}$

46. $6\sqrt[5]{11} - 8\sqrt[5]{11}$ **47.** $(\sqrt{8} + \sqrt{12})^2$ **48.** $\sqrt{8} \cdot \sqrt{15} \cdot \sqrt{21}$

49. $\frac{\sqrt{243}}{\sqrt{3}}$ **50.** $\frac{1}{3 + \sqrt{5}}$ **51.** $\frac{\sqrt{10}}{4 + \sqrt{2}}$

5-7 Radical Exponents

See pages 257–262.

Concept Summary

- For any nonzero real number b, and any integers m and n, with $n > 1$, $b^{\frac{m}{n}} = \sqrt[n]{b^m} = (\sqrt[n]{b})^m$

Examples

1 Write $32^{\frac{4}{5}} \cdot 32^{\frac{2}{5}}$ in radical form.

$32^{\frac{4}{5}} \cdot 32^{\frac{2}{5}} = 32^{\frac{4}{5} + \frac{2}{5}}$	Product of powers
$= 32^{\frac{6}{5}}$	Add.
$= (2^5)^{\frac{6}{5}}$	$32 = 2^5$
$= 2^6$ or 64	Power of a power

2 Simplify $\frac{3x}{\sqrt[3]{z}}$.

$\frac{3x}{\sqrt[3]{z}} = \frac{3x}{z^{\frac{1}{3}}}$	Rational exponents
$= \frac{3x}{z^{\frac{1}{3}}} \cdot \frac{z^{\frac{2}{3}}}{z^{\frac{2}{3}}}$	Rationalize the denominator.
$= \frac{3xz^{\frac{2}{3}}}{z}$ or $\frac{3x\sqrt[3]{z^2}}{z}$	Rewrite in radical form.

Exercises **Evaluate.** *See Examples 3 and 5 on pages 258 and 259.*

52. $27^{-\frac{2}{3}}$ **53.** $9^{\frac{1}{3}} \cdot 9^{\frac{5}{3}}$ **54.** $\left(\frac{8}{27}\right)^{-\frac{2}{3}}$

Simplify. *See Example 5 on page 259.*

55. $\frac{1}{y^{\frac{2}{5}}}$ **56.** $\frac{xy}{\sqrt[3]{z}}$ **57.** $\frac{3x + 4x^2}{x^{-\frac{2}{3}}}$

• Extra Practice, see pages 836–839.
• Mixed Problem Solving, see page 866.

5-8 Radical Equations and Inequalities

See pages 263–267.

Concept Summary

- To solve a radical equation, isolate the radical. Then raise each side of the equation to a power equal to the index of the radical.

Example **Solve $\sqrt{3x - 8} + 1 = 3$.**

$\sqrt{3x - 8} + 1 = 3$	Original equation
$\sqrt{3x - 8} = 2$	Subtract 1 from each side.
$\left(\sqrt{3x - 8}\right)^2 = 2^2$	Square each side.
$3x - 8 = 4$	Evaluate the squares.
$x = 4$	Solve for x.

Exercises **Solve each equation.** *See Examples 1–3 on pages 263 and 264.*

58. $\sqrt{x} = 6$
59. $y^{\frac{1}{3}} - 7 = 0$
60. $(x - 2)^{\frac{3}{2}} = -8$
61. $\sqrt{x + 5} - 3 = 0$
62. $\sqrt{3t - 5} - 3 = 4$
63. $\sqrt{2x - 1} = 3$
64. $\sqrt[4]{2x - 1} = 2$
65. $\sqrt{y + 5} = \sqrt{2y - 3}$
66. $\sqrt{y + 1} + \sqrt{y - 4} = 5$

5-9 Complex Numbers

See pages 270–275.

Concept Summary

- $i^2 = -1$ and $i = \sqrt{-1}$
- Complex conjugates can be used to simplify quotients of complex numbers.

Examples

1 Simplify $(15 - 2i) + (-11 + 5i)$.

$(15 - 2i) + (-11 + 5i) = [15 + (-11)] + (-2 + 5)i$ Group the real and imaginary parts.

$= 4 + 3i$ Add.

2 Simplify $\frac{7i}{2 + 3i}$.

$\frac{7i}{2 + 3i} = \frac{7i}{2 + 3i} \cdot \frac{2 - 3i}{2 - 3i}$ $2 + 3i$ and $2 - 3i$ are conjugates.

$= \frac{14i - 21i^2}{4 - 9i^2}$ Multiply.

$= \frac{21 + 14i}{13}$ or $\frac{21}{13} + \frac{14}{13}i$ $i^2 = -1$

Exercises **Simplify.** *See Examples 1–3 and 6–8 on pages 270, 272, and 273.*

67. $\sqrt{-64m^{12}}$
68. $(7 - 4i) - (-3 + 6i)$
69. $-6\sqrt{-9} \cdot 2\sqrt{-4}$
70. i^6
71. $(3 + 4i)(5 - 2i)$
72. $\left(\sqrt{6} + i\right)\left(\sqrt{6} - i\right)$
73. $\frac{1 + i}{1 - i}$
74. $\frac{4 - 3i}{1 + 2i}$
75. $\frac{3 - 9i}{4 + 2i}$

Chapter 5 Practice Test

Vocabulary and Concepts

Choose the term that best describes the shaded part of each trinomial.

1. $\boxed{2}x^2 - 3x + 4$
2. $4x^{\boxed{2}} - 6x - 3$
3. $9x^2 + 2x + \boxed{7}$

a. degree
b. constant term
c. coefficient

Skills and Applications

Simplify.

4. $(5b)^4(6c)^2$
5. $(13x - 1)(x + 3)$
6. $(2h - 6)^3$

Evaluate. Express the result in scientific notation.

7. $(3.16 \times 10^3)(24 \times 10^2)$
8. $\frac{7{,}200{,}000 \cdot 0.0011}{0.018}$

Simplify.

9. $(x^4 - x^3 - 10x^2 + 4x + 24) \div (x - 2)$
10. $(2x^3 + 9x^2 - 2x + 7) \div (x + 2)$

Factor completely. If the polynomial is not factorable, write *prime*.

11. $x^2 - 14x + 45$
12. $2r^2 + 3pr - 2p^2$
13. $x^2 + 2\sqrt{3}x + 3$

Simplify.

14. $\sqrt{175}$
15. $(5 + \sqrt{3})(7 - 2\sqrt{3})$
16. $3\sqrt{6} + 5\sqrt{54}$
17. $\frac{9}{5 - \sqrt{3}}$
18. $\left(9^{\frac{1}{2}} \cdot 9^{\frac{2}{3}}\right)^{\frac{1}{6}}$
19. $11^{\frac{1}{2}} \cdot 11^{\frac{7}{3}} \cdot 11^{\frac{1}{6}}$
20. $\sqrt[6]{256s^{11}t^{18}}$
21. $v^{-\frac{7}{11}}$
22. $\frac{b^{\frac{1}{2}}}{b^{\frac{3}{2}} - b^{\frac{1}{2}}}$

Solve each equation.

23. $\sqrt{b + 15} = \sqrt{3b + 1}$
24. $\sqrt{2x} = \sqrt{x - 4}$
25. $\sqrt[4]{y + 2} + 9 = 14$
26. $\sqrt[3]{2w - 1} + 11 = 18$
27. $\sqrt{4x + 28} = \sqrt{6x + 38}$
28. $1 + \sqrt{x + 5} = \sqrt{x + 12}$

Simplify.

29. $(5 - 2i) - (8 - 11i)$
30. $(14 - 5i)^2$

31. **SKYDIVING** The approximate time t in seconds that it takes an object to fall a distance of d feet is given by $t = \sqrt{\frac{d}{16}}$. Suppose a parachutist falls 11 seconds before the parachute opens. How far does the parachutist fall during this time period?

32. **GEOMETRY** The area of a triangle with sides of length a, b, and c is given by $\sqrt{s(s - a)(s - b)(s - c)}$, where $s = \frac{1}{2}(a + b + c)$. If the lengths of the sides of a triangle are 6, 9, and 12 feet, what is the area of the triangle expressed in radical form?

33. **STANDARDIZED TEST PRACTICE** $2 + \left(x + \frac{1}{x}\right)^2 =$

(A) 2
(B) 4
(C) $x^2 + \frac{1}{x^2}$
(D) $x^2 + \frac{1}{x^2} + 4$

Chapter 5 Standardized Test Practice

Standards Practice

Part 1 Multiple Choice

Record your answers on the answer sheet provided by your teacher or on a sheet of paper.

1. If $x^3 = 30$ and x is a real number, then x lies between which two consecutive integers?

(A) 2 and 3

(B) 3 and 4

(C) 4 and 5

(D) 5 and 6

2. If $12x + 7y = 19$ and $4x - y = 3$, then what is the value of $8x + 8y$?

(A) 2

(B) 8

(C) 16

(D) 22

3. For all positive integers n,

$\boxed{n} = n - 1$, if n is even and

$\boxed{n} = \frac{1}{2}(n + 1)$, if n is odd.

What is $\boxed{8} \times \boxed{13}$?

(A) 42

(B) 49

(C) 56

(D) 82

4. Let $x \circledast y = xy - y$ for all integers x and y. If $x \circledast y = 0$ and $y \neq 0$, what must x equal?

(A) -2

(B) -1

(C) 0

(D) 1

5. The sum of a number and its square is three times the number. What is the number?

(A) 0 only

(B) -2 only

(C) 2 only

(D) 0 or 2

6. In rectangle $ABCD$, $\overline{AD}$ is 8 units long. What is the length of $\overline{AB}$ in units?

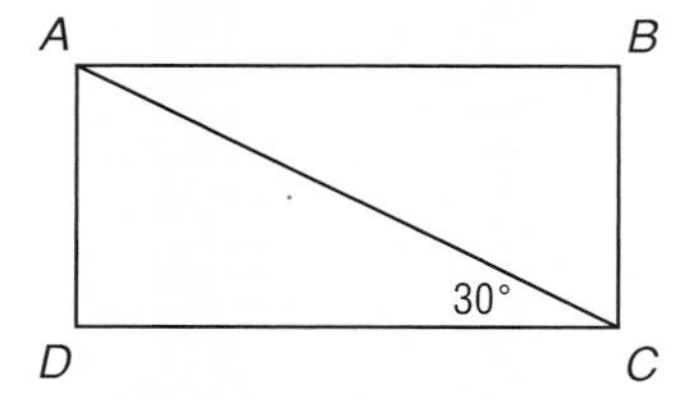

(A) 4

(B) 8

(C) $8\sqrt{3}$

(D) 16

7. The sum of two positive consecutive integers is s. In terms of s, what is the value of the greater integer?

(A) $\frac{s}{2} - 1$

(B) $\frac{s - 1}{2}$

(C) $\frac{s}{2}$

(D) $\frac{s + 1}{2}$

8. Latha, Renee, and Cindy scored a total of 30 goals for their soccer team this season. Latha scored three times as many goals as Renee. The combined number of goals scored by Latha and Cindy is four times the number scored by Renee. How many goals did Latha score?

(A) 5

(B) 6

(C) 18

(D) 20

9. If $s = t + 1$ and $t \geq 1$, then which of the following must be equal to $s^2 - t^2$?

(A) $(s - t)^2$

(B) $t^2 - 1$

(C) $s^2 - 1$

(D) $s + t$

Test-Taking Tip

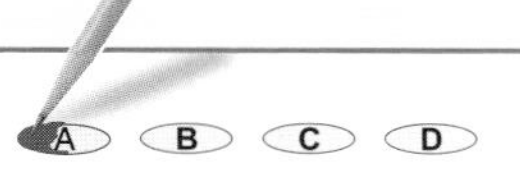

Question 9

If you simplify an expression and do not find your answer among the given answer choices, follow these steps. First, check your answer. Then, compare your answer with each of the given answer choices to determine whether it is equivalent to any of the answer choices.

Preparing for Standardized Tests
For test-taking strategies and more practice, see pages 877–892.

Part 2 Short Response/Grid In

Record your answers on the answer sheet provided by your teacher or on a sheet of paper.

10. Let $a ✿ b = a + \frac{1}{b}$, where $b \neq 0$. What is the value of $3 ✿ 4$?

11. If $3x^2 = 27$, what is the value of $3x^4$?

12. In the figure, if $x = 25$ and $z = 50$, what is the value of y?

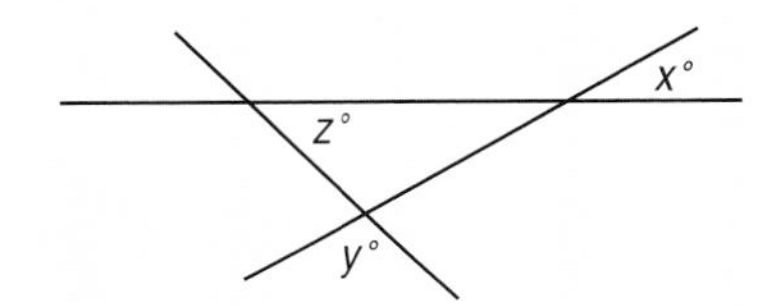

13. For all positive integers n, let ⬡n⬡ equal the greatest prime number that is a divisor of n. What does $\frac{⬡70⬡}{⬡27⬡}$ equal?

14. If $3x + 2y = 36$ and $\frac{5y}{3x} = 5$, then $x =$ __?__.

15. In the figure, a square with side of length $2\sqrt{2}$ is inscribed in a circle. If the area of the circle is $k\pi$, what is the exact value of k?

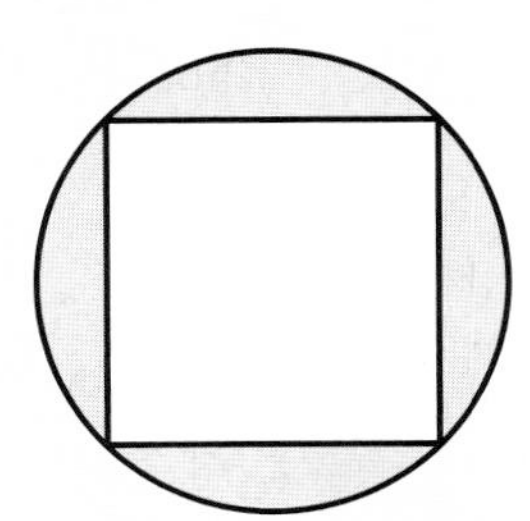

16. For all nonnegative numbers n, let $\boxed{n}$ be defined by $\boxed{n} = \frac{\sqrt{n}}{2}$. If $\boxed{n} = 4$, what is the value of n?

17. For the numbers a, b, and c, the average (arithmetic mean) is twice the median. If $a = 0$, and $a < b < c$, what is the value of $\frac{c}{b}$?

18. Write the numbers $\{\sqrt[3]{64}, \sqrt{18}, 4.1, 16^{\frac{1}{4}}\}$ in order from least to greatest. Show or describe your method.

19. Divide $x^3 - 7x^2 + 18x - 18$ by $x - 3$. Show each step of the process.

20. Simplify $\frac{x^2 - 2x - 8}{2x^2 - 9x + 4}$. Assume that the denominator is not equal to zero.

Part 3 Extended Response

Record your answers on a sheet of paper. Show your work.

For Exercises 21–23, use the information below.

The period of a pendulum is the time it takes for the pendulum to make one complete swing back and forth. The formula $T = 2\pi\sqrt{\frac{L}{32}}$ gives the period T in seconds for a pendulum L feet long.

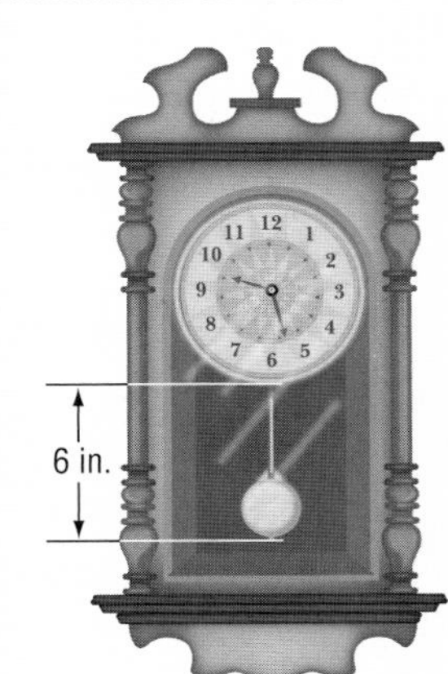

21. What is the period of the pendulum in the wall clock shown? Round to the nearest hundredth of a second.

22. Solve the formula for the length of the pendulum L in terms of the time T. Show each step of the process.

23. If you are building a grandfather clock and you want the pendulum to have a period of 2 seconds, how long should you make the pendulum? Round to the nearest tenth of a foot.

Chapter 6

Quadratic Functions and Inequalities

What You'll Learn

- **Lesson 6-1** Graph quadratic functions.
- **Lessons 6-2 through 6-5** Solve quadratic equations.
- **Lesson 6-3** Write quadratic equations and functions.
- **Lesson 6-6** Analyze graphs of quadratic functions.
- **Lesson 6-7** Graph and solve quadratic inequalities.

Key Vocabulary

- root (p. 294)
- zero (p. 294)
- completing the square (p. 307)
- Quadratic Formula (p. 313)
- discriminant (p. 316)

Why It's Important

Quadratic functions can be used to model real-world phenomena like the motion of a falling object. They can also be used to model the shape of architectural structures such as the supporting cables of a suspension bridge. *You will learn to calculate the value of the discriminant of a quadratic equation in order to describe the position of the supporting cables of the Golden Gate Bridge in Lesson 6-5.*

Getting Started

Prerequisite Skills To be successful in this chapter, you'll need to master these skills and be able to apply them in problem-solving situations. Review these skills before beginning Chapter 6.

For Lessons 6-1 and 6-2 — Graph Functions

Graph each equation by making a table of values. *(For review, see Lesson 2-1.)*

1. $y = 2x + 3$ **2.** $y = -x - 5$ **3.** $y = x^2 + 4$ **4.** $y = -x^2 - 2x + 1$

For Lessons 6-1, 6-2, and 6-5 — Multiply Polynomials

Find each product. *(For review, see Lesson 5-2.)*

5. $(x - 4)(7x + 12)$ **6.** $(x + 5)^2$ **7.** $(3x - 1)^2$ **8.** $(3x - 4)(2x - 9)$

For Lessons 6-3 and 6-4 — Factor Polynomials

Factor completely. If the polynomial is not factorable, write *prime*. *(For review, see Lesson 5-4.)*

9. $x^2 + 11x + 30$ **10.** $x^2 - 13x + 36$ **11.** $x^2 - x - 56$ **12.** $x^2 - 5x - 14$

13. $x^2 + x + 2$ **14.** $x^2 + 10x + 25$ **15.** $x^2 - 22x + 121$ **16.** $x^2 - 9$

For Lessons 6-4 and 6-5 — Simplify Radical Expressions

Simplify. *(For review, see Lessons 5-6 and 5-9.)*

17. $\sqrt{225}$ **18.** $\sqrt{48}$ **19.** $\sqrt{180}$ **20.** $\sqrt{68}$

21. $\sqrt{-25}$ **22.** $\sqrt{-32}$ **23.** $\sqrt{-270}$ **24.** $\sqrt{-15}$

Quadratic Functions and Inequalities Make this Foldable to help you organize your notes. Begin with one sheet of 11" by 17" paper.

Step 1 Fold and Cut

Fold in half lengthwise. Then fold in fourths crosswise. Cut along the middle fold from the edge to the last crease as shown.

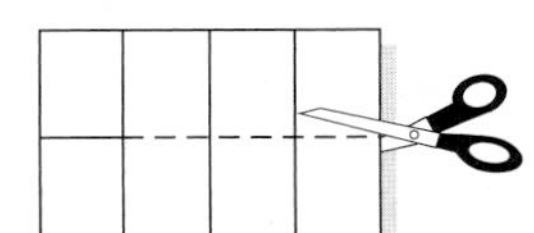

Step 2 Refold and Label

Refold along the lengthwise fold and staple the uncut section at the top. Label each section with a lesson number and close to form a booklet.

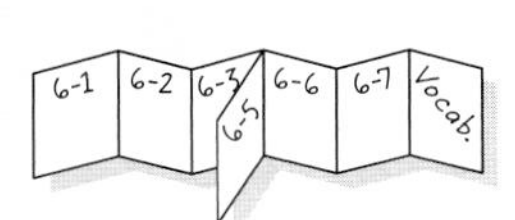

Reading and Writing As you read and study the chapter, fill the journal with notes, diagrams, and examples for each lesson.

6-1 Graphing Quadratic Functions

California Standards Standard 10.0 Students graph quadratic functions and determine the maxima, minima, and zeros of the function. (Key)

What You'll Learn

- Graph quadratic functions.
- Find and interpret the maximum and minimum values of a quadratic function.

Vocabulary

- quadratic function
- quadratic term
- linear term
- constant term
- parabola
- axis of symmetry
- vertex
- maximum value
- minimum value

How can income from a rock concert be maximized?

Rock music managers handle publicity and other business issues for the artists they manage. One group's manager has found that based on past concerts, the predicted income for a performance is $P(x) = -50x^2 + 4000x - 7500$, where x is the price per ticket in dollars.

The graph of this quadratic function is shown at the right. At first the income increases as the price per ticket increases, but as the price continues to increase, the income declines.

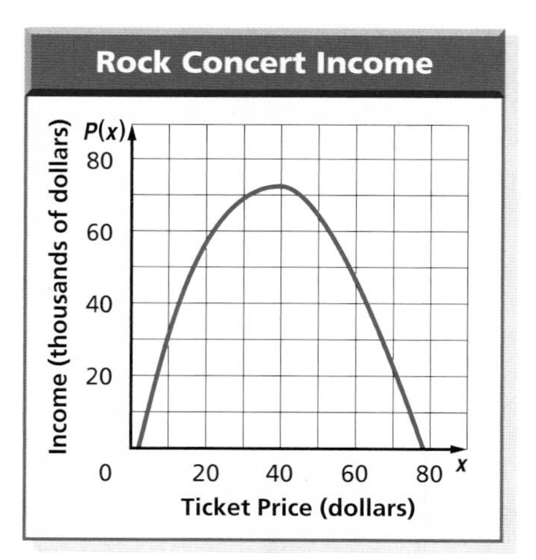

GRAPH QUADRATIC FUNCTIONS A **quadratic function** is described by an equation of the following form.

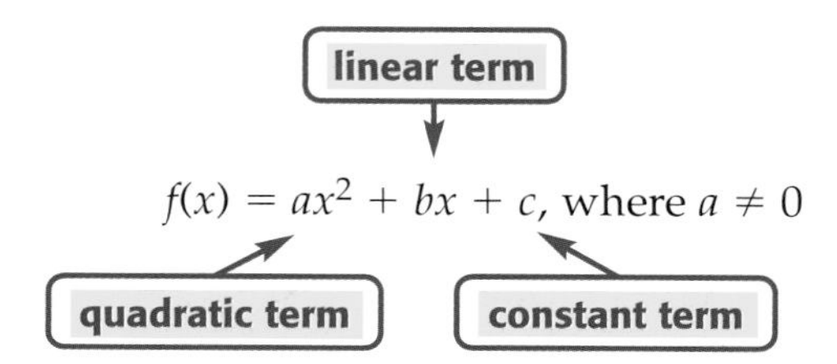

The graph of any quadratic function is called a **parabola**. One way to graph a quadratic function is to graph ordered pairs that satisfy the function.

Example 1 Graph a Quadratic Function

Graph $f(x) = 2x^2 - 8x + 9$ by making a table of values.

First, choose integer values for x. Then, evaluate the function for each x value. Graph the resulting coordinate pairs and connect the points with a smooth curve.

x	$2x^2 - 8x + 9$	$f(x)$	$(x, f(x))$
0	$2(0)^2 - 8(0) + 9$	9	(0, 9)
1	$2(1)^2 - 8(1) + 9$	3	(1, 3)
2	$2(2)^2 - 8(2) + 9$	1	(2, 1)
3	$2(3)^2 - 8(3) + 9$	3	(3, 3)
4	$2(4)^2 - 8(4) + 9$	9	(4, 9)

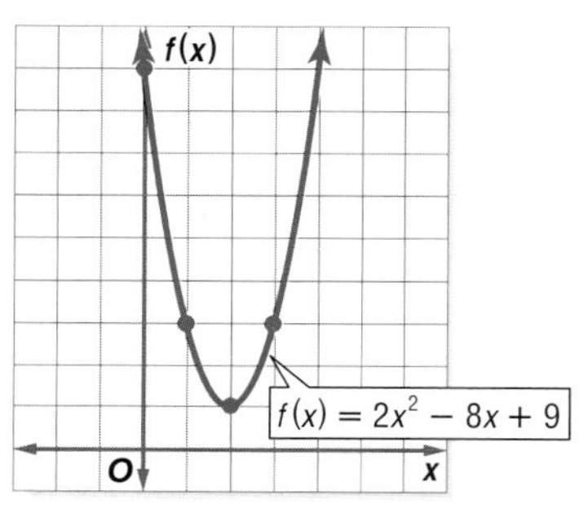

All parabolas have an **axis of symmetry**. If you were to fold a parabola along its axis of symmetry, the portions of the parabola on either side of this line would match.

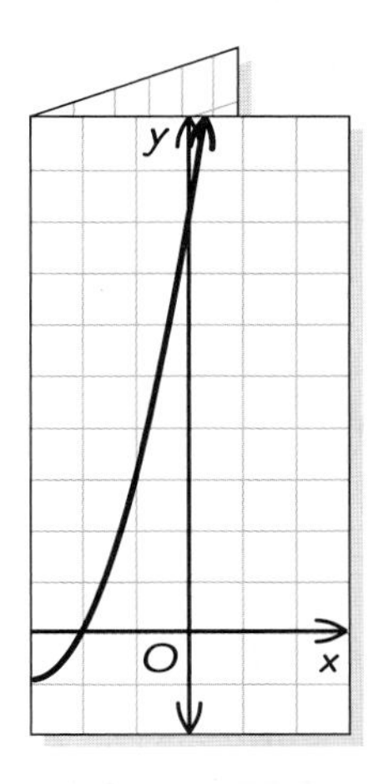

The point at which the axis of symmetry intersects a parabola is called the **vertex**. The y-intercept of a quadratic function, the equation of the axis of symmetry, and the x-coordinate of the vertex are related to the equation of the function as shown below.

Key Concept — Graph of a Quadratic Function

- **Words** Consider the graph of $y = ax^2 + bx + c$, where $a \neq 0$.
  - The y-intercept is $a(0)^2 + b(0) + c$ or c.
  - The equation of the axis of symmetry is $x = -\frac{b}{2a}$.
  - The x-coordinate of the vertex is $-\frac{b}{2a}$.
- **Model**

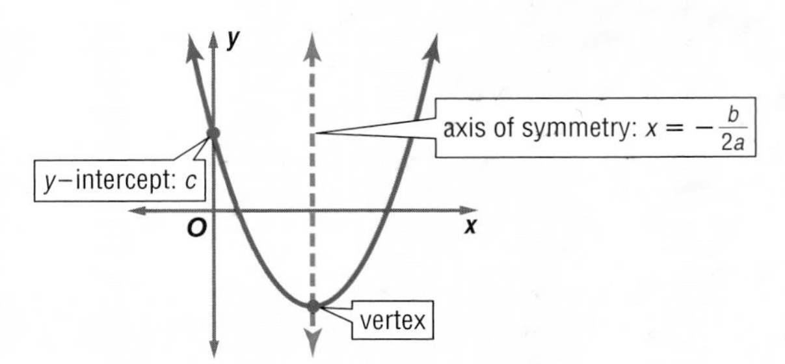

Knowing the location of the axis of symmetry, y-intercept, and vertex can help you graph a quadratic function.

Example 2 Axis of Symmetry, y-Intercept, and Vertex

Consider the quadratic function $f(x) = x^2 + 9 + 8x$.

a. Find the y-intercept, the equation of the axis of symmetry, and the x-coordinate of the vertex.

Begin by rearranging the terms of the function so that the quadratic term is first, the linear term is second, and the constant term is last. Then identify a, b, and c.

$$f(x) = ax^2 + bx + c$$

$$f(x) = x^2 + 9 + 8x \rightarrow f(x) = 1x^2 + 8x + 9$$

So, $a = 1$, $b = 8$, and $c = 9$.

The y-intercept is 9. You can find the equation of the axis of symmetry using a and b.

$x = -\frac{b}{2a}$ Equation of the axis of symmetry

$x = -\frac{8}{2(1)}$ $a = 1, b = 8$

$x = -4$ Simplify.

The equation of the axis of symmetry is $x = -4$. Therefore, the x-coordinate of the vertex is -4.

Study Tip

Symmetry

Sometimes it is convenient to use symmetry to help find other points on the graph of a parabola. Each point on a parabola has a mirror image located the same distance from the axis of symmetry on the other side of the parabola.

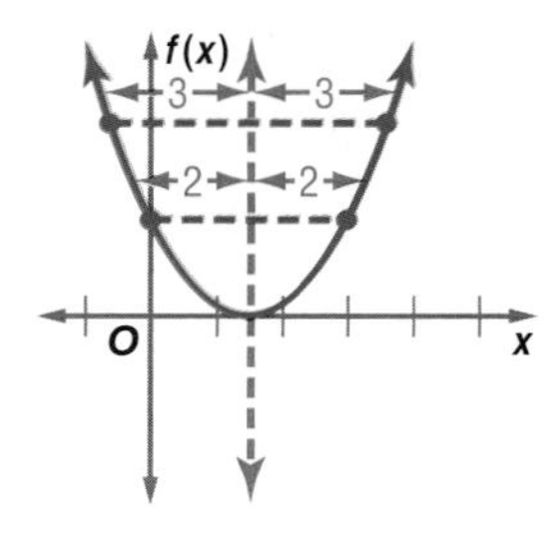

b. Make a table of values that includes the vertex.

Choose some values for x that are less than -4 and some that are greater than -4. This ensures that points on each side of the axis of symmetry are graphed.

x	$x^2 + 8x + 9$	$f(x)$	$(x, f(x))$
-6	$(-6)^2 + 8(-6) + 9$	-3	$(-6, -3)$
-5	$(-5)^2 + 8(-5) + 9$	-6	$(-5, -6)$
-4	$(-4)^2 + 8(-4) + 9$	-7	$(-4, -7)$ ← Vertex
-3	$(-3)^2 + 8(-3) + 9$	-6	$(-3, -6)$
-2	$(-2)^2 + 8(-2) + 9$	-3	$(-2, -3)$

c. Use this information to graph the function.

Graph the vertex and y-intercept. Then graph the points from your table connecting them and the y-intercept with a smooth curve. As a check, draw the axis of symmetry, $x = -4$, as a dashed line. The graph of the function should be symmetrical about this line.

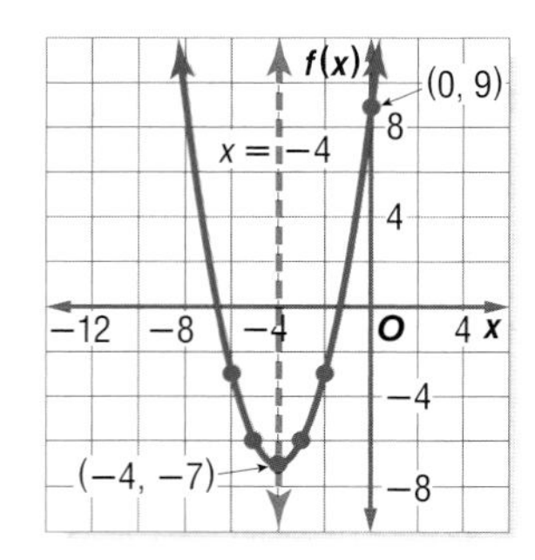

MAXIMUM AND MINIMUM VALUES The y-coordinate of the vertex of a quadratic function is the **maximum value** or **minimum value** obtained by the function.

Key Concept — *Maximum and Minimum Value*

- **Words** The graph of $f(x) = ax^2 + bx + c$, where $a \neq 0$,
  - opens up and has a minimum value when $a > 0$, and
  - opens down and has a maximum value when $a < 0$.
- **Models**

***a* is positive.**

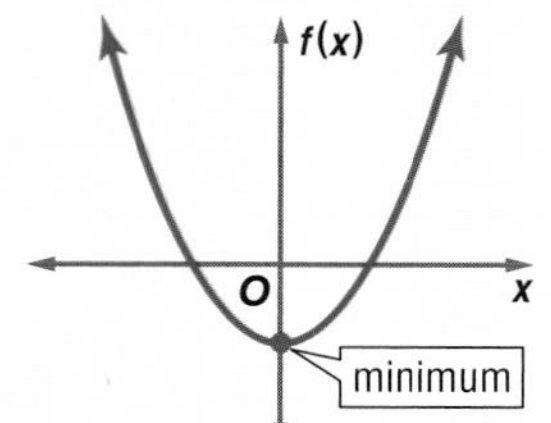

***a* is negative.**

f(x)

maximum

O

x

Example 3 *Maximum or Minimum Value*

Consider the function $f(x) = x^2 - 4x + 9$.

a. Determine whether the function has a maximum or a minimum value.

For this function, $a = 1$, $b = -4$, and $c = 9$. Since $a > 0$, the graph opens up and the function has a minimum value.

Study Tip

Common Misconception
The terms minimum *point* and minimum *value* are not interchangeable. The minimum point on the graph of a quadratic function is the set of coordinates that describe the location of the vertex. The minimum value of a function is the *y*-coordinate of the minimum point. It is the smallest value obtained when $f(x)$ is evaluated for all values of x.

b. State the maximum or minimum value of the function.

The minimum value of the function is the y-coordinate of the vertex.

The x-coordinate of the vertex is $-\frac{-4}{2(1)}$ or 2.

Find the y-coordinate of the vertex by evaluating the function for $x = 2$.

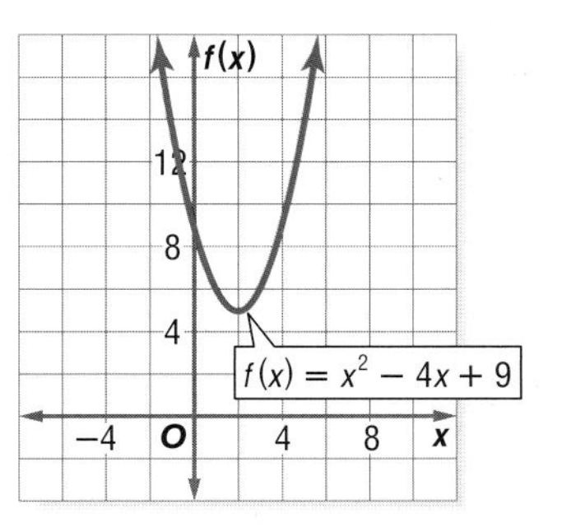

$f(x) = x^2 - 4x + 9$ Original function

$f(2) = (2)^2 - 4(2) + 9$ or 5 $x = 2$

Therefore, the minimum value of the function is 5.

When quadratic functions are used to model real-world situations, their maximum or minimum values can have real-world meaning.

Example 4 Find a Maximum Value

FUND-RAISING Four hundred people came to last year's winter play at Sunnybrook High School. The ticket price was $5. This year, the Drama Club is hoping to earn enough money to take a trip to a Broadway play. They estimate that for each $0.50 increase in the price, 10 fewer people will attend their play.

a. How much should the tickets cost in order to maximize the income from this year's play?

Words The income is the number of tickets multiplied by the price per ticket.

Variables Let x = the number of $0.50 price increases.
Then $5 + 0.50x$ = the price per ticket and
$400 - 10x$ = the number of tickets sold.

Let $I(x)$ = income as a function of x.

Equation

The income	is	the number of tickets	multiplied by	the price per ticket.
$I(x)$	$=$	$(400 - 10x)$	$\cdot$	$(5 + 0.50x)$

$= 400(5) + 400(0.50x) - 10x(5) - 10x(0.50x)$

$= 2000 + 200x - 50x - 5x^2$ Multiply.

$= 2000 + 150x - 5x^2$ Simplify.

$= -5x^2 + 150x + 2000$ Rewrite in $ax^2 + bx + c$ form.

$I(x)$ is a quadratic function with $a = -5$, $b = 150$, and $c = 2000$. Since $a < 0$, the function has a maximum value at the vertex of the graph. Use the formula to find the x-coordinate of the vertex.

x-coordinate of the vertex $= -\frac{b}{2a}$ Formula for the x-coordinate of the vertex

$= -\frac{150}{2(-5)}$ $a = -5, b = 150$

$= 15$ Simplify.

This means the Drama Club should make 15 price increases of $0.50 to maximize their income. Thus, the ticket price should be $5 + 0.50(15)$ or $12.50.

(continued on the next page)

b. What is the maximum income the Drama Club can expect to make?

To determine maximum income, find the maximum value of the function by evaluating $I(x)$ for $x = 15$.

$I(x) = -5x^2 + 150x + 2000$ Income function

$I(15) = -5(15)^2 + 150(15) + 2000$ $x = 15$

$= 3125$ Use a calculator.

Thus, the maximum income the Drama Club can expect is $3125.

CHECK Graph this function on a graphing calculator, and use the CALC menu to confirm this solution.

KEYSTROKES: [2nd] [CALC] 4

0 [ENTER] 25 [ENTER] [ENTER]

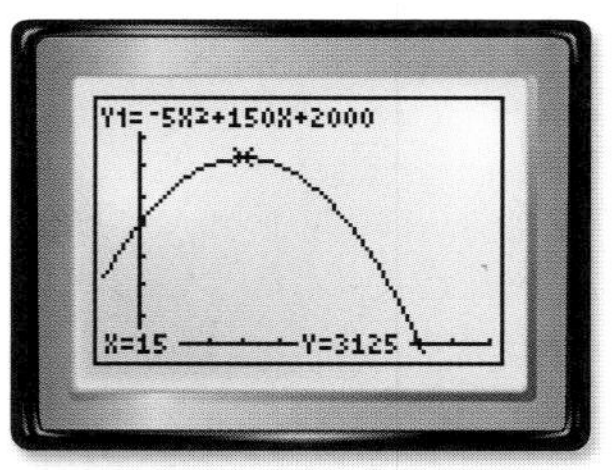

[−5, 50] scl: 5 by [−100, 4000] scl: 500

At the bottom of the display are the coordinates of the maximum point on the graph of $y = -5x^2 + 150x + 2000$. The y value of these coordinates is the maximum value of the function, or 3125. ✓

Check for Understanding

Concept Check

1. **OPEN ENDED** Give an example of a quadratic function. Identify its quadratic term, linear term, and constant term.

2. **Identify** the vertex and the equation of the axis of symmetry for each function graphed below.

a.

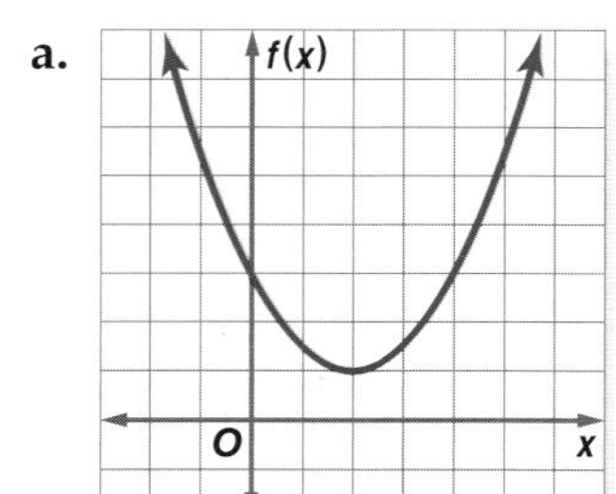

b.

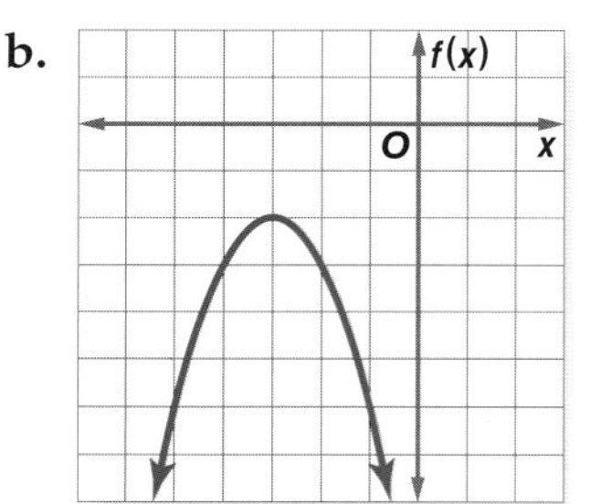

3. **State** whether the graph of each quadratic function opens *up* or *down*. Then state whether the function has a *maximum* or *minimum* value.

a. $f(x) = 3x^2 + 4x - 5$

b. $f(x) = -2x^2 + 9$

c. $f(x) = -5x^2 - 8x + 2$

d. $f(x) = 6x^2 - 5x$

Guided Practice

Complete parts a–c for each quadratic function.

a. Find the y-intercept, the equation of the axis of symmetry, and the x-coordinate of the vertex.

b. Make a table of values that includes the vertex.

c. Use this information to graph the function.

4. $f(x) = -4x^2$

5. $f(x) = x^2 + 2x$

6. $f(x) = -x^2 + 4x - 1$

7. $f(x) = x^2 + 8x + 3$

8. $f(x) = 2x^2 - 4x + 1$

9. $f(x) = 3x^2 + 10x$

Determine whether each function has a maximum or a minimum value. Then find the maximum or minimum value of each function.

10. $f(x) = -x^2 + 7$

11. $f(x) = x^2 - x - 6$

12. $f(x) = 4x^2 + 12x + 9$

Application

13. NEWSPAPERS Due to increased production costs, the Daily News must increase its subscription rate. According to a recent survey, the number of subscriptions will decrease by about 1250 for each 25¢ increase in the subscription rate. What weekly subscription rate will maximize the newspaper's income from subscriptions?

Daily News

Subscription Rate $7.50/wk

Current Circulation 50,000

Practice and Apply

Homework Help

For Exercises	See Examples
14–19	1
20–31	2
32–43, 54	3
44–53	4

Extra Practice
See page 839.

Complete parts a–c for each quadratic function.

a. **Find the y-intercept, the equation of the axis of symmetry, and the x-coordinate of the vertex.**

b. **Make a table of values that includes the vertex.**

c. **Use this information to graph the function.**

14. $f(x) = 2x^2$

15. $f(x) = -5x^2$

16. $f(x) = x^2 + 4$

17. $f(x) = x^2 - 9$

18. $f(x) = 2x^2 - 4$

19. $f(x) = 3x^2 + 1$

20. $f(x) = x^2 - 4x + 4$

21. $f(x) = x^2 - 9x + 9$

22. $f(x) = x^2 - 4x - 5$

23. $f(x) = x^2 + 12x + 36$

24. $f(x) = 3x^2 + 6x - 1$

25. $f(x) = -2x^2 + 8x - 3$

26. $f(x) = -3x^2 - 4x$

27. $f(x) = 2x^2 + 5x$

28. $f(x) = 0.5x^2 - 1$

29. $f(x) = -0.25x^2 - 3x$

30. $f(x) = \frac{1}{2}x^2 + 3x + \frac{9}{2}$

31. $f(x) = x^2 - \frac{2}{3}x - \frac{8}{9}$

Determine whether each function has a maximum or a minimum value. Then find the maximum or minimum value of each function.

32. $f(x) = 3x^2$

33. $f(x) = -x^2 - 9$

34. $f(x) = x^2 - 8x + 2$

35. $f(x) = x^2 + 6x - 2$

36. $f(x) = 4x - x^2 + 1$

37. $f(x) = 3 - x^2 - 6x$

38. $f(x) = 2x + 2x^2 + 5$

39. $f(x) = x - 2x^2 - 1$

40. $f(x) = -7 - 3x^2 + 12x$

41. $f(x) = -20x + 5x^2 + 9$

42. $f(x) = -\frac{1}{2}x^2 - 2x + 3$

43. $f(x) = \frac{3}{4}x^2 - 5x - 2$

More About. . .

Architecture

The Exchange House in London, England, is supported by two interior and two exterior steel arches. V-shaped braces add stability to the structure.

Source: Council on Tall Buildings and Urban Habitat

ARCHITECTURE **For Exercises 44 and 45, use the following information.**
The shape of each arch supporting the Exchange House can be modeled by $h(x) = -0.025x^2 + 2x$, where $h(x)$ represents the height of the arch and x represents the horizontal distance from one end of the base in meters.

44. Write the equation of the axis of symmetry, and find the coordinates of the vertex of the graph of $h(x)$.

45. According to this model, what is the maximum height of the arch?

PHYSICS For Exercises 46 and 47, use the following information.
An object is fired straight up from the top of a 200-foot tower at a velocity of 80 feet per second. The height $h(t)$ of the object t seconds after firing is given by $h(t) = -16t^2 + 80t + 200$.

46. Find the maximum height reached by the object and the time that the height is reached.

47. Interpret the meaning of the y-intercept in the context of this problem.

CONSTRUCTION For Exercises 48–50, use the following information.
Steve has 120 feet of fence to make a rectangular kennel for his dogs. He will use his house as one side.

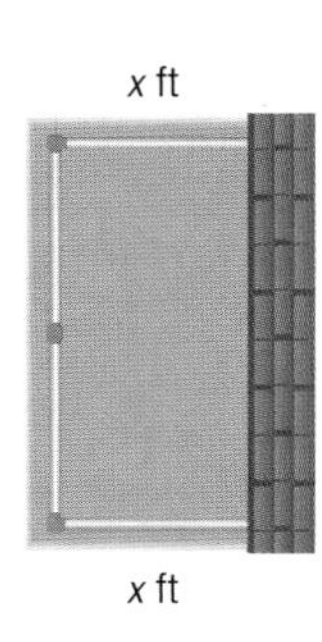

48. Write an algebraic expression for the kennel's length.

49. What dimensions produce a kennel with the greatest area?

50. Find the maximum area of the kennel.

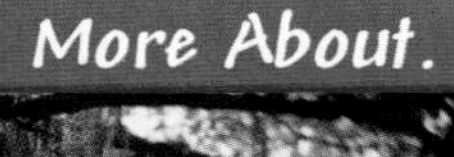

Tourism
Known as the Hostess City of the South, Savannah, Georgia, is a popular tourist destination. One of the first planned cities in the Americas, Savannah's Historic District is based on a grid-like pattern of streets and alleys surrounding open spaces called squares.
Source: savannah-online.com

TOURISM For Exercises 51 and 52, use the following information.
A tour bus in the historic district of Savannah, Georgia, serves 300 customers a day. The charge is $8 per person. The owner estimates that the company would lose 20 passengers a day for each $1 fare increase.

51. What charge would give the most income for the company?

52. If the company raised their fare to this price, how much daily income should they expect to bring in?

53. **GEOMETRY** A rectangle is inscribed in an isosceles triangle as shown. Find the dimensions of the inscribed rectangle with maximum area. (*Hint:* Use similar triangles.)

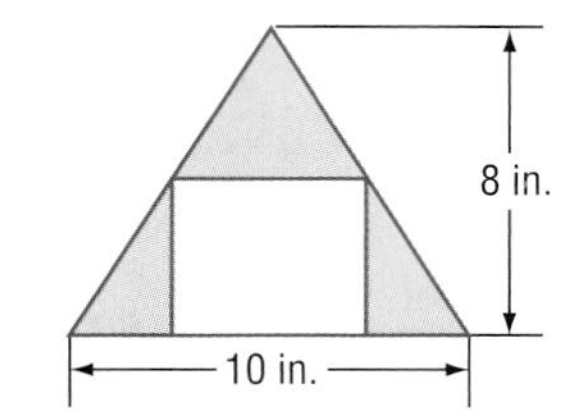

54. **CRITICAL THINKING** Write an expression for the minimum value of a function of the form $y = ax^2 + c$, where $a > 0$. Explain your reasoning. Then use this function to find the minimum value of $y = 8.6x^2 - 12.5$.

55. **WRITING IN MATH** Answer the question that was posed at the beginning of the lesson.

How can income from a rock concert be maximized?

Include the following in your answer:

- an explanation of why income increases and then declines as the ticket price increases, and
- an explanation of how to algebraically and graphically determine what ticket price should be charged to achieve maximum income.

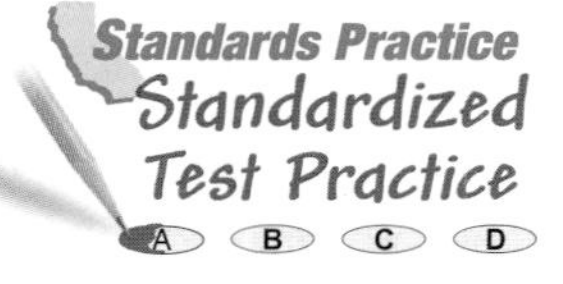

56. The graph of which of the following equations is symmetrical about the y-axis?

   (A) $y = x^2 + 3x - 1$ (B) $y = -x^2 + x$
   (C) $y = 6x^2 + 9$ (D) $y = 3x^2 - 3x + 1$

57. Which of the following tables represents a quadratic relationship between the two variables x and y?

(A)

x	1	2	3	4	5
y	3	3	3	3	3

(B)

x	1	2	3	4	5
y	5	4	3	2	1

(C)

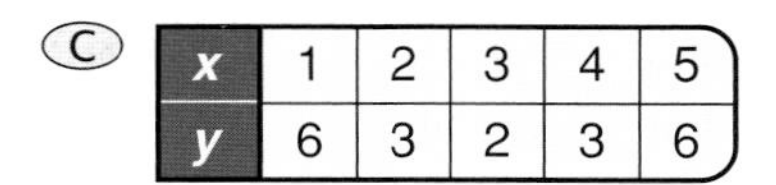

x	1	2	3	4	5
y	6	3	2	3	6

(D)

x	1	2	3	4	5
y	−4	−3	−4	−3	−4

Graphing Calculator

MAXIMA AND MINIMA You can use the MINIMUM or MAXIMUM feature on a graphing calculator to find the minimum or maximum value of a quadratic function. This involves defining an interval that includes the vertex of the parabola. A lower bound is an x value left of the vertex, and an upper bound is an x value right of the vertex.

Step 1 Graph the function so that the vertex of the parabola is visible.

Step 2 Select 3:minimum or 4:maximum from the CALC menu.

Step 3 Using the arrow keys, locate a left bound and press ENTER.

Step 4 Locate a right bound and press ENTER twice. The cursor appears on the maximum or minimum point of the function. The maximum or minimum value is the y-coordinate of that point.

Find the maximum or minimum value of each quadratic function to the nearest hundredth.

58. $f(x) = 3x^2 - 7x + 2$
59. $f(x) = -5x^2 + 8x$
60. $f(x) = 2x^2 - 3x + 2$
61. $f(x) = -6x^2 + 9x$
62. $f(x) = 7x^2 + 4x + 1$
63. $f(x) = -4x^2 + 5x$

Maintain Your Skills

Mixed Review

Simplify. *(Lesson 5-9)*

64. i^{14}
65. $(4 - 3i) - (5 - 6i)$
66. $(7 + 2i)(1 - i)$

Solve each equation. *(Lesson 5-8)*

67. $5 - \sqrt{b + 2} = 0$
68. $\sqrt[3]{x + 5} + 6 = 4$
69. $\sqrt{n + 12} - \sqrt{n} = 2$

Perform the indicated operations. *(Lesson 4-2)*

70. $[4 \quad 1 \quad -3] + [6 \quad -5 \quad 8]$
71. $[2 \quad -5 \quad 7] - [-3 \quad 8 \quad -1]$
72. $4\begin{bmatrix} -7 & 5 & -11 \\ 2 & -4 & 9 \end{bmatrix}$
73. $-2\begin{bmatrix} -3 & 0 & 12 \\ -7 & \frac{1}{3} & 4 \end{bmatrix}$

74. Graph the system of equations $y = -3x$ and $y - x = 4$. State the solution. Is the system of equations *consistent* and *independent*, *consistent* and *dependent*, or *inconsistent*? *(Lesson 3-1)*

Getting Ready for the Next Lesson

PREREQUISITE SKILL Evaluate each function for the given value.
*(To review **evaluating functions**, see Lesson 2-1.)*

75. $f(x) = x^2 + 2x - 3, x = 2$
76. $f(x) = -x^2 - 4x + 5, x = -3$
77. $f(x) = 3x^2 + 7x, x = -2$
78. $f(x) = \frac{2}{3}x^2 + 2x - 1, x = -3$

6-2 Solving Quadratic Equations by Graphing

California Standards Standard 10.0 Students graph quadratic functions and determine the maxima, minima, and zeros of the function. (Key)

What You'll Learn

- Solve quadratic equations by graphing.
- Estimate solutions of quadratic equations by graphing.

Vocabulary

- quadratic equation
- root
- zero

How does a quadratic function model a free-fall ride?

As you speed to the top of a free-fall ride, you are pressed against your seat so that you feel like you're being pushed downward. Then as you free-fall, you fall at the same rate as your seat. Without the force of your seat pressing on you, you *feel* weightless. The height above the ground (in feet) of an object in free-fall can be determined by the quadratic function $h(t) = -16t^2 + h_0$, where t is the time in seconds and the initial height is h_0 feet.

SOLVE QUADRATIC EQUATIONS When a quadratic function is set equal to a value, the result is a quadratic equation. A **quadratic equation** can be written in the form $ax^2 + bx + c = 0$, where $a \neq 0$.

Study Tip

Reading Math
In general, equations have roots, functions have zeros, and graphs of functions have x-intercepts.

The solutions of a quadratic equation are called the **roots** of the equation. One method for finding the roots of a quadratic equation is to find the **zeros** of the related quadratic function. The zeros of the function are the x-intercepts of its graph. These are the solutions of the related equation because $f(x) = 0$ at those points. The zeros of the function graphed at the right are 1 and 3.

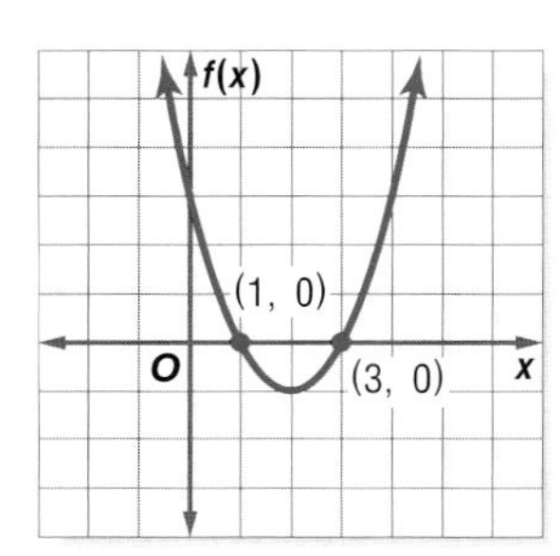

Example 1 Two Real Solutions

Solve $x^2 + 6x + 8 = 0$ by graphing.

Graph the related quadratic function $f(x) = x^2 + 6x + 8$. The equation of the axis of symmetry is $x = -\frac{6}{2(1)}$ or -3. Make a table using x values around -3. Then, graph each point.

x	-5	-4	-3	-2	-1
$f(x)$	3	0	-1	0	3

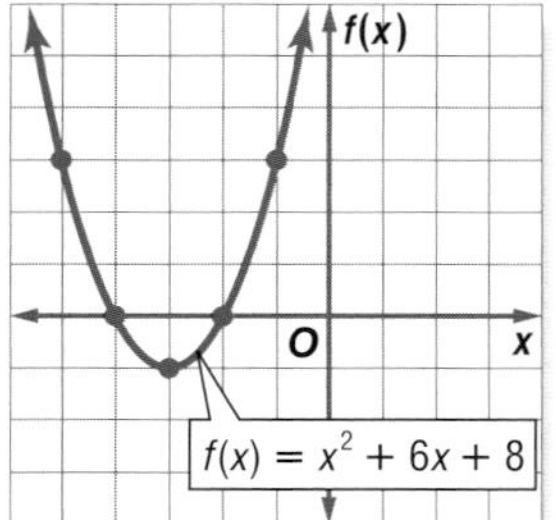

From the table and the graph, we can see that the zeros of the function are -4 and -2. Therefore, the solutions of the equation are -4 and -2.

CHECK Check the solutions by substituting each solution into the equation to see if it is satisfied.

$$x^2 + 6x + 8 = 0 \qquad\qquad x^2 + 6x + 8 = 0$$
$$(-4)^2 + 6(-4) + 8 \stackrel{?}{=} 0 \qquad\qquad (-2)^2 + 6(-2) + 8 \stackrel{?}{=} 0$$
$$0 = 0 \checkmark \qquad\qquad 0 = 0 \checkmark$$

The graph of the related function in Example 1 had two zeros; therefore, the quadratic equation had two real solutions. This is one of the three possible outcomes when solving a quadratic equation.

Key Concept — Solutions of a Quadratic Equation

- **Words** A quadratic equation can have one real solution, two real solutions, or no real solution.
- **Models**

One Real Solution	Two Real Solutions	No Real Solution
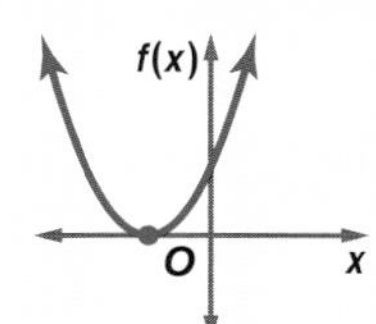	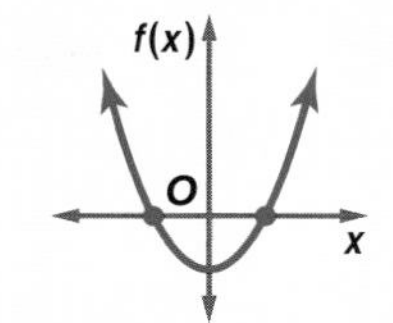	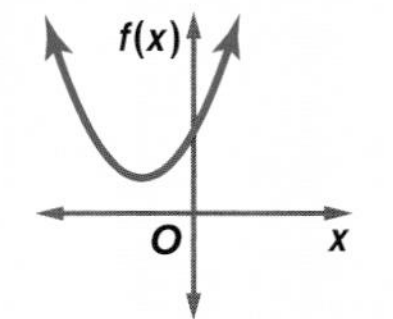

Study Tip

One Real Solution
When a quadratic equation has one real solution, it really has two solutions that are the same number.

Example 2 One Real Solution

Solve $8x - x^2 = 16$ by graphing.

Write the equation in $ax^2 + bx + c = 0$ form.

$8x - x^2 = 16 \rightarrow -x^2 + 8x - 16 = 0$ Subtract 16 from each side.

Graph the related quadratic function $f(x) = -x^2 + 8x - 16$.

x	2	3	4	5	6
$f(x)$	−4	−1	0	−1	−4

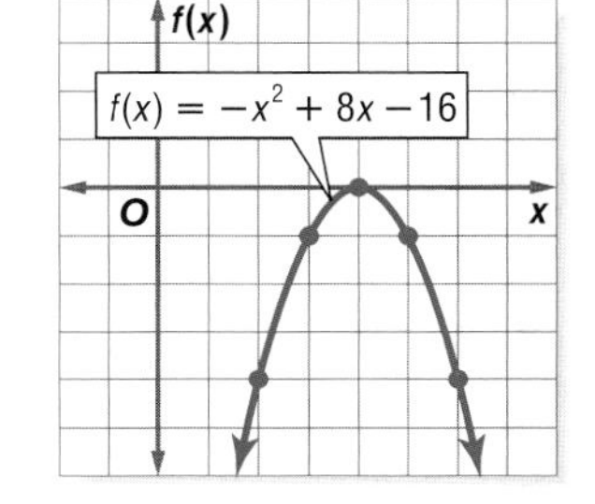

Notice that the graph has only one x-intercept, 4. Thus, the equation's only solution is 4.

Example 3 No Real Solution

NUMBER THEORY **Find two real numbers whose sum is 6 and whose product is 10 or show that no such numbers exist.**

Explore Let x = one of the numbers. Then $6 - x$ = the other number.

Plan Since the product of the two numbers is 10, you know that $x(6 - x) = 10$.

$x(6 - x) = 10$ Original equation

$6x - x^2 = 10$ Distributive Property

$-x^2 + 6x - 10 = 0$ Subtract 10 from each side.

Solve You can solve $-x^2 + 6x - 10 = 0$ by graphing the related function $f(x) = -x^2 + 6x - 10$.

x	1	2	3	4	5
$f(x)$	−5	−2	−1	−2	−5

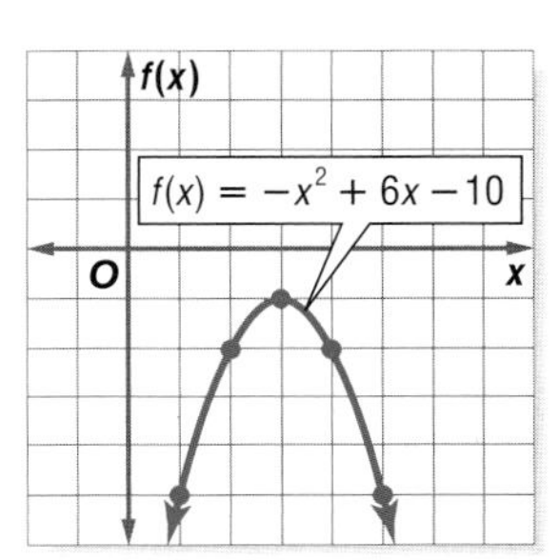

Notice that the graph has no x-intercepts. This means that the original equation has no real solution. Thus, it is *not* possible for two numbers to have a sum of 6 and a product of 10.

Examine Try finding the product of several pairs of numbers whose sum is 6. Is the product of each pair less than 10 as the graph suggests?

ESTIMATE SOLUTIONS Often exact roots cannot be found by graphing. In this case, you can estimate solutions by stating the consecutive integers between which the roots are located.

Example 4 Estimate Roots

Solve $-x^2 + 4x - 1 = 0$ by graphing. If exact roots cannot be found, state the consecutive integers between which the roots are located.

The equation of the axis of symmetry of the related function is $x = -\frac{4}{2(-1)}$ or 2.

x	0	1	2	3	4
$f(x)$	−1	2	3	2	−1

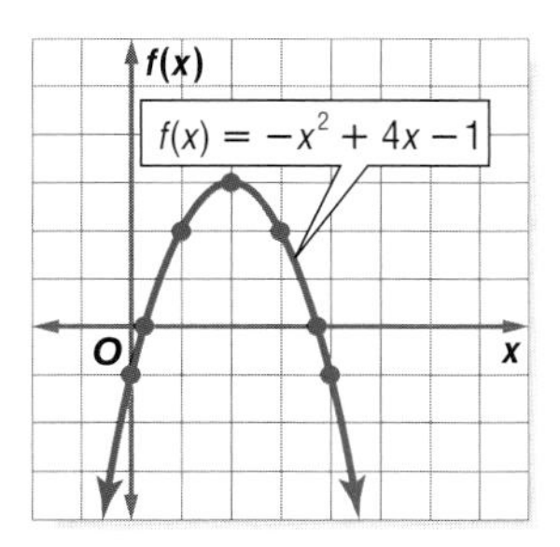

The x-intercepts of the graph are between 0 and 1 and between 3 and 4. So, one solution is between 0 and 1, and the other is between 3 and 4.

Study Tip

Location of Roots
Notice in the table of values that the value of the function changes from negative to positive between the x values of 0 and 1, and 3 and 4.

For many applications, an exact answer is not required, and approximate solutions are adequate. Another way to estimate the solutions of a quadratic equation is by using a graphing calculator.

Example 5 Write and Solve an Equation

EXTREME SPORTS On March 12, 1999, Adrian Nicholas broke the world record for the longest human flight. He flew 10 miles from his drop point in 4 minutes 55 seconds using a specially designed, aerodynamic suit. Using the information at the right and ignoring air resistance, how long would Mr. Nicholas have been in free-fall had he not used this special suit? Use the formula $h(t) = -16t^2 + h_0$, where the time t is in seconds and the initial height h_0 is in feet.

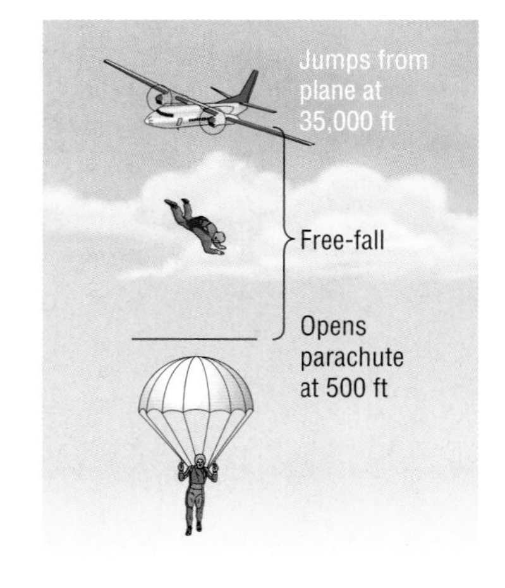

We need to find t when $h_0 = 35{,}000$ and $h(t) = 500$. Solve $500 = -16t^2 + 35{,}000$.

$500 = -16t^2 + 35{,}000$ Original equation

$0 = -16t^2 + 34{,}500$ Subtract 500 from each side.

Graph the related function $y = -16t^2 + 34{,}500$ using a graphing calculator. Adjust your window so that the x-intercepts of the graph are visible.

Use the ZERO feature, 2nd [CALC], to find the positive zero of the function, since time cannot be negative. Use the arrow keys to locate a left bound for the zero and press ENTER.

Then, locate a right bound and press ENTER twice. The positive zero of the function is approximately 46.4. Mr. Nicholas would have been in free-fall for about 46 seconds.

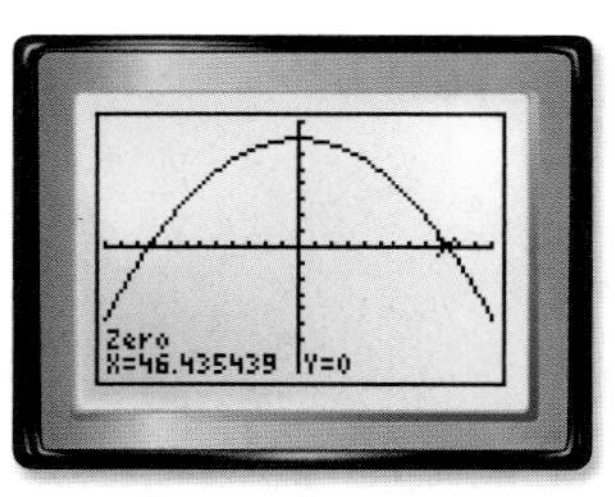

[−60, 60] scl: 5 by [−40000, 40000] scl: 5000

Check for Understanding

Concept Check

1. **Define** each term and explain how they are related.
   a. solution b. root c. zero of a function d. x-intercept
2. **OPEN ENDED** Give an example of a quadratic function and state its related quadratic equation.
3. **Explain** how you can estimate the solutions of a quadratic equation by examining the graph of its related function.

Guided Practice

Use the related graph of each equation to determine its solutions.

4. $x^2 + 3x - 4 = 0$

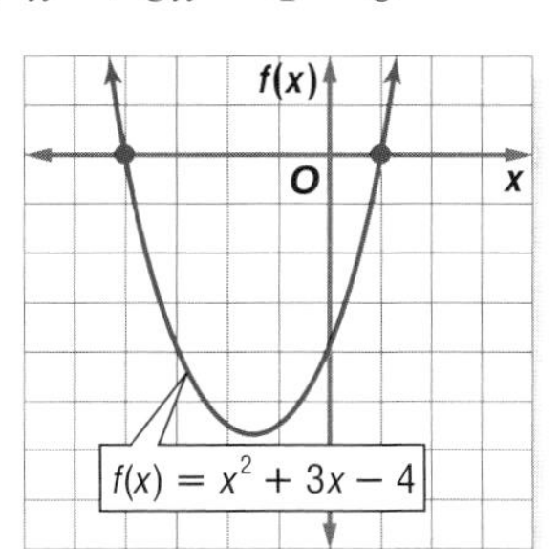

5. $2x^2 + 2x - 4 = 0$

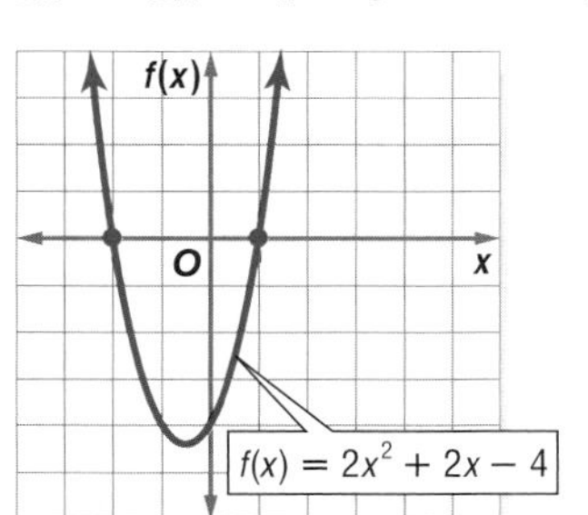

6. $x^2 + 8x + 16 = 0$

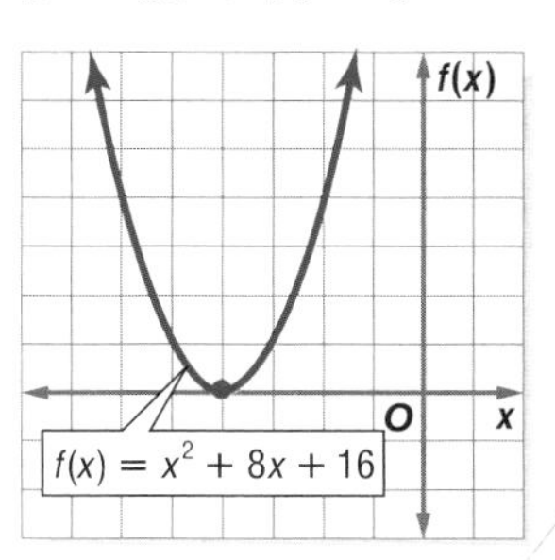

Solve each equation by graphing. If exact roots cannot be found, state the consecutive integers between which the roots are located.

7. $-x^2 - 7x = 0$
8. $x^2 - 2x - 24 = 0$
9. $x^2 + 3x = 28$
10. $25 + x^2 + 10x = 0$
11. $4x^2 - 7x - 15 = 0$
12. $2x^2 - 2x - 3 = 0$

Application

13. **NUMBER THEORY** Use a quadratic equation to find two real numbers whose sum is 5 and whose product is -14, or show that no such numbers exist.

Practice and Apply

Homework Help	
For Exercises	See Examples
14–19	1–3
20–37	1–4
38–41	3
42–46	5

Extra Practice
See page 840.

Use the related graph of each equation to determine its solutions.

14. $x^2 - 6x = 0$

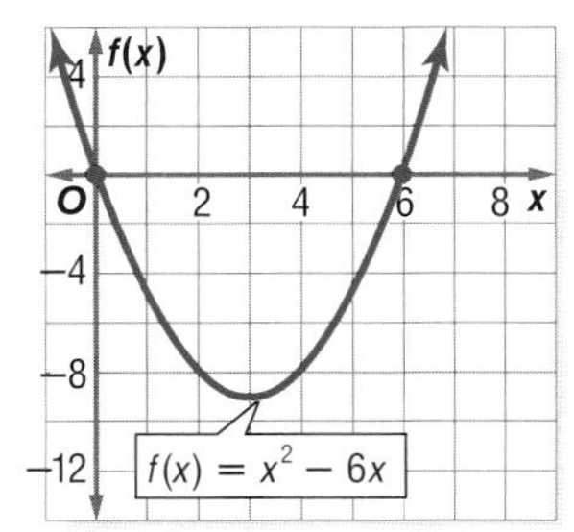

15. $x^2 - 6x + 9 = 0$

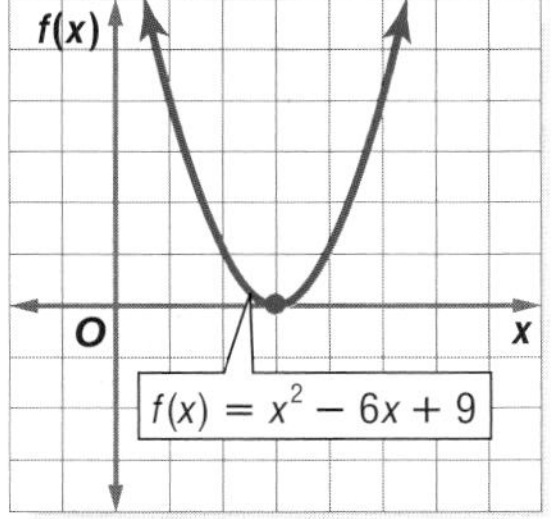

16. $-2x^2 - x + 6 = 0$

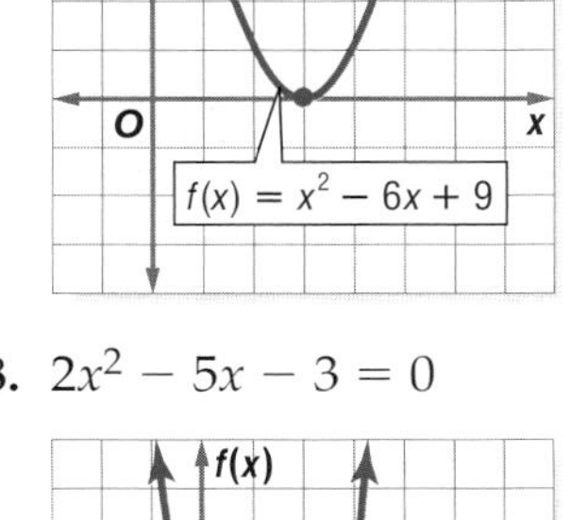

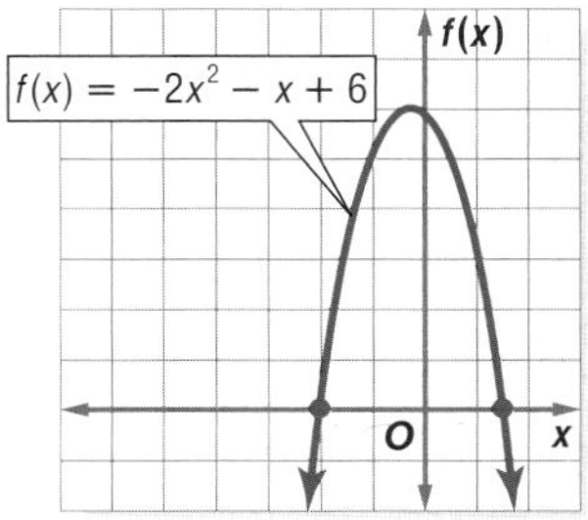

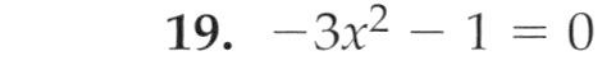

17. $-0.5x^2 = 0$

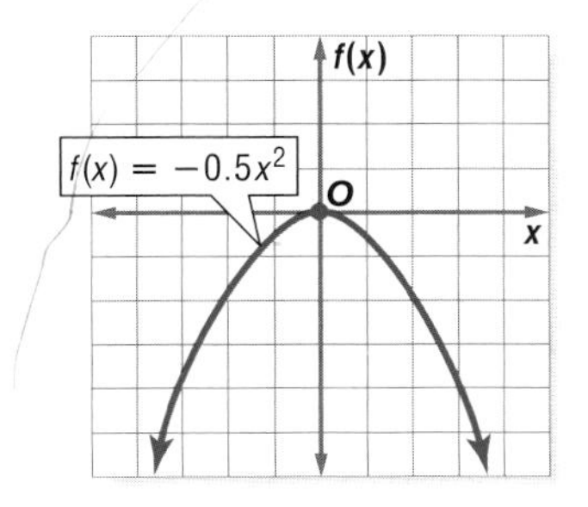

18. $2x^2 - 5x - 3 = 0$

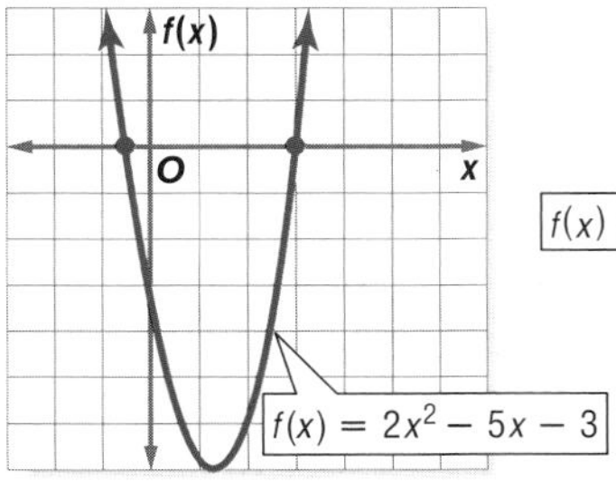

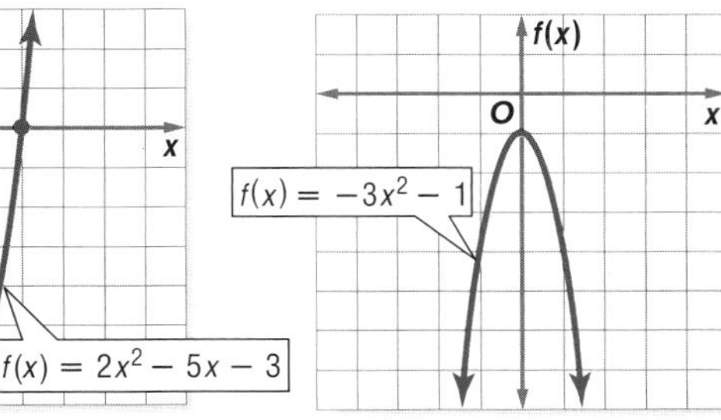

19. $-3x^2 - 1 = 0$

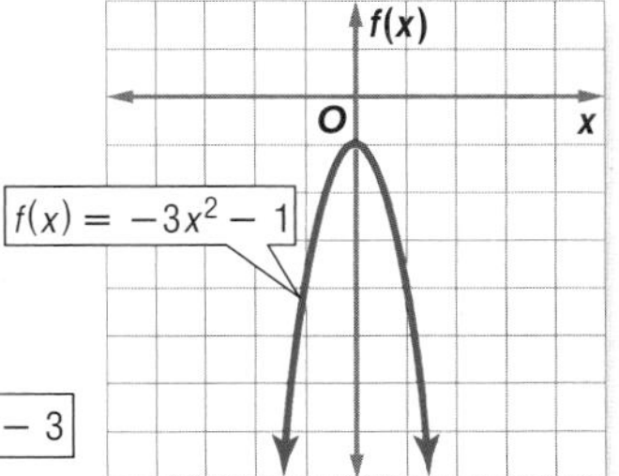

Solve each equation by graphing. If exact roots cannot be found, state the consecutive integers between which the roots are located.

20. $x^2 - 3x = 0$
21. $-x^2 + 4x = 0$
22. $x^2 + 4x - 4 = 0$
23. $x^2 - 2x - 1 = 0$
24. $-x^2 + x = -20$
25. $x^2 - 9x = -18$
26. $14x + x^2 + 49 = 0$
27. $-12x + x^2 = -36$
28. $2x^2 - 3x = 9$
29. $4x^2 - 8x = 5$
30. $2x^2 = -5x + 12$
31. $2x^2 = x + 15$
32. $x^2 + 3x - 2 = 0$
33. $x^2 - 4x + 2 = 0$
34. $-2x^2 + 3x + 3 = 0$
35. $0.5x^2 - 3 = 0$
36. $x^2 + 2x + 5 = 0$
37. $-x^2 + 4x - 6 = 0$

NUMBER THEORY Use a quadratic equation to find two real numbers that satisfy each situation, or show that no such numbers exist.

38. Their sum is −17, and their product is 72.
39. Their sum is 7, and their product is 14.
40. Their sum is −9, and their product is 24.
41. Their sum is 12, and their product is −28.

For Exercises 42–44, use the formula $h(t) = v_0t - 16t^2$ where $h(t)$ is the height of an object in feet, v_0 is the object's initial velocity in feet per second, and t is the time in seconds.

42. **ARCHERY** An arrow is shot upward with a velocity of 64 feet per second. Ignoring the height of the archer, how long after the arrow is released does it hit the ground?

43. **TENNIS** A tennis ball is hit upward with a velocity of 48 feet per second. Ignoring the height of the tennis player, how long does it take for the ball to fall to the ground?

44. **BOATING** A boat in distress launches a flare straight up with a velocity of 190 feet per second. Ignoring the height of the boat, how many seconds will it take for the flare to hit the water?

45. **LAW ENFORCEMENT** Police officers can use the length of skid marks to help determine the speed of a vehicle before the brakes were applied. If the skid marks are on dry concrete, the formula $\frac{s^2}{24} = d$ can be used. In the formula, s represents the speed in miles per hour, and d represents the length of the skid marks in feet. If the length of the skid marks on dry concrete are 50 feet, how fast was the car traveling?

46. **EMPIRE STATE BUILDING** Suppose you could conduct an experiment by dropping a small object from the Observatory of the Empire State Building. How long would it take for the object to reach the ground, assuming there is no air resistance? Use the information at the left and the formula $h(t) = -16t^2 + h_0$, where t is the time in seconds and the initial height h_0 is in feet.

47. **CRITICAL THINKING** A quadratic function has values $f(-4) = -11$, $f(-2) = 9$, and $f(0) = 5$. Between which two x values must $f(x)$ have a zero? Explain your reasoning.

More About. . .

Empire State Building

Located on the 86th floor, 1050 feet (320 meters) above the streets of New York City, the Observatory offers panoramic views from within a glass-enclosed pavilion and from the surrounding open-air promenade.

Source: www.esbnyc.com

48. **WRITING IN MATH** Answer the question that was posed at the beginning of the lesson.

How does a quadratic function model a free-fall ride?

Include the following in your answer:

- a graph showing the height at any given time of a free-fall ride that lifts riders to a height of 185 feet, and
- an explanation of how to use this graph to estimate how long the riders would be in free-fall if the ride were allowed to hit the ground before stopping.

Standards Practice

Standardized Test Practice

49. If one of the roots of the equation $x^2 + kx - 12 = 0$ is 4, what is the value of k?

Ⓐ -1 Ⓑ 0 Ⓒ 1 Ⓓ 3

50. For what value of x does $f(x) = x^2 + 5x + 6$ reach its minimum value?

Ⓐ -3 Ⓑ $-\frac{5}{2}$ Ⓒ -2 Ⓓ -5

Extending the Lesson

SOLVE ABSOLUTE VALUE EQUATIONS BY GRAPHING Similar to quadratic equations, you can solve absolute value equations by graphing. Graph the related absolute value function for each equation using a graphing calculator. Then use the ZERO feature, 2nd [CALC], to find its real solutions, if any, rounded to the nearest hundredth.

51. $|x + 1| = 0$

52. $|x| - 3 = 0$

53. $|x - 4| - 1 = 0$

54. $-|x + 4| + 5 = 0$

55. $2|3x| - 8 = 0$

56. $|2x - 3| + 1 = 0$

Maintain Your Skills

Mixed Review

Find the y-intercept, the equation of the axis of symmetry, and the x-coordinate of the vertex for each quadratic function. Then graph the function by making a table of values. *(Lesson 6-1)*

57. $f(x) = x^2 - 6x + 4$

58. $f(x) = -4x^2 + 8x - 1$

59. $f(x) = \frac{1}{4}x^2 + 3x + 4$

Simplify. *(Lesson 5-9)*

60. $\frac{2i}{3 + i}$

61. $\frac{4}{5 - i}$

62. $\frac{1 + i}{3 - 2i}$

Evaluate the determinant of each matrix. *(Lesson 4-3)*

63. $\begin{bmatrix} 6 & 4 \\ -3 & 2 \end{bmatrix}$

64. $\begin{bmatrix} 2 & -1 & -6 \\ 5 & 0 & 3 \\ -3 & 2 & 11 \end{bmatrix}$

65. $\begin{bmatrix} 6 & 5 & -2 \\ -3 & 0 & 6 \\ 1 & 4 & 2 \end{bmatrix}$

66. **COMMUNITY SERVICE** A drug awareness program is being presented at a theater that seats 300 people. Proceeds will be donated to a local drug information center. If every two adults must bring at least one student, what is the maximum amount of money that can be raised? *(Lesson 3-4)*

Getting Ready for the Next Lesson

PREREQUISITE SKILL Factor completely.
*(To review **factoring trinomials**, see Lesson 5-4.)*

67. $x^2 + 5x$

68. $x^2 - 100$

69. $x^2 - 11x + 28$

70. $x^2 - 18x + 81$

71. $3x^2 + 8x + 4$

72. $6x^2 - 14x - 12$

Graphing Calculator Investigation

A Follow-Up of Lesson 6-2

Modeling Real-World Data

You can use a TI-83 Plus to model data points whose curve of best fit is quadratic.

FALLING WATER Water is allowed to drain from a hole made in a 2-liter bottle. The table shows the level of the water y measured in centimeters from the bottom of the bottle after x seconds. Find and graph a linear regression equation and a quadratic regression equation. Determine which equation is a better fit for the data.

Time (s)	0	20	40	60	80	100	120	140	160	180	200	220
Water level (cm)	42.6	40.7	38.9	37.2	35.8	34.3	33.3	32.3	31.5	30.8	30.4	30.1

Step 1 *Find a linear regression equation.*

- Enter the times in L1 and the water levels in L2. Then find a linear regression equation.

  KEYSTROKES: *Review lists and finding a linear regression equation on page 87.*

- Graph a scatter plot and the regression equation.

  KEYSTROKES: *Review graphing a regression equation on page 87.*

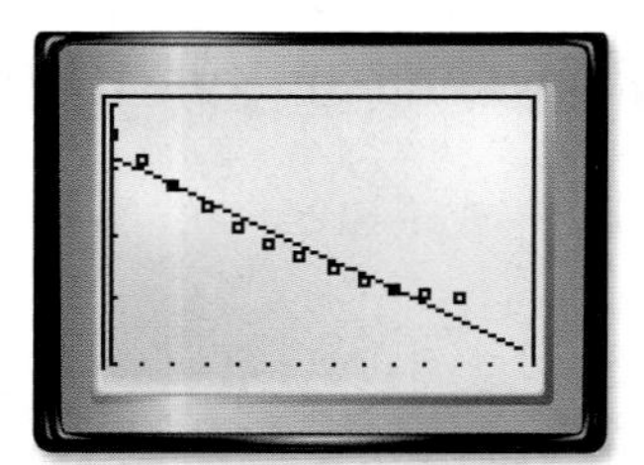

[0, 260] scl: 20 by [25, 45] scl: 5

Step 2 *Find a quadratic regression equation.*

- Find the quadratic regression equation. Then copy the equation to the Y= list and graph.

  KEYSTROKES: STAT ▶ 5 ENTER Y= VARS 5 ▶ ▶ ENTER GRAPH

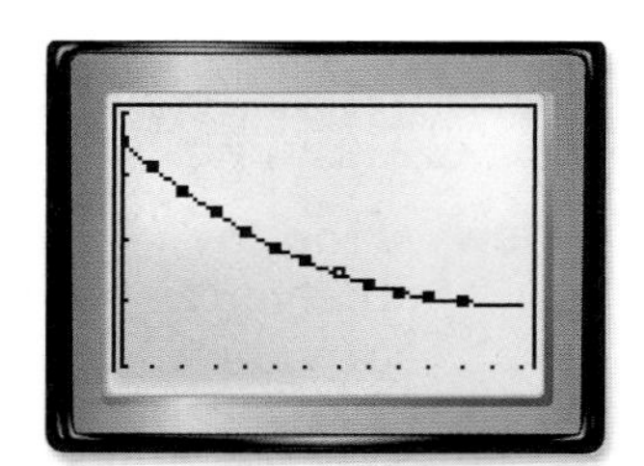

[0, 260] scl: 20 by [25, 45] scl: 5

The graph of the linear regression equation appears to pass through just two data points. However, the graph of the quadratic regression equation fits the data very well.

Exercises

For Exercises 1–4, use the graph of the braking distances for dry pavement.

1. Find and graph a linear regression equation and a quadratic regression equation for the data. Determine which equation is a better fit for the data.
2. Use the CALC menu with each regression equation to estimate the braking distance at speeds of 100 and 150 miles per hour.
3. How do the estimates found in Exercise 2 compare?
4. How might choosing a regression equation that does not fit the data well affect predictions made by using the equation?

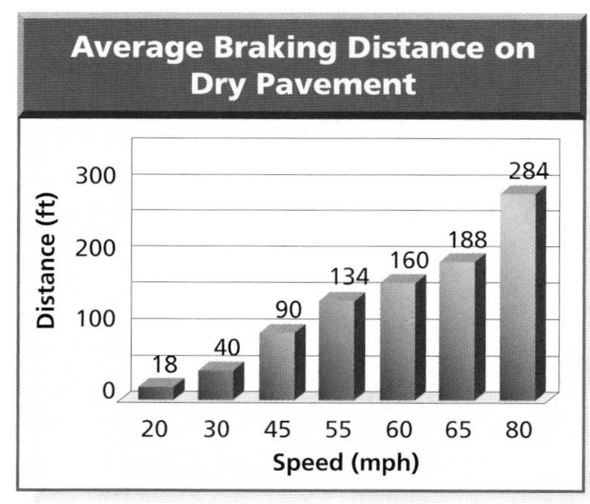

Source: Missouri Department of Revenue

www.algebra2.com/other_calculator_keystrokes

6-3 Solving Quadratic Equations by Factoring

California Standards **Standard 8.0 Students solve and graph quadratic equations by factoring,** completing the square, or using the quadratic formula. **Students apply these techniques in solving word problems.** (Key)

What You'll Learn

- Solve quadratic equations by factoring.
- Write a quadratic equation with given roots.

How is the Zero Product Property used in geometry?

The length of a rectangle is 5 inches more than its width, and the area of the rectangle is 24 square inches. To find the dimensions of the rectangle you need to solve the equation $x(x + 5) = 24$ or $x^2 + 5x = 24$.

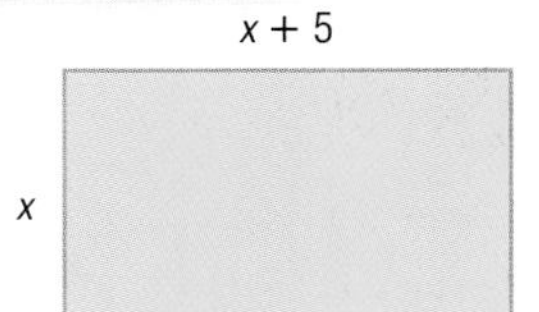

SOLVE EQUATIONS BY FACTORING In the last lesson, you learned to solve a quadratic equation like the one above by graphing. Another way to solve this equation is by factoring. Consider the following products.

$$7(0) = 0 \qquad 0(-2) = 0$$
$$(6 - 6)(0) = 0 \qquad -4(-5 + 5) = 0$$

Notice that in each case, *at least one* of the factors is zero. These examples illustrate the **Zero Product Property**.

Key Concept — *Zero Product Property*

- **Words** For any real numbers a and b, if $ab = 0$, then either $a = 0$, $b = 0$, or both a and b equal zero.
- **Example** If $(x + 5)(x - 7) = 0$, then $x + 5 = 0$ and/or $x - 7 = 0$.

Example 1 *Two Roots*

Solve each equation by factoring.

a. $x^2 = 6x$

$x^2 = 6x$	Original equation
$x^2 - 6x = 0$	Subtract $6x$ from each side.
$x(x - 6) = 0$	Factor the binomial.
$x = 0$ or $x - 6 = 0$	Zero Product Property
$x = 6$	Solve the second equation.

The solution set is $\{0, 6\}$.

CHECK Substitute 0 and 6 for x in the original equation.

$$x^2 = 6x \qquad x^2 = 6x$$
$$(0)^2 \stackrel{?}{=} 6(0) \qquad (6)^2 \stackrel{?}{=} 6(6)$$
$$0 = 0 \checkmark \qquad 36 = 36 \checkmark$$

b. $2x^2 + 7x = 15$

$2x^2 + 7x = 15$	Original equation
$2x^2 + 7x - 15 = 0$	Subtract 15 from each side.
$(2x - 3)(x + 5) = 0$	Factor the trinomial.
$2x - 3 = 0$ or $x + 5 = 0$	Zero Product Property
$2x = 3$ $\quad x = -5$	Solve each equation.
$x = \frac{3}{2}$	

The solution set is $\left\{-5, \frac{3}{2}\right\}$. Check each solution.

Study Tip

Double Roots
The application of the Zero Product Property produced two identical equations, $x - 8 = 0$, both of which have a root of 8. For this reason, 8 is called the *double root* of the equation.

Example 2 Double Root

Solve $x^2 - 16x + 64 = 0$ by factoring.

$x^2 - 16x + 64 = 0$	Original equation
$(x - 8)(x - 8) = 0$	Factor.
$x - 8 = 0$ or $x - 8 = 0$	Zero Product Property
$x = 8$ $\quad x = 8$	Solve each equation.

The solution set is {8}.

CHECK The graph of the related function, $f(x) = x^2 - 16x + 64$, intersects the x-axis only once. Since the zero of the function is 8, the solution of the related equation is 8.

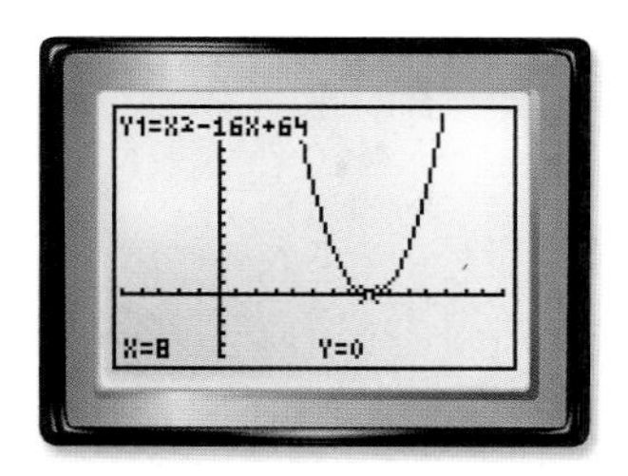

Standards Practice

Standardized Test Practice

A B C D

Example 3 Greatest Common Factor

Multiple-Choice Test Item

What is the positive solution of the equation $3x^2 - 3x - 60 = 0$?

Ⓐ -4 Ⓑ 2 Ⓒ 5 Ⓓ 10

Read the Test Item
You are asked to find the *positive* solution of the given quadratic equation. This implies that the equation also has a solution that is not positive. Since a quadratic equation can either have one, two, or no solutions, we should expect to find two solutions to this equation.

Solve the Test Item
Solve this equation by factoring. But before trying to factor $3x^2 - 3x - 60$ into two binomials, look for a greatest common factor. Notice that each term is divisible by 3.

$3x^2 - 3x - 60 = 0$	Original equation
$3(x^2 - x - 20) = 0$	Factor.
$x^2 - x - 20 = 0$	Divide each side by 3.
$(x + 4)(x - 5) = 0$	Factor.
$x + 4 = 0$ or $x - 5 = 0$	Zero Product Property
$x = -4$ $\quad x = 5$	Solve each equation.

Both solutions, -4 and 5, are listed among the answer choices. Since the question asked for the positive solution, the answer is C.

Test-Taking Tip
Because the problem asked for a *positive* solution, choice A could have been eliminated even before the expression was factored.

WRITE QUADRATIC EQUATIONS You have seen that a quadratic equation of the form $(x - p)(x - q) = 0$ has roots p and q. You can use this pattern to find a quadratic equation for a given pair of roots.

Study Tip

Writing an Equation

The pattern $(x - p)(x - q) = 0$ produces one equation with roots p and q. In fact, there are an infinite number of equations that have these same roots.

Example 4 Write an Equation Given Roots

Write a quadratic equation with $\frac{1}{2}$ and -5 as its roots. Write the equation in the form $ax^2 + bx + c = 0$, where a, b, and c are integers.

$(x - p)(x - q) = 0$ Write the pattern.

$\left(x - \frac{1}{2}\right)[x - (-5)] = 0$ Replace p with $\frac{1}{2}$ and q with -5.

$\left(x - \frac{1}{2}\right)(x + 5) = 0$ Simplify.

$x^2 + \frac{9}{2}x - \frac{5}{2} = 0$ Use FOIL.

$2x^2 + 9x - 5 = 0$ Multiply each side by 2 so that b and c are integers.

A quadratic equation with roots $\frac{1}{2}$ and -5 and integral coefficients is $2x^2 + 9x - 5 = 0$. You can check this result by graphing the related function.

Check for Understanding

Concept Check

1. **Write** the meaning of the Zero Product Property.
2. **OPEN ENDED** Choose two integers. Then, write an equation with those roots in the form $ax^2 + bx + c = 0$, where a, b, and c are integers.
3. **FIND THE ERROR** Lina and Kristin are solving $x^2 + 2x = 8$.

Lina

$x^2 + 2x = 8$
$x(x + 2) = 8$
$x = 8$ or $x + 2 = 8$
$x = 6$

Kristin

$x^2 + 2x = 8$
$x^2 + 2x - 8 = 0$
$(x + 4)(x - 2) = 0$
$x + 4 = 0$ or $x - 2 = 0$
$x = -4$ $x = 2$

Who is correct? Explain your reasoning.

Guided Practice

Solve each equation by factoring.

4. $x^2 - 11x = 0$
5. $x^2 + 6x - 16 = 0$
6. $x^2 = 49$
7. $x^2 + 9 = 6x$
8. $4x^2 - 13x = 12$
9. $5x^2 - 5x - 60 = 0$

Write a quadratic equation with the given roots. Write the equation in the form $ax^2 + bx + c = 0$, where a, b, and c are integers.

10. $-4, 7$
11. $\frac{1}{2}, \frac{4}{3}$
12. $-\frac{3}{5}, -\frac{1}{3}$

Standards Practice

Standardized Test Practice

13. Which of the following is the sum of the solutions of $x^2 - 2x - 8 = 0$?

(A) -6 (B) -4 (C) -2 (D) 2

Practice and Apply

Homework Help

For Exercises	See Examples
14–33, 42–46	1, 2
34–41	4
51–52	3

Extra Practice
See page 840.

Solve each equation by factoring.

14. $x^2 + 5x - 24 = 0$
15. $x^2 - 3x - 28 = 0$
16. $x^2 = 25$
17. $x^2 = 81$
18. $x^2 + 3x = 18$
19. $x^2 - 4x = 21$
20. $3x^2 = 5x$
21. $4x^2 = -3x$
22. $x^2 + 36 = 12x$
23. $x^2 + 64 = 16x$
24. $4x^2 + 7x = 2$
25. $4x^2 - 17x = -4$
26. $4x^2 + 8x = -3$
27. $6x^2 + 6 = -13x$
28. $9x^2 + 30x = -16$
29. $16x^2 - 48x = -27$
30. $-2x^2 + 12x - 16 = 0$
31. $-3x^2 - 6x + 9 = 0$

32. Find the roots of $x(x + 6)(x - 5) = 0$.

33. Solve $x^3 = 9x$ by factoring.

Write a quadratic equation with the given roots. Write the equation in the form $ax^2 + bx + c = 0$, where a, b, and c are integers.

34. 4, 5
35. $-2, 7$
36. $4, -5$
37. $-6, -8$
38. $\frac{1}{2}, 3$
39. $\frac{1}{3}, 5$
40. $-\frac{2}{3}, \frac{3}{4}$
41. $-\frac{3}{2}, -\frac{4}{5}$

42. **DIVING** To avoid hitting any rocks below, a cliff diver jumps up and out. The equation $h = -16t^2 + 4t + 26$ describes her height h in feet t seconds after jumping. Find the time at which she returns to a height of 26 feet.

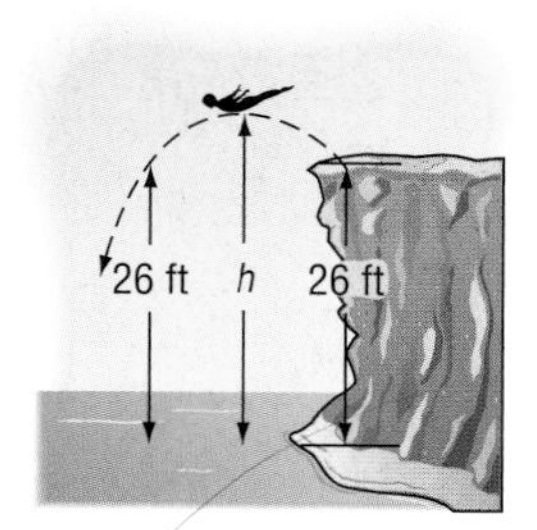

43. **NUMBER THEORY** Find two consecutive even integers whose product is 224.

44. **PHOTOGRAPHY** A rectangular photograph is 8 centimeters wide and 12 centimeters long. The photograph is enlarged by increasing the length and width by an equal amount in order to double its area. What are the dimensions of the new photograph?

More About. . .

Forestry

A board foot is a measure of lumber volume. One piece of lumber 1 foot long by 1 foot wide by 1 inch thick measures one board foot.

Source: www.wood-worker.com

FORESTRY For Exercises 45 and 46, use the following information.
Lumber companies need to be able to estimate the number of board feet that a given log will yield. One of the most commonly used formulas for estimating board feet is the *Doyle Log Rule*, $B = \frac{L}{16}(D^2 - 8D + 16)$, where B is the number of board feet, D is the diameter in inches, and L is the length of the log in feet.

45. Rewrite Doyle's formula for logs that are 16 feet long.

46. Find the root(s) of the quadratic equation you wrote in Exercise 45. What do the root(s) tell you about the kinds of logs for which Doyle's rule makes sense?

47. **CRITICAL THINKING** For a quadratic equation of the form $(x - p)(x - q) = 0$, show that the axis of symmetry of the related quadratic function is located halfway between the x-intercepts p and q.

CRITICAL THINKING Find a value of k that makes each statement true.

48. -3 is a root of $2x^2 + kx - 21 = 0$.

49. $\frac{1}{2}$ is a root of $2x^2 + 11x = -k$.

50. **WRITING IN MATH** Answer the question that was posed at the beginning of the lesson.

How is the Zero Product Property used in geometry?

Include the following in your answer:

- an explanation of how to find the dimensions of the rectangle using the Zero Product Property, and
- why the equation $x(x + 5) = 24$ is not solved by using $x = 24$ and $x + 5 = 24$.

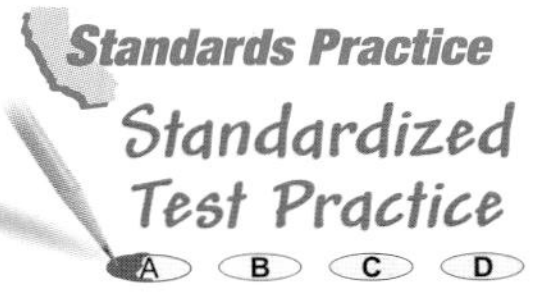

51. Which quadratic equation has roots $\frac{1}{2}$ and $\frac{1}{3}$?

(A) $5x^2 - 5x - 2 = 0$ (B) $5x^2 - 5x + 1 = 0$
(C) $6x^2 + 5x - 1 = 0$ (D) $6x^2 - 5x + 1 = 0$

52. If the roots of a quadratic equation are 6 and -3, what is the equation of the axis of symmetry?

(A) $x = 1$ (B) $x = \frac{3}{2}$ (C) $x = \frac{1}{2}$ (D) $x = -2$

Maintain Your Skills

Mixed Review

Solve each equation by graphing. If exact roots cannot be found, state the consecutive integers between which the roots are located. *(Lesson 6-2)*

53. $f(x) = -x^2 - 4x + 5$ 54. $f(x) = 4x^2 + 4x + 1$ 55. $f(x) = 3x^2 - 10x - 4$

56. Determine whether $f(x) = 3x^2 - 12x - 7$ has a maximum or a minimum value. Then find the maximum or minimum value. *(Lesson 6-1)*

Simplify. *(Lesson 5-6)*

57. $\sqrt{3}\left(\sqrt{6} - 2\right)$ 58. $\sqrt{108} - \sqrt{48} + \left(\sqrt{3}\right)^3$ 59. $\left(5 + \sqrt{8}\right)^2$

Solve each system of equations. *(Lesson 3-2)*

60. $4a - 3b = -4$
$3a - 2b = -4$

61. $2r + s = 1$
$r - s = 8$

62. $3x - 2y = -3$
$3x + y = 3$

Getting Ready for the Next Lesson

PREREQUISITE SKILL Simplify. *(To review* ***simplifying radicals****, see Lesson 5-5.)*

63. $\sqrt{8}$ 64. $\sqrt{20}$ 65. $\sqrt{27}$
66. $\sqrt{-50}$ 67. $\sqrt{-12}$ 68. $\sqrt{-48}$

Practice Quiz 1

Lessons 6-1 through 6-3

1. Find the y-intercept, the equation of the axis of symmetry, and the x-coordinate of the vertex for $f(x) = 3x^2 - 12x + 4$. Then graph the function by making a table of values. *(Lesson 6-1)*

2. Determine whether $f(x) = 3 - x^2 + 5x$ has a maximum or minimum value. Then find this maximum or minimum value. *(Lesson 6-1)*

3. Solve $2x^2 - 11x + 12 = 0$ by graphing. If exact roots cannot be found, state the consecutive integers between which the roots are located. *(Lesson 6-2)*

4. Solve $2x^2 - 5x - 3 = 0$ by factoring. *(Lesson 6-3)*

5. Write a quadratic equation with roots -4 and $\frac{1}{3}$. Write the equation in the form $ax^2 + bx + c = 0$, where a, b, and c are integers. *(Lesson 6-3)*

6-4 Completing the Square

California Standards Standard 8.0 **Students solve and graph quadratic equations by** factoring, **completing the square,** or using the quadratic formula. **Students apply these techniques in solving word problems.** (Key)

What You'll Learn

- Solve quadratic equations by using the Square Root Property.
- Solve quadratic equations by completing the square.

Vocabulary

- completing the square

How can you find the time it takes an accelerating race car to reach the finish line?

Under a yellow caution flag, race car drivers slow to a speed of 60 miles per hour. When the green flag is waved, the drivers can increase their speed.

Suppose the driver of one car is 500 feet from the finish line. If the driver accelerates at a constant rate of 8 feet per second squared, the equation $t^2 + 22t + 121 = 246$ represents the time t it takes the driver to reach this line. To solve this equation, you can use the Square Root Property.

SQUARE ROOT PROPERTY You have solved equations like $x^2 - 25 = 0$ by factoring. You can also use the **Square Root Property** to solve such an equation. This method is useful with equations like the one above that describes the race car's speed. In this case, the quadratic equation contains a perfect square trinomial set equal to a constant.

Study Tip

Reading Math
$\pm\sqrt{n}$ is read *plus or minus the square root of n.*

Key Concept — *Square Root Property*

For any real number n, if $x^2 = n$, then $x = \pm\sqrt{n}$.

Example 1 Equation with Rational Roots

Solve $x^2 + 10x + 25 = 49$ by using the Square Root Property.

$x^2 + 10x + 25 = 49$	Original equation
$(x + 5)^2 = 49$	Factor the perfect square trinomial.
$x + 5 = \pm\sqrt{49}$	Square Root Property
$x + 5 = \pm 7$	$\sqrt{49} = 7$
$x = -5 \pm 7$	Add -5 to each side.
$x = -5 + 7$ or $x = -5 - 7$	Write as two equations.
$x = 2$ $\quad$ $x = -12$	Solve each equation.

The solution set is $\{2, -12\}$. You can check this result by using factoring to solve the original equation.

Roots that are irrational numbers may be written as exact answers in radical form or as *approximate* answers in decimal form when a calculator is used.

Example 2 Equation with Irrational Roots

Solve $x^2 - 6x + 9 = 32$ by using the Square Root Property.

$x^2 - 6x + 9 = 32$	Original equation
$(x - 3)^2 = 32$	Factor the perfect square trinomial.
$x - 3 = \pm\sqrt{32}$	Square Root Property
$x = 3 \pm 4\sqrt{2}$	Add 3 to each side; $\sqrt{32} = 4\sqrt{2}$
$x = 3 + 4\sqrt{2}$ or $x = 3 - 4\sqrt{2}$	Write as two equations.
$x \approx 8.7$ $\quad$ $x \approx -2.7$	Use a calculator.

The exact solutions of this equation are $3 - 4\sqrt{2}$ and $3 + 4\sqrt{2}$. The approximate solutions are -2.7 and 8.7. Check these results by finding and graphing the related quadratic function.

$x^2 - 6x + 9 = 32$	Original equation
$x^2 - 6x - 23 = 0$	Subtract 32 from each side.
$y = x^2 - 6x - 23$	Related quadratic function

CHECK Use the ZERO function of a graphing calculator. The approximate zeros of the related function are -2.7 and 8.7.

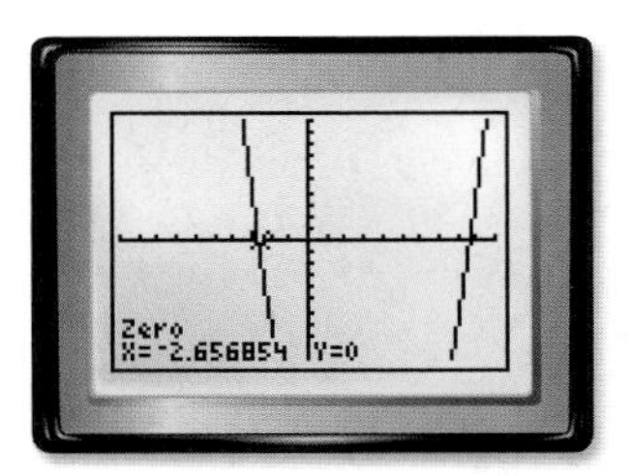

COMPLETE THE SQUARE The Square Root Property can only be used to solve quadratic equations when the side containing the quadratic expression is a perfect square. However, few quadratic expressions are perfect squares. To make a quadratic expression a perfect square, a method called **completing the square** may be used.

In a perfect square trinomial, there is a relationship between the coefficient of the linear term and the constant term. Consider the pattern for squaring a sum.

$(x + 7)^2 = x^2 + 2(7)x + 7^2$	Square of a sum pattern
$= x^2 + 14x + 49$	Simplify.
$\left(\frac{14}{2}\right)^2 \longrightarrow 7^2$	Notice that 49 is 7^2 and 7 is one-half of 14.

You can use this pattern of coefficients to complete the square of a quadratic expression.

Key Concept — Completing the Square

- **Words** To complete the square for any quadratic expression of the form $x^2 + bx$, follow the steps below.

  Step 1 Find one half of b, the coefficient of x.

  Step 2 Square the result in Step 1.

  Step 3 Add the result of Step 2 to $x^2 + bx$.

- **Symbols** $x^2 + bx + \left(\frac{b}{2}\right)^2 = \left(x + \frac{b}{2}\right)^2$

Example 3 Complete the Square

Find the value of c that makes $x^2 + 12x + c$ a perfect square. Then write the trinomial as a perfect square.

Step 1 Find one half of 12. $\frac{12}{2} = 6$

Step 2 Square the result of Step 1. $6^2 = 36$

Step 3 Add the result of Step 2 to $x^2 + 12x$. $x^2 + 12x + 36$

The trinomial $x^2 + 12x + 36$ can be written as $(x + 6)^2$.

You can solve any quadratic equation by completing the square. Because you are solving an equation, add the value you use to complete the square to each side.

Algebra Activity

Completing the Square

Use algebra tiles to complete the square for the equation $x^2 + 2x - 3 = 0$.

Step 1 Represent $x^2 + 2x - 3 = 0$ on an equation mat.

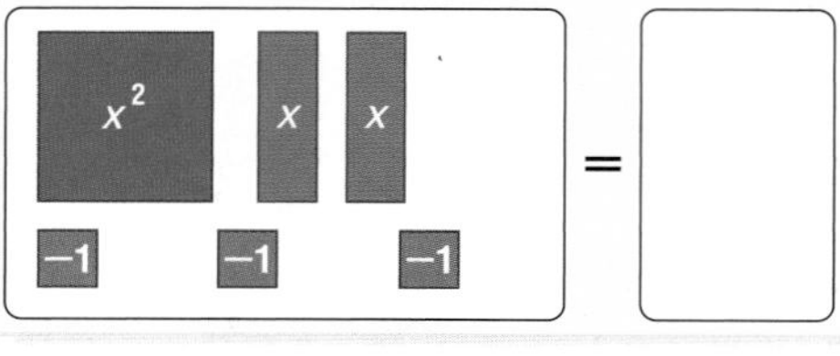

$x^2 + 2x - 3 = 0$

Step 2 Add 3 to each side of the mat. Remove the zero pairs.

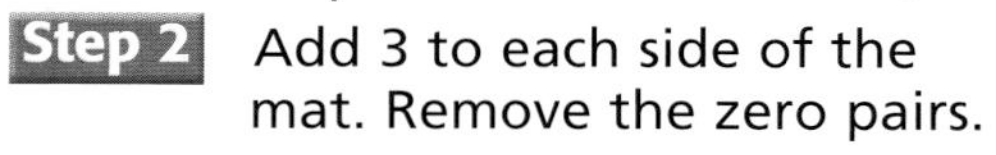
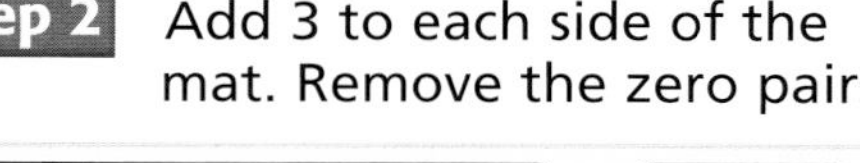
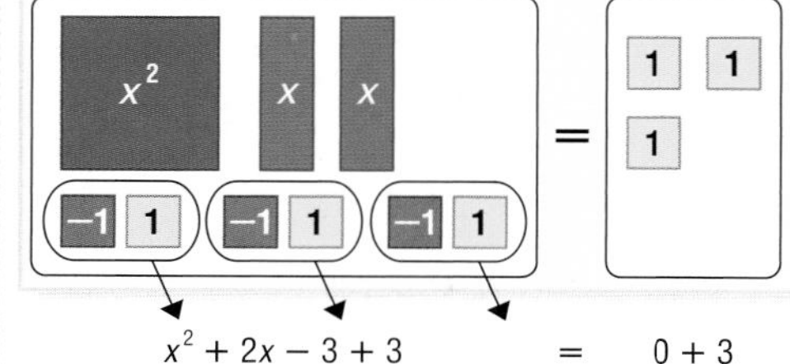

$x^2 + 2x - 3 + 3 = 0 + 3$

Step 3 Begin to arrange the x^2 and x tiles into a square.

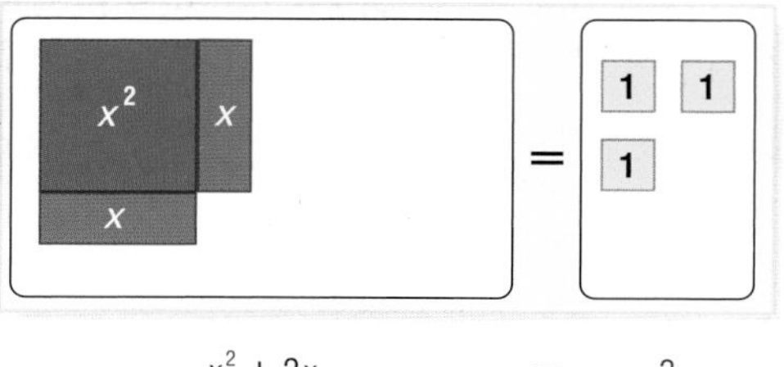

$x^2 + 2x = 3$

Step 4 To complete the square, add 1 yellow 1 tile to each side. The completed equation is $x^2 + 2x + 1 = 4$ or $(x + 1)^2 = 4$.

$x^2 + 2x + 1 = 3 + 1$

Model

Use algebra tiles to complete the square for each equation.

1. $x^2 + 2x - 4 = 0$
2. $x^2 + 4x + 1 = 0$
3. $x^2 - 6x = -5$
4. $x^2 - 2x = -1$

Study Tip

Common Misconception

When solving equations by completing the square, don't forget to add $\left(\frac{b}{2}\right)^2$ to *each* side of the equation.

Example 4 Solve an Equation by Completing the Square

Solve $x^2 + 8x - 20 = 0$ by completing the square.

$x^2 + 8x - 20 = 0$ Notice that $x^2 + 8x - 20$ is not a perfect square.

$x^2 + 8x = 20$ Rewrite so the left side is of the form $x^2 + bx$.

$x^2 + 8x + 16 = 20 + 16$ Since $\left(\frac{8}{2}\right)^2 = 16$, add 16 to each side.

$(x + 4)^2 = 36$ Write the left side as a perfect square by factoring.

$x + 4 = \pm 6$	Square Root Property
$x = -4 \pm 6$	Add -4 to each side.
$x = -4 + 6$ or $x = -4 - 6$	Write as two equations.
$x = 2$ $\quad$ $x = -10$	The solution set is $\{-10, 2\}$.

You can check this result by using factoring to solve the original equation.

When the coefficient of the quadratic term is not 1, you must first divide the equation by that coefficient before completing the square.

Example 5 Equation with $a \neq 1$

Solve $2x^2 - 5x + 3 = 0$ by completing the square.

$2x^2 - 5x + 3 = 0$	Notice that $2x^2 - 5x + 3$ is not a perfect square.
$x^2 - \frac{5}{2}x + \frac{3}{2} = 0$	Divide by the coefficient of quadratic term, 2.
$x^2 - \frac{5}{2}x = -\frac{3}{2}$	Subtract $\frac{3}{2}$ from each side.
$x^2 - \frac{5}{2}x + \frac{25}{16} = -\frac{3}{2} + \frac{25}{16}$	Since $\left(-\frac{5}{2} \div 2\right)^2 = \frac{25}{16}$, add $\frac{25}{16}$ to each side.
$\left(x - \frac{5}{4}\right)^2 = \frac{1}{16}$	Write the left side as a perfect square by factoring. Simplify the right side.
$x - \frac{5}{4} = \pm\frac{1}{4}$	Square Root Property
$x = \frac{5}{4} \pm \frac{1}{4}$	Add $\frac{5}{4}$ to each side.
$x = \frac{5}{4} + \frac{1}{4}$ or $x = \frac{5}{4} - \frac{1}{4}$	Write as two equations.
$x = \frac{3}{2}$ $\quad$ $x = 1$	The solution set is $\left\{1, \frac{3}{2}\right\}$.

Not all solutions of quadratic equations are real numbers. In some cases, the solutions are complex numbers of the form $a + bi$, where $b \neq 0$.

Example 6 Equation with Complex Solutions

Solve $x^2 + 4x + 11 = 0$ by completing the square.

$x^2 + 4x + 11 = 0$	Notice that $x^2 + 4x + 11$ is not a perfect square.
$x^2 + 4x = -11$	Rewrite so the left side is of the form $x^2 + bx$.
$x^2 + 4x + 4 = -11 + 4$	Since $\left(\frac{4}{2}\right)^2 = 4$, add 4 to each side.
$(x + 2)^2 = -7$	Write the left side as a perfect square by factoring.
$x + 2 = \pm\sqrt{-7}$	Square Root Property
$x + 2 = \pm i\sqrt{7}$	$\sqrt{-1} = i$
$x = -2 \pm i\sqrt{7}$	Subtract 2 from each side.

The solution set is $\{-2 + i\sqrt{7}, -2 - i\sqrt{7}\}$. Notice that these are imaginary solutions.

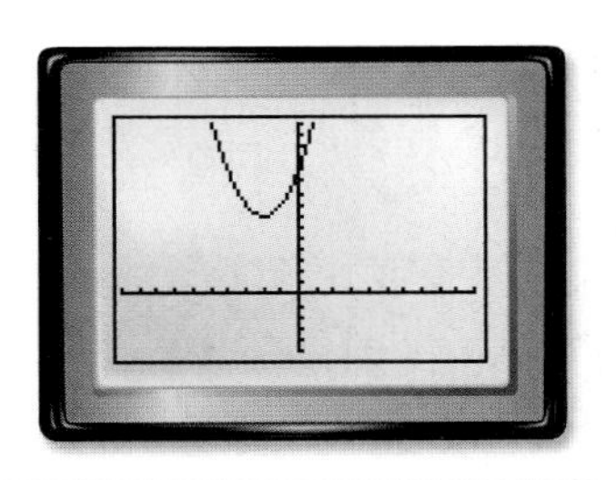

CHECK A graph of the related function shows that the equation has no real solutions since the graph has no x-intercepts. Imaginary solutions must be checked algebraically by substituting them in the original equation.

Check for Understanding

Concept Check

1. **Explain** what it means to *complete the square.*
2. **Determine** whether the value of c that makes $ax^2 + bx + c$ a perfect square trinomial is *sometimes, always,* or *never* negative. Explain your reasoning.
3. **FIND THE ERROR** Rashid and Tia are solving $2x^2 - 8x + 10 = 0$ by completing the square. Who is correct? Explain your reasoning.

Rashid	Tia
$2x^2 - 8x + 10 = 0$	$2x^2 - 8x + 10 = 0$
$2x^2 - 8x = -10$	$x^2 - 4x = 0 - 5$
$2x^2 - 8x + 16 = -10 + 16$	$x^2 - 4x + 4 = -5 + 4$
$(x - 4)^2 = 6$	$(x - 2)^2 = -1$
$x - 4 = \pm\sqrt{6}$	$x - 2 = \pm i$
$x = 4 \pm \sqrt{6}$	$x = 2 \pm i$

Guided Practice

Solve each equation by using the Square Root Property.

4. $x^2 + 14x + 49 = 9$
5. $9x^2 - 24x + 16 = 2$

Find the value of c that makes each trinomial a perfect square. Then write the trinomial as a perfect square.

6. $x^2 - 12x + c$
7. $x^2 - 3x + c$

Solve each equation by completing the square.

8. $x^2 + 3x - 18 = 0$
9. $x^2 - 8x + 11 = 0$
10. $x^2 + 2x + 6 = 0$
11. $2x^2 - 3x - 3 = 0$

Application

ASTRONOMY For Exercises 12 and 13, use the following information.

The height h of an object t seconds after it is dropped is given by $h = -\frac{1}{2}gt^2 + h_0$, where h_0 is the initial height and g is the acceleration due to gravity. The acceleration due to gravity near Earth's surface is 9.8 m/s^2, while on Jupiter it is 23.1 m/s^2. Suppose an object is dropped from an initial height of 100 meters from the surface of each planet.

12. On which planet should the object reach the ground first?
13. Find the time it takes for the object to reach the ground on each planet to the nearest tenth of a second.

Practice and Apply

Homework Help

For Exercises	See Examples
14–23, 48	1, 2
24–31	3
32–47, 49–50, 53	4–6

Extra Practice
See page 840.

Solve each equation by using the Square Root Property.

14. $x^2 + 4x + 4 = 25$
15. $x^2 - 10x + 25 = 49$
16. $x^2 + 8x + 16 = 7$
17. $x^2 - 6x + 9 = 8$
18. $4x^2 - 28x + 49 = 5$
19. $9x^2 + 30x + 25 = 11$
20. $x^2 + x + \frac{1}{4} = \frac{9}{16}$
21. $x^2 + 1.4x + 0.49 = 0.81$

22. **MOVIE SCREENS** The area A in square feet of a projected picture on a movie screen is given by $A = 0.16d^2$, where d is the distance from the projector to the screen in feet. At what distance will the projected picture have an area of 100 square feet?

23. **ENGINEERING** In an engineering test, a rocket sled is propelled into a target. The sled's distance d in meters from the target is given by the formula $d = -1.5t^2 + 120$, where t is the number of seconds after rocket ignition. How many seconds have passed since rocket ignition when the sled is 10 meters from the target?

More About. . .

Engineering

Reverse ballistic testing—accelerating a target on a sled to impact a stationary test item at the end of the track—was pioneered at the Sandia National Laboratories' Rocket Sled Track Facility in Albuquerque, New Mexico. This facility provides a 10,000-foot track for testing items at very high speeds.

Source: www.sandia.gov

Find the value of c that makes each trinomial a perfect square. Then write the trinomial as a perfect square.

24. $x^2 + 16x + c$
25. $x^2 - 18x + c$
26. $x^2 - 15x + c$
27. $x^2 + 7x + c$
28. $x^2 + 0.6x + c$
29. $x^2 - 2.4x + c$
30. $x^2 - \frac{8}{3}x + c$
31. $x^2 + \frac{5}{2}x + c$

Solve each equation by completing the square.

32. $x^2 - 8x + 15 = 0$
33. $x^2 + 2x - 120 = 0$
34. $x^2 + 2x - 6 = 0$
35. $x^2 - 4x + 1 = 0$
36. $x^2 - 4x + 5 = 0$
37. $x^2 + 6x + 13 = 0$
38. $2x^2 + 3x - 5 = 0$
39. $2x^2 - 3x + 1 = 0$
40. $3x^2 - 5x + 1 = 0$
41. $3x^2 - 4x - 2 = 0$
42. $2x^2 - 7x + 12 = 0$
43. $3x^2 + 5x + 4 = 0$
44. $x^2 + 1.4x = 1.2$
45. $x^2 - 4.7x = -2.8$
46. $x^2 - \frac{2}{3}x - \frac{26}{9} = 0$
47. $x^2 - \frac{3}{2}x - \frac{23}{16} = 0$

48. **FRAMING** A picture has a square frame that is 2 inches wide. The area of the picture is one-third of the total area of the picture and frame. What are the dimensions of the picture to the nearest quarter of an inch?

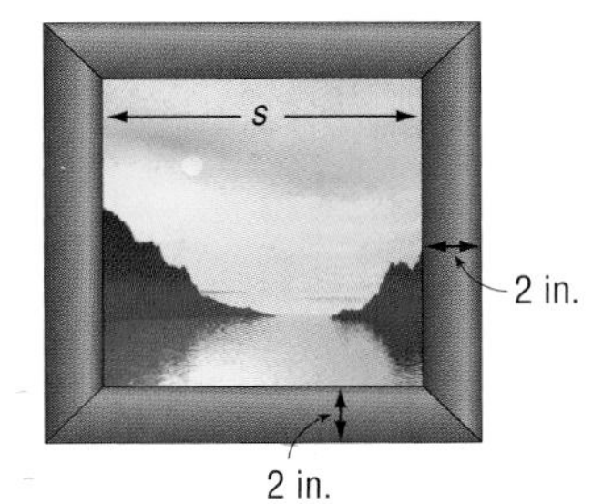

GOLDEN RECTANGLE **For Exercises 49–51, use the following information.**
A *golden rectangle* is one that can be divided into a square and a second rectangle that is geometrically similar to the original rectangle. The ratio of the length of the longer side to the shorter side of a golden rectangle is called the *golden ratio*.

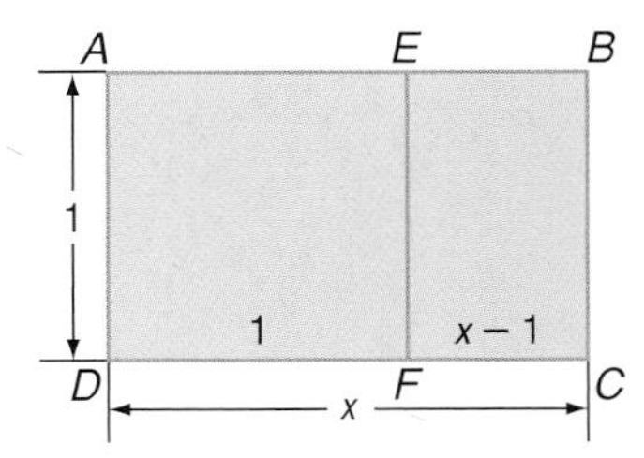

49. Find the ratio of the length of the longer side to the length of the shorter side for rectangle $ABCD$ and for rectangle $EBCF$.

50. Find the exact value of the golden ratio by setting the two ratios in Exercise 49 equal and solving for x. (*Hint*: The golden ratio is a positive value.)

51. **RESEARCH** Use the Internet or other reference to find examples of the golden rectangle in architecture. What applications does the reciprocal of the golden ratio have in music?

52. **CRITICAL THINKING** Find all values of n such that $x^2 + bx + \left(\frac{b}{2}\right)^2 = n$ has
   a. one real root.
   b. two real roots.
   c. two imaginary roots.

53. **KENNEL** A kennel owner has 164 feet of fencing with which to enclose a rectangular region. He wants to subdivide this region into three smaller rectangles of equal length, as shown. If the total area to be enclosed is 576 square feet, find the dimensions of the entire enclosed region. (*Hint*: Write an expression for ℓ in terms of w.)

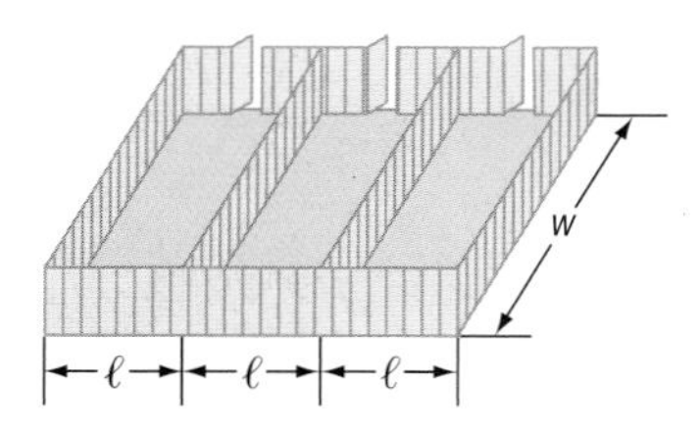

54. **WRITING IN MATH** Answer the question that was posed at the beginning of the lesson.

How can you find the time it takes an accelerating race car to reach the finish line?

Include the following in your answer:

- an explanation of why $t^2 + 22t + 121 = 246$ cannot be solved by factoring, and
- a description of the steps you would take to solve the equation $t^2 + 22t + 121 = 246$.

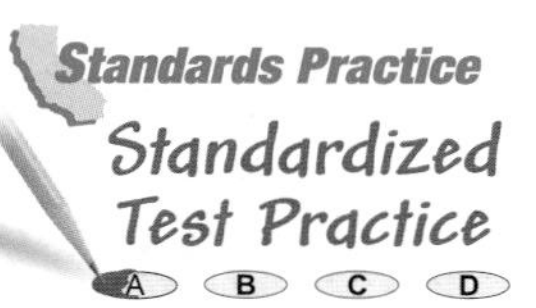

55. What is the absolute value of the product of the two solutions for x in $x^2 - 2x - 2 = 0$?

Ⓐ -1 Ⓑ 0 Ⓒ 1 Ⓓ 2

56. For which value of c will the roots of $x^2 + 4x + c = 0$ be real and equal?

Ⓐ 1 Ⓑ 2 Ⓒ 3 Ⓓ 4 Ⓔ 5

Maintain Your Skills

Mixed Review

Write a quadratic equation with the given root(s). Write the equation in the form $ax^2 + bx + c = 0$, where a, b, and c are integers. *(Lesson 6-3)*

57. 2, 1
58. $-3, 9$
59. $6, \frac{1}{3}$
60. $-\frac{1}{3}, -\frac{3}{4}$

Solve each equation by graphing. If exact roots cannot be found, state the consecutive integers between which the roots are located. *(Lesson 6-2)*

61. $3x^2 = 4 - 8x$
62. $x^2 + 48 = 14x$
63. $2x^2 + 11x = -12$

64. Write *the seventh root of 5 cubed* using exponents. *(Lesson 5-7)*

Solve each system of equations by using inverse matrices. *(Lesson 4-8)*

65. $5x + 3y = -5$
$7x + 5y = -11$

66. $6x + 5y = 8$
$3x - y = 7$

CHEMISTRY **For Exercises 67 and 68, use the following information.**
For hydrogen to be a liquid, its temperature must be within 2°C of −257°C. *(Lesson 1-4)*

67. Write an equation to determine the greatest and least temperatures for this substance.

68. Solve the equation.

Getting Ready for the Next Lesson

PREREQUISITE SKILL **Evaluate $b^2 - 4ac$ for the given values of a, b, and c.**
*(To review **evaluating expressions**, see Lesson 1-1.)*

69. $a = 1, b = 7, c = 3$
70. $a = 1, b = 2, c = 5$
71. $a = 2, b = -9, c = -5$
72. $a = 4, b = -12, c = 9$

6-5 The Quadratic Formula and the Discriminant

What You'll Learn

- Solve quadratic equations by using the Quadratic Formula.
- Use the discriminant to determine the number and type of roots of a quadratic equation.

Vocabulary

- Quadratic Formula
- discriminant

California Standards
Standard 8.0
Students solve and graph quadratic equations by factoring, completing the square, or **using the quadratic formula. Students apply these techniques in solving word problems. They also solve quadratic equations in the complex number system.** (Key)

How is blood pressure related to age?

As people age, their arteries lose their elasticity, which causes blood pressure to increase. For healthy women, average systolic blood pressure is estimated by $P = 0.01A^2 + 0.05A + 107$, where P is the average blood pressure in millimeters of mercury (mm Hg) and A is the person's age. For healthy men, average systolic blood pressure is estimated by $P = 0.006A^2 - 0.02A + 120$.

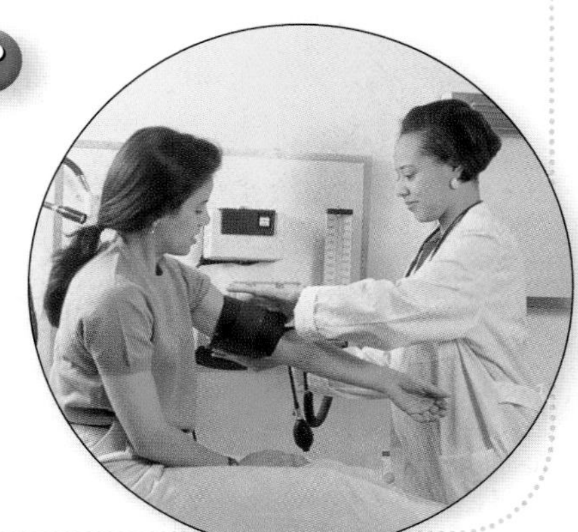

QUADRATIC FORMULA You have seen that exact solutions to some quadratic equations can be found by graphing, by factoring, or by using the Square Root Property. While completing the square can be used to solve any quadratic equation, the process can be tedious if the equation contains fractions or decimals. Fortunately, a formula exists that can be used to solve any quadratic equation of the form $ax^2 + bx + c = 0$. This formula can be derived by solving the general form of a quadratic equation.

$ax^2 + bx + c = 0$	General quadratic equation
$x^2 + \frac{b}{a}x + \frac{c}{a} = 0$	Divide each side by a.
$x^2 + \frac{b}{a}x = -\frac{c}{a}$	Subtract $\frac{c}{a}$ from each side.
$x^2 + \frac{b}{a}x + \frac{b^2}{4a^2} = -\frac{c}{a} + \frac{b^2}{4a^2}$	Complete the square.
$\left(x + \frac{b}{2a}\right)^2 = \frac{b^2 - 4ac}{4a^2}$	Factor the left side. Simplify the right side.
$x + \frac{b}{2a} = \pm\frac{\sqrt{b^2 - 4ac}}{2a}$	Square Root Property
$x = -\frac{b}{2a} \pm \frac{\sqrt{b^2 - 4ac}}{2a}$	Subtract $\frac{b}{2a}$ from each side.
$x = \frac{-b \pm \sqrt{b^2 - 4ac}}{2a}$	Simplify.

This equation is known as the **Quadratic Formula**.

Study Tip

Reading Math
The Quadratic Formula is read *x equals the opposite of b, plus or minus the square root of b squared minus 4ac, all divided by 2a.*

Key Concept — Quadratic Formula

The solutions of a quadratic equation of the form $ax^2 + bx + c = 0$, where $a \neq 0$, are given by the following formula.

$$x = \frac{-b \pm \sqrt{b^2 - 4ac}}{2a}$$

Example 1 Two Rational Roots

Solve $x^2 - 12x = 28$ by using the Quadratic Formula.

First, write the equation in the form $ax^2 + bx + c = 0$ and identify a, b, and c.

$$x^2 - 12x = 28 \longrightarrow 1x^2 - 12x - 28 = 0 \quad (ax^2 + bx + c = 0)$$

Then, substitute these values into the Quadratic Formula.

$$x = \frac{-b \pm \sqrt{b^2 - 4ac}}{2a}$$ Quadratic Formula

$$x = \frac{-(-12) \pm \sqrt{(-12)^2 - 4(1)(-28)}}{2(1)}$$ Replace a with 1, b with -12, and c with -28.

$$x = \frac{12 \pm \sqrt{144 + 112}}{2}$$ Simplify.

$$x = \frac{12 \pm \sqrt{256}}{2}$$ Simplify.

$$x = \frac{12 \pm 16}{2}$$ $\sqrt{256} = 16$

$$x = \frac{12 + 16}{2} \quad \text{or} \quad x = \frac{12 - 16}{2}$$ Write as two equations.

$$= 14 \qquad\qquad = -2$$ Simplify.

The solutions are -2 and 14. Check by substituting each of these values into the original equation.

Study Tip

Quadratic Formula
Although factoring may be an easier method to solve the equations in Examples 1 and 2, the Quadratic Formula can be used to solve any quadratic equation.

When the value of the radicand in the Quadratic Formula is 0, the quadratic equation has exactly one rational root.

Example 2 One Rational Root

Solve $x^2 + 22x + 121 = 0$ by using the Quadratic Formula.

Identify a, b, and c. Then, substitute these values into the Quadratic Formula.

$$x = \frac{-b \pm \sqrt{b^2 - 4ac}}{2a}$$ Quadratic Formula

$$x = \frac{-(22) \pm \sqrt{(22)^2 - 4(1)(121)}}{2(1)}$$ Replace a with 1, b with 22, and c with 121.

$$x = \frac{-22 \pm \sqrt{0}}{2}$$ Simplify.

$$x = \frac{-22}{2} \text{ or } -11$$ $\sqrt{0} = 0$

The solution is -11.

CHECK A graph of the related function shows that there is one solution at $x = -11$.

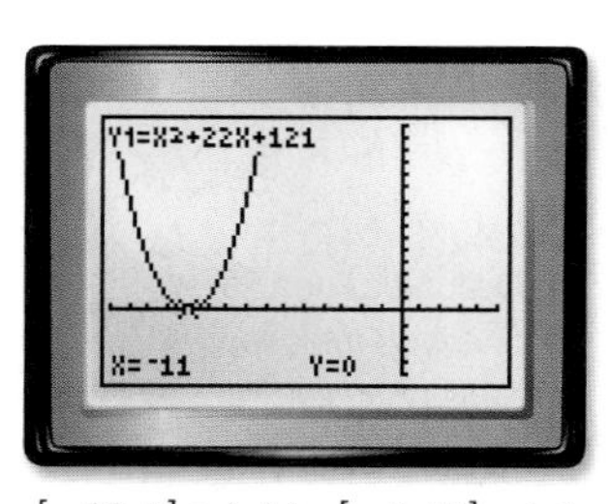

[−15, 5] scl: 1 by [−5, 15] scl: 1

You can express irrational roots exactly by writing them in radical form.

Example 3 Irrational Roots

Solve $2x^2 + 4x - 5 = 0$ by using the Quadratic Formula.

$x = \frac{-b \pm \sqrt{b^2 - 4ac}}{2a}$ Quadratic Formula

$x = \frac{-(4) \pm \sqrt{(4)^2 - 4(2)(-5)}}{2(2)}$ Replace a with 2, b with 4, and c with -5.

$x = \frac{-4 \pm \sqrt{56}}{4}$ Simplify.

$x = \frac{-4 \pm 2\sqrt{14}}{4}$ or $\frac{-2 \pm \sqrt{14}}{2}$ $\sqrt{56} = \sqrt{4 \cdot 14}$ or $2\sqrt{14}$

The exact solutions are $\frac{-2 - \sqrt{14}}{2}$ and $\frac{-2 + \sqrt{14}}{2}$. The approximate solutions are -2.9 and 0.9.

CHECK Check these results by graphing the related quadratic function, $y = 2x^2 + 4x - 5$. Using the ZERO function of a graphing calculator, the approximate zeros of the related function are -2.9 and 0.9.

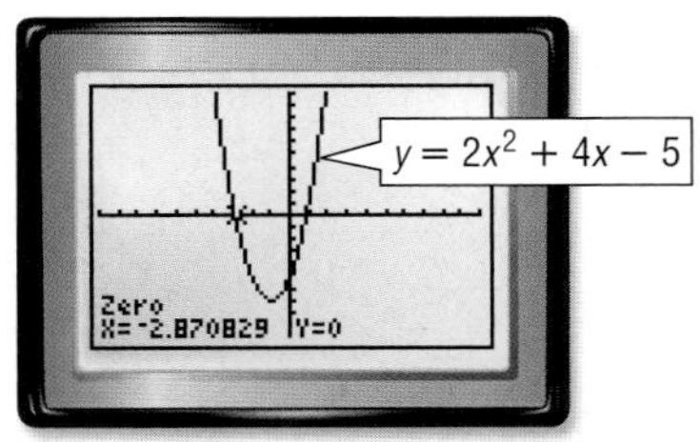

[−10, 10] scl: 1 by [−10, 10] scl: 1

When using the Quadratic Formula, if the radical contains a negative value, the solutions will be complex. Complex solutions always appear in conjugate pairs.

Example 4 Complex Roots

Solve $x^2 - 4x = -13$ by using the Quadratic Formula.

$x = \frac{-b \pm \sqrt{b^2 - 4ac}}{2a}$ Quadratic Formula

$x = \frac{-(-4) \pm \sqrt{(-4)^2 - 4(1)(13)}}{2(1)}$ Replace a with 1, b with -4, and c with 13.

$x = \frac{4 \pm \sqrt{-36}}{2}$ Simplify.

$x = \frac{4 \pm 6i}{2}$ $\sqrt{-36} = \sqrt{36(-1)}$ or $6i$

$x = 2 \pm 3i$ Simplify.

The solutions are the complex numbers $2 + 3i$ and $2 - 3i$.

A graph of the related function shows that the solutions are complex, but it cannot help you find them.

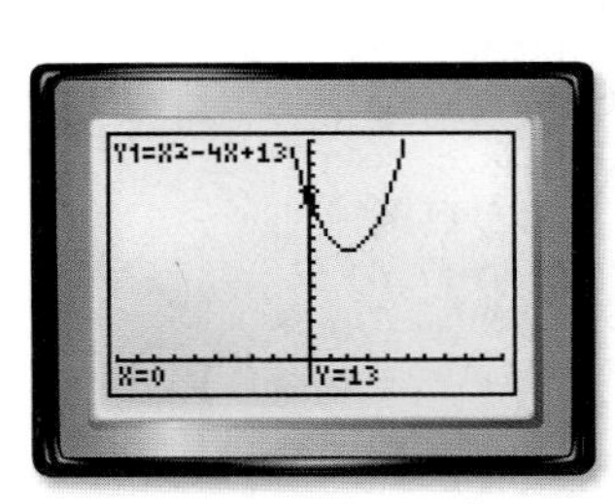

[−15, 5] scl: 1 by [−2, 18] scl: 1

Study Tip

Using the Quadratic Formula
Remember that to correctly identify a, b, and c for use in the Quadratic Formula, the equation must be written in the form $ax^2 + bx + c = 0$.

CHECK To check complex solutions, you must substitute them into the original equation. The check for $2 + 3i$ is shown below.

$x^2 - 4x = -13$	Original equation
$(2 + 3i)^2 - 4(2 + 3i) \stackrel{?}{=} -13$	$x = 2 + 3i$
$4 + 12i + 9i^2 - 8 - 12i \stackrel{?}{=} -13$	Sum of a square; Distributive Property
$-4 + 9i^2 \stackrel{?}{=} -13$	Simplify.
$-4 - 9 = -13$ ✓	$i^2 = -1$

Study Tip

Reading Math
Remember that the solutions of an equation are called *roots*.

ROOTS AND THE DISCRIMINANT

In Examples 1, 2, 3, and 4, observe the relationship between the value of the expression under the radical and the roots of the quadratic equation. The expression $b^2 - 4ac$ is called the **discriminant**.

$$x = \frac{-b \pm \sqrt{b^2 - 4ac}}{2a} \longleftarrow \text{discriminant}$$

The value of the discriminant can be used to determine the number and type of roots of a quadratic equation.

Key Concept — Discriminant

Consider $ax^2 + bx + c = 0$.

Value of Discriminant	Type and Number of Roots	Example of Graph of Related Function
$b^2 - 4ac > 0$; $b^2 - 4ac$ is a perfect square.	2 real, rational roots	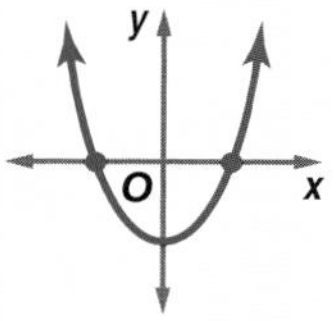
$b^2 - 4ac > 0$; $b^2 - 4ac$ is *not* a perfect square.	2 real, irrational roots	
$b^2 - 4ac = 0$	1 real, rational root	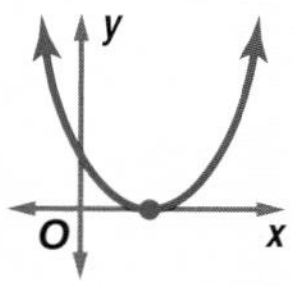
$b^2 - 4ac < 0$	2 complex roots	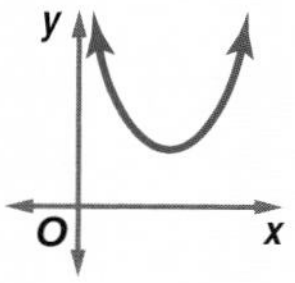

Study Tip

Using the Discriminant
The discriminant can help you check the solutions of a quadratic equation. Your solutions must match in number and in type to those determined by the discriminant.

Example 5 Describe Roots

Find the value of the discriminant for each quadratic equation. Then describe the number and type of roots for the equation.

a. $9x^2 - 12x + 4 = 0$

$a = 9, b = -12, c = 4$

$b^2 - 4ac = (-12)^2 - 4(9)(4)$

$= 144 - 144$

$= 0$

The discriminant is 0, so there is one rational root.

b. $2x^2 + 16x + 33 = 0$

$a = 2, b = 16, c = 33$

$b^2 - 4ac = (16)^2 - 4(2)(33)$

$= 256 - 264$

$= -8$

The discriminant is negative, so there are two complex roots.

c. $-5x^2 + 8x - 1 = 0$

$a = -5, b = 8, c = -1$

$$b^2 - 4ac = (8)^2 - 4(-5)(-1)$$
$$= 64 - 20$$
$$= 44$$

The discriminant is 44, which is not a perfect square. Therefore, there are two irrational roots.

d. $-7x + 15x^2 - 4 = 0$

$a = 15, b = -7, c = -4$

$$b^2 - 4ac = (-7)^2 - 4(15)(-4)$$
$$= 49 + 240$$
$$= 289 \text{ or } 17^2$$

The discriminant is 289, which is a perfect square. Therefore, there are two rational roots.

You have studied a variety of methods for solving quadratic equations. The table below summarizes these methods.

Concept Summary — Solving Quadratic Equations

Method	Can be Used	When to Use
Graphing	sometimes	Use only if an exact answer is not required. Best used to check the reasonableness of solutions found algebraically.
Factoring	sometimes	Use if the constant term is 0 or if the factors are easily determined. **Example** $x^2 - 3x = 0$
Square Root Property	sometimes	Use for equations in which a perfect square is equal to a constant. **Example** $(x + 13)^2 = 9$
Completing the Square	always	Useful for equations of the form $x^2 + bx + c = 0$, where b is even. **Example** $x^2 + 14x - 9 = 0$
Quadratic Formula	always	Useful when other methods fail or are too tedious. **Example** $3.4x^2 - 2.5x + 7.9 = 0$

Check for Understanding

Concept Check

1. **OPEN ENDED** Sketch the graph of a quadratic equation whose discriminant is
   a. positive. **b.** negative. **c.** zero.
2. **Explain** why the roots of a quadratic equation are complex if the value of the discriminant is less than 0.
3. **Describe** the relationship that must exist between a, b, and c in the equation $ax^2 + bx + c = 0$ in order for the equation to have exactly one solution.

Guided Practice

Complete parts a–c for each quadratic equation.

a. Find the value of the discriminant.

b. Describe the number and type of roots.

c. Find the exact solutions by using the Quadratic Formula.

4. $8x^2 + 18x - 5 = 0$
5. $2x^2 - 4x + 1 = 0$
6. $4x^2 + 4x + 1 = 0$
7. $x^2 + 3x + 8 = 5$

Solve each equation using the method of your choice. Find exact solutions.

8. $x^2 + 8x = 0$

9. $x^2 + 5x + 6 = 0$

10. $x^2 - 2x - 2 = 0$

11. $4x^2 + 20x + 25 = -2$

Application PHYSICS **For Exercises 12 and 13, use the following information.**
The height $h(t)$ in feet of an object t seconds after it is propelled straight up from the ground with an initial velocity of 85 feet per second is modeled by $h(t) = -16t^2 + 85t$.

12. When will the object be at a height of 50 feet?

13. Will the object ever reach a height of 120 feet? Explain your reasoning.

Practice and Apply

Homework Help	
For Exercises	See Examples
14–27	1–5
28–39, 42–44	1–4
40–41	5

Extra Practice
See page 841.

Complete parts a–c for each quadratic equation.

a. Find the value of the discriminant.

b. Describe the number and type of roots.

c. Find the exact solutions by using the Quadratic Formula.

14. $x^2 + 3x - 3 = 0$

15. $x^2 - 16x + 4 = 0$

16. $x^2 - 2x + 5 = 0$

17. $x^2 - x + 6 = 0$

18. $-12x^2 + 5x + 2 = 0$

19. $-3x^2 - 5x + 2 = 0$

20. $x^2 + 4x + 3 = 4$

21. $2x - 5 = -x^2$

22. $9x^2 - 6x - 4 = -5$

23. $25 + 4x^2 = -20x$

24. $4x^2 + 7 = 9x$

25. $3x + 6 = -6x^2$

26. $\frac{3}{4}x^2 - \frac{1}{3}x - 1 = 0$

27. $0.4x^2 + x - 0.3 = 0$

Solve each equation by using the method of your choice. Find exact solutions.

28. $x^2 - 30x - 64 = 0$

29. $7x^2 + 3 = 0$

30. $x^2 - 4x + 7 = 0$

31. $2x^2 + 6x - 3 = 0$

32. $4x^2 - 8 = 0$

33. $4x^2 + 81 = 36x$

34. $-4(x + 3)^2 = 28$

35. $3x^2 - 10x = 7$

36. $x^2 + 9 = 8x$

37. $10x^2 + 3x = 0$

38. $2x^2 - 12x + 7 = 5$

39. $21 = (x - 2)^2 + 5$

More About. . .

Bridges

The Golden Gate, located in San Francisco, California, is the tallest bridge in the world, with its towers extending 746 feet above the water and the floor of the bridge extending 220 feet above water.

Source: www.goldengatebridge.org

BRIDGES **For Exercises 40 and 41, use the following information.**
The supporting cables of the Golden Gate Bridge approximate the shape of a parabola. The parabola can be modeled by the quadratic function $y = 0.00012x^2 + 6$, where x represents the distance from the axis of symmetry and y represents the height of the cables. The related quadratic equation is $0.00012x^2 + 6 = 0$.

40. Calculate the value of the discriminant.

41. What does the discriminant tell you about the supporting cables of the Golden Gate Bridge?

FOOTBALL **For Exercises 42 and 43, use the following information.**
The average NFL salary $A(t)$ (in thousands of dollars) from 1975 to 2000 can be estimated using the function $A(t) = 2.3t^2 - 12.4t + 73.7$, where t is the number of years since 1975.

42. Determine a domain and range for which this function makes sense.

43. According to this model, in what year did the average salary first exceed 1 million dollars?

Online Research **Data Update** What is the current average NFL salary? How does this average compare with the average given by the function used in Exercises 42 and 43? Visit www.algebra2.com/data_update to learn more.

44. **HIGHWAY SAFETY** Highway safety engineers can use the formula $d = 0.05s^2 + 1.1s$ to estimate the minimum stopping distance d in feet for a vehicle traveling s miles per hour. If a car is able to stop after 125 feet, what is the fastest it could have been traveling when the driver first applied the brakes?

45. **CRITICAL THINKING** Find all values of k such that $x^2 - kx + 9 = 0$ has

a. one real root. **b.** two real roots. **c.** no real roots.

46. **WRITING IN MATH** Answer the question that was posed at the beginning of the lesson.

How is blood pressure related to age?

Include the following in your answer:

- an expression giving the average systolic blood pressure for a person of your age, and
- an example showing how you could determine A in either formula given a specific value of P.

47. If $2x^2 - 5x - 9 = 0$, then x could equal which of the following?

(A) -1.12 (B) 1.54 (C) 2.63 (D) 3.71

48. Which best describes the nature of the roots of the equation $x^2 - 3x + 4 = 0$?

(A) real and equal (B) real and unequal

(C) complex (D) real and complex

Maintain Your Skills

Mixed Review

Solve each equation by using the Square Root Property. *(Lesson 6-4)*

49. $x^2 + 18x + 81 = 25$ 50. $x^2 - 8x + 16 = 7$ 51. $4x^2 - 4x + 1 = 8$

Solve each equation by factoring. *(Lesson 6-3)*

52. $4x^2 + 8x = 0$ 53. $x^2 - 5x = 14$ 54. $3x^2 + 10 = 17x$

Simplify. *(Lesson 5-5)*

55. $\sqrt{a^8b^{20}}$ 56. $\sqrt{100p^{12}q^2}$ 57. $\sqrt[3]{64b^6c^6}$

58. **ANIMALS** The fastest-recorded physical action of any living thing is the wing beat of the common midge. This tiny insect normally beats its wings at a rate of 133,000 times per minute. At this rate, how many times would the midge beat its wings in an hour? Write your answer in scientific notation. *(Lesson 5-1)*

Solve each system of inequalities. *(Lesson 3-3)*

59. $x + y \leq 9$
$x - y \leq 3$
$y - x \geq 4$

60. $x \geq 1$
$y \leq -1$
$y \leq x$

Getting Ready for the Next Lesson

PREREQUISITE SKILL State whether each trinomial is a perfect square. If it is, factor it. *(To review **perfect square trinomials**, see Lesson 5-4.)*

61. $x^2 - 5x - 10$ 62. $x^2 - 14x + 49$

63. $4x^2 + 12x + 9$ 64. $25x^2 + 20x + 4$

65. $9x^2 - 12x + 16$ 66. $36x^2 - 60x + 25$

Graphing Calculator
A Preview of Lesson 6-6

California Standards Standard 9.0 Students demonstrate and explain the effect that changing a coefficient has on the graph of quadratic functions;. . . (Key)

Families of Parabolas

The general form of a quadratic equation is $y = a(x - h)^2 + k$. Changing the values of a, h, and k results in a different parabola in the family of quadratic functions. The parent graph of the family of parabolas is the graph of $y = x^2$.

You can use a TI-83 Plus graphing calculator to analyze the effects that result from changing each of the parameters a, h, and k.

Example 1

Graph each set of equations on the same screen in the standard viewing window. Describe any similarities and differences among the graphs.

$$y = x^2,\ y = x^2 + 3,\ y = x^2 - 5$$

The graphs have the same shape, and all open up. The vertex of each graph is on the y-axis. However, the graphs have different vertical positions.

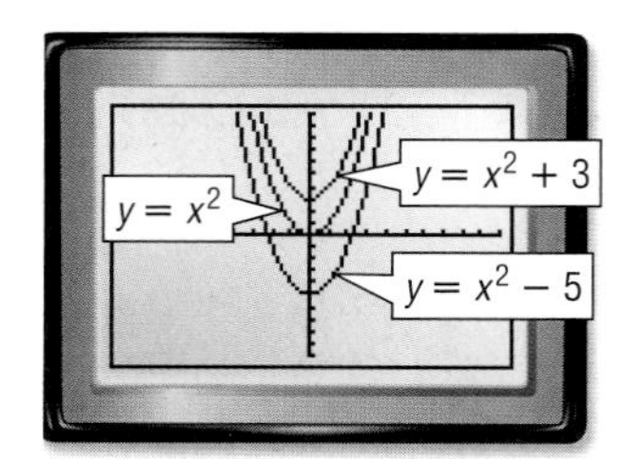

Example 1 shows how changing the value of k in the equation $y = a(x - h)^2 + k$ *translates* the parabola along the y-axis. If $k > 0$, the parabola is translated k units up, and if $k < 0$, it is translated k units down.

How do you think changing the value of h will affect the graph of $y = (x - h)^2$ as compared to the graph of $y = x^2$?

Example 2

Graph each set of equations on the same screen in the standard viewing window. Describe any similarities and differences among the graphs.

$$y = x^2,\ y = (x + 3)^2,\ y = (x - 5)^2$$

These three graphs all open up and have the same shape. The vertex of each graph is on the x-axis. However, the graphs have different horizontal positions.

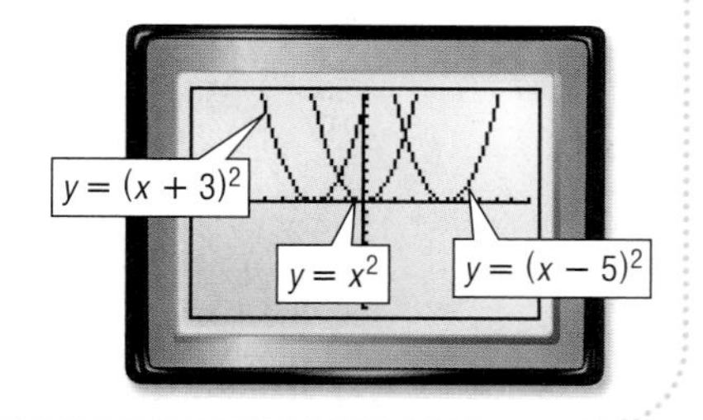

Example 2 shows how changing the value of h in the equation $y = a(x - h)^2 + k$ *translates* the graph horizontally. If $h > 0$, the graph translates to the right h units. If $h < 0$, the graph translates to the left h units.

www.algebra2.com/other_calculator_keystrokes

Investigation

How does the value a affect the graph of $y = ax^2$?

Example 3

Graph each set of equations on the same screen in the standard viewing window. Describe any similarities and differences among the graphs.

a. $y = x^2$, $y = -x^2$

The graphs have the same vertex and the same shape. However, the graph of $y = x^2$ opens up and the graph of $y = -x^2$ opens down.

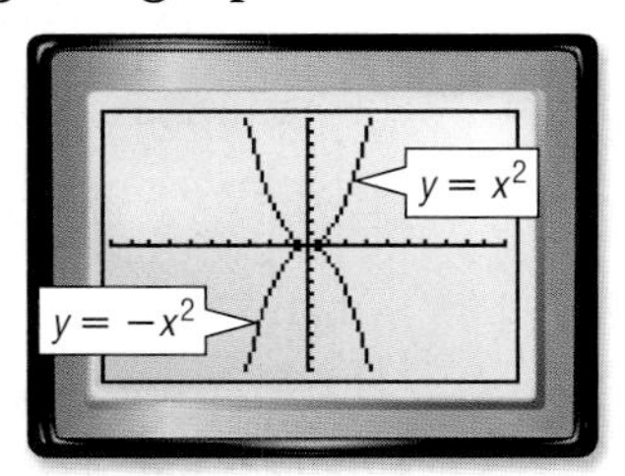

b. $y = x^2$, $y = 4x^2$, $y = \frac{1}{4}x^2$

The graphs have the same vertex, (0, 0), but each has a different shape. The graph of $y = 4x^2$ is narrower than the graph of $y = x^2$. The graph of $y = \frac{1}{4}x^2$ is wider than the graph of $y = x^2$.

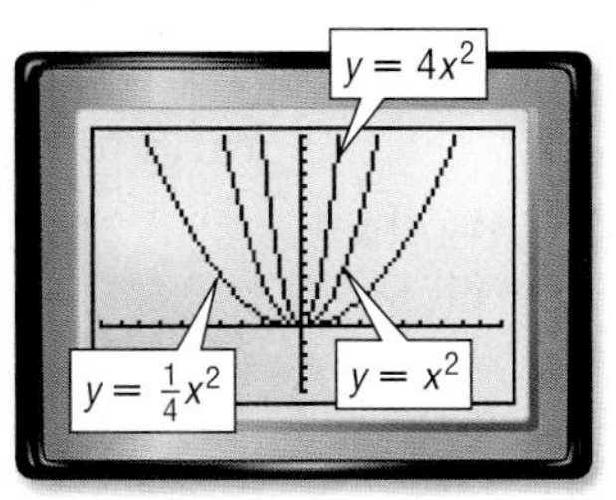

[−10, 10] scl: 1 by [−5, 15] scl: 1

Changing the value of a in the equation $y = a(x - h)^2 + k$ can affect the direction of the opening and the shape of the graph. If $a > 0$, the graph opens up, and if $a < 0$, the graph opens down or is *reflected* over the x-axis. If $|a| > 1$, the graph is narrower than the graph of $y = x^2$. If $|a| < 1$, the graph is wider than the graph of $y = x^2$. Thus, a change in the absolute value of a results in a *dilation* of the graph of $y = x^2$.

Exercises

Consider $y = a(x - h)^2 + k$.

1. How does changing the value of h affect the graph? Give an example.

2. How does changing the value of k affect the graph? Give an example.

3. How does using $-a$ instead of a affect the graph? Give an example.

Examine each pair of equations and predict the similarities and differences in their graphs. Use a graphing calculator to confirm your predictions. Write a sentence or two comparing the two graphs.

4. $y = x^2$, $y = x^2 + 2.5$

5. $y = -x^2$, $y = x^2 - 9$

6. $y = x^2$, $y = 3x^2$

7. $y = x^2$, $y = -6x^2$

8. $y = x^2$, $y = (x + 3)^2$

9. $y = -\frac{1}{3}x^2$, $y = -\frac{1}{3}x^2 + 2$

10. $y = x^2$, $y = (x - 7)^2$

11. $y = x^2$, $y = 3(x + 4)^2 - 7$

12. $y = x^2$, $y = -\frac{1}{4}x^2 + 1$

13. $y = (x + 3)^2 - 2$, $y = (x + 3)^2 + 5$

14. $y = 3(x + 2)^2 - 1$, $y = 6(x + 2)^2 - 1$

15. $y = 4(x - 2)^2 - 3$, $y = \frac{1}{4}(x - 2)^2 - 1$

6-6 Analyzing Graphs of Quadratic Functions

California Standards Standard 9.0 Students demonstrate and explain the effect that changing a coefficient has on the graph of quadratic functions;. . . (Key)

What You'll Learn

- Analyze quadratic functions of the form $y = a(x - h)^2 + k$.
- Write a quadratic function in the form $y = a(x - h)^2 + k$.

Vocabulary

- vertex form

How can the graph of $y = x^2$ be used to graph any quadratic function?

A *family of graphs* is a group of graphs that displays one or more similar characteristics. The graph of $y = x^2$ is called the *parent graph* of the family of quadratic functions.

Study the graphs of $y = x^2$, $y = x^2 + 2$, and $y = (x - 3)^2$. Notice that adding a constant to x^2 moves the graph up. Subtracting a constant from x before squaring it moves the graph to the right.

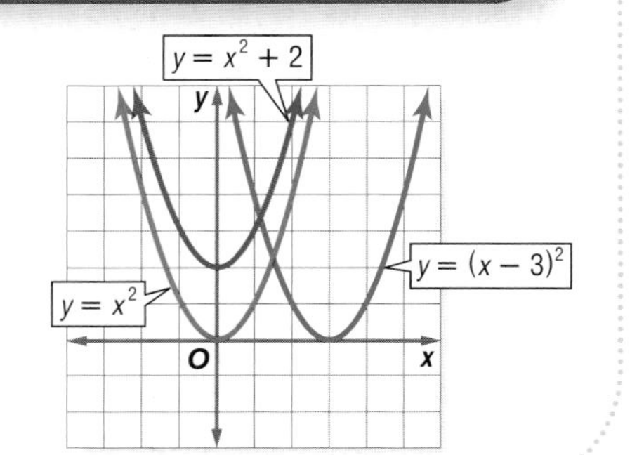

ANALYZE QUADRATIC FUNCTIONS Each function above can be written in the form $y = (x - h)^2 + k$, where (h, k) is the vertex of the parabola and $x = h$ is its axis of symmetry. This is often referred to as the **vertex form** of a quadratic function.

Equation	Vertex	Axis of Symmetry
$y = x^2$ or $y = (x - 0)^2 + 0$	(0, 0)	$x = 0$
$y = x^2 + 2$ or $y = (x - 0)^2 + 2$	(0, 2)	$x = 0$
$y = (x - 3)^2$ or $y = (x - 3)^2 + 0$	(3, 0)	$x = 3$

Recall that a translation slides a figure on the coordinate plane without changing its shape or size. As the values of h and k change, the graph of $y = a(x - h)^2 + k$ is the graph of $y = x^2$ translated:

- $|h|$ units *left* if h is negative or $|h|$ units *right* if h is positive, and
- $|k|$ units *up* if k is positive or $|k|$ units *down* if k is negative.

Example 1 Graph a Quadratic Function in Vertex Form

Analyze $y = (x + 2)^2 + 1$. Then draw its graph.

This function can be rewritten as $y = [x - (-2)]^2 + 1$. Then $h = -2$ and $k = 1$.

The vertex is at (h, k) or $(-2, 1)$, and the axis of symmetry is $x = -2$. The graph has the same shape as the graph of $y = x^2$, but is translated 2 units left and 1 unit up.

Now use this information to draw the graph.

Step 1 Plot the vertex, $(-2, 1)$.

Step 2 Draw the axis of symmetry, $x = -2$.

Step 3 Find and plot two points on one side of the axis of symmetry, such as $(-1, 2)$ and $(0, 5)$.

Step 4 Use symmetry to complete the graph.

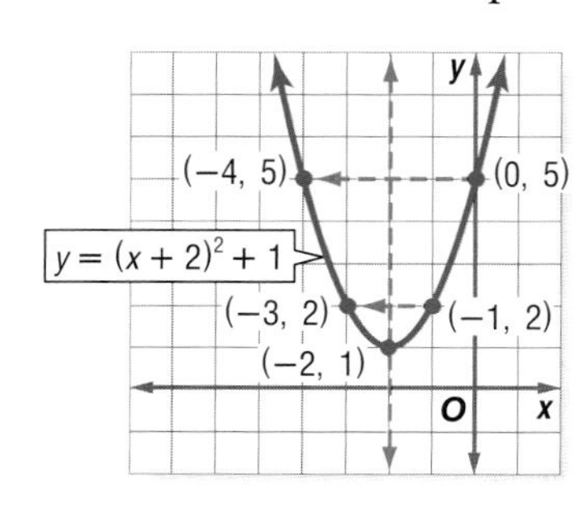

How does the value of a in the general form $y = a(x - h)^2 + k$ affect a parabola? Compare the graphs of the following functions to the parent function, $y = x^2$.

a. $y = 2x^2$ **b.** $y = \frac{1}{2}x^2$

c. $y = -2x^2$ **d.** $y = -\frac{1}{2}x^2$

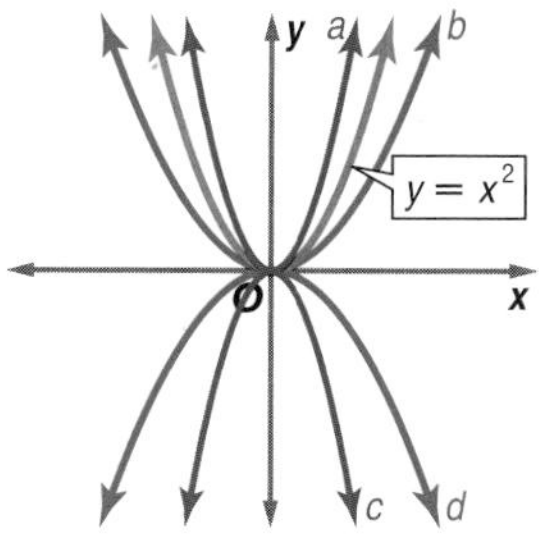

All of the graphs have the vertex (0, 0) and axis of symmetry $x = 0$.

Notice that the graphs of $y = 2x^2$ and $y = \frac{1}{2}x^2$ are *dilations* of the graph of $y = x^2$. The graph of $y = 2x^2$ is narrower than the graph of $y = x^2$, while the graph of $y = \frac{1}{2}x^2$ is wider. The graphs of $y = -2x^2$ and $y = 2x^2$ are *reflections* of each other over the x-axis, as are the graphs of $y = -\frac{1}{2}x^2$ and $y = \frac{1}{2}x^2$.

Changing the value of a in the equation $y = a(x - h)^2 + k$ can affect the direction of the opening and the shape of the graph.

- If $a > 0$, the graph opens up.
- If $a < 0$, the graph opens down.
- If $|a| > 1$, the graph is narrower than the graph of $y = x^2$.
- If $|a| < 1$, the graph is wider than the graph of $y = x^2$.

Study Tip

Reading Math
$|a| < 1$ means that a is a rational number between 0 and 1, such as $\frac{2}{5}$, or a rational number between -1 and 0, such as -0.3.

Concept Summary — *Quadratic Functions in Vertex Form*

The vertex form of a quadratic function is $y = a(x - h)^2 + k$.

h and *k*	*k*
Vertex and Axis of Symmetry	**Vertical Translation**

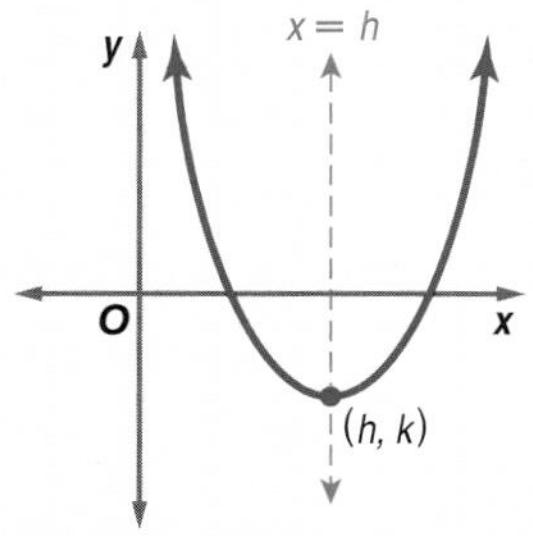

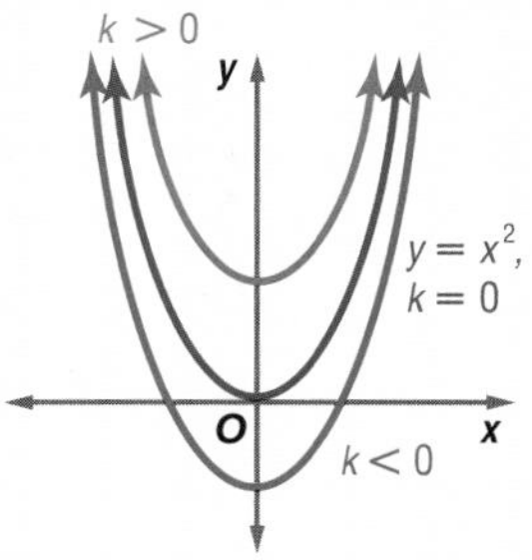

h	*a*
Horizontal Translation	**Direction of Opening and Shape of Parabola**

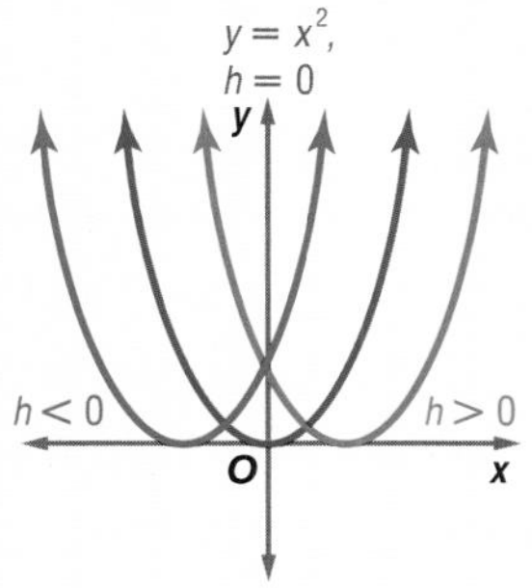

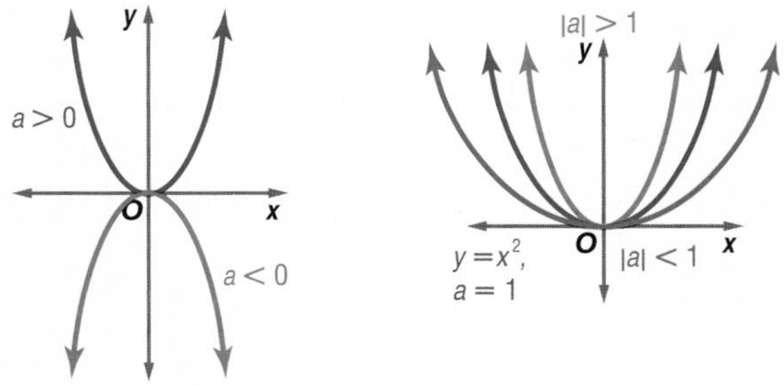

WRITE QUADRATIC FUNCTIONS IN VERTEX FORM Given a function of the form $y = ax^2 + bx + c$, you can complete the square to write the function in vertex form.

Example 2 Write $y = x^2 + bx + c$ in Vertex Form

Write $y = x^2 + 8x - 5$ in vertex form. Then analyze the function.

$y = x^2 + 8x - 5$	Notice that $x^2 + 8x - 5$ is not a perfect square.
$y = (x^2 + 8x + 16) - 5 - 16$	Complete the square by adding $\left(\frac{8}{2}\right)^2$ or 16. Balance this addition by subtracting 16.
$y = (x + 4)^2 - 21$	Write $x^2 + 8x + 16$ as a perfect square.

This function can be rewritten as $y = [x - (-4)]^2 + (-21)$. Written in this way, you can see that $h = -4$ and $k = -21$.

The vertex is at $(-4, -21)$, and the axis of symmetry is $x = -4$. Since $a = 1$, the graph opens up and has the same shape as the graph of $y = x^2$, but it is translated 4 units left and 21 units down.

CHECK You can check the vertex and axis of symmetry using the formula $x = -\frac{b}{2a}$. In the original equation, $a = 1$ and $b = 8$, so the axis of symmetry is $x = -\frac{8}{2(1)}$ or -4. Thus, the x-coordinate of the vertex is -4, and the y-coordinate of the vertex is $y = (-4)^2 + 8(-4) - 5$ or -21.

Study Tip

Check
As an additional check, graph the function in Example 2 to verify the location of its vertex and axis of symmetry.

When writing a quadratic function in which the coefficient of the quadratic term is not 1 in vertex form, the first step is to factor out that coefficient from the quadratic and linear terms. Then you can complete the square and write in vertex form.

Example 3 Write $y = ax^2 + bx + c$ in Vertex Form, $a \neq 1$

Write $y = -3x^2 + 6x - 1$ in vertex form. Then analyze and graph the function.

$y = -3x^2 + 6x - 1$	Original equation
$y = -3(x^2 - 2x) - 1$	Group $ax^2 + bx$ and factor, dividing by a.
$y = -3(x - 2x + 1) - 1 - (-3)(1)$	Complete the square by adding 1 inside the parentheses. Notice that this is an overall addition of $-3(1)$. Balance this addition by subtracting $-3(1)$.
$y = -3(x - 1)^2 + 2$	Write $x^2 - 2x + 1$ as a perfect square.

The vertex form of this function is $y = -3(x - 1)^2 + 2$. So, $h = 1$ and $k = 2$.

The vertex is at $(1, 2)$, and the axis of symmetry is $x = 1$. Since $a = -3$, the graph opens downward and is narrower than the graph of $y = x^2$. It is also translated 1 unit right and 2 units up.

Now graph the function. Two points on the graph to the right of $x = 1$ are $(1.5, 1.25)$ and $(2, -1)$. Use symmetry to complete the graph.

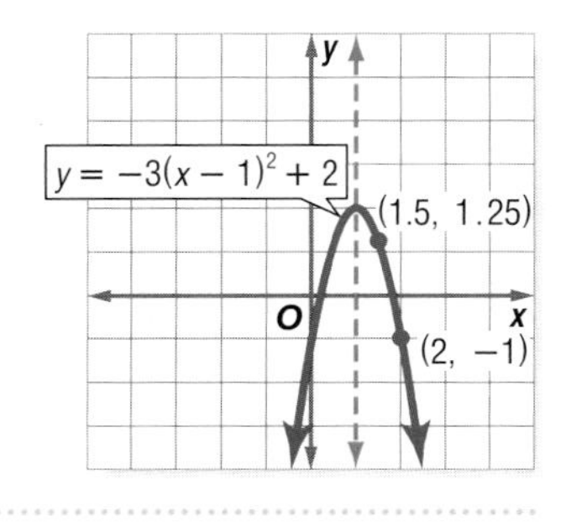

1. It moves it to the left or right. $y=(x-3)^2+2$ will be more to the right than $y=(x+2)^2+2$.

2. It moves it up or down on the graph. $y=(x-3)^2+3$

If the vertex and one other point on the graph of a parabola are known, you can write the equation of the parabola in vertex form.

Example 4 Write an Equation Given Points

Write an equation for the parabola whose vertex is at (−1, 4) and passes through (2, 1).

The vertex of the parabola is at (−1, 4), so $h = -1$ and $k = 4$. Since (2, 1) is a point on the graph of the parabola, let $x = 2$ and $y = 1$. Substitute these values into the vertex form of the equation and solve for a.

$y = a(x - h)^2 + k$	Vertex form
$1 = a[2 - (-1)]^2 + 4$	Substitute 1 for y, 2 for x, −1 for h, and 4 for k.
$1 = a(9) + 4$	Simplify.
$-3 = 9a$	Subtract 4 from each side.
$-\frac{1}{3} = a$	Divide each side by 9.

The equation of the parabola in vertex form is $y = -\frac{1}{3}(x + 1)^2 + 4$.

CHECK A graph of $y = -\frac{1}{3}(x + 1)^2 + 4$ verifies that the parabola passes through the point at (2, 1).

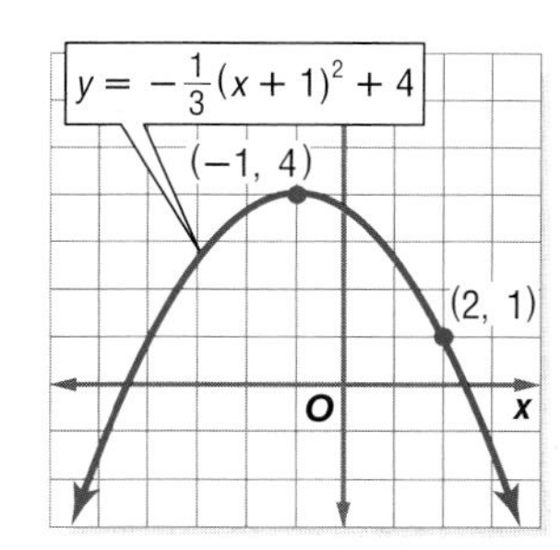

Check for Understanding

Concept Check

1. **Write** a quadratic equation that transforms the graph of $y = 2(x + 1)^2 + 3$ so that it is:
   - **a.** 2 units up.
   - **b.** 3 units down.
   - **c.** 2 units to the left.
   - **d.** 3 units to the right.
   - **e.** narrower.
   - **f.** wider.
   - **g.** opening in the opposite direction.
2. **Explain** how you can find an equation of a parabola using its vertex and one other point on its graph.
3. **OPEN ENDED** Write the equation of a parabola with a vertex of (2, −1).
4. **FIND THE ERROR** Jenny and Ruben are writing $y = x^2 - 2x + 5$ in vertex form.

Jenny	Ruben
$y = x^2 - 2x + 5$	$y = x^2 - 2x + 5$
$y = (x^2 - 2x + 1) + 5 - 1$	$y = (x^2 - 2x + 1) + 5 + 1$
$y = (x - 1)^2 + 4$	$y = (x - 1)^2 + 6$

Who is correct? Explain your reasoning.

Guided Practice

Write each quadratic function in vertex form, if not already in that form. Then identify the vertex, axis of symmetry, and direction of opening.

5. $y = 5(x + 3)^2 - 1$
6. $y = x^2 + 8x - 3$
7. $y = -3x^2 - 18x + 11$

Graph each function.

8. $y = 3(x + 3)^2$
9. $y = \frac{1}{3}(x - 1)^2 + 3$
10. $y = -2x^2 + 16x - 31$

Write an equation for the parabola with the given vertex that passes through the given point.

11. vertex: (2, 0)
    point: (1, 4)
12. vertex: (−3, 6)
    point: (−5, 2)
13. vertex: (−2, −3)
    point: (−4, −5)

Application

14. **FOUNTAINS** The height of a fountain's water stream can be modeled by a quadratic function. Suppose the water from a jet reaches a maximum height of 8 feet at a distance 1 foot away from the jet. If the water lands 3 feet away from the jet, find a quadratic function that models the height $h(d)$ of the water at any given distance d feet from the jet.

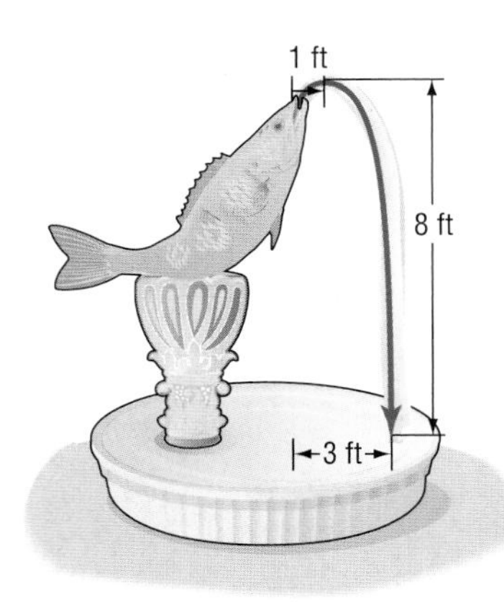

Practice and Apply

Homework Help

For Exercises	See Examples
15–26	2
27–38, 47, 48, 50–52	1, 3
39–46, 49	4

Extra Practice
See page 841.

Write each quadratic function in vertex form, if not already in that form. Then identify the vertex, axis of symmetry, and direction of opening.

15. $y = -2(x + 3)^2$
16. $y = \frac{1}{3}(x - 1)^2 + 2$
17. $y = 5x^2 - 6$
18. $y = -8x^2 + 3$
19. $y = -x^2 - 4x + 8$
20. $y = x^2 - 6x + 1$
21. $y = -3x^2 + 12x$
22. $y = 4x^2 + 24x$
23. $y = 4x^2 + 8x - 3$
24. $y = -2x^2 + 20x - 35$
25. $y = 3x^2 + 3x - 1$
26. $y = 4x^2 - 12x - 11$

Graph each function.

27. $y = 4(x + 3)^2 + 1$
28. $y = -(x - 5)^2 - 3$
29. $y = \frac{1}{4}(x - 2)^2 + 4$
30. $y = \frac{1}{2}(x - 3)^2 - 5$
31. $y = x^2 + 6x + 2$
32. $y = x^2 - 8x + 18$
33. $y = -4x^2 + 16x - 11$
34. $y = -5x^2 - 40x - 80$
35. $y = -\frac{1}{2}x^2 + 5x - \frac{27}{2}$
36. $y = \frac{1}{3}x^2 - 4x + 15$

37. Write one sentence that compares the graphs of $y = 0.2(x + 3)^2 + 1$ and $y = 0.4(x + 3)^2 + 1$.

38. Compare the graphs of $y = 2(x - 5)^2 + 4$ and $y = 2(x - 4)^2 - 1$.

WebQuest

You can use a quadratic function to model the world population. Visit www.algebra2.com/webquest to continue work on your WebQuest project.

Write an equation for the parabola with the given vertex that passes through the given point.

39. vertex: (6, 1)
    point: (5, 10)
40. vertex: (−4, 3)
    point: (−3, 6)
41. vertex: (3, 0)
    point: (6, −6)
42. vertex: (5, 4)
    point: (3, −8)
43. vertex: (0, 5)
    point: (3, 8)
44. vertex: (−3, −2)
    point: (−1, 8)

45. Write an equation for a parabola whose vertex is at the origin and passes through $(2, -8)$.

46. Write an equation for a parabola with vertex at $(-3, -4)$ and y-intercept 8.

Aerospace

The KC135A has the nickname "Vomit Comet." It starts its ascent at 24,000 feet. As it approaches maximum height, the engines are stopped, and the aircraft is allowed to free-fall at a determined angle. Zero gravity is achieved for 25 seconds as the plane reaches the top of its flight and begins its descent.

Source: NASA

47. **AEROSPACE** NASA's KC135A aircraft flies in parabolic arcs to simulate the weightlessness experienced by astronauts in space. The height h of the aircraft (in feet) t seconds after it begins its parabolic flight can be modeled by the equation $h(t) = -9.09(t - 32.5)^2 + 34{,}000$. What is the maximum height of the aircraft during this maneuver and when does it occur?

DIVING For Exercises 48–50, use the following information.

The distance of a diver above the water $d(t)$ (in feet) t seconds after diving off a platform is modeled by the equation $d(t) = -16t^2 + 8t + 30$.

48. Find the time it will take for the diver to hit the water.

49. Write an equation that models the diver's distance above the water if the platform were 20 feet higher.

50. Find the time it would take for the diver to hit the water from this new height.

LAWN CARE For Exercises 51 and 52, use the following information.

The path of water from a sprinkler can be modeled by a quadratic function. The three functions below model paths for three different angles of the water.

Angle A: $y = -0.28(x - 3.09)^2 + 3.27$

Angle B: $y = -0.14(x - 3.57)^2 + 2.39$

Angle C: $y = -0.09(x - 3.22)^2 + 1.53$

51. Which sprinkler angle will send water the highest? Explain your reasoning.

52. Which sprinkler angle will send water the farthest? Explain your reasoning.

53. **CRITICAL THINKING** Given $y = ax^2 + bx + c$ with $a \neq 0$, derive the equation for the axis of symmetry by completing the square and rewriting the equation in the form $y = a(x - h)^2 + k$.

54. **WRITING IN MATH** Answer the question that was posed at the beginning of the lesson.

How can the graph $y = x^2$ be used to graph any quadratic function?

Include the following in your answer:

- a description of the effects produced by changing a, h, and k in the equation $y = a(x - h)^2 + k$, and
- a comparison of the graph of $y = x^2$ and the graph of $y = a(x - h)^2 + k$ using values of your own choosing for a, h, and k.

55. If $f(x) = x^2 - 5x$ and $f(n) = -4$, then which of the following could be n?

Ⓐ -5 Ⓑ -4 Ⓒ -1 Ⓓ 1

56. The vertex of the graph of $y = 2(x - 6)^2 + 3$ is located at which of the following points?

Ⓐ $(2, 3)$ Ⓑ $(6, 3)$ Ⓒ $(6, -3)$ Ⓓ $(-2, 3)$

Maintain Your Skills

Mixed Review

Find the value of the discriminant for each quadratic equation. Then describe the number and type of roots for the equation. *(Lesson 6-5)*

57. $3x^2 - 6x + 2 = 0$ **58.** $4x^2 + 7x = 11$ **59.** $2x^2 - 5x + 6 = 0$

Solve each equation by completing the square. *(Lesson 6-4)*

60. $x^2 + 10x + 17 = 0$ **61.** $x^2 - 6x + 18 = 0$ **62.** $4x^2 + 8x = 9$

Find each quotient. *(Lesson 5-3)*

63. $(2t^3 - 2t - 3) \div (t - 1)$ **64.** $(t^3 - 3t + 2) \div (t + 2)$

65. $(n^4 - 8n^3 + 54n + 105) \div (n - 5)$ **66.** $(y^4 + 3y^3 + y - 1) \div (y + 3)$

67. EDUCATION The graph shows the number of U.S. students in study-abroad programs. *(Lesson 2-5)*

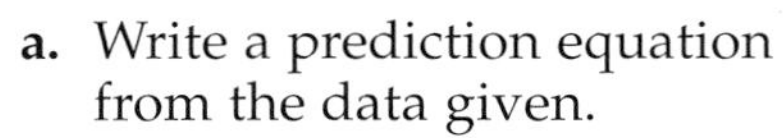

a. Write a prediction equation from the data given.

b. Use your equation to predict the number of students in these programs in 2005.

Getting Ready for the Next Lesson

PREREQUISITE SKILL Determine whether the given value satisfies the inequality.
*(To review **inequalities**, see Lesson 1-6.)*

68. $-2x^2 + 3 < 0; x = 5$

69. $4x^2 + 2x - 3 \geq 0; x = -1$

70. $4x^2 - 4x + 1 \leq 10; x = 2$

71. $6x^2 + 3x > 8; x = 0$

Practice Quiz 2

Lessons 6-4 through 6-6

Solve each equation by completing the square. *(Lesson 6-4)*

1. $x^2 + 14x + 37 = 0$ **2.** $2x^2 - 2x + 5 = 0$

Find the value of the discriminant for each quadratic equation. Then describe the number and type of roots for the equation. *(Lesson 6-5)*

3. $5x^2 - 3x + 1 = 0$ **4.** $3x^2 + 4x - 7 = 0$

Solve each equation by using the Quadratic Formula. *(Lesson 6-5)*

5. $x^2 + 9x - 11 = 0$ **6.** $-3x^2 + 4x = 4$

7. Write an equation for a parabola with vertex at $(2, -5)$ that passes through $(-1, 1)$. *(Lesson 6-6)*

Write each equation in vertex form. Then identify the vertex, axis of symmetry, and direction of opening. *(Lesson 6-6)*

8. $y = x^2 + 8x + 18$ **9.** $y = -x^2 + 12x - 36$ **10.** $y = 2x^2 + 12x + 13$

6-7 Graphing and Solving Quadratic Inequalities

What You'll Learn

- Graph quadratic inequalities in two variables.
- Solve quadratic inequalities in one variable.

Vocabulary

- quadratic inequality

How can you find the time a trampolinist spends above a certain height?

Trampolining was first featured as an Olympic sport at the 2000 Olympics in Sydney, Australia. Suppose the height $h(t)$ in feet of a trampolinist above the ground during one bounce is modeled by the quadratic function $h(t) = -16t^2 + 42t + 3.75$. We can solve a quadratic inequality to determine how long this performer is more than a certain distance above the ground.

Study Tip

Look Back
For review of **graphing linear inequalities,** see Lesson 2-7.

GRAPH QUADRATIC INEQUALITIES You can graph **quadratic inequalities** in two variables using the same techniques you used to graph linear inequalities in two variables.

Step 1 Graph the related quadratic equation, $y = ax^2 + bx + c$. Decide if the parabola should be solid or dashed.

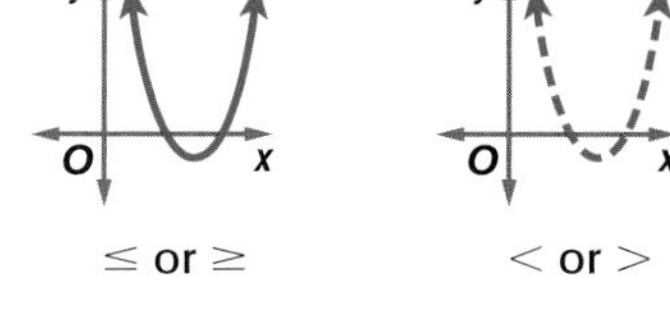

Step 2 Test a point (x_1, y_1) inside the parabola. Check to see if this point is a solution of the inequality.

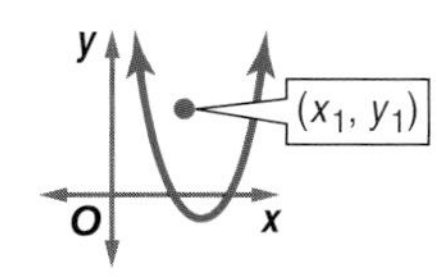

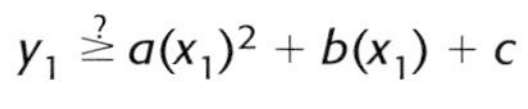

Step 3 If (x_1, y_1) is a solution, shade the region *inside* the parabola. If (x_1, y_1) is *not* a solution, shade the region *outside* the parabola.

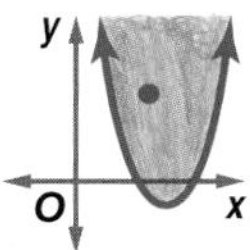

(x_1, y_1) is a solution.

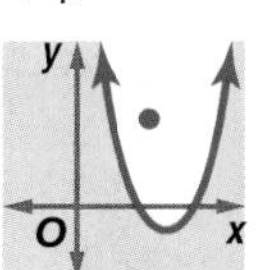

(x_1, y_1) is not a solution.

Example 1 Graph a Quadratic Inequality

Graph $y > -x^2 - 6x - 7$.

Step 1 Graph the related quadratic equation, $y = -x^2 - 6x - 7$.

Since the inequality symbol is >, the parabola should be dashed.

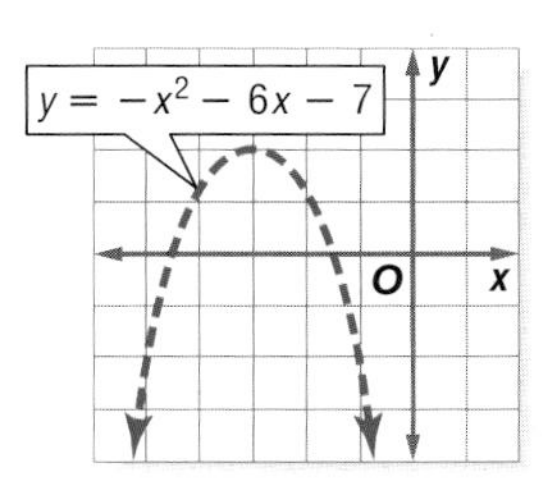

(continued on the next page)

Step 2 Test a point inside the parabola, such as $(-3, 0)$.

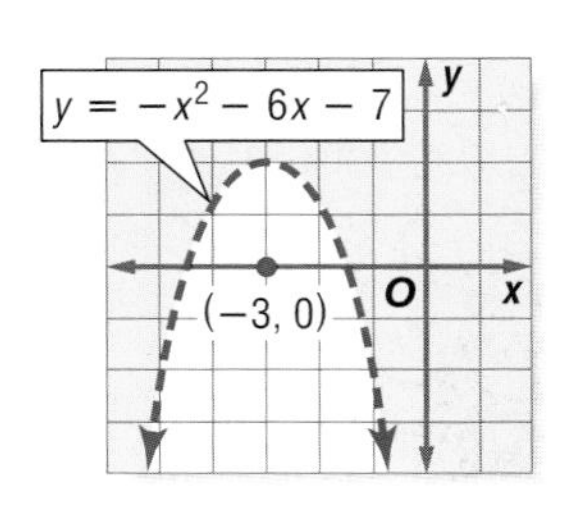

$$y > -x^2 - 6x - 7$$
$$0 \overset{?}{>} -(-3)^2 - 6(-3) - 7$$
$$0 \overset{?}{>} -9 + 18 - 7$$
$$0 \overset{?}{>} 2 \quad \times$$

So, $(-3, 0)$ is *not* a solution of the inequality.

Step 3 Shade the region outside the parabola.

SOLVE QUADRATIC INEQUALITIES To solve a quadratic inequality in one variable, you can use the graph of the related quadratic function.

To solve $ax^2 + bx + c < 0$, graph $y = ax^2 + bx + c$. Identify the x values for which the graph lies *below* the x-axis.

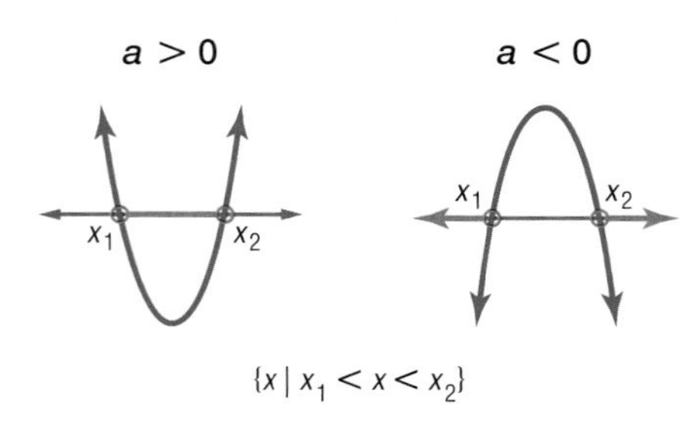

For $\leq$, include the x-intercepts in the solution.

To solve $ax^2 + bx + c > 0$, graph $y = ax^2 + bx + c$. Identify the x values for which the graph lies *above* the x-axis.

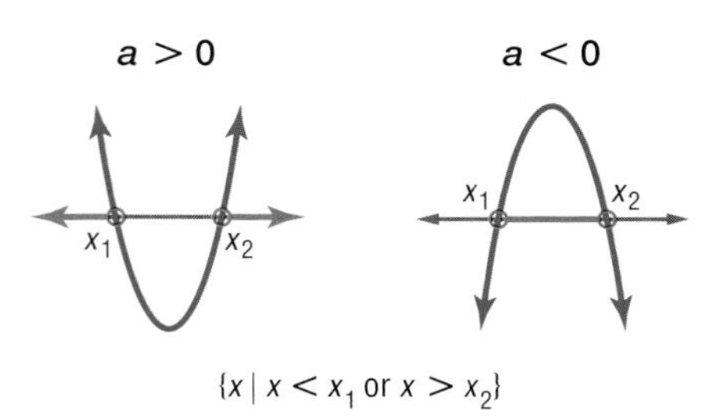

For $\geq$, include the x-intercepts in the solution.

Example 2 Solve $ax^2 + bx + c > 0$

Study Tip

Solving Quadratic Inequalities by Graphing

A precise graph of the related quadratic function is not necessary since the zeros of the function were found algebraically.

Solve $x^2 + 2x - 3 > 0$ by graphing.

The solution consists of the x values for which the graph of the related quadratic function lies *above* the x-axis. Begin by finding the roots of the related equation.

$x^2 + 2x - 3 = 0$	Related equation
$(x + 3)(x - 1) = 0$	Factor.
$x + 3 = 0$ or $x - 1 = 0$	Zero Product Property
$x = -3$ $\quad$ $x = 1$	Solve each equation.

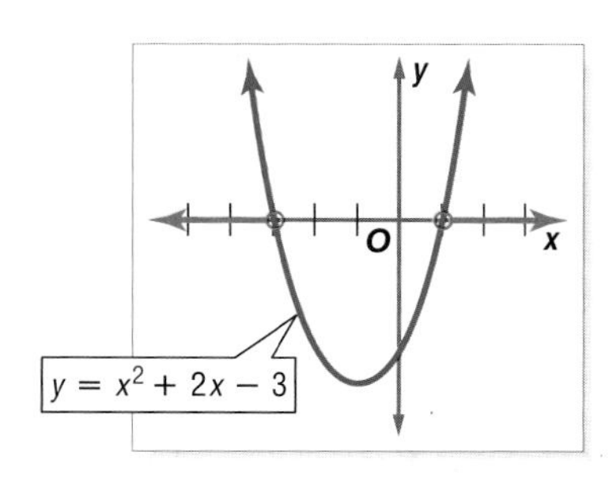

Sketch the graph of a parabola that has x-intercepts at -3 and 1. The graph should open up since $a > 0$.

The graph lies above the x-axis to the left of $x = -3$ and to the right of $x = 1$. Therefore, the solution set is $\{x \mid x < -3 \text{ or } x > 1\}$.

Example 3 Solve $ax^2 + bx + c \leq 0$

Solve $0 \geq 3x^2 - 7x - 1$ by graphing.

This inequality can be rewritten as $3x^2 - 7x - 1 \leq 0$. The solution consists of the x values for which the graph of the related quadratic function lies *on and below* the x-axis. Begin by finding the roots of the related equation.

$3x^2 - 7x - 1 = 0$ Related equation

$x = \frac{-b \pm \sqrt{b^2 - 4ac}}{2a}$ Use the Quadratic Formula.

$x = \frac{-(-7) \pm \sqrt{(-7)^2 - 4(3)(-1)}}{2(3)}$ Replace a with 3, b with -7, and c with -1.

$x = \frac{7 + \sqrt{61}}{6}$ or $x = \frac{7 - \sqrt{61}}{6}$ Simplify and write as two equations.

$x \approx 2.47$ $\quad x \approx -0.14$ Simplify.

Sketch the graph of a parabola that has x-intercepts of 2.47 and -0.14. The graph should open up since $a > 0$.

The graph lies on and below the x-axis at $x = -0.14$ and $x = 2.47$ and between these two values. Therefore, the solution set of the inequality is approximately $\{x \mid -0.14 \leq x \leq 2.47\}$.

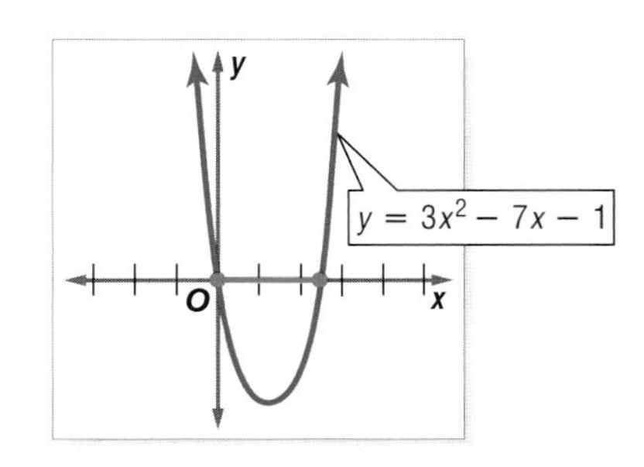

CHECK Test one value of x less than -0.14, one between -0.14 and 2.47, and one greater than 2.47 in the original inequality.

Test $x = -1$.	Test $x = 0$.	Test $x = 3$.
$0 \geq 3x^2 - 7x - 1$	$0 \geq 3x^2 - 7x - 1$	$0 \geq 3x^2 - 7x - 1$
$0 \stackrel{?}{\geq} 3(-1)^2 - 7(-1) - 1$	$0 \stackrel{?}{\geq} 3(0)^2 - 7(0) - 1$	$0 \stackrel{?}{\geq} 3(3)^2 - 7(3) - 1$
$0 \geq 9$ ✗	$0 \geq -1$ ✓	$0 \geq 5$ ✗

Real-world problems that involve vertical motion can often be solved by using a quadratic inequality.

More About. . .

Football

A long hang time allows the kicking team time to provide good coverage on a punt return. The suggested hang time for high school and college punters is 4.5–4.6 seconds.

Source: www.takeaknee.com

Example 4 Write an Inequality

FOOTBALL The height of a punted football can be modeled by the function $H(x) = -4.9x^2 + 20x + 1$, where the height $H(x)$ is given in meters and the time x is in seconds. At what time in its flight is the ball within 5 meters of the ground?

The function $H(x)$ describes the height of the football. Therefore, you want to find the values of x for which $H(x) \leq 5$.

$H(x) \leq 5$ Original inequality

$-4.9x^2 + 20x + 1 \leq 5$ $\quad H(x) = -4.9x^2 + 20x + 1$

$-4.9x^2 + 20x - 4 \leq 0$ Subtract 5 from each side.

Graph the related function $y = -4.9x^2 + 20x - 4$ using a graphing calculator. The zeros of the function are about 0.21 and 3.87, and the graph lies below the x-axis when $x < 0.21$ or $x > 3.87$.

Thus, the ball is within 5 meters of the ground for the first 0.21 second of its flight and again after 3.87 seconds until the ball hits the ground at 4.13 seconds.

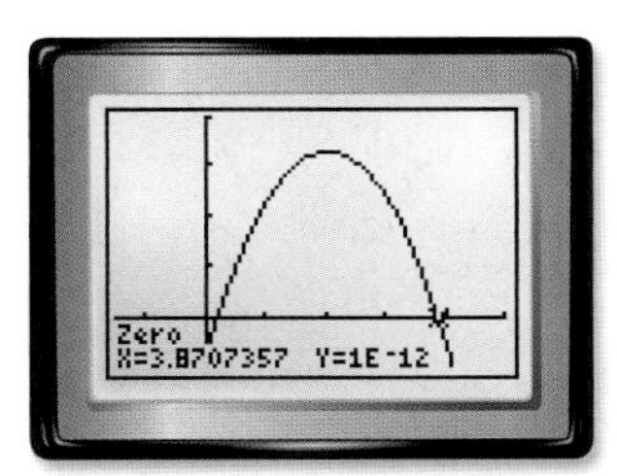

[−1.5, 5] scl: 1 by [−5, 20] scl: 5

You can also solve quadratic inequalities algebraically.

Example 5 Solve a Quadratic Inequality

Solve $x^2 + x > 6$ algebraically.

First solve the related quadratic equation $x^2 + x = 6$.

$x^2 + x = 6$	Related quadratic equation
$x^2 + x - 6 = 0$	Subtract 6 from each side.
$(x + 3)(x - 2) = 0$	Factor.
$x + 3 = 0$ or $x - 2 = 0$	Zero Product Property
$x = -3$ $\quad$ $x = 2$	Solve each equation.

Plot –3 and 2 on a number line. Use circles since these values are not solutions of the original inequality. Notice that the number line is now separated into three intervals.

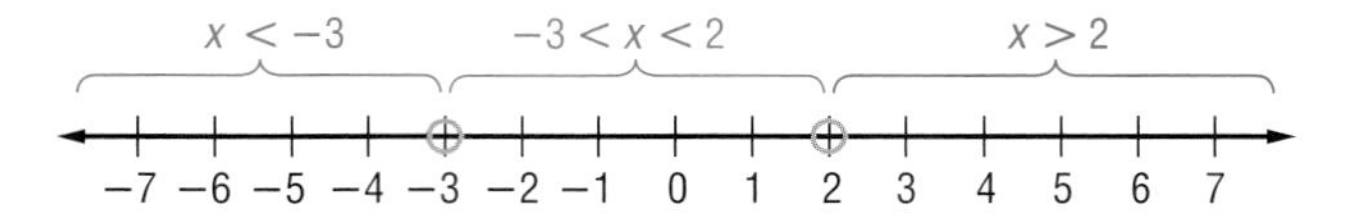

Test a value in each interval to see if it satisfies the original inequality.

$x < -3$	$-3 < x < 2$	$x > 2$
Test $x = -4$.	Test $x = 0$.	Test $x = 4$.
$x^2 + x > 6$	$x^2 + x > 6$	$x^2 + x > 6$
$(-4)^2 + (-4) \overset{?}{>} 6$	$0^2 + 0 \overset{?}{>} 6$	$4^2 + 4 \overset{?}{>} 6$
$12 > 6$ ✓	$0 > 6$ ✗	$20 > 6$ ✓

The solution set is $\{x \mid x < -3 \text{ or } x > 2\}$. This is shown on the number line below.

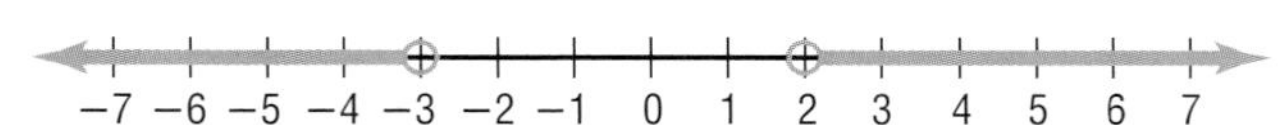

Study Tip

Solving Quadratic Inequalities Algebraically

As with linear inequalities, the solution set of a quadratic inequality can be all real numbers or the empty set, $\varnothing$. The solution is all real numbers when all three test points satisfy the inequality. It is the empty set when none of the tests points satisfy the inequality.

Check for Understanding

Concept Check

1. **Determine** which inequality, $y \geq (x - 3)^2 - 1$ or $y \leq (x - 3)^2 - 1$, describes the graph at the right.

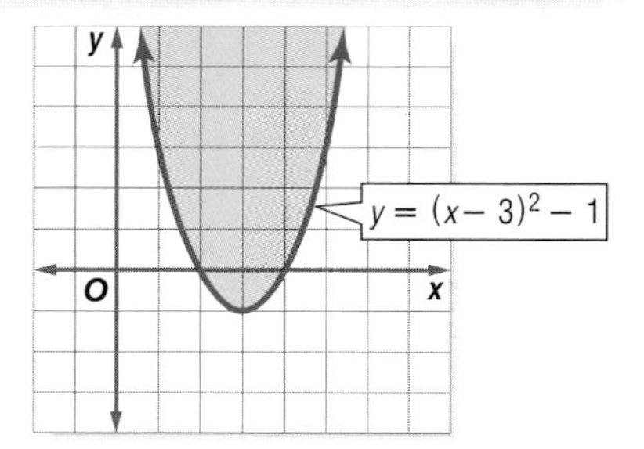

2. **OPEN ENDED** List three points you might test to find the solution of $(x + 3)(x - 5) < 0$.

3. **Examine** the graph of $y = x^2 - 4x - 5$ at the right.
   a. What are the solutions of $0 = x^2 - 4x - 5$?
   b. What are the solutions of $x^2 - 4x - 5 \geq 0$?
   c. What are the solutions of $x^2 - 4x - 5 \leq 0$?

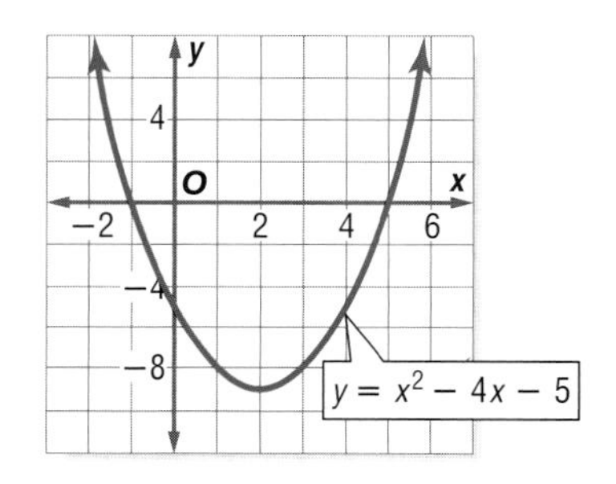

Guided Practice

Graph each inequality.

4. $y \geq x^2 - 10x + 25$
5. $y < x^2 - 16$
6. $y > -2x^2 - 4x + 3$
7. $y \leq -x^2 + 5x + 6$

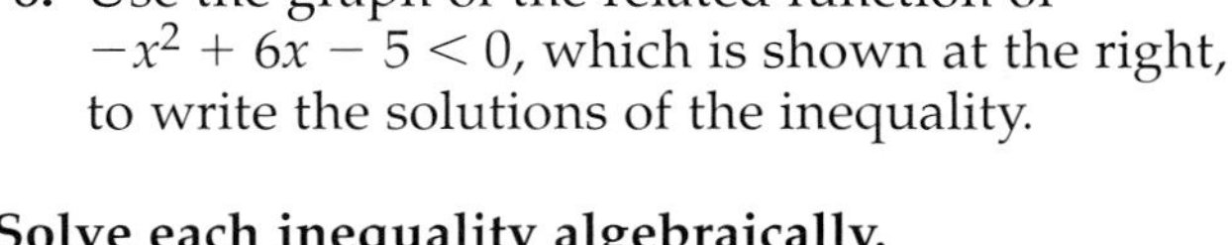

8. Use the graph of the related function of $-x^2 + 6x - 5 < 0$, which is shown at the right, to write the solutions of the inequality.

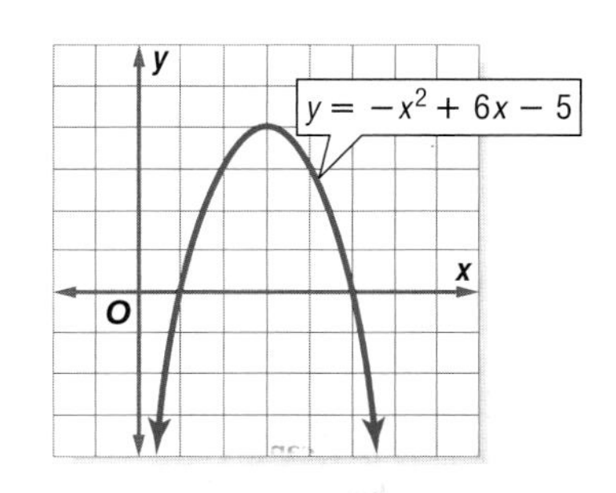

Solve each inequality algebraically.

9. $x^2 - 6x - 7 < 0$
10. $x^2 - x - 12 > 0$
11. $x^2 < 10x - 25$
12. $x^2 \leq 3$

Application

13. **BASEBALL** A baseball player hits a high pop-up with an initial upward velocity of 30 meters per second, 1.4 meters above the ground. The height $h(t)$ of the ball in meters t seconds after being hit is modeled by $h(t) = -4.9t^2 + 30t + 1.4$. How long does a player on the opposing team have to catch the ball if he catches it 1.7 meters above the ground?

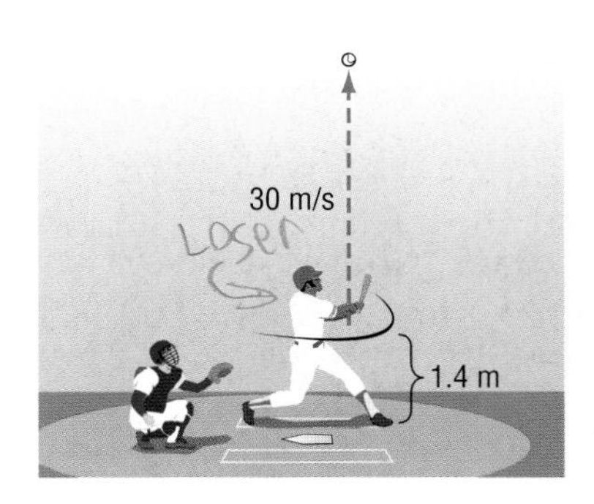

Practice and Apply

Homework Help

For Exercises	See Examples
14–25	1
26–29	2, 3
30–42	2, 3, 5
43–48	4

Extra Practice
See page 841.

Graph each inequality.

14. $y \geq x^2 + 3x - 18$
15. $y < -x^2 + 7x + 8$
16. $y \leq x^2 + 4x + 4$
17. $y \leq x^2 + 4x$
18. $y > x^2 - 36$
19. $y > x^2 + 6x + 5$
20. $y \leq -x^2 - 3x + 10$
21. $y \geq -x^2 - 7x + 10$
22. $y > -x^2 + 10x - 23$
23. $y < -x^2 + 13x - 36$
24. $y < 2x^2 + 3x - 5$
25. $y \geq 2x^2 + x - 3$

Use the graph of its related function to write the solutions of each inequality.

26. $-x^2 + 10x - 25 \geq 0$

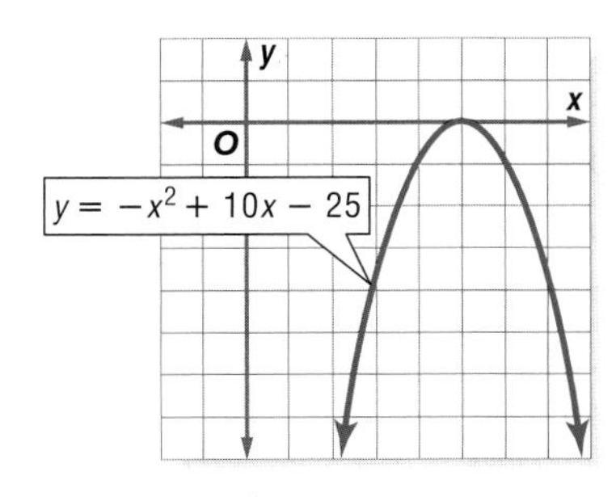

27. $x^2 - 4x - 12 \leq 0$

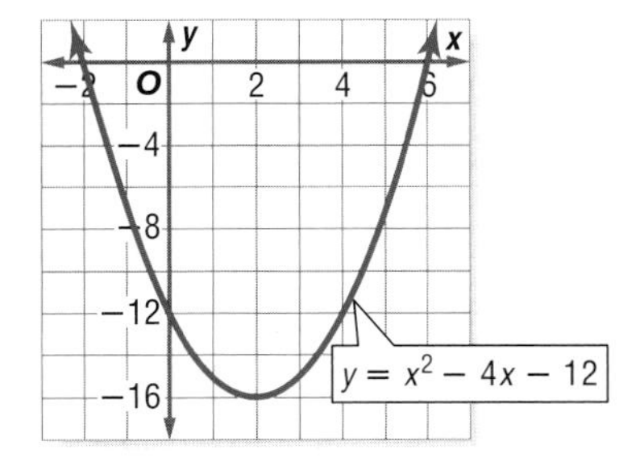

28. $x^2 - 9 > 0$

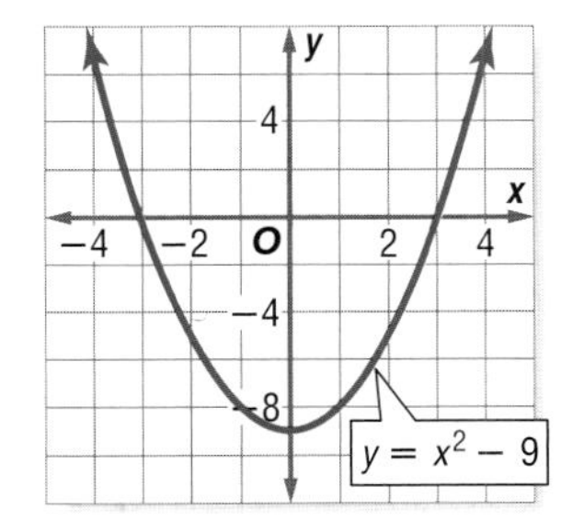

29. $-x^2 - 10x - 21 < 0$

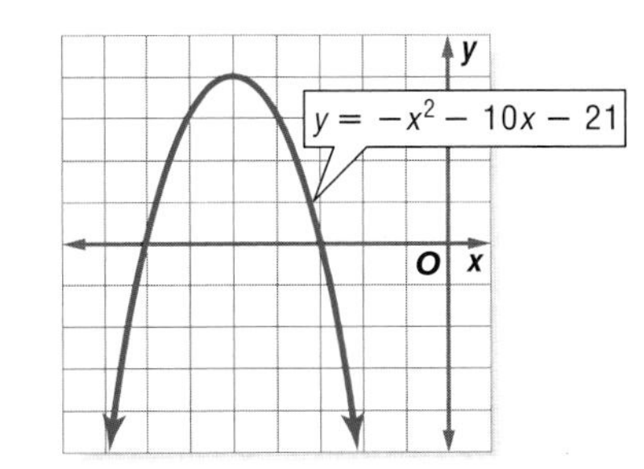

Solve each inequality algebraically.

30. $x^2 - 3x - 18 > 0$
31. $x^2 + 3x - 28 < 0$
32. $x^2 - 4x \leq 5$
33. $x^2 + 2x \geq 24$
34. $-x^2 - x + 12 \geq 0$
35. $-x^2 - 6x + 7 \leq 0$
36. $9x^2 - 6x + 1 \leq 0$
37. $4x^2 + 20x + 25 \geq 0$
38. $x^2 + 12x < -36$
39. $-x^2 + 14x - 49 \geq 0$
40. $18x - x^2 \leq 81$
41. $16x^2 + 9 < 24x$

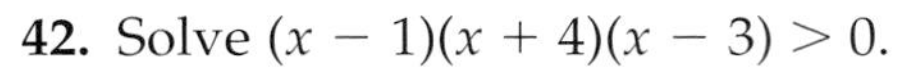

42. Solve $(x - 1)(x + 4)(x - 3) > 0$.

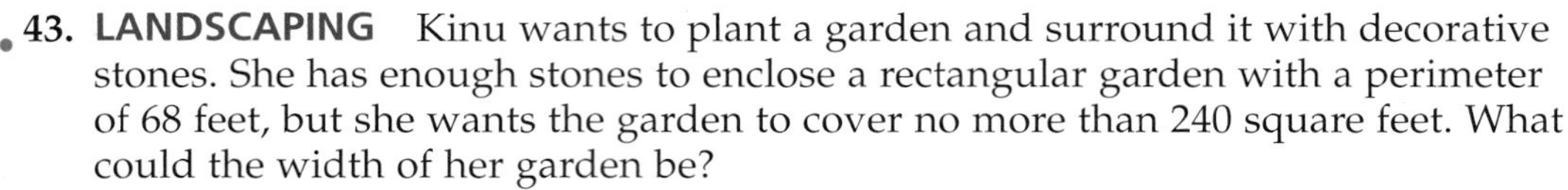

43. **LANDSCAPING** Kinu wants to plant a garden and surround it with decorative stones. She has enough stones to enclose a rectangular garden with a perimeter of 68 feet, but she wants the garden to cover no more than 240 square feet. What could the width of her garden be?

44. **BUSINESS** A mall owner has determined that the relationship between monthly rent charged for store space r (in dollars per square foot) and monthly profit $P(r)$ (in thousands of dollars) can be approximated by the function $P(r) = -8.1r^2 + 46.9r - 38.2$. Solve each quadratic equation or inequality. Explain what each answer tells about the relationship between monthly rent and profit for this mall.

 a. $-8.1r^2 + 46.9r - 38.2 = 0$
 b. $-8.1r^2 + 46.9r - 38.2 > 0$
 c. $-8.1r^2 + 46.9r - 38.2 > 10$
 d. $-8.1r^2 + 46.9r - 38.2 < 10$

45. **GEOMETRY** A rectangle is 6 centimeters longer than it is wide. Find the possible dimensions if the area of the rectangle is more than 216 square centimeters.

FUND-RAISING For Exercises 46–48, use the following information.
The girls' softball team is sponsoring a fund-raising trip to see a professional baseball game. They charter a 60-passenger bus for $525. In order to make a profit, they will charge $15 per person if all seats on the bus are sold, but for each empty seat, they will increase the price by $1.50 per person.

46. Write a quadratic function giving the softball team's profit $P(n)$ from this fund-raiser as a function of the number of passengers n.
47. What is the minimum number of passengers needed in order for the softball team not to lose money?
48. What is the maximum profit the team can make with this fund-raiser, and how many passengers will it take to achieve this maximum?

49. **CRITICAL THINKING** Graph the intersection of the graphs of $y \leq -x^2 + 4$ and $y \geq x^2 - 4$.

50. **WRITING IN MATH** Answer the question that was posed at the beginning of the lesson.

How can you find the time a trampolinist spends above a certain height?

Include the following in your answer:

- a quadratic inequality that describes the time the performer spends more than 10 feet above the ground, and
- two approaches to solving this quadratic inequality.

Career Choices

Landscape Architect

Landscape architects design outdoor spaces so that they are not only functional, but beautiful and compatible with the natural environment.

Online Research
For information about a career as a landscape architect, visit: www.algebra2.com/careers

Standardized Test Practice

Standards Practice

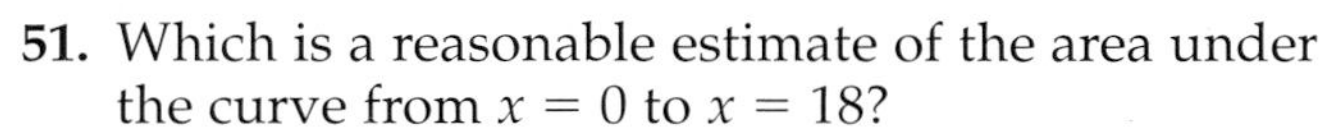

51. Which is a reasonable estimate of the area under the curve from $x = 0$ to $x = 18$?

    (A) 29 square units

    (B) 58 square units

    (C) 116 square units

    (D) 232 square units

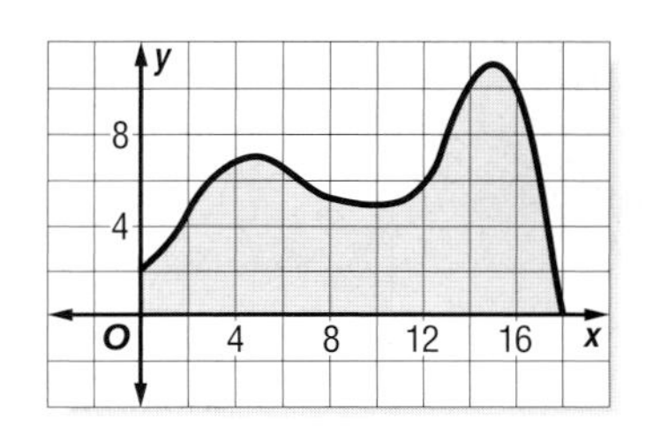

52. If $(x + 1)(x - 2)$ is positive, then

    (A) $x < -1$ or $x > 2$.

    (B) $x > -1$ or $x < 2$.

    (C) $-1 < x < 2$.

    (D) $-2 < x < 1$.

Extending the Lesson

SOLVE ABSOLUTE VALUE INEQUALITIES BY GRAPHING Similar to quadratic inequalities, you can solve absolute value inequalities by graphing.

Graph the related absolute value function for each inequality using a graphing calculator. For $>$ and $\geq$, identify the x values, if any, for which the graph lies *below* the x-axis. For $<$ and $\leq$, identify the x values, if any, for which the graph lies *above* the x-axis.

53. $|x - 2| > 0$

54. $|x| - 7 < 0$

55. $-|x + 3| + 6 < 0$

56. $2|x + 3| - 1 \geq 0$

57. $|5x + 4| - 2 \leq 0$

58. $|4x - 1| + 3 < 0$

Maintain Your Skills

Mixed Review

Write each equation in vertex form. Then identify the vertex, axis of symmetry, and direction of opening. *(Lesson 6-6)*

59. $y = x^2 - 2x + 9$

60. $y = -2x^2 + 16x - 32$

61. $y = \frac{1}{2}x^2 + 6x + 18$

Solve each equation using the method of your choice. Find exact solutions. *(Lesson 6-5)*

62. $x^2 + 12x + 32 = 0$

63. $x^2 + 7 = -5x$

64. $3x^2 + 6x - 2 = 3$

Simplify. *(Lesson 5-2)*

65. $(2a^2b - 3ab^2 + 5a - 6b) + (4a^2b^2 + 7ab^2 - b + 7a)$

66. $(x^3 - 3x^2y + 4xy^2 + y^3) - (7x^3 + x^2y - 9xy^2 + y^3)$

67. $x^{-3}y^2(x^4y + x^3y^{-1} + x^2y^{-2})$

68. $(5a - 3)(1 - 3a)$

Find each product, if possible. *(Lesson 4-3)*

69. $\begin{bmatrix} -6 & 3 \\ 4 & 7 \end{bmatrix} \cdot \begin{bmatrix} 2 & -5 \\ -3 & 6 \end{bmatrix}$

70. $[2 \quad -6 \quad 3] \cdot \begin{bmatrix} 3 & -3 \\ 9 & 0 \\ -2 & 4 \end{bmatrix}$

71. **LAW ENFORCEMENT** Thirty-four states classify drivers having at least a 0.1 blood alcohol content (BAC) as intoxicated. An infrared device measures a person's BAC through an analysis of his or her breath. A certain detector measures BAC to within 0.002. If a person's actual blood alcohol content is 0.08, write and solve an absolute value equation to describe the range of BACs that might register on this device. *(Lesson 1-6)*

Chapter 6 Study Guide and Review

Vocabulary and Concept Check

axis of symmetry (p. 287)
completing the square (p. 307)
constant term (p. 286)
discriminant (p. 316)
linear term (p. 286)
maximum value (p. 288)
minimum value (p. 288)
parabola (p. 286)
quadratic equation (p. 294)
Quadratic Formula (p. 313)
quadratic function (p. 286)
quadratic inequality (p. 329)
quadratic term (p. 286)
roots (p. 294)
Square Root Property (p. 306)
vertex (p. 287)
vertex form (p. 322)
Zero Product Property (p. 301)
zeros (p. 294)

Choose the letter of the term that best matches each phrase.

1. the graph of any quadratic function
2. process used to create a perfect square trinomial
3. the line passing through the vertex of a parabola and dividing the parabola into two mirror images
4. a function described by an equation of the form $f(x) = ax^2 + bx + c$, where $a \neq 0$
5. the solutions of an equation
6. $y = a(x - h)^2 + k$
7. in the Quadratic Formula, the expression under the radical sign, $b^2 - 4ac$
8. $x = \dfrac{-b \pm \sqrt{b^2 - 4ac}}{2a}$

a. axis of symmetry
b. completing the square
c. discriminant
d. constant term
e. linear term
f. parabola
g. Quadratic Formula
h. quadratic function
i. roots
j. vertex form

Lesson-by-Lesson Review

6-1 Graphing Quadratic Functions

See pages 286–293.

Concept Summary

The graph of $y = ax^2 + bx + c$, $a \neq 0$,

- opens up, and the function has a minimum value when $a > 0$, and
- opens down, and the function has a maximum value when $a < 0$.

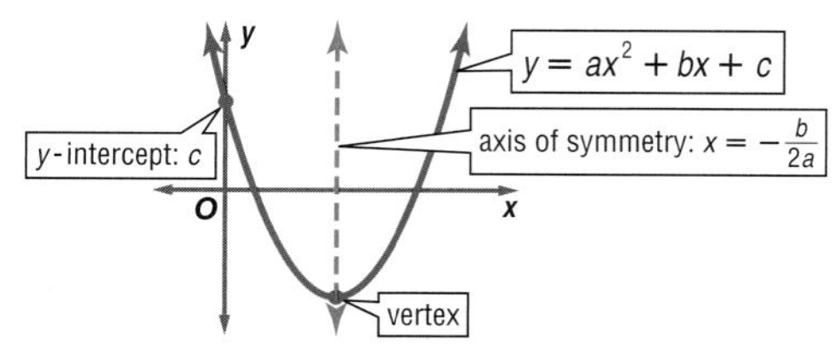

Example **Find the maximum or minimum value of $f(x) = -x^2 + 4x - 12$.**

Since $a < 0$, the graph opens down and the function has a maximum value. The maximum value of the function is the y-coordinate of the vertex. The x-coordinate of the vertex is $x = -\dfrac{4}{2(-1)}$ or 2. Find the y-coordinate by evaluating the function for $x = 2$.

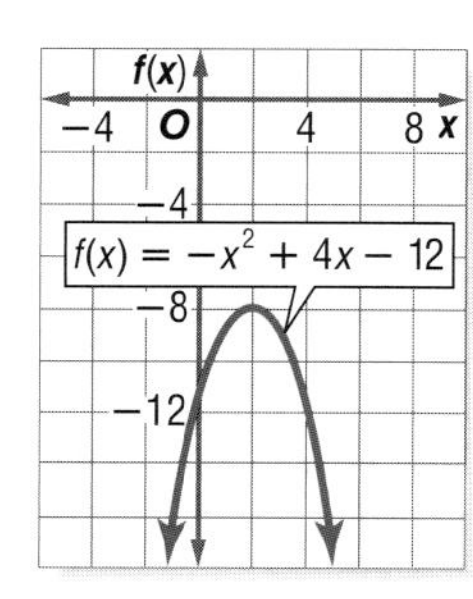

$f(x) = -x^2 + 4x - 12$ Original function

$f(2) = -(2)^2 + 4(2) - 12$ or -8 Replace x with 2.

Therefore, the maximum value of the function is -8.

www.algebra2.com/vocabulary_review

Exercises **Complete parts a–c for each quadratic function.**

a. Find the y-intercept, the equation of the axis of symmetry, and the x-coordinate of the vertex.

b. Make a table of values that includes the vertex.

c. Use this information to graph the function. *(See Example 2 on pages 287 and 288.)*

9. $f(x) = x^2 + 6x + 20$ **10.** $f(x) = x^2 - 2x - 15$ **11.** $f(x) = x^2 - 8x + 7$

12. $f(x) = -2x^2 + 12x - 9$ **13.** $f(x) = -x^2 - 4x - 3$ **14.** $f(x) = 3x^2 + 9x + 6$

Determine whether each function has a maximum or a minimum value. Then find the maximum or minimum value of each function.
(See Example 3 on pages 288 and 289.)

15. $f(x) = 4x^2 - 3x - 5$ **16.** $f(x) = -3x^2 + 2x - 2$ **17.** $f(x) = -2x^2 + 7$

6-2 Solving Quadratic Equations by Graphing

See pages 294–299.

Concept Summary

- The solutions, or roots, of a quadratic equation are the zeros of the related quadratic function. You can find the zeros of a quadratic function by finding the x-intercepts of its graph.
- A quadratic equation can have one real solution, two real solutions, or no real solution.

One Real Solution

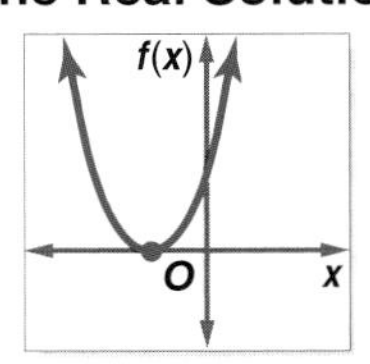

Two Real Solutions

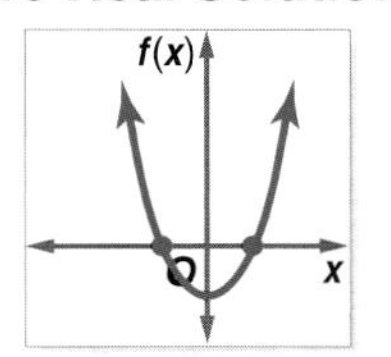

No Real Solution

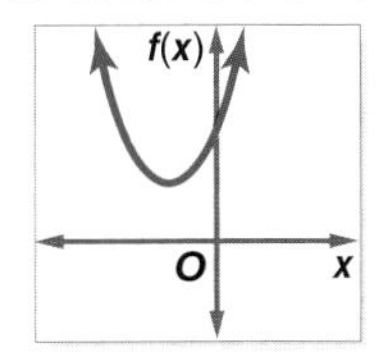

Example **Solve $2x^2 - 5x + 2 = 0$ by graphing.**

The equation of the axis of symmetry is $x = -\frac{-5}{2(2)}$ or $x = \frac{5}{4}$.

x	0	$\frac{1}{2}$	$\frac{5}{4}$	2	$\frac{5}{2}$
$f(x)$	2	0	$-\frac{9}{8}$	0	2

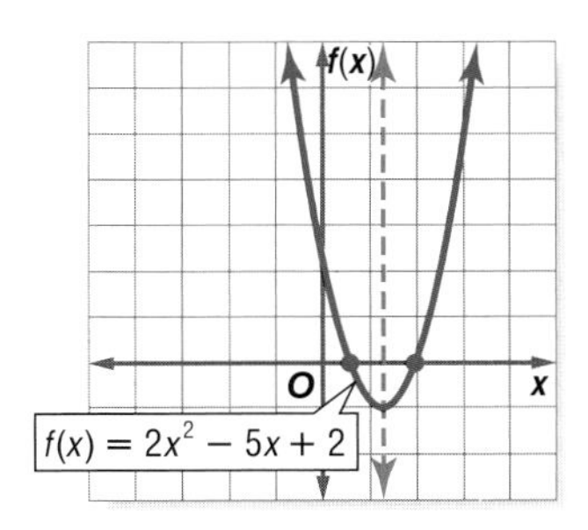

The zeros of the related function are $\frac{1}{2}$ and 2. Therefore, the solutions of the equation are $\frac{1}{2}$ and 2.

Exercises **Solve each equation by graphing. If exact roots cannot be found, state the consecutive integers between which the roots are located.**
(See Examples 1–3 on pages 294 and 295.)

18. $x^2 - 36 = 0$ **19.** $-x^2 - 3x + 10 = 0$ **20.** $2x^2 + x - 3 = 0$

21. $-x^2 - 40x - 80 = 0$ **22.** $-3x^2 - 6x - 2 = 0$ **23.** $\frac{1}{5}(x + 3)^2 - 5 = 0$

6-3 Solving Quadratic Equations by Factoring

See pages 301–305.

Concept Summary

- Zero Product Property: For any real numbers a and b, if $ab = 0$, then either $a = 0$, $b = 0$, or both a and $b = 0$.

Example **Solve $x^2 + 9x + 20 = 0$ by factoring.**

$x^2 + 9x + 20 = 0$	Original equation
$(x + 4)(x + 5) = 0$	Factor the trinomial.
$x + 4 = 0$ or $x + 5 = 0$	Zero Product Property
$x = -4$ $\quad$ $x = -5$	The solution set is $\{-5, -4\}$.

Exercises **Solve each equation by factoring.** *(See Examples 1–3 on pages 301 and 302.)*

24. $x^2 - 4x - 32 = 0$ **25.** $3x^2 + 6x + 3 = 0$ **26.** $5y^2 = 80$

27. $2c^2 + 18c - 44 = 0$ **28.** $25x^2 - 30x = -9$ **29.** $6x^2 + 7x = 3$

Write a quadratic equation with the given root(s). Write the equation in the form $ax^2 + bx + c$, where a, b, and c are integers. *(See Example 4 on page 303.)*

30. $-4, -25$ **31.** $10, -7$ **32.** $\frac{1}{3}, 2$

6-4 Completing the Square

See pages 306–312.

Concept Summary

- To complete the square for any quadratic expression $x^2 + bx$:

Step 1 Find one half of b, the coefficient of x.

Step 2 Square the result in Step 1.

Step 3 Add the result of Step 2 to $x^2 + bx$. $\quad x^2 + bx + \left(\frac{b}{2}\right)^2 = \left(x + \frac{b}{2}\right)^2$

Example **Solve $x^2 + 10x - 39 = 0$ by completing the square.**

$x^2 + 10x - 39 = 0$	Notice that $x^2 + 10x - 39 = 0$ is not a perfect square.
$x^2 + 10x = 39$	Rewrite so the left side is of the form $x^2 + bx$.
$x^2 + 10x + 25 = 39 + 25$	Since $\left(\frac{10}{2}\right)^2 = 25$, add 25 to each side.
$(x + 5)^2 = 64$	Write the left side as a perfect square by factoring.
$x + 5 = \pm 8$	Square Root Property
$x + 5 = 8$ or $x + 5 = -8$	Rewrite as two equations.
$x = 3$ $\quad$ $x = -13$	The solution set is $\{-13, 3\}$.

Exercises **Find the value of c that makes each trinomial a perfect square. Then write the trinomial as a perfect square.** *(See Example 3 on page 307.)*

33. $x^2 + 34x + c$ **34.** $x^2 - 11x + c$ **35.** $x^2 + \frac{7}{2}x + c$

Solve each equation by completing the square. *(See Examples 4–6 on pages 308 and 309.)*

36. $2x^2 - 7x - 15 = 0$ **37.** $2n^2 - 12n - 22 = 0$ **38.** $2x^2 - 5x + 7 = 3$

6-5 The Quadratic Formula and the Discriminant

See pages 313–319.

Concept Summary

- Quadratic Formula: $x = \frac{-b \pm \sqrt{b^2 - 4ac}}{2a}$ where $a \neq 0$

Solve $x^2 - 5x - 66 = 0$ by using the Quadratic Formula.

$x = \frac{-b \pm \sqrt{b^2 - 4ac}}{2a}$ — Quadratic Formula

$= \frac{-(-5) \pm \sqrt{(-5)^2 - 4(1)(-66)}}{2(1)}$ — Replace a with 1, b with -5, and c with -66.

$= \frac{5 \pm 17}{2}$ — Simplify.

$x = \frac{5 + 17}{2}$ or $x = \frac{5 - 17}{2}$ — Write as two equations.

$= 11$ $\quad = -6$ — The solution set is $\{11, -6\}$.

Exercises **Complete parts a–c for each quadratic equation.**

a. Find the value of the discriminant.

b. Describe the number and type of roots.

c. Find the exact solutions by using the Quadratic Formula.

(See Examples 1–4 on pages 314–316.)

39. $x^2 + 2x + 7 = 0$ **40.** $-2x^2 + 12x - 5 = 0$ **41.** $3x^2 + 7x - 2 = 0$

6-6 Analyzing Graphs of Quadratic Functions

See pages 322–328.

Concept Summary

- As the values of h and k change, the graph of $y = (x - h)^2 + k$ is the graph of $y = x^2$ translated
  - $|h|$ units left if h is negative or $|h|$ units right if h is positive.
  - $|k|$ units up if k is positive or $|k|$ units down if k is negative.
- Consider the equation $y = a(x - h)^2 + k$.
  - If $a > 0$, the graph opens up; if $a < 0$ the graph opens down.
  - If $|a| > 1$, the graph is narrower than the graph of $y = x^2$.
  - If $|a| < 1$, the graph is wider than the graph of $y = x^2$.

Example **Write the quadratic function $y = 3x^2 + 42x + 142$ in vertex form. Then identify the vertex, axis of symmetry, and direction of opening.**

$y = 3x^2 + 42x + 142$ — Original equation

$y = 3(x^2 + 14x) + 142$ — Group $ax^2 + bx$ and factor, dividing by a.

$y = 3(x^2 + 14x + 49) + 142 - 3(49)$ — Complete the square by adding $3\left(\frac{14}{2}\right)^2$. Balance this with a subtraction of 3(49).

$y = 3(x + 7)^2 - 5$ — Write $x^2 + 14x + 49$ as a perfect square.

So, $a = 3$, $h = -7$, and $k = -5$. The vertex is at $(-7, -5)$, and the axis of symmetry is $x = -7$. Since a is positive, the graph opens up.

• Extra Practice, see pages 839–841.
• Mixed Problem Solving, see page 867.

Exercises **Write each equation in vertex form, if not already in that form. Then identify the vertex, axis of symmetry, and direction of opening.** *(See Examples 1 and 3 on pages 322 and 324.)*

42. $y = -6(x + 2)^2 + 3$ **43.** $y = 5x^2 + 35x + 58$ **44.** $y = -\frac{1}{3}x^2 + 8x$

Graph each function. *(See Examples 1–3 on pages 322 and 324.)*

45. $y = (x - 2)^2 - 2$ **46.** $y = 2x^2 + 8x + 10$ **47.** $y = -9x^2 - 18x - 6$

Write an equation for the parabola with the given vertex that passes through the given point. *(See Example 4 on page 325.)*

48. vertex: $(4, 1)$
point: $(2, 13)$

49. vertex: $(-2, 3)$
point: $(-6, 11)$

50. vertex: $(-3, -5)$
point: $(0, -14)$

6-7 Graphing and Solving Quadratic Inequalities

See pages 329–335.

Concept Summary

- Graph quadratic inequalities in two variables as follows.

  Step 1 Graph the related quadratic equation, $y = ax^2 + bx + c$. Decide if the parabola should be solid or dashed.

  Step 2 Test a point (x_1, y_1) inside the parabola. Check to see if this point is a solution of the inequality.

  Step 3 If (x_1, y_1) *is* a solution, shade the region *inside* the parabola. If (x_1, y_1) is *not* a solution, shade the region *outside* the parabola.

- To solve a quadratic inequality in one variable, graph the related quadratic function. Identify the x values for which the graph lies *below* the x-axis for $<$ and $\leq$. Identify the x values for which the graph lies *above* the x-axis for $>$ and $\geq$.

Example **Solve $x^2 + 3x - 10 < 0$ by graphing.**

Find the roots of the related equation.

$0 = x^2 + 3x - 10$ Related equation

$0 = (x + 5)(x - 2)$ Factor.

$x + 5 = 0$ or $x - 2 = 0$ Zero Product Property

$x = -5$ $\quad x = 2$ Solve each equation.

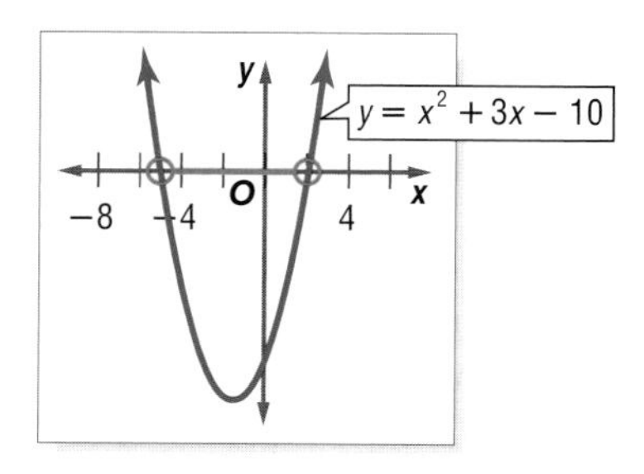

Sketch the graph of the parabola that has x-intercepts at -5 and 2. The graph should open up since $a > 0$. The graph lies below the x-axis between $x = -5$ and $x = 2$. Therefore, the solution set is $\{x \mid -5 < x < 2\}$.

Exercises **Graph each inequality.** *(See Example 1 on pages 329 and 330.)*

51. $y > x^2 - 5x + 15$ **52.** $y \leq 4x^2 - 36x + 17$ **53.** $y \geq -x^2 + 7x - 11$

Solve each inequality. *(See Examples 2, 3, and 5 on pages 330–332.)*

54. $6x^2 + 5x > 4$ **55.** $8x + x^2 \geq -16$ **56.** $2x^2 + 5x < 12$

57. $2x^2 - 5x > 3$ **58.** $4x^2 - 9 \leq -4x$ **59.** $3x^2 - 5 > 6x$

Chapter 6

Practice Test

Vocabulary and Concepts

Choose the word or term that best completes each statement.

1. The y-coordinate of the vertex of the graph of $y = ax^2 + bx + c$ is the (*maximum, minimum*) value obtained by the function when a is positive.
2. (*The Square Root Property, Completing the square*) can be used to solve any quadratic equation.

Skills and Applications

Complete parts a–c for each quadratic function.

a. **Find the y-intercept, the equation of the axis of symmetry, and the x-coordinate of the vertex.**
b. **Make a table of values that includes the vertex.**
c. **Use this information to graph the function.**

3. $f(x) = x^2 - 2x + 5$
4. $f(x) = -3x^2 + 8x$
5. $f(x) = -2x^2 - 7x - 1$

Determine whether each function has a maximum or a minimum value. Then find the maximum or minimum value of each function.

6. $f(x) = x^2 + 6x + 9$
7. $f(x) = 3x^2 - 12x - 24$
8. $f(x) = -x^2 + 4x$

9. Write a quadratic equation with roots -4 and 5. Write the equation in the form $ax^2 + bx + c = 0$, where a, b, and c are integers.

Solve each equation using the method of your choice. Find exact solutions.

10. $x^2 + x - 42 = 0$
11. $-1.6x^2 - 3.2x + 18 = 0$
12. $15x^2 + 16x - 7 = 0$
13. $x^2 + 8x - 48 = 0$
14. $x^2 + 12x + 11 = 0$
15. $x^2 - 9x - \frac{19}{4} = 0$
16. $3x^2 + 7x - 31 = 0$
17. $10x^2 + 3x = 1$
18. $-11x^2 - 174x + 221 = 0$

19. **BALLOONING** At a hot-air balloon festival, you throw a weighted marker straight down from an altitude of 250 feet toward a bull's eye below. The initial velocity of the marker when it leaves your hand is 28 feet per second. Find how long it will take the marker to hit the target by solving the equation $-16t^2 - 28t + 250 = 0$.

Write each equation in vertex form, if not already in that form. Then identify the vertex, axis of symmetry, and direction of opening.

20. $y = (x + 2)^2 - 3$
21. $y = x^2 + 10x + 27$
22. $y = -9x^2 + 54x - 8$

Graph each inequality.

23. $y \leq x^2 + 6x - 7$
24. $y > -2x^2 + 9$
25. $y \geq -\frac{1}{2}x^2 - 3x + 1$

Solve each inequality.

26. $(x - 5)(x + 7) < 0$
27. $3x^2 \geq 16$
28. $-5x^2 + x + 2 < 0$

29. **PETS** A rectangular turtle pen is 6 feet long by 4 feet wide. The pen is enlarged by increasing the length and width by an equal amount in order to double its area. What are the dimensions of the new pen?

30. **STANDARDIZED TEST PRACTICE** Which of the following is the sum of both solutions of the equation $x^2 + 8x - 48 = 0$?

(A) -16 (B) -8 (C) -4 (D) 12

Chapter 6 Standardized Test Practice

Standards Practice

Part 1 Multiple Choice

Record your answers on the answer sheet provided by your teacher or on a sheet of paper.

1. In a class of 30 students, half are girls and 24 ride the bus to school. If 4 of the girls do not ride the bus to school, how many boys in this class ride the bus to school?

(A) 2 (B) 11

(C) 13 (D) 15

2. In the figure below, the measures of $\angle m + \angle n + \angle p =$ ___?___

(A) 90 (B) 180

(C) 270 (D) 360

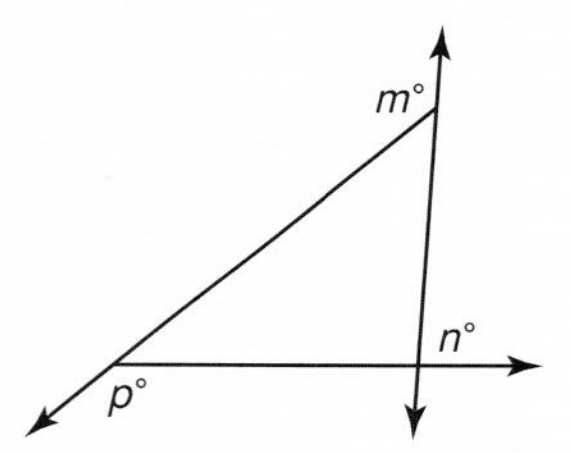

3. Of the points $(-4, -2)$, $(1, -3)$, $(-1, 3)$, $(3, 1)$, and $(-2, 1)$, which three lie on the same side of the line $y - x = 0$?

(A) $(-4, -2)$, $(1, -3)$, $(-2, 1)$

(B) $(-4, -2)$, $(1, -3)$, $(3, 1)$

(C) $(-4, -2)$, $(-1, 3)$, $(-2, 1)$

(D) $(1, -3)$, $(-1, 3)$, $(3, 1)$

4. If k is an integer, then which of the following must also be integers?

I. $\frac{5k + 5}{5k}$ II. $\frac{5k + 5}{k + 1}$ III. $\frac{5k^2 + k}{5k}$

(A) I only (B) II only

(C) I and II (D) II and III

5. Which of the following is a factor of $x^2 - 7x - 8$?

(A) $x + 2$ (B) $x - 1$

(C) $x - 4$ (D) $x - 8$

6. If $x > 0$, then $\frac{\sqrt{16x^2 + 64x + 64}}{x + 2} =$ ___?___

(A) 2 (B) 4

(C) 8 (D) 16

7. If x and p are both greater than zero and $4x^2p^2 + xp - 33 = 0$, then what is the value of p in terms of x?

(A) $-\frac{3}{x}$ (B) $-\frac{11}{4x}$

(C) $\frac{3}{4x}$ (D) $\frac{11}{4x}$

8. For all positive integers n, $\langle n \rangle = 3\sqrt{n}$. Which of the following equals 12?

(A) $\langle 4 \rangle$ (B) $\langle 8 \rangle$

(C) $\langle 16 \rangle$ (D) $\langle 32 \rangle$

9. Which number is the sum of both solutions of the equation $x^2 - 3x - 18 = 0$?

(A) -6 (B) -3

(C) 3 (D) 6

10. One of the roots of the polynomial $6x^2 + kx + 20 = 0$ is $-\frac{5}{2}$. What is the value of k?

(A) -23 (B) $-\frac{4}{3}$

(C) 23 (D) 7

Test-Taking Tip

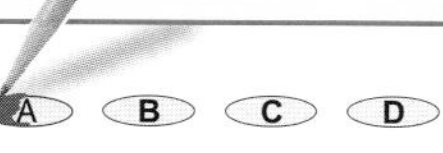

Questions 8, 11, 13, 16, 21, and 27

Be sure to use the information that describes the variables in any standardized test item. For example, if an item says that $x > 0$, check to be sure that your solution for x is not a negative number.

Preparing for Standardized Tests
For test-taking strategies and more practice, see pages 877–892.

Part 2 Short Response/Grid In

Record your answers on the answer sheet provided by your teacher or on a sheet of paper.

11. If n is a three-digit number that can be expressed as the product of three consecutive *even* integers, what is one possible value of n?

12. If x and y are *different* positive integers and $x + y = 6$, what is one possible value of $3x + 5y$?

13. If a circle of radius 12 inches has its radius decreased by 6 inches, by what percent is its area decreased?

14. What is the least positive integer k for which $12k$ is the cube of an integer?

15. If $AB = BC$ in the figure, what is the y-coordinate of point B?

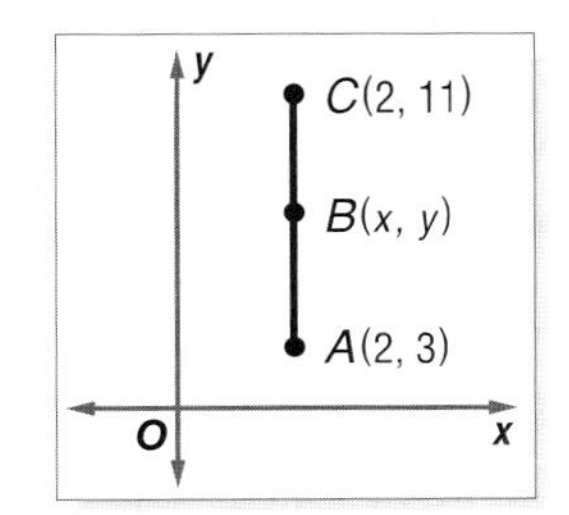

16. In the figure, if O is the center of the circle, what is the value of x?

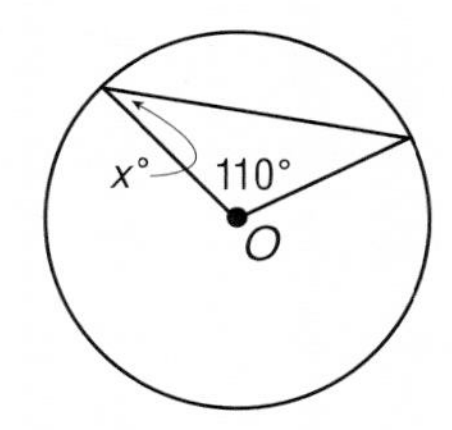

17. Let $a \blacklozenge b$ be defined as the sum of all integers greater than a and less than b. For example, $6 \blacklozenge 10 = 7 + 8 + 9$ or 24. What is the value of $(75 \blacklozenge 90) - (76 \blacklozenge 89)$?

18. If $x^2 - y^2 = 42$ and $x + y = 6$, what is the value of $x - y$?

19. By what amount does the sum of the roots exceed the product of the roots of the equation $(x - 7)(x + 3) = 0$?

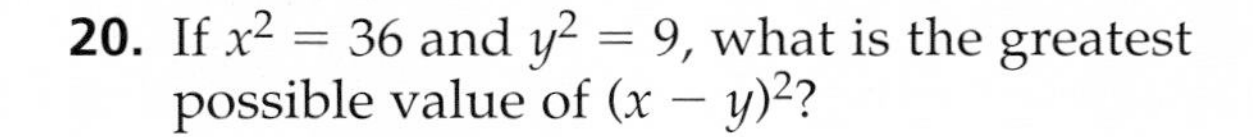

20. If $x^2 = 36$ and $y^2 = 9$, what is the greatest possible value of $(x - y)^2$?

Part 3 Extended Response

Record your answers on a sheet of paper. Show your work.

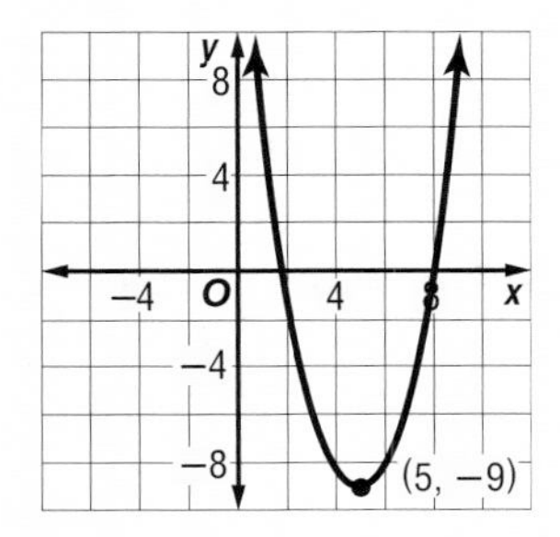

For Exercises 21–23, use the graph below.

21. What are the solutions of $f(x) = 0$? Explain how you found the solutions.

22. Does the function have a minimum value or a maximum value? Find the minimum or maximum value of the function.

23. Write the function whose graph is shown. Explain your method.

For Exercises 24 and 25, use the information below.

Scott launches a model rocket from ground level. The rocket's height h in meters is given by the equation $h = -4.9t^2 + 56t$, where t is the time in seconds after the launch.

24. What is the maximum height the rocket will reach? Round to the nearest tenth of a meter. Show each step and explain your method.

25. How long after it is launched will the rocket reach its maximum height? Round to the nearest tenth of a second.

Chapter 7 Polynomial Functions

What You'll Learn

- **Lessons 7-1 and 7-3** Evaluate polynomial functions and solve polynomial equations.
- **Lessons 7-2 and 7-9** Graph polynomial and square root functions.
- **Lessons 7-4, 7-5, and 7-6** Find factors and zeros of polynomial functions.
- **Lesson 7-7** Find the composition of functions.
- **Lesson 7-8** Determine the inverses of functions or relations.

Key Vocabulary

- polynomial function (p. 347)
- synthetic substitution (p. 365)
- Fundamental Theorem of Algebra (p. 371)
- composition of functions (p. 384)
- inverse function (p. 391)

Why It's Important

According to the Fundamental Theorem of Algebra, every polynomial equation has at least one root. Sometimes the roots have real-world meaning. Many real-world situations that cannot be modeled using a linear function can be approximated using a polynomial function.

You will learn how the power generated by a windmill can be modeled by a polynomial function in Lesson 7-1.

Getting Started

Prerequisite Skills To be successful in this chapter, you'll need to master these skills and be able to apply them in problem-solving situations. Review these skills before beginning Chapter 7.

For Lesson 7-2 — Solve Equations by Graphing

Use the related graph of each equation to determine its roots. If exact roots cannot be found, state the consecutive integers between which the roots are located.

(For review, see Lesson 6-2.)

1. $x^2 - 5x + 2 = 0$

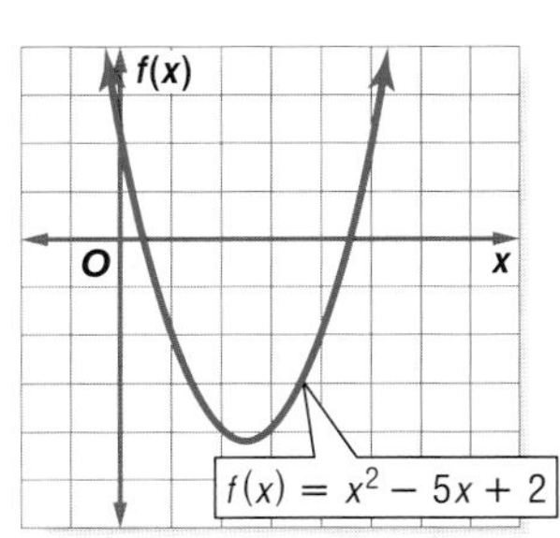

2. $3x^2 + x - 4 = 0$

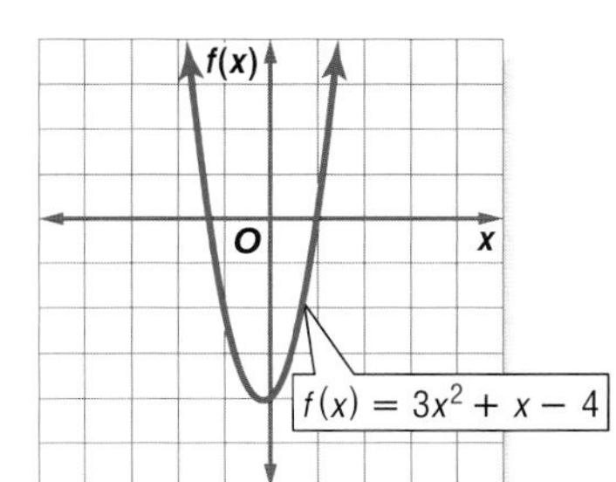

3. $\frac{2}{3}x^2 + 3x - 1 = 0$

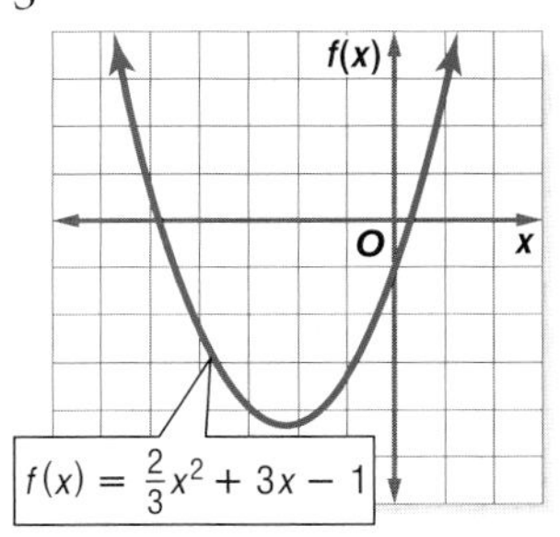

For Lesson 7-3 — Quadratic Formula

Solve each equation. *(For review, see Lesson 6-5.)*

4. $x^2 - 17x + 60 = 0$

5. $14x^2 + 23x + 3 = 0$

6. $2x^2 + 5x + 1 = 0$

For Lessons 7-4 through 7-6 — Synthetic Division

Simplify each expression using synthetic division. *(For review, see Lesson 5-3.)*

7. $(3x^2 - 14x - 24) \div (x - 6)$

8. $(a^2 - 2a - 30) \div (a + 7)$

For Lessons 7-1 and 7-7 — Evaluating Functions

Find each value if $f(x) = 4x - 7$ and $g(x) = 2x^2 - 3x + 1$. *(For review, see Lesson 2-1.)*

9. $f(-3)$

10. $g(2a)$

11. $f(4b^2) + g(b)$

Polynomial Functions Make this Foldable to help you organize your notes. Begin with five sheets of plain $8\frac{1}{2}$" by 11" paper.

Step 1 Stack and Fold

Stack sheets of paper with edges $\frac{3}{4}$-inch apart. Fold up the the bottom edges to create equal tabs.

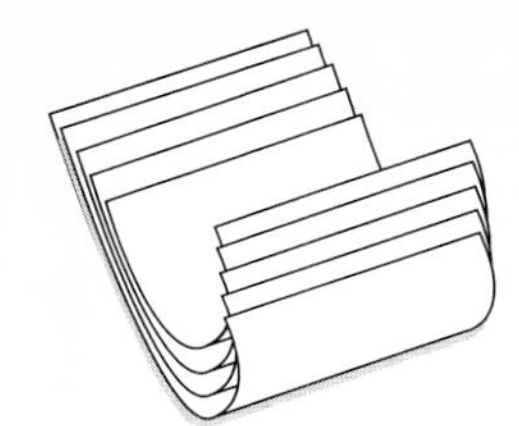

Step 2 Staple and Label

Staple along the fold. Label the tabs with lesson numbers.

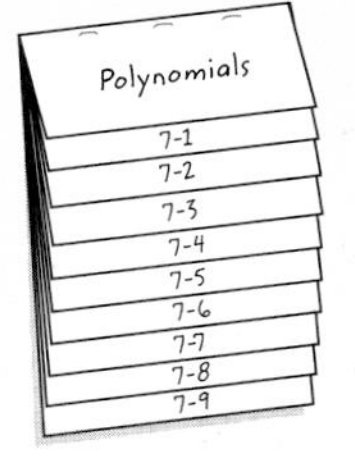

Reading and Writing As you read and study the chapter, use each page to write notes and examples.

7-1 Polynomial Functions

What You'll Learn

- Evaluate polynomial functions.
- Identify general shapes of graphs of polynomial functions.

Vocabulary

- polynomial in one variable
- degree of a polynomial
- leading coefficients
- polynomial function
- end behavior

Where are polynomial functions found in nature?

If you look at a cross section of a honeycomb, you see a pattern of hexagons. This pattern has one hexagon surrounded by six more hexagons. Surrounding these is a third ring of 12 hexagons, and so on. The total number of hexagons in a honeycomb can be modeled by the function $f(r) = 3r^2 - 3r + 1$, where r is the number of rings and $f(r)$ is the number of hexagons.

POLYNOMIAL FUNCTIONS Recall that a polynomial is a monomial or a sum of monomials. The expression $3r^2 - 3r + 1$ is a **polynomial in one variable** since it only contains one variable, r.

Key Concept — *Polynomial in One Variable*

- **Words** A polynomial of degree n in one variable x is an expression of the form $a_0x^n + a_1x^{n-1} + \ldots + a_{n-2}x^2 + a_{n-1}x + a_n$, where the coefficients $a_0, a_1, a_2, \ldots, a_n$, represent real numbers, a_0 is not zero, and n represents a nonnegative integer.
- **Examples** $3x^5 + 2x^4 - 5x^3 + x^2 + 1$

  $n = 5, a_0 = 3, a_1 = 2, a_2 = -5, a_3 = 1, a_4 = 0$, and $a_5 = 1$

The **degree of a polynomial** in one variable is the greatest exponent of its variable. The **leading coefficient** is the coefficient of the term with the highest degree.

Polynomial	Expression	Degree	Leading Coefficient
Constant	9	0	9
Linear	$x - 2$	1	1
Quadratic	$3x^2 + 4x - 5$	2	3
Cubic	$4x^3 - 6$	3	4
General	$a_0x^n + a_1x^{n-1} + \ldots + a_{n-2}x^2 + a_{n-1}x + a_n$	n	a_0

Example 1 Find Degrees and Leading Coefficients

State the degree and leading coefficient of each polynomial in one variable. If it is not a polynomial in one variable, explain why.

a. $7x^4 + 5x^2 + x - 9$

This is a polynomial in one variable.

The degree is 4, and the leading coefficient is 7.

b. $8x^2 + 3xy - 2y^2$

This is not a polynomial in one variable. It contains two variables, x and y.

c. $7x^6 - 4x^3 + \frac{1}{x}$

This is not a polynomial. The term $\frac{1}{x}$ cannot be written in the form x^n, where n is a nonnegative integer.

d. $\frac{1}{2}x^2 + 2x^3 - x^5$

Rewrite the expression so the powers of x are in decreasing order.

$$-x^5 + 2x^3 + \frac{1}{2}x^2$$

This is a polynomial in one variable with degree of 5 and leading coefficient of -1.

Study Tip

Power Function

A common type of function is a **power function**, which has an equation in the form $f(x) = ax^b$, where a and b are real numbers. When b is a positive integer, $f(x) = ax^b$ is a polynomial function.

A polynomial equation used to represent a function is called a **polynomial function**. For example, the equation $f(x) = 4x^2 - 5x + 2$ is a quadratic polynomial function, and the equation $p(x) = 2x^3 + 4x^2 - 5x + 7$ is a cubic polynomial function. Other polynomial functions can be defined by the following general rule.

Key Concept — *Definition of a Polynomial Function*

- **Words** A polynomial function of degree n can be described by an equation of the form $P(x) = a_0x^n + a_1x^{n-1} + \ldots + a_{n-2}x^2 + a_{n-1}x + a_n$, where the coefficients $a_0, a_1, a_2, \ldots, a_n$, represent real numbers, a_0 is not zero, and n represents a nonnegative integer.
- **Examples** $f(x) = 4x^2 - 3x + 2$
  $n = 2, a_0 = 4, a_1 = -3, a_2 = 2$

If you know an element in the domain of any polynomial function, you can find the corresponding value in the range. Recall that $f(3)$ can be found by evaluating the function for $x = 3$.

Example 2 *Evaluate a Polynomial Function*

NATURE Refer to the application at the beginning of the lesson.

a. Show that the polynomial function $f(r) = 3r^2 - 3r + 1$ gives the total number of hexagons when $r = 1, 2,$ and 3.

Find the values of $f(1)$, $f(2)$, and $f(3)$.

$f(r) = 3r^2 - 3r + 1$
$f(1) = 3(1)^2 - 3(1) + 1$
$= 3 - 3 + 1$ or 1

$f(r) = 3r^2 - 3r + 1$
$f(2) = 3(2)^2 - 3(2) + 1$
$= 12 - 6 + 1$ or 7

$f(r) = 3r^2 - 3r + 1$
$f(3) = 3(3)^2 - 3(3) + 1$
$= 27 - 9 + 1$ or 19

From the information given, you know the number of hexagons in the first ring is 1, the number of hexagons in the second ring is 6, and the number of hexagons in the third ring is 12. So, the total number of hexagons with one ring is 1, two rings is $6 + 1$ or 7, and three rings is $12 + 6 + 1$ or 19. These match the functional values for $r = 1, 2,$ and 3, respectively.

ring 3
ring 2
ring 1

Rings of a Honeycomb

b. Find the total number of hexagons in a honeycomb with 12 rings.

$f(r) = 3r^2 - 3r + 1$	Original function
$f(12) = 3(12)^2 - 3(12) + 1$	Replace r with 12.
$= 432 - 36 + 1$ or 397	Simplify.

You can also evaluate functions for variables and algebraic expressions.

Example 3 Functional Values of Variables

a. Find $p(a^2)$ if $p(x) = x^3 + 4x^2 - 5x$.

$p(x) = x^3 + 4x^2 - 5x$	Original function
$p(a^2) = (a^2)^3 + 4(a^2)^2 - 5(a^2)$	Replace x with a^2.
$= a^6 + 4a^4 - 5a^2$	Property of powers

b. Find $q(a + 1) - 2q(a)$ if $q(x) = x^2 + 3x + 4$.

To evaluate $q(a + 1)$, replace x in $q(x)$ with $a + 1$.

$q(x) = x^2 + 3x + 4$	Original function
$q(a + 1) = (a + 1)^2 + 3(a + 1) + 4$	Replace x with $a + 1$.
$= a^2 + 2a + 1 + 3a + 3 + 4$	Evaluate $(a + 1)^2$ and $3(a + 1)$.
$= a^2 + 5a + 8$	Simplify.

To evaluate $2q(a)$, replace x with a in $q(x)$, then multiply the expression by 2.

$q(x) = x^2 + 3x + 4$	Original function
$2q(a) = 2(a^2 + 3a + 4)$	Replace x with a.
$= 2a^2 + 6a + 8$	Distributive Property

Now evaluate $q(a + 1) - 2q(a)$.

$q(a + 1) - 2q(a) = a^2 + 5a + 8 - (2a^2 + 6a + 8)$	Replace $q(a + 1)$ and $2q(a)$ with evaluated expressions.
$= a^2 + 5a + 8 - 2a^2 - 6a - 8$	
$= -a^2 - a$	Simplify.

GRAPHS OF POLYNOMIAL FUNCTIONS The general shapes of the graphs of several polynomial functions are shown below. These graphs show the *maximum* number of times the graph of each type of polynomial may intersect the x-axis. Recall that the x-coordinate of the point at which the graph intersects the x-axis is called a *zero* of a function. How does the degree compare to the maximum number of real zeros?

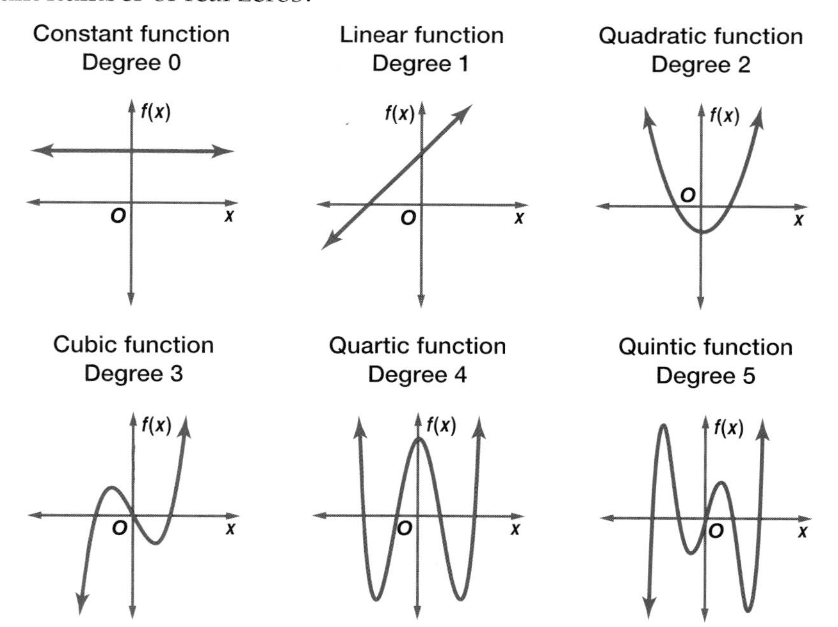

Notice the shapes of the graphs for even-degree polynomial functions and odd-degree polynomial functions. The degree and leading coefficient of a polynomial function determine the graph's end behavior.

The **end behavior** is the behavior of the graph as x approaches positive infinity ($+\infty$) or negative infinity ($-\infty$). This is represented as $x \to +\infty$ and $x \to -\infty$, respectively. $x \to +\infty$ is read *x approaches positive infinity.*

Concept Summary — End Behavior of a Polynomial Function

Degree: even
Leading Coefficient: positive
End Behavior:

$f(x) \to +\infty$ as $x \to -\infty$ $f(x) \to +\infty$ as $x \to +\infty$

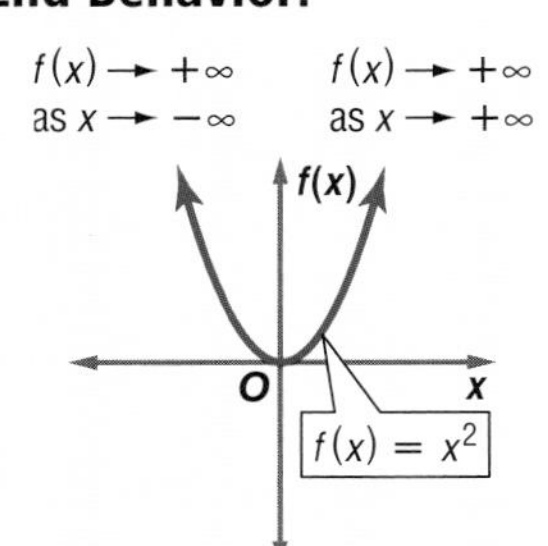

Degree: odd
Leading Coefficient: positive
End Behavior:

$f(x) \to +\infty$ as $x \to +\infty$

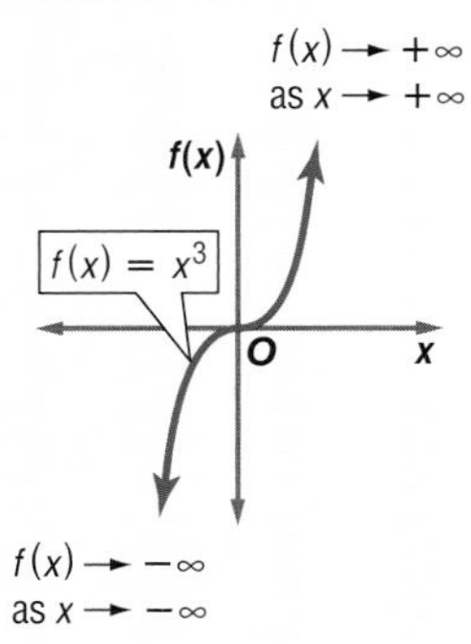

$f(x) \to -\infty$ as $x \to -\infty$

Degree: even
Leading Coefficient: negative
End Behavior:

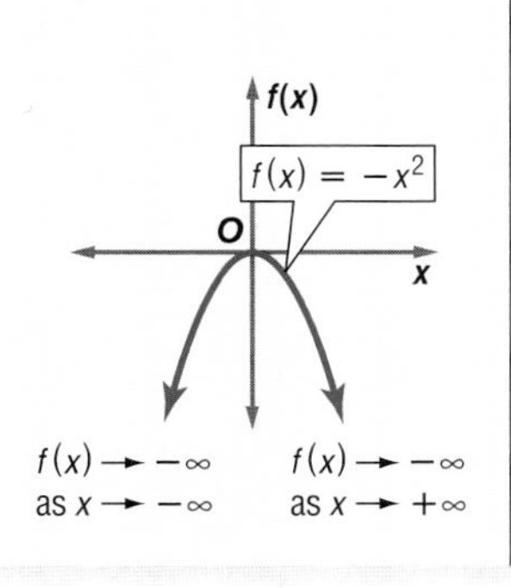

$f(x) \to -\infty$ as $x \to -\infty$ $f(x) \to -\infty$ as $x \to +\infty$

Degree: odd
Leading Coefficient: negative
End Behavior:

$f(x) \to +\infty$ as $x \to -\infty$

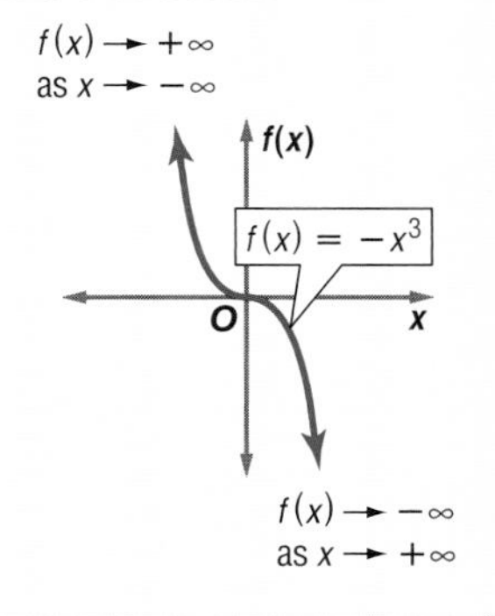

$f(x) \to -\infty$ as $x \to +\infty$

The graph of an even-degree function may or may not intersect the x-axis, depending on its location in the coordinate plane. If it intersects the x-axis in two places, the function has two real zeros. If it does not intersect the x-axis, the roots of the related equation are imaginary and cannot be determined from the graph. If the graph is tangent to the x-axis, as shown above, there are two zeros that are the same number. The graph of an odd-degree function always crosses the x-axis at least once, and thus the function always has at least one real zero.

Study Tip

Number of Zeros
The number of zeros of an odd-degree function may be less than the maximum by a multiple of 2. For example, the graph of a quintic function may only cross the x-axis 3 times.

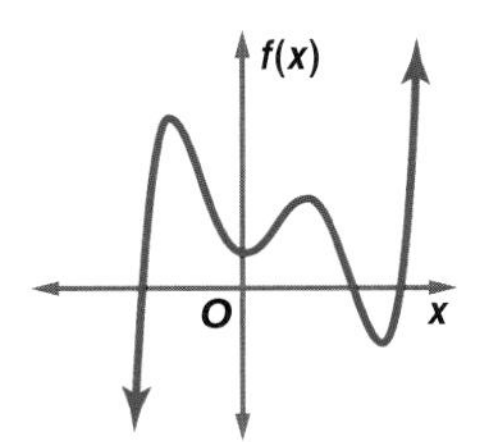

The same is true for an even-degree function. One exception is when the graph of $f(x)$ touches the x-axis.

Example 4 Graphs of Polynomial Functions

For each graph,

- **describe the end behavior,**
- **determine whether it represents an odd-degree or an even-degree polynomial function, and**
- **state the number of real zeros.**

a.

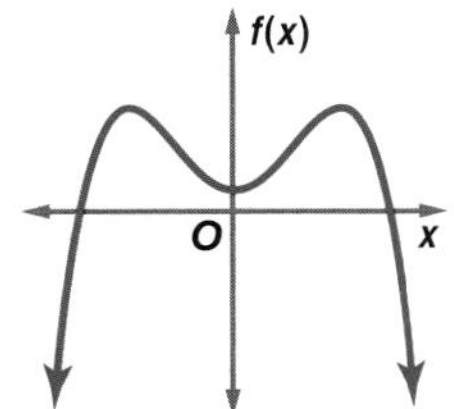

b.

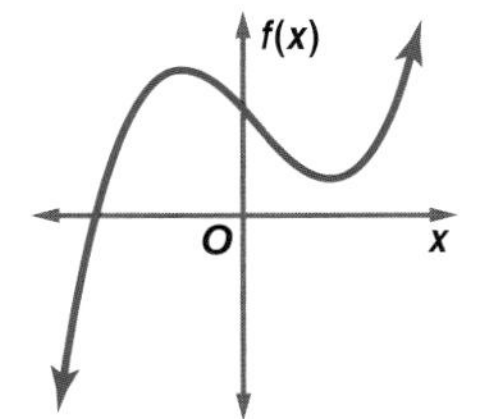

c.

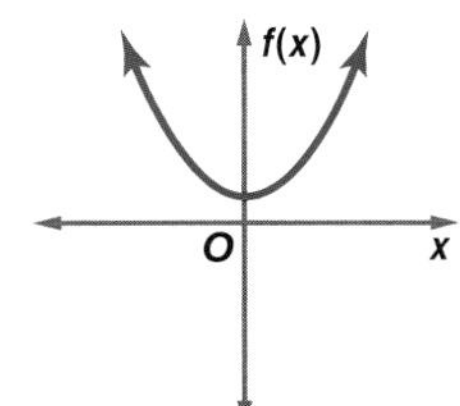

a.
- $f(x) \to -\infty$ as $x \to +\infty$. $f(x) \to -\infty$ as $x \to -\infty$.
- It is an even-degree polynomial function.
- The graph intersects the x-axis at two points, so the function has two real zeros.

b.
- $f(x) \to +\infty$ as $x \to +\infty$. $f(x) \to -\infty$ as $x \to -\infty$.
- It is an odd-degree polynomial function.
- The graph has one real zero.

c.
- $f(x) \to +\infty$ as $x \to +\infty$. $f(x) \to +\infty$ as $x \to -\infty$.
- It is an even-degree polynomial function.
- This graph does not intersect the x-axis, so the function has no real zeros.

Check for Understanding

Concept Check

1. **Explain** why a constant polynomial such as $f(x) = 4$ has degree 0 and a linear polynomial such as $f(x) = x + 5$ has degree 1.
2. **Describe** the characteristics of the graphs of odd-degree and even-degree polynomial functions whose leading coefficients are positive.
3. **OPEN ENDED** Sketch the graph of an odd-degree polynomial function with a negative leading coefficient and three real roots.
4. **Tell** whether the following statement is *always, sometimes* or *never* true. Explain.
   A polynomial function that has four real roots is a fourth-degree polynomial.

Guided Practice

State the degree and leading coefficient of each polynomial in one variable. If it is not a polynomial in one variable, explain why.

5. $5x^6 - 8x^2$
6. $2b + 4b^3 - 3b^5 - 7$

Find $p(3)$ and $p(-1)$ for each function.

7. $p(x) = -x^3 + x^2 - x$
8. $p(x) = x^4 - 3x^3 + 2x^2 - 5x + 1$

If $p(x) = 2x^3 + 6x - 12$ and $q(x) = 5x^2 + 4$, find each value.

9. $p(a^3)$
10. $5[q(2a)]$
11. $3p(a) - q(a + 1)$

For each graph,

a. describe the end behavior,

b. determine whether it represents an odd-degree or an even-degree polynomial function, and

c. state the number of real zeros.

12.
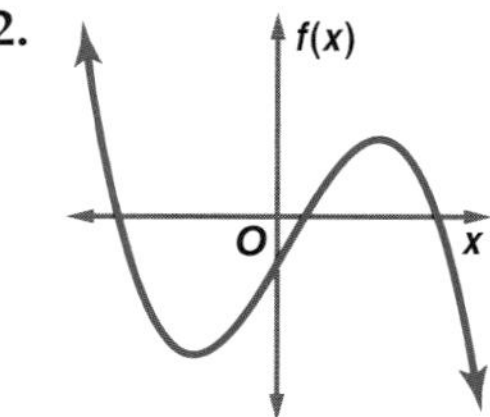

13.
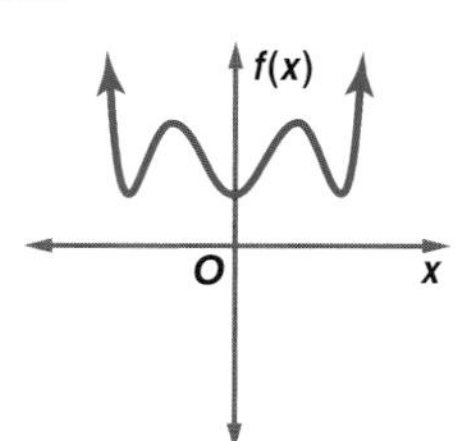

14.
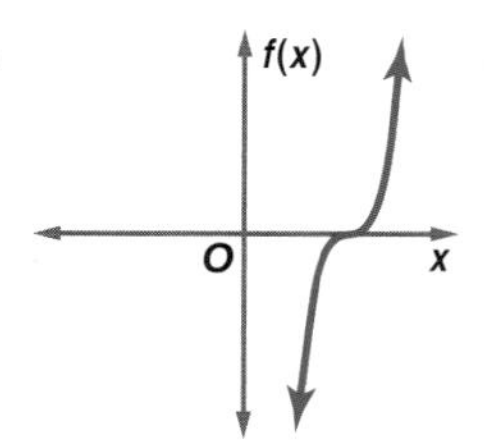

Application

15. **BIOLOGY** The intensity of light emitted by a firefly can be determined by $L(t) = 10 + 0.3t + 0.4t^2 - 0.01t^3$, where t is temperature in degrees Celsius and $L(t)$ is light intensity in lumens. If the temperature is 30°C, find the light intensity.

Practice and Apply

Homework Help

For Exercises	See Examples
16–21	1
22–29, 45	2
30–38	3
39–44, 46–48	4

Extra Practice
See page 842.

State the degree and leading coefficient of each polynomial in one variable. If it is not a polynomial in one variable, explain why.

16. $7 - x$
17. $(a + 1)(a^2 - 4)$
18. $a^2 + 2ab + b^2$
19. $6x^4 + 3x^2 + 4x - 8$
20. $7 + 3x^2 - 5x^3 + 6x^2 - 2x$
21. $c^2 + c - \frac{1}{c}$

Find $p(4)$ and $p(-2)$ for each function.

22. $p(x) = 2 - x$
23. $p(x) = x^2 - 3x + 8$
24. $p(x) = 2x^3 - x^2 + 5x - 7$
25. $p(x) = x^5 - x^2$
26. $p(x) = x^4 - 7x^3 + 8x - 6$
27. $p(x) = 7x^2 - 9x + 10$
28. $p(x) = \frac{1}{2}x^4 - 2x^2 + 4$
29. $p(x) = \frac{1}{8}x^3 - \frac{1}{4}x^2 - \frac{1}{2}x + 5$

If $p(x) = 3x^2 - 2x + 5$ and $r(x) = x^3 + x + 1$, find each value.

30. $r(3a)$

31. $4p(a)$

32. $p(a^2)$

33. $p(2a^3)$

34. $r(x + 1)$

35. $p(x^2 + 3)$

36. $2[p(x + 4)]$

37. $r(x + 1) - r(x^2)$

38. $3[p(x^2 - 1)] + 4p(x)$

For each graph,

a. describe the end behavior,

b. determine whether it represents an odd-degree or an even-degree polynomial function, and

c. state the number of real zeros.

39.

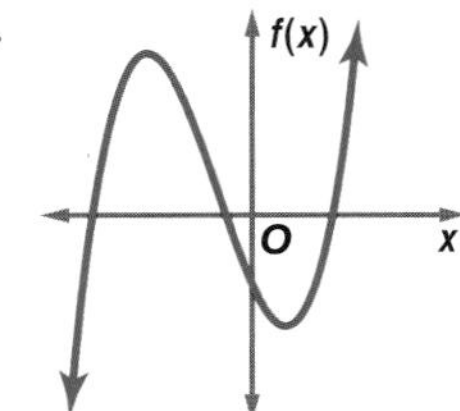

40.

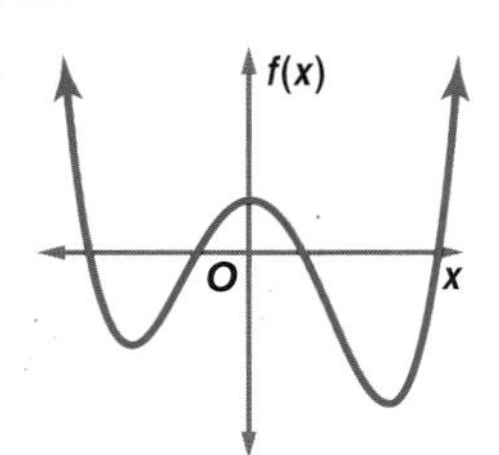

41.

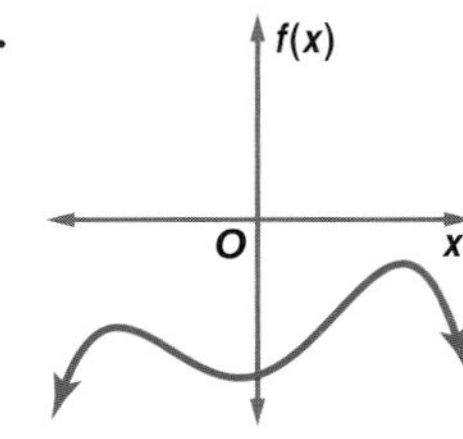

42.

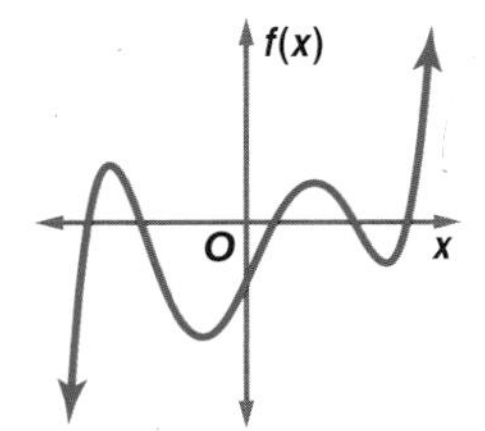

43.

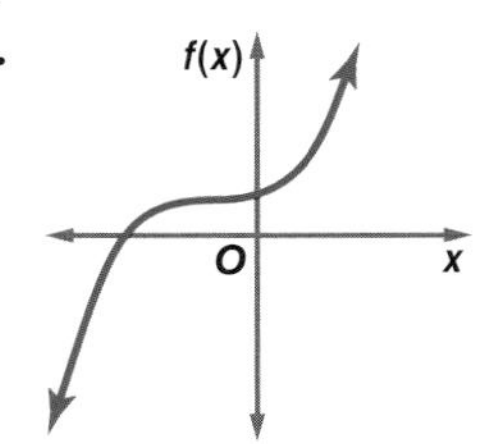

44.

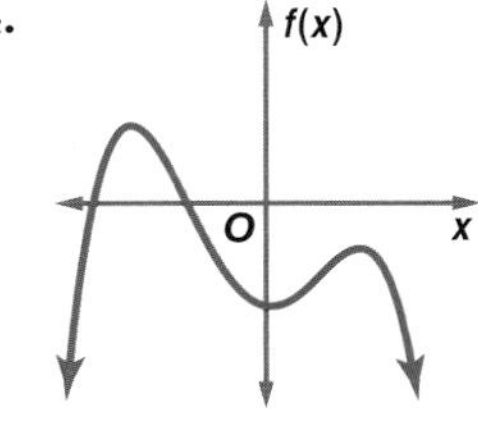

More About. . .

Theater

In 1997, *Cats* surpassed *A Chorus Line* as the longest-running Broadway show.

Source: www.newsherald.com

45. ENERGY The power generated by a windmill is a function of the speed of the wind. The approximate power is given by the function $P(s) = \frac{s^3}{1000}$, where s represents the speed of the wind in kilometers per hour. Find the units of power $P(s)$ generated by a windmill when the wind speed is 18 kilometers per hour.

THEATER For Exercises 46–48, use the graph that models the attendance to Broadway plays (in millions) from 1970–2000.

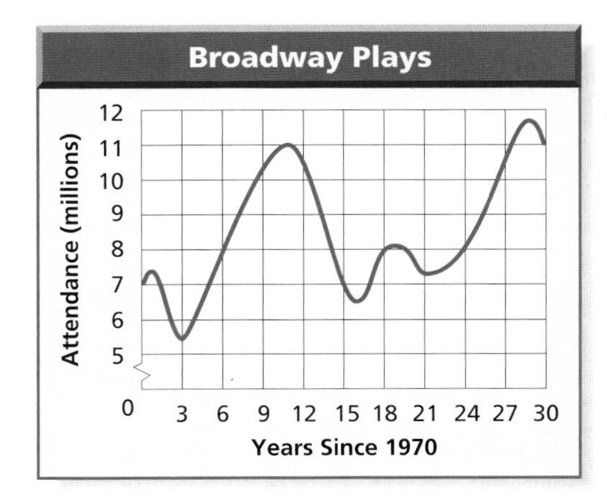

46. Is the graph an odd-degree or even-degree function?

47. Discuss the end behavior of the graph.

48. Do you think attendance at Broadway plays will increase or decrease after 2000? Explain your reasoning.

CRITICAL THINKING For Exercises 49–52, use the following information.

The graph of the polynomial function $f(x) = ax(x - 4)(x + 1)$ goes through the point at (5, 15).

49. Find the value of a.

50. For what value(s) of x will $f(x) = 0$?

51. Rewrite the function as a cubic function.

52. Sketch the graph of the function.

PATTERNS For Exercises 53–55, use the diagrams below that show the maximum number of regions formed by connecting points on a circle.

1 point, 1 region

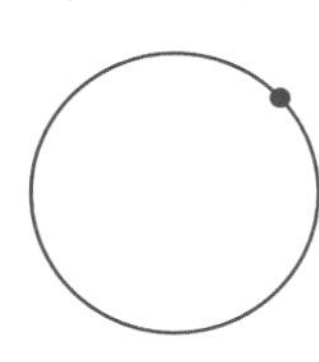

2 points, 2 regions

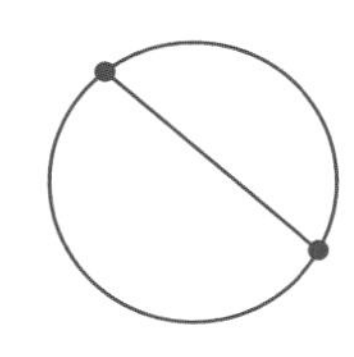

3 points, 4 regions

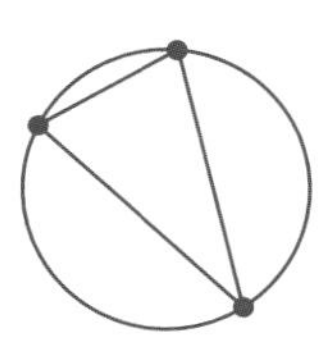

4 points, 8 regions

53. The maximum number of regions formed by connecting n points of a circle can be described by the function $f(n) = \frac{1}{24}(n^4 - 6n^3 + 23n^2 - 18n + 24)$. What is the degree of this polynomial function?

54. Find the maximum number of regions formed by connecting 5 points of a circle. Draw a diagram to verify your solution.

55. How many points would you have to connect to form 99 regions?

56. WRITING IN MATH Answer the question that was posed at the beginning of the lesson.

Where are polynomial functions found in nature?

Include the following in your answer:

- an explanation of how you could use the equation to find the number of hexagons in the tenth ring, and
- any other examples of patterns found in nature that might be modeled by a polynomial equation.

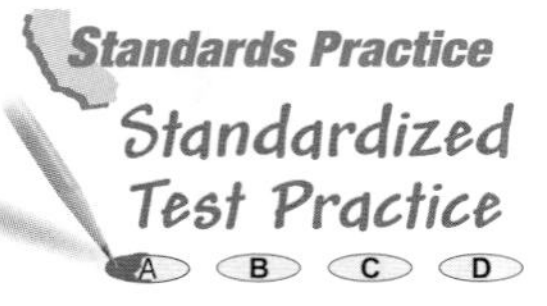

57. The figure at the right shows the graph of the polynomial function $f(x)$. Which of the following could be the degree of $f(x)$?

Ⓐ 2 Ⓑ 3 Ⓒ 4 Ⓓ 5

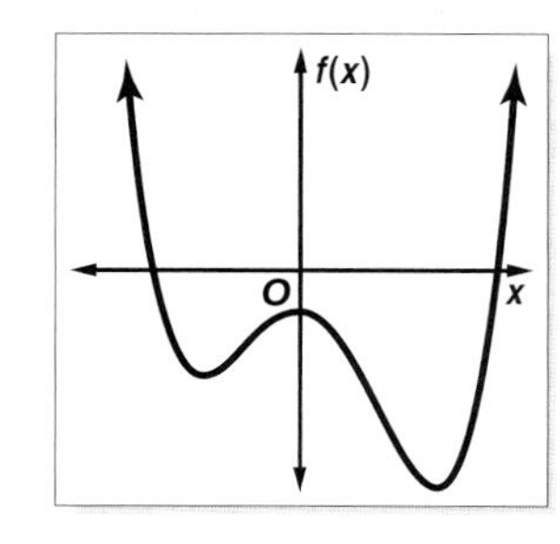

58. If $\frac{1}{2}x^2 - 6x + 2 = 0$, then x could equal which of the following?

Ⓐ −1.84 Ⓑ −0.81 Ⓒ 0.34 Ⓓ 2.37

Maintain Your Skills

Mixed Review

Solve each inequality algebraically. *(Lesson 6-7)*

59. $x^2 - 8x + 12 < 0$

60. $x^2 + 2x - 86 \geq -23$

61. $15x^2 + 3x - 12 \leq 0$

Graph each function. *(Lesson 6-6)*

62. $y = -2(x - 2)^2 + 3$

63. $y = \frac{1}{3}(x + 5)^2 - 1$

64. $y = \frac{1}{2}x^2 + x + \frac{3}{2}$

Solve each equation by completing the square. *(Lesson 6-4)*

65. $x^2 - 8x - 2 = 0$

66. $x^2 + \frac{1}{3}x - \frac{35}{36} = 0$

67. **BUSINESS** Becca is writing a computer program to find the salaries of her employees after their annual raise. The percent of increase is represented by p. Marty's salary is $23,450 now. Write a polynomial to represent Marty's salary after one year and another to represent Marty's salary after three years. Assume that the rate of increase will be the same for each of the three years. *(Lesson 5-2)*

Getting Ready for the Next Lesson

PREREQUISITE SKILL Graph each equation by making a table of values.
*(To review **graphing quadratic functions**, see Lesson 6-1.)*

68. $y = x^2 + 4$

69. $y = -x^2 + 6x - 5$

70. $y = \frac{1}{2}x^2 + 2x - 6$

7-2 Graphing Polynomial Functions

What You'll Learn

- Graph polynomial functions and locate their real zeros.
- Find the maxima and minima of polynomial functions.

How can graphs of polynomial functions show trends in data?

The percent of the United States population that was foreign-born since 1900 can be modeled by $P(t) = 0.00006t^3 - 0.007t^2 + 0.05t + 14$, where $t = 0$ in 1900. Notice that the graph is decreasing from $t = 5$ to $t = 75$ and then it begins to increase. The points at $t = 5$ and $t = 75$ are turning points in the graph.

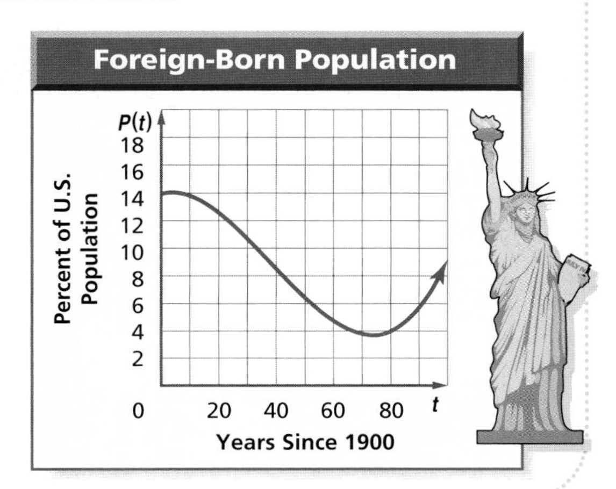

Vocabulary

- Location Principle
- relative maximum
- relative minimum

GRAPH POLYNOMIAL FUNCTIONS To graph a polynomial function, make a table of values to find several points and then connect them to make a smooth curve. Knowing the end behavior of the graph will assist you in completing the sketch of the graph.

Example 1 Graph a Polynomial Function

Graph $f(x) = x^4 + x^3 - 4x^2 - 4x$ by making a table of values.

x	$f(x)$
−2.5	≈ 8.4
−2.0	0.0
−1.5	≈ −1.3
−1.0	0.0
−0.5	≈ 0.9

x	$f(x)$
0.0	0.0
0.5	≈ −2.8
1.0	−6.0
1.5	≈ −6.6
2.0	0.0

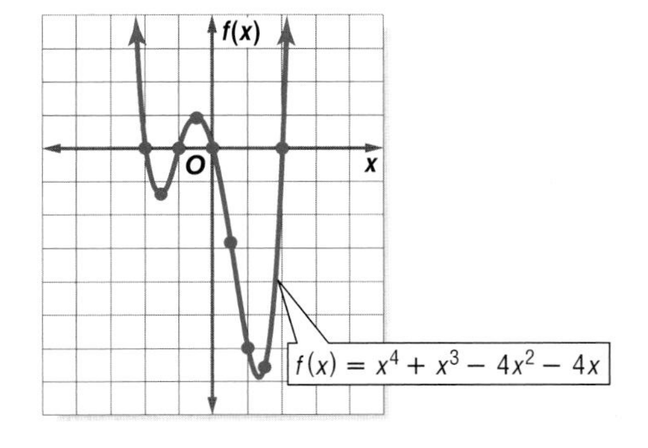

This is an even-degree polynomial with a positive leading coefficient, so $f(x) \to +\infty$ as $x \to +\infty$, and $f(x) \to +\infty$ as $x \to -\infty$. Notice that the graph intersects the x-axis at four points, indicating there are four real zeros of this function.

Study Tip

Graphing Polynomial Functions

To graph polynomial functions it will often be necessary to include x values that are not integers.

In Example 1, the zeros occur at integral values that can be seen in the table used to plot the function. Notice that the values of the function before and after each zero are different in sign. In general, the graph of a polynomial function will cross the x-axis somewhere between pairs of x values at which the corresponding $f(x)$ values change signs. Since zeros of the function are located at x-intercepts, there is a zero between each pair of these x values. This property for locating zeros is called the **Location Principle**.

Key Concept — Location Principle

- **Words** Suppose $y = f(x)$ represents a polynomial function and a and b are two numbers such that $f(a) < 0$ and $f(b) > 0$. Then the function has at least one real zero between a and b.
- **Model**

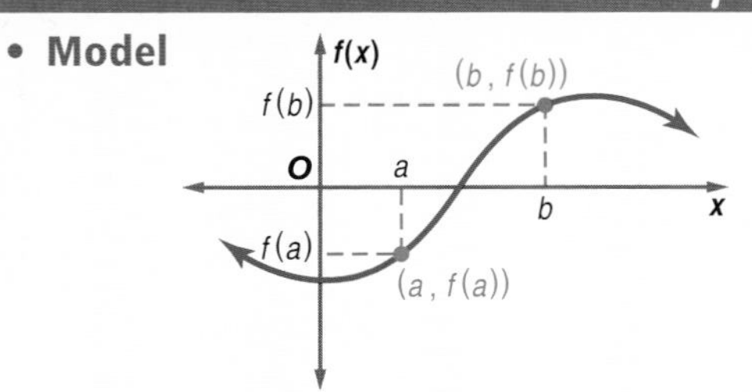

Example 2 Locate Zeros of a Function

Determine consecutive values of x between which each real zero of the function $f(x) = x^3 - 5x^2 + 3x + 2$ is located. Then draw the graph.

Make a table of values. Since $f(x)$ is a third-degree polynomial function, it will have either 1, 2, or 3 real zeros. Look at the values of $f(x)$ to locate the zeros. Then use the points to sketch a graph of the function.

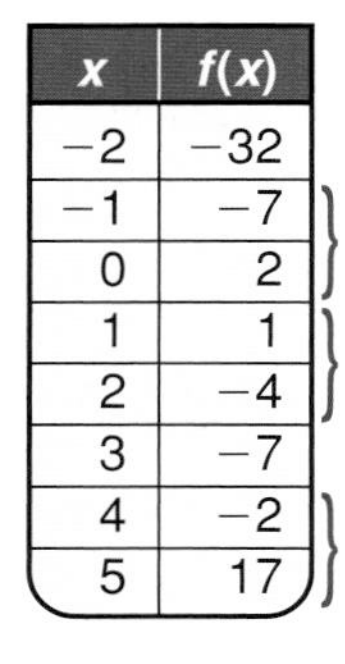

x	f(x)	
−2	−32	
−1	−7	change in signs
0	2	
1	1	change in signs
2	−4	
3	−7	
4	−2	change in signs
5	17	

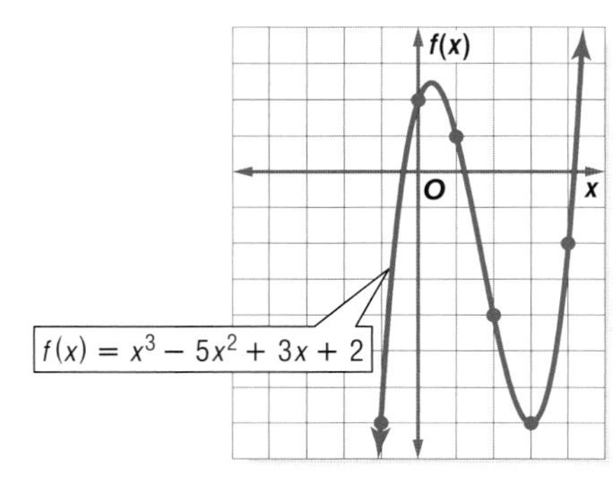

The changes in sign indicate that there are zeros between $x = -1$ and $x = 0$, between $x = 1$ and $x = 2$, and between $x = 4$ and $x = 5$.

MAXIMUM AND MINIMUM POINTS The graph at the right shows the shape of a general third-degree polynomial function.

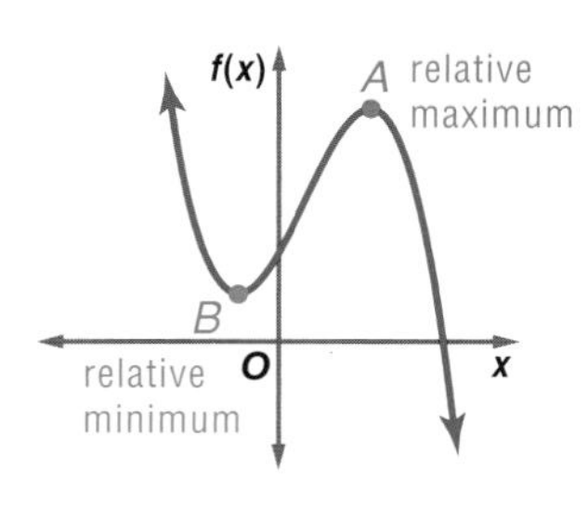

Point A on the graph is a **relative maximum** of the cubic function since no other nearby points have a greater y-coordinate. Likewise, point B is a **relative minimum** since no other nearby points have a lesser y-coordinate. These points are often referred to as *turning points*. The graph of a polynomial function of degree n has at most $n - 1$ turning points.

Study Tip

Reading Math
The plurals of maximum and minimum are *maxima* and *minima*.

Example 3 Maximum and Minimum Points

Graph $f(x) = x^3 - 3x^2 + 5$. Estimate the x-coordinates at which the relative maxima and relative minima occur.

Make a table of values and graph the equation.

x	f(x)	
−2	−15	zero between $x = -2$ and $x = -1$
−1	1	
0	5	← indicates a relative maximum
1	3	
2	1	← indicates a relative minimum
3	5	

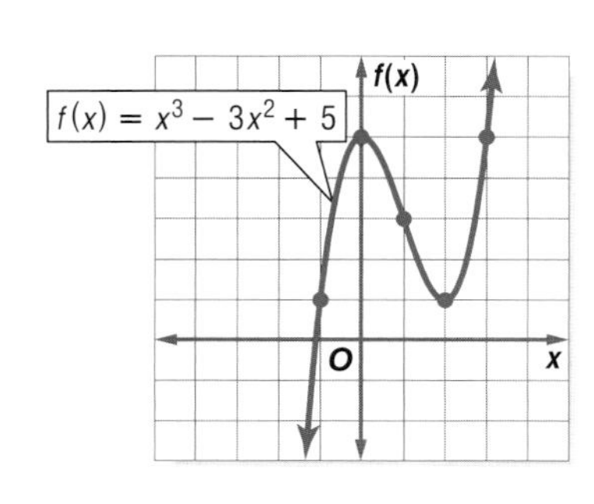

Look at the table of values and the graph.

- The values of $f(x)$ change signs between $x = -2$ and $x = -1$, indicating a zero of the function.
- The value of $f(x)$ at $x = 0$ is greater than the surrounding points, so it is a relative maximum.
- The value of $f(x)$ at $x = 2$ is less than the surrounding points, so it is a relative minimum.

The graph of a polynomial function can reveal trends in real-world data.

Example 4 Graph a Polynomial Model

ENERGY The average fuel (in gallons) consumed by individual vehicles in the United States from 1960 to 2000 is modeled by the cubic equation $F(t) = 0.025t^3 - 1.5t^2 + 18.25t + 654$, where t is the number of years since 1960.

a. Graph the equation.

Make a table of values for the years 1960–2000. Plot the points and connect with a smooth curve. Finding and plotting the points for every fifth year gives a good approximation of the graph.

t	$F(t)$
0	654
5	710.88
10	711.5
15	674.63
20	619
25	563.38
30	526.5
35	527.13
40	584

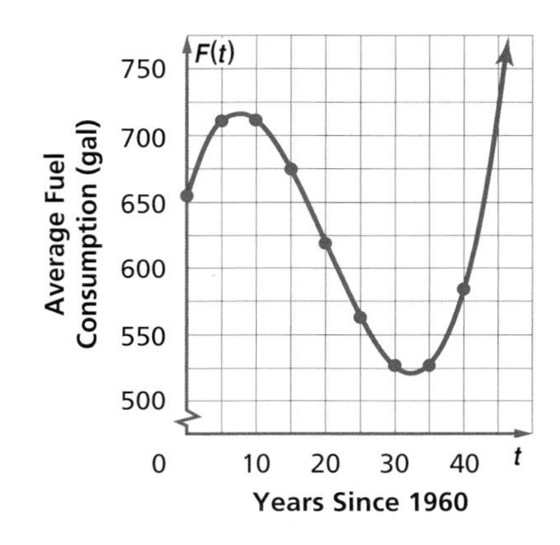

b. Describe the turning points of the graph and its end behavior.

There is a relative maximum between 1965 and 1970 and a relative minimum between 1990 and 1995. For the end behavior, as t increases, $F(t)$ increases.

c. What trends in fuel consumption does the graph suggest?

Average fuel consumption hit a maximum point around 1970 and then started to decline until 1990. Since 1990, fuel consumption has risen and continues to rise.

More About. . .

Energy

Gasoline and diesel fuels are the most familiar transportation fuels in this country, but other energy sources are available, including ethanol, a grain alcohol that can be produced from corn or other crops.

Source: U.S. Environmental Protection Agency

A graphing calculator can be helpful in finding the relative maximum and relative minimum of a function.

Graphing Calculator Investigation

Maximum and Minimum Points

You can use a TI-83 Plus to find the coordinates of relative maxima and relative minima. Enter the polynomial function in the Y= list and graph the function. Make sure that all the turning points are visible in the viewing window. Find the coordinates of the minimum and maximum points, respectively.

KEYSTROKES: *Refer to page 293 to review finding maxima and minima.*

(continued on the next page)

Think and Discuss

1. Graph $f(x) = x^3 - 3x^2 + 4$. Estimate the x-coordinates of the relative maximum and relative minimum points from the graph.
2. Use the maximum and minimum options from the CALC menu to find the exact coordinates of these points. You will need to use the arrow keys to select points to the left and to the right of the point.
3. Graph $f(x) = \frac{1}{2}x^4 - 4x^3 + 7x^2 - 8$. How many relative maximum and relative minimum points does the graph contain? What are the coordinates?

Check for Understanding

Concept Check

1. **Explain** the Location Principle in your own words.
2. **State** the number of turning points of the graph of a fifth-degree polynomial if it has five distinct real zeros.
3. **OPEN ENDED** Sketch a graph of a function that has one relative maximum point and two relative minimum points.

Guided Practice

Graph each polynomial function by making a table of values.

4. $f(x) = x^3 - x^2 - 4x + 4$
5. $f(x) = x^4 - 7x^2 + x + 5$

Determine consecutive values of x between which each real zero of each function is located. Then draw the graph.

6. $f(x) = x^3 - x^2 + 1$
7. $f(x) = x^4 - 4x^2 + 2$

Graph each polynomial function. Estimate the x-coordinates at which the relative maxima and relative minima occur.

8. $f(x) = x^3 + 2x^2 - 3x - 5$
9. $f(x) = x^4 - 8x^2 + 10$

Application

CABLE TV For Exercises 10–12, use the following information.
The number of cable TV systems after 1985 can be modeled by the function $C(t) = -43.2t^2 + 1343t + 790$, where t represents the number of years since 1985.

10. Graph this equation for the years 1985 to 2005.
11. Describe the turning points of the graph and its end behavior.
12. What trends in cable TV subscriptions does the graph suggest?

Practice and Apply

Homework Help

For Exercises	See Examples
13–26	1, 2, 3
27–35	4

Extra Practice
See page 842.

For Exercises 13–26, complete each of the following.

a. **Graph each function by making a table of values.**

b. **Determine consecutive values of x between which each real zero is located.**

c. **Estimate the x-coordinates at which the relative maxima and relative minima occur.**

13. $f(x) = -x^3 - 4x^2$
14. $f(x) = x^3 - 2x^2 + 6$
15. $f(x) = x^3 - 3x^2 + 2$
16. $f(x) = x^3 + 5x^2 - 9$
17. $f(x) = -3x^3 + 20x^2 - 36x + 16$
18. $f(x) = x^3 - 4x^2 + 2x - 1$
19. $f(x) = x^4 - 8$
20. $f(x) = x^4 - 10x^2 + 9$
21. $f(x) = -x^4 + 5x^2 - 2x - 1$
22. $f(x) = -x^4 + x^3 + 8x^2 - 3$
23. $f(x) = x^4 - 9x^3 + 25x^2 - 24x + 6$
24. $f(x) = 2x^4 - 4x^3 - 2x^2 + 3x - 5$
25. $f(x) = x^5 + 4x^4 - x^3 - 9x^2 + 3$
26. $f(x) = x^5 - 6x^4 + 4x^3 + 17x^2 - 5x - 6$

EMPLOYMENT For Exercises 27–30, use the graph that models the unemployment rates from 1975–2000.

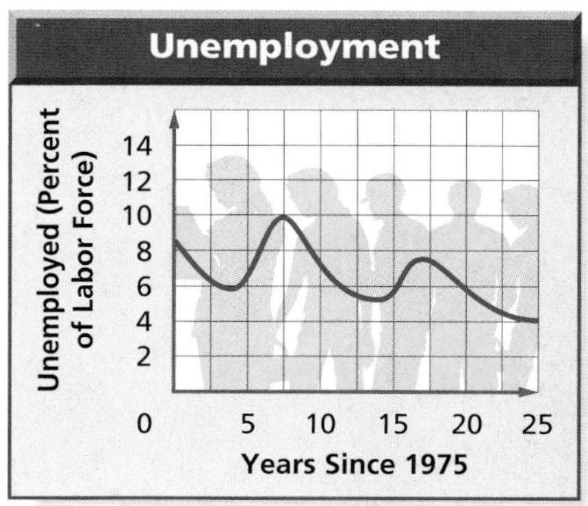

27. In what year was the unemployment rate the highest? the lowest?
28. Describe the turning points and end behavior of the graph.
29. If this graph was modeled by a polynomial equation, what is the least degree the equation could have?
30. Do you expect the unemployment rate to increase or decrease from 2001 to 2005? Explain your reasoning.

Online Research **Data Update** What is the current unemployment rate? Visit www.algebra2.com/data_update to learn more.

CHILD DEVELOPMENT For Exercises 31 and 32, use the following information. The average height (in inches) for boys ages 1 to 20 can be modeled by the equation $B(x) = -0.001x^4 + 0.04x^3 - 0.56x^2 + 5.5x + 25$, where x is the age (in years). The average height for girls ages 1 to 20 is modeled by the equation $G(x) = -0.0002x^4 + 0.006x^3 - 0.14x^2 + 3.7x + 26$.

31. Graph both equations by making a table of values. Use $x = \{0, 2, 4, 6, 8, 10, 12, 14, 16, 18, 20\}$ as the domain. Round values to the nearest inch.
32. Compare the graphs. What do the graphs suggest about the growth rate for both boys and girls?

More About. . .

Child Devolpment

As children develop, their sleeping needs change. Infants sleep about 16–18 hours a day. Toddlers usually sleep 10–12 hours at night and take one or two daytime naps. School-age children need 9–11 hours of sleep, and teens need at least 9 hours of sleep.

Source: www.kidshealth.org

PHYSIOLOGY For Exercises 33–35, use the following information. During a regular respiratory cycle, the volume of air in liters in the human lungs can be described by the function $V(t) = 0.173t + 0.152t^2 - 0.035t^3$, where t is the time in seconds.

33. Estimate the real zeros of the function by graphing.
34. About how long does a regular respiratory cycle last?
35. Estimate the time in seconds from the beginning of this respiratory cycle for the lungs to fill to their maximum volume of air.

CRITICAL THINKING For Exercises 36–39, sketch a graph of each polynomial.

36. even-degree polynomial function with one relative maximum and two relative minima
37. odd-degree polynomial function with one relative maximum and one relative minimum; the leading coefficient is negative
38. even-degree polynomial function with four relative maxima and three relative minima
39. odd-degree polynomial function with three relative maxima and three relative minima; the leftmost points are negative

40. **WRITING IN MATH** Answer the question that was posed at the beginning of the lesson.

How can graphs of polynomial functions show trends in data?

Include the following in your answer:

- a description of the types of data that are best modeled by polynomial equations rather than linear equations, and
- an explanation of how you would determine when the percent of foreign-born citizens was at its highest and when the percent was at its lowest since 1900.

Standardized Test Practice

Standards Practice

41. Which of the following could be the graph of $f(x) = x^3 + x^2 - 3x$?

(A)
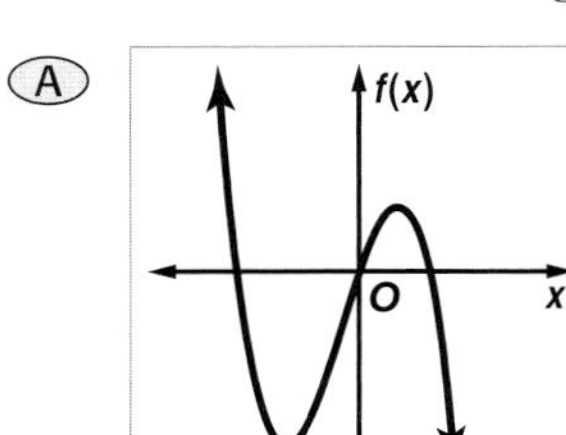

(B)
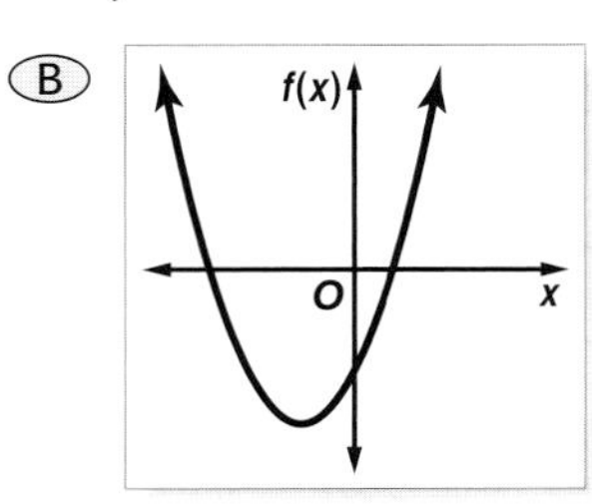

(C)
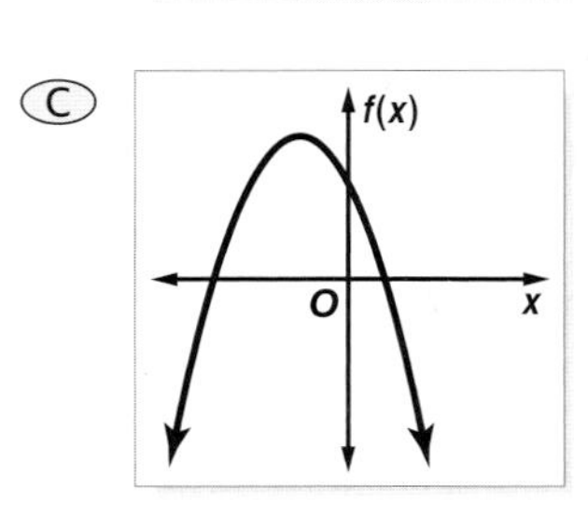

(D)
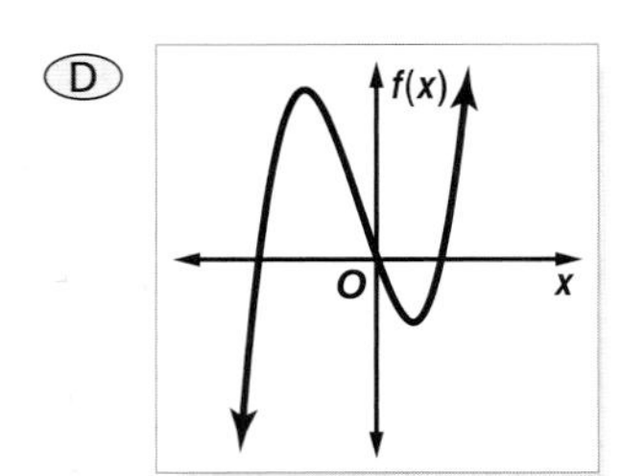

42. The function $f(x) = x^2 - 4x + 3$ has a relative minimum located at which of the following x values?

(A) -2 (B) 2 (C) 3 (D) 4

Graphing Calculator

Use a graphing calculator to estimate the x-coordinates at which the maxima and minima of each function occur. Round to the nearest hundredth.

43. $f(x) = x^3 + x^2 - 7x - 3$

44. $f(x) = -x^3 + 6x^2 - 6x - 5$

45. $f(x) = -x^4 + 3x^2 - 8$

46. $f(x) = 3x^4 - 7x^3 + 4x - 5$

Maintain Your Skills

Mixed Review

If $p(x) = 2x^2 - 5x + 4$ and $r(x) = 3x^3 - x^2 - 2$, find each value. *(Lesson 7-1)*

47. $r(2a)$

48. $5p(c)$

49. $p(2a^2)$

50. $r(x - 1)$

51. $p(x^2 + 4)$

52. $2[p(x^2 + 1)] - 3r(x - 1)$

Graph each inequality. *(Lesson 6-7)*

53. $y > x^2 - 4x + 6$

54. $y \leq -x^2 + 6x - 3$

55. $y < x^2 - 2x$

Solve each matrix equation or system of equations by using inverse matrices. *(Lesson 4-8)*

56. $\begin{bmatrix} 3 & 6 \\ 2 & -1 \end{bmatrix} \cdot \begin{bmatrix} a \\ b \end{bmatrix} = \begin{bmatrix} -3 \\ 18 \end{bmatrix}$

57. $\begin{bmatrix} 5 & -7 \\ -3 & 4 \end{bmatrix} \cdot \begin{bmatrix} m \\ n \end{bmatrix} = \begin{bmatrix} -1 \\ 1 \end{bmatrix}$

58. $3j + 2k = 8$
$j - 7k = 18$

59. $5y + 2z = 11$
$10y - 4z = -2$

60. **SPORTS** Bob and Minya want to build a ramp that they can use while rollerblading. If they want the ramp to have a slope of $\frac{1}{4}$, how tall should they make the ramp? *(Lesson 2-3)*

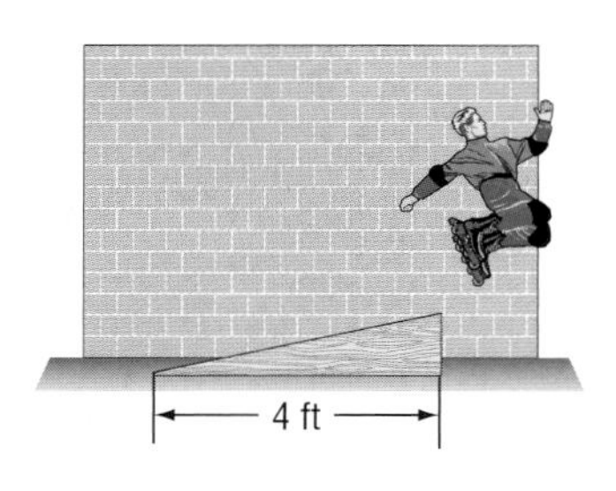

Getting Ready for the Next Lesson

PREREQUISITE SKILL Factor each polynomial.
*(To review **factoring polynomials**, see Lesson 5-4.)*

61. $x^2 - x - 30$

62. $2b^2 - 9b + 4$

63. $6a^2 + 17a + 5$

64. $4m^2 - 9$

65. $t^3 - 27$

66. $r^4 - 1$

Graphing Calculator Investigation

A Follow-Up of Lesson 7-2

Modeling Real-World Data

You can use a TI-83 Plus to model data whose curve of best fit is a polynomial function.

Example

The table shows the distance a seismic wave can travel based on its distance from an earthquake's epicenter. Draw a scatter plot and a curve of best fit that relates distance to travel time. Then determine approximately how far from the epicenter the wave will be felt 8.5 minutes after the earthquake occurs.

Source: University of Arizona

Travel Time (min)	1	2	5	7	10	12	13
Distance (km)	400	800	2500	3900	6250	8400	10,000

Step 1 Enter the travel times in L1 and the distances in L2.

KEYSTROKES: *Refer to page 87 to review how to enter lists.*

Step 2 Graph the scatter plot.

KEYSTROKES: *Refer to page 87 to review how to graph a scatter plot.*

Step 3 Compute and graph the equation for the curve of best fit. A quartic curve is the best fit for these data.

KEYSTROKES: [STAT] [▶] 7 [2nd] [L1] [,] [2nd] [L2] [ENTER] [Y=] [VARS] 5 [▶] [▶] 1 [GRAPH]

Step 4 Use the [CALC] feature to find the value of the function for $x = 8.5$.

KEYSTROKES: *Refer to page 87 to review how to find function values.*

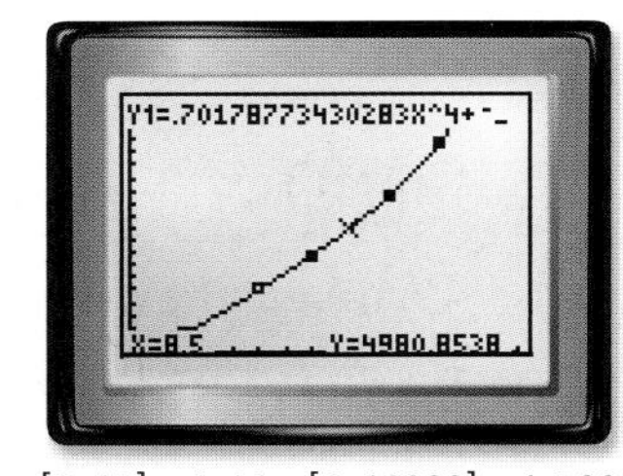

[0, 15] scl: 1 by [0, 10000] scl: 500

After 8.5 minutes, you would expect the wave to be felt approximately 5000 kilometers away.

Exercises

Use the table that shows how many minutes out of each eight-hour work day are used to pay one day's worth of taxes.

Year	Minutes
1940	83
1950	117
1960	130
1970	141
1980	145
1990	145
2000	160

Source: Tax Foundation

1. Draw a scatter plot of the data. Then graph several curves of best fit that relate the number of minutes to the year. Try LinReg, QuadReg, and CubicReg.
2. Write the equation for the curve that best fits the data.
3. Based on this equation, how many minutes should you expect to work each day in the year 2010 to pay one day's taxes?

www.algebra2.com/other_calculator_keystrokes

7-3 Solving Equations Using Quadratic Techniques

What You'll Learn

- Write expressions in quadratic form.
- Use quadratic techniques to solve equations.

Vocabulary

- quadratic form

How can solving polynomial equations help you to find dimensions?

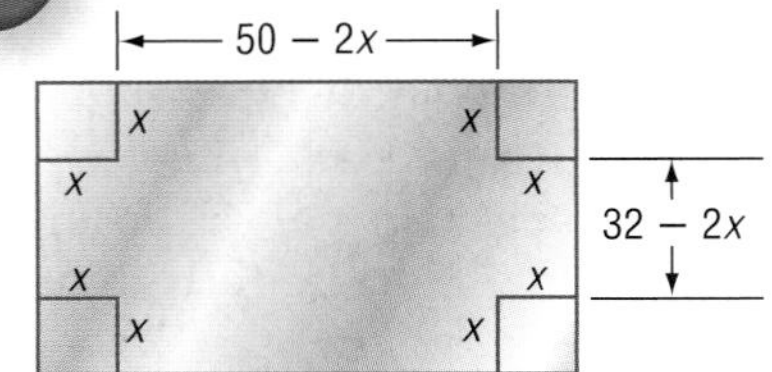

The Taylor Manufacturing Company makes open metal boxes of various sizes. Each sheet of metal is 50 inches long and 32 inches wide. To make a box, a square is cut from each corner. The volume of the box depends on the side length x of the cut squares. It is given by $V(x) = 4x^3 - 164x^2 + 1600x$. You can solve a polynomial equation to find the dimensions of the square to cut for a box with specific volume.

QUADRATIC FORM In some cases, you can rewrite a polynomial in x in the form $au^2 + bu + c$. For example, by letting $u = x^2$ the expression $x^4 - 16x^2 + 60$ can be written as $(x^2)^2 - 16(x^2) + 60$ or $u^2 - 16u + 60$. This new, but equivalent, expression is said to be in **quadratic form**.

Key Concept — *Quadratic Form*

An expression that is quadratic in form can be written as $au^2 + bu + c$ for any numbers a, b, and c, $a \neq 0$, where u is some expression in x. The expression $au^2 + bu + c$ is called the quadratic form of the original expression.

Example 1 Write Expressions in Quadratic Form

Write each expression in quadratic form, if possible.

a. $x^4 + 13x^2 + 36$

$x^4 + 13x^2 + 36 = (x^2)^2 + 13(x^2) + 36$ $\quad (x^2)^2 = x^4$

b. $16x^6 - 625$

$16x^6 - 625 = (4x^3)^2 - 625$ $\quad (x^3)^2 = x^6$

c. $12x^8 - x^2 + 10$

This cannot be written in quadratic form since $x^8 \neq (x^2)^2$.

d. $x - 9x^{\frac{1}{2}} + 8$

$x - 9x^{\frac{1}{2}} + 8 = \left(x^{\frac{1}{2}}\right)^2 - 9\left(x^{\frac{1}{2}}\right) + 8$ $\quad x^1 = \left(x^{\frac{1}{2}}\right)^2$

SOLVE EQUATIONS USING QUADRATIC FORM In Chapter 6, you learned to solve quadratic equations by using the Zero Product Property and the Quadratic Formula. You can extend these techniques to solve higher-degree polynomial equations that can be written using quadratic form or have an expression that contains a quadratic factor.

Example 2 Solve Polynomial Equations

Solve each equation.

a. $x^4 - 13x^2 + 36 = 0$

$$x^4 - 13x^2 + 36 = 0 \quad \text{Original equation}$$
$$(x^2)^2 - 13(x^2) + 36 = 0 \quad \text{Write the expression on the left in quadratic form.}$$
$$(x^2 - 9)(x^2 - 4) = 0 \quad \text{Factor the trinomial.}$$
$$(x - 3)(x + 3)(x - 2)(x + 2) = 0 \quad \text{Factor each difference of squares.}$$

Use the Zero Product Property.

$x - 3 = 0$ or $x + 3 = 0$ or $x - 2 = 0$ or $x + 2 = 0$

$x = 3$ $\quad$ $x = -3$ $\quad$ $x = 2$ $\quad$ $x = -2$

The solutions are -3, -2, 2, and 3.

CHECK The graph of $f(x) = x^4 - 13x^2 + 36$ shows that the graph intersects the x-axis at -3, -2, 2, and 3. ✓

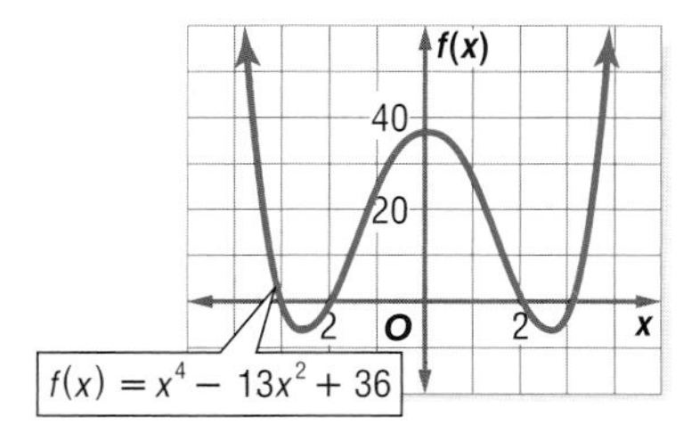

Study Tip

Look Back
To review the formula for factoring the **sum of two cubes**, see Lesson 5-4.

b. $x^3 + 343 = 0$

$$x^3 + 343 = 0 \quad \text{Original equation}$$
$$(x)^3 + 7^3 = 0 \quad \text{This is the sum of two cubes.}$$
$$(x + 7)[x^2 - x(7) + 7^2] = 0 \quad \text{Sum of two cubes formula with } a = x \text{ and } b = 7$$
$$(x + 7)(x^2 - 7x + 49) = 0 \quad \text{Simplify.}$$
$$x + 7 = 0 \text{ or } x^2 - 7x + 49 = 0 \quad \text{Zero Product Property}$$

The solution of the first equation is -7. The second equation can be solved by using the Quadratic Formula.

$$x = \frac{-b \pm \sqrt{b^2 - 4ac}}{2a} \quad \text{Quadratic Formula}$$
$$= \frac{-(-7) \pm \sqrt{(-7)^2 - 4(1)(49)}}{2(1)} \quad \text{Replace } a \text{ with 1, } b \text{ with } -7 \text{, and } c \text{ with 49.}$$
$$= \frac{7 \pm \sqrt{-147}}{2} \quad \text{Simplify.}$$
$$= \frac{7 \pm i\sqrt{147}}{2} \text{ or } \frac{7 \pm 7i\sqrt{3}}{2} \quad \sqrt{147} \times \sqrt{-1} = 7i\sqrt{3}$$

Thus, the solutions of the original equation are -7, $\frac{7 + 7i\sqrt{3}}{2}$, and $\frac{7 - 7i\sqrt{3}}{2}$.

Study Tip

Substitution
To avoid confusion, you can substitute another variable for the expression in parentheses.

For example, $\left(x^{\frac{1}{3}}\right)^2 - 6\left(x^{\frac{1}{3}}\right) + 5 = 0$ could be written as $u^2 - 6u + 5 = 0$. Then, once you have solved the equation for u, substitute $x^{\frac{1}{3}}$ for u and solve for x.

Some equations involving rational exponents can be solved by using a quadratic technique.

Example 3 Solve Equations with Rational Exponents

Solve $x^{\frac{2}{3}} - 6x^{\frac{1}{3}} + 5 = 0$.

$$x^{\frac{2}{3}} - 6x^{\frac{1}{3}} + 5 = 0 \quad \text{Original equation}$$
$$\left(x^{\frac{1}{3}}\right)^2 - 6\left(x^{\frac{1}{3}}\right) + 5 = 0 \quad \text{Write the expression on the left in quadratic form.}$$

(continued on the next page)

$\left(x^{\frac{1}{3}} - 1\right)\left(x^{\frac{1}{3}} - 5\right) = 0$ Factor the trinomial.

$x^{\frac{1}{3}} - 1 = 0$ or $x^{\frac{1}{3}} - 5 = 0$ Zero Product Property

$x^{\frac{1}{3}} = 1$ $\quad x^{\frac{1}{3}} = 5$ Isolate x on one side of the equation.

$\left(x^{\frac{1}{3}}\right)^3 = 1^3$ $\quad \left(x^{\frac{1}{3}}\right)^3 = 5^3$ Cube each side.

$x = 1$ $\quad x = 125$ Simplify.

CHECK Substitute each value into the original equation.

$x^{\frac{2}{3}} - 6x^{\frac{1}{3}} + 5 = 0$ $\qquad x^{\frac{2}{3}} - 6x^{\frac{1}{3}} + 5 = 0$

$1^{\frac{2}{3}} - 6(1)^{\frac{1}{3}} + 5 \stackrel{?}{=} 0$ $\qquad 125^{\frac{2}{3}} - 6(125)^{\frac{1}{3}} + 5 \stackrel{?}{=} 0$

$1 - 6 + 5 \stackrel{?}{=} 0$ $\qquad 25 - 30 + 5 \stackrel{?}{=} 0$

$0 = 0 \ \checkmark$ $\qquad 0 = 0 \ \checkmark$

The solutions are 1 and 125.

To use a quadratic technique, rewrite the equation so one side is equal to zero.

Example 4 Solve Radical Equations

Solve $x - 6\sqrt{x} = 7$.

$x - 6\sqrt{x} = 7$ Original equation

$x - 6\sqrt{x} - 7 = 0$ Rewrite so that one side is zero.

$(\sqrt{x})^2 - 6(\sqrt{x}) - 7 = 0$ Write the expression on the left in quadratic form.

You can use the Quadratic Formula to solve this equation.

$\sqrt{x} = \frac{-b \pm \sqrt{b^2 - 4ac}}{2a}$ Quadratic Formula

$\sqrt{x} = \frac{-(-6) \pm \sqrt{(-6)^2 - 4(1)(-7)}}{2(1)}$ Replace a with 1, b with -6, and c with -7.

$\sqrt{x} = \frac{6 \pm 8}{2}$ Simplify.

$\sqrt{x} = \frac{6 + 8}{2}$ or $\sqrt{x} = \frac{6 - 8}{2}$ Write as two equations.

$\sqrt{x} = 7$ $\quad \sqrt{x} = -1$ Simplify.

$x = 49$

Since the principal square root of a number cannot be negative, the equation $\sqrt{x} = -1$ has no solution. Thus, the only solution of the original equation is 49.

Study Tip

Look Back
To review **principal roots**, see Lesson 5-5.

Check for Understanding

Concept Check

1. **OPEN ENDED** Give an example of an equation that is not quadratic but can be written in quadratic form. Then write it in quadratic form.
2. **Explain** how the graph of the related polynomial function can help you verify the solution to a polynomial equation.
3. **Describe** how to solve $x^5 - 2x^3 + x = 0$.

Guided Practice Write each expression in quadratic form, if possible.

4. $5y^4 + 7y^3 - 8$
5. $84n^4 - 62n^2$

Solve each equation.

6. $x^3 + 9x^2 + 20x = 0$
7. $x^4 - 17x^2 + 16 = 0$
8. $x^3 - 216 = 0$
9. $x - 16x^{\frac{1}{2}} = -64$

Application

10. **POOL** The Shelby University swimming pool is in the shape of a rectangular prism and has a volume of 28,000 cubic feet. The dimensions of the pool are x feet deep by $7x - 6$ feet wide by $9x - 2$ feet long. How deep is the pool?

Practice and Apply

Homework Help

For Exercises	See Examples
11–16	1
17–28	2–4
29–36	2

Extra Practice See page 842.

Write each expression in quadratic form, if possible.

11. $2x^4 + 6x^2 - 10$
12. $a^8 + 10a^2 - 16$
13. $11n^6 + 44n^3$
14. $7b^5 - 4b^3 + 2b$
15. $7x^{\frac{1}{9}} - 3x^{\frac{1}{3}} + 4$
16. $6x^{\frac{2}{5}} - 4x^{\frac{1}{5}} - 16$

Solve each equation.

17. $m^4 + 7m^3 + 12m^2 = 0$
18. $a^5 + 6a^4 + 5a^3 = 0$
19. $b^4 = 9$
20. $t^5 - 256t = 0$
21. $d^4 + 32 = 12d^2$
22. $x^4 + 18 = 11x^2$
23. $x^3 + 729 = 0$
24. $y^3 - 512 = 0$
25. $x^{\frac{1}{2}} - 8x^{\frac{1}{4}} + 15 = 0$
26. $p^{\frac{2}{3}} + 11p^{\frac{1}{3}} + 28 = 0$
27. $y - 19\sqrt{y} = -60$
28. $z = 8\sqrt{z} + 240$
29. $s^3 + 4s^2 - s - 4 = 0$
30. $h^3 - 8h^2 + 3h - 24 = 0$

31. **GEOMETRY** The width of a rectangular prism is w centimeters. The height is 2 centimeters less than the width. The length is 4 centimeters more than the width. If the volume of the prism is 8 times the measure of the length, find the dimensions of the prism.

Career Choices

Designer

Designers combine practical knowledge with artistic ability to turn abstract ideas into formal designs. Designers usually specialize in a particular area, such as clothing, or home interiors.

Online Research For information about a career as a designer, visit: www.algebra2.com/careers

DESIGN For Exercises 32–34, use the following information.
Jill is designing a picture frame for an art project. She plans to have a square piece of glass in the center and surround it with a decorated ceramic frame, which will also be a square. The dimensions of the glass and frame are shown in the diagram at the right. Jill determines that she needs 27 square inches of material for the frame.

x

$x^2 - 3$ in.

32. Write a polynomial equation that models the area of the frame.
33. What are the dimensions of the glass piece?
34. What are the dimensions of the frame?

PACKAGING For Exercises 35 and 36, use the following information.
A computer manufacturer needs to change the dimensions of its foam packaging for a new model of computer. The width of the original piece is three times the height, and the length is equal to the height squared. The volume of the new piece can be represented by the equation $V(h) = 3h^4 + 11h^3 + 18h^2 + 44h + 24$, where h is the height of the original piece.

35. Factor the equation for the volume of the new piece to determine three expressions that represent the height, length, and width of the new piece.
36. How much did each dimension of the packaging increase for the new foam piece?

37. **CRITICAL THINKING** Explain how you would solve $|a - 3|^2 - 9|a - 3| = -8$. Then solve the equation.

38. **WRITING IN MATH** Answer the question that was posed at the beginning of the lesson.

How can solving polynomial equations help you to find dimensions?

Include the following items in your answer:

- an explanation of how you could determine the dimensions of the cut square if the desired volume was 3600 cubic inches, and
- an explanation of why there can be more than one square that can be cut to produce the same volume.

39. Which of the following is a solution of $x^4 - 2x^2 - 3 = 0$?

(A) $\sqrt[4]{2}$ (B) 1 (C) -3 (D) $\sqrt{3}$

40. **EXTENDED RESPONSE** Solve $18x + 9\sqrt{2x} - 4 = 0$ by first rewriting it in quadratic form. Show your work.

Maintain Your Skills

Mixed Review

Graph each function by making a table of values. *(Lesson 7-2)*

41. $f(x) = x^3 - 4x^2 + x + 5$

42. $f(x) = x^4 - 6x^3 + 10x^2 - x - 3$

Find $p(7)$ and $p(-3)$ for each function. *(Lesson 7-1)*

43. $p(x) = x^2 - 5x + 3$

44. $p(x) = x^3 - 11x - 4$

45. $p(x) = \frac{2}{3}x^4 - 3x^3$

For Exercises 46–48, use the following information.

Triangle ABC with vertices $A(-2, 1)$, $B(-3, -3)$, and $C(3, -1)$ is rotated 90° counterclockwise about the origin. *(Lesson 4-4)*

46. Write the coordinates of the triangle in a vertex matrix.

47. Find the coordinates of $\triangle A'B'C'$.

48. Graph the preimage and the image.

Getting Ready for the Next Lesson

PREREQUISITE SKILL Find each quotient.

*(To review **dividing polynomials**, see Lesson 5-3.)*

49. $(x^3 + 4x^2 - 9x + 4) \div (x - 1)$

50. $(4x^3 - 8x^2 - 5x - 10) \div (x + 2)$

51. $(x^4 - 9x^2 - 2x + 6) \div (x - 3)$

52. $(x^4 + 3x^3 - 8x^2 + 5x - 6) \div (x + 1)$

Practice Quiz 1

Lessons 7-1 through 7-3

1. If $p(x) = 2x^3 - x$, find $p(a - 1)$. *(Lesson 7-1)*

2. Describe the end behavior of the graph at the right. Then determine whether it represents an odd-degree or an even-degree polynomial function and state the number of real zeros. *(Lesson 7-1)*

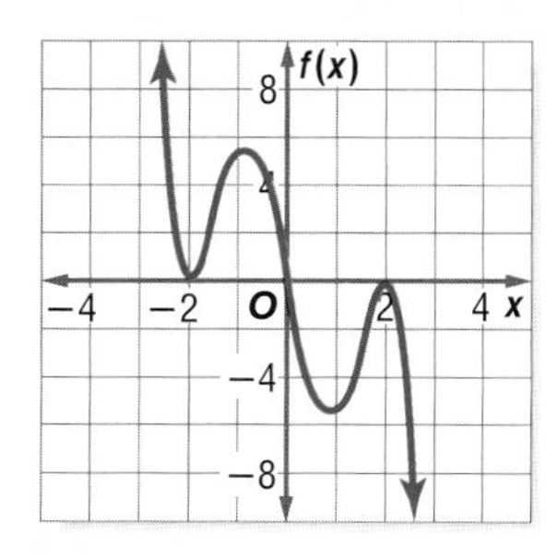

3. Graph $y = x^3 + 2x^2 - 4x - 6$. Estimate the x-coordinates at which the relative maxima and relative minima occur. *(Lesson 7-2)*

4. Write the expression $18x^{\frac{1}{3}} + 36x^{\frac{2}{3}} + 5$ in quadratic form. *(Lesson 7-3)*

5. Solve $a^4 = 6a^2 + 27$. *(Lesson 7-3)*

7-4 The Remainder and Factor Theorems

California Standards Standard 3.0 Students are adept at operations on polynomials, including long division. (Key)

What You'll Learn

- Evaluate functions using synthetic substitution.
- Determine whether a binomial is a factor of a polynomial by using synthetic substitution.

Vocabulary

- synthetic substitution
- depressed polynomial

How can you use the Remainder Theorem to evaluate polynomials?

The number of international travelers to the United States since 1986 can be modeled by the equation $T(x) = 0.02x^3 - 0.6x^2 + 6x + 25.9$, where x is the number of years since 1986 and $T(x)$ is the number of travelers in millions. To estimate the number of travelers in 2006, you can evaluate the function for $x = 20$, or you can use synthetic substitution.

Study Tip

Look Back
To review **dividing polynomials** and **synthetic division**, see Lesson 5-3.

SYNTHETIC SUBSTITUTION Synthetic division is a shorthand method of long division. It can also be used to find the value of a function. Consider the polynomial function $f(a) = 4a^2 - 3a + 6$. Divide the polynomial by $a - 2$.

Method 1 Long Division

$$\begin{array}{r} 4a + 5 \\ a - 2\overline{)4a^2 - 3a + 6} \\ \underline{4a^2 - 8a} \\ 5a + 6 \\ \underline{5a - 10} \\ 16 \end{array}$$

Method 2 Synthetic Division

$$\begin{array}{r|rrr} 2 & 4 & -3 & 6 \\ & & 8 & 10 \\ \hline & 4 & 5 & |\,16 \end{array}$$

Compare the remainder of 16 to $f(2)$.

$$\begin{aligned} f(2) &= 4(2)^2 - 3(2) + 6 && \text{Replace } a \text{ with 2.} \\ &= 16 - 6 + 6 && \text{Multiply.} \\ &= 16 && \text{Simplify.} \end{aligned}$$

Notice that the value of $f(2)$ is the same as the remainder when the polynomial is divided by $a - 2$. This illustrates the **Remainder Theorem**.

Key Concept *Remainder Theorem*

If a polynomial $f(x)$ is divided by $x - a$, the remainder is the constant $f(a)$, and

$$\underbrace{f(x)}_{\text{Dividend}} \underbrace{=}_{\text{equals}} \underbrace{q(x)}_{\text{quotient}} \underbrace{\cdot}_{\text{times}} \underbrace{(x-a)}_{\text{divisor}} \underbrace{+}_{\text{plus}} \underbrace{f(a),}_{\text{remainder.}}$$

where $q(x)$ is a polynomial with degree one less than the degree of $f(x)$.

When synthetic division is used to evaluate a function, it is called **synthetic substitution**. It is a convenient way of finding the value of a function, especially when the degree of the polynomial is greater than 2.

Example 1 Synthetic Substitution

If $f(x) = 2x^4 - 5x^2 + 8x - 7$, find $f(6)$.

Method 1 Synthetic Substitution

By the Remainder Theorem, $f(6)$ should be the remainder when you divide the polynomial by $x - 6$.

$$\begin{array}{r|rrrrr} 6 & 2 & 0 & -5 & 8 & -7 \\ & & 12 & 72 & 402 & 2460 \\ \hline & 2 & 12 & 67 & 410 & 2453 \end{array}$$

Notice that there is no x^3 term. A zero is placed in this position as a placeholder.

The remainder is 2453. Thus, by using synthetic substitution, $f(6) = 2453$.

Method 2 Direct Substitution

Replace x with 6.

$f(x) = 2x^4 - 5x^2 + 8x - 7$ Original function

$f(6) = 2(6)^4 - 5(6)^2 + 8(6) - 7$ Replace x with 6.

$= 2592 - 180 + 48 - 7$ or 2453 Simplify.

By using direct substitution, $f(6) = 2453$.

FACTORS OF POLYNOMIALS Divide $f(x) = x^4 + x^3 - 17x^2 - 20x + 32$ by $x - 4$.

$$\begin{array}{r|rrrrr} 4 & 1 & 1 & -17 & -20 & 32 \\ & & 4 & 20 & 12 & -32 \\ \hline & 1 & 5 & 3 & -8 & 0 \end{array}$$

Study Tip

Depressed Polynomial

A *depressed polynomial* has a degree that is one less than the original polynomial.

The quotient of $f(x)$ and $x - 4$ is $x^3 + 5x^2 + 3x - 8$. When you divide a polynomial by one of its binomial factors, the quotient is called a **depressed polynomial**. From the results of the division and by using the Remainder Theorem, we can make the following statement.

$$\underbrace{x^4 + x^3 - 17x^2 - 20x + 32}_{\text{Dividend}} \underbrace{=}_{\text{equals}} \underbrace{(x^3 + 5x^2 + 3x - 8)}_{\text{quotient}} \underbrace{\cdot}_{\text{times}} \underbrace{(x - 4)}_{\text{divisor}} \underbrace{+}_{\text{plus}} \underbrace{0}_{\text{remainder.}}$$

Since the remainder is 0, $f(4) = 0$. This means that $x - 4$ is a factor of $x^4 + x^3 - 17x^2 - 20x + 32$. This illustrates the **Factor Theorem**, which is a special case of the Remainder Theorem.

Key Concept *Factor Theorem*

The binomial $x - a$ is a factor of the polynomial $f(x)$ if and only if $f(a) = 0$.

Suppose you wanted to find the factors of $x^3 - 3x^2 - 6x + 8$. One approach is to graph the related function, $f(x) = x^3 - 3x^2 - 6x + 8$. From the graph at the right, you can see that the graph of $f(x)$ crosses the x-axis at -2, 1, and 4. These are the zeros of the function. Using these zeros and the Zero Product Property, we can express the polynomial in factored form.

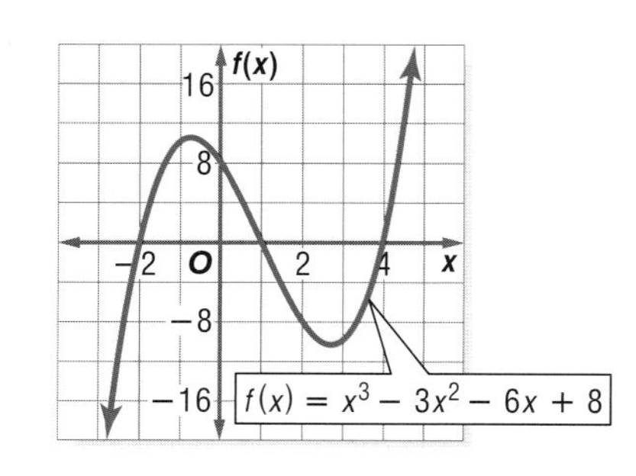

$$f(x) = [x - (-2)](x - 1)(x - 4)$$
$$= (x + 2)(x - 1)(x - 4)$$

This method of factoring a polynomial has its limitations. Most polynomial functions are not easily graphed and once graphed, the exact zeros are often difficult to determine.

The Factor Theorem can help you find all factors of a polynomial.

Example 2 Use the Factor Theorem

Show that $x + 3$ is a factor of $x^3 + 6x^2 - x - 30$. Then find the remaining factors of the polynomial.

The binomial $x + 3$ is a factor of the polynomial if -3 is a zero of the related polynomial function. Use the Factor Theorem and synthetic division.

$$\begin{array}{r|rrrr} -3 & 1 & 6 & -1 & -30 \\ & & -3 & -9 & 30 \\ \hline & 1 & 3 & -10 & 0 \end{array}$$

> **Study Tip**
>
> *Factoring*
> The factors of a polynomial do not have to be binomials. For example, the factors of $x^3 + x^2 - x + 15$ are $x + 3$ and $x^2 - 2x + 5$.

Since the remainder is 0, $x + 3$ is a factor of the polynomial. The polynomial $x^3 + 6x^2 - x - 30$ can be factored as $(x + 3)(x^2 + 3x - 10)$. The polynomial $x^2 + 3x - 10$ is the depressed polynomial. Check to see if this polynomial can be factored.

$x^2 + 3x - 10 = (x - 2)(x + 5)$ Factor the trinomial.

So, $x^3 + 6x^2 - x - 30 = (x + 3)(x - 2)(x + 5)$.

CHECK You can see that the graph of the related function $f(x) = x^3 + 6x^2 - x - 30$ crosses the x-axis at -5, -3, and 2. Thus, $f(x) = [x - (-5)][x - (-3)](x - 2)$. ✓

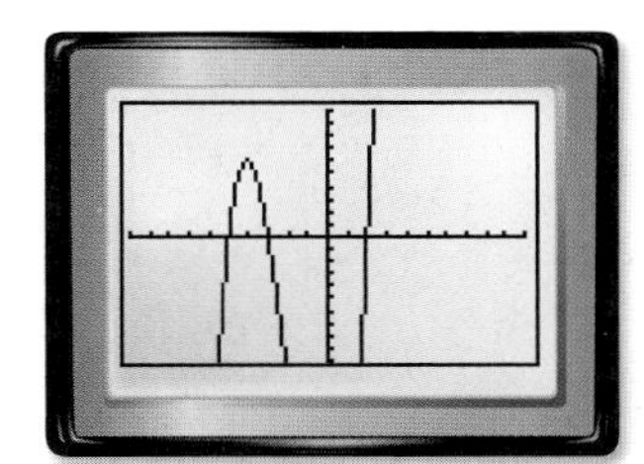

Example 3 Find All Factors of a Polynomial

GEOMETRY The volume of the rectangular prism is given by $V(x) = x^3 + 3x^2 - 36x + 32$. Find the missing measures.

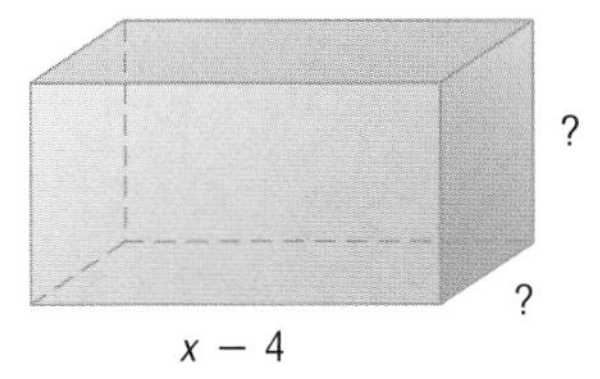

The volume of a rectangular prism is $\ell \times w \times h$.

You know that one measure is $x - 4$, so $x - 4$ is a factor of $V(x)$.

$$\begin{array}{r|rrrr} 4 & 1 & 3 & -36 & 32 \\ & & 4 & 28 & -32 \\ \hline & 1 & 7 & -8 & 0 \end{array}$$

The quotient is $x^2 + 7x - 8$. Use this to factor $V(x)$.

$V(x) = x^3 + 3x^2 - 36x + 32$ Volume function

$= (x - 4)(x^2 + 7x - 8)$ Factor.

$= (x - 4)(x + 8)(x - 1)$ Factor the trinomial $x^2 + 7x - 8$.

So, the missing measures of the prism are $x + 8$ and $x - 1$.

Check for Understanding

Concept Check

1. **OPEN ENDED** Give an example of a polynomial function that has a remainder of 5 when divided by $x - 4$.

2. **State** the degree of the depressed polynomial that is the result of dividing $x^5 + 3x^4 - 16x - 48$ by one of its first-degree binomial factors.

3. **Write** the dividend, divisor, quotient, and remainder represented by the synthetic division at the right.

$$\begin{array}{r|rrrr} -2 & 1 & 0 & 6 & 32 \\ & & -2 & 4 & -20 \\ \hline & 1 & -2 & 10 & \vert\; 12 \end{array}$$

Guided Practice

Use synthetic substitution to find $f(3)$ and $f(-4)$ for each function.

4. $f(x) = x^3 - 2x^2 - x + 1$

5. $f(x) = 5x^4 - 6x^2 + 2$

Given a polynomial and one of its factors, find the remaining factors of the polynomial. Some factors may not be binomials.

6. $x^3 - x^2 - 5x - 3;\ x + 1$

7. $x^3 - 3x + 2;\ x - 1$

8. $6x^3 - 25x^2 + 2x + 8;\ 3x - 2$

9. $x^4 + 2x^3 - 8x - 16;\ x + 2$

Application

For Exercises 10–12, use the graph at the right. The projected sales of e-books can be modeled by the function $S(x) = -17x^3 + 200x^2 - 113x + 44$, where x is the number of years since 2000.

10. Use synthetic substitution to estimate the sales for 2006 in billions of dollars.

11. Evaluate $S(6)$.

12. Which method—synthetic division or direct substitution—do you prefer to use to evaluate polynomials? Explain your answer.

Practice and Apply

Homework Help

For Exercises	See Examples
13–20	1
21–36	2
37–44	3

Extra Practice
See page 843.

Use synthetic substitution to find $g(3)$ and $g(-4)$ for each function.

13. $g(x) = x^2 - 8x + 6$

14. $g(x) = x^3 + 2x^2 - 3x + 1$

15. $g(x) = x^3 - 5x + 2$

16. $g(x) = x^4 - 6x - 8$

17. $g(x) = 2x^3 - 8x^2 - 2x + 5$

18. $g(x) = 3x^4 + x^3 - 2x^2 + x + 12$

19. $g(x) = x^5 + 8x^3 + 2x - 15$

20. $g(x) = x^6 - 4x^4 + 3x^2 - 10$

Given a polynomial and one of its factors, find the remaining factors of the polynomial. Some factors may not be binomials.

21. $x^3 + 2x^2 - x - 2;\ x - 1$

22. $x^3 - x^2 - 10x - 8;\ x + 1$

23. $x^3 + x^2 - 16x - 16;\ x + 4$

24. $x^3 - 6x^2 + 11x - 6;\ x - 2$

25. $2x^3 - 5x^2 - 28x + 15; x - 5$

26. $3x^3 + 10x^2 - x - 12; x + 3$

27. $2x^3 + 7x^2 - 53x - 28; 2x + 1$

28. $2x^3 + 17x^2 + 23x - 42; 2x + 7$

29. $x^4 + 2x^3 + 2x^2 - 2x - 3; x + 1$

30. $16x^5 - 32x^4 - 81x + 162; x - 2$

Changes in world population can be modeled by a polynomial function. Visit www.algebra2.com/webquest to continue work on your WebQuest project.

31. Use the graph of the polynomial function at the right to determine at least one binomial factor of the polynomial. Then find all the factors of the polynomial.

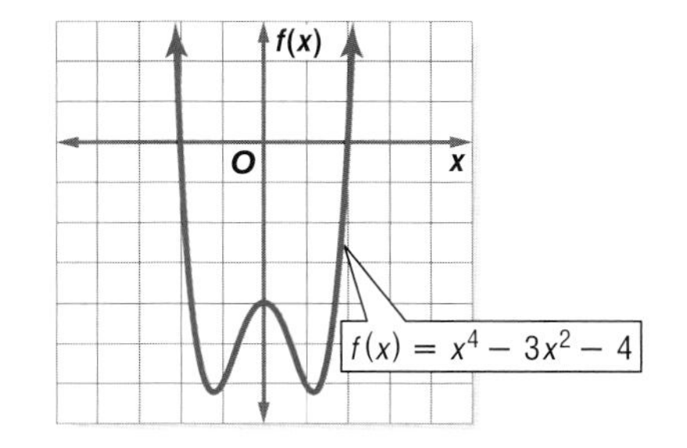

32. Use synthetic substitution to show that $x - 8$ is a factor of $x^3 - 4x^2 - 29x - 24$. Then find any remaining factors.

Find values of k so that each remainder is 3.

33. $(x^2 - x + k) \div (x - 1)$

34. $(x^2 + kx - 17) \div (x - 2)$

35. $(x^2 + 5x + 7) \div (x + k)$

36. $(x^3 + 4x^2 + x + k) \div (x + 2)$

ENGINEERING **For Exercises 37 and 38, use the following information.**
When a certain type of plastic is cut into sections, the length of each section determines its strength. The function $f(x) = x^4 - 14x^3 + 69x^2 - 140x + 100$ can describe the relative strength of a section of length x feet. Sections of plastic x feet long, where $f(x) = 0$, are extremely weak. After testing the plastic, engineers discovered that sections 5 feet long were extremely weak.

37. Show that $x - 5$ is a factor of the polynomial function.

38. Are there other lengths of plastic that are extremely weak? Explain your reasoning.

ARCHITECTURE **For Exercises 39 and 40, use the following information.**
Elevators traveling from one floor to the next do not travel at a constant speed. Suppose the speed of an elevator in feet per second is given by the function $f(t) = -0.5t^4 + 4t^3 - 12t^2 + 16t$, where t is the time in seconds.

39. Find the speed of the elevator at 1, 2, and 3 seconds.

40. It takes 4 seconds for the elevator to go from one floor to the next. Use synthetic substitution to find $f(4)$. Explain what this means.

41. **CRITICAL THINKING** Consider the polynomial $f(x) = ax^4 + bx^3 + cx^2 + dx + e$, where $a + b + c + d + e = 0$. Show that this polynomial is divisible by $x - 1$.

PERSONAL FINANCE **For Exercises 42–45, use the following information.**
Zach has purchased some home theater equipment for $2000, which he is financing through the store. He plans to pay $340 per month and wants to have the balance paid off after six months. The formula $B(x) = 2000x^6 - 340(x^5 + x^4 + x^3 + x^2 + x + 1)$ represents his balance after six months if x represents 1 plus the monthly interest rate (expressed as a decimal).

42. Find his balance after 6 months if the annual interest rate is 12%. (*Hint*: The monthly interest rate is the annual rate divided by 12, so $x = 1.01$.)

43. Find his balance after 6 months if the annual interest rate is 9.6%.

44. How would the formula change if Zach wanted to pay the balance in five months?

45. Suppose he finances his purchase at 10.8% and plans to pay $410 every month. Will his balance be paid in full after five months?

46. **WRITING IN MATH** Answer the question that was posed at the beginning of the lesson.

How can you use the Remainder Theorem to evaluate polynomials?

Include the following items in your answer:

- an explanation of when it is easier to use the Remainder Theorem to evaluate a polynomial rather than substitution, and
- evaluate the expression for the number of international travelers to the U.S. for $x = 20$.

47. Determine the zeros of the function $f(x) = x^2 + 7x + 12$ by factoring.

Ⓐ 7, 12 Ⓑ 3, 4 Ⓒ −5, 5 Ⓓ −4, −3

48. **SHORT RESPONSE** Using the graph of the polynomial function at the right, find all the factors of the polynomial $x^5 + x^4 - 3x^3 - 3x^2 - 4x - 4$.

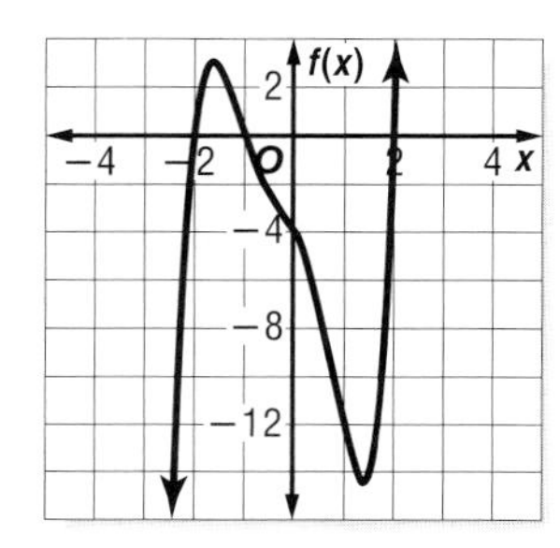

Maintain Your Skills

Mixed Review

Write each expression in quadratic form, if possible. *(Lesson 7-3)*

49. $x^4 - 8x^2 + 4$
50. $9d^6 + 5d^3 - 2$
51. $r^4 - 5r^3 + 18r$

Graph each polynomial function. Estimate the x-coordinates at which the relative maxima and relative minima occur. *(Lesson 7-2)*

52. $f(x) = x^3 - 6x^2 + 4x + 3$
53. $f(x) = -x^4 + 2x^3 + 3x^2 - 7x + 4$

54. **PHYSICS** A model airplane is fixed on a string so that it flies around in a circle. The formula $F_c = m\left(\frac{4\pi^2 r}{T^2}\right)$ describes the force required to keep the airplane going in a circle, where m represents the mass of the plane, r represents the radius of the circle, and T represents the time for a revolution. Solve this formula for T. Write in simplest radical form. *(Lesson 5-8)*

Solve each matrix equation. *(Lesson 4-1)*

55. $\begin{bmatrix} 7x \\ 12 \end{bmatrix} = \begin{bmatrix} 28 \\ -6y \end{bmatrix}$
56. $\begin{bmatrix} 5a + 2b \\ a - 7b \end{bmatrix} = \begin{bmatrix} -17 \\ 4 \end{bmatrix}$

Identify each function as S for step, C for constant, A for absolute value, or P for piecewise. *(Lesson 2-6)*

57.
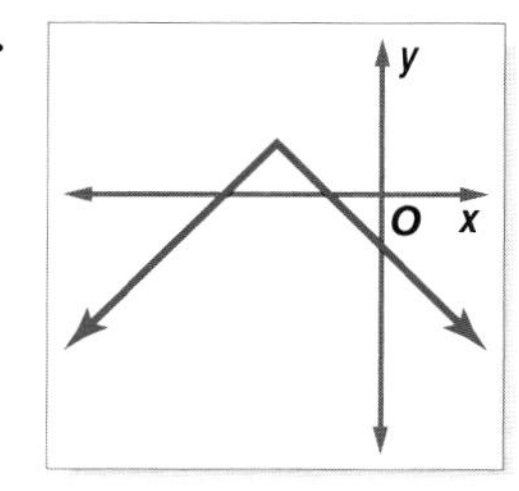

58.
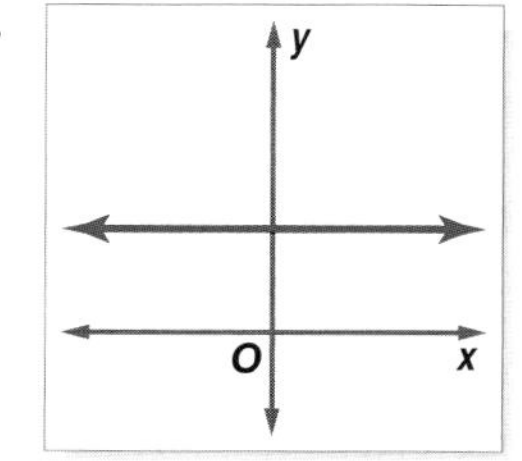

59.
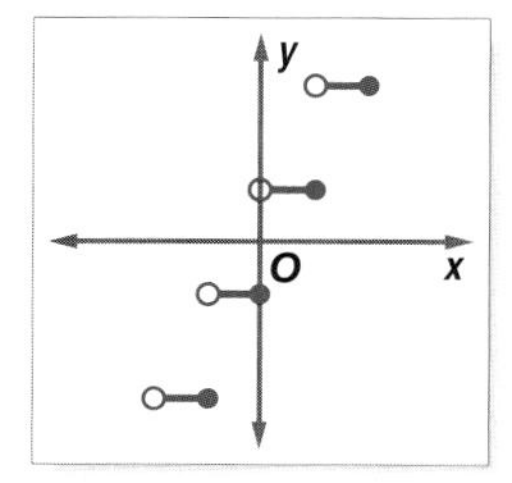

Getting Ready for the Next Lesson

PREREQUISITE SKILL Find the exact solutions of each equation by using the Quadratic Formula. *(For review of the **Quadratic Formula**, see Lesson 6-5.)*

60. $x^2 + 7x + 8 = 0$
61. $3x^2 - 9x + 2 = 0$
62. $2x^2 + 3x + 2 = 0$

7-5 Roots and Zeros

What You'll Learn

- Determine the number and type of roots for a polynomial equation.
- Find the zeros of a polynomial function.

How can the roots of an equation be used in pharmacology?

When doctors prescribe medication, they give patients instructions as to how much to take and how often it should be taken. The amount of medication in your body varies with time.

Suppose the equation $M(t) = 0.5t^4 + 3.5t^3 - 100t^2 + 350t$ models the number of milligrams of a certain medication in the bloodstream t hours after it has been taken. The doctor can use the roots of this equation to determine how often the patient should take the medication to maintain a certain concentration in the body.

TYPES OF ROOTS You have already learned that a zero of a function $f(x)$ is any value c such that $f(c) = 0$. When the function is graphed, the real zeros of the function are the x-intercepts of the graph.

Concept Summary — *Zeros, Factors, and Roots*

Let $f(x) = a_nx^n + \ldots + a_1x + a_0$ be a polynomial function. Then

- c is a zero of the polynomial function $f(x)$,
- $x - c$ is a factor of the polynomial $f(x)$, and
- c is a root or solution of the polynomial equation $f(x) = 0$.

In addition, if c is a real number, then $(c, 0)$ is an intercept of the graph of $f(x)$.

Study Tip

Look Back
For review of **complex numbers**, see Lesson 5-9.

When you solve a polynomial equation with degree greater than zero, it may have one or more real roots, or no real roots (the roots are imaginary numbers). Since real numbers and imaginary numbers both belong to the set of complex numbers, all polynomial equations with degree greater than zero will have at least one root in the set of complex numbers. This is the **Fundamental Theorem of Algebra**.

Key Concept — *Fundamental Theorem of Algebra*

Every polynomial equation with degree greater than zero has at least one root in the set of complex numbers.

Example 1 Determine Number and Type of Roots

Solve each equation. State the number and type of roots.

a. $x + 3 = 0$

$x + 3 = 0$ Original equation

$x = -3$ Subtract 3 from each side.

This equation has exactly one real root, -3.

b. $x^2 - 8x + 16 = 0$

$x^2 - 8x + 16 = 0$ Original equation

$(x - 4)^2 = 0$ Factor the left side as a perfect square trinomial.

$x = 4$ Solve for x using the Square Root Property.

Since $x - 4$ is twice a factor of $x^2 - 8x + 16$, 4 is a double root. So this equation has two real roots, 4 and 4.

Study Tip

Reading Math

In addition to double roots, equations can have triple or quadruple roots. In general, these roots are referred to as *repeated roots*.

c. $x^3 + 2x = 0$

$x^3 + 2x = 0$ Original equation

$x(x^2 + 2) = 0$ Factor out the GCF.

Use the Zero Product Property.

$x = 0$ or $x^2 + 2 = 0$

$x^2 = -2$ Subtract two from each side.

$x = \pm\sqrt{-2}$ or $\pm i\sqrt{2}$ Square Root Property

This equation has one real root, 0, and two imaginary roots, $i\sqrt{2}$ and $-i\sqrt{2}$.

d. $x^4 - 1 = 0$

$$x^4 - 1 = 0$$
$$(x^2 + 1)(x^2 - 1) = 0$$
$$(x^2 + 1)(x + 1)(x - 1) = 0$$

$x^2 + 1 = 0$ or $x + 1 = 0$ or $x - 1 = 0$

$x^2 = -1$ $\quad x = -1 \quad x = 1$

$x = \pm\sqrt{-1}$ or $\pm i$

This equation has two real roots, 1 and -1, and two imaginary roots, i and $-i$.

Compare the degree of each equation and the number of roots of each equation in Example 1. The following corollary of the Fundamental Theorem of Algebra is an even more powerful tool for problem solving.

Key Concept — Corollary

A polynomial equation of the form $P(x) = 0$ of degree n with complex coefficients has exactly n roots in the set of complex numbers.

Similarly, a polynomial function of nth degree has exactly n zeros.

More About...

Descartes

René Descartes (1596–1650) was a French mathematician and philosopher. One of his best-known quotations comes from his *Discourse on Method*: "I think, therefore I am."

Source: *A History of Mathematics*

French mathematician René Descartes made more discoveries about zeros of polynomial functions. His rule of signs is given below.

Key Concept — Descartes' Rule of Signs

If $P(x)$ is a polynomial with real coefficients whose terms are arranged in descending powers of the variable,

- the number of positive real zeros of $y = P(x)$ is the same as the number of changes in sign of the coefficients of the terms, or is less than this by an even number, and
- the number of negative real zeros of $y = P(x)$ is the same as the number of changes in sign of the coefficients of the terms of $P(-x)$, or is less than this number by an even number.

Example 2 Find Numbers of Positive and Negative Zeros

State the possible number of positive real zeros, negative real zeros, and imaginary zeros of $p(x) = x^5 - 6x^4 - 3x^3 + 7x^2 - 8x + 1$.

Since $p(x)$ has degree 5, it has five zeros. However, some of them may be imaginary. Use Descartes' Rule of Signs to determine the number and type of real zeros. Count the number of changes in sign for the coefficients of $p(x)$.

$$p(x) = x^5 \quad - \quad 6x^4 \quad - \quad 3x^3 \quad + \quad 7x^2 \quad - \quad 8x \quad + \quad 1$$

yes (+ to −), no (− to −), yes (− to +), yes (+ to −), yes (− to +)

Since there are 4 sign changes, there are 4, 2, or 0 positive real zeros.

Find $p(-x)$ and count the number of changes in signs for its coefficients.

$$p(x) = (-x)^5 - 6(-x)^4 - 3(-x)^3 + 7(-x)^2 - 8(-x) + 1$$
$$= -x^5 - 6x^4 + 3x^3 + 7x^2 + 8x + 1$$

no (− to −), yes (− to +), no (+ to +), no (+ to +), no (+ to +)

Since there is 1 sign change, there is exactly 1 negative real zero.

Thus, the function $p(x)$ has either 4, 2, or 0 positive real zeros and exactly 1 negative real zero. Make a chart of the possible combinations of real and imaginary zeros.

Number of Positive Real Zeros	Number of Negative Real Zeros	Number of Imaginary Zeros	Total Number of Zeros
4	1	0	4 + 1 + 0 = 5
2	1	2	2 + 1 + 2 = 5
0	1	4	0 + 1 + 4 = 5

Study Tip

Zero at the Origin

Recall that the number 0 has no sign. Therefore, if 0 is a zero of a function, the sum of the number of positive real zeros, negative real zeros, and imaginary zeros is reduced by how many times 0 is a zero of the function.

FIND ZEROS We can find all of the zeros of a function using some of the strategies you have already learned.

Example 3 Use Synthetic Substitution to Find Zeros

Find all of the zeros of $f(x) = x^3 - 4x^2 + 6x - 4$.

Since $f(x)$ has degree 3, the function has three zeros. To determine the possible number and type of real zeros, examine the number of sign changes for $f(x)$ and $f(-x)$.

$f(x) = x^3 - 4x^2 + 6x - 4$ — yes, yes, yes

$f(-x) = -x^3 - 4x^2 - 6x - 4$ — no, no, no

Since there are 3 sign changes for the coefficients of $f(x)$, the function has 3 or 1 positive real zeros. Since there are no sign changes for the coefficient of $f(-x)$, $f(x)$ has no negative real zeros. Thus, $f(x)$ has either 3 real zeros, or 1 real zero and 2 imaginary zeros.

To find these zeros, first list some possibilities and then eliminate those that are not zeros. Since none of the zeros are negative and $f(0)$ is -4, begin by evaluating $f(x)$ for positive integral values from 1 to 4. You can use a shortened form of synthetic substitution to find $f(a)$ for several values of a.

(continued on the next page)

Study Tip

Finding Zeros
While direct substitution could be used to find each real zero of a polynomial, using synthetic substitution provides you with a depressed polynomial that can be used to find any imaginary zeros.

x	1	−4	6	−4
1	1	−3	3	−1
2	1	−2	2	**0**
3	1	−1	3	5
4	1	0	6	20

Each row in the table shows the coefficients of the depressed polynomial and the remainder.

From the table, we can see that one zero occurs at $x = 2$. Since the depressed polynomial of this zero, $x^2 - 2x + 2$, is quadratic, use the Quadratic Formula to find the roots of the related quadratic equation, $x^2 - 2x + 2 = 0$.

$x = \frac{-b \pm \sqrt{b^2 - 4ac}}{2a}$ Quadratic Formula

$= \frac{-(-2) \pm \sqrt{(-2)^2 - 4(1)(2)}}{2(1)}$ Replace a with 1, b with −2, and c with 2.

$= \frac{2 \pm \sqrt{-4}}{2}$ Simplify.

$= \frac{2 \pm 2i}{2}$ $\sqrt{4} \times \sqrt{-1} = 2i$

$= 1 \pm i$ Simplify.

Thus, the function has one real zero at $x = 2$ and two imaginary zeros at $x = 1 + i$ and $x = 1 - i$. The graph of the function verifies that there is only one real zero.

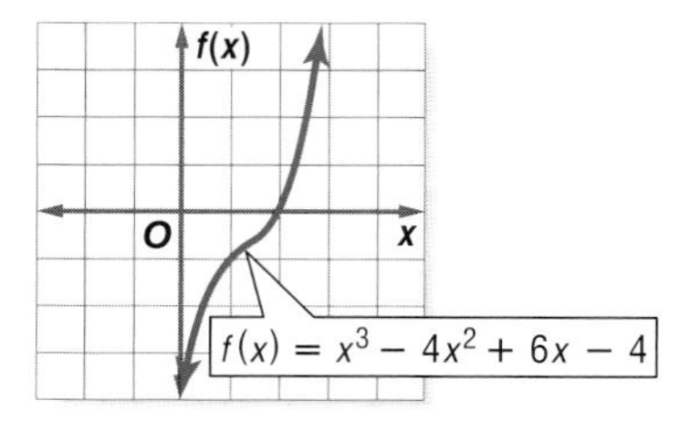

In Chapter 6, you learned that solutions of a quadratic equation that contains imaginary numbers come in pairs. This applies to the zeros of polynomial functions as well. For any polynomial function, if an imaginary number is a zero of that function, its conjugate is also a zero. This is called the **Complex Conjugates Theorem**.

Key Concept *Complex Conjugates Theorem*

Suppose a and b are real numbers with $b \neq 0$. If $a + bi$ is a zero of a polynomial function with real coefficients, then $a - bi$ is also a zero of the function.

Standardized Test Practice

Example 4 Use Zeros to Write a Polynomial Function

Short-Response Test Item

Write a polynomial function of least degree with integral coefficients whose zeros include 3 and $2 - i$.

Read the Test Item

- If $2 - i$ is a zero, then $2 + i$ is also a zero according to the Complex Conjugates Theorem. So, $x - 3$, $x - (2 - i)$, and $x - (2 + i)$ are factors of the polynomial function.

Solve the Test Item

- Write the polynomial function as a product of its factors.

  $f(x) = (x - 3)[x - (2 - i)][x - (2 + i)]$

- Multiply the factors to find the polynomial function.

$$\begin{aligned} f(x) &= (x-3)[x-(2-i)][x-(2+i)] && \text{Write an equation.} \\ &= (x-3)[(x-2)+i][(x-2)-i] && \text{Regroup terms.} \\ &= (x-3)[(x-2)^2-i^2] && \text{Rewrite as the difference of two squares.} \\ &= (x-3)[x^2-4x+4-(-1)] && \text{Square } x-2 \text{ and replace } i^2 \text{ with } -1. \\ &= (x-3)(x^2-4x+5) && \text{Simplify.} \\ &= x^3-4x^2+5x-3x^2+12x-15 && \text{Multiply using the Distributive Property.} \\ &= x^3-7x^2+17x-15 && \text{Combine like terms.} \end{aligned}$$

$f(x) = x^3 - 7x^2 + 17x - 15$ is a polynomial function of least degree with integral coefficients whose zeros are 3, $2 - i$, and $2 + i$.

Check for Understanding

Concept Check

1. **OPEN ENDED** Write a polynomial function $p(x)$ whose coefficients have two sign changes. Then describe the nature of its zeros.
2. **Explain** why an odd-degree function must always have at least one real root.
3. **State** the least degree a polynomial equation with real coefficients can have if it has roots at $x = 5 + i$, $x = 3 - 2i$, and a double root at $x = 0$.

Guided Practice

Solve each equation. State the number and type of roots.

4. $x^2 + 4 = 0$
5. $x^3 + 4x^2 - 21x = 0$

State the possible number of positive real zeros, negative real zeros, and imaginary zeros of each function.

6. $f(x) = 5x^3 + 8x^2 - 4x + 3$
7. $r(x) = x^5 - x^3 - x + 1$

Find all of the zeros of each function.

8. $p(x) = x^3 + 2x^2 - 3x + 20$
9. $f(x) = x^3 - 4x^2 + 6x - 4$
10. $v(x) = x^3 - 3x^2 + 4x - 12$
11. $f(x) = x^3 - 3x^2 + 9x + 13$

Standardized Test Practice

A B C D

12. **SHORT RESPONSE** Write a polynomial function of least degree with integral coefficients whose zeros include 2 and $4i$.

Practice and Apply

Homework Help

For Exercises	See Examples
13–18	1
19–24, 41	2
25–34, 44–48	3
35–40, 42, 43	4

Extra Practice
See page 843.

Solve each equation. State the number and type of roots.

13. $3x + 8 = 0$
14. $2x^2 - 5x + 12 = 0$
15. $x^3 + 9x = 0$
16. $x^4 - 81 = 0$
17. $x^4 - 16 = 0$
18. $x^5 - 8x^3 + 16x = 0$

State the possible number of positive real zeros, negative real zeros, and imaginary zeros of each function.

19. $f(x) = x^3 - 6x^2 + 1$
20. $g(x) = 5x^3 + 8x^2 - 4x + 3$
21. $h(x) = 4x^3 - 6x^2 + 8x - 5$
22. $q(x) = x^4 + 5x^3 + 2x^2 - 7x - 9$
23. $p(x) = x^5 - 6x^4 - 3x^3 + 7x^2 - 8x + 1$
24. $f(x) = x^{10} - x^8 + x^6 - x^4 + x^2 - 1$

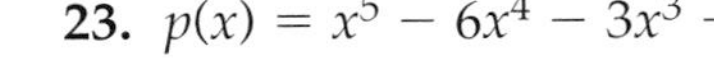
www.algebra2.com/self_check_quiz/ca

Find all of the zeros of each function.

25. $g(x) = x^3 + 6x^2 + 21x + 26$

26. $h(x) = x^3 - 6x^2 + 10x - 8$

27. $h(x) = 4x^4 + 17x^2 + 4$

28. $f(x) = x^3 - 7x^2 + 25x - 175$

29. $g(x) = 2x^3 - x^2 + 28x + 51$

30. $q(x) = 2x^3 - 17x^2 + 90x - 41$

31. $f(x) = x^3 - 5x^2 - 7x + 51$

32. $p(x) = x^4 - 9x^3 + 24x^2 - 6x - 40$

33. $r(x) = x^4 - 6x^3 + 12x^2 + 6x - 13$

34. $h(x) = x^4 - 15x^3 + 70x^2 - 70x - 156$

Write a polynomial function of least degree with integral coefficients that has the given zeros.

35. $-4, 1, 5$

36. $-2, 2, 4, 6$

37. $4i, 3, -3$

38. $2i, 3i, 1$

39. $9, 1 + 2i$

40. $6, 2 + 2i$

41. Sketch the graph of a polynomial function that has the indicated number and type of zeros.

a. 3 real, 2 imaginary **b.** 4 real **c.** 2 imaginary

More About. . .

Space Exploration

A space shuttle is a reusable vehicle, launched like a rocket, which can put people and equipment in orbit around Earth. The first space shuttle was launched in 1981.

Source: kidsastronomy.about.com

SCULPTING **For Exercises 42 and 43, use the following information.**
Antonio is preparing to make an ice sculpture. He has a block of ice that he wants to reduce in size by shaving off the same amount from the length, width, and height. He wants to reduce the volume of the ice block to 24 cubic feet.

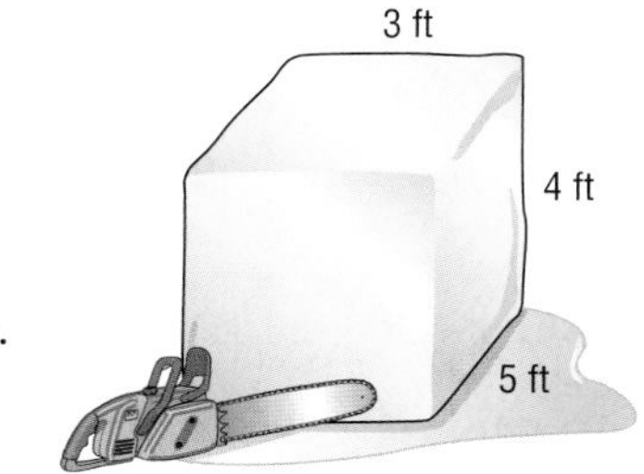

42. Write a polynomial equation to model this situation.

43. How much should he take from each dimension?

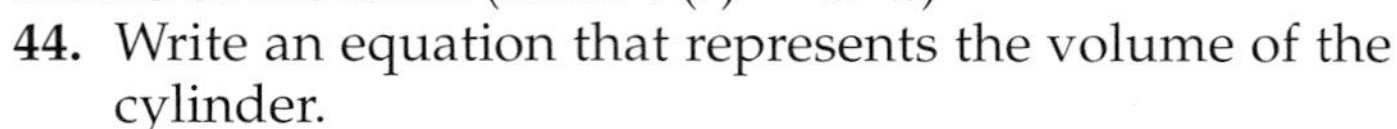

SPACE EXPLORATION **For Exercises 44 and 45, use the following information.**
The space shuttle has an external tank for the fuel that the main engines need for the launch. This tank is shaped like a capsule, a cylinder with a hemispherical dome at either end. The cylindrical part of the tank has an approximate volume of 336π cubic meters and a height of 17 meters more than the radius of the tank. (*Hint:* $V(r) = \pi r^2 h$)

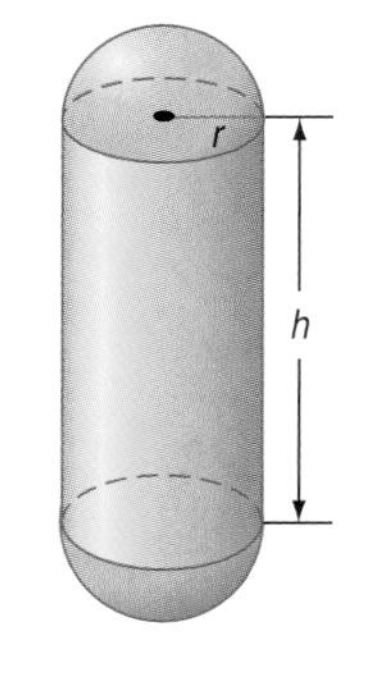

44. Write an equation that represents the volume of the cylinder.

45. What are the dimensions of the tank?

MEDICINE **For Exercises 46–48, use the following information.**
Doctors can measure cardiac output in patients at high risk for a heart attack by monitoring the concentration of dye injected into a vein near the heart. A normal heart's dye concentration is given by $d(x) = -0.006x^4 - 0.15x^3 - 0.05x^2 + 1.8x$, where x is the time in seconds.

46. How many positive real zeros, negative real zeros, and imaginary zeros exist for this function? (*Hint:* Notice that 0, which is neither positive nor negative, is a zero of this function since $d(0) = 0$.)

47. Approximate all real zeros to the nearest tenth by graphing the function using a graphing calculator.

48. What is the meaning of the roots in this problem?

49. CRITICAL THINKING Find a counterexample to disprove the following statement.

The polynomial function of least degree with integral coefficients with zeros at $x = 4$, $x = -1$, *and* $x = 3$, *is unique.*

50. **CRITICAL THINKING** If a sixth-degree polynomial equation has exactly five distinct real roots, what can be said of one of its roots? Draw a graph of this situation.

51. **WRITING IN MATH** Answer the question that was posed at the beginning of the lesson.

How can the roots of an equation be used in pharmacology?

Include the following items in your answer:

- an explanation of what the roots of this equation represent, and
- an explanation of what the roots of this equation reveal about how often a patient should take this medication.

52. The equation $x^4 - 1 = 0$ has exactly ___?___ complex root(s).

(A) 4 (B) 0 (C) 2 (D) 1

53. How many negative real zeros does $f(x) = x^5 - 2x^4 - 4x^3 + 4x^2 - 5x + 6$ have?

(A) 3 (B) 2 (C) 1 (D) 0

Maintain Your Skills

Mixed Review

Use synthetic substitution to find $f(-3)$ and $f(4)$ for each function. *(Lesson 7-4)*

54. $f(x) = x^3 - 5x^2 + 16x - 7$

55. $f(x) = x^4 + 11x^3 - 3x^2 + 2x - 5$

56. **RETAIL** The store Bunches of Boxes and Bags assembles boxes for mailing. The store manager found that the volume of a box made from a rectangular piece of cardboard with a square of length x inches cut from each corner is $4x^3 - 168x^2 + 1728x$ cubic inches. If the piece of cardboard is 48 inches long, what is the width? *(Lesson 7-3)*

Determine whether each function has a maximum or a minimum value. Then find the maximum or minimum value of each function. *(Lesson 6-1)*

57. $f(x) = x^2 - 8x + 3$

58. $f(x) = -3x^2 - 18x + 5$

59. $f(x) = -7 + 4x^2$

Factor completely. If the polynomial is not factorable, write *prime*. *(Lesson 5-4)*

60. $15a^2b^2 - 5ab^2c^2$

61. $12p^2 - 64p + 45$

62. $4y^3 + 24y^2 + 36y$

Use matrices A, B, C, and D to find the following. *(Lesson 4-2)*

$$A = \begin{bmatrix} -4 & 4 \\ 2 & -3 \\ 1 & 5 \end{bmatrix} \quad B = \begin{bmatrix} 7 & 0 \\ 4 & 1 \\ 6 & -2 \end{bmatrix} \quad C = \begin{bmatrix} -4 & -5 \\ -3 & 1 \\ 2 & 3 \end{bmatrix} \quad D = \begin{bmatrix} 1 & -2 \\ 1 & -1 \\ -3 & 4 \end{bmatrix}$$

63. $A + D$

64. $B - C$

65. $3B - 2A$

66. Write an inequality for the graph at the right. *(Lesson 2-7)*

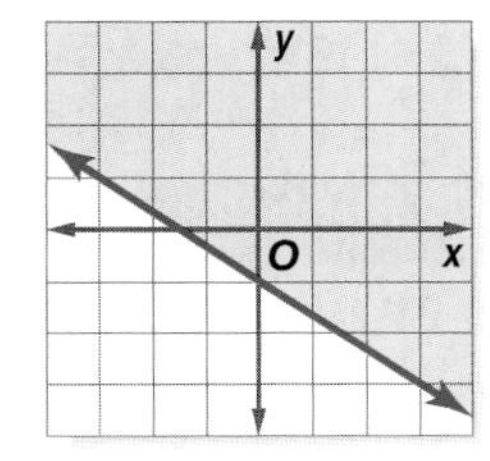

Getting Ready for the Next Lesson

BASIC SKILL **Find all values of $\pm\frac{a}{b}$ given each replacement set.**

67. $a = \{1, 5\}; b = \{1, 2\}$

68. $a = \{1, 2\}; b = \{1, 2, 7, 14\}$

69. $a = \{1, 3\}; b = \{1, 3, 9\}$

70. $a = \{1, 2, 4\}; b = \{1, 2, 4, 8, 16\}$

7-6 Rational Zero Theorem

What You'll Learn

- Identify the possible rational zeros of a polynomial function.
- Find all the rational zeros of a polynomial function.

How can the Rational Zero Theorem solve problems involving large numbers?

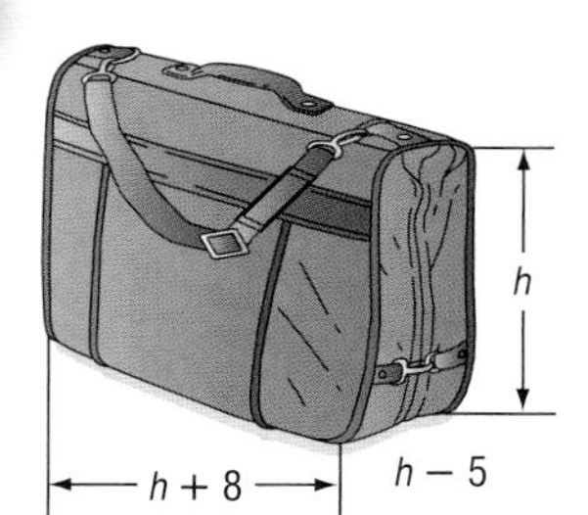

On an airplane, carry-on baggage must fit into the overhead compartment above the passenger's seat. The length of the compartment is 8 inches longer than the height, and the width is 5 inches shorter than the height. The volume of the compartment is 2772 cubic inches. You can solve the polynomial equation $h(h + 8)(h - 5) = 2772$, where h is the height, $h + 8$ is the length, and $h - 5$ is the width, to find the dimensions of the overhead compartment in which your luggage must fit.

IDENTIFY RATIONAL ZEROS Usually it is not practical to test all possible zeros of a polynomial function using only synthetic substitution. The **Rational Zero Theorem** can help you choose some possible zeros to test.

Key Concept — Rational Zero Theorem

- **Words** Let $f(x) = a_0x^n + a_1x^{n-1} + \ldots + a_{n-2}x^2 + a_{n-1}x + a_n$ represent a polynomial function with integral coefficients. If $\frac{p}{q}$ is a rational number in simplest form and is a zero of $y = f(x)$, then p is a factor of a_n and q is a factor of a_0.
- **Example** Let $f(x) = 2x^3 + 3x^2 - 17x + 12$. If $\frac{3}{2}$ is a zero of $f(x)$, then 3 is a factor of 12 and 2 is a factor of 2.

In addition, if the coefficient of the x term with the highest degree is 1, we have the following corollary.

Key Concept — Corollary (Integral Zero Theorem)

If the coefficients of a polynomial function are integers such that $a_0 = 1$ and $a_n \neq 0$, any rational zeros of the function must be factors of a_n.

Example 1 Identify Possible Zeros

List all of the possible rational zeros of each function.

a. $f(x) = 2x^3 - 11x^2 + 12x + 9$

If $\frac{p}{q}$ is a rational zero, then p is a factor of 9 and q is a factor of 2. The possible values of p are ±1, ±3, and ±9. The possible values for q are ±1 and ±2. So, $\frac{p}{q} = \pm1, \pm3, \pm9, \pm\frac{1}{2}, \pm\frac{3}{2}$, and $\pm\frac{9}{2}$.

b. $f(x) = x^3 - 9x^2 - x + 105$

Since the coefficient of x^3 is 1, the possible rational zeros must be a factor of the constant term 105. So, the possible rational zeros are the integers ±1, ±3, ±5, ±7, ±15, ±21, ±35, and ±105.

FIND RATIONAL ZEROS Once you have written the possible rational zeros, you can test each number using synthetic substitution.

Example 2 Use the Rational Zero Theorem

GEOMETRY The volume of a rectangular solid is 675 cubic centimeters. The width is 4 centimeters less than the height, and the length is 6 centimeters more than the height. Find the dimensions of the solid.

Study Tip

Descartes' Rule of Signs

Examine the signs of the coefficients in the equation, + + − −. There is one change of sign, so there is only one positive real zero.

Let x = the height, $x - 4$ = the width, and $x + 6$ = the length.

Write an equation for the volume.

$x(x - 4)(x + 6) = 675$ Formula for volume

$x^3 + 2x^2 - 24x = 675$ Multiply.

$x^3 + 2x^2 - 24x - 675 = 0$ Subtract 675.

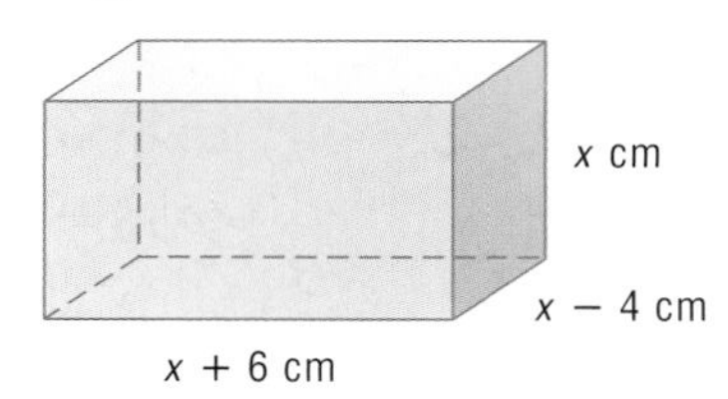

The leading coefficient is 1, so the possible integer zeros are factors of 675, ±1, ±3, ±5, ±9, ±15, ±25, ±27, ±45, ±75, ±135, ±225, and ±675. Since length can only be positive, we only need to check positive zeros. From Descartes' Rule of Signs, we also know there is only one positive real zero. Make a table and test possible real zeros.

p	1	2	−24	−675
1	1	3	−21	−696
3	1	5	−9	−702
5	1	7	11	−620
9	1	11	75	0

One zero is 9. Since there is only one positive real zero, we do not have to test the other numbers. The other dimensions are 9 − 4 or 5 centimeters and 9 + 6 or 15 centimeters.

CHECK Verify that the dimensions are correct. $5 \times 9 \times 15 = 675$ ✓

You usually do not need to test all of the possible zeros. Once you find a zero, you can try to factor the depressed polynomial to find any other zeros.

Example 3 Find All Zeros

Find all of the zeros of $f(x) = 2x^4 - 13x^3 + 23x^2 - 52x + 60$.

From the corollary to the Fundamental Theorem of Algebra, we know there are exactly 4 complex roots. According to Descartes' Rule of Signs, there are 4, 2, or 0 positive real roots and 0 negative real roots. The possible rational zeros are ±1, ±2, ±3, ±4, ±5, ±6, ±10, ±12, ±15, ±20, ±30, ±60, $\pm\frac{1}{2}$, $\pm\frac{3}{2}$, $\pm\frac{5}{2}$, and $\pm\frac{15}{2}$. Make a table and test some possible rational zeros.

$\frac{p}{q}$	2	−13	23	−52	60
1	2	−11	12	−40	20
2	2	−9	5	−42	−24
3	2	−7	2	−46	−78
5	2	−3	8	−12	0

(continued on the next page)

Since $f(5) = 0$, you know that $x = 5$ is a zero. The depressed polynomial is $2x^3 - 3x^2 + 8x - 12$.

Factor $2x^3 - 3x^2 + 8x - 12$.

$2x^3 - 3x^2 + 8x - 12 = 0$ Write the depressed polynomial.

$2x^3 + 8x - 3x^2 - 12 = 0$ Regroup terms.

$2x(x^2 + 4) - 3(x^2 + 4) = 0$ Factor by grouping.

$(x^2 + 4)(2x - 3) = 0$ Distributive Property

$x^2 + 4 = 0$ or $2x - 3 = 0$ Zero Product Property

$x^2 = -4$ $\quad 2x = 3$

$x = \pm 2i$ $\quad x = \frac{3}{2}$

There is another real zero at $x = \frac{3}{2}$ and two imaginary zeros at $x = 2i$ and $x = -2i$.

The zeros of this function are 5, $\frac{3}{2}$, $2i$ and $-2i$.

Check for Understanding

Concept Check

1. **Explain** why it is useful to use the Rational Zero Theorem when finding the zeros of a polynomial function.

2. **OPEN ENDED** Write a polynomial function that has possible rational zeros of $\pm 1, \pm 3, \pm \frac{1}{2}, \pm \frac{3}{2}$.

3. **FIND THE ERROR** Lauren and Luis are listing the possible rational zeros of $f(x) = 4x^5 + 4x^4 - 3x^3 + 2x^2 - 5x + 6$.

Lauren	Luis
$\pm 1, \pm\frac{1}{2}, \pm\frac{1}{3}, \pm\frac{1}{6}, \pm 2, \pm\frac{2}{3}, \pm 4, \pm\frac{4}{3}$	$\pm 1, \pm\frac{1}{2}, \pm\frac{1}{4}, \pm 2, \pm 3, \pm\frac{3}{2}, \pm\frac{3}{4}, \pm 6$

Who is correct? Explain your reasoning.

Guided Practice

List all of the possible rational zeros of each function.

4. $p(x) = x^4 - 10$

5. $d(x) = 6x^3 + 6x^2 - 15x - 2$

Find all of the rational zeros of each function.

6. $p(x) = x^3 - 5x^2 - 22x + 56$

7. $f(x) = x^3 - x^2 - 34x - 56$

8. $t(x) = x^4 - 13x^2 + 36$

9. $f(x) = 2x^3 - 7x^2 - 8x + 28$

10. Find all of the zeros of $f(x) = 6x^3 + 5x^2 - 9x + 2$.

Application

11. **GEOMETRY** The volume of the rectangular solid is 1430 cubic centimeters. Find the dimensions of the solid.

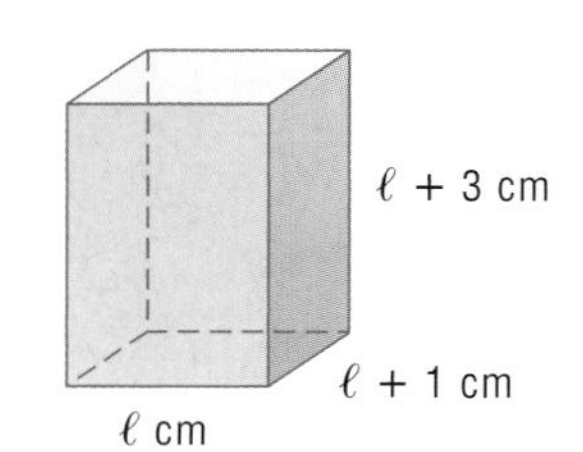

Practice and Apply

Homework Help

For Exercises	See Examples
12–17	1
18–29, 34–41	2
30–33	3

Extra Practice
See page 843.

List all of the possible rational zeros of each function.

12. $f(x) = x^3 + 6x + 2$
13. $h(x) = x^3 + 8x + 6$
14. $f(x) = 3x^4 + 15$
15. $n(x) = x^5 + 6x^3 - 12x + 18$
16. $p(x) = 3x^3 - 5x^2 - 11x + 3$
17. $h(x) = 9x^6 - 5x^3 + 27$

Find all of the rational zeros of each function.

18. $f(x) = x^3 + x^2 - 80x - 300$
19. $p(x) = x^3 - 3x - 2$
20. $h(x) = x^4 + x^2 - 2$
21. $g(x) = x^4 - 3x^3 - 53x^2 - 9x$
22. $f(x) = 2x^5 - x^4 - 2x + 1$
23. $f(x) = x^5 - 6x^3 + 8x$
24. $g(x) = x^4 - 3x^3 + x^2 - 3x$
25. $p(x) = x^4 + 10x^3 + 33x^2 + 38x + 8$
26. $p(x) = x^3 + 3x^2 - 25x + 21$
27. $h(x) = 6x^3 + 11x^2 - 3x - 2$
28. $h(x) = 10x^3 - 17x^2 - 7x + 2$
29. $g(x) = 48x^4 - 52x^3 + 13x - 3$

Find all of the zeros of each function.

30. $p(x) = 6x^4 + 22x^3 + 11x^2 - 38x - 40$
31. $g(x) = 5x^4 - 29x^3 + 55x^2 - 28x$
32. $h(x) = 9x^5 - 94x^3 + 27x^2 + 40x - 12$
33. $p(x) = x^5 - 2x^4 - 12x^3 - 12x^2 - 13x - 10$

FOOD **For Exercises 34–36, use the following information.**
Terri's Ice Cream Parlor makes gourmet ice cream cones. The volume of each cone is 8π cubic inches. The height is 4 inches more than the radius of the cone's opening.

34. Write a polynomial equation that represents the volume of an ice cream cone. Use the formula for the volume of a cone, $V = \frac{1}{3}\pi r^2 h$.
35. What are the possible values of r? Which of these values are reasonable?
36. Find the dimensions of the cone.

More About. . .

Food

The largest ice cream sundae, weighing 24.91 tons, was made in Edmonton, Alberta, in July 1988.

Source: *The Guinness Book of Records.*

AUTOMOBILES **For Exercises 37 and 38, use the following information.**
The length of the cargo space in a sport-utility vehicle is 4 inches greater than the height of the space. The width is sixteen inches less than twice the height. The cargo space has a total volume of 55,296 cubic inches.

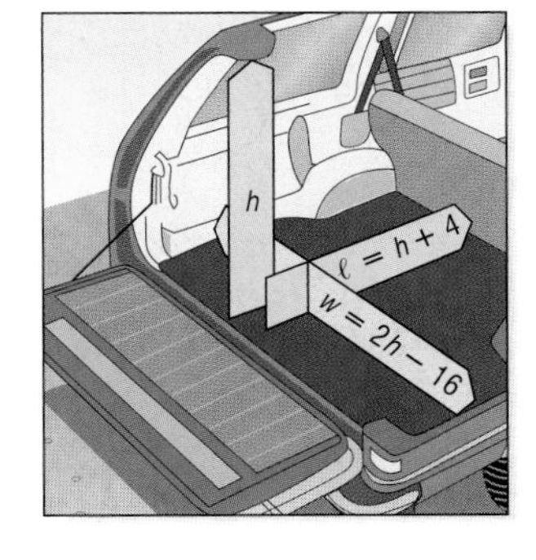

37. Write a polynomial function that represents the volume of the cargo space.
38. Find the dimensions of the cargo space.

AMUSEMENT PARKS **For Exercises 39–41, use the following information.**
An amusement park owner wants to add a new wilderness water ride that includes a mountain that is shaped roughly like a pyramid. Before building the new attraction, engineers must build and test a scale model.

39. If the height of the scale model is 9 inches less than its length and its base is a square, write a polynomial function that describes the volume of the model in terms of its length. Use the formula for the volume of a pyramid, $V = \frac{1}{3}Bh$.
40. If the volume of the model is 6300 cubic inches, write an equation for the situation.
41. What are the dimensions of the scale model?

42. **CRITICAL THINKING** Suppose k and $2k$ are zeros of $f(x) = x^3 + 4x^2 + 9kx - 90$. Find k and all three zeros of $f(x)$.

43. **WRITING IN MATH** Answer the question that was posed at the beginning of the lesson.

How can the Rational Zero Theorem solve problems involving large numbers?

Include the following items in your answer:

- the polynomial equation that represents the volume of the compartment, and
- a list of all reasonable measures of the width of the compartment, assuming that the width is a whole number.

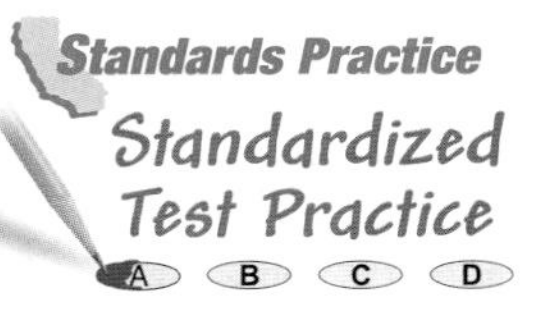

44. Using the Rational Zero Theorem, determine which of the following is a zero of the function $f(x) = 12x^5 - 5x^3 + 2x - 9$.

Ⓐ -6 Ⓑ $\frac{3}{8}$ Ⓒ $-\frac{2}{3}$ Ⓓ 1

45. **OPEN ENDED** Write a polynomial with -5, -2, 1, 3, and 4 as roots.

Maintain Your Skills

Mixed Review

Given a function and one of its zeros, find all of the zeros of the function. *(Lesson 7-5)*

46. $g(x) = x^3 + 4x^2 - 27x - 90; -3$

47. $h(x) = x^3 - 11x + 20; 2 + i$

48. $f(x) = x^3 + 5x^2 + 9x + 45; -5$

49. $g(x) = x^3 - 3x^2 - 41x + 203; -7$

Given a polynomial and one of its factors, find the remaining factors of the polynomial. Some factors may not be binomials. *(Lesson 7-4)*

50. $20x^3 - 29x^2 - 25x + 6; x - 2$

51. $3x^4 - 21x^3 + 38x^2 - 14x + 24; x - 3$

Simplify. *(Lesson 5-5)*

52. $\sqrt{245}$

53. $\pm\sqrt{18x^3y^2}$

54. $\sqrt{16x^2 - 40x + 25}$

55. **GEOMETRY** The perimeter of a right triangle is 24 centimeters. Three times the length of the longer leg minus two times the length of the shorter leg exceeds the hypotenuse by 2 centimeters. What are the lengths of all three sides? *(Lesson 3-5)*

Getting Ready for the Next Lesson

PREREQUISITE SKILL **Simplify.**

*(To review **operations with polynomials**, see Lessons 5-2 and 5-3.)*

56. $(x^2 - 7) + (x^3 + 3x^2 + 1)$

57. $(8x^2 - 3x) - (4x^2 + 5x - 3)$

58. $(x + 2)(x^2 + 3x - 5)$

59. $(x^3 + 3x^2 - 3x + 1)(x - 5)^2$

60. $(x^2 - 2x - 30) \div (x + 7)$

61. $(x^3 + 2x^2 - 3x + 1) \div (x + 1)$

Practice Quiz 2

Lessons 7-4 through 7-6

Use synthetic substitution to find $f(-2)$ and $f(3)$ for each function. *(Lesson 7-4)*

1. $f(x) = 7x^5 - 25x^4 + 17x^3 - 32x^2 + 10x - 22$

2. $f(x) = 3x^4 - 12x^3 - 21x^2 + 30x$

3. Write the polynomial equation of degree 4 with leading coefficient 1 that has roots at -2, -1, 3, and 4. *(Lesson 7-5)*

Find all of the rational zeros of each function. *(Lesson 7-6)*

4. $f(x) = 5x^3 - 29x^2 + 55x - 28$

5. $g(x) = 4x^3 + 16x^2 - x - 24$

7-7 Operations on Functions

California Standards Standard 25.0 Students use properties from number systems to justify steps in combining and simplifying functions.

What You'll Learn

- Find the sum, difference, product, and quotient of functions.
- Find the composition of functions.

Vocabulary

- composition of functions

Why is it important to combine functions in business?

Carol Coffmon owns a garden store where she sells birdhouses. The revenue from the sale of the birdhouses is given by $r(x) = 125x$. The function for the cost of making the birdhouses is given by $c(x) = 65x + 5400$. Her profit p is the revenue minus the cost or $p = r - c$. So the profit function $p(x)$ can be defined as $p(x) = (r - c)(x)$. If you have two functions, you can form a new function by performing arithmetic operations on them.

ARITHMETIC OPERATIONS Let $f(x)$ and $g(x)$ be any two functions. You can add, subtract, multiply, and divide functions according to the following rules.

Key Concept — Operations with Function

Operation	Definition	Examples if $f(x) = x + 2$, $g(x) = 3x$
Sum	$(f + g)(x) = f(x) + g(x)$	$(x + 2) + 3x = 4x + 2$
Difference	$(f - g)(x) = f(x) - g(x)$	$(x + 2) - 3x = -2x + 2$
Product	$(f \cdot g)(x) = f(x) \cdot g(x)$	$(x + 2)3x = 3x^2 + 6x$
Quotient	$\left(\frac{f}{g}\right)(x) = \frac{f(x)}{g(x)}, g(x) \neq 0$	$\frac{x + 2}{3x}$

Example 1 Add and Subtract Functions

Given $f(x) = x^2 - 3x + 1$ and $g(x) = 4x + 5$, find each function.

a. $(f + g)(x)$

$(f + g)(x) = f(x) + g(x)$ — Addition of functions

$= (x^2 - 3x + 1) + (4x + 5)$ — $f(x) = x^2 - 3x + 1$ and $g(x) = 4x + 5$

$= x^2 + x + 6$ — Simplify.

b. $(f - g)(x)$

$(f - g)(x) = f(x) - g(x)$ — Subtraction of functions

$= (x^2 - 3x + 1) - (4x + 5)$ — $f(x) = x^2 - 3x + 1$ and $g(x) = 4x + 5$

$= x^2 - 7x - 4$ — Simplify.

Notice that the functions f and g have the same domain of all real numbers. The functions $f + g$ and $f - g$ also have domains that include all real numbers. For each new function, the domain consists of the intersection of the domains of $f(x)$ and $g(x)$. The domain of the quotient function is further restricted by excluded values that make the denominator equal to zero.

Example 2 Multiply and Divide Functions

Given $f(x) = x^2 + 5x - 1$ and $g(x) = 3x - 2$, find each function.

a. $(f \cdot g)(x)$

$(f \cdot g)(x) = f(x) \cdot g(x)$	Product of functions
$= (x^2 + 5x - 1)(3x - 2)$	$f(x) = x^2 + 5x - 1$ and $g(x) = 3x - 2$
$= x^2(3x - 2) + 5x(3x - 2) - 1(3x - 2)$	Distributive Property
$= 3x^3 - 2x^2 + 15x^2 - 10x - 3x + 2$	Distributive Property
$= 3x^3 + 13x^2 - 13x + 2$	Simplify.

b. $\left(\frac{f}{g}\right)(x)$

$\left(\frac{f}{g}\right)(x) = \frac{f(x)}{g(x)}$	Division of functions
$= \frac{x^2 + 5x - 1}{3x - 2}, x \neq \frac{2}{3}$	$f(x) = x^2 + 5x - 1$ and $g(x) = 3x - 2$

Because $x = \frac{2}{3}$ makes $3x - 2 = 0$, $\frac{2}{3}$ is excluded from the domain of $\left(\frac{f}{g}\right)(x)$.

Study Tip

Reading Math

$[f \circ g](x)$ and $f[g(x)]$ are both read *f of g of x*.

COMPOSITION OF FUNCTIONS Functions can also be combined using **composition of functions**. In a composition, a function is performed, and then a second function is performed on the result of the first function. The composition of f and g is denoted by $f \circ g$.

Key Concept *Composition of Functions*

Suppose f and g are functions such that the range of g is a subset of the domain of f. Then the composite function $f \circ g$ can be described by the equation

$$[f \circ g](x) = f[g(x)].$$

The composition of functions can be shown by mappings. Suppose $f = \{(3, 4), (2, 3), (-5, 0)\}$ and $g = \{(3, -5), (4, 3), (0, 2)\}$. The composition of these functions is shown below.

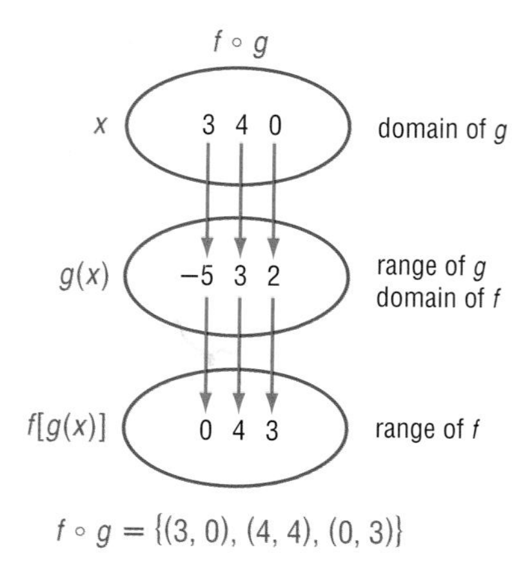

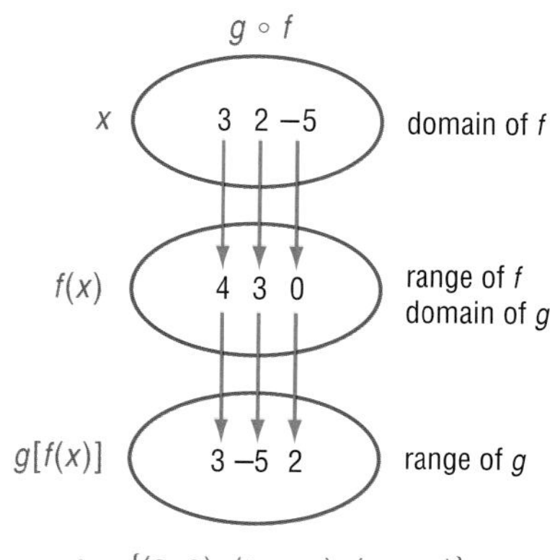

The composition of two functions may not exist. Given two functions f and g, $[f \circ g](x)$ is defined only if the range of $g(x)$ is a subset of the domain of $f(x)$. Similarly, $[g \circ f](x)$ is defined only if the range of $f(x)$ is a subset of the domain of $g(x)$.

Example 3 Evaluate Composition of Relations

If $f(x) = \{(7, 8), (5, 3), (9, 8), (11, 4)\}$ and $g(x) = \{(5, 7), (3, 5), (7, 9), (9, 11)\}$, find $f \circ g$ and $g \circ f$.

To find $f \circ g$, evaluate $g(x)$ first. Then use the range of g as the domain of f and evaluate $f(x)$.

$f[g(5)] = f(7)$ or 8 $\quad g(5) = 7$

$f[g(3)] = f(5)$ or 3 $\quad g(3) = 5$

$f[g(7)] = f(9)$ or 8 $\quad g(7) = 9$

$f[g(9)] = f(11)$ or 4 $\quad g(9) = 11$

$f \circ g = \{(5, 8), (3, 3), (7, 8), (9, 4)\}$

To find $g \circ f$, evaluate $f(x)$ first. Then use the range of f as the domain of g and evaluate $g(x)$.

$g[f(7)] = g(8)$ $\quad g(8)$ is undefined.

$g[f(5)] = g(3)$ or 5 $\quad f(5) = 3$

$g[f(9)] = g(8)$ $\quad g(8)$ is undefined.

$g[f(11)] = g(4)$ $\quad g(4)$ is undefined.

Since 8 and 4 are not in the domain of g, $g \circ f$ is undefined for $x = 7$, $x = 9$, and $x = 11$. However, $g[f(5)] = 5$ so $g \circ f = \{(5, 5)\}$.

Notice that in most instances $f \circ g \neq g \circ f$. Therefore, the order in which you compose two functions is very important.

Example 4 Simplify Composition of Functions

a. Find $[f \circ g](x)$ and $[g \circ f](x)$ for $f(x) = x + 3$ and $g(x) = x^2 + x - 1$.

$[f \circ g](x) = f[g(x)]$	Composition of functions
$= f(x^2 + x - 1)$	Replace $g(x)$ with $x^2 + x - 1$.
$= (x^2 + x - 1) + 3$	Substitute $x^2 + x - 1$ for x in $f(x)$.
$= x^2 + x + 2$	Simplify.

$[g \circ f](x) = g[f(x)]$	Composition of functions
$= g(x + 3)$	Replace $f(x)$ with $x + 3$.
$= (x + 3)^2 + (x + 3) - 1$	Substitute $x + 3$ for x in $g(x)$.
$= x^2 + 6x + 9 + x + 3 - 1$	Evaluate $(x + 3)^2$.
$= x^2 + 7x + 11$	Simplify.

So, $[f \circ g](x) = x^2 + x + 2$ and $[g \circ f](x) = x^2 + 7x + 11$.

b. Evaluate $[f \circ g](x)$ and $[g \circ f](x)$ for $x = 2$.

$[f \circ g](x) = x^2 + x + 2$	Function from part a
$[f \circ g](2) = (2)^2 + 2 + 2$	Replace x with 2.
$= 8$	Simplify.

$[g \circ f](x) = x^2 + 7x + 11$	Function from part a
$[g \circ f](2) = (2)^2 + 7(2) + 11$	Replace x with 2.
$= 29$	Simplify.

So, $[f \circ g](2) = 8$ and $[g \circ f](2) = 29$.

Example 5 Use Composition of Functions

TAXES Tyrone Davis has $180 deducted from every paycheck for retirement. He can have these deductions taken before taxes are applied, which reduces his taxable income. His federal income tax rate is 18%. If Tyrone earns $2200 every pay period, find the difference in his net income if he has the retirement deduction taken before taxes or after taxes.

Study Tip

Combining Functions
By combining functions, you can make the evaluation of the functions more efficient.

Explore Let x = Tyrone's income per paycheck, $r(x)$ = his income after the deduction for retirement, and $t(x)$ = his income after the deduction for federal income tax.

Plan Write equations for $r(x)$ and $t(x)$.
$180 is deducted from every paycheck for retirement: $r(x) = x - 180$.

Tyrone's tax rate is 18%: $t(x) = x - 0.18x$.

Solve If Tyrone has his retirement deducted *before* taxes, then his net income is represented by $[t \circ r](2200)$.

$[t \circ r](2200) = t(2200 - 180)$ Replace x with 2200 in $r(x) = x - 180$.
$= t(2020)$
$= 2020 - 0.18(2020)$ Replace x with 2020 in $t(x) = x - 0.18x$.
$= 1656.40$

If Tyrone has his retirement deducted *after* taxes, then his net income is represented by $[r \circ t](2200)$.

$[r \circ t](2200) = r[2200 - 0.18(2200)]$ Replace x with 2200 in $t(x) = x - 0.18x$.
$= r(1804)$
$= 1804 - 180$ Replace x with 1804 in $r(x) = x - 180$.
$= 1624$

$[t \circ r](2200) = 1656.40$ and $[r \circ t](2200) = 1624$. The difference is $1656.40 − $1624 or $32.40. So, his net pay is $32.40 more by having his retirement deducted before taxes.

Examine The answer makes sense. Since the taxes are being applied to a smaller amount, less taxes will be deducted from his paycheck.

Check for Understanding

Concept Check

1. **Determine** whether the following statement is *always*, *sometimes*, or *never* true. Support your answer with an example.
Given two functions f and g, $f \circ g = g \circ f$.

2. **OPEN ENDED** Write a set of ordered pairs for functions f and g, given that $f \circ g = \{(4, 3), (-1, 9), (-2, 7)\}$.

3. **FIND THE ERROR** Danette and Marquan are finding $[g \circ f](3)$ for $f(x) = x^2 + 4x + 5$ and $g(x) = x - 7$. Who is correct? Explain your reasoning.

Danette	Marquan
$[g \circ f](3) = g[(3)^2 + 4(3) + 5]$	$[g \circ f](3) = f(3 - 7)$
$= g(26)$	$= f(-4)$
$= 26 - 7$	$= (-4)^2 + 4(-4) + 5$
$= 19$	$= 5$

Guided Practice

Find $(f + g)(x)$, $(f - g)(x)$, $(f \cdot g)(x)$, and $\left(\frac{f}{g}\right)(x)$ for each $f(x)$ and $g(x)$. Justify each step.

4. $f(x) = 3x + 4$
   $g(x) = 5 + x$

5. $f(x) = x^2 + 3$
   $g(x) = x - 4$

For each set of ordered pairs, find $f \circ g$ and $g \circ f$, if they exist.

6. $f = \{(-1, 9), (4, 7)\}$
   $g = \{(-5, 4), (7, 12), (4, -1)\}$

7. $f = \{(0, -7), (1, 2), (2, -1)\}$
   $g = \{(-1, 10), (2, 0)\}$

Find $[g \circ h](x)$ and $[h \circ g](x)$. Justify each step.

8. $g(x) = 2x$
   $h(x) = 3x - 4$

9. $g(x) = x + 5$
   $h(x) = x^2 + 6$

If $f(x) = 3x$, $g(x) = x + 7$, and $h(x) = x^2$, find each value.

10. $f[g(3)]$

11. $g[h(-2)]$

12. $h[h(1)]$

Application

SHOPPING For Exercises 13–16, use the following information.
Mai-Lin is shopping for computer software. She finds a CD-ROM program that costs \$49.99, but is on sale at a 25% discount. She also has a \$5 coupon for the product.

13. Express the price of the CD after the discount and the price of the CD after the coupon using function notation. Let x represent the price of the CD, $p(x)$ represent the price after the 25% discount, and $c(x)$ represent the price after the coupon.

14. Find $c[p(x)]$ and explain what this value represents.

15. Find $p[c(x)]$ and explain what this value represents.

16. Which method results in the lower sale price? Explain your reasoning.

Practice and Apply

Homework Help

For Exercises	See Examples
17–22	1, 2
23–28	3
29–46	4
47–55	5

Extra Practice
See page 844.

Find $(f + g)(x)$, $(f - g)(x)$, $(f \cdot g)(x)$, and $\left(\frac{f}{g}\right)(x)$ for each $f(x)$ and $g(x)$. Justify each step.

17. $f(x) = x + 9$
    $g(x) = x - 9$

18. $f(x) = 2x - 3$
    $g(x) = 4x + 9$

19. $f(x) = 2x^2$
    $g(x) = 8 - x$

20. $f(x) = x^2 + 6x + 9$
    $g(x) = 2x + 6$

21. $f(x) = x^2 - 1$
    $g(x) = \frac{x}{x + 1}$

22. $f(x) = x^2 - x - 6$
    $g(x) = \frac{x - 3}{x + 2}$

For each set of ordered pairs, find $f \circ g$ and $g \circ f$ if they exist.

23. $f = \{(1, 1), (0, -3)\}$
    $g = \{(1, 0), (-3, 1), (2, 1)\}$

24. $f = \{(1, 2), (3, 4), (5, 4)\}$
    $g = \{(2, 5), (4, 3)\}$

25. $f = \{(3, 8), (4, 0), (6, 3), (7, -1)\}$
    $g = \{(0, 4), (8, 6), (3, 6), (-1, 8)\}$

26. $f = \{(4, 5), (6, 5), (8, 12), (10, 12)\}$
    $g = \{4, 6), (2, 4), (6, 8), (8, 10)\}$

27. $f = \{(2, 5), (3, 9), (-4, 1)\}$
    $g = \{(5, -4), (8, 3), (2, -2)\}$

28. $f = \{(7, 0), (-5, 3), (8, 3), (-9, 2)\}$
    $g = \{(2, -5), (1, 0), (2, -9), (3, 6)\}$

Find $[g \circ h](x)$ and $[h \circ g](x)$. Justify each step.

29. $g(x) = 4x$
    $h(x) = 2x - 1$

30. $g(x) = -5x$
    $h(x) = -3x + 1$

31. $g(x) = x + 2$
    $h(x) = x^2$

32. $g(x) = x - 4$
    $h(x) = 3x^2$

33. $g(x) = 2x$
    $h(x) = x^3 + x^2 + x + 1$

34. $g(x) = x + 1$
    $h(x) = 2x^2 - 5x + 8$

If $f(x) = 4x$, $g(x) = 2x - 1$, and $h(x) = x^2 + 1$, find each value.

35. $f[g(-1)]$
36. $h[g(4)]$
37. $g[f(5)]$
38. $f[h(-4)]$
39. $g[g(7)]$
40. $f[f(-3)]$
41. $h\left[f\left(\frac{1}{4}\right)\right]$
42. $g\left[h\left(-\frac{1}{2}\right)\right]$
43. $[g \circ (f \circ h)](3)$
44. $[f \circ (h \circ g)](3)$
45. $[h \circ (g \circ f)](2)$
46. $[f \circ (g \circ h)](2)$

More About. . .

Shopping

Americans spent over $500 million on inline skates and equipment in 2000.

Source: National Sporting Goods Association

POPULATION GROWTH **For Exercises 47 and 48, use the following information.**
From 1990 to 1999, the number of births $b(x)$ in the U.S. can be modeled by the function $b(x) = -27x + 4103$, and the number of deaths $d(x)$ can be modeled by the function $d(x) = 23x + 2164$, where x is the number of years since 1990 and $b(x)$ and $d(x)$ are in thousands.

47. The net increase in population P is the number of births per year minus the number of deaths per year or $P = b - d$. Write an expression that can be used to model the population increase in the U.S. from 1990 to 1999 in function notation.

48. Assume that births and deaths continue at the same rates. Estimate the net increase in population in 2010.

SHOPPING **For Exercises 49–51, use the following information.**
Liluye wants to buy a pair of inline skates that are on sale for 30% off the original price of $149. The sales tax is 5.75%.

49. Express the price of the inline skates after the discount and the price of the inline skates after the sales tax using function notation. Let x represent the price of the inline skates, $p(x)$ represent the price after the 30% discount, and $s(x)$ represent the price after the sales tax.

50. Which composition of functions represents the price of the inline skates, $p[s(x)]$ or $s[p(x)]$? Explain your reasoning.

51. How much will Liluye pay for the inline skates?

TEMPERATURE **For Exercises 52–54, use the following information.**
There are three temperature scales: Fahrenheit (°F), Celsius (°C), and Kelvin (K). The function $K(C) = C + 273$ can be used to convert Celsius temperatures to Kelvin. The function $C(F) = \frac{5}{9}(F - 32)$ can be used to convert Fahrenheit temperatures to Celsius.

52. Write a composition of functions that could be used to convert Fahrenheit temperatures to Kelvin.

53. Find the temperature in Kelvin for the boiling point of water and the freezing point of water if water boils at 212°F and freezes at 32°F.

54. While performing an experiment, Kimi found the temperature of a solution at different intervals. She needs to record the change in temperature in degrees Kelvin, but only has a thermometer with a Fahrenheit scale. What will she record when the temperature of the solution goes from 158°F to 256°F?

55. **FINANCE** Kachina pays $50 each month on a credit card that charges 1.6% interest monthly. She has a balance of $700. The balance at the beginning of the nth month is given by $f(n) = f(n - 1) + 0.016\,f(n - 1) - 50$. Find the balance at the beginning of the first five months. No additional charges are made on the card. (*Hint:* $f(1) = 700$)

56. **CRITICAL THINKING** If $f(0) = 4$ and $f(x + 1) = 3f(x) - 2$, find $f(4)$.

57. **WRITING IN MATH** Answer the question that was posed at the beginning of the lesson.

Why is it important to combine functions in business?

Include the following in your answer:

- a description of how to write a new function that represents the profit, using the revenue and cost functions, and
- an explanation of the benefits of combining two functions into one function.

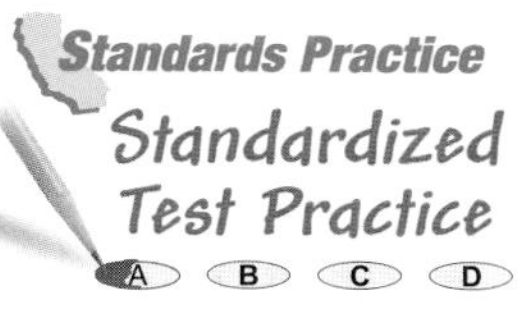

58. If $h(x) = 7x - 5$ and $g[h(x)] = 2x + 3$, then $g(x) =$

Ⓐ $\frac{2x + 31}{7}$.
Ⓑ $-5x + 8$.
Ⓒ $5x - 8$.
Ⓓ $\frac{2x + 26}{7}$.

59. If $f(x) = 4x^4 + 5x^3 - 3x^2 - 14x + 31$ and $g(x) = 7x^3 - 4x^2 + 5x - 42$, then $(f - g)(x) =$

Ⓐ $4x^4 + 12x^3 - 7x^2 - 9x - 11$.
Ⓑ $4x^4 - 2x^3 - 7x^2 - 19x - 11$.
Ⓒ $4x^4 - 2x^3 + x^2 - 19x + 73$.
Ⓓ $-3x^4 - 2x^3 - 7x^2 - 19x + 73$.

Maintain Your Skills

Mixed Review

List all of the possible rational zeros of each function. *(Lesson 7-6)*

60. $r(x) = x^2 - 6x + 8$
61. $f(x) = 4x^3 - 2x^2 + 6$
62. $g(x) = 9x^2 - 1$

Write a polynomial function of least degree with integral coefficients that has the given zeros. *(Lesson 7-5)*

63. 5, 3, −4
64. −3, −2, 8
65. $1, \frac{1}{2}, \frac{2}{3}$
66. $6, 2i$
67. $3, 3 - 2i$
68. $-5, 2, 1 - i$

69. **ELECTRONICS** There are three basic things to be considered in an electrical circuit: the flow of the electrical current I, the resistance to the flow Z called impedance, and electromotive force E called voltage. These quantities are related in the formula $E = I \cdot Z$. The current of a circuit is to be $35 - 40j$ amperes. Electrical engineers use the letter j to represent the imaginary unit. Find the impedance of the circuit if the voltage is to be $430 - 330j$ volts. *(Lesson 5-9)*

Find the inverse of each matrix, if it exists. *(Lesson 4-7)*

70. $\begin{bmatrix} 8 & 6 \\ 7 & 5 \end{bmatrix}$
71. $\begin{bmatrix} 1 & 2 \\ 1 & 3 \end{bmatrix}$
72. $\begin{bmatrix} 8 & 4 \\ 6 & 3 \end{bmatrix}$
73. $\begin{bmatrix} -4 & 2 \\ 3 & -1 \end{bmatrix}$
74. $\begin{bmatrix} 6 & -2 \\ 9 & -3 \end{bmatrix}$
75. $\begin{bmatrix} 2 & 2 \\ 3 & -5 \end{bmatrix}$

Getting Ready for the Next Lesson

PREREQUISITE SKILL Solve each equation or formula for the specified variable. *(To review **solving equations for a variable**, see Lesson 1-3.)*

76. $2x - 3y = 6$, for x
77. $4x^2 - 5xy + 2 = 3$, for y
78. $3x + 7xy = -2$, for x
79. $I = prt$, for t
80. $C = \frac{5}{9}(F - 32)$, for F
81. $F = G\frac{Mm}{r^2}$, for m

7-8 Inverse Functions and Relations

California Standards Standard 24.0 Students solve problems involving functional concepts, such as composition, defining the inverse function and performing arithmetic operations on functions.

What You'll Learn

- Find the inverse of a function or relation.
- Determine whether two functions or relations are inverses.

Vocabulary

- inverse relation
- inverse function
- identity function
- one-to-one

How are inverse functions related to measurement conversions?

Most scientific formulas involve measurements given in SI (International System) units. The SI units for speed are meters per second. However, the United States uses customary measurements such as miles per hour. To convert x miles per hour to an approximate equivalent in meters per second, you can evaluate

$f(x) = \frac{x \text{ miles}}{1 \text{ hour}} \cdot \frac{1600 \text{ meters}}{1 \text{ mile}} \cdot \frac{1 \text{ hour}}{3600 \text{ seconds}}$ or $f(x) = \frac{4}{9}x$. To convert x meters per second to an approximate equivalent in miles per hour, you can evaluate

$g(x) = \frac{x \text{ meters}}{1 \text{ second}} \cdot \frac{3600 \text{ seconds}}{1 \text{ hour}} \cdot \frac{1 \text{ mile}}{1600 \text{ meters}}$ or $g(x) = \frac{9}{4}x$.

Notice that $f(x)$ multiplies a number by 4 and divides it by 9. The function $g(x)$ does the inverse operation of $f(x)$. It divides a number by 4 and multiplies it by 9. The functions $f(x) = \frac{4}{9}x$ and $g(x) = \frac{9}{4}x$ are inverses.

FIND INVERSES Recall that a relation is a set of ordered pairs. The **inverse relation** is the set of ordered pairs obtained by reversing the coordinates of each original ordered pair. The domain of a relation becomes the range of the inverse, and the range of a relation becomes the domain of the inverse.

Key Concept — Inverse Relations

- **Words** Two relations are inverse relations if and only if whenever one relation contains the element (a, b), the other relation contains the element (b, a).
- **Example** $Q = \{(1, 2), (3, 4), (5, 6)\}$ $\quad S = \{(2, 1), (4, 3), (6, 5)\}$
  Q and S are inverse relations.

Example 1 Find an Inverse Relation

GEOMETRY The ordered pairs of the relation $\{(2, 1), (5, 1), (2, -4)\}$ are the coordinates of the vertices of a right triangle. Find the inverse of this relation and determine whether the resulting ordered pairs are also the vertices of a right triangle.

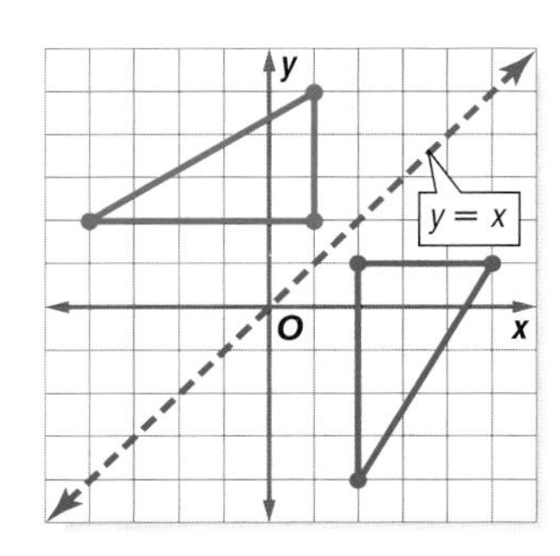

To find the inverse of this relation, reverse the coordinates of the ordered pairs.

The inverse of the relation is $\{(1, 2), (1, 5), (-4, 2)\}$.

Plotting the points shows that the ordered pairs also describe the vertices of a right triangle. Notice that the graphs of the relation and the inverse relation are reflections over the graph of $y = x$.

Study Tip

Reading Math

f^{-1} is read *f inverse* or *the inverse of f*. Note that -1 is not an exponent.

The ordered pairs of **inverse functions** are also related. We can write the inverse of function $f(x)$ as $f^{-1}(x)$.

Key Concept — *Property of Inverse Functions*

Suppose f and f^{-1} are inverse functions. Then, $f(a) = b$ if and only if $f^{-1}(b) = a$.

Let's look at the inverse functions $f(x) = x + 2$ and $f^{-1}(x) = x - 2$.

Evaluate $f(5)$.	Now, evaluate $f^{-1}(7)$.
$f(x) = x + 2$	$f^{-1}(x) = x - 2$
$f(5) = 5 + 2$ or 7	$f^{-1}(7) = 7 - 2$ or 5

Since $f(x)$ and $f^{-1}(x)$ are inverses, $f(5) = 7$ and $f^{-1}(7) = 5$. The inverse function can be found by exchanging the domain and range of the function.

Example 2 Find an Inverse Function

a. Find the inverse of $f(x) = \frac{x+6}{2}$.

Step 1 Replace $f(x)$ with y in the original equation.

$f(x) = \frac{x+6}{2} \qquad y = \frac{x+6}{2}$

Step 2 Interchange x and y.

$x = \frac{y+6}{2}$

Step 3 Solve for y.

$x = \frac{y+6}{2}$ Inverse

$2x = y + 6$ Multiply each side by 2.

$2x - 6 = y$ Subtract 6 from each side.

Step 4 Replace y with $f^{-1}(x)$.

$y = 2x - 6 \qquad f^{-1}(x) = 2x - 6$

The inverse of $f(x) = \frac{x+6}{2}$ is $f^{-1}(x) = 2x - 6$.

b. Graph the function and its inverse.

Graph both functions on the coordinate plane. The graph of $f^{-1}(x) = 2x - 6$ is the reflection of the graph of $f(x) = \frac{x+6}{2}$ over the line $y = x$.

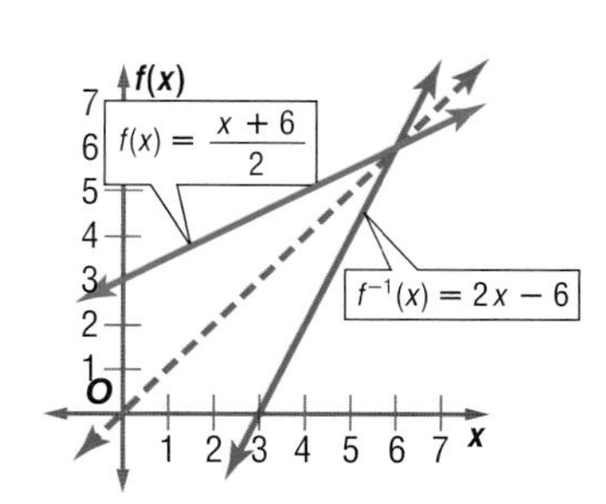

INVERSES OF RELATIONS AND FUNCTIONS You can determine whether two functions are inverses by finding both of their compositions. If both equal the $I(x) = x$, then the functions are inverse functions.

Key Concept — *Inverse Functions*

- **Words** Two functions f and g are inverse functions if and only if both of their compositions are the identity function.
- **Symbols** $[f \circ g](x) = x$ and $[g \circ f](x) = x$

Study Tip

Inverse Functions
Both compositions of $f(x)$ and $g(x)$ must be the identity function for $f(x)$ and $g(x)$ to be inverses. It is necessary to check them both.

Example 3 Verify Two Functions are Inverses

Determine whether $f(x) = 5x + 10$ and $g(x) = \frac{1}{5}x - 2$ are inverse functions.

Check to see if the compositions of $f(x)$ and $g(x)$ are identity functions.

$$[f \circ g](x) = f[g(x)] = f\left(\frac{1}{5}x - 2\right) = 5\left(\frac{1}{5}x - 2\right) + 10 = x - 10 + 10 = x$$

$$[g \circ f](x) = g[f(x)] = g(5x + 10) = \frac{1}{5}(5x + 10) - 2 = x + 2 - 2 = x$$

The functions are inverses since both $[f \circ g](x)$ and $[g \circ f](x)$ equal x.

You can also determine whether two functions are inverse functions by graphing. The graphs of a function and its inverse are mirror images with respect to the graph of the identity function $I(x) = x$.

Algebra Activity

Inverses of Functions

- Use a full sheet of grid paper. Draw and label the x- and y-axes.
- Graph $y = 2x - 3$.
- On the same coordinate plane, graph $y = x$ as a dashed line.
- Place a geomirror so that the drawing edge is on the line $y = x$. Carefully plot the points that are part of the reflection of the original line. Draw a line through the points.

Analyze

1. What is the equation of the drawn line?
2. What is the relationship between the line $y = 2x - 3$ and the line that you drew? Justify your answer.
3. Try this activity with the function $y = |x|$. Is the inverse also a function? Explain.

When the inverse of a function is a function, then the original function is said to be **one-to-one**. To determine if the inverse of a function is a function, you can use the *horizontal line test*.

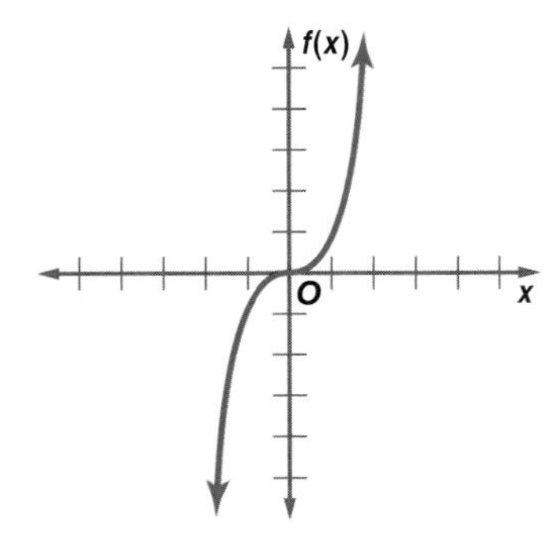

No horizontal line can be drawn so that it passes through more than one point. The inverse of this function is a function.

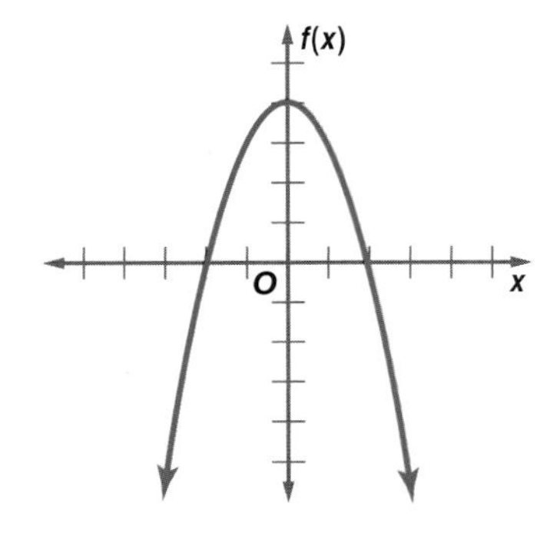

A horizontal line can be drawn that passes through more than one point. The inverse of this function is not a function.

Check for Understanding

Concept Check

1. **Determine** whether $f(x) = 3x + 6$ and $g(x) = x - 2$ are inverses.
2. **Explain** the steps you would take to find an inverse function.
3. **OPEN ENDED** Give an example of a function and its inverse. Verify that the two functions are inverses.
4. **Determine** the values of n for which $f(x) = x^n$ has an inverse that is a function. Assume that n is a whole number.

Guided Practice

Find the inverse of each relation.

5. $\{(2, 4), (-3, 1), (2, 8)\}$
6. $\{(1, 3), (1, -1), (1, -3), (1, 1)\}$

Find the inverse of each function. Then graph the function and its inverse.

7. $f(x) = -x$
8. $g(x) = 3x + 1$
9. $y = \frac{1}{2}x + 5$

Determine whether each pair of functions are inverse functions.

10. $f(x) = x + 7$
    $g(x) = x - 7$
11. $g(x) = 3x - 2$
    $f(x) = \frac{x - 2}{3}$

Application

PHYSICS **For Exercises 12 and 13, use the following information.**
The acceleration due to gravity is 9.8 meters per second squared (m/s^2). To convert to feet per second squared, you can use the following chain of operations:
$\frac{9.8 \text{ m}}{s^2} \times \frac{100 \text{ cm}}{1 \text{ m}} \times \frac{1 \text{ in.}}{2.54 \text{ cm}} \times \frac{1 \text{ ft}}{12 \text{ in.}}$.

12. Find the value of the acceleration due to gravity in feet per second squared.
13. An object is accelerating at 50 feet per second squared. How fast is it accelerating in meters per second squared?

Practice and Apply

Homework Help

For Exercises	See Examples
14–19	1
20–31, 38–43	2
32–37	3

Extra Practice
See page 844.

Find the inverse of each relation.

14. $\{(2, 6), (4, 5), (-3, -1)\}$
15. $\{(3, 8), (4, -2), (5, -3)\}$
16. $\{(7, -4), (3, 5), (-1, 4), (7, 5)\}$
17. $\{(-1, -2), (-3, -2), (-1, -4), (0, 6)\}$
18. $\{(6, 11), (-2, 7), (0, 3), (-5, 3)\}$
19. $\{(2, 8), (-6, 5), (8, 2), (5, -6)\}$

Find the inverse of each function. Then graph the function and its inverse.

20. $y = -3$
21. $g(x) = -2x$
22. $f(x) = x - 5$
23. $g(x) = x + 4$
24. $f(x) = 3x + 3$
25. $y = -2x - 1$
26. $y = \frac{1}{3}x$
27. $f(x) = \frac{5}{8}x$
28. $f(x) = \frac{1}{3}x + 4$
29. $f(x) = \frac{4}{5}x - 7$
30. $g(x) = \frac{2x + 3}{6}$
31. $f(x) = \frac{7x - 4}{8}$

Determine whether each pair of functions are inverse functions.

32. $f(x) = x - 5$
    $g(x) = x + 5$
33. $f(x) = 3x + 4$
    $g(x) = 3x - 4$
34. $f(x) = 6x + 2$
    $g(x) = x - \frac{1}{3}$
35. $g(x) = 2x + 8$
    $f(x) = \frac{1}{2}x - 4$
36. $h(x) = 5x - 7$
    $g(x) = \frac{1}{5}(x + 7)$
37. $g(x) = 2x + 1$
    $f(x) = \frac{x - 1}{2}$

NUMBER GAMES **For Exercises 38–40, use the following information.**
Damaso asked Sophia to choose a number between 1 and 20. He told her to add 7 to that number, multiply by 4, subtract 6, and divide by 2.

38. Write an equation that models this problem.

39. Find the inverse.

40. Sophia's final number was 35. What was her original number?

41. SALES Sales associates at Electronics Unlimited earn \$8 an hour plus a 4% commission on the merchandise they sell. Write a function to describe their income, and find how much merchandise they must sell in order to earn \$500 in a 40-hour week.

More About. . .

Temperature
The Fahrenheit temperature scale was established in 1724 by a physicist named Gabriel Daniel Fahrenheit. The Celsius temperature scale was established in the same year by an astronomer named Anders Celsius.
Source: www.infoplease.com

TEMPERATURE **For Exercises 42 and 43, use the following information.**
A formula for converting degrees Fahrenheit to Celsius is $C(x) = \frac{5}{9}(x - 32)$.

42. Find the inverse $C^{-1}(x)$. Show that $C(x)$ and $C^{-1}(x)$ are inverses.

43. Explain what purpose $C^{-1}(x)$ serves.

44. CRITICAL THINKING Give an example of a function that is its own inverse.

45. WRITING IN MATH Answer the question that was posed at the beginning of the lesson.

How are inverse functions related to measurement conversions?

Include the following items in your answer:
- an explanation of why you might want to know the customary units if you are given metric units even if it is not necessary for you to perform additional calculations, and
- a demonstration of how to convert the speed of light $c = 3.0 \times 10^8$ meters per second to miles per hour.

Standards Practice
Standardized Test Practice

46. Which of the following is the inverse of the function $f(x) = \frac{3x - 5}{2}$?

Ⓐ $g(x) = \frac{2x + 5}{3}$ Ⓑ $g(x) = \frac{3x + 5}{2}$ Ⓒ $g(x) = 2x + 5$ Ⓓ $g(x) = \frac{2x - 5}{3}$

47. For which of the following functions is the inverse also a function?

I. $f(x) = x^3$ **II.** $f(x) = x^4$ **III.** $f(x) = -|x|$

Ⓐ I and II only Ⓑ I only Ⓒ I, II, and III Ⓓ III only

Maintain Your Skills

Mixed Review

Find $[g \circ h](x)$ and $[h \circ g](x)$. *(Lesson 7-7)*

48. $g(x) = 4x$, $h(x) = x + 5$

49. $g(x) = 3x + 2$, $h(x) = 2x - 4$

50. $g(x) = x + 4$, $h(x) = x^2 - 3x - 28$

Find all of the rational zeros of each function. *(Lesson 7-6)*

51. $f(x) = x^3 + 6x^2 - 13x - 42$

52. $h(x) = 24x^3 - 86x^2 + 57x + 20$

Evaluate each expression. *(Lesson 5-7)*

53. $16^{\frac{3}{2}}$

54. $64^{\frac{1}{3}} \cdot 64^{\frac{1}{2}}$

55. $\frac{3^{\frac{4}{3}}}{81^{\frac{1}{12}}}$

Getting Ready for the Next Lesson

PREREQUISITE SKILL **Solve each equation.**
*(To review **solving radical equations**, see Lesson 5-8.)*

56. $\sqrt{x} - 5 = -3$

57. $\sqrt{x + 4} = 11$

58. $12 - \sqrt{x} = -2$

59. $\sqrt{x - 5} = \sqrt{2x + 2}$

60. $\sqrt{x - 3} = \sqrt{2} - \sqrt{x}$

61. $3 - \sqrt{x} = \sqrt{x - 6}$

7-9 Square Root Functions and Inequalities

What You'll Learn

- Graph and analyze square root functions.
- Graph square root inequalities.

How are square root functions used in bridge design?

The Sunshine Skyway Bridge across Tampa Bay, Florida, is supported by 21 steel cables, each 9 inches in diameter. The amount of weight that a steel cable can support is given by $w = 8d^2$, where d is the diameter of the cable in inches and w is the weight in tons. If you need to know what diameter a steel cable should have to support a given weight, you can use the equation $d = \sqrt{\frac{w}{8}}$.

Vocabulary

- square root function
- square root inequality

SQUARE ROOT FUNCTIONS If a function contains a square root of a variable, it is called a **square root function**. The inverse of a quadratic function is a square root function only if the range is restricted to nonnegative numbers.

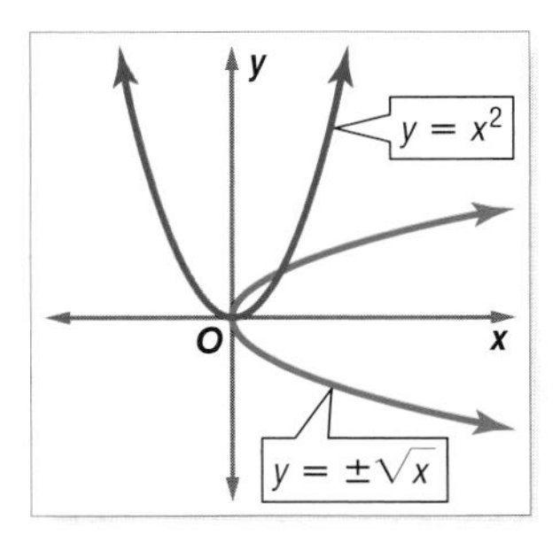

$y = \pm\sqrt{x}$ is not a function.

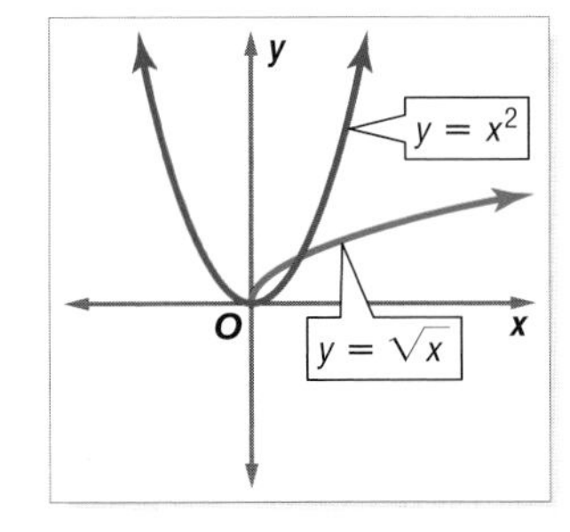

$y = \sqrt{x}$ is a function.

In order for a square root to be a real number, the radicand cannot be negative. When graphing a square root function, determine when the radicand would be negative and exclude those values from the domain.

Example 1 Graph a Square Root Function

Graph $y = \sqrt{3x + 4}$. State the domain, range, and x- and y-intercepts.

Since the radicand cannot be negative, identify the domain.

$3x + 4 \geq 0$ Write the expression inside the radicand as ≥ 0.

$x \geq -\frac{4}{3}$ Solve for x.

The x-intercept is $-\frac{4}{3}$.

Make a table of values and graph the function. From the graph, you can see that the domain is $x \geq -\frac{4}{3}$, and the range is $y \geq 0$. The y-intercept is 2.

x	y
$-\frac{4}{3}$	0
-1	1
0	2
2	3.2
4	4

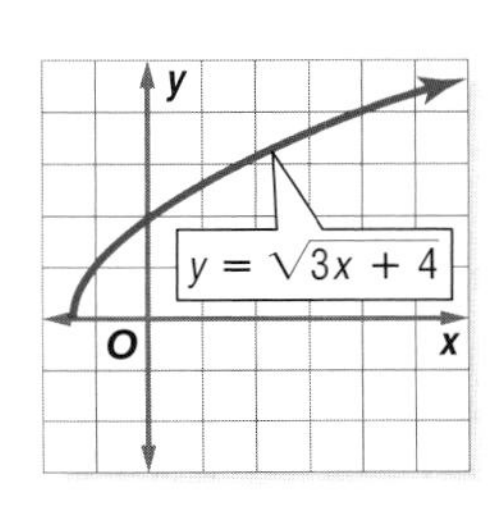

Example 2 Solve a Square Root Problem

SUBMARINES **A lookout on a submarine is h feet above the surface of the water. The greatest distance d in miles that the lookout can see on a clear day is given by the square root of the quantity h multiplied by $\frac{3}{2}$.**

a. Graph the function. State the domain and range.

The function is $d = \sqrt{\frac{3h}{2}}$. Make a table of values and graph the function.

h	d
0	0
2	$\sqrt{3}$ or 1.73
4	$\sqrt{6}$ or 2.45
6	$\sqrt{9}$ or 3.00
8	$\sqrt{12}$ or 3.46
10	$\sqrt{15}$ or 3.87

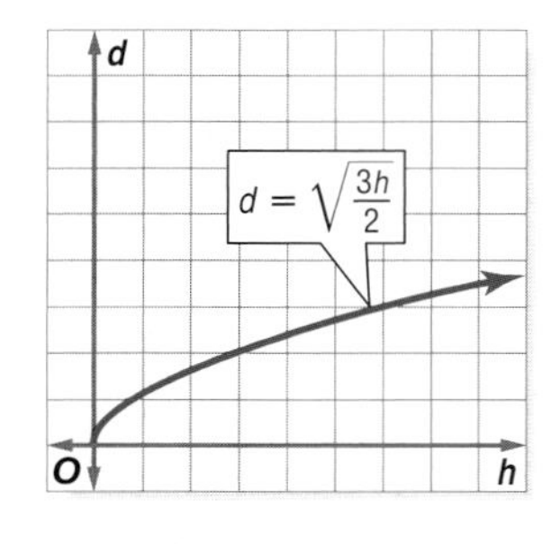

The domain is $h \geq 0$, and the range is $d \geq 0$.

b. A ship is 3 miles from a submarine. How high would the submarine have to raise its periscope in order to see the ship?

$d = \sqrt{\frac{3h}{2}}$ Original equation

$3 = \sqrt{\frac{3h}{2}}$ Replace d with 3.

$9 = \frac{3h}{2}$ Square each side.

$18 = 3h$ Multiply each side by 2.

$6 = h$ Divide each side by 3.

The periscope would have to be 6 feet above the water. Check this result on the graph.

More About...

Submarines

Submarines were first used by The United States in 1776 during the Revolutionary War.

Source: www.infoplease.com

Graphs of square root functions can be transformed just like quadratic functions.

Graphing Calculator Investigation

Square Root Functions

You can use a TI-83 Plus graphing calculator to graph square root functions. Use [2nd] [$\sqrt{\ }$] to enter the functions in the Y= list.

Think and Discuss

1. Graph $y = \sqrt{x}$, $y = \sqrt{x} + 1$, and $y = \sqrt{x} - 2$ in the viewing window $[-2, 8]$ by $[-4, 6]$. State the domain and range of each function and describe the similarities and differences among the graphs.
2. Graph $y = \sqrt{x}$, $y = \sqrt{2x}$, and $y = \sqrt{8x}$ in the viewing window $[0, 10]$ by $[0, 10]$. State the domain and range of each function and describe the similarities and differences among the graphs.
3. Make a conjecture on how you could write an equation that translates the parent graph $y = \sqrt{x}$ to the left three units. Test your conjecture with the graphing calculator.

SQUARE ROOT INEQUALITIES A **square root inequality** is an inequality involving square roots. You can use what you know about square root functions to graph square root inequalities.

Example 3 Graph a Square Root Inequality

a. Graph $y < \sqrt{2x - 6}$.

Graph the related equation $y = \sqrt{2x - 6}$. Since the boundary should not be included, the graph should be dashed.

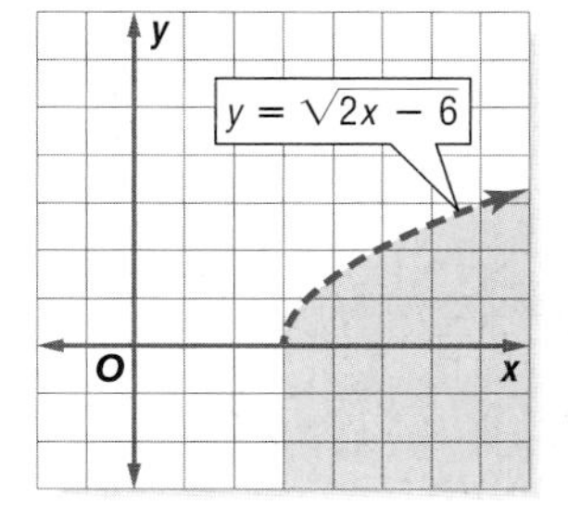

The domain includes values for $x \geq 3$, so the graph is to the right of $x = 3$. Select a point and test its ordered pair.

Test (4, 1).

$1 < \sqrt{2(4) - 6}$

$1 < \sqrt{2}$ true

Shade the region that includes the point (4, 1).

b. Graph $y \geq \sqrt{x + 1}$.

Graph the related equation $y = \sqrt{x + 1}$.

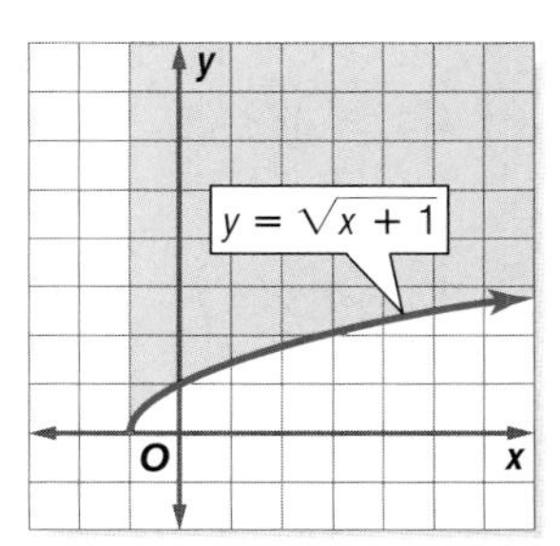

The domain includes values for $x \geq -1$, so the graph includes $x = -1$ and the values of x to the right of $x = -1$. Select a point and test its ordered pair.

Test (2, 1).

$y \geq \sqrt{x + 1}.$

$1 \geq \sqrt{2 + 1}$

$1 \geq \sqrt{3}$ false

Shade the region that does not include (2, 1).

Check for Understanding

Concept Check

1. **Explain** why the inverse of $y = 3x^2$ is not a square root function.
2. **Describe** the difference between the graphs of $y = \sqrt{x} - 4$ and $y = \sqrt{x - 4}$.
3. **OPEN ENDED** **Write** a square root function with a domain of $\{x \mid x \geq 2\}$.

Guided Practice

Graph each function. State the domain and range of the function.

4. $y = \sqrt{x} + 2$
5. $y = \sqrt{4x}$
6. $y = 3 - \sqrt{x}$
7. $y = \sqrt{x - 1} + 3$

Graph each inequality.

8. $y \leq \sqrt{x - 4} + 1$
9. $y > \sqrt{2x + 4}$
10. $y < 3 - \sqrt{5x + 1}$
11. $y \geq \sqrt{x + 2} - 1$

Application **FIREFIGHTING** **For Exercises 12 and 13, use the following information.**
When fighting a fire, the velocity v of water being pumped into the air is the square root of twice the product of the maximum height h and g, the acceleration due to gravity (32 ft/s^2).

12. Determine an equation that will give the maximum height of the water as a function of its velocity.

13. The Coolville Fire Department must purchase a pump that is powerful enough to propel water 80 feet into the air. Will a pump that is advertised to project water with a velocity of 75 ft/s meet the fire department's need? Explain.

Practice and Apply

Homework Help

For Exercises	See Examples
14–25	1
26–31	3
32–34	2

Extra Practice
See page 844.

Graph each function. State the domain and range of each function.

14. $y = \sqrt{3x}$
15. $y = -\sqrt{5x}$
16. $y = -4\sqrt{x}$
17. $y = \frac{1}{2}\sqrt{x}$
18. $y = \sqrt{x + 2}$
19. $y = \sqrt{x - 7}$
20. $y = -\sqrt{2x + 1}$
21. $y = \sqrt{5x - 3}$
22. $y = \sqrt{x + 6} - 3$
23. $y = 5 - \sqrt{x + 4}$
24. $y = \sqrt{3x - 6} + 4$
25. $y = 2\sqrt{3 - 4x} + 3$

Graph each inequality.

26. $y \leq -6\sqrt{x}$
27. $y < \sqrt{x + 5}$
28. $y > \sqrt{2x + 8}$
29. $y \geq \sqrt{5x - 8}$
30. $y \geq \sqrt{x - 3} + 4$
31. $y < \sqrt{6x - 2} + 1$

32. **ROLLER COASTERS** The velocity of a roller coaster as it moves down a hill is $v = \sqrt{v_0^2 + 64h}$, where v_0 is the initial velocity and h is the vertical drop in feet. An engineer wants a new coaster to have a velocity of 90 feet per second when it reaches the bottom of the hill. If the initial velocity of the coaster at the top of the hill is 10 feet per second, how high should the engineer make the hill?

AEROSPACE **For Exercises 33 and 34, use the following information.**
The force due to gravity decreases with the square of the distance from the center of Earth. So, as an object moves farther from Earth, its weight decreases. The radius of Earth is approximately 3960 miles. The formula relating weight and distance is $r = \sqrt{\frac{3960^2 W_E}{W_S}} - 3960$, where W_E represents the weight of a body on Earth, W_S represents the weight of a body a certain distance from the center of Earth, and r represents the distance of an object above Earth's surface.

33. An astronaut weighs 140 pounds on Earth and 120 pounds in space. How far is he above Earth's surface?

34. An astronaut weighs 125 pounds on Earth. What is her weight in space if she is 99 miles above the surface of Earth?

More About. . .

Aerospace

The weight of a person is equal to the product of the person's mass and the acceleration due to Earth's gravity. Thus, as a person moves away from Earth, the person's weight decreases. However, mass remains constant.

35. **RESEARCH** Use the Internet or another resource to find the weights, on Earth, of several space shuttle astronauts and the average distance they were from Earth during their missions. Use this information to calculate their weights while in orbit.

36. **CRITICAL THINKING** Recall how values of a, h, and k can affect the graph of a quadratic function of the form $y = a(x - h)^2 + k$. Describe how values of a, h, and k can affect the graph of a square root function of the form $y = a\sqrt{x - h} + k$.

37. **WRITING IN MATH** Answer the question that was posed at the beginning of the lesson.

How are square root functions used in bridge design?

Include the following in your answer:

- the weights for which a diameter less than 1 is reasonable, and
- the weight that the Sunshine Skyway Bridge can support.

38. What is the domain of $f(x) \geq \sqrt{5x - 3}$?

(A) $\left\{x \mid x > \frac{3}{5}\right\}$ (B) $\left\{x \mid x > -\frac{3}{5}\right\}$ (C) $\left\{x \mid x \geq \frac{3}{5}\right\}$ (D) $\left\{x \mid x \geq -\frac{3}{5}\right\}$

39. Given the graph of the square root function at the right, which of the following must be true?

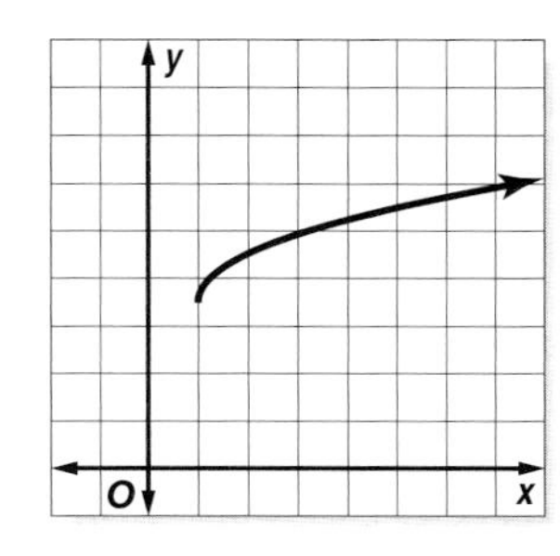

I. The domain is all real numbers.

II. The function is $y = \sqrt{x} + 3.5$.

III. The range is $\{y \mid y \geq 3.5\}$.

(A) I only (B) I, II, and III

(C) II and III (D) III only

Maintain Your Skills

Mixed Review **Determine whether each pair of functions are inverse functions.** *(Lesson 7-8)*

40. $f(x) = 3x$
$g(x) = \frac{1}{3}x$

41. $f(x) = 4x - 5$
$g(x) = \frac{1}{4}x - \frac{5}{16}$

42. $f(x) = \frac{3x + 2}{7}$
$g(x) = \frac{7x - 2}{3}$

Find $(f + g)(x)$, $(f - g)(x)$, $(f \cdot g)(x)$, and $\left(\frac{f}{g}\right)(x)$ for each $f(x)$ and $g(x)$. *(Lesson 7-7)*

43. $f(x) = x + 5$
$g(x) = x - 3$

44. $f(x) = 10x - 20$
$g(x) = x - 2$

45. $f(x) = 4x^2 - 9$
$g(x) = \frac{1}{2x + 3}$

46. **ENTERTAINMENT** A magician asked a member of his audience to choose any number. He said, "Multiply your number by 3. Add the sum of your number and 8 to that result. Now divide by the sum of your number and 2." The magician announced the final answer without asking the original number. What was the final answer? How did he know what it was? *(Lesson 5-4)*

Simplify. *(Lesson 5-2)*

47. $(x + 2)(2x - 8)$

48. $(3p + 5)(2p - 4)$

49. $(a^2 + a + 1)(a - 1)$

WebQuest Internet Project

Population Explosion

It is time to complete your project. Use the information and data you have gathered about the population to prepare a Web page. Be sure to include graphs, tables, and equations in the presentation.

www.algebra2.com/webquest

Chapter 7 Study Guide and Review

Vocabulary and Concept Check

- Complex Conjugates Theorem (p. 374)
- composition of functions (p. 384)
- degree of a polynomial (p. 346)
- depressed polynomial (p. 366)
- Descartes' Rule of Signs (p. 372)
- end behavior (p. 349)
- Factor Theorem (p. 366)
- Fundamental Theorem of Algebra (p. 371)
- identity function (p. 391)
- Integral Zero Theorem (p. 378)
- inverse function (p. 391)
- inverse relation (p. 390)
- leading coefficients (p. 346)
- Location Principle (p. 353)
- one-to-one (p. 392)
- polynomial function (p. 347)
- polynomial in one variable (p. 346)
- quadratic form (p. 360)
- Rational Zero Theorem (p. 378)
- relative maximum (p. 354)
- relative minimum (p. 354)
- Remainder Theorem (p. 365)
- square root function (p. 395)
- square root inequality (p. 397)
- synthetic substitution (p. 365)

Choose the letter that best matches each statement or phrase.

1. A point on the graph of a polynomial function that has no other nearby points with lesser y-coordinates is a ______.
2. The ______ is the coefficient of the term in a polynomial function with the highest degree.
3. The ______ says that in any polynomial function, if an imaginary number is a zero of that function, then its conjugate is also a zero.
4. When a polynomial is divided by one of its binomial factors, the quotient is called a(n) ______.
5. $(x^2)^2 - 17(x^2) + 16 = 0$ is written in ______.
6. $f(x) = 6x - 2$ and $g(x) = \frac{x + 2}{6}$ are ______ since $[f \circ g](x)$ and $[g \circ f](x) = x$.

a. Complex Conjugates Theorem
b. depressed polynomial
c. inverse functions
d. leading coefficient
e. quadratic form
f. relative minimum

Lesson-by-Lesson Review

7-1 Polynomial Functions

See pages 346–352.

Concept Summary

- The degree of a polynomial function in one variable is determined by the greatest exponent of its variable.

Example Find $p(a + 1)$ if $p(x) = 5x - x^2 + 3x^3$.

$p(a + 1) = 5(a + 1) - (a + 1)^2 + 3(a + 1)^3$ Replace x with $a + 1$.

$= 5a + 5 - (a^2 + 2a + 1) + 3(a^3 + 3a^2 + 3a + 1)$ Evaluate $5(a + 1)$, $(a + 1)^2$, and $3(a + 1)^3$.

$= 5a + 5 - a^2 - 2a - 1 + 3a^3 + 9a^2 + 9a + 3$

$= 3a^3 + 8a^2 + 12a + 7$ Simplify.

Exercises **Find $p(-4)$ and $p(x + h)$ for each function.**

See Examples 2 and 3 on pages 347 and 348.

7. $p(x) = x - 2$
8. $p(x) = -x + 4$
9. $p(x) = 6x + 3$
10. $p(x) = x^2 + 5$
11. $p(x) = x^2 - x$
12. $p(x) = 2x^3 - 1$

7-2 Graphing Polynomial Functions

See pages 353–358.

Concept Summary

- The Location Principle: Since zeros of a function are located at x-intercepts, there is also a zero between each pair of these zeros.
- Turning points of a function are called relative maxima and relative minima.

Example **Graph $f(x) = x^4 - 2x^2 + 10x - 2$ by making a table of values.**

Make a table of values for several values of x and plot the points. Connect the points with a smooth curve.

x	$f(x)$
−3	31
−2	−14
−1	−13
0	−2
1	7
2	26

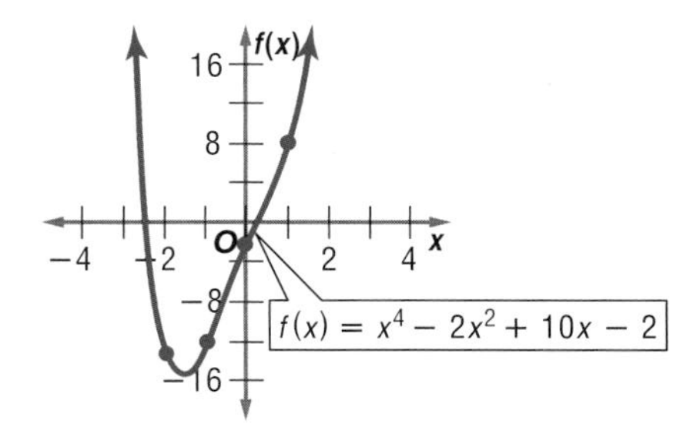

Exercises **For Exercises 13–18, complete each of the following.**

a. Graph each function by making a table of values.

b. Determine consecutive values of x between which each real zero is located.

c. Estimate the x-coordinates at which the relative maxima and relative minima occur. *See Example 1 on page 353.*

13. $h(x) = x^3 - 6x - 9$

14. $f(x) = x^4 + 7x + 1$

15. $p(x) = x^5 + x^4 - 2x^3 + 1$

16. $g(x) = x^3 - x^2 + 1$

17. $r(x) = 4x^3 + x^2 - 11x + 3$

18. $f(x) = x^3 + 4x^2 + x - 2$

7-3 Solving Equations Using Quadratic Techniques

See pages 360–364.

Concept Summary

- Solve polynomial equations by using quadratic techniques.

Example **Solve $x^3 - 3x^2 - 54x = 0$.**

$x^3 - 3x^2 - 54x = 0$ Original equation

$x(x^2 - 3x - 54) = 0$ Factor out the GCF.

$x(x - 9)(x + 6) = 0$ Factor the trinomial.

$x = 0$ or $x - 9 = 0$ or $x + 6 = 0$ Zero Product Property

$x = 0$ $\quad x = 9$ $\quad x = -6$

Exercises **Solve each equation.** *See Example 2 on page 361.*

19. $3x^3 + 4x^2 - 15x = 0$

20. $m^4 + 3m^3 = 40m^2$

21. $a^3 - 64 = 0$

22. $r + 9\sqrt{r} = -8$

23. $x^4 - 8x^2 + 16 = 0$

24. $x^{\frac{2}{3}} - 9x^{\frac{1}{3}} + 20 = 0$

7-4 The Remainder and Factor Theorems

See pages 365–370.

Concept Summary

- Remainder Theorem: If a polynomial $f(x)$ is divided by $x - a$, the remainder is the constant $f(a)$ and $f(x) = q(x) \cdot (x - a) + f(a)$ where $q(x)$ is a polynomial with degree one less than the degree of $f(x)$.
- Factor Theorem: $x - a$ is a factor of polynomial $f(x)$ if and only if $f(a) = 0$.

Example **Show that $x + 2$ is a factor of $x^3 - 2x^2 - 5x + 6$. Then find any remaining factors of the polynomial.**

$$\begin{array}{r|rrrr} -2 & 1 & -2 & -5 & 6 \\ & & -2 & 8 & -6 \\ \hline & 1 & -4 & 3 & 0 \end{array}$$

The remainder is 0, so $x + 2$ is a factor of $x^3 - 2x^2 - 5x + 6$. Since $x^3 - 2x^2 - 5x + 6 = (x + 2)(x^2 - 4x + 3)$, the factors of $x^3 - 2x^2 - 5x + 6$ are $(x + 2)(x - 3)(x - 1)$.

Exercises **Use synthetic substitution to find $f(3)$ and $f(-2)$ for each function.** *See Example 2 on page 367.*

25. $f(x) = x^2 - 5$ **26.** $f(x) = x^2 - 4x + 4$ **27.** $f(x) = x^3 - 3x^2 + 4x + 8$

Given a polynomial and one of its factors, find the remaining factors of the polynomial. Some factors may not be binomials. *See Example 3 on page 367.*

28. $x^3 + 5x^2 + 8x + 4$; $x + 1$ **29.** $x^3 + 4x^2 + 7x + 6$; $x + 2$

7-5 Roots and Zeros

See pages 371–377.

Concept Summary

- Fundamental Theorem of Algebra: Every polynomial equation with degree greater than zero has at least one root in the set of complex numbers.
- Use Descartes' Rule of Signs to determine types of zeros of polynomial functions.
- Complex Conjugates Theorem: If $a + bi$ is a zero of a polynomial function, then $a - bi$ is also a zero of the function.

Example **State the possible number of positive real zeros, negative real zeros, and imaginary zeros of $f(x) = 5x^4 + 6x^3 - 8x + 12$.**

Since $f(x)$ has two sign changes, there are 2 or 0 real positive zeros.

$f(-x) = 5x^4 - 6x^3 + 8x + 12$ Two sign changes $\rightarrow$ 0 or 2 negative real zeros

There are 0, 2, or 4 imaginary zeros.

Exercises **State the possible number of positive real zeros, negative real zeros, and imaginary zeros of each function.** *See Example 2 on page 373.*

30. $f(x) = 2x^4 - x^3 + 5x^2 + 3x - 9$ **31.** $f(x) = 7x^3 + 5x - 1$

32. $f(x) = -4x^4 - x^2 - x + 1$ **33.** $f(x) = 3x^4 - x^3 + 8x^2 + x - 7$

34. $f(x) = x^4 + x^3 - 7x + 1$ **35.** $f(x) = 2x^4 - 3x^3 - 2x^2 + 3$

7-6 Rational Zero Theorem

See pages 378–382.

Concept Summary

- Use the Rational Zero Theorem to find possible zeros of a polynomial function.
- Integral Zero Theorem: If the coefficients of a polynomial function are integers such that $a_0 = 1$ and $a_n \neq 0$, any rational zeros of the function must be factors of a_n.

Examples **Find all of the zeros of $f(x) = x^3 + 7x^2 - 36$.**

There are exactly three complex zeros.

There is exactly one positive real zero and two or zero negative real zeros.

The possible rational zeros are $\pm1, \pm2, \pm3, \pm4, \pm6, \pm9, \pm12, \pm18, \pm36$.

$$\begin{array}{r|rrrr} 2 & 1 & 7 & 0 & -36 \\ & & 2 & 18 & 36 \\ \hline & 1 & 9 & 18 & 0 \end{array}$$

$$\begin{aligned} x^3 + 7x^2 - 36 &= (x - 2)(x^2 + 9x + 18) \\ &= (x - 2)(x + 3)(x + 6) \end{aligned}$$

Therefore, the zeros are 2, −3, and −6.

Exercises **Find all of the rational zeros of each function.** *See Example 3 on page 379.*

36. $f(x) = 2x^3 - 13x^2 + 17x + 12$
37. $f(x) = x^4 + 5x^3 + 15x^2 + 19x + 8$
38. $f(x) = x^3 - 3x^2 - 10x + 24$
39. $f(x) = x^4 - 4x^3 - 7x^2 + 34x - 24$
40. $f(x) = 2x^3 - 5x^2 - 28x + 15$
41. $f(x) = 2x^4 - 9x^3 + 2x^2 + 21x - 10$

7-7 Operations of Functions

See pages 383–389.

Concept Summary

Operation	Definition	Operation	Definition
Sum	$(f + g)(x) = f(x) + g(x)$	Quotient	$\left(\frac{f}{g}\right)(x) = \frac{f(x)}{g(x)}, g(x) \neq 0$
Difference	$(f - g)(x) = f(x) - g(x)$	Composition	$[f \circ g](x) = f[g(x)]$
Product	$(f \cdot g)(x) = f(x) \cdot g(x)$	—	—

Example **If $f(x) = x^2 - 2$ and $g(x) = 8x - 1$. Find $g[f(x)]$ and $f[g(x)]$.**

$$\begin{aligned} g[f(x)] &= 8(x^2 - 2) - 1 && \text{Replace } f(x) \text{ with } x^2 - 2. \\ &= 8x^2 - 16 - 1 && \text{Multiply.} \\ &= 8x^2 - 17 && \text{Simplify.} \end{aligned}$$

$$\begin{aligned} f[g(x)] &= (8x - 1)^2 - 2 && \text{Replace } g(x) \text{ with } 8x - 1. \\ &= 64x^2 - 16x + 1 - 2 && \text{Expand the binomial.} \\ &= 64x^2 - 16x - 1 && \text{Simplify.} \end{aligned}$$

Exercises **Find $[g \circ h](x)$ and $[h \circ g](x)$.** *See Example 4 on page 385.*

42. $h(x) = 2x - 1$, $g(x) = 3x + 4$
43. $h(x) = x^2 + 2$, $g(x) = x - 3$
44. $h(x) = x^2 + 1$, $g(x) = -2x + 1$
45. $h(x) = -5x$, $g(x) = 3x - 5$
46. $h(x) = x^3$, $g(x) = x - 2$
47. $h(x) = x + 4$, $g(x) = |x|$

Chapter 7 **For More ...**

- Extra Practice, see pages 842–844.
- Mixed Problem Solving, see page 868.

7-8 Inverse Functions and Relations

See pages 390–394.

Concept Summary

- Reverse the coordinates of ordered pairs to find the inverse of a relation.
- Two functions are inverses if and only if both of their compositions are the identity function. $[f \circ g](x) = x$ and $[g \circ f](x) = x$
- A function is one-to-one when the inverse of the function is a function.

Example **Find the inverse of $f(x) = -3x + 1$.**

Rewrite $f(x)$ as $y = -3x + 1$. Then interchange the variables and solve for y.

$x = -3y + 1$ Interchange the variables.

$3y = -x + 1$ Solve for y.

$y = \frac{-x + 1}{3}$ Divide each side by 3.

$f^{-1}(x) = \frac{-x + 1}{3}$ Rewrite in function notation.

Exercises **Find the inverse of each function. Then graph the function and its inverse.** *See Example 2 on page 391.*

48. $f(x) = 3x - 4$ **49.** $f(x) = -2x - 3$ **50.** $g(x) = \frac{1}{3}x + 2$

51. $f(x) = \frac{-3x + 1}{2}$ **52.** $y = x^2$ **53.** $y = (2x + 3)^2$

7-9 Square Root Functions and Inequalities

See pages 395–399.

Concept Summary

- Graph square root inequalities in a similar manner as graphing square root equations.

Example **Graph $y = 2 + \sqrt{x - 1}$.**

x	y
1	2
2	3
3	$2 + \sqrt{2}$ or 3.4
4	$2 + \sqrt{3}$ or 3.7
5	4

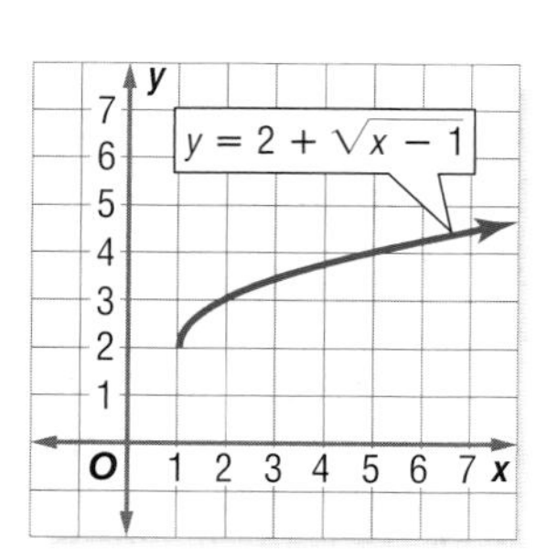

Exercises **Graph each function. State the domain and range of each function.** *See Examples 1 and 2 on pages 395 and 396.*

54. $y = \frac{1}{3}\sqrt{x + 2}$ **55.** $y = \sqrt{5x - 3}$ **56.** $y = 4 + 2\sqrt{x - 3}$

Graph each inequality. *See Example 3 on page 397.*

57. $y \geq \sqrt{x} - 2$ **58.** $y < \sqrt{4x - 5}$

Chapter 7

Practice Test

Vocabulary and Concepts

Match each statement with the term that it best describes.

1. $[f \circ g](x) = f[g(x)]$
2. $[f \circ g](x) = x$ and $[g \circ f](x) = x$
3. $(\sqrt{x})^2 - 2(\sqrt{x}) + 4 = 0$

a. quadratic form
b. composition of functions
c. inverse functions

Skills and Applications

For Exercises 4–7, complete each of the following.

a. Graph each function by making a table of values.
b. Determine consecutive values of x between which each real zero is located.
c. Estimate the x-coordinates at which the relative maxima and relative minima occur.

4. $g(x) = x^3 + 6x^2 + 6x - 4$
5. $h(x) = x^4 + 6x^3 + 8x^2 - x$
6. $f(x) = x^3 + 3x^2 - 2x + 1$
7. $g(x) = x^4 - 2x^3 - 6x^2 + 8x + 5$

Solve each equation.

8. $p^3 + 8p^2 = 18p$
9. $16x^4 - x^2 = 0$
10. $r^4 - 9r^2 + 18 = 0$
11. $p^{\frac{3}{2}} - 8 = 0$

Given a polynomial and one of its factors, find the remaining factors of the polynomial. Some factors may not be binomials.

12. $x^3 - x^2 - 5x - 3;\ x + 1$
13. $x^3 + 8x + 24;\ x + 2$

State the possible number of positive real zeros, negative real zeros, and imaginary zeros for each function.

14. $f(x) = x^3 - x^2 - 14x + 24$
15. $f(x) = 2x^3 - x^2 + 16x - 5$

Find all of the rational zeros of each function.

16. $g(x) = x^3 - 3x^2 - 53x - 9$
17. $h(x) = x^4 + 2x^3 - 23x^2 + 2x - 24$

Determine whether each pair of functions are inverse functions.

18. $f(x) = 4x - 9,\ g(x) = \frac{x - 9}{4}$
19. $f(x) = \frac{1}{x + 2},\ g(x) = \frac{1}{x} - 2$

If $f(x) = 2x - 4$ and $g(x) = x^2 + 3$, find each value.

20. $(f + g)(x)$
21. $(f - g)(x)$
22. $(f \cdot g)(x)$
23. $\left(\frac{f}{g}\right)(x)$

24. **FINANCIAL PLANNING** Toshi will start college in six years. According to their plan, Toshi's parents will save $1000 each year for the next three years. During the fourth and fifth years, they will save $1200 each year. During the last year before he starts college, they will save $2000.
    a. In the formula $A = P(1 + r)^t$, $A =$ the balance, $P =$ the amount invested, $r =$ the interest rate, and $t =$ the number of years the money has been invested. Use this formula to write a polynomial equation to describe the balance of the account when Toshi starts college.
    b. Find the balance of the account if the interest rate is 6%.

Standards Practice

25. **STANDARDIZED TEST PRACTICE** Which value is included in the graph of $y < \sqrt{2x}$?
    Ⓐ $(-2, -2)$ Ⓑ $(-1, -1)$ Ⓒ $(0, 0)$ Ⓓ None of these

www.algebra2.com/chapter_test/ca

Chapter 7

Standardized Test Practice

Standards Practice

Part 1 Multiple Choice

Record your answers on the answer sheet provided by your teacher or on a sheet of paper.

1. If $\frac{2}{p} - \frac{4}{p^2} = -\frac{2}{p^3}$, then what is the value of p?

Ⓐ -1 Ⓑ 1 Ⓒ $-\frac{1}{2}$ Ⓓ $\frac{1}{2}$

2. There are n gallons of liquid available to fill a tank. After k gallons of the liquid have filled the tank, how do you represent in terms of n and k the percent of liquid that has filled the tank?

Ⓐ $\frac{100k}{n}\%$ Ⓑ $\frac{n}{100k}\%$

Ⓒ $\frac{100n}{k}\%$ Ⓓ $\frac{n}{100(n-k)}\%$

3. How many different triangles have sides of lengths 4, 9 and s, where s is an integer and $4 < s < 9$?

Ⓐ 0 Ⓑ 1 Ⓒ 2 Ⓓ 3

4. Triangles ABC and DEF are similar. The area of $\triangle ABC$ is 36 square units. What is the perimeter of $\triangle DEF$?

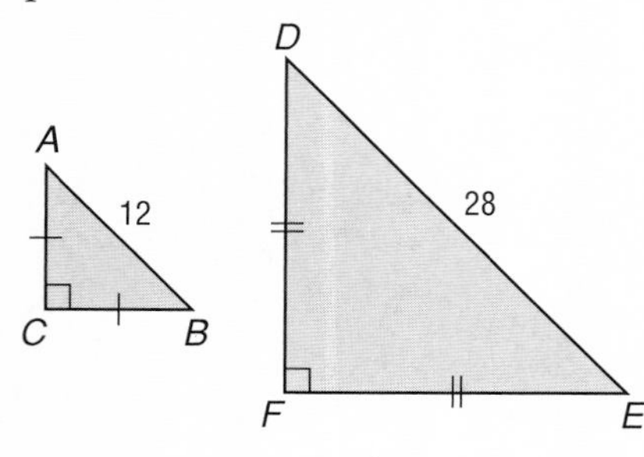

Ⓐ 56 units Ⓑ $28 + 28\sqrt{2}$ units

Ⓒ $56\sqrt{2}$ units Ⓓ $28 + 14\sqrt{2}$ units

5. If $2 - 3x > -1$ and $x + 5 > 0$, then x could equal each of the following *except*

Ⓐ -5. Ⓑ -4. Ⓒ -2. Ⓓ 0.

6. What is the midpoint of the line segment whose endpoints are represented on the coordinate grid by the points $(-5, -3)$ and $(-1, 4)$?

Ⓐ $\left(-3, -\frac{1}{2}\right)$ Ⓑ $\left(-3, \frac{1}{2}\right)$

Ⓒ $\left(-2, -\frac{7}{2}\right)$ Ⓓ $\left(-2, \frac{1}{2}\right)$

7. For all $n \neq 0$, what is the slope of the line passing through (n, k) and $(-n, -k)$?

Ⓐ 0 Ⓑ 1 Ⓒ $\frac{n}{k}$ Ⓓ $\frac{k}{n}$

8. Which of the following is a quadratic equation in one variable?

Ⓐ $3(x + 4) + 1 = 4x - 9$

Ⓑ $3x(x + 4) + 1 = 4x - 9$

Ⓒ $3x(x^2 + 4) + 1 = 4x - 9$

Ⓓ $y = 3x^2 + 8x + 10$

9. Simplify $\sqrt[4]{t^3} \cdot \sqrt[8]{t^2}$.

Ⓐ $t^{\frac{3}{16}}$ Ⓑ $t^{\frac{1}{2}}$ Ⓒ $t^{\frac{3}{4}}$ Ⓓ t

10. Which of the following is a quadratic equation that has roots of $2\frac{1}{2}$ and $\frac{2}{3}$?

Ⓐ $5x^2 + 11x - 7 = 0$

Ⓑ $5x^2 - 11x + 10 = 0$

Ⓒ $6x^2 - 19x + 10 = 0$

Ⓓ $6x^2 + 11x + 10 = 0$

11. If $f(x) = 3x - 5$ and $g(x) = 2 + x^2$, then what is equal to $f[g(2)]$?

Ⓐ 3 Ⓑ 6 Ⓒ 12 Ⓓ 13

12. Which of the following is a zero of $f(x) = x^3 - 7x + 6$?

Ⓐ -1 Ⓑ 2 Ⓒ 3 Ⓓ 6

Preparing for Standardized Tests
For test-taking strategies and more practice, see pages 877–892.

Part 2 Short Response/Grid In

Record your answers on the answer sheet provided by your teacher or on a sheet of paper.

13. A group of 34 people is to be divided into committees so that each person serves on exactly one committee. Each committee must have at least 3 members and not more than 5 members. If N represents the maximum number of committees that can be formed and n represents the minimum number of committees that can be formed, what is the value of $N - n$?

14. Raisins selling for \$2.00 per pound are to be mixed with peanuts selling for \$3.00 per pound. How many pounds of peanuts are needed to produce a 20-pound mixture that sells for \$2.75 per pound?

15. Jars X, Y, and Z each contain 10 marbles. What is the minimum number of marbles that must be transferred among the jars so that the ratio of the number of marbles in jar X to the number of marbles in jar Y to the number of marbles in jar Z is 1 : 2 : 3?

16. If the area of $\triangle BCD$ is 40% of the area of $\triangle ABC$, what is the measure of $\overline{AD}$?

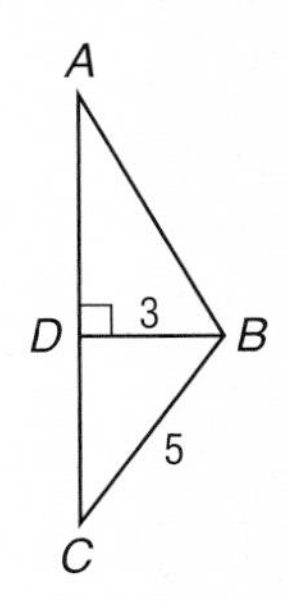

Test-Taking Tip

Questions 13, 15, and 18
Words such as *maximum*, *minimum*, *least*, and *greatest* indicate that a problem may involve an inequality. Take special care when simplifying inequalities that involve negative numbers.

17. The mean of 15 scores is 82. If the mean of 7 of these scores is 78, what is the mean of the remaining 8 scores?

18. If the measures of the sides of a triangle are 3, 8, and x and x is an integer, then what is the least possible perimeter of the triangle?

19. If the operation ❖ is defined by the equation x ❖ $y = 3x - y$, what is the value of w in the equation w ❖ $6 = 2$ ❖ w?

20. In the figure below, $\ell \parallel m$. Find b. Assume that the figure is not drawn to scale.

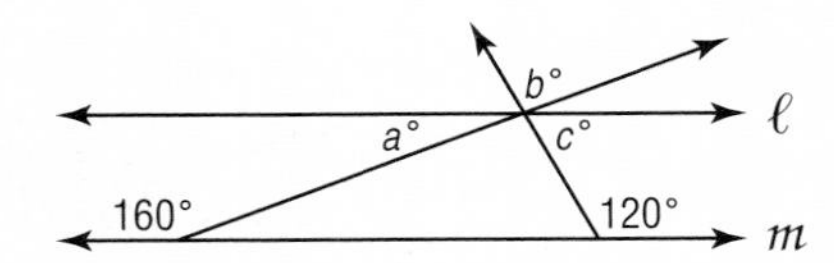

Part 3 Extended Response

Record your answers on a sheet of paper. Show your work.

For Exercises 21–25, use the polynomial function $f(x) = 3x^4 + 19x^3 + 7x^2 - 11x - 2$.

21. What is the degree of the function?

22. Evaluate $f(1)$, $f(-2)$, and $f(2a)$. Show your work.

23. State the number of possible positive real zeros, negative real zeros, and imaginary zeros of $f(x)$.

24. List all of the possible rational zeros of the function.

25. Find all of the zeros of the function.

26. Sketch the graphs of $f(x) = \frac{3x + 1}{2}$ and $g(x) = \frac{2x - 1}{3}$. Considering the graphs, describe the relationship between $f(x)$ and $g(x)$. Verify your conclusion.

UNIT 3

Advanced Functions and Relations

You can use functions and relations to investigate events like earthquakes. In this unit, you will learn about conic sections, rational expressions and equations, and exponential and logarithmic functions.

WebQuest Internet Project

On Quake Anniversary, Japan Still Worries

Source: *USA TODAY,* January 16, 2001

"As Japan marks the sixth anniversary of the devastating Kobe earthquake this week, a different seismic threat is worrying the country: Mount Fuji. Researchers have measured a sudden increase of small earthquakes on the volcano, indicating there is movement of magma underneath its snowcapped, nearly symmetrical cone about 65 miles from Tokyo." In this project, you will explore how functions and relations are related to locating, measuring, and classifying earthquakes.

Log on to www.algebra2.com/webquest. Begin your WebQuest by reading the Task.

Then continue working on your WebQuest as you study Unit 3.

Lesson	8-3	9-5	10-1
Page	429	502	529

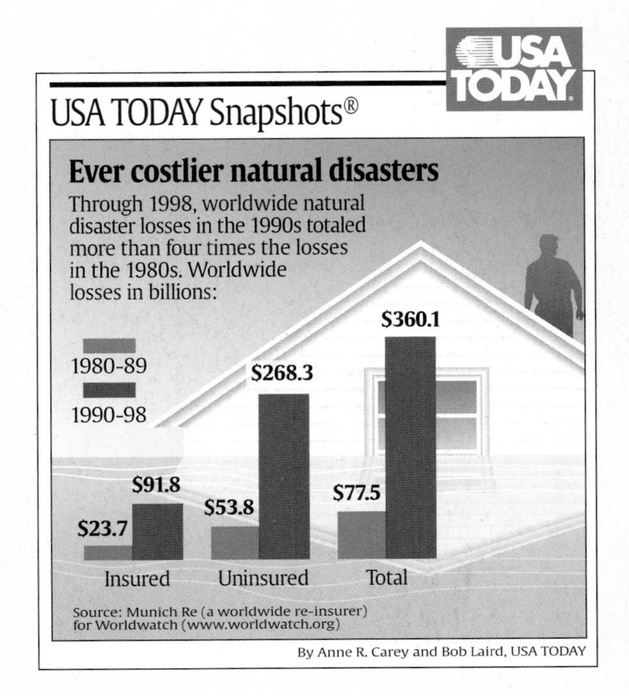

Chapter

8 Conic Sections

What You'll Learn

- **Lesson 8-1** Use the Midpoint and Distance Formulas.
- **Lessons 8-2 through 8-5** Write and graph equations of parabolas, circles, ellipses, and hyperbolas.
- **Lesson 8-6** Identify conic sections.
- **Lesson 8-7** Solve systems of quadratic equations and inequalities.

Key Vocabulary

- parabola (p. 419)
- conic section (p. 419)
- circle (p. 426)
- ellipse (p. 433)
- hyperbola (p. 441)

Why It's Important

Many planets, comets, and satellites have orbits in curves called *conic sections*. These curves include parabolas, circles, ellipses, and hyperbolas. The Moon's orbit is almost a perfect circle. *You will learn more about the orbits in Lessons 8-2 through 8-7.*

Getting Started

Prerequisite Skills To be successful in this chapter, you'll need to master these skills and be able to apply them in problem-solving situations. Review these skills before beginning Chapter 8.

For Lessons 8-2 through 8-6 — Completing the Square

Solve each equation by completing the square. *(For review, see Lesson 6-4.)*

1. $x^2 + 10x + 24 = 0$ **2.** $x^2 - 2x + 2 = 0$ **3.** $2x^2 + 5x - 12 = 0$

For Lessons 8-2 through 8-6 — Translation Matrices

A translation is given for each figure.

a. Write the vertex matrix for the given figure.

b. Write the translation matrix.

c. Find the coordinates in matrix form of the vertices of the translated figure.

(For review, see Lesson 4-4.)

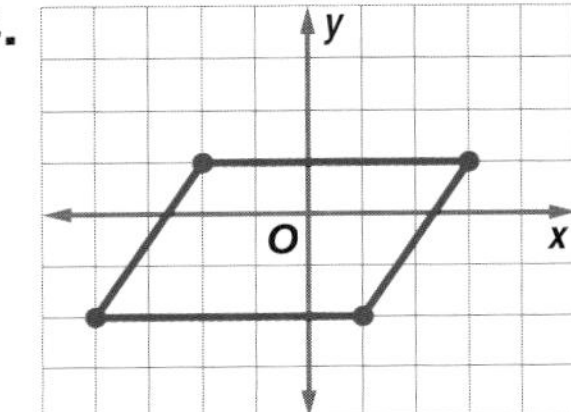

translated 4 units left and 2 units up

5.

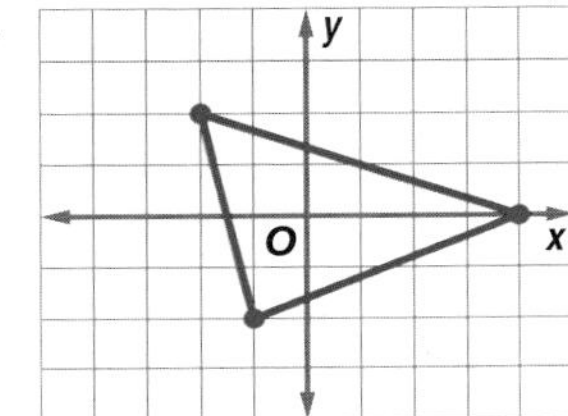

translated 5 units right and 3 units down

For Lesson 8-7 — Graph Linear Inequalities

Graph each inequality. *(For review, see Lesson 2-7.)*

6. $y < x + 2$ **7.** $x + y \leq 3$ **8.** $2x - 3y > 6$

Conic Sections Make this Foldable to help you organize your notes. Begin with four sheets of grid paper and one piece of construction paper.

Step 1 Fold and Staple

Stack sheets of grid paper with edges $\frac{1}{2}$ inch apart. Fold top edges back. Staple to construction paper at top.

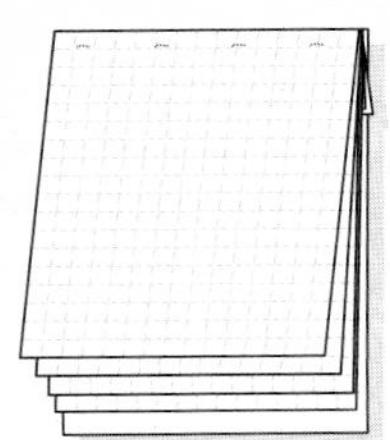

Step 2 Cut and Label

Cut grid paper in half lengthwise. Label tabs as shown.

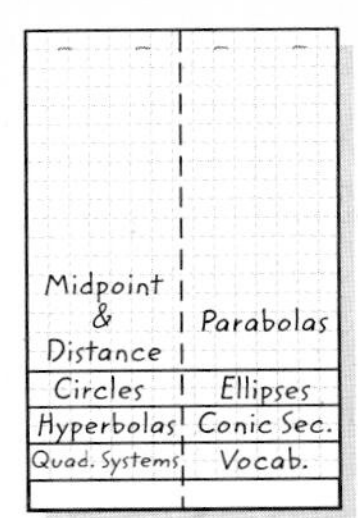

Reading and Writing As you read and study the chapter, use each tab to write notes, formulas, and examples for each conic section.

8-1 Midpoint and Distance Formulas

What You'll Learn

- Find the midpoint of a segment on the coordinate plane.
- Find the distance between two points on the coordinate plane.

How are the Midpoint and Distance Formulas used in emergency medicine?

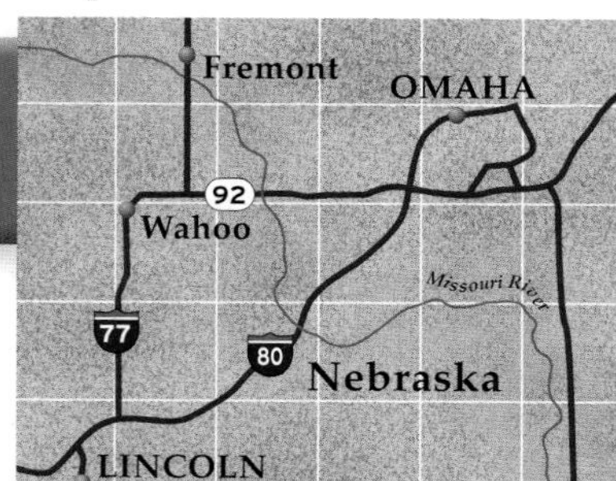

A square grid is superimposed on a map of eastern Nebraska where emergency medical assistance by helicopter is available from both Lincoln and Omaha. Each side of a square represents 10 miles. You can use the formulas in this lesson to determine whether the site of an emergency is closer to Lincoln or to Omaha.

THE MIDPOINT FORMULA Recall that point M is the midpoint of segment PQ if M is between P and Q and $PM = MQ$. There is a formula for the coordinates of the midpoint of a segment in terms of the coordinates of the endpoints. *You will show that this formula is correct in Exercise 41.*

Key Concept — Midpoint Formula

- **Words** If a line segment has endpoints at (x_1, y_1) and (x_2, y_2), then the midpoint of the segment has coordinates $\left(\frac{x_1 + x_2}{2}, \frac{y_1 + y_2}{2}\right)$.
- **Symbols** midpoint $= \left(\frac{x_1 + x_2}{2}, \frac{y_1 + y_2}{2}\right)$
- **Model**

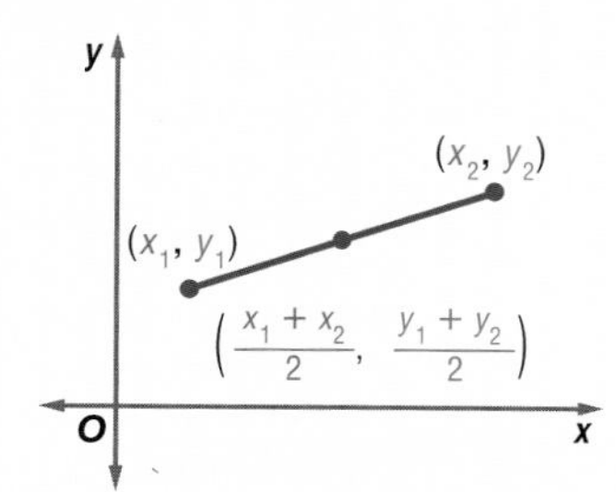

Study Tip

Midpoints
The coordinates of the midpoint are the means of the coordinates of the endpoints.

Example 1 Find a Midpoint

LANDSCAPING A landscape design includes two square flower beds and a sprinkler halfway between them. Find the coordinates of the sprinkler if the origin is at the lower left corner of the grid.

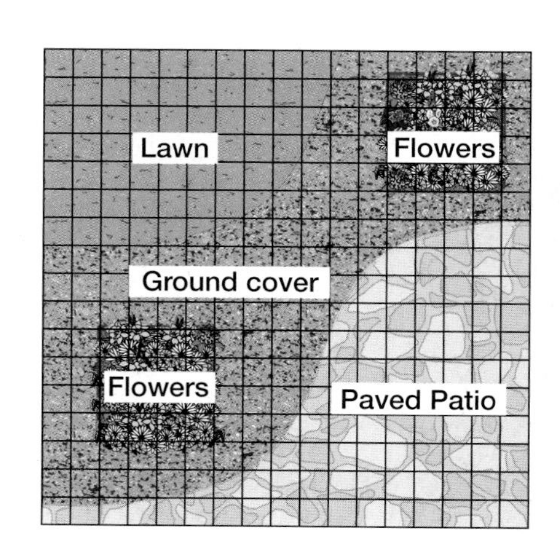

The centers of the flower beds are at (4, 5) and (14, 13). The sprinkler will be at the midpoint of the segment joining these points.

$$\left(\frac{x_1 + x_2}{2}, \frac{y_1 + y_2}{2}\right) = \left(\frac{4 + 14}{2}, \frac{5 + 13}{2}\right)$$

$$= \left(\frac{18}{2}, \frac{18}{2}\right) \text{ or } (9, 9)$$

The sprinkler will have coordinates (9, 9).

THE DISTANCE FORMULA Recall that the distance between two points on a number line whose coordinates are a and b is $|a - b|$ or $|b - a|$. You can use this fact and the Pythagorean Theorem to derive a formula for the distance between two points on a coordinate plane.

Suppose (x_1, y_1) and (x_2, y_2) name two points. Draw a right triangle with vertices at these points and the point (x_1, y_2). The lengths of the legs of the right triangle are $|x_2 - x_1|$ and $|y_2 - y_1|$. Let d represent the distance between (x_1, y_1) and (x_2, y_2). Now use the Pythagorean Theorem.

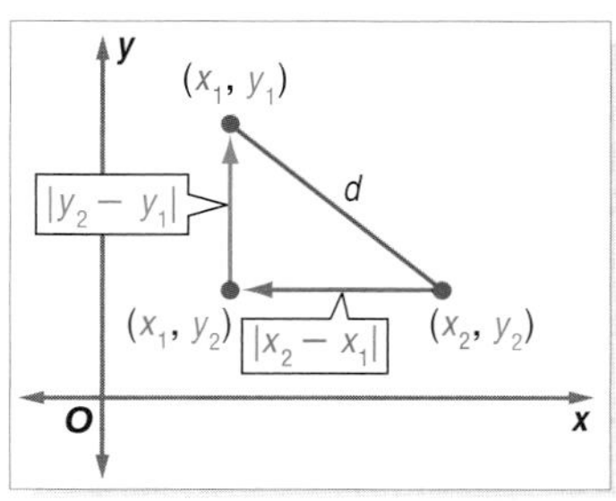

$c^2 = a^2 + b^2$	Pythagorean Theorem
$d^2 = \|x_2 - x_1\|^2 + \|y_2 - y_1\|^2$	Replace c with d, a with $\|x_2 - x_1\|$, and b with $\|y_2 - y_1\|$.
$d^2 = (x_2 - x_1)^2 + (y_2 - y_1)^2$	$\|x_2 - x_1\|^2 = (x_2 - x_1)^2$; $\|y_2 - y_1\|^2 = (y_2 - y_1)^2$
$d = \sqrt{(x_2 - x_1)^2 + (y_2 - y_1)^2}$	Find the nonnegative square root of each side.

Study Tip

Distance
In mathematics, distances are always positive.

Key Concept — Distance Formula

- **Words** The distance between two points with coordinates (x_1, y_1) and (x_2, y_2) is given by $d = \sqrt{(x_2 - x_1)^2 + (y_2 - y_1)^2}$.
- **Model**

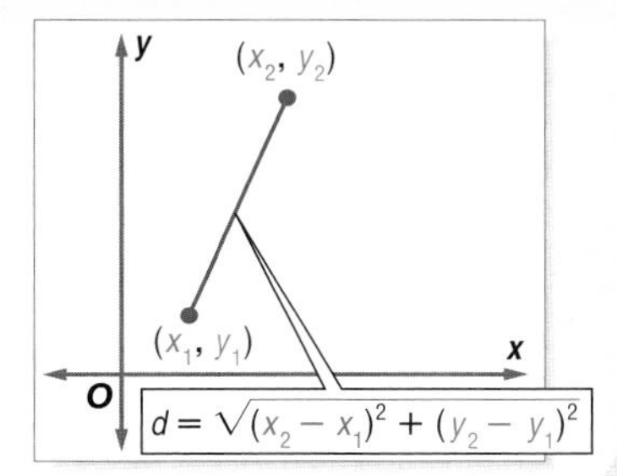

Example 2 Find the Distance Between Two Points

What is the distance between $A(-3, 6)$ and $B(4, -4)$?

$d = \sqrt{(x_2 - x_1)^2 + (y_2 - y_1)^2}$	Distance Formula
$= \sqrt{[4 - (-3)]^2 + (-4 - 6)^2}$	Let $(x_1, y_1) = (-3, 6)$ and $(x_2, y_2) = (4, -4)$.
$= \sqrt{7^2 + (-10)^2}$	Subtract.
$= \sqrt{49 + 100}$ or $\sqrt{149}$	Simplify.

The distance between the points is $\sqrt{149}$ units.

Example 3 Find the Farthest Point

Multiple-Choice Test Item

Which point is farthest from $(-1, 3)$?

(A) $(2, 4)$ (B) $(-4, 1)$ (C) $(0, 5)$ (D) $(3, -2)$

Read the Test Item

The word *farthest* refers to the greatest distance.

(continued on the next page)

Solve the Test Item

Use the Distance Formula to find the distance from (−1, 3) to each point.

Distance to (2, 4)

$$d = \sqrt{[2-(-1)]^2 + (4-3)^2}$$
$$= \sqrt{3^2 + 1^2} \text{ or } \sqrt{10}$$

Distance to (−4, 1)

$$d = \sqrt{[-4-(-1)]^2 + (1-3)^2}$$
$$= \sqrt{(-3)^2 + (-2)^2} \text{ or } \sqrt{13}$$

Distance to (0, 5)

$$d = \sqrt{[0-(-1)]^2 + (5-3)^2}$$
$$= \sqrt{1^2 + 2^2} \text{ or } \sqrt{5}$$

Distance to (3, −2)

$$d = \sqrt{[3-(-1)]^2 + (-2-3)^2}$$
$$= \sqrt{4^2 + (-5)^2} \text{ or } \sqrt{41}$$

The greatest distance is $\sqrt{41}$ units. So, the farthest point from (−1, 3) is (3, −2). The answer is D.

Check for Understanding

Concept Check

1. **Explain** how you can determine in which quadrant the midpoint of the segment with endpoints at (−6, 8) and (4, 3) lies without actually calculating the coordinates.
2. **Identify** all of the points that are equidistant from the endpoints of a given segment.
3. **OPEN ENDED** Find two points that are $\sqrt{29}$ units apart.

Guided Practice

Find the midpoint of each line segment with endpoints at the given coordinates.

4. (−5, 6), (1, 7)
5. (8, 9), (−3, −4.5)

Find the distance between each pair of points with the given coordinates.

6. (2, −4), (10, −10)
7. (7, 8), (−4, 9)
8. (0.5, 1.4), (1.1, 2.9)

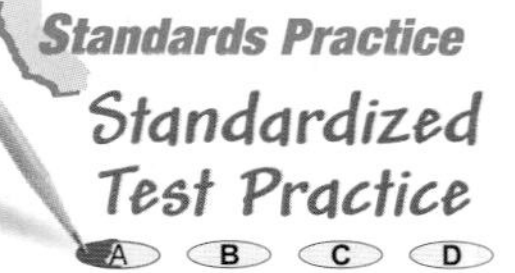

9. Which of the following points is closest to (2, −4)?

Ⓐ (3, 1) Ⓑ (−2, 0) Ⓒ (1, 5) Ⓓ (4, −2)

Practice and Apply

Homework Help	
For Exercises	See Examples
10–23	1
24–40	2, 3

Extra Practice
See page 845.

Find the midpoint of each line segment with endpoints at the given coordinates.

10. (8, 3), (16, 7)
11. (−5, 3), (−3, −7)
12. (6, −5), (−2, −7)
13. (5, 9), (12, 18)
14. (0.45, 7), (−0.3, −0.6)
15. (4.3, −2.1), (1.9, 7.5)
16. $\left(\frac{1}{2}, -\frac{2}{3}\right), \left(\frac{1}{3}, \frac{1}{4}\right)$
17. $\left(\frac{1}{3}, \frac{3}{4}\right), \left(-\frac{1}{4}, \frac{1}{2}\right)$

18. **GEOMETRY** Triangle *MNP* has vertices *M*(3, 5), *N*(−2, 8), and *P*(7, −4). Find the coordinates of the midpoint of each side.

19. **GEOMETRY** Circle *Q* has a diameter $\overline{AB}$. If *A* is at (−3, −5) and the center is at (2, 3), find the coordinates of *B*.

20. **REAL ESTATE** In John's town, the numbered streets and avenues form a grid. He belongs to a gym at the corner of 12th Street and 15th Avenue, and the deli where he works is at the corner of 4th Street and 5th Avenue. He wants to rent an apartment halfway between the two. In what area should he look?

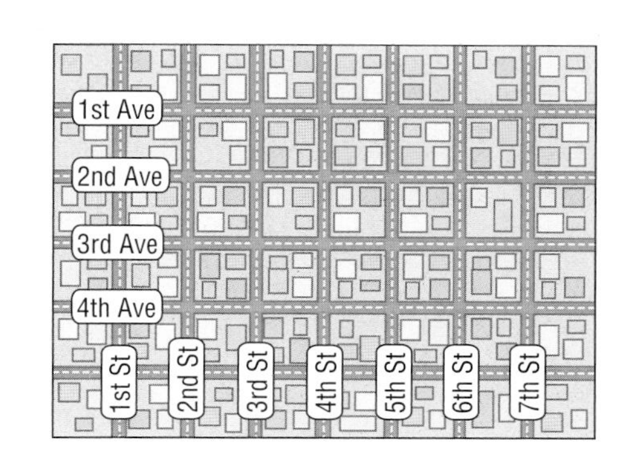

GEOGRAPHY For Exercises 21–23, use the following information.
The U.S. Geological Survey (USGS) has determined the official center of the continental United States.

21. Describe a method that might be used to approximate the geographical center of the continental United States.
22. **RESEARCH** Use the Internet or other reference to look up the USGS geographical center of the continental United States.
23. How does the location given by USGS compare to the result of your method?

Find the distance between each pair of points with the given coordinates.

24. $(-4, 9), (1, -3)$
25. $(1, -14), (-6, 10)$
26. $(-4, -10), (-3, -11)$
27. $(9, -2), (12, -14)$
28. $(0.23, 0.4), (0.68, -0.2)$
29. $(2.3, -1.2), (-4.5, 3.7)$
30. $\left(-3, -\frac{2}{11}\right), \left(5, \frac{9}{11}\right)$
31. $\left(0, \frac{1}{5}\right), \left(\frac{3}{5}, -\frac{3}{5}\right)$
32. $(2\sqrt{3}, -5), (-3\sqrt{3}, 9)$
33. $\left(\frac{2\sqrt{3}}{3}, \frac{\sqrt{5}}{4}\right), \left(-\frac{2\sqrt{3}}{3}, \frac{\sqrt{5}}{2}\right)$

34. **GEOMETRY** A circle has a radius with endpoints at $(2, 5)$ and $(-1, -4)$. Find the circumference and area of the circle.

35. **GEOMETRY** Find the perimeter and area of the triangle shown at the right.

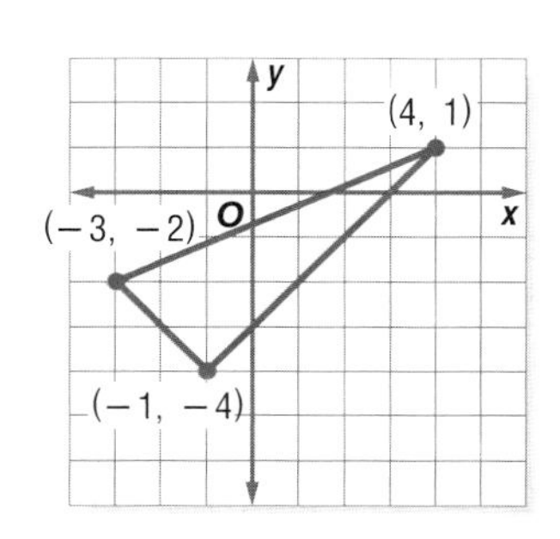

36. **GEOMETRY** Quadrilateral $RSTV$ has vertices $R(-4, 6)$, $S(4, 5)$, $T(6, 3)$, and $V(5, -8)$. Find the perimeter of the quadrilateral.

37. **GEOMETRY** Triangle CAT has vertices $C(4, 9)$, $A(8, -9)$, and $T(-6, 5)$. M is the midpoint of $\overline{TA}$. Find the length of median $\overline{CM}$. (*Hint*: A median connects a vertex of a triangle to the midpoint of the opposite side.)

TRAVEL For Exercises 38 and 39, use the figure at the right, where a grid is superimposed on a map of a portion of the state of Alabama.

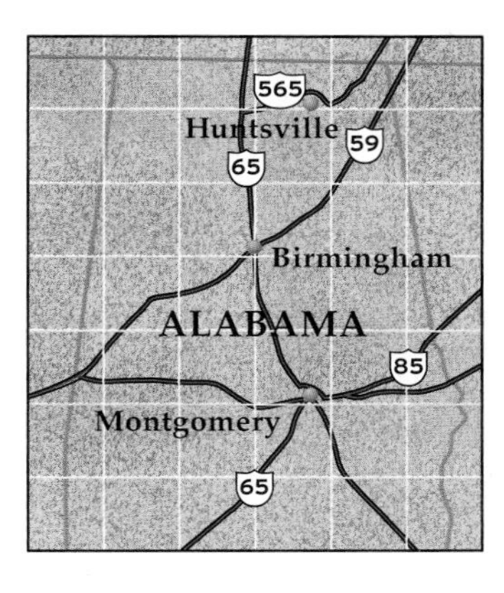

38. About how far is it from Birmingham to Montgomery if each unit on the grid represents 40 miles?
39. How long would it take a plane to fly from Huntsville to Montgomery if its average speed is 180 miles per hour?

40. WOODWORKING A stage crew is making the set for a children's play. They want to make some gingerbread shapes out of some leftover squares of wood with sides measuring 1 foot. They can make taller shapes by cutting them out of the wood diagonally. To the nearest inch, how tall is the gingerbread shape in the drawing at the right?

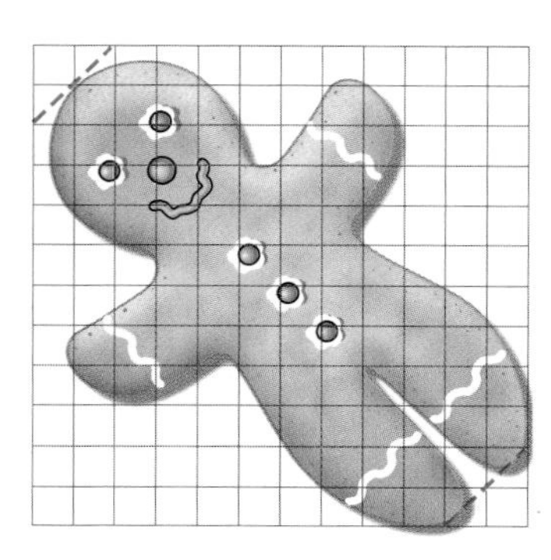

41. CRITICAL THINKING Verify the Midpoint Formula. (*Hint*: You must show that the formula gives the coordinates of a point on the line through the given endpoints and that the point is equidistant from the endpoints.)

42. WRITING IN MATH Answer the question that was posed at the beginning of the lesson.

How are the Midpoint and Distance Formulas used in emergency medicine?

Include the following in your answer:

- a few sentences explaining how to use the Distance Formula to approximate the distance between two cities on a map, and
- which city, Lincoln or Omaha, an emergency medical helicopter should be dispatched from to pick up a patient in Fremont.

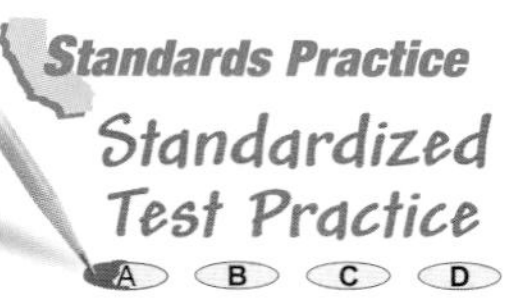

43. What is the distance between the points $A(4, -2)$ and $B(-4, -8)$?

Ⓐ 6 Ⓑ 8 Ⓒ 10 Ⓓ 14

44. Point $D(5, -1)$ is the midpoint of segment $\overline{CE}$. If point C has coordinates (3, 2), then what are the coordinates of point E?

Ⓐ (8, 1) Ⓑ (7, −4) Ⓒ (2, −3) Ⓓ $\left(4, \frac{1}{2}\right)$

Extending the Lesson

For Exercises 45 and 46, use the following information.
You can use midpoints and slope to describe some transformations. Suppose point A' is the image when point A is reflected over the line with equation $y = x$.

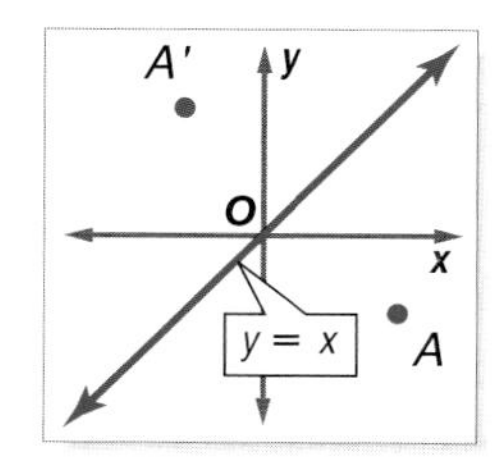

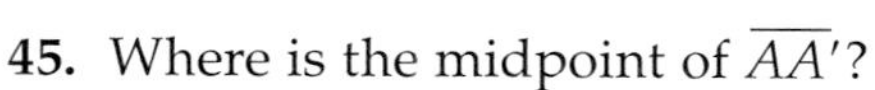

45. Where is the midpoint of $\overline{AA'}$?

46. What is the slope of $\overleftrightarrow{AA'}$? Explain.

Maintain Your Skills

Mixed Review

Graph each function. State the domain and range. *(Lesson 7-9)*

47. $y = \sqrt{x - 2}$ **48.** $y = \sqrt{x} - 1$ **49.** $y = 2\sqrt{x} + 1$

50. Determine whether the functions $f(x) = x - 2$ and $g(x) = 2x$ are inverse functions. *(Lesson 7-8)*

Simplify. *(Lesson 5-9)*

51. $(2 + 4i) + (-3 + 9i)$ **52.** $(4 - i) - (-2 + i)$ **53.** $(1 - 2i)(2 + i)$

Getting Ready for the Next Lesson

PREREQUISITE SKILL Write each equation in the form $y = a(x - h)^2 + k$.
*(To review **completing the square**, see Lesson 6-4.)*

54. $y = x^2 + 6x + 9$ **55.** $y = x^2 - 4x + 1$ **56.** $y = 2x^2 + 20x + 50$

57. $y = 3x^2 - 6x + 5$ **58.** $y = -x^2 - 4x + 6$ **59.** $y = -3x^2 - 18x - 10$

Algebra Activity

A Follow-Up of Lesson 8-1

Midpoint and Distance Formulas in Three Dimensions

You can derive a formula for distance in three-dimensional space. It may seem that the formula would involve a cube root, but it actually involves a square root, similar to the formula in two dimensions.

Suppose (x_1, y_1, z_1) and (x_2, y_2, z_2) name two points in space. Draw the rectangular box that has opposite vertices at these points. The dimensions of the box are $|x_2 - x_1|$, $|y_2 - y_1|$, and $|z_2 - z_1|$. Let a be the length of a diagonal of the bottom of the box. By the Pythagorean Theorem, $a^2 = |x_2 - x_1|^2 + |y_2 - y_1|^2$.

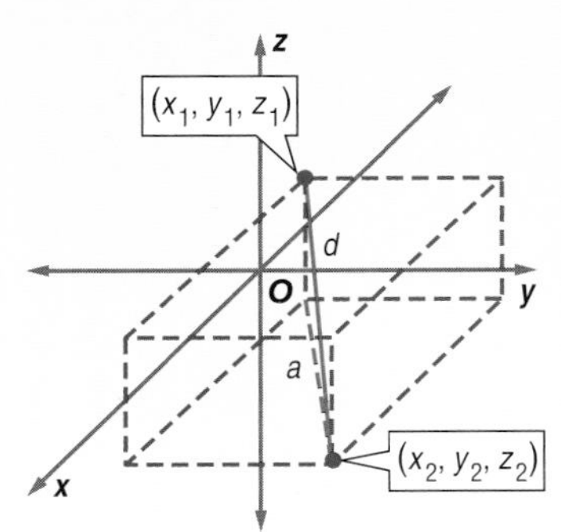

To find the distance d between (x_1, y_1, z_1) and (x_2, y_2, z_2), apply the Pythagorean Theorem to the right triangle whose legs are a diagonal of the bottom of the box and a vertical edge of the box.

$d^2 = a^2 + \lvert z_2 - z_1\rvert^2$	Pythagorean Theorem
$d^2 = \lvert x_2 - x_1\rvert^2 + \lvert y_2 - y_1\rvert^2 + \lvert z_2 - z_1\rvert^2$	$a^2 = \lvert x_2 - x_1\rvert^2 + \lvert y_2 - y_1\rvert^2$
$d^2 = (x_2 - x_1)^2 + (y_2 - y_1)^2 + (z_2 - z_1)^2$	$\lvert x_2 - x_1\rvert^2 = (x_2 - x_1)^2$, and so on
$d = \sqrt{(x_2 - x_1)^2 + (y_2 - y_1)^2 + (z_2 - z_1)^2}$	Take the square root of each side.

The distance d between the points with coordinates (x_1, y_1, z_1) and (x_2, y_2, z_2) is given by the formula $d = \sqrt{(x_2 - x_1)^2 + (y_2 - y_1)^2 + (z_2 - z_1)^2}$.

Example 1

Find the distance between (2, 0, −3) and (4, 2, 9).

$d = \sqrt{(x_2 - x_1)^2 + (y_2 - y_1)^2 + (z_2 - z_1)^2}$	Distance Formula
$= \sqrt{(4 - 2)^2 + (2 - 0)^2 + [9 - (-3)]^2}$	$(x_1, y_1, z_1) = (2, 0, -3)$ $(x_2, y_2, z_2) = (4, 2, 9)$
$= \sqrt{2^2 + 2^2 + 12^2}$	
$= \sqrt{152}$ or $2\sqrt{38}$	

The distance is $2\sqrt{38}$ or about 12.33 units.

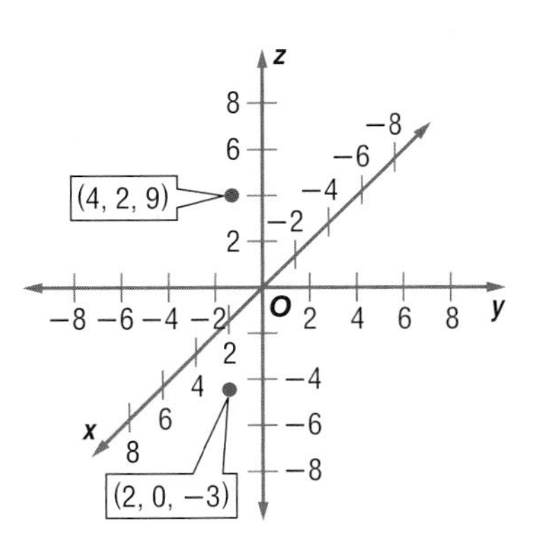

In three dimensions, the midpoint of the segment with coordinates (x_1, y_1, z_1) and (x_2, y_2, z_2) has coordinates $\left(\frac{x_1 + x_2}{2}, \frac{y_1 + y_2}{2}, \frac{z_1 + z_2}{2}\right)$. Notice how similar this is to the Midpoint Formula in two dimensions.

(continued on the next page)

Example 2

Find the coordinates of the midpoint of the segment with endpoints (6, −5, 1) and (−2, 4, 0).

$$\left(\frac{x_1 + x_2}{2}, \frac{y_1 + y_2}{2}, \frac{z_1 + z_2}{2}\right) = \left(\frac{6 + (-2)}{2}, \frac{-5 + 4}{2}, \frac{1 + 0}{2}\right) \quad (x_1, y_1, z_1) = (6, -5, 1),\ (x_2, y_2, z_2) = (-2, 4, 0)$$

$$= \left(\frac{4}{2}, \frac{-1}{2}, \frac{1}{2}\right) \quad \text{Add.}$$

$$= \left(2, -\frac{1}{2}, \frac{1}{2}\right) \quad \text{Simplify.}$$

The midpoint has coordinates $\left(2, -\frac{1}{2}, \frac{1}{2}\right)$.

Exercises

Find the distance between each pair of points with the given coordinates.

1. (2, 4, 5), (1, 2, 3)

2. (−1, 6, 2), (4, −3, 0)

3. (−2, 1, 7), (−2, 6, −3)

4. (0, 7, −1), (−4, 1, 3)

Find the midpoint of each line segment with endpoints at the given coordinates.

5. (2, 6, −1), (−4, 8, 5)

6. (4, −3, 2), (−2, 7, 6)

7. (1, 3, 7), (−4, 2, −1)

8. (2.3, −1.7, 0.6), (−2.7, 3.1, 1.8)

9. The coordinates of one endpoint of a segment are (4, −2, 3), and the coordinates of the midpoint are (3, 2, 5). Find the coordinates of the other endpoint.

10. Two of the opposite vertices of a rectangular solid are at (4, 1, −1) and (2, 3, 5). Find the coordinates of the other six vertices.

11. Determine whether a triangle with vertices at (2, −4, 2), (3, 1, 5), and (6, −3, −1) is a right triangle. Explain.

The vertices of a rectangular solid are at (−2, 3, 2), (3, 3, 2), (3, 1, 2), (−2, 1, 2), (−2, 3, 6), (3, 3, 6), (3, 1, 6), and (−2, 1, 6).

12. Find the volume of the solid.

13. Find the length of a diagonal of the solid.

14. Show that the point with coordinates $\left(\frac{x_1 + x_2}{2}, \frac{y_1 + y_2}{2}, \frac{z_1 + z_2}{2}\right)$ is equidistant from the points with coordinates (x_1, y_1, z_1) and (x_2, y_2, z_2).

15. Find the value of c so that the point with coordinates $(2, 3, c)$ is $3\sqrt{6}$ units from the point with coordinates $(-1, 0, 5)$.

The endpoints of a diameter of a sphere are at (2, −3, 2) and (−1, 1, −4).

16. Find the length of a radius of the sphere.

17. Find the coordinates of the center of the sphere.

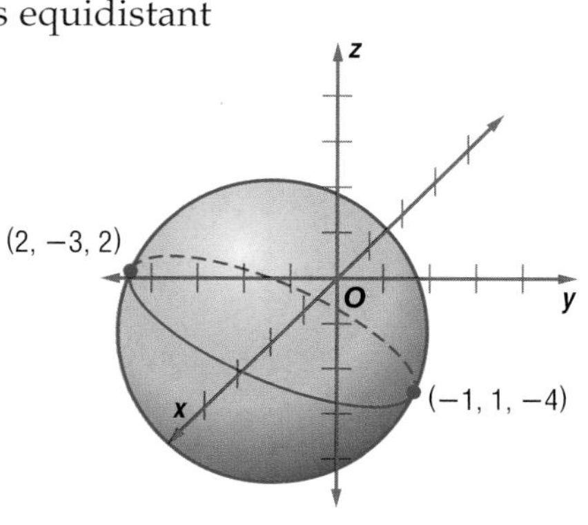

8-2 Parabolas

California Standards Standard 16.0 Students demonstrate and explain how the geometry of the graph of a conic section depends on the coefficients of the quadratic equation representing it.

What You'll Learn

- Write equations of parabolas in standard form.
- Graph parabolas.

How are parabolas used in manufacturing?

A mirror or other reflective object in the shape of a parabola has the property that parallel incoming rays are all reflected to the same point. Or, if that point is the source of rays, the rays become parallel when they are reflected.

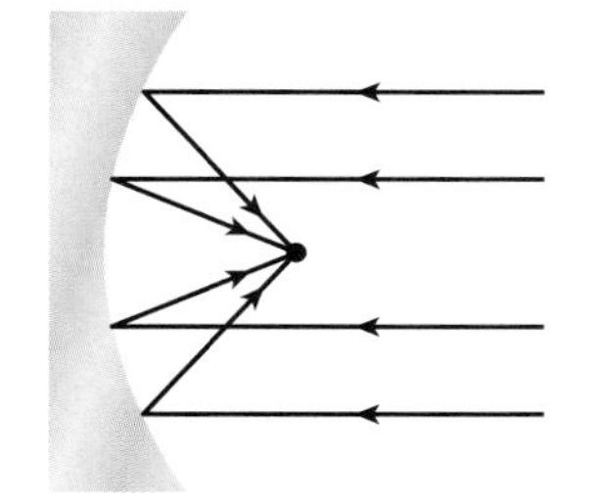

Vocabulary

- parabola
- conic section
- focus
- directrix
- latus rectum

EQUATIONS OF PARABOLAS In Chapter 6, you learned that the graph of an equation of the form $y = ax^2 + bx + c$ is a **parabola**. A parabola can also be obtained by slicing a double cone on a slant as shown below on the left. Any figure that can be obtained by slicing a double cone is called a **conic section**. Other conic sections are also shown below.

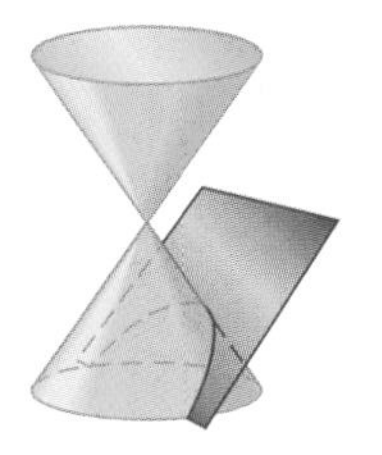

parabola

circle

ellipse

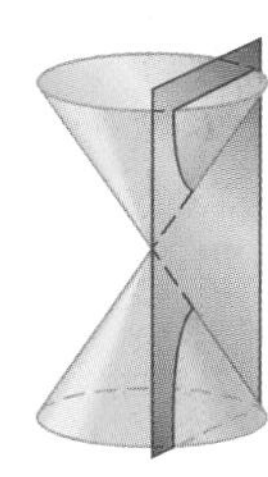

hyperbola

Study Tip

Focus of a Parabola
The focus is the special point referred to at the beginning of the lesson.

A parabola can also be defined as the set of all points in a plane that are the same distance from a given point called the **focus** and a given line called the **directrix**. The parabola at the right has its focus at (2, 3), and the equation of its directrix is $y = -1$. You can use the Distance Formula to find an equation of this parabola.

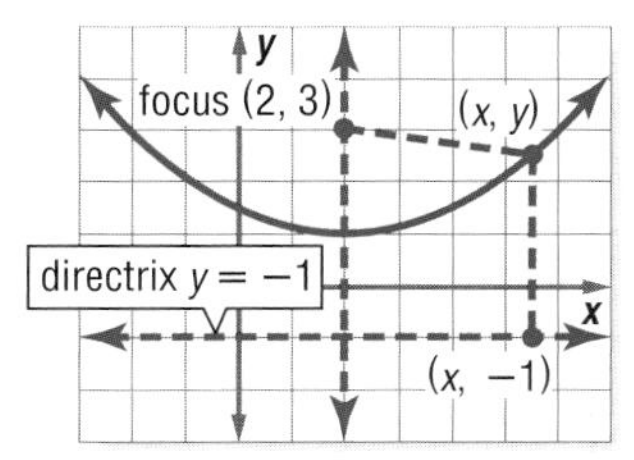

Let (x, y) be any point on this parabola. The distance from this point to the focus must be the same as the distance from this point to the directrix. The distance from a point to a line is measured along the perpendicular from the point to the line.

distance from (x, y) to $(2, 3)$ = distance from (x, y) to $(x, -1)$

$$\sqrt{(x-2)^2 + (y-3)^2} = \sqrt{(x-x)^2 + [y-(-1)]^2}$$

$$(x-2)^2 + (y-3)^2 = 0^2 + (y+1)^2$$ Square each side.

$$(x-2)^2 + y^2 - 6y + 9 = y^2 + 2y + 1$$ Square $y - 3$ and $y + 1$.

$$(x-2)^2 + 8 = 8y$$ Isolate the y-terms.

$$\frac{1}{8}(x-2)^2 + 1 = y$$ Divide each side by 8.

An equation of the parabola with focus at (2, 3) and directrix with equation $y = -1$ is $y = \frac{1}{8}(x - 2)^2 + 1$. The equation of the *axis of symmetry* for this parabola is $x = 2$. The axis of symmetry intersects the parabola at a point called the *vertex*. The vertex is the point where the graph turns. The vertex of this parabola is at (2, 1). Since $\frac{1}{8}$ is positive, the parabola opens upward.

Any equation of the form $y = ax^2 + bx + c$ can be written in standard form.

Key Concept — Equation of a Parabola

The standard form of the equation of a parabola with vertex (h, k) and axis of symmetry $x = h$ is $y = a(x - h)^2 + k$.

- If $a > 0$, k is the minimum value of the related function and the parabola opens upward.
- If $a < 0$, k is the maximum value of the related function and the parabola opens downward.

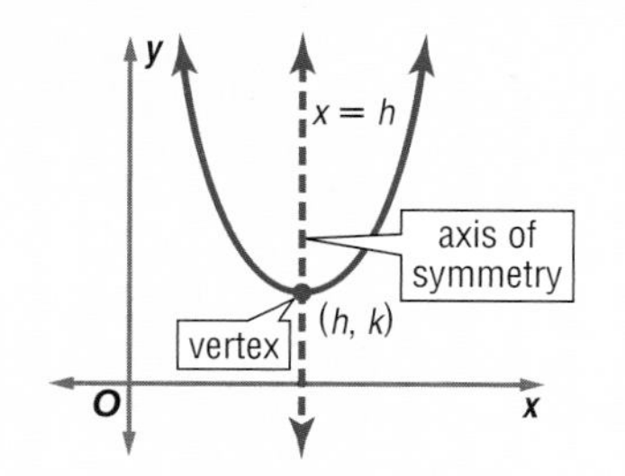

Example 1 Analyze the Equation of a Parabola

Write $y = 3x^2 + 24x + 50$ in standard form. Identify the vertex, axis of symmetry, and direction of opening of the parabola.

$y = 3x^2 + 24x + 50$	Original equation
$y = 3(x^2 + 8x) + 50$	Factor 3 from the x-terms.
$y = 3(x^2 + 8x + \blacksquare) + 50 - 3(\blacksquare)$	Complete the square on the right side.
$y = 3(x^2 + 8x + 16) + 50 - 3(16)$	The 16 added when you complete the square is multiplied by 3.
$y = 3(x + 4)^2 + 2$	
$y = 3[x - (-4)]^2 + 2$	$(h, k) = (-4, 2)$

The vertex of this parabola is located at $(-4, 2)$, and the equation of the axis of symmetry is $x = -4$. The parabola opens upward.

Study Tip

Look Back
To review **completing the square**, see Lesson 6-4.

GRAPH PARABOLAS You can use symmetry and translations to graph parabolas. The equation $y = a(x - h)^2 + k$ can be obtained from $y = ax^2$ by replacing x with $x - h$ and y with $y - k$. Therefore, the graph of $y = a(x - h)^2 + k$ is the graph of $y = ax^2$ translated h units to the right and k units up.

Example 2 Graph Parabolas

Graph each equation.

a. $y = -2x^2$

For this equation, $h = 0$ and $k = 0$. The vertex is at the origin. Since the equation of the axis of symmetry is $x = 0$, substitute some small positive integers for x and find the corresponding y-values.

x	y
1	−2
2	−8
3	−18

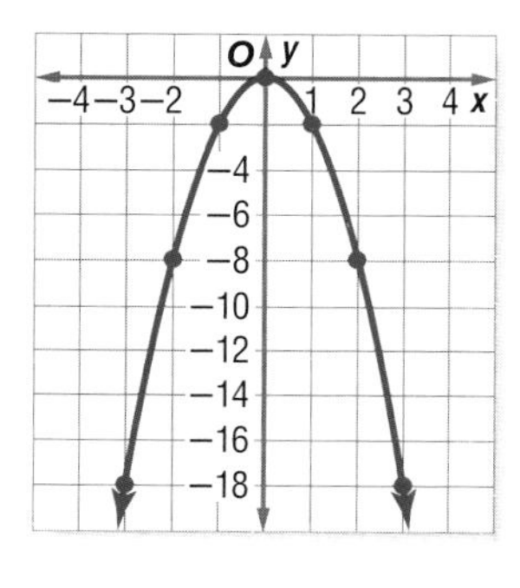

Since the graph is symmetric about the y-axis, the points at $(-1, -2)$, $(-2, -8)$, and $(-3, -18)$ are also on the parabola. Use all of these points to draw the graph.

Notice that each side of the graph is the reflection of the other side about the y-axis.

b. $y = -2(x - 2)^2 + 3$

The equation is of the form $y = a(x - h)^2 + k$, where $h = 2$ and $k = 3$. The graph of this equation is the graph of $y = -2x^2$ in part **a** translated 2 units to the right and up 3 units. The vertex is now at (2, 3).

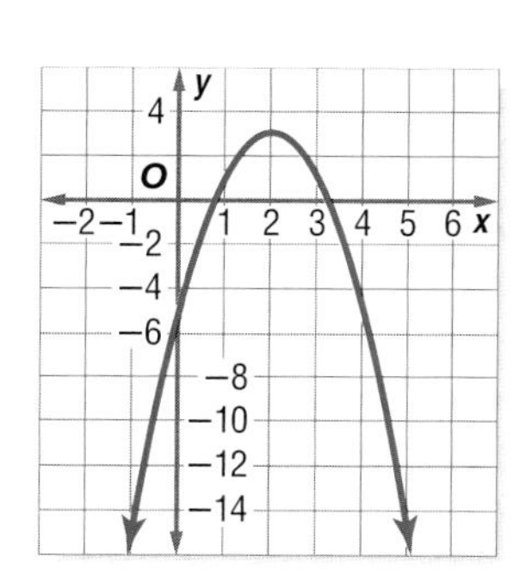

You can use paper folding to investigate the characteristics of a parabola.

Algebra Activity

Parabolas

Model

Step 1 Start with a sheet of wax paper that is about 15 inches long and 12 inches wide. Make a line that is perpendicular to the sides of the sheet by folding the sheet near one end. Open up the paper again. This line is the directrix. Mark a point about midway between the sides of the sheet so that the distance from the directrix is about 1 inch. This point is the focus.

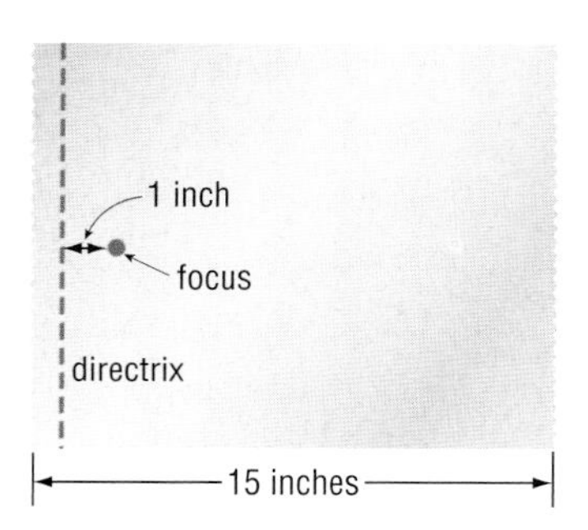

Put the focus on top of any point on the directrix and crease the paper. Make about 20 more creases by placing the focus on top of other points on the directrix. The lines form the outline of a parabola.

Step 2 Start with a new sheet of wax paper. Form another outline of a parabola with a focus that is about 3 inches from the directrix.

Step 3 On a new sheet of wax paper, form a third outline of a parabola with a focus that is about 5 inches from the directrix.

Analyze

Compare the shapes of the three parabolas. How does the distance between the focus and the directrix affect the shape of a parabola?

The shape of a parabola and the distance between the focus and directrix depend on the value of a in the equation. The line segment through the focus of a parabola and perpendicular to the axis of symmetry is called the **latus rectum**. The endpoints of the latus rectum lie on the parabola.

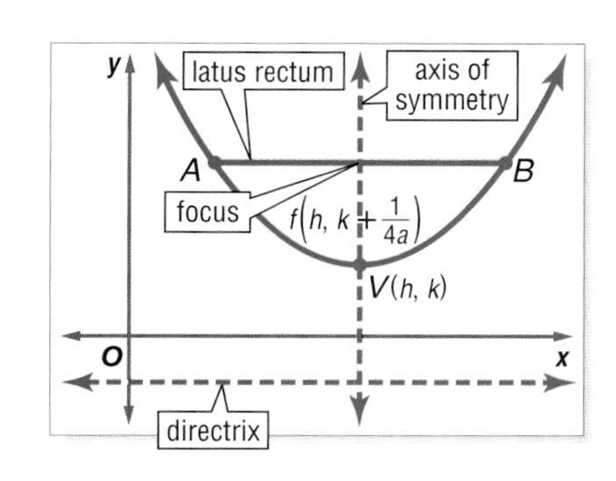

In the figure at the right, the latus rectum is $\overline{AB}$. The length of the latus rectum of the parabola with equation $y = a(x - h)^2 + k$ is $\left|\frac{1}{a}\right|$ units. The endpoints of the latus rectum are $\left|\frac{1}{2a}\right|$ units from the focus.

Equations of parabolas with vertical axes of symmetry are of the form $y = a(x - h)^2 + k$ and are functions. Equations of parabolas with horizontal axes of symmetry are of the form $x = a(y - k)^2 + h$ and are not functions.

Concept Summary *Information About Parabolas*

Form of Equation	$y = a(x - h)^2 + k$	$x = a(y - k)^2 + h$
Vertex	(h, k)	(h, k)
Axis of Symmetry	$x = h$	$y = k$
Focus	$\left(h, k + \frac{1}{4a}\right)$	$\left(h + \frac{1}{4a}, k\right)$
Directrix	$y = k - \frac{1}{4a}$	$x = h - \frac{1}{4a}$
Direction of Opening	upward if $a > 0$, downward if $a < 0$	right if $a > 0$, left if $a < 0$
Length of Latus Rectum	$\left\lvert\frac{1}{a}\right\rvert$ units	$\left\lvert\frac{1}{a}\right\rvert$ units

Example 3 Graph an Equation Not in Standard Form

Graph $4x - y^2 = 2y + 13$.

First, write the equation in the form $x = a(y - k)^2 + h$.

$4x - y^2 = 2y + 13$ — There is a y^2 term, so isolate the y and y^2 terms.

$4x = y^2 + 2y + 13$

$4x = (y^2 + 2y + ■) + 13 - ■$ — Complete the square.

$4x = (y^2 + 2y + 1) + 13 - 1$ — Add and subtract 1, since $\left(\frac{2}{2}\right)^2 = 1$.

$4x = (y + 1)^2 + 12$ — Write $y^2 + 2y + 1$ as a square.

$x = \frac{1}{4}(y + 1)^2 + 3$ — $(h, k) = (3, -1)$

Then use the following information to draw the graph.

vertex: $(3, -1)$

axis of symmetry: $y = -1$

focus: $\left(3 + \frac{1}{4\left(\frac{1}{4}\right)}, -1\right)$ or $(4, -1)$

directrix: $x = 3 - \frac{1}{4\left(\frac{1}{4}\right)}$ or 2

direction of opening: right, since $a > 0$

length of latus rectum: $\left\lvert\frac{1}{\frac{1}{4}}\right\rvert$ or 4 units

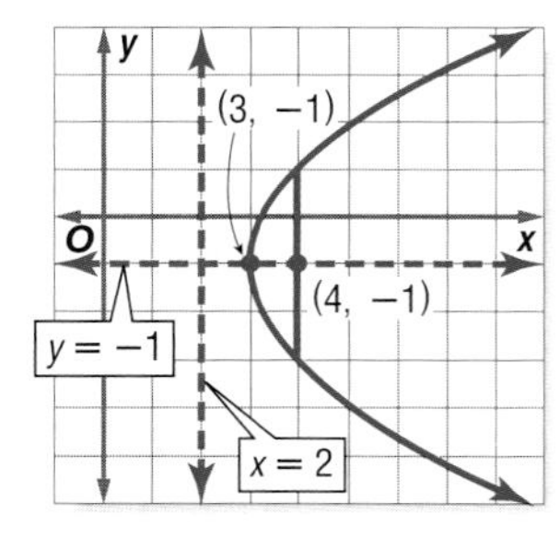

Remember that you can plot as many points as necessary to help you draw an accurate graph.

More About. . .

Satellite TV

The important characteristics of a satellite dish are the diameter D, depth d, and the ratio $\frac{f}{D}$, where f is the distance between the focus and the vertex. A typical dish has the values $D = 60$ cm, $d = 6.25$ cm, and $\frac{f}{D} = 0.6$.

Source: www.2000networks.com

Example 4 Write and Graph an Equation for a Parabola

SATELLITE TV Satellite dishes have parabolic cross sections.

a. Use the information at the left to write an equation that models a cross section of a satellite dish. Assume that the focus is at the origin and the parabola opens to the right.

First, solve for f. Since $\frac{f}{D} = 0.6$, and $D = 60$, $f = 0.6(60)$ or 36.

The focus is at $(0, 0)$, and the parabola opens to the right. So the vertex must be at $(-36, 0)$. Thus, $h = -36$ and $k = 0$. Use the x-coordinate of the focus to find a.

$$-36 + \frac{1}{4a} = 0 \quad h = -36\text{; The } x\text{-coordinate of the focus is 0.}$$

$$\frac{1}{4a} = 36 \quad \text{Add 36 to each side.}$$

$$1 = 144a \quad \text{Multiply each side by } 4a.$$

$$\frac{1}{144} = a \quad \text{Divide each side by 144.}$$

An equation of the parabola is $x = \frac{1}{144}y^2 - 36$.

b. Graph the equation.

The length of the latus rectum is $\left|\frac{1}{\frac{1}{144}}\right|$ or 144 units, so the graph must pass through (0, 72) and (0, −72). According to the diameter and depth of the dish, the graph must pass through (−29.75, 30) and (−29.75, −30). Use these points and the information from part **a** to draw the graph.

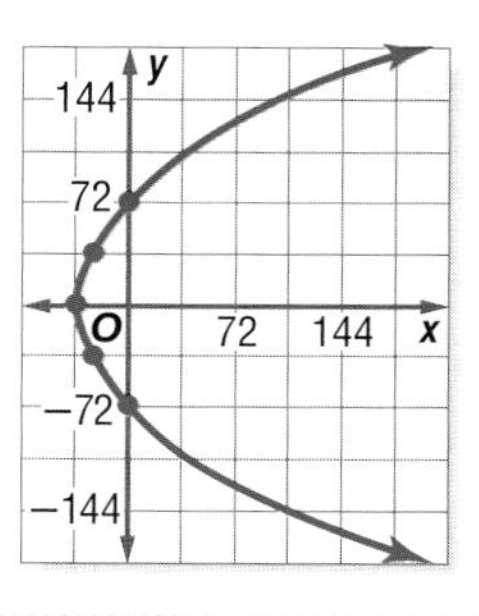

Check for Understanding

Concept Check

1. **Identify** the vertex, focus, axis of symmetry, and directrix of the graph of $y = 4(x - 3)^2 - 7$.
2. **OPEN ENDED** Write an equation for a parabola that opens to the left.
3. **FIND THE ERROR** Katie is finding the standard form of the equation $y = x^2 + 6x + 4$. What mistake did she make in her work?

$$y = x^2 + 6x + 4$$
$$y = x^2 + 6x + 9 + 4$$
$$y = (x + 3)^2 + 4$$

Guided Practice

4. Write $y = 2x^2 - 12x + 6$ in standard form.

Identify the coordinates of the vertex and focus, the equations of the axis of symmetry and directrix, and the direction of opening of the parabola with the given equation. Then find the length of the latus rectum and graph the parabola.

5. $y = (x - 3)^2 - 4$
6. $y = 2(x + 7)^2 + 3$
7. $y = -3x^2 - 8x - 6$
8. $x = \frac{2}{3}y^2 - 6y + 12$

Write an equation for each parabola described below. Then draw the graph.

9. focus (3, 8), directrix $y = 4$
10. vertex (5, −1), focus (3, −1)

Application

11. **COMMUNICATION** A microphone is placed at the focus of a parabolic reflector to collect sound for the television broadcast of a World Cup soccer game. Write an equation for the cross section, assuming that the focus is at the origin and the parabola opens to the right.

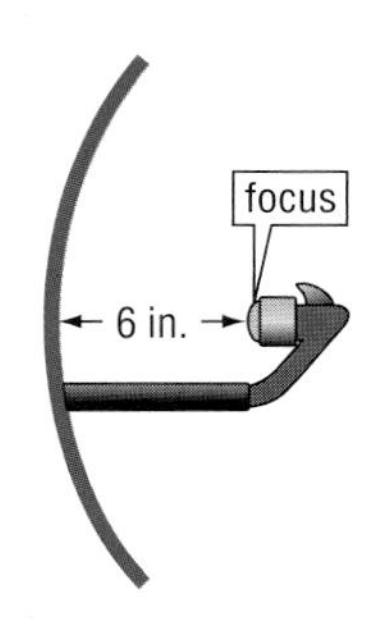

Practice and Apply

Homework Help

For Exercises	See Examples
12–15, 35	1
16–34	1–3
36–41	2–4
42–45	4a

Extra Practice
See page 845.

Write each equation in standard form.

12. $y = x^2 - 6x + 11$
13. $x = y^2 + 14y + 20$
14. $y = \frac{1}{2}x^2 + 12x - 8$
15. $x = 3y^2 + 5y - 9$

Identify the coordinates of the vertex and focus, the equations of the axis of symmetry and directrix, and the direction of opening of the parabola with the given equation. Then find the length of the latus rectum and graph the parabola.

16. $-6y = x^2$
17. $y^2 = 2x$
18. $3(y - 3) = (x + 6)^2$
19. $-2(y - 4) = (x - 1)^2$
20. $4(x - 2) = (y + 3)^2$
21. $(y - 8)^2 = -4(x - 4)$
22. $y = x^2 - 12x + 20$
23. $x = y^2 - 14y + 25$
24. $x = 5y^2 + 25y + 60$
25. $y = 3x^2 - 24x + 50$
26. $y = -2x^2 + 5x - 10$
27. $x = -4y^2 + 6y + 2$
28. $y = \frac{1}{2}x^2 - 3x + \frac{19}{2}$
29. $x = -\frac{1}{3}y^2 - 12y + 15$

For Exercises 30–34, use the equation $x = 3y^2 + 4y + 1$.

30. Draw the graph.
31. Find the x-intercept(s).
32. Find the y-intercept(s).
33. What is the equation of the axis of symmetry?
34. What are the coordinates of the vertex?

35. **MANUFACTURING** The reflective surface in a flashlight has a parabolic cross section that can be modeled by $y = \frac{1}{3}x^2$, where x and y are in centimeters. How far from the vertex should the filament of the light bulb be located?

Write an equation for each parabola described below. Then draw the graph.

36. vertex (0, 1), focus (0, 5)
37. vertex (8, 6), focus (2, 6)
38. focus (−4, −2), directrix $x = -8$
39. vertex (1, 7), directrix $y = 3$
40. vertex (−7, 4), axis of symmetry $x = -7$, measure of latus rectum 6, $a < 0$
41. vertex (4, 3), axis of symmetry $y = 3$, measure of latus rectum 4, $a > 0$

42. Write an equation for the graph at the right.

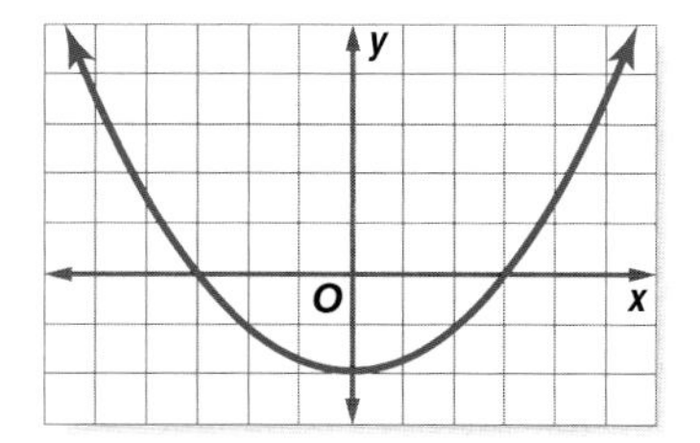

43. **BRIDGES** The Bayonne Bridge connects Staten Island, New York, to New Jersey. It has an arch in the shape of a parabola that opens downward. Write an equation of a parabola to model the arch, assuming that the origin is at the surface of the water, beneath the vertex of the arch.

44. SPORTS When a ball is thrown or kicked, the path it travels is shaped like a parabola. Suppose a football is kicked from ground level, reaches a maximum height of 25 feet, and hits the ground 100 feet from where it was kicked. Assuming that the ball was kicked at the origin, write an equation of the parabola that models the flight of the ball.

45. AEROSPACE A spacecraft is in a circular orbit 150 kilometers above Earth. Once it attains the velocity needed to escape Earth's gravity, the spacecraft will follow a parabolic path with the center of Earth as the focus. Suppose the spacecraft reaches escape velocity above the North Pole. Write an equation to model the parabolic path of the spacecraft, assuming that the center of Earth is at the origin and the radius of Earth is 6400 kilometers.

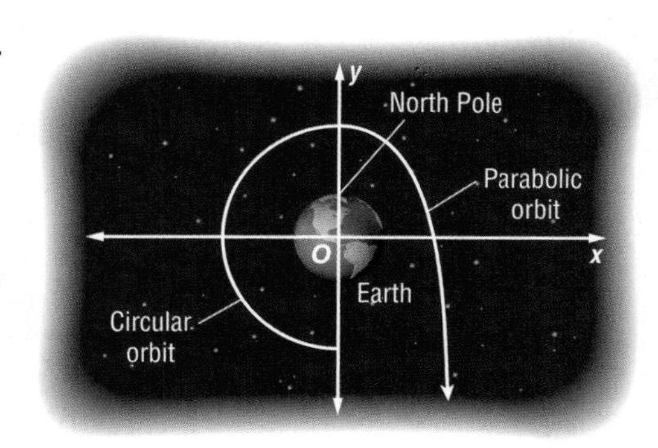

46. CRITICAL THINKING The parabola with equation $y = (x - 4)^2 + 3$ has its vertex at (4, 3) and passes through (5, 4). Find an equation of a different parabola with its vertex at (4, 3) that passes through (5, 4).

47. WRITING IN MATH Answer the question that was posed at the beginning of the lesson.

How are parabolas used in manufacturing?

Include the following in your answer:

- how you think the focus of a parabola got its name, and
- why a car headlight with a parabolic reflector is better than one with an unreflected light bulb.

48. Which equation has a graph that opens downward?

Ⓐ $y = 3x^2 - 2$ Ⓑ $y = 2 - 3x^2$ Ⓒ $x = 3y^2 - 2$ Ⓓ $x = 2 - 3y^2$

49. Find the vertex of the parabola with equation $y = x^2 - 10x + 8$.

Ⓐ (5, −17) Ⓑ (10, 8) Ⓒ (0, 8) Ⓓ (5, 8)

Maintain Your Skills

Mixed Review

Find the distance between each pair of points with the given coordinates. *(Lesson 8-1)*

50. (7, 3), (−5, 8) **51.** (4, −1), (−2, 7) **52.** (−3, 1), (0, 6)

53. Graph $y \leq \sqrt{x + 1}$. *(Lesson 7-9)*

54. HEALTH Ty's heart rate is usually 120 beats per minute when he runs. If he runs for 2 hours every day, about how many times will his heart beat during the amount of time he exercises in two weeks? Express the answer in scientific notation. *(Lesson 5-1)*

Getting Ready for the Next Lesson

PREREQUISITE SKILL Simplify each radical expression.
*(To review **simplifying radicals**, see Lessons 5-5 and 5-6.)*

55. $\sqrt{16}$ **56.** $\sqrt{25}$ **57.** $\sqrt{81}$ **58.** $\sqrt{144}$

59. $\sqrt{12}$ **60.** $\sqrt{18}$ **61.** $\sqrt{48}$ **62.** $\sqrt{72}$

8-3 Circles

California Standards Standard 16.0 Students demonstrate and explain how the geometry of the graph of a conic section depends on the coefficients of the quadratic equation representing it.

What You'll Learn

- Write equations of circles.
- Graph circles.

Vocabulary

- circle
- center
- tangent

Why are circles important in air traffic control?

Radar equipment can be used to detect and locate objects that are too far away to be seen by the human eye. The radar systems at major airports can typically detect and track aircraft up to 45 to 70 miles in any direction from the airport. The boundary of the region that a radar system can monitor can be modeled by a circle.

EQUATIONS OF CIRCLES A **circle** is the set of all points in a plane that are equidistant from a given point in the plane, called the **center**. Any segment whose endpoints are the center and a point on the circle is a *radius* of the circle.

Assume that (x, y) are the coordinates of a point on the circle at the right. The center is at (h, k), and the radius is r. You can find an equation of the circle by using the Distance Formula.

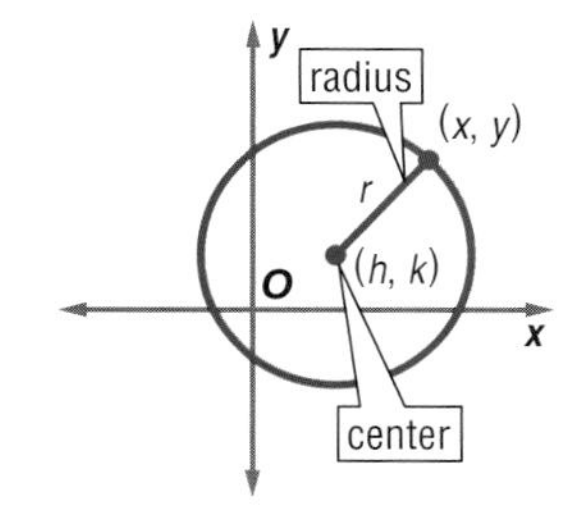

$$\sqrt{(x_2 - x_1)^2 + (y_2 - y_1)^2} = d \quad \text{Distance Formula}$$

$$\sqrt{(x - h)^2 + (y - k)^2} = r \quad (x_1, y_1) = (h, k),\ (x_2, y_2) = (x, y),\ d = r$$

$$(x - h)^2 + (y - k)^2 = r^2 \quad \text{Square each side.}$$

Key Concept — Equation of a Circle

The equation of a circle with center (h, k) and radius r units is $(x - h)^2 + (y - k)^2 = r^2$.

Example 1 Write an Equation Given the Center and Radius

NUCLEAR POWER In 1986, a nuclear reactor exploded at a power plant about 110 kilometers north and 15 kilometers west of Kiev. At first, officials evacuated people within 30 kilometers of the power plant. Write an equation to represent the boundary of the evacuated region if the origin of the coordinate system is at Kiev.

Since Kiev is at $(0, 0)$, the power plant is at $(-15, 110)$. The boundary of the evacuated region is the circle centered at $(-15, 110)$ with radius 30 kilometers.

$$(x - h)^2 + (y - k)^2 = r^2 \quad \text{Equation of a circle}$$

$$[x - (-15)]^2 + (y - 110)^2 = 30^2 \quad (h, k) = (-15, 110),\ r = 30$$

$$(x + 15)^2 + (y - 110)^2 = 900 \quad \text{Simplify.}$$

The equation is $(x + 15)^2 + (y - 110)^2 = 900$.

Example 2 Write an Equation Given a Diameter

Write an equation for a circle if the endpoints of a diameter are at (5, 4) and (−2, −6).

Explore To write an equation of a circle, you must know the center and the radius.

Plan You can find the center of the circle by finding the midpoint of the diameter. Then you can find the radius of the circle by finding the distance from the center to one of the given points.

Solve First find the center of the circle.

$$(h, k) = \left(\frac{x_1 + x_2}{2}, \frac{y_1 + y_2}{2}\right) \quad \text{Midpoint Formula}$$

$$= \left(\frac{5 + (-2)}{2}, \frac{4 + (-6)}{2}\right) \quad (x_1, y_1) = (5, 4), (x_2, y_2) = (-2, -6)$$

$$= \left(\frac{3}{2}, \frac{-2}{2}\right) \quad \text{Add.}$$

$$= \left(\frac{3}{2}, -1\right) \quad \text{Simplify.}$$

Now find the radius.

$$r = \sqrt{(x_2 - x_1)^2 + (y_2 - y_1)^2} \quad \text{Distance Formula}$$

$$= \sqrt{\left(\frac{3}{2} - 5\right)^2 + (-1 - 4)^2} \quad (x_1, y_1) = (5, 4), (x_2, y_2) = \left(\frac{3}{2}, -1\right)$$

$$= \sqrt{\left(-\frac{7}{2}\right)^2 + (-5)^2} \quad \text{Subtract.}$$

$$= \sqrt{\frac{149}{4}} \quad \text{Simplify.}$$

The radius of the circle is $\sqrt{\frac{149}{4}}$ units, so $r^2 = \frac{149}{4}$.

Substitute h, k, and r^2 into the standard form of the equation of a circle. An equation of the circle is $\left(x - \frac{3}{2}\right)^2 + [y - (-1)]^2 = \frac{149}{4}$ or $\left(x - \frac{3}{2}\right)^2 + (y + 1)^2 = \frac{149}{4}$.

Examine Each of the given points satisfies the equation, so the equation is reasonable.

A line in the plane of a circle can intersect the circle in zero, one, or two points. A line that intersects the circle in exactly one point is said to be **tangent** to the circle. The line and the circle are tangent to each other at this point.

Example 3 Write an Equation Given the Center and a Tangent

Write an equation for a circle with center at (−4, −3) that is tangent to the x-axis.

Sketch the circle. Since the circle is tangent to the x-axis, its radius is 3.

An equation of the circle is $(x + 4)^2 + (y + 3)^2 = 9$.

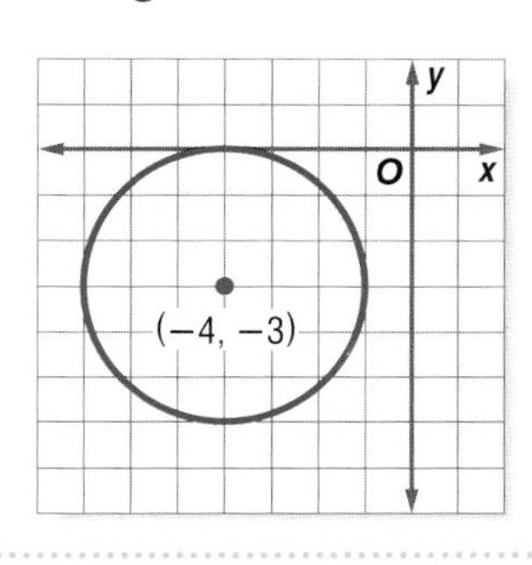

GRAPH CIRCLES You can use completing the square, symmetry, and transformations to help you graph circles. The equation $(x - h)^2 + (y - k)^2 = r^2$ is obtained from the equation $x^2 + y^2 = r^2$ by replacing x with $x - h$ and y with $y - k$. So, the graph of $(x - h)^2 + (y - k)^2 = r^2$ is the graph of $x^2 + y^2 = r^2$ translated h units to the right and k units up.

Example 4 Graph an Equation in Standard Form

Find the center and radius of the circle with equation $x^2 + y^2 = 25$. Then graph the circle.

The center of the circle is at (0, 0), and the radius is 5.

The table lists some integer values for x and y that satisfy the equation.

x	y
0	5
3	4
4	3
5	0

Since the circle is centered at the origin, it is symmetric about the y-axis. Therefore, the points at (−3, 4), (−4, 3) and (−5, 0) lie on the graph.

The circle is also symmetric about the x-axis, so the points at (−4, −3), (−3, −4), (0, −5), (3, −4), and (4, −3) lie on the graph.

Graph all of these points and draw the circle that passes through them.

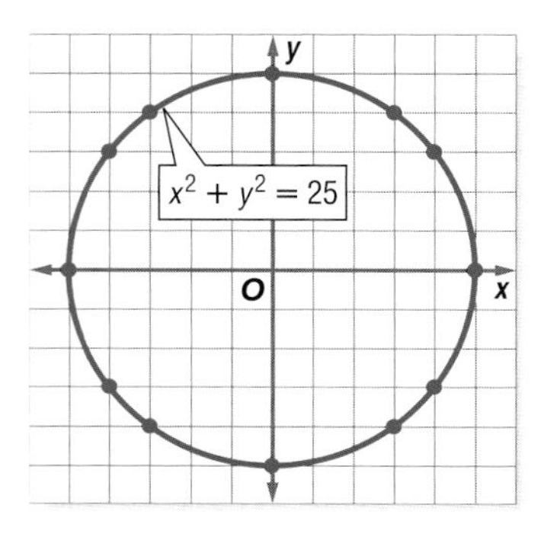

Example 5 Graph an Equation Not in Standard Form

Find the center and radius of the circle with equation $x^2 + y^2 - 4x + 8y - 5 = 0$. Then graph the circle.

Complete the squares.

$$x^2 + y^2 - 4x + 8y - 5 = 0$$
$$x^2 - 4x + \blacksquare + y^2 + 8y + \blacksquare = 5 + \blacksquare + \blacksquare$$
$$x^2 - 4x + 4 + y^2 + 8y + 16 = 5 + 4 + 16$$
$$(x - 2)^2 + (y + 4)^2 = 25$$

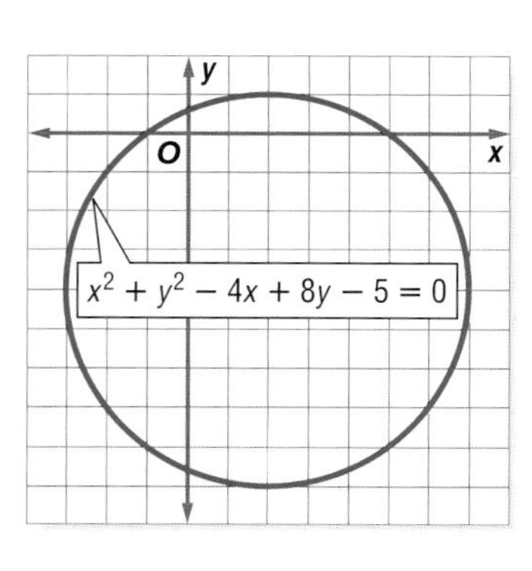

The center of the circle is at (2, −4), and the radius is 5. In the equation from Example 4, x has been replaced by $x - 2$, and y has been replaced by $y + 4$. The graph is the graph from Example 4 translated 2 units to the right and down 4 units.

Check for Understanding

Concept Check

1. **OPEN ENDED** Write an equation for a circle with center at (6, −2).
2. **Write** $x^2 + y^2 + 6x - 2y - 54 = 0$ in standard form by completing the square. Describe the transformation that can be applied to the graph of $x^2 + y^2 = 64$ to obtain the graph of the given equation.
3. **FIND THE ERROR** Juwan says that the circle with equation $(x - 4)^2 + y^2 = 36$ has radius 36 units. Lucy says that the radius is 6 units. Who is correct? Explain.

Guided Practice

4. Write an equation for the graph at the right.

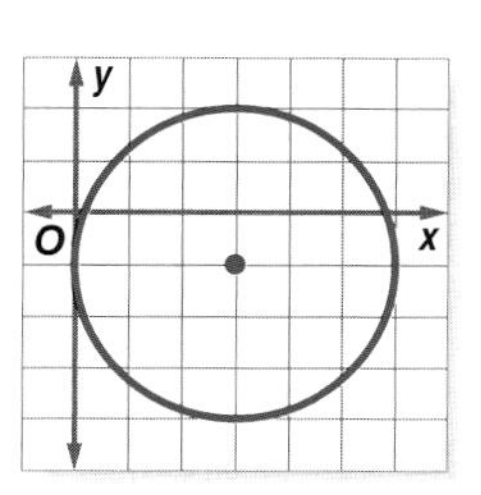

Write an equation for the circle that satisfies each set of conditions.

5. center $(-1, -5)$, radius 2 units
6. endpoints of a diameter at $(-4, 1)$ and $(4, -5)$
7. center $(3, -7)$, tangent to the y-axis

Find the center and radius of the circle with the given equation. Then graph the circle.

8. $(x - 4)^2 + (y - 1)^2 = 9$
9. $x^2 + (y - 14)^2 = 34$
10. $(x - 4)^2 + y^2 = \frac{16}{25}$
11. $\left(x + \frac{2}{3}\right)^2 + \left(y - \frac{1}{2}\right)^2 = \frac{8}{9}$
12. $x^2 + y^2 + 8x - 6y = 0$
13. $x^2 + y^2 + 4x - 8 = 0$

Application

AEROSPACE **For Exercises 14 and 15, use the following information.**
In order for a satellite to remain in a circular orbit above the same spot on Earth, the satellite must be 35,800 kilometers above the equator.

14. Write an equation for the orbit of the satellite. Use the center of Earth as the origin and 6400 kilometers for the radius of Earth.
15. Draw a labeled sketch of Earth and the orbit to scale.

Practice and Apply

Homework Help

For Exercises	See Examples
16–29	1–3
30–48	4, 5

Extra Practice
See page 845.

Write an equation for each graph.

16.

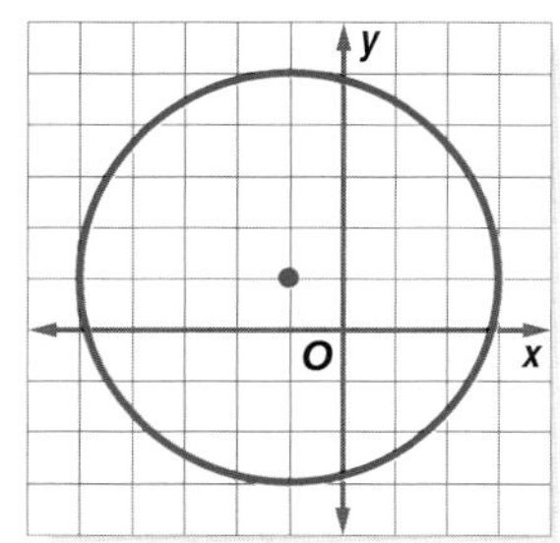

17.

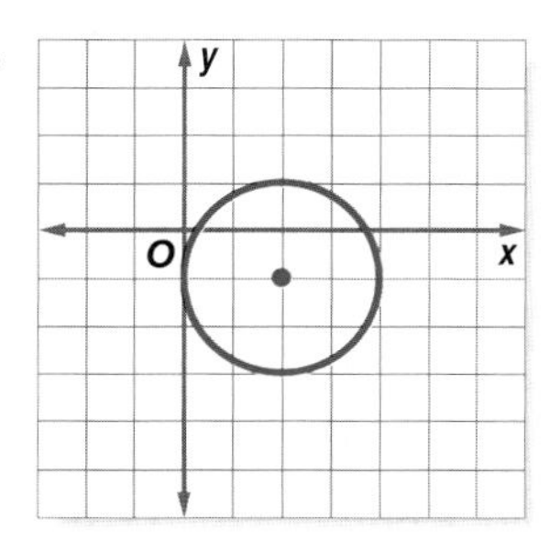

Write an equation for the circle that satisfies each set of conditions.

18. center $(0, 3)$, radius 7 units
19. center $(-8, 7)$, radius $\frac{1}{2}$ unit
20. endpoints of a diameter at $(-5, 2)$ and $(3, 6)$
21. endpoints of a diameter at $(11, 18)$ and $(-13, -19)$
22. center $(8, -9)$, passes through $(21, 22)$
23. center $\left(-\sqrt{13}, 42\right)$, passes through the origin
24. center at $(-8, -7)$, tangent to y-axis
25. center at $(4, 2)$, tangent to x-axis
26. center in the first quadrant; tangent to $x = -3$, $x = 5$, and the x-axis
27. center in the second quadrant; tangent to $y = -1$, $y = 9$, and the y-axis

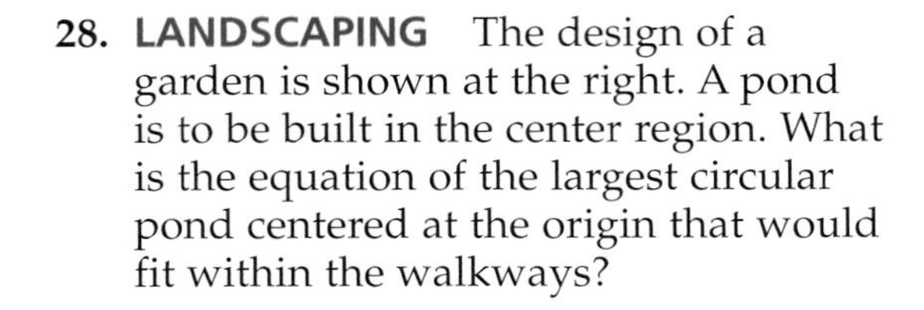

28. **LANDSCAPING** The design of a garden is shown at the right. A pond is to be built in the center region. What is the equation of the largest circular pond centered at the origin that would fit within the walkways?

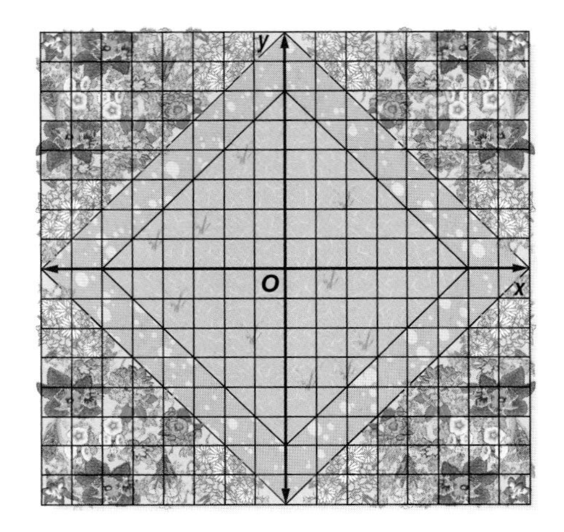

The epicenter of an earthquake can be located by using the equation of a circle. Visit www.algebra2.com/webquest to continue work on your WebQuest project.

29. **EARTHQUAKES** The University of Southern California is located about 2.5 miles west and about 2.8 miles south of downtown Los Angeles. Suppose an earthquake occurs with its epicenter about 40 miles from the university. Assume that the origin of a coordinate plane is located at the center of downtown Los Angeles. Write an equation for the set of points that could be the epicenter of the earthquake.

Find the center and radius of the circle with the given equation. Then graph the circle.

30. $x^2 + (y + 2)^2 = 4$
31. $x^2 + y^2 = 144$
32. $(x - 3)^2 + (y - 1)^2 = 25$
33. $(x + 3)^2 + (y + 7)^2 = 81$
34. $(x - 3)^2 + y^2 = 16$
35. $(x - 3)^2 + (y + 7)^2 = 50$
36. $\left(x + \sqrt{5}\right)^2 + y^2 - 8y = 9$
37. $x^2 + \left(y - \sqrt{3}\right)^2 + 4x = 25$
38. $x^2 + y^2 + 6y = -50 - 14x$
39. $x^2 + y^2 - 6y - 16 = 0$
40. $x^2 + y^2 + 2x - 10 = 0$
41. $x^2 + y^2 - 18x - 18y + 53 = 0$
42. $x^2 + y^2 + 9x - 8y + 4 = 0$
43. $x^2 + y^2 - 3x + 8y = 20$
44. $x^2 - 12x + 84 = -y^2 + 16y$
45. $x^2 + y^2 + 2x + 4y = 9$
46. $3x^2 + 3y^2 + 12x - 6y + 9 = 0$
47. $4x^2 + 4y^2 + 36y + 5 = 0$

48. **RADIO** The diagram at the right shows the relative locations of some cities in North Dakota. The x-axis represents Interstate 94. The scale is 1 unit = 30 miles. While driving west on the highway, Doralina is listening to a radio station in Minot. She estimates the range of the signal to be 120 miles. How far west of Bismarck will she be able to pick up the signal?

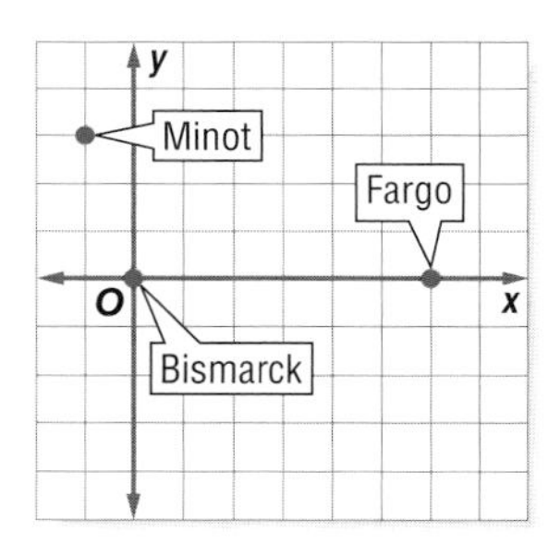

49. **CRITICAL THINKING** A circle has its center on the line with equation $y = 2x$. The circle passes through $(1, -3)$ and has a radius of $\sqrt{5}$ units. Write an equation of the circle.

50. **WRITING IN MATH** Answer the question that was posed at the beginning of the lesson.

Why are circles important in air traffic control?

Include the following in your answer:

- an equation of the circle that determines the boundary of the region where planes can be detected if the range of the radar is 50 miles and the radar is at the origin, and
- how an air traffic controller's job would be different for a region whose boundary is modeled by $x^2 + y^2 = 4900$ instead of $x^2 + y^2 = 1600$.

51. Find the radius of the circle with equation $x^2 + y^2 + 8x + 8y + 28 = 0$.

(A) 2 (B) 4 (C) 8 (D) 28

52. Find the center of the circle with equation $x^2 + y^2 - 10x + 6y + 27 = 0$.

(A) $(-10, 6)$ (B) $(1, 1)$ (C) $(10, -6)$ (D) $(5, -3)$

CIRCLES For Exercises 53–56, use the following information.
Since a circle is not the graph of a function, you cannot enter its equation directly into a graphing calculator. Instead, you must solve the equation for y. The result will contain a $\pm$ symbol, so you will have two functions.

53. Solve $(x + 3)^2 + y^2 = 16$ for y.

54. What two functions should you enter to graph the given equation?

55. Graph $(x + 3)^2 + y^2 = 16$ on a graphing calculator.

56. Solve $(x + 3)^2 + y^2 = 16$ for x. What parts of the circle do the two expressions for x represent?

Maintain Your Skills

Mixed Review

Identify the coordinates of the vertex and focus, the equations of the axis of symmetry and directrix, and the direction of opening of the parabola with the given equation. Then find the length of the latus rectum and graph the parabola. *(Lesson 8-2)*

57. $x = -3y^2 + 1$ 58. $y + 2 = -(x - 3)^2$ 59. $y = x^2 + 4x$

Find the midpoint of the line segment with endpoints at the given coordinates. *(Lesson 8-1)*

60. $(5, -7), (3, -1)$ 61. $(2, -9), (-4, 5)$ 62. $(8, 0), (-5, 12)$

Find all of the rational zeros for each function. *(Lesson 7-5)*

63. $f(x) = x^3 + 5x^2 + 2x - 8$ 64. $g(x) = 2x^3 - 9x^2 + 7x + 6$

65. **PHOTOGRAPHY** The perimeter of a rectangular picture is 86 inches. Twice the width exceeds the length by 2 inches. What are the dimensions of the picture? *(Lesson 3-2)*

Getting Ready for the Next Lesson

PREREQUISITE SKILL Solve each equation. Assume that all variables are positive. *(To review **solving quadratic equations**, see Lesson 6-4.)*

66. $c^2 = 13^2 - 5^2$ 67. $c^2 = 10^2 - 8^2$ 68. $(\sqrt{7})^2 = a^2 - 3^2$

69. $24^2 = a^2 - 7^2$ 70. $4^2 = 6^2 - b^2$ 71. $(2\sqrt{14})^2 = 8^2 - b^2$

Practice Quiz 1 — Lessons 8-1 through 8-3

Find the distance between each pair of points with the given coordinates. *(Lesson 8-1)*

1. $(9, 5), (4, -7)$ 2. $(0, -5), (10, -3)$

Identify the coordinates of the vertex and focus, the equations of the axis of symmetry and directrix, and the direction of opening of the parabola with the given equation. Then find the length of the latus rectum and graph the parabola. *(Lesson 8-2)*

3. $y^2 = 6x$ 4. $y = x^2 + 8x + 20$

5. Find the center and radius of the circle with equation $x^2 + (y - 4)^2 = 49$. Then graph the circle. *(Lesson 8-3)*

Algebra Activity

A Preview of Lesson 8-4

California Standards Standard 16.0 Students demonstrate and explain how the geometry of the graph of a conic section depends on the coefficients of the quadratic equation representing it.

Investigating Ellipses

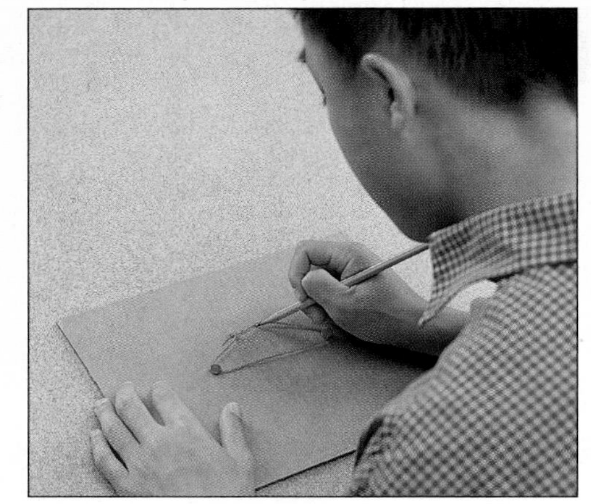

Follow the steps below to construct another type of conic section.

Step 1 Place two thumbtacks in a piece of cardboard, about 1 foot apart.

Step 2 Tie a knot in a piece of string and loop it around the thumbtacks.

Step 3 Place your pencil in the string. Keep the string tight and draw a curve.

Step 4 Continue drawing until you return to your starting point.

The curve you have drawn is called an **ellipse**. The points where the thumbtacks are located are called the **foci** of the ellipse. *Foci* is the plural of *focus*.

Model and Analyze

Place a large piece of grid paper on a piece of cardboard.

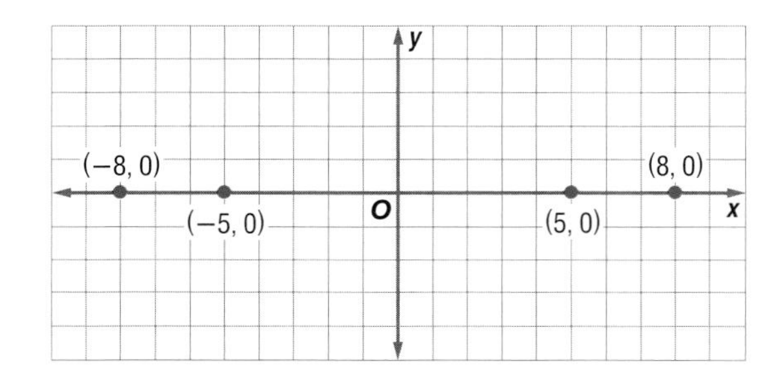

1. Place the thumbtacks at (8, 0) and (−8, 0). Choose a string long enough to loop around both thumbtacks. Draw an ellipse.

2. Repeat Exercise 1, but place the thumbtacks at (5, 0) and (−5, 0). Use the same loop of string and draw an ellipse. How does this ellipse compare to the one in Exercise 1?

Place the thumbtacks at each set of points and draw an ellipse.

You may change the length of the loop of string if you like.

3. (12, 0), (−12, 0) **4.** (2, 0), (−2, 0) **5.** (14, 4), (−10, 4)

Make a Conjecture

In Exercises 6–10, describe what happens to the shape of an ellipse when each change is made.

6. The thumbtacks are moved closer together.

7. The thumbtacks are moved farther apart.

8. The length of the loop of string is increased.

9. The thumbtacks are arranged vertically.

10. One thumbtack is removed, and the string is looped around the remaining thumbtack.

11. Pick a point on one of the ellipses you have drawn. Use a ruler to measure the distances from that point to the points where the thumbtacks were located. Add the distances. Repeat for other points on the same ellipse. What relationship do you notice?

12. Could this activity be done with a rubber band instead of a piece of string? Explain.

8-4 Ellipses

California Standards Standard 16.0 Students demonstrate and explain how the geometry of the graph of a conic section depends on the coefficients of the quadratic equation representing it.

What You'll Learn

- Write equations of ellipses.
- Graph ellipses.

Vocabulary

- ellipse
- foci
- major axis
- minor axis
- center

Why are ellipses important in the study of the solar system?

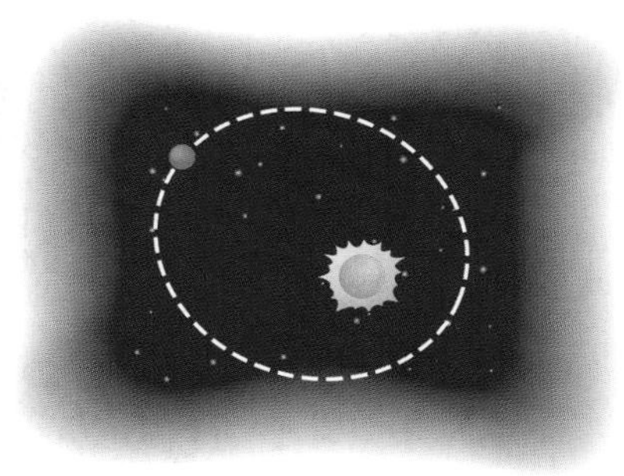

Fascination with the sky has caused people to wonder, observe, and make conjectures about the planets since the beginning of history. Since the early 1600s, the orbits of the planets have been known to be ellipses with the Sun at a focus.

EQUATIONS OF ELLIPSES As you discovered in the Algebra Activity on page 432, an **ellipse** is the set of all points in a plane such that the sum of the distances from two fixed points is constant. The two fixed points are called the **foci** of the ellipse.

The ellipse at the right has foci at (5, 0) and (−5, 0). The distances from either of the x-intercepts to the foci are 2 units and 12 units, so the sum of the distances from any point with coordinates (x, y) on the ellipse to the foci is 14 units.

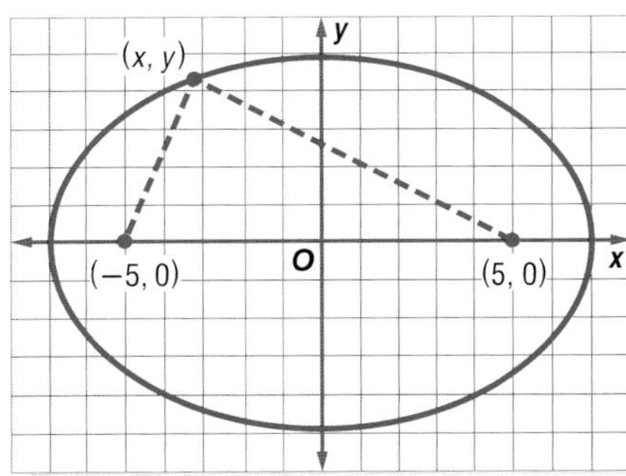

You can use the Distance Formula and the definition of an ellipse to find an equation of this ellipse.

$$\underbrace{\text{The distance between } (x, y) \text{ and } (-5, 0)} + \underbrace{\text{the distance between } (x, y) \text{ and } (5, 0)} = 14.$$

$$\sqrt{(x+5)^2 + y^2} + \sqrt{(x-5)^2 + y^2} = 14$$

$$\sqrt{(x+5)^2 + y^2} = 14 - \sqrt{(x-5)^2 + y^2} \quad \text{Isolate the radicals.}$$

$$(x+5)^2 + y^2 = 196 - 28\sqrt{(x-5)^2 + y^2} + (x-5)^2 + y^2 \quad \text{Square each side.}$$

$$x^2 + 10x + 25 + y^2 = 196 - 28\sqrt{(x-5)^2 + y^2} + x^2 - 10x + 25 + y^2$$

$$20x - 196 = -28\sqrt{(x-5)^2 + y^2} \quad \text{Simplify.}$$

$$5x - 49 = -7\sqrt{(x-5)^2 + y^2} \quad \text{Divide each side by 4.}$$

$$25x^2 - 490x + 2401 = 49[(x-5)^2 + y^2] \quad \text{Square each side.}$$

$$25x^2 - 490x + 2401 = 49x^2 - 490x + 1225 + 49y^2 \quad \text{Distributive Property}$$

$$-24x^2 - 49y^2 = -1176 \quad \text{Simplify.}$$

$$\frac{x^2}{49} + \frac{y^2}{24} = 1 \quad \text{Divide each side by } -1176.$$

An equation for this ellipse is $\frac{x^2}{49} + \frac{y^2}{24} = 1$.

Every ellipse has two axes of symmetry. The points at which the ellipse intersects its axes of symmetry determine two segments with endpoints on the ellipse called the **major axis** and the **minor axis**. The axes intersect at the **center** of the ellipse. The foci of an ellipse always lie on the major axis.

Study the ellipse at the right. The sum of the distances from the foci to any point on the ellipse is the same as the length of the major axis, or $2a$ units. The distance from the center to either focus is c units. By the Pythagorean Theorem, a, b, and c are related by the equation $c^2 = a^2 - b^2$. Notice that the x- and y-intercepts, $(\pm a, 0)$ and $(0, \pm b)$, satisfy the quadratic equation $\frac{x^2}{a^2} + \frac{y^2}{b^2} = 1$. This is the standard form of the equation of an ellipse with its center at the origin and a horizontal major axis.

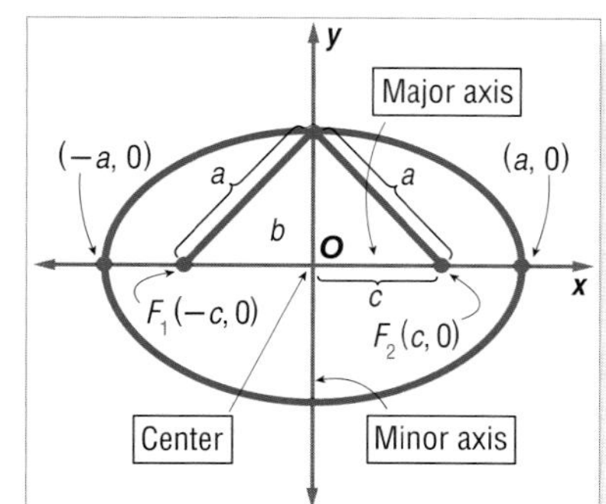

Study Tip

Vertices of Ellipses
The endpoints of each axis are called the *vertices* of the ellipse.

Key Concept — *Equations of Ellipses with Centers at the Origin*

Standard Form of Equation	$\frac{x^2}{a^2} + \frac{y^2}{b^2} = 1$	$\frac{y^2}{a^2} + \frac{x^2}{b^2} = 1$
Direction of Major Axis	horizontal	vertical
Foci	$(c, 0), (-c, 0)$	$(0, c), (0, -c)$
Length of Major Axis	$2a$ units	$2a$ units
Length of Minor Axis	$2b$ units	$2b$ units

In either case, $a^2 \geq b^2$ and $c^2 = a^2 - b^2$. You can determine if the foci are on the x-axis or the y-axis by looking at the equation. If the x^2 term has the greater denominator, the foci are on the x-axis. If the y^2 term has the greater denominator, the foci are on the y-axis.

Example 1 Write an Equation for a Graph

Write an equation for the ellipse shown at the right.

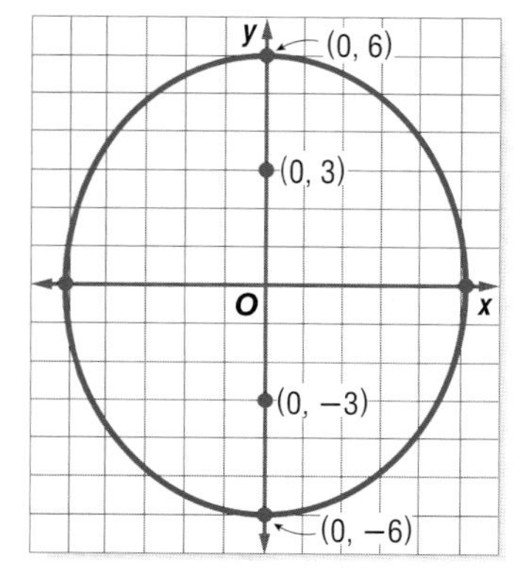

In order to write the equation for the ellipse, we need to find the values of a and b for the ellipse. We know that the length of the major axis of any ellipse is $2a$ units. In this ellipse, the length of the major axis is the distance between the points at (0, 6) and (0, −6). This distance is 12 units.

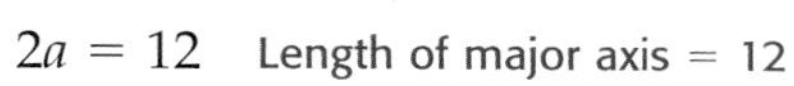

$2a = 12$ Length of major axis = 12

$a = 6$ Divide each side by 2.

The foci are located at (0, 3) and (0, −3), so $c = 3$. We can use the relationship between a, b, and c to determine the value of b.

$c^2 = a^2 - b^2$ Equation relating a, b, and c

$9 = 36 - b^2$ $c = 3$ and $a = 6$

$b^2 = 27$ Solve for b^2.

Since the major axis is vertical, substitute 36 for a^2 and 27 for b^2 in the form $\frac{y^2}{a^2} + \frac{x^2}{b^2} = 1$. An equation of the ellipse is $\frac{y^2}{36} + \frac{x^2}{27} = 1$.

Example 2 Write an Equation Given the Lengths of the Axes

MUSEUMS In an ellipse, sound or light coming from one focus is reflected to the other focus. In a whispering gallery, a person can hear another person whisper from across the room if the two people are standing at the foci. The whispering gallery at the Museum of Science and Industry in Chicago has an elliptical cross section that is 13 feet 6 inches by 47 feet 4 inches.

a. Write an equation to model this ellipse. Assume that the center is at the origin and the major axis is horizontal.

The length of the major axis is $47\frac{1}{3}$ or $\frac{142}{3}$ feet.

$2a = \frac{142}{3}$ Length of major axis $= \frac{142}{3}$

$a = \frac{71}{3}$ Divide each side by 2.

The length of the minor axis is $13\frac{1}{2}$ or $\frac{27}{2}$ feet.

$2b = \frac{27}{2}$ Length of minor axis $= \frac{27}{2}$

$b = \frac{27}{4}$ Divide each side by 2.

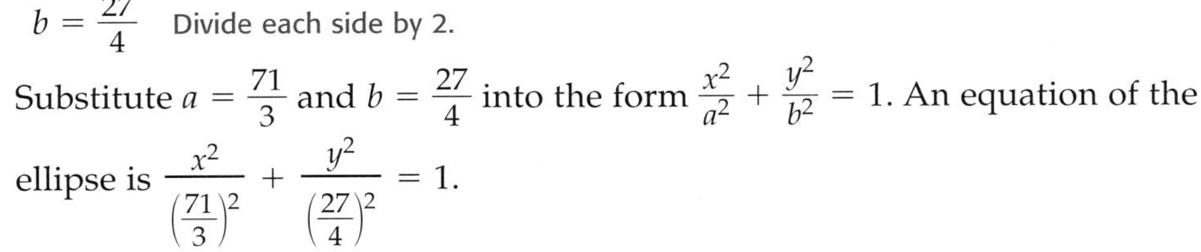

Substitute $a = \frac{71}{3}$ and $b = \frac{27}{4}$ into the form $\frac{x^2}{a^2} + \frac{y^2}{b^2} = 1$. An equation of the ellipse is $\frac{x^2}{\left(\frac{71}{3}\right)^2} + \frac{y^2}{\left(\frac{27}{4}\right)^2} = 1$.

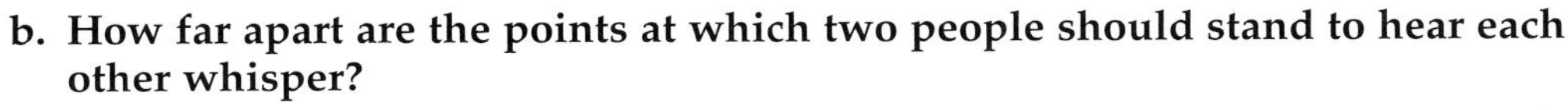

b. How far apart are the points at which two people should stand to hear each other whisper?

People should stand at the two foci of the ellipse. The distance between the foci is $2c$ units.

$c^2 = a^2 - b^2$ Equation relating a, b, and c

$c = \sqrt{a^2 - b^2}$ Take the square root of each side.

$2c = 2\sqrt{a^2 - b^2}$ Multiply each side by 2.

$2c = 2\sqrt{\left(\frac{71}{3}\right)^2 - \left(\frac{27}{4}\right)^2}$ Substitute $a = \frac{71}{3}$ and $b = \frac{27}{4}$.

$2c \approx 45.37$ Use a calculator.

The points where two people should stand to hear each other whisper are about 45.37 feet or 45 feet 4 inches apart.

More About. . .

Museums

The whispering gallery at Chicago's Museum of Science and Industry has a parabolic dish at each focus to help collect sound.

Source: www.msichicago.org

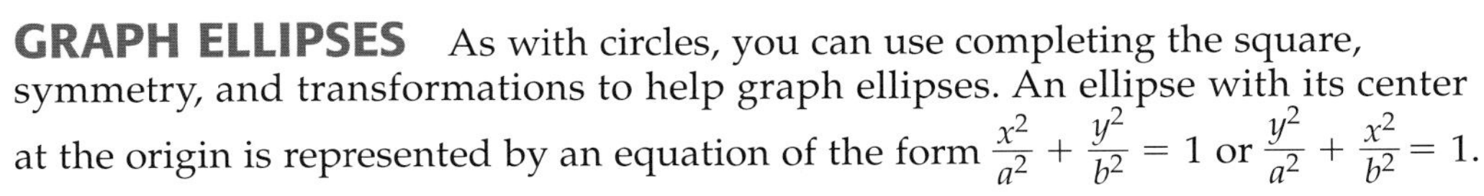

GRAPH ELLIPSES As with circles, you can use completing the square, symmetry, and transformations to help graph ellipses. An ellipse with its center at the origin is represented by an equation of the form $\frac{x^2}{a^2} + \frac{y^2}{b^2} = 1$ or $\frac{y^2}{a^2} + \frac{x^2}{b^2} = 1$. The ellipse could be translated h units to the right and k units up. This would move the center to the point (h, k). Such a move would be equivalent to replacing x with $x - h$ and replacing y with $y - k$.

Key Concept — Equations of Ellipses with Centers at (h, k)

Standard Form of Equation	$\frac{(x-h)^2}{a^2} + \frac{(y-k)^2}{b^2} = 1$	$\frac{(y-k)^2}{a^2} + \frac{(x-h)^2}{b^2} = 1$
Direction of Major Axis	horizontal	vertical
Foci	$(h \pm c, k)$	$(h, k \pm c)$

Example 3 Graph an Equation in Standard Form

Find the coordinates of the center and foci and the lengths of the major and minor axes of the ellipse with equation $\frac{x^2}{16} + \frac{y^2}{4} = 1$. Then graph the ellipse.

The center of this ellipse is at (0, 0).

Since $a^2 = 16$, $a = 4$. Since $b^2 = 4$, $b = 2$.

The length of the major axis is 2(4) or 8 units, and the length of the minor axis is 2(2) or 4 units. Since the x^2 term has the greater denominator, the major axis is horizontal.

$c^2 = a^2 - b^2$ Equation relating a, b, and c

$c^2 = 4^2 - 2^2$ or 12 $a = 4, b = 2$

$c = \sqrt{12}$ or $2\sqrt{3}$ Take the square root of each side.

The foci are at $(2\sqrt{3}, 0)$ and $(-2\sqrt{3}, 0)$.

You can use a calculator to find some approximate nonnegative values for x and y that satisfy the equation. Since the ellipse is centered at the origin, it is symmetric about the y-axis. Therefore, the points at (−4, 0), (−3, 1.3), (−2, 1.7), and (−1, 1.9) lie on the graph.

x	y
0	2.0
1	1.9
2	1.7
3	1.3
4	0.0

The ellipse is also symmetric about the x-axis, so the points at (−3, −1.3), (−2, −1.7), (−1, −1.9), (0, −2), (1, −1.9), (2, −1.7), and (3, −1.3) lie on the graph.

Graph the intercepts, (−4, 0), (4, 0), (0, 2), and (0, −2), and draw the ellipse that passes through them and the other points.

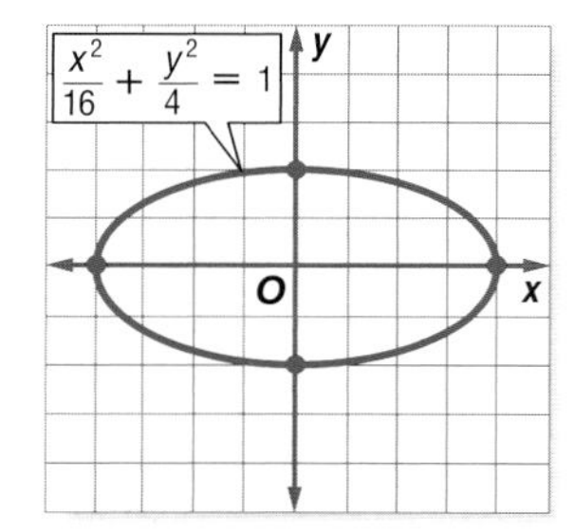

Study Tip

Graphing Calculator

You can graph an ellipse on a graphing calculator by first solving for y. Then graph the two equations that result on the same screen.

If you are given an equation of an ellipse that is not in standard form, write it in standard form first. This will make graphing the ellipse easier.

Example 4 Graph an Equation Not in Standard Form

Find the coordinates of the center and foci and the lengths of the major and minor axes of the ellipse with equation $x^2 + 4y^2 + 4x - 24y + 24 = 0$. Then graph the ellipse.

Complete the square for each variable to write this equation in standard form.

$x^2 + 4y^2 + 4x - 24y + 24 = 0$ Original equation

$(x^2 + 4x + ■) + 4(y^2 - 6y + ■) = -24 + ■ + 4(■)$ Complete the squares.

$(x^2 + 4x + 4) + 4(y^2 - 6y + 9) = -24 + 4 + 4(9)$ $\left(\frac{4}{2}\right)^2 = 4, \left(\frac{-6}{2}\right)^2 = 9$

$(x + 2)^2 + 4(y - 3)^2 = 16$ Write the trinomials as perfect squares.

$\frac{(x + 2)^2}{16} + \frac{(y - 3)^2}{4} = 1$ Divide each side by 16.

The graph of this ellipse is the graph from Example 3 translated 2 units to the left and up 3 units. The center is at $(-2, 3)$ and the foci are at $(-2 + 2\sqrt{3}, 0)$ and $(-2 - 2\sqrt{3}, 0)$. The length of the major axis is still 8 units, and the length of the minor axis is still 4 units.

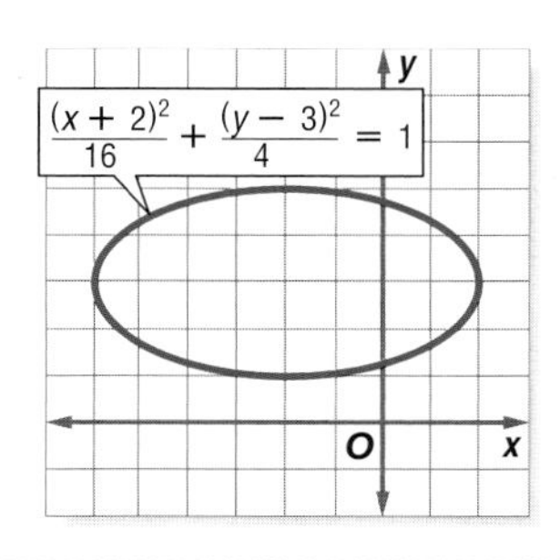

You can use a circle to locate the foci on the graph of a given ellipse.

Algebra Activity

Locating Foci

Step 1 Graph an ellipse so that its center is at the origin. Let the endpoints of the major axis be at $(-9, 0)$ and $(9, 0)$, and let the endpoints of the minor axis be at $(0, -5)$ and $(0, 5)$.

Step 2 Use a compass to draw a circle with center at $(0, 0)$ and radius 9 units.

Step 3 Draw the line with equation $y = 5$ and mark the points at which the line intersects the circle.

Step 4 Draw perpendicular lines from the points of intersection to the x-axis. The foci of the ellipse are located at the points where the perpendicular lines intersect the x-axis.

Make a Conjecture

Draw another ellipse and locate its foci. Why does this method work?

Check for Understanding

Concept Check

1. **Identify** the axes of symmetry of the ellipse at the right.

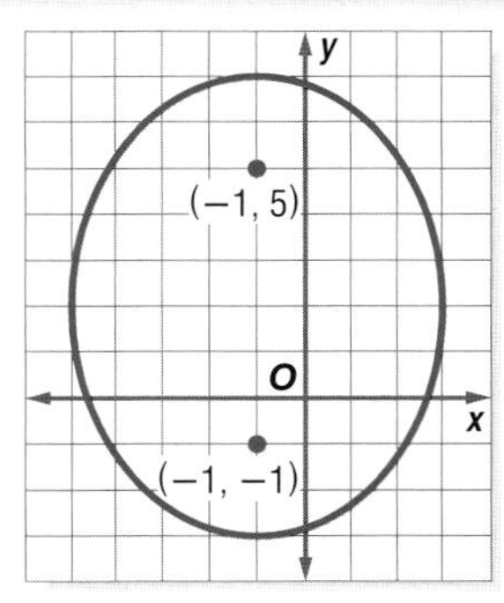

2. **Explain** why a circle is a special case of an ellipse.

3. **OPEN ENDED** Write an equation for an ellipse with its center at $(2, -5)$ and a horizontal major axis.

Guided Practice

4. Write an equation for the ellipse shown at the right.

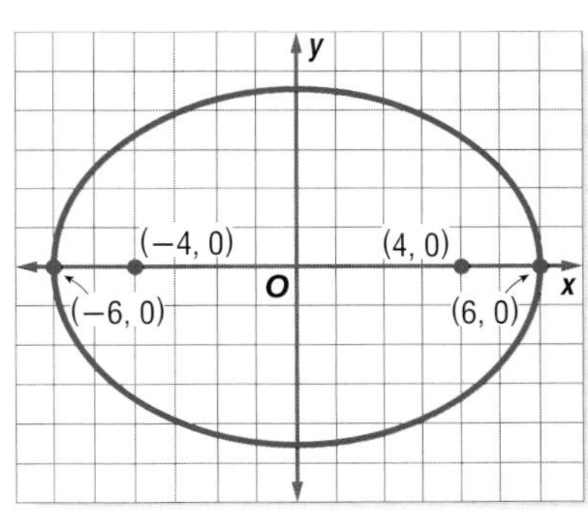

Write an equation for the ellipse that satisfies each set of conditions.

5. endpoints of major axis at $(2, 2)$ and $(2, -10)$, endpoints of minor axis at $(0, -4)$ and $(4, -4)$

6. endpoints of major axis at $(0, 10)$ and $(0, -10)$, foci at $(0, 8)$ and $(0, -8)$

Find the coordinates of the center and foci and the lengths of the major and minor axes for the ellipse with the given equation. Then graph the ellipse.

7. $\frac{y^2}{18} + \frac{x^2}{9} = 1$

8. $\frac{(x-1)^2}{20} + \frac{(y+2)^2}{4} = 1$

9. $4x^2 + 8y^2 = 32$

10. $x^2 + 25y^2 - 8x + 100y + 91 = 0$

Application

11. ASTRONOMY At its closest point, Mercury is 29.0 million miles from the center of the Sun. At its farthest point, Mercury is 43.8 million miles from the center of the Sun. Write an equation for the orbit of Mercury, assuming that the center of the orbit is the origin and the Sun lies on the x-axis.

Practice and Apply

Homework Help

For Exercises	See Examples
12–24	1, 2
25–38	3, 4

Extra Practice
See page 846.

Write an equation for each ellipse.

12.

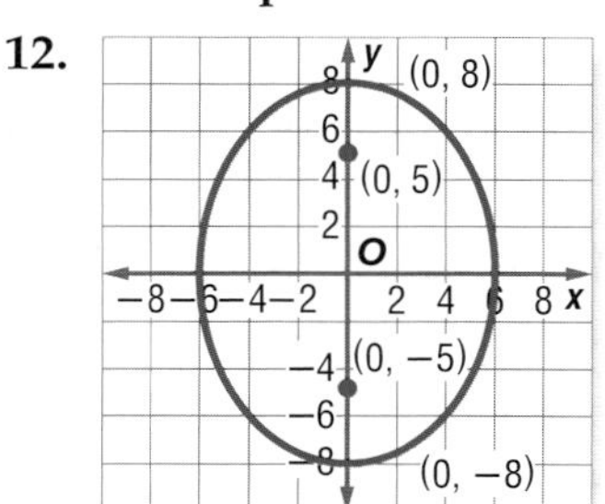

13.

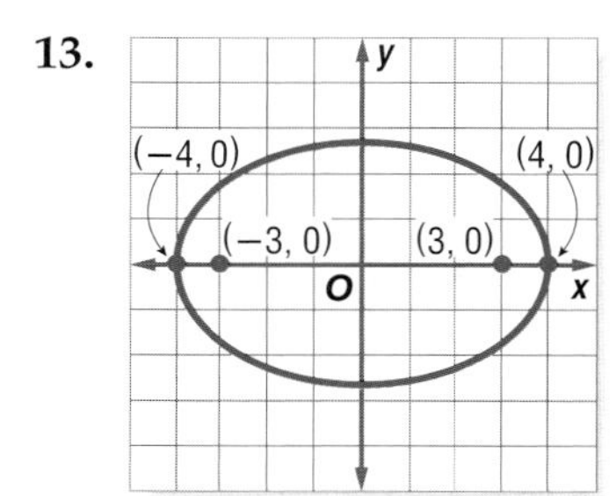

14.

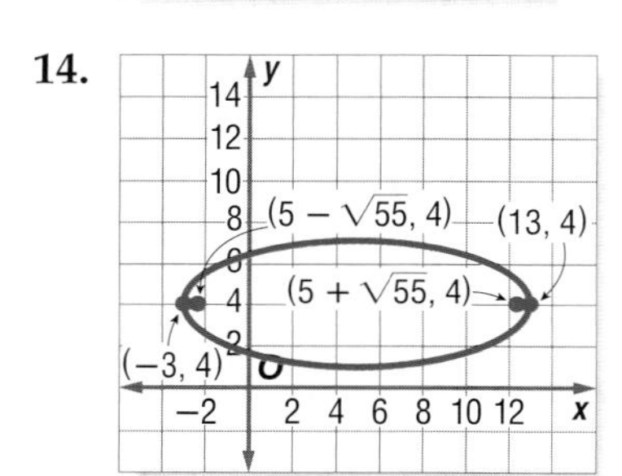

15.

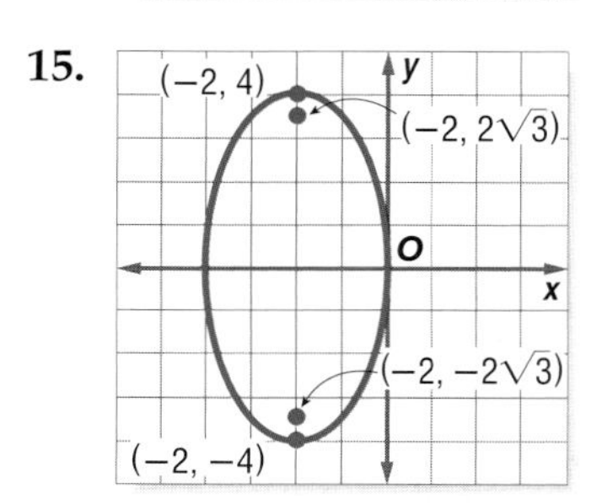

Write an equation for the ellipse that satisfies each set of conditions.

16. endpoints of major axis at $(-11, 5)$ and $(7, 5)$, endpoints of minor axis at $(-2, 9)$ and $(-2, 1)$

17. endpoints of major axis at $(2, 12)$ and $(2, -4)$, endpoints of minor axis at $(4, 4)$ and $(0, 4)$

18. major axis 20 units long and parallel to y-axis, minor axis 6 units long, center at $(4, 2)$

19. major axis 16 units long and parallel to x-axis, minor axis 9 units long, center at $(5, 4)$

20. endpoints of major axis at $(10, 2)$ and $(-8, 2)$, foci at $(6, 2)$ and $(-4, 2)$

21. endpoints of minor axis at $(0, 5)$ and $(0, -5)$, foci at $(12, 0)$ and $(-12, 0)$

22. INTERIOR DESIGN The rounded top of the window is the top half of an ellipse. Write an equation for the ellipse if the origin is at the midpoint of the bottom edge of the window.

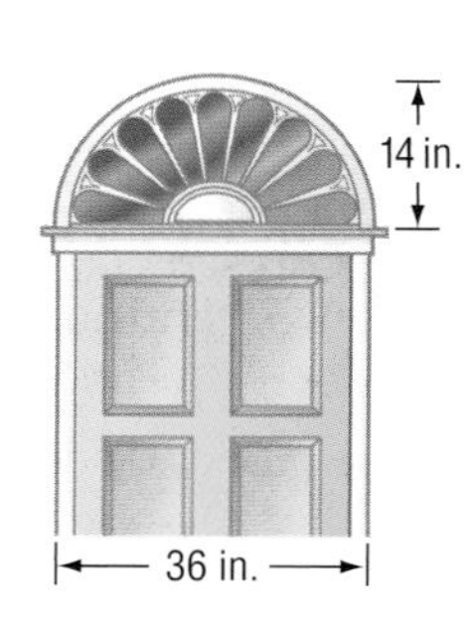

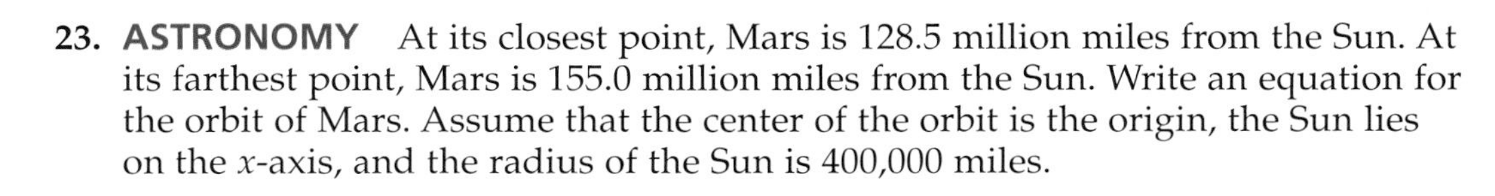

23. **ASTRONOMY** At its closest point, Mars is 128.5 million miles from the Sun. At its farthest point, Mars is 155.0 million miles from the Sun. Write an equation for the orbit of Mars. Assume that the center of the orbit is the origin, the Sun lies on the x-axis, and the radius of the Sun is 400,000 miles.

More About. . .

White House

The Ellipse, also known as President's Park South, has an area of about 16 acres.

Source: www.nps.gov

24. **WHITE HOUSE** There is an open area south of the White House known as the Ellipse. Write an equation to model the Ellipse. Assume that the origin is at the center of the Ellipse.

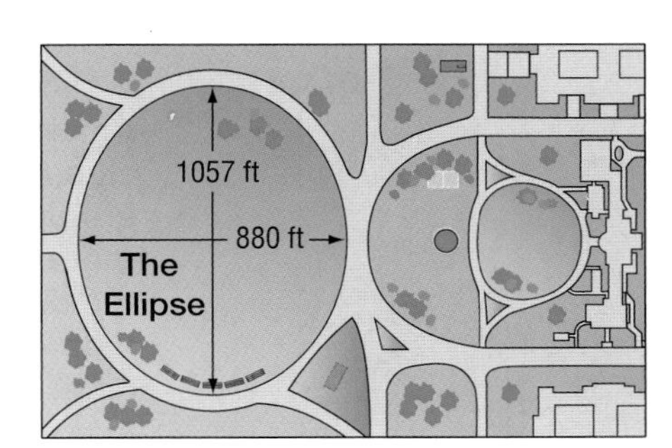

25. Write the equation $10x^2 + 2y^2 = 40$ in standard form.

26. What is the standard form of the equation $x^2 + 6y^2 - 2x + 12y - 23 = 0$?

Find the coordinates of the center and foci and the lengths of the major and minor axes for the ellipse with the given equation. Then graph the ellipse.

27. $\frac{y^2}{10} + \frac{x^2}{5} = 1$

28. $\frac{x^2}{25} + \frac{y^2}{9} = 1$

29. $\frac{(x+8)^2}{144} + \frac{(y-2)^2}{81} = 1$

30. $\frac{(y+11)^2}{144} + \frac{(x-5)^2}{121} = 1$

31. $3x^2 + 9y^2 = 27$

32. $27x^2 + 9y^2 = 81$

33. $16x^2 + 9y^2 = 144$

34. $36x^2 + 81y^2 = 2916$

35. $3x^2 + y^2 + 18x - 2y + 4 = 0$

36. $x^2 + 5y^2 + 4x - 70y + 209 = 0$

37. $7x^2 + 3y^2 - 28x - 12y = -19$

38. $16x^2 + 25y^2 + 32x - 150y = 159$

39. **CRITICAL THINKING** Find an equation for the ellipse with foci at $(\sqrt{3}, 0)$ and $(-\sqrt{3}, 0)$ that passes through $(0, 3)$.

40. **WRITING IN MATH** Answer the question that was posed at the beginning of the lesson.

Why are ellipses important in the study of the solar system?

Include the following in your answer:

- why an equation that is an accurate model of the path of a planet might be useful, and
- the distance from the center of Earth's orbit to the center of the Sun given that the Sun is at a focus of the orbit of Earth. Use the information in the figure at the right.

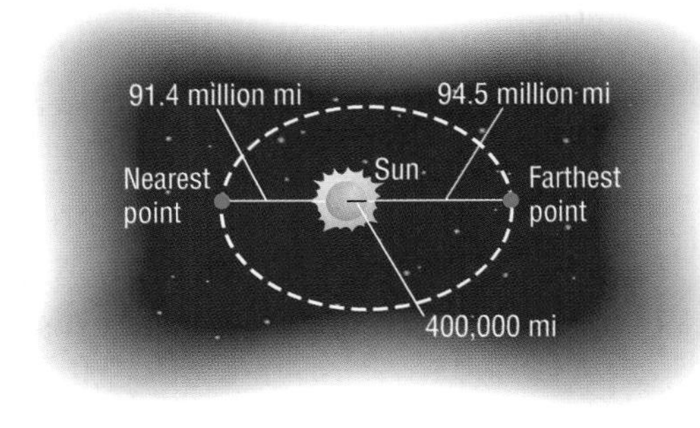

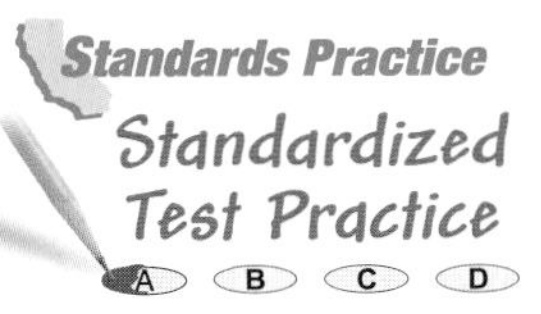

41. In the figure, A, B, and C are collinear. What is the measure of $\angle DBE$?

Ⓐ 40° Ⓑ 65°

Ⓒ 80° Ⓓ 100°

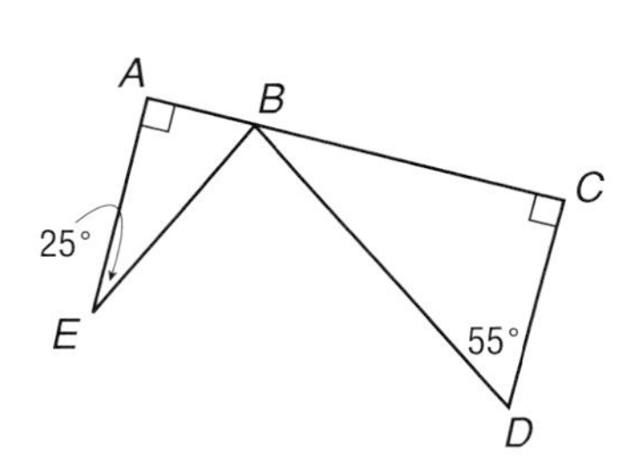

42. $\sqrt{25 + 144} =$

Ⓐ 7 Ⓑ 13 Ⓒ 17 Ⓓ 169

Extending the Lesson

43. **ASTRONOMY** In an ellipse, the ratio $\frac{c}{a}$ is called the **eccentricity** and is denoted by the letter e. Eccentricity measures the elongation of an ellipse. As shown in the graph at the right, the closer e is to 0, the more an ellipse looks like a circle. Pluto has the most eccentric orbit in our solar system with $e \approx 0.25$. Find an equation to model the orbit of Pluto, given that the length of the major axis is about 7.34 billion miles. Assume that the major axis is horizontal and that the center of the orbit is the origin.

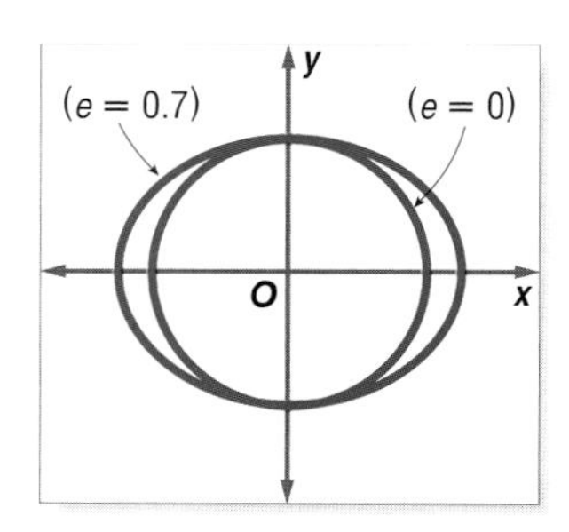

Maintain Your Skills

Mixed Review

Write an equation for the circle that satisfies each set of conditions. *(Lesson 8-3)*

44. center $(3, -2)$, radius 5 units

45. endpoints of a diameter at $(5, -9)$ and $(3, 11)$

46. center $(-1, 0)$, passes through $(2, -6)$

47. center $(4, -1)$, tangent to y-axis

48. Write an equation of a parabola with vertex $(3, 1)$ and focus $\left(3, 1\frac{1}{2}\right)$. Then draw the graph. *(Lesson 8-2)*

MARRIAGE For Exercises 49–51, use the table at the right that shows the number of married Americans over the last few decades. *(Lesson 2-5)*

49. Draw a scatter plot in which x is the number of years since 1980.

50. Find a prediction equation.

51. Predict the number of married Americans in 2010.

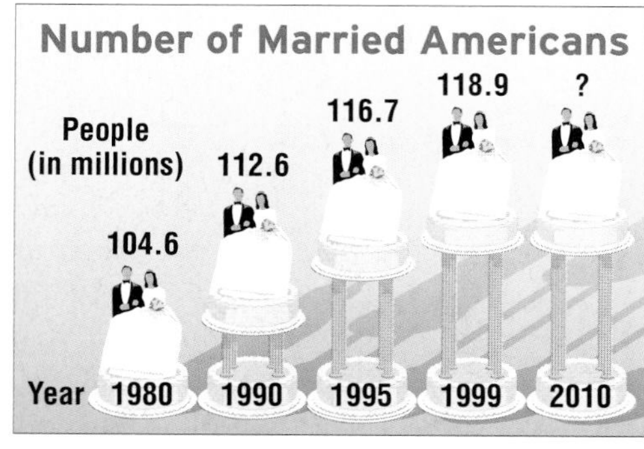

Source: U.S. Census Bureau

Online Research **Data Update** For the latest statistics on marriage and other characteristics of the population, visit www.algebra2.com/data_update to learn more.

Getting Ready for the Next Lesson

PREREQUISITE SKILL Graph the line with the given equation.
*(To review **graphing lines**, see Lessons 2-1, 2-2, and 2-3.)*

52. $y = 2x$

53. $y = -2x$

54. $y = -\frac{1}{2}x$

55. $y = \frac{1}{2}x$

56. $y + 2 = 2(x - 1)$

57. $y + 2 = -2(x - 1)$

8-5 Hyperbolas

California Standards Standard 16.0 Students demonstrate and explain how the geometry of the graph of a conic section depends on the coefficients of the quadratic equation representing it.

What You'll Learn

- Write equations of hyperbolas.
- Graph hyperbolas.

How are hyperbolas different from parabolas?

A hyperbola is a conic section with the property that rays directed toward one focus are reflected toward the other focus. Notice that, unlike the other conic sections, a hyperbola has two branches.

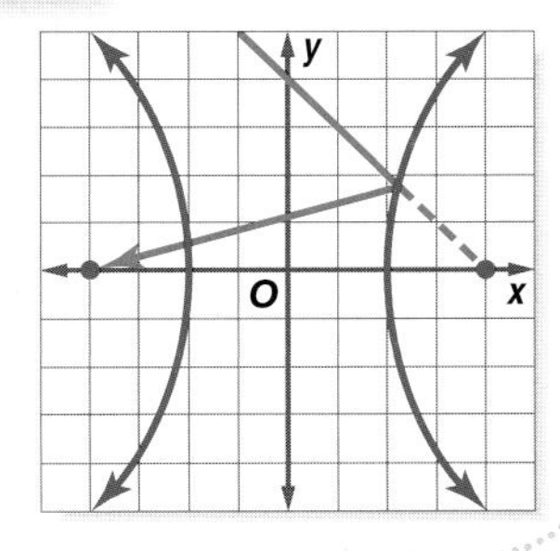

Vocabulary

- hyperbola
- foci
- center
- vertex
- asymptote
- transverse axis
- conjugate axis

EQUATIONS OF HYPERBOLAS A **hyperbola** is the set of all points in a plane such that the absolute value of the difference of the distances from two fixed points, called the **foci**, is constant.

The hyperbola at the right has foci at (0, 3) and (0, −3). The distances from either of the y-intercepts to the foci are 1 unit and 5 units, so the difference of the distances from any point with coordinates (x, y) on the hyperbola to the foci is 4 or −4 units, depending on the order in which you subtract.

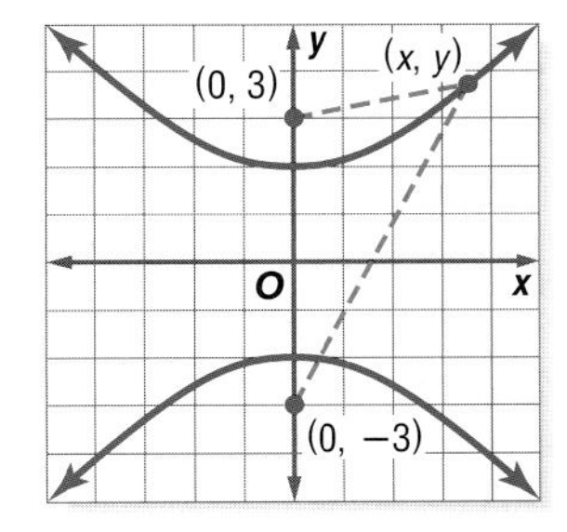

You can use the Distance Formula and the definition of a hyperbola to find an equation of this hyperbola.

The distance between (x, y) and $(0, 3)$ − the distance between (x, y) and $(0, -3)$ = ±4.

$$\sqrt{x^2 + (y - 3)^2} - \sqrt{x^2 + (y + 3)^2} = \pm 4$$

$$\sqrt{x^2 + (y - 3)^2} = \pm 4 + \sqrt{x^2 + (y + 3)^2} \quad \text{Isolate the radicals.}$$

$$x^2 + (y - 3)^2 = 16 \pm 8\sqrt{x^2 + (y + 3)^2} + x^2 + (y + 3)^2 \quad \text{Square each side.}$$

$$x^2 + y^2 - 6y + 9 = 16 \pm 8\sqrt{x^2 + (y + 3)^2} + x^2 + y^2 + 6y + 9$$

$$-12y - 16 = \pm 8\sqrt{x^2 + (y + 3)^2} \quad \text{Simplify.}$$

$$3y + 4 = \pm 2\sqrt{x^2 + (y + 3)^2} \quad \text{Divide each side by } -4.$$

$$9y^2 + 24y + 16 = 4[x^2 + (y + 3)^2] \quad \text{Square each side.}$$

$$9y^2 + 24y + 16 = 4x^2 + 4y^2 + 24y + 36 \quad \text{Distributive Property}$$

$$5y^2 - 4x^2 = 20 \quad \text{Simplify.}$$

$$\frac{y^2}{4} - \frac{x^2}{5} = 1 \quad \text{Divide each side by 20.}$$

An equation of this hyperbola is $\frac{y^2}{4} - \frac{x^2}{5} = 1$.

The diagram below shows the parts of a hyperbola.

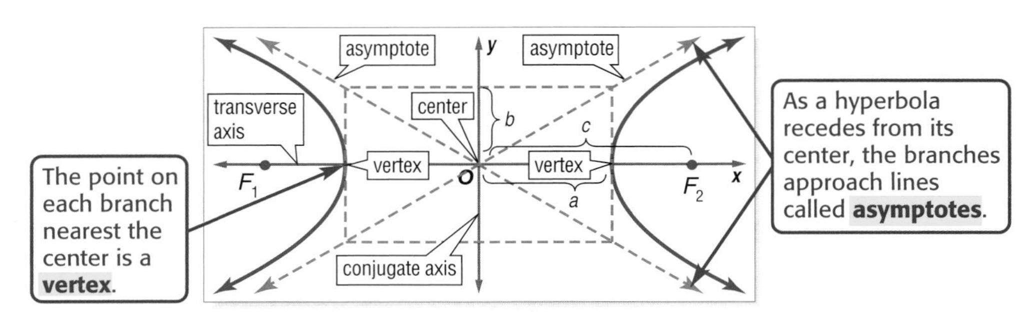

A hyperbola has some similarities to an ellipse. The distance from the **center** to a vertex is a units. The distance from the center to a focus is c units. There are two axes of symmetry. The **transverse axis** is a segment of length $2a$ whose endpoints are the vertices of the hyperbola. The **conjugate axis** is a segment of length $2b$ units that is perpendicular to the transverse axis at the center. The values of a, b, and c are related differently for a hyperbola than for an ellipse. For a hyperbola, $c^2 = a^2 + b^2$. The table below summarizes many of the properties of hyperbolas with centers at the origin.

Study Tip

Reading Math

In the standard form of a hyperbola, the squared terms are subtracted ($-$). For an ellipse, they are added ($+$).

Key Concept *Equations of Hyperbolas with Centers at the Origin*

Standard Form of Equation	$\frac{x^2}{a^2} - \frac{y^2}{b^2} = 1$	$\frac{y^2}{a^2} - \frac{x^2}{b^2} = 1$
Direction of Transverse Axis	horizontal	vertical
Foci	$(c, 0), (-c, 0)$	$(0, c), (0, -c)$
Vertices	$(a, 0), (-a, 0)$	$(0, a), (0, -a)$
Length of Transverse Axis	$2a$ units	$2a$ units
Length of Conjugate Axis	$2b$ units	$2b$ units
Equations of Asymptotes	$y = \pm\frac{b}{a}x$	$y = \pm\frac{a}{b}x$

Example 1 Write an Equation for a Graph

Write an equation for the hyperbola shown at the right.

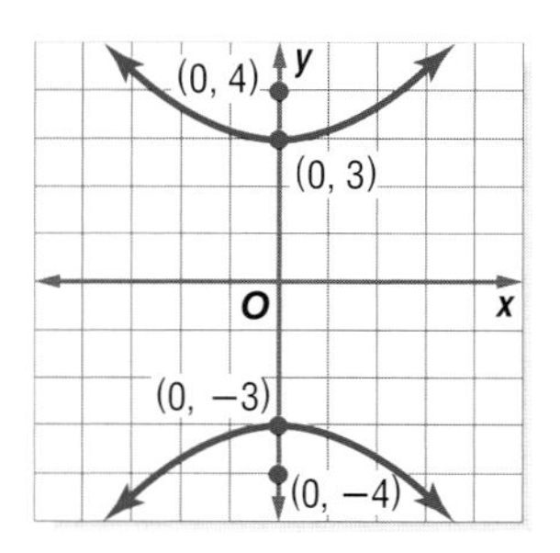

The center is the midpoint of the segment connecting the vertices, or (0, 0).

The value of a is the distance from the center to a vertex, or 3 units. The value of c is the distance from the center to a focus, or 4 units.

$c^2 = a^2 + b^2$	Equation relating a, b, and c for a hyperbola
$4^2 = 3^2 + b^2$	$c = 4$, $a = 3$
$16 = 9 + b^2$	Evaluate the squares.
$7 = b^2$	Solve for b^2.

Since the transverse axis is vertical, the equation is of the form $\frac{y^2}{a^2} - \frac{x^2}{b^2} = 1$.

Substitute the values for a^2 and b^2. An equation of the hyperbola is $\frac{y^2}{9} - \frac{x^2}{7} = 1$.

Example 2 Write an Equation Given the Foci and Transverse Axis

NAVIGATION The LORAN navigational system is based on hyperbolas. Two stations send out signals at the same time. A ship notes the difference in the times at which it receives the signals. The ship is on a hyperbola with the stations at the foci. Suppose a ship determines that the difference of its distances from two stations is 50 nautical miles. The stations are 100 nautical miles apart. Write an equation for a hyperbola on which the ship lies if the stations are at (−50, 0) and (50, 0).

More About. . .

Navigation

LORAN stands for *Long Range Navigation*. The LORAN system is generally accurate to within 0.25 nautical mile.

Source: U.S. Coast Guard

First, draw a figure. By studying either of the x-intercepts, you can see that the difference of the distances from any point on the hyperbola to the stations at the foci is the same as the length of the transverse axis, or $2a$. Therefore, $2a = 50$, or $a = 25$. According to the coordinates of the foci, $c = 50$.

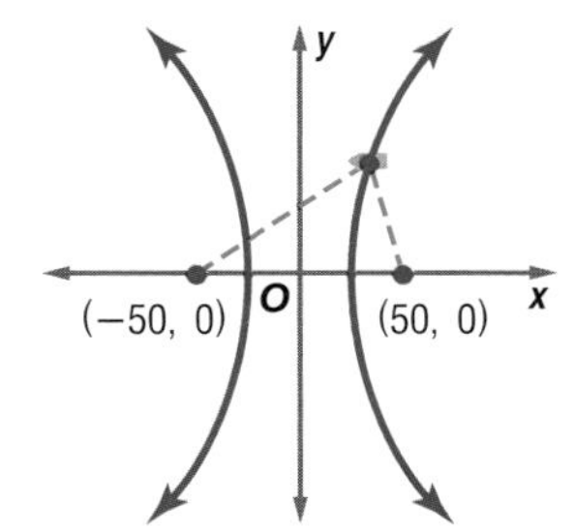

Use the values of a and c to determine the value of b for this hyperbola.

$c^2 = a^2 + b^2$	Equation relating a, b, and c for a hyperbola
$50^2 = 25^2 + b^2$	$c = 50$, $a = 25$
$2500 = 625 + b^2$	Evaluate the squares.
$1875 = b^2$	Solve for b^2.

Since the transverse axis is horizontal, the equation is of the form $\frac{x^2}{a^2} - \frac{y^2}{b^2} = 1$.

Substitute the values for a^2 and b^2. An equation of the hyperbola is $\frac{x^2}{625} - \frac{y^2}{1875} = 1$.

GRAPH HYPERBOLAS So far, you have studied hyperbolas that are centered at the origin. A hyperbola may be translated so that its center is at (h, k). This corresponds to replacing x by $x - h$ and y by $y - k$ in both the equation of the hyperbola and the equations of the asymptotes.

Key Concept Equations of Hyperbolas with Centers at (h, k)

Standard Form of Equation	$\frac{(x-h)^2}{a^2} - \frac{(y-k)^2}{b^2} = 1$	$\frac{(y-k)^2}{a^2} - \frac{(x-h)^2}{b^2} = 1$
Direction of Transverse Axis	horizontal	vertical
Equations of Asymptotes	$y - k = \pm\frac{b}{a}(x - h)$	$y - k = \pm\frac{a}{b}(x - h)$

It is easier to graph a hyperbola if the asymptotes are drawn first. To graph the asymptotes, use the values of a and b to draw a rectangle with dimensions $2a$ and $2b$. The diagonals of the rectangle should intersect at the center of the hyperbola. The asymptotes will contain the diagonals of the rectangle.

Example 3 Graph an Equation in Standard Form

Find the coordinates of the vertices and foci and the equations of the asymptotes for the hyperbola with equation $\frac{x^2}{9} - \frac{y^2}{4} = 1$. Then graph the hyperbola.

The center of this hyperbola is at the origin. According to the equation, $a^2 = 9$ and $b^2 = 4$, so $a = 3$ and $b = 2$. The coordinates of the vertices are (3, 0) and (−3, 0).

(continued on the next page)

$c^2 = a^2 + b^2$ Equation relating a, b, and c for a hyperbola

$c^2 = 3^2 + 2^2$ $a = 3$, $b = 2$

$c^2 = 13$ Simplify.

$c = \sqrt{13}$ Take the square root of each side.

The foci are at $(\sqrt{13}, 0)$ and $(-\sqrt{13}, 0)$.

The equations of the asymptotes are $y = \pm\frac{b}{a}x$ or $y = \pm\frac{2}{3}x$.

You can use a calculator to find some approximate nonnegative values for x and y that satisfy the equation. Since the hyperbola is centered at the origin, it is symmetric about the y-axis. Therefore, the points at $(-8, 4.9)$, $(-7, 4.2)$, $(-6, 3.5)$, $(-5, 2.7)$, $(-4, 1.8)$, and $(-3, 0)$ lie on the graph.

x	y
3	0
4	1.8
5	2.7
6	3.5
7	4.2
8	4.9

The hyperbola is also symmetric about the x-axis, so the points at $(-8, -4.9)$, $(-7, -4.2)$, $(-6, -3.5)$, $(-5, -2.7)$, $(-4, -1.8)$, $(4, -1.8)$, $(5, -2.7)$, $(6, -3.5)$, $(7, -4.2)$, and $(8, -4.9)$ also lie on the graph.

Study Tip

Graphing Calculator

You can graph a hyperbola on a graphing calculator. Similar to an ellipse, first solve the equation for y. Then graph the two equations that result on the same screen.

Draw a 6-unit by 4-unit rectangle. The asymptotes contain the diagonals of the rectangle. Graph the vertices, which, in this case, are the x-intercepts. Use the asymptotes as a guide to draw the hyperbola that passes through the vertices and the other points. The graph does not intersect the asymptotes.

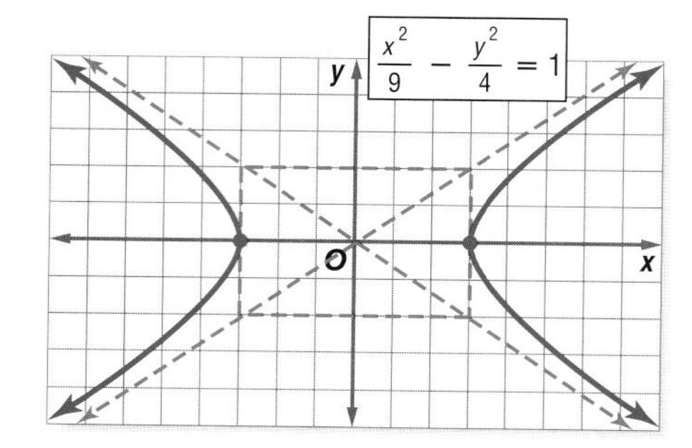

When graphing a hyperbola given an equation that is not in standard form, begin by rewriting the equation in standard form.

Example 4 Graph an Equation Not in Standard Form

Find the coordinates of the vertices and foci and the equations of the asymptotes for the hyperbola with equation $4x^2 - 9y^2 - 32x - 18y + 19 = 0$. Then graph the hyperbola.

Complete the square for each variable to write this equation in standard form.

$4x^2 - 9y^2 - 32x - 18y + 19 = 0$ Original equation

$4(x^2 - 8x + \blacksquare) - 9(y^2 + 2y + \blacksquare) = -19 + 4(\blacksquare) - 9(\blacksquare)$ Complete the squares.

$4(x^2 - 8x + 16) - 9(y^2 + 2y + 1) = -19 + 4(16) - 9(1)$

$4(x - 4)^2 - 9(y + 1)^2 = 36$ Write the trinomials as perfect squares.

$\frac{(x-4)^2}{9} - \frac{(y+1)^2}{4} = 1$ Divide each side by 36.

The graph of this hyperbola is the graph from Example 3 translated 4 units to the right and down 1 unit. The vertices are at $(7, -1)$ and $(1, -1)$, and the foci are at $(4 + \sqrt{13}, -1)$ and $(4 - \sqrt{13}, -1)$. The equations of the asymptotes are $y + 1 = \pm\frac{2}{3}(x - 4)$.

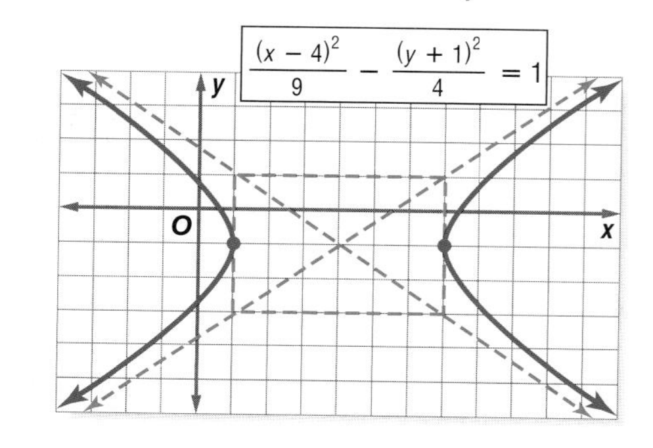

Check for Understanding

Concept Check

1. **Determine** whether the statement is *sometimes, always,* or *never* true. *The graph of a hyperbola is symmetric about the x-axis.*
2. **Describe** how the graph of $y^2 - \frac{x^2}{k^2} = 1$ changes as k increases.
3. **OPEN ENDED** Find a counterexample to the following statement. *If the equation of a hyperbola is* $\frac{x^2}{a^2} - \frac{y^2}{b^2} = 1$, *then* $a^2 \geq b^2$.

Guided Practice

4. Write an equation for the hyperbola shown at the right.
5. A hyperbola has foci at (4, 0) and (−4, 0). The value of a is 1. Write an equation for the hyperbola.

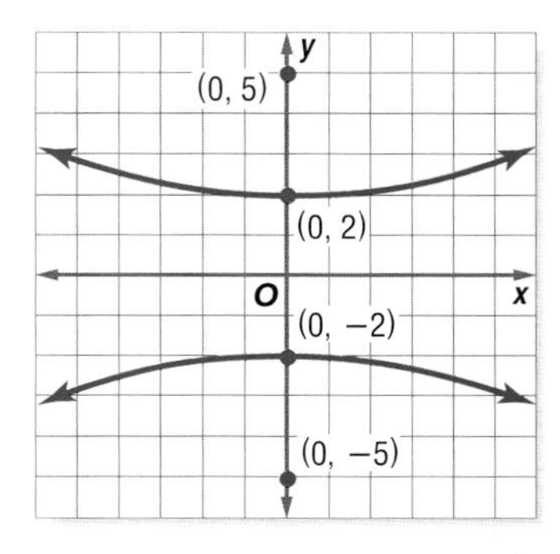

Find the coordinates of the vertices and foci and the equations of the asymptotes for the hyperbola with the given equation. Then graph the hyperbola.

6. $\frac{y^2}{18} - \frac{x^2}{20} = 1$
7. $\frac{(y+6)^2}{20} - \frac{(x-1)^2}{25} = 1$
8. $x^2 - 36y^2 = 36$
9. $5x^2 - 4y^2 - 40x - 16y - 36 = 0$

Application

10. **ASTRONOMY** Comets that pass by Earth only once may follow hyperbolic paths. Suppose a comet's path is modeled by a branch of the hyperbola with equation $\frac{y^2}{225} - \frac{x^2}{400} = 1$. Find the coordinates of the vertices and foci and the equations of the asymptotes for the hyperbola. Then graph the hyperbola.

Practice and Apply

Homework Help

For Exercises	See Examples
11–20, 35	1, 2
21–34	3, 4
36–38	

Extra Practice
See page 846.

Write an equation for each hyperbola.

11.

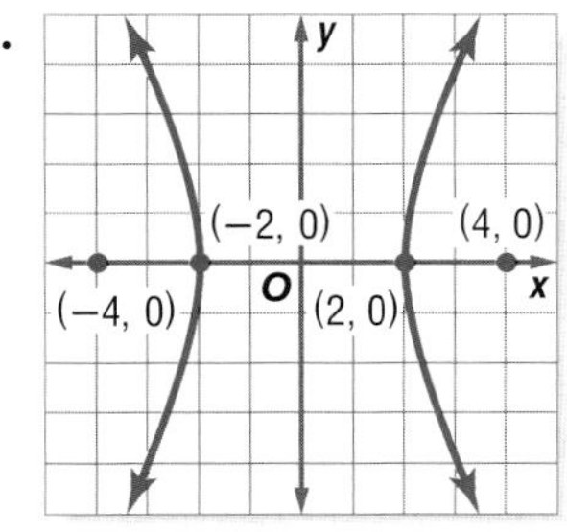

12.

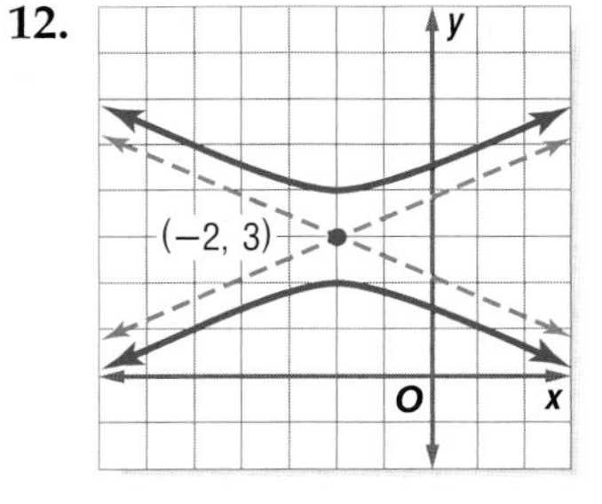

13.

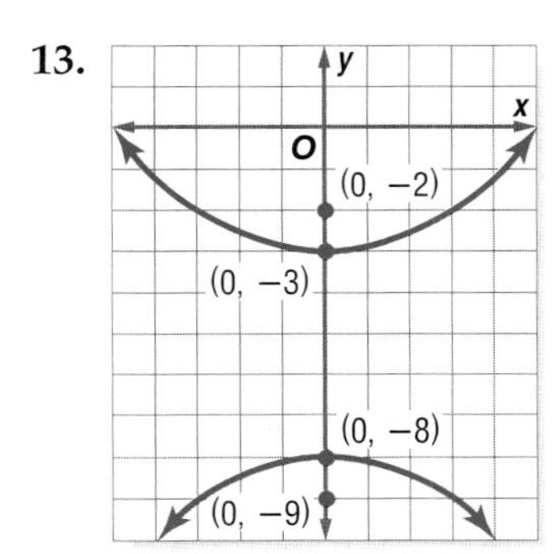

14.

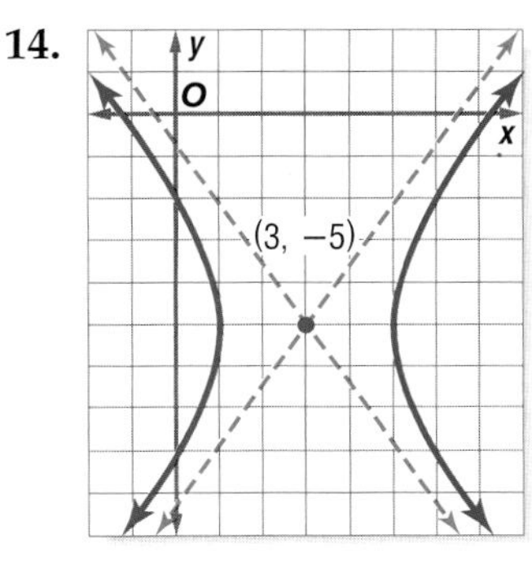

Write an equation for the hyperbola that satisfies each set of conditions.

15. vertices $(-5, 0)$ and $(5, 0)$, conjugate axis of length 12 units
16. vertices $(0, -4)$ and $(0, 4)$, conjugate axis of length 14 units
17. vertices $(9, -3)$ and $(-5, -3)$, foci $\left(2 \pm \sqrt{53}, -3\right)$
18. vertices $(-4, 1)$ and $(-4, 9)$, foci $\left(-4, 5 \pm \sqrt{97}\right)$

19. Find an equation for a hyperbola centered at the origin with a horizontal transverse axis of length 8 units and a conjugate axis of length 6 units.

20. What is an equation for the hyperbola centered at the origin with a vertical transverse axis of length 12 units and a conjugate axis of length 4 units?

Find the coordinates of the vertices and foci and the equations of the asymptotes for the hyperbola with the given equation. Then graph the hyperbola.

21. $\frac{x^2}{81} - \frac{y^2}{49} = 1$
22. $\frac{y^2}{36} - \frac{x^2}{4} = 1$
23. $\frac{y^2}{16} - \frac{x^2}{25} = 1$
24. $\frac{x^2}{9} - \frac{y^2}{25} = 1$
25. $x^2 - 2y^2 = 2$
26. $x^2 - y^2 = 4$
27. $y^2 = 36 + 4x^2$
28. $6y^2 = 2x^2 + 12$
29. $\frac{(y-4)^2}{16} - \frac{(x+2)^2}{9} = 1$
30. $\frac{(y-3)^2}{25} - \frac{(x-2)^2}{16} = 1$
31. $\frac{(x+1)^2}{4} - \frac{(y+3)^2}{9} = 1$
32. $\frac{(x+6)^2}{36} - \frac{(y+3)^2}{9} = 1$
33. $y^2 - 3x^2 + 6y + 6x - 18 = 0$
34. $4x^2 - 25y^2 - 8x - 96 = 0$

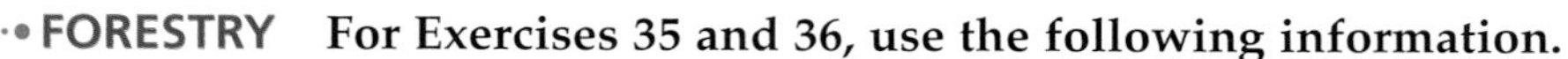

FORESTRY **For Exercises 35 and 36, use the following information.**
A forester at an outpost and another forester at the primary station both heard an explosion. The outpost and the primary station are 6 kilometers apart.

35. If one forester heard the explosion 6 seconds before the other, write an equation that describes all the possible locations of the explosion. Place the two forester stations on the x-axis with the midpoint between the stations at the origin. The transverse axis is horizontal. (*Hint*: The speed of sound is about 0.35 kilometer per second.)

36. Draw a sketch of the possible locations of the explosion. Include the ranger stations in the drawing.

Career Choices

Forester

Foresters work for private companies or governments to protect and manage forest land. They also supervise the planting of trees and use controlled burning to clear weeds, brush, and logging debris.

Online Research
For information about a career as a forester, visit:
www.algebra2.com/careers

37. **STRUCTURAL DESIGN** An architect's design for a building includes some large pillars with cross sections in the shape of hyperbolas. The curves can be modeled by the equation $\frac{x^2}{0.25} - \frac{y^2}{9} = 1$, where the units are in meters. If the pillars are 4 meters tall, find the width of the top of each pillar and the width of each pillar at the narrowest point in the middle. Round to the nearest centimeter.

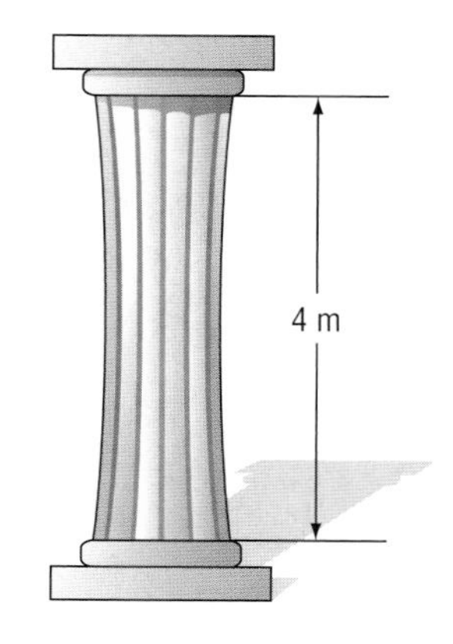

38. **CRITICAL THINKING** A hyperbola with a horizontal transverse axis contains the point at $(4, 3)$. The equations of the asymptotes are $y - x = 1$ and $y + x = 5$. Write the equation for the hyperbola.

39. **PHOTOGRAPHY** A curved mirror is placed in a store for a wide-angle view of the room. The right-hand branch of $\frac{x^2}{1} - \frac{y^2}{3} = 1$ models the curvature of the mirror. A small security camera is placed so that all of the 2-foot diameter of the mirror is visible. If the back of the room lies on $x = -18$, what width of the back of the room is visible to the camera? (*Hint*: Find the equations of the lines through the focus and each edge of the mirror.)

40. **WRITING IN MATH** Answer the question that was posed at the beginning of the lesson.

How are hyperbolas different from parabolas?

Include the following in your answer:
- differences in the graphs of hyperbolas and parabolas, and
- differences in the reflective properties of hyperbolas and parabolas.

Standards Practice

Standardized Test Practice

41. A leg of an isosceles right triangle has a length of 5 units. What is the length of the hypotenuse?

(A) $\frac{5\sqrt{2}}{2}$ units (B) 5 units (C) $5\sqrt{2}$ units (D) 10 units

42. In the figure, what is the sum of the slopes of $\overline{AB}$ and $\overline{AC}$?

(A) -1 (B) 0
(C) 1 (D) 8

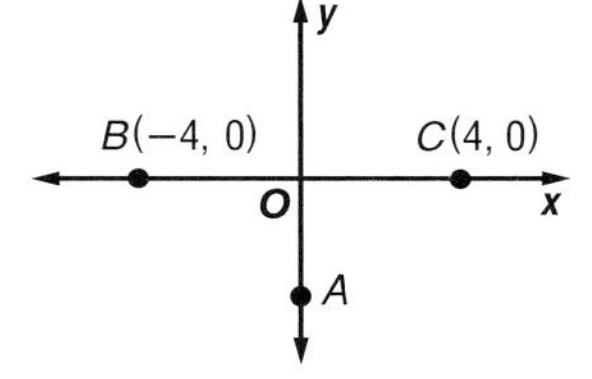

Extending the Lesson

A hyperbola with asymptotes that are not perpendicular is called a **nonrectangular hyperbola**. Most of the hyperbolas you have studied so far are nonrectangular. A **rectangular hyperbola** has perpendicular asymptotes. For example, the graph of $x^2 - y^2 = 1$ is a rectangular hyperbola. The graphs of equations of the form $xy = c$, where c is a constant, are rectangular hyperbolas with the coordinate axes as their asymptotes.

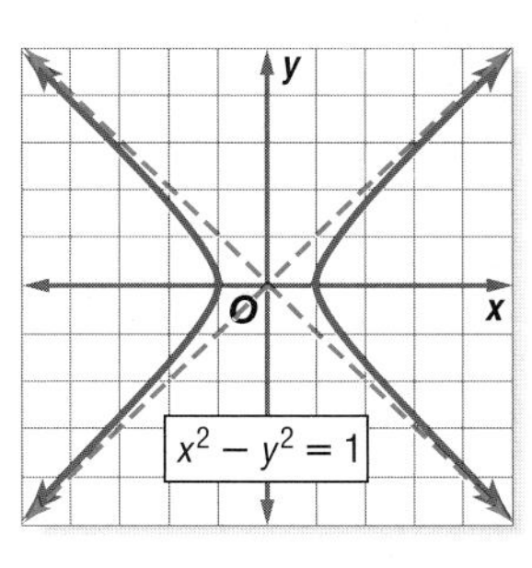

For Exercises 43 and 44, consider the equation $xy = 2$.

43. Plot some points and use them to graph the equation. Be sure to consider negative values for the variables.

44. Find the coordinates of the vertices of the graph of the equation.

45. Graph $xy = -2$.

46. Describe the transformations that can be applied to the graph of $xy = 2$ to obtain the graph of $xy = -2$.

Maintain Your Skills

Mixed Review

Write an equation for the ellipse that satisfies each set of conditions. *(Lesson 8-4)*

47. endpoints of major axis at (1, 2) and (9, 2), endpoints of minor axis at (5, 1) and (5, 3)

48. major axis 8 units long and parallel to y-axis, minor axis 6 units long, center at $(-3, 1)$

49. foci at (5, 4) and $(-3, 4)$, major axis 10 units long

50. Find the center and radius of the circle with equation $x^2 + y^2 - 10x + 2y + 22 = 0$. Then graph the circle. *(Lesson 8-3)*

Solve each equation by factoring. *(Lesson 6-2)*

51. $x^2 + 6x + 8 = 0$

52. $2q^2 + 11q = 21$

Perform the indicated operations, if possible. *(Lesson 4-5)*

53. $\begin{bmatrix} 2 & -1 \\ 0 & 5 \end{bmatrix} \cdot \begin{bmatrix} -3 & 2 \\ 1 & 4 \end{bmatrix}$

54. $[1 \quad -3] \cdot \begin{bmatrix} 4 & -2 & 1 \\ -3 & 2 & 0 \end{bmatrix}$

55. **PAGERS** Refer to the graph at the right. What was the average rate of change of the number of pager subscribers from 1996 to 1999? *(Lesson 2-3)*

56. Solve $|2x + 1| = 9$. *(Lesson 1-4)*

57. Simplify $7x + 8y + 9y - 5x$. *(Lesson 1-2)*

Getting Ready for the Next Lesson

PREREQUISITE SKILL **Each equation is of the form $Ax^2 + Bxy + Cy^2 + Dx + Ey + F = 0$. Identify the values of A, B, and C.**
(To review ***coefficients****, see Lesson 5-1.)*

58. $2x^2 + 3xy - 5y^2 = 0$

59. $x^2 - 2xy + 9y^2 = 0$

60. $-3x^2 + xy + 2y^2 + 4x - 7y = 0$

61. $5x^2 - 2y^2 + 5x - y = 0$

62. $x^2 - 4x + 5y + 2 = 0$

63. $xy - 2x - 3y + 6 = 0$

Practice Quiz 2

Lessons 8-4 and 8-5

1. Write an equation of the ellipse with foci at (3, 8) and (3, −6) and endpoints of the major axis at (3, −8) and (3, 10). *(Lesson 8-4)*

Find the coordinates of the center and foci and the lengths of the major and minor axes of the ellipse with the given equation. Then graph the ellipse. *(Lesson 8-4)*

2. $\frac{(x-4)^2}{9} + \frac{(y+2)^2}{1} = 1$

3. $16x^2 + 5y^2 + 32x - 10y - 59 = 0$

Write an equation for the hyperbola that satisfies each set of conditions. *(Lesson 8-5)*

4. vertices (−3, 0) and (3, 0), conjugate axis of length 8 units

5. vertices (−2, 2) and (6, 2), foci $(2 \pm \sqrt{21}, 2)$

8-6 Conic Sections

California Standards Standard 17.0 Given a quadratic equation of the form $ax^2 + by^2 + cx + dy + e = 0$, students can use the method for completing the square to put the equation into standard form and can recognize whether the graph of the equation is a circle, ellipse, parabola, or hyperbola. Students can then graph the equation.

What You'll Learn

- Write equations of conic sections in standard form.
- Identify conic sections from their equations.

How can you use a flashlight to make conic sections?

Recall that parabolas, circles, ellipses, and hyperbolas are called conic sections because they are the cross sections formed when a double cone is sliced by a plane. You can use a flashlight and a flat surface to make patterns in the shapes of conic sections.

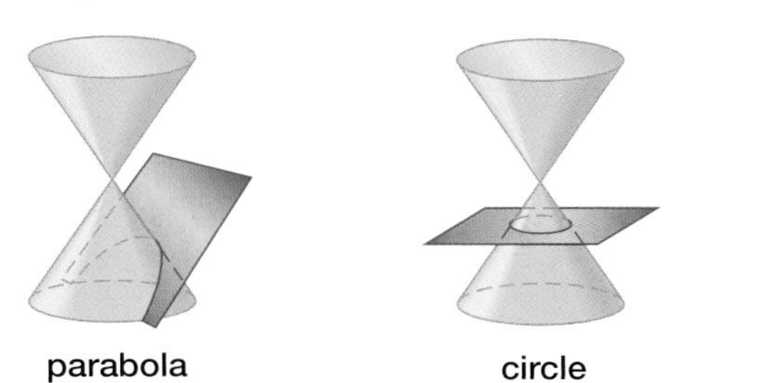

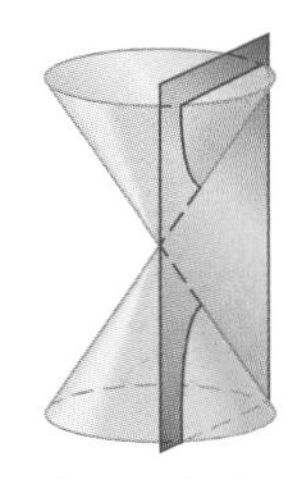

STANDARD FORM The equation of any conic section can be written in the form of the general quadratic equation $Ax^2 + Bxy + Cy^2 + Dx + Ey + F = 0$, where A, B, and C are not all zero. If you are given an equation in this general form, you can complete the square to write the equation in one of the standard forms you have learned.

Study Tip

Reading Math
In this lesson, the word *ellipse* means an ellipse that is not a circle.

Concept Summary — Standard Form of Conic Sections

Conic Section	Standard Form of Equation
Parabola	$y = a(x - h)^2 + k$ or $x = a(y - k)^2 + h$
Circle	$(x - h)^2 + (y - k)^2 = r^2$
Ellipse	$\frac{(x-h)^2}{a^2} + \frac{(y-k)^2}{b^2} = 1$ or $\frac{(y-k)^2}{a^2} + \frac{(x-h)^2}{b^2} = 1$, $a \neq b$
Hyperbola	$\frac{(x-h)^2}{a^2} - \frac{(y-k)^2}{b^2} = 1$ or $\frac{(y-k)^2}{a^2} - \frac{(x-h)^2}{b^2} = 1$

Example 1 Rewrite an Equation of a Conic Section

Write the equation $x^2 + 4y^2 - 6x - 7 = 0$ in standard form. State whether the graph of the equation is a *parabola, circle, ellipse,* or *hyperbola*. Then graph the equation.

Write the equation in standard form.

$x^2 + 4y^2 - 6x - 7 = 0$	Original equation
$x^2 - 6x + ■ + 4y^2 = 7 + ■$	Isolate terms.
$x^2 - 6x + 9 + 4y^2 = 7 + 9$	Complete the square.
$(x - 3)^2 + 4y^2 = 16$	$x^2 - 6x + 9 = (x - 3)^2$
$\frac{(x-3)^2}{16} + \frac{y^2}{4} = 1$	Divide each side by 16.

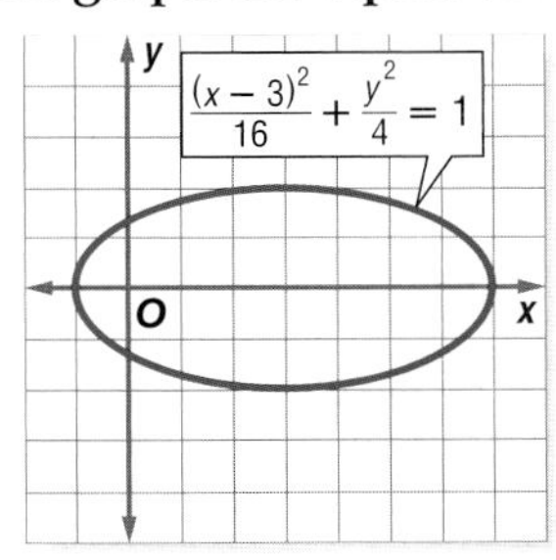

The graph of the equation is an ellipse with its center at (3, 0).

IDENTIFY CONIC SECTIONS Instead of writing the equation in standard form, you can determine what type of conic section an equation of the form $Ax^2 + Bxy + Cy^2 + Dx + Ey + F = 0$, where $B = 0$, represents by looking at A and C.

Concept Summary — *Identifying Conic Sections*

Conic Section	Relationship of A and C
Parabola	$A = 0$ or $C = 0$, but not both.
Circle	$A = C$
Ellipse	A and C have the same sign and $A \neq C$.
Hyperbola	A and C have opposite signs.

Example 2 *Analyze an Equation of a Conic Section*

Without writing the equation in standard form, state whether the graph of each equation is a *parabola, circle, ellipse,* or *hyperbola.*

a. $y^2 - 2x^2 - 4x - 4y - 4 = 0$

$A = -2$ and $C = 1$. Since A and C have opposite signs, the graph is a hyperbola.

b. $4x^2 + 4y^2 + 20x - 12y + 30 = 0$

$A = 4$ and $C = 4$. Since $A = C$, the graph is a circle.

c. $y^2 - 3x + 6y + 12 = 0$

$C = 1$. Since there is no x^2 term, $A = 0$. The graph is a parabola.

Check for Understanding

Concept Check

1. **OPEN ENDED** Write an equation of the form $Ax^2 + Bxy + Cy^2 + Dx + Ey + F = 0$, where $A = 2$, that represents a circle.
2. **Write** the general quadratic equation for which $A = 2$, $B = 0$, $C = 0$, $D = -4$, $E = 7$, and $F = 1$.
3. **Explain** why the graph of $x^2 + y^2 - 4x + 2y + 5 = 0$ is a single point.

Guided Practice

Write each equation in standard form. State whether the graph of the equation is a *parabola, circle, ellipse,* or *hyperbola.* Then graph the equation.

4. $y = x^2 + 3x + 1$
5. $y^2 - 2x^2 - 16 = 0$
6. $x^2 + y^2 = x + 2$
7. $x^2 + 4y^2 + 2x - 24y + 33 = 0$

Without writing the equation in standard form, state whether the graph of each equation is a *parabola, circle, ellipse,* or *hyperbola.*

8. $y^2 - x - 10y + 34 = 0$
9. $3x^2 + 2y^2 + 12x - 28y + 104 = 0$

Application

AVIATION For Exercises 10 and 11, use the following information.
When an airplane flies faster than the speed of sound, it produces a shock wave in the shape of a cone. Suppose the shock wave intersects the ground in a curve that can be modeled by $x^2 - 14x + 4 = 9y^2 - 36y$.

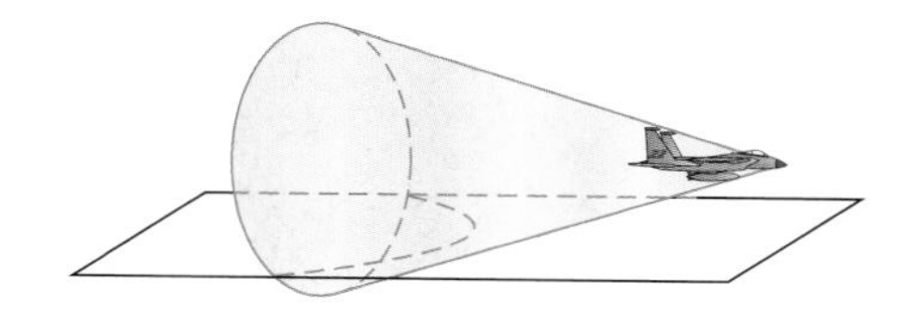

10. Identify the shape of the curve.

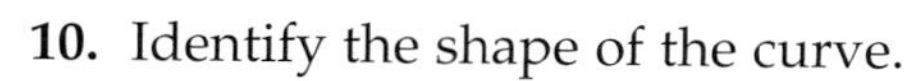

11. Graph the equation.

Practice and Apply

Homework Help

For Exercises	See Examples
12–32	1
33–43	2

Extra Practice
See page 846.

Write each equation in standard form. State whether the graph of the equation is a *parabola, circle, ellipse,* or *hyperbola.* Then graph the equation.

12. $6x^2 + 6y^2 = 162$
13. $4x^2 + 2y^2 = 8$
14. $x^2 = 8y$
15. $4y^2 - x^2 + 4 = 0$
16. $(x - 1)^2 - 9(y - 4)^2 = 36$
17. $y + 4 = (x - 2)^2$
18. $(y - 4)^2 = 9(x - 4)$
19. $x^2 + y^2 + 4x - 6y = -4$
20. $x^2 + y^2 + 6y + 13 = 40$
21. $x^2 - y^2 + 8x = 16$
22. $x^2 + 2y^2 = 2x + 8$
23. $x^2 - 8y + y^2 + 11 = 0$
24. $9y^2 + 18y = 25x^2 + 216$
25. $3x^2 + 4y^2 + 8y = 8$
26. $x^2 + 4y^2 - 11 = 2(4y - x)$
27. $y + x^2 = -(8x + 23)$
28. $6x^2 - 24x - 5y^2 - 10y - 11 = 0$
29. $25y^2 + 9x^2 - 50y - 54x = 119$

30. **ASTRONOMY** The orbits of comets follow paths in the shapes of conic sections. For example, Halley's Comet follows an elliptical orbit with the Sun located at one focus. What type(s) of orbit(s) pass by the Sun only once?

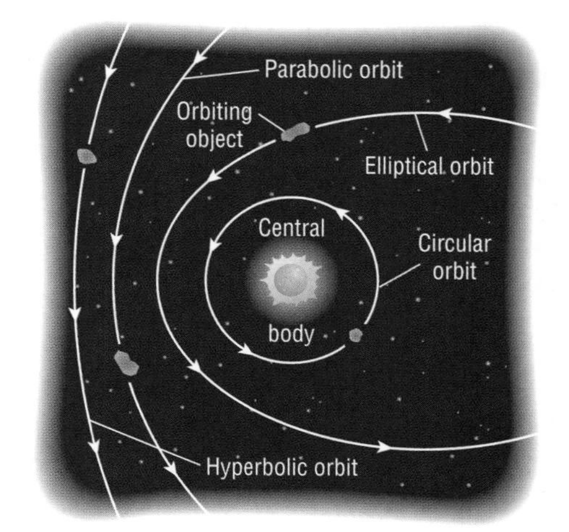

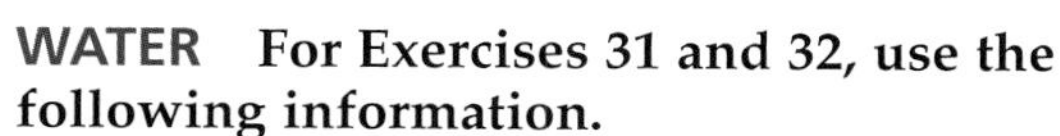

WATER For Exercises 31 and 32, use the following information.
If two stones are thrown into a lake at different points, the points of intersection of the resulting ripples will follow a conic section. Suppose the conic section has the equation $x^2 - 2y^2 - 2x - 5 = 0$.

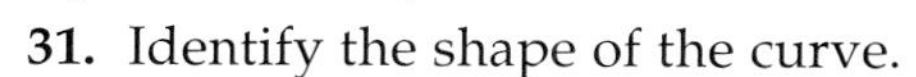

31. Identify the shape of the curve.
32. Graph the equation.

Without writing the equation in standard form, state whether the graph of each equation is a *parabola, circle, ellipse,* or *hyperbola.*

33. $x^2 + y^2 - 8x - 6y + 5 = 0$
34. $3x^2 - 2y^2 + 32y - 134 = 0$
35. $y^2 + 18y - 2x = -84$
36. $7x^2 - 28x + 4y^2 + 8y = -4$
37. $5x^2 + 6x - 4y = x^2 - y^2 - 2x$
38. $2x^2 + 12x + 18 - y^2 = 3(2 - y^2) + 4y$

39. Identify the shape of the graph of the equation $2x^2 + 3x - 4y + 2 = 0$.

40. What type of conic section is represented by the equation $y^2 - 6y = x^2 - 8$?

For Exercises 41–43, match each equation below with the situation that it could represent.

a. $9x^2 + 4y^2 - 36 = 0$
b. $0.004x^2 - x + y - 3 = 0$
c. $x^2 + y^2 - 20x + 30y - 75 = 0$

41. **SPORTS** the flight of a baseball
42. **PHOTOGRAPHY** the oval opening in a picture frame
43. **GEOGRAPHY** the set of all points that are 20 miles from a landmark

CRITICAL THINKING **For Exercises 44 and 45, use the following information.**

The graph of an equation of the form $\frac{x^2}{a^2} - \frac{y^2}{b^2} = 0$ is a special case of a hyperbola.

44. Identify the graph of such an equation.

45. Explain how to obtain such a set of points by slicing a double cone with a plane.

46. **WRITING IN MATH** Answer the question that was posed at the beginning of the lesson.

How can you use a flashlight to make conic sections?

Include the following in your answer:

- an explanation of how you could point the flashlight at a ceiling or wall to make a circle, and
- an explanation of how you could point the flashlight to make a branch of a hyperbola.

47. Which conic section is not symmetric about the y-axis?

Ⓐ $x^2 - y + 3 = 0$ Ⓑ $y^2 - x^2 - 1 = 0$

Ⓒ $6x^2 + y^2 - 6 = 0$ Ⓓ $x^2 + y^2 - 2x - 3 = 0$

48. What is the equation of the graph at the right?

Ⓐ $y = x^2 + 1$ Ⓑ $y - x = 1$

Ⓒ $y^2 - x^2 = 1$ Ⓓ $x^2 + y^2 = 1$

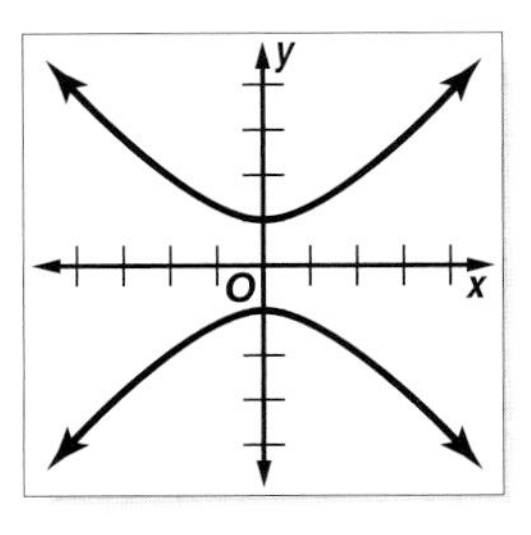

Extending the Lesson

49. Refer to Exercise 43 on page 440. Eccentricity can be studied for conic sections other than ellipses. The expression for the eccentricity of a hyperbola is $\frac{c}{a}$, just as for an ellipse. The eccentricity of a parabola is 1. Find inequalities for the eccentricities of noncircular ellipses and hyperbolas, respectively.

Maintain Your Skills

Mixed Review

Write an equation of the hyperbola that satisfies each set of conditions. *(Lesson 8-5)*

50. vertices (5, 10) and (5, −2), conjugate axis of length 8 units

51. vertices (6, −6) and (0, −6), foci $(3 \pm \sqrt{13}, -6)$

52. Find the coordinates of the center and foci and the lengths of the major and minor axes of the ellipse with equation $4x^2 + 9y^2 - 24x + 72y + 144 = 0$. Then graph the ellipse. *(Lesson 8-4)*

Simplify. Assume that no variable equals 0. *(Lesson 5-1)*

53. $(x^3)^4$ **54.** $(m^5n^{-3})^2m^2n^7$ **55.** $\frac{x^2y^{-3}}{x^{-5}y}$

56. **HEALTH** The prediction equation $y = 205 - 0.5x$ relates a person's maximum heart rate for exercise y and age x. Use the equation to find the maximum heart rate for an 18-year-old. *(Lesson 2-5)*

Getting Ready for the Next Lesson

PREREQUISITE SKILL **Solve each system of equations.**
(To review ***solving systems of linear equations****, see Lesson 3-2.)*

57. $y = x + 4$
$2x + y = 10$

58. $4x + y = 14$
$4x - y = 10$

59. $x + 5y = 10$
$3x - 2y = -4$

Algebra Activity

A Follow-Up of Lesson 8-6

Conic Sections

Recall that a parabola is the set of all points that are equidistant from the focus and the directrix.

You can draw a parabola based on this definition by using special conic graph paper. This graph paper contains a series of concentric circles equally spaced from each other and a series of parallel lines tangent to the circles.

Number the circles consecutively beginning with the smallest circle. Number the lines with consecutive integers as shown in the sample at the right. Be sure that line 1 is tangent to circle 1.

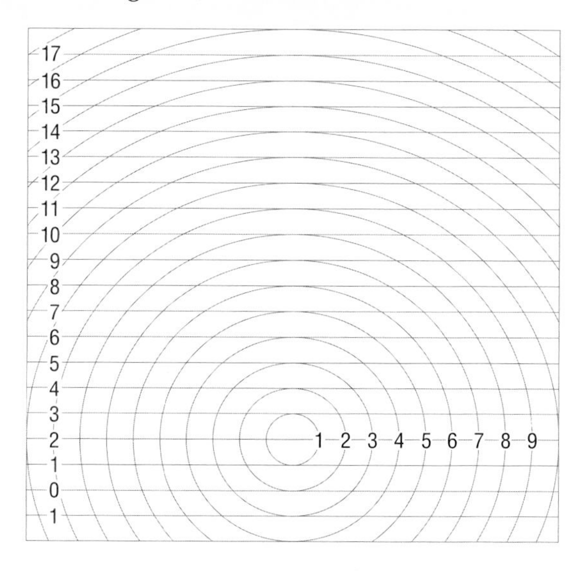

Activity 1

Mark the point at the intersection of circle 1 and line 1. Mark both points that are on line 2 and circle 2. Continue this process, marking both points on line 3 and circle 3, and so on. Then connect the points with a smooth curve.

Look at the diagram at the right. What shape is the graph? Note that every point on the graph is equidistant from the center of the small circle and the line labeled 0. The center of the small circle is the focus of the parabola, and line 0 is the directrix.

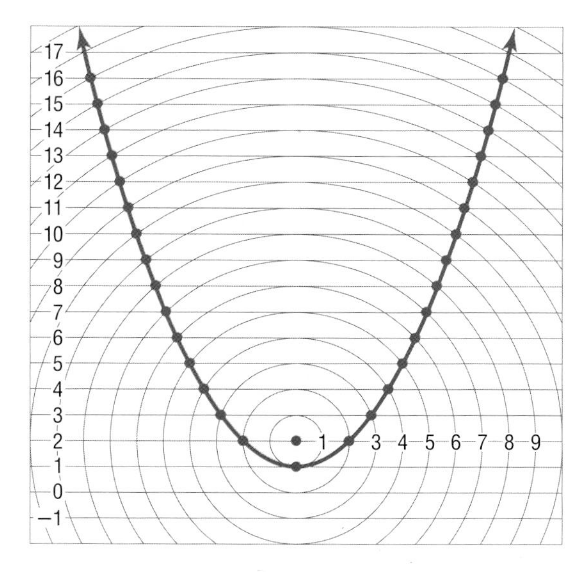

Activity 2

An ellipse is the set of points such that the sum of the distances from two fixed points is constant. The two fixed points are called the foci.

- Use graph paper like that shown. It contains two small circles and a series of concentric circles from each. The concentric circles are tangent to each other as shown.
- Choose the constant 13. Mark the points at the intersections of circle 9 and circle 4, because $9 + 4 = 13$. Continue this process until you have marked the intersection of all circles whose sum is 13.
- Connect the points to form a smooth curve. The curve is an ellipse whose foci are the centers of the two small circles on the graph paper.

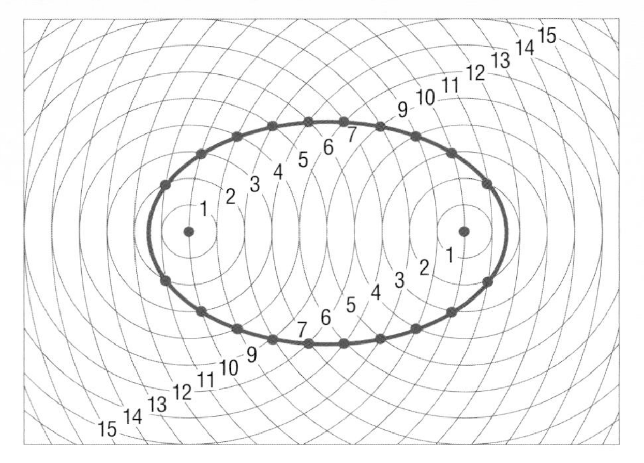

Activity 3

A hyperbola is the set of points such that the difference of the distances from two fixed points is constant. The two fixed points are called the foci.

- Use the same type of graph paper that you used for the ellipse in Activity 2. Choose the constant 7. Mark the points at the intersections of circle 9 and circle 2, because $9 - 2 = 7$. Continue this process until you have marked the intersections of all circles whose difference in radius is 7.
- Connect the points to form a hyperbola.

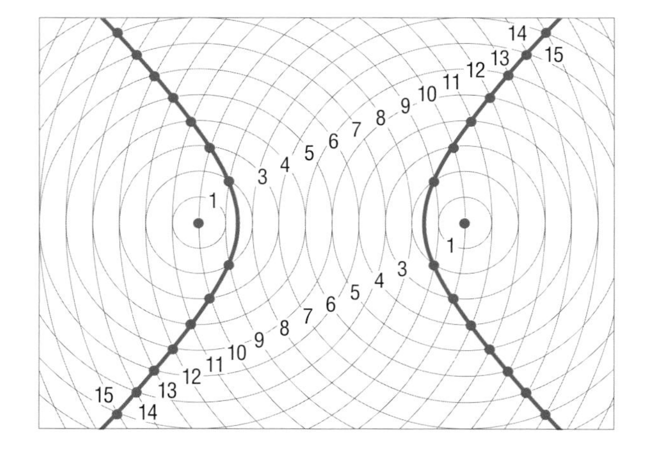

Model and Analyze

1. Use the type of graph paper you used in Activity 1. Mark the intersection of line 0 and circle 2. Then mark the two points on line 1 and circle 3, the two points on line 2 and circle 4, and so on. Draw the new parabola. Continue this process and make as many parabolas as you can on one sheet of the graph paper. The focus is always the center of the small circle. Why are the resulting graphs parabolas?

2. In Activity 2, you drew an ellipse such that the sum of the distances from two fixed points was 13. Choose 10, 11, 12, 14, and so on for that sum, and draw as many ellipses as you can on one piece of the graph paper.

a. Why can you not start with 9 as the sum?

b. What happens as the sum increases? decreases?

3. In Activity 3, you drew a hyperbola such that the difference of the distances from two fixed points was 7. Choose other numbers and draw as many hyperbolas as you can on one piece of graph paper. What happens as the difference increases? decreases?

8-7 Solving Quadratic Systems

What You'll Learn

- Solve systems of quadratic equations algebraically and graphically.
- Solve systems of quadratic inequalities graphically.

How do systems of equations apply to video games?

Computer software often uses a coordinate system to keep track of the locations of objects on the screen. Suppose an enemy space station is located at the center of the screen, which is the origin in a coordinate system. The space station is surrounded by a circular force field of radius 50 units. If the spaceship you control is flying toward the center along the line with equation $y = 3x$, the point where the ship hits the force field is a solution of a system of equations.

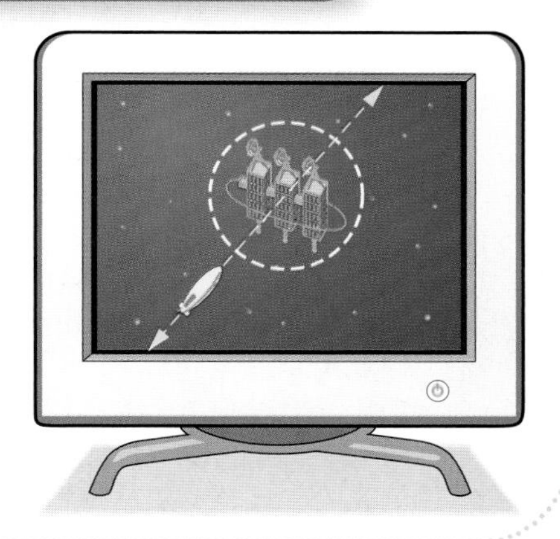

SYSTEMS OF QUADRATIC EQUATIONS If the graphs of a system of equations are a conic section and a line, the system may have zero, one, or two solutions. Some of the possible situations are shown below.

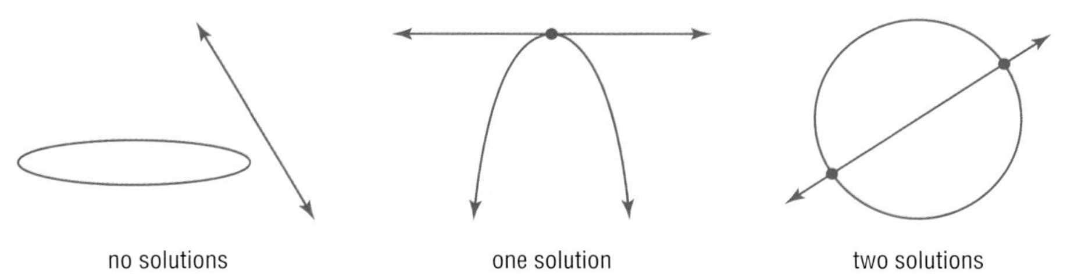

You have solved systems of linear equations graphically and algebraically. You can use similar methods to solve systems involving quadratic equations.

Example 1 Linear-Quadratic System

Solve the system of equations.

$x^2 - 4y^2 = 9$
$4y - x = 3$

You can use a graphing calculator to help visualize the relationships of the graphs of the equations and predict the number of solutions.

Solve each equation for y to obtain $y = \pm\frac{\sqrt{x^2 - 9}}{2}$ and $y = \frac{1}{4}x + \frac{3}{4}$. Enter the functions $y = \frac{\sqrt{x^2 - 9}}{2}$, $y = -\frac{\sqrt{x^2 - 9}}{2}$, and $y = \frac{1}{4}x + \frac{3}{4}$ on the Y= screen. The graph indicates that the hyperbola and line intersect in two points. So the system has two solutions.

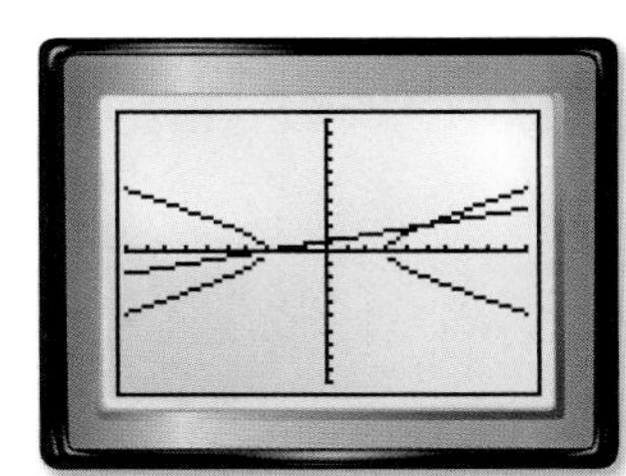

[−10, 10] scl: 1 by [−10, 10] scl: 1

(continued on the next page)

Use substitution to solve the system. First rewrite $4y - x = 3$ as $x = 4y - 3$.

$x^2 - 4y^2 = 9$	First equation in the system
$(4y - 3)^2 - 4y^2 = 9$	Substitute $4y - 3$ for x.
$12y^2 - 24y = 0$	Simplify.
$y^2 - 2y = 0$	Divide each side by 12.
$y(y - 2) = 0$	Factor.
$y = 0 \quad \text{or} \quad y - 2 = 0$	Zero Product Property
$y = 2$	Solve for y.

Now solve for x.

$x = 4y - 3$	$x = 4y - 3$	Equation for x in terms of y
$= 4(0) - 3$	$= 4(2) - 3$	Substitute the y values.
$= -3$	$= 5$	Simplify.

The solutions of the system are $(-3, 0)$ and $(5, 2)$. Based on the graph, these solutions are reasonable.

If the graphs of a system of equations are two conic sections, the system may have zero, one, two, three, or four solutions. Some of the possible situations are shown below.

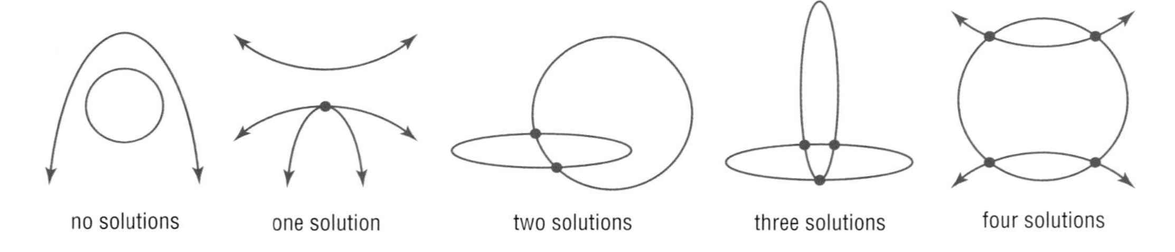

Example 2 Quadratic-Quadratic System

Solve the system of equations.

$$y^2 = 13 - x^2$$
$$x^2 + 4y^2 = 25$$

A graphing calculator indicates that the circle and ellipse intersect in four points. So, this system has four solutions.

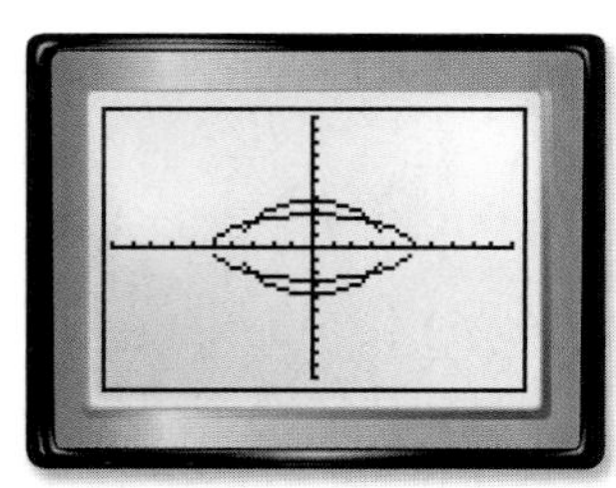

[−10, 10] scl: 1 by [−10, 10] scl: 1

Study Tip

Graphing Calculators

If you use ZSquare on the ZOOM menu, the graph of the first equation will look like a circle.

Use the elimination method to solve the system.

$$y^2 = 13 - x^2$$
$$x^2 + 4y^2 = 25$$

$-x^2 - y^2 = -13$	Rewrite the first original equation.
$(+)\ x^2 + 4y^2 = 25$	Second original equation
$3y^2 = 12$	Add.
$y^2 = 4$	Divide each side by 3.
$y = \pm 2$	Take the square root of each side.

Substitute 2 and -2 for y in either of the original equations and solve for x.

$x^2 + 4y^2 = 25$	$x^2 + 4y^2 = 25$	Second original equation
$x^2 + 4(2)^2 = 25$	$x^2 + 4(-2)^2 = 25$	Substitute for y.
$x^2 = 9$	$x^2 = 9$	Subtract 16 from each side.
$x = \pm 3$	$x = \pm 3$	Take the square root of each side.

The solutions are (3, 2), (−3, 2), (−3, −2), and (3, −2).

A graphing calculator can be used to approximate the solutions of a system of equations.

Graphing Calculator Investigation

Quadratic Systems

The calculator screen shows the graphs of two circles.

Think and Discuss

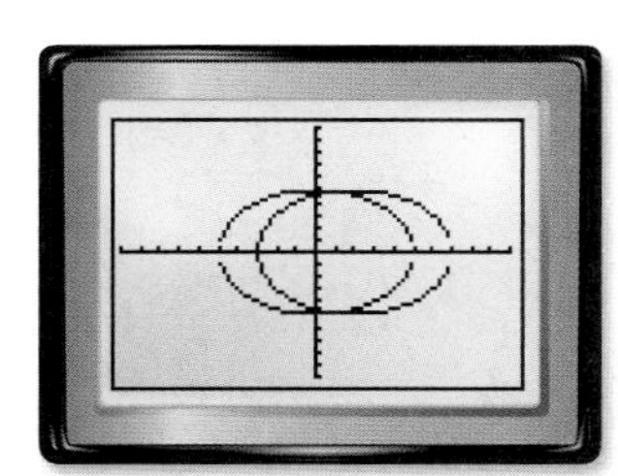

[−10, 10] scl: 1 by [−10, 10] scl: 1

1. Write the system of equations represented by the graph.
2. Enter the equations into a TI-83 Plus and use the intersect feature on the CALC menu to solve the system. Round to the nearest hundredth.
3. Solve the system algebraically.

4. Can you always find the exact solution of a system using a graphing calculator? Explain.

Use a graphing calculator to solve each system of equations. Round to the nearest hundredth.

5. $y = x + 2$
   $x^2 + y^2 = 9$

6. $3x^2 + y^2 = 11$
   $y = x^2 + x + 1$

SYSTEMS OF QUADRATIC INEQUALITIES You have learned how to solve systems of linear inequalities by graphing. Systems of quadratic inequalities are also solved by graphing.

The graph of an inequality involving a parabola, circle, or ellipse is either the interior or the exterior of the conic section. The graph of an inequality involving a hyperbola is either the region between the branches or the two regions inside the branches. As with linear inequalities, examine the inequality symbol to determine whether to include the boundary.

Example 3 System of Quadratic Inequalities

Solve the system of inequalities by graphing.

$y \leq x^2 - 2$
$x^2 + y^2 < 16$

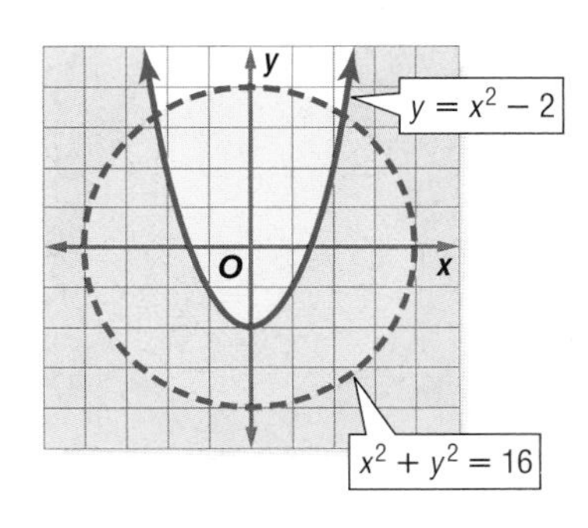

The graph of $y \leq x^2 - 2$ is the parabola $y = x^2 - 2$ and the region outside or below it. This region is shaded blue.

The graph of $x^2 + y^2 < 16$ is the interior of the circle $x^2 + y^2 = 16$. This region is shaded yellow.

The intersection of these regions, shaded green, represents the solution of the system of inequalities.

Study Tip

Graphing Quadratic Inequalities
If you are unsure about which region to shade, you can test one or more points, as you did with linear inequalities.

Check for Understanding

Concept Check

1. **Graph** each system of equations. Use the graph to solve the system.

   a. $4x - 3y = 0$
   $x^2 + y^2 = 25$

   b. $y = 5 - x^2$
   $y = 2x^2 + 2$

2. **Sketch** a parabola and an ellipse that intersect at exactly three points.

3. **OPEN ENDED** Write a system of quadratic equations for which (2, 6) is a solution.

Guided Practice

Find the exact solution(s) of each system of equations.

4. $y = 5$
   $y^2 = x^2 + 9$

5. $y - x = 1$
   $x^2 + y^2 = 25$

6. $3x = 8y^2$
   $8y^2 - 2x^2 = 16$

7. $5x^2 + y^2 = 30$
   $9x^2 - y^2 = -16$

Solve each system of inequalities by graphing.

8. $x + y < 4$
   $9x^2 - 4y^2 \geq 36$

9. $x^2 + y^2 < 25$
   $4x^2 - 9y^2 < 36$

Application

10. **EARTHQUAKES** In a coordinate system where a unit represents one mile, the epicenter of an earthquake was determined to be 50 miles from a station at the origin. It was also 40 miles from a station at (0, 30) and 13 miles from a station at (35, 18). Where was the epicenter located?

Practice and Apply

Homework Help

For Exercises	See Examples
11–16, 25 29	1
17–24, 26–28, 30, 31	2
32–37	3

Extra Practice See page 847.

Find the exact solution(s) of each system of equations.

11. $y = x + 2$
    $y = x^2$

12. $y = x + 3$
    $y = 2x^2$

13. $x^2 + y^2 = 36$
    $y = x + 2$

14. $y^2 + x^2 = 9$
    $y = 7 - x$

15. $\frac{x^2}{30} + \frac{y^2}{6} = 1$
    $x = y$

16. $\frac{x^2}{36} - \frac{y^2}{4} = 1$
    $x = y$

17. $4x + y^2 = 20$
    $4x^2 + y^2 = 100$

18. $y + x^2 = 3$
    $x^2 + 4y^2 = 36$

19. $x^2 + y^2 = 64$
    $x^2 + 64y^2 = 64$

20. $y^2 + x^2 = 25$
    $y^2 + 9x^2 = 25$

21. $y^2 = x^2 - 25$
    $x^2 - y^2 = 7$

22. $y^2 = x^2 - 7$
    $x^2 + y^2 = 25$

23. $2x^2 + 8y^2 + 8x - 48y + 30 = 0$
    $2x^2 - 8y^2 = -48y + 90$

24. $3x^2 - 20y^2 - 12x + 80y - 96 = 0$
    $3x^2 + 20y^2 = 80y + 48$

25. Where do the graphs of the equations $y = 2x + 1$ and $2x^2 + y^2 = 11$ intersect?

26. What are the coordinates of the points that lie on the graphs of both $x^2 + y^2 = 25$ and $2x^2 + 3y^2 = 66$?

27. **ROCKETS** Two rockets are launched at the same time, but from different heights. The height y in feet of one rocket after t seconds is given by $y = -16t^2 + 150t + 5$. The height of the other rocket is given by $y = -16t^2 + 160t$. After how many seconds are the rockets at the same height?

28. ADVERTISING The corporate logo for an automobile manufacturer is shown at the right. Write a system of three equations to model this logo.

29. MIRRORS A hyperbolic mirror is a mirror in the shape of one branch of a hyperbola. Such a mirror reflects light rays directed at one focus toward the other focus. Suppose a hyperbolic mirror is modeled by the upper branch of the hyperbola with equation $\frac{y^2}{9} - \frac{x^2}{16} = 1$. A light source is located at $(-10, 0)$. Where should the light from the source hit the mirror so that the light will be reflected to $(0, -5)$?

More About. . .

Astronomy

The astronomical unit (AU) is the mean distance between Earth and the Sun. One AU is about 93 million miles or 150 million kilometers.

Source: www.infoplease.com

ASTRONOMY **For Exercises 30 and 31, use the following information.**

The orbit of Pluto can be modeled by the equation $\frac{x^2}{39.5^2} + \frac{y^2}{38.3^2} = 1$, where the units are astronomical units. Suppose a comet is following a path modeled by the equation $x = y^2 + 20$.

30. Find the point(s) of intersection of the orbits of Pluto and the comet. Round to the nearest tenth.

31. Will the comet necessarily hit Pluto? Explain.

Solve each system of inequalities by graphing.

32. $x + 2y > 1$
$x^2 + y^2 \le 25$

33. $x + y \le 2$
$4x^2 - y^2 \ge 4$

34. $x^2 + y^2 \ge 4$
$4y^2 + 9x^2 \le 36$

35. $x^2 + y^2 < 36$
$4x^2 + 9y^2 > 36$

36. $y^2 < x$
$x^2 - 4y^2 < 16$

37. $x^2 \le y$
$y^2 - x^2 \ge 4$

CRITICAL THINKING **For Exercises 38–42, find all values of k for which the system of equations has the given number of solutions. If no values of k meet the condition, write *none*.**

$$x^2 + y^2 = k^2 \qquad \frac{x^2}{9} + \frac{y^2}{4} = 1$$

38. no solutions

39. one solution

40. two solutions

41. three solutions

42. four solutions

43. **WRITING IN MATH** Answer the question that was posed at the beginning of the lesson.

How do systems of equations apply to video games?

Include the following in your answer:

- a linear-quadratic system of equations that applies to this situation,
- an explanation of how you know that the spaceship is headed directly toward the center of the screen, and
- the coordinates of the point at which the spaceship will hit the force field, assuming that the spaceship moves from the bottom of the screen toward the center.

44. If $\boxed{x}$ is defined to be $x^2 - 4x$ for all numbers x, which of the following is the greatest?

Ⓐ $\boxed{0}$ Ⓑ $\boxed{1}$ Ⓒ $\boxed{2}$ Ⓓ $\boxed{3}$

45. How many three-digit numbers are divisible by 3?

Ⓐ 299 Ⓑ 300 Ⓒ 301 Ⓓ 302

SYSTEMS OF EQUATIONS Write a system of equations that satisfies each condition. Use a graphing calculator to verify that you are correct.

46. two parabolas that intersect in two points

47. a hyperbola and a circle that intersect in three points

48. a circle and an ellipse that do not intersect

49. a circle and an ellipse that intersect in four points

50. a hyperbola and an ellipse that intersect in two points

51. two circles that intersect in three points

Maintain Your Skills

Mixed Review

Write each equation in standard form. State whether the graph of the equation is a *parabola, circle, ellipse,* or *hyperbola*. Then graph the equation. *(Lesson 8-6)*

52. $x^2 + y^2 + 4x + 2y - 6 = 0$

53. $9x^2 + 4y^2 - 24y = 0$

54. Find the coordinates of the vertices and foci and the equations of the asymptotes of the hyperbola with the equation $6y^2 - 2x^2 = 24$. Then graph the hyperbola. *(Lesson 8-5)*

Solve each equation by factoring. *(Lesson 6-5)*

55. $x^2 + 7x = 0$

56. $x^2 - 3x = 0$

57. $21 = x^2 + 4x$

58. $35 = -2x + x^2$

59. $9x^2 + 24x = -16$

60. $8x^2 + 2x = 3$

For Exercises 61 and 62, complete parts a–c for each quadratic equation. *(Lesson 6-5)*

a. Find the value of the discriminant.

b. Describe the number and type of roots.

c. Find the exact solutions by using the Quadratic Formula.

61. $5x^2 = 2$

62. $-3x^2 + 6x - 7 = 0$

Simplify. *(Lesson 5-9)*

63. $(3 + 2i) - (1 - 7i)$

64. $(8 - i)(4 - 3i)$

65. $\frac{2 + 3i}{1 + 2i}$

66. **CHEMISTRY** The mass of a proton is about 1.67×10^{-27} kilogram. The mass of an electron is about 9.11×10^{-31} kilogram. About how many times as massive as an electron is a proton? *(Lesson 5-1)*

Evaluate each determinant. *(Lesson 4-3)*

67. $\begin{vmatrix} 2 & -3 \\ 2 & 0 \end{vmatrix}$

68. $\begin{vmatrix} -4 & -2 \\ 5 & 3 \end{vmatrix}$

69. $\begin{vmatrix} 2 & 1 & -2 \\ 4 & 0 & 3 \\ -3 & 1 & 7 \end{vmatrix}$

70. Solve the system of equations. *(Lesson 3-5)*

$r + s + t = 15$
$r + t = 12$
$s + t = 10$

Write an equation in slope-intercept form for each graph. *(Lesson 2-4)*

71.

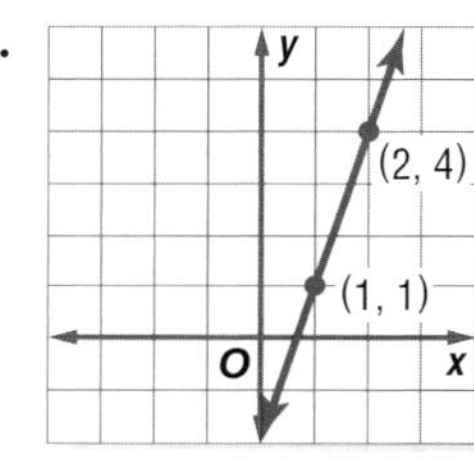

72.

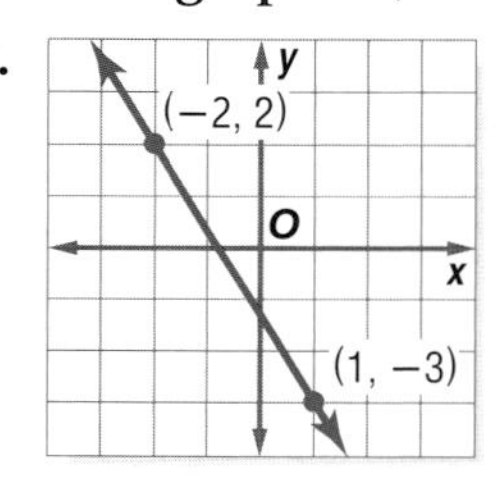

Study Guide and Review

Vocabulary and Concept Check

asymptote (p. 442)
center of a circle (p. 426)
center of a hyperbola (p. 442)
center of an ellipse (p. 434)
circle (p. 426)
conic section (p. 419)
conjugate axis (p. 442)
directrix (p. 419)
Distance Formula (p. 413)
ellipse (p. 433)
foci of a hyperbola (p. 441)
foci of an ellipse (p. 433)
focus of a parabola (p. 419)
hyperbola (p. 441)
latus rectum (p. 421)
major axis (p. 434)
Midpoint Formula (p. 412)
minor axis (p. 434)
parabola (p. 419)
tangent (p. 427)
transverse axis (p. 442)
vertex of a hyperbola (p. 442)

Tell whether each statement is *true* or *false*. If the statement is false, correct it to make it true.

1. An ellipse is the set of all points in a plane such that the sum of the distances from two given points in the plane, called the foci, is constant.
2. The major axis is the longer of the two axes of symmetry of an ellipse.
3. The formula used to find the distance between two points in a coordinate plane is $d = \sqrt{(x_2 - x_1)^2 + (y_2 - y_1)^2}$.
4. A parabola is the set of all points that are the same distance from a given point called the directrix and a given line called the focus.
5. The radius is the distance from the center of a circle to any point on the circle.
6. The conjugate axis of a hyperbola is a line segment parallel to the transverse axis.
7. A conic section is formed by slicing a double cone by a plane.
8. A hyperbola is the set of all points in a plane such that the absolute value of the sum of the distances from any point on the hyperbola to two given points is constant.
9. The midpoint formula is given by $\left(\frac{x_1 - x_2}{2}, \frac{y_1 - y_2}{2}\right)$.
10. The set of all points in a plane that are equidistant from a given point in a plane, called the center, forms a circle.

Lesson-by-Lesson Review

8-1 Midpoint and Distance Formulas

See pages 412–416.

Concept Summary

- Midpoint Formula: $M = \left(\frac{x_1 + x_2}{2}, \frac{y_1 + y_2}{2}\right)$
- Distance Formula: $d = \sqrt{(x_2 - x_1)^2 + (y_2 - y_1)^2}$

Examples 1 **Find the midpoint of a segment whose endpoints are at (−5, 9) and (11, −1).**

$\left(\frac{x_1 + x_2}{2}, \frac{y_1 + y_2}{2}\right) = \left(\frac{-5 + 11}{2}, \frac{9 + (-1)}{2}\right)$ Let $(x_1, y_1) = (-5, 9)$ and $(x_2, y_2) = (11, -1)$.

$= \left(\frac{6}{2}, \frac{8}{2}\right)$ or (3, 4) Simplify.

2 **Find the distance between $P(6, -4)$ and $Q(-3, 8)$.**

$d = \sqrt{(x_2 - x_1)^2 + (y_2 - y_1)^2}$ Distance Formula

$= \sqrt{(-3 - 6)^2 + [8 - (-4)]^2}$ Let $(x_1, y_1) = (6, -4)$ and $(x_2, y_2) = (-3, 8)$.

$= \sqrt{81 + 144}$ Subtract.

$= \sqrt{225}$ or 15 units Simplify.

Exercises **Find the midpoint of the line segment with endpoints at the given coordinates.** *See Example 1 on page 412.*

11. $(1, 2), (4, 6)$ **12.** $(-8, 0), (-2, 3)$ **13.** $\left(\frac{3}{5}, -\frac{7}{4}\right), \left(\frac{1}{4}, -\frac{2}{5}\right)$

Find the distance between each pair of points with the given coordinates. *See Examples 2 and 3 on pages 413 and 414.*

14. $(-2, 10), (-2, 13)$ **15.** $(8, 5), (-9, 4)$ **16.** $(7, -3), (1, 2)$

8-2 Parabolas

See pages 419–425.

Concept Summary

Parabolas		
Standard Form	$y = a(x - h)^2 + k$	$x = a(y - k)^2 + h$
Vertex	(h, k)	(h, k)
Axis of Symmetry	$x = h$	$y = k$
Focus	$\left(h, k + \frac{1}{4a}\right)$	$\left(h + \frac{1}{4a}, k\right)$
Directrix	$y = k - \frac{1}{4a}$	$x = h - \frac{1}{4a}$

Example **Graph $4y - x^2 = 14x - 27$.**

First write the equation in the form $y = a(x - h)^2 + k$.

$4y - x^2 = 14x - 27$ Original equation

$4y = x^2 + 14x - 27$ Isolate the terms with x.

$4y = (x^2 + 14x + ■) - 27 - ■$ Complete the square.

$4y = (x^2 + 14x + 49) - 27 - 49$ Add and subtract 49, since $\left(\frac{14}{2}\right)^2 = 49$.

$4y = (x + 7)^2 - 76$ $x^2 + 14x + 49 = (x + 7)^2$

$y = \frac{1}{4}(x + 7)^2 - 19$ Divide each side by 4.

vertex: $(-7, -19)$ axis of symmetry: $x = -7$

focus: $\left(-7, -19 + \frac{1}{4\left(\frac{1}{4}\right)}\right)$ or $(-7, -18)$

directrix: $y = -19 - \frac{1}{4\left(\frac{1}{4}\right)}$ or $y = -20$

direction of opening: upward since $a > 0$

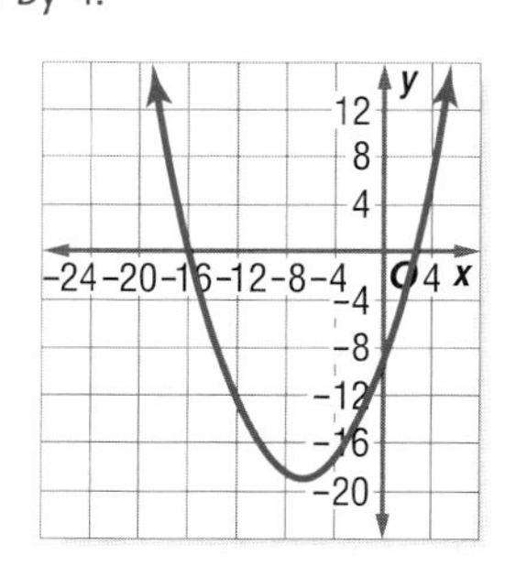

Exercises **Identify the coordinates of the vertex and focus, the equations of the axis of symmetry and directrix, and the direction of opening of the parabola with the given equation. Then find the length of the latus rectum and graph the parabola.** *See Examples 2–4 on pages 420–423.*

17. $(x - 1)^2 = 12(y - 1)$

18. $y + 6 = 16(x - 3)^2$

19. $x^2 - 8x + 8y + 32 = 0$

20. $x = 16y^2$

21. Write an equation for a parabola with vertex (0, 1) and focus (0, −1). Then graph the parabola. *See Example 4 on pages 422 and 423.*

8-3 Circles

See pages 426–431.

Concept Summary

- The equation of a circle with center (h, k) and radius r can be written in the form $(x - h)^2 + (y - k)^2 = r^2$.

Example **Graph $x^2 + y^2 + 8x - 24y + 16 = 0$.**

First write the equation in the form $(x - h)^2 + (y - k)^2 = r^2$.

$$x^2 + y^2 + 8x - 24y + 16 = 0 \quad \text{Original equation}$$

$$x^2 + 8x + \blacksquare + y^2 - 24y + \blacksquare = -16 + \blacksquare + \blacksquare \quad \text{Complete the squares.}$$

$$x^2 + 8x + 16 + y^2 - 24y + 144 = -16 + 16 + 144 \quad \left(\frac{8}{2}\right)^2 = 16, \left(\frac{-24}{2}\right)^2 = 144$$

$$(x + 4)^2 + (y - 12)^2 = 144 \quad \text{Write the trinomials as squares.}$$

The center of the circle is at (−4, 12) and the radius is 12.

Now draw the graph.

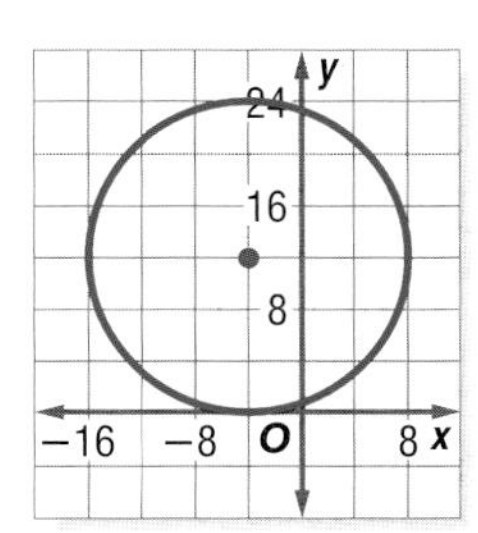

Exercises **Write an equation for the circle that satisfies each set of conditions.** *See Example 1 on page 426.*

22. center (2, −3), radius 5 units

23. center (−4, 0), radius $\frac{3}{4}$ unit

24. endpoints of a diameter at (9, 4) and (−3, −2)

25. center at (−1, 2), tangent to x-axis

Find the center and radius of the circle with the given equation. Then graph the circle. *See Examples 4 and 5 on page 428.*

26. $x^2 + y^2 = 169$

27. $(x + 5)^2 + (y - 11)^2 = 49$

28. $x^2 + y^2 - 6x + 16y - 152 = 0$

29. $x^2 + y^2 + 6x - 2y - 15 = 0$

8-4 Ellipses

See pages 433–440.

Concept Summary

Ellipses		
Standard Form of Equation	$\frac{(x-h)^2}{a^2} + \frac{(y-k)^2}{b^2} = 1$	$\frac{(y-k)^2}{a^2} + \frac{(x-h)^2}{b^2} = 1$
Direction of Major Axis	horizontal	vertical

Example **Graph $x^2 + 3y^2 - 16x + 24y + 31 = 0$.**

First write the equation in standard form by completing the squares.

$$x^2 + 3y^2 - 16x + 24y + 31 = 0 \quad \text{Original equation}$$

$$x^2 - 16x + \blacksquare + 3(y^2 + 8y + \blacksquare) = -31 + \blacksquare + 3(\blacksquare) \quad \text{Complete the squares.}$$

$$x^2 - 16x + 64 + 3(y^2 + 8y + 16) = -31 + 64 + 3(16) \quad \left(\frac{-16}{2}\right)^2 = 64, \left(\frac{8}{2}\right)^2 = 16$$

$$(x - 8)^2 + 3(y + 4)^2 = 81 \quad \text{Write the trinomials as squares.}$$

$$\frac{(x-8)^2}{81} + \frac{(y+4)^2}{27} = 1 \quad \text{Divide each side by 81.}$$

The center of the ellipse is at $(8, -4)$. The length of the major axis is 18, and the length of the minor axis is $6\sqrt{3}$.

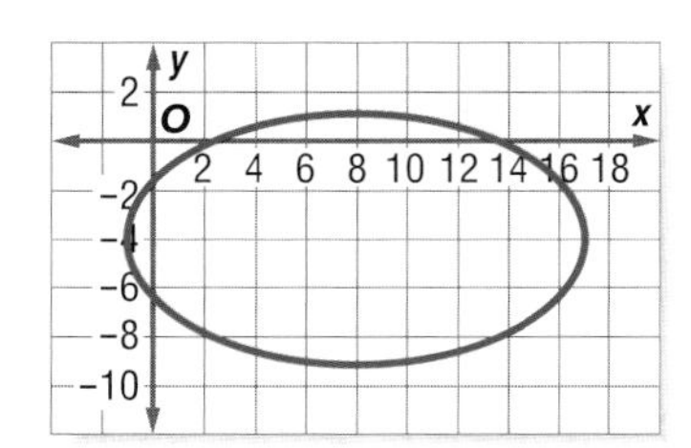

Exercises

30. Write an equation for the ellipse with endpoints of the major axis at $(4, 1)$ and $(-6, 1)$ and endpoints of the minor axis at $(-1, 3)$ and $(-1, -1)$.
See Examples 1 and 2 on pages 434 and 435.

Find the coordinates of the center and foci and the lengths of the major and minor axes for the ellipse with the given equation. Then graph the ellipse.
See Examples 3 and 4 on pages 436 and 437.

31. $\frac{x^2}{16} + \frac{y^2}{25} = 1$
32. $\frac{(x+2)^2}{16} + \frac{(y-3)^2}{9} = 1$
33. $x^2 + 4y^2 - 2x + 16y + 13 = 0$

8-5 Hyperbolas

See pages 441–448.

Concept Summary

Hyperbolas		
Standard Form	$\frac{(x-h)^2}{a^2} - \frac{(y-k)^2}{b^2} = 1$	$\frac{(y-k)^2}{a^2} - \frac{(x-h)^2}{b^2} = 1$
Transverse Axis	horizontal	vertical
Asymptotes	$y - k = \pm\frac{b}{a}(x - h)$	$y - k = \pm\frac{a}{b}(x - h)$

Example **Graph $9x^2 - 4y^2 + 18x + 32y - 91 = 0$.**

Complete the square for each variable to write this equation in standard form.

$$9x^2 - 4y^2 + 18x + 32y - 91 = 0 \quad \text{Original equation}$$
$$9(x^2 + 2x + \blacksquare) - 4(y^2 - 8y + \blacksquare) = 91 + 9(\blacksquare) - 4(\blacksquare) \quad \text{Complete the squares.}$$
$$9(x^2 + 2x + 1) - 4(y^2 - 8y + 16) = 91 + 9(1) - 4(16) \quad \left(\frac{2}{2}\right)^2 = 1, \left(\frac{-8}{2}\right)^2 = 16$$
$$9(x + 1)^2 - 4(y - 4)^2 = 36 \quad \text{Write the trinomials as squares.}$$
$$\frac{(x + 1)^2}{4} - \frac{(y - 4)^2}{9} = 1 \quad \text{Divide each side by 36.}$$

The center is at $(-1, 4)$. The vertices are at $(-3, 4)$ and $(1, 4)$ and the foci are at $(-1 \pm \sqrt{13}, 4)$. The equations of the asymptotes are $y - 4 = \pm\frac{3}{2}(x + 1)$.

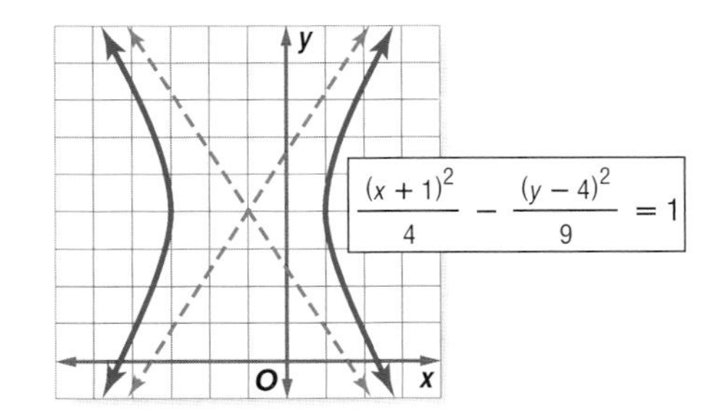

Exercises

34. Write an equation for a hyperbola that has vertices at (2, 5) and (2, 1) and a conjugate axis of length 6 units. *See Example 1 on page 442.*

Find the coordinates of the vertices and foci and the equations of the asymptotes for the hyperbola with the given equation. Then graph the hyperbola.
See Examples 3 and 4 on pages 443 and 444.

35. $\frac{y^2}{4} - \frac{x^2}{9} = 1$

36. $\frac{(x - 2)^2}{1} - \frac{(y + 1)^2}{9} = 1$

37. $9y^2 - 16x^2 = 144$

38. $16x^2 - 25y^2 - 64x - 336 = 0$

8-6 Conic Sections

See pages 449–452.

Concept Summary

- Conic sections can be identified directly from their equations of the form $Ax^2 + Bxy + Cy^2 + Dx + Ey + F = 0$, assuming $B = 0$.

Conic Section	Relationship of *A* and *C*
Parabola	$A = 0$ or $C = 0$, but not both.
Circle	$A = C$
Ellipse	*A* and *C* have the same sign and $A \neq C$.
Hyperbola	*A* and *C* have opposite signs.

Example **Without writing the equation in standard form, state whether the graph of $4x^2 + 9y^2 + 16x - 18y - 11 = 0$ is a *parabola, circle, ellipse,* or *hyperbola.***

In this equation, $A = 4$ and $C = 9$. Since *A* and *C* are both positive and $A \neq C$, the graph is an ellipse.

Chapter 8 **For More ...**

- Extra Practice, see pages 845–847.
- Mixed Problem Solving, see page 869.

Exercises **Write each equation in standard form. State whether the graph of the equation is a *parabola, circle, ellipse,* or *hyperbola*. Then graph the equation.**
See Example 1 on page 449.

39. $x^2 + 4x - y = 0$

40. $9x^2 + 4y^2 = 36$

41. $-4x^2 + y^2 + 8x - 8 = 0$

42. $x^2 + y^2 - 4x - 6y + 4 = 0$

Without writing the equation in standard form, state whether the graph of each equation is a *parabola, circle, ellipse,* or *hyperbola*. *See Example 2 on page 450.*

43. $7x^2 + 9y^2 = 63$

44. $x^2 - 8x + 16 = 6y$

45. $x^2 + 4x + y^2 - 285 = 0$

46. $5y^2 + 2y + 4x - 13x^2 = 81$

8-7 Solving Quadratic Systems

See pages 455–460.

Concept Summary

- Systems of quadratic equations can be solved using substitution and elimination.
- A system of quadratic equations can have zero, one, two, three, or four solutions.

Example **Solve the system of equations.**

$$x^2 + y^2 + 2x - 12y + 12 = 0$$
$$y + x = 0$$

Use substitution to solve the system.

First, rewrite $y + x = 0$ as $y = -x$.

$x^2 + y^2 + 2x - 12y + 12 = 0$	First original equation
$x^2 + (-x)^2 + 2x - 12(-x) + 12 = 0$	Substitute $-x$ for y.
$2x^2 + 14x + 12 = 0$	Simplify.
$x^2 + 7x + 6 = 0$	Divide each side by 2.
$(x + 6)(x + 1) = 0$	Factor.
$x + 6 = 0$ or $x + 1 = 0$	Zero Product Property.
$x = -6$ $\quad$ $x = -1$	Solve for x.

Now solve for y.

$y = -x$	$y = -x$	Equation for y in terms of x
$= -(-6)$ or 6	$= -(-1)$ or 1	Substitute the x values.

The solutions of the system are $(-6, 6)$ and $(-1, 1)$.

Exercises **Find the exact solution(s) of each system of equations.**
See Examples 1 and 2 on pages 455–457.

47. $x^2 + y^2 - 18x + 24y + 200 = 0$
$4x + 3y = 0$

48. $4x^2 + y^2 = 16$
$x^2 + 2y^2 = 4$

Solve each system of inequalities by graphing. *See Example 3 on page 457.*

49. $y < x$
$y > x^2 - 4$

50. $x^2 + y^2 \leq 9$
$x^2 + 4y^2 \leq 16$

Chapter 8

Practice Test

Vocabulary and Concepts

Choose the letter that best matches each description.

1. the set of all points in a plane that are the same distance from a given point, the focus, and a given line, the directrix
2. the set of all points in a plane such that the absolute value of the difference of the distances from two fixed points, the foci, is constant
3. the set of all points in a plane such that the sum of the distances from two fixed points, the foci, is constant

a. ellipse
b. parabola
c. hyperbola

Skills and Applications

Find the midpoint of the line segment with endpoints at the given coordinates.

4. $(7, 1), (-5, 9)$
5. $\left(\frac{3}{8}, -1\right), \left(-\frac{8}{5}, 2\right)$
6. $(-13, 0), (-1, -8)$

Find the distance between each pair of points with the given coordinates.

7. $(-6, 7), (3, 2)$
8. $\left(\frac{1}{2}, \frac{5}{2}\right), \left(-\frac{3}{4}, -\frac{11}{4}\right)$
9. $(8, -1), (8, -9)$

State whether the graph of each equation is a *parabola, circle, ellipse,* or *hyperbola*. Then graph the equation.

10. $x^2 + 4y^2 = 25$
11. $y = 4x^2 + 1$
12. $x^2 = 36 - y^2$
13. $(x + 4)^2 = 7(y + 5)$
14. $4x^2 - 26y^2 + 10 = 0$
15. $25x^2 + 49y^2 = 1225$
16. $-(y^2 - 24) = x^2 + 10x$
17. $5x^2 - y^2 = 49$
18. $\frac{1}{3}x^2 - 4 = y$
19. $\frac{y^2}{9} - \frac{x^2}{25} = 1$

20. **TUNNELS** The opening of a tunnel is in the shape of a semielliptical arch. The arch is 60 feet wide and 40 feet high. Find the height of the arch 12 feet from the edge of the tunnel.

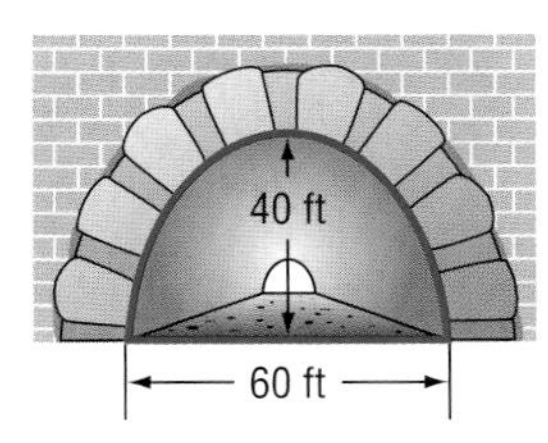

Find the exact solution(s) of each system of equations.

21. $x^2 + y^2 = 100$
    $y = 2 - x$
22. $x^2 + 2y^2 = 6$
    $x + y = 1$
23. $x^2 - y^2 - 12x + 12y = 36$
    $x^2 + y^2 - 12x - 12y + 36 = 0$

24. Solve the system of inequalities by graphing.
    $x^2 - y^2 \geq 1$
    $x^2 + y^2 \leq 16$

Standards Practice

25. **STANDARDIZED TEST PRACTICE** Which is *not* the equation of a parabola?

    (A) $y = 2x^2 + 4x - 9$
    (B) $3x + 2y^2 + y + 1 = 0$
    (C) $x^2 + 2y^2 + 8y = 8$
    (D) $x = \frac{1}{2}(y - 1)^2 + 5$

Chapter 8

Standardized Test Practice

Standards Practice

Part 1 Multiple Choice

Record your answers on the answer sheet provided by your teacher or on a sheet of paper.

1. The product of a prime number and a composite number must be

(A) prime. (B) composite.
(C) even. (D) negative.

2. In 1990, the population of Clayton was 54,200, and the population of Montrose was 47,500. By 2000, the population of each city had decreased by exactly 5%. How many more people lived in Clayton than in Montrose in 2000?

(A) 335 (B) 5085
(C) 6365 (D) 6700

3. If 4% of n is equal to 40% of p, then n is what percent of $10p$?

(A) $\frac{1}{1000}\%$ (B) 10%
(C) 100% (D) 1000%

4. Leroy bought m magazines at d dollars per magazine and p paperback books at $2d + 1$ dollars per book. Which of the following represents the total amount Leroy spent?

(A) $d(m + 2p) + p$ (B) $(m + p)(3d + 1)$
(C) $md + 2pd + 1$ (D) $pd(m + 2)$

Test-Taking Tip

Questions 3, 4
In problems with variables, you can substitute values to try to eliminate some of the answer choices. For example, in Question 3, choose a value for n and compute the corresponding value of p. Then find $\frac{n}{10p}$ to answer the question.

5. What is the midpoint of the line segment whose endpoints are at $(-5, -3)$ and $(-1, 4)$?

(A) $\left(-3, -\frac{1}{2}\right)$ (B) $\left(-3, \frac{1}{2}\right)$
(C) $\left(-2, \frac{7}{2}\right)$ (D) $\left(-2, \frac{1}{2}\right)$

6. Point $M(-2, 3)$ is the midpoint of line segment NP. If point N has coordinates $(-7, 1)$, then what are the coordinates of point P?

(A) $(-5, 2)$ (B) $(-4, 6)$
(C) $\left(-\frac{9}{2}, 2\right)$ (D) $(3, 5)$

7. Which equation's graph is a parabola?

(A) $3x^2 - 2y^2 = 10$
(B) $4x^2 + 3y^2 = 20$
(C) $2x^2 + 2y^2 = 15$
(D) $3x^2 + 4y = 8$

8. What is the center of the circle with equation $x^2 + y^2 - 4x + 6y - 9 = 0$?

(A) $(-4, 6)$ (B) $(-2, 3)$
(C) $(2, -3)$ (D) $(3, 3)$

9. What is the distance between the points shown in the graph?

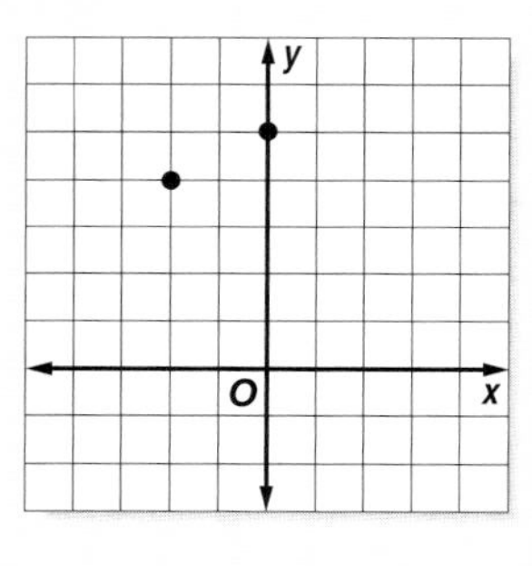

(A) $\sqrt{3}$ units
(B) $\sqrt{5}$ units
(C) 3 units
(D) $\sqrt{17}$ units

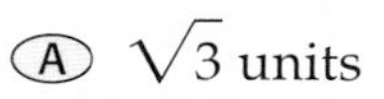

10. The median of seven test scores is 52, the mode is 64, the lowest score is 40, and the average is 53. If the scores are integers, what is the greatest possible test score?

(A) 68 (B) 72 (C) 76 (D) 84

Preparing for Standardized Tests
For test-taking strategies and more practice, see pages 877–892.

Part 2 Short Response/Grid In

Record your answers on the answer sheet provided by your teacher or on a sheet of paper.

11. What is the least positive integer p for which $2^{2p} + 3$ is not a prime number?

12. The ratio of cars to SUVs in a parking lot is 4 to 5. After 6 cars leave the parking lot, the ratio of cars to SUVs becomes 1 to 2. How many SUVs are in the parking lot?

13. Each dimension of a rectangular box is an integer greater than 1. If the area of one side of the box is 27 square units and the area of another side is 12 square units, what is the volume of the box in cubic units?

14. Let the operation * be defined as $a * b = 2ab - (a + b)$. If $4 * x = 10$, then what is the value of x?

15. If the slope of line PQ in the figure is $\frac{1}{4}$, what is the area of quadrilateral $OPQR$?

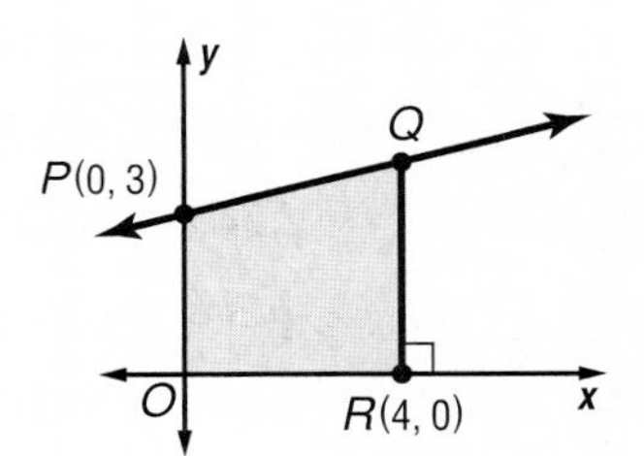

16. In the figure, the slope of line ℓ is $\frac{5}{4}$, and the slope of line k is $\frac{3}{8}$. What is the distance from point A to point B?

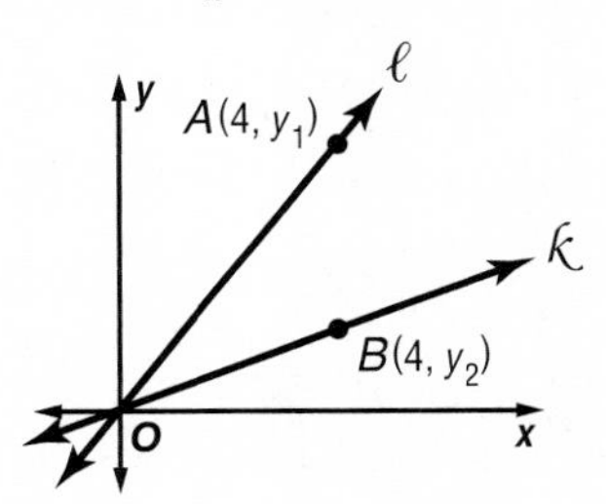

17. If $(2x - 3)(4x + n) = ax^2 + bx - 15$ for all values of x, what is the value of $a + b$?

Part 3 Extended Response

Record your answers on a sheet of paper. Show your work.

For Exercises 18–20, use the following information.

A certain map of New York City's Central Park shows the Loeb Boathouse at (4, 7), Belvedere Castle at (3, 9), and the center of the Great Lawn at (3, 12). Each grid on the map represents 0.1 mile in the park.

18. How far apart are the Loeb Boathouse and Belvedere Castle? Round to the nearest hundredth of a mile.

19. Loeb Boathouse is halfway between the center of the Great Lawn and the statue of Balto. Where is the statue of Balto on this map?

20. The Great Lawn is approximately shaped like an ellipse. The major axis is 0.2 mile and the minor axis is 0.14 mile. If the major axis is vertical, write an equation to model the Great Lawn.

21. A fountain sprays water from a jet at ground level. The water reaches a maximum height of 84 inches and hits the ground 18 inches away from the jet. Write an equation of the parabola that models the path of the water. Assume that the jet is at the origin.

For Exercises 22–25, use the following information.

The endpoints of a diameter of a circle are at $(-1, 0)$, and $(5, -8)$.

22. What are the coordinates of the center of the circle? Explain your method.

23. Find the radius of the circle. Explain your method.

24. Write an equation of the circle.

25. Find the point(s) at which the circle intersects the line with equation $y = x - 7$.

Chapter 9 Rational Expressions and Equations

What You'll Learn

- **Lessons 9-1 and 9-2** Simplify rational expressions.
- **Lesson 9-3** Graph rational functions.
- **Lesson 9-4** Solve direct, joint, and inverse variation problems.
- **Lesson 9-5** Identify graphs and equations as different types of functions.
- **Lesson 9-6** Solve rational equations and inequalities.

Key Vocabulary

- rational expression (p. 472)
- asymptote (p. 485)
- point discontinuity (p. 485)
- direct variation (p. 492)
- inverse variation (p. 493)

Why It's Important

Rational expressions, functions, and equations can be used to solve problems involving mixtures, photography, electricity, medicine, and travel, to name a few. Direct, joint, and inverse variation are important applications of rational expressions. For example, scuba divers can use direct variation to determine the amount of pressure at various depths. *You will learn how to determine the amount of pressure exerted on the ears of a diver in Lesson 9-4.*

Getting Started

Prerequisite Skills To be successful in this chapter, you'll need to master these skills and be able to apply them in problem-solving situations. Review these skills before beginning Chapter 9.

For Lesson 9-1 — Solve Equations with Rational Numbers

Solve each equation. Write your answer in simplest form. *(For review, see Lesson 1-3.)*

1. $\frac{8}{5}x = \frac{4}{15}$ **2.** $\frac{27}{14}t = \frac{6}{7}$ **3.** $\frac{3}{10} = \frac{12}{25}a$

4. $\frac{6}{7} = 9m$ **5.** $\frac{9}{8}b = 18$ **6.** $\frac{6}{7}s = \frac{3}{4}$

7. $\frac{1}{3}r = \frac{5}{6}$ **8.** $\frac{2}{3}n = 7$ **9.** $\frac{4}{5}r = \frac{5}{6}$

For Lesson 9-3 — Determine Asymptotes and Graph Equations

Draw the asymptotes and graph each hyperbola. *(For review, see Lesson 8-5.)*

10. $\frac{(x-3)^2}{4} - \frac{(y+5)^2}{9} = 1$ **11.** $\frac{y^2}{4} - \frac{(x+4)^2}{1} = 1$ **12.** $\frac{(x+2)^2}{4} - \frac{(y-3)^2}{25} = 1$

For Lesson 9-4 — Solve Proportions

Solve each proportion.

13. $\frac{3}{4} = \frac{r}{16}$ **14.** $\frac{8}{16} = \frac{5}{y}$ **15.** $\frac{6}{8} = \frac{m}{20}$

16. $\frac{t}{3} = \frac{5}{24}$ **17.** $\frac{5}{a} = \frac{6}{18}$ **18.** $\frac{3}{4} = \frac{b}{6}$

19. $\frac{v}{9} = \frac{12}{18}$ **20.** $\frac{7}{p} = \frac{1}{4}$ **21.** $\frac{2}{5} = \frac{3}{z}$

22. $\frac{16}{18} = \frac{6}{x}$ **23.** $\frac{12}{14} = \frac{3}{2}$ **24.** $\frac{5}{7} = \frac{1.5}{t}$

Rational Expressions and Equations Make this Foldable to help you organize your notes. Begin with a sheet of plain $8\frac{1}{2}$" by 11" paper.

Step 1 Fold

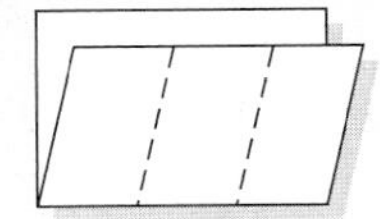

Fold in half lengthwise leaving a $1\frac{1}{2}$" margin at the top. Fold again in thirds.

Step 2 Cut and Label

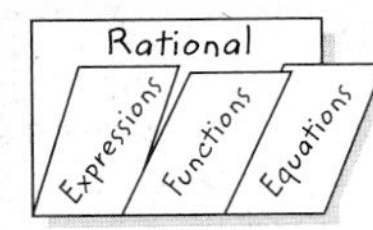

Open. Cut along the second folds to make three tabs. Label as shown.

Reading and Writing As you read and study the chapter, write notes and examples for each concept under the tabs.

9-1 Multiplying and Dividing Rational Expressions

What You'll Learn

- Simplify rational expressions.
- Simplify complex fractions.

How are rational expressions used in mixtures?

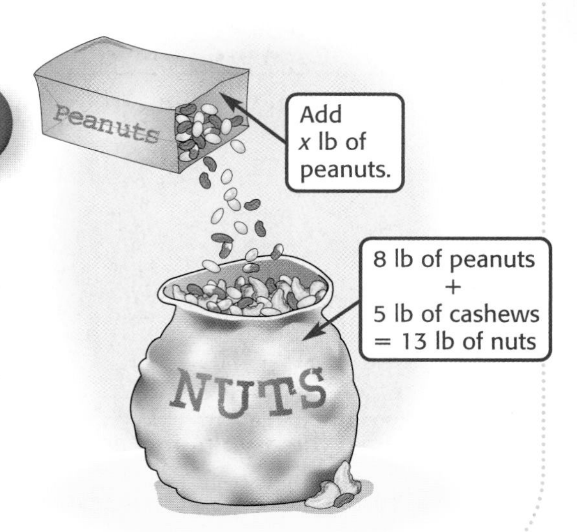

The Goodie Shoppe sells candy and nuts by the pound. One of their items is a mixture of peanuts and cashews. This mixture is made with 8 pounds of peanuts and 5 pounds of cashews. Therefore, $\frac{8}{8+5}$ or $\frac{8}{13}$ of the mixture is peanuts. If the store manager adds an additional x pounds of peanuts to the mixture, then $\frac{8+x}{13+x}$ of the mixture will be peanuts.

Vocabulary

- rational expression
- complex fraction

California Standards
Standard 7.0
Students add, subtract, **multiply, divide, reduce,** and evaluate **rational expressions with monomial and polynomial denominators and simplify complicated rational expressions, including those with negative exponents in the denominator.** (Key)

SIMPLIFY RATIONAL EXPRESSIONS A ratio of two polynomial expressions such as $\frac{8+x}{13+x}$ is called a **rational expression**. Because variables in algebra represent real numbers, operations with rational numbers and rational expressions are similar.

To write a fraction in simplest form, you divide both the numerator and denominator by their greatest common factor (GCF). To simplify a rational expression, you use similar properties.

Example 1 Simplify a Rational Expression

a. Simplify $\frac{2x(x-5)}{(x-5)(x^2-1)}$.

Look for common factors.

$$\frac{2x(x-5)}{(x-5)(x^2-1)} = \frac{2x}{x^2-1} \cdot \frac{\overset{1}{\cancel{x-5}}}{\underset{1}{\cancel{x-5}}}$$ How is this similar to simplifying $\frac{10}{15}$?

$$= \frac{2x}{x^2-1}$$ Simplify.

b. Under what conditions is this expression undefined?

Just as with a fraction, a rational expression is undefined if the denominator is equal to 0. To find when this expression is undefined, completely factor the original denominator.

$$\frac{2x(x-5)}{(x-5)(x^2-1)} = \frac{2x(x-5)}{(x-5)(x-1)(x+1)}$$ $x^2 - 1 = (x-1)(x+1)$

The values that would make the denominator equal to 0 are 5, 1, or -1. So the expression is undefined when $x = 5$, $x = 1$, or $x = -1$. These numbers are called *excluded values.*

Standardized Test Practice

Standards Practice

Example 2 Use the Process of Elimination

Multiple-Choice Test Item

For what value(s) of x is $\frac{x^2 + x - 12}{x^2 + 7x + 12}$ undefined?

(A) $-4, -3$ (B) -4 (C) 0 (D) $-4, 3$

Read the Test Item

You want to determine which values of x make the denominator equal to 0.

Solve the Test Item

Look at the possible answers. Notice that if x equals 0 or a positive number, $x^2 + 7x + 12$ must be greater than 0. Therefore, you can eliminate choices C and D. Since both choices A and B contain -4, determine whether the denominator equals 0 when $x = -3$.

$x^2 + 7x + 12 = (-3)^2 + 7(-3) + 12$ $x = -3$

$= 9 - 21 + 12$ Multiply.

$= 0$ Simplify.

Since the denominator equals 0 when $x = -3$, the answer is A.

Test-Taking Tip

Sometimes you can save time by looking at the possible answers and eliminating choices, rather than actually evaluating an expression or solving an equation.

Sometimes you can factor out -1 in the numerator or denominator to help simplify rational expressions.

Example 3 Simplify by Factoring Out -1

Simplify $\frac{z^2w - z^2}{z^3 - z^3w}$.

$\frac{z^2w - z^2}{z^3 - z^3w} = \frac{z^2(w - 1)}{z^3(1 - w)}$ Factor the numerator and the denominator.

$= \frac{\overset{1}{\cancel{z^2}}(-1)\overset{1}{\cancel{(1 - w)}}}{\underset{z}{\cancel{z^3}}\underset{1}{\cancel{(1 - w)}}}$ $w - 1 = -(-w + 1)$ or $-1(1 - w)$

$= \frac{-1}{z}$ or $-\frac{1}{z}$ Simplify.

Remember that to multiply two fractions, you first multiply the numerators and then multiply the denominators. To divide two fractions, you multiply by the multiplicative inverse, or reciprocal, of the divisor.

Multiplication

$\frac{5}{6} \cdot \frac{4}{15} = \frac{\overset{1}{\cancel{5}} \cdot \overset{1}{\cancel{2}} \cdot 2}{\underset{1}{\cancel{2}} \cdot 3 \cdot 3 \cdot \underset{1}{\cancel{5}}}$

$= \frac{2}{3 \cdot 3}$ or $\frac{2}{9}$

Division

$\frac{3}{7} \div \frac{9}{14} = \frac{3}{7} \cdot \frac{14}{9}$

$= \frac{\overset{1}{\cancel{3}} \cdot 2 \cdot \overset{1}{\cancel{7}}}{\underset{1}{\cancel{7}} \cdot \underset{1}{\cancel{3}} \cdot 3}$

$= \frac{2}{3}$

The same procedures are used for multiplying and dividing rational expressions.

Key Concept — Rational Expressions

Multiplying Rational Expressions

- **Words** To multiply two rational expressions, multiply the numerators and the denominators.
- **Symbols** For all rational expressions $\frac{a}{b}$ and $\frac{c}{d}$, $\frac{a}{b} \cdot \frac{c}{d} = \frac{ac}{bd}$, if $b \neq 0$ and $d \neq 0$.

Dividing Rational Expressions

- **Words** To divide two rational expressions, multiply by the reciprocal of the divisor.
- **Symbols** For all rational expressions $\frac{a}{b}$ and $\frac{c}{d}$, $\frac{a}{b} \div \frac{c}{d} = \frac{a}{b} \cdot \frac{d}{c} = \frac{ad}{bc}$, if $b \neq 0$, $c \neq 0$, and $d \neq 0$.

The following examples show how these rules are used with rational expressions.

Study Tip

Alternative Method

When multiplying rational expressions, you can multiply first and then divide by the common factors. For instance, in Example 4,

$\frac{4a}{5b} \cdot \frac{15b^2}{16a^3} = \frac{60ab^2}{80a^3b}$.

Now divide the numerator and denominator by the common factors.

$\frac{\overset{3}{\cancel{60}}\, \overset{1}{\cancel{a}}\, \overset{b}{\cancel{b^2}}}{\underset{4}{\cancel{80}}\, \underset{a^2}{\cancel{a^3}}\, \underset{1}{\cancel{b}}} = \frac{3b}{4a^2}$

Example 4 Multiply Rational Expressions

Simplify each expression.

a. $\frac{4a}{5b} \cdot \frac{15b^2}{16a^3}$

$$\frac{4a}{5b} \cdot \frac{15b^2}{16a^3} = \frac{\overset{1}{\cancel{2}} \cdot \overset{1}{\cancel{2}} \cdot \overset{1}{\cancel{a}} \cdot 3 \cdot \overset{1}{\cancel{5}} \cdot \overset{1}{\cancel{b}} \cdot b}{\underset{1}{\cancel{5}} \cdot \underset{1}{\cancel{b}} \cdot \underset{1}{\cancel{2}} \cdot \underset{1}{\cancel{2}} \cdot 2 \cdot 2 \cdot \underset{1}{\cancel{a}} \cdot a \cdot a}$$ Factor.

$$= \frac{3 \cdot b}{2 \cdot 2 \cdot a \cdot a}$$ Simplify.

$$= \frac{3b}{4a^2}$$ Simplify.

b. $\frac{8t^2s}{5r^2} \cdot \frac{15sr}{12t^3s^2}$

$$\frac{8t^2s}{5r^2} \cdot \frac{15sr}{12t^3s^2} = \frac{\overset{1}{\cancel{2}} \cdot \overset{1}{\cancel{2}} \cdot 2 \cdot \overset{1}{\cancel{t}} \cdot \overset{1}{\cancel{t}} \cdot \overset{1}{\cancel{s}} \cdot \overset{1}{\cancel{3}} \cdot \overset{1}{\cancel{5}} \cdot \overset{1}{\cancel{s}} \cdot \overset{1}{\cancel{r}}}{\underset{1}{\cancel{5}} \cdot \underset{1}{\cancel{r}} \cdot r \cdot \underset{1}{\cancel{2}} \cdot \underset{1}{\cancel{2}} \cdot \underset{1}{\cancel{3}} \cdot \underset{1}{\cancel{t}} \cdot \underset{1}{\cancel{t}} \cdot t \cdot \underset{1}{\cancel{s}} \cdot \underset{1}{\cancel{s}}}$$ Factor.

$$= \frac{2}{rt}$$ Simplify.

Example 5 Divide Rational Expressions

Simplify $\frac{4x^2y}{15a^3b^3} \div \frac{2xy^2}{5ab^3}$.

$$\frac{4x^2y}{15a^3b^3} \div \frac{2xy^2}{5ab^3} = \frac{4x^2y}{15a^3b^3} \cdot \frac{5ab^3}{2xy^2}$$ Multiply by the reciprocal of the divisor.

$$= \frac{\overset{1}{\cancel{2}} \cdot 2 \cdot \overset{1}{\cancel{x}} \cdot x \cdot \overset{1}{\cancel{y}} \cdot \overset{1}{\cancel{5}} \cdot \overset{1}{\cancel{a}} \cdot \overset{1}{\cancel{b}} \cdot \overset{1}{\cancel{b}} \cdot \overset{1}{\cancel{b}}}{3 \cdot \underset{1}{\cancel{5}} \cdot \underset{1}{\cancel{a}} \cdot a \cdot a \cdot \underset{1}{\cancel{b}} \cdot \underset{1}{\cancel{b}} \cdot \underset{1}{\cancel{b}} \cdot \underset{1}{\cancel{2}} \cdot \underset{1}{\cancel{x}} \cdot \underset{1}{\cancel{y}} \cdot y}$$ Factor.

$$= \frac{2 \cdot x}{3 \cdot a \cdot a \cdot y}$$ Simplify.

$$= \frac{2x}{3a^2y}$$ Simplify.

These same steps are followed when the rational expressions contain numerators and denominators that are polynomials.

Example 6 Polynomials in the Numerator and Denominator

Simplify each expression.

a. $\frac{x^2 + 2x - 8}{x^2 + 4x + 3} \cdot \frac{3x + 3}{x - 2}$

$$\frac{x^2 + 2x - 8}{x^2 + 4x + 3} \cdot \frac{3x + 3}{x - 2} = \frac{(x + 4)\overset{1}{\cancel{(x - 2)}}}{(x + 3)\underset{1}{\cancel{(x + 1)}}} \cdot \frac{3\overset{1}{\cancel{(x + 1)}}}{\underset{1}{\cancel{(x - 2)}}} \quad \text{Factor.}$$

$$= \frac{3(x + 4)}{(x + 3)} \quad \text{Simplify.}$$

$$= \frac{3x + 12}{x + 3} \quad \text{Simplify.}$$

b. $\frac{a + 2}{a + 3} \div \frac{a^2 + a - 12}{a^2 - 9}$

$$\frac{a + 2}{a + 3} \div \frac{a^2 + a - 12}{a^2 - 9} = \frac{a + 2}{a + 3} \cdot \frac{a^2 - 9}{a^2 + a - 12} \quad \text{Multiply by the reciprocal of the divisor.}$$

$$= \frac{(a + 2)\overset{1}{\cancel{(a + 3)}}\overset{1}{\cancel{(a - 3)}}}{\underset{1}{\cancel{(a + 3)}}(a + 4)\underset{1}{\cancel{(a - 3)}}} \quad \text{Factor.}$$

$$= \frac{a + 2}{a + 4} \quad \text{Simplify.}$$

Study Tip

Factor First
As in Example 6, sometimes you must factor the numerator and/or the denominator first before you can simplify a quotient of rational expressions.

SIMPLIFY COMPLEX FRACTIONS A **complex fraction** is a rational expression whose numerator and/or denominator contains a rational expression. The expressions below are complex fractions.

$$\frac{\frac{a}{5}}{3b} \qquad \frac{\frac{3}{t}}{t + 5} \qquad \frac{\frac{m^2 - 9}{8}}{\frac{3 - m}{12}} \qquad \frac{\frac{1}{p} + 2}{\frac{3}{p} - 4}$$

Remember that a fraction is nothing more than a way to express a division problem. For example, $\frac{2}{5}$ can be expressed as $2 \div 5$. So to simplify any complex fraction, rewrite it as a division expression and use the rules for division.

Example 7 Simplify a Complex Fraction

Simplify $\dfrac{\frac{r^2}{r^2 - 25s^2}}{\frac{r}{5s - r}}$.

$$\frac{\frac{r^2}{r^2 - 25s^2}}{\frac{r}{5s - r}} = \frac{r^2}{r^2 - 25s^2} \div \frac{r}{5s - r} \quad \text{Express as a division expression.}$$

$$= \frac{r^2}{r^2 - 25s^2} \cdot \frac{5s - r}{r} \quad \text{Multiply by the reciprocal of the divisor.}$$

$$= \frac{\overset{1}{\cancel{r}} \cdot r(-1)\overset{1}{\cancel{(r - 5s)}}}{(r + 5s)\underset{1}{\cancel{(r - 5s)}}\underset{1}{\cancel{r}}} \quad \text{Factor.}$$

$$= \frac{-r}{r + 5s} \text{ or } -\frac{r}{r + 5s} \quad \text{Simplify.}$$

Check for Understanding

Concept Check

1. **OPEN ENDED** Write two rational expressions that are equivalent.
2. **Explain** how multiplication and division of rational expressions are similar to multiplication and division of rational numbers.
3. **Determine** whether $\frac{2d+5}{3d+5} = \frac{2}{3}$ is *sometimes, always,* or *never* true. Explain.

Guided Practice

Simplify each expression.

4. $\frac{45mn^3}{20n^7}$
5. $\frac{a+b}{a^2-b^2}$
6. $\frac{6y^3-9y^2}{2y^2+5y-12}$
7. $\frac{2a^2}{5b^2c} \cdot \frac{3bc^2}{8a^2}$
8. $\frac{35}{16x^2} \div \frac{21}{4x}$
9. $\frac{3t+6}{7t-7} \cdot \frac{14t-14}{5t+10}$
10. $\frac{12p^2+6p-6}{4(p+1)^2} \div \frac{6p-3}{2p+10}$
11. $\frac{\frac{c^3d^3}{a}}{\frac{xc^2d}{ax^2}}$
12. $\frac{\frac{2y}{y^2-4}}{\frac{3}{y^2-4y+4}}$

Standards Practice

Standardized Test Practice

13. Identify all of the values of y for which the expression $\frac{y-4}{y^2-4y-12}$ is undefined.

Ⓐ $-2, 4, 6$ Ⓑ $-6, -4, 2$ Ⓒ $-2, 0, 6$ Ⓓ $-2, 6$

Practice and Apply

Homework Help

For Exercises	See Examples
14–21	1, 3
22–35	4–6
36–41	7
42, 43, 50	2

Extra Practice
See page 847.

Simplify each expression.

14. $\frac{30bc}{12b^2}$
15. $\frac{-3mn^4}{21m^2n^2}$
16. $\frac{(-3x^2y)^3}{9x^2y^2}$
17. $\frac{(-2rs^2)^2}{12r^2s^3}$
18. $\frac{5t-5}{t^2-1}$
19. $\frac{c+5}{2c+10}$
20. $\frac{y^2+4y+4}{3y^2+5y-2}$
21. $\frac{a^2+2a+1}{2a^2+3a+1}$
22. $\frac{3xyz}{4xz} \cdot \frac{6x^2}{3y^2}$
23. $\frac{-4ab}{21c} \cdot \frac{14c^2}{18a^2}$
24. $\frac{3}{5d} \div \left(\frac{-9}{15df}\right)$
25. $\frac{p^3}{2q} \div \frac{-p}{4q}$
26. $\frac{2x^3y}{z^5} \div \left(\frac{4xy}{z^3}\right)^2$
27. $\frac{xy}{a^3} \div \left(\frac{xy}{ab}\right)^3$
28. $\frac{3t^2}{t+2} \cdot \frac{t+2}{t^2}$
29. $\frac{4w+4}{3} \cdot \frac{1}{w+1}$
30. $\frac{4t^2-4}{9(t+1)^2} \cdot \frac{3t+3}{2t-2}$
31. $\frac{3p-21}{p^2-49} \cdot \frac{p^2+7p}{3p}$
32. $\frac{5x^2+10x-75}{4x^2-24x-28} \cdot \frac{2x^2-10x-28}{x^2+7x+10}$
33. $\frac{w^2-11w+24}{w^2-18w+80} \cdot \frac{w^2-15w+50}{w^2-9w+20}$
34. $\frac{r^2+2r-8}{r^2+4r+3} \div \frac{r-2}{3r+3}$
35. $\frac{a^2+2a-15}{a-3} \div \frac{a^2-4}{2}$

36. $\dfrac{\frac{m^3}{3n}}{-\frac{m^4}{9n^2}}$

37. $\dfrac{\frac{p^3}{2q}}{\frac{-p^2}{4q}}$

38. $\dfrac{\frac{m+n}{5}}{\frac{m^2+n^2}{5}}$

39. $\dfrac{\frac{x+y}{2x-y}}{\frac{x+y}{2x+y}}$

40. $\dfrac{\frac{6y^2-6}{8y^2+8y}}{\frac{3y-3}{4y^2+4y}}$

41. $\dfrac{\frac{5x^2-5x-30}{45-15x}}{\frac{6+x-x^2}{4x-12}}$

42. Under what conditions is $\dfrac{2d(d+1)}{(d+1)(d^2-4)}$ undefined?

43. Under what conditions is $\dfrac{a^2+ab+b^2}{a^2-b^2}$ undefined?

More About. . .

Basketball

After graduating from the U.S. Naval Academy, David Robinson became the NBA Rookie of the Year in 1990. He has played basketball in 3 different Olympic Games.

Source: NBA

BASKETBALL **For Exercises 44 and 45, use the following information.**
At the end of the 2000–2001 season, David Robinson had made 6827 field goals out of 13,129 attempts during his NBA career.

44. Write a fraction to represent the ratio of the number of career field goals made to career field goals attempted by David Robinson at the end of the 2000–2001 season.

45. Suppose David Robinson attempted a field goals and made m field goals during the 2001–2002 season. Write a rational expression to represent the number of career field goals made to the number of career field goals attempted at the end of the 2001–2002 season.

Online Research **Data Update** What are the current scoring statistics of your favorite NBA player? Visit www.algebra2.com/data_update to learn more.

46. **GEOMETRY** A parallelogram with an area of $6x^2 - 7x - 5$ square units has a base of $3x - 5$ units. Determine the height of the parallelogram.

47. **GEOMETRY** Parallelogram L has an area of $3x^2 + 10x + 3$ square meters and a height of $3x + 1$ meters. Parallelogram M has an area of $2x^2 - 13x + 20$ square meters and a height of $x - 4$ meters. Find the area of rectangle N.

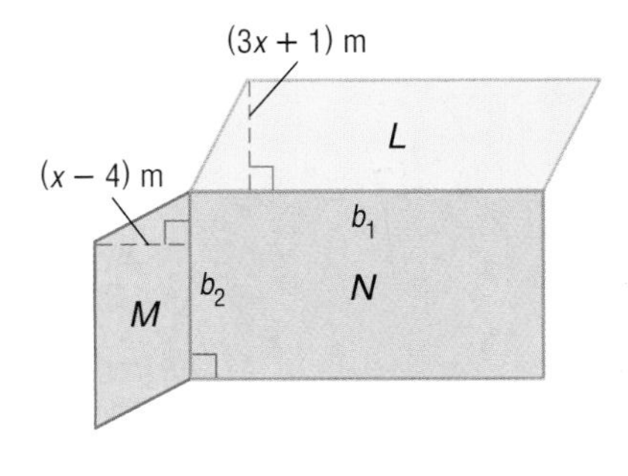

48. **CRITICAL THINKING** Simplify $\dfrac{(a^2 - 5a + 6)^{-1}}{(a-2)^{-2}} \div \dfrac{(a-3)^{-1}}{(a-2)^{-2}}$.

49. **WRITING IN MATH** Answer the question that was posed at the beginning of the lesson.

How are rational expressions used in mixtures?

Include the following in your answer:

- an explanation of how to determine whether the rational expression representing the nut mixture is in simplest form, and
- an example of a mixture problem that could be represented by $\dfrac{8+x}{13+x+y}$.

Standardized Test Practice

Standards Practice

50. For what value(s) of x is the expression $\frac{4x}{x^2 - x}$ undefined?

(A) $-1, 1$ (B) $-1, 0, 1$ (C) $0, 1$ (D) 0 (E) $1, 2$

51. **Compare the quantity in Column A and the quantity in Column B. Then determine whether:**

(A) **the quantity in Column A is greater,**

(B) **the quantity in Column B is greater,**

(C) **the two quantities are equal, or**

(D) **the relationship cannot be determined from the information given.**

Column A	Column B
$\frac{a^2 + 3a - 10}{a - 2}$	$\frac{a^2 + a - 6}{a + 3}$

Maintain Your Skills

Mixed Review

Find the exact solution(s) of each system of equations. *(Lesson 8–7)*

52. $x^2 + 2y^2 = 33$
$x^2 + y^2 - 19 = 2x$

53. $x^2 + 2y^2 = 33$
$x^2 - y^2 = 9$

Write each equation in standard form. State whether the graph of the equation is a *parabola, circle, ellipse,* or *hyperbola*. Then graph the equation. *(Lesson 8–6)*

54. $y^2 - 3x + 6y + 12 = 0$

55. $x^2 - 14x + 4 = 9y^2 - 36y$

Determine whether each graph represents an odd-degree function or an even-degree function. Then state how many real zeros each function has. *(Lesson 7–1)*

56.

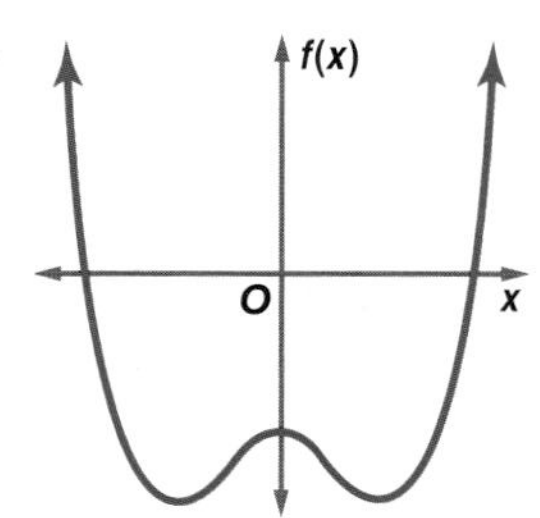

57.

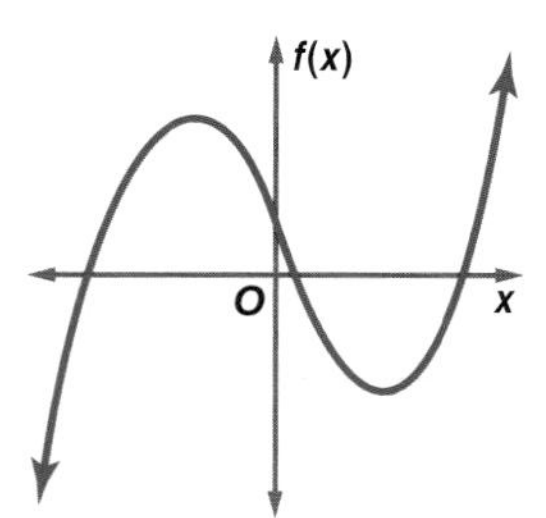

58.

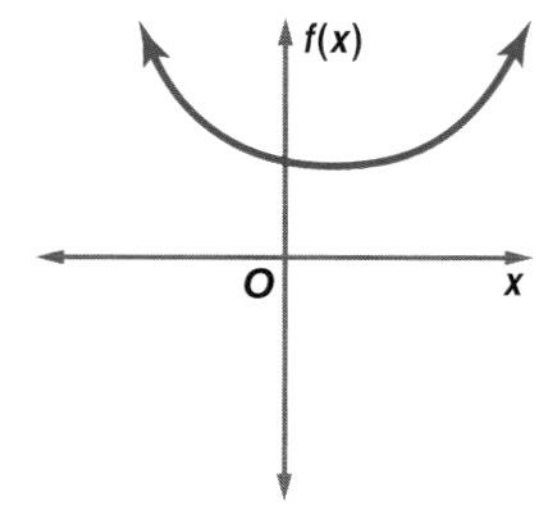

Solve each equation by factoring. *(Lesson 6–3)*

59. $r^2 - 3r = 4$

60. $18u^2 - 3u = 1$

61. $d^2 - 5d = 0$

62. **ASTRONOMY** Earth is an average 1.496×10^8 kilometers from the Sun. If light travels 3×10^5 kilometers per second, how long does it take sunlight to reach Earth? *(Lesson 5–1)*

Solve each equation. *(Lesson 1–4)*

63. $|2x + 7| + 5 = 0$

64. $5|3x - 4| = x + 1$

Getting Ready for the Next Lesson

PREREQUISITE SKILL **Solve each equation.** *(To review **solving equations**, see Lesson 1–3.)*

65. $\frac{2}{3} + x = -\frac{4}{9}$

66. $x + \frac{5}{8} = -\frac{5}{6}$

67. $x - \frac{3}{5} = \frac{2}{3}$

68. $x + \frac{3}{16} = -\frac{1}{2}$

69. $x - \frac{1}{6} = -\frac{7}{9}$

70. $x - \frac{3}{8} = -\frac{5}{24}$

9-2 Adding and Subtracting Rational Expressions

What You'll Learn

- Determine the LCM of polynomials.
- Add and subtract rational expressions.

California Standards
Standard 7.0
Students add, subtract, multiply, divide, **reduce,** and evaluate **rational expressions with monomial and polynomial denominators** and simplify complicated rational expressions, including those with negative exponents in the denominator. (Key)

How is subtraction of rational expressions used in photography?

To take sharp, clear pictures, a photographer must focus the camera precisely. The distance from the object to the lens p and the distance from the lens to the film q must be accurately calculated to ensure a sharp image. The focal length of the lens is f.

The formula $\frac{1}{q} = \frac{1}{f} - \frac{1}{p}$ can be used to determine how far the film should be placed from the lens to create a perfect photograph.

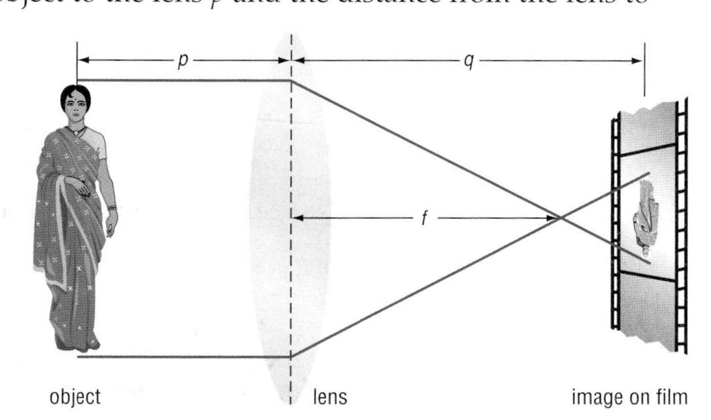

LCM OF POLYNOMIALS To find $\frac{5}{6} - \frac{1}{4}$ or $\frac{1}{f} - \frac{1}{p}$, you must first find the least common denominator (LCD). The LCD is the least common multiple (LCM) of the denominators.

To find the LCM of two or more numbers or polynomials, factor each number or polynomial. The LCM contains *each* factor the greatest number of times it appears as a factor.

LCM of 6 and 4

$6 = 2 \cdot 3$

$4 = 2^2$

$\text{LCM} = 2^2 \cdot 3 \text{ or } 12$

LCM of $a^2 - 6a + 9$ and $a^2 + a - 12$

$a^2 - 6a + 9 = (a - 3)^2$

$a^2 + a - 12 = (a - 3)(a + 4)$

$\text{LCM} = (a - 3)^2(a + 4)$

Example 1 LCM of Monomials

Find the LCM of $18r^2s^5$, $24r^3st^2$, and $15s^3t$.

$18r^2s^5 = 2 \cdot 3^2 \cdot r^2 \cdot s^5$ Factor the first monomial.

$24r^3st^2 = 2^3 \cdot 3 \cdot r^3 \cdot s \cdot t^2$ Factor the second monomial.

$15s^3t = 3 \cdot 5 \cdot s^3 \cdot t$ Factor the third monomial.

$\text{LCM} = 2^3 \cdot 3^2 \cdot 5 \cdot r^3 \cdot s^5 \cdot t^2$ Use each factor the greatest number of times it appears as a factor and simplify.

$= 360r^3s^5t^2$

Example 2 LCM of Polynomials

Find the LCM of $p^3 + 5p^2 + 6p$ and $p^2 + 6p + 9$.

$p^3 + 5p^2 + 6p = p(p + 2)(p + 3)$ Factor the first polynomial.

$p^2 + 6p + 9 = (p + 3)^2$ Factor the second polynomial.

$\text{LCM} = p(p + 2)(p + 3)^2$ Use each factor the greatest number of times it appears as a factor.

ADD AND SUBTRACT RATIONAL EXPRESSIONS As with fractions, to add or subtract rational expressions, you must have common denominators.

Specific Case		General Case
$\frac{2}{3} + \frac{3}{5} = \frac{2 \cdot 5}{3 \cdot 5} + \frac{3 \cdot 3}{5 \cdot 3}$	Find equivalent fractions that have a common denominator.	$\frac{a}{c} + \frac{b}{d} = \frac{a \cdot d}{c \cdot d} + \frac{b \cdot c}{d \cdot c}$
$= \frac{10}{15} + \frac{9}{15}$	Simplify each numerator and denominator.	$= \frac{ad}{cd} + \frac{bc}{cd}$
$= \frac{19}{15}$	Add the numerators.	$= \frac{ad + bc}{cd}$

Example 3 Monomial Denominators

Simplify $\frac{7x}{15y^2} + \frac{y}{18xy}$.

$\frac{7x}{15y^2} + \frac{y}{18xy} = \frac{7x \cdot 6x}{15y^2 \cdot 6x} + \frac{y \cdot 5y}{18xy \cdot 5y}$ The LCD is $90xy^2$. Find equivalent fractions that have this denominator.

$= \frac{42x^2}{90xy^2} + \frac{5y^2}{90xy^2}$ Simplify each numerator and denominator.

$= \frac{42x^2 + 5y^2}{90xy^2}$ Add the numerators.

Example 4 Polynomial Denominators

Simplify $\frac{w + 12}{4w - 16} - \frac{w + 4}{2w - 8}$.

$\frac{w + 12}{4w - 16} - \frac{w + 4}{2w - 8} = \frac{w + 12}{4(w - 4)} - \frac{w + 4}{2(w - 4)}$ Factor the denominators.

$= \frac{w + 12}{4(w - 4)} - \frac{(w + 4)(2)}{2(w - 4)(2)}$ The LCD is $4(w-4)$.

$= \frac{(w + 12) - (2)(w + 4)}{4(w - 4)}$ Subtract the numerators.

$= \frac{w + 12 - 2w - 8}{4(w - 4)}$ Distributive Property

$= \frac{-w + 4}{4(w - 4)}$ Combine like terms.

$= \frac{-1\overset{1}{\cancel{(w - 4)}}}{4\underset{1}{\cancel{(w - 4)}}}$ or $-\frac{1}{4}$ Simplify.

Study Tip

Common Factors

Sometimes when you simplify the numerator, the polynomial contains a factor common to the denominator. Thus, the rational expression can be further simplified.

Sometimes simplifying complex fractions involves adding or subtracting rational expressions. One way to simplify a complex fraction is to simplify the numerator and the denominator separately, and then simplify the resulting expressions.

Example 5 Simplify Complex Fractions

Simplify $\dfrac{\frac{1}{x} - \frac{1}{y}}{1 + \frac{1}{x}}$.

$$\frac{\frac{1}{x} - \frac{1}{y}}{1 + \frac{1}{x}} = \frac{\frac{y}{xy} - \frac{x}{xy}}{\frac{x}{x} + \frac{1}{x}}$$ The LCD of the numerator is xy. The LCD of the denominator is x.

$$= \frac{\frac{y - x}{xy}}{\frac{x + 1}{x}}$$ Simplify the numerator and denominator.

$$= \frac{y - x}{xy} \div \frac{x + 1}{x}$$ Write as a division expression.

$$= \frac{y - x}{\cancel{x}y} \cdot \frac{\cancel{x}}{x + 1}$$ Multiply by the reciprocal of the divisor.

$$= \frac{y - x}{y(x + 1)} \text{ or } \frac{y - x}{xy + y}$$ Simplify.

Example 6 Use a Complex Fraction to Solve a Problem

COORDINATE GEOMETRY **Find the slope of the line that passes through $A\left(\frac{2}{p}, \frac{1}{2}\right)$ and $B\left(\frac{1}{3}, \frac{3}{p}\right)$.**

$$m = \frac{y_2 - y_1}{x_2 - x_1}$$ Definition of slope

$$= \frac{\frac{3}{p} - \frac{1}{2}}{\frac{1}{3} - \frac{2}{p}}$$ $y_2 = \frac{3}{p}, y_1 = \frac{1}{2}, x_2 = \frac{1}{3}$, and $x_1 = \frac{2}{p}$

$$= \frac{\frac{6 - p}{2p}}{\frac{p - 6}{3p}}$$ The LCD of the numerator is $2p$. The LCD of the denominator is $3p$.

$$= \frac{6 - p}{2p} \div \frac{p - 6}{3p}$$ Write as a division expression.

$$= \frac{\cancel{6 - p}}{2\cancel{p}} \cdot \frac{3\cancel{p}}{\cancel{p - 6}} \text{ or } -\frac{3}{2}$$ The slope is $-\frac{3}{2}$.

Study Tip

Check Your Solution

You can check your answer be letting p equal any nonzero number, say 1. Use the definition of slope to find the slope of the line through the points.

Check for Understanding

Concept Check

1. **FIND THE ERROR** Catalina and Yong-Chan are simplifying $\frac{x}{a} - \frac{x}{b}$.

Catalina

$$\frac{x}{a} - \frac{x}{b} = \frac{bx}{ab} - \frac{ax}{ab} = \frac{bx - ax}{ab}$$

Yong-Chan

$$\frac{x}{a} - \frac{x}{b} = \frac{x}{a - b}$$

Who is correct? Explain your reasoning.

2. **OPEN ENDED** Write two polynomials that have a LCM of $d^3 - d$.

3. Consider $\frac{1}{a} + \frac{1}{b} + \frac{1}{c}$ if a, b, and c are real numbers. Determine whether each statement is *sometimes*, *always*, or *never* true. Explain your answer.
   a. abc is a common denominator.
   b. abc is the LCD.
   c. ab is the LCD.
   d. b is the LCD.
   e. The sum is $\frac{bc + ac + ab}{abc}$.

Guided Practice

Find the LCM of each set of polynomials.

4. $12y^2, 6x^2$
5. $16ab^3, 5b^2a^2, 20ac$
6. $x^2 - 2x, x^2 - 4$

Simplify each expression.

7. $\frac{2}{x^2y} - \frac{x}{y}$
8. $\frac{7a}{15b^2} + \frac{b}{18ab}$
9. $\frac{5}{3m} - \frac{2}{7m} - \frac{1}{2m}$
10. $\frac{6}{d^2 + 4d + 4} + \frac{5}{d + 2}$
11. $\frac{a}{a^2 - a - 20} + \frac{2}{a + 4}$
12. $\frac{x + \frac{x}{3}}{x - \frac{x}{6}}$

Application

13. **GEOMETRY** Find the perimeter of the quadrilateral. Express in simplest form.

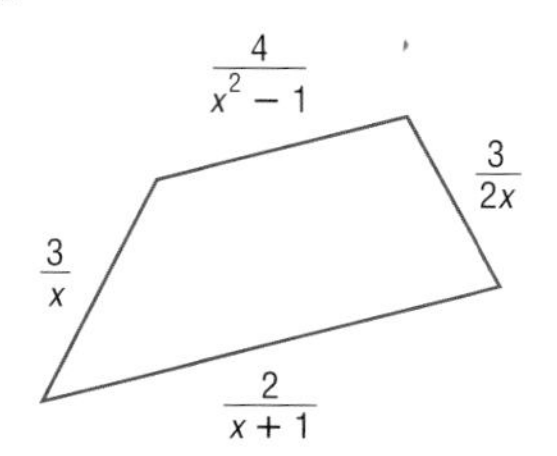

Practice and Apply

Homework Help

For Exercises	See Examples
14–21	1, 2
22–39	3, 4
40–43	5
44–49	6

Extra Practice
See page 847.

Find the LCM of each set of polynomials.

14. $10s^2, 35s^2t^2$
15. $36x^2y, 20xyz$
16. $14a^3, 15bc^3, 12b^3$
17. $9p^2q^3, 6pq^4, 4p^3$
18. $4w - 12, 2w - 6$
19. $x^2 - y^2, x^3 + x^2y$
20. $2t^2 + t - 3, 2t^2 + 5t + 3$
21. $n^2 - 7n + 12, n^2 - 2n - 8$

Simplify each expression.

22. $\frac{6}{ab} + \frac{8}{a}$
23. $\frac{5}{6v} + \frac{7}{4v}$
24. $\frac{5}{r} + 7$
25. $\frac{2x}{3y} + 5$
26. $\frac{3x}{4y^2} - \frac{y}{6x}$
27. $\frac{5}{a^2b} - \frac{7a}{5b^2}$
28. $\frac{3}{4q} - \frac{2}{5q} - \frac{1}{2q}$
29. $\frac{11}{9} - \frac{7}{2w} - \frac{6}{5w}$
30. $\frac{7}{y - 8} - \frac{6}{8 - y}$
31. $\frac{a}{a - 4} - \frac{3}{4 - a}$
32. $\frac{m}{m^2 - 4} + \frac{2}{3m + 6}$
33. $\frac{y}{y + 3} - \frac{6y}{y^2 - 9}$

34. $\dfrac{5}{x^2 - 3x - 28} + \dfrac{7}{2x - 14}$

35. $\dfrac{d - 4}{d^2 + 2d - 8} - \dfrac{d + 2}{d^2 - 16}$

36. $\dfrac{1}{h^2 - 9h + 20} - \dfrac{5}{h^2 - 10h + 25}$

37. $\dfrac{x}{x^2 + 5x + 6} - \dfrac{2}{x^2 + 4x + 4}$

38. $\dfrac{m^2 + n^2}{m^2 - n^2} + \dfrac{m}{n - m} + \dfrac{n}{m + n}$

39. $\dfrac{y + 1}{y - 1} + \dfrac{y + 2}{y - 2} + \dfrac{y}{y^2 - 3y + 2}$

40. $\dfrac{\dfrac{1}{b + 2} + \dfrac{1}{b - 5}}{\dfrac{2b^2 - b - 3}{b^2 - 3b - 10}}$

41. $\dfrac{(x + y)\left(\dfrac{1}{x} - \dfrac{1}{y}\right)}{(x - y)\left(\dfrac{1}{x} + \dfrac{1}{y}\right)}$

42. Write $\left(\dfrac{2s}{2s + 1} - 1\right) \div \left(1 + \dfrac{2s}{1 - 2s}\right)$ in simplest form.

43. What is the simplest form of $\left(3 + \dfrac{5}{a + 2}\right) \div \left(3 - \dfrac{10}{a + 7}\right)$?

ELECTRICITY For Exercises 44 and 45, use the following information.
In an electrical circuit, if two resistors with resistance R_1 and R_2 are connected in parallel as shown, the relationship between these resistances and the resulting combination resistance R is $\dfrac{1}{R} = \dfrac{1}{R_1} + \dfrac{1}{R_2}$.

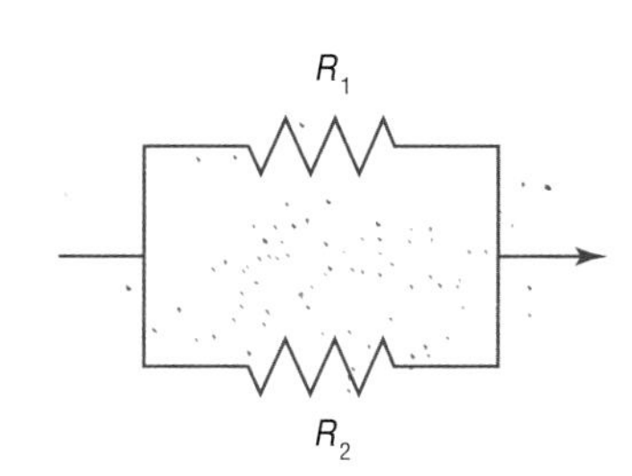

44. If R_1 is x ohms and R_2 is 4 ohms less than twice x ohms, write an expression for $\dfrac{1}{R}$.

45. Find the effective resistance of a 30-ohm resistor and a 20-ohm resistor that are connected in parallel.

More About. . .

Bicycling
The Tour de France is the most popular bicycle road race. It lasts 24 days and covers 2500 miles.
Source: *World Book Encyclopedia*

BICYCLING For Exercises 46–48, use the following information.
Jalisa is competing in a 48-mile bicycle race. She travels half the distance at one rate. The rest of the distance, she travels 4 miles per hour slower.

46. If x represents the faster pace in miles per hour, write an expression that represents the time spent at that pace.

47. Write an expression for the amount of time spent at the slower pace.

48. Write an expression for the amount of time Jalisa needed to complete the race.

49. **MAGNETS** For a bar magnet, the magnetic field strength H at a point P along the axis of the magnet is $H = \dfrac{m}{2L(d - L)^2} - \dfrac{m}{2L(d + L)^2}$. Write a simpler expression for H.

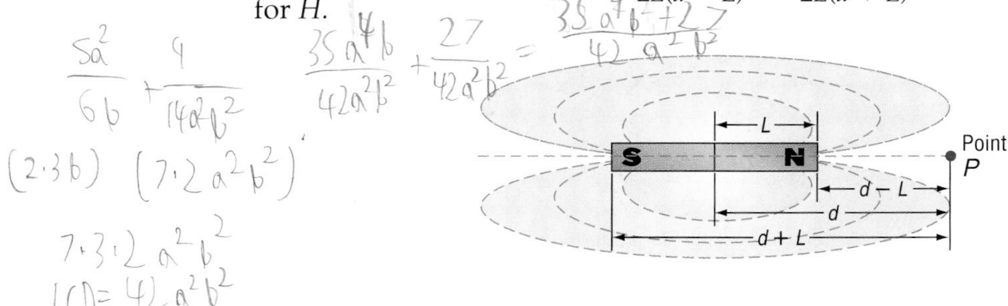

50. **CRITICAL THINKING** Find two rational expressions whose sum is $\dfrac{2x - 1}{(x + 1)(x - 2)}$.

51. **WRITING IN MATH** Answer the question that was posed at the beginning of the lesson.

How is subtraction of rational expressions used in photography?
Include the following in your answer:

- an explanation of how to subtract rational expressions, and
- an equation that could be used to find the distance between the lens and the film if the focal length of the lens is 10 centimeters and the distance between the lens and the object is 60 centimeters.

Standards Practice
Standardized Test Practice

52. For all $t \neq 5$, $\frac{t^2 - 25}{3t - 15} =$

(A) $\frac{t-5}{3}$. (B) $\frac{t+5}{3}$. (C) $t - 5$. (D) $t + 5$. (E) $\frac{t-5}{t-3}$.

53. What is the sum of $\frac{x-y}{5}$ and $\frac{x+y}{4}$?

(A) $\frac{9x+9y}{20}$ (B) $\frac{x+9y}{20}$ (C) $\frac{9x+y}{20}$ (D) $\frac{9x-y}{20}$ (E) $\frac{x-9y}{20}$

Maintain Your Skills

Mixed Review

Simplify each expression. *(Lesson 9-1)*

54. $\frac{9x^2y^3}{(5xyz)^2} \div \frac{(3xy)^3}{20x^2y}$

55. $\frac{5a^2 - 20}{2a + 2} \cdot \frac{4a}{10a - 20}$

Solve each system of inequalities by graphing. *(Lesson 8-7)*

56. $9x^2 + y^2 < 81$
$x^2 + y^2 \geq 16$

57. $(y - 3)^2 \geq x + 2$
$x^2 \leq y + 4$

58. **GARDENS** Helene Jonson has a rectangular garden 25 feet by 50 feet. She wants to increase the garden on all sides by an equal amount. If the area of the garden is to be increased by 400 square feet, by how much should each dimension be increased? *(Lesson 6-4)*

Getting Ready for the Next Lesson

PREREQUISITE SKILL Draw the asymptotes and graph each hyperbola.
*(To review **graphing hyperbolas**, see Lesson 8-5.)*

59. $\frac{x^2}{16} - \frac{y^2}{20} = 1$

60. $\frac{y^2}{49} - \frac{x^2}{25} = 1$

61. $\frac{(x+2)^2}{16} - \frac{(y-5)^2}{25} = 1$

Practice Quiz 1

Lessons 9-1 and 9-2

Simplify each expression. *(Lesson 9-1)*

1. $\frac{t^2 - t - 6}{t^2 - 6t + 9}$

2. $\frac{3ab^3}{8a^2b} \cdot \frac{4ac}{9b^4}$

3. $-\frac{4}{8x} \div \frac{16}{xy^2}$

4. $\frac{48}{6a + 42} \cdot \frac{7a + 49}{16}$

5. $\frac{w^2 + 5w + 4}{6} \div \frac{w + 1}{18w + 24}$

6. $\frac{\frac{x^2 + x}{x + 1}}{\frac{x}{x - 1}}$

Simplify each expression. *(Lesson 9-2)*

7. $\frac{4a + 2}{a + b} + \frac{1}{-b - a}$

8. $\frac{2x}{5ab^3} + \frac{4y}{3a^2b^2}$

9. $\frac{5}{n + 6} - \frac{4}{n - 1}$

10. $\frac{x - 5}{2x - 6} - \frac{x - 7}{4x - 12}$

9-3 Graphing Rational Functions

California Standards Standard 7.0 **Students** add, subtract, multiply, divide, **reduce,** and evaluate **rational expressions with** monomial and **polynomial denominators** and simplify complicated rational expressions, including those with negative exponents in the denominator. (Key)

What You'll Learn

- Determine the vertical asymptotes and the point discontinuity for the graphs of rational functions.
- Graph rational functions.

Vocabulary

- rational function
- continuity
- asymptote
- point discontinuity

How can rational functions be used when buying a group gift?

A group of students want to get their favorite teacher, Mr. Salgado, a retirement gift. They plan to get him a gift certificate for a weekend package at a lodge in a state park. The certificate costs \$150. If c represents the cost for each student and s represents the number of students, then $c = \frac{150}{s}$.

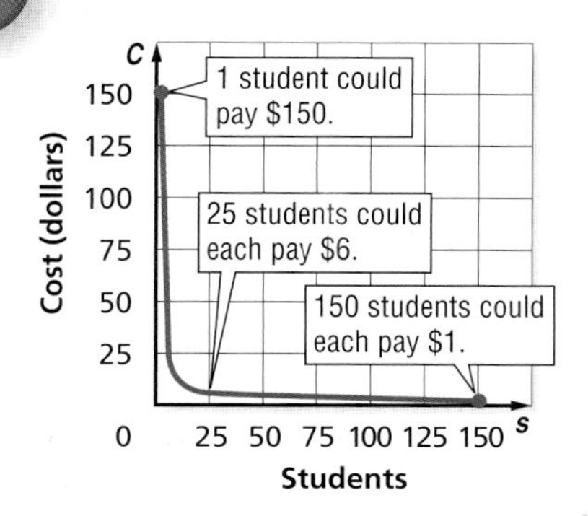

VERTICAL ASYMPTOTES AND POINT DISCONTINUITY The function $c = \frac{150}{s}$ is an example of a rational function. A **rational function** is an equation of the form $f(x) = \frac{p(x)}{q(x)}$, where $p(x)$ and $q(x)$ are polynomial functions and $q(x) \neq 0$. Here are other examples of rational functions.

$$f(x) = \frac{x}{x + 3} \qquad g(x) = \frac{5}{x - 6} \qquad h(x) = \frac{x + 4}{(x - 1)(x + 4)}$$

No denominator in a rational function can be zero because division by zero is not defined. In the examples above, the functions are not defined at $x = -3$, $x = 6$, and $x = 1$ and $x = -4$, respectively.

The graphs of rational functions may have breaks in **continuity**. This means that, unlike polynomial functions, which can be traced with a pencil never leaving the paper, not all rational functions are traceable. Breaks in continuity can appear as a vertical **asymptote** or as a **point discontinuity**. Recall that an asymptote is a line that the graph of the function approaches, but never crosses. Point discontinuity is like a hole in a graph.

Study Tip

Look Back
To review **asymptotes**, see Lesson 8-5.

Key Concept *Vertical Asymptotes*

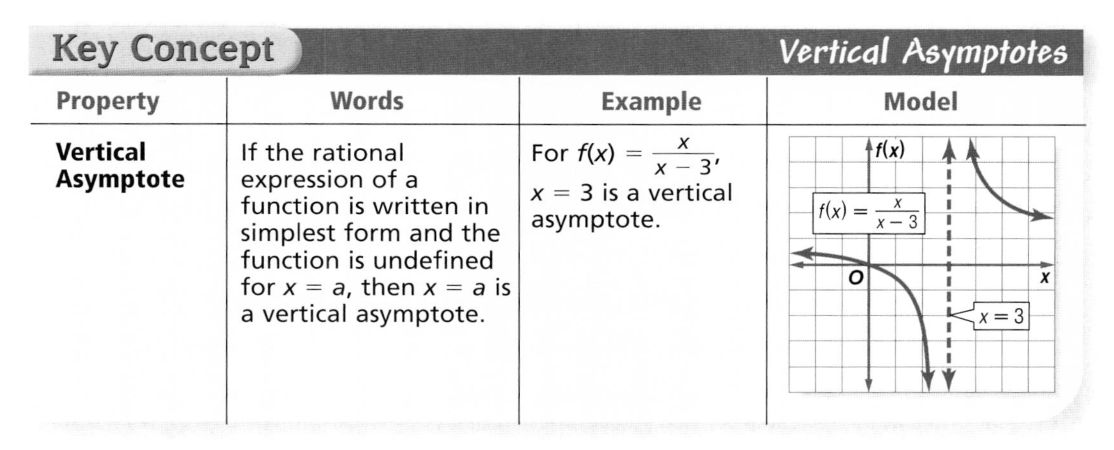

Property	Words	Example	Model
Vertical Asymptote	If the rational expression of a function is written in simplest form and the function is undefined for $x = a$, then $x = a$ is a vertical asymptote.	For $f(x) = \frac{x}{x - 3}$, $x = 3$ is a vertical asymptote.	$f(x)$; $f(x) = \frac{x}{x - 3}$; O; x; $x = 3$

Key Concept — Point Discontinuity

Property	Words	Example	Model
Point Discontinuity	If the original function is undefined for $x = a$ but the rational expression of the function in simplest form is defined for $x = a$, then there is a hole in the graph at $x = a$.	$f(x) = \frac{(x + 2)(x - 1)}{x + 2}$ can be simplified to $f(x) = x - 1$. So, $x = -2$ represents a hole in the graph.	

Example 1 Vertical Asymptotes and Point Discontinuity

Determine the equations of any vertical asymptotes and the values of x for any holes in the graph of $f(x) = \frac{x^2 - 1}{x^2 - 6x + 5}$.

First factor the numerator and denominator of the rational expression.

$$\frac{x^2 - 1}{x^2 - 6x + 5} = \frac{(x - 1)(x + 1)}{(x - 1)(x - 5)}$$

The function is undefined for $x = 1$ and $x = 5$. Since $\frac{\overset{1}{\cancel{(x - 1)}}(x + 1)}{\underset{1}{\cancel{(x - 1)}}(x - 5)} = \frac{x + 1}{x - 5}$,

$x = 5$ is a vertical asymptote, and $x = 1$ represents a hole in the graph.

GRAPH RATIONAL FUNCTIONS You can use what you know about vertical asymptotes and point discontinuity to graph rational functions.

Example 2 Graph with a Vertical Asymptote

Graph $f(x) = \frac{x}{x - 2}$.

The function is undefined for $x = 2$. Since $\frac{x}{x - 2}$ is in simplest form, $x = 2$ is a vertical asymptote. Draw the vertical asymptote. Make a table of values. Plot the points and draw the graph.

x	$f(x)$
−50	0.96154
−20	0.90909
−10	0.83333
−2	0.5
−1	0.33333
0	0
1	−1
3	3
4	2
5	1.6667
10	1.25
20	1.1111
50	1.0417

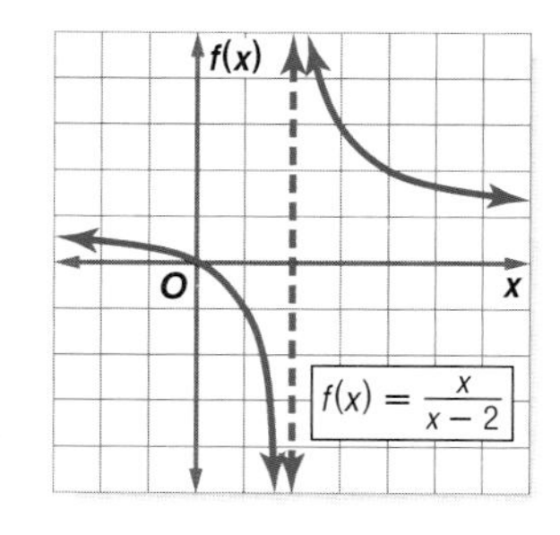

As $|x|$ increases, it appears that the y values of the function get closer and closer to 1. The line with the equation $f(x) = 1$ is a horizontal asymptote of the function.

Study Tip

Graphing Rational Functions

Finding the x- and y-intercepts is often useful when graphing rational functions.

As you have learned, graphs of rational functions may have point discontinuity rather than vertical asymptotes. The graphs of these functions appear to have holes. These holes are usually shown as circles on graphs.

Example 3 Graph with Point Discontinuity

Graph $f(x) = \frac{x^2 - 9}{x + 3}$.

Notice that $\frac{x^2 - 9}{x + 3} = \frac{(x + 3)(x - 3)}{x + 3}$ or $x - 3$.

Therefore, the graph of $f(x) = \frac{x^2 - 9}{x + 3}$ is the graph of $f(x) = x - 3$ with a hole at $x = -3$.

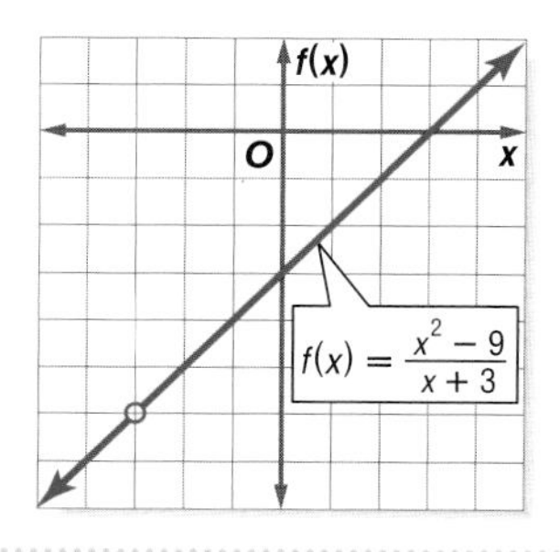

Many real-life situations can be described by using rational functions.

Algebra Activity

Rational Functions

The density of a material can be expressed as $D = \frac{m}{V}$, where m is the mass of the material in grams and V is the volume in cubic centimeters. By finding the volume and density of 200 grams of each liquid, you can draw a graph of the function $D = \frac{200}{V}$.

Collect the Data

- Use a balance and metric measuring cups to find the volumes of 200 grams of different liquids such as water, cooking oil, isopropyl alcohol, sugar water, and salt water.
- Use $D = \frac{m}{V}$ to find the density of each liquid.

Analyze the Data

1. Graph the data by plotting the points (volume, density) on a graph. Connect the points.
2. From the graph, find the asymptotes.

In the real world, sometimes values on the graph of a rational function are not meaningful.

Example 4 Use Graphs of Rational Functions

TRANSPORTATION **A train travels at one velocity V_1 for a given amount of time t_1 and then another velocity V_2 for a different amount of time t_2. The average velocity is given by $V = \frac{V_1 t_1 + V_2 t_2}{t_1 + t_2}$.**

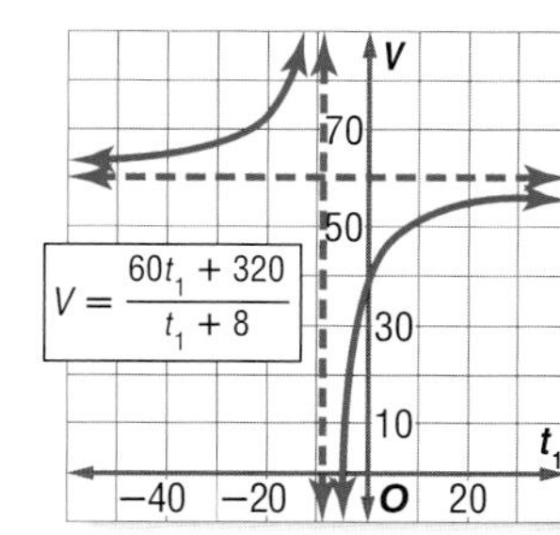

a. Let t_1 be the independent variable and let V be the dependent variable. Draw the graph if $V_1 = 60$ miles per hour, $V_2 = 40$ miles per hour, and $t_2 = 8$ hours.

The function is $V = \frac{60t_1 + 40(8)}{t_1 + 8}$ or $V = \frac{60t_1 + 320}{t_1 + 8}$. The vertical asymptote is $t_1 = -8$. Graph the vertical asymptote and the function. Notice that the horizontal asymptote is $V = 60$.

b. **What is the V-intercept of the graph?**

The V-intercept is 40.

c. **What values of t_1 and V are meaningful in the context of the problem?**

In the problem context, time and velocity are positive values. Therefore, only values of t_1 greater than 0 and values of V between 40 and 60 are meaningful.

Check for Understanding

Concept Check

1. **OPEN ENDED** Write a function whose graph has two vertical asymptotes located at $x = -5$ and $x = 2$.

2. **Compare and contrast** the graphs of $f(x) = \frac{(x-1)(x+5)}{x-1}$ and $g(x) = x + 5$.

3. **Describe** the graph at the right. Include the equations of any asymptotes, the x values of any holes, and the x- and y-intercepts.

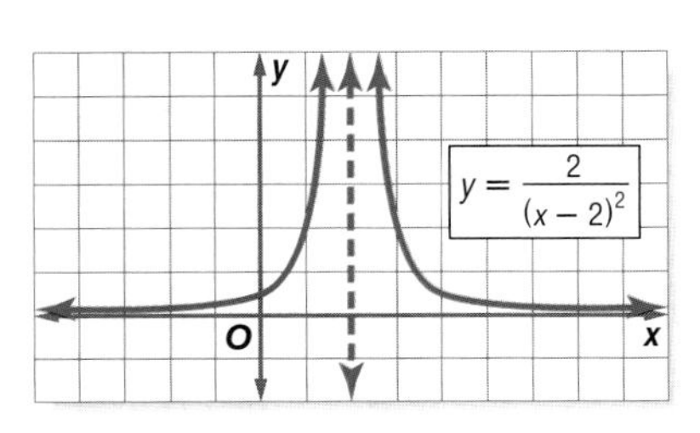

Guided Practice

Determine the equations of any vertical asymptotes and the values of x for any holes in the graph of each rational function.

4. $f(x) = \frac{3}{x^2 - 4x + 4}$

5. $f(x) = \frac{x - 1}{x^2 + 4x - 5}$

Graph each rational function.

6. $f(x) = \frac{x}{x + 1}$

7. $f(x) = \frac{6}{(x-2)(x+3)}$

8. $f(x) = \frac{x^2 - 25}{x - 5}$

9. $f(x) = \frac{x - 5}{x + 1}$

10. $f(x) = \frac{4}{(x-1)^2}$

11. $f(x) = \frac{x + 2}{x^2 - x - 6}$

Application

MEDICINE For Exercises 12–15, use the following information.

For certain medicines, health care professionals may use Young's Rule, $C = \frac{y}{y + 12} \cdot D$, to estimate the proper dosage for a child when the adult dosage is known. In this equation, C represents the child's dose, D represents the adult dose, and y represents the child's age in years.

12. Use Young's Rule to estimate the dosage of amoxicillin for an eight-year-old child if the adult dosage is 250 milligrams.

13. Graph $C = \frac{y}{y + 12}$.

14. Give the equations of any asymptotes and y- and C-intercepts of the graph.

15. What values of y and C are meaningful in the context of the problem?

Practice and Apply

Homework Help

For Exercises	See Examples
16–21	1
22–39	2, 3
40–50	4

Extra Practice See page 849.

Determine the equations of any vertical asymptotes and the values of x for any holes in the graph of each rational function.

16. $f(x) = \frac{2}{x^2 - 5x + 6}$
17. $f(x) = \frac{4}{x^2 + 2x - 8}$
18. $f(x) = \frac{x + 3}{x^2 + 7x + 12}$
19. $f(x) = \frac{x - 5}{x^2 - 4x - 5}$
20. $f(x) = \frac{x^2 - 8x + 16}{x - 4}$
21. $f(x) = \frac{x^2 - 3x + 2}{x - 1}$

Graph each rational function.

22. $f(x) = \frac{1}{x}$
23. $f(x) = \frac{3}{x}$
24. $f(x) = \frac{1}{x + 2}$
25. $f(x) = \frac{-5}{x + 1}$
26. $f(x) = \frac{x}{x - 3}$
27. $f(x) = \frac{5x}{x + 1}$
28. $f(x) = \frac{-3}{(x - 2)^2}$
29. $f(x) = \frac{1}{(x + 3)^2}$
30. $f(x) = \frac{x + 4}{x - 1}$
31. $f(x) = \frac{x - 1}{x - 3}$
32. $f(x) = \frac{x^2 - 36}{x + 6}$
33. $f(x) = \frac{x^2 - 1}{x - 1}$
34. $f(x) = \frac{3}{(x - 1)(x + 5)}$
35. $f(x) = \frac{-1}{(x + 2)(x - 3)}$
36. $f(x) = \frac{x}{x^2 - 1}$
37. $f(x) = \frac{x - 1}{x^2 - 4}$
38. $f(x) = \frac{6}{(x - 6)^2}$
39. $f(x) = \frac{1}{(x + 2)^2}$

More About. . .

History

Mathematician Maria Gaetana Agnesi was one of the greatest scholars of all time. Born in Milan, Italy, in 1718, she mastered Greek, Hebrew, and several modern languages by the age of 11.

Source: *A History of Mathematics*

HISTORY **For Exercises 40–42, use the following information.**
In Maria Gaetana Agnesi's book *Analytical Institutions*, Agnesi discussed the characteristics of the equation $x^2y = a^2(a - y)$, whose graph is called the "curve of Agnesi." This equation can be expressed as $y = \frac{a^3}{x^2 + a^2}$.

40. Graph $f(x) = \frac{a^3}{x^2 + a^2}$ if $a = 4$.
41. Describe the graph.
42. **Make a conjecture** about the shape of the graph of $f(x) = \frac{a^3}{x^2 + a^2}$ if $a = -4$. Explain your reasoning.

AUTO SAFETY **For Exercises 43–45, use the following information.**
When a car has a front-end collision, the objects in the car (including passengers) keep moving forward until the impact occurs. After impact, objects are repelled. Seat belts and airbags limit how far you are jolted forward. The formula for the velocity at which you are thrown backward is $V_f = \frac{(m_1 - m_2)v_i}{m_1 + m_2}$, where m_1 and m_2 are masses of the two objects meeting and v_i is the initial velocity.

43. Let m_1 be the independent variable, and let V_f be the dependent variable. Graph the function if $m_2 = 7$ kilograms and $v_i = 5$ meters per second.
44. Give the equation of the vertical asymptote and the m_1- and V_f-intercepts of the graph.
45. Find the value of V_f when the value of m_1 is 5 kilograms.

46. **CRITICAL THINKING** Write three rational functions that have a vertical asymptote at $x = 3$ and a hole at $x = -2$.

BASKETBALL **For Exercises 47–50, use the following information.**
Zonta plays basketball for Centerville High School. So far this season, she has made 6 out of 10 free throws. She is determined to improve her free-throw percentage. If she can make x consecutive free throws, her free-throw percentage can be determined using $P(x) = \frac{6 + x}{10 + x}$.

47. Graph the function.
48. What part of the graph is meaningful in the context of the problem?
49. Describe the meaning of the y-intercept.
50. What is the equation of the horizontal asymptote? Explain its meaning with respect to Zonta's shooting percentage.

51. **WRITING IN MATH** Answer the question that was posed at the beginning of the lesson.

How can rational functions be used when buying a group gift?

Include the following in your answer:
- a complete graph of the function $c = \frac{150}{s}$ with asymptotes, and
- an explanation of why only part of the graph is meaningful in the context of the problem.

52. Which set is the domain of the function graphed at the right?

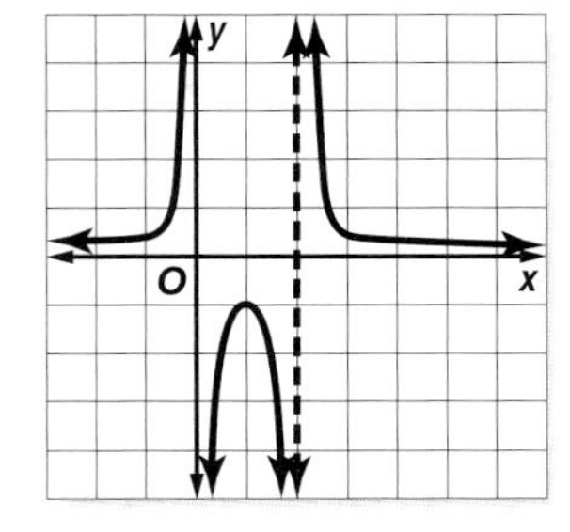

Ⓐ $\{x \mid x \neq 0, 2\}$
Ⓑ $\{x \mid x \neq -2, 0\}$
Ⓒ $\{x \mid x < 4\}$
Ⓓ $\{x \mid x > -4\}$

53. Which set is the range of the function $y = \frac{x^2 + 8}{2}$?

Ⓐ $\{y \mid y \neq \pm 2\sqrt{2}\}$
Ⓑ $\{y \mid y \geq 4\}$
Ⓒ $\{y \mid y \geq 0\}$
Ⓓ $\{y \mid y \leq 0\}$

Maintain Your Skills

Mixed Review

Simplify each expression. *(Lessons 9-2 and 9-1)*

54. $\frac{3m + 2}{m + n} + \frac{4}{2m + 2n}$
55. $\frac{5}{x + 3} - \frac{2}{x - 2}$
56. $\frac{2w - 4}{w + 3} \div \frac{2w + 6}{5}$

Find the coordinates of the center and the radius of the circle with the given equation. Then graph the circle. *(Lesson 8-3)*

57. $(x - 6)^2 + (y - 2)^2 = 25$
58. $x^2 + y^2 + 4x = 9$

59. **ART** Joyce Jackson purchases works of art for an art gallery. Two years ago, she bought a painting for \$20,000, and last year, she bought one for \$35,000. If paintings appreciate 14% per year, how much are the two paintings worth now? *(Lesson 7-1)*

Solve each equation by completing the square. *(Lesson 6-4)*

60. $x^2 + 8x + 20 = 0$
61. $x^2 + 2x - 120 = 0$
62. $x^2 + 7x - 17 = 0$

Getting Ready for the Next Lesson

BASIC SKILL **Solve each proportion.**

63. $\frac{16}{v} = \frac{32}{9}$
64. $\frac{7}{25} = \frac{a}{5}$
65. $\frac{6}{15} = \frac{8}{s}$
66. $\frac{b}{9} = \frac{40}{30}$

Graphing Calculator Investigation

A Follow-Up of Lesson 9-3

Graphing Rational Functions

A TI-83 Plus graphing calculator can be used to explore the graphs of rational functions. These graphs have some features that never appear in the graphs of polynomial functions.

Example 1 **Graph $y = \frac{8x - 5}{2x}$ in the standard viewing window. Find the equations of any asymptotes.**

- Enter the equation in the Y= list.

KEYSTROKES: [Y=] [(] 8 [X,T,θ,n] [−] 5 [)] [÷] [(] 2 [X,T,θ,n] [)] [ZOOM] 6

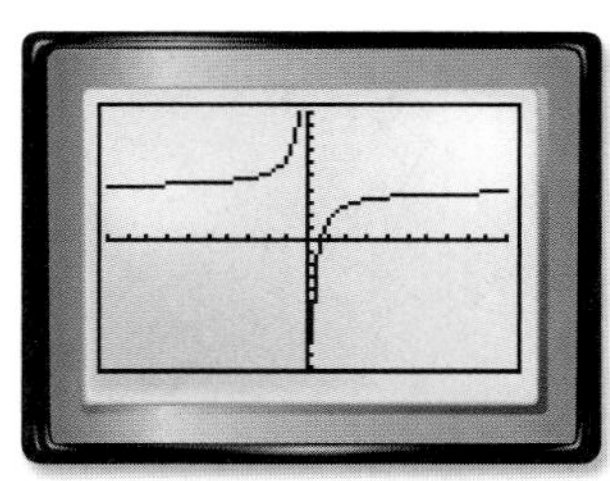
[−10, 10] scl: 1 by [−10, 10] scl: 1

By looking at the equation, we can determine that if $x = 0$, the function is undefined. The equation of the vertical asymptote is $x = 0$. Notice what happens to the y values as x grows larger and as x gets smaller. The y values approach 4. So, the equation for the horizontal asymptote is $y = 4$.

Example 2 **Graph $y = \frac{x^2 - 16}{x + 4}$ in the window [−5, 4.4] by [−10, 2] with scale factors of 1.**

- Because the function is not continuous, put the calculator in dot mode.

KEYSTROKES: [MODE] [▼] [▼] [▼] [▼] [▶] [ENTER]

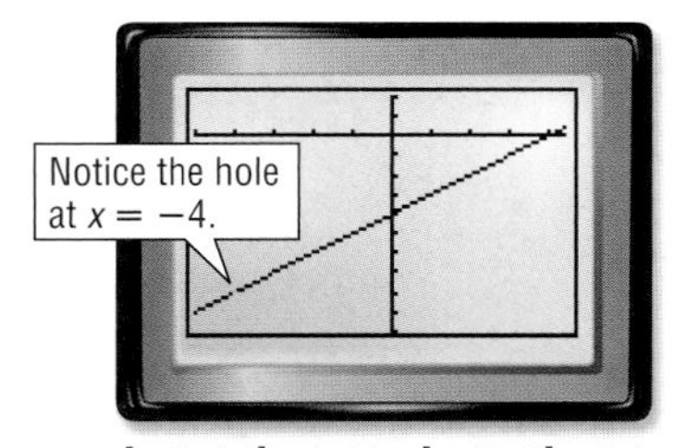

[−5, 4.4] scl: 1 by [−10, 2] scl: 1

This graph looks like a line with a break in continuity at $x = -4$. This happens because the denominator is 0 when $x = -4$. Therefore, the function is undefined when $x = -4$.

If you TRACE along the graph, when you come to $x = -4$, you will see that there is no corresponding y value.

Exercises

Use a graphing calculator to graph each function. Be sure to show a complete graph. Draw the graph on a sheet of paper. Write the x-coordinates of any points of discontinuity and/or the equations of any asymptotes.

1. $f(x) = \frac{1}{x}$

2. $f(x) = \frac{x}{x + 2}$

3. $f(x) = \frac{2}{x - 4}$

4. $f(x) = \frac{2x}{3x - 6}$

5. $f(x) = \frac{4x + 2}{x - 1}$

6. $f(x) = \frac{x^2 - 9}{x + 3}$

7. Which graph(s) has point discontinuity?

8. Describe functions that have point discontinuity.

www.algebra2.com/other_calculator_keystrokes

9-4 Direct, Joint, and Inverse Variation

What You'll Learn

- Recognize and solve direct and joint variation problems.
- Recognize and solve inverse variation problems.

Vocabulary

- direct variation
- constant of variation
- joint variation
- inverse variation

How is variation used to find the total cost given the unit cost?

The total high-tech spending t of an average public college can be found by using the equation $t = 149s$, where s is the number of students.

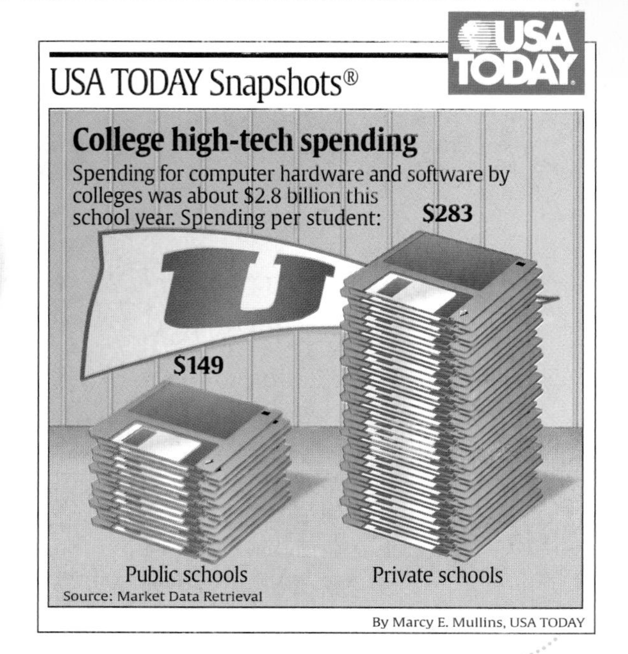

DIRECT VARIATION AND JOINT VARIATION The relationship given by $t = 149s$ is an example of direct variation. A **direct variation** can be expressed in the form $y = kx$. The k in this equation is a constant and is called the **constant of variation**.

Notice that the graph of $t = 149s$ is a straight line through the origin. An equation of a direct variation is a special case of an equation written in slope-intercept form, $y = mx + b$. When $m = k$ and $b = 0$, $y = mx + b$ becomes $y = kx$. So the slope of a direct variation equation is its constant of variation.

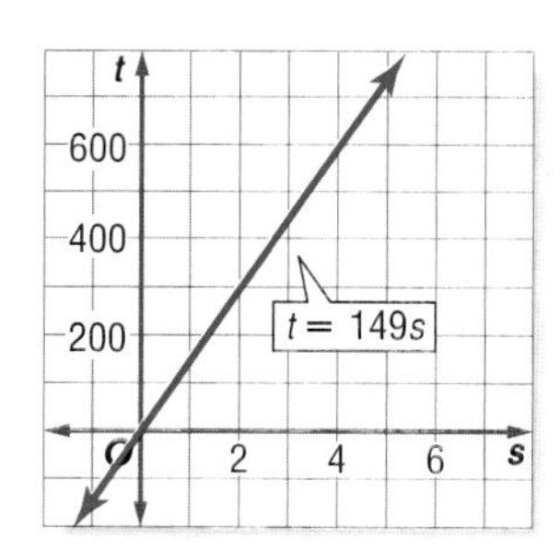

To express a direct variation, we say that y varies directly as x. In other words, as x increases, y increases or decreases at a constant rate.

Key Concept — Direct Variation

y varies directly as x if there is some nonzero constant k such that $y = kx$. k is called the constant of variation.

If you know that y varies directly as x and one set of values, you can use a proportion to find the other set of corresponding values.

$$y_1 = kx_1 \quad \text{and} \quad y_2 = kx_2$$

$$\frac{y_1}{x_1} = k \qquad \frac{y_2}{x_2} = k$$

Therefore, $\frac{y_1}{x_1} = \frac{y_2}{x_2}$.

Using the properties of equality, you can find many other proportions that relate these same x and y values.

Example 1 Direct Variation

If y varies directly as x and $y = 12$ when $x = -3$, find y when $x = 16$.

Use a proportion that relates the values.

$\frac{y_1}{x_1} = \frac{y_2}{x_2}$ Direct proportion

$\frac{12}{-3} = \frac{y_2}{16}$ $y_1 = 12$, $x_1 = -3$, and $x_2 = 16$

$16(12) = -3(y_2)$ Cross multiply.

$192 = -3y_2$ Simplify.

$-64 = y_2$ Divide each side by -3.

When $x = 16$, the value of y is -64.

Another type of variation is joint variation. **Joint variation** occurs when one quantity varies directly as the product of two or more other quantities.

Key Concept — Joint Variation

y varies jointly as x and z if there is some number k such that $y = kxz$, where $k \neq 0$, $x \neq 0$, and $z \neq 0$.

If you know y varies jointly as x and z and one set of values, you can use a proportion to find the other set of corresponding values.

$$y_1 = kx_1z_1 \quad \text{and} \quad y_2 = kx_2z_2$$

$$\frac{y_1}{x_1z_1} = k \qquad \frac{y_2}{x_2z_2} = k$$

Therefore, $\frac{y_1}{x_1z_1} = \frac{y_2}{x_2z_2}$.

Example 2 Joint Variation

Suppose y varies jointly as x and z. Find y when $x = 8$ and $z = 3$, if $y = 16$ when $z = 2$ and $x = 5$.

Use a proportion that relates the values.

$\frac{y_1}{x_1z_1} = \frac{y_2}{x_2z_2}$ Joint variation

$\frac{16}{5(2)} = \frac{y_2}{8(3)}$ $y_1 = 16$, $x_1 = 5$, $z_1 = 2$, $x_2 = 8$, and $z_2 = 3$

$8(3)(16) = 5(2)(y_2)$ Cross multiply.

$384 = 10y_2$ Simplify.

$38.4 = y_2$ Divide each side by 10.

When $x = 8$ and $z = 3$, the value of y is 38.4.

INVERSE VARIATION Another type of variation is inverse variation. For two quantities with **inverse variation**, as one quantity increases, the other quantity decreases. For example, speed and time for a fixed distance vary inversely with each other. When you travel to a particular location, as your speed increases, the time it takes to arrive at that location decreases.

Key Concept — Inverse Variation

y varies inversely as *x* if there is some nonzero constant *k* such that

$$xy = k \text{ or } y = \frac{k}{x}.$$

Suppose y varies inversely as x such that $xy = 6$ or $y = \frac{6}{x}$. The graph of this equation is shown at the right. Note that in this case, k is a positive value 6, so as the values of x increase, the values of y decrease.

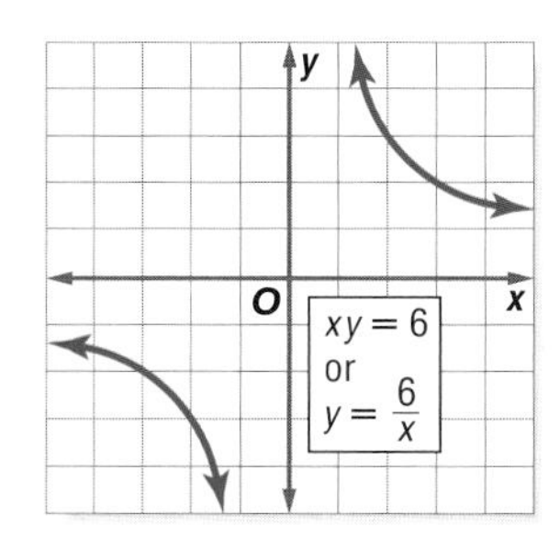

Just as with direct variation and joint variation, a proportion can be used with inverse variation to solve problems where some quantities are known. The following proportion is only one of several that can be formed.

$x_1y_1 = k$ and $x_2y_2 = k$

$x_1y_1 = x_2y_2$ Substitution Property of Equality

$\frac{x_1}{y_2} = \frac{x_2}{y_1}$ Divide each side by y_1y_2.

Example 3 Inverse Variation

If r varies inversely as t and $r = 18$ when $t = -3$, find r when $t = -11$.

Use a proportion that relates the values.

$\frac{r_1}{t_2} = \frac{r_2}{t_1}$ Inverse variation

$\frac{18}{-11} = \frac{r_2}{-3}$ $r_1 = 18$, $t_1 = -3$, and $t_2 = -11$

$18(-3) = -11(r_2)$ Cross multiply.

$-54 = -11r_2$ Simplify.

$4\frac{10}{11} = r_2$ Divide each side by -11.

When $t = -11$, the value of r is $4\frac{10}{11}$.

More About. . .

Space

Mercury is about 36 million miles from the Sun, making it the closest planet to the Sun. Its proximity to the Sun causes its temperature to be as high as 800°F.

Source: *World Book Encyclopedia*

Example 4 Use Inverse Variation

SPACE The apparent length of an object is inversely proportional to one's distance from the object. Earth is about 93 million miles from the Sun. Use the information at the left to find how much larger the diameter of the Sun would appear on Mercury than on Earth.

Explore You know that the apparent diameter of the Sun varies inversely with the distance from the Sun. You also know Mercury's distance from the Sun and Earth's distance from the Sun. You want to determine how much larger the diameter of the Sun appears on Mercury than on Earth.

Plan Let the apparent diameter of the Sun from Earth equal 1 unit and the apparent diameter of the Sun from Mercury equal m. Then use a proportion that relates the values.

Solve

$$\frac{\text{distance from Mercury}}{\text{apparent diameter from Earth}} = \frac{\text{distance from Earth}}{\text{apparent diameter from Mercury}} \quad \text{Inverse variation}$$

$$\frac{36 \text{ million miles}}{1 \text{ unit}} = \frac{93 \text{ million miles}}{m \text{ units}} \quad \text{Substitution}$$

$$(36 \text{ million miles})(m \text{ units}) = (93 \text{ million miles})(1 \text{ unit}) \quad \text{Cross multiply.}$$

$$m = \frac{(93 \text{ million miles})(1 \text{ unit})}{36 \text{ million miles}} \quad \text{Divide each side by 36 million miles.}$$

$$m \approx 2.58 \text{ units} \quad \text{Simplify.}$$

Examine Since the distance between the Sun and Earth is between 2 and 3 times the distance between the Sun and Mercury, the answer seems reasonable. From Mercury, the diameter of the Sun will appear about 2.58 times as large as it appears from Earth.

Check for Understanding

Concept Check

1. **Determine** whether each graph represents a *direct* or an *inverse* variation.

a.

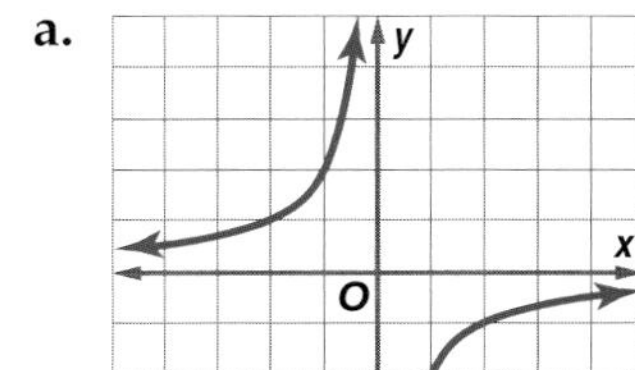

b.

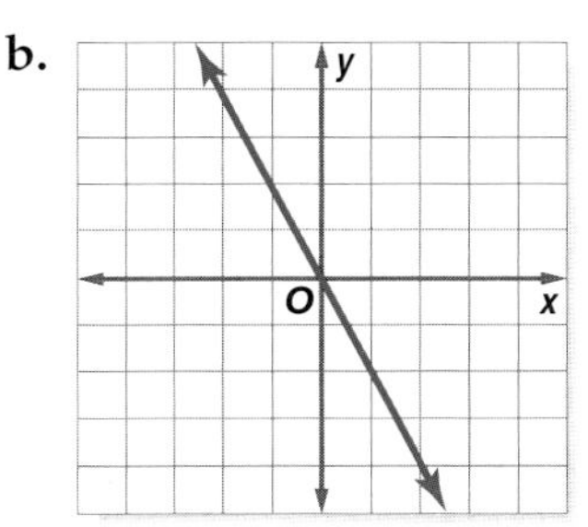

2. **Compare and contrast** $y = 5x$ and $y = -5x$.
3. **OPEN ENDED** Describe two quantities in real life that vary directly with each other and two quantities that vary inversely with each other.

Guided Practice

State whether each equation represents a *direct, joint,* or *inverse* variation. Then name the constant of variation.

4. $ab = 20$
5. $\frac{y}{x} = -0.5$
6. $A = \frac{1}{2}bh$

Find each value.

7. If y varies directly as x and $y = 18$ when $x = 15$, find y when $x = 20$.
8. Suppose y varies jointly as x and z. Find y when $x = 9$ and $z = -5$, if $y = -90$ when $z = 15$ and $x = -6$.
9. If y varies inversely as x and $y = -14$ when $x = 12$, find x when $y = 21$.

Application

SWIMMING For Exercises 10–13, use the following information.
When a person swims underwater, the pressure in his or her ears varies directly with the depth at which he or she is swimming.

4.3 pounds per square inch (psi)
10 ft

10. Write an equation of direct variation that represents this situation.
11. Find the pressure at 60 feet.
12. It is unsafe for amateur divers to swim where the water pressure is more than 65 pounds per square inch. How deep can an amateur diver safely swim?
13. Make a table showing the number of pounds of pressure at various depths of water. Use the data to draw a graph of pressure versus depth.

Practice and Apply

Homework Help

For Exercises	See Examples
14–37	1–3
38–53	4

Extra Practice
See page 848.

State whether each equation represents a *direct, joint,* or *inverse* variation. Then name the constant of variation.

14. $\frac{n}{m} = 1.5$ **15.** $a = 5bc$ **16.** $vw = -18$ **17.** $3 = \frac{a}{b}$

18. $p = \frac{12}{q}$ **19.** $y = -7x$ **20.** $V = \frac{1}{3}Bh$ **21.** $\frac{2.5}{t} = s$

22. CHEMISTRY Boyle's Law states that when a sample of gas is kept at a constant temperature, the volume varies inversely with the pressure exerted on it. Write an equation for Boyle's Law that expresses the variation in volume V as a function of pressure P.

23. CHEMISTRY Charles' Law states that when a sample of gas is kept at a constant pressure, its volume V will increase as the temperature t increases. Write an equation for Charles' Law that expresses volume as a function.

24. GEOMETRY How does the circumference of a circle vary with respect to its radius? What is the constant of variation?

25. TRAVEL A map is scaled so that 3 centimeters represents 45 kilometers. How far apart are two towns if they are 7.9 centimeters apart on the map?

Find each value.

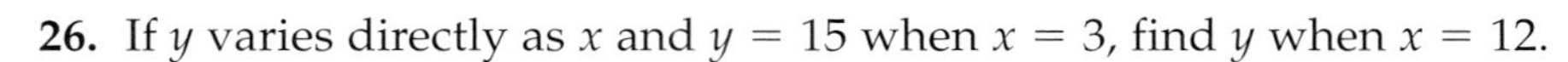

26. If y varies directly as x and $y = 15$ when $x = 3$, find y when $x = 12$.

27. If y varies directly as x and $y = 8$ when $x = 6$, find y when $x = 15$.

28. Suppose y varies jointly as x and z. Find y when $x = 2$ and $z = 27$, if $y = 192$ when $x = 8$ and $z = 6$.

29. If y varies jointly as x and z and $y = 80$ when $x = 5$ and $z = 8$, find y when $x = 16$ and $z = 2$.

30. If y varies inversely as x and $y = 5$ when $x = 10$, find y when $x = 2$.

31. If y varies inversely as x and $y = 16$ when $x = 5$, find y when $x = 20$.

32. If y varies inversely as x and $y = 2$ when $x = 25$, find x when $y = 40$.

33. If y varies inversely as x and $y = 4$ when $x = 12$, find y when $x = 5$.

34. If y varies directly as x and $y = 9$ when $x = -15$, find y when $x = 21$.

35. If y varies directly as x and $x = 6$ when $y = 0.5$, find y when $x = 10$.

36. Suppose y varies jointly as x and z. Find y when $x = \frac{1}{2}$ and $z = 6$, if $y = 45$ when $x = 6$ and $z = 10$.

37. If y varies jointly as x and z and $y = \frac{1}{8}$ when $x = \frac{1}{2}$ and $z = 3$, find y when $x = 6$ and $z = \frac{1}{3}$.

38. WORK Paul drove from his house to work at an average speed of 40 miles per hour. The drive took him 15 minutes. If the drive home took him 20 minutes and he used the same route in reverse, what was his average speed going home?

39. WATER SUPPLY Many areas of Northern California depend on the snowpack of the Sierra Nevada Mountains for their water supply. If 250 cubic centimeters of snow will melt to 28 cubic centimeters of water, how much water does 900 cubic centimeters of snow produce?

Career Choices

Travel Agent
Travel agents give advice and make arrangements for transportation, accommodations, and recreation. For international travel, they also provide information on customs and currency exchange.

Online Research
For information about a career as a travel agent, visit: www.algebra2.com/careers

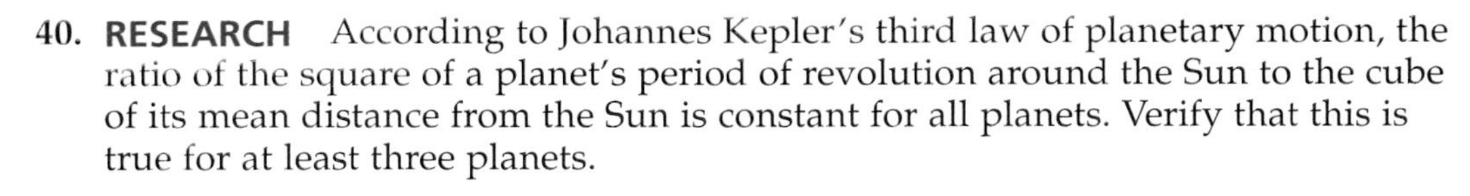

40. **RESEARCH** According to Johannes Kepler's third law of planetary motion, the ratio of the square of a planet's period of revolution around the Sun to the cube of its mean distance from the Sun is constant for all planets. Verify that this is true for at least three planets.

More About. . .

Biology

In order to sustain itself in its cold habitat, a Siberian tiger requires 20 pounds of meat per day.

Source: *Wildlife Fact File*

BIOLOGY **For Exercises 41–43, use the information at the left.**

41. Write an equation to represent the amount of meat needed to sustain s Siberian tigers for d days.

42. Is your equation in Exercise 41 a *direct*, *joint*, or *inverse* variation?

43. How much meat do three Siberian tigers need for the month of January?

LAUGHTER **For Exercises 44–46, use the following information.**
According to *The Columbus Dispatch*, the average American laughs 15 times per day.

44. Write an equation to represent the average number of laughs produced by m household members during a period of d days.

45. Is your equation in Exercise 44 a *direct*, *joint*, or *inverse* variation?

46. Assume that members of your household laugh the same number of times each day as the average American. How many times would the members of your household laugh in a week?

ARCHITECTURE **For Exercises 47–49, use the following information.**
When designing buildings such as theaters, auditoriums, or museums architects have to consider how sound travels. Sound intensity I is inversely proportional to the square of the distance from the sound source d.

47. Write an equation that represents this situation.

48. If d is the independent variable and I is the dependent variable, graph the equation from Exercise 47 when $k = 16$.

49. If a person in a theater moves to a seat twice as far from the speakers, compare the new sound intensity to that of the original.

Study Tip

Combined Variation

Many applied problems involve a combination of direct, inverse, and joint variation. This is called *combined variation.*

TELECOMMUNICATIONS **For Exercises 50–53, use the following information.**
It has been found that the average number of daily phone calls C between two cities is directly proportional to the product of the populations P_1 and P_2 of two cities and inversely proportional to the square of the distance d between the cities. That is, $C = \frac{kP_1P_2}{d^2}$.

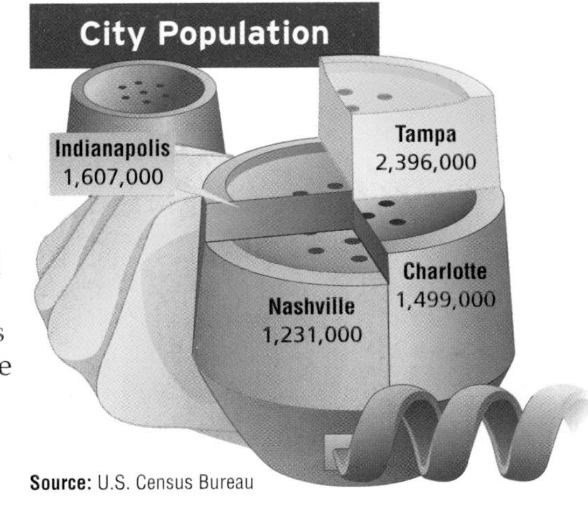

Source: U.S. Census Bureau

50. The distance between Nashville and Charlotte is about 425 miles. If the average number of daily phone calls between the cities is 204,000, find the value of k and write the equation of variation. Round to the nearest hundredth.

51. Nashville is about 680 miles from Tampa. Find the average number of daily phone calls between them.

52. The average daily phone calls between Indianapolis and Charlotte is 133,380. Find the distance between Indianapolis and Charlotte.

53. Could you use this formula to find the populations or the average number of phone calls between two adjoining cities? Explain.

54. **CRITICAL THINKING** Write a real-world problem that involves a joint variation. Solve the problem.

55. **WRITING IN MATH** Answer the question that was posed at the beginning of the lesson.

How is variation used to find the total cost given the unit cost?

Include the following in your answer:

- an explanation of why the equation for the total cost is a direct variation, and
- a problem involving unit cost and total cost of an item and its solution.

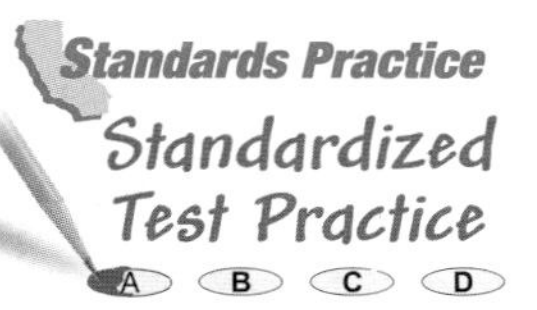

56. If the ratio of $2a$ to $3b$ is 4 to 5, what is the ratio of $5a$ to $4b$?

Ⓐ $\frac{4}{3}$ Ⓑ $\frac{3}{4}$ Ⓒ $\frac{9}{8}$ Ⓓ $\frac{3}{2}$

57. Suppose b varies inversely as the square of a. If a is multiplied by 9, which of the following is true for the value of b?

Ⓐ It is multiplied by $\frac{1}{3}$. Ⓑ It is multiplied by $\frac{1}{9}$.

Ⓒ It is multiplied by $\frac{1}{81}$. Ⓓ It is multiplied by 3.

Maintain Your Skills

Mixed Review

Determine the equations of any vertical asymptotes and the values of x for any holes in the graph of each rational function. *(Lesson 9-3)*

58. $f(x) = \frac{x + 1}{x^2 - 1}$ 59. $f(x) = \frac{x + 3}{x^2 + x - 12}$ 60. $f(x) = \frac{x^2 + 4x + 3}{x + 3}$

Simplify each expression. *(Lesson 9-2)*

61. $\frac{3x}{x - y} + \frac{4x}{y - x}$ 62. $\frac{t}{t + 2} - \frac{2}{t^2 - 4}$ 63. $\frac{m - \frac{1}{m}}{1 + \frac{4}{m} - \frac{5}{m^2}}$

64. **ASTRONOMY** The distance from Earth to the Sun is approximately 93,000,000 miles. Write this number in scientific notation. *(Lesson 5-1)*

State the slope and the y-intercept of the graph of each equation. *(Lesson 2-4)*

65. $y = 0.4x + 1.2$ 66. $2y = 6x + 14$ 67. $3x + 5y = 15$

Getting Ready for the Next Lesson

PREREQUISITE SKILL **Identify each function as S for step, C for constant, A for absolute value, or P for piecewise.** *(To review **special functions**, see Lesson 2-6.)*

68. $h(x) = \frac{2}{3}$ 69. $g(x) = 3|x|$ 70. $f(x) = [\![2x]\!]$

71. $f(x) = \begin{cases} 1 \text{ if } x > 0 \\ -1 \text{ if } x \leq 0 \end{cases}$ 72. $h(x) = |x - 2|$ 73. $g(x) = -3$

Practice Quiz 2 — *Lessons 9-3 and 9-4*

Graph each rational function. *(Lesson 9-3)*

1. $f(x) = \frac{x - 1}{x - 4}$ 2. $f(x) = \frac{-2}{x^2 - 6x + 9}$

Find each value. *(Lesson 9-4)*

3. If y varies inversely as x and $x = 14$ when $y = 7$, find x when $y = 2$.

4. If y varies directly as x and $y = 1$ when $x = 5$, find y when $x = 22$.

5. If y varies jointly as x and z and $y = 80$ when $x = 25$ and $z = 4$, find y when $x = 20$ and $z = 7$.

9-5 Classes of Functions

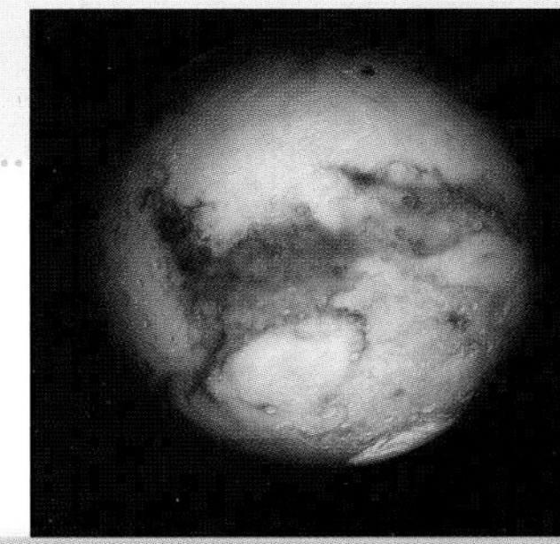

What You'll Learn

- Identify graphs as different types of functions.
- Identify equations as different types of functions.

How can graphs of functions be used to determine a person's weight on a different planet?

The purpose of the 2001 Mars Odyssey Mission is to study conditions on Mars. The findings will help NASA prepare for a possible mission with human explorers. The graph at the right compares a person's weight on Earth with his or her weight on Mars. This graph represents a direct variation, which you studied in the previous lesson.

Weight in Pounds

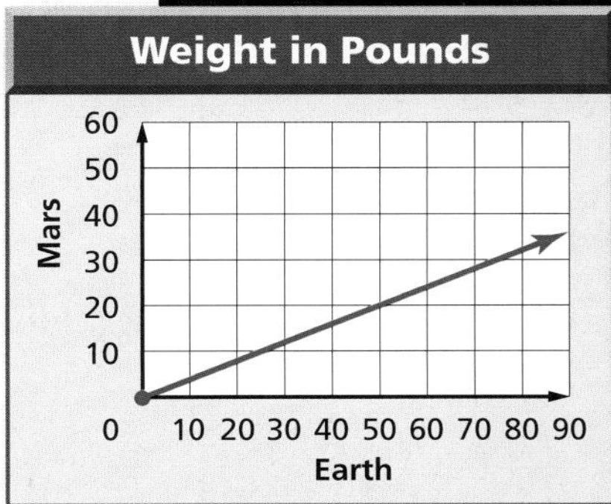

IDENTIFY GRAPHS In this book, you have studied several types of graphs representing special functions. The following is a summary of these graphs.

Concept Summary — *Special Functions*

Constant Function

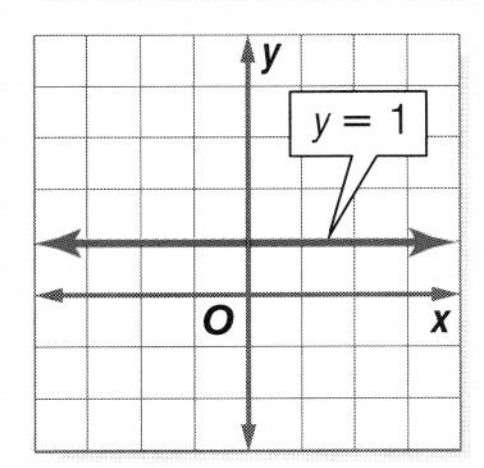

The general equation of a constant function is $y = a$, where a is any number. Its graph is a horizontal line that crosses the y-axis at a.

Direct Variation Function

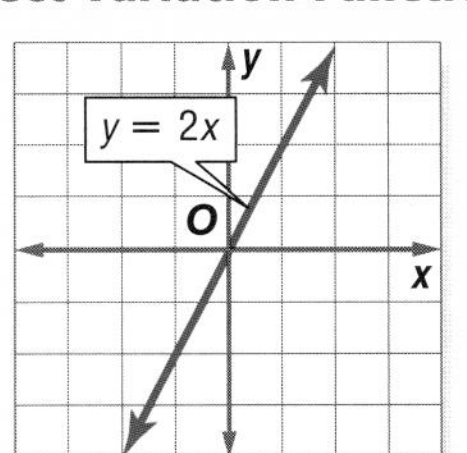

The general equation of a direct variation function is $y = ax$, where a is a nonzero constant. Its graph is a line that passes through the origin and is neither horizontal nor vertical.

Identity Function

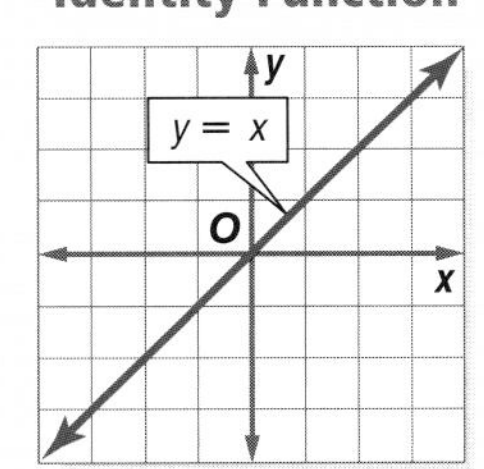

The identity function $y = x$ is a special case of the direct variation function in which the constant is 1. Its graph passes through all points with coordinates (a, a).

Greatest Integer Function

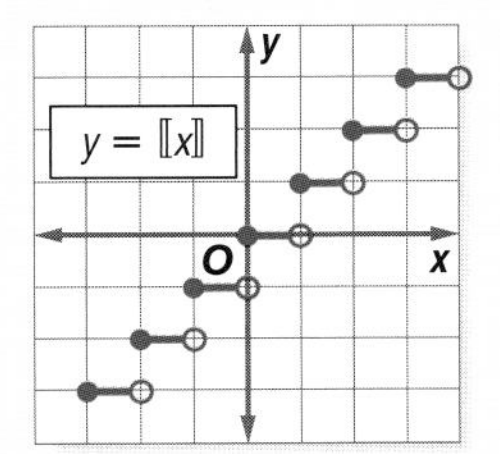

If an equation includes an expression inside the greatest integer symbol, the function is a greatest integer function. Its graph looks like steps.

Absolute Value Function

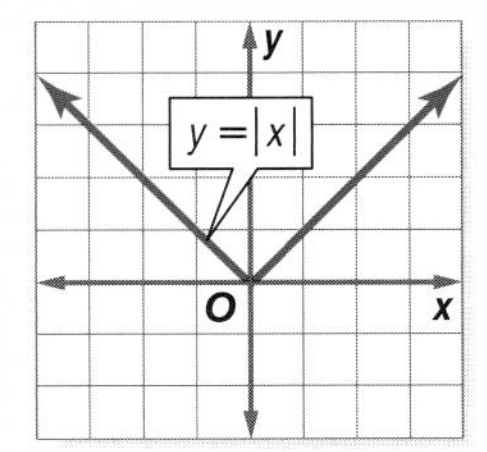

An equation with a direct variation expression inside absolute value symbols is an absolute value function. Its graph is in the shape of a V.

Quadratic Function

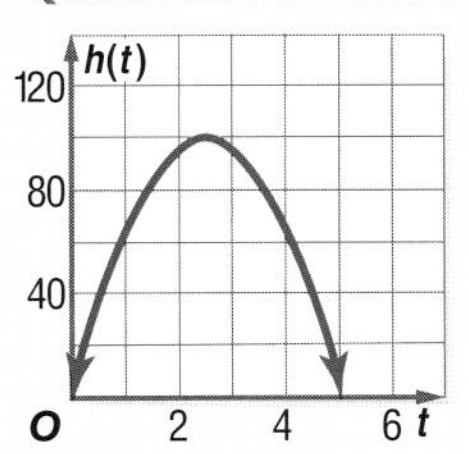

The general equation of a quadratic function is $y = ax^2 + bx + c$, where $a \neq 0$. Its graph is a parabola.

(continued on the next page)

Concept Summary — Special Functions

Square Root Function

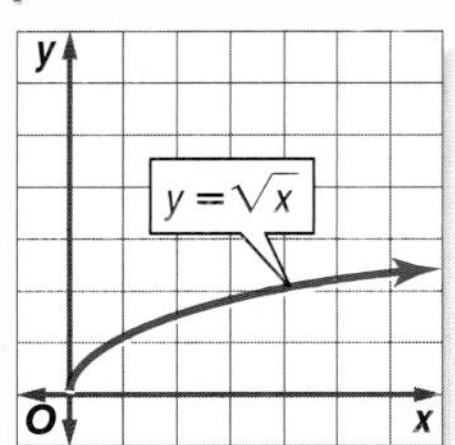

If an equation includes an expression inside the radical sign, the function is a square root function. Its graph is a curve that starts at a point and continues in only one direction.

Rational Function

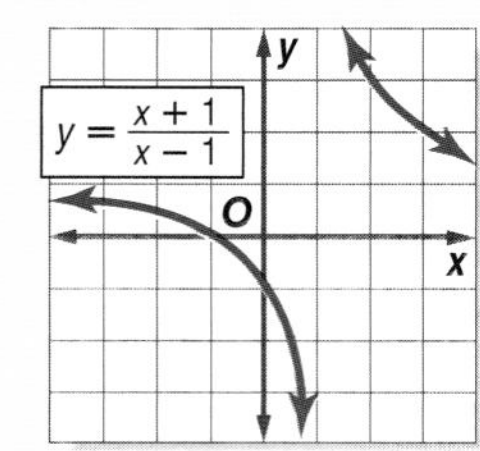

The general equation for a rational function is $y = \frac{p(x)}{q(x)}$, where $p(x)$ and $q(x)$ are polynomial functions. Its graph has one or more asymptotes and/or holes.

Inverse Variation Function

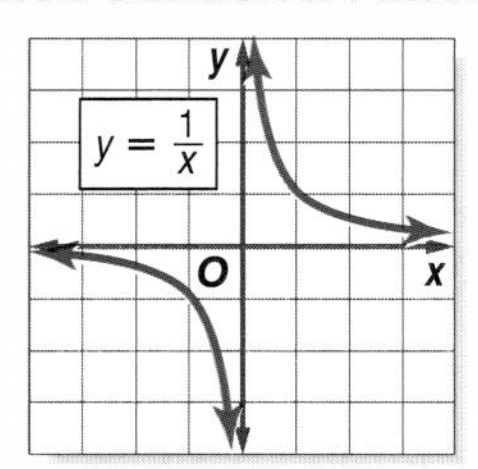

The inverse variation function $y = \frac{a}{x}$ is a special case of the rational function where $p(x)$ is a constant and $q(x) = x$. Its graph has two asymptotes, $x = 0$ and $y = 0$.

Example 1 Identify a Function Given the Graph

Identify the type of function represented by each graph.

a.

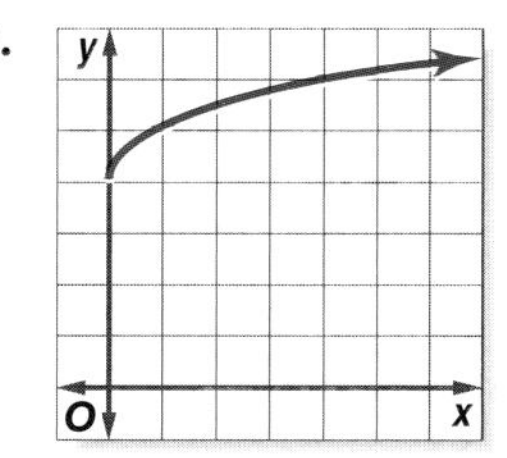

The graph has a starting point and curves in one direction. The graph represents a square root function.

b.

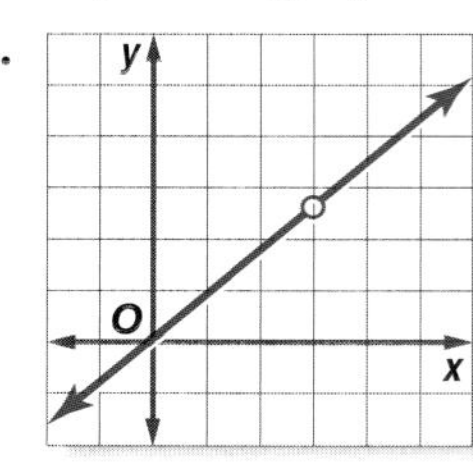

The graph appears to be a direct variation since it is a straight line passing through the origin. However, the hole indicates that it represents a rational function.

IDENTIFY EQUATIONS If you can identify an equation as a type of function, you can determine the shape of the graph.

Example 2 Match Equation with Graph

ROCKETRY **Emily launched a toy rocket from ground level. The height above the ground level h, in feet, after t seconds is given by the formula $h(t) = -16t^2 + 80t$. Which graph depicts the height of the rocket during its flight?**

a.

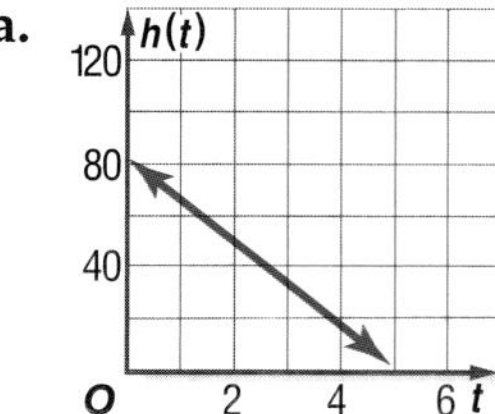

b.

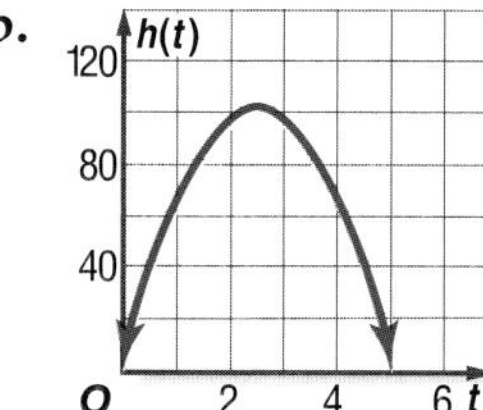

c.

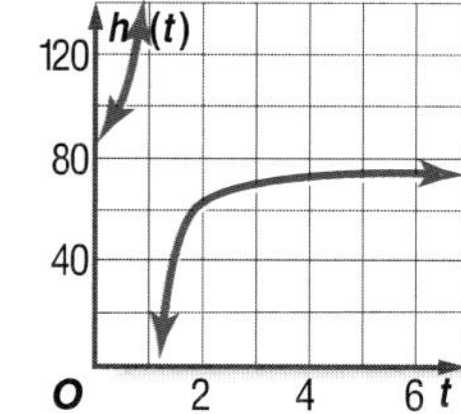

The function includes a second-degree polynomial. Therefore, it is a quadratic function, and its graph is a parabola. Graph b is the only parabola. Therefore, the answer is graph **b**.

Sometimes recognizing an equation as a specific type of function can help you graph the function.

Example 3 Identify a Function Given its Equation

Identify the type of function represented by each equation. Then graph the equation.

a. $y = |x| - 1$

Since the equation includes an expression inside absolute value symbols, it is an absolute value function. Therefore, the graph will be in the shape of a V. Determine some points on the graph and use what you know about graphs of absolute value functions to graph the function.

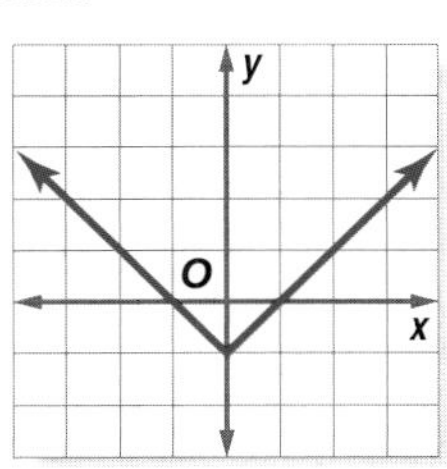

b. $y = -\frac{2}{3}x$

The function is in the form $y = ax$, where $a = -\frac{2}{3}$. Therefore, it is a direct variation function. The graph passes through the origin and has a slope of $-\frac{2}{3}$.

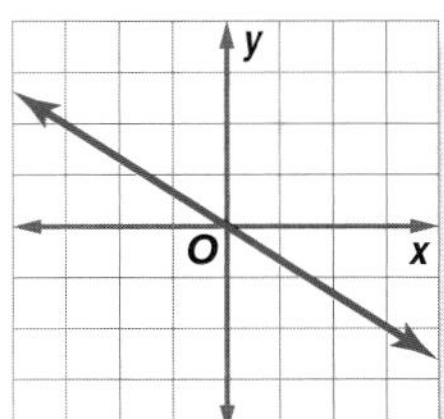

Check for Understanding

Concept Check

1. **OPEN ENDED** Find a counterexample to the statement *All functions are continuous*. Describe your function.
2. **Name** three special functions whose graphs are straight lines. Give an example of each function.
3. **Describe** the graph of $y = [\![x + 2]\!]$.

Guided Practice

Identify the type of function represented by each graph.

4.

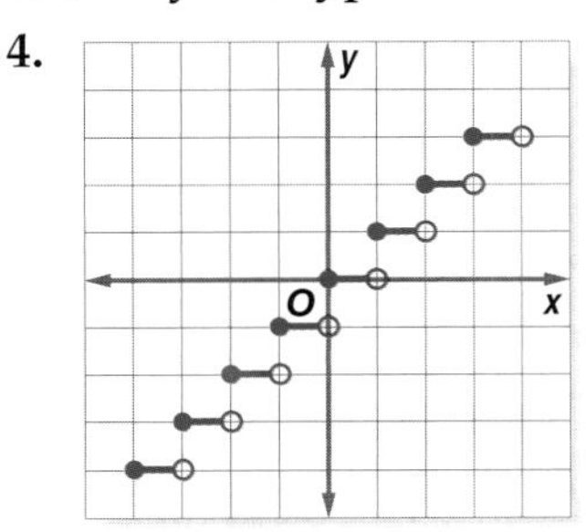

5.

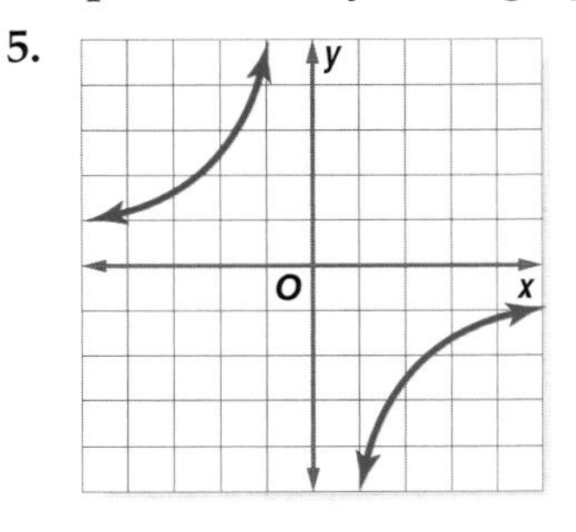

6.

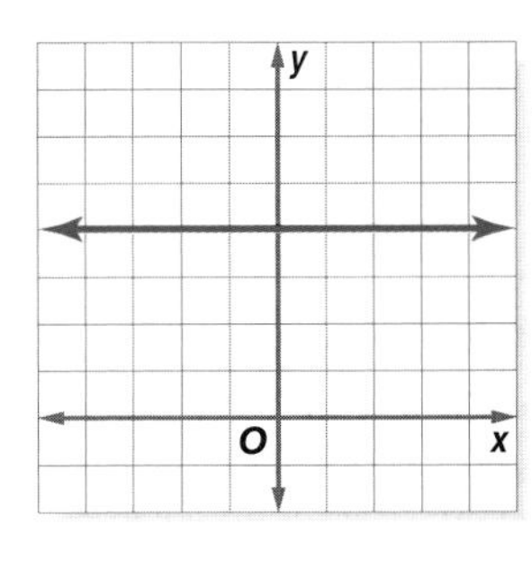

Match each graph with an equation at the right.

7.

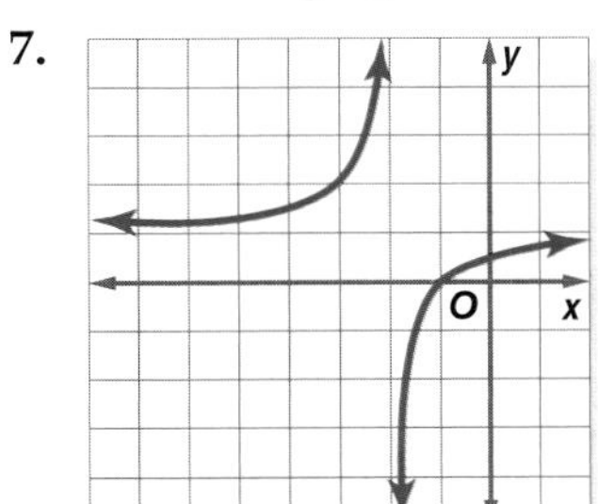

8.

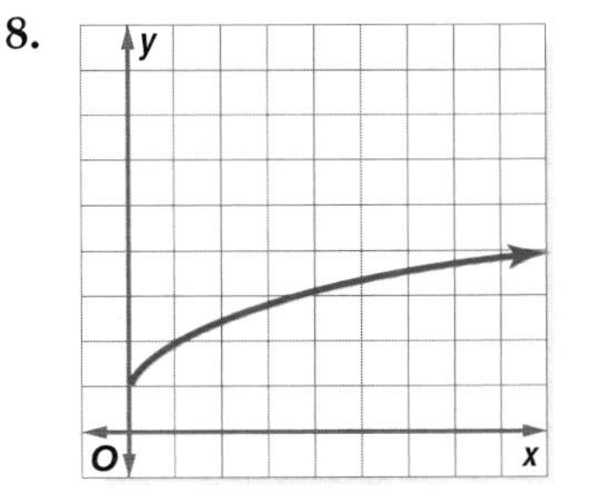

a. $y = x^2 + 2x + 3$
b. $y = \sqrt{x + 1}$
c. $y = \frac{x + 1}{x + 2}$
d. $y = [\![2x]\!]$

Identify the type of function represented by each equation. Then graph the equation.

9. $y = x$ **10.** $y = -x^2 + 2$ **11.** $y = |x + 2|$

Application

12. GEOMETRY Write the equation for the area of a circle. Identify the equation as a type of function. Describe the graph of the function.

Practice and Apply

Homework Help

For Exercises	See Examples
13–18	1
19–22, 31–36	2
23–30	3

Extra Practice
See page 848.

Identify the type of function represented by each graph.

13.
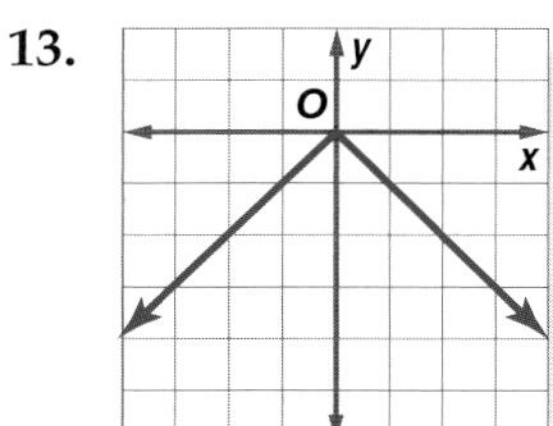

14.
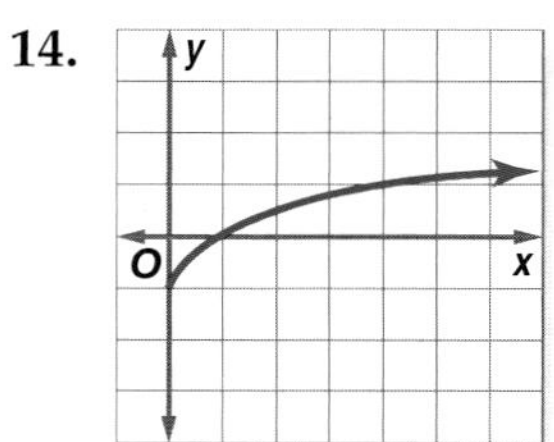

15.
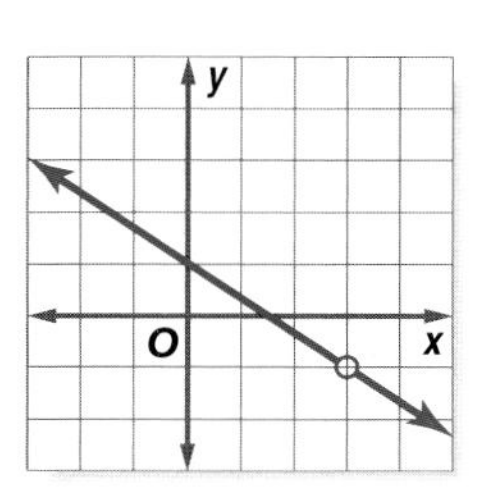

16.
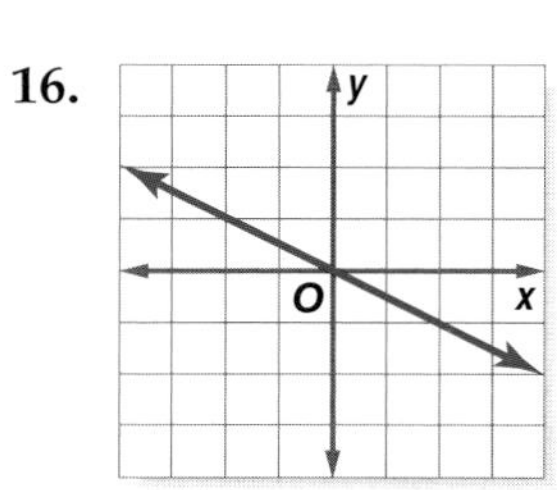

17.
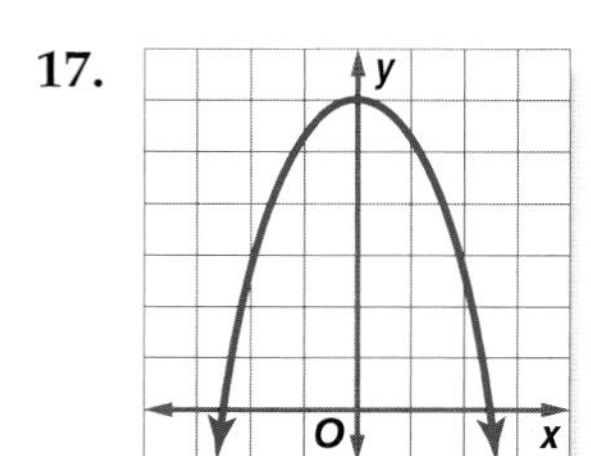

18.
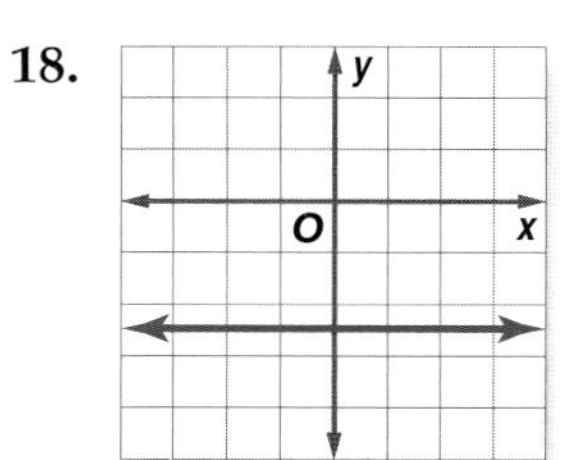

Match each graph with an equation at the right.

19.
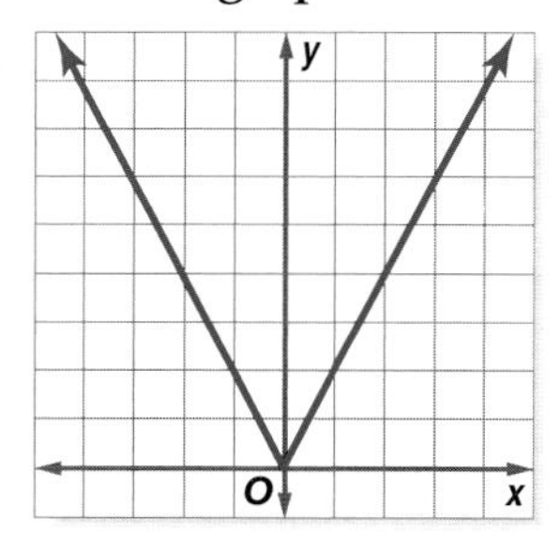

20.
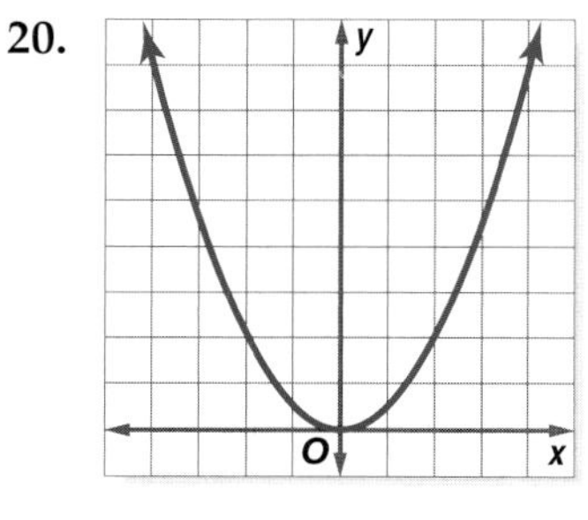

21.
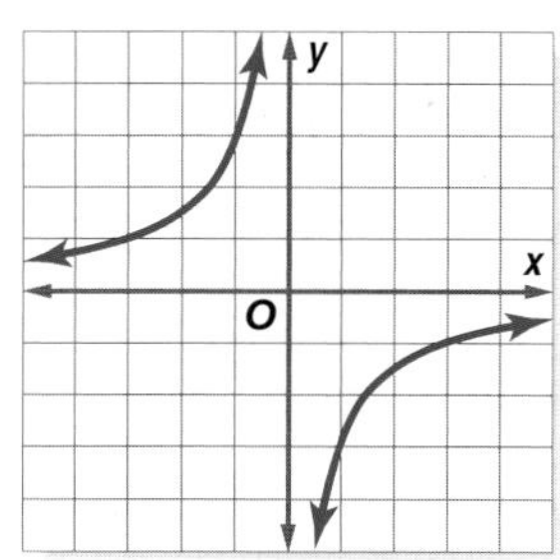

22.
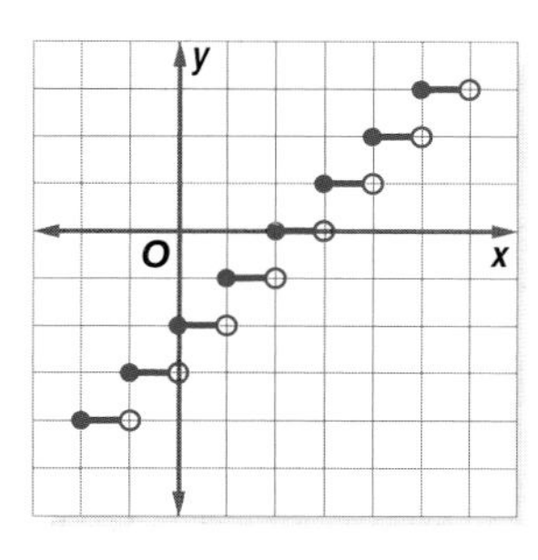

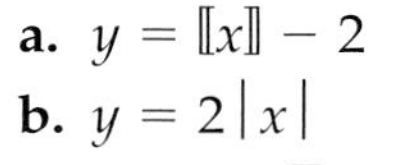

a. $y = [\![x]\!] - 2$
b. $y = 2|x|$
c. $y = 2\sqrt{x}$
d. $y = -3x$
e. $y = 0.5x^2$
f. $y = -\frac{3}{x + 1}$
g. $y = -\frac{3}{x}$

WebQuest

You can use functions to determine the relationship between primary and secondary earthquake waves. Visit www.algebra2.com/webquest to continue work on your WebQuest project.

Identify the type of function represented by each equation. Then graph the equation.

23. $y = -1.5$ **24.** $y = 2.5x$ **25.** $y = \sqrt{9x}$ **26.** $y = \frac{4}{x}$

27. $y = \frac{x^2 - 1}{x - 1}$ **28.** $y = 3[\![x]\!]$ **29.** $y = |2x|$ **30.** $y = 2x^2$

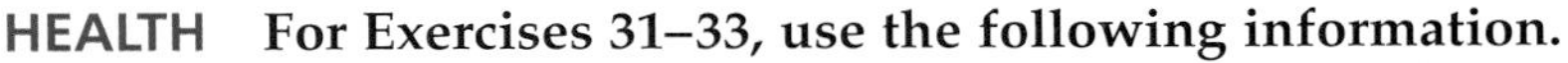

HEALTH For Exercises 31–33, use the following information.
A woman painting a room will burn an average of 4.5 Calories per minute.

31. Write an equation for the number of Calories burned in m minutes.

32. Identify the equation in Exercise 31 as a type of function.

33. Describe the graph of the function.

More About. . .

Architecture

The Gateway Arch is 630 feet high and is the tallest monument in the United States.

Source: *World Book Encyclopedia*

34. ARCHITECTURE The shape of the Gateway Arch of the Jefferson National Expansion Memorial in St. Louis, Missouri, resembles the graph of the function $f(x) = -0.00635x^2 + 4.0005x - 0.07875$, where x is in feet. Describe the shape of the Gateway Arch.

MAIL For Exercises 35 and 36, use the following information.
In 2001, the cost to mail a first-class letter was 34¢ for any weight up to and including 1 ounce. Each additional ounce or part of an ounce added 21¢ to the cost.

35. Make a graph showing the postal rates to mail any letter from 0 to 8 ounces.

36. Compare your graph in Exercise 35 to the graph of the greatest integer function.

37. CRITICAL THINKING Identify each table of values as a type of function.

a.

x	$f(x)$
−5	7
−3	5
−1	3
0	2
1	3
3	5
5	7
7	9

b.

x	$f(x)$
−5	24
−3	8
−1	0
0	−1
1	0
3	8
5	24
7	48

c.

x	$f(x)$
−1.3	−1
−1.7	−1
0	1
0.8	1
0.9	1
1	2
1.5	2
2.3	3

d.

x	$f(x)$
−5	undefined
−3	undefined
−1	undefined
0	0
1	1
4	2
9	3
16	4

38. WRITING IN MATH Answer the question that was posed at the beginning of the lesson.

How can graphs of functions be used to determine a person's weight on a different planet?

Include the following in your answer:
- an explanation of why the graph comparing weight on Earth and Mars represents a direct variation function, and
- an equation and a graph comparing a person's weight on Earth and Venus if a person's weight on Venus is 0.9 of his or her weight on Earth.

39. The curve at the right could be part of the graph of which function?

Ⓐ $y = \sqrt{x}$

Ⓑ $y = x^2 - 5x + 4$

Ⓒ $xy = 4$

Ⓓ $y = -x + 20$

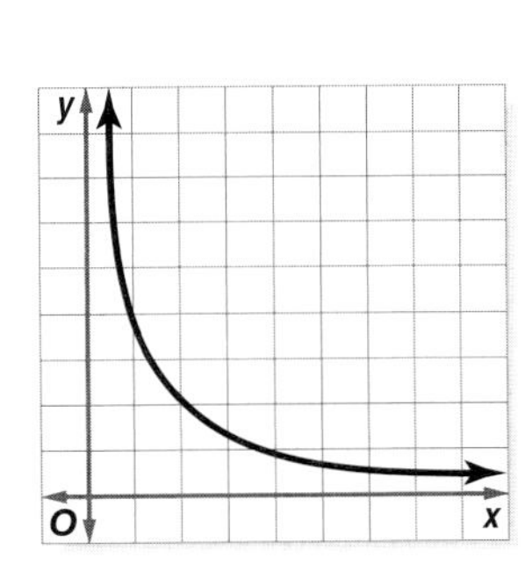

40. If $g(x) = [\![x]\!]$, which of the following is the graph of $g\left(\frac{x}{2}\right) + 2$?

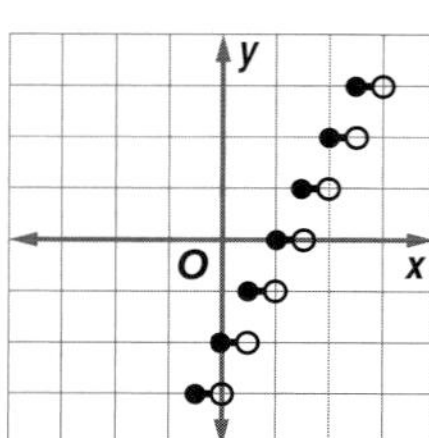

Ⓑ
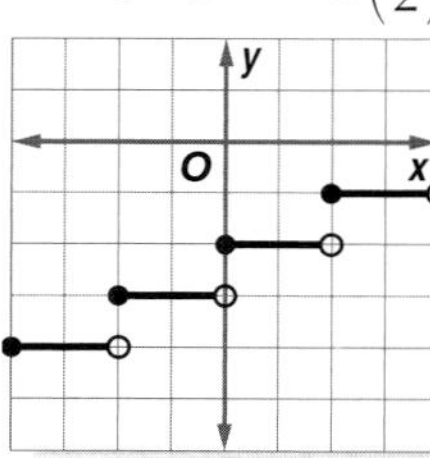

Ⓒ
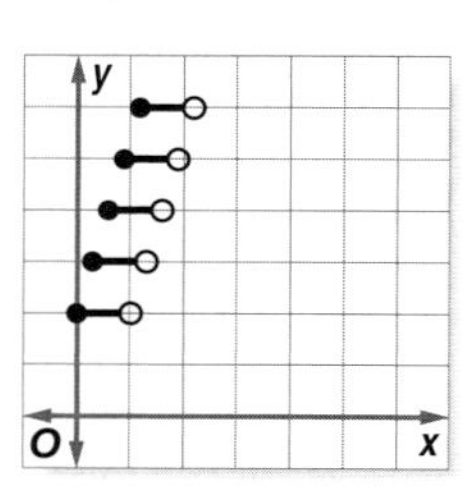

Ⓓ
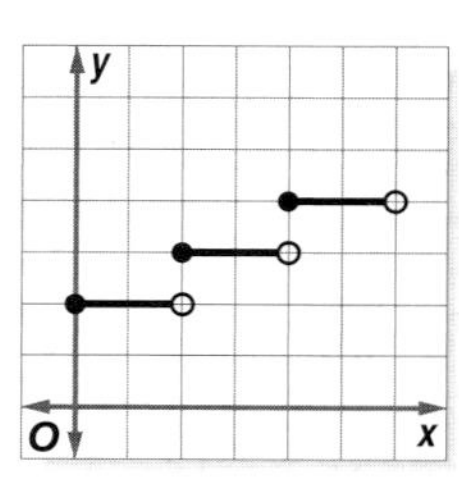

Maintain Your Skills

Mixed Review

41. If x varies directly as y and $y = \frac{1}{5}$ when $x = 11$, find x when $y = \frac{2}{5}$. *(Lesson 9-4)*

Graph each rational function. *(Lesson 9-3)*

42. $f(x) = \frac{3}{x + 2}$

43. $f(x) = \frac{8}{(x - 1)(x + 3)}$

44. $f(x) = \frac{x^2 - 5x + 4}{x - 4}$

Identify the coordinates of the vertex and focus, the equations of the axis of symmetry and directrix, and the direction of opening of the parabola with the given equation. Then find the length of the latus rectum and graph the parabola. *(Lesson 8-2)*

45. $\frac{1}{2}(y + 1) = (x - 8)^2$

46. $x = \frac{1}{4}y^2 - \frac{1}{2}y - 3$

47. $3x - y^2 = 8y + 31$

Find each product, if possible. *(Lesson 4-3)*

48. $\begin{bmatrix} 3 & -5 \\ 2 & 7 \end{bmatrix} \cdot \begin{bmatrix} 5 & 1 & -3 \\ 8 & -4 & 9 \end{bmatrix}$

49. $\begin{bmatrix} 4 & -1 & 6 \\ 1 & 5 & -8 \end{bmatrix} \cdot \begin{bmatrix} 1 & 3 \\ 9 & -6 \end{bmatrix}$

Solve each system of equations by using either substitution or elimination. *(Lesson 3-2)*

50. $3x + 5y = -4$
$2x - 3y = 29$

51. $3a - 2b = -3$
$3a + b = 3$

52. $3s - 2t = 10$
$4s + t = 6$

Determine the value of r so that a line through the points with the given coordinates has the given slope. *(Lesson 2-3)*

53. $(r, 2), (4, -6)$; slope $= \frac{-8}{3}$

54. $(r, 6), (8, 4)$; slope $= \frac{1}{2}$

55. Evaluate $[(-7 + 4) \times 5 - 2] \div 6$. *(Lesson 1-1)*

Getting Ready for the Next Lesson

PREREQUISITE SKILL **Find the LCM of each set of polynomials.**
*(To review **least common multiples of polynomials,** see Lesson 9-2.)*

56. $15ab^2c, 6a^3, 4bc^2$

57. $9x^3, 5xy^2, 15x^2y^3$

58. $5d - 10, 3d - 6$

59. $x^2 - y^2, 3x + 3y$

60. $a^2 - 2a - 3, a^2 - a - 6$

61. $2t^2 - 9t - 5, t^2 + t - 30$

9-6

Solving Rational Equations and Inequalities

California Standards Standard 4.0 Students factor polynomials representing the difference of squares, perfect square trinomials, and the sum and difference of two cubes. (Key)

What You'll Learn

- Solve rational equations.
- Solve rational inequalities.

Vocabulary

- rational equation
- rational inequality

How are rational equations used to solve problems involving unit price?

The Coast to Coast Phone Company advertises 5¢ a minute for long-distance calls. However, it also charges a monthly fee of \$5. If the customer has x minutes in long distance calls last month, the bill in cents will be $500 + 5x$. The actual cost per minute is $\frac{500 + 5x}{x}$. To find how many long-distance minutes a person would need to make the actual cost per minute 6¢, you would need to solve the equation $\frac{500 + 5x}{x} = 6$.

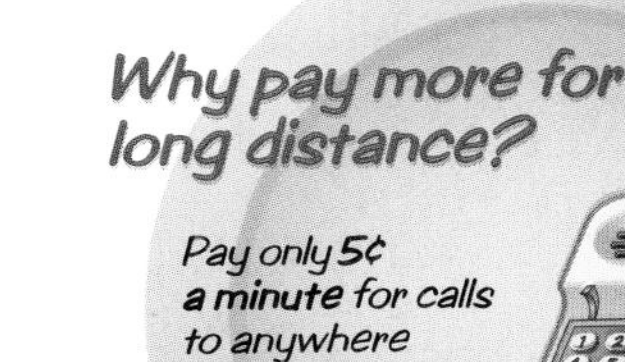

SOLVE RATIONAL EQUATIONS The equation $\frac{500 + 5x}{x} = 6$ is an example of a rational equation. In general, any equation that contains one or more rational expressions is called a **rational equation**.

Rational equations are easier to solve if the fractions are eliminated. You can eliminate the fractions by multiplying each side of the equation by the least common denominator (LCD). Remember that when you multiply each side by the LCD, each term on each side must be multiplied by the LCD.

Example 1 Solve a Rational Equation

Solve $\frac{9}{28} + \frac{3}{z + 2} = \frac{3}{4}$. Check your solution.

The LCD for the three denominators is $28(z + 2)$.

$$\frac{9}{28} + \frac{3}{z + 2} = \frac{3}{4}$$ Original equation

$$28(z + 2)\left(\frac{9}{28} + \frac{3}{z + 2}\right) = 28(z + 2)\left(\frac{3}{4}\right)$$ Multiply each side by $28(z + 2)$.

$$\overset{1}{\cancel{28}}(z + 2)\left(\frac{9}{\underset{1}{\cancel{28}}}\right) + 28\overset{1}{\cancel{(z + 2)}}\left(\frac{3}{\underset{1}{\cancel{z + 2}}}\right) = \overset{7}{\cancel{28}}(z + 2)\left(\frac{3}{\underset{1}{\cancel{4}}}\right)$$ Distributive Property

$$(9z + 18) + 84 = 21z + 42$$ Simplify.

$$9z + 102 = 21z + 42$$ Simplify.

$$60 = 12z$$ Subtract $9z$ and 42 from each side.

$$5 = z$$ Divide each side by 12.

CHECK $\frac{9}{28} + \frac{3}{z+2} = \frac{3}{4}$ Original equation

$\frac{9}{28} + \frac{3}{5+2} \stackrel{?}{=} \frac{3}{4}$ $z = 5$

$\frac{9}{28} + \frac{3}{7} \stackrel{?}{=} \frac{3}{4}$ Simplify.

$\frac{9}{28} + \frac{12}{28} \stackrel{?}{=} \frac{3}{4}$ Simplify.

$\frac{3}{4} = \frac{3}{4}$ ✓ The solution is correct.

The solution is 5.

When solving a rational equation, any possible solution that results in a zero in the denominator must be excluded from your list of solutions.

Example 2 Elimination of a Possible Solution

Solve $r + \frac{r^2 - 5}{r^2 - 1} = \frac{r^2 + r + 2}{r + 1}$. Check your solution.

The LCD is $(r^2 - 1)$.

$$r + \frac{r^2 - 5}{r^2 - 1} = \frac{r^2 + r + 2}{r + 1}$$ Original equation

$$(r^2 - 1)\left(r + \frac{r^2 - 5}{r^2 - 1}\right) = (r^2 - 1)\left(\frac{r^2 + r + 2}{r + 1}\right)$$ Multiply each side by the LCD, $(r^2 - 1)$.

$$(r^2 - 1)r + \overset{1}{\cancel{(r^2 - 1)}}\left(\frac{r^2 - 5}{\underset{1}{\cancel{r^2 - 1}}}\right) = \overset{(r-1)}{\cancel{(r^2 - 1)}}\left(\frac{r^2 + r + 2}{\underset{1}{\cancel{r + 1}}}\right)$$ Distributive Property

$$(r^3 - r) + (r^2 - 5) = (r - 1)(r^2 + r + 2)$$ Simplify.

$$r^3 + r^2 - r - 5 = r^3 + r - 2$$ Simplify.

$$r^2 - 2r - 3 = 0$$ Subtract $(r^3 + r - 2)$ from each side.

$$(r - 3)(r + 1) = 0$$ Factor.

$$r - 3 = 0 \quad \text{or} \quad r + 1 = 0$$ Zero Product Property

$$r = 3 \qquad\qquad r = -1$$

CHECK $r + \frac{r^2 - 5}{r^2 - 1} = \frac{r^2 + r + 2}{r + 1}$ Original equation

$3 + \frac{3^2 - 5}{3^2 - 1} \stackrel{?}{=} \frac{3^2 + 3 + 2}{3 + 1}$ $r = 3$

$3 + \frac{4}{8} \stackrel{?}{=} \frac{14}{4}$ Simplify.

$\frac{7}{2} = \frac{7}{2}$ ✓

$r + \frac{r^2 - 5}{r^2 - 1} = \frac{r^2 + r + 2}{r + 1}$ Original equation

$-1 + \frac{(-1)^2 - 5}{(-1)^2 - 1} \stackrel{?}{=} \frac{(-1)^2 + (-1) + 2}{-1 + 1}$ $r = -1$

$-1 + \frac{-4}{0} \stackrel{?}{=} \frac{2}{0}$ Simplify.

Since $r = -1$ results in a zero in the denominator, eliminate -1 from the list of solutions.

The solution is 3.

Study Tip

Extraneous Solutions
Multiplying each side of an equation by the LCD of rational expressions can yield results that are not solutions of the original equation. These solutions are called *extraneous solutions.*

Some real-world problems can be solved with rational equations.

Example 3 Work Problem

TUNNELS **When building the Chunnel, the English and French each started drilling on opposite sides of the English Channel. The two sections became one in 1990. The French used more advanced drilling machinery than the English. Suppose the English could drill the Chunnel in 6.2 years and the French could drill it in 5.8 years. How long would it have taken the two countries to drill the tunnel?**

In 1 year, the English could complete $\frac{1}{6.2}$ of the tunnel.

In 2 years, the English could complete $\frac{1}{6.2} \cdot 2$ or $\frac{2}{6.2}$ of the tunnel.

In t years, the English could complete $\frac{1}{6.2} \cdot t$ or $\frac{t}{6.2}$ of the tunnel.

Likewise, in t years, the French could complete $\frac{1}{5.8} \cdot t$ or $\frac{t}{5.8}$ of the tunnel.

Together, they completed the whole tunnel.

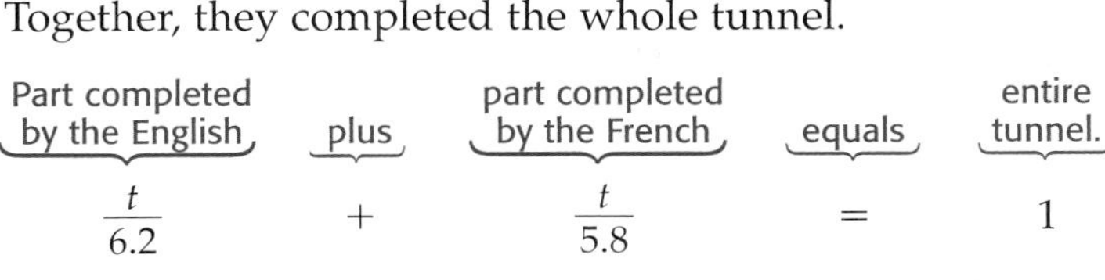

$$\frac{t}{6.2} + \frac{t}{5.8} = 1$$

Solve the equation.

$\frac{t}{6.2} + \frac{t}{5.8} = 1$	Original equation
$17.98\left(\frac{t}{6.2} + \frac{t}{5.8}\right) = 17.98(1)$	Multiply each side by 17.98.
$17.98\left(\frac{t}{6.2}\right) + 17.98\left(\frac{t}{5.8}\right) = 17.98$	Distributive Property
$2.9t + 3.1t = 17.98$	Simplify.
$6t = 17.98$	Simplify.
$t \approx 3.00$	Divide each side by 6.

It would have taken about 3 years to build the Chunnel.

More About. . .

Tunnels

The Chunnel is a tunnel under the English Channel that connects England with France. It is 32 miles long with 23 miles of the tunnel under water.

Source: www.pbs.org

Rate problems frequently involve rational equations.

Example 4 Rate Problem

NAVIGATION **The speed of the current in the Puget sound is 5 miles per hour. A barge travels 26 miles with the current and returns in $10\frac{2}{3}$ hours. What is the speed of the barge in still water?**

WORDS The formula that relates distance, time, and rate is $d = rt$ or $\frac{d}{r} = t$.

VARIABLES Let r be the speed of the barge in still water. Then the speed of the barge with the current is $r + 5$, and the speed of the barge against the current is $r - 5$.

Time going with the current plus time going against the current equals total time.

EQUATION $\frac{26}{r+5} + \frac{26}{r-5} = 10\frac{2}{3}$

(continued on the next page)

Solve the equation.

$$\frac{26}{r+5} + \frac{26}{r-5} = 10\frac{2}{3}$$ Original equation

$$3(r^2 - 25)\left(\frac{26}{r+5} + \frac{26}{r-5}\right) = 3(r^2 - 25)\left(10\frac{2}{3}\right)$$ Multiply each side by $3(r^2 - 25)$.

$$3\overset{(r-5)}{(r^2 - 25)}\left(\frac{26}{\underset{1}{r+5}}\right) + 3\overset{(r+5)}{(r^2 - 25)}\left(\frac{26}{\underset{1}{r-5}}\right) = \overset{1}{3}(r^2 - 25)\left(\frac{32}{\underset{1}{3}}\right)$$ Distributive Property

$$(78r - 390) + (78r + 390) = 32r^2 - 800$$ Simplify.

$$156r = 32r^2 - 800$$ Simplify.

$$0 = 32r^2 - 156r - 800$$ Subtract $156r$ from each side.

$$0 = 8r^2 - 39r - 200$$ Divide each side by 4.

Use the Quadratic Formula to solve for r.

$$x = \frac{-b \pm \sqrt{b^2 - 4ac}}{2a}$$ Quadratic Formula

$$r = \frac{-(-39) \pm \sqrt{(-39)^2 - 4(8)(-200)}}{2(8)}$$ $x = r$, $a = 8$, $b = -39$, and $c = -200$

$$r = \frac{39 \pm \sqrt{7921}}{16}$$ Simplify.

$$r = \frac{39 \pm 89}{16}$$ Simplify.

$$r = 8 \text{ or } -3.125$$ Simplify.

Since the speed must be positive, the answer is 8 miles per hour.

Study Tip

Look Back

To review the **Quadratic Formula,** see Lesson 6-5.

SOLVE RATIONAL INEQUALITIES Inequalities that contain one or more rational expressions are called **rational inequalities**. To solve rational inequalities, complete the following steps.

Step 1 State the excluded values.

Step 2 Solve the related equation.

Step 3 Use the values determined in Steps 1 and 2 to divide a number line into regions. Test a value in each region to determine which regions satisfy the original inequality.

Example 5 Solve a Rational Inequality

Solve $\frac{1}{4a} + \frac{5}{8a} > \frac{1}{2}$.

Step 1 Values that make a denominator equal to 0 are excluded from the domain. For this inequality, the excluded value is 0.

Step 2 Solve the related equation.

$$\frac{1}{4a} + \frac{5}{8a} = \frac{1}{2}$$ Related equation

$$8a\left(\frac{1}{4a} + \frac{5}{8a}\right) = 8a\left(\frac{1}{2}\right)$$ Multiply each side by $8a$.

$$2 + 5 = 4a$$ Simplify.

$$7 = 4a$$ Add.

$$1\frac{3}{4} = a$$ Divide each side by 4.

Step 3 Draw vertical lines at the excluded value and at the solution to separate the number line into regions.

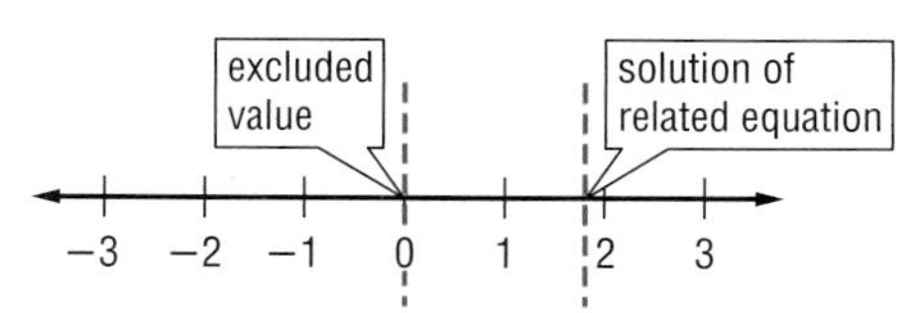

Now test a sample value in each region to determine if the values in the region satisfy the inequality.

Test $a = -1$.

$$\frac{1}{4(-1)} + \frac{5}{8(-1)} \stackrel{?}{>} \frac{1}{2}$$
$$-\frac{1}{4} - \frac{5}{8} \stackrel{?}{>} \frac{1}{2}$$
$$-\frac{7}{8} \not> \frac{1}{2}$$

$a < 0$ is *not* a solution.

Test $a = 1$.

$$\frac{1}{4(1)} + \frac{5}{8(1)} \stackrel{?}{>} \frac{1}{2}$$
$$\frac{1}{4} + \frac{5}{8} \stackrel{?}{>} \frac{1}{2}$$
$$\frac{7}{8} > \frac{1}{2} \ \checkmark$$

$0 < a < 1\frac{3}{4}$ is a solution.

Test $a = 2$.

$$\frac{1}{4(2)} + \frac{5}{8(2)} \stackrel{?}{>} \frac{1}{2}$$
$$\frac{1}{8} + \frac{5}{16} \stackrel{?}{>} \frac{1}{2}$$
$$\frac{7}{16} \not> \frac{1}{2}$$

$a > 1\frac{3}{4}$ is *not* a solution.

The solution is $0 < a < 1\frac{3}{4}$.

Check for Understanding

Concept Check

1. **OPEN ENDED** Write a rational equation that can be solved by first multiplying each side by $5(a + 2)$.

2. State the expression by which you would multiply each side of $\frac{x}{x+4} + \frac{1}{2} = 1$ in order to solve the equation. What value(s) of x cannot be a solution?

3. **FIND THE ERROR** Jeff and Dustin are solving $2 - \frac{3}{a} = \frac{2}{3}$.

Jeff

$$2 - \frac{3}{a} = \frac{2}{3}$$
$$6a - 9 = 2a$$
$$4a = 9$$
$$a = 2.25$$

Dustin

$$2 - \frac{3}{a} = \frac{2}{3}$$
$$2 - 9 = 2a$$
$$-7 = 2a$$
$$-3.5 = a$$

Who is correct? Explain your reasoning.

Guided Practice

Solve each equation or inequality. Check your solutions.

4. $\frac{2}{d} + \frac{1}{4} = \frac{11}{12}$

5. $t + \frac{12}{t} - 8 = 0$

6. $\frac{1}{x-1} + \frac{2}{x} = 0$

7. $\frac{12}{v^2 - 16} - \frac{24}{v-4} = 3$

8. $\frac{4}{c+2} > 1$

9. $\frac{1}{3v} + \frac{1}{4v} < \frac{1}{2}$

Application

10. **WORK** A bricklayer can build a wall of a certain size in 5 hours. Another bricklayer can do the same job in 4 hours. If the bricklayers work together, how long would it take to do the job?

Practice and Apply

Homework Help

For Exercises	See Examples
11–30	1, 2, 5
31–39	3, 4

Extra Practice
See page 849.

Solve each equation or inequality. Check your solutions.

11. $\frac{y}{y+1} = \frac{2}{3}$

12. $\frac{p}{p-2} = \frac{2}{5}$

13. $s + 5 = \frac{6}{s}$

14. $a + 1 = \frac{6}{a}$

15. $\frac{7}{a+1} > 7$

16. $\frac{10}{m+1} > 5$

17. $\frac{9}{t-3} = \frac{t-4}{t-3} + \frac{1}{4}$

18. $\frac{w}{w-1} + w = \frac{4w-3}{w-1}$

19. $5 + \frac{1}{t} > \frac{16}{t}$

20. $7 - \frac{2}{b} < \frac{5}{b}$

21. $\frac{2}{3y} + \frac{5}{6y} > \frac{3}{4}$

22. $\frac{1}{2p} + \frac{3}{4p} < \frac{1}{2}$

23. $\frac{b-4}{b-2} = \frac{b-2}{b+2} + \frac{1}{b-2}$

24. $\frac{4n^2}{n^2-9} - \frac{2n}{n+3} = \frac{3}{n-3}$

25. $\frac{1}{d+4} = \frac{2}{d^2+3d-4} - \frac{1}{1-d}$

26. $\frac{2}{y+2} - \frac{y}{2-y} = \frac{y^2+4}{y^2-4}$

27. $\frac{3}{b^2+5b+6} + \frac{b-1}{b+2} = \frac{7}{b+3}$

28. $\frac{1}{n-2} = \frac{2n+1}{n^2+2n-8} + \frac{2}{n+4}$

29. $\frac{2q}{2q+3} - \frac{2q}{2q-3} = 1$

30. $\frac{4}{z-2} - \frac{z+6}{z+1} = 1$

31. **NUMBER THEORY** The ratio of 8 less than a number to 28 more than that number is 2 to 5. What is the number?

32. **NUMBER THEORY** The sum of a number and 8 times its reciprocal is 6. Find the number(s).

33. **ACTIVITIES** The band has 30 more members than the school chorale. If each group had 10 more members, the ratio of their membership would be 3:2. How many members are in each group?

PHYSICS For Exercises 34 and 35, use the following information.
The distance a spring stretches is related to the mass attached to the spring. This is represented by $d = km$, where d is the distance, m is the mass, and k is the spring constant. When two springs with spring constants k_1 and k_2 are attached in a series, the resulting spring constant k is found by the equation $\frac{1}{k} = \frac{1}{k_1} + \frac{1}{k_2}$.

34. If one spring with constant of 12 centimeters per gram is attached in a series with another spring with constant of 8 centimeters per gram, find the resultant spring constant.

35. If a 5-gram object is hung from the series of springs, how far will the springs stretch?

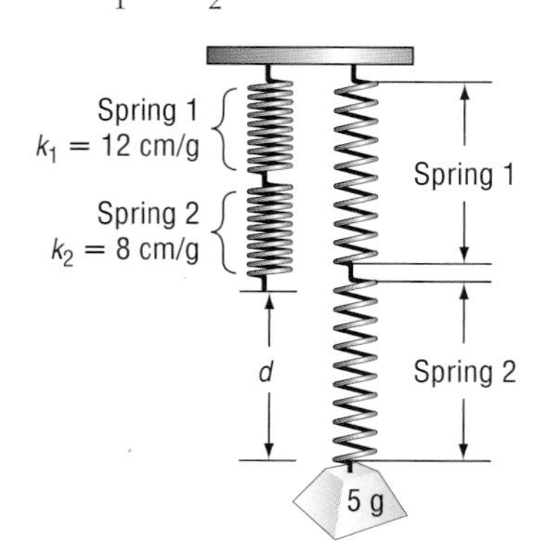

36. **CYCLING** On a particular day, the wind added 3 kilometers per hour to Alfonso's rate when he was cycling with the wind and subtracted 3 kilometers per hour from his rate on his return trip. Alfonso found that in the same amount of time he could cycle 36 kilometers with the wind, he could go only 24 kilometers against the wind. What is his normal bicycling speed with no wind?

Chemist
Many chemists work for manufacturers developing products or doing quality control to ensure the products meet industry and government standards.

Online Research
For information about a career as a chemist, visit: www.algebra2.com/careers

37. CHEMISTRY Kiara adds an 80% acid solution to 5 milliliters of solution that is 20% acid. The function that represents the percent of acid in the resulting solution is $f(x) = \frac{5(0.20) + x(0.80)}{5 + x}$, where x is the amount of 80% solution added. How much 80% solution should be added to create a solution that is 50% acid?

STATISTICS For Exercises 38 and 39, use the following information.
A number x is the *harmonic mean* of y and z if $\frac{1}{x}$ is the average of $\frac{1}{y}$ and $\frac{1}{z}$.

38. Find y if $x = 8$ and $z = 20$.

39. Find x if $y = 5$ and $z = 8$.

40. CRITICAL THINKING Solve for a if $\frac{1}{a} - \frac{1}{b} = c$.

41. **WRITING IN MATH** Answer the question that was posed at the beginning of the lesson.

How are rational equations used to solve problems involving unit price?

Include the following in your answer:

- an explanation of how to solve $\frac{500 + 5x}{x} = 6$, and
- the reason why the actual price per minute could never be 5¢.

42. If $T = \frac{4st}{s - t}$, what is the value of s when $t = 5$ and $T = 40$?

Ⓐ 20 Ⓑ 10 Ⓒ 5 Ⓓ 2

43. Amanda wanted to determine the average of her 6 test scores. She added the scores correctly to get T, but divided by 7 instead of 6. Her average was 12 less than the actual average. Which equation could be used to determine the value of T?

Ⓐ $6T + 12 = 7T$

Ⓑ $\frac{T}{7} = \frac{T - 12}{6}$

Ⓒ $\frac{T}{7} + 12 = \frac{T}{6}$

Ⓓ $\frac{T}{6} = \frac{T - 12}{7}$

Maintain Your Skills

Mixed Review **Identify the type of function represented by each equation. Then graph the equation.** *(Lesson 9-5)*

44. $y = 2x^2 + 1$ **45.** $y = 2\sqrt{x}$ **46.** $y = 0.8x$

47. If y varies inversely as x and $y = 24$ when $x = 9$, find y when $x = 6$. *(Lesson 9-4)*

48. If y varies directly as x and $y = 9$ when $x = 4$, find y when $x = 15$. *(Lesson 9-4)*

Find the distance between each pair of points with the given coordinates. *(Lesson 8-1)*

49. $(-5, 7), (9, -11)$ **50.** $(3, 5), (7, 3)$ **51.** $(-1, 3), (-5, -8)$

Solve each inequality. *(Lesson 6-7)*

52. $(x + 11)(x - 3) > 0$ **53.** $x^2 - 4x \leq 0$ **54.** $2b^2 - b < 6$

Graphing Calculator Investigation

A Follow-Up of Lesson 9-6

Solving Rational Equations by Graphing

You can use a graphing calculator to solve rational equations. You need to graph both sides of the equation and locate the point(s) of intersection. You can also use a graphing calculator to confirm solutions that you have found algebraically.

Example

Use a graphing calculator to solve $\frac{4}{x+1} = \frac{3}{2}$.

- First, rewrite as two functions, $y_1 = \frac{4}{x+1}$ and $y_2 = \frac{3}{2}$.
- Next, graph the two functions on your calculator.

  KEYSTROKES: [Y=] 4 [÷] [(] [X,T,θ,n] 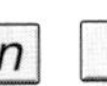 [+]  1 [)]  [▼] 3 [÷] 2 [ZOOM] 6

  Notice that because the calculator is in connected mode, a vertical line is shown connecting the two branches of the hyperbola. This line is not part of the graph.

[−10, 10] scl: 1 by [−10, 10] scl: 1

- Next, locate the point(s) of intersection.

  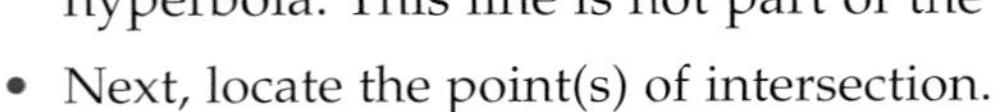

  KEYSTROKES: [2nd] [CALC] 5

  Select one graph and press [ENTER]. Select the other graph, press [ENTER], and press [ENTER] again. The solution is $1\frac{2}{3}$. Check this solution by substitution.

Exercises

Use a graphing calculator to solve each equation.

1. $\frac{1}{x} + \frac{1}{2} = \frac{2}{x}$

2. $\frac{1}{x-4} = \frac{2}{x-2}$

3. $\frac{4}{x} = \frac{6}{x^2}$

4. $\frac{1}{1-x} = 1 - \frac{x}{x-1}$

5. $\frac{1}{x+4} = \frac{2}{x^2+3x-4} - \frac{1}{1-x}$

6. $\frac{1}{x-1} + \frac{1}{x+2} = \frac{1}{2}$

Solve each equation algebraically. Then, confirm your solution(s) using a graphing calculator.

7. $\frac{3}{x} + \frac{7}{x} = 9$

8. $\frac{1}{x-1} + \frac{2}{x} = 0$

9. $1 + \frac{5}{x-1} = \frac{7}{6}$

10. $\frac{1}{x^2-1} = \frac{2}{x^2+x-2}$

11. $\frac{6}{x^2+2x} - \frac{x+1}{x+2} = \frac{2}{x}$

12. $\frac{3}{x^2+5x+6} + \frac{x-1}{x+2} = \frac{7}{x+3}$

www.algebra2.com/other_calculator_keystrokes

Chapter 9
Study Guide and Review

Vocabulary and Concept Check

asymptote (p. 485)	direct variation (p. 492)	rational equation (p. 505)
complex fraction (p. 475)	inverse variation (p. 493)	rational expression (p. 472)
constant of variation (p. 492)	joint variation (p. 493)	rational function (p. 485)
continuity (p. 485)	point discontinuity (p. 485)	rational inequality (p. 508)

State whether each sentence is *true* or *false*. If false, replace the underlined word or number to make a true sentence.

1. The equation $y = \frac{x^2 - 1}{x + 1}$ has a(n) *asymptote* at $x = -1$.
2. The equation $y = 3x$ is an example of a *direct* variation equation.
3. The equation $y = \frac{x^2}{x + 1}$ is a(n) *polynomial* equation.
4. The graph of $y = \frac{4}{x - 4}$ has a(n) *variation* at $x = 4$.
5. The equation $b = \frac{2}{a}$ is a(n) *inverse* variation equation.
6. On the graph of $y = \frac{x - 5}{x + 2}$, there is a break in continuity at $x = \underline{2}$.

Lesson-by-Lesson Review

9-1 Multiplying and Dividing Rational Expressions

See pages 472–478.

Concept Summary

- Multiplying and dividing rational expressions is similar to multiplying and dividing fractions.

Examples

1 Simplify $\frac{3x}{2y} \cdot \frac{8y^3}{6x^2}$.

$$\frac{3x}{2y} \cdot \frac{8y^3}{6x^2} = \frac{\overset{1}{\cancel{3}} \cdot \overset{1}{\cancel{x}} \cdot \overset{1}{\cancel{2}} \cdot \overset{1}{\cancel{2}} \cdot 2 \cdot \overset{1}{\cancel{y}} \cdot y \cdot y}{\underset{1}{\cancel{2}} \cdot \underset{1}{\cancel{y}} \cdot \underset{1}{\cancel{2}} \cdot \underset{1}{\cancel{3}} \cdot \underset{1}{\cancel{x}} \cdot x}$$

$$= \frac{2y^2}{x}$$

2 Simplify $\frac{p^2 + 7p}{3p} \div \frac{49 - p^2}{3p - 21}$.

$$\frac{p^2 + 7p}{3p} \div \frac{49 - p^2}{3p - 21} = \frac{p^2 + 7p}{3p} \cdot \frac{3p - 21}{49 - p^2}$$

$$= \frac{\overset{1}{\cancel{p}}\overset{1}{\cancel{(7 + p)}}}{\underset{1}{\cancel{3}}\underset{1}{\cancel{p}}} \cdot \frac{\overset{-1}{\cancel{-3}}\overset{1}{\cancel{(7 - p)}}}{\underset{1}{\cancel{(7 + p)}}\underset{1}{\cancel{(7 - p)}}}$$

$$= -1$$

Exercises **Simplify each expression.** *See Examples 4–7 on pages 474 and 475.*

7. $\frac{-4ab}{21c} \cdot \frac{14c^2}{22a^2}$
8. $\frac{a^2 - b^2}{6b} \div \frac{a + b}{36b^2}$
9. $\frac{y^2 - y - 12}{y + 2} \div \frac{y - 4}{y^2 - 4y - 12}$
10. $\dfrac{\frac{x^2 + 7x + 10}{x + 2}}{\frac{x^2 + 2x - 15}{x + 2}}$
11. $\dfrac{\frac{1}{n^2 - 6n + 9}}{\frac{n + 3}{2n^2 - 18}}$
12. $\frac{x^2 + 3x - 10}{x^2 + 8x + 15} \cdot \frac{x^2 + 5x + 6}{x^2 + 4x + 4}$

9-2 Adding and Subtracting Rational Expressions

See pages 479–484.

Concept Summary

- To add or subtract rational expressions, find a common denominator.
- To simplify complex fractions, simplify the numerator and the denominator separately, and then simplify the resulting expression.

Example Simplify $\frac{14}{x+y} - \frac{9x}{x^2 - y^2}$.

$$\frac{14}{x+y} - \frac{9x}{x^2-y^2} = \frac{14}{x+y} - \frac{9x}{(x+y)(x-y)} \quad \text{Factor the denominators.}$$

$$= \frac{14(x-y)}{(x+y)(x-y)} - \frac{9x}{(x+y)(x-y)} \quad \text{The LCD is } (x+y)(x-y).$$

$$= \frac{14(x-y) - 9x}{(x+y)(x-y)} \quad \text{Subtract the numerators.}$$

$$= \frac{14x - 14y - 9x}{(x+y)(x-y)} \quad \text{Distributive Property}$$

$$= \frac{5x - 14y}{(x+y)(x-y)} \quad \text{Simplify.}$$

Exercises **Simplify each expression.** *See Examples 3 and 4 on page 480.*

13. $\frac{x+2}{x-5} + 6$ **14.** $\frac{x-1}{x^2-1} + \frac{2}{5x+5}$ **15.** $\frac{7}{y} - \frac{2}{3y}$

16. $\frac{7}{y-2} - \frac{11}{2-y}$ **17.** $\frac{3}{4b} - \frac{2}{5b} - \frac{1}{2b}$ **18.** $\frac{m+3}{m^2-6m+9} - \frac{8m-24}{9-m^2}$

9-3 Graphing Rational Functions

See pages 485–490.

Concept Summary

- Functions are undefined at any x value where the denominator is zero.
- An asymptote is a line that the graph of the function approaches, but never crosses.

Example Graph $f(x) = \frac{5}{x(x+4)}$.

The function is undefined for $x = 0$ and $x = -4$. Since $\frac{5}{x(x+4)}$ is in simplest form, $x = 0$ and $x = -4$ are vertical asymptotes. Draw the two asymptotes and sketch the graph.

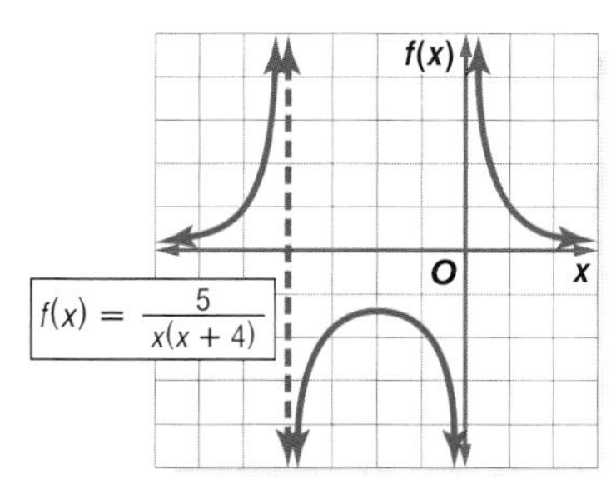

Exercises **Graph each rational function.** *See Examples 2–4 on pages 486–488.*

19. $f(x) = \frac{4}{x-2}$ **20.** $f(x) = \frac{x}{x+3}$ **21.** $f(x) = \frac{2}{x}$

22. $f(x) = \frac{x-4}{x+3}$ **23.** $f(x) = \frac{5}{(x+1)(x-3)}$ **24.** $f(x) = \frac{x^2+2x+1}{x+1}$

9-4 Direct, Joint, and Inverse Variation

See pages 492–498.

Concept Summary

- Direct Variation: There is a nonzero constant k such that $y = kx$.
- Joint Variation: There is a number k such that $y = kxz$, where $x \neq 0$ and $z \neq 0$.
- Inverse Variation: There is a nonzero constant k such that $xy = k$ or $y = \frac{k}{x}$.

Example **If y varies inversely as x and $x = 14$ when $y = -6$, find x when $y = -11$.**

$\frac{x_1}{y_2} = \frac{x_2}{y_1}$ Inverse variation

$\frac{14}{-11} = \frac{x_2}{-6}$ $x_1 = 14, y_1 = -6, y_2 = -11$

$14(-6) = -11(x_2)$ Cross multiply.

$-84 = -11x_2$ Simplify.

$7\frac{7}{11} = x_2$ When $y = -11$, the value of x is $7\frac{7}{11}$.

Exercises **Find each value.** *See Examples 1–3 on pages 493 and 494.*

25. If y varies directly as x and $y = 21$ when $x = 7$, find x when $y = -5$.

26. If y varies inversely as x and $y = 9$ when $x = 2.5$, find y when $x = -0.6$.

27. If y varies inversely as x and $x = 28$ when $y = 18$, find x when $y = 63$.

28. If y varies directly as x and $x = 28$ when $y = 18$, find x when $y = 63$.

29. If y varies jointly as x and z and $x = 2$ and $z = 4$ when $y = 16$, find y when $x = 5$ and $z = 8$.

9-5 Classes of Functions

See pages 499–504.

Concept Summary

The following is a list of special functions.

- constant function
- direct variation function
- identity function
- greatest integer function
- absolute value function
- quadratic function
- square root function
- rational function
- inverse variation function

Examples **Identify the type of function represented by each graph.**

1

The graph has a parabolic shape, therefore it is a quadratic function.

2

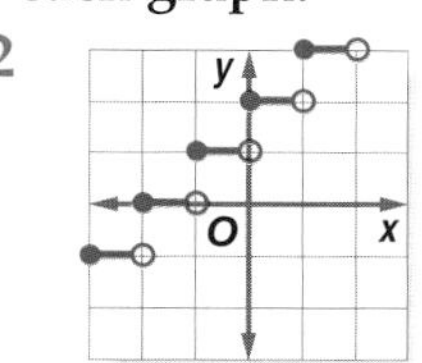

The graph has a stair-step pattern, therefore it is a greatest integer function.

Chapter 9 For More ...

- Extra Practice, see pages 847–849.
- Mixed Problem Solving, see page 870.

Exercises **Identify the type of function represented by each graph.**
See Example 1 on page 500.

30.

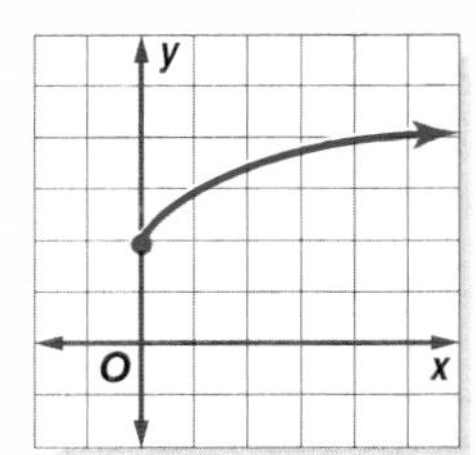

31.

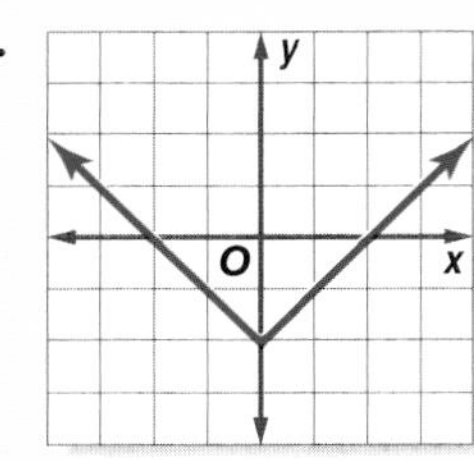

32.

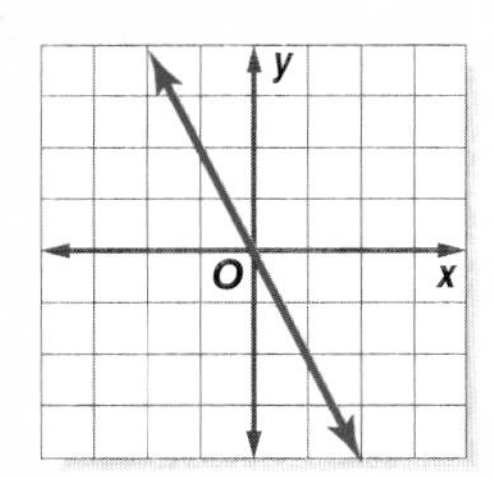

9-6 Solving Rational Equations and Inequalities

See pages 505–511.

Concept Summary

- Eliminate fractions in rational equations by multiplying each side of the equation by the LCD.
- Possible solutions to a rational equation must exclude values that result in zero in the denominator.
- To solve rational inequalities, find the excluded values, solve the related equation, and use these values to divide a number line into regions. Then test a value in each region to determine which regions satisfy the original inequality.

Example **Solve $\frac{1}{x-1} + \frac{2}{x} = 0$.**

The LCD is $x(x - 1)$.

$$\frac{1}{x-1} + \frac{2}{x} = 0 \quad \text{Original equation}$$

$$x(x-1)\left(\frac{1}{x-1} + \frac{2}{x}\right) = x(x-1)(0) \quad \text{Multiply each side by } x(x-1).$$

$$x(x-1)\left(\frac{1}{x-1}\right) + x(x-1)\left(\frac{2}{x}\right) = x(x-1)(0) \quad \text{Distributive Property}$$

$$1(x) + 2(x-1) = 0 \quad \text{Simplify.}$$

$$x + 2x - 2 = 0 \quad \text{Distributive Property}$$

$$3x - 2 = 0 \quad \text{Simplify.}$$

$$3x = 2 \quad \text{Add 2 to each side.}$$

$$x = \frac{2}{3} \quad \text{Divide each side by 3.}$$

The solution is $\frac{2}{3}$.

Exercises **Solve each equation or inequality. Check your solutions.**
See Examples 1, 2, and 5 on pages 505, 506, 508, and 509.

33. $\frac{3}{y} + \frac{7}{y} = 9$

34. $1 + \frac{5}{y-1} = \frac{7}{6}$

35. $\frac{3x+2}{4} = \frac{9}{4} - \frac{3-2x}{6}$

36. $\frac{1}{r^2-1} = \frac{2}{r^2+r-2}$

37. $\frac{x}{x^2-1} + \frac{2}{x+1} = 1 + \frac{1}{2x-2}$

38. $\frac{1}{3b} - \frac{3}{4b} > \frac{1}{6}$

Chapter 9 Practice Test

Vocabulary and Concepts

Match each example with the correct term.

1. $y = 4xz$
2. $y = 5x$
3. $y = \frac{7}{x}$

a. inverse variation equation
b. direct variation equation
c. joint variation equation

Skills and Applications

Simplify each expression.

4. $\frac{a^2 - ab}{3a} \div \frac{a - b}{15b^2}$

5. $\frac{x^2 - y^2}{y^2} \cdot \frac{y^3}{y - x}$

6. $\frac{x^2 - 2x + 1}{y - 5} \div \frac{x - 1}{y^2 - 25}$

7. $\dfrac{\frac{x^2 - 1}{x^2 - 3x - 10}}{\frac{x^2 + 3x + 2}{x^2 - 12x + 35}}$

8. $\frac{x - 2}{x - 1} + \frac{6}{7x - 7}$

9. $\frac{x}{x^2 - 9} + \frac{1}{2x + 6}$

Identify the type of function represented by each graph.

10. 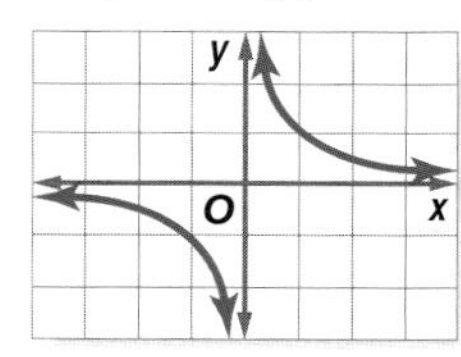

11. 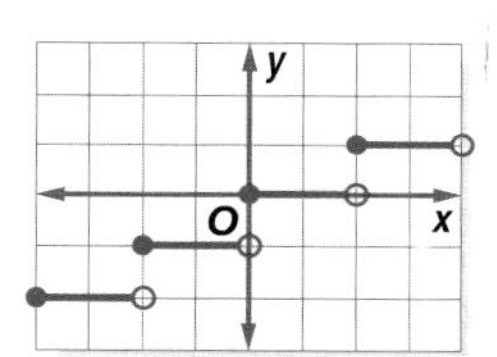

Graph each rational function.

12. $f(x) = \frac{-4}{x - 3}$

13. $f(x) = \frac{2}{(x - 2)(x + 1)}$

Solve each equation or inequality.

14. $\frac{2}{x - 1} = 4 - \frac{x}{x - 1}$

15. $\frac{9}{28} + \frac{3}{z + 2} = \frac{3}{4}$

16. $5 + \frac{3}{t} > -\frac{2}{t}$

17. $x + \frac{12}{x} - 8 = 0$

18. $\frac{5}{6} - \frac{2m}{2m + 3} = \frac{19}{6}$

19. $\frac{x - 3}{2x} = \frac{x - 2}{2x + 1} - \frac{1}{2}$

20. If y varies inversely as x and $y = 9$ when $x = -\frac{2}{3}$, find x when $y = -7$.

21. If g varies directly as w and $g = 10$ when $w = -3$, find w when $g = 4$.

22. Suppose y varies jointly as x and z. If $x = 10$ when $y = 250$ and $z = 5$, find x when $y = 2.5$ and $z = 4.5$.

23. **AUTO MAINTENANCE** When air is pumped into a tire, the pressure required varies inversely as the volume of the air. If the pressure is 30 pounds per square inch when the volume is 140 cubic inches, find the pressure when the volume is 100 cubic inches.

24. **ELECTRICITY** The current I in a circuit varies inversely with the resistance R.
    a. Use the table at the right to write an equation relating the current and the resistance.
    b. What is the constant of variation?

I	0.5	1.0	1.5	2.0	2.5	3.0	5.0
R	12.0	6.0	4.0	3.0	2.4	2.0	1.2

Standards Practice

25. **STANDARDIZED TEST PRACTICE** If $m = \frac{1}{x}$, $n = 7m$, $p = \frac{1}{n}$, $q = 14p$, and $r = \frac{1}{\frac{1}{2}q}$, find x.

Ⓐ r Ⓑ q Ⓒ p Ⓓ $\frac{1}{r}$ Ⓔ $\frac{1}{q}$

Chapter 9

Standardized Test Practice

Standards Practice

Part 1 Multiple Choice

Record your answers on the answer sheet provided by your teacher or on a sheet of paper.

1. Best Bikes has 5000 bikes in stock on May 1. By the end of May, 40 percent of the bikes have been sold. By the end of June, 40 percent of the remaining bikes have been sold. How many bikes remain unsold?

Ⓐ 1000 Ⓑ 1200
Ⓒ 1800 Ⓓ 2000

2. In $\triangle ABC$, if AB is equal to 8, then BC is equal to

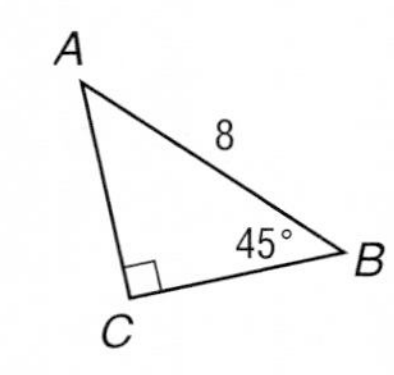

Ⓐ $\frac{\sqrt{2}}{8}$. Ⓑ 4.
Ⓒ $4\sqrt{2}$. Ⓓ 8.

3. In the figure, the slope of $\overline{AC}$ is $-\frac{1}{3}$ and $m\angle C = 30°$. What is the length of $\overline{BC}$?

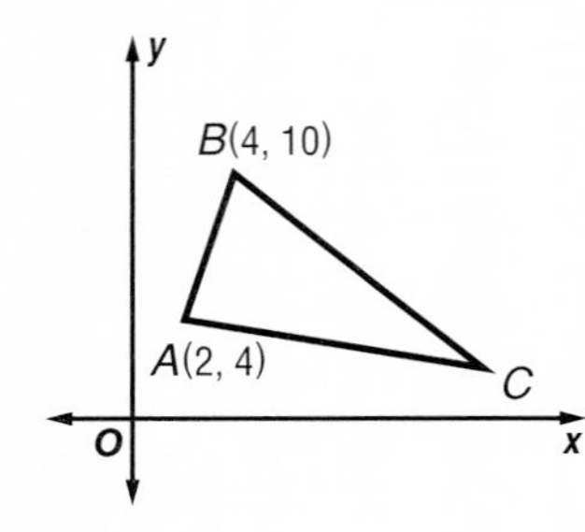

Ⓐ $\sqrt{10}$ Ⓑ $2\sqrt{10}$
Ⓒ $3\sqrt{10}$ Ⓓ $4\sqrt{10}$

4. Given that $-|2 - 4k| = -14$, which of the following could be k?

Ⓐ 5 Ⓑ 4
Ⓒ 3 Ⓓ 2

5. In a hardware store, n nails cost c cents. Which of the following expresses the cost of k nails?

Ⓐ nck Ⓑ $\frac{kc}{n}$
Ⓒ $n + \frac{k}{c}$ Ⓓ $n + \frac{c}{n}$

6. If $5w + 3 \leq w - 9$, then

Ⓐ $w \leq 3$. Ⓑ $w \geq 3$.
Ⓒ $w \leq 12$. Ⓓ $w \leq -3$.

7. The graphs show a driver's distance d from a designated point as a function of time t. The driver passed the designated point at 60 mph and continued at that speed for 2 hours. Then she slowed to 50 mph for 1 hour. She stopped for gas and lunch for 1 hour and then drove at 60 mph for 1 hour. Which graph best represents this trip?

Ⓐ
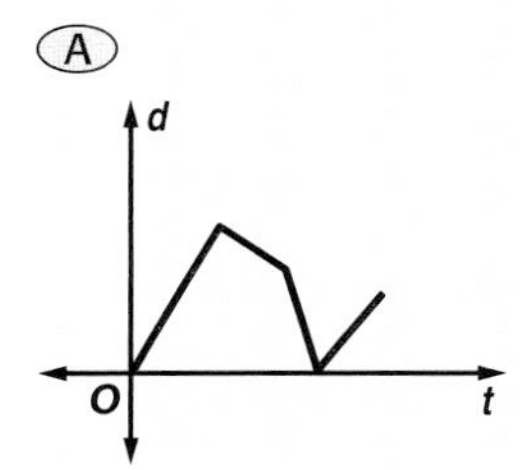

Ⓑ
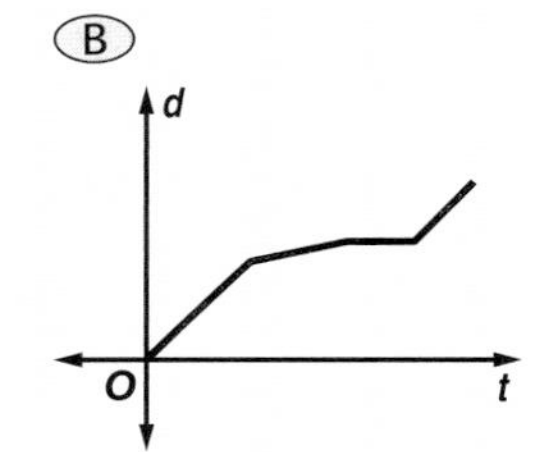

Ⓒ
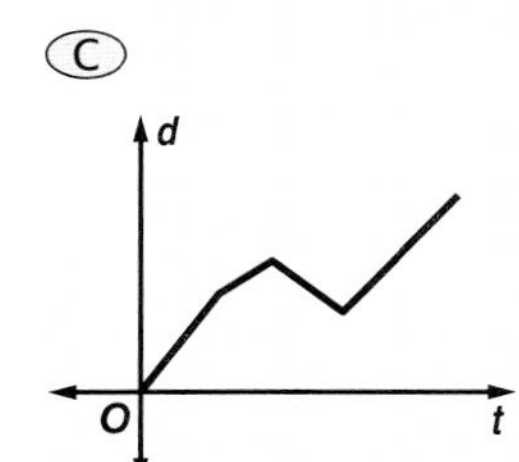

Ⓓ
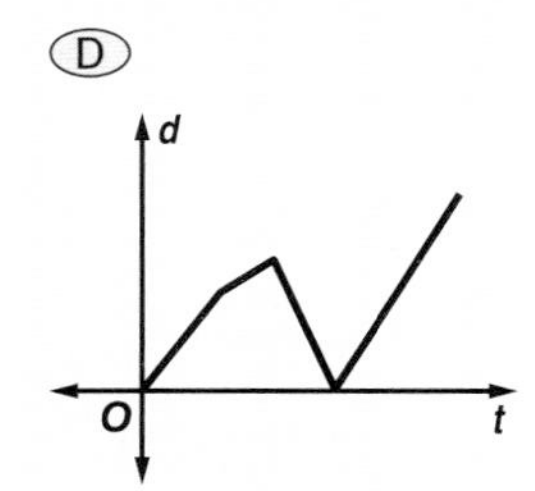

8. Which equation has roots of $-2n$, $2n$, and 2?

Ⓐ $2x^2 - 8n^2 = 0$
Ⓑ $8n^2 - 2x^2 = 0$
Ⓒ $x^3 - 2x^2 - 4n^2x - 8n^2 = 0$
Ⓓ $x^3 - 2x^2 - 4n^2x + 8n^2 = 0$

9. What point is on the graph of $y - x^2 = 2$ and has a y-coordinate of 5?

Ⓐ $(-\sqrt{3}, 5)$ Ⓑ $(\sqrt{7}, 5)$
Ⓒ $(5, \sqrt{3})$ Ⓓ $(3, 5)$

Preparing for Standardized Tests
For test-taking strategies and more practice, see pages 877–892.

Part 2 Short Response/Grid In

Record your answers on the answer sheet provided by your teacher or on a sheet of paper.

10. In the figure, what is the equation of the circle Q that is circumscribed around the square $ABCD$?

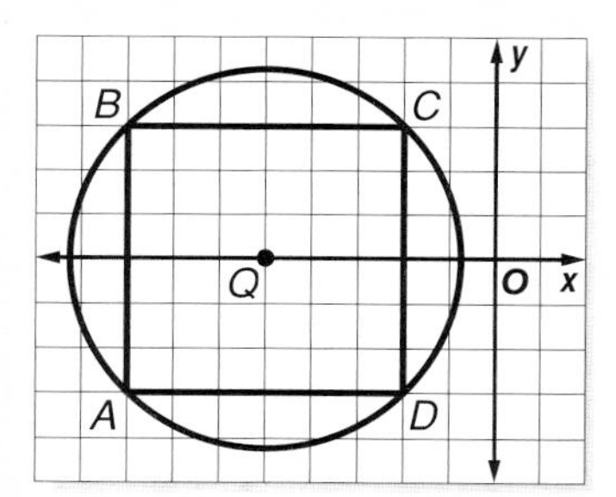

11. Find one possible value for k such that k is an integer between 20 and 40 that has a remainder of 2 when it is divided by 3 and that has a remainder of 2 when divided by 4.

12. The coordinates of the vertices of a triangle are $(2, -4)$, $(10, -4)$, and (a, b). If the area of the triangle is 36 square units, what is a possible value for b?

13. If $(x + 2)(x - 3) = 6$, what is a possible value of x?

14. If the average of five consecutive even integers is 76, what is the greatest of these integers?

15. In May, Hank's Camping Supply Store sold 45 tents. In June, it sold 90 tents. What is the percent increase in the number of tents sold?

16. If $2^{n-4} = 64$, what is the value of n?

17. If $xy = 5$ and $x^2 + y^2 = 20$, what is the value of $(x + y)^2$?

18. If $\frac{2}{a} - \frac{8}{a^2} = \frac{-8}{a^3}$, then what is the value of a?

19. If $\sqrt[x]{80} = 2\sqrt[x]{5}$, what is the value of x?

20. What is the y-intercept of the graph of $3x + 2 = 4y - 6$?

Part 3 Extended Response

Record your answers on a sheet of paper. Show your work.

21. Write an expression that is undefined when $x = 1$ or $x = -1$. Justify your answer.

22. Identify the type of function represented by the graph below. Then write an equation for the function. Explain your answer.

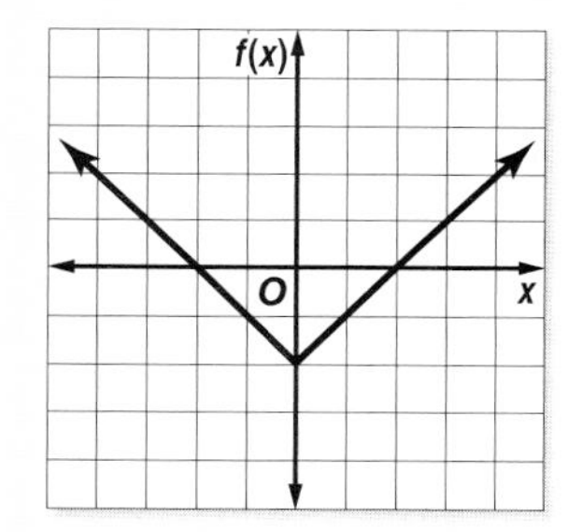

23. If 100 feet of a certain type of cable weighs 12 pounds, how much do 3 yards of the same cable weigh?

For Exercises 24 and 25, use the following information.

A gear that is 8 inches in diameter turns a smaller gear that is 3 inches in diameter.

24. Does this situation represent a direct or inverse variation? Explain your reasoning.

25. If the larger gear makes 36 revolutions, how many revolutions does the smaller gear make in that time?

Test-Taking Tip

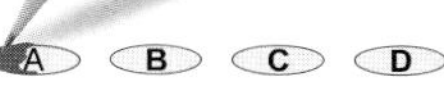

Questions 23–25
When solving problems involving variation, pay careful attention to the units of measure in your calculations. Check that the units in the fractions of a proportion for a direct variation problem are the same in the numerators and the denominators.

Chapter 10

Exponential and Logarithmic Relations

What You'll Learn

- **Lessons 10-1 through 10-3** Simplify exponential and logarithmic expressions.
- **Lessons 10-1, 10-4, and 10-5** Solve exponential equations and inequalities.
- **Lessons 10-2 and 10-3** Solve logarithmic equations and inequalities.
- **Lesson 10-6** Solve problems involving exponential growth and decay.

Key Vocabulary

- exponential growth (p. 524)
- exponential decay (p. 524)
- logarithm (p. 531)
- common logarithm (p. 547)
- natural logarithm (p. 554)

Why It's Important

Exponential functions are often used to model problems involving growth and decay. Logarithms can also be used to solve such problems. *You will learn how a declining farm population can be modeled by an exponential function in Lesson 10-1.*

Getting Started

Prerequisite Skills To be successful in this chapter, you'll need to master these skills and be able to apply them in problem-solving situations. Review these skills before beginning Chapter 10.

Lessons 10-1 through 10-3 — Multiply and Divide Monomials

Simplify. Assume that no variable equals 0. *(For review, see Lesson 5-1.)*

1. $x^5 \cdot x \cdot x^6$ **2.** $(3ab^4c^2)^3$ **3.** $\dfrac{-36x^7y^4z^3}{21x^4y^9z^4}$ **4.** $\left(\dfrac{4ab^2}{64b^3c}\right)^2$

Lessons 10-2 and 10-3 — Solve Inequalities

Solve each inequality. *(For review, see Lesson 1-5)*

5. $a + 4 < -10$ **6.** $-5n \leq 15$ **7.** $3y + 2 \geq -4$ **8.** $15 - x > 9$

Lessons 10-2 and 10-3 — Inverse Functions

Find the inverse of each function. Then graph the function and its inverse.
(For review, see Lesson 7-8.)

9. $f(x) = -2x$ **10.** $f(x) = 3x - 2$ **11.** $f(x) = -x + 1$ **12.** $f(x) = \dfrac{x - 4}{3}$

Lessons 10-2 and 10-3 — Composition of Functions

Find $g[h(x)]$ and $h[g(x)]$. *(For review, see Lesson 7-7.)*

13. $h(x) = 3x + 4$
$g(x) = x - 2$

14. $h(x) = 2x - 7$
$g(x) = 5x$

15. $h(x) = x - 4$
$g(x) = x^2$

16. $h(x) = 4x + 1$
$g(x) = -2x - 3$

Exponential and Logarithmic Relations Make this Foldable to help you organize your notes. Begin with four sheets of grid paper.

Step 1 Fold and Cut

Fold in half along the width. On the first two sheets, cut along the fold at the ends. On the second two sheets, cut in the center of the fold as shown.

First Sheets

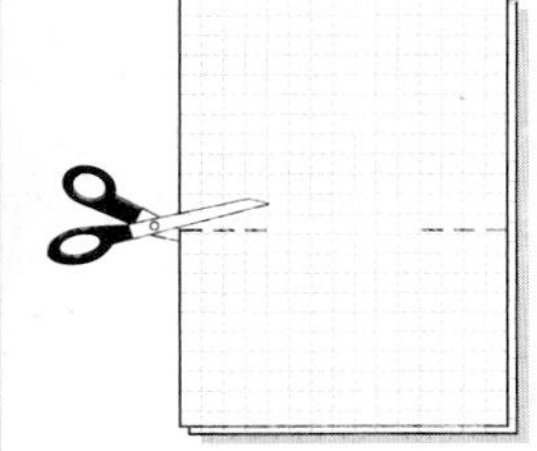

Second Sheets

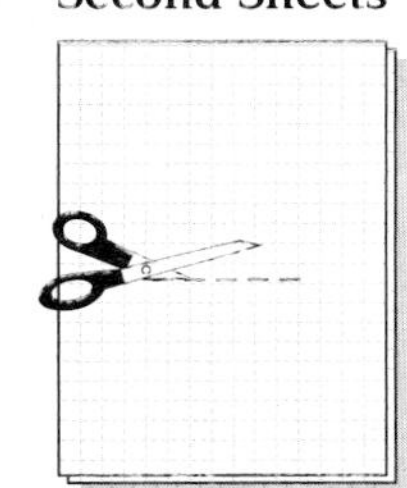

Step 2 Fold and Label

Insert first sheets through second sheets and align the folds. Label the pages with lesson numbers.

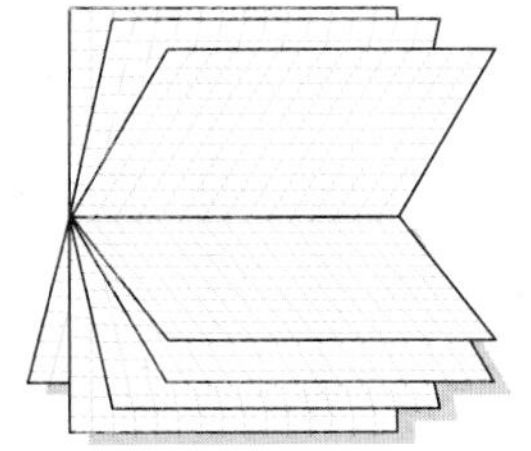

Reading and Writing As you read and study the chapter, fill the journal with notes, diagrams, and examples for each lesson.

Algebra Activity

A Preview of Lesson 10-1

California Standards **Standard 12.0** **Students** know the laws of fractional exponents, **understand exponential functions,** and use these functions in problems involving exponential growth and decay. (Key)

Investigating Exponential Functions

Collect the Data

Step 1 Cut a sheet of notebook paper in half.

Step 2 Stack the two halves, one on top of the other.

Step 3 Make a table like the one below and record the number of sheets of paper you have in the stack after one cut.

Number of Cuts	Number of Sheets
0	1
1	
2	

Step 4 Cut the two stacked sheets in half, placing the resulting pieces in a single stack. Record the number of sheets of paper in the new stack after 2 cuts.

Step 5 Continue cutting the stack in half, each time putting the resulting piles in a single stack and recording the number of sheets in the stack. Stop when the resulting stack is too thick to cut.

Analyze the Data

1. Write a list of ordered pairs (x, y), where x is the number of cuts and y is the number of sheets in the stack. Notice that the list starts with the ordered pair $(0, 1)$, which represents the single sheet of paper before any cuts were made.
2. Continue the list, beyond the point where you stopped cutting, until you reach the ordered pair for 7 cuts. Explain how you calculated the last y values for your list, after you had stopped cutting.
3. Plot the ordered pairs in your list on a coordinate grid. Be sure to choose a scale for the y-axis so that you can plot all of the points.
4. Describe the pattern of the points you have plotted. Do they lie on a straight line?

Make a Conjecture

5. Write a function that expresses y as a function of x.
6. Use a calculator to evaluate the function you wrote in Exercise 5 for $x = 8$ and $x = 9$. Does it give the correct number of sheets in the stack after 8 and 9 cuts?
7. Notebook paper usually stacks about 500 sheets to the inch. How thick would your stack of paper be if you had been able to make 9 cuts?
8. Suppose each cut takes about 5 seconds. If you had been able to keep cutting, you would have made 36 cuts in three minutes. At 500 sheets to the inch, make a conjecture as to how thick you think the stack would be after 36 cuts.
9. Use your function from Exercise 5 to calculate the thickness of your stack after 36 cuts. Write your answer in miles.

10-1 Exponential Functions

California Standards Standard 12.0 **Students** know the laws of fractional exponents, **understand exponential functions,** and use these functions in problems involving exponential growth and decay. (Key)

What You'll Learn

- Graph exponential functions.
- Solve exponential equations and inequalities.

How does an exponential function describe tournament play?

The NCAA women's basketball tournament begins with 64 teams and consists of 6 rounds of play. The winners of the first round play against each other in the second round. The winners then move from the Sweet Sixteen to the Elite Eight to the Final Four and finally to the Championship Game.

The number of teams y that compete in a tournament of x rounds is $y = 2^x$.

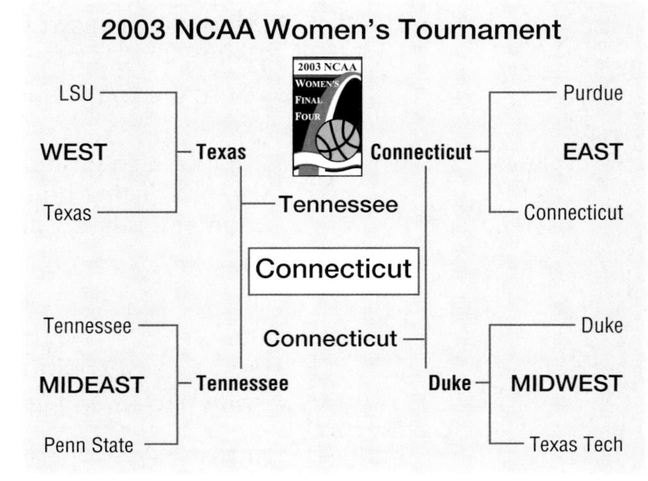

Vocabulary

- exponential function
- exponential growth
- exponential decay
- exponential equation
- exponential inequality

EXPONENTIAL FUNCTIONS In an exponential function like $y = 2^x$, the base is a constant, and the exponent is a variable. Let's examine the graph of $y = 2^x$.

Study Tip

Common Misconception
Be sure not to confuse polynomial functions and exponential functions. While $y = x^2$ and $y = 2^x$ each have an exponent, $y = x^2$ is a polynomial function and $y = 2^x$ is an exponential function.

Example 1 Graph an Exponential Function

Sketch the graph of $y = 2^x$. Then state the function's domain and range.

Make a table of values. Connect the points to sketch a smooth curve.

x	$y = 2^x$
-3	$2^{-3} = \frac{1}{8}$
-2	$2^{-2} = \frac{1}{4}$
-1	$2^{-1} = \frac{1}{2}$
0	$2^0 = 1$
$\frac{1}{2}$	$2^{\frac{1}{2}} = \sqrt{2}$
1	$2^1 = 2$
2	$2^2 = 4$
3	$2^3 = 8$

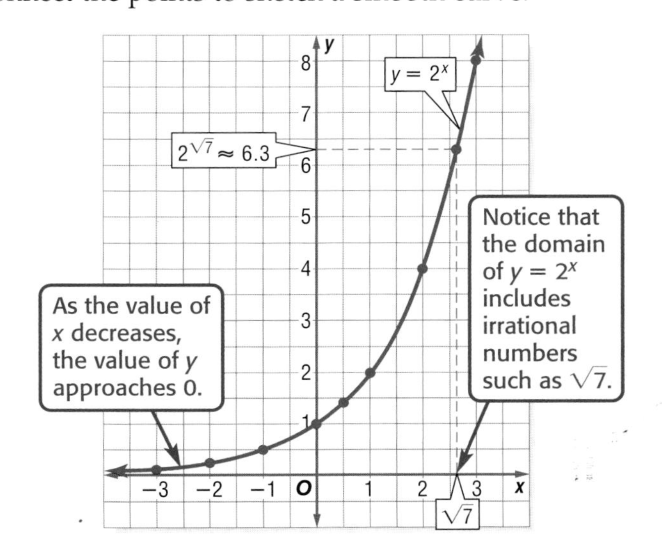

The domain is all real numbers, while the range is all positive numbers.

You can use a TI-83 Plus graphing calculator to look at the graph of two other exponential functions, $y = 3^x$ and $y = \left(\frac{1}{3}\right)^x$.

Graphing Calculator Investigation

Families of Exponential Functions

The calculator screen shows the graphs of $y = 3^x$ and $y = \left(\frac{1}{3}\right)^x$.

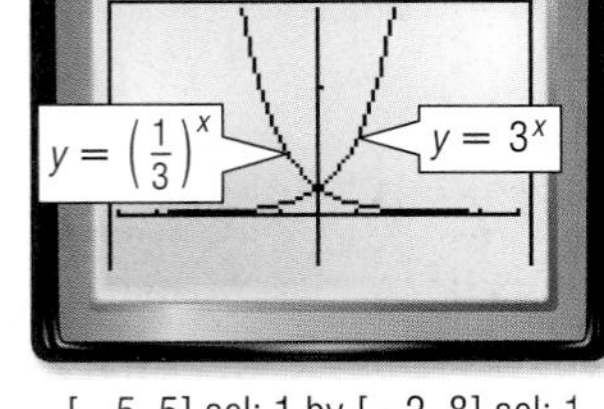

[−5, 5] scl: 1 by [−2, 8] scl: 1

Think and Discuss

1. How do the shapes of the graphs compare?
2. How do the asymptotes and y-intercepts of the graphs compare?
3. Describe the relationship between the graphs.

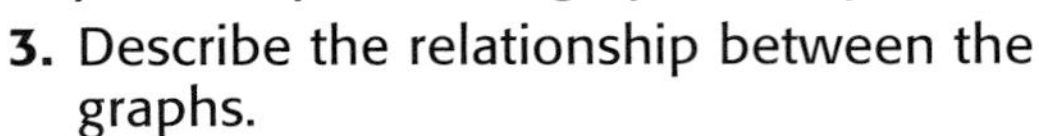

4. Graph each group of functions on the same screen. Then compare the graphs, listing both similarities and differences in shape, asymptotes, domain, range, and y-intercepts.
   a. $y = 2^x$, $y = 3^x$, and $y = 4^x$
   b. $y = \left(\frac{1}{2}\right)^x$, $y = \left(\frac{1}{3}\right)^x$, and $y = \left(\frac{1}{4}\right)^x$
   c. $y = -3(2)^x$ and $y = 3(2)^x$; $y = -1(2)^x$ and $y = 2^x$.
5. Describe the relationship between the graphs of $y = -1(2)^x$ and $y = 2^x$.

Study Tip

Look Back

To review **continuous functions**, see page 62, Exercises 60 and 61. To review **one-to-one functions**, see Lesson 2-1.

In general, an equation of the form $y = ab^x$, where $a \neq 0$, $b > 0$, and $b \neq 1$, is called an **exponential function** with base b. Exponential functions have the following characteristics.

1. The function is continuous and one-to-one.
2. The domain is the set of all real numbers.
3. The x-axis is an asymptote of the graph.
4. The range is the set of all positive numbers if $a > 0$ and all negative numbers if $a < 0$.
5. The graph contains the point $(0, a)$. That is, the y-intercept is a.
6. The graphs of $y = ab^x$ and $y = a\left(\frac{1}{b}\right)^x$ are reflections across the y-axis.

Study Tip

Exponential Growth and Decay

Notice that the graph of an exponential growth function *rises* from left to right. The graph of an exponential decay function *falls* from left to right.

There are two types of exponential functions: **exponential growth** and **exponential decay**. The base of an exponential growth function is a number greater than one. The base of an exponential decay function is a number between 0 and 1.

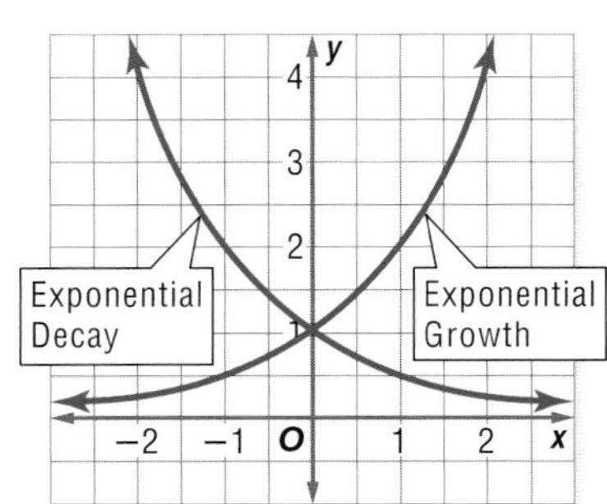

Key Concept — *Exponential Growth and Decay*

- If $a > 0$ and $b > 1$, the function $y = ab^x$ represents exponential growth.
- If $a > 0$ and $0 < b < 1$, the function $y = ab^x$ represents exponential decay.

Example 2 Identify Exponential Growth and Decay

Determine whether each function represents exponential *growth* or *decay*.

Function	Exponential Growth or Decay?
a. $y = \left(\frac{1}{5}\right)^x$	The function represents exponential decay, since the base, $\frac{1}{5}$, is between 0 and 1.
b. $y = 3(4)^x$	The function represents exponential growth, since the base, 4, is greater than 1.
c. $y = 7(1.2)^x$	The function represents exponential growth, since the base, 1.2, is greater than 1.

Exponential functions are frequently used to model the growth or decay of a population. You can use the y-intercept and one other point on the graph to write the equation of an exponential function.

Example 3 Write an Exponential Function

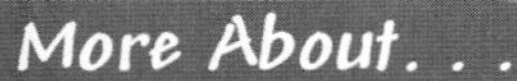

Farming

In 1999, 47% of the net farm income in the United States was from direct government payments. The USDA has set a goal of reducing this percent to 14% by 2005.

Source: USDA

FARMING In 1983, there were 102,000 farms in Minnesota, but by 1998, this number had dropped to 80,000.

a. Write an exponential function of the form $y = ab^x$ that could be used to model the farm population y of Minnesota. Write the function in terms of x, the number of years since 1983.

For 1983, the time x equals 0, and the initial population y is 102,000. Thus, the y-intercept, and value of a, is 102,000.

For 1998, the time x equals 1998 − 1983 or 15, and the population y is 80,000. Substitute these values and the value of a into an exponential function to approximate the value of b.

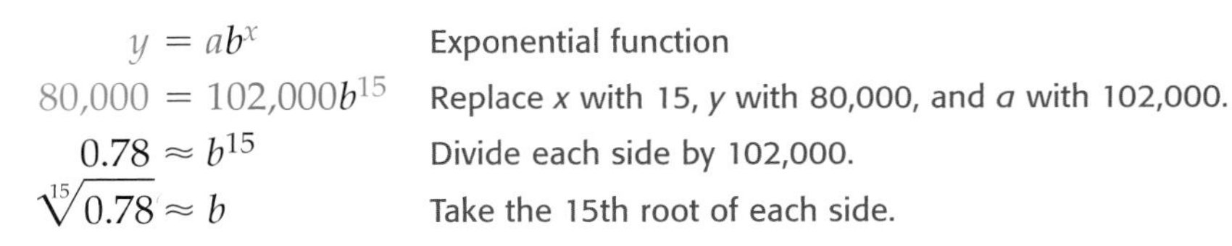

$y = ab^x$ Exponential function

$80{,}000 = 102{,}000b^{15}$ Replace x with 15, y with 80,000, and a with 102,000.

$0.78 \approx b^{15}$ Divide each side by 102,000.

$\sqrt[15]{0.78} \approx b$ Take the 15th root of each side.

To find the 15th root of 0.78, use selection 5: $\sqrt[x]{\ }$ under the MATH menu on the TI-83 Plus.

KEYSTROKES: 15 [MATH] 5 0.78 [ENTER] .9835723396

An equation that models the farm population of Minnesota from 1983 to 1998 is $y = 102{,}000(0.98)^x$.

b. Suppose the number of farms in Minnesota continues to decline at the same rate. Estimate the number of farms in 2010.

For 2010, the time x equals 2010 − 1983 or 27.

$y = 102{,}000(0.98)^x$ Modeling equation

$y = 102{,}000(0.98)^{27}$ Replace x with 27.

$y \approx 59{,}115$ Use a calculator.

The farm population in Minnesota will be about 59,115 in 2010.

Study Tip

Look Back
To review **Properties of Power**, see Lesson 5-1.

EXPONENTIAL EQUATIONS AND INEQUALITIES Since the domain of an exponential function includes irrational numbers such as $\sqrt{2}$, all the properties of rational exponents apply to irrational exponents.

Example 4 Simplify Expressions with Irrational Exponents

Simplify each expression.

a. $2^{\sqrt{5}} \cdot 2^{\sqrt{3}}$

$2^{\sqrt{5}} \cdot 2^{\sqrt{3}} = 2^{\sqrt{5} + \sqrt{3}}$ Product of Powers

b. $\left(7^{\sqrt{2}}\right)^{\sqrt{3}}$

$\left(7^{\sqrt{2}}\right)^{\sqrt{3}} = 7^{\sqrt{2} \cdot \sqrt{3}}$ Power of a Power

$= 7^{\sqrt{6}}$ Product of Radicals

The following property is useful for solving exponential equations. **Exponential equations** are equations in which variables occur as exponents.

Key Concept — Property of Equality for Exponential Functions

- **Symbols** If b is a positive number other than 1, then $b^x = b^y$ if and only if $x = y$.
- **Example** If $2^x = 2^8$, then $x = 8$.

Example 5 Solve Exponential Equations

Solve each equation.

a. $3^{2n + 1} = 81$

$3^{2n+1} = 81$ Original equation

$3^{2n+1} = 3^4$ Rewrite 81 as 3^4 so each side has the same base.

$2n + 1 = 4$ Property of Equality for Exponential Functions

$2n = 3$ Subtract 1 from each side.

$n = \frac{3}{2}$ Divide each side by 2.

The solution is $\frac{3}{2}$.

CHECK $3^{2n+1} = 81$ Original equation

$3^{2\left(\frac{3}{2}\right)+1} \stackrel{?}{=} 81$ Substitute $\frac{3}{2}$ for n.

$3^4 \stackrel{?}{=} 81$ Simplify.

$81 = 81$ ✓ Simplify.

b. $4^{2x} = 8^{x - 1}$

$4^{2x} = 8^{x-1}$ Original equation

$(2^2)^{2x} = (2^3)^{x-1}$ Rewrite each side with a base of 2.

$2^{4x} = 2^{3(x-1)}$ Power of a Power

$4x = 3(x - 1)$ Property of Equality for Exponential Functions

$4x = 3x - 3$ Distributive Property

$x = -3$ Subtract $3x$ from each side.

The solution is -3.

The following property is useful for solving inequalities involving exponential functions or **exponential inequalities**.

Key Concept *Property of Inequality for Exponential Functions*

- **Symbols** If $b > 1$, then $b^x > b^y$ if and only if $x > y$, and $b^x < b^y$ if and only if $x < y$.
- **Example** If $5^x < 5^4$, then $x < 4$.

This property also holds for $\leq$ and $\geq$.

Example 6 Solve Exponential Inequalities

Solve $4^{3p-1} > \frac{1}{256}$.

$4^{3p-1} > \frac{1}{256}$ Original inequality

$4^{3p-1} > 4^{-4}$ Rewrite $\frac{1}{256}$ as $\frac{1}{4^4}$ or 4^{-4} so each side has the same base.

$3p - 1 > -4$ Property of Inequality for Exponential Functions

$3p > -3$ Add 1 to each side.

$p > -1$ Divide each side by 3.

The solution set is $p > -1$.

CHECK Test a value of p greater than -1; for example, $p = 0$.

$4^{3p-1} > \frac{1}{256}$ Original inequality

$4^{3(0)-1} \overset{?}{>} \frac{1}{256}$ Replace p with 0.

$4^{-1} \overset{?}{>} \frac{1}{256}$ Simplify.

$\frac{1}{4} > \frac{1}{256}$ ✓ $a^{-1} = \frac{1}{a}$

Check for Understanding

Concept Check

1. **OPEN ENDED** Give an example of a value of b for which $y = b^x$ represents exponential decay.
2. **Identify** each function as *linear, quadratic,* or *exponential.*

   a. $y = 3x^2$ b. $y = 4(3)^x$ c. $y = 2x + 4$ d. $y = 4(0.2)^x + 1$

Match each function with its graph.

3. $y = 5^x$
4. $y = 2(5)^x$
5. $y = \left(\frac{1}{5}\right)^x$

a.

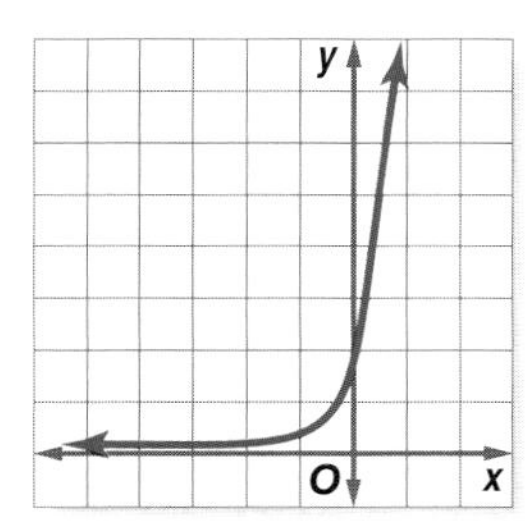

b.

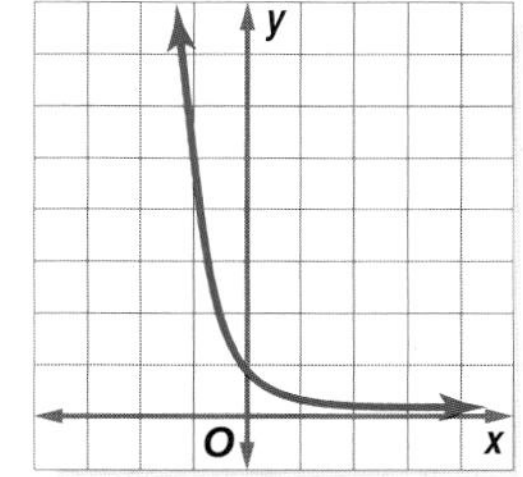

c.

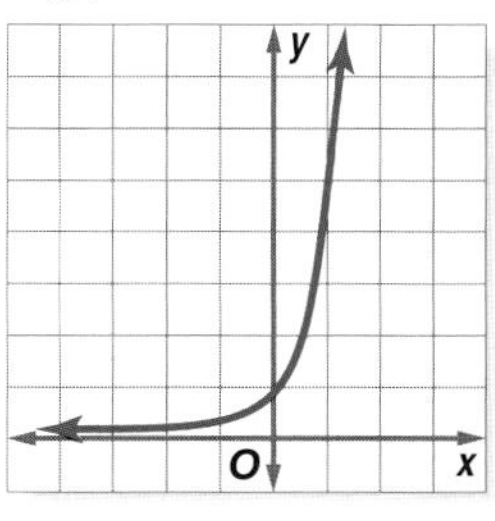

Guided Practice **Sketch the graph of each function. Then state the function's domain and range.**

6. $y = 3(4)^x$
7. $y = 2\left(\frac{1}{3}\right)^x$

Determine whether each function represents exponential *growth* or *decay*.

8. $y = 2(7)^x$ **9.** $y = (0.5)^x$ **10.** $y = 0.3(5)^x$

Write an exponential function whose graph passes through the given points.

11. (0, 3) and (−1, 6) **12.** (0, −18) and (−2, −2)

Simplify each expression.

13. $2^{\sqrt{7}} \cdot 2^{\sqrt{7}}$ **14.** $(a^{\pi})^4$ **15.** $81^{\sqrt{2}} \div 3^{\sqrt{2}}$

Solve each equation or inequality. Check your solution.

16. $2^{n+4} = \frac{1}{32}$ **17.** $5^{2x+3} \leq 125$ **18.** $9^{2y-1} = 27^y$

Application

ANIMAL CONTROL **For Exercises 19 and 20, use the following information.** During the 19th century, rabbits were brought to Australia. Since the rabbits had no natural enemies on that continent, their population increased rapidly. Suppose there were 65,000 rabbits in Australia in 1865 and 2,500,000 in 1867.

19. Write an exponential function that could be used to model the rabbit population y in Australia. Write the function in terms of x, the number of years since 1865.

20. Assume that the rabbit population continued to grow at that rate. Estimate the Australian rabbit population in 1872.

Practice and Apply

Homework Help

For Exercises	See Examples
21–26	1
27–32	2
33–38, 57–66	3
39–44	4
45–56	5, 6

Extra Practice
See page 849.

Sketch the graph of each function. Then state the function's domain and range.

21. $y = 2(3)^x$ **22.** $y = 5(2)^x$ **23.** $y = 0.5(4)^x$

24. $y = 4\left(\frac{1}{3}\right)^x$ **25.** $y = -\left(\frac{1}{5}\right)^x$ **26.** $y = -2.5(5)^x$

Determine whether each function represents exponential *growth* or *decay*.

27. $y = 10(3.5)^x$ **28.** $y = 2(4)^x$ **29.** $y = 0.4\left(\frac{1}{3}\right)^x$

30. $y = 3\left(\frac{5}{2}\right)^x$ **31.** $y = 30^{-x}$ **32.** $y = 0.2(5)^{-x}$

Write an exponential function whose graph passes through the given points.

33. (0, −2) and (−2, −32) **34.** (0, 3) and (1, 15)

35. (0, 7) and (2, 63) **36.** (0, −5) and (−3, −135)

37. (0, 0.2) and (4, 51.2) **38.** (0, −0.3) and (5, −9.6)

Simplify each expression.

39. $\left(5^{\sqrt{2}}\right)^{\sqrt{8}}$ **40.** $\left(x^{\sqrt{5}}\right)^{\sqrt{3}}$ **41.** $7^{\sqrt{2}} \cdot 7^{3\sqrt{2}}$

42. $y^{3\sqrt{3}} \div y^{\sqrt{3}}$ **43.** $n^2 \cdot n^{\pi}$ **44.** $64^{\pi} \div 2^{\pi}$

Solve each equation or inequality. Check your solution.

45. $3^{n-2} = 27$ **46.** $2^{3x+5} = 128$ **47.** $5^{n-3} = \frac{1}{25}$

48. $2^{2n} \leq \frac{1}{16}$ **49.** $\left(\frac{1}{9}\right)^m = 81^{m+4}$ **50.** $\left(\frac{1}{7}\right)^{y-3} = 343$

51. $16^n < 8^{n+1}$ **52.** $10^{x-1} = 100^{2x-3}$ **53.** $36^{2p} = 216^{p-1}$

54. $32^{5p+2} \geq 16^{5p}$ **55.** $3^{5x} \cdot 81^{1-x} = 9^{x-3}$ **56.** $49^x = 7^{x^2-15}$

The magnitude of an earthquake can be represented by an exponential equation. Visit www.algebra2.com/webquest to continue work on your WebQuest project.

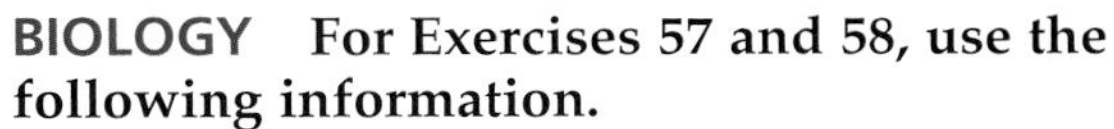

BIOLOGY For Exercises 57 and 58, use the following information.
The number of bacteria in a colony is growing exponentially.

57. Write an exponential function to model the population y of bacteria x hours after 2 P.M.

58. How many bacteria were there at 7 P.M. that day?

POPULATION For Exercises 59–61, use the following information.
Every ten years, the Bureau of the Census counts the number of people living in the United States. In 1790, the population of the U.S. was 3.93 million. By 1800, this number had grown to 5.31 million.

59. Write an exponential function that could be used to model the U.S. population y in millions for 1790 to 1800. Write the equation in terms of x, the number of decades x since 1790.

60. Assume that the U.S. population continued to grow at that rate. Estimate the population for the years 1820, 1840, and 1860. Then compare your estimates with the actual population for those years, which were 9.64, 17.06, and 31.44 million, respectively.

61. RESEARCH Estimate the population of the U.S. in 2000. Then use the Internet or other reference to find the actual population of the U.S. in 2000. Has the population of the U.S. continued to grow at the same rate at which it was growing in the early 1800s? Explain.

More About. . .

Computers

Since computers were invented, computational speed has multiplied by a factor of 4 about every three years.

Source: www.wired.com

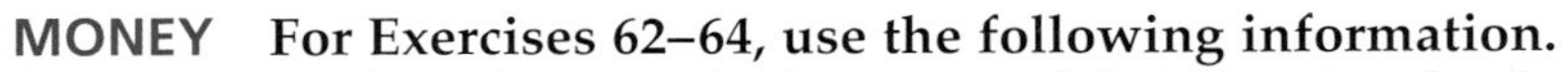

MONEY For Exercises 62–64, use the following information.
Suppose you deposit a principal amount of P dollars in a bank account that pays compound interest. If the annual interest rate is r (expressed as a decimal) and the bank makes interest payments n times every year, the amount of money A you would have after t years is given by $A(t) = P\left(1 + \frac{r}{n}\right)^{nt}$.

62. If the principal, interest rate, and number of interest payments are known, what type of function is $A(t) = P\left(1 + \frac{r}{n}\right)^{nt}$? Explain your reasoning.

63. Write an equation giving the amount of money you would have after t years if you deposit \$1000 into an account paying 4% annual interest compounded quarterly (four times per year).

64. Find the account balance after 20 years.

COMPUTERS For Exercises 65 and 66, use the information at the left.

65. If a typical computer operates with a computational speed s today, write an expression for the speed at which you can expect an equivalent computer to operate after x three-year periods.

66. Suppose your computer operates with a processor speed of 600 megahertz and you want a computer that can operate at 4800 megahertz. If a computer with that speed is currently unavailable for home use, how long can you expect to wait until you can buy such a computer?

67. CRITICAL THINKING Decide whether the following statement is *sometimes*, *always*, or *never* true. Explain your reasoning.

For a positive base b other than 1, $b^x > b^y$ if and only if $x > y$.

68. **WRITING IN MATH** Answer the question that was posed at the beginning of the lesson.

How does an exponential function describe tournament play?

Include the following in your answer:

- an explanation of how you could use the equation $y = 2^x$ to determine the number of rounds of tournament play for 128 teams, and
- an example of an inappropriate number of teams for tournament play with an explanation as to why this number would be inappropriate.

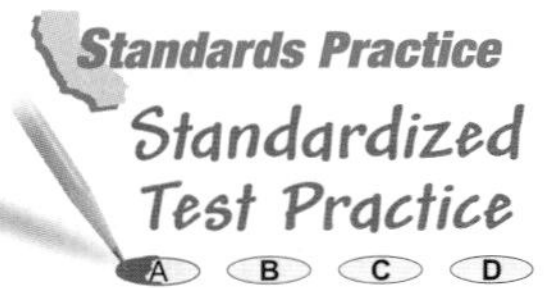

69. If $4^{x+2} = 48$, then $4^x =$

Ⓐ 3.0. Ⓑ 6.4. Ⓒ 6.9. Ⓓ 12.0. Ⓔ 24.0.

70. **GRID IN** Suppose you deposit \$500 in an account paying 4.5% interest compounded semiannually. Find the dollar value of the account rounded to the nearest penny after 10 years.

FAMILIES OF GRAPHS Graph each pair of functions on the same screen. Then compare the graphs, listing both similarities and differences in shape, asymptotes, domain, range, and y-intercepts.

71. $y = 2^x$ and $y = 2^x + 3$

72. $y = 3^x$ and $y = 3^{x+1}$

73. $y = \left(\frac{1}{5}\right)^x$ and $y = \left(\frac{1}{5}\right)^{x-2}$

74. $y = \left(\frac{1}{4}\right)^x$ and $y = \left(\frac{1}{4}\right)^x - 1$

75. Describe the effect of changing the values of h and k in the equation $y = 2^{x-h} + k$.

Maintain Your Skills

Mixed Review

Solve each equation or inequality. Check your solutions. *(Lesson 9-6)*

76. $\frac{15}{p} + p = 16$

77. $\frac{s-3}{s+4} = \frac{6}{s^2 - 16}$

78. $\frac{2a-5}{a-9} + \frac{a}{a+9} = \frac{-6}{a^2 - 81}$

79. $\frac{x-2}{x} < \frac{x-4}{x-6}$

Identify each equation as a type of function. Then graph the equation. *(Lesson 9-5)*

80. $y = \sqrt{x-2}$

81. $y = -2[\![x]\!]$

82. $y = 8$

Find the inverse of each matrix, if it exists. *(Lesson 4-7)*

83. $\begin{bmatrix} 1 & 0 \\ 0 & 1 \end{bmatrix}$

84. $\begin{bmatrix} 2 & 4 \\ 5 & 10 \end{bmatrix}$

85. $\begin{bmatrix} -5 & 6 \\ -11 & 3 \end{bmatrix}$

86. **ENERGY** A circular cell must deliver 18 watts of energy. If each square centimeter of the cell that is in sunlight produces 0.01 watt of energy, how long must the radius of the cell be? *(Lesson 5-8)*

Getting Ready for the Next Lesson

PREREQUISITE SKILL Find $g[h(x)]$ and $h[g(x)]$.

(To review ***composition of functions****, see Lesson 7-7.)*

87. $h(x) = 2x - 1$
$g(x) = x - 5$

88. $h(x) = x + 3$
$g(x) = x^2$

89. $h(x) = 2x + 5$
$g(x) = -x + 3$

10-2 Logarithms and Logarithmic Functions

What You'll Learn

- Evaluate logarithmic expressions.
- Solve logarithmic equations and inequalities.

Why is a logarithmic scale used to measure sound?

Many scientific measurements have such an enormous range of possible values that it makes sense to write them as powers of 10 and simply keep track of their exponents. For example, the loudness of sound is measured in units called *decibels*. The graph shows the relative intensities and decibel measures of common sounds.

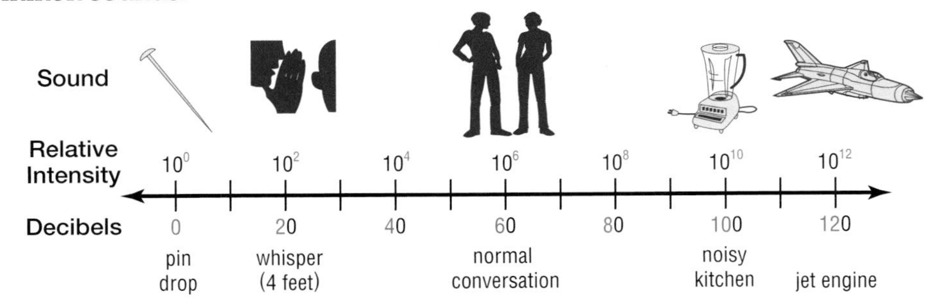

The decibel measure of the loudness of a sound is the exponent or logarithm of its relative intensity multiplied by 10.

Vocabulary

- logarithm
- logarithmic function
- logarithmic equation
- logarithmic inequality

California Standards
Standard 11.1
Students understand the inverse relationship between exponents and logarithms and use this relationship to solve problems involving logarithms and exponents. (Key)

LOGARITHMIC FUNCTIONS AND EXPRESSIONS To better understand what is meant by a logarithm, let's look at the graph of $y = 2^x$ and its inverse. Since exponential functions are one-to-one, the inverse of $y = 2^x$ exists and is also a function. Recall that you can graph the inverse of a function by interchanging the x and y values in the ordered pairs of the function.

Study Tip

Look Back
To review **inverse functions**, see Lesson 7-8.

$y = 2^x$

x	y
-3	$\frac{1}{8}$
-2	$\frac{1}{4}$
-1	$\frac{1}{2}$
0	1
1	2
2	4
3	8

$x = 2^y$

x	y
$\frac{1}{8}$	-3
$\frac{1}{4}$	-2
$\frac{1}{2}$	-1
1	0
2	1
4	2
8	3

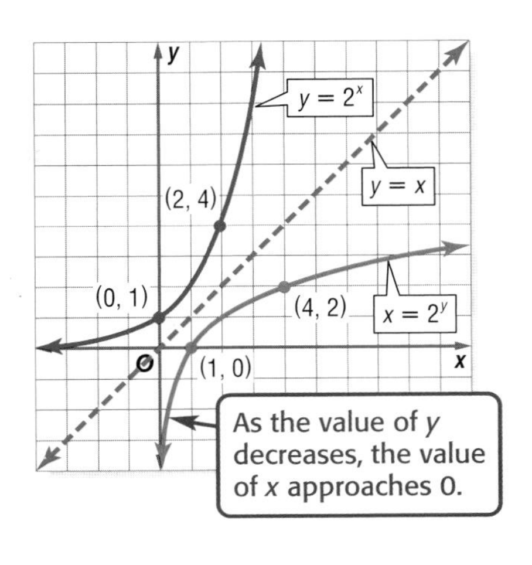

The inverse of $y = 2^x$ can be defined as $x = 2^y$. Notice that the graphs of these two functions are reflections of each other over the line $y = x$.

In general, the inverse of $y = b^x$ is $x = b^y$. In $x = b^y$, y is called the **logarithm** of x. It is usually written as $y = \log_b x$ and is read *y equals log base b of x*.

Key Concept — Logarithm with Base b

- **Words** Let b and x be positive numbers, $b \neq 1$. The *logarithm of x with base b* is denoted $\log_b x$ and is defined as the exponent y that makes the equation $b^y = x$ true.
- **Symbols** Suppose $b > 0$ and $b \neq 1$. For $x > 0$, there is a number y such that $\log_b x = y$ if and only if $b^y = x$.

Example 1 Logarithmic to Exponential Form

Write each equation in exponential form.

a. $\log_8 1 = 0$

$\log_8 1 = 0 \rightarrow 1 = 8^0$

b. $\log_2 \frac{1}{16} = -4$

$\log_2 \frac{1}{16} = -4 \rightarrow \frac{1}{16} = 2^{-4}$

Example 2 Exponential to Logarithmic Form

Write each equation in logarithmic form.

a. $10^3 = 1000$

$10^3 = 1000 \rightarrow \log_{10} 1000 = 3$

b. $9^{\frac{1}{2}} = 3$

$9^{\frac{1}{2}} = 3 \rightarrow \log_9 3 = \frac{1}{2}$

You can use the definition of logarithm to find the value of a logarithmic expression.

Example 3 Evaluate Logarithmic Expressions

Evaluate $\log_2 64$.

$\log_2 64 = y$	Let the logarithm equal y.
$64 = 2^y$	Definition of logarithm
$2^6 = 2^y$	$64 = 2^6$
$6 = y$	Property of Equality for Exponential Functions

So, $\log_2 64 = 6$.

The function $y = \log_b x$, where $b > 0$ and $b \neq 1$, is called a **logarithmic function**. As shown in the graph on the previous page, this function is the inverse of the exponential function $y = b^x$ and has the following characteristics.

1. The function is continuous and one-to-one.
2. The domain is the set of all positive real numbers.
3. The y-axis is an asymptote of the graph.
4. The range is the set of all real numbers.
5. The graph contains the point (1, 0). That is, the x-intercept is 1.

> **Study Tip**
> *Look Back*
> To review **composition of functions**, see Lesson 7-7.

Since the exponential function $f(x) = b^x$ and the logarithmic function $g(x) = \log_b x$ are inverses of each other, their composites are the identity function. That is, $f[g(x)] = x$ and $g[f(x)] = x$.

$$f[g(x)] = x \qquad g[f(x)] = x$$
$$f(\log_b x) = x \qquad g(b^x) = x$$
$$b^{\log_b x} = x \qquad \log_b b^x = x$$

Thus, if their bases are the same, exponential and logarithmic functions "undo" each other. You can use this inverse property of exponents and logarithms to simplify expressions.

Example 4 Inverse Property of Exponents and Logarithms

Evaluate each expression.

a. $\log_6 6^8$

$\log_6 6^8 = 8$ $\quad \log_b b^x = x$

b. $3^{\log_3 (4x - 1)}$

$3^{\log_3 (4x-1)} = 4x - 1$ $\quad b^{\log_b x} = x$

SOLVE LOGARITHMIC EQUATIONS AND INEQUALITIES

A **logarithmic equation** is an equation that contains one or more logarithms. You can use the definition of a logarithm to help you solve logarithmic equations.

Example 5 Solve a Logarithmic Equation

Solve $\log_4 n = \frac{5}{2}$.

$\log_4 n = \frac{5}{2}$	Original equation
$n = 4^{\frac{5}{2}}$	Definition of logarithm
$n = (2^2)^{\frac{5}{2}}$	$4 = 2^2$
$n = 2^5$	Power of a Power
$n = 32$	Simplify.

A **logarithmic inequality** is an inequality that involves logarithms. In the case of inequalities, the following property is helpful.

Key Concept — Logarithmic to Exponential Inequality

- **Symbols** If $b > 1$, $x > 0$, and $\log_b x > y$, then $x > b^y$.
  If $b > 1$, $x > 0$, and $\log_b x < y$, then $0 < x < b^y$.
- **Examples** $\log_2 x > 3$, $x > 2^3$ $\qquad$ $\log_3 x < 5$, $0 < x < 3^5$

Example 6 Solve a Logarithmic Inequality

Solve $\log_5 x < 2$. Check your solution.

$\log_5 x < 2$	Original inequality
$0 < x < 5^2$	Logarithmic to exponential inequality
$0 < x < 25$	Simplify.

The solution set is $\{x \mid 0 < x < 25\}$.

CHECK Try 5 to see if it satisfies the inequality.

$\log_5 x < 2$	Original inequality
$\log_5 5 \overset{?}{<} 2$	Substitute 5 for x.
$1 < 2$ ✓	$\log_5 5 = 1$ because $5^1 = 5$.

Study Tip

Special Values
If $b > 0$ and $b \neq 1$, then the following statements are true.
- $\log_b b = 1$ because $b^1 = b$.
- $\log_b 1 = 0$ because $b^0 = 1$.

Use the following property to solve logarithmic equations that have logarithms with the same base on each side.

Key Concept *Property of Equality for Logarithmic Functions*

- **Symbols** If b is a positive number other than 1, then $\log_b x = \log_b y$ if and only if $x = y$.
- **Example** If $\log_7 x = \log_7 3$, then $x = 3$.

Example 7 Solve Equations with Logarithms on Each Side

Solve $\log_5 (p^2 - 2) = \log_5 p$. Check your solution.

$\log_5 (p^2 - 2) = \log_5 p$	Original equation
$p^2 - 2 = p$	Property of Equality for Logarithmic Functions
$p^2 - p - 2 = 0$	Subtract p from each side.
$(p - 2)(p + 1) = 0$	Factor.
$p - 2 = 0$ or $p + 1 = 0$	Zero Product Property
$p = 2$ $\quad p = -1$	Solve each equation.

CHECK Substitute each value into the original equation.

$\log_5 (2^2 - 2) \stackrel{?}{=} \log_5 2$	Substitute 2 for p.
$\log_5 2 = \log_5 2$ ✓	Simplify.
$\log_5 [(-1)^2 - 2] \stackrel{?}{=} \log_5 (-1)$	Substitute -1 for p.

Since $\log_5 (-1)$ is undefined, -1 is an *extraneous* solution and must be eliminated. Thus, the solution is 2.

Study Tip

Extraneous Solutions
The domain of a logarithmic function does not include negative values. For this reason, be sure to check for extraneous solutions of logarithmic equations.

Use the following property to solve logarithmic inequalities that have the same base on each side. Exclude values from your solution set that would result in taking the logarithm of a number less than or equal to zero in the original inequality.

Key Concept *Property of Inequality for Logarithmic Functions*

- **Symbols** If $b > 1$, then $\log_b x > \log_b y$ if and only if $x > y$, and $\log_b x < \log_b y$ if and only if $x < y$.
- **Example** If $\log_2 x > \log_2 9$, then $x > 9$.

This property also holds for $\leq$ and $\geq$.

Example 8 Solve Inequalities with Logarithms on Each Side

Solve $\log_{10} (3x - 4) < \log_{10} (x + 6)$. Check your solution.

$\log_{10} (3x - 4) < \log_{10} (x + 6)$	Original inequality
$3x - 4 < x + 6$	Property of Inequality for Logarithmic Functions
$2x < 10$	Addition and Subtraction Properties of Inequalities
$x < 5$	Divide each side by 2.

We must exclude from this solution all values of x such that $3x - 4 \leq 0$ or $x + 6 \leq 0$. Thus, the solution set is $x > \frac{4}{3}$, $x > -6$, and $x < 5$. This compound inequality simplifies to $\frac{4}{3} < x < 5$.

Study Tip

Look back
To review **compound inequalities**, see Lesson 1-6.

Check for Understanding

Concept Check

1. **OPEN ENDED** Give an example of an exponential equation and its related logarithmic equation.
2. **Describe** the relationship between $y = 3^x$ and $y = \log_3 x$.
3. **FIND THE ERROR** Paul and Scott are solving $\log_3 x = 9$.

Paul	Scott
$\log_3 x = 9$	$\log_3 x = 9$
$3^x = 9$	$x = 3^9$
$3^x = 3^2$	$x = 19{,}683$
$x = 2$	

Who is correct? Explain your reasoning.

Guided Practice

Write each equation in logarithmic form.

4. $5^4 = 625$
5. $7^{-2} = \frac{1}{49}$

Write each equation in exponential form.

6. $\log_3 81 = 4$
7. $\log_{36} 6 = \frac{1}{2}$

Evaluate each expression.

8. $\log_4 256$
9. $\log_2 \frac{1}{8}$
10. $3^{\log_3 21}$
11. $\log_5 5^{-1}$

Solve each equation or inequality. Check your solutions.

12. $\log_9 x = \frac{3}{2}$
13. $\log_{\frac{1}{10}} x = -3$
14. $\log_3 (2x - 1) \leq 2$
15. $\log_5 (3x - 1) = \log_5 2x^2$
16. $\log_2 (3x - 5) > \log_2 (x + 7)$
17. $\log_b 9 = 2$

Application

SOUND For Exercises 18–20, use the following information.
An equation for loudness L, in decibels, is $L = 10 \log_{10} R$, where R is the relative intensity of the sound.

18. Solve $130 = 10 \log_{10} R$ to find the relative intensity of a fireworks display with a loudness of 130 decibels.
19. Solve $75 = 10 \log_{10} R$ to find the relative intensity of a concert with a loudness of 75 decibels.
20. How many times more intense is the fireworks display than the concert? In other words, find the ratio of their intensities.

Fireworks	130-190
Car racing	100-130
Parades	80-120
Yard work	95-115
Movies	90-110
Concerts	75-110

Note: Sounds listed by range of peak levels.
Source: National Campaign for Hearing Health
By Hilary Wasson and Sam Ward, USA TODAY

Practice and Apply

Homework Help

For Exercises	See Examples
21–26	1
27–32	2
33–46	3
47–62	4–7
63–65	4
68–70	5

Extra Practice
See page 849.

Write each equation in logarithmic form.

21. $8^3 = 512$ **22.** $3^3 = 27$ **23.** $5^{-3} = \frac{1}{125}$

24. $\left(\frac{1}{3}\right)^{-2} = 9$ **25.** $100^{\frac{1}{2}} = 10$ **26.** $2401^{\frac{1}{4}} = 7$

Write each equation in exponential form.

27. $\log_5 125 = 3$ **28.** $\log_{13} 169 = 2$ **29.** $\log_4 \frac{1}{4} = -1$

30. $\log_{100} \frac{1}{10} = -\frac{1}{2}$ **31.** $\log_8 4 = \frac{2}{3}$ **32.** $\log_{\frac{1}{5}} 25 = -2$

Evaluate each expression.

33. $\log_2 16$ **34.** $\log_{12} 144$ **35.** $\log_{16} 4$

36. $\log_9 243$ **37.** $\log_2 \frac{1}{32}$ **38.** $\log_3 \frac{1}{81}$

39. $\log_5 5^7$ **40.** $2^{\log_2 45}$ **41.** $\log_{11} 11^{(n-5)}$

42. $6^{\log_6 (3x+2)}$ **43.** $\log_{10} 0.001$ **44.** $\log_4 16^x$

WORLD RECORDS **For Exercises 45 and 46, use the information given for Exercises 18–20 to find the relative intensity of each sound.** **Source:** *The Guinness Book of Records*

45. The loudest animal sounds are the low-frequency pulses made by blue whales when they communicate. These pulses have been measured up to 188 decibels.

46. The loudest insect is the African cicada. It produces a calling song that measures 106.7 decibels at a distance of 50 centimeters.

Solve each equation or inequality. Check your solutions.

47. $\log_9 x = 2$ **48.** $\log_2 c > 8$

49. $\log_{64} y \leq \frac{1}{2}$ **50.** $\log_{25} n = \frac{3}{2}$

51. $\log_{\frac{1}{7}} x = -1$ **52.** $\log_{\frac{1}{3}} p < 0$

53. $\log_2 (3x - 8) \geq 6$ **54.** $\log_{10} (x^2 + 1) = 1$

55. $\log_b 64 = 3$ **56.** $\log_b 121 = 2$

57. $\log_5 5^{6n+1} = 13$ **58.** $\log_5 x = \frac{1}{2}$

59. $\log_6 (2x - 3) = \log_6 (x + 2)$ **60.** $\log_2 (4y - 10) \geq \log_2 (y - 1)$

61. $\log_{10} (a^2 - 6) > \log_{10} a$ **62.** $\log_7 (x^2 + 36) = \log_7 100$

Show that each statement is true.

63. $\log_5 25 = 2 \log_5 5$ **64.** $\log_{16} 2 \cdot \log_2 16 = 1$ **65.** $\log_7 [\log_3 (\log_2 8)] = 0$

66. **a.** Sketch the graphs of $y = \log_{\frac{1}{2}} x$ and $y = \left(\frac{1}{2}\right)^x$ on the same axes.

b. Describe the relationship between the graphs.

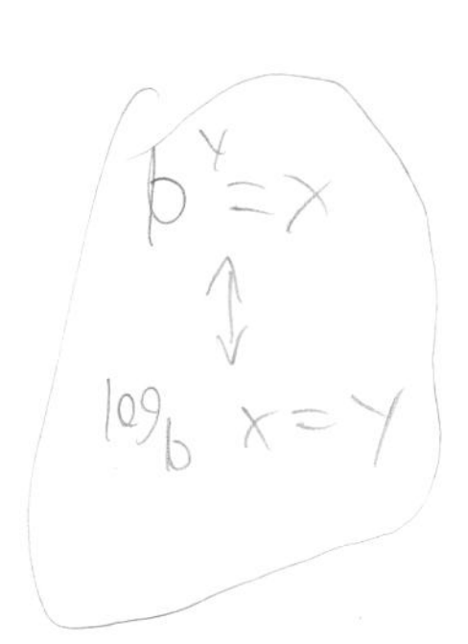

67. **a.** Sketch the graphs of $y = \log_2 x + 3$, $y = \log_2 x - 4$, $y = \log_2 (x - 1)$, and $y = \log_2 (x + 2)$.

b. Describe this family of graphs in terms of its parent graph $y = \log_2 x$.

EARTHQUAKE **For Exercises 68 and 69, use the following information.**
The magnitude of an earthquake is measured on a logarithmic scale called the Richter scale. The magnitude M is given by $M = \log_{10} x$, where x represents the amplitude of the seismic wave causing ground motion.

68. How many times as great is the amplitude caused by an earthquake with a Richter scale rating of 7 as an aftershock with a Richter scale rating of 4?

69. How many times as great was the motion caused by the 1906 San Francisco earthquake that measured 8.3 on the Richter scale as that caused by the 2001 Bhuj, India, earthquake that measured 6.9?

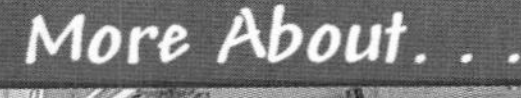

Earthquake

The Loma Prieta earthquake measured 7.1 on the Richter scale and interrupted the 1989 World Series in San Francisco.

Source: U.S. Geological Survey

70. **NOISE ORDINANCE** A proposed city ordinance will make it illegal to create sound in a residential area that exceeds 72 decibels during the day and 55 decibels during the night. How many times more intense is the noise level allowed during the day than at night?

71. **CRITICAL THINKING** The value of $\log_2 5$ is between two consecutive integers. Name these integers and explain how you determined them.

72. **CRITICAL THINKING** Using the definition of a logarithmic function where $y = \log_b x$, explain why the base b cannot equal 1.

73. **WRITING IN MATH** Answer the question that was posed at the beginning of the lesson.

Why is a logarithmic scale used to measure sound?

Include the following in your answer:

- the relative intensities of a pin drop, a whisper, normal conversation, kitchen noise, and a jet engine written in scientific notation,
- a plot of each of these relative intensities on the scale shown below, and

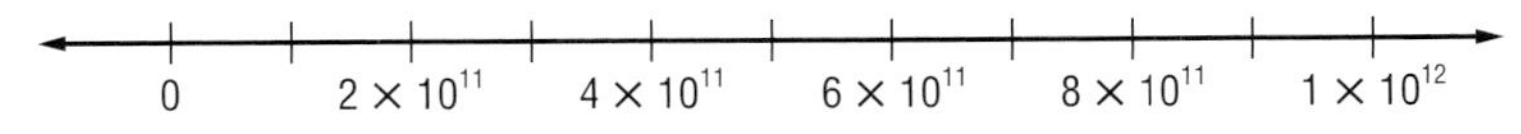

- an explanation as to why the logarithmic scale might be preferred over the scale shown above.

74. What is the equation of the function graphed at the right?

Ⓐ $y = 2(3)^x$

Ⓑ $y = 2\left(\frac{1}{3}\right)^x$

Ⓒ $y = 3\left(\frac{1}{2}\right)^x$

Ⓓ $y = 3(2)^x$

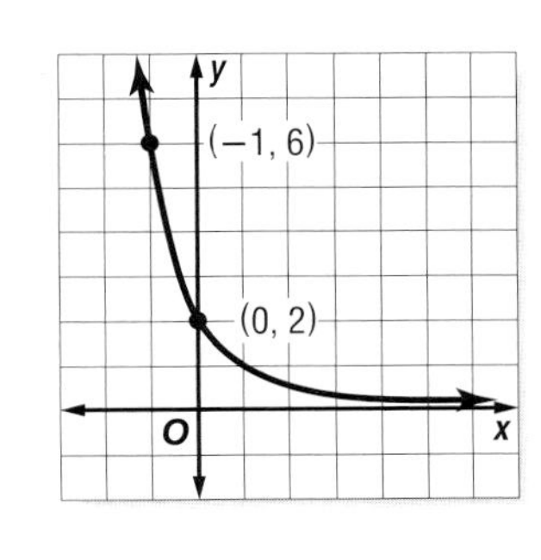

75. In the figure at the right, if $y = \frac{2}{7}x$ and $z = 3w$, then $x =$

Ⓐ 14. Ⓑ 20.

Ⓒ 28. Ⓓ 35.

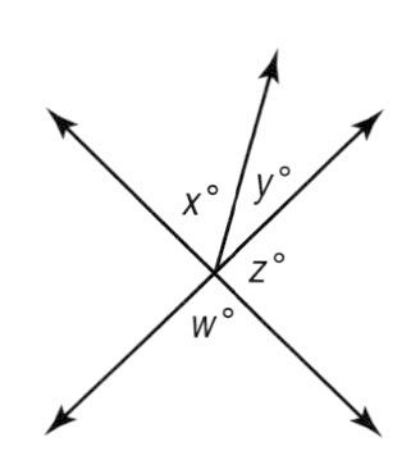

Maintain Your Skills

Mixed Review **Simplify each expression.** *(Lesson 10-1)*

76. $x^{\sqrt{6}} \cdot x^{\sqrt{6}}$

77. $\left(b^{\sqrt{6}}\right)^{\sqrt{24}}$

Solve each equation. Check your solutions. *(Lesson 9-6)*

78. $\frac{2x+1}{x} - \frac{x+1}{x-4} = \frac{-20}{x^2-4x}$

79. $\frac{2a-5}{a-9} - \frac{a-3}{3a+2} = \frac{5}{3a^2-25a-18}$

Solve each equation by using the method of your choice. Find exact solutions. *(Lesson 6-5)*

80. $9y^2 = 49$

81. $2p^2 = 5p + 6$

Simplify each expression. *(Lesson 9-2)*

82. $\frac{3}{2y} + \frac{4}{3y} - \frac{7}{5y}$

83. $\frac{x-7}{x^2-9} - \frac{x-3}{x^2+10x+21}$

84. **BANKING** Donna Bowers has $4000 she wants to save in the bank. A certificate of deposit (CD) earns 8% annual interest, while a regular savings account earns 3% annual interest. Ms. Bowers doesn't want to tie up all her money in a CD, but she has decided she wants to earn $240 in interest for the year. How much money should she put in to each type of account? (*Hint:* Use Cramer's Rule.) *(Lesson 4-4)*

Getting Ready for the Next Lesson **PREREQUISITE SKILL** **Simplify. Assume that no variable equals zero.**
*(To review **multiplying and dividing monomials**, see Lesson 5-1.)*

85. $x^4 \cdot x^6$

86. $(y^3)^8$

87. $(2a^2b)^3$

88. $\frac{a^4n^7}{a^3n}$

89. $\frac{x^5yz^2}{x^2y^3z^5}$

90. $\left(\frac{b^7}{a^4}\right)^0$

Practice Quiz 1

Lessons 10-1 and 10-2

1. Determine whether $5(1.2)^x$ represents exponential *growth* or *decay*. *(Lesson 10-1)*

2. Write an exponential function whose graph passes through (0, 2) and (2, 32).

3. Write an equivalent logarithmic equation for $4^6 = 4096$. *(Lesson 10-2)*

4. Write an equivalent exponential equation for $\log_9 27 = \frac{3}{2}$. *(Lesson 10-2)*

Evaluate each expression. *(Lesson 10-2)*

5. $\log_8 16$

6. $\log_4 4^{15}$

Solve each equation or inequality. Check your solutions. *(Lessons 10-1 and 10-2)*

7. $3^{4x} = 3^{3-x}$

8. $3^{2n} \leq \frac{1}{9}$

9. $\log_2 (x + 6) > 5$

10. $\log_5 (4x - 1) = \log_5 (3x + 2)$

Graphing Calculator Investigation

A Follow-Up of Lesson 10-2

Modeling Real-World Data: Curve Fitting

We are often confronted with data for which we need to find an equation that best fits the information. We can find exponential and logarithmic functions of best fit using a TI-83 Plus graphing calculator.

Example

The population per square mile in the United States has changed dramatically over a period of years. The table shows the number of people per square mile for several years.

U.S. Population Density			
Year	People per square mile	Year	People per square mile
1790	4.5	1900	21.5
1800	6.1	1910	26.0
1810	4.3	1920	29.9
1820	5.5	1930	34.7
1830	7.4	1940	37.2
1840	9.8	1950	42.6
1850	7.9	1960	50.6
1860	10.6	1970	57.5
1870	10.9	1980	64.0
1880	14.2	1990	70.3
1890	17.8	2000	80.0

Source: Northeast-Midwest Institute

a. Use a graphing calculator to enter the data and draw a scatter plot that shows how the number of people per square mile is related to the year.

Step 1 Enter the year into L1 and the people per square mile into L2.

KEYSTROKES: *See pages 87 and 88 to review how to enter lists.*

Be sure to clear the Y= list. Use the [▶] key to move the cursor from L1 to L2.

Step 2 Draw the scatter plot.

KEYSTROKES: *See pages 87 and 88 to review how to graph a scatter plot.*

Make sure that **Plot 1** is on, the scatter plot is chosen, **Xlist** is **L1**, and **Ylist** is **L2**. Use the viewing window [1780, 2020] with a scale factor of 10 by [0, 115] with a scale factor of 5.

We see from the graph that the equation that best fits the data is a curve. Based on the shape of the curve, try an exponential model.

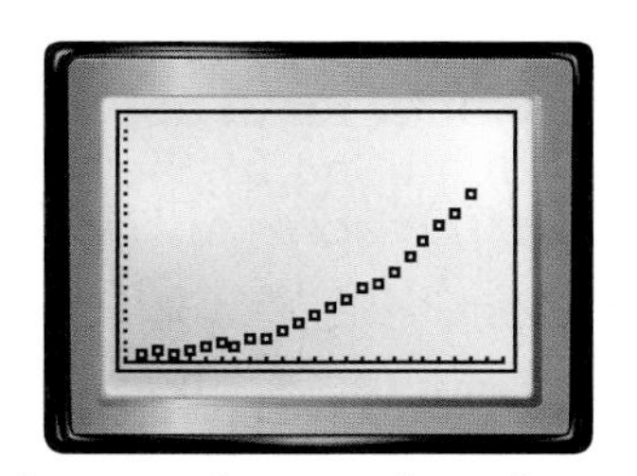

[1780, 2020] scl: 10 by [0, 115] scl: 5

Step 3 To determine the exponential equation that best fits the data, use the exponential regression feature of the calculator.

KEYSTROKES: [STAT] [▶] 0 [2nd] [L1] [,] [2nd] [L2] [ENTER]

The equation is $y = 1.835122 \times 10^{-11}(1.014700091)^x$.

(continued on the next page)

The calculator also reports an r value of 0.991887235. Recall that this number is a correlation coefficient that indicates how well the equation fits the data. A perfect fit would be $r = 1$. Therefore, we can conclude that this equation is a pretty good fit for the data.

To check this equation visually, overlap the graph of the equation with the scatter plot.

KEYSTROKES: [Y=] [VARS] 5 [▶] [▶] 1 [GRAPH]

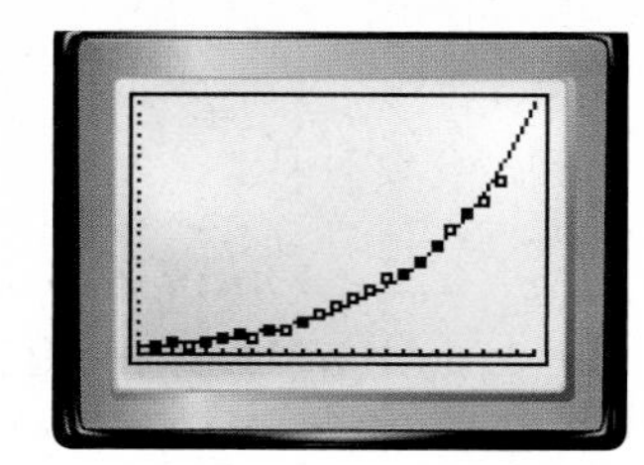

[1780, 2020] scl: 10 by [0, 115] scl: 5

The *residual* is the difference between actual and predicted data. The predicted population per square mile in 2000 using this model was 86.9 (To calculate, press [2nd] [CALC] 1 2000 [ENTER].) So the residual for 2000 was 80.0 − 86.9 or −6.9.

b. If this trend continues, what will be the population per square mile in 2010?

To determine the population per square mile in 2010, from the graphics screen, find the value of y when $x = 2010$.

KEYSTROKES: [2nd] [CALC] 1 2010 [ENTER]

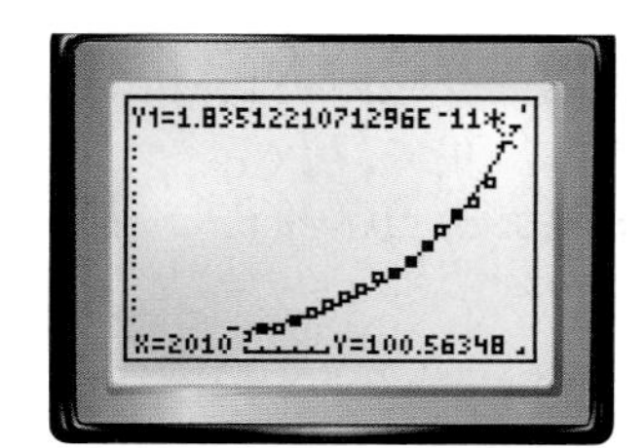

[1780, 2020] scl: 10 by [0, 115] scl: 5

The calculator returns a value of approximately 100.6. If this trend continues, in 2010, there will be approximately 100.6 people per square mile.

Exercises

In 1985, Erika received $30 from her aunt and uncle for her seventh birthday. Her father deposited it into a bank account for her. Both Erika and her father forgot about the money and made no further deposits or withdrawals. The table shows the account balance for several years.

Elapsed Time (years)	Balance
0	$30.00
5	$41.10
10	$56.31
15	$77.16
20	$105.71
25	$144.83
30	$198.43

1. Use a graphing calculator to draw a scatter plot for the data.

2. Calculate and graph the curve of best fit that shows how the elapsed time is related to the balance. Use ExpReg for this exercise.

3. Write the equation of best fit.

4. Write a sentence that describes the fit of the graph to the data.

5. Based on the graph, estimate the balance in 41 years. Check this using the CALC value.

6. Do you think there are any other types of equations that would be good models for these data? Why or why not?

10-3 Properties of Logarithms

What You'll Learn

- Simplify and evaluate expressions using the properties of logarithms.
- Solve logarithmic equations using the properties of logarithms.

California Standards
Standard 11.0 Students prove simple laws of logarithms. (Key)

Standard 14.0 Students understand and use the properties of logarithms to simplify logarithmic numeric expressions and to identify their approximate values.

How are the properties of exponents and logarithms related?

In Lesson 5-1, you learned that the product of powers is the sum of their exponents.

$$9 \cdot 81 = 3^2 \cdot 3^4 \text{ or } 3^{2+4}$$

In Lesson 10-2, you learned that logarithms *are* exponents, so you might expect that a similar property applies to logarithms. Let's consider a specific case. Does $\log_3 (9 \cdot 81) = \log_3 9 + \log_3 81$?

$\log_3 (9 \cdot 81) = \log_3 (3^2 \cdot 3^4)$ Replace 9 with 3^2 and 81 with 3^4.

$= \log_3 3^{(2+4)}$ Product of Powers

$= 2 + 4 \text{ or } 6$ Inverse property of exponents and logarithms

$\log_3 9 + \log_3 81 = \log_3 3^2 + \log_3 3^4$ Replace 9 with 3^2 and 81 with 3^4.

$= 2 + 4 \text{ or } 6$ Inverse property of exponents and logarithms

So, $\log_3 (9 \cdot 81) = \log_3 9 + \log_3 81$.

$\log_5 x^2 = \log_5 9$

$x^2 = 9$

$x = 3$

PROPERTIES OF LOGARITHMS Since logarithms are exponents, the properties of logarithms can be derived from the properties of exponents. The example above and other similar examples suggest the following property of logarithms.

Key Concept — Product Property of Logarithms

- **Words** The logarithm of a product is the sum of the logarithms of its factors.
- **Symbols** For all positive numbers m, n, and b, where $b \neq 1$, $\log_b mn = \log_b m + \log_b n$.
- **Example** $\log_3 (4)(7) = \log_3 4 + \log_3 7$

To show that this property is true, let $b^x = m$ and $b^y = n$. Then, using the definition of logarithm, $x = \log_b m$ and $y = \log_b n$.

$b^x b^y = mn$

$b^{x+y} = mn$ Product of Powers

$\log_b b^{x+y} = \log_b mn$ Property of Equality for Logarithmic Functions

$x + y = \log_b mn$ Inverse Property of Exponents and Logarithms

$\log_b m + \log_b n = \log_b mn$ Replace x with $\log_b m$ and y with $\log_b n$.

You can use the Product Property of Logarithms to approximate logarithmic expressions.

Example 1 Use the Product Property

Use $\log_2 3 \approx 1.5850$ to approximate the value of $\log_2 48$.

$\log_2 48 = \log_2 (2^4 \cdot 3)$ Replace 48 with $16 \cdot 3$ or $2^4 \cdot 3$.

$= \log_2 2^4 + \log_2 3$ Product Property

$= 4 + \log_2 3$ Inverse Property of Exponents and Logarithms

$\approx 4 + 1.5850$ or 5.5850 Replace $\log_2 3$ with 1.5850.

Thus, $\log_2 48$ is approximately 5.5850.

Recall that the quotient of powers is found by subtracting exponents. The property for the logarithm of a quotient is similar.

Key Concept — Quotient Property of Logarithms

- **Words** The logarithm of a quotient is the difference of the logarithms of the numerator and the denominator.
- **Symbols** For all positive numbers m, n, and b, where $b \neq 1$, $\log_b \frac{m}{n} = \log_b m - \log_b n$.

You will show that this property is true in Exercise 47.

Career Choices

Sound Technician

Sound technicians produce movie sound tracks in motion picture production studios, control the sound of live events such as concerts, or record music in a recording studio.

Online Research For information about a career as a sound technician, visit: www.algebra2.com/careers

Example 2 Use the Quotient Property

Use $\log_3 5 \approx 1.4650$ and $\log_3 20 \approx 2.7268$ to approximate $\log_3 4$.

$\log_3 4 = \log_3 \frac{20}{5}$ Replace 4 with the quotient $\frac{20}{5}$.

$= \log_3 20 - \log_3 5$ Quotient Property

$\approx 2.7268 - 1.4650$ or 1.2618 $\log_3 20 = 2.7268$ and $\log_3 5 = 1.4650$

Thus, $\log_3 4$ is approximately 1.2618.

CHECK Using the definition of logarithm and a calculator, $3^{1.2618} \approx 4$. ✓

Example 3 Use Properties of Logarithms

SOUND The loudness L of a sound in decibels is given by $L = 10 \log_{10} R$, where R is the sound's relative intensity. Suppose one person talks with a relative intensity of 10^6 or 60 decibels. Would the sound of ten people each talking at that same intensity be ten times as loud or 600 decibels? Explain your reasoning.

Let L_1 be the loudness of one person talking. → $L_1 = 10 \log_{10} 10^6$

Let L_2 be the loudness of ten people talking. → $L_2 = 10 \log_{10} (10 \cdot 10^6)$

Then the increase in loudness is $L_2 - L_1$.

$L_2 - L_1 = 10 \log_{10} (10 \cdot 10^6) - 10 \log_{10} 10^6$ Substitute for L_1 and L_2.

$= 10(\log_{10} 10 + \log_{10} 10^6) - 10 \log_{10} 10^6$ Product Property

$= 10 \log_{10} 10 + 10 \log_{10} 10^6 - 10 \log_{10} 10^6$ Distributive Property

$= 10 \log_{10} 10$ Subtract.

$= 10(1)$ or 10 Inverse Property of Exponents and Logarithms

The sound of ten people talking is perceived by the human ear to be only about 10 decibels louder than the sound of one person talking, or 70 decibels.

Recall that the power of a power is found by multiplying exponents. The property for the logarithm of a power is similar.

Key Concept — *Power Property of Logarithms*

- **Words** The logarithm of a power is the product of the logarithm and the exponent.
- **Symbols** For any real number p and positive numbers m and b, where $b \neq 1$, $\log_b m^p = p \log_b m$.

You will show that this property is true in Exercise 50.

Example 4 Power Property of Logarithms

Given $\log_4 6 \approx 1.2925$, approximate the value of $\log_4 36$.

$\log_4 36 = \log_4 6^2$	Replace 36 with 6^2.
$= 2 \log_4 6$	Power Property
$\approx 2(1.2925)$ or 2.585	Replace $\log_4 6$ with 1.2925.

SOLVE LOGARITHMIC EQUATIONS You can use the properties of logarithms to solve equations involving logarithms.

Example 5 Solve Equations Using Properties of Logarithms

Solve each equation.

a. $3 \log_5 x - \log_5 4 = \log_5 16$

$3 \log_5 x - \log_5 4 = \log_5 16$	Original equation
$\log_5 x^3 - \log_5 4 = \log_5 16$	Power Property
$\log_5 \frac{x^3}{4} = \log_5 16$	Quotient Property
$\frac{x^3}{4} = 16$	Property of Equality for Logarithmic Functions
$x^3 = 64$	Multiply each side by 4.
$x = 4$	Take the cube root of each side.

The solution is 4.

b. $\log_4 x + \log_4 (x - 6) = 2$

$\log_4 x + \log_4 (x - 6) = 2$	Original equation
$\log_4 x(x - 6) = 2$	Product Property
$x(x - 6) = 4^2$	Definition of logarithm
$x^2 - 6x - 16 = 0$	Subtract 16 from each side.
$(x - 8)(x + 2) = 0$	Factor.
$x - 8 = 0$ or $x + 2 = 0$	Zero Product Property
$x = 8$ $\quad$ $x = -2$	Solve each equation.

CHECK Substitute each value into the original equation.

$\log_4 8 + \log_4 (8 - 6) \stackrel{?}{=} 2$

$\log_4 8 + \log_4 2 \stackrel{?}{=} 2$

$\log_4 (8 \cdot 2) \stackrel{?}{=} 2$

$\log_4 16 \stackrel{?}{=} 2$

$2 = 2 \checkmark$

$\log_4 (-2) + \log_4 (-2 - 6) \stackrel{?}{=} 2$

$\log_4 (-2) + \log_4 (-8) \stackrel{?}{=} 2$

Since $\log_4 (-2)$ and $\log_4 (-8)$ are undefined, -2 is an extraneous solution and must be eliminated.

The only solution is 8.

Study Tip

Checking Solutions
It is wise to check all solutions to see if they are valid since the domain of a logarithmic function is not the complete set of real numbers.

Check for Understanding

Concept Check

1. **Name** the properties that are used to derive the properties of logarithms.
2. **OPEN ENDED** Write an expression that can be simplified by using two or more properties of logarithms. Then simplify it.
3. **FIND THE ERROR** Umeko and Clemente are simplifying $\log_7 6 + \log_7 3 - \log_7 2$.

Umeko	Clemente
$\log_7 6 + \log_7 3 - \log_7 2$	$\log_7 6 + \log_7 3 - \log_7 2$
$= \log_7 18 - \log_7 2$	$= \log_7 9 - \log_7 2$
$= \log_7 9$	$= \log_7 7$ or 1

Who is correct? Explain your reasoning.

Guided Practice

Use $\log_3 2 \approx 0.6310$ and $\log_3 7 \approx 1.7712$ to approximate the value of each expression.

4. $\log_3 \frac{7}{2}$
5. $\log_3 18$
6. $\log_3 \frac{2}{3}$

Solve each equation. Check your solutions.

7. $\log_3 42 - \log_3 n = \log_3 7$
8. $\log_2 3x + \log_2 5 = \log_2 30$
9. $2 \log_5 x = \log_5 9$
10. $\log_{10} a + \log_{10} (a + 21) = 2$

Application

MEDICINE **For Exercises 11 and 12, use the following information.**
The pH of a person's blood is given by $\text{pH} = 6.1 + \log_{10} B - \log_{10} C$, where B is the concentration of bicarbonate, which is a base, in the blood and C is the concentration of carbonic acid in the blood.

11. Use the Quotient Property of Logarithms to simplify the formula for blood pH.
12. Most people have a blood pH of 7.4. What is the approximate ratio of bicarbonate to carbonic acid for blood with this pH?

Practice and Apply

Homework Help

For Exercises	See Examples
13–20	1, 2, 4
21–34	5
37–45	3

Extra Practice
See page 850.

Use $\log_5 2 \approx 0.4307$ and $\log_5 3 \approx 0.6826$ to approximate the value of each expression.

13. $\log_5 9$
14. $\log_5 8$
15. $\log_5 \frac{2}{3}$
16. $\log_5 \frac{3}{2}$
17. $\log_5 50$
18. $\log_5 30$
19. $\log_5 0.5$
20. $\log_5 \frac{10}{9}$

Solve each equation. Check your solutions.

21. $\log_3 5 + \log_3 x = \log_3 10$
22. $\log_4 a + \log_4 9 = \log_4 27$
23. $\log_{10} 16 - \log_{10} 2t = \log_{10} 2$
24. $\log_7 24 - \log_7 (y + 5) = \log_7 8$
25. $\log_2 n = \frac{1}{4} \log_2 16 + \frac{1}{2} \log_2 49$
26. $2 \log_{10} 6 - \frac{1}{3} \log_{10} 27 = \log_{10} x$
27. $\log_{10} z + \log_{10} (z + 3) = 1$
28. $\log_6 (a^2 + 2) + \log_6 2 = 2$
29. $\log_2 (12b - 21) - \log_2 (b^2 - 3) = 2$
30. $\log_2 (y + 2) - \log_2 (y - 2) = 1$
31. $\log_3 0.1 + 2 \log_3 x = \log_3 2 + \log_3 5$
32. $\log_5 64 - \log_5 \frac{8}{3} + \log_5 2 = \log_5 4p$

Solve for n.

33. $\log_a 4n - 2\log_a x = \log_a x$

34. $\log_b 8 + 3\log_b n = 3\log_b (x - 1)$

CRITICAL THINKING Tell whether each statement is *true* or *false*. If true, show that it is true. If false, give a counterexample.

35. For all positive numbers m, n, and b, where $b \neq 1$, $\log_b (m + n) = \log_b m + \log_b n$.

36. For all positive numbers m, n, x, and b, where $b \neq 1$, $n \log_b x + m \log_b x = (n + m) \log_b x$.

37. **EARTHQUAKES** The great Alaskan earthquake in 1964 was about 100 times more intense than the Loma Prieta earthquake in San Francisco in 1989. Find the difference in the Richter scale magnitudes of the earthquakes.

BIOLOGY For Exercises 38–40, use the following information.

The energy E (in kilocalories per gram molecule) needed to transport a substance from the outside to the inside of a living cell is given by $E = 1.4(\log_{10} C_2 - \log_{10} C_1)$, where C_1 is the concentration of the substance outside the cell and C_2 is the concentration inside the cell.

38. Express the value of E as one logarithm.

39. Suppose the concentration of a substance inside the cell is twice the concentration outside the cell. How much energy is needed to transport the substance on the outside of the cell to the inside? (Use $\log_{10} 2 \approx 0.3010$.)

40. Suppose the concentration of a substance inside the cell is four times the concentration outside the cell. How much energy is needed to transport the substance from the outside of the cell to the inside?

More About. . .

Star Light

The Greek astronomer Hipparchus made the first known catalog of stars. He listed the brightness of each star on a scale of 1 to 6, the brightest being 1. With no telescope, he could only see stars as dim as the 6th magnitude.

Source: NASA

SOUND For Exercises 41–43, use the formula for the loudness of sound in Example 3 on page 542. Use $\log_{10} 2 \approx 0.3010$ and $\log_{10} 3 \approx 0.47712$.

41. A certain sound has a relative intensity of R. By how many decibels does the sound increase when the intensity is doubled?

42. A certain sound has a relative intensity of R. By how many decibels does the sound decrease when the intensity is halved?

43. A stadium containing 10,000 cheering people can produce a crowd noise of about 90 decibels. If every one cheers with the same relative intensity, how much noise, in decibels, is a crowd of 30,000 people capable of producing? Explain your reasoning.

STAR LIGHT For Exercises 44–46, use the following information.

The brightness, or apparent magnitude, m of a star or planet is given by the formula $m = 6 - 2.5 \log_{10} \frac{L}{L_0}$, where L is the amount of light coming to Earth from the star or planet and L_0 is the amount of light from a sixth magnitude star.

44. Find the difference in the magnitudes of Sirius and the crescent moon.

45. Find the difference in the magnitudes of Saturn and Neptune.

46. **RESEARCH** Use the Internet or other reference to find the magnitude of the dimmest stars that we can now see with ground-based telescopes.

47. CRITICAL THINKING Use the properties of exponents to prove the Quotient Property of Logarithms.

48. WRITING IN MATH Answer the question that was posed at the beginning of the lesson.

How are the properties of exponents and logarithms related?

Include the following in your answer:

- examples like the one shown at the beginning of the lesson illustrating the Quotient Property and Power Property of Logarithms, and
- an explanation of the similarity between one property of exponents and its related property of logarithms.

49. Simplify $2 \log_5 12 - \log_5 8 - 2 \log_5 3$.

(A) $\log_5 2$ (B) $\log_5 3$ (C) $\log_5 0.5$ (D) 1

50. SHORT RESPONSE Show that $\log_b m^p = p \log_b m$ for any real number p and positive number m and b, where $b \neq 1$.

Maintain Your Skills

Mixed Review

Evaluate each expression. *(Lesson 10-2)*

51. $\log_3 81$ **52.** $\log_9 \frac{1}{729}$ **53.** $\log_7 7^{2x}$

Solve each equation or inequality. Check your solutions. *(Lesson 10-1)*

54. $3^{5n+3} = 3^{33}$ **55.** $7^a = 49^{-4}$ **56.** $3^{d+4} > 9^d$

Determine whether each graph represents an odd-degree polynomial function or an even-degree polynomial function. Then state how many real zeros each function has. *(Lesson 7-1)*

57.

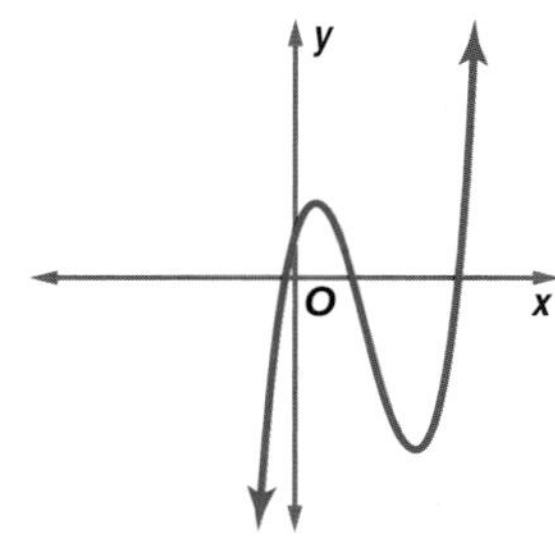

58.

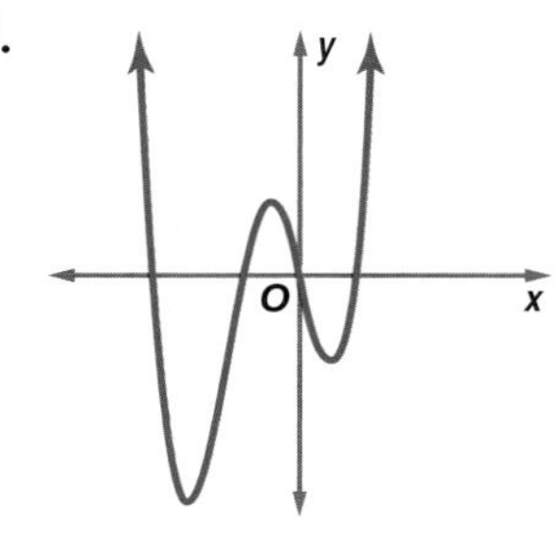

Simplify each expression. *(Lesson 9-1)*

59. $\frac{39a^3b^4}{13a^4b^3}$ **60.** $\frac{k+3}{5kl} \cdot \frac{10kl}{k+3}$ **61.** $\frac{5y - 15z}{42x^2} \div \frac{y - 3z}{14x}$

62. PHYSICS If a stone is dropped from a cliff, the equation $t = \frac{1}{4}\sqrt{d}$ represents the time t in seconds that it takes for the stone to reach the ground. If d represents the distance in feet that the stone falls, find how long it would take for a stone to fall from a 150-foot cliff. *(Lesson 5-6)*

Getting Ready for the Next Lesson

PREREQUISITE SKILL **Solve each equation or inequality. Check your solutions.**

(To review ***solving logarithmic equations and inequalities****, see Lesson 10-2.)*

63. $\log_3 x = \log_3 (2x - 1)$ **64.** $\log_{10} 2^x = \log_{10} 32$

65. $\log_2 3x > \log_2 5$ **66.** $\log_5 (4x + 3) < \log_5 11$

10-4 Common Logarithms

California Standards Standard 13.0 Students use the definition of logarithms to translate between logarithms in any base.

What You'll Learn

- Solve exponential equations and inequalities using common logarithms.
- Evaluate logarithmic expressions using the Change of Base Formula.

Vocabulary

- common logarithm
- Change of Base Formula

Why is a logarithmic scale used to measure acidity?

The pH level of a substance measures its acidity. A low pH indicates an acid solution while a high pH indicates a basic solution. The pH levels of some common substances are shown.

The pH level of a substance is given by $pH = -\log_{10}[H+]$, where $H+$ is the substance's hydrogen ion concentration in moles per liter. Another way of writing this formula is $pH = -\log[H+]$.

Acidity of Common Substances

Substance	pH Level
Battery acid	1.0
Sauerkraut	3.5
Tomatoes	4.2
Black Coffee	5.0
Milk	6.4
Distilled Water	7.0
Eggs	7.8
Milk of magnesia	10.0

COMMON LOGARITHMS You have seen that the base 10 logarithm function, $y = \log_{10} x$, is used in many applications. Base 10 logarithms are called **common logarithms**. Common logarithms are usually written without the subscript 10.

$$\log_{10} x = \log x, x > 0$$

Most calculators have a LOG key for evaluating common logarithms.

Study Tip

Technology
Nongraphing scientific calculators often require entering the number followed by the function, for example, 3 LOG.

Example 1 Find Common Logarithms

Use a calculator to evaluate each expression to four decimal places.

a. log 3 KEYSTROKES: LOG 3 ENTER .4771212547 about 0.4771

b. log 0.2 KEYSTROKES: LOG 0.2 ENTER -.6989700043 about −0.6990

Sometimes an application of logarithms requires that you use the inverse of logarithms, or exponentiation.

$$10^{\log x} = x$$

Example 2 Solve Logarithmic Equations Using Exponentiation

EARTHQUAKES The amount of energy E, in ergs, that an earthquake releases is related to its Richter scale magnitude M by the equation $\log E = 11.8 + 1.5M$. The Chilean earthquake of 1960 measured 8.5 on the Richter scale. How much energy was released?

$\log E = 11.8 + 1.5M$	Write the formula.
$\log E = 11.8 + 1.5(8.5)$	Replace M with 8.5.
$\log E = 24.55$	Simplify.
$10^{\log E} = 10^{24.55}$	Write each side using exponents and base 10.
$E = 10^{24.55}$	Inverse Property of Exponents and Logarithms
$E \approx 3.55 \times 10^{24}$	Use a calculator.

The amount of energy released by this earthquake was about 3.55×10^{24} ergs.

Example 3 Solve Exponential Equations Using Logarithms

Solve $3^x = 11$.

$3^x = 11$	Original equation
$\log 3^x = \log 11$	Property of Equality for Logarithmic Functions
$x \log 3 = \log 11$	Power Property of Logarithms
$x = \frac{\log 11}{\log 3}$	Divide each side by log 3.
$x \approx \frac{1.0414}{0.4771}$	Use a calculator.
$x \approx 2.1828$	The solution is approximately 2.1828.

CHECK You can check this answer using a calculator or by using estimation. Since $3^2 = 9$ and $3^3 = 27$, the value of x is between 2 and 3. In addition, the value of x should be closer to 2 than 3, since 11 is closer to 9 than 27. Thus, 2.1828 is a reasonable solution. ✓

Study Tip

Using Logarithms
When you use the Property for Logarithmic Functions as in the second step of Example 3, this is sometimes referred to as *taking the logarithm of each side.*

Example 4 Solve Exponential Inequalities Using Logarithms

Solve $5^{3y} < 8^{y-1}$.

$5^{3y} < 8^{y-1}$	Original inequality
$\log 5^{3y} < \log 8^{y-1}$	Property of Inequality for Logarithmic Functions
$3y \log 5 < (y - 1) \log 8$	Power Property of Logarithms
$3y \log 5 < y \log 8 - \log 8$	Distributive Property
$3y \log 5 - y \log 8 < -\log 8$	Subtract y log 8 from each side.
$y(3 \log 5 - \log 8) < -\log 8$	Distributive Property
$y < \frac{-\log 8}{3 \log 5 - \log 8}$	Divide each side by $3 \log 5 - \log 8$.
$y < \frac{-(0.9031)}{3(0.6990) - 0.9031}$	Use a calculator.
$y < -0.7564$	The solution set is $\{y \mid y < -0.7564\}$.

CHECK Test $y = -1$.

$5^{3y} < 8^{y-1}$	Original inequality
$5^{3(-1)} < 8^{(-1)-1}$	Replace y with 1.
$5^{-3} < 8^{-2}$	Simplify.
$\frac{1}{125} < \frac{1}{64}$ ✓	Negative Exponent Property

CHANGE OF BASE FORMULA The **Change of Base Formula** allows you to write equivalent logarithmic expressions that have different bases.

Key Concept — Change of Base Formula

- **Symbols** For all positive numbers, a, b and n, where $a \neq 1$ and $b \neq 1$,

$$\log_a n = \frac{\log_b n}{\log_b a}. \quad \begin{matrix} \leftarrow \text{log base } b \text{ of original number} \\ \leftarrow \text{log base } b \text{ of old base} \end{matrix}$$

- **Example** $\log_5 12 = \frac{\log_{10} 12}{\log_{10} 5}$

To prove this formula, let $\log_a n = x$.

$a^x = n$	Definition of logarithm
$\log_b a^x = \log_b n$	Property of Equality for Logarithms
$x \log_b a = \log_b n$	Power Property of Logarithms
$x = \frac{\log_b n}{\log_b a}$	Divide each side by $\log_b a$.
$\log_a n = \frac{\log_b n}{\log_b a}$	Replace x with $\log_a n$.

This formula makes it possible to evaluate a logarithmic expression of any base by translating the expression into one that involves common logarithms.

Example 5 Change of Base Formula

Express $\log_4 25$ in terms of common logarithms. Then approximate its value to four decimal places.

$\log_4 25 = \frac{\log_{10} 25}{\log_{10} 4}$ Change of Base Formula

≈ 2.3219 Use a calculator.

The value of $\log_4 25$ is approximately 2.3219.

Check for Understanding

Concept Check

1. **Name** the base used by the calculator LOG key. What are these logarithms called?
2. **OPEN ENDED** Give an example of an exponential equation requiring the use of logarithms to solve. Then solve your equation.
3. **Explain** why you must use the Change of Base Formula to find the value of $\log_2 7$ on a calculator.

Guided Practice

Use a calculator to evaluate each expression to four decimal places.

4. log 4
5. log 23
6. log 0.5

Solve each equation or inequality. Round to four decimal places.

7. $9^x = 45$
8. $4^{5n} > 30$
9. $3.1^{a-3} = 9.42$
10. $11^{x^2} = 25.4$
11. $7^{t-2} = 5^t$
12. $4^{p-1} \leq 3^p$

Express each logarithm in terms of common logarithms. Then approximate its value to four decimal places.

13. $\log_7 5$
14. $\log_3 42$
15. $\log_2 9$

Application

16. **DIET** Sandra's doctor has told her to avoid foods with a pH that is less than 4.5. What is the hydrogen ion concentration of foods Sandra is allowed to eat? Use the information at the beginning of the lesson.

Practice and Apply

Use a calculator to evaluate each expression to four decimal places.

17. log 5
18. log 12
19. log 7.2
20. log 2.3
21. log 0.8
22. log 0.03

Homework Help

For Exercises	See Examples
17–22	1
23–44, 53–57	3, 4
45–50	5
51–55	2

Extra Practice
See page 850.

ACIDITY For Exercises 23–26, use the information at the beginning of the lesson to find the pH of each substance given its concentration of hydrogen ions.

23. ammonia: $[H+] = 1 \times 10^{-11}$ mole per liter

24. vinegar: $[H+] = 6.3 \times 10^{-3}$ mole per liter

25. lemon juice: $[H+] = 7.9 \times 10^{-3}$ mole per liter

26. orange juice: $[H+] = 3.16 \times 10^{-4}$ mole per liter

Solve each equation or inequality. Round to four decimal places.

27. $6^x \geq 42$

28. $5^x = 52$

29. $8^{2a} < 124$

30. $4^{3p} = 10$

31. $3^{n+2} = 14.5$

32. $9^{z-4} = 6.28$

33. $8.2^{n-3} = 42.5$

34. $2.1^{t-5} = 9.32$

35. $20^{x^2} = 70$

36. $2^{x^2-3} = 15$

37. $8^{2n} > 52^{4n+3}$

38. $2^{2x+3} = 3^{3x}$

39. $16^{d-4} = 3^{3-d}$

40. $7^{p+2} \leq 13^{5-p}$

41. $5^{5y-2} = 2^{2y+1}$

42. $8^{2x-5} = 5^{x+1}$

43. $2^n = \sqrt{3^{n-2}}$

44. $4^x = \sqrt{5^{x+2}}$

Express each logarithm in terms of common logarithms. Then approximate its value to four decimal places.

45. $\log_2 13$

46. $\log_5 20$

47. $\log_7 3$

48. $\log_3 8$

49. $\log_4 (1.6)^2$

50. $\log_6 \sqrt{5}$

For Exercises 51 and 52, use the information presented at the beginning of the lesson.

51. POLLUTION The acidity of water determines the toxic effects of runoff into streams from industrial or agricultural areas. A pH range of 6.0 to 9.0 appears to provide protection for freshwater fish. What is this range in terms of the water's hydrogen ion concentration?

52. BUILDING DESIGN The 1971 Sylmar earthquake in Los Angeles had a Richter scale magnitude of 6.3. Suppose an architect has designed a building strong enough to withstand an earthquake 50 times as intense as the Sylmar quake. Find the magnitude of the strongest quake this building is designed to withstand.

ASTRONOMY For Exercises 53–55, use the following information.
Some stars appear bright only because they are very close to us. Absolute magnitude M is a measure of how bright a star would appear if it were 10 parsecs, about 32 light years, away from Earth. A lower magnitude indicates a brighter star. Absolute magnitude is given by $M = m + 5 - 5 \log d$, where d is the star's distance from Earth measured in parsecs and m is its apparent magnitude.

53. Sirius and Vega are two of the brightest stars in Earth's sky. The apparent magnitude of Sirius is -1.44 and of Vega is 0.03. Which star appears brighter?

54. Sirius is 2.64 parsecs from Earth while Vega is 7.76 parsecs from Earth. Find the absolute magnitude of each star.

55. Which star is actually brighter? That is, which has a lower absolute magnitude?

56. CRITICAL THINKING

a. Without using a calculator, find the value of $\log_2 8$ and $\log_8 2$.

b. Without using a calculator, find the value of $\log_9 27$ and $\log_{27} 9$.

c. Make and prove a conjecture as to the relationship between $\log_a b$ and $\log_b a$.

MONEY For Exercises 57 and 58, use the following information.
If you deposit P dollars into a bank account paying an annual interest rate r (expressed as a decimal), with n interest payments each year, the amount A you would have after t years is $A = P\left(1 + \frac{r}{n}\right)^{nt}$. Marta places $100 in a savings account earning 6% annual interest, compounded quarterly.

57. If Marta adds no more money to the account, how long will it take the money in the account to reach $125?

58. How long will it take for Marta's money to double?

59. **WRITING IN MATH** Answer the question that was posed at the beginning of the lesson.

Why is a logarithmic scale used to measure acidity?

Include the following in your answer:

- the hydrogen ion concentration of three substances listed in the table, and
- an explanation as to why it is important to be able to distinguish between a hydrogen ion concentration of 0.00001 mole per liter and 0.0001 mole per liter.

Standards Practice
Standardized Test Practice

60. The expression log $(x - 5)$ is undefined for all values of x such that

(A) $x \leq 5$
(B) $x > 5$
(C) $x \leq 0$
(D) $x > 1$

61. If $2^4 = 3^x$, then what is the value of x?

(A) 0.63 (B) 2.34
(C) 2.52 (D) 4

Maintain Your Skills

Mixed Review

Use $\log_7 2 \approx 0.3562$ and $\log_7 3 \approx 0.5646$ to approximate the value of each expression. *(Lesson 10-3)*

62. $\log_7 16$ **63.** $\log_7 27$ **64.** $\log_7 36$

Solve each equation or inequality. Check your solutions. *(Lesson 10-2)*

65. $\log_4 r = 3$ **66.** $\log_8 z \leq -2$ **67.** $\log_3 (4x - 5) = 5$

68. Use synthetic substitution to find $f(-2)$ for $f(x) = x^3 + 6x - 2$. *(Lesson 7-4)*

Factor completely. If the polynomial is not factorable, write prime. *(Lesson 5-4)*

69. $3d^2 + 2d - 8$ **70.** $42pq - 35p + 18q - 15$ **71.** $13xyz + 3x^2z + 4k$

Getting Ready for the Next Lesson

PREREQUISITE SKILLS Write an equivalent exponential equation.
*(For review of **logarithmic equations**, see Lesson 10-2.)*

72. $\log_2 3 = x$ **73.** $\log_3 x = 2$ **74.** $\log_5 125 = 3$

Write an equivalent logarithmic equation.
*(For review of **logarithmic equations**, see Lesson 10-2.)*

75. $5^x = 45$ **76.** $7^3 = x$ **77.** $b^y = x$

Graphing Calculator

A Follow-Up of Lesson 10-4

Solving Exponential and Logarithmic Equations and Inequalities

You can use a TI-83 Plus graphing calculator to solve exponential and logarithmic equations and inequalities. This can be done by graphing each side of the equation separately and using the **intersect** feature on the calculator.

Example 1

Solve $2^{3x-9} = \left(\frac{1}{2}\right)^{x-3}$ by graphing.

Step 1 *Graph each side of the equation.*

- Graph each side of the equation as a separate function. Enter 2^{3x-9} as Y1. Enter $\left(\frac{1}{2}\right)^{x-3}$ as Y2. Be sure to include the added parentheses around each exponent. Then graph the two equations.

KEYSTROKES: *See pages 87 and 88 to review graphing equations.*

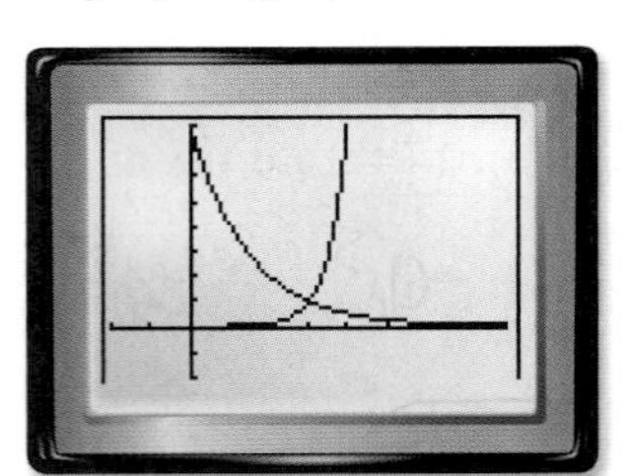

[−2, 8] scl: 1 by [−2, 8] scl: 1

Step 2 *Use the intersect feature.*

- You can use the **intersect** feature on the **CALC** menu to approximate the ordered pair of the point at which the curves cross.

KEYSTROKES: *See page 115 to review how to use the intersect feature.*

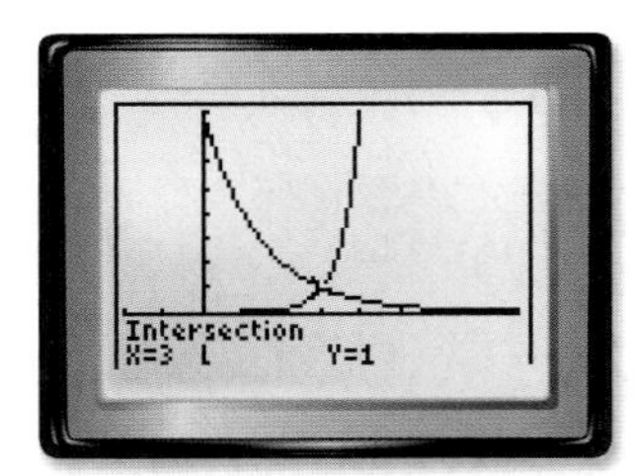

[−2, 8] scl: 1 by [−2, 8] scl: 1

The calculator screen shows that the x-coordinate of the point at which the curves cross is 3. Therefore, the solution of the equation is 3.

The TI-83 Plus has $y = \log_{10} x$ as a built-in function. Enter [Y=] [LOG] [X,T,θ,n] [GRAPH] to view this graph. To graph logarithmic functions with bases other than 10, you must use the Change of Base Formula,

$$\log_a n = \frac{\log_b n}{\log_b a}.$$

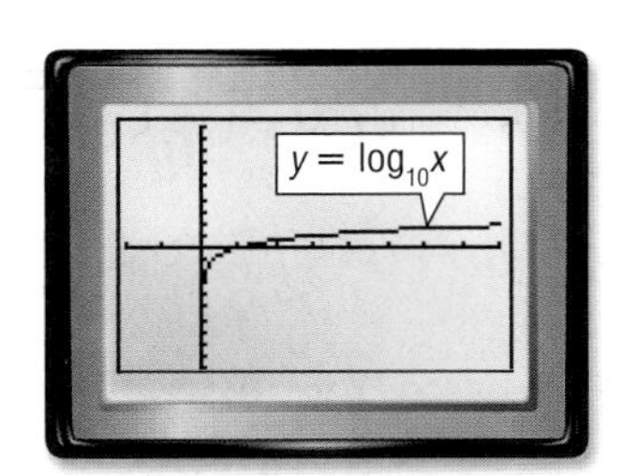

[−2, 8] scl: 1 by [−5, 5] scl: 1

For example, $\log_3 x = \frac{\log_{10} x}{\log_{10} 3}$, so to graph $y = \log_3 x$ you must enter [LOG] [X,T,θ,n] [)] [÷] [LOG] 3 [)] as Y1.

www.algebra2.com/other_calculator_keystrokes

Investigation

Example 2

Solve $\log_2 2x \geq \log_{\frac{1}{2}} 2x$ by graphing.

Step 1 *Rewrite the problem as a system of common logarithmic inequalities.*

- The first inequality is $\log_2 2x \geq y$ or $y \leq \log_2 2x$. The second inequality is $y \geq \log_{\frac{1}{2}} 2x$.
- Use the Change of Base Formula to create equations that can be entered into the calculator.

$$\log_2 2x = \frac{\log 2x}{\log 2} \qquad \log_{\frac{1}{2}} 2x = \frac{\log 2x}{\log \frac{1}{2}}$$

Thus, the two inequalities are $y \leq \frac{\log 2x}{\log 2}$ and $y \geq \frac{\log 2x}{\log \frac{1}{2}}$.

Step 2 *Enter the first inequality.*

- Enter $y \leq \frac{\log 2x}{\log 2}$ as Y1. Since the inequality includes *less than*, shade below the curve.

KEYSTROKES: [Y=] [LOG] 2 [X,T,θ,n] [)] [÷] [LOG] 2 [)]

Use the arrow and [ENTER] keys to choose the shade below icon, ▙.

Step 3 *Enter the second inequality.*

- Enter $y \geq \frac{\log 2x}{\log \frac{1}{2}}$ as Y2. Since the inequality includes *greater than*, shade above the curve.

KEYSTROKES: [LOG] 2 [X,T,θ,n] [)] [÷] [LOG] 1 [÷] 2 [)] [GRAPH]

Use the arrow and [ENTER] keys to choose the shade above icon, ▜.

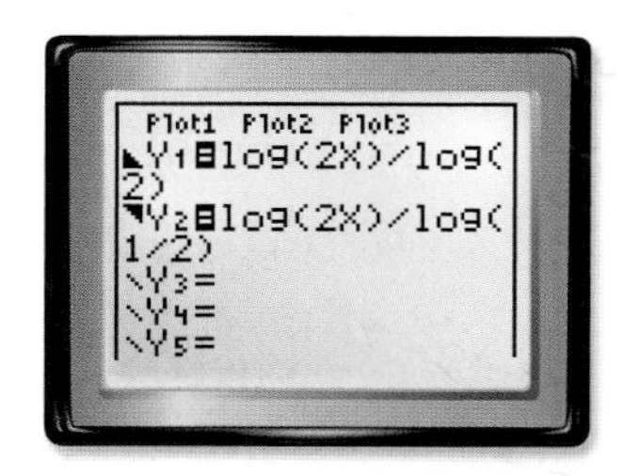

Step 4 *Graph the inequalities.*

KEYSTROKES: [GRAPH]

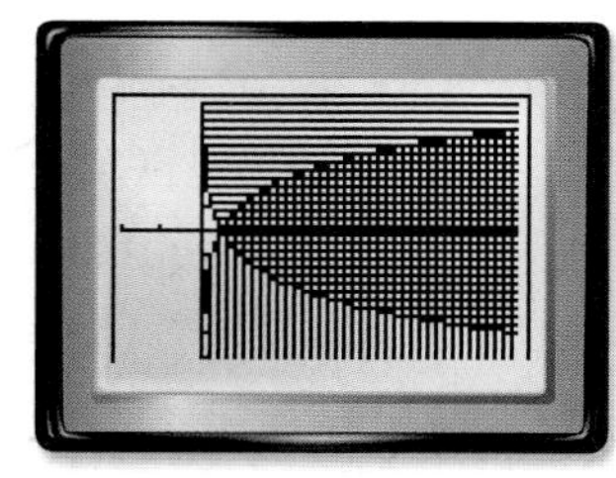

[−2, 8] scl: 1 by [−5, 5] scl: 1

The x values of the points in the region where the shadings overlap is the solution set of the original inequality. Using the calculator's intersect feature, you can conclude that the solution set is $\{x \mid x \geq 0.5\}$.

Exercises **Solve each equation or inequality by graphing.**

1. $3.5^{x+2} = 1.75^{x+3}$
2. $-3^{x+4} = -0.5^{2x+3}$
3. $6^{2-x} - 4 = -0.25^{x-2.5}$
4. $3^x - 4 = 5^{\frac{x}{2}}$
5. $\log_2 3x = \log_3 (2x + 2)$
6. $2^{x-2} \geq 0.5^{x-3}$
7. $\log_3 (3x - 5) \geq \log_3 (x + 7)$
8. $5^{x+3} \leq 2^{x+4}$
9. $\log_2 2x \leq \log_4 (x + 3)$

10-5 Base e and Natural Logarithms

California Standards Standard 11.1 Students understand the inverse relationship between exponents and logarithms and use this relationship to solve problems involving logarithms and exponents. (Key)

What You'll Learn

- Evaluate expressions involving the natural base and natural logarithms.
- Solve exponential equations and inequalities using natural logarithms.

How is the natural base e used in banking?

Suppose a bank compounds interest on accounts *continuously*, that is, with no waiting time between interest payments.

To develop an equation to determine continuously compounded interest, examine what happens to the value A of an account for increasingly larger numbers of compounding periods n. Use a principal P of \$1, an interest rate r of 100% or 1, and time t of 1 year.

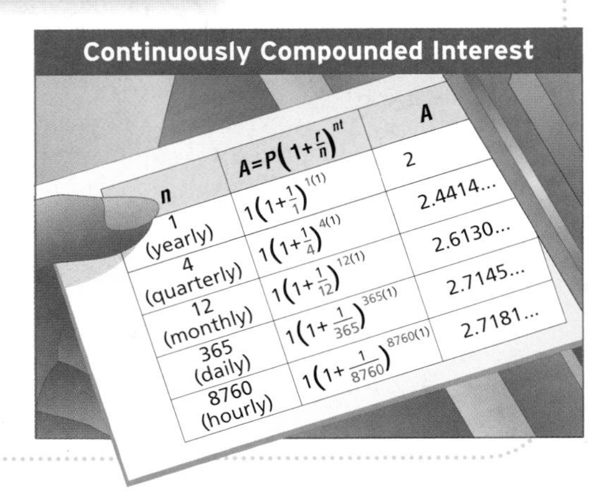

Continuously Compounded Interest

n	$A = P\left(1+\frac{r}{n}\right)^{nt}$	A
1 (yearly)	$1\left(1+\frac{1}{1}\right)^{1(1)}$	2
4 (quarterly)	$1\left(1+\frac{1}{4}\right)^{4(1)}$	2.4414...
12 (monthly)	$1\left(1+\frac{1}{12}\right)^{12(1)}$	2.6130...
365 (daily)	$1\left(1+\frac{1}{365}\right)^{365(1)}$	2.7145...
8760 (hourly)	$1\left(1+\frac{1}{8760}\right)^{8760(1)}$	2.7181...

Vocabulary

- natural base, e
- natural base exponential function
- natural logarithm
- natural logarithmic function

BASE e AND NATURAL LOGARITHMS In the table above, as n increases, the expression $1\left(1 + \frac{1}{n}\right)^{n(1)}$ or $\left(1 + \frac{1}{n}\right)^{n}$ approaches the irrational number 2.71828... . This number is referred to as the **natural base, e**.

An exponential function with base e is called a **natural base exponential function**. The graph of $y = e^x$ is shown at the right. Natural base exponential functions are used extensively in science to model quantities that grow and decay continuously.

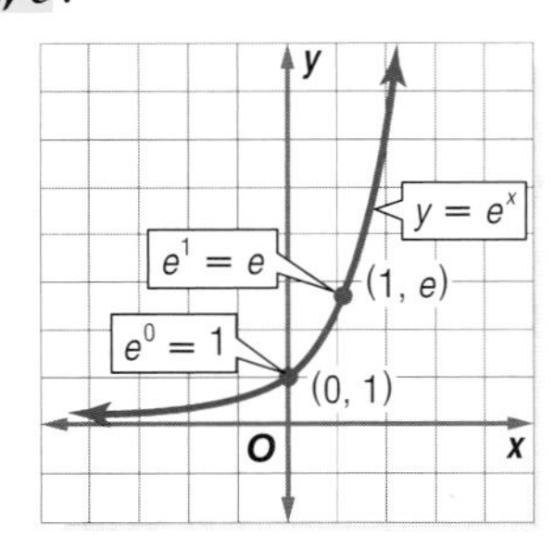

Most calculators have an e^x function for evaluating natural base expressions.

Study Tip

Simplifying Expressions with e

You can simplify expressions involving e in the same manner in which you simplify expressions involving π.

Examples:

- $\pi^2 \cdot \pi^3 = \pi^5$
- $e^2 \cdot e^3 = e^5$

Example 1 Evaluate Natural Base Expressions

Use a calculator to evaluate each expression to four decimal places.

a. e^2 KEYSTROKES: [2nd] [e^x] 2 [ENTER] 7.389056099 about 7.3891

b. $e^{-1.3}$ KEYSTROKES: [2nd] [e^x] -1.3 [ENTER] .272531793 about 0.2725

The logarithm with base e is called the **natural logarithm**, sometimes denoted by $\log_e x$, but more often abbreviated $\ln x$. The **natural logarithmic function**, $y = \ln x$, is the inverse of the natural base exponential function, $y = e^x$. The graph of these two functions shows that $\ln 1 = 0$ and $\ln e = 1$.

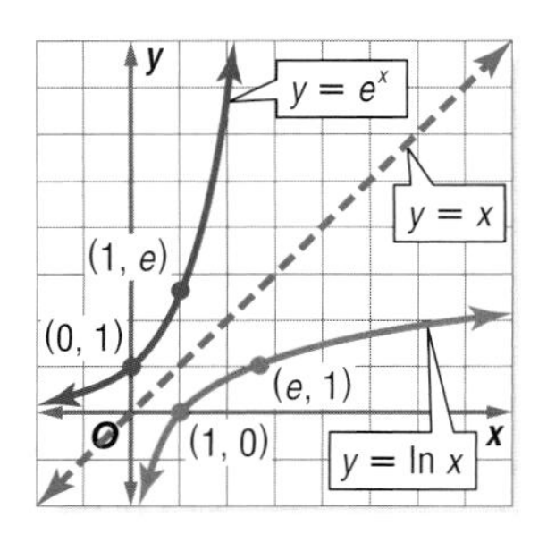

Most calculators have an LN key for evaluating natural logarithms.

Example 2 Evaluate Natural Logarithmic Expressions

Use a calculator to evaluate each expression to four decimal places.

a. $\ln 4$ KEYSTROKES: LN 4 ENTER 1.386294361 about 1.3863

b. $\ln 0.05$ KEYSTROKES: LN 0.05 ENTER −2.995732274 about −2.9957

You can write an equivalent base e exponential equation for a natural logarithmic equation and vice versa by using the fact that $\ln x = \log_e x$.

Example 3 Write Equivalent Expressions

Write an equivalent exponential or logarithmic equation.

a. $e^x = 5$

$e^x = 5 \rightarrow \log_e 5 = x$

$\ln 5 = x$

b. $\ln x \approx 0.6931$

$\ln x \approx 0.6931 \rightarrow \log_e x \approx 0.6931$

$x \approx e^{0.6931}$

Since the natural base function and the natural logarithmic function are inverses, these two functions can be used to "undo" each other.

$$e^{\ln x} = x \qquad \ln e^x = x$$

Example 4 Inverse Property of Base e and Natural Logarithms

Evaluate each expression.

a. $e^{\ln 7}$

$e^{\ln 7} = 7$

b. $\ln e^{4x + 3}$

$\ln e^{4x + 3} = 4x + 3$

EQUATIONS AND INEQUALITIES WITH e AND ln Equations and inequalities involving base e are easier to solve using natural logarithms than using common logarithms. All of the properties of logarithms that you have learned apply to natural logarithms as well.

Example 5 Solve Base e Equations

Solve $5e^{-x} - 7 = 2$.

$5e^{-x} - 7 = 2$	Original equation
$5e^{-x} = 9$	Add 7 to each side.
$e^{-x} = \frac{9}{5}$	Divide each side by 5.
$\ln e^{-x} = \ln \frac{9}{5}$	Property of Equality for Logarithms
$-x = \ln \frac{9}{5}$	Inverse Property of Exponents and Logarithms
$x = -\ln \frac{9}{5}$	Divide each side by −1.
$x \approx -0.5878$	Use a calculator.

The solution is about −0.5878.

CHECK You can check this value by substituting −0.5878 into the original equation or by finding the intersection of the graphs of $y = 5e^{-x} - 7$ and $y = 2$.

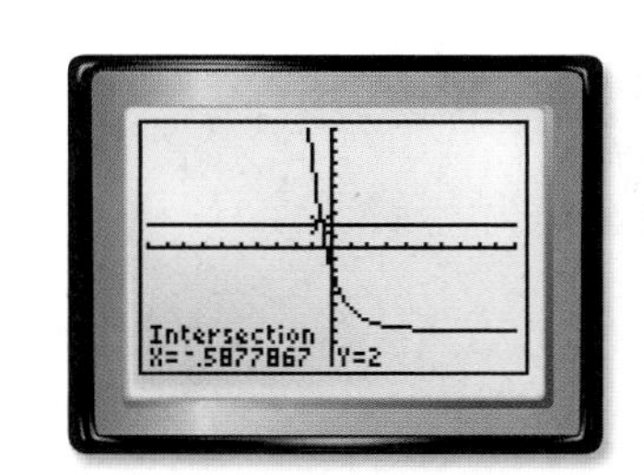

Study Tip

Continuously Compounded Interest
Although no banks actually pay interest compounded continuously, the equation $A = Pe^{rt}$ is so accurate in computing the amount of money for quarterly compounding, or daily compounding, that it is often used for this purpose.

When interest is compounded continuously, the amount A in an account after t years is found using the formula $A = Pe^{rt}$, where P is the amount of principal and r is the annual interest rate.

Example 6 Solve Base e Inequalities

SAVINGS Suppose you deposit $1000 in an account paying 5% annual interest, compounded continuously.

a. What is the balance after 10 years?

$A = Pe^{rt}$	Continuous compounding formula
$= 1000e^{(0.05)(10)}$	Replace P with 1000, r with 0.05, and t with 10.
$= 1000e^{0.5}$	Simplify.
≈ 1648.72	Use a calculator.

The balance after 10 years would be $1648.72.

b. How long will it take for the balance in your account to reach at least $1500?

$$\underbrace{\text{The balance}}_{A}\ \underbrace{\text{is at least}}_{\geq}\ \underbrace{\$1500.}_{1500}$$

$A \geq 1500$	Write an inequality.
$1000e^{(0.05)t} \geq 1500$	Replace A with $1000e^{(0.05)t}$.
$e^{(0.05)t} \geq 1.5$	Divide each side by 1000.
$\ln e^{(0.05)t} \geq \ln 1.5$	Property of Equality for Logarithms
$0.05t \geq \ln 1.5$	Inverse Property of Exponents and Logarithms
$t \geq \frac{\ln 1.5}{0.05}$	Divide each side by 0.05.
$t \geq 8.11$	Use a calculator.

It will take at least 8.11 years for the balance to reach $1500.

Study Tip

Equations with ln
As with other logarithmic equations, remember to check for extraneous solutions.

Example 7 Solve Natural Log Equations and Inequalities

Solve each equation or inequality.

a. $\ln 5x = 4$

$\ln 5x = 4$	Original equation
$e^{\ln 5x} = e^4$	Write each side using exponents and base e.
$5x = e^4$	Inverse Property of Exponents and Logarithms
$x = \frac{e^4}{5}$	Divide each side by 5.
$x \approx 10.9196$	Use a calculator.

The solution is 10.9196. Check this solution using substitution or graphing.

b. $\ln (x - 1) > -2$

$\ln (x - 1) > -2$	Original inequality
$e^{\ln (x - 1)} > e^{-2}$	Write each side using exponents and base e.
$x - 1 > e^{-2}$	Inverse Property of Exponents and Logarithms
$x > e^{-2} + 1$	Add 1 to each side.
$x > 1.1353$	Use a calculator.

The solution is all numbers greater than about 1.1353. Check this solution using substitution.

Check for Understanding

Concept Check

1. **Name** the base of natural logarithms.
2. **OPEN ENDED** Give an example of an exponential equation that requires using natural logarithms instead of common logarithms to solve.
3. **FIND THE ERROR** Colby and Elsu are solving $\ln 4x = 5$.

Colby	Elsu
$\ln 4x = 5$	$\ln 4x = 5$
$10^{\ln 4x} = 10^5$	$e^{\ln 4x} = e^5$
$4x = 100{,}000$	$4x = e^5$
$x = 25{,}000$	$x = \frac{e^5}{4}$
	$x \approx 37.1033$

Who is correct? Explain your reasoning.

Guided Practice

Use a calculator to evaluate each expression to four decimal places.

4. e^6
5. $e^{-3.4}$
6. $\ln 1.2$
7. $\ln 0.1$

Write an equivalent exponential or logarithmic equation.

8. $e^x = 4$
9. $\ln 1 = 0$

Evaluate each expression.

10. $e^{\ln 3}$
11. $\ln e^{5x}$

Solve each equation or inequality.

12. $e^x > 30$
13. $2e^x - 5 = 1$
14. $3 + e^{-2x} = 8$
15. $\ln x < 6$
16. $2 \ln 3x + 1 = 5$
17. $\ln x^2 = 9$

Application

ALTITUDE **For Exercises 18 and 19, use the following information.**
The altimeter in an airplane gives the altitude or height h (in feet) of a plane above sea level by measuring the outside air pressure P (in kilopascals). The height and air pressure are related by the model $P = 101.3\, e^{-\frac{h}{26{,}200}}$

18. Find a formula for the height in terms of the outside air pressure.
19. Use the formula you found in Exercise 18 to approximate the height of a plane above sea level when the outside air pressure is 57 kilopascals.

Practice and Apply

Homework Help

For Exercises	See Examples
20–29	1, 2
30–33	3
34–37	4
38–53	5–7
54–57	6
58–61	3, 5

Extra Practice
See page 850.

Use a calculator to evaluate each expression to four decimal places.

20. e^4
21. e^5
22. $e^{-1.2}$
23. $e^{0.5}$
24. $\ln 3$
25. $\ln 10$
26. $\ln 5.42$
27. $\ln 0.03$

28. **SAVINGS** If you deposit \$150 in a savings account paying 4% interest compounded continuously, how much money will you have after 5 years? Use the formula presented in Example 6.
29. **PHYSICS** The equation $\ln \frac{I_0}{I} = 0.014d$ relates the intensity of light at a depth of d centimeters of water I with the intensity in the atmosphere I_0. Find the depth of the water where the intensity of light is half the intensity of the light in the atmosphere.

Write an equivalent exponential or logarithmic equation.

30. $e^{-x} = 5$ **31.** $e^2 = 6x$ **32.** $\ln e = 1$ **33.** $\ln 5.2 = x$

Evaluate each expression.

34. $e^{\ln 0.2}$ **35.** $e^{\ln y}$ **36.** $\ln e^{-4x}$ **37.** $\ln e^{45}$

Solve each equation or inequality.

38. $3e^x + 1 = 5$ **39.** $2e^x - 1 = 0$ **40.** $e^x < 4.5$

41. $e^x > 1.6$ **42.** $-3e^{4x} + 11 = 2$ **43.** $8 + 3e^{3x} = 26$

44. $e^{5x} \geq 25$ **45.** $e^{-2x} \leq 7$ **46.** $\ln 2x = 4$

47. $\ln 3x = 5$ **48.** $\ln (x + 1) = 1$ **49.** $\ln (x - 7) = 2$

50. $\ln x + \ln 3x = 12$ **51.** $\ln 4x + \ln x = 9$

52. $\ln (x^2 + 12) = \ln x + \ln 8$ **53.** $\ln x + \ln (x + 4) = \ln 5$

More About. . .

Money

To determine the doubling time on an account paying an interest rate r that is compounded *annually*, investors use the "Rule of 72." Thus, the amount of time needed for the money in an account paying 6% interest compounded annually to double is $\frac{72}{6}$ or 12 years.

Source: www.datachimp.com

MONEY **For Exercises 54–57, use the formula for continuously compounded interest found in Example 6.**

54. If you deposit \$100 in an account paying 3.5% interest compounded continuously, how long will it take for your money to double?

55. Suppose you deposit A dollars in an account paying an interest rate r as a percent, compounded continuously. Write an equation giving the time t needed for your money to double, or the *doubling time.*

56. Explain why the equation you found in Exercise 55 might be referred to as the "Rule of 70."

57. **MAKE A CONJECTURE** State a rule that could be used to approximate the amount of time t needed to triple the amount of money in a savings account paying r percent interest compounded continuously.

POPULATION **For Exercises 58 and 59, use the following information.**
In 2000, the world's population was about 6 billion. If the world's population continues to grow at a constant rate, the future population P, in billions, can be predicted by $P = 6e^{0.02t}$, where t is the time in years since 2000.

58. According to this model, what will the world's population be in 2010?

59. Some experts have estimated that the world's food supply can support a population of, at most, 18 billion. According to this model, for how many more years will the world's population remain at 18 billion or less?

Online Research **Data Update** What is the current world population? Visit www.algebra2.com/data_update to learn more.

RUMORS **For Exercises 60 and 61, use the following information.**
The number of people H who have heard a rumor can be approximated by $H = \frac{P}{1 + (P - S)e^{-0.35t}}$, where P is the total population, S is the number of people who start the rumor, and t is the time in minutes. Suppose two students start a rumor that the principal will let everyone out of school one hour early that day.

60. If there are 1600 students in the school, how many students will have heard the rumor after 10 minutes?

61. How much time will pass before half of the students have heard the rumor?

62. **CRITICAL THINKING** Determine whether the following statement is *sometimes*, *always*, or *never* true. Explain your reasoning.

For all positive numbers x and y, $\frac{\log x}{\log y} = \frac{\ln x}{\ln y}$.

63. **WRITING IN MATH** Answer the question that was posed at the beginning of the lesson.

How is the natural base e used in banking?

Include the following in your answer:

- an explanation of how to calculate the value of an account whose interest is compounded continuously, and
- an explanation of how to use natural logarithms to find when the account will have a specified value.

64. If $e^x \neq 1$ and $e^{x^2} = \frac{1}{(\sqrt{2})^x}$, what is the value of x?

Ⓐ -1.41 Ⓑ -0.35 Ⓒ 1.00 Ⓓ 1.10

65. **SHORT RESPONSE** The population of a certain country can be modeled by the equation $P(t) = 40\,e^{0.02t}$, where P is the population in millions and t is the number of years since 1900. When will the population be 100 million, 200 million, and 400 million? What do you notice about these time periods?

Maintain Your Skills

Mixed Review

Express each logarithm in terms of common logarithms. Then approximate its value to four decimal places. *(Lesson 10-4)*

66. $\log_4 68$ 67. $\log_6 0.047$ 68. $\log_{50} 23$

Solve each equation. Check your solutions. *(Lesson 10-3)*

69. $\log_3 (a + 3) + \log_3 (a - 3) = \log_3 16$ 70. $\log_{11} 2 + 2 \log_{11} x = \log_{11} 32$

State whether each equation represents a *direct, joint,* or *inverse* variation. Then name the constant of variation. *(Lesson 9-4)*

71. $mn = 4$ 72. $\frac{a}{b} = c$ 73. $y = -7x$

74. **COMMUNICATION** A microphone is placed at the focus of a parabolic reflector to collect sounds for the television broadcast of a football game. The focus of the parabola that is the cross section of the reflector is 5 inches from the vertex. The latus rectum is 20 inches long. Assuming that the focus is at the origin and the parabola opens to the right, write the equation of the cross section. *(Lesson 8-2)*

Getting Ready for the Next Lesson

PREREQUISITE SKILL **Solve each equation or inequality.**
*(To review **exponential equations and inequalities**, see Lesson 10-1.)*

75. $2^x = 10$ 76. $5^x = 12$ 77. $6^x = 13$

78. $2(1 + 0.1)^x = 50$ 79. $10(1 + 0.25)^x = 200$ 80. $400(1 - 0.2)^x = 50$

Practice Quiz 2 — Lessons 10-3 through 10-5

1. Express $\log_4 5$ in terms of common logarithms. Then approximate its value to four decimal places. *(Lesson 10-4)*

2. Write an equivalent exponential equation for $\ln 3x = 2$. *(Lesson 10-5)*

Solve each equation or inequality. *(Lesson 10-3 through 10-5)*

3. $\log_2 (9x + 5) = 2 + \log_2 (x^2 - 1)$ 4. $2^{x-3} > 5$ 5. $2e^x - 1 = 7$

10-6 Exponential Growth and Decay

California Standards Standard 12.0 **Students** know the laws of fractional exponents, **understand exponential functions, and use these functions in problems involving exponential growth and decay.** (Key)

What You'll Learn

- Use logarithms to solve problems involving exponential decay.
- Use logarithms to solve problems involving exponential growth.

Vocabulary

- rate of decay
- rate of growth

How can you determine the current value of your car?

Certain assets, like homes, can *appreciate* or increase in value over time. Others, like cars, *depreciate* or decrease in value with time. Suppose you buy a car for \$22,000 and the value of the car decreases by 16% each year. The table shows the value of the car each year for up to 5 years after it was purchased.

Years after Purchase	Value of Car (\$)
0	22,000.00
1	18,480.00
2	15,523.20
3	13,039.49
4	10,953.17
5	9200.66

EXPONENTIAL DECAY The depreciation of the value of a car is an example of exponential decay. When a quantity *decreases* by a fixed percent each year, or other period of time, the amount y of that quantity after t years is given by $y = a(1 - r)^t$, where a is the initial amount and r is the percent of decrease expressed as a decimal. The percent of decrease r is also referred to as the **rate of decay**.

Example 1 Exponential Decay of the Form $y = a(1 - r)^t$

CAFFEINE A cup of coffee contains 130 milligrams of caffeine. If caffeine is eliminated from the body at a rate of 11% per hour, how long will it take for half of this caffeine to be eliminated from a person's body?

Study Tip

Rate of Change
Remember to rewrite the rate of change as a decimal before using it in the formula.

Explore The problem gives the amount of caffeine consumed and the rate at which the caffeine is eliminated. It asks you to find the time it will take for half of the caffeine to be eliminated from a person's body.

Plan Use the formula $y = a(1 - r)^t$. Let t be the number of hours since drinking the coffee. The amount remaining y is half of 130 or 65.

Solve

$y = a(1 - r)^t$ Exponential decay formula

$65 = 130(1 - 0.11)^t$ Replace y with 65, a with 130, and r with 11% or 0.11.

$0.5 = (0.89)^t$ Divide each side by 130.

$\log 0.5 = \log (0.89)^t$ Property of Equality for Logarithms

$\log 0.5 = t \log (0.89)$ Power Property for Logarithms

$\frac{\log 0.5}{\log 0.89} = t$ Divide each side by log 0.89.

$5.9480 \approx t$ Use a calculator.

It will take approximately 6 hours for half of the caffeine to be eliminated from a person's body.

Examine Use the formula to find how much of the original 130 milligrams of caffeine would remain after 6 hours.

$y = a(1 - r)^t$	Exponential decay formula
$y = 130(1 - 0.11)^6$	Replace a with 130, r with 0.11, and t with 6.
$y \approx 64.6$	Use a calculator.

Half of 130 is 65, so the answer seems reasonable.

Another model for exponential decay is given by $y = ae^{-kt}$, where k is a constant. This is the model preferred by scientists. Use this model to solve problems involving radioactive decay.

Example 2 Exponential Decay of the Form $y = ae^{-kt}$

PALEONTOLOGY The *half-life* of a radioactive substance is the time it takes for half of the atoms of the substance to become disintegrated. All life on Earth contains the radioactive element Carbon-14, which decays continuously at a fixed rate. The half-life of Carbon-14 is 5760 years. That is, every 5760 years half of a mass of Carbon-14 decays away.

a. What is the value of k for Carbon-14?

To determine the constant k for Carbon-14, let a be the initial amount of the substance. The amount y that remains after 5760 years is then represented by $\frac{1}{2}a$ or $0.5a$.

$y = ae^{-kt}$	Exponential decay formula
$0.5a = ae^{-k(5760)}$	Replace y with $0.5a$ and t with 5760.
$0.5 = e^{-5760k}$	Divide each side by a.
$\ln 0.5 = \ln e^{-5760k}$	Property of Equality for Logarithmic Functions
$\ln 0.5 = -5760k$	Inverse Property of Exponents and Logarithms
$\frac{\ln 0.5}{-5760} = k$	Divide each side by -5760.
$0.00012 \approx k$	Use a calculator.

The constant for Carbon-14 is 0.00012. Thus, the equation for the decay of Carbon-14 is $y = ae^{-0.00012t}$, where t is given in years.

b. A paleontologist examining the bones of a woolly mammoth estimates that they contain only 3% as much Carbon-14 as they would have contained when the animal was alive. How long ago did the mammoth die?

Let a be the initial amount of Carbon-14 in the animal's body. Then the amount y that remains after t years is 3% of a or $0.03a$.

$y = ae^{-0.00012t}$	Formula for the decay of Carbon-14
$0.03a = ae^{-0.00012t}$	Replace y with $0.03a$.
$0.03 = e^{-0.00012t}$	Divide each side by a.
$\ln 0.03 = \ln e^{-0.00012t}$	Property of Equality for Logarithms
$\ln 0.03 = -0.00012t$	Inverse Property of Exponents and Logarithms
$\frac{\ln 0.03}{-0.00012} = t$	Divide each side by -0.00012.
$29{,}221 \approx t$	Use a calculator.

The mammoth lived about 29,000 years ago.

Career Choices

Paleontologist

Paleontologists study fossils found in geological formations. They use these fossils to trace the evolution of plant and animal life and the geologic history of Earth.

Online Research For information about a career as a paleontologist, visit: www.algebra2.com/careers

Source: U.S. Department of Labor

EXPONENTIAL GROWTH When a quantity *increases* by a fixed percent each time period, the amount y of that quantity after t time periods is given by $y = a(1 + r)^t$, where a is the initial amount and r is the percent of increase expressed as a decimal. The percent of increase r is also referred to as the **rate of growth**.

Example 3 Exponential Growth of the Form $y = a(1 + r)^t$

Multiple-Choice Test Item

In 1910, the population of a city was 120,000. Since then, the population has increased by exactly 1.5% per year. If the population continues to grow at this rate, what will the population be in 2010?

Ⓐ 138,000
Ⓑ 531,845
Ⓒ 1,063,690
Ⓓ 1.4×10^{11}

Test-Taking Tip

To change a percent to a decimal, drop the percent symbol and move the decimal point two places to the left.
1.5% = 0.015

Read the Test Item

You need to find the population of the city 2010 − 1910 or 100 years later. Since the population is growing at a fixed percent each year, use the formula $y = a(1 + r)^t$.

Solve the Test Item

$y = a(1 + r)^t$	Exponential growth formula
$y = 120{,}000(1 + 0.015)^{100}$	Replace $a = 120{,}000$, r with 0.015, and t with 2010 − 1910 or 100.
$y = 120{,}000(1.015)^{100}$	Simplify.
$y \approx 531{,}845.48$	Use a calculator.

The answer is B.

Another model for exponential growth, preferred by scientists, is $y = ae^{kt}$, where k is a constant. Use this model to find the constant k.

Example 4 Exponential Growth of the Form $y = ae^{kt}$

POPULATION As of 2000, China was the world's most populous country, with an estimated population of 1.26 billion people. The second most populous country was India, with 1.01 billion. The populations of India and China can be modeled by $I(t) = 1.01e^{0.015t}$ and $C(t) = 1.26e^{0.009t}$, respectively. According to these models, when will India's population be more than China's?

You want to find t such that $I(t) > C(t)$.

$I(t) > C(t)$	
$1.01e^{0.015t} > 1.26e^{0.009t}$	Replace $I(t)$ with $1.01e^{0.015t}$ and $C(t)$ with $1.26e^{0.009t}$
$\ln 1.01e^{0.015t} > \ln 1.26e^{0.009t}$	Property of Inequality for Logarithms
$\ln 1.01 + \ln e^{0.015t} > \ln 1.26 + \ln e^{0.009t}$	Product Property of Logarithms
$\ln 1.01 + 0.015t > \ln 1.26 + 0.009t$	Inverse Property of Exponents and Logarithms
$0.006t > \ln 1.26 - \ln 1.01$	Subtract $0.009t$ from each side.
$t > \dfrac{\ln 1.26 - \ln 1.01}{0.006}$	Divide each side by 0.006.
$t > 36.86$	Use a calculator.

After 37 years or in 2037, India will be the most populous country in the world.

Check for Understanding

Concept Check

1. **Write** a general formula for exponential growth and decay where r is the percent of change.
2. **Explain** how to solve $y = (1 + r)^t$ for t.
3. **OPEN ENDED** Give an example of a quantity that grows or decays at a fixed rate.

Guided Practice

SPACE For Exercises 4–6, use the following information.
A radioisotope is used as a power source for a satellite. The power output P (in watts) is given by $P = 50e^{-\frac{t}{250}}$, where t is the time in days.

4. Is the formula for power output an example of exponential growth or decay? Explain your reasoning.
5. Find the power available after 100 days.
6. Ten watts of power are required to operate the equipment in the satellite. How long can the satellite continue to operate?

POPULATION GROWTH For Exercises 7 and 8, use the following information.
The city of Raleigh, North Carolina, grew from a population of 212,000 in 1990 to a population of 259,000 in 1998.

7. Write an exponential growth equation of the form $y = ae^{kt}$ for Raleigh, where t is the number of years after 1990.
8. Use your equation to predict the population of Raleigh in 2010.

9. Suppose the weight of a bar of soap decreases by 2.5% each time it is used. If the bar weighs 95 grams when it is new, what is its weight to the nearest gram after 15 uses?

   (A) 57.5 g (B) 59.4 g (C) 65 g (D) 93 g

Practice and Apply

Homework Help

For Exercises	See Examples
10	1
12–14,	2
11, 17–20	3
15, 16	4

Extra Practice
See page 851.

10. **COMPUTERS** Zeus Industries bought a computer for $2500. It is expected to depreciate at a rate of 20% per year. What will the value of the computer be in 2 years?

11. **REAL ESTATE** The Martins bought a condominium for $85,000. Assuming that the value of the condo will appreciate at most 5% a year, how much will the condo be worth in 5 years?

12. **MEDICINE** Radioactive iodine is used to determine the health of the thyroid gland. It decays according to the equation $y = ae^{-0.0856t}$, where t is in days. Find the half-life of this substance.

13. **PALEONTOLOGY** A paleontologist finds a bone that might be a dinosaur bone. In the laboratory, she finds that the Carbon-14 found in the bone is $\frac{1}{12}$ of that found in living bone tissue. Could this bone have belonged to a dinosaur? Explain your reasoning. (*Hint:* The dinosaurs lived from 220 million years ago to 63 million years ago.)

14. **ANTHROPOLOGY** An anthropologist finds there is so little remaining Carbon-14 in a prehistoric bone that instruments cannot measure it. This means that there is less than 0.5% of the amount of Carbon-14 the bones would have contained when the person was alive. How long ago did the person die?

Olympics

The women's high jump competition first took place in the USA in 1895, but it did not become an Olympic event until 1928.

Source: www.princeton.edu

BIOLOGY For Exercises 15 and 16, use the following information.
Bacteria usually reproduce by a process known as *binary fission*. In this type of reproduction, one bacterium divides, forming two bacteria. Under ideal conditions, some bacteria reproduce every 20 minutes.

15. Find the constant k for this type of bacteria under ideal conditions.

16. Write the equation for modeling the exponential growth of this bacterium.

ECONOMICS For Exercises 17 and 18, use the following information.
The annual Gross Domestic Product (GDP) of a country is the value of all of the goods and services produced in the country during a year. During the period 1985–1999, the Gross Domestic Product of the United States grew about 3.2% per year, measured in 1996 dollars. In 1985, the GDP was $5717 billion.

17. Assuming this rate of growth continues, what will the GDP of the United States be in the year 2010?

18. In what year will the GDP reach $20 trillion?

19. OLYMPICS In 1928, when the high jump was first introduced as a women's sport at the Olympic Games, the winning women's jump was 62.5 inches, while the winning men's jump was 76.5 inches. Since then, the winning jump for women has increased by about 0.38% per year, while the winning jump for men has increased at a slower rate, 0.3%. If these rates continue, when will the women's winning high jump be higher than the men's?

20. HOME OWNERSHIP The Mendes family bought a new house 10 years ago for $120,000. The house is now worth $191,000. Assuming a steady rate of growth, what was the yearly rate of appreciation?

21. CRITICAL THINKING The half-life of Radium is 1620 years. When will a 20-gram sample of Radium be completely gone? Explain your reasoning.

22. WRITING IN MATH Answer the question that was posed at the beginning of the lesson.

How can you determine the current value of your car?

Include the following in your answer:
- a description of how to find the percent decrease in the value of the car each year, and
- a description of how to find the value of a car for any given year when the rate of depreciation is known.

23. SHORT RESPONSE An artist creates a sculpture out of salt that weighs 2000 pounds. If the sculpture loses 3.5% of its mass each year to erosion, after how many years will the statue weigh less than 1000 pounds?

24. The curve shown at the right represents a portion of the graph of which function?

(A) $y = 50 - x$ (B) $y = \log x$

(C) $y = e^{-x}$ (D) $xy = 5$

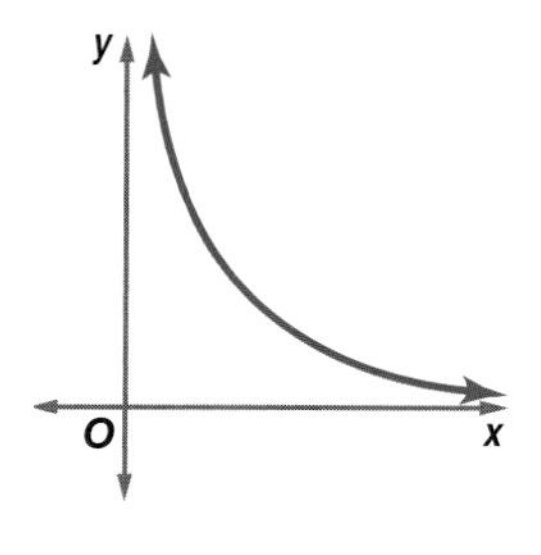

Maintain Your Skills

Mixed Review **Write an equivalent exponential or logarithmic equation.** *(Lesson 10-5)*

25. $e^3 = y$ **26.** $e^{4n-2} = 29$ **27.** $\ln 4 + 2 \ln x = 8$

Solve each equation or inequality. Round to four decimal places. *(Lesson 10-4)*

28. $16^x = 70$ **29.** $2^{3p} > 1000$ **30.** $\log_b 81 = 2$

BUSINESS **For Exercises 31–33, use the following information.**
The board of a small corporation decided that 8% of the annual profits would be divided among the six managers of the corporation. There are two sales managers and four nonsales managers. Fifty percent of the amount would be split equally among all six managers. The other 50% would be split among the four nonsales managers. Let p represent the annual profits of the corporation. *(Lesson 9-2)*

31. Write an expression to represent the share of the profits each nonsales manager will receive.

32. Simplify this expression.

33. Write an expression in simplest form to represent the share of the profits each sales manager will receive.

Without writing the equation in standard form, state whether the graph of each equation is a *parabola, circle, ellipse,* or *hyperbola.* *(Lesson 8-6)*

34. $4y^2 - 3x^2 + 8y - 24x = 50$ **35.** $7x^2 - 42x + 6y^2 - 24y = -45$

36. $y^2 + 3x - 8y = 4$ **37.** $x^2 + y^2 - 6x + 2y + 5 = 0$

AGRICULTURE **For Exercises 38–40, use the graph at the right.** *(Lesson 5-1)*

38. Write the number of pounds of pecans produced by U.S. growers in 2000 in scientific notation.

39. Write the number of pounds of pecans produced by the state of Georgia in 2000 in scientific notation.

40. What percent of the overall pecan production for 2000 can be attributed to Georgia?

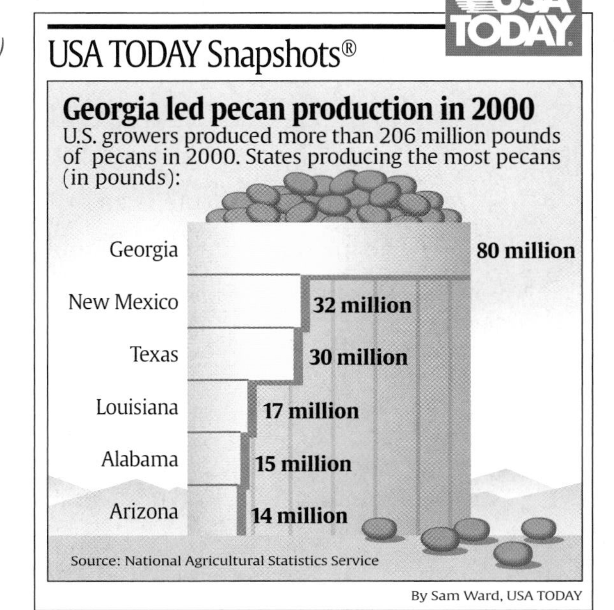

WebQuest **Internet Project**

On Quake Anniversary, Japan Still Worries

It is time to complete your project. Use the information and data you have gathered about earthquakes to prepare a research report or Web page. Be sure to include graphs, tables, diagrams, and any calculations you need for the earthquake you chose.

www.algebra2.com/webquest

Chapter 10

Study Guide and Review

Vocabulary and Concept Check

Change of Base Formula (p. 548)
common logarithm (p. 547)
exponential decay (p. 524)
exponential equation (p. 526)
exponential function (p. 524)
exponential growth (p. 524)
exponential inequality (p. 527)
logarithm (p. 531)
logarithmic equation (p. 533)
logarithmic function (p. 532)
logarithmic inequality (p. 533)
natural base, e (p. 554)
natural base exponential function (p. 554)
natural logarithm (p. 554)
natural logarithmic function (p. 554)
Power Property of Logarithms (p. 543)
Product Property of Logarithms (p. 541)
Property of Equality for Exponential Functions (p. 526)
Property of Equality for Logarithmic Functions (p. 534)
Property of Inequality for Exponential Functions (p. 527)
Property of Inequality for Logarithmic Functions (p. 534)
Quotient Property of Logarithms (p. 542)
rate of decay (p. 560)
rate of growth (p. 562)

State whether each sentence is *true* or *false*. If false, replace the underlined word(s) to make a true statement.

1. If $24^{2y+3} = 24^{y-4}$, then $2y + 3 = y - 4$ by the *Property of Equality for Exponential Functions* .
2. The number of bacteria in a petri dish over time is an example of *exponential decay*.
3. The *natural logarithm* is the inverse of the exponential function with base 10.
4. The *Power Property of Logarithms* shows that $\ln 9 < \ln 81$.
5. If a savings account yields 2% interest per year, then 2% is the *rate of growth*.
6. Radioactive half-life is an example of *exponential decay*.
7. The inverse of an exponential function is a *composite function*.
8. The *Quotient Property of Logarithms* is shown by $\log_4 2x = \log_4 2 + \log_4 x$.
9. The function $f(x) = 2(5)^x$ is an example of a *quadratic function*.

Lesson-by-Lesson Review

10-1 Exponential Functions

See pages 523–530.

Concept Summary

- An exponential function is in the form $y = ab^x$, where $a \neq 0$, $b > 0$, and $b \neq 1$.
- The function $y = ab^x$ represents exponential growth for $a > 0$ and $b > 1$, and exponential decay for $a > 0$ and $0 < b < 1$.
- Property of Equality for Exponential Functions: If b is a positive number other than 1, then $b^x = b^y$ if and only if $x = y$.
- Property of Inequality for Exponential Functions: If $b > 1$, then $b^x > b^y$ if and only if $x > y$, and $b^x < b^y$ if and only if $x < y$.

Example **Solve $64 = 2^{3n+1}$ for n.**

$64 = 2^{3n+1}$ Original equation

$2^6 = 2^{3n+1}$ Rewrite 64 as 2^6 so each side has the same base.

$6 = 3n + 1$ Property of Equality for Exponential Functions

$\frac{5}{3} = n$ The solution is $\frac{5}{3}$.

Exercises **Determine whether each function represents exponential *growth* or *decay*.** *See Example 2 on page 525.*

10. $y = 5(0.7)^x$

11. $y = \frac{1}{3}(4)^x$

Write an exponential function whose graph passes through the given points.
See Example 3 on page 525.

12. $(0, -2)$ and $(3, -54)$

13. $(0, 7)$ and $(1, 1.4)$

Solve each equation or inequality. *See Examples 5 and 6 on pages 526 and 527.*

14. $9^x = \frac{1}{81}$

15. $2^{6x} = 4^{5x+2}$

16. $49^{3p+1} = 7^{2p-5}$

17. $9^{x^2} \leq 27^{x^2-2}$

10-2 Logarithms and Logarithmic Functions

See pages 531–538.

Concept Summary

- Suppose $b > 0$ and $b \neq 1$. For $x > 0$, there is a number y such that $\log_b x = y$ if and only if $b^y = x$.
- Logarithmic to exponential inequality:
  If $b > 1$, $x > 0$, and $\log_b x > y$, then $x > b^y$.

  If $b > 1$, $x > 0$, and $\log_b x < y$, then $0 < x < b^y$.
- Property of Equality for Logarithmic Functions:
  If b is a positive number other than 1, then $\log_b x = \log_b y$ if and only if $x = y$.
- Property of Inequality for Logarithmic Functions:
  If $b > 1$, then $\log_b x > \log_b y$ if and only if $x > y$, and $\log_b x < \log_b y$ if and only if $x < y$.

Examples

$\log_7 x = 2 \rightarrow 7^2 = x$

$\log_2 x > 5 \rightarrow x > 2^5$

$\log_3 x < 4 \rightarrow 0 < x < 3^4$

If $\log_5 x = \log_5 6$, then $x = 6$.

If $\log_4 x > \log_4 10$, then $x > 10$.

Examples **1** **Solve $\log_9 n > \frac{3}{2}$.**

$\log_9 n > \frac{3}{2}$ Original inequality

$n > 9^{\frac{3}{2}}$ Logarithmic to exponential inequality

$n > (3^2)^{\frac{3}{2}}$ $9 = 3^2$

$n > 3^3$ Power of a Power

$n > 27$ Simplify.

2 **Solve $\log_3 12 = \log_3 2x$.**

$\log_3 12 = \log_3 2x$	Original equation
$12 = 2x$	Property of Equality for Logarithmic Functions
$6 = x$	Divide each side by 2.

Exercises **Write each equation in logarithmic form.** *See Example 1 on page 532.*

18. $7^3 = 343$ **19.** $5^{-2} = \frac{1}{25}$ **20.** $4^{\frac{3}{2}} = 8$

Write each equation in exponential form. *See Example 2 on page 532.*

21. $\log_4 64 = 3$ **22.** $\log_8 2 = \frac{1}{3}$ **23.** $\log_6 \frac{1}{36} = -2$

Evaluate each expression. *See Examples 3 and 4 on pages 532 and 533.*

24. $4^{\log_4 9}$ **25.** $\log_7 7^{-5}$ **26.** $\log_{81} 3$ **27.** $\log_{13} 169$

Solve each equation or inequality. *See Examples 5–8 on pages 533 and 534.*

28. $\log_4 x = \frac{1}{2}$
29. $\log_{81} 729 = x$
30. $\log_b 9 = 2$
31. $\log_8 (3y - 1) < \log_8 (y + 5)$
32. $\log_5 12 < \log_5 (5x - 3)$
33. $\log_8 (x^2 + x) = \log_8 12$

10-3 Properties of Logarithms

See pages 541–546.

Concept Summary

- The logarithm of a product is the sum of the logarithms of its factors.
- The logarithm of a quotient is the difference of the logarithms of the numerator and the denominator.
- The logarithm of a power is the product of the logarithm and the exponent.

Use $\log_{12} 9 \approx 0.884$ and $\log_{12} 18 \approx 1.163$ to approximate the value of $\log_{12} 2$.

$\log_{12} 2 = \log_{12} \frac{18}{9}$	Replace 2 with $\frac{18}{9}$.
$= \log_{12} 18 - \log_{12} 9$	Quotient Property
$\approx 1.163 - 0.884$ or 0.279	Replace $\log_{12} 9$ with 0.884 and $\log_{12} 18$ with 1.163.

Exercises **Use $\log_9 7 \approx 0.8856$ and $\log_9 4 \approx 0.6309$ to approximate the value of each expression.** *See Examples 1 and 2 on page 542.*

34. $\log_9 28$ **35.** $\log_9 49$ **36.** $\log_9 144$

Solve each equation. *See Example 5 on page 543.*

37. $\log_2 y = \frac{1}{3} \log_2 27$
38. $\log_5 7 + \frac{1}{2} \log_5 4 = \log_5 x$
39. $2 \log_2 x - \log_2 (x + 3) = 2$
40. $\log_3 x - \log_3 4 = \log_3 12$
41. $\log_6 48 - \log_6 \frac{16}{5} + \log_6 5 = \log_6 5x$
42. $\log_7 m = \frac{1}{3} \log_7 64 + \frac{1}{2} \log_7 121$

10-4 Common Logarithms

See pages 547–551.

Concept Summary

- Base 10 logarithms are called common logarithms and are usually written without the subscript 10: $\log_{10} x = \log x$.
- You use the inverse of logarithms, or exponentiation, to solve equations or inequalities involving common logarithms: $10^{\log x} = x$.
- The Change of Base Formula: $\log_a n = \frac{\log_b n}{\log_b a}$ ← log base b original number; ← log base b old base

Example **Solve $5^x = 7$.**

$5^x = 7$	Original equation
$\log 5^x = \log 7$	Property of Equality for Logarithmic Functions
$x \log 5 = \log 7$	Power Property of Logarithms
$x = \frac{\log 7}{\log 5}$	Divide each side by log 5.
$x \approx \frac{0.8451}{0.6990}$ or 1.2090	Use a calculator.

Exercises **Solve each equation or inequality. Round to four decimal places.** *See Examples 3 and 4 on page 548.*

43. $2^x = 53$ **44.** $2.3^{x^2} = 66.6$ **45.** $3^{4x-7} < 4^{2x+3}$

46. $6^{3y} = 8^{y-1}$ **47.** $12^{x-5} \geq 9.32$ **48.** $2.1^{x-5} = 9.32$

Express each logarithm in terms of common logarithms. Then approximate its value to four decimal places. *See Example 5 on page 549.*

49. $\log_4 11$ **50.** $\log_2 15$ **51.** $\log_{20} 1000$

10-5 Base e and Natural Logarithms

See pages 554–559.

Concept Summary

- You can write an equivalent base e exponential equation for a natural logarithmic equation and vice versa by using the fact that $\ln x = \log_e x$.
- Since the natural base function and the natural logarithmic function are inverses, these two functions can be used to "undo" each other.
  $e^{\ln x} = x$ and $\ln e^x = x$

Example **Solve $\ln (x + 4) > 5$.**

$\ln (x + 4) > 5$	Original inequality
$e^{\ln (x + 4)} > e^5$	Write each side using exponents and base e.
$x + 4 > e^5$	Inverse Property of Exponents and Logarithms
$x > e^5 - 4$	Subtract 4 from each side.
$x > 144.4132$	Use a calculator.

• Extra Practice, see pages 849–851.
• Mixed Problem Solving, see page 871.

Exercises **Write an equivalent exponential or logarithmic equation.**
See Example 3 on page 555.

52. $e^x = 6$ **53.** $\ln 7.4 = x$

Evaluate each expression. *See Example 4 on page 555.*

54. $e^{\ln 12}$ **55.** $\ln e^{7x}$

Solve each equation or inequality.
See Examples 5 and 7 on pages 555 and 556.

56. $2e^x - 4 = 1$ **57.** $e^x > 3.2$ **58.** $-4e^{2x} + 15 = 7$
59. $\ln 3x \leq 5$ **60.** $\ln (x - 10) = 0.5$ **61.** $\ln x + \ln 4x = 10$

10-6 Exponential Growth and Decay

See pages 560–565.

Concept Summary

- Exponential decay: $y = a(1 - r)^t$ or $y = ae^{-kt}$
- Exponential growth: $y = a(1 + r)^t$ or $y = ae^{kt}$

Example BIOLOGY **A certain culture of bacteria will grow from 500 to 4000 bacteria in 1.5 hours. Find the constant k for the growth formula. Use $y = ae^{kt}$.**

$y = ae^{kt}$ Exponential growth formula

$4000 = 500e^{k(1.5)}$ Replace y with 4000, a with 500, and t with 1.5.

$8 = e^{1.5k}$ Divide each side by 500.

$\ln 8 = \ln e^{1.5k}$ Property of Equality for Logarithmic Functions

$\ln 8 = 1.5k$ Inverse Property of Exponents and Logarithms

$\frac{\ln 8}{1.5} = k$ Divide each side by 1.5.

$1.3863 \approx k$ Use a calculator.

Exercises *See Examples 1–4 on pages 560–562.*

62. BUSINESS Able Industries bought a fax machine for $250. It is expected to depreciate at a rate of 25% per year. What will be the value of the fax machine in 3 years?

63. BIOLOGY For a certain strain of bacteria, k is 0.872 when t is measured in days. How long will it take 9 bacteria to increase to 738 bacteria?

64. CHEMISTRY Radium-226 decomposes radioactively. Its half-life, the time it takes for half of the sample to decompose, is 1800 years. Find the constant k in the decay formula for this compound.

65. POPULATION The population of a city 10 years ago was 45,600. Since then, the population has increased at a steady rate each year. If the population is currently 64,800, find the annual rate of growth for this city.

Chapter 10 Practice Test

Vocabulary and Concepts

Choose the term that best completes each sentence.

1. The equation $y = 0.3(4)^x$ is an exponential (*growth, decay*) function.
2. The logarithm of a quotient is the (*sum, difference*) of the logarithms of the numerator and the denominator.
3. The base of a natural logarithm is (*10, e*).

Skills and Applications

4. Write $3^7 = 2187$ in logarithmic form.
5. Write $\log_8 16 = \frac{4}{3}$ in exponential form.
6. Write an exponential function whose graph passes through (0, 0.4) and (2, 6.4).
7. Express $\log_3 5$ in terms of common logarithms.
8. Evaluate $\log_2 \frac{1}{32}$.

Use $\log_4 7 \approx 1.4037$ and $\log_4 3 \approx 0.7925$ to approximate the value of each expression.

9. $\log_4 21$
10. $\log_4 \frac{7}{12}$

Simplify each expression.

11. $\left(3^{\sqrt{8}}\right)^{\sqrt{2}}$
12. $81^{\sqrt{5}} \div 3^{\sqrt{5}}$

Solve each equation or inequality. Round to four decimal places if necessary.

13. $2^{x-3} = \frac{1}{16}$
14. $27^{2p+1} = 3^{4p-1}$
15. $\log_2 x < 7$
16. $\log_m 144 = -2$
17. $\log_3 x - 2\log_3 2 = 3\log_3 3$
18. $\log_9 (x + 4) + \log_9 (x - 4) = 1$
19. $\log_5 (8y - 7) = \log_5 (y^2 + 5)$
20. $\log_3 3^{(4x-1)} = 15$
21. $7.6^{x-1} = 431$
22. $\log_2 5 + \frac{1}{3}\log_2 27 = \log_2 x$
23. $3^x = 5^{x-1}$
24. $4^{2x-3} = 9^{x+3}$
25. $e^{3y} > 6$
26. $2e^{3x} + 5 = 11$
27. $\ln 3x - \ln 15 = 2$

COINS **For Exercises 28 and 29, use the following information.**
You buy a commemorative coin for \$25. The value of the coin increases 3.25% per year.

28. How much will the coin be worth in 15 years?
29. After how many years will the coin have doubled in value?

Standards Practice

30. **STANDARDIZED TEST PRACTICE** Which equation represents the graph at the right?

Ⓐ $y = x^2 + 2$

Ⓑ $y = 2^x$

Ⓒ $y = \log_2 x$

Ⓓ $y = 2^x + 1$

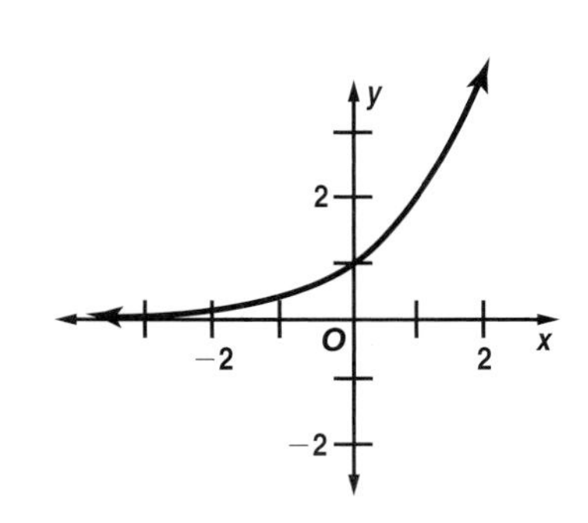

Chapter 10 Standardized Test Practice

Standards Practice

Part 1 Multiple Choice

Record your answers on the answer sheet provided by your teacher or on a sheet of paper.

1. The arc shown is part of a circle. Find the area of the shaded region.

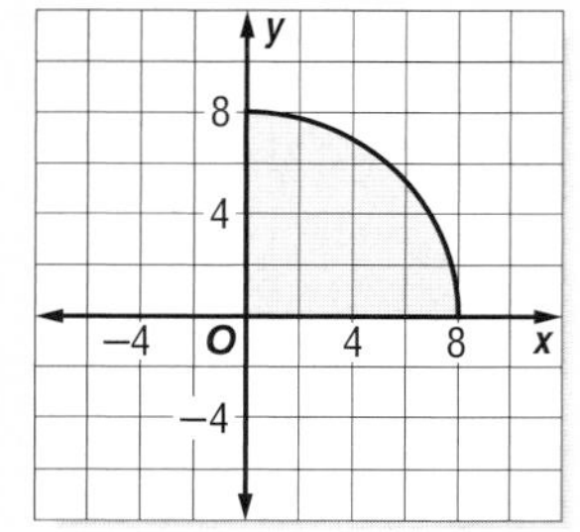

(A) 8π units^2

(B) 16π units^2

(C) 32π units^2

(D) 64π units^2

2. If line ℓ is parallel to line m in the figure below, what is the value of x?

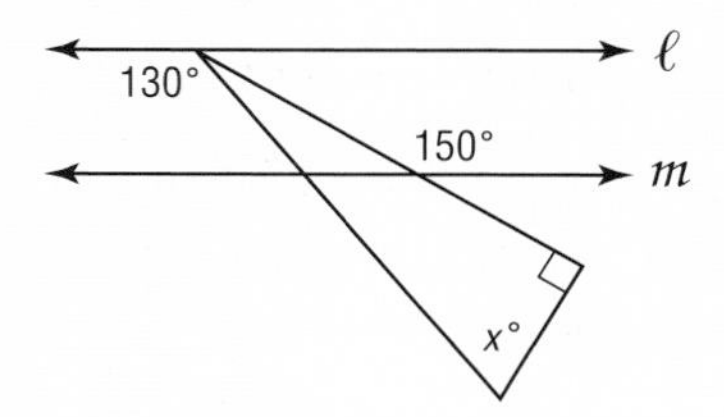

(A) 40 (B) 50

(C) 60 (D) 70

3. According to the graph, what was the percent of increase in sales from 1998 to 2000?

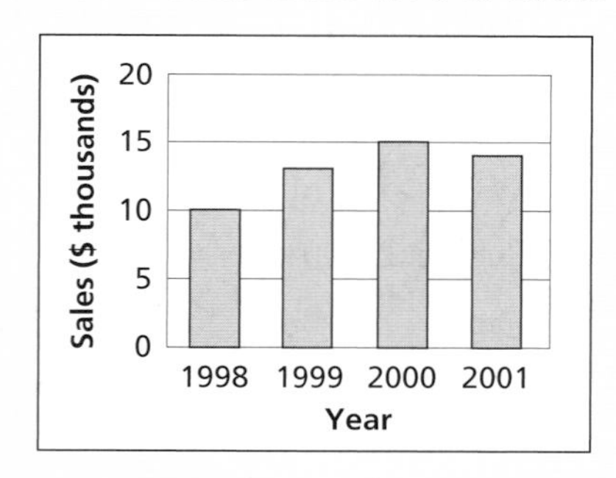

(A) 5% (B) 15% (C) 25% (D) 50%

4. What is the x-intercept of the line described by the equation $y = 2x + 5$?

(A) -5 (B) $-\frac{5}{2}$

(C) 0 (D) $\frac{5}{2}$

5. $\frac{(xy)^2z^0}{y^2x^3} =$

(A) $\frac{1}{x^2y}$ (B) $\frac{z}{x^2}$

(C) $\frac{z}{x}$ (D) $\frac{1}{x}$

6. If $\frac{v^2 - 36}{6 - v} = 10$, then $v =$

(A) -16. (B) -4.

(C) 4. (D) 8.

7. The expression $\frac{1}{3}\sqrt{45}$ is equivalent to

(A) $\sqrt{5}$. (B) $3\sqrt{5}$.

(C) 5. (D) 15.

8. What are all the values for x such that $x^2 < 3x + 18$?

(A) $x < -3$ (B) $-3 < x < 6$

(C) $x > -3$ (D) $x < 6$

9. If $f(x) = 2x^3 - 18x$, what are all the values of x at which $f(x) = 0$?

(A) 0, 3 (B) -3, 0, 3

(C) -6, 0, 6 (D) -3, 2, 3

10. Which of the following is equal to $\frac{17.5(10^{-2})}{500(10^{-4})}$?

(A) $0.035(10^{-2})$ (B) $0.35(10^{-2})$

(C) $0.0035(10^2)$ (D) $0.035(10^2)$

Test-Taking Tip

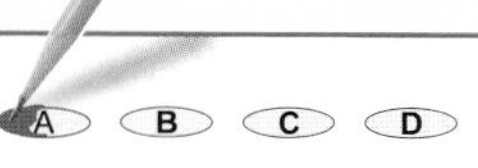

Question 7

You can use estimates to help you eliminate answer choices. For example, in Question 7, you can estimate that $\frac{1}{3}\sqrt{45}$ is less than $\frac{1}{3}\sqrt{49}$, which is $\frac{7}{3}$ or $2\frac{1}{3}$. Eliminate choices C and D.

Preparing for Standardized Tests
For test-taking strategies and more practice, see pages 877–892.

Part 2 Short Response/Grid In

Record your answers on the answer sheet provided by your teacher or on a sheet of paper.

11. If the outer diameter of a cylindrical tank is 62.46 centimeters and the inner diameter is 53.32 centimeters, what is the thickness of the tank?

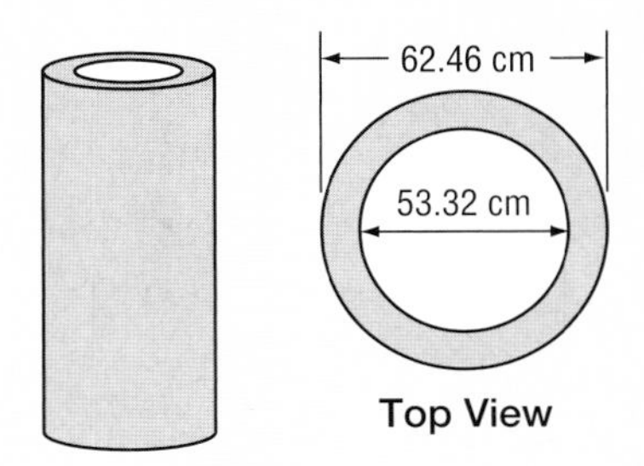

12. What number added to 80% of itself is equal to 45?

13. Of 200 families surveyed, 95% have at least one TV and 60% of those with TVs have more than 2 TVs. If 50 families have exactly 2 TVs, how many families have exactly 1 TV?

14. In the figure, if $ED = 8$, what is the measure of line segment AE?

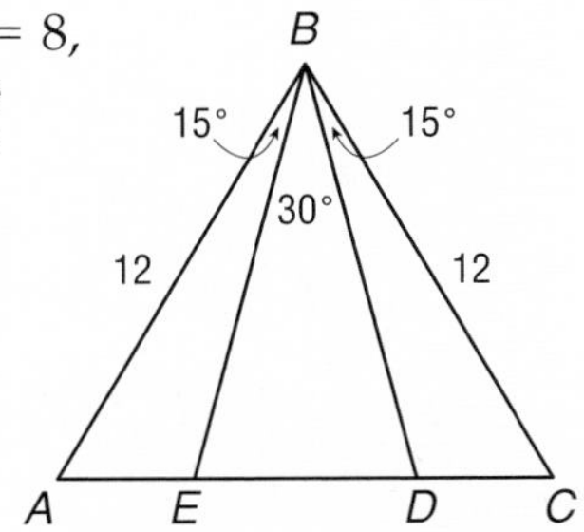

15. If $a \leftrightarrow b$ is defined as $a - b + ab$, find the value of $4 \leftrightarrow 2$.

16. If $6(m + k) = 26 + 4(m + k)$, what is the value of $m + k$?

17. If $y = 1 - x^2$ and $-3 \leq x \leq 1$, what number is found by subtracting the *least* possible value of y from the *greatest* possible value of y?

18. If $f(x) = (x - \pi)(x - 3)(x - e)$, what is the difference between the greatest and least roots of $f(x)$? Round to the nearest hundredth.

19. Suppose you deposit \$700 in an account that pays 2.5% interest compounded quarterly. What will the dollar value of the account be in 5 years? Round to the nearest penny.

20. Solve $128^{x+1} = 64^{3x}$ for x.

21. State the domain and range of the function $y = \log_5 (x + 1)$.

Part 3 Extended Response

Record your answers on a sheet of paper. Show your work.

For Exercises 22–25, use the following information.

In $y = 8.430\,(1.058)^x$, x represents the number of years since 1990, and y represents the approximate number of millions of Americans 7 years of age and older who participated in backpacking two or more times that year.

22. Describe how the number of millions of Americans who go backpacking is changing over time.

23. What do the numbers 8.430 and 1.058 represent?

24. About how many Americans went backpacking in 1996?

25. In what year would you expect the number of Americans who participated in backpacking to reach 20 million for the first time?

26. Sketch the graphs of $f(x) = 2^x$ and $g(x) = \log_2 x$.

    a. Study the graphs and describe the relationship between $f(x)$ and $g(x)$.

    b. Specify the domain and range of $g(x)$.

UNIT 4

Discrete Mathematics

Discrete mathematics is the branch of mathematics that involves finite or discontinuous quantities. In this unit, you will learn about sequences, series, probability, and statistics.

Richard Kaye
Professor of Mathematics
University of Birmingham

Chapter 11
Sequences and Series

Chapter 12
Probability and Statistics

WebQuest Internet Project

'Minesweeper': Secret to Age-Old Puzzle?

Source: *USA TODAY,* November 3, 2000

"*Minesweeper*, a seemingly simple game included on most personal computers, could help mathematicians crack one of the field's most intriguing problems. The buzz began after Richard Kaye, a mathematics professor at the University of Birmingham in England, started playing *Minesweeper*. After playing the game steadily for a few weeks, Kaye realized that *Minesweeper*, if played on a much larger grid, has the same mathematical characteristics as other problems deemed insolvable." In this project, you will research a mathematician of the past and his or her role in the development of discrete mathematics.

Log on to www.algebra2.com/webquest. Begin your WebQuest by reading the Task.

Then continue working on your WebQuest as you study Unit 4.

Lesson	11-7	12-1
Page	616	635

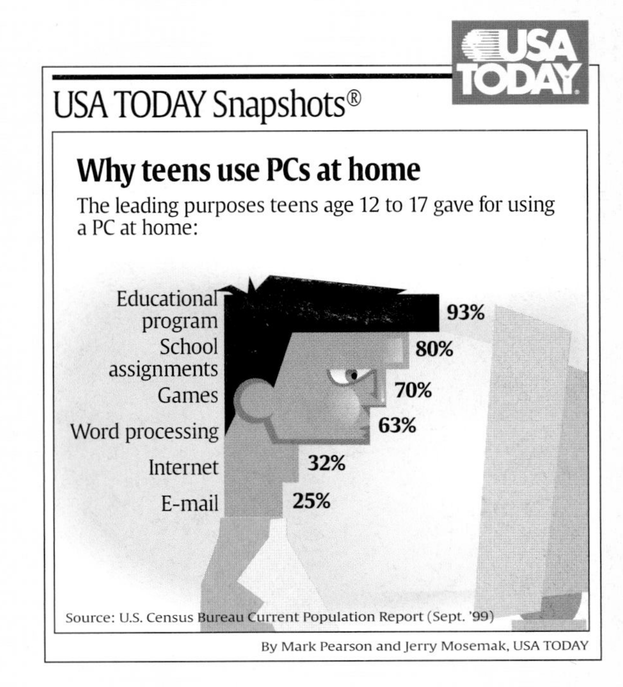

Chapter 11 Sequences and Series

What You'll Learn

- **Lessons 11-1 through 11-5** Use arithmetic and geometric sequences and series.
- **Lesson 11-6** Use special sequences and iterate functions.
- **Lesson 11-7** Expand powers by using the Binomial Theorem.
- **Lesson 11-8** Prove statements by using mathematical induction.

Key Vocabulary

- arithmetic sequence (p. 578)
- arithmetic series (p. 583)
- sigma notation (p. 585)
- geometric sequence (p. 588)
- geometric series (p. 594)

Why It's Important

Many number patterns found in nature and used in business can be modeled by sequences, which are lists of numbers. Some sequences are classified by the method used to predict the next term from the previous term(s). When the terms of a sequence are added, a series is formed. *In Lesson 11-2, you will learn how the number of seats in the rows of an amphitheater can be modeled using a series.*

Getting Started

Prerequisite Skills To be successful in this chapter, you'll need to master these skills and be able to apply them in problem-solving situations. Review these skills before beginning Chapter 11.

For Lessons 11-1 and 11-3 — Solve Equations

Solve each equation. *(For review, see Lessons 1-3 and 5-5.)*

1. $36 = 12 + 4x$

2. $-40 = 10 + 5x$

3. $12 - 3x = 27$

4. $162 = 2x^4$

5. $\frac{1}{8} = 4x^5$

6. $3x^3 + 4 = -20$

For Lessons 11-1 and 11-5 — Graph Functions

Graph each function. *(For review, see Lesson 2-1.)*

7. $\{(1, 1), (2, 3), (3, 5), (4, 7), (5, 9)\}$

8. $\{(1, -20), (2, -16), (3, -12), (4, -8), (5, -4)\}$

9. $\left\{(1, 64), (2, 16), (3, 4), (4, 1), \left(5, \frac{1}{4}\right)\right\}$

10. $\left\{(1, 2), (2, 3), \left(3, \frac{7}{2}\right), \left(4, \frac{15}{4}\right), \left(5, \frac{31}{8}\right)\right\}$

For Lessons 11-1 through 11-5, 11-8 — Evaluate Expressions

Evaluate each expression for the given value(s) of the variable(s). *(For review, see Lesson 1-1.)*

11. $x + (y - 1)z$ if $x = 3$, $y = 8$, and $z = 2$

12. $\frac{x}{2}(y + z)$ if $x = 10$, $y = 3$, and $z = 25$

13. $a \cdot b^{c-1}$ if $a = 2$, $b = \frac{1}{2}$, and $c = 7$

14. $\frac{a(1 - bc)^2}{1 - b}$ if $a = -2$, $b = 3$, and $c = 5$

15. $\frac{a}{1 - b}$ if $a = \frac{1}{2}$, and $b = \frac{1}{6}$

16. $\frac{n(n + 1)}{2}$ if $n = 10$

Sequences and Series Make this Foldable to help you organize your notes. Begin with one sheet of 11" by 17" paper and four sheets of notebook paper.

Step 1 Fold and Cut

Fold the short sides of the 11" by 17" paper to meet in the middle.

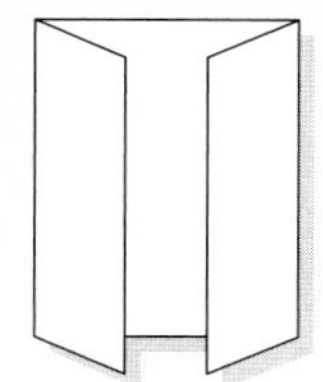

Step 2 Staple and Label

Fold the notebook paper in half lengthwise. Insert two sheets of notebook paper in each tab and staple the edges. Label with lesson numbers.

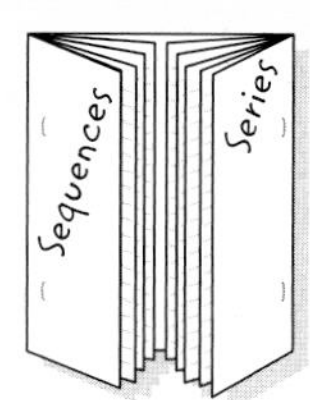

Reading and Writing As you read and study the chapter, fill the journal with examples for each lesson.

11-1 Arithmetic Sequences

California Standards **Standard 22.0 Students find the general term** and the sums **of arithmetic series** and of both finite and infinite geometric series.

What You'll Learn

- Use arithmetic sequences.
- Find arithmetic means.

How are arithmetic sequences related to roofing?

A roofer is nailing shingles to the roof of a house in overlapping rows. There are three shingles in the top row. Since the roof widens from top to bottom, one additional shingle is needed in each successive row.

Row	1	2	3	4	5	6	7
Shingles	3	4	5	6	7	8	9

Vocabulary

- sequence
- term
- arithmetic sequence
- common difference
- arithmetic means

ARITHMETIC SEQUENCES The numbers 3, 4, 5, 6, ..., representing the number of shingles in each row, are an example of a sequence of numbers. A **sequence** is a list of numbers in a particular order. Each number in a sequence is called a **term**. The first term is symbolized by a_1, the second term is symbolized by a_2, and so on.

The graph represents the information from the table above. A sequence is a discrete function whose domain is the set of positive integers.

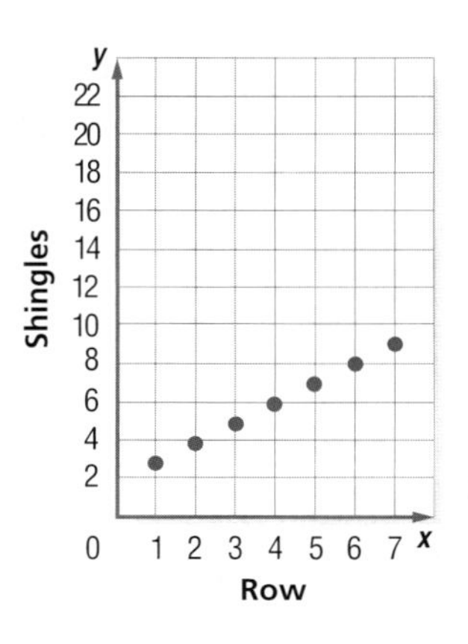

Many sequences have patterns. For example, in the sequence above for the number of shingles, each term can be found by adding 1 to the previous term. A sequence of this type is called an arithmetic sequence. An **arithmetic sequence** is a sequence in which each term after the first is found by adding a constant, called the **common difference** d, to the previous term.

Study Tip

Sequences

The numbers in a sequence may not be ordered. For example, the numbers 33, 25, 36, 40, 36, 66, 63, 50, ... are a sequence that represents the number of home runs Sammy Sosa hit in each year beginning with 1993.

Example 1 Find the Next Terms

Find the next four terms of the arithmetic sequence 55, 49, 43, ...

Find the common difference d by subtracting two consecutive terms.

$49 - 55 = -6$ and $43 - 49 = -6$ So, $d = -6$.

Now add -6 to the third term of the sequence, and then continue adding -6 until the next four terms are found.

$$43 \xrightarrow{+(-6)} 37 \xrightarrow{+(-6)} 31 \xrightarrow{+(-6)} 25 \xrightarrow{+(-6)} 19$$

The next four terms of the sequence are 37, 31, 25, and 19.

It is possible to develop a formula for each term of an arithmetic sequence in terms of the first term a_1 and the common difference d. Consider Example 1.

Sequence	numbers	55	49	43	37	...	
	symbols	a_1	a_2	a_3	a_4	...	a_n
Expressed in Terms of d and the First Term	numbers	$55 + 0(-6)$	$55 + 1(-6)$	$55 + 2(-6)$	$55 + 3(-6)$	...	$55 + (n - 1)(-6)$
	symbols	$a_1 + 0 \cdot d$	$a_1 + 1 \cdot d$	$a_1 + 2 \cdot d$	$a_1 + 3 \cdot d$	...	$a_1 + (n - 1)d$

The following formula generalizes this pattern for any arithmetic sequence.

Key Concept *nth Term of an Arithmetic Sequence*

The nth term a_n of an arithmetic sequence with first term a_1 and common difference d is given by

$$a_n = a_1 + (n - 1)d,$$

where n is any positive integer.

More About. . .

Construction

The table below shows typical costs for a construction company to rent a crane for one, two, three, or four months.

Months	Cost ($)
1	75,000
2	90,000
3	105,000
4	120,000

Source: www.howstuffworks.com

Example 2 Find a Particular Term

CONSTRUCTION **Refer to the information at the left. Assuming that the arithmetic sequence continues, how much would it cost to rent the crane for twelve months?**

Explore Since the difference between any two successive costs is $15,000, the costs form an arithmetic sequence with common difference 15,000.

Plan You can use the formula for the nth term of an arithmetic sequence with $a_1 = 75{,}000$ and $d = 15{,}000$ to find a_{12}, the cost for twelve months.

Solve

$a_n = a_1 + (n - 1)d$ Formula for nth term

$a_{12} = 75{,}000 + (12 - 1)15{,}000$ $n = 12$, $a_1 = 75{,}000$, $d = 15{,}000$

$a_{12} = 240{,}000$ Simplify.

It would cost $240,000 to rent the crane for twelve months.

Examine You can find terms of the sequence by adding 15,000. a_5 through a_{12} are 135,000, 150,000, 165,000, 180,000, 195,000, 210,000, 225,000, and 240,000. Therefore, $240,000 is correct.

Example 3 Write an Equation for the nth Term

Write an equation for the nth term of the arithmetic sequence 8, 17, 26, 35, ...

In this sequence, $a_1 = 8$ and $d = 9$. Use the nth term formula to write an equation.

$a_n = a_1 + (n - 1)d$ Formula for nth term

$a_n = 8 + (n - 1)9$ $a_1 = 8$, $d = 9$

$a_n = 8 + 9n - 9$ Distributive Property

$a_n = 9n - 1$ Simplify.

An equation is $a_n = 9n - 1$.

Algebra Activity

Arithmetic Sequences

Study the figures below. The length of an edge of each cube is 1 centimeter.

 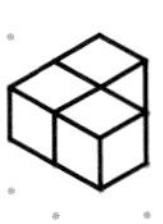 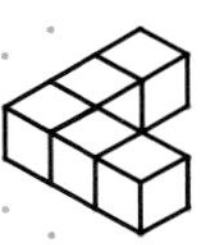

Model and Analyze

1. Based on the pattern, draw the fourth figure on a piece of isometric dot paper.
2. Find the volumes of the four figures.
3. Suppose the number of cubes in the pattern continues. Write an equation that gives the volume of Figure n.
4. What would the volume of the twelfth figure be?

ARITHMETIC MEANS Sometimes you are given two terms of a sequence, but they are not successive terms of that sequence. The terms between any two nonsuccessive terms of an arithmetic sequence are called **arithmetic means**. In the sequence below, 41, 52, and 63 are the three arithmetic means between 30 and 74.

$$19, 30, \underbrace{41, 52, 63}, 74, 85, 96, \ldots$$

3 arithmetic means between 30 and 74

Example 4 Find Arithmetic Means

Find the four arithmetic means between 16 and 91.

You can use the nth term formula to find the common difference. In the sequence 16, __?__, __?__, __?__, __?__, 91, ..., a_1 is 16 and a_6 is 91.

$a_n = a_1 + (n - 1)d$	Formula for the nth term
$a_6 = 16 + (6 - 1)d$	$n = 6$, $a_1 = 16$
$91 = 16 + 5d$	$a_6 = 91$
$75 = 5d$	Subtract 16 from each side.
$15 = d$	Divide each side by 5.

Now use the value of d to find the four arithmetic means.

16 → (+ 15) → 31 → (+ 15) → 46 → (+ 15) → 61 → (+ 15) → 76

The arithmetic means are 31, 46, 61, and 76. **CHECK** $76 + 15 = 91$ ✓

Study Tip

Alternate Method

You may prefer this method. The four means will be $16 + d$, $16 + 2d$, $16 + 3d$, and $16 + 4d$. The common difference is $d = 91 - (16 + 4d)$ or $d = 15$.

Check for Understanding

Concept Check

1. **Explain** why the sequence 4, 5, 7, 10, 14, ... is not arithmetic.
2. **Find** the 15th term in the arithmetic sequence $-3, 4, 11, 18, \ldots$.
3. **OPEN ENDED** Write an arithmetic sequence with common difference -5.

Guided Practice

Find the next four terms of each arithmetic sequence.

4. 12, 16, 20, ...
5. $3, 1, -1, \ldots$

Find the first five terms of each arithmetic sequence described.

6. $a_1 = 5, d = 3$
7. $a_1 = 14, d = -2$

8. Find a_{13} for the arithmetic sequence $-17, -12, -7, \ldots$.

Find the indicated term of each arithmetic sequence.

9. $a_1 = 3, d = -5, n = 24$

10. $a_1 = -5, d = 7, n = 13$

11. Complete: 68 is the __?__ th term of the arithmetic sequence $-2, 3, 8, \ldots$.

12. Write an equation for the nth term of the arithmetic sequence $-26, -15, -4, 7, \ldots$.

13. Find the three arithmetic means between 44 and 92.

Application

14. **ENTERTAINMENT** A basketball team has a halftime promotion where a fan gets to shoot a 3-pointer to try to win a jackpot. The jackpot starts at $5000 for the first game and increases $500 each time there is no winner. Ken has tickets to the fifteenth game of the season. How much will the jackpot be for that game if no one wins by then?

Practice and Apply

Homework Help

For Exercises	See Examples
15–28, 49	1
29–45, 51	2
46–48, 50	3
52–55	4

Extra Practice
See page 851.

Find the next four terms of each arithmetic sequence.

15. $9, 16, 23, \ldots$

16. $31, 24, 17, \ldots$

17. $-6, -2, 2, \ldots$

18. $-8, -5, -2, \ldots$

19. $\frac{1}{3}, 1, \frac{5}{3}, \ldots$

20. $\frac{18}{5}, \frac{16}{5}, \frac{14}{5}, \ldots$

21. $6.7, 6.3, 5.9, \ldots$

22. $1.3, 3.8, 6.3, \ldots$

Find the first five terms of each arithmetic sequence described.

23. $a_1 = 2, d = 13$

24. $a_1 = 41, d = 5$

25. $a_1 = 6, d = -4$

26. $a_1 = 12, d = -3$

27. $a_1 = \frac{4}{3}, d = -\frac{1}{3}$

28. $a_1 = \frac{5}{8}, d = \frac{3}{8}$

29. Find a_8 if $a_n = 4 + 3n$.

30. If $a_n = 1 - 5n$, what is a_{10}?

Find the indicated term of each arithmetic sequence.

31. $a_1 = 3, d = 7, n = 14$

32. $a_1 = -4, d = -9, n = 20$

33. $a_1 = 35, d = 3, n = 101$

34. $a_1 = 20, d = 4, n = 81$

35. $a_1 = 5, d = \frac{1}{3}, n = 12$

36. $a_1 = \frac{5}{2}, d = -\frac{3}{2}, n = 11$

37. a_{12} for $-17, -13, -9, \ldots$

38. a_{12} for $8, 3, -2, \ldots$

39. a_{21} for $121, 118, 115, \ldots$

40. a_{43} for $5, 9, 13, 17, \ldots$

41. **GEOLOGY** Geologists estimate that the continents of Europe and North America are drifting apart at a rate of an average of 12 miles every 1 million years, or about 0.75 inch per year. If the continents continue to drift apart at that rate, how many inches will they drift in 50 years? (*Hint*: $a_1 = 0.75$)

42. **TOWER OF PISA** To prove that objects of different weights fall at the same rate, Galileo dropped two objects with different weights from the Leaning Tower of Pisa in Italy. The objects hit the ground at the same time. When an object is dropped from a tall building, it falls about 16 feet in the first second, 48 feet in the second second, and 80 feet in the third second, regardless of its weight. How many feet would an object fall in the tenth second?

Complete the statement for each arithmetic sequence.

43. 170 is the ___?___ term of $-4, 2, 8, \ldots$.

44. 124 is the ___?___ term of $-2, 5, 12, \ldots$.

45. -14 is the ___?___ term of $2\frac{1}{5}, 2, 1\frac{4}{5}, \ldots$.

Write an equation for the *n*th term of each arithmetic sequence.

46. 7, 16, 25, 34, ...

47. 18, 11, 4, -3, ...

48. $-3, -5, -7, -9, \ldots$

GEOMETRY **For Exercises 49–51, refer to the first three arrays of numbers below.**

49. Make drawings to find the next three numbers in this pattern.

50. Write an equation representing the *n*th number in this pattern.

51. Is 397 a number in this pattern? Explain.

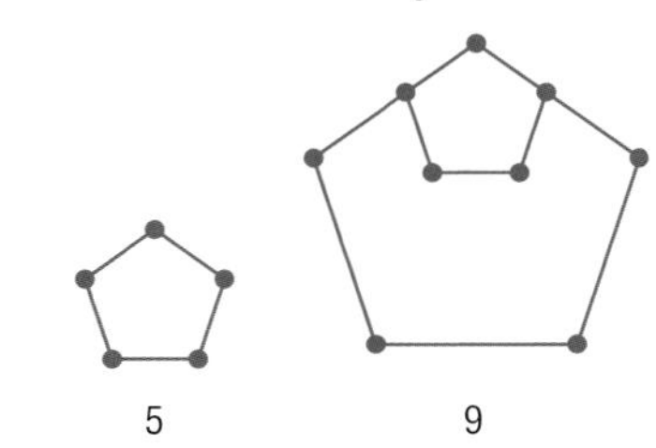

Find the arithmetic means in each sequence.

52. 55, ___?___, ___?___, ___?___, 115

53. 10, ___?___, ___?___, -8

54. -8, ___?___, ___?___, ___?___, ___?___, 7

55. 3, ___?___, ___?___, ___?___, ___?___, ___?___, 27

56. **CRITICAL THINKING** The numbers x, y, and z are the first three terms of an arithmetic sequence. Express z in terms of x and y.

57. **WRITING IN MATH** Answer the question that was posed at the beginning of the lesson.

How are arithmetic sequences related to roofing?

Include the following in your answer:

- the words that indicate that the numbers of shingles in the rows form an arithmetic sequence, and
- explanations of at least two ways to find the number of shingles in the fifteenth row.

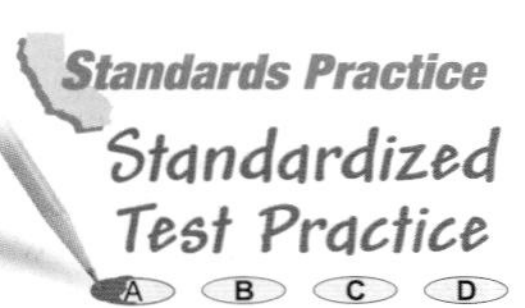

58. What number follows 20 in this arithmetic sequence?

8, 11, 14, 17, 20, ...

Ⓐ 5 Ⓑ 23 Ⓒ 26 Ⓓ 29

59. Find the first term in the arithmetic sequence.

_____, $8\frac{1}{3}, 7, 5\frac{2}{3}, 4\frac{1}{3}, \ldots$

Ⓐ 3 Ⓑ $9\frac{2}{3}$ Ⓒ $10\frac{1}{3}$ Ⓓ 11

Maintain Your Skills

Mixed Review

60. **COMPUTERS** Suppose a computer that costs \$3000 new is only worth \$600 after 3 years. What is the average annual rate of depreciation? *(Lesson 10-6)*

Solve each equation. *(Lesson 10-5)*

61. $3e^x - 2 = 0$

62. $e^{3x} = 4$

63. $\ln(x + 2) = 5$

64. If y varies directly as x and $y = 5$ when $x = 2$, find y when $x = 6$. *(Lesson 9-4)*

Getting Ready for the Next Lesson

PREREQUISITE SKILL **Evaluate each expression for the given values of the variable.** *(To review **evaluating expressions**, see Lesson 1-1.)*

65. $3n - 1$; $n = 1, 2, 3, 4$

66. $6 - j$; $j = 1, 2, 3, 4$

67. $4m + 7$; $m = 1, 2, 3, 4, 5$

11-2 Arithmetic Series

California Standards Standard 23.0 Students derive the summation formulas for arithmetic series and for both finite and infinite geometric series. (Key)

What You'll Learn

- Find sums of arithmetic series.
- Use sigma notation.

Vocabulary

- series
- arithmetic series
- sigma notation
- index of summation

How do arithmetic series apply to amphitheaters?

The first amphitheaters were built for contests between gladiators. Modern amphitheaters are usually used for the performing arts. Amphitheaters generally get wider as the distance from the stage increases.

Suppose a small amphitheater can seat 18 people in the first row and each row can seat 4 more people than the previous row.

Study Tip

Indicated Sum
The sum of a series is the result when the terms of the series are added. An *indicated sum* is the expression that illustrates the series, which includes the terms + or −.

ARITHMETIC SERIES The numbers of seats in the rows of the amphitheater form an arithmetic sequence. To find the number of people who could sit in the first four rows, add the first four terms of the sequence. That sum is 18 + 22 + 26 + 30 or 96. A **series** is an indicated sum of the terms of a sequence. Since 18, 22, 26, 30 is an arithmetic sequence, 18 + 22 + 26 + 30 is an **arithmetic series**. Below are some more arithmetic sequences and the corresponding arithmetic series.

Arithmetic Sequence	Arithmetic Series
$-9, -3, 3$	$-9 + (-3) + 3$
$\frac{3}{8}, \frac{8}{8}, \frac{13}{8}, \frac{18}{8}$	$\frac{3}{8} + \frac{8}{8} + \frac{13}{8} + \frac{18}{8}$

S_n represents the sum of the first n terms of a series. For example, S_4 is the sum of the first four terms.

To develop a formula for the sum of any arithmetic series, consider the series below.

$$S_9 = 4 + 11 + 18 + 25 + 32 + 39 + 46 + 53 + 60$$

Write S_9 in two different orders and add the two equations.

$$\begin{aligned} S_9 &= 4 + 11 + 18 + 25 + 32 + 39 + 46 + 53 + 60 \\ (+)\ S_9 &= 60 + 53 + 46 + 39 + 32 + 25 + 18 + 11 + 4 \\ \hline 2S_9 &= 64 + 64 + 64 + 64 + 64 + 64 + 64 + 64 + 64 \\ 2S_9 &= 9(64) \\ S_9 &= \frac{9}{2}(64) \end{aligned}$$

Note that the sum had 9 terms.

The first and last terms of the sum are 64.

An arithmetic series S_n has n terms, and the sum of the first and last terms is $a_1 + a_n$. Thus, the formula $S_n = \frac{n}{2}(a_1 + a_n)$ represents the sum of any arithmetic series.

Key Concept — Sum of an Arithmetic Series

The sum S_n of the first n terms of an arithmetic series is given by

$$S_n = \frac{n}{2}[2a_1 + (n - 1)d] \text{ or } S_n = \frac{n}{2}(a_1 + a_n).$$

Example 1 Find the Sum of an Arithmetic Series

Find the sum of the first 100 positive integers.

The series is $1 + 2 + 3 + \ldots + 100$. Since you can see that $a_1 = 1$, $a_{100} = 100$, and $d = 1$, you can use either sum formula for this series.

Method 1

$S_n = \frac{n}{2}(a_1 + a_n)$	Sum formula
$S_{100} = \frac{100}{2}(1 + 100)$	$n = 100, a_1 = 1, a_{100} = 100, d = 1$
$S_{100} = 50(101)$	Simplify.
$S_{100} = 5050$	Multiply.

Method 2

$S_n = \frac{n}{2}[2a_1 + (n - 1)d]$

$S_{100} = \frac{100}{2}[2(1) + (100 - 1)1]$

$S_{100} = 50(101)$

$S_{100} = 5050$

The sum of the first 100 positive integers is 5050.

Example 2 Find the First Term

RADIO A radio station considered giving away \$4000 every day in the month of August for a total of \$124,000. Instead, they decided to increase the amount given away every day while still giving away the same total amount. If they want to increase the amount by \$100 each day, how much should they give away the first day?

You know the values of n, S_n, and d. Use the sum formula that contains d.

$S_n = \frac{n}{2}[2a_1 + (n - 1)d]$	Sum formula
$S_{31} = \frac{31}{2}[2a_1 + (31 - 1)100]$	$n = 31, d = 100$
$124{,}000 = \frac{31}{2}(2a_1 + 3000)$	$S_{31} = 124{,}000$
$8000 = 2a_1 + 3000$	Multiply each side by $\frac{2}{31}$.
$5000 = 2a_1$	Subtract 3000 from each side.
$2500 = a_1$	Divide each side by 2.

The radio station should give away \$2500 the first day.

More About. . .

Radio

99.0% of teens ages 12–17 listen to the radio at least once a week. 79.1% listen at least once a day.

Source: Radio Advertising Bureau

Sometimes it is necessary to use both a sum formula and the formula for the nth term to solve a problem.

Example 3 Find the First Three Terms

Find the first three terms of an arithmetic series in which $a_1 = 9$, $a_n = 105$, and $S_n = 741$.

Step 1 Since you know a_1, a_n, and S_n, use $S_n = \frac{n}{2}(a_1 + a_n)$ to find n.

$S_n = \frac{n}{2}(a_1 + a_n)$

$741 = \frac{n}{2}(9 + 105)$

$741 = 57n$

$13 = n$

Step 2 Find d.

$a_n = a_1 + (n - 1)d$

$105 = 9 + (13 - 1)d$

$96 = 12d$

$8 = d$

Step 3 Use d to determine a_2 and a_3.

$a_2 = 9 + 8$ or 17 $\qquad a_3 = 17 + 8$ or 25

The first three terms are 9, 17, and 25.

Study Tip

Sigma Notation

There are many ways to represent a given series.

$\sum_{r=4}^{9}(r-3)$

$=\sum_{s=2}^{7}(s-1)$

$=\sum_{j=0}^{5}(j+1)$

SIGMA NOTATION Writing out a series can be time-consuming and lengthy. For convenience, there is a more concise notation called **sigma notation**. The series $3+6+9+12+\ldots+30$ can be expressed as $\sum_{n=1}^{10}3n$. This expression is read *the sum of 3n as n goes from 1 to 10.*

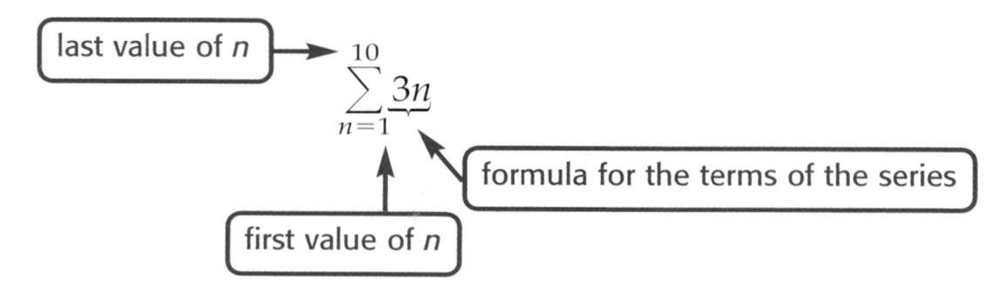

The variable, in this case n, is called the **index of summation**.

To generate the terms of a series given in sigma notation, successively replace the index of summation with consecutive integers between the first and last values of the index, inclusive. For the series above, the values of n are 1, 2, 3, and so on, through 10.

Example 4 Evaluate a Sum in Sigma Notation

Evaluate $\sum_{j=5}^{8}(3j-4)$.

Method 1

Find the terms by replacing j with 5, 6, 7, and 8. Then add.

$$\sum_{j=5}^{8}(3j-4) = [3(5)-4]+[3(6)-4]+[3(7)-4]+[3(8)-4]$$

$$= 11+14+17+20$$

$$= 62$$

Method 2

Since the sum is an arithmetic series, use the formula $S_n=\frac{n}{2}(a_1+a_n)$.

There are 4 terms, $a_1=3(5)-4$ or 11, and $a_4=3(8)-4$ or 20.

$$S_4=\frac{4}{2}(11+20)$$

$$S_4=62$$

The sum of the series is 62.

You can use the sum and sequence features on a graphing calculator to find the sum of a series.

Study Tip

Graphing Calculators

On the TI-83 Plus, sum(is located on the LIST MATH menu. The function seq(is located on the LIST OPS menu.

Graphing Calculator Investigation

Sums of Series

The calculator screen shows the evaluation of $\sum_{N=2}^{10}(5N-2)$. The first four entries for seq(are

```
sum(seq(5N-2,N,2
,10,1))
             252
```

- the formula for the general term of the series,
- the index of summation,
- the first value of the index, and
- the last value of the index, respectively.

The last entry is always 1 for the types of series that we are considering.

Think and Discuss

1. Explain why you can use any letter for the index of summation.
2. Evaluate $\sum_{n=1}^{8}(2n-1)$ and $\sum_{j=5}^{12}(2j-9)$. **Make a conjecture** as to their relationship and explain why you think it is true.

Check for Understanding

Concept Check

1. **Explain** the difference between a sequence and a series.
2. **OPEN ENDED** Write an arithmetic series for which $S_5 = 10$.
3. **OPEN ENDED** Write the series $7 + 10 + 13 + 16$ using sigma notation.

Guided Practice

Find S_n for each arithmetic series described.

4. $a_1 = 4, a_n = 100, n = 25$
5. $a_1 = 40, n = 20, d = -3$
6. $a_1 = 132, d = -4, a_n = 52$
7. $d = 5, n = 16, a_n = 72$

Find the sum of each arithmetic series.

8. $5 + 11 + 17 + \ldots + 95$
9. $38 + 35 + 32 + \ldots + 2$
10. $\sum_{n=1}^{7} (2n + 1)$
11. $\sum_{k=3}^{7} (3k + 4)$

Find the first three terms of each arithmetic series described.

12. $a_1 = 11, a_n = 110, S_n = 726$
13. $n = 8, a_n = 36, S_n = 120$

Application

14. **WORLD CULTURES** The African-American festival of *Kwanzaa* includes a ritual involving candles. The first night, a candle is lit and then blown out. The second night, a new candle and the candle from the previous night are lit and blown out. This pattern of lighting a new candle and relighting all the candles from the previous nights is continued for seven nights. Use a formula from this lesson to find the total number of candle lightings during the festival.

Practice and Apply

Homework Help

For Exercises	See Examples
15–32, 39, 40, 45	1, 2
33–38	4
31–43	3

Extra Practice
See page 851.

Find S_n for each arithmetic series described.

15. $a_1 = 7, a_n = 79, n = 8$
16. $a_1 = 58, a_n = -7, n = 26$
17. $a_1 = 43, n = 19, a_n = 115$
18. $a_1 = 76, n = 21, a_n = 176$
19. $a_1 = 7, d = -2, n = 9$
20. $a_1 = 3, d = -4, n = 8$
21. $a_1 = 5, d = \frac{1}{2}, n = 13$
22. $a_1 = 12, d = \frac{1}{3}, n = 13$
23. $d = -3, n = 21, a_n = -64$
24. $d = 7, n = 18, a_n = 72$
25. $d = \frac{1}{5}, n = 10, a_n = \frac{23}{10}$
26. $d = -\frac{1}{4}, n = 20, a_n = -\frac{53}{12}$

27. **TOYS** Jamila is making a triangular wall with building blocks. The top row has one block, the second row has three, the third has five, and so on. How many rows can she make with a set of 100 blocks?

28. **CONSTRUCTION** A construction company will be fined for each day it is late completing its current project. The daily fine will be \$4000 for the first day and will increase by \$1000 each day. Based on its budget, the company can only afford \$60,000 in total fines. What is the maximum number of days it can be late?

Find the sum of each arithmetic series.

29. $6 + 13 + 20 + 27 + \ldots + 97$
30. $7 + 14 + 21 + 28 + \ldots + 98$
31. $34 + 30 + 26 + \ldots + 2$
32. $16 + 10 + 4 + \ldots + (-50)$
33. $\sum_{n=1}^{6} (2n + 11)$
34. $\sum_{n=1}^{5} (2 - 3n)$
35. $\sum_{k=7}^{11} (42 - 9k)$
36. $\sum_{t=19}^{23} (5t - 3)$
37. $\sum_{i=1}^{300} (7i - 3)$
38. $\sum_{k=1}^{150} (11 + 2k)$

39. Find the sum of the first 1000 positive even integers.

40. What is the sum of the multiples of 3 between 3 and 999, inclusive?

Find the first three terms of each arithmetic series described.

41. $a_1 = 17, a_n = 197, S_n = 2247$

42. $a_1 = -13, a_n = 427, S_n = 18{,}423$

43. $n = 31, a_n = 78, S_n = 1023$

44. $n = 19, a_n = 103, S_n = 1102$

45. **AEROSPACE** On the Moon, a falling object falls just 2.65 feet in the first second after being dropped. Each second it falls 5.3 feet farther than in the previous second. How far would an object fall in the first ten seconds after being dropped?

CRITICAL THINKING State whether each statement is *true* or *false*. Explain.

46. Doubling each term in an arithmetic series will double the sum.

47. Doubling the number of terms in an arithmetic series, but keeping the first term and common difference the same, will double the sum.

48. **WRITING IN MATH** Answer the question that was posed at the beginning of the lesson.

How do arithmetic series apply to amphitheaters?

Include the following in your answer:

- explanations of what the sequence and the series that can be formed from the given numbers represent, and
- two ways to find the amphitheater capacity if it has ten rows of seats.

Standards Practice

Standardized Test Practice

49. $18 + 22 + 26 + 30 + \ldots + 50 = ?$

Ⓐ 146 Ⓑ 272 Ⓒ 306 Ⓓ 340

50. The angles of a triangle form an arithmetic sequence. If the smallest angle measures 36°, what is the measure of the largest angle?

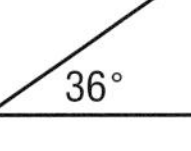

Ⓐ 60° Ⓑ 72°

Ⓒ 84° Ⓓ 144°

Graphing Calculator

Use a graphing calculator to find the sum of each arithmetic series.

51. $\sum_{n=21}^{75} (2n + 5)$

52. $\sum_{n=10}^{50} (3n - 1)$

53. $\sum_{n=20}^{60} (4n + 3)$

Maintain Your Skills

Mixed Review

Find the indicated term of each arithmetic sequence. *(Lesson 11-1)*

54. $a_1 = 46, d = 5, n = 14$

55. $a_1 = 12, d = -7, n = 22$

56. **RADIOACTIVITY** The decay of Radon-222 can be modeled by the equation $y = ae^{-0.1813t}$, where t is measured in days. What is the half-life of Radon-222? *(Lesson 10-6)*

Solve each equation by completing the square. *(Lesson 6-4)*

57. $x^2 + 9x + 20.25 = 0$

58. $9x^2 + 96x + 256 = 0$

59. $x^2 - 3x - 20 = 0$

Simplify. *(Lesson 5-6)*

60. $5\sqrt{3} - 4\sqrt{3}$

61. $\sqrt{26} \cdot \sqrt{39} \cdot \sqrt{14}$

62. $(\sqrt{10} - \sqrt{6})(\sqrt{5} + \sqrt{3})$

Getting Ready for the Next Lesson

PREREQUISITE SKILL Evaluate the expression $a \cdot b^{n-1}$ for the given values of a, b, and n. *(To review **evaluating expressions**, see Lesson 1-1.)*

63. $a = 1, b = 2, n = 5$

64. $a = 2, b = -3, n = 4$

65. $a = 18, b = \frac{1}{3}, n = 6$

11-3 Geometric Sequences

California Standards Standard 22.0 **Students find the general term** and the sums of arithmetic series and **of** both **finite** and infinite **geometric series.**

What You'll Learn

- Use geometric sequences.
- Find geometric means.

Vocabulary

- geometric sequence
- common ratio
- geometric means

How do geometric sequences apply to a bouncing ball?

If you have ever bounced a ball, you know that when you drop it, it never rebounds to the height from which you dropped it. Suppose a ball is dropped from a height of three feet, and each time it falls, it rebounds to 60% of the height from which it fell. The heights of the ball's rebounds form a sequence.

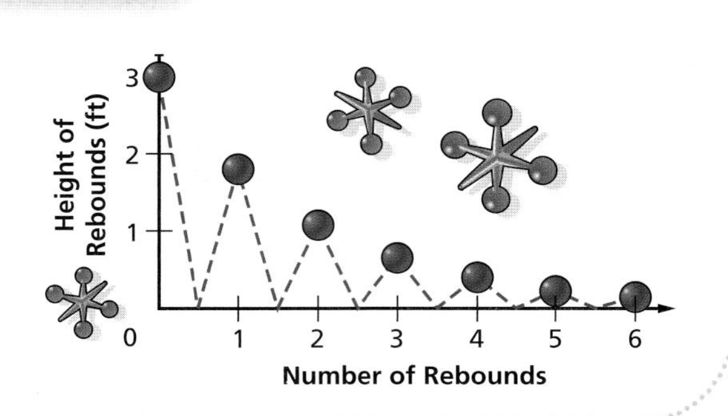

GEOMETRIC SEQUENCES The height of the first rebound of the ball is 3(0.6) or 1.8 feet. The height of the second rebound is 1.8(0.6) or 1.08 feet. The height of the third rebound is 1.08(0.6) or 0.648 feet. The sequence of heights, 1.8, 1.08, 0.648, …, is an example of a **geometric sequence**. A geometric sequence is a sequence in which each term after the first is found by multiplying the previous term by a constant r called the **common ratio**.

As with an arithmetic sequence, you can label the terms of a geometric sequence as a_1, a_2, a_3, and so on. The nth term is a_n and the previous term is a_{n-1}. So, $a_n = r(a_{n-1})$. Thus, $r = \frac{a_n}{a_{n-1}}$. That is, the common ratio can be found by dividing any term by its previous term.

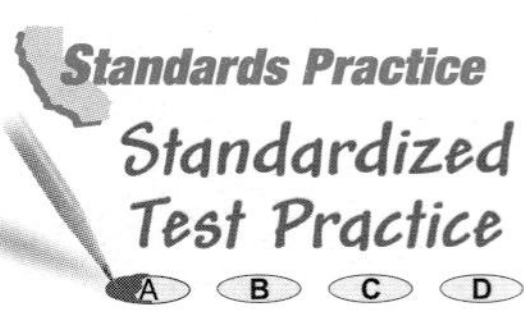

Example 1 Find the Next Term

Multiple-Choice Test Item

Find the missing term in the geometric sequence: 8, 20, 50, 125, ___.

Ⓐ 75 Ⓑ 200 Ⓒ 250 Ⓓ 312.5

Test-Taking Tip

Since the terms of this sequence are increasing, the missing term must be greater than 125. You can immediately eliminate 75 as a possible answer.

Read the Test Item

Since $\frac{20}{8} = 2.5$, $\frac{50}{20} = 2.5$, and $\frac{125}{50} = 2.5$, the sequence has a common ratio of 2.5.

Solve the Test Item

To find the missing term, multiply the last given term by 2.5: $125(2.5) = 312.5$.

The answer is D.

You have seen that each term of a geometric sequence can be expressed in terms of r and its previous term. It is also possible to develop a formula that expresses each term of a geometric sequence in terms of r and the first term a_1. Study the patterns shown in the table on the next page for the sequence 2, 6, 18, 54, … .

Sequence							
Sequence	numbers	2	6	18	54	...	
	symbols	a_1	a_2	a_3	a_4	...	a_n
Expressed in Terms of r and the Previous Term	numbers	2	2(3)	6(3)	18(3)	...	
	symbols	a_1	$a_1 \cdot r$	$a_2 \cdot r$	$a_3 \cdot r$	...	$a_{n-1} \cdot r$
Expressed in Terms of r and the First Term	numbers	2	2(3)	2(9)	2(27)	...	
		$2(3^0)$	$2(3^1)$	$2(3^2)$	$2(3^3)$	...	
	symbols	$a_1 \cdot r^0$	$a_1 \cdot r^1$	$a_1 \cdot r^2$	$a_1 \cdot r^3$	...	$a_1 \cdot r^{n-1}$

The three entries in the last column of the table all describe the nth term of a geometric sequence. This leads us to the following formula for finding the nth term of a geometric sequence.

Key Concept — *n*th Term of a Geometric Sequence

The nth term a_n of a geometric sequence with first term a_1 and common ratio r is given by

$$a_n = a_1 \cdot r^{n-1},$$

where n is any positive integer.

Example 2 Find a Particular Term

Find the eighth term of a geometric sequence for which $a_1 = -3$ and $r = -2$.

$a_n = a_1 \cdot r^{n-1}$ Formula for nth term

$a_8 = (-3) \cdot (-2)^{8-1}$ $n = 8, a_1 = -3, r = -2$

$a_8 = (-3) \cdot (-128)$ $(-2)^7 = -128$

$a_8 = 384$ Multiply.

The eighth term is 384.

Example 3 Write an Equation for the nth Term

Write an equation for the nth term of the geometric sequence 3, 12, 48, 192, ...

In this sequence, $a_1 = 3$ and $r = 4$. Use the nth term formula to write an equation.

$a_n = a_1 \cdot r^{n-1}$ Formula for nth term

$a_n = 3 \cdot 4^{n-1}$ $a_1 = 3, r = 4$

An equation is $a_n = 3 \cdot 4^{n-1}$.

You can also use the formula for the nth term if you know the common ratio and one term of a geometric sequence, but not the first term.

Example 4 Find a Term Given the Fourth Term and the Ratio

Find the tenth term of a geometric sequence for which $a_4 = 108$ and $r = 3$.

First, find the value of a_1.

$a_n = a_1 \cdot r^{n-1}$ Formula for nth term

$a_4 = a_1 \cdot 3^{4-1}$ $n = 4, r = 3$

$108 = 27a_1$ $a_4 = 108$

$4 = a_1$ Divide each side by 27.

Now find a_{10}.

$a_n = a_1 \cdot r^{n-1}$ Formula for nth term

$a_{10} = 4 \cdot 3^{10-1}$ $n = 10, a_1 = 4, r = 3$

$a_{10} = 78{,}732$ Use a calculator.

The tenth term is 78,732.

GEOMETRIC MEANS In Lesson 11-1, you learned that missing terms between two nonsuccessive terms in an arithmetic sequence are called *arithmetic means*. Similarly, the missing terms(s) between two nonsuccessive terms of a geometric sequence are called **geometric means**. For example, 6, 18, and 54 are three geometric means between 2 and 162 in the sequence 2, 6, 18, 54, 162, … . You can use the common ratio to find the geometric means in a given sequence.

Example 5 Find Geometric Means

Find three geometric means between 2.25 and 576.

Use the nth term formula to find the value of r. In the sequence 2.25, __?__, __?__, __?__, 576, a_1 is 2.25 and a_5 is 576.

$a_n = a_1 \cdot r^{n-1}$	Formula for nth term
$a_5 = 2.25 \cdot r^{5-1}$	$n = 5$, $a_1 = 2.25$
$576 = 2.25r^4$	$a_5 = 576$
$256 = r^4$	Divide each side by 2.25.
$\pm 4 = r$	Take the fourth root of each side.

There are two possible common ratios, so there are two possible sets of geometric means. Use each value of r to find three geometric means.

$r = 4$	$r = -4$
$a_2 = 2.25(4)$ or 9	$a_2 = 2.25(-4)$ or -9
$a_3 = 9(4)$ or 36	$a_3 = -9(-4)$ or 36
$a_4 = 36(4)$ or 144	$a_4 = 36(-4)$ or -144

The geometric means are 9, 36, and 144, or -9, 36, and -144.

Study Tip

Alternate Method

You may prefer this method. The three means will be $2.25r$, $2.25r^2$, and $2.25r^3$. Then the common ratio is $r = \frac{576}{2.25r^3}$ or $r^4 = \frac{576}{2.25}$. Thus, $r = 4$.

Check for Understanding

Concept Check

1. **Decide** whether each sequence is *arithmetic* or *geometric*. Explain.
   a. 1, −2, 4, −8, … **b.** 1, −2, −5, −8, …
2. **OPEN ENDED** Write a geometric sequence with a common ratio of $\frac{2}{3}$.
3. **FIND THE ERROR** Marika and Lori are finding the seventh term of the geometric sequence 9, 3, 1, … .

Marika	Lori
$r = \frac{3}{9}$ or $\frac{1}{3}$	$r = \frac{9}{3}$ or 3
$a_7 = 9\left(\frac{1}{3}\right)^{7-1}$	$a_7 = 9 \cdot 3^{7-1}$
$= \frac{1}{81}$	$= 6561$

Who is correct? Explain your reasoning.

Guided Practice

Find the next two terms of each geometric sequence.

4. 20, 30, 45, …
5. $-\frac{1}{4}, \frac{1}{2}, -1, \ldots$
6. Find the first five terms of the geometric sequence for which $a_1 = -2$ and $r = 3$.

7. Find a_9 for the geometric sequence 60, 30, 15, ...

Find the indicated term of each geometric sequence.

8. $a_1 = 7, r = 2, n = 4$

9. $a_3 = 32, r = -0.5, n = 6$

10. Write an equation for the nth term of the geometric sequence 4, 8, 16, ...

11. Find two geometric means between 1 and 27.

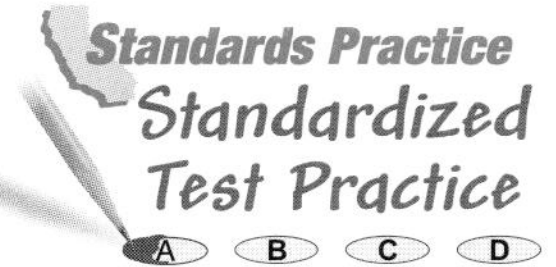

12. Find the missing term in the geometric sequence: $\frac{9}{4}, \frac{3}{4}, \frac{1}{4}, \frac{1}{12}$, ____.

Ⓐ $\frac{1}{36}$ Ⓑ $\frac{1}{20}$ Ⓒ $\frac{1}{6}$ Ⓓ $\frac{1}{3}$

Practice and Apply

Homework Help

For Exercises	See Examples
13–24	1
25–30, 33–38, 47, 48	2
31, 32	4
39–42	3
43–46	5

Extra Practice
See page 852.

Find the next two terms of each geometric sequence.

13. 405, 135, 45, ...

14. 81, 108, 144, ...

15. 16, 24, 36, ...

16. 162, 108, 72, ...

17. $\frac{5}{2}, \frac{5}{3}, \frac{10}{9}, \ldots$

18. $\frac{1}{3}, \frac{5}{6}, \frac{25}{12}, \ldots$

19. 1.25, −1.5, 1.8, ...

20. 1.4, −3.5, 8.75, ...

Find the first five terms of each geometric sequence described.

21. $a_1 = 2, r = -3$

22. $a_1 = 1, r = 4$

23. $a_1 = 243, r = \frac{1}{3}$

24. $a_1 = 576, r = -\frac{1}{2}$

25. Find a_7 if $a_n = 12\left(\frac{1}{2}\right)^{n-1}$.

26. If $a_n = \frac{1}{3} \cdot 6^{n-1}$, what is a_6?

Find the indicated term of each geometric sequence.

27. $a_1 = \frac{1}{3}, r = 3, n = 8$

28. $a_1 = \frac{1}{64}, r = 4, n = 9$

29. $a_1 = 16{,}807, r = \frac{3}{7}, n = 6$

30. $a_1 = 4096, r = \frac{1}{4}, n = 8$

31. $a_4 = 16, r = 0.5, n = 8$

32. $a_6 = 3, r = 2, n = 12$

33. a_9 for $\frac{1}{5}, 1, 5, \ldots$

34. a_7 for $\frac{1}{32}, \frac{1}{16}, \frac{1}{8}, \ldots$

35. a_8 for 4, −12, 36, ...

36. a_6 for 540, 90, 15, ...

37. **ART** A one-ton ice sculpture is melting so that it loses one-fifth of its weight per hour. How much of the sculpture will be left after five hours? Write the answer in pounds.

38. **SALARIES** Geraldo's current salary is $40,000 per year. His annual pay raise is always a percent of his salary at the time. What would his salary be if he got four consecutive 4% increases?

Write an equation for the *n*th term of each geometric sequence.

39. 36, 12, 4, ...

40. 64, 16, 4, ...

41. −2, 10, −50, ...

42. 4, −12, 36, ...

Find the geometric means in each sequence.

43. 9, __?__, __?__, __?__, 144

44. 4, __?__, __?__, __?__, 324

45. 32, __?__, __?__, __?__, __?__, 1

46. 3, __?__, __?__, __?__, __?__, 96

MEDICINE For Exercises 47 and 48, use the following information.
Iodine-131 is a radioactive element used to study the thyroid gland.

47. **RESEARCH** Use the Internet or other resource to find the *half-life* of Iodine-131, rounded to the nearest day. This is the amount of time it takes for half of a sample of Iodine-131 to decay into another element.

48. How much of an 80-milligram sample of Iodine-131 would be left after 32 days?

CRITICAL THINKING Determine whether each statement is *true* or *false*. If true, explain. If false, provide a counterexample.

49. Every sequence is either arithmetic or geometric.

50. There is no sequence that is both arithmetic and geometric.

51. **WRITING IN MATH** Answer the question that was posed at the beginning of the lesson.

How do geometric sequences apply to a bouncing ball?

Include the following in your answer:

- the first five terms of the sequence of heights from which the ball falls, and
- any similarities or differences in the sequences for the heights the ball rebounds and the heights from which the ball falls.

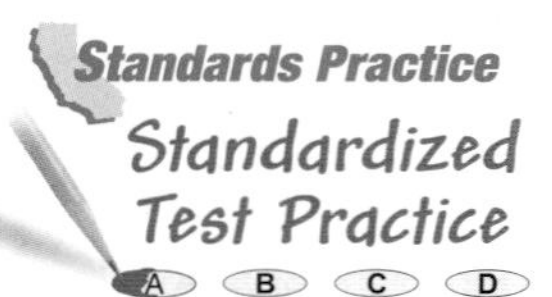

52. Find the missing term in the geometric sequence: $-5, 10, -20, 40,$ ____.

(A) -80 (B) -35 (C) 80 (D) 100

53. What is the tenth term in the geometric sequence: $144, 72, 36, 18, \ldots$?

(A) 0 (B) $\frac{9}{64}$ (C) $\frac{9}{32}$ (D) $\frac{9}{16}$

Maintain Your Skills

Mixed Review

Find S_n for each arithmetic series described. *(Lesson 11-2)*

54. $a_1 = 11, a_n = 44, n = 23$

55. $a_1 = -5, d = 3, n = 14$

Find the arithmetic means in each sequence. *(Lesson 11-1)*

56. $15,$? , ? , 27

57. $-8,$? , ? , ? , -24

58. **GEOMETRY** Find the perimeter of a triangle with vertices at $(2, 4)$, $(-1, 3)$ and $(1, -3)$. *(Lesson 8-1)*

Getting Ready for the Next Lesson

PREREQUISITE SKILL Evaluate each expression. *(To review **expressions**, see Lesson 1-1.)*

59. $\dfrac{1 - 2^7}{1 - 2}$

60. $\dfrac{1 - \left(\frac{1}{2}\right)^6}{1 - \left(\frac{1}{2}\right)}$

61. $\dfrac{1 - \left(-\frac{1}{3}\right)^5}{1 - \left(-\frac{1}{3}\right)}$

Practice Quiz 1

Lessons 11-1 through 11-3

Find the indicated term of each arithmetic sequence. *(Lesson 11-1)*

1. $a_1 = 7, d = 3, n = 14$

2. $a_1 = 2, d = \frac{1}{2}, n = 8$

Find the sum of each arithmetic series described. *(Lesson 11-2)*

3. $a_1 = 5, a_n = 29, n = 11$

4. $6 + 12 + 18 + \ldots + 96$

5. Find a_7 for the geometric sequence $729, -243, 81, \ldots$. *(Lesson 11-3)*

Graphing Calculator Investigation

A Preview of Lesson 11-4

Limits

You may have noticed that in some geometric sequences, the later the term in the sequence, the closer the value is to 0. Another way to describe this is that as n increases, a_n approaches 0. The value that the terms of a sequence approach, in this case 0, is called the **limit** of the sequence. Other types of infinite sequences may also have limits. If the terms of a sequence do not approach a unique value, we say that the limit of the sequence does not exist.

Find the limit of the geometric sequence $1, \frac{1}{3}, \frac{1}{9}, \ldots$.

Step 1 *Enter the sequence.*

- The formula for this sequence is $a_n = \left(\frac{1}{3}\right)^{n-1}$.
- Position the cursor on L1 in the STAT EDIT Edit … screen and enter the formula seq(N,N,1,10,1). This generates the values 1, 2, …, 10 of the index N.
- Position the cursor on L2 and enter the formula seq((1/3)^(N-1),N,1,10,1). This generates the first ten terms of the sequence.

  KEYSTROKES: *Review sequences in the Graphing Calculator Investigation on page 585.*

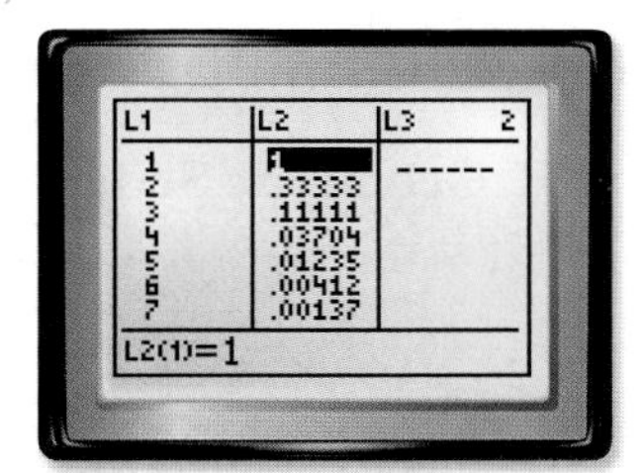

Notice that as n increases, the terms of the given sequence get closer and closer to 0. If you scroll down, you can see that for $n \geq 8$ the terms are so close to 0 that the calculator expresses them in scientific notation. This suggests that the limit of the sequence is 0.

Step 2 *Graph the sequence.*

- Use a STAT PLOT to graph the sequence. Use L1 as the Xlist and L2 as the Ylist.

  KEYSTROKES: *Review STAT PLOTs on page 87.*

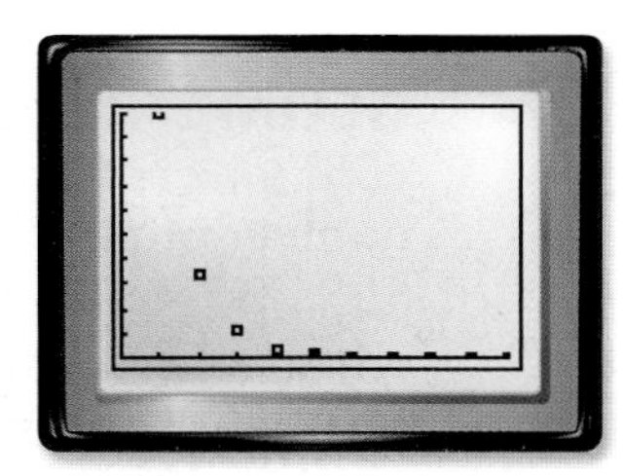

[0, 10] scl: 1 by [0, 1] scl: 0.1

The graph also shows that, as n increases, the terms approach 0. In fact, for $n \geq 6$, the marks appear to lie on the horizontal axis. This strongly suggests that the limit of the sequence is 0.

Exercises

Use a graphing calculator to find the limit, if it exists, of each sequence.

1. $a_n = \left(\frac{1}{2}\right)^n$

2. $a_n = \left(-\frac{1}{2}\right)^n$

3. $a_n = 4^n$

4. $a_n = \frac{1}{n^2}$

5. $a_n = \frac{2^n}{2^n + 1}$

6. $a_n = \frac{n^2}{n + 1}$

11-4 Geometric Series

California Standards Standard 23.0 **Students derive the summation formulas for** arithmetic series and for both **finite** and infinite **geometric series.** (Key)

What You'll Learn

- Find sums of geometric series.
- Find specific terms of geometric series.

Vocabulary

- geometric series

How is e-mailing a joke like a geometric series?

Suppose you e-mail a joke to three friends on Monday. Each of those friends sends the joke on to three of their friends on Tuesday. Each person who receives the joke on Tuesday sends it to three more people on Wednesday, and so on.

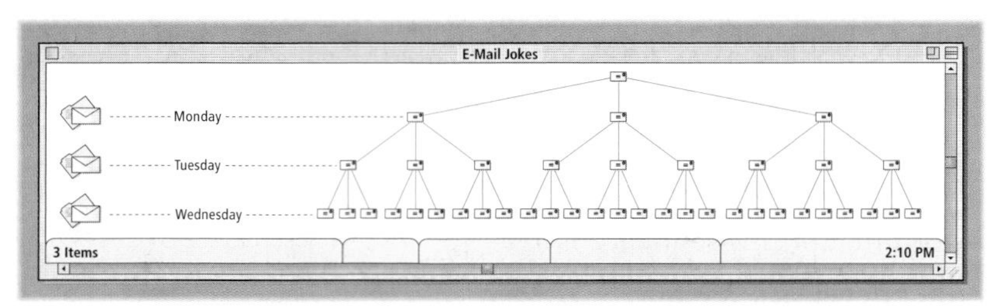

GEOMETRIC SERIES Notice that every day, the number of people who read your joke is three times the number that read it the day before. By Sunday, the number of people, including yourself, who have read the joke is 1 + 3 + 9 + 27 + 81 + 243 + 729 + 2187 or 3280!

The numbers 1, 3, 9, 27, 81, 243, 729, and 2187 form a geometric sequence in which $a_1 = 1$ and $r = 3$. Since 1, 3, 9, 27, 81, 243, 729, 2187 is a geometric sequence, 1 + 3 + 9 + 27 + 81 + 243 + 729 + 2187 is called a **geometric series**. Below are some more examples of geometric sequences and their corresponding geometric series.

Geometric Sequences	Geometric Series
1, 2, 4, 8, 16	1 + 2 + 4 + 8 + 16
4, −12, 36	4 + (−12) + 36
$5, 1, \frac{1}{5}, \frac{1}{25}$	$5 + 1 + \frac{1}{5} + \frac{1}{25}$

To develop a formula for the sum of a geometric series, consider the series given in the e-mail situation above.

$$\begin{aligned} S_8 &= 1 + 3 + 9 + 27 + 81 + 243 + 729 + 2187 \\ (-)\ 3S_8 &= \phantom{1 + {}} 3 + 9 + 27 + 81 + 243 + 729 + 2187 + 6561 \\ \hline (1-3)S_8 &= 1 + 0 + 0 + 0 + 0 + 0 + 0 + 0 - 6561 \end{aligned}$$

$$S_8 = \frac{1 - 6561}{1 - 3} \text{ or } 3280$$

The expression for S_8 can be written as $S_8 = \frac{a_1 - a_1 r^8}{1 - r}$. A rational expression like this can be used to find the sum of any geometric series.

Study Tip

Terms of Geometric Sequences

Remember that a_9 can also be written as $a_1 r^8$.

Key Concept — Sum of a Geometric Series

The sum S_n of the first n terms of a geometric series is given by

$$S_n = \frac{a_1 - a_1 r^n}{1 - r} \text{ or } S_n = \frac{a_1(1 - r^n)}{1 - r}, \text{ where } r \neq 1.$$

You cannot use the formula for the sum with a geometric series for which $r = 1$ because division by 0 would result. In a geometric series with $r = 1$, the terms are constant. For example, $4 + 4 + 4 + \ldots + 4$ is such a series. In general, the sum of n terms of a geometric series with $r = 1$ is $n \cdot a^1$.

Example 1 Find the Sum of the First n Terms

GENEALOGY In the book *Roots*, author Alex Haley traced his family history back many generations to the time one of his ancestors was brought to America from Africa. If you could trace your family back for 15 generations, starting with your parents, how many ancestors would there be?

Counting your two parents, four grandparents, eight great-grandparents, and so on gives you a geometric series with $a_1 = 2$, $r = 2$, and $n = 15$.

$S_n = \frac{a_1(1 - r^n)}{1 - r}$ Sum formula

$S_{15} = \frac{2(1 - 2^{15})}{1 - 2}$ $n = 15, a_1 = 2, r = 2$

$S_{15} = 65{,}534$ Use a calculator.

Going back 15 generations, you have 65,534 ancestors.

More About. . .

Genealogy

When he died in 1992, Samuel Must of Fryburg, Pennsylvania, had a record 824 living descendants.

Source: *The Guinness Book of Records*

As with arithmetic series, you can use sigma notation to represent geometric series.

Example 2 Evaluate a Sum Written in Sigma Notation

Evaluate $\sum_{n=1}^{6} 5 \cdot 2^{n-1}$.

Method 1

Find the terms by replacing n with 1, 2, 3, 4, 5, and 6. Then add.

$$\begin{aligned}\sum_{n=1}^{6} 5 \cdot 2^{n-1} &= 5(2^{1-1}) + 5(2^{2-1}) \\ &\quad + 5(2^{3-1}) + 5(2^{4-1}) \\ &\quad + 5(2^{5-1}) + 5(2^{6-1}) \\ &= 5(1) + 5(2) + 5(4) + 5(8) \\ &\quad + 5(16) + 5(32) \\ &= 5 + 10 + 20 + 40 + 80 \\ &\quad + 160 \\ &= 315\end{aligned}$$

The sum of the series is 315.

Method 2

Since the sum is a geometric series, you can use the formula

$S_n = \frac{a_1(1 - r^n)}{1 - r}$.

$S_6 = \frac{5(1 - 2^6)}{1 - 2}$ $n = 6, a_1 = 5, r = 2$

$S_6 = \frac{5(-63)}{-1}$ $2^6 = 64$

$S_6 = 315$ Simplify.

How can you find the sum of a geometric series if you know the first and last terms and the common ratio, but not the number of terms? Remember the formula for the nth term of a geometric sequence or series, $a_n = a_1 \cdot r^{n-1}$. You can use this formula to find an expression involving r^n.

$a_n = a_1 \cdot r^{n-1}$ Formula for nth term

$a_n \cdot r = a_1 \cdot r^{n-1} \cdot r$ Multiply each side by r.

$a_n \cdot r = a_1 \cdot r^n$ $r^{n-1} \cdot r^1 = r^{n-1+1}$ or r^n

Now substitute $a_n \cdot r$ for $a_1 \cdot r^n$ in the formula for the sum of a geometric series. The result is $S_n = \frac{a_1 - a_n r}{1 - r}$.

Example 3 Use the Alternate Formula for a Sum

Find the sum of a geometric series for which $a_1 = 15{,}625$, $a_n = -5$, and $r = -\frac{1}{5}$.

Since you do not know the value of n, use the formula derived above.

$S_n = \frac{a_1 - a_n r}{1 - r}$ Alternate sum formula

$= \frac{15{,}625 - (-5)\left(-\frac{1}{5}\right)}{1 - \left(-\frac{1}{5}\right)}$ $a_1 = 15{,}625$, $a_n = -5$, $r = -\frac{1}{5}$

$= \frac{15{,}624}{\frac{6}{5}}$ or 13,020 Simplify.

SPECIFIC TERMS You can use the formula for the sum of a geometric series to help find a particular term of the series.

Example 4 Find the First Term of a Series

Find a_1 in a geometric series for which $S_8 = 39{,}360$ and $r = 3$.

$S_n = \frac{a_1(1 - r^n)}{1 - r}$ Sum formula

$39{,}360 = \frac{a_1(1 - 3^8)}{1 - 3}$ $S_8 = 39{,}360$; $r = 3$; $n = 8$

$39{,}360 = \frac{-6560a_1}{-2}$ Subtract.

$39{,}360 = 3280a_1$ Divide.

$12 = a_1$ Divide each side by 3280.

The first term of the series is 12.

Check for Understanding

Concept Check

1. **OPEN ENDED** Write a geometric series for which $r = \frac{1}{2}$ and $n = 4$.
2. **Explain**, using geometric series, why the polynomial $1 + x + x^2 + x^3$ can be written as $\frac{x^4 - 1}{x - 1}$, assuming $x \neq 1$.
3. **Explain** how to write the series 2 + 12 + 72 + 432 + 2592 using sigma notation.

Guided Practice

Find S_n for each geometric series described.

4. $a_1 = 12$, $a_5 = 972$, $r = -3$
5. $a_1 = 3$, $a_n = 46{,}875$, $r = -5$
6. $a_1 = 5$, $r = 2$, $n = 14$
7. $a_1 = 243$, $r = -\frac{2}{3}$, $n = 5$

Find the sum of each geometric series.

8. 54 + 36 + 24 + 16 + … to 6 terms
9. 3 − 6 + 12 − … to 7 terms
10. $\sum_{n=1}^{5} \frac{1}{4} \cdot 2^{n-1}$
11. $\sum_{n=1}^{7} 81\left(\frac{1}{3}\right)^{n-1}$

Find the indicated term for each geometric series described.

12. $S_n = \frac{381}{64}, r = \frac{1}{2}, n = 7; a_1$

13. $S_n = 33, a_n = 48, r = -2; a_1$

Application

14. **WEATHER** Heavy rain caused a river to rise. The river rose three inches the first day, and each additional day it rose twice as much as the previous day. How much did the river rise in five days?

Practice and Apply

Homework Help

For Exercises	See Examples
15–34, 47	1, 3
35–40	2
41–46	4

Extra Practice
See page 852.

Find S_n for each geometric series described.

15. $a_1 = 2, a_6 = 486, r = 3$

16. $a_1 = 3, a_8 = 384, r = 2$

17. $a_1 = 1296, a_n = 1, r = -\frac{1}{6}$

18. $a_1 = 343, a_n = -1, r = -\frac{1}{7}$

19. $a_1 = 4, r = -3, n = 5$

20. $a_1 = 5, r = 3, n = 12$

21. $a_1 = 2401, r = -\frac{1}{7}, n = 5$

22. $a_1 = 625, r = \frac{3}{5}, n = 5$

23. $a_1 = 162, r = \frac{1}{3}, n = 6$

24. $a_1 = 80, r = -\frac{1}{2}, n = 7$

25. $a_1 = 625, r = 0.4, n = 8$

26. $a_1 = 4, r = 0.5, n = 8$

27. $a_2 = -36, a_5 = 972, n = 7$

28. $a_3 = -36, a_6 = -972, n = 10$

29. **HEALTH** Contagious diseases can spread very quickly. Suppose five people are ill during the first week of an epidemic and that each person who is ill spreads the disease to four people by the end of the next week. By the end of the tenth week of the epidemic, how many people have been affected by the illness?

30. **LEGENDS** There is a legend of a king who wanted to reward a boy for a good deed. The king gave the boy a choice. He could have $1,000,000 at once, or he could be rewarded daily for a 30-day month, with one penny on the first day, two pennies on the second day, and so on, receiving twice as many pennies each day as the previous day. How much would the second option be worth?

Find the sum of each geometric series.

31. $4096 - 512 + 64 - \ldots$ to 5 terms

32. $7 + 21 + 63 + \ldots$ to 10 terms

33. $\frac{1}{16} + \frac{1}{4} + 1 + \ldots$ to 7 terms

34. $\frac{1}{9} - \frac{1}{3} + 1 - \ldots$ to 6 terms

35. $\sum_{n=1}^{9} 5 \cdot 2^{n-1}$

36. $\sum_{n=1}^{6} 2(-3)^{n-1}$

37. $\sum_{n=1}^{7} 144\left(-\frac{1}{2}\right)^{n-1}$

38. $\sum_{n=1}^{8} 64\left(\frac{3}{4}\right)^{n-1}$

39. $\sum_{n=1}^{20} 3 \cdot 2^{n-1}$

40. $\sum_{n=1}^{16} 4 \cdot 3^{n-1}$

Find the indicated term for each geometric series described.

41. $S_n = 165, a_n = 48, r = -\frac{2}{3}; a_1$

42. $S_n = 688, a_n = 16, r = -\frac{1}{2}; a_1$

43. $S_n = -364, r = -3, n = 6; a_1$

44. $S_n = 1530, r = 2, n = 8; a_1$

45. $S_n = 315, r = 0.5, n = 6; a_2$

46. $S_n = 249.92, r = 0.2, n = 5, a_3$

47. **LANDSCAPING** Rob is helping his dad install a fence. He is using a sledgehammer to drive the pointed fence posts into the ground. On his first swing, he drives a post five inches into the ground. Since the soil is denser the deeper he drives, on each swing after the first, he can only drive the post 30% as far into the ground as he did on the previous swing. How far has he driven the post into the ground after five swings?

48. **CRITICAL THINKING** If a_1 and r are integers, explain why the value of $\frac{a_1 - a_1 r^n}{1 - r}$ must also be an integer.

49. **WRITING IN MATH** Answer the question that was posed at the beginning of the lesson.

How is e-mailing a joke like a geometric series?

Include the following in your answer:

- how the related geometric series would change if each person e-mailed the joke on to four people instead of three, and
- how the situation could be changed to make it better to use a formula than to add terms.

50. The first term of a geometric series is -1, and the common ratio is -3. How many terms are in the series if its sum is 182?

(A) 6 (B) 7 (C) 8 (D) 9

51. What is the first term in a geometric series with ten terms, a common ratio of 0.5, and a sum of 511.5?

(A) 64 (B) 128 (C) 256 (D) 512

Graphing Calculator

Use a graphing calculator to find the sum of each geometric series.

52. $\sum_{n=1}^{20} 3(-2)^{n-1}$

53. $\sum_{n=1}^{15} 2\left(\frac{1}{2}\right)^{n-1}$

54. $\sum_{n=1}^{10} 5(0.2)^{n-1}$

Maintain Your Skills

Mixed Review

Find the geometric means in each sequence. *(Lesson 11-3)*

55. $\frac{1}{24}$, ___?___, ___?___, ___?___, 54

56. -2, ___?___, ___?___, ___?___, ___?___, $-\frac{243}{16}$

Find the sum of each arithmetic series. *(Lesson 11-2)*

57. $50 + 44 + 38 + \ldots + 8$

58. $\sum_{n=1}^{12} (2n + 3)$

ENTERTAINMENT **For Exercises 59–61, use the table that shows the number of drive-in movie screens in the United States for 1995–2000.** *(Lesson 2-5)*

Year	1995	1996	1997	1998	1999	2000
Screens	848	826	815	750	737	637

Source: National Association of Theatre Owners

59. Draw a scatter plot, in which x is the number of years since 1995.

60. Find a prediction equation.

61. Predict the number of screens in 2010.

Online Research **Data Update** For the latest statistics on the movie industry, visit: www.algebra2.com/data_update

Getting Ready for the Next Lesson

PREREQUISITE SKILL **Evaluate $\frac{a}{1 - b}$ for the given values of a and b.**
(To review ***evaluating expressions****, see Lesson 1-1.)*

62. $a = 1, b = \frac{1}{2}$

63. $a = 3, b = -\frac{1}{2}$

64. $a = \frac{1}{3}, b = -\frac{1}{3}$

65. $a = \frac{1}{2}, b = \frac{1}{4}$

66. $a = -1, b = 0.5$

67. $a = 0.9, b = -0.5$

11-5 Infinite Geometric Series

California Standards Standard 22.0 **Students find the general term and the sums of** arithmetic series and of both finite and **infinite geometric series.**

What You'll Learn

- Find the sum of an infinite geometric series.
- Write repeating decimals as fractions.

Vocabulary

- infinite geometric series
- partial sum
- convergent series

How does an infinite geometric series apply to a bouncing ball?

Refer to the beginning of Lesson 11-3. Suppose you wrote a geometric series to find the sum of the heights of the rebounds of the ball. The series would have no last term because theoretically there is no last bounce of the ball. For every rebound of the ball, there is another rebound, 60% as high. Such a geometric series is called an **infinite geometric series**.

In the Bleachers By Steve Moore

"And that, ladies and gentlemen, is the way the ball bounces."

INFINITE GEOMETRIC SERIES Consider the infinite geometric series $\frac{1}{2} + \frac{1}{4} + \frac{1}{8} + \frac{1}{16} + \ldots$. You have already learned how to find the sum S_n of the first n terms of a geometric series. For an infinite series, S_n is called a **partial sum** of the series. The table and graph show some values of S_n.

n	S_n
1	$\frac{1}{2}$ or 0.5
2	$\frac{3}{4}$ or 0.75
3	$\frac{7}{8}$ or 0.875
4	$\frac{15}{16}$ or 0.9375
5	$\frac{31}{32}$ or 0.96875
6	$\frac{63}{64}$ or 0.984375
7	$\frac{127}{128}$ or 0.9921875

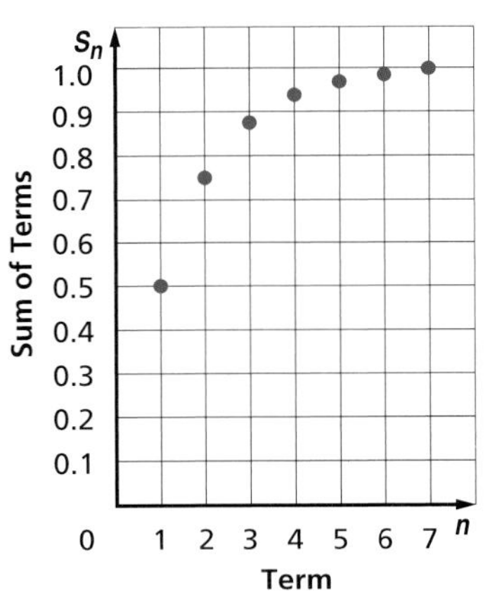

Study Tip

Absolute Value Recall that $|r| < 1$ means $-1 < r < 1$.

Notice that as n increases, the partial sums level off and approach a limit of 1. This leveling-off behavior is characteristic of infinite geometric series for which $|r| < 1$.

Let's look at the formula for the sum of a finite geometric series and use it to find a formula for the sum of an infinite geometric series.

$$S_n = \frac{a_1 - a_1r^n}{1 - r}$$ Sum of first n terms

$$= \frac{a_1}{1 - r} - \frac{a_1r^n}{1 - r}$$ Write the fraction as a difference of fractions.

If $-1 < r < 1$, the value of r^n will approach 0 as n increases. Therefore, the partial sums of an infinite geometric series will approach $\frac{a_1}{1 - r} - \frac{a_1(0)}{1 - r}$ or $\frac{a_1}{1 - r}$. This expression gives the sum of an infinite geometric series. An infinite series that has a sum is called a **convergent series**.

Key Concept — Sum of an Infinite Geometric Series

The sum S of an infinite geometric series with $-1 < r < 1$ is given by

$$S = \frac{a_1}{1 - r}.$$

Study Tip

Formula for Sum if $-1 < r < 1$

To convince yourself of this formula, make a table of the first ten partial sums of the geometric series with $r = \frac{1}{2}$ and $a_1 = 100$.

Term Number	Term	Partial Sum
1	100	100
2	50	150
3	25	175
⋮	⋮	⋮
10		

Complete the table and compare the sum that the series is approaching to that obtained by using the formula.

An infinite geometric series for which $|r| \geq 1$ does not have a sum. Consider the series $1 + 3 + 9 + 27 + 81 + \ldots$. In this series, $a_1 = 1$ and $r = 3$. The table shows some of the partial sums of this series. As n increases, S_n rapidly increases and has no limit. That is, the partial sums do not approach a particular value.

n	S_n
5	121
10	29,524
15	7,174,453
20	1,743,392,200

Example 1 Sum of an Infinite Geometric Series

Find the sum of each infinite geometric series, if it exists.

a. $\frac{1}{2} + \frac{3}{8} + \frac{9}{32} + \ldots$

First, find the value of r to determine if the sum exists.

$a_1 = \frac{1}{2}$ and $a_2 = \frac{3}{8}$, so $r = \frac{\frac{3}{8}}{\frac{1}{2}}$ or $\frac{3}{4}$. Since $\left|\frac{3}{4}\right| < 1$, the sum exists.

Now use the formula for the sum of an infinite geometric series.

$$S = \frac{a_1}{1 - r}$$ Sum formula

$$= \frac{\frac{1}{2}}{1 - \frac{3}{4}}$$ $a_1 = \frac{1}{2}, r = \frac{3}{4}$

$$= \frac{\frac{1}{2}}{\frac{1}{4}} \text{ or } 2$$ Simplify.

The sum of the series is 2.

b. $1 - 2 + 4 - 8 + \ldots$

$a_1 = 1$ and $a_2 = -2$, so $r = \frac{-2}{1}$ or -2. Since $|-2| \geq 1$, the sum does not exist.

In Lessons 11-2 and 11-4, we used sigma notation to represent finite series. You can also use sigma notation to represent infinite series. An *infinity symbol* ∞ is placed above the Σ to indicate that a series is infinite.

Example 2 Infinite Series in Sigma Notation

Evaluate $\sum_{n=1}^{\infty} 24\left(-\frac{1}{5}\right)^{n-1}$.

In this infinite geometric series, $a_1 = 24$ and $r = -\frac{1}{5}$.

$S = \frac{a_1}{1 - r}$ Sum formula

$= \frac{24}{1 - \left(-\frac{1}{5}\right)}$ $a_1 = 24, r = -\frac{1}{5}$

$= \frac{24}{\frac{6}{5}}$ or 20 Simplify.

Thus, $\sum_{n=1}^{\infty} 24\left(-\frac{1}{5}\right)^{n-1} = 20$.

REPEATING DECIMALS The formula for the sum of an infinite geometric series can be used to write a repeating decimal as a fraction. Remember that decimals with bar notation such as $0.\overline{2}$ and $0.\overline{47}$ represent 0.222222… and 0.474747…, respectively. Each of these expressions can be written as an infinite geometric series.

Example 3 Write a Repeating Decimal as a Fraction

Write $0.\overline{39}$ as a fraction.

Method 1

Write the repeating decimal as a sum.

$0.\overline{39} = 0.393939\ldots$

$= 0.39 + 0.0039 + 0.000039 + \ldots$

$= \frac{39}{100} + \frac{39}{10,000} + \frac{39}{1,000,000} + \ldots$

In this series, $a_1 = \frac{39}{100}$ and $r = \frac{1}{100}$.

$S = \frac{a_1}{1 - r}$ Sum formula

$= \frac{\frac{39}{100}}{1 - \frac{1}{100}}$ $a_1 = \frac{39}{100}, r = \frac{1}{100}$

$= \frac{\frac{39}{100}}{\frac{99}{100}}$ Subtract.

$= \frac{39}{99}$ or $\frac{13}{33}$ Simplify.

Thus, $0.\overline{39} = \frac{13}{33}$.

Method 2

$S = 0.\overline{39}$ Label the given decimal.

$S = 0.393939\ldots$ Repeating decimal

$100S = 39.393939\ldots$ Multiply each side by 100.

$99S = 39$ Subtract the second equation from the third.

$S = \frac{39}{99}$ or $\frac{13}{33}$ Divide each side by 99.

Check for Understanding

Concept Check

1. **OPEN ENDED** Write the series $\frac{1}{2} + \frac{1}{4} + \frac{1}{8} + \frac{1}{16} + \dots$ using sigma notation.
2. **Explain** why $0.999999\dots = 1$.
3. **FIND THE ERROR** Miguel and Beth are discussing the series $-\frac{1}{3} + \frac{4}{9} - \frac{16}{27} + \dots$. Miguel says that the sum of the series is $-\frac{1}{7}$. Beth says that the series does not have a sum. Who is correct? Explain your reasoning.

Miguel

$$S = \frac{-\frac{1}{3}}{1 - \left(-\frac{4}{3}\right)}$$

$$= -\frac{1}{7}$$

Guided Practice

Find the sum of each infinite geometric series, if it exists.

4. $a_1 = 36, r = \frac{2}{3}$
5. $a_1 = 18, r = -1.5$
6. $16 + 24 + 36 + \dots$
7. $\frac{1}{4} + \frac{1}{6} + \frac{2}{18} + \dots$
8. $6 - 2.4 + 0.96 - \dots$
9. $\sum_{n=1}^{\infty} 40\left(\frac{3}{5}\right)^{n-1}$

Write each repeating decimal as a fraction.

10. $0.\overline{5}$
11. $0.\overline{73}$
12. $0.\overline{175}$

Application

13. **CLOCKS** Jasmine's old grandfather clock is broken. When she tries to set the pendulum in motion by holding it against the side of the clock and letting it go, it first swings 24 centimeters to the other side, then 18 centimeters back, then 13.5 centimeters, and so on. What is the total distance that the pendulum swings?

Practice and Apply

Homework Help

For Exercises	See Examples
14–27, 32–39	1
28–31	2
40–47	3

Extra Practice
See page 852.

Find the sum of each infinite geometric series, if it exists.

14. $a_1 = 4, r = \frac{5}{7}$
15. $a_1 = 14, r = \frac{7}{3}$
16. $a_1 = 12, r = -0.6$
17. $a_1 = 18, r = 0.6$
18. $16 + 12 + 9 + \dots$
19. $-8 - 4 - 2 - \dots$
20. $12 - 18 + 24 - \dots$
21. $18 - 12 + 8 - \dots$
22. $1 + \frac{2}{3} + \frac{4}{9} + \dots$
23. $\frac{5}{3} + \frac{25}{3} + \frac{125}{3} + \dots$
24. $\frac{5}{3} - \frac{10}{9} + \frac{20}{27} - \dots$
25. $\frac{3}{2} - \frac{3}{4} + \frac{3}{8} - \dots$
26. $3 + 1.8 + 1.08 + \dots$
27. $1 - 0.5 + 0.25 - \dots$
28. $\sum_{n=1}^{\infty} 48\left(\frac{2}{3}\right)^{n-1}$
29. $\sum_{n=1}^{\infty} \left(\frac{3}{8}\right)\left(\frac{3}{4}\right)^{n-1}$
30. $\sum_{n=1}^{\infty} 3(0.5)^{n-1}$
31. $\sum_{n=1}^{\infty} (1.5)(0.25)^{n-1}$

32. **CHILD'S PLAY** Kimimela's little sister likes to swing at the playground. Yesterday, Kimimela pulled the swing back and let it go. The swing traveled a distance of 9 feet before heading back the other way. Each swing afterward was only 70% as long as the previous one. Find the total distance the swing traveled.

GEOMETRY **For Exercises 33 and 34, refer to square *ABCD*, which has a perimeter of 40 centimeters.**

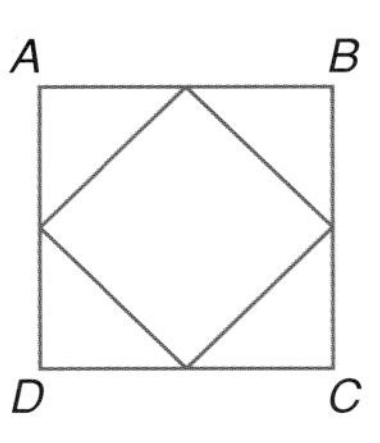

If the midpoints of the sides are connected, a smaller square results. Suppose the process of connecting midpoints of sides and drawing new squares is continued indefinitely.

33. Write an infinite geometric series to represent the sum of the perimeters of all of the squares.

34. Find the sum of the perimeters of all of the squares.

Aviation

The largest hot-air balloon ever flown had a capacity of 2.6 million cubic feet.

Source: *The Guinness Book of Records*

35. **AVIATION** A hot-air balloon rises 90 feet in its first minute of flight. In each succeeding minute, it rises only 90% as far as it did during the preceding minute. What is the final height of the balloon?

36. The sum of an infinite geometric series is 81, and its common ratio is $\frac{2}{3}$. Find the first three terms of the series.

37. The sum of an infinite geometric series is 125, and the value of r is 0.4. Find the first three terms of the series.

38. The common ratio of an infinite geometric series is $\frac{11}{16}$, and its sum is $76\frac{4}{5}$. Find the first four terms of the series.

39. The first term of an infinite geometric series is -8, and its sum is $-13\frac{1}{3}$. Find the first four terms of the series.

Write each repeating decimal as a fraction.

40. $0.\overline{7}$
41. $0.\overline{1}$
42. $0.\overline{36}$
43. $0.\overline{82}$
44. $0.\overline{246}$
45. $0.\overline{427}$
46. $0.4\overline{5}$
47. $0.2\overline{31}$

48. **CRITICAL THINKING** Derive the formula for the sum of an infinite geometric series by using the technique in Lessons 11-2 and 11-4. That is, write an equation for the sum S of a general infinite geometric series, multiply each side of the equation by r, and subtract equations.

49. **WRITING IN MATH** Answer the question that was posed at the beginning of the lesson.

How does an infinite geometric series apply to a bouncing ball?

Include the following in your answer:
- some formulas you might expect to see on the chalkboard if the character in the comic strip really was discussing a bouncing ball, and
- an explanation of how to find the total distance traveled, both up and down, by the bouncing ball described at the beginning of Lesson 11-3.

50. What is the sum of an infinite geometric series with a first term of 6 and a common ratio of $\frac{1}{2}$?

(A) 3 (B) 4 (C) 9 (D) 12

51. $2 + \frac{2}{3} + \frac{2}{9} + \frac{2}{27} + \ldots =$

(A) $\frac{3}{2}$ (B) $\frac{80}{27}$ (C) 3 (D) does not exist

Maintain Your Skills

Mixed Review

Find S_n for each geometric series described. *(Lesson 11-4)*

52. $a_1 = 1, a_6 = -243, r = -3$

53. $a_1 = 72, r = \frac{1}{3}, n = 7$

54. **PHYSICS** A vacuum pump removes 20% of the air from a container with each stroke of its piston. What percent of the original air remains after five strokes of the piston? *(Lesson 11-3)*

Solve each equation or inequality. Check your solution. *(Lesson 10-1)*

55. $6^x = 216$

56. $2^{2x} = \frac{1}{8}$

57. $3^{x-2} \geq 27$

Simplify each expression. *(Lesson 9-2)*

58. $\frac{-2}{ab} + \frac{5}{a^2}$

59. $\frac{1}{x-3} - \frac{2}{x+1}$

60. $\frac{1}{x^2 + 6x + 8} + \frac{3}{x+4}$

Write an equation for the circle that satisfies each set of conditions. *(Lesson 8-3)*

61. center (2, 4), radius 6

62. endpoints of a diameter at (7, 3) and (−1, −5)

Find all the zeros of each function. *(Lesson 7-5)*

63. $f(x) = 8x^3 - 36x^2 + 22x + 21$

64. $g(x) = 12x^4 + 4x^3 - 3x^2 - x$

Write a quadratic equation with the given roots. Write the equation in the form $ax^2 + bx + c = 0$, where a, b, and c are integers. *(Lesson 6-3)*

65. 6, −6

66. −2, −7

67. 6, 4

RECREATION For Exercises 68 and 69, refer to the graph at the right. *(Lesson 2-3)*

68. Find the average rate of change of the number of visitors to Yosemite National Park from 1996 to 1999.

69. Was the number of visitors increasing or decreasing from 1996 to 1999?

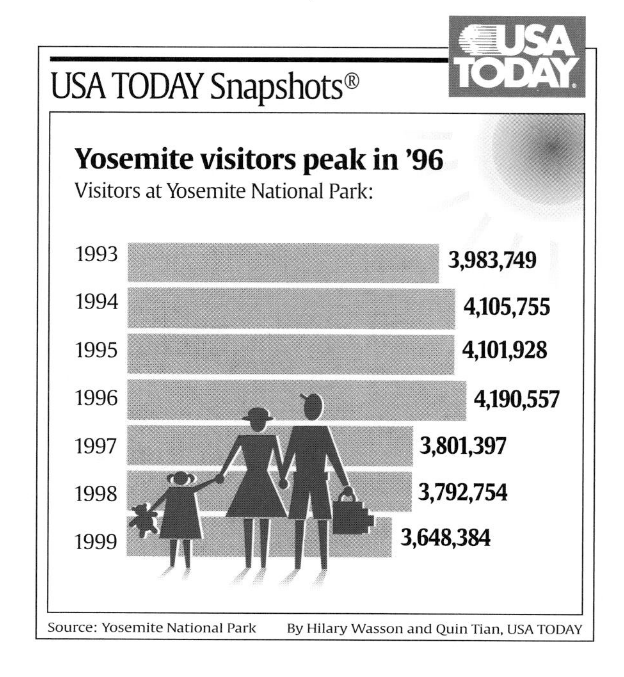

Getting Ready for the Next Lesson

PREREQUISITE SKILL Find each function value.
*(To review **evaluating functions**, see Lesson 2-1.)*

70. $f(x) = 2x, f(1)$

71. $g(x) = 3x - 3, g(2)$

72. $h(x) = -2x + 2, h(0)$

73. $f(x) = 3x - 1, f\left(\frac{1}{2}\right)$

74. $g(x) = x^2, g(2)$

75. $h(x) = 2x^2 - 4, h(0)$

Spreadsheet Investigation

A Preview of Lesson 11-6

Amortizing Loans

When a payment is made on a loan, part of the payment is used to cover the interest that has accumulated since the last payment. The rest is used to reduce the *principal*, or original amount of the loan. This process is called *amortization*. You can use a spreadsheet to analyze the payments, interest, and balance on a loan. A table that shows this kind of information is called an *amortization schedule*.

Example

Marisela just bought a new sofa for $495. The store is letting her make monthly payments of $43.29 at an interest rate of 9% for one year. How much will she still owe after six months?

Every month, the interest on the remaining balance will be $\frac{9\%}{12}$ or 0.75%. You can find the balance after a payment by multiplying the balance after the previous payment by 1 + 0.0075 or 1.0075 and then subtracting 43.29.

In a spreadsheet, use the column of numbers for the number of payments and use column B for the balance. Enter the interest rate and monthly payment in cells in column A so that they can be easily updated if the information changes.

The spreadsheet at the right shows the formulas for the balances after each of the first six payments. After six months, Marisela still owes $253.04.

Loans

	A	B
1	Interest rate	=495*(1+A2)-A5
2	0.0075	=B1*(1+A2)-A5
3		=B2*(1+A2)-A5
4	Monthly payment	=B3*(1+A2)-A5
5	43.29	=B4*(1+A2)-A5
6		=B5*(1+A2)-A5
7		

Sheet1 / Sh

Exercises

1. Let b_n be the balance left on Marisela's loan after n months. Write an equation relating b_n and b_{n+1}.
2. Payments at the beginning of a loan go more toward interest than payments at the end. What percent of Marisela's loan remains to be paid after half a year?
3. Extend the spreadsheet to the whole year. What is the balance after 12 payments? Why is it not 0?
4. Suppose Marisela decides to pay $50 every month. How long would it take her to pay off the loan?
5. Suppose that, based on how much she can afford, Marisela will pay a variable amount each month in addition to the $43.29. Explain how the flexibility of a spreadsheet can be used to adapt to this situation.
6. Jamie has a three-year, $12,000 car loan. The annual interest rate is 6%, and his monthly payment is $365.06. After twelve months, he receives an inheritance which he wants to use to pay off the loan. How much does he owe at that point?

11-6 Recursion and Special Sequences

What You'll Learn

- Recognize and use special sequences.
- Iterate functions.

Vocabulary

- Fibonacci sequence
- recursive formula
- iteration

How is the Fibonacci sequence illustrated in nature?

A shoot on a sneezewort plant must grow for two months before it is strong enough to put out another shoot. After that, it puts out at least one shoot every month.

Month	1	2	3	4	5
Shoots	1	1	2	3	5

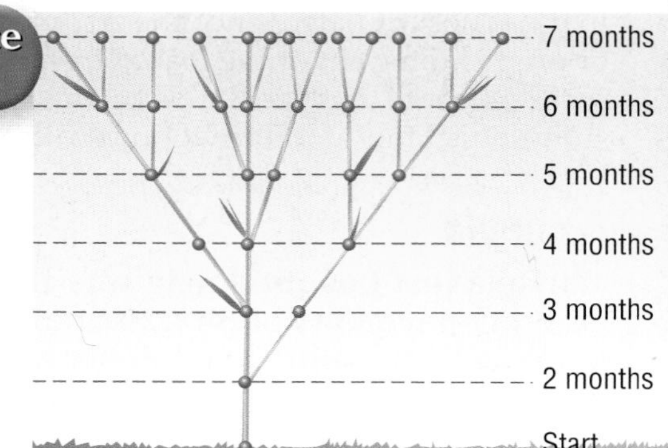

SPECIAL SEQUENCES Notice that the sequence 1, 1, 2, 3, 5, 8, 13, … has a pattern. Each term in the sequence is the sum of the two previous terms. For example, $8 = 3 + 5$ and $13 = 5 + 8$. This sequence is called the **Fibonacci sequence**, and it is found in many places in nature.

first term	a_1		1
second term	a_2		1
third term	a_3	$a_1 + a_2$	$1 + 1 = 2$
fourth term	a_4	$a_2 + a_3$	$1 + 2 = 3$
fifth term	a_5	$a_3 + a_4$	$2 + 3 = 5$
⋮	⋮	⋮	
nth term	a_n	$a_{n-2} + a_{n-1}$	

Study Tip

Reading Math
A recursive formula is often called a *recursive relation* or a *recurrence relation.*

The formula $a_n = a_{n-2} + a_{n-1}$ is an example of a **recursive formula**. This means that each term is formulated from one or more previous terms. To be able to use a recursive formula, you must be given the value(s) of the first term(s) so that you can start the sequence and then use the formula to generate the rest of the terms.

Example 1 Use a Recursive Formula

Find the first five terms of the sequence in which $a_1 = 4$ and $a_{n+1} = 3a_n - 2$, $n \geq 1$.

$a_{n+1} = 3a_n - 2$ Recursive formula

$a_{1+1} = 3a_1 - 2$ $n = 1$
$a_2 = 3(4) - 2$ or 10 $a_1 = 4$

$a_{2+1} = 3a_2 - 2$ $n = 2$
$a_3 = 3(10) - 2$ or 28 $a_2 = 10$

$a_{3+1} = 3a_3 - 2$ $n = 3$
$a_4 = 3(28) - 2$ or 82 $a_3 = 28$

$a_{4+1} = 3a_4 - 2$ $n = 4$
$a_5 = 3(82) - 2$ or 244 $a_4 = 82$

The first five terms of the sequence are 4, 10, 28, 82, and 244.

Example 2 Find and Use a Recursive Formula

GARDENING Mr. Yazaki discovered that there were 225 dandelions in his garden on the first Saturday of spring. He had time to pull out 100, but by the next Saturday, there were twice as many as he had left. Each Saturday in spring, he removed 100 dandelions, only to find that the number of remaining dandelions had doubled by the following Saturday.

a. Write a recursive formula for the number of dandelions Mr. Yazaki finds in his garden each Saturday.

Let d_n represent the number of dandelions at the beginning of the nth Saturday. Mr. Yazaki will pull 100 of these out of his garden, leaving $d_n - 100$. The number d_{n+1} of dandelions the next Saturday will be twice this number. So, $d_{n+1} = 2(d_n - 100)$ or $2d_n - 200$.

b. Find the number of dandelions Mr. Yazaki would find on the fifth Saturday.

On the first Saturday, there were 225 dandelions, so $d_1 = 225$.

$d_{n+1} = 2d_n - 200$ Recursive formula

$d_{1+1} = 2d_1 - 200$ $n = 1$
$d_2 = 2(225) - 200$ or 250

$d_{2+1} = 2d_2 - 200$ $n = 2$
$d_3 = 2(250) - 200$ or 300

$d_{3+1} = 2d_3 - 200$ $n = 3$
$d_4 = 2(300) - 200$ or 400

$d_{4+1} = 2d_4 - 200$ $n = 4$
$d_5 = 2(400) - 200$ or 600

On the fifth Saturday, there would be 600 dandelions in Mr. Yazaki's garden.

You can use sequences to analyze some games.

Algebra Activity

Special Sequences

The object of the *Towers of Hanoi* game is to move a stack of n coins from one position to another in the fewest number a_n of moves with these rules.

- You may only move one coin at a time.
- A coin must be placed on top of another coin, not underneath.
- A smaller coin may be placed on top of a larger coin, but not vice versa. For example, a penny may not be placed on top of a dime.

Model and Analyze

1. Draw three circles on a sheet of paper, as shown above. Place a penny on the first circle. What is the least number of moves required to get the penny to the second circle?
2. Place a nickel and a penny on the first circle, with the penny on top. What is the least number of moves that you can make to get the stack to another circle? (Remember, a nickel cannot be placed on top of a penny.)
3. Place a nickel, penny, and dime on the first circle. What is the least number of moves that you can take to get the stack to another circle?

Make a Conjecture

4. Place a quarter, nickel, penny, and dime on the first circle. Experiment to find the least number of moves needed to get the stack to another circle. Make a conjecture about a formula for the minimum number a_n of moves required to move a stack of n coins.

Study Tip

Look Back

To review **composition of functions**, see Lesson 7-7.

ITERATION **Iteration** is the process of composing a function with itself repeatedly. For example, if you compose a function with itself once, the result is $f \circ f(x)$ or $f(f(x))$. If you compose a function with itself two times, the result is $f \circ f \circ f\,(x)$ or $f(f(f(x)))$, and so on.

You can use iteration to recursively generate a sequence. Start with an initial value x_0. Let $x_1 = f(x_0)$, $x_2 = f(x_1)$ or $f(f(x_0))$, $x_3 = f(\,x_2)$ or $f(f(f(x_0)))$, and so on.

Example 3 Iterate a Function

Find the first three iterates x_1, x_2, x_3 of the function $f(x) = 2x + 3$ for an initial value of $x_0 = 1$.

To find the first iterate x_1, find the value of the function for $x_0 = 1$.

$x_1 = f(x_0)$ Iterate the function.

$= f(1)$ $x_0 = 1$

$= 2(1) + 3$ or 5 Simplify.

To find the second iterate x_2, substitute x_1 for x.

$x_2 = f(x_1)$ Iterate the function.

$= f(5)$ $x_1 = 5$

$= 2(5) + 3$ or 13 Simplify.

Substitute x_2 for x to find the third iterate.

$x_3 = f(x_2)$ Iterate the function.

$= f(13)$ $x_2 = 13$

$= 2(13) + 3$ or 29 Simplify.

Therefore, 1, 5, 13, 29 is an example of a sequence generated using iteration.

Check for Understanding

Concept Check

1. **Write** recursive formulas for the nth terms of arithmetic and geometric sequences.
2. **OPEN ENDED** Write a recursive formula for a sequence whose first three terms are 1, 1, and 3.
3. **State** whether the statement $x_n \neq x_{n-1}$ is *sometimes*, *always*, or *never* true if $x_n = f(x_{n-1})$. Explain.

Guided Practice

Find the first five terms of each sequence.

4. $a_1 = 12, a_{n+1} = a_n - 3$
5. $a_1 = -3, a_{n+1} = a_n + n$
6. $a_1 = 0, a_{n+1} = -2a_n - 4$
7. $a_1 = 1, a_2 = 2, a_{n+2} = 4a_{n+1} - 3a_n$

Find the first three iterates of each function for the given initial value.

8. $f(x) = 3x - 4, x_0 = 3$
9. $f(x) = -2x + 5, x_0 = 2$
10. $f(x) = x^2 + 2, x_0 = -1$

Application

BANKING For Exercises 11 and 12, use the following information.
Rita has deposited \$1000 in a bank account. At the end of each year, the bank posts interest to her account in the amount of 5% of the balance, but then takes out a \$10 annual fee.

11. Let b_0 be the amount Rita deposited. Write a recursive equation for the balance b_n in her account at the end of n years.
12. Find the balance in the account after four years.

Practice and Apply

Homework Help	
For Exercises	See Examples
13–30	1–2
31–39	3

Extra Practice
See page 853.

Find the first five terms of each sequence.

13. $a_1 = -6, a_{n+1} = a_n + 3$

14. $a_1 = 13, a_{n+1} = a_n + 5$

15. $a_1 = 2, a_{n+1} = a_n - n$

16. $a_1 = 6, a_{n+1} = a_n + n + 3$

17. $a_1 = 9, a_{n+1} = 2\,a_n - 4$

18. $a_1 = 4, a_{n+1} = 3a_n - 6$

19. $a_1 = -1, a_2 = 5, a_{n+1} = a_n + a_{n-1}$

20. $a_1 = 4, a_2 = -3, a_{n+2} = a_{n+1} + 2a_n$

21. $a_1 = \frac{7}{2}, a_{n+1} = \frac{n}{n+1} \cdot a_n$

22. $a_1 = \frac{3}{4}, a_{n+1} = \frac{n^2+1}{n} \cdot a_n$

23. If $a_0 = 7$ and $a_{n+1} = a_n + 12$ for $n \geq 0$, find the value of a_5.

24. If $a_0 = 1$ and $a_{n+1} = -2.1$ for $n \geq 0$, then what is the value of a_4?

GEOMETRY **For Exercises 25 and 26, use the following information.**
Join two 1-unit by 1-unit squares to form a rectangle. Next, draw a larger square along a long side of the rectangle. Continue this process of drawing a square along a long side of the rectangle formed at the previous step.

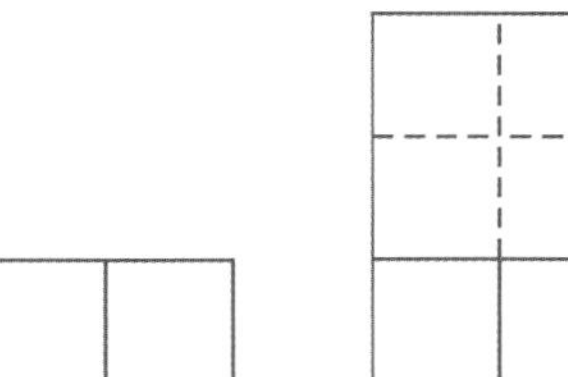
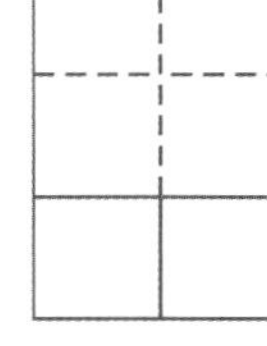
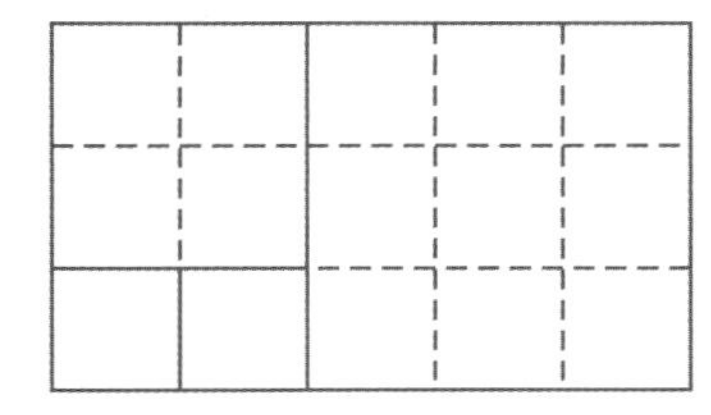

Step 1 Step 2 Step 3

25. Write the sequence of the lengths of the sides of the squares you added at each step. Begin the sequence with the lengths of the sides of the two original squares.

26. Identify the sequence in Exercise 25.

27. LOANS The Cruz family is taking out a mortgage loan for \$100,000 to buy a house. Their monthly payment is \$678.79. The recursive formula $b_n = 1.006\,b_{n-1} - 678.79$ describes the balance left on the loan after n payments. Find the balances of the loan after each of the first eight payments.

GEOMETRY **For Exercises 28–30, study the triangular numbers shown below.**

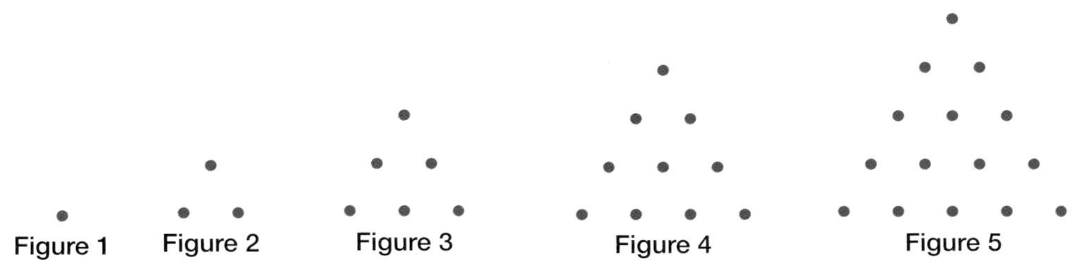

28. Write a sequence of the first five triangular numbers.

29. Write a recursive formula for the nth triangular number t_n.

30. What is the 200th triangular number?

Career Choices

Real Estate Agent
Most real estate agents are independent business-people who earn their income from commission.

Online Research
To learn more about a career in real estate, visit: www.algebra2.com/careers

Find the first three iterates of each function for the given initial value.

31. $f(x) = 9x - 2, x_0 = 2$
32. $f(x) = 4x - 3, x_0 = 2$
33. $f(x) = 3x + 5, x_0 = -4$
34. $f(x) = 5x + 1, x_0 = -1$
35. $f(x) = 2x^2 - 5, x_0 = -1$
36. $f(x) = 3x^2 - 4, x_0 = 1$
37. $f(x) = 2x^2 + 2x + 1, x_0 = \frac{1}{2}$
38. $f(x) = 3x^2 - 3x + 2, x_0 = \frac{1}{3}$

39. **ECONOMICS** If the rate of inflation is 2%, the cost of an item in future years can be found by iterating the function $c(x) = 1.02x$. Find the cost of a $70 portable stereo in four years if the rate of inflation remains constant.

40. **CRITICAL THINKING** Are there a function $f(x)$ and an initial value x_0 such that the first three iterates, in order, are 4, 4, and 7? If so, state such a function and initial value. If not, explain.

41. **WRITING IN MATH** Answer the question that was posed at the beginning of the lesson.

How is the Fibonacci sequence illustrated in nature?

Include the following in your answer:

- the 13th term in the Fibonacci sequence, with an explanation of what it tells you about the plant described, and
- an explanation of why the Fibonacci sequence is neither arithmetic nor geometric.

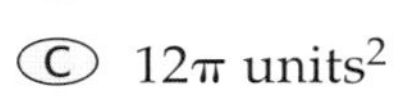

42. If a is positive, what percent of $4a$ is 8?

(A) $\frac{a}{100}\%$ (B) $\frac{a}{2}\%$ (C) $\frac{8}{a}\%$ (D) $\frac{200}{a}\%$

43. The figure at the right is made of three concentric semicircles. What is the total area of the shaded regions?

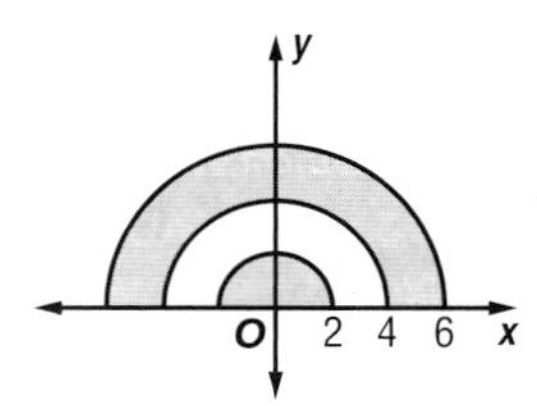

(A) 4π units^2 (B) 10π units^2

(C) 12π units^2 (D) 20π units^2

Maintain Your Skills

Mixed Review

Find the sum of each infinite geometric series, if it exists. *(Lesson 11-5)*

44. $9 + 6 + 4 + \ldots$
45. $\frac{1}{8} + \frac{1}{32} + \frac{1}{128} + \ldots$
46. $4 - \frac{8}{3} + \frac{16}{9} + \ldots$

Find the sum of each geometric series. *(Lesson 11-4)*

47. $2 - 10 + 50 - \ldots$ to 6 terms
48. $3 + 1 + \frac{1}{3} + \ldots$ to 7 terms

49. **GEOMETRY** The area of rectangle $ABCD$ is $6x^2 + 38x + 56$ square units. Its width is $2x + 8$ units. What is the length of the rectangle? *(Lesson 5-3)*

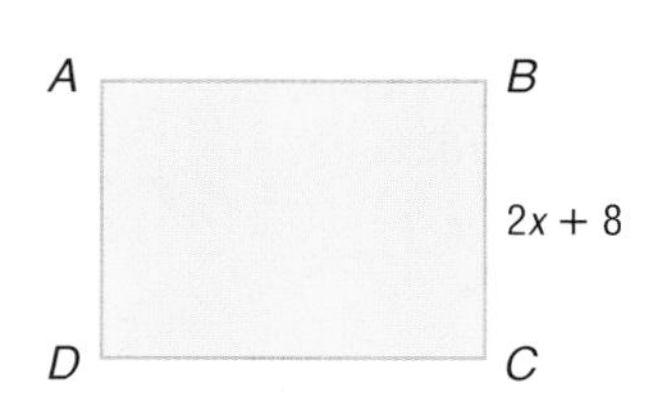

Getting Ready for the Next Lesson

BASIC SKILL Evaluate each expression.

50. $5 \cdot 4 \cdot 3 \cdot 2 \cdot 1$
51. $7 \cdot 6 \cdot 5 \cdot 4 \cdot 3 \cdot 2 \cdot 1$
52. $\frac{4 \cdot 3}{2 \cdot 1}$
53. $\frac{6 \cdot 5 \cdot 4}{3 \cdot 2 \cdot 1}$
54. $\frac{9 \cdot 8 \cdot 7 \cdot 6}{4 \cdot 3 \cdot 2 \cdot 1}$
55. $\frac{10 \cdot 9 \cdot 8 \cdot 7 \cdot 6 \cdot 5}{6 \cdot 5 \cdot 4 \cdot 3 \cdot 2 \cdot 1}$

Algebra Activity

A Follow-Up of Lesson 11-6

Fractals

Fractals are sets of points that often involve intricate geometric shapes. Many fractals have the property that when small parts are magnified, the detail of the fractal is not lost. In other words, the magnified part is made up of smaller copies of itself. Such fractals can be constructed recursively.

You can use isometric dot paper to draw stages of the construction of a fractal called the *von Koch snowflake.*

Stage 1 Draw an equilateral triangle with sides of length 9 units on the dot paper.

Stage 2 Now remove the middle third of each side of the triangle from Stage 1 and draw the other two sides of an equilateral triangle pointing outward.

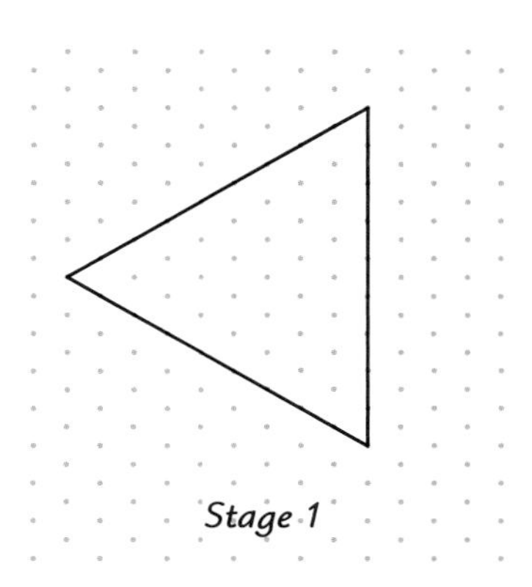

Stage 1

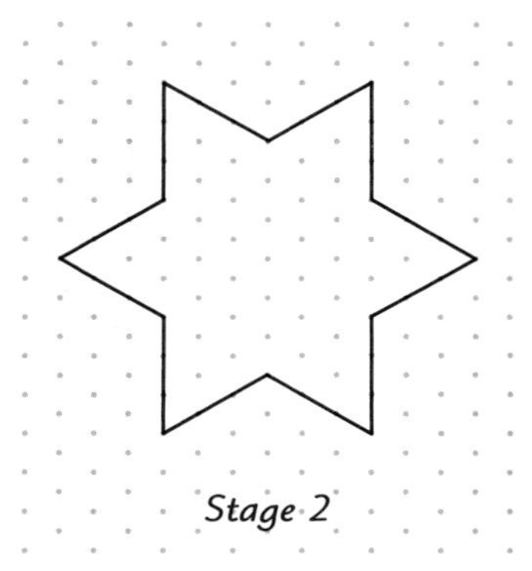

Stage 2

Imagine continuing this process indefinitely. The von Koch snowflake is the shape that these stages approach.

Model and Analyze

1. Copy and complete the table. Draw Stage 3, if necessary.

Stage	1	2	3	4
Number of Segments	3	12		
Length of each Segment	9	3		
Perimeter	27	36		

2. Write recursive formulas for the number s_n of segments in Stage n, the length ℓ_n of each segment in Stage n, and the perimeter P_n of Stage n.

3. Write nonrecursive formulas for s_n, ℓ_n, and P_n.

4. What is the perimeter of the von Koch snowflake? Explain.

5. Explain why the area of the von Koch snowflake can be represented by the infinite series $\frac{81\sqrt{3}}{4} + \frac{27\sqrt{3}}{4} + 3\sqrt{3} + \frac{4\sqrt{3}}{3} + \dots$.

6. Find the sum of the series in Exercise 5. Explain your steps.

7. Do you think the results of Exercises 4 and 6 are contradictory? Explain.

11-7 The Binomial Theorem

California Standards Standard 20.0 Students know the binomial theorem and use it to expand binomial expressions that are raised to positive integer powers. (Key)

What You'll Learn

- Use Pascal's triangle to expand powers of binomials.
- Use the Binomial Theorem to expand powers of binomials.

How does a power of a binomial describe the numbers of boys and girls in a family?

According to the U.S. Census Bureau, ten percent of families have three or more children. If a family has four children, there are six sequences of births of boys and girls that result in two boys and two girls. These sequences are listed below.

BBGG BGBG BGGB GBBG GBGB GGBB

Vocabulary

- Pascal's triangle
- Binomial Theorem
- factorial

Pascal's Triangle Although he did not discover it, Pascal's triangle is named for the French mathematician Blaise Pascal (1623–1662).

PASCAL'S TRIANGLE You can use the coefficients in powers of binomials to count the number of possible sequences in situations such as the one above. Remember that a binomial is a polynomial with two terms. Expand a few powers of the binomial $b + g$.

$$(b + g)^0 = 1b^0g^0$$
$$(b + g)^1 = 1b^1g^0 + 1b^0g^1$$
$$(b + g)^2 = 1b^2g^0 + 2b^1g^1 + 1b^0g^1$$
$$(b + g)^3 = 1b^3g^0 + 3b^2g^1 + 3b^1g^2 + 1b^0g^3$$
$$(b + g)^4 = 1b^4g^0 + 4b^3g^1 + 6b^2g^2 + 4b^1g^3 + 1b^0g^4$$

The coefficient 6 of the b^2g^2 term in the expansion of $(b + g)^4$ gives the number of sequences of births that result in two boys and two girls. As another example, the coefficient 4 of the b^1g^3 term gives the number of sequences with one boy and 3 girls.

Here are some patterns that can be seen in any binomial expansion of the form $(a + b)^n$.

1. There are $n + 1$ terms.
2. The exponent n of $(a + b)^n$ is the exponent of a in the first term and the exponent of b in the last term.
3. In successive terms, the exponent of a decreases by one, and the exponent of b increases by one.
4. The sum of the exponents in each term is n.
5. The coefficients are symmetric. They increase at the beginning of the expansion and decrease at the end.

The coefficients form a pattern that is often displayed in a triangular formation. This is known as **Pascal's triangle**. Notice that each row begins and ends with 1. Each coefficient is the sum of the two coefficients above it in the previous row.

$(a + b)^0$						1					
$(a + b)^1$					1		1				
$(a + b)^2$				1		2		1			
$(a + b)^3$			1		3		3		1		
$(a + b)^4$		1		4		6		4		1	
$(a + b)^5$	1		5		10		10		5		1

Example 1 Use Pascal's Triangle

Expand $(x + y)^7$.

Write two more rows of Pascal's triangle.

1 6 15 20 15 6 1

1 7 21 35 35 21 7 1

Use the patterns of a binomial expansion and the coefficients to write the expansion of $(x + y)^7$.

$$(x + y)^7 = 1x^7y^0 + 7x^6y^1 + 21x^5y^2 + 35x^4y^3 + 35x^3y^4 + 21x^2y^5 + 7x^1y^6 + 1x^0y^7$$
$$= x^7 + 7x^6y + 21x^5y^2 + 35x^4y^3 + 35x^3y^4 + 21x^2y^5 + 7xy^6 + y^7$$

THE BINOMIAL THEOREM Another way to show the coefficients in a binomial expansion is to write them in terms of the previous coefficients.

$(a+b)^0$	1				
$(a+b)^1$	1	$\frac{1}{1}$			
$(a+b)^2$	1	$\frac{2}{1}$	$\frac{2\cdot 1}{1\cdot 2}$		
$(a+b)^3$	1	$\frac{3}{1}$	$\frac{3\cdot 2}{1\cdot 2}$	$\frac{3\cdot 2\cdot 1}{1\cdot 2\cdot 3}$	
$(a+b)^4$	1	$\frac{4}{1}$	$\frac{4\cdot 3}{1\cdot 2}$	$\frac{4\cdot 3\cdot 2}{1\cdot 2\cdot 3}$	$\frac{4\cdot 3\cdot 2\cdot 1}{1\cdot 2\cdot 3\cdot 4}$

Eliminate common factors that are shown in color.

This pattern provides the coefficients of $(a + b)^n$ for any nonnegative integer n. The pattern is summarized in the **Binomial Theorem**.

Key Concept — Binomial Theorem

If n is a nonnegative integer, then

$$(a + b)^n = 1a^nb^0 + \frac{n}{1}a^{n-1}b^1 + \frac{n(n-1)}{1\cdot 2}a^{n-2}b^2 + \frac{n(n-1)(n-2)}{1\cdot 2\cdot 3}a^{n-3}b^3 + \ldots + 1a^0b^n.$$

Example 2 Use the Binomial Theorem

Expand $(a - b)^6$.

The expansion will have seven terms. Use the sequence $1, \frac{6}{1}, \frac{6\cdot 5}{1\cdot 2}, \frac{6\cdot 5\cdot 4}{1\cdot 2\cdot 3}$ to find the coefficients for the first four terms. Then use symmetry to find the remaining coefficients.

$$(a - b)^6 = 1a^6(-b)^0 + \frac{6}{1}a^5(-b)1 + \frac{6\cdot 5}{1\cdot 2}a^4(-b)^2 + \frac{6\cdot 5\cdot 4}{1\cdot 2\cdot 3}a^3(-b)^3 + \ldots + 1a^0(-b)^6$$
$$= a^6 - 6a^5b + 15a^4b^2 - 20a^3b^3 + 15a^2b^4 - 6ab^5 + b^6$$

Notice that in terms having the same coefficients, the exponents are reversed, as in $15a^4b^2$ and $15a^2b^4$.

Study Tip

Graphing Calculators

On a TI-83 Plus, the factorial symbol, !, is located on the MATH PRB menu.

The factors in the coefficients of binomial expansions involve special products called **factorials**. For example, the product $4 \cdot 3 \cdot 2 \cdot 1$ is written $4!$ and is read *4 factorial*. In general, if n is a positive integer, then $n! = n(n-1)(n-2)(n-3) \ldots 2 \cdot 1$. *By definition, $0! = 1$.*

Example 3 Factorials

Evaluate $\frac{8!}{3!5!}$.

$$\frac{8!}{3!5!} = \frac{8 \cdot 7 \cdot 6 \cdot \overset{1}{\cancel{5 \cdot 4 \cdot 3 \cdot 2 \cdot 1}}}{3 \cdot 2 \cdot 1 \cdot \underset{1}{\cancel{5 \cdot 4 \cdot 3 \cdot 2 \cdot 1}}}$$

Note that $8! = 8 \cdot 7 \cdot 6 \cdot 5!$, so $\frac{8!}{3!5!} = \frac{8 \cdot 7 \cdot 6 \cdot 5!}{3!5!}$ or $\frac{8 \cdot 7 \cdot 6}{3 \cdot 2 \cdot 1}$

$$= \frac{8 \cdot 7 \cdot 6}{3 \cdot 2 \cdot 1} \text{ or } 56$$

Study Tip

Missing Steps

If you don't understand a step like $\frac{6 \cdot 5 \cdot 4}{1 \cdot 2 \cdot 3} = \frac{6!}{3!3!}$, work it out on a piece of scrap paper.

$$\frac{6 \cdot 5 \cdot 4}{1 \cdot 2 \cdot 3} = \frac{6 \cdot 5 \cdot 4 \cdot 3!}{1 \cdot 2 \cdot 3 \cdot 3!} = \frac{6!}{3!3!}$$

An expression such as $\frac{6 \cdot 5 \cdot 4}{1 \cdot 2 \cdot 3}$ in Example 2 can be written as a quotient of factorials. In this case, $\frac{6 \cdot 5 \cdot 4}{1 \cdot 2 \cdot 3} = \frac{6!}{3!3!}$. Using this idea, you can rewrite the expansion of $(a + b)^6$ using factorials.

$$(a + b)^6 = \frac{6!}{6!0!}a^6b^0 + \frac{6!}{5!1!}a^5b^1 + \frac{6!}{4!2!}a^4b^2 + \frac{6!}{3!3!}a^3b^3 + \frac{6!}{2!4!}a^2b^4 + \frac{6!}{1!5!}a^1b^5 + \frac{6!}{0!6!}a^0b^6$$

You can also write this series using sigma notation.

$$(a + b)^6 = \sum_{k=0}^{6} \frac{6!}{(6 - k)!k!}a^{6-k}b^k$$

In general, the Binomial Theorem can be written both in factorial notation and in sigma notation.

Key Concept — Binomial Theorem, Factorial Form

$$(a + b)^n = \frac{n!}{n!0!}a^nb^0 + \frac{n!}{(n - 1)!1!}a^{n-1}b^1 + \frac{n!}{(n - 2)!2!}a^{n-2}b^2 + \ldots + \frac{n!}{0!n!}a^0b^n$$

$$= \sum_{k=0}^{n} \frac{n!}{(n - k)!k!}a^{n-k}b^k$$

Example 4 Use a Factorial Form of the Binomial Theorem

Expand $(2x + y)^5$.

$$(2x + y)^5 = \sum_{k=0}^{5} \frac{5!}{(5 - k)!k!}(2x)^{5-k}y^k$$ Binomial Theorem, factorial form

$$= \frac{5!}{5!0!}(2x)^5y^0 + \frac{5!}{4!1!}(2x)^4y^1 + \frac{5!}{3!2!}(2x)^3y^2 + \frac{5!}{2!3!}(2x)^2y^3 + \frac{5!}{1!4!}(2x)^1y^4 + \frac{5!}{0!5!}(2x)^0y^5$$ Let $k = 0, 1, 2, 3, 4$, and 5.

$$= \frac{5 \cdot 4 \cdot 3 \cdot 2 \cdot 1}{5 \cdot 4 \cdot 3 \cdot 2 \cdot 1 \cdot 1}(2x)^5 + \frac{5 \cdot 4 \cdot 3 \cdot 2 \cdot 1}{4 \cdot 3 \cdot 2 \cdot 1 \cdot 1}(2x)^4 y + \frac{5 \cdot 4 \cdot 3 \cdot 2 \cdot 1}{3 \cdot 2 \cdot 1 \cdot 2 \cdot 1}(2x)^3 y^2 + \frac{5 \cdot 4 \cdot 3 \cdot 2 \cdot 1}{2 \cdot 1 \cdot 3 \cdot 2 \cdot 1}(2x)^2 y^3 + \frac{5 \cdot 4 \cdot 3 \cdot 2 \cdot 1}{1 \cdot 4 \cdot 3 \cdot 2 \cdot 1}(2x)y^4 + \frac{5 \cdot 4 \cdot 3 \cdot 2 \cdot 1}{1 \cdot 5 \cdot 4 \cdot 3 \cdot 2 \cdot 1}y^5$$

$$= 32x^5 + 80x^4y + 80x^3y^2 + 40x^2y^3 + 10xy^4 + y^5$$ Simplify.

Sometimes you need to know only a particular term of a binomial expansion. Note that when the Binomial Theorem is written in sigma notation, $k = 0$ for the first term, $k = 1$ for the second term, and so on. In general, the value of k is always one less than the number of the term you are finding.

Example 5 Find a Particular Term

Find the fifth term in the expansion of $(p + q)^{10}$.

First, use the Binomial Theorem to write the expansion in sigma notation.

$$(p + q)^{10} = \sum_{k=0}^{10} \frac{10!}{(10 - k)!k!} p^{10-k} q^k$$

In the fifth term, $k = 4$.

$$\frac{10!}{(10-k)!k!} p^{10-k} q^k = \frac{10!}{(10-4)!4!} p^{10-4} q^4 \quad k = 4$$

$$= \frac{10 \cdot 9 \cdot 8 \cdot 7}{4 \cdot 3 \cdot 2 \cdot 1} p^6 q^4 \quad \frac{10!}{6!4!} = \frac{10 \cdot 9 \cdot 8 \cdot 7 \cdot 6!}{6!4!} \text{ or } \frac{10 \cdot 9 \cdot 8 \cdot 7}{4 \cdot 3 \cdot 2 \cdot 1}$$

$$= 210p^6q^4 \quad \text{Simplify.}$$

Check for Understanding

Concept Check

1. **List** the coefficients in the row of Pascal's triangle corresponding to $n = 8$.
2. **Identify** the coefficient of $a^{n-1}b$ in the expansion of $(a + b)^n$.
3. **OPEN ENDED** Write a power of a binomial for which the first term of the expansion is $625x^4$.

Guided Practice

Evaluate each expression.

4. $8!$
5. $\frac{13!}{9!}$
6. $\frac{12!}{2!10!}$

Expand each power.

7. $(p + q)^5$
8. $(t + 2)^6$
9. $(x - 3y)^4$

Find the indicated term of each expansion.

10. fourth term of $(a + b)^8$
11. fifth term of $(2a + 3b)^{10}$

Application

12. **SCHOOL** Mr. Hopkins is giving a five-question true-false quiz. How many ways could a student answer the questions with three trues and two falses?

Practice and Apply

Homework Help

For Exercises	See Examples
13–18	3
19–33	1, 2, 4
34–41	5

Extra Practice See page 853.

Evaluate each expression.

13. $9!$
14. $13!$
15. $\frac{9!}{7!}$
16. $\frac{7!}{4!}$
17. $\frac{12!}{8!4!}$
18. $\frac{14!}{5!9!}$

Expand each power.

19. $(a - b)^3$
20. $(m + n)^4$
21. $(r + s)^8$
22. $(m - a)^5$
23. $(x + 3)^5$
24. $(a - 2)^4$
25. $(2b - x)^4$
26. $(2a + b)^6$
27. $(3x - 2y)^5$
28. $(3x + 2y)^4$
29. $\left(\frac{a}{2} + 2\right)^5$
30. $\left(3 + \frac{m}{3}\right)^5$

31. **GEOMETRY** Write an expanded expression for the volume of the cube at the right.

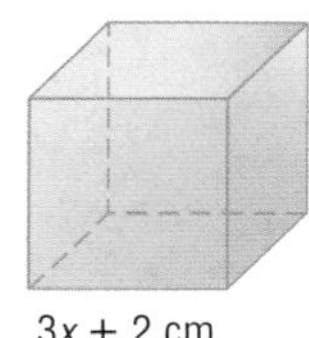

Pascal's triangle displays many patterns. Visit www.algebra2.com/webquest to continue work on your WebQuest project.

32. **GAMES** The diagram shows the board for a game in which ball bearings are dropped down a chute. A pattern of nails and dividers causes the bearings to take various paths to the sections at the bottom. For each section, how many paths through the board lead to that section?

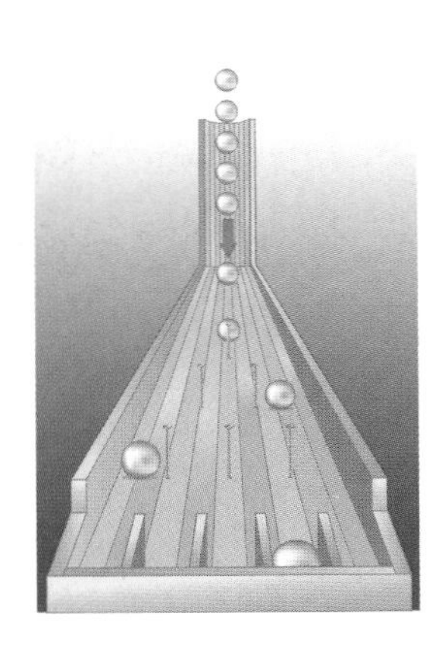

33. **INTRAMURALS** Ofelia is taking ten shots in the intramural free-throw shooting competition. How many sequences of makes and misses are there that result in her making eight shots and missing two?

Find the indicated term of each expansion.

34. sixth term of $(x - y)^9$
35. seventh term of $(x + y)^{12}$
36. fourth term of $(x + 2)^7$
37. fifth term of $(a - 3)^8$
38. fifth term of $(2a + 3b)^{10}$
39. fourth term of $(2x + 3y)^9$
40. fourth term of $\left(x + \frac{1}{3}\right)^7$
41. sixth term of $\left(x - \frac{1}{2}\right)^{10}$

42. **CRITICAL THINKING** Explain why $\frac{12!}{7!5!} + \frac{12!}{6!6!} = \frac{13!}{7!6!}$ without finding the value of any of the expressions.

43. **WRITING IN MATH** Answer the question that was posed at the beginning of the lesson.

How does a power of a binomial describe the numbers of boys and girls in a family?

Include the following in your answer:
- the expansion of $(b + g)^5$ and what it tells you about sequences of births of boys and girls in families with five children, and
- an explanation of how to find a formula for the number of sequences of births that have exactly k girls in a family of n children.

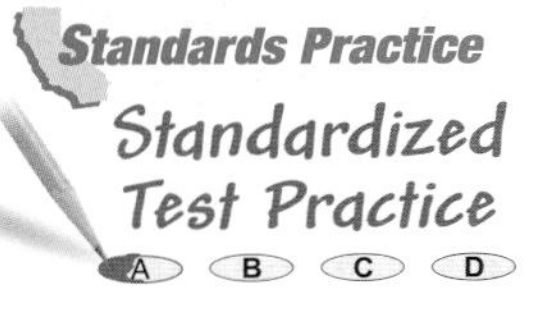

44. Which of the following represents the values of x that are solutions of the inequality $x^2 < x + 20$?

Ⓐ $x > -4$

Ⓑ $x < 5$

Ⓒ $-5 < x < 4$

Ⓓ $-4 < x < 5$

45. If four lines intersect as shown in the figure at the right, $x + y =$

Ⓐ 70.

Ⓑ 115.

Ⓒ 140.

Ⓓ It cannot be determined from the information given.

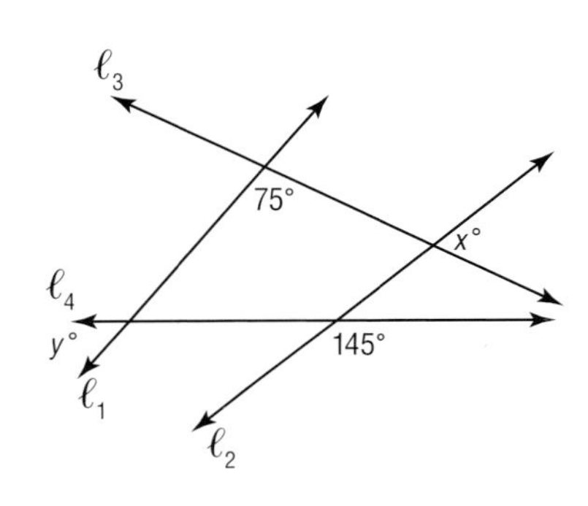

Maintain Your Skills

Mixed Review

Find the first five terms of each sequence. *(Lesson 11-6)*

46. $a_1 = 7, a_{n+1} = a_n - 2$

47. $a_1 = 3, a_{n+1} = 2a_n - 1$

48. **CLOCKS** The spring in Juanita's old grandfather clock is broken. When you try to set the pendulum in motion by holding it against the wall of the clock and letting go, it follows a swing pattern of 25 centimeters, 20 centimeters, 16 centimeters, and so on until it comes to rest. What is the total distance the pendulum swings before coming to rest? *(Lesson 11-5)*

Express each logarithm in terms of common logarithms. Then approximate its value to four decimal places. *(Lesson 10-4)*

49. $\log_2 5$

50. $\log_3 10$

51. $\log_5 8$

Determine any vertical asymptotes and holes in the graph of each rational function. *(Lesson 9-3)*

52. $f(x) = \dfrac{1}{x^2 + 5x + 6}$

53. $f(x) = \dfrac{x + 2}{x^2 + 3x - 4}$

54. $f(x) = \dfrac{x^2 + 4x + 3}{x + 3}$

Without writing the equation in standard form, state whether the graph of each equation is a *parabola, circle, ellipse,* or *hyperbola.* *(Lesson 8-6)*

55. $x^2 - 6x - y^2 - 3 = 0$

56. $4y - x + y^2 = 1$

Determine whether each pair of functions are inverse functions. *(Lesson 7-8)*

57. $f(x) = x + 3$
$g(x) = x - 3$

58. $f(x) = 2x + 1$
$g(x) = \dfrac{x + 1}{2}$

Getting Ready for the Next Lesson

PREREQUISITE SKILL **State whether each statement is *true* or *false* when $n = 1$. Explain.** *(To review* ***evaluating expressions****, see Lesson 1-1.)*

59. $1 = \dfrac{n(n + 1)}{2}$

60. $1 = \dfrac{(n + 1)(2n + 1)}{2}$

61. $1 = \dfrac{n^2(n + 1)^2}{4}$

62. $3^n - 1$ is even.

Practice Quiz 2

Lessons 11-4 through 11-7

Find the sum of each geometric series. *(Lessons 11-4 and 11-5)*

1. $a_1 = 5, r = 3, n = 12$

2. $\sum_{n-1}^{6} 2(-3)^{n-1}$

3. $\sum_{n=1}^{\infty} 8\left(\frac{2}{3}\right)^{n-1}$

4. $5 + 1 + \frac{1}{5} + \ldots$

Find the first five terms of each sequence. *(Lesson 11-6)*

5. $a_1 = 1, a_{n+1} = 2a_n + 3$

6. $a_1 = 2, a_{n+1} = a_n + 2n$

7. Find the first three iterates of the function $f(x) = -3x + 2$ for an initial value of $x_0 = -1$. *(Lesson 11-6)*

Expand each power. *(Lesson 11-7)*

8. $(3x + y)^5$

9. $(a + 2)^6$

10. Find the fifth term of the expansion of $(2a + b)^9$. *(Lesson 11-7)*

11-8 Proof and Mathematical Induction

California Standards
Standard 21.0 Students apply the method of mathematical induction to prove general statements about the positive integers.

What You'll Learn

- Prove statements by using mathematical induction.
- Disprove statements by finding a counterexample.

How does the concept of a ladder help you prove statements about numbers?

Imagine the positive integers as a ladder that goes upward forever. You know that you cannot leap to the top of the ladder, but you can stand on the first step, and no matter which step you are on, you can always climb one step higher. Is there any step you cannot reach?

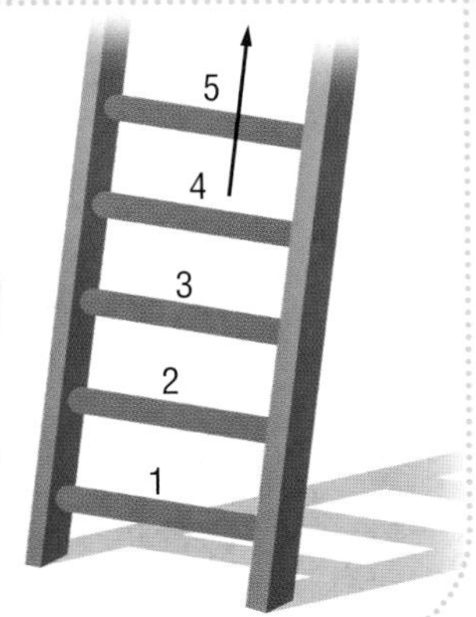

Vocabulary

- mathematical induction
- inductive hypothesis

MATHEMATICAL INDUCTION **Mathematical induction** is used to prove statements about positive integers. An induction proof consists of three steps.

Study Tip

Step 1
In many cases, it will be helpful to let $n = 1$.

Key Concept — Mathematical Induction

Step 1 Show that the statement is true for some integer n.

Step 2 Assume that the statement is true for some positive integer k, where $k \geq n$. This assumption is called the **inductive hypothesis**.

Step 3 Show that the statement is true for the next integer $k + 1$.

Example 1 Summation Formula

Prove that the sum of the squares of the first n positive integers is $\frac{n(n+1)(2n+1)}{6}$. That is, prove that $1^2 + 2^2 + 3^2 + \ldots + n^2 = \frac{n(n+1)(2n+1)}{6}$.

Step 1 When $n = 1$, the left side of the given equation is 1^2 or 1. The right side is $\frac{1(1+1)[2(1)+1]}{6}$ or 1. Thus, the equation is true for $n = 1$.

Step 2 Assume $1^2 + 2^2 + 3^2 + \ldots + k^2 = \frac{k(k+1)(2k+1)}{6}$ for a positive integer k.

Step 3 Show that the given equation is true for $n = k + 1$.

$$1^2 + 2^2 + 3^2 + \ldots + k^2 + (k+1)^2 = \frac{k(k+1)(2k+1)}{6} + (k+1)^2 \quad \text{Add } (k+1)^2 \text{ to each side.}$$

$$= \frac{k(k+1)(2k+1) + 6(k+1)^2}{6} \quad \text{Add.}$$

$$= \frac{(k+1)[k(2k+1) + 6(k+1)]}{6} \quad \text{Factor.}$$

$$= \frac{(k+1)[2k^2 + 7k + 6]}{6} \quad \text{Simplify.}$$

$$= \frac{(k+1)(k+2)(2k+3)}{6} \quad \text{Factor.}$$

$$= \frac{(k+1)[(k+1)+1][2(k+1)+1]}{6}$$

The last expression on page 618 is the right side of the equation to be proved, where n has been replaced by $k + 1$. Thus, the equation is true for $n = k + 1$.

This proves that $1^2 + 2^2 + 3^2 + \ldots + n^2 = \frac{n(n + 1)(2n + 1)}{6}$ for all positive integers n.

Example 2 Divisibility

Prove that $7^n - 1$ is divisible by 6 for all positive integers n.

Step 1 When $n = 1$, $7^n - 1 = 7^1 - 1$ or 6. Since 6 is divisible by 6, the statement is true for $n = 1$.

Step 2 Assume that $7^k - 1$ is divisible by 6 for some positive integer k. This means that there is a whole number r such that $7^k - 1 = 6r$.

Step 3 Show that the statement is true for $n = k + 1$.

$7^k - 1 = 6r$ Inductive hypothesis

$7^k = 6r + 1$ Add 1 to each side.

$7(7^k) = 7(6r + 1)$ Multiply each side by 7.

$7^{k+1} = 42r + 7$ Simplify.

$7^{k+1} - 1 = 42r + 6$ Subtract 1 from each side.

$7^{k+1} - 1 = 6(7r + 1)$ Factor.

Since r is a whole number, $7r + 1$ is a whole number. Therefore, $7^{k+1} - 1$ is divisible by 6. Thus, the statement is true for $n = k + 1$.

This proves that $7^n - 1$ is divisible by 6 for all positive integers n.

Study Tip

Reading Math
One of the meanings of *counter* is *to oppose*, so a counterexample is an example that opposes a hypothesis.

COUNTEREXAMPLES Of course, not every formula that you can write is true. A formula that works for a few positive integers may not work for *every* positive integer. You can show that a formula is not true by finding a *counterexample*. This often involves trial and error.

Example 3 Counterexample

Find a counterexample for the formula $1^4 + 2^4 + 3^4 + \ldots + n^4 = 1 + (4n - 4)^2$.

Check the first few positive integers.

n	Left Side of Formula	Right Side of Formula	
1	1^4 or 1	$1 + [4(1) - 4]^2 = 1 + 0^2$ or 1	true
2	$1^4 + 2^4 = 1 + 16$ or 17	$1 + [4(2) - 4]^2 = 1 + 4^2$ or 17	true
3	$1^4 + 2^4 + 3^4 = 1 + 16 + 81$ or 98	$1 + [4(3) - 4]^2 = 1 + 64$ or 65	false

The value $n = 3$ is a counterexample for the formula.

Check for Understanding

Concept Check

1. **Describe** some of the types of statements that can be proved by using mathematical induction.

2. **Explain** the difference between mathematical induction and a counterexample.

3. **OPEN ENDED** Write an expression of the form $b^n - 1$ that is divisible by 2 for all positive integers n.

Guided Practice **Prove that each statement is true for all positive integers.**

4. $1 + 2 + 3 + \ldots + n = \frac{n(n + 1)}{2}$

5. $\frac{1}{2} + \frac{1}{2^2} + \frac{1}{2^3} + \ldots + \frac{1}{2^n} = 1 - \frac{1}{2^n}$

6. $4^n - 1$ is divisible by 3.

7. $5^n + 3$ is divisible by 4.

Find a counterexample for each statement.

8. $1 + 2 + 3 + \ldots + n = n^2$

9. $2^n + 2n$ is divisible by 4.

Application

10. **PARTIES** Suppose that each time a new guest arrives at a party, he or she shakes hands with each person already at the party. Prove that after n guests have arrived, a total of $\frac{n(n - 1)}{2}$ handshakes have taken place.

Practice and Apply

Homework Help	
For Exercises	See Examples
11–23, 31	1
24	1, 2
25–30	3

Extra Practice
See page 853.

Prove that each statement is true for all positive integers.

11. $1 + 5 + 9 + \ldots + (4n - 3) = n(2n - 1)$

12. $2 + 5 + 8 + \ldots + (3n - 1) = \frac{n(3n + 1)}{2}$

13. $1^3 + 2^3 + 3^3 + \ldots + n^3 = \frac{n^2(n + 1)^2}{4}$

14. $1^2 + 3^2 + 5^2 + \ldots + (2n - 1)^2 = \frac{n(2n - 1)(2n + 1)}{3}$

15. $\frac{1}{3} + \frac{1}{3^2} + \frac{1}{3^3} + \ldots + \frac{1}{3^n} = \frac{1}{2}\left(1 - \frac{1}{3^n}\right)$

16. $\frac{1}{4} + \frac{1}{4^2} + \frac{1}{4^3} + \ldots + \frac{1}{4^n} = \frac{1}{3}\left(1 - \frac{1}{4^n}\right)$

17. $8^n - 1$ is divisible by 7.

18. $9^n - 1$ is divisible by 8.

19. $12^n + 10$ is divisible by 11.

20. $13^n + 11$ is divisible by 12.

21. **ARCHITECTURE** A memorial being constructed in a city park will be a brick wall, with a top row of six gold-plated bricks engraved with the names of six local war veterans. Each row has two more bricks than the row above it. Prove that the number of bricks in the top n rows is $n^2 + 5n$.

22. **GEOMETRIC SERIES** Use mathematical induction to prove the formula $a_1 + a_1r + a_1r^2 + \ldots + a_1r^{n-1} = \frac{a_1(1 - r^n)}{1 - r}$ for the sum of a finite geometric series.

23. **ARITHMETIC SERIES** Use mathematical induction to prove the formula $a_1 + (a_1 + d) + (a_1 + 2d) + \ldots + [a_1 + (n - 1)d] = \frac{n}{2}[2\,a_1 + (n - 1)d]$ for the sum of an arithmetic series.

24. **PUZZLES** Show that a 2^n by 2^n checkerboard with the top right square missing can always be covered by nonoverlapping L-shaped tiles like the one at the right.

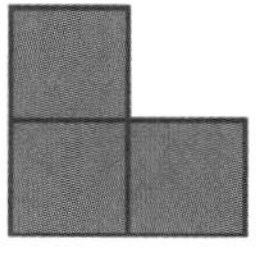

Find a counterexample for each statement.

25. $1^2 + 2^2 + 3^2 + \ldots + n^2 = \frac{n(3n - 1)}{2}$

26. $1^3 + 3^3 + 5^3 + \ldots + (2n - 1)^3 = 12n^3 - 23n^2 + 12n$

27. $3^n + 1$ is divisible by 4.

28. $2^n + 2n^2$ is divisible by 4.

29. $n^2 - n + 11$ is prime.

30. $n^2 + n + 41$ is prime.

31. **CRITICAL THINKING** Refer to Example 2. Explain how to use the Binomial Theorem to show that $7^n - 1$ is divisible by 6 for all positive integers n.

32. **WRITING IN MATH** Answer the question that was posed at the beginning of the lesson.

How does the concept of a ladder help you prove statements about numbers?

Include the following in your answer:

- an explanation of which part of an inductive proof corresponds to stepping onto the bottom step of the ladder, and
- an explanation of which part of an inductive proof corresponds to climbing from one step on the ladder to the next.

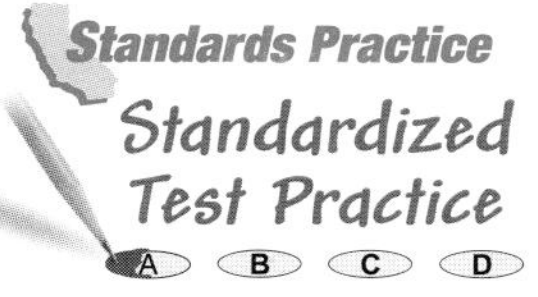

33. $\dfrac{x - \frac{4}{x}}{1 - \frac{4}{x} + \frac{4}{x^2}} =$

Ⓐ $\frac{x}{x - 2}$ Ⓑ $\frac{x^2 + 2}{x - 2}$ Ⓒ $\frac{x^2 + 2x}{x - 2}$ Ⓓ $\frac{x^2 + 2x}{(x - 2)^2}$

34. *PQRS* is a square. Find the ratio of the length of diagonal $\overline{QS}$ to the length of side $\overline{RS}$.

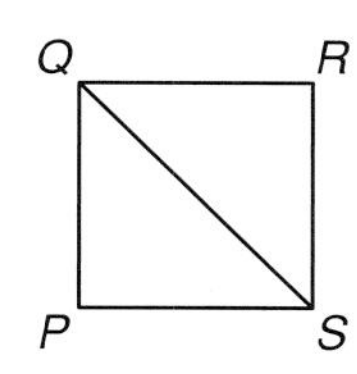

Ⓐ $\sqrt{2}$

Ⓑ 1

Ⓒ $\frac{\sqrt{2}}{2}$

Ⓓ The relationship cannot be determined from the given information.

Maintain Your Skills

Mixed Review

Expand each power. *(Lesson 11-7)*

35. $(x + y)^6$ **36.** $(a - b)^7$ **37.** $(2x + y)^8$

Find the first three iterates of each function for the given initial value. *(Lesson 11-6)*

38. $f(x) = 3x - 2, x_0 = 2$ **39.** $f(x) = 4x^2 - 2, x_0 = 1$

40. **BIOLOGY** Suppose an amoeba divides into two amoebas once every hour. How long would it take for a single amoeba to become a colony of 4096 amoebas? *(Lesson 10-2)*

Solve each equation. Check your solutions. *(Lesson 9-6)*

41. $\frac{1}{y + 1} - \frac{3}{y - 3} = 2$ **42.** $\frac{6}{a - 7} = \frac{a - 49}{a^2 - 7a} + \frac{1}{a}$

Chapter 11 Study Guide and Review

Vocabulary and Concept Check

arithmetic means (p. 580)
arithmetic sequence (p. 578)
arithmetic series (p. 583)
Binomial Theorem (p. 613)
common difference (p. 578)
common ratio (p. 588)
convergent series (p. 600)
factorial (p. 613)
Fibonacci sequence (p. 606)
geometric means (p. 590)
geometric sequence (p. 588)
geometric series (p. 594)
index of summation (p. 585)
inductive hypothesis (p. 618)
infinite geometric series (p. 599)
iteration (p. 608)
mathematical induction (p. 618)
partial sum (p. 599)
Pascal's triangle (p. 612)
recursive formula (p. 606)
sequence (p. 578)
series (p. 583)
sigma notation (p. 585)
term (p. 578)

Choose the term from the list above that best completes each statement.

1. A(n) ______________ of an infinite series is the sum of a certain number of terms.
2. If a sequence has a common ratio, then it is a(n) ______________.
3. Using __________, the series 2 + 5 + 8 + 11 + 14 can be written as $\sum_{n=1}^{5}(3n - 1)$.
4. Eleven and 17 are the two ____ between 5 and 23 in the sequence 5, 11, 17, 23.
5. Using the __________, $(a - 2)^4$ can be expanded to $a^4 - 8a^3 + 24a^2 - 32a + 16$.
6. The __________ of the sequence 3, 2, $\frac{4}{3}, \frac{8}{9}, \frac{16}{27}$ is $\frac{2}{3}$.
7. The ________ 11 + 16.5 + 22 + 27.5 + 33 has a sum of 110.
8. A(n) ____ is expressed as $n! = n(n - 1)(n - 2) \ldots 2 \cdot 1$.

Lesson-by-Lesson Review

11-1 Arithmetic Sequences

See pages 578–582.

Concept Summary

- An arithmetic sequence is formed by adding a constant to each term to get the next term.
- The nth term a_n of an arithmetic sequence with first term a_1 and common difference d is given by $a_n = a_1 + (n - 1)d$, where n is any positive integer.

Examples

1 Find the 12th term of an arithmetic sequence if $a_1 = -17$ and $d = 4$.

$a_n = a_1 + (n - 1)d$ — Formula for the nth term

$a_{12} = -17 + (12 - 1)4$ — $n = 12, a_1 = -17, d = 4$

$a_{12} = 27$ — Simplify.

2 Find the two arithmetic means between 4 and 25.

$a_n = a_1 + (n - 1)d$ — Formula for the nth term

$a_4 = 4 + (4 - 1)d$ — $n = 4, a_1 = 4$

$25 = 4 + 3d$ — $a_4 = 25$

$7 = d$ — The arithmetic means are 4 + 7 or 11 and 11 + 7 or 18.

www.algebra2.com/vocabulary_review

Exercises **Find the indicated term of each arithmetic sequence.** *See Example 2 on p. 579.*

9. $a_1 = 6, d = 8, n = 5$

10. $a_1 = -5, d = 7, n = 22$

11. $a_1 = 5, d = -2, n = 9$

12. $a_1 = -2, d = -3, n = 15$

Find the arithmetic means in each sequence. *See Example 4 on page 580.*

13. -7, __?__, __?__, __?__, 9

14. 12, __?__, __?__, 4

15. 9, __?__, __?__, __?__, __?__, -6

16. 56, __?__, __?__, __?__, 28

11-2 Arithmetic Series

See pages 583–587.

Concept Summary

- The sum S_n of the first n terms of an arithmetic series is given by $S_n = \frac{n}{2}[2a_1 + (n - 1)d]$ or $S_n = \frac{n}{2}(a_1 + a_n)$.

Example **Find S_n for the arithmetic series with $a_1 = 34$, $a_n = 2$, and $n = 9$.**

$S_n = \frac{n}{2}(a_1 + a_n)$ Sum formula

$S_9 = \frac{9}{2}(34 + 2)$ $n = 9, a_1 = 34, a_n = 2$

$S_9 = 162$ Simplify.

Exercises **Find S_n for each arithmetic series.** *See Examples on pages 584 and 585.*

17. $a_1 = 12, a_n = 117, n = 36$

18. $4 + 10 + 16 + \ldots + 106$

19. $10 + 4 + (-2) + \ldots + (-50)$

20. $\sum_{n=2}^{13} (3n + 1)$

11-3 Geometric Sequences

See pages 588–592.

Concept Summary

- A geometric sequence is one in which each term after the first is found by multiplying the previous term by a common ratio.
- The nth term a_n of a geometric sequence with first term a_1 and common ratio r is given by $a_n = a_1 \cdot r^{n-1}$, where n is any positive integer.

Examples

1 **Find the fifth term of a geometric sequence for which $a_1 = 7$ and $r = 3$.**

$a_n = a_1 \cdot r^{n-1}$ Formula for nth term

$a_5 = 7 \cdot 3^{5-1}$ $n = 5, a_1 = 7, r = 3$

$a_5 = 567$ The fifth term is 567.

2 **Find two geometric means between 1 and 8.**

$a_n = a_1 \cdot r^{n-1}$ Formula for nth term

$a_4 = 1 \cdot r^{4-1}$ $n = 4$ and $a_1 = 1$

$8 = r^3$ $a_4 = 8$

$2 = r$ The geometric means are 1(2) or 2 and 2(2) or 4.

Exercises **Find the indicated term of each geometric sequence.**
See Example 2 on page 589.

21. $a_1 = 2, r = 2, n = 5$

22. $a_1 = 7, r = 2, n = 4$

23. $a_1 = 243, r = -\frac{1}{3}, n = 5$

24. a_6 for $\frac{2}{3}, \frac{4}{3}, \frac{8}{3}, \ldots$

Find the geometric means in each sequence. *See Example 5 on page 590.*

25. 3, ___?___, ___?___, 24

26. 7.5, ___?___, ___?___, ___?___, 120

27. 8, ___?___, ___?___, ___?___, ___?___, $\frac{1}{4}$

28. 5, ___?___, ___?___, ___?___, 80

11-4 Geometric Series

See pages 594–598.

Concept Summary

- The sum S_n of the first n terms of a geometric series is given by

$$S_n = \frac{a_1(1 - r^n)}{1 - r} \text{ or } S_n = \frac{a_1 - a_1 r^n}{1 - r}, \text{ where } r \neq 1.$$

Example **Find the sum of a geometric series for which $a_1 = 7$, $r = 3$, and $n = 14$.**

$S_n = \frac{a_1 - a_1 r^n}{1 - r}$ Sum formula

$S_{14} = \frac{7 - 7 \cdot 3^{14}}{1 - 3}$ $n = 14, a_1 = 7, r = 3$

$S_{14} = 16{,}740{,}388$ Use a calculator.

Exercises **Find S_n for each geometric series.** *See Examples 1 and 3 on pages 595 and 596.*

29. $a_1 = 12, r = 3, n = 5$

30. $4 - 2 + 1 - \ldots$ to 6 terms

31. $256 + 192 + 144 + \ldots$ to 7 terms

32. $\sum_{n=1}^{5} \left(-\frac{1}{2}\right)^{n-1}$

11-5 Infinite Geometric Series

See pages 599–604.

Concept Summary

- The sum S of an infinite geometric series with $-1 < r < 1$ is given by $S = \frac{a_1}{1 - r}$.

Example **Find the sum of the infinite geometric series for which $a_1 = 18$ and $r = -\frac{2}{7}$.**

$S = \frac{a_1}{1 - r}$ Sum formula

$= \frac{18}{1 - \left(-\frac{2}{7}\right)}$ $a_1 = 18, r = -\frac{2}{7}$

$= \frac{18}{\frac{9}{7}}$ or 14 Simplify.

Exercises **Find the sum of each infinite geometric series, if it exists.**
See Example 1 on page 600.

33. $a_1 = 6, r = \frac{11}{12}$ **34.** $\frac{1}{8} - \frac{3}{16} + \frac{9}{32} - \frac{27}{64} + \ldots$ **35.** $\sum_{n=1}^{\infty} -2\left(-\frac{5}{8}\right)^{n-1}$

11-6 Recursion and Special Sequences

See pages 606–610.

Concept Summary

- In a recursive formula, each term is formulated from one or more previous terms.
- Iteration is the process of composing a function with itself repeatedly.

Examples

1 **Find the first five terms of the sequence in which $a_1 = 2$ and $a_{n+1} = 2a_n - 1$.**

$a_{n+1} = 2a_n - 1$ Recursive formula

$a_{1+1} = 2a_1 - 1$ $n = 1$
$a_2 = 2(2) - 1$ or 3 $a_1 = 2$
$a_{2+1} = 2a_2 - 1$ $n = 2$
$a_3 = 2(3) - 1$ or 5 $a_2 = 3$

$a_{3+1} = 2a_3 - 1$ $n = 3$
$a_4 = 2(5) - 1$ or 9 $a_3 = 5$
$a_{4+1} = 2a_4 - 1$ $n = 4$
$a_5 = 2(9) - 1$ or 17 $a_4 = 9$

The first five terms of the sequence are 2, 3, 5, 9, and 17.

2 **Find the first three iterates of $f(x) = -5x - 1$ for an initial value of $x_0 = -1$.**

$x_1 = f(x_0)$
$= f(-1)$
$= -5(-1) - 1$ or 4

$x_2 = f(x_1)$
$= f(4)$
$= -5(4) - 1$ or -21

$x_3 = f(x_2)$
$= f(-21)$
$= -5(-21) - 1$ or 104

The first three iterates are 4, -21, and 104.

Exercises **Find the first five terms of each sequence.** *See Example 1 on page 606.*

36. $a_1 = -2, a_{n+1} = a_n + 5$
37. $a_1 = 3, a_{n+1} = 4a_n - 10$
38. $a_1 = 2, a_{n+1} = a_n + 3n$
39. $a_1 = 1, a_2 = 3, a_{n+2} = a_{n+1} + a_n$

Find the first three iterates of each function for the given initial value.
See Example 3 on page 608.

40. $f(x) = -2x + 3, x_0 = 1$
41. $f(x) = 7x - 4, x_0 = 2$
42. $f(x) = x^2 - 6, x_0 = -1$
43. $f(x) = -2x^2 - x + 5, x_0 = -2$

11-7 The Binomial Theorem

See pages 612–617.

Concept Summary

- Pascal's triangle can be used to find the coefficients in a binomial expansion.
- The Binomial Theorem: $(a + b)^n = \sum_{k=0}^{n} \frac{n!}{(n-k)!k!} a^{n-k} b^k$

• Extra Practice, see pages 851–855.
• Mixed Problem Solving, see page 872.

Example **Expand $(a - 2b)^4$.**

$$(a - 2b)^4 = \sum_{k=0}^{4} \frac{4!}{(4-k)!k!} a^{4-k}(-2b)^k \quad \text{Binomial Theorem}$$

$$= \frac{4!}{4!0!}a^4(-2b)^0 + \frac{4!}{3!1!}a^3(-2b)^1 + \frac{4!}{2!2!}a^2(-2b)^2 + \frac{4!}{1!3!}a^1(-2b)^3 + \frac{4!}{0!4!}a^0(-2b)^4$$

$$= a^4 - 8a^3b + 24a^2b^2 - 32ab^3 + 16b^4 \quad \text{Simplify.}$$

Exercises **Expand each power.** *See Examples 1, 2, and 4 on pages 613 and 614.*

44. $(x + y)^3$ **45.** $(x - 2)^4$ **46.** $(3r + s)^5$

Find the indicated term of each expansion. *See Example 5 on page 615.*

47. fourth term of $(x + 2y)^6$ **48.** second term of $(4x - 5)^{10}$

11-8 Proof and Mathematical Induction

See pages 618–621.

Concept Summary

- Mathematical induction is a method of proof used to prove statements about the positive integers.

Example **Prove $1 + 5 + 25 + \ldots + 5^{n-1} = \frac{1}{4}(5^n - 1)$ for all positive integers n.**

Step 1 When $n = 1$, the left side of the given equation is 1. The right side is $\frac{1}{4}(5^1 - 1)$ or 1. Thus, the equation is true for $n = 1$.

Step 2 Assume that $1 + 5 + 25 + \ldots + 5^{k-1} = \frac{1}{4}(5^k - 1)$ for some positive integer k.

Step 3 Show that the given equation is true for $n = k + 1$.

$$1 + 5 + 25 + \ldots + 5^{k-1} + 5^{(k+1)-1} = \frac{1}{4}(5^k - 1) + 5^{(k+1)-1} \quad \text{Add } 5^{(k+1)-1} \text{ to each side.}$$

$$= \frac{1}{4}(5^k - 1) + 5^k \quad \text{Simplify the exponent.}$$

$$= \frac{5^k - 1 + 4 \cdot 5^k}{4} \quad \text{Common denominator}$$

$$= \frac{5 \cdot 5^k - 1}{4} \quad \text{Distributive Property}$$

$$= \frac{1}{4}(5^{k+1} - 1) \quad 5 \cdot 5^k = 5^{k+1}$$

The last expression above is the right side of the equation to be proved, where n has been replaced by $k + 1$. Thus, the equation is true for $n = k + 1$.

This proves that $1 + 5 + 25 + \ldots + 5^{n-1} = \frac{1}{4}(5^n - 1)$ for all positive integers n.

Exercises **Prove that each statement is true for all positive integers.**
See Examples 1 and 2 on pages 618 and 619.

49. $1 + 2 + 4 + \ldots + 2^{n-1} = 2^n - 1$ **50.** $6^n - 1$ is divisible by 5.

Practice Test

Vocabulary and Concepts

Choose the correct term to complete each sentence.

1. A sequence in which each term after the first is found by adding a constant to the previous term is called a(n) (*arithmetic*, *geometric*) sequence.
2. A (*Fibonacci sequence, series*) is a sum of terms of a sequence.
3. (*Pascal's triangle*, *Recursive formulas*) and the Binomial Theorem can be used to expand powers of binomials.

Skills and Applications

4. Find the next four terms of the arithmetic sequence 42, 37, 32, … .
5. Find the 27th term of an arithmetic sequence for which $a_1 = 2$ and $d = 6$.
6. Find the three arithmetic means between -4 and 16.
7. Find the sum of the arithmetic series for which $a_1 = 7$, $n = 31$, and $a_n = 127$.
8. Find the next two terms of the geometric sequence $\frac{1}{81}, \frac{1}{27}, \frac{1}{9}, \ldots$.
9. Find the sixth term of the geometric sequence for which $a_1 = 5$ and $r = -2$.
10. Find the two geometric means between 7 and 189.
11. Find the sum of the geometric series for which $a_1 = 125$, $r = \frac{2}{5}$, and $n = 4$.

Find the sum of each series, if it exists.

12. $\sum_{k=3}^{15}(14 - 2k)$
13. $\sum_{n=1}^{\infty}\frac{1}{3}(-2)^{n-1}$
14. $91 + 85 + 79 + \ldots + (-29)$
15. $12 + (-6) + 3 - \frac{3}{2} + \ldots$

Find the first five terms of each sequence.

16. $a_1 = 1, a_{n+1} = a_n + 3$
17. $a_1 = -3, a_{n+1} = a_n + n^2$
18. Find the first three iterates of $f(x) = x^2 - 3x$ for an initial value of $x_0 = 1$.
19. Expand $(2s - 3t)^5$.
20. Find the third term of the expansion of $(x + y)^{10}$.

Prove that each statement is true for all positive integers.

21. $1 + 3 + 5 + \ldots + (2n - 1) = n^2$
22. $14^n - 1$ is divisible by 13.
23. **DESIGN** A landscaper is designing a wall of white brick and red brick. The pattern starts with 20 red bricks on the bottom row. Each row above it contains 3 fewer red bricks than the preceding row. If the top row contains no red bricks, how many rows are there and how many red bricks were used?
24. **RECREATION** One minute after it is released, a gas-filled balloon has risen 100 feet. In each succeeding minute, the balloon rises only 50% as far as it rose in the previous minute. How far will the balloon rise in 5 minutes?

Standards Practice

25. **STANDARDIZED TEST PRACTICE** Find the next term in the geometric sequence $8, 6, \frac{9}{2}, \frac{27}{8}, \ldots$.

Ⓐ $\frac{11}{8}$ Ⓑ $\frac{27}{16}$ Ⓒ $\frac{9}{4}$ Ⓓ $\frac{81}{32}$

Chapter 11 Standardized Test Practice

Standards Practice

Part 1 Multiple Choice

Record your answers on the answer sheet provided by your teacher or on a sheet of paper.

1. For all positive integers, let $\boxed{n} = n + g$, where g is the greatest factor of n, and $g < n$. If $\boxed{18} = x$, then $\boxed{x} =$

Ⓐ 9. Ⓑ 8.

Ⓒ 27. Ⓓ 36.

2. If p is positive, what percent of $6p$ is 12?

Ⓐ $\frac{p}{100}\%$ Ⓑ $\frac{p}{2}\%$

Ⓒ $\frac{12}{p}\%$ Ⓓ $\frac{200}{p}\%$

3. A box is 12 units tall, 6 units long, and 8 units wide. A designer is creating a new box that must have the same volume as the first box. If the length and width of the new box are each 50% greater than the length and width of the first box, about how many units tall will the new box be?

Ⓐ 5.3 Ⓑ 6.8

Ⓒ 7.1 Ⓓ 8.5

4. Which of the following statements must be true when $0 < m < 1$?

I $\frac{\sqrt{m}}{m} > 1$ **II** $4m < 1$ **III** $m^2 - m^3 < 0$

Ⓐ I only

Ⓑ III only

Ⓒ I and II only

Ⓓ I, II, and III

5. If $3kx - \frac{4s}{t} = 3ky$, then $x - y = ?$

Ⓐ $-\frac{4s}{3kt}$ Ⓑ $\frac{-4s}{t} + \frac{1}{3k}$

Ⓒ $\frac{4s}{3t} - k$ Ⓓ $\frac{4s}{3kt}$

6. For all $n \neq 0$, what is the slope of the line passing through $(3n, -k)$ and $(-n, -k)$?

Ⓐ 0 Ⓑ $\frac{k}{2n}$

Ⓒ $\frac{2n}{k}$ Ⓓ undefined

7. Which is the graph of the equation $x^2 + (y - 4)^2 = 20$?

Ⓐ line Ⓑ parabola

Ⓒ circle Ⓓ ellipse

8. $\dfrac{x - \frac{9}{x}}{1 - \frac{6}{x} + \frac{9}{x^2}} =$

Ⓐ $\frac{x}{x - 3}$ Ⓑ $\frac{x^2 + 3}{x - 3}$

Ⓒ $\frac{x^2 + 3x}{x - 3}$ Ⓓ $\frac{x^2 + 3x}{(x - 3)^2}$

9. What is the sum of the positive even factors of 30?

Ⓐ 18 Ⓑ 30

Ⓒ 48 Ⓓ 72

10. If ℓ_1 is parallel to ℓ_2 in the figure, what is the value of x?

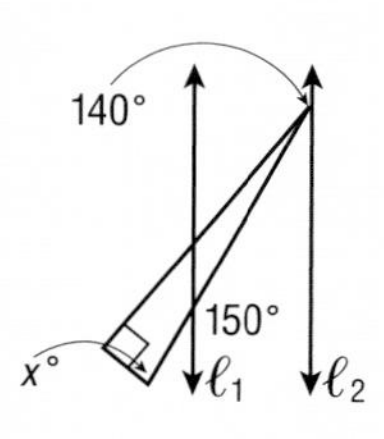

Ⓐ 30

Ⓑ 40

Ⓒ 70

Ⓓ 80

Test-Taking Tip

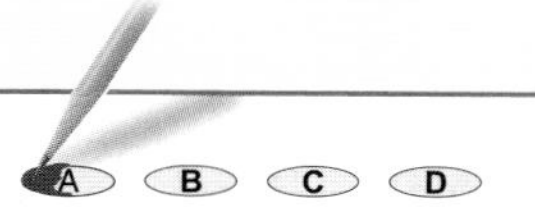

Question 5

Some questions ask you to find the value of an expression. It is often not necessary to find the value of each variable in the expression. For example, to answer Question 5, it is not necessary to find the values of x and y. Isolate the expression $x - y$ on one side of the equation.

Preparing for Standardized Tests
For test-taking strategies and more practice, see pages 877–892.

Part 2 Short Response/Grid In

Record your answers on the answer sheet provided by your teacher or on a sheet of paper.

11. $\begin{array}{r} AA \\ +\ BB \\ \hline CC \end{array}$

If A, B, and C are each digits and $A = 3B$, then what is one possible value of C?

12. In the figure, each arc is a semicircle. If B is the midpoint of $\overline{AD}$ and C is the midpoint of $\overline{BD}$, what is the ratio of the area of the semicircle $\widehat{CD}$ to the area of the semicircle $\widehat{AD}$?

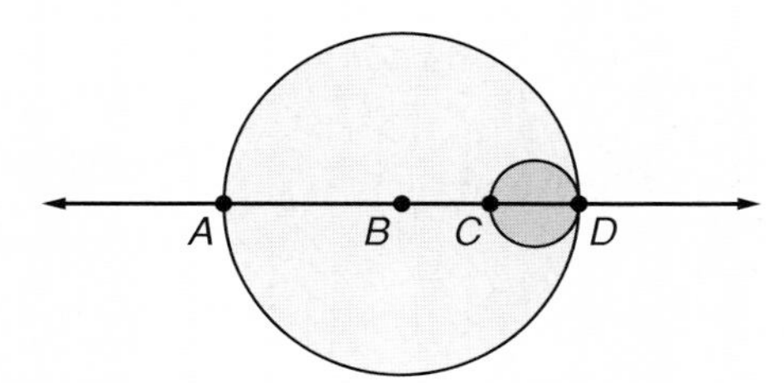

13. Two people are 17.5 miles apart. They begin to walk toward each other along a straight line at the same time. One walks at the rate of 4 miles per hour, and the other walks at the rate of 3 miles per hour. In how many hours will they meet?

14. If $\frac{x + y}{x} = \frac{5}{4}$, then $\frac{y}{x} =$

15. A car's gasoline tank is $\frac{1}{2}$ full. After adding 7 gallons of gas, the gauge shows that the tank is $\frac{3}{4}$ full. How many gallons does the tank hold?

16. If $a = 15 - b$, what is the value of $3a + 3b$?

17. If $x^9 = \frac{45}{y}$ and $x^7 = \frac{1}{5y}$, and $x > 0$, what is the value of x?

18. The rates at a parking garage are shown. How much will it cost to park in the garage for $4\frac{1}{2}$ hours?

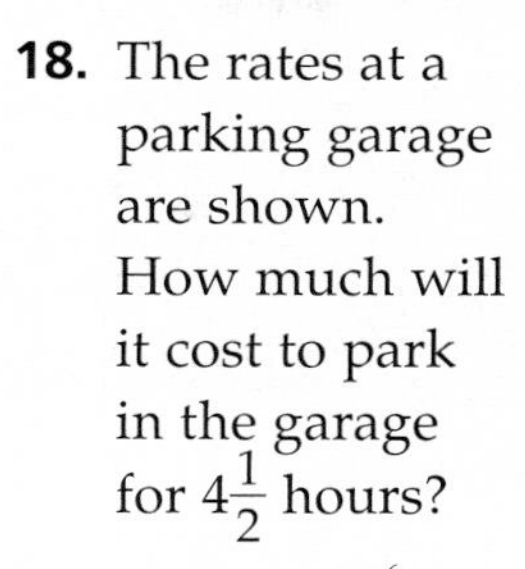

19. Evaluate. $\sum_{n=1}^{6}(1 - 4n)$.

20. Does an infinite geometric series with a first term of 36 and a common ratio of $\frac{1}{4}$ have a sum? If the series has a sum, find the sum.

21. What is the last term in the expansion of $(x + 2y)^6$?

Part 3 Extended Response

Record your answers on a sheet of paper. Show your work.

For Exercises 22–25, use the following information.

The amount of a certain medication remaining in the body is reduced by 20% each hour after the medication is given. The prescription that Dr. Barr gave Kim called for a 100-milligram dose of medication.

22. Is the amount of medication in Kim's body each hour after she takes it modeled by an arithmetic sequence or a geometric sequence? Explain.

23. Write an equation to find the amount of medication A in Kim's body each hour h after she takes it.

24. How many milligrams of medication remain in Kim's body 4 hours after she takes a 100-milligram dose? If necessary, round to the nearest tenth of a milligram.

25. If Dr. Barr wants Kim to have at least 30 milligrams of medication in her body at all times, after how many hours should Kim take a second dose of medication? Explain your method of solution.

Chapter

12 Probability and Statistics

What You'll Learn

- **Lessons 12-1 and 12-2** Solve problems involving independent events, dependent events, permutations, and combinations.
- **Lessons 12-3, 12-4, 12-5, and 12-8** Find probability and odds.
- **Lesson 12-6** Find statistical measures.
- **Lesson 12-7** Use the normal distribution.
- **Lesson 12-9** Determine whether a sample is unbiased.

Key Vocabulary

- **permutation** (p. 638)
- **combination** (p. 640)
- **probability** (p. 644)
- **measures of central tendency** (p. 664)
- **measures of variation** (p. 665)

Why It's Important

Being able to analyze data is an important skill for every citizen. Business decision-makers rely on statistical measures to ensure quality products, medical researchers test and design new treatments by performing experiments with sample populations, and sports coaches use probabilities to design a winning team.

Each day during a presidential election campaign, journalists report the results of public opinion polls. Pollsters must make sure that the sample they choose accurately represents all of the voters. *You will investigate how opinion polls are used in political campaigns in Lesson 12-9.*

Getting Started

Prerequisite Skills To be successful in this chapter, you'll need to master these skills and be able to apply them in problem-solving situations. Review these skills before beginning Chapter 12.

For Lesson 12-3 — Find Simple Probability

Find each probability if a die is rolled once.

1. $P(2)$ **2.** $P(5)$ **3.** $P(\text{even number})$

4. $P(\text{odd number})$ **5.** $P(\text{numbers less than 5})$ **6.** $P(\text{numbers greater than 1})$

For Lesson 12-6 — Box-and-Whisker Plots

Make a box-and-whisker plot for each set of data. *(For review, see pages 826 and 827.)*

7. {24, 32, 38, 38, 26, 33, 37, 39, 23, 31, 40, 21}

8. {25, 46, 31, 53, 39, 59, 48, 43, 68, 64, 29}

9. {51, 69, 46, 27, 60, 53, 55, 39, 81, 54, 46, 23}

10. {13.6, 15.1, 14.9, 15.7, 16.0, 14.1, 16.3, 14.3, 13.8}

For Lesson 12-6 — Evaluate Expressions

Evaluate $\sqrt{\dfrac{(a-b)^2 + (c-b)^2}{d}}$ for each set of values. *(For review, see Lesson 5-6.)*

11. $a = 4, b = 7, c = 1, d = 5$ **12.** $a = 2, b = 6, c = 9, d = 5$

13. $a = 5, b = 1, c = 7, d = 4$ **14.** $a = 3, b = 4, c = 11, d = 10$

For Lesson 12-8 — Expand Binomials

Expand each binomial. *(For review, see Lesson 5-2.)*

15. $(a + b)^3$ **16.** $(c + d)^4$ **17.** $(m - n)^5$ **18.** $(x + y)^6$

Probability and Statistics Make this Foldable to help you organize your notes. Begin with one sheet of 11" by 17" paper.

Step 1 Fold

Fold 2" tabs on each of the short sides.

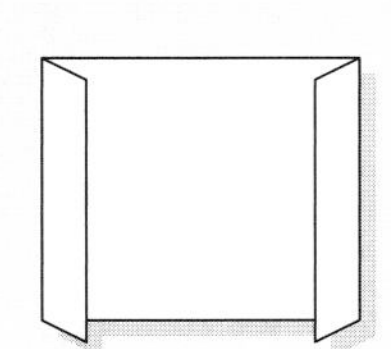

Step 2 Fold and Cut

Then fold in half in both directions. Open and cut as shown.

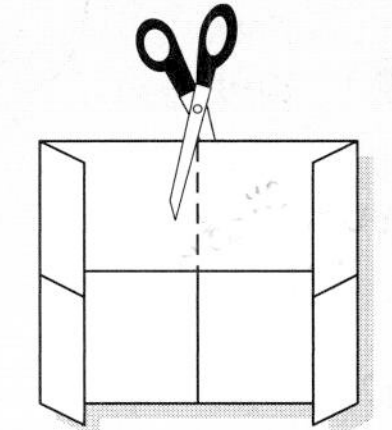

Step 3 Staple and Label

Refold along the width. Staple each pocket. Label pockets as *The Counting Principle*, *Permutations and Combinations*, *Probability*, and *Statistics*.

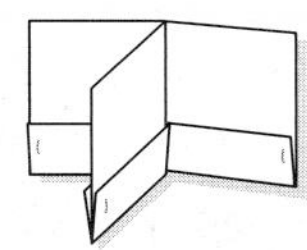

Reading and Writing As you read and study the chapter, you can write notes and examples on index cards and store the cards in the Foldable pockets.

12-1 The Counting Principle

California Standards Standard 18.0 **Students use fundamental counting principles** to compute combinations and permutations. (Key)

What You'll Learn

- Solve problems involving independent events.
- Solve problems involving dependent events.

Vocabulary

- outcomes
- sample space
- event
- independent events
- Fundamental Counting Principle
- dependent events

How can you count the maximum number of license plates a state can issue?

Most states have letters and digits on their license plates. The number of possible plates is too great to count by listing all of the possibilities. It is much more efficient to count the number of possibilities by using the Fundamental Counting Principle.

INDEPENDENT EVENTS An **outcome** is the result of a single trial. For example, the trial of flipping a coin once has two outcomes: head or tail. The set of all possible outcomes is called the **sample space**. An **event** consists of one or more outcomes of a trial. The choices of letters and digits to be put on a license plate are called **independent events** because each letter or digit chosen does *not* affect the choices for the others.

For situations in which the number of choices leads to a small number of total possibilities, you can use a tree diagram or a table to count them.

Example 1 Independent Events

FOOD A sandwich cart offers customers a choice of hamburger, chicken, or fish on either a plain or a sesame seed bun. How many different combinations of meat and a bun are possible?

First, note that the choice of the type of meat does not affect the choice of the type of bun, so these events are independent.

Method 1 Tree Diagram

Let H represent hamburger, C, chicken, F, fish, P, plain, and S, sesame seed. Make a tree diagram in which the first row shows the choice of meat and the second row shows the choice of bun.

Meat	H		C		F	
Bun	P	S	P	S	P	S
Possible Combinations	HP	HS	CP	CS	FP	FS

There are six possible outcomes.

Method 2 Make a Table

Make a table in which each row represents a type of meat and each column represents a type of bun.

This method also shows that there are six outcomes.

		Bun	
		Plain	**Sesame**
Meat	**Hamburger**	HP	HS
	Chicken	CP	CS
	Fish	FP	FS

Notice that in Example 1, there are 3 ways to choose the type of meat, 2 ways to choose the type of bun, and $3 \cdot 2$ or 6 total ways to choose a combination of the two. This illustrates the **Fundamental Counting Principle**.

Key Concept — *Fundamental Counting Principle*

- **Words** If event M can occur in m ways and event N can occur in n ways, then event M followed by event N can occur in $m \cdot n$ ways.
- **Example** If event M can occur in 2 ways and event N can occur in 3 ways, then M followed by N can occur in $2 \cdot 3$ or 6 ways.

This rule can be extended to any number of events.

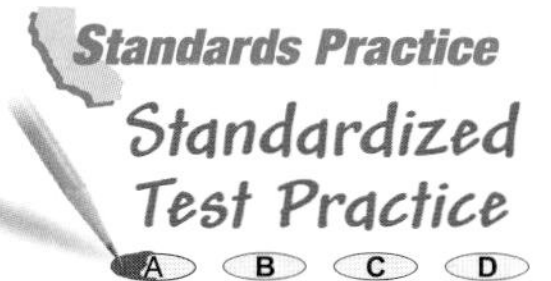

Example 2 *Fundamental Counting Principle*

Multiple-Choice Test Item

Kim won a contest on a radio station. The prize was a restaurant gift certificate and tickets to a sporting event. She can select one of three different restaurants and tickets to a football, baseball, basketball, or hockey game. How many different ways can she select a restaurant followed by a sporting event?

(A) 7 (B) 12 (C) 15 (D) 16

Read the Test Item

Her choice of a restaurant does not affect her choice of a sporting event, so these events are independent.

Solve the Test Item

There are 3 ways she can choose a restaurant and there are 4 ways she can choose the sporting event. By the Fundamental Counting Principle, there are $3 \cdot 4$ or 12 total ways she can choose her two prizes. The answer is B.

Test-Taking Tip

Remember that you can check your answer by making a tree diagram or a table showing the outcomes.

The Fundamental Counting Principle can be used to count the number of outcomes possible for any number of successive events.

Example 3 *More than Two Independent Events*

COMMUNICATION **Many answering machines allow owners to call home and get their messages by entering a 3-digit code. How many codes are possible?**

The choice of any digit does not affect the other two digits, so the choices of the digits are independent events.

There are 10 possible first digits in the code, 10 possible second digits, and 10 possible third digits. So, there are $10 \cdot 10 \cdot 10$ or 1000 possible different code numbers.

DEPENDENT EVENTS Some situations involve dependent events. With **dependent events**, the outcome of one event *does* affect the outcome of another event. The Fundamental Counting Principle applies to dependent events as well as independent events.

Example 4 Dependent Events

SCHOOL Charlita wants to take 6 different classes next year. Assuming that each class is offered each period, how many different schedules could she have?

When Charlita schedules a given class for a given period, she cannot schedule that class for any other period. Therefore, the choices of which class to schedule each period are dependent events.

There are 6 classes Charlita can take during first period. That leaves 5 classes she can take second period. After she chooses which classes to take the first two periods, there are 4 remaining choices for third period, and so on.

Period	1st	2nd	3rd	4th	5th	6th
Number of Choices	6	5	4	3	2	1

There are $6 \cdot 5 \cdot 4 \cdot 3 \cdot 2 \cdot 1$ or 720 schedules that Charlita could have.
Note that $6 \cdot 5 \cdot 4 \cdot 3 \cdot 2 \cdot 1 = 6!$.

Study Tip

Look Back
To review **factorials**, see Lesson 11-7.

Concept Summary — *Independent and Dependent Events*

- **Words** If the outcome of an event does *not* affect the outcome of another event, the two events are *independent*.
- **Example** Tossing a coin and rolling a die are independent events.

- **Words** If the outcome of an event *does* affect the outcome of another event, the two events are *dependent*.
- **Example** Taking a piece of candy from a jar and then taking a second piece without replacing the first are dependent events because taking the first piece affects what is available to be taken next.

Check for Understanding

Concept Check

1. **List** the possible outcomes when a coin is tossed three times. Use H for heads and T for tails.
2. **OPEN ENDED** Describe a situation in which you can use the Fundamental Counting Principle to show that there are 18 total possibilities.
3. **Explain** how choosing to buy a car or a pickup truck and then selecting the color of the vehicle could be dependent events.

Guided Practice

State whether the events are *independent* or *dependent*.

4. choosing the color and size of a pair of shoes
5. choosing the winner and runner-up at a dog show

Solve each problem.

6. An ice cream shop offers a choice of two types of cones and 15 flavors of ice cream. How many different 1-scoop ice cream cones can a customer order?
7. Lance's math quiz has eight true-false questions. How many different choices for giving answers to the eight questions are possible?
8. For a college application, Macawi must select one of five topics on which to write a short essay. She must also select a different topic from the list for a longer essay. How many ways can she choose the topics for the two essays?

9. A bookshelf holds 4 different biographies and 5 different mystery novels. How many ways can one book of each type be selected?

(A) 1 (B) 9 (C) 10 (D) 20

Practice and Apply

Homework Help

For Exercises	See Examples
10–23 25–27	1–4

Extra Practice
See page 854.

State whether the events are *independent* or *dependent*.

10. choosing a president, vice president, secretary, and treasurer for Student Council, assuming that a person can hold only one office

11. selecting a fiction book and a nonfiction book at the library

12. Each of six people guess the total number of points scored in a basketball game. Each person writes down his or her guess without telling what it is.

13. The letters A through Z are written on pieces of paper and placed in a jar. Four of them are selected one after the other without replacing any of them.

Solve each problem.

14. Tim wants to buy one of three different albums he sees in a music store. Each is available on tape and on CD. From how many combinations of album and format does he have to choose?

15. A video store has 8 new releases this week. Each is available on videotape and on DVD. How many ways can a customer choose a new release and a format to rent?

16. Carlos has homework to do in math, chemistry, and English. How many ways can he choose the order in which to do his homework?

17. The menu for a banquet has a choice of 2 types of salad, 5 main courses, and 3 desserts. How many ways can a salad, main course, and dessert be selected to form a meal?

18. A golf club manufacturer makes drivers with 4 different shaft lengths, 3 different lofts, 2 different grips, and 2 different club head materials. How many different combinations are possible?

19. Each question on a five-question multiple-choice quiz has answer choices labeled A, B, C, and D. How many different ways can a student answer the five questions?

20. How many ways can six different books be arranged on a shelf if one of the books is a dictionary and it must be on an end?

21. In how many orders can eight actors be listed in the opening credits of a movie if the leading actor must be listed first or last?

You can use the Fundamental Counting Principle to list possible outcomes in games. Visit www.algebra2.com/webquest to continue work on your WebQuest project.

22. PASSWORDS Abby is registering at a Web site. She must select a password containing 6 numerals to be able to use the site. How many passwords are allowed if no digit may be used more than once?

23. ENTERTAINMENT Solve the problem in the comic strip below. Assume that the books are all different.

Peanuts®

24. CRITICAL THINKING The members of the Math Club need to elect a president and a vice-president. They determine that there are a total of 272 ways that they can fill the positions with two different members. How many people are in the Math Club?

25. **HOME SECURITY** How many different 5-digit codes are possible using the keypad shown at the right if the first digit cannot be 0 and no digit may be used more than once?

More About. . .

Area Codes

Before 1995, area codes had the following format.

(XYZ)

X = 2, 3, …, or 9

Y = 0 or 1

Z = 0, 1, 2, …, or 9

Source: www.nanpa.com

AREA CODES **For Exercises 26 and 27, refer to the information about telephone area codes at the left.**

26. How many area codes were possible before 1995?

27. In 1995, the restriction on the middle digit was removed, allowing any digit in that position. How many total codes were possible after this change was made?

28. **RESEARCH** Use the Internet or other resource to find the configuration of letters and numbers on license plates in your state. Then find the number of possible plates.

29. **WRITING IN MATH** Answer the question that was posed at the beginning of the lesson.

How can you count the maximum number of license plates a state can issue?

Include the following in your answer:

- an explanation of how to use the Fundamental Counting Principle to find the number of different license plates in a state such as Florida, which has 3 letters followed by 3 numbers, and
- a way that a state can increase the number of possible plates without increasing the length of the plate number.

Standards Practice

Standardized Test Practice

30. How many numbers between 100 and 999, inclusive, have 7 in the tens place?

(A) 90 (B) 100 (C) 110 (D) 120

31. A coin is tossed four times. How many possible sequences of heads or tails are possible?

(A) 4 (B) 8 (C) 16 (D) 32

Extending the Lesson

For Exercises 32 and 33, use the following information.
A **finite graph** is a collection of points, called **vertices**, and segments, called **edges**, connecting the vertices. For example, the graph shown at the right has 4 vertices and 2 edges.

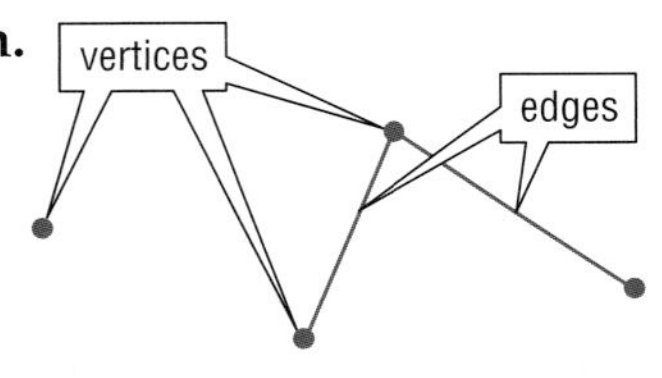

32. Suppose a graph has 10 vertices and each pair of vertices is connected by exactly one edge. Find the number of edges in the graph. (*Hint:* If you use the Fundamental Counting Principle, be sure to count each edge only once.)

33. **TRANSPORTATION** The table shows the distances in miles of the roads between some towns. Draw a graph in which the vertices represent the towns and the edges are labeled with the lengths of the roads. Use your graph to find the length of the shortest route from Greenville to Red Rock.

Route	Miles
Greenville to Roseburg	14
Greenville to Bluemont	12
Greenville to Whiteston	9
Roseburg to Bluemont	8
Bluemont to Whiteston	5
Roseburg to Red Rock	7
Bluemont to Red Rock	9
Whiteston to Red Rock	11

Maintain Your Skills

Mixed Review

34. Prove that $4 + 7 + 10 \cdots + (3n + 1) = \frac{n(3n+5)}{2}$ for all positive integers n. *(Lesson 11-8)*

Find the indicated term of each expansion. *(Lesson 11-7)*

35. third term of $(x + y)^8$

36. fifth term of $(2a - b)^7$

Evaluate each expression. *(Lesson 10-2)*

37. $\log_2 128$

38. $\log_3 243$

39. $\log_9 3$

Simplify each expression. *(Lesson 9-1)*

40. $-\frac{x^2 - y^2}{x + y} \cdot \frac{1}{x - y}$

41. $\dfrac{\frac{x^2}{x^2 - 25y^2}}{\frac{x}{5y - x}}$

42. **CARTOGRAPHY** Edison is located at (9, 3) in the coordinate system on a road map. Kettering is located at (12, 5) on the same map. Each side of a square on the map represents 10 miles. To the nearest mile, what is the distance between Edison and Kettering? *(Lesson 8-1)*

Solve each equation. *(Lesson 7-3)*

43. $x^4 - 5x^2 + 4 = 0$

44. $y^4 + 4y^3 + 4y^2 = 0$

Write an equation of the form $y = a(x - h)^2 + k$ for the parabola with the given vertex that passes through the given point. *(Lesson 6-6)*

45. vertex (3, 2)
point (5, 6)

46. vertex (−1, 4)
point (−2, 2)

47. vertex (0, 8)
point (4, 0)

Solve each equation. *(Lesson 5-8)*

48. $\sqrt{2x + 1} = 3$

49. $3 + \sqrt{x + 1} = 5$

50. $\sqrt{x} + \sqrt{x + 5} = 5$

Find the inverse of each matrix, if it exists. *(Lesson 4-7)*

51. $\begin{bmatrix} 3 & 1 \\ -4 & 1 \end{bmatrix}$

52. $\begin{bmatrix} 4 & -5 \\ 2 & -1 \end{bmatrix}$

53. $\begin{bmatrix} -3 & 2 \\ -6 & 4 \end{bmatrix}$

Write an equation in slope-intercept form for each graph. *(Lesson 2-4)*

54.

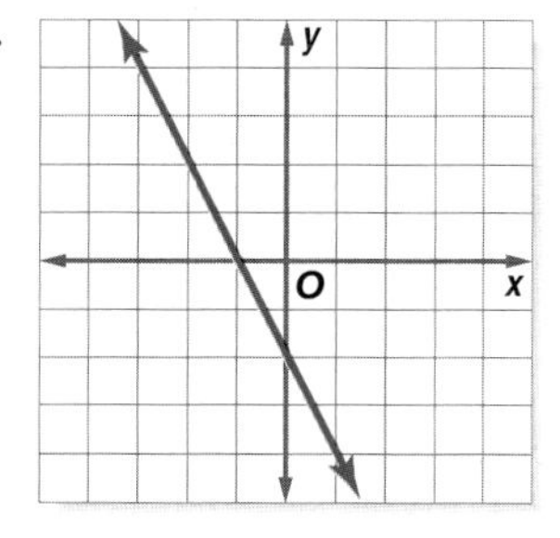

55.

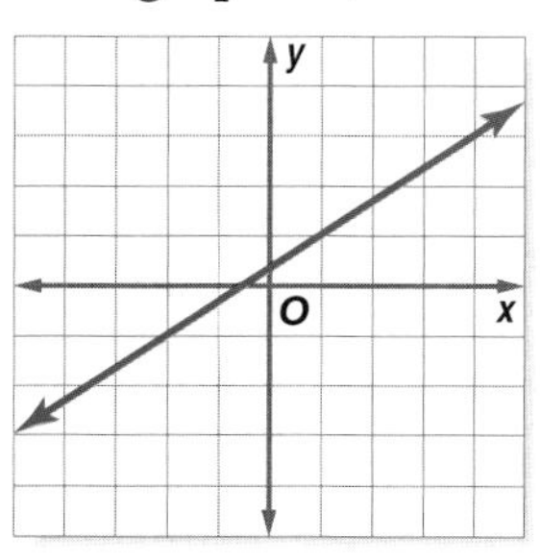

Getting Ready for the Next Lesson

PREREQUISITE SKILL **Evaluate each expression.**
*(To review **factorials**, see Lesson 11-7.)*

56. $\frac{5!}{2!}$

57. $\frac{6!}{4!}$

58. $\frac{7!}{3!}$

59. $\frac{6!}{1!}$

60. $\frac{4!}{2!2!}$

61. $\frac{6!}{2!4!}$

62. $\frac{8!}{3!5!}$

63. $\frac{5!}{5!0!}$

12-2 Permutations and Combinations

California Standards Standard 18.0 Students use fundamental counting principles to compute combinations and permutations. (Key)

What You'll Learn

- Solve problems involving linear permutations.
- Solve problems involving combinations.

How do permutations and combinations apply to softball?

When the manager of a softball team fills out her team's lineup card before the game, the order in which she fills in the names is important because it determines the order in which the players will bat.

Suppose she has 7 possible players in mind for the top 4 spots in the lineup. You know from the Fundamental Counting Principle that there are $7 \cdot 6 \cdot 5 \cdot 4$ or 840 ways that she could assign players to the top 4 spots.

Vocabulary

- permutation
- linear permutation
- combination

PERMUTATIONS When a group of objects or people are arranged in a certain order, the arrangement is called a **permutation**. In a permutation, the *order* of the objects is very important. The arrangement of objects or people in a line is called a **linear permutation**.

Notice that $7 \cdot 6 \cdot 5 \cdot 4$ is the product of the first 4 factors of 7!. You can rewrite this product in terms of 7!.

$$7 \cdot 6 \cdot 5 \cdot 4 = 7 \cdot 6 \cdot 5 \cdot 4 \cdot \frac{3 \cdot 2 \cdot 1}{3 \cdot 2 \cdot 1}$$ Multiply by $\frac{3 \cdot 2 \cdot 1}{3 \cdot 2 \cdot 1}$ or 1.

$$= \frac{7 \cdot 6 \cdot 5 \cdot 4 \cdot 3 \cdot 2 \cdot 1}{3 \cdot 2 \cdot 1} \text{ or } \frac{7!}{3!}$$ $7! = 7 \cdot 6 \cdot 5 \cdot 4 \cdot 3 \cdot 2 \cdot 1$ and $3! = 3 \cdot 2 \cdot 1$

Notice that 3! is the same as $(7 - 4)!$.

The number of ways to arrange 7 people or objects taken 4 at a time is written $P(7, 4)$. The expression for the softball lineup above is a case of the following formula.

Study Tip

Reading Math
The expression $P(n, r)$ is read *the number of permutations of n objects taken r at a time*. It is sometimes written as $_nP_r$.

Key Concept — Permutations

The number of permutations of n distinct objects taken r at a time is given by

$$P(n, r) = \frac{n!}{(n - r)!}.$$

Example 1 Permutation

FIGURE SKATING There are 10 finalists in a figure skating competition. How many ways can gold, silver, and bronze medals be awarded?

Since each winner will receive a different medal, order is important. You must find the number of permutations of 10 things taken 3 at a time.

$$P(n, r) = \frac{n!}{(n-r)!}$$ Permutation formula

$$P(10, 3) = \frac{10!}{(10-3!)}$$ $n = 10, r = 3$

$$= \frac{10!}{7!}$$ Simplify.

$$= \frac{10 \cdot 9 \cdot 8 \cdot \overset{1}{\cancel{7}} \cdot \overset{1}{\cancel{6}} \cdot \overset{1}{\cancel{5}} \cdot \overset{1}{\cancel{4}} \cdot \overset{1}{\cancel{3}} \cdot \overset{1}{\cancel{2}} \cdot \overset{1}{\cancel{1}}}{\underset{1}{\cancel{7}} \cdot \underset{1}{\cancel{6}} \cdot \underset{1}{\cancel{5}} \cdot \underset{1}{\cancel{4}} \cdot \underset{1}{\cancel{3}} \cdot \underset{1}{\cancel{2}} \cdot \underset{1}{\cancel{1}}} \text{ or } 720$$ Divide by common factors.

The gold, silver, and bronze medals can be awarded in 720 ways.

Notice that in Example 1, all of the factors of $(n - r)!$ are also factors of $n!$. Instead of writing all of the factors, you can also evaluate the expression in the following way.

$$\frac{10!}{(10-3)!} = \frac{10!}{7!}$$ Simplify.

$$= \frac{10 \cdot 9 \cdot 8 \cdot 7!}{7!}$$ $\frac{7!}{7!} = 1$

$$= 10 \cdot 9 \cdot 8 \text{ or } 720$$ Multiply.

Suppose you want to rearrange the letters of the word *geometry* to see if you can make a different word. If the two *e*'s were not identical, the eight letters in the word could be arranged in $P(8, 8)$ or $8!$ ways. To account for the identical *e*'s, divide $P(8, 8)$ or 40,320 by the number of arrangements of *e*. The two *e*'s can be arranged in $P(2, 2)$ or $2!$ ways.

$$\frac{P(8, 8)}{P(2, 2)} = \frac{8!}{2!}$$ Divide.

$$= \frac{8 \cdot 7 \cdot 6 \cdot 5 \cdot 4 \cdot 3 \cdot 2!}{2!} \text{ or } 20{,}160$$ Simplify.

Thus, there are 20,160 ways to arrange the letters in *geometry*.

When some letters or objects are alike, use the rule below to find the number of permutations.

Key Concept — Permutations with Repetitions

The number of permutations of n objects of which p are alike and q are alike is $\frac{n!}{p!q!}$.

This rule can be extended to any number of objects that are repeated.

Example 2 Permutation with Repetition

How many different ways can the letters of the word *MISSISSIPPI* be arranged?

The second, fifth, eighth, and eleventh letters are each I.

The third, fourth, sixth, and seventh letters are each S.

The ninth and tenth letters are each P.

You need to find the number of permutations of 11 letters of which 4 of one letter, 4 of another letter, and 2 of another letter are the same.

$$\frac{11!}{4!4!2!} = \frac{11 \cdot 10 \cdot 9 \cdot 8 \cdot 7 \cdot 6 \cdot 5 \cdot 4!}{4!4!2!} \text{ or } 34{,}650$$

There are 34,650 ways to arrange the letters.

www.algebra2.com/extra_examples/ca

COMBINATIONS

An arrangement or selection of objects in which order is *not* important is called a **combination**. The number of combinations of n objects taken r at a time is written $C(n, r)$. *It is sometimes written* $_nC_r$.

You know that there are $P(n, r)$ ways to select r objects from a group of n if the order is important. There are $r!$ ways to order the r objects that are selected, so there are $r!$ permutations that are all the same combination. Therefore,

$$C(n, r) = \frac{P(n, r)}{r!} \text{ or } \frac{n!}{(n - r)!r!}.$$

Study Tip

Permutations and Combinations

- If order in an arrangement *is* important, the arrangement is a *permutation*.
- If order is *not* important, the arrangement is a *combination*.

Key Concept — Combinations

The number of combinations of n distinct objects taken r at a time is given by

$$C(n, r) = \frac{n!}{(n - r)!r!}.$$

Example 3 Combination

A group of seven students working on a project needs to choose two from their group to present the group's report to the class. How many ways can they choose the two students?

Since the order they choose the students is not important, you must find the number of combinations of 7 students taken 2 at a time.

$C(n, r) = \frac{n!}{(n - r)!r!}$ Combination formula

$C(7, 2) = \frac{7!}{(7 - 2)!2!}$ $n = 7$ and $r = 2$

$= \frac{7!}{5!2!}$ or 21 Simplify.

There are 21 possible ways to choose the two students.

In more complicated situations, you may need to multiply combinations and/or permutations.

Study Tip

Deck of Cards

In this text, a *standard deck of cards* always means a deck of 52 playing cards. There are 4 suits—clubs (black), diamonds (red), hearts (red), and spades (black)—with 13 cards in each suit.

Example 4 Multiple Events

Five cards are drawn from a standard deck of cards. How many hands consist of three clubs and two diamonds?

By the Fundamental Counting Principle, you can multiply the number of ways to select three clubs and the number of ways to select two diamonds.

Only the cards in the hand matter, not the order in which they were drawn, so use combinations.

$C(13, 3)$ Three of 13 clubs are to be drawn.

$C(13, 2)$ Two of 13 diamonds are to be drawn.

$C(13, 3) \cdot C(13, 2) = \frac{13!}{(13 - 3)!3!} \cdot \frac{13!}{(13 - 2)!2!}$ Combination formula

$= \frac{13!}{10!3!} \cdot \frac{13!}{11!2!}$ Subtract.

$= 286 \cdot 78$ or 22,308 Simplify.

There are 22,308 hands consisting of 3 clubs and 2 diamonds.

Check for Understanding

Concept Check

1. **OPEN ENDED** Describe a situation in which the number of outcomes is given by $P(6, 3)$.
2. **Show** that $C(n, n - r) = C(n, r)$.
3. **Determine** whether the statement $C(n, r) = P(n, r)$ is *sometimes*, *always*, or *never* true. Explain your reasoning.

Guided Practice

Evaluate each expression.

4. $P(5, 3)$
5. $P(6, 3)$
6. $C(4, 2)$
7. $C(6, 1)$

Determine whether each situation involves a *permutation* or a *combination*. Then find the number of possibilities.

8. choosing 2 different pizza toppings from a list of 6
9. seven shoppers in line at a checkout counter
10. an arrangement of the letters in the word *intercept*

Application

11. **SCHOOL** The principal at Cobb County High School wants to start a mentoring group. He needs to narrow his choice of students to be mentored to six from a group of nine. How many ways can a group of six be selected?

Practice and Apply

Homework Help

For Exercises	See Examples
12–15	1
16–19	3
20, 21, 32–35	4
22–31	1–3

Extra Practice
See page 854.

Evaluate each expression.

12. $P(8, 2)$
13. $P(9, 1)$
14. $P(7, 5)$
15. $P(12, 6)$
16. $C(5, 2)$
17. $C(8, 4)$
18. $C(12, 7)$
19. $C(10, 4)$
20. $C(12, 4) \cdot C(8, 3)$
21. $C(9, 3) \cdot C(6, 2)$

Determine whether each situation involves a *permutation* or a *combination*. Then find the number of possibilities.

22. the winner and first, second, and third runners-up in a contest with 10 finalists
23. selecting two of eight employees to attend a business seminar
24. an arrangement of the letters in the word *algebra*
25. placing an algebra book, a geometry book, a chemistry book, an English book, and a health book on a shelf
26. selecting nine books to check out of the library from a reading list of twelve
27. an arrangement of the letters in the word *parallel*
28. choosing two CDs to buy from ten that are on sale
29. selecting three of fifteen flavors of ice cream at the grocery store

30. **MOVIES** The manager of a four-screen movie theater is deciding which of 12 available movies to show. The screens are in rooms with different seating capacities. How many ways can he show four different movies on the screens?

31. **LANGUAGES** How many different arrangements of the letters of the Hawaiian word *aloha* are possible?

32. **GOVERNMENT** How many ways can five members of the 100-member United States Senate be chosen to be put on a committee?

33. How many ways can a hand of five cards consisting of four cards from one suit and one card from another suit be drawn from a standard deck of cards?

34. How many ways can a hand of five cards consisting of three cards from one suit and two cards from another suit be drawn from a standard deck of cards?

35. **LOTTERIES** In a multi-state lottery, the player must guess which five of forty nine white balls numbered from 1 to 49 will be drawn. The order in which the balls are drawn does not matter. The player must also guess which one of forty-two red balls numbered from 1 to 42 will be drawn. How many ways can the player fill out a lottery ticket?

More About. . .

Card Games

Hanafuda cards are often called "flower cards" because each suit is depicted by a different flower. Each flower is representative of the calendar month in which the flower blooms.

Source: www.gamesdomain.com

36. **CARD GAMES** *Hanafuda* is a Japanese game that uses a deck of cards made up of 12 suits, with each suit having four cards. How many 7-card hands can be formed so that 3 are from one suit and 4 are from another?

37. **CRITICAL THINKING** Show that $C(n-1, r) + C(n-1, r-1) = C(n, r)$.

38. **WRITING IN MATH** Answer the question that was posed at the beginning of the lesson.

How do permutations and combinations apply to softball?

Include the following in your answer:

- an explanation of how to find the number of 9-person lineups that are possible, and
- an explanation of how many ways there are to choose 9 players if 16 players show up for a game.

39. How many ways can eight runners in an Olympic race finish in first, second, and third places?

(A) 8 (B) 24 (C) 56 (D) 336

40. How many diagonals can be drawn in the pentagon?

(A) 5 (B) 10

(C) 15 (D) 20

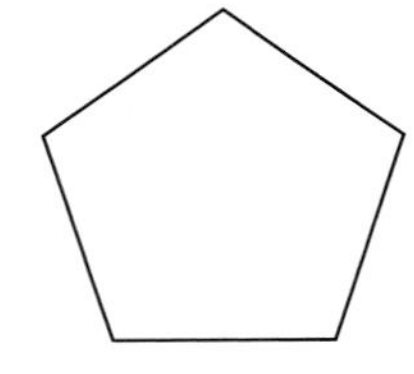

Extending the Lesson

When n distinct objects are arranged in a circle, there are n ways that the arrangement can be rotated to obtain an arrangement that is really the same as the original. For example, the two arrangements of three objects shown below are the same. Therefore, the number of **circular permutations** of n distinct objects is $\frac{n!}{n}$ or $(n-1)!$ *Note that the keys are not turned over.*

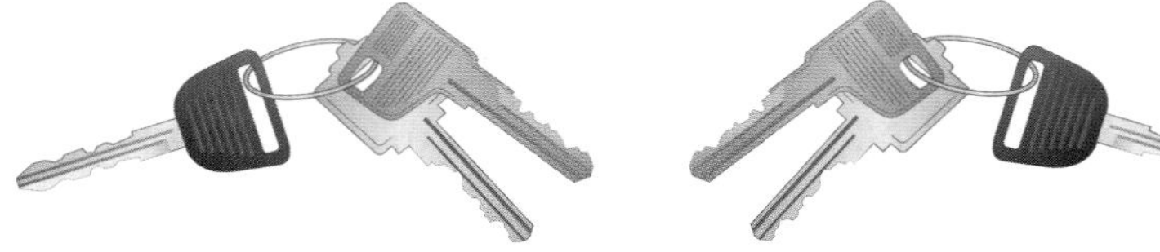

Find the number of possibilities for each situation.

41. a basketball huddle of 5 players

42. four different dishes on a revolving tray in the middle of a table at a Chinese restaurant

43. six quarters with designs from six different states arranged in a circle on top of your desk

Maintain Your Skills

Mixed Review

44. Darius can do his homework in pencil or pen, using lined or unlined paper, and on one or both sides of each page. How many ways can he prepare his homework? *(Lesson 12-1)*

45. A customer in an ice cream shop can order a sundae with a choice of 10 flavors of ice cream, a choice of 4 flavors of sauce, and with or without a cherry on top. How many different sundaes are possible? *(Lesson 12-1)*

Find a counterexample to each statement. *(Lesson 11-8)*

46. $1 + 2 + 3 + \ldots + n = 2n - 1$

47. $5^n + 1$ is divisible by 6.

Solve each equation or inequality. *(Lesson 10-5)*

48. $3e^x + 1 = 2$

49. $e^{2x} > 5$

50. $\ln (x - 1) = 3$

51. CONSTRUCTION A painter works on a job for 10 days and is then joined by an associate. Together they finish the job in 6 more days. The associate could have done the job in 30 days. How long would it have taken the painter to do the job alone? *(Lesson 9-6)*

Write an equation for each ellipse. *(Lesson 8-4)*

52.

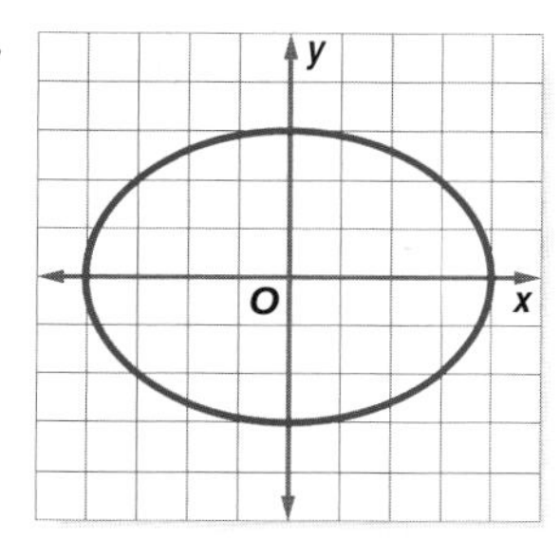

53.

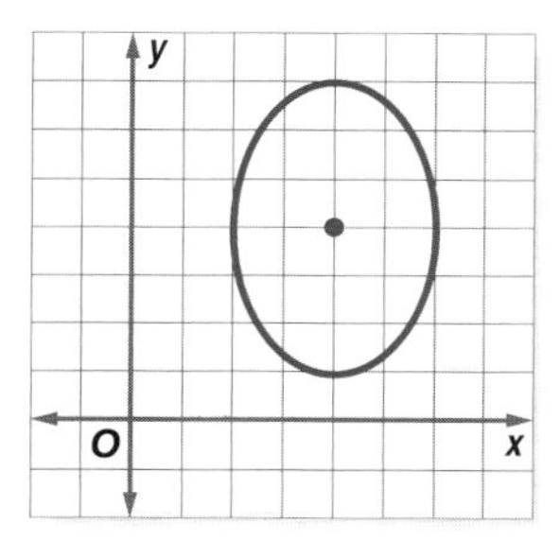

Find $p(-1)$ and $p(5)$ for each function. *(Lesson 7-1)*

54. $p(x) = \frac{1}{2}x^2 + 3x - 1$

55. $p(x) = x^4 - 4x^3 + 2x - 7$

Solve each equation by factoring. *(Lesson 6-3)*

56. $x^2 - 16 = 0$

57. $x^2 - 3x - 10 = 0$

58. $3x^2 + 8x - 3 = 0$

Simplify. *(Lesson 5-6)*

59. $\sqrt{128}$

60. $\sqrt{3x^6y^4}$

61. $\sqrt{20} + 2\sqrt{45} - \sqrt{80}$

Solve each system of equations by using inverse matrices. *(Lesson 4-8)*

62. $x + 2y = 5$
$3x - 3y = -12$

63. $5a + 2b = 4$
$-3a + b = 2$

Find the slope of the line that passes through each pair of points. *(Lesson 2-3)*

64. $(2, 1), (5, -3)$

65. $(0, 4), (7, -2)$

66. $(5, 3), (2, 3)$

Solve each equation. Check your solutions. *(Lesson 1-4)*

67. $|x - 4| = 11$

68. $|2x + 2| = -3$

Getting Ready for the Next Lesson

PREREQUISITE SKILL Evaluate the expression $\frac{x}{x + y}$ for the given values of x and y. *(To review **evaluating expressions**, see Lesson 1-1.)*

69. $x = 3, y = 2$

70. $x = 4, y = 4$

71. $x = 2, y = 8$

72. $x = 5, y = 10$

12-3 Probability

California Standards Standard 19.0 Students use combinations and permutations to compute probabilities. (Key)

What You'll Learn

- Find the probability and odds of events.
- Create and use graphs of probability distributions.

What do probability and odds tell you about life's risks?

The risk of getting struck by lightning in any given year is 1 in 750,000. The chances of surviving a lightning strike are 3 in 4. These risks and chances are a way of describing the probability of an event. The **probability** of an event is a ratio that measures the chances of the event occurring.

Vocabulary

- probability
- success
- failure
- random
- odds
- random variable
- probability distribution
- uniform distribution
- relative-frequency histogram

PROBABILITY AND ODDS Mathematicians often use tossing of coins and rolling of dice to illustrate probability. When you toss a coin, there are only two possible outcomes—heads or tails. A desired outcome is called a **success**. Any other outcome is called a **failure**.

Key Concept — *Probability of Success and Failure*

If an event can succeed in s ways and fail in f ways, then the probabilities of success, $P(S)$, and of failure, $P(F)$, are as follows.

$$P(S) = \frac{s}{s + f} \qquad P(F) = \frac{f}{s + f}$$

The probability of an event occurring is always between 0 and 1, inclusive. The closer the probability of an event is to 1, the more likely the event is to occur. The closer the probability of an event is to 0, the less likely the event is to occur.

Example 1 Probability

When two coins are tossed, what is the probability that both are tails?

You can use a tree diagram to find the sample space.

First coin	H		T	
Second coin	H	T	H	T
Possible outcomes	HH	HT	TH	TT

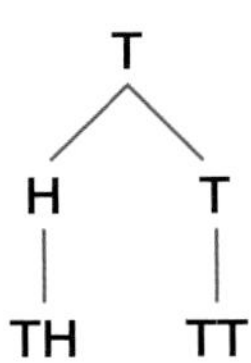

There are 4 possible outcomes. You can confirm this using the Fundamental Counting Principle. There are 2 possible results for the first coin and 2 for the second coin, so there are $2 \cdot 2$ or 4 possible outcomes. Only one of these outcomes, TT, is a success, so $s = 1$. The other three outcomes are failures, so $f = 3$.

$$P(\text{two tails}) = \frac{s}{s + f} \quad \text{Probability formula}$$

$$= \frac{1}{1 + 3} \text{ or } \frac{1}{4} \quad s = 1, f = 3$$

The probability of tossing two heads is $\frac{1}{4}$. *This probability can also be written as a decimal, 0.25, or as a percent, 25%.*

Study Tip

Reading Math
When P is followed by an event in parentheses, P stands for *probability*. When there are two numbers in parentheses, P stands for *permutations*.

In more complicated situations, you may need to use permutations and/or combinations to count the outcomes. When all outcomes have an equally likely chance of occurring, we say that the outcomes occur at **random**.

Example 2 Probability with Combinations

Monifa has a collection of 32 CDs—18 R&B and 14 rap. As she is leaving for a trip, she randomly chooses 6 CDs to take with her. What is the probability that she selects 3 R&B and 3 rap?

Step 1 Determine how many 6-CD selections meet the conditions.

$C(18, 3)$ Select 3 R&B CDs. Their order does not matter.
$C(14, 3)$ Select 3 rap CDs.

Step 2 Use the Fundamental Counting Principle to find the number of successes.

$C(18, 3) \cdot C(14, 3) = \frac{18!}{15!3!} \cdot \frac{14!}{11!3!}$ or 297,024

Step 3 Find the total number, $s + f$, of possible 6-CD selections.

$C(32, 6) = \frac{32!}{26!6!}$ or 906,192 $\quad s + f = 906,192$

Step 4 Determine the probability.

$P(\text{3 R\&B CDs and 3 rap CDs}) = \frac{s}{s + f}$ Probability formula

$= \frac{297,024}{906,192}$ Substitute.

≈ 0.32777 Use a calculator.

The probability of selecting 3 R&B CDs and 3 rap CDs is about 0.32777 or 33%.

Another way to measure the chance of an event occurring is with odds. The **odds** that an event will occur can be expressed as the ratio of the number of successes to the number of failures.

Key Concept — Odds

The odds that an event will occur can be expressed as the ratio of the number of ways it can succeed to the number of ways it can fail. If an event can succeed in s ways and fail in f ways, then the odds of success and of failure are as follows.

Odds of success $= s{:}f$ Odds of failure $= f{:}s$

Example 3 Odds

LIFE EXPECTANCY According to the U.S. National Center for Health Statistics, the chances of a male born in 1990 living to be at least 65 years of age are about 3 in 4. For females, the chances are about 17 in 20.

a. What are the odds of a male living to be at least 65?

Three out of four males will live to be at least 65, so the number of successes (living to 65) is 3. The number of failures is $4 - 3$ or 1.

odds of a male living to 65 $= s{:}f$ Odds formula

$= 3{:}1$ $\quad s = 3, f = 1$

The odds of a male living to at least 65 are 3:1.

b. What are the odds of a female living to be at least 65?

Seventeen out of twenty females will live to be at least 65, so the number of successes in this case is 17. The number of failures is 20 − 17 or 3.

odds of a female living to be 65 $= s{:}f$ — Odds formula

$= 17{:}3$ — $s = 17, f = 3$

The odds of a female living to at least 65 are 17:3.

PROBABILITY DISTRIBUTIONS Many experiments, such as rolling a die, have numerical outcomes. A **random variable** is a variable whose value is the numerical outcome of a random event. For example, when rolling a die we can let the random variable D represent the number showing on the die. Then D can equal 1, 2, 3, 4, 5, or 6. A **probability distribution** for a particular random variable is a function that maps the sample space to the probabilities of the outcomes in the sample space. The table below illustrates the probability distribution for rolling a die. *A distribution like this one where all of the probabilities are the same is called a* ***uniform distribution***.

Study Tip

Reading Math

The notation $P(X = n)$ is used with random variables. $P(D = 4) = \frac{1}{6}$ is read *the probability that D equals 4 is one sixth.*

D = Roll	1	2	3	4	5	6
Probability	$\frac{1}{6}$	$\frac{1}{6}$	$\frac{1}{6}$	$\frac{1}{6}$	$\frac{1}{6}$	$\frac{1}{6}$

$P(D = 4) = \frac{1}{6}$

To help visualize a probability distribution, you can use a table of probabilities or a graph, called a **relative-frequency histogram**.

Example 4 Probability Distribution

Suppose two dice are rolled. The table and the relative-frequency histogram show the distribution of the sum of the numbers rolled.

S = Sum	2	3	4	5	6	7	8	9	10	11	12
Probability	$\frac{1}{36}$	$\frac{1}{18}$	$\frac{1}{12}$	$\frac{1}{9}$	$\frac{5}{36}$	$\frac{1}{6}$	$\frac{5}{36}$	$\frac{1}{9}$	$\frac{1}{12}$	$\frac{1}{18}$	$\frac{1}{36}$

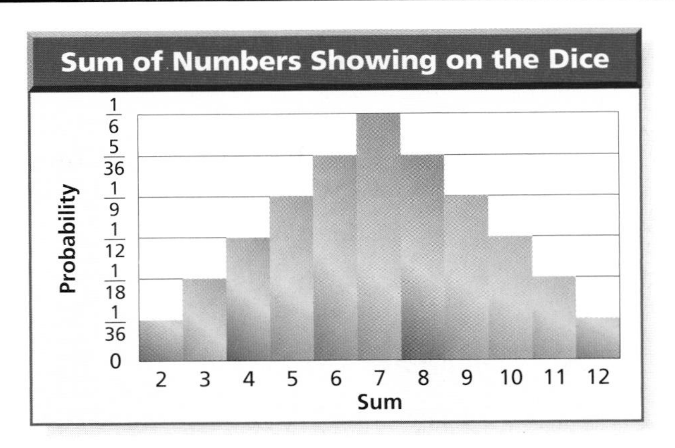

a. Use the graph to determine which outcome is most likely. What is its probability?

The most likely outcome is a sum of 7, and its probability is $\frac{1}{6}$.

b. Use the table to find $P(S = 9)$. What other sum has the same probability?

According to the table, the probability of a sum of 9 is $\frac{1}{9}$. The other outcome with a probability of $\frac{1}{9}$ is 5.

c. What are the odds of rolling a sum of 7?

Step 1 Identify s and f.

$$P(\text{rolling a } 7) = \frac{1}{6}$$

$$= \frac{s}{s+f} \quad s = 1, f = 5$$

Step 2 Find the odds.

$$\text{Odds} = s{:}f$$

$$= 1{:}5$$

So, the odds of rolling a sum of 7 are 1:5.

Check for Understanding

Concept Check

1. **OPEN ENDED** Describe an event that has a probability of 0 and an event that has a probability of 1.
2. **Write** the probability of an event whose odds are 3:2.
3. **Verify** the probabilities given for sums of 2 and 3 in Example 4.

Guided Practice

Suppose you select 2 letters at random from the word *compute*. Find each probability.

4. P(2 vowels)
5. P(2 consonants)
6. P(1 vowel, 1 consonant)

Find the odds of an event occurring, given the probability of the event.

7. $\frac{8}{9}$
8. $\frac{1}{6}$
9. $\frac{2}{9}$

Find the probability of an event occurring, given the odds of the event.

10. 6:5
11. 10:1
12. 2:5

The table and the relative-frequency histogram show the distribution of the number of heads when 3 coins are tossed. Find each probability.

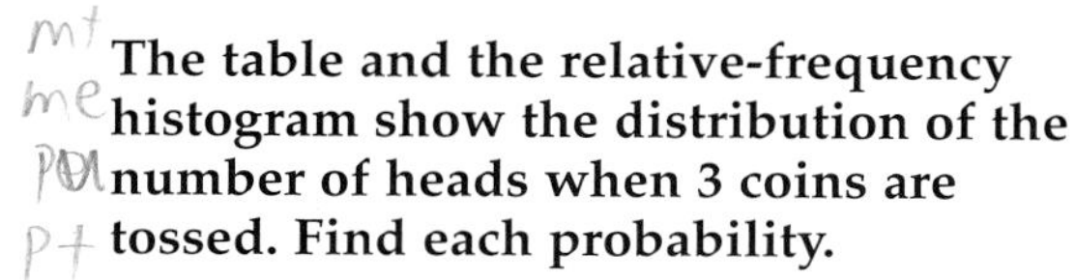

H = Heads	0	1	2	3
Probability	$\frac{1}{8}$	$\frac{3}{8}$	$\frac{3}{8}$	$\frac{1}{8}$

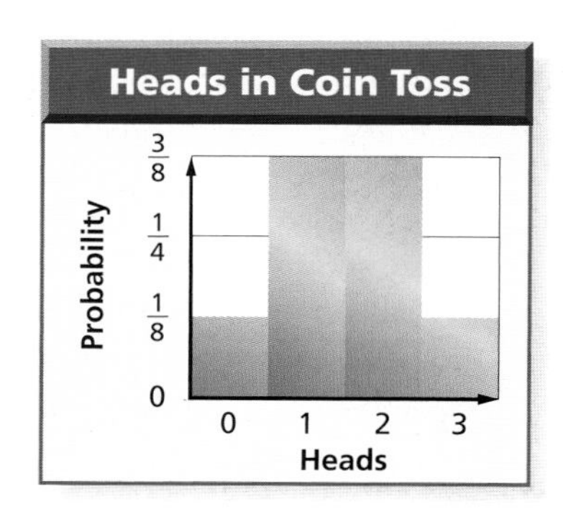

13. $P(H = 0)$
14. $P(H = 2)$

Application

GEOGRAPHY **For Exercises 15–18, find each probability if a state is chosen at random from the 50 states.**

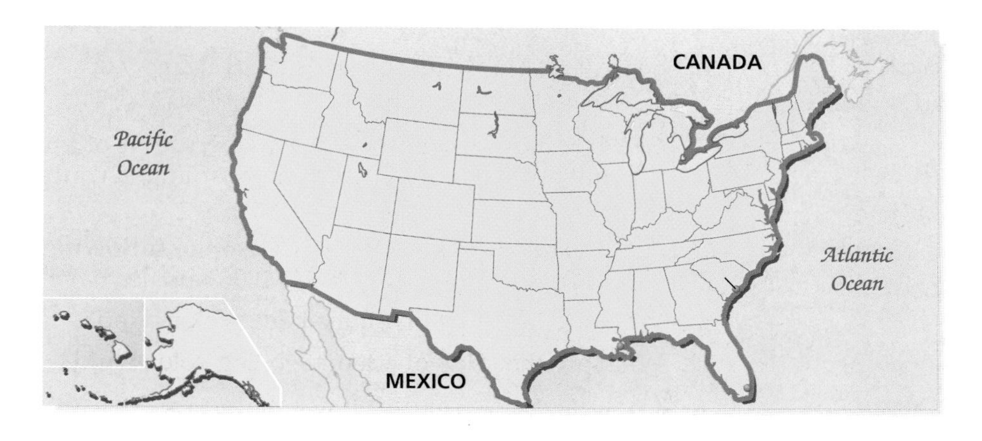

15. P(next to the Pacific Ocean)
16. P(has at least five neighboring states)
17. P(borders Mexico)
18. P(is surrounded by water)

Practice and Apply

Homework Help

For Exercises	See Examples
19–33, 54	1, 2
34–53	3
55–60	4

Extra Practice
See page 854.

Ebony has 4 male kittens and 7 female kittens. She picks up 2 kittens to give to a friend. Find the probability of each selection.

19. P(2 male)
20. P(2 female)
21. P(1 of each)

Bob is moving and all of his CDs are mixed up in a box. Twelve CDs are rock, eight are jazz, and five are classical. If he reaches in the box and selects them at random, find each probability.

22. P(3 jazz)
23. P(3 rock)
24. P(1 classical, 2 jazz)
25. P(2 classical, 1 rock)
26. P(1 jazz, 2 rock)
27. P(1 classical, 1 jazz, 1 rock)
28. P(2 rock, 2 classical)
29. P(2 jazz, 1 reggae)

30. **LOTTERIES** The state of Florida has a lottery in which 6 numbers out of 53 are drawn at random. What is the probability of a given ticket matching all 6 numbers in any order?

ENTRANCE TESTS For Exercises 31–33, use the table that shows the college majors of the students who took the Medical College Admission Test (MCAT) in April 2000.
If a student taking the test were randomly selected, find each probability. Express as decimals rounded to the nearest thousandth.

31. P(math or statistics)
32. P(biological sciences)
33. P(physical sciences)

Major	Students
biological sciences	15,819
humanities	963
math or statistics	179
physical sciences	2770
social sciences	2482
specialized health sciences	1431
other	1761

More About. . .

Entrance Tests

In addition to the MCAT, most medical schools require applicants to have had one year each of biology, physics, and English, and two years of chemistry in college.

Find the odds of an event occurring, given the probability of the event.

34. $\frac{1}{2}$
35. $\frac{3}{8}$
36. $\frac{11}{12}$
37. $\frac{5}{8}$
38. $\frac{4}{7}$
39. $\frac{1}{5}$
40. $\frac{4}{11}$
41. $\frac{3}{4}$

Find the probability of an event occurring, given the odds of the event.

42. 6:1
43. 3:7
44. 5:6
45. 4:5
46. 9:8
47. 1:8
48. 7:9
49. 3:2

50. **GENEOLOGY** The odds that an American is of English ancestry are 1:9. What is the probability that an American is of English ancestry?

GENETICS For Exercises 51 and 52, use the following information.
Eight out of 100 males and 1 out of 1000 females have some form of color blindness.

51. What are the odds of a male being color-blind?
52. What are the odds of a female being color-blind?

53. **EDUCATION** Josefina's guidance counselor estimates that the probability she will get a college scholarship is $\frac{4}{5}$. What are the odds that she will *not* earn a scholarship?

54. **CARD GAMES** The game of euchre is played using only the 9s, 10s, jacks, queens, kings, and aces from a standard deck of cards. Find the probability of being dealt a 5-card euchre hand containing all four suits.

Three students are selected at random from a group of 3 sophomores and 3 juniors. The table and relative-frequency histogram show the distribution of the number of sophomores chosen. Find each probability.

Sophomores	0	1	2	3
Probability	$\frac{1}{20}$	$\frac{9}{20}$	$\frac{9}{20}$	$\frac{1}{20}$

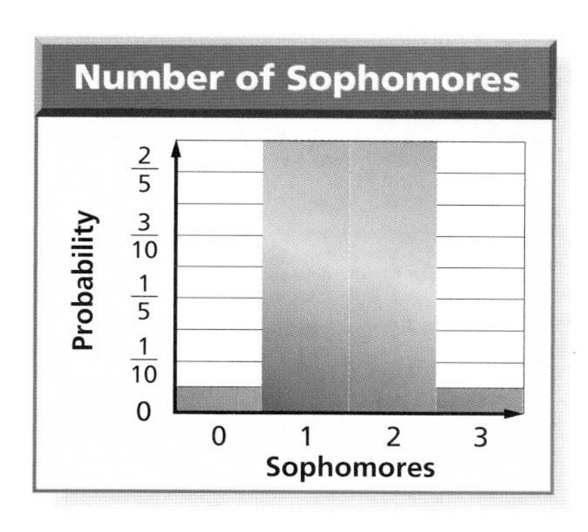

55. P(0 sophomores)
56. P(1 sophomore)
57. P(2 sophomores)
58. P(3 sophomores)
59. P(2 juniors)
60. P(1 junior)

61. **WRITING** Josh types the 5 entries in the bibliography of his term paper in random order, forgetting that they should be in alphabetical order by author. What is the probability that he actually typed them in alphabetical order?

62. **CRITICAL THINKING** Find the probability that a point chosen at random in the figure is in the shaded region. Write your answer in terms of π.

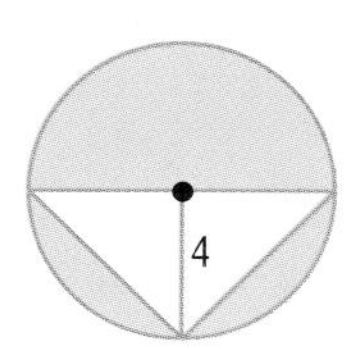

63. **WRITING IN MATH** Answer the question that was posed at the beginning of the lesson.

What do probability and odds tell you about life's risks?

Include the following in your answer:

- the odds of being struck by lightning and surviving the lightning strike, and
- a description of the meaning of *success* and *failure* in this case.

64. $\frac{6!}{2!} = ?$

Ⓐ 3 Ⓑ 60 Ⓒ 360 Ⓓ 720

65. A jar contains 4 red marbles, 3 green marbles, and 2 blue marbles. If a marble is drawn at random, what is the probability that it is not green?

Ⓐ $\frac{2}{9}$ Ⓑ $\frac{1}{3}$ Ⓒ $\frac{4}{9}$ Ⓓ $\frac{2}{3}$

Extending the Lesson

Theoretical probability is determined using mathematical methods and assumptions about the fairness of coins, dice, and so on. **Experimental probability** is determined by performing experiments and observing the outcomes.

Determine whether each probability is *theoretical* or *experimental*. Then find the probability.

66. Two dice are rolled. What is the probability that the sum will be 12?
67. A baseball player has 126 hits in 410 at-bats this season. What is the probability that he gets a hit in his next at-bat?
68. A bird watcher observes that 5 out of 25 birds in a garden are red. What is the probability that the next bird to fly into the garden will be red?
69. A hand of 2 cards is dealt from a standard deck of cards. What is the probability that both cards are clubs?

Maintain Your Skills

Mixed Review **Determine whether each situation involves a *permutation* or a *combination*. Then find the number of possibilities.** *(Lesson 12-2)*

70. arranging 5 different books on a shelf

71. arranging the letters of the word *arrange*

72. picking 3 apples from the last 7 remaining at the grocery store

73. A mail-order computer company offers a choice of 4 amounts of memory, 2 sizes of hard drives, and 2 sizes of monitors. How many different systems are available to a customer? *(Lesson 12-1)*

74. How many ways can 4 different gifts be placed into 4 different gift bags if each bag gets exactly 1 gift? *(Lesson 12-1)*

Identify the type of function represented by each graph. *(Lesson 9-5)*

75.

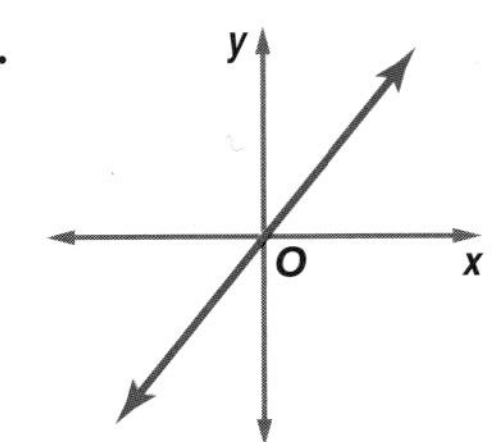

76.

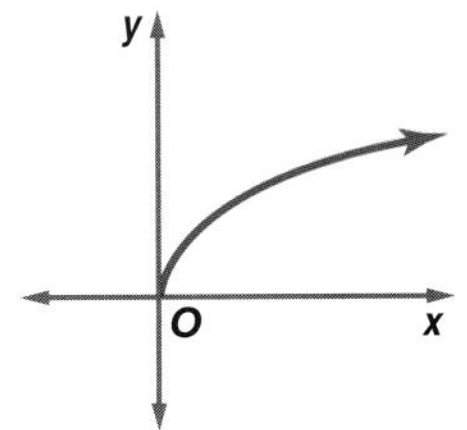

Solve each matrix equation. *(Lesson 4-1)*

77. $[x \quad y] = [y \quad 4]$

78. $\begin{bmatrix} 3y \\ 2x \end{bmatrix} = \begin{bmatrix} x + 8 \\ y - x \end{bmatrix}$

Getting Ready for the Next Lesson BASIC SKILL **Find each product if $a = \frac{3}{5}$, $b = \frac{2}{7}$, $c = \frac{3}{4}$, and $d = \frac{1}{3}$.**

79. ab **80.** bc **81.** cd **82.** bd **83.** ac

Practice Quiz 1

Lessons 12-1 through 12-3

1. At the Burger Bungalow, you can order your hamburger with or without cheese, with or without onions or pickles, and either rare, medium, or well-done. How many different ways can you order your hamburger? *(Lesson 12-1)*

2. For a particular model of car, a dealer offers 3 sizes of engines, 2 types of stereos, 18 body colors, and 7 upholstery colors. How many different possibilities are available for that model? *(Lesson 12-1)*

3. How many codes consisting of a letter followed by 3 digits can be made if no digit can be used more than once? *(Lesson 12-1)*

Evaluate each expression. *(Lesson 12-2)*

4. $P(12, 3)$ **5.** $C(8, 3)$

Determine whether each situation involves a *permutation* or a *combination*. Then find the number of possibilities. *(Lesson 12-2)*

6. 8 cars in a row parked next to a curb

7. a hand of 6 cards from a standard deck of cards

Two cards are drawn from a standard deck of cards. Find each probability. *(Lesson 12-3)*

8. P(2 aces) **9.** P(1 heart, 1 club) **10.** P(1 queen, 1 king)

12-4 Multiplying Probabilities

What You'll Learn

- Find the probability of two independent events.
- Find the probability of two dependent events.

Vocabulary

- area diagram

How does probability apply to basketball?

Reggie Miller of the Indiana Pacers is one of the best free-throw shooters in the National Basketball Association. The table shows the five highest season free-throw statistics of his career. For any year, you can determine the probability that Miller will make two free throws in a row based on the probability of his making one free throw.

Season	FT%
1990–91	91.8
1993–94	90.8
1998–99	91.5
1999–00	92.9
2000–01	92.8

Source: *Sporting News*

PROBABILITY OF INDEPENDENT EVENTS In a situation with two events like shooting a free throw and then shooting another one, you can find the probability of *both* events occurring if you know the probability of each event occurring. You can use an **area diagram** to model the probability of the two events occurring at the same time.

Algebra Activity

Area Diagrams

Suppose there are 1 red and 3 blue paper clips in one drawer and 1 gold and 2 silver paper clips in another drawer. The area diagram represents the probabilities of choosing one colored paper clip and one metallic paper clip if one of each is chosen at random. For example, rectangle A represents drawing 1 silver clip and 1 blue clip.

		Colored: blue $\frac{3}{4}$	Colored: red $\frac{1}{4}$
Metallic	silver $\frac{2}{3}$	A	B
Metallic	gold $\frac{1}{3}$	C	D

Model and Analyze

1. Find the areas of rectangles A, B, C, and D, and explain what each area represents.
2. What is the probability of choosing a red paper clip and a silver paper clip?
3. What are the length and width of the whole square? What is the area? Why does the area need to have this value?
4. Make an area diagram that represents the probability of each outcome if you spin each spinner once. Label the diagram and describe what the area of each rectangle represents.

In Exercise 4 of the activity, spinning one spinner has no effect on the second spinner. These events are independent.

Key Concept *Probability of Two Independent Events*

If two events, A and B, are independent, then the probability of both events occurring is $P(A \text{ and } B) = P(A) \cdot P(B)$.

This formula can be applied to any number of independent events.

Example 1 Two Independent Events

At a picnic, Julio reaches into an ice-filled cooler containing 8 regular soft drinks and 5 diet soft drinks. He removes a can, then decides he is not really thirsty, and puts it back. What is the probability that Julio and the next person to reach into the cooler both randomly select a regular soft drink?

Study Tip

Alternative Method
You could use the Fundamental Counting Principle to find the number of successes and the number of total outcomes.
both regular $= 8 \cdot 8$ or 64
total outcomes $= 13 \cdot 13$ or 169
So, $P(\text{both reg.}) = \frac{64}{169}$.

Explore These events are independent since Julio replaced the can that he removed. The outcome of the second person's selection is not affected by Julio's selection.

Plan Since there are 13 cans, the probability of each person's getting a regular soft drink is $\frac{8}{13}$.

Solve

$P(\text{both regular}) = P(\text{regular}) \cdot P(\text{regular})$ — Probability of independent events

$= \frac{8}{13} \cdot \frac{8}{13}$ or $\frac{64}{169}$ — Substitute and multiply.

The probability that both people select a regular soft drink is $\frac{64}{169}$ or about 0.38.

Examine You can verify this result by making a tree diagram that includes probabilities. Let R stand for regular and D stand for diet.

$P(R, R) = \frac{8}{13} \cdot \frac{8}{13}$

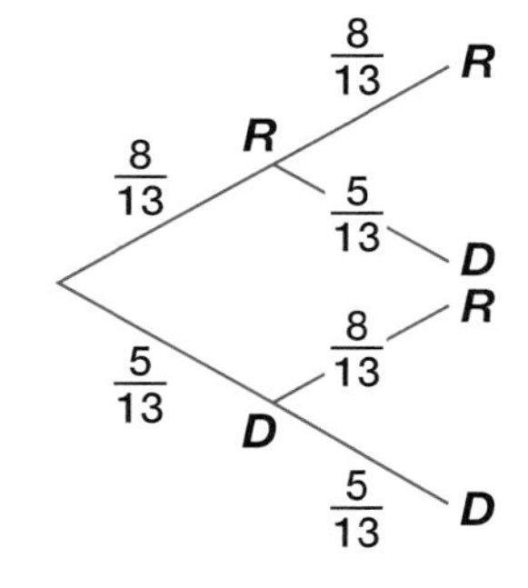

The formula for the probability of independent events can be extended to any number of independent events.

Example 2 Three Independent Events

In a board game, three dice are rolled to determine the number of moves for the players. What is the probability that the first die shows a 6, the second die shows a 6, and the third die does not?

Let A be the event that the first die shows a 6. → $P(A) = \frac{1}{6}$

Let B be the event that the second die shows a 6. → $P(B) = \frac{1}{6}$

Let C be the event that the third die does *not* show a 6. → $P(C) = \frac{5}{6}$

$$P(A, B, \text{and } C) = P(A) \cdot P(B) \cdot P(C)$$ Probability of independent events

$$= \frac{1}{6} \cdot \frac{1}{6} \cdot \frac{5}{6} \text{ or } \frac{5}{216}$$ Substitute and multiply.

The probability that the first and second dice show a 6 and the third die does not is $\frac{5}{36}$.

PROBABILITY OF DEPENDENT EVENTS In Example 1, what is the probability that both people select a regular soft drink if Julio does not put his back in the cooler? In this case, the two events are dependent because the outcome of the first event affects the outcome of the second event.

Study Tip

Conditional Probability
The event of getting a regular soft drink the second time *given* that Julio got a regular soft drink the first time is called a *conditional probability.*

First selection	Second selection	
$P(\text{regular}) = \frac{8}{13}$	$P(\text{regular}) = \frac{7}{12}$	Notice that when Julio removes his can, there is not only one fewer regular soft drink but also one fewer drink in the cooler.

$$P(\text{both regular}) = P(\text{regular}) \cdot P(\text{regular following regular})$$

$$= \frac{8}{13} \cdot \frac{7}{12} \text{ or } \frac{14}{39}$$ Substitute and multiply.

The probability that both people select a regular soft drink is $\frac{14}{39}$ or about 0.36.

Key Concept — Probability of Two Dependent Events

If two events, A and B, are dependent, then the probability of both events occurring is $P(A \text{ and } B) = P(A) \cdot P(B \text{ following } A)$.

This formula can be extended to any number of dependent events.

Example 3 Two Dependent Events

The host of a game show is drawing chips from a bag to determine the prizes for which contestants will play. Of the 10 chips in the bag, 6 show *television,* 3 show *vacation,* and 1 shows *car.* If the host draws the chips at random and does not replace them, find each probability.

Because the first chip is not replaced, the events are dependent. Let T represent a television, V a vacation, and C a car.

a. a vacation, then a car

$$P(V, \text{then } C) = P(V) \cdot P(C \text{ following } V)$$ Dependent events

$$= \frac{3}{10} \cdot \frac{1}{9} \text{ or } \frac{1}{30}$$ After the first chip is drawn, there are 9 left.

The probability of a vacation and then a car is $\frac{1}{30}$ or about 0.03.

b. two televisions

$$P(T, \text{then } T) = P(T) \cdot P(T \text{ following } T)$$ Dependent events

$$= \frac{6}{10} \cdot \frac{5}{9} \text{ or } \frac{1}{3}$$ If the first chip shows television, then 5 of the remaining 9 show television.

The probability of the host drawing two televisions is $\frac{1}{3}$.

Example 4 Three Dependent Events

Three cards are drawn from a standard deck of cards without replacement. Find the probability of drawing a diamond, a club, and another diamond in that order.

Since the cards are not replaced, the events are dependent. Let D represent a diamond and C a club.

$$P(D, C, D) = P(D) \cdot P(C \text{ following } D) \cdot P(D \text{ following } D \text{ and } C)$$

$$= \frac{13}{52} \cdot \frac{13}{51} \cdot \frac{12}{50} \text{ or } \frac{13}{850}$$ If the first two cards are a diamond and a club, then 12 of the remaining cards are diamonds.

The probability is $\frac{13}{850}$ or about 0.015.

Check for Understanding

Concept Check

1. **OPEN ENDED** Describe two real-life events that are dependent.
2. **Write** a formula for $P(A, B, C, \text{and } D)$ if A, B, C, and D are independent.
3. **FIND THE ERROR** Mario and Tabitha are calculating the probability of getting a 4 and then a 2 if they roll a die twice.

Mario

$P(4, \text{then } 2) = \frac{1}{6} \cdot \frac{1}{6}$

$= \frac{1}{36}$

Tabitha

$P(4, \text{then } 2) = \frac{1}{6} \cdot \frac{1}{5}$

$= \frac{1}{30}$

Who is correct? Explain your reasoning.

Guided Practice

A die is rolled twice. Find each probability.

4. $P(5, \text{then } 1)$
5. $P(\text{two even numbers})$

Two cards are drawn from a standard deck of cards. Find each probability if no replacement occurs.

6. $P(\text{two hearts})$
7. $P(\text{ace, then king})$

There are 8 action, 3 romantic comedy, and 5 children's DVDs on a shelf. Suppose two DVDs are selected at random from the shelf. Find each probability.

8. $P(2 \text{ action DVDs})$, if no replacement occurs
9. $P(2 \text{ action DVDs})$, if replacement occurs
10. $P(\text{a romantic comedy DVD, then a children's DVD})$, if no replacement occurs

Determine whether the events are *independent* or *dependent*. Then find the probability.

11. Yana has 7 blue pens, 3 black pens, and 2 red pens in his desk drawer. If he selects three pens at random with no replacement, what is the probability that he will first select a blue pen, then a black pen, and then another blue pen?
12. A black die and a white die are rolled. What is the probability that a 3 shows on the black die and a 5 shows on the white die?

Application

13. **ELECTIONS** Tami, Sonia, Malik, and Roger are the four candidates for student council president. If their names are placed in random order on the ballot, what is the probability that Malik's name will be first on the ballot followed by Sonia's name second?

Practice and Apply

Homework Help

For Exercises	See Examples
14–19, 36–39, 44–46	1, 2
20–29	1, 3
30–35	1–4
40–43	3

Extra Practice
See page 855.

A die is rolled twice. Find each probability.

14. P(2, then 3)
15. P(no 6s)
16. P(two 4s)
17. P(1, then any number)
18. P(two of the same number)
19. P(two different numbers)

The tiles *A, B, G, I, M, R,* and *S* of a word game are placed face down in the lid of the game. If two tiles are chosen at random, find each probability.

20. $P(R$, then $S)$, if no replacement occurs
21. $P(A$, then $M)$, if replacement occurs
22. P(2 consonants), if replacement occurs
23. P(2 consonants), if no replacement occurs
24. $P(B$, then $D)$, if replacement occurs
25. P(selecting the same letter twice), if no replacement occurs

Ashley takes her 3-year-old brother Alex into an antique shop. There are 4 statues, 3 picture frames, and 3 vases on a shelf. Alex accidentally knocks 2 items off the shelf and breaks them. Find each probability.

26. P(breaking 2 vases)
27. P(breaking 2 statues)
28. P(breaking a picture frame, then a vase)
29. P(breaking a statue, then a picture frame)

Determine whether the events are *independent* or *dependent*. Then find the probability.

30. There are 3 miniature chocolate bars and 5 peanut butter cups in a candy dish. Judie chooses 2 of them at random. What is the probability that she chooses 2 miniature chocolate bars?
31. A bowl contains 4 peaches and 5 apricots. Maxine randomly selects one, puts it back, and then randomly selects another. What is the probability that both selections were apricots?
32. A bag contains 7 red, 4 blue, and 6 yellow marbles. If 3 marbles are selected in succession, what is the probability of selecting blue, then yellow, then red, if replacement occurs each time?
33. Joe's wallet contains three \$1 bills, four \$5 bills, and two \$10 bills. If he selects three bills in succession, find the probability of selecting a \$10 bill, then a \$5 bill, and then a \$1 bill if the bills are not replaced.
34. What is the probability of getting heads each time if a coin is tossed 5 times?
35. When Diego plays his favorite video game, the odds are 3 to 4 that he will reach the highest level of the game. What is the probability that he will reach the highest level each of the next four times he plays?

For Exercises 36–39, suppose you spin the spinner twice.

36. Sketch a tree diagram showing all of the possibilities. Use it to find the probability of spinning red and then blue.

37. Sketch an area diagram of the outcomes. Shade the region on your area diagram corresponding to getting the same color twice.

38. What is the probability that you get the same color on both spins?

39. If you spin the same color twice, what is the probability that the color is red?

Spelling

The National Spelling Bee has been held every year since 1925, except for 1943-1945. Of the first 76 champions, 42 were girls and 34 were boys.

Source: www.spellingbee.com

Find each probability if 13 cards are drawn from a standard deck of cards and no replacement occurs.

40. P(all clubs)

41. P(all black cards)

42. P(all one suit)

43. P(no aces)

44. **UTILITIES** A city water system includes a sequence of 4 pumps as shown below. Water enters the system at point A, is pumped through the system by pumps at locations 1, 2, 3, and 4, and exits the system at point B.

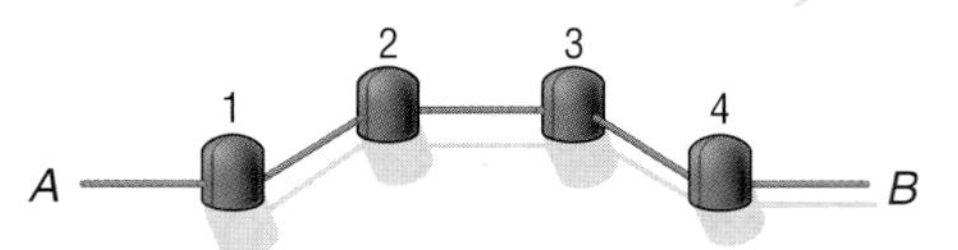

If the probability of failure for any one pump is $\frac{1}{100}$, what is the probability that water will flow all the way through the system from A to B?

45. **SPELLING** Suppose a contestant in a spelling bee has a 93% chance of spelling any given word correctly. What is the probability that he or she spells the first five words in a bee correctly and then misspells the sixth word?

46. **LITERATURE** The following quote is from *The Mirror Crack'd*, which was written by Agatha Christie in 1962.

> "I think you're begging the question," said Haydock, "and I can see looming ahead one of those terrible exercises in probability where six men have white hats and six men have black hats and you have to work it out by mathematics how likely it is that the hats will get mixed up and in what proportion. If you start thinking about things like that, you would go round the bend. Let me assure you of that!"

If the twelve hats are all mixed up and each man randomly chooses a hat, what is the probability that the first three men get their own hats? Assume that no replacement occurs.

For Exercises 47–49, use the following information.
You have a bag containing 10 marbles. In this problem, a *cycle* means that you draw a marble, record its color, and put it back.

47. You go through the cycle 10 times. If you do not record any black marbles, can you conclude that there are no black marbles in the bag?

48. Can you conclude that there are none if you repeat the cycle 50 times?

49. How many times do you have to repeat the cycle to be certain that there are no black marbles in the bag? Explain your reasoning.

50. **CRITICAL THINKING** If one bulb in a string of holiday lights fails to work, the whole string will not light. If each bulb in a set has a 99.5% chance of working, what is the maximum number of lights that can be strung together with at least a 90% chance of the whole string lighting?

51. **WRITING IN MATH** Answer the question that was posed at the beginning of the lesson.

How does probability apply to basketball?

Include the following in your answer:

- an explanation of how a value such as one of those in the table at the beginning of the lesson could be used to find the chances of Reggie Miller making 0, 1, or 2 of 2 successive free throws, assuming the 2 free throws are independent, and
- a possible psychological reason why 2 free throws on the same trip to the foul line might not be independent.

52. The spinner is spun four times. What is the probability that the spinner lands on 2 each time?

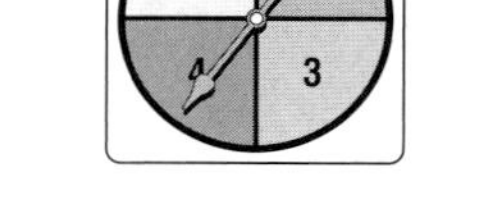

(A) $\frac{1}{2}$ (B) $\frac{1}{4}$

(C) $\frac{1}{16}$ (D) $\frac{1}{256}$

53. A coin is tossed and a die is rolled. What is the probability of a head and a 3?

(A) $\frac{1}{4}$ (B) $\frac{1}{8}$ (C) $\frac{1}{12}$ (D) $\frac{1}{24}$

Maintain Your Skills

Mixed Review

A gumball machine contains 7 red, 8 orange, 9 purple, 7 white, and 5 yellow gumballs. Tyson buys 3 gumballs. Find each probability, assuming that the machine dispenses the gumballs at random. *(Lesson 12-3)*

54. P(3 red)
55. P(2 white, 1 purple)
56. P(1 purple, 1 orange, 1 yellow)

57. **PHOTOGRAPHY** A photographer is taking a picture of a bride and groom together with 6 attendants. How many ways can he arrange the 8 people in a row if the bride and groom stand in the middle? *(Lesson 12-2)*

Solve each equation. Check your solutions. *(Lesson 10-3)*

58. $\log_5 5 + \log_5 x = \log_5 30$
59. $\log_{16} c - 2\log_{16} 3 = \log_{16} 4$

Given a polynomial and one of its factors, find the remaining factors of the polynomial. Some factors may not be binomials. *(Lesson 7-4)*

60. $x^3 - x^2 - 10x + 6; x + 3$
61. $x^3 - 7x^2 + 12x; x - 3$

Graph each inequality. *(Lesson 6-7)*

62. $y \leq x^2 + x - 2$
63. $y < x^2 - 4$
64. $y > x^2 - 3x$

Simplify. *(Lesson 5-5)*

65. $\sqrt{(153)^2}$
66. $\sqrt[3]{-729}$
67. $\sqrt[16]{b^{16}}$
68. $\sqrt{25a^8b^6}$

Solve each system of equations. *(Lesson 3-2)*

69. $z = 4y - 2$
$z = -y + 3$

70. $j - k = 4$
$2j + k = 35$

71. $3x + 1 = -y - 1$
$2y = -4x$

Getting Ready for the Next Lesson

BASIC SKILL **Find each sum if $a = \frac{1}{2}$, $b = \frac{1}{6}$, $c = \frac{2}{3}$, and $d = \frac{3}{4}$.**

72. $a + b$
73. $b + c$
74. $a + d$
75. $b + d$
76. $c + a$
77. $c + d$

12-5 Adding Probabilities

California Standards Standard 19.0 **Students use combinations** and permutations **to compute probabilities.** (Key)

What You'll Learn

- Find the probability of mutually exclusive events.
- Find the probability of inclusive events.

How does probability apply to your personal habits?

The graph shows the results of a survey about bedtime rituals. Determining the probability that a randomly selected person reads a book or brushes his or her teeth before going to bed requires adding probabilities.

Vocabulary

- simple event
- compound event
- mutually exclusive events
- inclusive events

MUTUALLY EXCLUSIVE EVENTS When you roll a die, an event such as rolling a 1 is called a **simple event** because it consists of only one event. An event that consists of two or more simple events is called a **compound event**. For example, the event of rolling an odd number or a number greater than 5 is a compound event because it consists of the simple events rolling a 1, rolling a 3, rolling a 5, or rolling a 6.

When there are two events, it is important to understand how they are related before finding the probability of one or the other event occurring. Suppose you draw a card from a standard deck of cards. What is the probability of drawing a 2 or an ace? Since a card cannot be both a 2 *and* an ace, these are called **mutually exclusive events**. That is, the two events cannot occur at the same time. The probability of drawing a 2 or an ace is found by adding their individual probabilities.

$P(2 \text{ or ace}) = P(2) + P(\text{ace})$ Add probabilities.

$= \frac{4}{52} + \frac{4}{52}$ There are 4 twos and 4 aces in a deck.

$= \frac{8}{52} \text{ or } \frac{2}{13}$ Simplify.

The probability of drawing a 2 or an ace is $\frac{2}{13}$.

Key Concept: Probability of Mutually Exclusive Events

- **Words** If two events, A and B, are mutually exclusive, then the probability that A or B occurs is the sum of their probabilities.
- **Symbols** $P(A \text{ or } B) = P(A) + P(B)$

This formula can be extended to any number of mutually exclusive events.

Example 1 Two Mutually Exclusive Events

Keisha has a stack of 8 baseball cards, 5 basketball cards, and 6 soccer cards. If she selects a card at random from the stack, what is the probability that it is a baseball or a soccer card?

These are mutually exclusive events, since the card cannot be both a baseball card *and* a soccer card. Note that there is a total of 19 cards.

$P(\text{baseball or soccer}) = P(\text{baseball}) + P(\text{soccer})$ Mutually exclusive events

$= \frac{8}{19} + \frac{6}{19} \text{ or } \frac{14}{19}$ Substitute and add.

The probability that Keisha selects a baseball or a soccer card is $\frac{14}{19}$.

Example 2 Three Mutually Exclusive Events

There are 7 girls and 6 boys on the junior class homecoming committee. A subcommittee of 4 people is being chosen at random to decide the theme for the class float. What is the probability that the subcommittee will have at least 2 girls?

At least 2 girls means that the subcommittee may have 2, 3, or 4 girls. It is not possible to select a group of 2 girls, a group of 3 girls, and a group of 4 girls all in the same 4-member subcommittee, so the events are mutually exclusive. Add the probabilities of each type of committee.

Study Tip

Choosing a Committee
C(13, 4) refers to choosing 4 subcommittee members from 13 committee members. Since order does not matter, the number of combinations is found.

$P(\text{at least 2 girls}) = P(\text{2 girls}) + P(\text{3 girls}) + P(\text{4 girls})$

2 girls, 2 boys; 3 girls, 1 boy; 4 girls, 0 boys

$= \frac{C(7, 2) \cdot C(6, 2)}{C(13, 4)} + \frac{C(7, 3) \cdot C(6, 1)}{C(13, 4)} + \frac{C(7, 4) \cdot C(6, 0)}{C(13, 4)}$

$= \frac{315}{715} + \frac{210}{715} + \frac{35}{715} \text{ or } \frac{112}{143}$ Simplify.

The probability of at least 2 girls on the subcommittee is $\frac{112}{143}$ or about 0.78.

INCLUSIVE EVENTS What is the probability of drawing a queen or a diamond from a standard deck of cards? Since it is possible to draw a card that is both a queen and a diamond, these events are *not* mutually exclusive. These are called **inclusive events**.

P(queen)	**P(diamond)**	**P(diamond, queen)**
$\frac{4}{52}$	$\frac{13}{52}$	$\frac{1}{52}$
1 queen in each suit	diamonds	queen of diamonds

Study Tip

Common Misconception
In mathematics, unlike everyday language, the expression *A* or *B* allows the possibility of both *A* and *B* occurring.

In the first two fractions above, the probability of drawing the queen of diamonds is counted twice, once for a queen and once for a diamond. To find the correct probability, you must subtract P(queen of diamonds) from the sum of the first two probabilities.

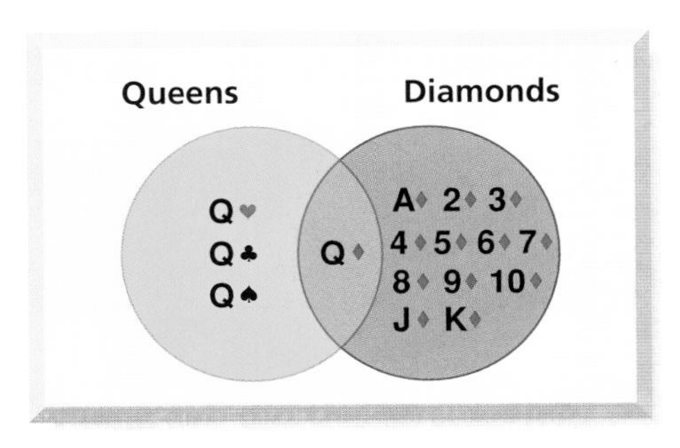

$$P(\text{queen or diamond}) = P(\text{queen}) + P(\text{diamond}) - P(\text{queen of diamonds})$$
$$= \frac{4}{52} + \frac{13}{52} - \frac{1}{52} \text{ or } \frac{4}{13}$$

The probability of drawing a queen or a diamond is $\frac{4}{13}$.

Key Concept — *Probability of Inclusive Events*

- **Words** If two events, A and B, are inclusive, then the probability that A or B occurs is the sum of their probabilities decreased by the probability of both occurring.
- **Symbols** $P(A \text{ or } B) = P(A) + P(B) - P(A \text{ and } B)$

Example 3 *Inclusive Events*

EDUCATION The enrollment at Southburg High School is 1400. Suppose 550 students take French, 700 take algebra, and 400 take both French and algebra. What is the probability that a student selected at random takes French or algebra?

Since some students take both French and algebra, the events are inclusive.

$P(\text{French}) = \frac{550}{1400}$ $\quad P(\text{algebra}) = \frac{700}{1400}$ $\quad P(\text{French and algebra}) = \frac{400}{1400}$

$$P(\text{French or algebra}) = P(\text{French}) + P(\text{algebra}) - P(\text{French and algebra})$$
$$= \frac{550}{1400} + \frac{700}{1400} - \frac{400}{1400} \text{ or } \frac{17}{28} \quad \text{Substitute and simplify.}$$

The probability that a student selected at random takes French or algebra is $\frac{17}{28}$.

Check for Understanding

Concept Check

1. **OPEN ENDED** Describe two mutually exclusive events and two inclusive events.
2. **Draw** a Venn diagram to illustrate Example 3.
3. **FIND THE ERROR** Refer to the comic below.

The Born Loser®

Why is the weather forecaster's prediction incorrect?

Guided Practice

A die is rolled. Find each probability.

4. $P(1 \text{ or } 6)$
5. $P(\text{at least } 5)$
6. $P(\text{less than } 3)$
7. $P(\text{prime})$
8. $P(\text{even or prime})$
9. $P(\text{multiple of 2 or 3})$

A card is drawn from a standard deck of cards. Determine whether the events are *mutually exclusive* or *inclusive*. Then find the probability.

10. P(6 or king)

11. P(queen or spade)

Application

12. **SCHOOL** There are 8 girls and 8 boys on the student senate. Three of the students are seniors. What is the probability that a person selected from the student senate is not a senior?

Practice and Apply

Homework Help

For Exercises	See Examples
13–22, 33–42	1, 2
23–26	1–3
27–32, 43–46	3

Extra Practice
See page 855.

Lisa has 9 rings in her jewelry box. Five are gold and 4 are silver. If she randomly selects 3 rings to wear to a party, find each probability.

13. P(2 silver or 2 gold)

14. P(all gold or all silver)

15. P(at least 2 gold)

16. P(at least 1 silver)

Seven girls and six boys walk into a video store at the same time. There are five salespeople available to help them. Find the probability that the salespeople will first help the given numbers of girls and boys.

17. P(4 girls or 4 boys)

18. P(3 girls or 3 boys)

19. P(all girls or all boys)

20. P(at least 3 girls)

21. P(at least 4 girls or at least 4 boys)

22. P(at least 2 boys)

For Exercises 23–26, determine whether the events are *mutually exclusive* or *inclusive*. Then find the probability.

23. There are 3 literature books, 4 algebra books, and 2 biology books on a shelf. If a book is randomly selected, what is the probability of selecting a literature book or an algebra book?

24. A die is rolled. What is the probability of rolling a 5 or a number greater than 3?

25. In the Math Club, 7 of the 20 girls are seniors, and 4 of the 14 boys are seniors. What is the probability of randomly selecting a boy or a senior to represent the Math Club at a statewide math contest?

26. A card is drawn from a standard deck of cards. What is the probability of drawing an ace or a face card? (*Hint:* A face card is a jack, queen, or king.)

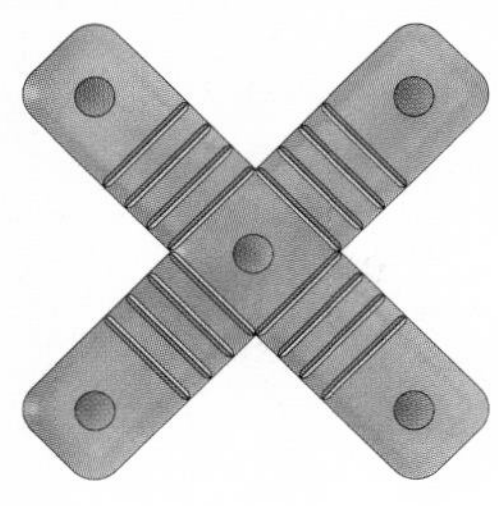

World Cultures

Totolospi is a Hopi game of chance. The players use cane dice, which have both a flat side and a round side, and a counting board inscribed in stone.

27. One tile with each letter of the alphabet is placed in a bag, and one is drawn at random. What is the probability of selecting a vowel or a letter from the word *equation*?

28. Each of the numbers from 1 to 30 is written on a card and placed in a bag. If one card is drawn at random, what is the probability that the number is a multiple of 2 or a multiple of 3?

Two cards are drawn from a standard deck of cards. Find each probability.

29. P(both kings or both black)

30. P(both kings or both face cards)

31. P(both face cards or both red)

32. P(both either red or a king)

WORLD CULTURES **For Exercises 33–36, refer to the information at the left.** When tossing 3 cane dice, if three round sides land up, the player advances 2 lines. If three flat sides land up, the player advances 1 line. If a combination is thrown, the player loses a turn. Find each probability.

33. P(advancing 2 lines)

34. P(advancing 1 line)

35. P(advancing at least 1 line)

36. P(losing a turn)

For Exercises 37–42, use the following information.
Each of the numbers 1 through 30 is written on a table tennis ball and placed in a wire cage. Each of the numbers 20 through 45 is written on a table tennis ball and placed in a different wire cage. One ball is chosen at random from each spinning cage. Find each probability.

37. P(each is a 25)

38. P(neither is a 20)

39. P(exactly one is a 30)

40. P(exactly one is a 40)

41. P(the numbers are equal)

42. P(the sum is 30)

More About. . .

Recycling

The United States recycles 28% of its waste.

Source: The U.S. Environmental Protection Agency

43. RECYCLING In one community, 300 people were surveyed to see if they would participate in a curbside recycling program. Of those surveyed, 134 said they would recycle aluminum cans, and 108 said they would recycle glass. Of those, 62 said they would recycle both. What is the probability that a randomly selected member of the community would recycle aluminum or glass?

SCHOOL For Exercises 44–46, use the Venn diagram that shows the number of participants in extracurricular activities for a junior class of 324 students. Determine each probability if a student is selected at random from the class.

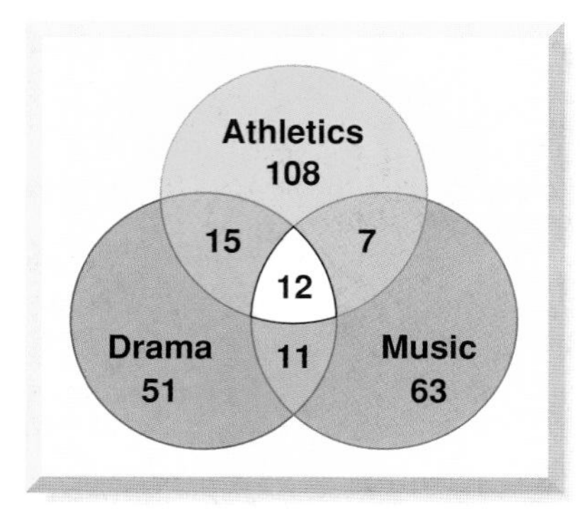

44. P(drama or music)

45. P(drama or athletics)

46. P(athletics and drama, or music and athletics)

47. CRITICAL THINKING Consider the following probability equation.

$$P(A \text{ and } B) = P(A) + P(B) - P(A \text{ or } B)$$

A textbook gives this equation for events A and B that are mutually exclusive or inclusive. Is this correct? Explain.

48. WRITING IN MATH Answer the question that was posed at the beginning of the lesson.

How does probability apply to your personal habits?

Include the following in your answer:

- an explanation of whether the events listed in the graphic are mutually exclusive or inclusive, and
- an explanation of how to determine the probability that a randomly selected person reads a book or brushes his or her teeth before going to bed if in a survey of 2000 people, 600 said that they do both.

49. In a jar of red and white gumballs, the ratio of white gumballs to red gumballs is 5:4. If the jar contains a total of 180 gumballs, how many of them are red?

Ⓐ 45 Ⓑ 64 Ⓒ 80 Ⓓ 100

50. $\langle x \rangle = \frac{1}{2}x$ if x is composite. $\langle x \rangle = 2x$ if x is prime. What is the value of $\langle 7 \rangle + \langle 18 \rangle$?

Ⓐ 23 Ⓑ 46 Ⓒ 50 Ⓓ 64

Maintain Your Skills

Mixed Review

A die is rolled three times. Find each probability. *(Lesson 12-4)*

51. $P(1, \text{then } 2, \text{then } 3)$

52. $P(\text{no 4s})$

53. $P(\text{three 1s})$

54. $P(\text{three even numbers})$

Find the odds of an event occurring, given the probability of the event. *(Lesson 12-3)*

55. $\frac{4}{5}$

56. $\frac{1}{9}$

57. $\frac{2}{7}$

58. $\frac{5}{8}$

Find the sum of each series. *(Lessons 11-2 and 11-4)*

59. $2 + 4 + 8 + \cdots + 128$

60. $\sum_{n=1}^{3} (5n - 2)$

Find the exact solution(s) of each system of equations. *(Lesson 8-7)*

61. $y = -10$
$y^2 = x^2 + 36$

62. $x^2 = 144$
$x^2 + y^2 = 169$

63. Use the graph of the polynomial function at the right to determine at least one binomial factor of the polynomial. Then find all factors of the polynomial. *(Lesson 7-4)*

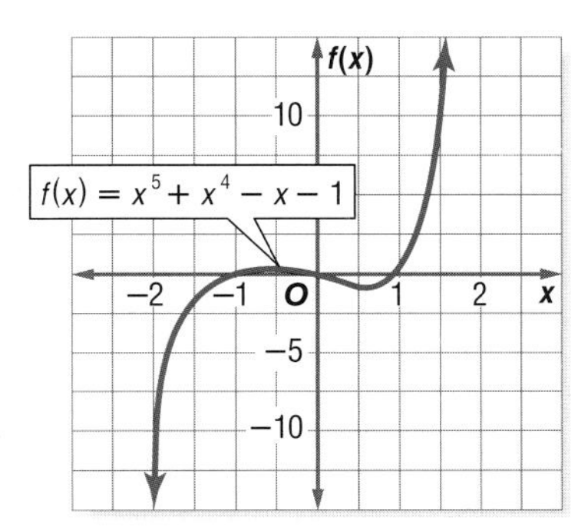

Find the maxima and minima of each function. Round to the nearest hundredth. *(Lesson 6-2)*

64. $f(x) = x^3 + 2x^2 - 5$

65. $f(x) = x^3 + 3x^2 + 2x + 1$

Graph each system of inequalities. Name the coordinates of the vertices of the feasible region. Find the maximum and minimum values of the given function for this region. *(Lesson 3-4)*

66. $y \geq x - 2$
$x \geq 0$
$y \leq 2 - x$
$f(x, y) = 3x + y$

67. $y \geq 2x - 3$
$1 \leq x \leq 3$
$y \leq x + 2$
$f(x, y) = x + 4y$

SPEED SKATING For Exercises 68 and 69, use the following information.
In the 1988 Winter Olympics, Bonnie Blair set a world record for women's speed skating by skating approximately 12.79 meters per second in the 500-meter race. *(Lesson 2-6)*

68. Suppose she could maintain that speed. Write an equation that represents how far she could travel in t seconds.

69. What type of equation is the one in Exercise 68?

Getting Ready for the Next Lesson

PREREQUISITE SKILL Find the mean, median, mode, and range for each set of data. Round to the nearest hundredth, if necessary.
*(To review **mean**, **median**, **mode**, and **range**, see pages 822 and 823.)*

70. 298, 256, 399, 388, 276

71. 3, 75, 58, 7, 34

72. 4.8, 5.7, 2.1, 2.1, 4.8, 2.1

73. 80, 50, 65, 55, 70, 65, 75, 50

74. 61, 89, 93, 102, 45, 89

75. 13.3, 15.4, 12.5, 10.7

12-6 Statistical Measures

What You'll Learn

- Use measures of central tendency to represent a set of data.
- Find measures of variation for a set of data.

Vocabulary

- univariate data
- measure of central tendency
- measure of variation
- dispersion
- variance
- standard deviation

What statistics should a teacher tell the class after a test?

On Mr. Dent's most recent Algebra 2 test, his students earned these scores.

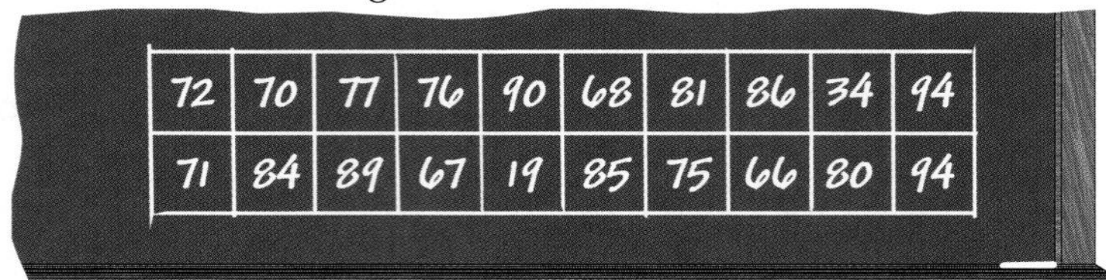

72	70	77	76	90	68	81	86	34	94
71	84	89	67	19	85	75	66	80	94

When his students ask how they did on the test, which measure of central tendency should Mr. Dent use to describe the scores?

MEASURES OF CENTRAL TENDENCY Data with one variable, such as the test scores, is called **univariate data**. Sometimes it is convenient to have one number that describes a set of univariate data. This number is called a **measure of central tendency**, because it represents the center or middle of the data. The most commonly used measures of central tendency are the *mean*, *median*, and *mode*.

When deciding which measure of central tendency to use to represent a set of data, look closely at the data itself.

Study Tip

Look Back
To review **outliers**, see Lesson 2-5.

Concept Summary — Measures of Tendency

Use	When . . .
mean	the data are spread out, and you want an average of the values.
median	the data contain outliers.
mode	the data are tightly clustered around one or two values.

Example 1 Choose a Measure of Central Tendency

SWEEPSTAKES A sweepstakes offers a first prize of $10,000, two second prizes of $100, and one hundred third prizes of $10.

a. Which measure of central tendency best represents the available prizes?

Since 100 of the 103 prizes are $10, the mode ($10) best represents the available prizes. Notice that in this case the median is the same as the mode.

b. Which measure of central tendency would the organizers of the sweepstakes be most likely to use in their advertising?

The organizers would be most likely to use the mean (about $109) to make people think they had a better chance of winning more money.

Study Tip

Reading Math

The symbol σ is the lower case Greek letter *sigma*. $\bar{x}$ is read *x bar*.

MEASURES OF VARIATION

Measures of variation or **dispersion** measure how spread out or scattered a set of data is. The simplest measure of variation to calculate is the *range,* the difference between the greatest and the least values in a set of data. Variance and standard deviation are measures of variation that indicate how much the data values differ from the mean.

To find the **variance** σ^2 of a set of data, follow these steps.

1. Find the mean, $\bar{x}$.
2. Find the difference between each value in the set of data and the mean.
3. Square each difference.
4. Find the mean of the squares.

The **standard deviation** σ is the square root of the variance.

Key Concept — *Standard Deviation*

If a set of data consists of the n values $x_1, x_2, \ldots, x_n$ and has mean $\bar{x}$, then the standard deviation σ is given by the following formula.

$$\sigma = \sqrt{\frac{(x_1 - \bar{x})^2 + (x_2 - \bar{x})^2 + \cdots + (x_n - \bar{x})^2}{n}}$$

Example 2 *Standard Deviation*

STATES The table shows the populations in millions of 11 eastern states as of the 2000 Census. Find the variance and standard deviation of the data to the nearest tenth.

State	Population	State	Population	State	Population
NY	19.0	MD	5.3	RI	1.0
PA	12.3	CT	3.4	DE	0.8
NJ	8.4	ME	1.3	VT	0.6
MA	6.3	NH	1.2	—	—

Source: U.S. Census Bureau

Step 1 Find the mean. Add the data and divide by the number of items.

$$\bar{x} = \frac{19.0 + 12.3 + 8.4 + 6.3 + 5.3 + 3.4 + 1.3 + 1.2 + 1.0 + 0.8 + 0.6}{11}$$

$\approx 5.4\overline{18}$ The mean is about 5.4 people.

Step 2 Find the variance.

$$\sigma^2 = \frac{(x_1 - \bar{x})^2 + (x_2 - \bar{x})^2 + \cdots + (x_n - \bar{x})^2}{n}$$ Variance formula

$$\approx \frac{(19.0 - 5.4)^2 + (12.3 - 5.4)^2 + \cdots + (0.8 - 5.4)^2 + (0.6 - 5.4)^2}{11}$$

$\approx \frac{344.4}{11}$ Simplify.

$\approx 31.3\overline{09}$ The variance is about 31.3 people.

Step 3 Find the standard deviation.

$\sigma^2 \approx 31.3$ Take the square root of each side.

$\sigma \approx 5.594640292$ The standard deviation is about 5.6 people.

Most of the members of a set of data are within 1 standard deviation of the mean. The populations of the states in Example 2 can be broken down as shown below.

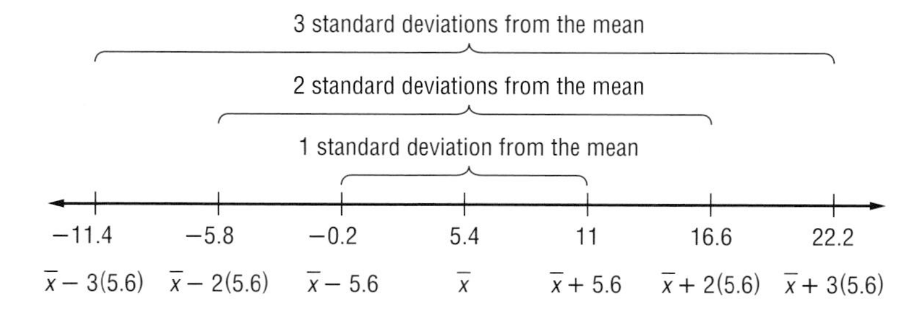

Looking at the original data, you can see that most of the states' populations were between 2.4 million and 20.2 million. That is, the majority of members of the data set were within 1 standard deviation of the mean.

You can use a TI-83 Plus graphing calculator to find statistics for the data in Example 2.

Graphing Calculator Investigation

One-Variable Statistics

The TI-83 Plus can compute a set of one-variable statistics from a list of data. These statistics include the mean, variance, and standard deviation. Enter the data into L1.

KEYSTROKES: [STAT] [ENTER] 19.0 [ENTER] 12.3 [ENTER] ...

Then use [STAT] [▶] 1 [ENTER] to show the statistics. The mean $\bar{x}$ is about 5.4, the sum of the values $\sum x$ is 59.6, the standard deviation σx is about 5.6, and there are $n = 11$ data items. If you scroll down, you will see the least value (minX = .6), the three quartiles (1, 3.4, and 8.4), and the greatest value (maxX = 19).

Think and Discuss

1. Find the variance of the data set.
2. Enter the data set in list L1 but without the outlier 19.0. What are the new mean, median, and standard deviation?
3. Did the mean or median change less when the outlier was deleted?

Check for Understanding

Concept Check

1. **OPEN ENDED** Give a sample set of data with a variance and standard deviation of 0.
2. **Find a counterexample** for the following statement.
   The standard deviation of a set of data is always less than the variance.
3. **Write** the formula for standard deviation using sigma notation. (*Hint:* To review sigma notation, see Lesson 11-5.)

Guided Practice

Find the variance and standard deviation of each set of data to the nearest tenth.

4. {48, 36, 40, 29, 45, 51, 38, 47, 39, 37}
5. {321, 322, 323, 324, 325, 326, 327, 328, 329, 330}
6. {43, 56, 78, 81, 47, 42, 34, 22, 78, 98, 38, 46, 54, 67, 58, 92, 55}

Application

EDUCATION For Exercises 7 and 8, use the following information.
The table below shows the amounts of money spent on education per student in 1998 in two regions of the United States.

Pacific States		Southwest Central States	
State	Expenditures per Student ($)	State	Expenditures per Student ($)
Alaska	10,650	Texas	6291
California	5345	Arkansas	5222
Washington	6488	Louisiana	5194
Oregon	6719	Oklahoma	4634

Source: National Education Association

7. Find the mean for each region.
8. For which region is the mean more representative of the data? Explain.

Practice and Apply

Homework Help

For Exercises	See Examples
17–26	1
9–16, 27–33	2

Extra Practice
See page 855.

Find the variance and standard deviation of each set of data to the nearest tenth.

9. {400, 300, 325, 275, 425, 375, 350}
10. {5, 4, 5, 5, 5, 5, 6, 6, 6, 6, 7, 7, 7, 7, 8, 9}
11. {2.4, 5.6, 1.9, 7.1, 4.3, 2.7, 4.6, 1.8, 2.4}
12. {4.3, 6.4, 2.9, 3.1, 8.7, 2.8, 3.6, 1.9, 7.2}
13. {234, 345, 123, 368, 279, 876, 456, 235, 333, 444}
14. {13, 14, 15, 16, 17, 18, 19, 20, 21, 23, 67, 56, 34, 99, 44, 55}

15.

Stem	Leaf
4	4 5 6 7 7
5	3 5 6 7 8 9
6	7 7 8 9 9 9

4|5 = 45

16.

Stem	Leaf
5	7 7 7 8 9
6	3 4 5 5 6 7
7	2 3 4 5 6

6|3 = 63

More About. . .

Basketball

Chamique Holdsclaw of the Washington Mystics led the Women's National Basketball Association in rebounding in 2003 with 284 rebounds in 26 games, an average of about 10.9 rebounds per game.

Source: WNBA

BASKETBALL For Exercises 17 and 18, use the following information.
The table below shows the rebounding totals for the 2003 Los Angeles Sparks.

141	231	220	126	175	55	29	68	19	32	19	21	6	23	1

Source: WNBA

17. Find the mean, median, and mode of the data to the nearest tenth.
18. Which measure of central tendency best represents the data? Explain.

Online Research **Data Update** For the latest rebounding statistics for both women's and men's professional basketball, visit: www.algebra2.com/data_update

EDUCATION For Exercises 19 and 20, use the following information. The Millersburg school board is negotiating a pay raise with the teacher's union. Three of the administrators have salaries of $80,000 each. However, a majority of the teachers have salaries of about $35,000 per year.

19. You are a member of the school board and would like to show that the current salaries are reasonable. Would you quote the mean, median, or mode as the "average" salary to justify your claim? Explain.
20. You are the head of the teacher's union and maintain that a pay raise is in order. Which of the mean, median, or mode would you quote to justify your claim? Explain your reasoning.

ADVERTISING For Exercises 21–23, use the following information.
A camera store placed an ad in the newspaper showing five digital cameras for sale. The ad says, "Our digital cameras average $695." The prices of the digital cameras are $1200, $999, $1499, $895, $695, $1100, $1300, and $695.

21. Find the mean, median, and mode of the prices.
22. Which measure is the store using in its ad? Why did they choose this measure?
23. As a consumer, which measure would you want to see advertised? Explain.

SHOPPING MALLS For Exercises 24–26, use the following information.
The table lists the areas of some large shopping malls in the United States.

	Mall	Gross Leasable Area (ft^2)
1	Del Amo Fashion Center, Torrance, CA	3,000,000
2	South Coast Plaza/Crystal Court, Costa Mesa, CA	2,918,236
3	Mall of America, Bloomington, MN	2,472,500
4	Lakewood Center Mall, Lakewood, CA	2,390,000
5	Roosevelt Field Mall, Garden City, NY	2,300,000
6	Gurnee Mills, Gurnee, IL	2,200,000
7	The Galleria, Houston, TX	2,100,000
8	Randall Park Mall, North Randall, OH	2,097,416
9	Oakbrook Shopping Center, Oak Brook, IL	2,006,688
10	Sawgrass Mills, Sunrise, FL	2,000,000
10	The Woodlands Mall, The Woodlands, TX	2,000,000
10	Woodfield, Schaumburg, IL	2,000,000

Source: Blackburn Marketing Service

More About. . .

Shopping
While the Mall of America does not have the most gross leasable area, it is the largest fully enclosed retail and entertainment complex in the United States.
Source: Mall of America

24. Find the mean, median, and mode of the gross leasable areas.
25. You are a realtor who is trying to lease mall space in different areas of the country to a large retailer. Which measure would you talk about if the customer felt that the malls were too large for his store? Explain.
26. Which measure would you talk about if the customer had a large inventory? Explain.

FOOTBALL For Exercises 27–30, use the weights in pounds of the starting offensive linemen of the football teams from three high schools.

Jackson	Washington	King
170, 165, 140, 188, 195	144, 177, 215, 225, 197	166, 175, 196, 206, 219

27. Find the standard deviation of the weights for Jackson High.
28. Find the standard deviation of the weights for Washington High.
29. Find the standard deviation of the weights for King High
30. Which team had the most variation in weights? How do you think this variation will impact their play?

SCHOOL For Exercises 31–33, use the frequency table at the right that shows the scores on a multiple-choice test.

31. Find the variance and standard deviation of the scores.
32. What percent of the scores are within one standard deviation of the mean?
33. What percent of the scores are within two standard deviations of the mean?

Score	Frequency
90	3
85	2
80	3
75	7
70	6
65	4

For Exercises 34–36, consider the two graphs below.

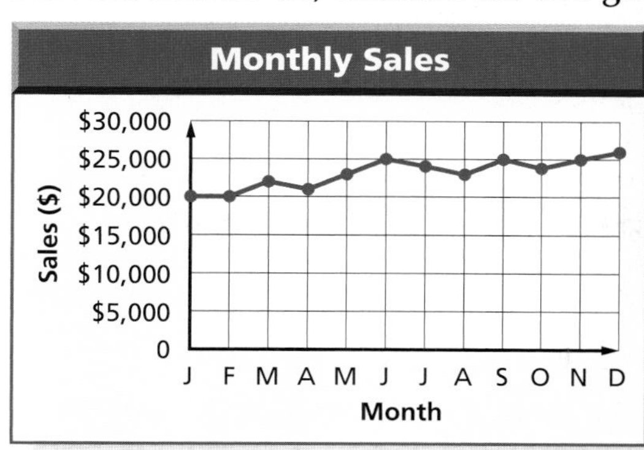

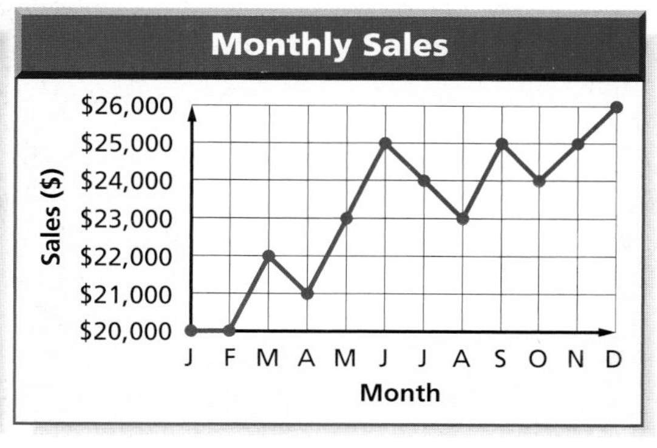

34. Explain why the graphs made from the same data look different.

35. Describe a situation where the first graph might be used.

36. Describe a situation where the second graph might be used.

CRITICAL THINKING **For Exercises 37 and 38, consider the two sets of data.**

A = {1, 2, 2, 2, 2, 3, 3, 3, 3, 4}, B = {1, 1, 2, 2, 2, 3, 3, 3, 4, 4}

37. Find the mean, median, variance, and standard deviation of each set of data.

38. Explain how you can tell which histogram below goes with each data set without counting the frequencies in the sets.

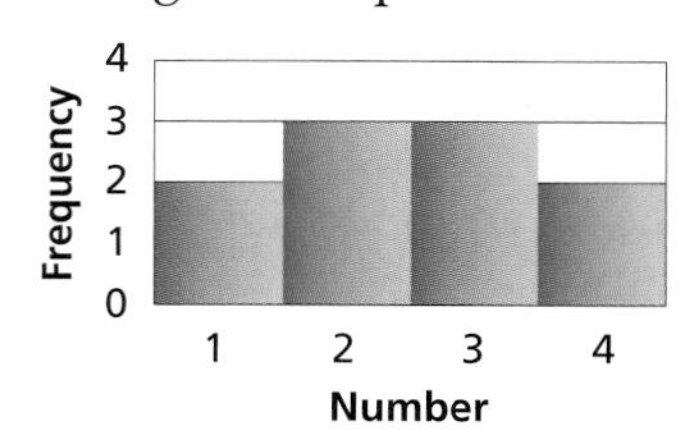

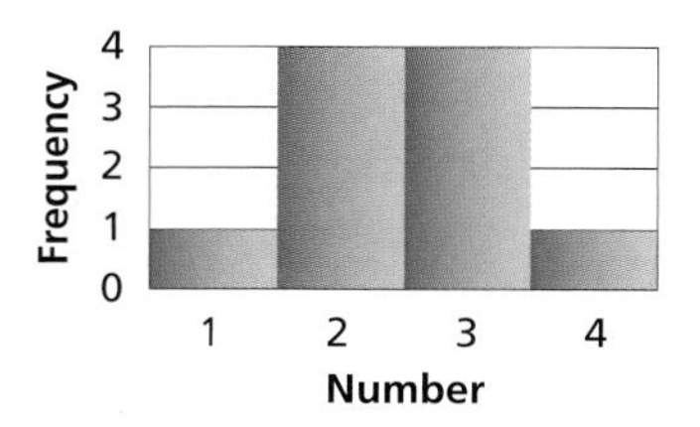

39. WRITING IN MATH Answer the question that was posed at the beginning of the lesson.

What statistics should a teacher tell the class after a test?

Include the following in your answer:

- the mean, median, and mode of the given data set,
- which measure of central tendency best represents the test scores and why, and
- how the measures of central tendency are affected if Mr. Dent adds 5 points to each score.

Standards Practice

Standardized Test Practice

40. What is the mean of the numbers represented by $x + 1$, $3x - 2$, and $2x - 5$?

(A) $2x - 2$ (B) $\frac{6x - 7}{3}$ (C) $\frac{x + 1}{3}$ (D) $x + 4$

41. Manuel got scores of 92, 85, and 84 on three successive tests. What score must he get on a fourth test in order to have an average of 90?

(A) 96 (B) 97 (C) 98 (D) 99

Extending the Lesson

Mean deviation is another method of dispersion. It is the mean of the deviations of the data from the mean of the data. If a set of data consists of n values $x_1, x_2, \ldots, x_n$ and has mean $\bar{x}$, then the mean deviation is given by the following formula.

$$\text{MD} = \frac{|x_1 - \bar{x}| + |x_2 - \bar{x}| + \cdots + |x_n - \bar{x}|}{n} \text{ or } \frac{1}{n}\sum_{i=1}^{n}|x_i - \bar{x}|$$

Study Tip

Reading Math
Mean deviation is also sometimes called *mean absolute deviation*.

Find the mean deviation of each set of data to the nearest tenth.

42. {95, 91, 88, 86}

43. {10.4, 11.4, 16.2, 14.9, 13.5}

44. Suppose two sets of data have the same mean and different standard deviations. Describe their mean deviations.

Maintain Your Skills

Mixed Review

Determine whether the events are mutually exclusive or inclusive. Then find the probability. *(Lesson 12-5)*

45. A card is drawn from a standard deck of cards. What is the probability that it is a 5 or a spade?

46. A jar of change contains 5 quarters, 8 dimes, 10 nickels, and 19 pennies. If a coin is pulled from the jar at random, what is the probability that it is a nickel or a dime?

Two cards are drawn from a standard deck of cards. Find each probability. *(Lesson 12-4)*

47. P(ace, then king) if replacement occurs

48. P(ace, then king) if no replacement occurs

49. P(heart, then club) if no replacement occurs

50. P(heart, then club) if replacement occurs

51. Find the coordinates of the vertices and foci and the slopes of the asymptotes for the hyperbola given by $\frac{y^2}{81} - \frac{x^2}{25} = 1$. *(Lesson 8-5)*

If $f(x) = x - 7$, $g(x) = 4x^2$, and $h(x) = 2x + 1$, find each value. *(Lesson 7-7)*

52. $f[g(-1)]$ **53.** $h[f(15)]$ **54.** $f \circ h(2)$

55. BUSINESS The Energy Booster Company keeps their stock of Health Aid liquid in a rectangular tank whose sides measure $x - 1$ centimeters, $x + 3$ centimeters, and $x - 2$ centimeters. Suppose they would like to bottle their Health Aid in $x - 3$ containers of the same size. How much liquid in cubic centimeters will remain unbottled? *(Lesson 7-2)*

Use Cramer's Rule to solve each system of equations. *(Lesson 4-6)*

56. $2x + 6y = 28$
$-x - 4y = -20$

57. $7c - 3d = -8$
$4c + d = 9$

58. $m - 2n = -7$
$-3m + n = -4$

Getting Ready for the Next Lesson

BASIC SKILL Find each percent.

59. 68% of 200 **60.** 68% of 500 **61.** 95% of 400

62. 95% of 500 **63.** 99% of 400 **64.** 99% of 500

Practice Quiz 2 — Lessons 12-4 through 12-6

A bag contains 5 red marbles, 3 green marbles, and 2 blue marbles. Two marbles are drawn at random from the bag. Find each probability. *(Lesson 12-4)*

1. P(red, then green) if replacement occurs

2. P(red, then green) if no replacement occurs

3. P(2 red) if no replacement occurs

4. P(2 red) if replacement occurs

A twelve-sided die has sides numbered 1 through 12. The die is rolled once. Find each probability. *(Lesson 12-5)*

5. P(4 or 5) **6.** P(even or a multiple of 3) **7.** P(odd or a multiple of 4)

Find the variance and standard deviation of each set of data to the nearest tenth. *(Lesson 12-6)*

8. {5, 8, 2, 9, 4} **9.** {16, 22, 18, 31, 25, 22} **10.** {425, 400, 395, 415, 420}

12-7 The Normal Distribution

What You'll Learn

- Determine whether a set of data appears to be normally distributed or skewed.
- Solve problems involving normally distributed data.

Vocabulary

- discrete probability distribution
- continuous probability distribution
- normal distribution
- skewed distribution

How are the heights of professional athletes distributed?

The frequency table below lists the heights of the 2001 Baltimore Ravens. The table shows the heights of the players, but it does not show how these heights compare to the height of an average player. To make that comparison, you can determine how the heights are distributed.

Height (in.)	67	69	70	71	72	73	74	75	76	77	80
Frequency	1	1	4	4	10	6	6	8	7	5	1

Source: www.ravenszone.net

NORMAL AND SKEWED DISTRIBUTIONS The probability distributions you have studied thus far are **discrete probability distributions** because they have only a finite number of possible values. A discrete probability distribution can be represented by a histogram. For a **continuous probability distribution**, the outcome can be any value in an interval of real numbers. Continuous probability distributions are represented by curves instead of histograms.

The curve at the right represents a continuous probability distribution. Notice that the curve is symmetric. Such a curve is often called a *bell curve*. Many distributions with symmetric curves or histograms are **normal distributions**.

Normal Distribution

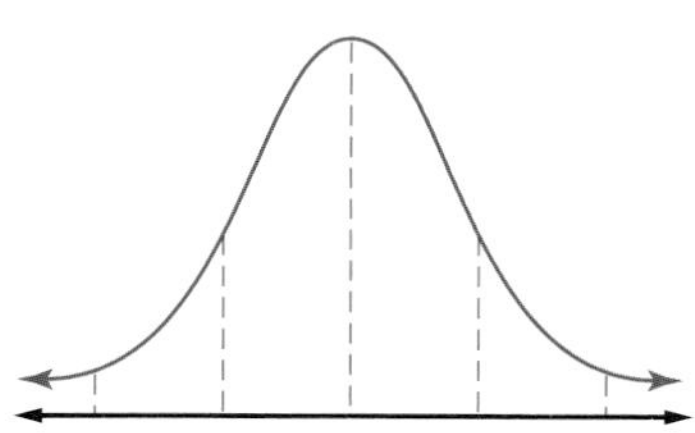

Study Tip

Skewed Distributions

In a positively skewed distribution, the long tail is in the positive direction. These are sometimes said to be *skewed to the right*. In a negatively skewed distribution, the long tail is in the negative directon. These are sometimes said to be *skewed to the left*.

A curve or histogram that is not symmetric represents a **skewed distribution**. For example, the distribution for a curve that is high at the left and has a tail to the right is said to be *positively skewed*. Similarly, the distribution for a curve that is high at the right and has a tail to the left is said to be *negatively skewed*.

Positively Skewed

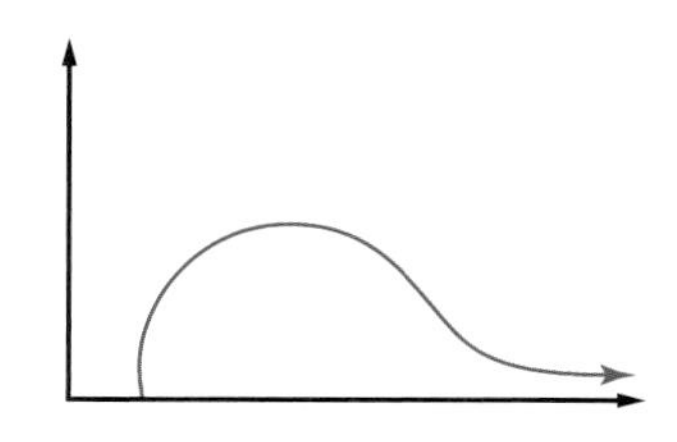

Negatively Skewed

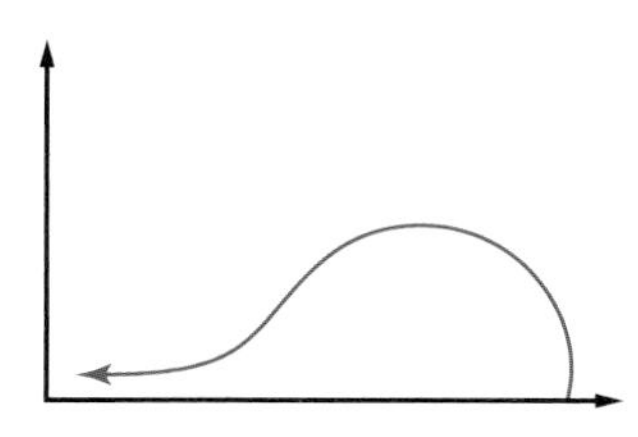

Example 1 Classify a Data Distribution

Determine whether the data {14, 15, 11, 13, 13, 14, 15, 14, 12, 13, 14, 15} appear to be *positively skewed, negatively skewed,* or *normally distributed.*

Make a frequency table for the data. Then use the table to make a histogram.

Value	11	12	13	14	15
Frequency	1	1	3	4	3

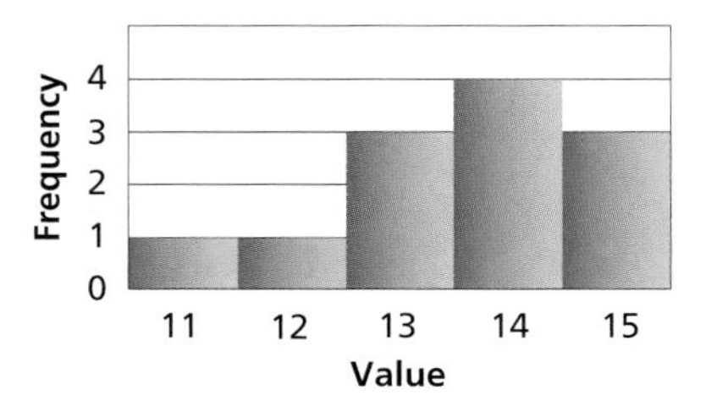

Since the histogram is high at the right and has a tail to the left, the data are negatively skewed.

USE NORMAL DISTRIBUTIONS Normal distributions occur quite frequently in real life. Standardized test scores, the lengths of newborn babies, the useful life and size of manufactured items, and production levels can all be represented by normal distributions. In all of these cases, the number of data values must be large for the distribution to be approximately normal.

Study Tip

Normal Distribution

If you randomly select an item from data that are normally distributed, the probability that the one you pick will be within one standard deviation of the mean is 0.68. If you do this 1000 times, about 680 of those picked will be within one standard deviation of the mean.

Key Concept — Normal Distribution

Normal distributions have these properties.

The graph is maximized at the mean.

The mean, median, and mode are about equal.

0.5% 2% 13.5% 34% 34% 13.5% 2% 0.5%

About 68% of the values are within one standard deviation of the mean.

About 95% of the values are within two standard deviations of the mean.

About 99% of the values are within three standard deviations of the mean.

Example 2 Normal Distribution

PHYSIOLOGY The reaction times for a hand-eye coordination test administered to 1800 teenagers are normally distributed with a mean of 0.35 second and a standard deviation of 0.05 second.

a. About how many teens had reaction times between 0.25 and 0.45 second?

Draw a normal curve. Label the mean and the mean plus or minus multiples of the standard deviation.

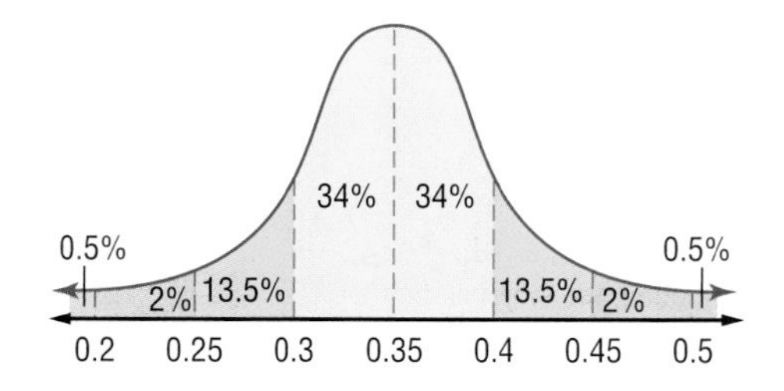

The values 0.25 and 0.45 are 2 standard deviations *below and above* the mean, respectively. Therefore, about 95% of the data are between 0.25 and 0.45.

$1800 \times 95\% = 1710$ Multiply 1800 by 0.95.

About 1710 of the teenagers had reaction times between 0.25 and 0.45 second.

b. What is the probability that a teenager selected at random had a reaction time greater than 0.4 second?

The value 0.4 is one standard deviation above the mean. You know that about 100% − 68% or 32% of the data are more than one standard deviation away from the mean. By the symmetry of the normal curve, half of 32%, or 16%, of the data are more than one standard deviation above the mean.

The probability that a teenager selected at random had a reaction time greater than 0.4 second is about 16% or 0.16.

Check for Understanding

Concept Check

1. **OPEN ENDED** Sketch a positively skewed graph. Describe a situation in which you would expect data to be distributed this way.
2. **Compare and contrast** the means and standard deviations of the graphs.

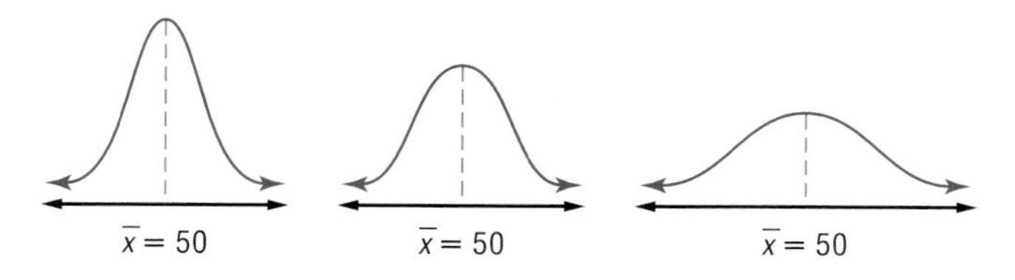

3. **Explain** how to find what percent of a set of normally distributed data is more than 3 standard deviations above the mean.

Guided Practice

4. The table at the right shows female mathematics SAT scores in 2000. Determine whether the data appear to be *positively skewed*, *negatively skewed*, or *normally distributed*.

Score	Percent of Females
200–299	3
300–399	14
400–499	33
500–599	31
600–699	15
700–800	4

Source: www.collegeboard.org

For Exercises 5–7, use the following information.
Mrs. Sung gave a test in her trigonometry class. The scores were normally distributed with a mean of 85 and a standard deviation of 3.

5. What percent would you expect to score between 82 and 88?
6. What percent would you expect to score between 88 and 91?
7. What is the probability that a student chosen at random scored between 79 and 91?

Application

QUALITY CONTROL **For Exercises 8–11, use the following information.**
The useful life of a radial tire is normally distributed with a mean of 30,000 miles and a standard deviation of 5000 miles. The company makes 10,000 tires a month.

8. About how many tires will last between 25,000 and 35,000 miles?
9. About how many tires will last more than 40,000 miles?
10. About how many tires will last less than 25,000 miles?
11. What is the probability that if you buy a radial tire at random, it will last between 20,000 and 35,000 miles?

Practice and Apply

Homework Help

For Exercises	See Examples
12–14	1
15–26	2

Extra Practice
See page 856.

Determine whether the data in each table appear to be *positively skewed*, *negatively skewed*, or *normally distributed*.

12.

U.S. Population	
Age	Percent
0–19	28.7
20–39	29.3
40–59	25.5
60–79	13.3
80–99	3.2
100+	0.0

Source: U.S. Census Bureau

13.

Record Low Temperatures in the 50 States	
Temperature (°F)	Number of States
−80 to −65	4
−64 to −49	12
−48 to −33	19
−32 to −17	12
−16 to −1	2
0 to 15	1

Source: *The World Almanac*

14. SCHOOL The frequency table at the right shows the grade-point averages (GPAs) of the juniors at Stanhope High School. Do the data appear to be *positively skewed*, *negatively skewed*, or *normally distributed*? Explain.

GPA	Frequency
0.0–0.4	4
0.5–0.9	4
1.0–1.4	2
1.5–1.9	32
2.0–2.4	96
2.5–2.9	91
3.0–3.4	110
3.5–4.0	75

FOOD **For Exercises 15–18, use the following information.**
The shelf life of a particular dairy product is normally distributed with a mean of 12 days and a standard deviation of 3.0 days.

15. About what percent of the products last between 9 and 15 days?
16. About what percent of the products last between 12 and 15 days?
17. About what percent of the products last less than 3 days?
18. About what percent of the products last more than 15 days?

VENDING **For Exercises 19–21, use the following information.**
The vending machine in the school cafeteria usually dispenses about 6 ounces of soft drink. Lately, it is not working properly, and the variability of how much of the soft drink it dispenses has been getting greater. The amounts are normally distributed with a standard deviation of 0.2 ounce.

19. What percent of the time will you get more than 6 ounces of soft drink?
20. What percent of the time will you get less than 6 ounces of soft drink?
21. What percent of the time will you get between 5.6 and 6.4 ounces of soft drink?

MANUFACTURING **For Exercises 22–24, use the following information.**
A company manufactures 1000 CDs per hour that are supposed to be 120 millimeters in diameter. These CDs are made for drives 122 millimeters wide. The sizes of CDs made by this company are normally distributed with a standard deviation of 1 millimeter.

22. What percent of the CDs would you expect to be greater than 120 millimeters?
23. In one hour, how many CDs would you expect to be between 119 and 122 millimeters?
24. About how many CDs per hour will be too large to fit in the drives?

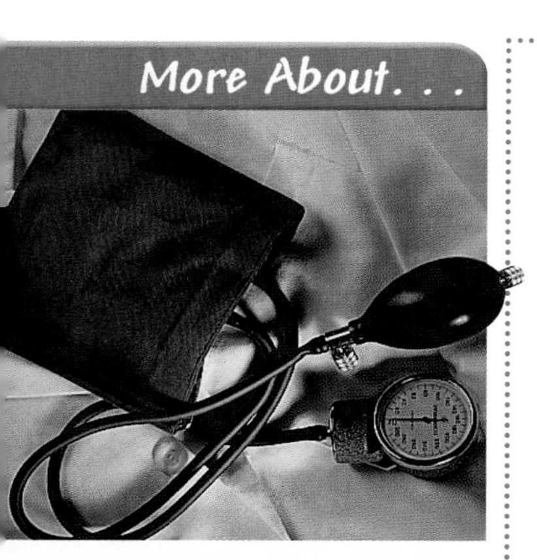

Health

A systolic blood pressure below 130 is normal and between 130 and 139 is "high normal."

Source: National Institutes of Health

HEALTH For Exercises 25 and 26, use the following information.
A recent study showed that the systolic blood pressure of high school students ages 14–17 is normally distributed with a mean of 120 and a standard deviation of 12. Suppose a high school has 800 students.

25. About what percent of the students have blood pressures below 108?

26. About how many students have blood pressures between 108 and 144?

27. CRITICAL THINKING The graphing calculator screen shows the graph of a normal distribution for a large set of test scores whose mean is 500 and whose standard deviation is 100. If every test score in the data set were increased by 25 points, describe how the mean, standard deviation, and graph of the data would change.

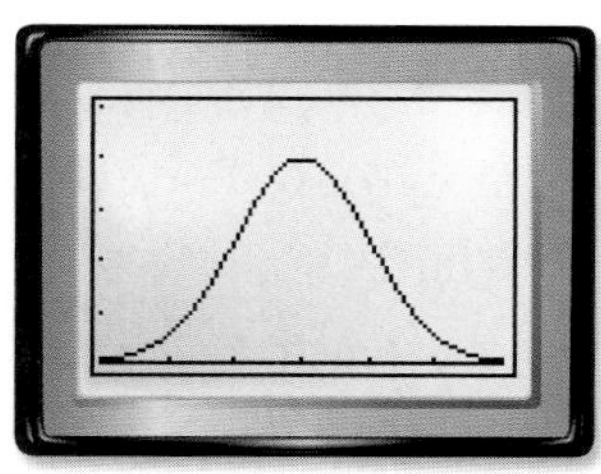

[200, 800] scl: 100 by [0, 0.005] scl: 0.001

28. WRITING IN MATH Answer the question that was posed at the beginning of the lesson.

How are the heights of professional athletes distributed?

Include the following items in your answer:
- a histogram of the given data, and
- an explanation of whether you think the data are normally distributed.

Standards Practice

Standardized Test Practice

29. If $x + y = 5$ and $xy = 6$, what is the value of $x^2 + y^2$?

Ⓐ 13 Ⓑ 17 Ⓒ 25 Ⓓ 37

30. Which of the following is not the square of a rational number?

Ⓐ 0.04 Ⓑ 0.16 Ⓒ $\frac{4}{9}$ Ⓓ $\frac{2}{3}$

Maintain Your Skills

Mixed Review

Find the variance and standard deviation of each set of data to the nearest tenth. *(Lesson 12-6)*

31. {7, 16, 9, 4, 12, 3, 9, 4}

32. {12, 14, 28, 19, 11, 7, 10}

A card is drawn from a standard deck of cards. Find each probability. *(Lesson 12-5)*

33. P(jack or queen)

34. P(ace or heart)

35. P(2 or face card)

Find all of the rational zeros for each function. *(Lesson 7-6)*

36. $f(x) = x^3 + 4x^2 - 5x$

37. $p(x) = x^3 - 3x^2 - 10x + 24$

38. $h(x) = x^4 - 2x^2 + 1$

39. $f(x) = 4x^4 - 13x^3 - 13x^2 + 28x - 6$

METEOROLOGY For excercises 40 and 41, use the following information.
Weather forecasters can determine the approximate time that a thunderstorm will last if they know the diameter d of the storm in miles. The time t in hours can be found by using the formula $216t^2 = d^3$. *(Lesson 6-2)*

40. Graph $y = 216t^2 - 5^3$ and use it to estimate how long a thunderstorm will last if its diameter is 5 miles.

41. Find how long a thunderstorm will last if its diameter is 5 miles and compare this time with your estimate in Exercise 40.

Getting Ready for the Next Lesson

PREREQUISITE SKILL Find the indicated term of each expression.
*(For review of **binomial expansions**, see Lesson 5-2.)*

42. third term of $(a + b)^7$

43. fourth term of $(c + d)^8$

44. fifth term of $(x + y)^9$

12-8 Binomial Experiments

California Standards Standard 20.0 Students know the binomial theorem and use it to expand binomial expressions that are raised to positive integer powers. (Key)

What You'll Learn

- Use binomial expansions to find probabilities.
- Find probabilities for binomial experiments.

Vocabulary

- binomial experiment

How can you determine whether guessing is worth it?

What is the probability of getting exactly 4 questions correct on a 5-question multiple-choice quiz if you guess at every question?

Study Tip

Look Back
To review the **Binomial Theorem**, see Lesson 11-7.

BINOMIAL EXPANSIONS You can use the Binomial Theorem to find probabilities in certain situations where there are two possible outcomes. The 5 possible ways of getting 4 questions right r and 1 question wrong w are shown at the right. This chart shows the combination of 5 things (answer choices) taken 4 at a time (right answers) or $C(5, 4)$.

w	r	r	r	r
r	w	r	r	r
r	r	w	r	r
r	r	r	w	r
r	r	r	r	w

The terms of the binomial expansion of $(r + w)^5$ can be used to find the probabilities of each combination of right and wrong.

$$(r + w)^5 = r^5 + 5r^4w + 10r^3w^2 + 10r^2w^3 + 5rw^4 + w^5$$

Coefficient	Term	Meaning
$C(5, 5) = 1$	r^5	1 way to get all 5 questions right
$C(5, 4) = 5$	$5r^4w$	5 ways to get 4 questions right and 1 question wrong
$C(5, 3) = 10$	$10r^3w^2$	10 ways to get 3 questions right and 2 questions wrong
$C(5, 2) = 10$	$10r^2w^3$	10 ways to get 2 questions right and 3 questions wrong
$C(5, 1) = 5$	$5rw^4$	5 ways to get 1 question right and 4 questions wrong
$C(5, 0) = 1$	w^5	1 way to get all 5 questions wrong

The probability of getting a question right that you guessed on is $\frac{1}{4}$. So, the probability of getting the question wrong is $\frac{3}{4}$. To find the probability of getting 4 questions right and 1 question wrong, substitute $\frac{1}{4}$ for r and $\frac{3}{4}$ for w in the term $5r^4w$.

$$P(\text{4 right, 1 wrong}) = 5r^4w$$

$$= 5\left(\frac{1}{4}\right)^4\left(\frac{3}{4}\right) \quad r = \frac{1}{4}, w = \frac{3}{4}$$

$$= \frac{15}{1024} \quad \text{Multiply.}$$

The probability of getting exactly 4 questions correct is $\frac{15}{1024}$ or about 1.5%.

Example 1 Binomial Theorem

If a family has 4 children, what is the probability that they have 3 boys and 1 girl?

There are two possible outcomes for the gender of each of their children: boy or girl. The probability of a boy b is $\frac{1}{2}$, and the probability of a girl g is $\frac{1}{2}$.

$(b + g)^4 = b^4 + 4b^3g + 6b^2g^2 + 4bg^3 + g^4$

The term $4b^3g$ represents 3 boys and 1 girl.

$$P(3 \text{ boys, } 1 \text{ girl}) = 4b^3g$$
$$= 4\left(\frac{1}{2}\right)^3\left(\frac{1}{2}\right) \quad b = \frac{1}{2}, g = \frac{1}{2}$$
$$= \frac{1}{4} \quad \text{Multiply.}$$

The probability of 3 boys and 1 girl is $\frac{1}{4}$ or 25%.

BINOMIAL EXPERIMENTS Problems like Example 1 that can be solved using binomial expansion are called **binomial experiments**.

Key Concept — Binomial Experiments

A binomial experiment exists if and only if all of these conditions occur.

- There are exactly two possible outcomes for each trial.
- There is a fixed number of trials.
- The trials are independent.
- The probabilities for each trial are the same.

A binomial experiment is sometimes called a *Bernoulli experiment*.

Suppose that in the application at the beginning of the lesson, the first 3 questions are answered correctly. Then the last 2 are answered incorrectly. The probability of this occurring is $\frac{1}{4} \cdot \frac{1}{4} \cdot \frac{1}{4} \cdot \frac{3}{4} \cdot \frac{3}{4}$ or $\left(\frac{1}{4}\right)^3\left(\frac{3}{4}\right)^2$. In general, there are $C(5, 3)$ ways to arrange 3 correct answers among the 5 questions, so the probability of exactly 3 correct answers is given by $C(5, 3)\left(\frac{1}{4}\right)^3\left(\frac{3}{4}\right)^2$.

More About. . .

Sports

The National Hockey League record for most goals in a game by one player is seven. A player has scored five or more goals in a game 53 times in league history.

Source: NHL

Example 2 Binomial Experiment

SPORTS Suppose that when hockey star Jaromir Jagr takes a shot, he has a $\frac{1}{7}$ probability of scoring a goal. He takes 6 shots in a game one night.

a. What is the probability that he will score exactly 2 goals?

The probability that he scores a goal on a given shot is $\frac{1}{7}$. The probability that he does not score on a given shot is $\frac{6}{7}$. There are $C(6, 2)$ ways to choose the 2 shots that score.

$$P(2 \text{ goals}) = C(6, 2)\left(\frac{1}{7}\right)^2\left(\frac{6}{7}\right)^4 \quad \text{If he scores on 2 shots, he fails to score on 4 shots.}$$
$$= \frac{6 \cdot 5}{2}\left(\frac{1}{7}\right)^2\left(\frac{6}{7}\right)^4 \quad C(6, 2) = \frac{6!}{4!2!}$$
$$= \frac{19{,}440}{117{,}649} \quad \text{Simplify.}$$

The probability that Jagr will score exactly 2 goals is $\frac{19{,}440}{117{,}649}$ or about 0.17.

b. What is the probability that he will score at least 2 goals?

Instead of adding the probabilities of getting exactly 2, 3, 4, 5, and 6 goals, it is easier to subtract the probabilities of getting exactly 0 or 1 goal from 1.

$$P(\text{at least 2 goals}) = 1 - P(0 \text{ goals}) - P(1 \text{ goal})$$

$$= 1 - C(6, 0)\left(\frac{1}{7}\right)^0\left(\frac{6}{7}\right)^6 - C(6, 1)\left(\frac{1}{7}\right)^1\left(\frac{6}{7}\right)^5$$

$$= 1 - \frac{46{,}656}{117{,}649} - \frac{46{,}656}{117{,}649} \quad \text{Simplify.}$$

$$= \frac{24{,}337}{117{,}649} \quad \text{Subtract.}$$

The probability that Jagr will score at least 2 goals is $\frac{24{,}337}{117{,}649}$ or about 0.21.

Check for Understanding

Concept Check

1. **OPEN ENDED** Describe a situation for which the P(2 or more) can be found by using a binomial expansion.
2. Refer to the application at the beginning of the lesson. List the possible sequences of 3 right answers and 2 wrong answers.
3. **Explain** why each experiment is not binomial.
   a. rolling a die and recording whether a 1, 2, 3, 4, 5, or 6 comes up
   b. tossing a coin repeatedly until it comes up heads
   c. removing marbles from a bag and recording whether each one is black or white, if no replacement occurs

Guided Practice

Find each probability if a coin is tossed 3 times.

4. P(exactly 2 heads)
5. P(0 heads)
6. P(at least 1 head)

Four cards are drawn from a standard deck of cards. Each card is replaced before the next one is drawn. Find each probability.

7. P(4 jacks)
8. P(exactly 3 jacks)
9. P(at most 1 jack)

Application

SPORTS For Exercises 10 and 11, use the following information.
Jessica Mendoza of Stanford University was the 2000 NCAA women's softball batting leader with an average of .475. This means that the probability of her getting a hit in a given at-bat was 0.475.

10. Find the probability of her getting 4 hits in 4 at-bats.
11. Find the probability of her getting exactly 2 hits in 4 at-bats.

Practice and Apply

Find each probability if a coin is tossed 4 times.

12. P(4 tails)
13. P(0 tails)
14. P(exactly 2 tails)
15. P(exactly 1 tail)
16. P(at least 3 tails)
17. P(at most 2 tails)

Find each probability if a die is rolled 5 times.

18. P(exactly one 5)
19. P(exactly three 5s)
20. P(at most two 5s)
21. P(at least three 5s)

As an apartment manager, Jackie Thomas is responsible for showing prospective renters different models of apartments. When showing a model, the probability that she selects the correct key from her set is $\frac{1}{4}$. If she shows 5 models in a day, find each probability.

22. P(never the correct key)
23. P(always the correct key)
24. P(correct exactly 4 times)
25. P(correct exactly 2 times)
26. P(no more than 2 times correct)
27. P(at least 3 times correct)

Prisana guesses at all 10 true/false questions on her history test. Find each probability.

28. P(exactly 6 correct)
29. P(exactly 4 correct)
30. P(at most half correct)
31. P(at least half correct)

If a thumbtack is dropped, the probability of it landing point-up is 0.4. If 12 tacks are dropped, find each probability.

32. P(at least 9 points up)
33. P(at most 4 points up)

34. **CARS** According to a recent survey, about 1 in 3 new cars is leased rather than bought. What is the probability that 3 of 7 randomly-selected new cars are leased?

35. **INTERNET** In 2001, it was estimated that 32.5% of U.S. adults use the Internet. What is the probability that exactly 2 out of 5 randomly-selected U.S. adults use the Internet?

More About. . .

Internet

The word *Internet* was virtually unknown until the mid-1980s. By 1997, 19 million Americans were using the Internet. That number tripled in 1998 and passed 100 million in 1999.

Source: UCLA

WORLD CULTURES **For Exercises 36 and 37, use the following information.**
The Cayuga Indians played a game of chance called *Dish*, in which they used 6 flattened peach stones blackened on one side. They placed the peach stones in a wooden bowl and tossed them. The winner was the first person to get a prearranged number of points. The table below shows the points that were given for each toss. Assume that each face (black or neutral) of each stone has an equal chance of showing up.

36. Copy and complete the table by finding the probability of each outcome.
37. Find the probability that a player gets at least 1 point for a toss.

Outcome	Points	Probability
6 black	5	
5 black, 1 neutral	1	
4 black, 2 neutral	0	
3 black, 3 neutral	0	
2 black, 4 neutral	0	
1 black, 5 neutral	1	
6 neutral	5	

38. **CRITICAL THINKING** Write an expression for the probability of exactly m successes in n trials of a binomial experiment where the probability of success in a given trial is p.

39. **WRITING IN MATH** Answer the question that was posed at the beginning of the lesson.

    How can you determine whether guessing is worth it?

    Include the following in your answer:
    - an explanation of how to find the probability of getting any number of questions right on a 5-question multiple-choice quiz, and
    - the probability of each score.

40. GRID IN In the figure, if $DE = 2$, what is the sum of the area of $\triangle ABE$ and the area of $\triangle BCD$?

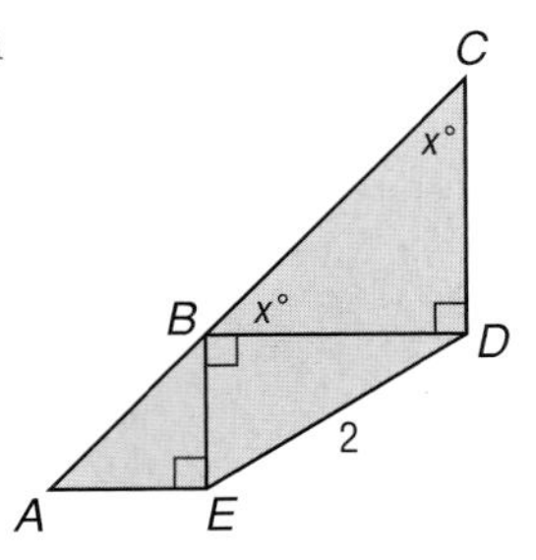

41. What is the net result if a discount of 5% is applied to a bill of \$340.60?

Ⓐ \$306.54　　Ⓑ \$323.57

Ⓒ \$335.60　　Ⓓ \$357.63

BINOMIAL DISTRIBUTION You can use a TI-83 Plus to investigate the graph of a binomial distribution.

Step 1 Enter the number of trials in L1. Start with 10 trials.

KEYSTROKES: [STAT] 1 [▲] [2nd] [LIST] [▶] 5 [X,T,θ,n] [,] [X,T,θ,n] [,] 0 [,] 10 [)] [ENTER]

Step 2 Calculate the probability of success for each trial in L2.

KEYSTROKES: [▶] [▲] [2nd] [DISTR] 0 10 [,] .5 [,] [2nd] [L1] [)] [ENTER]

Step 3 Graph the histogram.

KEYSTROKES: [2nd] [STATPLOT]

Use the arrow and [ENTER] keys to choose ON, the histogram, L1 as the Xlist, and L2 as the frequency. Use the window [0, 10] scl:1 by [0, 0.5] scl: 0.1.

42. Replace the 10 in the keystrokes for steps 1 and 2 to graph the binomial distribution for several values of n less than or equal to 47. You may have to adjust your viewing window to see all of the histogram. Make sure Xscl is 1.

43. What type of distribution does the binomial distribution start to resemble as n increases?

Maintain Your Skills

Mixed Review

For Exercises 44–46, use the following information.
A set of 400 test scores is normally distributed with a mean of 75 and a standard deviation of 8. *(Lesson 12-7)*

44. What percent of the test scores lie between 67 and 83?

45. How many of the test scores are greater than 91?

46. What is the probability that a randomly-selected score is less than 67?

47. A salesperson had sales of \$11,000, \$15,000, \$11,000, \$16,000, \$12,000, and \$12,000 in the last six months. Which measure of central tendency would he be likely to use to represent these data when he talks with his supervisor? Explain. *(Lesson 12-6)*

Graph each inequality. *(Lesson 2-7)*

48. $x \geq -3$　　**49.** $x + y \leq 4$　　**50.** $y > |5x|$

Getting Ready for the Next Lesson

PREREQUISITE SKILL **Evaluate $2\sqrt{\frac{p(1-p)}{n}}$ for the given values of p and n. Round to the nearest thousandth, if necessary.** *(For review of **radical expressions**, see Lesson 5-6.)*

51. $p = 0.5, n = 100$　　**52.** $p = 0.5, n = 400$

53. $p = 0.25, n = 500$　　**54.** $p = 0.75, n = 1000$

55. $p = 0.3, n = 500$　　**56.** $p = 0.6, n = 1000$

Algebra Activity

A Follow-Up of Lesson 12-8

Simulations

A **simulation** uses a probability experiment to mimic a real-life situation. You can use a simulation to solve the following problem about **expected value**.

A brand of cereal is offering one of six different prizes in every box. If the prizes are equally and randomly distributed within the cereal boxes, how many boxes, on average, would you have to buy in order to get a complete set of the six prizes?

Collect the Data

Work in pairs or small groups to complete steps 1 through 4.

Step 1 Use the six numbers on a die to represent the six different prizes.

Step 2 Roll the die and record which prize was in the first box of cereal. Use a tally sheet like the one shown.

Step 3 Continue to roll the die and record the prize number until you have a complete set of prizes. Stop as soon as you have a complete set. This is the end of one trial in your simulation. Record the number of boxes required for this trial.

Step 4 Repeat steps 1, 2, and 3 until your group has carried out 25 trials. Use a new tally sheet for each trial.

Simulation Tally Sheet	
Prize Number	**Boxes Purchased**
1	
2	
3	
4	
5	
6	
Total Needed	

Analyze the Data

1. Create two different statistical graphs of the data collected for 25 trials.
2. Determine the mean, median, maximum, minimum, and standard deviation of the total number of boxes needed in the 25 trials.
3. Combine the small-group results and determine the mean, median, maximum, minimum, and standard deviation of the number of boxes required for all the trials conducted by the class.

Make a Conjecture

4. If you carry out 25 additional trials, will your results be the same as in the first 25 trials? Explain.
5. Should the small-group results or the class results give a better idea of the average number of boxes required to get a complete set of superheroes? Explain.
6. If there were 8 superheroes instead of 6, would you need to buy more boxes of cereal or fewer boxes of cereal on average?
7. What if one of the 6 prizes was more common than the other 5? For instance, suppose that one prize, Amazing Amy, appears in 25% of all the boxes and the other 5 prizes are equally and randomly distributed among the remaining 75% of the boxes? Design and carry out a new simulation to predict the average number of boxes you would need to buy to get a complete set. Include some measures of central tendency and dispersion with your data.

12-9 Sampling and Error

What You'll Learn

- Determine whether a sample is unbiased.
- Find margins of sampling error.

Vocabulary

- unbiased sample
- margin of sampling error

How are opinion polls used in political campaigns?

About a month before the 2000 presidential election, Mason-Dixon Polling & Research surveyed the preferences of Florida voters. The results shown were published in the *Orlando Sentinel*.

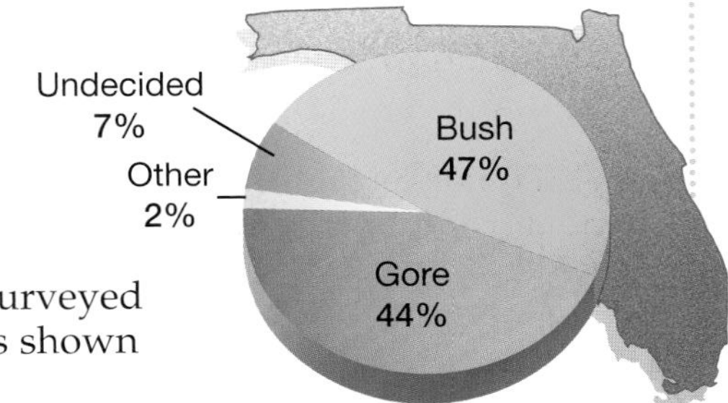

BIAS When polling organizations want to find how the public feels about an issue, they do not have the time or money to ask everyone. Instead, they obtain their results by polling a small portion of the population. To be sure that the results are representative of the population, they need to make sure that this portion is a random or **unbiased sample** of the population. A sample of size n is random when every possible sample of size n has an equal chance of being selected.

Example 1 Biased and Unbiased Samples

State whether each method would produce a random sample. Explain.

a. asking every tenth person coming out of a health club how many times a week they exercise to determine how often people in the city exercise

This would not result in a random sample because the people surveyed would probably exercise more often than the average person.

b. surveying people going into an Italian restaurant to find out people's favorite type of food

This would probably not result in a random sample because the people surveyed would probably be more likely than others to prefer Italian food.

Study Tip

Law of Large Numbers
The principle that as sample size increases, the closer the experimental probability is to the theoretical probability is called the *Law of Large Numbers.*

MARGIN OF ERROR As the size of a sample increases, it more accurately reflects the population. If you sampled only three people and two prefer Brand A, you could say, "Two out of three people chose Brand A over any other brand," but you may not be giving a true picture of how the total population would respond. The **margin of sampling error (ME)** gives a limit on the difference between how a sample responds and how the total population would respond.

Key Concept — Margin of Sampling Error

If the percent of people in a sample responding in a certain way is p and the size of the sample is n, then 95% of the time, the percent of the population responding in that same way will be between $p - ME$ and $p + ME$, where

$$ME = 2\sqrt{\frac{p(1-p)}{n}}.$$

That is, the probability is 0.95 that $p \pm ME$ will contain the true population results.

Example 2 Find a Margin of Error

In a survey of 1000 randomly selected adults, 37% answered "yes" to a particular question. What is the margin of error?

$ME = 2\sqrt{\frac{p(1-p)}{n}}$ Formula for margin of sampling error

$= 2\sqrt{\frac{0.37(1-0.37)}{1000}}$ $p = 37\%$ or 0.37, $n = 1000$

≈ 0.030535 Use a calculator.

The margin of error is about 3%. This means that there is a 95% chance that the percent of people in the whole population who would answer "yes" is between 37 − 3 or 34% and 37 + 3 or 40%.

Published survey results often include the margin of error for the data. You can use this information to determinine the sample size.

Example 3 Analyze a Margin of Error

HEALTH **In a recent Gallup Poll, 25% of the people surveyed said they had smoked cigarettes in the past week. The margin of error was 3%.**

a. What does the 3% indicate about the results?

The 3% means that the probability is 95% that the percent of people in the population who had smoked cigarettes in the past week was between 25 − 3 or 22% and 25 + 3 or 28%.

b. How many people were surveyed?

$ME = 2\sqrt{\frac{p(1-p)}{n}}$ Formula for margin of sampling error

$0.03 = 2\sqrt{\frac{0.25(1-0.25)}{n}}$ $ME = 0.03$, $p = 0.25$

$0.015 = \sqrt{\frac{0.25(0.75)}{n}}$ Divide each side by 2.

$0.000225 = \frac{0.25(0.75)}{n}$ Square each side.

$n = \frac{0.25(0.75)}{0.000225}$ Multiply by n and divide by 0.000225.

$n \approx 833.33$ Use a calculator.

About 833 people were surveyed.

More About. . .

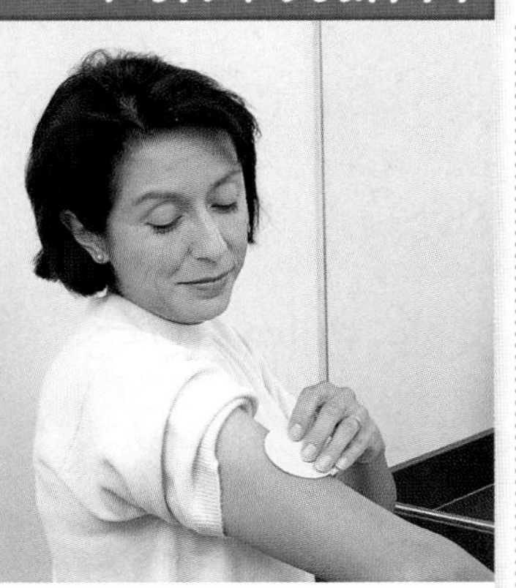

Health

The percent of smokers in the United States population declined from 38.7% in 1985 to 25.8% in 1999. New therapies, like the nicotine patch, are helping more people to quit.

Source: U.S. Department of Health and Human Services

Check for Understanding

Concept Check

1. **Describe** how sampling techniques can influence the results of a survey.
2. **OPEN ENDED** Give an example of a good sample and a bad sample. Explain your reasoning.
3. **Explain** what happens to the margin of sampling error when the size of the sample n increases. Why does this happen?

Guided Practice

Determine whether each situation would produce a random sample. Write *yes* or *no* and explain your answer.

4. the government sending a tax survey to everyone whose social security number ends in a particular digit

5. surveying students in the honors chemistry classes to determine the average time students in your school study each week

For Exercises 6–8, find the margin of sampling error to the nearest percent.

6. $p = 72\%$, $n = 100$

7. $p = 31\%$, $n = 500$

8. In a survey of 520 randomly-selected high school students, 68% of those surveyed stated that they were involved in extracurricular activities at their school.

Application

MEDIA **For Exercises 9 and 10, use the following information.**
According to a survey in *American Demographics*, 77% of Americans age 12 or older said they listen to the radio every day. Suppose the survey had a margin of error of 5%.

9. What does the 5% indicate about the results?

10. How many people were surveyed?

Practice and Apply

Homework Help

For Exercises	See Examples
11–14	1
15–26	2
27–28	3

Extra Practice
See page 856.

Determine whether each situation would produce a random sample. Write *yes* or *no* and explain your answer.

11. pointing with your pencil at a class list with your eyes closed as a way to find a sample of students in your class

12. putting the names of all seniors in a hat, then drawing names from the hat to select a sample of seniors

13. calling every twentieth person listed in the telephone book to determine which political candidate is favored

14. finding the heights of all the boys in a freshman physical education class to determine the average height of all the boys in your school

For Exercises 15–24, find the margin of sampling error to the nearest percent.

15. $p = 81\%$, $n = 100$

16. $p = 16\%$, $n = 400$

17. $p = 54\%$, $n = 500$

18. $p = 48\%$, $n = 1000$

19. $p = 33\%$, $n = 1000$

20. $p = 67\%$, $n = 1500$

21. A poll asked people to name the most serious problem facing the country. Forty-six percent of the 800 randomly selected people said crime.

22. Although skim milk has as much calcium as whole milk, only 33% of 2406 adults surveyed in *Shape* magazine said skim milk is a good calcium source.

23. Three hundred sixty-seven of 425 high school students said pizza was their favorite food in the school cafeteria.

24. Nine hundred thirty-four of 2150 subscribers to a particular newspaper said their favorite sport was football.

25. **ECONOMICS** In a poll conducted by ABC News, 83% of the 1020 people surveyed said they supported raising the minimum wage. What was the margin of error?

Physician

Physicians diagnose illnesses and prescribe and administer treatment.

Online Research For information about a career as a physician, visit: www.algebra2.com/careers

26. **PHYSICIANS** In a recent Harris Poll, 61% of the 1010 people surveyed said they considered being a physician to be a very prestigious occupation. What was the margin of error?

27. **SHOPPING** According to a Gallup Poll, 33% of shoppers planned to spend $1000 or more during a recent holiday season. The margin of error was 3%. How many people were surveyed?

28. **CRITICAL THINKING** One hundred people were asked a yes-or-no question in an opinion poll. How many said "yes" if the margin of error was 9.6%?

29. **WRITING IN MATH** Answer the question that was posed at the beginning of the lesson.

How are opinion polls used in political campaigns?

Include the following in your answer:

- a description of how a candidate could use statistics from opinion polls to determine where to make campaign stops,
- the margin of error for Bush if 807 people were surveyed, and
- an explanation of how to use the margin of error to determine the range of percent of Florida voters who favored Bush.

30. In rectangle $ABCD$, what is $x + y$ in terms of z?

(A) $90 + z$ (B) $190 - z$
(C) $180 + z$ (D) $270 - z$

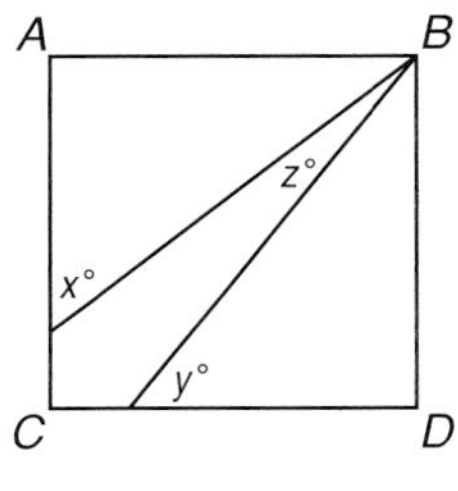

31. If $xy^{-2} + y^{-1} = y^{-2}$, then the value of x *cannot* equal which of the following?

(A) -1 (B) 0 (C) 1 (D) 2

Maintain Your Skills

Mixed Review

A student guesses at all 5 questions on a true-false quiz. Find each probability. *(Lesson 12-8)*

32. P(all 5 correct)
33. P(exactly 4 correct)
34. P(at least 3 correct)

A set of 250 data values is normally distributed with a mean of 50 and a standard deviation of 5.5. *(Lesson 12-7)*

35. What percent of the data lies between 39 and 61?

36. How many data values are less than 55.5?

37. What is the probability that a data value selected at random is greater than 39?

38. Given $x^3 - 3x^2 - 4x + 12$ and one of its factors $x + 2$, find the remaining factors of the polynomial. *(Lesson 7-4)*

'Minesweeper': Secret to Age-Old Puzzle?

It is time to complete your project. Use the information and data you have gathered about the history of mathematics to prepare a presentation or web page. Be sure to include transparencies and a sample mathematics problem or idea in the presentation.

www.algebra2.com/webquest

Algebra Activity

A Follow-Up of Lesson 12-9

Testing Hypotheses

A **hypothesis** is a statement to be tested. Testing a hypothesis to determine whether it is supported by the data involves five steps.

Step 1 State the hypothesis. The statement should include a *null hypothesis*, which is the hypothesis to be tested, and an *alternative hypothesis*.

Step 2 Design the experiment.

Step 3 Conduct the experiment and collect the data.

Step 4 Evaluate the data. Decide whether to reject the null hypothesis.

Step 5 Summarize the results.

Test the following hypothesis.

People react to sound and touch at the same rate.

You can measure reaction time by having someone drop a ruler and then having someone else catch it between their fingers. The distance the ruler falls will depend on their reaction time. Half of the class will investigate the time it takes to react when someone is told the ruler has dropped. The other half will measure the time it takes to react when the catcher is alerted by touch.

Step 1 The null hypothesis H_0 and alternative hypothesis H_1 are as follows.
These statements often use $=$, $\neq$, $<$, $>$, $\geq$, and $\leq$.

- H_0: reaction time to sound = reaction time to touch
- H_1: reaction time to sound $\neq$ reaction time to touch

Step 2 You will need to decide the height from which the ruler is dropped, the position of the person catching the ruler, the number of practice runs, and whether to use one try or the average of several tries.

Step 3 Conduct the experiment in each group and record the results.

Step 4 Organize the results so that they can be compared.

Step 5 Based on the results, do you think the hypothesis is true? Explain.

Analyze

State the null and alternative hypotheses for each conjecture.

1. A teacher feels that playing classical music during a math test will cause the test scores to change (either up or down). In the past, the average test score was 73.
2. An engineer thinks that the mean number of defects can be decreased by using robots on an assembly line. Currently, there are 18 defects for every 1000 items.
3. A researcher is concerned that a new medicine will cause pulse rates to rise dangerously. The mean pulse rate for the population is 82 beats per minute.
4. **MAKE A CONJECTURE** Design and conduct an experiment to test the following hypothesis. Interpret the data and present your results. *Pulse rates increase 20% after moderate exercise.*

Chapter 12

Study Guide and Review

Vocabulary and Concept Check

Choose the letter of the term that best matches each statement or phrase.

1. the ratio of the number of ways an event can succeed to the number of possible outcomes
2. an arrangement of objects in which order does not matter
3. two or more events in which the outcome of one event affects the outcome of another event
4. a sample in which every member of the population has an equal chance to be selected
5. an arrangement of objects in which order matters
6. two events in which the outcome can never be the same
7. the ratio of the number of ways an event can succeed to the number of ways it can fail

a. dependent events
b. combination
c. probability
d. permutation
e. mutually exclusive events
f. odds
g. unbiased sample

Lesson-by-Lesson Review

12-1 The Counting Principle

See pages 632–637.

Concept Summary

- Fundamental Counting Principle: If event M can occur in m ways and event N can occur in n ways, then event M followed by event N can occur in $m \cdot n$ ways.
- Independent Events: The outcome of one event does *not* affect the outcome of another.
- Dependent Events: The outcome of one event *does* affect the outcome of another.

Example

How many different license plates are possible with two letters followed by three digits?

There are 26 possibilities for each letter. There are 10 possibilities, the digits 0–9, for each number. Thus, the number of possible license plates is as follows.

$26 \cdot 26 \cdot 10 \cdot 10 \cdot 10 = 26^2 \cdot 10^3$ or 676,000

Exercises **Solve each problem.** *See Examples 2 and 3 on page 633.*

8. The letters a, c, e, g, i, and k are used to form 6-letter passwords for a movie theater security system. How many passwords can be formed if the letters can be used more than once in any given password?

9. How many 4-digit personal identification codes can be formed if each numeral can only be used once?

12-2 Permutations and Combinations

See pages 638–643.

Concept Summary

- In a permutation, the order of objects is important.
- In a combination, the order of objects is not important.

A basket contains 3 apples, 6 oranges, 7 pears, and 9 peaches. How many ways can 1 apple, 2 oranges, 6 pears, and 2 peaches be selected?

This involves the product of four combinations, one for each type of fruit.

$$C(3, 1) \cdot C(6, 2) \cdot C(7, 6) \cdot C(9, 2) = \frac{3!}{(3-1)!1!} \cdot \frac{6!}{(6-2)!2!} \cdot \frac{7!}{(7-6)!6!} \cdot \frac{9!}{(9-2)!2!}$$

$$= 3 \cdot 15 \cdot 7 \cdot 36 \text{ or } 11{,}340$$

There are 11,340 different ways to choose the fruit from the basket.

Exercises **Solve each problem.** *See Example 4 on page 640.*

10. A committee of 3 is selected from Jillian, Miles, Mark, and Nikia. How many committees contain 2 boys and 1 girl?

11. Five cards are drawn from a standard deck of cards. How many different hands consist of four queens and one king?

12. A box of pencils contains 4 red, 2 white, and 3 blue pencils. How many different ways can 2 red, 1 white, and 1 blue pencil be selected?

12-3 Probability

See pages 644–650.

Concept Summary

- $P(\text{success}) = \frac{s}{s+f}$; $P(\text{failure}) = \frac{f}{s+f}$
- odds of success = $s:f$; odds of failure = $f:s$

Example

A bag of golf tees contains 23 red, 19 blue, 16 yellow, 21 green, 11 orange, 19 white, and 17 black tees. What is the probability that if you choose a tee from the bag at random, you will choose a green tee?

There are 21 ways to choose a green tee and 23 + 19 + 16 + 11 + 19 + 17 or 105 ways not to choose a green tee. So, s is 21 and f is 105.

$$P(\text{green tee}) = \frac{s}{s+f}$$

$$= \frac{21}{21+105} \text{ or } \frac{1}{6}$$ The probability is 1 out of 6 or about 16.7%.

Exercises **Find the odds of an event occurring, given the probability of the event.**
See Example 3 on pages 645 and 646.

13. $\frac{1}{4}$ 14. $\frac{5}{8}$ 15. $\frac{7}{12}$ 16. $\frac{3}{7}$ 17. $\frac{2}{5}$

18. The table shows the distribution of the number of heads occurring when four coins are tossed. Find $P(H = 3)$.
See Example 4 on page 646.

H = Heads	0	1	2	3	4
Probability	$\frac{1}{16}$	$\frac{1}{4}$	$\frac{3}{8}$	$\frac{1}{4}$	$\frac{1}{16}$

12-4 Multiplying Probabilities

See pages 651–657.

Concept Summary

- Probability of two independent events: $P(A \text{ and } B) = P(A) \cdot P(B)$
- Probability of two dependent events: $P(A \text{ and } B) = P(A) \cdot P(B \text{ following } A)$

There are 3 dimes, 2 quarters, and 5 nickels in Langston's pocket. If he reaches in and selects three coins at random without replacing any of them, what is the probability that he will choose a dime d, then a quarter q, then a nickel n?

Because the outcomes of the first and second choices affect the later choices, these are dependent events.

$P(d, \text{ then } q, \text{ then } n) = \frac{3}{10} \cdot \frac{2}{9} \cdot \frac{5}{8}$ or $\frac{1}{24}$ The probability is $\frac{1}{24}$ or about 4.2%.

Exercises **Determine whether the events are *independent* or *dependent*. Then find the probability.** *See Examples 1–4 on pages 652 and 654.*

19. Two dice are rolled. What is the probability that each die shows a 4?
20. Two cards are drawn from a standard deck of cards without replacement. Find the probability of drawing a heart and a club, in that order.
21. Luz has 2 red, 2 white, and 3 blue marbles in a cup. If she draws two marbles at random and does not replace the first one, find the probability of a white marble and then a blue marble.

12-5 Adding Probabilities

See pages 658–663.

Concept Summary

- Probability of mutually exclusive events: $P(A \text{ or } B) = P(A) + P(B)$
- Probability of inclusive events: $P(A \text{ or } B) = P(A) + P(B) - P(A \text{ and } B)$

Example

Trish has four \$1 bills and six \$5 bills. She takes three bills from her wallet at random. What is the probability that Trish will select at least two \$1 bills?

$$P(\text{at least two \$1 bills}) = P(\text{two \$1, one \$5}) + P(\text{three \$1, no \$5})$$

$$= \frac{C(4, 2) \cdot C(6, 1)}{C(10, 3)} + \frac{C(4, 3) \cdot C(6, 0)}{C(10, 3)}$$

$$= \frac{\frac{4! \cdot 6!}{(4-2)!2!(6-1)!1!}}{C(10, 3)} + \frac{\frac{4! \cdot 6!}{(4-3)!3!(6-0)!0!}}{C(10, 3)}$$

$$= \frac{36}{120} + \frac{4}{120} \text{ or } \frac{1}{3}$$

The probability is $\frac{1}{3}$ or about 0.333.

Exercises **Determine whether the events are *mutually exclusive* or *inclusive*. Then find the probability.** *See Examples 1–3 on pages 659 and 660.*

22. There are 5 English, 2 math, and 3 chemistry books on a shelf. If a book is randomly selected, what is the probability of selecting a math book or a chemistry book?

23. A die is rolled. What is the probability of rolling a 6 or a number less than 4?

24. A die is rolled. What is the probability of rolling a 6 or a number greater than 4?

25. A card is drawn from a standard deck of cards. What is the probability of drawing a king or a red card?

12-6 Statistical Measures

See pages 664–670.

Concept Summary

- To represent a set of data, use the mean if the data are spread out and you want an average of the values, the median when the data contain outliers, or the mode when the data are tightly clustered around one or two values.
- Standard deviation for n values:

$$\sigma = \sqrt{\frac{(x_1 - \overline{x})^2 + (x_2 - \overline{x})^2 + \cdots + (x_n - \overline{x})^2}{n}}, \overline{x} \text{ is the mean}$$

Example **Find the variance and standard deviation for {100, 156, 158, 159, 162, 165, 170, 190}.**

Step 1 Find the mean.

$$\overline{x} = \frac{100 + 156 + 158 + 159 + 162 + 165 + 170 + 190}{8}$$ Add the data and divide by the number of items.

$$= \frac{1260}{8}$$

$$= 157.5$$

Step 2 Find the variance.

$$\sigma^2 = \frac{(x_1 - \overline{x})^2 + (x_2 - \overline{x})^2 + \cdots + (x_n - \overline{x})^2}{n}$$

$$= \frac{(100 - 157.5)^2 + (156 - 157.5)^2 + \cdots + (170 - 157.5)^2 + (190 - 157.5)^2}{8}$$

$$= \frac{4600}{8}$$ Simplify.

$$= 575$$ Use a calculator.

Step 3 Find the standard deviation.

$\sigma^2 = 575$ Take the square root of each side.

$\sigma \approx 23.98$ Use a calculator.

Exercises **Find the variance and standard deviation of each set of data to the nearest tenth.** *See Examples 1 and 2 on pages 664 and 665.*

26. {56, 56, 57, 58, 58, 58, 59, 61}

27. {302, 310, 331, 298, 348, 305, 314, 284, 321, 337}

28. {3.4, 4.2, 8.6, 5.1, 3.6, 2.8, 7.1, 4.4, 5.2, 5.6}

12-7 The Normal Distribution

See pages 671–675.

Concept Summary

Normal distributions have these properties.

- The graph is maximized and the data are symmetric at the mean.
- The mean, median, and mode are about equal.
- About 68% of the values are within one standard deviation of the mean.
- About 95% of the values are within two standard deviations of the mean.
- About 99% of the values are within three standard deviations of the mean.

Example **Mr. Byrum gave an exam to his 30 Algebra 2 students at the end of the first semester. The scores were normally distributed with a mean score of 78 and a standard deviation of 6.**

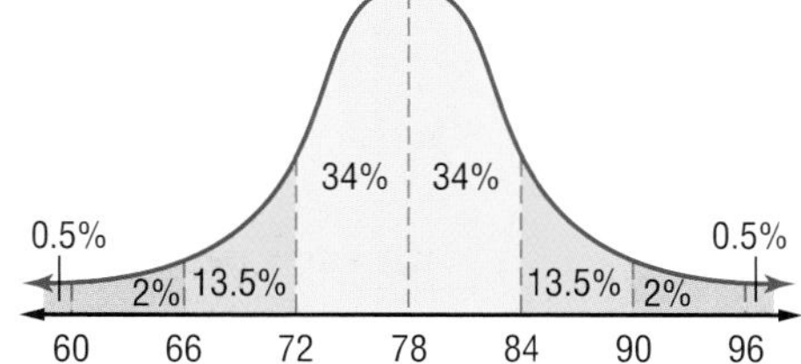

a. What percent of the class would you expect to have scored between 72 and 84?

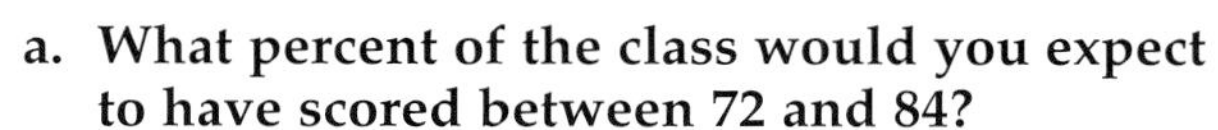

Since 72 and 84 are 1 standard deviation to the left and right of the mean, respectively, 34% + 34% or 68% of the students scored within this range.

b. What percent of the class would you expect to have scored between 90 and 96?

90 to 96 on the test includes 2% of the students.

c. Approximately how many students scored between 84 and 90?

84 to 90 on the test includes 13.5% of the students. $0.135 \times 30 = 4$ students

d. Approximately how many students scored between 72 and 84?

34% + 34% or 68% of the students scored between 72 and 84.

$0.68 \times 30 = 20$ students

Exercises **For Exercises 29–32, use the following information.**

The utility bills in a city of 5000 households are normally distributed with a mean of \$180 and a standard deviation of \$16. *See Example 2 on pages 672 and 673.*

29. About how many utility bills were between \$164 and \$196?

30. About how many bills were more than \$212?

31. About how many bills were less than \$164?

32. What is the probability that a household selected at random will have a utility bill between \$164 and \$180?

12-8 Binomial Experiments

See pages 676–680.

Concept Summary

A binomial experiment exists if and only if all of these conditions occur.

- There are exactly two possible outcomes for each trial.
- There is a fixed number of trials.
- The trials are independent.
- The possibilities for each trial are the same.

• Extra Practice, see pages 854–856.
• Mixed Problem Solving, see page 837.

Example **To practice for a jigsaw puzzle competition, Laura and Julian completed four jigsaw puzzles. The probability that Laura places the last piece is $\frac{3}{5}$, and the probability that Julian places the last piece is $\frac{2}{5}$. What is the probability that Laura will place the last piece of at least two puzzles?**

$P = L^4 + 4L^3J + 6L^2J^2$ — P(last piece in 4) + P(last piece in 3) + P(last piece in 2)

$= \left(\frac{3}{5}\right)^4 + 4\left(\frac{3}{5}\right)^3\left(\frac{2}{5}\right) + 6\left(\frac{3}{5}\right)^2\left(\frac{2}{5}\right)^2$ — $L = \frac{3}{5}, J = \frac{2}{5}$

$= \frac{81}{625} + \frac{216}{625} + \frac{216}{625}$ or 0.8208 — The probability is 82.08%.

Exercises *See Example 2 on pages 677 and 678.*

33. Find the probability of getting 7 heads in 8 tosses of a coin.

34. Find the probability that a family with seven children has exactly five boys.

Find each probability if a die is rolled twelve times.

35. P(twelve 3s) **36.** P(exactly one 3) **37.** P(six 3s)

12-9 Sampling and Error

See pages 682–685.

Concept Summary

- Margin of sampling error: $ME = 2\sqrt{\frac{p(1-p)}{n}}$ if the percent of people in a sample responding in a certain way is p and the size of the sample is n

Example **In a survey taken at a local high school, 75% of the student body stated that they thought school lunches should be free. This survey had a margin of error of 2%. How many people were surveyed?**

$ME = 2\sqrt{\frac{p(1-p)}{n}}$ — Formula for margin of sampling error

$0.02 = 2\sqrt{\frac{0.75(1-0.75)}{n}}$ — $ME = 0.02, p = 0.75$

$0.01 = \sqrt{\frac{0.75(1-0.75)}{n}}$ — Divide each side by 2.

$0.0001 = \frac{0.75(0.25)}{n}$ — Square each side of the equation.

$n = \frac{0.75(0.25)}{0.0001}$ — Multiply each side by n and divide each side by 0.0001.

$n = 1875$ — There were about 1875 people in the survey.

Exercises

38. In a poll asking people to name their most valued freedom, 51% of the randomly selected people said it was the freedom of speech. Find the margin of sampling error if 625 people were randomly selected. *See Example 2 on page 683.*

39. According to a recent survey of mothers with children who play sports, 63% of them would prefer that their children not play football. Suppose the margin of error is 4.5%. How many mothers were surveyed? *See Example 3 on page 683.*

Chapter 12

Practice Test

Vocabulary and Concepts

Match the following terms and descriptions.

1. data are symmetric about the mean
2. variance and standard deviation
3. mode, median, mean

- **a.** measures of central tendency
- **b.** measures of variation
- **c.** normal distribution

Skills and Applications

Evaluate each expression.

4. $P(7, 3)$
5. $C(7, 3)$
6. $P(13, 5)$

Solve each problem.

7. How many ways can 9 bowling balls be arranged on the upper rack of a bowling ball rack?
8. How many different outfits can be made if you choose 1 each from 11 skirts, 9 blouses, 3 belts, and 7 pairs of shoes?
9. How many ways can the letters of the word *probability* be arranged?
10. How many different soccer teams consisting of 11 players can be formed from 18 players?
11. In a row of 10 parking spaces in a parking lot, how many ways can 4 cars park?
12. Eleven points are equally spaced on a circle. How many ways can 5 of these points be chosen as the vertices of a pentagon?
13. A number is drawn at random from a hat that contains all the numbers from 1 to 100. What is the probability that the number is less than sixteen?
14. Two cards are drawn in succession from a standard deck of cards without replacement. What is the probability that both cards are greater than 2 and less than 9?
15. A shipment of ten television sets contains 3 defective sets. How many ways can a hospital purchase 4 of these sets and receive at least 2 of the defective sets?
16. While shooting arrows, William Tell can hit an apple 9 out of 10 times. What is the probability that he will hit it exactly 4 out of 7 times?
17. Ten people are going on a camping trip in 3 cars that hold 5, 2, and 4 passengers, respectively. How many ways is it possible to transport the 10 people to their campsite?
18. From a box containing 5 white golf balls and 3 red golf balls, 3 golf balls are drawn in succession, each being replaced in the box before the next draw is made. What is the probability that all 3 golf balls are the same color?

For Exercises 19–21, use the following information.
In a ten-question multiple-choice test with four choices for each question, a student who was not prepared guesses on each item. Find each probability.

19. six questions correct
20. at least eight questions correct
21. fewer than eight questions correct

Standards Practice

22. **STANDARDIZED TEST PRACTICE** Lila throws a die and writes down the number showing. If she throws the number cube again, what is the probability that the second throw will have the same number showing as the first throw?

Ⓐ $\frac{1}{2}$ Ⓑ $\frac{1}{3}$ Ⓒ $\frac{1}{4}$ Ⓓ $\frac{1}{6}$

www.algebra2.com/chapter_test/ca

Chapter 12 Standardized Test Practice

Standards Practice

Part 1 Multiple Choice

Record your answers on the answer sheet provided by your teacher or on a sheet of paper.

1. In a jar of red and green gumdrops, the ratio of red gumdrops to green gumdrops is 7 to 3. If the jar contains a total of 150 gumdrops, how many gumdrops are green?

(A) 21 (B) 30

(C) 45 (D) 105

2. ⟨x⟩ $= \frac{1}{2}x$ if x is composite and ⟨x⟩ $= 2x$ if x is prime. What is the value of ⟨16⟩ + ⟨11⟩?

(A) 10 (B) 30

(C) 54 (D) 60

3. In rhombus $ABCD$, which of the following are true?

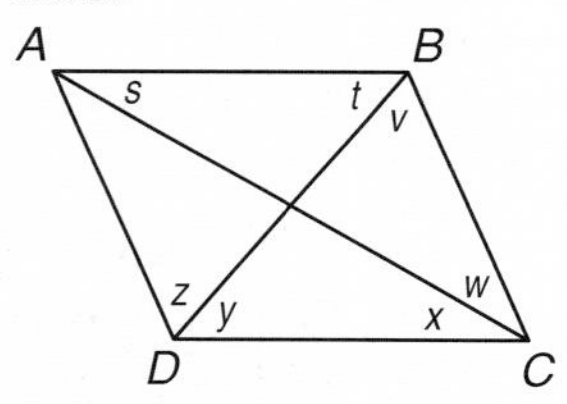

I. $\angle s$ and $\angle x$ are congruent.

II. $\angle t$ and $\angle v$ are congruent.

III. $\angle z$ and $\angle t$ are congruent.

(A) I only

(B) II only

(C) I and II only

(D) I, II, and III

4. What is the area of an isosceles right triangle with hypotenuse $3\sqrt{2}$ units?

(A) $1.5\sqrt{2}$ units^2

(B) 4.5 units^2

(C) 9 units^2

(D) $6 + 3\sqrt{2}$ units^2

5. What is the solution set for $t(t + 7) = 18$?

(A) $\{-2, 9\}$

(B) $\{-3, 6\}$

(C) $\{0, 18\}$

(D) $\{-9, 2\}$

6. The equation $3x - 8 = 5x^2 - y$ represents which of the following conic sections?

(A) hyperbola

(B) parabola

(C) circle

(D) ellipse

7. If the equations $x^2 + y^2 = 16$ and $y = x^2 + 4$ are graphed on the same coordinate plane, how many points of intersection exist?

(A) none

(B) one

(C) two

(D) three

8. A number is chosen at random from the set $\{1, 2, 3, \ldots 20\}$. What is the probability that the number is odd and divisible by 3?

(A) $\frac{3}{20}$ (B) $\frac{3}{10}$

(C) $\frac{7}{20}$ (D) $\frac{13}{20}$

9. What is the least positive integer that is divisible by 3, 4, 5, and 6?

(A) 60 (B) 180

(C) 240 (D) 360

10. If $4y - 5x + 6xy - 50 = 0$ and $x + 7 = 13$, then what is $y + 5$?

(A) 2 (B) 6

(C) 7 (D) 11

Preparing for Standardized Tests
For test-taking strategies and more practice, see pages 877–892.

Part 2 Short Response/Grid In

Record your answers on the answer sheet provided by your teacher or on a sheet of paper.

11. In a high school, 250 students take math and 50 students take art. If there are 280 students enrolled in the school and they all take at least one of these courses, how many students take both math and art?

12. If $20 < y < 30$ and x and y are both integers, what is the greatest possible value for x?

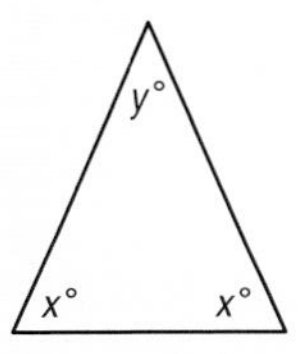

13. Four numbers are selected at random. Their average (arithmetic mean) is 45. The fourth number selected is 34. What is the sum of the other three numbers?

14. If one half of an even positive integer and three fourths of the next greater even integer have a sum of 24, what is the mean of the two integers?

15. Shane has six tiles, each of which has one of the letters A, B, C, D, E, or F on it. If one of the letters must be A and the last letter must be F, how many different arrangements of three letters (such as ADF) can Shane create with these titles?

Test-Taking Tip (A) (B) (C) (D)

Question 10
When answering questions, read carefully and make sure that you know exactly what the question is asking you to find. For example, if you only find the value of y in Question 10, you have not solved the problem. You need to find the value of $y + 5$.

16. The lunch special at Wally's Hot Dog Stand offers a sandwich, snack, and a drink for $3.99. How many different lunch combinations can be ordered?

Wally's lunch specials
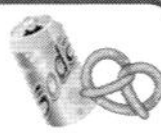

Sandwiches: hotdog, a bratwurst, or Italian sausage
Snacks: potato chips, pretzels, or corn chips
Drinks: diet soda, regular cola, lemon-lime bottled soda, or bottled water

17. A gumball machine contains 84 gumballs. Of the gumballs, 19 are yellow, 32 are red, and 33 are green. Each gumball sold is selected at random. If a yellow gumball is sold, what is the probability that the next gumball is also yellow?

18. Esteban is the place kicker on his high school football team. Based on past experience, the probability that Esteban will succeed in kicking an extra point is $\frac{9}{10}$. What is the probability that he will succeed on exactly three of the four extra point attempts in a game?

Part 3 Extended Response

Record your answers on a sheet of paper. Show your work.

For Exercises 19-21, use the following information.

Twenty-five students took an English examination and received the following scores.

76, 62, 78, 69, 88, 82, 79, 70, 81, 87, 90, 71, 75, 78, 88, 83, 91, 93, 95, 78, 80, 82, 86, 89, 79

19. Find the mean, median, mode, and standard deviation of the data. If necessary, round to the nearest hundredth.

20. How many students scored within one standard deviation of the mean?

21. Do the results of the examination approximate a normal distribution? Justify your answer.

UNIT 5

Trigonometry

Trigonometry is used in navigation, physics, and construction, among other fields. In this unit, you will learn about trigonometric functions, graphs, and identities.

WebQuest Internet Project

Trig Class Angles for Lessons in Lit

Source: *USA TODAY,* November 21, 2000

"The groans from the trigonometry students immediately told teacher Michael Buchanan what the class thought of his idea to read Homer Hickam's *October Sky*. In the story, in order to accomplish what they would like, the kids had to teach themselves trig, calculus, and physics." In this project, you will research applications of trigonometry as it applies to a possible career for you.

Log on to www.algebra2.com/webquest. Begin your WebQuest by reading the Task.

Then continue working on your WebQuest as you study Unit 5.

Lesson	13-1	14-2
Page	708	775

Chapter 13

Trigonometric Functions

What You'll Learn

- **Lessons 13-1, 13-2, 13-3, 13-6, and 13-7** Find values of trigonometric functions.
- **Lessons 13-1, 13-4, and 13-5** Solve problems by using right triangle trigonometry.
- **Lessons 13-4 and 13-5** Solve triangles by using the Law of Sines and Law of Cosines.

Key Vocabulary

- solve a right triangle (p. 704)
- radian (p. 710)
- Law of Sines (p. 726)
- Law of Cosines (p. 733)
- circular function (p. 740)

Why It's Important

Trigonometry is the study of the relationships among the angles and sides of right triangles. One of the many real-world applications of trigonometric functions involves solving problems using indirect measurement. For example, surveyors use a trigonometric function to find the heights of buildings. *You will learn how architects who design fountains use a trigonometric function to aim the water jets in Lesson 13-7.*

Getting Started

Prerequisite Skills To be successful in this chapter, you'll need to master these skills and be able to apply them in problem-solving situations. Review these skills before beginning Chapter 13.

For Lessons 13-1 and 13-3 — Pythagorean Theorem

Find the value of x to the nearest tenth. *(For review, see pages 820 and 821.)*

1.

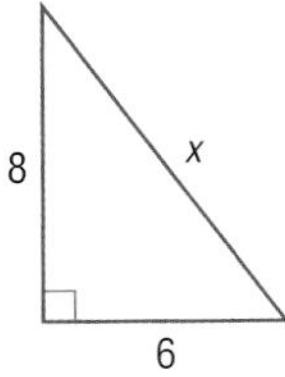

2.

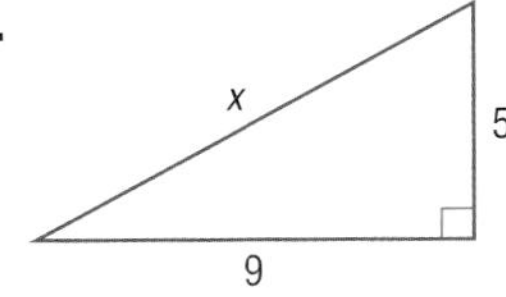

3.

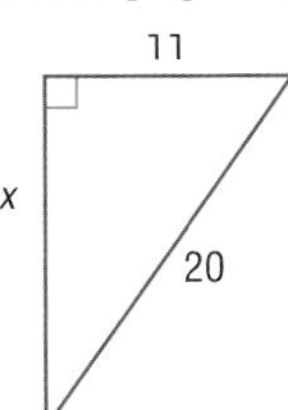

4.

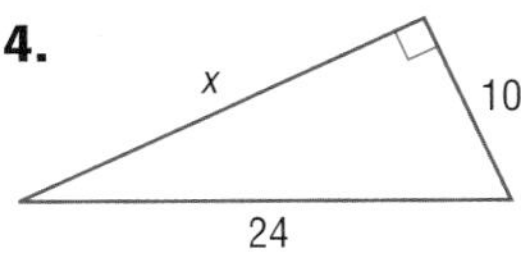

For Lesson 13-1 — 45°-45°-90° and 30°-60°-90° Triangles

Find each missing measure. Write all radicals in simplest form.

5.

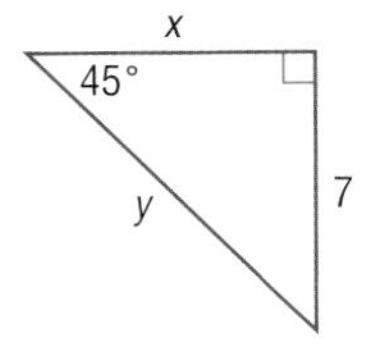

6.

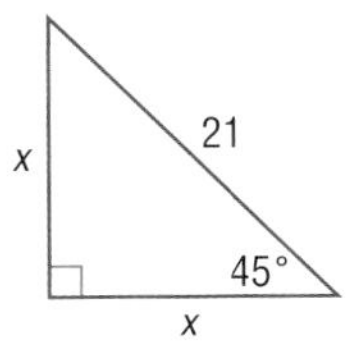

7.

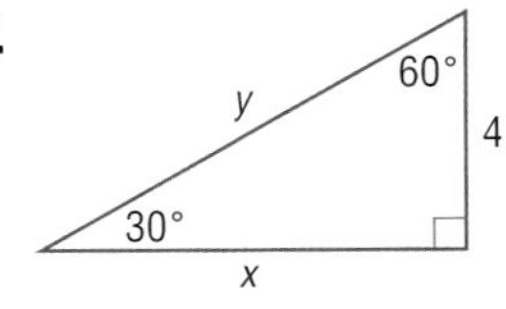

8.

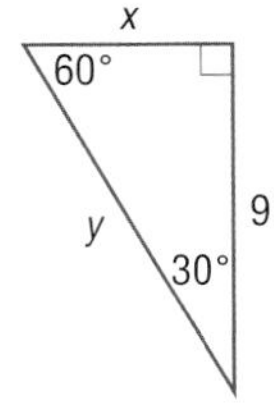

For Lesson 13-7 — Inverse Functions

Find the inverse of each function. Then graph the function and its inverse.
(For review, see Lesson 7-8.)

9. $f(x) = x + 3$

10. $f(x) = \frac{x - 2}{5}$

11. $f(x) = x^2 - 4$

12. $f(x) = -7x - 9$

FOLDABLES™ Study Organizer

Trigonometric Functions Make this Foldable to help you organize your notes. Begin with one sheet of construction paper and two pieces of grid paper.

Step 1 Fold and Cut

Stack and fold on the diagonal. Cut to form a triangular stack.

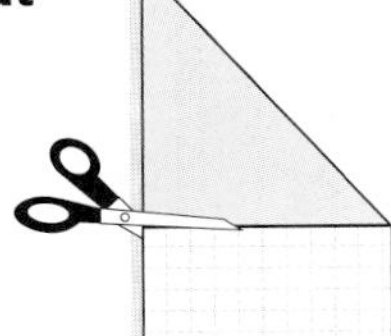

Step 2 Staple and Label

Staple the edge to form a booklet.

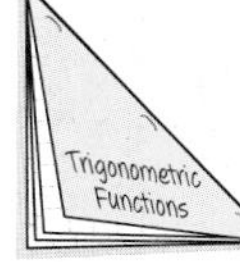

Reading and Writing As you read and study the chapter, you can write notes, draw diagrams, and record formulas on the grid paper.

Spreadsheet Investigation

A Preview of Lesson 13-1

Special Right Triangles

You can use a computer spreadsheet program to investigate the relationships among the ratios of the side measures of special right triangles.

Example

The legs of a 45°-45°-90° triangle, *a* and *b*, are equal in measure. Use a spreadsheet to investigate the dimensions of 45°-45°-90° triangles. What patterns do you observe in the ratios of the side measures of these triangles?

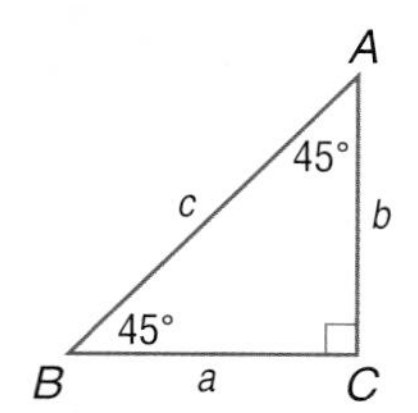

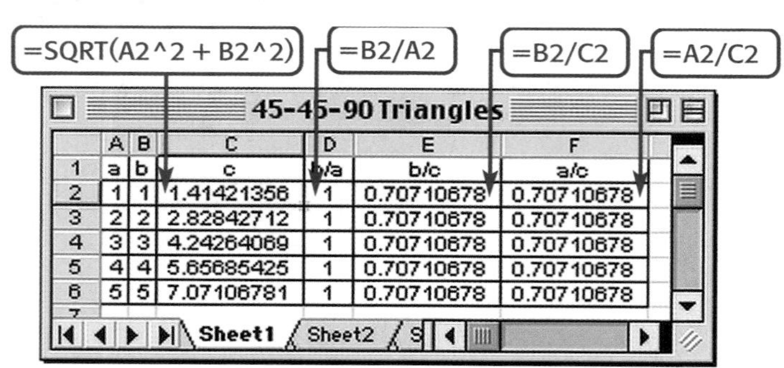

45-45-90 Triangles

	A	B	C	D	E	F
1	a	b	c	b/a	b/c	a/c
2	1	1	1.41421356	1	0.70710678	0.70710678
3	2	2	2.82842712	1	0.70710678	0.70710678
4	3	3	4.24264069	1	0.70710678	0.70710678
5	4	4	5.65685425	1	0.70710678	0.70710678
6	5	5	7.07106781	1	0.70710678	0.70710678
7						

Sheet1 / Sheet2 / S

The spreadsheet shows the formula that will calculate the length of side c. The formula uses the Pythagorean Theorem in the form $c = \sqrt{a^2 + b^2}$. Since 45°-45°-90° triangles share the same angle measures, the triangles listed in the spreadsheet are all similar triangles. Notice that all of the ratios of side b to side a are 1. All of the ratios of side b to side c and of side a to side c are approximately 0.71.

Exercises

For Exercises 1–3, use the spreadsheet below for 30°-60°-90° triangles.

If the measure of one leg of a right triangle and the measure of the hypotenuse are in a ratio of 1 to 2, then the acute angles of the triangle measure 30° and 60°.

B, 60°, c, a, 30°, A, b, C

30-60-90 Triangles

	A	B	C	D	E	F
1	a	b	c	b/a	b/c	a/c
2	1		2			
3	2		4			
4	3		6			
5	4		8			
6	5		10			
7						

Sheet1 / Sheet2 / Sheet3

1. Copy and complete the spreadsheet above.
2. Describe the relationship among the 30°-60°-90° triangles whose dimensions are given.
3. What patterns do you observe in the ratios of the side measures of these triangles?

13-1 Right Triangle Trigonometry

What You'll Learn

- Find values of trigonometric functions for acute angles.
- Solve problems involving right triangles.

Vocabulary

- trigonometry
- trigonometric functions
- sine
- cosine
- tangent
- cosecant
- secant
- cotangent
- solve a right triangle
- angle of elevation
- angle of depression

How is trigonometry used in building construction?

The Americans with Disabilities Act (ADA) provides regulations designed to make public buildings accessible to all. Under this act, the slope of an entrance ramp designed for those with mobility disabilities must not exceed a ratio of 1 to 12. This means that for every 12 units of horizontal run, the ramp can rise or fall no more than 1 unit.

When viewed from the side, a ramp forms a right triangle. The slope of the ramp can be described by the *tangent* of the angle the ramp makes with the ground. In this example, the tangent of angle A is $\frac{1}{12}$.

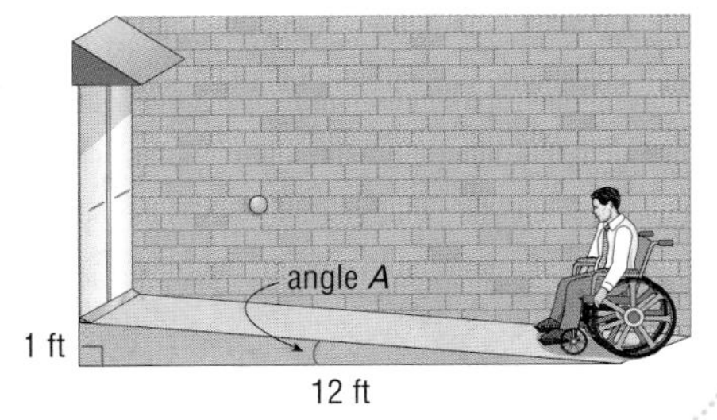

Study Tip

Reading Math
The word *trigonometry* is derived from two Greek words—*trigon* meaning triangle and *metra* meaning measurement.

TRIGONOMETRIC VALUES The tangent of an angle is one of the ratios used in trigonometry. **Trigonometry** is the study of the relationships among the angles and sides of a right triangle.

Consider right triangle ABC in which the measure of acute angle A is identified by the Greek letter *theta*, θ. The sides of the triangle are the *hypotenuse*, the *leg opposite* θ, and the *leg adjacent to* θ.

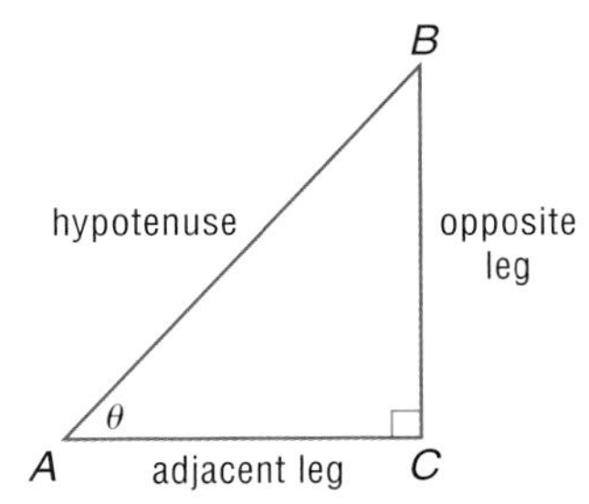

Using these sides, you can define six **trigonometric functions**: **sine**, **cosine**, **tangent**, **cosecant**, **secant**, and **cotangent**. These functions are abbreviated sin, cos, tan, sec, csc, and cot, respectively.

Key Concept — Trigonometric Functions

If θ is the measure of an acute angle of a right triangle, *opp* is the measure of the leg opposite θ, *adj* is the measure of the leg adjacent to θ, and *hyp* is the measure of the hypotenuse, then the following are true.

$$\sin\theta = \frac{\text{opp}}{\text{hyp}} \qquad \cos\theta = \frac{\text{adj}}{\text{hyp}} \qquad \tan\theta = \frac{\text{opp}}{\text{adj}}$$

$$\csc\theta = \frac{\text{hyp}}{\text{opp}} \qquad \sec\theta = \frac{\text{hyp}}{\text{adj}} \qquad \cot\theta = \frac{\text{adj}}{\text{opp}}$$

Notice that the sine, cosine, and tangent functions are reciprocals of the cosecant, secant, and cotangent functions, respectively. Thus, the following are also true.

$$\csc\theta = \frac{1}{\sin\theta} \qquad \sec\theta = \frac{1}{\cos\theta} \qquad \cot\theta = \frac{1}{\tan\theta}$$

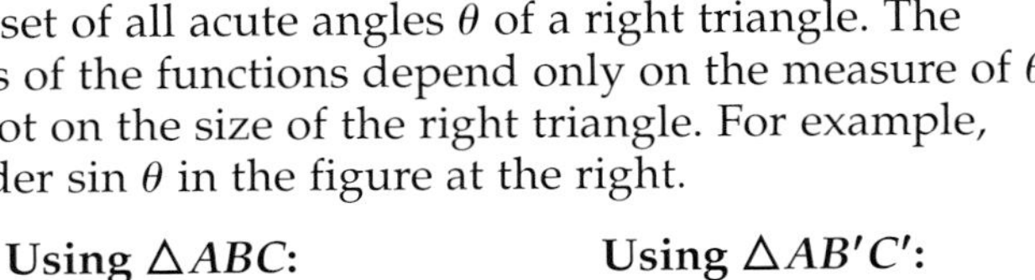

The domain of each of these trigonometric functions is the set of all acute angles θ of a right triangle. The values of the functions depend only on the measure of θ and not on the size of the right triangle. For example, consider $\sin \theta$ in the figure at the right.

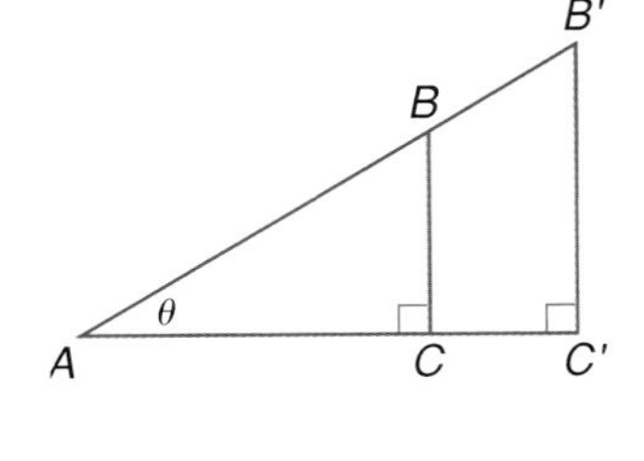

Using $\triangle ABC$: $\sin \theta = \frac{BC}{AB}$

Using $\triangle AB'C'$: $\sin \theta = \frac{B'C'}{AB'}$

The right triangles are similar because they share angle θ. Since they are similar, the ratios of corresponding sides are equal. That is, $\frac{BC}{AB} = \frac{B'C'}{AB'}$. Therefore, you will find the same value for $\sin \theta$ regardless of which triangle you use.

Study Tip

Memorize Trigonometric Ratios

SOH-CAH-TOA is a mnemonic device for remembering the first letter of each word in the ratios for sine, cosine, and tangent.

$\sin \theta = \frac{\text{opp}}{\text{hyp}}$

$\cos \theta = \frac{\text{adj}}{\text{hyp}}$

$\tan \theta = \frac{\text{opp}}{\text{adj}}$

Example 1 Find Trigonometric Values

Find the values of the six trigonometric functions for angle θ.

For this triangle, the leg opposite θ is $\overline{AB}$, and the leg adjacent to θ is $\overline{CB}$. Recall that the hypotenuse is always the longest side of a right triangle, in this case $\overline{AC}$.

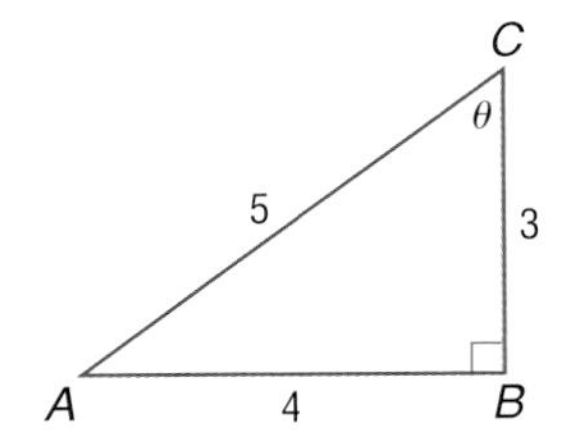

Use opp = 4, adj = 3, and hyp = 5 to write each trigonometric ratio.

$\sin \theta = \frac{\text{opp}}{\text{hyp}} = \frac{4}{5}$ $\quad \cos \theta = \frac{\text{adj}}{\text{hyp}} = \frac{3}{5}$ $\quad \tan \theta = \frac{\text{opp}}{\text{adj}} = \frac{4}{3}$

$\csc \theta = \frac{\text{hyp}}{\text{opp}} = \frac{5}{4}$ $\quad \sec \theta = \frac{\text{hyp}}{\text{adj}} = \frac{5}{3}$ $\quad \cot \theta = \frac{\text{adj}}{\text{opp}} = \frac{3}{4}$

Throughout Unit 5, a capital letter will be used to denote both a vertex of a triangle and the measure of the angle at that vertex. The same letter in lowercase will be used to denote the side opposite that angle and its measure.

Example 2 Use One Trigonometric Ratio to Find Another

Multiple-Choice Test Item

If $\cos A = \frac{2}{5}$, find the value of $\tan A$.

(A) $\frac{5}{2}$ (B) $\frac{2\sqrt{21}}{21}$ (C) $\frac{\sqrt{21}}{2}$ (D) $\sqrt{21}$

Test-Taking Tip

Whenever necessary or helpful, draw a diagram of the situation.

Read the Test Item

Begin by drawing a right triangle and labeling one acute angle A. Since $\cos \theta = \frac{\text{adj}}{\text{hyp}}$ and $\cos A = \frac{2}{5}$ in this case, label the adjacent leg 2 and the hypotenuse 5.

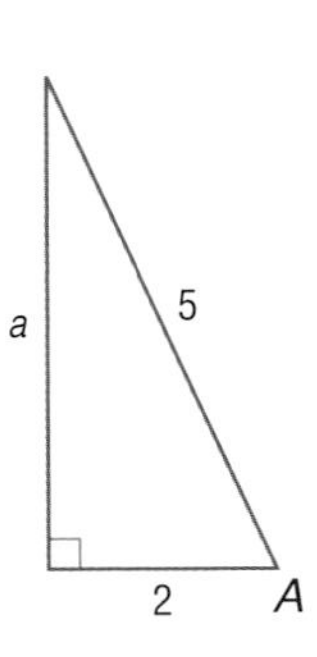

Solve the Test Item

Use the Pythagorean Theorem to find a.

$a^2 + b^2 = c^2$	Pythagorean Theorem
$a^2 + 2^2 = 5^2$	Replace b with 2 and c with 5.
$a^2 + 4 = 25$	Simplify.
$a^2 = 21$	Subtract 4 from each side.
$a = \sqrt{21}$	Take the square root of each side.

Now find tan A.

$\tan A = \frac{\text{opp}}{\text{adj}}$ Tangent ratio

$= \frac{\sqrt{21}}{2}$ Replace *opp* with $\sqrt{21}$ and *adj* with 2.

The answer is C.

Angles that measure 30°, 45°, and 60° occur frequently in trigonometry. The table below gives the values of the six trigonometric functions for these angles. To remember these values, use the properties of 30°-60°-90° and 45°-45°-90° triangles.

Key Concept — Trigonometric Values for Special Angles

30°-60°-90° Triangle

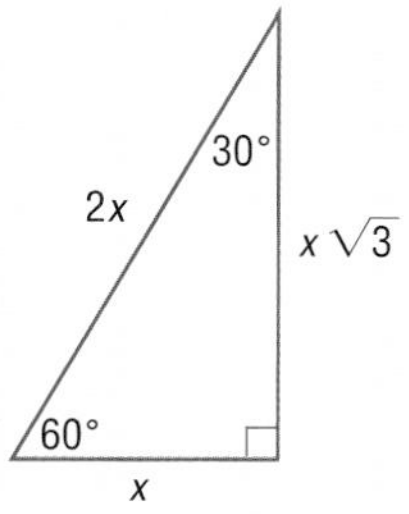

45°-45°-90° Triangle

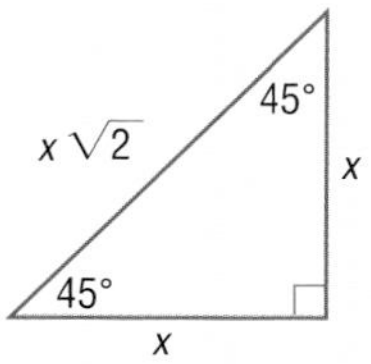

θ	$\sin\theta$	$\cos\theta$	$\tan\theta$	$\csc\theta$	$\sec\theta$	$\cot\theta$
30°	$\frac{1}{2}$	$\frac{\sqrt{3}}{2}$	$\frac{\sqrt{3}}{3}$	2	$\frac{2\sqrt{3}}{3}$	$\sqrt{3}$
45°	$\frac{\sqrt{2}}{2}$	$\frac{\sqrt{2}}{2}$	1	$\sqrt{2}$	$\sqrt{2}$	1
60°	$\frac{\sqrt{3}}{2}$	$\frac{1}{2}$	$\sqrt{3}$	$\frac{2\sqrt{3}}{3}$	2	$\frac{\sqrt{3}}{3}$

You will verify some of these values in Exercises 27 and 28.

RIGHT TRIANGLE PROBLEMS You can use trigonometric functions to solve problems involving right triangles.

Example 3 Find a Missing Side Length of a Right Triangle

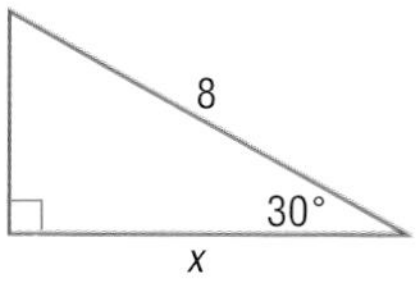

Write an equation involving sin, cos, or tan that can be used to find the value of x. Then solve the equation. Round to the nearest tenth.

The measure of the hypotenuse is 8. The side with the missing length is *adjacent* to the angle measuring 30°. The trigonometric function relating the adjacent side of a right triangle and the hypotenuse is the cosine function.

$\cos\theta = \frac{\text{adj}}{\text{hyp}}$ cosine ratio

$\cos 30° = \frac{x}{8}$ Replace θ with 30°, *adj* with x, and *hyp* with 8.

$\frac{\sqrt{3}}{2} = \frac{x}{8}$ $\cos 30° = \frac{\sqrt{3}}{2}$.

$4\sqrt{3} = x$ Multiply each side by 8.

The value of x is $4\sqrt{3}$ or about 6.9.

Study Tip

Common Misconception

The $\cos^{-1} x$ on a graphing calculator does *not* find $\frac{1}{\cos x}$. To find sec x or $\frac{1}{\cos x}$, find cos x and then use the x^{-1} key.

A calculator can be used to find the value of trigonometric functions for *any* angle, not just the special angles mentioned. Use SIN, COS, and TAN for sine, cosine, and tangent. Use these keys and the reciprocal key, x^{-1}, for cosecant, secant, and cotangent. Be sure your calculator is in degree mode.

Here are some calculator examples.

cos 46° **KEYSTROKES:** [COS] 46 [ENTER] .6946583705

cot 20° **KEYSTROKES:** [TAN] 20 [ENTER] [x^{-1}] [ENTER] 2.747477419

If you know the measures of any two sides of a right triangle or the measures of one side and one acute angle, you can determine the measures of all the sides and angles of the triangle. This process of finding the missing measures is known as **solving a right triangle**.

Example 4 Solve a Right Triangle

Solve $\triangle XYZ$. Round measures of sides to the nearest tenth and measures of angles to the nearest degree.

You know the measures of one side, one acute angle, and the right angle. You need to find x, z, and Y.

Find x and z.

$$\tan 35° = \frac{x}{10} \qquad \sec 35° = \frac{z}{10}$$

$$10 \tan 35° = x \qquad \frac{1}{\cos 35°} = \frac{z}{10}$$

$$7.0 \approx x \qquad \frac{10}{\cos 35°} = z$$

$$12.2 \approx z$$

Find Y.

$35° + Y = 90°$ Angles X and Y are complementary.

$Y = 55°$ Solve for Y.

Therefore, $Y = 55°$, $x \approx 7.0$, and $z \approx 12.2$.

Study Tip

Error in Measurement

The value of z in Example 4 is found using the secant instead of using the Pythagorean Theorem. This is because the secant uses values given in the problem rather than calculated values.

Use the inverse capabilities of your calculator to find the measure of an angle when one of its trigonometric ratios is known. For example, use the SIN^{-1} function to find the measure of an angle when the sine of the angle is known. *You will learn more about inverses of trigonometric functions in Lesson 13-7.*

Example 5 Find Missing Angle Measures of Right Triangles

Solve $\triangle ABC$. Round measures of sides to the nearest tenth and measures of angles to the nearest degree.

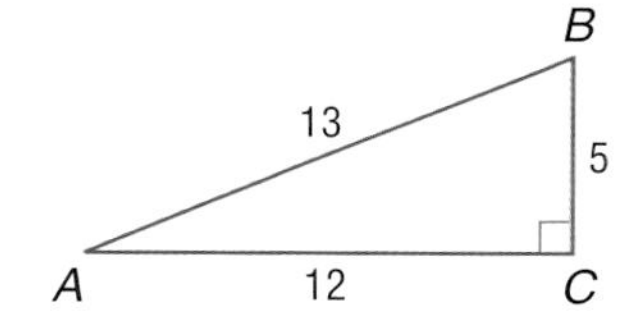

You know the measures of the sides. You need to find A and B.

Find A. $\sin A = \frac{5}{13}$ $\sin A = \frac{\text{opp}}{\text{hyp}}$

Use a calculator and the SIN^{-1} function to find the angle whose sine is $\frac{5}{13}$.

KEYSTROKES: [2nd] [SIN^{-1}] 5 [÷] 13 [)] [ENTER] 22.61986495

To the nearest degree, $A \approx 23°$.

Find B. $23° + B \approx 90°$ Angles A and B are complementary.

$B \approx 67°$ Solve for B.

Therefore, $A \approx 23°$ and $B \approx 67°$.

Trigonometry has many practical applications. Among the most important is the ability to find distances or lengths that either cannot be measured directly or are not easily measured directly.

Example 6 Indirect Measurement

BRIDGE CONSTRUCTION In order to construct a bridge across a river, the width of the river at that location must be determined. Suppose a stake is planted on one side of the river directly across from a second stake on the opposite side. At a distance 50 meters to the left of the stake, an angle of 82° is measured between the two stakes. Find the width of the river.

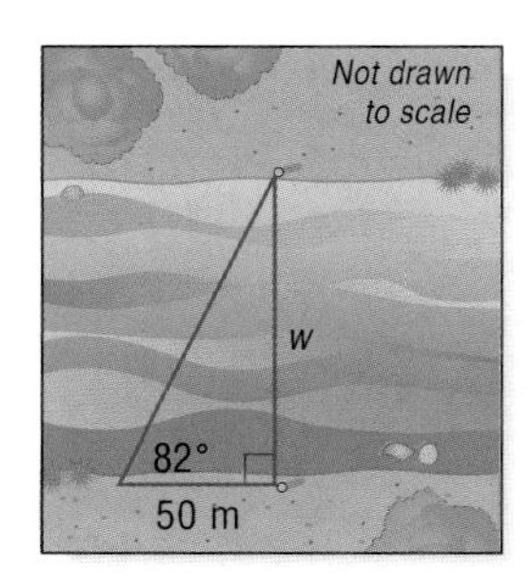

Let w represent the width of the river at that location. Write an equation using a trigonometric function that involves the ratio of the distance w and 50.

$\tan 82° = \frac{w}{50}$ — $\tan \theta = \frac{\text{opp}}{\text{adj}}$

$50 \tan 82° = w$ — Multiply each side by 50.

$355.8 \approx w$ — Use a calculator.

The width of the river is about 355.8 meters.

Some applications of trigonometry use an angle of elevation or depression. In the figure at the right, the angle formed by the line of sight from the observer and a line parallel to the ground is called the **angle of elevation**. The angle formed by the line of sight from the plane and a line parallel to the ground is called the **angle of depression**.

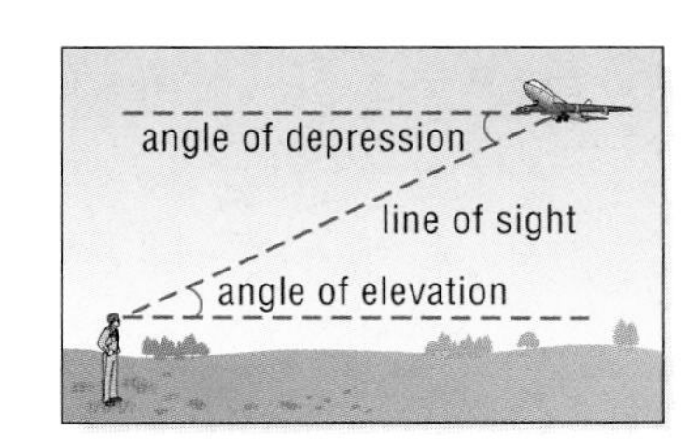

The angle of elevation and the angle of depression are congruent since they are alternate interior angles of parallel lines.

Example 7 Use an Angle of Elevation

SKIING The Aerial run in Snowbird, Utah, has an angle of elevation of 20.2°. Its vertical drop is 2900 feet. Estimate the length of this run.

Let ℓ represent the length of the run. Write an equation using a trigonometric function that involves the ratio of ℓ and 2900.

$\sin 20.2° = \frac{2900}{\ell}$ — $\sin \theta = \frac{\text{opp}}{\text{hyp}}$

$\ell = \frac{2900}{\sin 20.2°}$ — Solve for ℓ.

$\ell \approx 8398.5$ — Use a calculator.

The length of the run is about 8399 feet.

More About. . .

Skiing

The average annual snowfall in Snowbird, Utah, is 500 inches. The longest designated run there is Chip's Run, at 2.5 miles.

Source: www.utahskiing.com

Check for Understanding

Concept Check

1. **Define** the word *trigonometry*.
2. **OPEN ENDED** Draw a right triangle. Label one of its acute angles θ. Then, label the hypotenuse, the leg adjacent to θ, and the leg opposite θ.
3. **Find a counterexample** to the following statement.
   It is always possible to solve a right triangle.

Guided Practice

Find the values of the six trigonometric functions for angle θ.

4.
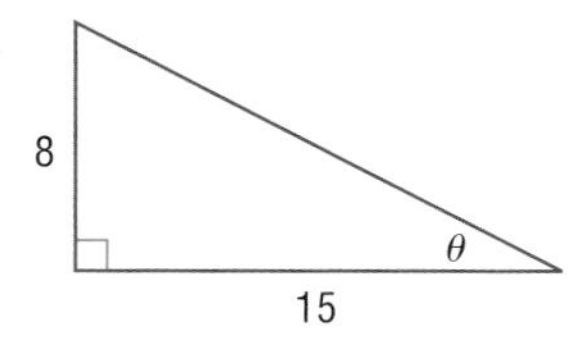

5.
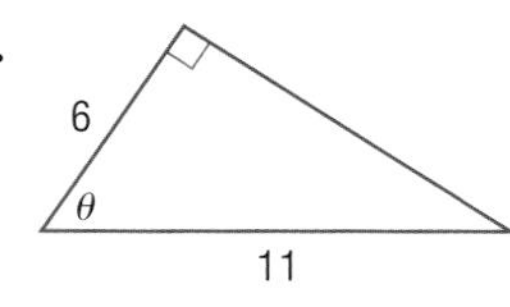

6.
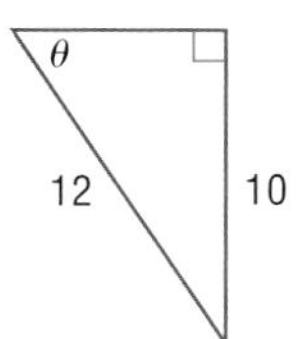

Write an equation involving sin, cos, or tan that can be used to find x. Then solve the equation. Round measures of sides to the nearest tenth and angles to the nearest degree.

7.
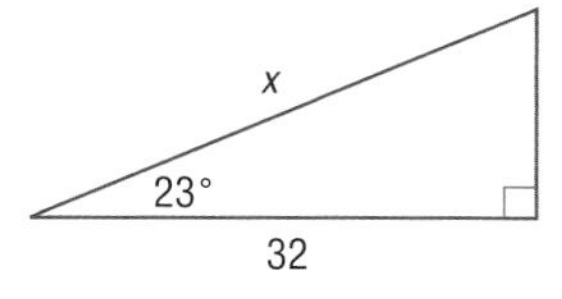

8.
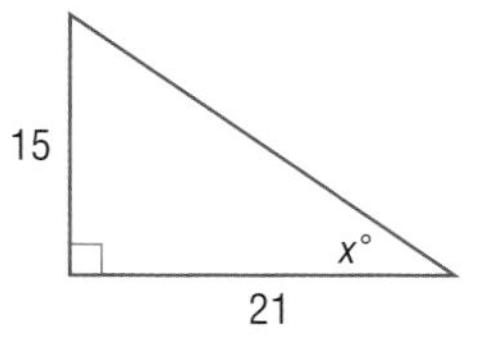

Solve $\triangle ABC$ by using the given measurements. Round measures of sides to the nearest tenth and measures of angles to the nearest degree.

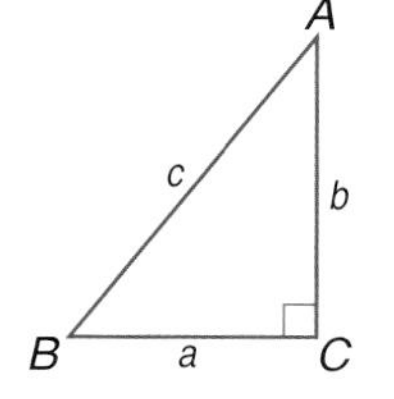

9. $A = 45°, b = 6$
10. $B = 56°, c = 16$
11. $b = 7, c = 18$
12. $a = 14, b = 13$

13. **AVIATION** When landing, a jet will average a $3°$ angle of descent. What is the altitude x, to the nearest foot, of a jet on final descent as it passes over an airport beacon 6 miles from the start of the runway?

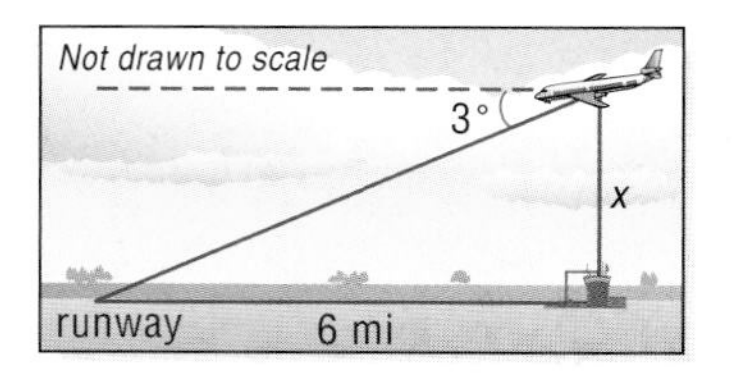

14. If $\tan \theta = 3$, find the value of $\sin \theta$.

   Ⓐ $\frac{3}{10}$ Ⓑ $\frac{3\sqrt{10}}{10}$ Ⓒ $\frac{10}{3}$ Ⓓ $\frac{1}{3}$

Practice and Apply

Homework Help

For Exercises	See Examples
15–20	1
21–24, 27, 28	3
25, 26, 37–40	5
29–36	4
41–46	6, 7
49	2

Extra Practice
See page 857.

Find the values of the six trigonometric functions for angle θ.

15.
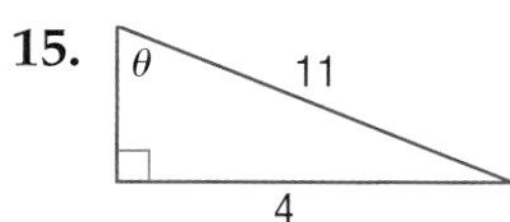

16.
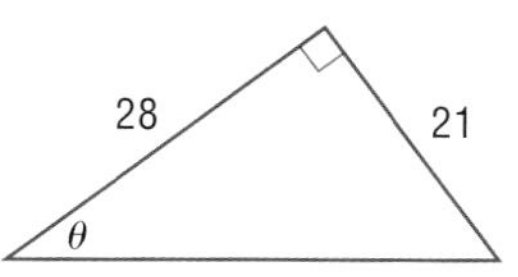

17.
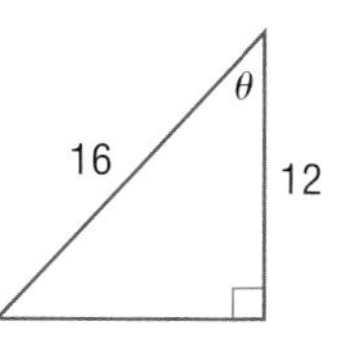

18.
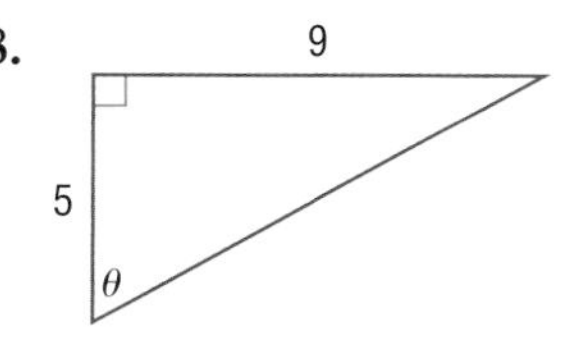

19.
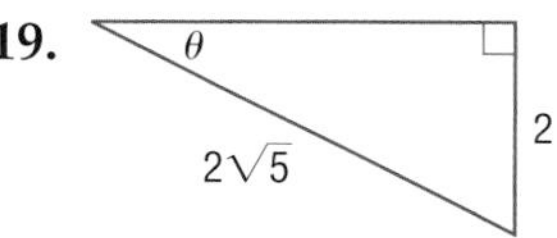

20.
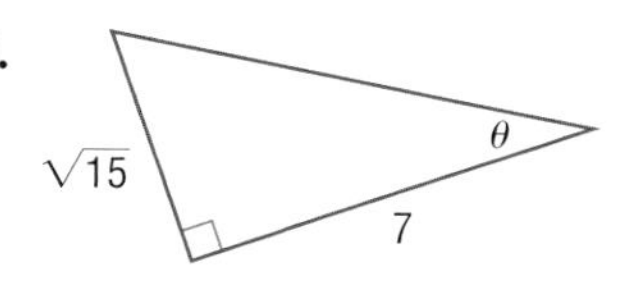

Write an equation involving sin, cos, or tan that can be used to find x. Then solve the equation. Round measures of sides to the nearest tenth and measures of angles to the nearest degree.

21. 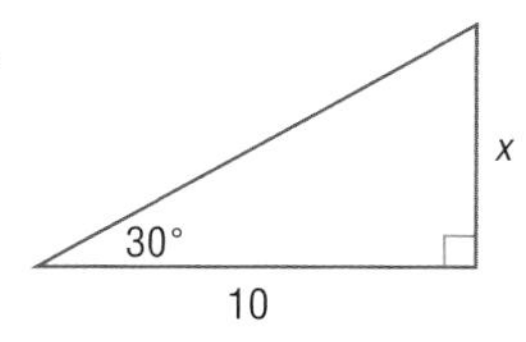

22. 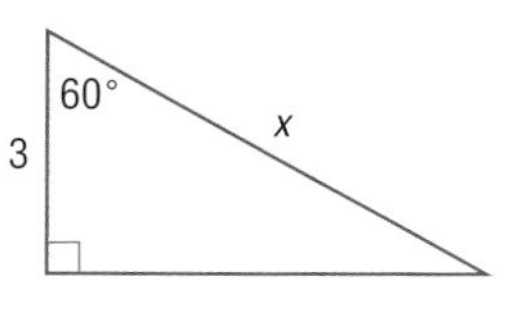

23. 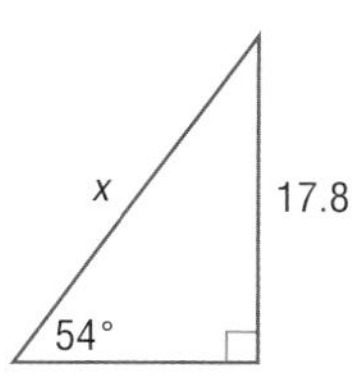

24. 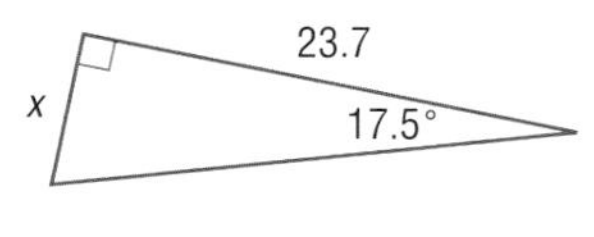

25. 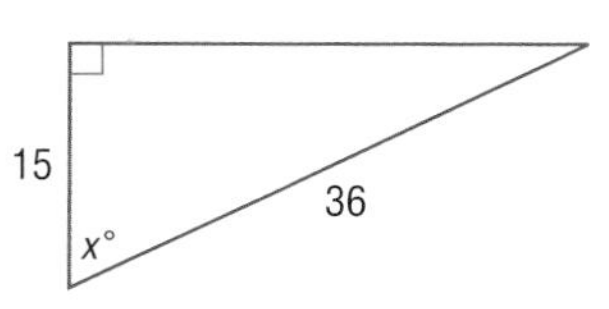

26. 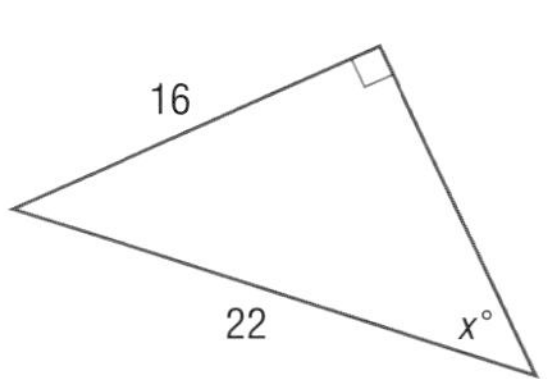

27. Using the 30°-60°-90° triangle shown on page 703, verify each value.

a. $\sin 30° = \frac{1}{2}$ **b.** $\cos 30° = \frac{\sqrt{3}}{2}$ **c.** $\sin 60° = \frac{\sqrt{3}}{2}$

28. Using the 45°-45°-90° triangle shown on page 703, verify each value.

a. $\sin 45° = \frac{\sqrt{2}}{2}$ **b.** $\cos 45° = \frac{\sqrt{2}}{2}$ **c.** $\tan 45° = 1$

Career Choices

Surveyor

Land surveyors manage survey parties that measure distances, directions, and angles between points, lines, and contours on Earth's surface.

Online Research For information about a career as a surveyor, visit: www.algebra2.com/careers

Solve $\triangle ABC$ by using the given measurements. Round measures of sides to the nearest tenth and measures of angles to the nearest degree.

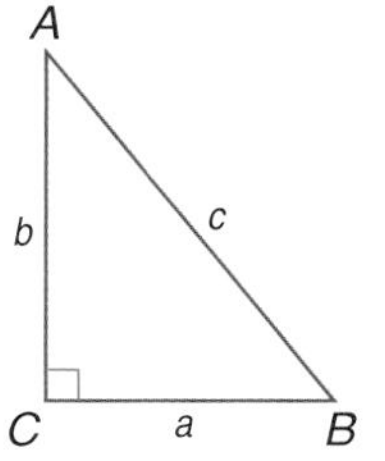

29. $A = 16°, c = 14$ **30.** $B = 27°, b = 7$

31. $A = 34°, a = 10$ **32.** $B = 15°, c = 25$

33. $B = 30°, b = 11$ **34.** $A = 45°, c = 7\sqrt{2}$

35. $B = 18°, a = \sqrt{15}$ **36.** $A = 10°, b = 15$

37. $b = 6, c = 13$ **38.** $a = 4, c = 9$

39. $\tan B = \frac{7}{8}, b = 7$ **40.** $\sin A = \frac{1}{3}, a = 5$

41. TRAVEL In a sightseeing boat near the base of the Horseshoe Falls at Niagara Falls, a passenger estimates the angle of elevation to the top of the falls to be 30°. If the Horseshoe Falls are 173 feet high, what is the distance from the boat to the base of the falls?

42. SURVEYING A surveyor stands 100 feet from a building and sights the top of the building at a 55° angle of elevation. Find the height of the building.

EXERCISE For Exercises 43 and 44, use the following information.
A preprogrammed workout on a treadmill consists of intervals walking at various rates and angles of incline. A 1% incline means 1 unit of vertical rise for every 100 units of horizontal run.

43. At what angle, with respect to the horizontal, is the treadmill bed when set at a 10% incline? Round to the nearest degree.

44. If the treadmill bed is 40 inches long, what is the vertical rise when set at an 8% incline?

45. GEOMETRY Find the area of the regular hexagon with point O as its center. (*Hint*: First find the value of x.)

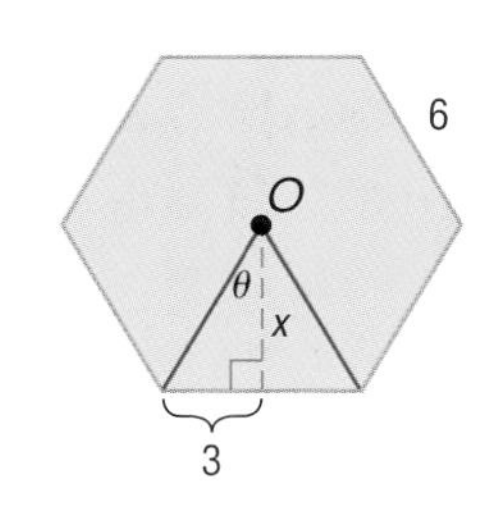

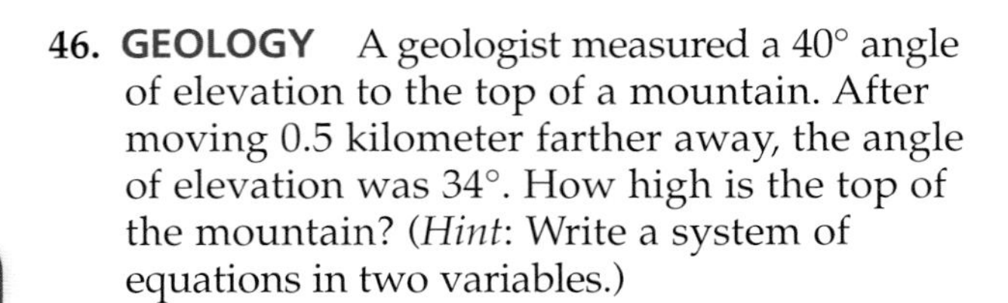

46. GEOLOGY A geologist measured a 40° angle of elevation to the top of a mountain. After moving 0.5 kilometer farther away, the angle of elevation was 34°. How high is the top of the mountain? (*Hint*: Write a system of equations in two variables.)

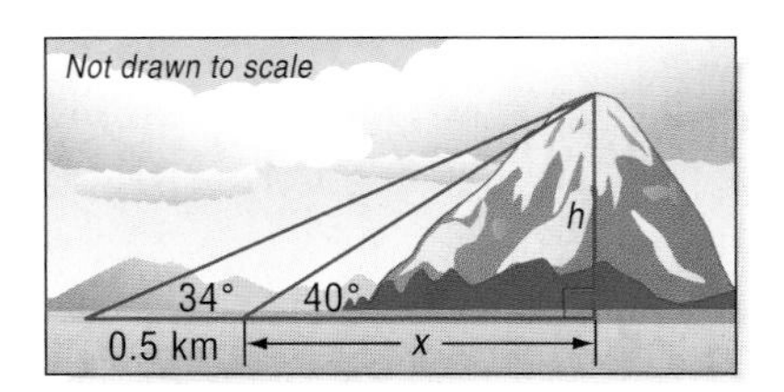

You can use the tangent ratio to determine the maximum height of a rocket. Visit www.algebra2.com/webquest to continue work on your WebQuest project.

47. CRITICAL THINKING Explain why the sine and cosine of an acute angle are never greater than 1, but the tangent of an acute angle may be greater than 1.

48. WRITING IN MATH Answer the question that was posed at the beginning of the lesson.

How is trigonometry used in building construction?

Include the following in your answer:

- an explanation as to why the ratio of vertical rise to horizontal run on an entrance ramp is the tangent of the angle the ramp makes with the horizontal, and
- an explanation of how an architect can use the tangent ratio to ensure that all the ramps he or she designs meet the ADA requirement.

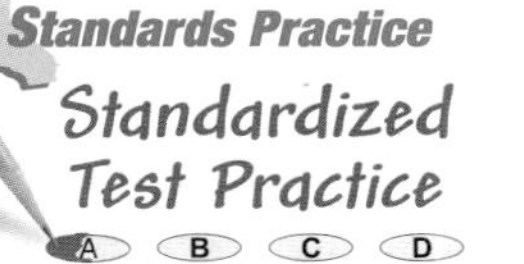

49. If the secant of an angle θ is $\frac{25}{7}$, what is the sine of angle θ?

(A) $\frac{5}{25}$ (B) $\frac{7}{25}$ (C) $\frac{24}{25}$ (D) $\frac{25}{7}$

50. GRID IN The tailgate of a moving truck is 2 feet above the ground. The incline of the ramp used for loading the truck is 15° as shown. Find the length of the ramp to the nearest tenth of a foot.

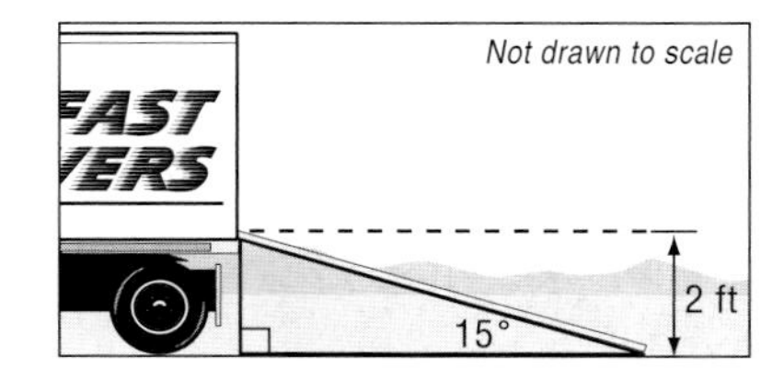

Maintain Your Skills

Mixed Review

Determine whether each situation would produce a random sample. Write *yes* or *no* and explain your answer. *(Lesson 12-9)*

51. surveying band members to find the most popular type of music at your school

52. surveying people coming into a post office to find out what color cars are most popular

Find each probability if a coin is tossed 4 times. *(Lesson 12-8)*

53. P(exactly 2 heads) **54.** P(4 heads) **55.** P(at least 1 head)

Solve each equation. *(Lesson 7-3)*

56. $y^4 - 64 = 0$ **57.** $x^5 - 5x^3 + 4x = 0$ **58.** $d + \sqrt{d} - 132 = 0$

Getting Ready for the Next Lesson

PREREQUISITE SKILL Find each product. Include the appropriate units with your answer. *(To review **dimensional analysis**, see Lesson 5-1.)*

59. $5 \text{ gallons}\left(\frac{4 \text{ quarts}}{1 \text{ gallon}}\right)$ **60.** $6.8 \text{ miles}\left(\frac{5280 \text{ feet}}{1 \text{ mile}}\right)$

61. $\left(\frac{2 \text{ square meters}}{5 \text{ dollars}}\right) 30 \text{ dollars}$ **62.** $\left(\frac{4 \text{ liters}}{5 \text{ minutes}}\right) 60 \text{ minutes}$

13-2 Angles and Angle Measure

What You'll Learn

- Change radian measure to degree measure and vice versa.
- Identify coterminal angles.

Vocabulary

- initial side
- terminal side
- standard position
- unit circle
- radian
- coterminal angles

How can angles be used to describe circular motion?

The Ferris wheel at Navy Pier in Chicago has a 140-foot diameter and 40 gondolas equally spaced around its circumference. The average angular velocity ω of one of the gondolas is given by $\omega = \frac{\theta}{t}$, where θ is the angle through which the gondola has revolved after a specified amount of time t. For example, if a gondola revolves through an angle of 225° in 40 seconds, then its average angular velocity is $225° \div 40$ or about 5.6° per second.

ANGLE MEASUREMENT What does an angle measuring 225° look like? In Lesson 13-1, you worked only with acute angles, those measuring between 0° and 90°, but angles can have *any* real number measurement.

Study Tip

Reading Math

In trigonometry, an angle is sometimes referred to as an *angle of rotation*.

On a coordinate plane, an angle may be generated by the rotation of two rays that share a fixed endpoint at the origin. One ray, called the **initial side** of the angle, is fixed along the positive x-axis. The other ray, called the **terminal side** of the angle, can rotate about the center. An angle positioned so that its vertex is at the origin and its initial side is along the positive x-axis is said to be in **standard position**.

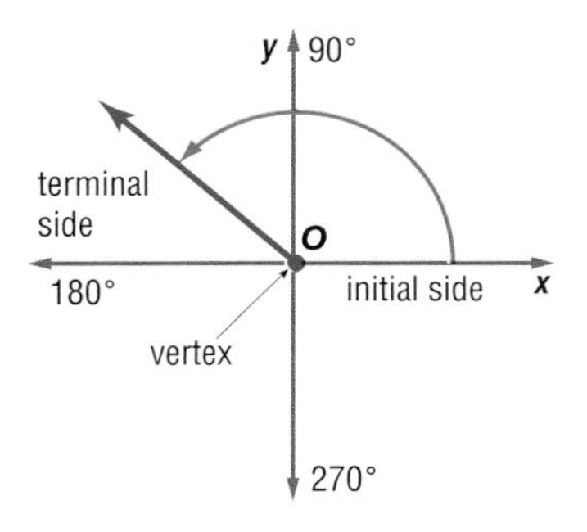

The measure of an angle is determined by the amount and direction of rotation from the initial side to the terminal side.

Positive Angle Measure
counterclockwise

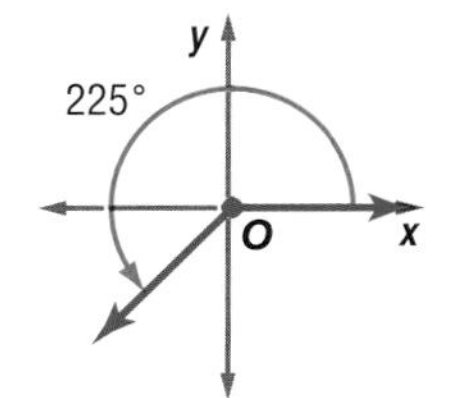

Negative Angle Measure
clockwise

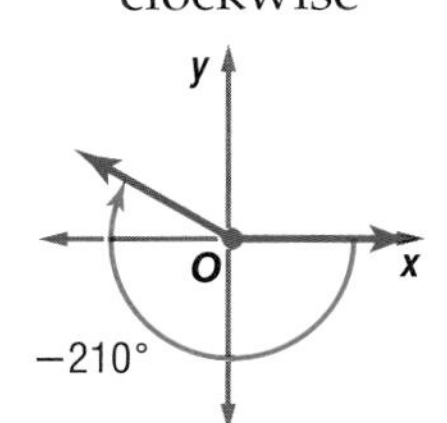

When terminal sides rotate, they may sometimes make one or more revolutions. An angle whose terminal side has made exactly one revolution has a measure of 360°.

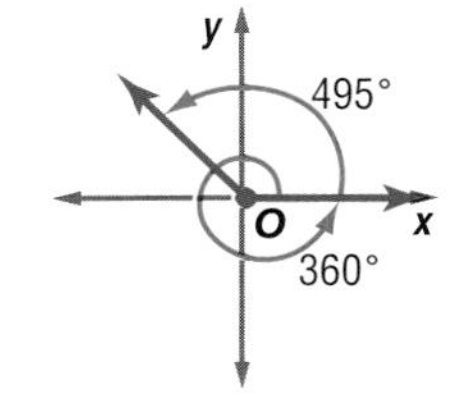

Example 1 Draw an Angle in Standard Position

Draw an angle with the given measure in standard position.

a. 240° $240° = 180° + 60°$
Draw the terminal side of the angle 60° counterclockwise past the negative x-axis.

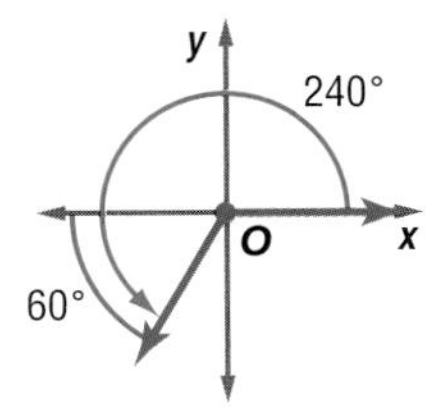

b. −30° The angle is negative.
Draw the terminal side of the angle 30° clockwise from the positive x-axis.

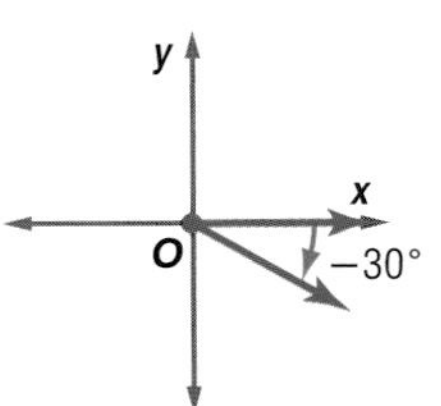

c. 450° $450° = 360° + 90°$
Draw the terminal side of the angle 90° counterclockwise past the positive x-axis.

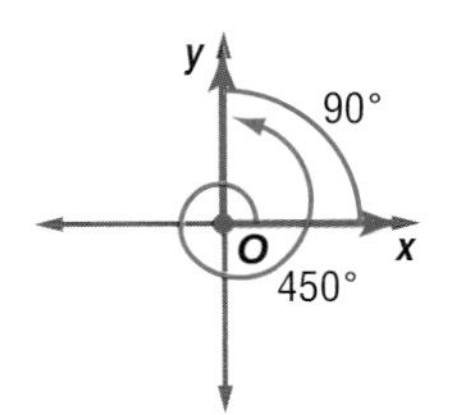

Another unit used to measure angles is a radian. The definition of a radian is based on the concept of a **unit circle**, which is a circle of radius 1 unit whose center is at the origin of a coordinate system. One **radian** is the measure of an angle θ in standard position whose rays intercept an arc of length 1 unit on the unit circle.

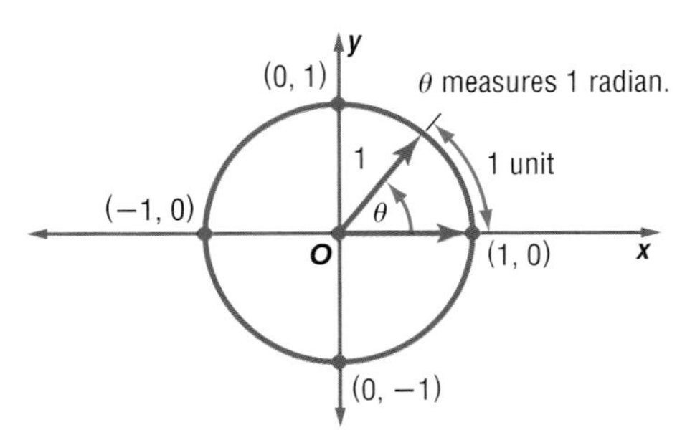

Study Tip

Radian Measure
As with degrees, the measure of an angle in radians is positive if its rotation is counterclockwise. The measure is negative if the rotation is clockwise.

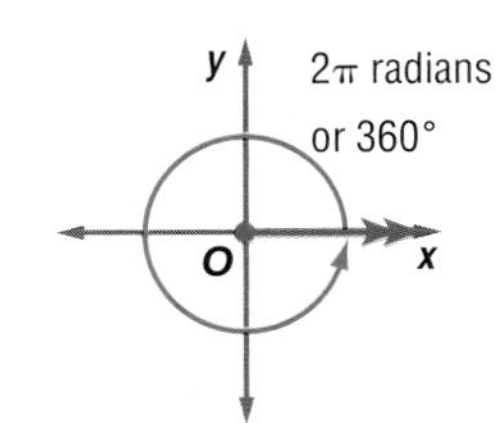

The circumference of any circle is $2\pi r$, where r is the radius measure. So the circumference of a unit circle is $2\pi(1)$ or 2π units. Therefore, an angle representing one complete revolution of the circle measures 2π radians. This same angle measures 360°. Therefore, the following equation is true.

$$2\pi \text{ radians} = 360°$$

To change angle measures from radians to degrees or vice versa, solve the equation above in terms of both units.

$$2\pi \text{ radians} = 360°$$
$$\frac{2\pi \text{ radians}}{2\pi} = \frac{360°}{2\pi}$$
$$1 \text{ radian} = \frac{180°}{\pi}$$

1 radian is about 57 degrees.

$$2\pi \text{ radians} = 360°$$
$$\frac{2\pi \text{ radians}}{360} = \frac{360°}{360}$$
$$\frac{\pi \text{ radians}}{180} = 1°$$

1 degree is about 0.0175 radian.

These equations suggest a method for converting between radian and degree measure.

Key Concept — Radian and Degree Measure

- To rewrite the radian measure of an angle in degrees, multiply the number of radians by $\frac{180°}{\pi \text{ radians}}$.
- To rewrite the degree measure of an angle in radians, multiply the number of degrees by $\frac{\pi \text{ radians}}{180°}$.

Example 2 Convert Between Degree and Radian Measure

Rewrite the degree measure in radians and the radian measure in degrees.

a. 60°

$$60° = 60°\left(\frac{\pi \text{ radians}}{180°}\right)$$
$$= \frac{60\pi}{180} \text{ radians or } \frac{\pi}{3}$$

b. $-\frac{7\pi}{4}$

$$-\frac{7\pi}{4} = \left(-\frac{7\pi}{4} \text{ radians}\right)\left(\frac{180°}{\pi \text{ radians}}\right)$$
$$= -\frac{1260°}{4} \text{ or } -315°$$

Study Tip

Reading Math

The word *radian* is usually omitted when angles are expressed in radian measure. Thus, when no units are given for an angle measure, radian measure is implied.

You will find it useful to learn equivalent degree and radian measures for the special angles shown in the diagram at the right. This diagram is more easily learned by memorizing the equivalent degree and radian measures for the first quadrant and for 90°. All of the other special angles are multiples of these angles.

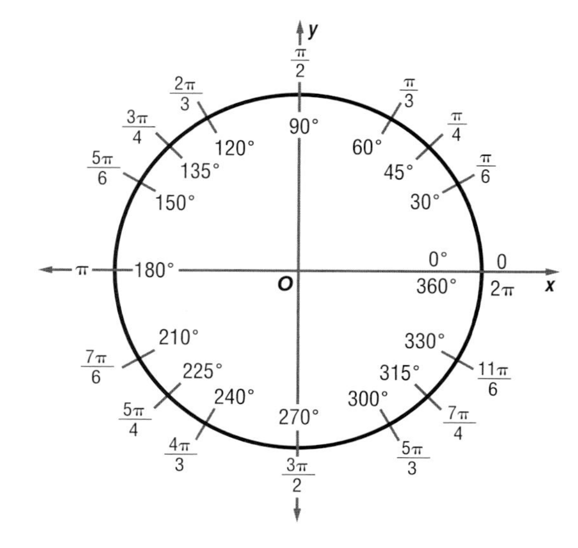

Example 3 Measure an Angle in Degrees and Radians

TIME **Find both the degree and radian measures of the angle through which the hour hand on a clock rotates from 1:00 P.M. to 3:00 P.M.**

The numbers on a clock divide it into 12 equal parts with 12 equal angles. The angle from 1 to 3 on the clock represents $\frac{2}{12}$ or $\frac{1}{6}$ of a complete rotation of 360°. $\frac{1}{6}$ of 360° is 60°.

Since the rotation is clockwise, the angle through which the hour hand rotates is negative. Therefore, the angle measures −60°.

60° has an equivalent radian measure of $\frac{\pi}{3}$. So the equivalent radian measure of −60° is $-\frac{\pi}{3}$.

COTERMINAL ANGLES If you graph a 405° angle and a 45° angle in standard position on the same coordinate plane, you will notice that the terminal side of the 405° angle is the same as the terminal side of the 45° angle. When two angles in standard position have the same terminal sides, they are called **coterminal angles**.

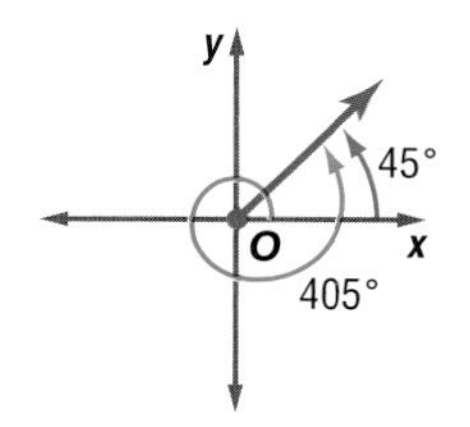

Notice that $405° - 45° = 360°$. In degree measure, coterminal angles differ by an integral multiple of 360°. You can find an angle that is coterminal to a given angle by adding or subtracting a multiple of 360°. In radian measure, a coterminal angle is found by adding or subtracting a multiple of 2π.

Example 4 Find Coterminal Angles

Find one angle with positive measure and one angle with negative measure coterminal with each angle.

a. 240°

A positive angle is $240° + 360°$ or $600°$.

A negative angle is $240° - 360°$ or $-120°$.

Study Tip

Coterminal Angles
Notice in Example 4b that it is necessary to subtract a multiple of 2π to find a coterminal angle with negative measure.

b. $\frac{9\pi}{4}$

A positive angle is $\frac{9\pi}{4} + 2\pi$ or $\frac{17\pi}{4}$. $\quad \frac{9\pi}{4} + \frac{8\pi}{4} = \frac{17\pi}{4}$

A negative angle is $\frac{9\pi}{4} - 2(2\pi)$ or $-\frac{7\pi}{4}$. $\quad \frac{9\pi}{4} + \left(-\frac{16\pi}{4}\right) = -\frac{7\pi}{4}$

Check for Understanding

Concept Check

1. **Name** the set of numbers to which angle measures belong.
2. **Define** the term radian.
3. **OPEN ENDED** Draw and label an example of an angle with negative measure in standard position. Then find an angle with positive measure that is coterminal with this angle.

Guided Practice

Draw an angle with the given measure in standard position.

4. 70°
5. 300°
6. 570°
7. $-45°$

Rewrite each degree measure in radians and each radian measure in degrees.

8. 130°
9. $-10°$
10. 485°
11. $\frac{3\pi}{4}$
12. $-\frac{\pi}{6}$
13. $\frac{19\pi}{3}$

Find one angle with positive measure and one angle with negative measure coterminal with each angle.

14. 60°
15. 425°
16. $\frac{\pi}{3}$

Application

ASTRONOMY For Exercises 17 and 18, use the following information.
Earth rotates on its axis once every 24 hours.

17. How long does it take Earth to rotate through an angle of 315°?
18. How long does it take Earth to rotate through an angle of $\frac{\pi}{6}$?

Practice and Apply

Draw an angle with the given measure in standard position.

19. 235°
20. 270°
21. 790°
22. 380°
23. $-150°$
24. $-50°$
25. π
26. $-\frac{2\pi}{3}$

Homework Help

For Exercises	See Examples
19–26	1
27–42	2
43–54	4
55–59	3

Extra Practice
See page 857.

Rewrite each degree measure in radians and each radian measure in degrees.

27. $120°$ **28.** $60°$ **29.** $-15°$ **30.** $-225°$

31. $660°$ **32.** $570°$ **33.** $158°$ **34.** $260°$

35. $\frac{5\pi}{6}$ **36.** $\frac{11\pi}{4}$ **37.** $-\frac{\pi}{4}$ **38.** $-\frac{\pi}{3}$

39. $\frac{29\pi}{4}$ **40.** $\frac{17\pi}{6}$ **41.** 9 **42.** 3

Find one angle with positive measure and one angle with negative measure coterminal with each angle.

43. $225°$ **44.** $30°$ **45.** $-15°$

46. $-140°$ **47.** $368°$ **48.** $760°$

49. $\frac{3\pi}{4}$ **50.** $\frac{7\pi}{6}$ **51.** $-\frac{5\pi}{4}$

52. $-\frac{2\pi}{3}$ **53.** $\frac{9\pi}{2}$ **54.** $\frac{17\pi}{4}$

55. DRIVING Some sport-utility vehicles (SUVs) use 15-inch radius wheels. When driven 40 miles per hour, determine the measure of the angle through which a point on the wheel travels every second. Round to both the nearest degree and nearest radian.

More About. . .

Driving

A leading U.S. automaker plans to build a hybrid sport-utility vehicle in the near future that will use an electric motor to boost fuel efficiency and reduce polluting emissions.

Source: *The Dallas Morning News*

GEOMETRY For Exercises 56 and 57, use the following information.

A *sector* is a region of a circle that is bounded by a central angle θ and its intercepted arc. The area A of a sector with radius r and central angle θ is given by $A = \frac{1}{2}r^2\theta$, where θ is measured in radians.

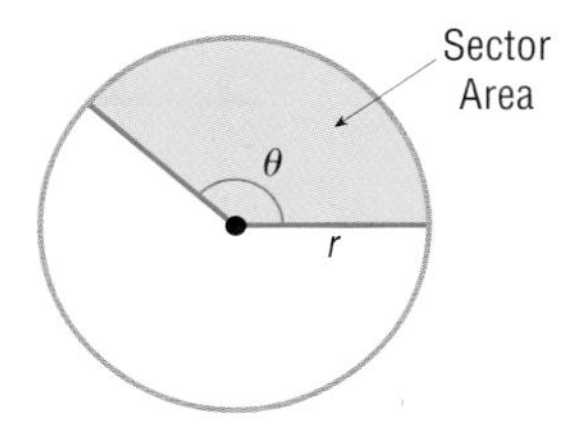

56. Find the area of a sector with a central angle of $\frac{4\pi}{3}$ radians in a circle whose radius measures 10 inches.

57. Find the area of a sector with a central angle of $150°$ in a circle whose radius measures 12 meters.

58. ENTERTAINMENT Suppose the gondolas on the Navy Pier Ferris wheel were numbered from 1 through 40 consecutively in a counterclockwise fashion. If you were sitting in gondola number 3 and the wheel were to rotate counterclockwise through $\frac{47\pi}{10}$ radians, which gondola used to be in the position that you are in now?

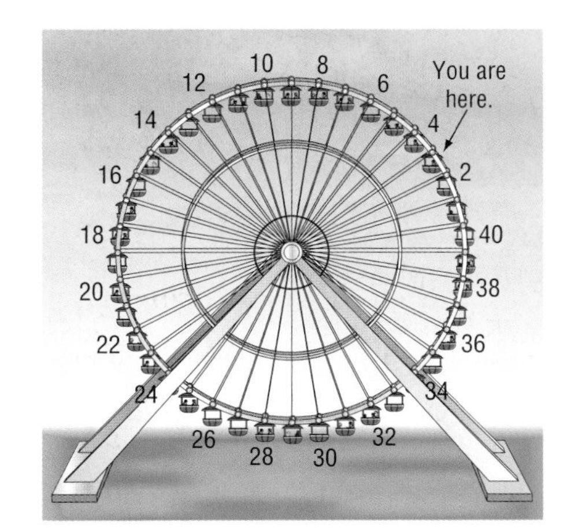

59. CARS Use the Area of a Sector Formula in Exercises 56 and 57 to find the area swept by the rear windshield wiper of the car shown at the right.

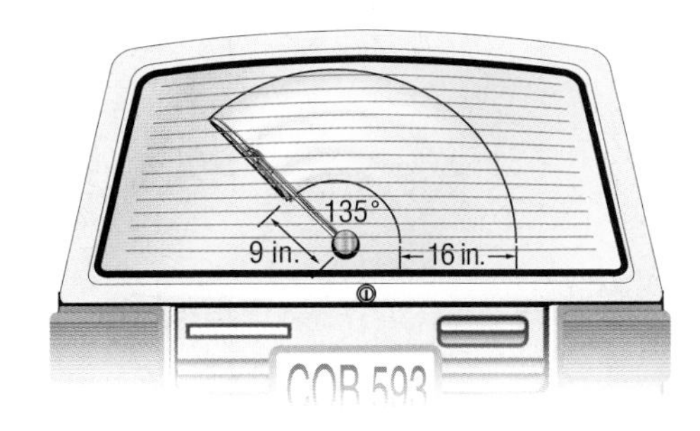

60. CRITICAL THINKING If (a, b) is on a circle that has radius r and center at the origin, prove that each of the following points is also on this circle.

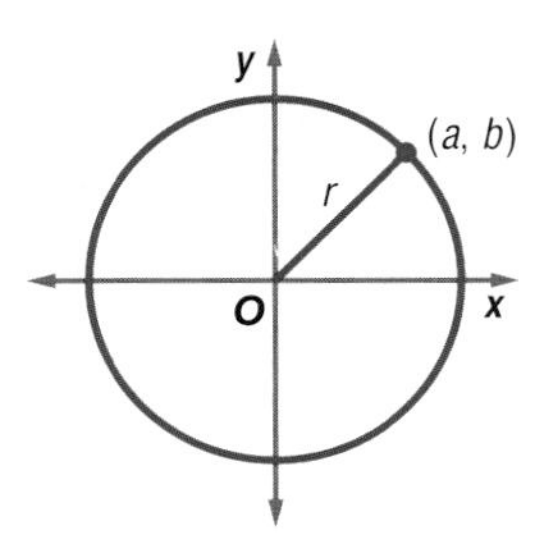

a. $(a, -b)$

b. (b, a)

c. $(b, -a)$

61. **WRITING IN MATH** Answer the question that was posed at the beginning of the lesson.

How can angles be used to describe circular motion?

Include the following in your answer:

- an explanation of the significance of angles of more than 180° in terms of circular motion,
- an explanation of the significance of angles with negative measure in terms of circular motion, and
- an interpretation of a rate of more than 360° per minute.

Standards Practice

Standardized Test Practice

62. Choose the radian measure that is equal to 56°.

(A) $\frac{\pi}{15}$ (B) $\frac{7\pi}{45}$

(C) $\frac{14\pi}{45}$ (D) $\frac{\pi}{3}$

63. Angular velocity is defined by the equation $\omega = \frac{\theta}{t}$, where θ is usually expressed in radians and t represents time. Find the angular velocity in radians per second of a point on a bicycle tire if it completes 2 revolutions in 3 seconds.

(A) $\frac{\pi}{3}$ (B) $\frac{\pi}{2}$

(C) $\frac{2\pi}{3}$ (D) $\frac{4\pi}{3}$

Maintain Your Skills

Mixed Review

Solve $\triangle ABC$ by using the given measurements. Round measures of sides to the nearest tenth and measures of angles to the nearest degree. *(Lesson 13-1)*

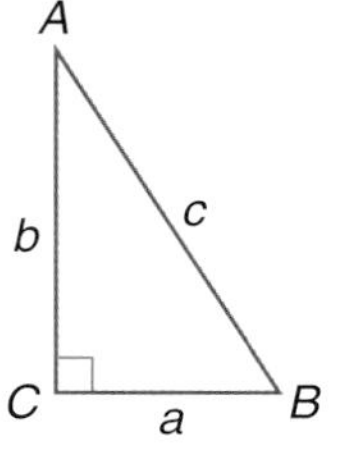

64. $A = 34°, b = 5$

65. $B = 68°, b = 14.7$

66. $B = 55°, c = 16$

67. $a = 0.4, b = 0.4\sqrt{3}$

Find the margin of sampling error. *(Lesson 12-9)*

68. $p = 72\%, n = 100$

69. $p = 50\%, n = 200$

Determine whether each situation involves a *permutation* or a *combination*. Then find the number of possibilities. *(Lesson 12-2)*

70. choosing an arrangement of 5 CDs from your 30 favorite CDs

71. choosing 3 different types of snack foods out of 7 at the store to take on a trip

Find $[g \circ h](x)$ and $[h \circ g](x)$. *(Lesson 7-7)*

72. $g(x) = 2x$
$h(x) = 3x - 4$

73. $g(x) = 2x + 5$
$h(x) = 2x^2 - 3x + 9$

For Exercises 74 and 75, use the graph at the right.
The number of sports radio stations can be modeled by $R(x) = 7.8x^2 + 16.6x + 95.8$, where x is the number of years since 1996. *(Lesson 7-5)*

74. Use synthetic substitution to estimate the number of sports radio stations for 2006.

75. Evaluate $R(12)$. What does this value represent?

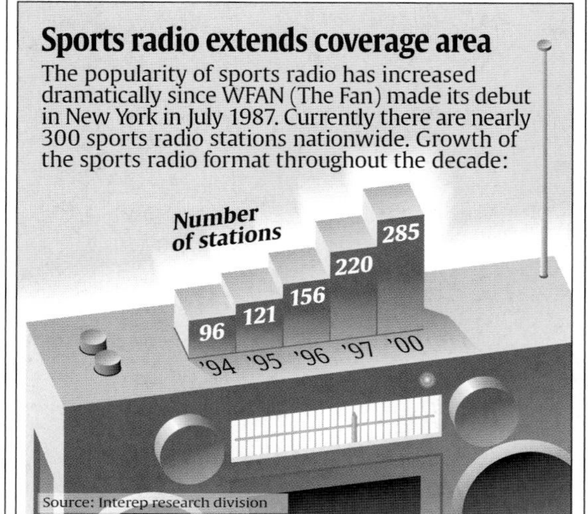

Getting Ready for the Next Lesson

PREREQUISITE SKILL Simplify each expression.
*(To review **rationalizing denominators**, see Lesson 5-6.)*

76. $\frac{2}{\sqrt{3}}$ **77.** $\frac{3}{\sqrt{5}}$ **78.** $\frac{4}{\sqrt{6}}$

79. $\frac{5}{\sqrt{10}}$ **80.** $\frac{\sqrt{7}}{\sqrt{2}}$ **81.** $\frac{\sqrt{5}}{\sqrt{8}}$

Practice Quiz 1 — Lessons 13-1 and 13-2

Solve $\triangle ABC$ by using the given measurements. Round measures of sides to the nearest tenth and measures of angles to the nearest degree. *(Lesson 13-1)*

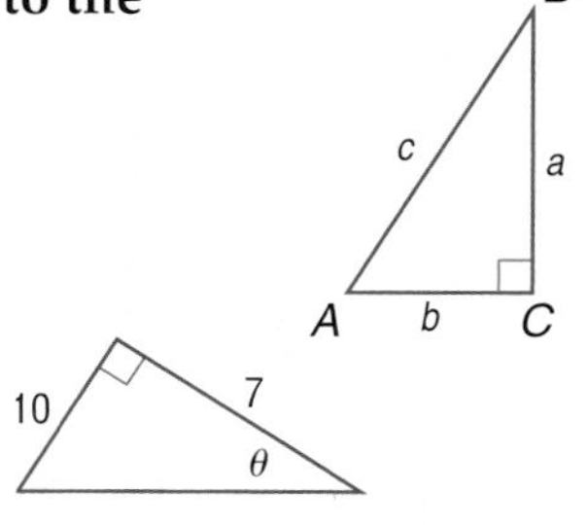

1. $A = 48°, b = 12$ **2.** $a = 18, c = 21$

3. Draw an angle measuring $-60°$ in standard position. *(Lesson 13-1)*

4. Find the values of the six trigonometric functions for angle θ in the triangle at the right. *(Lesson 13-1)*

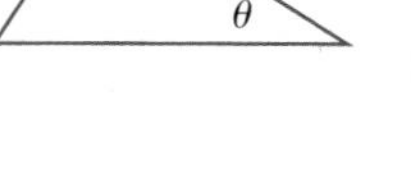

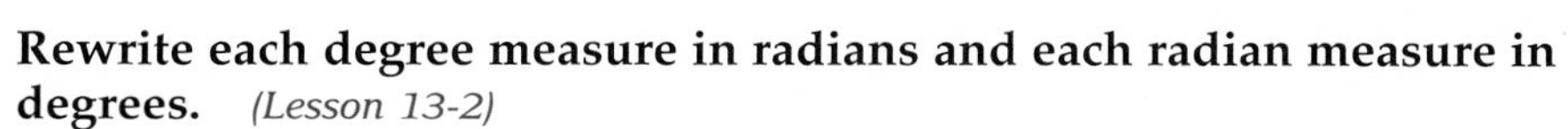

Rewrite each degree measure in radians and each radian measure in degrees. *(Lesson 13-2)*

5. $190°$ **6.** $450°$ **7.** $\frac{7\pi}{6}$ **8.** $-\frac{11\pi}{5}$

Find one angle with positive measure and one angle with negative measure coterminal with each angle. *(Lesson 13-2)*

9. $-55°$ **10.** $\frac{11\pi}{3}$

Algebra Activity

A Follow-Up of Lesson 13-2

Investigating Regular Polygons Using Trigonometry

Collect the Data

- Use a compass to draw a circle with a radius of one inch. Inscribe an equilateral triangle inside of the circle. To do this, use a protractor to measure three angles of 120° at the center of the circle, since $\frac{360°}{3} = 120°$. Then connect the points where the sides of the angles intersect the circle using a straightedge.
- The **apothem** of a regular polygon is a segment that is drawn from the center of the polygon perpendicular to a side of the polygon. Use the cosine of angle θ to find the length of an apothem, labeled a in the diagram below.

Analyze the Data

1. Make a table like the one shown below and record the length of the apothem of the equilateral triangle.

Number of Sides, n	θ	a
3	60	
4	45	
5		
6		
7		
8		
9		
10		

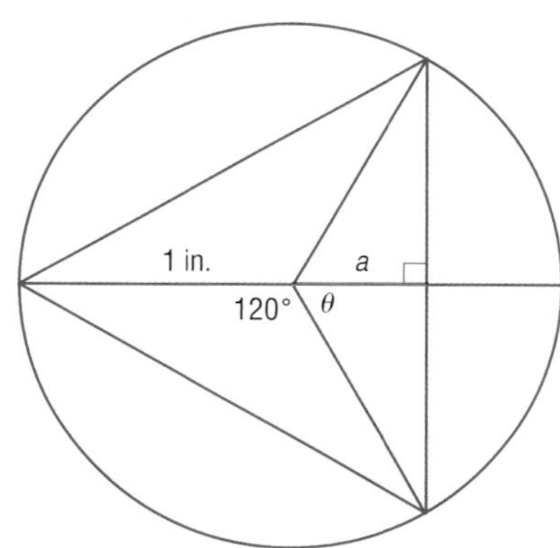

Inscribe each regular polygon named in the table in a circle of radius one inch. Copy and complete the table.

2. What do you notice about the measure of θ as the number of sides of the inscribed polygon increases?

3. What do you notice about the values of a?

Make a Conjecture

4. Suppose you inscribe a 20-sided regular polygon inside a circle. Find the measure of angle θ.

5. Write a formula that gives the measure of angle θ for a polygon with n sides.

6. Write a formula that gives the length of the apothem of a regular polygon inscribed in a circle of radius one inch.

7. How would the formula you wrote in Exercise 6 change if the radius of the circle was not one inch?

13-3 Trigonometric Functions of General Angles

What You'll Learn

- Find values of trigonometric functions for general angles.
- Use reference angles to find values of trigonometric functions.

Vocabulary

- quadrantal angle
- reference angle

How can you model the position of riders on a skycoaster?

A skycoaster consists of a large arch from which two steel cables hang and are attached to riders suited together in a harness. A third cable, coming from a larger tower behind the arch, is attached with a ripcord. Riders are hoisted to the top of the larger tower, pull the ripcord, and then plunge toward Earth. They swing through the arch, reaching speeds of more than 60 miles per hour. After the first several swings of a certain skycoaster, the angle θ of the riders from the center of the arch is given by $\theta = 0.2 \cos (1.6t)$, where t is the time in seconds after leaving the bottom of their swing.

TRIGONOMETRIC FUNCTIONS AND GENERAL ANGLES In Lesson 13-1, you found values of trigonometric functions whose domains were the set of all acute angles, angles between 0 and $\frac{\pi}{2}$, of a right triangle. For $t > 0$ in the equation above, you must find the cosine of an angle greater than $\frac{\pi}{2}$. In this lesson, we will extend the domain of trigonometric functions to include angles of *any* measure.

Key Concept — Trigonometric Functions, θ in Standard Position

Let θ be an angle in standard position and let $P(x, y)$ be a point on the terminal side of θ. Using the Pythagorean Theorem, the distance r from the origin to P is given by $r = \sqrt{x^2 + y^2}$. The trigonometric functions of an angle in standard position may be defined as follows.

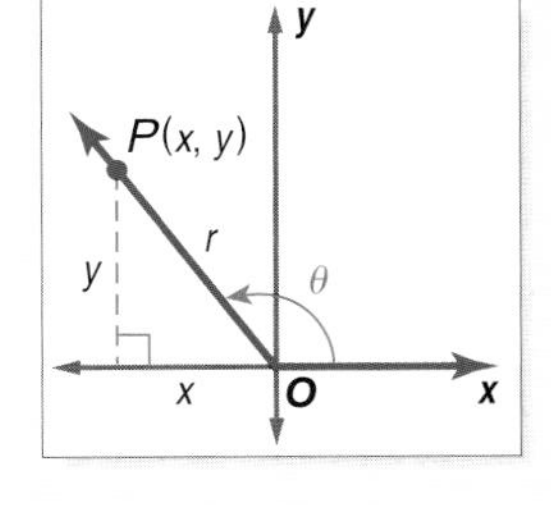

$\sin \theta = \frac{y}{r}$ $\qquad \cos \theta = \frac{x}{r}$ $\qquad \tan \theta = \frac{y}{x}, x \neq 0$

$\csc \theta = \frac{r}{y}, y \neq 0$ $\qquad \sec \theta = \frac{r}{x}, x \neq 0$ $\qquad \cot \theta = \frac{x}{y}, y \neq 0$

Example 1 Evaluate Trigonometric Functions for a Given Point

Find the exact values of the six trigonometric functions of θ if the terminal side of θ contains the point $(5, -12)$.

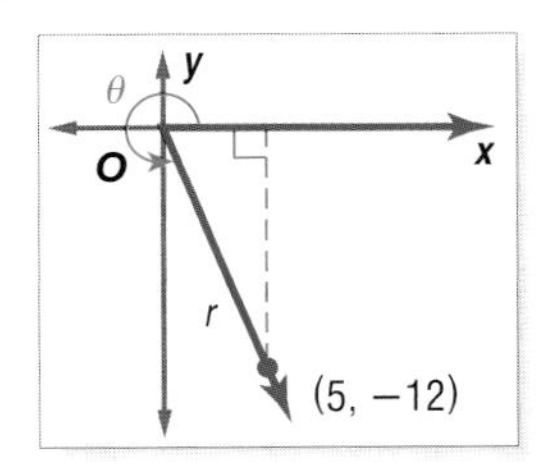

From the coordinates given, you know that $x = 5$ and $y = -12$. Use the Pythagorean Theorem to find r.

(continued on the next page)

$$r = \sqrt{x^2 + y^2}$$ Pythagorean Theorem

$$= \sqrt{5^2 + (-12)^2}$$ Replace x with 5 and y with -12.

$$= \sqrt{169} \text{ or } 13$$ Simplify.

Now, use $x = 5$, $y = -12$, and $r = 13$ to write the ratios.

$\sin \theta = \frac{y}{r}$
$= \frac{-12}{13}$ or $-\frac{12}{13}$

$\cos \theta = \frac{x}{r}$
$= \frac{5}{13}$

$\tan \theta = \frac{y}{x}$
$= \frac{-12}{5}$ or $-\frac{12}{5}$

$\csc \theta = \frac{r}{y}$
$= \frac{13}{-12}$ or $-\frac{13}{12}$

$\sec \theta = \frac{r}{x}$
$= \frac{13}{5}$

$\cot \theta = \frac{x}{y}$
$= \frac{5}{-12}$ or $-\frac{5}{12}$

If the terminal side of angle θ lies on one of the axes, θ is called a **quadrantal angle**. The quadrantal angles are 0°, 90°, 180°, and 270°. Notice that for these angles either x or y is equal to 0. Since division by zero is undefined, two of the trigonometric values are undefined for each quadrantal angle.

Key Concept — Quadrantal Angles

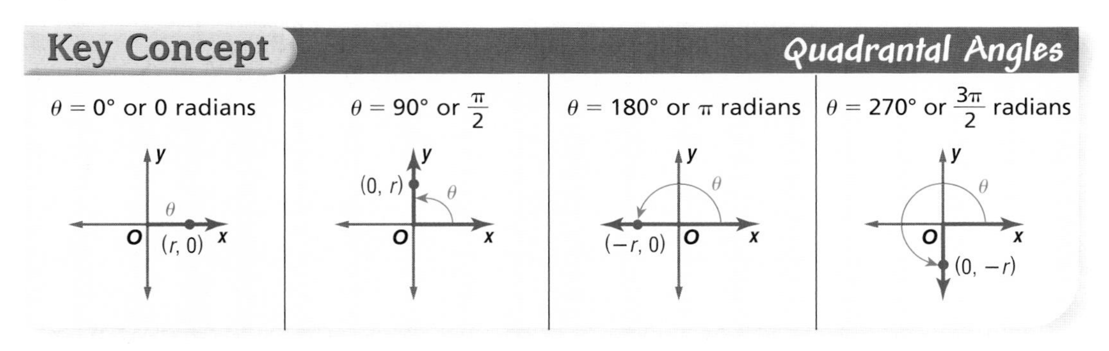

Example 2 Quadrantal Angles

Find the values of the six trigonometric functions for an angle in standard position that measures 270°.

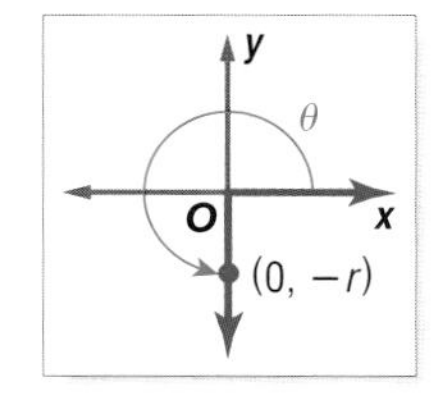

When $\theta = 270°$, $x = 0$ and $y = -r$.

$\sin \theta = \frac{y}{r}$
$= \frac{-r}{r}$ or -1

$\cos \theta = \frac{x}{r}$
$= \frac{0}{r}$ or 0

$\tan \theta = \frac{y}{x}$
$= \frac{-r}{0}$ or undefined

$\csc \theta = \frac{r}{y}$
$= \frac{r}{-r}$ or -1

$\sec \theta = \frac{r}{x}$
$= \frac{r}{0}$ or undefined

$\cot \theta = \frac{x}{y}$
$= \frac{0}{-r}$ or 0

REFERENCE ANGLES To find the values of trigonometric functions of angles greater than 90° (or less than 0°), you need to know how to find the measures of reference angles. If θ is a nonquadrantal angle in standard position, its **reference angle**, θ', is defined as the acute angle formed by the terminal side of θ and the x-axis.

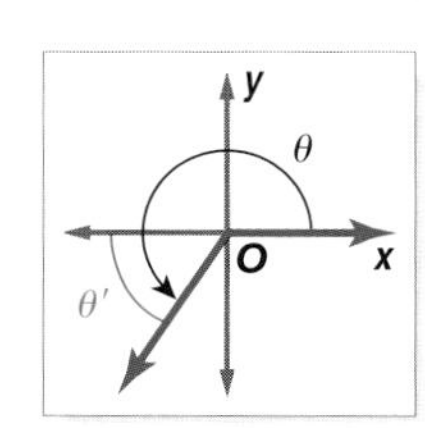

Study Tip

Reading Math
θ' is read *theta prime*.

You can use the rule below to find the reference angle for any nonquadrantal angle θ where $0° < \theta < 360°$ (or $0 < \theta < 2\pi$).

Key Concept — Reference Angle Rule

For any nonquadrantal angle θ, $0° < \theta < 360°$ (or $0 < \theta < 2\pi$), its reference angle θ' is defined as follows.

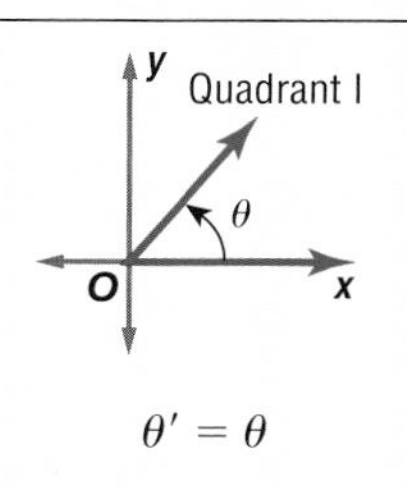

$\theta' = \theta$

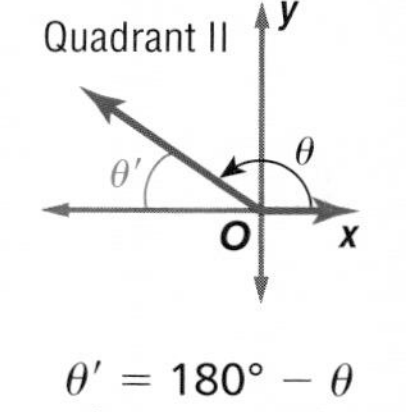

$\theta' = 180° - \theta$
$(\theta' = \pi - \theta)$

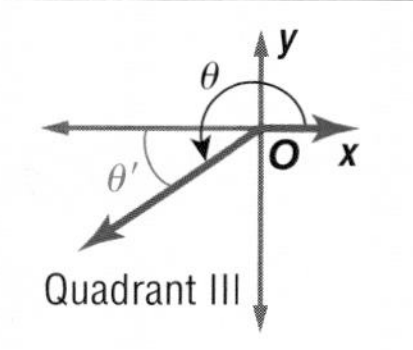

$\theta' = \theta - 180°$
$(\theta' = \theta - \pi)$

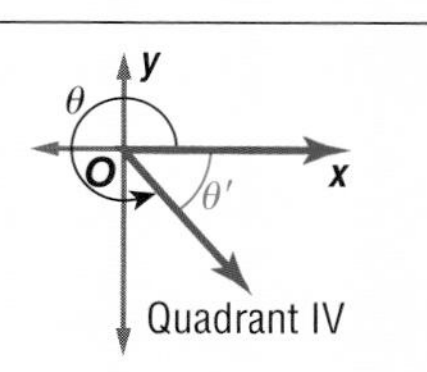

$\theta' = 360° - \theta$
$(\theta' = 2\pi - \theta)$

If the measure of θ is greater than 360° or less than 0°, its reference angle can be found by associating it with a coterminal angle of positive measure between 0° and 360°.

Example 3 Find the Reference Angle for a Given Angle

Sketch each angle. Then find its reference angle.

a. 300°

Because the terminal side of 300° lies in Quadrant IV, the reference angle is $360° - 300°$ or $60°$.

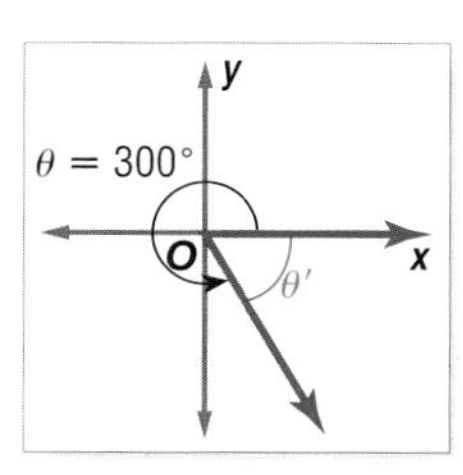

b. $-\frac{2\pi}{3}$

A coterminal angle of $-\frac{2\pi}{3}$ is $2\pi - \frac{2\pi}{3}$ or $\frac{4\pi}{3}$.

Because the terminal side of this angle lies in Quadrant III, the reference angle is $\frac{4\pi}{3} - \pi$ or $\frac{\pi}{3}$.

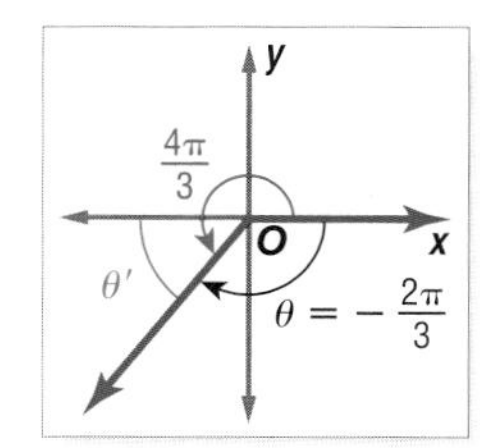

To use the reference angle θ' to find a trigonometric value of θ, you need to know the sign of that function for an angle θ. From the function definitions, these signs are determined by x and y, since r is always positive. Thus, the sign of each trigonometric function is determined by the quadrant in which the terminal side of θ lies.

The chart below summarizes the signs of the trigonometric functions for each quadrant.

	Quadrant			
Function	**I**	**II**	**III**	**IV**
sin θ or csc θ	+	+	−	−
cos θ or sec θ	+	−	−	+
tan θ or cot θ	+	−	+	−

Use the following steps to find the value of a trigonometric function of any angle θ.

Step 1 Find the reference angle θ'.

Step 2 Find the value of the trigonometric function for θ'.

Step 3 Using the quadrant in which the terminal side of θ lies, determine the sign of the trigonometric function value of θ.

Example 4 Use a Reference Angle to Find a Trigonometric Value

Study Tip

Look Back
To review **trigonometric values of angles measuring 30°, 45°, and 60°**, see Lesson 13-1.

Find the exact value of each trigonometric function.

a. sin 120°

Because the terminal side of 120° lies in Quadrant II, the reference angle θ' is $180° - 120°$ or 60°. The sine function is positive in Quadrant II, so $\sin 120° = \sin 60°$ or $\frac{\sqrt{3}}{2}$.

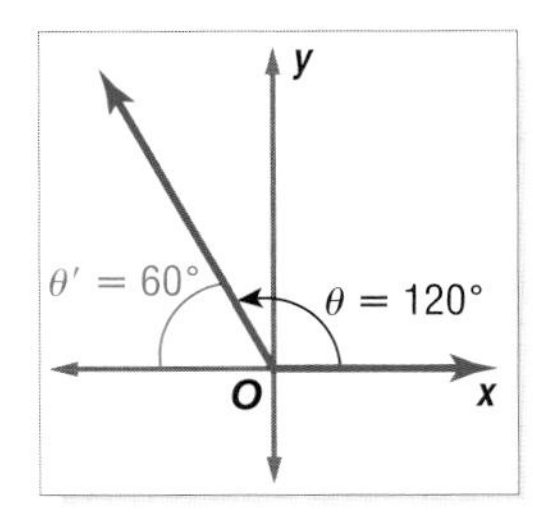

b. $\cot \frac{7\pi}{4}$

Because the terminal side of $\frac{7\pi}{4}$ lies in Quadrant IV, the reference angle θ' is $2\pi - \frac{7\pi}{4}$ or $\frac{\pi}{4}$. The cotangent function is negative in Quadrant IV.

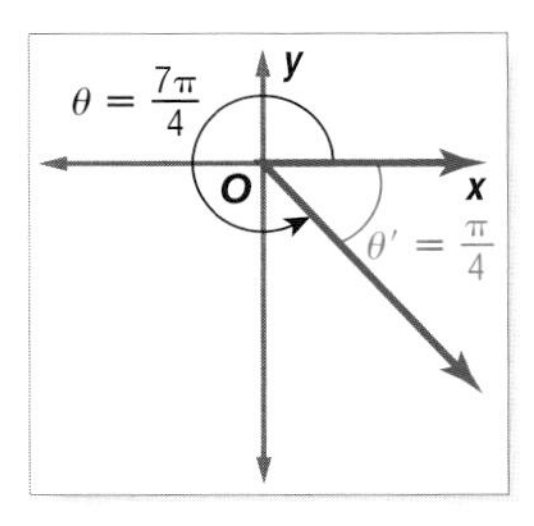

$$\cot \frac{7\pi}{4} = -\cot \frac{\pi}{4}$$

$$= -\cot 45° \quad \frac{\pi}{4} \text{ radians} = 45°$$

$$= -1 \quad \cot 45° = 1$$

If you know the quadrant that contains the terminal side of θ in standard position and the exact value of one trigonometric function of θ, you can find the values of the other trigonometric functions of θ using the function definitions.

Example 5 Quadrant and One Trigonometric Value of θ

Suppose θ is an angle in standard position whose terminal side is in Quadrant III and $\sec \theta = -\frac{4}{3}$. Find the exact values of the remaining five trigonometric functions of θ.

Draw a diagram of this angle, labeling a point $P(x, y)$ on the terminal side of θ. Use the definition of secant to find the values of x and r.

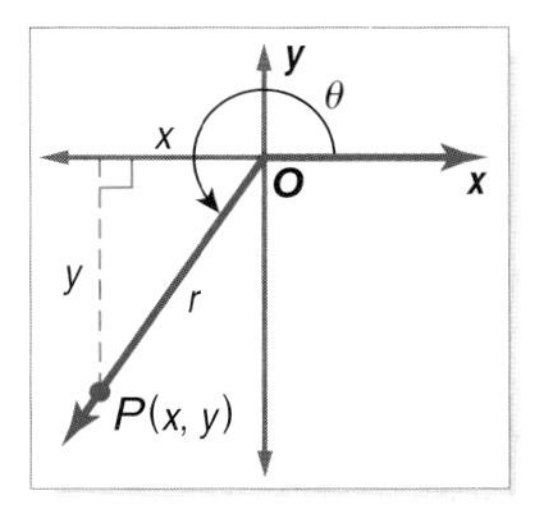

$$\sec \theta = -\frac{4}{3} \quad \text{Given}$$

$$\frac{r}{x} = -\frac{4}{3} \quad \text{Definition of secant}$$

Since x is negative in Quadrant III and r is always positive, $x = -3$ and $r = 4$. Use these values and the Pythagorean Theorem to find y.

$$x^2 + y^2 = r^2 \quad \text{Pythagorean Theorem}$$
$$(-3)^2 + y^2 = 4^2 \quad \text{Replace } x \text{ with } -3 \text{ and } r \text{ with 4.}$$
$$y^2 = 16 - 9 \quad \text{Simplify. Then subtract 9 from each side.}$$
$$y = \pm\sqrt{7} \quad \text{Simplify. Then take the square root of each side.}$$
$$y = -\sqrt{7} \quad y \text{ is negative in Quadrant III.}$$

Use $x = -3$, $y = -\sqrt{7}$, and $r = 4$ to write the remaining trigonometric ratios.

$$\sin \theta = \frac{y}{r} = \frac{-\sqrt{7}}{4} \qquad \cos \theta = \frac{x}{r} = \frac{-3}{4} \qquad \tan \theta = \frac{y}{x} = \frac{-\sqrt{7}}{-3} \text{ or } \frac{\sqrt{7}}{3}$$

$$\csc \theta = \frac{r}{y} = -\frac{4}{\sqrt{7}} \text{ or } -\frac{4\sqrt{7}}{7} \qquad \cot \theta = \frac{x}{y} = \frac{-3}{-\sqrt{7}} \text{ or } \frac{3\sqrt{7}}{7}$$

Just as an exact point on the terminal side of an angle can be used to find trigonometric function values, trigonometric function values can be used to find the exact coordinates of a point on the terminal side of an angle.

Example 6 Find Coordinates Given a Radius and an Angle

ROBOTICS In a robotics competition, a robotic arm 4 meters long is to pick up an object at point A and release it into a container at point B. The robot's owner programs the arm to rotate through an angle of precisely 135° to accomplish this task. What is the new position of the object relative to the pivot point O?

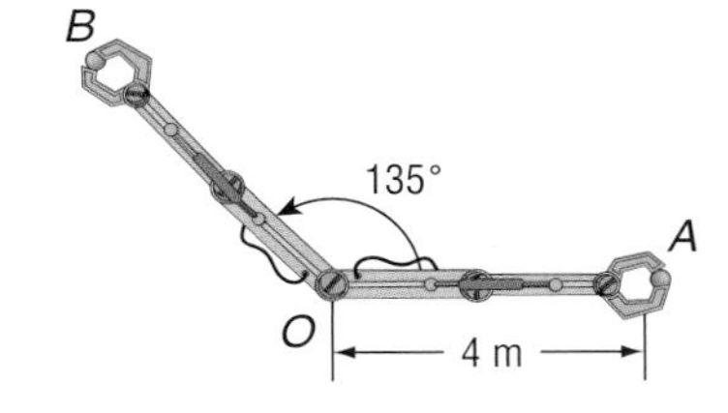

With the pivot point at the origin and the angle through which the arm rotates in standard position, point A has coordinates (4, 0). The reference angle θ' for 135° is $180° - 135°$ or 45°.

Let the position of point B have coordinates (x, y). Then, use the definitions of sine and cosine to find the value of x and y. The value of r is the length of the robotic arm, 4 meters. Because B is in Quadrant II, the cosine of 135° is negative.

$$\cos 135° = \frac{x}{r} \quad \text{cosine ratio}$$
$$-\cos 45° = \frac{x}{4} \quad 180° - 135° = 45°$$
$$-\frac{\sqrt{2}}{2} = \frac{x}{4} \quad \cos 45° = \frac{\sqrt{2}}{2}$$
$$-2\sqrt{2} = x \quad \text{Solve for } x.$$

$$\sin 135° = \frac{y}{r} \quad \text{sine ratio}$$
$$\sin 45° = \frac{y}{4} \quad 180° - 35° = 45°$$
$$\frac{\sqrt{2}}{2} = \frac{y}{4} \quad \sin 45° = \frac{\sqrt{2}}{2}$$
$$2\sqrt{2} = y \quad \text{Solve for } y.$$

The exact coordinates of B are $(-2\sqrt{2}, 2\sqrt{2})$. Since $2\sqrt{2}$ is about 2.83, the object is about 2.83 meters to the left of the pivot point and about 2.83 meters in front of the pivot point.

More About. . .

Robotics

RoboCup is an annual event in which teams from all over the world compete in a series of soccer matches in various classes according to the size and intellectual capacity of their robot. The robots are programmed to react to the ball and communicate with each other.

Source: www.robocup.org

Check for Understanding

Concept Check

1. **Determine** whether the following statement is *true* or *false*. If true, explain your reasoning. If false, give a counterexample.

   The values of the secant and tangent functions for any quadrantal angle are undefined.

2. **OPEN ENDED** Give an example of an angle whose sine is negative.

3. **Explain** how the reference angle θ' is used to find the value of a trigonometric function of θ, where θ is greater than 90°.

Guided Practice

Find the exact values of the six trigonometric functions of θ if the terminal side of θ in standard position contains the given point.

4. $(-15, 8)$
5. $(-3, 0)$
6. $(4, 4)$

Sketch each angle. Then find its reference angle.

7. 235°
8. $\frac{7\pi}{4}$
9. $-240°$

Find the exact value of each trigonometric function.

10. $\sin 300°$
11. $\cos 180°$
12. $\tan \frac{5\pi}{3}$
13. $\sec \frac{7\pi}{6}$

Suppose θ is an angle in standard position whose terminal side is in the given quadrant. For each function, find the exact values of the remaining five trigonometric functions of θ.

14. $\cos \theta = -\frac{1}{2}$, Quadrant II
15. $\cot \theta = -\frac{\sqrt{2}}{2}$, Quadrant IV

Application

16. **BASKETBALL** The maximum height H in feet that a basketball reaches after being shot is given by the formula $H = \frac{V_0^2 (\sin \theta)^2}{64}$, where V_0 represents the initial velocity in feet per second, θ represents the degree measure of the angle that the path of the basketball makes with the ground. Find the maximum height reached by a ball shot with an initial velocity of 30 feet per second at an angle of 70°.

Practice and Apply

Homework Help

For Exercises	See Examples
17–24	1
25–32	3
33–46	2, 4
47–52	5
53–55	6

Extra Practice
See page 857.

Find the exact values of the six trigonometric functions of θ if the terminal side of θ in standard position contains the given point.

17. $(7, 24)$
18. $(2, 1)$
19. $(5, -8)$
20. $(4, -3)$
21. $(0, -6)$
22. $(-1, 0)$
23. $(\sqrt{2}, -\sqrt{2})$
24. $(-\sqrt{3}, -\sqrt{6})$

Sketch each angle. Then find its reference angle.

25. 315°
26. 240°
27. $-210°$
28. $-125°$
29. $\frac{5\pi}{4}$
30. $\frac{5\pi}{6}$
31. $\frac{13\pi}{7}$
32. $-\frac{2\pi}{3}$

Find the exact value of each trigonometric function.

33. $\sin 240°$
34. $\sec 120°$
35. $\tan 300°$
36. $\cot 510°$
37. $\csc 5400°$
38. $\cos \frac{11\pi}{3}$
39. $\cot \left(-\frac{5\pi}{6}\right)$
40. $\sin \frac{3\pi}{4}$
41. $\sec \frac{3\pi}{2}$
42. $\csc \frac{17\pi}{6}$
43. $\cos (-30°)$
44. $\tan \left(-\frac{5\pi}{4}\right)$

45. **SKYCOASTING** Mikhail and Anya visit a local amusement park to ride a skycoaster. After the first several swings, the angle the skycoaster makes with the vertical is modeled by $\theta = 0.2 \cos \pi t$, with θ measured in radians and t measured in seconds. Determine the measure of the angle for $t = 0, 0.5, 1, 1.5, 2, 2.5$, and 3 in both radians and degrees.

46. **NAVIGATION** Ships and airplanes measure distance in nautical miles. The formula 1 nautical mile $= 6077 - 31 \cos 2\theta$ feet, where θ is the latitude in degrees, can be used to find the approximate length of a nautical mile at a certain latitude. Find the length of a nautical mile where the latitude is 60°.

More About. . .

Baseball

If a major league pitcher throws a pitch at 95 miles per hour, it takes only about 4 tenths of a second for the ball to travel the 60 feet, 6 inches from the pitcher's mound to home plate. In that time, the hitter must decide whether to swing at the ball and if so, when to swing.

Source: www.exploratorium.edu

Suppose θ is an angle in standard position whose terminal side is in the given quadrant. For each function, find the exact values of the remaining five trigonometric functions of θ.

47. $\cos \theta = \frac{3}{5}$, Quadrant IV
48. $\tan \theta = -\frac{1}{5}$, Quadrant II
49. $\sin \theta = \frac{1}{3}$, Quadrant II
50. $\cot \theta = \frac{1}{2}$, Quadrant III
51. $\sec \theta = -\sqrt{10}$, Quadrant III
52. $\csc \theta = -5$, Quadrant IV

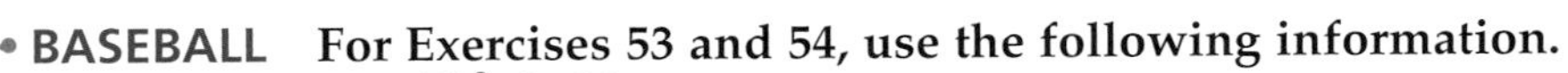

BASEBALL For Exercises 53 and 54, use the following information.

The formula $R = \frac{V_0^2 \sin 2\theta}{32}$ gives the distance of a baseball that is hit at an initial velocity of V_0 feet per second at an angle of θ with the ground.

53. If the ball was hit with an initial velocity of 80 feet per second at an angle of 30°, how far was it hit?

54. Which angle will result in the greatest distance? Explain your reasoning.

55. **CAROUSELS** Anthony's little brother gets on a carousel that is 8 meters in diameter. At the start of the ride, his brother is 3 meters from the fence to the ride. How far will his brother be from the fence after the carousel rotates 240°?

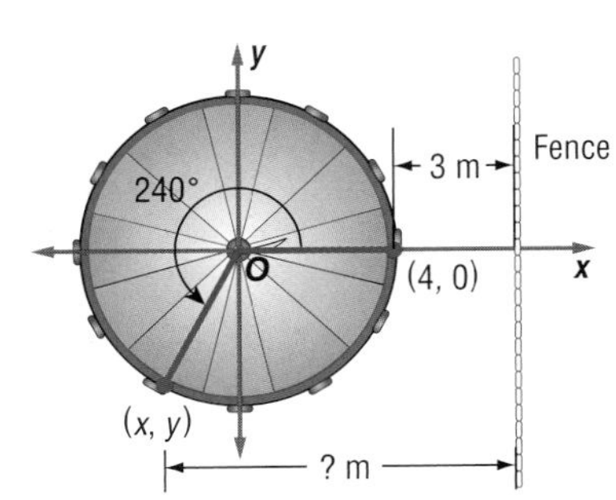

Online Research **Data Update** What is the diameter of the world's largest carousel? Visit www.algebra2.com/data_update to learn more.

CRITICAL THINKING Suppose θ is an angle in standard position with the given conditions. State the quadrant(s) in which the terminal side of θ lies.

56. $\sin \theta > 0$
57. $\sin \theta > 0$, $\cos \theta < 0$
58. $\tan \theta > 0$, $\cos \theta < 0$

59. **WRITING IN MATH** Answer the question that was posed at the beginning of the lesson.

How can you model the position of riders on a skycoaster?

Include the following in your answer:

- an explanation of how you could use the cosine of the angle θ and the length of the cable from which they swing to find the horizontal position of a person on a skycoaster relative to the center of the arch, and
- an explanation of how you would use the angle θ, the height of the tower, and the length of the cable to find the height of riders from the ground.

60. If the cotangent of angle θ is 1, then the tangent of angle θ is

Ⓐ -1. Ⓑ 0. Ⓒ 1. Ⓓ 3.

61. **SHORT RESPONSE** Find the exact coordinates of point P, which is located at the intersection of a circle of radius 5 and the terminal side of angle θ measuring $\frac{5\pi}{3}$.

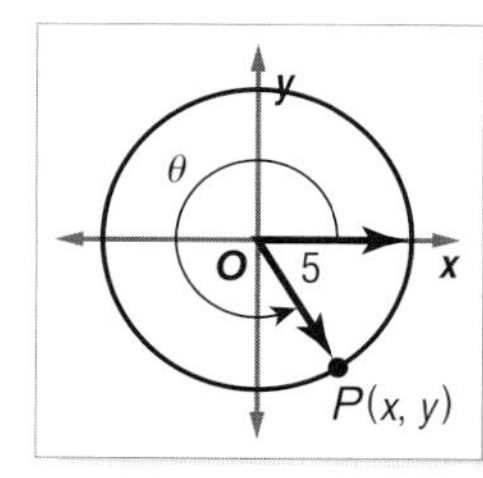

Maintain Your Skills

Mixed Review

Rewrite each degree measure in radians and each radian measure in degrees. *(Lesson 13-2)*

62. $90°$ 63. $\frac{5\pi}{3}$ 64. 5

Write an equation involving sin, cos, or tan that can be used to find x. Then solve the equation. Round measures of sides to the nearest tenth and measures of angles to the nearest degree. *(Lesson 13-1)*

65.
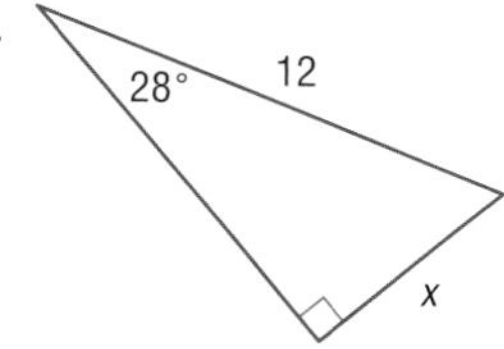

66.
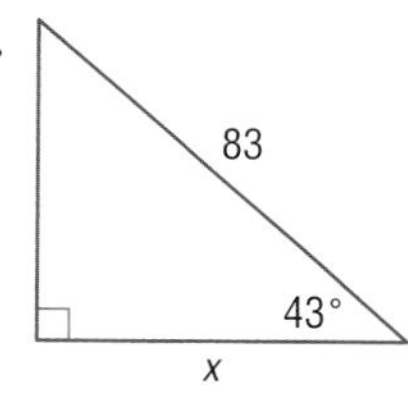

67.
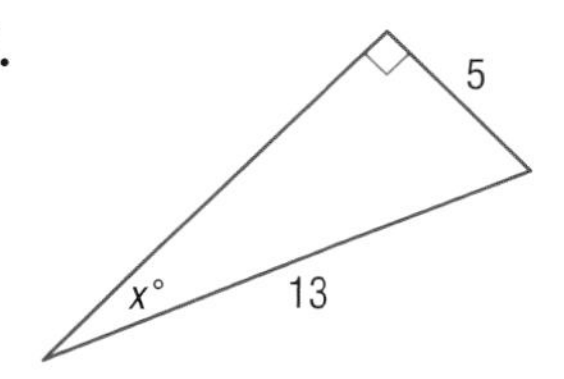

68. **LITERATURE** In one of *Grimm's Fairy Tales*, Rumpelstiltskin has the ability to spin straw into gold. Suppose on the first day, he spun 5 pieces of straw into gold, and each day thereafter he spun twice as much. How many pieces of straw would he have spun into gold by the end of the week? *(Lesson 11-4)*

Use Cramer's Rule to solve each system of equations. *(Lesson 4-6)*

69. $3x - 4y = 13$
$-2x + 5y = -4$

70. $5x + 7y = 1$
$3x + 5y = 3$

71. $2x + 3y = -2$
$-6x + y = -34$

Getting Ready for the Next Lesson

PREREQUISITE SKILL Solve each equation. Round to the nearest tenth.
(To review ***solving equations with trigonometric functions****, see Lesson 13-1.)*

72. $\frac{a}{\sin 32°} = \frac{8}{\sin 65°}$ 73. $\frac{b}{\sin 45°} = \frac{21}{\sin 100°}$ 74. $\frac{c}{\sin 60°} = \frac{3}{\sin 75°}$

75. $\frac{\sin A}{14} = \frac{\sin 104°}{25}$ 76. $\frac{\sin B}{3} = \frac{\sin 55°}{7}$ 77. $\frac{\sin C}{10} = \frac{\sin 35°}{9}$

13-4 Law of Sines

What You'll Learn

- Solve problems by using the Law of Sines.
- Determine whether a triangle has one, two, or no solutions.

Vocabulary

- Law of Sines

How can trigonometry be used to find the area of a triangle?

You know how to find the area of a triangle when the base and the height are known. Using this formula, the area of $\triangle ABC$ below is $\frac{1}{2}ch$. If the height h of this triangle were not known, you could still find the area given the measures of angle A and the length of side b.

$$\sin A = \frac{h}{b} \rightarrow h = b \sin A$$

By combining this equation with the area formula, you can find a new formula for the area of the triangle.

$$\text{Area} = \frac{1}{2}ch \rightarrow \text{Area} = \frac{1}{2}c(b \sin A)$$

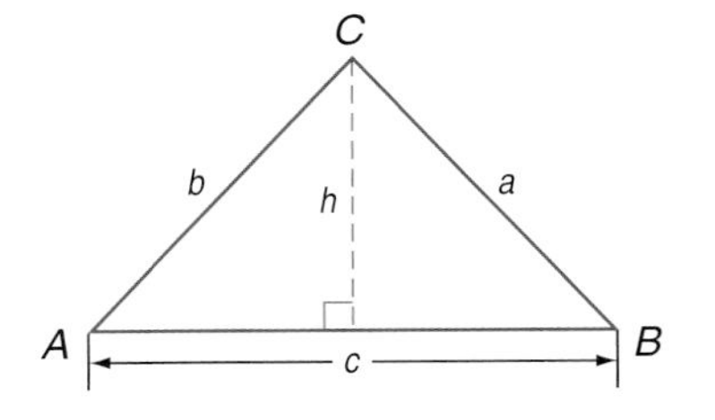

LAW OF SINES You can find two other formulas for the area of the triangle above in a similar way. These formulas, summarized below, allow you to find the area of any triangle when you know the measures of two sides and the included angle.

Key Concept — Area of a Triangle

- **Words** The area of a triangle is one half the product of the lengths of two sides and the sine of their included angle.
- **Symbols** $\text{area} = \frac{1}{2}bc \sin A$

  $\text{area} = \frac{1}{2}ac \sin B$

  $\text{area} = \frac{1}{2}ab \sin C$

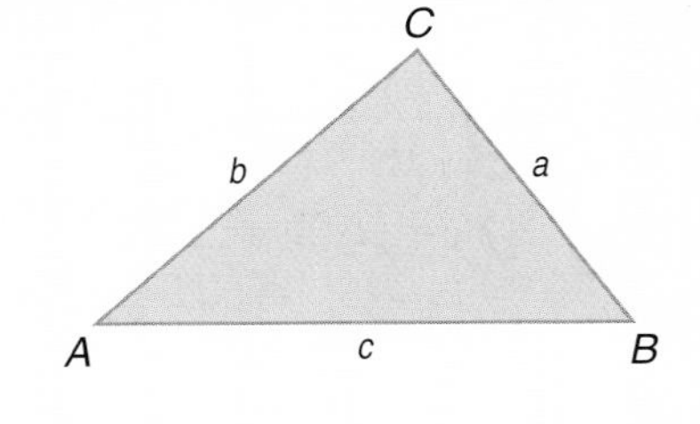

Example 1 Find the Area of a Triangle

Find the area of $\triangle ABC$ to the nearest tenth.

In this triangle, $a = 5$, $c = 6$, and $B = 112°$. Choose the second formula because you know the values of its variables.

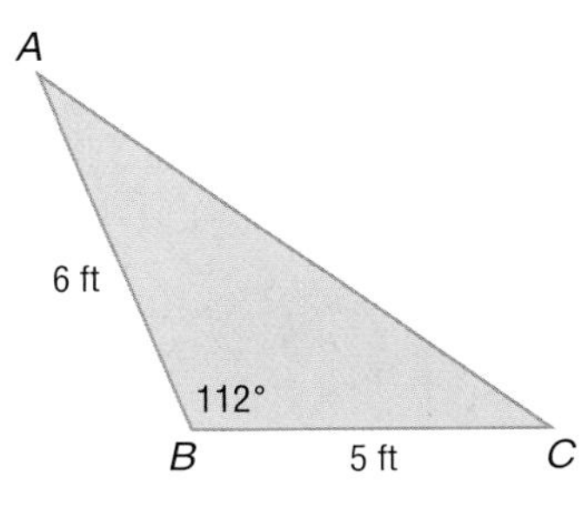

$\text{Area} = \frac{1}{2}ac \sin B$	Area formula
$= \frac{1}{2}(5)(6) \sin 112°$	Replace a with 5, c with 6, and B with 112°.
≈ 13.9	Use a calculator.

To the nearest tenth, the area is 13.9 square feet.

All of the area formulas for $\triangle ABC$ represent the area of the same triangle. So, $\frac{1}{2}bc \sin A$, $\frac{1}{2}ac \sin B$, and $\frac{1}{2}ab \sin C$ are all equal. You can use this fact to derive the **Law of Sines**.

$$\frac{1}{2}bc \sin A = \frac{1}{2}ac \sin B = \frac{1}{2}ab \sin C$$ Set area formulas equal to each other.

$$\frac{\frac{1}{2}bc \sin A}{\frac{1}{2}abc} = \frac{\frac{1}{2}ac \sin B}{\frac{1}{2}abc} = \frac{\frac{1}{2}ab \sin C}{\frac{1}{2}abc}$$ Divide each expression by $\frac{1}{2}abc$.

$$\frac{\sin A}{a} = \frac{\sin B}{b} = \frac{\sin C}{c}$$ Simplify.

Study Tip

Alternate Representations
The Law of Sines may also be written as $\frac{a}{\sin A} = \frac{b}{\sin B} = \frac{c}{\sin C}$.

Key Concept — *Law of Sines*

Let $\triangle ABC$ be any triangle with a, b, and c representing the measures of sides opposite angles with measurements A, B, and C respectively. Then,

$$\frac{\sin A}{a} = \frac{\sin B}{b} = \frac{\sin C}{c}.$$

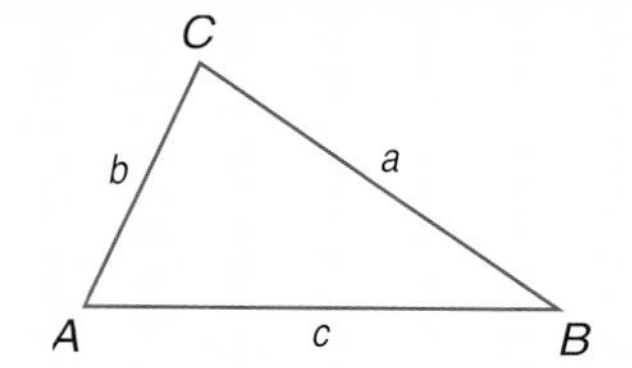

The Law of Sines can be used to write three different equations.

$$\frac{\sin A}{a} = \frac{\sin B}{b} \quad \text{or} \quad \frac{\sin B}{b} = \frac{\sin C}{c} \quad \text{or} \quad \frac{\sin A}{a} = \frac{\sin C}{c}$$

In Lesson 13-1, you learned how to solve right triangles. To solve *any* triangle, you can apply the Law of Sines if you know

- the measures of two angles and any side or
- the measures of two sides and the angle opposite one of them.

Example 2 Solve a Triangle Given Two Angles and a Side

Solve $\triangle ABC$.

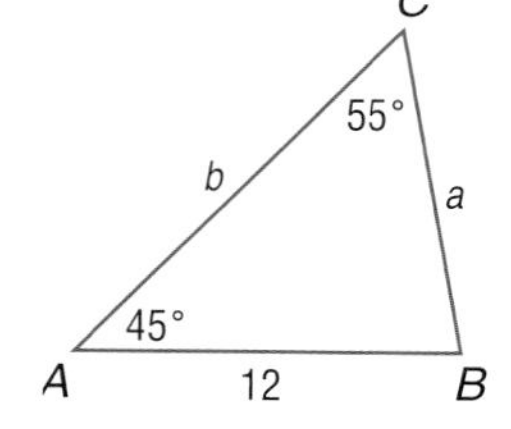

You are given the measures of two angles and a side. First, find the measure of the third angle.

$45° + 55° + B = 180°$ The sum of the angle measures of a triangle is 180°.

$B = 80°$ $180 - (45 + 55) = 80$

Now use the Law of Sines to find a and b. Write two equations, each with one variable.

$\frac{\sin A}{a} = \frac{\sin C}{c}$	Law of Sines	$\frac{\sin B}{b} = \frac{\sin C}{c}$
$\frac{\sin 45°}{a} = \frac{\sin 55°}{12}$	Replace A with 45°, B with 80°, C with 55°, and c with 12.	$\frac{\sin 80°}{b} = \frac{\sin 55°}{12}$
$a = \frac{12 \sin 45°}{\sin 55°}$	Solve for the variable.	$b = \frac{12 \sin 80°}{\sin 55°}$
$a \approx 10.4$	Use a calculator.	$b \approx 14.4$

Therefore, $B = 80°$, $a \approx 10.4$, and $b \approx 14.4$.

ONE, TWO, OR NO SOLUTIONS When solving a triangle, you must analyze the data you are given to determine whether there is a solution. For example, if you are given the measures of two angles and a side, as in Example 2, the triangle has a unique solution. However, if you are given the measures of two sides and the angle opposite one of them, a single solution may not exist. One of the following will be true.

- No triangle exists, and there is no solution.
- Exactly one triangle exists, and there is one solution.
- Two triangles exist, and there are two solutions.

Key Concept — Possible Triangles Given Two Sides and One Opposite Angle

Suppose you are given a, b, and A for a triangle.

A Is Acute ($A < 90°$).

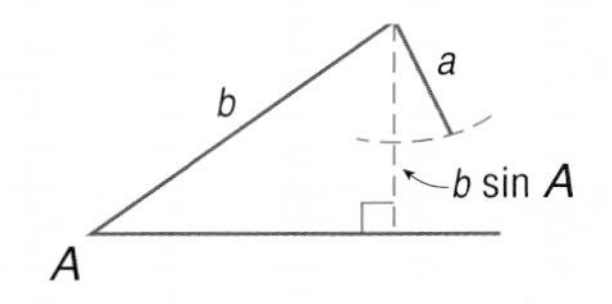

$a < b \sin A$
no solution

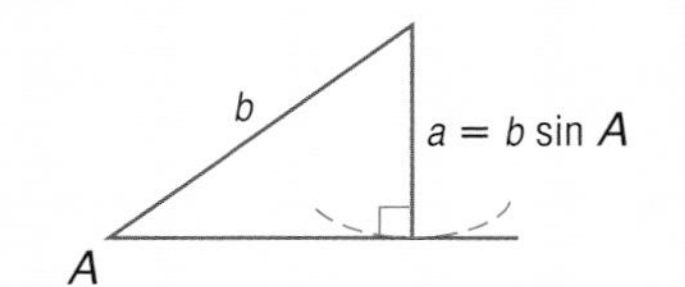

$a = b \sin A$
one solution

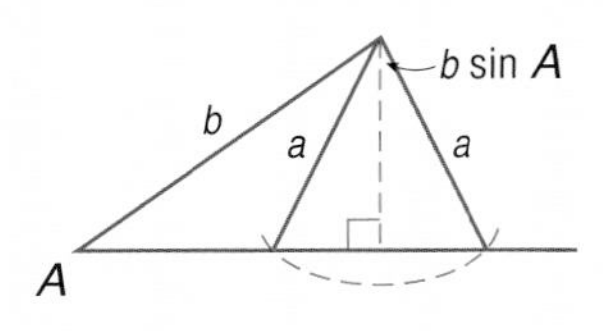

$b > a > b \sin A$
two solutions

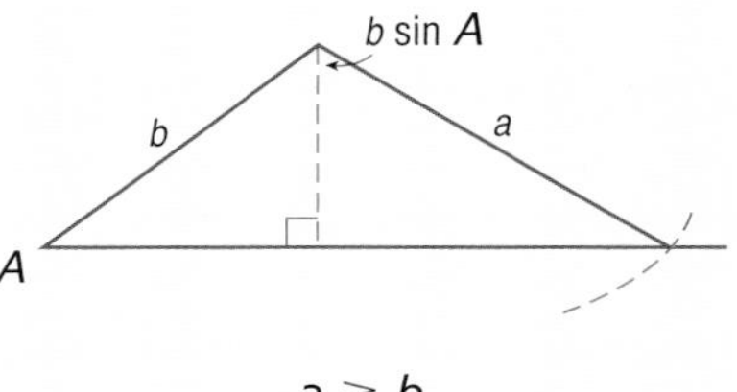

$a \geq b$
one solution

A Is Right or Obtuse ($A \geq 90°$).

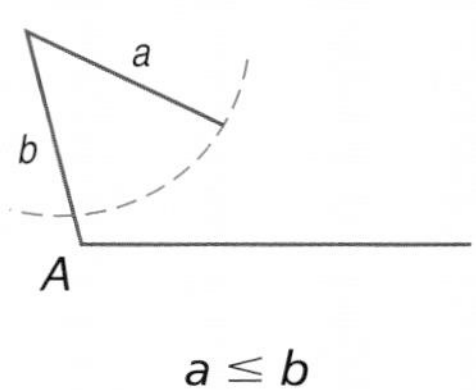

$a \leq b$
no solution

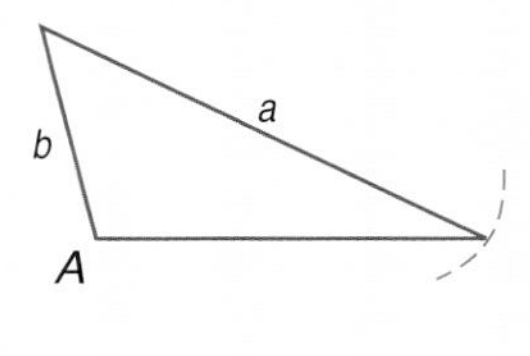

$a > b$
one solution

Example 3 One Solution

In $\triangle ABC$, $A = 118°$, $a = 20$, and $b = 17$. Determine whether $\triangle ABC$ has *no* solution, *one* solution, or *two* solutions. Then solve $\triangle ABC$.

Because angle A is obtuse and $a > b$, you know that one solution exists.

Make a sketch and then use the Law of Sines to find B.

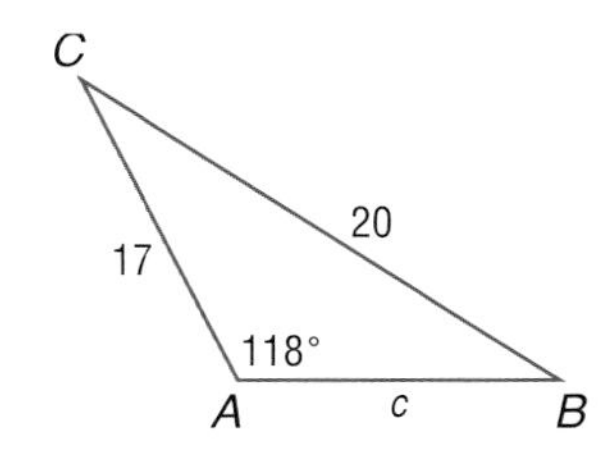

$\frac{\sin B}{17} = \frac{\sin 118°}{20}$ Law of Sines

$\sin B = \frac{17 \sin 118°}{20}$ Multiply each side by 17.

$\sin B \approx 0.7505$ Use a calculator.

$B \approx 49°$ Use the $\sin^{-1}$ function.

The measure of angle C is approximately $180 - (118 + 49)$ or $13°$.

Use the Law of Sines again to find c.

$\frac{\sin 13}{c} = \frac{\sin 118°}{20}$ Law of Sines

$c = \frac{20 \sin 13°}{\sin 118°}$ or about 5.1 Use a calculator.

Therefore, $B \approx 49°$, $C \approx 13°$, and $c \approx 5.1$.

Example 4 No Solution

In $\triangle ABC$, $A = 50°$, $a = 5$, and $b = 9$. Determine whether $\triangle ABC$ has *no* solution, *one* solution, or *two* solutions. Then solve $\triangle ABC$.

Since angle A is acute, find $b \sin A$ and compare it with a.

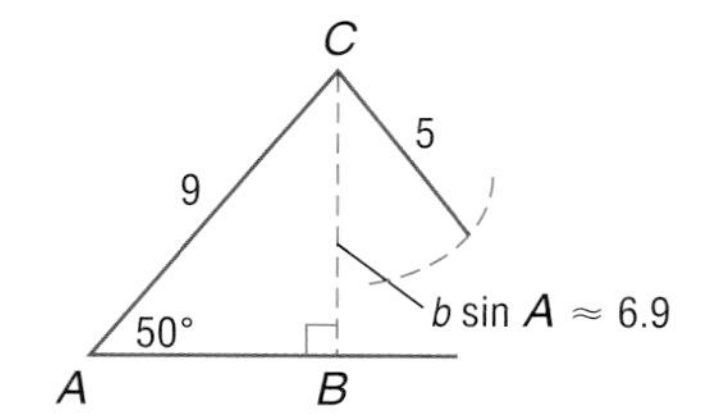

$b \sin A = 9 \sin 50°$ Replace b with 9 and A with 50°.

≈ 6.9 Use a calculator.

Since $5 < 6.9$, there is no solution.

Study Tip

A Is Acute

We compare $b \sin A$ to a because $b \sin A$ is the minimum distance from C to $\overline{AB}$ when A is acute.

When two solutions for a triangle exist, it is called the *ambiguous case.*

Example 5 Two Solutions

In $\triangle ABC$, $A = 39°$, $a = 10$, and $b = 14$. Determine whether $\triangle ABC$ has *no* solution, *one* solution, or *two* solutions. Then solve $\triangle ABC$.

Since angle A is acute, find $b \sin A$ and compare it with a.

$b \sin A = 14 \sin 39°$ Replace b with 14 and A with 39°.

≈ 8.81 Use a calculator.

Since $14 > 10 > 8.81$, there are two solutions. Thus, there are two possible triangles to be solved.

Case 1 Acute Angle B

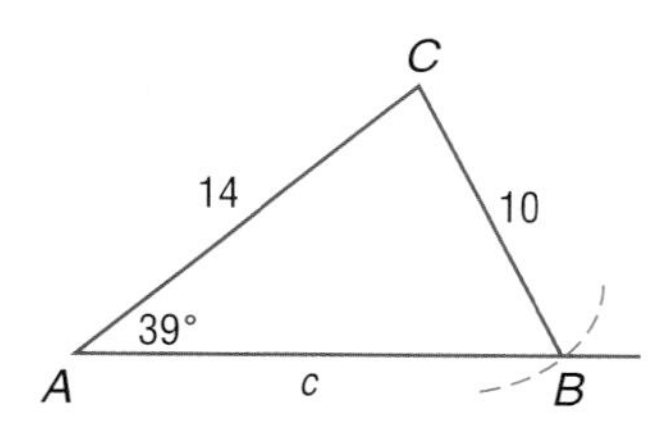

First, use the Law of Sines to find B.

$$\frac{\sin B}{14} = \frac{\sin 39°}{10}$$

$$\sin B = \frac{14 \sin 39°}{10}$$

$$\sin B = 0.8810$$

$$B \approx 62°$$

The measure of angle C is approximately $180 - (39 + 62)$ or $79°$.

Use the Law of Sines again to find c.

$$\frac{\sin 79°}{c} = \frac{\sin 39°}{10}$$

$$c = \frac{10 \sin 79°}{\sin 39°}$$

$$c \approx 15.6$$

Therefore, $B \approx 62°$, $C \approx 79°$, and $c \approx 15.6$.

Case 2 Obtuse Angle B

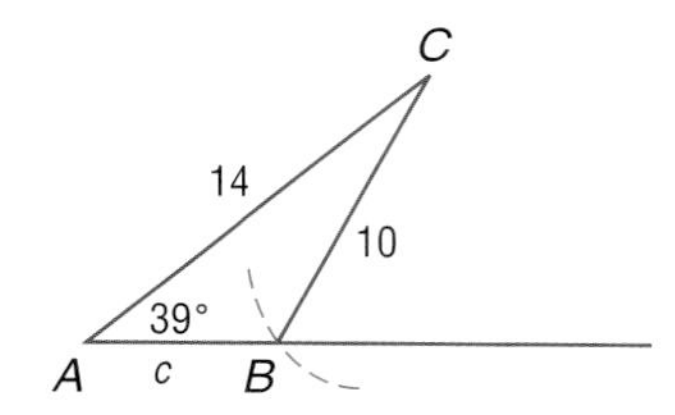

To find B, you need to find an obtuse angle whose sine is also 0.8810. To do this, subtract the angle given by your calculator, 62°, from 180°. So B is approximately $180 - 62$ or $118°$.

The measure of angle C is approximately $180 - (39 + 118)$ or $23°$.

Use the Law of Sines to find c.

$$\frac{\sin 23°}{c} = \frac{\sin 39°}{10}$$

$$c = \frac{10 \sin 23°}{\sin 39°}$$

$$c \approx 6.2$$

Therefore, $B \approx 118°$, $C \approx 23°$, and $c \approx 6.2$.

Study Tip

Alternate Method

Another way to find the obtuse angle in Case 2 of Example 5 is to notice in the figure below that $\triangle CBB'$ is isosceles. Since the base angles of an isosceles triangle are always congruent and $m\angle B' = 62°$, $m\angle CBB' = 62°$. Also, $\angle ABC$ and $m\angle CBB'$ are supplementary. Therefore, $m\angle ABC = 180° - 62°$ or $118°$.

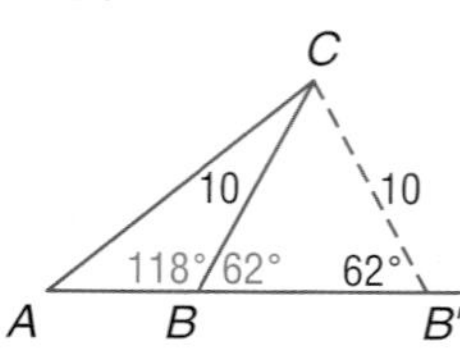

Example 6 Use the Law of Sines to Solve a Problem

LIGHTHOUSES A lighthouse is located on a rock at a certain distance from a straight shore. The light revolves counterclockwise at a steady rate of one revolution per minute. As the beam revolves, it strikes a point on the shore that is 2000 feet from the lighthouse. Three seconds later, the light strikes a point 750 feet further down the shore. To the nearest foot, how far is the lighthouse from the shore?

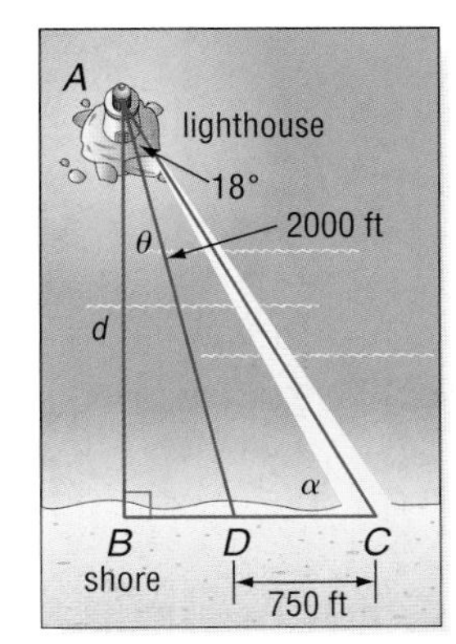

Because the lighthouse makes one revolution every 60 seconds, the angle through which the light revolves in 3 seconds is $\frac{3}{60}(360°)$ or 18°.

Use the Law of Sines to find the measure of angle α.

$\frac{\sin \alpha}{2000} = \frac{\sin 18°}{750}$	Law of Sines
$\sin \alpha = \frac{2000 \sin 18°}{750}$	Multiply each side by 2000.
$\sin \alpha \approx 0.8240$	Use a calculator.
$\alpha \approx 55°$	Use the $\sin^{-1}$ function.

Use this angle measure to find the measure of angle θ. Since $\triangle ABC$ is a right triangle, the measures of angle α and $\angle BAC$ are complementary.

$\alpha + m\angle BAC = 90°$	Angles α and $\angle BAC$ are complementary.
$55° + (\theta + 18°) \approx 90°$	$\alpha \approx 55°$ and $m\angle BAC = \theta + 18°$
$\theta + 73° \approx 90°$	Simplify.
$\theta \approx 17°$	Solve for θ.

To find the distance from the lighthouse to the shore, solve $\triangle ABD$ for d.

$\cos \theta = \frac{AB}{AD}$	Cosine ratio
$\cos 17° \approx \frac{d}{2000}$	$\theta = 17°$ and $AD = 2000$
$d \approx 2000 \cos 17°$	Solve for d.
$d \approx 1913$	Use a calculator.

The distance from the lighthouse to the shore, to the nearest foot, is 1913 feet. This answer is reasonable since 1913 is less than 2000.

More About. . .

Lighthouses

Standing 208 feet tall, the Cape Hatteras Lighthouse in North Carolina is the tallest lighthouse in the United States.

Source: www.oldcapehatteraslighthouse.com

Check for Understanding

Concept Check

1. **Determine** whether the following statement is *sometimes, always* or *never* true. Explain your reasoning.

   If given the measure of two sides of a triangle and the angle opposite one of them, you will be able to find a unique solution.

2. **OPEN ENDED** Give an example of a triangle that has two solutions by listing measures for A, a, and b, where a and b are in centimeters. Then draw both cases using a ruler and protractor.

3. **FIND THE ERROR** Dulce and Gabe are finding the area of $\triangle ABC$ for $A = 64°$, $a = 15$ meters, and $b = 8$ meters using the sine function.

Dulce	Gabe
$Area = \frac{1}{2}(15)(8)\sin 64°$	$Area = \frac{1}{2}(15)(8)\sin 87.4°$
$\approx 53.9\ m^2$	$\approx 59.9\ m^2$

Who is correct? Explain your reasoning.

Guided Practice

Find the area of $\triangle ABC$ to the nearest tenth.

4.

B
10 in.
50°
A
15 in.
C

5.

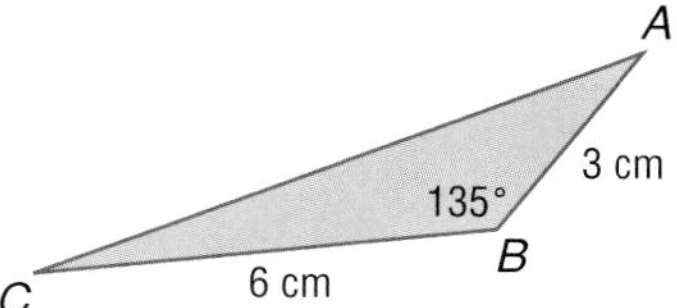

Solve each triangle. Round measures of sides to the nearest tenth and measures of angles to the nearest degree.

6.

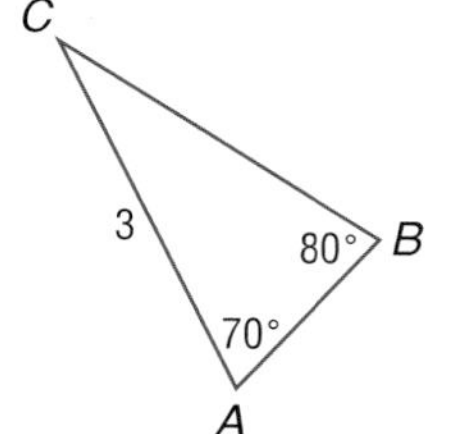

7.

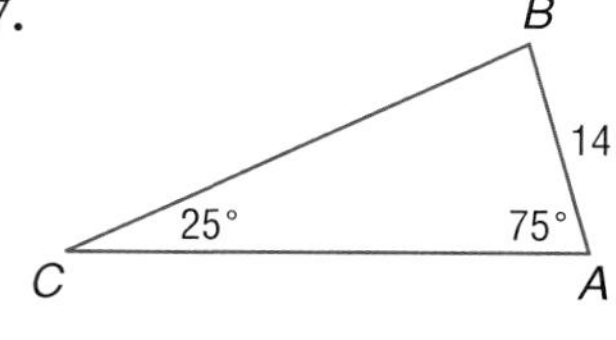

8.

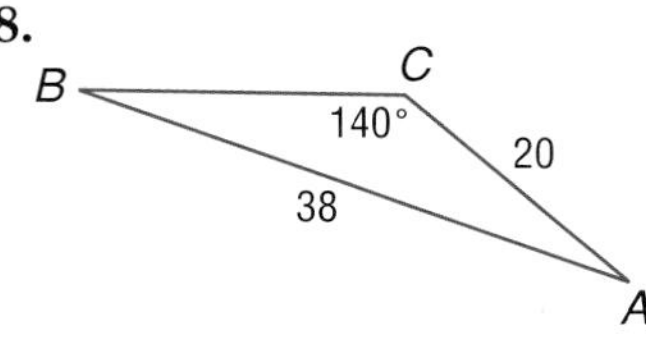

Determine whether each triangle has *no* solution, *one* solution, or *two* solutions. Then solve each triangle. Round measures of sides to the nearest tenth and measures of angles to the nearest degree.

9. $A = 123°, a = 12, b = 23$

10. $A = 30°, a = 3, b = 4$

11. $A = 55°, a = 10, b = 5$

12. $A = 145°, a = 18, b = 10$

Application

13. **WOODWORKING** Latisha is constructing a triangular brace from three beams of wood. She is to join the 6-meter beam to the 7-meter beam so the angle opposite the 7-meter beam measures 75°. To what length should Latisha cut the third beam in order to form the triangular brace? Round to the nearest tenth.

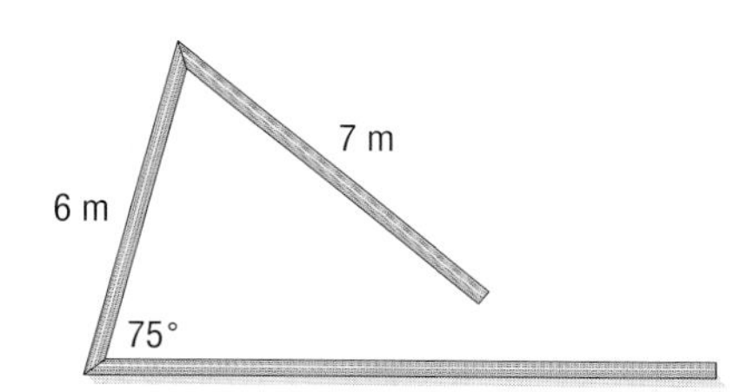

Practice and Apply

Homework Help

For Exercises	See Examples
14–19	1
20–37	2–5
38–41	6

Extra Practice
See page 858.

Find the area of $\triangle ABC$ to the nearest tenth.

14.

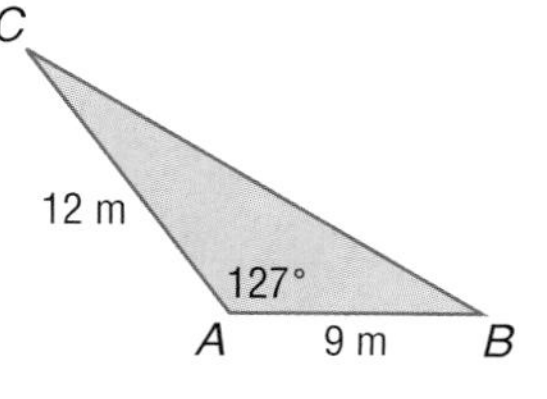

15.

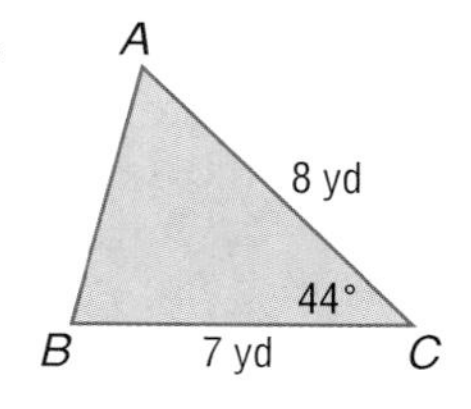

16. $B = 85°, c = 23$ ft, $a = 50$ ft

17. $A = 60°, b = 12$ cm, $c = 12$ cm

18. $C = 136°, a = 3$ m, $b = 4$ m

19. $B = 32°, a = 11$ mi, $c = 5$ mi

Solve each triangle. Round measures of sides to the nearest tenth and measures of angles to the nearest degree.

20.

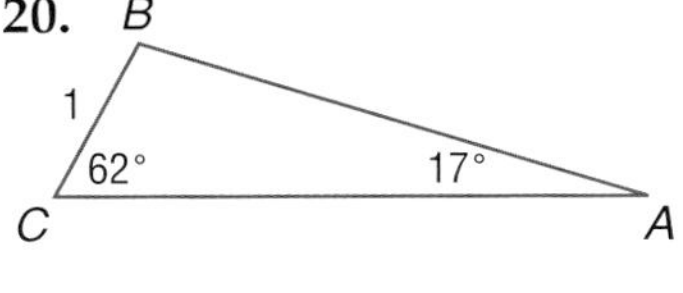

21.

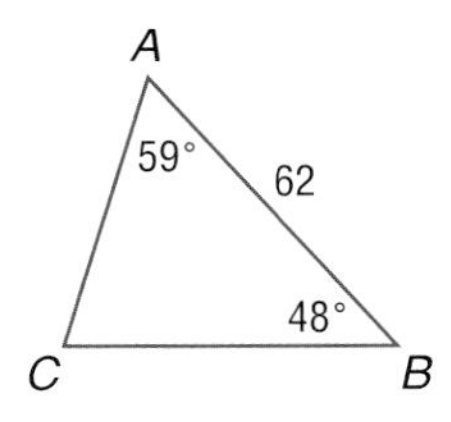

22.

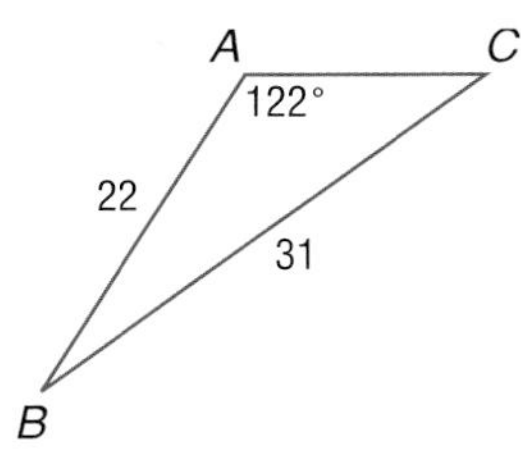

23.

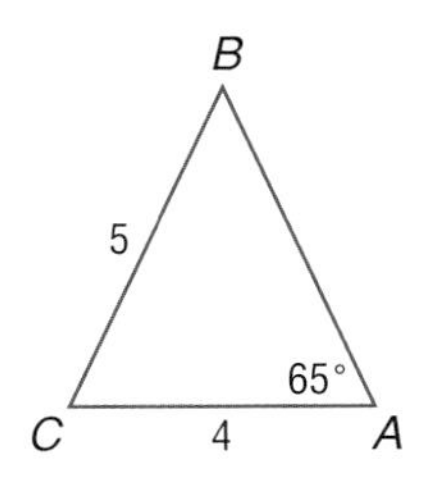

24.

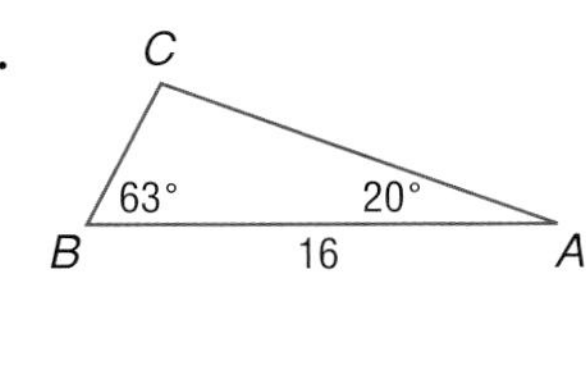

25.

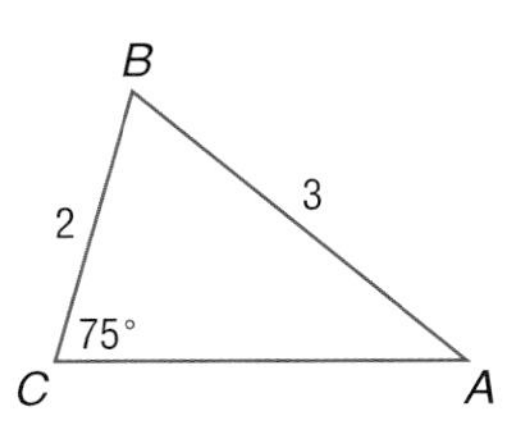

26. $A = 50°, a = 2.5, c = 3$

27. $B = 18°, C = 142°, b = 20$

Determine whether each triangle has *no* solution, *one* solution, or *two* solutions. Then solve each triangle. Round measures of sides to the nearest tenth and measures of angles to the nearest degree.

28. $A = 124°, a = 1, b = 2$

29. $A = 99°, a = 2.5, b = 1.5$

30. $A = 33°, a = 2, b = 3.5$

31. $A = 68°, a = 3, b = 5$

32. $A = 30°, a = 14, b = 28$

33. $A = 61°, a = 23, b = 8$

34. $A = 52°, a = 190, b = 200$

35. $A = 80°, a = 9, b = 9.1$

36. $A = 28°, a = 8.5, b = 7.2$

37. $A = 47°, a = 67, b = 83$

38. RADIO A radio station providing local tourist information has its transmitter on Beacon Road, 8 miles from where it intersects with the interstate highway. If the radio station has a range of 5 miles, between what two distances from the intersection can cars on the interstate tune in to hear this information?

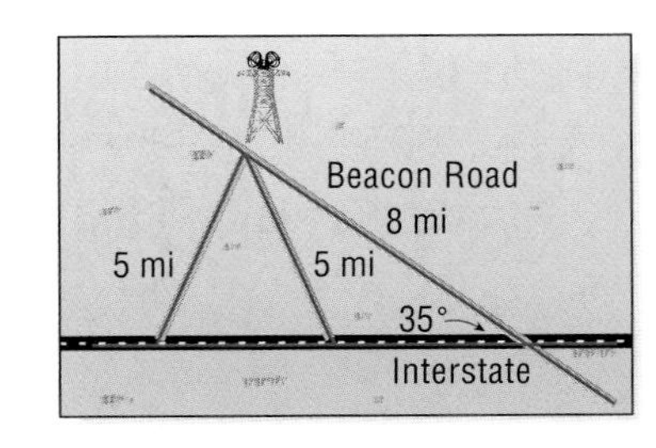

39. FORESTRY Two forest rangers, 12 miles from each other on a straight service road, both sight an illegal bonfire away from the road. Using their radios to communicate with each other, they determine that the fire is between them. The first ranger's line of sight to the fire makes an angle of 38° with the road, and the second ranger's line of sight to the fire makes a 63° angle with the road. How far is the fire from each ranger?

More About. . .

Ballooning

Hot-air balloons range in size from approximately 54,000 cubic feet to over 250,000 cubic feet.

Source: www.unicorn-ballon.com

40. BALLOONING As a hot-air balloon crosses over a straight portion of interstate highway, its pilot eyes two consecutive mileposts on the same side of the balloon. When viewing the mileposts the angles of depression are 64° and 7°. How high is the balloon to the nearest foot?

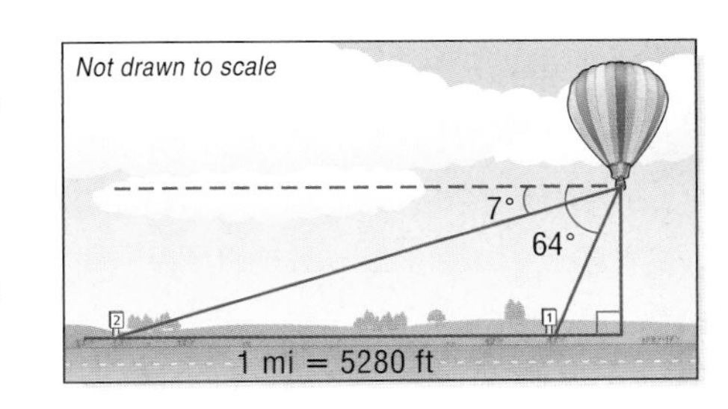

41. **NAVIGATION** Two fishing boats, A and B, are anchored 4500 feet apart in open water. A plane flies at a constant speed in a straight path directly over the two boats, maintaining a constant altitude. At one point during the flight, the angle of depression to A is 85°, and the angle of depression to B is 25°. Ten seconds later the plane has passed over A and spots B at a 35° angle of depression. How fast is the plane flying?

42. **CRITICAL THINKING** Given $\triangle ABC$, if $a = 20$ and $B = 47°$, then determine all possible values of b so that the triangle has

a. two solutions. b. one solution. c. no solutions.

43. **WRITING IN MATH** Answer the question that was posed at the beginning of the lesson.

How can trigonometry be used to find the area of a triangle?

Include the following in your answer:

- the conditions that would indicate that trigonometry is needed to find the area of a triangle,
- an example of a real-world situation in which you would need trigonometry to find the area of a triangle, and
- a derivation of one of the other two area formulas.

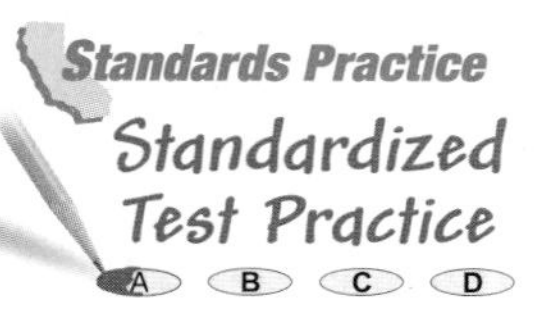

44. Which of the following is the perimeter of the triangle shown?

(A) 49.0 cm (B) 66.0 cm

(C) 91.4 cm (D) 93.2 cm

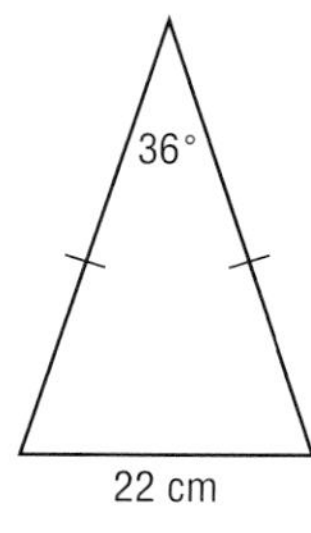

45. **SHORT RESPONSE** The longest side of a triangle is 67 inches. Two angles have measures of 47° and 55°. Solve the triangle.

Maintain Your Skills

Mixed Review

Find the exact value of each trigonometric function. *(Lesson 13-3)*

46. cos 30° 47. $\cot\left(\frac{\pi}{3}\right)$ 48. $\csc\left(\frac{\pi}{4}\right)$

Find one angle with positive measure and one angle with negative measure coterminal with each angle. *(Lesson 13-2)*

49. 300° 50. 47° 51. $\frac{5\pi}{6}$

Two cards are drawn from a deck of cards. Find each probability. *(Lesson 12-5)*

52. P(both 5s or both spades) 53. P(both 7s or both red)

54. **AERONAUTICS** A rocket rises 20 feet in the first second, 60 feet in the second second, and 100 feet in the third second. If it continues at this rate, how many feet will it rise in the 20th second? *(Lesson 11-1)*

Getting Ready for the Next Lesson

PREREQUISITE SKILL Solve each equation. Round to the nearest tenth.
*(To review **solving equations with trigonometric functions**, see Lesson 13-1.)*

55. $a^2 = 3^2 + 5^2 - 2(3)(5) \cos 85°$ 56. $c^2 = 12^2 + 10^2 - 2(12)(10) \cos 40°$

57. $7^2 = 11^2 + 9^2 - 2(11)(9) \cos B°$ 58. $13^2 = 8^2 + 6^2 - 2(8)(6) \cos A°$

13-5 Law of Cosines

What You'll Learn

- Solve problems by using the Law of Cosines.
- Determine whether a triangle can be solved by first using the Law of Sines or the Law of Cosines.

Vocabulary

- Law of Cosines

How can you determine the angle at which to install a satellite dish?

The GE-3 satellite is in a *geosynchronous orbit* about Earth, meaning that it circles Earth once each day. As a result, the satellite appears to remain stationary over one point on the equator. A receiving dish for the satellite can be directed at one spot in the sky. The satellite orbits 35,786 kilometers above the equator at 87°W longitude. The city of Valparaiso, Indiana, is located at approximately 87°W longitude and 41.5°N latitude.

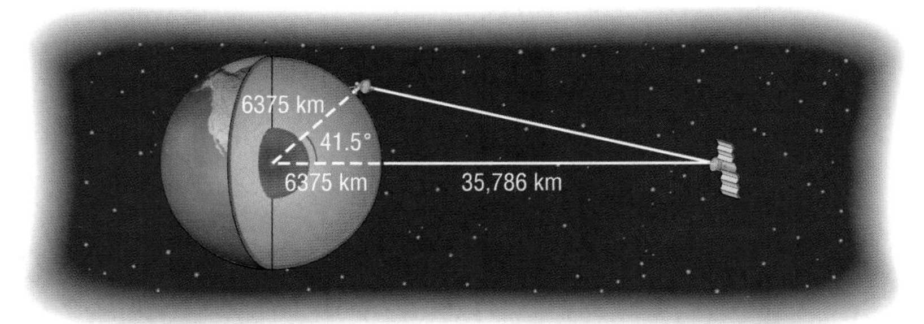

Knowing the radius of Earth to be about 6375 kilometers, a satellite dish installer can use trigonometry to determine the angle at which to direct the receiver.

LAW OF COSINES Problems such as this, in which you know the measures of two sides and the included angle of a triangle, cannot be solved using the Law of Sines. You can solve problems such as this by using the **Law of Cosines**.

To derive the Law of Cosines, consider $\triangle ABC$. What relationship exists between a, b, c, and A?

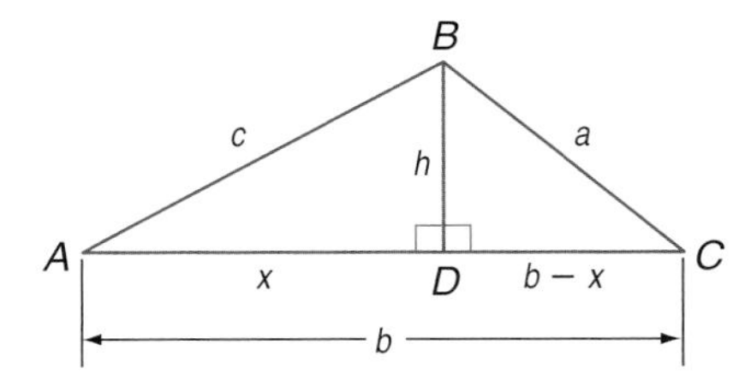

$a^2 = (b - x)^2 + h^2$	Use the Pythagorean Theorem for $\triangle DBC$.
$= b^2 - 2bx + x^2 + h^2$	Expand $(b - x)^2$.
$= b^2 - 2bx + c^2$	In $\triangle ADB$, $c^2 = x^2 + h^2$.
$= b^2 - 2b(c \cos A) + c^2$	$\cos A = \frac{x}{c}$, so $x = c \cos A$.
$= b^2 + c^2 - 2bc \cos A$	Commutative Property

Key Concept — Law of Cosines

Let $\triangle ABC$ be any triangle with a, b, and c representing the measures of sides, and opposite angles with measures A, B, and C, respectively. Then the following equations are true.

$$a^2 = b^2 + c^2 - 2bc \cos A$$
$$b^2 = a^2 + c^2 - 2ac \cos B$$
$$c^2 = a^2 + b^2 - 2ab \cos C$$

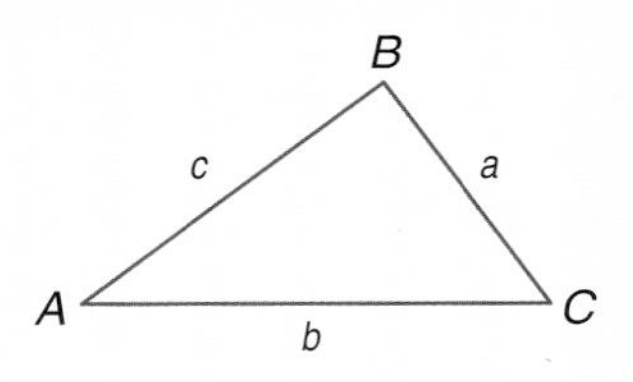

You can apply the Law of Cosines to a triangle if you know

- the measures of two sides and the included angle, or
- the measures of three sides.

Example 1 Solve a Triangle Given Two Sides and Included Angle

Solve $\triangle ABC$.

You are given the measures of two sides and the included angle. Begin by using the Law of Cosines to determine c.

$c^2 = a^2 + b^2 - 2ab \cos C$ — Law of Cosines

$c^2 = 18^2 + 24^2 - 2(18)(24) \cos 57°$ — $a = 18$, $b = 24$, and $C = 57°$

$c^2 \approx 429.4$ — Simplify using a calculator.

$c \approx 20.7$ — Take the square root of each side.

Next, you can use the Law of Sines to find the measure of angle A.

$\frac{\sin A}{a} = \frac{\sin C}{c}$ — Law of Sines

$\frac{\sin A}{18} \approx \frac{\sin 57°}{20.7}$ — $a = 18$, $C = 57°$, and $c \approx 20.7$

$\sin A \approx \frac{18 \sin 57°}{20.7}$ — Multiply each side by 18.

$\sin A \approx 0.7293$ — Use a calculator.

$A \approx 47°$ — Use the $\sin^{-1}$ function.

The measure of the angle B is approximately $180° - (57° + 47°)$ or $76°$. Therefore, $c \approx 20.7$, $A \approx 47°$, and $B \approx 76°$.

Alternate Method
After finding the measure of c in Example 1, the Law of Cosines could be used again to find a second angle.

Example 2 Solve a Triangle Given Three Sides

Solve $\triangle ABC$.

You are given the measures of three sides. Use the Law of Cosines to find the measure of the largest angle first, angle A.

$a^2 = b^2 + c^2 - 2bc \cos A$ — Law of Cosines

$15^2 = 9^2 + 7^2 - 2(9)(7) \cos A$ — $a = 15$, $b = 9$, and $c = 7$

$15^2 - 9^2 - 7^2 = -2(9)(7) \cos A$ — Subtract 9^2 and 7^2 from each side.

$\frac{15^2 - 9^2 - 7^2}{-2(9)(7)} = \cos A$ — Divide each side by $-2(9)(7)$.

$-0.7540 \approx \cos A$ — Use a calculator.

$139° \approx A$ — Use the $\cos^{-1}$ function.

You can use the Law of Sines to find the measure of angle B.

$\frac{\sin B}{b} = \frac{\sin A}{a}$ — Law of Sines

$\frac{\sin B}{9} \approx \frac{\sin 139°}{15}$ — $b = 9$, $A \approx 139°$, and $a = 15$

$\sin B \approx \frac{9 \sin 139°}{15}$ — Multiply each side by 9.

$\sin B \approx 0.3936$ — Use a calculator.

$B \approx 23°$ — Use the $\sin^{-1}$ function.

The measure of the angle C is approximately $180° - (139° + 23°)$ or $18°$. Therefore, $A \approx 139°$, $B \approx 23°$, and $C \approx 18°$.

Sides and Angles
When solving triangles, remember that the angle with the greatest measure is always opposite the longest side. The angle with the least measure is always opposite the shortest side.

CHOOSE THE METHOD To solve a triangle that is *oblique*, or having no right angle, you need to know the measure of at least one side and any two other parts. If the triangle has a solution, then you must decide whether to begin solving by using the Law of Sines or by using the Law of Cosines. Use the chart below to help you choose.

Concept Summary — Solving an Oblique Triangle

Given	Begin by Using
two angles and any side	Law of Sines
two sides and an angle opposite one of them	Law of Sines
two sides and their included angle	Law of Cosines
three sides	Law of Cosines

More About. . .

Emergency Medicine

Medical evacuation (Medevac) helicopters provide quick transportation from areas that are difficult to reach by any other means. These helicopters can cover long distances and are primary emergency vehicles in locations where there are few hospitals.

Source: The Helicopter Education Center

Example 3 Apply the Law of Cosines

EMERGENCY MEDICINE A medical rescue helicopter has flown from its home base at point C to pick up an accident victim at point A and then from there to the hospital at point B. The pilot needs to know how far he is now from his home base so he can decide whether to refuel before returning. How far is the hospital from the helicopter's base?

You are given the measures of two sides and their included angle, so use the Law of Cosines to find a.

$a^2 = b^2 + c^2 - 2bc \cos A$ Law of Cosines

$a^2 = 50^2 + 45^2 - 2(50)(45) \cos 130°$ $b = 50$, $c = 45$, and $A = 130°$

$a^2 \approx 7417.5$ Use a calculator to simplify.

$a \approx 86.1$ Take the square root of each side.

The distance between the hospital and the helicopter base is approximately 86.1 miles.

Check for Understanding

Concept Check

1. **FIND THE ERROR** Mateo and Amy are deciding which method, the Law of Sines or the Law of Cosines, should be used first to solve $\triangle ABC$.

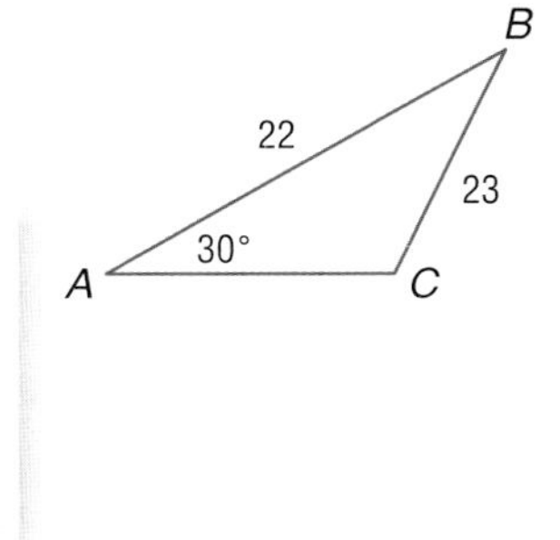

Mateo	Amy
Begin by using the Law of Sines, since you are given two sides and an angle opposite one of them.	Begin by using the Law of Cosines, since you are given two sides and their included angle.

Who is correct? Explain your reasoning.

2. **Explain** how to solve a triangle by using the Law of Cosines if the lengths of
   a. three sides are known.
   b. two sides and the measure of the angle between them are known.
3. **OPEN ENDED** Give an example of a triangle that can be solved by first using the Law of Cosines.

Guided Practice

Determine whether each triangle should be solved by beginning with the Law of Sines or Law of Cosines. Then solve each triangle. Round measures of sides to the nearest tenth and measures of angles to the nearest degree.

4.

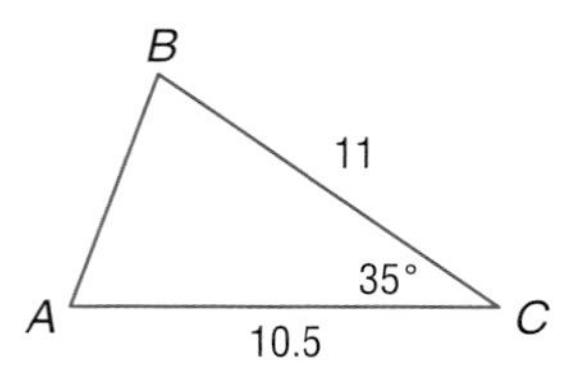

5.

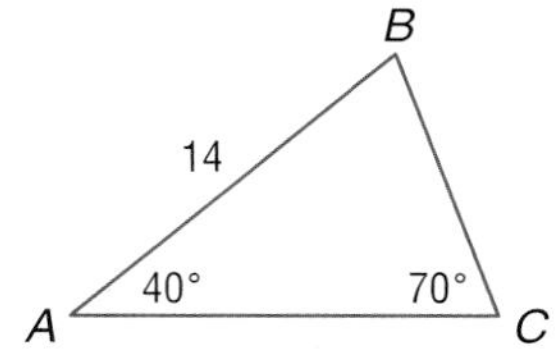

6. $A = 42°, b = 57, a = 63$

7. $a = 5, b = 12, c = 13$

Application

BASEBALL For Exercises 8 and 9, use the following information.

In Australian baseball, the bases lie at the vertices of a square 27.5 meters on a side and the pitcher's mound is 18 meters from home plate.

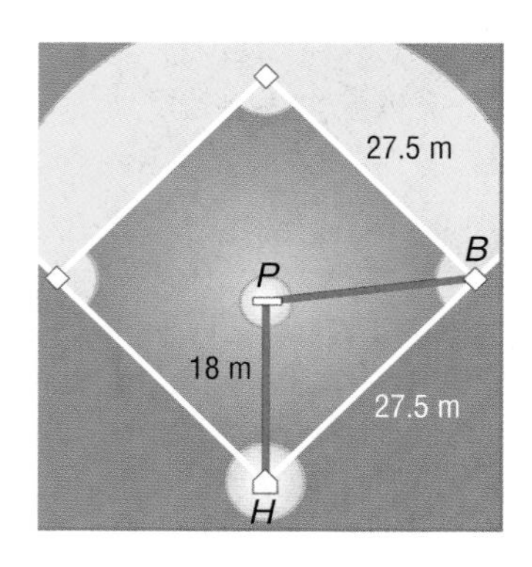

8. Find the distance from the pitcher's mound to first base.
9. Find the angle between home plate, the pitcher's mound, and first base.

Practice and Apply

Homework Help

For Exercises	See Examples
10–27	1, 2
28–33	3

Extra Practice
See page 858.

Determine whether each triangle should be solved by beginning with the Law of Sines or Law of Cosines. Then solve each triangle. Round measures of sides to the nearest tenth and measures of angles to the nearest degree.

10.

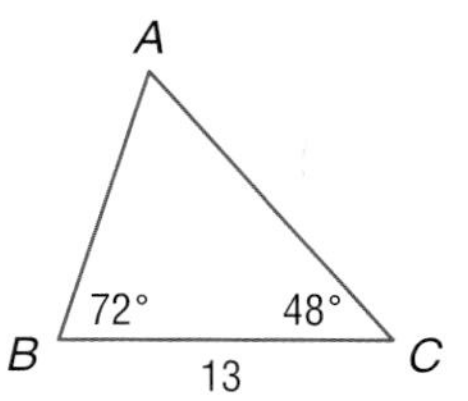

11.

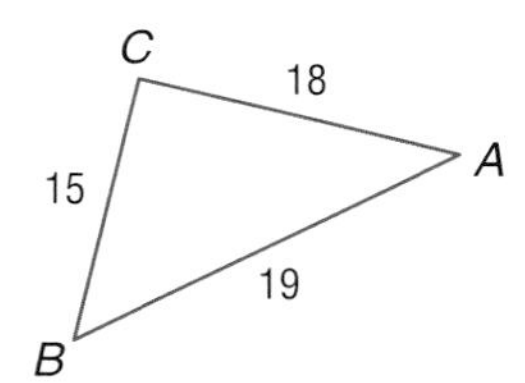

12.

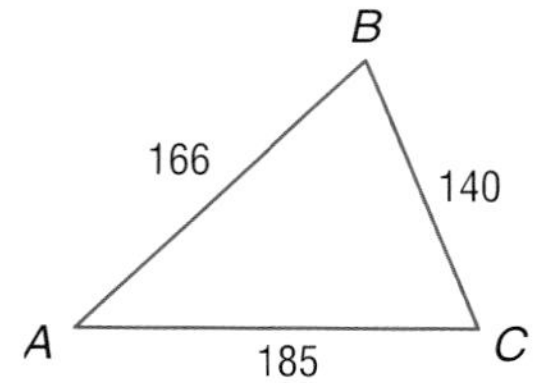

13.

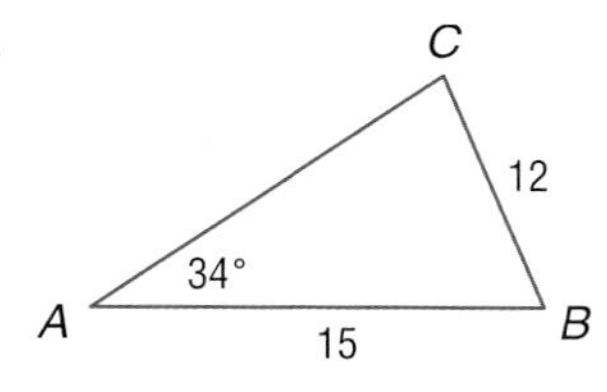

14.

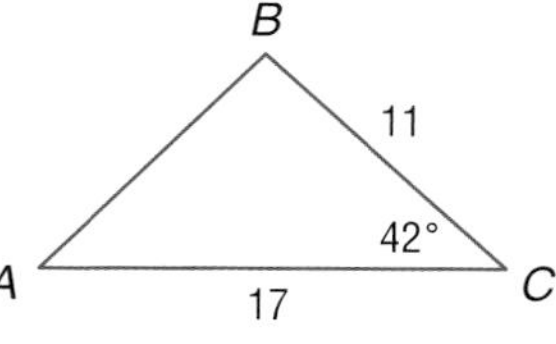

15.

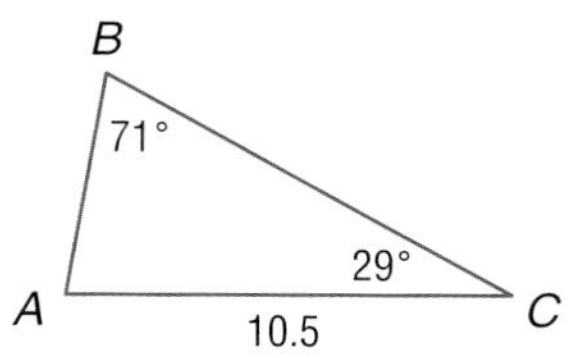

16. $a = 20, c = 24, B = 47°$

17. $a = 345, b = 648, c = 442$

18. $A = 36°, a = 10, b = 19$

19. $A = 25°, B = 78°, a = 13.7$

20. $a = 21.5, b = 16.7, c = 10.3$

21. $a = 16, b = 24, c = 41$

22. $a = 8, b = 24, c = 18$

23. $B = 19°, a = 51, c = 61$

24. $A = 56°, B = 22°, a = 12.2$

25. $a = 4, b = 8, c = 5$

26. $a = 21.5, b = 13, C = 38°$

27. $A = 40°, b = 7, a = 6$

More About. . .

Dinosaurs

At digs such as the one at the Glen Rose formation in Texas, anthropologists study the footprints made by dinosaurs millions of years ago. *Locomoter* parameters, such as pace and stride, taken from these prints can be used to describe how a dinosaur once moved.

Source: Mid-America Paleontology Society

DINOSAURS For Exercises 28–30, use the diagram at the right.

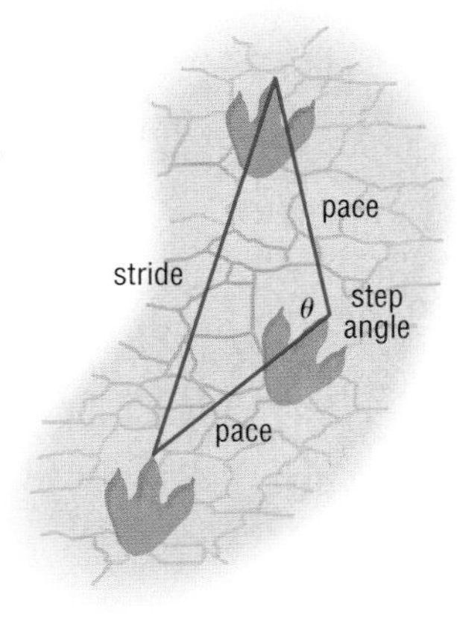

28. An anthropologist examining the footprints made by a bipedal (two-footed) dinosaur finds that the dinosaur's average pace was about 1.60 meters and average stride was about 3.15 meters. Find the step angle θ for this dinosaur.

29. Find the step angle θ made by the hindfeet of a herbivorous dinosaur whose pace averages 1.78 meters and stride averages 2.73 meters.

30. An efficient walker has a step angle that approaches 180°, meaning that the animal minimizes "zig-zag" motion while maximizing forward motion. What can you tell about the motion of each dinosaur from its step angle?

31. GEOMETRY In rhombus $ABCD$, the measure of $\angle ADC$ is 52°. Find the measures of diagonals $\overline{AC}$ and $\overline{DB}$ to the nearest tenth.

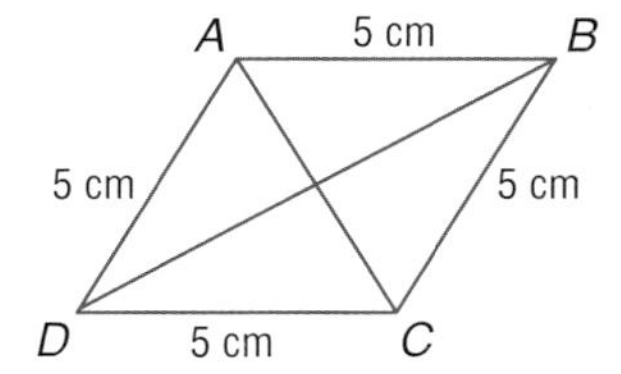

32. SURVEYING Two sides of a triangular plot of land have lengths of 425 feet and 550 feet. The measure of the angle between those sides is 44.5°. Find the perimeter and area of the plot.

33. AVIATION A pilot typically flies a route from Bloomington to Rockford, covering a distance of 117 miles. In order to avoid a storm, the pilot first flies from Bloomington to Peoria, a distance of 42 miles, then turns the plane and flies 108 miles on to Rockford. Through what angle did the pilot turn the plane over Peoria?

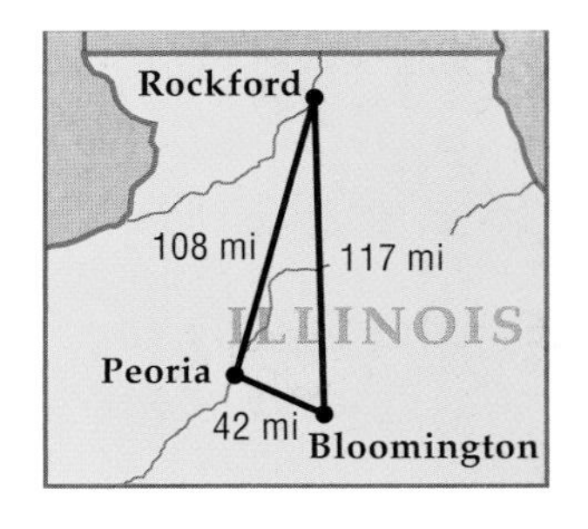

34. CRITICAL THINKING Explain how the Pythagorean Theorem is a special case of the Law of Cosines.

35. WRITING IN MATH Answer the question that was posed at the beginning of the lesson.

How can you determine the angle at which to install a satellite dish?

Include the following in your answer:

- a description of the conditions under which you can use the Law of Cosines to solve a triangle, and
- given the latitude of a point on Earth's surface, an explanation of how you can determine the angle at which to install a satellite dish at the same longitude.

36. In $\triangle DEF$, what is the value of θ to the nearest degree?

(A) 26° (B) 74°

(C) 80° (D) 141°

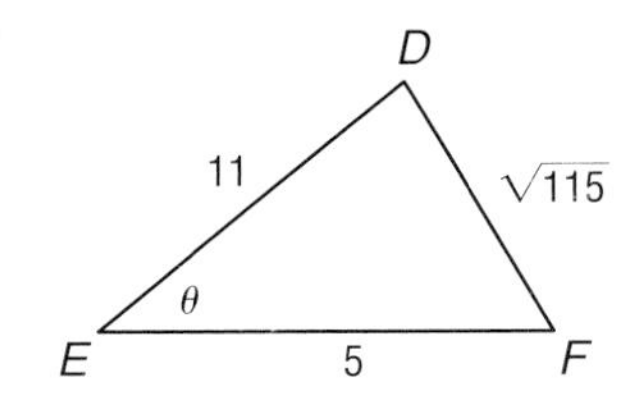

37. Two trucks, A and B, start from the intersection C of two straight roads at the same time. Truck A is traveling twice as fast as truck B and after 4 hours, the two trucks are 350 miles apart. Find the approximate speed of truck B in miles per hour.

Ⓐ 35 Ⓑ 37 Ⓒ 57 Ⓓ 73

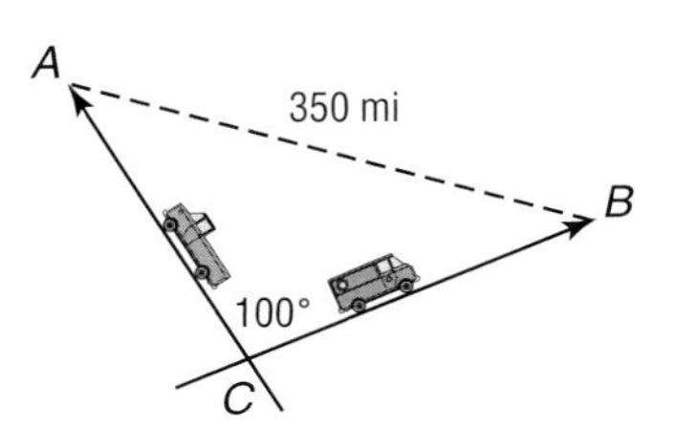

Extending the Lesson

ERROR IN MEASUREMENT For Exercises 38–40, use the following information. Consider $\triangle ABC$, in which $a = 17$, $b = 8$, and $c = 20$.

38. Find the measure of angle C in one step using the Law of Cosines. Round to the nearest tenth.

39. Find the measure of angle C in two steps using the Law of Cosines and then the Law of Sines. Round to the nearest tenth.

40. Explain why your answers for Exercises 38 and 39 are different. Which answer gives you the better approximation for the measure of angle C?

Maintain Your Skills

Mixed Review

Determine whether each triangle has *no* solution, *one* solution, or *two* solutions. Then solve each triangle. Round measures of sides to the nearest tenth and measures of angles to the nearest degree. *(Lesson 13-4)*

41. $A = 55°, a = 8, b = 7$

42. $A = 70°, a = 7, b = 10$

Find the exact values of the six trigonometric functions of θ if the terminal side of θ in standard position contains the given point. *(Lesson 13-3)*

43. $(5, 12)$

44. $(4, 7)$

45. $(\sqrt{10}, \sqrt{6})$

Solve each equation or inequality. *(Lesson 10-5)*

46. $e^x + 5 = 9$

47. $4e^x - 3 > -1$

48. $\ln (x + 3) = 2$

Getting Ready for the Next Lesson

PREREQUISITE SKILL Find one angle with positive measure and one angle with negative measure coterminal with each angle.
*(To review **coterminal angles**, see Lesson 13-2.)*

49. $45°$

50. $30°$

51. $180°$

52. $\frac{\pi}{2}$

53. $\frac{7\pi}{6}$

54. $\frac{4\pi}{3}$

Practice Quiz 2

Lessons 13-3 through 13-5

1. Find the exact value of the six trigonometric functions of θ if the terminal side of θ in standard position contains the point $(-2, 3)$. *(Lesson 13-3)*

2. Find the exact value of csc $\frac{5\pi}{3}$. *(Lesson 13-3)*

3. Find the area of $\triangle DEF$ to the nearest tenth. *(Lesson 13-4)*

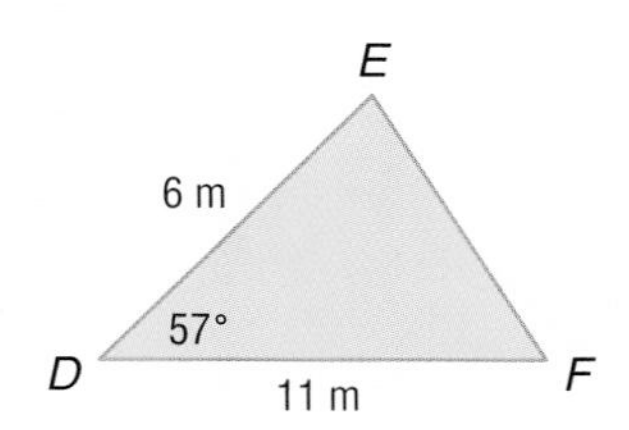

4. Determine whether $\triangle ABC$, with $A = 22°$, $a = 15$, and $b = 18$, has *no* solution, *one* solution, or *two* solutions. Then solve the triangle, if possible. Round measures of sides to the nearest tenth and measures of angles to the nearest degree. *(Lesson 13-4)*

5. Determine whether $\triangle ABC$, with $b = 11$, $c = 14$, and $A = 78°$, should be solved by beginning with the Law of Sines or Law of Cosines. Then solve the triangle. Round measures of sides to the nearest tenth and measures of angles to the nearest degree. *(Lesson 13-5)*

13-6 Circular Functions

What You'll Learn

- Define and use the trigonometric functions based on the unit circle.
- Find the exact values of trigonometric functions of angles.

Vocabulary

- circular function
- periodic
- period

How can you model annual temperature fluctuations?

The average high temperatures, in degrees Fahrenheit, for Barrow, Alaska, are given in the table at the right. With January assigned a value of 1, February a value of 2, March a value of 3, and so on, these data can be graphed as shown below. This pattern of temperature fluctuations repeats after a period of 12 months.

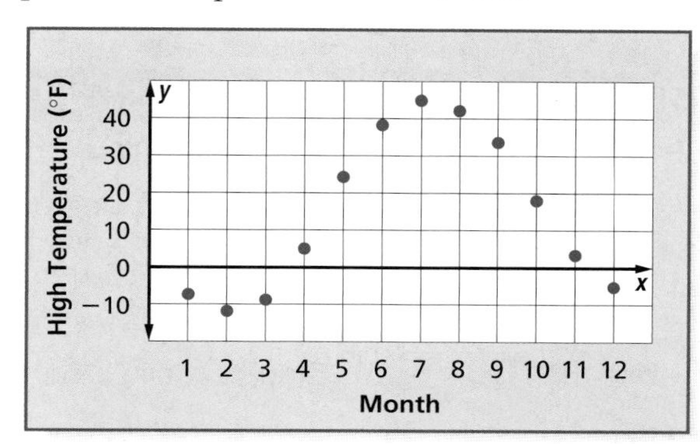

BARROW, ALASKA

MONTH	HIGH TEMP. (°F)
Jan	-7.4
Feb	-11.8
March	-9.0
April	4.7
May	24.2
June	38.3
July	45.0
Aug	42.3
Sept	33.8
Oct	18.1
Nov	3.5
Dec	-5.2

Source: www.met.utah.edu

UNIT CIRCLE DEFINITIONS From your work with reference angles, you know that the values of trigonometric functions also repeat. For example, sin 30° and sin 150° have the same value, $\frac{1}{2}$. In this lesson, we will further generalize the trigonometric functions by defining them in terms of the unit circle.

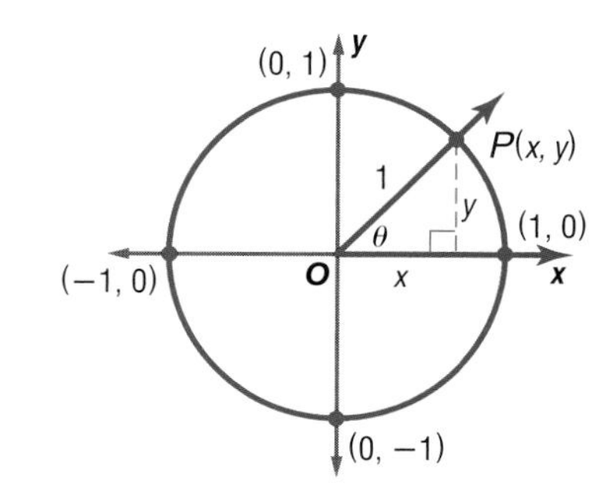

Consider an angle θ in standard position. The terminal side of the angle intersects the unit circle at a unique point, $P(x, y)$. Recall that $\sin \theta = \frac{y}{r}$ and $\cos \theta = \frac{x}{r}$. Since $P(x, y)$ is on the unit circle, $r = 1$. Therefore, $\sin \theta = y$ and $\cos \theta = x$.

Key Concept — Definition of Sine and Cosine

- **Words** If the terminal side of an angle θ in standard position intersects the unit circle at $P(x, y)$, then $\cos \theta = x$ and $\sin \theta = y$. Therefore, the coordinates of P can be written as $P(\cos \theta, \sin \theta)$.

- **Model**

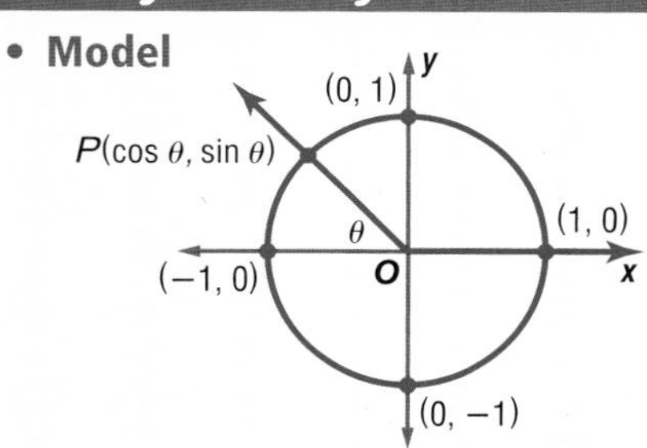

Since there is exactly one point $P(x, y)$ for any angle θ, the relations $\cos \theta = x$ and $\sin \theta = y$ are functions of θ. Because they are both defined using a unit circle, they are often called **circular functions**.

Study Tip

Reading Math
To help you remember that $x = \cos \theta$ and $y = \sin \theta$, notice that alphabetically x comes before y and cosine comes before sine.

Example 1 Find Sine and Cosine Given Point on Unit Circle

Given an angle θ in standard position, if $P\left(\frac{2\sqrt{2}}{3}, -\frac{1}{3}\right)$ lies on the terminal side and on the unit circle, find $\sin \theta$ and $\cos \theta$.

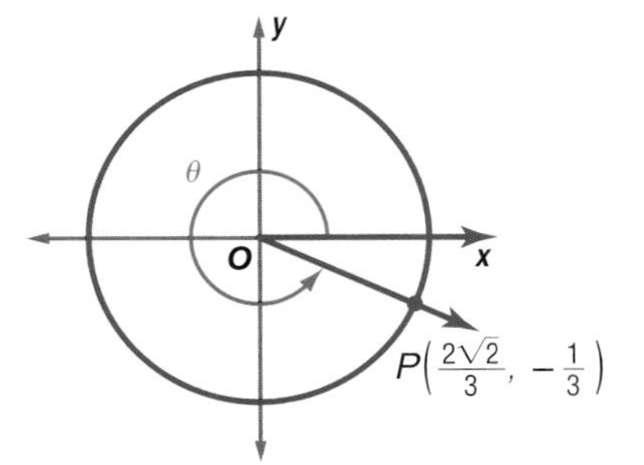

$P\left(\frac{2\sqrt{2}}{3}, -\frac{1}{3}\right) = P(\cos \theta, \sin \theta)$,

so $\sin \theta = -\frac{1}{3}$ and $\cos \theta = \frac{2\sqrt{2}}{3}$.

In the Investigation below, you will explore the behavior of the sine and cosine functions on the unit circle.

Graphing Calculator Investigation

Sine and Cosine on the Unit Circle

Press [MODE] on a TI-83 Plus and highlight Degree and Par. Then use the following range values to set up a viewing window: TMIN = 0, TMAX = 360, TSTEP = 15, XMIN = −2.4, XMAX = 2.35, XSCL = 0.5, YMIN = −1.5, YMAX = 1.55, YSCL = 0.5. Press [Y=] to define the unit circle with $X_{1T} = \cos T$ and $Y_{1T} = \sin T$. Press [GRAPH]. Use the [TRACE] function to move around the circle.

Think and Discuss

1. What does T represent? What does the x value represent? What does the y value represent?
2. Determine the sine and cosine of the angles whose terminal sides lie at 0°, 90°, 180°, and 270°.
3. How do the values of sine change as you move around the unit circle? How do the values of cosine change?

The exact values of the sine and cosine functions for specific angles are summarized using the definition of sine and cosine on the unit circle below.

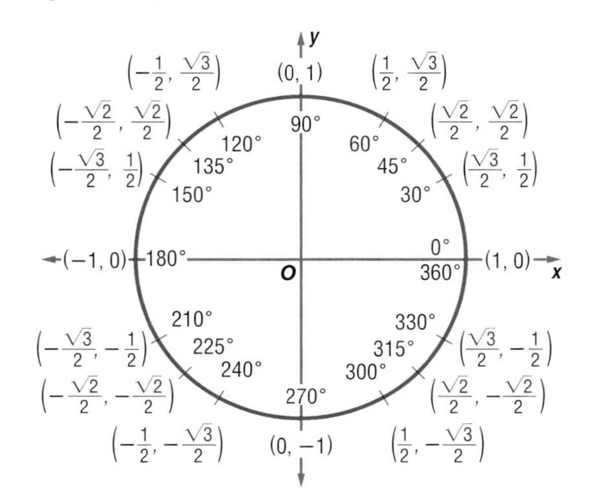

This same information is presented on the graphs of the sine and cosine functions below, where the horizontal axis shows the values of θ and the vertical axis shows the values of $\sin \theta$ or $\cos \theta$.

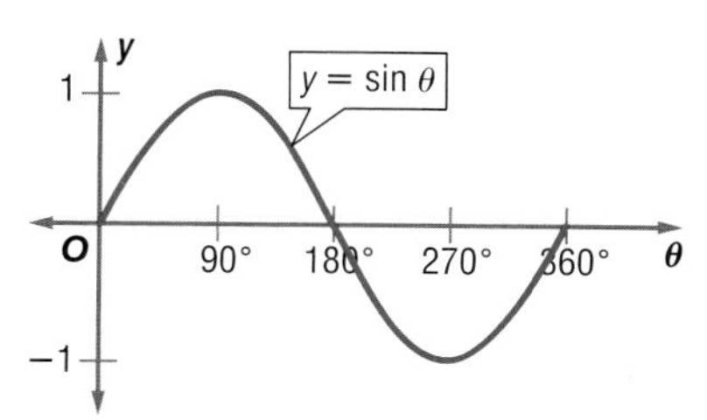

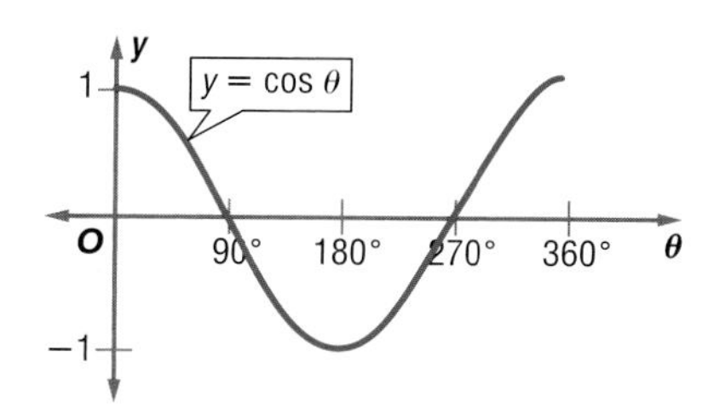

PERIODIC FUNCTIONS Notice in the graph above that the values of sine for the coterminal angles 0° and 360° are both 0. The values of cosine for these angles are both 1. Every 360° or 2π radians, the sine and cosine functions repeat their values. So, we can say that the sine and cosine functions are **periodic**, each having a **period** of 360° or 2π radians.

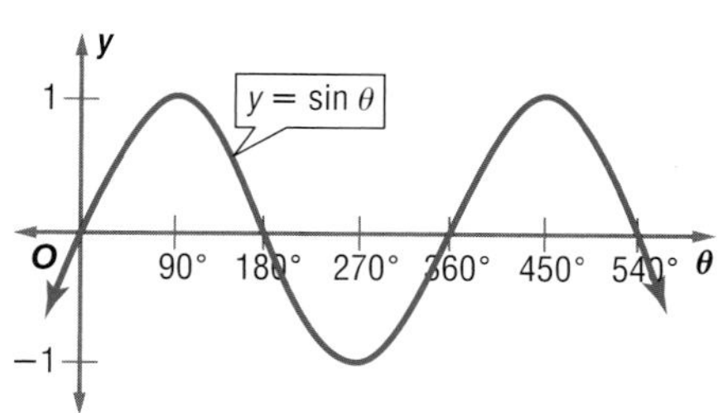

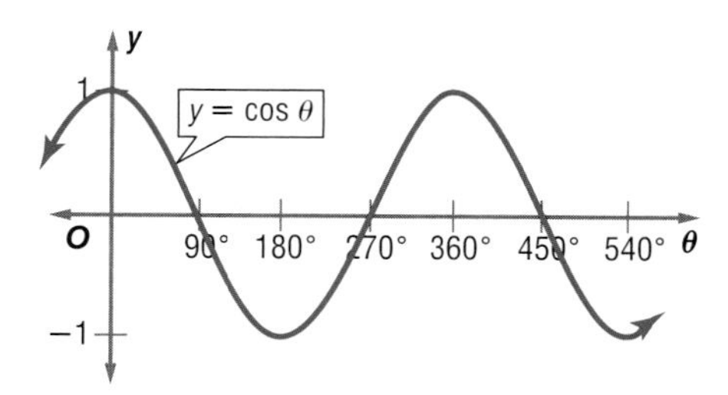

Key Concept — *Periodic Function*

A function is called periodic if there is a number a such that $f(x) = f(x + a)$ for all x in the domain of the function. The least positive value of a for which $f(x) = f(x + a)$ is called the period of the function.

For the sine and cosine functions, $\cos (x + 360°) = \cos x$, and $\sin (x + 360°) = \sin x$. In radian measure, $\cos (x + 2\pi) = \cos x$, and $\sin (x + 2\pi) = \sin x$. Therefore, the period of the sine and cosine functions is 360° or 2π.

Example 2 *Find the Value of a Trigonometric Function*

Find the exact value of each function.

a. $\mathbf{\cos 675°}$

$$\cos 675° = \cos (315° + 360°)$$
$$= \cos 315°$$
$$= \frac{\sqrt{2}}{2}$$

b. $\mathbf{\sin \left(-\frac{5\pi}{6}\right)}$

$$\sin \left(-\frac{5\pi}{6}\right) = \sin \left(-\frac{5\pi}{6} + 2\pi\right)$$
$$= \sin \frac{7\pi}{6}$$
$$= -\frac{1}{2}$$

When you look at the graph of a periodic function, you will see a repeating pattern: a shape that repeats over and over as you move to the right on the x-axis. The period is the distance along the x-axis from the beginning of the pattern to the point at which it begins again.

Many real-world situations have characteristics that can be described with periodic functions.

Example 3 Find the Value of a Trigonometric Function

FERRIS WHEEL As you ride a Ferris wheel, the height that you are above the ground varies periodically as a function of time. Consider the height of the center of the wheel to be the starting point. A particular wheel has a diameter of 38 feet and travels at a rate of 4 revolutions per minute.

a. Identify the period of this function.

Since the wheel makes 4 complete counterclockwise rotations every minute, the period is the time it takes to complete one rotation, which is $\frac{1}{4}$ of a minute or 15 seconds.

b. Make a graph in which the horizontal axis represents the time *t* in seconds and the vertical axis represents the height *h* in feet in relation to the starting point.

Since the diameter of the wheel is 38 feet, the wheel reaches a maximum height of $\frac{38}{2}$ or 19 feet above the starting point and a minimum of 19 feet below the starting point.

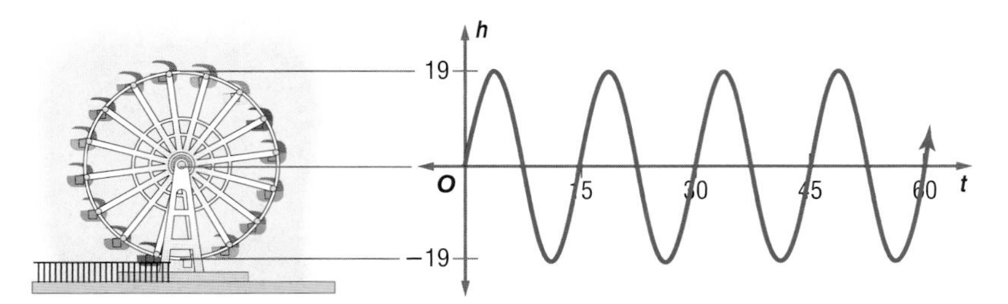

Check for Understanding

Concept Check

1. **State** the conditions under which $\cos \theta = x$ and $\sin \theta = y$.
2. **OPEN ENDED** Give an example of a situation that could be described by a periodic function. Then state the period of the function.
3. **Compare and contrast** the graphs of the sine and cosine functions on page 741.

Guided Practice

If the given point *P* is located on the unit circle, find $\sin \theta$ and $\cos \theta$.

4. $P\left(\frac{5}{13}, -\frac{12}{13}\right)$
5. $P\left(\frac{\sqrt{2}}{2}, \frac{\sqrt{2}}{2}\right)$

Find the exact value of each function.

6. $\sin -240°$
7. $\cos \frac{10\pi}{3}$

8. Determine the period of the function that is graphed below.

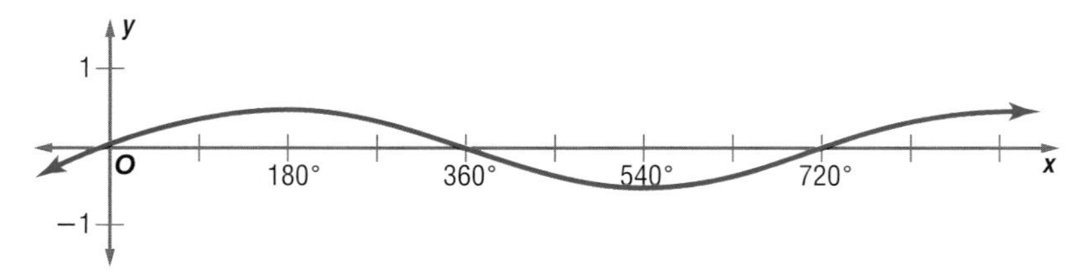

Application **PHYSICS For Exercises 9 and 10, use the following information.**

The motion of a weight on a spring varies periodically as a function of time. Suppose you pull the weight down 3 inches from its equilibrium point and then release it. It bounces above the equilibrium point and then returns below the equilibrium point in 2 seconds.

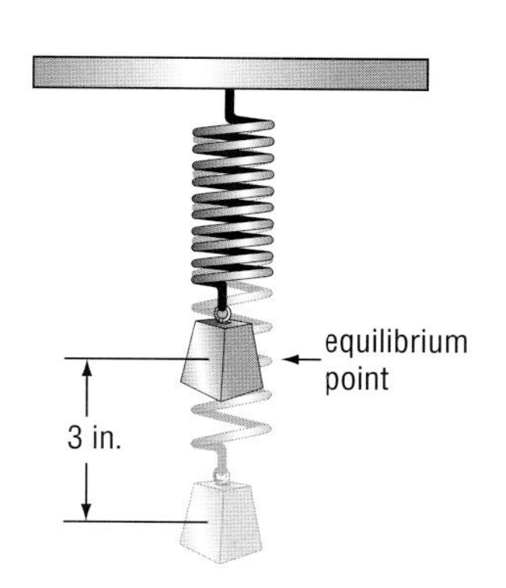

9. Find the period of this function.

10. Graph the height of the spring as a function of time.

Practice and Apply

Homework Help

For Exercises	See Examples
11–16	1
17–28	2
29–42	3

Extra Practice
See page 858.

The given point P is located on the unit circle. Find $\sin\theta$ and $\cos\theta$.

11. $P\left(-\frac{3}{5}, \frac{4}{5}\right)$

12. $P\left(-\frac{12}{13}, -\frac{5}{13}\right)$

13. $P\left(\frac{8}{17}, \frac{15}{17}\right)$

14. $P\left(\frac{\sqrt{3}}{2}, -\frac{1}{2}\right)$

15. $P\left(-\frac{1}{2}, \frac{\sqrt{3}}{2}\right)$

16. $P(0.6, 0.8)$

Find the exact value of each function.

17. $\sin 690°$

18. $\cos 750°$

19. $\cos 5\pi$

20. $\sin\left(\frac{14\pi}{6}\right)$

21. $\sin\left(-\frac{3\pi}{2}\right)$

22. $\cos(-225°)$

23. $\frac{\cos 60° + \sin 30°}{4}$

24. $3(\sin 60°)(\cos 30°)$

25. $\sin 30° - \sin 60°$

26. $\frac{4 \cos 330° + 2 \sin 60°}{3}$

27. $12(\sin 150°)(\cos 150°)$

28. $(\sin 30°)^2 + (\cos 30°)^2$

Determine the period of each function.

29.

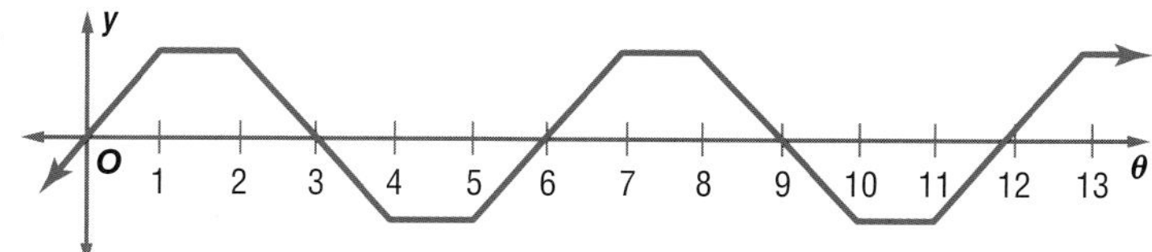

30.

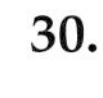

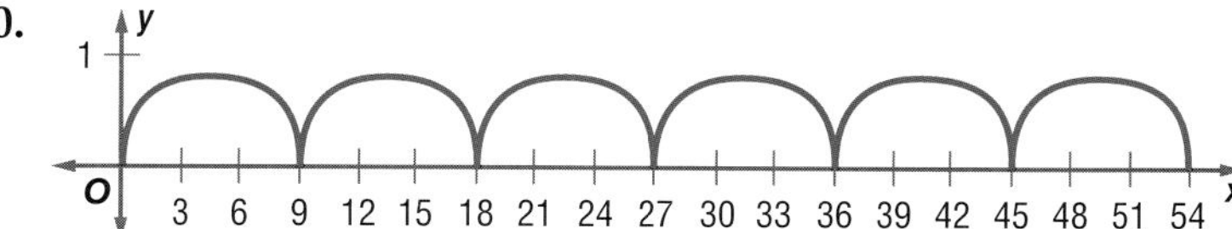

31.

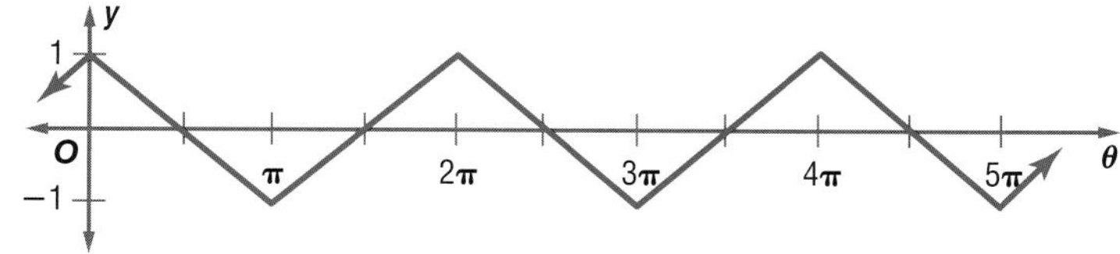

32.

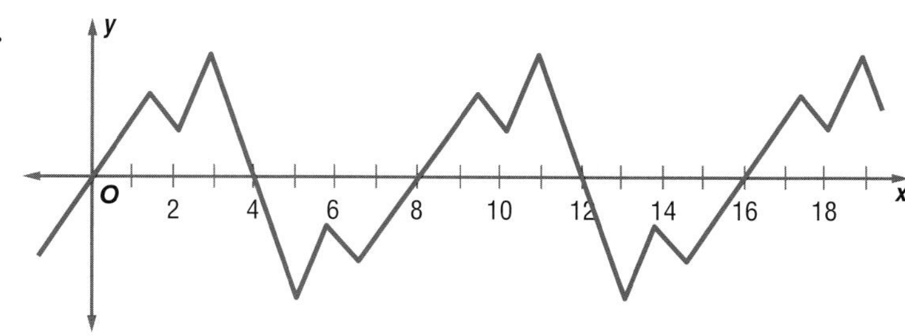

GUITAR **For Exercises 33 and 34, use the following information.**
When a guitar string is plucked, it is displaced from a fixed point in the middle of the string and vibrates back and forth, producing a musical tone. The exact tone depends on the frequency, or number of cycles per second, that the string vibrates. To produce an A, the frequency is 440 cycles per second, or 440 hertz (Hz).

33. Find the period of this function.

34. Graph the height of the fixed point on the string from its resting position as a function of time. Let the maximum distance above the resting position have a value of 1 unit and the minimum distance below this position have a value of 1 unit.

More About. . .

Guitar

Most guitars have six strings. The frequency at which one of these strings vibrates is controlled by the length of the string, the amount of tension on the string, the weight of the string, and springiness of the strings' material.

Source: www.howstuffworks.com

35. GEOMETRY A regular hexagon is inscribed in a unit circle centered at the origin. If one vertex of the hexagon is at (1, 0), find the exact coordinates of the remaining vertices.

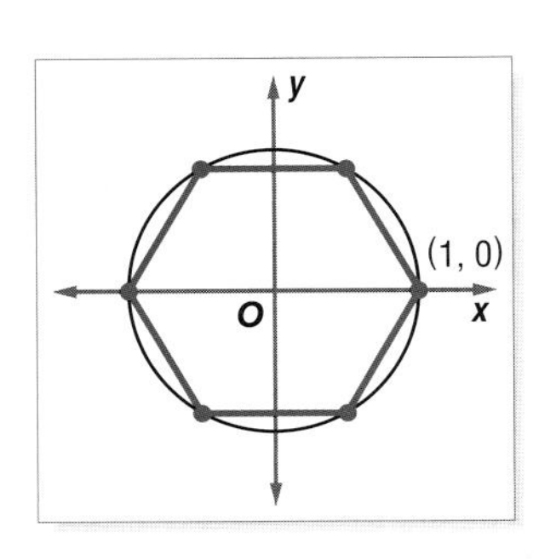

36. BIOLOGY In a certain area of forested land, the population of rabbits R increases and decreases periodically throughout the year. If the population can be modeled by $R = 425 + 200 \sin\left[\frac{\pi}{365}(d - 60)\right]$, where d represents the dth day of the year, describe what happens to the population throughout the year.

SLOPE **For Exercises 37–42, use the following information.**
Suppose the terminal side of an angle θ in standard position intersects the unit circle at $P(x, y)$.

37. What is the slope of $\overline{OP}$?

38. Which of the six trigonometric functions is equal to the slope of $\overline{OP}$?

39. What is the slope of any line perpendicular to $\overline{OP}$?

40. Which of the six trigonometric functions is equal to the slope of any line perpendicular to $\overline{OP}$?

41. Find the slope of $\overline{OP}$ when $\theta = 60°$.

42. If $\theta = 60°$, find the slope of the line tangent to circle O at point P.

43. CRITICAL THINKING Determine the domain and range of the functions $y = \sin \theta$ and $y = \cos \theta$.

44. WRITING IN MATH Answer the question that was posed at the beginning of the lesson.

How can you model annual temperature fluctuations?

Include the following in your answer:

- a description of how the sine and cosine functions are similar to annual temperature fluctuations, and
- if the formula for the temperature T in degrees Fahrenheit of a city t months into the year is given by $T = 50 + 25 \sin\left(\frac{\pi}{6}t\right)$, explain how to find the average temperature and the maximum and minimum predicted over the year.

Standardized Test Practice

Standards Practice

45. If $\triangle ABC$ is an equilateral triangle, what is the length of $\overline{AD}$, in units?

(A) $5\sqrt{2}$

(B) 5

(C) $10\sqrt{2}$

(D) 10

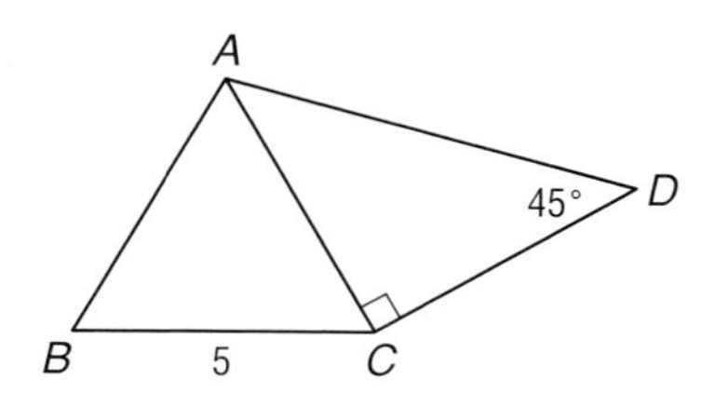

46. SHORT RESPONSE What is the exact value of tan 1830°?

Maintain Your Skills

Mixed Review

Determine whether each triangle should be solved by beginning with the Law of Sines or Law of Cosines. Then solve each triangle. Round measures of sides to the nearest tenth and measures of angles to the nearest degree. *(Lesson 13-5)*

47.

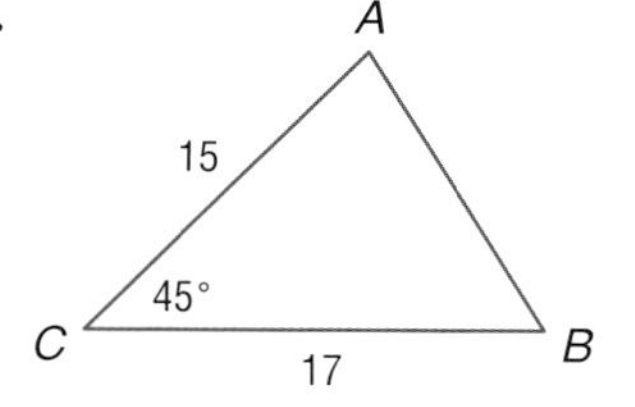

48.

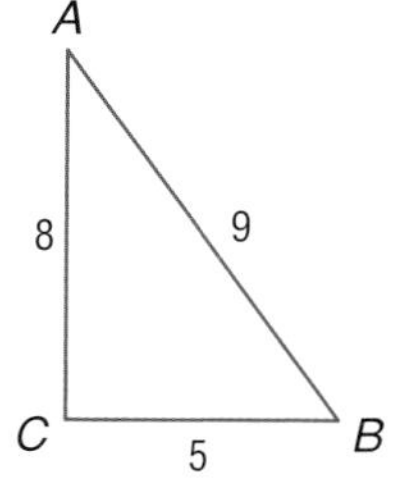

Find the area of $\triangle ABC$. Round to the nearest tenth. *(Lesson 13-4)*

49. $a = 11$ in., $c = 5$ in., $B = 79°$

50. $b = 4$ m, $c = 7$ m, $A = 63°$

BULBS For Exercises 51–56, use the following information.
The lifetimes of 10,000 light bulbs are normally distributed. The mean lifetime is 300 days, and the standard deviation is 40 days. *(Lesson 12-7)*

51. How many light bulbs will last between 260 and 340 days?

52. How many light bulbs will last between 220 and 380 days?

53. How many light bulbs will last fewer than 300 days?

54. How many light bulbs will last more than 300 days?

55. How many light bulbs will last more than 380 days?

56. How many light bulbs will last fewer than 180 days?

Find the sum of each infinite geometric series, if it exists. *(Lesson 11-5)*

57. $a_1 = 3, r = 1.2$

58. $16, 4, 1, \frac{1}{4}, \ldots$

59. $\sum_{n=1}^{\infty} 13(-0.625)^{n-1}$

Use synthetic division to find each quotient. *(Lesson 5-3)*

60. $(4x^2 - 13x + 10) \div (x - 2)$

61. $(2x^2 + 21x + 54) \div (x + 6)$

62. $(5y^3 + y^2 - 7) \div (y + 1)$

63. $(2y^2 + y - 16) \div (y - 3)$

Getting Ready for the Next Lesson

PREREQUISITE SKILL Find each value of θ. Round to the nearest degree.
*(To review **finding angle measures**, see Lesson 13-1.)*

64. $\sin \theta = 0.3420$

65. $\cos \theta = -0.3420$

66. $\tan \theta = 3.2709$

67. $\tan \theta = 5.6713$

68. $\sin \theta = 0.8290$

69. $\cos \theta = 0.0175$

13-7 Inverse Trigonometric Functions

What You'll Learn

- Solve equations by using inverse trigonometric functions.
- Find values of expressions involving trigonometric functions.

How are inverse trigonometric functions used in road design?

Vocabulary

- principal values
- Arcsine function
- Arccosine function
- Arctangent function

When a car travels a curve on a horizontal road, the friction between the tires and the road keeps the car on the road. Above a certain speed, however, the force of friction will not be great enough to hold the car in the curve. For this reason, civil engineers design banked curves.

The proper banking angle θ for a car making a turn of radius r feet at a velocity v in feet per second is given by the equation $\tan \theta = \frac{v^2}{32r}$. In order to determine the appropriate value of θ for a specific curve, you need to know the radius of the curve, the maximum allowable velocity of cars making the curve, and how to determine the angle θ given the value of its tangent.

SOLVE EQUATIONS USING INVERSES Sometimes the value of a trigonometric function for an angle is known and it is necessary to find the measure of the angle. The concept of inverse functions can be applied to find the inverse of trigonometric functions.

In Lesson 8-8, you learned that the inverse of a function is the relation in which all the values of x and y are reversed. The graphs of $y = \sin x$ and its inverse, $x = \sin y$, are shown below.

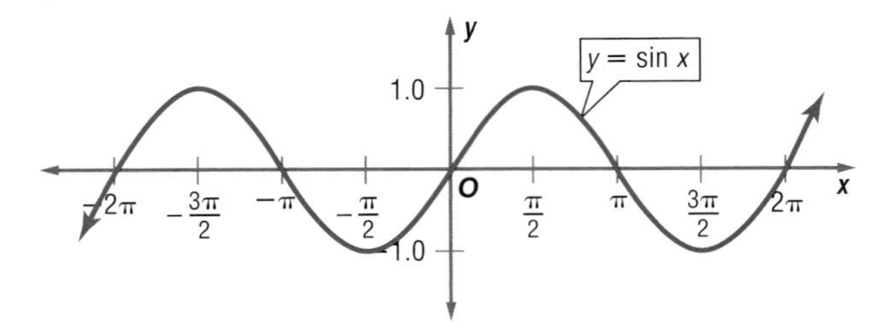

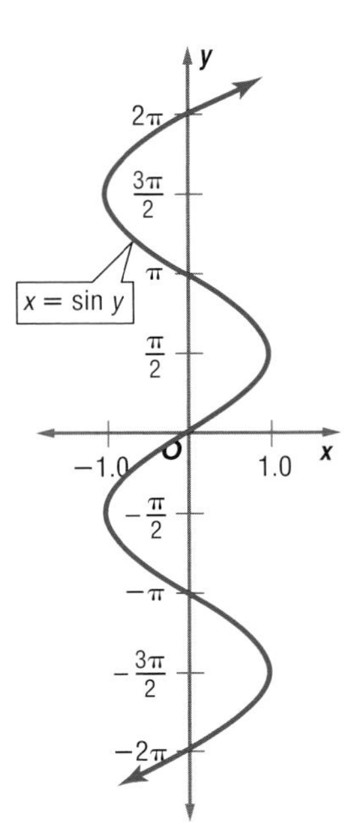

Notice that the inverse is not a function, since it fails the vertical line test. None of the inverses of the trigonometric functions are functions.

We must restrict the domain of trigonometric functions so that their inverses are functions. The values in these restricted domains are called **principal values**. Capital letters are used to distinguish trigonometric functions with restricted domains from the usual trigonometric functions.

Key Concept — *Principal Values of Sine, Cosine, and Tangent*

$y = \text{Sin } x$ if and only if $y = \sin x$ and $-\frac{\pi}{2} \leq x \leq \frac{\pi}{2}$.

$y = \text{Cos } x$ if and only if $y = \cos x$ and $0 \leq x \leq \pi$.

$y = \text{Tan } x$ if and only if $y = \tan x$ and $-\frac{\pi}{2} < x < \frac{\pi}{2}$.

The inverse of the Sine function is called the **Arcsine function** and is symbolized by **Sin^{-1}** or **Arcsin**. The Arcsine function has the following characteristics.

- Its domain is the set of real numbers from -1 to 1.
- Its range is the set of angle measures from $-\frac{\pi}{2} \leq x \leq \frac{\pi}{2}$.
- $\text{Sin } x = y$ if and only if $\text{Sin}^{-1} y = x$.
- $[\text{Sin}^{-1} \circ \text{Sin}](x) = [\text{Sin} \circ \text{Sin}^{-1}](x) = x$.

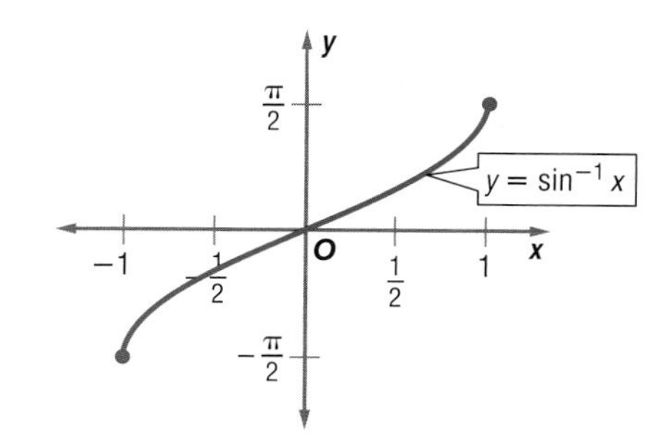

Study Tip

Look Back
To review **composition of functions**, see Lesson 8-7.

The definitions of the Arccosine and Arctangent functions are similar to the definition of the Arcsine function.

Concept Summary — *Inverse Sine, Cosine, and Tangent*

- Given $y = \text{Sin } x$, the inverse Sine function is defined by $y = \text{Sin}^{-1} x$ or $y = \text{Arcsin } x$.
- Given $y = \text{Cos } x$, the inverse Cosine function is defined by $y = \text{Cos}^{-1} x$ or $y = \text{Arccos } x$.
- Given $y = \text{Tan } x$, the inverse Tangent function is defined by $y = \text{Tan}^{-1} x$ or $y = \text{Arctan } x$.

The expressions in each row of the table below are equivalent. You can use these expressions to rewrite and solve trigonometric equations.

$y = \text{Sin } x$	$x = \text{Sin}^{-1} y$	$x = \text{Arcsin } y$
$y = \text{Cos } x$	$x = \text{Cos}^{-1} y$	$x = \text{Arccos } y$
$y = \text{Tan } x$	$x = \text{Tan}^{-1} y$	$x = \text{Arctan } y$

Example 1 *Solve an Equation*

Solve $\text{Sin } x = \frac{\sqrt{3}}{2}$ by finding the value of x to the nearest degree.

If $\text{Sin } x = \frac{\sqrt{3}}{2}$, then x is the least value whose sine is $\frac{\sqrt{3}}{2}$. So, $x = \text{Arcsin } \frac{\sqrt{3}}{2}$.

Use a calculator to find x.

KEYSTROKES: [2nd] [SIN⁻¹] [2nd] [√] 3 [)] [÷] 2 [)] [ENTER] 60

Therefore, $x = 60°$.

Many application problems involve finding the inverse of a trigonometric function.

Example 2 Apply an Inverse to Solve a Problem

DRAWBRIDGE Each leaf of a certain double-leaf drawbridge is 130 feet long. If an 80-foot wide ship needs to pass through the bridge, what is the minimum angle θ, to the nearest degree, which each leaf of the bridge should open so that the ship will fit?

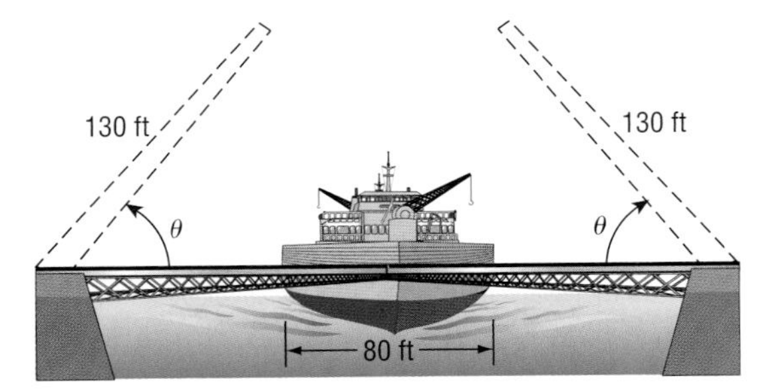

More About. . .

Drawbridges

Bascule bridges have spans (leaves) that pivot upward utilizing gears, motors, and counterweights.

Source: www.multnomah.lib.or.us

When the two parts of the bridge are in their lowered position, the bridge spans 130 + 130 or 260 feet. In order for the ship to fit, the distance between the leaves must be at least 80 feet.

This leaves a horizontal distance of $\frac{260 - 80}{2}$ or 90 feet from the pivot point of each leaf to the ship as shown in the diagram at the right.

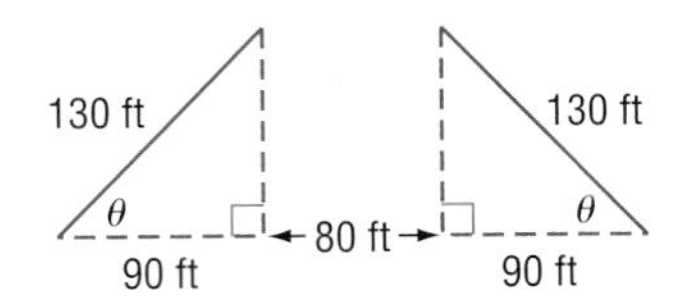

To find the measure of angle θ, use the cosine ratio for right triangles.

$\cos\theta = \frac{\text{adj}}{\text{hyp}}$	Cosine ratio
$\cos\theta = \frac{90}{130}$	Replace *adj* with 90 and *hyp* with 130.
$\theta = \cos^{-1}\left(\frac{90}{130}\right)$	Inverse cosine function
$\theta \approx 46.2°$	Use a calculator.

Thus, the minimum angle through which each leaf of the bridge should open is 47°.

TRIGONOMETRIC VALUES You can use a calculator to find the values of trigonometric expressions.

Example 3 Find a Trigonometric Value

Find each value. Write angle measures in radians. Round to the nearest hundredth.

Study Tip

Angle Measure

Remember that when evaluating an inverse trigonometric function the result is an angle measure.

a. $\text{ArcSin}\ \frac{\sqrt{3}}{2}$

KEYSTROKES: [2nd] [SIN⁻¹] [2nd] [√] 3 [)] [÷] 2 [)] [ENTER] 1.047197551

Therefore, $\text{ArcSin}\ \frac{\sqrt{3}}{2} \approx 1.05$ radians.

b. $\tan\left(\text{Cos}^{-1}\frac{6}{7}\right)$

KEYSTROKES: [TAN] [2nd] [COS⁻¹] 6 [÷] 7 [)] [ENTER] .6009252126

Therefore, $\tan\left(\text{Cos}^{-1}\frac{6}{7}\right) \approx 0.60$.

Check for Understanding

Concept Check

1. **Explain** how you know when the domain of a trigonometric function is restricted.
2. **OPEN ENDED** Write an equation giving the value of the Cosine function for an angle measure in its domain. Then, write your equation in the form of an inverse function.
3. Describe how $y = \text{Cos } x$ and $y = \text{Arccos } x$ are related.

Guided Practice

Write each equation in the form of an inverse function.

4. $\text{Tan } \theta = x$
5. $\text{Cos } \alpha = 0.5$

Solve each equation by finding the value of x to the nearest degree.

6. $x = \text{Cos}^{-1} \frac{\sqrt{2}}{2}$
7. $\text{Arctan } 0 = x$

Find each value. Write angle measures in radians. Round to the nearest hundredth.

8. $\text{Tan}^{-1} \left(-\frac{\sqrt{3}}{3}\right)$
9. $\text{Cos}^{-1} (-1)$
10. $\cos \left(\text{Cos}^{-1} \frac{2}{9}\right)$
11. $\sin \left(\text{Sin}^{-1} \frac{3}{4}\right)$
12. $\sin \left(\text{Cos}^{-1} \frac{3}{4}\right)$
13. $\tan \left(\text{Sin}^{-1} \frac{1}{2}\right)$

Application

14. **ARCHITECTURE** The support for a roof is shaped like two right triangles as shown at the right. Find θ.

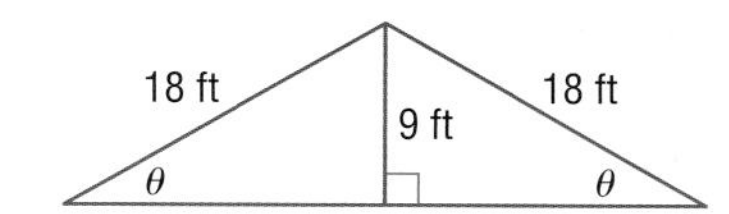

Practice and Apply

Homework Help	
For Exercises	See Examples
15–26	1
27–42	3
43–48	2

Extra Practice
See page 859.

Write each equation in the form of an inverse function.

15. $\alpha = \text{Sin } \beta$
16. $\text{Tan } a = b$
17. $\text{Cos } y = x$
18. $\text{Sin } 30° = \frac{1}{2}$
19. $\text{Cos } 45° = y$
20. $-\frac{4}{3} = \text{Tan } x$

Solve each equation by finding the value of x to the nearest degree.

21. $x = \text{Cos}^{-1} \frac{1}{2}$
22. $\text{Sin}^{-1} \frac{1}{2} = x$
23. $\text{Arctan } 1 = x$
24. $x = \text{Arctan } \frac{\sqrt{3}}{3}$
25. $x = \text{Sin}^{-1}$
26. $x = \text{Cos}^{-1} 0$

Find each value. Write angle measures in radians. Round to the nearest hundredth.

27. $\text{Cos}^{-1} \left(-\frac{1}{2}\right)$
28. $\text{Sin}^{-1} \frac{\pi}{2}$
29. $\text{Arctan } \frac{\sqrt{3}}{3}$
30. $\text{Arccos } \frac{\sqrt{3}}{2}$
31. $\sin \left(\text{Sin}^{-1} \frac{1}{2}\right)$
32. $\cot \left(\text{Sin}^{-1} \frac{5}{6}\right)$
33. $\tan \left(\text{Cos}^{-1} \frac{6}{7}\right)$
34. $\sin \left(\text{Arctan } \frac{\sqrt{3}}{3}\right)$
35. $\cos \left(\text{Arcsin } \frac{3}{5}\right)$
36. $\cot \left(\text{Sin}^{-1} \frac{7}{9}\right)$
37. $\cos \left(\text{Tan}^{-1} \sqrt{3}\right)$
38. $\tan (\text{Arctan } 3)$
39. $\cos \left[\text{Arccos} \left(-\frac{1}{2}\right)\right]$
40. $\text{Sin}^{-1} \left(\tan \frac{\pi}{4}\right)$
41. $\cos \left(\text{Cos}^{-1} \frac{\sqrt{2}}{2} - \frac{\pi}{2}\right)$
42. $\text{Cos}^{-1} (\text{Sin}^{-1} 90)$
43. $\sin \left(2 \text{ Cos}^{-1} \frac{3}{5}\right)$
44. $\sin \left(2 \text{ Sin}^{-1} \frac{1}{2}\right)$

45. TRAVEL The cruise ship *Reno* sailed due west 24 miles before turning south. When the *Reno* became disabled and radioed for help, the rescue boat found that the fastest route to her covered a distance of 48 miles. The cosine of the angle at which the rescue boat should sail is 0.5. Find the angle θ, to the nearest tenth of a degree, at which the rescue boat should travel to aid the *Reno*.

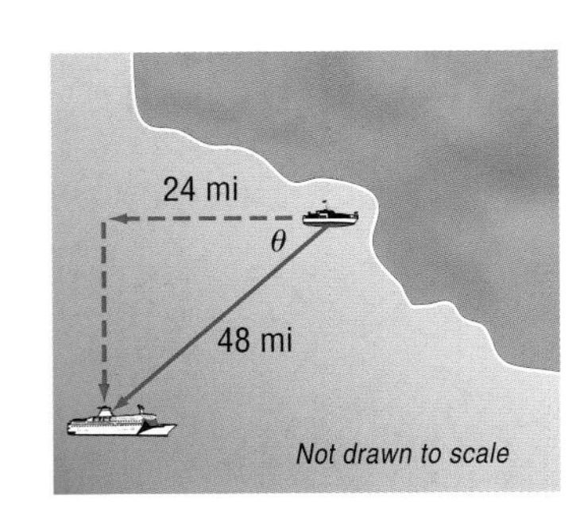

46. FOUNTAINS Architects who design fountains know that both the height and distance that a water jet will project is dependent on the angle θ at which the water is aimed. For a given angle θ, the ratio of the maximum height H of the parabolic arc to the horizontal distance D it travels is given by $\frac{H}{D} = \frac{1}{4}\tan\theta$. Find the value of θ, to the nearest degree, that will cause the arc to go twice as high as it travels horizontally.

More About. . .

Track and Field

The shot is a metal sphere that can be made out of solid iron. Shot putters stand inside a seven-foot circle and must "put" the shot from the shoulder with one hand.

Source: www.coolrunning.com.au

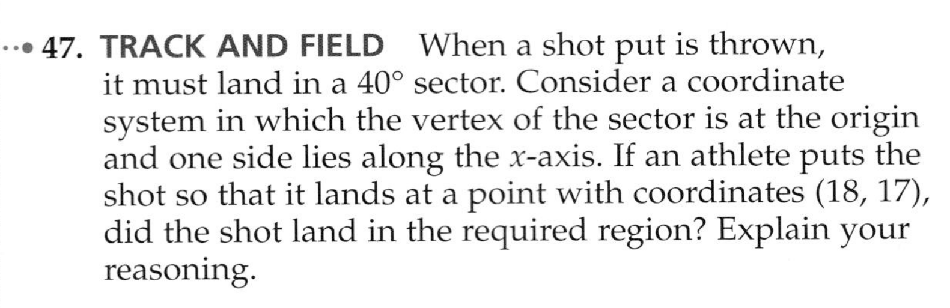

47. TRACK AND FIELD When a shot put is thrown, it must land in a 40° sector. Consider a coordinate system in which the vertex of the sector is at the origin and one side lies along the x-axis. If an athlete puts the shot so that it lands at a point with coordinates (18, 17), did the shot land in the required region? Explain your reasoning.

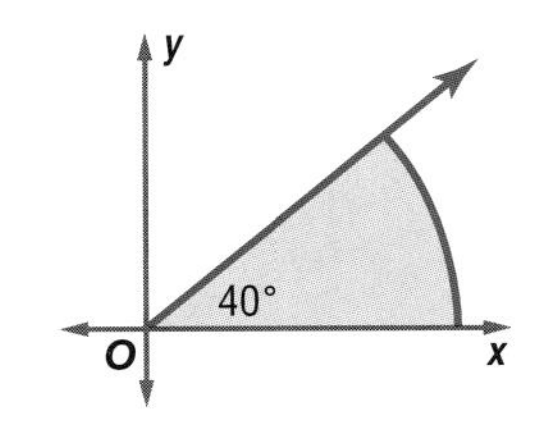

48. OPTICS You may have polarized sunglasses that eliminate glare by polarizing the light. When light is polarized, all of the waves are traveling in parallel planes. Suppose horizontally-polarized light with intensity I_0 strikes a polarizing filter with its axis at an angle of θ with the horizontal. The intensity of the transmitted light I_t and θ are related by the equation $\cos\theta = \sqrt{\frac{I_t}{I_0}}$. If one fourth of the polarized light is transmitted through the lens, what angle does the transmission axis of the filter make with the horizontal?

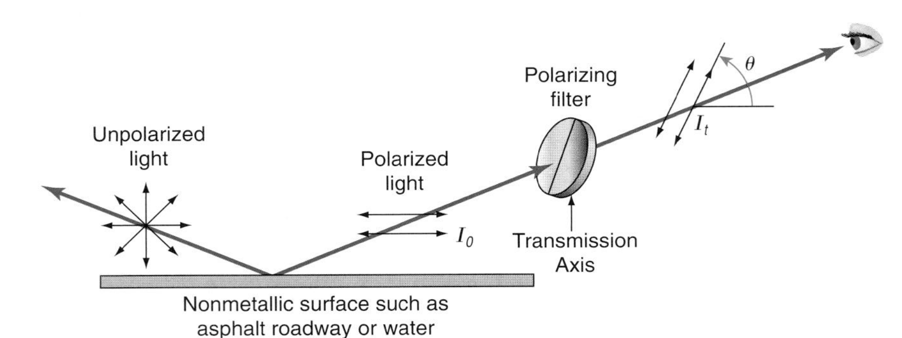

CRITICAL THINKING For Exercises 49–51, use the following information.
If the graph of the line $y = mx + b$ intersects the x-axis such that an angle of θ is formed with the positive x-axis, then $\tan\theta = m$.

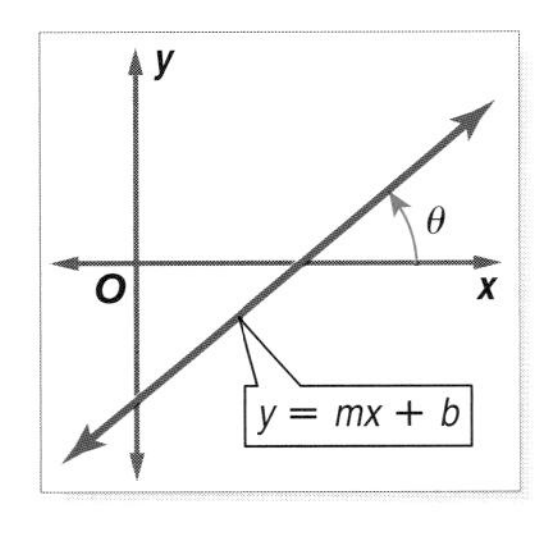

49. Find the acute angle that the graph of $3x + 5y = 7$ makes with the positive x-axis to the nearest degree.

50. Determine the obtuse angle formed at the intersection of the graphs of $2x + 5y = 8$ and $6x - y = -8$. State the measure of the angle to the nearest degree.

51. Explain why this relationship, $\tan\theta = m$, holds true.

52. **WRITING IN MATH** Answer the question that was posed at the beginning of the lesson.

How are inverse trigonometric functions used in road design?

Include the following in your answer:

- a few sentences describing how to determine the banking angle for a road, and
- a description of what would have to be done to a road if the speed limit were increased and the banking angle was not changed.

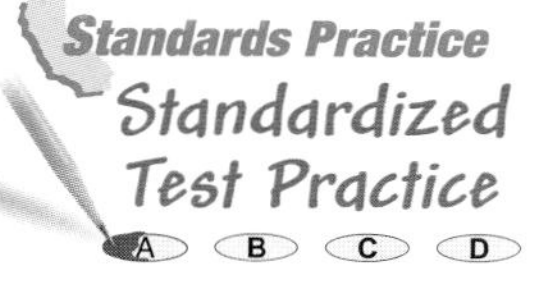

53. **GRID IN** Find the angle of depression θ between the shallow end and the deep end of the swimming pool to the nearest degree.

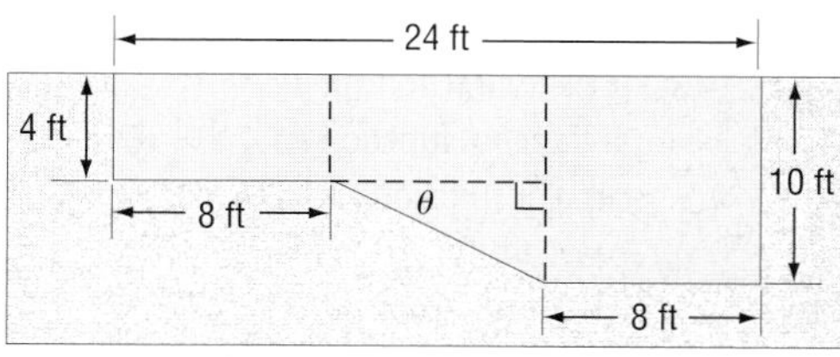

Side View of Swimming Pool

54. If $\sin \theta = \frac{2}{3}$ and $-90° \leq \theta \leq 90°$, then $\cos 2\theta =$

Ⓐ $-\frac{1}{9}$. Ⓑ $-\frac{1}{3}$. Ⓒ $\frac{1}{3}$. Ⓓ $\frac{1}{9}$. Ⓔ 1.

ADDITION OF TRIGONOMETRIC INVERSES Consider the function $y = \text{Sin}^{-1}\, x + \text{Cos}^{-1}\, x$.

55. Copy and complete the table below by evaluating y for each value of x.

x	0	$\frac{1}{2}$	$\frac{\sqrt{2}}{2}$	$\frac{\sqrt{3}}{2}$	1	$-\frac{1}{2}$	$-\frac{\sqrt{2}}{2}$	$-\frac{\sqrt{3}}{2}$	-1
y									

56. Make a conjecture about the function $y = \text{Sin}^{-1}\, x + \text{Cos}^{-1}\, x$.

57. Considering only positive values of x, provide an explanation of why your conjecture might be true.

Maintain Your Skills

Mixed Review

Find the exact value of each function. *(Lesson 13-6)*

58. $\sin -660°$

59. $\cos 25\pi$

60. $(\sin 135°)^2 + (\cos -675°)^2$

Determine whether each triangle should be solved by beginning with the Law of Sines or Law of Cosines. Then solve each triangle. Round measures of sides to the nearest tenth and measures of angles to the nearest degree. *(Lesson 13-5)*

61. $a = 3.1, b = 5.8, A = 30°$

62. $a = 9, b = 40, c = 41$

Use synthetic substitution to find $f(3)$ and $f(-4)$ for each function. *(Lesson 7-4)*

63. $f(x) = 5x^2 + 6x - 17$

64. $f(x) = -3x^2 + 2x - 1$

65. $f(x) = 4x^2 - 10x + 5$

66. **PHYSICS** A toy rocket is fired upward from the top of a 200-foot tower at a velocity of 80 feet per second. The height of the rocket t seconds after firing is given by the formula $h(t) = -16t^2 + 80t + 200$. Find the time at which the rocket reaches its maximum height of 300 feet. *(Lesson 6-5)*

Chapter 13 Study Guide and Review

Vocabulary and Concept Check

State whether each sentence is *true* or *false*. If false, replace the underlined word(s) or number to make a true sentence.

1. When two angles in standard position have the same terminal side, they are called quadrantal angles.
2. The Law of Sines is used to solve a triangle when the measure of two angles and the measure of any side are known.
3. Trigonometric functions can be defined by using a unit circle.
4. For all values of θ, $\csc \theta = \frac{1}{\cos \theta}$.
5. A radian is the measure of an angle on the unit circle where the rays of the angle intercept an arc with length 1 unit.
6. If the measures of three sides of a triangle are known, then the Law of Sines can be used to solve the triangle.
7. An angle measuring 60° is a quadrantal angle.
8. For all values of x, $\cos (x + 180°) = \cos x$.
9. In a coordinate plane, the initial side of an angle is the ray that rotates about the center.

Lesson-by-Lesson Review

13-1 Right Triangle Trigonometry

See pages 701–708.

Concept Summary

θ: measure of an acute angle of a right triangle
opp: measure of the leg opposite θ
adj: measure of the leg adjacent to θ
hyp: measure of the hypotenuse

$\sin \theta = \frac{\text{opp}}{\text{hyp}}$ $\quad \cos \theta = \frac{\text{adj}}{\text{hyp}}$ $\quad \tan \theta = \frac{\text{opp}}{\text{adj}}$

$\csc \theta = \frac{\text{hyp}}{\text{opp}}$ $\quad \sec \theta = \frac{\text{hyp}}{\text{adj}}$ $\quad \cot \theta = \frac{\text{adj}}{\text{opp}}$

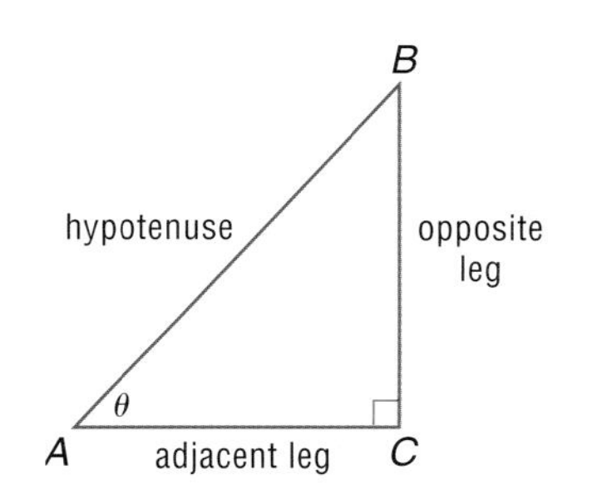

Example **Solve $\triangle ABC$. Round measures of sides to the nearest tenth and measures of angles to the nearest degree.**

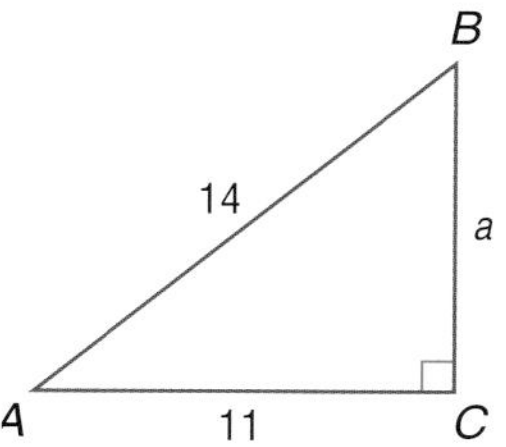

Find a.

$a^2 + b^2 = c^2$ Pythagorean Theorem

$a^2 + 11^2 = 14^2$ $b = 11$ and $c = 14$

$a = \sqrt{14^2 - 11^2}$ Solve for a.

$a \approx 8.7$ Use a calculator.

Find A. $\cos A = \frac{11}{14}$ $\cos A = \frac{\text{adj}}{\text{hyp}}$

Use a calculator to find the angle whose cosine is $\frac{11}{14}$.

KEYSTROKES: [2nd] [COS^{-1}] 11 [÷] 14 [)] [ENTER] 38.2132107

To the nearest degree, $A \approx 38°$.

Find B. $38° + B \approx 90°$ Angles A and B are complementary.

$B \approx 52°$ Solve for B.

Therefore, $a \approx 8.7$, $A \approx 38°$, and $B \approx 52°$.

Exercises **Solve $\triangle ABC$ by using the given measurements. Round measures of sides to the nearest tenth and measures of angles to the nearest degree.** *See Examples 4 and 5 on page 704.*

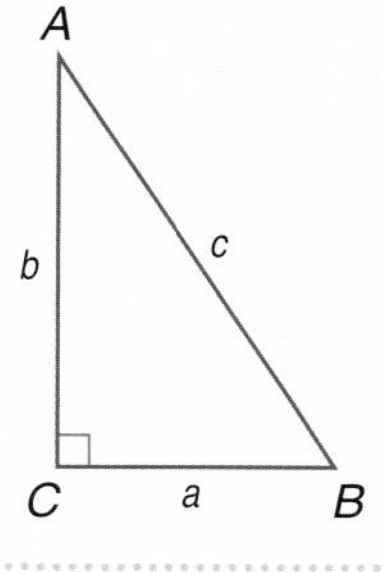

10. $c = 16, a = 7$

11. $A = 25°, c = 6$

12. $B = 45°, c = 12$

13. $B = 83°, b = \sqrt{31}$

14. $a = 9, B = 49°$

15. $\cos A = \frac{1}{4}, a = 4$

13-2 Angles and Angle Measure

See pages 709–715.

Concept Summary

- An angle in standard position has its vertex at the origin and its initial side along the positive x-axis.
- The measure of an angle is determined by the amount of rotation from the initial side to the terminal side. If the rotation is in a counterclockwise direction, the measure of the angle is positive. If the rotation is in a clockwise direction, the measure of the angle is negative.

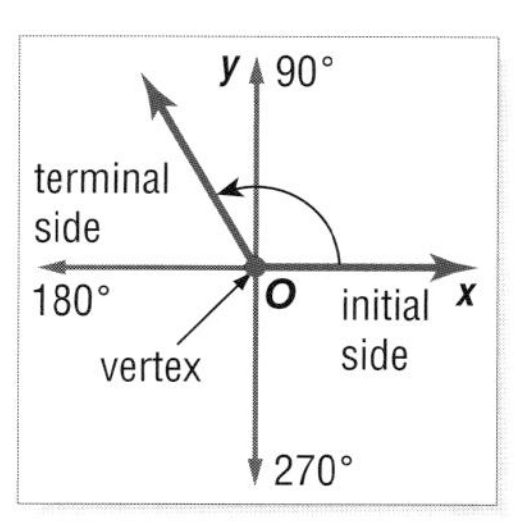

Examples **Rewrite the degree measure in radians and the radian measure in degrees.**

1 **240°**

$$240° = 240°\left(\frac{\pi \text{ radians}}{180°}\right)$$

$$= \frac{240\pi}{180} \text{ radians or } \frac{4\pi}{3}$$

2 $\frac{\pi}{12}$

$$\frac{\pi}{12} = \left(\frac{\pi}{12} \text{ radians}\right)\left(\frac{180°}{\pi \text{ radians}}\right)$$

$$= \frac{180°}{12} \text{ or } 15°$$

Exercises **Rewrite each degree measure in radians and each radian measure in degrees.** *See Example 2 on page 711.*

16. $255°$ **17.** $-210°$ **18.** $\frac{7\pi}{4}$ **19.** -4π

Find one angle with positive measure and one angle with negative measure coterminal with each angle. *See Example 4 on page 712.*

20. $205°$ **21.** $-40°$ **22.** $\frac{4\pi}{3}$ **23.** $-\frac{7\pi}{4}$

13-3 Trigonometric Functions of General Angles

See pages 717–724.

Concept Summary

- You can find the exact values of the six trigonometric functions of θ given the coordinates of a point $P(x, y)$ on the terminal side of the angle.

$\sin \theta = \frac{y}{r}$ $\cos \theta = \frac{x}{r}$ $\tan \theta = \frac{y}{x}, x \neq 0$

$\csc \theta = \frac{r}{y}, y \neq 0$ $\sec \theta = \frac{r}{x}, x \neq 0$ $\cot \theta = \frac{x}{y}, y \neq 0$

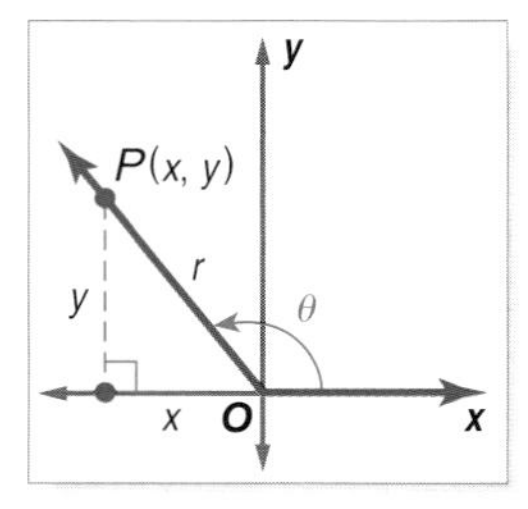

Example **Find the exact value of cos 150°.**

Because the terminal side of 150° lies in Quadrant II, the reference angle θ' is $180° - 150°$ or $30°$. The cosine function is negative in Quadrant II, so $\cos 150° = -\cos 30°$ or $-\frac{\sqrt{3}}{2}$.

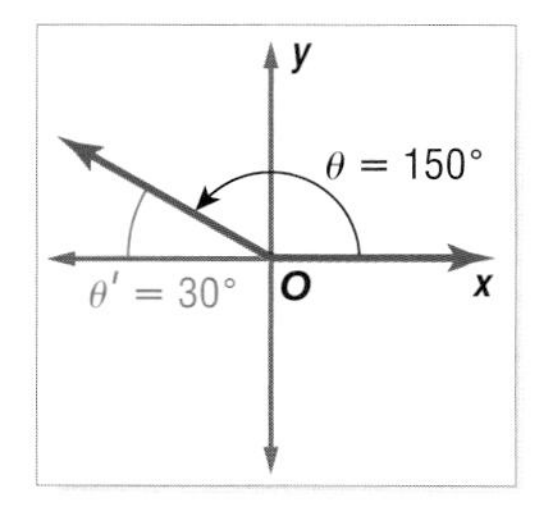

Exercises **Find the exact value of the six trigonometric functions of θ if the terminal side of θ in standard position contains the given point.**
See Example 1 on pages 717 and 718.

24. $P(2, 5)$ **25.** $P(15, -8)$

Find the exact value of each trigonometric function. *See Example 4 on page 720.*

26. $\cos 3\pi$ **27.** $\tan 120°$ **28.** $\sin \frac{5\pi}{4}$ **29.** $\sec (-30°)$

13-4 Law of Sines

See pages 725–732.

Concept Summary

- You can find the area of $\triangle ABC$ if the measures of two sides and their included angle are known.

area $= \frac{1}{2}bc \sin A$ area $= \frac{1}{2}ac \sin B$ area $= \frac{1}{2}ab \sin C$

- Law of Sines: $\frac{\sin A}{a} = \frac{\sin B}{b} = \frac{\sin C}{c}$

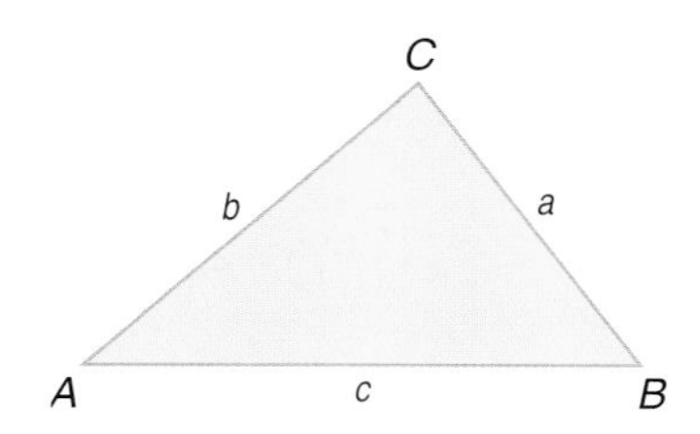

Example **Solve $\triangle ABC$.**

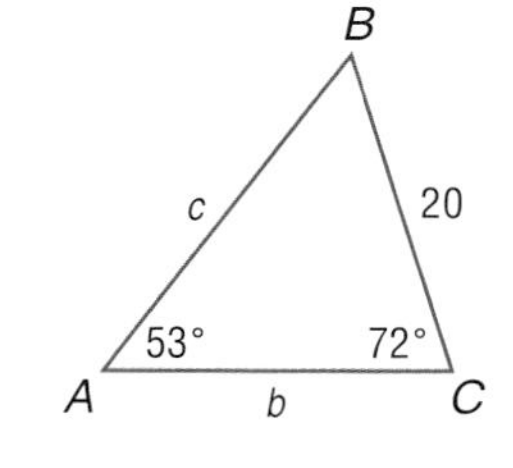

First, find the measure of the third angle.

$53° + 72° + B = 180°$ The sum of the angle measures is 180°.

$B = 55°$ $180 - (53 + 72) = 55$

Now use the Law of Sines to find b and c. Write two equations, each with one variable.

$\frac{\sin A}{a} = \frac{\sin C}{c}$	Law of Sines	$\frac{\sin B}{b} = \frac{\sin A}{a}$
$\frac{\sin 53°}{20} = \frac{\sin 72°}{c}$	Replace A with 53°, B with 55°, C with 72°, and a with 20.	$\frac{\sin 55°}{b} = \frac{\sin 53°}{20}$
$c = \frac{20 \sin 72°}{\sin 53°}$	Solve for the variable.	$b = \frac{20 \sin 55°}{\sin 53°}$
$c \approx 23.8$	Use a calculator.	$b \approx 20.5$

Therefore, $B = 55°$, $b \approx 20.5$, and $c \approx 23.8$.

Exercises **Determine whether each triangle has *no* solution, *one* solution, or *two* solutions. Then solve each triangle. Round measures of sides to the nearest tenth and measures of angles to the nearest degree.** *See Examples 3–5 on pages 727 and 728.*

30. $a = 24, b = 36, A = 64°$

31. $A = 40°, b = 10, a = 8$

32. $b = 10, c = 15, C = 66°$

33. $A = 82°, a = 9, b = 12$

34. $A = 105°, a = 18, b = 14$

35. $B = 46°, C = 83°, b = 65$

13-5 Law of Cosines

See pages 733–738.

Concept Summary

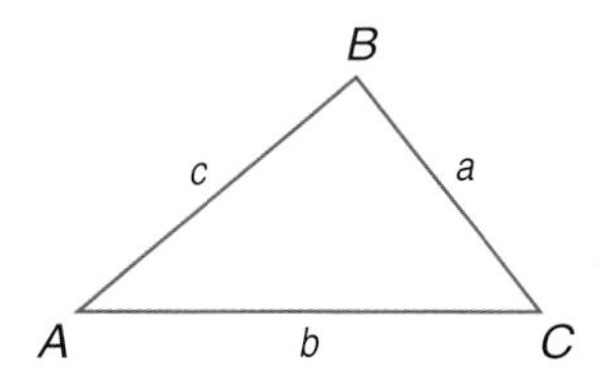

- Law of Cosines: $a^2 = b^2 + c^2 - 2bc \cos A$
  $b^2 = a^2 + c^2 - 2ac \cos B$
  $c^2 = a^2 + b^2 - 2ab \cos C$

Example **Solve $\triangle ABC$ for $A = 62°$, $b = 15$, and $c = 12$.**

You are given the measure of two sides and the included angle. Begin by drawing a diagram and using the Law of Cosines to determine a.

$a^2 = b^2 + c^2 - 2bc \cos A$	Law of Cosines
$a^2 = 15^2 + 12^2 - 2(15)(12) \cos 62°$	$b = 15$, $c = 12$, and $A = 62°$
$a^2 \approx 200$	Simplify.
$a \approx 14.1$	Take the square root of each side.

Next, you can use the Law of Sines to find the measure of angle C.

$\frac{\sin 62°}{14.1} \approx \frac{\sin C}{12}$	Law of Sines
$\sin C \approx \frac{12 \sin 62°}{14.1}$ or about 48.7°	Use a calculator.

The measure of the angle B is approximately $180 - (62 + 48.7)$ or 69.3°.

Therefore, $a \approx 14.1$, $C \approx 48.7°$, $B \approx 69.3°$.

• Extra Practice, see pages 857–859.
• Mixed Problem Solving, see page 874.

Exercises **Determine whether each triangle should be solved by beginning with the Law of Sines or Law of Cosines. Then solve each triangle. Round measures of sides to the nearest tenth and measures of angles to the nearest degree.** *See Examples 1 and 2 on pages 734 and 735.*

36.

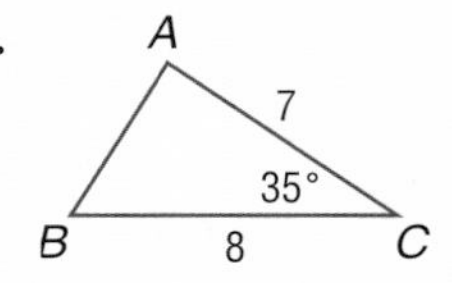

37.

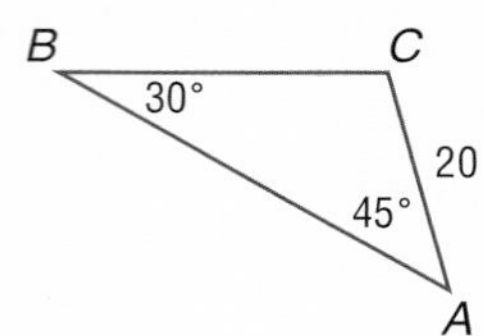

38.

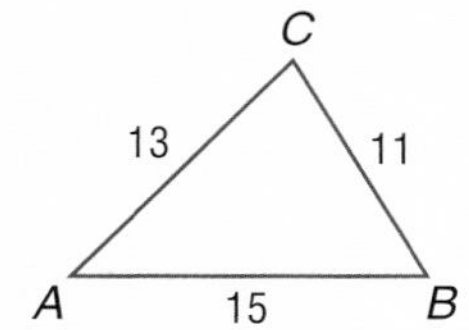

39. $C = 65°, a = 4, b = 7$ **40.** $A = 36°, a = 6, b = 8$ **41.** $b = 7.6, c = 14.1, A = 29°$

13-6 Circular Functions

See pages 739–745.

Concept Summary

- If the terminal side of an angle θ in standard position intersects the unit circle at $P(x, y)$, then $\cos \theta = x$ and $\sin \theta = y$. Therefore, the coordinates of P can be written as $P(\cos \theta, \sin \theta)$.

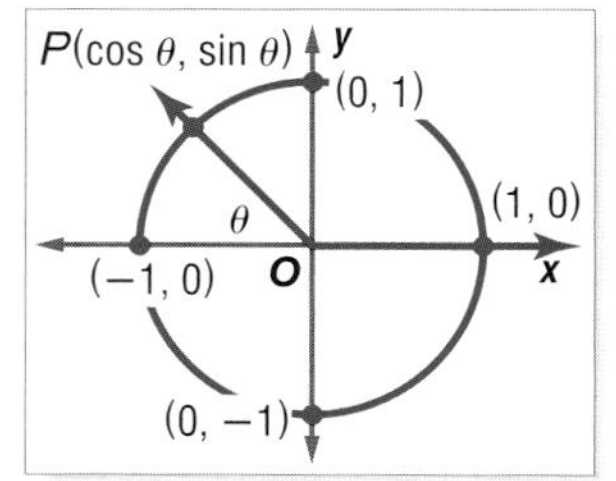

Example **Find the exact value of $\cos\left(-\frac{7\pi}{4}\right)$.**

$$\cos\left(-\frac{7\pi}{4}\right) = \cos\left(-\frac{7\pi}{4} + 2\pi\right) = \cos\frac{\pi}{4} \text{ or } \frac{\sqrt{2}}{2}$$

Exercises **Find the exact value of each function.** *See Example 2 on page 741.*

42. $\sin(-150°)$ **43.** $\cos 300°$ **44.** $(\sin 45°)(\sin 225°)$

45. $\sin \frac{5\pi}{4}$ **46.** $(\sin 30°)^2 + (\cos 30°)^2$ **47.** $\frac{4 \cos 150° + 2 \sin 300°}{3}$

13-7 Inverse Trigonometric Functions

See pages 746–751.

Concept Summary

- $y = \text{Sin } x$ if and only if $y = \sin x$ and $-\frac{\pi}{2} \le x \le \frac{\pi}{2}$.
- $y = \text{Cos } x$ if and only if $y = \cos x$ and $0 \le x \le \pi$.
- $y = \text{Tan } x$ if and only if $y = \tan x$ and $-\frac{\pi}{2} < x < \frac{\pi}{2}$.

Example **Find the value of $\text{Cos}^{-1}\left[\tan\left(-\frac{\pi}{6}\right)\right]$ in radians. Round to the nearest hundredth.**

KEYSTROKES: 2nd [COS⁻¹] TAN (–) 2nd [π] ÷ 6)) ENTER 2.186276035

Therefore, $\text{Cos}^{-1}\left[\tan\left(-\frac{\pi}{6}\right)\right] \approx 2.19$ radians.

Exercises **Find each value. Write angle measures in radians. Round to the nearest hundredth.** *See Example 3 on page 748.*

48. $\text{Sin}^{-1}(-1)$ **49.** $\text{Tan}^{-1}\sqrt{3}$ **50.** $\tan\left(\text{Arcsin}\frac{3}{5}\right)$ **51.** $\cos(\text{Sin}^{-1} 1)$

Chapter 13

Practice Test

Vocabulary and Concepts

1. **Draw** a right triangle and label one of the acute angles θ. Then label the hypotenuse *hyp*, the side opposite θ *opp*, and the side adjacent θ *adj*.
2. **State** the Law of Sines for $\triangle ABC$.
3. **Describe** a situation in which you would solve a triangle by first applying the Law of Cosines.

Skills and Applications

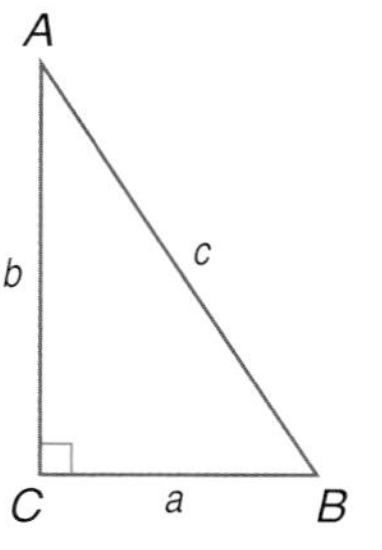

Solve $\triangle ABC$ by using the given measurements. Round measures of sides to the nearest tenth and measures of angles to the nearest degree.

4. $a = 7, A = 49°$
5. $B = 75°, b = 6$
6. $A = 22°, c = 8$
7. $a = 7, c = 16$

Rewrite each degree measure in radians and each radian measure in degrees.

8. $275°$
9. $-\frac{\pi}{6}$
10. $\frac{11\pi}{2}$
11. $330°$
12. $-600°$
13. $-\frac{7\pi}{4}$

Find the exact value of each expression. Write angle measures in degrees.

14. $\cos(-120°)$
15. $\sin \frac{7\pi}{4}$
16. $\cot 300°$
17. $\sec\left(-\frac{7\pi}{6}\right)$
18. $\text{Sin}^{-1}\left(-\frac{\sqrt{3}}{2}\right)$
19. Arctan 1
20. $\tan 135°$
21. $\csc \frac{5\pi}{6}$

22. Determine the number of possible solutions for a triangle in which $A = 40°$, $b = 10$, and $a = 14$. If a solution exists, solve the triangle. Round measures of sides to the nearest tenth and measures of angles to the nearest degree.

23. Suppose θ is an angle in standard position whose terminal side lies in Quadrant II. Find the exact values of the remaining five trigonometric functions for θ for $\cos\theta = -\frac{\sqrt{3}}{2}$.

24. **GEOLOGY** From the top of the cliff, a geologist spots a dry riverbed. The measurement of the angle of depression to the riverbed is 70°. The cliff is 50 meters high. How far is the riverbed from the base of the cliff?

Standards Practice

25. **STANDARDIZED TEST PRACTICE** Triangle *ABC* has a right angle at *C*, angle $B = 30°$, and $BC = 6$. Find the area of triangle *ABC*.

(A) 6 units^2 (B) $\sqrt{3}$ units^2 (C) $6\sqrt{3}$ units^2 (D) 12 units^2

Chapter 13 Standardized Test Practice

Standards Practice

Part 1 Multiple Choice

Record your answers on the answer sheet provided by your teacher or on a sheet of paper.

1. If $3n + k = 30$ and n is a positive even integer, then which of the following statements must be true?
I. k is divisible by 3.
II. k is an even integer.
III. k is less than 20.

(A) I only (B) II only
(C) I and II only (D) I, II, and III

2. If $4x^2 + 5x = 80$ and $4x^2 - 5y = 30$, then what is the value of $6x + 6y$?

(A) 10 (B) 50 (C) 60 (D) 110

3. If $a = b + cb$, then what does $\frac{b}{a}$ equal in terms of c?

(A) $\frac{1}{c}$ (B) $\frac{1}{1 + c}$
(C) $1 - c$ (D) $1 + c$

4. What is the value of $\sum_{n=1}^{5} 3n^2$?

(A) 55 (B) 58
(C) 75 (D) 165

5. There are 16 green marbles, 2 red marbles, and 6 yellow marbles in a jar. How many yellow marbles need to be added to the jar in order to double the probability of selecting a yellow marble?

(A) 4 (B) 6 (C) 8 (D) 12

> **Test-Taking Tip** (A) (B) (C) (D)
> **Questions 1–10**
> The answer choices to multiple-choice questions can provide clues to help you solve a problem. In Question 5, you can add the values in the answer choices to the number of yellow marbles and the total number of marbles to find which is the correct answer.

6. From a lookout point on a cliff above a lake, the angle of depression to a boat on the water is 12°. The boat is 3 kilometers from the shore just below the cliff. What is the height of the cliff from the surface of the water to the lookout point?

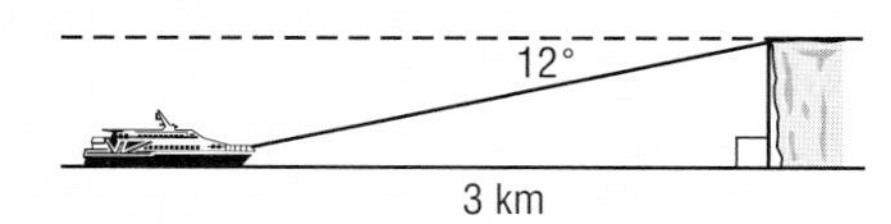

(A) $\frac{3}{\sin 12°}$ (B) $\frac{3}{\tan 12°}$
(C) $\frac{3}{\cos 12°}$ (D) $3 \tan 12°$

7. If $x + y = 90°$ and x and y are positive, then $\frac{\cos x}{\sin y} =$

(A) 0. (B) $\frac{1}{2}$.
(C) 1. (D) cannot be determined

8. A child flying a kite holds the string 4 feet above the ground. The taut string is 40 feet long and makes an angle of 35° with the horizontal. How high is the kite off the ground?

(A) $4 + 40 \sin 35°$ (B) $4 + 40 \cos 35°$
(C) $4 + 40 \tan 35°$ (D) $4 + \frac{40}{\sin 35°}$

9. If $\sin \theta = -\frac{1}{2}$ and $180° < \theta < 270°$, then $\theta =$

(A) 200°. (B) 210°.
(C) 225°. (D) 240°.

10. If $\cos \theta = \frac{8}{17}$ and the terminal side of the angle is in quadrant IV, then $\sin \theta =$

(A) $-\frac{15}{8}$. (B) $-\frac{17}{15}$.
(C) $-\frac{15}{17}$. (D) $\frac{15}{17}$.

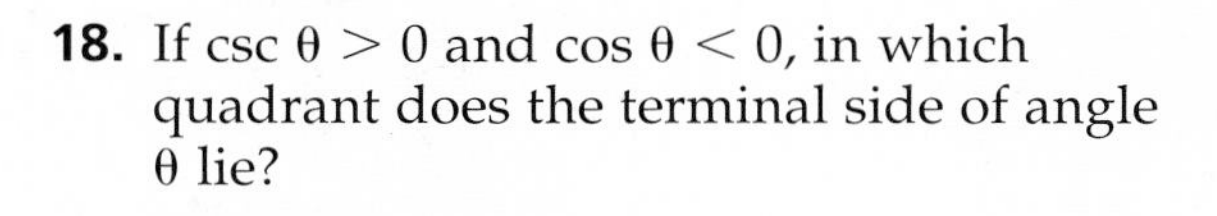

Part 2 Short Response/Grid In

Record your answers on the answer sheet provided by your teacher or on a sheet of paper.

11. The length, width, and height of the rectangular box illustrated below are each integers greater than 1. If the area of *ABCD* is 18 square units and the area of *CDEF* is 21 square units, what is the volume of the box?

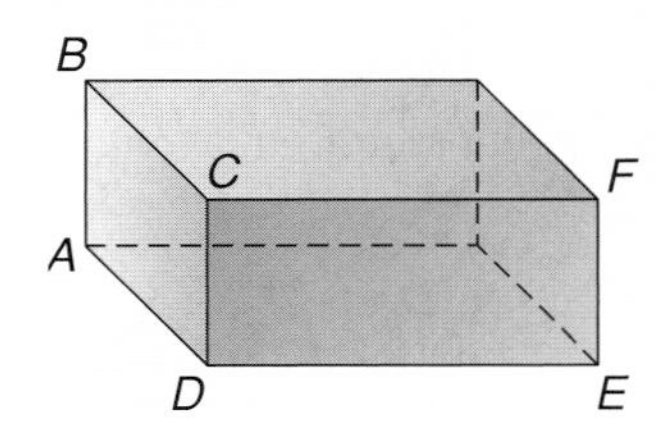

12. When six consecutive integers are multiplied, their product is 0. What is their greatest possible sum?

13. The average (arithmetic mean) score for the 25 players on a team is n. Their scores range from 60 to 100, inclusive. The average score of 20 of the players is 70. What is the difference between the greatest and least possible values of n?

14. The variables a, b, c, d, and e are integers in a sequence, where $a = 2$ and $b = 12$. To find the next term, double the last term and add that result to one less than the next-to-last term. For example, $c = 25$, because $2(12) = 24$, $2 - 1 = 1$, and $24 + 1 = 25$. What is the value of e?

15. In the figure, if $t = 2v$, what is the value of x?

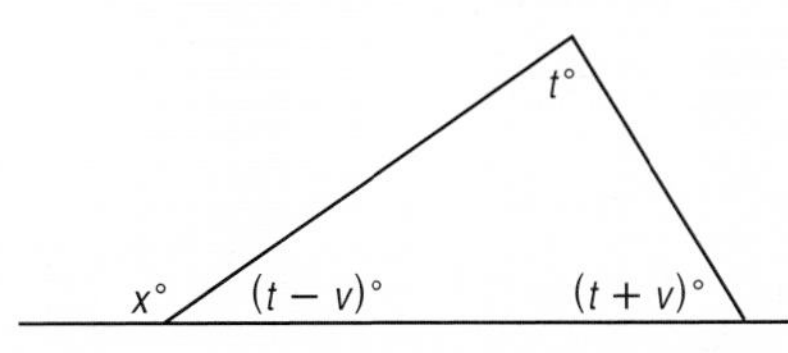

16. If $b = 4$, then what is the value of a in the equations below?
$3a + 4b + 2c = 33$
$2b + 4c = 12$

17. At the head table at a banquet, 3 men and 3 women sit in a row. In how many ways can the row be arranged so that the men and women alternate?

18. If $\csc \theta > 0$ and $\cos \theta < 0$, in which quadrant does the terminal side of angle θ lie?

19. In $\triangle ABC$, if $AC = 15$, $BC = 10$, and $m\angle A = 35°$, what type of angle could angle C be?

20. Through how many radians does the minute hand of a clock turn in 18 minutes?

21. The terminal side of an angle θ passes through the point at $\left(-\frac{1}{2}, \frac{\sqrt{3}}{2}\right)$ on a unit circle. If θ is in standard position, name a possible value of θ.

Part 3 Extended Response

Record your answers on a sheet of paper. Show your work.

For Exercises 22–24, use the following information.

A boater at sea uses a sextant to sight an angle of elevation of 42° to the top of a lighthouse. After the ship travels 330 feet directly toward the lighthouse, she makes another sighting and the new angle of elevation is 49°. The ship's charts show that water becomes too shallow for the boat 125 feet from the base of the lighthouse.

22. Tell whether you can find the distance from the boat to the lighthouse by using the Law of Sines or the Law of Cosines. Justify your answer.

23. Find the distance from the boat to the base of the lighthouse. Show each step and round to the nearest foot.

24. If the boat is traveling at 44 feet per second, how long does the boater have to change course before the boat runs aground in the shallow water? Round to the nearest tenth of a second.

Chapter 14 Trigonometric Graphs and Identities

What You'll Learn

- **Lessons 14-1 and 14-2** Graph trigonometric functions and determine period, amplitude, phase shifts, and vertical shifts.
- **Lessons 14-3 and 14-4** Use and verify trigonometric identities.
- **Lessons 14-5 and 14-6** Use sum and difference formulas and double- and half-angle formulas.
- **Lesson 14-7** Solve trigonometric equations.

Key Vocabulary

- amplitude (p. 763)
- phase shift (p. 769)
- vertical shift (p. 771)
- trigonometric identity (p. 777)
- trigonometric equation (p. 799)

Why It's Important

Some equations contain one or more trigonometric functions. It is important to know how to simplify trigonometric expressions to solve these equations. Trigonometric functions can be used to model many real-world applications, such as music. *You will learn how a trigonometric function can be used to describe music in Lesson 14-6.*

Getting Started

Prerequisite Skills To be successful in this chapter, you'll need to master these skills and be able to apply them in problem-solving situations. Review these skills before beginning Chapter 14.

For Lessons 14-1 and 14-2 — Trigonometric Values

Find the exact value of each trigonometric function. *(For review, see Lesson 13-3.)*

1. $\sin 135°$ **2.** $\tan 315°$ **3.** $\cos 90°$ **4.** $\tan 45°$

5. $\sin \frac{5\pi}{4}$ **6.** $\cos \frac{7\pi}{6}$ **7.** $\sin \frac{11\pi}{6}$ **8.** $\tan \frac{3\pi}{2}$

For Lessons 14-3, 14-5, and 14-6 — Circular Functions

Find the exact value of each trigonometric function. *(For review, see Lesson 13-6.)*

9. $\cos(-150°)$ **10.** $\sin 510°$ **11.** $\cot \frac{9\pi}{4}$ **12.** $\sec \frac{13\pi}{6}$

13. $\tan\left(-\frac{3\pi}{2}\right)$ **14.** $\csc(-720°)$ **15.** $\cos \frac{7\pi}{3}$ **16.** $\tan \frac{8\pi}{3}$

For Lesson 14-4 — Factor Polynomials

Factor completely. If the polynomial is not factorable, write *prime*. *(For review, see Lesson 5-4.)*

17. $-15x^2 - 5x$ **18.** $2x^4 - 4x^2$ **19.** $x^3 + 4$

20. $x^2 - 6x + 8$ **21.** $2x^2 - 3x - 2$ **22.** $3x^3 - 2x^2 - x$

For Lesson 14-7 — Solve Quadratic Equations

Solve each equation by factoring. *(For review, see Lesson 6-3.)*

23. $x^2 - 5x - 24 = 0$ **24.** $x^2 - 2x - 48 = 0$ **25.** $x^2 + 3x - 40 = 0$

26. $x^2 - 12x = 0$ **27.** $-2x^2 - 11x - 12 = 0$ **28.** $x^2 - 16 = 0$

Trigonometric Graphs and Identities Make this Foldable to help you organize your notes. Begin with eight sheets of grid paper.

Step 1 Staple

Staple the stack of grid paper along the top to form a booklet.

Step 2 Cut and Label

Cut seven lines from the bottom of the top sheet, six lines from the second sheet, and so on. Label with lesson numbers as shown.

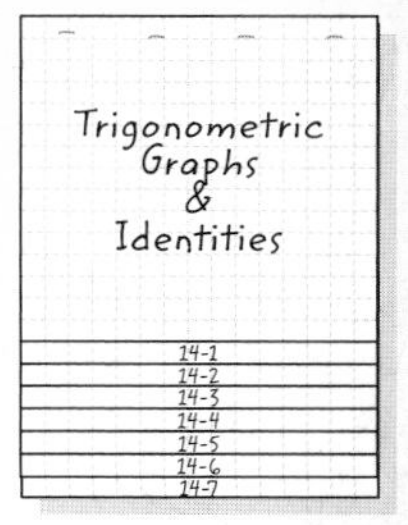

Reading and Writing As you read and study the chapter, use each page to write notes and to graph examples for each lesson.

14-1 Graphing Trigonometric Functions

What You'll Learn

- Graph trigonometric functions.
- Find the amplitude and period of variation of the sine, cosine, and tangent functions.

Vocabulary
- amplitude

Why can you predict the behavior of tides?

The rise and fall of tides can have great impact on the communities and ecosystems that depend upon them. One type of tide is a semidiurnal tide. This means that bodies of water, like the Atlantic Ocean, have two high tides and two low tides a day. Because tides are periodic, they behave the same way each day.

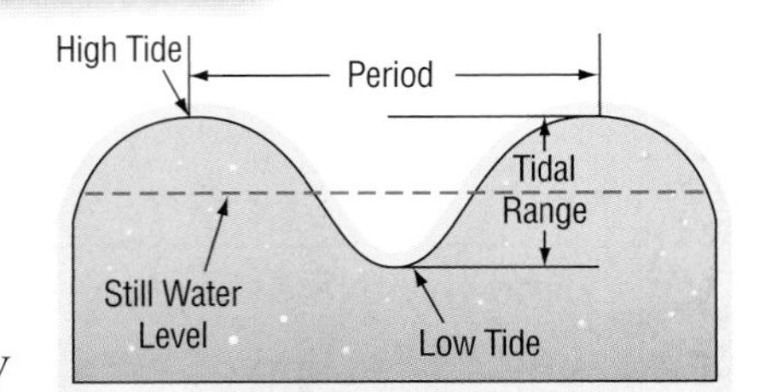

GRAPH TRIGONOMETRIC FUNCTIONS The diagram below illustrates the water level as a function of time for a body of water with semidiurnal tides.

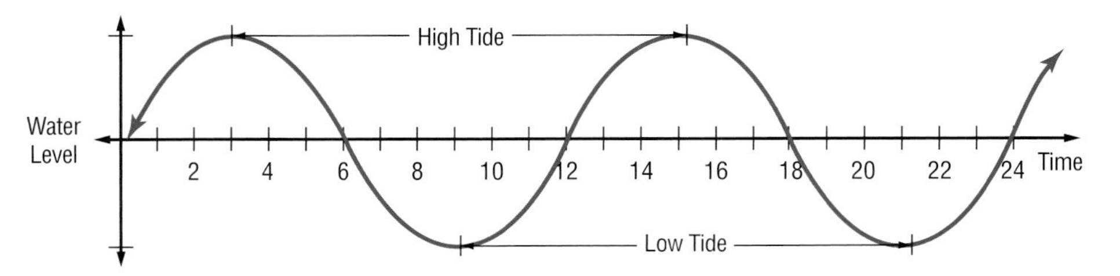

Study Tip

Look Back
To review **period** and **periodic functions**, see Lesson 13-6.

In each cycle of high and low tides, the pattern repeats itself. Recall that a function whose graph repeats a basic pattern is called a *periodic function.*

To find the period, start from any point on the graph and proceed to the right until the pattern begins to repeat. The simplest approach is to begin at the origin. Notice that after about 12 hours the graph begins to repeat. Thus, the period of the function is about 12 hours.

To graph the periodic functions $y = \sin \theta$, $y = \cos \theta$, or $y = \tan \theta$, use values of θ expressed either in degrees or radians. Ordered pairs for points on these graphs are of the form $(\theta, \sin \theta)$, $(\theta, \cos \theta)$, and $(\theta, \tan \theta)$, respectively.

θ	0°	30°	45°	60°	90°	120°	135°	150°	180°	210°	225°	240°	270°	300°	315°	330°	360°
$\sin \theta$	0	$\frac{1}{2}$	$\frac{\sqrt{2}}{2}$	$\frac{\sqrt{3}}{2}$	1	$\frac{\sqrt{3}}{2}$	$\frac{\sqrt{2}}{2}$	$\frac{1}{2}$	0	$-\frac{1}{2}$	$-\frac{\sqrt{2}}{2}$	$-\frac{\sqrt{3}}{2}$	−1	$-\frac{\sqrt{3}}{2}$	$-\frac{\sqrt{2}}{2}$	$-\frac{1}{2}$	0
nearest tenth	0	0.5	0.7	0.9	1	0.9	0.7	0.5	0	−0.5	−0.7	−0.9	−1	−0.9	−0.7	−0.5	0
$\cos \theta$	1	$\frac{\sqrt{3}}{2}$	$\frac{\sqrt{2}}{2}$	$\frac{1}{2}$	0	$-\frac{1}{2}$	$-\frac{\sqrt{2}}{2}$	$-\frac{\sqrt{3}}{2}$	−1	$-\frac{\sqrt{3}}{2}$	$-\frac{\sqrt{2}}{2}$	$-\frac{1}{2}$	0	$\frac{1}{2}$	$\frac{\sqrt{2}}{2}$	$\frac{\sqrt{3}}{2}$	1
nearest tenth	1	0.9	0.7	0.5	0	−0.5	−0.7	−0.9	−1	−0.9	−0.7	−0.5	0	0.5	0.7	0.9	1
$\tan \theta$	0	$\frac{\sqrt{3}}{3}$	1	$\sqrt{3}$	nd	$-\sqrt{3}$	−1	$-\frac{\sqrt{3}}{3}$	0	$\frac{\sqrt{3}}{3}$	1	$\sqrt{3}$	nd	$-\sqrt{3}$	−1	$-\frac{\sqrt{3}}{3}$	0
nearest tenth	0	0.6	1	1.7	nd	−1.7	−1	−0.6	0	0.6	1	1.7	nd	−1.7	−1	−0.6	0
θ	0	$\frac{\pi}{6}$	$\frac{\pi}{4}$	$\frac{\pi}{3}$	$\frac{\pi}{2}$	$\frac{2\pi}{3}$	$\frac{3\pi}{4}$	$\frac{5\pi}{6}$	π	$\frac{7\pi}{6}$	$\frac{5\pi}{4}$	$\frac{4\pi}{3}$	$\frac{3\pi}{2}$	$\frac{5\pi}{3}$	$\frac{7\pi}{4}$	$\frac{11\pi}{6}$	2π

nd = not defined

After plotting several points, complete the graphs of $y = \sin \theta$ and $y = \cos \theta$ by connecting the points with a smooth, continuous curve. Recall from Chapter 13 that each of these functions has a period of 360° or 2π radians. That is, the graph of each function repeats itself every 360° or 2π radians.

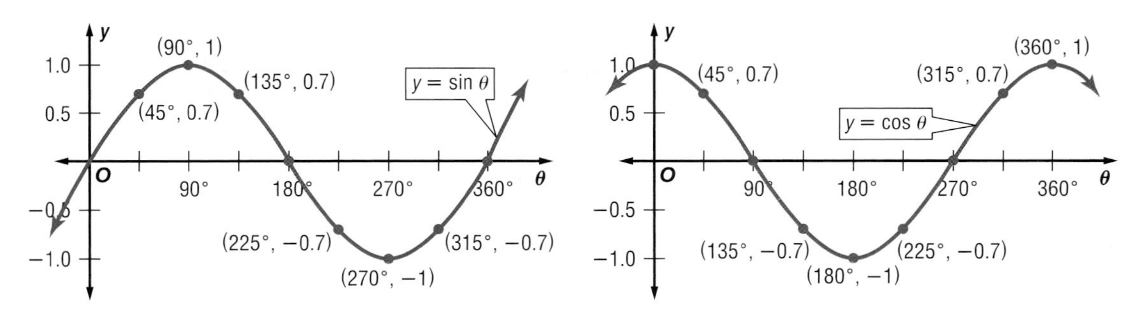

Notice that both the sine and cosine have a maximum value of 1 and a minimum value of −1. The **amplitude** of the graph of a periodic function is the absolute value of half the difference between its maximum value and its minimum value. So, for both the sine and cosine functions, the amplitude of their graphs is $\left|\frac{1-(-1)}{2}\right|$ or 1.

The graph of the tangent function can also be drawn by plotting points. By examining the values for $\tan \theta$ in the table, you can see that the tangent function is not defined for 90°, 270°, …, $90° + k \cdot 180°$, where k is an integer. The graph is separated by vertical asymptotes whose x-intercepts are the values for which $y = \tan \theta$ is not defined.

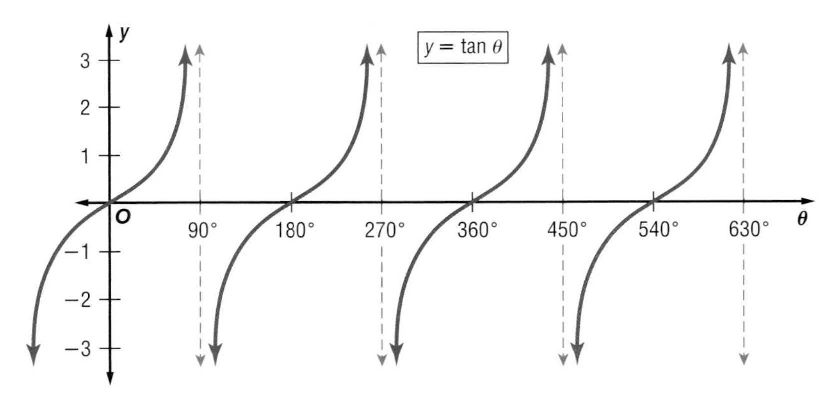

The period of the tangent function is 180° or π radians. Since the tangent function has no maximum or minimum value, it has no amplitude.

The graphs of the secant, cosecant, and cotangent functions are shown below. Compare them to the graphs of the cosine, sine, and tangent functions, which are shown in red.

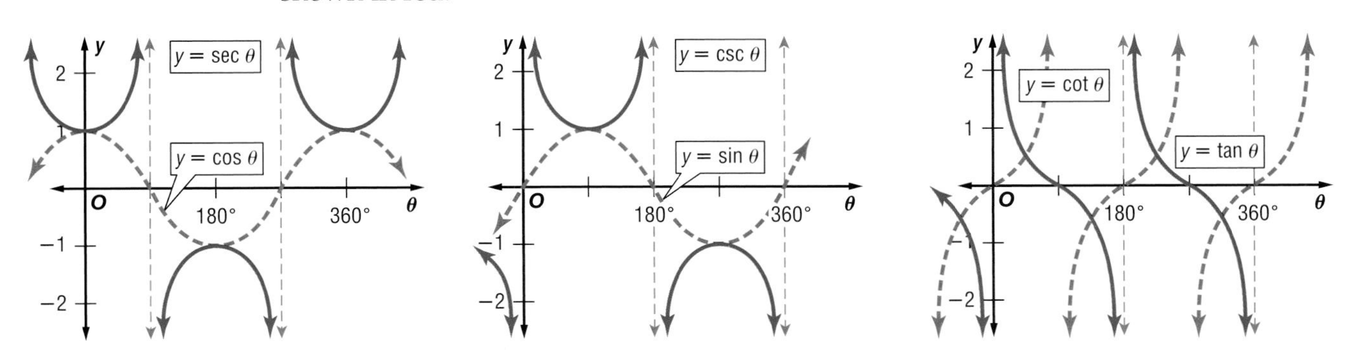

Notice that the period of the secant and cosecant functions is 360° or 2π radians. The period of the cotangent is 180° or π radians. Since none of these functions have a maximum or minimum value, they have no amplitude.

VARIATIONS OF TRIGONOMETRIC FUNCTIONS Just as with other functions, a trigonometric function can be used to form a family of graphs by changing the period and amplitude.

Graphing Calculator Investigation

Period and Amplitude

On a TI-83 Plus graphing calculator, set the MODE to degrees.

Think and Discuss

1. Graph $y = \sin x$ and $y = \sin 2x$. What is the maximum value of each function?

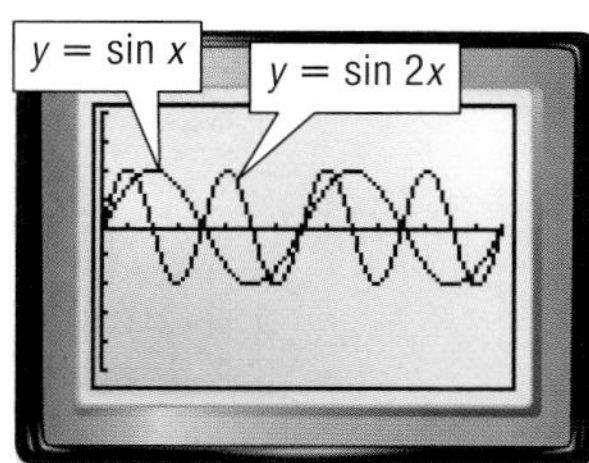

[0, 720] scl: 45 by [−2.5, 2.5] scl: 0.5

2. How many times does each function reach a maximum value?
3. Graph $y = \sin\left(\frac{x}{2}\right)$. What is the maximum value of this function? How many times does this function reach its maximum value?

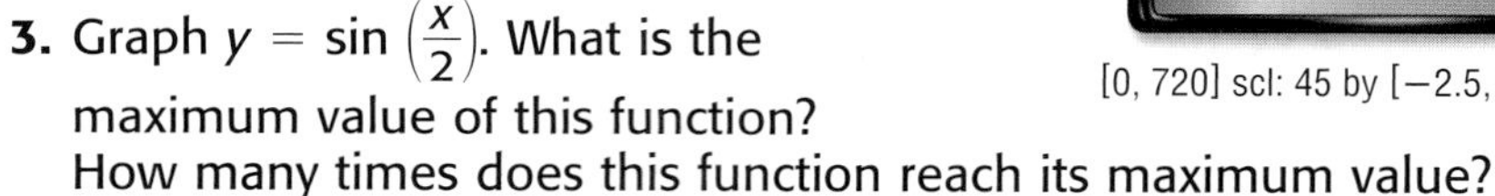

4. Use the equations $y = \sin bx$ and $y = \cos bx$. Repeat Exercises 1–3 for maximum values and the other values of b. What conjecture can you make about the effect of b on the maximum values and the periods of these functions?
5. Graph $y = \sin x$ and $y = 2 \sin x$. What is the maximum value of each function? What is the period of each function?

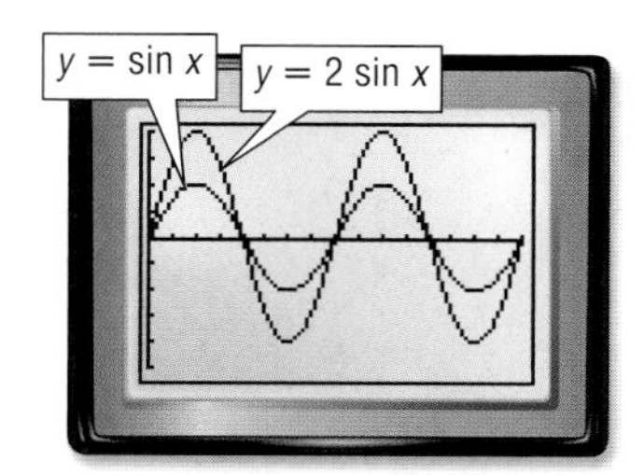

[0, 720] scl: 45 by [−2.5, 2.5] scl: 0.5

6. Graph $y = \frac{1}{2} \sin x$. What is the maximum value of this function? What is the period of this function?
7. Use the equations $y = a \sin x$ and $y = a \cos x$. Repeat Exercises 5 and 6 for other values of a. What conjecture can you make about the effect of a on the amplitudes and periods of $y = a \sin x$ and $y = a \cos x$?

Study Tip

Amplitude and Period
Note that the amplitude affects the graph along the vertical axis and the period affects it along the horizontal axis.

The results of the investigation suggest the following generalization.

Key Concept — *Amplitudes and Periods*

- **Words** For functions of the form $y = a \sin b\theta$ and $y = a \cos b\theta$, the amplitude is $|a|$, and the period is $\frac{360°}{|b|}$ or $\frac{2\pi}{|b|}$.
  For functions of the form $y = a \tan b\theta$, the amplitude is not defined, and the period is $\frac{180°}{|b|}$ or $\frac{\pi}{|b|}$.
- **Examples**

$y = 3 \sin 4\theta$	amplitude 3 and period $\frac{360°}{4}$ or 90°
$y = -6 \cos 5\theta$	amplitude $\lvert -6 \rvert$ or 6 and period $\frac{2\pi}{5}$
$y = 2 \tan \frac{1}{3}\theta$	no amplitude and period 3π

You can use the amplitude and period of a trigonometric function to help you graph the function.

Example 1 Graph Trigonometric Functions

Find the amplitude and period of each function. Then graph the function.

a. $y = \cos 3\theta$

First, find the amplitude.

$|a| = |1|$ The coefficient of $\cos 3\theta$ is 1.

Next, find the period.

$\frac{360°}{|b|} = \frac{360°}{|3|}$ $b = 3$

$= 120°$

Use the amplitude and period to graph the function.

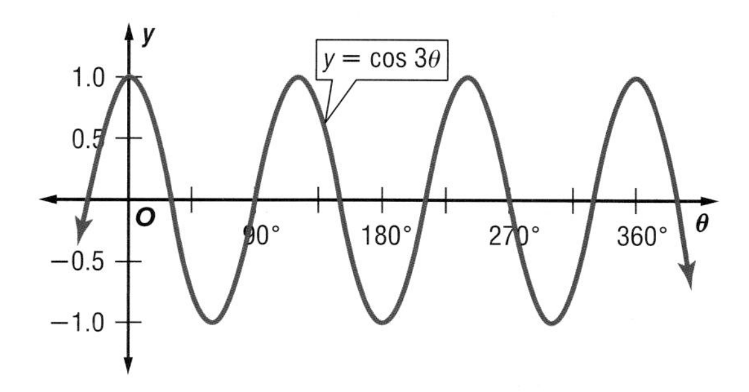

b. $y = \frac{1}{4} \sin \theta$

Amplitude: $|a| = \left|\frac{1}{4}\right|$

$= \frac{1}{4}$

Period: $\frac{360°}{|b|} = \frac{360°}{|1|}$

$= 360°$

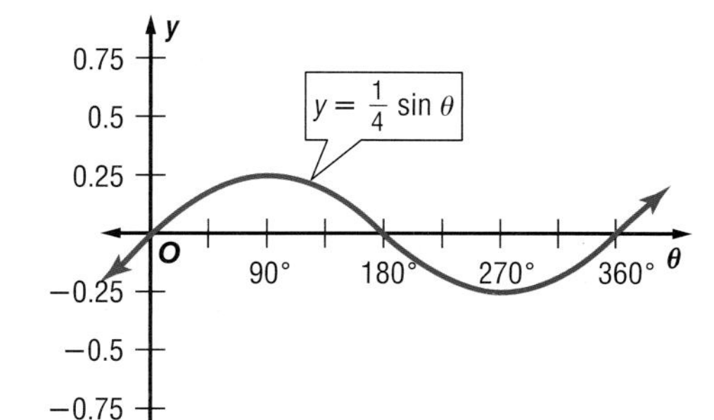

c. $y = \frac{1}{2} \sin \left(-\frac{1}{3}\theta\right)$

Amplitude: $|a| = \left|\frac{1}{2}\right|$

$= \frac{1}{2}$

Period: $\frac{2\pi}{|b|} = \frac{2\pi}{\left|-\frac{1}{3}\right|}$

$= 6\pi$

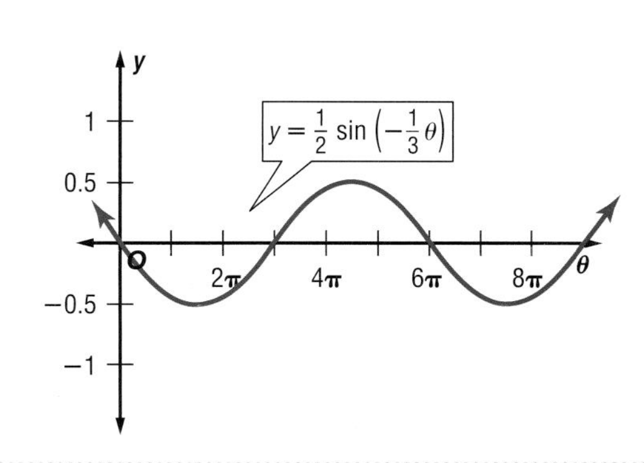

You can use trigonometric functions to describe real-world situations.

Example 2 Use Trigonometric Functions

OCEANOGRAPHY Refer to the application at the beginning of the lesson. Suppose the tidal range of a city on the Atlantic coast is 18 feet. A tide is at *equilibrium* when it is at its normal level, halfway between its highest and lowest points.

a. Write a function to represent the height h of the tide. Assume that the tide is at equilibrium at $t = 0$ and that the high tide is beginning.

Since the height of the tide is 0 at $t = 0$, use the sine function $h = a \sin bt$, where a is the amplitude of the tide and t is the time in hours.

Find the amplitude. The difference between high tide and low tide is the tidal range or 18 feet.

$a = \frac{18}{2}$ or 9

Find the value of b. Each tide cycle lasts about 12 hours.

$\frac{2\pi}{|b|} = 12$ period $= \frac{2\pi}{|b|}$

$b = \frac{2\pi}{12}$ or $\frac{\pi}{6}$ Solve for b.

Thus, an equation to represent the height of the tide is $h = 9 \sin \frac{\pi}{6}t$.

b. Graph the tide function.

More About. . .

Oceanography

Lake Superior has one of the smallest tidal ranges. It can be measured in inches, while the tidal range in the Bay of Funday in Canada measures up to 50 feet.

Source: Office of Naval Research

Check for Understanding

Concept Check

1. **OPEN ENDED** Explain why $y = \tan \theta$ has no amplitude.
2. **Explain** what it means to say that the period of a function is 180°.
3. **FIND THE ERROR** Dante and Jamile graphed $y = 3 \cos \frac{2}{3}\theta$.

Dante

Jamile

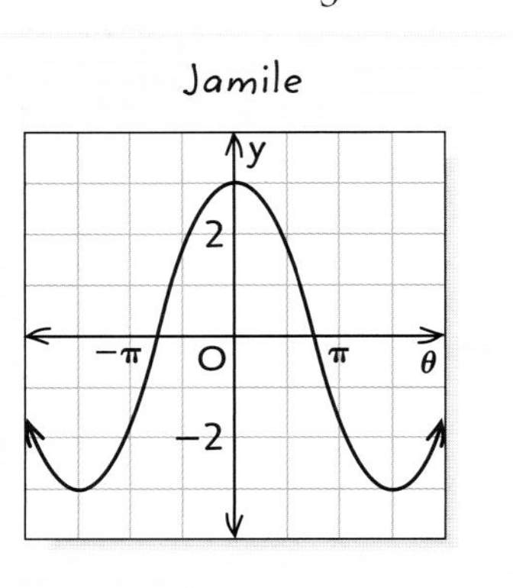

Who is correct? Explain your reasoning.

Guided Practice **Find the amplitude, if it exists, and period of each function. Then graph each function.**

4. $y = \frac{1}{2} \sin \theta$
5. $y = 2 \sin \theta$
6. $y = \frac{2}{3} \cos \theta$
7. $y = \frac{1}{4} \tan \theta$
8. $y = \csc 2\theta$
9. $y = 4 \sin 2\theta$
10. $y = 4 \cos \frac{3}{4}\theta$
11. $y = \frac{1}{2} \sec 3\theta$
12. $y = \frac{3}{4} \cos \frac{1}{2}\theta$

Application BIOLOGY **For Exercises 13 and 14, use the following information.**
In a certain wildlife refuge, the population of field mice can be modeled by $y = 3000 + 1250 \sin \frac{\pi}{6}t$, where y represents the number of mice and t represents the number of months past March 1 of a given year.

13. Determine the period of the function. What does this period represent?
14. What is the maximum number of mice and when does this occur?

Practice and Apply

Homework Help	
For Exercises	See Examples
15–35	1
36–41	2

Extra Practice
See page 859.

Find the amplitude, if it exists, and period of each function. Then graph each function.

15. $y = 3 \sin \theta$
16. $y = 5 \cos \theta$
17. $y = 2 \csc \theta$
18. $y = 2 \tan \theta$
19. $y = \frac{1}{5} \sin \theta$
20. $y = \frac{1}{3} \sec \theta$
21. $y = \sin 4\theta$
22. $y = \sin 2\theta$
23. $y = \sec 3\theta$
24. $y = \cot 5\theta$
25. $y = 4 \tan \frac{1}{3}\theta$
26. $y = 2 \cot \frac{1}{2}\theta$
27. $y = 6 \sin \frac{2}{3}\theta$
28. $y = 3 \cos \frac{1}{2}\theta$
29. $y = 3 \csc \frac{1}{2}\theta$
30. $y = \frac{1}{2} \cot 2\theta$
31. $2y = \tan \theta$
32. $\frac{3}{4}y = \frac{2}{3} \sin \frac{3}{5}\theta$

33. Draw a graph of a sine function with an amplitude of $\frac{3}{5}$ and a period of 90°. Then write an equation for the function.

34. Draw a graph of a cosine function with an amplitude of $\frac{7}{8}$ and a period of $\frac{2\pi}{5}$. Then write an equation for the function.

35. COMMUNICATIONS The carrier wave for a certain FM radio station can be modeled by the equation $y = A \sin (10^7 \cdot 2\pi t)$, where A is the amplitude of the wave and t is the time in seconds. Determine the period of the carrier wave.

PHYSICS **For Exercises 36 and 37, use the following information.**
The sound wave produced by a tuning fork can be modeled by a sine function.

36. If the amplitude of the sine function is 0.25, write the equations for tuning forks that resonate with a frequency of 64, 256, and 512 Hertz.

37. How do the periods of the tuning forks compare?

38. CRITICAL THINKING A function is called *even* if the graphs of $y = f(x)$ and $y = f(-x)$ are the same. A function is *odd* if $f(-x) = -f(x)$. The graphs of even functions have reflection symmetry with respect to the y-axis and the graphs of odd functions have rotation symmetry with respect to the origin. Which of the six trigonometric functions are even and which are odd? Justify your answer with a graph of each function.

BOATING **For Exercises 39–41, use the following information.**
A marker buoy off the coast of Gulfport, Mississippi, bobs up and down with the waves. The distance between the highest and lowest point is 4 feet. The buoy moves from its highest point to its lowest point and back to its highest point every 10 seconds.

39. Write an equation for the motion of the buoy. Assume that it is at equilibrium at $t = 0$ and that it is on the way up from the normal water level.

40. Draw a graph showing the height of the buoy as a function of time.

41. What is the height of the buoy after 12 seconds?

42. WRITING IN MATH Answer the question that was posed at the beginning of the lesson.

Why can you predict the behavior of tides?

Include the following in your answer:

- an explanation of why certain tidal characteristics follow the patterns seen in the graph of the sine function, and
- a description of how to determine the amplitude of a function using the maximum and minimum values.

43. What is the period of $f(x) = \frac{1}{2}\cos 3x$?

Ⓐ 120° Ⓑ 180° Ⓒ 360° Ⓓ 720°

44. Identify the equation of the graphed function.

Ⓐ $y = \frac{1}{2}\sin 4\theta$

Ⓑ $y = 2\sin\frac{1}{4}\theta$

Ⓒ $y = \frac{1}{4}\sin 2\theta$

Ⓓ $y = 4\sin\frac{1}{2}\theta$

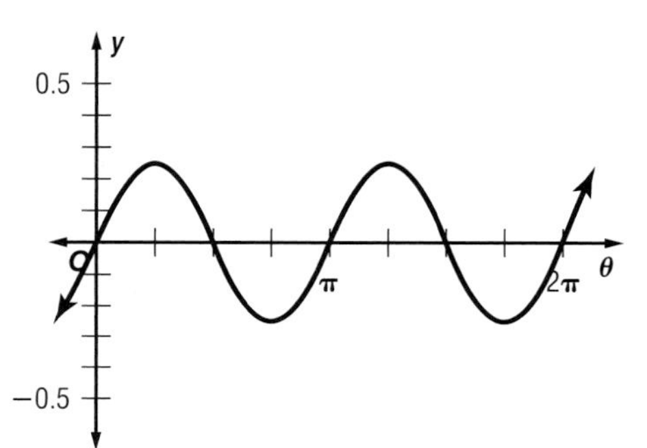

Maintain Your Skills

Mixed Review

Solve each equation. *(Lesson 13-7)*

45. $x = \text{Sin}^{-1} 1$ **46.** $\text{Arcsin}(-1) = y$ **47.** $\text{Arccos}\frac{\sqrt{2}}{2} = x$

Find the exact value of each function. *(Lesson 13-6)*

48. $\sin 390°$ **49.** $\sin(-315°)$ **50.** $\cos 405°$

51. **PROBABILITY** There are 8 girls and 8 boys on the Faculty Advisory Board. Three are juniors. Find the probability of selecting a boy or a girl from the committee who is not a junior. *(Lesson 12-5)*

52. Find the first five terms of the sequence in which $a_1 = 3$, $a_{n+1} = 2a_n + 5$. *(Lesson 11-5)*

Getting Ready for the Next Lesson

PREREQUISITE SKILL **Graph each pair of functions on the same set of axes.**
(To review ***graphs of quadratic functions****, see Lesson 6-6.)*

53. $y = x^2, y = 3x^2$ **54.** $y = 3x^2, y = 3x^2 - 4$

55. $y = 2x^2, y = 2(x + 1)^2$ **56.** $y = x^2 + 2, y = (x - 3)^2 + 2$

14-2 Translations of Trigonometric Graphs

What You'll Learn

- Graph horizontal translations of trigonometric graphs and find phase shifts.
- Graph vertical translations of trigonometric graphs.

Vocabulary

- phase shift
- vertical shift
- midline

How can translations of trigonometric graphs be used to show animal populations?

In predator-prey ecosystems, the number of predators and the number of prey tend to vary in a periodic manner. In a certain region with coyotes as predators and rabbits as prey, the rabbit population R can be modeled by the equation $R = 1200 + 250 \sin \frac{1}{2}\pi t$, where t is the time in years since January 1, 2001.

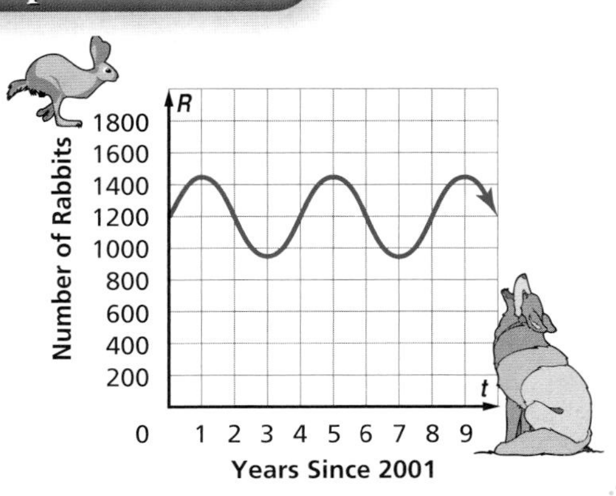

HORIZONTAL TRANSLATIONS Recall that a translation is a type of transformation in which the image is identical to the preimage in all aspects except its location on the coordinate plane. A horizontal translation shifts to the left or right, and not upward or downward.

Graphing Calculator Investigation

Horizontal Translations

On a TI-83 Plus, set the MODE to degrees.

Think and Discuss

1. Graph $y = \sin x$ and $y = \sin (x - 30)$. How do the two graphs compare?
2. Graph $y = \sin (x + 60)$. How does this graph compare to the other two?
3. What conjecture can you make about the effect of h in the function $y = \sin (x - h)$?
4. Test your conjecture on the following pairs of graphs.
   - $y = \cos x$ and $y = \cos (x + 30)$
   - $y = \tan x$ and $y = \tan (x - 45)$
   - $y = \sec x$ and $y = \sec (x + 75)$

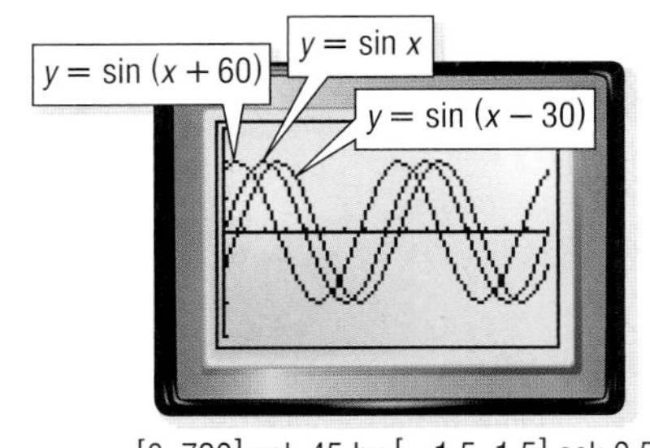

[0, 720] scl: 45 by [−1.5, 1.5] scl: 0.5

Notice that when a constant is added to an angle measure in a trigonometric function, the graph is shifted to the left or to the right. If (x, y) are coordinates of $y = \sin x$, then $(x \pm h, y)$ are coordinates of $y = \sin (x \mp h)$. A horizontal translation of a trigonometric function is called a **phase shift**.

Key Concept — Phase Shift

- **Words:** The phase shift of the functions $y = a \sin b(\theta - h)$, $y = a \cos b(\theta - h)$, and $y = a \tan b(\theta - h)$ is h, where $b > 0$.

  If $h > 0$, the shift is to the right. If $h < 0$, the shift is to the left.

- **Models:**

Sine

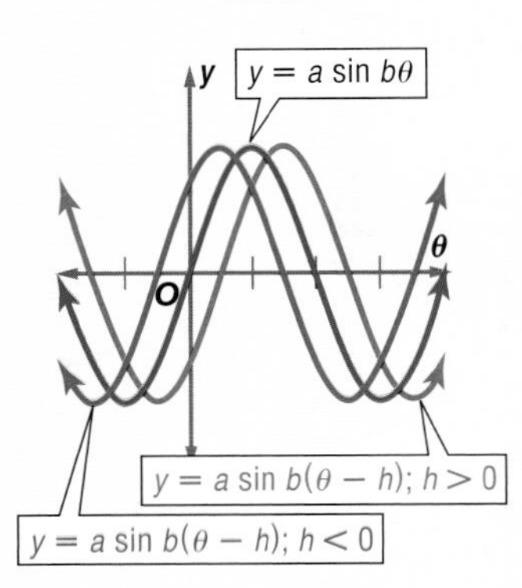

Cosine

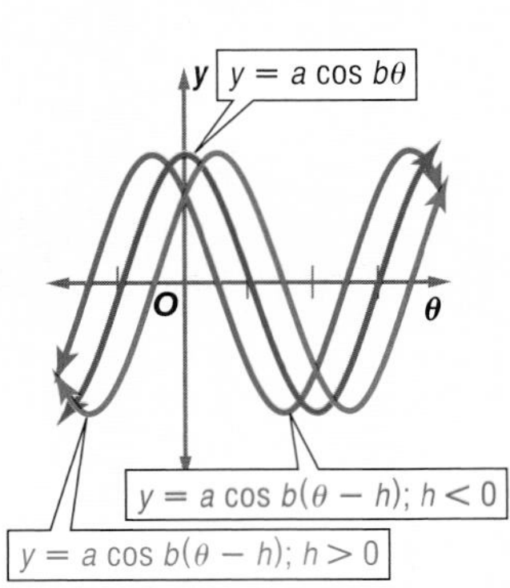

Tangent

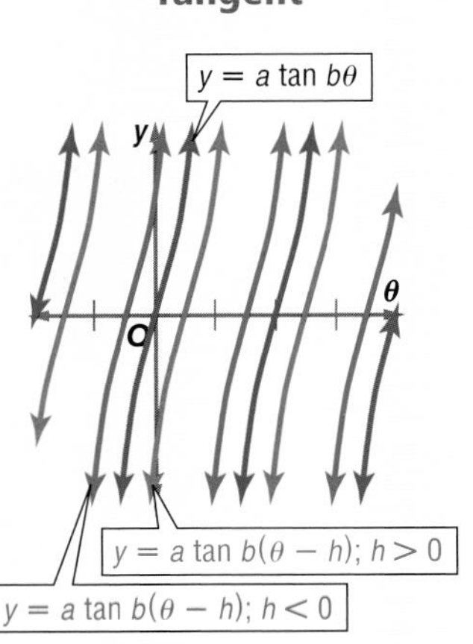

The secant, cosecant, and cotangent can be graphed using the same rules.

Example 1 Graph Horizontal Translations

State the amplitude, period, and phase shift for each function. Then graph the function.

a. $y = \cos(\theta - 60°)$

Since $a = 1$ and $b = 1$, the amplitude and period of the function are the same as $y = \cos \theta$. However, $h = 60°$, so the phase shift is $60°$. Because $h > 0$, the parent graph is shifted to the right.

To graph $y = \cos(\theta - 60°)$, consider the graph of $y = \cos \theta$. Graph this function and then shift the graph $60°$ to the right. The graph $y = \cos(\theta - 60°)$ is the graph of $y = \cos \theta$ shifted to the right.

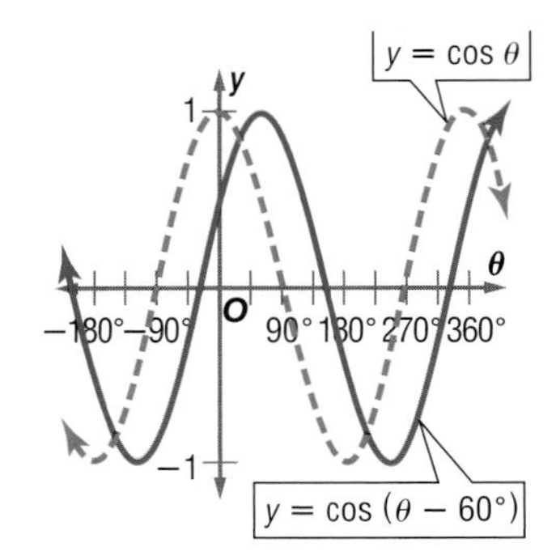

Study Tip

Verifying a Graph
After drawing the graph of a trigonometric function, select values of θ and evaluate them in the equation to verify your graph.

b. $y = 2 \sin\left(\theta + \frac{\pi}{4}\right)$

Amplitude: $a = |2|$ or 2

Period: $\frac{2\pi}{|b|} = \frac{2\pi}{|1|}$ or 2π

Phase Shift: $h = -\frac{\pi}{4}$ $\left(\theta + \frac{\pi}{4}\right) = \theta - \left(-\frac{\pi}{4}\right)$

The phase shift is to the left since $-\frac{\pi}{4} < 0$.

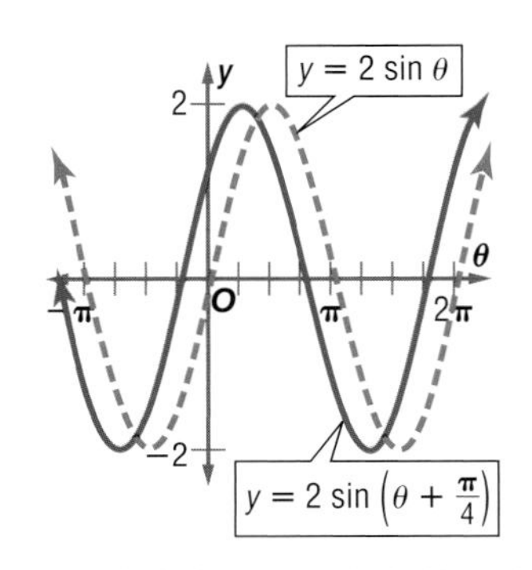

VERTICAL TRANSLATIONS

In Chapter 6, you learned that the graph of $y = x^2 + 4$ is a vertical translation of the parent graph of $y = x^2$. Similarly, graphs of trigonometric functions can be translated vertically through a **vertical shift**.

When a constant is added to a trigonometric function, the graph is shifted upward or downward. If (x, y) are coordinates of $y = \sin x$, then $(x, y \pm k)$ are coordinates of $y = \sin x \pm k$.

A new horizontal axis called the **midline** becomes the reference line about which the graph oscillates. For the graph of $y = \sin \theta + k$, the midline is the graph of $y = k$.

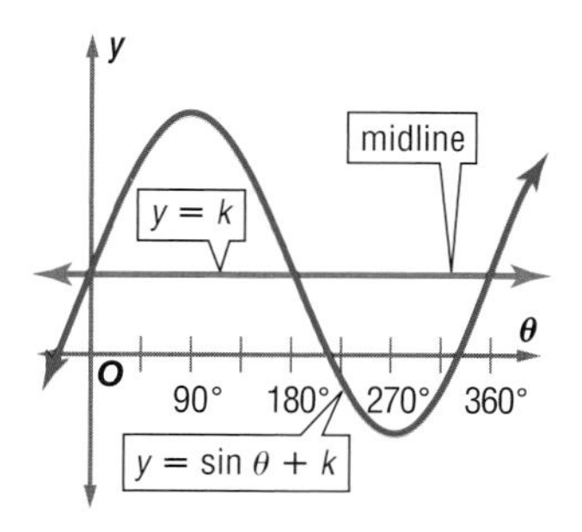

Study Tip

Look Back

Pay close attention to trigonometric functions for the placement of parentheses. Note that $\sin(\theta + x) \neq \sin \theta + x$. The expression on the left represents a phase shift while the expression on the right represents a vertical shift.

Key Concept — Vertical Shift

- **Words** The vertical shift of the functions $y = a \sin b(\theta - h) + k$, $y = a \cos b(\theta - h) + k$, and $y = a \tan b(\theta - h) + k$ is k.

  If $k > 0$, the shift is up. If $k < 0$, the shift is down. The midline is $y = k$.

- **Models:**

Sine

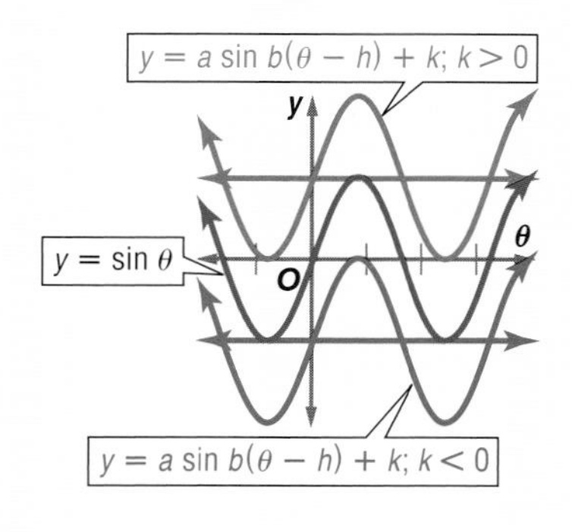

Cosine

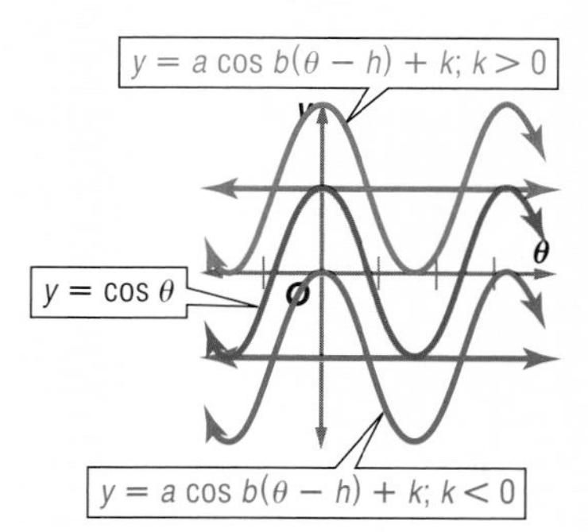

Tangent

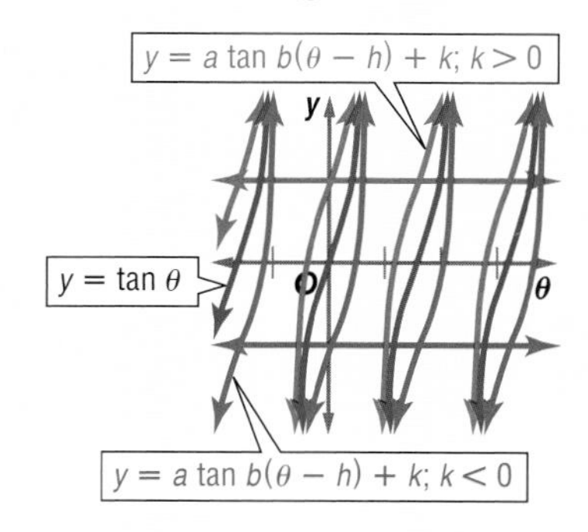

The secant, cosecant, and cotangent can be graphed using the same rules.

Example 2 Graph Vertical Translations

State the vertical shift, equation of the midline, amplitude, and period for each function. Then graph the function.

a. $y = \tan \theta - 2$

Since $\tan \theta - 2 = \tan \theta + (-2)$, $k = -2$, and the vertical shift is -2. Draw the midline, $y = -2$. The tangent function has no amplitude and the period is the same as that of $\tan \theta$.

Draw the graph of the function relative to the midline.

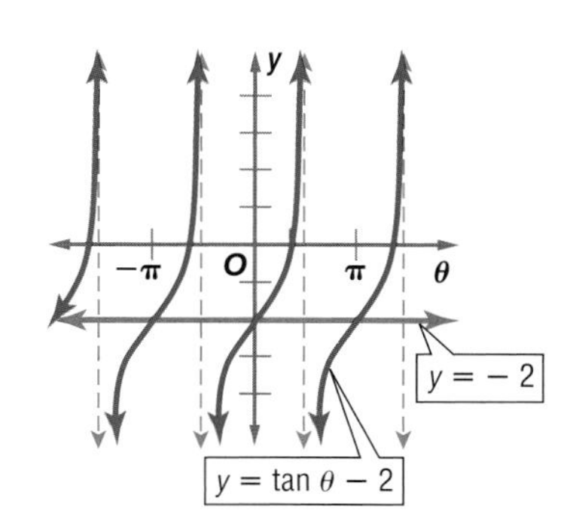

Study Tip

Look Back

It may be helpful to first graph the parent graph $y = \sin \theta$ in one color. Then apply the vertical shift and graph the function in another color. Then apply the change in amplitude and graph the function in the final color.

b. $y = \frac{1}{2} \sin \theta + 1$

Vertical shift: $k = 1$, so the midline is the graph of $y = 1$.

Amplitude: $|a| = \left|\frac{1}{2}\right|$ or $\frac{1}{2}$

Period: $\frac{2\pi}{|b|} = 2\pi$

Since the amplitude of the function is $\frac{1}{2}$, draw dashed lines parallel to the midline that are $\frac{1}{2}$ unit above and below the midline. Then draw the sine curve.

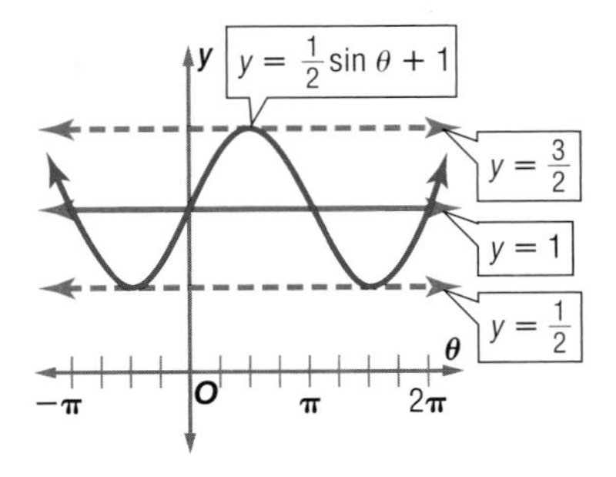

In general, use the following steps to graph any trigonometric function.

Concept Summary — Graphing Trigonometric Functions

Step 1 Determine the vertical shift, and graph the midline.

Step 2 Determine the amplitude, if it exists. Use dashed lines to indicate the maximum and minimum values of the function.

Step 3 Determine the period of the function and graph the appropriate function.

Step 4 Determine the phase shift and translate the graph accordingly.

Example 3 Graph Transformations

State the vertical shift, amplitude, period, and phase shift of $y = 4 \cos \left[\frac{1}{2}\left(\theta - \frac{\pi}{3}\right)\right] - 6$. Then graph the function.

The function is written in the form $y = a \cos [b(\theta - h)] + k$. Identify the values of k, a, b, and h.

$k = -6$, so the vertical shift is -6.

$a = 4$, so the amplitude is $|4|$ or 4.

$b = \frac{1}{2}$, so the period is $\frac{2\pi}{\left|\frac{1}{2}\right|}$ or 4π.

$h = \frac{\pi}{3}$, so the phase shift is $\frac{\pi}{3}$ to the right.

Then graph the function.

Step 1 The vertical shift is -6. Graph the midline $y = -6$.

Step 2 The amplitude is 4. Draw dashed lines 4 units above and below the midline at $y = -2$ and $y = -10$.

Step 3 The period is 4π, so the graph will be stretched. Graph $y = 4 \cos \frac{1}{2}\theta - 6$ using the midline as a reference.

Step 4 Shift the graph $\frac{\pi}{3}$ to the right.

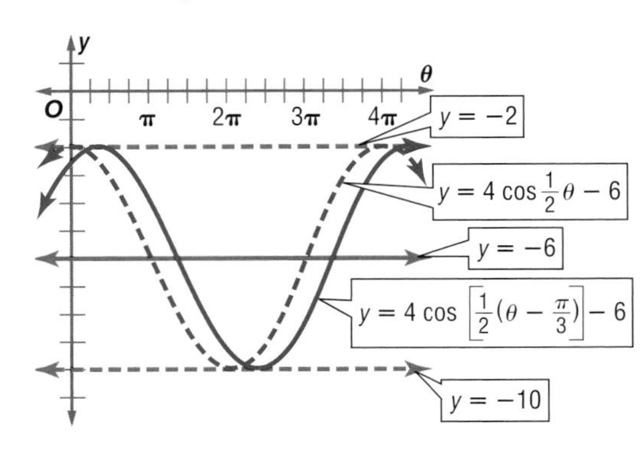

You can use information about amplitude, period, and translations of trigonometric functions to model real-world applications.

Example 4 Use Translations to Solve a Problem

HEALTH Suppose a person's resting blood pressure is 120 over 80. This means that the blood pressure oscillates between a maximum of 120 and a minimum of 80. If this person's resting heart rate is 60 beats per minute, write a sine function that represents the blood pressure for *t* seconds. Then graph the function.

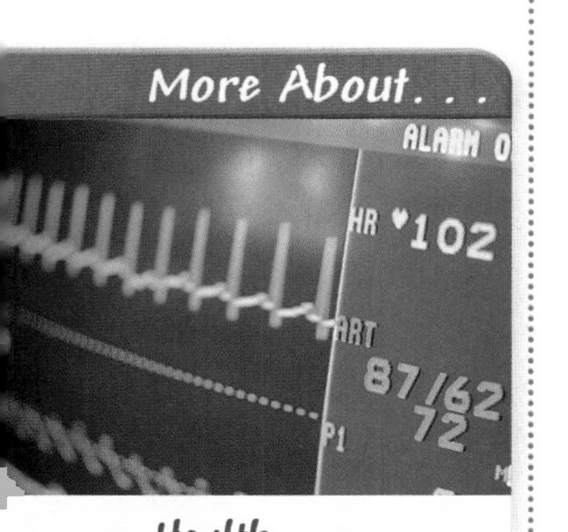

More About. . .

Health

Blood pressure can change from minute to minute and can be affected by the slightest of movements, such as tapping your fingers or crossing your arms.

Source: American Heart Association

Explore You know that the function is periodic and can be modeled using sine.

Plan Let P represent blood pressure and let t represent time in seconds. Use the equation $P = a \sin [b(t - h)] + k$.

Solve

- Write the equation for the midline. Since the maximum is 120 and the minimum is 80, the midline lies halfway between these values.

  $P = \frac{120 + 80}{2}$ or 100

- Determine the amplitude by finding the difference between the midline value and the maximum and minimum values.

  $a = |120 - 100|$ $\quad$ $a = |80 - 100|$

  $= |20|$ or 20 $\quad$ $= |-20|$ or 20

  Thus, $a = 20$.

- Determine the period of the function and solve for b. Recall that the period of a function can be found using the expression $\frac{2\pi}{|b|}$. Since the heart rate is 60 beats per minute, there is one heartbeat, or cycle, per second. So, the period is 1 second.

  $1 = \frac{2\pi}{|b|}$ Write an equation.

  $|b| = 2\pi$ Multiply each side by $|b|$.

  $b = \pm 2\pi$ Solve.

  For this example, let $b = 2\pi$. The use of the positive or negative value depends upon whether you begin a cycle with a maximum value (positive) or a minimum value (negative).

- There is no phase shift, so $h = 0$. So, the equation is $P = 20 \sin 2\pi t + 100$.

- Graph the function.

  Step 1 Draw the midline $P = 100$.

  Step 2 Draw maximum and minimum reference lines.

  Step 3 Use the period to draw the graph of the function.

  Step 4 There is no phase shift.

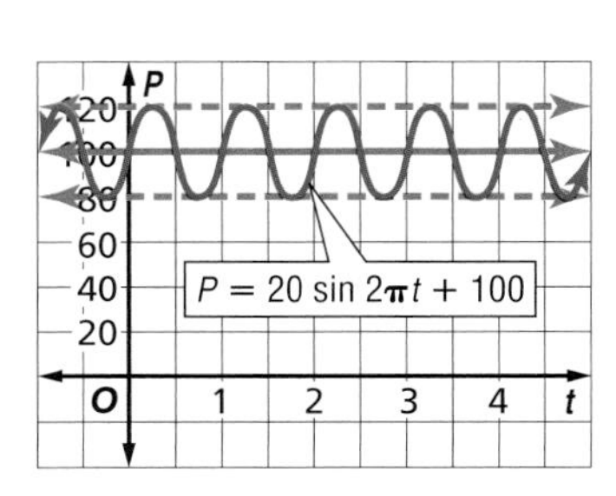

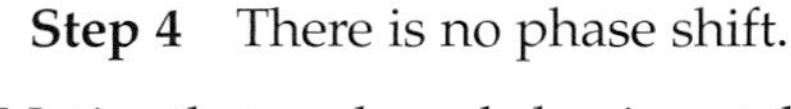

Examine Notice that each cycle begins at the midline, rises to 120, drops to 80, and then returns to the midline. This represents the blood pressure of 120 over 80 for one heartbeat. Since each cycle lasts 1 second, there will be 60 cycles, or heartbeats, in 1 minute. Therefore, the graph accurately represents the information.

Check for Understanding

Concept Check

1. **Identify** the vertical shift, amplitude, period, and phase shift of the graph of $y = 3\cos(2x - 90°) + 15$.
2. **Define** the midline of a trigonometric graph.
3. **OPEN ENDED** Write the equation of a trigonometric function with a phase shift of $-45°$.

Guided Practice

State the amplitude, period, and phase shift for each function. Then graph the function.

4. $y = \sin\left(\theta - \frac{\pi}{2}\right)$
5. $y = \tan(\theta + 60°)$
6. $y = \cos(\theta - 45°)$
7. $y = \sec\left(\theta + \frac{\pi}{3}\right)$

State the vertical shift, equation of the midline, amplitude, and period for each function. Then graph the function.

8. $y = \cos\theta + \frac{1}{4}$
9. $y = \sec\theta - 5$
10. $y = \tan\theta + 4$
11. $y = \sin\theta + 0.25$

State the vertical shift, amplitude, period, and phase shift of each function. Then graph the function.

12. $y = 3\sin[2(\theta - 30°)] + 10$
13. $y = 2\cot(3\theta + 135°) - 6$
14. $y = \frac{1}{2}\sec\left[4\left(\theta - \frac{\pi}{4}\right)\right] + 1$
15. $y = \frac{2}{3}\cos\left[\frac{1}{2}\left(\theta + \frac{\pi}{6}\right)\right] - 2$

Application

PHYSICS For Exercises 16–18, use the following information.
A weight is attached to a spring and suspended from the ceiling. At equilibrium, the weight is located 4 feet above the floor. The weight is pulled down 1 foot and released.

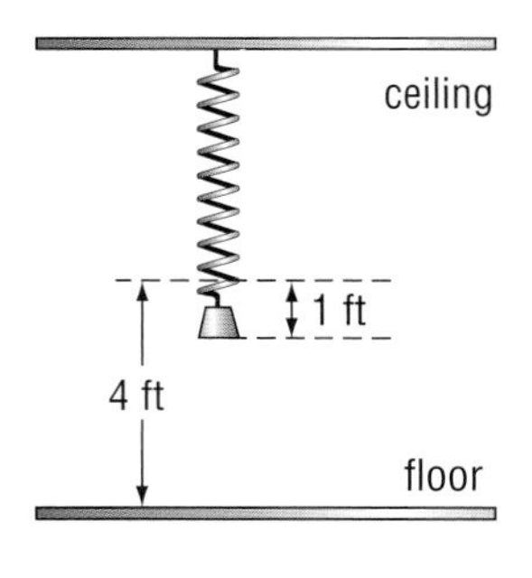

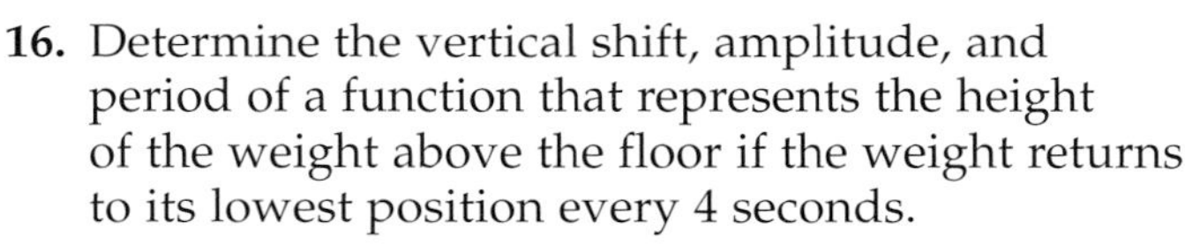

16. Determine the vertical shift, amplitude, and period of a function that represents the height of the weight above the floor if the weight returns to its lowest position every 4 seconds.
17. Write the equation for the height h of the weight above the floor as a function of time t seconds.
18. Draw a graph of the function you wrote in Exercise 17.

Practice and Apply

State the amplitude, period, and phase shift for each function. Then graph the function.

19. $y = \cos(\theta + 90°)$
20. $y = \cot(\theta - 30°)$
21. $y = \sin\left(\theta - \frac{\pi}{4}\right)$
22. $y = \cos\left(\theta + \frac{\pi}{3}\right)$
23. $y = \frac{1}{4}\tan(\theta + 22.5°)$
24. $y = 3\sin(\theta - 75°)$

Homework Help

For Exercises	See Examples
19–24	1
25–30	2
31, 32	1, 2
33–42	3
37–40	4

Extra Practice
See page 859.

State the vertical shift, equation of the midline, amplitude, and period for each function. Then graph the function.

25. $y = \sin \theta - 1$
26. $y = \sec \theta + 2$
27. $y = \cos \theta - 5$
28. $y = \csc \theta - \frac{3}{4}$
29. $y = \frac{1}{2} \sin \theta + \frac{1}{2}$
30. $y = 6 \cos \theta + 1.5$

31. Graph $y = 5 + \tan\left(\theta + \frac{\pi}{4}\right)$. Describe the transformation to the parent graph $y = \tan \theta$.

32. Draw a graph of the function $y = \frac{2}{3} \cos (\theta - 50°) + 2$. How does this graph compare to the graph of $y = \cos \theta$?

State the vertical shift, amplitude, period, and phase shift of each function. Then graph the function.

33. $y = 2 \sin [3(\theta - 45°)] + 1$
34. $y = 4 \cos [2(\theta + 30°)] - 5$
35. $y = 3 \csc \left[\frac{1}{2}(\theta + 60°)\right] - 3.5$
36. $y = 6 \cot \left[\frac{2}{3}(\theta - 90°)\right] + 0.75$
37. $y = \frac{1}{4} \cos (2\theta - 150°) + 1$
38. $y = \frac{2}{5} \tan (6\theta + 135°) - 4$
39. $y = 3 + 2 \sin \left[2\left(\theta + \frac{\pi}{4}\right)\right]$
40. $y = 4 + 5 \sec \left[\frac{1}{3}\left(\theta + \frac{2\pi}{3}\right)\right]$

41. Graph $y = 3 - \frac{1}{2} \cos \theta$ and $y = 3 - \frac{1}{2} \cos (\theta + \pi)$. How do the graphs compare?

42. Compare the graphs of $y = -\sin \left[\frac{1}{4}\left(\theta - \frac{\pi}{2}\right)\right]$ and $y = \cos \left[\frac{1}{4}\left(\theta + \frac{3\pi}{2}\right)\right]$.

WebQuest

Translations of trigonometric graphs can be used to describe temperature trends. Visit www.algebra2.com/webquest to continue work on your WebQuest project.

43. **MUSIC** When represented on an oscilloscope, the note A above middle C has period of $\frac{1}{440}$. Which of the following can be an equation for an oscilloscope graph of this note? The amplitude of the graph is K.
   a. $y = K \sin 220\pi t$ **b.** $y = K \sin 440\pi t$ **c.** $y = K \sin 880\pi t$

ZOOLOGY For Exercises 44–46, use the following information.
The population of predators and prey in a closed ecological system tends to vary periodically over time. In a certain system, the population of owls O can be represented by $O = 150 + 30 \sin\left(\frac{\pi}{10}t\right)$ where t is the time in years since January 1, 2001. In that same system, the population of mice M can be represented by $M = 600 + 300 \sin\left(\frac{\pi}{10}t + \frac{\pi}{20}\right)$.

44. Find the maximum number of owls. After how many years does this occur?
45. What is the minimum number of mice? How long does it take for the population of mice to reach this level?
46. Why would the maximum owl population follow behind the population of mice?

47. **TIDES** The height of the water in a harbor rose to a maximum height of 15 feet at 6:00 P.M. and then dropped to a minimum level of 3 feet by 3:00 A.M. Assume that the water level can be modeled by the sine function. Write an equation that represents the height h of the water t hours after noon on the first day.

Online Research **Data Update** Use the Internet or another resource to find tide data for a location of your choice. Write a sine function to represent your data. Then graph the function. Visit www.algebra2.com/data_update to learn more.

48. **CRITICAL THINKING** The graph of $y = \cot \theta$ is a transformation of the graph of $y = \tan \theta$. Determine a, b, and h so that $\cot \theta = a \tan [b(\theta - h)]$ for all values of θ for which each function is defined.

49. **WRITING IN MATH** Answer the question that was posed at the beginning of the lesson.

How can translations of trigonometric graphs be used to show animal populations?

Include the following in your answer:
- a description of what each number in the equation $R = 1200 + 250 \sin \frac{1}{2}\pi t$ represents, and
- a comparison of the graphs of $y = a \cos bx$, $y = a \cos bx + k$, and $y = a \cos [b(x - h)]$.

Standards Practice

Standardized Test Practice

50. Which equation is represented by the graph?

Ⓐ $y = \cot (\theta + 45°)$

Ⓑ $y = \cot (\theta - 45°)$

Ⓒ $y = \tan (\theta + 45°)$

Ⓓ $y = \tan (\theta - 45°)$

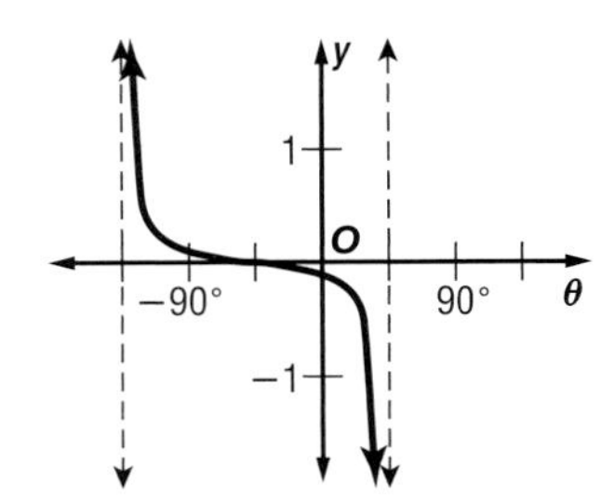

51. Identify the equation for a sine function of period 90°, after a phase shift 20° to the left.

Ⓐ $y = \sin [0.25(\theta - 20°)]$

Ⓑ $y = \sin [4(\theta - 20°)]$

Ⓒ $y = \sin [0.25(\theta + 20°)]$

Ⓓ $y = \sin [4(\theta + 20°)]$

Maintain Your Skills

Mixed Review

Find the amplitude, if it exists, and period of each function. Then graph each function. *(Lesson 14-1)*

52. $y = 3 \csc \theta$

53. $y = \sin \frac{\theta}{2}$

54. $y = 3 \tan \frac{2}{3}\theta$

Find each value. *(Lesson 13-7)*

55. $\sin \left(\text{Cos}^{-1} \frac{2}{3}\right)$

56. $\cos \left(\text{Cos}^{-1} \frac{4}{7}\right)$

57. $\text{Sin}^{-1} \left(\sin \frac{5}{6}\right)$

58. $\cos \left(\text{Tan}^{-1} \frac{3}{4}\right)$

59. **GEOMETRY** Find the total number of diagonals that can be drawn in a decagon. *(Lesson 12-2)*

Solve each equation. Round to the nearest hundredth. *(Lesson 10-4)*

60. $4^x = 24$

61. $4.3^{3x+1} = 78.5$

62. $7^{x-2} = 53^{-x}$

Simplify each expression. *(Lesson 9-4)*

63. $\frac{3}{a - 2} + \frac{2}{a - 3}$

64. $\frac{w + 12}{4w - 16} - \frac{w + 4}{2w - 8}$

65. $\frac{3y + 1}{2y - 10} + \frac{1}{y^2 - 2y - 15}$

Getting Ready for the Next Lesson

PREREQUISITE SKILL **Find the value of each function.**
*(To review **reference angles**, see Lesson 13-3.)*

66. $\cos 150°$

67. $\tan 135°$

68. $\sin \frac{3\pi}{2}$

69. $\cos \left(-\frac{\pi}{3}\right)$

70. $\sin (-\pi)$

71. $\tan \left(-\frac{5\pi}{6}\right)$

72. $\cos 225°$

73. $\tan 405°$

14-3 Trigonometric Identities

What You'll Learn

- Use identities to find trigonometric values.
- Use trigonometric identities to simplify expressions.

Vocabulary

- trigonometric identity

How can trigonometry be used to model the path of a baseball?

A model for the height of a baseball after it is hit as a function of time can be determined using trigonometry. If the ball is hit with an initial velocity of v feet per second at an angle of θ from the horizontal, then the height h of the ball after t seconds can be represented by $h = \left(\frac{-16}{v^2\cos^2\theta}\right)t^2 + \left(\frac{\sin\theta}{\cos\theta}\right)t + h_0$, where h_0 is the height of the ball in feet the moment it is hit.

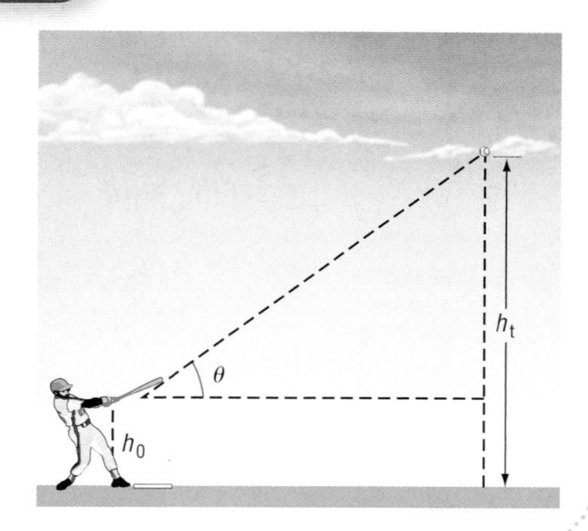

FIND TRIGONOMETRIC VALUES In the equation above, the second term $\left(\frac{\sin\theta}{\cos\theta}\right)t$ can also be written as $(\tan\theta)t$. $\left(\frac{\sin\theta}{\cos\theta}\right)t = (\tan\theta)t$ is an example of a trigonometric identity. A **trigonometric identity** is an equation involving trigonometric functions that is true for all values for which every expression in the equation is defined.

The identity $\tan\theta = \frac{\sin\theta}{\cos\theta}$ is true except for angle measures such as 90°, 270°, 450°, …, $90° + 180° \cdot k$. The cosine of each of these angle measures is 0, so none of the expressions tan 90°, tan 270°, tan 450°, and so on, are defined. An identity similar to this is $\cot\theta = \frac{\cos\theta}{\sin\theta}$.

These identities are sometimes called *quotient identities*. These and other basic trigonometric identities are listed below.

Key Concept *Basic Trigonometric Identities*

Quotient Identities	$\tan\theta = \frac{\sin\theta}{\cos\theta}$	$\cot\theta = \frac{\cos\theta}{\sin\theta}$	
Reciprocal Identities	$\csc\theta = \frac{1}{\sin\theta}$	$\sec\theta = \frac{1}{\cos\theta}$	$\cot\theta = \frac{1}{\tan\theta}$
Pythagorean Identities	$\cos^2\theta + \sin^2\theta = 1$ $\tan^2\theta + 1 = \sec^2\theta$ $\cot^2\theta + 1 = \csc^2\theta$		

You can use trigonometric identities to find values of trigonometric functions.

Example 1 Find a Value of a Trigonometric Function

a. Find $\cos\theta$ if $\sin\theta = -\frac{3}{5}$ and $90° < \theta < 180°$.

$\cos^2\theta + \sin^2\theta = 1$	Trigonometric identity
$\cos^2\theta = 1 - \sin^2\theta$	Subtract $\sin^2\theta$ from each side.
$\cos^2\theta = 1 - \left(\frac{3}{5}\right)^2$	Substitute $\frac{3}{5}$ for $\sin\theta$.
$\cos^2\theta = 1 - \frac{9}{25}$	Square $\frac{3}{5}$.
$\cos^2\theta = \frac{16}{25}$	Subtract.
$\cos\theta = \pm\frac{4}{5}$	Take the square root of each side.

Since θ is in the second quadrant, $\cos\theta$ is negative. Thus, $\cos\theta = -\frac{4}{5}$.

b. Find $\csc\theta$ if $\cot\theta = -\frac{1}{4}$ and $270° < \theta < 360°$.

$\cot^2\theta + 1 = \csc^2\theta$	Trigonometric identity
$\left(-\frac{1}{4}\right)^2 + 1 = \csc^2\theta$	Substitute $-\frac{1}{4}$ for $\cot\theta$.
$\frac{1}{16} + 1 = \csc^2\theta$	Square $-\frac{1}{4}$.
$\frac{17}{16} = \csc^2\theta$	Add.
$\pm\frac{\sqrt{17}}{4} = \csc\theta$	Take the square root of each side.

Since θ is in the fourth quadrant, $\csc\theta$ is negative. Thus, $\csc\theta = -\frac{\sqrt{17}}{4}$.

SIMPLIFY EXPRESSIONS Trigonometric identities can also be used to simplify expressions containing trigonometric functions. Simplifying an expression that contains trigonometric functions means that the expression is written as a numerical value or in terms of a single trigonometric function, if possible.

Example 2 Simplify an Expression

Simplify $\frac{\csc^2\theta - \cot^2\theta}{\cos\theta}$.

$\frac{\csc^2\theta - \cot^2\theta}{\cos\theta} = \frac{\frac{1}{\sin^2\theta} - \frac{\cos^2\theta}{\sin^2\theta}}{\cos\theta}$	$\csc^2\theta = \frac{1}{\sin^2\theta}$, $\cot^2\theta = \frac{\cos^2\theta}{\sin^2\theta}$
$= \frac{\frac{1 - \cos^2\theta}{\sin^2\theta}}{\cos\theta}$	Add.
$= \frac{\frac{\sin^2\theta}{\sin^2\theta}}{\cos\theta}$	$1 - \cos^2\theta = \sin^2\theta$
$= \frac{1}{\cos\theta}$	$\frac{\sin^2\theta}{\sin^2\theta} = 1$
$= \sec\theta$	$\frac{1}{\cos\theta} = \sec\theta$

Example 3 Simplify and Use an Expression

BASEBALL Refer to the application at the beginning of the lesson. Rewrite the equation in terms of tan θ.

$h = \left(\frac{-16}{v^2 \cos^2 \theta}\right)t^2 + \left(\frac{\sin \theta}{\cos \theta}\right)t + h_0$ — Original equation

$= \frac{-16}{v^2}\left(\frac{1}{\cos^2 \theta}\right)t^2 + \left(\frac{\sin \theta}{\cos \theta}\right)t + h_0$ — Factor.

$= \frac{-16}{v^2}\left(\frac{1}{\cos^2 \theta}\right)t^2 + (\tan \theta)t + h_0$ — $\frac{\sin \theta}{\cos \theta} = \tan \theta$

$= \frac{-16}{v^2}(\sec^2 \theta)t^2 + (\tan \theta)t + h_0$ — Since $\frac{1}{\cos \theta} = \sec \theta$, $\frac{1}{\cos^2 \theta} = \sec^2 \theta$.

$= \frac{-16}{v^2}(1 + \tan^2 \theta)t^2 + (\tan \theta)t + h_0$ — $\sec^2 \theta = 1 + \tan^2 \theta$

Thus, $\left(\frac{-16}{v^2 \cos^2 \theta}\right)t^2 + \left(\frac{\sin \theta}{\cos \theta}\right)t + h_0 = \frac{-16}{v^2}(1 + \tan^2 \theta)t^2 + (\tan \theta)t + h_0$.

Check for Understanding

Concept Check

1. **Describe** how you can determine the quadrant in which the terminal side of angle α lies if $\sin \alpha = -\frac{1}{4}$.
2. **Explain** why the Pythagorean identities are so named.
3. **OPEN ENDED** Explain what it means to simplify a trigonometric expression.

Guided Practice

Find the value of each expression.

4. $\tan \theta$, if $\sin \theta = \frac{1}{2}$; $90° \leq \theta < 180°$
5. $\csc \theta$, if $\cos \theta = -\frac{3}{5}$; $180° \leq \theta < 270°$
6. $\cos \theta$, if $\sin \theta = \frac{4}{5}$; $0° \leq \theta < 90°$
7. $\sec \theta$, if $\tan \theta = -1$; $270° < \theta < 360°$

Simplify each expression.

8. $\csc \theta \cos \theta \tan \theta$
9. $\sec^2 \theta - 1$
10. $\frac{\tan \theta}{\sin \theta}$
11. $\sin \theta (1 + \cot^2 \theta)$

Application

12. **PHYSICAL SCIENCE** When a person moves along a circular path, the body leans away from a vertical position. The nonnegative acute angle that the body makes with the vertical is called the *angle of inclination* and is represented by the equation $\tan \theta = \frac{v^2}{gR}$, where R is the radius of the circular path, v is the speed of the person in meters per second, and g is the acceleration due to gravity, 9.8 meters per second squared. Write an equivalent expression using $\sin \theta$ and $\cos \theta$.

Practice and Apply

Practice and Apply

Find the value of each expression.

13. $\tan \theta$, if $\cot \theta = 2$; $0° \leq \theta < 90°$
14. $\sin \theta$, if $\cos \theta = \frac{2}{3}$; $0° \leq \theta < 90°$
15. $\sec \theta$, if $\tan \theta = -2$; $90° < \theta < 180°$
16. $\tan \theta$, if $\sec \theta = -3$; $180° < \theta < 270°$

Homework Help	
For Exercises	See Examples
13–24	1
25–36	2
37–43	3

Extra Practice
See page 860.

Find the value of each expression.

17. $\csc \theta$, if $\cos \theta = -\frac{3}{5}$; $90° < \theta < 180°$
18. $\cos \theta$, if $\sec \theta = \frac{5}{3}$; $270° < \theta < 360°$
19. $\cos \theta$, if $\sin \theta = \frac{1}{2}$; $0° \leq \theta < 90°$
20. $\csc \theta$, if $\cos \theta = -\frac{2}{3}$; $180° < \theta < 270°$
21. $\tan \theta$, if $\cos \theta = \frac{4}{5}$; $0° \leq \theta < 90°$
22. $\cos \theta$, if $\csc \theta = -\frac{5}{3}$; $270° < \theta < 360°$
23. $\sec \theta$, if $\sin \theta = \frac{3}{4}$; $90° < \theta < 180°$
24. $\sin \theta$, if $\tan \theta = 4$; $180° < \theta < 270°$

Simplify each expression.

25. $\cos \theta \csc \theta$
26. $\tan \theta \cot \theta$
27. $\sin \theta \cot \theta$
28. $\cos \theta \tan \theta$
29. $2(\csc^2 \theta - \cot^2 \theta)$
30. $3(\tan^2 \theta - \sec^2 \theta)$
31. $\dfrac{\cos \theta \csc \theta}{\tan \theta}$
32. $\dfrac{\sin \theta \csc \theta}{\cot \theta}$
33. $\dfrac{1 - \cos^2 \theta}{\sin^2 \theta}$
34. $\dfrac{1 - \sin^2 \theta}{\sin^2 \theta}$
35. $\dfrac{\sin^2 \theta + \cos^2 \theta}{\sin^2 \theta}$
36. $\dfrac{\tan^2 \theta - \sin^2 \theta}{\tan^2 \theta \sin^2 \theta}$

AMUSEMENT PARKS **For Exercises 37–39, use the following information.**
Suppose a child is riding on a merry-go-round and is seated on an outside horse. The diameter of the merry-go-round is 16 meters.

37. If the sine of the angle of inclination of the child is $\frac{1}{5}$, what is the angle of inclination made by the child? Refer to Exercise 12 for information on the angle of inclination.
38. What is the velocity of the merry-go-round?
39. If the speed of the merry-go-round is 3.6 meters per second, what is the value of the angle of inclination of a rider?

More About. . .

Amusement Parks

The oldest operational carousel in the United States is the Flying Horse Carousel at Martha's Vineyard, Massachusetts.

Source: Martha's Vineyard Preservation Trust

LIGHTING **For Exercises 40 and 41, use the following information.**
The amount of light that a source provides to a surface is called the illuminance. The illuminance E in foot candles on a surface is related to the distance R in feet from the light source. The formula $\sec \theta = \frac{I}{ER^2}$, where I is the intensity of the light source measured in candles and θ is the angle between the light beam and a line perpendicular to the surface, can be used in situations in which lighting is important.

40. Solve the formula in terms of E.
41. Is the equation in Exercise 40 equivalent to $R^2 = \frac{I \tan \theta \cos \theta}{E}$? Explain.

ELECTRONICS **For Exercises 42 and 43, use the following information.**
When an alternating current of frequency f and a peak current I pass through a resistance R, then the power delivered to the resistance at time t seconds is $P = I^2R - I^2R \cos^2 (2ft\pi)$.

42. Write an expression for the power in terms of $\sin^2 2ft\pi$.
43. Write an expression for the power in terms of $\tan^2 2ft\pi$.

44. **CRITICAL THINKING** If $\tan \beta = \frac{3}{4}$, find $\dfrac{\sin \beta \sec \beta}{\cot \beta}$.

45. **WRITING IN MATH** Answer the question that was posed at the beginning of the lesson.

How can trigonometry be used to model the path of a baseball?

Include the following in your answer:

- an explanation of why the equation at the beginning of the lesson is the same as $y = \frac{-16 \sec^2 \theta}{v^2}x^2 + (\tan \theta)x + h_0$, and
- examples of how you might use this equation for other situations.

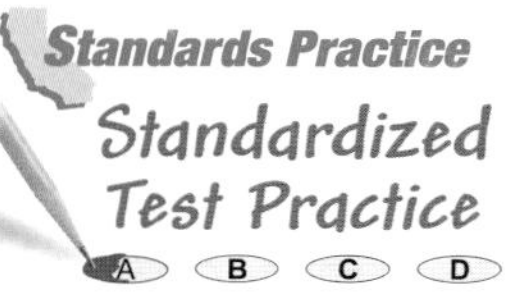

Standards Practice

Standardized Test Practice

46. If $\sin x = m$ and $0 < x < 90°$, then $\tan x =$

Ⓐ $\frac{1}{m^2}$.　Ⓑ $\frac{m}{\sqrt{1 - m^2}}$.　Ⓒ $\frac{1 - m^2}{m}$.　Ⓓ $\frac{m}{1 - m^2}$.

47. $\frac{1}{1 + \sin x} + \frac{1}{1 - \sin x} =$

Ⓐ $2 \sec^2 x$　Ⓑ $-\sec^2 x$　Ⓒ $2 \csc^2 x$　Ⓓ $-\csc^2 x$

Maintain Your Skills

Mixed Review

State the vertical shift, equation of the midline, amplitude, and period for each function. Then graph the function. *(Lesson 14-2)*

48. $y = \sin \theta - 1$

49. $y = \tan \theta + 12$

Find the amplitude, if it exists, and period of each function. Then graph each function. *(Lesson 14-1)*

50. $y = \csc 2\theta$

51. $y = \cos 3\theta$

52. $y = \frac{1}{3} \cot 5\theta$

53. Find the sum of a geometric series for which $a_1 = 48$, $a_n = 3$, and $r = \frac{1}{2}$. *(Lesson 11-4)*

54. Write an equation of a parabola with focus at $(11, -1)$ and whose directrix is $y = 2$. *(Lesson 8-2)*

Getting Ready for the Next Lesson

PREREQUISITE SKILL **Name the property illustrated by each statement.**
*(To review **properties of equality**, see Lesson 1-3.)*

55. If $4 + 8 = 12$, then $12 = 4 + 8$.

56. If $7 + s = 21$, then $s = 14$.

57. If $4x = 16$, then $12x = 48$.

58. If $q + (8 + 5) = 32$, then $q + 13 = 32$.

Practice Quiz 1

Lessons 14-1 through 14-3

1. Find the amplitude and period of $y = \frac{3}{4} \sin \frac{1}{2}\theta$. Then graph the function. *(Lesson 14-1)*

2. State the vertical shift, amplitude, period, and phase shift for $y = 2 \cos \left[\frac{1}{4}\left(\theta - \frac{\pi}{4}\right)\right] - 5$. Then graph the function. *(Lesson 14-2)*

Find the value of each expression. *(Lesson 14-3)*

3. $\cos \theta$, if $\sin \theta = \frac{4}{5}$; $90° < \theta < 180°$

4. $\csc \theta$, if $\cot \theta = -\frac{2}{3}$; $270° < \theta < 360°$

5. $\sec \theta$, if $\tan \theta = \frac{1}{2}$; $0° < \theta < 90°$

14-4

Verifying Trigonometric Identities

What You'll Learn

- Verify trigonometric identities by transforming one side of an equation into the form of the other side.
- Verify trigonometric identities by transforming each side of the equation into the same form.

How can you verify trigonometric identities?

Examine the graphs of $y = \tan^2 \theta - \sin^2 \theta$ and $y = \tan^2 \theta \sin^2 \theta$. Recall that when the graphs of two functions coincide, the functions are equivalent. However, the graphs only show a limited range of solutions. It is not sufficient to show some values of θ and conclude that the statement is true for all values of θ. In order to show that the equation $\tan^2 \theta - \sin^2 \theta = \tan^2 \theta \sin^2 \theta$ for all values of θ, you must consider the general case.

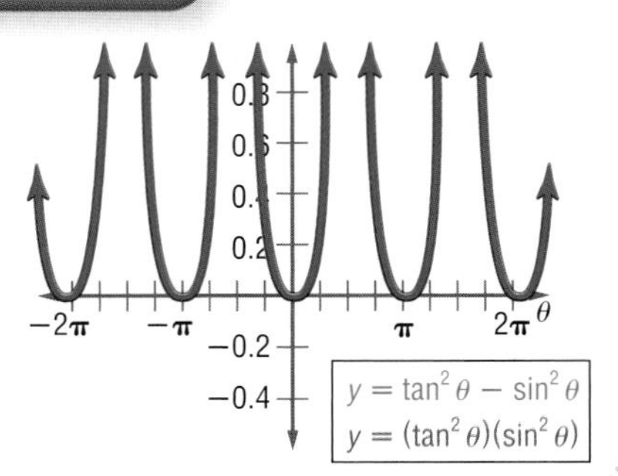

Study Tip

Common Misconception
You cannot perform operations to the quantities from each side of an unverified identity as you do with equations. Until an identity is verified it is not considered an equation, so the properties of equality do not apply.

TRANSFORM ONE SIDE OF AN EQUATION You can use the basic trigonometric identities along with the definitions of the trigonometric functions to verify identities. For example, if you wish to show that $\tan^2 \theta - \sin^2 \theta = \tan^2 \theta \sin^2 \theta$ is an identity, you need to show that it is true for all values of θ.

Verifying an identity is like checking the solution of an equation. You must simplify one or both sides of an equation *separately* until they are the same. In many cases, it is easier to work with only one side of an equation. You may choose either side, but it is often easier to begin with the more complicated side of the equation. Transform that expression into the form of the simpler side.

Example 1 Transform One Side of an Equation

Verify that $\tan^2 \theta - \sin^2 \theta = \tan^2 \theta \sin^2 \theta$ is an identity.

Transform the left side.

$\tan^2 \theta - \sin^2 \theta \stackrel{?}{=} \tan^2 \theta \sin^2 \theta$ Original equation

$\frac{\sin^2 \theta}{\cos^2 \theta} - \sin^2 \theta \stackrel{?}{=} \tan^2 \theta \sin^2 \theta$ $\tan^2 \theta = \frac{\sin^2 \theta}{\cos^2 \theta}$

$\frac{\sin^2 \theta}{\cos^2 \theta} - \frac{\sin^2 \theta \cos^2 \theta}{\cos^2 \theta} \stackrel{?}{=} \tan^2 \theta \sin^2 \theta$ Rewrite using the LCD, $\cos^2 \theta$.

$\frac{\sin^2 \theta - \sin^2 \theta \cos^2 \theta}{\cos^2 \theta} \stackrel{?}{=} \tan^2 \theta \sin^2 \theta$ Subtract.

$\frac{\sin^2 \theta (1 - \cos^2 \theta)}{\cos^2 \theta} \stackrel{?}{=} \tan^2 \theta \sin^2 \theta$ Factor.

$\frac{\sin^2 \theta \sin^2 \theta}{\cos^2 \theta} \stackrel{?}{=} \tan^2 \theta \sin^2 \theta$ $1 - \cos^2 \theta = \sin^2 \theta$

$\frac{\sin^2 \theta}{\cos^2 \theta} \cdot \frac{\sin^2 \theta}{1} \stackrel{?}{=} \tan^2 \theta \sin^2 \theta$ $\frac{ab}{c} = \frac{a}{c} \cdot \frac{b}{1}$

$\tan^2 \theta \sin^2 \theta = \tan^2 \theta \sin^2 \theta$ $\frac{\sin^2 \theta}{\cos^2 \theta} = \tan \theta$

Standardized Test Practice

Ⓐ Ⓑ Ⓒ Ⓓ

Standards Practice

Example 2 Find an Equivalent Expression

Multiple-Choice Test Item

$\sin\theta\left(\frac{1}{\sin\theta} - \frac{\cos\theta}{\cot\theta}\right) =$

Ⓐ $\cos\theta$ Ⓑ $\sin\theta$ Ⓒ $\cos^2\theta$ Ⓓ $\sin^2\theta$

Read the Test Item

Find an expression that is equal to the given expression.

Solve the Test Item

Write a trigonometric identity by using the basic trigonometric identities and the definitions of trigonometric functions to transform the given expression to match one of the choices.

$$\sin\theta\left(\frac{1}{\sin\theta} - \frac{\cos\theta}{\cot\theta}\right) = \sin\theta\left(\frac{1}{\sin\theta} - \frac{\cos\theta}{\frac{\cos\theta}{\sin\theta}}\right) \quad \cot\theta = \frac{\cos\theta}{\sin\theta}$$

$$= \sin\theta\left(\frac{1}{\sin\theta} - \frac{\cos\theta\sin\theta}{\cos\theta}\right) \quad \text{Simplify.}$$

$$= \sin\theta\left(\frac{1}{\sin\theta} - \sin\theta\right) \quad \text{Simplify.}$$

$$= 1 - \sin^2\theta \quad \text{Distributive property.}$$

$$= \cos^2\theta \quad 1 - \sin^2\theta = \cos^2\theta$$

Since $\sin\theta\left(\frac{1}{\sin\theta} - \frac{\cos\theta}{\cot\theta}\right) = \cos^2\theta$, the answer is C.

Test-Taking Tip

Verify your answer by choosing values for θ. Then evaluate the original expression and compare to your answer choice.

TRANSFORM BOTH SIDES OF AN EQUATION Sometimes it is easier to transform both sides of an equation separately into a common form. The following suggestions may be helpful as you verify trigonometric identities.

- Substitute one or more basic trigonometric identities to simplify an expression.
- Factor or multiply to simplify an expression.
- Multiply both the numerator and denominator by the same trigonometric expression.
- Write both sides of the identity in terms of sine and cosine only. Then simplify each side as much as possible.

Example 3 Verify by Transforming Both Sides

Verify that $\sec^2\theta - \tan^2\theta = \tan\theta\cot\theta$ is an identity.

$$\sec^2\theta - \tan^2\theta \stackrel{?}{=} \tan\theta\cot\theta \quad \text{Original equation}$$

$$\frac{1}{\cos^2\theta} - \frac{\sin^2\theta}{\cos^2\theta} \stackrel{?}{=} \frac{\sin\theta}{\cos\theta}\cdot\frac{\cos\theta}{\sin\theta} \quad \text{Express all terms using sine and cosine.}$$

$$\frac{1-\sin^2\theta}{\cos^2\theta} \stackrel{?}{=} 1 \quad \text{Subtract on the left. Multiply on the right.}$$

$$\frac{\cos^2\theta}{\cos^2\theta} \stackrel{?}{=} 1 \quad 1 - \sin^2\theta = \cos^2\theta$$

$$1 = 1 \quad \text{Simplify the left side.}$$

Check for Understanding

Concept Check

1. **Explain** the steps used to verify the identity $\sin \theta \tan \theta = \sec \theta - \cos \theta$.
2. **Describe** the various methods you can use to show that two trigonometric expressions form an identity.
3. **OPEN ENDED** Write a trigonometric equation that is not an identity. Explain how you know it is not an identity.

Guided Practice

Verify that each of the following is an identity.

4. $\tan \theta (\cot \theta + \tan \theta) = \sec^2 \theta$
5. $\tan^2 \theta \cos^2 \theta = 1 - \cos^2 \theta$
6. $\frac{\cos^2 \theta}{1 - \sin \theta} = 1 + \sin \theta$
7. $\frac{1 + \tan^2 \theta}{\csc^2 \theta} = \tan^2 \theta$
8. $\frac{\sin \theta}{\sec \theta} = \frac{1}{\tan \theta + \cot \theta}$
9. $\frac{\sec \theta + 1}{\tan \theta} = \frac{\tan \theta}{\sec \theta - 1}$

Standards Practice

Standardized Test Practice

10. Which expression is equivalent to $\frac{\sec \theta + \csc \theta}{1 + \tan \theta}$?

(A) $\sin \theta$ (B) $\cos \theta$ (C) $\tan \theta$ (D) $\csc \theta$

Practice and Apply

Homework Help

For Exercises	See Examples
11–24, 26–32	1, 2
25	3

Extra Practice
See page 860.

Verify that each of the following is an identity.

11. $\cos^2 \theta + \tan^2 \theta \cos^2 \theta = 1$
12. $\cot \theta (\cot \theta + \tan \theta) = \csc^2 \theta$
13. $1 + \sec^2 \theta \sin^2 \theta = \sec^2 \theta$
14. $\sin \theta \sec \theta \cot \theta = 1$
15. $\frac{1 - \cos \theta}{1 + \cos \theta} = (\csc \theta - \cot \theta)^2$
16. $\frac{1 - 2\cos^2 \theta}{\sin \theta \cos \theta} = \tan \theta - \cot \theta$
17. $\cot \theta \csc \theta = \frac{\cot \theta + \csc \theta}{\sin \theta + \tan \theta}$
18. $\sin \theta + \cos \theta = \frac{1 + \tan \theta}{\sec \theta}$
19. $\frac{\sec \theta}{\sin \theta} - \frac{\sin \theta}{\cos \theta} = \cot \theta$
20. $\frac{\sin \theta}{1 - \cos \theta} + \frac{1 - \cos \theta}{\sin \theta} = 2 \csc \theta$
21. $\frac{1 + \sin \theta}{\sin \theta} = \frac{\cot^2 \theta}{\csc \theta - 1}$
22. $\frac{1 + \tan \theta}{1 + \cot \theta} = \frac{\sin \theta}{\cos \theta}$
23. $\frac{1}{\sec^2 \theta} + \frac{1}{\csc^2 \theta} = 1$
24. $1 + \frac{1}{\cos \theta} = \frac{\tan^2 \theta}{\sec \theta - 1}$
25. $1 - \tan^4 \theta = 2\sec^2 \theta - \sec^4 \theta$
26. $\cos^4 \theta - \sin^4 \theta = \cos^2 \theta - \sin^2 \theta$
27. $\frac{1 - \cos \theta}{\sin \theta} = \frac{\sin \theta}{1 + \cos \theta}$
28. $\frac{\cos \theta}{1 + \sin \theta} + \frac{\cos \theta}{1 - \sin \theta} = 2 \sec \theta$
29. Verify that $\tan \theta \sin \theta \cos \theta \csc^2 \theta = 1$ is an identity.
30. Show that $1 + \cos \theta$ and $\frac{\sin^2 \theta}{1 - \cos \theta}$ form an identity.

PHYSICS **For Exercises 31 and 32, use the following information.**
If an object is propelled from ground level, the maximum height that it reaches is given by $h = \frac{v^2 \sin^2 \theta}{2g}$, where θ is the angle between the ground and the initial path of the object, v is the object's initial velocity, and g is the acceleration due to gravity, 9.8 meters per second squared.

31. Verify the identity $\frac{v^2 \sin^2 \theta}{2g} = \frac{v^2 \tan^2 \theta}{2g \sec^2 \theta}$.
32. A model rocket is launched with an initial velocity of 110 meters per second at an angle of 80° with the ground. Find the maximum height of the rocket.

33. **CRITICAL THINKING** Present a logical argument for why the identity $\sin^{-1} x + \cos^{-1} x = \frac{\pi}{2}$ is true when $0 \leq x \leq 1$.

34. **WRITING IN MATH** Answer the question that was posed at the beginning of the lesson.

How can you verify trigonometric identities?

Include the following in your answer:

- an explanation of why you cannot perform operations to each side of an unverified identity,
- an explanation of how you can tell if two expressions are equivalent, and
- an explanation of why you cannot use the graphs of two equations to verify an identity.

35. Which of the following is not equivalent to $\cos \theta$?

(A) $\frac{\cos \theta}{\cos^2 \theta + \sin^2 \theta}$ (B) $\frac{1 - \sin^2 \theta}{\cos \theta}$ (C) $\cot \theta \sin \theta$ (D) $\tan \theta \csc \theta$

36. Which of the following is equivalent to $\sin \theta + \cot \theta \cos \theta$?

(A) $2 \sin \theta$ (B) $\frac{1}{\sin \theta}$ (C) $\cos^2 \theta$ (D) $\frac{\sin \theta + \cos \theta}{\sin^2 \theta}$

VERIFYING TRIGONOMETRIC IDENTITIES You can determine whether or not an equation may be a trigonometric identity by graphing the expressions on either side of the equals sign as two separate functions. If the graphs do not match, then the equation is not an identity. If the two graphs do coincide, the equation *might* be an identity. The equation has to be verified algebraically to ensure that it is an identity.

Determine whether each of the following *may be* or *is not* an identity.

37. $\cot x + \tan x = \csc x \cot x$

38. $\sec^2 x - 1 = \sin^2 x \sec^2 x$

39. $(1 + \sin x)(1 - \sin x) = \cos^2 x$

40. $\frac{1}{\sec x \tan x} = \csc x - \sin x$

41. $\frac{\sec^2 x}{\tan x} = \sec x \csc x$

42. $\frac{1}{\sec x} + \frac{1}{\csc x} = 1$

Maintain Your Skills

Mixed Review

Find the value of each expression. *(Lesson 14-3)*

43. $\sec \theta$, if $\tan \theta = \frac{1}{2}$; $0° < \theta < 90°$

44. $\cos \theta$, if $\sin \theta = -\frac{2}{3}$; $180° < \theta < 270°$

45. $\csc \theta$, if $\cot \theta = -\frac{7}{12}$; $90° < \theta < 180°$

46. $\sin \theta$, if $\cos \theta = \frac{3}{4}$; $270° < \theta < 360°$

State the amplitude, period, and phase shift of each function. Then graph each function. *(Lesson 14-2)*

47. $y = \cos(\theta - 30°)$

48. $y = \sin(\theta - 45°)$

49. $y = 3 \cos\left(\theta + \frac{\pi}{2}\right)$

50. What is the probability that an event occurs if the odds of the event occurring are 5:1? *(Lesson 12-4)*

Getting Ready for the Next Lesson

PREREQUISITE SKILL **Simplify each expression.**
*(To review **simplifying radical expressions**, see Lesson 5-6.)*

51. $\frac{\sqrt{3}}{2} \cdot \frac{\sqrt{2}}{2}$

52. $\frac{1}{2} \cdot \frac{\sqrt{2}}{2}$

53. $\frac{\sqrt{6}}{4} + \frac{\sqrt{2}}{2}$

54. $\frac{1}{2} - \frac{\sqrt{3}}{4}$

14-5

Sum and Difference of Angles Formulas

What You'll Learn

- Find values of sine and cosine involving sum and difference formulas.
- Verify identities by using sum and difference formulas.

How are the sum and difference formulas used to describe communication interference?

Have you ever been talking on a cell phone and temporarily lost the signal? Radio waves that pass through the same place at the same time cause interference. *Constructive interference* occurs when two waves combine to have a greater amplitude than either of the component waves. *Destructive interference* occurs when the component waves combine to have a smaller amplitude.

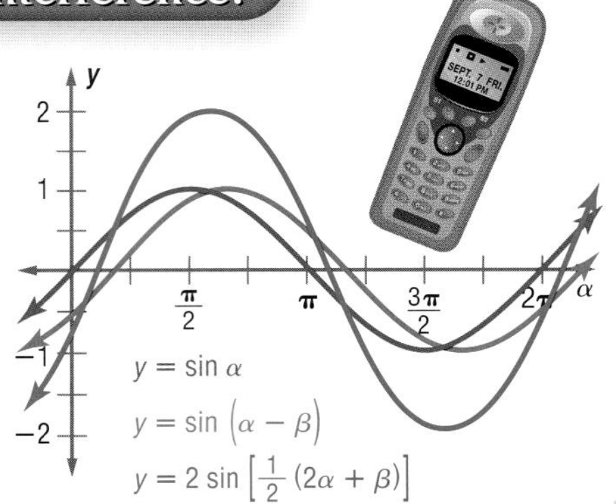

Study Tips

Reading Math

The Greek letter *beta*, β, can be used to denote the measure of an angle.

It is important to realize that $\sin(\alpha \pm \beta)$ is not the same as $\sin \alpha \pm \sin \beta$.

SUM AND DIFFERENCE FORMULAS Notice that the third equation shown above involves the sum of α and β. It is often helpful to use formulas for the trigonometric values of the difference or sum of two angles. For example, you could find sin 15° by evaluating $\sin(60° - 45°)$. Formulas can be developed that can be used to evaluate expressions like $\sin(\alpha - \beta)$ or $\cos(\alpha + \beta)$.

The figure at the right shows two angles α and β in standard position on the unit circle. Use the Distance Formula to find d, where $(x_1, y_1) = (\cos \beta, \sin \beta)$ and $(x_2, y_2) = (\cos \alpha, \sin \alpha)$.

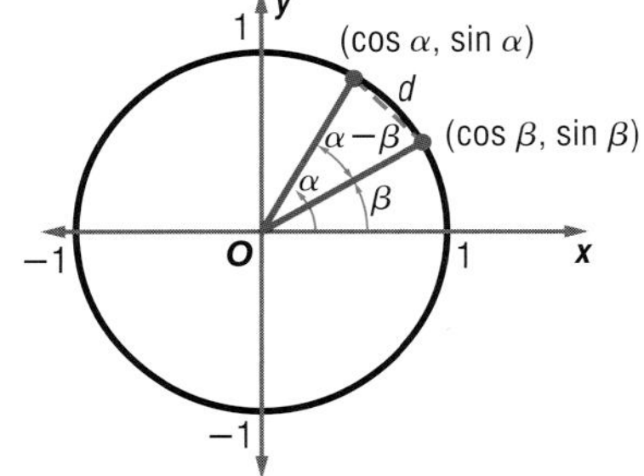

$$d = \sqrt{(\cos \alpha - \cos \beta)^2 + (\sin \alpha - \sin \beta)^2}$$
$$d^2 = (\cos \alpha - \cos \beta)^2 + (\sin \alpha - \sin \beta)^2$$
$$d^2 = (\cos^2 \alpha - 2\cos \alpha \cos \beta + \cos^2 \beta) + (\sin^2 \alpha - 2\sin \alpha \sin \beta + \sin^2 \beta)$$
$$d^2 = \cos^2 \alpha + \sin^2 \alpha + \cos^2 \beta + \sin^2 \beta - 2\cos \alpha \cos \beta - 2\sin \alpha \sin \beta$$
$$d^2 = 1 + 1 - 2\cos \alpha \cos \beta - 2\sin \alpha \sin \beta \qquad \sin^2 \alpha + \cos^2 \alpha = 1 \text{ and}$$
$$d^2 = 2 - 2\cos \alpha \cos \beta - 2\sin \alpha \sin \beta \qquad \sin^2 \beta + \cos^2 \beta = 1$$

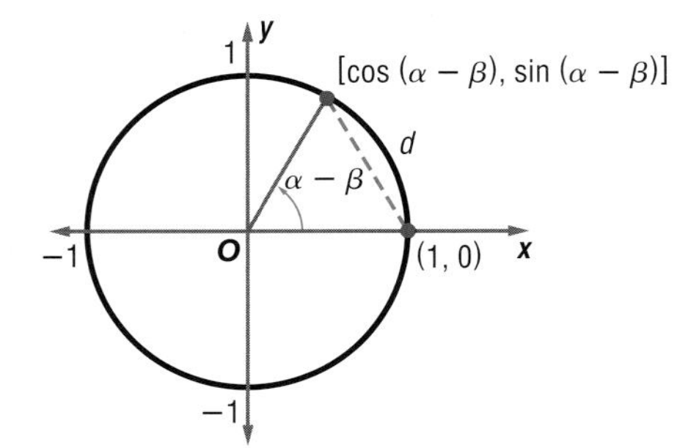

Now find the value of d^2 when the angle having measure $\alpha - \beta$ is in standard position on the unit circle, as shown in the figure at the left.

$$d = \sqrt{[\cos(\alpha - \beta) - 1]^2 + [\sin(\alpha - \beta) - 0]^2}$$
$$d^2 = [\cos(\alpha - \beta) - 1]^2 + [\sin(\alpha - \beta) - 0]^2$$
$$= [\cos^2(\alpha - \beta) - 2\cos(\alpha - \beta) + 1] + \sin^2(\alpha - \beta)$$
$$= \cos^2(\alpha - \beta) + \sin^2(\alpha - \beta) - 2\cos(\alpha - \beta) + 1$$
$$= 1 - 2\cos(\alpha - \beta) + 1$$
$$= 2 - 2\cos(\alpha - \beta)$$

By equating the two expressions for d^2, you can find a formula for $\cos(\alpha - \beta)$.

$$d^2 = d^2$$

$$2 - 2\cos(\alpha - \beta) = 2 - 2\cos\alpha\cos\beta - 2\sin\alpha\sin\beta$$

$$-1 + \cos(\alpha - \beta) = -1 + \cos\alpha\cos\beta + \sin\alpha\sin\beta$$ Divide each side by -2.

$$\cos(\alpha - \beta) = \cos\alpha\cos\beta + \sin\alpha\sin\beta$$ Add 1 to each side.

Use the formula for $\cos(\alpha - \beta)$ to find a formula for $\cos(\alpha + \beta)$.

$$\cos(\alpha - \beta) = \cos[\alpha - (-\beta)]$$

$$= \cos\alpha\cos(-\beta) + \sin\alpha\sin(-\beta)$$

$$= \cos\alpha\cos\beta - \sin\alpha\sin\beta$$ $\cos(-\beta) = \cos\beta$; $\sin(-\beta) = -\sin\beta$

You can use a similar method to find formulas for $\sin(\alpha + \beta)$ and $\sin(\alpha - \beta)$.

Key Concept — Sum and Difference of Angles Formulas

The following identities hold true for all values of α and β.

$$\cos(\alpha \pm \beta) = \cos\alpha\cos\beta \mp \sin\alpha\sin\beta$$

$$\sin(\alpha \pm \beta) = \sin\alpha\cos\beta \pm \cos\alpha\sin\beta$$

Notice the symbol $\mp$ in the formula for $\cos(\alpha \pm \beta)$. It means "minus or plus." In the cosine formula, when the sign on the left side of the equation is plus, the sign on the right side is minus; when the sign on the left side is minus, the sign on the right side is plus. The signs match each other in the sine formula.

Example 1 Use Sum and Difference of Angles Formulas

Find the exact value of each expression.

a. cos 75°

Use the formula $\cos(\alpha + \beta) = \cos\alpha\cos\beta - \sin\alpha\sin\beta$.

$$\cos 75° = \cos(30° + 45°)$$ $\alpha = 30°, \beta = 45°$

$$= \cos 30°\cos 45° - \sin 30°\sin 45°$$

$$= \left(\frac{\sqrt{3}}{2} \cdot \frac{\sqrt{2}}{2}\right) - \left(\frac{1}{2} \cdot \frac{\sqrt{2}}{2}\right)$$ Evaluate each expression.

$$= \frac{\sqrt{6}}{4} - \frac{\sqrt{2}}{4}$$ Multiply.

$$= \frac{\sqrt{6} - \sqrt{2}}{4}$$ Simplify.

b. sin (−210°)

Use the formula $\sin(\alpha - \beta) = \sin\alpha\cos\beta - \cos\alpha\sin\beta$.

$$\sin(-210°) = \sin(60° - 270°)$$ $\alpha = 60°, \beta = 270°$

$$= \sin 60°\cos 270° - \cos 60°\sin 270°$$

$$= \left(\frac{\sqrt{3}}{2}\right)(0) - \left(\frac{1}{2}\right)(-1)$$ Evaluate each expression.

$$= 0 - \left(-\frac{1}{2}\right)$$ Multiply.

$$= \frac{1}{2}$$ Simplify.

Example 2 Use Sum and Difference Formulas to Solve a Problem

Study Tip

Reading Math
The symbol ϕ is the lowercase Greek letter *phi*.

PHYSICS On June 22, the maximum amount of light energy falling on a square foot of ground at a location in the northern hemisphere is given by $E \sin (113.5° - \phi)$, where ϕ is the latitude of the location and E is the amount of light energy when the Sun is directly overhead. Use the difference of angles formula to determine the amount of light energy in Rochester, New York, located at a latitude of 43.1° N.

Use the difference formula for sine.

$$\begin{aligned}\sin (113.5° - \phi) &= \sin 113.5° \cos \phi - \cos 113.5° \sin \phi \\ &= \sin 113.5° \cos 43.1° - \cos 113.5° \sin 43.1° \\ &= 0.9171 \cdot 0.7302 - (-0.3987) \cdot 0.6833 \\ &= 0.9420\end{aligned}$$

In Rochester, New York, the maximum light energy per square foot is $0.9420E$.

VERIFY IDENTITIES You can also use the sum and difference formulas to verify identities.

Example 3 Verify Identities

Verify that each of the following is an identity.

a. $\sin (180° + \theta) = -\sin \theta$

$\sin (180° + \theta) \stackrel{?}{=} -\sin \theta$	Original equation
$\sin 180° \cos \theta + \cos 180° \sin \theta \stackrel{?}{=} -\sin \theta$	Sum of angles formula
$0 \cos \theta + (-1) \sin \theta \stackrel{?}{=} -\sin \theta$	Evaluate each expression.
$-\sin \theta = -\sin \theta$	Simplify.

b. $\cos (180° + \theta) = -\cos \theta$

$\cos (180° + \theta) \stackrel{?}{=} -\cos \theta$	Original equation
$\cos 180° \cos \theta - \sin 180° \sin \theta \stackrel{?}{=} -\cos \theta$	Sum of angles formula
$(-1) \cos \theta - 0 \sin \theta \stackrel{?}{=} -\cos \theta$	Evaluate each expression.
$-\cos \theta = -\cos \theta$	Simplify.

Check for Understanding

Concept Check

1. **Determine** whether $\sin (\alpha + \beta) = \sin \alpha + \sin \beta$ is an identity.
2. **Describe** a method for finding the exact value of sin 105°. Then find the value.
3. **OPEN ENDED** Determine whether $\cos (\alpha - \beta) < 1$ is *sometimes, always,* or *never* true. Explain your reasoning.

Guided Practice

Find the exact value of each expression.

4. sin 75°
5. sin 165°
6. cos 255°
7. cos (−30°)
8. sin (−240°)
9. cos (−120°)

Verify that each of the following is an identity.

10. $\cos (270° - \theta) = -\sin \theta$
11. $\sin \left(\theta + \frac{\pi}{2}\right) = \cos \theta$
12. $\sin (\theta + 30°) + \cos (\theta + 60°) = \cos \theta$

Application **13. GEOMETRY** Determine the exact value of $\tan \alpha$.

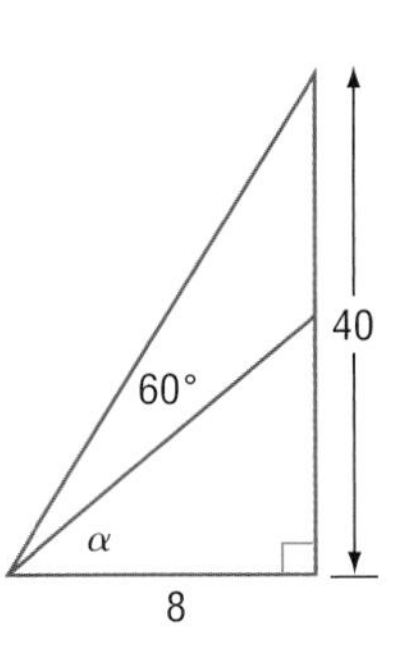

Practice and Apply

Homework Help

For Exercises	See Examples
14–27	1
28–39	3
40, 41, 43–46	2

Extra Practice
See page 860.

Find the exact value of each expression.

14. $\sin 135°$ **15.** $\cos 105°$ **16.** $\sin 285°$

17. $\cos 165°$ **18.** $\cos 195°$ **19.** $\sin 255°$

20. $\cos 225°$ **21.** $\sin 315°$ **22.** $\sin(-15°)$

23. $\cos(-45°)$ **24.** $\cos(-150°)$ **25.** $\sin(-165°)$

26. What is the exact value of $\sin 75° - \sin 15°$?

27. Find the exact value of $\cos 105° + \cos 225°$.

Verify that each of the following is an identity.

28. $\sin(270° - \theta) = -\cos \theta$

29. $\cos(90° + \theta) = -\sin \theta$

30. $\cos(90° - \theta) = \sin \theta$

31. $\sin(90° - \theta) = \cos \theta$

32. $\sin\left(\theta + \frac{3\pi}{2}\right) = -\cos \theta$

33. $\cos(\pi - \theta) = -\cos \theta$

34. $\cos(2\pi + \theta) = \cos \theta$

35. $\sin(\pi - \theta) = \sin \theta$

36. $\sin(60° + \theta) + \sin(60° - \theta) = \sqrt{3} \cos \theta$

37. $\sin\left(\theta + \frac{\pi}{3}\right) - \cos\left(\theta + \frac{\pi}{6}\right) = \sin \theta$

38. $\sin(\alpha + \beta) \sin(\alpha - \beta) = \sin^2 \alpha - \sin^2 \beta$

39. $\cos(\alpha + \beta) = \frac{1 - \tan \alpha \tan \beta}{\sec \alpha \sec \beta}$

COMMUNICATION **For Exercises 40 and 41, use the following information.**
A radio transmitter sends out two signals, one for voice communication and another for data. Suppose the equation of the voice wave is $v = 10 \sin(2t - 30°)$ and the equation of the data wave is $d = 10 \cos(2t + 60°)$.

40. Draw a graph of the waves when they are combined.

41. Refer to the application at the beginning of the lesson. What type of interference results? Explain.

More About. . .

Physics

In the northern hemisphere, the day with the least number of hours of daylight is December 21 or 22, the first day of winter.

Source: www.infoplease.com

PHYSICS **For Exercises 42–45, use the following information.**
On December 22, the maximum amount of light energy that falls on a square foot of ground at a certain location is given by $E \sin(113.5° + \phi)$, where ϕ is the latitude of the location. Use the sum of angles formula to find the amount of light energy, in terms of E, for each location.

42. Salem, OR (Latitude: 44.9° N)

43. Chicago, IL (Latitude: 41.8° N)

44. Charleston, SC (Latitude: 28.5°N)

45. San Diego, CA (Latitude 32.7° N)

46. CRITICAL THINKING Use the sum and difference formulas for sine and cosine to derive formulas for $\tan(\alpha + \beta)$ and $\tan(\alpha - \beta)$.

47. **WRITING IN MATH** Answer the question that was posed at the beginning of the lesson.

How are the sum and difference formulas used to describe communication interference?

Include the following in your answer:

- an explanation of the difference between constructive and destructive interference, and
- a description of how you would explain wave interference to a friend.

Standards Practice

Standardized Test Practice

48. Find the exact value of $\sin \theta$.

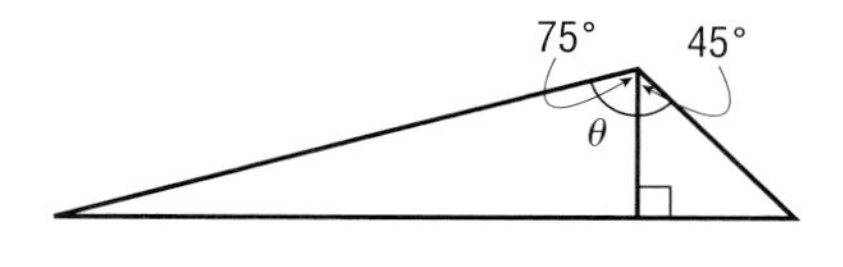

Ⓐ $\frac{\sqrt{3}}{2}$ Ⓑ $\frac{\sqrt{2}}{2}$

Ⓒ $\frac{1}{2}$ Ⓓ $\frac{\sqrt{3}}{3}$

49. Find the exact value of $\cos(-210°)$.

Ⓐ $\frac{\sqrt{3}}{2}$ Ⓑ 0.5 Ⓒ $-\frac{\sqrt{3}}{2}$ Ⓓ -0.5

Maintain Your Skills

Mixed Review

Verify that each of the following is an identity. *(Lesson 14-4)*

50. $\cot \theta + \sec \theta = \frac{\cos^2 \theta + \sin \theta}{\sin \theta \cos \theta}$

51. $\sin^2 \theta + \tan^2 \theta = (1 - \cos^2 \theta) + \frac{\sec^2 \theta}{\csc^2 \theta}$

52. $\sin \theta (\sin \theta + \csc \theta) = 2 - \cos^2 \theta$

53. $\frac{\sec \theta}{\tan \theta} = \csc \theta$

Simplify each expression. *(Lesson 14-3)*

54. $\frac{\tan \theta \csc \theta}{\sec \theta}$

55. $4\left(\sec^2 \theta - \frac{\sin^2 \theta}{\cos^2 \theta}\right)$

56. $(\cot \theta + \tan \theta)\sin \theta$

57. $\csc \theta \tan \theta + \sec \theta$

Find the exact values of the six trigonometric functions of θ if the terminal side of θ in standard position contains the given point. *(Lesson 13-3)*

58. $(5, -3)$ 59. $(-3, -4)$ 60. $(0, 2)$

Evaluate each expression. *(Lesson 12-2)*

61. $P(6, 4)$

62. $P(12, 7)$

63. $C(8, 3)$

64. $C(10, 4)$

65. **AVIATION** A pilot is flying from Chicago to Columbus, a distance of 300 miles. In order to avoid an area of thunderstorms, she alters her initial course by 15° and flies on this course for 75 miles. How far is she from Columbus? *(Lesson 13-5)*

66. Write $6y^2 - 34x^2 = 204$ in standard form. *(Lesson 8-5)*

Getting Ready for the Next Lesson

PREREQUISITE SKILL **Solve each equation.**
*(To review **solving equations using the Square Root Property**, see Lesson 6-4.)*

67. $x^2 = \frac{20}{16}$ 68. $x^2 = \frac{9}{25}$ 69. $x^2 = \frac{5}{25}$ 70. $x^2 = \frac{18}{32}$

71. $x^2 - 1 = \frac{1}{2}$ 72. $x^2 - 1 = \frac{4}{5}$ 73. $x^2 = \frac{\sqrt{3}}{2} - \frac{1}{2}$ 74. $x^2 = \frac{\sqrt{2}}{2} - 1$

14-6 Double-Angle and Half-Angle Formulas

What You'll Learn

- Find values of sine and cosine involving double-angle formulas.
- Find values of sine and cosine involving half-angle formulas.

How can trigonometric functions be used to describe music?

Stringed instruments such as a piano, guitar, or violin rely on waves to produce the tones we hear. When the strings are struck or plucked, they vibrate. If the motion of the strings were observed in slow motion, you could see that there are places on the string, called *nodes*, that do not move under the vibration. Halfway between each pair of consecutive nodes are *antinodes* that undergo the maximum vibration. The nodes and antinodes form *harmonics*. These harmonics can be represented using variations of the equations $y = \sin 2\theta$ and $y = \sin \frac{1}{2}\theta$.

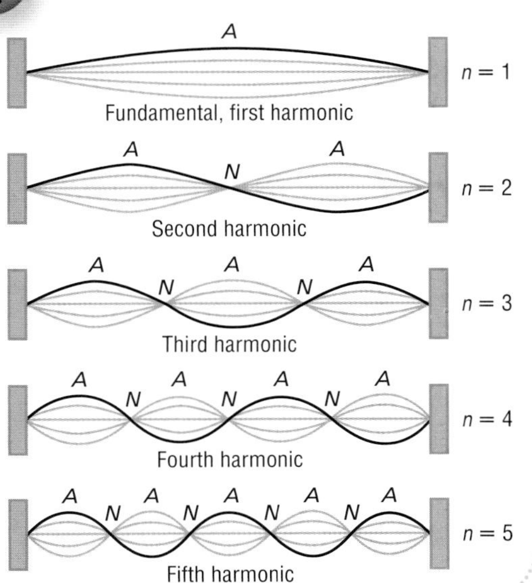

DOUBLE-ANGLE FORMULAS You can use the formula for $\sin(\alpha + \beta)$ to find the sine of twice an angle θ, $\sin 2\theta$, and the formula for $\cos(\alpha + \beta)$ to find the cosine of twice an angle θ, $\cos 2\theta$.

$$\begin{aligned} \sin 2\theta &= \sin(\theta + \theta) \\ &= \sin\theta\cos\theta + \cos\theta\sin\theta \\ &= 2\sin\theta\cos\theta \end{aligned}$$

$$\begin{aligned} \cos 2\theta &= \cos(\theta + \theta) \\ &= \cos\theta\cos\theta - \sin\theta\sin\theta \\ &= \cos^2\theta - \sin^2\theta \end{aligned}$$

You can find alternate forms for $\cos 2\theta$ by making substitutions into the expression $\cos^2\theta - \sin^2\theta$.

$\cos^2\theta - \sin^2\theta = (1 - \sin^2\theta) - \sin^2\theta$ Substitute $1 - \sin^2\theta$ for $\cos^2\theta$.

$= 1 - 2\sin^2\theta$ Simplify.

$\cos^2\theta - \sin^2\theta = \cos^2\theta - (1 - \cos^2\theta)$ Substitute $1 - \cos^2\theta$ for $\sin^2\theta$.

$= 2\cos^2\theta - 1$ Simplify.

These formulas are called the **double-angle formulas**.

Key Concept — Double-Angle Formulas

The following identities hold true for all values of θ.

$\sin 2\theta = 2\sin\theta\cos\theta$

$\cos 2\theta = \cos^2\theta - \sin^2\theta$

$\cos 2\theta = 1 - 2\sin^2\theta$

$\cos 2\theta = 2\cos^2\theta - 1$

Example 1 Double-Angle Formulas

Find the exact value of each expression if $\sin \theta = \frac{4}{5}$ and θ is between 90° and 180°.

a. $\sin 2\theta$

Use the identity $\sin 2\theta = 2 \sin \theta \cos \theta$.

First, find the value of $\cos \theta$.

$\cos^2 \theta = 1 - \sin^2 \theta$ — $\cos^2 \theta + \sin^2 \theta = 1$

$\cos^2 \theta = 1 - \left(\frac{4}{5}\right)^2$ — $\sin \theta = \frac{4}{5}$

$\cos^2 \theta = \frac{9}{25}$ — Subtract.

$\cos \theta = \pm\frac{3}{5}$ — Find the square root of each side.

Since θ is in the second quadrant, cosine is negative. Thus, $\cos \theta = -\frac{3}{5}$.

Now find $\sin 2\theta$.

$\sin 2\theta = 2 \sin \theta \cos \theta$ — Double-angle formula

$\sin 2\theta = 2\left(\frac{4}{5}\right)\left(-\frac{3}{5}\right)$ — $\sin \theta = \frac{4}{5}$, $\cos \theta = -\frac{3}{5}$

$= -\frac{24}{25}$ — The value of $\sin 2\theta$ is $-\frac{24}{25}$.

b. $\cos 2\theta$

Use the identity $\cos 2\theta = 1 - 2 \sin^2 \theta$.

$\cos 2\theta = 1 - 2 \sin^2 \theta$ — Double-angle formula

$= 1 - 2\left(\frac{4}{5}\right)^2$ — $\sin \theta = \frac{4}{5}$

$= -\frac{7}{25}$ — The value of $\cos 2\theta$ is $-\frac{7}{25}$.

HALF-ANGLE FORMULAS You can derive formulas for the sine and cosine of half a given angle using the double-angle formulas.

Find $\sin \frac{\alpha}{2}$.

$1 - 2 \sin^2 \theta = \cos 2\theta$ — Double-angle formula

$1 - 2 \sin^2 \frac{\alpha}{2} = \cos \alpha$ — Substitute $\frac{\alpha}{2}$ for θ and α for 2θ.

$\sin^2 \frac{\alpha}{2} = \frac{1 - \cos \alpha}{2}$ — Solve for $\sin^2 \frac{\alpha}{2}$.

$\sin \frac{\alpha}{2} = \pm\sqrt{\frac{1 - \cos \alpha}{2}}$ — Take the square root of each side.

Find $\cos \frac{\alpha}{2}$.

$2 \cos^2 \theta - 1 = \cos 2\theta$ — Double-angle formula

$2 \cos^2 \frac{\alpha}{2} - 1 = \cos \alpha$ — Substitute $\frac{\alpha}{2}$ for θ and α for 2θ.

$\cos^2 \frac{\alpha}{2} = \frac{1 + \cos \alpha}{2}$ — Solve for $\cos^2 \frac{\alpha}{2}$.

$\cos \frac{\alpha}{2} = \pm\sqrt{\frac{1 + \cos \alpha}{2}}$ — Take the square root of each side.

These are called the **half-angle formulas**. The signs are determined by the function of $\frac{\alpha}{2}$.

Key Concept — *Half-Angle Formulas*

The following identities hold true for all values of α.

$$\sin \frac{\alpha}{2} = \pm\sqrt{\frac{1 - \cos \alpha}{2}} \qquad \cos \frac{\alpha}{2} = \pm\sqrt{\frac{1 + \cos \alpha}{2}}$$

Example 2 Half-Angle Formulas

Find $\cos \frac{\alpha}{2}$ if $\sin \alpha = -\frac{3}{4}$ and α is in the third quadrant.

Since $\cos \frac{\alpha}{2} = \pm\sqrt{\frac{1 + \cos \alpha}{2}}$, we must find $\cos \alpha$ first.

$\cos^2 \alpha = 1 - \sin^2 \alpha$ — $\cos^2 \alpha + \sin^2 \alpha = 1$

$\cos^2 \alpha = 1 - \left(-\frac{3}{4}\right)^2$ — $\sin \alpha = -\frac{3}{4}$

$\cos^2 \alpha = \frac{7}{16}$ — Simplify.

$\cos \alpha = \pm\frac{\sqrt{7}}{4}$ — Take the square root of each side.

Since α is in the third quadrant, $\cos \alpha = -\frac{\sqrt{7}}{4}$.

$\cos \frac{\alpha}{2} = \pm\sqrt{\frac{1 + \cos \alpha}{2}}$ — Half-angle formula

$= \pm\sqrt{\frac{1 - \frac{\sqrt{7}}{4}}{2}}$ — $\cos \alpha = -\frac{\sqrt{7}}{4}$

$= \pm\sqrt{\frac{4 - \sqrt{7}}{8}}$ — Simplify the radicand.

$= \pm\frac{\sqrt{4 - \sqrt{7}}}{2\sqrt{2}} \cdot \frac{\sqrt{2}}{\sqrt{2}}$ — Rationalize.

$= \pm\frac{\sqrt{8 - 2\sqrt{7}}}{4}$ — Multiply.

Since α is between 180° and 270°, $\frac{\alpha}{2}$ is between 90° and 135°. Thus, $\cos \frac{\alpha}{2}$ is negative and equals $-\frac{\sqrt{8 - 2\sqrt{7}}}{4}$.

Study Tip

Choosing the Sign

You may want to determine the quadrant in which the terminal side of $\frac{\alpha}{2}$ will lie in the first step of the solution. Then you can use the correct sign from the beginning.

Example 3 Evaluate Using Half-Angle Formulas

Find the exact value of each expression by using the half-angle formulas.

a. sin 105°

$\sin 105° = \sin \frac{210°}{2}$

$= \sqrt{\frac{1 - \cos 210°}{2}}$ — $\sin \frac{\alpha}{2} = \pm\sqrt{\frac{1 - \cos \alpha}{2}}$

(continued on the next page)

$$= \sqrt{\frac{1 - \left(-\frac{\sqrt{3}}{2}\right)}{2}} \quad \cos 210° = -\frac{\sqrt{3}}{2}$$

$$= \sqrt{\frac{2 + \sqrt{3}}{4}} \quad \text{Simplify the radicand.}$$

$$= \frac{\sqrt{2 + \sqrt{3}}}{2} \quad \text{Simplify the denominator.}$$

b. $\cos \frac{\pi}{8}$

$$\cos \frac{\pi}{8} = \frac{\frac{\pi}{4}}{2}$$

$$= \sqrt{\frac{1 + \cos \frac{\pi}{4}}{2}} \quad \cos \frac{\alpha}{2} = \pm\sqrt{\frac{1 + \cos \alpha}{2}}$$

$$= \sqrt{\frac{1 + \frac{\sqrt{2}}{2}}{2}} \quad \cos \frac{\pi}{4} = \frac{\sqrt{2}}{2}$$

$$= \sqrt{\frac{2 + \sqrt{2}}{4}} \quad \text{Simplify the radicand.}$$

$$= \frac{\sqrt{2 + \sqrt{2}}}{2} \quad \text{Simplify the denominator.}$$

Recall that you can use the sum and difference formulas to verify identities. Double- and half-angle formulas can also be used to verify identities.

Example 4 Verify Identities

Verify that $(\sin \theta + \cos \theta)^2 = 1 + \sin 2\theta$ is an identity.

$$(\sin \theta + \cos \theta)^2 \stackrel{?}{=} 1 + \sin 2\theta \quad \text{Original equation}$$

$$\sin^2 \theta + 2 \sin \theta \cos \theta + \cos^2 \theta \stackrel{?}{=} 1 + \sin 2\theta \quad \text{Multiply.}$$

$$1 + 2 \sin \theta \cos \theta \stackrel{?}{=} 1 + \sin 2\theta \quad \sin^2 \theta + \cos^2 \theta = 1$$

$$1 + \sin 2\theta = 1 + \sin 2\theta \quad \text{Double-angle formula}$$

Check for Understanding

Concept Check

1. **Explain** how to find $\cos \frac{x}{2}$ if x is in the third quadrant.
2. **Find a counterexample** to show that $\cos 2\theta = 2 \cos \theta$ is not an identity.
3. **OPEN ENDED** Describe the conditions under which you would use each of the three identities for $\cos 2\theta$.

Guided Practice

Find the exact values of $\sin 2\theta$, $\cos 2\theta$, $\sin \frac{\theta}{2}$, and $\cos \frac{\theta}{2}$ for each of the following.

4. $\cos \theta = \frac{3}{5}$; $0° < \theta < 90°$
5. $\cos \theta = -\frac{2}{3}$; $180° < \theta < 270°$
6. $\sin \theta = \frac{1}{2}$; $0° < \theta < 90°$
7. $\sin \theta = -\frac{3}{4}$; $270° < \theta < 360°$

Find the exact value of each expression by using the half-angle formulas.

8. $\sin 195°$
9. $\cos \frac{19\pi}{12}$

Verify that each of the following is an identity.

10. $\cot x = \dfrac{\sin 2x}{1 - \cos 2x}$

11. $\cos^2 2x + 4 \sin^2 x \cos^2 x = 1$

Application

12. AVIATION When a jet travels at speeds greater than the speed of sound, a sonic boom is created by the sound waves forming a cone behind the jet. If θ is the measure of the angle at the vertex of the cone, then the Mach number M can be determined using the formula $\sin \dfrac{\theta}{2} = \dfrac{1}{M}$. Find the Mach number of a jet if a sonic boom is created by a cone with a vertex angle of 75°.

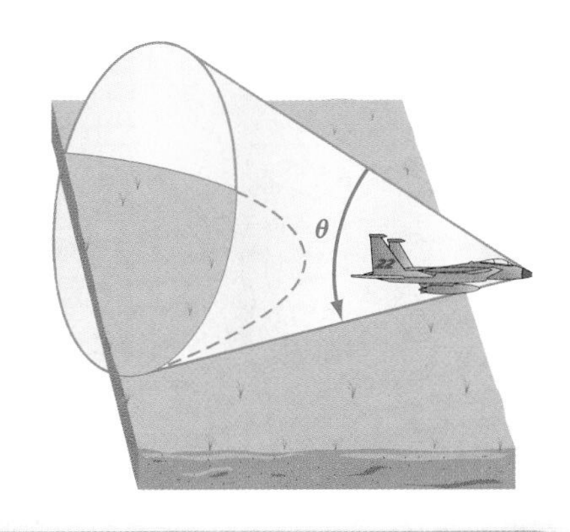

Practice and Apply

Homework Help

For Exercises	See Examples
13–24, 38, 39	1, 2
25–30, 37	3
31–36	4

Extra Practice
See page 861.

Find the exact values of $\sin 2\theta$, $\cos 2\theta$, $\sin \frac{\theta}{2}$, and $\cos \frac{\theta}{2}$ for each of the following.

13. $\sin \theta = \frac{5}{13}$; $90° < \theta < 180°$

14. $\cos \theta = \frac{1}{5}$; $270° < \theta < 360°$

15. $\cos \theta = -\frac{1}{3}$; $180° < \theta < 270°$

16. $\sin \theta = -\frac{3}{5}$; $180° < \theta < 270°$

17. $\sin \theta = -\frac{3}{8}$; $270° < \theta < 360°$

18. $\cos \theta = -\frac{1}{4}$; $90° < \theta < 180°$

19. $\cos \theta = \frac{1}{6}$; $0° < \theta < 90°$

20. $\cos \theta = -\frac{12}{13}$; $180° < \theta < 270°$

21. $\sin \theta = -\frac{1}{3}$; $270° < \theta < 360°$

22. $\sin \theta = -\frac{1}{4}$; $180° < \theta < 270°$

23. $\cos \theta = \frac{2}{3}$; $0° < \theta < 90°$

24. $\sin \theta = \frac{2}{5}$; $90° < \theta < 180°$

Find the exact value of each expression by using the half-angle formulas.

25. $\cos 165°$

26. $\sin 22\frac{1}{2}°$

27. $\cos 157\frac{1}{2}°$

28. $\sin 345°$

29. $\sin \frac{7\pi}{8}$

30. $\cos \frac{7\pi}{12}$

Verify that each of the following is an identity.

31. $\sin 2x = 2 \cot x \sin^2 x$

32. $2 \cos^2 \frac{x}{2} = 1 + \cos x$

33. $\sin^4 x - \cos^4 x = 2 \sin^2 x - 1$

34. $\sin^2 x = \frac{1}{2}(1 - \cos 2x)$

35. $\tan^2 \frac{x}{2} = \dfrac{1 - \cos x}{1 + \cos x}$

36. $\dfrac{1}{\sin x \cos x} - \dfrac{\cos x}{\sin x} = \tan x$

More About. . .

Optics

A rainbow appears when the sun shines through water droplets that act as a prism.

37. OPTICS If a glass prism has an apex angle of measure α and an angle of deviation of measure β, then the index of refraction n of the prism is given by $n = \dfrac{\sin\left[\frac{1}{2}(\alpha + \beta)\right]}{\sin \frac{\alpha}{2}}$.

What is the angle of deviation of a prism with an apex angle of 40° and an index of refraction of 2?

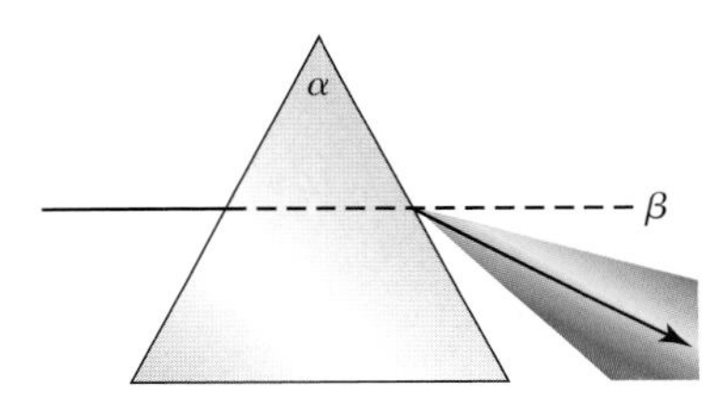

GEOGRAPHY **For Excercises 38 and 39, use the following information.**
A Mercator projection map uses a flat projection of Earth in which the distance between the lines of latitude increases with their distance from the equator. The calculation of the location of a point on this projection uses the expression $\tan\left(45° + \frac{L}{2}\right)$, where L is the latitude of the point.

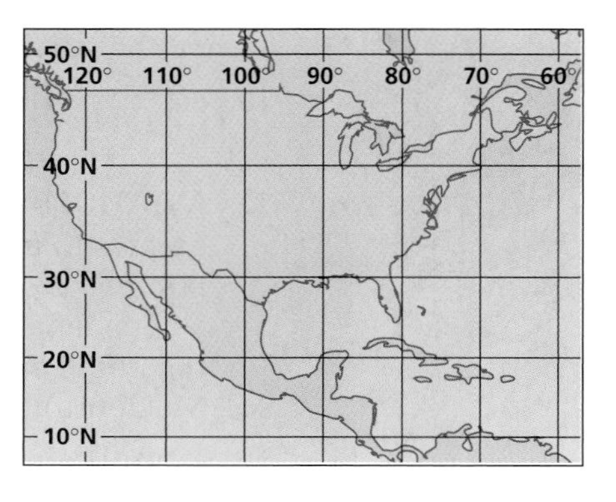

38. Write this expression in terms of a trigonometric function of L.

39. Find the exact value of the expression if $L = 60°$.

PHYSICS **For Exercises 40 and 41, use the following information.**
An object is propelled from ground level with an initial velocity of v at an angle of elevation θ.

40. The horizontal distance d it will travel can be determined using $d = \frac{v^2 \sin 2\theta}{g}$, where g is acceleration due to gravity. Verify that this expression is the same as $\frac{2}{g}v^2(\tan\theta - \tan\theta\sin^2\theta)$.

41. The maximum height h the object will reach can be determined using the formula $h = \frac{v^2 \sin^2\theta}{2g}$. Find the ratio of the maximum height attained to the horizontal distance traveled.

CRITICAL THINKING **For Exercises 42–46, use the following information.**
Consider the functions $f(x) = \sin 2x$, $g(x) = \sin^2 x$, $h(x) = -\cos^2 x$, and $k(x) = -\frac{1}{2}\cos 2x$.

42. Draw the graphs of $y = g(x)$, $y = h(x)$, and $y = k(x)$ on the same coordinate plane on the interval from $x = -2\pi$ to $x = 2\pi$. What do you notice about the graphs?

43. Where do the maxima and minima of g, h, and k occur?

44. Draw the graph of $y = f(x)$ on a separate coordinate plane.

45. What is the behavior of the graph of $f(x)$ at the locations found in Exercise 43?

46. Use what you know about transformations to determine c and d so that $g(x) = h(x) + c = k(x) + d$.

47. WRITING IN MATH Answer the question that was posed at the beginning of the lesson.

How can trigonometric functions be used to describe music?

Include the following in your answer:

- a description of what happens to the graph of the function of a vibrating string as it moves from one harmonic to the next, and
- an explanation of what happens to the period of the function as you move from the nth harmonic to the $(n + 1)$th harmonic.

48. Find the exact value of $\cos 2\theta$ if $\sin\theta = \frac{-\sqrt{5}}{3}$ and $180° < \theta < 270°$.

Ⓐ $\frac{-\sqrt{6}}{6}$ Ⓑ $\frac{-\sqrt{30}}{6}$ Ⓒ $\frac{-4\sqrt{5}}{9}$ Ⓓ $\frac{-1}{9}$

49. Find the exact value of $\sin\frac{\theta}{2}$ if $\cos\theta = \frac{\sqrt{3}}{2}$ and $0° < \theta < 90°$.

Ⓐ $\frac{\sqrt{3}}{2}$ Ⓑ $\frac{\sqrt{2 - \sqrt{3}}}{2}$ Ⓒ $\frac{\sqrt{2 + \sqrt{3}}}{2}$ Ⓓ $\frac{1}{2}$

Maintain Your Skills

Mixed Review

Find the exact value of each expression. *(Lesson 14-5)*

50. $\cos 15°$

51. $\sin 15°$

52. $\sin(-135°)$

53. $\cos 150°$

54. $\sin 105°$

55. $\cos(-300°)$

Verify that each of the following is an identity. *(Lesson 14-4)*

56. $\cot^2 \theta - \sin^2 \theta = \frac{\cos^2 \theta \csc^2 \theta - \sin^2 \theta}{\sin^2 \theta \csc^2 \theta}$

57. $\cos \theta (\cos \theta + \cot \theta) = \cot \theta \cos \theta (\sin \theta + 1)$

EARTHQUAKE For Exercises 58 and 59, use the following information. The magnitude of an earthquake M measured on the Richter scale is given by $M = \log_{10} x$, where x represents the amplitude of the seismic wave causing ground motion. *(Lesson 10-2)*

58. How many times as great was the 1960 Chile earthquake as the 1938 Indonesia earthquake?

59. The largest aftershock of the 1964 Alaskan earthquake was 6.7 on the Richter scale. How many times as great was the main earthquake as this aftershock?

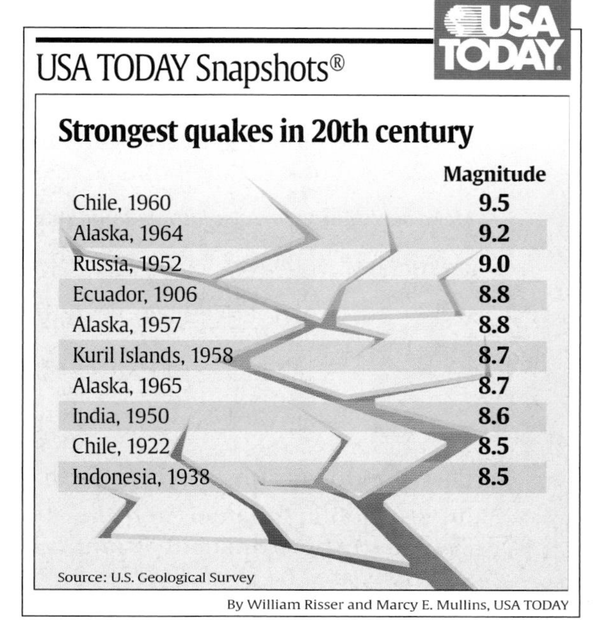

USA TODAY Snapshots®

Strongest quakes in 20th century

	Magnitude
Chile, 1960	9.5
Alaska, 1964	9.2
Russia, 1952	9.0
Ecuador, 1906	8.8
Alaska, 1957	8.8
Kuril Islands, 1958	8.7
Alaska, 1965	8.7
India, 1950	8.6
Chile, 1922	8.5
Indonesia, 1938	8.5

Source: U.S. Geological Survey

By William Risser and Marcy E. Mullins, USA TODAY

Getting Ready for the Next Lesson

PREREQUISITE SKILL Solve each equation.
*(To review **solving equations using the Zero Product Property**, see Lesson 6-3.)*

60. $(x + 6)(x - 5) = 0$

61. $(x - 1)(x + 1) = 0$

62. $x(x + 2) = 0$

63. $(2x - 5)(x + 2) = 0$

64. $(2x + 1)(2x - 1) = 0$

65. $x^2(2x + 1) = 0$

Practice Quiz 2

Lessons 14-4 through 14-6

Verify that each of the following is an identity. *(Lessons 14-5)*

1. $\sin \theta \sec \theta = \tan \theta$

2. $\sec \theta - \cos \theta = \sin \theta \tan \theta$

3. $\sin \theta + \tan \theta = \frac{\sin \theta (\cos \theta + 1)}{\cos \theta}$

Verify that each of the following is an identity. *(Lessons 14-4 and 14-5)*

4. $\sin(90° + \theta) = \cos \theta$

5. $\cos\left(\frac{3\pi}{2} - \theta\right) = -\sin \theta$

6. $\sin(\theta + 30°) + \cos(\theta + 60°) = \cos \theta$

Find the exact value of each expression by using the double-angle or half-angle formulas. *(Lesson 14-6)*

7. $\sin 2\theta$ if $\cos \theta = -\frac{\sqrt{3}}{2}$; $180° < \theta < 270°$

8. $\cos \frac{\theta}{2}$ if $\sin \theta = -\frac{9}{41}$; $270° \le \theta < 360°$

9. $\sin 165°$

10. $\cos \frac{5\pi}{8}$

Graphing Calculator Investigation

A Preview of Lesson 14-7

Solving Trigonometric Equations

The graph of a trigonometric function is made up of points that represent all values that satisfy the function. To solve a trigonometric equation, you need to find all values of the variable that satisfy the equation. You can use a TI-83 Plus to solve trigonometric equations by graphing each side of the equation as a function and then locating the points of intersection.

Example 1 **Use a graphing calculator to solve $\sin x = 0.2$ if $0° \leq x < 360°$.**

Rewrite the equation as two functions, $y = \sin x$ and $y = 0.2$. Then graph the two functions. Look for the point of intersection.

Make sure that your calculator is in degree mode to get the correct viewing window.

KEYSTROKES: [MODE] [▼] [▼] [▶] [ENTER] [WINDOW] 0 [ENTER] 360 [ENTER] 90 [ENTER] −2 [ENTER] 1 [ENTER] 1 [ENTER] [Y=] [SIN] [X,T,θ,n] [ENTER] 0.2 [ENTER] [GRAPH]

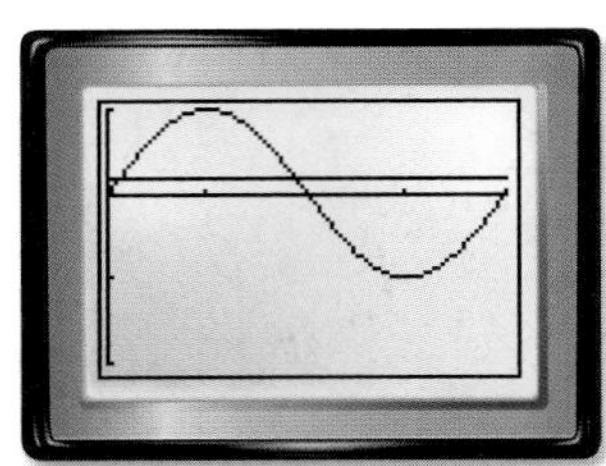

[0, 360] scl: 90 by [−2, 1] scl: 1

Based on the graph, you can see that there are two points of intersection in the interval $0° \leq x < 360°$. Use [Zoom] or [2nd] [CALC] 5 to approximate the solutions. The approximate solutions are 168.5° and 11.5°.

Like other equations you have studied, some trigonometric equations have no real solutions. Carefully examine the graphs over their respective periods for points of intersection. If there are no points of intersection, then the trigonometric equation has no real solutions.

Example 2 **Use a graphing calculator to solve $\tan^2 x \cos x + 5 \cos x = 0$ if $0° \leq x < 360°$.**

Because the tangent function is not continuous, place the calculator in Dot mode. The related functions to be graphed are $y = \tan^2 x \cos x + 5 \cos x$ and $y = 0$.

These two functions do not intersect. Therefore, the equation $\tan^2 x \cos x + 5 \cos x = 0$ has no real solutions.

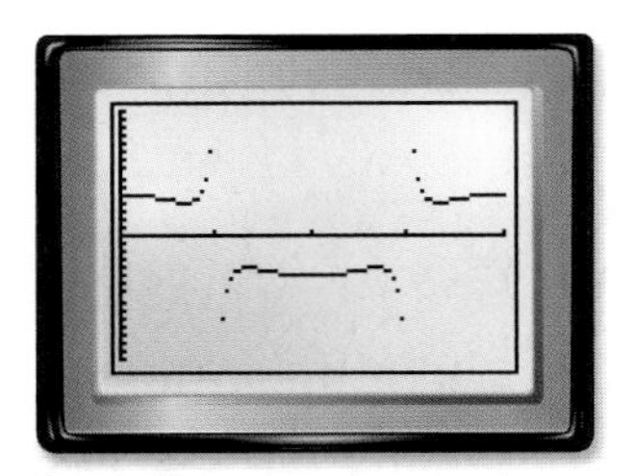

[0, 360] scl: 90 by [−15, 15] scl: 1

Exercises

Use a graphing calculator to solve each equation for the values of x indicated.

1. $\sin x = 0.8$ if $0° \leq x < 360°$
2. $\tan x = \sin x$ if $0° \leq x < 360°$
3. $2 \cos x + 3 = 0$ if $0° \leq x < 360°$
4. $0.5 \cos x = 1.4$ if $-720° \leq x < 720°$
5. $\sin 2x = \sin x$ if $0° \leq x < 360°$
6. $\sin 2x - 3 \sin x = 0$ if $-360° \leq x < 360°$

14-7 Solving Trigonometric Equations

What You'll Learn

- Solve trigonometric equations.
- Use trigonometric equations to solve real-world problems.

Vocabulary

- trigonometric equation

How can trigonometric equations be used to predict temperature?

The average daily high temperature for a region can be described by a trigonometric function. For example, the average daily high temperature for each month in Orlando, Florida, can be modeled by the function $T = 11.56 \sin (0.4516x - 1.641) + 80.89$, where T represents the average daily high temperature in degrees Fahrenheit and x represents the month of the year. This equation can be used to predict the months in which the average temperature in Orlando will be at or above a desired temperature.

SOLVE TRIGONOMETRIC EQUATIONS You have seen that trigonometric identities are true for *all* values of the variable for which the equation is defined. However, most **trigonometric equations** like some algebraic equations, are true for *some* but not *all* values of the variable.

Example 1 Solve Equations for a Given Interval

Find all solutions of each equation for the given interval.

a. $\cos^2 \theta = 1; 0° \leq \theta < 360°$

$$\cos^2 \theta = 1 \quad \text{Original equation}$$
$$\cos^2 \theta - 1 = 0 \quad \text{Solve for 0.}$$
$$(\cos \theta + 1)(\cos \theta - 1) = 0 \quad \text{Factor.}$$

Now use the Zero Product Property.

$$\cos \theta + 1 = 0 \quad \text{or} \quad \cos \theta - 1 = 0$$
$$\cos \theta = -1 \qquad \cos \theta = 1$$
$$\theta = 180° \qquad \theta = 0°$$

The solutions are 0° and 180°.

b. $\sin 2\theta = 2 \cos \theta; 0 \leq \theta < 2\pi$

$$\sin 2\theta = 2 \cos \theta \quad \text{Original equation}$$
$$2 \sin \theta \cos \theta = 2 \cos \theta \quad \sin 2\theta = 2 \sin \theta \cos \theta$$
$$2 \sin \theta \cos \theta - 2 \cos \theta = 0 \quad \text{Solve for 0.}$$
$$2 \cos \theta (\sin \theta - 1) = 0 \quad \text{Factor.}$$

(continued on the next page)

Use the Zero Product Property.

$2\cos\theta = 0$ or $\sin\theta - 1 = 0$

$\cos\theta = 0$ $\quad$ $\sin\theta = 1$

$\theta = \frac{\pi}{2}$ or $\frac{3\pi}{2}$ $\quad$ $\theta = \frac{\pi}{2}$

The solutions are $\frac{\pi}{2}$ and $\frac{3\pi}{2}$.

Trigonometric equations are usually solved for values of the variable between 0° and 360° or 0 radians and 2π radians. There are solutions outside that interval. These other solutions differ by integral multiples of the period of the function.

Example 2 Solve Trigonometric Equations

a. Solve $2\sin\theta = -1$ for all values of θ if θ is measured in radians.

$2\sin\theta = -1$ Original equation

$\sin\theta = -\frac{1}{2}$ Divide each side by 2.

Look at the graph of $y = \sin\theta$ to find solutions of $\sin\theta = -\frac{1}{2}$.

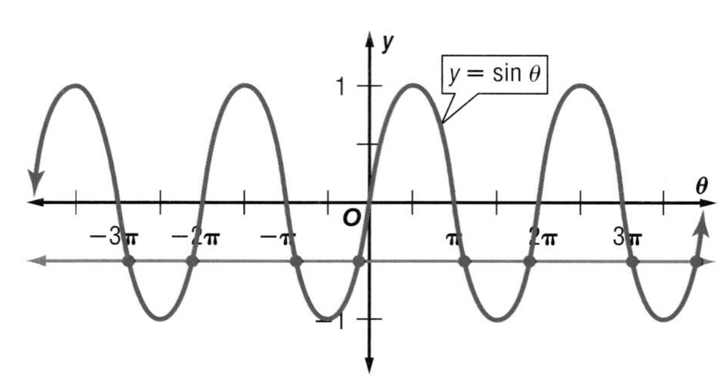

The solutions are $\frac{7\pi}{6}, \frac{11\pi}{6}, \frac{19\pi}{6}, \frac{23\pi}{6}$, and so on, and $-\frac{7\pi}{6}, -\frac{11\pi}{6}, -\frac{19\pi}{6}, -\frac{23\pi}{6}$, and so on. The only solutions in the interval 0 to 2π are $\frac{7\pi}{6}$ and $\frac{11\pi}{6}$. The period of the sine function is 2π radians. So the solutions can be written as $\frac{7\pi}{6} + 2k\pi$ and $\frac{11\pi}{6} + 2k\pi$, where k is any integer.

b. Solve $\cos 2\theta + \cos\theta + 1 = 0$ for all values of θ if θ is measured in degrees.

$\cos 2\theta + \cos\theta + 1 = 0$ Original equation

$2\cos^2\theta - 1 + \cos\theta + 1 = 0$ $\cos 2\theta = 2\cos^2\theta - 1$

$2\cos^2\theta + \cos\theta = 0$ Simplify.

$\cos\theta\,(2\cos\theta + 1) = 0$ Factor.

Solve for θ in the interval 0° to 360°.

$\cos\theta = 0$ or $2\cos\theta + 1 = 0$

$\theta = 90°$ or $270°$ $\quad$ $2\cos\theta = -1$

$\cos\theta = -\frac{1}{2}$

$\theta = 120°$ or $240°$

The solutions are $90° + k \cdot 180°$, $120° + k \cdot 360°$, and $240° + k \cdot 360°$.

Study Tip

Expressing Solutions as Multiples

The expression $90° + k \cdot 180°$ includes 270° and its multiples, so it is not necessary to list them separately.

If an equation cannot be solved easily by factoring, try rewriting the expression using trigonometric identities. However, using identities and some algebraic operations, such as squaring, may result in extraneous solutions. So, it is necessary to check your solutions using the original equation.

Example 3 Solve Trigonometric Equations Using Identities

Solve $\cos\theta\tan\theta - \sin^2\theta = 0$.

$\cos\theta\tan\theta - \sin^2\theta = 0$ Original equation

$\cos\theta\left(\frac{\sin\theta}{\cos\theta}\right) - \sin^2\theta = 0$ $\tan\theta = \frac{\sin\theta}{\cos\theta}$

$\sin\theta - \sin^2\theta = 0$ Multiply.

$\sin\theta(1 - \sin\theta) = 0$ Factor.

$\sin\theta = 0$ or $1 - \sin\theta = 0$

$\theta = 0°, 180°, \text{ or } 360°$ $\quad\sin\theta = 1$

$\theta = 90°$

CHECK

$\cos\theta\tan\theta - \sin^2\theta = 0$

$\cos 0°\tan 0° - \sin^2 0° \stackrel{?}{=} 0$ $\theta = 0°$

$1 \cdot 0 - 0 \stackrel{?}{=} 0$

$0 = 0$ ✓

$\cos\theta\tan\theta - \sin^2\theta = 0$

$\cos 180°\tan 180° - \sin^2 180° \stackrel{?}{=} 0$ $\theta = 180°$

$-1 \cdot 0 - 0 \stackrel{?}{=} 0$

$0 = 0$ ✓

$\cos\theta\tan\theta - \sin^2\theta = 0$

$\cos 360°\tan 360° - \sin^2 360° \stackrel{?}{=} 0$ $\theta = 360°$

$1 \cdot 0 - 0 \stackrel{?}{=} 0$

$0 = 0$ ✓

$\cos\theta\tan\theta - \sin^2\theta = 0$

$\cos 90°\tan 90° - \sin^2 90° \stackrel{?}{=} 0$ $\theta = 90°$

$\tan 90°$ is undefined.

Thus, 90° is not a solution.

The solution is $0° + k \cdot 180°$.

Some trigonometric equations have no solution. For example, the equation $\cos x = 4$ has no solution since all values of $\cos x$ are between -1 and 1, inclusive. Thus, the solution set for $\cos x = 4$ is empty.

Example 4 Determine Whether a Solution Exists

Solve $3\cos 2\theta - 5\cos\theta = 1$.

$3\cos 2\theta - 5\cos\theta = 1$ Original equation

$3(2\cos^2\theta - 1) - 5\cos\theta = 1$ $\cos 2\theta = 2\cos^2\theta - 1$

$6\cos^2\theta - 3 - 5\cos\theta = 1$ Multiply.

$6\cos^2\theta - 5\cos\theta - 4 = 0$ Subtract 1 from each side.

$(3\cos\theta - 4)(2\cos\theta + 1) = 0$ Factor.

$3\cos\theta - 4 = 0$ or $2\cos\theta + 1 = 0$

$3\cos\theta = 4$ $\quad 2\cos\theta = -1$

$\cos\theta = \frac{4}{3}$ $\quad \cos\theta = -\frac{1}{2}$

Not possible since $\cos\theta$ cannot be greater than 1. $\quad \theta = 120° \text{ or } 240°$

Thus, the solutions are $120° + k \cdot 360°$ and $240° + k \cdot 360°$.

USE TRIGONOMETRIC EQUATIONS Trigonometric equations are often used to solve real-world situations.

Example 5 Use a Trigonometric Equation

GARDENING Rhonda wants to wait to plant her flowers until there are at least 14 hours of daylight. The number of hours of daylight H in her town can be represented by $H = 11.45 + 6.5 \sin(0.0168d - 1.333)$, where d is the day of the year and angle measures are in radians. On what day is it safe for Rhonda to plant her flowers?

$H = 11.45 + 6.5 \sin(0.0168d - 1.333)$	Original equation
$14 = 11.45 + 6.5 \sin(0.0168d - 1.333)$	$H = 14$
$2.55 = 6.5 \sin(0.0168d - 1.333)$	Subtract 11.45 from each side.
$0.392 = \sin(0.0168d - 1.333)$	Divide each side by 6.5.
$0.403 = 0.0168d - 1.333$	$\sin^{-1} 0.392 = 0.403$
$1.736 = 0.0168d$	Add 1.333 to each side.
$103.333 = d$	Divide each side by 0.0168.

Rhonda can safely plant her flowers around the 104th day of the year, or around April 14.

Check for Understanding

Concept Check

1. **Tell** why the equation $\sec \theta = 0$ has no solutions.
2. **Explain** why the number of solutions to the equation $\sin \theta = \frac{\sqrt{3}}{2}$ is infinite.
3. **OPEN ENDED** Write an example of a trigonometric equation that has no solution.

Guided Practice

Find all solutions of each equation for the given interval.

4. $4 \cos^2 \theta = 1; 0° \leq \theta < 360°$
5. $2 \sin^2 \theta - 1 = 0; 90° < \theta < 270°$
6. $\sin 2\theta = \cos \theta; 0 \leq \theta < 2\pi$
7. $3 \sin^2 \theta - \cos^2 \theta = 0; 0 \leq \theta < \frac{\pi}{2}$

Solve each equation for all values of θ if θ is measured in radians.

8. $\cos 2\theta = \cos \theta$
9. $\sin \theta + \sin \theta \cos \theta = 0$

Solve each equation for all values of θ if θ is measured in degrees.

10. $\sin \theta = 1 + \cos \theta$
11. $2 \cos^2 \theta + 2 = 5 \cos \theta$

Solve each equation for all values of θ.

12. $2 \sin^2 \theta - 3 \sin \theta - 2 = 0$
13. $2 \cos^2 \theta + 3 \sin \theta - 3 = 0$

Application

14. **PHYSICS** According to Snell's law, the angle at which light enters water α is related to the angle at which light travels in water β by the equation $\sin \alpha = 1.33 \sin \beta$. At what angle does a beam of light enter the water if the beam travels at an angle of 23° through the water?

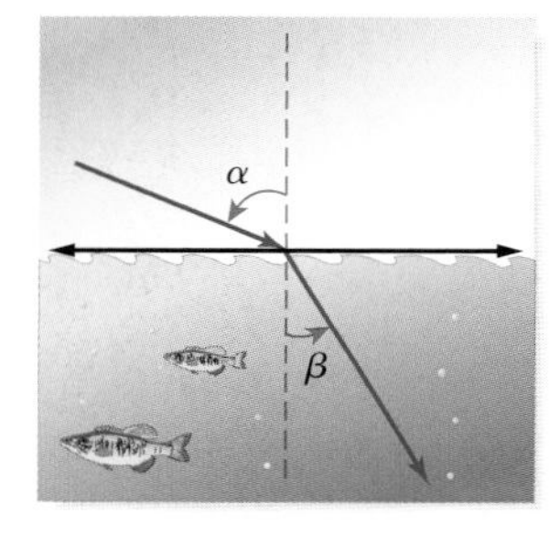

Practice and Apply

Homework Help

For Exercises	See Examples
15–22	1
23–34	2
35–40	3, 4
41–43	5

Extra Practice
See page 861.

Find all solutions of each equation for the given interval.

15. $2 \cos \theta - 1 = 0; 0° \leq \theta < 360°$

16. $2 \sin \theta = -\sqrt{3}; 180° < \theta < 360°$

17. $4 \sin^2 \theta = 1; 180° < \theta < 360°$

18. $4 \cos^2 \theta = 3; 0° \leq \theta < 360°$

19. $2 \cos^2 \theta = \sin \theta + 1; 0 \leq \theta < 2\pi$

20. $\sin^2 \theta - 1 = \cos^2 \theta; 0 \leq \theta < \pi$

21. $2 \sin^2 \theta + \sin \theta = 0; \pi < \theta < 2\pi$

22. $2 \cos^2 \theta = -\cos \theta; 0 \leq \theta < 2\pi$

Solve each equation for all values of θ if θ is measured in radians.

23. $\cos 2\theta + 3 \cos \theta - 1 = 0$

24. $2 \sin^2 \theta - \cos \theta - 1 = 0$

25. $\cos^2 \theta - \frac{5}{2} \cos \theta - \frac{3}{2} = 0$

26. $\cos \theta = 3 \cos \theta - 2$

27. $4 \cos^2 \theta - 4 \cos \theta + 1 = 0$

28. $\cos 2\theta = 1 - \sin \theta$

Solve each equation for all values of θ if θ is measured in degrees.

29. $\sin \theta = \cos \theta$

30. $\tan \theta = \sin \theta$

31. $\sin^2 \theta - 2 \sin \theta - 3 = 0$

32. $4 \sin^2 \theta - 4 \sin \theta + 1 = 0$

33. $\tan^2 \theta - \sqrt{3} \tan \theta = 0$

34. $\cos^2 \theta - \frac{7}{2} \cos \theta - 2 = 0$

Solve each equation for all values of θ.

35. $\sin^2 \theta + \cos 2\theta - \cos \theta = 0$

36. $2 \sin^2 \theta - 3 \sin \theta - 2 = 0$

37. $\sin^2 \theta = \cos^2 \theta - 1$

38. $2 \cos^2 \theta + \cos \theta = 0$

39. $\sin \frac{\theta}{2} + \cos \theta = 1$

40. $\sin \frac{\theta}{2} + \cos \frac{\theta}{2} = \sqrt{2}$

LIGHT **For Exercises 41 and 42, use the information shown.**

41. The length of the shadow S of the International Peace Memorial at Put-In-Bay, Ohio, depends upon the angle of inclination of the Sun, θ. Express S as a function of θ.

42. Find the angle of inclination θ that will produce a shadow 560 feet long.

More About. . .

Waves
In the oceans, the height and period of water waves are determined by wind velocity, the duration of the wind, and the distance the wind has blown across the water.

Source: www.infoplease.com

WAVES **For Exercises 43 and 44, use the following information.**
For a short time after a wave is created by a boat, the height of the wave can be modeled using $y = \frac{1}{2}h + \frac{1}{2}h \sin \frac{2\pi t}{P}$, where h is the maximum height of the wave in feet, P is the period in seconds, and t is the propagation of the wave in seconds.

43. If $h = 3$ and $P = 2$ seconds, write the equation for the wave and draw its graph over a 10-second interval.

44. How many times over the first 10 seconds does the graph predict the wave to be one foot high?

45. **CRITICAL THINKING** Computer games often use transformations to distort images on the screen. In one such transformation, an image is rotated counterclockwise using the equations $x' = x \cos \theta - y \sin \theta$ and $y' = x \sin \theta + y \cos \theta$. If the coordinates of an image point are (3, 4) after a 60° rotation, what are the coordinates of the preimage point?

46. **WRITING IN MATH** Answer the question that was posed at the beginning of the lesson.

How can trigonometric equations be used to predict temperature?

Include the following in your answer:

- an explanation of why the sine function can be used to model the average daily temperature, and
- an explanation of why, during one period, you might find a specific average temperature twice.

Standards Practice

Standardized Test Practice

47. Which of the following is *not* a possible solution of $0 = \sin \theta + \cos \theta \tan^2 \theta$?

(A) $\frac{3\pi}{4}$ (B) $\frac{7\pi}{4}$ (C) 2π (D) $\frac{5\pi}{2}$

48. The graph of the equation $y = 2 \cos \theta$ is shown. Which is a solution for $2 \cos \theta = 1$?

(A) $\frac{8\pi}{3}$ (B) $\frac{13\pi}{3}$

(C) $\frac{10\pi}{3}$ (D) $\frac{15\pi}{3}$

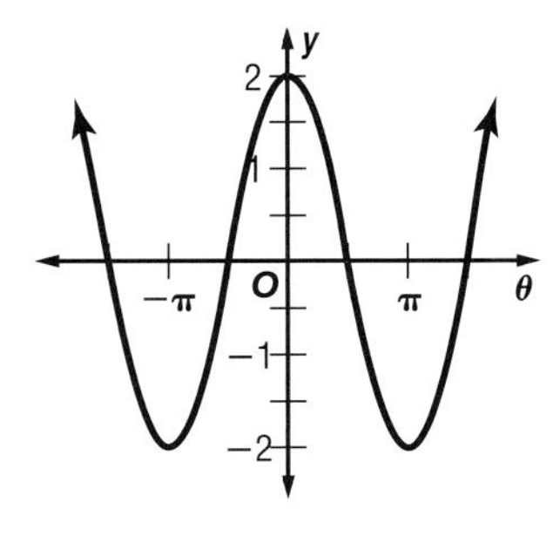

Maintain Your Skills

Mixed Review **Find the exact value of $\sin 2\theta$, $\cos 2\theta$, $\sin \frac{\theta}{2}$, and $\cos \frac{\theta}{2}$ for each of the following.** *(Lesson 14-6)*

49. $\sin \theta = \frac{3}{5}$; $0° < \theta < 90°$

50. $\cos \theta = \frac{1}{2}$; $0° < \theta < 90°$

51. $\cos \theta = \frac{5}{6}$; $0° < \theta < 90°$

52. $\sin \theta = \frac{4}{5}$; $0° < \theta < 90°$

Find the exact value of each expression. *(Lesson 14-5)*

53. $\sin 240°$

54. $\cos 315°$

55. Solve $\triangle ABC$. Round measures of sides and angles to the nearest tenth. *(Lesson 13-4)*

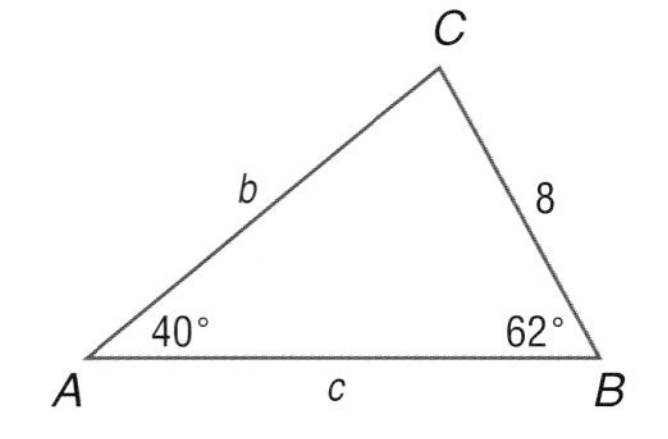

WebQuest Internet Project

Trig Class Angles for Lessons in Lit

It is time to complete your project. Use the information and data you have gathered about the applications of trigonometry to prepare a poster, report, or Web page. Be sure to include graphs, tables, or diagrams in the presentation.

www.algebra2.com/webquest

Chapter 14 Study Guide and Review

Vocabulary and Concept Check

amplitude (p. 763)
double-angle formula (p. 791)
half-angle formula (p. 793)
midline (p. 771)
phase shift (p. 769)
trigonometric equation (p. 799)
trigonometric identity (p. 777)
vertical shift (p. 771)

Choose the correct letter that best matches each phrase.

1. horizontal translation of a trigonometric function
2. a reference line about which a graph oscillates
3. vertical translation of a trigonometric function
4. the formula used to find $\cos 22\frac{1}{2}°$
5. $\sin 2\theta = 2 \sin \theta \cos \theta$
6. a measure of how long it takes for a graph to repeat itself
7. $\cos (\alpha - \beta) = \cos \alpha \cos \beta + \sin \alpha \sin \beta$
8. the absolute value of half the difference between the maximum and minimum values of a periodic function

a. amplitude
b. midline
c. period
d. vertical shift
e. double-angle formula
f. half-angle formula
g. difference of angles formula
h. phase shift

Lesson-by-Lesson Review

14-1 Graphing Trigonometric Functions

See pages 762–768.

Concept Summary

- For trigonometric functions of the form $y = a \sin b\theta$ and $y = a \cos b\theta$, the amplitude is $|a|$, and the period is $\frac{360°}{|b|}$ or $\frac{2\pi}{|b|}$.
- The period of $y = a \tan b\theta$ is $\frac{180°}{|b|}$ or $\frac{\pi}{|b|}$.

Example **Find the amplitude and period of $y = 2 \cos 4\theta$. Then graph the function.**

The amplitude is $|2|$ or 2.

The period is $\frac{360°}{|4|}$ or 90°.

Use the amplitude and period to graph the function.

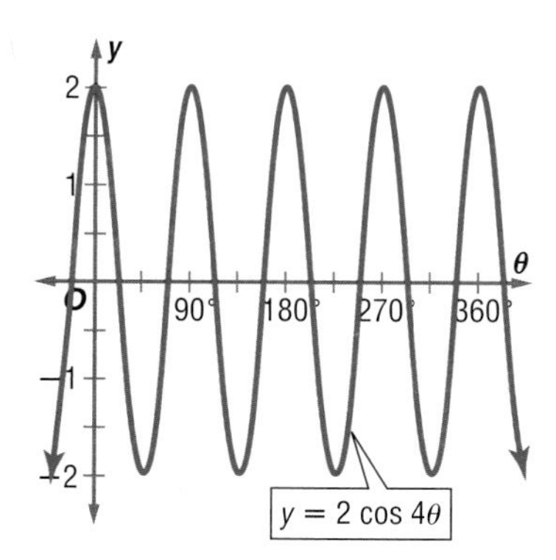

Exercises **Find the amplitude, if it exists, and period of each function. Then graph each function.** *See Example 1 on page 765.*

9. $y = -\frac{1}{2} \cos \theta$
10. $y = 4 \sin 2\theta$
11. $y = \sin \frac{1}{2}\theta$
12. $y = 5 \sec \theta$
13. $y = \frac{1}{2} \csc \frac{2}{3}\theta$
14. $y = \tan 4\theta$

14-2 Translations of Trigonometric Graphs

See pages 769–776.

Concept Summary

- For trigonometric functions of the form $y = a \sin b(\theta - h)$, $y = a \cos(\theta - h)$, and $y = a \tan(\theta - h)$, the phase shift is to the right when $h > 0$ and to the left when $h < 0$.
- For trigonometric functions of the form $y = a \sin b(\theta - h) + k$, $y = a \cos(\theta - h) + k$, and $y = a \tan(\theta - h) + k$, the vertical shift is up when $k > 0$ and down when $k < 0$.

Example **State the vertical shift, amplitude, period, and phase shift of $y = 3 \sin\left[2\left(\theta - \frac{\pi}{2}\right)\right] - 2$. Then graph the function.**

Identify the values of k, a, b, and h.

$k = -2$, so the vertical shift is –2.

$a = 3$, so the amplitude is 3.

$b = 2$, so the period is $\frac{2\pi}{|2|}$ or π.

$h = \frac{\pi}{2}$, so the phase shift is $\frac{\pi}{2}$ to the right.

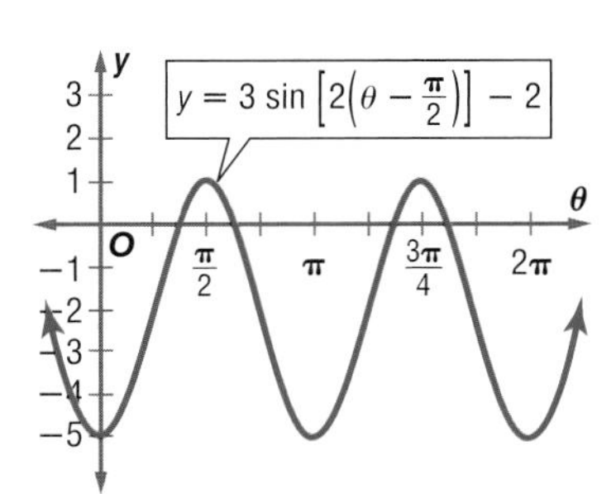

Exercises **State the vertical shift, amplitude, period, and phase shift of each function. Then graph the function.** *See Example 3 on page 772.*

15. $y = \frac{1}{2} \sin[2(\theta - 60°)] - 1$

16. $y = 2 \tan\left[\frac{1}{4}(\theta - 90°)\right] + 3$

17. $y = 3 \sec\left[\frac{1}{2}\left(\theta + \frac{\pi}{4}\right)\right] + 1$

18. $y = \frac{1}{3} \cos\left[\frac{1}{3}\left(\theta - \frac{2\pi}{3}\right)\right] - 2$

14-3 Trigonometric Identities

See pages 777–781.

Concept Summary

- Quotient Identities: $\tan\theta = \frac{\sin\theta}{\cos\theta}$, $\cot\theta = \frac{\cos\theta}{\sin\theta}$
- Reciprocal Identities: $\csc\theta = \frac{1}{\sin\theta}$, $\sec\theta = \frac{1}{\cos\theta}$, $\cot\theta = \frac{1}{\tan\theta}$
- Pythagorean Identities: $\cos^2\theta + \sin^2\theta = 1$, $\tan^2\theta + 1 = \sec^2\theta$, $\cot^2\theta + 1 = \csc^2\theta$

Example **Simplify $\sin\theta \cot\theta \cos\theta$.**

$\sin\theta \cot\theta \cos\theta = \frac{\sin\theta}{1} \cdot \frac{\cos\theta}{\sin\theta} \cdot \frac{\cos\theta}{1}$ $\cot\theta = \frac{\cos\theta}{\sin\theta}$

$= \cos^2\theta$ Multiply.

Exercises **Find the value of each expression.** *See Example 1 on page 778.*

19. $\cot\theta$, if $\csc\theta = -\frac{5}{3}$; $270° < \theta < 360°$

20. $\sec\theta$, if $\sin\theta = \frac{1}{2}$; $0° \le \theta < 90°$

Simplify each expression. *See Example 2 on page 778.*

21. $\sin\theta \csc\theta - \cos^2\theta$

22. $\cos^2\theta \sec\theta \csc\theta$

23. $\cos\theta + \sin\theta \tan\theta$

14-4 Verifying Trigonometric Identities

See pages 782–785.

Concept Summary

- Use the basic trigonometric identities to transform one or both sides of a trigonometric equation into the same form.

Example **Verify that $\tan \theta + \cot \theta = \sec \theta \csc \theta$.**

$\tan \theta + \cot \theta \stackrel{?}{=} \sec \theta \csc \theta$	Original equation
$\frac{\sin \theta}{\cos \theta} + \frac{\cos \theta}{\sin \theta} \stackrel{?}{=} \sec \theta \csc \theta$	$\tan \theta = \frac{\sin \theta}{\cos \theta}$, $\cot \theta = \frac{\cos \theta}{\sin \theta}$
$\frac{\sin^2 \theta + \cos^2 \theta}{\cos \theta \sin \theta} \stackrel{?}{=} \sec \theta \csc \theta$	Rewrite using the LCD, $\cos \theta \sin \theta$.
$\frac{1}{\cos \theta \sin \theta} \stackrel{?}{=} \sec \theta \csc \theta$	$\sin^2 \theta + \cos^2 \theta = 1$
$\frac{1}{\cos \theta} \cdot \frac{1}{\sin \theta} \stackrel{?}{=} \sec \theta \csc \theta$	Rewrite as the product of two expressions.
$\sec \theta \csc \theta = \sec \theta \csc \theta$	$\frac{1}{\cos \theta} = \sec \theta$, $\frac{1}{\sin \theta} = \csc \theta$

Exercises **Verify that each of the following is an identity.**
See Examples 1–3 on pages 782–783.

24. $\frac{\sin \theta}{\tan \theta} + \frac{\cos \theta}{\cot \theta} = \cos \theta + \sin \theta$

25. $\frac{\sin \theta}{1 - \cos \theta} = \csc \theta + \cot \theta$

26. $\cot^2 \theta \sec^2 \theta = 1 + \cot^2 \theta$

27. $\sec \theta\,(\sec \theta - \cos \theta) = \tan^2 \theta$

14-5 Sum and Difference of Angles Formulas

See pages 786–790.

Concept Summary

- For all values of α and β: $\cos (\alpha \pm \beta) = \cos \alpha \cos \beta \mp \sin \alpha \sin \beta$
  $\sin (\alpha \pm \beta) = \sin \alpha \cos \beta \pm \cos \alpha \sin \beta$

Example **Find the exact value of sin 195°.**

$\sin 195° = \sin (150° + 45°)$	$195° = 150° + 45°$
$= \sin 150° \cos 45° + \cos 150° \sin 45°$	$\alpha = 150°, \beta = 45°$
$= \left(\frac{1}{2}\right)\left(\frac{\sqrt{2}}{2}\right) + \left(-\frac{\sqrt{3}}{2}\right)\left(\frac{\sqrt{2}}{2}\right)$	Evaluate each expression.
$= \frac{\sqrt{2} - \sqrt{6}}{4}$	Simplify.

Exercises **Find the exact value of each expression.** *See Example 1 on page 787.*

28. $\cos 15°$ **29.** $\cos 285°$ **30.** $\sin 150°$

31. $\sin 195°$ **32.** $\cos (-210°)$ **33.** $\sin (-105°)$

Verify that each of the following is an identity. *See Example 3 on page 788.*

34. $\cos (90° + \theta) = -\sin \theta$

35. $\sin (30° - \theta) = \cos (60° + \theta)$

36. $\sin (\theta + \pi) = -\sin \theta$

37. $-\cos \theta = \cos (\pi + \theta)$

• Extra Practice, see pages 859–861.
• Mixed Problem Solving, see page 875.

14-6 Double-Angle and Half-Angle Formulas

See pages 791–797.

Concept Summary

- Double-angle formulas: $\sin 2\theta = 2 \sin \theta \cos \theta$, $\cos 2\theta = \cos^2 \theta - \sin^2 \theta$, $\cos 2\theta = 1 - 2 \sin^2 \theta$, $\cos 2\theta = 2 \cos^2 \theta - 1$
- Half-angle formulas: $\sin \frac{\alpha}{2} = \pm\sqrt{\frac{1 - \cos \alpha}{2}}$, $\cos \frac{\alpha}{2} = \pm\sqrt{\frac{1 + \cos \alpha}{2}}$

Example **Verify that $\csc 2\theta = \frac{\sec \theta}{2 \sin \theta}$ is an identity.**

$\csc 2\theta \stackrel{?}{=} \frac{\sec \theta}{2 \sin \theta}$ Original equation

$\frac{1}{\sin 2\theta} \stackrel{?}{=} \frac{\frac{1}{\cos \theta}}{2 \sin \theta}$ $\csc \theta = \frac{1}{\sin \theta}$, $\sec \theta = \frac{1}{\cos \theta}$

$\frac{1}{\sin 2\theta} \stackrel{?}{=} \frac{1}{2 \sin \theta \cos \theta}$ Simplify the complex fraction.

$\frac{1}{\sin 2\theta} = \frac{1}{\sin 2\theta}$ $2 \sin \theta \cos \theta = \sin 2\theta$

Exercises **Find the exact values of $\sin 2\theta$, $\cos 2\theta$, $\sin \frac{\theta}{2}$, and $\cos \frac{\theta}{2}$ for each of the following.** *See Examples 1 and 2 on pages 792 and 793.*

38. $\sin \theta = \frac{1}{4}$; $0° < \theta < 90°$

39. $\sin \theta = -\frac{5}{13}$; $180° < \theta < 270°$

40. $\cos \theta = -\frac{5}{17}$; $90° < \theta < 180°$

41. $\cos \theta = \frac{12}{13}$; $270° < \theta < 360°$

14-7 Solving Trigonometric Equations

See pages 799–804.

Concept Summary

- Solve trigonometric equations by factoring or by using trigonometric identities.

Example **Solve $\sin 2\theta + \sin \theta = 0$ if $0° \leq \theta < 360°$.**

$\sin 2\theta + \sin \theta = 0$ Original equation

$2 \sin \theta \cos \theta + \sin \theta = 0$ $\sin 2\theta = 2 \sin \theta \cos \theta$

$\sin \theta (2 \cos \theta + 1) = 0$ Factor.

$\sin \theta = 0$ or $2 \cos \theta + 1 = 0$

$\theta = 0°$ or $180°$ $\theta = 120°$ or $240°$

Exercises **Find all solutions of each equation for the interval $0° \leq \theta < 360°$.** *See Example 1 on page 799.*

42. $2 \sin 2\theta = 1$

43. $2 \cos^2 \theta + \sin^2 \theta = 2 \cos \theta$

Solve each equation for all values of θ if θ is measured in radians. *See Example 2 on page 800.*

44. $6 \sin^2 \theta - 5 \sin \theta - 4 = 0$

45. $2 \cos^2 \theta = 3 \sin \theta$

Chapter 14 Practice Test

Vocabulary and Concepts

Choose the correct term to complete each sentence.

1. The (*period*, *phase shift*) of $y = 3 \sin 2(\theta - 60°) + 2$ is 180°.
2. A midline is used with a (*phase shift*, *vertical shift*) of a trigonometric function.
3. The amplitude of $y = \frac{1}{3} \cos [3(\theta + 4)] - 1$ is $\left(\frac{1}{3}, 3\right)$.
4. The (*cosine*, *cosecant*) has no amplitude.

Skills and Applications

State the vertical shift, amplitude, period, and phase shift of each function. Then graph the function.

5. $y = \frac{2}{3} \sin 2\theta + 5$
6. $y = 4 \cos \left[\frac{1}{2}(\theta + 30°)\right] - 1$

Find the value of each expression.

7. $\tan \theta$, if $\sin \theta = \frac{1}{2}$; $90° < \theta < 180°$
8. $\sec \theta$, if $\cot \theta = \frac{3}{4}$; $180° < \theta < 270°$

Verify that each of the following is an identity.

9. $(\sin \theta - \cos \theta)^2 = 1 - \sin 2\theta$
10. $\frac{\cos \theta}{1 - \sin^2 \theta} = \sec \theta$
11. $\frac{\sec \theta}{\sin \theta} - \frac{\sin \theta}{\cos \theta} = \cot \theta$
12. $\frac{1 + \tan^2 \theta}{\cos^2 \theta} = \sec^4 \theta$

Find the exact value of each expression.

13. $\cos 285°$
14. $\sin 345°$
15. $\sin (-225°)$
16. $\cos 480°$
17. $\cos 67.5°$
18. $\sin 75°$

Solve each equation for all values of θ if θ is measured in degrees.

19. $\sec \theta = 1 + \tan \theta$
20. $\cos 2\theta = \cos \theta$
21. $\cos 2\theta + \sin \theta = 1$
22. $\sin \theta = \tan \theta$

GOLF For Exercises 23 and 24, use the following information.
A golf ball is hit with an initial velocity of 100 feet per second. The distance the ball travels is found by the formula $d = \frac{v_0^2}{g} \sin 2\theta$, where v_0 is the initial velocity, g is the acceleration due to gravity, 32 feet per second squared, and θ is the measurement of the angle that the path of the ball makes with the ground.

23. Find the distance that the ball travels if the angle between the path of the ball and the ground measures 60°.
24. If a ball travels 312.5 feet, what was the angle the path of the ball made with the ground to the nearest degree?

Standards Practice

25. **STANDARDIZED TEST PRACTICE** Identify the equation of the graphed function.

    (A) $y = 3 \cos 2\theta$
    (B) $y = \frac{1}{3} \cos 2\theta$
    (C) $y = 3 \cos \frac{1}{2}\theta$
    (D) $y = \frac{1}{3} \cos \frac{1}{2}\theta$

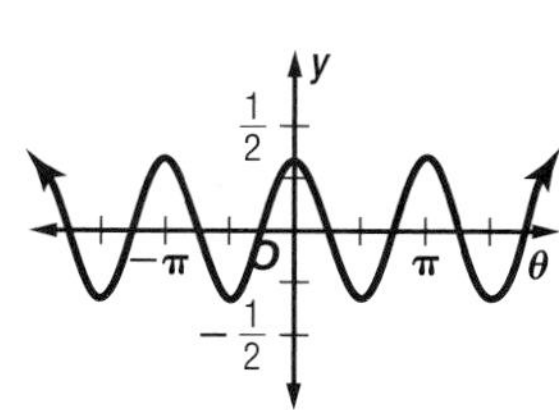

Chapter 14 Standardized Test Practice

Standards Practice

Part 1 Multiple Choice

Record your answers on the answer sheet provided by your teacher or on a sheet of paper.

1. Which of the following is *not* equal to 3.5×10^{-2}?

(A) $\frac{35}{1000}$ (B) 0.035

(C) $\frac{7}{200}$ (D) (0.5)(0.007)

2. The sum of five consecutive odd integers is 55. What is the sum of the greatest and least of these integers?

(A) 11 (B) 22 (C) 26 (D) 30

3. If 8 bananas cost a cents and 6 oranges cost b cents, what is the cost of 2 bananas and 2 oranges in terms of a and b?

(A) $\frac{ab}{12}$ (B) $3a + \frac{b}{3}$

(C) $3a + 4b$ (D) $\frac{3a + 4b}{12}$

4. A bag contains 16 peppermint candies, 10 butterscotch candies, and 8 cherry candies. Emma chooses one piece at random, puts it in her pocket, and then repeats the process. If she has chosen 3 peppermint candies, 2 butterscotch candies, and 1 cherry candy, what is the probability that the next piece of candy she chooses will be cherry?

(A) $\frac{7}{34}$ (B) $\frac{8}{34}$ (C) $\frac{1}{4}$ (D) $\frac{3}{4}$

5. What is the value of $\frac{\sin \frac{\pi}{6}}{\cos \frac{2\pi}{3}}$?

(A) $-\sqrt{3}$ (B) -1 (C) $-\frac{\sqrt{3}}{3}$ (D) 1

6. In right triangle QRS, what is the value of $\tan R$?

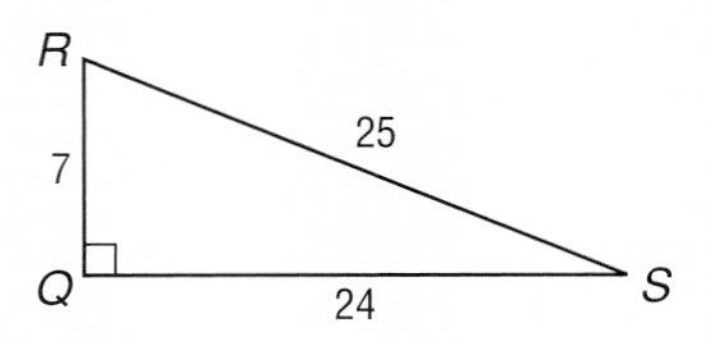

(A) $\frac{7}{25}$ (B) $\frac{7}{24}$

(C) $\frac{25}{24}$ (D) $\frac{24}{7}$

7. What is the value of $\sin\left(\cos^{-1}\frac{1}{3}\right)$ in Quadrant I?

(A) $\frac{2}{3}$ (B) $\frac{2\sqrt{2}}{3}$

(C) $\frac{\sqrt{2}}{3}$ (D) $\frac{\sqrt{6}}{3}$

8. What is the least positive value for x where $y = \sin 2x$ reaches its minimum?

(A) $\frac{\pi}{2}$ (B) π

(C) $\frac{3\pi}{4}$ (D) $\frac{3\pi}{2}$

9. Which of the following is equivalent to $\frac{\sin^2 \theta + \cos^2 \theta}{\sec^2 \theta}$?

(A) $\cos^2 \theta$ (B) $\sin^2 \theta$

(C) $\tan^2 \theta$ (D) $\sin^2 \theta + 1$

10. If $\cos \theta = -\frac{1}{2}$ and θ is in Quadrant II, what is the value of $\sin 2\theta$?

(A) $\frac{1}{2}$ (B) $-\frac{1}{2}$

(C) $\frac{\sqrt{3}}{2}$ (D) $-\frac{\sqrt{3}}{2}$

Preparing for Standardized Tests
For test-taking strategies and more practice, see pages 877–892.

Part 2 Short Response/Grid In

Record your answers on the answer sheet provided by your teacher or on a sheet of paper.

11. If k is a positive integer, and $7k + 3$ equals a prime number that is less than 50, then what is one possible value of $7k + 3$?

12. It costs \$8 to make a book. The selling price will include an additional 200%. What will be the selling price?

13. The mean of seven numbers is 0. The sum of three of the numbers is -9. What is the sum of the remaining four numbers?

14. If $4a - 6b = 0$ and $c = 9b$, what is the ratio of a to c?

15. What is the value of x if $\frac{3^3 \cdot 3}{\sqrt{81}} = 3^x$?

16. The ages of children at a party are 6, 7, 6, 6, 7, 7, 8, 6, 7, 8, 9, 7, and 7. Let N represent the median of their ages and m represent the mode. What is $N - m$?

17. In the figure below, $CEFG$ is a square, ABD is a right triangle, D is the midpoint of side CE, H is the midpoint of side CG, and C is the midpoint of side BD. $BCDE$ is a line segment, and AHD is a line segment. If the measure of the area of square $CEFG$ is 16, what is the measure of the area of quadrilateral $ABCH$?

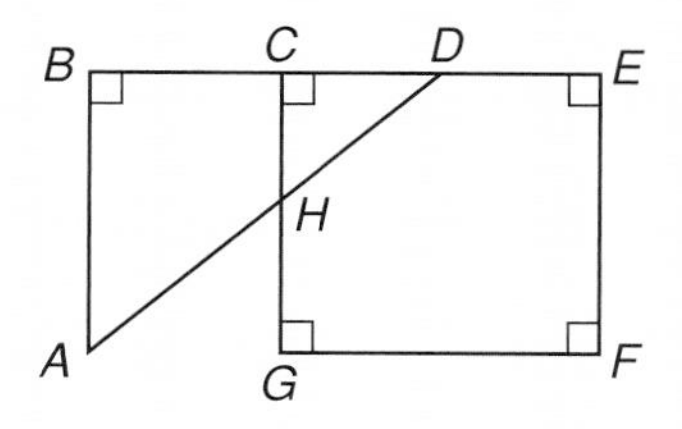

Test-Taking Tip

Always write down your calculations on scrap paper or in the test booklet, even if you think you can do the calculations in your head. Writing down your calculations will help you avoid making simple mistakes.

18. A line with a slope of $\frac{3}{8}$ passes through $(6, 4n)$ and $(0, n)$. What is the value of n?

19. If $\sin 60° = \frac{\sqrt{3}}{2}$, what is the value of $\sin^2 30° + \cos^2 30°$?

20. A and B are both positive acute angles. If $\cos A = \frac{12}{13}$ and $\sin B = \frac{9}{41}$, find the value of $\sin (A + B)$.

21. The diagram below shows a portion of a graph of a trigonometric function. Write an equation that could represent the graph.

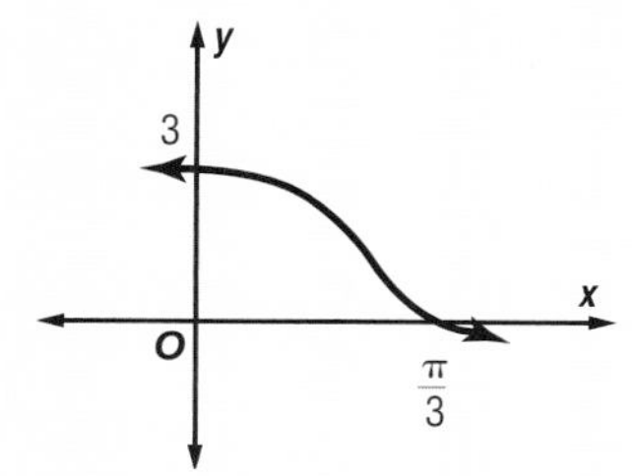

Part 3 Extended Response

Record your answers on a sheet of paper. Show your work.

For Exercises 22–24, use the following information.

The outdoor temperature T varies with the time of day. During the course of a certain day, the outdoor temperature can be modeled by the equation $T = 16.9 \sin 6.9t + 59.9$, where t is the number of hours since midnight.

22. Use the equation to find the maximum temperature of the day. At about what time was that maximum reached?

23. Use the equation to find the minimum temperature of the day. At about what time was that minimum reached?

24. The air conditioning in a home operates when the outside temperature is 75°F or above. Determine, to the nearest tenth of an hour, the amount of time in this day that the air conditioning is on in the home.

www.algebra2.com/standardized_test/ca

Student Handbook

Skills

Prerequisite Skills

Reference

Prerequisite Skills

1 Comparing and Ordering Real Numbers

- To determine which of two real numbers is greater, express each number as a decimal. Then compare the numbers.

Example 1 **Replace each ● with <, >, or = to make a true sentence.**

a. $\frac{4}{7}$ ● $0.\overline{5}$

$\frac{4}{7} \approx 0.57$ Round to the nearest hundredth.

$0.\overline{5} \approx 0.56$

Since $0.57 > 0.56$, $\frac{4}{7} > 0.\overline{5}$.

b. $\frac{1}{8}$ ● $\frac{1}{\sqrt{18}}$

$\frac{1}{8} = 0.125$ Use a calculator to find a rational approximation of $\frac{1}{\sqrt{18}}$.

$\frac{1}{\sqrt{18}} \approx 0.236$

Since $0.125 < 0.236$, $\frac{1}{8} < \frac{1}{\sqrt{18}}$.

- To order real numbers, first express each number as a decimal. Then write the decimals in order from least to greatest, and write the corresponding real numbers in the same order.

Example 2 **Order $2.\overline{54}$, $-\sqrt{6}$, $\frac{9}{4}$, $\frac{22}{-9}$ from least to greatest.**

$2.\overline{54} = 2.545454\ldots$ or about 2.55

$-\sqrt{6} = -2.44948974\ldots$ or about -2.45

$\frac{9}{4} = 2.25$

$\frac{22}{-9} = -2.4444\ldots$ or about -2.44

$-2.45 < -2.44 < 2.25 < 2.55$

Thus, the order from least to greatest is $-\sqrt{6}$, $\frac{22}{-9}$, $\frac{9}{4}$, $2.\overline{54}$.

Exercises **Replace each ● with <, >, or = to make a true sentence.**

1. $\frac{2}{3}$ ● 0.6
2. 0.35 ● $\frac{3}{8}$
3. $-\frac{3}{5}$ ● $-\frac{5}{7}$
4. $\frac{1}{39}$ ● $-\frac{2}{15}$
5. $0.\overline{4}$ ● $\frac{4}{9}$
6. $\frac{7}{11}$ ● $0.\overline{63}$
7. $\sqrt{5}$ ● $2\frac{1}{4}$
8. $\frac{1}{\sqrt{3}}$ ● $\frac{5}{9}$
9. $\frac{1}{\sqrt{2}}$ ● $\frac{11}{12}$

Order each set of numbers from least to greatest.

10. 0.1, 0.01, $\frac{3}{10}$, 0.2
11. $\frac{1}{3}$, 0.3, 0.4, $\frac{1}{4}$
12. 26.1, 26, 25.9, $\frac{181}{7}$
13. $\frac{8}{9}$, 0.89, $0.8\overline{9}$, $\frac{8}{11}$
14. 1.32, $-\sqrt{3}$, $\frac{4}{3}$, $\frac{-15}{11}$
15. $\frac{9}{2}$, $\frac{40}{-9}$, $-4.\overline{05}$, $\sqrt{18}$
16. $7.\overline{8}$, $8.\overline{7}$, $8.\overline{8}$, $8.\overline{78}$
17. 3.04, $4.0\overline{3}$, $3.0\overline{4}$, 4.03

Explain the difference between the following numbers and arrange them in order from least to greatest.

18. $3.\overline{54}$, $3.5\overline{4}$, 3.5, 3.54, $3.\overline{5}$
19. $2.\overline{987}$, $2.98\overline{7}$, 2.987, $2.9\overline{87}$, $2.\overline{98}$
20. 8.6, $8.6\overline{7}$, $8\frac{2}{3}$, 8.7676…
21. $-4.\overline{10}$, $-4\frac{1}{9}$, $-4.121231234\ldots$, $-4.1\overline{2}$

2 Factoring Polynomials

- Some polynomials can be factored using the Distributive Property.

Example 1 **Factor $4a^2 + 8a$.**

Find the GCF of $4a^2$ and $8a$.

$4a^2 = 2 \cdot 2 \cdot a \cdot a$

$8a = 2 \cdot 2 \cdot 2 \cdot a$

GCF: $2 \cdot 2 \cdot a$ or $4a$

$4a^2 + 8a = 4a(a) + 4a(2)$ Rewrite each term using the GCF.

$= 4a(a + 2)$ Distributive Property

Thus, the completely factored form of $4a^2 + 8a$ is $4a(a + 2)$.

- To factor quadratic trinomials of the form $x^2 + bx + c$, find two integers m and n whose product is equal to c and whose sum is equal to b. Then write $x^2 + bx + c$ using the pattern $(x + m)(x + n)$.

Example 2 **Factor each polynomial.**

a. $x^2 + 5x + 6$ ← Both b and c are positive.

In this trinomial, b is 5 and c is 6.

Find two numbers whose product is 6 and whose sum is 5.

Factors of 6	Sum of Factors
1, 6	7
2, 3	5

The correct factors are 2 and 3.

$x^2 + 5x + 6 = (x + m)(x + n)$ Write the pattern.

$= (x + 2)(x + 3)$ $m = 2$ and $n = 3$

CHECK Multiply the binomials to check the factorization.

$(x + 2)(x + 3) = x^2 + 3x + 2x + 2(3)$ FOIL

$= x^2 + 5x + 6$ ✓

b. $x^2 - 8x + 12$ ← b is negative and c is positive.

In this trinomial, $b = -8$ and $c = 12$. This means that $m + n$ is negative and mn is positive. So m and n must both be negative.

Factors of 12	Sum of Factors
−1, −12	−13
−2, −6	−8

The correct factors are −2 and −6.

$x^2 - 8x + 12 = (x + m)(x + n)$ Write the pattern.

$= [x + (-2)][x + (-6)]$ $m = -2$ and $n = -6$

$= (x - 2)(x - 6)$ Simplify.

c. $x^2 + 14x - 15$ ← b is positive and c is negative.

In this trinomial, $b = 14$ and $c = -15$. This means that $m + n$ is positive and mn is negative. So either m or n must be negative, but not both.

Factors of −15	Sum of Factors
1, −15	−14
−1, 15	14

The correct factors are −1 and 15.

$x^2 + 14x - 15 = (x + m)(x + n)$ Write the pattern.

$= [x + (-1)](x + 15)$ $m = -1$ and $n = 15$

$= (x - 1)(x + 15)$ Simplify.

- To factor quadratic trinomials of the form $ax^2 + bx + c$, find two integers m and n whose product is equal to ac and whose sum is equal to b. Write $ax^2 + bx + c$ using the pattern $ax^2 + mx + nx + c$. Then factor by grouping.

Example 3 **Factor $6x^2 + 7x - 3$.**

In this trinomial, $a = 6$, $b = 7$ and $c = -3$.
Find two numbers whose product is $6 \cdot (-3)$ or -18 and whose sum is 7.

Factors of −18	Sum of Factors
1, −18	−17
−1, 18	17
2, −9	−7
−2, 9	7

The correct factors are −2 and 9.

$6x^2 + 7x - 3 = 6x^2 + mx + nx - 3$ Write the pattern.
$= 6x^2 + (-2)x + 9x - 3$ $m = -2$ and $n = 9$
$= (6x^2 - 2x) + (9x - 3)$ Group terms with common factors.
$= 2x(3x - 1) + 3(3x - 1)$ Factor the GCF from each group.
$= (2x + 3)(3x - 1)$ Distributive Property

- Here are some special products.

Perfect Square Trinomials

$(a + b)^2 = (a + b)(a + b)$
$= a^2 + 2ab + b^2$

$(a - b)^2 = (a - b)(a - b)$
$= a^2 - 2ab + b^2$

Difference of Squares

$a^2 - b^2 = (a + b)(a - b)$

Example 4 **Factor each polynomial.**

a. $4x^2 + 20x + 25$

The first and last terms are perfect squares.
The middle term is equal to $2(2x)(5)$.
This is a perfect square trinomial of the form $(a + b)^2$.

$4x^2 + 20x + 25 = (2x)^2 + 2(2x)(5) + 5^2$ Write as $a^2 + 2ab + b^2$.
$= (2x + 5)^2$ Factor using the pattern.

b. $x^2 - 4$

This is a difference of squares.

$x^2 - 4 = x^2 - (2)^2$ Write in the form $a^2 - b^2$.
$= (x + 2)(x - 2)$ Factor the difference of squares.

Exercises **Factor the following polynomials.**

1. $12x^2 + 4x$
2. $6x^2y + 2x$
3. $8ab^2 - 12ab$
4. $x^2 + 5x + 4$
5. $y^2 + 12y + 27$
6. $x^2 + 6x + 8$
7. $3y^2 + 13y + 4$
8. $7x^2 + 51x + 14$
9. $3x^2 + 28x + 32$
10. $x^2 - 5x + 6$
11. $y^2 - 5y + 4$
12. $6x^2 - 13x + 5$
13. $6a^2 - 50ab + 16b^2$
14. $11x^2 - 78x + 7$
15. $18x^2 - 31xy + 6y^2$
16. $x^2 + 4xy + 4y^2$
17. $9x^2 - 24x + 16$
18. $4a^2 + 12ab + 9b^2$
19. $x^2 - 144$
20. $4c^2 - 9$
21. $16y^2 - 1$
22. $25x^2 - 4y^2$
23. $36y^2 - 16$
24. $9a^2 - 49b^2$

3 Congruent and Similar Figures

Congruent figures have the same size and the same shape.

- Two polygons are congruent if their corresponding sides are congruent and their corresponding angles are congruent.

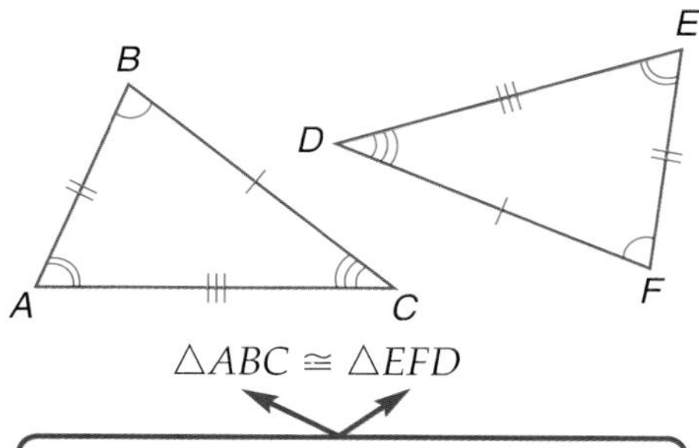

Congruent Angles	Congruent Sides
$\angle A \cong \angle E$	$\overline{AB} \cong \overline{EF}$
$\angle B \cong \angle F$	$\overline{BC} \cong \overline{FD}$
$\angle C \cong \angle D$	$\overline{AC} \cong \overline{ED}$

$\triangle ABC \cong \triangle EFD$

The order of the vertices indicates the corresponding parts.

Read the symbol $\cong$ as *is congruent to.*

Example 1 **If $\triangle XYZ \cong \triangle PQR$, name the congruent angles and sides.**

Name the pairs of congruent angles by looking at the order of the vertices in the statement $\triangle XYZ \cong \triangle PQR$.

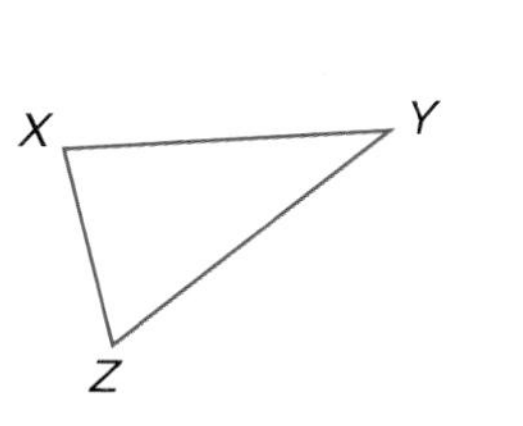

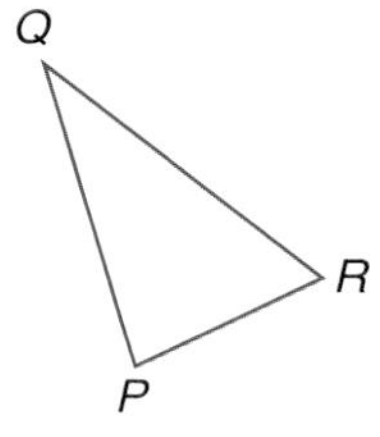

So, $\angle X \cong \angle P$, $\angle Y \cong \angle Q$, and $\angle Z \cong \angle R$.

Since X corresponds to P, and Y corresponds to Q, $\overline{XY} \cong \overline{PQ}$.

Since Y corresponds to Q, and Z corresponds to R, $\overline{YZ} \cong \overline{QR}$.

Since Z corresponds to R, and X corresponds to P, $\overline{ZX} \cong \overline{RP}$.

Example 2 **The corresponding parts of two congruent triangles are marked on the figure. Write a congruence statement for the two triangles.**

List the congruent angles and sides.

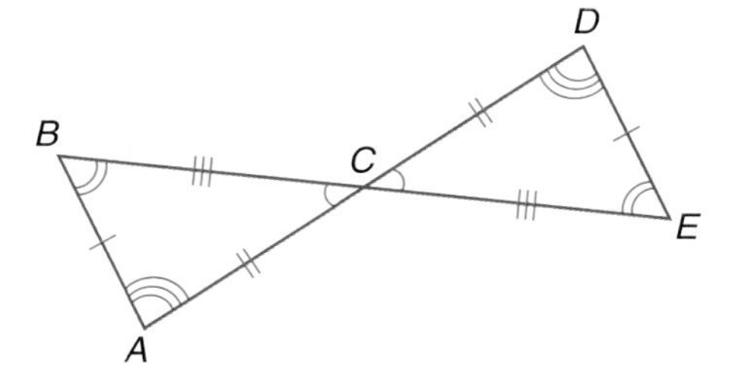

$\angle A \cong \angle D$	$\overline{AB} \cong \overline{DE}$
$\angle B \cong \angle E$	$\overline{AC} \cong \overline{DC}$
$\angle ACB \cong \angle DCE$	$\overline{BC} \cong \overline{EC}$

Match the vertices of the congruent angles. Therefore, $\triangle ABC \cong \triangle DEC$.

Similar figures have the same shape, but not necessarily the same size.

- In similar figures, corresponding angles are congruent, and the measures of corresponding sides are proportional. (They have equivalent ratios.)

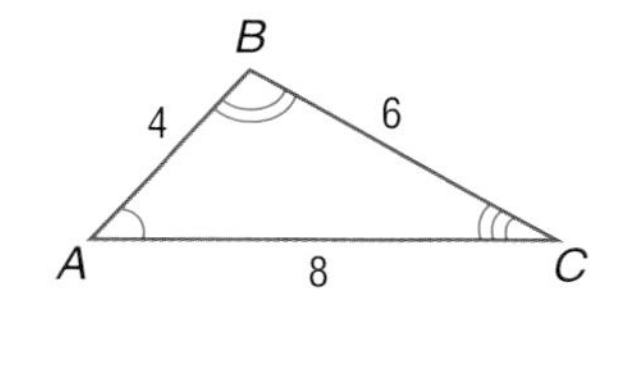

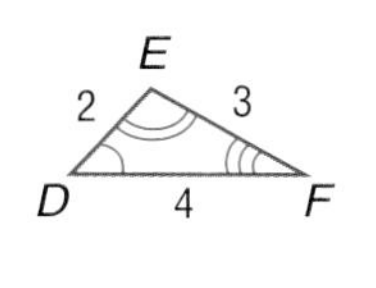

Congruent Angles

$\angle A \cong \angle D$, $\angle B \cong \angle E$, $\angle C \cong \angle F$

Proportional Sides

$\frac{AB}{DE} = \frac{BC}{EF} = \frac{AC}{DF}$

$\triangle ABC \sim \triangle DEF$

Read the symbol $\sim$ as *is similar to.*

Example 3 **Determine whether the polygons are similar. Justify your answer.**

a.

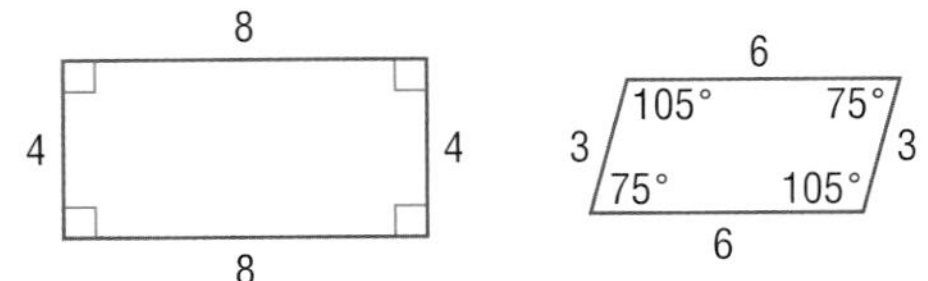

Since $\frac{4}{3} = \frac{8}{6} = \frac{4}{3} = \frac{8}{6}$, the measures of the sides of the polygons are proportional. However, the corresponding angles are not congruent. The polygons are not similar.

b.

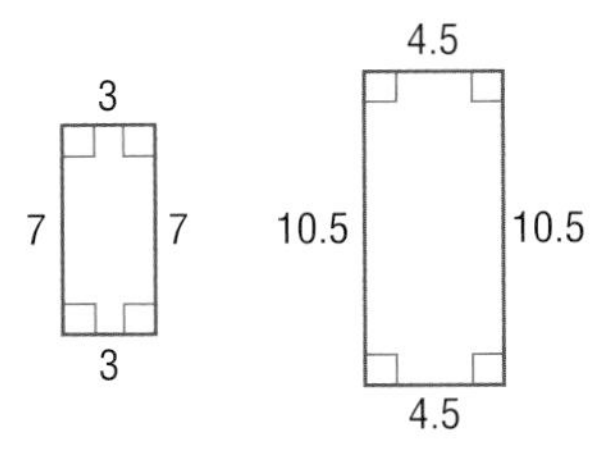

Since $\frac{7}{10.5} = \frac{3}{4.5} = \frac{7}{10.5} = \frac{3}{4.5}$, the measures of the sides of the polygons are proportional. The corresponding angles are congruent. Therefore, the polygons are similar.

Example 4 **The triangles are similar. Find the values of x and y.**

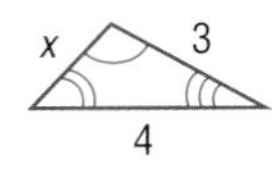

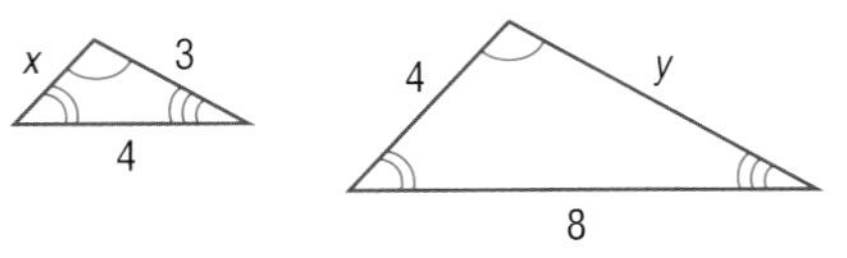

Write proportions using corresponding parts. Then solve to find the missing measures.

$\frac{x}{4} = \frac{4}{8}$	Definition of similar polygons	$\frac{3}{y} = \frac{4}{8}$	Definition of similar polygons
$x(8) = 4(4)$	Cross products	$3(8) = y(4)$	Cross products
$8x = 16$	Simplify.	$24 = 4y$	Simplify.
$\frac{8x}{8} = \frac{16}{8}$	Divide each side by 8.	$\frac{24}{4} = \frac{4y}{4}$	Divide each side by 4.
$x = 2$	Simplify.	$6 = y$	Simplify.

Example 5 CIVIL ENGINEERING **The city of Mansfield plans to build a bridge across Pine Lake. Use the information in the diagram to find the distance across Pine Lake.**

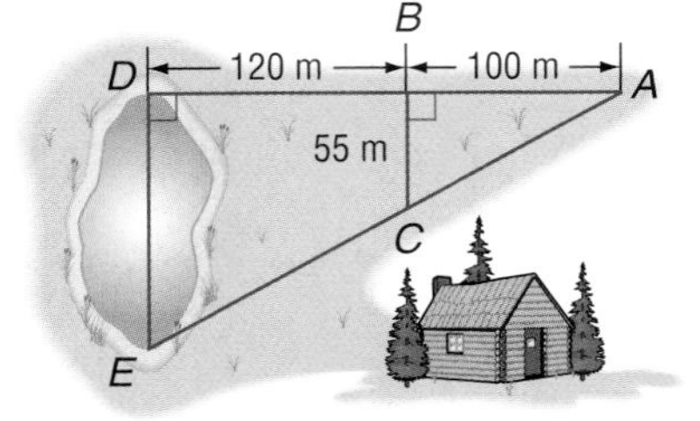

$\triangle ABC \sim \triangle ADE$

$\frac{AB}{AD} = \frac{BC}{DE}$	Definition of similar polygons
$\frac{100}{220} = \frac{55}{DE}$	$AB = 100$, $AD = 100 + 120 = 220$, $BC = 55$
$100DE = 220(55)$	Cross products
$100DE = 12{,}100$	Simplify.
$DE = 121$	Divide each side by 100.

The distance across the lake is 121 meters.

Exercises **If $\triangle GHI \cong \triangle JKL$, name the part that is congruent to each angle or segment.**

1. $\angle K$
2. $\overline{HI}$
3. $\angle I$
4. $\angle J$
5. $\overline{JK}$
6. $\overline{GI}$

Complete each congruence statement.

7.

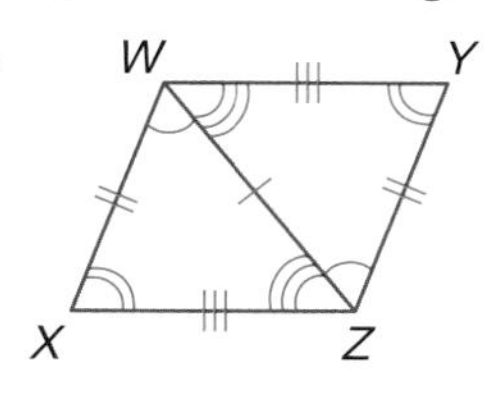

$\triangle XWZ \cong \triangle$ __?__

8.

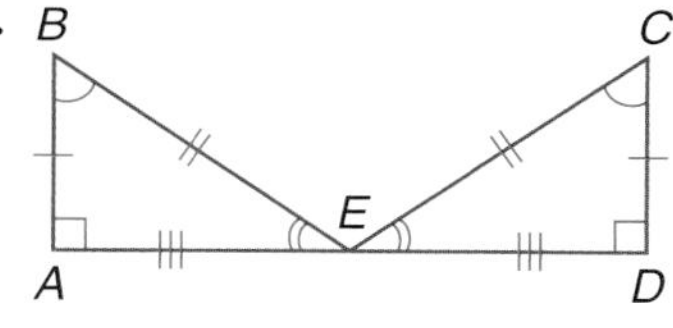

$\triangle ABE \cong \triangle$ __?__

9.

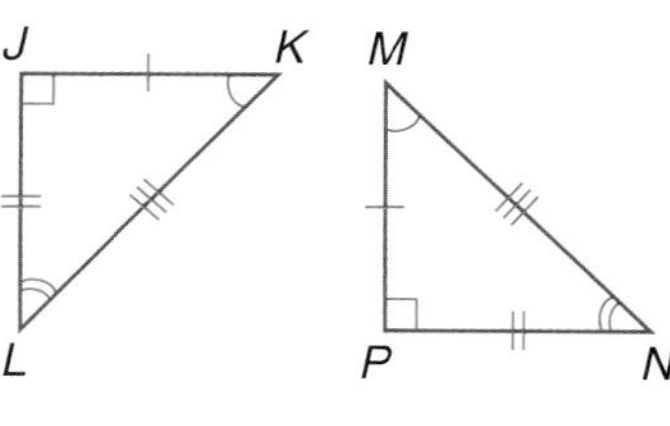

$\triangle JKL \cong \triangle$ __?__

Determine whether each pair of figures is *similar*, *congruent*, or *neither*.

10.

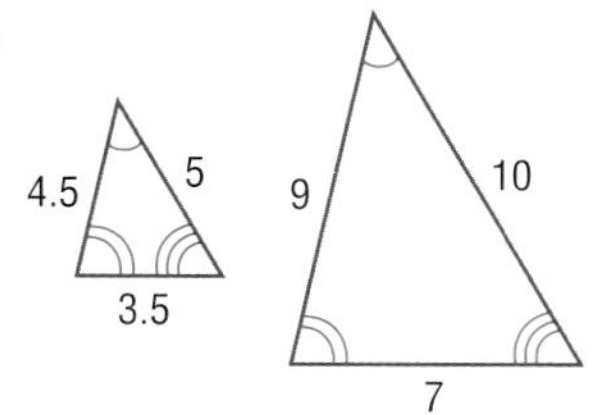

11.

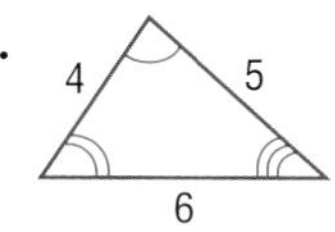

12.

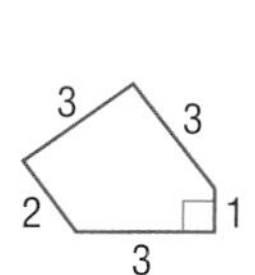

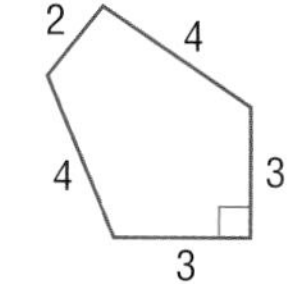

13.

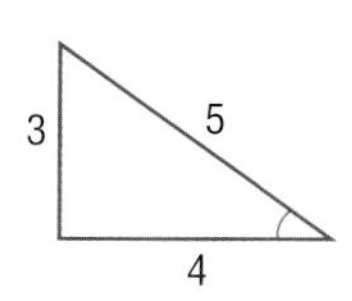

14.

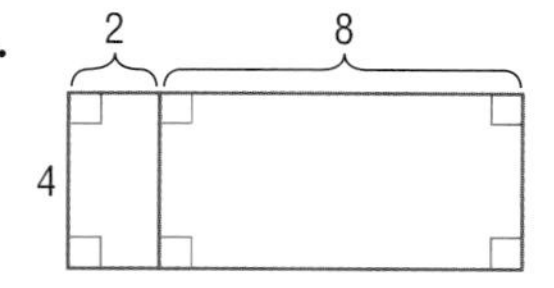

15.

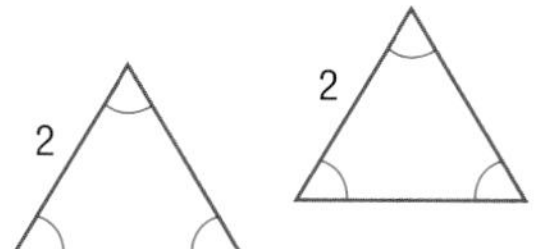

16.

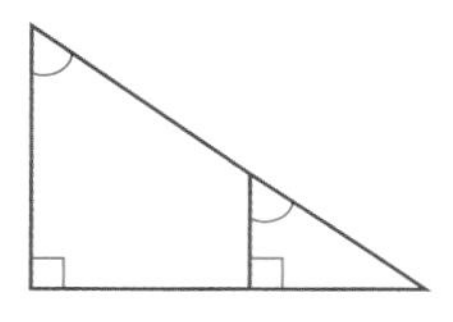

17.

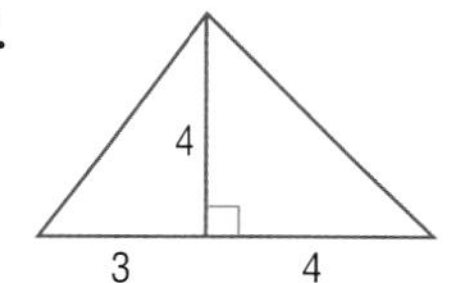

18.

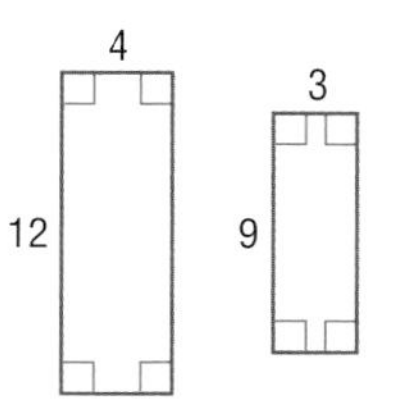

Each pair of polygons is similar. Find the values of x and y.

19.

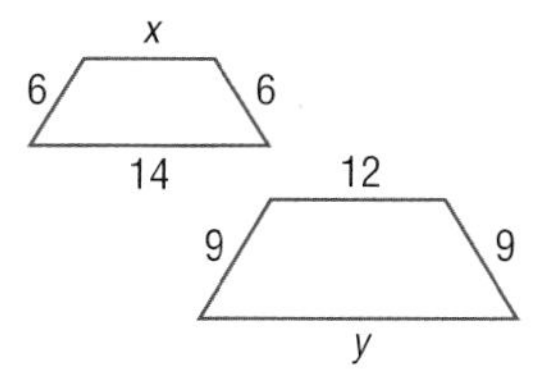

20.

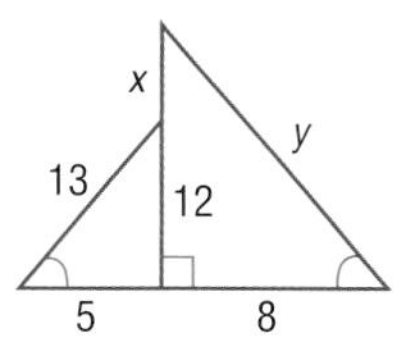

21.

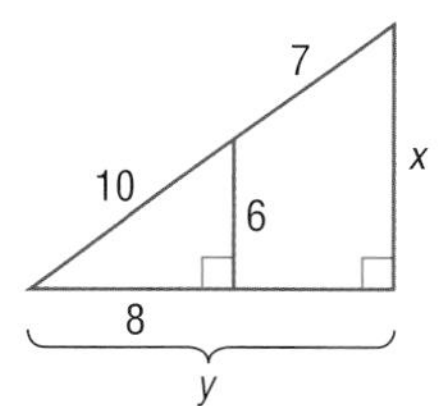

22. What are the dimensions of a scale drawing of a room that measures 10 feet by 12 feet if $\frac{1}{2}$ inch represents 1 foot?

23. SHADOWS On a sunny day, Jason measures the length of his shadow and the length of a tree's shadow. Use the figures at the right to find the height of the tree.

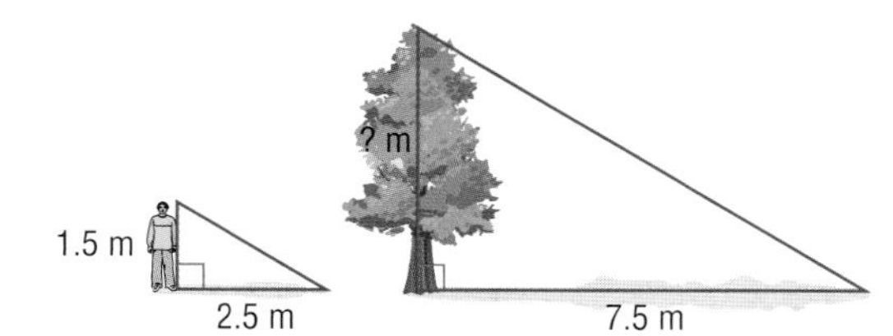

24. PHOTOGRAPHY A photo that is 4 inches wide by 6 inches long must be reduced to fit in a space 3 inches wide. How long will the reduced photo be?

25. SURVEYING Surveyors use instruments to measure objects that are too large or too far away to measure by hand. They can use the shadows that objects cast to find the height of the objects without measuring them. A surveyor finds that a telephone pole that is 25 feet tall is casting a shadow 20 feet long. A nearby building is casting a shadow 52 feet long. What is the height of the building?

4 Pythagorean Theorem

The **Pythagorean Theorem** states that in a right triangle, the square of the length of the hypotenuse c is equal to the sum of the squares of the lengths of the legs a and b.

That is, in any right triangle, $c^2 = a^2 + b^2$.

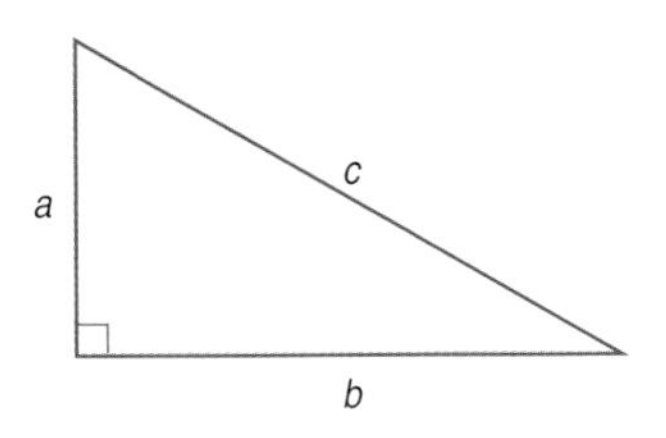

Example 1 **Find the length of the hypotenuse of each right triangle.**

a.

5 in.
c in.
12 in.

$c^2 = a^2 + b^2$ Pythagorean Theorem

$c^2 = 5^2 + 12^2$ Replace a with 5 and b with 12.

$c^2 = 25 + 144$ Simplify.

$c^2 = 169$ Add.

$c = \sqrt{169}$ Take the square root of each side.

$c = 13$ Simplify.

The length of the hypotenuse is 13 inches.

b.

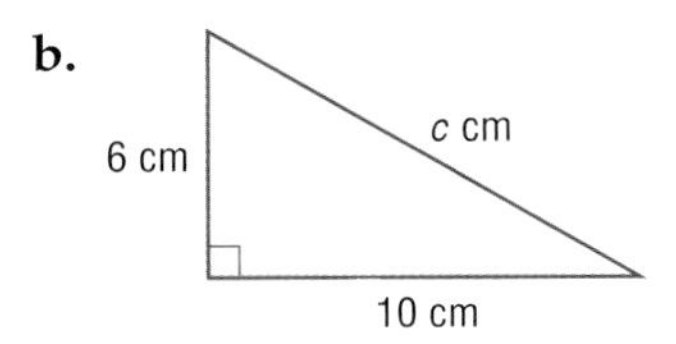

$c^2 = a^2 + b^2$ Pythagorean Theorem

$c^2 = 6^2 + 10^2$ Replace a with 6 and b with 10.

$c^2 = 36 + 100$ Simplify.

$c^2 = 136$ Add.

$c = \sqrt{136}$ Take the square root of each side.

$c \approx 11.7$ Use a calculator to find the square root of 136. Round to the nearest tenth.

To the nearest tenth, the length of the hypotenuse is 11.7 centimeters.

Example 2 **Find the length of the missing leg in each right triangle.**

a.

7 ft
25 ft
a ft

$c^2 = a^2 + b^2$ Pythagorean Theorem

$25^2 = a^2 + 7^2$ Replace c with 25 and b with 7.

$625 = a^2 + 49$ Simplify.

$625 - 49 = a^2 + 49 - 49$ Subtract 49 from each side.

$576 = a^2$ Simplify.

$\sqrt{576} = a$ Take the square root of each side.

$24 = a$ Simplify.

The length of the leg is 24 feet.

b.

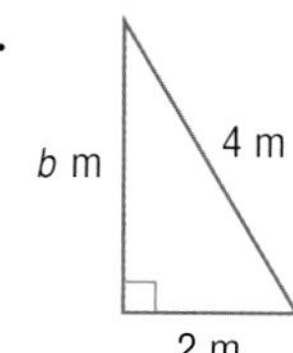

$c^2 = a^2 + b^2$	Pythagorean Theorem
$4^2 = 2^2 + b^2$	Replace *c* with 4 and *a* with 2.
$16 = 4 + b^2$	Simplify.
$12 = b^2$	Subtract 4 from each side.
$\sqrt{12} = b$	Take the square root of each side.
$3.5 \approx b$	Use a calculator to find the square root of 12. Round to the nearest tenth.

To the nearest tenth, the length of the leg is 3.5 meters.

Example 3 **The lengths of the three sides of a triangle are 5, 7, and 9 inches. Determine whether this triangle is a right triangle.**

Since the longest side is 9 inches, use 9 as *c*, the measure of the hypotenuse.

$c^2 = a^2 + b^2$	Pythagorean Theorem
$9^2 \stackrel{?}{=} 5^2 + 7^2$	Replace *c* with 9, *a* with 5, and *b* with 7.
$81 \stackrel{?}{=} 25 + 49$	Evaluate 9^2, 5^2, and 7^2.
$81 \neq 74$	Simplify.

Since $c^2 \neq a^2 + b^2$, the triangle is *not* a right triangle.

Exercises **Find each missing measure. Round to the nearest tenth, if necessary.**

1.

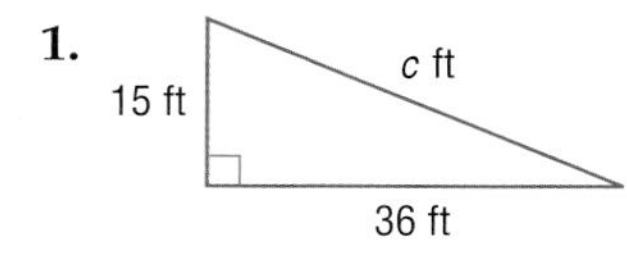

2.

3.

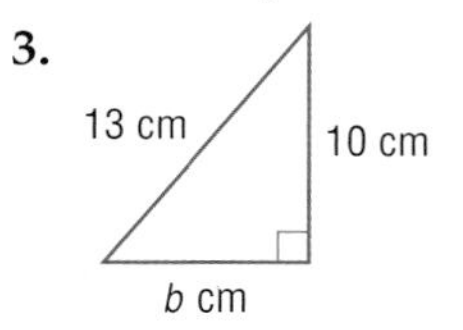

4.

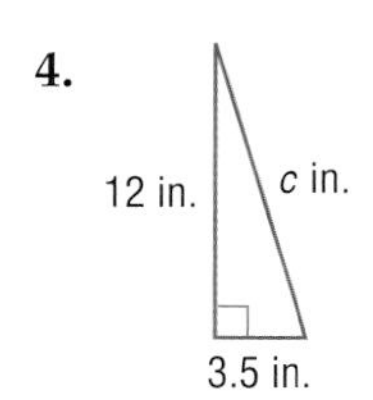

5.

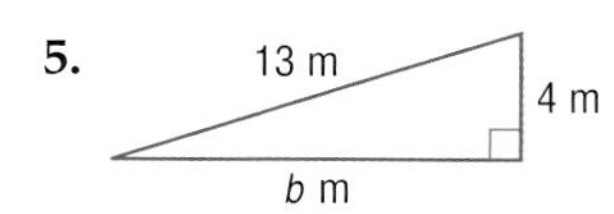

6.

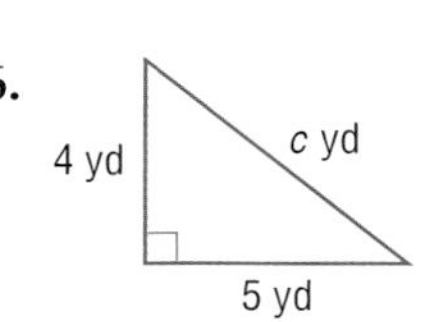

7. $a = 3, b = 4, c = ?$

8. $a = ?, b = 12, c = 13$

9. $a = 14, b = ?, c = 50$

10. $a = 2, b = 9, c = ?$

11. $a = 6, b = ?, c = 13$

12. $a = ?, b = 7, c = 11$

The lengths of three sides of a triangle are given. Determine whether each triangle is a right triangle.

13. 5 in., 7 in., 8 in.

14. 9 m, 12 m, 15 m

15. 6 cm, 7 cm, 12 cm

16. 11 ft, 12 ft, 16 ft

17. 10 yd, 24 yd, 26 yd

18. 11 km, 60 km, 61 km

19. **FLAGPOLES** Mai-Lin wants to find the distance from her feet to the top of the flagpole. If the flagpole is 30 feet tall and Mai-Lin is standing a distance of 15 feet from the flagpole, what is the distance from her feet to the top of the flagpole?

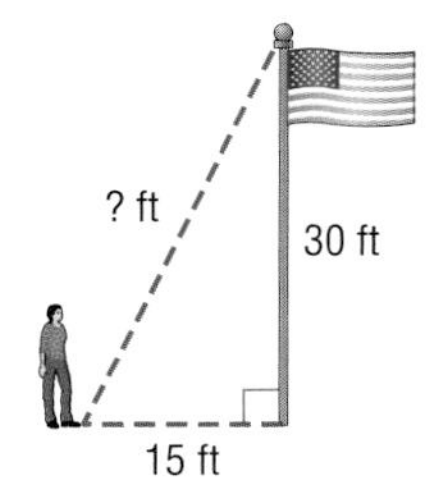

20. **CONSTRUCTION** The walls of the Downtown Recreation Center are being covered with paneling. The doorway into one room is 0.9 meters wide and 2.5 meters high. What is the width of the widest rectangular panel that can be taken through this doorway?

5 Mean, Median, and Mode

Mean, median, and mode are measures of central tendency that are often used to represent a set of data.

- To find the **mean**, find the sum of the data and divide by the number of items in the data set. (The mean is often called the average.)
- To find the **median**, arrange the data in numerical order. The median is the middle number. If there is an even number of data, the median is the mean of the two middle numbers.
- The **mode** is the number (or numbers) that appears most often in a set of data. If no item appears most often, the set has no mode.

Example 1 **Michelle is saving to buy a car. She saved $200 in June, $300 in July, $400 in August, and $150 in September. What was her mean (or average) monthly savings?**

$$\text{mean} = \frac{\text{sum of monthly savings}}{\text{number of months}}$$

$$= \frac{\$200 + \$300 + \$400 + \$150}{4}$$

$$= \frac{\$1050}{4} \text{ or } \$262.50$$

Michelle's mean monthly savings was $262.50.

Example 2 **Find the median of the data.**

Peter's Best Running Times	
Week	**Minutes to Run a Mile**
1	4.5
2	3.7
3	4.1
4	4.1
5	3.6
6	3.4

To find the median, order the numbers from least to greatest.
The median is in the middle.
3.4, 3.6, $\underbrace{3.7, 4.1}$, 4.1, 4.5

$$\frac{3.7 + 4.1}{2} = 3.9$$

There is an even number of data. Find the mean of the middle two.

Example 3 GOLF **Four players tied for first in the 2001 PGA Tour Championship. The scores for each player for each round are shown in the table below. What is the mode score?**

Player	Round 1	Round 2	Round 3	Round 4
Mike Weir	68	66	68	68
David Toms	73	66	64	67
Sergio Garcia	69	67	66	68
Ernie Els	69	68	65	68

Source: ESPN

The mode is the score that occurred most often. Since the score of 68 occurred 6 times, it is the mode of these data.

- The **range** of a set of data is the difference between the greatest and the least values of the set. It describes how a set of data varies.

Example 4 **Find the range of the data.**
{6, 11, 18, 4, 9, 15, 6, 3}

The greatest value is 18 and the least value is 3.
So, the range is 18 − 3 or 15.

Exercises **Find the mean, median, mode, and range for each set of data. Round to the nearest tenth if necessary.**

1. {2, 8, 12, 13, 15}
2. {66, 78, 78, 64, 34, 88}
3. {87, 95, 84, 89, 100, 82}
4. {99, 100, 85, 96, 94, 99}
5. {9.9, 9.9, 10, 9.9, 8.8, 9.5, 9.5}
6. {501, 503, 502, 502, 502, 504, 503, 503}
7. {7, 19, 15, 13, 11, 17, 9}
8. {6, 12, 21, 43, 1, 3, 13, 8}
9. {0.8, 0.04, 0.9, 1.1, 0.25}
10. $\left\{2\frac{1}{2}, 1\frac{7}{8}, 2\frac{5}{8}, 2\frac{3}{4}, 2\frac{1}{8}\right\}$

11. **CHARITY** The table shows the amounts collected by classes at Jackson High School. Find the mean, median, mode, and range of the data.

Amounts Collected for Charity	
Class	**Amount**
A	$150
B	$300
C	$55
D	$40
E	$10
F	$25
G	$200
H	$100

12. **SCHOOL** The table shows Pilar's grades in chemistry class for the semester. Find her mean, median, and mode scores, and the range of her scores.

Chemistry Grades	
Assignment	**Grade (out of 100)**
Homework	100
Electron Project	98
Test I	87
Atomic Mass Project	95
Test II	88
Phase Change Project	90
Test III	95

13. **WEATHER** The table shows the precipitation for the month of July in Cape Hatteras, North Carolina in various years. Find the mean, median, mode, and range of the data.

July Precipitation in Cape Hatteras, North Carolina												
Year	1990	1991	1992	1993	1994	1995	1996	1997	1998	1999	2000	2001
Inches	4.23	8.58	5.28	2.03	3.93	1.08	9.54	4.94	10.85	2.66	6.04	3.26

Source: National Climatic Data Center

14. **SCHOOL** Kaitlyn's scores on her first five algebra tests are 88, 90, 91, 89, and 92. What test score must Kaitlyn earn on the sixth test so that her mean score will be at least 90?
15. **GOLF** Colin's average for three rounds of golf is 94. What is the highest score he can receive for the fourth round to have an average (mean) of 92?
16. **SCHOOL** Mika has a mean score of 21 on his first four Spanish quizzes. If each quiz is worth 25 points, what is the highest possible mean score he can have after the fifth quiz?
17. **SCHOOL** To earn a grade of B in math, Latisha must have an average (mean) score of at least 84 on five math tests. Her scores on the first three tests are 85, 89, and 82. What is the lowest total score that Latisha must have on the last two tests to earn a B test average?

6 Bar and Line Graphs

A **bar graph** compares different categories of data by showing each as a bar whose length is related to the frequency. A **double bar graph** compares two sets of data. Another way to represent data is by using a **line graph**. A line graph usually shows how data changes over a period of time.

Example 1 MARRIAGE **The table shows the average age at which Americans marry for the first time. Make a double bar graph to display the data.**

Average Age to Marry		
Year	1990	1998
Men	26	27
Women	22	25

Source: U.S. Census Bureau

Step 1 Draw a horizontal and a vertical axis and label them as shown.

Step 2 Draw side-by-side bars to represent each category.

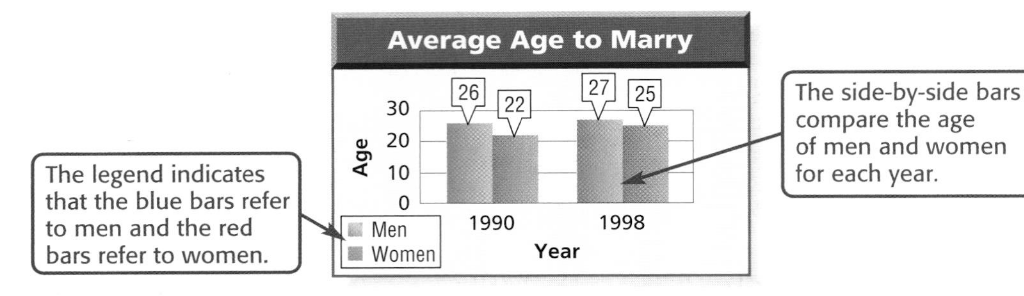

Example 2 HEALTH **The table shows Mark's height at 2-year intervals. Make a line graph to display the data.**

Age	2	4	6	8	10	12	14	16
Height (feet)	2.8	3.5	4.0	4.6	4.9	5.2	5.8	6

Step 1 Draw a horizontal and a vertical axis. Label them as shown.

Step 2 Plot the points.

Step 3 Draw a line connecting each pair of consecutive points.

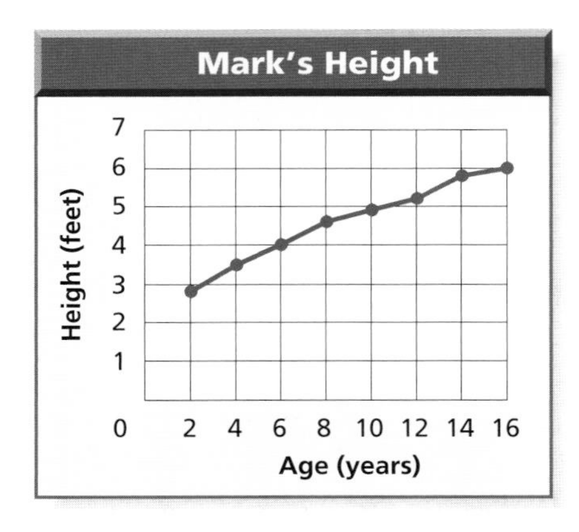

Exercises

1. **HEALTH** The table below shows the life expectancy for Americans born in each year listed. Make a double-bar graph to display the data.

Life Expectancy		
Year of Birth	Male	Female
1980	70.0	77.5
1985	71.2	78.2
1990	71.8	78.8
1995	72.5	78.9
1998	73.9	79.4

2. **MONEY** The amount of money in Becky's savings account from August through March is shown in the table below. Make a line graph to display the data.

Month	Amount	Month	Amount
August	$300	December	$780
September	$400	January	$800
October	$700	February	$950
November	$780	March	$900

7 Stem-and-Leaf Plots

In a **stem-and-leaf plot**, data are organized in two columns. The greatest place value of the data is used for the stems. The next greatest place value forms the leaves. Stem-and-leaf plots are useful for organizing long lists of numbers.

Example

SCHOOL **Isabella has collected data on the GPAs (grade point average) of the 16 students in the art club. Display the data in a stem-and-leaf plot.**
{4.0, 3.9, 3.1, 3.9, 3.8, 3.7, 1.8, 2.6, 4.0, 3.9, 3.5, 3.3, 2.9, 2.5, 1.1, 3.5}

Step 1 Find the least and the greatest number. Then identify the greatest place-value digit in each number. In this case, ones.

least data: 1.1 — The least number has 1 in the ones place.

greatest data: 4.0 — The greatest number has 4 in the ones place.

Step 2 Draw a vertical line and write the stems from 1 to 4 to the left of the line.

Step 3 Write the leaves to the right of the line, with the corresponding stem. For example, write 0 to the right of 4 for 4.0.

Stem	Leaf
1	8 1
2	5 6 9
3	9 1 9 8 7 9 5 3 5
4	0 0

Step 4 Rearrange the leaves so they are ordered from least to greatest.

Step 5 Include a key or an explanation.

Stem	Leaf
1	1 8
2	5 6 9
3	1 3 5 5 7 8 9 9 9
4	0 0 *3|1 = 3.1*

Exercises

GAMES **For Exercises 1–4, use the following information.**
The stem-and-leaf plot at the right shows Charmaine's scores for her favorite computer game.

Stem	Leaf
9	0 0 0 1 3 4 5 5 7 8 8 8 9 9
10	0 3 4 4 5 6 9
11	0 3 9 9
12	1 2 6
13	0 *12|6 = 126*

1. What are Charmaine's highest and lowest scores?
2. Which score(s) occurred most frequently?
3. How many scores were above 115?
4. Has Charmaine ever scored 123?
5. **SCHOOL** The class scores on a 50-item test are shown in the table at the right. Make a stem-and-leaf plot of the data.

Test Scores					
45	15	30	40	28	35
39	29	38	18	43	49
46	44	48	35	36	30

6. **GEOGRAPHY** The table shows the land area of each county in Wyoming. Round each area to the nearest hundred square miles and organize the data in a stem-and-leaf plot.

County	Area (mi²)	County	Area (mi²)	County	Area (mi²)
Albany	4273	Hot Springs	2004	Sheridan	2523
Big Horn	3137	Johnson	4166	Sublette	4883
Campbell	4797	Laramie	2686	Sweetwater	10,425
Carbon	7896	Lincoln	4069	Teton	4008
Converse	4255	Natrona	5340	Unita	2082
Crook	2859	Niobrara	2626	Washakie	2240
Fremont	9182	Park	6942	Weston	2398
Goshen	2225	Platte	2085		

Source: *The World Almanac*

8 Box-and-Whisker Plots

In a set of data, **quartiles** are values that divide the data into four equal parts.

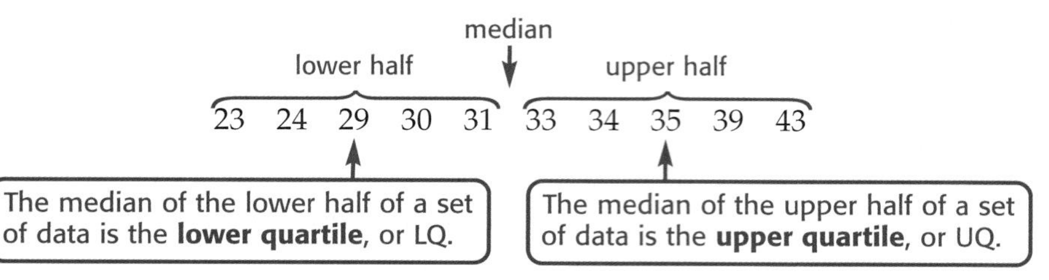

- To make a **box-and-whisker plot**, draw a box around the quartile values, and lines or *whiskers* to represent the values in the lower fourth of the data and the upper fourth of the data.

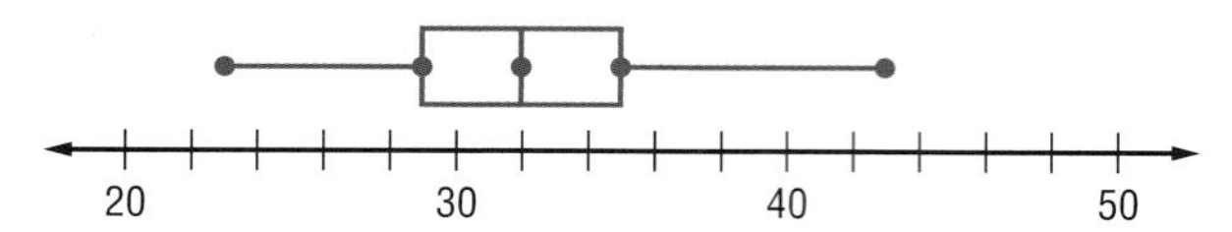

Example 1 MONEY **The amount spent in the cafeteria by 20 students is shown. Display the data in a box-and-whisker plot.**

Amount Spent			
$2.00	$2.00	$1.00	$4.00
$1.00	$2.50	$2.50	$2.00
$2.50	$1.00	$4.00	$2.50
$3.50	$2.00	$3.00	$2.50
$4.00	$4.00	$5.50	$1.50

Step 1 Find the least and greatest number. Then draw a number line that covers the range of the data. In this case, the least value is 1 and the greatest value is 5.5.

$1 $2 $3 $4 $5 $6

Step 2 Find the median, the extreme values, and the upper and lower quartiles. Mark these points above the number line.

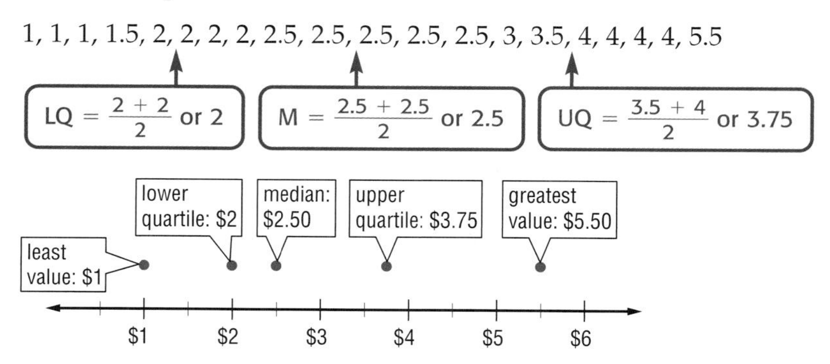

Step 3 Draw a box and the whiskers.

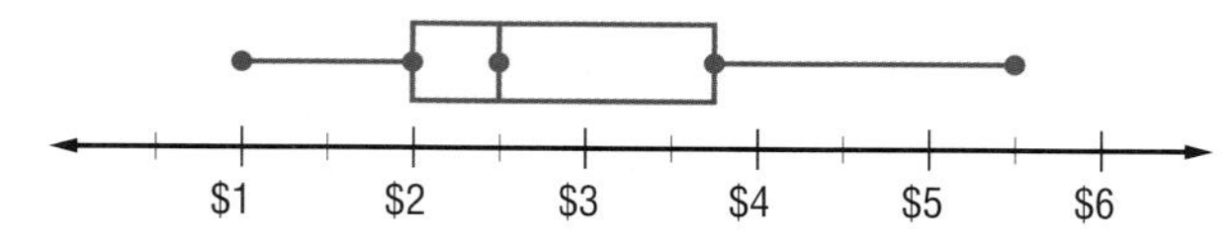

- The **interquartile range (IQR)** is the range of the middle half of the data and contains 50% of the data in the set.

$$\text{Interquartile range} = \text{UQ} - \text{LQ}$$

The interquartile range of the data in Example 1 is $3.75 - 2$ or 1.75.

- An **outlier** is any element of a set that is at least 1.5 interquartile ranges less than the lower quartile or greater than the upper quartile. The whisker representing the data is drawn from the box to the least or greatest value that is not an outlier.

Example 2 SCHOOL **The number of hours José studied each day for the last month is shown in the box-and-whiskers plot below.**

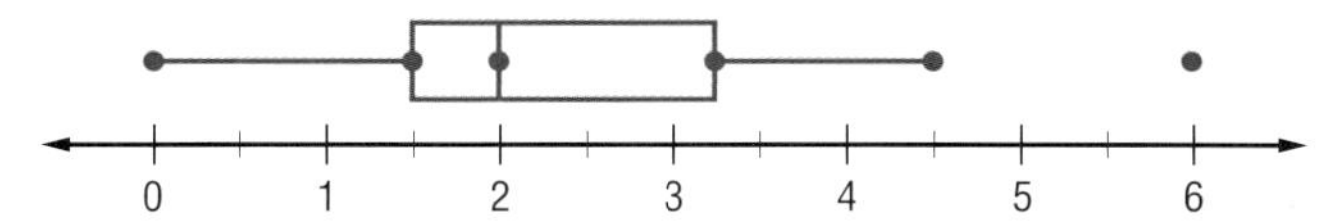

a. What percent of the data lies between 1.5 and 3.25?
The value 1.5 is the lower quartile and 3.25 is the upper quartile. The values between the lower and upper quartiles represent 50% of the data.

b. What was the greatest amount of time José studied in a day?
The greatest value in the plot is 6, so the greatest amount of time José studied in a day was 6 hours.

c. What is the interquartile range of this box-and-whisker plot?
The interquartile range is UQ − LQ. For this plot, the interquartile range is $3.25 - 1.5$ or 1.75 hours.

d. Identify any outliers in the data.
An outlier is at least 1.5(1.75) less than the lower quartile or more than the upper quartile. Since $3.25 + (1.5)(1.75) = 5.875$, and $6 > 5.875$, the value 6 is an outlier, and was not included in the whisker.

Exercises

DRIVING For Exercises 1–3, use the following information.
Tyler surveyed 20 randomly chosen students at his school about how many miles they drive in an average day. The results are shown in the box-and-whisker plot.

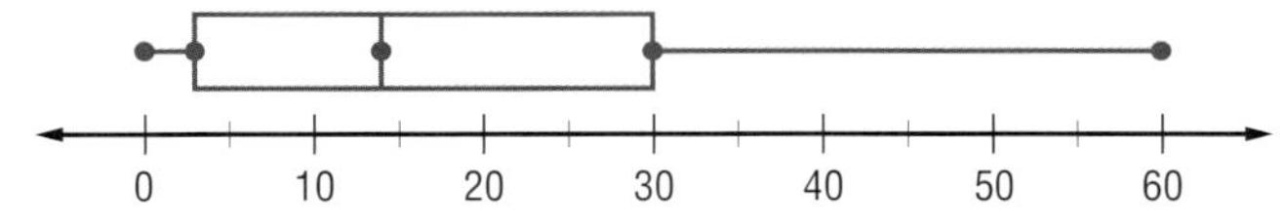

1. What percent of the students drive more than 30 miles in a day?
2. What is the interquartile range of the box-and-whisker plot?
3. Does a student at Tyler's school have a better chance to meet someone who drives the same mileage they do if they drive 50 miles in a day or 15 miles in a day? Why?
4. **SOFT DRINKS** Carlos surveyed his friends to find the number of cans of soft drink they drink in an average week. Make a box-and-whisker plot of the data.
   {0, 0, 0, 1, 1, 1, 2, 2, 3, 4, 4, 5, 5, 7, 10, 10, 10, 11, 11}
5. **ANIMALS** The average life span of some animals commonly found in a zoo are given below. Make a box-and-whisker plot of the data.
   {1, 7, 7, 10, 12, 12, 15, 15, 18, 20, 20, 20, 25, 40, 100}
6. **BASEBALL** The table shows the number of sacrifice hits made by teams in the National Baseball League in the 2001 season. Make a box-and-whisker plot of the data.

Team	Home Runs	Team	Home Runs
Arizona	71	Milwaukee	65
Atlanta	64	Montreal	64
Chicago	117	New York	52
Cincinnati	66	Philadelphia	67
Colorado	81	Pittsburgh	60
Florida	60	San Diego	29
Houston	71	San Francisco	67
Los Angeles	57	St. Louis	83

Source: ESPN

Extra Practice

Lesson 1-1

(pages 6–10)

Find the value of each expression.

1. $2(3 + 8) - 3$
2. $(5 + 3) - 16 \div 4$
3. $4 + 8(4) \div 2 - 10$
4. $15 \div 3 \cdot 5 + 1$
5. $3(2^2 + 3)$
6. $5 + 3^2 - 16 + 4$
7. $[(4 + 8)^2 \div 9] \cdot 5$
8. $5 + 8^2 \div 4 \cdot 3$
9. $5 \cdot 7 - 2(5 + 1) \div 3$
10. $3 + 7^2 - 16 \div 2$
11. $12 + 20 \div 4 - 5$
12. $0.5[7 - (8 - 6)^2] - 1$
13. $\frac{1}{2}(3^2 + 5 \cdot 7) - 8$
14. $\frac{3 \cdot 5 + 3^2}{2^3}$
15. $\frac{6^2 + 4(2^4)}{28 + 9 \cdot 8}$
16. $\frac{2^3 - 8(4^2)}{18 + 2^5}$

Evaluate each expression if $a = -0.5$, $b = 4$, $c = 5$, and $d = -3$.

17. $3b + 4d$
18. $ab^2 + c$
19. $bc + d \div a$
20. $7ab - 3d$
21. $ad + b^2 - c$
22. $\frac{4a + 3c}{3b}$
23. $\frac{3ab^2 - d^3}{a}$
24. $\frac{5a + ad}{bc}$

Lesson 1-2

(pages 11–18)

Name the sets of numbers to which each number belongs. (Use N, W, Z, Q, I, and R.)

1. 8.2
2. -9
3. $\sqrt{36}$
4. $-\frac{1}{3}$
5. $\sqrt{2}$
6. $-0.\overline{24}$

Name the property illustrated by each equation.

7. $(4 + 9a)2b = 2b(4 + 9a)$
8. $3\left(\frac{1}{3}\right) = 1$
9. $a(3 - 2) = a \cdot 3 - a \cdot 2$
10. $(-3b) + 3b = 0$
11. $jk + 0 = jk$
12. $(2a)b = 2(ab)$

Name the additive inverse and multiplicative inverse for each number.

13. 3
14. $-\frac{1}{8}$
15. 0.2
16. $-2\frac{2}{7}$

Simplify each expression.

17. $7s + 9t + 2s - 7t$
18. $6(2a + 3b) + 5(3a - 4b)$
19. $4(3x - 5y) - 8(2x + y)$
20. $0.2(5m - 8) + 0.3(6 - 2m)$
21. $\frac{1}{2}(7p + 3q) + \frac{3}{4}(6p - 4q)$
22. $\frac{4}{5}(3v - 2w) - \frac{1}{5}(7v - 2w)$

Lesson 1-3

(pages 20–27)

Write an algebraic expression to represent each verbal expression.

1. twelve decreased by the square of a number
2. twice the sum of a number and negative nine
3. the product of the square of a number and 6
4. the square of the sum of a number and 11

Name the property illustrated by each statement.

5. If $a + 1 = 6$, then $3(a + 1) = 3(6)$
6. If $x + (4 + 5) = 21$, then $x + 9 = 21$
7. If $7x = 42$, then $7x - 5 = 42 - 5$
8. If $3 + 5 = 8$ and $8 = 2 \cdot 4$, then $3 + 5 = 2 \cdot 4$.

Solve each equation.

9. $5t + 8 = 88$
10. $27 - x = -4$
11. $\frac{3}{4}y = \frac{2}{3}y + 5$
12. $8s - 3 = 5(2s + 1)$
13. $3(k - 2) = k + 4$
14. $0.5z + 10 = z + 4$
15. $8q - \frac{q}{3} = 46$
16. $-\frac{2}{7}r + \frac{3}{7} = 5$
17. $d - 1 = \frac{1}{2}(d - 2)$

Solve each equation or formula for the specified variable.

18. $C = \pi r$; for r
19. $I = Prt$, for t
20. $m = \frac{n - 2}{n}$, for n

Lesson 1-4

(pages 28–32)

Evaluate each expression if $x = -5$, $y = 3$, and $z = -2.5$.

1. $|2x|$
2. $|-3y|$
3. $|2x + y|$
4. $|y + 5z|$
5. $-|x + z|$
6. $8 - |5y - 3|$
7. $2|x| - 4|2 + y|$
8. $|x + y| - 6|z|$

Solve each equation.

9. $|d + 1| = 7$
10. $|a - 6| = 10$
11. $2|x - 5| = 22$
12. $|t + 9| - 8 = 5$
13. $|p + 1| + 10 = 5$
14. $6|g - 3| = 42$
15. $2|y + 4| = 14$
16. $|3b - 10| = 2b$
17. $|3x + 7| + 4 = 0$
18. $|2c + 3| - 15 = 0$
19. $7 - |m - 1| = 3$
20. $3 + |z + 5| = 10$
21. $4|h + 1| = 32$
22. $2|2x + 3| = 34$
23. $3|a - 5| - 4 = 14$
24. $2|2d - 7| + 1 = 35$
25. $|3t + 6| + 9 = 30$
26. $|d - 3| = 2d + 9$
27. $|4y - 5| + 4 = 7y + 8$
28. $|2b + 4| - 3 = 6b + 1$
29. $|5t| + 2 = 3t + 18$

Lesson 1-5

(pages 33–39)

Solve each inequality. Describe the solution set using set-builder or interval notation. Then, graph the solution set on a number line.

1. $2z + 5 \leq 7$
2. $3r - 8 > 7$
3. $0.75b < 3$
4. $-3x > 6$
5. $2(3f + 5) \geq 28$
6. $-33 > 5g + 7$
7. $-3(y - 2) \geq -9$
8. $7a + 5 > 4a - 7$
9. $5(b - 3) \leq b - 7$
10. $3(2x - 5) < 5(x - 4)$
11. $2(4m - 1) + 3(m + 4) \geq 6m$
12. $8(2c - 1) > 11c + 22$
13. $5y - 4(2y + 1) \leq 2(0.5 - 2y)$
14. $2(d + 4) - 5 \geq 5(d + 3)$
15. $8 - 3t < 4(3 - t)$
16. $-x \geq \frac{x + 4}{7}$
17. $\frac{a + 8}{4} \leq \frac{7 + a}{3}$
18. $-y < \frac{y + 5}{2}$
19. $2 + 4(d - 2) \leq 3 - (d - 1)$
20. $5(x - 1) - 4x \geq 3(3 - x)$
21. $6s - (4s + 7) > 5 - s$

Define a variable and write an inequality for each problem. Then solve the resulting inequality.

22. The product of 7 and a number is greater than 42.
23. The difference of twice a number and 3 is at most 11.
24. The product of −10 and a number is greater than or equal to 20.
25. Thirty increased by a number is less than twice the number plus three.

Lesson 1-6

(pages 40–46)

Write an absolute value inequality for each of the following. Then graph the solution set on a number line.

1. all numbers less than −9 and greater than 9
2. all numbers between −5.5 and 5.5
3. all numbers greater than or equal to −2 and less than or equal to 2

Solve each inequality. Graph the solution set on a number line.

4. $3m - 2 < 7$ or $2m + 1 > 13$
5. $2 < n + 4 < 7$
6. $-3 \leq s - 2 \leq 5$
7. $5t + 3 \leq -7$ or $5t - 2 \geq 8$
8. $7 \leq 4x + 3 \leq 19$
9. $4x + 7 < 5$ or $2x - 4 > 12$
10. $|7x| \geq 21$
11. $|8p| \leq 16$
12. $|7d| \geq -42$
13. $|a + 3| < 1$
14. $|t - 4| > 1$
15. $|2y - 5| < 3$
16. $|3d + 6| \geq 3$
17. $|4x - 1| < 5$
18. $|6v + 12| > 18$
19. $|2r + 4| < 6$
20. $|5w - 3| \geq 9$
21. $|z + 2| \geq 0$
22. $12 + |2q| < 0$
23. $|3h| + 15 < 0$
24. $|5n| - 16 \geq 4$

Lesson 2-1

(pages 56–62)

Determine whether each relation is a function. Write *yes* or *no*.

1.

Year	Population
1970	11,605
1980	13,468
1990	15,630
2000	18,140

2.

x	y
1	5
2	5
3	5
4	5

3.

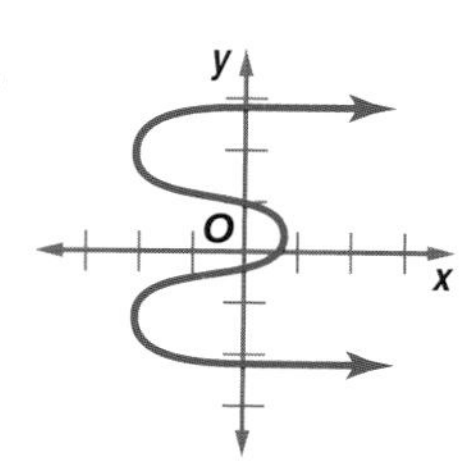

Graph each relation or equation and find the domain and range. Then determine whether the relation or equation is a function.

4. $\{(1, 2), (2, 3), (3, 4), (4, 5)\}$
5. $\{(0, 3), (0, 2), (0, 1), (0, 0)\}$
6. $y = -x$
7. $y = 2x - 1$
8. $y = 2x^2$
9. $y = -x^2$

Find each value if $f(x) = x + 7$ and $g(x) = (x + 1)^2$.

10. $f(2)$
11. $f(-4)$
12. $f(a + 2)$
13. $g(4)$
14. $g(-2)$
15. $f(0.5)$
16. $g(b - 1)$
17. $g(3c)$

Lesson 2-2

(pages 63–67)

State whether each equation or function is linear. Write *yes* or *no*. If no, explain.

1. $\frac{x}{2} - y = 7$
2. $\sqrt{x} = y + 5$
3. $g(x) = \frac{2}{x - 3}$
4. $x = 3 + y$
5. $f(x) = 7$
6. $\frac{3}{x} - \frac{1}{4} = \frac{4}{3}$

Write each equation in standard form. Identify *A*, *B*, and *C*.

7. $x + 7 = y$
8. $x = -3y$
9. $5x = 7y + 3$
10. $y = \frac{2}{3}x + 8$
11. $-0.4x = 10$
12. $0.75y = -6$

Find the *x*-intercept and the *y*-intercept of the graph of each equation. Then graph the equation.

13. $2x + y = 6$
14. $3x - 2y = -12$
15. $y = -x$
16. $x = 3y$
17. $\frac{3}{4}y - x = 1$
18. $y = -3$

Lesson 2-3

(pages 68–74)

Find the slope of the line that passes through each pair of points.

1. $(0, 3), (5, 0)$
2. $(2, 3), (5, 7)$
3. $(2, 8), (2, -8)$
4. $(1.5, -1), (3, 1.5)$
5. $\left(-\frac{1}{2}, \frac{3}{5}\right), \left(\frac{3}{10}, -\frac{1}{4}\right)$
6. $(-3, c), (4, c)$

Graph the line passing through the given point with the given slope.

7. $(0, 3); 1$
8. $(2, 3); 0$
9. $(-1, 1); -\frac{1}{3}$

Graph the line that satisfies each set of conditions.

10. passes through $(0, 1)$, parallel to a line whose slope is -2
11. passes through $(2, -3)$, perpendicular to a line whose slope is 5
12. passes through $(1, -1)$, parallel to the graph of $x + y = 3$
13. passes through $(4, -5)$, perpendicular to the graph of $-2x + 5y = 1$

Lesson 2-4

(pages 75–80)

State the slope and y-intercept of the graph of each equation.

1. $y = -3x + 4$
2. $x - y = 5$
3. $x = -4$

Write an equation in slope-intercept form for each graph.

4.

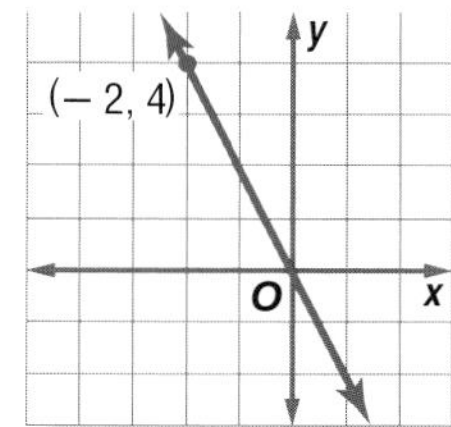

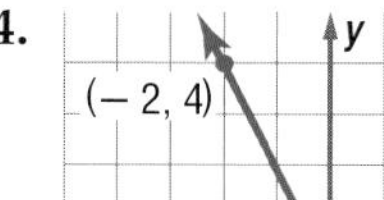

5.

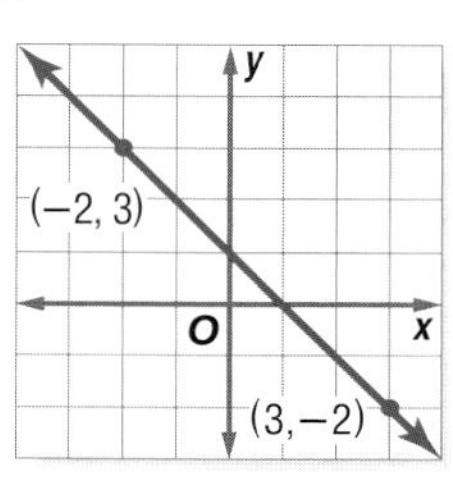

Write an equation in slope-intercept form for the line that satisfies each set of conditions.

6. slope -1, passes through $(7, 2)$
7. slope $\frac{3}{4}$, passes through the origin
8. passes through $(1, -3)$ and $(-1, 2)$
9. x-intercept -5, y-intercept 2
10. passes through $(1, 1)$, parallel to the graph of $2x + 3y = 5$
11. passes through $(6, -2)$, perpendicular to the graph of $-x + 8y = 5$
12. passes through $(0, 0)$, perpendicular to the graph of $2y + 3x = 4$

Lesson 2-5

(pages 81–86)

Complete parts a–c for each set of data in Exercises 1–3.

a. Draw a scatter plot.

b. Use two ordered pairs to write a prediction equation.

c. Use your prediction equation to predict the missing value.

1.

Telephone Costs	
Minutes	**Cost ($)**
1	0.20
3	0.52
4	0.68
6	1.00
9	1.48
15	?

2.

Washington	
Year	**Population**
1960	2,853,214
1970	3,413,244
1980	4,132,353
1990	4,866,669
2000	5,894,121
2010	?

Source: *The World Almanac*

3.

Federal Minimum Wage	
Year	**Wage**
1981	$3.35
1990	$3.80
1991	$4.25
1996	$4.75
1997	$5.15
2005	?

Source: *The World Almanac*

Lesson 2-6

(pages 89–95)

Identify each function as S for step, C for constant, A for absolute value, or P for piecewise.

1.

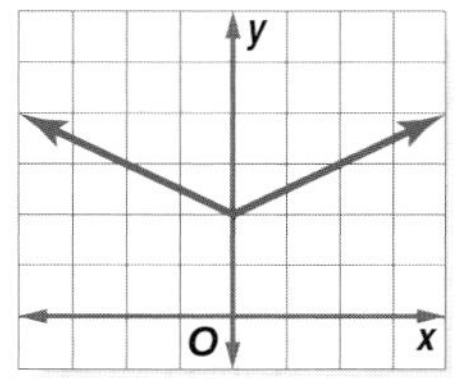

2.

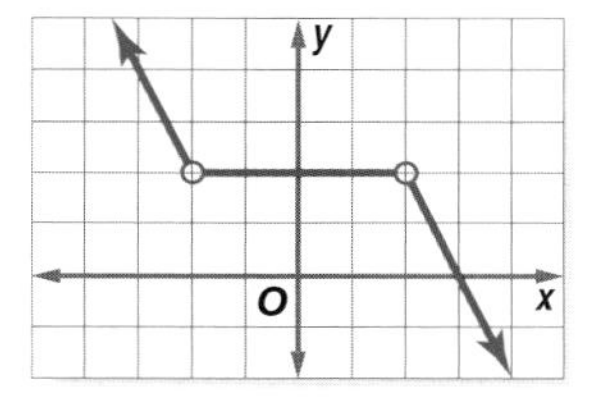

Graph each function. Identify the domain and range.

3. $f(x) = [\![x + 5]\!]$
4. $g(x) = [\![x]\!] - 2$
5. $f(x) = -2[\![x]\!]$
6. $h(x) = |x| - 3$
7. $h(x) = |x - 1|$
8. $g(x) = |2x| + 2$
9. $h(x) = \begin{cases} x \text{ if } x < -2 \\ 4 \text{ if } x \geq -2 \end{cases}$
10. $f(x) = \begin{cases} -3 \text{ if } x \leq 1 \\ -x \text{ if } x > 1 \end{cases}$

Extra Practice

Lesson 2-7

(pages 96–99)

Graph each inequality.

1. $y \geq x - 2$
2. $y < -3x - 1$
3. $4y \leq -3x + 8$
4. $3x > y$
5. $x + 2 \geq y - 7$
6. $2x < 5 - y$
7. $y > \frac{1}{5}x - 8$
8. $2y - 5x \leq 8$
9. $-2x + 5 \leq \frac{2}{3}y$
10. $3x + 2y \geq 0$
11. $x \leq 2$
12. $\frac{y}{2} \leq x - 1$
13. $y - 3 < 5$
14. $y \geq -|x|$
15. $|x| \leq y + 3$
16. $y > |5x - 3|$
17. $y \leq |8 - x|$
18. $y < |x + 3| - 1$
19. $y + |2x| \geq 4$
20. $y \geq |2x - 1| + 5$
21. $y < \left|\frac{2x}{3}\right| - 1$

Lesson 3-1

(pages 110–115)

Solve each system of equations by graphing.

1. $x + 3y = 18$, $-x + 2y = 7$
2. $x - y = 2$, $2x - 2y = 10$
3. $2x + 6y = 6$, $\frac{1}{3}x + y = 1$
4. $x + 3y = 0$, $2x + 6y = 5$
5. $2x - y = 7$, $\frac{2}{5}x - \frac{4}{3}y = -2$
6. $y = \frac{1}{3}x + 1$, $y = 4x + 1$

Graph each system of equations and describe it as *consistent and independent*, *consistent and dependent*, or *inconsistent*.

7. $2x + 3y = 5$, $-6x - 9y = -15$
8. $x - 2y = 4$, $y = x - 2$
9. $y = 0.5x$, $2y = x + 4$
10. $9x - 5 = 7y$, $4.5x - 3.5y = 2.5$
11. $\frac{3}{4}x - y = 0$, $\frac{1}{3}y + \frac{1}{2}x = 6$
12. $\frac{2}{3}x = \frac{5}{3}y$, $2x - 5y = 0$

Lesson 3-2

(pages 116–122)

Solve each system of equations by using substitution.

1. $2x + 3y = 10$, $x + 6y = 32$
2. $x = 4y - 10$, $5x + 3y = -4$
3. $3x - 4y = -27$, $2x + y = -7$

Solve each system of equations by using elimination.

4. $7x + y = 9$, $5x - y = 15$
5. $r + 5s = -17$, $2r - 6s = -2$
6. $6p + 8q = 20$, $5p - 4q = -26$

Solve each system of equations by using either substitution or elimination.

7. $2x - 3y = 7$, $3x + 6y = 42$
8. $2a + 5b = -13$, $3a - 4b = 38$
9. $3c + 4d = -1$, $6c - 2d = 3$
10. $7x - y = 35$, $y = 5x - 19$
11. $3m + 4n = 28$, $5m - 3n = -21$
12. $x = 2y - 1$, $4x - 3y = 21$
13. $2.5x + 1.5y = -2$, $3.5x - 0.5y = 18$
14. $\frac{5}{2}x + \frac{1}{3}y = 13$, $\frac{1}{2}x - y = -7$
15. $\frac{2}{7}c - \frac{4}{3}d = 16$, $\frac{4}{7}c + \frac{8}{3}d = -16$

Lesson 3-3

(pages 123–127)

Solve each system of inequalities by graphing.

1. $x \le 5$
   $y \ge -3$
2. $y < 3$
   $y - x \ge -1$
3. $x + y < 5$
   $x < 2$
4. $y + x < 2$
   $y \ge x$
5. $x + y \le 2$
   $y - x \le 4$
6. $y \le x + 4$
   $y - x \ge 1$
7. $y < \frac{1}{3}x + 5$
   $y > 2x + 1$
8. $y + x \ge 1$
   $y - x \ge -1$
9. $|x| > 2$
   $|y| \le 5$
10. $|x - 3| \le 3$
    $4y - 2x \le 6$
11. $4x + 3y \ge 12$
    $2y - x \ge -1$
12. $y \le -1$
    $3x - 2y \ge 6$

Find the coordinates of the vertices of the figure formed by each system of inequalities.

13. $y \le 3$
    $x \le 2$
    $y \ge -\frac{3}{2}x + 3$
14. $y \ge -1$
    $y \le x$
    $y \le -x + 4$
15. $y \le \frac{1}{3}x + \frac{7}{3}$
    $4x - y \le 5$
    $y \ge -\frac{3}{2}x + \frac{1}{2}$

Lesson 3-4

(pages 129–135)

A feasible region has vertices at (−3, 2), (1, 3), (6, 1), and (2, −2). Find the maximum and minimum values of each function.

1. $f(x, y) = 2x - y$
2. $f(x, y) = x + 5y$
3. $f(x, y) = y - 4x$
4. $f(x, y) = -x + 3y$
5. $f(x, y) = 3x - y$
6. $f(x, y) = 2y - 2x$

Graph each system of inequalities. Name the coordinates of the vertices of the feasible region. Find the maximum and minimum values of the given function for this region.

7. $4x - 5y \le -10$
   $y \le 6$
   $2x + y \ge 2$
   $f(x, y) = x + y$
8. $x \le 5$
   $y \ge 2$
   $2x - 5y \ge -10$
   $f(x, y) = 3x + y$
9. $x - 2y \ge -7$
   $x + y \le 8$
   $y \ge 5x + 8$
   $f(x, y) = 3x - 4y$
10. $y \le 4x + 6$
    $x + 4y \ge 7$
    $2x + y \le 7$
    $f(x, y) = 2x - y$
11. $y \ge 0$
    $y \le 5$
    $y \le -x + 7$
    $5x + 3y \ge 20$
    $f(x, y) = x + 2y$
12. $y \ge 0$
    $3x - 2y \ge 0$
    $x + 3y \le 11$
    $2x + 3y \le 16$
    $f(x, y) = 4x + y$

Lesson 3-5

(pages 138–144)

For each system of equations, an ordered triple is given. Determine whether or not it is a solution of the system.

1. $4x + 2y - 6z = -38$
   $5x - 4y + z = -18$
   $x + 3y + 7z = 38$; $(-3, 2, 5)$
2. $u + 3v + w = 14$
   $2u - v + 3w = -9$
   $4u - 5v - 2w = -2$; $(1, 5, -2)$
3. $x + y = -6$
   $x + z = -2$
   $y + z = 2$; $(-4, -2, 2)$

Solve each system of equations.

4. $5a = 5$
   $6b - 3c = 15$
   $2a + 7c = -5$
5. $s + 2t = 5$
   $7r - 3s + t = 20$
   $2t = 8$
6. $2u - 3v = 13$
   $3v + w = -3$
   $4u - w = 2$
7. $4a + 2b - c = 5$
   $2a + b - 5c = -11$
   $a - 2b + 3c = 6$
8. $x + 2y - z = 1$
   $x + 3y + 2z = 7$
   $2x + 6y + z = 8$
9. $2x + y - z = 7$
   $3x - y + 2z = 15$
   $x - 4y + z = 2$

Lesson 4-1

(pages 154–158)

State the dimensions of each matrix.

1. $\begin{bmatrix} 8 & 4 & 3 \end{bmatrix}$
2. $\begin{bmatrix} 2 & -1 \\ 0 & 6 \end{bmatrix}$
3. $\begin{bmatrix} 4 \\ -3 \end{bmatrix}$

Solve each matrix equation.

4. $\begin{bmatrix} 2x & 3y & -z \end{bmatrix} = \begin{bmatrix} 2y & -z & 15 \end{bmatrix}$
5. $\begin{bmatrix} x + y \\ 4x - 3y \end{bmatrix} = \begin{bmatrix} 1 \\ 11 \end{bmatrix}$
6. $-2\begin{bmatrix} w + 5 & x - z \\ 3y & 8 \end{bmatrix} = \begin{bmatrix} -16 & -4 \\ 6 & 2x + 8z \end{bmatrix}$
7. $y\begin{bmatrix} 2 & x \\ 5 & 1 \end{bmatrix} = \begin{bmatrix} 4 & -10 \\ 10 & 2z \end{bmatrix}$
8. $\begin{bmatrix} 2x \\ -y \\ 3z \end{bmatrix} = \begin{bmatrix} 16 \\ 18 \\ -21 \end{bmatrix}$
9. $\begin{bmatrix} x - 3y \\ 4y - 3x \end{bmatrix} = -5\begin{bmatrix} 2 \\ x \end{bmatrix}$
10. $\begin{bmatrix} x^2 + 4 & y + 6 \\ x - y & 2 - y \end{bmatrix} = \begin{bmatrix} 5 & 7 \\ 0 & 1 \end{bmatrix}$
11. $\begin{bmatrix} x + y & 3 \\ y & 6 \end{bmatrix} = \begin{bmatrix} 0 & 2y - x \\ z & 4 - 2x \end{bmatrix}$

Lesson 4-2

(pages 160–166)

Perform the indicated matrix operations. If the matrix does not exist, write impossible.

1. $\begin{bmatrix} 3 & 5 \\ -7 & 2 \end{bmatrix} + \begin{bmatrix} -2 & 6 \\ 8 & -1 \end{bmatrix}$
2. $\begin{bmatrix} 0 & -1 & 3 \end{bmatrix} + \begin{bmatrix} 5 \\ -2 \\ -3 \end{bmatrix}$
3. $\begin{bmatrix} 45 & 36 & 18 \\ 63 & 29 & 5 \end{bmatrix} - \begin{bmatrix} 45 & -2 & 36 \\ 18 & 9 & -10 \end{bmatrix}$
4. $4\begin{bmatrix} -8 & 2 & 9 \end{bmatrix} - 3\begin{bmatrix} 2 & -7 & 6 \end{bmatrix}$
5. $\begin{bmatrix} -3 & 6 & -9 \\ 4 & -3 & 0 \\ 8 & -2 & 3 \end{bmatrix} - \begin{bmatrix} 1 & 5 & 7 \\ 5 & 2 & -6 \\ 3 & 0 & -2 \end{bmatrix}$
6. $5\begin{bmatrix} 3 & 1 & 0 \\ 0 & 0 & 2 \\ 1 & -1 & -1 \end{bmatrix} + \begin{bmatrix} 2 & 0 & 3 \\ 1 & 1 & 2 \\ 2 & 1 & -1 \end{bmatrix}$
7. $5\begin{bmatrix} 6 & -2 \\ 5 & 4 \end{bmatrix} - 2\begin{bmatrix} 6 & -2 \\ 5 & 4 \end{bmatrix} + 4\begin{bmatrix} 7 & -6 \\ -4 & 2 \end{bmatrix}$
8. $1.3\begin{bmatrix} 3.7 \\ -5.4 \end{bmatrix} + 4.1\begin{bmatrix} 6.4 \\ -3.7 \end{bmatrix} - 6.2\begin{bmatrix} -0.8 \\ 7.4 \end{bmatrix}$

Use the matrices A, B, C, D, and E to find the following.

$A = \begin{bmatrix} 1 & 0 \\ 0 & 1 \end{bmatrix}$, $B = \begin{bmatrix} -1 & 0 \\ 0 & -1 \end{bmatrix}$, $C = \begin{bmatrix} 2 & -2 \\ -3 & 3 \end{bmatrix}$, $D = \begin{bmatrix} -2 & 2 \\ 3 & -3 \end{bmatrix}$, $E = \begin{bmatrix} 5 & -3 \\ -2 & 4 \end{bmatrix}$

9. $A + B$
10. $C + D$
11. $A - B$
12. $4B$
13. $D - C$
14. $E + 2A$
15. $D - 2B$
16. $2A + 3E - D$

Lesson 4-3

(pages 167–174)

Determine whether each matrix product is defined. If so, state the dimensions of the product.

1. $A_{4 \times 3} \cdot B_{3 \times 4}$
2. $R_{2 \times 5} \cdot S_{2 \times 5}$
3. $G_{2 \times 5} \cdot H_{2 \times 3}$
4. $M_{3 \times 8} \cdot N_{8 \times 2}$
5. $X_{4 \times 3} \cdot Y_{2 \times 6}$
6. $J_{5 \times 3} \cdot K_{3 \times 6}$
7. $C_{m \times n} \cdot D_{n \times p}$
8. $X_{2 \times 9} \cdot Y_{9 \times 7}$

Find each product, if possible.

9. $\begin{bmatrix} -3 & 4 \end{bmatrix} \cdot \begin{bmatrix} -1 \\ 2 \end{bmatrix}$
10. $\begin{bmatrix} 2 & -4 \\ 0 & 5 \end{bmatrix} \cdot \begin{bmatrix} 1 & 3 \\ -2 & -1 \end{bmatrix}$
11. $\begin{bmatrix} 1 & 3 \\ -2 & -1 \end{bmatrix} \cdot \begin{bmatrix} 2 & -4 \\ 0 & 5 \end{bmatrix}$
12. $\begin{bmatrix} 3 & 2 \\ 5 & 2 \end{bmatrix} \cdot \begin{bmatrix} -8 \\ 15 \end{bmatrix}$
13. $\begin{bmatrix} -1 \\ 2 \\ 1 \end{bmatrix} \cdot \begin{bmatrix} 7 & 6 & 1 \\ 2 & -4 & 0 \end{bmatrix}$
14. $\begin{bmatrix} 0 & 1 & -2 \\ 5 & 3 & -4 \\ -1 & 0 & 0 \end{bmatrix} \cdot \begin{bmatrix} 1 & -3 & 0 \\ 2 & 0 & -1 \\ 0 & 1 & -2 \end{bmatrix}$
15. $\begin{bmatrix} 3 & -2 \\ 4 & 5 \end{bmatrix} \cdot \begin{bmatrix} 1 & 0 \\ 0 & 1 \end{bmatrix}$
16. $\begin{bmatrix} -1 & 0 & 2 \\ -6 & 5 & -3 \end{bmatrix} \cdot \begin{bmatrix} -2 \\ 1 \\ 7 \end{bmatrix}$

Lesson 4-4

(pages 175–181)

For Exercises 1–3, use the following information.
Triangle XYZ with vertices $X(2, 5)$, $Y(-3, 1)$, and $Z(1, -4)$ is translated 3 units right and 2 units down.

1. Write the translation matrix.
2. Find the coordinates of $\triangle X'Y'Z'$.
3. Graph the preimage and the image.

For Exercises 4–6, use the following information.
The vertices of quadrilateral $ABCD$ are $A(1, 1)$, $B(-2, 3)$, $C(-4, -1)$, and $D(2, -3)$. The quadrilateral is dilated so that its perimeter is 2 times the original perimeter.

4. Write the coordinates for $ABCD$ in a vertex matrix.
5. Find the coordinates of the image $A'B'C'D'$.
6. Graph $ABCD$ and $A'B'C'D'$.

For Exercises 7–13, use the following information.
The vertices of $\triangle MQN$ are $M(2, 4)$, $Q(3, -5)$, and $N(1, -1)$.

7. Write the coordinates of $\triangle MQN$ in a vertex matrix.
8. Write the reflection matrix for reflecting over the line $y = x$.
9. Find the coordinates of $\triangle M'Q'N'$ after the reflection.
10. Graph $\triangle MQN$ and $\triangle M'Q'N'$.
11. Write a rotation matrix for rotating $\triangle MQN$ 90° counterclockwise about the origin.
12. Find the coordinates of $\triangle M'Q'N'$ after the rotation.
13. Graph $\triangle MQN$ and $\triangle M'Q'N'$.

Lesson 4-5

(pages 182–188)

Find the value of each determinant.

1. $\begin{vmatrix} 1 & -5 \\ 3 & 4 \end{vmatrix}$
2. $\begin{vmatrix} 1 & 1 \\ 1 & 1 \end{vmatrix}$
3. $\begin{vmatrix} 2 & -2 \\ 3 & -3 \end{vmatrix}$
4. $\begin{vmatrix} 2 & -3 \\ -3 & -2 \end{vmatrix}$

Evaluate each determinant using expansion by minors.

5. $\begin{vmatrix} 2 & -3 & 5 \\ 1 & -2 & -7 \\ -1 & 4 & -3 \end{vmatrix}$
6. $\begin{vmatrix} 0 & -1 & 2 \\ -2 & 1 & 0 \\ 2 & 0 & -1 \end{vmatrix}$
7. $\begin{vmatrix} 4 & 3 & -2 \\ 2 & 5 & -8 \\ 6 & 4 & -1 \end{vmatrix}$
8. $\begin{vmatrix} -3 & 0 & 2 \\ 1 & -2 & -1 \\ 0 & 5 & 0 \end{vmatrix}$

Evaluate each determinant using diagonals.

9. $\begin{vmatrix} 3 & 2 & -1 \\ 2 & 3 & 0 \\ -1 & 0 & 3 \end{vmatrix}$
10. $\begin{vmatrix} 1 & 0 & 0 \\ 0 & 1 & 0 \\ 0 & 0 & 1 \end{vmatrix}$
11. $\begin{vmatrix} 6 & 4 & -1 \\ 2 & 5 & -8 \\ 4 & 3 & -2 \end{vmatrix}$
12. $\begin{vmatrix} 6 & 12 & 15 \\ 9 & 3 & 14 \\ 5 & 6 & 3 \end{vmatrix}$

Lesson 4-6

(pages 189–194)

Use Cramer's Rule to solve each system of equations.

1. $5x - 3y = 19$
   $7x + 2y = 8$
2. $4p - 3q = 22$
   $2p + 8q = 30$
3. $-x + y = 5$
   $2x + 4y = 38$
4. $\frac{1}{3}x - \frac{1}{2}y = -8$
   $\frac{3}{5}x + \frac{5}{6}y = -4$
5. $\frac{1}{4}c + \frac{2}{3}d = 6$
   $\frac{3}{4}c - \frac{5}{3}d = -4$
6. $0.3a + 1.6b = 0.44$
   $0.4a + 2.5b = 0.66$
7. $x + y + z = 6$
   $2x - y - z = -3$
   $3x + y - 2z = -1$
8. $2a + b - c = -6$
   $a - 2b + c = 8$
   $-a - 3b + 2c = 14$
9. $r + 2s - t = 10$
   $-2r + 3s + t = 6$
   $3r - 2s + 2t = -19$

Extra Practice

Lesson 4-7

(pages 195–201)

Determine whether each pair of matrices are inverses.

1. $A = \begin{bmatrix} -7 & -6 \\ 8 & 7 \end{bmatrix}, B = \begin{bmatrix} -7 & -6 \\ 8 & 7 \end{bmatrix}$

2. $C = \begin{bmatrix} -3 & 4 \\ 2 & -2 \end{bmatrix}, D = \begin{bmatrix} -2 & -2 \\ -4 & -3 \end{bmatrix}$

3. $X = \begin{bmatrix} 1 & 0 \\ 0 & 1 \end{bmatrix}, Y = \begin{bmatrix} 1 & 0 \\ 0 & 1 \end{bmatrix}$

4. $N = \begin{bmatrix} 1 & 0 \\ 0 & 1 \end{bmatrix}, M = \begin{bmatrix} 1 & 1 \\ 1 & 1 \end{bmatrix}$

Find the inverse of each matrix, if it exists.

5. $\begin{bmatrix} 2 & 3 \\ 1 & 1 \end{bmatrix}$

6. $\begin{bmatrix} 3 & 2 \\ 0 & -4 \end{bmatrix}$

7. $\begin{bmatrix} 3 & 8 \\ 0 & -1 \end{bmatrix}$

8. $\begin{bmatrix} 3 & -6 \\ 2 & -4 \end{bmatrix}$

9. $\begin{bmatrix} 2 & 4 \\ 2 & 3 \end{bmatrix}$

10. $\begin{bmatrix} 8 & -5 \\ -6 & 4 \end{bmatrix}$

11. $\begin{bmatrix} 10 & 3 \\ 5 & -2 \end{bmatrix}$

12. $\begin{bmatrix} -3 & 4 \\ -4 & 8 \end{bmatrix}$

13. $\begin{bmatrix} -1 & -3 \\ -2 & -4 \end{bmatrix}$

14. $\begin{bmatrix} -1 & 0 \\ 0 & -1 \end{bmatrix}$

15. $\begin{bmatrix} 3 & -3 \\ 3 & -3 \end{bmatrix}$

16. $\begin{bmatrix} \frac{1}{3} & -\frac{1}{2} \\ -\frac{1}{4} & \frac{1}{5} \end{bmatrix}$

Lesson 4-8

(pages 202–207)

Write a matrix equation for each system of equations.

1. $5a + 3b = 6$; $2a - b = 9$

2. $3x + 4y = -8$; $2x - 3y = 6$

3. $m + 3n = 1$; $4m - n = -22$

4. $4c - 3d = -1$; $5c - 2d = 39$

5. $x + 2y - z = 6$; $-2x + 3y + z = 1$; $x + y + 3z = 8$

6. $2a - 3b - c = 4$; $4a + b + c = 15$; $a - b - c = -2$

Solve each matrix equation or system of equations by using inverse matrices.

7. $\begin{bmatrix} 3 & 4 \\ 2 & -5 \end{bmatrix} \cdot \begin{bmatrix} x \\ y \end{bmatrix} = \begin{bmatrix} 33 \\ -1 \end{bmatrix}$

8. $\begin{bmatrix} -1 & 1 \\ 7 & -6 \end{bmatrix} \cdot \begin{bmatrix} x \\ y \end{bmatrix} = \begin{bmatrix} 0 \\ 3 \end{bmatrix}$

9. $\begin{bmatrix} 1 & 0 \\ 0 & 1 \end{bmatrix} \cdot \begin{bmatrix} x \\ y \end{bmatrix} = \begin{bmatrix} -29 \\ 52 \end{bmatrix}$

10. $5x - y = 7$; $8x + 2y = 4$

11. $3m + n = 4$; $2m + 2n = 3$

12. $6c + 5d = 7$; $3c - 10d = -4$

13. $3a - 5b = 1$; $a + 3b = 5$

14. $2r - 7s = 24$; $-r + 8s = -21$

15. $x + y = -3$; $3x - 10y = 43$

16. $2m - 3n = 3$; $-4m + 9n = -8$

17. $x + y = 1$; $2x - 2y = -12$

Lesson 5-1

(pages 222–228)

Simplify. Assume that no variable equals 0.

1. $x^7 \cdot x^3 \cdot x$
2. $m^8 \cdot m \cdot m^{10}$
3. $7^5 \cdot 7^2$
4. $(-3)^4(-3)$
5. $\frac{t^{12}}{t}$
6. $-\frac{16x^8}{8x^2}$
7. $\frac{6^5}{6^3}$
8. $\frac{p^5q^7}{p^2q^5}$
9. $-(m^3)^8$
10. $\frac{x}{x^7}$
11. $(3^5)^7$
12. -3^4
13. $(abc)^3$
14. $(x^2)^5$
15. $-\left(\frac{x}{5}\right)^2$
16. $(b^4)^6$
17. $(-2y^5)^2$
18. $3x^0$
19. $(5x^4)^{-2}$
20. $\left(\frac{5a^7}{2b^5c}\right)^3$
21. $(-3)^{-2}$
22. -3^{-2}
23. $\frac{1}{x^{-3}}$
24. $\frac{5^6a^{x+y}}{5^4a^{x-y}}$

Evaluate. Express the result in scientific notation.

25. $(8.95 \times 10^9)(1.82 \times 10^7)$
26. $(3.1 \times 10^5)(7.9 \times 10^{-8})$
27. $\frac{(2.38 \times 10^{13})(7.56 \times 10^{-5})}{(4.2 \times 10^{18})}$

Lesson 5-2
(pages 229–232)

Determine whether each expression is a polynomial. If it is a polynomial, state the degree of the polynomial.

1. $5x - 3x^2 + 7$
2. $\sqrt{7} - u^3$
3. $5r^3 + 7r^2s - 3rs^2 + 6s^3$
4. $\frac{xy}{3}$
5. $2 + \frac{7}{w}$
6. $\sqrt{a - 1}$

Simplify.

7. $(4x^3 + 5x - 7x^2) + (-2x^3 + 5x^2 - 7y^2)$
8. $(2x^2 - 3x + 11) + (7x^2 + 2x - 8)$
9. $(-3x^2 + 7x + 23) + (-8x^2 - 5x + 13)$
10. $(-3x^2 + 7x + 23) - (-8x^2 - 5x + 13)$
11. $(5x^2 + 7x + 23) - (x^2 - 9x - 5) - (-3x^2 + 4x + 2)$
12. $5a^2b(4a - 3b)$
13. $\frac{7}{uw}\left(4u^2w^3 - 5uw + \frac{w}{7u}\right)$
14. $-4x^5(-3x^4 - x^3 + x + 7)$
15. $(2x - 3)(4x + 7)$
16. $(3x - 5)(-2x - 1)$
17. $(3x - 5)(2x - 1)$
18. $(2x + 5)(2x - 5)$
19. $(3x - 7)(3x + 7)$
20. $(5 + 2w)(5 - 2w)$
21. $(2a^2 + 8)(2a^2 - 8)$
22. $(-5x + 10)(-5x - 10)$
23. $(4x - 3)^2$
24. $(5x + 6)^2$
25. $(-x + 1)^2$
26. $\frac{3}{4}x(x^2 + 4x + 14)$
27. $-\frac{1}{2}a^2(a^3 - 6a^2 + 5a)$

Lesson 5-3
(pages 233–238)

Simplify.

1. $\frac{18r^3s^2 + 36r^2s^3}{9r^2s^2}$
2. $\frac{15v^3w^2 - 5v^4w^3}{-5v^4w^3}$
3. $\frac{x^2 - x + 1}{x}$
4. $(5bh + 5ch) \div (b + c)$
5. $(25c^4d + 10c^3d^2 - cd) \div 5cd$
6. $(16f^{18} + 20f^9 - 8f^6) \div 4f^3$
7. $(33m^5 + 55mn^5 - 11m^3)(11m)^{-1}$
8. $(8g^3 + 19g^2 - 12g + 9) \div (g + 3)$
9. $(p^{21} + 3p^{14} + p^7 - 2)(p^7 + 2)^{-1}$
10. $(15x^3 + 19x^2y - 7xy^2 + 5y^3) \div (3x + 5y)$
11. $(8k^2 - 56k + 98) \div (2k - 7)$
12. $(2r^2 + 5r - 3) \div (r + 3)$
13. $(n^3 + 125) \div (n + 5)$

Use synthetic division to find each quotient.

14. $(10y^4 + 3y^2 - 7) \div (2y^2 - 1)$
15. $(q^4 + 8q^3 + 3q + 17) \div (q + 8)$
16. $(15v^3 + 8v^2 - 21v + 6) \div (5v - 4)$
17. $(-2x^3 + 15x^2 - 10x + 3) \div (x + 3)$
18. $(6a^4 - 22a^3 - 9a^2 + 9a - 17) \div (a - 4)$
19. $(5s^3 + s^2 - 7) \div (s + 1)$
20. $(t^4 - 2t^3 + t^2 - 3t + 2) \div (t - 2)$
21. $(z^4 - 3z^3 - z^2 - 11z - 4) \div (z - 4)$
22. $(3r^4 - 6r^3 - 2r^2 + r - 6) \div (r + 1)$
23. $(2b^3 - 11b^2 + 12b + 9) \div (b - 3)$

Lesson 5-4
(pages 239–244)

Factor completely. If the polynomial is not factorable, write *prime*.

1. $14a^3b^3c - 21a^2b^4c + 7a^2b^3c$
2. $10ax - 2xy - 15ab + 3by$
3. $x^2 + x - 42$
4. $2x^2 + 5x + 3$
5. $6x^2 + 71x - 12$
6. $6x^4 - 12x^3 + 3x^2$
7. $x^2 - 6x + 2$
8. $x^2 - 2x - 15$
9. $6x^2 + 23x + 20$
10. $24x^2 - 76x + 40$
11. $6p^2 - 13pq - 28q^2$
12. $2x^2 - 6x + 3$
13. $x^2 + 49 - 14x$
14. $9x^2 - 64$
15. $36 - t^{10}$
16. $x^2 + 16$
17. $a^4 - 81b^4$
18. $3a^3 + 12a^2 - 63a$
19. $x^3 - 8x^2 + 15x$
20. $x^2 + 6x + 9$
21. $18x^3 - 8x$
22. $3x^2 - 42x + 40$
23. $2x^2 + 4x - 1$
24. $2x^3 + 6x^2 + x + 3$
25. $35ac - 3bd - 7ad + 15bc$
26. $5h^2 - 10hj + h - 2j$

Simplify. Assume that no denominator is equal to 0.

27. $\frac{x^2 + 8x + 15}{x^2 + 4x + 3}$
28. $\frac{x^2 + x - 2}{x^2 - 6x + 5}$
29. $\frac{x^2 - 15x + 56}{x^2 - 4x - 21}$
30. $\frac{x^2 + x - 6}{x^3 + 9x^2 + 27x + 27}$

Lesson 5-5

(pages 245–249)

Use a calculator to approximate each value to three decimal places.

1. $\sqrt{289}$
2. $\sqrt{7832}$
3. $\sqrt[4]{0.0625}$
4. $\sqrt[3]{-343}$
5. $\sqrt[10]{32^4}$
6. $\sqrt[3]{49}$
7. $\sqrt[5]{5}$
8. $-\sqrt[4]{25}$

Simplify.

9. $\sqrt{9h^{22}}$
10. $\sqrt[5]{0}$
11. $\sqrt{\frac{16}{9}}$
12. $\sqrt{\left(-\frac{2}{3}\right)^4}$
13. $\sqrt[5]{-32}$
14. $-\sqrt{-144}$
15. $\sqrt[4]{a^{16}b^8}$
16. $\pm\sqrt[4]{81x^4}$
17. $\sqrt[5]{\frac{1}{100{,}000}}$
18. $\sqrt[3]{-d^6}$
19. $\sqrt[5]{p^{25}q^{15}r^5s^{20}}$
20. $\sqrt[4]{(2x^2-y^8)^8}$
21. $\pm\sqrt{16m^6n^2}$
22. $-\sqrt[3]{(2x-y)^3}$
23. $\sqrt[4]{(r+s)^4}$
24. $\sqrt{9a^2+6a+1}$
25. $\sqrt{4y^2+12y+9}$
26. $-\sqrt{x^2-2x+1}$
27. $\pm\sqrt{x^2+2x+1}$
28. $\sqrt[3]{a^3+6a^2+12a+8}$

Lesson 5-6

(pages 250–256)

Simplify.

1. $\sqrt{75}$
2. $7\sqrt{12}$
3. $\sqrt[3]{81}$
4. $\sqrt{5r^5}$
5. $\sqrt[4]{7^8x^5y^6}$
6. $3\sqrt{5}+6\sqrt{5}$
7. $\sqrt{18}-\sqrt{50}$
8. $4\sqrt[3]{32}+\sqrt[3]{500}$
9. $\sqrt{12}\sqrt{27}$
10. $3\sqrt{12}+2\sqrt{300}$
11. $\sqrt[3]{54}-\sqrt[3]{24}$
12. $\sqrt{10}(2-\sqrt{5})$
13. $-\sqrt{3}(2\sqrt{6}-\sqrt{63})$
14. $(5+\sqrt{2})(3+\sqrt{3})$
15. $(2+\sqrt{5})(2-\sqrt{5})$
16. $(8+\sqrt{11})^2$
17. $(\sqrt{3}+\sqrt{6})(\sqrt{3}-\sqrt{6})$
18. $(\sqrt{8}+\sqrt{13})^2$
19. $(1-\sqrt{7})(4+\sqrt{7})$
20. $(5-2\sqrt{7})^2$
21. $\sqrt{\frac{3m^3}{24n^5}}$
22. $\frac{\sqrt{18}}{\sqrt{32}}$
23. $2\sqrt[3]{\frac{r^5}{2s^2t}}$
24. $\sqrt[3]{\frac{4}{7}}$
25. $\sqrt[5]{\frac{32}{a^4}}$
26. $\sqrt{\frac{2}{3}}-\sqrt{\frac{3}{8}}$
27. $\frac{5}{3-\sqrt{10}}$
28. $\frac{\sqrt{5}}{1+\sqrt{3}}$
29. $\frac{-2+\sqrt{7}}{2+\sqrt{7}}$
30. $\frac{1-\sqrt{3}}{1+\sqrt{8}}$
31. $\frac{\sqrt{2}+\sqrt{3}}{\sqrt{2}-\sqrt{3}}$
32. $\frac{x+\sqrt{5}}{x-\sqrt{5}}$

Lesson 5-7

(pages 257–262)

Write each expression in radical form.

1. $10^{\frac{1}{3}}$
2. $8^{\frac{1}{4}}$
3. $a^{\frac{2}{3}}$
4. $(b^2)^{\frac{3}{4}}$

Write each radical using rational exponents.

5. $\sqrt{35}$
6. $\sqrt[4]{32}$
7. $\sqrt[3]{27a^2x}$
8. $\sqrt[5]{25ab^3c^4}$

Evaluate each expression.

9. $2401^{\frac{1}{4}}$
10. $27^{\frac{4}{3}}$
11. $(-32)^{\frac{2}{5}}$
12. $-81^{\frac{3}{4}}$
13. $(-125)^{-\frac{2}{3}}$
14. $16^{\frac{5}{2}}\cdot 16^{\frac{1}{2}}$
15. $8^{-\frac{2}{3}}\cdot 64^{\frac{1}{6}}$
16. $\left(\frac{48}{1875}\right)^{-\frac{5}{4}}$

Simplify each expression.

17. $7^{\frac{5}{9}}\cdot 7^{\frac{4}{9}}$
18. $32^{\frac{2}{3}}\cdot 32^{\frac{3}{5}}$
19. $\left(k^{\frac{8}{5}}\right)^5$
20. $x^{\frac{2}{5}}\cdot x^{\frac{8}{5}}$
21. $m^{\frac{2}{5}}\cdot m^{\frac{4}{5}}$
22. $\left(p^{\frac{5}{4}}\cdot q^{\frac{7}{2}}\right)^{\frac{8}{3}}$
23. $\left(4^{\frac{9}{2}}c^{\frac{3}{2}}\right)^2$
24. $\frac{7^{\frac{3}{4}}}{7^{\frac{5}{3}}}$
25. $\frac{1}{t^{\frac{9}{5}}}$
26. $a^{\frac{-8}{7}}$
27. $\frac{r}{r^{\frac{7}{5}}}$
28. $\sqrt[4]{36}$
29. $\sqrt[4]{9a^2}$
30. $\sqrt[3]{\sqrt{81}}$
31. $\frac{v^{\frac{11}{7}}-v^{\frac{4}{7}}}{v^{\frac{4}{7}}}$
32. $\frac{1}{5^{\frac{1}{2}}+3^{\frac{1}{2}}}$

Lesson 5-8

(pages 263–267)

Solve each equation or inequality.

1. $\sqrt{x} = 16$
2. $\sqrt{z + 3} = 7$
3. $\sqrt[3]{a + 5} = 1$
4. $5\sqrt{s} - 8 = 3$
5. $\sqrt[4]{m + 7} + 11 = 9$
6. $d + \sqrt{d^2 - 8} = 4$
7. $g\sqrt{5} + 4 = g + 4$
8. $\sqrt{x - 8} = \sqrt{13 + x}$
9. $\sqrt{3x + 10} = 1 + \sqrt{2x + 5}$
10. $\sqrt{3x + 9} > 2$
11. $\sqrt{3n - 1} \leq 5$
12. $2 - 4\sqrt{21 - 6c} < -6$
13. $\sqrt{5y + 4} > 8$
14. $\sqrt{2w + 3} + 5 \geq 7$
15. $\sqrt{x + 29} - 3 = \sqrt{x - 16}$
16. $\sqrt{3x + 25} + \sqrt{10 - 2x} = 0$
17. $\sqrt{2c + 3} - 7 > 0$
18. $\sqrt{3z - 5} - 3 = 1$
19. $\sqrt{5y + 1} + 6 < 10$
20. $\sqrt{3n + 1} - 2 \leq 6$
21. $\sqrt{y - 5} - \sqrt{y} \geq 1$
22. $(5n - 1)^{\frac{1}{2}} = 0$
23. $(7x - 6)^{\frac{1}{3}} + 1 = 3$
24. $(6a - 8)^{\frac{1}{4}} + 9 \geq 10$

Lesson 5-9

(pages 270–275)

Simplify.

1. $\sqrt{-289}$
2. $\sqrt{-\frac{25}{121}}$
3. $\sqrt{-625b^8}$
4. $\sqrt{-\frac{28t^6}{27s^5}}$
5. $(7i)^2$
6. $(6i)(-2i)(11i)$
7. $(\sqrt{-8})(\sqrt{-12})$
8. $-i^{22}$
9. $i^{17} \cdot i^{12} \cdot i^{26}$
10. $(14 - 5i) + (-8 + 19i)$
11. $(7i) - (2 + 3i)$
12. $(2 + 2i) - (5 + i)$
13. $(7 + 3i)(7 - 3i)$
14. $(8 - 2i)(5 + i)$
15. $(6 + 8i)^2$
16. $\frac{3}{6 - 2i}$
17. $\frac{5i}{3 + 4i}$
18. $\frac{3 - 7i}{5 + 4i}$

Solve each equation.

19. $x^2 + 8 = 3$
20. $\frac{4x^2}{49} + 6 = 3$
21. $8x^2 + 5 = 1$
22. $12 - 9x^2 = 38$
23. $9x^2 + 7 = 4$
24. $\frac{1}{2}x^2 + 1 = 0$

Lesson 6-1

(pages 286–293)

For Exercises 1–12, complete parts a–c for each quadratic function.

a. Find the y-intercept, the equation of the axis of symmetry, and the x-coordinate of the vertex.

b. Make a table of values that includes the vertex.

c. Use this information to graph the function.

1. $f(x) = 6x^2$
2. $f(x) = -x^2$
3. $f(x) = x^2 + 5$
4. $f(x) = -x^2 - 2$
5. $f(x) = 2x^2 + 1$
6. $f(x) = -3x^2 + 6x$
7. $f(x) = x^2 + 6x - 3$
8. $f(x) = x^2 - 2x - 8$
9. $f(x) = -3x^2 - 6x + 12$
10. $f(x) = x^2 + 5x - 6$
11. $f(x) = 2x^2 + 7x - 4$
12. $f(x) = -5x^2 + 10x + 1$

Determine whether each function has a maximum or a minimum value. Then find the maximum or minimum value of each function.

13. $f(x) = 9x^2$
14. $f(x) = 9 - x^2$
15. $f(x) = x^2 - 5x + 6$
16. $f(x) = 2 + 7x - 6x^2$
17. $f(x) = 4x^2 - 9$
18. $f(x) = x^2 + 2x + 1$
19. $f(x) = 8 - 3x - 4x^2$
20. $f(x) = x^2 - x + \frac{5}{4}$
21. $f(x) = -x^2 + \frac{14}{3}x + \frac{5}{3}$

Lesson 6-2

(pages 294–299)

Use the related graph of each equation to determine its solutions.

1. $x^2 + x - 6 = 0$

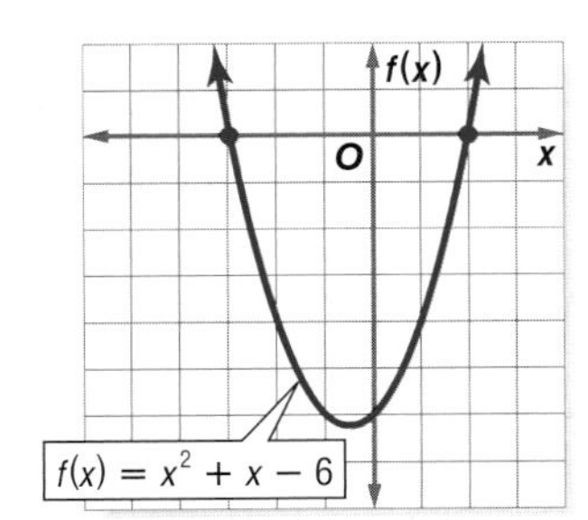

2. $-2x^2 = 0$

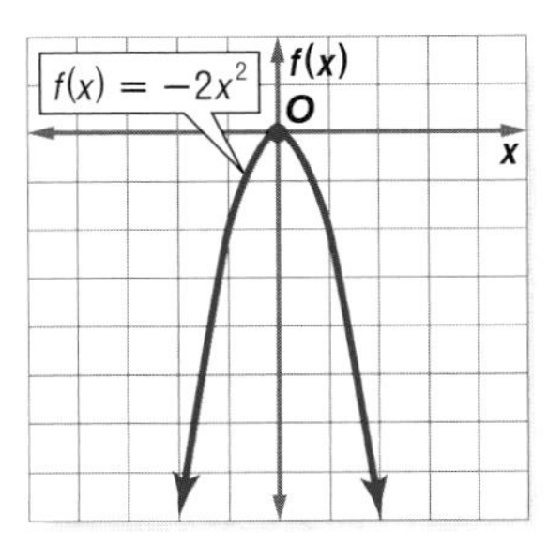

3. $x^2 - 4x - 5 = 0$

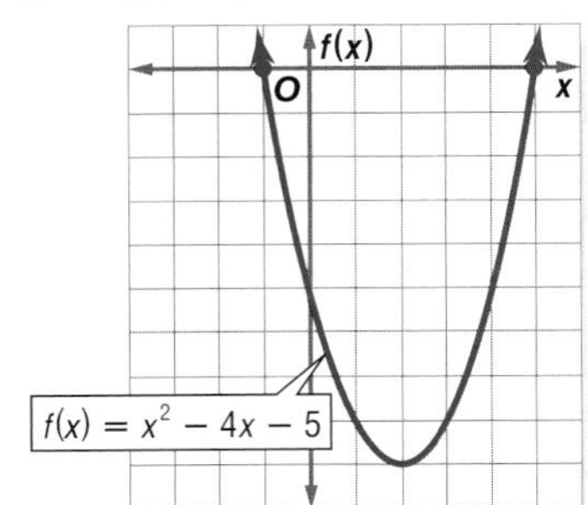

Solve each equation by graphing. If exact roots cannot be found, state the consecutive integers between which the roots are located.

4. $x^2 - 2x = 0$
5. $x^2 + 8x - 20 = 0$
6. $-2x^2 + 10x - 5 = 0$
7. $-5x + 2x^2 - 3 = 0$
8. $3x^2 - x + 8 = 0$
9. $-x^2 + 2 = 7x$
10. $4x^2 - 4x + 1 = 0$
11. $4x + 1 = 3x^2$
12. $x^2 = -9x$
13. $x^2 + 6x - 27 = 0$
14. $0.4x^2 + 1 = 0$
15. $0.5x^2 + 3x - 2 = 0$

Lesson 6-3

(pages 301–305)

Solve each equation by factoring.

1. $x^2 + 7x + 10 = 0$
2. $3x^2 = 75x$
3. $2x^2 + 7x = 9$
4. $8x^2 = 48 - 40x$
5. $5x^2 = 20x$
6. $12x^2 - 71x - 6 = 0$
7. $16x^2 - 64 = 0$
8. $5x^2 - 45x + 90 = 0$
9. $24x^2 - 15 = 2x$
10. $x^2 = 72 - x$
11. $2x^2 + 5x + 3 = 0$
12. $4x^2 + 9 = 12x$
13. $2x^2 - 8x = 0$
14. $8x^2 + 10x = 3$
15. $12x^2 - 5x = 3$
16. $x^2 + 8x + 12 = 0$
17. $x^2 + 9x + 14 = 0$
18. $9x^2 + 1 = 6x$
19. $6x^2 + 7x = 3$
20. $x^2 - 4x = 21$

Write a quadratic equation with the given roots. Write the equation in the form $ax^2 + bx + c = 0$, where a, b, and c are integers.

21. 2, 1
22. −3, 4
23. −1, −7
24. $-1, \frac{1}{2}$
25. $-5, \frac{1}{4}$
26. $-\frac{1}{3}, -\frac{1}{2}$

Lesson 6-4

(pages 306–312)

Find the value of c that makes each trinomial a perfect square. Then write the trinomial as a perfect square.

1. $x^2 - 4x + c$
2. $x^2 + 20x + c$
3. $x^2 - 11x + c$
4. $x^2 - \frac{2}{3}x + c$
5. $x^2 + 30x + c$
6. $x^2 + \frac{3}{8}x + c$
7. $x^2 - \frac{2}{5}x + c$
8. $x^2 - 3x + c$

Solve each equation by completing the square.

9. $x^2 + 3x - 4 = 0$
10. $x^2 + 5x = 0$
11. $x^2 + 2x - 63 = 0$
12. $3x^2 - 16x - 35 = 0$
13. $x^2 + 7x + 13 = 0$
14. $5x^2 - 8x + 2 = 0$
15. $x^2 - 6x + 11 = 0$
16. $x^2 - 12x + 36 = 0$
17. $8x^2 + 13x - 4 = 0$
18. $3x^2 + 5x + 6 = 0$
19. $x^2 + 14x - 1 = 0$
20. $4x^2 - 32x + 15 = 0$
21. $3x^2 - 11x - 4 = 0$
22. $x^2 + 8x - 84 = 0$
23. $x^2 - 7x + 5 = 0$
24. $x^2 + 3x - 8 = 0$
25. $x^2 - 5x - 10 = 0$
26. $3x^2 - 12x + 4 = 0$
27. $x^2 + 20x + 75 = 0$
28. $x^2 - 5x - 24 = 0$
29. $2x^2 + x - 21 = 0$

Lesson 6-5

(pages 313–319)

For Exercises 1–16, complete parts a–c for each quadratic equation.

a. Find the value of the discriminant.

b. Describe the number and type of roots.

c. Find the exact solutions by using the Quadratic Formula.

1. $x^2 + 7x + 13 = 0$ **2.** $6x^2 + 6x - 21 = 0$ **3.** $5x^2 - 5x + 4 = 0$ **4.** $9x^2 + 42x + 49 = 0$

5. $4x^2 - 16x + 3 = 0$ **6.** $2x^2 = 5x + 3$ **7.** $x^2 + 81 = 18x$ **8.** $3x^2 - 30x + 75 = 0$

9. $24x^2 + 10x = 43$ **10.** $9x^2 + 4 = 2x$ **11.** $7x = 8x^2$ **12.** $18x^2 = 9x + 45$

13. $x^2 - 4x + 4 = 0$ **14.** $4x^2 + 16x + 15 = 0$ **15.** $x^2 - 6x + 13 = 0$ **16.** $3x^2 = 108x$

Solve each equation by using the method of your choice. Find the exact solutions.

17. $x^2 + 4x + 29 = 0$ **18.** $4x^2 + 3x - 2 = 0$ **19.** $2x^2 + 5x = 9$ **20.** $x^2 = 8x - 16$

21. $7x^2 = 4x$ **22.** $2x^2 + 6x + 5 = 0$ **23.** $9x^2 - 30x + 25 = 0$ **24.** $3x^2 - 4x + 2 = 0$

Lesson 6-6

(pages 322–328)

Write each quadratic function in vertex form, if not already in that form. Then identify the vertex, axis of symmetry, and direction of opening.

1. $y = (x + 6)^2 - 1$ **2.** $y = 2(x - 8)^2 - 5$ **3.** $y = -(x + 1)^2 + 7$ **4.** $y = -9(x - 7)^2 + 3$

5. $y = -x^2 + 10x - 3$ **6.** $y = -2x^2 + 16x + 7$ **7.** $y = 3x^2 + 9x + 8$ **8.** $y = \frac{3}{4}x^2 - 6x - 5$

Graph each function.

9. $y = x^2 - 2x + 4$ **10.** $y = -3x^2 + 18x$ **11.** $y = -2x^2 - 4x + 1$ **12.** $y = 2x^2 - 8x + 9$

13. $y = \frac{1}{3}x^2 + 2x + 7$ **14.** $y = x^2 + 6x + 9$ **15.** $y = x^2 + 3x + 6$ **16.** $y = 2x^2 + 8x + 9$

17. $y = x^2 - 8x + 9$ **18.** $y = -x^2 - x + 10$ **19.** $y = -0.5x^2 + 4x - 3$ **20.** $y = -2x^2 - 8x - 1$

Write an equation for the parabola with the given vertex that passes through the given point.

21. vertex: $(-1, 5)$ point: $(2, -4)$

22. vertex: $(2, -1)$ point: $(-2, 7)$

23. vertex: $(-5, -3)$ point: $(-1, 5)$

24. vertex: $(0, -8)$ point: $(2, -2)$

Lesson 6-7

(pages 329–335)

Graph each inequality.

1. $y \leq 5x^2 + 3x - 2$ **2.** $y > -3x^2 + 2$ **3.** $y \geq x^2 - 8x$ **4.** $y \geq -x^2 - x + 3$

5. $y \leq 3x^2 + 4x - 8$ **6.** $y \leq -5x^2 + 2x - 3$ **7.** $y > 4x^2 + x$ **8.** $y \geq -x^2 - 3$

Use the graph of its related function to write the solutions of each inequality.

9. $x^2 - 4 \leq 0$

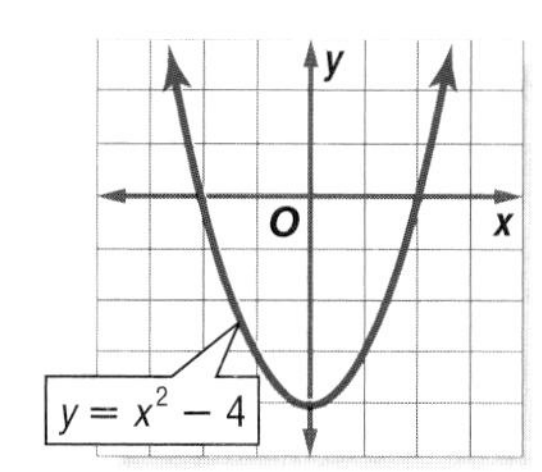

10. $-x^2 + 6x - 9 \geq 0$

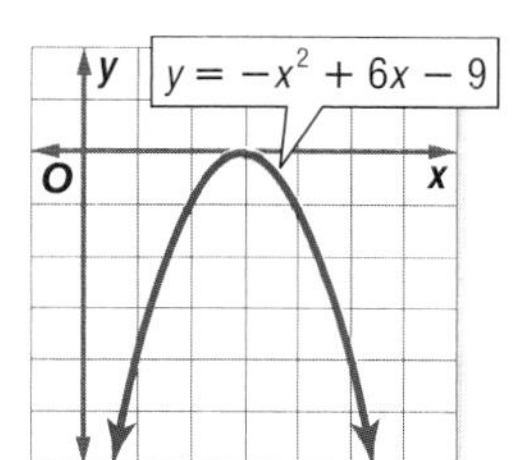

11. $x^2 + 4x - 5 < 0$

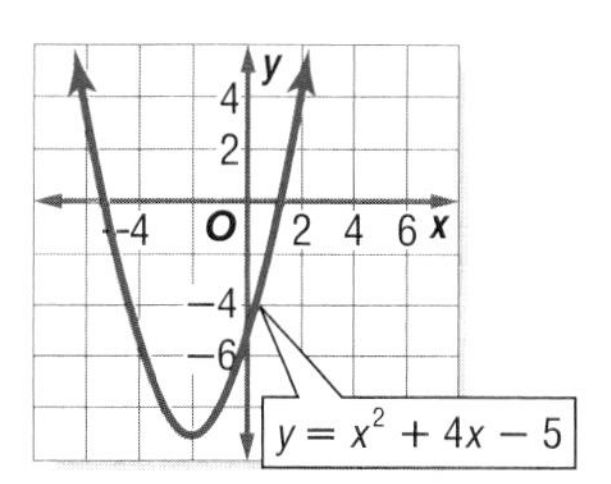

Solve each inequality algebraically.

12. $x^2 - 1 < 0$ **13.** $10x^2 - x - 2 \geq 0$ **14.** $-x^2 - 5x - 6 > 0$ **15.** $-3x^2 \geq 5$

16. $x^2 - 2x - 8 \leq 0$ **17.** $2x^2 \geq 5x + 12$ **18.** $x^2 + 3x - 4 > 0$ **19.** $2x - x^2 \leq -15$

Lesson 7-1

(pages 346–352)

State the degree and leading coefficient of each polynomial in one variable. If it is not a polynomial in one variable, explain why.

1. $2n^2 + 2m^2$
2. $5 - 3a^2$
3. $(x^2 + 2)(x^3 - 5)$
4. $-3a^3 + 5a - \frac{2}{a}$
5. $5c + c^2 - \frac{3}{5}$
6. $-4m - 7m^6 + 6m^5 + 3m^2$

Find $p(5)$ and $p(-1)$ for each function.

7. $p(x) = 7x - 3$
8. $p(x) = -3x^2 + 5x - 4$
9. $p(x) = 5x^4 + 2x^2 - 2x$
10. $p(x) = -13x^3 + 5x^2 - 3x + 2$
11. $p(x) = x^6 - 2$
12. $p(x) = \frac{2}{3}x^2 + 5x$
13. $p(x) = x^3 + x^2 - x + 1$
14. $p(x) = x^4 - x^2 - 1$
15. $p(x) = 1 - x^3$

If $p(x) = -2x^2 + 5x + 1$ and $q(x) = x^3 - 1$, find each value.

16. $q(n)$
17. $p(2b)$
18. $q(z^3)$
19. $p(3m^2)$
20. $q(x + 1)$
21. $p(3 - x)$
22. $q(a^2 - 2)$
23. $3q(h - 3)$
24. $5[p(c - 4)]$
25. $q(n - 2) + q(n^2)$
26. $-3p(4a) - p(a)$
27. $2[q(d^2 + 1)] + 3q(d)$

Lesson 7-2

(pages 353–358)

For Exercises 1–16, complete each of the following.
a. Graph each function by making a table of values.
b. Determine the values of x between which the real zeros are located.
c. Estimate the x-coordinates at which the relative maxima and relative minima occur.

1. $f(x) = x^3 + x^2 - 3x$
2. $f(x) = -x^4 + x^3 + 5$
3. $f(x) = x^3 - 3x^2 + 8x - 7$
4. $f(x) = 2x^5 + 3x^4 - 8x^2 + x + 4$
5. $f(x) = x^4 - 5x^3 + 6x^2 - x - 2$
6. $f(x) = 2x^6 + 5x^4 - 3x^2 - 5$
7. $f(x) = -x^3 - 8x^2 + 3x - 7$
8. $f(x) = -x^4 - 3x^3 + 5x$
9. $f(x) = x^5 - 7x^4 - 3x^3 + 2x^2 - 4x + 9$
10. $f(x) = x^4 - 5x^3 + x^2 - x - 3$
11. $f(x) = x^4 - 128x^2 + 960$
12. $f(x) = -x^5 + x^4 - 208x^2 + 145x + 9$
13. $f(x) = x^5 - x^3 - x + 1$
14. $f(x) = x^3 - 2x^2 - x + 5$
15. $f(x) = 2x^4 - x^3 + x^2 - x + 1$
16. $f(x) = -x^3 - x^2 - x - 1$

Lesson 7-3

(pages 360–364)

Write each polynomial in quadratic form, if possible.

1. $5x^{10} - 6x^5 - 3$
2. $2y^6 + 3y^4 + 10$
3. $z^6 - 8z^3$
4. $y^5 - 6y^3 + 4$
5. $x - 10x^{\frac{1}{2}} + 25$
6. $x^4 - 7x + 12$
7. $3r + 2r^{\frac{1}{2}} - 7$
8. $r^{\frac{2}{3}} - 5r^{\frac{1}{3}} + 6$
9. $x^{\frac{1}{2}} + 9x^{\frac{1}{4}} + 18$

Solve each equation.

10. $8x^3 - 27 = 0$
11. $2m^4 = 3m^2 + 5$
12. $3b^5 = 7b^3$
13. $x^3 + 10x^2 + 16x = 0$
14. $y^4 - 3y^2 + 2 = 0$
15. $a^3 = 125$
16. $5z^{\frac{2}{3}} - z^{\frac{1}{3}} = 4$
17. $x^{\frac{1}{2}} - 6x^{\frac{1}{4}} + 8 = 0$
18. $3m + m^{\frac{1}{2}} - 2 = 0$
19. $m - 9\sqrt{m} + 8 = 0$
20. $r^2 - 12r + 20 = 0$
21. $\sqrt[3]{x^2} - 8\sqrt[3]{x} + 15 = 0$
22. $m - 11\sqrt{m} + 30 = 0$
23. $y^3 - 8\sqrt{y^3} + 16 = 0$
24. $g^{\frac{2}{3}} - 2g^{\frac{1}{3}} - 8 = 0$

Lesson 7-4

(pages 365–370)

Use synthetic substitution to find $f(3)$ and $f(-4)$ for each function.

1. $f(x) = x^2 - 6x + 2$
2. $f(x) = x^3 + 5x - 6$
3. $f(x) = x^3 - x^2 - 3x + 1$
4. $f(x) = -3x^3 + 5x^2 + 7x - 3$
5. $f(x) = 3x^5 - 5x^3 + 2x - 8$
6. $f(x) = -2x^4 + 7x^3 + 8x^2 - 3x + 5$
7. $f(x) = 10x^3 + 2$
8. $f(x) = x^5 + x^4 + x^3 + x^2 + x + 1$

Given a polynomial and one of its factors, find the remaining factors of the polynomial. Some factors may not be binomials.

9. $(x^3 - x^2 + x + 14); (x + 2)$
10. $(5x^3 - 17x^2 + 6x); (x - 3)$
11. $(2x^3 + x^2 - 41x + 20); (x - 4)$
12. $(x^3 - 8); (x - 2)$
13. $(x^2 + 6x + 5); (x + 1)$
14. $(x^4 + x^3 + x^2 + x); (x + 1)$
15. $(x^3 - 8x^2 + x + 42); (x - 7)$
16. $(6x^4 + 13x^3 - 36x^2 - 43x + 30); (x - 2)$
17. $(x^4 + 5x^3 - 27x - 135); (x - 3)$
18. $(2x^3 - 15x^2 - 2x + 120); (2x + 5)$
19. $(6x^3 - 17x^2 + 6x + 8); (3x - 4)$
20. $(30x^3 - 68x^2 + 10x + 12); (5x - 3)$
21. $(10x^3 + x^2 - 46x + 35); (5x - 7)$
22. $(x^3 + 9x^2 + 23x + 15); (x + 1)$

Lesson 7-5

(pages 371–377)

Solve each equation and state the number and type of roots.

1. $-5x - 7 = 0$
2. $3x^2 + 10 = 0$
3. $x^4 - 2x^3 - 7x^2 - 2x - 8 = 0$
4. $x^4 - 2x^3 = 23x^2 - 60x$

State the number of positive real zeros, negative real zeros, and imaginary zeros for each function.

5. $f(x) = 5x^8 - x^6 + 7x^4 - 8x^2 - 3$
6. $f(x) = 6x^5 - 7x^2 + 5$
7. $f(x) = -2x^6 - 5x^5 + 8x^2 - 3x + 1$
8. $f(x) = 4x^3 + x^2 - 38x + 56$
9. $f(x) = 3x^8 - 15x^5 - 7x^4 - 8x^3 - 3$
10. $f(x) = -x^6 - 8x^5 - 5x^4 - 11x^3 - 2x^2 - 5x - 1$
11. $f(x) = 3x^4 - 5x^3 + 2x^2 - 7x + 5$
12. $f(x) = x^5 - x^4 + 7x^3 - 25x^2 + 8x - 13$

Find all of the zeros of the function.

13. $f(x) = x^3 - 7x^2 + 16x - 10$
14. $f(x) = 10x^3 + 7x^2 - 82x + 56$
15. $f(x) = x^3 - 16x^2 + 79x - 114$
16. $f(x) = -3x^3 + 6x^2 + 5x - 8$
17. $f(x) = 6x^4 + 13x^3 - 18x^2 - 7x + 6$
18. $f(x) = 4x^4 + 36x^3 + 57x^2 + 225x + 200$
19. $f(x) = 24x^3 + 64x^2 + 6x - 10$
20. $f(x) = 2x^3 + 2x^2 - 34x + 30$

Write a polynomial function of least degree with integral coefficients that has the given zeros.

21. $-3, 1, 2$
22. $-5, -3, 3, 5$
23. $-6, 6, -5i, 5i$
24. $3, \pm i\sqrt{7}$

Lesson 7-6

(pages 378–382)

List all of the possible rational zeros for each function.

1. $f(x) = 3x^5 - 7x^3 - 8x + 6$
2. $f(x) = 4x^3 + 2x^2 - 5x + 8$
3. $f(x) = 6x^9 - 7$

Find all of the rational zeros for each function.

4. $f(x) = x^4 + 3x^3 - 7x^2 - 27x - 18$
5. $f(x) = 6x^4 - 31x^3 - 119x^2 + 214x + 560$
6. $f(x) = 20x^4 - 16x^3 + 11x^2 - 12x - 3$
7. $f(x) = 2x^4 - 30x^3 + 117x^2 - 75x + 280$
8. $f(x) = 3x^4 + 8x^3 + 9x^2 + 32x - 12$
9. $f(x) = 2x^6 - 12x^5 + 17x^4 + 6x^3 - 10x^2 + 6x - 9$
10. $f(x) = 2x^5 - 6x^4 + 18x^3 - 54x^2 - 20x + 60$
11. $f(x) = x^5 - x^4 + x^3 + 3x^2 - x$

Find all of the zeros of each function.

12. $f(x) = x^4 + 8x^2 - 9$
13. $f(x) = 2x^4 - 6x^3 - 70x^2 - 30x - 400$
14. $f(x) = 3x^4 - 9x^2 - 12$
15. $f(x) = 4x^4 + 19x^2 - 63$

Lesson 7-7

(pages 383–389)

Find $(f+g)(x)$, $(f-g)(x)$, $(f \cdot g)(x)$, and $\left(\frac{f}{g}\right)(x)$ for each $f(x)$ and $g(x)$.

1. $f(x) = 3x + 5$, $g(x) = x - 3$
2. $f(x) = \sqrt{x}$, $g(x) = x^2$
3. $f(x) = 2x^2 - 5x + 8$, $g(x) = \frac{x-8}{3}$
4. $f(x) = x^2 - 5$, $g(x) = x^2 + 5$
5. $f(x) = x + 2$, $g(x) = x^2 + 4x + 4$
6. $f(x) = x^2 + 1$, $g(x) = x + 1$

For each set of ordered pairs, find $f \circ g$ and $g \circ f$ if they exist.

7. $f = \{(-1, 1), (2, -1), (-3, 5)\}$, $g = \{(1, -1), (-1, 2), (5, -3)\}$
8. $f = \{(0, 6), (5, -8), (-9, 2)\}$, $g = \{(-8, 3), (6, 4), (2, 1)\}$
9. $f = \{(8, 2), (6, 5), (-3, 4), (1, 0)\}$, $g = \{(2, 8), (5, 6), (4, -3), (0, 1)\}$
10. $f = \{(10, 4), (-1, 2), (5, 6), (-1, 0)\}$, $g = \{(-4, 10), (2, -9), (-7, 5), (-2, -1)\}$

Find $[g \circ h](x)$ and $[h \circ g](x)$.

11. $g(x) = 8 - 2x$, $h(x) = 3x$
12. $g(x) = x^2 - 7$, $h(x) = 3x + 2$
13. $g(x) = 2x + 7$, $h(x) = \frac{x-7}{2}$
14. $g(x) = 3x + 2$, $h(x) = 5 - 3x$

If $f(x) = x^2 + 1$, $g(x) = 2x$, and $h(x) = x - 1$, find each value.

15. $g[f(1)]$
16. $[f \circ h](3)$
17. $[h \circ f](3)$
18. $[g \circ f](-2)$
19. $g[h(-20)]$
20. $f[h(-3)]$
21. $g[f(a)]$
22. $[f \circ (g \circ f)](c)$

Lesson 7-8

(pages 390–394)

Find the inverse of each relation.

1. $\{(-2, 7), (3, 0), (5, -8)\}$
2. $\{(-3, 9), (-2, 4), (3, 9), (-1, 1)\}$
3. $\{(1, 5), (2, 3), (4, 3), (-1, 5)\}$

Find the inverse of each function. Then graph the function and its inverse.

4. $f(x) = x - 7$
5. $y = 2x + 8$
6. $g(x) = 3x - 8$
7. $h(x) = \frac{x}{5} + 1$
8. $y = -2$
9. $g(x) = 5 - 2x$
10. $y = -5x - 6$
11. $h(x) = -\frac{2}{3}x$
12. $y = \frac{x-5}{3}$
13. $y = \frac{1}{2}x - 1$
14. $f(x) = \frac{3x+8}{4}$
15. $g(x) = \frac{2x-1}{3}$

Determine whether each pair of functions are inverse functions.

16. $f(x) = \frac{2x-3}{5}$, $g(x) = \frac{3x-5}{3}$
17. $f(x) = 5x - 6$, $g(x) = \frac{x+6}{5}$
18. $f(x) = 6 - 3x$, $g(x) = 2 - \frac{1}{3}x$
19. $f(x) = 3x - 7$, $g(x) = \frac{1}{3}x + 7$

Lesson 7-9

(pages 395–399)

Graph each function. State the domain and range of each function.

1. $y = \sqrt{x - 4}$
2. $y = \sqrt{x + 3} - 1$
3. $y = \frac{1}{3}\sqrt{x + 2}$
4. $y = \sqrt{2x + 5}$
5. $y = -\sqrt{4x}$
6. $y = 2\sqrt{x}$
7. $y = -3\sqrt{x}$
8. $y = \sqrt{x} + 5$
9. $y = \sqrt{2x} - 1$
10. $y = 5\sqrt{x} + 1$
11. $y = \sqrt{x + 1} - 2$
12. $y = 6 - \sqrt{x + 3}$

Graph each inequality.

13. $y > \sqrt{2x}$
14. $y \leq \sqrt{-5x}$
15. $y \geq \sqrt{x + 6} + 6$
16. $y < \sqrt{3x + 1} + 2$
17. $y \geq \sqrt{8x - 3} + 1$
18. $y < \sqrt{5x - 1} + 3$

Lesson 8-1

(pages 412–416)

Find the midpoint of the line segment with endpoints at the given coordinates.

1. $(7, -3), (-11, 13)$
2. $(16, 29), (-7, 2)$
3. $(43, -18), (-78, -32)$
4. $(-7.54, 3.42), (4.89, -9.28)$
5. $\left(\frac{1}{2}, \frac{1}{4}\right), \left(\frac{2}{3}, \frac{3}{5}\right)$
6. $\left(-\frac{1}{4}, \frac{2}{3}\right), \left(-\frac{1}{2}, -\frac{1}{2}\right)$

Find the distance between each pair of points with the given coordinates.

7. $(5, 7), (3, 19)$
8. $(-2, -1), (5, 3)$
9. $(-3, 15), (7, -8)$
10. $(6, -3), (-4, -9)$
11. $(3.89, -0.38), (4.04, -0.18)$
12. $(5\sqrt{3}, 2\sqrt{2}), (-11\sqrt{3}, -4\sqrt{2})$
13. $\left(\frac{1}{4}, 0\right), \left(-\frac{2}{3}, \frac{1}{2}\right)$
14. $\left(4, -\frac{5}{6}\right), \left(-2, \frac{1}{6}\right)$
15. A circle has a radius with endpoints at $(-3, 1)$ and $(2, -5)$. Find the circumference and area of the circle. Write the answer in terms of π.
16. Triangle ABC has vertices $A(0, 0)$, $B(-3, 4)$, and $C(2, 6)$. Find the perimeter of the triangle.

Lesson 8-2

(pages 419–425)

Write each equation in standard form.

1. $y = x^2 - 4x + 7$
2. $y = 2x^2 + 12x + 17$
3. $x = 3y^2 - 6y + 5$

Identify the coordinates of the vertex and focus, the equations of the axis of symmetry and directrix, and the direction of opening of the parabola with the given equation. Then find the length of the latus rectum and graph the parabola.

4. $y + 4 = x^2$
5. $y = 5(x + 2)^2$
6. $4(y + 2) = 3(x - 1)^2$
7. $5x + 3y^2 = 15$
8. $y = 2x^2 - 8x + 7$
9. $x = 2y^2 - 8y + 7$
10. $3(x - 8)^2 = 5(y + 3)$
11. $x = 3(y + 4)^2 + 1$
12. $8y + 5x^2 + 30x + 101 = 0$
13. $x = -\frac{1}{5}y^2 + \frac{8}{5}y - 7$
14. $6x = y^2 - 6y + 39$
15. $-8y = x^2$
16. $y = 4x^2 + 24x + 38$
17. $y = x^2 - 6x + 3$
18. $y = x^2 + 4x + 1$

Write an equation for each parabola described below. Then draw the graph.

19. focus $(1, 1)$, directrix $y = -1$
20. vertex $(-1, 2)$, directrix $y = -4$
21. vertex $(2, -3)$, focus $(0, -3)$

Lesson 8-3

(pages 426–431)

Write an equation for the circle that satisfies each set of conditions.

1. center $(3, 2)$, $r = 5$ units
2. center $(-5, 8)$, $r = 3$ units
3. center $(1, -6)$, $r = \frac{2}{3}$ units
4. center $(0, 7)$, tangent to x-axis
5. center $(-2, -4)$, tangent to y-axis
6. endpoints of a diameter at $(-9, 0)$ and $(2, -5)$
7. endpoints of a diameter at $(4, 1)$ and $(-3, 2)$
8. center $(6, -10)$, passes through origin
9. center $(0.8, 0.5)$, passes through $(2, 2)$

Find the center and radius of the circle with the given equation. Then graph the circle.

10. $x^2 + y^2 = 36$
11. $\left(x - \frac{3}{4}\right)^2 + \left(y + \frac{2}{3}\right)^2 = \frac{32}{49}$
12. $(x - 5)^2 + (y + 4)^2 = 1$
13. $x^2 + 3x + y^2 - 5y = 0.5$
14. $x^2 + y^2 = 14x - 24$
15. $x^2 + y^2 = 2(y - x)$
16. $x^2 + 10x + (y - \sqrt{3})^2 = 11$
17. $x^2 + y^2 = 4x + 9$
18. $x^2 + y^2 + 12x - 10y + 45 = 0$
19. $x^2 + y^2 - 6x + 4y = 156$
20. $x^2 + y^2 - 2x + 7y = 1$
21. $16(x^2 + y^2) - 8(3x + 5y) + 33 = 0$

Lesson 8-4

(pages 433–440)

Write an equation for the ellipse that satisfies each set of conditions.

1. endpoints of major axis at $(-2, 7)$ and $(4, 7)$, endpoints of minor axis at $(1, 5)$ and $(1, 9)$
2. endpoints of minor axis at $(1, -4)$ and $(1, 5)$, endpoints of major axis at $(-4, 0.5)$ and $(6, 0.5)$
3. major axis 24 units long and parallel to the y-axis, minor axis 4 units long, center at $(0, 3)$

Find the coordinates of the center and foci and the lengths of the major and minor axes for the ellipse with the given equation. Then graph the ellipse.

4. $\frac{x^2}{36} + \frac{y^2}{81} = 1$
5. $\frac{x^2}{121} + \frac{(y-5)^2}{16} = 1$
6. $\frac{(x+2)^2}{12} + \frac{(y+1)^2}{16} = 1$
7. $\frac{(x+2)^2}{36} + \frac{(y-4)^2}{40} = 1$
8. $\frac{(x+8)^2}{121} + \frac{(y-7)^2}{64} = 1$
9. $\frac{(x-4)^2}{16} + \frac{(y+1)^2}{9} = 1$
10. $8x^2 + 2y^2 = 32$
11. $7x^2 + 3y^2 = 84$
12. $9x^2 + 16y^2 = 144$
13. $169x^2 - 338x + 169 + 25y^2 = 4225$
14. $x^2 + 4y^2 + 8x - 64y = -128$
15. $4x^2 + 5y^2 = 6(6x + 5y) + 658$
16. $9x^2 + 16y^2 - 54x + 64y + 1 = 0$

Lesson 8-5

(pages 441–448)

Find the coordinates of the vertices and foci and the equations of the asymptotes for the hyperbola with the given equation. Then graph the hyperbola.

1. $\frac{y^2}{25} - \frac{x^2}{9} = 1$
2. $\frac{x^2}{4} - \frac{y^2}{9} = 1$
3. $\frac{x^2}{81} - \frac{y^2}{36} = 1$
4. $\frac{x^2}{9} - \frac{y^2}{16} = 1$
5. $\frac{y^2}{100} - \frac{x^2}{144} = 1$
6. $\frac{x^2}{16} - \frac{y^2}{4} = 1$
7. $\frac{(x-4)^2}{64} - \frac{(y+1)^2}{16} = 1$
8. $\frac{(y-7)^2}{2.25} - \frac{(x-3)^2}{4} = 1$
9. $(x+5)^2 - \frac{(y+3)^2}{48} = 1$
10. $x^2 - 9y^2 = 36$
11. $4x^2 - 9y^2 = 72$
12. $49x^2 - 16y^2 = 784$
13. $144x^2 + 1152x - 25y^2 - 100y = 1396$
14. $576y^2 = 49x^2 + 490x + 29449$
15. $23.04y^2 - 46.08y - 1.96x^2 - 3.92x = 24.0784$
16. $25(y+5)^2 - 20(x-1)^2 = 500$

Write an equation for the hyperbola that satisfies each set of conditions.

17. vertices $(-3, 0)$ and $(3, 0)$; conjugate axis of length 8 units
18. vertices $(0, -7)$ and $(0, 7)$; conjugate axis of length 25 units
19. center $(0, 0)$; horizontal transverse axis of length 12 units and a conjugate axis of length 10 units

Lesson 8-6

(pages 449–452)

Write each equation in standard form. State whether the graph of the equation is a *parabola, circle, ellipse,* or *hyperbola*. Then graph the equation.

1. $9x^2 - 36x + 36 = 4y^2 + 24y + 72$
2. $x^2 + 4x + 2y^2 + 16y + 32 = 0$
3. $x^2 + 6x + y^2 - 6y + 9 = 0$
4. $9y^2 = 25x^2 + 400x + 1825$
5. $2y^2 + 12y - x + 6 = 0$
6. $x^2 + y^2 = 10x + 2y + 23$
7. $3x^2 + y = 12x - 17$
8. $9x^2 - 18x + 16y^2 + 160y = -265$
9. $x^2 + 10x + 5 = 4y^2 + 16$
10. $\frac{(y-5)^2}{4} - (x+1)^2 = 4$
11. $9x^2 + 49y^2 = 441$
12. $4x^2 - y^2 = 4$

Without writing the equation in standard form, state whether the graph of each equation is a *parabola, circle, ellipse,* or *hyperbola*.

13. $(x+3)^2 = 8(y+2)$
14. $x^2 + 4x + y^2 - 8y = 2$
15. $9x^2 + 9y^2 = 9$
16. $y - x^2 = x + 3$
17. $2x^2 - 13y^2 + 5 = 0$
18. $16(x-3)^2 + 81(y+4)^2 = 1296$
19. $x^2 + 5y^2 = 16$
20. $4x^2 - y^2 = 16$

Lesson 8-7

(pages 455–460)

Solve each system of inequalities by graphing.

1. $\frac{x^2}{16} - \frac{y^2}{1} \geq 1$
   $x^2 + y^2 \leq 49$
2. $\frac{x^2}{25} + \frac{y^2}{16} \leq 1$
   $y \leq x - 2$
3. $y \geq x + 3$
   $x^2 + y^2 < 25$
4. $4x^2 + (y - 3)^2 \leq 16$
   $x + 2y \geq 4$

Find the exact solution(s) of each system of equations.

5. $\frac{x^2}{16} + \frac{y^2}{16} = 1$
   $y = x + 3$
6. $x = y^2$
   $(x + 3)^2 + y^2 = 53$
7. $\frac{x^2}{3} - \frac{(y + 2)^2}{4} = 1$
   $x^2 = y^2 + 11$
8. $\frac{(x - 1)^2}{5} + \frac{y^2}{2} = 1$
   $y = x + 1$
9. $x^2 + y^2 = 13$
   $x^2 - y^2 = -5$
10. $\frac{x^2}{25} - \frac{y^2}{5} = 1$
    $y = x - 4$
11. $x^2 + y = 0$
    $x + y = -2$
12. $x^2 - 9y^2 = 36$
    $x = y$
13. $4x^2 + 6y^2 = 360$
    $y = x$

Lesson 9-1

(pages 472–478)

Simplify each expression.

1. $\frac{25xy^2}{15y}$
2. $\frac{-4a^2b^3}{28ab^4}$
3. $\frac{(-2cd^3)^2}{8c^2d^5}$
4. $\frac{3x^3}{-2} \cdot \frac{-4}{9x}$
5. $\frac{21x^2}{-5} \cdot \frac{10}{7x^3}$
6. $\frac{2u^2}{3} \div \frac{6u^3}{5}$
7. $\frac{15x^3}{14} \div \frac{18x}{7}$
8. $\frac{xy^2}{2} \cdot \frac{x^2}{2y} \cdot \frac{2}{x^2y}$
9. $axy \div \frac{ax}{y}$
10. $\frac{9u^2}{28v} \div \frac{27u^2}{8v^2}$
11. $\frac{x^2 - 4}{4x^2 - 1} \cdot \frac{2x - 1}{x + 2}$
12. $\frac{x^2 - 1}{2x^2 - x - 1} \div \frac{x^2 - 4}{2x^2 - 3x - 2}$
13. $\frac{2x^2 + x - 1}{2x^2 + 3x - 2} \div \frac{x^2 - 2x + 1}{x^2 + x - 2}$
14. $\frac{\frac{(ab)^2}{c}}{\frac{xa^3b}{cx^2}}$
15. $\frac{\frac{x^4 - y^4}{x^3 + y^3}}{\frac{x^3 - y^3}{x + y}}$

Lesson 9-2

(pages 479–484)

Find the LCM of each set of polynomials.

1. $2a^2b$, $4ab^2$, $20a$
2. $48c^2d$, $72cd^2$
3. $x^2 - 4x - 12$, $x^2 + 7x + 10$

Simplify each expression.

4. $\frac{12}{7d} - \frac{3}{14d}$
5. $\frac{x + 1}{x} - \frac{x - 1}{x^2}$
6. $\frac{2x + 1}{4x^2} - \frac{x + 3}{6x}$
7. $\frac{7x}{13y^2} + \frac{4y}{6x^2}$
8. $\frac{x}{x - 1} + \frac{1}{1 - x}$
9. $\frac{1}{3v^2} + \frac{1}{uv} + \frac{3}{4u^2}$
10. $\frac{1}{x^2 - x} + \frac{1}{x^2 + x}$
11. $\frac{1}{x^2 - 1} - \frac{1}{(x - 1)^2}$
12. $\frac{5}{x} - \frac{3}{x + 5}$
13. $y - 1 + \frac{1}{y - 1}$
14. $3m + 1 - \frac{2m}{3m + 1}$
15. $\frac{3x}{x - y} + \frac{4x}{y - x}$
16. $\frac{6}{4m^2 - 12mn + 9n^2} + \frac{2}{2mn - 3n^2}$
17. $\frac{3}{x^2 + 5ax + 6a^2} + \frac{2}{x^2 - 4a^2}$
18. $\frac{x}{x^2 + 5x + 6} - \frac{2}{x^2 + 4x + 4}$
19. $\frac{4}{a^2 - 4} - \frac{3}{a^2 + 4a + 4}$
20. $\frac{4}{3 - 3z^2} - \frac{2}{z^2 + 5z + 4}$
21. $\frac{2c}{c^2 - 9} - \frac{1}{c^2 + 6c + 9}$
22. $\frac{\frac{1}{x + y}}{\frac{1}{x} + \frac{1}{y}}$
23. $\frac{1 - \frac{1}{x + 1}}{1 + \frac{1}{x - 1}}$
24. $\frac{4 + \frac{1}{x - 2}}{3 - \frac{1}{x - 2}}$

Extra Practice

Lesson 9-3

(pages 485–490)

Determine the equations of any vertical asymptotes and the values of x for any holes in the graph of each rational function.

1. $f(x) = \frac{1}{x+4}$
2. $f(x) = \frac{x-2}{x+3}$
3. $f(x) = \frac{5}{(x+1)(x-8)}$
4. $f(x) = \frac{x}{x+2}$
5. $f(x) = \frac{x^2-4}{x+2}$
6. $f(x) = \frac{x^2+x-6}{x^2+8x+15}$

Graph each rational function.

7. $f(x) = \frac{1}{x-5}$
8. $f(x) = \frac{3x}{x+1}$
9. $f(x) = \frac{x^2-16}{x-4}$
10. $f(x) = \frac{x}{x-6}$
11. $f(x) = \frac{1}{(x-3)^2}$
12. $f(x) = \frac{2}{(x+3)(x-4)}$
13. $f(x) = \frac{x+4}{x^2-1}$
14. $f(x) = \frac{x+2}{x+3}$
15. $f(x) = \frac{x^2+5x-14}{x^2+9x+14}$

Lesson 9-4

(pages 492–498)

State whether each equation represents a *direct*, *joint*, or *inverse* variation. Then name the constant of variation.

1. $xy = 10$
2. $x = 6y$
3. $\frac{x}{7} = y$
4. $\frac{x}{y} = -6$
5. $10x = y$
6. $x = \frac{2}{y}$
7. $A = lw$
8. $\frac{1}{4}b = -\frac{3}{5}c$
9. $D = rt$

Find each value.

10. If y varies directly as x and $y = 16$ when $x = 4$, find y when $x = 12$.
11. If x varies inversely as y and $x = 12$ when $y = -3$, find x when $y = -18$.
12. If m varies directly as w and $m = -15$ when $w = 2.5$, find m when $w = 12.5$.
13. If y varies jointly as x and z and $y = 10$ when $z = 4$ and $x = 5$, find y when $x = 4$ and $z = 2$.
14. If y varies inversely as x and $y = \frac{1}{4}$ when $x = 24$, find y when $x = \frac{3}{4}$.
15. If y varies jointly as x and z and $y = 45$ when $x = 9$ and $z = 15$, find y when $x = 25$ and $z = 12$.

Lesson 9-5

(pages 499–504)

Identify the type of function represented by each graph.

1. 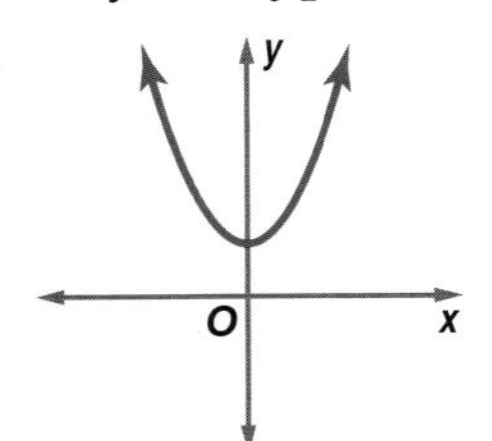

2. 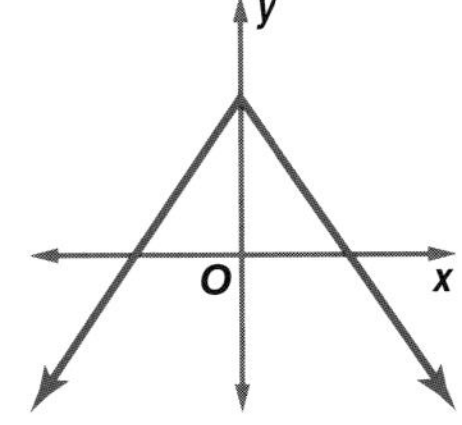

3. 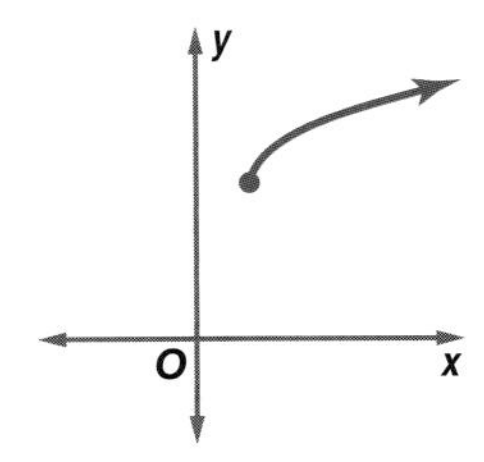

Identify the function represented by each equation. Then graph the equation.

4. $y = \sqrt{5x}$
5. $y = \frac{3}{4}x$
6. $y = |x| + 3$
7. $y = x^2 - 2$
8. $y = \frac{2}{x}$
9. $y = 2[\![x]\!]$
10. $y = -2x^2 + 1$
11. $y = \frac{x^2+2x-3}{x^2+7x+12}$
12. $y = -3$

Lesson 9-6

(pages 505–511)

Solve each equation or inequality. Check your solutions.

1. $\frac{x}{x-3} = \frac{1}{4}$
2. $\frac{5}{x} + \frac{3}{5} = \frac{2}{x}$
3. $\frac{5}{b-2} < 5$
4. $\frac{4}{a+3} > 2$
5. $\frac{x-2}{x} = \frac{x-4}{x-6}$
6. $-6 - \frac{8}{n} < n$
7. $\frac{2}{d} + \frac{1}{d-2} = 1$
8. $\frac{1}{2+3x} + \frac{2}{2-3x} = 0$
9. $\frac{1}{n+1} + \frac{1}{n-1} = \frac{2}{n^2-1}$
10. $\frac{1}{x-3} + \frac{1}{x+5} = \frac{x+1}{x-3}$
11. $\frac{4}{x^2-2x-3} = \frac{-x}{3-x} - \frac{1}{x+1}$
12. $\frac{p}{p+1} + \frac{3}{p-3} + 1 = 0$
13. $\frac{3x}{x^2+2x-8} = \frac{1}{x-2} + \frac{x}{x+4}$
14. $\frac{5z+2}{z^2-4} = \frac{-5z}{2-z} + \frac{2}{z+2}$
15. $\frac{1}{x-3} + \frac{2}{x^2-9} = \frac{5}{x+3}$
16. $\frac{1}{m^2-1} = \frac{2}{m^2+m-2}$
17. $\frac{12}{x^2-16} - \frac{24}{x-4} = 3$
18. $n + \frac{1}{n+3} = \frac{n^2}{n-1}$

Lesson 10-1

(pages 523–530)

Sketch the graph of each function. Then state the function's domain and range.

1. $y = 3(5)^x$
2. $y = 0.5(2)^x$
3. $y = 3\left(\frac{1}{4}\right)^x$
4. $y = 2(1.5)^x$

Determine whether each function represents exponential growth or decay.

5. $y = 4(3)^x$
6. $y = 10^{-x}$
7. $y = 5\left(\frac{1}{2}\right)^x$
8. $y = 2\left(\frac{5}{4}\right)^x$

Write an exponential function whose graph passes through the given points.

9. (0, 6) and (2, 54)
10. (0, −4) and (−4, −64)
11. (0, 1.5) and (3, 40.5)
12. (0, −3.7) and (5, −118.4)

Simplify each expression.

13. $4^{\sqrt{2}} \cdot 4^{\sqrt{8}}$
14. $(5^{\sqrt{5}})^{\sqrt{45}}$
15. $(w^{\sqrt{6}})^{\sqrt{3}}$
16. $27^{\sqrt{5}} \div 3^{\sqrt{5}}$
17. $8^{2\sqrt{3}} \cdot 4^{\sqrt{3}}$
18. $5^{\sqrt{2}} \cdot 5^{\sqrt{3}}$
19. $7^{\sqrt{3}} \cdot 7^{2\sqrt{3}}$
20. $(y^{\sqrt{3}})^{\sqrt{27}}$

Solve each equation or inequality. Check your solution.

21. $27^{2x-1} = 3$
22. $8^{2+x} \geq 2$
23. $4^{2x+5} < 8^{x+1}$
24. $6^{x+1} = 36^{x-1}$
25. $10^{x-1} > 100^{4-x}$
26. $\left(\frac{1}{5}\right)^{x-3} = 125$
27. $2^{x^2+1} = 32$
28. $36^x = 6^{x^2-3}$

Lesson 10-2

(pages 531–538)

Write each equation in logarithmic form.

1. $3^5 = 243$
2. $10^3 = 1000$
3. $4^{-3} = \frac{1}{64}$

Write each equation in exponential form.

4. $\log_2 \frac{1}{8} = -3$
5. $\log_{25} 5 = \frac{1}{2}$
6. $\log_7 \frac{1}{7} = -1$

Evaluate each expression.

7. $\log_4 16$
8. $\log_{10} 10{,}000$
9. $\log_3 \frac{1}{9}$
10. $\log_2 1024$
11. $\log_6 6^5$
12. $\log_{\frac{1}{2}} 8$
13. $\log_{11} 121$
14. $5^{\log_5 10}$

Solve each equation or inequality. Check your solution.

15. $\log_8 b = 2$
16. $\log_4 x < 3$
17. $\log_{\frac{1}{9}} n = -\frac{1}{2}$
18. $\log_x 7 = 1$
19. $\log_{\frac{2}{3}} a < 3$
20. $\log_2 (x^2 - 9) = 4$

Lesson 10-3

(pages 541–546)

Use $\log_3 5 \approx 1.4651$ and $\log_3 7 \approx 1.7712$ to approximate the value of each expression.

1. $\log_3 \frac{7}{5}$
2. $\log_3 245$
3. $\log_3 35$

Solve each equation. Check your solutions.

4. $\log_2 x + \log_2 (x - 2) = \log_2 3$
5. $\log_3 x = 2 \log_3 3 + \log_3 5$
6. $\log_5 (x^2 + 7) = \frac{2}{3} \log_5 64$
7. $\log_7 (3x + 5) - \log_7 (x - 5) = \log_7 8$
8. $\log_2 (x^2 - 9) = 4$
9. $\log_3 (x + 2) + \log_3 6 = 3$
10. $\log_6 x + \log_6 (x - 5) = 2$
11. $\log_5 (x + 3) = \log_5 8 - \log_5 2$
12. $2 \log_3 x - \log_3 (x - 2) = 2$
13. $\log_6 x = \frac{3}{2} \log_6 9 + \log_6 2$
14. $\log_8 (x + 6) + \log_8 (x - 6) = 2$
15. $\frac{1}{2}\log_4 (x + 2) + \frac{1}{2}\log_4 (x + 2) = \frac{2}{3}\log_4 27$
16. $\log_3 14 + \log_3 x = \log_3 42$
17. $\log_{10} x = \frac{1}{2}\log_{10} 81$

Lesson 10-4

(pages 547–551)

Use a calculator to evaluate each expression to four decimal places.

1. log 55
2. log 6.7
3. log 3.3
4. log 0.08
5. log 9.9
6. log 0.6

Solve each equation or inequality. Round to four decimal places.

7. $2^x = 15$
8. $4^{2a} > 45$
9. $7^{2x} = 35$
10. $11^{x + 4} > 57$
11. $1.5^{a - 7} = 9.6$
12. $3^{b^2} = 64$
13. $7^{3c} < 35^{2c - 1}$
14. $5^{m^2 + 1} = 30$
15. $7^{3y - 1} < 2^{2y + 4}$
16. $9^{n - 3} = 2^{n + 3}$
17. $11^{t + 1} \leq 22^{t + 3}$
18. $2^{3a - 1} = 3^{a + 2}$

Express each logarithm in terms of common logarithms. Then approximate its value to four decimal places.

19. $\log_3 21$
20. $\log_4 62$
21. $\log_5 28$
22. $\log_2 25$
23. $\log_{12} 30$
24. $\log_4 63$
25. $\log_7 35$
26. $\log_6 100$

Lesson 10-5

(pages 554–559)

Use a calculator to evaluate each expression to four decimal places.

1. e^3
2. $e^{0.75}$
3. e^{-4}
4. $e^{-2.5}$
5. ln 5
6. ln 8
7. ln 8.4
8. ln 0.6

Write an equivalent exponential or logarithmic equation.

9. $e^x = 10$
10. $\ln x \approx 2.3026$
11. $e^3 = 9x$
12. $\ln 0.2 = x$

Evaluate each expression.

13. $e^{\ln 5}$
14. $\ln e^{x - 2}$
15. $\ln e^{-3n}$
16. $e^{\ln 2.5}$

Solve each equation or inequality.

17. $25e^x = 1000$
18. $e^{0.075x} > 25$
19. $e^x < 3.8$
20. $-2e^x + 5 = 1$
21. $5 + 4e^{2x} = 17$
22. $e^{-3x} \leq 15$
23. $\ln 7x = 10$
24. $\ln 4x = 8$
25. $3 \ln 2x \geq 9$
26. $\ln (x + 2) = 4$
27. $\ln (2x + 3) > 0$
28. $\ln (3x - 1) = 5$

Lesson 10-6

(pages 560–565)

1. Mr. Rogers purchased a combine for $175,000 for his farming operation. It is expected to depreciate at a rate of 18% per year. What will be the value of the combine in 3 years?
2. The Jacksons bought a house for $65,000 in 1992. Houses in the neighborhood have appreciated at the rate of 4.5% a year. How much is the house worth in 2003?
3. In 1950, the population of a city was 50,000. Since then, the population has increased by 2.25% per year. If it continues to grow at this rate, what will the population be in 2005?
4. In a particular state, the population of black bears has been decreasing at the rate of 0.75% per year. In 1990, it was estimated that there were 400 black bears in the state. If the population continues to decline at the same rate, what will the population be in 2010?
5. Scott invests $8500 in a certificate of deposit (CD). The interest is calculated on the money once per year. If the interest rate for the CD is 6.875% and he invests in a 4-year CD, how much money will he have at the end of the investment period?
6. Mrs. Nguyen takes a job as a teacher with a starting salary of $20,050. Over the past few years in this district, teachers have received a 4% increase in pay each year. If the pay continues to increase at the same rate, what will be her salary in 5 years?

Lesson 11-1

(pages 578–582)

Find the next four terms of each arithmetic sequence.

1. 9, 7, 5, ...
2. 3, 4.5, 6, ...
3. 40, 35, 30, ...
4. 2, 5, 8, ...

Find the first five terms of each arithmetic sequence described.

5. $a_1 = 1, d = 7$
6. $a_1 = -5, d = 2$
7. $a_1 = 1.2, d = 3.7$
8. $a_1 = -\frac{5}{4}, d = -\frac{1}{2}$

Find the indicated term of each arithmetic sequence.

9. $a_1 = 4, d = 5, n = 10$
10. $a_1 = -30, d = -6, n = 5$
11. $a_1 = -3, d = 32, n = 8$
12. $a_1 = \frac{3}{4}, d = -\frac{1}{4}, n = 72$
13. $a_1 = -\frac{1}{5}, d = \frac{3}{5}, n = 17$
14. $a_1 = 20, d = -3, n = 16$

Write an equation for the *n*th term of each arithmetic sequence.

15. 3, 5, 7, 9, ...
16. 2, −1, −4, −7, ...
17. 20, 28, 36, 44, ...

Find the arithmetic means in each sequence.

18. 2, ___?___, ___?___, ___?___, 34
19. 0, ___?___, ___?___, ___?___, −28
20. −10, ___?___, ___?___, ___?___, 14

Lesson 11-2

(pages 583–587)

Find S_n for each arithmetic series described.

1. $a_1 = 3, a_n = 20, n = 6$
2. $a_1 = 15, a_n = -12, n = 30$
3. $a_1 = 90, a_n = -4, n = 10$
4. $a_1 = 16, a_n = 14, n = 12$
5. $a_1 = -80, a_n = 120, n = 18$
6. $a_1 = -3, a_n = -72, n = 14$
7. $a_1 = -1, d = 10, n = 30$
8. $a_1 = 4, d = -5, n = 11$
9. $a_1 = 5, d = -\frac{1}{2}, n = 17$

Find the sum of each arithmetic series.

10. $3 + 12 + 21 + 30 + \ldots + 57$
11. $1 + 4 + 7 + 10 + \ldots + 31$
12. $8 + 16 + 24 + \ldots + 80$
13. $\sum_{n=1}^{6} (n + 2)$
14. $\sum_{n=5}^{10} (2n - 5)$
15. $\sum_{k=1}^{5} (40 - 2k)$
16. $\sum_{k=8}^{12} (6 - 3k)$
17. $\sum_{n=1}^{4} (10n + 2)$
18. $\sum_{n=6}^{10} (2 + 3n)$

Find the first three terms of each arithmetic series described.

19. $a_1 = 11, a_n = 38, S_n = 245$
20. $a_1 = -6, a_n = -34, S_n = -300$
21. $a_1 = 0, a_n = 165, S_n = 990$
22. $n = 12, a_n = 13, S_n = -42$
23. $n = 11, a_n = 5, S_n = 0$
24. $a_1 = -25, a_n = 15, S_n = -45$

Lesson 11-3 *(pages 588–592)*

Find the next two terms of each geometric sequence.

1. 5, 15, 45, … **2.** 2, 10, 50, … **3.** 64, 16, 4, …

4. −9, 27, −81, … **5.** 0.5, 0.75, 1.125, … **6.** $\frac{1}{2}, -\frac{3}{8}, \frac{9}{32}, \ldots$

Find the first five terms of each geometric sequence described.

7. $a_1 = -2, r = 6$ **8.** $a_1 = 4, r = -5$ **9.** $a_1 = 1, r = -3$ **10.** $a_1 = 1, r = 0.8$

11. $a_1 = 0.8, r = 2.5$ **12.** $a_1 = 1.5, r = -3$ **13.** $a_1 = \frac{2}{5}, r = -3$ **14.** $a_1 = -\frac{1}{3}, r = -\frac{3}{5}$

Find the indicated term of each geometric sequence.

15. $a_1 = 5, r = 7, n = 6$ **16.** $a_1 = 200, r = -\frac{1}{2}, n = 10$ **17.** $a_1 = 60, r = -2, n = 4$

18. $a_1 = 300, r = \frac{1}{4}, n = 6$ **19.** $a_1 = 8, r = -2, n = 8$ **20.** $a_1 = 1, r = -1, n = 30$

Write an equation for the *n*th term of each geometric sequence.

21. 20, 40, 80, … **22.** $-\frac{1}{2}, -\frac{1}{8}, -\frac{1}{32}, \ldots$

Find the geometric means in each sequence.

23. 1, _?_, _?_, _?_, 81 **24.** 5, _?_, _?_, _?_, 6480

Lesson 11-4 *(pages 594–598)*

Find S_n for each geometric series described.

1. $a_1 = \frac{1}{81}, r = 3, n = 6$ **2.** $a_1 = 1, r = -2, n = 7$ **3.** $a_1 = 5, r = 4, n = 5$

4. $a_1 = -27, r = -\frac{1}{3}, n = 6$ **5.** $a_1 = 1000, r = \frac{1}{2}, n = 7$ **6.** $a_1 = 125, r = -\frac{2}{5}, n = 5$

7. $a_1 = 10, r = 3, n = 6$ **8.** $a_1 = 1250, r = -\frac{1}{5}, n = 5$ **9.** $a_1 = 1215, r = \frac{1}{3}, n = 5$

10. $a_1 = 16, r = \frac{3}{2}, n = 5$ **11.** $a_1 = 7, r = 2, n = 7$ **12.** $a_1 = -\frac{3}{2}, r = -\frac{1}{2}, n = 6$

Find the sum of each geometric series.

13. $\sum_{k=1}^{5} 2^k$ **14.** $\sum_{n=0}^{3} 3^{-n}$ **15.** $\sum_{n=0}^{3} 2(5^n)$ **16.** $\sum_{k=2}^{5} -(-3)^{k-1}$

Find the indicated term for each geometric series described.

17. $S_n = 300, a_n = 160, r = 2; a_1$ **18.** $S_n = -171, n = 9, r = -2\,; a_5$ **19.** $S_n = -4372, a_n = -2916, r = 3\,; a_4$

Lesson 11-5 *(pages 599–604)*

Find the sum of each infinite geometric series, if it exists.

1. $a_1 = 54, r = \frac{1}{3}$ **2.** $a_1 = 2, r = -1$ **3.** $a_1 = 1000, r = -0.2$ **4.** $a_1 = 7, r = \frac{3}{7}$

5. $49 + 14 + 4 + \ldots$ **6.** $\frac{3}{4} + \frac{1}{2} + \frac{1}{3} + \ldots$ **7.** $12 - 4 + \frac{4}{3} - \ldots$ **8.** $3 - 9 + 27 - \ldots$

9. $3 - 2 + \frac{4}{3} - \ldots$ **10.** $\sum_{n=1}^{\infty} 3\left(\frac{1}{4}\right)^{n-1}$ **11.** $\sum_{n=1}^{\infty} 5\left(-\frac{1}{10}\right)^{n-1}$ **12.** $\sum_{n=1}^{\infty} -\frac{2}{3}\left(-\frac{3}{4}\right)^{n-1}$

Write each repeating decimal as a fraction.

13. $0.\overline{4}$ **14.** $0.\overline{27}$ **15.** $0.\overline{123}$ **16.** $0.\overline{645}$ **17.** $0.6\overline{7}$ **18.** $0.8\overline{53}$

Lesson 11-6

(pages 606–610)

Find the first five terms of each sequence.

1. $a_1 = 4, a_{n+1} = 2a_n + 1$
2. $a_1 = 6, a_{n+1} = a_n + 7$
3. $a_1 = 16, a_{n+1} = a_n + (n + 4)$
4. $a_1 = 1, a_{n+1} = \frac{n}{n+2} \cdot a_n$
5. $a_1 = -\frac{1}{2}, a_{n+1} = 2a_n + \frac{1}{4}$
6. $a_1 = \frac{1}{3}, a_2 = \frac{1}{4}, a_{n+1} = a_n + a_{n-1}$

Find the first three iterates of each function for the given initial value.

7. $f(x) = 3x - 1, x_0 = 3$
8. $f(x) = 2x^2 - 8, x_0 = -1$
9. $f(x) = 4x + 5, x_0 = 0$
10. $f(x) = 3x^2 + 1, x_0 = 1$
11. $f(x) = x^2 + 4x + 4, x_0 = 1$
12. $f(x) = x^2 + 9, x_0 = 2$
13. $f(x) = 2x^2 + x + 1, x_0 = -\frac{1}{2}$
14. $f(x) = 3x^2 + 2x - 1, x_0 = \frac{2}{3}$

Lesson 11-7

(pages 612–617)

Evaluate each expression.

1. $6!$
2. $4!$
3. $\frac{13!}{6!}$
4. $\frac{10!}{3!7!}$
5. $\frac{14!}{4!10!}$
6. $\frac{7!}{2!5!}$
7. $\frac{12!}{5!}$
8. $\frac{9!}{8!}$
9. $\frac{10!}{10!0!}$

Expand each power.

10. $(z - 3)^5$
11. $(m + 1)^4$
12. $(x + 6)^4$
13. $(z - y)^2$
14. $(m + n)^5$
15. $(a - b)^4$
16. $(2n + 1)^4$
17. $(3n - 4)^3$
18. $(2n - m)^0$
19. $(4x - a)^4$
20. $(3r - 4s)^5$
21. $\left(\frac{b}{2} - 1\right)^4$

Find the indicated term of each expansion.

22. sixth term of $(x + 3)^8$
23. fourth term of $(x - 2)^7$
24. fifth term of $(a + b)^6$
25. fourth term of $(x - y)^9$
26. sixth term of $(x + 4y)^7$
27. fifth term of $(3x + 5y)^{10}$

Lesson 11-8

(pages 618–621)

Prove that each statement is true for all positive integers.

1. $2 + 4 + 6 + \ldots + 2n = n^2 + n$
2. $1^3 + 3^3 + 5^3 + \ldots + (2n - 1)^3 = n^2(2n^2 - 1)$
3. $1 \cdot 3 + 3 \cdot 5 + 5 \cdot 7 + \ldots + (2n - 1)(2n + 1) = \frac{n(4n^2 + 6n - 1)}{3}$
4. $\frac{1}{1 \cdot 3} + \frac{1}{2 \cdot 4} + \frac{1}{3 \cdot 5} + \ldots + \frac{1}{n(n + 2)} = \frac{n(3n + 5)}{4(n + 1)(n + 2)}$
5. $1 \cdot 3 + 2 \cdot 4 + 3 \cdot 5 + \ldots + n(n + 2) = \frac{n(n + 1)(2n + 7)}{6}$
6. $\frac{5}{1 \cdot 2} \cdot \frac{1}{3} + \frac{7}{2 \cdot 3} \cdot \frac{1}{3^2} + \frac{9}{3 \cdot 4} \cdot \frac{1}{3^3} + \ldots + \frac{2n + 3}{n(n + 1)} \cdot \frac{1}{3^n} = 1 - \frac{1}{3^n(n + 1)}$

Find a counterexample for each statement.

7. $n^2 + 2n - 1$ is divisible by 2.
8. $2^n + 3^n$ is prime.
9. $2^{n-1} + n = 2^n + 2 - n$ for all integers $n \geq 2$.
10. $3^n - 2n = 3^n - 2^n$ for all integers $n \geq 1$

Extra Practice

Lesson 12-1

(pages 632–637)

List the possible outcomes for each situation.

1. tossing a penny and rolling a number cube
2. choosing a denim jacket that comes in dark blue, stone washed, or black that has buttons or snaps
3. ordering a large pizza with thin or thick crust, and one topping of either pepperoni, sausage, or vegetables, and either jack or mozzarella cheese

State whether the events are *independent* or *dependent*.

4. A comedy video and an action video are selected from the video store.
5. The numbers 1–10 are written on pieces of paper and are placed in a hat. Three of them are selected one after the other without replacing any of the pieces of paper.

Solve each problem.

6. On a bookshelf there are 10 different algebra books, 6 different geometry books, and 4 different calculus books. In how many ways can you choose 3 books, one of each kind?
7. In how many different ways can a 10-question true-false test be answered?
8. A student council has 6 seniors, 5 juniors, and 1 sophomore as members. In how many ways can a 3-member council committee be formed that includes one member of each class?
9. How many license plates of 5 symbols can be made using a letter for the first symbol and digits for the remaining 4 symbols?

Lesson 12-2

(pages 638–643)

Evaluate each expression.

1. $P(3, 2)$
2. $P(5, 2)$
3. $P(10, 6)$
4. $P(4, 3)$
5. $P(12, 2)$
6. $P(7, 2)$
7. $C(8, 6)$
8. $C(20, 17)$
9. $C(9, 4) \cdot C(5, 3)$
10. $C(6, 1) \cdot C(4, 1)$
11. $C(10, 5) \cdot C(8, 4)$
12. $C(7, 6) \cdot C(3, 1)$

Determine whether each situation involves a *permutation* or a *combination*. Then find the number of possibilities.

13. choosing a team of 9 players from a group of 20
14. selecting the batting order of 9 players in a baseball game
15. arranging the order of 8 songs on a CD
16. finding the number of 5-card hands that include 4 diamonds and 1 club

Lesson 12-3

(pages 644–650)

A jar contains 3 red, 4 green, and 5 orange marbles. If three marbles are drawn at random and not replaced, find each probability.

1. P(all green)
2. P(1 red, then 2 not red)
3. P(2 orange, then 1 not orange)

Find the odds of an event occurring, given the probability of the event.

4. $\frac{5}{9}$
5. $\frac{4}{8}$
6. $\frac{3}{10}$

Find the probability of an event occurring, given the odds of the event.

7. $\frac{2}{7}$
8. $\frac{6}{13}$
9. $\frac{1}{19}$

The table shows the number of ways to achieve each product when two dice are tossed. Find each probability.

Product	1	2	3	4	5	6	8	9	10	12	15	16	18	20	24	25	30	36
Ways	1	2	2	3	2	4	2	1	2	4	2	1	2	2	2	1	2	1

10. $P(6)$
11. $P(12)$
12. P(not 36)
13. P(not 12)

Lesson 12-4

(pages 651–657)

An octahedral die is rolled twice. The sides are numbered 1–8. Find each probability.

1. P(1, then 8)
2. P(two 7s)
3. P(8, then any number)
4. P(two of the same number)
5. P(two different numbers)
6. P(no 8s)

Two cards are drawn from a standard deck of cards. Find each probability if no replacement occurs.

7. P(jack, jack)
8. P(heart, club)
9. P(two diamonds)
10. P(2 of hearts, diamond)
11. P(2 red cards)
12. P(2 black aces)

Determine whether the events are *independent* or *dependent*. Then find the probability.

13. According to the weather reports, the probability of rain on a certain day is 70% in Yellow Falls and 50% in Copper Creek. What is the probability that it will rain in both cities?
14. A contestant on a game show reaches into a container without looking and picks two paper bills. There are 2 $100 bills, 4 $50 bills, 10 $20 bills, and 20 $10 bills. What is the probability that the contestant draws 2 $100 bills one after the other without replacement?
15. The odds of winning a carnival game are 1 to 5. What is the probability that a player will win the game three consecutive times?

Lesson 12-5

(pages 658–663)

An octahedral die is rolled. The sides are numbered 1–8. Find each probability.

1. P(7 or 8)
2. P(less than 4)
3. P(greater than 6)
4. P(not prime)
5. P(odd or prime)
6. P(multiple of 5 or odd)

Ten slips of paper are placed in a container. Each is labeled with a number from 1 through 10. Determine whether the events are *mutually exclusive* or *inclusive*. Then find the probability.

7. P(1 or 10)
8. P(3 or odd)
9. P(6 or less than 7)

Find each probability.

10. Two letters are chosen at random from the word GEESE and two are chosen at random from the word PLEASE. What is the probability that all four letters are Es or none of the letters is an E?
11. Three dice are thrown. What is the probability that all three dice show the same number?
12. Two marbles are simultaneously drawn at random from a bag containing 3 red, 5 blue, and 6 green marbles.
    a. P(at least one red marble)
    b. P(at least one green marble)
    c. P(two marbles of the same color)
    d. P(two marbles of different colors)

Lesson 12-6

(pages 664–670)

Find the mean, median, mode, and standard deviation of each set of data. Round to the nearest hundredth, if necessary.

1. {4, 1, 2, 1, 1}
2. {86, 71, 74, 65, 45, 42, 76}
3. {16, 20, 15, 14, 24, 23, 25, 10, 19}
4. {18, 24, 16, 24, 22, 24, 22, 22, 24, 13, 17, 18, 16, 20, 16, 7, 22, 5, 4, 24}
5. {55, 50, 50, 55, 65, 50, 45, 35, 50, 40, 70, 40, 70, 50, 90, 30, 35, 55, 55, 40, 75, 35, 40, 45, 65, 50, 60}
6. {364, 305, 217, 331, 305, 311, 352, 319, 272, 238, 311, 226, 220, 226, 215, 160, 123, 4, 24, 238, 99}
7. {25.5, 26.7, 20.9, 23.4, 26.8, 24.0, 25.7}
8. The following high temperatures were recorded during a cold spell in Cleveland lasting thirty-eight days.
   29° 26° 17° 12° 5° 4° 25° 17° 23° 18° 13° 6° 25° 20° 27° 22° 26° 30° 31°
   2° 12° 27° 16° 27° 16° 30° 6° 16° 5° 0° 5° 29° 18° 16° 22° 29° 8° 23°

Extra Practice

Lesson 12-7

(pages 671–675)

1. Determine whether the data in the table appear to be *positively skewed*, *negatively skewed*, or *normally distributed*. The average size of a farm in each U.S. state was determined.

Acres	85–559	560–1034	1035–1509	1510–1984	1985–2459	2460–2934	2935–3409	3410–3884
States	37	4	3	1	2	1	0	2

Source: *The World Almanac*

2. The diameters of metal fittings made by a machine are normally distributed. The diameters have a mean of 7.5 centimeters and a standard deviation of 0.5 centimeters.
   a. What percent of the fittings have diameters between 7.0 and 8.0 centimeters?
   b. What percent of the fittings have diameters between 7.5 and 8.0 centimeters?
   c. What percent of the fittings have diameters greater than 6.5 centimeters?
   d. Of 100 fittings, how many will have a diameter between 6.0 and 8.5 centimeters?
3. A college entrance exam was administered at a state university. The scores were normally distributed with a mean of 510, and a standard deviation of 80.
   a. What percent would you expect to score above 510?
   b. What percent would you expect to score between 430 and 590?
   c. What is the probability that a student chosen at random scored between 350 and 670?

Lesson 12-8

(pages 676–680)

Find each probability if a coin is tossed 5 times.

1. P(0 heads)
2. P(exactly 4 heads)
3. P(exactly 3 tails)

Find each probability.

4. Ten percent of a batch of toothpaste is defective. Five tubes of toothpaste are selected at random from this batch.
   a. P(0 defective)
   b. P(exactly one defective)
   c. P(at least three defective)
   d. P(less than three defective)
5. On a 20-question true-false test, you guess at every question.
   a. P(all answers correct)
   b. P(exactly 10 correct)

Lesson 12-9

(pages 682–685)

Determine whether each situation would produce a random sample. Write *yes* or *no* and explain your answer.

1. finding the most often prescribed pain reliever by asking all of the doctors at a hospital
2. taking a poll of the most popular baby girl names this year by studying birth announcements in newspapers from different cities across the country
3. polling people who are leaving a pizza parlor about their favorite restaurant in the city

Find the margin of sampling error.

4. $p = 45\%, n = 125$
5. $p = 62\%, n = 240$
6. $p = 24\%, n = 600$
7. $p = 67\%, n = 180$
8. $p = 82\%, n = 1000$
9. $p = 15\%, n = 2500$
10. A poll conducted on the favorite breakfast choice of students in your school showed that 75% of the 2250 students asked indicated oatmeal as their favorite breakfast.
11. Of the 420 people polled at a supermarket, 56% felt that they were easily swayed by the sample items in the aisles and purchased those items, even though they were not intending to when they arrived at the store.
12. Of 3000 women between the ages of 25 and 35 polled, only 45% felt that they consume the recommended daily allowance of calcium by the National Institute of Health.

Lesson 13-1

(pages 701–708)

Find the values of the six trigonometric functions for angle θ.

1. 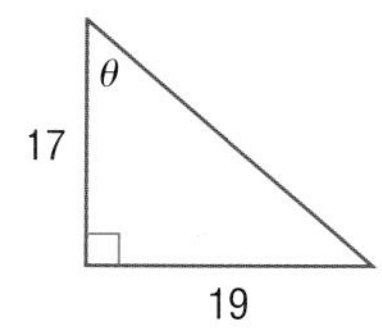

2. 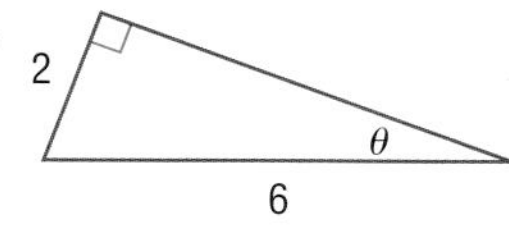

3. 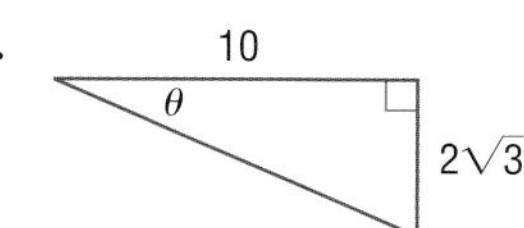

Solve $\triangle ABC$ using the diagram at the right and the given measurements. Round measures of sides to the nearest tenth and measures of angles to the nearest degree.

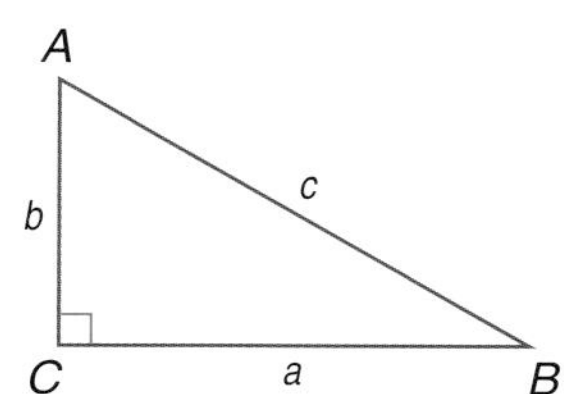

4. $B = 42°, c = 30$
5. $A = 84°, a = 4$
6. $B = 19°, b = 34$
7. $A = 75°, c = 55$
8. $b = 24, c = 36$
9. $a = 51, c = 115$
10. $\cos B = \frac{2}{5}, a = 12$
11. $\tan A = \frac{3}{2}, b = 22$

Lesson 13-2

(pages 709–715)

Draw an angle with the given measure in standard position.

1. $60°$
2. $250°$
3. $315°$
4. $150°$

Rewrite each degree measure in radians and each radian measure in degrees.

5. $-135°$
6. $-315°$
7. $45°$
8. $80°$
9. $24°$
10. $-54°$
11. $-\pi$
12. $\frac{9\pi}{4}$
13. $\frac{3\pi}{2}$
14. $-\frac{7\pi}{2}$
15. $\frac{9\pi}{10}$
16. $\frac{17\pi}{30}$
17. $\frac{7\pi}{12}$
18. 1
19. $-2\frac{1}{3}$

Find one angle with positive measure and one angle with negative measure coterminal with each angle.

20. $50°$
21. $-75°$
22. $125°$
23. $-400°$
24. $550°$
25. 3π
26. -2π
27. $\frac{2\pi}{3}$
28. $\frac{12\pi}{5}$
29. 0

Lesson 13-3

(pages 717–724)

Find the exact values of the six trigonometric functions of θ if the terminal side of θ in standard position contains the given point.

1. $P(3, -4)$
2. $P(1, \sqrt{3})$
3. $P(0, -4)$
4. $P(-5, -5)$
5. $P(-\sqrt{2}, -\sqrt{2})$

Find the exact value of each trigonometric function.

6. $\cos 225°$
7. $\sin\left(-\frac{5\pi}{3}\right)$
8. $\tan \frac{7\pi}{6}$
9. $\tan(-300°)$
10. $\cos \frac{7\pi}{4}$

Suppose θ is an angle in standard position whose terminal side is in the given quadrant. For each function, find the exact values of the remaining five trigonometric functions of θ.

11. $\cos \theta = -\frac{1}{3}$; Quadrant III
12. $\sec \theta = 2$; Quadrant IV
13. $\sin \theta = \frac{2}{3}$; Quadrant II
14. $\tan \theta = -4$; Quadrant IV
15. $\csc \theta = -5$; Quadrant III
16. $\cot \theta = -2$; Quadrant II
17. $\tan \theta = \frac{1}{3}$; Quadrant III
18. $\cos \theta = \frac{1}{4}$; Quadrant I
19. $\csc \theta = -\frac{5}{2}$; Quadrant IV

Lesson 13-4
(pages 725–732)

Find the area of $\triangle ABC$. Round to the nearest tenth.

1. $a = 11$ m, $b = 13$ m, $C = 31°$ **2.** $a = 15$ ft, $b = 22$ ft, $C = 90°$ **3.** $a = 12$ cm, $b = 12$ cm, $C = 50°$

Solve each triangle. Round to the nearest tenth.

4. $A = 18°, B = 37°, a = 15$ **5.** $A = 60°, C = 25°, c = 3$ **6.** $B = 40°, C = 32°, b = 10$
7. $B = 10°, C = 23°, c = 8$ **8.** $A = 12°, B = 60°, b = 5$ **9.** $A = 35°, C = 45°, a = 30$

Determine whether each triangle has *no* solution, *one* solution, or *two* solutions. Then solve each triangle. Round to the nearest tenth.

10. $A = 40°, B = 60°, c = 20$ **11.** $B = 70°, C = 58°, a = 84$ **12.** $A = 40°, a = 5, b = 12$
13. $A = 58°, a = 26, b = 29$ **14.** $A = 38°, B = 63°, c = 15$ **15.** $A = 150°, a = 6, b = 8$
16. $A = 57°, a = 12, b = 19$ **17.** $A = 25°, a = 125, b = 150$ **18.** $C = 98°, a = 64, c = 90$
19. $A = 40°, B = 60°, c = 20$ **20.** $A = 132°, a = 33, b = 50$ **21.** $A = 45°, a = 83, b = 79$

Lesson 13-5
(pages 733–738)

Determine whether each triangle should be solved by beginning with the Law of Sines or Law of Cosines. Then solve each triangle.

1.
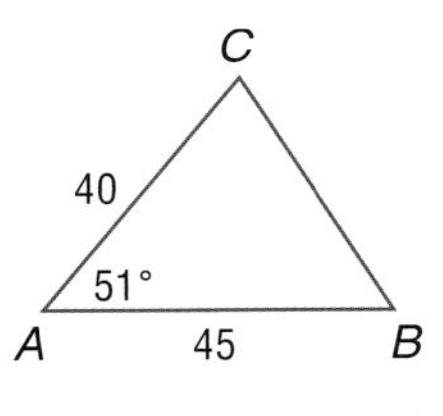

2.
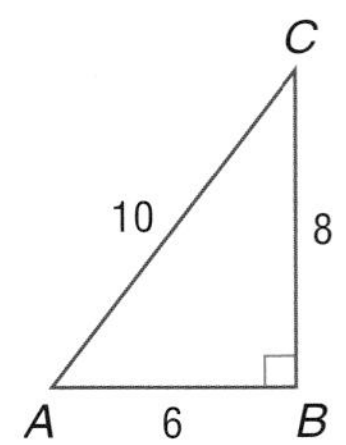

3.
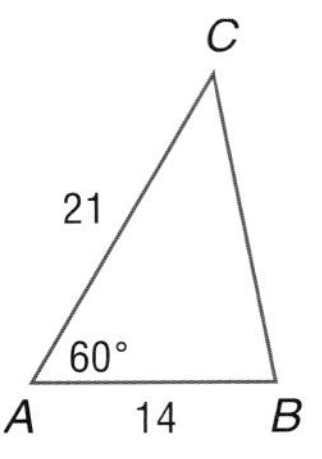

4. $a = 14, b = 15, c = 16$ **5.** $B = 41°, C = 52°, c = 27$ **6.** $a = 19, b = 24.3, c = 21.8$
7. $A = 112°, a = 32, c = 20$ **8.** $b = 8, c = 7, A = 28°$ **9.** $a = 5, b = 6, c = 7$
10. $C = 25°, a = 12, b = 9$ **11.** $a = 8, A = 49°, B = 58°$ **12.** $A = 42°, b = 120, c = 160$
13. $c = 10, A = 35°, C = 65°$ **14.** $a = 10, b = 16, c = 19$ **15.** $B = 45°, a = 40, c = 48$
16. $B = 100°, a = 10, c = 8$ **17.** $A = 40°, B = 45°, c = 4$ **18.** $A = 20°, b = 100, c = 84$

Lesson 13-6
(pages 739–745)

The given point P is located on the unit circle. Find $\sin \theta$ and $\cos \theta$.

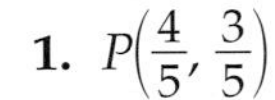

1. $P\left(\frac{4}{5}, \frac{3}{5}\right)$ **2.** $P\left(\frac{12}{13}, -\frac{5}{13}\right)$ **3.** $P\left(-\frac{8}{17}, -\frac{15}{17}\right)$ **4.** $P\left(\frac{3}{7}, \frac{2\sqrt{10}}{7}\right)$ **5.** $P\left(-\frac{2}{3}, \frac{\sqrt{5}}{3}\right)$

Find the exact value of each function.

6. $\sin 210°$ **7.** $\cos 150°$ **8.** $\cos(-135°)$ **9.** $\cos \frac{3\pi}{4}$
10. $\sin 570°$ **11.** $\sin 390°$ **12.** $\sin \frac{4\pi}{3}$ **13.** $\cos\left(-\frac{7\pi}{3}\right)$
14. $\cos 30° + \cos 60°$ **15.** $5(\sin 45°)(\cos 45°)$ **16.** $\frac{\sin 210° + \cos 240°}{2}$ **17.** $\frac{6 \cos 120° + 4 \sin 150°}{5}$

Determine the period of each function.

18.
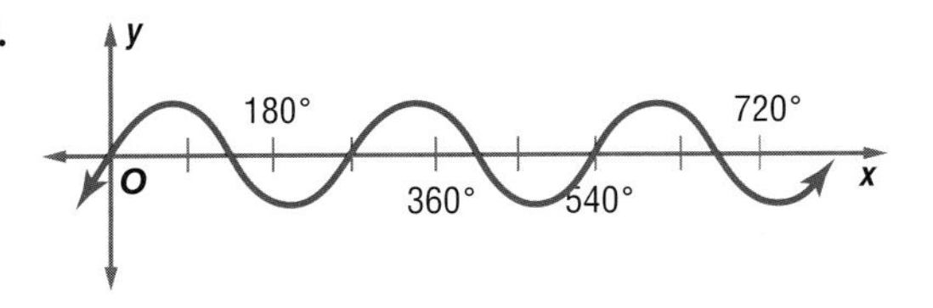

19.
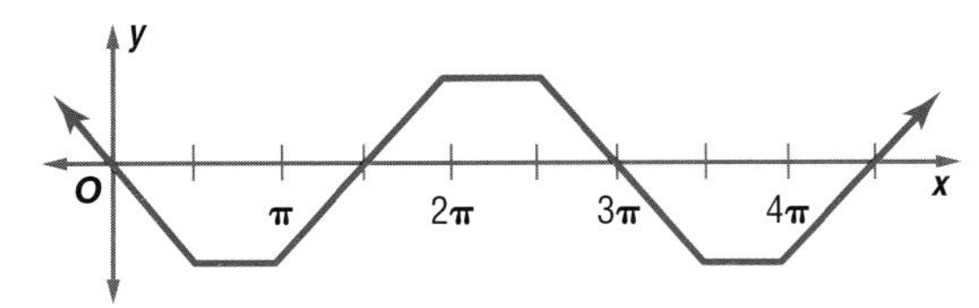

Lesson 13–7

(pages 746–751)

Write each equation in the form of an inverse function.

1. $\text{Sin } m = n$
2. $\text{Tan } 45° = 1$
3. $\text{Cos } x = \frac{1}{2}$
4. $\text{Sin } 65° = a$
5. $\text{Tan } 60° = \sqrt{3}$
6. $\text{Sin } x = \frac{\sqrt{2}}{2}$

Solve each equation.

7. $y = \text{Sin}^{-1}\left(-\frac{\sqrt{2}}{2}\right)$
8. $\text{Tan}^{-1}(1) = x$
9. $a = \text{Arccos}\left(\frac{\sqrt{3}}{2}\right)$
10. $\text{Arcsin}(0) = x$
11. $y = \text{Cos}^{-1}\left(\frac{1}{2}\right)$
12. $y = \text{Sin}^{-1}(1)$

Find each value. Round to the nearest hundredth.

13. $\text{Arccos}\left(-\frac{\sqrt{2}}{2}\right)$
14. $\text{Sin}^{-1}(-1)$
15. $\cos\left[\text{Arcsin}\left(\frac{\sqrt{2}}{2}\right)\right]$
16. $\tan\left[\text{Sin}^{-1}\left(\frac{5}{13}\right)\right]$
17. $\sin 2\left[\text{Arccos}\left(\frac{1}{2}\right)\right]$
18. $\sin\left[\text{Arccos}\left(\frac{5}{17}\right)\right]$
19. $\sin\left[\text{Tan}^{-1}\left(\frac{5}{12}\right)\right]$
20. $\tan\left[\text{Arccos}\left(-\frac{\sqrt{3}}{2}\right)\right]$
21. $\sin^{-1}[\text{Cos}^{-1}(1) - 1]$
22. $\text{Cos}^{-1}\left[\tan\left(\frac{\pi}{4}\right)\right]$
23. $\cos\left[\text{Sin}^{-1}\left(\frac{1}{2}\right)\right]$
24. $\sin[\text{Cos}^{-1}(0)]$

Lesson 14-1

(pages 762–768)

Find the amplitude, if it exists, and period of each function. Then graph each function.

1. $y = 2\cos\theta$
2. $y = \frac{1}{3}\sin\theta$
3. $y = \sin 3\theta$
4. $y = 3\sec\theta$
5. $y = \sec\frac{1}{3}\theta$
6. $y = 2\csc\theta$
7. $y = 3\tan\theta$
8. $y = 3\sin\frac{2}{3}\theta$
9. $y = 2\sin\frac{1}{5}\theta$
10. $y = 3\sin 2\theta$
11. $y = \frac{1}{2}\cos\frac{3}{4}\theta$
12. $y = 5\csc 3\theta$
13. $y = 2\cot 6\theta$
14. $y = 2\csc 6\theta$
15. $y = 3\tan\frac{1}{3}\theta$

Lesson 14-2

(pages 769–776)

State the phase shift for each function. Then graph the function.

1. $y = \sin(\theta + 60°)$
2. $y = \cos(\theta - 90°)$
3. $y = \tan\left(\theta + \frac{\pi}{2}\right)$
4. $y = \sin\left(\theta + \frac{\pi}{6}\right)$

State the vertical shift and the equation of the midline for each function. Then graph the function.

5. $y = \cos\theta + 3$
6. $y = \sin\theta - 2$
7. $y = \sec\theta + 5$
8. $y = \csc\theta - 6$
9. $y = 2\sin\theta - 4$
10. $y = \frac{1}{3}\sin\theta + 7$

State the vertical shift, amplitude, period, and phase shift of each function. Then graph the function.

11. $y = 3\cos[2(\theta + 30°)] + 4$
12. $y = 2\tan[3(\theta - 60°)] - 2$
13. $y = \frac{1}{2}\sin[4(\theta - 45°)] + 1$
14. $y = \frac{2}{5}\cos[6(\theta + 45°)] - 5$
15. $y = 6 + 2\sin\left[3\left(\theta + \frac{\pi}{2}\right)\right]$
16. $y = 3 + 3\cos\left[2\left(\theta - \frac{\pi}{3}\right)\right]$

Lesson 14-3

(pages 777–781)

Find the value of each expression.

1. $\sin \theta$, if $\cos \theta = \frac{4}{5}$; $0° \leq \theta \leq 90°$
2. $\tan \theta$, if $\sin \theta = \frac{1}{2}$; $0° \leq \theta \leq 90°$
3. $\csc \theta$, if $\sin \theta = \frac{3}{4}$; $90° \leq \theta \leq 180°$
4. $\cos \theta$, if $\tan \theta = -4$; $90° \leq \theta \leq 180°$
5. $\sec \theta$, if $\tan \theta = -4$; $90° \leq \theta \leq 180°$
6. $\sin \theta$, if $\cot \theta = -\frac{1}{4}$; $270° \leq \theta \leq 360°$
7. $\tan \theta$, if $\sec \theta = -3$; $90° \leq \theta \leq 180°$
8. $\sin \theta$, if $\cos \theta = \frac{3}{5}$; $270° \leq \theta \leq 360°$
9. $\cos \theta$, if $\sin \theta = -\frac{1}{2}$; $270° \leq \theta \leq 360°$
10. $\csc \theta$, if $\cot \theta = -\frac{1}{4}$; $90° \leq \theta \leq 180°$
11. $\csc \theta$, if $\sec \theta = -\frac{5}{3}$; $180° \leq \theta \leq 270°$
12. $\cos \theta$, if $\tan \theta = 5$; $180° \leq \theta \leq 270°$

Simplify each expression.

13. $\csc^2 \theta - \cot^2 \theta$
14. $\sin \theta \tan \theta \csc \theta$
15. $\tan \theta \csc \theta$
16. $\sec \theta \cot \theta \cos \theta$
17. $\cos \theta (1 - \cos^2 \theta)$
18. $\frac{1 - \sin^2 \theta}{\cos^2 \theta}$
19. $\frac{\sin^2 \theta + \cos^2 \theta}{\cos^2 \theta}$
20. $\frac{1 + \tan^2 \theta}{1 + \cot^2 \theta}$
21. $\frac{1}{1 + \sin \theta} + \frac{1}{1 - \sin \theta}$

Lesson 14-4

(pages 782–785)

Verify that each of the following is an identity.

1. $\sin^2 \theta + \cos^2 \theta + \tan^2 \theta = \sec^2 \theta$
2. $\frac{\tan \theta}{\sin \theta} = \sec \theta$
3. $\frac{\tan \theta}{\cot \theta} = \tan^2 \theta$
4. $\csc^2 \theta (1 - \cos^2 \theta) = 1$
5. $1 - \cot^4 \theta = 2 \csc^2 \theta - \csc^4 \theta$
6. $\sin^4 \theta - \cos^4 \theta = \sin^2 \theta - \cos^2 \theta$
7. $\sin^2 \theta + \cot^2 \theta \sin^2 \theta = 1$
8. $\frac{\cos \theta}{\csc \theta} - \frac{\csc \theta}{\sec \theta} = -\frac{\cos^3 \theta}{\sin \theta}$
9. $\frac{\cos \theta}{\sec \theta - 1} + \frac{\cos \theta}{\sec \theta + 1} = 2 \cot^2 \theta$
10. $\frac{1 + \cos \theta}{\sin \theta} = \frac{\sin \theta}{1 - \cos \theta}$
11. $\sec \theta + \tan \theta = \frac{\cos \theta}{1 - \sin \theta}$
12. $\tan \theta + \cot \theta = \csc \theta \sec \theta$
13. $\frac{\cot^2 \theta}{1 + \cot^2 \theta} = 1 - \sin^2\theta$
14. $\frac{\tan \theta - \sin \theta}{\sec \theta} = \frac{\sin^3 \theta}{1 + \cos \theta}$
15. $\sin^2 \theta(1 - \cos^2 \theta) = \sin^4 \theta$
16. $\sin^2 \theta + \sin^2 \theta \tan^2 \theta = \tan^2 \theta$
17. $\frac{\sec \theta - 1}{\sec \theta + 1} + \frac{\cos \theta - 1}{\cos \theta + 1} = 0$
18. $\tan^2 \theta(1 - \sin^2 \theta) = \sin^2 \theta$
19. $\tan \theta + \frac{\cos \theta}{1 + \sin \theta} = \sec \theta$
20. $\frac{\tan \theta}{\sec \theta + 1} = \frac{1 - \cos \theta}{\sin \theta}$
21. $\csc \theta - \frac{\sin \theta}{1 + \cos \theta} = \cot \theta$

Lesson 14-5

(pages 786–790)

Find the exact value of each expression.

1. $\sin 195°$
2. $\cos 285°$
3. $\sin 255°$
4. $\sin 105°$
5. $\cos 15°$
6. $\sin 15°$
7. $\cos 375°$
8. $\sin 165°$
9. $\sin (-225°)$
10. $\cos (-210°)$
11. $\cos (-225°)$
12. $\sin (-30°)$
13. $\sin 120°$
14. $\sin 225°$
15. $\cos (-30°)$

Verify that each of the following is an identity.

16. $\sin (90° + \theta) = \cos \theta$
17. $\cos (180° - \theta) = -\cos \theta$
18. $\sin (\pi + \theta) = -\sin \theta$
19. $\sin (\theta + 30°) + \sin (\theta + 60°) = \frac{\sqrt{3} + 1}{2} (\sin \theta + \cos \theta)$
20. $\cos (30° - \theta) + \cos (30° + \theta) = \sqrt{3} \cos \theta$

Lesson 14-6

(pages 791–797)

Find the exact value of $\sin 2\theta$, $\cos 2\theta$, $\sin \frac{\theta}{2}$, and $\cos \frac{\theta}{2}$ for each of the following.

1. $\cos \theta = \frac{7}{25}$; $0 < \theta < 90°$
2. $\sin \theta = \frac{2}{7}$; $0 < \theta < 90°$
3. $\cos \theta = -\frac{1}{8}$; $180 < \theta < 270°$
4. $\sin \theta = -\frac{5}{13}$; $270 < \theta < 360°$
5. $\sin \theta = \frac{\sqrt{35}}{6}$; $0 < \theta < 90°$
6. $\cos \theta = -\frac{17}{18}$; $90 < \theta < 180°$

Find the exact value of each expression by using the half-angle formulas.

7. $\sin 75°$
8. $\cos 75°$
9. $\sin \frac{\pi}{8}$
10. $\cos \frac{13\pi}{12}$
11. $\cos 22.5°$
12. $\cos \frac{\pi}{4}$

Verify that each of the following is an identity.

13. $\frac{\sin 2\theta}{2 \sin^2 \theta} = \cot \theta$
14. $1 + \cos 2\theta = \frac{2}{1 + \tan^2 \theta}$
15. $\csc \theta \sec \theta = 2 \csc 2\theta$
16. $\sin 2\theta (\cot \theta + \tan \theta) = 2$
17. $\frac{1 - \tan^2 \theta}{1 + \tan^2 \theta} = \cos 2\theta$
18. $\frac{\cos \theta + \sin \theta}{\cos \theta - \sin \theta} = \frac{1 + \sin 2\theta}{\cos 2\theta}$

Lesson 14-7

(pages 799–804)

Find all the solutions for each equation for $0° \leq \theta < 360°$.

1. $\cos \theta = -\frac{\sqrt{3}}{2}$
2. $\sin 2\theta = -\frac{\sqrt{3}}{2}$
3. $\cos 2\theta = 8 - 15 \sin \theta$
4. $\sin \theta + \cos \theta = 1$
5. $2 \sin^2 \theta + \sin \theta = 0$
6. $\sin 2\theta = \cos \theta$

Solve each equation for all values of θ if θ is measured in radians.

7. $\cos 2\theta \sin \theta = 1$
8. $\sin \frac{\theta}{2} + \cos \frac{\theta}{2} = \sqrt{2}$
9. $\cos 2\theta + 4 \cos \theta = -3$
10. $\sin \frac{\theta}{2} + \cos \theta = 1$
11. $3 \tan^2 \theta - \sqrt{3} \tan \theta = 0$
12. $4 \sin \theta \cos \theta = -\sqrt{3}$

Solve each equation for all values of θ if θ is measured in degrees.

13. $2 \sin^2 \theta - 1 = 0$
14. $\cos \theta - 2 \cos \theta \sin \theta = 0$
15. $\cos 2\theta \sin \theta = 1$
16. $(\tan \theta - 1)(2 \cos \theta + 1) = 0$
17. $2 \cos^2 \theta = 0.5$
18. $\sin \theta \tan \theta - \tan \theta = 0$

Solve each equation for all values of θ.

19. $\tan \theta = 1$
20. $\cos 8\theta = 1$
21. $\sin \theta + 1 = \cos 2\theta$
22. $8 \sin \theta \cos \theta = 2\sqrt{3}$
23. $\cos \theta = 1 + \sin \theta$
24. $2 \cos^2 \theta = \cos \theta$

Mixed Problem Solving

Chapter 1 Solving Equations and Inequalities

(pages 4–53)

GEOMETRY For Exercises 1 and 2, use the following information.
The formula for the surface area of a sphere is $SA = 4\pi r^2$, and the formula for the volume of a sphere is $V = \frac{4}{3}\pi r^3$. *(Lesson 1-1)*

1. Find the volume and surface area of a sphere with radius 2 inches. Write your answer in terms of π.
2. Is it possible for a sphere to have the same numerical value for the surface area and volume? If so, find the radius of such a sphere.

3. **CONSTRUCTION** The Birtic family is building a family room on their house. The dimensions of the room are 26 feet by 28 feet. Show how to use the Distributive Property to mentally calculate the area of the room. *(Lesson 1-2)*

GEOMETRY For Exercises 4–6, use the following information.
The formula for the surface area of a cylinder is $SA = 2\pi r^2 + 2\pi rh$. *(Lesson 1-2)*

4. Use the Distributive Property to rewrite the formula by factoring out the greatest common factor of the two terms.
5. Find the surface area for a cylinder with radius 3 centimeters and height 10 centimeters using both formulas. Leave the answer in terms of π.
6. Which formula do you prefer? Explain your reasoning.

POPULATION For Exercises 7 and 8, use the following information.
In 1990, the population of Mankato, Minnesota, was 31,460. For each of the next eight years, the population decreased by an average of 85 people per year. *(Lesson 1-3)*

7. What was the population in 1998?
8. If the population continues to decline at the same rate as from 1990 to 1998, what would you expect the population to be in 2005?

9. **WEATHER** The average yearly temperature for a particular coastal city in California is 64°F. The temperature seldom varies more than 7 degrees from the average. Write and solve an equation describing the maximum and minimum temperatures for this city. *(Lesson 1-4)*

ASTRONOMY For Exercises 10 and 11, use the following information.
The planets in our solar system travel in orbits that are not circular. For example, Pluto's farthest distance from the Sun is 4539 million miles, and its closest distance is 2756 million miles. *(Lesson 1-4)*

10. What is the average of the two distances?
11. Write an equation that can be solved to find the minimum and maximum distances from the Sun to Pluto.

HEALTH For Exercises 12 and 13, use the following information.
The National Heart Association recommends that less than 30% of a person's total daily Caloric intake come from fat. One gram of fat yields nine Calories. Jason is a healthy 21-year old male whose average daily Caloric intake is between 2500 and 3300 Calories. *(Lesson 1-5)*

12. Write an inequality that represents the suggested fat intake for Jason.
13. What is the greatest suggested fat intake for Jason?

TRAVEL For Exercises 14 and 15, use the following information.
Bonnie is planning a 5-day trip to a convention. She wants to spend no more than \$1000. The plane ticket is \$375, and the hotel is \$85 per night. *(Lesson 1-5)*

14. Let f represent the cost of food for one day. Write an inequality to represent this situation.
15. Solve the inequality and interpret the solution.

16. **PAINTING** Phil owns and operates a home remodeling business. He estimates that he will need 12–15 gallons of paint for a particular project. If each gallon of paint costs \$18.99, write and solve a compound inequality to determine what the cost c of the paint could be. *(Lesson 1-6)*

CONSTRUCTION For Exercises 17 and 18, use the following information.
A new playground is to be built in the shape of a rectangle. The length must be 1.5 times the width. The playground must be more than 3750 square feet, but less than 15,000 square feet. *(Lesson 1-6)*

17. Write and solve a compound inequality to determine possible dimensions for the playground.
18. Give three possible dimensions for the playground.

AGRICULTURE For Exercises 1–3, use the following information.
The table shows the average prices received by farmers for a bushel of corn from 1940–1999. *(Lesson 2-1)*

Year	Price	Year	Price
1940	$0.62	1980	$3.11
1950	$1.52	1990	$2.28
1960	$1.00	1999	$1.90
1970	$1.33		

Source: *The World Almanac*

1. Write a relation to represent the data.
2. Graph the relation.
3. Is the relation a function? Explain your reasoning.

MEASUREMENT For Exercises 4 and 5, use the following information.
The equation $y = 0.3937x$ can be used to convert any number of centimeters x to inches y. *(Lesson 2-2)*

4. Find the number of inches in 100 centimeters.
5. Find the number of centimeters in 12 inches.

POPULATION For Exercises 6–8, use the following information.
The table shows the population of Miami, Florida, for various years since 1950. *(Lesson 2-3)*

Year	Population	Year	Population
1950	249,276	1990	358,648
1970	334,859	1998	368,624
1980	346,681	2001	365,127

Source: *The World Almanac*

6. Graph the data in the table.
7. Find the average rate of change in population from 1950 to 2001.
8. Find the rate of change from 1998 to 2001. What does your answer mean?

HEALTH For Exercises 9–11, use the following information.
In 1985, 39% of people in the United States age 12 and over reported using cigarettes. The percent of people using cigarettes has decreased about 1% per year following 1985. **Source:** *The World Almanac* *(Lesson 2-4)*

9. Write an equation that represents how many people use cigarettes in x years.
10. If the percent of people using cigarettes continues to decrease at the same rate, what percent of people would you predict to be using cigarettes in 2005?
11. If the trend continues, when would you predict there to be no people using cigarettes in the U.S.? How accurate is your prediction?

EMPLOYMENT For Exercises 12–16, use the following information.
The table shows the number of unemployed people in the United States and the percent of the population unemployed for 1993 to 1999. *(Lesson 2-5)*

Year	Number Unemployed	Percent Unemployed
1993	8,940,000	6.9
1994	7,996,000	6.1
1995	7,404,000	5.6
1996	7,236,000	5.4
1997	6,739,000	4.9
1998	6,210,000	4.5
1999	5,880,000	4.2

Source: *The World Almanac*

12. Draw two scatter plots of the data: one with year and number unemployed and the other with year and percent unemployed.
13. Use two ordered pairs to write a prediction equation for each scatter plot.
14. Compare the two equations.
15. Predict the percent of people that will be unemployed in 2005.
16. In 1999, what was the total number of people in the United States?
17. **EDUCATION** At Madison Elementary, each classroom of students can have a maximum of 25 students. Draw a graph of a step function that shows the relationship between the number of students x and the number of classrooms y that are needed. *(Lesson 2-6)*

CRAFTS For Exercises 18–20, use the following information.
Priscilla makes stuffed animals and plans to sell them at a local craft show. She charges $10 for the small animals and $15 for the large animals. To cover expenses at the craft show, she needs to sell at least $350 worth of animals. *(Lesson 2-7)*

18. Write an inequality that describes this situation.
19. Graph the inequality.
20. If she sells 10 small and 15 large animals, will she cover her expenses?

EXERCISE **For Exercises 1–4, use the following information.**
At Everybody's Gym, you have two options for becoming a member. You can pay $400 per year or you can pay $150 per year plus $5 per visit. *(Lesson 3-1)*

1. For each option, write an equation that represents the cost of belonging to the gym.
2. Graph the equations. Estimate the break-even point for the gym memberships.
3. Explain what the break-even point means.
4. If you plan to visit the gym at least once per week during the year, which option should you choose?

POPULATION **For Exercises 5–8, use the following information.**
In 1994, there were about 66.5 million men in the United States work force and 56.6 million women. Over the years, the number of men has increased by about 0.98 million per year, and the number of women has increased by about 1.08 million per year.
Source: *The World Almanac* *(Lesson 3-2)*

5. Write a system of equations that represents the number of men and women in the work force y for any number of years x.
6. Solve the system to determine the year in which the number of men and women in the work force will be the same.
7. What would be the number of men in the work force in the year in Exercise 6?
8. Do you think the solution to this system makes sense for this situation? Explain.

9. **GEOMETRY** Find the coordinates of the vertices of the parallelogram whose sides are contained in the lines whose equations are $y = 3$, $y = 7$, $y = 2x$, and $y = 2x - 13$. *(Lesson 3-2)*

EDUCATION **For Exercises 10–13, use the following information.**
Mr. Gunlikson needs to purchase equipment for his physical education classes. His budget for the year is $4250. He decides to purchase cross-country ski equipment. He is able to find skis for $75 per pair and boots for $40 per pair. He knows that he should buy more boots than skis because the skis are adjustable to several sizes of boots. *(Lesson 3-3)*

10. Let y be the number of pairs of boots and x be the number of pairs of skis. Write a system of inequalities for this situation. (Remember that the number of pairs of boots and skis must be positive.)
11. Graph the region that shows how many pairs of boots and skis he can buy.
12. Give an example of three different purchases that Mr. Gunlikson can make.
13. Suppose Mr. Gunlikson wants to spend all of the money. What combination of skis and boots should he buy? Explain.

MANUFACTURING **For Exercises 14–18, use the following information.**
A shoe manufacturer makes outdoor and indoor soccer shoes. There is a two-step process for both kinds of shoes. Each pair of outdoor shoes requires 2 hours in step one, 1 hour in step two, and produces a profit of $20. Each pair of indoor shoes requires 1 hour in step one, 3 hours in step two, and produces a profit of $15. The company has 40 hours of labor per day available for step one and 60 hours available for step 2. *(Lesson 3-4)*

14. Let x represent the number of pairs of outdoor shoes and let y represent the number of indoor shoes that can be produced per day. Write a system of inequalities to represent the number of pairs of outdoor and indoor soccer shoes that can be produced in one day.
15. Draw the graph showing the feasible region.
16. List the coordinates of the vertices of the feasible region.
17. Write a function for the total profit on the shoes.
18. What is the maximum profit? What is the combination of shoes for this profit?

GEOMETRY **For Exercises 19–21, use the following information.**
An isosceles trapezoid has shorter base of measure a, longer base of measure c, and congruent legs of measure b. The perimeter of the trapezoid is 58 inches. The average of the bases is 19 inches and the longer base is twice the leg plus 7. *(Lesson 3-5)*

19. Write a system of three equations that represents this situation.
20. Find the lengths of the sides of the trapezoid.
21. Find the area of the trapezoid.

22. **EDUCATION** The three American universities with the greatest endowments in 2000 were Harvard, Yale, and Stanford. Their combined endowments are $38.1 billion. Harvard had $0.1 billion more in endowments than Yale and Stanford together. Stanford's endowments trailed Harvard's by $10.2 billion. What were the endowments of each of these universities? *(Lesson 3-5)*

AGRICULTURE For Exercises 1 and 2, use the following information.
In 1999, the United States produced 62,662,000 metric tons of wheat, 9,546,000 metric tons of rice, and 239,719,000 metric tons of corn. In that same year, Russia produced 30,960,000 metric tons of wheat, 444,000 metric tons of rice, and 1,070,000 metric tons of corn. **Source:** *The World Almanac* *(Lesson 4-1)*

1. Organize the data in two different matrices.
2. What are the dimensions of the matrices in Exercise 1?

LIFE EXPECTANCY For Exercises 3–5, use the following information.
The table shows the life expectancy for males and females for the given years. **Source:** *The World Almanac* *(Lesson 4-2)*

Year	1910	1930	1950	1970	1990
Male	48.4	58.1	65.6	67.1	71.8
Female	51.8	61.6	71.1	74.7	78.8

3. Organize all the data in a matrix.
4. Show how to organize the data in two matrices in such a way that you can find the difference between the life expectancies of males and females for the given years. Find the difference.
5. Does addition of any two matrices you can write for the data make sense? Explain.

CRAFTS For Exercises 6 and 7, use the following information.
Mrs. Long is selling crocheted items at a craft fair. She sells large afghans for \$60, baby blankets for \$40, doilies for \$25, and pot holders for \$5. She takes the following number of items to the fair: 12 afghans, 25 baby blankets, 45 doilies, and 50 pot holders. *(Lesson 4-3)*

6. Write an inventory matrix for the number of each item and a cost matrix for the price of each item.
7. Suppose Mrs. Long sells all of the items. Find her total income expressed as a matrix.

GEOMETRY For Exercises 8–11, use the following information.
A trapezoid has vertices $T(3, 3)$, $R(-1, 3)$, $A(-2, -4)$, and $P(5, -4)$. *(Lesson 4-4)*

8. Show how to use a reflection matrix to find the vertices of $TRAP$ after a reflection over the x-axis.
9. The area of a trapezoid is found by multiplying one-half the sum of the bases by the height. Find the areas of $TRAP$ and $T'R'A'P'$. How do they compare?
10. Show how to use a matrix and scalar multiplication to find the vertices of $TRAP$ after a dilation that triples the perimeter of $TRAP$.
11. Find the areas of $TRAP$ and $T'R'A'P'$ in Exercise 10. How do they compare?

AGRICULTURE For Exercises 12 and 13, use the following information.
A farm has a triangular plot planted with alfalfa defined by the coordinates $\left(-\frac{1}{2}, -\frac{1}{4}\right)$, $\left(\frac{1}{3}, \frac{1}{2}\right)$, and $\left(\frac{2}{3}, -\frac{1}{2}\right)$, where units are in square miles. *(Lesson 4-5)*

12. Find the area of the region in square miles.
13. One square miles equals 640 acres. To the nearest acre, how many acres are in the triangular plot?

ART For Exercises 14 and 15, use the following information.
Alberda sells beads to use in making Native American jewelry. Small beads sell for \$5.80 per pound, and large beads sell for \$4.60 per pound. Bernadette bought a bag of beads for \$33.00 that contained 3 times as many pounds of the small beads as the large beads. *(Lesson 4-6)*

14. Write a system of equations using the information given.
15. How many pounds of small and large beads did Bernadette buy?

MATRICES For Exercises 16 and 17, use the following information.
Two 2×2 inverse matrices have a sum of $\begin{bmatrix} -2 & 0 \\ 0 & -2 \end{bmatrix}$. The value of each entry is no less than -3 and no greater than 2. *(Lesson 4-7)*

16. Find the two matrices that satisfy the conditions.
17. Explain your method for finding the matrices.

18. **CONSTRUCTION** Alan enjoys building decks during the summer. For labor he charges \$750 to build a small deck and \$1250 to build a large deck. During the spring and summer of 2001, he built 5 more small decks than large decks. If he earned \$11,750, how many of each type of deck did he build? *(Lesson 4-8)*

EDUCATION For Exercises 1–3, use the following information.
In 1998 in the United States, there were 2,826,146 classroom teachers and 46,534,687 students. An average of $6662 was spent per student.
Source: *The World Almanac* *(Lesson 5-1)*

1. Write the numbers of teachers and students in scientific notation.
2. Find the number of students per teacher. Write the answer in standard notation rounded to the nearest whole number.
3. Find the total amount of money spent for students in 1998. Write the answer in both scientific and standard notation.

POPULATION For Exercises 4–6, use the following information.
In 2000, the population of Mexico City, Mexico, was 18,131,000, and the population of Bombay, India, was 18,066,000. It is projected that, until the year 2015, the population of Mexico City will increase at the rate of 0.4% per year and the population of Bombay will increase at the rate of 3% per year. **Source:** *The World Almanac* *(Lesson 5-2)*

4. Let r represent the rate of increase in population for each city. Write a polynomial to represent the population of each city in 2002.
5. Predict the population of each city in 2015.
6. If the projected rates are accurate, in what year will the two cities have approximately the same population?

GEOMETRY For Exercises 7 and 8, use the following information.
A rectangular box for a new product is designed in such a way that the three dimensions always have a particular relationship defined by the variable x. The volume of the box can be written as $6x^3 + 31x^2 + 53x + 30$, and the height is always $x + 2$. *(Lesson 5-3)*

7. What are the width and length of the box in terms of x?
8. Will the ratio of the dimensions of the box always be the same regardless of the value of x? Explain.

GEOMETRY For Exercises 9 and 10, use the following information.
Hero's formula for the area of a triangle is given by $A = \sqrt{s(s - a)(s - b)(s - c)}$, where a, b, and c are the lengths of the sides of the triangle and $s = 0.5(a + b + c)$. *(Lesson 5-4)*

9. Find the lengths of the sides of the triangle given in this application of Hero's formula: $A = \sqrt{s^4 - 12s^3 + 47s^2 - 60s}$.
10. What type of triangle is this?

11. **PHYSICS** The speed of sound in a liquid is $s = \sqrt{\frac{B}{d}}$, where B is known as the bulk modulus of the liquid and d is the density of the liquid. For water $B = 2.1 \cdot 10^9$ N/m^2 and $d = 10^3$ kg/m^3. Find the speed of sound in water to the nearest meter per second. *(Lesson 5-5)*

LAW ENFORCEMENT For Exercises 12 and 13, use the following information.
The approximate speed s in miles per hour that a car was traveling if it skidded d feet is given by the formula $s = 5.5\sqrt{kd}$, where k is the coefficient of friction. *(Lesson 5-6)*

12. For a dry concrete road, $k = 0.8$. If a car skids 110 feet on a dry concrete road, find its speed in miles per hour to the nearest whole number.
13. Another formula using the same variables is $s = 2\sqrt{5kd}$. Compare the results using the two formulas. Explain any variations in the answers.

PHYSICS For Exercises 14–16, use the following information.
Kepler's Third Law of planetary motion states that the square of the orbital period of any planet, in Earth years, is equal to the cube of the planet's distance from the Sun in astronomical units (AU).
Source: *The World Almanac* *(Lesson 5-7)*

14. The orbital period of Mercury is 87.97 Earth days. What is Mercury's distance from the Sun in AU?
15. Pluto's period of revolution is 247.66 Earth years. What is Pluto's distance from the Sun?
16. What is Earth's distance from the Sun in AU? Explain your result.

PHYSICS For Exercises 17 and 18, use the following information.
The time T in seconds that it takes a pendulum to make a complete swing back and forth is given by the formula $T = 2\pi\sqrt{\frac{L}{g}}$, where L is the length of the pendulum in feet and g is the acceleration due to gravity, 32 feet per second squared. *(Lesson 5-8)*

17. In Tokyo, Japan, a huge pendulum in the Shinjuku building measures 73 feet 9.75 inches. How long does it take for the pendulum to make a complete swing? **Source:** *The Guinness Book of Records*
18. A clockmaker wants to build a pendulum that takes 20 seconds to swing back and forth. How long should the pendulum be?

NUMBER THEORY For Exercises 19 and 20, use the following information.
Two complex conjugate numbers have a sum of 12 and a product of 40. *(Lesson 5-9)*

19. Find the two numbers.
20. Explain the method you used to find the numbers.

PHYSICS For Exercises 1–3, use the following information.

A model rocket is shot straight up from the top of a 100-foot building at a velocity of 800 feet per second. *(Lesson 6-1)*

1. The height $h(t)$ of the model rocket t seconds after firing is given by $h(t) = -16t^2 + at + b$ where a is the initial velocity in feet per second and b is the initial height of the rocket above the ground. Write an equation for the model rocket.
2. Find the maximum height reached by the rocket and the time that the height is reached.
3. Suppose a rocket is fired from the ground (initial height is 0). Find values for a, initial velocity, and t, time, such that the rocket reaches a height of 32,000 feet at time t.

RIDES For Exercises 4 and 5, use the following information.

An amusement park ride carries riders to the top of a 225-foot tower. The riders then free-fall in their seats until they reach 30 feet above the ground. *(Lesson 6-2)*

4. Use the formula $h(t) = -16t^2 + h_0$, where the time t is in seconds and the initial height h_0 is in feet to find how long the riders are in free-fall.
5. Suppose the designer of the ride wants the riders to experience free-fall for 5 seconds before stopping 30 feet above the ground. What should be the height of the tower?

CONSTRUCTION For Exercises 6 and 7, use the following information.

Nicole's new house has a small deck that measures 6 feet by 12 feet. She would like to build a larger deck. *(Lesson 6-3)*

6. By what amount must each dimension be increased to triple the area of the original deck?
7. What are the new dimensions of the deck?

CONSTRUCTION For Exercises 8 and 9, use the following information.

A contractor wants to construct a rectangular pool with a length that is twice the width. He plans to build a two-meter-wide walkway around the pool. He wants the area of the walkway to equal the surface area of the pool. *(Lesson 6-4)*

8. Find the dimensions of the pool to the nearest tenth of a meter.
9. What is the surface area of the pool to the nearest square meter?

PHYSICS For Exercises 10–12, use the following information.

A ball is thrown vertically into the air with a velocity of 112 feet per second. The ball was released 6 feet above the ground. The height above the ground t seconds after release is modeled by the equation $h(t) = -16t^2 + 112t + 6$. *(Lesson 6-5)*

10. When will the ball reach a height of 130 feet?
11. Will the ball ever reach 250 feet? Explain your reasoning.
12. In how many seconds after its release will the ball hit the ground?

WEATHER For Exercises 13–15, use the following information.

The table shows the normal high temperatures for Albany, New York.

Source: *The World Almanac* *(Lesson 6-6)*

Month	Temperature (°F)
January	21
February	24
March	34
April	46
May	58
June	67
July	72
August	70
September	61
October	50
November	40
December	27

13. Suppose the months are numbered such that January = 1, February = 2, and so on. A graphing calculator gave the following function as a model for the data: $y = -1.5x^2 + 21.2x - 8.5$. Graph the points in the table and the function on the same coordinate plane.
14. Identify the vertex, axis of symmetry, and direction of opening for the calculator function.
15. Discuss how well you think the function models the actual temperature data.

16. **MODELS** John is building a display table for model cars. He wants the perimeter of the table to be 26 feet, but he wants the area of the table to be no more than 30 square feet. What could the width of the table be? *(Lesson 6-7)*

WEATHER **For Exercises 1 and 2, use the following information.**
The graph models the average monthly temperature of Boise, Idaho. The values of x are $x = 1$ for January, $x = 2$ for February, and so on. The temperatures are in °F. **Source:** *The World Almanac* *(Lesson 7-1)*

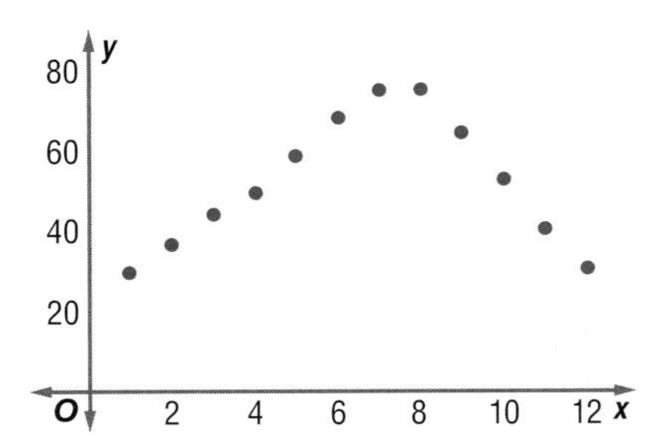

1. Is the function odd or even?
2. What do the relative maxima and minima represent?

POPULATION **For Exercises 3–5, use the following information.**
The table shows the percent of the U.S. population that was foreign-born during various years. The x values are years since 1900 and the y values are the percent of the population. **Source:** *The World Almanac* *(Lesson 7-2)*

U.S. Foreign-Born Population

x	y	x	y
0	13.6	60	5.4
10	14.7	70	4.7
20	13.2	80	6.2
30	11.6	90	8.0
40	8.8	99	9.7
50	6.9		

3. Graph the function.
4. Describe the turning points of the graph and its end behavior.
5. If this graph was modeled by a polynomial equation, what is the least degree the equation could have?

DESIGN **For Exercises 6 and 7, use the following information.**
A cylindrical can has a volume of approximately 628 cubic inches and a height of 9 inches. The volume of the can is represented by $V = \pi(a + 3)^2(9)$. *(Lesson 7-3)*

6. Use $\pi = 3.14$ to find a to the nearest tenth.
7. What is the radius of the can?

SALES **For Exercises 8 and 9, use the following information.**
The sales of items related to information technology can be modeled by $S(x) = -1.7x^3 + 18x^2 + 26.4x + 678$, where x is the number of years since 1996 and y is billions of dollars. **Source:** *Wall Street Almanac* *(Lesson 7-4)*

8. Use synthetic substitution to estimate the sales for 2003 and 2006.
9. Do you think this model is useful in estimating the sales of this industry in the future? Explain.

MANUFACTURING **For Exercises 10 and 11, use the following information.**
A box measures 12 inches by 16 inches by 18 inches. The manufacturer will increase each dimension of the box by the same number of inches and have a new volume of 5985 cubic inches. *(Lesson 7-5)*

10. Write a polynomial equation to model this situation.
11. How much should be added to each dimension?
12. **CONSTRUCTION** A picnic area has the shape of a trapezoid. The longer base is 8 more than 3 times the length of the shorter base and the height is 1 more than 3 times the shorter base. What are the dimensions of the trapezoid if the area is 4104 square feet? *(Lesson 7-6)*

EMPLOYMENT **For Exercises 13 and 14, use the following information.**
From 1994 to 1999, the number of employed women and men in the United States, age 16 and over, can be modeled by the following equations where x is the number of years since 1994 and y is the number of people in thousands. **Source:** *The World Almanac* *(Lesson 7-7)*

women: $y = 1086.4x + 56{,}610$
men: $y = 999.2x + 66{,}450$

13. Write a function that models the total number of men and women employed in the United States.
14. If f is the function for the number of men and g is the function for the number of women, what does $(f - g)(x)$ represent?
15. **HEALTH** The average weight of a baby born at a certain hospital is $7\frac{1}{2}$ pounds, and the average length is 19.5 inches. One kilogram is about 2.2 pounds, and 1 centimeter is about 0.3937 inches. Find the average weight in kilograms and the length in centimeters. *(Lesson 7-8)*

SAFETY **For Exercises 16 and 17, use the following information.**
The table shows the total stopping distance x, in meters, of a vehicle and the speed y, in meters per second. *(Lesson 7-9)*

Distance	92	68	49	32	18
Speed	29	25	20	16	11

16. Graph the data in the table.
17. Graph the function $y = 2\sqrt{2x}$ on the same coordinate plane. How well do you think this function models the given data? Explain.

GEOMETRY For Exercises 1–4, use the following information.
Triangle ABC has vertices $A(2, 1)$, $B(-6, 5)$, and $C(-2, -3)$. *(Lesson 8-1)*

1. An isosceles triangle has two sides with equal length. Is triangle ABC isosceles? Explain your reasoning.
2. An equilateral triangle has three sides of equal length. Is triangle ABC equilateral? Explain your reasoning.
3. Triangle EFG is formed by joining the midpoints of the sides of triangle ABC. What type of triangle is $\triangle EFG$? Explain your reasoning.
4. Describe any relationship between the lengths of the sides of the two triangles.

ENERGY For Exercises 5–8, use the following information.
Solar energy plays an important role in space satellites. As they become more efficient, solar batteries are being used for more purposes on Earth. A parabolic mirror can be used to collect solar energy. The mirrors reflect the rays from the Sun to the focus of the parabola. The latus rectum of a particular mirror is 40 feet long. *(Lesson 8-2)*

5. Write an equation for the parabola formed by the mirror if the vertex of the mirror is 9.75 feet below the origin.
6. One foot is exactly 0.3048 meter. Rewrite the equation for the mirror in terms of meters.
7. Graph one of the equations for the mirror.
8. Which equation did you choose to graph? Explain.

COMMUNICATION For Exercises 9–11, use the following information.
Radio waves carry information from the transmitter to the receivers. The radio tower for KCGM, Voice of the Prairies, has a circular radius for broadcasting of 65 miles. The radio tower for KVCK has a circular radius for broadcasting of 85 miles. *(Lesson 8-3)*

9. Let the radio tower for KCGM be located at the origin of a coordinate system. Write an equation for the set of points at the maximum broadcast distance from the tower.
10. The radio tower for KVCK is 50 miles south and 15 miles west of the KCGM tower. Let each mile represent one unit on the coordinate system. Write an equation for the set of points at the maximum broadcast distance from the KVCK tower.
11. Graph the two equations and show the area where the radio signals overlap.

ASTRONOMY For Exercises 12–14, use the following information.
The table shows the closest and farthest distances of Venus and Jupiter from the Sun in millions of miles. **Source:** *The World Almanac* *(Lesson 8-4)*

Planet	Closest	Farthest
Venus	66.8	67.7
Jupiter	460.1	507.4

12. Write an equation for the orbit of each planet, assuming that the center of the orbit is the origin, the center of the Sun is a focus of the ellipse, and the Sun lies on the x-axis.
13. Find the eccentricity e, or the ratio $\frac{c}{a}$, for each planet.
14. Which planet has an orbit that is closer to looking like a circle? Explain your reasoning.
15. A comet follows a path that is one branch of a hyperbola. Suppose Earth is the center of the hyperbolic curve and has coordinates (0, 0). Write an equation for the path of the comet if $c = 5{,}225{,}000$ miles and $a = 2{,}500{,}000$ miles. Let the x-axis be the transverse axis. *(Lesson 8-5)*

AVIATION For Exercises 16–18, use the following information.
A military jet performs for an air show. The path of the plane during one trick can be modeled by a conic section with equation $24x^2 + 1000y - 31{,}680x - 45{,}600 = 0$, where distances are represented in feet.

16. Identify the shape of the curved path of the jet. Write the equation in standard form.
17. If the jet begins its path upward or ascent at (0, 0), what is the horizontal distance traveled by the jet from the beginning of the ascent to the end of the descent?
18. What is the maximum height of the jet?

SATELLITES For Exercises 19–21, use the following information.
Two satellites are placed in orbit about Earth. The equations of the two orbits are $\frac{x^2}{(300)^2} + \frac{y^2}{(900)^2} = 1$ and $\frac{x^2}{(600)^2} + \frac{y^2}{(690)^2} = 1$, where distances are in km and Earth is the center of each curve. *(Lesson 8-7)*

19. Solve each equation for y.
20. Use a graphing calculator to estimate the intersection points of the two orbits.
21. Compare the orbits of the two satellites.

MANUFACTURING For Exercises 1–3, use the following information.
A shipping container in the shape of a rectangular prism is designed using any value for x such that the volume can be represented by the polynomial $6x^3 + 11x^2 + 4x$, where the height is x. *(Lesson 9-1)*

1. Find the length and width of the container in terms of x.
2. Find the ratio of the three dimensions of the container when $x = 2$.
3. Will the ratio of the three dimensions be the same for all values of x?

PHOTOGRAPHY For Exercises 4–6, use the following information.
The formula $\frac{1}{q} = \frac{1}{f} - \frac{1}{p}$ can be used to determine how far the film should be placed from the lens of a camera to create a perfect photograph. The variable q represents the distance from the lens to the film, f represents the focal length of the lens, and p represents the distance from the object to the lens. *(Lesson 9-2)*

4. Solve the formula for $\frac{1}{p}$.
5. Write the expression containing f and q as a single rational expression.
6. If a camera has a focal length of 8 centimeters and the lens is 10 centimeters from the film, how far should an object be from the lens so that the picture will be in focus?

PHYSICS For Exercises 7–9, use the following information.
Isaac Newton's law of universal gravitation depends upon the Inverse Square Law. It states that the relationship between two variables is related to the equation $y = \frac{1}{x^2}$. *(Lesson 9-3)*

7. Graph $y = \frac{1}{x^2}$.
8. Give the equations of any asymptotes of the graph.
9. If x represents distance in an application of the Inverse Square Law, what values of x make sense for the situation?

PHYSICS For Exercises 10 and 11, use the following information.
In 1798, Henry Cavendish found a value for the universal gravitational constant G that Newton used in his formula $F = G\frac{m_A m_B}{d^2}$ for finding the gravitational force between two objects. The variables in the formula are defined as follows: F is the gravitational force between the objects, G is the universal constant, m_A is the mass of the first object, m_B is the mass of the second object, and d is the distance between the centers of the objects. *(Lesson 9-4)*

10. If the mass of object A is constant, does Newton's formula represent a *direct* or *inverse* variation between the mass of object B and the distance?
11. Cavendish found the value of G to be 6.67×10^{-11} N · m^2/kg^2. If two objects each weighing 5 kilograms are placed so that their centers are 0.5 meter apart, what is the gravitational force between the two objects?

EDUCATION For Exercises 12–14, use the following information.
The table shows the average number of students per computer in United States public schools for various years. *(Lesson 9-5)*

Year	Students	Year	Students
1984	125	1992	18
1985	75	1993	16
1986	50	1994	14
1987	37	1995	10.5
1988	32	1996	10
1989	25	1997	7.8
1990	22	1998	6.1
1991	20	1999	5.7

Source: *The World Almanac*

12. Let x represent years where 1984 = 1, 1985 = 2, and so on. Let y represent the number of students. Graph the data.
13. What type of function does the graph most closely resemble?
14. Use a graphing calculator to find an equation that models the data.

TRAVEL For Exercises 15–17, use the following information.
A trip between two towns in Montana takes 4 hours under ideal conditions. The first 150 miles of the trip is on an interstate, and the second 130 miles is on a highway with a speed limit that is 10 miles per hour less than on the interstate. *(Lesson 9-6)*

15. If x represents the speed limit on the interstate, write an expression that represents the time spent at that speed.
16. Write an expression for the time spent on the other highway.
17. Write and solve an equation to find the speed limit on the interstate and on the highway.

POPULATION For Exercises 1–4, use the following information.
The population of the world has been growing rapidly. In 1950, the world population was about 2.556 billion. By 1980, it had increased to about 4.458 billion. **Source:** *The World Almanac* *(Lesson 10-1)*

1. Write an exponential function of the form $y = ab^x$ that could be used to model the world population y in billions for 1950 to 1980. Write the equation in terms of x, the number of years since 1950. (Round the value of b to the nearest ten thousandth.)
2. Suppose the world population continued to grow at that rate. Estimate the population for the year 2000.
3. In 2000, the population of the world was about 6.08 billion. Compare your estimate to the actual population.
4. Use the equation you wrote in Exercise 1 to estimate the world population in the year 2020. How accurate do you think the estimate is? Explain your reasoning.

EARTHQUAKES For Exercises 5–8, use the following information.
The table shows the Richter scale that measures earthquake intensity. Column 2 shows the increase in intensity between each number. For example, an earthquake that measures 7 is 10 times more intense than one measuring 6. *(Lesson 10-2)*

Richter Number	Increase in Magnitude
1	1
2	10
3	100
4	1000
5	10,000
6	100,000
7	1,000,000
8	10,000,000

Source: *The New York Public Library*

5. Graph this function where x is the Richter number and y is the increase in magnitude.
6. Write an equation of the form $y = b^{x-c}$ for the function in Exercise 5. (*Hint*: Write the values in the second column as powers of 10 to see a pattern and find the value of c.)
7. Graph the inverse of the function in Exercise 6.
8. Write an equation of the form $y = \log_{10}x + c$ for the function in Exercise 7.

EARTHQUAKES For Exercises 9–11, use the following information.
The table shows the magnitude on the Richter scale of some major earthquakes. *(Lesson 10-3)*

Year/Location	Magnitude
1939/Turkey	8.0
1963/Yugoslavia	6.0
1970/Peru	7.8
1988/Armenia	7.0
1995/Japan	6.9

Source: *The World Almanac*

9. Name two earthquakes such that the intensity of one was 10 times the intensity of the other.
10. Name two earthquakes such that the intensity of one was 100 times the intensity of the other.
11. What would be the magnitude of an earthquake that is 1000 times as intense as the 1963 earthquake in Yugoslavia?
12. Suppose you know that $\log_7 2 \approx 0.3562$ and $\log_7 3 \approx 0.5646$. Describe two different methods that you could use to approximate $\log_7 2.5$. (You can use a calculator, of course.) Then describe how you can check your result. *(Lesson 10-4)*

WEATHER For Exercises 13–15, use the following information.
The atmospheric pressure P, in bars, of a given height on Earth can be found by using the formula $P = s \cdot e^{-\frac{h}{H}}$. In the formula, s is the surface pressure on Earth, which is approximately 1 bar, h is the altitude for which you want to find the pressure in kilometers, and H is always 7 kilometers. *(Lesson 10-5)*

13. Find the pressure for 2 kilometers, 4 kilometers, and 7 kilometers.
14. What do you notice about the pressure as altitude increases?
15. What is the pressure on top of Mount Everest at an altitude of 8700 meters?

AGRICULTURE For Exercises 16–19, use the following information.
An equation that models the decline in the number of U.S. farms is $y = 3{,}962{,}520(0.98)^x$, where x is years since 1960 and y is the number of farms.
"**Source:** *Wall Street Journal* *(Lesson 10-6)*

16. By examining the equation, how can you determine that the number of farms is declining?
17. By what rate per year is the number of farms declining?
18. How many farms were there in 1960?
19. Predict when the number of farms will be less than 1.5 million.

CLUBS For Exercises 1 and 2, use the following information.
Kim and her mother are in a quilting club consisting of 9 members. Each week, the club meets, and each member must bring one completed quilt square. *(Lesson 11-1)*

1. Find the first eight terms of the sequence that describes the total number of squares that have been made for the quilt after each meeting.
2. One particular quilt measures 72 inches by 84 inches and is being designed with 4-inch squares. After how many meetings will the quilt be complete?

ART For Exercises 3 and 4, use the following information.
Alberta is making a Native American beadwork design consisting of rows of colored beads. The first row consists of 10 beads, and each consecutive row will have 15 more beads than the previous row. *(Lesson 11-2)*

3. Write an equation for the number of beads in the nth row.
4. Find the number of beads in the design if it contains 25 rows.

GAMES For Exercises 5 and 6, use the following information.
An audition is held for a TV game show. At the end of each round, one-half of the prospective contestants are eliminated from the competition. On a particular day, 524 contestants begin the audition. *(Lesson 11-3)*

5. Write an equation for finding the number of contestants that are left after n rounds.
6. Using this method, will the number of contestants that are to be eliminated always be a whole number? Explain.

SPORTS For Exercises 7–9, use the following information.
Caitlin is training for a marathon (about 26 miles). Her trainer advises her to begin by running 2 miles. Then she is to run every other day. During each session, she is to multiply the distance she ran the previous session by one and a half. *(Lesson 11-4)*

7. Write the first five terms of a sequence describing the number of miles she is to run during consecutive training sessions.
8. When will she exceed 26 miles in one training session?
9. When will she have run at least 100 total miles?

GEOMETRY For Exercises 10–12, use the following information.
You can illustrate the sum of an infinite geometric series by using a square of paper. *(Lesson 11-5)*

10. Cut a square of paper at least 8 inches on a side. Let the square be one unit. Cut away one-half of the square. Call this piece Term 1. Next, cut away one-half of the remaining sheet of paper. Call this piece Term 2. Continue cutting the remaining paper in half and labeling the pieces with a term number as long as possible. List the fractions represented by the pieces.
11. If you could cut squares indefinitely, you would have an infinite series. Find the sum of the series.
12. How does the sum of the series relate to the original square of paper?

BIOLOGY For Exercises 13–15, use the following information.
In a particular forest, scientists are interested in how the population of wolves will change over the next two years. One mathematical model for animal population is the Verhulst population model. The formula for this model is $p_{n+1} = p_n + rp_n(1 - p_n)$, where n represents the number of time periods that have passed, p_n represents the percent of the maximum sustainable population that exists at time n, and r is the growth factor. *(Lesson 11-6)*

13. To find the population of the wolves after one year, you must evaluate the expression $p_1 = 0.45 + 1.5(0.45)(1 - 0.45)$. What is the value of the expression?
14. Explain what each number in the expression in Exercise 13 represents.
15. The current population of wolves is 165. Find the new population by multiplying 165 by the value in Exercise 13.

16. **PASCAL'S TRIANGLE** Study the first eight rows of Pascal's triangle. Write the sum of the terms in each row as a list. Make a conjecture about the sums of the rows of Pascal's triangle. *(Lesson 11-7)*

17. **NUMBER THEORY** Two statements that can be proved using mathematical induction are $\frac{1}{3} + \frac{1}{3^2} + \frac{1}{3^3} + \cdots + \frac{1}{3^n} = \frac{1}{2}\left(1 - \frac{1}{3^n}\right)$ and $\frac{1}{4} + \frac{1}{4^2} + \frac{1}{4^3} + \cdots + \frac{1}{4^n} = \frac{1}{3}\left(1 - \frac{1}{4^n}\right)$.
Write and prove a conjecture involving $\frac{1}{5}$ that is similar to the two given statements. *(Lesson 11-8)*

NUMBER THEORY For Exercises 1–3, use the following information.
According to the Rational Zero Theorem, if $\frac{p}{q}$ is a rational root, then p is a factor of the constant of the polynomial, and q is a factor of the leading coefficient. *(Lesson 12-1)*

1. What is the maximum number of possible rational roots that you may need to check for the polynomial $3x^4 - 5x^3 + 2x^2 - 7x + 10$? Explain your answer.
2. Why may you not need to check the maximum number of possible roots?
3. Are choosing the numerator and the denominator for a possible rational root independent or dependent events?

4. **GARDENING** A gardener is selecting plants for a special display. There are 15 varieties of pansies from which to choose. The gardener can only use 9 varieties in the display. How many ways can 9 varieties be chosen from the 15 varieties? *(Lesson 12-2)*

SPEED LIMITS For Exercises 5–7, use the following information.
In 1995, states were allowed to set their own highway speed limits. The table shows the number of states having each speed limit for their rural interstates. *(Lesson 12-3)*

Speed Limit	Number of States
55	1
65	20
70	18
75	11

Source: *The World Almanac*

5. If a state is randomly selected, what is the probability that its speed limit is 75?
6. If a state is randomly selected, what is the probability that its speed limit is 55?
7. If a state is randomly selected, what is the probability that its speed limit is 55 or greater?

8. **LOTTERIES** A lottery number for a particular lottery has seven digits which can be any digit from 0 to 9. It is advertised that the odds of winning the lottery are 1 to 10,000,000. Is this statement about the odds correct? Explain your reasoning. *(Lesson 12-4)*

For Exercises 9 and 10, use the following information.
The table shows the results of a survey of the most popular colors for luxury cars in 1999. *(Lesson 12-5)*

Color	% of cars	Color	% of cars
white	16.1	gold	7.0
silver	14.8	green	6.1
lt. brown	12.9	red	6.0
black	9.4	blue	4.9
gray	8.3	other	14.5

Source: *The World Almanac*

9. If a car sold in 1999 is randomly selected, what is the probability that it is white or silver?
10. In a parking lot of 1000 cars sold in 1999, how many cars would you expect to be blue or green?

EDUCATION For Exercises 11–13, use the following information.
The list of numbers given are the average scores for each state for the ACT for 1999–2000. *(Lesson 12-6)*
20.2, 21.3, 21.5, 20.3, 21.4, 21.5, 21.3, 20.6, 20.6, 19.9, 21.6, 21.4, 21.5, 21.4, 22.0, 21.6, 20.1, 19.6, 21.9, 20.7, 21.9, 21.3, 22.0, 18.7, 21.6, 21.8, 21.7, 21.5, 22.5, 20.7, 20.1, 22.2, 19.5, 21.4, 21.4, 20.8, 22.7, 21.4, 21.1, 19.3, 21.5, 20.0, 20.3, 21.5, 22.2, 20.5, 22.4, 20.2, 22.2, 21.6

11. Compare the mean and median of the data.
12. Find the standard deviation of the data. Round to the nearest hundredth.
13. Suppose the state with an average score of 20.0 incorrectly reported the results. The score for the state is actually 22.5. How are the mean and median of the data affected by this data change?
14. **HEALTH** The heights of students at Madison High School are normally distributed with a mean of 66 inches and a standard deviation of 2 inches. Of the 1080 students in the school, how many would you expect to be less than 62 inches tall? *(Lesson 12-7)*

EDUCATION For Exercises 15 and 16, use the following information.
A mathematics contest consists of four tests. Each test has ten multiple-choice questions with four answer options. *(Lesson 12-8)*

15. What is the probability that a student who randomly answers all questions on a test will get a score of 5 correct?
16. What is the probability that a student who randomly answers all questions on a test will get every question incorrect?
17. **SURVEY** A poll of 1750 people shows that 78% enjoy travel. Find the margin of the sampling error for the survey. *(Lesson 12-9)*

CABLE CARS **For Exercises 1 and 2, use the following information.**
The longest cable car route in the world is located in Venezuela. It begins at an altitude of 5379 feet and ends at an altitude of 15,629 feet. The ride is 8 miles long. **Source:** *The Guinness Book of Records* *(Lesson 13-1)*

1. Draw a diagram to represent this situation.
2. To the nearest degree, what is the average angle of elevation of the cable car ride?

RIDES **For Exercises 3–5, use the following information.**
In 2000, a gigantic Ferris wheel, the London Eye, opened in England. The wheel has 32 cars evenly spaced around the circumference. *(Lesson 13-2)*

3. What is the measure, in degrees, of the angle between any two consecutive cars?
4. What is the measure, in radians, of the angle between any two consecutive cars?
5. If a car is located such that the measure in standard position is $-60°$, what are the measures of one angle with positive measure and one angle with negative measure coterminal with the angle of this car?

BASKETBALL **For Exercises 6 and 7, use the following information.**
During the halftime of a basketball game, a person is selected to try to make a shot at a distance of 12 feet from the basket. The formula $R = \frac{V_0^2 \sin 2\theta}{32}$ gives the distance of a basketball shot with an initial velocity of V_0 feet per second at an angle of θ with the ground. *(Lesson 13-3)*

6. If the basketball was shot with an initial velocity of 24 feet per second at an angle of 75°, how far will the basketball travel?
7. Find an initial velocity and angle of release that would result in the ball traveling approximately 12 feet to the basket.
8. **COMMUNICATIONS** A telecommunications tower needs to be supported by two wires. The angle between the ground and the tower on one side must be 35° and the angle between the ground and the second tower must be 72°. The distance between the two wires is 110 feet.

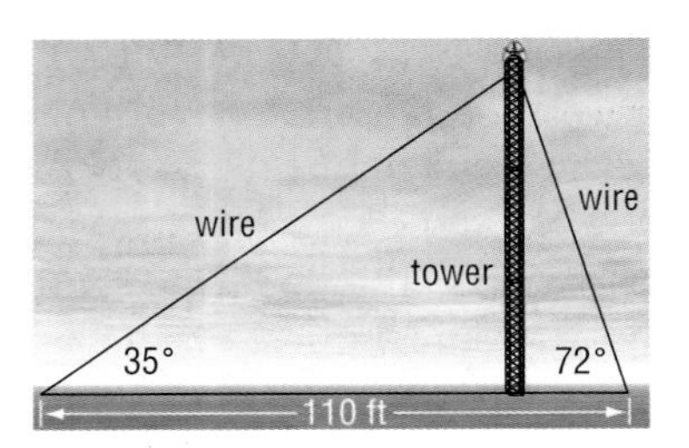

To the nearest foot, what should be the lengths of the two wires? *(Lesson 13-4)*

SURVEYING **For Exercises 9–11, use the following information.**
A triangular plot of farm land measures 0.9 by 0.5 by 1.25 miles. *(Lesson 13-5)*

9. If the plot of land is fenced on the border, what will be the angles at which the fences of the three sides meet? Round to the nearest degree.
10. What is the area of the plot of land? (*Hint*: Use the area formula in Lesson 13-4.)
11. One square mile is 640 acres. How many acres are in the plot of land?

WEATHER **For Exercises 12 and 13, use the following information.**
The monthly normal temperatures, in degrees Fahrenheit, for New York City are given in the table. January is assigned a value of 1, February a value of 2, and so on. *(Lesson 13-6)*

Month	Temperature	Month	Temperature
1	32	7	77
2	34	8	76
3	42	9	68
4	53	10	58
5	63	11	48
6	72	12	37

Source: *The World Almanac*

12. Graph the data in a scatter plot.
13. A trigonometric model for the temperature T in degrees Fahrenheit of New York City at t months is given by $T = 22.5 \sin\left(\frac{\pi}{6}x - 2.25\right) + 54.3$. A quadratic model for the same situation is $T = -1.34x^2 + 18.84x + 5$. Which model do you think best fits the data? Explain your reasoning.

PHYSICS **For Exercises 14–16, use the following information.**
When light passes from one substance to another, it may be reflected and refracted. Snell's law can be used to find the angle of refraction as a beam of light passes from one substance to another. One form of the formula for Snell's law is $n_1 \sin \theta_1 = n_2 \sin \theta_2$, where n_1 and n_2 are the indices of refraction for the two substances and θ_1 and θ_2 are the angles of the light rays passing through the two substances. *(Lesson 13-7)*

14. Solve the equation for $\sin \theta_1$.
15. Write an equation in the form of an inverse function that allows you to find θ_1.
16. If a light beam in air with index of refraction of 1.00 hits a diamond with index of 2.42 at an angle of 30°, find the angle of refraction.

Mixed Problem Solving

TIDES For Exercises 1–3, use the following information.
The world's record for the highest tide is held by the Minas Basin in Nova Scotia, Canada, with a tidal range of 54.6 feet. A tide is at equilibrium when it is at its normal level halfway between its highest and lowest points. *(Lesson 14-1)*

1. Write an equation to represent the height h of the tide. Assume that the tide is at equilibrium at $t = 0$, that the high tide is beginning, and that the tide completes one cycle in 12 hours.
2. Mobile, Alabama, has a very small tidal range at only one foot six inches. Write an equation to represent the height h of the tide in Mobile.
3. Graph the functions for the tides in Minas Basin and Mobile on the same axis system. How do the graphs compare?

RIDES For Exercises 4–7, use the following information.
The Cosmoclock 21 is a huge Ferris wheel in Yokohama City, Japan. The diameter is 328 feet. Suppose that a rider enters the ride at 0 feet and then rotates in 90° increments counterclockwise. The table shows the angle measures of rotation and the height above the ground of the rider. *(Lesson 14-2)*

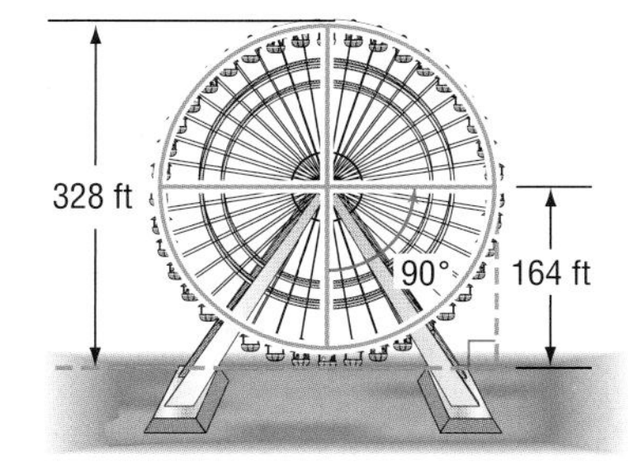

Angle	Height	Angle	Height
0°	0	450°	
90°	164	540°	
180°	328	630°	
270°		720°	
360°			

4. Copy and complete the table. Then graph the points (angle, height).
5. A function that models the data is $y = 164 \cdot (\sin(x - 90°)) + 164$. Identify the vertical shift, amplitude, period, and phase shift of the graph.
6. Write an equation using the sine that models the position of a rider on the Vienna Giant Ferris Wheel in Vienna, Austria, with a diameter of 200 feet. Check your equation by plotting the points and the equation with a graphing calculator.
7. Write an equation using the cosine that models the position of a rider on the Vienna Giant Ferris Wheel.
8. **TRIGONOMETRY** Using the exact values for the sine and cosine functions, show that the identity $\cos^2\theta + \sin^2\theta = 1$ is true for angles of measure 30°, 45°, 60°, 90°, and 180°. *(Lesson 14-3)*
9. **ROCKETS** In the formula $h = \frac{v^2 \sin^2\theta}{2g}$, h is the maximum height reached by a rocket, θ is the angle between the ground and the initial path of the object, v is the rocket's initial velocity, and g is the acceleration due to gravity. Verify the identity $\frac{v^2 \sin^2\theta}{2g} = \frac{v^2 \cos^2\theta}{2g \cot^2\theta}$. *(Lesson 14-4)*

WEATHER For Exercises 10–12, use the following information.
The monthly high temperatures for Minneapolis, Minnesota, can be modeled by the equation $y = 31.65 \sin\left(\frac{\pi}{6}x - 2.09\right) + 52.35$, where the months x are January $= 1$, February $= 2$, and so on. The monthly low temperatures for Minneapolis can be modeled by the equation $y = 30.15 \sin\left(\frac{\pi}{6}x - 2.09\right) + 32.95$. *(Lesson 14-5)*

10. Write a new function by adding the expressions on the right side of each equation and dividing the result by 2.
11. Graph the two original functions and the new function on the same calculator screen. What do you notice?
12. What is the meaning of the function you wrote in Exercise 10?
13. Begin with one of the Pythagorean Identities. Perform equivalent operations on each side to create a new trigonometric identity. Then show that the identity is true. *(Lesson 14-6)*
14. **TELEVISION** The tallest structure in the world is a television transmitting tower located near Fargo, North Dakota, with a height of 2064 feet.

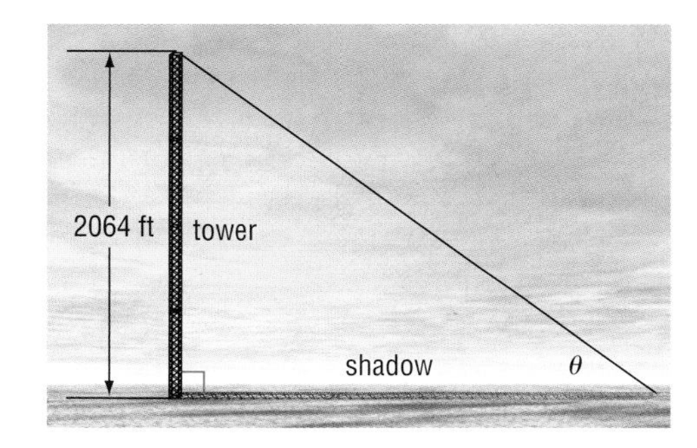

What is the measure of θ if the length of the shadow is 1 mile? **Source:** *The Guinness Book of Records* *(Lesson 14–7)*

Mixed Problem Solving

Preparing For Standardized Tests 

Becoming a Better Test-Taker

At some time in your life, you will probably have to take a standardized test. Sometimes this test may determine if you go on to the next grade or course, or even if you will graduate from high school. This section of your textbook is dedicated to making you a better test-taker.

TYPES OF TEST QUESTIONS In the following pages, you will see examples of four types of questions commonly seen on standardized tests. A description of each type of question is shown in the table below.

Type of Question	Description	See Pages
multiple choice	Four or five possible answer choices are given from which you choose the best answer.	878–879
gridded response	You solve the problem. Then you enter the answer in a special grid and shade in the corresponding circles.	880–883
short response	You solve the problem, showing your work and/or explaining your reasoning.	884–887
extended response	You solve a multi-part problem, showing yourwork and/or explaining your reasoning	888–892

PRACTICE After being introduced to each type of question, you can practice that type of question. Each set of practice questions is divided into five sections that represent the concepts most commonly assessed on standardized tests.

- Number and Operations
- Algebra
- Geometry
- Measurement
- Data Analysis and Probability

USING A CALCULATOR On some tests, you are permitted to use a calculator. You should check with your teacher to determine if calculator use is permitted on the test you will be taking, and if so, what type of calculator can be used.

TEST-TAKING TIPS In addition to Test-Taking Tips like the one shown at the right, here are some additional thoughts that might help you.

- Get a good night's rest before the test. Cramming the night before does not improve your results.
- Budget your time when taking a test. Don't dwell on problems that you cannot solve. Just make sure to leave that question blank on your answer sheet.
- Watch for key words like NOT and EXCEPT. Also look for order words like LEAST, GREATEST, FIRST, and LAST.

Test-Taking Tip

If you are allowed to use a calculator, make sure you are familiar with how it works so that you won't waste time trying to figure out the calculator when taking the test.

Multiple-Choice Questions

Multiple-choice questions are the most common type of questions on standardized tests. These questions are sometimes called *selected-response questions.* You are asked to choose the best answer from four or five possible answers. To record a multiple-choice answer, you may be asked to shade in a bubble that is a circle or an oval or to just write the letter of your choice. Always make sure that your shading is dark enough and completely covers the bubble.

The answer to a multiple-choice question is usually not immediately obvious from the choices, but you may be able to eliminate some of the possibilities by using your knowledge of mathematics. Another answer choice might be that the correct answer is not given.

Incomplete Shading
Ⓐ Ⓑ Ⓒ Ⓓ
Too light shading
Ⓐ Ⓑ Ⓒ Ⓓ
Correct shading
Ⓐ Ⓑ Ⓒ Ⓓ

Example 1

White chocolate pieces sell for \$3.25 per pound while dark chocolate pieces sell for \$2.50 per pound. How many pounds of white chocolate are needed to produce a 10-pound mixture of both kinds that sells for \$2.80 per pound?

Strategy

Reasonableness
Check to see that your answer is reasonable with the given information.

Ⓐ 2 lb **Ⓑ 4 lb** **Ⓒ 6 lb** **Ⓓ 10 lb**

The question asks you to find the number of pounds of the white chocolate. Let w be the number of pounds of white chocolate and let d be the number of pounds of dark chocolate. Write a system of equations.

$w + d = 10$ There is a total of 10 pounds of chocolate.

$3.25w + 2.50d = 2.80(10)$ The price is \$2.80 × 10 for the mixed chocolate.

Use substitution to solve.

$3.25w + 2.50d = 2.80(10)$	Original equation
$3.25w + 2.50(10 - w) = 28$	Solve the first equation for d and substitute.
$3.25w + 25 - 2.5w = 28$	Distributive Property
$0.75w = 3$	Simplify.
$w = 4$	Divide each side by 0.75.

The answer is B.

Example 2

Josh throws a baseball upward at a velocity of 105 feet per second, releasing the baseball when it is 5 feet above the ground. The height of the baseball t seconds after being thrown is given by the formula $h(t) = -16t^2 + 105t + 5$. Find the time at which the baseball reaches its maximum height. Round to the nearest tenth of a second.

Strategy

Diagrams
Drawing a diagram for a situation may help you to answer the question.

Ⓕ 1.0 s **Ⓖ 3.3 s** **Ⓗ 6.6 s** **Ⓙ 177.3 s**

Graph the equation. The graph is a parabola. Make sure to label the horizontal axis as t (time in seconds) and the vertical axis as h for height in feet. The ball is at its maximum height at the vertex of the graph.

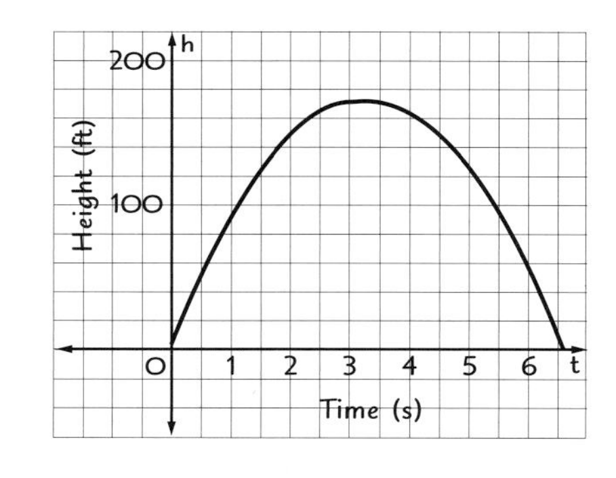

The graph indicates that the maximum height is achieved between 3 and 4 seconds after launch.

The answer is G.

Multiple-Choice Practice

Choose the best answer.

Number and Operations

1. In 2002, 1.8123×10^8 people in the United States and Canada used the Internet while 5.442×10^8 people worldwide used the Internet. What percent of users were from the United States and Canada?

Ⓐ 33% Ⓑ 35% Ⓒ 37% Ⓓ 50%

2. Serena has 6 plants to put in her garden. How many different ways can she arrange the plants?

Ⓐ 21 Ⓑ 30 Ⓒ 360 Ⓓ 720

Algebra

3. The sum of Kevin's, Anna's, and Tia's ages is 40. Anna is 1 year more than twice as old as Tia. Kevin is 3 years older than Anna. How old is Anna?

Ⓐ 7 Ⓑ 14 Ⓒ 15 Ⓓ 18

4. Rafael's Theatre Company sells tickets for $10. At this price, they sell 400 tickets. Rafael estimates that they would sell 40 less tickets for each $2 price increase. What charge would give the most income?

Ⓐ 10 Ⓑ 13 Ⓒ 15 Ⓓ 20

Geometry

5. Hai stands 75 feet from the base of a building and sights the top at a 35° angle. What is the height of the building to the nearest tenth of a foot?

Ⓐ 0.0 ft Ⓑ 43.0 ft

Ⓒ 52.5 ft Ⓓ 61.4 ft

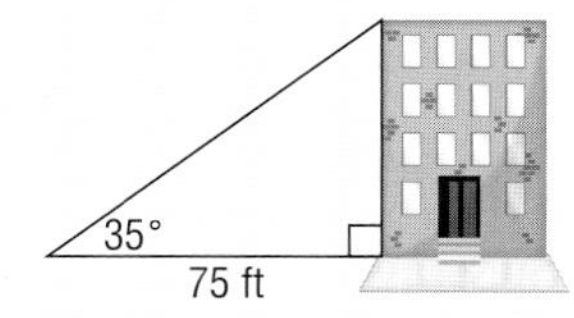

6. Samone draws $\triangle QRS$ on grid paper to use for a design in her art class. She needs to rotate the triangle 180° counterclockwise. What will be the y-coordinate of the image of S?

Ⓐ −6 Ⓑ −2 Ⓒ −1 Ⓓ 2

Measurement

7. Lakeisha is teaching a summer art class for children. For one project, she estimates that she will need $\frac{2}{3}$ yard of string for each 3 students. How many yards will she need for 16 students?

Ⓐ 3 yd Ⓑ $3\frac{5}{9}$ yd

Ⓒ $10\frac{2}{3}$ yd Ⓓ 16 yd

8. Kari works at night so she needs to make her room as dark as possible during the day to sleep. How much black paper will she need to cover the window in her room, which is shaped as shown. Use 3.14 for π. Round to the nearest tenth of a square foot.

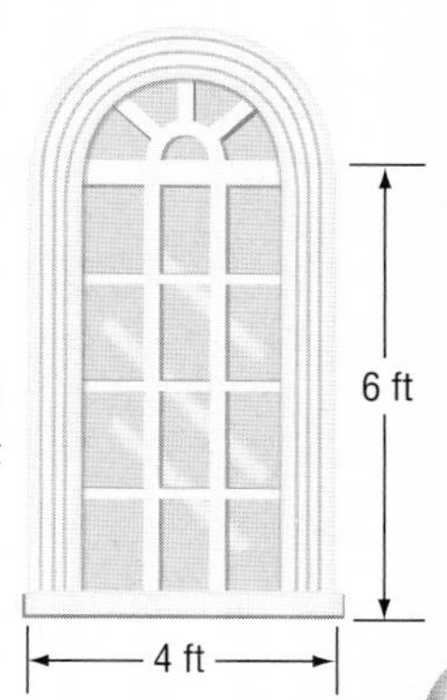

Ⓐ 24.0 ft² Ⓑ 24.6 ft²

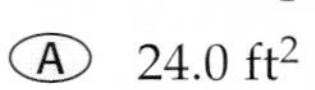

Ⓒ 30.3 ft² Ⓓ 36.6 ft²

Data Analysis and Probability

9. A card is drawn from a standard deck of 52 cards. If one card is drawn, what is the probability that it is a heart or a 2?

Ⓐ $\frac{1}{52}$ Ⓑ $\frac{1}{13}$ Ⓒ $\frac{1}{4}$ Ⓓ $\frac{4}{13}$

10. The weight of candy in boxes is normally distributed with a mean of 12 ounces and a standard deviation of 0.5 ounce. About what percent of the time will you get a box that weighs over 12.5 ounces?

Ⓐ 13.5% Ⓑ 16% Ⓒ 50% Ⓓ 68%

Test-Taking Tip

Question 8

Many standardized tests include a reference sheet with common formulas that you may use during the test. If it is available before the test, familiarize yourself with the reference sheet for quick reference during the test.

Gridded-Response Questions

Gridded-response questions are another type of question on standardized tests. These questions are sometimes called *student-produced response* or *grid-in,* because you must create the answer yourself, not just choose from four or five possible answers.

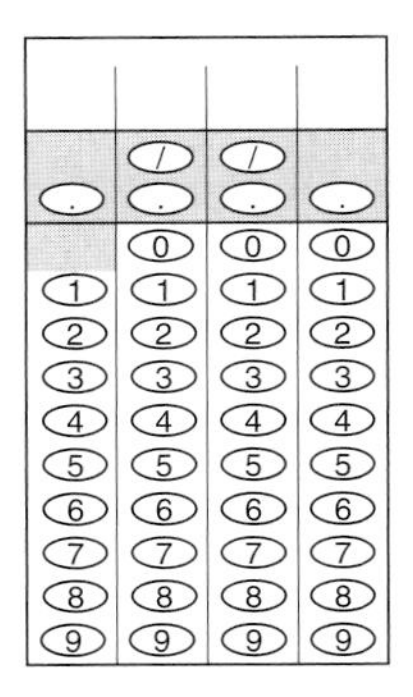

For gridded response, you must mark your answer on a grid printed on an answer sheet. The grid contains a row of four or five boxes at the top, two rows of ovals or circles with decimal and fraction symbols, and four or five columns of ovals, numbered 0–9. At the right is an example of a grid from an answer sheet.

Example 1 **Find the x-coordinate for the solution of the given system of equations.**

$$4x - y = 14$$
$$-3x + y = -11$$

What value do you need to find?

You need to find only the x-coordinate of the point where the graphs of the two equations intersect. You could graph the system, but that takes time. The easiest method is probably the substitution method since the second equation can be solved easily for y.

$-3x + y = -11$	Second equation
$y = -11 + 3x$	Solve the second equation for y.

$4x - y = 14$	First equation
$4x - (-11 + 3x) = 14$	Substitute for y.
$4x + 11 - 3x = 14$	Distributive Property
$x = 3$	Simplify.

The answer to be filled in on the grid is 3.

How do you fill in the grid for the answer?

- Write your answer in the answer boxes.
- Write only one digit or symbol in each answer box.
- Do not write any digits or symbols outside the answer boxes.
- You may write your answer with the first digit in the left answer box, or with the last digit in the right answer box. You may leave blank any boxes you do not need on the right or the left side of your answer.
- Fill in only one bubble for every answer box that you have written in. Be sure not to fill in a bubble under a blank answer box.

Many gridded response questions result in an answer that is a fraction or a decimal. These values can also be filled in on the grid.

Example 2 **Zuri is solving a problem about the area of a room. The equation she needs to solve is $4x^2 + 11x - 3 = 0$. Since the answer will be a length, she only needs to find the positive root. What is the solution?**

Since you can see the equation is not easily factorable, use the Quadratic Formula.

$$x = \frac{-b \pm \sqrt{b^2 - 4ac}}{2a}$$

$$= \frac{-11 \pm \sqrt{11^2 - 4(4)(-3)}}{2(4)}$$

$$= \frac{-11 + 13}{8}$$ There are two roots, but you only need the positive one.

$$= \frac{2}{8} \text{ or } \frac{1}{4}$$

How do you grid the answer?

You can either grid the fraction $\frac{1}{4}$, or rewrite it as 0.25 and grid the decimal. Be sure to write the decimal point or fraction bar in the answer box. The following are acceptable answer responses that represent $\frac{1}{4}$ and 0.25.

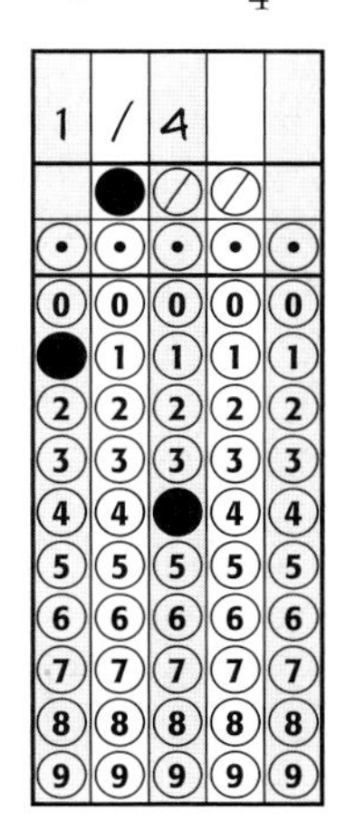

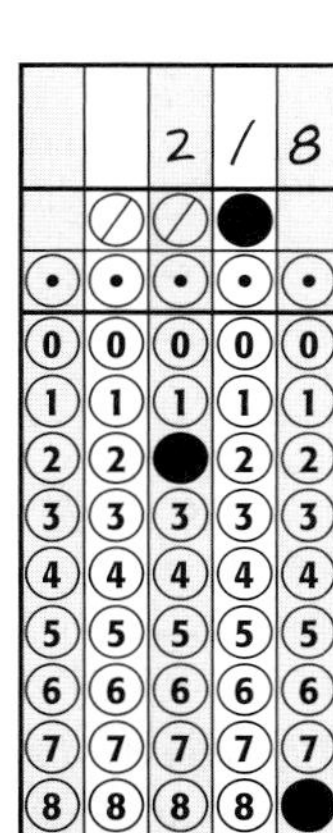

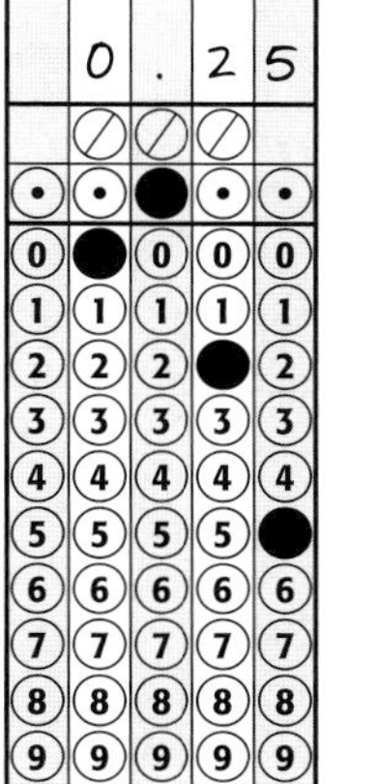

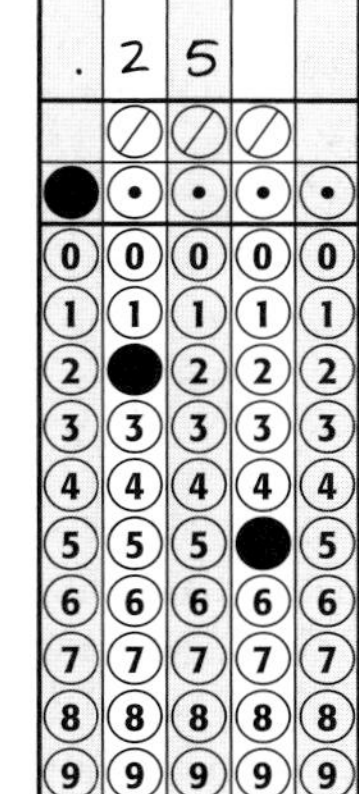

Do not leave a blank answer box in the middle of an answer.

Some problems may result in an answer that is a mixed number. Before filling in the grid, change the mixed number to an equivalent improper fraction or decimal. For example, if the answer is $1\frac{1}{2}$, do not enter 1 1/2 as this will be interpreted as $\frac{11}{2}$. Instead, either enter 3/2 or 1.5.

Example 3 **José is using this figure for a computer graphics design. He wants to dilate the figure by a scale factor of $\frac{7}{4}$. What will be the y-coordinate of the image of D?**

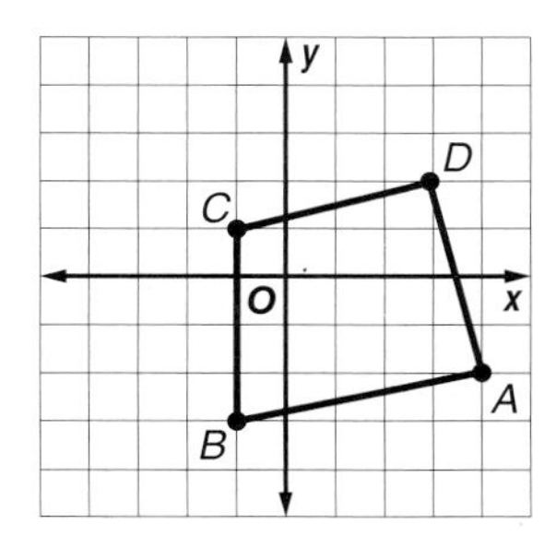

To find the y-coordinate of the image of D, multiply the x-coordinate by the scale factor of $\frac{7}{4}$.

$$2 \cdot \frac{7}{4} = \frac{7}{2}$$

Grid in 7/2 or 3.5. Do not grid 3 1/2.

Gridded-Response Practice

Solve each problem. Then copy and complete a grid.

Number and Operations

1. Rewrite 16^3 as a power of 2. What is the value of the exponent for 2?

2. Wolf 359 is the fourth closest star to Earth. It is 45,531,250 million miles from Earth. A light-year, the distance light travels in a year, is 5.88×10^{12} miles. What is the distance from Earth to Wolf 359 in light-years? Round to the nearest tenth.

3. A store received a shipment of coats. The coats were marked up 50% to sell to the customers. At the end of the season, the coats were discounted 60%. Find the ratio of the discounted price of a coat to the original cost of the coat to the store.

4. Find the value of the determinant $\begin{vmatrix} -1 & 4 \\ -3 & 0 \end{vmatrix}$.

5. Kendra is displaying eight sweaters in a store window. There are four identical red sweaters, three identical brown sweaters, and one white sweater. How many different arrangements of the eight sweaters are possible?

6. An electronics store reduced the price of a DVD player by 10% because it was used as a display model. If the reduced price was \$107.10, what was the cost in dollars before it was reduced? Round to the nearest cent if necessary.

Algebra

7. The box shown can be purchased to ship merchandise at the Pack 'n Ship Store. The volume of the box is 945 cubic inches. What is the measure of the greatest dimension of the box in inches?

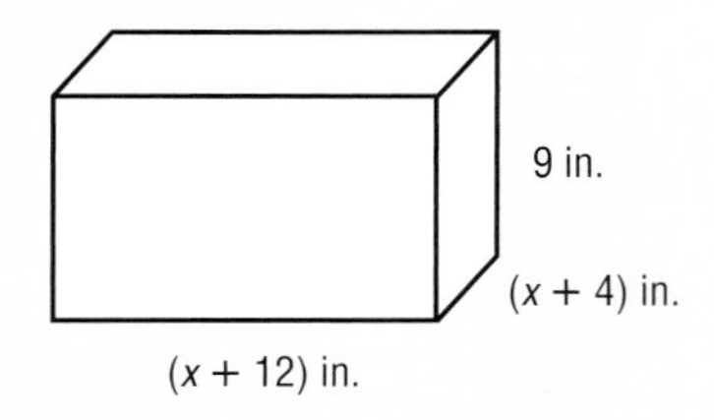

8. If $f(x) = 2x^2 - 3x + 10$, find $f(-1)$.

Test-Taking Tip

Question 22
Fractions do not have to be written in lowest terms. Any equivalent fraction that fits the grid is correct.

9. Find the positive root of $\frac{10}{a^2 - 16} - \frac{6}{a + 4} = \frac{4}{9}$.

10. The graph shows the retail price per gallon of unleaded gasoline in the U.S. from 1980 to 2000. What is the slope if a line is drawn through the points for 15 and 20 years since 1980?

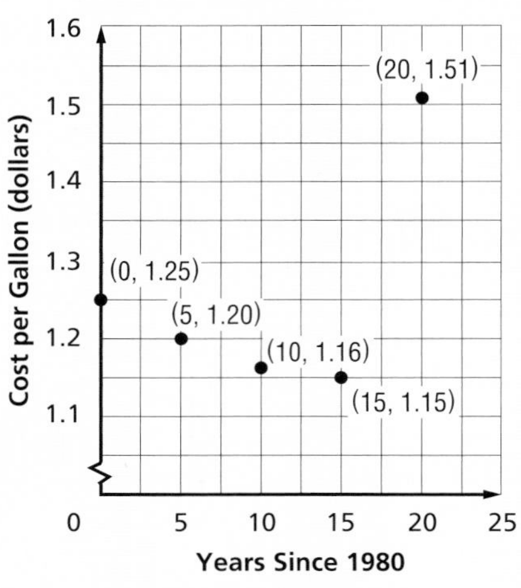

Source: U.S. Dept. of Energy

11. Solve $\sqrt{x + 11} - 9 = \sqrt{x} - 8$.

Geometry

12. Polygon $DABC$ is rotated 90° counterclockwise and then reflected over the line $y = x$. What is the x-coordinate of the final image of A?

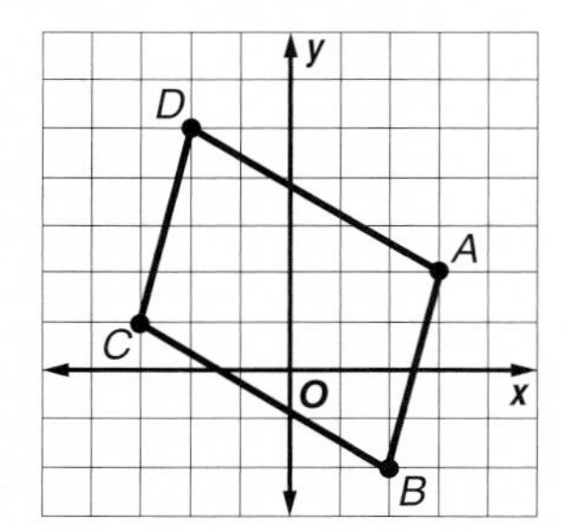

13. A garden is shaped as shown below. What is the measure of $\angle A$ to the nearest degree?

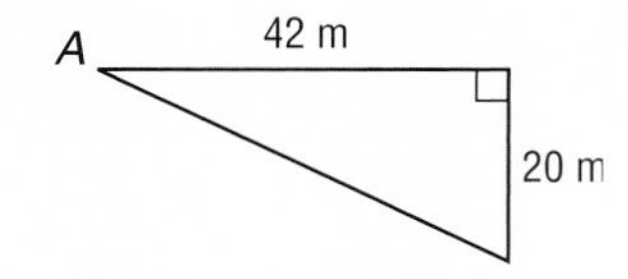

14. A circle of radius r is circumscribed about a square. What is the ratio of the area of the circle to the area of the square? Express the ratio as a decimal rounded to the nearest hundredth.

15. Find the degree measure of $\angle 1$.

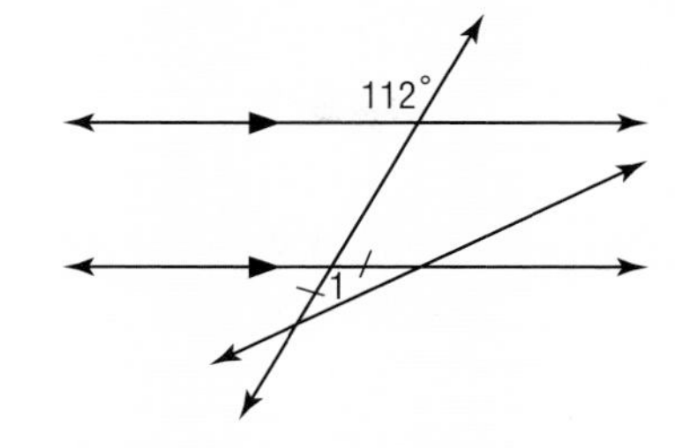

16. Each angle of a regular polygon measures 150°. How many sides does the polygon have?

Measurement

17. The Pasacal (P) is a measure of pressure that is equivalent to 1 Newton per square meter. The typical pressure in an automobile tire is 2×10^5 P while typical blood pressure is 1.6×10^4 P. How many times greater is the pressure in a tire than typical blood pressure?

18. A circular ride at an amusement park rotated $\frac{7\pi}{4}$ radians while loading riders. What is the degree measure of the rotation?

19. Find the value of x in the triangle. Round to the nearest tenth of a foot.

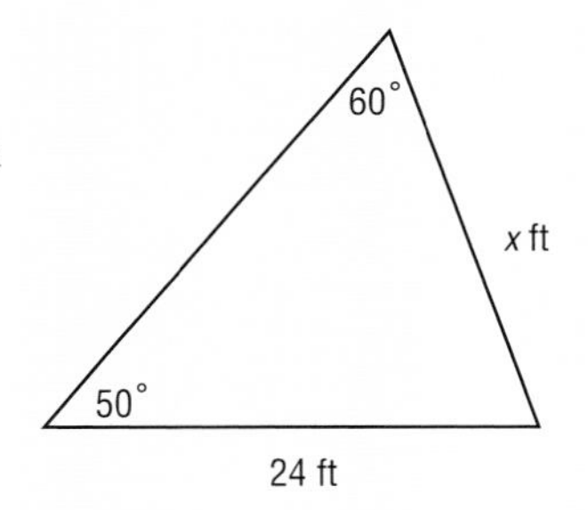

20. Four equal-sized cylindrical juice cans are packed tightly in the box shown. What is the volume of space in the box that is not occupied by the cans in cubic inches? Use 3.14 for π and round to the nearest cubic inch.

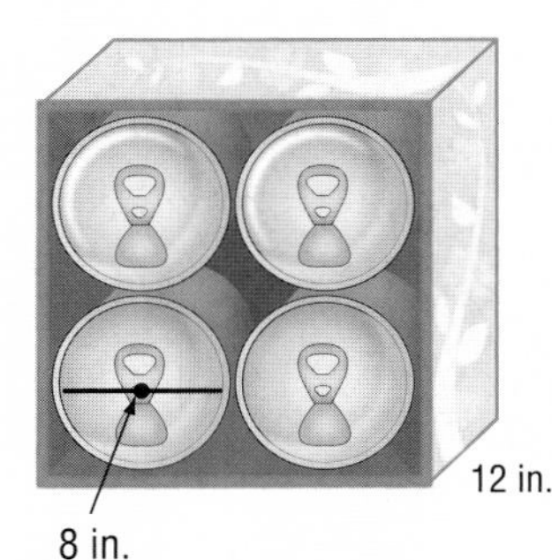

21. Caroline is making a quilt. The diagram shows a piece of cloth she will cut for a portion of the pattern. Find the area of the entire hexagonal piece to the nearest tenth of a square inch.

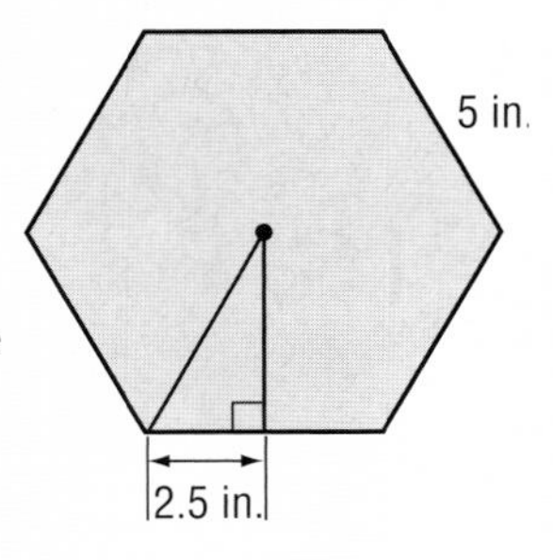

Data Analysis and Probability

22. Of ten girls on a team, three have blue eyes. If two girls are chosen at random, what is the probability that neither has blue eyes?

23. In order to win a game, Miguel needs to advance his game piece 4 spaces. What is the probability that the sum of the numbers on the two dice he rolls will be 4?

24. The table shows the number of televisions owned per 1000 people in each country. What is the absolute value of the difference between the mean and the median of the data?

Country	Televisions
United States	844
Latvia	741
Japan	719
Canada	715
Australia	706
United Kingdom	652
Norway	648
Finland	643
France	623

Source: International Telecommunication Union

25. The table shows the amount of breakfast cereal eaten per person each year by the ten countries that eat the most. Find the standard deviation of the data set. Round to the nearest tenth of a pound.

Country	Cereal (lb)
Sweden	23
Canada	17
Australia	16
United Kingdom	15
Nauru	14
New Zealand	14
Ireland	12
United States	11
Finland	10
Denmark	7

Source: Euromonitor

26. Two number cubes are rolled. If the two numbers appearing on the faces of the number cubes are different, find the probability that the sum is 6. Round to the nearest hundredth.

Preparing for Standardized Tests

Short-Response Questions

Short-response questions require you to provide a solution to the problem, as well as any method, explanation, and/or justification you used to arrive at the solution. These are sometimes called *constructed-response, open-response, open-ended, free-response,* or *student-produced questions.* The following is a sample rubric, or scoring guide, for scoring short-response questions.

Criteria	Score
Full credit: The answer is correct and a full explanation is provided that shows each step in arriving at the final answer.	2
Partial credit: There are two different ways to receive partial credit. • The answer is correct, but the explanation provided is incomplete or incorrect. • The answer is incorrect, but the explanation and method of solving the problem is correct.	1
No credit: Either an answer is not provided or the answer does not make sense.	0

On some standardized tests, no credit is given for a correct answer if your work is not shown.

Example

Mr. Youngblood has a fish pond in his backyard. It is circular with a diameter of 10 feet. He wants to build a walkway of equal width around the pond. He wants the total area of the pond and walkway to be about 201 square feet. To the nearest foot, what should be the width of the walkway?

Full Credit Solution

Strategy

Diagrams

Draw a diagram of the pond and the walkway. Label important information.

First draw a diagram to represent the situation.

10 ft

x ft

5 ft

Since the diameter of the pond is 10 feet, the radius is 5 ft. Let the width of the walkway be x feet.

$A = \pi r^2$

$201 = \pi(x + 5)^2$ $\quad r = x + 5, A = 201$

$201 = \pi(x^2 + 10x + 25)$ $\quad$ Multiply.

$\frac{201}{\pi} = \frac{\pi(x^2 + 10x + 25)}{\pi}$ $\quad$ Divide by π, using 3.14 for π.

$64 = x^2 + 10x + 25$ or $x^2 + 10x - 39 = 0$

Use the Quadratic Formula.

$x = \frac{-b \pm \sqrt{b^2 - 4ac}}{2a}$

$= \frac{-10 \pm \sqrt{(-10)^2 - 4(1)(-39)}}{2(1)}$ $\quad a = 1, b = 10, c = -39$

$= \frac{-10 + \sqrt{256}}{2}$ or 3

The width of the walkway should be 3 feet.

The steps, calculations, and reasoning are clearly stated.

Before taking a standardized test, memorize common formulas, like the Quadratic Formula, to save time.

Since length must be positive, eliminate the negative solution.

Partial Credit Solution

In this sample solution, the equation that can be used to solve the problem is correct. However, there is no justification for any of the calculations.

There is no explanation of how the quadratic equation was found.

$x^2 + 10x - 39 = 0$

$x = \frac{-10 + \sqrt{256}}{2}$

$= -13 \text{ or } 3$

The walkway should be 3 feet wide.

Partial Credit Solution

In this sample solution, the answer is incorrect because the wrong root was chosen.

Since the diameter of the pond is 10 feet, the radius is 5 ft. Let the width of the walkway be x feet. Use the formula for the area of a circle.

$A = \pi r^2$

$201 = \pi(x + 5)^2$

$201 = \pi(x^2 + 10x + 25)$

$\frac{201}{\pi} = \frac{\pi(x^2 + 10x + 25)}{\pi}$

$64 = x^2 \text{ a } 10x \text{ a } 25 \text{ or } x^2 \text{ a } 10x - 39 = 0$

Use the Quadratic Formula.

$x = \frac{-b \pm \sqrt{b^2 - 4ac}}{2a}$

$= -13 \text{ or } 3$

The walkway should be 13 feet wide.

The negative root was chosen as the solution.

No Credit Solution

Use the formula for the area of a circle.

$A = \pi r^2 x$

$201 = \pi(5)^2 x$

$201 = 3.14(25)x$

$201 = 78.5x$

$x = 2.56$

Build the walkway 3 feet wide.

The width of the walkway x is used incorrectly in the area formula for a circle. However, when the student rounds the value for the width of the walkway, the answer is correct. No credit is given for an answer achieved using faulty reasoning.

Preparing for Standardized Tests

Short-Response Practice

Solve each problem. Show all your work.

Number and Operations

1. An earthquake that measures a value of 1 on the Richter scale releases the same amount of energy as 170 grams of TNT, while one that measures 4 on the scale releases the energy of 5 metric tons of TNT. One metric ton is 1000 kilograms and 1000 grams is 1 kilograms. How many times more energy is released by an earthquake measuring 4 than one measuring 1?

2. In 2000, Cook County, Illinois was the second largest county in the U.S with a population of about 5,377,000. This was about 43.3% of the population of Illinois. What was the approximate population of Illinois in 2000?

3. Show why $\begin{bmatrix} 1 & 0 \\ 0 & 1 \end{bmatrix}$ is the identity matrix for multiplication for 2×2 matrices.

4. The total volume of the oceans on Earth is 3.24×10^8 cubic miles. The total surface area of the water of the oceans is 139.8 million square miles. What is the average depth of the oceans?

5. At the Blaine County Fair, there are 12 finalists in the technology project competition. How many ways can 1st, 2nd, 3rd, and 4th place be awarded?

Algebra

6. Factor $3x^2a^2 - 3x^2b^2$. Explain each step.

7. Solve and graph $7 - 2a > \frac{15 - 2a}{6}$.

8. Solve the system of equations.
$x^2 + 9y^2 = 25 \quad y - x = -5$

9. The table shows what Miranda Richards charges for landscaping services for various numbers of hours. Write an equation to find the charge for any amount of time, where y is the total charge in dollars and x is the amount of time in hours. Explain the meaning of the slope and y-intercept of the graph of the equation.

Hours	Charge (dollars)	Hours	Charge (dollars)
0	17.50	3	64.00
1	33.00	4	79.50
2	48.50	5	95.00

10. Write an equation that fits the data in the table.

x	y
−3	12
−1	4
0	3
2	7
4	19

Geometry

11. A sledding hill at the local park has an angle of elevation of 15°. Its vertical drop is 400 feet. What is the length of the sledding path?

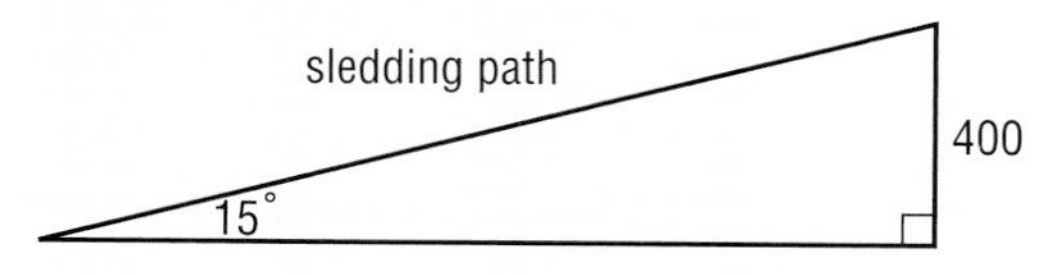

12. Polygon $BDFH$ is transformed using the matrix $\begin{bmatrix} 0 & 1 \\ 1 & 0 \end{bmatrix}$. Graph $B'D'F'H'$ and identify the type of transformation.

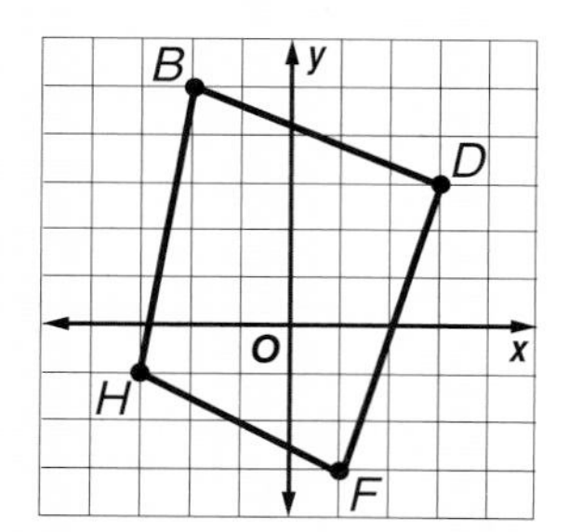

13. The map shows the trails that connect three hiking destinations. If Amparo hikes from Deer Ridge to Egg Mountain to Lookout Point and back to Deer Ridge, what is the distance she will have traveled?

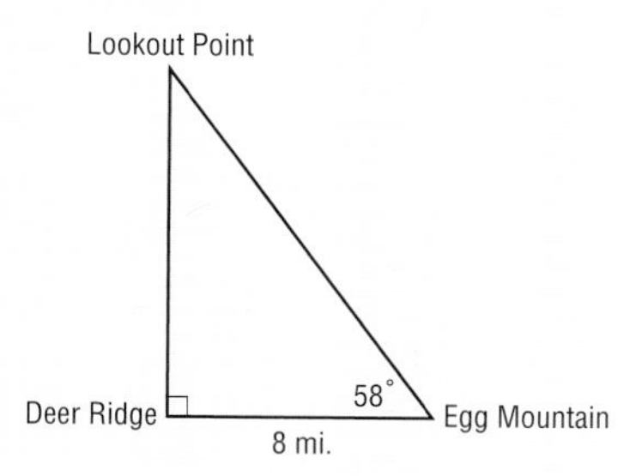

14. Mr Washington is making a cement table for his backyard. The tabletop will be circular with a diameter of 6 feet and a depth of 6 inches. How much cement will Mr. Washington need to make the top of the table? Use 3.14 for π and round to the nearest cubic foot.

15. A triangular garden is plotted on grid paper, where each unit is 1 meter. Its sides are segments that are parts of the lines with equations $y = -\frac{5}{4}x + 2$, $2y - 5x = 4$, and $y = -3$. Graph the triangle and find its area.

16. Dylan is flying a kite. He wants to know how high above the ground it is. He knows that he has let out 75 feet of string and that it is flying directly over a nearby fence post. If he is 50 feet from the fence post, how high is the kite? Round to the nearest tenth of a foot.

Measurement

17. The temperature of the Sun can reach 27,000,000°F. The relationship between Fahrenheit F and Celsius C temperatures is given by the equation $F = 1.8C + 32$. Find the temperature of the Sun in degrees Celsius.

18. In 2003, Monaco was the most densely populated country in the world. There were about 32,130 people occupying the country at the rate of 16,477 people per square kilometer. What is the area of Monaco?

19. A box containing laundry soap is a cylinder with a diameter of 10.5 inches and a height of 16 inches. What is the surface area of the box?

20. Light travels at 186,291 miles per second or 299,792 kilometers per second. What is the relationship between miles and kilometers?

21. Home Place Hardware sells storage buildings for your backyard. The front of the building is a trapezoid as shown. The store manager wants to advertise the total volume of the building. Find the volume in cubic feet.

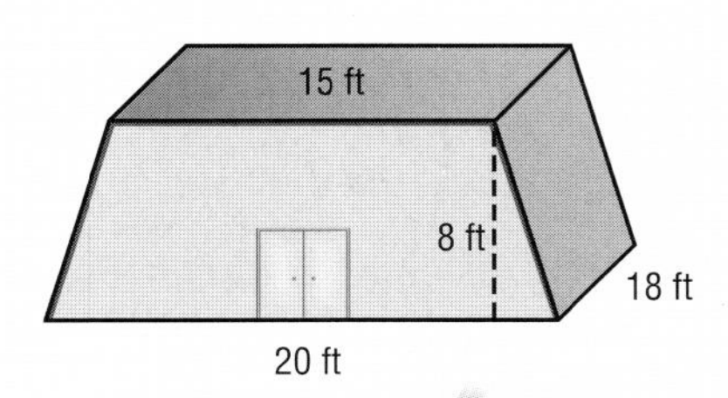

Test-Taking Tip
A B C D

Questions 15, 16, and 22

Be sure to read the instructions of each problem carefully. Some questions ask for more than one solution, specify how to round answers, or require an explanation.

Data Analysis and Probability

22. The table shows the 2000 populations of the six largest cities in Tennessee. Which measure, mean or median, do you think best represents the data? Explain your answer.

City	Population
Chattanooga	155,404
Clarksville	105,898
Jackson	60,635
Knoxville	173,661
Memphis	648,882
Nashville-Davidson	545,915

Source: International Telecommunication Union

23. The scatter plot shows the number of hours worked per week for U.S. production workers from 1970 through 2000. Let y be the hours worked per week and x be the years since 1970. Write an equation that you think best models the data.

Average Hours Worked per Week for Production Workers

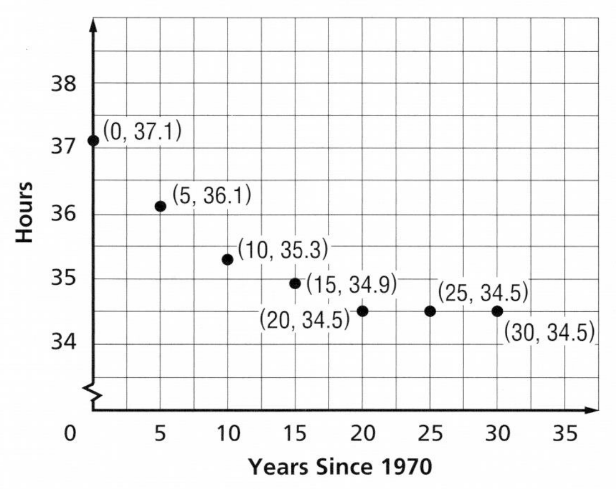

Source: Bureau of Labor Statistics

24. A day camp has 240 participants. Children can sign up for various activities. Suppose 135 children take swimming, 160 take soccer, and 75 take both swimming and soccer. What is the probability that a child selected at random takes swimming or soccer?

25. In how many different ways can seven members of a student government committee sit around a circular table?

26. Illinois residents can choose to buy an environmental license plate to support Illinois parks. Each environmental license plate displays 3 or 4 letters followed by a number 1 thru 99. How many different environmental license plates can be issued?

Extended-Response Questions

Extended-response questions are often called *open-ended* or *constructed-response questions.* Most extended-response questions have multiple parts. You must answer all parts to receive full credit.

Extended-response questions are similar to short-response questions in that you must show all of your work in solving the problem and a rubric is used to determine whether you receive full, partial, or no credit. The following is a sample rubric for scoring extended-response questions.

Criteria	Credit
Full credit: The answer is correct and a full explanation is provided that shows each step in arriving at the final answer.	2
Partial credit: There are two different ways to receive partial credit. • The answer is correct, but the explanation provided is incomplete or incorrect. • The answer is incorrect, but the explanation and method of solving the problem is correct.	1
No credit: Either an answer is not provided or the answer does not make sense.	0

On some standardized tests, no credit is given for a correct answer if your work is not shown.

Make sure that when the problem says to *Show your work,* you show every aspect of your solution including figures, sketches of graphing calculator screens, or reasoning behind computations.

Example

Libby throws a ball into the air with a velocity of 64 feet per second. She releases the ball 5 feet above the ground. The height of the ball above the ground t seconds after release is modeled by an equation of the form $h(t) = -16t^2 + v_o t + h_o$ where v_o is the initial velocity in feet per second and h_o is the height at which the ball is released.

a. Write an equation for the flight of the ball. Sketch the graph of the equation.

b. Find the maximum height that the ball reaches and the time that this height is reached.

c. Change only the speed of the release of the ball such that the ball will reach a maximum height greater than 100 feet. Write an equation for the flight of the ball.

Full Credit Solution

Part a A complete graph includes appropriate scales and labels for the axes, and points that are correctly graphed.

- A complete graph also shows the basic characteristics of the graph. The student should realize that the graph of this equation is a parabola opening downward with a maximum point reached at the vertex.
- The student should choose appropriate points to show the important characteristics of the graph.
- Students should realize that t and x and $h(t)$ and y are interchangeable on a graph on the coordinate plane.

To write the equation for the ball, I substituted $v_o = 64$ and $h_o = 5$ into the equation $h(t) = -16t^2 + v_o t + h_o$, so the equation is $h(t) = -16t^2 + 64t + 5$. To graph the equation, I found the equation of the axis of symmetry and the vertex.

$$x = -\frac{b}{2a}$$
$$= -\frac{64}{2(-16)} \text{ or } 2$$

The equation of the axis of symmetry is $x = 2$, so the x-coordinate of the vertex is 2. You let $t = x$ and then $h(t) = -16t^2 + 64t + 5 = -16(2)^2 + 64(2) + 5$ or 69. The vertex is (2, 69). I found some other points and sketched the graph. I graphed points $(t, h(t))$ as (x, y).

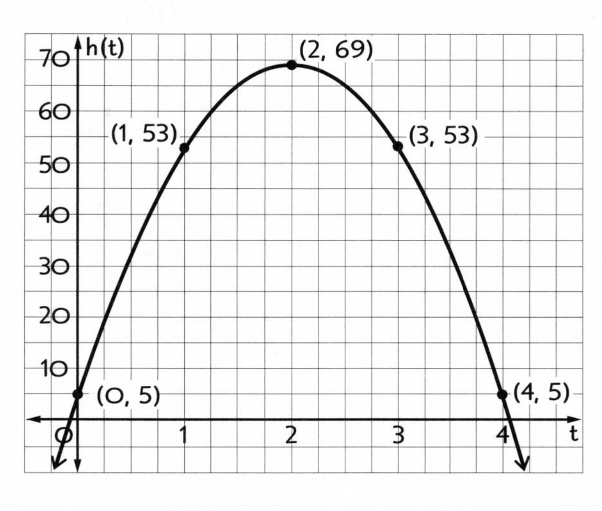

Part b

The maximum height of the ball is reached at the vertex of the parabola. So, the maximum height is 69 feet and the time it takes to reach the maximum height is 2 seconds.

Part c

In part c, any equation whose graph has a vertex with *y*-coordinate greater than 100 would be a correct answer earning full credit.

Since I have a graphing calculator, I changed the value of v_o until I found a graph in which the y or $h(t)$ coordinate was greater than 100. The equation I used was $h(t) = -16t^2 + 80t + 5$.

Partial Credit Solution

Part a This sample answer does not earn full credit because it includes no explanation of how the equation was written or the vertex was found.

$h(t) = -16t^2 + 64t + 5$; (2, 69)

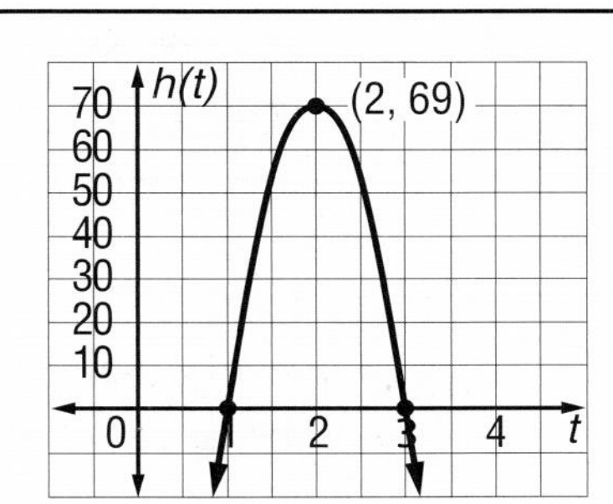

Part b Full credit is given because the vertex is correct and is interpreted correctly.

The vertex shows the maximum height of the ball. The time it takes to reach the maximum height of 69 feet is 2 seconds.

Part c Partial credit is given for part c since no explanation is given for using this equation. The student did not mention that the vertex would have a y-coordinate greater than 100.

I will write the equation $h(t) = -16t^2 + 100t + 5$.

This sample answer might have received a score of 2 or 1, depending on the judgment of the scorer. Had the student sketched a more accurate graph and given more complete explanations for Parts a and c, the score would probably have been a 3.

No Credit Solution

Part a No credit is given because the equation is incorrect with no explanation and the sketch of the graph has no labels, making it impossible to determine whether the student understands the relationship between the equation for a parabola and the graph.

$h(t) = -16t^2 + 5t + 64$

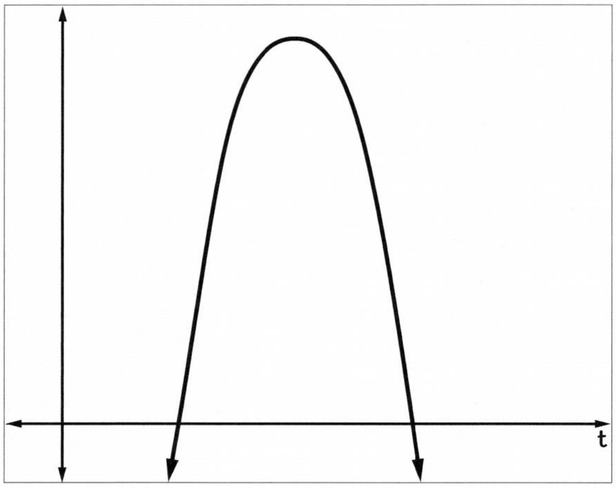

t

Part b

It reaches about 10 feet.

Part c

A good equation for the ball is $h(t) = -16t^2 + 5t + 100$.

In this sample answer, the student does not understand how to substitute the given information into the equation, graph a parabola, or interpret the vertex of a parabola.

Extended Response Practice

Solve each problem. Show all your work.

Number and Operations

1. Mrs. Ebbrect is assigning identification (ID) numbers to freshman students. She plans to use only the digits 2, 3, 5, 6, 7, and 9. The ID numbers will consist of three digits with no repetitions.

a. How many 3-digit ID numbers can be formed?

b. How many more ID numbers can Mrs. Ebbrect make if she allows repetitions?

c. What type of system could Mrs. Ebbrect use to choose the numbers if there are at least 400 students who need ID numbers?

2. Use these four matrices.

$A = \begin{bmatrix} -1 & 0 \\ 3 & -2 \end{bmatrix}$ $B = \begin{bmatrix} 4 & -6 & 5 \\ 0 & 1 & -3 \end{bmatrix}$

$C = \begin{bmatrix} -7 & 3 \\ -6 & 2 \end{bmatrix}$ $D = \begin{bmatrix} -3 & 1 \end{bmatrix}$

a. Find $A + C$.

b. Compare the dimensions of AB and DB.

c. Compare the matrices BC and CB.

Algebra

3. Roger is using the graph showing the gross cash income for all farms in the U.S. from 1930 through 2000 to make some predictions for the future.

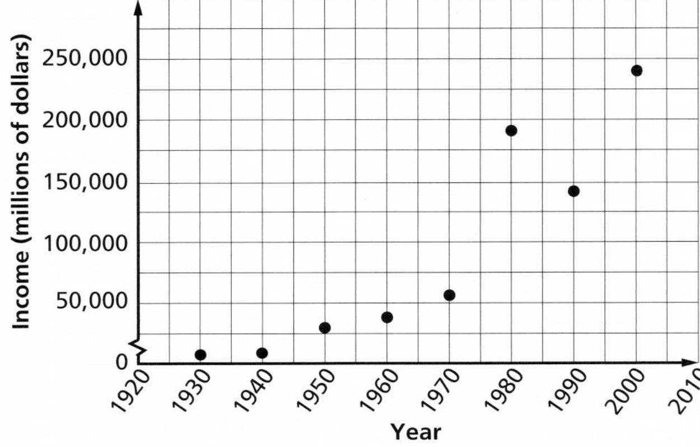

Source: U.S. Department of Agriculture

a. Write an equation in slope-intercept form for the line passing through the point for 1930 and the point for 1970.

b. Write an equation in slope-intercept form for the line passing through the point for 1980 and the point for 2000. Compare the slope of this line to the slope of the line in part a.

c. Which equation, if any, do you think Roger should use to model the data? Explain.

d. Suggest an equation that is not linear for Roger to use.

4. Brad is coaching the bantam age division (8 years old and younger) swim team. On the first day of practice, he has the team swim 4 laps of the 25-meter pool. For each of the next practices, he increases the laps by 3. In other words, the children swim 4 laps the first day, 7 laps the second day, 10 laps the third day, and so on.

a. Write a formula for the nth term of the sequence of the number of laps each day. Explain how you found the formula.

b. How many laps will the children swim on the 10th day?

c. Brad's goal is to have the children swim at least one mile during practice on the 20th day. If one mile is approximately 1.6 kilometers, will Brad reach his goal?

Geometry

5. Alejandra is planning to use a star shape in a galaxy-themed mural on her wall. The pentagon in the center is regular, and the triangles forming the points are isosceles.

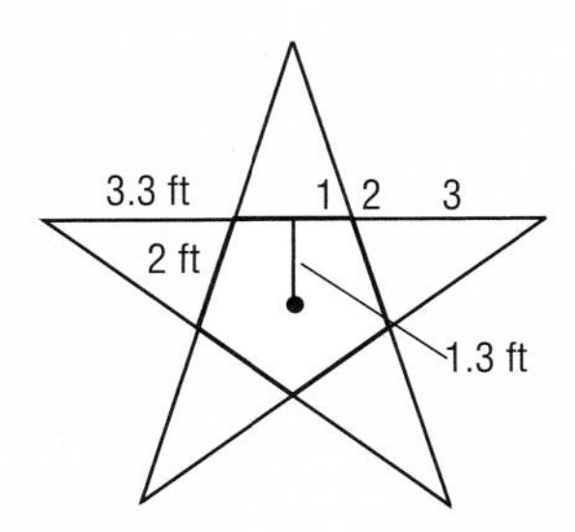

a. Find the measures of $\angle 1$, $\angle 2$, and $\angle 3$. Explain your method.

b. The approximate dimensions of the design are given. The segment of length 1.3 feet is the apothem of the pentagon. Find the approximate area of the design.

c. If Alejandra circumscribes a circle about the star, what is the area of the circle?

6. Kareem is using polygon $ABCDE$, shown on a coordinate plane, as a basis for a computer graphics design. He plans to perform various transformations on the polygon to produce a variety of interesting designs.

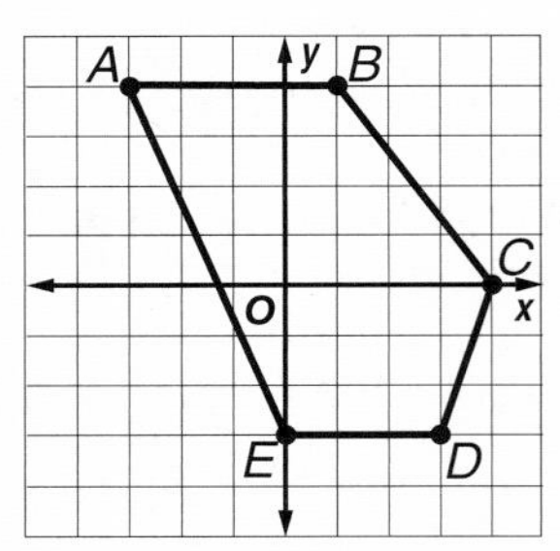

a. First, Kareem creates polygon $A'B'C'D'E'$ by rotating $ABCDE$ counterclockwise about the origin 270°. Graph polygon $A'B'C'D'E'$ and describe the relationship between the coordinates of $ABCDE$ and $A'B'C'D'E'$.

b. Next, Kareem reflects polygon $A'B'C'D'E'$ in the line $y = x$ to produce polygon $A''B''C''D''E''$. Graph $A''B''C''D''E''$ and describe the relationship between the coordinates of $A'B'C'D'E'$ and $A''B''C''D''E''$.

c. Describe how Kareem could transform polygon $ABCDE$ to polygon $A''B''C''D''E''$ with only one transformation.

Measurement

7. The speed of a satellite orbiting Earth can be found using the equation $v = \sqrt{\dfrac{\frac{Gm_E}{r}}{m_E}}$. G is the gravitational constant for Earth, m_E is the mass of Earth, and r is the radius of the orbit which includes the radius of Earth and the height of the satellite.

a. The radius of Earth is 6.38×10^6 meters. The distance of a particular satellite above Earth is 350 kilometers. What is the value of r? (*Hint:* The center of the orbit is the center of Earth.)

b. The gravitational constant for Earth is 6.67×10^{-11} N • m^2/kg^2. The mass of Earth is 5.97×10^{24} kg. Find the speed of the satellite in part a.

c. As a satellite increases in distance from the Earth, what is the effect on the speed of the orbit? Explain your reasoning.

Test-Taking Tip

Question 6

When questions require graphing, make sure your graph is accurate to receive full credit for your correct solution.

8. A cylindrical cooler has a diameter of 9 inches and a height of 11 inches. Scott plans to use it for soda cans that have a diameter of 2.5 inches and a height of 4.75 inches.

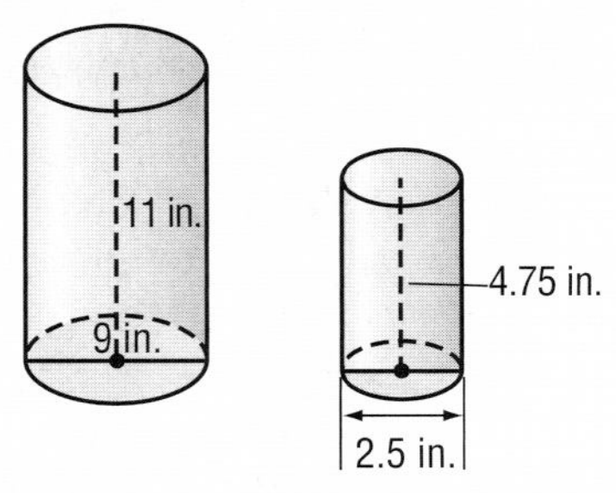

a. Scott plans to place two layers consisting of 9 cans each into the cooler. What is the volume of the space that will not be filled with cans?

b. Find the ratio of the volume of the cooler to the volume of the cans in part b.

Data Analysis and Probability

9. The table shows the total world population from 1950 through 2000.

Year	Population	Year	Population
1950	2,566,000,053	1980	4,453,831,714
1960	3,039,451,023	1990	5,278,639,789
1970	3,706,618,163	2000	6,082,966,429

a. Between which two decades was the percent increase in population the greatest?

b. Make a scatter plot of the data.

c. Find a function that models the data.

d. Predict the world population for 2030.

10. Each year, a university sponsors a conference for women. The Venn diagram shows the number of participants in three activities for the 680 women that attended. Suppose women who attended are selected at random for a survey.

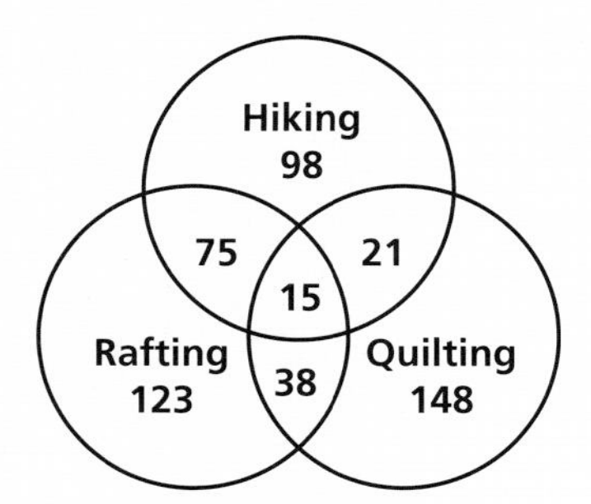

a. What is the probability that a woman selected participated in hiking or sculpting?

b. Describe a set of women such that the probability of their being selected is about 0.39.

Glossary/Glosario

Cómo usar el glosario en español:
1. Busca el término en inglés que desees encontrar.
2. El término en español, junto con la definición, se encuentran en la columna de la derecha.

English	Español
A	
absolute value (28) A number's distance from zero on the number line, represented by $\|x\|$.	**valor absoluto** Distancia entre un número y cero en una recta numérica; se denota con $\|x\|$.
absolute value function (90) A function written as $f(x) = \|x\|$, where $f(x) \geq 0$ for all values of x.	**función del valor absoluto** Una función que se escribe $f(x) = \|x\|$, donde $f(x) \geq 0$, para todos los valores de x.
absolute value inequalities (42) For all real numbers a and b, $b > 0$, the following statements are true. **1.** If $\|a\| < b$, then $-b < a < b$ **2.** If $\|a\| > b$, then $a > b$ or $a < -b$.	**desigualdades con valor absoluto** Para todo número real a y b, $b > 0$, se cumple lo siguiente. **1.** Si $\|a\| < b$, entonces $-b < a < b$ **2.** Si $\|a\| > b$, entonces $a > b$ o $a < -b$.
algebraic expression (7) An expression that contains at least one variable.	**expresión algebraica** Expresión que contiene al menos una variable.
amplitude (763) For functions in the form $y = a \sin b\theta$ or $y = a \cos b\theta$, the amplitude is $\|a\|$.	**amplitud** Para funciones de la forma $y = a$ sen $b\theta$ o $y = a \cos b\theta$, la amplitud es $\|a\|$.
angle of depression (705) The angle between a horizontal line and the line of sight from the observer to an object at a lower level.	**ángulo de depresión** Ángulo entre una recta horizontal y la línea visual de un observador a una figura en un nivel inferior.
angle of elevation (705) The angle between a horizontal line and the line of sight from the observer to an object at a higher level.	**ángulo de elevación** Ángulo entre una recta horizontal y la línea visual de un observador a una figura en un nivel superior.
arccosine (747) The inverse of $y = \cos x$, written as $x = \arccos y$.	**arcocoseno** La inversa de $y = \cos x$, que se escribe como $x = \arccos y$.
arcsine (747) The inverse of $y = \sin x$, written as $x = \arcsin y$.	**arcoseno** La inversa de $y =$ sen x, que se escribe como $x =$ arcsen y.
arctangent (747) The inverse of $y = \tan x$ written as $x = \arctan y$.	**arcotangente** La inversa de $y = \tan x$ que se escribe como $x = \arctan y$.
arithmetic mean (580) The terms between any two nonconsecutive terms of an arithmetic sequence.	**media aritmética** Cualquier término entre dos términos no consecutivos de una sucesión aritmética.
arithmetic sequence (578) A sequence in which each term after the first is found by adding a constant, the common difference d, to the previous term.	**sucesión aritmética** Sucesión en que cualquier término después del primero puede hallarse sumando una constante, la diferencia común d, al término anterior.
arithmetic series (583) The indicated sum of the terms of an arithmetic sequence.	**serie aritmética** Suma específica de los términos de una sucesión aritmética.
asymptote (442, 485) A line that a graph approaches but never crosses.	**asíntota** Recta a la que se aproxima una gráfica, sin jamás cruzarla.
augmented matrix (208) A coefficient matrix with an extra column containing the constant terms.	**matriz ampliada** Matriz coeficiente con una columna extra que contiene los términos constantes.
axis of symmetry (287) A line about which a figure is symmetric.	**eje de simetría** Recta respecto a la cual una figura es simétrica.

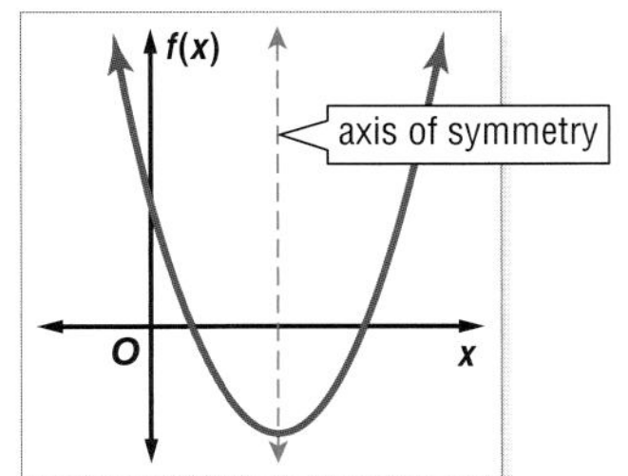

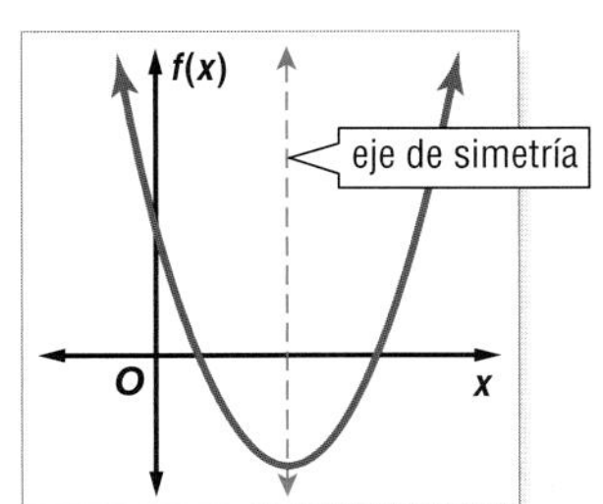

Glossary/Glosario

B

$b^{\frac{1}{n}}$ (257) For any real number b and for any positive integer n, $b^{\frac{1}{n}} = \sqrt[n]{b}$, except when $b < 0$ and n is even.

$b^{\frac{1}{n}}$ Para cualquier número real b y para cualquier entero positivo n, $b^{\frac{1}{n}} = \sqrt[n]{b}$, excepto cuando $b < 0$ y n es par.

binomial (229) A polynomial that has two unlike terms.

binomio Polinomio con dos términos diferentes.

binomial experiment (677) An experiment in which there are exactly two possible outcomes for each trial, a fixed number of independent trials, and the probabilities for each trial are the same.

experimento binomial Experimento con exactamente dos resultados posibles para cada prueba, un número fijo de pruebas independientes y en el cual cada prueba tiene igual probabilidad.

Binomial Theorem (613) If n is a nonnegative integer, then $(a + b)^n =$
$1a^nb^0 + \frac{n}{1}a^{n-1}b^1 + \frac{n(n-1)}{1 \cdot 2}a^{n-2}b^2 + \ldots + 1a^0b^n$.

Teorema del binomio Si n es un entero no negativo, entonces $(a + b)^n =$
$1a^nb^0 + \frac{n}{1}a^{n-1}b^1 + \frac{n(n-1)}{1 \cdot 2}a^{n-2}b^2 + \ldots + 1a^0b^n$.

boundary (96) A line or curve that separates the coordinate plane into two regions.

frontera Recta o curva que divide un plano de coordenadas en dos regiones.

bounded (129) A region is bounded when the graph of a system of constraints is a polygonal region.

acotada Una región está acotada cuando la gráfica de un sistema de restricciones es una región poligonal.

C

Cartesian coordinate plane (56) A plane divided into four quadrants by the intersection of the x-axis and the y-axis at the origin.

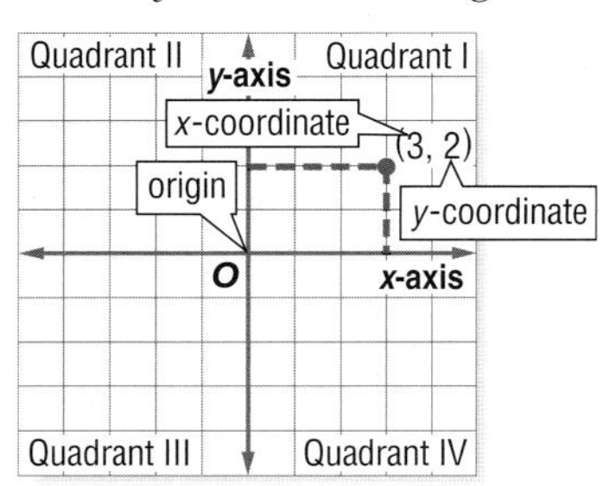

plano de coordenadas cartesiano Plano dividido en cuatro cuadrantes mediante la intersección en el origen de los ejes x y y.

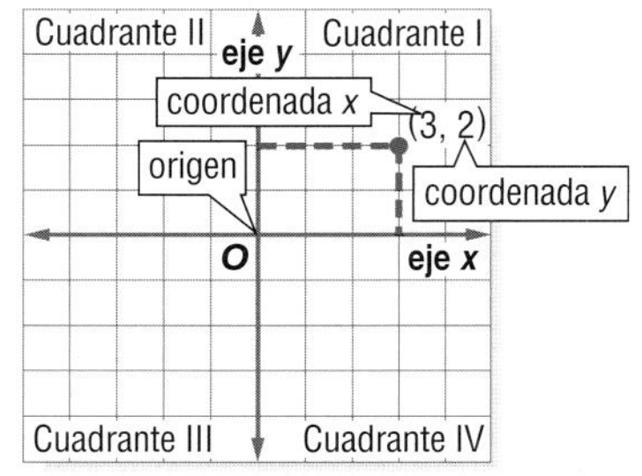

center of a circle (426) The point from which all points on a circle are equidistant.

centro de un círculo El punto desde el cual todos los puntos de un círculo están equidistantes.

center of an ellipse (434) The point at which the major axis and minor axis of an ellipse intersect.

centro de una elipse Punto de intersección de los ejes mayor y menor de una elipse.

center of a hyperbola (442) The midpoint of the segment whose endpoints are the foci.

centro de una hipérbola Punto medio del segmento cuyos extremos son los focos.

change of base formula (548) For all positive numbers a, b, and n, where $a \neq 1$ and $b \neq 1$, $\log_a n = \frac{\log_b n}{\log_b a}$.

fórmula del cambio de base Para todo número positivo a, b y n, donde $a \neq 1$ y $b \neq 1$, $\log_b n = \frac{\log_b n}{\log_b a}$.

circle (426) The set of all points in a plane that are equidistant from a given point in the plane, called the center.

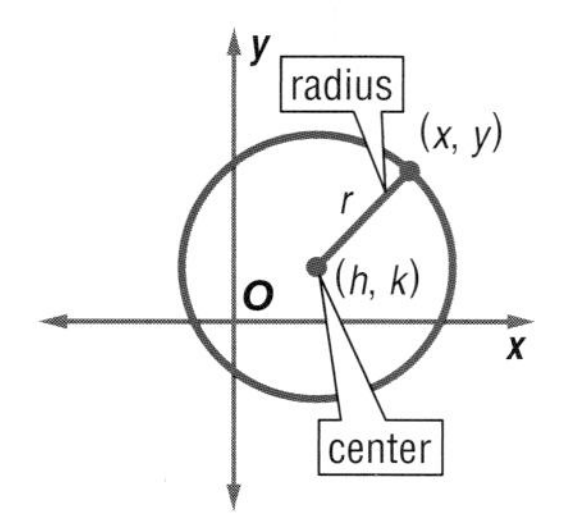

círculo Conjunto de todos los puntos en un plano que equidistan de un punto dado del plano llamado centro.

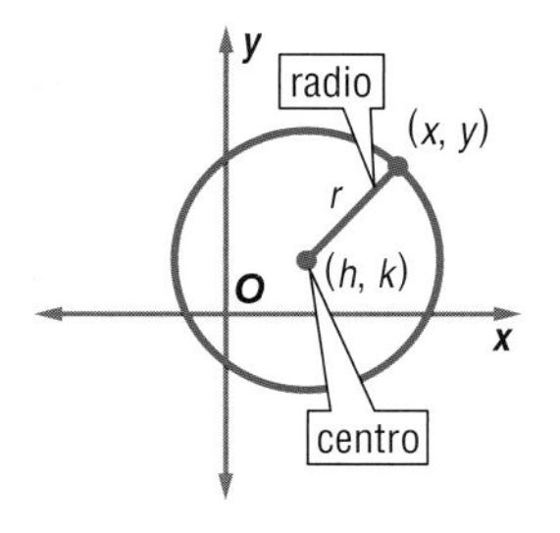

circular functions (740) Functions defined using a unit circle.

funciones circulares Funciones definidas en un círculo unitario.

coefficient (222) The numerical factor of a monomial.

coeficiente Factor numérico de un monomio.

column matrix (155) A matrix that has only one column.

matriz columna Matriz que sólo tiene una columna.

combination (640) An arrangement of objects in which order is not important.

combinación Arreglo de elementos en que el orden no es importante.

common difference (578) The difference between the successive terms of an arithmetic sequence.

diferencia común Diferencia entre términos consecutivos de una sucesión aritmética.

common logarithms (547) Logarithms that use 10 as the base.

logaritmos comunes El logaritmo de base 10.

common ratio (588) The ratio of successive terms of a geometric sequence.

razón común Razón entre términos consecutivos de una sucesión geométrica.

Commutative Property of Addition (12) For any real numbers a and b, $a + b = b + a$.

Propiedad conmutativa de la adición Para cualquier número real a y b, $a + b = b + a$.

Commutative Property of Multiplication (12) For any real numbers a and b, $a \cdot b = b \cdot a$.

Propiedad conmutativa de la multiplicación Para cualquier número real a y b, $a \cdot b = b \cdot a$.

completing the square (307) A process used to make a quadratic expression into a perfect square trinomial.

completar el cuadrado Proceso mediante el cual una expresión cuadrática se transforma en un trinomio cuadrado perfecto.

complex conjugates (273) Two complex numbers of the form a + $b\mathbf{i}$ and a − $b\mathbf{i}$.

conjugados complejos Dos números complejos de la forma a + $b\mathbf{i}$ y a − $b\mathbf{i}$.

complex fraction (475) A rational expression whose numerator and/or denominator contains a rational expression.

fracción compleja Expresión racional cuyo numerador o denominador contiene una expresión racional.

complex number (271) Any number that can be written in the form $a + b\mathbf{i}$, where a and b are real numbers and $\mathbf{i}$ is the imaginary unit.

número complejo Cualquier número que puede escribirse de la forma $a + b\mathbf{i}$, donde a y b son números reales e $\mathbf{i}$ es la unidad imaginaria.

composition of functions (384) A function is performed, and then a second function is performed on the result of the first function. The composition of f and g is denoted by $f \circ g$, and $[f \circ g](x) = f[g(x)]$.

composición de funciones Se evalúa una función y luego se evalúa una segunda función en el resultado de la primera función. La composición de f y g se define con $f \circ g$ y $[f \circ g](x) = f[g(x)]$.

compound event (658) Two or more simple events.

evento compuesto Dos o más eventos simples.

compound inequality (40) Two inequalities joined by the word *and* or *or*.

desigualdad compuesta Dos desigualdades unidas por las palabras *y* u *o*.

conic section (419) Any figure that can be obtained by slicing a double cone.

sección cónica Cualquier figura obtenida mediante el corte de un cono doble.

conjugate axis (442) The segment of length $2b$ units that is perpendicular to the transverse axis at the center.

eje conjugado El segmento de $2b$ unidades de longitud que es perpendicular al eje transversal en el centro.

conjugates (253) Binomials of the form $a\sqrt{b} + c\sqrt{d}$ and $a\sqrt{b} - c\sqrt{d}$, where a, b, c, and d are rational numbers.

conjugados Binomios de la forma $a\sqrt{b} + c\sqrt{d}$ y $a\sqrt{b} - c\sqrt{d}$, donde a, b, c y d son números racionales.

consistent (111) A system of equations that has at least one solution.

consistente Sistema de ecuaciones que posee por lo menos una solución.

constant (222) Monomials that contain no variables.

constante Monomios que carecen de variables.

constant function (90) A linear function of the form $f(x) = b$.

función constante Función lineal de la forma $f(x) = b$.

constant of variation (492) The constant k used with direct or inverse variation.

constante de variación La constante k que se usa en variación directa o inversa.

Glossary/Glosario

constant term (286) In $f(x) = ax^2 + bx + c$, c is the constant term.

constraints (129) Conditions given to variables, often expressed as linear inequalities.

continuity (485) A graph of a function that can be traced with a pencil that never leaves the paper.

continuous probability distribution (671) The outcome can be any value in an interval of real numbers, represented by curves.

cosecant (701) For any angle, with measure α, a point $P(x, y)$ on its terminal side, $r = \sqrt{x^2 + y^2}$, $\csc \alpha = \frac{r}{y}$.

cosine (701) For any angle, with measure α, a point $P(x, y)$ on its terminal side, $r = \sqrt{x^2 + y^2}$, $\cos \alpha = \frac{x}{r}$.

cotangent (701) For any angle, with measure α, a point $P(x, y)$ on its terminal side, $r = \sqrt{x^2 + y^2}$, $\cot \alpha = \frac{x}{y}$.

coterminal angles (711) Two angles in standard position that have the same terminal side.

Cramer's Rule (189) A method that uses determinants to solve a system of linear equations.

término constante En $f(x) = ax^2 + bx + c$, c es el término constante.

restricciones Condiciones a que están sujetas las variables, a menudo escritas como desigualdades lineales.

continuidad La gráfica de una función que se puede calcar sin levantar nunca el lápiz del papel.

distribución de probabilidad continua El resultado puede ser cualquier valor de un intervalo de números reales, representados por curvas.

cosecante Para cualquier ángulo de medida α, un punto $P(x, y)$ en su lado terminal, $r = \sqrt{x^2 + y^2}$, $\csc \alpha = \frac{r}{y}$.

coseno Para cualquier ángulo de medida α, un punto $P(x, y)$ en su lado terminal, $r = \sqrt{x^2 + y^2}$, $\cos \alpha = \frac{x}{r}$.

cotangente Para cualquier ángulo de medida α, un punto $P(x, y)$ en su lado terminal, $r = \sqrt{x^2 + y^2}$, $\cot \alpha = \frac{x}{y}$.

ángulos coterminales Dos ángulos en posición estándar que tienen el mismo lado terminal.

Regla de Crámer Método que usa determinantes para resolver un sistema de ecuaciones lineales.

D

degree (222) The sum of the exponents of the variables of a monomial.

degree of a polynomial in one variable (346) The greatest exponent of the variable of the polynomial.

dependent events (633) The outcome of one event does affect the outcome of another event.

dependent system (111) A consistent system of equations that has an infinite number of solutions.

dependent variable (59) The other variable in a function, usually y, whose values depend on x.

depressed polynomial (366) The quotient when a polynomial is divided by one of its binomial factors.

determinant (182) A square array of numbers or variables enclosed between two parallel lines.

dilation (176) A transformation in which a geometric figure is enlarged or reduced.

dimensional analysis (225) Performing operations with units.

dimensions of a matrix (155) The number of rows, m, and the number of columns, n, of the matrix written as $m \times n$.

grado Suma de los exponentes de las variables de un monomio.

grado de un polinomio de una variable El exponente máximo de la variable del polinomio.

eventos dependientes El resultado de un evento afecta el resultado de otro evento.

sistema dependiente Sistema de ecuaciones que posee un número infinito de soluciones.

variable dependiente La otra variable de una función, por lo general y, cuyo valor depende de x.

polinomio reducido El cociente cuando se divide un polinomio entre uno de sus factores binomiales.

determinante Arreglo cuadrado de números o variables encerrados entre dos rectas paralelas.

dilatación Transformación en que se amplía o reduce una figura geométrica.

análisis dimensional Realizar operaciones con unidades.

tamaño de una matriz El número de filas, m, y columnas, n, de una matriz, lo que se escribe $m \times n$.

directrix (419) See parabola.

direct variation (492) y varies directly as x if there is some nonzero constant k such that $y = kx$. k is called the constant of variation.

discrete probability distributions (671) Probabilities that have a finite number of possible values.

discriminant (316) In the Quadratic Formula, the expression $b^2 - 4ac$.

Distance Formula (413) The distance between two points with coordinates (x_1, y_1) and (x_2, y_2) is given by $d = \sqrt{(x_2 - x_1)^2 + (y_2 - y_1)^2}$.

domain (56) The set of all x-coordinates of the ordered pairs of a relation.

directriz Véase parábola.

variación directa y varía directamente con x si hay una constante no nula k tal que $y = kx$. k se llama la constante de variación.

distribución de probabilidad discreta Probabilidades que tienen un número finito de valores posibles.

discriminante En la fórmula cuadrática, la expresión $b^2 - 4ac$.

Fórmula de la distancia La distancia entre dos puntos (x_1, y_1) y (x_2, y_2) viene dada por $d = \sqrt{(x_2 - x_1)^2 + (y_2 - y_1)^2}$.

dominio El conjunto de todas las coordenadas x de los pares ordenados de una relación.

E

e (554) The irrational number 2.71828.... e is the base of the natural logarithms.

element (155) Each value in a matrix.

elimination method (118) Eliminate one of the variables in a system of equations by adding or subtracting the equations.

ellipse (433) The set of all points in a plane such that the sum of the distances from two given points in the plane, called foci, is constant.

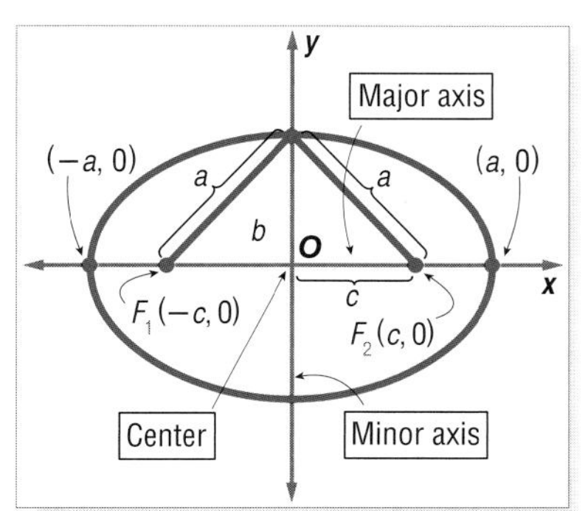

empty set (29) The solution set for an equation that has no solution, symbolized by { } or $\varnothing$.

end behavior (349) The behavior of the graph as x approaches positive infinity $(+\infty)$ or negative infinity $(-\infty)$.

equal matrices (155) Two matrices that have the same dimensions and each element of one matrix is equal to the corresponding element of the other matrix.

equation (20) A mathematical sentence stating that two mathematical expressions are equal.

expansion by minors (183) A method of evaluating a third or high order determinant by using determinants of lower order.

e El número irracional 2.71828.... e es la base de los logaritmos naturales.

elemento Cada valor de una matriz.

método de eliminación Eliminar una de las variables de un sistema de ecuaciones sumando o restando las ecuaciones.

elipse Conjunto de todos los puntos de un plano en los que la suma de sus distancias a dos puntos dados del plano, llamados focos, es constante.

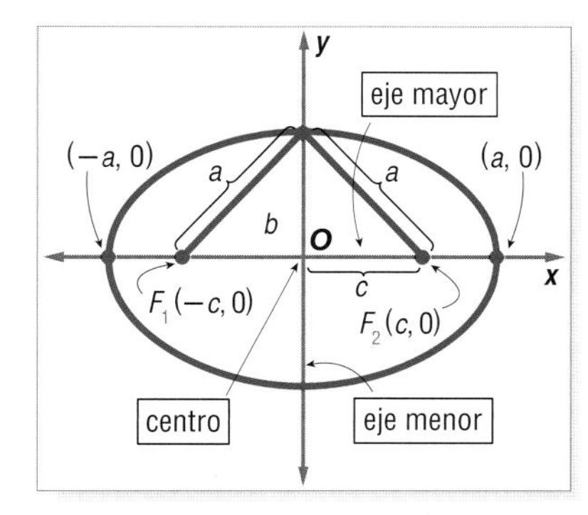

conjunto vacío Conjunto solución de una ecuación que no tiene solución, denotado por { } o $\varnothing$.

comportamiento final El comportamiento de una gráfica a medida que x tiende a más infinito $(+\infty)$ o menos infinito $(-\infty)$.

matrices iguales Dos matrices que tienen las mismas dimensiones y en las que cada elemento de una de ellas es igual al elemento correspondiente en la otra matriz.

ecuación Enunciado matemático que afirma la igualdad de dos expresiones matemáticas.

expansión por determinantes menores Un método de calcular el determinante de tercer orden o mayor mediante el uso de determinantes de orden más bajo.

exponential decay (524) Exponential decay occurs when a quantity decreases exponentially over time.

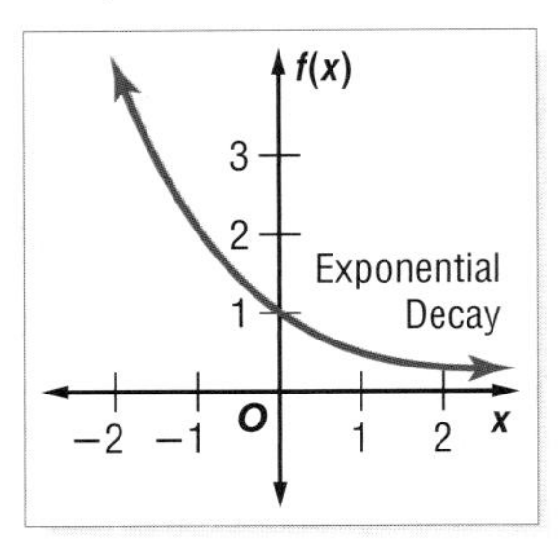

desintegración exponencial Ocurre cuando una cantidad disminuye exponencialmente con el tiempo.

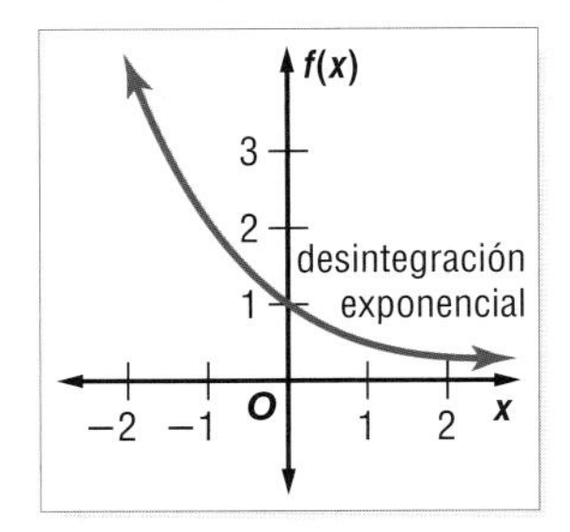

exponential equation (526) An equation in which the variables occur as exponents.

ecuación exponencial Ecuación en que las variables aparecen en los exponentes.

exponential function (524) A function of the form $y = ab^x$, where $a \neq 0$, $b > 0$, and $b \neq 1$.

función exponencial Una función de la forma $y = ab^x$, donde $a \neq 0$, $b > 0$, y $b \neq 1$.

exponential growth (524) Exponential growth occurs when a quantity increases exponentially over time.

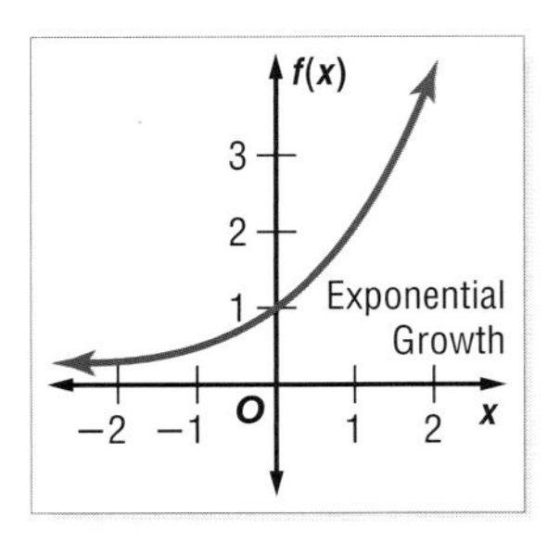

crecimiento exponencial El que ocurre cuando una cantidad aumenta exponencialmente con el tiempo.

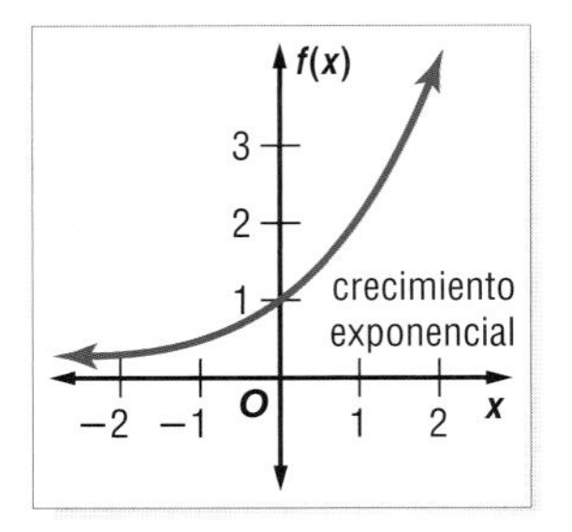

extraneous solution (263) A number that does not satisfy the original equation.

solución extraña Número que no satisface la ecuación original.

extrapolation (82) Predicting for an x-value greater than any in the data set.

extrapolación Predicción para un valor de x mayor que cualquiera de los de un conjunto de datos.

F

factorial (613) If n is a positive integer, then $n! = n(n - 1)(n - 2) \ldots 2 \cdot 1$.

factorial Si n es un entero positivo, entonces $n! = n(n - 1)(n - 2) \ldots 2 \cdot 1$.

failure (644) Any outcome other than the desired outcome.

fracaso Cualquier resultado distinto del deseado.

family of graphs (70) A group of graphs that displays one or more similar characteristics.

familia de gráficas Grupo de gráficas que presentan una o más características similares.

feasible region (129) The intersection of the graphs in a system of constraints.

región viable Intersección de las gráficas de un sistema de restricciones.

Fibonacci sequence (606) A sequence in which the first two terms are 1 and each of the additional terms is the sum of the two previous terms.

sucesión de Fibonacci Sucesión en que los dos primeros términos son iguales a 1 y cada término que sigue es igual a la suma de los dos anteriores.

focus (419, 433, 441) See parabola, ellipse, hyperbola.

foco Véase parábola, elipse, hipérbola.

FOIL method (230) The product of two binomials is the sum of the products of **F** the *first* terms, **O** the *outer* terms, **I** the *inner* terms, and **L** the *last* terms.

método FOIL El producto de dos binomios es la suma de los productos de los primeros (***F****irst)* términos, los términos exteriores (***O****uter)*, los términos interiores (***I****nner)* y los últimos (***L****ast)* términos.

formula (8) A mathematical sentence that expresses the relationship between certain quantities.

fórmula Enunciado matemático que describe la relación entre ciertas cantidades.

function (57) A relation in which each element of the domain is paired with exactly one element in the range.

function notation (59) An equation of y in terms of x can be rewritten so that $y = f(x)$. For example, $y = 2x + 1$ can be written as $f(x) = 2x + 1$.

función Relación en que a cada elemento del dominio le corresponde un solo elemento del rango.

notación funcional Una ecuación de y en términos de x puede escribirse en la forma $y = f(x)$. Por ejemplo, $y = 2x + 1$ puede escribirse como $f(x) = 2x + 1$.

G

geometric mean (590) The terms between any two nonsuccessive terms of a geometric sequence.

geometric sequence (588) A sequence in which each term after the first is found by multiplying the previous term by a constant r, called the common ratio.

geometric series (594) The sum of the terms of a geometric sequence.

greatest integer function (89) A step function, written as $f(x) = [\![x]\!]$, where $f(x)$ is the greatest integer less than or equal to x.

media geométrica Cualquier término entre dos términos no consecutivos de una sucesión geométrica.

sucesión geométrica Sucesión en que cualquier término después del primero puede hallarse multiplicando el término anterior por una constante r, llamada razón común .

serie geométrica La suma de los términos de una sucesión geométrica.

función del máximo entero Una función etapa que se escribe $f(x) = [x]$, donde $f(x)$ es el meaximo entero que es menor que o igual a x.

H

hyperbola (441) The set of all points in the plane such that the absolute value of the difference of the distances from two given points in the plane, called foci, is constant.

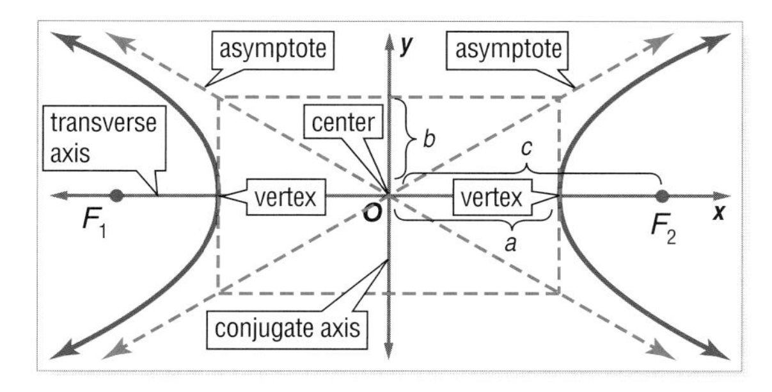

hypothesis (686) A statement to be tested.

hipérbola Conjunto de todos los puntos de un plano en los que el valor absoluto de la diferencia de sus distancias a dos puntos dados del plano, llamados focos, es constante.

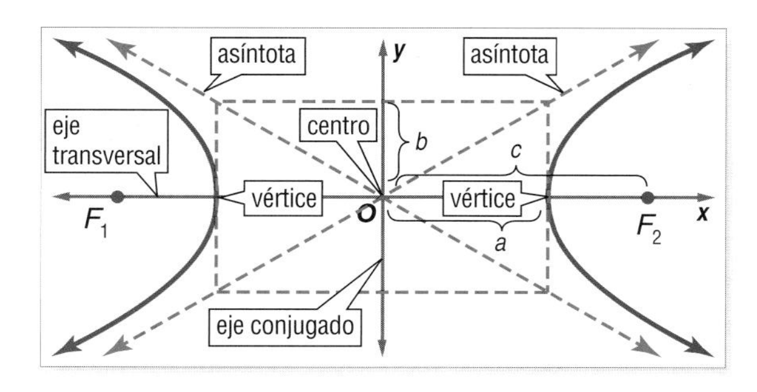

hipótesis Proposición que debe ser verificada.

I

identity function (90, 391) The function $I(x) = x$.

identity matrix (195) A square matrix that, when multiplied by another matrix, equals that same matrix. If A is any $n \times n$ matrix and I is the $n \times n$ identity matrix, then $A \cdot I = A$ and $I \cdot A = A$.

image (175) The graph of an object after a transformation.

imaginary unit (270) i, or the principal square root of -1.

inclusive (659) Two events whose outcomes may be the same.

función identidad La función $I(x) = x$.

matriz identidad Matriz cuadrada que al multiplicarse por otra matriz, es igual a la misma matriz. Si A es una matriz de $n \times n$ e I es la matriz identidad de $n \times n$, entonces $A \cdot I = A$ y $I \cdot A = A$.

imagen Gráfica de una figura después de una transformación.

unidad imaginaria i, o la raíz cuadrada principal de -1.

inclusivo Dos eventos que pueden tener los mismos resultados.

Glossary/Glosario

inconsistent (111) A system of equations that has no solutions.

independent (111) A system of equations that has exactly one solution.

independent events (632) Events that do not affect each other.

independent variable (59) In a function, the variable, usually x, whose values make up the domain.

index of summation (585) The variable used with the summation symbol. In the expression below, the index of summation is n.

$$\sum_{n=1}^{3} 4n$$

inductive hypothesis (618) The assumption that a statement is true for some positive integer k, where $k \geq n$.

infinite geometric series (599) A geometric series with an infinite number of terms.

initial side of an angle (709) The fixed ray of an angle.

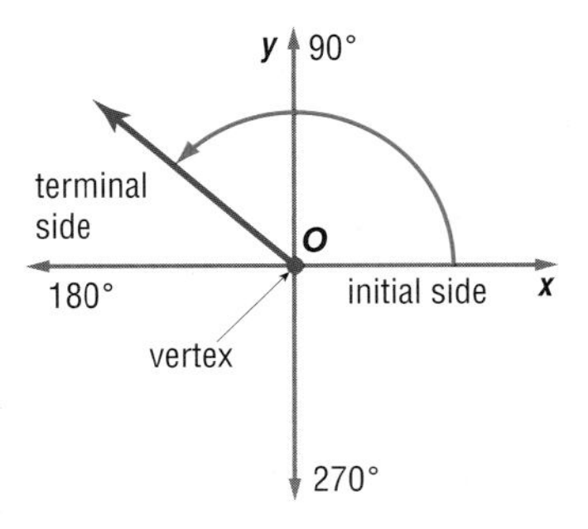

interpolation (82) Predicting for an x-value between the least and greatest values of the set.

intersection (40) The graph of a compound inequality containing *and*.

interval notation (35) Using the infinity symbols, $+\infty$ and $-\infty$, to indicate that the solution set of an inequality is unbounded in the positive or negative direction, respectively.

inverse (195) Two $n \times n$ matrices are inverses of each other if their product is the identity matrix.

inverse function (391) Two functions f and g are inverse functions if and only if both of their compositions are the identity function.

inverse of a trigonometric function (746) The arccosine, arcsine, and arctangent relations.

inverse relations (390) Two relations are inverse relations if and only if whenever one relation contains the element (a, b) the other relation contains the element (b, a).

inverse variation (493) y varies inversely as x if there is some nonzero constant k such that $xy = k$ or $y = \frac{k}{x}$.

inconsistente Sistema de ecuaciones que no tiene solución alguna.

independiente Sistema de ecuaciones que sólo tiene una solución.

eventos independientes Eventos que no se afectan mutuamente.

variable independiente En una función, la variable, por lo general x, cuyos valores forman el dominio.

índice de suma Variable que se usa con el símbolo de suma. En la siguiente expresión, el índice de suma es n.

$$\sum_{n=1}^{3} 4n$$

hipótesis inductiva El suponer que un enunciado es verdadero para algún entero positivo k, donde $k \geq n$.

serie geométrica infinita Serie geométrica con un número infinito de términos.

lado inicial de un ángulo El rayo fijo de un ángulo.

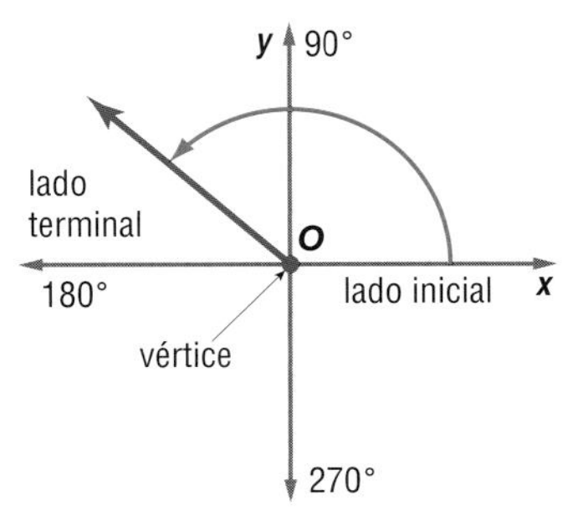

interpolación Predecir un valor de x entre los valores máximo y mínimo del conjunto de datos.

intersección Gráfica de una desigualdad compuesta que contiene la palabra *y*.

notación de intervalo Uso de los símbolos de infinito, $+\infty$ y $-\infty$, para indicar que el conjunto solución de una desigualdad no es acotado en la dirección positiva o negativa, respectivamente.

inversa Dos matrices de $n \times n$ son inversas mutuas si su producto es la matriz identidad.

función inversa Dos funciones f y g son inversas mutuas si y sólo si las composiciones de ambas son la función identidad.

inversa de una función trigonométrica Las relaciones arcocoseno, arcoseno y arcotangente.

relaciones inversas Dos relaciones son relaciones inversas mutuas si y sólo si cada vez que una de las relaciones contiene el elemento (a, b), la otra contiene el elemento (b, a).

variación inversa y varía inversamente con x *si* hay una constante no nula k tal que $xy = k$ o $y = \frac{k}{x}$.

irrational number (11) A real number that is not rational. The decimal form neither terminates nor repeats.

número irracional Número que no es racional. Su expansión decimal no es ni terminal ni periódica.

isometry (175) A transformation in which the image and preimage are congruent figures.

isometría Transformación en que la imagen y la preimagen son figuras congruentes.

iteration (608) The process of composing a function with itself repeatedly.

iteración Proceso de componer una función consigo misma repetidamente.

J

joint variation (493) y varies jointly as x and z if there is some nonzero constant k such that $y = kxz$, where $x \neq 0$ and $z \neq 0$.

variación conjunta y varía conjuntamente con x y z si hay una constante no nula k tal que $y = kxz$, donde $x \neq 0$ y $z \neq 0$.

L

latus rectum (421) The line segment through the focus of a parabola and perpendicular to the axis of symmetry.

latus rectum El segmento de recta que pasa por el foco de una parábola y que es perpendicular a su eje de simetría.

Law of Cosines (733–734) Let $\triangle ABC$ be any triangle with a, b, and c representing the measures of sides, and opposite angles with measures A, B, and C, respectively. Then the following equations are true.

$a^2 = b^2 + c^2 - 2bc \cos A$

$b^2 = a^2 + c^2 - 2ac \cos B$

$c^2 = a^2 + b^2 - 2ab \cos C$

Ley de los cosenos Sea $\triangle ABC$ un triángulo cualquiera, con a, b y c las longitudes de los lados y con ángulos opuestos de medidas A, B y C, respectivamente. Entonces se cumplen las siguientes ecuaciones.

$a^2 = b^2 + c^2 - 2bc \cos A$

$b^2 = a^2 + c^2 - 2ac \cos B$

$c^2 = a^2 + b^2 - 2ab \cos C$

Law of Sines (726) Let $\triangle ABC$ be any triangle with a, b, and c representing the measures of sides opposite angles with measurements A, B, and C, respectively. Then $\frac{\sin A}{a} = \frac{\sin B}{b} = \frac{\sin C}{c}$.

Ley de los senos Sea $\triangle ABC$ cualquier triángulo con a, b y c las longitudes de los lados y con ángulos opuestos de medidas A, B y C, respectivamente. Entonces $\frac{\sin A}{a} = \frac{\sin B}{b} = \frac{\sin C}{c}$.

leading coefficient (346) The coefficient of the term with the highest degree.

coeficiente líder Coeficiente del término de mayor grado.

like radical expressions (252) Two radical expressions in which both the radicands and indices are alike.

expresiones radicales semejantes Dos expresiones radicales en que tanto los radicandos como los índices son semejantes.

like terms (229) Monomials that can be combined.

términos semejantes Monomios que pueden combinarse.

limit (593) The value that the terms of a sequence approach.

límite El valor al que tienden los términos de una sucesión.

linear equation (63) An equation that has no operations other than addition, subtraction, and multiplication of a variable by a constant.

ecuación lineal Ecuación sin otras operaciones que las de adición, sustracción y multiplicación de una variable por una constante.

linear function (63) A function whose ordered pairs satisfy a linear equation.

función lineal Función cuyos pares ordenados satisfacen una ecuación lineal.

linear permutation (638) The arrangement of objects or people in a line.

permutación lineal Arreglo de personas o figuras en una línea.

linear programming (130) The process of finding the maximum or minimum values of a function for a region defined by inequalities.

programación lineal Proceso de hallar los valores máximo o mínimo de una función lineal en una región definida por las desigualdades.

linear term (286) In the equation $f(x) = ax^2 + bx + c$, bx is the linear term.

término lineal En la ecuación $f(x) = ax^2 + bx + c$, el término lineal es bx.

line of fit (81) A line that closely approximates a set of data.

recta de ajuste Recta que se aproxima estrechamente a un conjunto de datos.

logarithm (531) In the function $x = b^y$, y is called the logarithm, base b, of x. Usually written as $y = \log_b x$ and is read "y equals log base b of x."

logaritmo En la función $x = b^y$, y es el logaritmo en base b, de x. Generalmente escrito como $y = \log_b x$ y se lee "y es igual al logaritmo en base b de x."

logarithmic equation (533) An equation that contains one or more logarithms.

ecuación logarítmica Ecuación que contiene uno o más logaritmos.

logarithmic function (532) The function $y = \log_b x$, where $b > 0$ and $b \neq 1$, which is the inverse of the exponential function $y = b^x$.

función logarítmica La función $y = \log_b x$, donde $b > 0$ y $b \neq 1$, inversa de la función exponencial $y = b^x$.

M

$m \times n$ matrix (155) A matrix with m rows and n columns.

matriz de $m \times n$ Matriz de m filas y n columnas.

major axis (434) The longer of the two line segments that form the axes of symmetry of an ellipse.

eje mayor El más largo de dos segmentos de recta que forman los ejes de simetría de una elipse.

mapping (57) How each member of the domain is paired with each member of the range.

transformaciones La correspondencia entre cada miembro del dominio con cada miembro del rango.

margin of sampling error (ME) (682) The limit on the difference between how a sample responds and how the total population would respond.

margen de error muestral (EM) Límite en la diferencia entre las respuestas obtenidas con una muestra y cómo pudiera responder la población entera.

mathematical induction (618) A method of proof used to prove statements about positive integers.

inducción matemática Método de demostrar enunciados sobre los enteros positivos.

matrix (154) Any rectangular array of variables or constants in horizontal rows and vertical columns.

matriz Arreglo rectangular de variables o constantes en filas horizontales y columnas verticales.

maximum value (288) The y-coordinate of the vertex of the quadratic function $f(x) = ax^2 + bx + c$, where $a < 0$.

valor máximo La coordenada y del vértice de la función cuadrática $f(x) = ax^2 + bx + c$, donde $a < 0$.

measure of central tendency (665) A number that represents the center or middle of a set of data.

medida de tendencia central Número que representa el centro o medio de un conjunto de datos.

measure of variation (664) A representation of how spread out or scattered a set of data is.

medida de variación Número que representa la dispersión de un conjunto de datos.

midline (771) A horizontal axis used as the reference line about which the graph of a periodic function oscillates.

recta central Eje horizontal que se usa como recta de referencia alrededor de la cual oscila la gráfica de una función periódica.

minimum value (288) The y-coordinate of the vertex of the quadratic function $f(x) = ax^2 + bx + c$, where $a > 0$.

valor mínimo La coordenada y del vértice de la función cuadrática $f(x) = ax^2 + bx + c$, donde $a > 0$.

minor (183) The determinant formed when the row and column containing that element are deleted.

determinante menor El que se forma cuando se descartan la fila y columna que contienen dicho elemento.

minor axis (434) The shorter of the two line segments that form the axes of symmetry of an ellipse.

eje menor El más corto de los dos segmentos de recta de los ejes de simetría de una elipse.

monomial (222) An expression that is a number, a variable, or the product of a number and one or more variables.

monomio Expresión que es un número, una variable o el producto de un número por una o más variables.

mutually exclusive (658) Two events that cannot occur at the same time.

mutuamente exclusivos Dos eventos que no pueden ocurrir simultáneamente.

N

n^{th} root (245) For any real numbers a and b, and any positive integer n, if $a^n = b$, then a is an nth root of b.

raíz *enésima* Para cualquier número real a y b y cualquier entero positivo n, si $a^n = b$, entonces a se llama una raíz *enésima* de b.

natural base exponential function (554) An exponential function with base e, $y = e^x$.

función exponencial natural La función exponencial de base e, $y = e^x$.

natural logarithm (554) Logarithms with base e, written $\ln x$.

logaritmo natural Logaritmo de base e, el que se escribe $\ln x$.

natural logarithmic function (554) $y = \ln x$, the inverse of the natural base exponential function $y = e^x$.

función logarítmica natural $y = \ln x$, la inversa de la función exponencial natural $y = e^x$.

negative exponent (222) For any real number $a \neq 0$ and any integer n, $a^{-n} = \frac{1}{a^n}$ and $\frac{1}{a^{-n}} = a^n$.

exponente negativo Para cualquier número real $a \neq 0$ cualquier entero positivo n, $a^{-n} = \frac{1}{a^n}$ y $\frac{1}{a^{-n}} = a^n$.

normal distribution (671) A frequency distribution that often occurs when there is a large number of values in a set of data: about 68% of the values are within one standard deviation of the mean, 95% of the values are within two standard deviations from the mean, and 99% of the values are within three standard deviations.

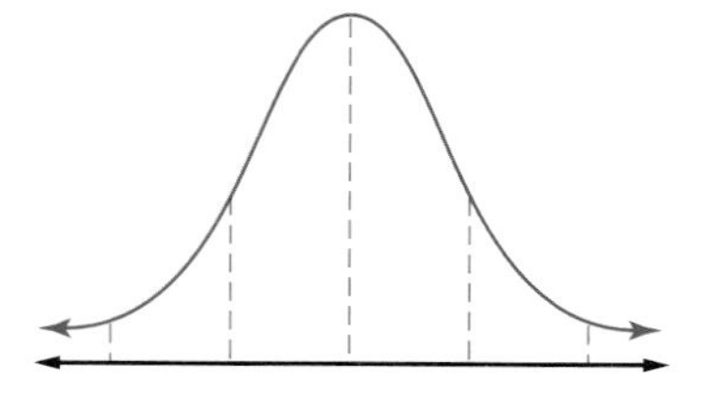

distribución normal Distribución de frecuencia que aparece a menudo cuando hay un número grande de datos: cerca del 68% de los datos están dentro de una desviación estándar de la media, 95% están dentro de dos desviaciones estándar de la media y 99% están dentro de tres desviaciones estándar de la media.

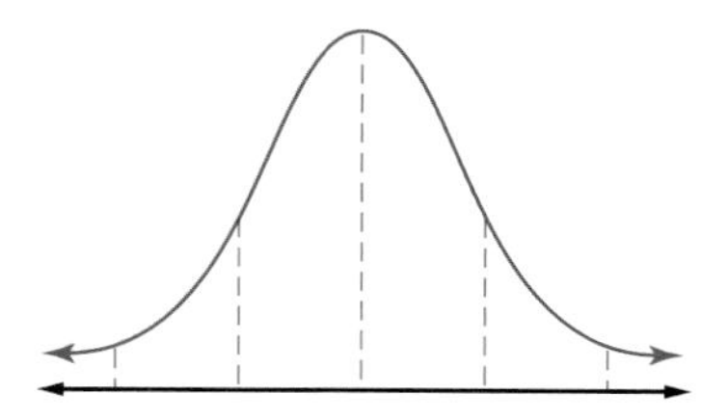

O

octants (136) The eight regions of three-dimensional space.

octantes Las ocho regiones del espacio tridimensional.

odds (645) The ratio of the number of the successes of an event to the number of failures.

posibilidades Razón del número de éxitos de un evento a su número de fracasos.

one-to-one function (57, 392) **1.** A function where each element of the range is paired with exactly one element of the domain **2.** A function whose inverse is a function.

función biunívoca **1.** Función en la que a cada elemento del rango le corresponde sólo un elemento del dominio. **2.** Función cuya inversa es una función.

open sentence (20) A mathematical sentence containing one or more variables.

enunciado abierto Enunciado matemático que contiene una o más variables.

ordered pair (56) A pair of coordinates, written in the form (x, y), used to locate any point on a coordinate plane.

par ordenado Un par de números, escrito en la forma (x, y), que se usa para ubicar cualquier punto en un plano de coordenadas.

ordered triple (136, 139) **1.** The coordinates of a point in space **2.** The solution of a system of equations in three variables x, y, and z.

triple ordenado **1.** Las coordenadas de un punto en el espacio **2.** Solución de un sistema de ecuaciones en tres variables x, y y z.

Order of Operations (6)

Step 1 Evaluate expressions inside grouping symbols.

Step 2 Evaluate all powers.

Step 3 Do all multiplications and/or divisions from left to right.

Step 4 Do all additions and subtractions from left to right.

Orden de las operaciones

Paso 1 Evalúa las expresiones dentro de símbolos de agrupamiento.

Paso 2 Evalúa todas las potencias.

Paso 3 Ejecuta todas las multiplicaciones y divisiones de izquierda a derecha.

Paso 4 Ejecuta todas las adiciones y sustracciones de izquierda a derecha.

outcomes (632) The results of a probability experiment/an event.

resultados Lo que produce un experimento o evento probabilístico.

outlier (826) A data point that does not appear to belong to the rest of the set.

valor atípico Dato que no parece pertenecer al resto el conjunto.

P

parabola (286, 419) The set of all points in a plane that are the same distance from a given point, called the focus, and a given line, called the directrix.

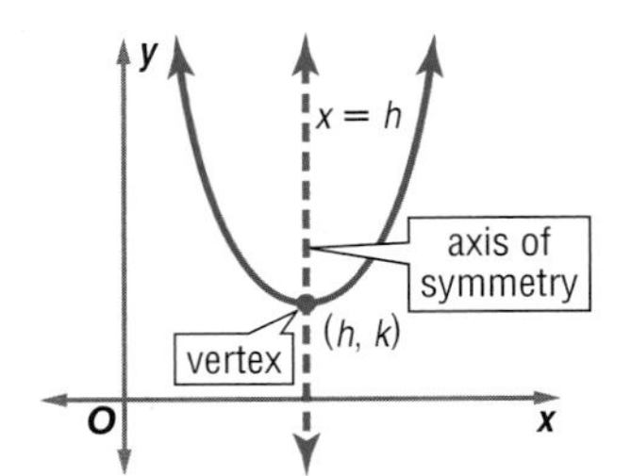

parábola Conjunto de todos los puntos de un plano que están a la misma distancia de un punto dado, llamado foco, y de una recta dada, llamada directriz.

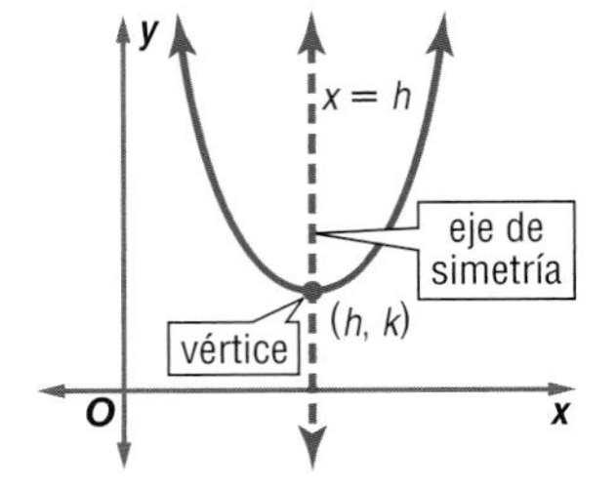

parallel lines (70) Nonvertical coplanar lines with the same slope.

rectas paralelas Rectas coplanares no verticales con la misma pendiente.

parent graph (70) The simplest of graphs in a family.

gráfica madre La gráfica más sencilla en una familia de gráficas.

partial sum (599) The sum of the first n terms of a series.

suma parcial La suma de los primeros n términos de una serie.

Pascal's triangle (612) A triangular array of numbers such that the $(n + 1)^{th}$ row is the coefficient of the terms of the expansion $(x + y)^n$ for $n = 0, 1, 2$ ...

Triángulo de Pascal Arreglo triangular de números en el que la fila $(n + 1)^n$ proporciona los coeficientes de los términos de la expansión de $(x + y)^n$ para $n = 0, 1, 2$ ...

period (741) The least possible value of a for which $f(x) = f(x + a)$.

período El menor valor positivo posible para a, para el cual $f(x) = f(x + a)$.

periodic function (741) A function is called periodic if there is a number a such that $f(x) = f(x + a)$ for all x in the domain of the function.

función periódica Función para la cual hay un número a tal que $f(x) = f(x + a)$ para todo x en el dominio de la función .

permutation (638) An arrangement of objects in which order is important.

permutación Arreglo de elementos en que el orden es importante.

perpendicular lines (71) In a plane, any two oblique lines the product of whose slopes is -1.

rectas perpendiculares En un plano, dos rectas oblicuas cualesquiera cuyas pendientes tienen un producto igual a -1.

phase shift (769) A horizontal translation of a trigonometric function.

desvío de fase Traslación horizontal de una función trigonométrica.

piecewise function (91) A function that is written using two or more expressions.

función a intervalos Función que se escribe usando dos o más expresiones.

point discontinuity (485) If the original function is undefined for $x = a$ but the related rational expression of the function in simplest form is defined for $x = a$, then there is a hole in the graph at $x = a$.

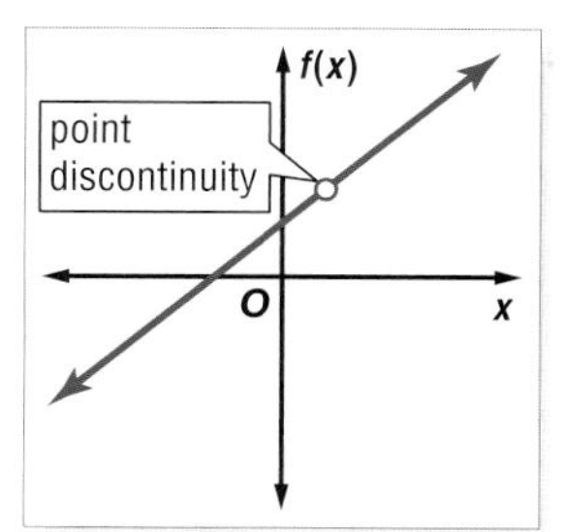

discontinuidad evitable Si la función original no está definida en $x = a$ pero la expresión racional reducida correspondiente de la función está definida en $x = a$, entonces la gráfica tiene una ruptura o corte en $x = a$.

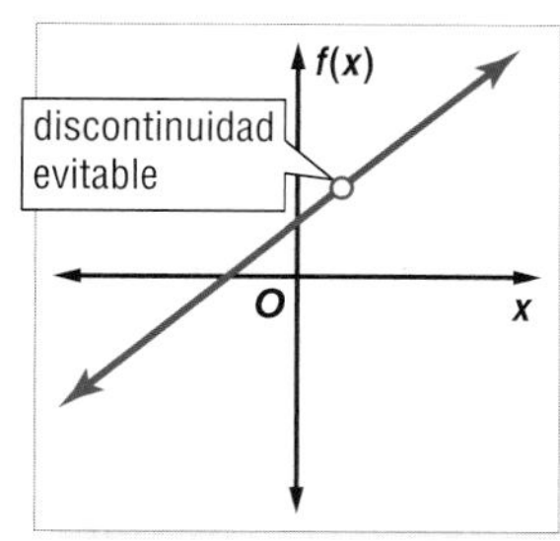

point-slope form (76) An equation in the form $y - y_1 = m(x - x_1)$ where (x_1, y_1) are the coordinates of a point on the line and m is the slope of the line.

forma punto-pendiente Ecuación de la forma $y - y_1 = m(x - x_1)$ donde (x_1, y_1) es un punto en la recta y m es la pendiente de la recta.

polynomial (229) A monomial or a sum of monomials.

polinomio Monomio o suma de monomios.

polynomial function (347) A function that is represented by a polynomial equation.

función polinomial Función representada por una ecuación polinomial.

polynomial in one variable (346) $a_0x^n + a_1x^{n-1} + \ldots + a_{n-2}x^2 + a_{n-1}x + a_n$, where the coefficients $a_0, a_1, \ldots, a_n$ represent real numbers, and a_0 is not zero and n is a nonnegative integer.

polinomio de una variable $a_0x^n + a_1x^{n-1} + \ldots + a_{n-2}x^2 + a_{n-1}x + a_n$, donde los coeficientes $a_0, a_1, \ldots, a_n$ son números reales, a_0 no es nulo y n es un entero no negativo.

power (222) An expression of the form x^n.

potencia Expresión de la forma x^n.

power function (704) An equation in the form $f(x) = ax^b$, where a and b are real numbers.

función potencia Ecuación de la forma $f(x) = ax^b$, donde a y b son números reales.

prediction equation (81) An equation suggested by the points of a scatter plot that is used to predict other points.

ecuación de predicción Ecuación sugerida por los puntos de una gráfica de dispersión y que se usa para predecir otros puntos.

preimage (175) The graph of an object before a transformation.

preimagen Gráfica de una figura antes de una transformación.

principal root (246) The nonnegative root.

raíz principal La raíz no negativa.

principal values (746) The values in the restricted domains of trigonometric functions.

valores principales Valores en los dominios restringidos de las funciones trigonométricas.

probability (644) A ratio that measures the chances of an event occurring.

probabilidad Razón que mide la posibilidad de que ocurra un evento.

probability distribution (646) A function that maps the sample space to the probabilities of the outcomes in the sample space for a particular random variable.

distribución de probabilidad Función que aplica el espacio muestral a las probabilidades de los resultados en el espacio muestral obtenidos para una variable aleatoria particular.

pure imaginary number (270) The square roots of negative real numbers. For any positive real number b, $\sqrt{-b^2} = \sqrt{b^2} \cdot \sqrt{-1}$, or bi.

número imaginario puro Raíz cuadrada de un número real negativo. Para cualquier número real positivo b, $\sqrt{-b^2} = \sqrt{b^2} \cdot \sqrt{-1}$ ó bi.

Q

quadrantal angle (718) An angle in standard position whose terminal side coincides with one of the axes.

ángulo de cuadrante Ángulo en posición estándar cuyo lado terminal coincide con uno de los ejes.

quadrants (56) The four areas of a Cartesian coordinate plane.

cuadrantes Las cuatro regiones de un plano de coordenadas cartesiano.

quadratic equation (294) A quadratic function set equal to a value, in the form $ax^2 + bx + c$, where $a \neq 0$.

quadratic form (360) For any numbers a, b, and c, except for $a = 0$, an equation that can be written in the form $a[f(x)^2] + b[f(x)] + c = 0$, where $f(x)$ is some expression in x.

Quadratic Formula (313) The solutions of a quadratic equation of the form $ax^2 + bx + c = 0$, where $a \neq 0$, are given by the Quadratic Formula, which is $x = \frac{-b \pm \sqrt{b^2 - 4ac}}{2a}$.

quadratic function (286) A function described by the equation $f(x) = ax^2 + bx + c$, where $a \neq 0$.

quadratic term (286) In the equation $f(x) = ax^2 + bx + c$, ax^2 is the quadratic term.

ecuación cuadrática Función cuadrática igual a un valor, de la forma $ax^2 + bx + c$, donde $a \neq 0$.

forma de ecuación cuadrática Para cualquier número a, b y c, excepto $a = 0$, una ecuación que puede escribirse de la forma $a[f(x)^2] + b[f(x)] + c = 0$, donde $f(x)$ es una expresión en x.

Fórmula cuadrática Las soluciones de una ecuación cuadrática de la forma $ax^2 + bx + c = 0$, donde $a \neq 0$, se dan por la fórmula cuadrática, que es $x = \frac{-b \pm \sqrt{b^2 - 4ac}}{2a}$.

función cuadrática Función descrita por la ecuación $f(x) = ax^2 + bx + c$, donde $a \neq 0$.

término cuadrático En la ecuación $f(x) = ax^2 + bx + c$, el término cuadrático es ax^2.

R

radian (710) The measure of an angle θ in standard position whose rays intercept an arc of length 1 unit on the unit circle.

radical equation (263) An equation with radicals that have variables in the radicands.

radical inequality (264) An inequality that has a variable in the radicand.

random (645) All outcomes have an equally likely chance of happening.

random variable (646) The outcome of a random process that has a numerical value.

range (56) The set of all y-coordinates of a relation.

rate of change (69) How much a quantity changes on average, relative to the change in another quantity, often time.

rate of decay (560) The percent decrease r in the equation $y = a(1 - r)^t$.

rate of growth (562) The percent increase r in the equation $y = a(1 + r)^t$.

rational equation (505) Any equation that contains one or more rational expressions.

rational exponent (258) For any nonzero real number b, and any integers m and n, with $n > 1$, $b^{\frac{m}{n}} = \sqrt[n]{b^m} = \left(\sqrt[n]{b}\right)^m$, except when $b < 0$ and n is even.

rational expression (472) A ratio of two polynomial expressions.

rational function (472) An equation of the form $f(x) = \frac{p(x)}{q(x)}$, where $p(x)$ and $q(x)$ are polynomial functions, and $q(x) \neq 0$.

radián Medida de un ángulo θ en posición normal cuyos rayos intersecan un arco de 1 unidad de longitud en el círculo unitario.

ecuación radical Ecuación con radicales que tienen variables en el radicando.

desigualdad radical Desigualdad que tiene una variable en el radicando.

aleatorio Todos los resultados son equiprobables.

variable aleatoria El resultado de un proceso aleatorio que tiene un valor numérico.

rango Conjunto de todas las coordenadas y de una relación.

tasa de cambio Lo que cambia una cantidad en promedio, respecto al cambio en otra cantidad, por lo general el tiempo.

tasa de desintegración Disminución porcentual r en la ecuación $y = a(1 - r)^t$.

tasa de crecimiento Aumento porcentual r en la ecuación $y = a(1 + r)^t$.

ecuación racional Cualquier ecuación que contiene una o más expresiones racionales.

exponente racional Para cualquier número real no nulo b y cualquier entero m y n, con $n > 1$, $b^{\frac{m}{n}} = \sqrt[n]{b^m} = \left(\sqrt[n]{b}\right)^m$, excepto cuando $b < 0$ y n es par.

expresión racional Razón de dos expresiones polinomiales.

función racional Ecuación de la forma $f(x) = \frac{p(x)}{q(x)}$, donde $p(x)$ y $q(x)$ son funciones polinomiales y $q(x) \neq 0$.

rational inequality (508) Any inequality that contains one or more rational expressions.

desigualdad racional Cualquier desigualdad que contiene una o más expresiones racionales.

rationalizing the denominator (251) To eliminate radicals from a denominator or fractions from a radicand.

racionalizar el denominador La eliminación de radicales de un denominador o de fracciones de un radicando.

rational number (11) Any number $\frac{m}{n}$, where m and n are integers and n is not zero. The decimal form is either a terminating or repeating decimal.

número racional Cualquier número $\frac{m}{n}$, donde m y n son enteros y n no es cero. Su expansión decimal es o terminal o periódica.

real numbers (11) All numbers used in everyday life; the set of all rational and irrational numbers.

números reales Todos los números que se usan en la vida cotidiana; el conjunto de los todos los números racionales e irracionales.

recursive formula (606) Each term is formulated from one or more previous terms.

fórmula recursiva Cada término proviene de uno o más términos anteriores.

reference angle (718) The acute angle formed by the terminal side of an angle in standard position and the x-axis.

ángulo de referencia El ángulo agudo formado por el lado terminal de un ángulo en posición estándar y el eje x.

reflection (177) A transformation in which every point of a figure is mapped to a corresponding image across a line of symmetry.

reflexión Transformación en que cada punto de una figura se aplica a través de una recta de simetría a su imagen correspondiente.

reflection matrix (177) A matrix used to reflect an object over a line or plane.

matriz de reflexión Matriz que se usa para reflejar una figura sobre una recta o plano.

regression line (87) A line of best fit.

recta de regresión Una recta de óptimo ajuste.

relation (56) A set of ordered pairs.

relación Conjunto de pares ordenados.

relative frequency histogram (646) A table of probabilities or a graph to help visualize a probability distribution.

histograma de frecuencia relativa Tabla de probabilidades o gráfica para asistir en la visualización de una distribución de probabilidad.

relative maximum (354) A point on the graph of a function where no other nearby points have a greater y-coordinate.

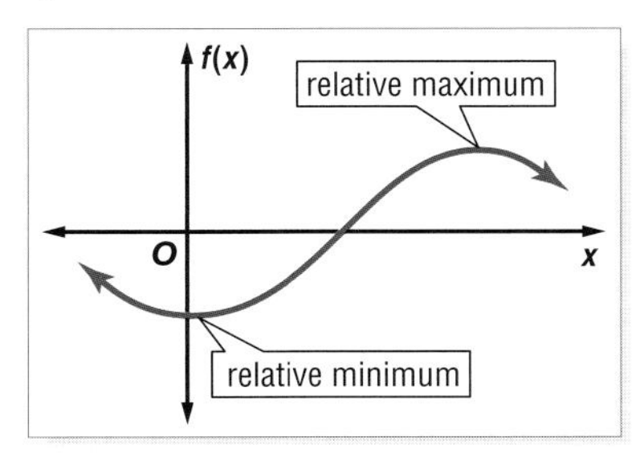

máximo relativo Punto en la gráfica de una función en donde ningún otro punto cercano tiene una coordenada y mayor.

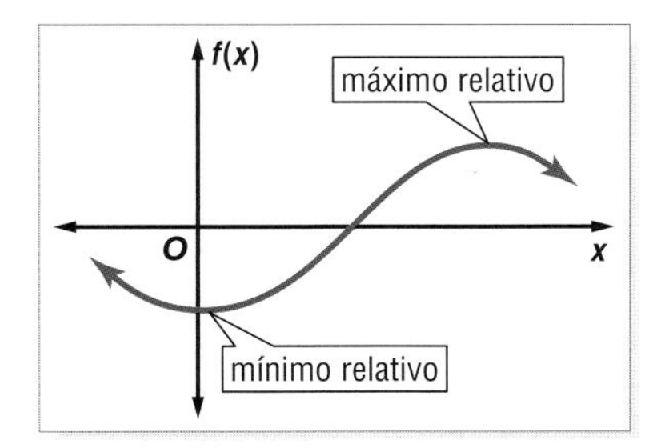

relative minimum (354) A point on the graph of a function where no other nearby points have a lesser y-coordinate.

mínimo relativo Punto en la gráfica de una función en donde ningún otro punto cercano tiene una coordenada y menor.

root (294) The solutions of a quadratic equation.

raíz Las soluciones de una ecuación cuadrática.

rotation (178) A transformation in which an object is moved around a center point, usually the origin.

rotación Transformación en que una figura se hace girar alrededor de un punto central, generalmente el origen.

rotation matrix (178) A matrix used to rotate an object.

matriz de rotación Matriz que se usa para hacer girar un objeto.

row matrix (155) A matrix that has only one row.

matriz fila Matriz que sólo tiene una fila.

S

sample space (632) The set of all possible outcomes of an event.

espacio muestral Conjunto de todos los resultados posibles de un experimento probabilístico.

scalar (162) A constant.

escalar Una constante.

scalar multiplication (162) Multiplying any matrix by a constant called a scalar; the product of a scalar k and an $m \times n$ matrix.

multiplicación por escalares Multiplicación de una matriz por una constante llamada escalar; producto de un escalar k y una matriz de $m \times n$.

scatter plot (81) A set of data graphed as ordered pairs in a coordinate plane.

gráfica de dispersión Conjuntos de datos graficados como pares ordenados en un plano de coordenadas.

scientific notation (225) The expression of a number in the form $a \times 10^n$, where $1 \leq a < 10$ and n is an integer.

notación científica Escritura de un número en la forma $a \times 10^n$, donde $1 \leq a < 10$ y n es un entero.

secant (701) For any angle, with measure α, a point $P(x, y)$ on its terminal side, $r = \sqrt{x^2 + y^2}$, $\sec \alpha = \frac{r}{x}$.

secante Para cualquier ángulo de medida α, un punto $P(x, y)$ en su lado terminal, $r = \sqrt{x^2 + y^2}$, $\sec \alpha = \frac{r}{x}$.

second-order determinant (182) The determinant of a 2×2 matrix.

determinante de segundo orden El determinante de una matriz de 2×2.

sequence (578) A list of numbers in a particular order.

sucesión Lista de números en un orden particular.

series (583) The sum of the terms of a sequence.

serie Suma específica de los términos de una sucesión.

set-builder notation (34) The expression of the solution set of an inequality, for example $\{x \mid x > 9\}$.

notación de construcción de conjuntos Escritura del conjunto solución de una desigualdad, por ejemplo, $\{x \mid x > 9\}$.

sigma notation (585) For any sequence $a_1, a_2, a_3, \ldots$, the sum of the first k terms may be written $\sum_{n=1}^{k} a_n$, which is read "the summation from $n = 1$ to k of a_n." Thus, $\sum_{n=1}^{k} a_n = a_1 + a_2 + a_3 + \ldots + a_k$, where k is an integer value.

notación de suma Para cualquier sucesión $a_1, a_2, a_3, \ldots$, la suma de los k primeros términos puede escribirse $\sum_{n=1}^{k} a_n$, lo que se lee "la suma de $n = 1$ a k de los a_n." Así, $\sum_{n=1}^{k} a_n = a_1 + a_2 + a_3 + \ldots + a_k$, donde k es un valor entero.

simple event (658) One event.

evento simple Un solo evento.

simplify (222) To rewrite an expression without parentheses or negative exponents.

reducir Escribir una expresión sin paréntesis o exponentes negativos.

simulation (681) The use of a probability experiment to mimic a real-life situation.

simulación Uso de un experimento probabilístico para imitar una situación de la vida real.

sine (701) For any angle, with measure α, a point $P(x, y)$ on its terminal side, $r = \sqrt{x^2 + y^2}$, $\sin \alpha = \frac{y}{r}$.

seno Para cualquier ángulo, de medida α, un punto $P(x, y)$ en su lado terminal, $r = \sqrt{x^2 + y^2}$, $\sin \alpha = \frac{y}{r}$.

skewed distribution (671) A curve or histogram that is not symmetric.

Positively Skewed

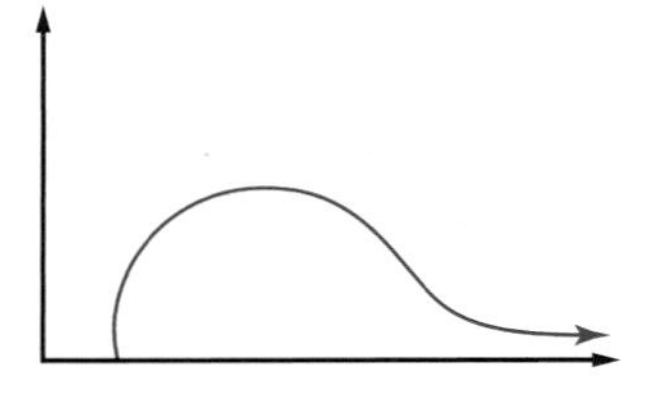

Negatively Skewed

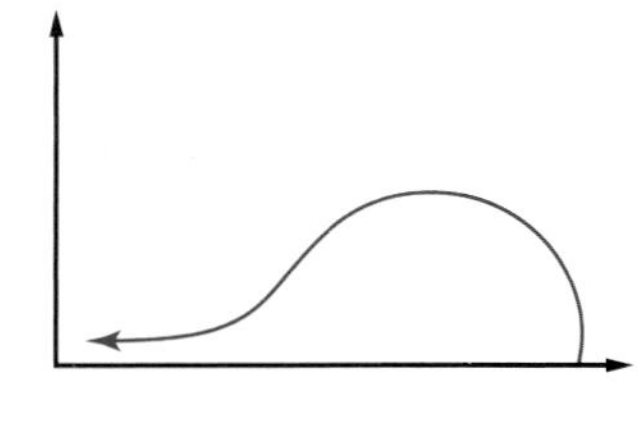

distribución asimétrica Curva o histograma que no es simétrico.

Positivamente Alabeada

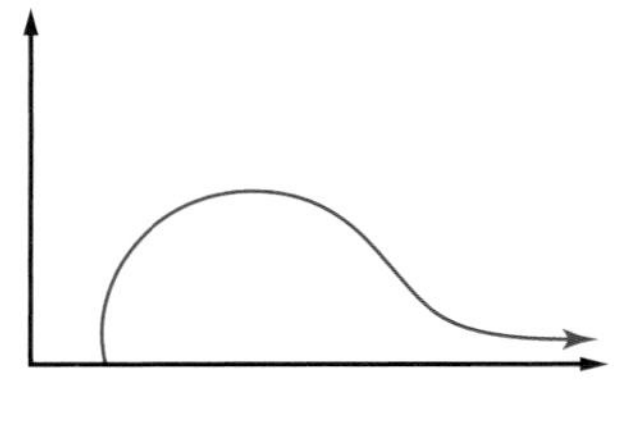

Negativamente Alabeada

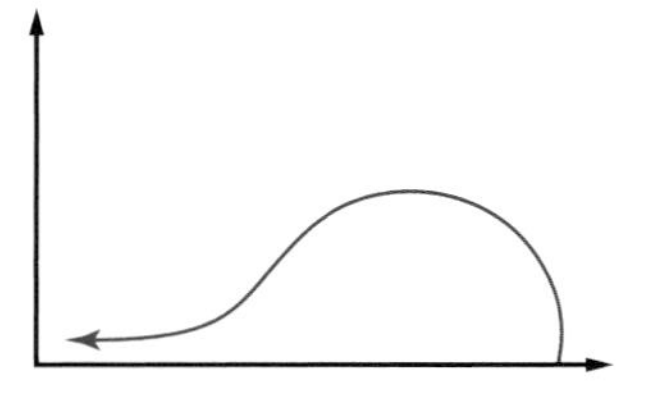

slope (68) The ratio of the change in y-coordinates to the change in x-coordinates.

pendiente La razón del cambio en coordenadas y al cambio en coordenadas x.

slope-intercept form (75) The equation of a line in the form $y = mx + b$, where m is the slope and b is the y-intercept.

forma pendiente-intersección Ecuación de una recta de la forma $y = mx + b$, donde m es la pendiente y b la intersección.

solution (20) A replacement for the variable in an open sentence that results in a true sentence.

solución Sustitución de la variable de un enunciado abierto que resulta en un enunciado verdadero.

solving a right triangle (704) The process of finding the measures of all of the sides and angles of a right triangle.

resolver un triángulo rectángulo Proceso de hallar las medidas de todos los lados y ángulos de un triángulo rectángulo.

square matrix (155) A matrix with the same number of rows and columns.

matriz cuadrada Matriz con el mismo número de filas y columnas.

square root (245) For any real numbers a and b, if $a^2 = b$, then a is a square root of b.

raíz cuadrada Para cualquier número real a y b, si $a^2 = b$, entonces a es una raíz cuadrada de b.

square root function (395) A function that contains a square root of a variable.

función radical Función que contiene la raíz cuadrada de una variable.

Square Root Property (306) For any real number n, if $x^2 = n$, then $x = \pm\sqrt{n}$.

Propiedad de la raíz cuadrada Para cualquier número real n, si $x^2 = n$, entonces $x = \pm\sqrt{n}$.

standard deviation (665) The square root of the variance, represented by α.

desviación estándar La raíz cuadrada de la varianza, la que se escribe α.

standard form (64) A linear equation written in the form $Ax + By = C$, where A, B, and C are real numbers and A and B are not both zero.

forma estándar Ecuación lineal escrita de la forma $Ax + By = C$, donde A, B, y C son números reales y A y B no son cero simultáneamente.

standard position (709) An angle positioned so that its vertex is at the origin and its initial side is along the positive x-axis.

posición estándar Ángulo en posición tal que su vértice está en el origen y su lado inicial está a lo largo del eje x positivo.

step function (89) A function whose graph is a series of line segments.

función etapa Función cuya gráfica es una serie de segmentos de recta.

substitution method (116) A method of solving a system of equations in which one equation is solved for one variable in terms of the other.

método de sustitución Método para resolver un sistema de ecuaciones en que una de las ecuaciones se resuelve en una de las variables en términos de la otra.

success (644) The desired outcome of an event.

éxito El resultado deseado de un evento.

synthetic division (234) A method used to divide a polynomial by a binomial.

división sintética Método que se usa para dividir un polinomio entre un binomio.

synthetic substitution (365) The use of synthetic division to evaluate a function.

sustitución sintética Uso de la división sintética para evaluar una función polinomial.

system of equations (110) A set of equations with the same variables.

sistema de ecuaciones Conjunto de ecuaciones con las mismas variables.

system of inequalities (123) A set of inequalities with the same variables.

sistema de desigualdades Conjunto de desigualdades con las mismas variables.

T

tangent (427, 701) **1.** A line that intersects a circle at exactly one point. **2.** For any angle, with measure α, a point $P(x, y)$ on its terminal side, $r = \sqrt{x^2 + y^2}$, $\tan \alpha = \frac{y}{x}$.

tangente **1.** Recta que interseca un círculo en un solo punto. **2.** Para cualquier ángulo, de medida α, un punto $P(x, y)$ en su lado terminal, $r = \sqrt{x^2 + y^2}$, $\tan \alpha = \frac{y}{x}$.

term (229, 578) **1.** The monomials that make up a polynomial. **2.** Each number in a sequence or series.

término **1.** Los monomios que constituyen un polinomio. **2.** Cada número de una sucesión o serie.

terminal side of an angle (709) A ray of an angle that rotates about the center.

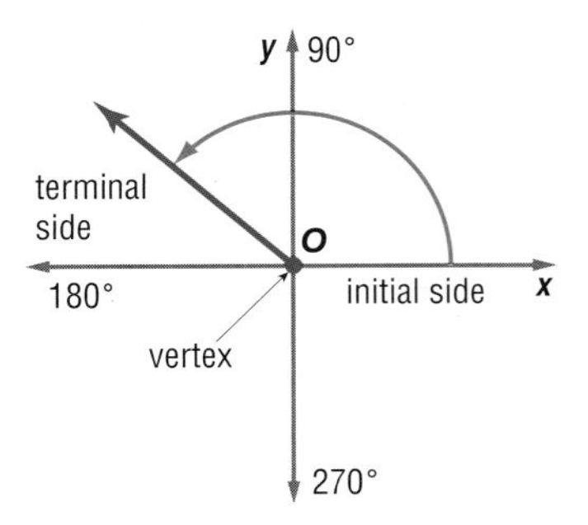

lado terminal de un ángulo Rayo de un ángulo que gira alrededor de un centro.

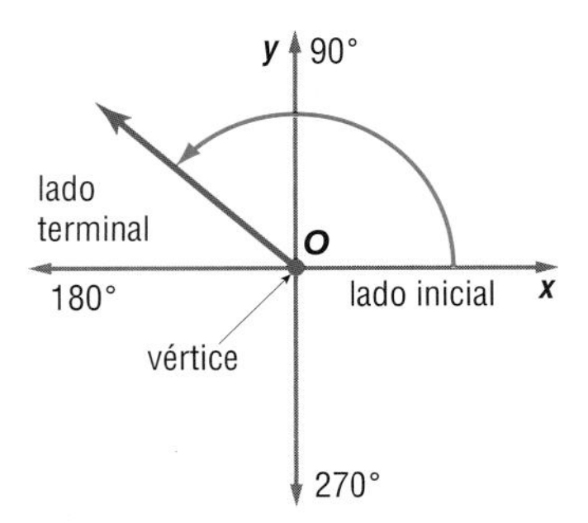

third-order determinant (183) Determinants of a 3×3 matrix.

determinante de tercer orden Determinante de una matriz de 3×3.

transformation (175) Functions that map points of a pre-image onto its image.

transformación Funciones que aplican puntos de una preimagen en su imagen.

translation (175) A figure is moved from one location to another on the coordinate plane without changing its size, shape, or orientation.

traslación Se mueve una figura de un lugar a otro en un plano de coordenadas sin cambiar su tamaño, forma u orientación.

translation matrix (175) A matrix that represents a translated figure.

matriz de traslación Matriz que representa una figura trasladada.

transverse axis (442) The segment of length $2a$ whose endpoints are the vertices of a hyperbola.

eje transversal El segmento de longitud $2a$ cuyos extremos son los vértices de una hipérbola.

trigonometric equation (799) An equation containing at least one trigonometric function that is true for some but not all values of the variable.

ecuación trigonométrica Ecuación que contiene por lo menos una función trigonométrica y que sólo se cumple para algunos valores de la variable.

trigonometric functions (701, 717) For any angle, with measure α, a point $P(x, y)$ on its terminal side, $r = \sqrt{x^2 + y^2}$, the trigonometric functions of α are as follows.

$\sin \alpha = \frac{y}{r}$ $\cos \alpha = \frac{x}{r}$ $\tan \alpha = \frac{y}{x}$

$\csc \alpha = \frac{r}{y}$ $\sec \alpha = \frac{r}{x}$ $\cot \alpha = \frac{x}{y}$

funciones trigonométricas Para cualquier ángulo, de medida α, un punto $P(x, y)$ en su lado terminal, $r = \sqrt{x^2 + y^2}$, las funciones trigonométricas de α son las siguientes.

$\text{sen } \alpha = \frac{y}{r}$ $\cos \alpha = \frac{x}{r}$ $\tan \alpha = \frac{y}{x}$

$\csc \alpha = \frac{r}{y}$ $\sec \alpha = \frac{r}{x}$ $\cot \alpha = \frac{x}{y}$

trigonometric identity (777) An equation involving a trigonometric function that is true for all values of the variable.

identidad trigonométrica Ecuación que involucra una o más funciones trigonométricas y que se cumple para todos los valores de la variable.

trigonometry (701) The study of the relationships between the angles and sides of a right triangle.

trigonometría Estudio de las relaciones entre los lados y ángulos de un triángulo rectángulo.

trinomial (229) A polynomial with three unlike terms.

trinomio Polinomio con tres términos diferentes.

U

unbiased sample (682) A sample in which every possible sample has an equal chance of being selected.

muestra no sesgada Muestra en que cualquier muestra posible tiene la misma posibilidad de seleccionarse.

unbounded (130) A system of inequalities that forms a region that is open.

no acotado Sistema de desigualdades que forma una región abierta.

union (41) The graph of a compound inequality containing *or*.

unión Gráfica de una desigualdad compuesta que contiene la palabra *o*.

unit circle (710) A circle of radius 1 unit whose center is at the origin of a coordinate system.

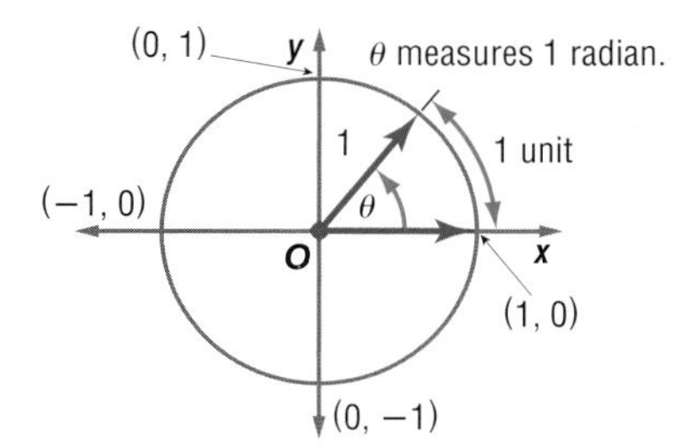

círculo unitario Círculo de radio 1 cuyo centro es el origen de un sistema de coordenadas.

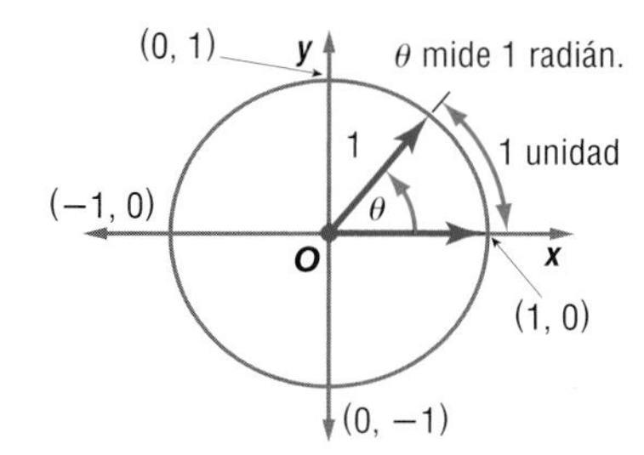

V

variables (7) Symbols, usually letters, used to represent unknown quantities.

variables Símbolos, por lo general letras, que se usan para representar cantidades desconocidas.

variance (665) The mean of the squares of the deviations from the arithmetic mean.

varianza Media de los cuadrados de las desviaciones de la media aritmética.

vertex (287, 442) **1.** The point at which the axis of symmetry intersects a parabola. **2.** The point on each branch nearest the center of a hyperbola.

vértice **1.** Punto en el que el eje de simetría interseca una parábola. **2.** El punto en cada rama más cercano al centro de una hipérbola.

vertex form (322) A quadratic function in the form $y = a(x - h)^2 + k$, where (h, k) is the vertex of the parabola and $x = h$ is its axis of symmetry.

forma de vértice Función cuadrática de la forma $y = a(x - h)^2 + k$, donde (h, k) es el vértice de la parábola y $x = h$ es su eje de simetría.

vertex matrix (175) A matrix used to represent the coordinates of the vertices of a polygon.

matriz de vértice Matriz que se usa para escribir las coordenadas de los vértices de un polígono.

vertical asymptote (485) If the related rational expression of a function is written in simplest form and is undefined for $x = a$, then $x = a$ is a vertical asymptote.

asíntota vertical Si la expresión racional que corresponde a una función racional se reduce y está no definida en $x = a$, entonces $x = a$ es una asíntota vertical.

vertical line test (57) If no vertical line intersects a graph in more than one point, then the graph represents a function.

prueba de la recta vertical Si ninguna recta vertical interseca una gráfica en más de un punto, entonces la gráfica representa una función.

vertices (129) The maximum or minimum value that a linear function has for the points in a feasible region.

vértices El valor máximo o mínimo que una función lineal tiene para los puntos en una región viable.

X

***x*-intercept** (65) The x-coordinate of the point at which a graph crosses the x-axis.

intersección *x* La coordenada x del punto o puntos en que una gráfica interseca o cruza el eje x.

Y

***y*-intercept** (65) The y-coordinate of the point at which a graph crosses the y-axis.

intersección *y* La coordenada y del punto o puntos en que una gráfica interseca o cruza el eje y.

Z

zeros (294) The x-intercepts of the graph of a quadratic equation; the points for which $f(x) = 0$.

ceros Las intersecciones x de la gráfica de una ecuación cuadrática; los puntos x para los que $f(x) = 0$.

zero matrix (155) A matrix in which every element is zero.

matriz nula Matriz cuyos elementos son todos igual a cero.

Selected Answers

Chapter 1 Solving Equations and Inequalities

Page 5 Chapter 1 Getting Started

1. 19.84 **3.** −17.51 **5.** $-\frac{5}{12}$ **7.** $-2\frac{1}{6}$ **9.** 0.48 **11.** 1.1 **13.** $-2\frac{2}{3}$ **15.** $8\frac{4}{5}$ **17.** 8 **19.** 49 **21.** 0.64 **23.** $\frac{4}{9}$ **25.** false **27.** true **29.** false **31.** true

Pages 8–10 Lesson 1-1

1. First, find the sum of c and d. Divide this sum by e. Multiply the quotient by b. Finally, add a. **3.** b; The sum of the cost of adult and children tickets should be subtracted from 50. Therefore parentheses need to be inserted around this sum to insure that this addition is done before subtraction. **5.** 6 **7.** 1 **9.** 119 **11.** −23 **13.** $432 **15.** $1162.50 **17.** 3 **19.** 25 **21.** −34 **23.** 5 **25.** −31 **27.** 14 **29.** −3 **31.** 162 **33.** 2.56 **35.** $25\frac{1}{3}$ **37.** 31.25 drops per min **39.** 2 **41.** −4.2 **43.** −4 **45.** 1.4 **47.** −8 **49.** $2\frac{1}{6}$ **51.** −16 **53.** $8266.03 **55.** Sample answer: $4 - 4 + 4 \div 4 = 1$; $4 \div 4 + 4 \div 4 = 2$; $(4 + 4 + 4) \div 4 = 3$; $4 \times (4 - 4) + 4 = 4$; $(4 \times 4 + 4) \div 4 = 5$; $(4 + 4) \div 4 + 4 = 6$; $44 \div 4 - 4 = 7$; $(4 + 4) \times (4 \div 4) = 8$; $4 + 4 + 4 \div 4 = 9$; $(44 - 4) \div 4 = 10$ **57.** C **59.** 3 **61.** 10 **63.** −2 **65.** $\frac{2}{3}$

Pages 14–17 Lesson 1-2

1a. Sample answer: 2 **1b.** Sample answer: 5 **1c.** Sample answer: −11 **1d.** Sample answer: 1.3 **1e.** Sample answer: $\sqrt{2}$ **1f.** Sample answer: −1.3 **3.** 0; Zero does not have a multiplicative inverse since $\frac{1}{0}$ is undefined. **5.** N, W, Z, Q, R **7.** Multiplicative Inverse **9.** Additive Identity **11.** $-\frac{1}{3}, 3$ **13.** $-2x + 4y$ **15.** $3c + 18d$ **17.** 1.5(10 + 15 + 12 + 8 + 19 + 22 + 31) or 1.5(10) + 1.5(15) + 1.5(12) + 1.5(8) + 1.5(19) + 1.5(22) + 1.5(31) **19.** W, Z, Q, R **21.** N, W, Z, Q, R **23.** I, R **25.** N, W, Z, Q, R **27.** Q, R; 2.4, 2.49, $2.\overline{49}$, $2.4\overline{9}$, $2.\overline{9}$ **29.** Associative (×) **31.** Associative (+) **33.** Multiplicative Inverse **35.** Multiplicative Identity **37.** $-m$; Additive Inverse **39.** 1 **41.** $\sqrt{2}$ units **43.** 10; $-\frac{1}{10}$ **45.** 0.125; −8 **47.** $-\frac{4}{3}, \frac{3}{4}$ **49.** $3a - 2b$ **51.** $40x - 7y$ **53.** $-12r + 4t$ **55.** $-3.4m + 1.8n$ **57.** $-8 + 9y$ **59.** true **61.** false; 6 **63.** 6.5(4.5 + 4.25 + 5.25 + 6.5 + 5) or 6.5(4.5) + 6.5(4.25) + (6.5)5.25 + 6.5(6.5) + 6.5(5)

65. $3\left(2\frac{1}{4}\right) + 2\left(1\frac{1}{8}\right)$

$= 3\left(2 + \frac{1}{4}\right) + 2\left(1 + \frac{1}{8}\right)$	*Definition of a mixed number*
$= 3(2) + 3\left(\frac{1}{4}\right) + 2(1) + 2\left(\frac{1}{8}\right)$	*Distributive Property*
$= 6 + \frac{3}{4} + 2 + \frac{1}{4}$	*Multiply.*
$= 6 + 2 + \frac{3}{4} + \frac{1}{4}$	*Commutative Property (+)*
$= 8 + \frac{3}{4} + \frac{1}{4}$	*Add.*
$= 8 + \left(\frac{3}{4} + \frac{1}{4}\right)$	*Associative Property (+)*
$= 8 + 1$ or 9	*Add.*

67. 4700 ft^2 **69.** $62.15

71. Answers should include the following.

- Instead of doubling each coupon value and then adding these values together, the Distributive Property could be applied allowing you to add the coupon values first and then double the sum.
- If a store had a 25% off sale on all merchandise, the Distributive Property could be used to calculate these savings. For example, the savings on a $15 shirt, $40 pair of jeans, and $25 pair of slacks could be calculated as 0.25(15) + 0.25(40) + 0.25(25) or as 0.25(15 + 40 + 25) using the Distributive Property.

73. C **75.** False; 0 − 1 = −1, which is not a whole number. **77.** False; $2 \div 3 = \frac{2}{3}$, which is not a whole number. **79.** 6 **81.** −2.75 **83.** −11 **85.** −4.3

Page 17 Practice Quiz 1

1. 14 **3.** 6 **5.** 2 amperes **7.** N, W, Z, Q, R **9.** $-\frac{6}{7}, \frac{7}{6}$

Pages 24–27 Lesson 1-3

1. Sample answer: $2x = -14$
3. Jamal; his method can be confirmed by solving the equation using an alternative method.

$$C = \frac{5}{9}(F - 32)$$
$$C = \frac{5}{9}F - \frac{5}{9}(32)$$
$$C + \frac{5}{9}(32) = \frac{5}{9}F$$
$$\frac{9}{5}\left[C + \frac{5}{9}(32)\right] = F$$
$$\frac{9}{5}C + 32 = F$$

5. $2n - n^3$ **7.** Sample answer: 5 plus 3 times the square of a number is twice that number. **9.** Addition (=) **11.** 14 **13.** −4.8 **15.** 16 **17.** $p = \frac{I}{rt}$ **19.** $5 + 3n$ **21.** $n^2 - 4$ **23.** $5(9 + n)$ **25.** $\left(\frac{n}{4}\right)^2$ **27.** $2\pi rh + 2\pi r^2$ **29.** Sample answer: 5 less than a number is 12. **31.** Sample answer: A number squared is equal to 4 times the number. **33.** Sample answer: A number divided by 4 is equal to twice the sum of that number and 1. **35.** Substitution (=) **37.** Transitive (=) **39.** Symmetric (=) **41.** 7 **43.** 3.2 **45.** $\frac{1}{12}$ **47.** −8 **49.** −7 **51.** 1 **53.** $\frac{1}{4}$ **55.** $-\frac{55}{2}$ **57.** $\frac{d}{t} = r$ **59.** $\frac{3V}{\pi r^2} = h$ **61.** $b = \frac{x(c-3)}{a} + 2$ **63.** n = number of games; $2(1.50) + n(2.50) = 16.75$; 5 **65.** x = cost of gasoline per mile; $972 + 114 + 105 + 7600x = 1837$; 8.5¢/mi **67.** a = Chun-Wei's age; $a + (2a + 8) + (2a + 8 + 3) = 94$; Chun-Wei: 15 yrs old, mother: 38 yrs old, father: 41 yrs old **69.** n = number of lamps broken; $12(125) - 45n = 1365$; 3 lamps **71.** 15.1 mi/month **73.** The Central Pacific had to lay their track through the Rocky Mountains, while the Union Pacific mainly built track over flat prairie. **75.** the product of 3 and the difference of a number and 5 added to the product of four times the number and the sum of the number and 1 **77.** B **79.** $-6x + 8y + 4z$ **81.** 6.6 **83.** 105 cm^2 **85.** 3 **87.** $-\frac{1}{4}$ **89.** $-5 + 6y$

Pages 30–32 Lesson 1-4

1. $|a| = -a$ when a is a negative number and the negative of a negative number is positive. **3.** Always; since the opposite of 0 is still 0, this equation has only one case, $ax + b = 0$. The solution is $\frac{-b}{a}$. **5.** 8 **7.** −17 **9.** $\{-18, -12\}$ **11.** $\{-32, 36\}$ **13.** $\{8\}$ **15.** least: 158°F; greatest: 162°F **17.** 15 **19.** 0 **21.** 3 **23.** −4 **25.** −9.4 **27.** 55 **29.** $\{8, 42\}$ **31.** $\{-45, 21\}$ **33.** $\{-2, 16\}$ **35.** $\left\{\frac{3}{2}\right\}$ **37.** $\left\{2, \frac{9}{2}\right\}$ **39.** $\varnothing$ **41.** $\{-5, 11\}$ **43.** $\left\{-\frac{11}{3}, -3\right\}$ **45.** $\{8\}$ **47.** $|x - 200| = 5$; maximum: 205°F; minimum: 195°F **49.** $|x - 13| = 5$; maximum: 18 km, minimum: 8 km **51.** sometimes; true only if $c \geq 0$ **53.** B **55.** $|x + 1| + 2 = x + 4$; $|x + 1| + 2 = -(x + 4)$ **57.** $\{-1.5\}$ **59.** $2(n - 11)$ **61.** $\frac{16}{3}$ **63.** 14 **65.** Distributive **67.** Additive Identity **69.** true **71.** false; 1.2 **73.** 364 ft^2 **75.** 8 **77.** $\frac{2}{3}$ **79.** $-\frac{3}{4}$

Pages 37–39 Lesson 1-5

1. Dividing by a number is the same as multiplying by its inverse. **3.** Sample answer: $x + 2 < x + 1$

5. $\left\{x \mid x \leq \frac{5}{3}\right\}$ or $\left(-\infty, \frac{5}{3}\right]$

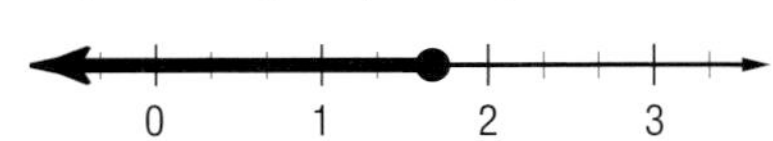

7. $\{y \mid y > 6\}$ or $(6, +\infty)$

9. $\{p \mid p > 15\}$ or $(15, +\infty)$

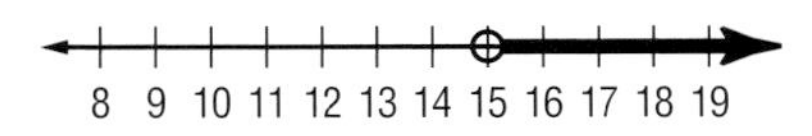

11. all real numbers or $(-\infty, +\infty)$

13. $2n - 3 \leq 5$; $n \leq 4$

15. $\{n \mid n \geq -11\}$ or $[-11, +\infty)$

17. $\{x \mid x < 7\}$ or $(-\infty, 7)$

19. $\{g \mid g \leq 27\}$ or $(-\infty, 27]$

21. $\{k \mid k \geq -3.5\}$ or $[-3.5, +\infty)$

23. $\{m \mid m > -4\}$ or $(-4, +\infty)$

25. $\{t \mid t \leq 0\}$ or $(-\infty, 0]$

27. $\{n \mid n \geq 1.75\}$ or $[1.75, +\infty)$

29. $\{x \mid x < -279\}$ or $(-\infty, -279)$

31. $\{d \mid d \geq -5\}$ or $[-5, +\infty)$

33. $\{g \mid g < 2\}$ or $(-\infty, 2)$

35. $\left\{y \mid y < \frac{1}{5}\right\}$ or $\left(-\infty, \frac{1}{5}\right)$

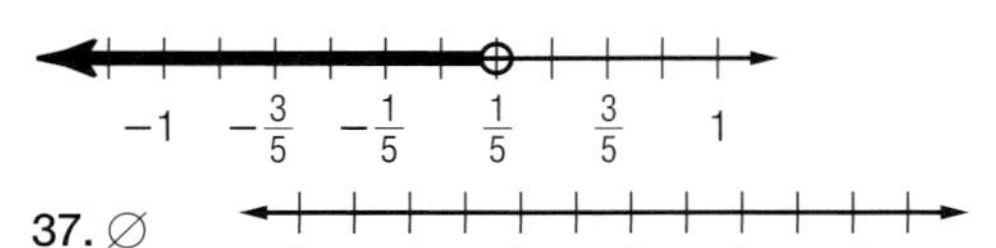

37. $\varnothing$

39. at least 25 h **41.** $n + 8 > 2$; $n > -6$ **43.** $\frac{1}{2}n - 7 \geq 5$; $n \geq 24$ **45.** $2(n + 5) \leq 3n + 11$; $n \geq -1$ **47.** $2(7m) \geq 17$; $m \geq \frac{17}{14}$; at least 2 child care staff members

49. $n \geq 34.97$; She must sell at least 35 cars. **51.** $s \geq 91$; Ahmik must score at least 91 on her next test to have an A test average. **53.** Answers should include the following.

- $150 < 400$
- Let n equal the number of minutes used. Write an expression representing the cost of Plan 1 and for Plan 2 for n minutes. The cost for Plan 1 would include a monthly access fee of \$35 plus 40¢ for each minute over 150 minutes or $35 + 0.4(n - 150)$. The cost for Plan 2 for 400 minutes or less would be \$55. To find where Plan 2 would cost less than Plan 1 solve $55 < 35 + 0.4(n - 150)$ for n. The solution set is $\{n \mid n > 200\}$, which means that for more than 200 minutes of calls, Plan 2 is cheaper.

55. D **57.** $x \geq -2$ **59.** $\{-14, 20\}$ **61.** $\varnothing$ **63.** N, W, Z, Q, R **65.** I, R **67.** $\{-7, 7\}$ **69.** $\left\{4, -\frac{4}{5}\right\}$ **71.** $\{-11, -1\}$

Page 39 Practice Quiz 2

1. 0.5 **3.** 14 **5.** $\left\{m \mid m > \frac{4}{9}\right\}$ or $\left(\frac{4}{9}, +\infty\right)$

Pages 43–46 Lesson 1-6

1. $5 \leq c \leq 15$ **3.** Sabrina; an absolute value inequality of the form $|a| > b$ should be rewritten as an *or* compound inequality, $a > b$ or $a < -b$.

5. $|n| > 3$

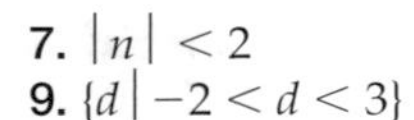

7. $|n| < 2$
9. $\{d \mid -2 < d < 3\}$

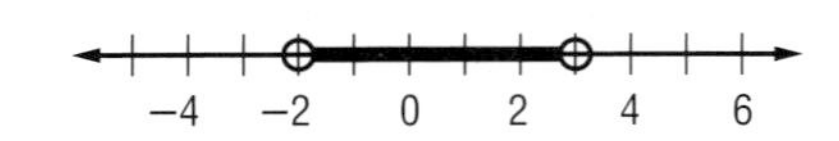

11. $\{g \mid -13 \leq g \leq 5\}$

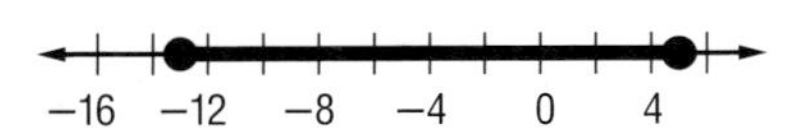

13. all real numbers

15. $|n| \geq 5$

17. $|n| < 4$

19. $|n| > 8$

21. $|n| > 1$ 23. $|n| \geq 1.5$ 25. $|n + 1| > 1$
27. $\{p \mid p \leq 2 \text{ or } p \geq 8\}$

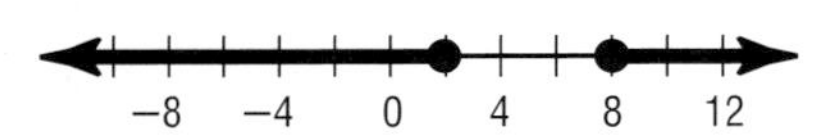

29. $\{x \mid -2 < x < 4\}$

31. $\{f \mid -7 < f < -5\}$

33. $\{g \mid -9 \leq g \leq 9\}$

35. $\varnothing$

37. $\{b \mid b > 10 \text{ or } b < -2\}$

39. $\left\{w \mid -\frac{7}{3} \leq w \leq 1\right\}$

41. all real numbers

43. $\left\{n \mid n = \frac{7}{2}\right\}$

45. $6.8 < x < 7.4$ 47. $45 \leq s \leq 55$
49. 108 in. $< L + D \leq$ 130 in.
51. $a + b > c, a + c > b, b + c > a$

53a.

53b.

53c.

53d. $3 < |x + 2| \leq 8$ can be rewritten as $|x + 2| > 3$ and $|x + 2| \leq 8$. The solution of $|x + 2| > 3$ is $x > 1$ or $x < -5$. The solution of $|x + 2| \leq 8$ is $-10 \leq x \leq 6$. Therefore, the union of these two sets is ($x > 1$ or $x < -5$) and ($-10 \leq x \leq 6$). The union of the graph of $x > 1$ or $x < -5$ and the graph of $-10 \leq x \leq 6$ is shown below. From this we can see that solution can be rewritten as ($-10 \leq x < -5$) or ($1 < x \leq 6$).

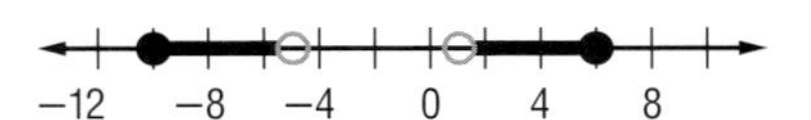

55. $x > -5$ or $x < -6$

57.

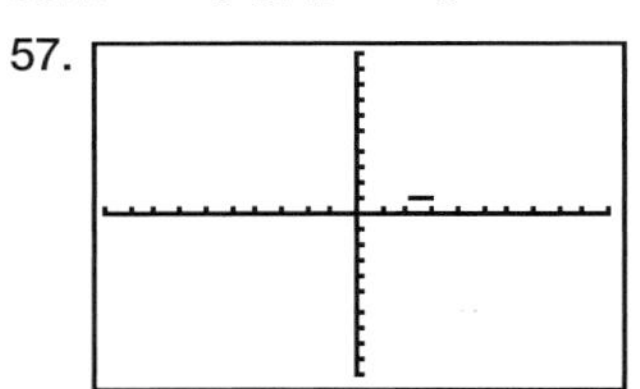

59. $(5x + 2 \geq 3)$ or $(5x + 2 \leq -3)$; $\{x \mid x \geq 0.2 \text{ or } x \leq -1\}$
61. $\{d \mid d \geq -6\}$ or $[-6, +\infty)$

63. $\{n \mid n < -1\}$ or $(-\infty, -1)$

65. $\{-10, 16\}$ 67. $\varnothing$ 69. Symmetric (=) 71. $3a + 7b$ 73. 2 75. -7

Pages 47–50 Chapter 1 Study Guide and Review

1. compound inequality 3. Commutative (×)
5. Reflexive (=) 7. Multiplicative Inverse 9. absolute value
11. 22 13. -49 15. -23 17. 37.5 19. Q, R 21. I, R
23. $-5a + 24b$ 25. -14 27. -13 29. -4 31. $x = \frac{C - By}{A}$
33. $p = \frac{A}{1 + rt}$ 35. $\{6, -18\}$ 37. $\{6\}$ 39. $\left\{-\frac{3}{2}, -1\right\}$
41. $\{x \mid x \geq 5\}$ or $[5, +\infty)$

43. $\{a \mid a > 2\}$ or $(2, +\infty)$

45. $\{x \mid x > -1.8\}$ or $(-1.8, +\infty)$

47. $\left\{y \mid \frac{5}{3} < y \leq 5\right\}$

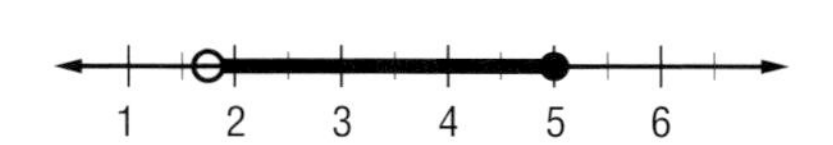

49. $\{y \mid -9 \le y \le 18\}$

51. $\left\{b \mid b < -4 \text{ or } b > -\frac{10}{3}\right\}$

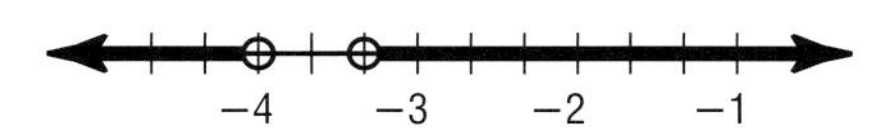

Chapter 2 Linear Relations and Functions

Page 55 Chapter 2 Getting Started

1. $(-3, 3)$ **3.** $(-3, -1)$ **5.** $(0, -4)$ **7.** -2 **9.** 9 **11.** 2
13. $x + 1$ **15.** $2x + 6$ **17.** $\frac{1}{2}x + 2$ **19.** 3 **21.** 15 **23.** 2.5

Pages 60–62 Lesson 2-1

1. Sample answer: $\{(-4, 3), (-2, 3), (1, 5), (-2, 1)\}$

3. Molly; to find $g(2a)$, replace x with $2a$. Teisha found $2g(a)$, not $g(2a)$. **5.** yes

7. D = {7}, R = {−1, 2, 5, 8}, no

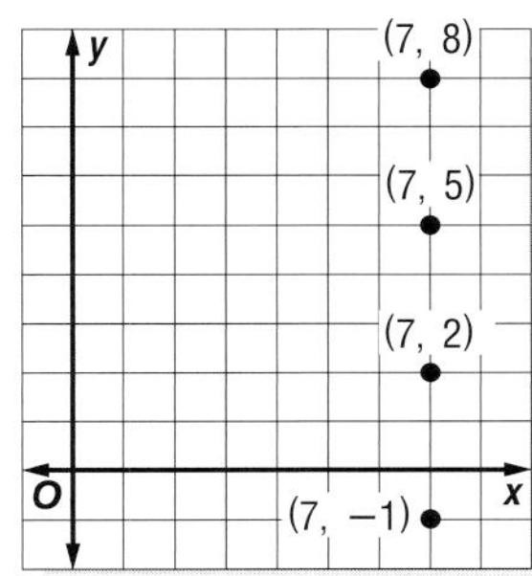

9. D = all reals, R = all reals, yes

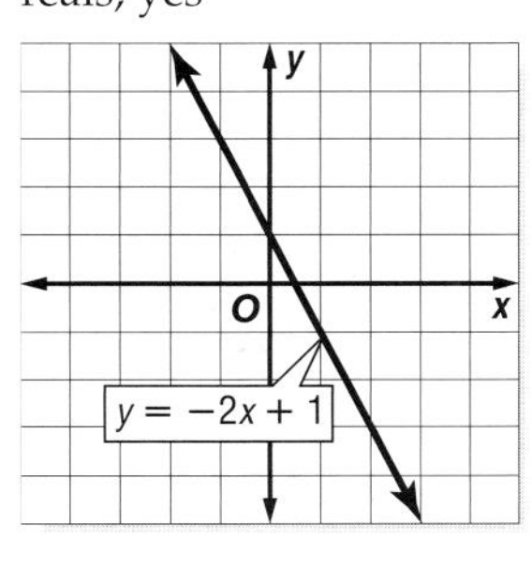

11. 10 **13.** D = {70, 72, 88}, R = {95, 97, 105, 114}

15.

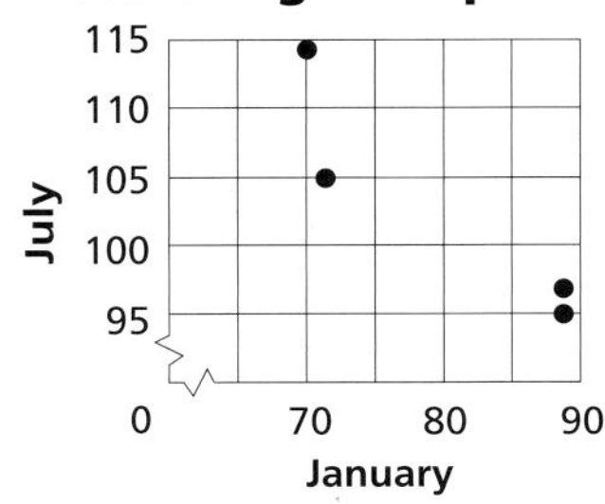

17. yes **19.** no **21.** yes

23. D = {−3, 1, 2}, R = {0, 1, 5}; yes

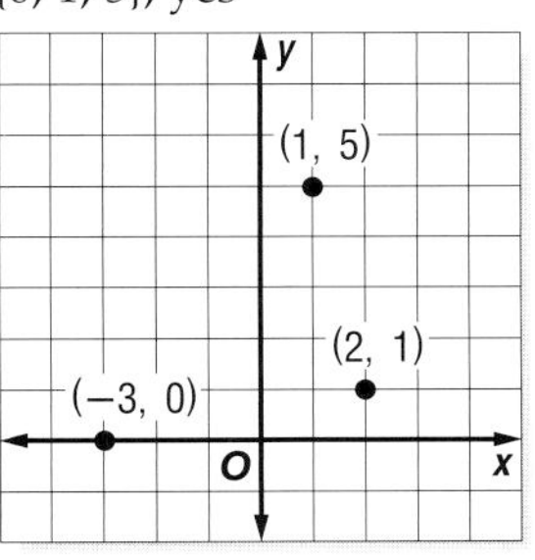

25. D = {−2, 3}, R = {5, 7, 8}; no

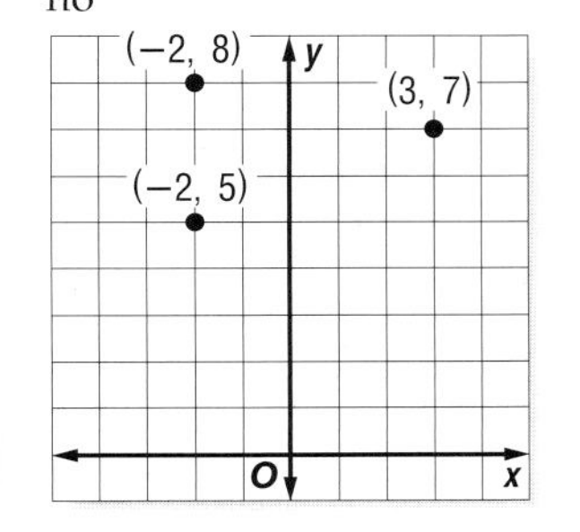

27. D = {−3.6, 0, 1.4, 2}, R = {−3, −1.1, 2, 8}; yes

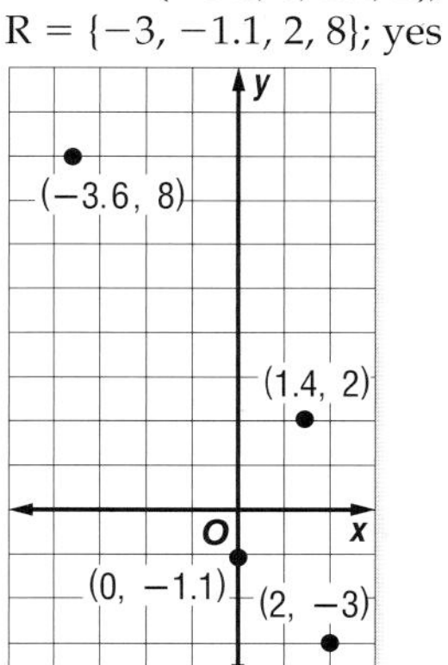

29. D = all reals, R = all reals; yes

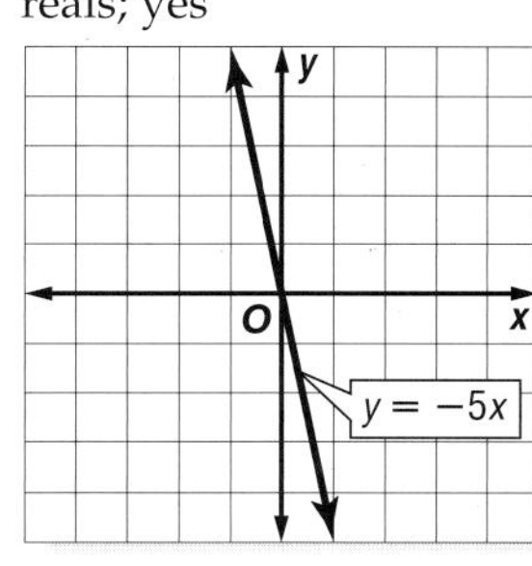

31. D = all reals, R = all reals; yes

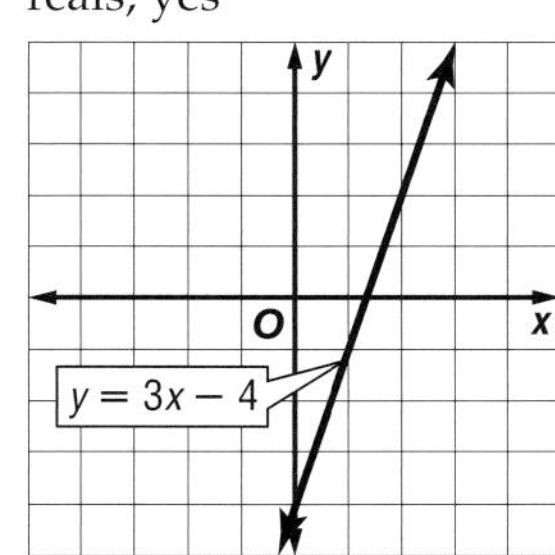

33. D = all reals, R = $\{y \mid y \ge 0\}$; yes

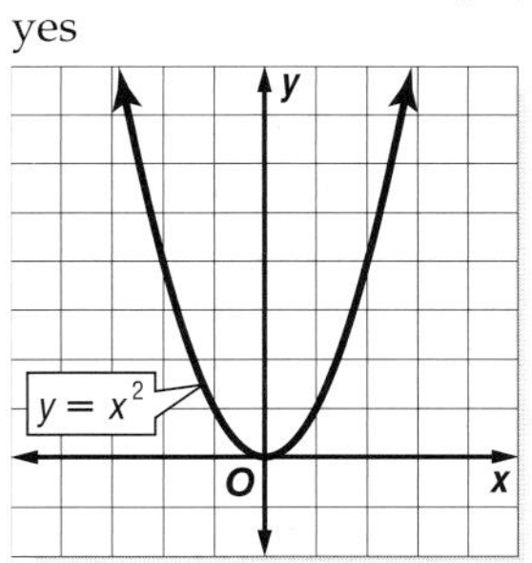

35.

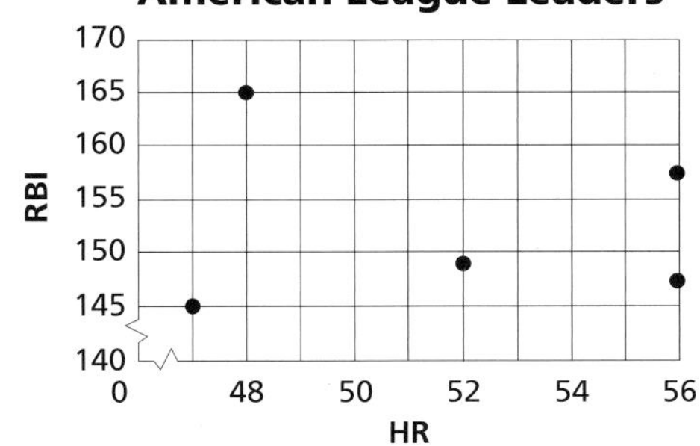

37. No; the domain value 56 is paired with two different range values.

39.

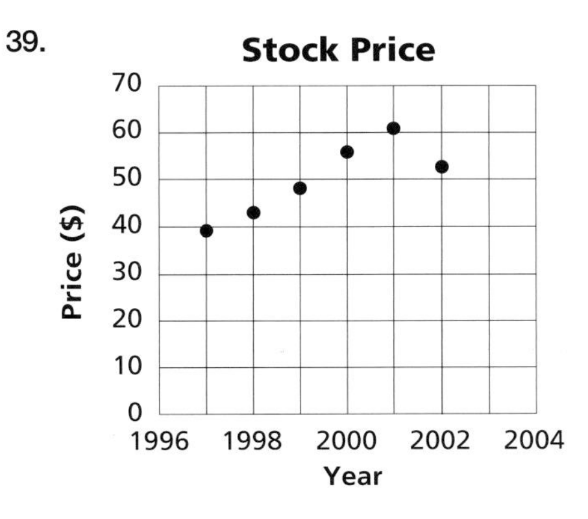

41. Yes; each domain value is paired with only one range value.

43.

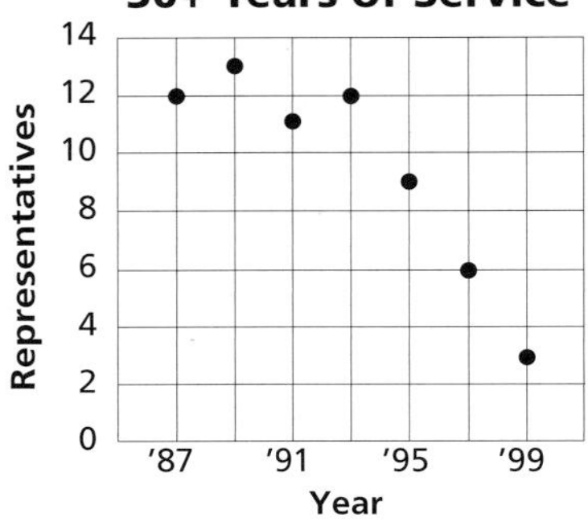

45. Yes; no; each domain value is paired with only one range value so the relation is a function, but the range value 12 is paired with two domain values so the function is not one-to-one. **47.** 6 **49.** −3 **51.** $25n^2 - 5n$ **53.** 11 **55.** $f(x) = 4x - 3$ **57.** B **59.** discrete **61.** discrete **63.** $\{y \mid -8 < y < 6\}$ **65.** $\{x \mid x < 5.1\}$ **67.** $29.82 **69.** $31a + 10b$ **71.** 2 **73.** 15

Pages 65–67 Lesson 2-2

1. The function can be written as $f(x) = \frac{1}{2}x + 1$, so it is of the form $f(x) = mx + b$, where $m = \frac{1}{2}$ and $b = 1$. **3.** Sample answer: $x + y = 2$ **5.** yes **7.** $2x - 5y = 3$; 2, −5, 3

9. $-\frac{5}{3}$, −5

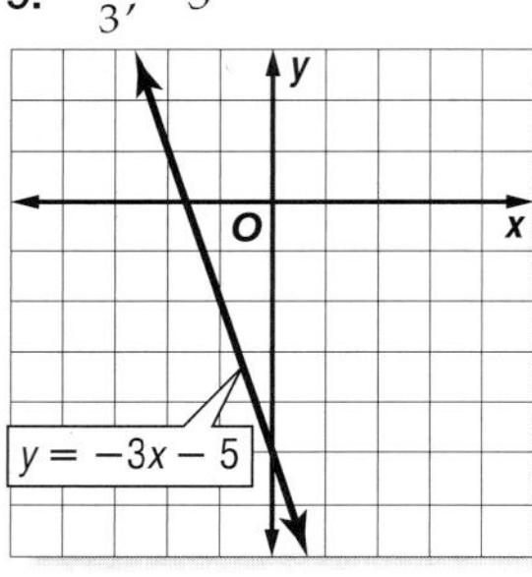

11. 2, 3

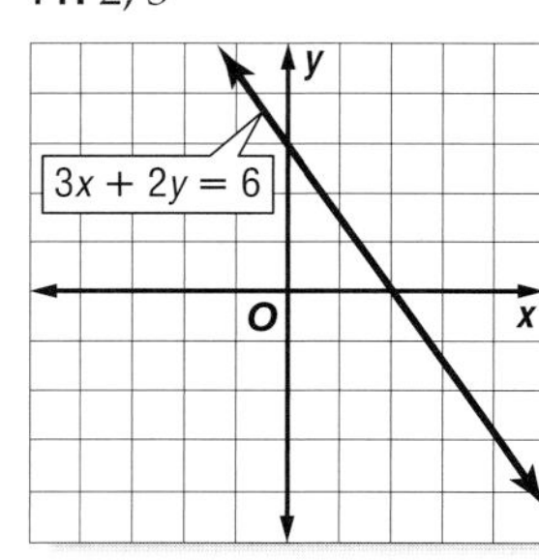

13. $177.62 **15.** yes **17.** No; y is inside a square root. **19.** No; x appears in a denominator. **21.** No; x has an exponent other than 1. **23.** $x^2 + 5y = 0$ **25.** 7200 m **27.** $3x + y = 4$; 3, 1, 4 **29.** $x - 4y = -5$; 1, −4, −5 **31.** $2x - y = 5$; 2, −1, 5 **33.** $x + y = 12$; 1, 1, 12 **35.** $x = 6$; 1, 0, 6 **37.** $25x + 2y = 9$; 25, 2, 9

39. 3, 5

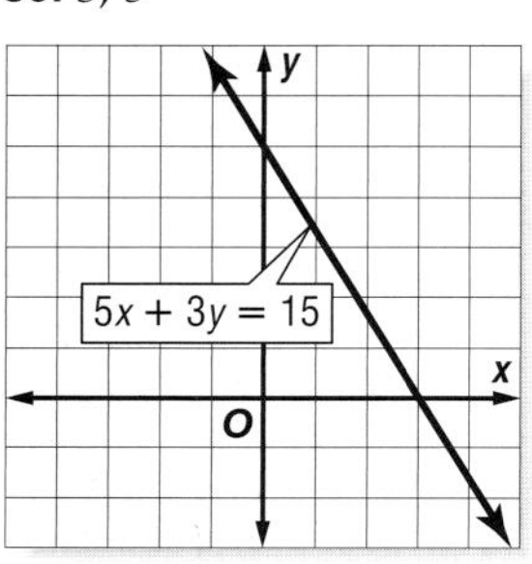

41. $\frac{10}{3}$, $-\frac{5}{2}$

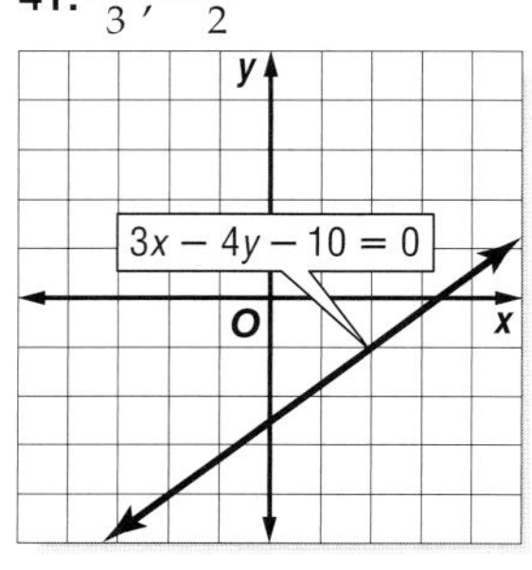

43. 0, 0

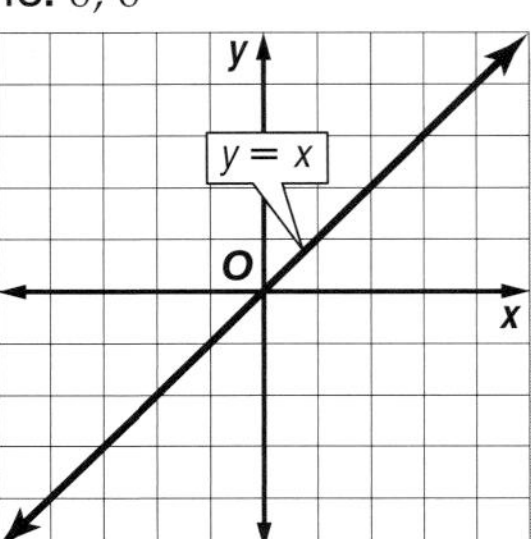

45. none, −2

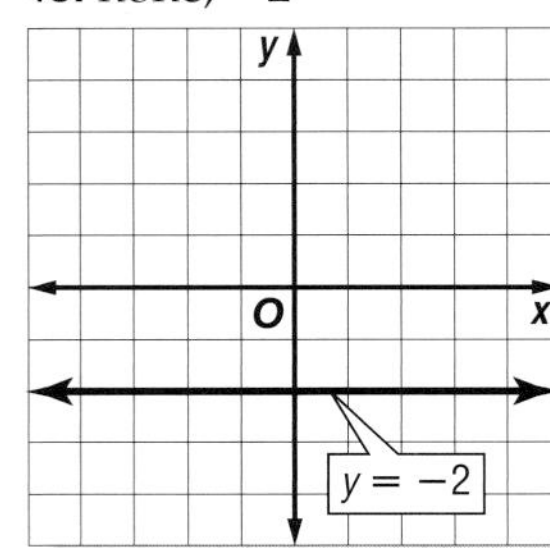

47. 8, none

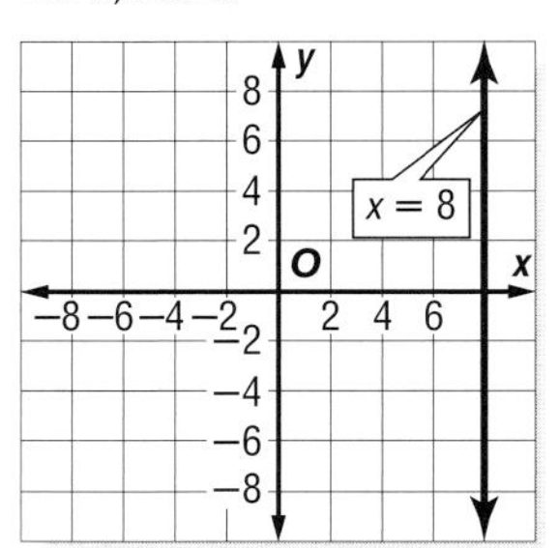

49. $\frac{1}{4}$, −1

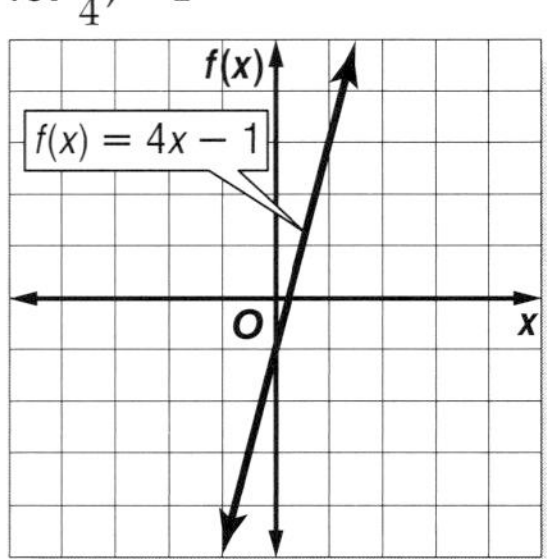

51. y; x + y = 5; x; O; x + y = 0; x + y = −5

The lines are parallel but have different y-intercepts.

53. 90°C

55.

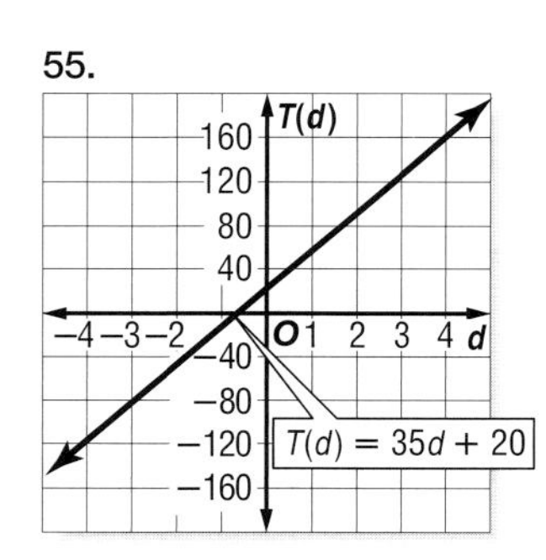

57.

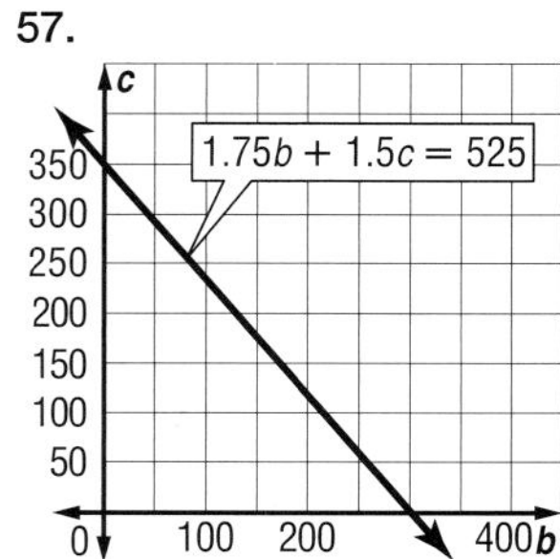

59. no **61.** A linear equation can be used to relate the amounts of time that a student spends on each of two subjects if the total amount of time is fixed. Answers should include the following.

- x and y must be nonnegative because Lolita cannot spend a negative amount of time studying a subject.
- The intercepts represent Lolita spending all of her time on one subject. The x-intercept represents her spending all of her time on math, and the y-intercept represents her spending all of her time on chemistry.

63. B **65.** D = {0, 1, 2}, R = {−1, 0, 2, 3}; no

67. $\{x \mid x < -6 \text{ or } x > -2\}$ **69.** $3s + 14$ **71.** $\frac{1}{3}$ **73.** 2 **75.** −5 **77.** 0.4

Pages 71–74 Lesson 2-3

1. Sample answer: $y = 1$ **3.** Luisa; Mark did not subtract in a consistent manner when using the slope formula. If $y_2 = 5$ and $y_1 = 4$, then x_2 must be −1 and x_1 must be 2, not vice versa. **5.** $-\frac{1}{2}$

7.

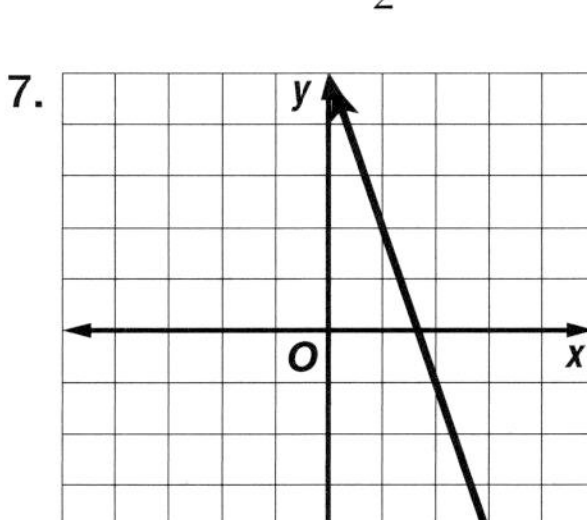

9.

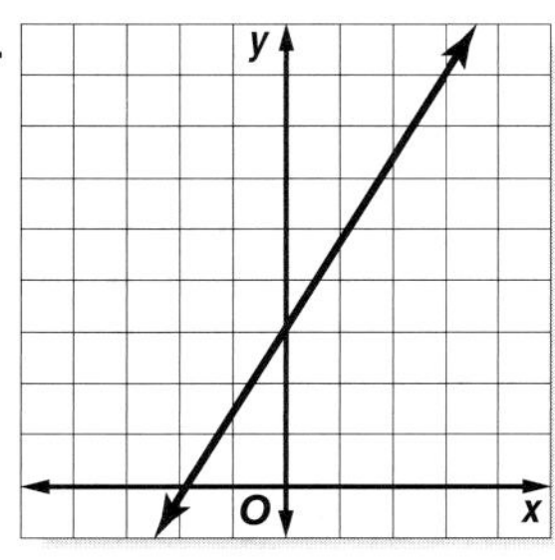

11.

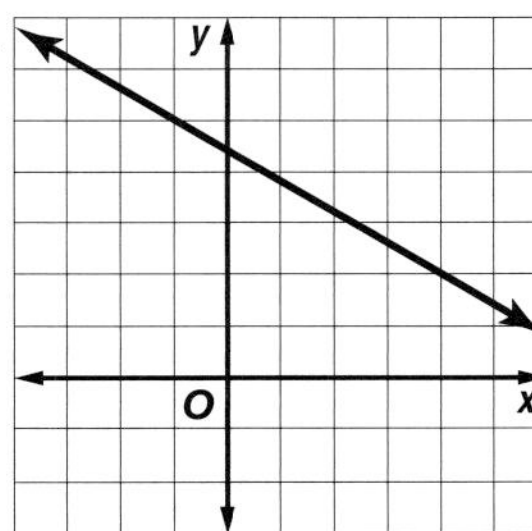

13. 1.25°/hr **15.** $-\frac{5}{2}$ **17.** $\frac{3}{5}$ **19.** 0 **21.** 8 **23.** −4 **25.** undefined **27.** 1 **29.** about 0.6

31.

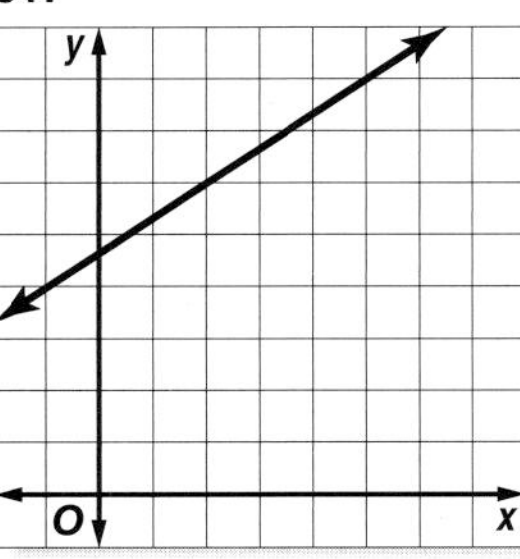

33.

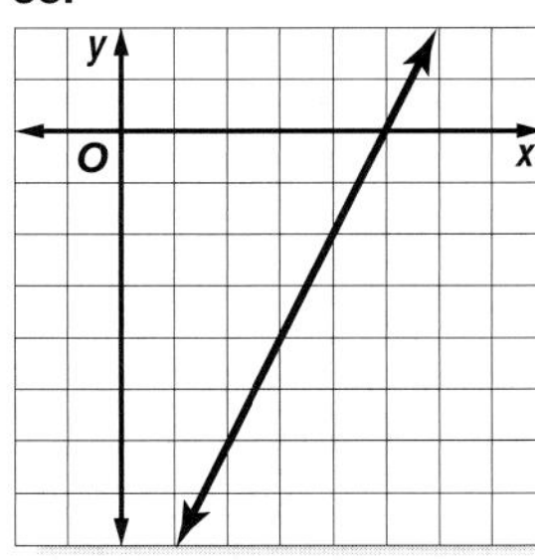

35.

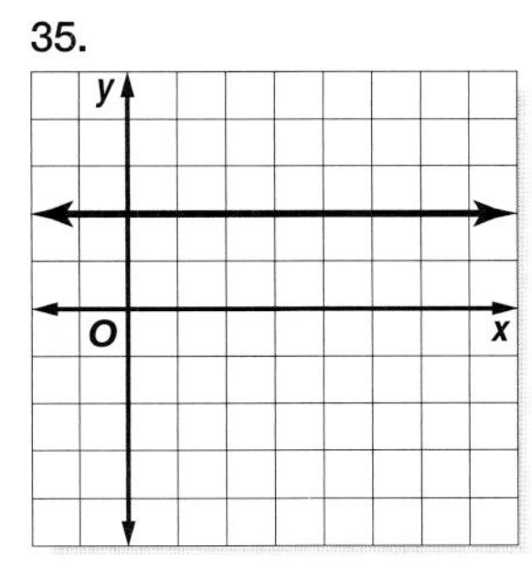

37. about 68 million per year **39.** The number of cassette tapes shipped has been decreasing. **41.** 45 mph

43.

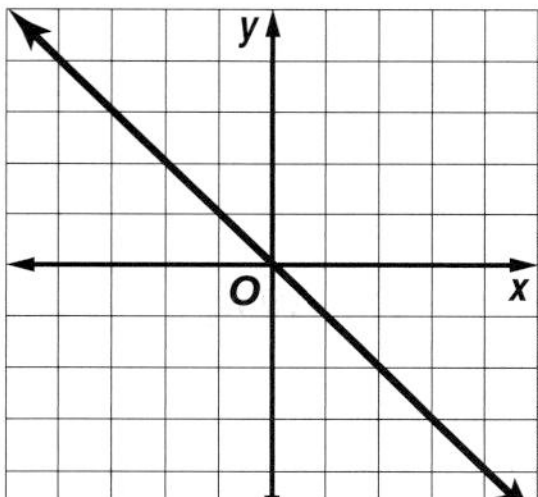

45.

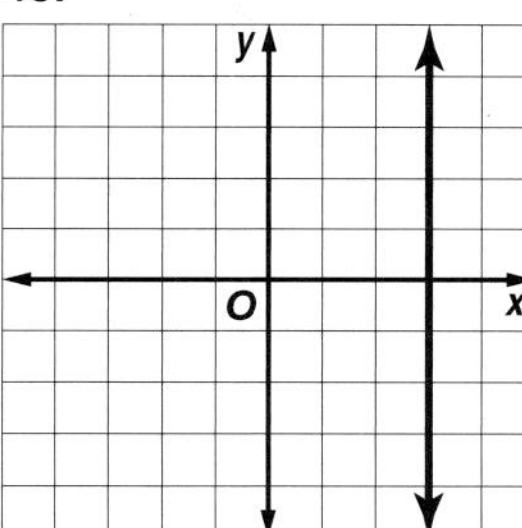

47.

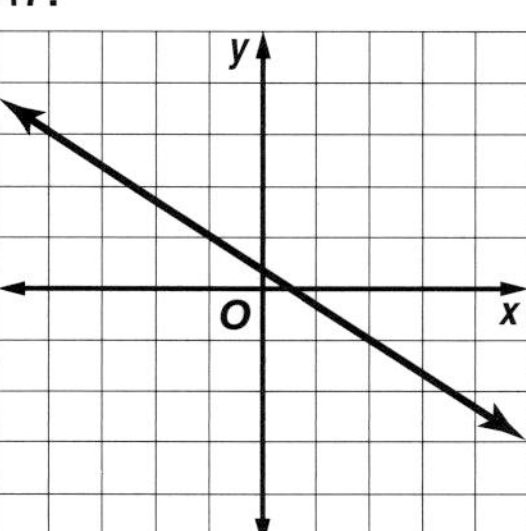

49.

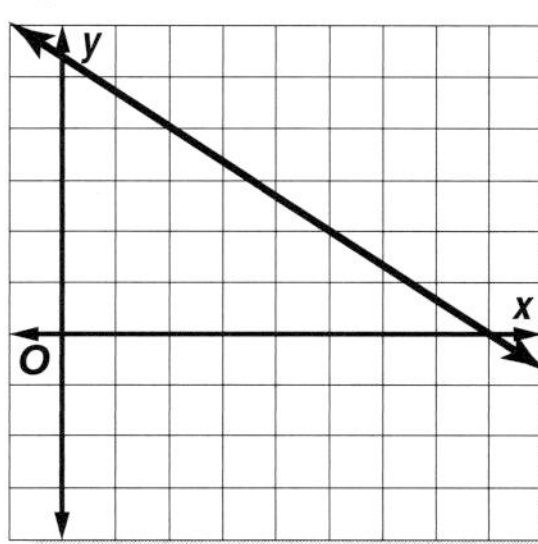

51. Yes; slopes show that adjacent sides are perpendicular.

53. The grade or steepness of a road can be interpreted mathematically as a slope. Answers should include the following.

- Think of the diagram at the beginning of the lesson as being in a coordinate plane. Then the rise is a change in y-coordinates and the horizontal distance is a change in x-coordinates. Thus, the grade is a slope expressed as a percent.
- 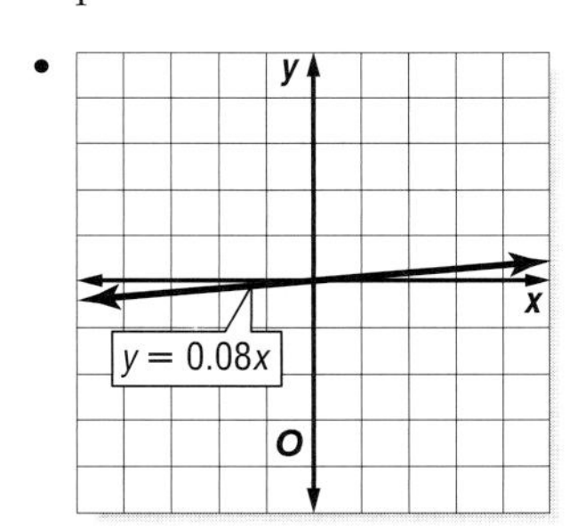

55. D **57.** The graphs have the same y-intercept. As the slopes become more negative, the lines get steeper.

59. $-2, \frac{8}{3}$

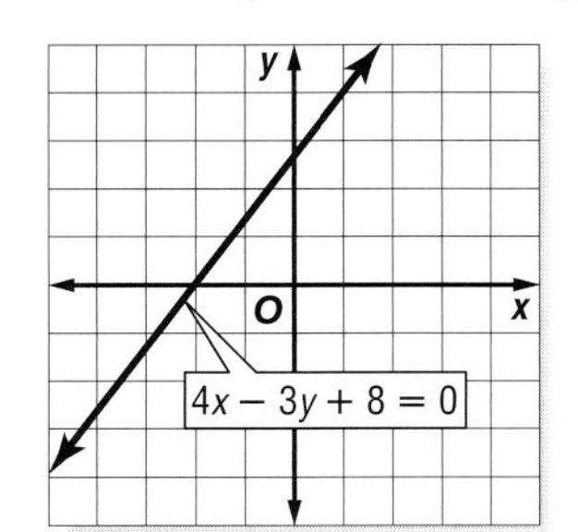

61. −7 **63.** $-\frac{5}{2}$ **65.** $\{x \mid -1 < x < 3\}$ **67.** at least 8 **69.** 9

71. $y = -4x + 2$ **73.** $y = \frac{5}{2}x - \frac{1}{2}$ **75.** $y = -\frac{2}{3}x + \frac{11}{3}$

Page 74 Practice Quiz 1

1. D = {−7, −3, 0, 2}, R = {−2, 1, 2, 4, 5} **3.** $6x + y = 4$

5. 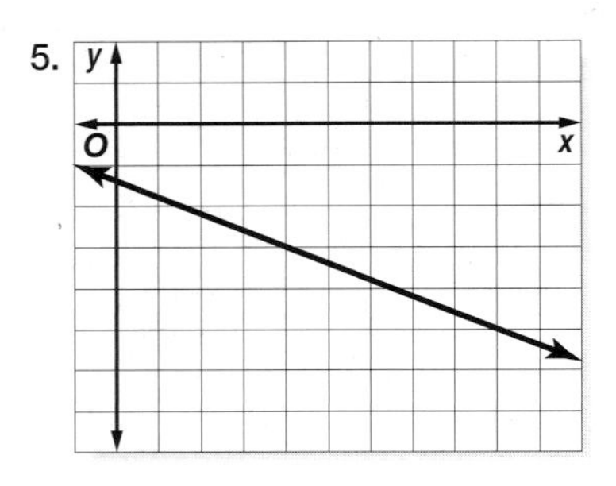

Pages 78–80 Lesson 2-4

1. Sample answer: $y = 3x + 2$ **3.** Solve the equation for y to get $y = \frac{3}{5}x - \frac{2}{5}$. The slope of this line is $\frac{3}{5}$. The slope of a parallel line is the same. **5.** $-\frac{3}{2}, 5$ **7.** $y = -\frac{3}{4}x + 2$ **9.** $y = -\frac{3}{5}x + \frac{16}{5}$ **11.** $y = \frac{5}{4}x + 7$ **13.** $-\frac{2}{3}, -4$ **15.** $\frac{1}{2}, -\frac{5}{2}$ **17.** undefined, none **19.** $y = 0.8x$ **21.** $y = -4$ **23.** $y = 3x - 6$ **25.** $y = -\frac{1}{2}x + \frac{7}{2}$ **27.** $y = -0.5x - 2$ **29.** $y = -\frac{4}{5}x + \frac{17}{5}$ **31.** $y = 0$ **33.** $y = x + 4$ **35.** $y = \frac{2}{3}x + \frac{10}{3}$ **37.** $y = -\frac{1}{15}x - \frac{23}{5}$ **39.** $y = 3x - 2$ **41.** $d = 180c - 360$ **43.** 540° **45.** 10 mi **47.** 68°F **49.** $y = 0.35x + 1.25$ **51.** $y = 2x + 4$ **53.** C **55.** $\frac{x}{\frac{5}{2}} - \frac{y}{5} = 1$ **57.** -2 **59.** 0 **61.** $\varnothing$ **63.** $\{r \mid r \geq 6\}$ **65.** 6.5 **67.** 5.85

Pages 83–86 Lesson 2-5

1. d **3.** Sample answer using (4, 130.0) and (6, 140.0): $y = 5x + 110$

5a. **Cable Television**

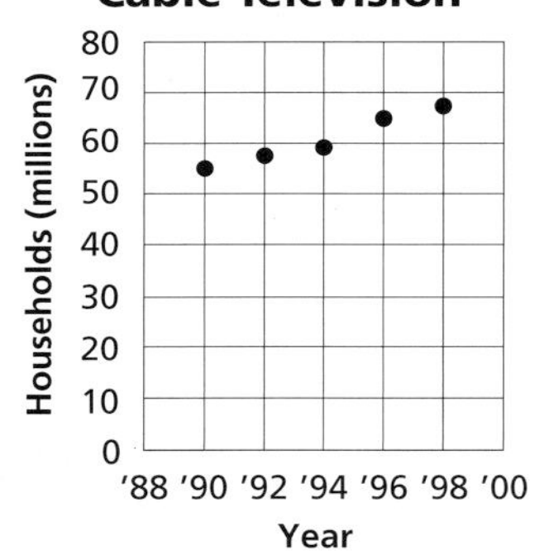

5b. Sample answer using (1992, 57) and (1998, 67): $y = 1.67x - 3269.64$ **5c.** Sample answer: about 87 million

7a. **2000–2001 Detroit Red Wings**

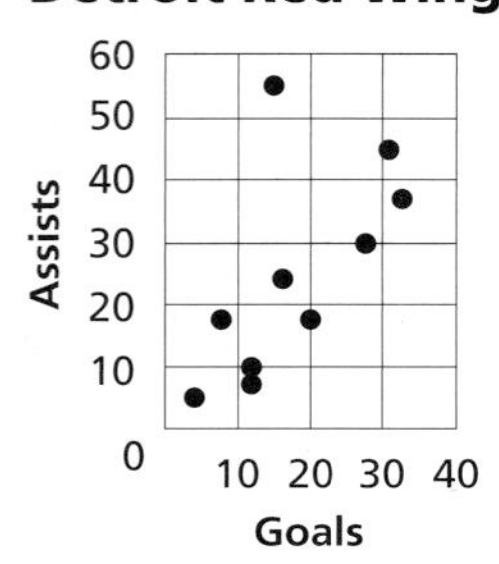

7b. Sample answer using (4, 5) and (32, 37): $y = 1.14x + 0.44$
7c. Sample answer: about 13

9a. **Broadway Play Revenue**

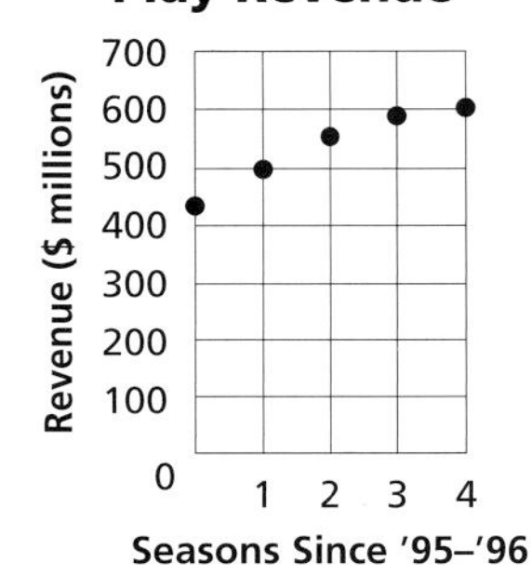

9b. Sample answer using (1, 499) and (3, 588): $y = 44.5x + 454.5$, where x is the number of seasons since 1995–1996 **9c.** Sample answer: about $1078 million or $1.1 billion **11.** Sample answer: $1091 **13.** Sample answer: Using the data for August and November, a prediction equation for Company 1 is $y = -0.86x + 25.13$, where x is the number of months since August. The negative slope suggests that the value of Company 1's stock is going down. Using the data for October and November, a prediction equation for Company 2 is $y = 0.38x + 31.3$, where x is the number of months since August. The positive slope suggests that the value of Company 2's stock is going up. Since the value of Company 1's stock appears to be going down, and the value of Company 2's stock appears to be going up, Della should buy Company 2.

15. **World Cities**

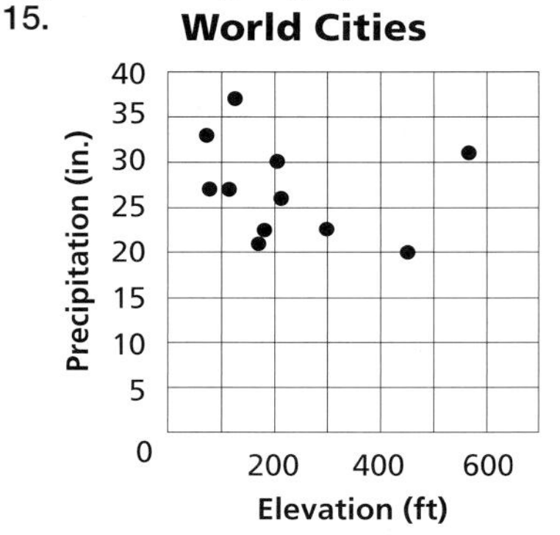

17. Sample answer: about 23 in. **19.** Sample answer: Using (1975, 62.5) and (1995, 81.7): 96.1% **23.** D **25.** 1988, 1993, 1998; 247, 360.5, 461 **27.** 354 **29.** $y = 21.4x - 42{,}294.03$ **31.** $y = 4x + 6$ **33.** 3 **35.** $\frac{29}{3}$ **37.** $\{x \mid x < -7 \text{ or } x > -1\}$ **39.** 11 **41.** $\frac{2}{3}$

Pages 92–95 Lesson 2-6

1. Sample answer: $[\![1.9]\!] = 1$ **3.** Sample answer: $f(x) = |x - 1|$ **5.** S

7. D = all reals, R = all integers

$g(x) = [\![2x]\!]$

9. D = all reals, R = all nonnegative reals

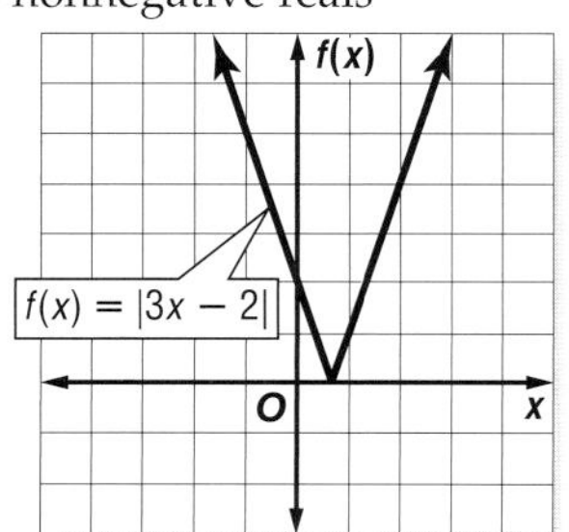

11. D = all reals, R = all reals

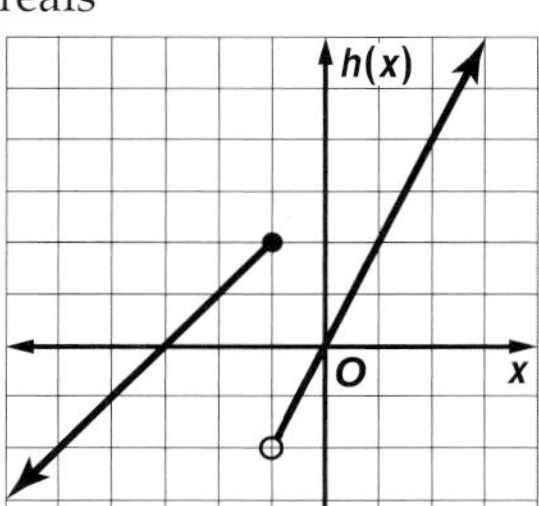

13.

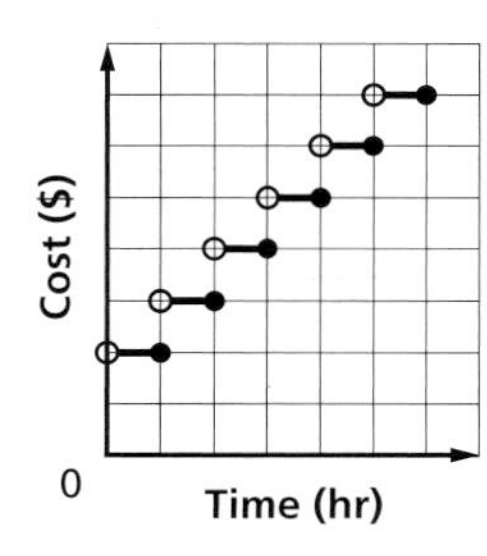

15. C **17.** S **19.** A **21.**

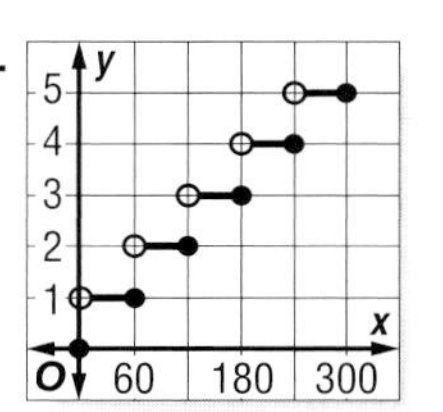
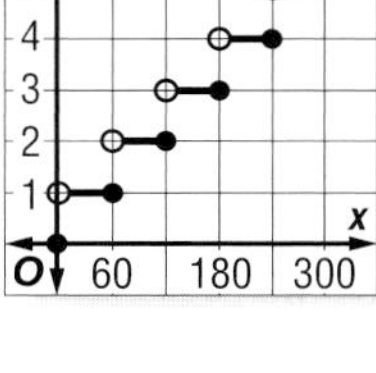

23. $1.00

25. D = all reals, R = all integers

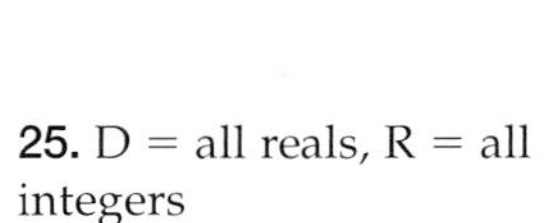

27. D = all reals, R = $\{3a \mid a$ is an integer.$\}$

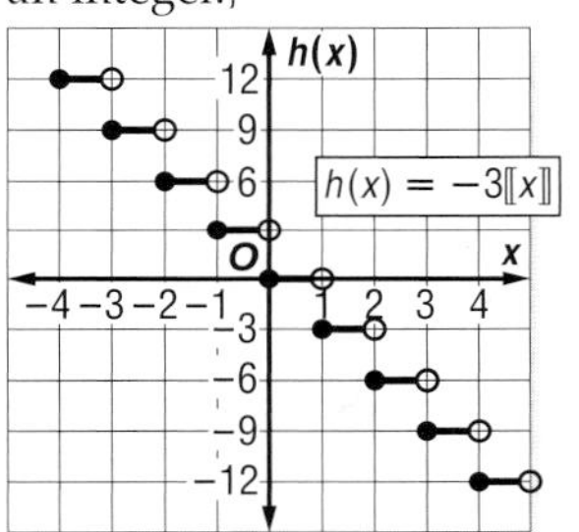

29. D = all reals, R = all integers

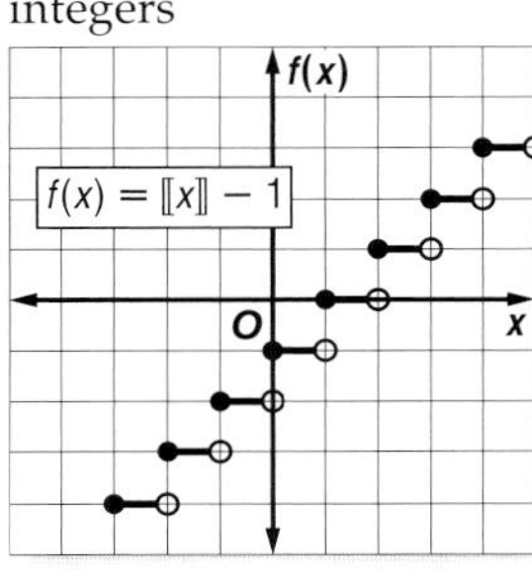

31. D = all reals, R = all nonnegative reals

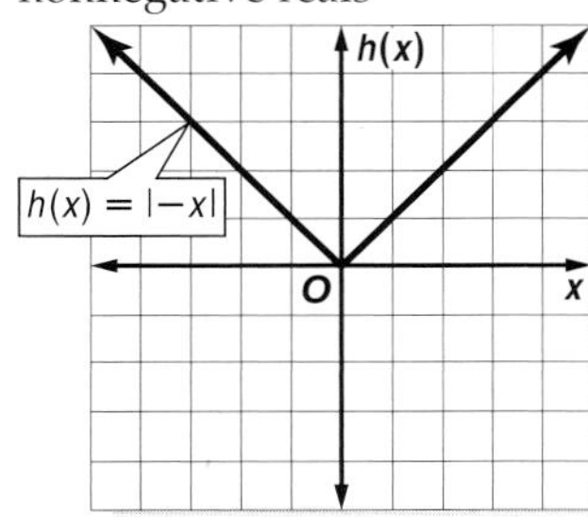

33. D = all reals, R = $\{y \mid y \geq -4\}$

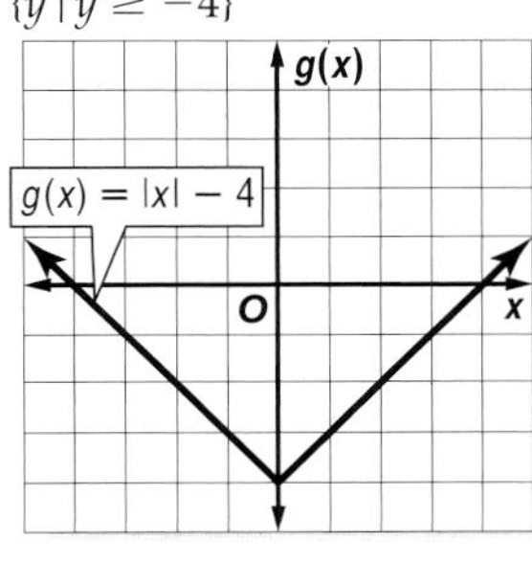

35. D = all reals, R = all nonnegative reals

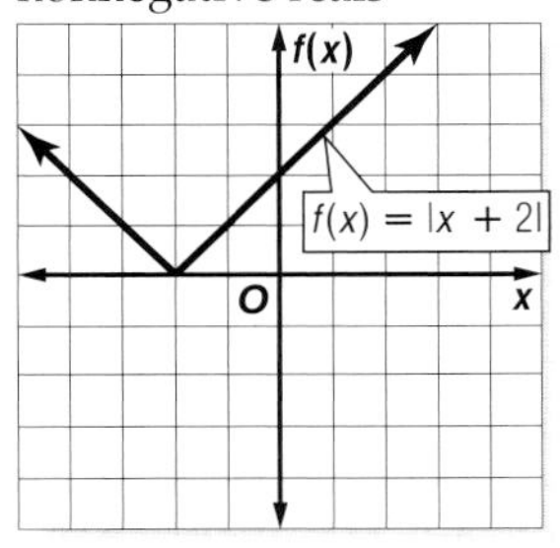

37. D = all reals, R = all nonnegative reals

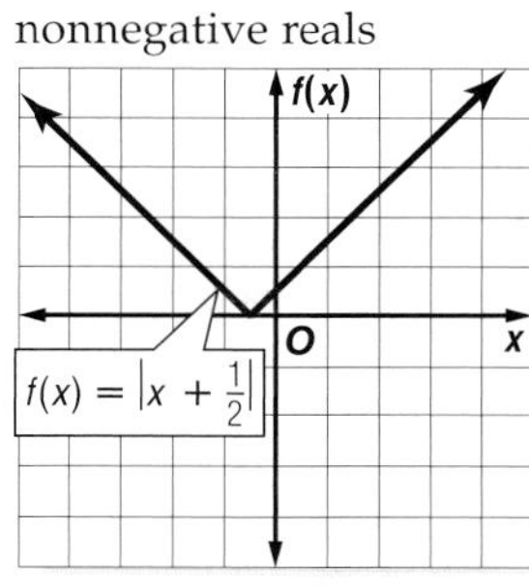

39. D = $\{x \mid x < -2$ or $x > 2\}$, R = $\{-1, 1\}$

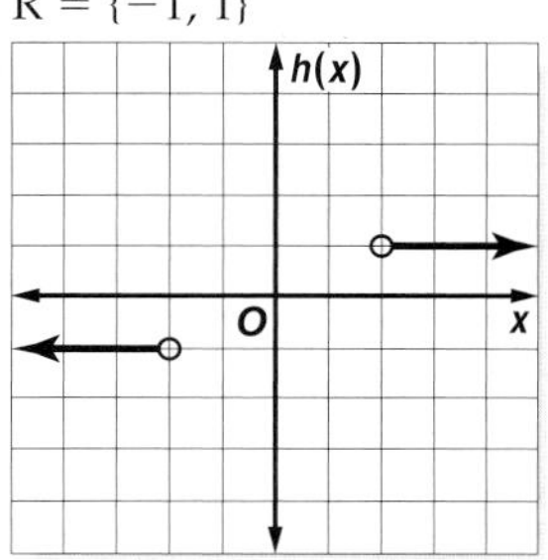

41. D = all reals, R = $\{y \mid y < 2\}$

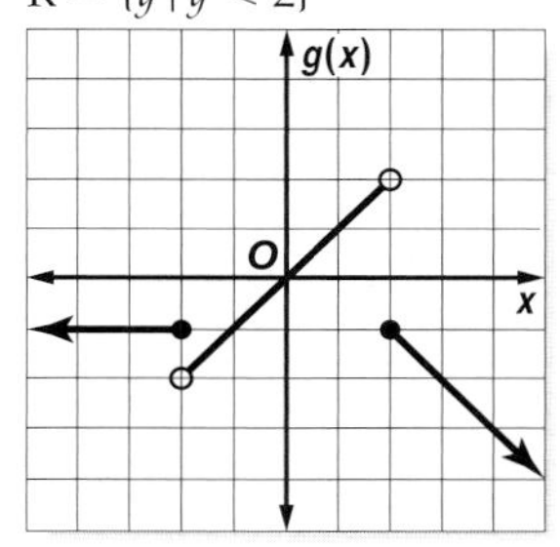

43. D = all reals, R = all whole numbers

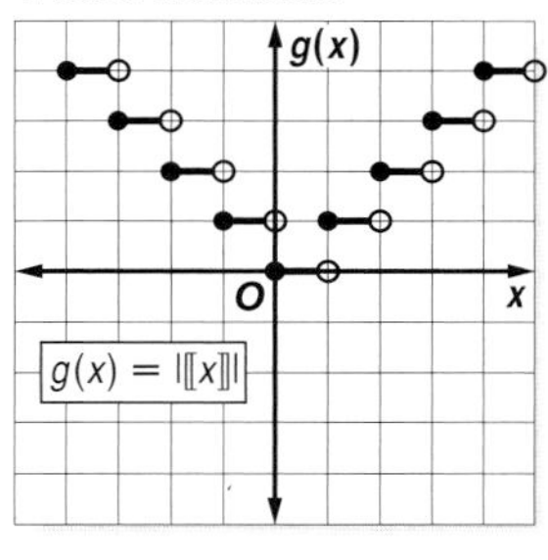

45. $f(x) = |x - 2|$

47.

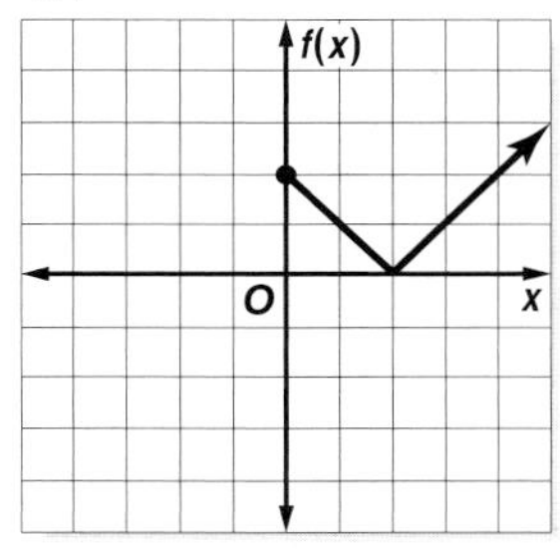

49.

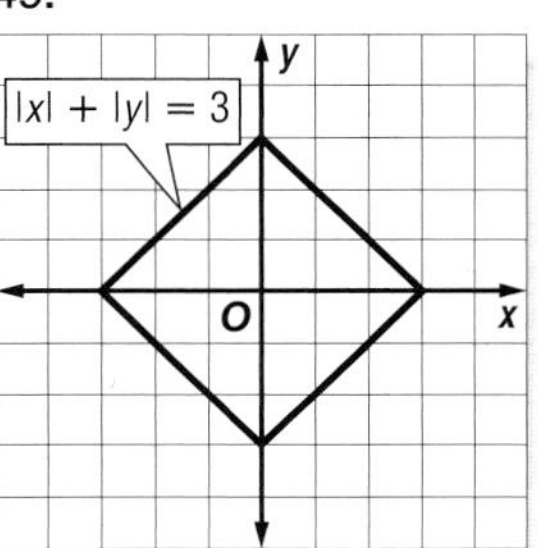

51. B **53.**

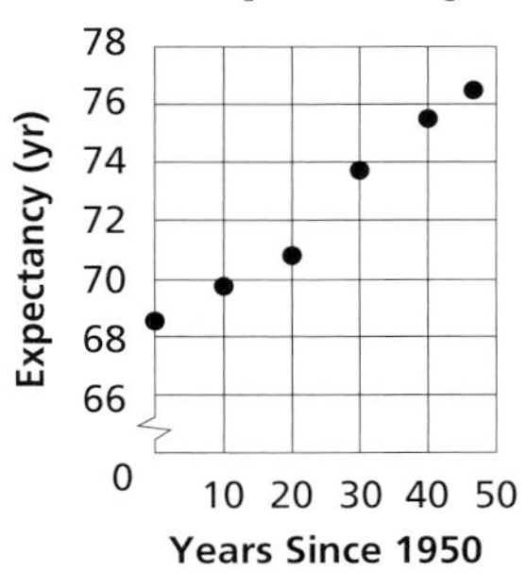

55. Sample answer: 78.7 yr **57.** $y = x - 2$

59. $\left\{y \mid y > \frac{5}{6}\right\}$ −3 −2 −1 0 1 2 3

61. no **63.** yes **65.** yes

Page 95 Practice Quiz 2

1. $y = -\frac{2}{3}x + \frac{11}{3}$ **3.** Sample answer using (66, 138) and (74, 178): $y = 5x - 192$ **5.** D = all reals, R = nonnegative reals

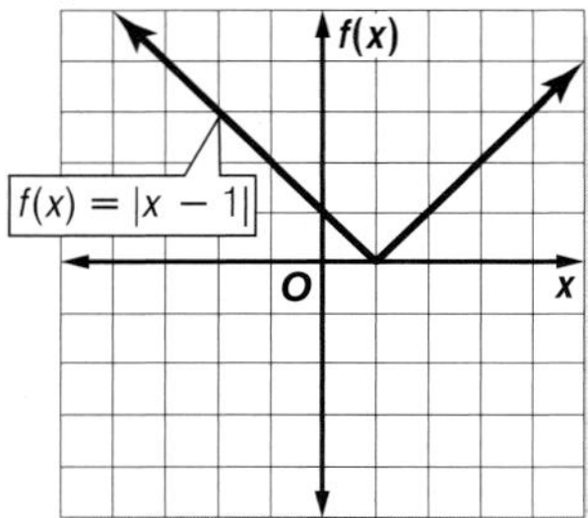

Pages 98–99 Lesson 2-7

1. $y \leq -3x + 4$ **3.** Sample answer: $y \geq |x|$

5.

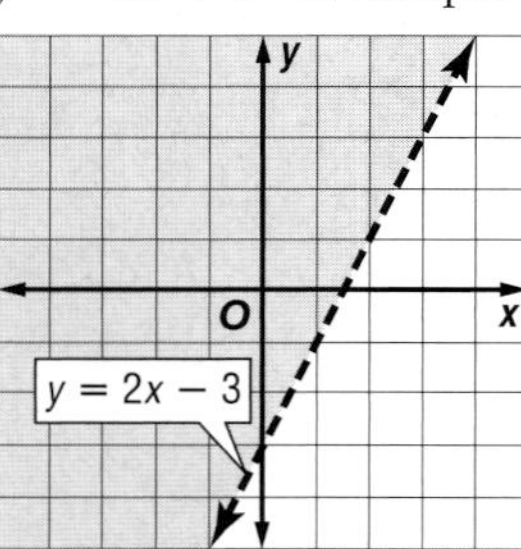

7.

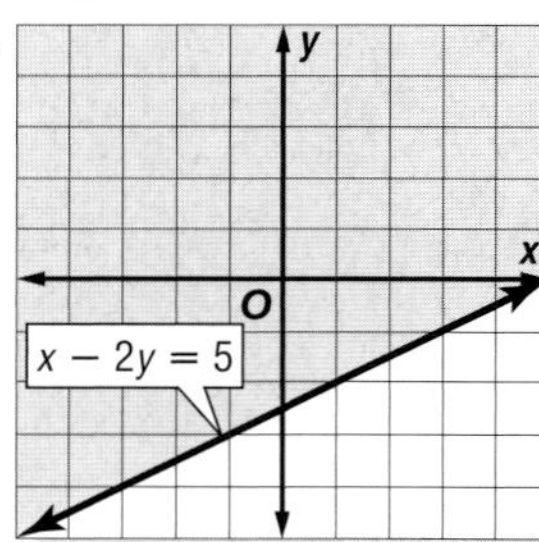

9.

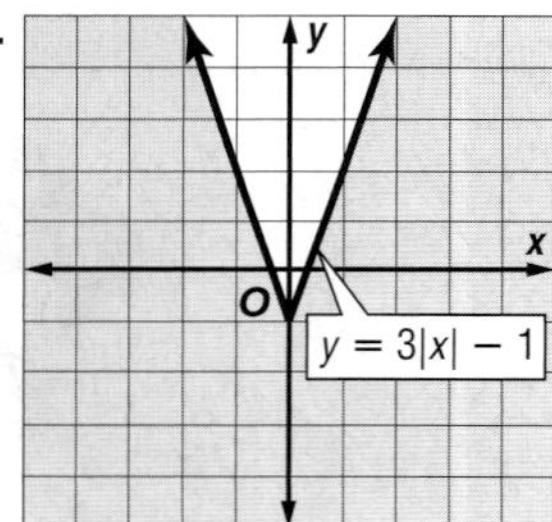

11.

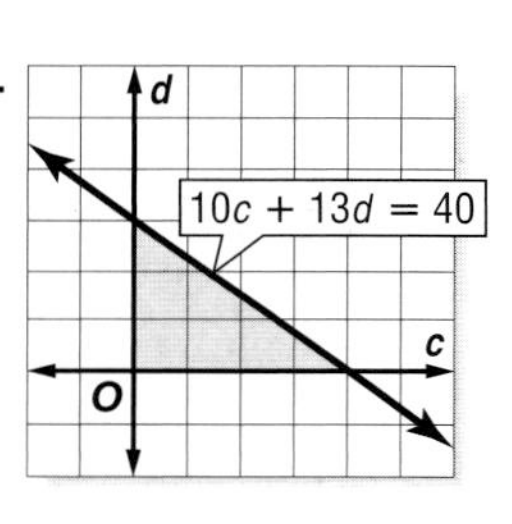

13.

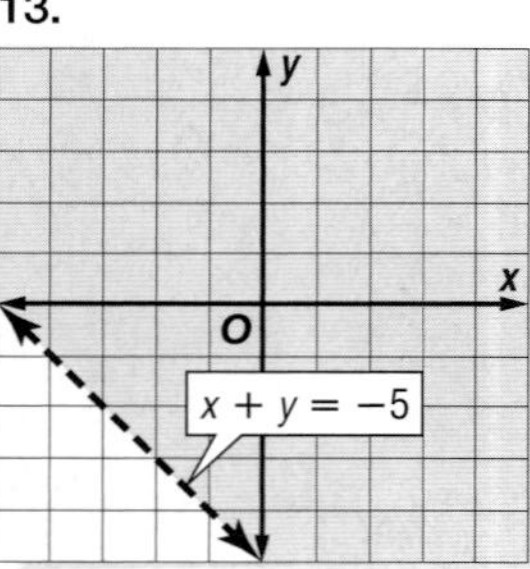

15.

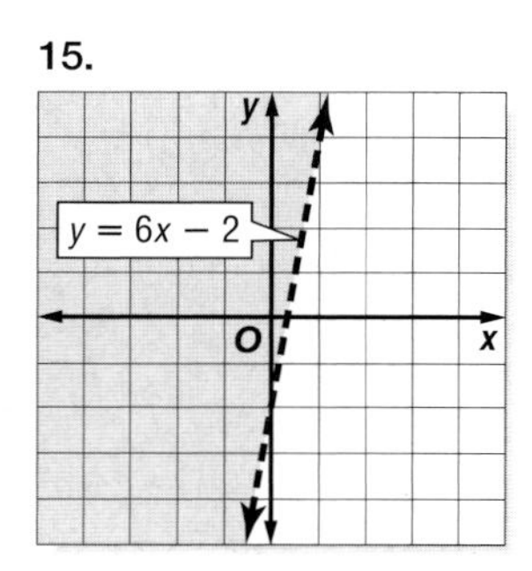

17.

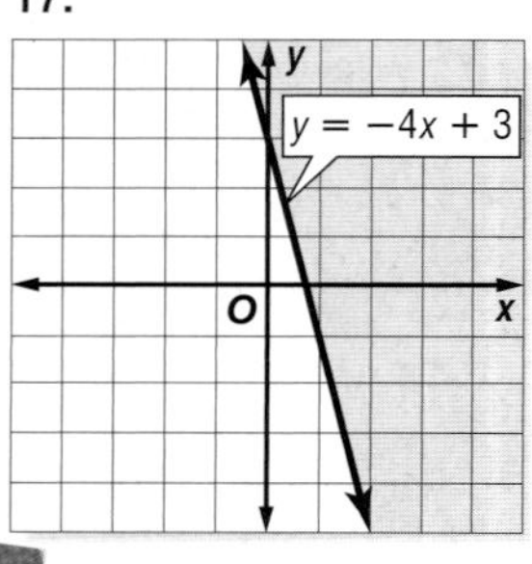

19.

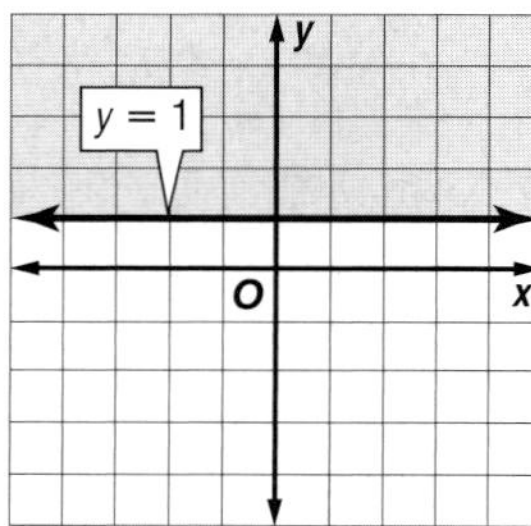
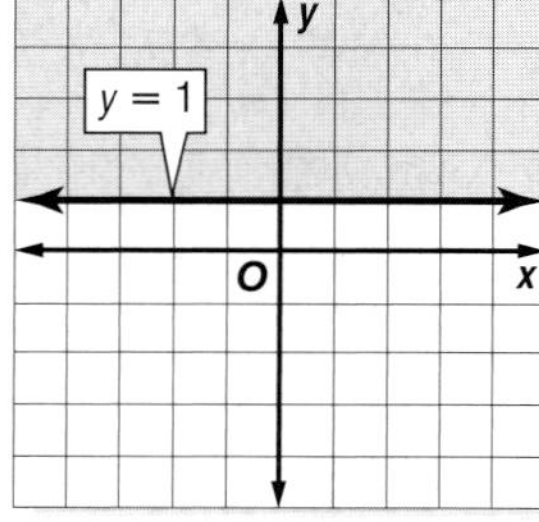

21.

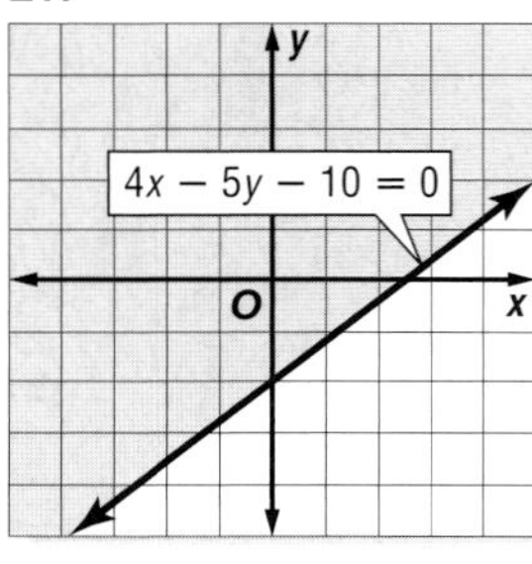

23.

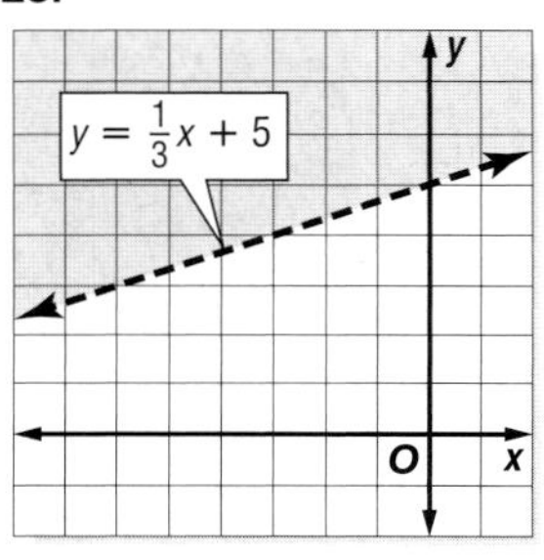

25.

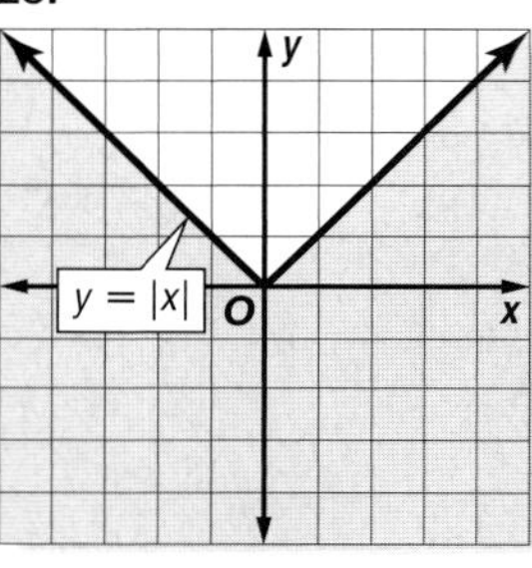

27.

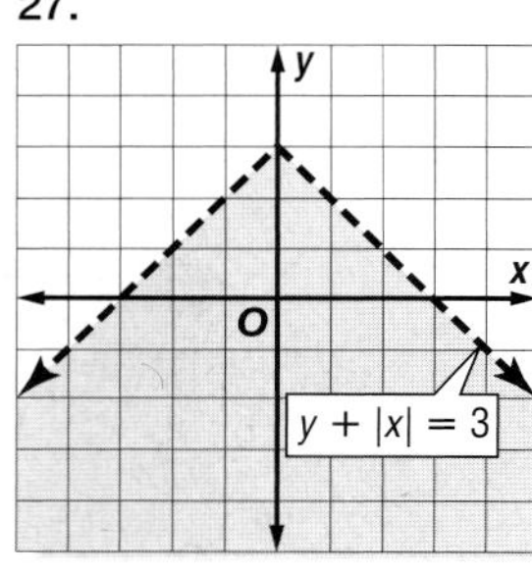

29.

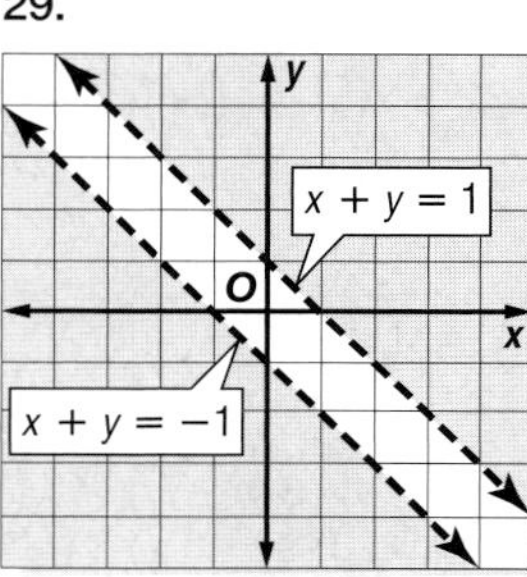

31. $x < -2$

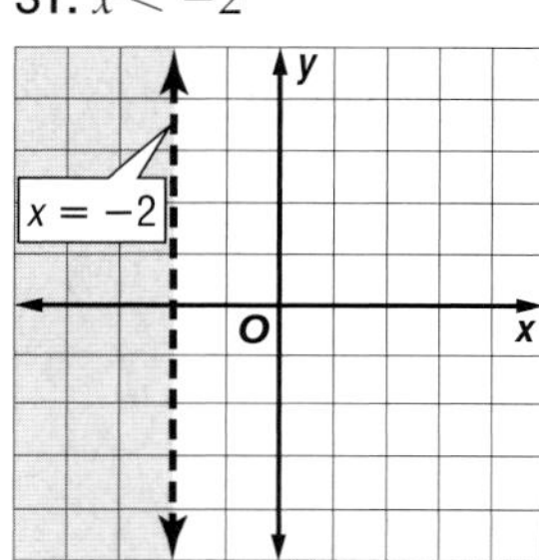

33.

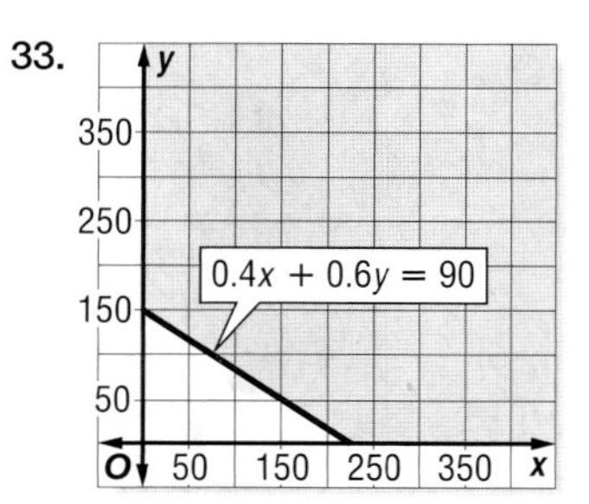

35. $4a + 3s \geq 2000$ **37.** yes **39.** yes **41.** Linear inequalities can be used to track the performance of players in fantasy football leagues. Answers should include the following.

- Let x be the number of receiving yards and let y be the number of touchdowns. The number of points Dana gets from receiving yards is $5x$ and the number of points he gets from touchdowns is $100y$. His total number of points is $5x + 100y$. He wants at least 1000 points, so the inequality $5x + 100y \geq 1000$ represents the situation.

- $5x + 100y = 1000$

- the first one

43. B **45.**

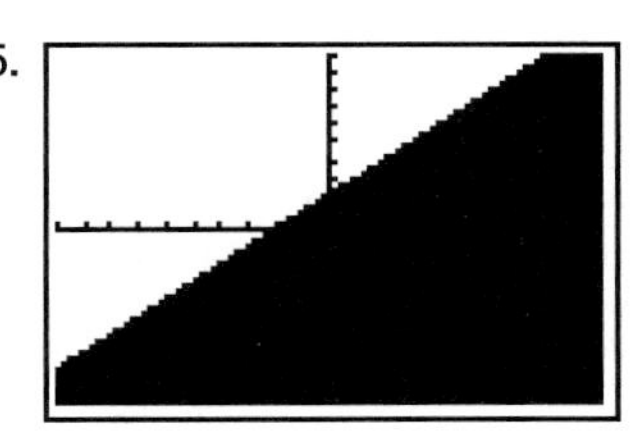

[−10, 10] scl: 1 by [−10, 10] scl: 1

47.

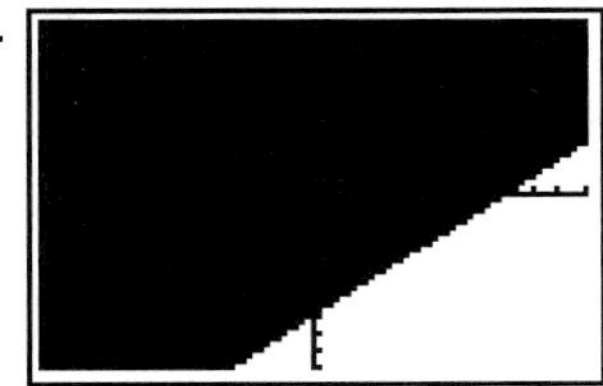

[−10, 10] scl: 1 by [−10, 10] scl: 1

49. D = all reals, R = $\{y \mid y \geq -1\}$

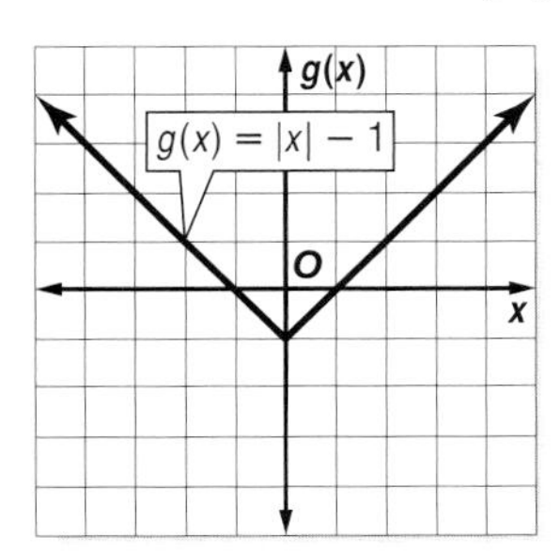

51.

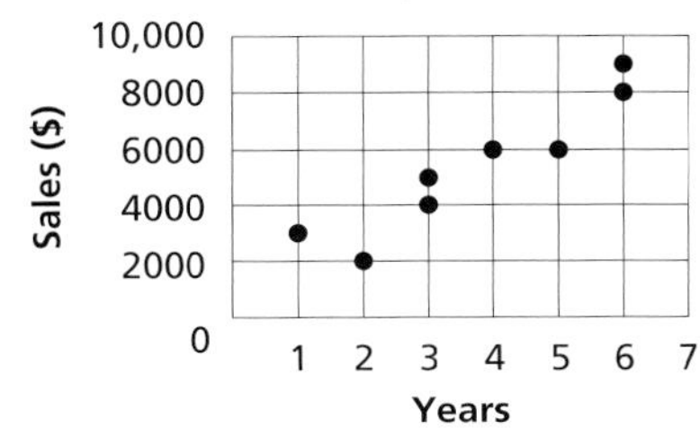

53. Sample answer: $10,000 **55.** 3

Pages 100–104 Chapter 2 Study Guide and Review

1. identity **3.** standard **5.** domain **7.** slope

9. D = {−2, 2, 6}, R = {1, 3}; yes

(−2, 3) (6, 3) (2, 1)

11. D = all reals, R = all reals; yes

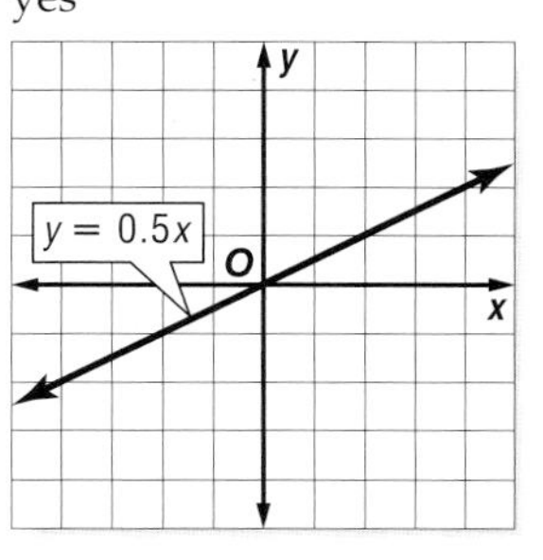

13. 21 **15.** $5y - 9$ **17.** No; x has an exponent other than 1. **19.** No; x is inside a square root. **21.** $5x + 2y = -4$; 5, 2, −4

23. −4, −20

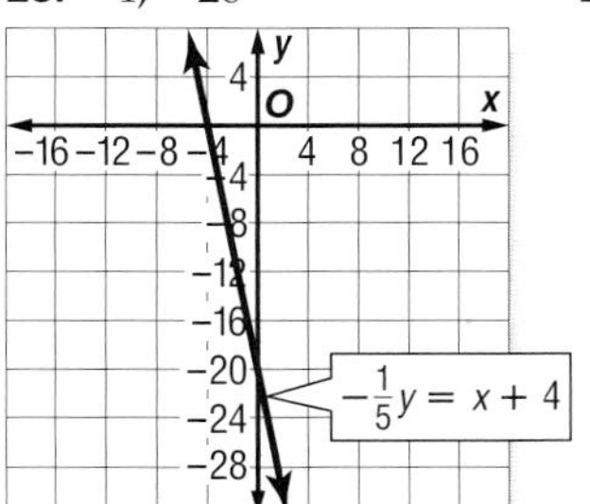

25. 9, −9

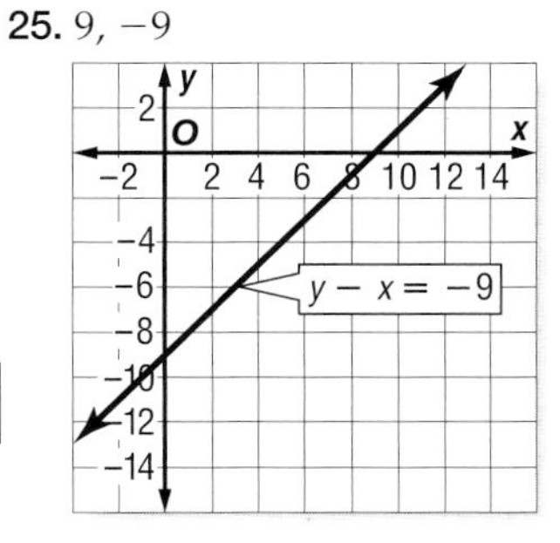

27. $-\frac{3}{11}$

29.

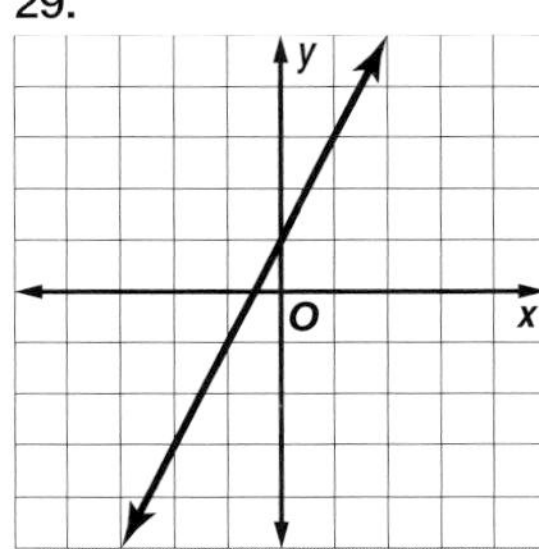

31.

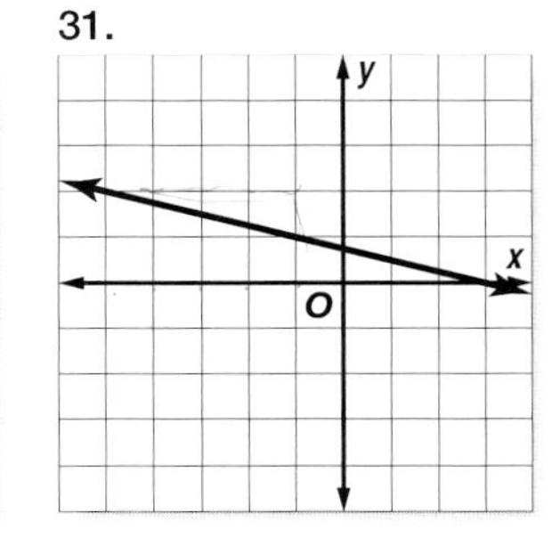

33.

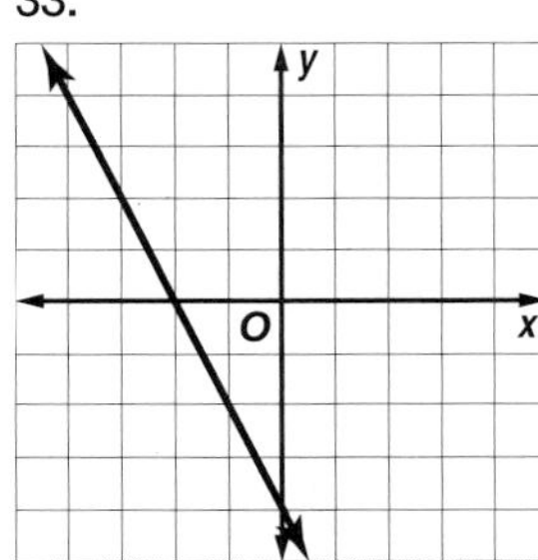

35.

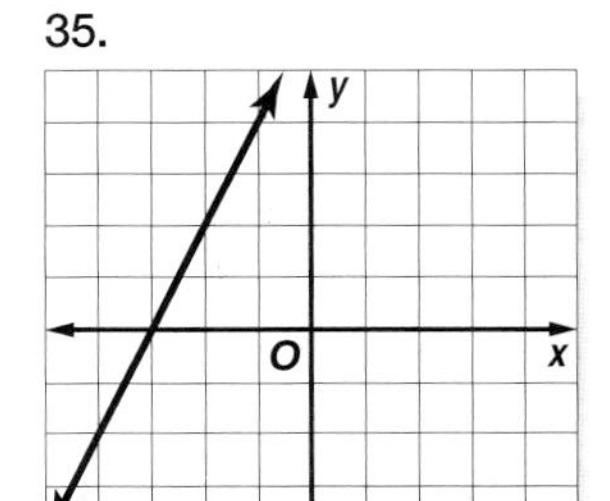

37. $y = -\frac{5}{3}x - 3$ **39.** $y = -\frac{3}{4}x + \frac{17}{4}$ **41.** Sample answer using (1980, 29.3) and (1990, 33.6): $y = 0.43x - 822.1$

43. D = all reals, R = all integers

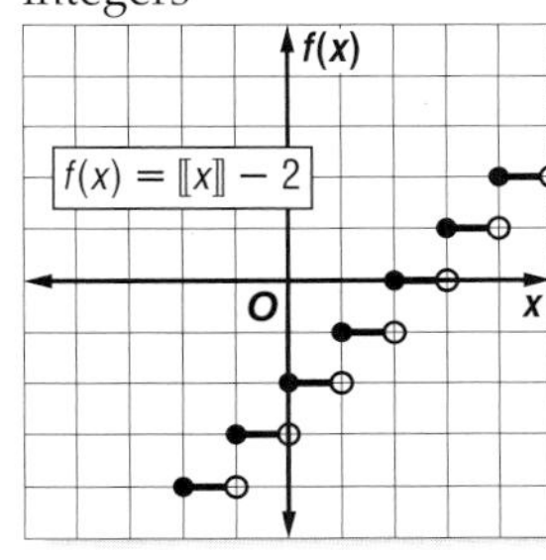

45. D = all reals, R = $\{y \mid y \geq 4\}$

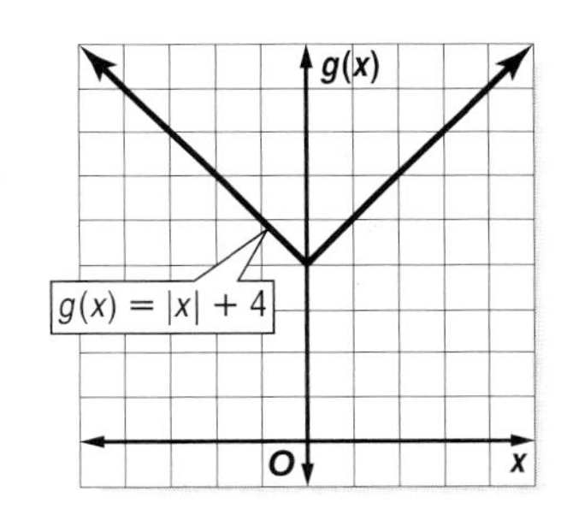

47. D = all reals, R = $\{y \mid y \leq 0 \text{ or } y = 2\}$

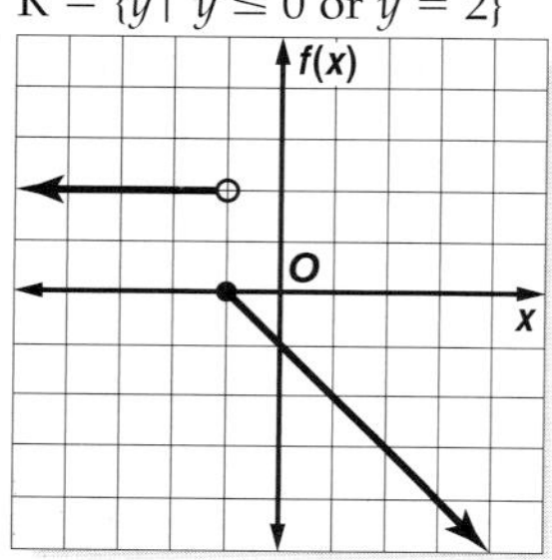

49.

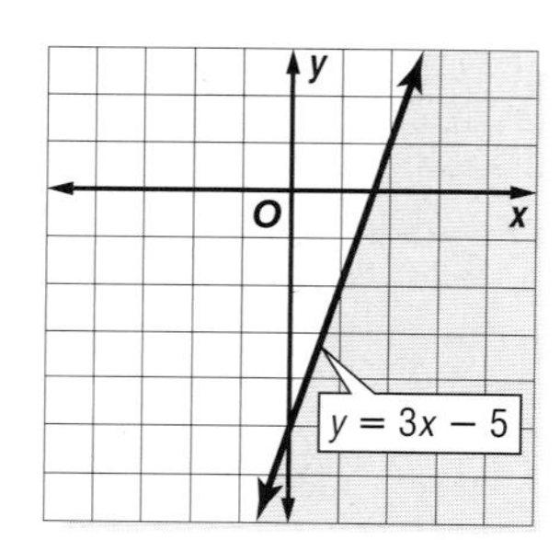

51.

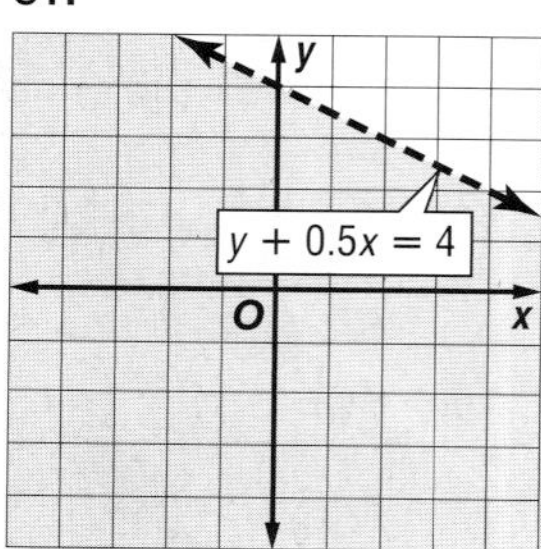

53.

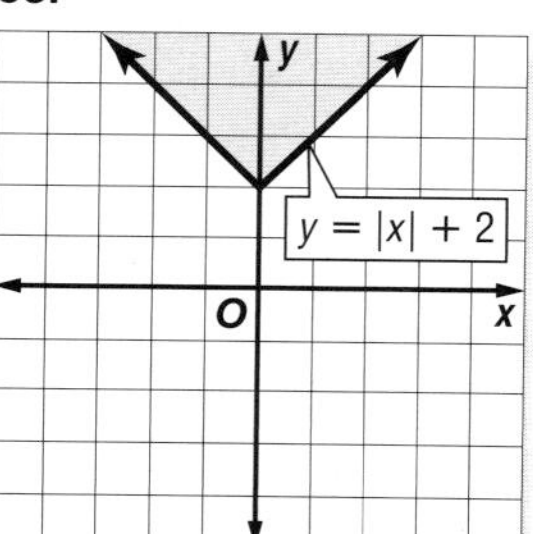

Chapter 3 Systems of Equations and Inequalities

Page 109 Chapter 3 Getting Started

1.

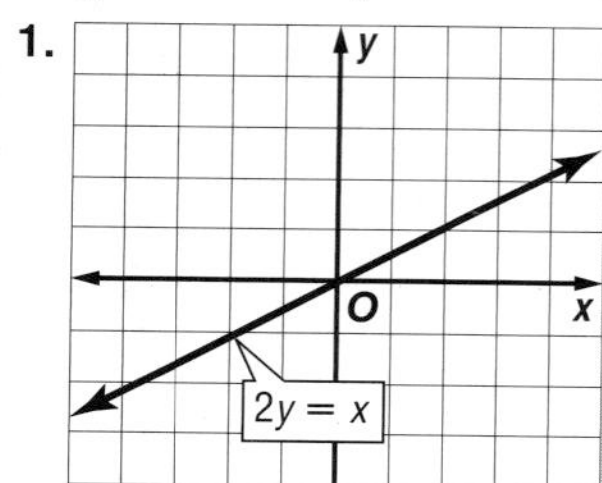

3.

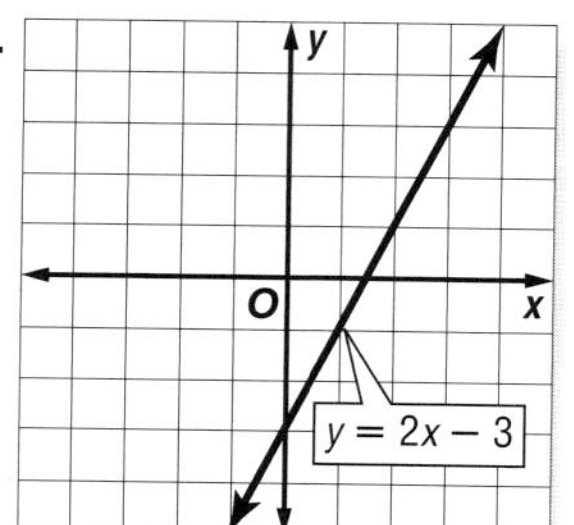

5.

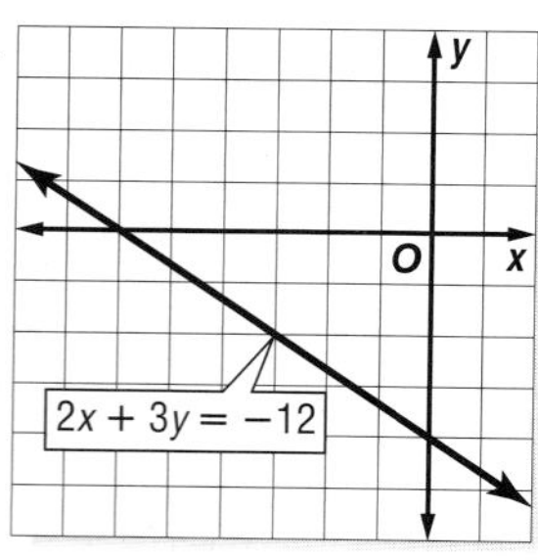

7. $y = -2x$ **9.** $y = 6 - 3x$
11. $y = 2 - 6x$

13.

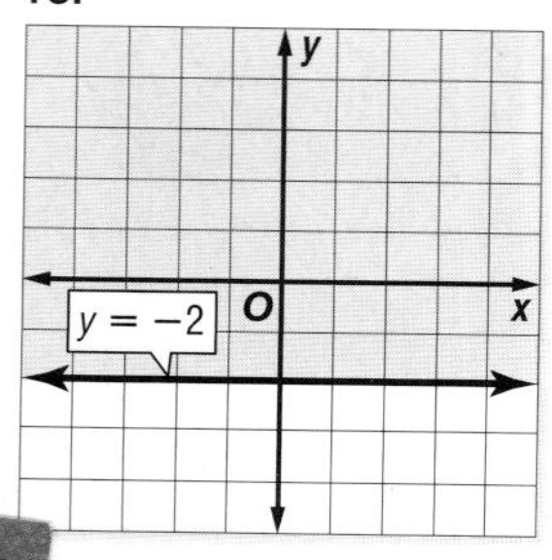

15.

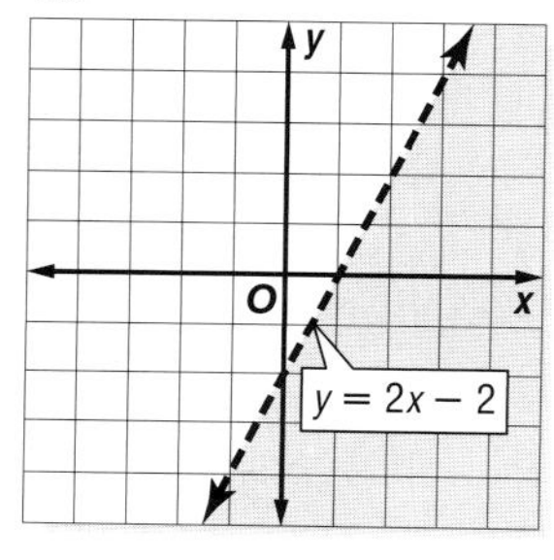

17.

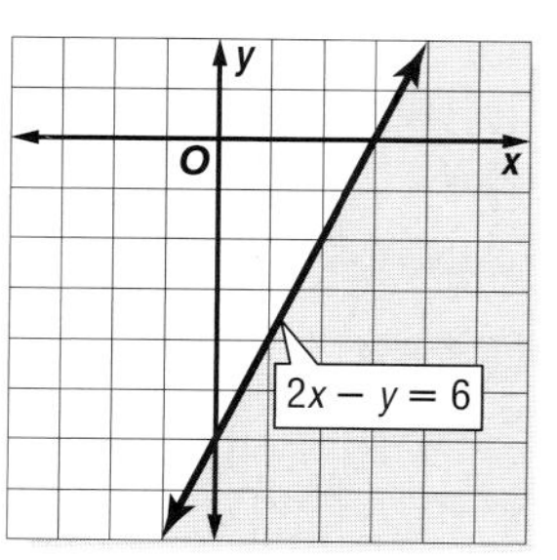

19. −9 **21.** 0 **23.** −22

Pages 112–115 Lesson 3-1

1. Two lines cannot intersect in exactly two points.
3. A graph is used to estimate the solution. To determine that the point lies on both lines, you must check that it satisfies both equations.

5.

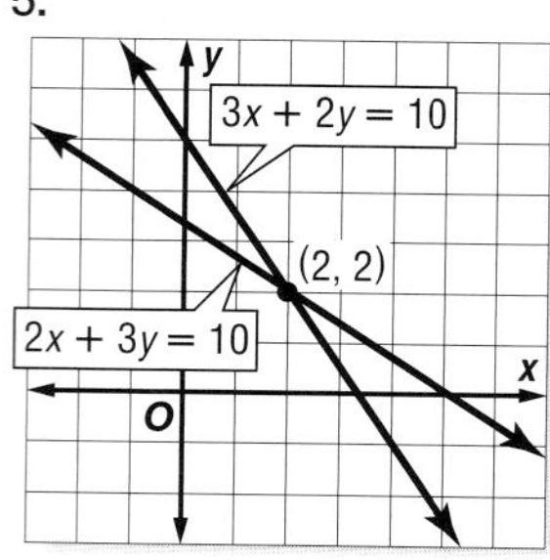

7. consistent and independent

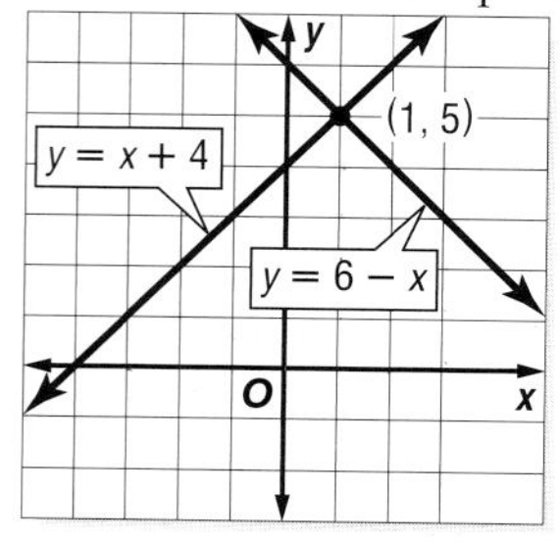

9. consistent and dependent

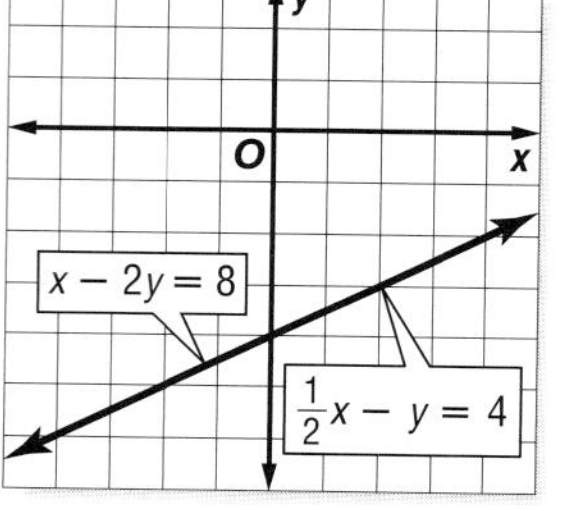

11. The cost is $5.60 for both stores to develop 30 prints.

13.

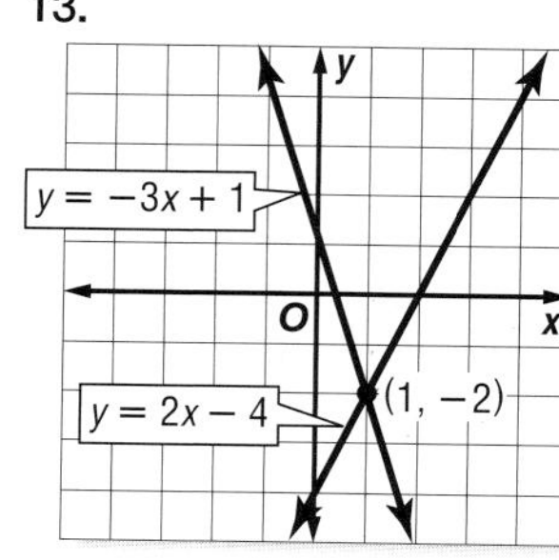

15.

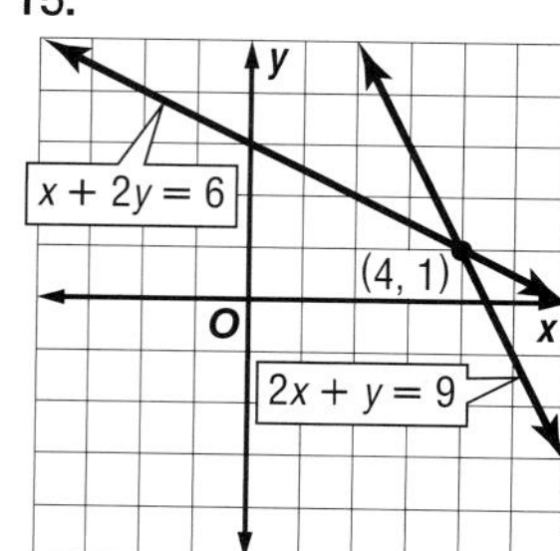

17.

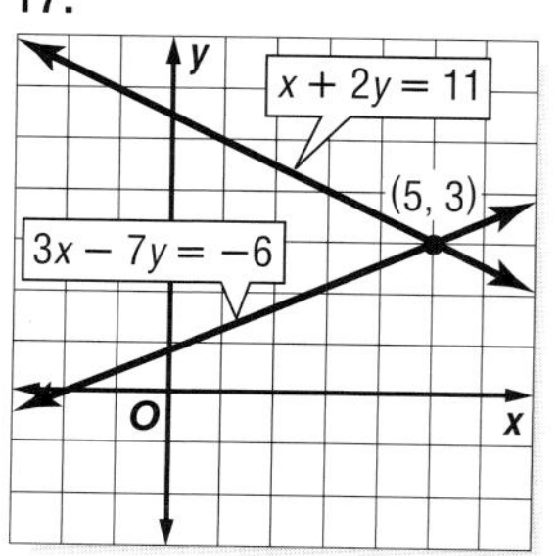

19.

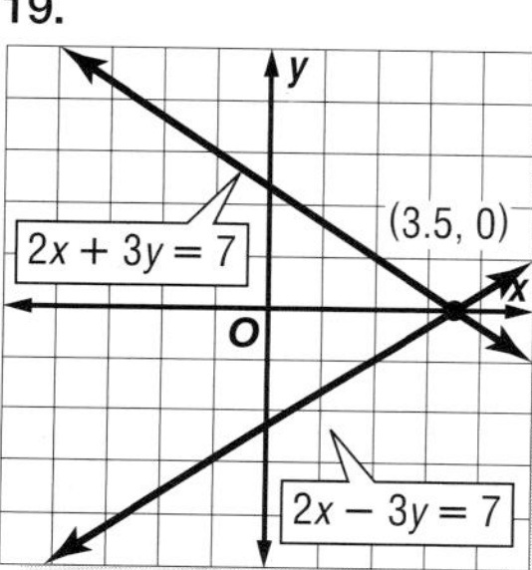

21.

23.

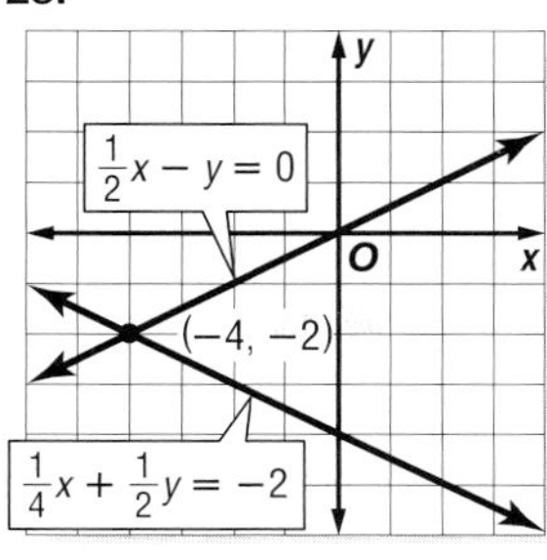

25. inconsistent

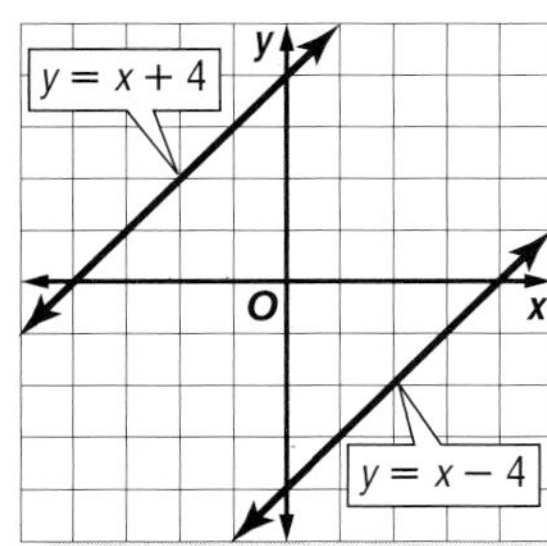

27. consistent and independent

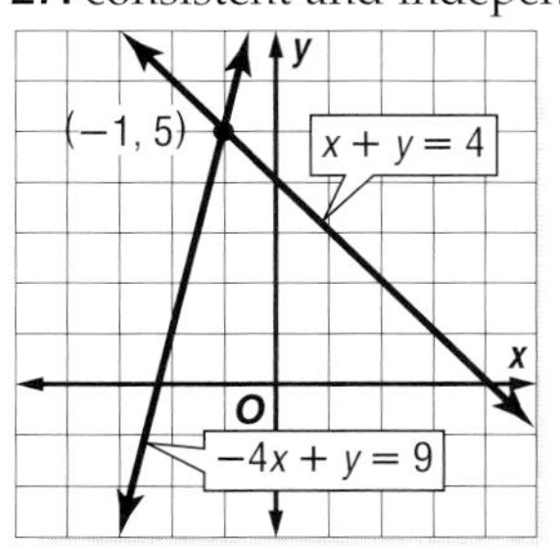

29. inconsistent

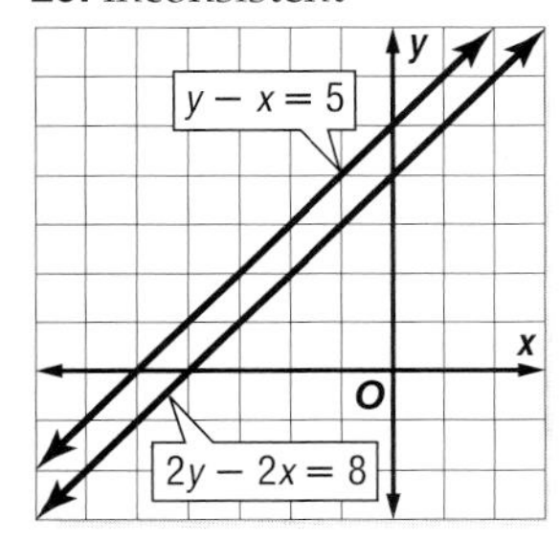

31. consistent and independent

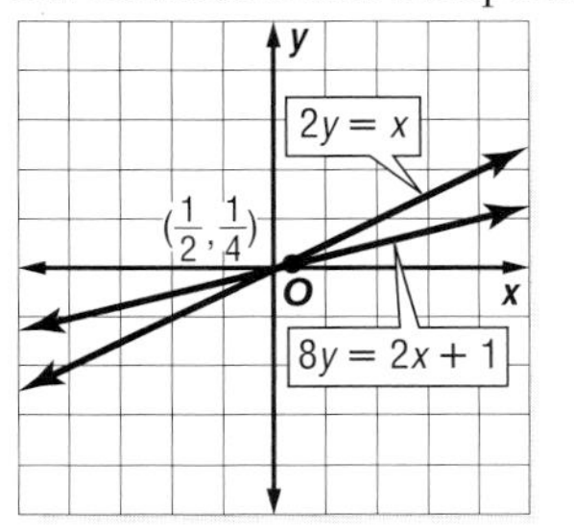

33. consistent and independent

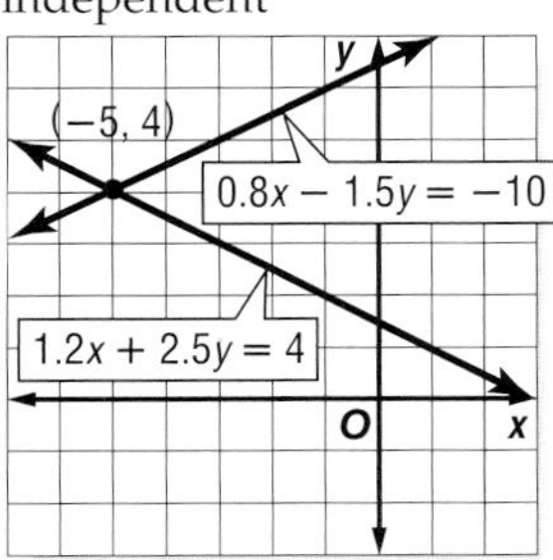

35. inconsistent

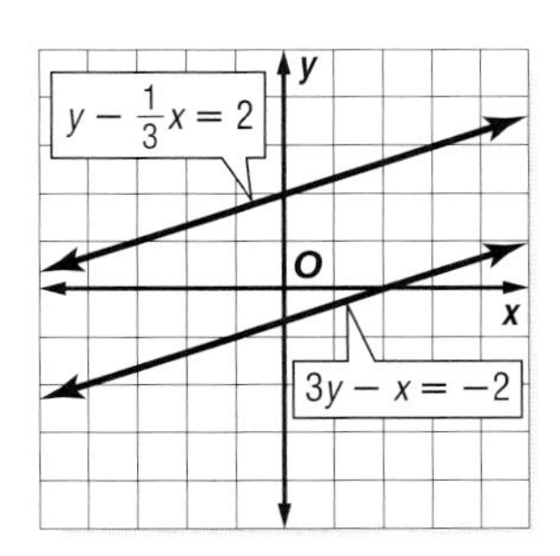

37. $(-3, 1)$ **39.** $y = 52 + 0.23x$, $y = 80$ **41.** Deluxe Plan **43.** Supply, 300,000; demand, 200,000; prices will tend to fall. **45.** $y = 304x + 15{,}982$, $y = 98.6x + 18{,}976$ **47.** FL will probably be ranked third by 2020. The graphs intersect in the year 2015, so NY will still have a higher population in 2010, but FL will have a higher population in 2020.

49. You can use a system of equations to track sales and make predictions about future growth based on past performance and trends in the graphs. Answers should include the following.

- The coordinates (6, 54) represent that 6 years after 1999 both the in-store sales and online sales will be $54,000.
- The in-store sales and the online sales will never be equal and in-store sales will continue to be higher than online sales.

51. C **53.** $(-5.56, 12)$ **55.** no solution **57.** $(2.64, 42.43)$

59.

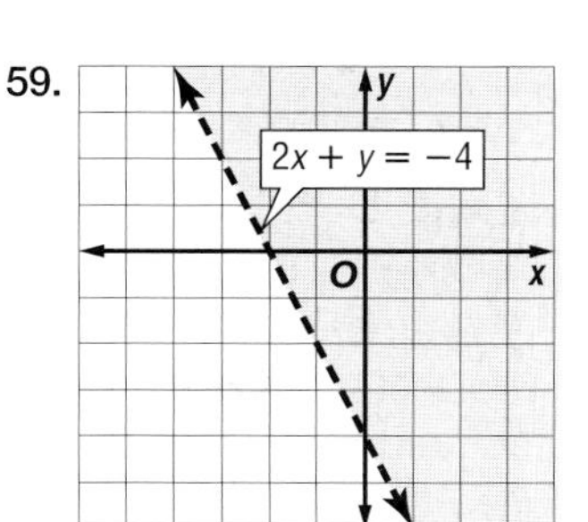

61. A **63.** P **65.** $\{-15, 9\}$ **67.** $\{-2, 3\}$ **69.** $\{9\}$ **71.** $x^2 - 6$ **73.** $\frac{z}{3} + 1$ **75.** $9y + 1$ **77.** $12x + 18y - 6$ **79.** $x + 4y$

Pages 119–122 Lesson 3-2

3. Vincent; Juanita subtracted the two equations incorrectly; $-y - y = -2y$, not 0. **5.** $(1, 3)$ **7.** $(5, 2)$ **9.** $(6, -20)$ **11.** $\left(3\frac{1}{3}, 2\frac{2}{3}\right)$ **13.** $(9, 5)$ **15.** $(3, -2)$ **17.** no solution **19.** $(4, 3)$ **21.** $(2, 0)$ **23.** $(10, -1)$ **25.** $(4, -3)$ **27.** $(-8, -3)$ **29.** no solution **31.** $\left(-\frac{1}{2}, \frac{3}{2}\right)$ **33.** $(-6, 11)$ **35.** $(1.5, 0.5)$ **37.** 8, 6 **39.** $x + y = 28$, $16x + 19y = 478$ **41.** 4 2-bedroom, 2 3-bedroom **43.** $x + y = 30$, $700x + 200y = 15{,}000$ **45.** $2x + 4y = 100$, $y = 2x$ **47.** Yes; they should finish the test within 40 minutes. **49.** 25 min of step aerobics, 15 min of stretching **51.** You can use a system of equations to find the monthly fee and rate per minute charged during the months of January and February. Answers should include the following.

- The coordinates of the point of intersection are (0.08, 3.5).
- Currently, Yolanda is paying a monthly fee of $3.50 and an additional 8¢ per minute. If she graphs $y = 0.08x + 3.5$ (to represent what she is paying currently) and $y = 0.10x + 3$ (to represent the other long-distance plan) and finds the intersection, she can identify which plan would be better for a person with her level of usage.

53. A

55. consistent and dependent

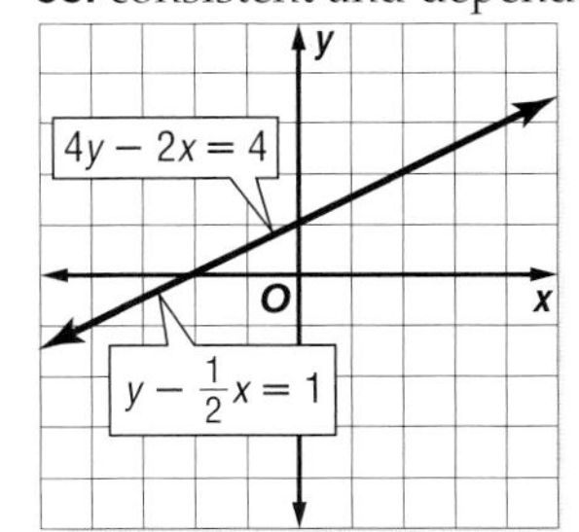

57.

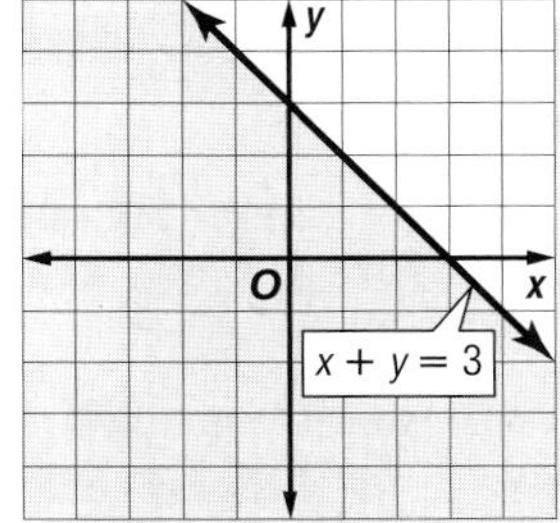

59.

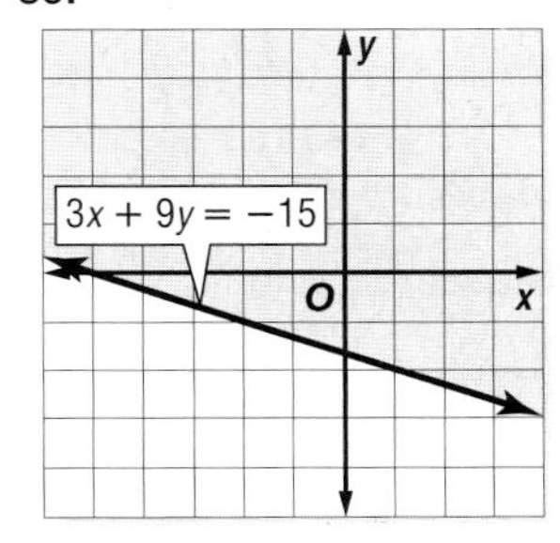

61. $x - y = 0$; 1, −1, 0 **63.** $2x - y = -3$; 2, −1, −3 **65.** $3x + 2y = 21$; 3, 2, 21 **67.** yes **69.** no

Selected Answers

Page 122 Practice Quiz 1

1.

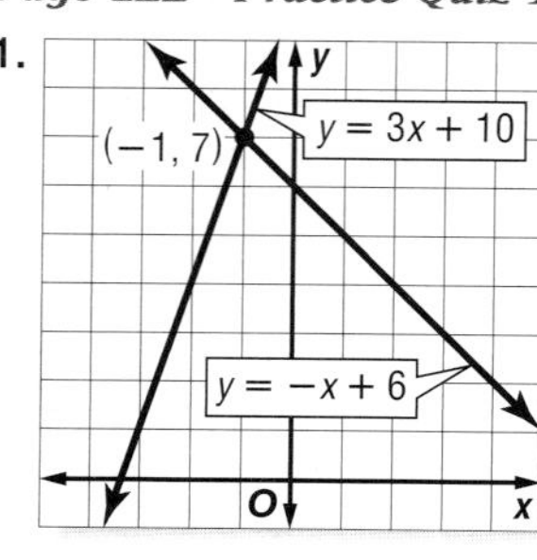

3. (2, 7) 5. Hartsfield, 78 million; O'Hare, 72.5 million

Pages 125–127 Lesson 3-3

1. Sample answer: $y > x + 3, y < x - 2$ 3a. 4 3b. 2 3c. 1 3d. 3

5.

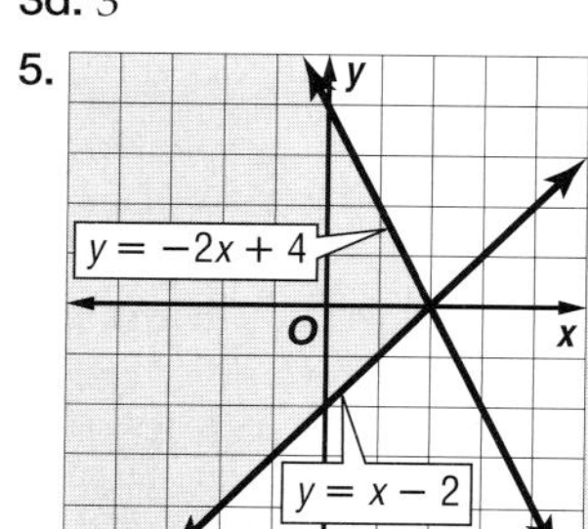

7.

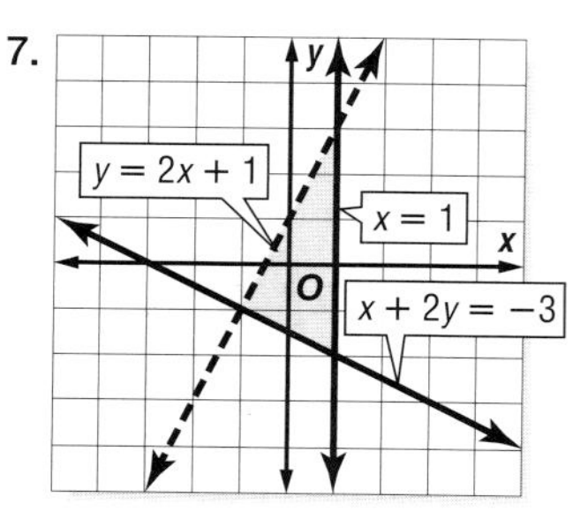

9. (−4, 3), (1, −2), (2, 9), (7, 4) 11. Sample answer: 3 packages of bagels, 4 packages of muffins; 4 packages of bagels, 4 packages of muffins; 3 packages of bagels, 5 packages of muffins

13.

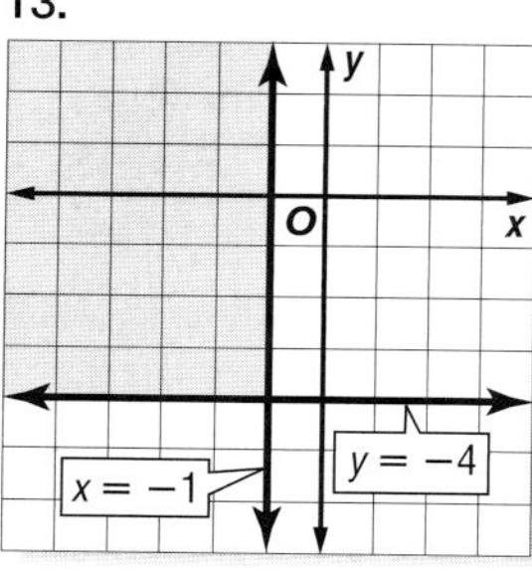

15.

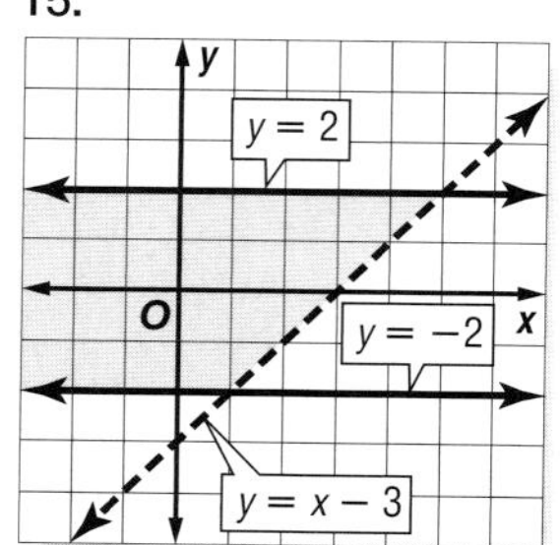

17.

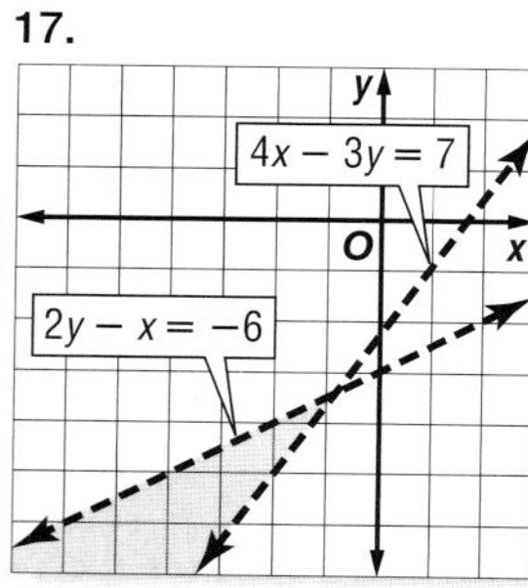

19. no solution

21.

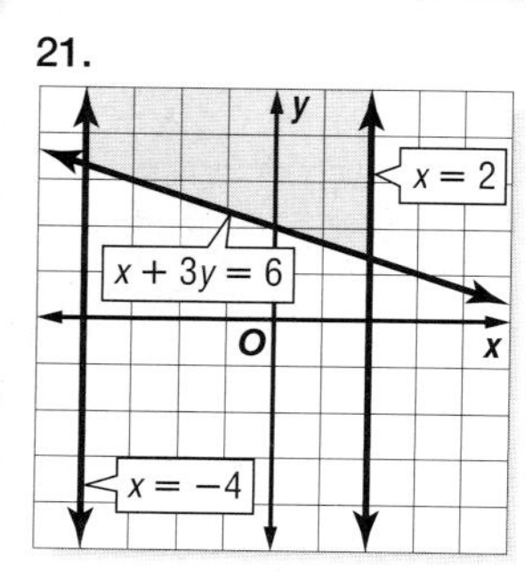

23.

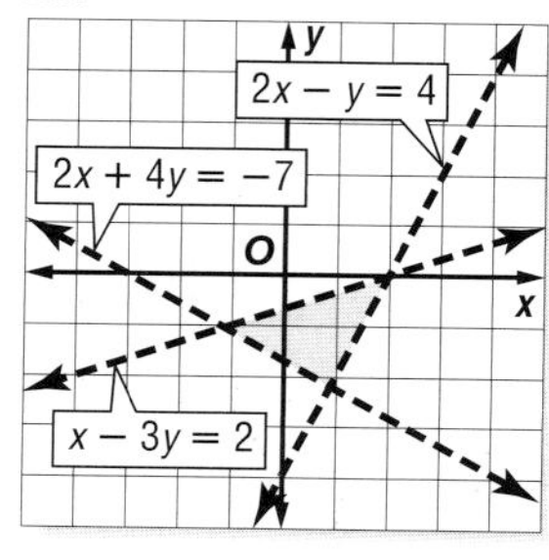

25. (−3, −4), (5, −4), (1, 4) 27. (−6, −9), (2, 7), (10, −1)

29. (−4, 3), (−2, 7), (4, −1), $\left(7\frac{1}{3}, 2\frac{1}{3}\right)$ 31. 64 units²

33. $s \geq 111, s \leq 130, h \geq 9, h \leq 12$

35.

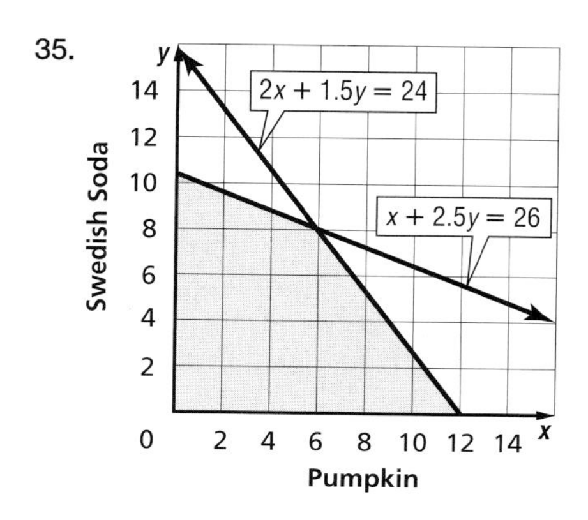

37. 6 pumpkin, 8 soda

39. The range for normal blood pressure satisfies four inequalities that can be graphed to find their intersection. Answers should include the following.

- Graph the blood pressure as an ordered pair; if the point lies in the shaded region, it is in the normal range.
- High systolic pressure is represented by the region to the right of $x = 140$ and high diastolic pressure is represented by the region above $y = 90$.

41. Sample answer: $y \leq 6, y \geq 2, x \leq 5, x \geq 1$ 43. (6, 5)

45.

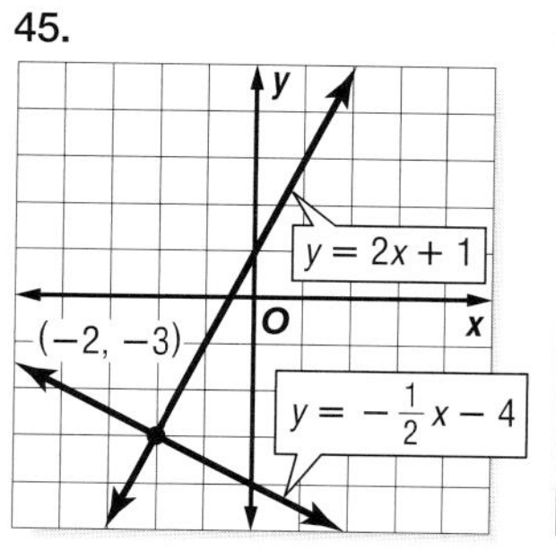

47.

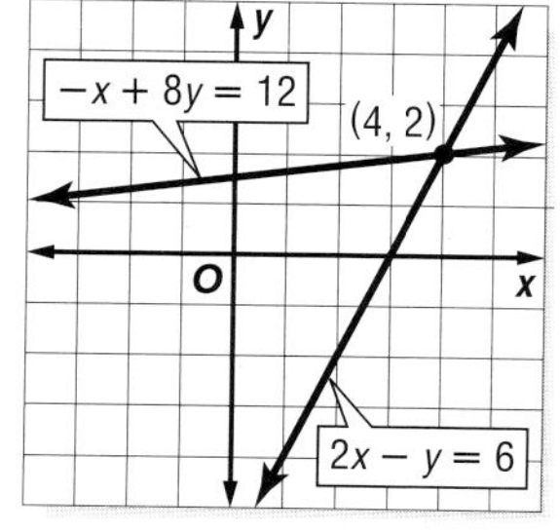

49. −5 51. 8 53. 5

Pages 132–135 Lesson 3-4

1. sometimes

3. 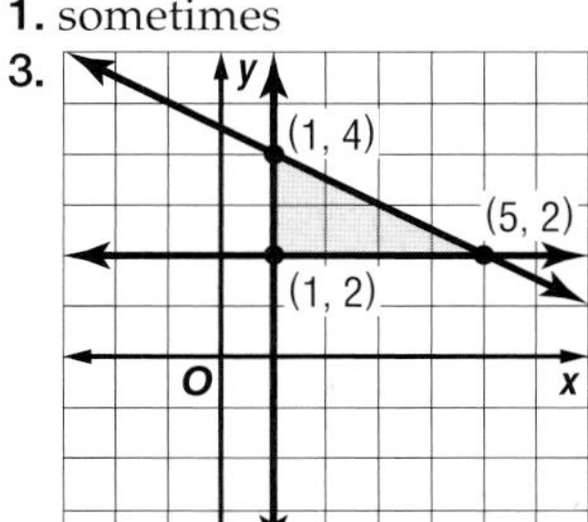

vertices: (1, 2), (1, 4), (5, 2); max: $f(5, 2) = 4$, min: $f(1, 4) = -10$

5. 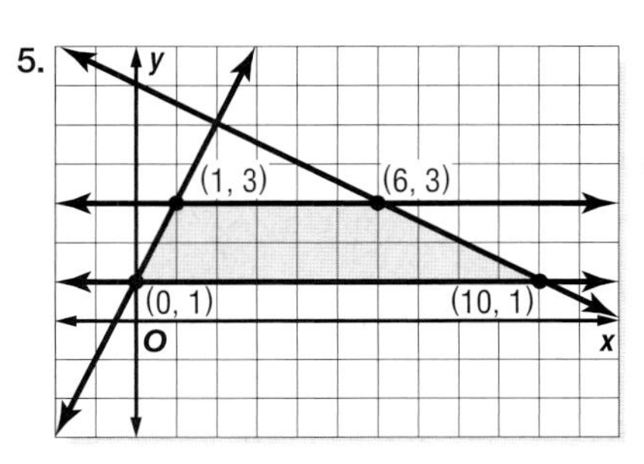

vertices: (0, 1), (1, 3), (6, 3), (10, 1); max: $f(10, 1) = 31$, min: $f(0, 1) = 1$

7. 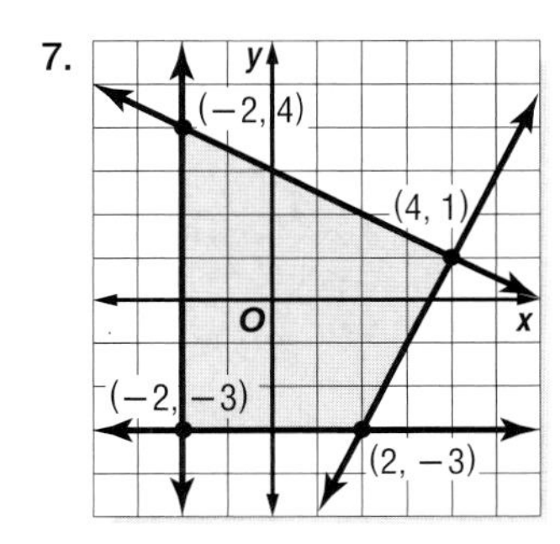

vertices: (−2, 4), (−2, −3), (2, −3), (4, 1); max: $f(2, -3) = 5$; min: $f(-2, 4) = -6$

9. $c \geq 0$, $\ell \geq 0$, $c + 3\ell \leq 56$, $4c + 2\ell \leq 104$ **11.** (0, 0), (26, 0), (20, 12), $\left(0, 18\frac{2}{3}\right)$

13. 20 canvas tote bags and 12 leather tote bags

15. 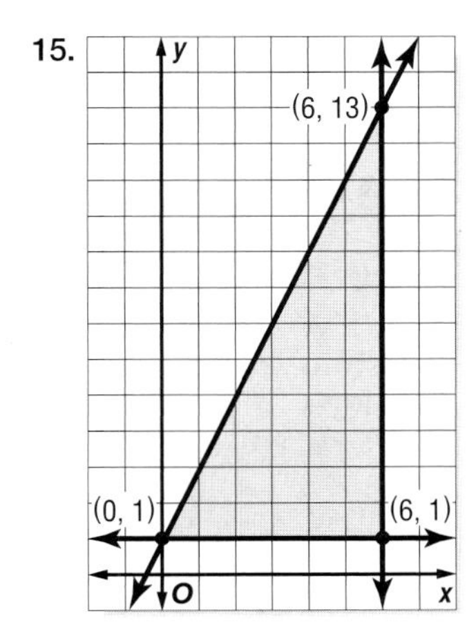

vertices: (0, 1), (6, 1), (6, 13); max: $f(6, 13) = 19$; min: $f(0, 1) = 1$

17. 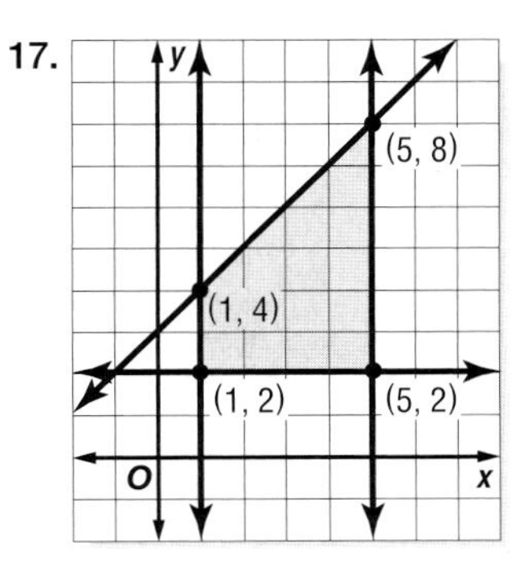

vertices: (1, 4), (5, 8), (5, 2), (1, 2); max: $f(5, 2) = 11$, min: $f(1, 4) = -5$

19. 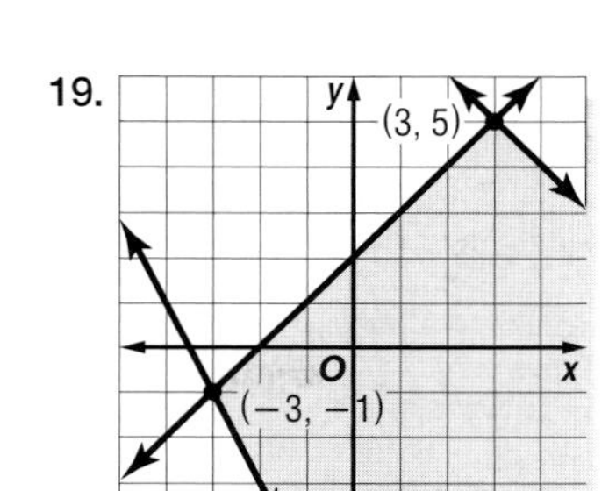

vertices: (−3, −1), (3, 5); min: $f(-3, -1) = -9$; no maximum

21. 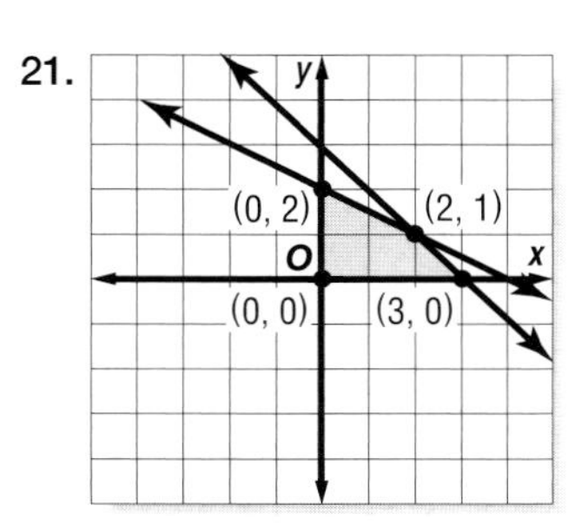

vertices: (0, 0), (0, 2), (2, 1), (3, 0); max: $f(0, 2) = 6$; min: $f(3, 0) = -12$

23. 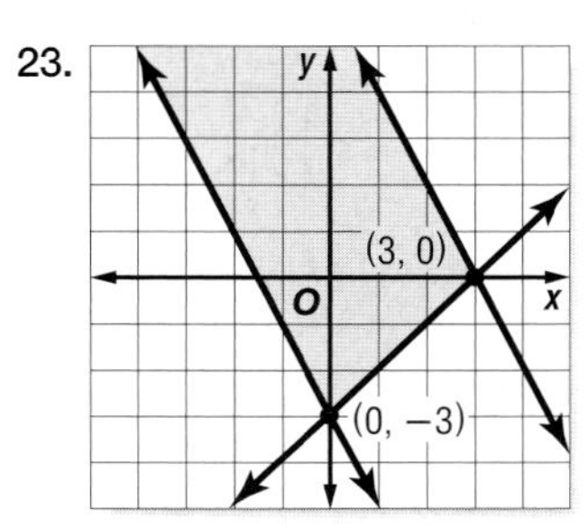

vertices: (3, 0), (0, −3); min: $f(0, -3) = -12$; no maximum

25. 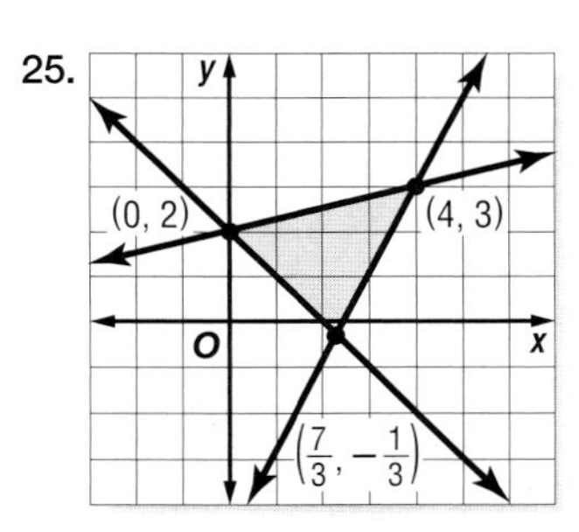

vertices: (0, 2), (4, 3), $\left(\frac{7}{3}, -\frac{1}{3}\right)$; max: $f(4, 3) = 25$, min: $f(0, 2) = 6$

27. 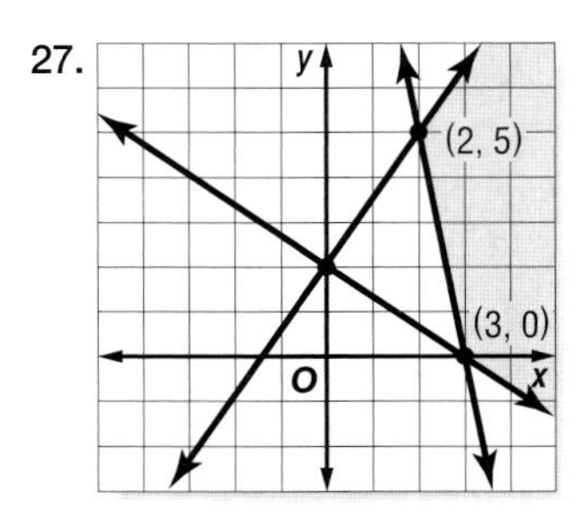

vertices: (2, 5), (3, 0); no maximum; no minimum

29. 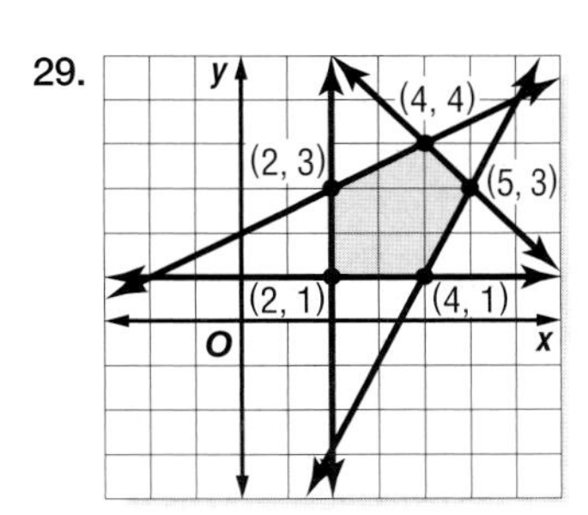

vertices: (2, 1), (2, 3), (4, 1), (4, 4), (5, 3); max: $f(4, 1) = 0$, min: $f(4, 4) = -12$

31. $g \geq 0, c \geq 0, 1.5g + c \leq 85, 2g + 0.5c \leq 40$ **33.** (0, 0), (0, 20), (80, 0) **35.** 0 graphing calculators, 80 CAS calculators

39. 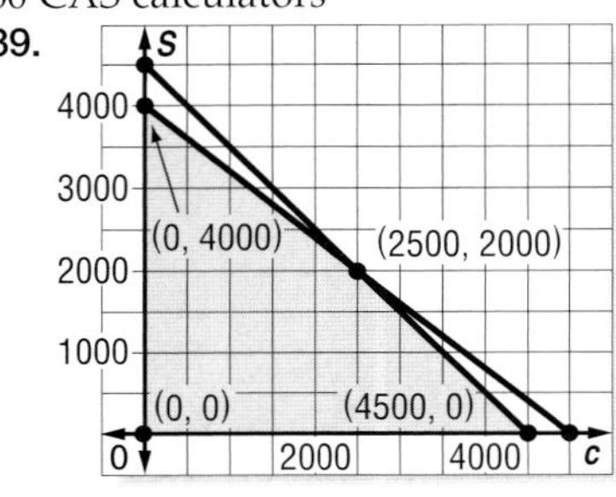

(0, 0), (0, 4000), (2500, 2000), (4500, 0)

41. 4500 acres corn, 0 acres soybeans; $130,500

43. There are many variables in scheduling tasks. Linear programming can help make sure that all the requirements are met. Answers should include the following.

- Let x = the number of buoy replacements and let y = the number of buoy repairs. Then, $x \geq 0, y \geq 0, x \leq 8$ and $2.5x + y \leq 24$.
- The captain would want to maximize the number of buoys that a crew could repair and replace so $f(x, y) = x + y$.
- Graph the inequalities and find the vertices of the intersection of the graphs. The coordinate (0, 24) maximizes the function. So the crew can service the maximum number of buoys if they replace 0 and repair 24 buoys.

45. C **47.** 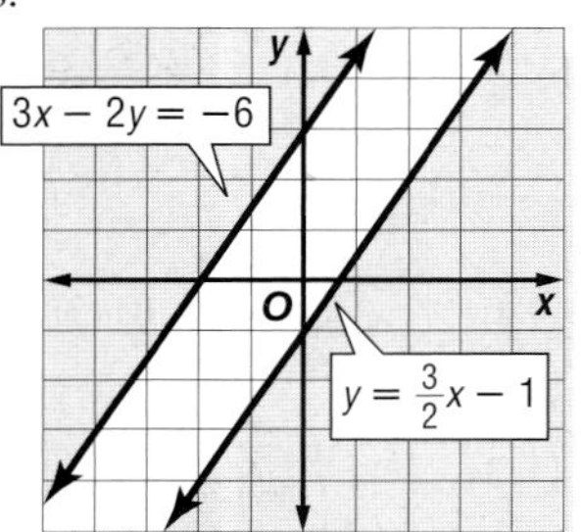

$\varnothing$

49. (2, 3) **51.** c = average cost each year; $15c + 3479 = 7489$ **53.** Additive Inverse **55.** Multiplicative Inverse **57.** 9 **59.** 16 **61.** 8

Page 135 Practice Quiz 2

1. 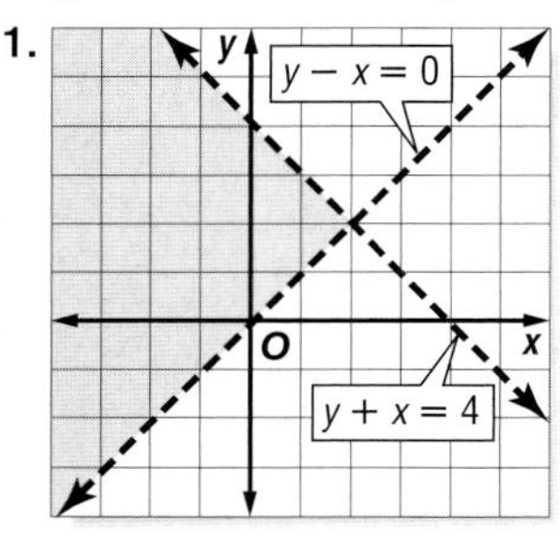

3. 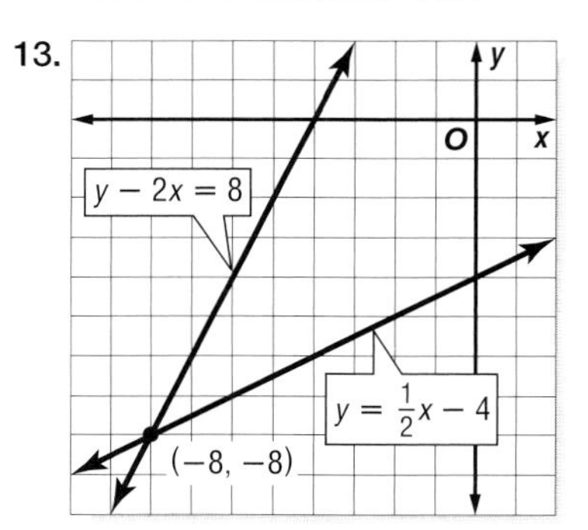

5. 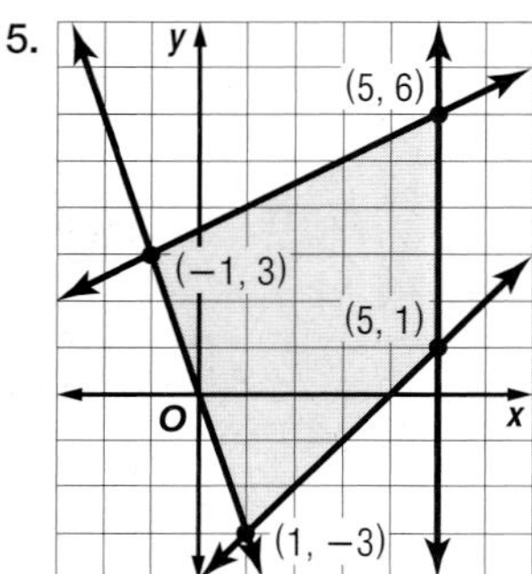

vertices: (1, −3), (−1, 3), (5, 6), (5, 1);
max: $f(5, 1) = 17$,
min: $f(-1, 3) = -13$

Pages 142–144 Lesson 3-5

1. You can use elimination or substitution to eliminate one of the variables. Then you can solve two equations in two variables.
3. Sample answer: $x + y + z = 4$, $2x - y + z = -9$, $x + 2y - z = 5$; $-3 + 5 + 2 = 4$, $2(-3) - 5 + 2 = -9$, $-3 + 2(5) - 2 = 5$ **5.** (−1, −3, 7) **7.** (5, 2, −1) **9.** (4, 0, 8) **11.** $4\frac{1}{2}$ lb chicken, 3 lb sausage, 6 lb rice **13.** (−2, 1, 5) **15.** (4, 0, −1) **17.** (1, 5, 7) **19.** infinitely many **21.** $\left(\frac{1}{3}, -\frac{1}{2}, \frac{1}{4}\right)$ **23.** (−5, 9, 4) **25.** 8, 1, 3 **27.** enchilada, $2.50; taco, $1.95; burrito, $2.65 **29.** $x + y + z = 355$, $x + 2y + 3z = 646$, $y = z + 27$ **31.** $a = \frac{3}{2}, b = 0, c = 3$; $y = \frac{3}{2}x^2 + 0x + 3$ or $y = \frac{3}{2}x^2 + 3$ **33.** D **35.** 120 units of notebook paper and 80 units of newsprint

37. 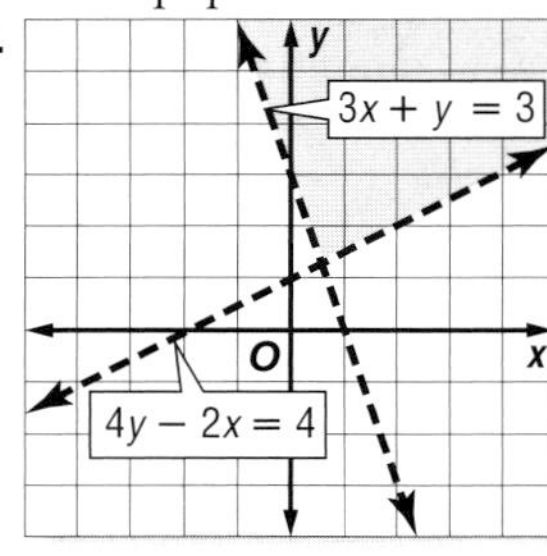

39. Sample answer using (7, 15) and (14, 22): $y = x + 8$ **41.** $x + 3y$ **43.** $9s + 4t$

Pages 145–148 Study Guide and Review

1. c **3.** f **5.** a **7.** h **9.** d

11. 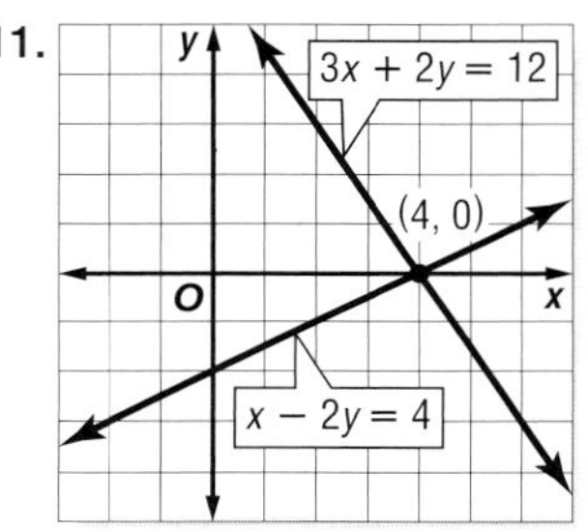

13. 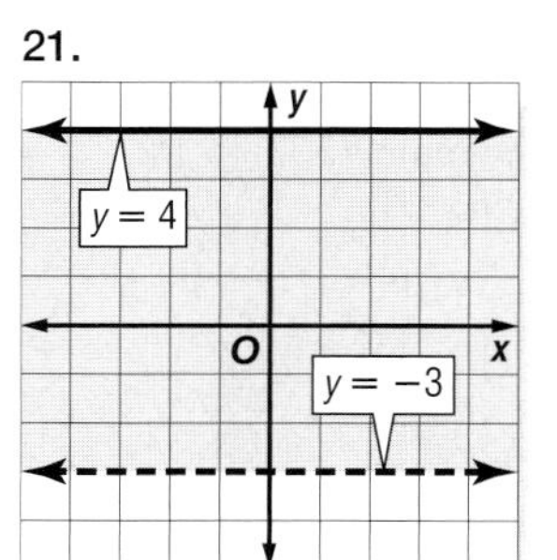

15. (3, 2) **17.** (9, 4) **19.** (−1, 2)

21. 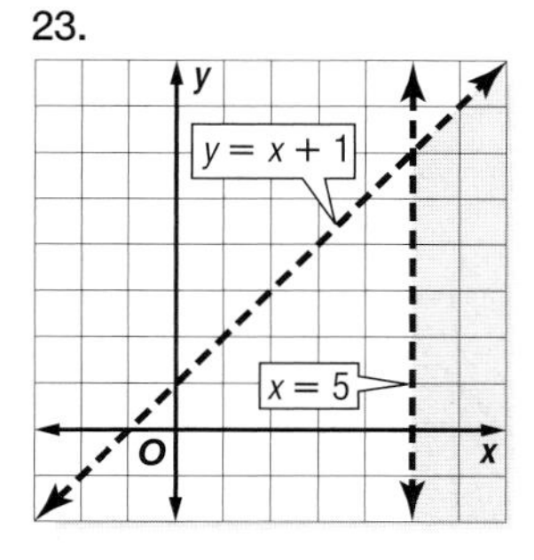

23.

25. 160 My Real Babies, 320 My First Babies **27.** (4, −2, 1)

Chapter 4 Matrices

Page 153 Chapter 4 Getting Started

1. 6 **3.** $4\frac{3}{4}$ **5.** −13 **7.** $-3; \frac{1}{3}$ **9.** $-8; \frac{1}{8}$ **11.** −1.25; 0.8

13. $\frac{8}{3}; -\frac{3}{8}$

15.

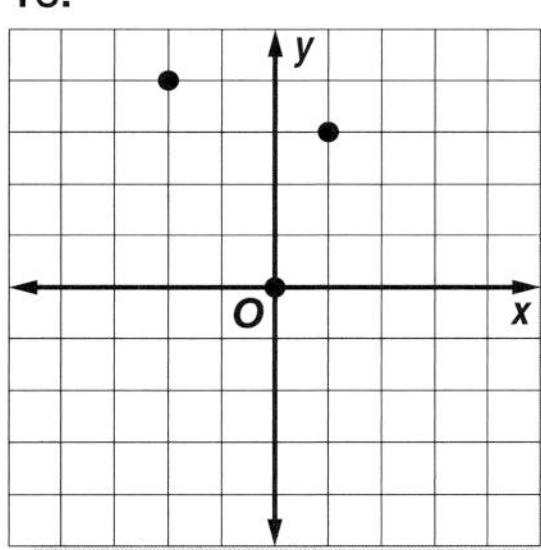

17.

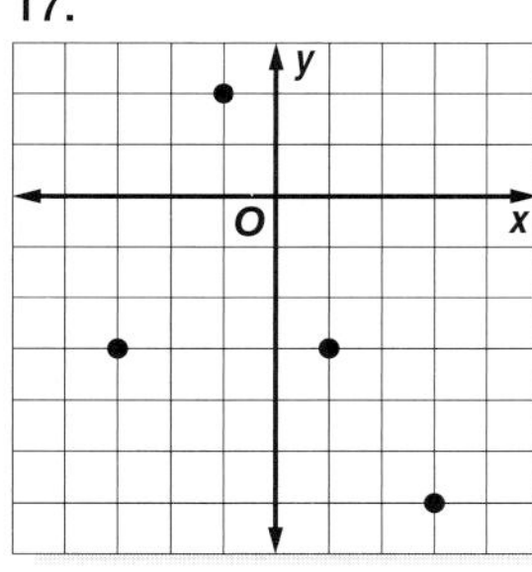

19. (6, 1) **21.** (8, −5) **23.** (2, −2)

Pages 156–158 Lesson 4-1

1. The matrices must have the same dimensions and each element of one matrix must be equal to the corresponding element of the other matrix. **3.** Corresponding elements are elements in the same row and column positions.
5. 3 × 4 **7.** (3, 3) **9.** 2 × 5 **11.** 3 × 1 **13.** 3 × 3
15. 3 × 2 **17.** $\left(3, -\frac{1}{3}\right)$ **19.** (3, −5, 6) **21.** (4, −3)
23. (14, 15) **25.** (5, 3, 2) **27.** 3 × 3 **29.** Sample answer: Mason's Steakhouse; it was given the highest rating possible for service and atmosphere, location was given one of the highest ratings, and it is moderately priced.

31.

	Single	Double	Suite
Weekday	60	70	75
Weekend	79	89	95

33. row 6, column 9 **35.** B **37.** (7, 5, 4) **39.** $\left(-\frac{4}{5}, \frac{3}{5}, -11\right)$

41.

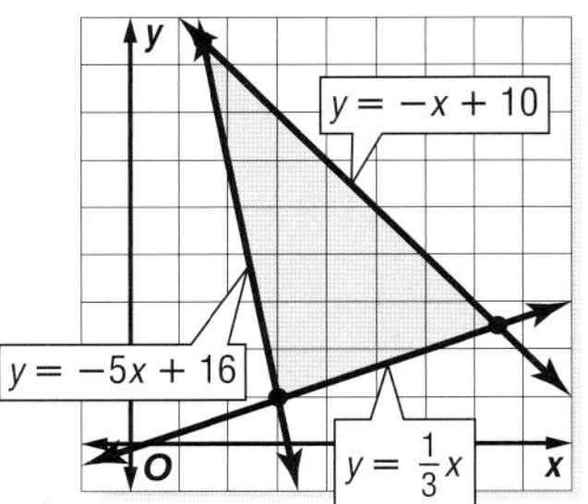

vertices: (3, 1), $\left(\frac{15}{2}, \frac{5}{2}\right)$, $\left(\frac{3}{2}, \frac{17}{2}\right)$; max: $f\left(\frac{15}{2}, \frac{5}{2}\right) =$ 35, min: $f\left(\frac{3}{2}, \frac{17}{2}\right) = -1$

43.

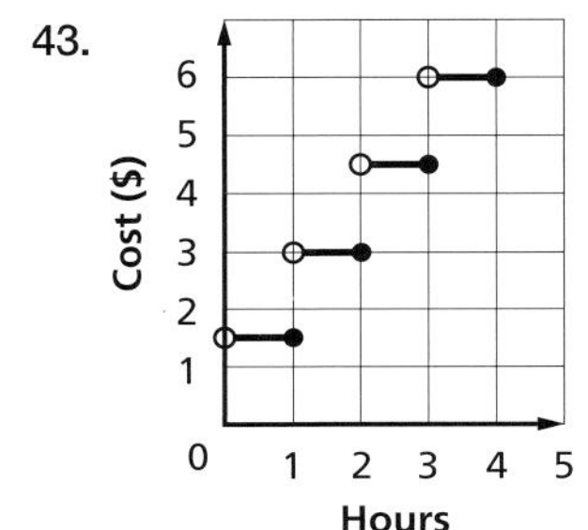

45. $4.50 **47.** 2 **49.** 20
51. −10 **53.** −18 **55.** −3
57. $\frac{3}{2}$

Pages 163–166 Lesson 4-2

1. They must have the same dimensions.

3. $\begin{bmatrix} 4 & 4 \\ 4 & 4 \\ 4 & 4 \end{bmatrix}$ **5.** $\begin{bmatrix} 1 & 10 \\ -7 & 5 \end{bmatrix}$ **7.** $\begin{bmatrix} -22 & 8 \\ 3 & 24 \end{bmatrix}$ **9.** $\begin{bmatrix} -21 & 29 \\ 12 & -22 \end{bmatrix}$

11. Males = $\begin{bmatrix} 16{,}763 & 549{,}499 \\ 14{,}620 & 477{,}960 \\ 14{,}486 & 455{,}305 \\ 9041 & 321{,}416 \\ 5234 & 83{,}411 \end{bmatrix}$, Females = $\begin{bmatrix} 16{,}439 & 456{,}873 \\ 14{,}545 & 405{,}163 \\ 12{,}679 & 340{,}480 \\ 7931 & 257{,}586 \\ 5450 & 133{,}235 \end{bmatrix}$

13. No; many schools offer the same sport for males and females, so those schools would be counted twice.

15. impossible **17.** $\begin{bmatrix} -4 & 8 & -2 \\ 6 & -10 & -16 \\ -14 & -12 & 4 \end{bmatrix}$ **19.** $\begin{bmatrix} -13 \\ -3 \\ 23 \end{bmatrix}$

21. $\begin{bmatrix} 1.5 & 3 \\ 4.5 & 9 \end{bmatrix}$ **23.** $\begin{bmatrix} -5\frac{1}{2} & 3 & 9 \\ 10\frac{2}{3} & 1\frac{2}{3} & -2\frac{1}{2} \end{bmatrix}$ **25.** $\begin{bmatrix} -2 & -1 \\ 4 & -1 \\ -7 & -4 \end{bmatrix}$

27. $\begin{bmatrix} 38 & 4 \\ 32 & -6 \\ 18 & 42 \end{bmatrix}$ **29.** $\begin{bmatrix} 2 & 4\frac{2}{3} \\ 1 & 5 \\ 6 & -1 \end{bmatrix}$ **31.** $\begin{bmatrix} 232 & 184 & 120 & 149 \\ 164 & 124 & 75 & 130 \\ 160 & 182 & 72 & 108 \end{bmatrix}$

33. $\begin{bmatrix} 245 \\ 228 \\ 319 \\ 227 \\ 117 \end{bmatrix}$ **35.** 1996, floods; 1997, floods; 1998, floods; 1999, tornadoes; 2000, lightning

37. $\begin{bmatrix} 1.50 & 2.25 \\ 1.00 & 1.75 \end{bmatrix}$ **39.** $\begin{bmatrix} 1.00 & 1.00 \\ 1.50 & 1.50 \end{bmatrix}$

41. You can use matrices to track dietary requirements and add them to find the total each day or each week. Answers should include the following.

- Breakfast = $\begin{bmatrix} 566 & 18 & 7 \\ 482 & 12 & 17 \\ 530 & 10 & 11 \end{bmatrix}$, Lunch = $\begin{bmatrix} 785 & 22 & 19 \\ 622 & 23 & 20 \\ 710 & 26 & 12 \end{bmatrix}$,

  Dinner = $\begin{bmatrix} 1257 & 40 & 26 \\ 987 & 32 & 45 \\ 1380 & 29 & 38 \end{bmatrix}$
- Add the three matrices: $\begin{bmatrix} 2608 & 80 & 52 \\ 2091 & 67 & 82 \\ 2620 & 65 & 61 \end{bmatrix}$.

43. A **45.** 1 × 4 **47.** 3 × 3 **49.** 4 × 3 **51.** (5, 3, 7)
53. (2, 5) **55.** (6, −1)

57.

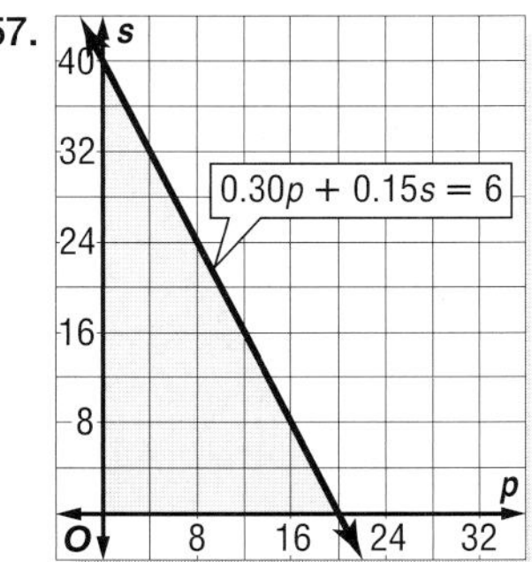

59. Multiplicative Inverse
61. Distributive

Pages 171–174 Lesson 4-3

1. Sample answer: $\begin{bmatrix} 1 & 2 \\ 3 & 4 \\ 5 & 6 \end{bmatrix} \cdot \begin{bmatrix} 7 & 8 \\ 9 & 10 \end{bmatrix}$ **3.** The Right Distributive Property says that $(A + B)C = AC + BC$, but $AC + BC \neq CA + CB$ since the Commutative Property does not hold for matrix multiplication in most cases. **5.** undefined

7. $\begin{bmatrix} 15 & -5 & 20 \\ 24 & -8 & 32 \end{bmatrix}$ **9.** $\begin{bmatrix} 24 \\ 41 \end{bmatrix}$ **11.** $\begin{bmatrix} 45 & 55 & 65 \end{bmatrix}$, $\begin{bmatrix} 350 & 280 \\ 320 & 165 \\ 180 & 120 \end{bmatrix}$

13. 4×2 **15.** undefined **17.** undefined

19. [6] **21.** not possible **23.** $\begin{bmatrix} 1 & -25 & 2 \\ 29 & 1 & -30 \end{bmatrix}$

25. $\begin{bmatrix} 24 & 16 \\ -32 & -5 \\ -48 & -11 \end{bmatrix}$

27. yes

$$AC + BC = \begin{bmatrix} 1 & -2 \\ 4 & 3 \end{bmatrix} \cdot \begin{bmatrix} 5 & 1 \\ 2 & -4 \end{bmatrix} + \begin{bmatrix} -5 & 2 \\ 4 & 3 \end{bmatrix} \cdot \begin{bmatrix} 5 & 1 \\ 2 & -4 \end{bmatrix}$$
$$= \begin{bmatrix} 1 & 9 \\ 26 & -8 \end{bmatrix} + \begin{bmatrix} -21 & -13 \\ 26 & -8 \end{bmatrix}$$
$$= \begin{bmatrix} -20 & -4 \\ 52 & -16 \end{bmatrix}$$

$$(A + B)C = \left(\begin{bmatrix} 1 & -2 \\ 4 & 3 \end{bmatrix} + \begin{bmatrix} -5 & 2 \\ 4 & 3 \end{bmatrix}\right) \cdot \begin{bmatrix} 5 & 1 \\ 2 & -4 \end{bmatrix}$$
$$= \begin{bmatrix} -4 & 0 \\ 8 & 6 \end{bmatrix} \cdot \begin{bmatrix} 5 & 1 \\ 2 & -4 \end{bmatrix}$$
$$= \begin{bmatrix} -20 & -4 \\ 52 & -16 \end{bmatrix}$$

29. no

$$C(A + B) = \begin{bmatrix} 5 & 1 \\ 2 & -4 \end{bmatrix} \cdot \left(\begin{bmatrix} 1 & -2 \\ 4 & 3 \end{bmatrix} + \begin{bmatrix} -5 & 2 \\ 4 & 3 \end{bmatrix}\right)$$
$$= \begin{bmatrix} 5 & 1 \\ 2 & -4 \end{bmatrix} \cdot \begin{bmatrix} -4 & 0 \\ 8 & 6 \end{bmatrix}$$
$$= \begin{bmatrix} -12 & 6 \\ -40 & -24 \end{bmatrix}$$

$$AC + BC = \begin{bmatrix} 1 & -2 \\ 4 & 3 \end{bmatrix} \cdot \begin{bmatrix} 5 & 1 \\ 2 & -4 \end{bmatrix} + \begin{bmatrix} -5 & 2 \\ 4 & 3 \end{bmatrix} \cdot \begin{bmatrix} 5 & 1 \\ 2 & -4 \end{bmatrix}$$
$$= \begin{bmatrix} 1 & 9 \\ 26 & -8 \end{bmatrix} + \begin{bmatrix} -21 & -13 \\ 26 & -8 \end{bmatrix}$$
$$= \begin{bmatrix} -20 & -4 \\ 52 & -16 \end{bmatrix}$$

31. $\begin{bmatrix} 290 & 165 & 210 \\ 175 & 240 & 190 \\ 110 & 75 & 0 \end{bmatrix}$ **33.** $\begin{bmatrix} 14{,}285 \\ 13{,}270 \\ 4295 \end{bmatrix}$

35. any two matrices $\begin{bmatrix} a & b \\ c & d \end{bmatrix}$ and $\begin{bmatrix} e & f \\ g & h \end{bmatrix}$ where $bg = cf$, $a = d$, and $e = h$ **37.** $\begin{bmatrix} 96.50 \\ 99.50 \\ 118 \\ 117 \end{bmatrix}$ **39.** \$431 **41.** \$26,360

43. $a = 1, b = 0, c = 0, d = 1$; the original matrix **45.** B

47. $\begin{bmatrix} 12 & -6 \\ -3 & 21 \end{bmatrix}$ **49.** $\begin{bmatrix} -20 & 2 \\ -28 & 12 \end{bmatrix}$ **51.** $(5, -9)$ **53.** \$2.50; \$1.50

55. 8; -16

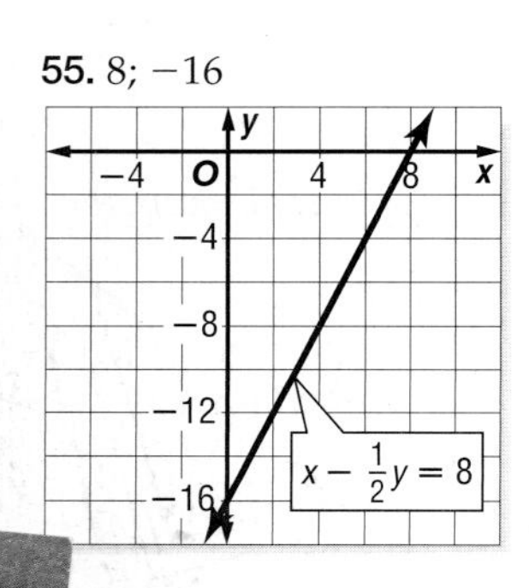

57.

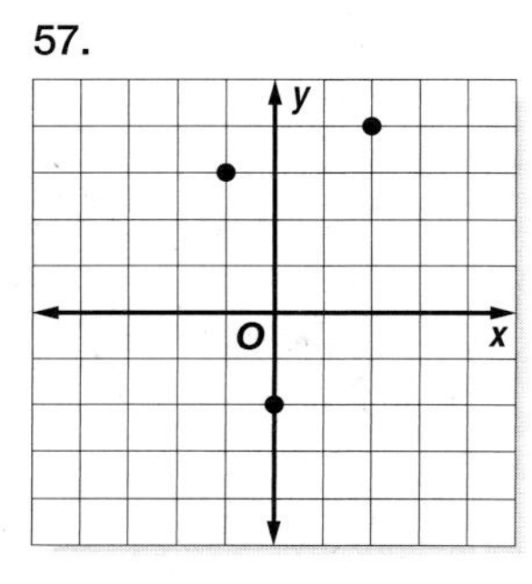

59.

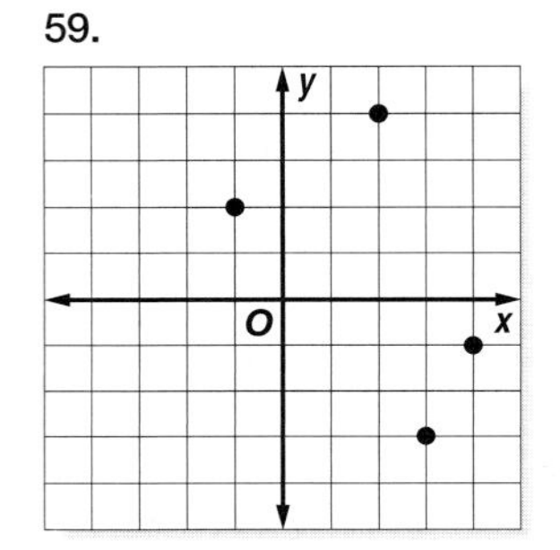

Page 174 Practice Quiz 1

1. $(6, 3)$ **3.** $(1, 3, 5)$ **5.** $\begin{bmatrix} 232 & 159 & 120 & 149 \\ 134 & 200 & 159 & 103 \end{bmatrix}$ **7.** $\begin{bmatrix} 4 & 3 \\ 1 & 3 \end{bmatrix}$

9. not possible

Pages 178–181 Lesson 4-4

1.

Transformation	Size	Shape	Isometry
reflection	same	same	yes
rotation	same	same	yes
translation	same	same	yes
dilation	changes	same	no

3. Sample answer: $\begin{bmatrix} -4 & -4 & -4 \\ 1 & 1 & 1 \end{bmatrix}$ **5.** $A'(4, 3)$, $B'(5, -6)$, $C'(-3, -7)$ **7.** $\begin{bmatrix} 0 & 5 & 5 & 0 \\ 4 & 4 & 0 & 0 \end{bmatrix}$ **9.** $A'(0, -4)$, $B'(5, -4)$, $C'(5, 0)$, $D'(0, 0)$ **11.** B **13.** $D'(-3, 6)$, $E'(-2, -3)$, $F'(-10, -4)$

15. $\begin{bmatrix} 0 & 1.5 & -2.5 \\ 2 & -1.5 & 0 \end{bmatrix}$

17.

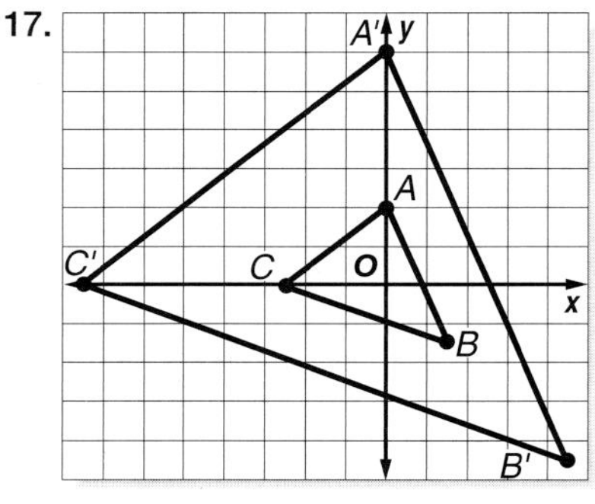

19. $X'(-1, 1)$, $Y'(-4, 2)$, $Z'(-1, 7)$

21. $\begin{bmatrix} 2 & 5 & 4 & 1 \\ 4 & 4 & 1 & 1 \end{bmatrix}$

23.

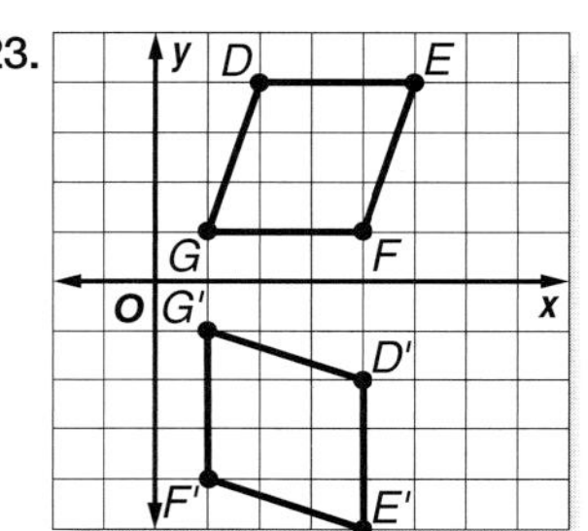

25. $J(-5, 3)$, $K(7, 2)$, $L(4, -1)$ **27.**

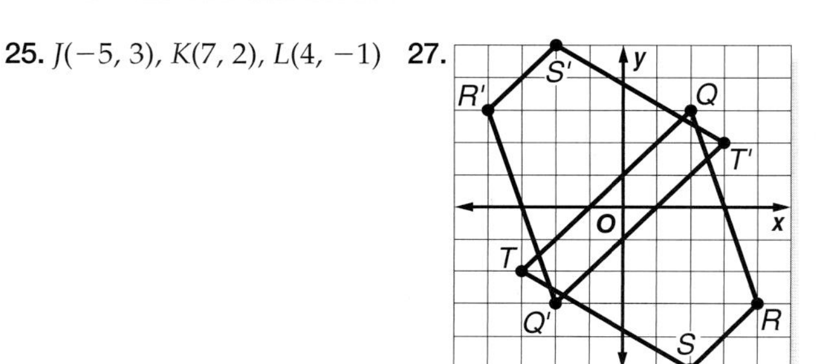

Chapter 4 Matrices

Page 153 Chapter 4 Getting Started

1. 6 **3.** $4\frac{3}{4}$ **5.** −13 **7.** $-3; \frac{1}{3}$ **9.** $-8; \frac{1}{8}$ **11.** −1.25; 0.8
13. $\frac{8}{3}; -\frac{3}{8}$

15.

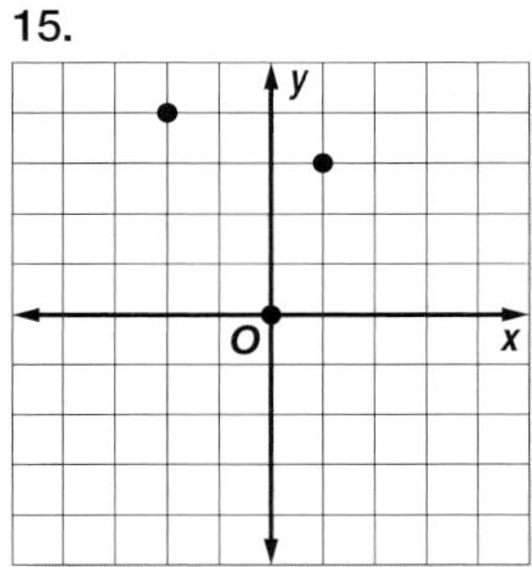

17.

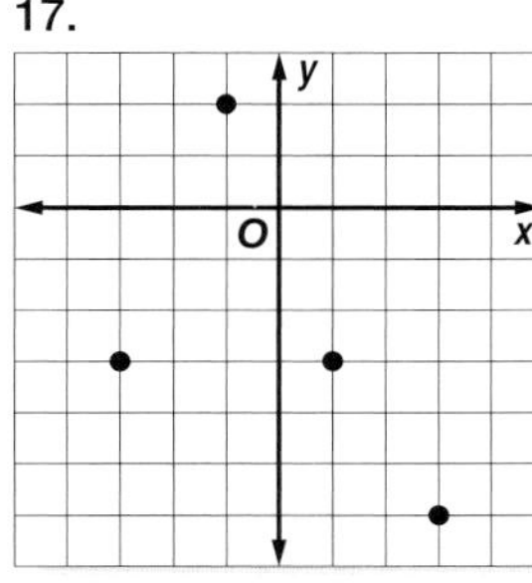

19. (6, 1) **21.** (8, −5) **23.** (2, −2)

Pages 156–158 Lesson 4-1

1. The matrices must have the same dimensions and each element of one matrix must be equal to the corresponding element of the other matrix. **3.** Corresponding elements are elements in the same row and column positions.
5. 3 × 4 **7.** (3, 3) **9.** 2 × 5 **11.** 3 × 1 **13.** 3 × 3
15. 3 × 2 **17.** $\left(3, -\frac{1}{3}\right)$ **19.** (3, −5, 6) **21.** (4, −3)
23. (14, 15) **25.** (5, 3, 2) **27.** 3 × 3 **29.** Sample answer: Mason's Steakhouse; it was given the highest rating possible for service and atmosphere, location was given one of the highest ratings, and it is moderately priced.

31.

	Single	Double	Suite
Weekday	60	70	75
Weekend	79	89	95

33. row 6, column 9 **35.** B **37.** (7, 5, 4) **39.** $\left(-\frac{4}{5}, \frac{3}{5}, -11\right)$

41.

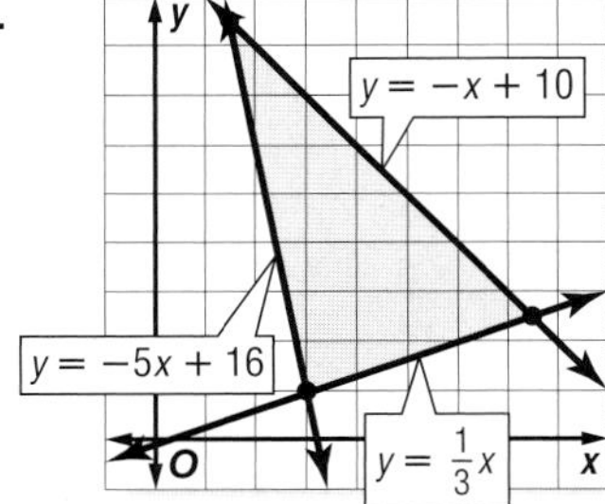

vertices: (3, 1), $\left(\frac{15}{2}, \frac{5}{2}\right)$, $\left(\frac{3}{2}, \frac{17}{2}\right)$; max: $f\left(\frac{15}{2}, \frac{5}{2}\right) = 35$, min: $f\left(\frac{3}{2}, \frac{17}{2}\right) = -1$

43.

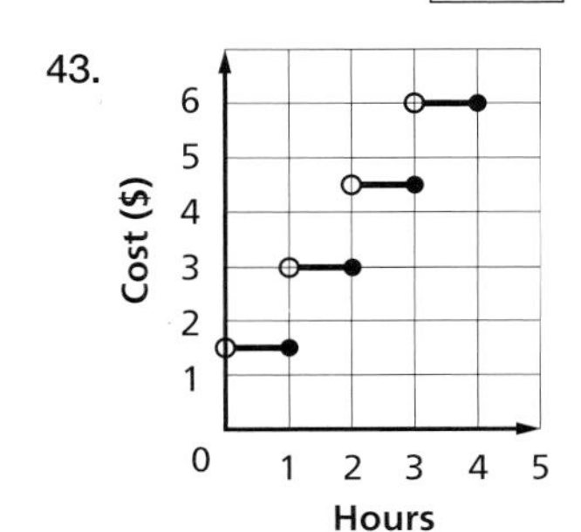

45. $4.50 **47.** 2 **49.** 20
51. −10 **53.** −18 **55.** −3
57. $\frac{3}{2}$

Pages 163–166 Lesson 4-2

1. They must have the same dimensions.

3. $\begin{bmatrix} 4 & 4 \\ 4 & 4 \\ 4 & 4 \end{bmatrix}$ **5.** $\begin{bmatrix} 1 & 10 \\ -7 & 5 \end{bmatrix}$ **7.** $\begin{bmatrix} -22 & 8 \\ 3 & 24 \end{bmatrix}$ **9.** $\begin{bmatrix} -21 & 29 \\ 12 & -22 \end{bmatrix}$

11. Males = $\begin{bmatrix} 16{,}763 & 549{,}499 \\ 14{,}620 & 477{,}960 \\ 14{,}486 & 455{,}305 \\ 9041 & 321{,}416 \\ 5234 & 83{,}411 \end{bmatrix}$, Females = $\begin{bmatrix} 16{,}439 & 456{,}873 \\ 14{,}545 & 405{,}163 \\ 12{,}679 & 340{,}480 \\ 7931 & 257{,}586 \\ 5450 & 133{,}235 \end{bmatrix}$

13. No; many schools offer the same sport for males and females, so those schools would be counted twice.

15. impossible **17.** $\begin{bmatrix} -4 & 8 & -2 \\ 6 & -10 & -16 \\ -14 & -12 & 4 \end{bmatrix}$ **19.** $\begin{bmatrix} -13 \\ -3 \\ 23 \end{bmatrix}$

21. $\begin{bmatrix} 1.5 & 3 \\ 4.5 & 9 \end{bmatrix}$ **23.** $\begin{bmatrix} -5\frac{1}{2} & 3 & 9 \\ 10\frac{2}{3} & 1\frac{2}{3} & -2\frac{1}{2} \end{bmatrix}$ **25.** $\begin{bmatrix} -2 & -1 \\ 4 & -1 \\ -7 & -4 \end{bmatrix}$

27. $\begin{bmatrix} 38 & 4 \\ 32 & -6 \\ 18 & 42 \end{bmatrix}$ **29.** $\begin{bmatrix} 2 & 4\frac{2}{3} \\ 1 & 5 \\ 6 & -1 \end{bmatrix}$ **31.** $\begin{bmatrix} 232 & 184 & 120 & 149 \\ 164 & 124 & 75 & 130 \\ 160 & 182 & 72 & 108 \end{bmatrix}$

33. $\begin{bmatrix} 245 \\ 228 \\ 319 \\ 227 \\ 117 \end{bmatrix}$ **35.** 1996, floods; 1997, floods; 1998, floods; 1999, tornadoes; 2000, lightning

37. $\begin{bmatrix} 1.50 & 2.25 \\ 1.00 & 1.75 \end{bmatrix}$ **39.** $\begin{bmatrix} 1.00 & 1.00 \\ 1.50 & 1.50 \end{bmatrix}$

41. You can use matrices to track dietary requirements and add them to find the total each day or each week. Answers should include the following.

- Breakfast = $\begin{bmatrix} 566 & 18 & 7 \\ 482 & 12 & 17 \\ 530 & 10 & 11 \end{bmatrix}$, Lunch = $\begin{bmatrix} 785 & 22 & 19 \\ 622 & 23 & 20 \\ 710 & 26 & 12 \end{bmatrix}$,

  Dinner = $\begin{bmatrix} 1257 & 40 & 26 \\ 987 & 32 & 45 \\ 1380 & 29 & 38 \end{bmatrix}$

- Add the three matrices: $\begin{bmatrix} 2608 & 80 & 52 \\ 2091 & 67 & 82 \\ 2620 & 65 & 61 \end{bmatrix}$.

43. A **45.** 1 × 4 **47.** 3 × 3 **49.** 4 × 3 **51.** (5, 3, 7)
53. (2, 5) **55.** (6, −1)

57.

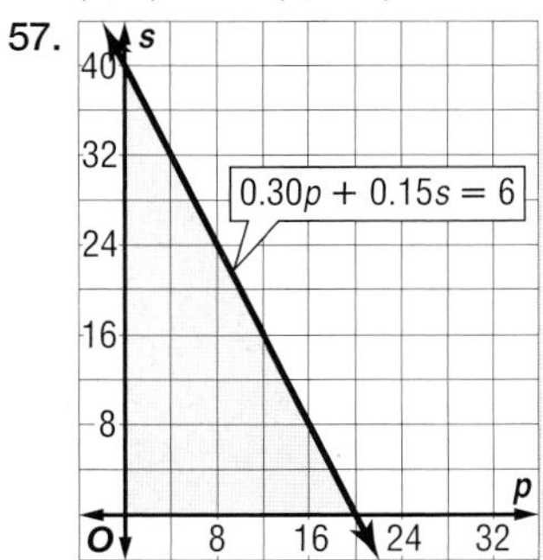

59. Multiplicative Inverse
61. Distributive

Pages 171–174 Lesson 4-3

1. Sample answer: $\begin{bmatrix} 1 & 2 \\ 3 & 4 \\ 5 & 6 \end{bmatrix} \cdot \begin{bmatrix} 7 & 8 \\ 9 & 10 \end{bmatrix}$ **3.** The Right Distributive Property says that $(A + B)C = AC + BC$, but $AC + BC \neq CA + CB$ since the Commutative Property does not hold for matrix multiplication in most cases. **5.** undefined

7. $\begin{bmatrix} 15 & -5 & 20 \\ 24 & -8 & 32 \end{bmatrix}$ **9.** $\begin{bmatrix} 24 \\ 41 \end{bmatrix}$ **11.** [45 55 65], $\begin{bmatrix} 350 & 280 \\ 320 & 165 \\ 180 & 120 \end{bmatrix}$

Selected Answers

13. 4×2 **15.** undefined **17.** undefined

19. [6] **21.** not possible **23.** $\begin{bmatrix} 1 & -25 & 2 \\ 29 & 1 & -30 \end{bmatrix}$

25. $\begin{bmatrix} 24 & 16 \\ -32 & -5 \\ -48 & -11 \end{bmatrix}$

27. yes

$$AC + BC = \begin{bmatrix} 1 & -2 \\ 4 & 3 \end{bmatrix} \cdot \begin{bmatrix} 5 & 1 \\ 2 & -4 \end{bmatrix} + \begin{bmatrix} -5 & 2 \\ 4 & 3 \end{bmatrix} \cdot \begin{bmatrix} 5 & 1 \\ 2 & -4 \end{bmatrix}$$
$$= \begin{bmatrix} 1 & 9 \\ 26 & -8 \end{bmatrix} + \begin{bmatrix} -21 & -13 \\ 26 & -8 \end{bmatrix}$$
$$= \begin{bmatrix} -20 & -4 \\ 52 & -16 \end{bmatrix}$$

$$(A + B)C = \left(\begin{bmatrix} 1 & -2 \\ 4 & 3 \end{bmatrix} + \begin{bmatrix} -5 & 2 \\ 4 & 3 \end{bmatrix}\right) \cdot \begin{bmatrix} 5 & 1 \\ 2 & -4 \end{bmatrix}$$
$$= \begin{bmatrix} -4 & 0 \\ 8 & 6 \end{bmatrix} \cdot \begin{bmatrix} 5 & 1 \\ 2 & -4 \end{bmatrix}$$
$$= \begin{bmatrix} -20 & -4 \\ 52 & -16 \end{bmatrix}$$

29. no

$$C(A + B) = \begin{bmatrix} 5 & 1 \\ 2 & -4 \end{bmatrix} \cdot \left(\begin{bmatrix} 1 & -2 \\ 4 & 3 \end{bmatrix} + \begin{bmatrix} -5 & 2 \\ 4 & 3 \end{bmatrix}\right)$$
$$= \begin{bmatrix} 5 & 1 \\ 2 & -4 \end{bmatrix} \cdot \begin{bmatrix} -4 & 0 \\ 8 & 6 \end{bmatrix}$$
$$= \begin{bmatrix} -12 & 6 \\ -40 & -24 \end{bmatrix}$$

$$AC + BC = \begin{bmatrix} 1 & -2 \\ 4 & 3 \end{bmatrix} \cdot \begin{bmatrix} 5 & 1 \\ 2 & -4 \end{bmatrix} + \begin{bmatrix} -5 & 2 \\ 4 & 3 \end{bmatrix} \cdot \begin{bmatrix} 5 & 1 \\ 2 & -4 \end{bmatrix}$$
$$= \begin{bmatrix} 1 & 9 \\ 26 & -8 \end{bmatrix} + \begin{bmatrix} -21 & -13 \\ 26 & -8 \end{bmatrix}$$
$$= \begin{bmatrix} -20 & -4 \\ 52 & -16 \end{bmatrix}$$

31. $\begin{bmatrix} 290 & 165 & 210 \\ 175 & 240 & 190 \\ 110 & 75 & 0 \end{bmatrix}$ **33.** $\begin{bmatrix} 14{,}285 \\ 13{,}270 \\ 4295 \end{bmatrix}$

35. any two matrices $\begin{bmatrix} a & b \\ c & d \end{bmatrix}$ and $\begin{bmatrix} e & f \\ g & h \end{bmatrix}$ where $bg = cf$, $a = d$, and $e = h$ **37.** $\begin{bmatrix} 96.50 \\ 99.50 \\ 118 \\ 117 \end{bmatrix}$ **39.** \$431 **41.** \$26,360

43. $a = 1$, $b = 0$, $c = 0$, $d = 1$; the original matrix **45.** B

47. $\begin{bmatrix} 12 & -6 \\ -3 & 21 \end{bmatrix}$ **49.** $\begin{bmatrix} -20 & 2 \\ -28 & 12 \end{bmatrix}$ **51.** $(5, -9)$ **53.** \$2.50; \$1.50

55. 8; −16

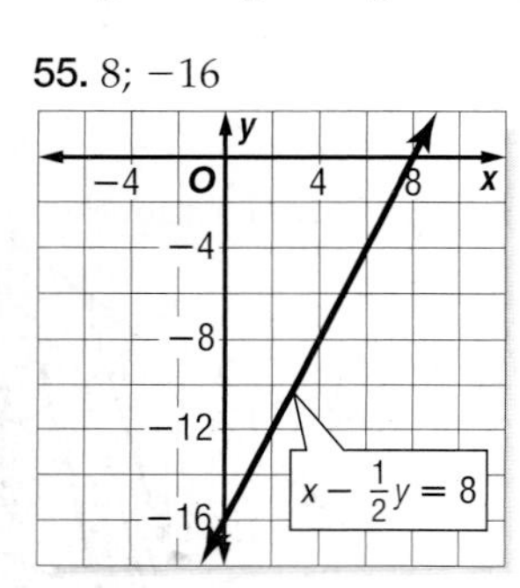

57.

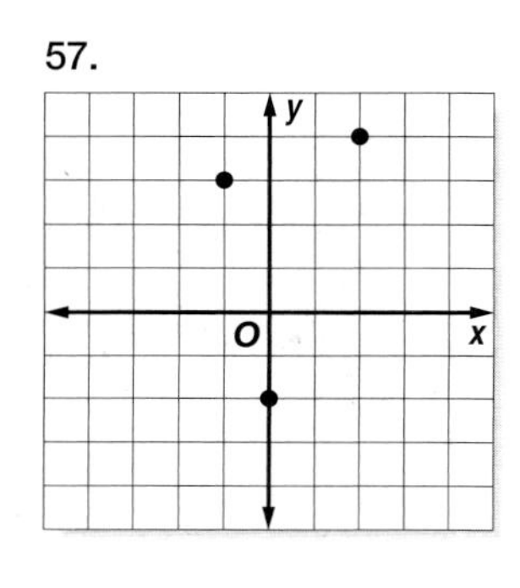

59.

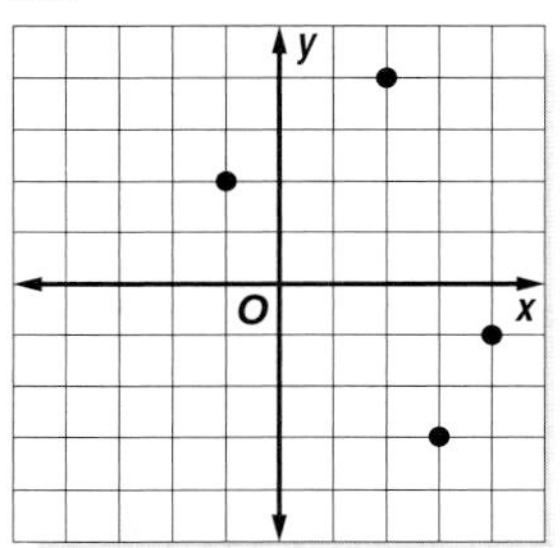

Page 174 Practice Quiz 1

1. (6, 3) **3.** (1, 3, 5) **5.** $\begin{bmatrix} 232 & 159 & 120 & 149 \\ 134 & 200 & 159 & 103 \end{bmatrix}$ **7.** $\begin{bmatrix} 4 & 3 \\ 1 & 3 \end{bmatrix}$

9. not possible

Pages 178–181 Lesson 4-4

1.

Transformation	Size	Shape	Isometry
reflection	same	same	yes
rotation	same	same	yes
translation	same	same	yes
dilation	changes	same	no

3. Sample answer: $\begin{bmatrix} -4 & -4 & -4 \\ 1 & 1 & 1 \end{bmatrix}$ **5.** $A'(4, 3)$, $B'(5, -6)$, $C'(-3, -7)$ **7.** $\begin{bmatrix} 0 & 5 & 5 & 0 \\ 4 & 4 & 0 & 0 \end{bmatrix}$ **9.** $A'(0, -4)$, $B'(5, -4)$, $C'(5, 0)$, $D'(0, 0)$ **11.** B **13.** $D'(-3, 6)$, $E'(-2, -3)$, $F'(-10, -4)$

15. $\begin{bmatrix} 0 & 1.5 & -2.5 \\ 2 & -1.5 & 0 \end{bmatrix}$

17.

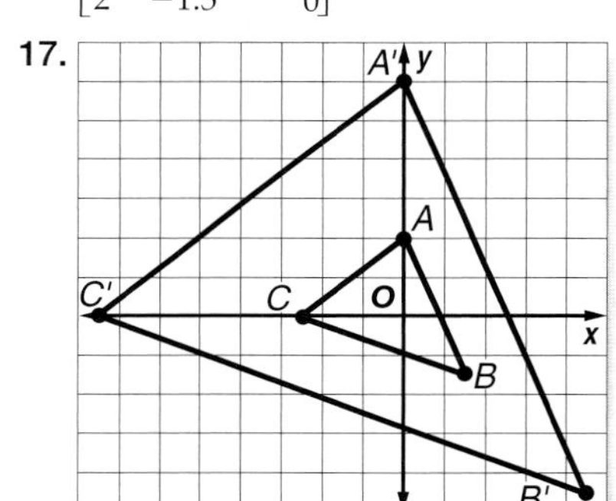

19. $X'(-1, 1)$, $Y'(-4, 2)$, $Z'(-1, 7)$

21. $\begin{bmatrix} 2 & 5 & 4 & 1 \\ 4 & 4 & 1 & 1 \end{bmatrix}$

23.

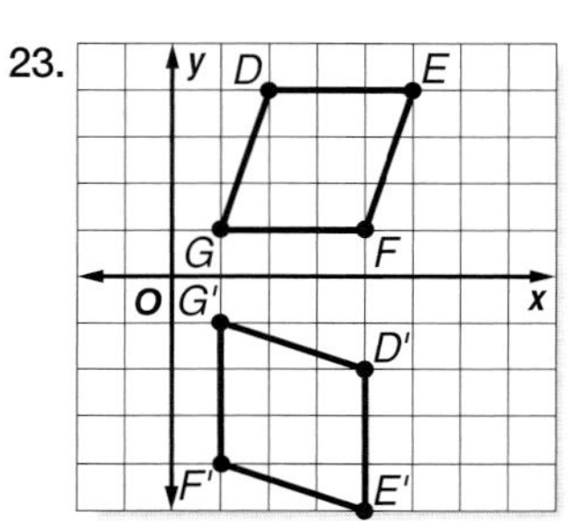

25. $J(-5, 3)$, $K(7, 2)$, $L(4, -1)$ **27.**

29. $\begin{bmatrix} 4 & -4 & -4 & 4 \\ -4 & -4 & 4 & 4 \end{bmatrix}$ **31.** $\begin{bmatrix} 4 & 4 & -4 & -4 \\ -4 & 4 & 4 & -4 \end{bmatrix}$

33. $(-1.5, -1.5), (-4.5, -1.5), (-6, -3.75), (-3, -3.75)$

35. $\begin{bmatrix} 3 \\ 4 \end{bmatrix}$ **37.** $(-8, 7), (-7, -8)$, and $(8, -7)$ **39.** Multiply the coordinates by $\begin{bmatrix} 1 & 0 \\ 0 & -1 \end{bmatrix}$, then add the result to $\begin{bmatrix} 6 \\ 0 \end{bmatrix}$.

41. $(17, -2), (23, 2)$

43. Transformations are used in computer graphics to create special effects. You can simulate the movement of an object, like in space, which you wouldn't be able to recreate otherwise. Answers should include the following.

- A figure with points $(a, b), (c, d), (e, f), (g, h)$, and (i, j) could be written in a 2×5 matrix $\begin{bmatrix} a & c & e & g & i \\ b & d & f & h & j \end{bmatrix}$ and multiplied on the left by the 2×2 rotation matrix.
- The object would get smaller and appear to be moving away from you.

45. A **47.** undefined **49.** $\begin{bmatrix} 11 & 24 & -7 \\ 18 & -13 & 8 \\ 33 & -8 & 21 \end{bmatrix}$

51.

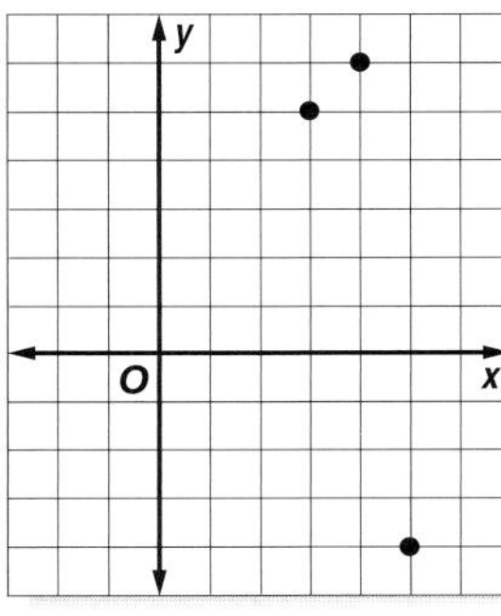

$D = \{3, 4, 5\}$,
$R = \{-4, 5, 6\}$; yes

53.

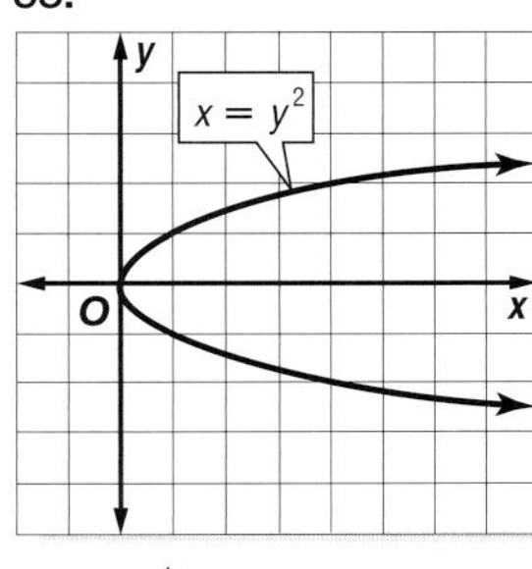

$D = \{x \mid x \geq 0\}$,
$R = \{\text{all real numbers}\}$; no

55. $|x| < 2.8$ **57.** $|x - 1| < 1$ **59.** 6 **61.** 28 **63.** $\frac{9}{4}$

Pages 185–188 Lesson 4-5

1. Sample answer: $\begin{bmatrix} 2 & 1 \\ 8 & 4 \end{bmatrix}$ **3.** It is not a square matrix.

5. Cross out the column and row that contains 6. The minor is the remaining 2×2 matrix. **7.** -38 **9.** -40 **11.** -43 **13.** 45 **15.** 20 **17.** -22 **19.** -29 **21.** 63 **23.** 32 **25.** 32 **27.** -58 **29.** 62 **31.** 172 **33.** -22 **35.** -5 **37.** -141 **39.** -6 **41.** 14.5 units^2 **43.** about 26 ft^2

45. Sample answer: $\begin{vmatrix} 1 & 1 & 1 \\ 1 & 1 & 1 \\ 1 & 1 & 1 \end{vmatrix}$

47. If you know the coordinates of the vertices of a triangle, you can use a determinant to find the area. This is convenient since you don't need to know any additional information such as the measure of the angles. Answers should include the following.

- You could place a coordinate grid over a map of the Bermuda Triangle with one vertex at the origin. By using the scale of the map, you could determine coordinates to represent the other two vertices and use a determinant to estimate the area.
- The determinant method is advantageous since you don't need to physically measure the lengths of each side or the measure of the angles between the vertices.

49. C **51.** -36.9 **53.** -493 **55.** -3252 **57.** $A'(-5, 2.5)$, $B'(2.5, 5)$, $C'(5, -7.5)$ **59.** $[-4]$ **61.** undefined **63.** $[14 \quad -8]$ **65.** 138,435 ft **67.** $y = -\frac{4}{3}x$ **69.** $y = \frac{1}{2}x + 5$ **71.** $(1, 9)$ **73.** $(-1, 1)$ **75.** $(4, 7)$

Pages 192–194 Lesson 4-6

1. The determinant of the coefficient matrix cannot be zero. **3.** $3x + 5y = -6, 4x - 2y = 30$ **5.** $(0.75, 0.5)$ **7.** no solution **9.** $\left(6, -\frac{1}{2}, 2\right)$ **11.** savings account, \$1500; certificate of deposit, \$2500 **13.** $(-12, 4)$ **15.** $(6, 3)$ **17.** $(-0.75, 3)$ **19.** $(-8.5625, -19.0625)$ **21.** $(4, -8)$ **23.** $\left(\frac{2}{3}, \frac{5}{6}\right)$ **25.** $(3, -4)$ **27.** $(2, -1, 3)$ **29.** $\left(\frac{141}{29}, -\frac{102}{29}, \frac{244}{29}\right)$ **31.** $\left(-\frac{155}{28}, \frac{143}{70}, \frac{673}{140}\right)$ **33.** race car, 5 plays; snowboard, 3 plays **35.** silk, \$34.99; cotton, \$24.99 **37.** peanuts, 2 lb; raisins, 1 lb; pretzels, 2 lb **39.** Cramer's Rule is a formula for the variables x and y where (x, y) is a solution for a system of equations. Answers should include the following.

- Cramer's Rule uses determinants composed of the coefficients and constants in a system of linear equations to solve the system.
- Cramer's Rule is convenient when coefficients are large or involve fractions or decimals. Finding the value of the determinant is sometimes easier than trying to find a greatest common factor if you are solving by using elimination or substituting complicated numbers.

41. 111°, 69° **43.** 40 **45.** $\begin{bmatrix} 1 & 1 & 1 \\ 3 & 3 & 3 \end{bmatrix}$

47.

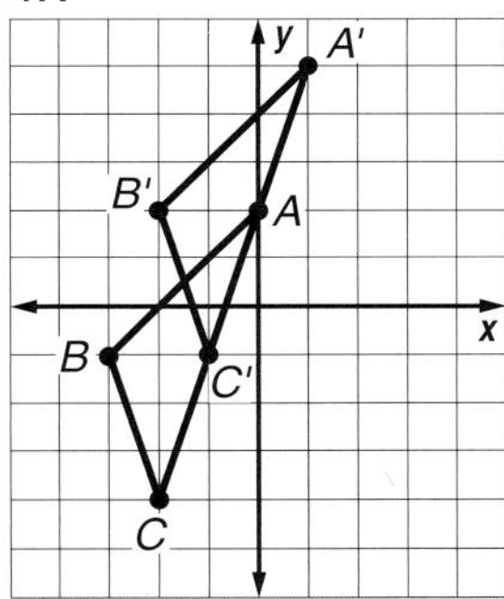

49.

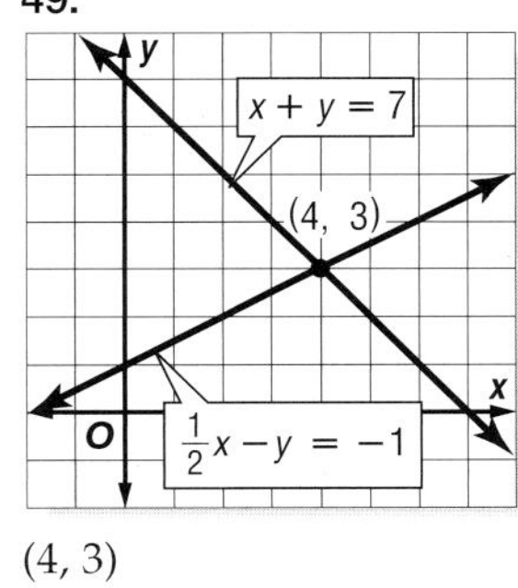

(4, 3)

51. $c = 10h + 35$ **53.** $\begin{bmatrix} 72 & 9 \\ 66 & -23 \end{bmatrix}$

Page 194 Practice Quiz 2

1. $\begin{bmatrix} 1 & 4 & 1 & -2 \\ 2 & -1 & -4 & -1 \end{bmatrix}$

3.

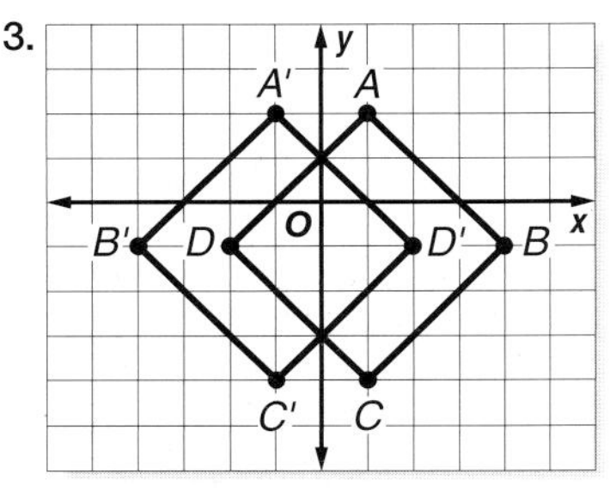

5. -58 **7.** 26 **9.** $(4, -5)$

Selected Answers

Pages 198–201 Lesson 4-7

1. $\begin{bmatrix} 1 & 0 & 0 & 0 \\ 0 & 1 & 0 & 0 \\ 0 & 0 & 1 & 0 \\ 0 & 0 & 0 & 1 \end{bmatrix}$ **3.** Sample answer: $\begin{bmatrix} 3 & 3 \\ 3 & 3 \end{bmatrix}$ **5.** yes

7. no inverse exists **11.** yes **13.** no **15.** yes **17.** true **19.** false **21.** no inverse exists **23.** $\frac{1}{7}\begin{bmatrix} 1 & -1 \\ 4 & 3 \end{bmatrix}$

25. $\frac{1}{4}\begin{bmatrix} -6 & -7 \\ -2 & -3 \end{bmatrix}$ **27.** $-\frac{1}{12}\begin{bmatrix} 6 & 0 \\ -5 & -2 \end{bmatrix}$ **29.** $\frac{1}{32}\begin{bmatrix} 1 & 5 \\ -6 & 2 \end{bmatrix}$

31. $10\begin{bmatrix} \frac{3}{4} & -\frac{5}{8} \\ -\frac{1}{5} & \frac{3}{10} \end{bmatrix}$ **33a.** yes

33b. Sample answer:

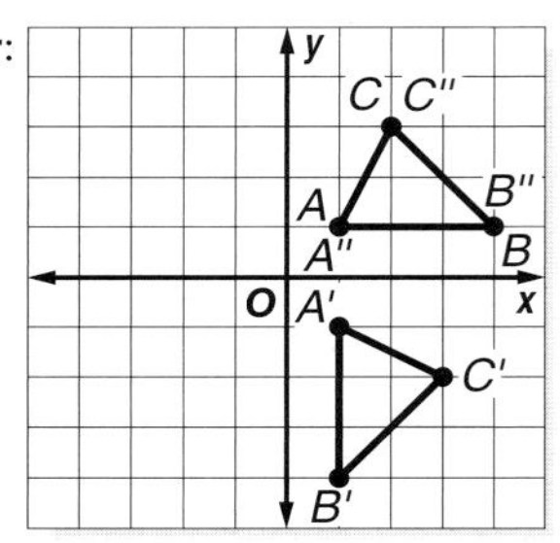

35. $\begin{bmatrix} 0 & -4 & 4 & 8 \\ 0 & 4 & 12 & 8 \end{bmatrix}$ **37.** dilation by a scale factor of $\frac{1}{2}$

39. MEET_IN_THE_LIBRARY **41.** BRING_YOUR_BOOK

43. $a = \pm 1, d = \pm 1, b = c = 0$ **45.** A **47.** $\begin{bmatrix} -5 & -9 \\ -6 & -11 \end{bmatrix}$

49. $\begin{bmatrix} \frac{3}{5} & -\frac{1}{5} \\ \frac{1}{5} & -\frac{2}{5} \end{bmatrix}$ **51.** $\begin{bmatrix} -1 & 1 & \frac{1}{3} \\ -\frac{1}{3} & \frac{2}{3} & 0 \\ \frac{7}{3} & -\frac{8}{3} & -\frac{1}{3} \end{bmatrix}$ **53.** $(2, -4)$

55. $(-5, 4, 1)$ **57.** -14 **59.** 1 **61.** -5 **63.** $\frac{5}{2}$ **65.** 7.82 tons/in^2 **67.** $5\frac{1}{2}$ **69.** 3 **71.** 300 **73.** -2 **75.** 4 **77.** -34

Pages 205–207 Lesson 4-8

1. $2r - 3s = 4, r + 4s = -2$ **3.** Tommy; a 2×1 matrix cannot be multiplied by a 2×2 matrix.

5. $\begin{bmatrix} 2 & 3 \\ -4 & -7 \end{bmatrix} \cdot \begin{bmatrix} g \\ h \end{bmatrix} = \begin{bmatrix} 8 \\ -5 \end{bmatrix}$ **7.** $(5, -2)$ **9.** $(-3, 5)$

11. $h = 1, c = 12$ **13.** $\begin{bmatrix} 4 & -7 \\ 3 & 5 \end{bmatrix} \cdot \begin{bmatrix} x \\ y \end{bmatrix} = \begin{bmatrix} 2 \\ 9 \end{bmatrix}$

15. $\begin{bmatrix} 3 & -7 \\ 6 & 5 \end{bmatrix} \cdot \begin{bmatrix} m \\ n \end{bmatrix} = \begin{bmatrix} -43 \\ -10 \end{bmatrix}$ **17.** $\begin{bmatrix} 3 & -5 & 2 \\ 1 & -7 & 3 \\ 4 & 0 & -3 \end{bmatrix} \cdot \begin{bmatrix} x \\ y \\ z \end{bmatrix} = \begin{bmatrix} 9 \\ 11 \\ -1 \end{bmatrix}$

19. $\begin{bmatrix} 3 & -5 & 6 \\ 11 & -12 & 16 \\ -5 & 8 & -3 \end{bmatrix} \cdot \begin{bmatrix} r \\ s \\ t \end{bmatrix} = \begin{bmatrix} 21 \\ 15 \\ -7 \end{bmatrix}$ **21.** $(3, 4)$ **23.** $(6, 1)$

25. $\left(-\frac{1}{3}, 4\right)$ **27.** $(-2, -2)$ **29.** $(0, 9)$ **31.** $\left(\frac{3}{2}, \frac{1}{3}\right)$ **33.** 2010 **35.** The solution set is the empty set or infinite solutions.

37. D **39.** $(-6, 2, 5)$ **41.** $(0, -1, 3)$ **43.** $\begin{bmatrix} 4 & -5 \\ -7 & 9 \end{bmatrix}$

45. $(4, -2)$ **47.** $(-6, -8)$ **49.** $\{-4, 10\}$ **51.** $\{2, 7\}$

Pages 209–214 Chapter 4 Study Guide and Review

1. identity matrix **3.** Scalar multiplication **5.** determinant **7.** dimensions **9.** equal matrices **11.** $(-5, -1)$ **13.** $(-1, 0)$

15. $\begin{bmatrix} -3 & 0 \\ -2 & -6 \end{bmatrix}$ **17.** $\begin{bmatrix} 1 & -2 \\ -14 & 9 \end{bmatrix}$ **19.** $[-18]$ **21.** not possible

23. $A'(1, 0), B'(8, -2), C'(3, -7)$ **25.** $A'(3, 5), B'(-4, 3), C'(1, -2)$ **27.** 109 **29.** 0 **31.** -52 **33.** $\left(\frac{2}{3}, 5\right)$ **35.** $(-1, -3)$

37. $(1, 2, -1)$ **39.** $-\frac{1}{14}\begin{bmatrix} -2 & -2 \\ -4 & 3 \end{bmatrix}$ **41.** $\frac{1}{24}\begin{bmatrix} 6 & -4 \\ 3 & 2 \end{bmatrix}$

43. $-\frac{1}{10}\begin{bmatrix} -4 & -2 \\ -5 & 0 \end{bmatrix}$ **45.** $(4, 2)$ **47.** $(-3, 1)$

Chapter 5 Polynomials

Page 221 Chapter 5 Getting Started

1. $2 + (-7)$ **3.** $x + (-y)$ **5.** $2xy + (-6yz)$ **7.** $-8x^3 - 2x + 6$ **9.** $-x + 3$ **11.** $-\frac{3}{2}a - 1$ **13.** 6.3; reals, rationals **15.** 17; reals, rationals, integers, whole numbers, natural numbers **17.** 4; reals, rationals, integers, whole numbers, natural numbers

Pages 226–228 Lesson 5-1

1. Sample answer: $(2x^2)^3 = 8x^6$ since $(2x^2)^3 = (2x^2)^3 \cdot (2x^2)^3 \cdot (2x^2)^3 = 2x^2 \cdot 2x^2 \cdot 2x^2 = 2x \cdot x \cdot 2x \cdot x \cdot 2x \cdot x = 8x^6$

3. Alejandra; when Kyle used the Power of a Product property in his first step, he forgot to put an exponent of -2 on a. Also, in his second step, $(-2)^{-2}$ should be $\frac{1}{4}$, not 4.

5. $16b^4$ **7.** $-6y^2$ **9.** $9p^2q^3$ **11.** $\frac{9}{c^2d^2}$ **13.** 4.21×10^5 **15.** 3.762×10^3 **17.** about 1.28 s **19.** b^4 **21.** z^{10} **23.** $-8c^3$ **25.** $-y^3z^2$ **27.** $-21b^5c^3$ **29.** $-24r^7s^5$ **31.** $90a^4b^4$

33. $\frac{a^2c^2}{3b^4}$ **35.** $-\frac{m^4n^9}{3}$ **37.** $\frac{8y^3}{x^6}$ **39.** $\frac{1}{v^3w^6}$ **41.** $\frac{2x^3y^2}{5z^7}$ **43.** 7

45. 4.32×10^4 **47.** 6.81×10^{-3} **49.** 6.754×10^8 **51.** 6.02×10^{-5} **53.** 6.2×10^{10} **55.** 1.681×10^{-7} **57.** 2×10^{-7} m **59.** about 330,000 times

61. Definition of an exponent

63. Economics often involves large amounts of money. Answers should include the following.

- The national debt in 2000 was five trillion, six hundred seventy-four billion, two hundred million or 5.6742×10^{12} dollars. The population was two hundred eighty-one million or 2.81×10^8.
- Divide the national debt by the population. $\frac{5.6742 \times 10^{12}}{2.81 \times 10^8} \approx \$2.0193 \times 10^4$ or about \$20,193 per person.

65. B **67.** $(-3, 3)$ **69.** $\begin{bmatrix} -\frac{1}{2} & \frac{3}{2} \\ 1 & -2 \end{bmatrix}$ **71.** 7 **73.** $(2, 0, 4)$

75. Sample answer using (0, 4.9) and (28, 8.3): $y = 0.12x + 4.9$ **77.** 7 **79.** $2x + 2y$ **81.** $4x + 8$ **83.** $-5x + 10y$

Pages 231–232 Lesson 5-2

1. Sample answer: $x^5 + x^4 + x^3$

3.

	x	x	x
x	x^2	x^2	x^2
2	x	x	x
	x	x	x

5. yes, 3 **7.** $10a - 2b$ **9.** $6xy + 18x$ **11.** $y^2 - 3y - 70$ **13.** $4z^2 - 1$ **15.** $7.5x^2 + 12.5x$ ft^2 **17.** yes, 3 **19.** no **21.** yes, 7 **23.** $-3y - 3y^2$ **25.** $10m^2 + 5m - 15$

27. $7x^2 - 8xy + 4y^2$ **29.** $12a^3 + 4ab$
31. $6x^2y^4 - 8x^2y^2 + 4xy^5$ **33.** $2a^4 - 3a^3b + 4a^4b^4$
35. $-0.001x^2 + 5x - 500$ **37.** $p^2 + 2p - 24$ **39.** $b^2 - 25$
41. $6x^2 + 34x + 48$ **43.** $a^6 - b^2$ **45.** $x^2 - 6xy + 9y^2$
47. $d^2 - 2 + \frac{1}{d^4}$ **49.** $27b^3 - 27b^2c + 9bc^2 - c^3$
51. $9c^2 - 12cd + 7d^2$ **53.** $R^2 + 2RW + W^2$
55. The expression for how much an amount of money will grow to is a polynomial in terms of the interest rate. Answers should include the following.
- If an amount A grows by r percent for n years, the amount will be $A(1 + r)^n$ after n years. When this expression is expanded, a polynomial results.
- $8820(1 + r)^3$, $8820r^3 + 26{,}460r^2 + 26{,}460r + 8820$
- Evaluate one of the expressions when $r = 0.04$. For example, $8820(1 + r)^3 = 8820(1.04)^3$ or \$9921.30 to the nearest cent. The value given in the table is \$9921 rounded to the nearest dollar.

57. B **59.** $20r^3t^4$ **61.** $\frac{b^2}{4a^2}$

63.

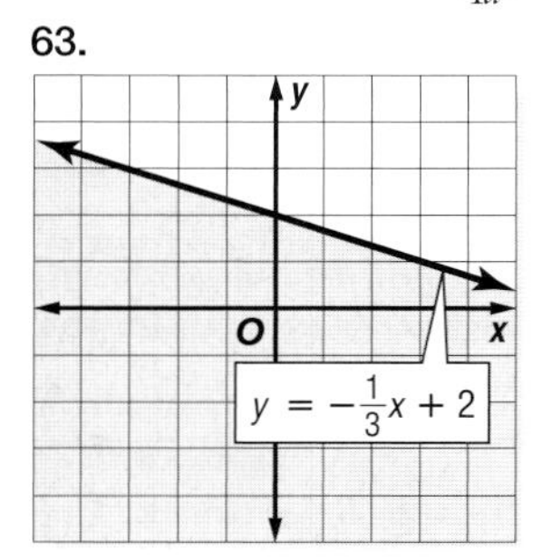

65.

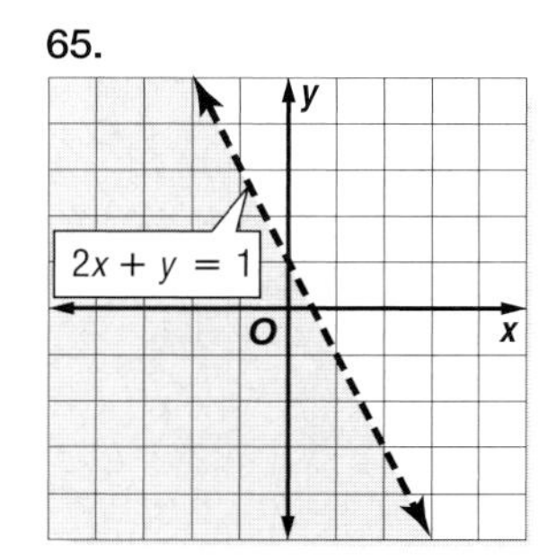

67. $2y^3$ **69.** $3a^2$

Pages 236–238 Lesson 5-3

1. Sample answer: $(x^2 + x + 5) \div (x + 1)$ **3.** Jorge; Shelly is subtracting in the columns instead of adding.
5. $5b - 4 + 7a$ **7.** $3a^3 - 9a^2 + 7a - 6$ **9.** $x^2 - xy + y^2$
11. $b^3 + b - 1$ **13.** $3b + 5$ **15.** $3ab - 6b^2$ **17.** $2c^2 - 3d + 4d^2$ **19.** $2y^2 + 4yz - 8y^3z^4$ **21.** $b^2 + 10b$ **23.** $n^2 - 2n + 3$
25. $x^3 - 5x^2 + 11x - 22 + \frac{39}{x + 2}$ **27.** x^2 **29.** $y^2 - y - 1$
31. $a^3 - 6a^2 - 7a + 7 + \frac{3}{a + 1}$
33. $x^4 - 3x^3 + 2x^2 - 6x + 19 - \frac{56}{x + 3}$ **35.** $g + 5$
37. $t^4 + 2t^3 + 4t^2 + 5t + 10$ **39.** $3t^2 - 2t + 3$
41. $3d^2 + 2d + 3 - \frac{2}{3d - 2}$ **43.** $x^3 - x - \frac{6}{2x + 3}$ **45.** $x - 3$
47. $x + 2$ **49.** $x^2 - x + 3$ **51.** $\$0.03x + 4 + \frac{1000}{x}$
53. $170 - \frac{170}{t^2 + 1}$ **55.** $x^3 + x^2 + 6x - 24$ ft
57. $x^2 + 3x + 12$ ft/s **59.** Division of polynomials can be used to solve for unknown quantities in geometric formulas that apply to manufacturing situations. Answers should include the following.
- $8x$ in. by $4x + s$ in.
- The area of a rectangle is equal to the length times the width. That is, $A = \ell w$.
- Substitute $32x^2 + x$ for A, $8x$ for ℓ, and $4x + s$ for w. Solving for s involves dividing $32x^2 + x$ by $8x$.

$$A = \ell w$$
$$32x^2 + x = 8x(4x + s)$$
$$\frac{32x^2 + x}{8x} = 4x + s$$
$$4x + \frac{1}{8} = 4x + s$$
$$\frac{1}{8} = s$$

The seam is $\frac{1}{8}$ inch.

61. D **63.** $y^4z^4 - y^3z^3 + 3y^2z$ **65.** $a^2 - 2ab + b^2$
67. $y = -x + 2$ **69.** 9 **71.** 4 **73.** 6

Page 238 Practice Quiz 1

1. 6.53×10^8 **3.** $-108x^8y^3$ **5.** $\frac{x^2}{z^6}$ **7.** $3t^2 + 2t - 8$
9. $m^2 - 3 - \frac{19}{m - 4}$

Pages 242–244 Lesson 5-4

1. Sample answer: $x^2 + 2x + 1$ **3.** sometimes
5. $a(a + 5 + b)$ **7.** $(y - 2)(y - 4)$ **9.** $3(b - 4)(b + 4)$
11. $(h + 20)(h^2 - 20h + 400)$ **13.** $\frac{2y}{y - 4}$ **15.** $2x(y^3 - 5)$
17. $2cd^2(6d - 4c + 5c^4d)$ **19.** $(2z - 3)(4y - 3)$
21. $(x + 1)(x + 6)$ **23.** $(2a + 1)(a + 1)$ **25.** $(2c + 3)(3c + 2)$
27. $3(n + 8)(n - 1)$ **29.** $(x + 6)^2$ **31.** prime
33. $(y^2 + z)(y^2 - z)$ **35.** $(z + 5)(z^2 - 5z + 25)$
37. $(p^2 + 1)(p + 1)(p - 1)$ **39.** $(7a + 2b)(c + d)(c - d)$
41. $(a - b)(5ax + 4by + 3cz)$ **43.** $(3x - 2)(x + 1)$
45. 30 ft by 40 ft **47.** $\frac{x + 5}{x - 6}$ **49.** $\frac{x - 4}{x^2 + 2x + 4}$ **51.** $x + 2$
53. $16x + 16$ ft/s **55.** $(8p^n + 1)^2$ **57.** B **59.** yes **61.** no; $(2x + 1)(x - 3)$ **63.** $t^2 - 2t + 1$ **65.** $x^2 + 2$
67. $4x^2 + 3xy - 3y^2$ **69.** $[-2]$ **71.** 15 in. by 28 in. **73.** no
75. Associative Property (+) **77.** irrational **79.** rational
81. rational

Pages 247–249 Lesson 5-5

1. Sample answer: 64 **3.** Sometimes; it is true when $x > 0$.
5. -2.668 **7.** 4 **9.** -3 **11.** x **13.** $6|a|b^2$ **15.** about 3.01 mi
17. -12.124 **19.** 2.066 **21.** -7.830 **23.** 3.890 **25.** 4.647
27. 59.161 **29.** ± 13 **31.** 18 **33.** -2 **35.** $\frac{1}{5}$ **37.** -0.4
39. $-|x|$ **41.** $8a^4$ **43.** $-c^2$ **45.** $4z^2$ **47.** $6x^2z^2$
49. $3p^6|q^3|$ **51.** $-3c^3d^4$ **53.** $p + q$ **55.** $|z + 4|$ **57.** not a real number **59.** -5 **61.** about 1.35 m **63.** $x = 0$ and $y \geq 0$, or $y = 0$ and $x \geq 0$ **65.** B **67.** $7xy^2(y - 2xy^3 + 4x^2)$
69. $(2x + 5)(x + 5)$ **71.** $4x^2 + x + 5 + \frac{8}{x - 2}$
73. $\begin{bmatrix} 810 & 2320 \\ 1418 & 2504 \end{bmatrix}$ **75.** $(1, -3)$ **77.** $x^2 + 11x + 24$
79. $a^2 - 7a - 18$ **81.** $x^2 - 9y^2$

Pages 254–256 Lesson 5-6

1. Sometimes; $\frac{1}{\sqrt[n]{a}} = \sqrt[n]{a}$ only when $a = 1$. **3.** The product of two conjugates yields a difference of two squares. Each square produces a rational number and the difference of two rational numbers is a rational number. **5.** $2x|y|\sqrt[4]{x}$
7. $-24\sqrt{35}$ **9.** $2a^2b^2\sqrt{3}$ **11.** $22\sqrt[3]{2}$ **13.** $2 + \sqrt{5}$
15. $9\sqrt{3}$ **17.** $3\sqrt[3]{2}$ **19.** $5x^2\sqrt{2}$ **21.** $3|x|y\sqrt{2y}$
23. $6y^2z\sqrt[3]{7}$ **25.** $\frac{1}{3}c|d|\sqrt[4]{c}$ **27.** $\frac{\sqrt[3]{6}}{2}$ **29.** $\frac{a^2\sqrt{b}}{b^2}$ **31.** $36\sqrt{7}$
33. $\frac{\sqrt{6}}{2}$ **35.** $3\sqrt{3}$ **37.** $7\sqrt{3} - 2\sqrt{2}$
39. $25 - 5\sqrt{2} + 5\sqrt{6} - 2\sqrt{3}$ **41.** $13 - 2\sqrt{22}$
43. $\frac{28 + 7\sqrt{3}}{13}$ **45.** $\frac{-1 - \sqrt{3}}{2}$ **47.** $\frac{\sqrt{x^2 - 1}}{x - 1}$ **49.** $6 + 16\sqrt{2}$ yd, $24 + 6\sqrt{2}$ yd^2 **51.** 0 ft/s **53.** about 18.18 m **55.** x and y are nonnegative. **57.** B **59.** $12z^4$ **61.** $|y + 2|$
63. $\frac{x + 1}{x + 4}$ **65.** $\begin{bmatrix} 1 & 4 \\ -5 & -4 \end{bmatrix}$ **67.** consistent and independent
69. -5 **71.** $-2, 4$ **73.** $\{x \mid x > 6\}$ **75.** $\frac{1}{4}$ **77.** $\frac{5}{6}$ **79.** $\frac{13}{24}$
81. $\frac{3}{8}$

Selected Answers

Page 256 Practice Quiz 2

1. $x^2y(3x + y + 1)$ **3.** $a(x + 3)^2$ **5.** $6|x||y^3|$ **7.** $|2n + 3|$ **9.** $-1 - \sqrt{7}$

Pages 260–262 Lesson 5-7

1. Sample answer: 64 **3.** In exponential form $\sqrt[n]{b^m}$ is equal to $(b^m)^{\frac{1}{n}}$. By the Power of a Power Property, $(b^m)^{\frac{1}{n}} = b^{\frac{m}{n}}$. But, $b^{\frac{m}{n}}$ is also equal to $\left(b^{\frac{1}{n}}\right)^m$ by the Power of a Power Property. This last expression is equal to $\left(\sqrt[n]{b}\right)^m$. Thus, $\sqrt[n]{b^m} = \left(\sqrt[n]{b}\right)^m$. **5.** $\sqrt[3]{x^2}$ or $\left(\sqrt[3]{x}\right)^2$ **7.** $6^{\frac{1}{3}}x^{\frac{5}{3}}y^{\frac{7}{3}}$ **9.** $\frac{1}{3}$ **11.** 2 **13.** $x^{\frac{2}{3}}$ **15.** $a^{\frac{3}{2}}b^{\frac{2}{3}}$ **17.** $\frac{z(x - 2y)^{\frac{1}{2}}}{x - 2y}$ **19.** $\sqrt{3}$ **21.** $\sqrt[5]{6}$ **23.** $\sqrt[5]{c^2}$ or $\left(\sqrt[5]{c}\right)^2$ **25.** $23^{\frac{1}{2}}$ **27.** $2z^{\frac{1}{2}}$ **29.** 2 **31.** $\frac{1}{5}$ **33.** $\frac{1}{9}$ **35.** 81 **37.** $\frac{2}{3}$ **39.** $\frac{4}{3}$ **41.** y^4 **43.** $b^{\frac{1}{5}}$ **45.** $\frac{w^{\frac{1}{5}}}{w}$ **47.** $t^{\frac{1}{4}}$ **49.** $\frac{a^{\frac{5}{12}}}{6a}$ **51.** $\frac{y^2 - 2y^{\frac{3}{2}}}{y - 4}$ **53.** $\sqrt{5}$ **55.** $17\sqrt[6]{17}$ **57.** $\sqrt[4]{5x^2y^2}$ **59.** $\frac{xy\sqrt{z}}{z}$ **61.** $\sqrt{2}$ **63.** $2\sqrt{6} - 5$ **65.** $2^{\frac{3}{2}} + 3^{\frac{1}{2}}$

67. 880 vibrations per second **69.** about 336

71. The equation that determines the size of the region around a planet where the planet's gravity is stronger than the Sun's can be written in terms of a fractional exponent. Answers should include the following.

- The radical form of the equation is $r = D\sqrt[5]{\left(\frac{M_p}{M_s}\right)^2}$ or $r = D\sqrt[5]{\frac{M_p^2}{M_s^2}}$. Multiply the fraction under the radical by $\frac{M_s^3}{M_s^3}$.

$$r = D\sqrt[5]{\frac{M_p^2}{M_s^2} \cdot \frac{M_s^3}{M_s^3}}$$

$$= D\sqrt[5]{\frac{M_p^2 M_s^3}{M_s^5}}$$

$$= D\,\frac{\sqrt[5]{M_p^2 M_s^3}}{\sqrt[5]{M_s^5}}$$

$$= \frac{D\sqrt[5]{M_p^2 M_s^3}}{M_s}$$

The simplified radical form is $r = \frac{D\sqrt[5]{M_p^2 M_s^3}}{M_s}$.

- If M_p and M_s are constant, then r increases as D increases because r is a linear function of D with positive slope.

73. C **75.** $36\sqrt{2}$ **77.** 8 **79.** $\frac{1}{2}x^2$ **81.** $x - 2$ **83.** $x + 2\sqrt{x} + 1$

Pages 265–267 Lesson 5-8

1. Since x is not under the radical, the equation is a linear equation, not a radical equation. The solution is $x = \frac{\sqrt{3} - 1}{2}$. **3.** Sample answer: $\sqrt{x} + \sqrt{x + 3} = 3$ **5.** -9 **7.** 15 **9.** 31 **11.** $0 \leq b < 4$ **13.** 16 **15.** no solution **17.** 9 **19.** -1 **21.** -20 **23.** no solution **25.** $x > 1$ **27.** $x \leq -11$ **29.** no solution **31.** 3 **33.** $0 \leq x \leq 2$ **35.** $b \geq 5$ **37.** 3 **39.** 1152 lb **41.** 34 ft **43.** Since $\sqrt{x + 2} \geq 0$ and $\sqrt{2x - 3} \geq 0$, the left side of the equation is nonnegative. Therefore, the left side of the equation cannot equal -1. Thus, the equation has no solution. **45.** D **47.** $5^{\frac{3}{7}}$

49. $(x^2 + 1)^{\frac{2}{3}}$ **51.** $\frac{\sqrt[3]{100}}{10}$

53. $x + y = 7$, $30x + 20y = 160$; $(2, 5)$

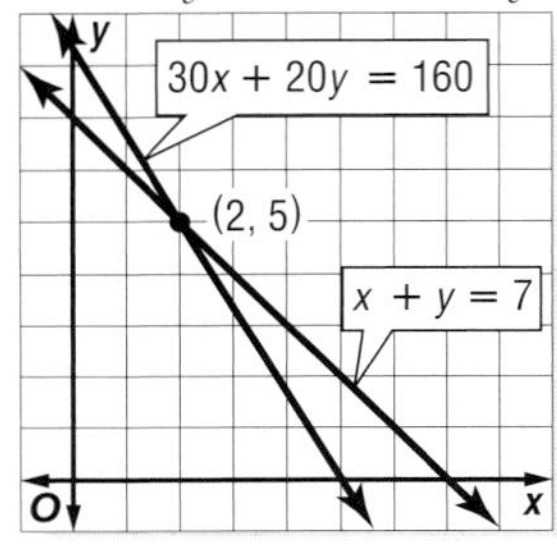

55. $1 - y$ **57.** -11 **59.** $-3 - 10x - 8x^2$

Pages 273–275 Lesson 5-9

1a. true **1b.** true **3.** Sample answer: $1 + 3i$ and $1 - 3i$ **5.** $5i|xy|\sqrt{2}$ **7.** $-180\sqrt{3}$ **9.** $6 + 3i$ **11.** $\frac{7}{17} - \frac{11}{17}i$ **13.** $\pm 2i\sqrt{2}$ **15.** 3, -3 **17.** $10 + 3j$ amps **19.** $9i$ **21.** $10a^2|b|i$ **23.** -12 **25.** $-75i$ **27.** 1 **29.** $-i$ **31.** 6 **33.** $4 - 5i$ **35.** $6 - 7i$ **37.** $-8 + 4i$ **39.** $\frac{10}{17} - \frac{6}{17}i$ **41.** $\frac{2}{5} + \frac{1}{5}i$ **43.** $20 + 15i$ **45.** $-\frac{1}{3} - \frac{2\sqrt{2}}{3}i$ **47.** $(5 - 2i)x^2 + (-1 + i)x + 7 + i$ **49.** $\pm 4i$ **51.** $\pm 2i\sqrt{3}$ **53.** $\pm 2i\sqrt{10}$ **55.** $\pm\frac{\sqrt{5}}{2}i$ **57.** 4, -3 **59.** $\frac{5}{3}$, 4 **61.** $\frac{67}{11}$, $\frac{19}{11}$ **63.** $13 + 18j$ volts

65. Case 1: $i > 0$

Multiply each side by i to get $i^2 > 0 \cdot i$ or $-1 > 0$. This is a contradiction.

Case 2: $i < 0$

Since you are assuming i is negative in this case, you must change the inequality symbol when you multiply each side by i. The result is again $i^2 > 0 \cdot i$ or $-1 > 0$, a contradiction.

Since both possible cases result in contradictions, the order relation "<" cannot be applied to the complex numbers.

67. C **69.** $-1, -i, 1, i, -1, -i, 1, i, -1$ **71.** 12 **73.** 4 **75.** $y^{\frac{1}{3}}$ **77.** $\begin{bmatrix} 2 & 1 & -2 \\ 3 & -2 & 1 \end{bmatrix}$ **79.** $\begin{bmatrix} 2 & 1 & -2 \\ -3 & 2 & -1 \end{bmatrix}$

81. sofa: \$1200, love seat: \$600, coffee table: \$250

83.

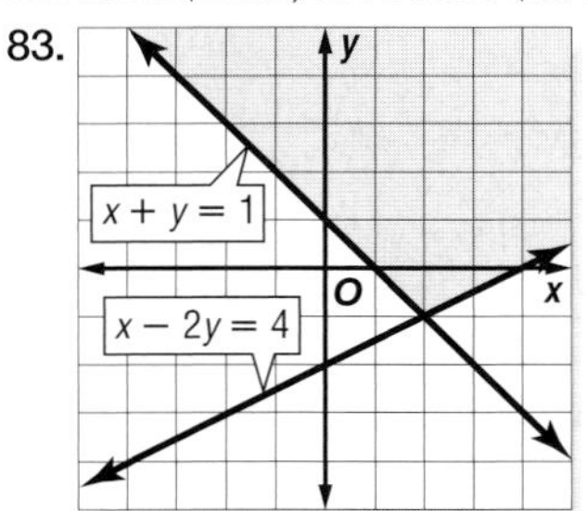

85. 0

Pages 276–280 Chapter 5 Study Guide and Review

1. scientific notation **3.** FOIL method **5.** extraneous solution **7.** square root **9.** principal root **11.** $\frac{1}{f^3}$ **13.** $8xy^4$ **15.** 1.7×10^8 **17.** 9×10^2 **19.** $4x^2 + 22x - 34$ **21.** $x^3y + x^2y^4$ **23.** $4a^4 + 24a^2 + 36$ **25.** $2x^3 + x - \frac{3}{x - 3}$ **27.** $x - 4$ **29.** $50(2x + 1)(2x - 1)$ **31.** $(5w^2 + 3)(w - 4)$ **33.** $(s + 8)(s^2 - 8s + 64)$ **35.** ± 16 **37.** 8 **39.** $|x^4 - 3|$ **41.** $2m^2$ **43.** $2\sqrt{2}$ **45.** $-5\sqrt{3}$ **47.** $20 + 8\sqrt{6}$ **49.** 9

51. $\frac{2\sqrt{10} - \sqrt{5}}{7}$ 53. 81 55. $\frac{y^{\frac{3}{5}}}{y}$ 57. $3x^{\frac{5}{3}} + 4x^{\frac{8}{3}}$ 59. 343 61. 4 63. 5 65. 8 67. $8m^6i$ 69. 72 71. $23 + 14i$ 73. i 75. $\frac{-3 - 21i}{10}$

Chapter 6 Quadratic Functions and Inequalities

Page 284 Chapter 6 Getting Started

1.

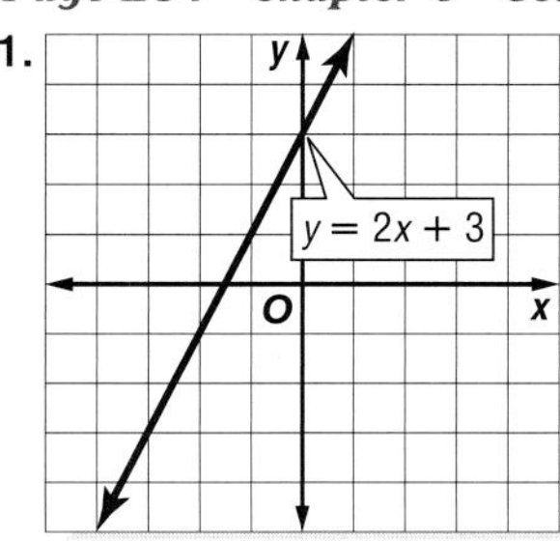

3.

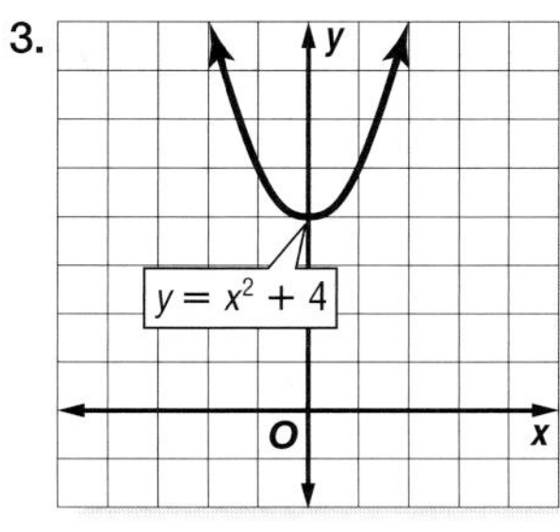

5. $7x^2 - 16x - 48$ 7. $9x^2 - 6x + 1$ 9. $(x + 6)(x + 5)$ 11. $(x - 8)(x + 7)$ 13. prime 15. $(x - 11)^2$ 17. 15 19. $6\sqrt{5}$ 21. $5i$ 23. $3i\sqrt{30}$

Pages 290–293 Lesson 6-1

1. Sample answer: $f(x) = 3x^2 + 5x - 6$; $3x^2, 5x, -6$
3a. up; min. 3b. down; max. 3c. down; max.
3d. up; min. 5a. 0; $x = -1$; -1

5b.

x	$f(x)$
−3	3
−2	0
−1	−1
0	0
1	3

5c.

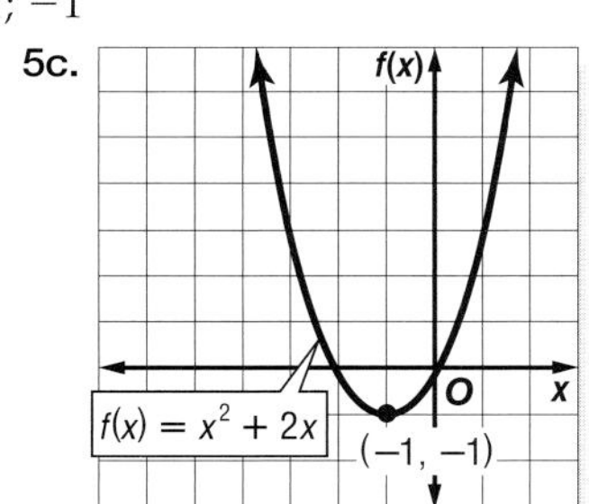

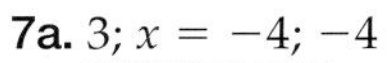

7a. 3; $x = -4$; -4

7b.

x	$f(x)$
−6	−9
−5	−12
−4	−13
−3	−12
−2	−9

7c.

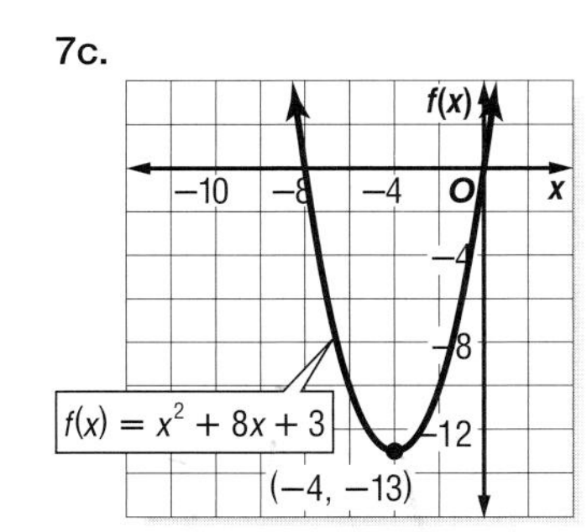

9a. 0; $x = -\frac{5}{3}$; $-\frac{5}{3}$

9b.

x	$f(x)$
−3	−3
−2	−8
$-\frac{5}{3}$	$-\frac{25}{3}$
−1	−7
0	0

9c.

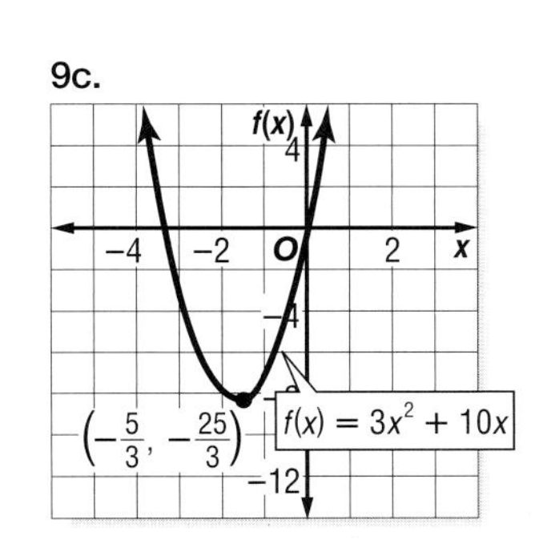

11. min.; $-\frac{25}{4}$ 13. $8.75 15a. 0; $x = 0$; 0

15b.

x	$f(x)$
−2	−20
−1	−5
0	0
1	−5
2	−20

15c.

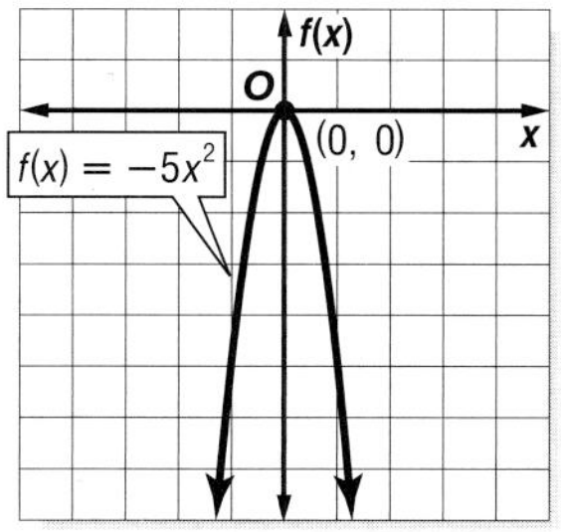

17a. −9; $x = 0$; 0

17b.

x	$f(x)$
−2	−5
−1	−8
0	−9
1	−8
2	−5

17c.

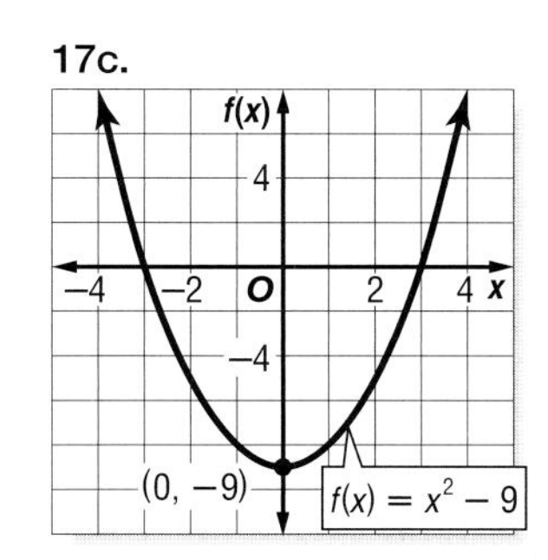

19a. 1; $x = 0$; 0

19b.

x	$f(x)$
−2	13
−1	4
0	1
1	4
2	13

19c.

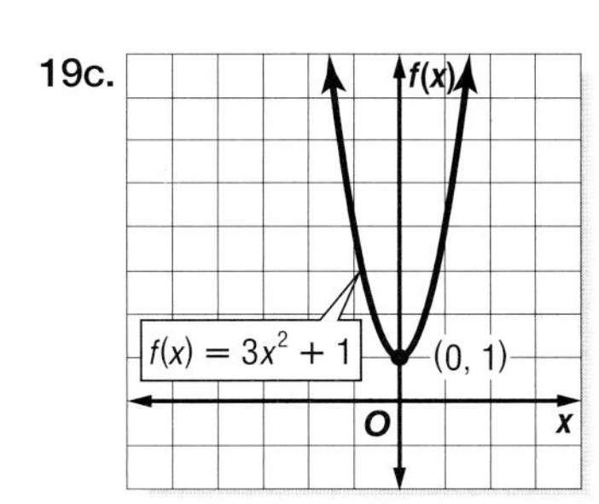

21a. 9; $x = 4.5$; 4.5

21b.

x	$f(x)$
3	−9
4	−11
4.5	−11.25
5	−11
6	−9

21c.

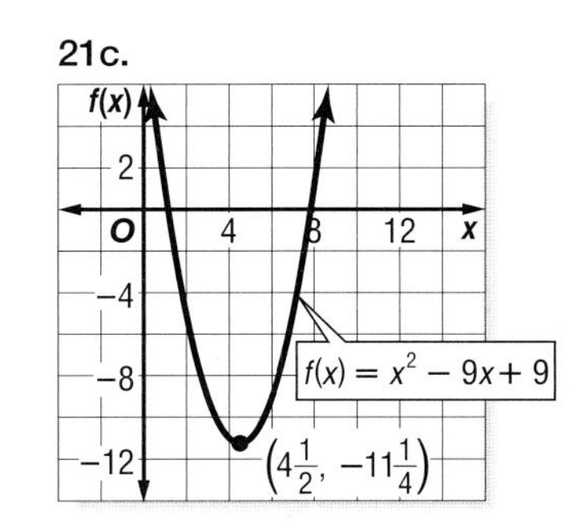

23a. 36; $x = -6$; -6

23b.

x	$f(x)$
−8	4
−7	1
−6	0
−5	1
−4	4

23c.

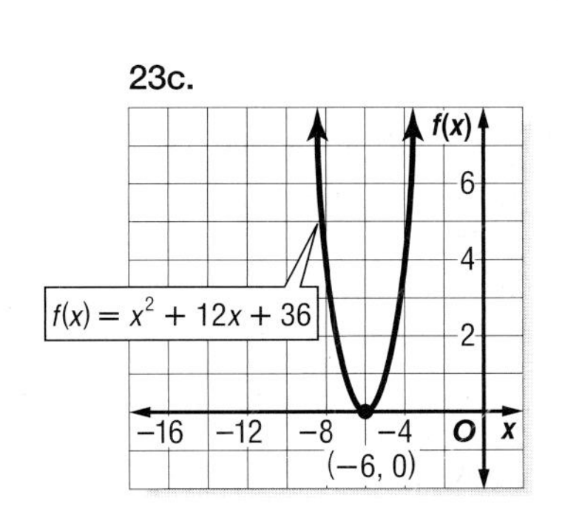

25a. -3; $x = 2$, 2

25b.

x	$f(x)$
0	-3
1	3
2	5
3	3
4	-3

25c.

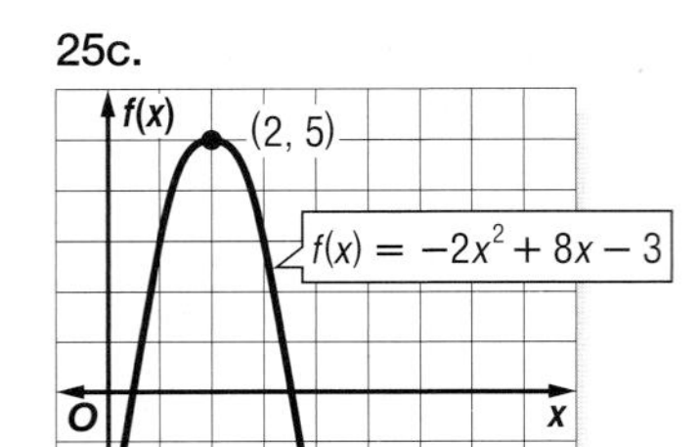

27a. 0; $x = -\frac{5}{4}$; $-\frac{5}{4}$

27b.

x	$f(x)$
-3	-3
-2	-2
$-\frac{5}{4}$	$-\frac{25}{8}$
-1	-3
0	0

27c.

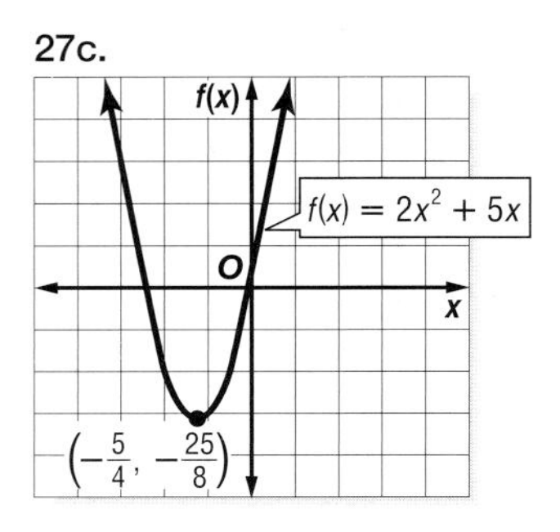

29a. 0; $x = -6$; -6

29b.

x	$f(x)$
-8	8
-7	8.75
-6	9
-5	8.75
-4	8

29c.

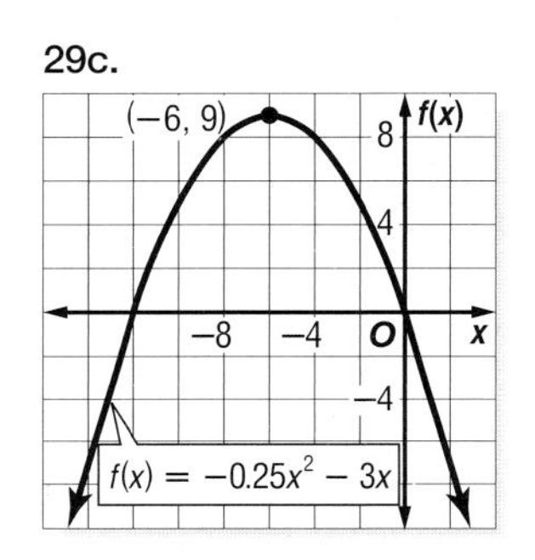

31a. $-\frac{8}{9}$; $x = \frac{1}{3}$; $\frac{1}{3}$

31b.

x	$f(x)$
-1	$\frac{7}{9}$
0	$-\frac{8}{9}$
$\frac{1}{3}$	-1
1	$-\frac{5}{9}$
2	$1\frac{7}{9}$

31c.

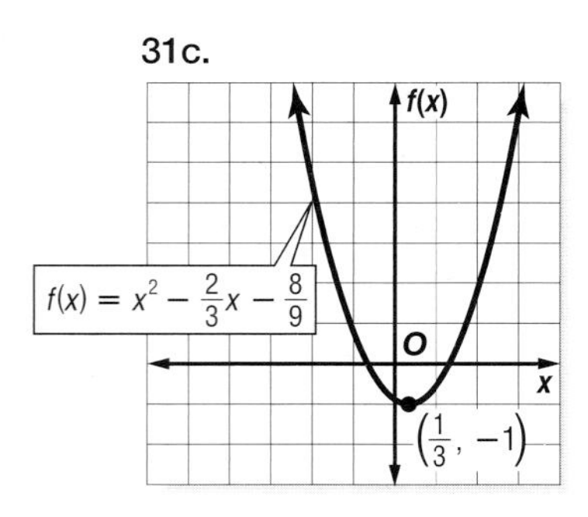

33. max.; -9 **35.** min.; -11 **37.** max.; 12 **39.** max.; $-\frac{7}{8}$ **41.** min.; -11 **43.** min.; $-10\frac{1}{3}$ **45.** 40 m
47. The y-intercept is the initial height of the object.
49. 60 ft by 30 ft **51.** \$11.50 **53.** 5 in. by 4 in.
55. If a quadratic function can be used to model ticket price versus profit, then by finding the x-coordinate of the vertex of the parabola you can determine the price per ticket that should be charged to achieve maximum profit. Answers should include the following.

- If the price of a ticket is too low, then you won't make enough money to cover your costs, but if the ticket price is too high fewer people will buy them.
- You can locate the vertex of the parabola on the graph of the function. It occurs when $x = 40$. Algebraically, this is found by calculating $x = -\frac{b}{2a}$ which, for this case, is $x = \frac{-4000}{2(-50)}$ or 40. Thus the ticket price should be set at \$40 each to achieve maximum profit.

57. C **59.** 3.20 **61.** 3.38 **63.** 1.56 **65.** $-1 + 3i$ **67.** 23
69. 4 **71.** $[5 \quad -13 \quad 8]$ **73.** $\begin{bmatrix} 6 & 0 & -24 \\ 14 & -\frac{2}{3} & -8 \end{bmatrix}$ **75.** 5 **77.** -2

Pages 297–299 Lesson 6-2

1a. The solution is the value that satisfies an equation.
1b. A root is a solution of an equation. **1c.** A zero is the x value of a function that makes the function equal to 0.
1d. An x-intercept is the point at which a graph crosses the x-axis. The solutions, or roots, of a quadratic equation are the zeros of the related quadratic function. You can find the zeros of a quadratic function by finding the x-intercepts of its graph. **3.** The x-intercepts of the related function are the solutions to the equation. You can estimate the solutions by stating the consecutive integers between which the x-intercepts are located. **5.** $-2, 1$ **7.** $-7, 0$ **9.** $-7, 4$
11. between -2 and -1, 3 **13.** $-2, 7$ **15.** 3 **17.** 0
19. no real solutions **21.** 0, 4 **23.** between -1 and 0; between 2 and 3 **25.** 3, 6 **27.** 6 **29.** $-\frac{1}{2}, 2\frac{1}{2}$ **31.** $-2\frac{1}{2}, 3$
33. between 0 and 1; between 3 and 4 **35.** between -3 and -2; between 2 and 3 **37.** no real solutions
39. Let x be the first number.
Then, $7 - x$ is the other number.

$$x(7 - x) = 14$$
$$-x^2 + 7x - 14 = 0$$

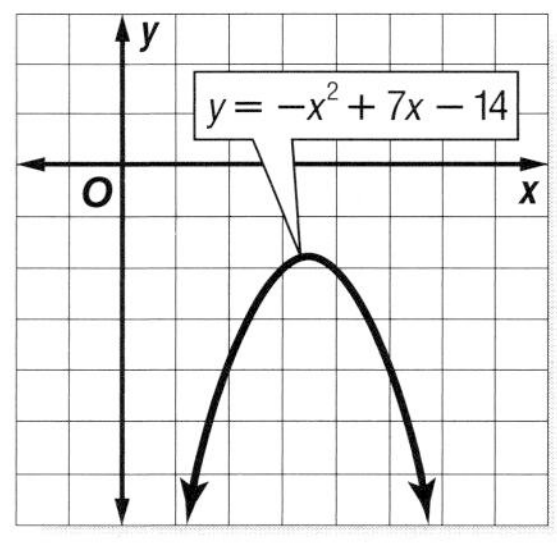

Since the graph of the related function does not intersect the x-axis, this equation has no real solutions. Therefore no such numbers exist. **41.** $-2, 14$ **43.** 3 s **45.** about 35 mph **47.** -4 and -2; The value of the function changes from negative to positive, therefore the value of the function is zero between these two numbers. **49.** A **51.** -1 **53.** 3, 5 **55.** ± 1.33

57. 4, $x = 3$; 3

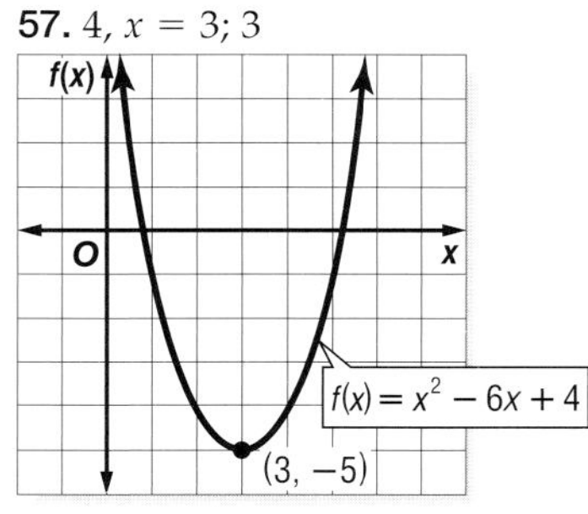

59. 4; $x = -6$; -6

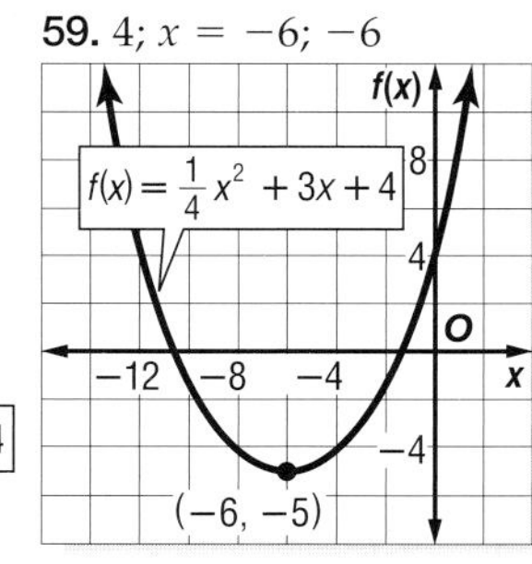

61. $\frac{10}{13} + \frac{2}{13}i$ **63.** 24 **65.** -60 **67.** $x(x + 5)$
69. $(x - 7)(x - 4)$ **71.** $(3x + 2)(x + 2)$

Pages 303–305 Lesson 6-3

1. Sample answer: If the product of two factors is zero, then at least one of the factors must be zero. **3.** Kristin; the Zero Product Property applies only when one side of the equation is 0. **5.** $\{-8, 2\}$ **7.** $\{3\}$ **9.** $\{-3, 4\}$

11. $6x^2 - 11x + 4 = 0$ **13.** D **15.** $\{-4, 7\}$ **17.** $\{-9, 9\}$
19. $\{-3, 7\}$ **21.** $\left\{0, -\frac{3}{4}\right\}$ **23.** $\{8\}$ **25.** $\left\{\frac{1}{4}, 4\right\}$ **27.** $\left\{-\frac{2}{3}, -\frac{3}{2}\right\}$
29. $\left\{\frac{3}{4}, \frac{9}{4}\right\}$ **31.** $\{-3, 1\}$ **33.** $0, -3, 3$ **35.** $x^2 - 5x - 14 = 0$
37. $x^2 + 14x + 48 = 0$ **39.** $3x^2 - 16x + 5 = 0$
41. $10x^2 + 23x + 12 = 0$ **43.** 14, 16 or −14, −16
45. $B = D^2 - 8D + 16$
47. $y = (x - p)(x - q)$
$y = x^2 - px - qx + pq$
$y = x^2 - (p + q)x + pq$
$a = 1, b = -(p + q), c = +pq$
axis of symmetry: $x = -\frac{b}{2a}$
$x = -\frac{-(p + q)}{2(1)}$
$x = \frac{p + q}{2}$
The axis of symmetry is the average of the x-intercepts. Therefore the axis of symmetry is located halfway between the x-intercepts. **49.** −6 **51.** D **53.** −5, 1 **55.** between −1 and 0; between 3 and 4 **57.** $3\sqrt{2} - 2\sqrt{3}$
59. $33 + 20\sqrt{2}$ **61.** $(3, -5)$ **63.** $2\sqrt{2}$ **65.** $3\sqrt{3}$
67. $2i\sqrt{3}$

Page 305 Practice Quiz 1

1. 4; $x = 2$; 2

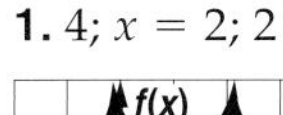

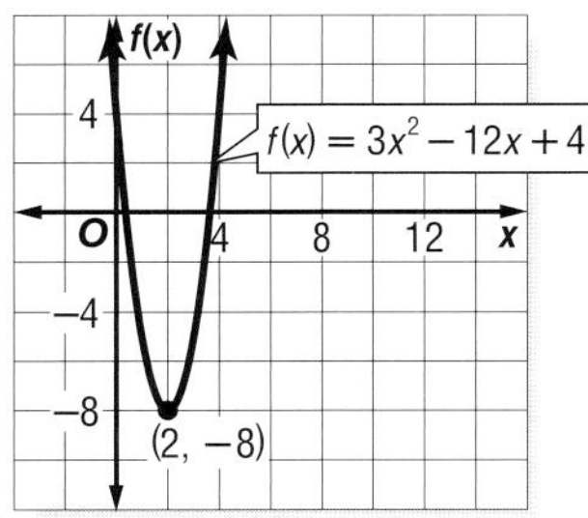

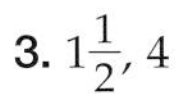

3. $1\frac{1}{2}$, 4
5. $3x^2 + 11x - 4 = 0$

Pages 310–312 Lesson 6-4

1. Completing the square allows you to rewrite one side of a quadratic equation in the form of a perfect square. Once in this form, the equation is solved by using the Square Root Property. **3.** Tia; before completing the square, you must first check to see that the coefficient of the quadratic term is 1. If it is not, you must first divide the equation by that coefficient. **5.** $\left\{\frac{4 \pm \sqrt{2}}{3}\right\}$ **7.** $\frac{9}{4}$; $\left(x - \frac{3}{2}\right)^2$ **9.** $\{4 \pm \sqrt{5}\}$
11. $\left\{\frac{3 \pm \sqrt{33}}{4}\right\}$ **13.** Earth: 4.5 s, Jupiter: 2.9 s
15. $\{-2, 12\}$ **17.** $\{3 \pm 2\sqrt{2}\}$ **19.** $\left\{\frac{-5 \pm \sqrt{11}}{3}\right\}$ **21.** $\{-1.6, 0.2\}$
23. about 8.56 s **25.** 81; $(x - 9)^2$ **27.** $\frac{49}{4}$; $\left(x + \frac{7}{2}\right)^2$
29. 1.44; $(x - 1.2)^2$ **31.** $\frac{25}{16}$; $\left(x + \frac{5}{4}\right)^2$ **33.** $\{-12, 10\}$
35. $\{2 \pm \sqrt{3}\}$ **37.** $\{-3 \pm 2i\}$ **39.** $\left\{\frac{1}{2}, 1\right\}$ **41.** $\left\{\frac{2 \pm \sqrt{10}}{3}\right\}$
43. $\left\{\frac{-5 \pm i\sqrt{23}}{6}\right\}$ **45.** $\{0.7, 4\}$ **47.** $\left\{\frac{3}{4} \pm \sqrt{2}\right\}$ **49.** $\frac{x}{1}, \frac{1}{x - 1}$
51. Sample answers: The golden rectangle is found in much of ancient Greek architecture, such as the Parthenon, as well as in modern architecture, such as in the windows of the United Nations building. Many songs have their climax at a point occurring 61.8% of the way through the piece, with 0.618 being about the reciprocal of the golden ratio. The reciprocal of the golden ratio is also used in the design of some violins. **53.** 18 ft by 32 ft or 64 ft by 9 ft

55. D **57.** $x^2 - 3x + 2 = 0$ **59.** $3x^2 - 19x + 6 = 0$
61. between −4 and −3; between 0 and 1 **63.** $-4, -1\frac{1}{2}$
65. $(2, -5)$ **67.** $|x - (-257)| = 2$ **69.** 37 **71.** 121

Pages 317–319 Lesson 6-5

1a. Sample answer:

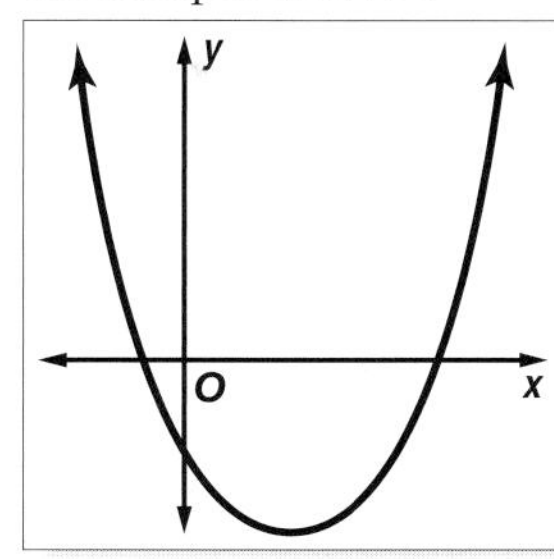

1b. Sample answer:

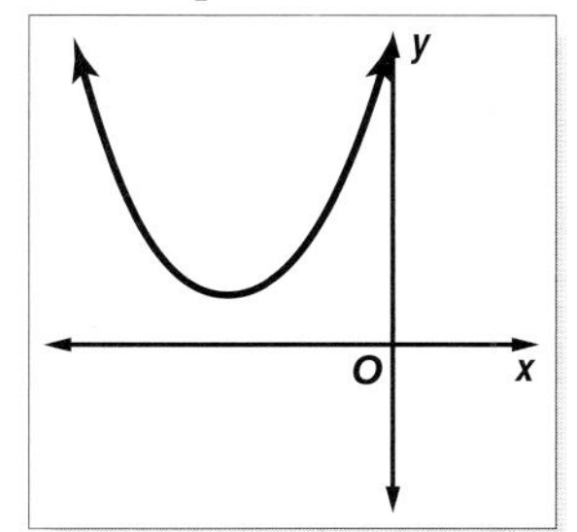

1c. Sample answer:

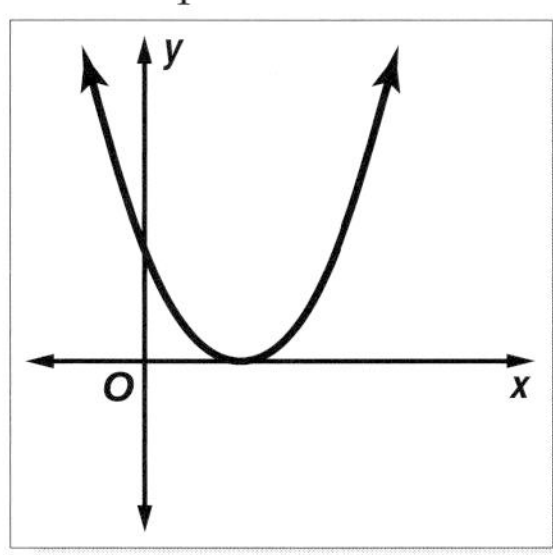

3. $b^2 - 4ac$ must equal 0. **5a.** 8 **5b.** 2 irrational
5c. $\frac{2 \pm \sqrt{2}}{2}$ **7a.** −3 **7b.** two complex **7c.** $\frac{-3 \pm i\sqrt{3}}{2}$
9. −3, −2 **11.** $\frac{-5 \pm i\sqrt{2}}{2}$ **13.** No; the discriminant of $-16t^2 + 85t = 120$ is −455, indicating that the equation has no real solutions. **15a.** 240 **15b.** 2 irrational
15c. $8 \pm 2\sqrt{15}$ **17a.** −23 **17b.** 2 complex **17c.** $\frac{1 \pm i\sqrt{23}}{2}$
19a. 49 **19b.** 2 rational **19c.** $-2, \frac{1}{3}$ **21a.** 24
21b. 2 irrational **21c.** $-1 \pm \sqrt{6}$ **23a.** 0 **23b.** one rational
23c. $-\frac{5}{2}$ **25a.** −135 **25b.** 2 complex **25c.** $\frac{-1 \pm i\sqrt{15}}{4}$
27a. 1.48 **27b.** 2 irrational **27c.** $\frac{-1 \pm 2\sqrt{0.37}}{0.8}$ **29.** $\pm i\frac{\sqrt{21}}{7}$
31. $\frac{-3 \pm \sqrt{15}}{2}$ **33.** $\frac{9}{2}$ **35.** $\frac{5 \pm \sqrt{46}}{3}$ **37.** $0, -\frac{3}{10}$ **39.** −2, 6
41. This means that the cables do not touch the floor of the bridge, since the graph does not intersect the x-axis and the roots are imaginary. **43.** 1998 **45a.** $k = \pm 6$ **45b.** $k < -6$ or $k > 6$ **45c.** $-6 < k < 6$ **47.** D **49.** −14, −4
51. $\frac{1 \pm 2\sqrt{2}}{2}$ **53.** −2, 7 **55.** a^4b^{10} **57.** $4b^2c^2$

59.

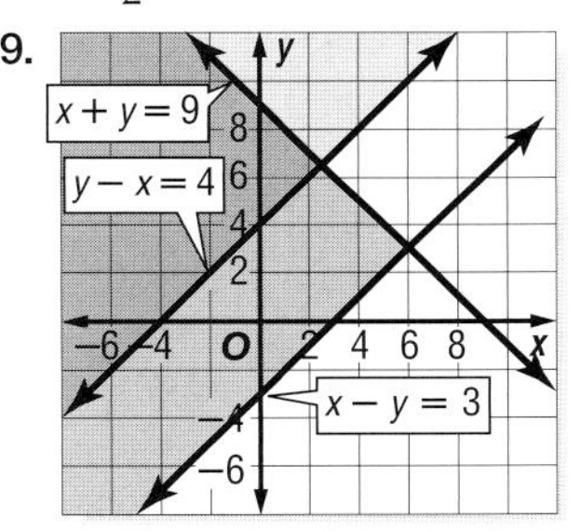

61. no **63.** yes; $(2x + 3)^2$ **65.** no

Selected Answers

Pages 325–328 Lesson 6-6

1a. $y = 2(x + 1)^2 + 5$ **1b.** $y = 2(x + 1)^2$
1c. $y = 2(x + 3)^2 + 3$ **1d.** $y = 2(x - 2)^2 + 3$
1e. Sample answer: $y = 4(x + 1)^2 + 3$ **1f.** Sample answer: $y = (x + 1)^2 + 3$ **1g.** $y = -2(x + 1)^2 + 3$ **3.** Sample answer: $y = 2(x - 2)^2 - 1$ **5.** $(-3, -1)$; $x = -3$; up
7. $y = -3(x + 3)^2 + 38$; $(-3, 38)$; $x = -3$; down

9.

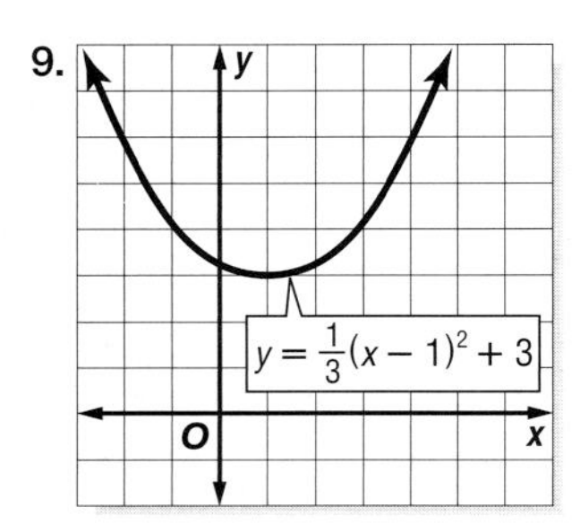

11. $y = 4(x - 2)^2$ **13.** $y = -\frac{1}{2}(x + 2)^2 - 3$ **15.** $(-3, 0)$; $x = -3$; down **17.** $(0, -6)$; $x = 0$; up
19. $y = -(x + 2)^2 + 12$; $(-2, 12)$; $x = -2$; down
21. $y = -3(x - 2)^2 + 12$; $(2, 12)$; $x = 2$; down
23. $y = 4(x + 1)^2 - 7$; $(-1, -7)$; $x = -1$; up
25. $y = 3\left(x + \frac{1}{2}\right)^2 - \frac{7}{4}$; $\left(-\frac{1}{2}, -\frac{7}{4}\right)$; $x = -\frac{1}{2}$; up

27.

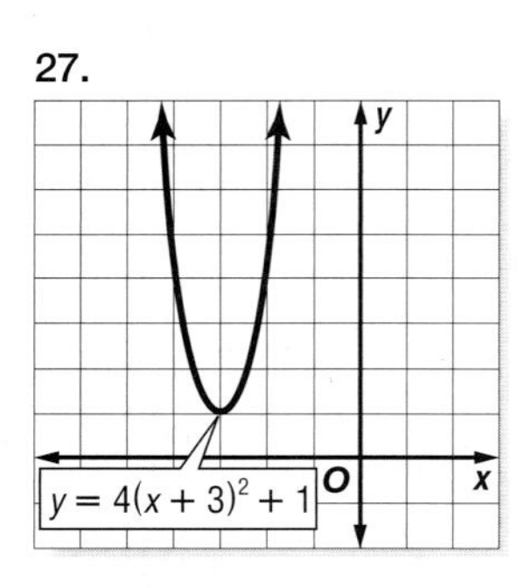

29.

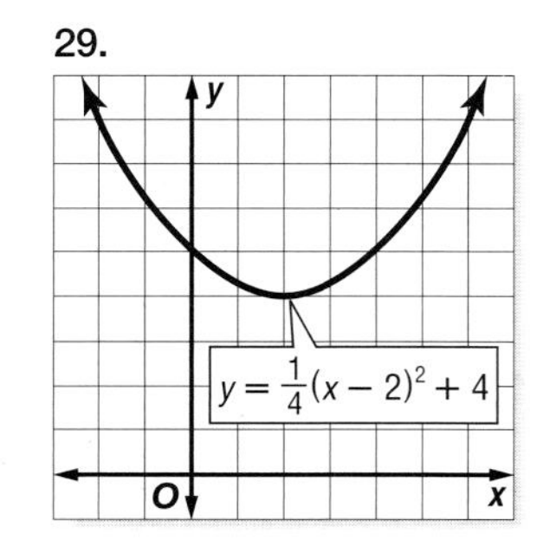

31.

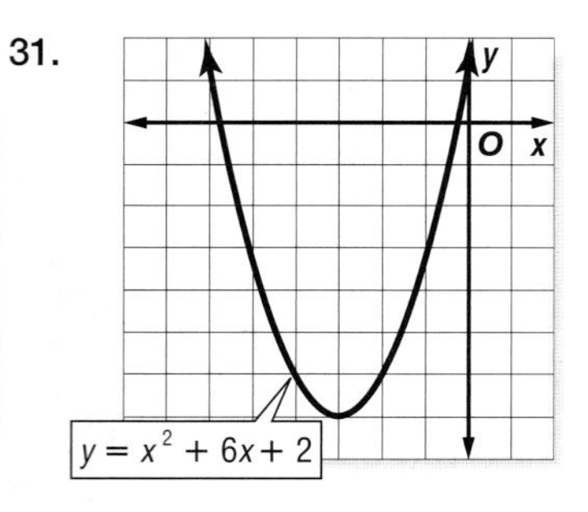

33.

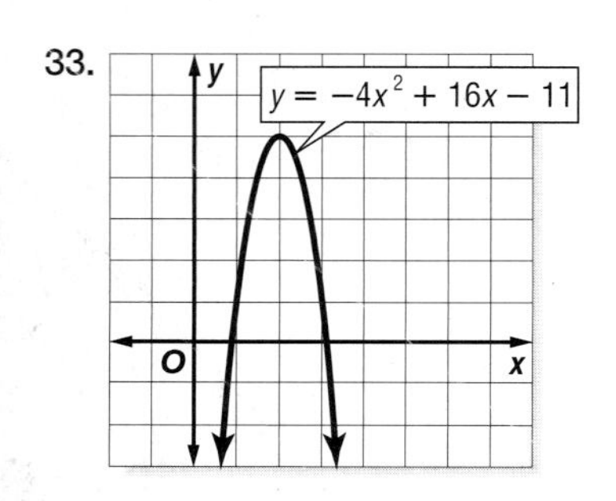

35.

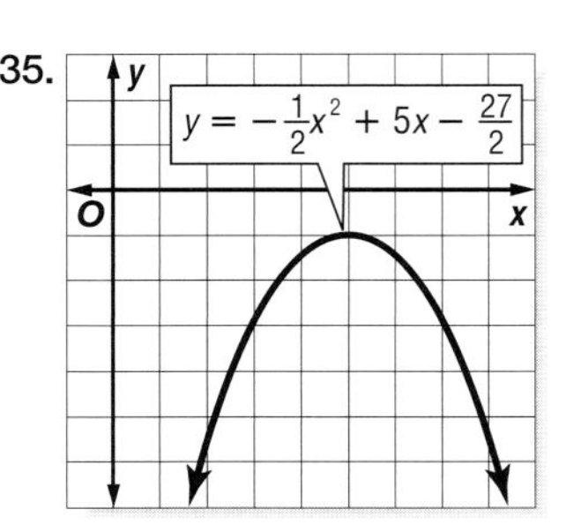

37. Sample answer: the graph of $y = 0.4(x + 3)^2 + 1$ is narrower than the graph of $y = 0.2(x + 3)^2 + 1$.
39. $y = 9(x - 6)^2 + 1$ **41.** $y = -\frac{2}{3}(x - 3)^2$ **43.** $y = \frac{1}{3}x^2 + 5$
45. $y = -2x^2$ **47.** 34,000 feet; 32.5 s after the aircraft begins its parabolic flight **49.** $d(t) = -16t^2 + 8t + 50$
51. Angle A; the graph of the equation for angle A is higher than the other two since 3.27 is greater than 2.39 or 1.53.
53. $y = ax^2 + bx + c$

$y = a\left(x^2 + \frac{b}{a}x\right) + c$

$y = a\left[x^2 + \frac{b}{a}x + \left(\frac{b}{2a}\right)^2\right] + c - a\left(\frac{b}{2a}\right)^2$

$y = a\left(x + \frac{b}{2a}\right)^2 + c - \frac{b^2}{4a}$

The axis of symmetry is $x = h$ or $-\frac{b}{2a}$. **55.** D
57. 12; 2 irrational **59.** -23; 2 complex **61.** $\{3 \pm 3i\}$
63. $2t^2 + 2t - \frac{3}{t - 1}$ **65.** $n^3 - 3n^2 - 15n - 21$
67a. Sample answer using (1994, 76,302) and (1997, 99,448): $y = 7715x - 15,307,408$ **67b.** 161,167 **69.** no **71.** no

Page 328 Practice Quiz 2

1. $\{-7 \pm 2\sqrt{3}\}$ **3.** -11; 2 complex **5.** $\left\{\frac{-9 \pm 5\sqrt{5}}{2}\right\}$
7. $y = \frac{2}{3}(x - 2)^2 - 5$ **9.** $y = -(x - 6)^2$; $(6, 0)$, $x = 6$; down

Pages 332–335 Lesson 6-7

1. $y \geq (x - 3)^2 - 1$ **3a.** $x = -1, 5$ **3b.** $x \leq -1$ or $x \geq 5$
3c. $-1 \leq x \leq 5$

5.

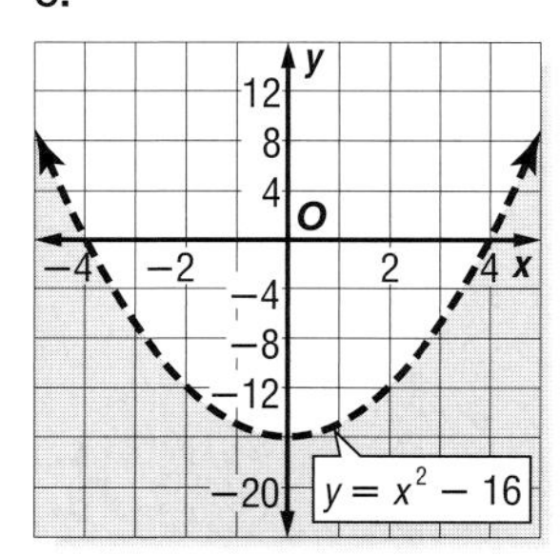

7.

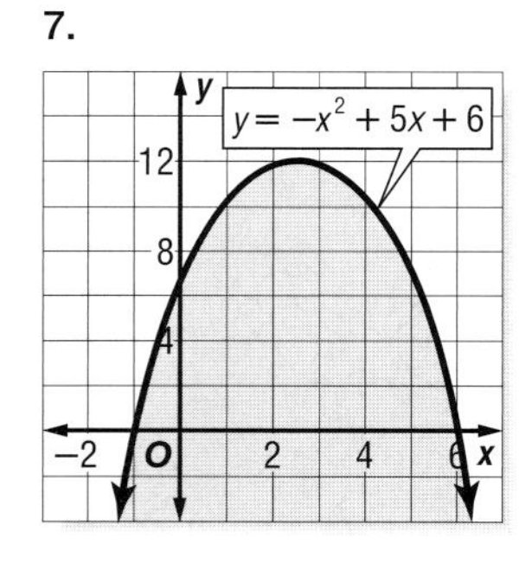

9. $\{x \mid -1 < x < 7\}$ **11.** $\varnothing$ **13.** about 6.1 s

15.

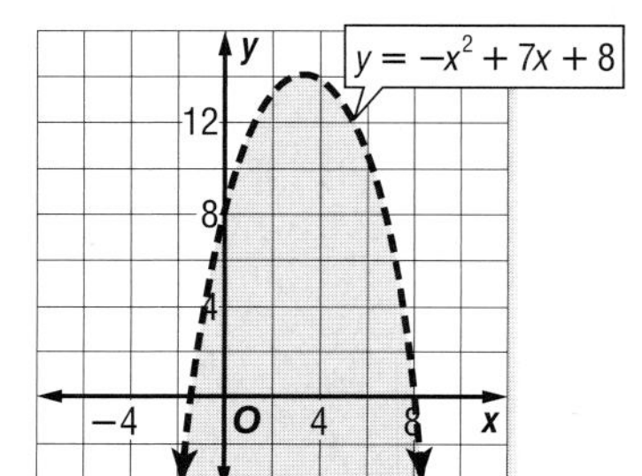

17.

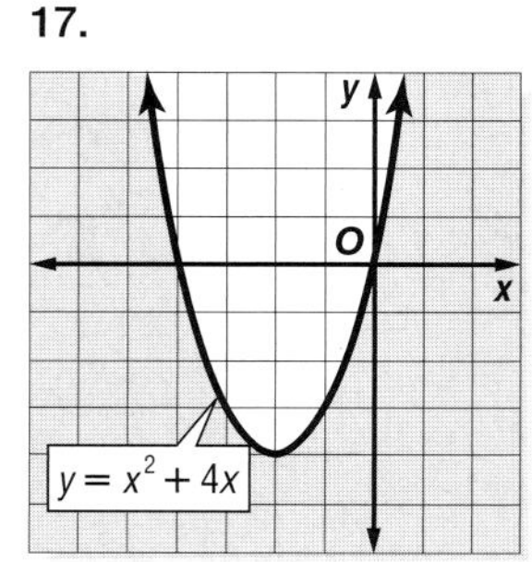

19.

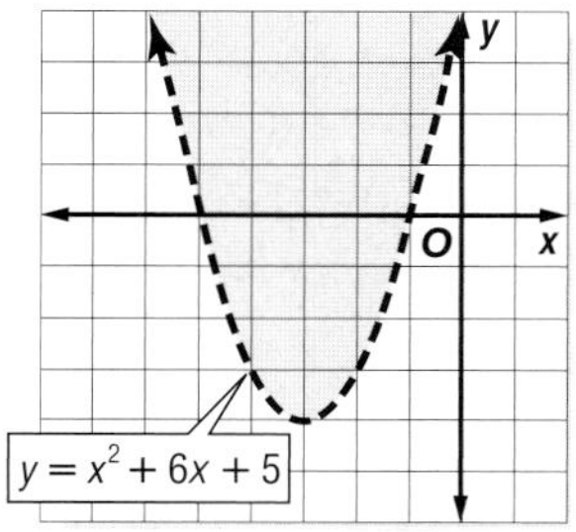

$y = x^2 + 6x + 5$

21.

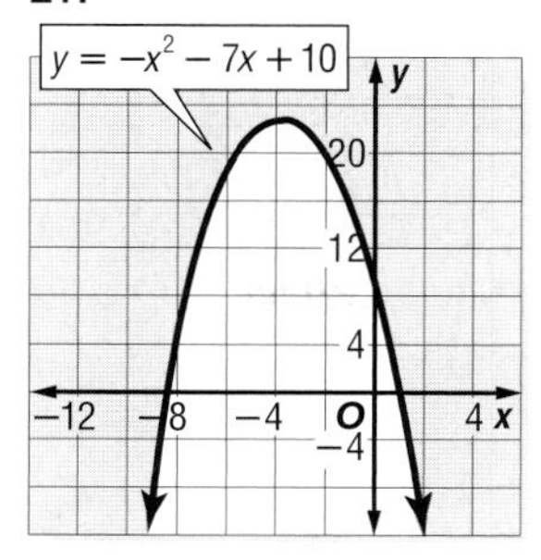

23.

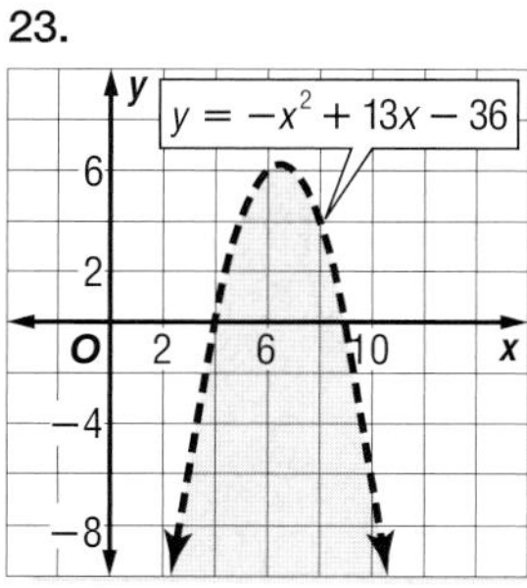

25.

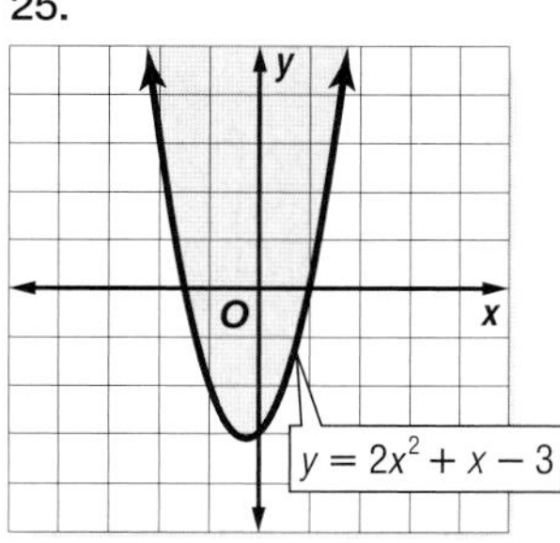

27. $-2 \le x \le 6$ **29.** $x < -7$ or $x > -3$ **31.** $\{x \mid -7 < x < 4\}$ **33.** $\{x \mid x \le -6 \text{ or } x \ge 4\}$ **35.** $\{x \mid x \le -7 \text{ or } x \ge 1\}$ **37.** all reals **39.** $\{x \mid x = 7\}$ **41.** $\varnothing$ **43.** 0 to 10 ft or 24 to 34 ft **45.** The width should be greater than 12 cm and the length should be greater than 18 cm **47.** 6

49.

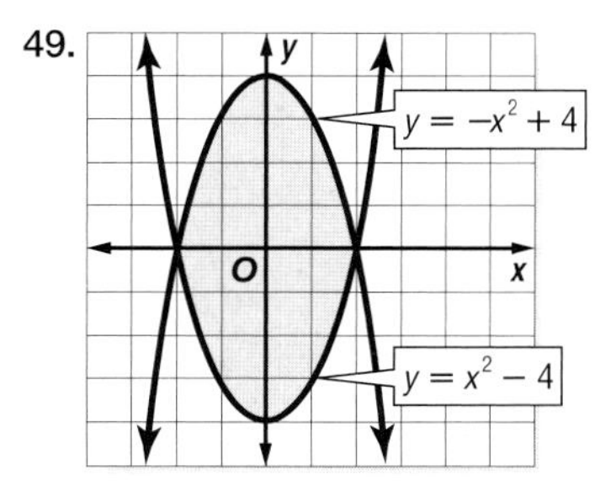

51. C **53.** $\{x \mid \text{all reals}, x \ne 2\}$ **55.** $\{x \mid x < -9 \text{ or } x > 3\}$ **57.** $\{x \mid -1.2 \le x \le -0.4\}$ **59.** $y = (x - 1)^2 + 8$; $(1, 8)$, $x = 1$; up **61.** $y = \frac{1}{2}(x + 6)^2$; $(-6, 0)$, $x = -6$; up **63.** $\frac{-5 \pm i\sqrt{3}}{2}$ **65.** $4a^2b^2 + 2a^2b + 4ab^2 + 12a - 7b$ **67.** $xy^3 + y + \frac{1}{x}$ **69.** $\begin{bmatrix} -21 & 48 \\ -13 & 22 \end{bmatrix}$ **71.** $|x - 0.08| \le 0.002$; $0.078 \le x \le 0.082$

Pages 336–340 Chapter 6 Study Guide and Review

1. f **3.** a **5.** i **7.** c **9a.** 20; $x = -3$; -3

9b.

x	f(x)
−5	15
−4	12
−3	11
−2	12
−1	15

9c.

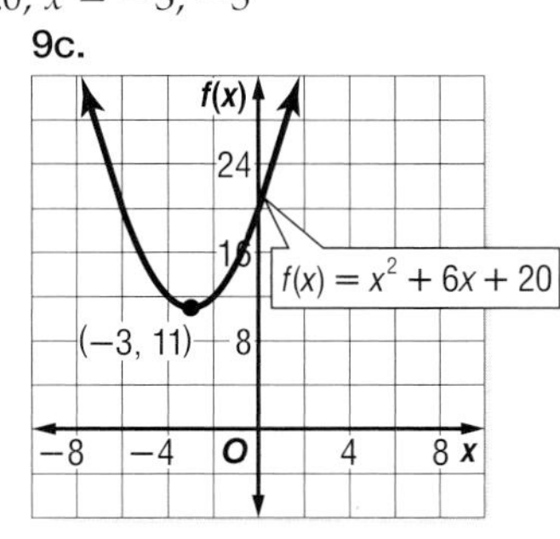

11a. 7; $x = 4$; 4

11b.

x	f(x)
2	−5
3	−8
4	−9
5	−8
6	−5

11c.

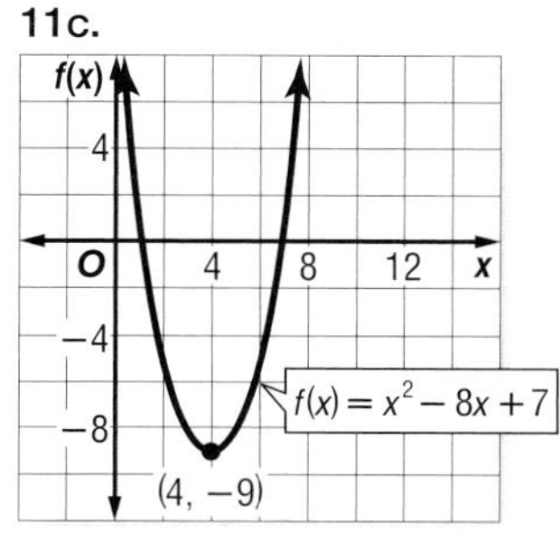

13a. -3; $x = -2$; -2

13b.

x	f(x)
−4	−3
−3	0
−2	1
−1	0
0	−3

13c.

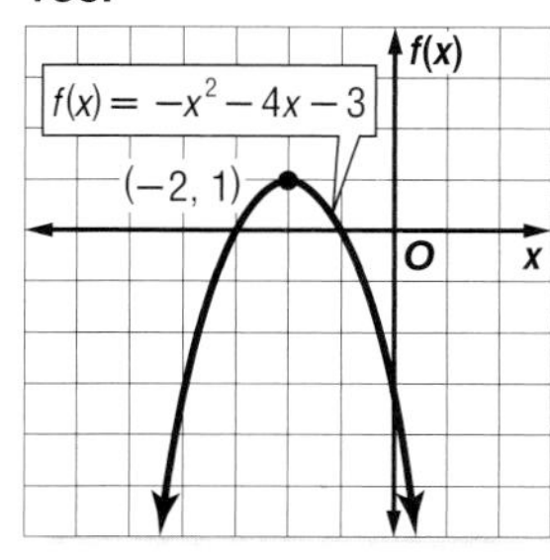

15. min.; $-\frac{89}{16}$ **17.** max.; 7 **19.** 2, −5 **21.** between −3 and −2; between −38 and −37 **23.** 2, −8 **25.** $\{-1\}$ **27.** $\{-11, 2\}$ **29.** $\left\{\frac{1}{3}, -\frac{3}{2}\right\}$ **31.** $x^2 - 3x - 70 = 0$ **33.** 289; $(x + 17)^2$ **35.** $\frac{49}{16}$; $\left(x + \frac{7}{4}\right)^2$ **37.** $3 \pm 2\sqrt{5}$ **39a.** −24 **39b.** 2 complex **39c.** $-1 \pm \sqrt{6}i$ **41a.** 73 **41b.** 2 irrational **41c.** $\frac{-7 \pm \sqrt{73}}{6}$ **43.** $y = 5\left(x + \frac{7}{2}\right)^2 - \frac{13}{4}$; $\left(-\frac{7}{2}, -\frac{13}{4}\right)$; $x = -\frac{7}{2}$; up

45.

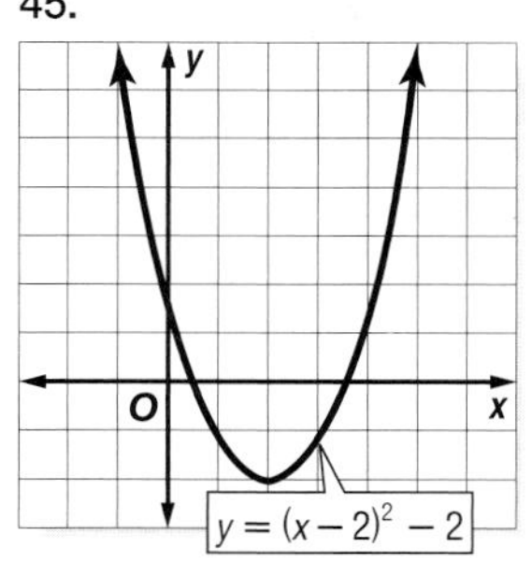

47.

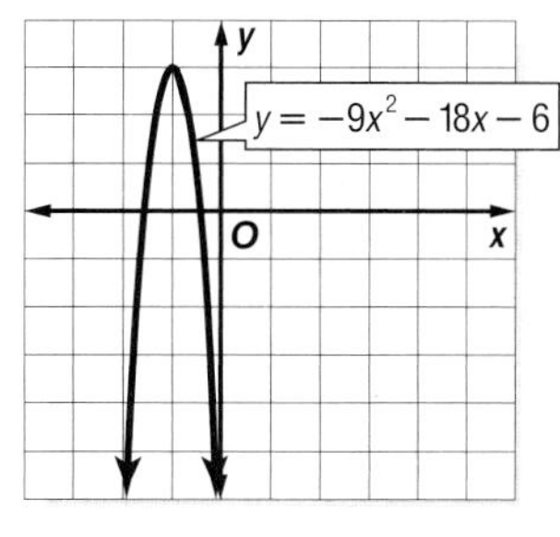

49. $y = \frac{1}{2}(x + 2)^2 + 3$

51.

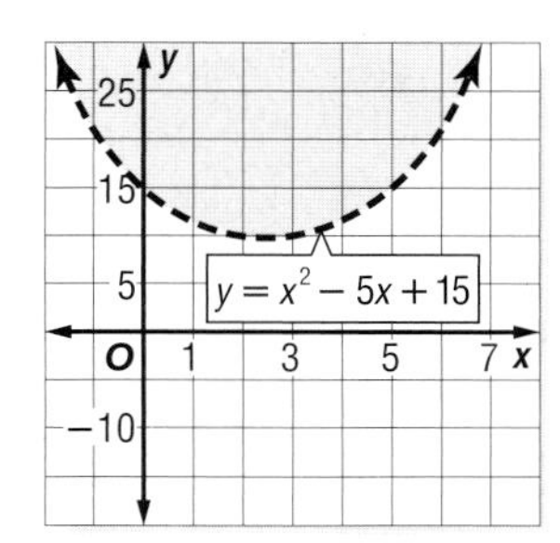

53.

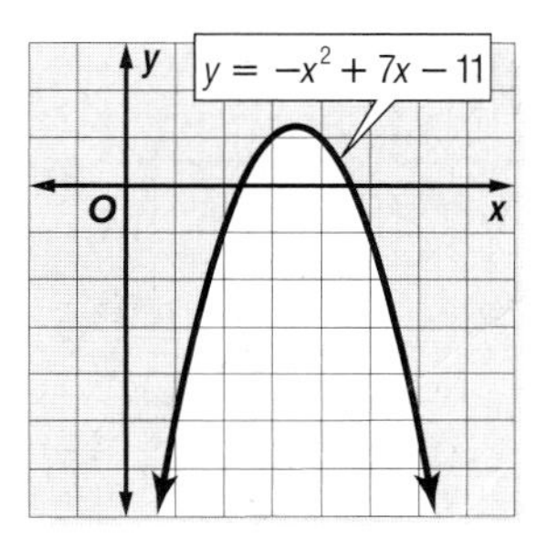

55. all reals **57.** $\left\{x \mid x < -\frac{1}{2} \text{ or } x > 3\right\}$

59. $\left\{x \mid x < \frac{3 - 2\sqrt{6}}{3} \text{ or } x > \frac{3 + 2\sqrt{6}}{3}\right\}$

Chapter 7 Polynomial Functions

Page 345 Chapter 7 Getting Started

1. between 0 and 1, between 4 and 5 **3.** between -5 and -4, between 0 and 1 **5.** $-\frac{3}{2}, -\frac{1}{7}$ **7.** $3x + 4$ **9.** -19 **11.** $18b^2 - 3b - 6$

Pages 350–352 Lesson 7-1

1. $4 = 4x^0$; $x = x^1$ **3.** Sample answer given.

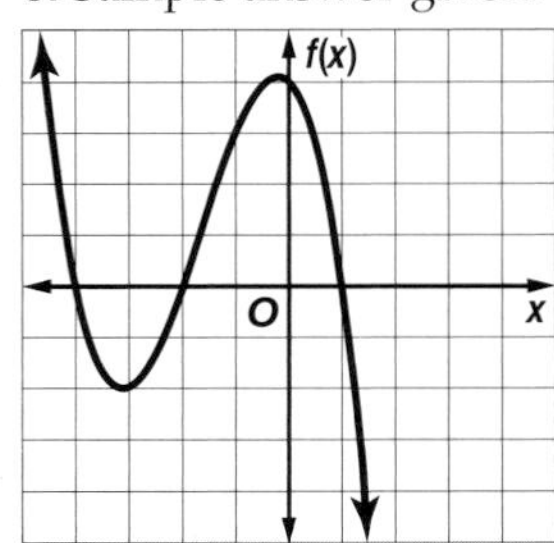

5. 6; 5 **7.** -21; 3 **9.** $2a^9 + 6a^3 - 12$ **11.** $6a^3 - 5a^2 + 8a - 45$ **13a.** $f(x) \to +\infty$ as $x \to +\infty$, $f(x) \to +\infty$ as $x \to -\infty$ **13b.** even **13c.** 0 **15.** 109 lumens **17.** 3; 1 **19.** 4; 6 **21.** No, this is not a polynomial because the term $\frac{1}{c}$ cannot be written in the form x^n, where n is a nonnegative integer. **23.** 12; 18 **25.** 1008; -36 **27.** 86; 56 **29.** 7; 4 **31.** $12a^2 - 8a + 20$ **33.** $12a^6 - 4a^3 + 5$ **35.** $3x^4 + 16x^2 + 26$ **37.** $-x^6 + x^3 + 2x^2 + 4x + 2$ **39a.** $f(x) \to +\infty$ as $x \to +\infty$, $f(x) \to -\infty$ as $x \to -\infty$ **39b.** odd **39c.** 3 **41a.** $f(x) \to -\infty$ as $x \to +\infty$, $f(x) \to -\infty$ as $x \to -\infty$ **41b.** even **41c.** 0 **43a.** $f(x) \to +\infty$ as $x \to +\infty$, $f(x) \to -\infty$ as $x \to -\infty$ **43b.** odd **43c.** 1 **45.** 5.832 units **47.** $f(x) \to -\infty$ as $x \to +\infty$; $f(x) \to -\infty$ as $x \to -\infty$ **49.** $\frac{1}{2}$ **51.** $f(x)\ \frac{1}{2}x^3 - \frac{3}{2}x^2 - 2x$ **53.** 4 **55.** 8 points **57.** C **59.** $\{x \mid 2 < x < 6\}$ **61.** $\left\{x \mid -1 \leq x \leq \frac{4}{5}\right\}$

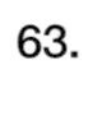

63.

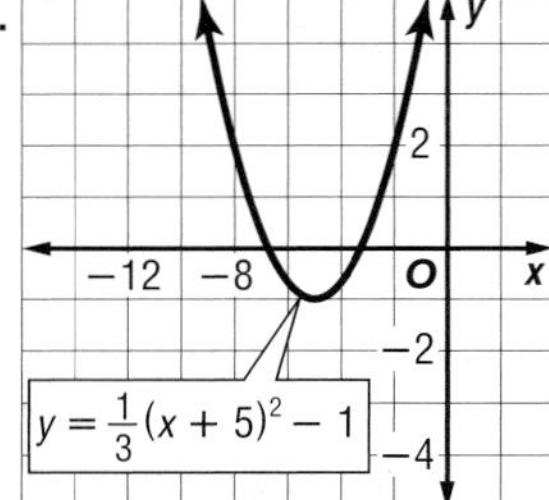

65. $\{4 \pm 3\sqrt{2}\}$ **67.** $23{,}450(1 + p)$; $23{,}450(1 + p)^3$

69.

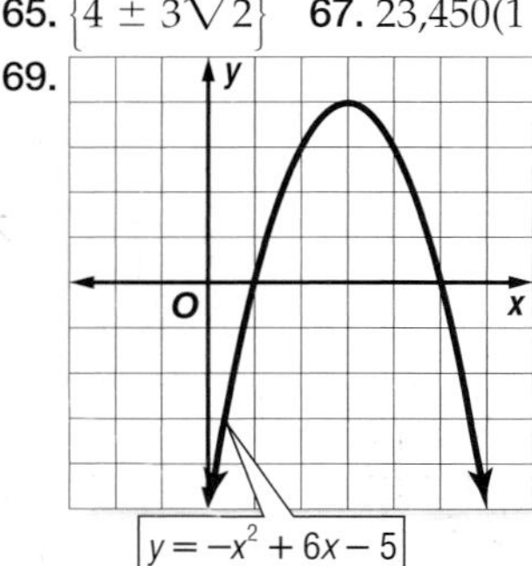

Pages 356–358 Lesson 7-2

1. There must be at least one real zero between two points on a graph when one of the points lies below the x-axis and the other point lies above the x-axis.

3.

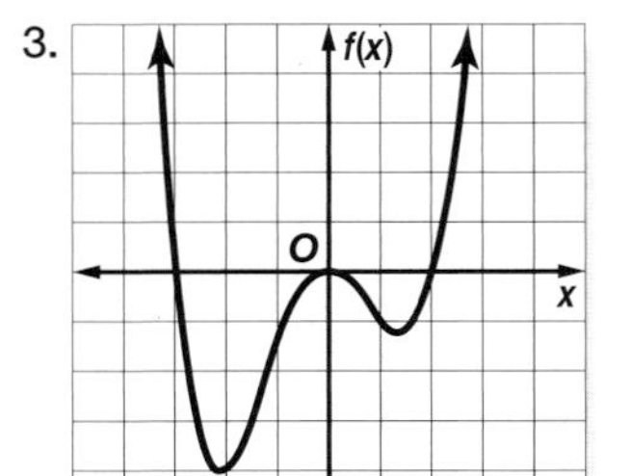

5.

x	$f(x)$
-3	20
-2	-9
-1	-2
0	5
1	0
2	-5
3	26

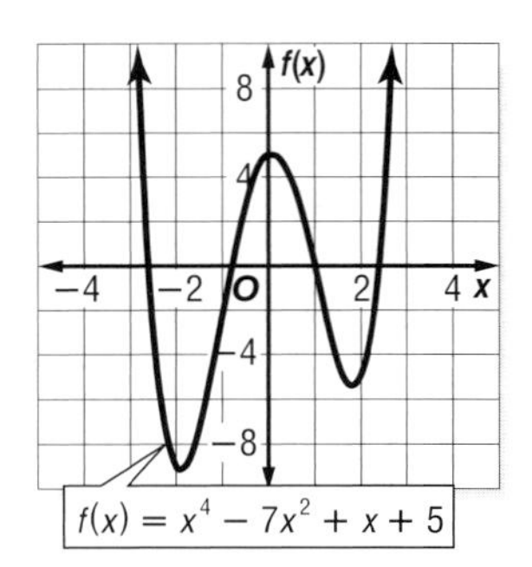

7. between -2 and -1, between -1 and 0, between 0 and 1, and between 1 and 2

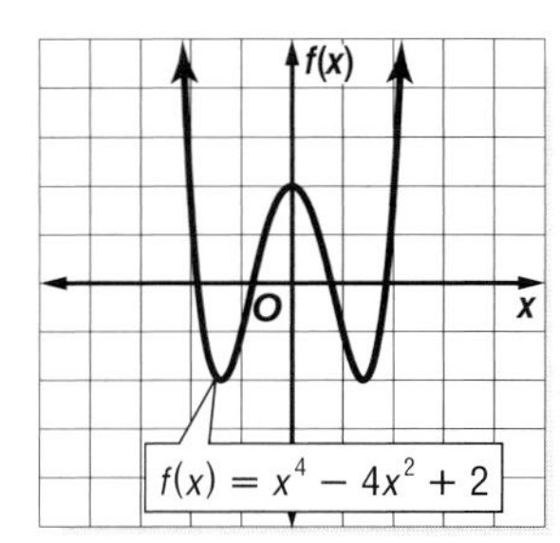

9.

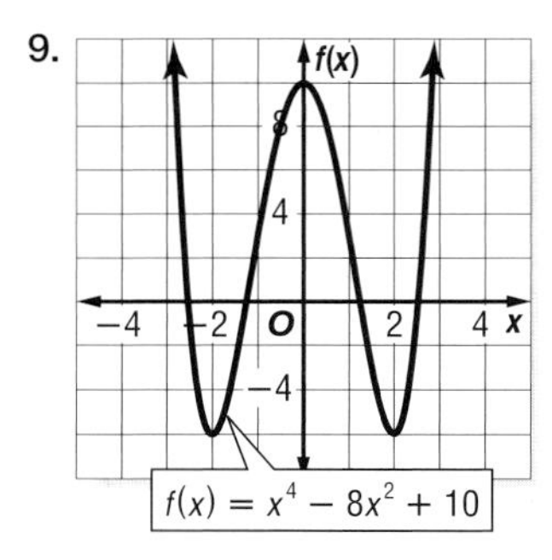

Sample answer: rel. max. at $x = 0$, rel. min. at $x = -2$ and at $x = 2$

11. rel. max. between $x = 15$ and $x = 16$, and no rel. min.; $f(x) \to -\infty$ as $x \to -\infty$, $f(x) \to -\infty$ as $x \to +\infty$.

13a.

x	$f(x)$
-5	25
-4	0
-3	-9
-2	-8
-1	-3
0	0
1	-5
2	-24

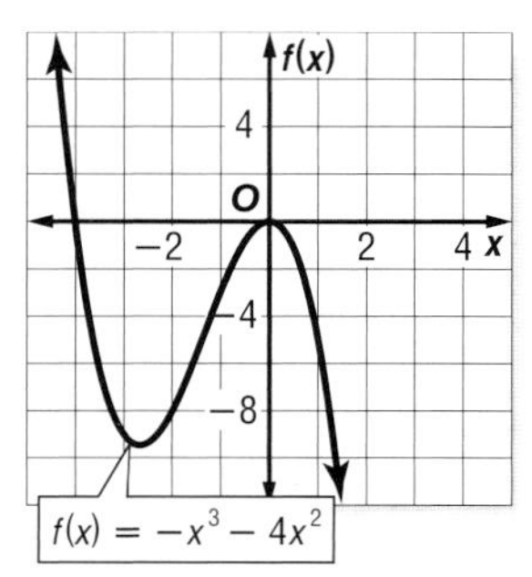

13b. at $x = -4$ and $x = 0$ **13c.** Sample answer: rel. max. at $x = 0$, rel. min. at $x = -3$

15a.

x	$f(x)$
−2	−18
−1	−2
0	2
1	0
2	−2
3	2
4	18

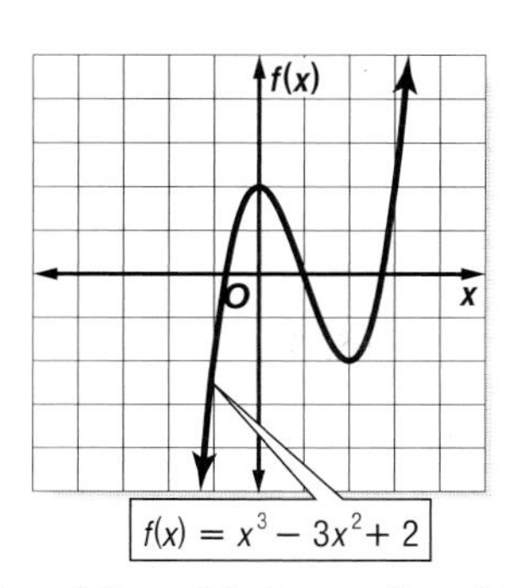

15b. at $x = 1$, between -1 and 0, and between 2 and 3
15c. Sample answer: rel. max. at $x = 0$, rel. min. at $x = 2$

17a.

x	$f(x)$
−1	75
0	16
1	−3
2	0
3	7
4	0
5	−39

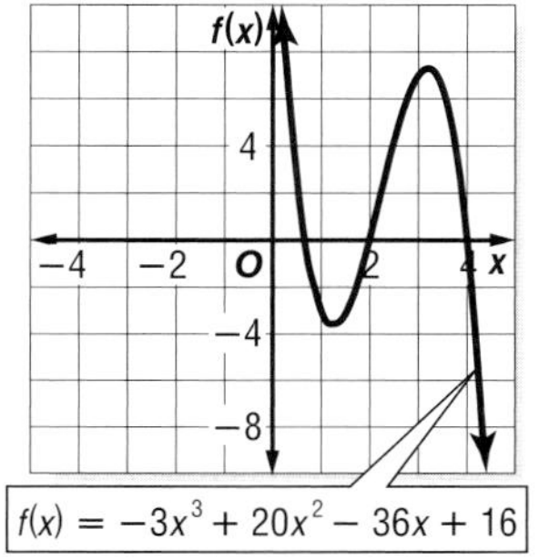

17b. between 0 and 1, at $x = 2$, and at $x = 4$
17c. Sample answer: rel. max. at $x = 3$, rel. min. at $x = 1$

19a.

x	$f(x)$
−3	73
−2	8
−1	−7
0	−8
1	−7
2	8
3	73

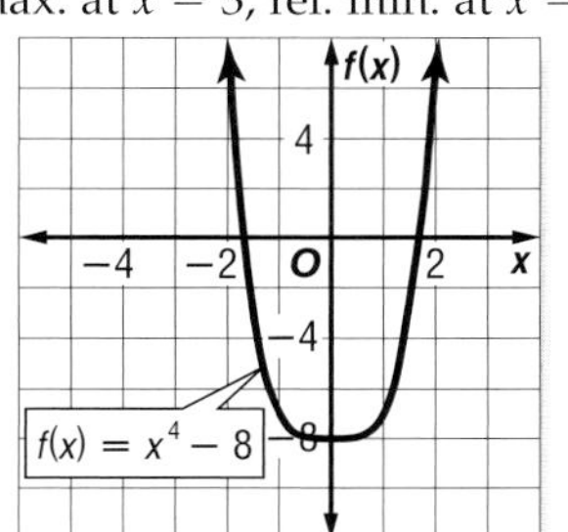

19b. between -2 and -1 and between 1 and 2
19c. Sample answer: no rel. max., rel. min. at $x = 0$

21a.

x	$f(x)$
−4	−169
−3	−31
−2	7
−1	5
0	−1
1	1
2	−1
3	−43

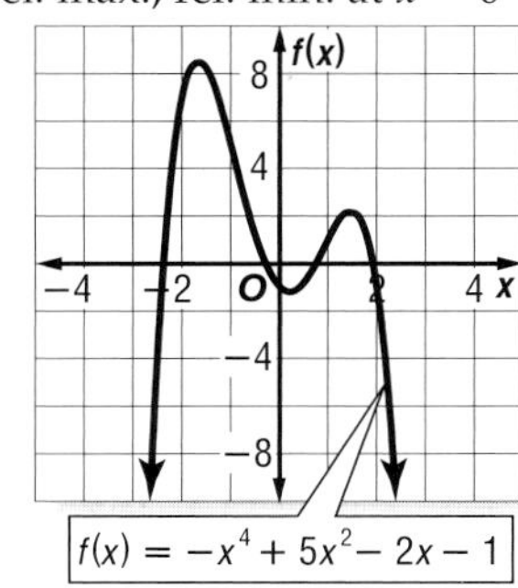

21b. between -3 and -2, between -1 and 0, between 0 and 1, and between 1 and 2 **21c.** Sample answer: rel. max. at $x = -2$ and at $x = 1.5$, rel. min. at $x = 0$

23a.

x	$f(x)$
−1	65
0	6
1	−1
2	2
3	−3
4	−10
5	11

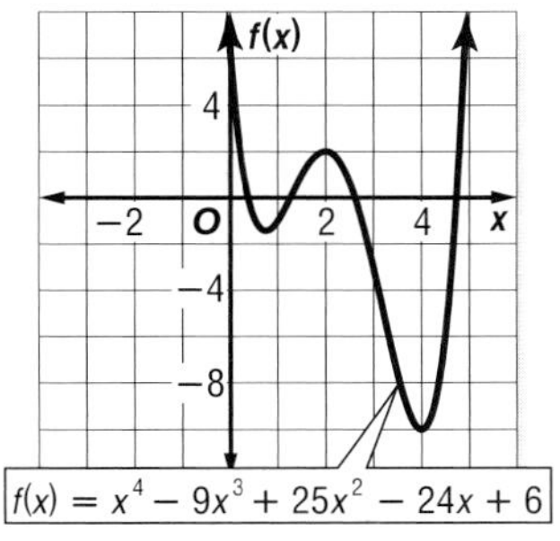

23b. between 0 and 1, between 1 and 2, between 2 and 3, and between 4 and 5 **23c.** Sample answer: rel. max. at $x = 2$, rel. min. at $x = 0.5$ and at $x = 4$

25a.

x	$f(x)$
−4	−77
−3	30
−2	7
−1	−2
0	3
1	−2
2	55

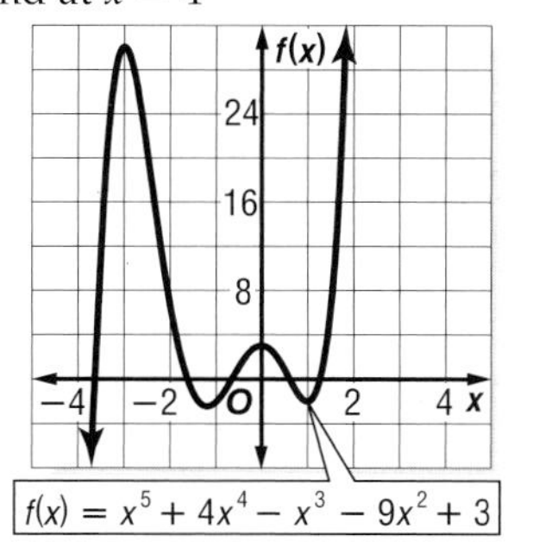

25b. between -4 and -3, between -2 and -1, between -1 and 0, between 0 and 1, and between 1 and 2 **25c.** Sample answer: rel. max. at $x = -3$ and at $x = 0$, rel. min. at $x = -1$ and at $x = 1$ **27.** highest: 1982; lowest: 2000 **29.** 5

31.

x	0	2	4	6	8	10	12	14	16	18	20
$B(x)$	25	34	40	45	50	54	59	64	68	71	71
$G(x)$	26	33	39	44	49	53	56	59	61	61	60

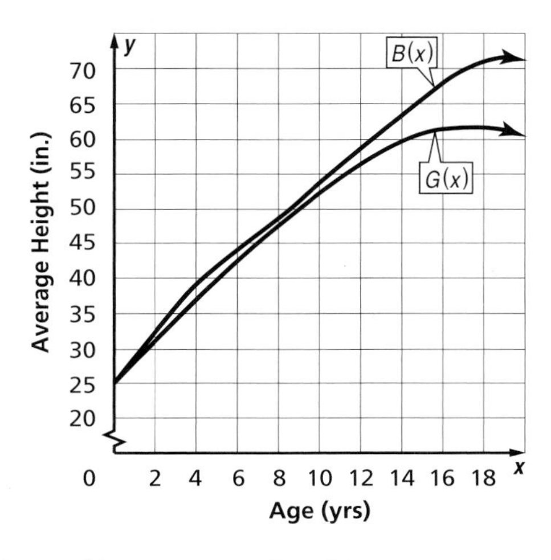

33. 0 and between 5 and 6 **35.** 3.4 s

37.

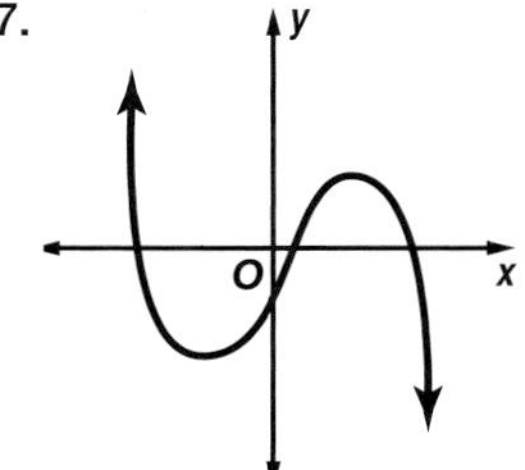

39.

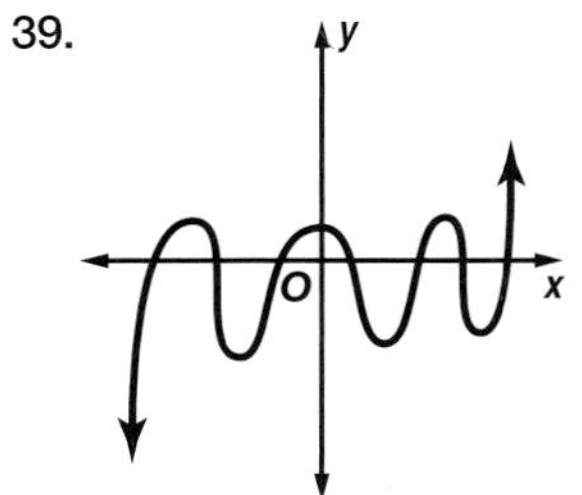

41. D **43.** -1.90; 1.23 **45.** 0; -1.22, 1.22 **47.** $24a^3 - 4a^2 - 2$ **49.** $8a^4 - 10a^2 + 4$ **51.** $2x^4 + 11x^2 + 16$

53.

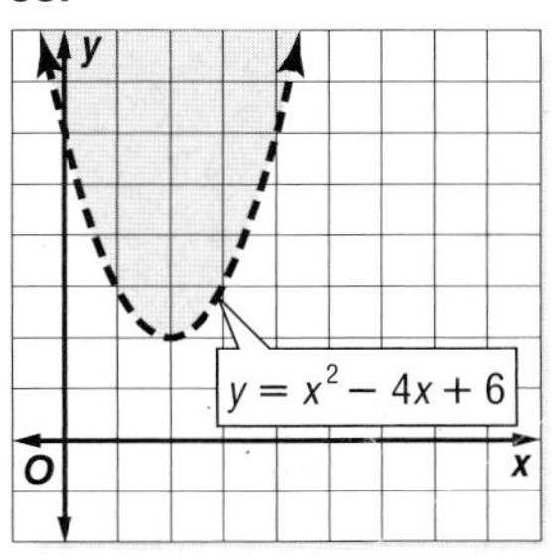

55.

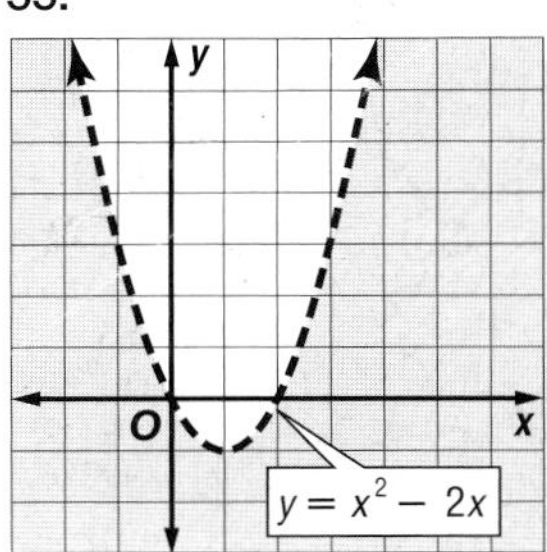

57. $(-3, -2)$ **59.** $(1, 3)$ **61.** $(x + 5)(x - 6)$
63. $(3a + 1)(2a + 5)$ **65.** $(t - 3)(t^2 + 3t + 9)$

Pages 362–364 Lesson 7-3

1. Sample answer: $16x^4 - 12x^2 = 0$; $4[4(x^2)^2 - 3x^2] = 0$
3. Factor out an x and write the equation in quadratic form so you have $x[(x^2)^2 - 2(x^2) + 1] = 0$. Factor the trinomial and solve for x using the Zero Product Property. The solutions are -1, 0, and 1. **5.** $84(n^2)^2 - 62(n^2)$ **7.** $-4, -1, 4, 1$ **9.** 64 **11.** $2(x^2)^2 + 6(x^2) - 10$ **13.** $11(n^3)^2 + 44(n^3)$ **15.** not possible **17.** $0, -4, -3$ **19.** $-\sqrt{3}, \sqrt{3}, -i\sqrt{3}, i\sqrt{3}$
21. $2, -2, 2\sqrt{2}, -2\sqrt{2}$ **23.** $-9, \frac{9 + 9i\sqrt{3}}{2}, \frac{9 - 9i\sqrt{3}}{2}$
25. 81, 625 **27.** 225, 16 **29.** $1, -1, 4$ **31.** $w = 4$ cm, $\ell = 8$ cm, $h = 2$ cm **33.** 3×3 in. **35.** $h^2 + 4, 3h + 2, h + 3$
37. Write the equation in quadratic form, $u^2 - 9x - 8 = 0$, where $u = |a - 3|$. Then factor and use the Zero Product Property to solve for a; 11, 4, 2, and -5. **39.** D
41.

x	$f(x)$
−2	−21
−1	−1
0	5
1	3
2	−1
3	−1
4	9
5	35

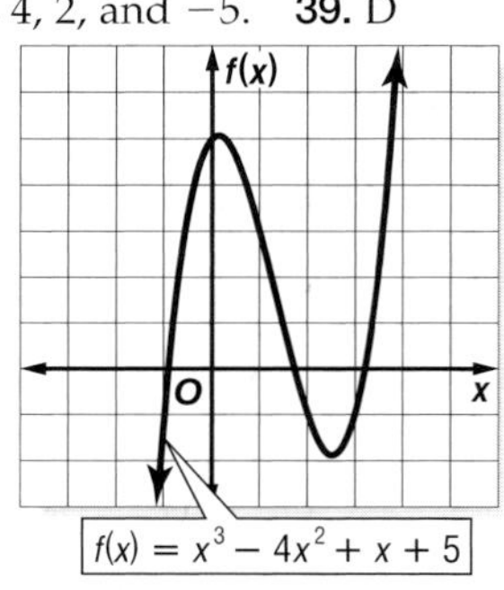

43. 17; 27 **45.** $\frac{1715}{3}$; 135 **47.** $A'(-1, -2)$, $B'(3, -3)$, $C'(1, 3)$
49. $x^2 + 5x - 4$ **51.** $x^3 - 6x - 20 - \frac{54}{x - 3}$

Page 364 Practice Quiz 1

1. $2a^3 - 6a^2 + 5a - 1$
3. Sample answer: maximum at $x = -2$, minimum at $x = 0.5$ **5.** $-3, 3, -i\sqrt{3}, i\sqrt{3}$

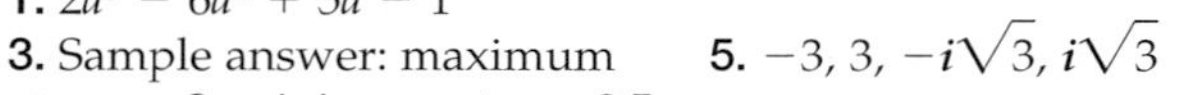

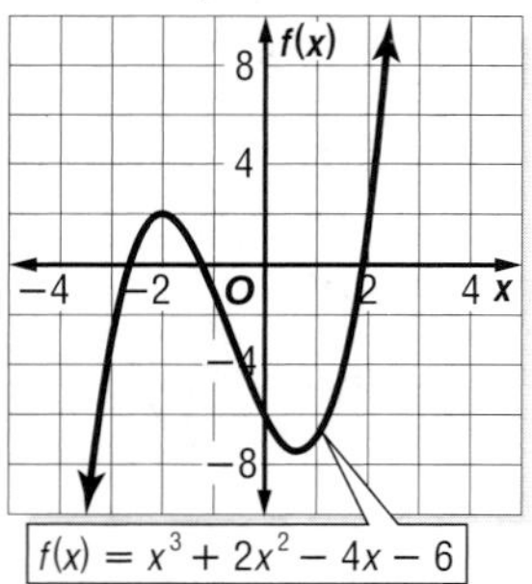

Pages 368–370 Lesson 7-4

1. Sample answer: $f(x) = x^2 - 2x - 3$ **3.** dividend: $x^3 + 6x + 32$; divisor: $x + 2$; quotient: $x^2 - 2x + 10$; remainder: 12 **5.** 353, 1186 **7.** $x - 1, x + 2$ **9.** $x - 2, x^2 + 2x + 4$
11. \$2.894 billion **13.** $-9, 54$ **15.** $14, -42$ **17.** $-19, -243$
19. $450, -1559$ **21.** $x + 1, x + 2$ **23.** $x - 4, x + 1$
25. $x + 3, x - \frac{1}{2}$ or $2x - 1$ **27.** $x + 7, x - 4$
29. $x - 1, x^2 + 2x + 3$ **31.** $x - 2, x + 2, x^2 + 1$
33. 3 **35.** 1, 4 **37.**

$$\begin{array}{r|rrrrr} 5 & 1 & -14 & 69 & -140 & 100 \\ & & 5 & -45 & 120 & -100 \\ \hline & 1 & -9 & 24 & -20 & 0 \end{array}$$

39. 7.5 ft/s, 8 ft/s, 7.5 ft/s **41.** By the Remainder Theorem, the remainder when $f(x)$ is divided by $x - 1$ is equivalent to $f(1)$, or $a + b + c + d + e$. Since $a + b + c + d + e = 0$, the remainder when $f(x)$ is divided by $x - 1$ is 0. Therefore, $x - 1$ is a factor of $f(x)$. **43.** \$16.70 **45.** No, he will still owe \$4.40. **47.** D **49.** $(x^2)^2 - 8(x^2) + 4$ **51.** not possible **53.** Sample answer: rel. max. and $x = -1$ and $x = 1.5$, rel. min. at $x = 1$

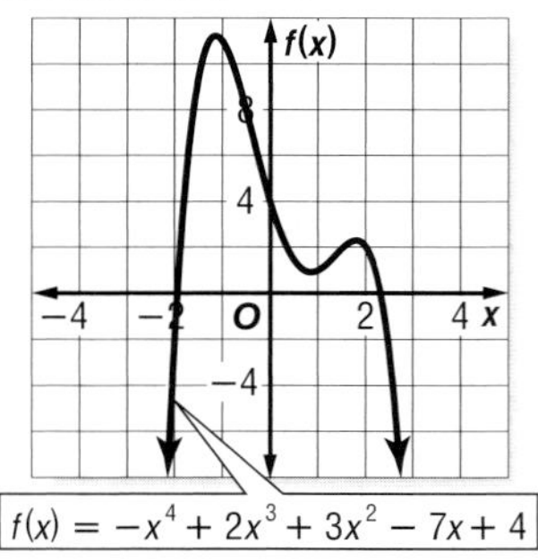

55. $(4, -2)$ **57.** A
59. S **61.** $\frac{9 \pm \sqrt{57}}{6}$

Pages 375–377 Lesson 7-5

1. Sample answer: $p(x) = x^3 - 6x^2 + x + 1$; $p(x)$ has either 2 or 0 positive real zeros, 1 negative real zero, and 2 or 0 imaginary zeros. **3.** 6 **5.** -7, 0, and 3; 3 real **7.** 2 or 0; 1; 2 or 4 **9.** $2, 1 + i, 1 - i$ **11.** $2 + 3i, 2 - 3i, -1$ **13.** $-\frac{8}{3}$; 1 real **15.** $0, 3i, -3i$; 1 real, 2 imaginary **17.** $2, -2, 2i$, and $-2i$; 2 real, 2 imaginary **19.** 2 or 0; 1; 2 or 0 **21.** 3 or 1; 0; 2 or 0 **23.** 4, 2, or 0; 1; 4, 2, or 0 **25.** $-2, -2 + 3i, -2 - 3i$
27. $2i, -2i, \frac{i}{2}, -\frac{i}{2}$ **29.** $-\frac{3}{2}, 1 + 4i, 1 - 4i$ **31.** $4 - i$, $4 + i, -3$ **33.** $3 - 2i, 3 + 2i, -1, 1$ **35.** $f(x) = x^3 - 2x^2 - 19x + 20$ **37.** $f(x) = x^4 + 7x^2 - 144$ **39.** $f(x) = x^3 - 11x^2 + 23x - 45$

41a.

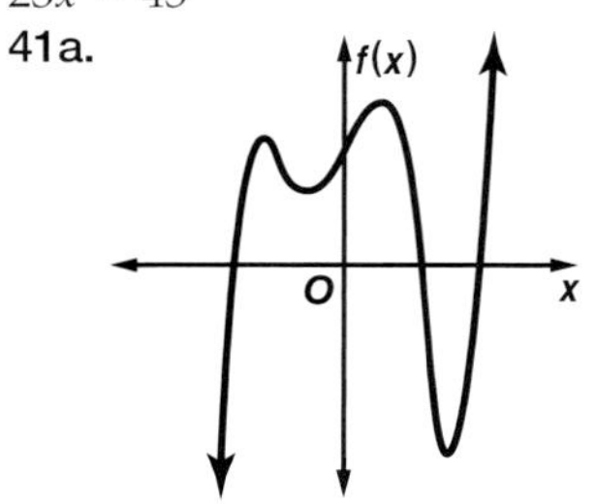

41b.

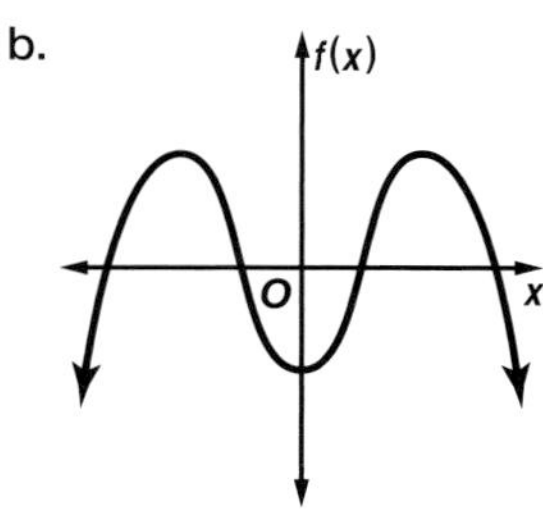

41c.

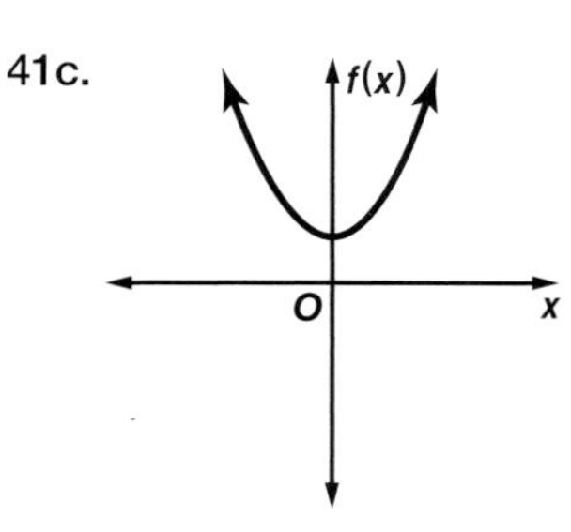

43. 1 ft **45.** radius = 4 m, height = 21 m **47.** $-24.1, -4.0$, 0, and 3.1

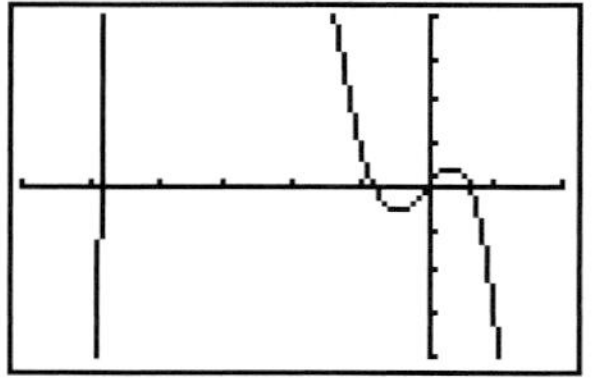

[−30, 10] scl: 5 by [−20, 20] scl: 5

49. Sample answer: $f(x) = x^3 - 6x^2 + 5x + 12$ and $g(x) = 2x^3 - 12x^2 + 10x + 24$; each have zeros at $x = 4$, $x = -2$, and $x = 3$.
51. If the equation models the level of a medication in a patient's bloodstream, a doctor can use the roots of the equation to determine how often the patient should take the medication to maintain the necessary concentration in the body. Answers should include the following.
- A graph of this equation reveals that only the first positive real root of the equation, 5, has meaning for this situation, since the next positive real root occurs after the medication level in the bloodstream has dropped below 0 mg. Thus according to this model, after 5 hours there is no significant amount of medicine left in the bloodstream.
- The patient should not go more than 5 hours before taking their next dose of medication.

53. C **55.** $-254, 915$ **57.** min.; -13 **59.** min.; -7
61. $(6p - 5)(2p - 9)$ **63.** $\begin{bmatrix} -3 & 2 \\ 3 & -4 \\ -2 & 9 \end{bmatrix}$ **65.** $\begin{bmatrix} 29 & -8 \\ 8 & 9 \\ 16 & -16 \end{bmatrix}$
67. $\pm\frac{1}{2}, \pm1, \pm\frac{5}{2}, \pm5$ **69.** $\pm\frac{1}{9}, \pm\frac{1}{3}, \pm1, \pm3$

Pages 380–382 Lesson 7-6

1. Sample answer: You limit the number of possible solutions.
3. Luis; Lauren found numbers in the form $\frac{q}{p}$, not $\frac{p}{q}$ as Luis did according to the Rational Zero Theorem. **5.** $\pm1, \pm2, \pm\frac{1}{2}, \pm\frac{1}{3}, \pm\frac{1}{6}, \pm\frac{2}{3}$ **7.** $-2, -4, 7$ **9.** $-2, 2, \frac{7}{2}$ **11.** 10 cm × 11 cm × 13 cm **13.** $\pm1, \pm2, \pm3, \pm6$ **15.** $\pm1, \pm2, \pm3, \pm6, \pm9, \pm18$ **17.** $\pm1, \pm\frac{1}{3}, \pm\frac{1}{9}, \pm3, \pm9, \pm27$ **19.** $-1, -1, 2$
21. 0, 9 **23.** $0, 2, -2$ **25.** $-2, -4$ **27.** $\frac{1}{2}, -\frac{1}{3}, -2$
29. $-\frac{1}{2}, \frac{1}{3}, \frac{1}{2}, \frac{3}{4}$ **31.** $\frac{4}{5}, 0, \frac{5 \pm i\sqrt{3}}{2}$ **33.** $-1, -2, 5, i, -i$
35. $2, -3 \pm i\sqrt{3}; 2$ **37.** $V = 2h^3 - 8h^2 - 64h$
39. $V = \frac{1}{3}\ell^3 - 3\ell^2$ **41.** $\ell = 30$ in., $w = 30$ in., $h = 21$ in.
43. The Rational Zero Theorem helps factor large numbers by eliminating some possible zeros because it is not practical to test all of them using synthetic substitution. Answers should include the following.
- The polynomial equation that represents the volume of the compartment is $V = w^3 + 3w^2 - 40w$.
- Reasonable measures of the width of the compartment are, in inches, 1, 2, 3, 4, 6, 7, 9, 12, 14, 18, 21, 22, 28, 33, 36, 42, 44, 63, 66, 77, and 84. The solution shows that $w = 14$ in., $\ell = 22$ in., and $d = 9$ in.

45. Sample answer $x^5 - x^4 - 27x^3 + 41x^2 + 106x - 120$
47. $-4, 2 + i, 2 - i$ **49.** $-7, 5 + 2i, 5 - 2i$ **51.** $x - 4$, $3x^2 + 2$ **53.** $\pm3xy\sqrt{2x}$ **55.** 6 cm, 8 cm, 10 cm
57. $4x^2 - 8x + 3$ **59.** $x^5 - 7x^4 - 8x^3 + 106x^2 - 85x + 25$
61. $x^2 + x - 4 + \frac{5}{x + 1}$

Page 382 Practice Quiz 2

1. $-930, -145$ **3.** $x^4 - 4x^3 - 7x^2 + 22x + 24 = 0$ **5.** $-\frac{3}{2}$

Pages 386–389 Lesson 7-7

1. Sometimes; sample answer: If $f(x) = x - 2$, $g(x) = x + 8$, then $f \circ g = x + 6$ and $g \circ f = x + 6$.
3. Danette; $[g \circ f](x) = g[f(x)]$ means to evaluate the f function first and then the g function. Marquan evaluated the functions in the wrong order. **5.** $x^2 + x - 1$; $x^2 - x + 7$; $x^3 - 4x^2 + 3x - 12$; $\frac{x^2 + 3}{x - 4}$, $x \neq 4$ **7.** $\{(2, -7)\}$; $\{(1, 0), (2, 10)\}$
9. $x^2 + 11$; $x^2 + 10x + 31$ **11.** 11 **13.** $p(x) = \frac{3}{4}x$; $c(x) = x - 5$
15. \$33.74; price of CD when coupon is subtracted and then 25% discount is taken **17.** $2x$; 18; $x^2 - 81$; $\frac{x + 9}{x - 9}$, $x \neq 9$
19. $2x^2 - x + 8$; $2x^2 + x - 8$; $-2x^3 + 16x^2$; $\frac{2x^2}{8 - x}$, $x \neq 8$
21. $\frac{x^3 + x^2 - 1}{x + 1}$, $x \neq -1$; $\frac{x^3 + x^2 - 2x - 1}{x + 1}$, $x \neq -1$; $x^2 - x$, $x \neq -1$; $\frac{x^3 + x^2 - x - 1}{x}$, $x \neq 0$ **23.** $\{(1, -3), (-3, 1), (2, 1)\}$; $\{(1, 0), (0, 1)\}$ **25.** $\{(0, 0), (8, 3), (3, 3)\}$; $\{(3, 6), (4, 4), (6, 6), (7, 8)\}$ **27.** $\{(5, 1), (8, 9)\}$; $\{(2, -4)\}$ **29.** $8x - 4$; $8x - 1$
31. $x^2 + 2$; $x^2 + 4x + 4$ **33.** $2x^3 + 2x^2 + 2x + 2$; $8x^3 + 4x^2 + 2x + 1$ **35.** -12 **37.** 39 **39.** 25 **41.** 2 **43.** 79 **45.** 226
47. $P(x) = -50x + 1939$ **49.** $p(x) = 0.70x$; $s(x) = 1.0575x$
51. \$110.30 **53.** 373 K; 273 K **55.** \$700, \$661.20, \$621.78, \$581.73, \$541.04 **57.** Answers should include the following.
- Using the revenue and cost functions, a new function that represents the profit is $p(x) = r(c(x))$.
- The benefit of combining two functions into one function is that there are fewer steps to compute and it is less confusing to the general population of people reading the formulas.

59. C **61.** $\pm1, \pm\frac{1}{2}, \pm\frac{1}{4}, \pm2, \pm3, \pm\frac{3}{2}, \pm\frac{3}{4}, \pm6$ **63.** $x^3 - 4x^2 - 17x + 60$ **65.** $6x^3 - 13x^2 + 9x - 2$ **67.** $x^3 - 9x^2 + 31x - 39$
69. $10 + 2j$ **71.** $\begin{bmatrix} 3 & -2 \\ -1 & 1 \end{bmatrix}$ **73.** $-\frac{1}{2}\begin{bmatrix} -1 & -2 \\ -3 & -4 \end{bmatrix}$
75. $-\frac{1}{16}\begin{bmatrix} -5 & -2 \\ -3 & 2 \end{bmatrix}$ **77.** $y = \frac{1 - 4x^2}{-5x}$ **79.** $t = \frac{I}{pr}$
81. $m = \frac{Fr^2}{GM}$

Pages 393–394 Lesson 7-8

1. no **3.** Sample answer: $f(x) = 2x$, $f^{-1}(x) = 0.5x$; $f[f^{-1}(x)] = f^{-1}[f(x)] = x$ **5.** $\{(4, 2), (1, -3), (8, 2)\}$
7. $f^{-1}(x) = -x$ **9.** $y = 2x - 10$

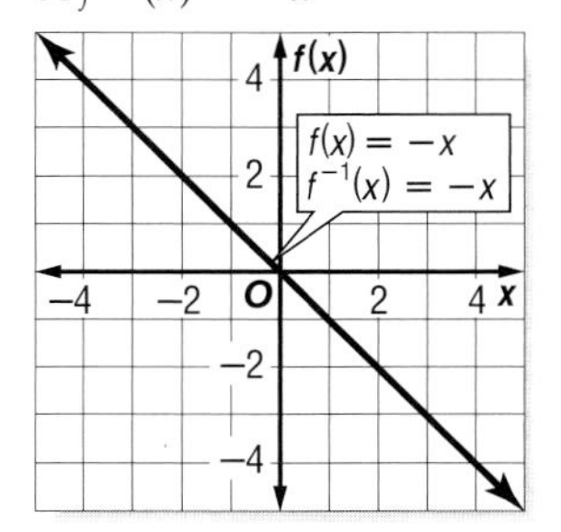

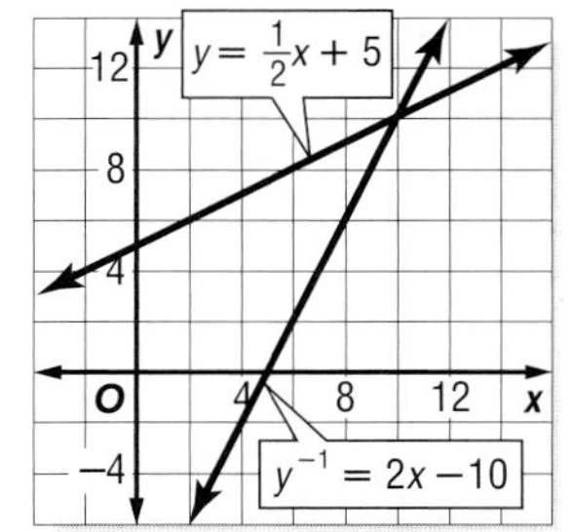

11. no **13.** 15.24 m/s² **15.** $\{(8, 3), (-2, 4), (-3, 5)\}$
17. $\{(-2, -1), (-2, -3), (-4, -1), (6, 0)\}$
19. $\{(8, 2), (5, -6), (2, 8), (-6, 5)\}$
21. $g^{-1}(x) = -\frac{1}{2}x$ **23.** $g^{-1}(x) = x - 4$

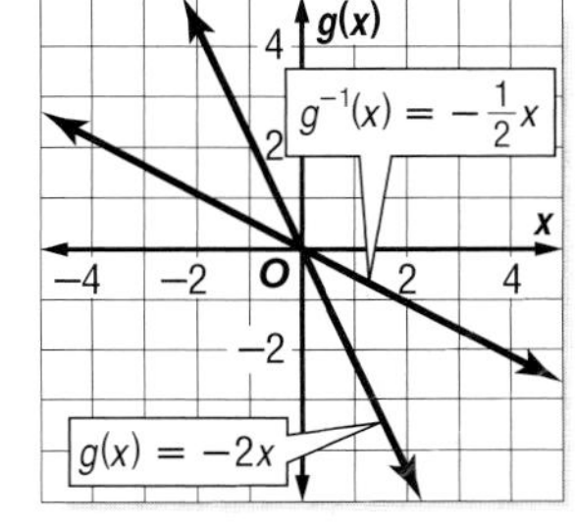

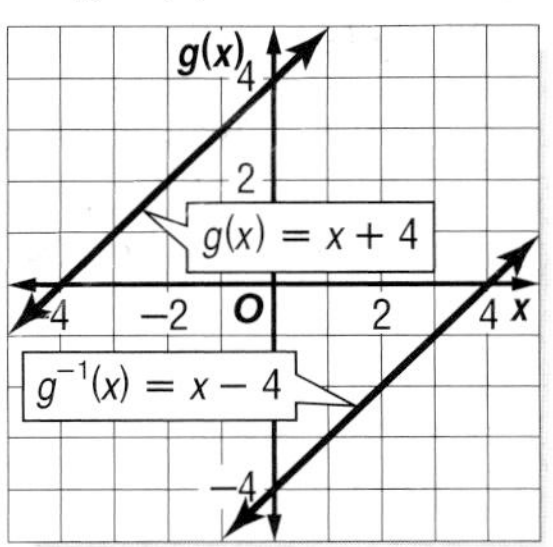

25. $y = -\frac{1}{2}x - \frac{1}{2}$

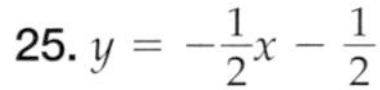

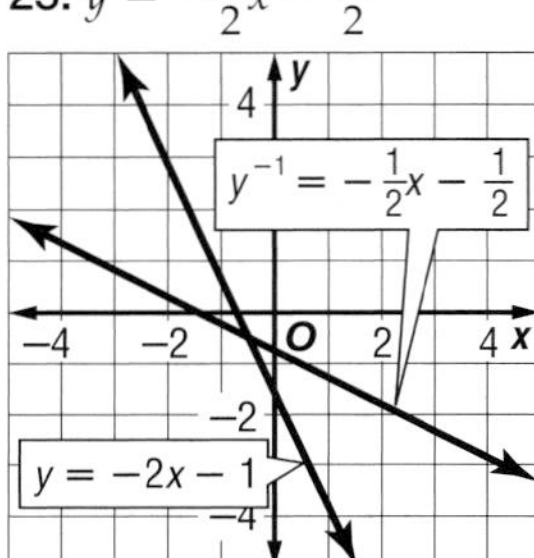

27. $f^{-1}(x) = \frac{8}{5}x$

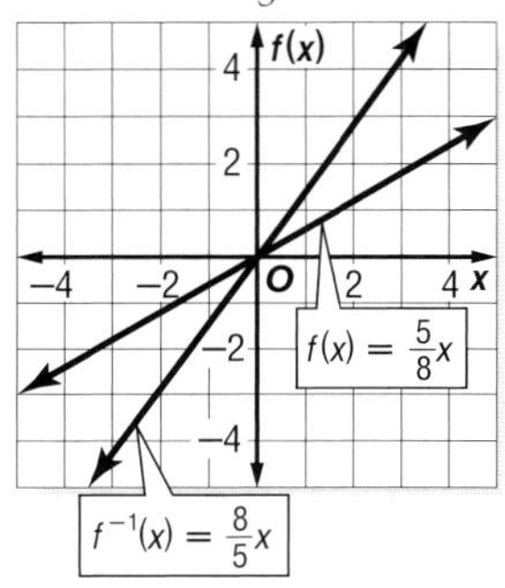

29. $f^{-1}(x) = \frac{5}{4}x + \frac{35}{4}$

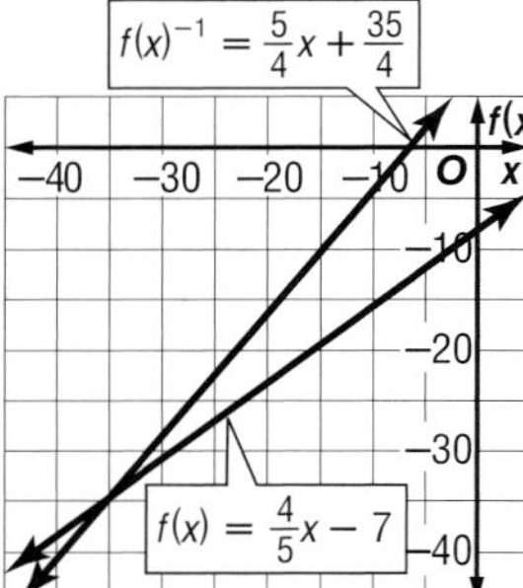

31. $f^{-1}(x) = \frac{8}{7}x + \frac{4}{7}$

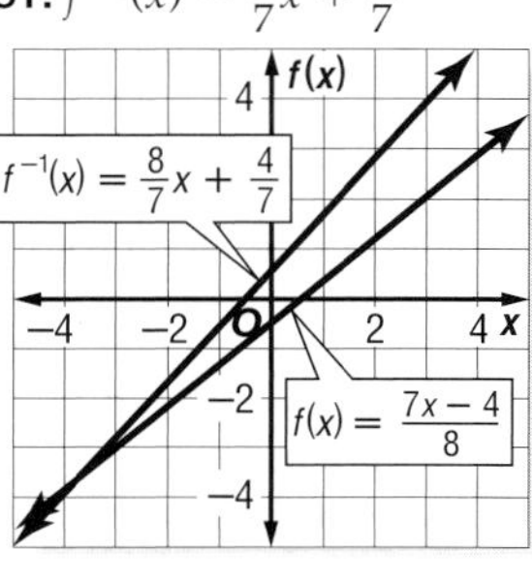

33. no 35. yes 37. yes 39. $y = \frac{1}{2}x - \frac{11}{2}$ 41. $I(m) = 320 + 0.04m$; \$4500 43. It can be used to convert Celsius to Fahrenheit.

45. Inverses are used to convert between two units of measurement. Answers should include the following.

- Even if it is not necessary, it is helpful to know the imperial units when given the metric units because most measurements in the U.S. are given in imperial units so it is easier to understand the quantities using our system.
- To convert the speed of light from meters per second to miles per hour,

$$f(x) \approx \frac{3.0 \times 10^8 \text{ meters}}{1 \text{ second}} \cdot \frac{3600 \text{ seconds}}{1 \text{ hour}} \cdot \frac{1 \text{ mile}}{1600 \text{ meters}}$$
$$\approx 675{,}000{,}000 \text{ mi/hr}$$

47. B 49. $g[h(x)] = 6x - 10$; $h[g(x)] = 6x$ 51. −7, −2, 3 53. 64 55. 3 57. 117 59. −7 61. $\frac{25}{4}$

Pages 397–399 Lesson 7-9

1. In order for it to be a square root function, only the nonnegative range can be considered. 3. Sample answer: $y = \sqrt{2x - 4}$

5.

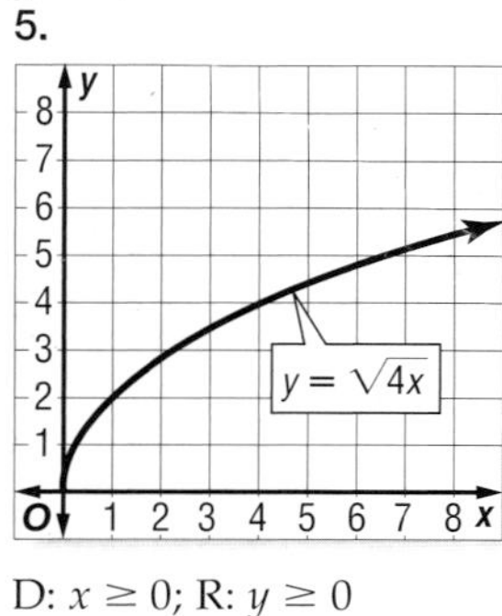

D: $x \geq 0$; R: $y \geq 0$

7.

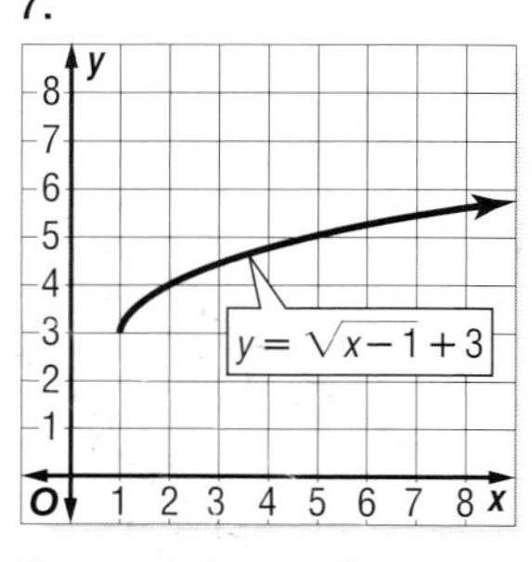

D: $x \geq 1$; R: $y \geq 3$

9.

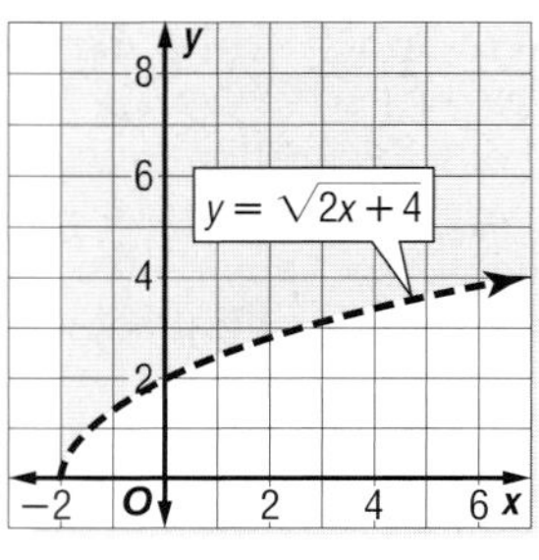

11.

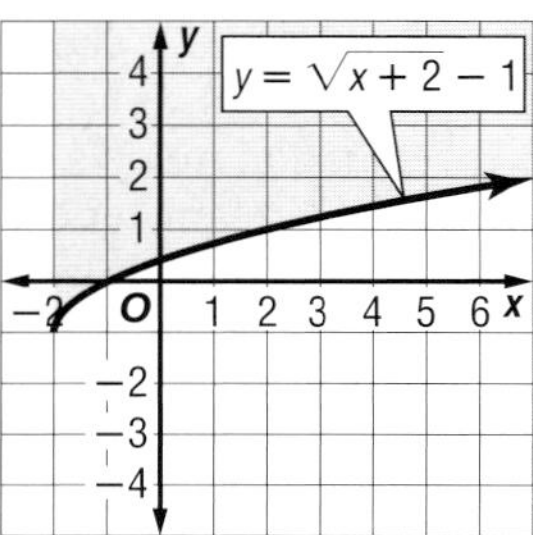

13. Yes; sample answer: The advertised pump will reach a maximum height of 87.9 ft.

15.

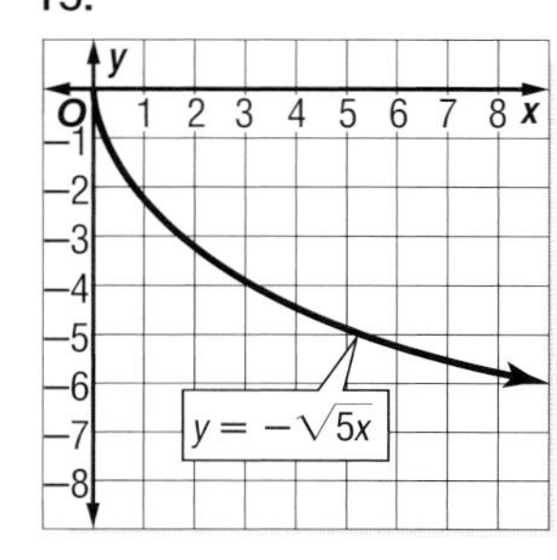

D: $x \geq 0$, R: $y \leq 0$

17.

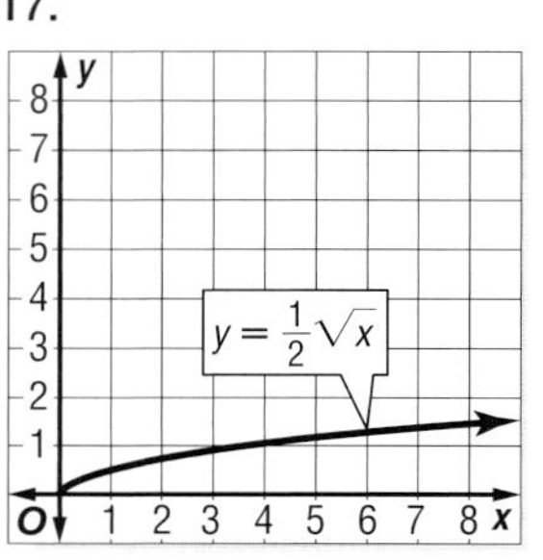

D: $x \geq 0$, R: $y \geq 0$

19.

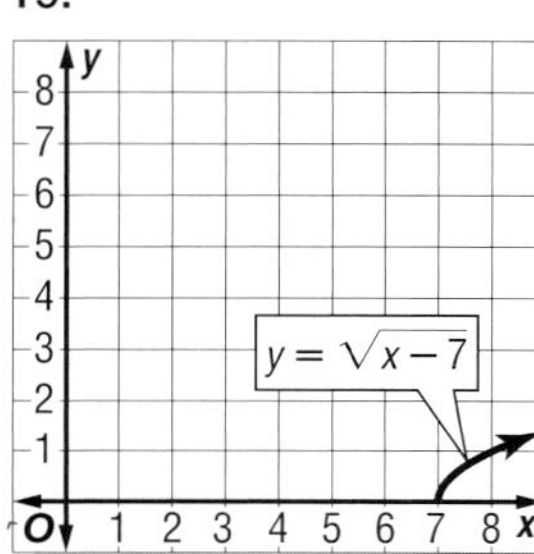

D: $x \geq 7$, R: $y \geq 0$

21.

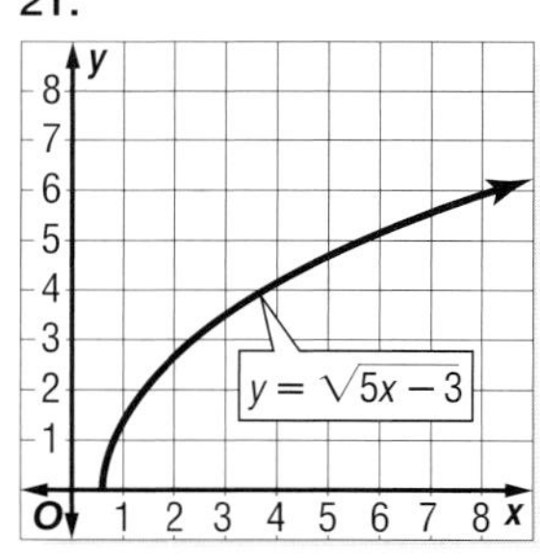

D: $x \geq 0.6$, R: $y \geq 0$

23.

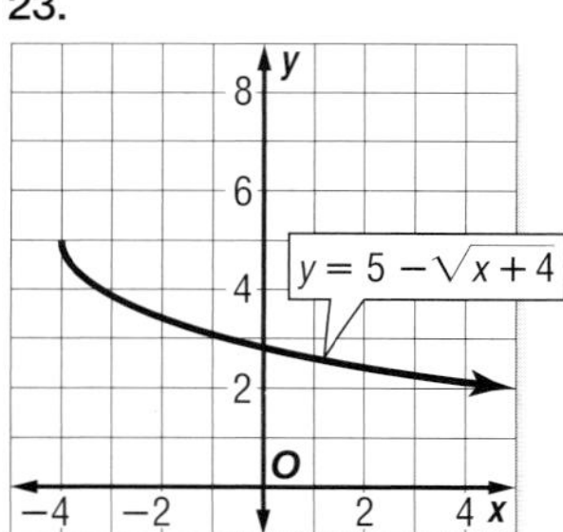

D: $x \geq -4$, R: $y \leq 5$

25.

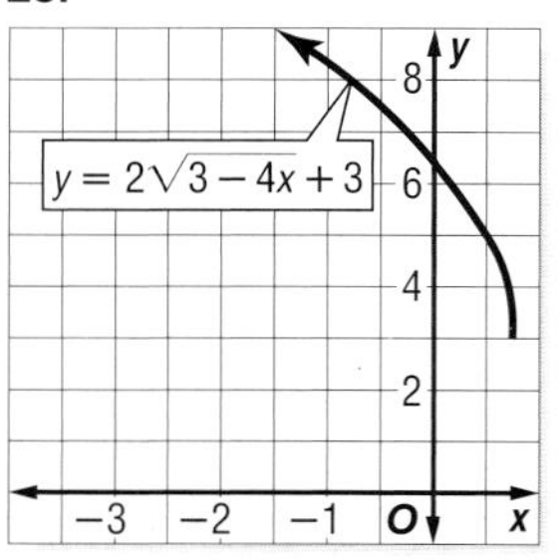

D: $x \leq 0.75$, R: $y \geq 3$

27.

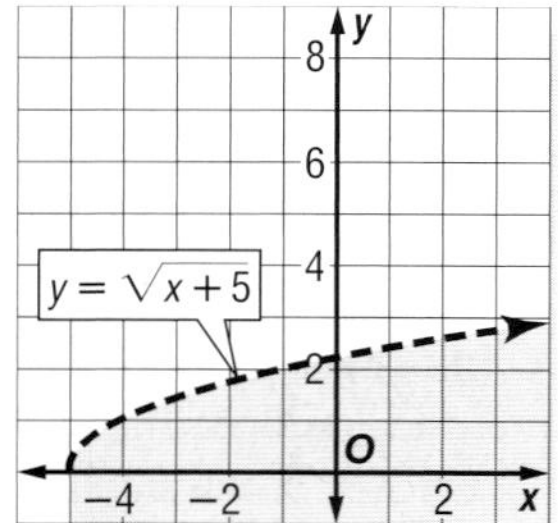

29.

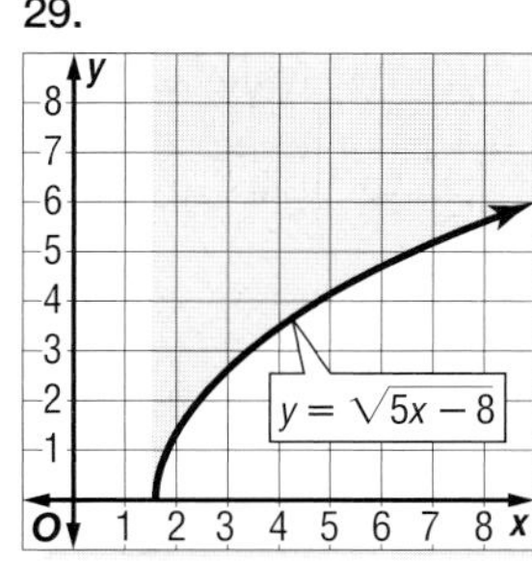

31.

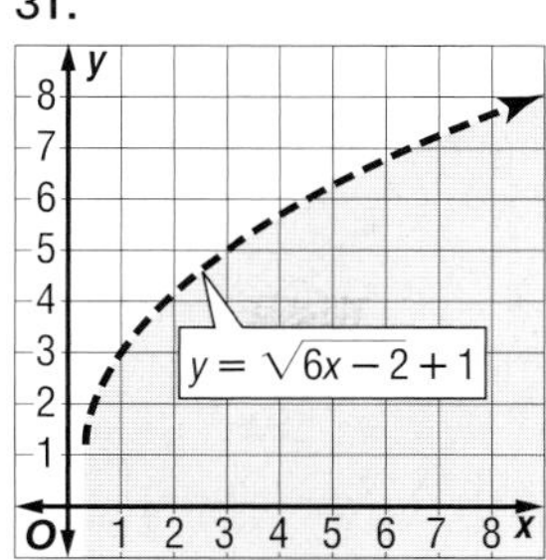

33. 317.29 mi

37. Square root functions are used in bridge design because the engineers must determine what diameter of steel cable needs to be used to support a bridge based on its weight. Answers should include the following.

- Sample answer: When the weight to be supported is less than 8 tons.
- 13,608 tons

39. D **41.** no **43.** $2x + 2$; 8; $x^2 + 2x - 15$; $\frac{x+5}{x-3}$, $x \neq 3$

45. $\frac{8x^3 + 12x^2 - 18x - 26}{2x + 3}$, $x \neq -\frac{3}{2}$; $\frac{8x^3 + 12x^2 - 18x - 28}{2x + 3}$, $x \neq -\frac{3}{2}$; $2x - 3$, $x \neq -\frac{3}{2}$; $8x^3 + 12x^2 - 18x - 27$, $x \neq -\frac{3}{2}$

47. $2x^2 - 4x - 16$ **49.** $a^3 - 1$

Pages 400–404 Chapter 7 Study Guide and Review

1. f **3.** a **5.** e **7.** -6; $x + h - 2$ **9.** -21; $6x + 6h + 3$

11. 20; $x^2 + 2xh + h^2 - x - h$

13a.

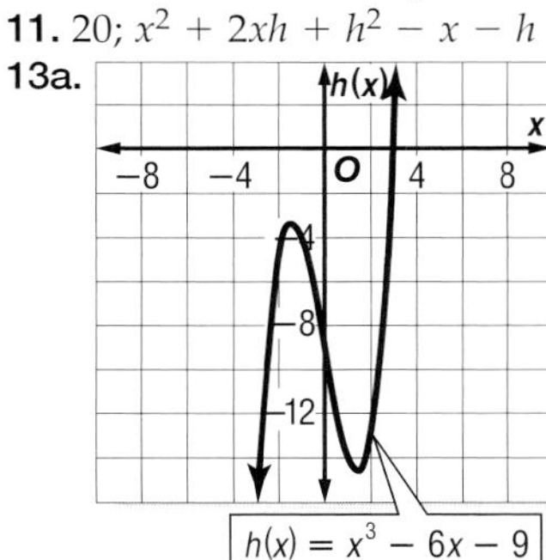

13b. at $x = 3$

13c. Sample answer: rel. max. at $x = -1.4$, rel. min. at $x = 1.4$

15a.

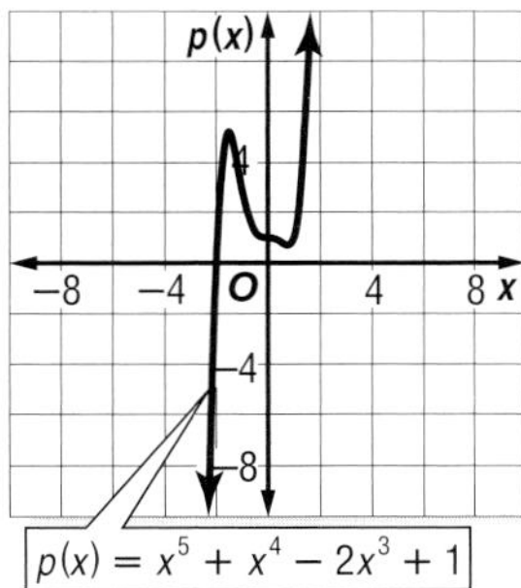

15b. between -2 and -3

15c. Sample answer: rel. max. at $x = -1.6$, rel. min. at $x = 0.8$

17a.

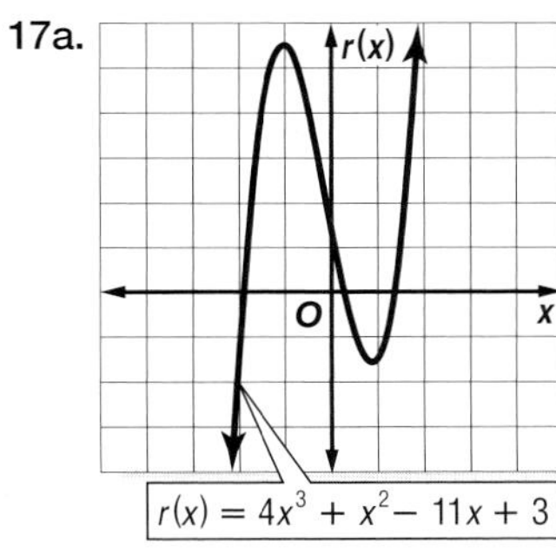

17b. between -2 and -1, between 0 and 1, and between 1 and 2

17c. Sample answer: rel. max. at $x = -1$, rel. min. at $x = 0.9$

19. $\frac{5}{3}$, -3, 0 **21.** 4, $-2 \pm 2i\sqrt{3}$ **23.** 2, -2 **25.** 4, -1

27. 20, -20 **29.** $x^2 + 2x + 3$ **31.** 1; 0; 2 **33.** 3 or 1; 1; 0 or 2 **35.** 2 or 0; 2 or 0; 4, 2, or 0 **37.** -1, -1

39. 1, 2, 4, -3 **41.** $\frac{1}{2}$, 2 **43.** $x^2 - 1$; $x^2 - 6x + 11$

45. $-15x - 5$; $-15x + 25$

47. $|x + 4|$; $|x| + 4$ **49.** $f^{-1}(x) = \frac{-x - 3}{2}$

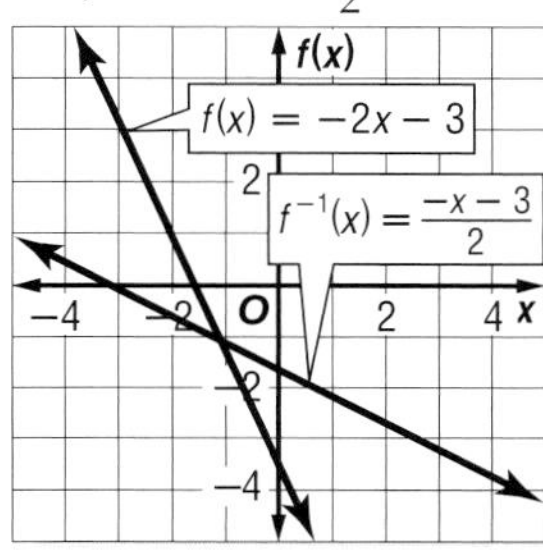

51. $f^{-1}(x) = \frac{2x - 1}{-3}$

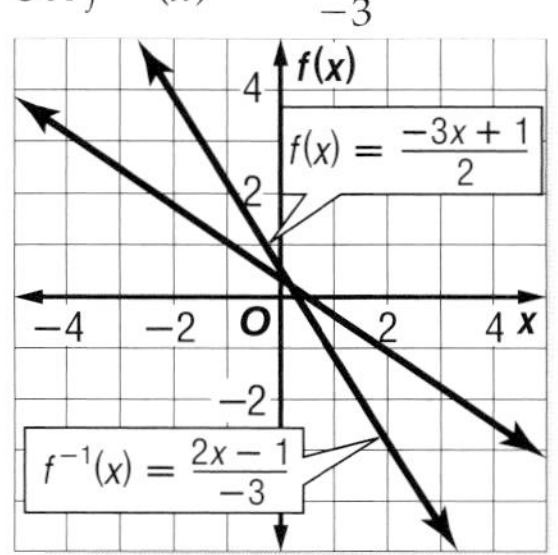

53. $y^{-1} = \pm\frac{1}{2}\sqrt{x} - \frac{3}{2}$

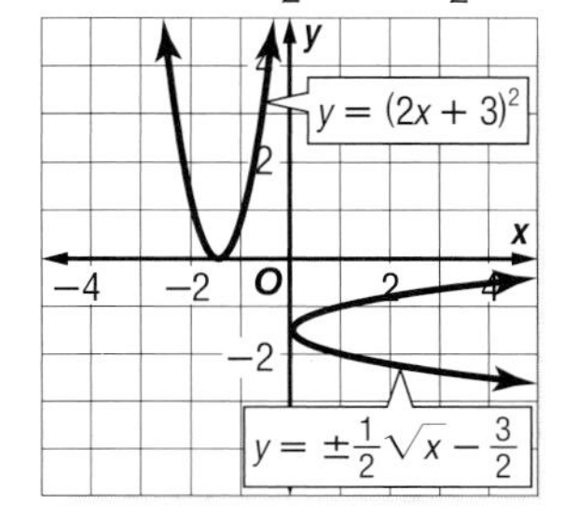

55. D: $x \geq \frac{3}{5}$, R: $y \geq 0$

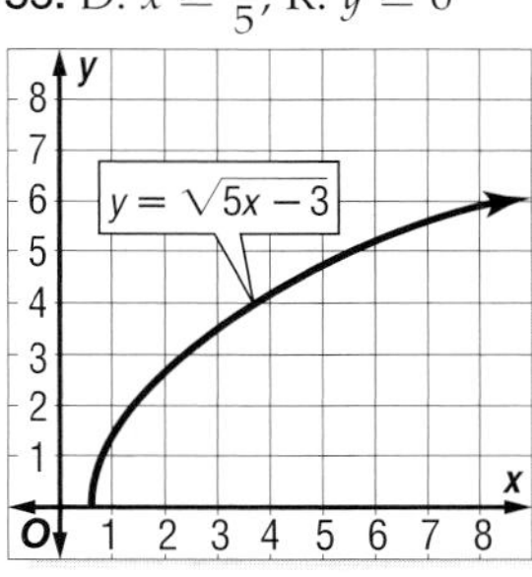

57.

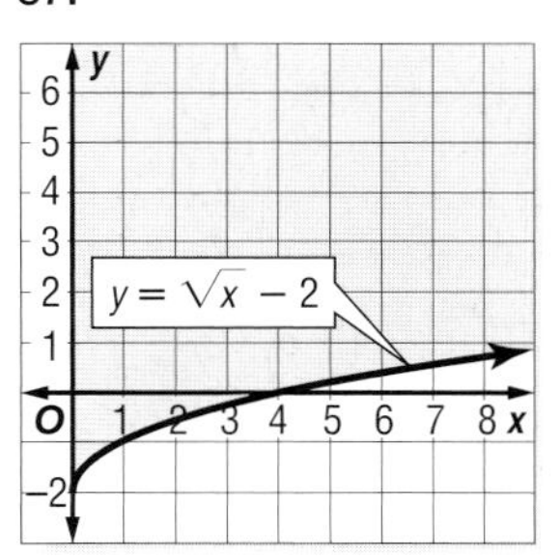

Chapter 8 Conic Sections

Page 411 Chapter 8 Getting Started

1. $\{-4, -6\}$ **3.** $\left\{\frac{3}{2}, -4\right\}$ **5a.** $\begin{bmatrix} -2 & 4 & -1 \\ 2 & 0 & -2 \end{bmatrix}$

5b. $\begin{bmatrix} 5 & 5 & 5 \\ -3 & -3 & -3 \end{bmatrix}$ **5c.** $\begin{bmatrix} 3 & 9 & 4 \\ -1 & -3 & -5 \end{bmatrix}$

7.

$x + y = 3$

Pages 414–416 Lesson 8-1

1. Since the sum of the x-coordinates of the given points is negative, the x-coordinate of the midpoint is negative. Since the sum of the y-coordinates of the given points is positive, the y-coordinate of the midpoint is positive. Therefore, the midpoint is in Quadrant II. **3.** Sample answer: (0, 0) and (5, 2) **5.** (2.5, 2.25) **7.** $\sqrt{122}$ units **9.** D **11.** $(-4, -2)$

13. $\left(\frac{17}{2}, \frac{27}{2}\right)$ **15.** (3.1, 2.7) **17.** $\left(\frac{1}{24}, \frac{5}{8}\right)$ **19.** (7, 11)

21. Sample answer: Draw several line segments across the U.S. One should go from the northeast corner to the southwest corner; another should go from the southeast corner to the northwest corner; another should go across the middle of the U.S. from east to west; and so on. Find the midpoints of these segments. Locate a point to represent all of these midpoints.
25. 25 units **27.** $3\sqrt{17}$ units **29.** $\sqrt{70.25}$ units **31.** 1 unit **33.** $\frac{\sqrt{813}}{12}$ units **35.** $7\sqrt{2} + \sqrt{58}$ units, 10 units^2

37. $\sqrt{130}$ units **39.** about 0.9 h **41.** The slope of the line through (x_1, y_1) and (x_2, y_2) is $\frac{y_2 - y_1}{x_2 - x_1}$ and the point-slope form of the equation of the line is $y - y_1 = \frac{y_2 - y_1}{x_2 - x_1}(x - x_1)$. Substitute $\left(\frac{x_1 + x_2}{2}, \frac{y_1 + y_2}{2}\right)$ into this equation. The left side is $\frac{y_1 + y_2}{2} - y_1$ or $\frac{y_2 - y_1}{2}$. The right side is $\frac{y_2 - y_1}{x_2 - x_1}\left(\frac{x_1 + x_2}{2} - x_1\right) = \frac{y_2 - y_1}{x_2 - x_1}\left(\frac{x_2 - x_1}{2}\right)$ or $\frac{y_2 - y_1}{2}$. Therefore, the point with coordinates $\left(\frac{x_1 + x_2}{2}, \frac{y_1 + y_2}{2}\right)$ lies on the line through (x_1, y_1) and (x_2, y_2).
The distance from $\left(\frac{x_1 + x_2}{2}, \frac{y_1 + y_2}{2}\right)$ to (x_1, y_1) is $\sqrt{\left(x_1 - \frac{x_1 + x_2}{2}\right)^2 + \left(y_1 - \frac{y_1 + y_2}{2}\right)^2}$ or $\sqrt{\left(\frac{x_1 - x_2}{2}\right)^2 + \left(\frac{y_1 - y_2}{2}\right)^2}$. The distance from $\left(\frac{x_1 + x_2}{2}, \frac{y_1 + y_2}{2}\right)$ to (x_2, y_2) is
$\sqrt{\left(x_2 - \frac{x_1 + x_2}{2}\right)^2 + \left(y_2 - \frac{y_1 + y_2}{2}\right)^2} = \sqrt{\left(\frac{x_2 - x_1}{2}\right)^2 + \left(\frac{y_2 - y_1}{2}\right)^2}$ or $\sqrt{\left(\frac{x_1 - x_2}{2}\right)^2 + \left(\frac{y_1 - y_2}{2}\right)^2}$. Therefore the point with coordinates $\left(\frac{x_1 + x_2}{2}, \frac{y_1 + y_2}{2}\right)$ is equidistant from (x_1, y_1) and (x_2, y_2). **43.** C **45.** on the line with equation $y = x$

47. D = $\{x \mid x \geq 2\}$, R = $\{y \mid y \geq 0\}$

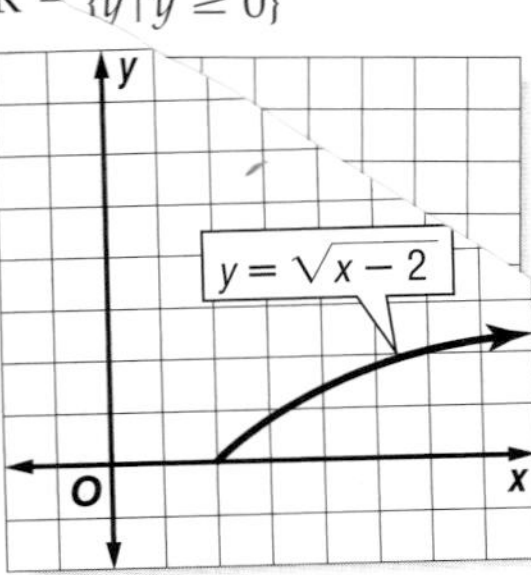

49. D = $\{x \mid x \geq 0\}$, R = $\{y \mid y \geq 1\}$

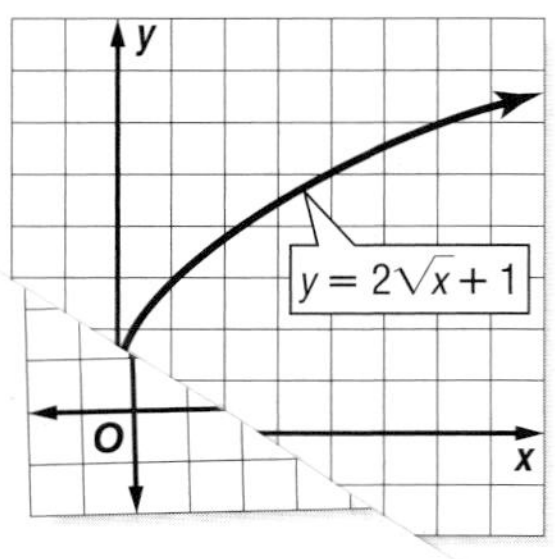

51. $-1 + 13i$ **53.** $4 - 3i$ **55.** $y = (x - 2)^2 - 3$ **57.** $y = 3(x - 1)^2 + 2$ **59.** $y = -3(x + 3)^2 + 17$

Pages 423–425 Lesson 8-2

1. $(3, -7)$, $\left(3, -6\frac{15}{16}\right)$, $x = 3$, $y = -7\frac{1}{16}$ **3.** When she added 9 to complete the square, she forgot to also subtract 9. The standard form is $y = (x + 3)^2 - 9 + 4$ or $y = (x + 3)^2 - 5$.

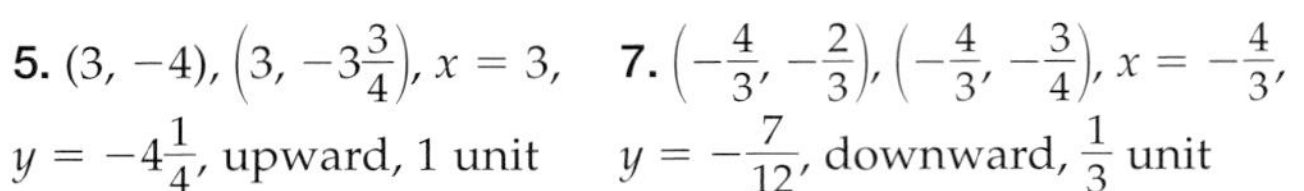

5. $(3, -4)$, $\left(3, -3\frac{3}{4}\right)$, $x = 3$, $y = -4\frac{1}{4}$, upward, 1 unit

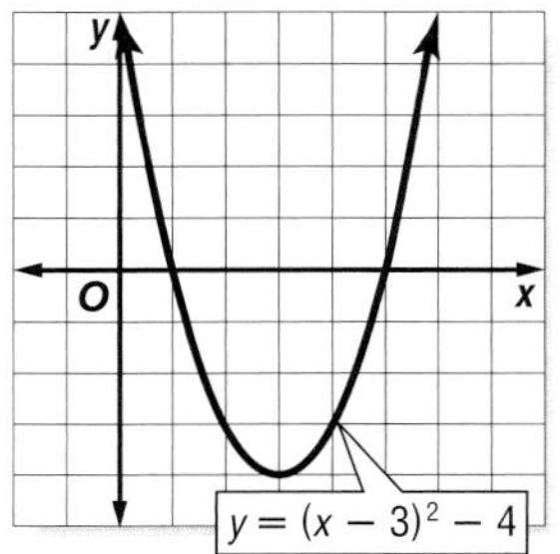

7. $\left(-\frac{4}{3}, -\frac{2}{3}\right)$, $\left(-\frac{4}{3}, -\frac{3}{4}\right)$, $x = -\frac{4}{3}$, $y = -\frac{7}{12}$, downward, $\frac{1}{3}$ unit

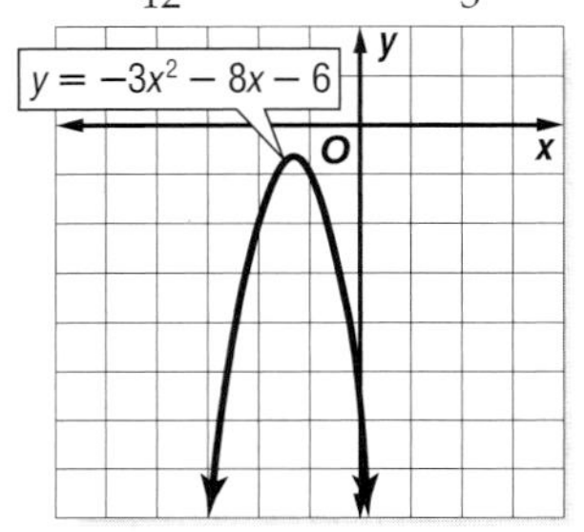

9. $y = \frac{1}{8}(x - 3)^2 + 6$

11. $x = \frac{1}{24}y^2 - 6$
13. $x = (y + 7)^2 - 29$
15. $x = 3\left(y + \frac{5}{6}\right)^2 - 11\frac{1}{12}$

17. $(0, 0)$, $\left(\frac{1}{2}, 0\right)$, $y = 0$, $x = -\frac{1}{2}$, right, 2 units

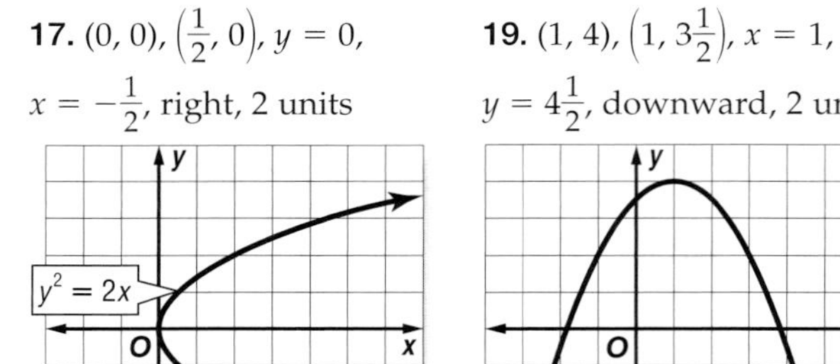

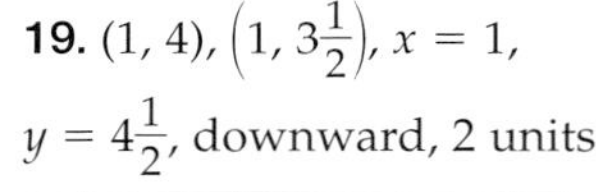

19. $(1, 4)$, $\left(1, 3\frac{1}{2}\right)$, $x = 1$, $y = 4\frac{1}{2}$, downward, 2 units

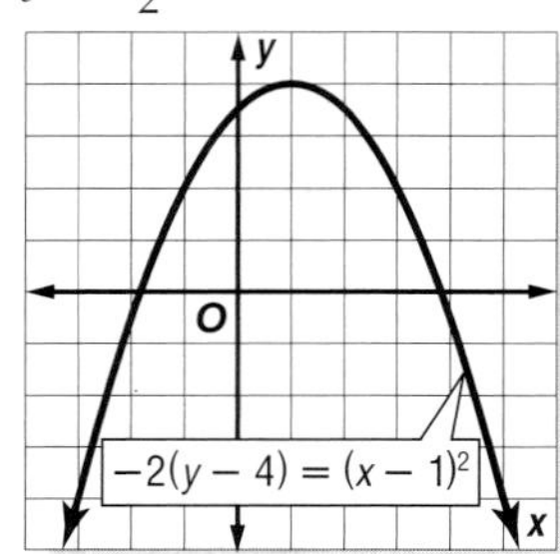

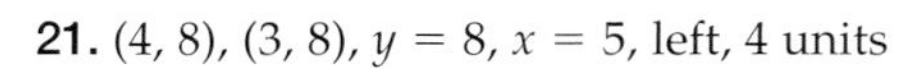

21. $(4, 8)$, $(3, 8)$, $y = 8$, $x = 5$, left, 4 units

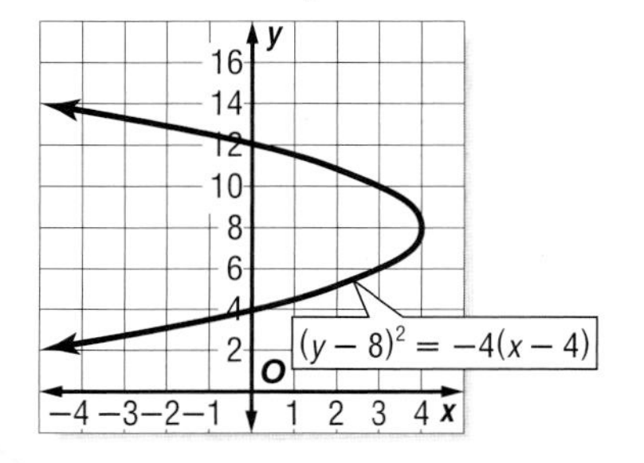

Selected Answers

23. $(-24, 7), \left(-23\frac{3}{4}, 7\right), y = 7, x = -24\frac{1}{4}$, right, 1 unit

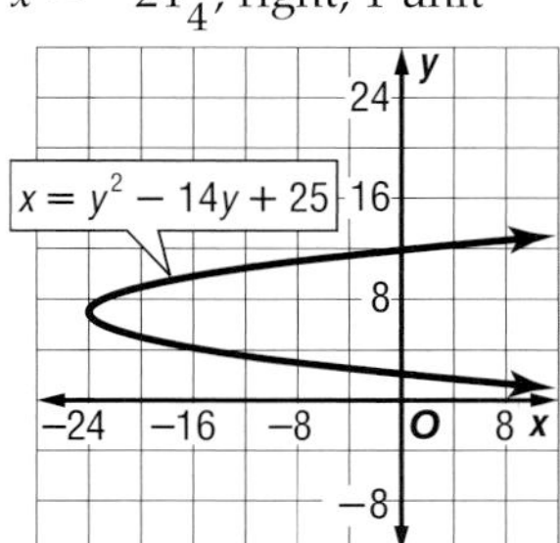

25. $(4, 2), \left(4, 2\frac{1}{12}\right), x = 4, y = 1\frac{11}{12}$, upward, $\frac{1}{3}$ unit

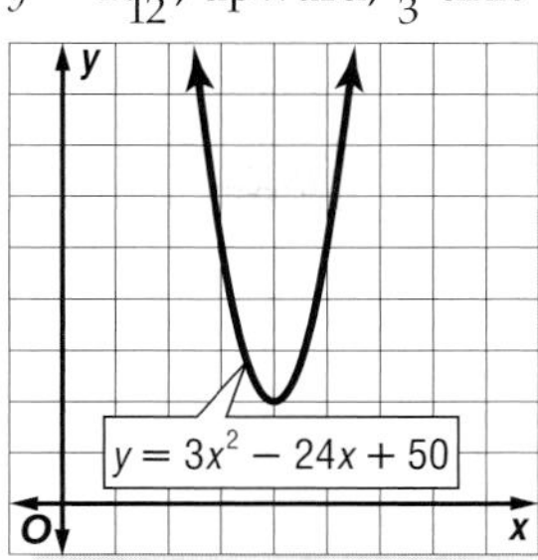

27. $\left(\frac{17}{4}, \frac{3}{4}\right), \left(\frac{67}{16}, \frac{3}{4}\right), y = \frac{3}{4}, x = \frac{69}{16}$, left, $\frac{1}{4}$ unit

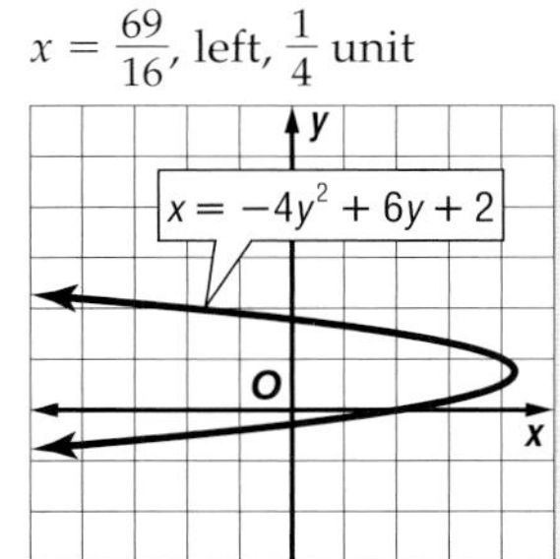

29. $(123, -18), \left(122\frac{1}{4}, -18\right), y = -18, x = 123\frac{3}{4}$, left, 3 units

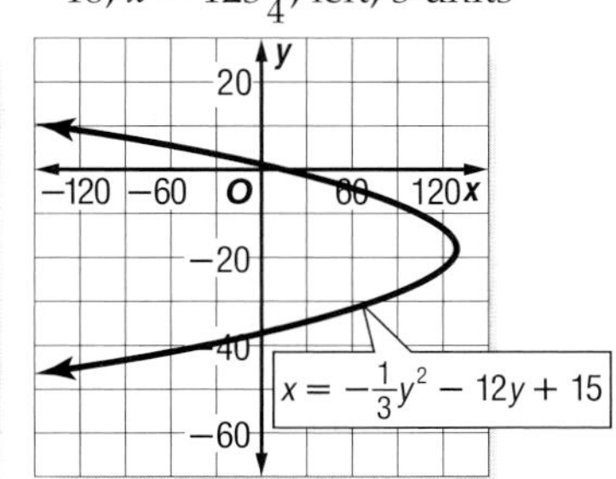

31. 1 **33.** $y = -\frac{2}{3}$ **35.** 0.75 cm

37. $x = -\frac{1}{24}(y - 6)^2 + 8$

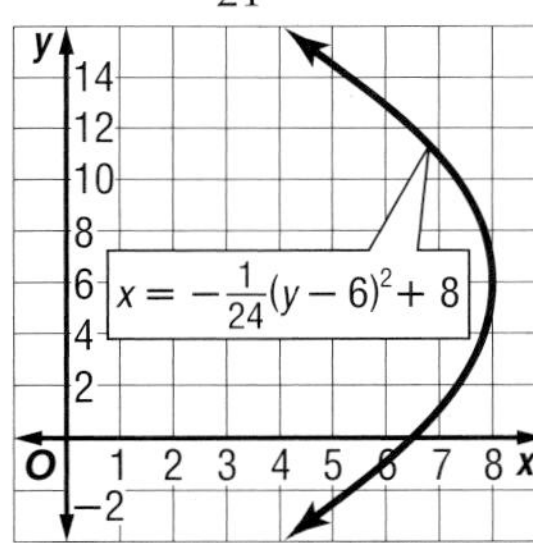

39. $y = \frac{1}{16}(x - 1)^2 + 7$

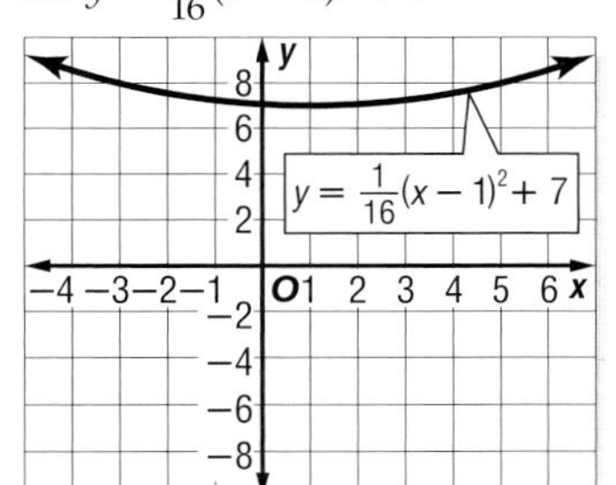

41. $x = \frac{1}{4}(y - 3)^2 + 4$

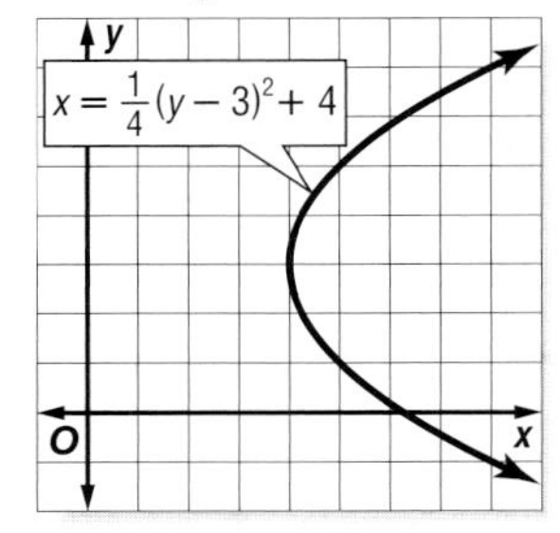

43. about $y = -0.00046x^2 + 325$ **45.** $y = -\frac{1}{26{,}200}x^2 + 6550$

47. A parabolic reflector can be used to make a car headlight more effective. Answers should include the following.

- Reflected rays are *focused* at that point.
- The light from an unreflected bulb would shine in all directions. With a parabolic reflector, most of the light can be directed forward toward the road.

49. A **51.** 10 units

53.

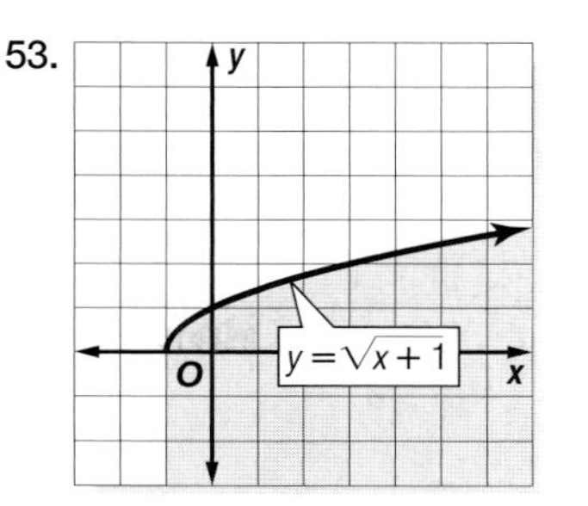

55. 4 **57.** 9 **59.** $2\sqrt{3}$ **61.** $4\sqrt{3}$

Pages 428–431 Lesson 8-3

1. Sample answer: $(x - 6)^2 + (y + 2)^2 = 16$ **3.** Lucy; 36 is the *square* of the radius, so the radius is 6 units.
5. $(x + 1)^2 + (y + 5)^2 = 4$ **7.** $(x - 3)^2 + (y + 7)^2 = 9$

9. $(0, 14)$, $\sqrt{34}$ units

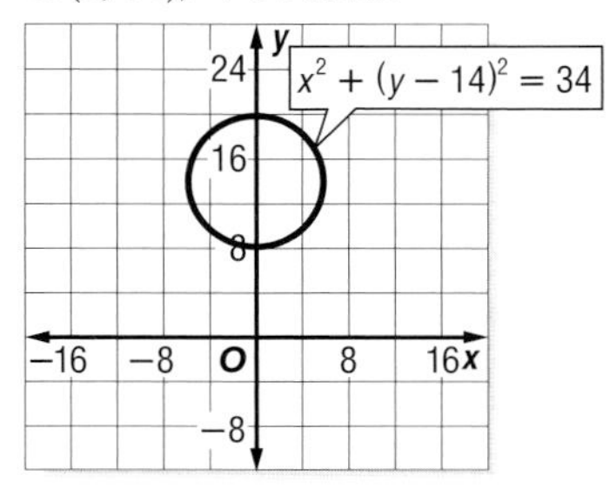

11. $\left(-\frac{2}{3}, \frac{1}{2}\right), \frac{2\sqrt{2}}{3}$ unit

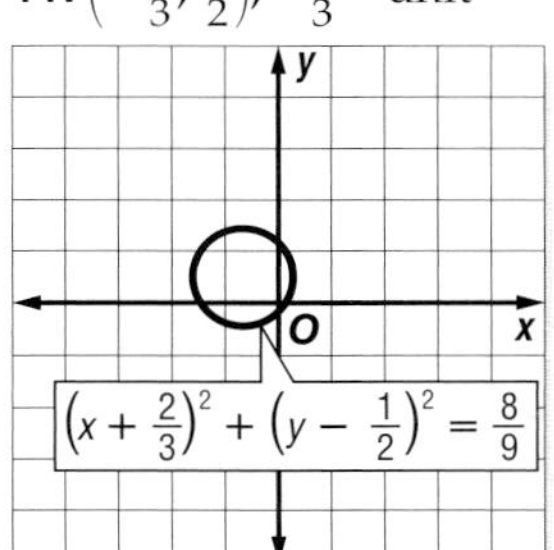

13. $(-2, 0)$, $2\sqrt{3}$ units

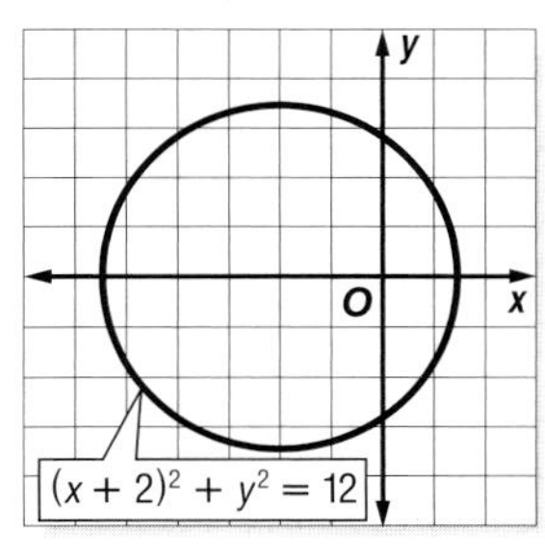

15.

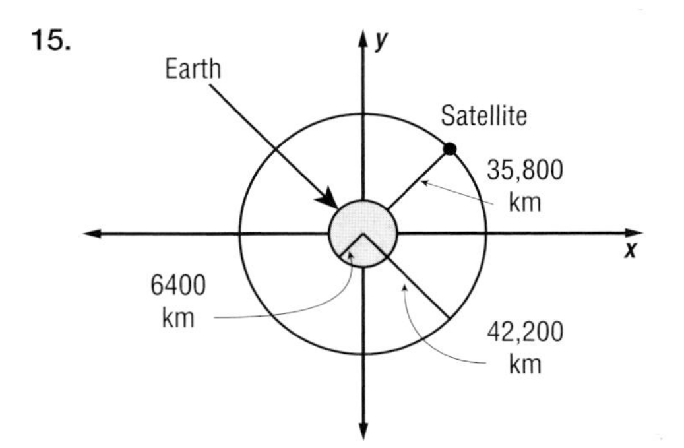

17. $(x - 2)^2 + (y + 1)^2 = 4$ **19.** $(x + 8)^2 + (y - 7)^2 = \frac{1}{4}$

21. $(x + 1)^2 + \left(y + \frac{1}{2}\right)^2 = \frac{1945}{4}$

23. $\left(x + \sqrt{13}\right)^2 + (y - 42)^2 = 1777$
25. $(x - 4)^2 + (y - 2)^2 = 4$ **27.** $(x + 5)^2 + (y - 4)^2 = 25$
29. $(x + 2.5)^2 + (y + 2.8)^2 = 1600$

Selected Answers

31. (0, 0), 12 units

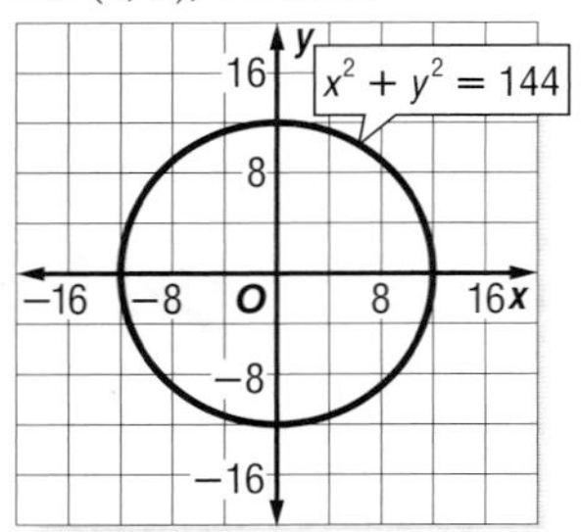

33. (−3, −7), 9 units

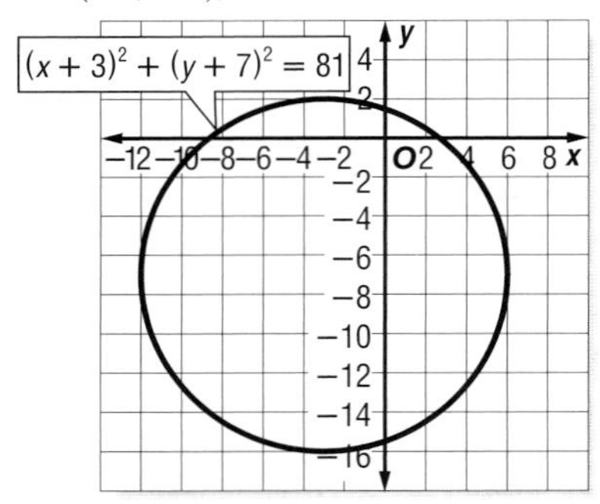

35. (3, −7), $5\sqrt{2}$ units

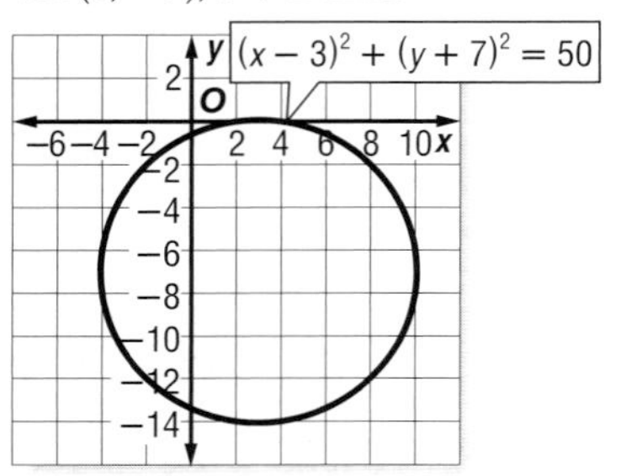

37. $(-2, \sqrt{3})$, $\sqrt{29}$ units

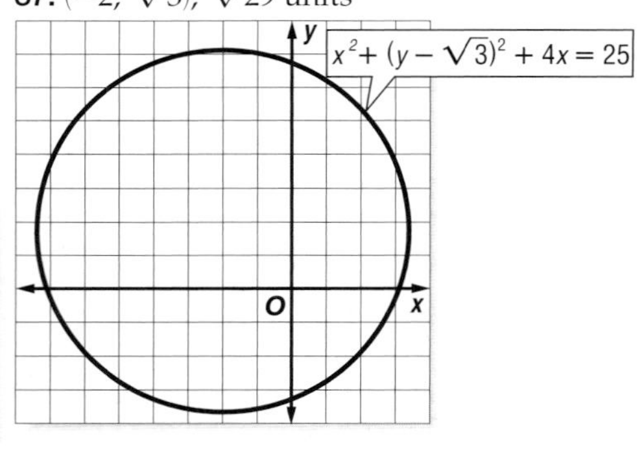

39. (0, 3), 5 units

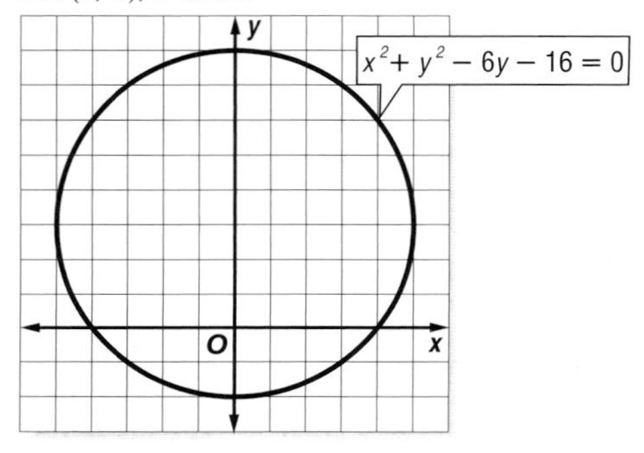

41. (9, 9), $\sqrt{109}$ units

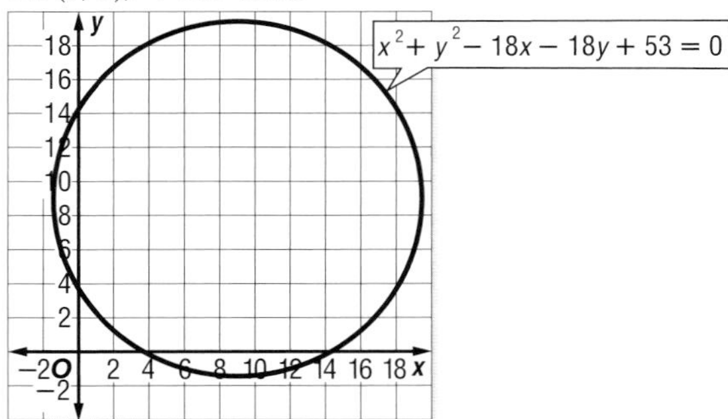

43. $\left(\frac{3}{2}, -4\right)$, $\frac{3\sqrt{17}}{2}$ units **45.** (−1, −2), $\sqrt{14}$ units

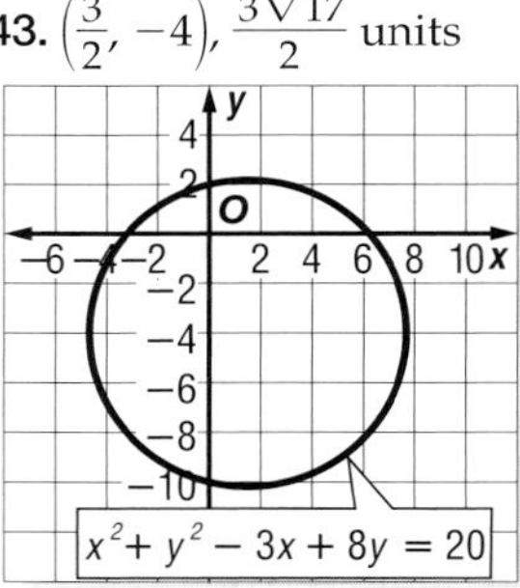

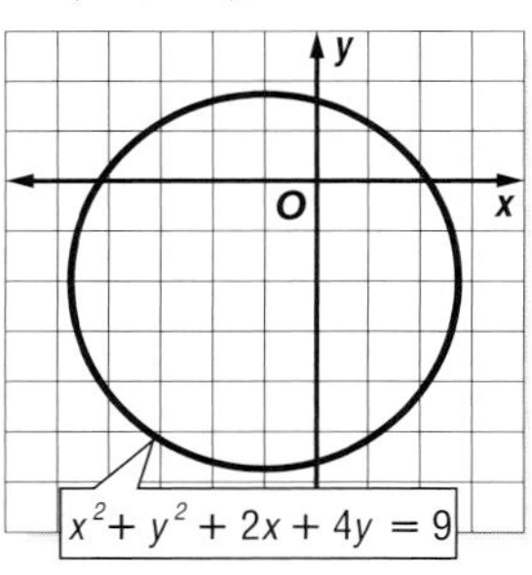

47. $\left(0, -\frac{9}{2}\right)$, $\sqrt{19}$ units

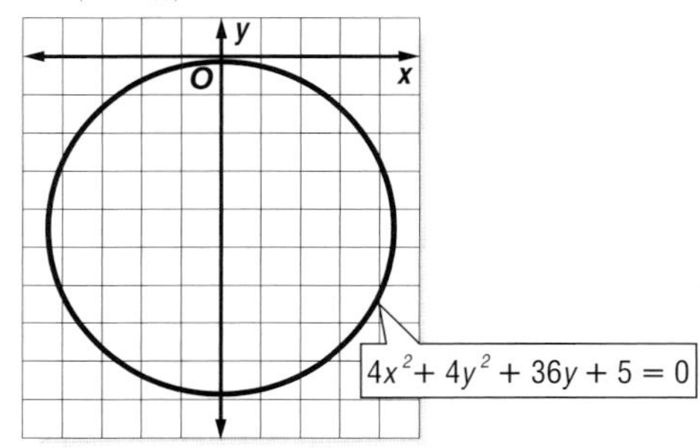

49. $(x + 1)^2 + (y + 2)^2 = 5$ **51.** A **53.** $y = \pm\sqrt{16 - (x + 3)^2}$

55.

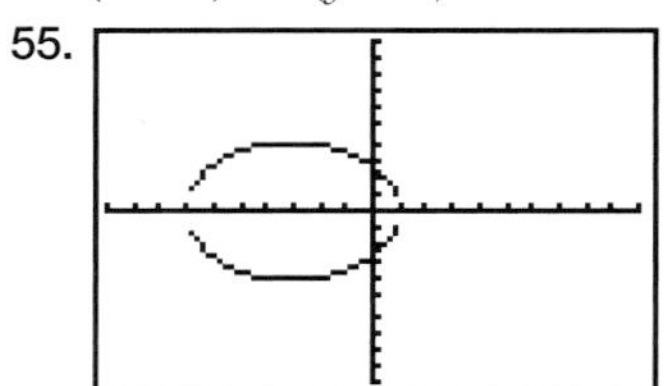

[−10, 10] scl:1 by [−10, 10] scl:1

57. (1, 0), $\left(\frac{11}{12}, 0\right)$, $y = 0$, $x = 1\frac{1}{12}$, left, $\frac{1}{3}$ unit **59.** (−2, −4), $\left(-2, -3\frac{3}{4}\right)$, $x = -2$, $y = -4\frac{1}{4}$, upward, 1 unit

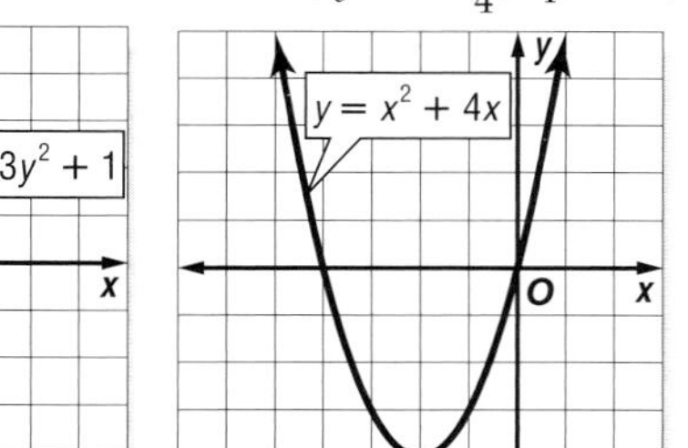

61. (−1, −2) **63.** −4, −2, 1 **65.** 28 in. by 15 in. **67.** 6 **69.** 25 **71.** $2\sqrt{2}$

Page 431 Practice Quiz 1

1. 13 units

3. $(0, 0)$, $\left(1\frac{1}{2}, 0\right)$, $y = 0$, $x = -1\frac{1}{2}$, right, 6 units

5. $(0, 4)$, 7 units

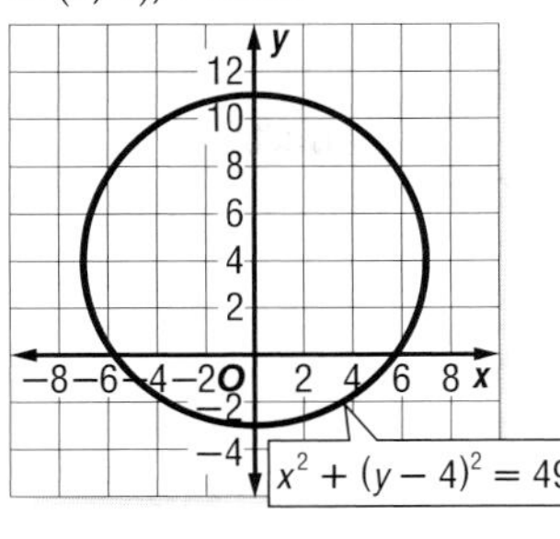

Pages 437–440 Lesson 8-4

1. $x = -1, y = 2$ **3.** Sample answer: $\frac{(x-2)^2}{4} + \frac{(y+5)^2}{1} = 1$

5. $\frac{(y+4)^2}{36} + \frac{(x-2)^2}{4} = 1$

7. $(0, 0)$: $(0, \pm 3)$; $6\sqrt{2}$; 6

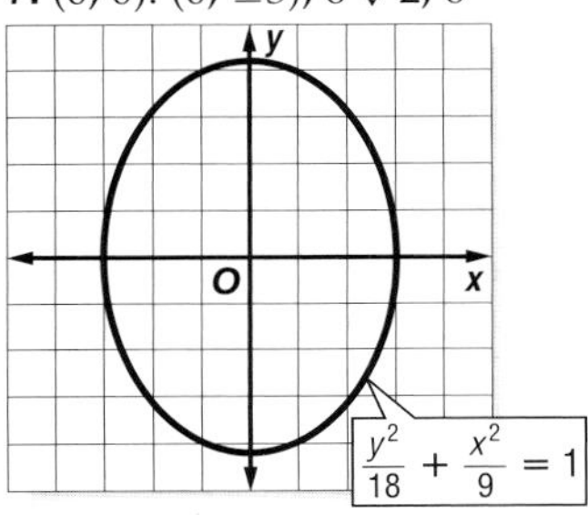

9. $(0, 0)$; $(\pm 2, 0)$; $4\sqrt{2}$; 4

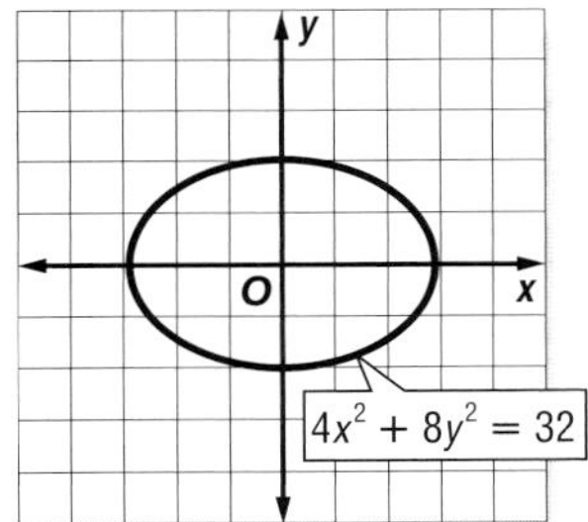

11. about $\frac{x^2}{1.32 \times 10^{15}} + \frac{y^2}{1.27 \times 10^{15}} = 1$ **13.** $\frac{x^2}{16} + \frac{y^2}{7} = 1$

15. $\frac{y^2}{16} + \frac{(x+2)^2}{4} = 1$ **17.** $\frac{(y-4)^2}{64} + \frac{(x-2)^2}{4} = 1$

19. $\frac{(x-5)^2}{64} + \frac{(y-4)^2}{\frac{81}{4}} = 1$ **21.** $\frac{x^2}{169} + \frac{y^2}{25} = 1$

23. about $\frac{x^2}{2.02 \times 10^{16}} + \frac{y^2}{2.00 \times 10^{16}} = 1$ **25.** $\frac{y^2}{20} + \frac{x^2}{4} = 1$

27. $(0, 0)$; $\left(0, \pm\sqrt{5}\right)$; $2\sqrt{10}$; $2\sqrt{5}$

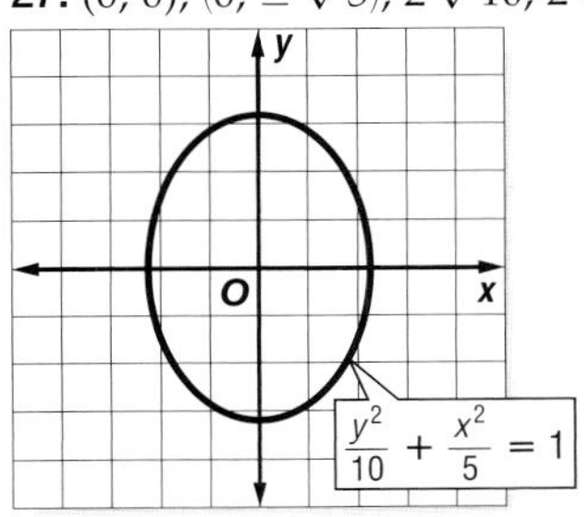

29. $(-8, 2)$; $\left(-8 \pm 3\sqrt{7}, 2\right)$; 24; 18

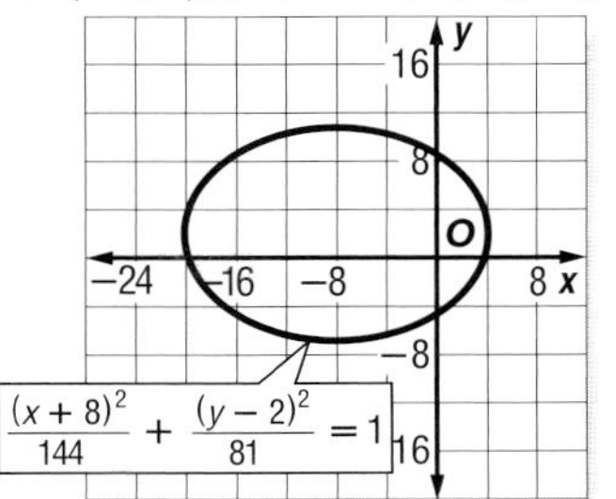

31. $(0, 0)$; $\left(\pm\sqrt{6}, 0\right)$; 6; $2\sqrt{3}$

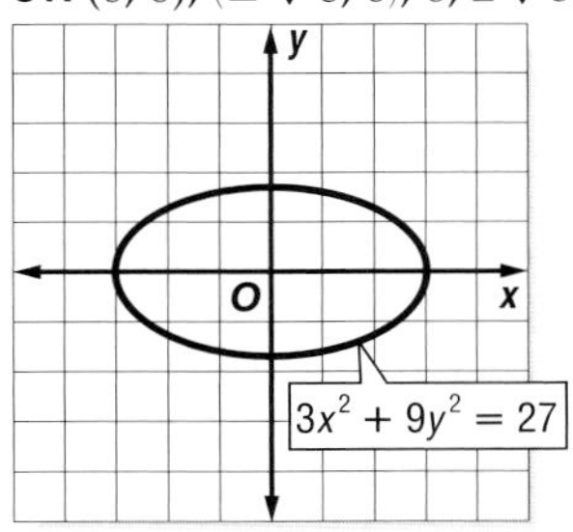

33. $(0, 0)$; $\left(0, \pm\sqrt{7}\right)$; 8; 6

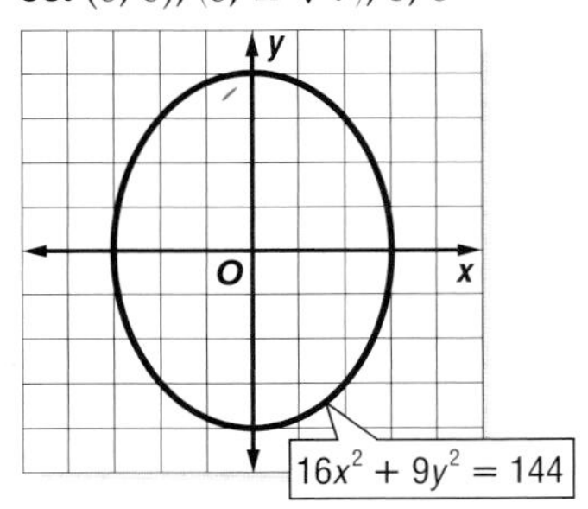

35. $(-3, 1)$; $(-3, 5)$, $(-3, -3)$; $4\sqrt{6}$; $4\sqrt{2}$

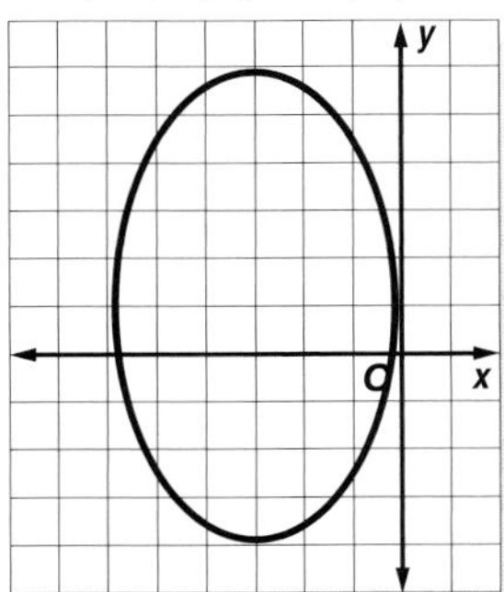

37. $(2, 2)$; $(2, 4)$, $(2, 0)$; $2\sqrt{7}$; $2\sqrt{3}$

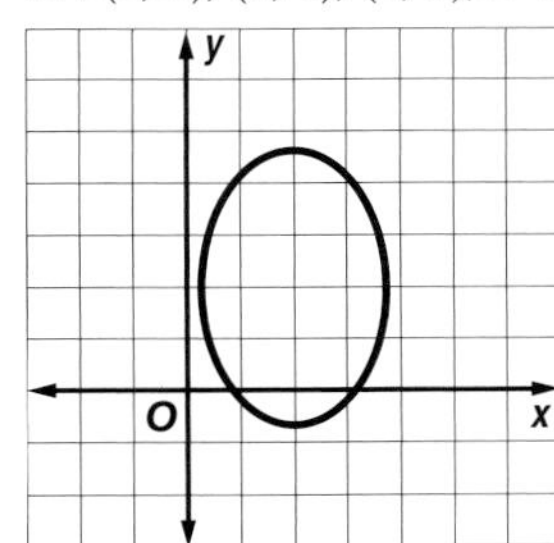

39. $\frac{x^2}{12} + \frac{y^2}{9} = 1$ **41.** C **43.** about $\frac{x^2}{1.35 \times 10^{19}} + \frac{y^2}{1.26 \times 10^{19}} = 1$

45. $(x-4)^2 + (y-1)^2 = 101$ **47.** $(x-4)^2 + (y+1)^2 = 16$

49.

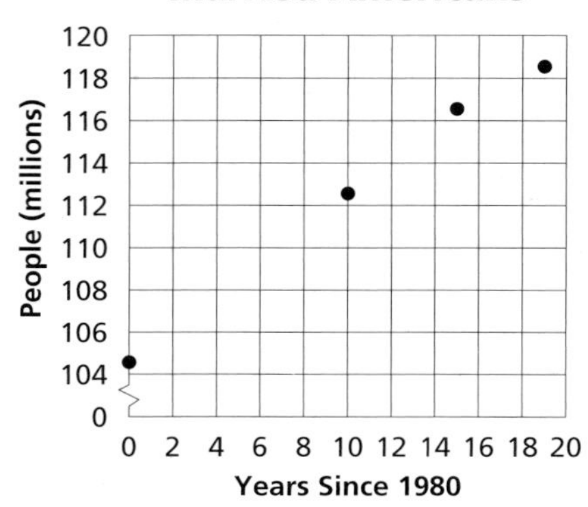

51. Sample answer: 128,600,000

53.

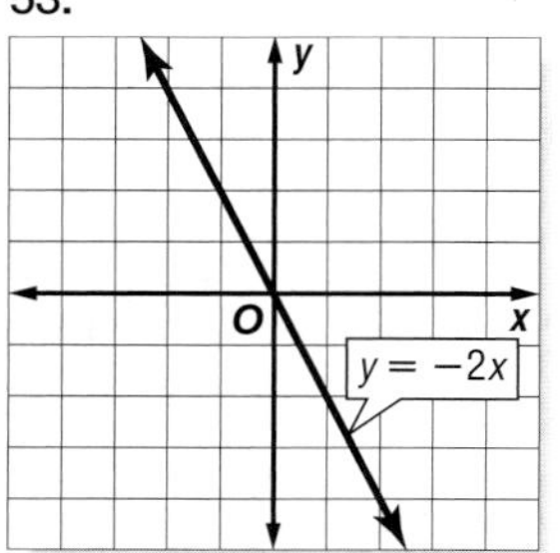

55.

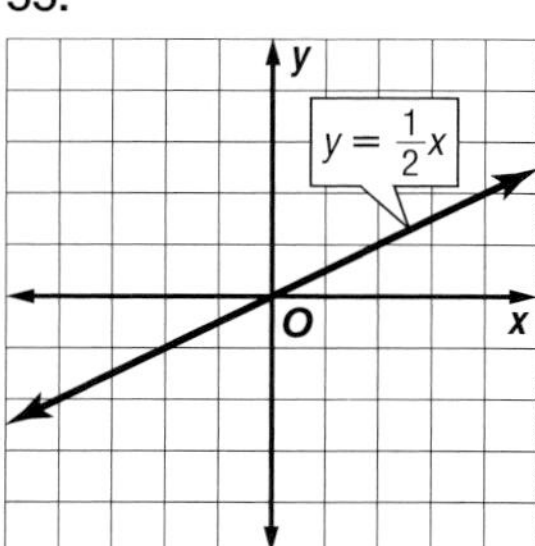

57.

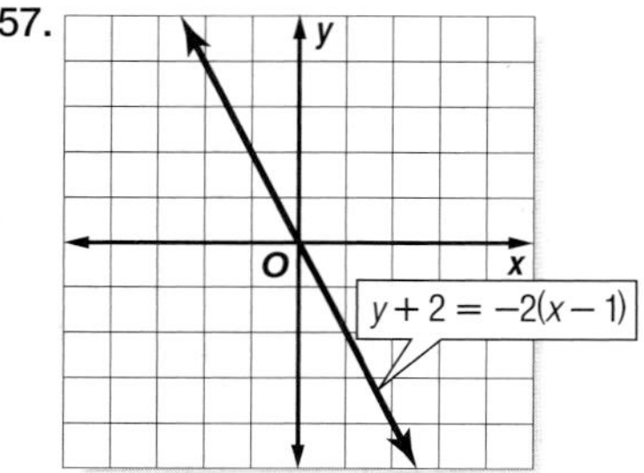

Pages 445–448 Lesson 8-5

1. sometimes **3.** Sample answer: $\frac{x^2}{4} - \frac{y^2}{9} = 1$

5. $\frac{x^2}{1} - \frac{y^2}{15} = 1$

7. $\left(1, -6 \pm 2\sqrt{5}\right)$; $\left(1, -6 \pm 3\sqrt{5}\right)$; $y + 6 = \pm\frac{2\sqrt{5}}{5}(x - 1)$

9. $\left(4 \pm 2\sqrt{5}, -2\right)$; $\left(4 \pm 3\sqrt{5}, -2\right)$; $y + 2 = \pm\frac{\sqrt{5}}{2}(x - 4)$

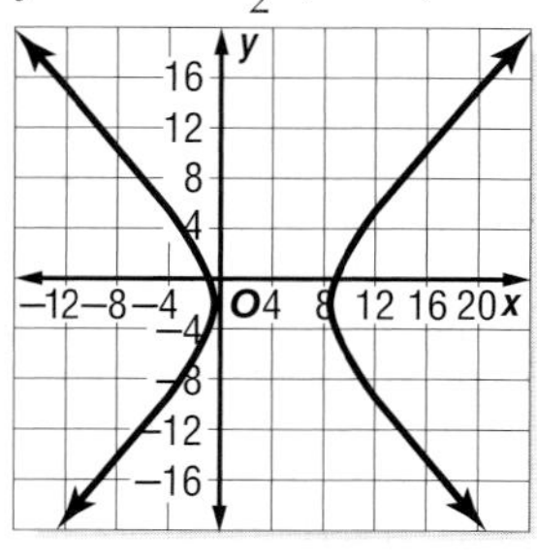
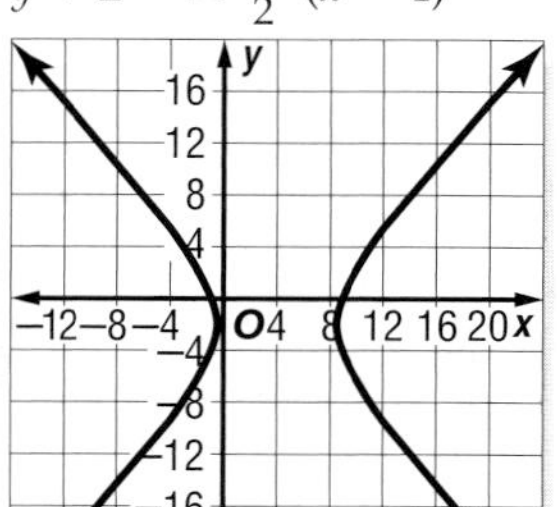

11. $\frac{x^2}{4} - \frac{y^2}{12} = 1$ **13.** $\frac{\left(y + \frac{11}{2}\right)^2}{\frac{25}{4}} - \frac{x^2}{6} = 1$

15. $\frac{x^2}{25} - \frac{y^2}{36} = 1$ **17.** $\frac{(x - 2)^2}{49} - \frac{(y + 3)^2}{4} = 1$

19. $\frac{x^2}{16} - \frac{y^2}{9} = 1$

21. $(\pm 9, 0)$; $\left(\pm\sqrt{130}, 0\right)$; $y = \pm\frac{7}{9}x$

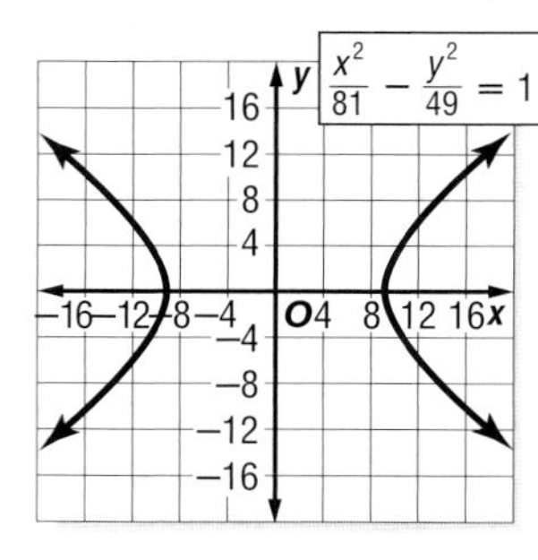

23. $(0, \pm 4)$; $\left(0, \pm\sqrt{41}\right)$; $y = \pm\frac{4}{5}x$

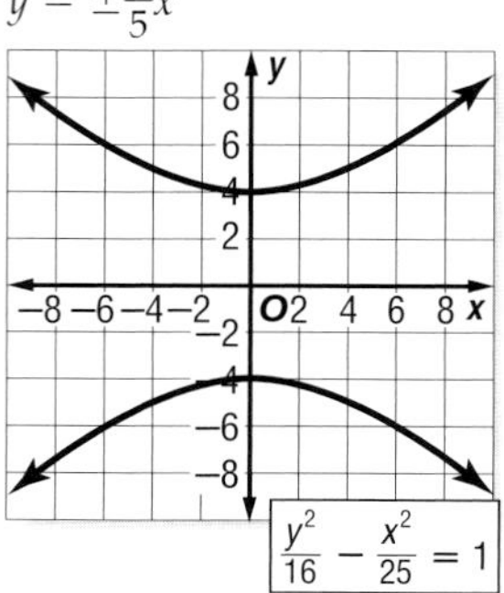

25. $\left(\pm\sqrt{2}, 0\right)$; $\left(\pm\sqrt{3}, 0\right)$; $y = \pm\frac{\sqrt{2}}{2}x$

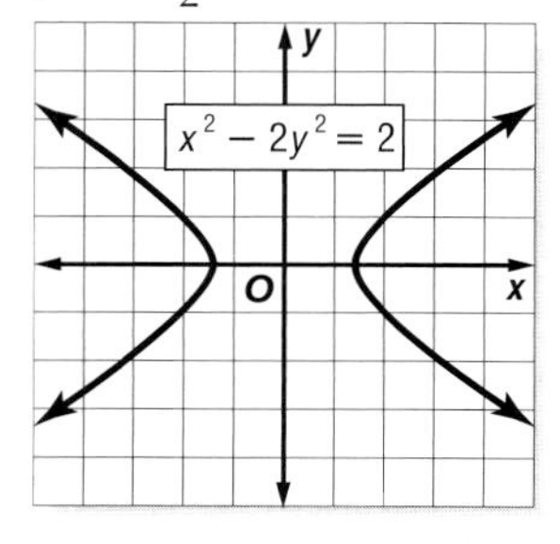

27. $(0, \pm 6)$; $\left(0, \pm 3\sqrt{5}\right)$; $y = \pm 2x$

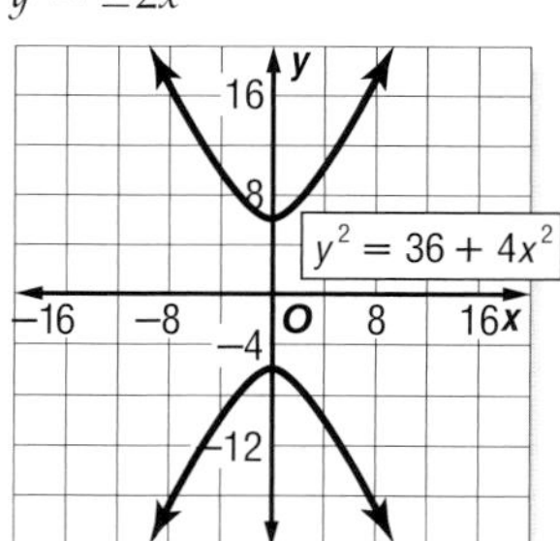

29. $(-2, 0)$, $(-2, 8)$; $(-2, -1)$, $(-2, 9)$; $y - 4 = \pm\frac{4}{3}(x + 2)$

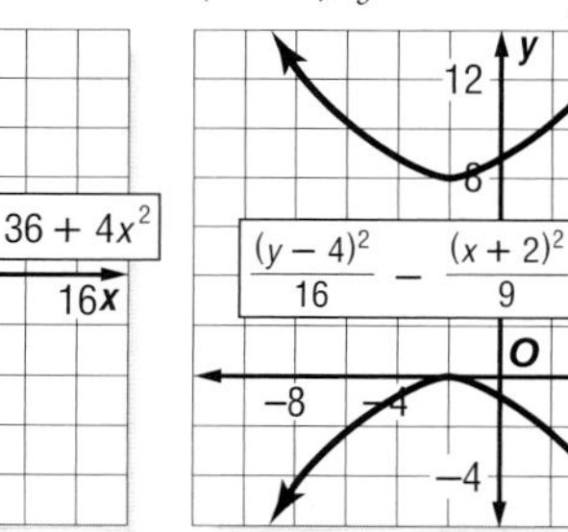

31. $(-3, -3)$, $(1, -3)$; $\left(-1 \pm \sqrt{13}, -3\right)$; $y + 3 = \pm\frac{3}{2}(x + 1)$

33. $\left(1, -3 \pm 2\sqrt{6}\right)$; $\left(1, -3 \pm 4\sqrt{2}\right)$; $y + 3 = \pm\sqrt{3}(x - 1)$

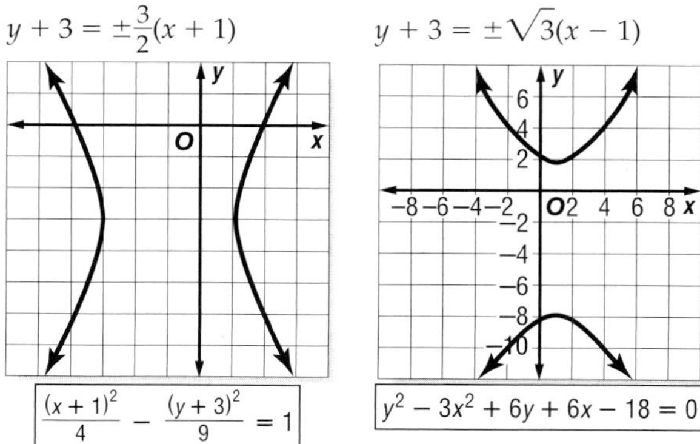

35. $\frac{x^2}{1.1025} - \frac{y^2}{7.8975} = 1$ 37. 120 cm, 100 cm

39. about 47.32 ft 41. C

43.

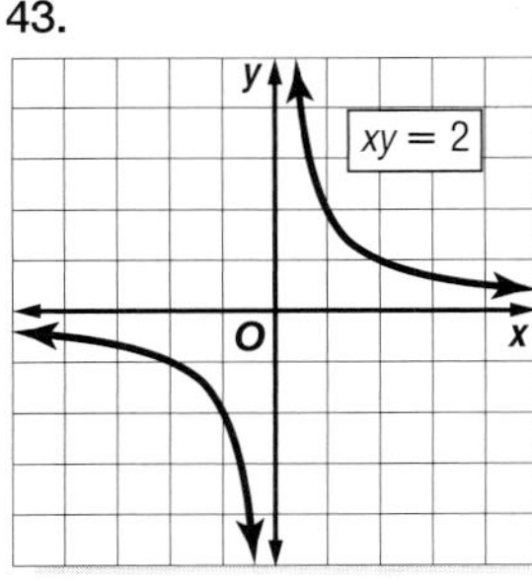

45.

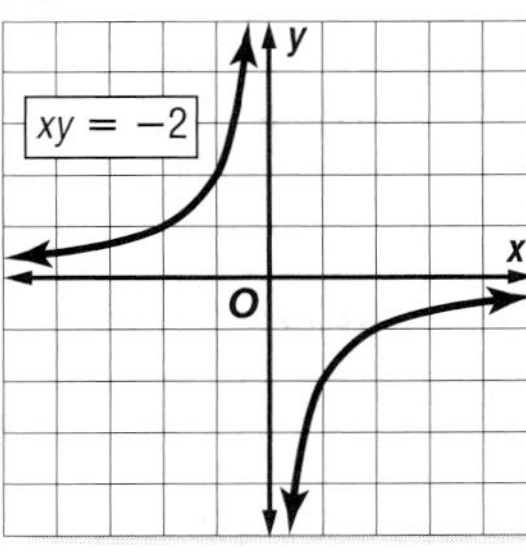

47. $\frac{(x-5)^2}{16} + \frac{(y-2)^2}{1} = 1$ 49. $\frac{(x-1)^2}{25} + \frac{(y-4)^2}{9} = 1$

51. −4, −2 53. $\begin{bmatrix} -7 & 0 \\ 5 & 20 \end{bmatrix}$ 55. about 5,330,000 subscribers per year 57. $2x + 17y$ 59. 1, −2, 9 61. 5, 0, −2

63. 0, 1, 0

Page 448 Practice Quiz 2

1. $\frac{(y-1)^2}{81} + \frac{(x-3)^2}{32} = 1$

3. $(-1, 1)$; $(-1, 1 \pm \sqrt{11})$; 8; $2\sqrt{5}$

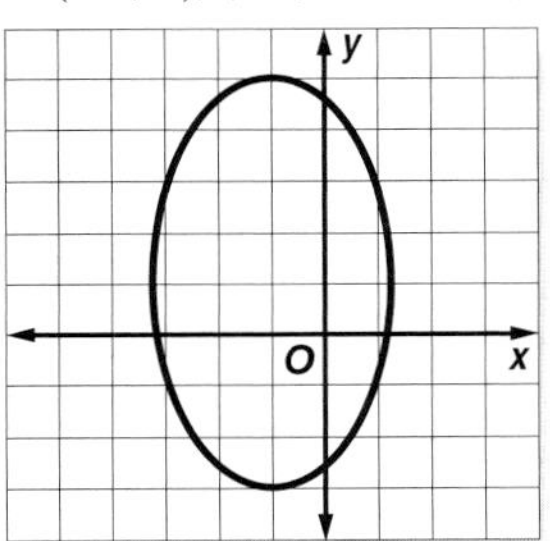

5. $\frac{(x-2)^2}{16} - \frac{(y-2)^2}{5} = 1$

Pages 450–452 Lesson 8-6

1. Sample answer: $2x^2 + 2y^2 - 1 = 0$ 3. The standard form of the equation is $(x-2)^2 + (y+1)^2 = 0$. This is an equation of a circle centered at $(2, -1)$ with radius 0. In other words, $(2, -1)$ is the only point that satisfies the equation.

5. $\frac{y^2}{16} - \frac{x^2}{8} = 1$, hyperbola 7. $\frac{(x+1)^2}{4} + \frac{(y-3)^2}{1} = 1$, ellipse

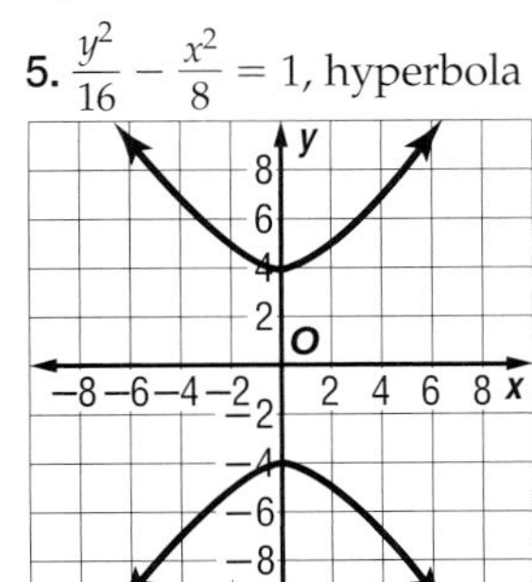

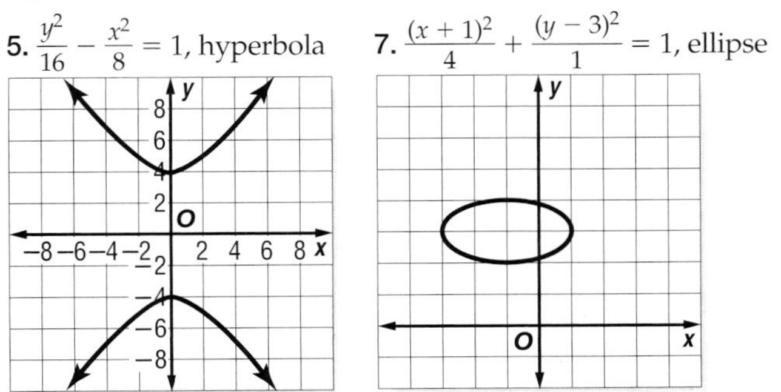

9. ellipse

11.

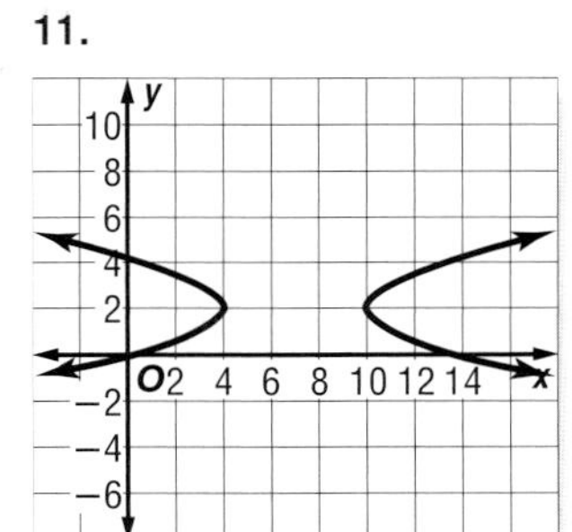

13. $\frac{y^2}{4} + \frac{x^2}{2} = 1$, ellipse

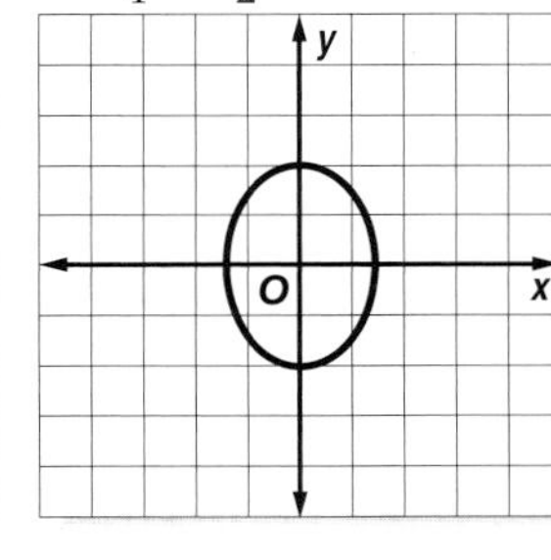

15. $\frac{x^2}{4} - \frac{y^2}{1} = 1$, hyperbola 17. $y = (x-2)^2 - 4$, parabola

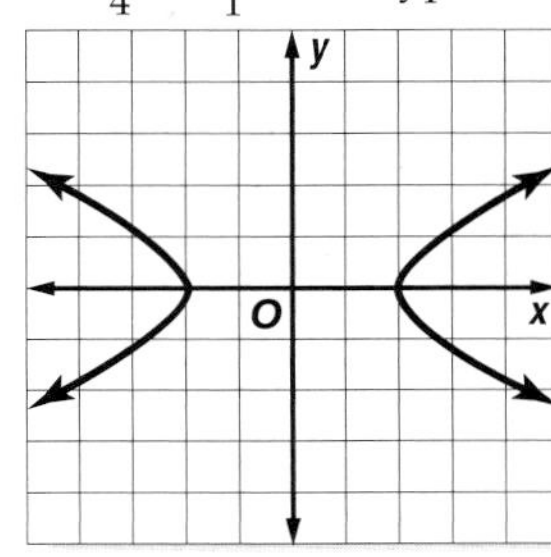

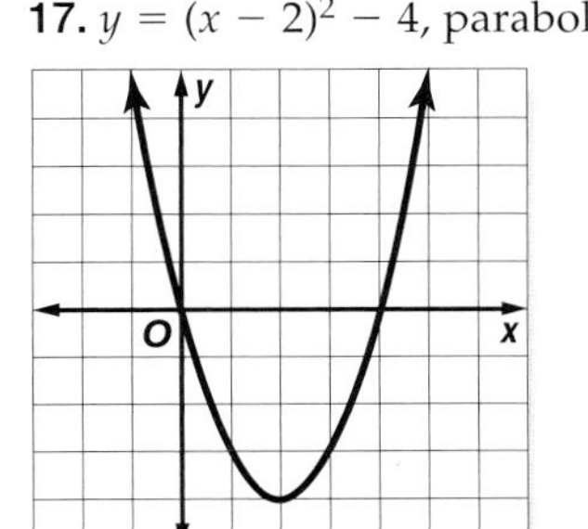

19. $(x+2)^2 + (y-3)^2 = 9$, circle

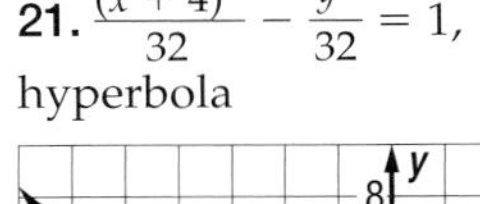

21. $\frac{(x+4)^2}{32} - \frac{y^2}{32} = 1$, hyperbola

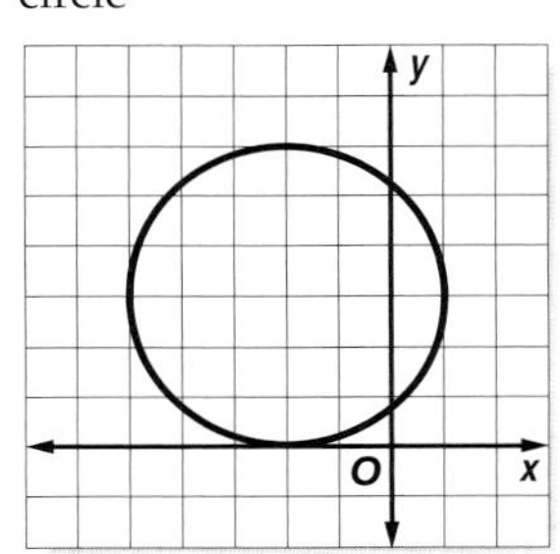

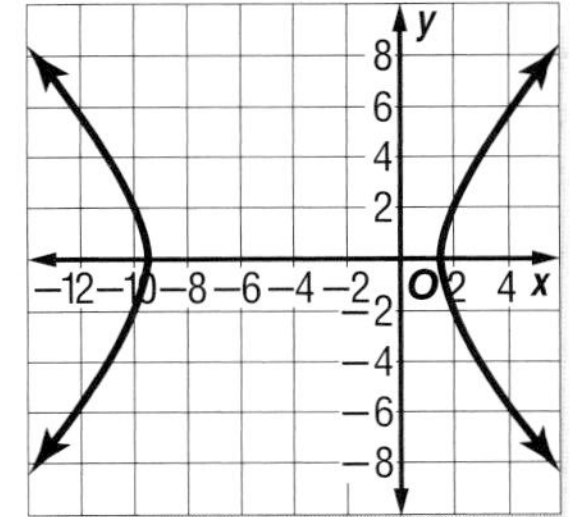

23. $x^2 + (y-4)^2 = 5$, circle 25. $\frac{x^2}{4} + \frac{(y+1)^2}{3} = 1$, ellipse

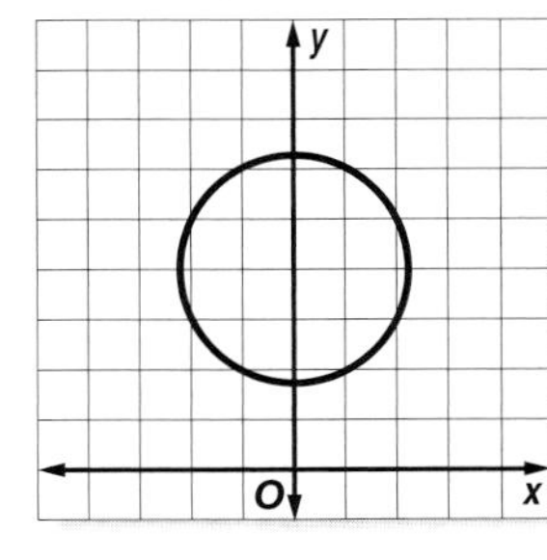

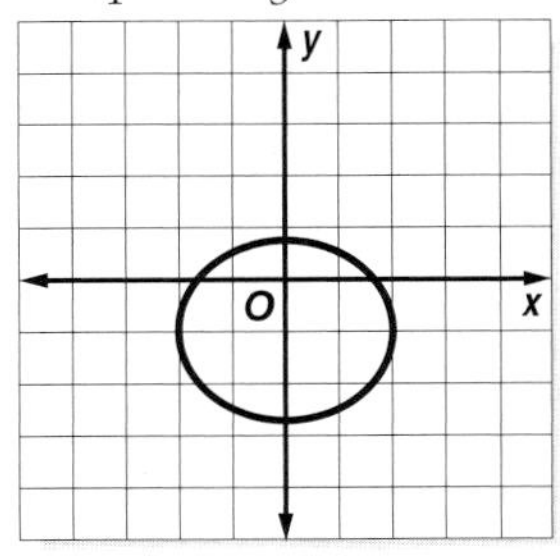

27. $y = -(x+4)^2 - 7$, parabola 29. $\frac{(x-3)^2}{25} + \frac{(y-1)^2}{9} = 1$, ellipse

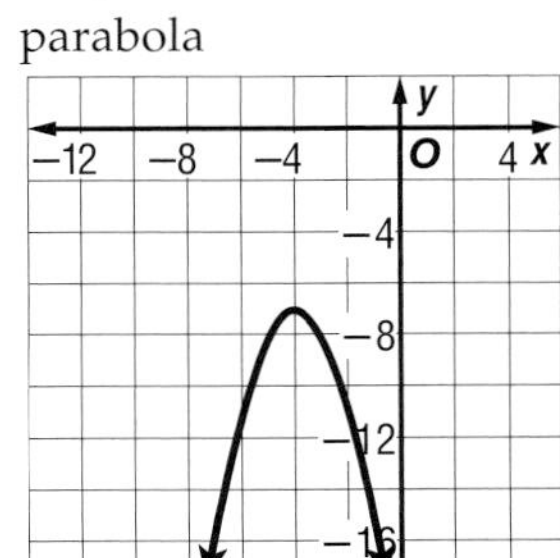

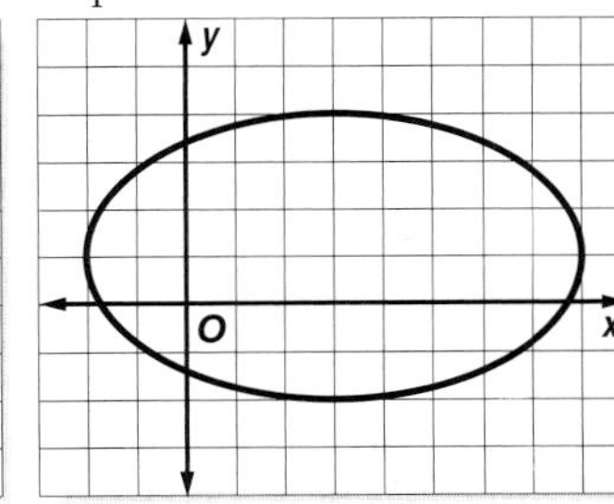

31. hyperbola 33. circle 35. parabola 37. ellipse

Selected Answers

39. parabola **41.** b **43.** c **45.** The plane should be vertical and contain the axis of the double cone.

47. D **49.** $0 < e < 1, e > 1$ **51.** $\frac{(x-3)^2}{9} - \frac{(y+6)^2}{4} = 1$

53. x^{12} **55.** $\frac{x^7}{y^4}$ **57.** (2, 6) **59.** (0, 2)

Pages 458–460 Lesson 8-7

1a. (−3, −4), (3, 4)

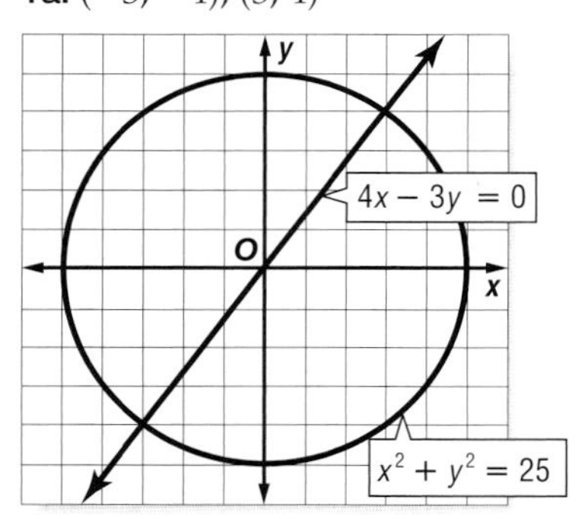

1b. (±1, 4)

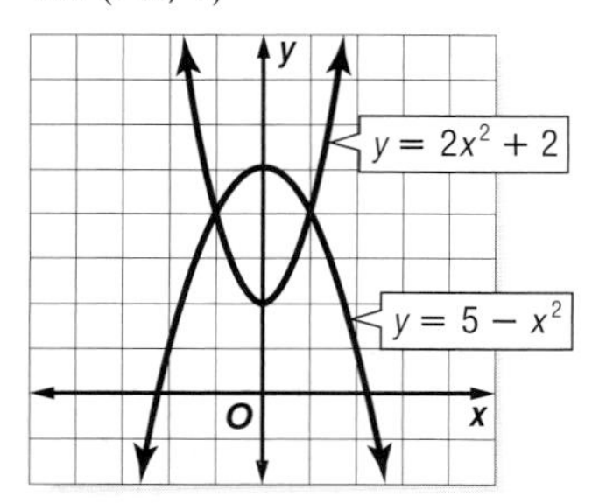

3. Sample answer: $x^2 + y^2 = 40, y = x^2 + x$

5. (−4, −3), (3, 4) **7.** (1, ±5), (−1, ±5)

9.

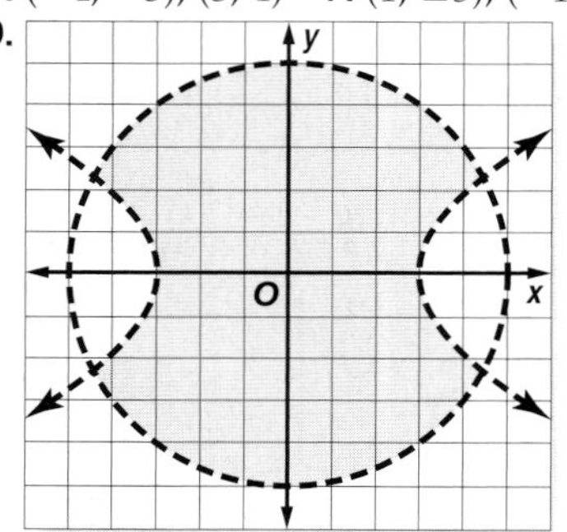

11. (2, 4), (−1, 1) **13.** $(-1 + \sqrt{17}, 1 + \sqrt{17})$, $(-1 - \sqrt{17}, 1 - \sqrt{17})$ **15.** $(\sqrt{5}, \sqrt{5})$, $(-\sqrt{5}, -\sqrt{5})$

17. (5, 0), (−4, ±6) **19.** (±8, 0) **21.** no solution

23. (−5, 5), (−5, 1), (3, 3) **25.** $\left(-\frac{5}{3}, -\frac{7}{3}\right)$, (1, 3) **27.** 0.5 s

29. $\left(\frac{40 - 24\sqrt{5}}{5}, \frac{45 - 12\sqrt{5}}{5}\right)$ **31.** No; the comet and Pluto may not be at either point of intersection at the same time.

33.

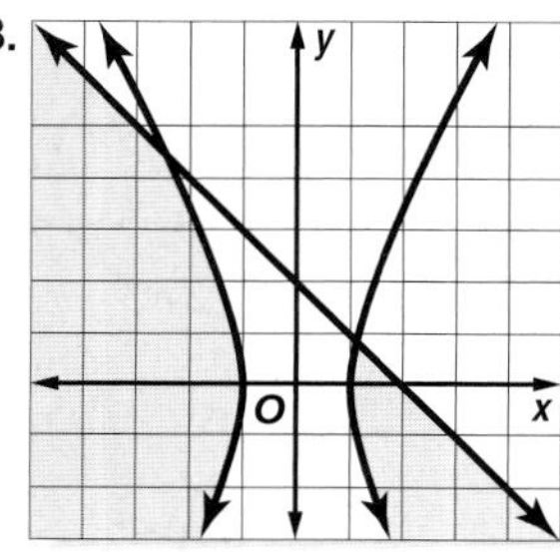

35.

37.

39. none **41.** none

43. Systems of equations can be used to represent the locations and/or paths of objects on the screen. Answers should include the following.

- $y = 3x, x^2 + y^2 = 2500$
- The y-intercept of the graph of the equation $y = 3x$ is 0, so the path of the spaceship contains the origin.
- $(-5\sqrt{10}, -15\sqrt{10})$ or about (−15.81, −47.43)

45. B **47.** Sample answer: $x^2 + y^2 = 36, \frac{(x+2)^2}{16} - \frac{y^2}{4} = 1$

49. Sample answer: $x^2 + y^2 = 81, \frac{x^2}{4} + \frac{y^2}{100} = 1$

51. impossible

53. $\frac{(y-3)^2}{9} + \frac{x^2}{4} = 1$, ellipse

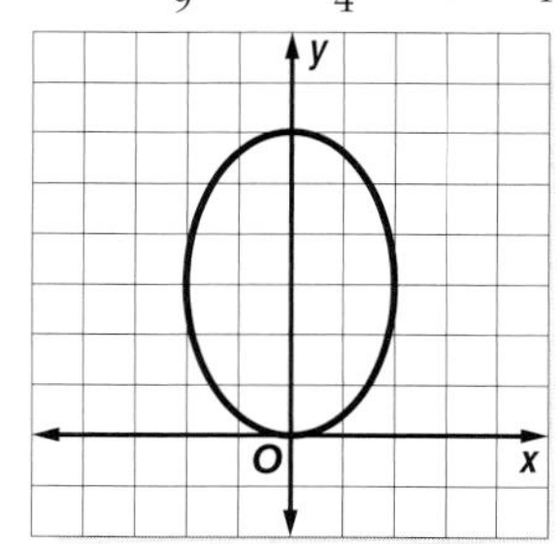

55. −7, 0 **57.** −7, 3 **59.** $-\frac{4}{3}$ **61a.** 40

61b. two real, irrational **61c.** $\pm\frac{\sqrt{10}}{5}$ **63.** $2 + 9i$

65. $\frac{8}{5} - \frac{1}{5}i$ **67.** 6 **69.** −51 **71.** $y = 3x - 2$

Pages 461–466 Chapter 8 Study Guide and Review

1. true **3.** true **5.** true **7.** true **9.** False; the midpoint formula is given by $\left(\frac{x_1 + x_2}{2}, \frac{y_1 + y_2}{2}\right)$. **11.** $\left(\frac{5}{2}, 4\right)$

13. $\left(\frac{17}{40}, -\frac{43}{40}\right)$ **15.** $\sqrt{290}$ units

17. (1, 1); (1, 4); $x = 1$; $y = -2$; upward; 12 units

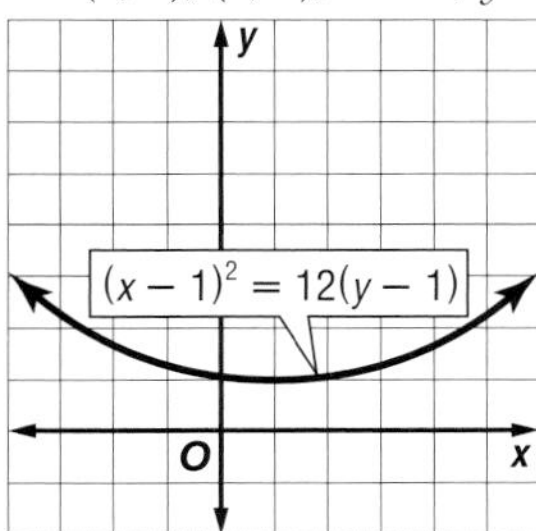

19. (4, −2); (4, −4); $x = 4$; $y = 0$; downward; 8 units

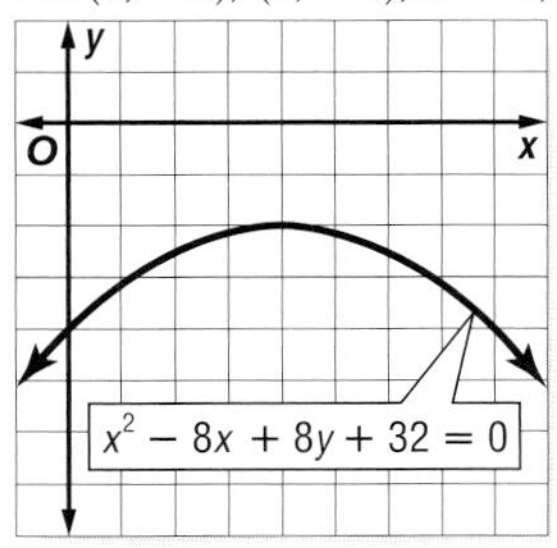

21. $y = -\frac{1}{8}x^2 + 1$

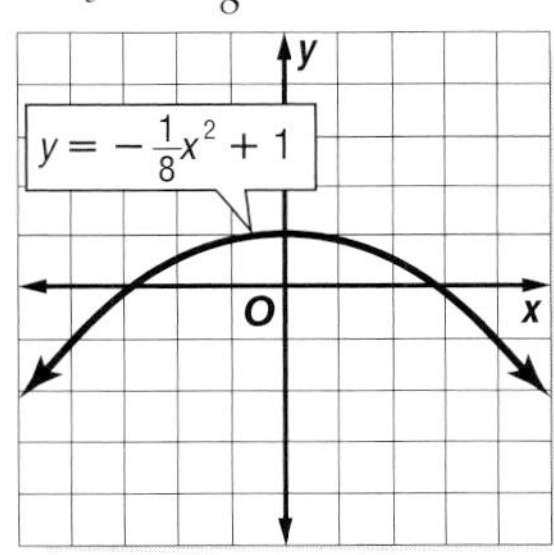

23. $(x + 4)^2 + y^2 = \frac{9}{16}$ **25.** $(x + 1)^2 + (y - 2)^2 = 4$

27. (−5, 11); 7 units

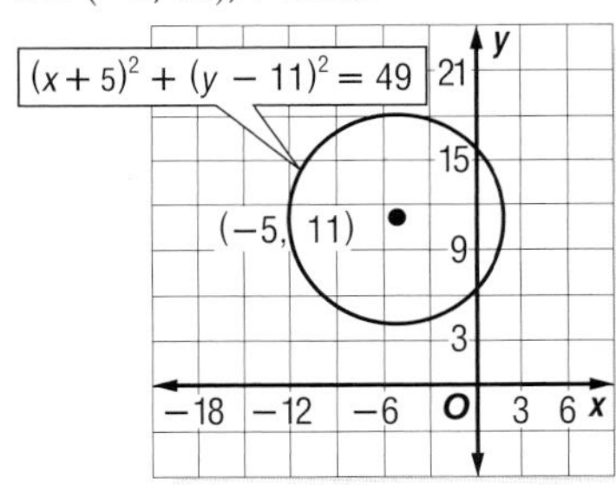

29. (−3, 1); 5 units

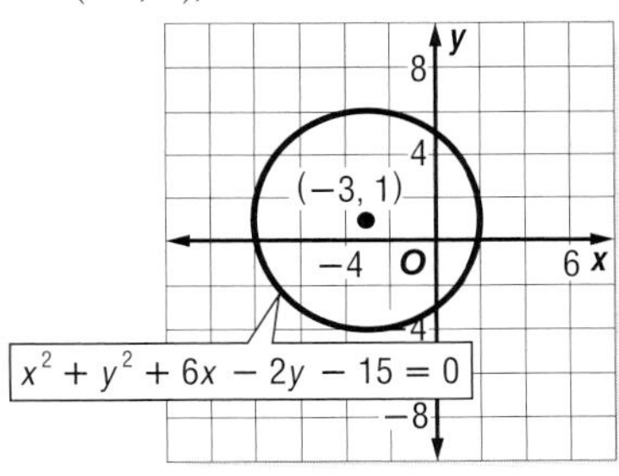

31. (0, 0); (0, ±3); 10; 8

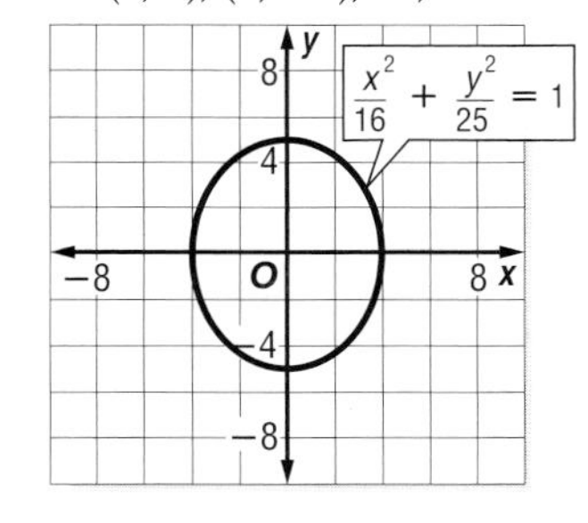

33. (1, −2); $(1 \pm \sqrt{3}, -2)$; 4; 2

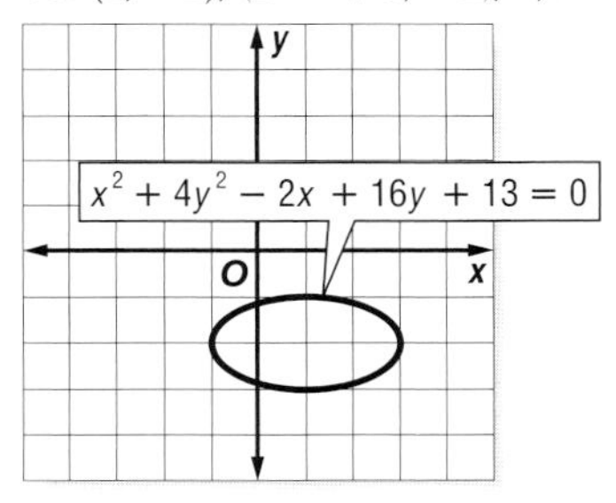

35. (0, ±2); $(0, \pm\sqrt{13})$; $y = \pm\frac{2}{3}x$

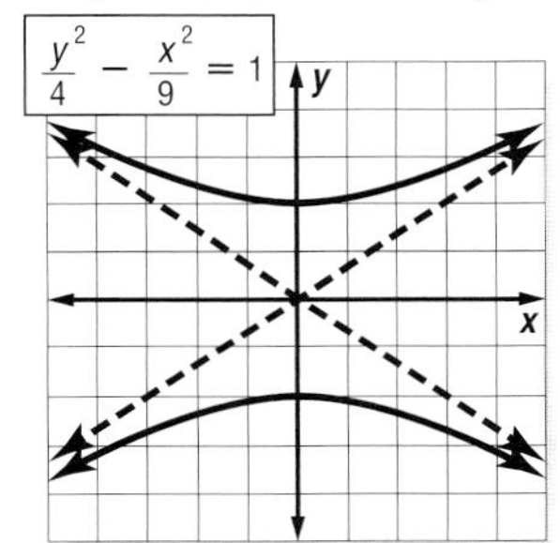

37. (0, ±4); (0, ±5); $y = \pm\frac{4}{3}x$

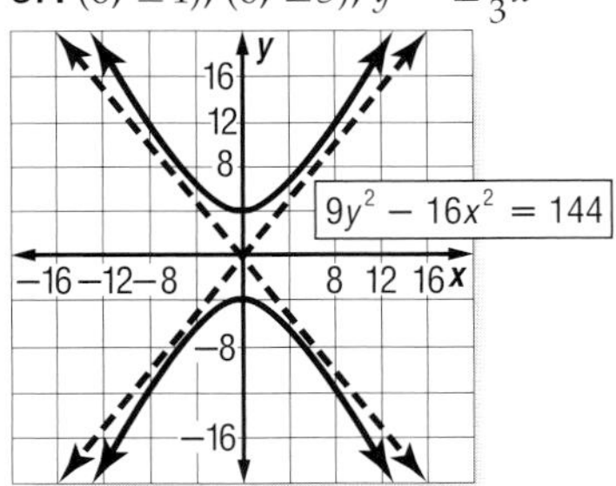

39. $y = (x + 2)^2 - 4$; parabola

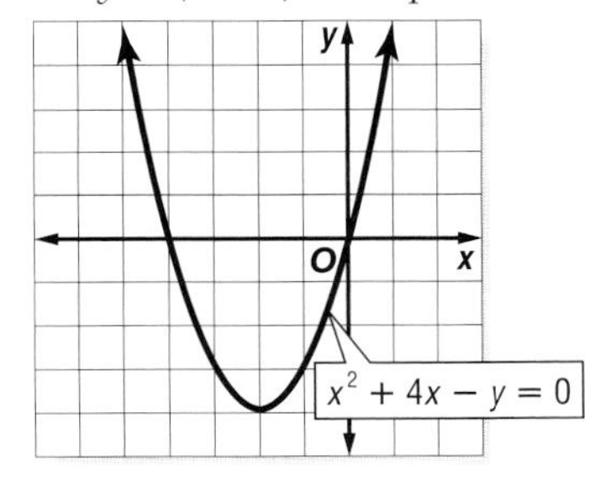

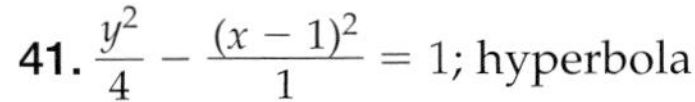

41. $\frac{y^2}{4} - \frac{(x-1)^2}{1} = 1$; hyperbola

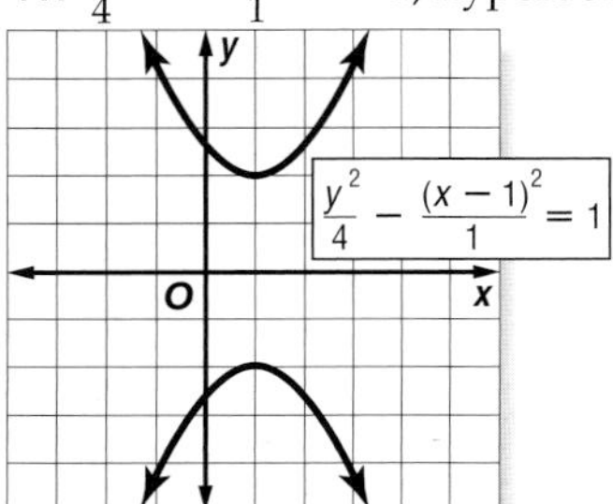

43. ellipse **45.** circle **47.** (6, −8), (12, −16)

49.

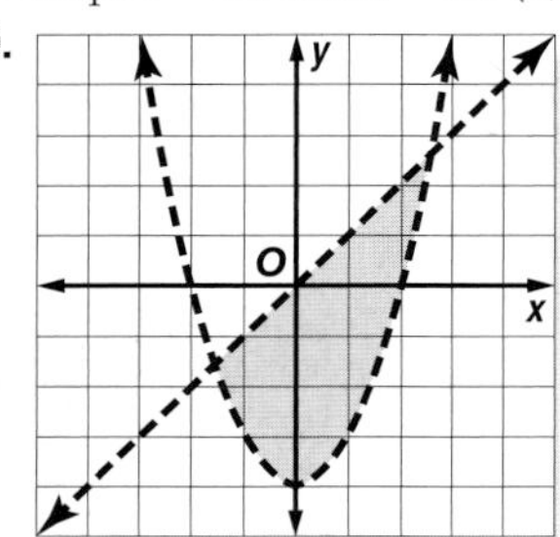

Chapter 9 Rational Expressions and Equations

Page 471 Chapter 9 Getting Started

1. $\frac{1}{6}$ **3.** $\frac{5}{8}$ **5.** 16 **7.** $2\frac{1}{2}$ **9.** $1\frac{1}{24}$

11.

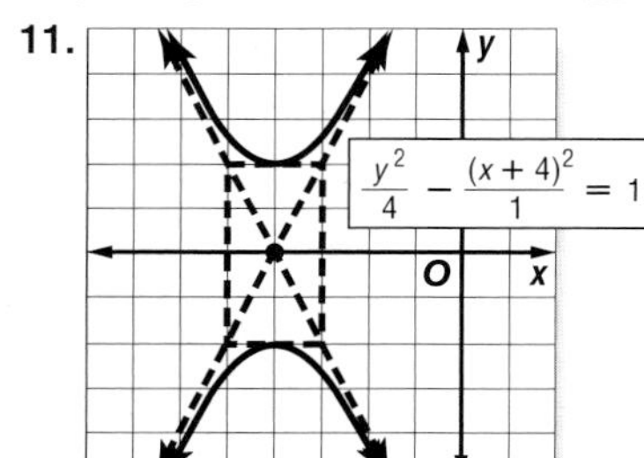

13. 12 **15.** 15 **17.** 15 **19.** 6 **21.** $7\frac{1}{2}$

Pages 476–478 Lesson 9-1

1. Sample answer: $\frac{4}{6}, \frac{4(x+2)}{6(x+2)}$ **3.** Never; solving the equation using cross products leads to 15 = 10, which is never true. **5.** $\frac{1}{a-b}$ **7.** $\frac{3c}{20b}$ **9.** $\frac{6}{5}$ **11.** cd^2x **13.** D **15.** $-\frac{n^2}{7m}$ **17.** $\frac{s}{3}$ **19.** $\frac{1}{2}$ **21.** $\frac{a+1}{2a+1}$ **23.** $-\frac{4bc}{27a}$ **25.** $-2p^2$ **27.** $\frac{b^3}{x^2y^2}$ **29.** $\frac{4}{3}$ **31.** 1 **33.** $\frac{w-3}{w-4}$ **35.** $\frac{2(a+5)}{(a-2)(a+2)}$ **37.** $-2p$ **39.** $\frac{2x+y}{2x-y}$ **41.** $\frac{4}{3}$ **43.** $a = -b$ or b **45.** $\frac{6827+m}{13{,}129+a}$ **47.** $(2x^2 + x - 15)$ m^2

49. A rational expression can be used to express the fraction of a nut mixture that is peanuts. Answers should include the following.

- The rational expression $\frac{8+x}{13+x}$ is in simplest form because the numerator and the denominator have no common factors.
- Sample answer: $\frac{8+x}{13+x+y}$ could be used to represent the fraction that is peanuts if x pounds of peanuts and y pounds of cashews were added to the original mixture.

51. A **53.** $(\pm\sqrt{17}, \pm 2\sqrt{2})$

55. $\frac{(x-7)^2}{9} - \frac{(y-2)^2}{1} = 1$; hyperbola

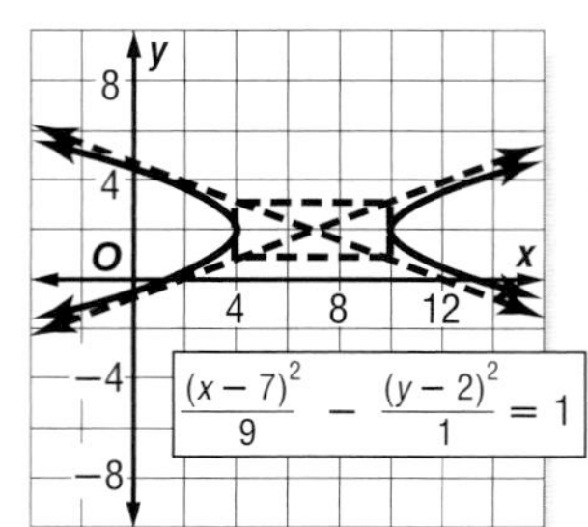

57. odd; 3 **59.** {−1, 4} **61.** {0, 5} **63.** ∅ **65.** $-1\frac{1}{9}$ **67.** $1\frac{4}{15}$ **69.** $-\frac{11}{18}$

Pages 481–484 Lesson 9-2

1. Catalina: you need a common denominator, not a common numerator, to subtract two rational expressions. **3a.** Always; since a, b, and c are factors of abc, abc is always a common denominator of $\frac{1}{a} + \frac{1}{b} + \frac{1}{c}$. **3b.** Sometimes; if a, b, and c have no common factors, then abc is the LCD of $\frac{1}{a} + \frac{1}{b} + \frac{1}{c}$. **3c.** Sometimes; if a and b have no common factors and c is a factor of ab, then ab is the LCD of $\frac{1}{a} + \frac{1}{b} + \frac{1}{c}$. **3d.** Sometimes; if a and c are factors of b, then b is the LCD of $\frac{1}{a} + \frac{1}{b} + \frac{1}{c}$. **3e.** Always; since $\frac{1}{a} + \frac{1}{b} + \frac{1}{c} = \frac{bc}{abc} + \frac{ac}{abc} + \frac{ab}{abc}$, the sum is always $\frac{bc+ac+ab}{abc}$. **5.** $80a^2b^3c$ **7.** $\frac{2-x^3}{x^2y}$ **9.** $\frac{37}{42m}$ **11.** $\frac{3a-10}{(a-5)(a+4)}$ **13.** $\frac{13x^2+4x-9}{2x(x-1)(x+1)}$ units **15.** $180x^2yz$ **17.** $36p^3q^4$ **19.** $x^2(x-y)(x+y)$ **21.** $(n-4)(n-3)(n+2)$ **23.** $\frac{31}{12v}$ **25.** $\frac{2x+15y}{3y}$ **27.** $\frac{25b-7a^3}{5a^2b^2}$ **29.** $\frac{110w-423}{90w}$ **31.** $\frac{a+3}{a-4}$ **33.** $\frac{y(y-9)}{(y+3)(y-3)}$ **35.** $\frac{-8d+20}{(d-4)(d+4)(d-2)}$ **37.** $\frac{x^2-6}{(x+2)^2(x+3)}$ **39.** $\frac{2y^2+y-4}{(y-1)(y-2)}$ **41.** −1 **43.** $\frac{a+7}{a+2}$ **45.** 12 ohms **47.** $\frac{24}{x-4}$ h **49.** $\frac{2md}{(d-L)^2(d+L)^2}$ or $\frac{2md}{(d^2-L^2)^2}$

51. Subtraction of rational expressions can be used to determine the distance between the lens and the film if the focal length of the lens and the distance between the lens and the object are known. Answers should include the following.

- To subtract rational expressions, first find a common denominator. Then, write each fraction as an equivalent fraction with the common denominator. Subtract the numerators and place the difference over the common denominator. If possible, reduce the answer.
- $\frac{1}{q} = \frac{1}{10} - \frac{1}{60}$ could be used to determine the distance between the lens and the film if the focal length of the lens is 10 cm and the distance between the lens and the object is 60 cm.

53. C **55.** $\frac{a(a+2)}{a+1}$

Selected Answers

57.

$(y - 3)^2 = x + 2$

$x^2 = y + 4$

59.

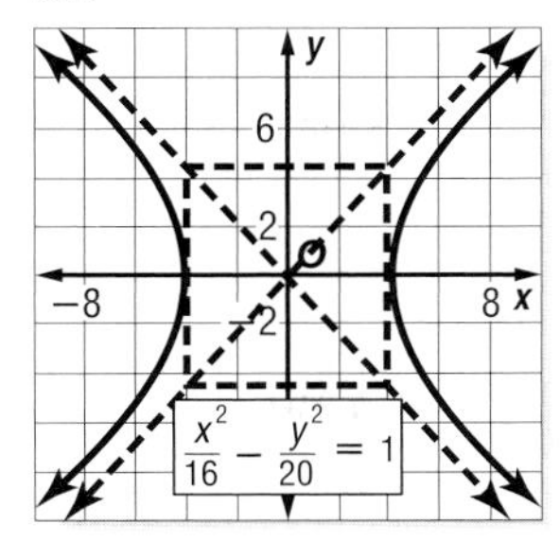

61.

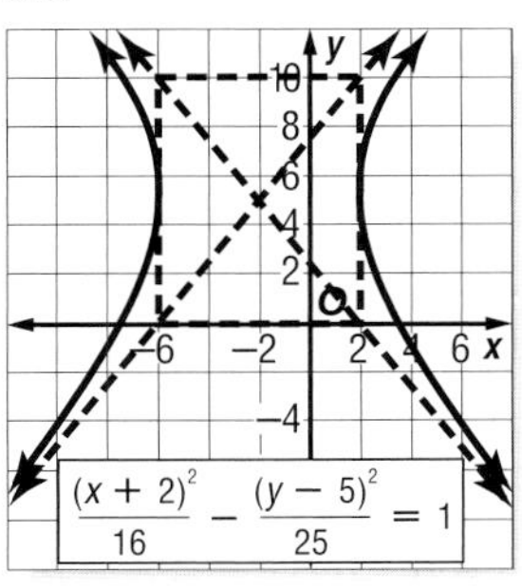

Page 484 Practice Quiz 1

1. $\frac{t + 2}{t - 3}$ **3.** $-\frac{y^2}{32}$ **5.** $(w + 4)(3w + 4)$ **7.** $\frac{4a + 1}{a + b}$
9. $\frac{n - 29}{(n + 6)(n - 1)}$

Pages 488–490 Lesson 9-3

1. Sample answer: $f(x) = \frac{1}{(x + 5)(x - 2)}$ **3.** $x = 2$ and $y = 0$ are asymptotes of the graph. The y-intercept is 0.5 and there is no x-intercept because $y = 0$ is an asymptote.
5. asymptote: $x = -5$; hole: $x = 1$

7.

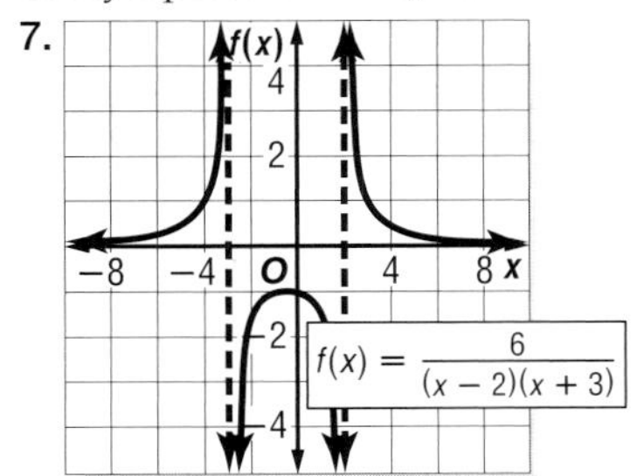

9.

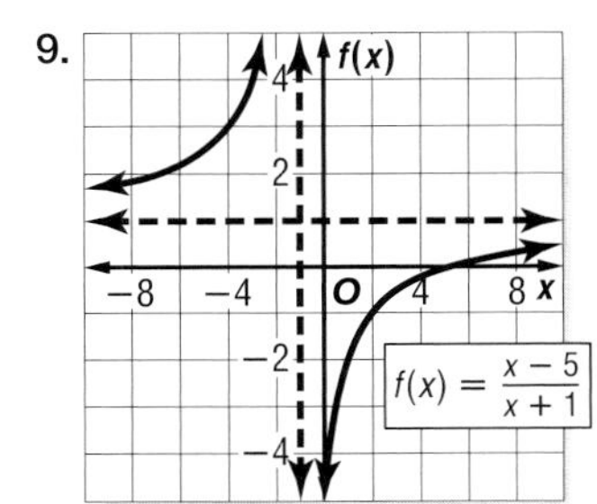

11.

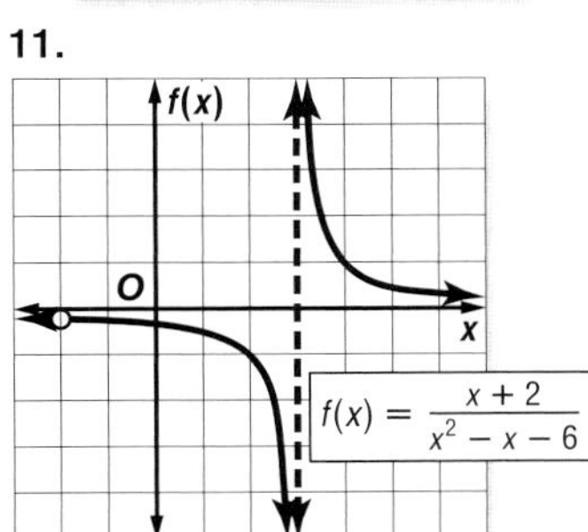

13.

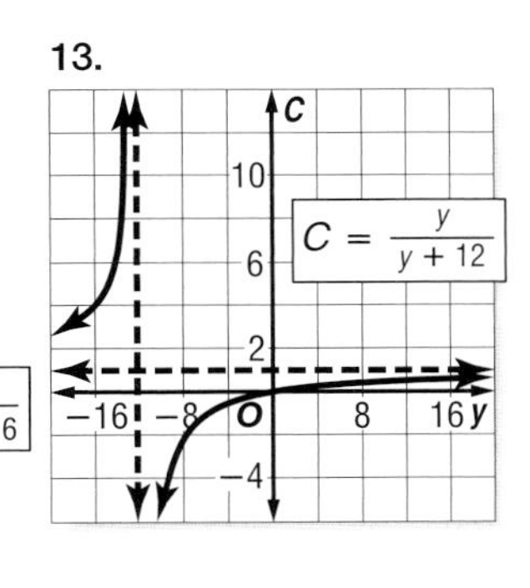

15. $y > 0$ and $0 < C < 1$ **17.** asymptotes: $x = -4$, $x = 2$
19. asymptotes: $x = -1$, hole: $x = 5$ **21.** hole: $x = 1$

23.

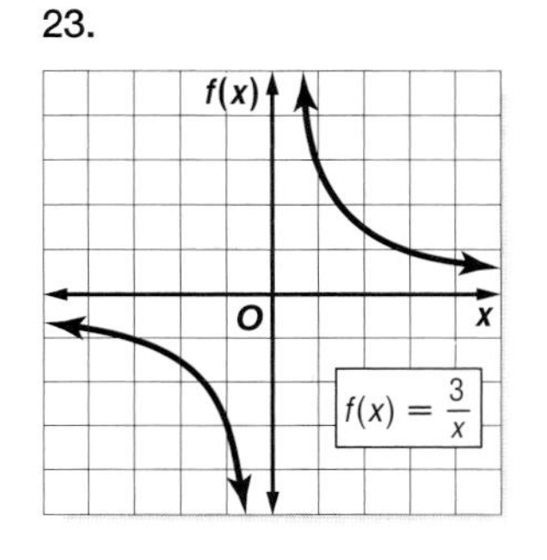

25.

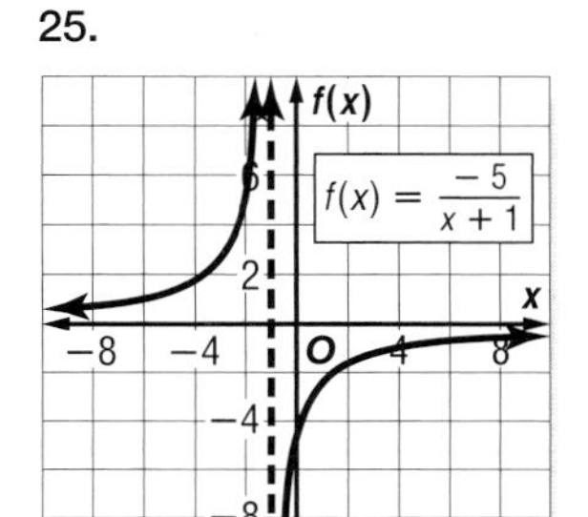

27.

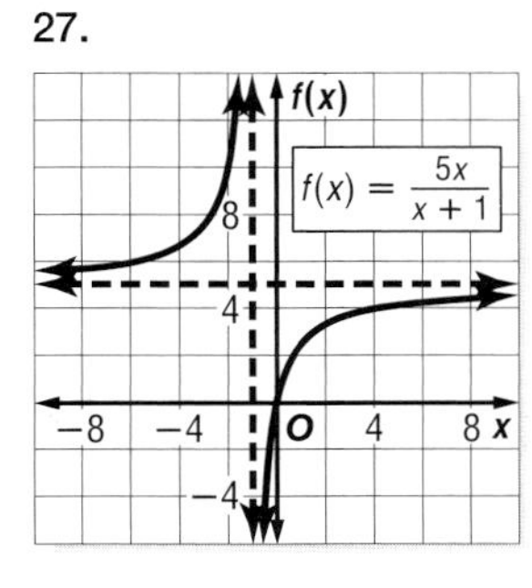

29.

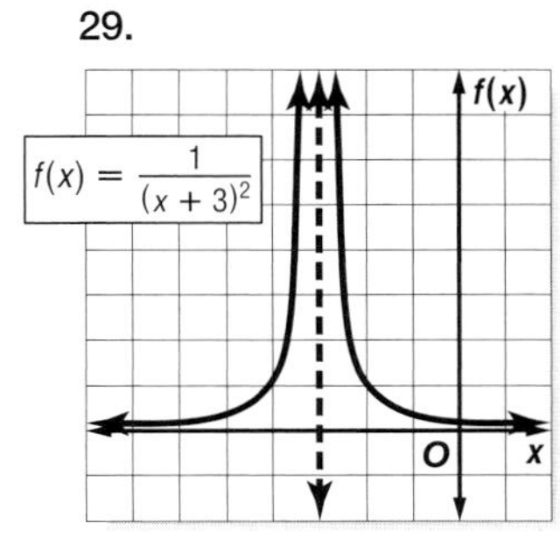

31.

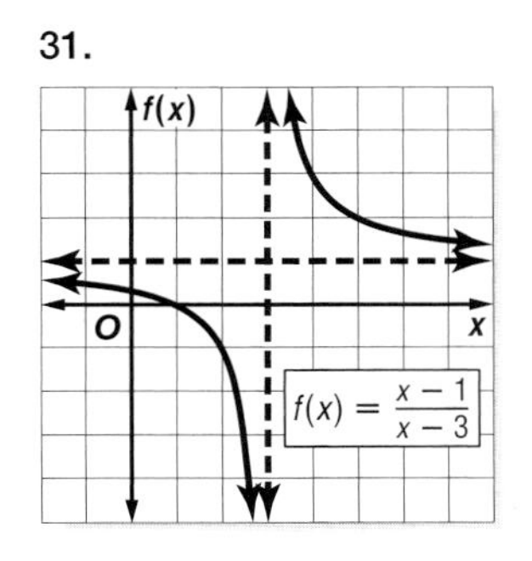

33.

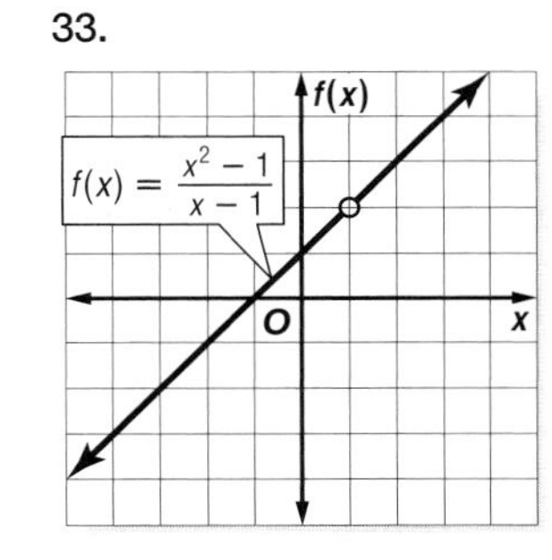

35.

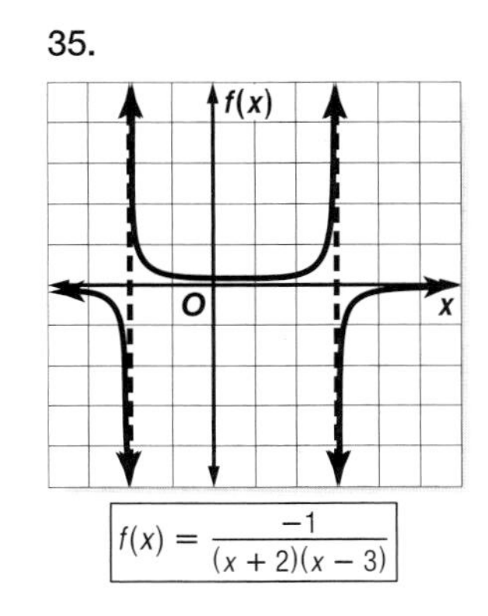

37.

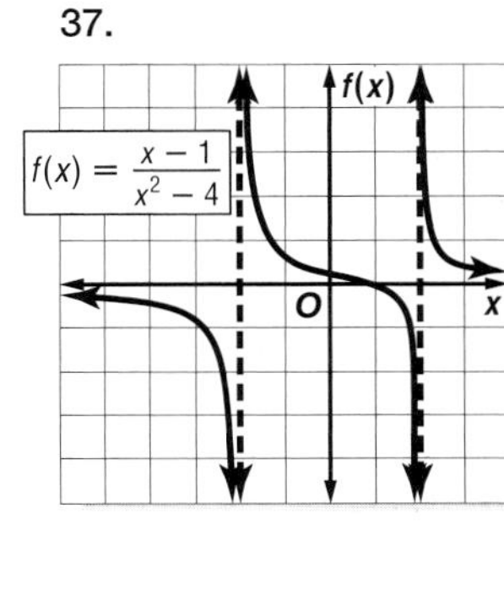

39.

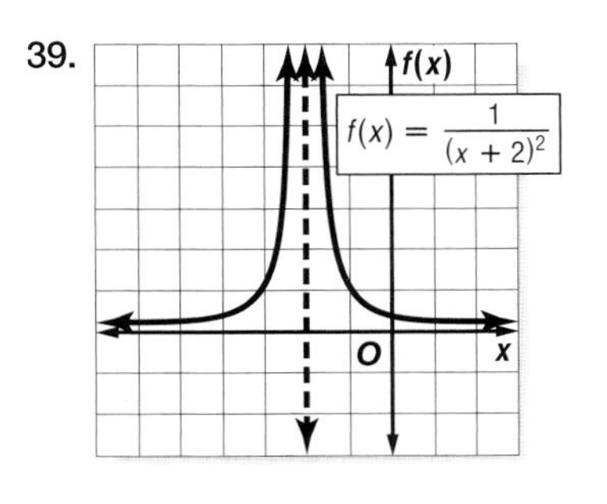

Selected Answers

41. The graph is bell-shaped with a horizontal asymptote at $f(x) = 0$.

43. 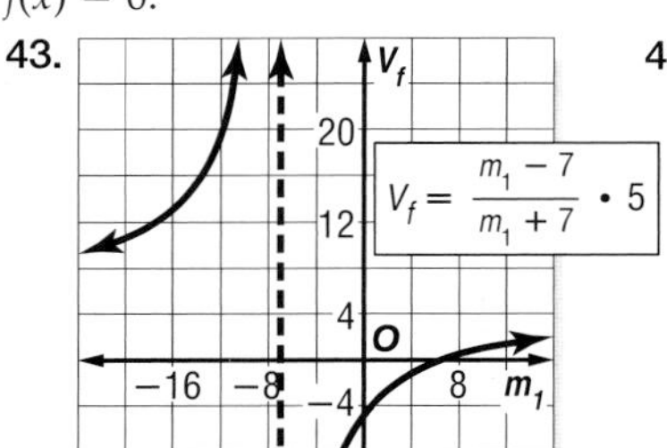

45. about −0.83 m/s

47. 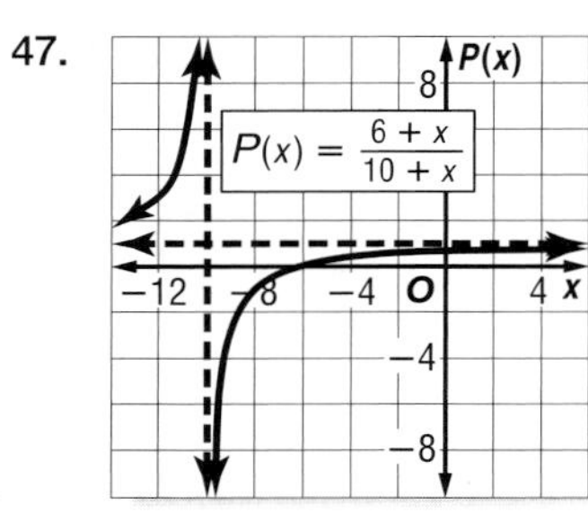

49. It represents her original free-throw percentage of 60%.

51. A rational function can be used to determine how much each person owes if the cost of the gift is known and the number of people sharing the cost is s. Answers should include the following.

- 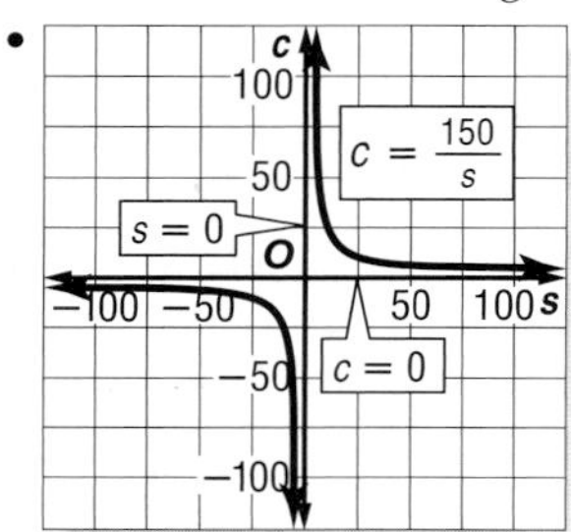

- Only the portion in the first quadrant is significant in the real world because there cannot be a negative number of people nor a negative amount of money owed for the gift.

53. B 55. $\frac{3x - 16}{(x + 3)(x - 2)}$

57. (6, 2); 5 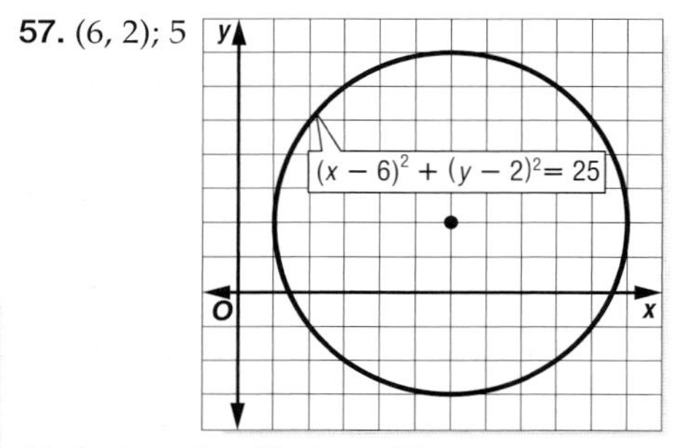

59. $65,892

61. {−12, 10} 63. 4.5 65. 20

Pages 495–498 Lesson 9-4

1a. inverse 1b. direct 3. Sample answers: wages and hours worked, total cost and number of pounds of apples; distances traveled and amount of gas remaining in the tank, distance of an object and the size it appears 5. direct; −0.5 7. 24 9. −8 11. 25.8 psi

13.

Depth (ft)	Pressure (psi)
0	0
1	0.43
2	0.86
3	1.29
4	1.72

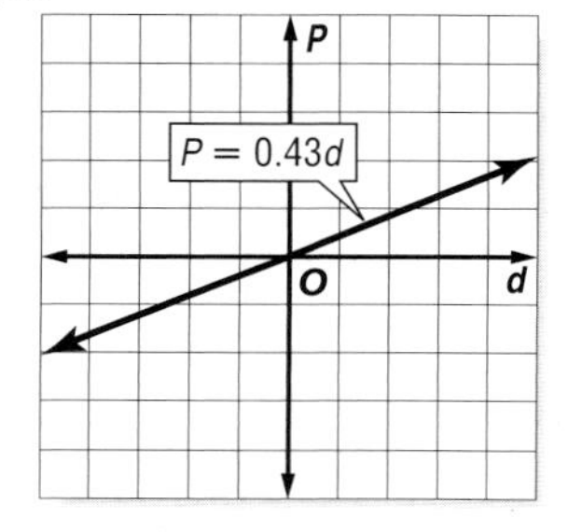

15. joint; 5 17. direct; 3 19. direct; −7 21. inverse; 2.5 23. $V = kt$ 25. 118.5 km 27. 20 29. 64 31. 4 33. 9.6 35. 0.83 37. $\frac{1}{6}$ 39. 100.8 cm^3 41. $m = 20sd$ 43. 1860 lb 45. joint 47. $I = \frac{k}{d^2}$ 49. The sound will be heard $\frac{1}{4}$ as intensely. 51. about 127,572 calls 53. no; $d \neq 0$

55. A direct variation can be used to determine the total cost when the cost per unit is known. Answers should include the following.

- Since the total cost T is the cost per unit u times the number of units n or $T = un$, the relationship is a direct variation. In this equation u is the constant of variation.
- Sample answer: The school store sells pencils for 20¢ each. John wants to buy 5 pencils. What is the total cost of the pencils? ($1.00)

57. C 59. asymptotes: $x = -4$, $x = 3$ 61. $\frac{x}{y - x}$ 63. $\frac{m(m + 1)}{m + 5}$ 65. 0.4; 1.2 67. $-\frac{3}{5}$; 3 69. A 71. P 73. C

Page 498 Practice Quiz 2

1. 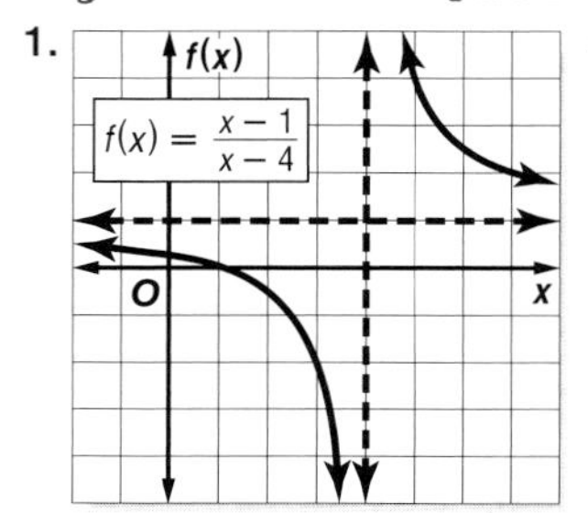

3. 49 5. 112

Pages 501–504 Lesson 9-5

1. Sample answer:

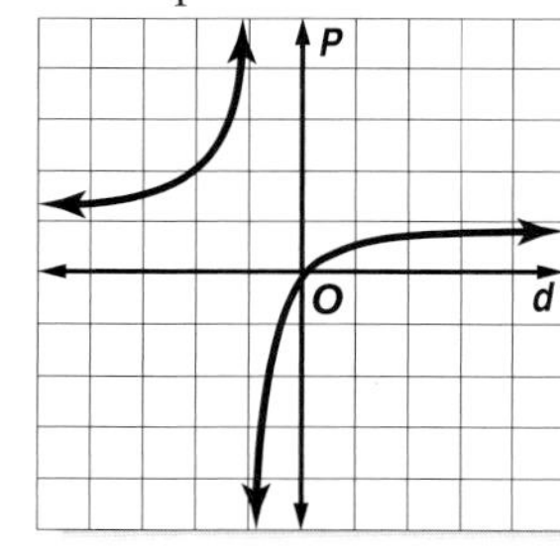

This graph is a rational function. It has an asymptote at $x = -1$.

3. The equation is a greatest integer function. The graph looks like a series of steps. 5. inverse variation or rational 7. c

9. identity or direct variation

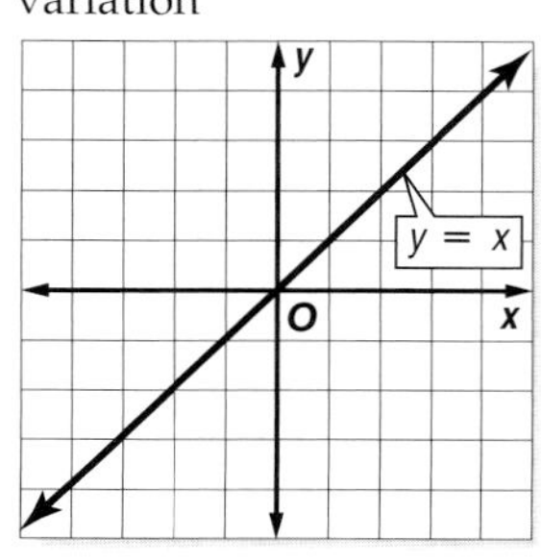

11. absolute value

y
$y = |x + 2|$
O
x

13. absolute value 15. rational 17. quadratic 19. b 21. g

23. constant

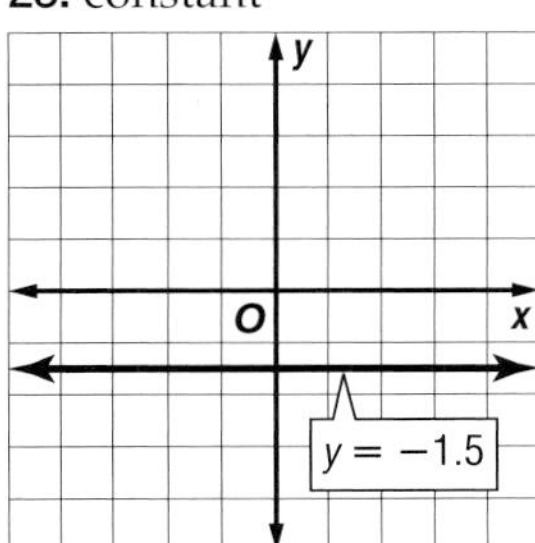

25. square root

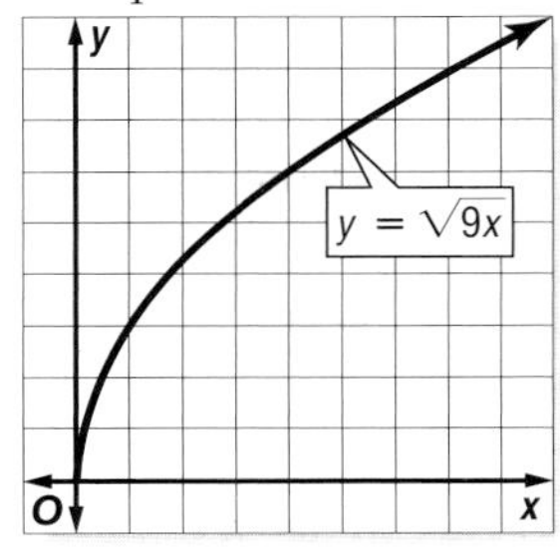

27. rational

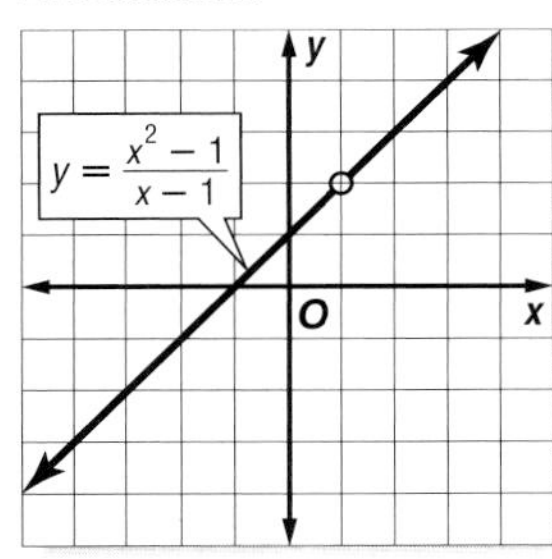

29. absolute value

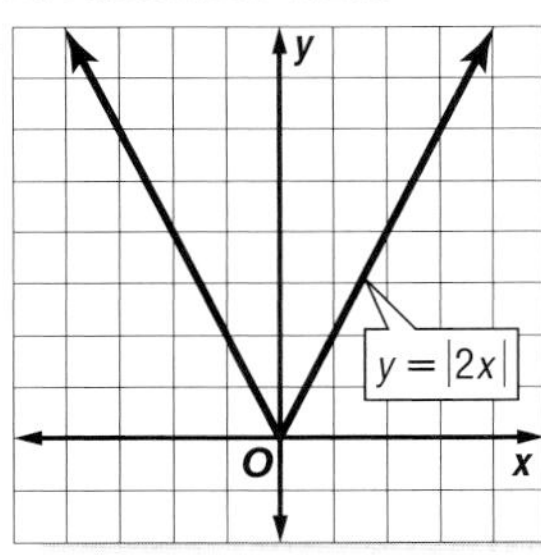

31. $C = 4.5m$ **33.** a line slanting to the right and passing through the origin

35.

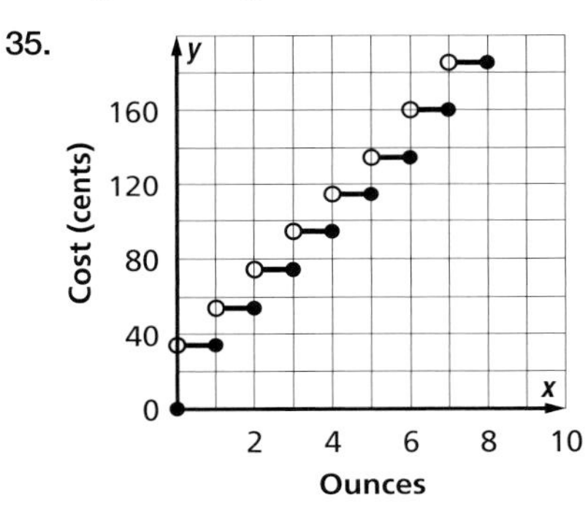

37a. absolute value **37b.** quadratic **37c.** greatest integer **37d.** square root **39.** C **41.** 22

43.

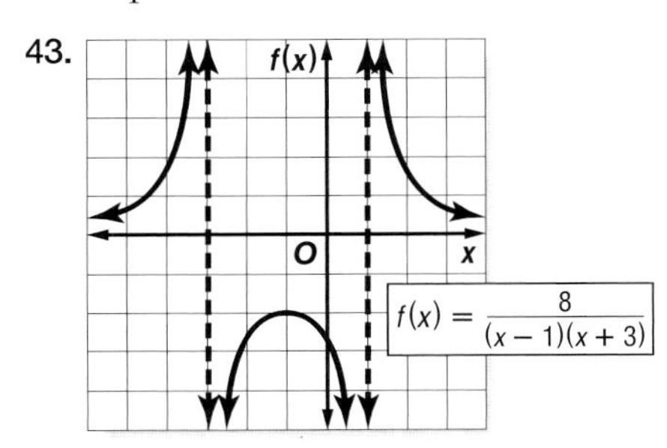

45. $(8, -1)$; $\left(8, -\frac{7}{8}\right)$; $x = 8$; $y = -1\frac{1}{8}$; up; $\frac{1}{2}$ unit

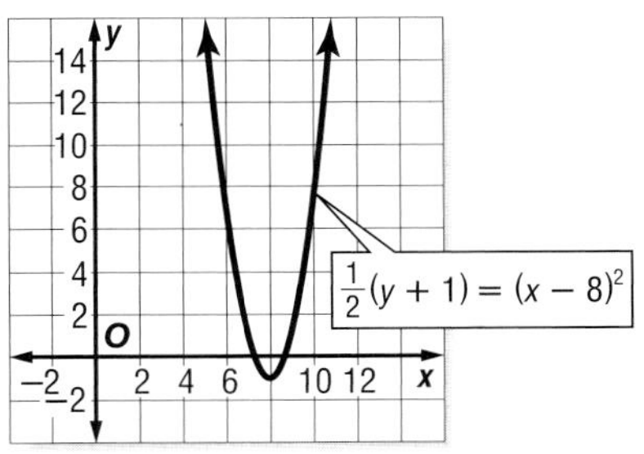

47. $(5, -4)$; $\left(5\frac{3}{4}, -4\right)$; $y = -4$; $x = 4\frac{1}{4}$; right; 3 units

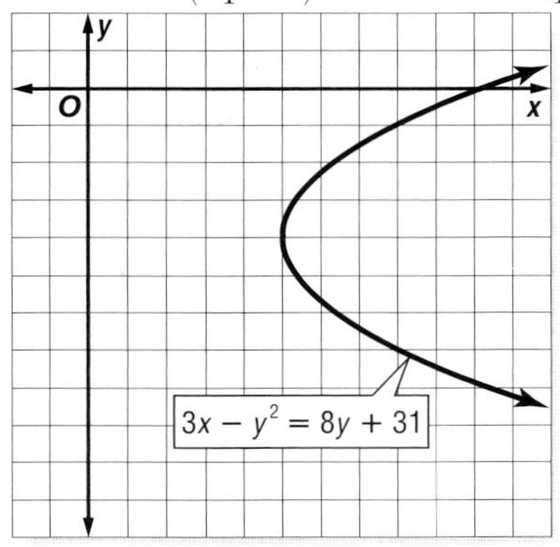

49. impossible **51.** $\left(\frac{1}{3}, 2\right)$ **53.** 1 **55.** $-\frac{17}{6}$ **57.** $45x^3y^3$ **59.** $3(x - y)(x + y)$ **61.** $(t - 5)(t + 6)(2t + 1)$

Pages 509–511 Lesson 9-6

1. Sample answer: $\frac{1}{5} + \frac{2}{a + 2} = 1$ **3.** Jeff; when Dustin multiplied by $3a$, he forgot to multiply the 2 by $3a$. **5.** 2, 6 **7.** −6, −2 **9.** $v < 0$ or $v > 1\frac{1}{6}$ **11.** 2 **13.** −6, 1 **15.** $-1 < a < 0$ **17.** 11 **19.** $t < 0$ or $t > 3$ **21.** $0 < y < 2$ **23.** 14 **25.** $\varnothing$ **27.** 7 **29.** $\frac{-3 \pm 3\sqrt{2}}{2}$ **31.** 32 **33.** band, 80 members; chorale, 50 members **35.** 24 cm **37.** 5 mL **39.** 6.15

41. If something has a general fee and cost per unit, rational equations can be used to determine how many units a person must buy in order for the actual unit price to be a given number. Answers should include the following.

- To solve $\frac{500 + 5x}{x} = 6$, multiply each side of the equation by x to eliminate the rational expression. Then subtract $5x$ from each side. Therefore, $500 = x$. A person would need to make 500 minutes of long distance minutes to make the actual unit price 6¢.
- Since the cost is 5¢ per minute plus \$5.00 per month, the actual cost per minute could never be 5¢ or less.

43. C **45.** square root

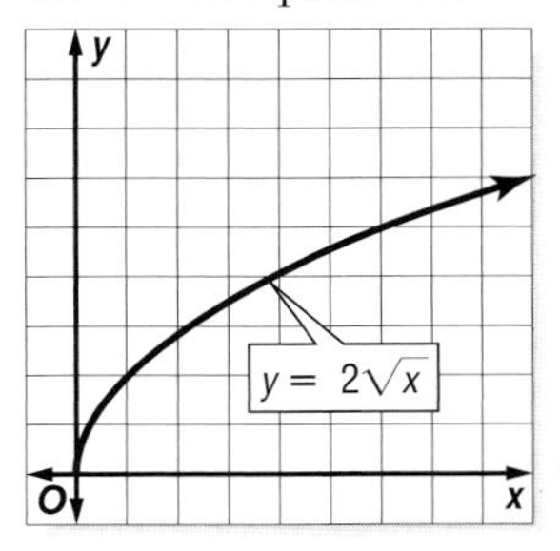

47. 36 **49.** $2\sqrt{130}$ **51.** $\sqrt{137}$ **53.** $\{x \mid 0 \le x \le 4\}$

Pages 513–516 Chapter 9 Study Guide and Review

1. false; point discontinuity **3.** false; rational **5.** true **7.** $\frac{-4bc}{33a}$ **9.** $(y + 3)(y - 6)$ **11.** $\frac{2}{n - 3}$ **13.** $\frac{7(x - 4)}{x - 5}$ **15.** $\frac{19}{3y}$ **17.** $-\frac{3}{20b}$

Selected Answers

19.

21.

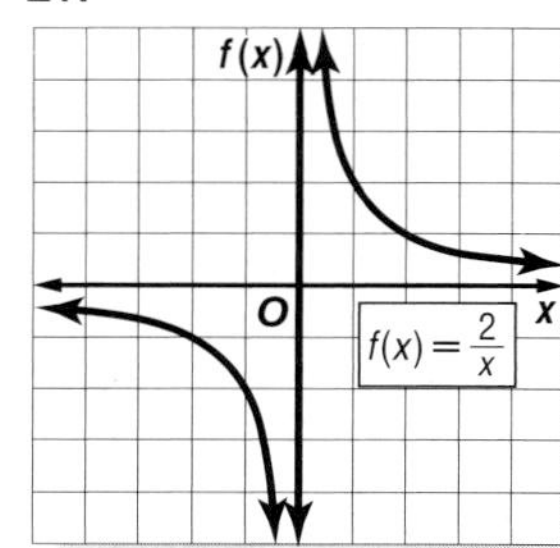

23.

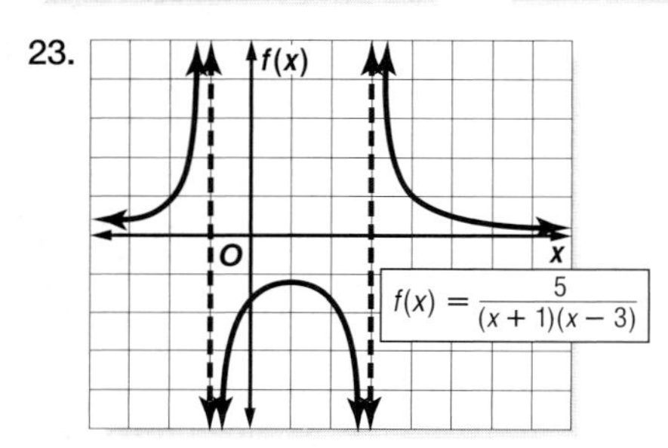

25. $-1\frac{2}{3}$ **27.** 8 **29.** 80 **31.** absolute value **33.** $1\frac{1}{9}$ **35.** 3 **37.** $1\frac{1}{2}$

Chapter 10 Exponential and Logarithmic Relations

Page 521 Chapter 10 Getting Started

1. x^{12} **3.** $-\frac{12x^3}{7y^5z}$ **5.** $a < -14$ **7.** $y \geq -2$

9. $f^{-1}(x) = -\frac{1}{2}x$

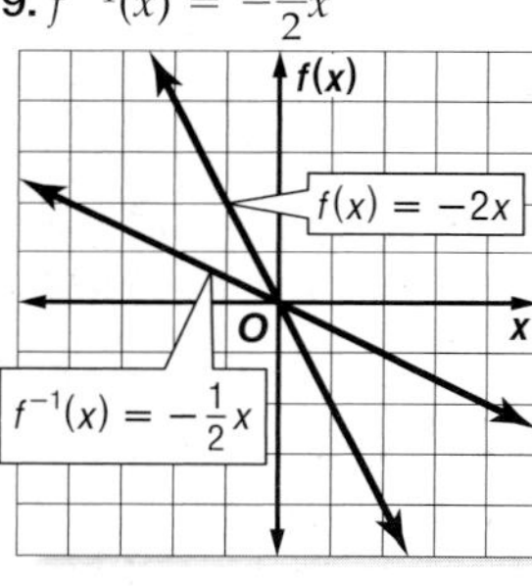

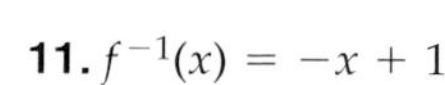

11. $f^{-1}(x) = -x + 1$

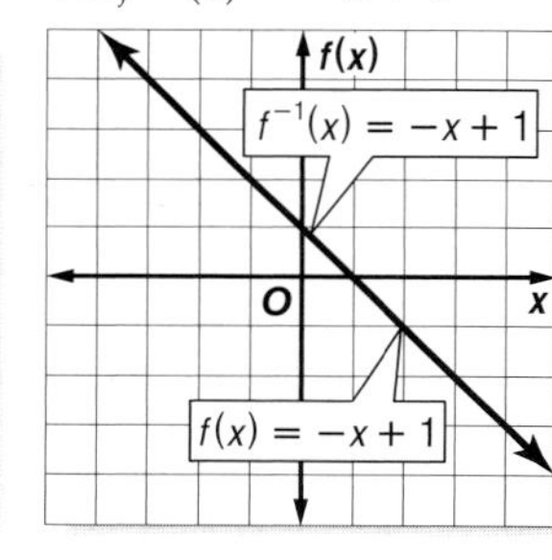

13. $g[h(x)] = 3x + 2$; $h[g(x)] = 3x - 2$
15. $g[h(x)] = x^2 - 8x + 16$; $h[g(x)] = x^2 - 4$

Pages 527–530 Lesson 10-1

1. Sample answer: 0.8 **3.** c **5.** b
7. D = $\{x \mid x$ is all real numbers.$\}$, R = $\{y \mid y > 0\}$

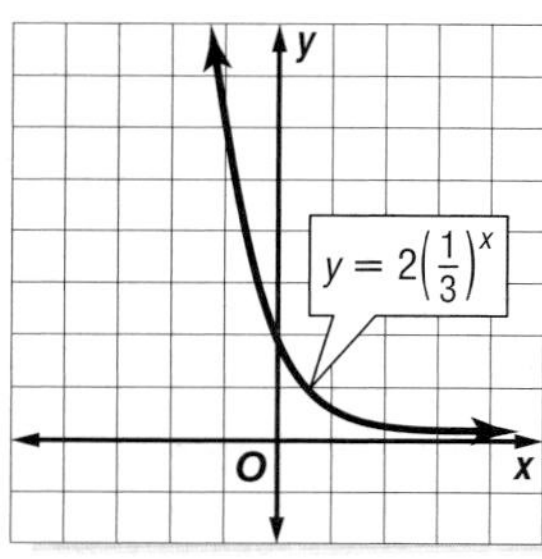

9. decay **11.** $y = 3\left(\frac{1}{2}\right)^x$ **13.** $2^{2\sqrt{7}}$ or $4^{\sqrt{7}}$
15. $3^{3\sqrt{2}}$ or $27^{\sqrt{2}}$ **17.** $x \leq 0$ **19.** $y = 65{,}000(6.20)^x$

21. D = $\{x \mid x$ is all real numbers.$\}$, R = $\{y \mid y > 0\}$

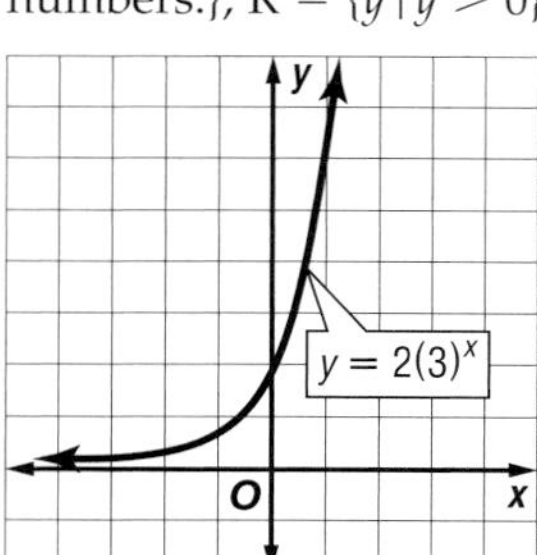

23. D = $\{x \mid x$ is all real numbers.$\}$, R = $\{y \mid y > 0\}$

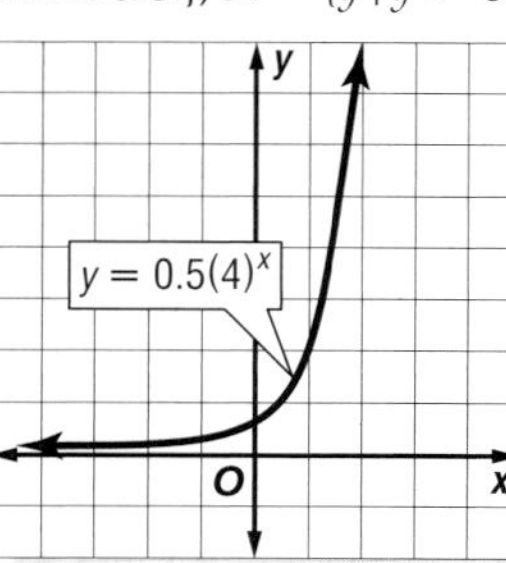

25. D = $\{x \mid x$ is all real numbers.$\}$, R = $\{y \mid y < 0\}$

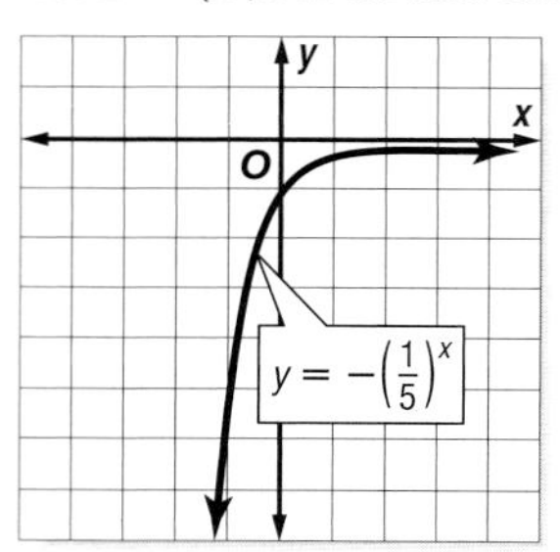

27. growth **29.** decay **31.** decay **33.** $y = -2\left(\frac{1}{4}\right)^x$
35. $y = 7(3)^x$ **37.** $y = 0.2(4)^x$ **39.** 5^4 or 625 **41.** $7^{4\sqrt{2}}$
43. $n^{2+\pi}$ **45.** $n = 5$ **47.** 1 **49.** $-\frac{8}{3}$ **51.** $n < 3$ **53.** -3
55. 10 **57.** $y = 100(6.32)^x$ **59.** $y = 3.93(1.35)^x$
61. 2144.97 million; 281.42 million; No, the growth rate has slowed considerably. The population in 2000 was much smaller than the equation predicts it would be.
63. $A(t) = 1000(1.01)^{4t}$ **65.** $s \cdot 4^x$ **67.** Sometimes; true when $b > 1$, but false when $b < 1$. **69.** A

71.

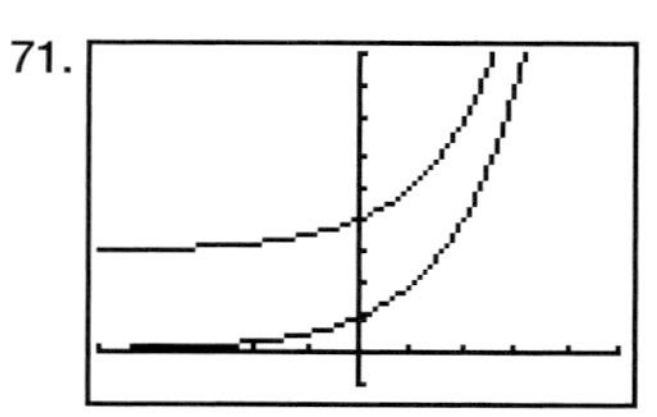

[−5, 5] scl: 1 by [−1, 9] scl: 1

The graphs have the same shape. The graph of $y = 2^x + 3$ is the graph of $y = 2^x$ translated three units up. The asymptote for the graph of $y = 2^x$ is the line $y = 0$ and for $y = 2^x + 3$ is the line $y = 3$. The graphs have the same domain, all real numbers, but the range of $y = 2^x$ is $y > 0$ and the range of $y = 2^x + 3$ is $y > 3$. The y-intercept of the graph of $y = 2^x$ is 1 and for the graph of $y = 2^x + 3$ is 4.

73.

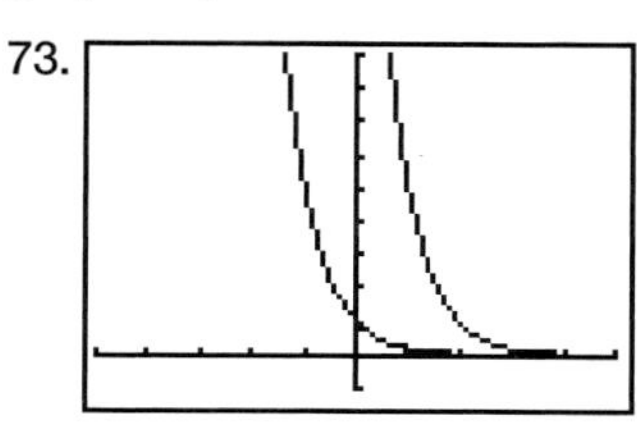

[−5, 5] scl: 1 by [−1, 9] scl: 1

The graphs have the same shape. The graph of $y = \left(\frac{1}{5}\right)^{x-2}$ is the graph of $y = \left(\frac{1}{5}\right)^x$ translated two units to the right. The asymptote for the graph of $y = \left(\frac{1}{5}\right)^x$ and for $y = \left(\frac{1}{5}\right)^{x-2}$ is

the line $y = 0$. The graphs have the same domain, all real numbers, and range, $y > 0$. The y-intercept of the graph of $y = \left(\frac{1}{5}\right)^x$ is 1 and for the graph of $y = \left(\frac{1}{5}\right)^{x-2}$ is 25. **75.** For $h > 0$, the graph of $y = 2^x$ is translated $|h|$ units to the right. For $h < 0$, the graph of $y = 2^x$ is translated $|h|$ units to the left. For $k > 0$, the graph of $y = 2^x$ is translated $|k|$ units up. For $k < 0$, the graph of $y = 2^x$ is translated $|k|$ units down. **77.** 1, 6 **79.** $0 < x < 3$ or $x > 6$
81. greatest integer

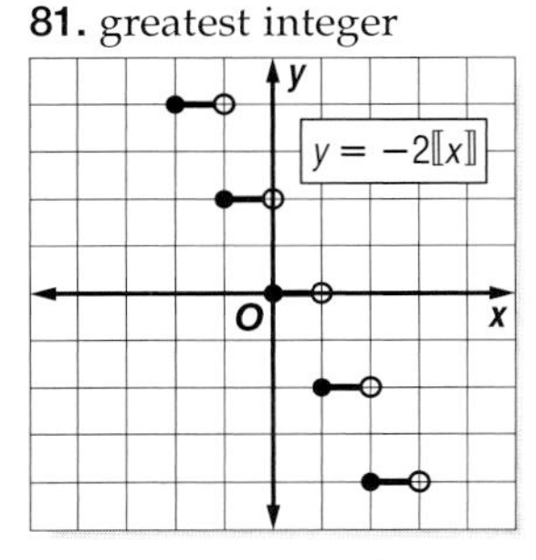

83. $\begin{bmatrix} 1 & 0 \\ 0 & 1 \end{bmatrix}$ **85.** $\frac{1}{51}\begin{bmatrix} 3 & -6 \\ 11 & -5 \end{bmatrix}$ **87.** $g[h(x)] = 2x - 6$; $h[g(x)] = 2x - 11$ **89.** $g[h(x)] = -2x - 2$; $h[g(x)] = -2x + 11$

Pages 535–538 Lesson 10-2

1. Sample answer: $x = 5^y$ and $y = \log_5 x$ **3.** Scott; the value of a logarithmic equation, 9, is the exponent of the equivalent exponential equation, and the base of the logarithmic expression, 3, is the base of the exponential equation. Thus, $x = 3^9$ or 19,683. **5.** $\log_7 \frac{1}{49} = -2$
7. $36^{\frac{1}{2}} = 6$ **9.** -3 **11.** -1 **13.** 1000 **15.** $\frac{1}{2}$, 1 **17.** 3
19. $10^{7.5}$ **21.** $\log_8 512 = 3$ **23.** $\log_5 \frac{1}{125} = -3$
25. $\log_{100} 10 = \frac{1}{2}$ **27.** $5^3 = 125$ **29.** $4^{-1} = \frac{1}{4}$ **31.** $8^{\frac{2}{3}} = 4$
33. 4 **35.** $\frac{1}{2}$ **37.** -5 **39.** 7 **41.** $n - 5$ **43.** -3 **45.** $10^{18.8}$
47. 81 **49.** $0 < y \leq 8$ **51.** 7 **53.** $x \geq 24$ **55.** 4 **57.** 2
59. 5 **61.** $a > 3$

63.

$\log_5 25 \stackrel{?}{=} 2 \log_5 5$	*Original equation*
$\log_5 5^2 \stackrel{?}{=} 2 \log_5 5^1$	*$25 = 5^2$ and $5 = 5^1$*
$2 \stackrel{?}{=} 2(1)$	*Inverse Property of Exponents and Logarithms*
$2 = 2\ \checkmark$	*Simplify.*

65.

$\log_7 [\log_3 (\log_2 8)] \stackrel{?}{=} 0$	*Original equation*
$\log_7 [\log_3 (\log_2 2^3)] \stackrel{?}{=} 0$	*$8 = 2^3$*
$\log_7 (\log_3 3) \stackrel{?}{=} 0$	*Inverse Property of Exponents and Logarithms*
$\log_7 (\log_3 3^1) \stackrel{?}{=} 0$	*$3 = 3^1$*
$\log_7 1 \stackrel{?}{=} 0$	*Inverse Property of Exponents and Logarithms*
$\log_7 7^0 \stackrel{?}{=} 0$	*$1 = 7^0$*
$0 = 0\ \checkmark$	*Inverse Property of Exponents and Logarithms*

67a.

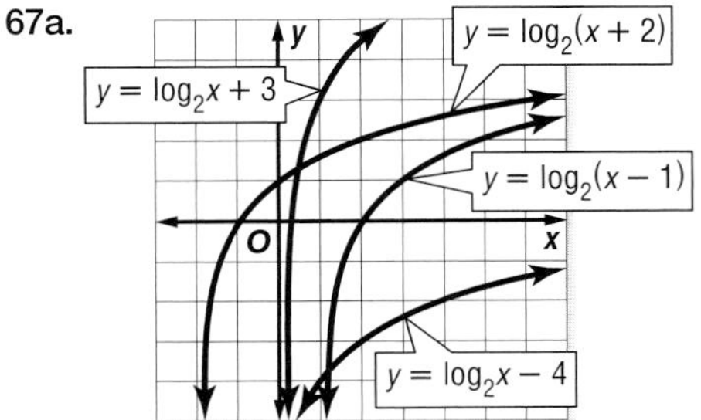

67b. The graph of $y = \log_2 x + 3$ is the graph of $y = \log_2 x$ translated 3 units up. The graph of $y = \log_2 x - 4$ is the graph of $y = \log_2 x$ translated 4 units down. The graph of $\log_2 (x - 1)$ is the graph of $y = \log_2 x$ translated 1 unit to the right. The graph of $\log_2 (x + 2)$ is the graph of $y = \log_2 x$ translated 2 units to the left. **69.** $10^{1.4}$ or about 25 times as great **71.** 2 and 3; Sample answer: 5 is between 2^2 and 2^3. **73.** A logarithmic scale illustrates that values next to each other vary by a factor of 10. Answers should include the following.

- Pin drop: 1×10^0; Whisper: 1×10^2; Normal conversation: 1×10^6; Kitchen noise: 1×10^{10}; Jet engine: 1×10^{12}
- 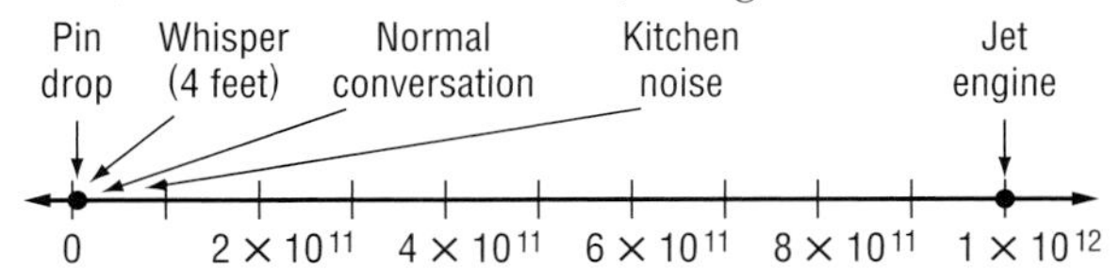

- On the scale shown above, the sound of a pin drop and the sound of normal conversation appear not to differ by much at all, when in fact they do differ in terms of the loudness we perceive. The first scale shows this difference more clearly.

75. D **77.** b^{12} **79.** -3, $\frac{14}{5}$ **81.** $\frac{5 \pm \sqrt{73}}{4}$
83. $\frac{6x - 58}{(x - 3)(x + 3)(x + 7)}$ **85.** x^{10} **87.** $8a^6b^3$ **89.** $\frac{x^3}{y^2z^3}$

Page 538 Practice Quiz 1

1. growth **3.** $\log_4 4096 = 6$ **5.** $\frac{4}{3}$ **7.** $\frac{3}{5}$ **9.** $x > 26$

Page 544–546 Lesson 10-3

1. properties of exponents **3.** Umeko; Clemente incorrectly applied the product and quotient properties of logarithms.
$\log_7 6 + \log_7 3 = \log_7 (6 \cdot 3)$ or $\log_7 18$
Product Property of Logarithms
$\log_7 18 - \log_7 2 = \log_7 (18 \div 2)$ or $\log_7 9$
Quotient Property of Logarithms

5. 2.6310 **7.** 6 **9.** 3 **11.** $\text{pH} = 6.1 + \log_{10} \frac{B}{C}$ **13.** 1.3652
15. -0.2519 **17.** 2.4307 **19.** -0.4307 **21.** 2 **23.** 4 **25.** 14
27. 2 **29.** $\varnothing$ **31.** 10 **33.** $\frac{x^3}{4}$ **35.** False; $\log_2 (2^2 + 2^3) = \log_2 12$, $\log_2 2^2 + \log_2 2^3 = 2 + 3$ or 5, and $\log_2 12 \neq 5$, since $2^5 \neq 12$. **37.** 2 **39.** about 0.4214 kilocalories per gram
41. 3 **43.** About 95 decibels; $L = 10 \log_{10} R$, where L is the loudness of the sound in decibels and R is the relative intensity of the sound. Since the crowd increased by a factor of 3, we assume that the intensity also increases by a factor of 3. Thus, we need to find the loudness of $3R$.
$L = 10 \log_{10} 3R$
$L = 10 (\log_{10} 3 + \log_{10} R)$
$L = 10 \log_{10} 3 + 10 \log_{10} R$
$L \approx 10(0.4771) + 90$
$L \approx 4.771 + 90$ or about 95
45. 7.5
47. Let $b^x = m$ and $b^y = n$. Then $\log_b m = x$ and $\log_b n = y$.

$\frac{b^x}{b^y} = \frac{m}{n}$	
$b^{x-y} = \frac{m}{n}$	*Quotient Property*
$\log_b b^{x-y} = \log_b \frac{m}{n}$	*Property of Equality for Logarithmic Equations*
$x - y = \log_b \frac{m}{n}$	*Inverse Property of Exponents and Logarithms*
$\log_b m - \log_b n = \log_b \frac{m}{n}$	*Replace x with $\log_b m$ and y with $\log_b n$.*

49. A **51.** 4 **53.** $2x$ **55.** -8

Selected Answers

57. odd; 3 59. $\frac{3b}{a}$ 61. $\frac{5}{3x}$ 63. 1 65. $x > \frac{5}{3}$

Pages 549–551 Lesson 10-4

1. 10; common logarithms 3. A calculator is not programmed to find base 2 logarithms. 5. 1.3617 7. 1.7325 9. 4.9824 11. 11.5665 13. $\frac{\log 5}{\log 7}$; 0.8271 15. $\frac{\log 9}{\log 2}$; 3.1699 17. 0.6990 19. 0.8573 21. −0.0969 23. 11 25. 2.1 27. $\{x \mid x \geq 2.0860\}$ 29. $\{a \mid a < 1.1590\}$ 31. 0.4341 33. 4.7820 35. ±1.1909 37. $\{n \mid n < -1.0178\}$ 39. 3.7162 41. 0.5873 43. −7.6377 45. $\frac{\log 13}{\log 2} \approx 3.7004$ 47. $\frac{\log 3}{\log 7} \approx 0.5646$ 49. $\frac{2 \log 1.6}{\log 4} \approx 0.6781$ 51. between 0.000000001 and 0.000001 mole per liter 53. Sirius 55. Vega 57. about 3.75 yr or 3 yr 9 mo 59. Comparisons between substances of different acidities are more easily distinguished on a logarithmic scale. Answers should include the following.
Sample Answer:

- Tomatoes: 6.3×10^{-5} mole per liter
  Milk: 3.98×10^{-7} mole per liter
  Eggs: 1.58×10^{-8} mole per liter
- Those measurements correspond to pH measurements of 5 and 4, indicating a weak acid and a stronger acid. On the logarithmic scale we can see the difference in these acids, whereas on a normal scale, these hydrogen ion concentrations would appear nearly the same. For someone who has to watch the acidity of the foods they eat, this could be the difference between an enjoyable meal and heartburn.

61. C 63. 1.6938 65. 64 67. 62 69. $(d + 2)(3d - 4)$ 71. prime 73. $3^2 = x$ 75. $\log_5 45 = x$ 77. $\log_b x = y$

Pages 557–559 Lesson 10-5

1. the number e 3. Elsu; Colby tried to write each side as a power of 10. Since the base of the natural logarithmic function is e, he should have written each side as a power of e; $10^{\ln 4x} \neq 4x$. 5. 0.0334 7. −2.3026 9. $e^0 = 1$ 11. $5x$ 13. 1.0986 15. $0 < x < 403.4288$ 17. ±90.0171 19. about 15,066 ft 21. 148.4132 23. 1.6487 25. 2.3026 27. −3.5066 29. about 49.5 cm 31. $2 = \ln 6x$ 33. $e^x = 5.2$ 35. y 37. 45 39. −0.6931 41. $x > 0.4700$ 43. 0.5973 45. $x \geq -0.9730$ 47. 49.4711 49. 14.3891 51. 45.0086 53. 1 55. $t = \frac{100 \ln 2}{r}$ 57. $t = \frac{110}{r}$ 59. about 55 yr 61. about 21 min 63. The number e is used in the formula for continuously compounded interest, $A = Pe^{rt}$. Although no banks actually pay interest compounded continually, the equation is so accurate in computing the amount of money for quarterly compounding, or daily compounding, that it is often used for this purpose. Answers should include the following.

- If you know the annual interest rate r and the principal P, the value of the account after t years is calculated by multiplying P times e raised to the r times t power. Use a calculator to find the value of e^{rt}.
- If you know the value A you wish the account to achieve, the principal P, and the annual interest rate r, the time t needed to achieve this value is found by first taking the natural logarithm of A minus the natural logarithm of P. Then, divide this quantity by r.

65. 1946, 1981, 2015; It takes between 34 and 35 years for the population to double. 67. $\frac{\log 0.047}{\log 6} = -1.7065$ 69. 5 71. inverse; 4 73. direct; −7 75. 3.32 77. 1.43 79. 13.43

Page 559 Practice Quiz 2

1. $\frac{\log 5}{\log 4}$; 1.1610 3. 3 5. 1.3863

Pages 563–565 Lesson 10-6

1. $y = a(1 + r)^t$, where $r > 0$ represents exponential growth and $r < 0$ represents exponential decay 3. Sample answer: money in a bank 5. about 33.5 watts 7. $y = 212{,}000e^{0.025t}$ 9. C 11. at most \$108,484.93 13. No; the bone is only about 21,000 years old, and dinosaurs died out 63,000,000 years ago. 15. about 0.0347 17. \$12,565 billion 19. after the year 2182 21. Never; theoretically, the amount left will always be half of the previous amount. 23. about 19.5 yr 25. $\ln y = 3$ 27. $4x^2 = e^8$ 29. $p > 3.3219$ 31. $\frac{0.5(0.08p)}{6} + \frac{0.5(0.08p)}{4}$ 33. $\frac{p}{150}$ 35. ellipse 37. circle 39. 8×10^7

Pages 566–570 Chapter 10 Study Guide and Review

1. true 3. false; common logarithm 5. true 7. false; logarithmic function 9. false; exponential function 11. growth 13. $y = 7\left(\frac{1}{5}\right)^x$ 15. −1 17. $x \leq -\sqrt{6}$ or $x \geq \sqrt{6}$ 19. $\log_5 \frac{1}{25} = -2$ 21. $4^3 = 64$ 23. $6^{-2} = \frac{1}{36}$ 25. −5 27. 2 29. $\frac{3}{2}$ 31. $\frac{1}{3} < y < 3$ 33. −4, 3 35. 1.7712 37. 3 39. 6 41. 15 43. 5.7279 45. $x < 7.3059$ 47. $x \geq 5.8983$ 49. $\frac{\log 11}{\log 4}$; 1.7297 51. $\frac{\log 1000}{\log 20}$; 2.3059 53. $e^x = 7.4$ 55. $7x$ 57. $x > 1.1632$ 59. $0 < x \leq 49.4711$ 61. 74.2066 63. 5.05 days 65. about 3.6%

Chapter 11 Sequences and Series

Page 577 Chapter 11 Getting Started

1. 6 3. −5 5. $\frac{1}{2}$

7.

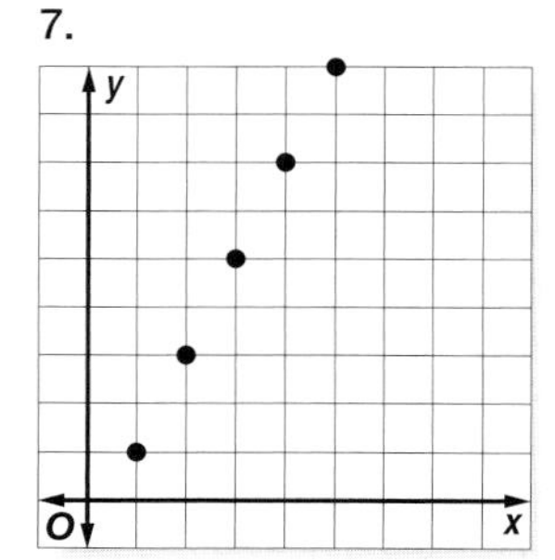

9.

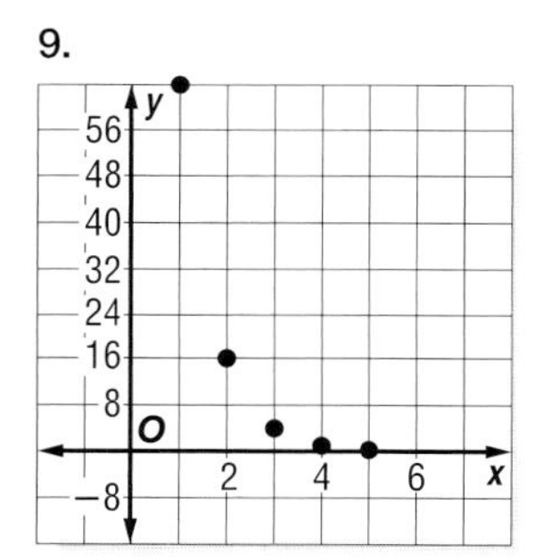

11. 17 13. $\frac{1}{32}$ 15. $\frac{3}{5}$

Pages 580–582 Lesson 11-1

1. The differences between the terms are not constant. 3. Sample answer: 1, −4, −9, −14, … 5. −3, −5, −7, −9 7. 14, 12, 10, 8, 6 9. −112 11. 15 13. 56, 68, 80 15. 30, 37, 44, 51 17. 6, 10, 14, 18 19. $\frac{7}{3}, 3, \frac{11}{3}, \frac{13}{3}$ 21. 5.5, 5.1, 4.7, 4.3 23. 2, 15, 28, 41, 54 25. 6, 2, −2, −6, −10 27. $\frac{4}{3}, 1, \frac{2}{3}, \frac{1}{3}, 0$ 29. 28 31. 94

33. 335 **35.** $\frac{26}{3}$ **37.** 27 **39.** 61 **41.** 37.5 in. **43.** 30th
45. 82nd **47.** $a_n = -7n + 25$
49. 13, 17, 21

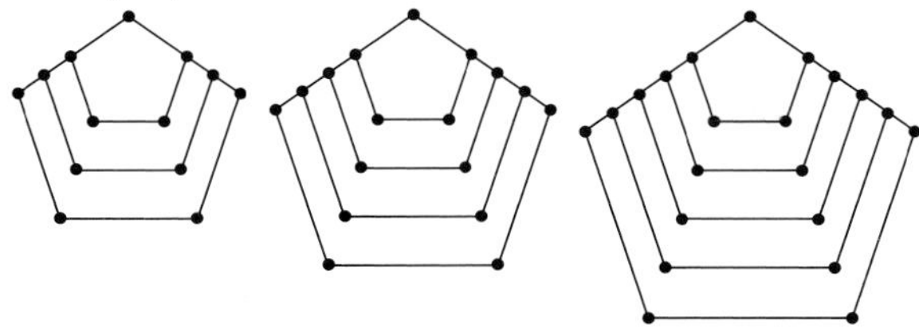

51. Yes; it corresponds to $n = 100$. **53.** 4, −2
55. 7, 11, 15, 19, 23
57. Arithmetic sequences can be used to model the numbers of shingles in the rows on a section of roof. Answers should include the following.
- One additional shingle is needed in each successive row.
- One method is to successively add 1 to the terms of the sequence: $a_8 = 9 + 1$ or 10, $a_9 = 10 + 1$ or 11, $a_{10} = 11 + 1$ or 12, $a_{11} = 12 + 1$ or 13, $a_{12} = 13 + 1$ or 14, $a_{13} = 14 + 1$ or 15, $a_{14} = 15 + 1$ or 16, $a_{15} = 16 + 1$ or 17. Another method is to use the formula for the nth term: $a_{15} = 3 + (15 - 1)1$ or 17.

59. B **61.** −0.4055 **63.** 146.4132 **65.** 2, 5, 8, 11
67. 11, 15, 19, 23, 27

Pages 586–587 Lesson 11-2
1. In a series, the terms are added. In a sequence, they are not. **3.** Sample answer: $\sum_{n=1}^{4}(3n + 4)$ **5.** 230 **7.** 552
9. 260 **11.** 95 **13.** −6, 0, 6 **15.** 344 **17.** 1501 **19.** −9
21. 104 **23.** −714 **25.** 14 **27.** 10 rows **29.** 721 **31.** 162
33. 108 **35.** −195 **37.** 315,150 **39.** 1,001,000 **41.** 17, 26, 35 **43.** −12, −9, −6 **45.** 265 ft **47.** False; for example, 7 + 10 + 13 + 16 = 46, but 7 + 10 + 13 + 16 + 19 + 22 + 25 + 28 = 140. **49.** C **51.** 5555 **53.** 6683 **55.** −135
57. $-\frac{9}{2}$ **59.** $\frac{3 \pm \sqrt{89}}{2}$ **61.** $26\sqrt{21}$ **63.** 16 **65.** $\frac{2}{27}$

Pages 590–592 Lesson 11-3
1a. Geometric; the terms have a common ratio of −2.
1b. Arithmetic; the terms have a common difference of −3.
3. Marika; Lori divided in the wrong order when finding r.
5. 2, −4 **7.** $\frac{15}{64}$ **9.** −4 **11.** 3, 9 **13.** 15, 5 **15.** 54, 81
17. $\frac{20}{27}, \frac{40}{81}$ **19.** −2.16, 2.592 **21.** 2, −6, 18, −54, 162
23. 243, 81, 27, 9, 3 **25.** $\frac{3}{16}$ **27.** 729 **29.** 243 **31.** 1
33. 78,125 **35.** −8748 **37.** 655.36 lb **39.** $a_n = 36\left(\frac{1}{3}\right)^{n-1}$
41. $a_n = -2(-5)^{n-1}$ **43.** ±18, 36, ±72 **45.** 16, 8, 4, 2
47. 8 days **49.** False; the sequence 1, 4, 9, 16, …, for example, is neither arithmetic nor geometric.
51. The heights of the bounces of a ball and the heights from which a bouncing ball falls each form geometric sequences. Answers should include the following.
- 3, 1.8, 1.08, 0.648, 0.3888
- The common ratios are the same, but the first terms are different. The sequence of heights from which the ball falls is the sequence of heights of the bounces with the term 3 inserted at the beginning.

53. C **55.** 203 **57.** −12, −16, −20 **59.** 127 **61.** $\frac{61}{81}$

Page 592 Practice Quiz 1
1. 46 **3.** 187 **5.** 1

Pages 596–598 Lesson 11-4
1. Sample answer: $4 + 2 + 1 + \frac{1}{2}$
3. Sample answer: The first term is $a_1 = 2$. Divide the second term by the first to find that the common ratio is $r = 6$. Therefore, the nth term of the series is given by $2 \cdot 6^{n-1}$. There are five terms, so the series can be written as $\sum_{n=1}^{5} 2 \cdot 6^{n-1}$. **5.** 39,063 **7.** 165 **9.** 129 **11.** $\frac{1093}{9}$
13. 3 **15.** 728 **17.** 1111 **19.** 244 **21.** 2101 **23.** $\frac{728}{3}$
25. 1040.984 **27.** 6564 **29.** 1,747,625 **31.** 3641 **33.** $\frac{5461}{16}$
35. 2555 **37.** $\frac{387}{4}$ **39.** 3,145,725 **41.** 243 **43.** 2 **45.** 80
47. about 7.13 in. **49.** If the number of people that each person sends the joke to is constant, then the total number of people who have seen the joke is the sum of a geometric series. Answers should include the following.
- The common ratio would change from 3 to 4.
- Increase the number of days that the joke circulates so that it is inconvenient to find and add all the terms of the series.

51. C **53.** 3.99987793 **55.** $\pm\frac{1}{4}, \frac{3}{2}, \pm 9$ **57.** 232
59. **Drive-In Movie Screens**

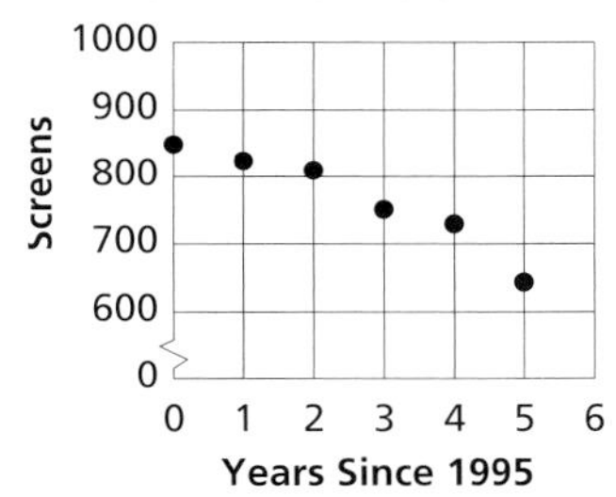

61. Sample answer: 294 **63.** 2 **65.** $\frac{2}{3}$ **67.** 0.6

Pages 602–604 Lesson 11-5
1. Sample answer: $\sum_{n=1}^{\infty}\left(\frac{1}{2}\right)^n$ **3.** Beth; the common ratio for the infinite geometric series is $-\frac{4}{3}$. Since $\left|-\frac{4}{3}\right| \geq 1$, the series does not have a sum and the formula $S = \frac{a_1}{1 - r}$ does not apply. **5.** does not exist **7.** $\frac{3}{4}$ **9.** 100 **11.** $\frac{73}{99}$
13. 96 cm **15.** does not exist **17.** 45 **19.** −16 **21.** $\frac{54}{5}$
23. does not exist **25.** 1 **27.** $\frac{2}{3}$ **29.** $\frac{3}{2}$ **31.** 2
33. $40 + 20\sqrt{2} + 20 + \ldots$ **35.** 900 ft **37.** 75, 30, 12
39. $-8, -3\frac{1}{5}, -1\frac{7}{25}, -\frac{64}{125}$ **41.** $\frac{1}{9}$ **43.** $\frac{82}{99}$ **45.** $\frac{427}{999}$ **47.** $\frac{229}{990}$
49. The total distance that a ball bounces, both up and down, can be found by adding the sums of two infinite geometric series. Answers should include the following.
- $a_n = a_1 \cdot r^{n-1}$, $S_n = \frac{a_1(1 - r^n)}{1 - r}$, or $S = \frac{a_1}{1 - r}$
- The total distance the ball falls is given by the infinite geometric series $3 + 3(0.6) + 3(0.6)^2 + \ldots$. The sum of this series is $\frac{3}{1 - 0.6}$ or 7.5. The total distance the ball bounces up is given by the infinite geometric series $1.8(0.6) + 1.8(0.6)^2 + 1.8(0.6)^3 + \ldots$. The sum of this

series is $\frac{1.8(0.6)}{1 - 0.6}$ or 2.7. Thus, the total distance the ball travels is 7.5 + 2.7 or 10.2 feet.

51. C **53.** $\frac{8744}{81}$ **55.** 3 **57.** $x \geq 5$ **59.** $\frac{-x + 7}{(x - 3)(x + 1)}$

61. $(x - 2)^2 + (y - 4)^2 = 36$ **63.** $-\frac{1}{2}, \frac{3}{2}, \frac{7}{2}$ **65.** $x^2 - 36 = 0$

67. $x^2 - 10x + 24 = 0$ **69.** The number of visitors was decreasing. **71.** 3 **73.** $\frac{1}{2}$ **75.** -4

Pages 608–610 Lesson 11-6

1. $a_n = a_{n-1} + d$; $a_n = r \cdot a_{n-1}$ **3.** Sometimes; if $f(x) = x^2$ and $x_1 = 2$, then $x_2 = 2^2$ or 4, so $x_2 \neq x_1$. But, if $x_1 = 1$, then $x_2 = 1$, so $x_2 = x_1$. **5.** $-3, -2, 0, 3, 7$ **7.** 1, 2, 5, 14, 41
9. 1, 3, -1 **11.** $b_n = 1.05b_{n-1} - 10$ **13.** $-6, -3, 0, 3, 6$
15. 2, 1, $-1, -4, -8$ **17.** 9, 14, 24, 44, 84 **19.** -1, 5, 4, 9, 13

21. $\frac{7}{2}, \frac{7}{4}, \frac{7}{6}, \frac{7}{8}, \frac{7}{10}$ **23.** 67 **25.** 1, 1, 2, 3, 5, ...

27. \$99,921.21, \$99,841.95, \$99,762.21, \$99,681.99, \$99,601.29, \$99,520.11, \$99,438.44, \$99,356.28 **29.** $t_n = t_{n-1} + n$
31. 16, 142, 1276 **33.** $-7, -16, -43$ **35.** -3, 13, 333

37. $\frac{5}{2}, \frac{37}{2}, \frac{1445}{2}$ **39.** \$75.78

41. Under certain conditions, the Fibonacci sequence can be used to model the number of shoots on a plant. Answers should include the following.

- The 13th term of the sequence is 233, so there are 233 shoots on the plant during the 13th month.
- The Fibonacci sequence is not arithmetic because the differences $(0, 1, 1, 2, \ldots)$ of the terms are not constant. The Fibonacci sequence is not geometric because the ratios $\left(1, 2, \frac{3}{2}, \ldots\right)$ of the terms are not constant.

43. C **45.** $\frac{1}{6}$ **47.** -5208 **49.** $3x + 7$ units **51.** 5040

53. 20 **55.** 210

Pages 615–617 Lesson 11-7

1. 1, 8, 28, 56, 70, 56, 28, 8, 1 **3.** Sample answer: $(5x + y)^4$
5. 17,160 **7.** $p^5 + 5p^4q + 10p^3q^2 + 10p^2q^3 + 5pq^4 + q^5$
9. $x^4 - 12x^3y + 54x^2y^2 - 108xy^3 + 81y^4$ **11.** $1{,}088{,}640a^6b^4$
13. 362,880 **15.** 72 **17.** 495 **19.** $a^3 - 3a^2b + 3ab^2 - b^3$
21. $r^8 + 8r^7s + 28r^6s^2 + 56r^5s^3 + 70r^4s^4 + 56r^3s^5 + 28r^2s^6 + 8rs^7 + s^8$ **23.** $x^5 + 15x^4 + 90x^3 + 270x^2 + 405x + 243$
25. $16b^4 - 32b^3x + 24b^2x^2 - 8bx^3 + x^4$ **27.** $243x^5 - 810x^4y + 1080x^3y^2 - 720x^2y^3 + 240xy^4 - 32y^5$ **29.** $\frac{a^5}{32} + \frac{5a^4}{8} + 5a^3 + 20a^2 + 40a + 32$ **31.** $27x^3 + 54x^2 + 36x + 8$ cm^3 **33.** 45
35. $924x^6y^6$ **37.** $5670a^4$ **39.** $145{,}152x^6y^3$ **41.** $-\frac{63}{8}x^5$

43. The coefficients in a binomial expansion give the numbers of sequences of births resulting in given numbers of boys and girls. Answers should include the following.

- $(b + g)^5 = b^5 + 5b^4g + 10b^3g^2 + 10b^2g^3 + 5bg^4 + g^5$; There is one sequence of births with all five boys, five sequences with four boys and one girl, ten sequences with three boys and two girls, ten sequences with two boys and three girls, five sequences with one boy and four girls, and one sequence with all five girls.
- The number of sequences of births that have exactly k girls in a family of n children is the coefficient of $b^{n-k}g^k$ in the expansion of $(b + g)^n$. According to the Binomial Theorem, this coefficient is $\frac{n!}{(n-k)!k!}$.

45. C **47.** 3, 5, 9, 17, 33 **49.** $\frac{\log 5}{\log 2}$; 2.3219 **51.** $\frac{\log 8}{\log 5}$; 1.2920 **53.** asymptotes: $x = -4$, $x = 1$ **55.** hyperbola

57. yes **59.** True; $\frac{1(1+1)}{2} = \frac{1(2)}{2}$ or 1. **61.** True; $\frac{1^2(1+1)^2}{4} = \frac{1(4)}{4}$ or 1.

Page 617 Practice Quiz 2

1. 1,328,600 **3.** 24 **5.** 1, 5, 13, 29, 61 **7.** 5, -13, 41
9. $a^6 + 12a^5 + 60a^4 + 160a^3 + 240a^2 + 192a + 64$

Pages 619–621 Lesson 11-8

1. Sample answers: formulas for the sums of powers of the first n positive integers and statements that expressions involving exponents of n are divisible by certain numbers
3. Sample answer: $3^n - 1$
5. Step 1: When $n = 1$, the left side of the given equation is $\frac{1}{2}$. The right side is $1 - \frac{1}{2}$ or $\frac{1}{2}$, so the equation is true for $n = 1$.

Step 2: Assume $\frac{1}{2} + \frac{1}{2^2} + \frac{1}{2^3} + \ldots + \frac{1}{2^k} = 1 - \frac{1}{2^k}$ for some positive integer k.

Step 3:
$$\frac{1}{2} + \frac{1}{2^2} + \frac{1}{2^3} + \ldots + \frac{1}{2^k} + \frac{1}{2^{k+1}} = 1 - \frac{1}{2^k} + \frac{1}{2^{k+1}}$$
$$= 1 - \frac{2}{2^{k+1}} + \frac{1}{2^{k+1}}$$
$$= 1 - \frac{1}{2^{k+1}}$$

The last expression is the right side of the equation to be proved, where $n = k + 1$. Thus, the equation is true for $n = k + 1$.

Therefore, $\frac{1}{2} + \frac{1}{2^2} + \frac{1}{2^3} + \ldots + \frac{1}{2^n} = 1 - \frac{1}{2^n}$ for all positive integers n.

7. Step 1: $5^1 + 3 = 8$, which is divisible by 4. The statement is true for $n = 1$.
Step 2: Assume that $5^k + 3$ is divisible by 4 for some positive integer k. This means that $5^k + 3 = 4r$ for some positive integer r.
Step 3:
$$5^k + 3 = 4r$$
$$5^k = 4r - 3$$
$$5^{k+1} = 20r - 15$$
$$5^{k+1} + 3 = 20r - 12$$
$$5^{k+1} + 3 = 4(5r - 3)$$

Since r is a positive integer, $5r - 3$ is a positive integer. Thus, $5^{k+1} + 3$ is divisible by 4, so the statement is true for $n = k + 1$.

Therefore, $5^n + 3$ is divisible by 4 for all positive integers n.
9. Sample answer: $n = 3$
11. Step 1: When $n = 1$, the left side of the given equation is 1. The right side is $1[2(1) - 1]$ or 1, so the equation is true for $n = 1$.
Step 2: Assume $1 + 5 + 9 + \ldots + (4k - 3) = k(2k - 1)$ for some positive integer k.
Step 3: $1 + 5 + 9 + \ldots + (4k - 3) + [4(k + 1) - 3]$
$$= k(2k - 1) + [4(k + 1) - 3]$$
$$= 2k^2 - k + 4k + 4 - 3$$
$$= 2k^2 + 3k + 1$$
$$= (k + 1)(2k + 1)$$
$$= (k + 1)[2(k + 1) - 1]$$

The last expression is the right side of the equation to be proved, where $n = k + 1$. Thus, the equation is true for $n = k + 1$.

Therefore, $1 + 5 + 9 + \ldots + (4n - 3) = n(2n - 1)$ for all positive integers n.
13. Step 1: When $n = 1$, the left side of the given equation is 1^3 or 1. The right side is $\frac{1^2(1+1)^2}{4}$ or 1, so the equation is true for $n = 1$.
Step 2: Assume $1^3 + 2^3 + 3^3 + \ldots + k^3 = \frac{k^2(k+1)^2}{4}$ for some

positive integer k.

Step 3: $1^3 + 2^3 + 3^3 + \ldots + k^3 + (k+1)^3$

$$= \frac{k^2(k+1)^2}{4} + (k+1)^3$$
$$= \frac{k^2(k+1)^2 + 4(k+1)^3}{4}$$
$$= \frac{(k+1)^2[k^2 + 4(k+1)]}{4}$$
$$= \frac{(k+1)^2(k^2 + 4k + 4)}{4}$$
$$= \frac{(k+1)^2(k+2)^2}{4}$$
$$= \frac{(k+1)^2[(k+1)+1]^2}{4}$$

The last expression is the right side of the equation to be proved, where $n = k + 1$. Thus, the equation is true for $n = k + 1$.

Therefore, $1^3 + 2^3 + 3^3 + \ldots + n^3 = \frac{n^2(n+1)^2}{4}$ for all positive integers n.

15. Step 1: When $n = 1$, the left side of the given equation is $\frac{1}{3}$. The right side is $\frac{1}{2}\left(1 - \frac{1}{3}\right)$ or $\frac{1}{3}$, so the equation is true for $n = 1$.

Step 2: Assume $\frac{1}{3} + \frac{1}{3^2} + \frac{1}{3^3} + \ldots + \frac{1}{3^k} = \frac{1}{2}\left(1 - \frac{1}{3^k}\right)$ for some positive integer k.

Step 3: $\frac{1}{3} + \frac{1}{3^2} + \frac{1}{3^3} + \ldots + \frac{1}{3^k} + \frac{1}{3^{k+1}} = \frac{1}{2}\left(1 - \frac{1}{3^k}\right) + \frac{1}{3^{k+1}}$

$$= \frac{1}{2} - \frac{1}{2 \cdot 3^k} + \frac{1}{3^{k+1}}$$
$$= \frac{3^{k+1} - 3 + 2}{2 \cdot 3^{k+1}}$$
$$= \frac{3^{k+1} - 1}{2 \cdot 3^{k+1}}$$
$$= \frac{1}{2}\left(\frac{3^{k+1} - 1}{3^{k+1}}\right)$$
$$= \frac{1}{2}\left(1 - \frac{1}{3^{k+1}}\right)$$

The last expression is the right side of the equation to be proved, where $n = k + 1$. Thus, the equation is true for $n = k + 1$.

Therefore, $\frac{1}{3} + \frac{1}{3^2} + \frac{1}{3^3} + \ldots + \frac{1}{3^n} = \frac{1}{2}\left(1 - \frac{1}{3^n}\right)$ for all positive integers n.

17. Step 1: $8^1 - 1 = 7$, which is divisible by 7. The statement is true for $n = 1$.

Step 2: Assume that $8^k - 1$ is divisible by 7 for some positive integer k. This means that $8^k - 1 = 7r$ for some whole number r.

Step 3:
$$8^k - 1 = 7r$$
$$8^k = 7r + 1$$
$$8^{k+1} = 56r + 8$$
$$8^{k+1} - 1 = 56r + 7$$
$$8^{k+1} - 1 = 7(8r + 1)$$

Since r is a whole number, $8r + 1$ is a whole number. Thus, $8^{k+1} - 1$ is divisible by 7, so the statement is true for $n = k + 1$.

Therefore, $8^n - 1$ is divisible by 7 for all positive integers n.

19. Step 1: $12^1 + 10 = 22$, which is divisible by 11. The statement is true for $n = 1$.

Step 2: Assume that $12^k + 10$ is divisible by 11 for some positive integer k. This means that $12^k + 10 = 11r$ for some positive integer r.

Step 3:
$$12^k + 10 = 11r$$
$$12^k = 11r - 10$$
$$12^{k+1} = 132r - 120$$
$$12^{k+1} + 10 = 132r - 110$$
$$12^{k+1} + 10 = 11(12r - 10)$$

Since r is a positive integer, $12r - 10$ is a positive integer. Thus, $12^{k+1} + 10$ is divisible by 11, so the statement is true for $n = k + 1$.

Therefore, $12^n + 10$ is divisible by 11 for all positive integers n.

21. Step 1: There are 6 bricks in the top row, and $1^2 + 5(1) = 6$, so the formula is true for $n = 1$.

Step 2: Assume that there are $k^2 + 5k$ bricks in the top k rows for some positive integer k.

Step 3: Since each row has 2 more bricks than the one above, the numbers of bricks in the rows form an arithmetic sequence. The number of bricks in the $(k + 1)$st row is $6 + [(k+1) - 1](2)$ or $2k + 6$. Then the number of bricks in the top $k + 1$ rows is $k^2 + 5k + (2k + 6)$ or $k^2 + 7k + 6$. $k^2 + 7k + 6 = (k+1)^2 + 5(k+1)$, which is the formula to be proved, where $n = k + 1$. Thus, the formula is true for $n = k + 1$.

Therefore, the number of bricks in the top n rows is $n^2 + 5n$ for all positive integers n.

23. Step 1: When $n = 1$, the left side of the given equation is a_1. The right side is $\frac{1}{2}[2a_1 + (1 - 1)d]$ or a_1, so the equation is true for $n = 1$.

Step 2: Assume $a_1 + (a_1 + d) + (a_1 + 2d) + \ldots + [a_1 + (k-1)d] = \frac{k}{2}[2a_1 + (k-1)d]$ for some positive integer k.

Step 3: $a_1 + (a_1 + d) + (a_1 + 2d) + \ldots + [a_1 + (k-1)d] + [a_1 + (k + 1 - 1)d]$

$$= \frac{k}{2}[2a_1 + (k-1)d] + [a_1 + (k + 1 - 1)d]$$
$$= \frac{k}{2}[2a_1 + (k-1)d] + a_1 + kd$$
$$= \frac{k[2a_1 + (k-1)d] + 2(a_1 + kd)}{2}$$
$$= \frac{k \cdot 2a_1 + (k^2 - k)d + 2a_1 + 2kd}{2}$$
$$= \frac{(k+1)2a_1 + (k^2 - k + 2k)d}{2}$$
$$= \frac{(k+1)2a_1 + k(k+1)d}{2}$$
$$= \frac{k+1}{2}(2a_1 + kd)$$
$$= \frac{k+1}{2}[2a_1 + (k + 1 - 1)d]$$

The last expression is the right side of the formula to be proved, where $n = k + 1$. Thus, the formula is true for $n = k + 1$.

Therefore, $a_1 + (a_1 + d) + (a_1 + 2d) + \ldots + [a_1 + (n-1)d] = \frac{n}{2}[2a_1 + (n-1)d]$ for all positive integers n.

25. Sample answer: $n = 3$ **27.** Sample answer: $n = 2$ **29.** Sample answer: $n = 11$ **31.** Write 7^n as $(6 + 1)^n$. Then use the Binomial Theorem.

$$7^n - 1 = (6+1)^n - 1$$
$$= 6^n + n \cdot 6^{n-1} + \frac{n(n-1)}{2}6^{n-2} + \ldots + n \cdot 6 + 1 - 1$$
$$= 6^n + n \cdot 6^{n-1} + \frac{n(n-1)}{2}6^{n-2} + \ldots + n \cdot 6$$

Since each term in the last expression is divisible by 6, the whole expression is divisible by 6. Thus, $7^n - 1$ is divisible by 6. **33.** C **35.** $x^6 + 6x^5y + 15x^4y^2 + 20x^3y^3 + 15x^2y^4 + 6xy^5 + y^6$ **37.** $256x^8 + 1024x^7y + 1792x^6y^2 + 1792x^5y^3 + 1120x^4y^4 + 448x^3y^5 + 112x^2y^6 + 16xy^7 + y^8$ **39.** 2, 14, 782 **41.** 0, 1

Pages 622–626 Chapter 11 Study Guide and Review

1. partial sum **3.** sigma notation **5.** Binomial Theorem **7.** arithmetic series **9.** 38 **11.** −11 **13.** −3, 1, 5 **15.** 6, 3, 0, −3 **17.** 2322 **19.** −220 **21.** 32 **23.** 3 **25.** 6, 12 **27.** 4, 2, 1, $\frac{1}{2}$ **29.** 1452 **31.** $\frac{14{,}197}{16}$ **33.** 72 **35.** $-\frac{16}{13}$ **37.** 3, 2, −2, −18, −82 **39.** 1, 3, 4, 7, 11 **41.** 10, 66, 458 **43.** −1, 4, −31 **45.** $x^4 - 8x^3 + 24x^2 - 32x + 16$ **47.** $160x^3y^3$

49. Step 1: When $n = 1$, the left side of the given equation is 1. The right side is $2^1 - 1$ or 1, so the equation is true for $n = 1$.

Step 2: Assume $1 + 2 + 4 + \ldots + 2^{k-1} = 2^k - 1$ for some positive integer k.

Step 3:
$$\begin{aligned} 1 + 2 + 4 + \ldots + 2^{k-1} + 2^{(k+1)-1} &= 2^k - 1 + 2k \\ &= 2 \cdot 2^k - 1 \\ &= 2^{k+1} - 1 \end{aligned}$$

The last expression is the right side of the equation to be proved, where $n = k + 1$. Thus, the equation is true for $n = k + 1$.

Therefore, $1 + 2 + 4 + \ldots + 2^{n-1} = 2^n - 1$ for all positive integers n.

Chapter 12 Probability and Statistics

Page 631 Chapter 12 Getting Started

1. $\frac{1}{6}$ **3.** $\frac{1}{2}$ **5.** $\frac{2}{3}$

7.

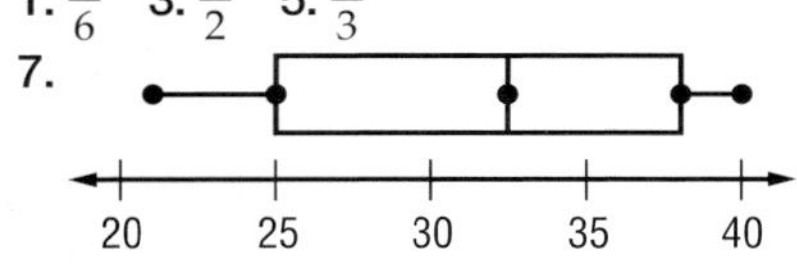

9.

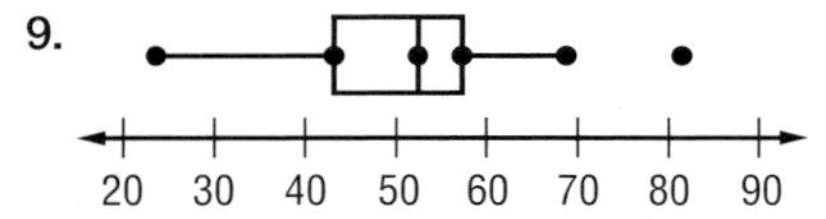

11. 3 **13.** $\sqrt{13}$ **15.** $a^3 + 3a^2b + 3ab^2 + b^3$ **17.** $m^5 - 5m^4n + 10m^3n^2 - 10m^2n^3 + 5mn^4 - n^5$

Pages 634–637 Lesson 12-1

1. HHH, HHT, HTH, HTT, THH, THT, TTH, TTT **3.** The available colors for the car could be different from those for the truck. **5.** dependent **7.** 256 **9.** D **11.** independent **13.** dependent **15.** 16 **17.** 30 **19.** 1024 **21.** 10,080 **23.** 362,880 **25.** 27,216 **27.** 800

29. The maximum number of license plates is a product with factors of 26s and 10s, depending on how many letters are used and how many digits are used. Answers should include the following.

- There are 26 choices for the first letter, 26 for the second, and 26 for the third. There are 10 choices for the first number, 10 for the second, and 10 for the third. By the Fundamental Counting Principle, there are $26^3 \cdot 10^3$ or 17,576,000 possible license plates.
- Replace positions containing numbers with letters.

31. C **33.** 20 mi **35.** $28x^6y^2$ **37.** 7 **39.** $\frac{1}{2}$ **41.** $-\frac{x}{x + 5y}$ **43.** ±1, ±2 **45.** $y = (x - 3)^2 + 2$ **47.** $y = -\frac{1}{2}x^2 + 8$ **49.** 3 **51.** $\frac{1}{7}\begin{bmatrix} 1 & -1 \\ 4 & 3 \end{bmatrix}$ **53.** no inverse exists **55.** $y = \frac{2}{3}x + \frac{1}{3}$ **57.** 30 **59.** 720 **61.** 15 **63.** 1

Pages 641–643 Lesson 12-2

1. Sample answer: There are six people in a contest. How many ways can the first, second, and third prizes be awarded? **3.** Sometimes; the statement is only true when $r = 1$. **5.** 120 **7.** 6 **9.** permutation; 5040 **11.** 84 **13.** 9 **15.** 665,280 **17.** 70 **19.** 210 **21.** 1260 **23.** combination; 28 **25.** permutation; 120 **27.** permutation; 3360 **29.** combination; 455 **31.** 60 **33.** 111,540 **35.** 80,089,128

37. $C(n - 1, r) + C(n - 1, r - 1)$
$$= \frac{(n-1)!}{(n-1-r)!r!} + \frac{(n-1)!}{[n-1-(r-1)]!(r-1)!}$$
$$= \frac{(n-1)!}{(n-r-1)!r!} + \frac{(n-1)!}{(n-r)!(r-1)!}$$
$$= \frac{(n-1)!}{(n-r-1)!r!} \cdot \frac{n-r}{n-r} + \frac{(n-1)!}{(n-r)!(r-1)!} \cdot \frac{r}{r}$$
$$= \frac{(n-1)!(n-r)}{(n-r)!r!} + \frac{(n-1)!r}{(n-r)!r!}$$
$$= \frac{(n-1)!(n-r+r)}{(n-r)!r!}$$
$$= \frac{(n-1)!n}{(n-r)!r!}$$
$$= \frac{n!}{(n-r)!r!}$$
$$= C(n, r)$$

39. D **41.** 24 **43.** 120 **45.** 80 **47.** Sample answer: $n = 2$ **49.** $x > 0.8047$ **51.** 20 days **53.** $\frac{(y-4)^2}{9} + \frac{(x-4)^2}{4} = 1$ **55.** −4; 128 **57.** {−2, 5} **59.** $8\sqrt{2}$ **61.** $4\sqrt{5}$ **63.** (0, 2) **65.** $-\frac{6}{7}$ **67.** {−7, 15} **69.** $\frac{3}{5}$ **71.** $\frac{1}{5}$

Pages 647–650 Lesson 12-3

1. Sample answer: The event *July comes before June* has a probability of 0. The event *June comes before July* has a probability of 1. **3.** There are $6 \cdot 6$ or 36 possible outcomes for the two dice. Only 1 outcome, 1 and 1, results in a sum of 2, so $P(2) = \frac{1}{36}$. There are 2 outcomes, 1 and 2 as well as 2 and 1, that result in a sum of 3, so $P(3) = \frac{2}{36}$ or $\frac{1}{18}$. **5.** $\frac{2}{7}$ **7.** 8:1 **9.** 2:7 **11.** $\frac{10}{11}$ **13.** $\frac{1}{8}$ **15.** $\frac{1}{10}$ **17.** $\frac{2}{25}$ **19.** $\frac{6}{55}$ **21.** $\frac{28}{55}$ **23.** $\frac{11}{115}$ **25.** $\frac{6}{115}$ **27.** $\frac{24}{115}$ **29.** 0 **31.** 0.007 **33.** 0.109 **35.** 3:5 **37.** 5:3 **39.** 1:4 **41.** 3:1 **43.** $\frac{3}{10}$ **45.** $\frac{4}{9}$ **47.** $\frac{1}{9}$ **49.** $\frac{3}{5}$ **51.** 2:23 **53.** 1:4 **55.** $\frac{1}{20}$ **57.** $\frac{9}{20}$ **59.** $\frac{9}{20}$ **61.** $\frac{1}{120}$ **63.** Probability and odds are good tools for assessing risk. Answers should include the following.

- $P(\text{struck by lightning}) = \frac{s}{s+f} = \frac{1}{750{,}000}$, so Odds = 1:(750,000 − 1) or 1:749,999. $P(\text{surviving a lightning strike}) = \frac{s}{s+f} = \frac{3}{4}$, so Odds = 3:(4 − 3) or 3:1.
- In this case, success is being struck by lightning or surviving the lightning strike. Failure is not being struck by lightning or not surviving the lightning strike.

65. D **67.** experimental; about 0.307 **69.** theoretical; $\frac{1}{17}$ **71.** permutation; 1260 **73.** 16 **75.** direct variation **77.** (4, 4) **79.** $\frac{6}{35}$ **81.** $\frac{1}{4}$ **83.** $\frac{9}{20}$

Page 650 Practice Quiz 1

1. 24 **3.** 18,720 **5.** 56 **7.** combination; 20,358,520 **9.** $\frac{13}{102}$

Pages 654–657 Lesson 12-4

1. Sample answer: putting on your socks, and then your shoes **3.** Mario; the probabilities of rolling a 4 and rolling

a 2 are both $\frac{1}{6}$. **5.** $\frac{1}{4}$ **7.** $\frac{4}{663}$ **9.** $\frac{1}{4}$ **11.** dependent; $\frac{21}{220}$
13. $\frac{1}{12}$ **15.** $\frac{25}{36}$ **17.** $\frac{1}{6}$ **19.** $\frac{5}{6}$ **21.** $\frac{1}{49}$ **23.** $\frac{10}{21}$ **25.** 0 **27.** $\frac{2}{15}$
29. $\frac{2}{15}$ **31.** independent; $\frac{25}{81}$ **33.** dependent; $\frac{1}{21}$
35. dependent; $\frac{81}{2401}$
37.

		First Spin: Blue $\frac{1}{3}$	First Spin: Yellow $\frac{1}{3}$	First Spin: Red $\frac{1}{3}$
Second Spin	Blue $\frac{1}{3}$	BB $\frac{1}{9}$	BY $\frac{1}{9}$	BR $\frac{1}{9}$
	Yellow $\frac{1}{3}$	YB $\frac{1}{9}$	YY $\frac{1}{9}$	YR $\frac{1}{9}$
	Red $\frac{1}{3}$	RB $\frac{1}{9}$	RY $\frac{1}{9}$	RR $\frac{1}{9}$

39. $\frac{1}{3}$ **41.** $\frac{19}{1{,}160{,}054}$ **43.** $\frac{6327}{20{,}825}$ **45.** about 4.87% **47.** no
49. Sample answer: As the number of trials increases, the results become more reliable. However, you cannot be absolutely certain that there are no black marbles in the bag without looking at all of the marbles. **51.** Probability can be used to analyze the chances of a player making 0, 1, or 2 free throws when he or she goes to the foul line to shoot 2 free throws. Answers should include the following.

- One of the decimals in the table could be used as the value of p, the probability that a player makes a given free throw. The probability that a player misses both free throws is $(1 - p)(1 - p)$ or $(1 - p)^2$. The probability that a player makes both free throws is $p \cdot p$ or p^2. Since the sum of the probabilities of all the possible outcomes is 1, the probability that a player makes exactly 1 of the 2 free throws is $1 - (1 - p)^2 - p^2$ or $2p(1 - p)$.
- The result of the first free throw could affect the player's confidence on the second free throw. For example, if the player makes the first free throw, the probability of he or she making the second free throw might increase. Or, if the player misses the first free throw, the probability that he or she makes the second free throw might decrease.

53. C **55.** $\frac{3}{340}$ **57.** 1440 ways **59.** 36 **61.** x, $x - 4$

63.

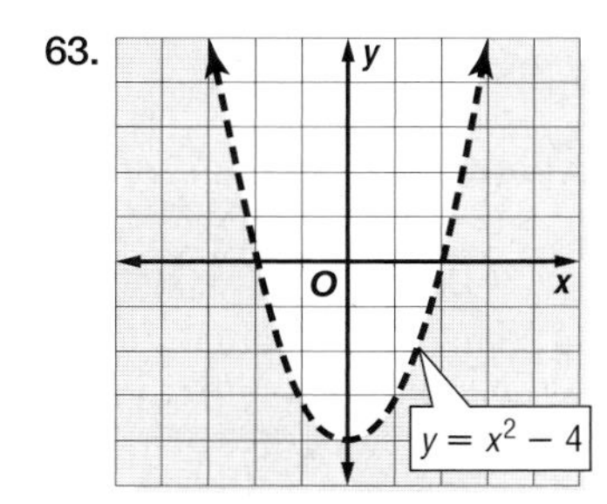

65. 153 **67.** $|b|$ **69.** (1, 2)
71. (−2, 4) **73.** $\frac{5}{6}$ **75.** $\frac{11}{12}$
77. $1\frac{5}{12}$

Pages 660–663 Lesson 12-5
1. Sample answer: mutually exclusive events: tossing a coin and rolling a die; inclusive events: drawing a 7 and a diamond from a standard deck of cards **3.** The events are not mutually exclusive, so the chance of rain is less than 100%. **5.** $\frac{1}{3}$ **7.** $\frac{1}{2}$ **9.** $\frac{2}{3}$ **11.** inclusive; $\frac{4}{13}$ **13.** $\frac{5}{6}$
15. $\frac{25}{42}$ **17.** $\frac{35}{143}$ **19.** $\frac{3}{143}$ **21.** $\frac{38}{143}$ **23.** mutually exclusive; $\frac{7}{9}$ **25.** inclusive; $\frac{21}{34}$ **27.** $\frac{4}{13}$ **29.** $\frac{55}{221}$ **31.** $\frac{188}{663}$

33. $\frac{1}{8}$ **35.** $\frac{1}{4}$ **37.** $\frac{1}{780}$ **39.** $\frac{9}{130}$ **41.** $\frac{11}{780}$ **43.** $\frac{3}{5}$ **45.** $\frac{17}{27}$

47. Subtracting $P(A \text{ and } B)$ from each side and adding $P(A \text{ or } B)$ to each side results in the equation $P(A \text{ or } B) = P(A) + P(B) - P(A \text{ and } B)$. This is the equation for the probability of inclusive events. If A and B are mutually exclusive, then $P(A \text{ and } B) = 0$, so the equation simplifies to $P(A \text{ or } B) = P(A) + P(B)$, which is the equation for the probability of mutually exclusive events. Therefore, the equation is correct in either case. **49.** C **51.** $\frac{1}{216}$ **53.** $\frac{1}{216}$

55. 4:1 **57.** 2:5 **59.** 254 **61.** (±8, −10) **63.** $(x + 1)^2(x - 1)(x^2 + 1)$ **65.** min: (−0.42, 0.62); max: (−1.58, 1.38)
67.

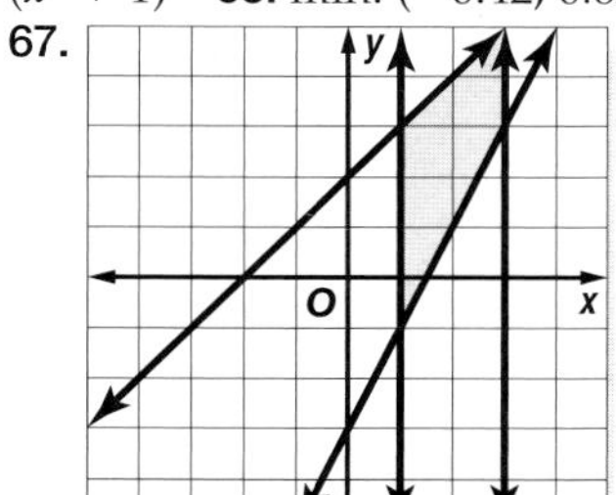

(1, 3), (1, −1), (3, 3), (3, 5); max: $f(3, 5) = 23$; min: $f(1, -1) = -3$
69. direct variation
71. 35.4, 34, no mode, 72
73. 63.75, 65, 50 and 65, 30
75. 12.98, 12.9, no mode, 4.7

Pages 666–670 Lesson 12-6
1. Sample answer: {10, 10, 10, 10, 10, 10}
3. $\sigma = \sqrt{\frac{1}{n}\sum_{i=1}^{n}(x_i - \bar{x})^2}$ **5.** 8.3, 2.9 **7.** \$7300.50, \$5335.25
9. 2500, 50 **11.** 3.1, 1.7 **13.** 37,691.2, 194.1 **15.** 82.9, 9.1
17. 77.7; 32; 19 **19.** Mean; it is highest. **21.** \$1047.88, \$1049.50, \$695 **23.** Mean or median; they are nearly equal and are more representative of the prices than the mode.
25. Mode; it is lowest. **27.** 19.3 **29.** 19.5 **31.** 59.8, 7.7
33. 100% **35.** Sample answer: The first graph might be used by a sales manager to show a salesperson that he or she does not deserve a big raise. It appears that sales are steady but not increasing fast enough to warrant a big raise.
37. A: 2.5, 2.5, 0.7, 0.8; B: 2.5, 2.5, 1.1, 1.0
39. The statistic(s) that best represent a set of test scores depends on the distribution of the particular set of scores. Answers should include the following.

- mean, 73.9; median, 76.5; mode, 94
- The mode is not representative at all because it is the highest score. The median is more representative than the mean because it is influenced less than the mean by the two very low scores of 34 and 19.
- Each measure is increased by 5.

41. D **43.** 1.9 **45.** inclusive; $\frac{4}{13}$ **47.** $\frac{1}{169}$ **49.** $\frac{13}{204}$
51. (0, ±9); $(0, \pm\sqrt{106})$; $\pm\frac{9}{5}$ **53.** 17 **55.** 12 cm³ **57.** (1, 5)
59. 136 **61.** 380 **63.** 396

Page 670 Practice Quiz 2
1. $\frac{3}{20}$ **3.** $\frac{2}{9}$ **5.** $\frac{1}{6}$ **7.** $\frac{3}{4}$ **9.** 23.6, 4.9

Pages 673–675 Lesson 12-7
1. Sample answer:

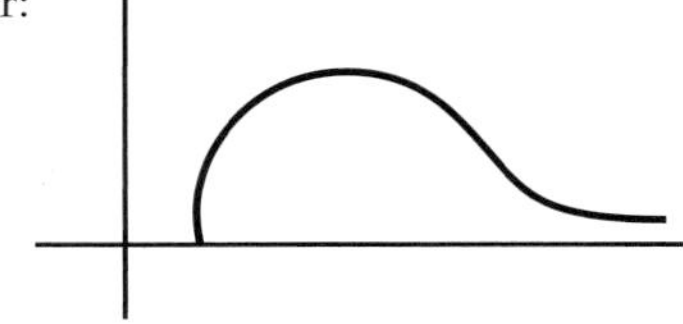

The use of cassettes since CDs were introduced.

Selected Answers

3. Since 99% of the data is within 3 standard deviations of the mean, 1% of the data is more than 3 standard deviations from the mean. By symmetry, half of this, or 0.5%, is more than 3 standard deviations above the mean. **5.** 68% **7.** 95% **9.** 250 **11.** 81.5% **13.** normally distributed **15.** 68% **17.** 0.5% **19.** 50% **21.** 95% **23.** 815 **25.** 16% **27.** The mean would increase by 25; the standard deviation would not change; and the graph would be translated 25 units to the right. **29.** A **31.** 17.5, 4.2 **33.** $\frac{2}{13}$ **35.** $\frac{4}{13}$ **37.** $-3, 2, 4$ **39.** $\frac{1}{4}$, 1 **41.** 0.76 h **43.** $56c^5d^3$

Pages 678–680 Lesson 12-8

1. Sample answer: In a 5-card hand, what is the probability that at least 2 cards are hearts? **3a.** Each trial has more than two possible outcomes. **3b.** The number of trials is not fixed. **3c.** The trials are not independent. **5.** $\frac{1}{8}$ **7.** $\frac{1}{28,561}$ **9.** $\frac{27,648}{28,561}$ **11.** about 0.37 **13.** $\frac{1}{16}$ **15.** $\frac{1}{4}$ **17.** $\frac{11}{16}$ **19.** $\frac{125}{3888}$ **21.** $\frac{23}{648}$ **23.** $\frac{1}{1024}$ **25.** $\frac{135}{512}$ **27.** $\frac{53}{512}$ **29.** $\frac{105}{512}$ **31.** $\frac{319}{512}$ **33.** about 0.44 **35.** about 0.32 **37.** $\frac{7}{32}$

39. Getting a right answer and a wrong answer are the outcomes of a binomial experiment. The probability is far greater that guessing will result in a low grade than in a high grade. Answers should include the following.
- Use $(r + w)^5 = r^5 + 5r^4w + 10r^3w^2 + 10r^2w^3 + 5rw^4 + w^5$ and the chart on page 48 to determine the probabilities of each combination of right and wrong.
- P(5 right): $r^5 = \left(\frac{1}{4}\right)^5 = \frac{1}{1024}$ or about 0.098%;

  P(4 right, 1 wrong): $\frac{15}{1024}$ or about 1.5%;

  P(3 right, 2 wrong): $10r^3w^2 = 10\left(\frac{1}{4}\right)^3\left(\frac{3}{4}\right)^2 = \frac{45}{512}$ or about 8.8%; P(3 wrong, 2 right): $10r^2w^3 = 10\left(\frac{1}{4}\right)^2\left(\frac{3}{4}\right)^3 = \frac{135}{512}$ or about 26.4%; P(4 wrong, 1 right): $5rw^4 = 5\left(\frac{1}{4}\right)\left(\frac{3}{4}\right)^4 = \frac{405}{1024}$ or about 39.6%; P(5 wrong): $w^5 = \left(\frac{3}{4}\right)^5 = \frac{243}{1024}$ or about 23.7%.

41. B **43.** normal distribution **45.** 10
47. Mean; it is highest.
49.

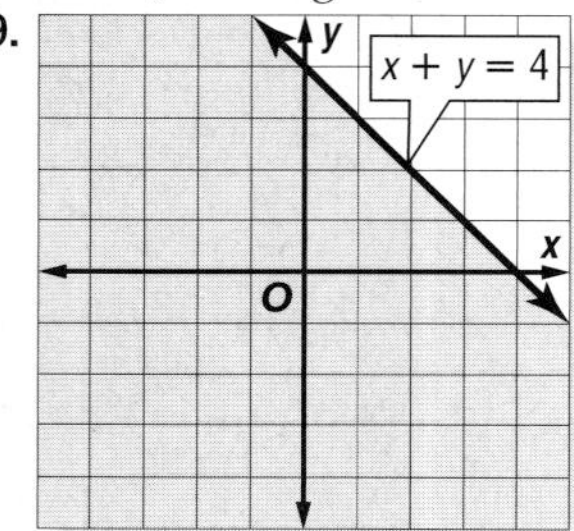

51. 0.1 **53.** 0.039 **55.** 0.041

Pages 684–685 Lesson 12-9

1. Sample answer: If a sample is not random, the results of a survey may not be valid. **3.** The margin of sampling error decreases when the size of the sample n increases. As n increases, $\frac{p(1-p)}{n}$ decreases. **5.** No; these students probably study more than average. **7.** about 4% **9.** The probability is 0.95 that the percent of Americans ages 12 and older who listen to the radio every day is between 72% and 82% **11.** No; you would tend to point toward the middle of the page. **13.** Yes; a wide variety of people would be called since almost everyone has a phone. **15.** about 8% **17.** about 4% **19.** about 3% **21.** about 4% **23.** about 3% **25.** about 2% **27.** about 983 **29.** A political candidate can use the statistics from an opinion poll to analyze his or her standing and to help plan the rest of the campaign. Answers should include the following.
- The candidate could decide to skip areas where he or she is way ahead or way behind, and concentrate on areas where the polls indicate the race is close.
- about 3.5%
- The margin of error indicates that with a probability of 0.95 the percent of the Florida population that favored Bush was between 43.5% and 50.5%. The margin of error for Gore was also about 3.5%, so with probability 0.95 the percent that favored Gore was between 40.5% and 47.5%. Therefore, it was possible that the percent of the Florida population that favored Bush was less than the percent that favored Gore.

31. C **33.** $\frac{5}{32}$ **35.** 95% **37.** 97.5%

Pages 687–692 Chapter 12 Study Guide and Review

1. c **3.** a **5.** d **7.** f **9.** 5040 codes **11.** 4 **13.** 1:3 **15.** 7:5 **17.** 2:3 **19.** independent; $\frac{1}{36}$ **21.** dependent; $\frac{1}{7}$ **23.** mutually exclusive; $\frac{2}{3}$ **25.** inclusive; $\frac{7}{13}$ **27.** 341.0, 18.5 **29.** 3400 **31.** 800 **33.** $\frac{1}{32}$ **35.** $\frac{1}{2,176,782,336}$ **37.** $\frac{14,437,500}{2,176,782,336}$ **39.** 460 mothers

Chapter 13 Trigonometric Functions

Page 699 Chapter 13 Getting Started

1. 10 **3.** 16.7 **5.** $x = 7, y = 7\sqrt{2}$ **7.** $x = 4\sqrt{3}, y = 8$
9. $f^{-1}(x) = x - 3$ **11.** $f^{-1}(x) = \pm\sqrt{x+4}$

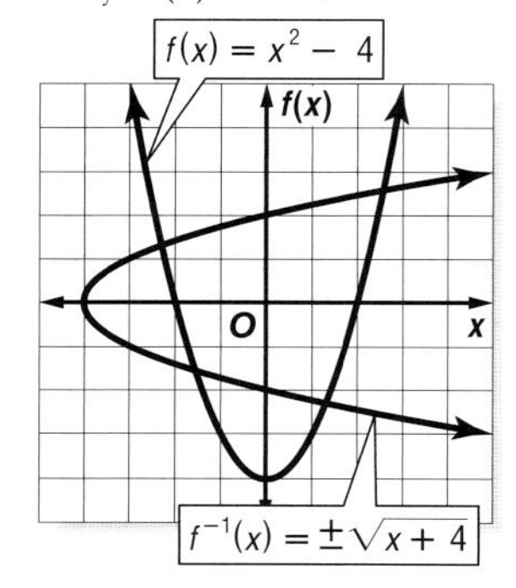

Pages 706–708 Lesson 13-1

1. Trigonometry is the study of the relationships between the angles and sides of a right triangle. **3.** Given only the measures of the angles of a right triangle, you cannot find the measures of its sides. **5.** $\sin\theta = \frac{\sqrt{85}}{11}$; $\cos\theta = \frac{6}{11}$; $\tan\theta = \frac{\sqrt{85}}{6}$; $\csc\theta = \frac{11\sqrt{85}}{85}$; $\sec\theta = \frac{11}{6}$; $\cot\theta = \frac{6\sqrt{85}}{85}$

7. $\cos 23° = \frac{32}{x}$; $x \approx 34.8$ **9.** $B = 45°, a = 6, c \approx 8.5$ **11.** $a \approx 16.6, A \approx 67°, B \approx 23°$ **13.** 1660 ft **15.** $\sin\theta = \frac{4}{11}$; $\cos\theta = \frac{\sqrt{105}}{11}$; $\tan\theta = \frac{4\sqrt{105}}{105}$; $\csc\theta = \frac{11}{4}$; $\sec\theta = \frac{11\sqrt{105}}{105}$; $\cot\theta = \frac{\sqrt{105}}{4}$ **17.** $\sin\theta = \frac{\sqrt{7}}{4}$; $\cos\theta = \frac{3}{4}$; $\tan\theta = \frac{\sqrt{7}}{3}$; $\csc\theta = \frac{4\sqrt{7}}{7}$; $\sec\theta = \frac{4}{3}$; $\cot\theta = \frac{3\sqrt{7}}{7}$ **19.** $\sin\theta = \frac{\sqrt{5}}{5}$;

$\cos\theta = \frac{2\sqrt{5}}{5}$; $\tan\theta = \frac{1}{2}$; $\csc\theta = \sqrt{5}$; $\sec\theta = \frac{\sqrt{5}}{2}$; $\cot\theta = 2$ **21.** $\tan 30° = \frac{x}{10}$, $x \approx 5.8$ **23.** $\sin 54° = \frac{17.8}{x}$, $x \approx 22.0$ **25.** $\cos x° = \frac{15}{36}$, $x \approx 65$

27a. $\sin 30° = \frac{\text{opp}}{\text{hyp}}$ *sine ratio*

$\sin 30° = \frac{x}{2x}$ *Replace opp with x and hyp with 2x.*

$\sin 30° = \frac{1}{2}$ *Simplify.*

27b. $\cos 30° = \frac{\text{adj}}{\text{hyp}}$ *cosine ratio*

$\cos 30° = \frac{\sqrt{3}x}{2x}$ *Replace adj with $\sqrt{3}x$ and hyp with 2x.*

$\cos 30° = \frac{\sqrt{3}}{2}$ *Simplify.*

27c. $\sin 60° = \frac{\text{opp}}{\text{hyp}}$ *sine ratio*

$\sin 60° = \frac{\sqrt{3}x}{2x}$ *Replace opp with $\sqrt{3}x$ and hyp with 2x.*

$\sin 60° = \frac{\sqrt{3}}{2}$ *Simplify.*

29. $B = 74°, a \approx 3.9, b \approx 13.5$ **31.** $B = 56°, b \approx 14.8, c \approx 17.9$ **33.** $A = 60°, a \approx 19.1, c = 22$ **35.** $A = 72°, b \approx 1.3, c \approx 4.1$ **37.** $A \approx 63°, B \approx 27°, a \approx 11.5$ **39.** $A \approx 49°, B \approx 41°, a = 8, c \approx 10.6$ **41.** about 300 ft **43.** about 6° **45.** 93.53 units^2 **47.** The sine and cosine ratios of acute angles of right triangles each have the longest measure of the triangle, the hypotenuse, as their denominator. A fraction whose denominator is greater than its numerator is less than 1. The tangent ratio of an acute angle of a right triangle does not involve the measure of the hypotenuse, $\frac{\text{opp}}{\text{adj}}$. If the measure of the opposite side is greater than the measure of the adjacent side, the tangent ratio is greater than 1. If the measure of the opposite side is less than the measure of the adjacent side, the tangent ratio is less than 1. **49.** C **51.** No; band members may be more likely to like the same kinds of music. **53.** $\frac{3}{8}$ **55.** $\frac{15}{16}$ **57.** $\{-2, -1, 0, 1, 2\}$ **59.** 20 qt **61.** 12 m^2

Pages 712–715 Lesson 13-2

1. reals

3.

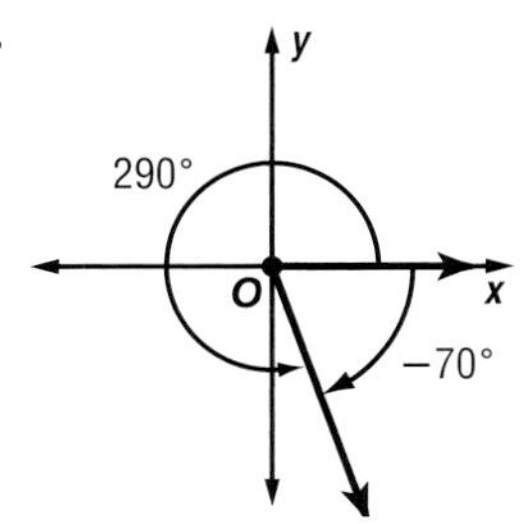

5.

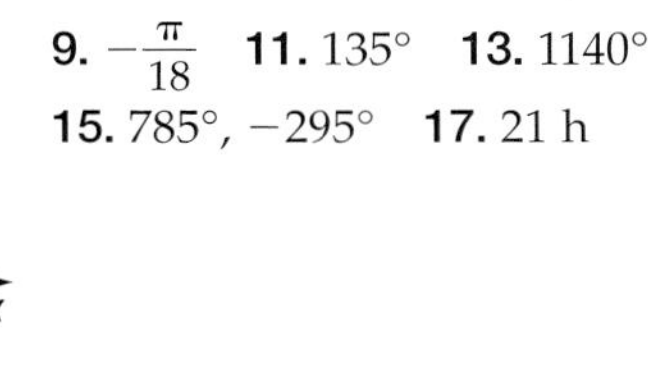

7.

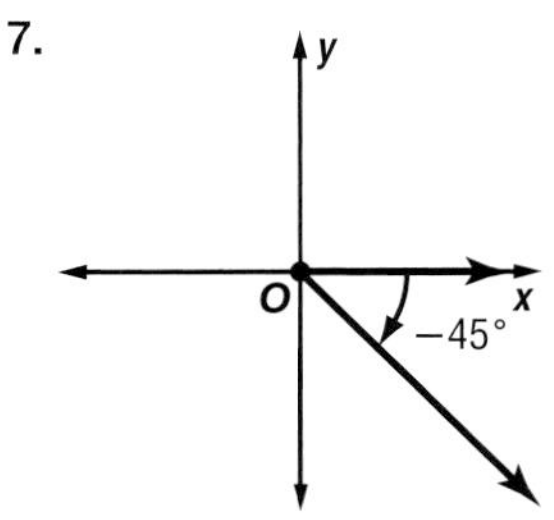

9. $-\frac{\pi}{18}$ **11.** 135° **13.** 1140° **15.** 785°, −295° **17.** 21 h

19.

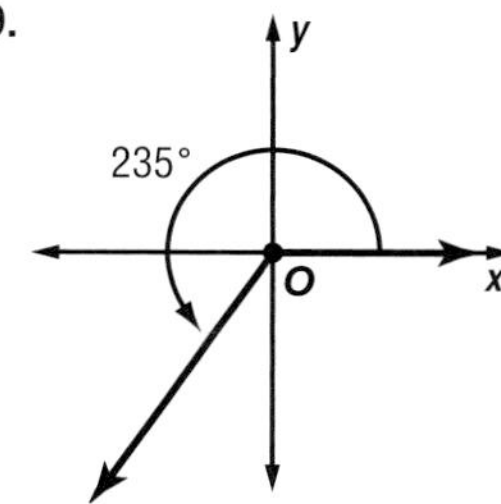

21. (790°)

23.

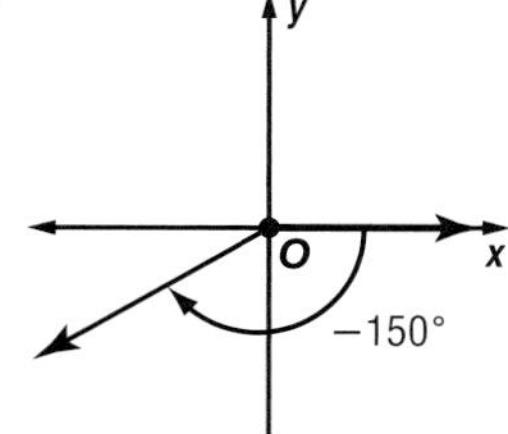

25. (π)

27. $\frac{2\pi}{3}$ **29.** $-\frac{\pi}{12}$ **31.** $\frac{11\pi}{3}$ **33.** $\frac{79\pi}{90}$ **35.** 150° **37.** −45° **39.** 1305° **41.** $\frac{1620}{\pi} \approx 515.7°$ **43.** Sample answer: 585°, −135° **45.** Sample answer: 345°, −375° **47.** Sample answer: 8°, −352° **49.** Sample answer: $\frac{11\pi}{4}, -\frac{5\pi}{4}$ **51.** Sample answer: $\frac{3\pi}{4}, -\frac{13\pi}{4}$ **53.** Sample answer: $\frac{13\pi}{2}, -\frac{3\pi}{2}$ **55.** 2689° per second; 47 radians per second **57.** about 188.5 m^2 **59.** about 640.88 in^2 **61.** Student answers should include the following.

- An angle with a measure of more than 180° gives an indication of motion in a circular path that ended at a point more than halfway around the circle from where it started.
- Negative angles convey the same meaning as positive angles, but in an opposite direction. The standard convention is that negative angles represent rotations in a clockwise direction.
- Rates over 360° per minute indicate that an object is rotating or revolving more than one revolution per minute.

63. D **65.** $A = 22°, a \approx 5.9, c \approx 15.9$ **67.** $c = 0.8, A = 30°, B = 60°$ **69.** about 7.07% **71.** combination, 35 **73.** $[g \circ h](x) = 4x^2 - 6x + 23$, $[h \circ g](x) = 8x^2 + 34x + 44$ **75.** 1418.2 or about 1418; the number of sports radio stations in 2008 **77.** $\frac{3\sqrt{5}}{5}$ **79.** $\frac{\sqrt{10}}{2}$ **81.** $\frac{\sqrt{10}}{4}$

Page 715 Practice Quiz 1

1. $B = 42°, a \approx 13.3, c \approx 17.9$

3. (−60°)

5. $\frac{19\pi}{18}$ **7.** 210° **9.** 305°; −415°

Pages 722–724 Lesson 13-3

1. False; $\sec 0° = \frac{r}{r}$ or 1 and $\tan 0° = \frac{0}{r}$ or 0.
3. To find the value of a trigonometric function of θ, where θ is greater than 90°, find the value of the trigonometric function for θ', then use the quadrant in which the terminal

Selected Answers

side of θ lies to determine the sign of the trigonometric function value of θ. **5.** $\sin\theta = 0$, $\cos\theta = -1$, $\tan\theta = 0$, $\csc\theta$ = undefined, $\sec\theta = -1$, $\cot\theta$ = undefined

7. 55°

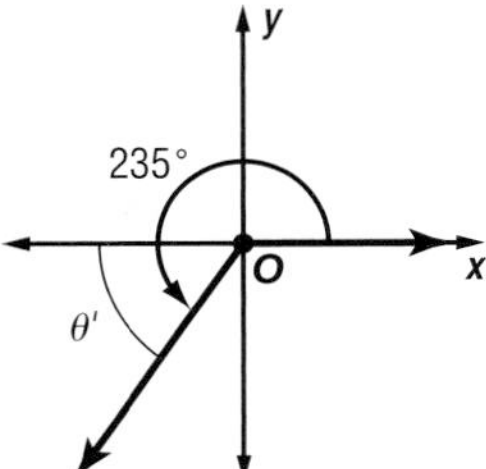

9. 60º

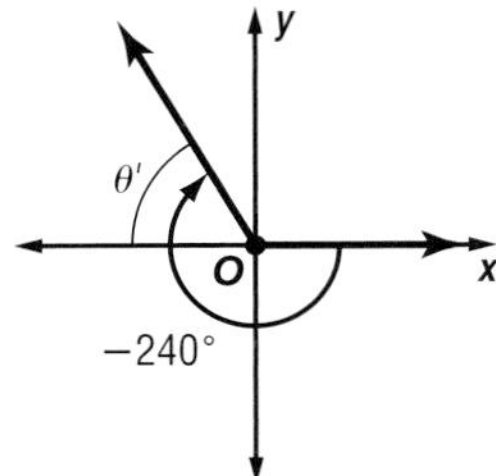

11. −1 **13.** $-\frac{2\sqrt{3}}{3}$ **15.** $\sin\theta = -\frac{\sqrt{6}}{3}$, $\cos\theta = \frac{\sqrt{3}}{3}$, $\tan\theta = -\sqrt{2}$, $\csc\theta = -\frac{\sqrt{6}}{2}$, $\sec\theta = \sqrt{3}$ **17.** $\sin\theta = \frac{24}{25}$, $\cos\theta = \frac{7}{25}$, $\tan\theta = \frac{24}{7}$, $\csc\theta = \frac{25}{24}$, $\sec\theta = \frac{25}{7}$, $\cot\theta = \frac{7}{24}$ **19.** $\sin\theta = -\frac{8\sqrt{89}}{89}$, $\cos\theta = \frac{5\sqrt{89}}{89}$, $\tan\theta = -\frac{8}{5}$, $\csc\theta = -\frac{\sqrt{89}}{8}$, $\sec\theta = \frac{\sqrt{89}}{5}$, $\cot\theta = -\frac{5}{8}$ **21.** $\sin\theta = -1$, $\cos\theta = 0$, $\tan\theta$ = undefined, $\csc\theta = -1$, $\sec\theta$ = undefined, $\cot\theta = 0$ **23.** $\sin\theta = -\frac{\sqrt{2}}{2}$, $\cos\theta = \frac{\sqrt{2}}{2}$, $\tan\theta = -1$, $\csc\theta = -\sqrt{2}$, $\sec\theta = \sqrt{2}$, $\cot\theta = -1$

25. 45°

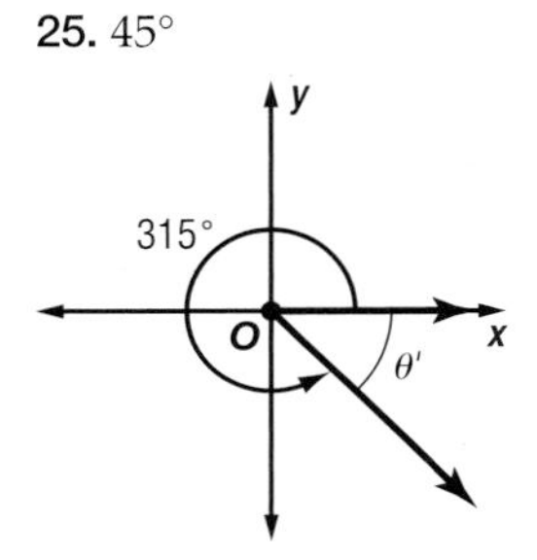

27. 30°

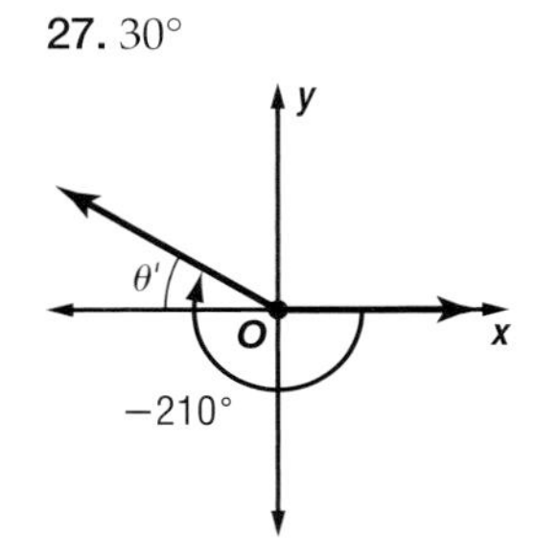

29. $\frac{\pi}{4}$

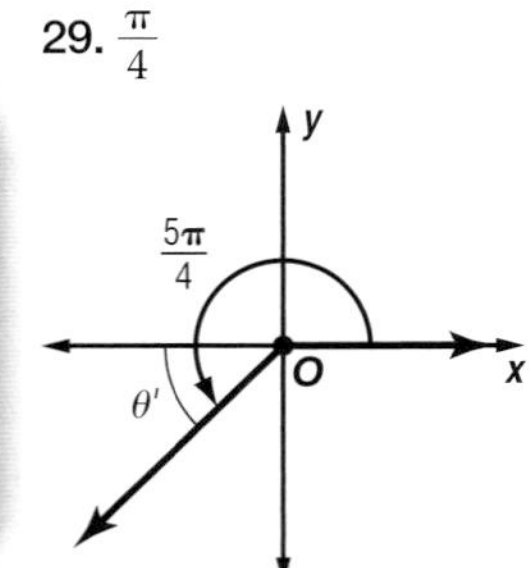

31. $\frac{\pi}{7}$

33. $-\frac{\sqrt{3}}{2}$ **35.** $-\sqrt{3}$ **37.** undefined **39.** $\sqrt{3}$ **41.** undefined **43.** $\frac{\sqrt{3}}{2}$ **45.** 0.2, 0, −0.2, 0, 0.2, 0, and −0.2; or about 11.5°, 0°, −11.5°, 0°, 11.5°, 0°, and −11.5° **47.** $\sin\theta = -\frac{4}{5}$, $\tan\theta = -\frac{4}{3}$, $\csc\theta = -\frac{5}{4}$, $\sec\theta = \frac{5}{3}$, $\cot\theta = -\frac{3}{4}$ **49.** $\cos\theta = -\frac{2\sqrt{2}}{3}$, $\tan\theta = -\frac{\sqrt{2}}{4}$, $\csc\theta = 3$, $\sec\theta = -\frac{3\sqrt{2}}{4}$, $\cot\theta = -2\sqrt{2}$ **51.** $\sin\theta = -\frac{3\sqrt{10}}{10}$, $\cos\theta = -\frac{\sqrt{10}}{10}$, $\tan\theta = 3$, $\csc\theta = -\frac{\sqrt{10}}{3}$, $\cot\theta = \frac{1}{3}$

53. about 173.2 ft **55.** 9 meters **57.** II

59. Answers should include the following.

- The cosine of any angle is defined as $\frac{x}{r}$, where x is the x-coordinate of any point on the terminal ray of the angle and r is the distance from the origin to that point. This means that for angles with terminal sides to the left of the y-axis, the cosine is negative, and those with terminal sides to the right of the y-axis, the cosine is positive. Therefore the cosine function can be used to model real-world data that oscillate between being positive and negative.
- If we knew the length of the cable we could find the vertical distance from the top of the tower to the rider. Then if we knew the height of the tower we could subtract from it the vertical distance calculated previously. This will leave the height of the rider from the ground.

61. $\left(\frac{5}{2}, -\frac{5\sqrt{3}}{2}\right)$ **63.** 300° **65.** $\sin 28° = \frac{x}{12}$, 5.6 **67.** $\sin x° = \frac{5}{13}$, 23 **69.** (7, 2) **71.** (5, −4) **73.** 15.1 **75.** 32.9° **77.** 39.6°

Pages 729–732 Lesson 13-4

1. Sometimes; only when A is acute, $a = b \sin A$ or $a > b$ and when A is obtuse, $a > b$.

3. Gabe;

$$\frac{\sin 64°}{15} = \frac{\sin B}{8}$$
$$\sin B = \frac{8 \sin 64°}{15}$$
$$B \approx 28.6°$$

$$m\angle C \approx 180° - (64° + 28.6°) \approx 87.4°$$
$$\text{Area} = \frac{1}{2}ab \sin C \approx \frac{1}{2}(15)(8) \sin 87.4° \approx 59.9 \text{ m}^2$$

5. 6.4 cm^2 **7.** $B = 80°$, $a \approx 32.0$, $b \approx 32.6$ **9.** no solution **11.** one; $B \approx 24°$, $C \approx 101°$, $c \approx 12.0$ **13.** 5.5 m **15.** 19.5 yd^2 **17.** 62.4 cm^2 **19.** 14.6 mi^2 **21.** $C = 73°$, $a \approx 55.6$, $b \approx 48.2$ **23.** $B \approx 46°$, $C \approx 69°$, $c \approx 5.1$ **25.** $A \approx 40°$, $B \approx 65°$, $b \approx 2.8$ **27.** $A = 20°$, $a \approx 22.1$, $c \approx 39.8$ **29.** one; $B \approx 36°$, $C \approx 45°$, $c \approx 1.8$ **31.** no **33.** one; $B \approx 18°$, $C \approx 101°$, $c \approx 25.8$ **35.** two; $B \approx 85°$, $C \approx 15°$, $c \approx 2.4$; $B \approx 95°$, $C \approx 5°$, $c \approx 0.8$ **37.** two; $B \approx 65°$, $C \approx 68°$, $c \approx 84.9$; $B \approx 115°$, $C \approx 18°$, $c \approx 28.3$

39. 7.5 mi from Ranger B, 10.9 mi from Ranger A

41. 107 mph **43.** Answers should include the following.

- If the height of the triangle is not given, but the measure of two sides and their included angle are given, then the formula for the area of a triangle using the sine function should be used.
- You might use this formula to find the area of a triangular piece of land, since it might be easier to measure two sides and use surveying equipment to measure the included angle than to measure the perpendicular distance from one vertex to its opposite side.
- The area of $\triangle ABC$ is $\frac{1}{2}ah$.

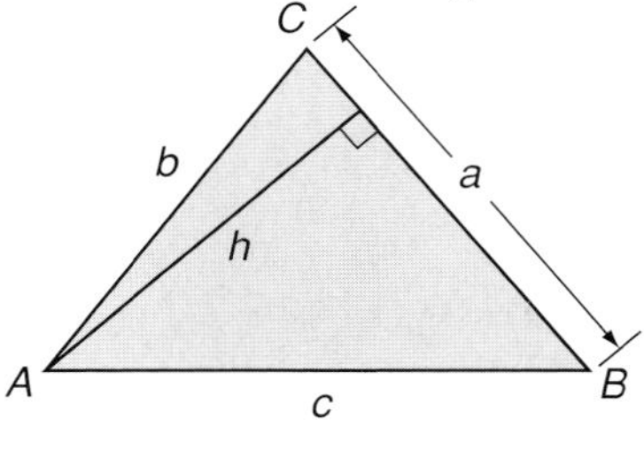

$\sin B = \frac{h}{c}$ or $h = c \sin B$

Selected Answers

• Area $= \frac{1}{2}ah$ or Area $= \frac{1}{2}a(c \sin B)$

45. $B = 78°$, $a \approx 50.1$, $c \approx 56.1$ **47.** $\frac{\sqrt{3}}{3}$ **49.** 660°, −60°
51. $\frac{17\pi}{6}$, $-\frac{7\pi}{6}$ **53.** $\frac{55}{221}$ **55.** 5.6 **57.** 39.4°

Pages 735–738 Lesson 13-5

1. Mateo; the angle given is not between the two sides; therefore the Law of Sines should be used.
3. Sample answer:

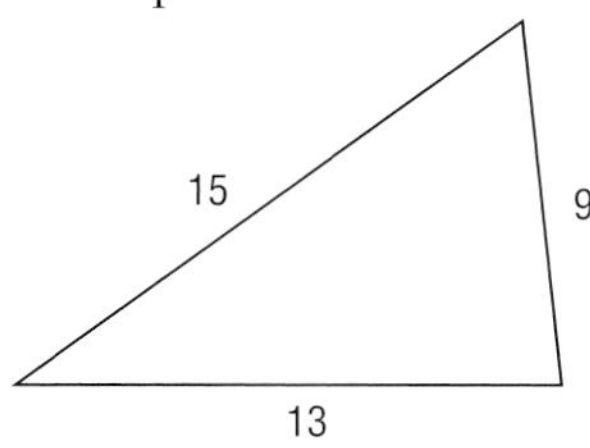

5. sines; $B = 70°$, $a \approx 9.6$, $b = 14$ **7.** cosines; $A \approx 23°$, $B \approx 67°$, $C = 90°$ **9.** 94.3° **11.** cosines; $A \approx 48°$, $B \approx 63°$, $C \approx 70°$ **13.** sines; $B \approx 102°$, $C \approx 44°$, $b \approx 21.0$
15. $A = 80°$, $a \approx 10.9$, $c \approx 5.4$ **17.** cosines; $A \approx 30°$, $B \approx 110°$, $C \approx 40°$ **19.** sines; $C \approx 77°$, $b \approx 31.7$, $c \approx 31.6$
21. no **23.** cosines; $A \approx 52°$, $C \approx 109°$, $b \approx 21.0$
25. cosines; $A \approx 24°$, $B \approx 125°$, $C \approx 31°$ **27.** sines; $B \approx 49°$, $C \approx 91°$, $c \approx 9.3$ **29.** about 100.1° **31.** 4.4 cm, 9.0 cm **33.** 91.6°
35. Answers should include the following.
• The Law of Cosines can be used when you know all three sides of a triangle or when you know two sides and the included angle. It can even be used with two sides and the nonincluded angle. This set of conditions leaves a quadratic equation to be solved. It may have one, two, or no solution just like the SSA case with the Law of Sines.
• Given the latitude of a point on the surface of Earth, you can use the radius of the Earth and the orbiting height of a satellite in geosynchronous orbit to create a triangle. This triangle will have two known sides and the measure of the included angle. Find the third side using the Law of Cosines and then use the Law of Sines to determine the angles of the triangle. Subtract 90 degrees from the angle with its vertex on Earth's surface to find the angle at which to aim the receiver dish.
37. A **39.** Sample answer: 100.2° **41.** one; $B \approx 46°$, $C \approx 79°$, $c \approx 9.6$ **43.** $\sin\theta = \frac{12}{13}$, $\cos\theta = \frac{5}{13}$, $\tan\theta = \frac{12}{5}$, $\csc\theta = \frac{13}{12}$, $\sec\theta = \frac{13}{5}$, $\cot\theta = \frac{5}{12}$ **45.** $\sin\theta = \frac{\sqrt{6}}{4}$, $\cos\theta = \frac{\sqrt{10}}{4}$, $\tan\theta = \frac{\sqrt{15}}{5}$, $\csc\theta = \frac{2\sqrt{6}}{3}$, $\sec\theta = \frac{2\sqrt{10}}{5}$, $\cot\theta = \frac{\sqrt{15}}{3}$ **47.** $\{x \mid x > -0.6931\}$ **49.** 405°, −315°

51. 540°, −180° **53.** $\frac{19\pi}{6}$, $-\frac{5\pi}{6}$

Page 738 Practice Quiz 2

1. $\sin\theta = \frac{3\sqrt{13}}{13}$; $\cos\theta = -\frac{2\sqrt{13}}{13}$; $\tan\theta = -\frac{3}{2}$; $\csc\theta = \frac{\sqrt{13}}{3}$; $\sec\theta = -\frac{\sqrt{13}}{2}$; $\cot\theta = -\frac{2}{3}$ **3.** 27.7 m²

5. cosines; $c \approx 15.9$, $C \approx 59°$, $B \approx 43°$

Pages 742–745 Lesson 13-6

1. The terminal side of the angle θ in standard position must intersect the unit circle at $P(x, y)$. **3.** Sample answer: The graphs have the same shape, but cross the x-axis at different points. **5.** $\sin\theta = \frac{\sqrt{2}}{2}$; $\cos\theta = \frac{\sqrt{2}}{2}$ **7.** $-\frac{1}{2}$
9. 2 s **11.** $\sin\theta = \frac{4}{5}$; $\cos\theta = -\frac{3}{5}$ **13.** $\sin\theta = \frac{15}{17}$; $\cos\theta = \frac{8}{17}$
15. $\sin\theta = \frac{\sqrt{3}}{2}$; $\cos\theta = -\frac{1}{2}$ **17.** $-\frac{1}{2}$ **19.** −1 **21.** 1
23. $\frac{1}{4}$ **25.** $\frac{1-\sqrt{3}}{2}$ **27.** $-3\sqrt{3}$ **29.** 6 **31.** 2π **33.** $\frac{1}{440}$ s
35. $\left(\frac{1}{2}, \frac{\sqrt{3}}{2}\right)$, $\left(-\frac{1}{2}, \frac{\sqrt{3}}{2}\right)$, $(-1, 0)$, $\left(-\frac{1}{2}, -\frac{\sqrt{3}}{2}\right)$, $\left(\frac{1}{2}, -\frac{\sqrt{3}}{2}\right)$
37. $\frac{y}{x}$ **39.** $-\frac{x}{y}$ **41.** $\sqrt{3}$ **43.** sine: D = {all reals}, R = $\{-1 \le y \le 1\}$; cosine: D = {all reals}, R = $\{-1 \le y \le 1\}$ **45.** A
47. cosines; $c \approx 12.4$, $B \approx 59°$, $A \approx 76°$ **49.** 27.0 in²
51. 6800 **53.** 5000 **55.** 250 **57.** does not exist **59.** 8
61. $2x + 9$ **63.** $2y + 7 + \frac{5}{y-3}$ **65.** 110° **67.** 80° **69.** 89°

Pages 749–751 Lesson 13-7

1. Restricted domains are denoted with a capital letter.
3. They are inverses of each other. **5.** α = Arccos 0.5 **7.** 0°
9. $\pi \approx 3.14$ **11.** 0.75 **13.** 0.58 **15.** β = Arcsin α
17. y = Arccos x **19.** Arccos y = 45° **21.** 60° **23.** 45°
25. 45° **27.** 2.09 **29.** 0.52 **31.** 0.5 **33.** 0.60 **35.** 0.8
37. 0.5 **39.** −0.5 **41.** 0.71 **43.** 0.96 **45.** 60° south of west
47. No; with this point on the terminal side of the throwing angle θ, the measure of θ is found by solving the equation $\tan\theta = \frac{17}{18}$. Thus $\theta = \tan^{-1}\frac{17}{18}$ or about 43.4°, which is greater than the 40° requirement. **49.** 31° **51.** Suppose $P(x_1, y_1)$ and $Q(x_2, y_2)$ lie on the line $y = mx + b$. Then $m = \frac{y_2 - y_1}{x_2 - x_1}$. The tangent of the angle θ the line makes with the positive x-axis is equal to the ratio $\frac{\text{opp}}{\text{adj}}$ or $\frac{y_2 - y_1}{x_2 - x_1}$. Thus $\tan\theta = m$.

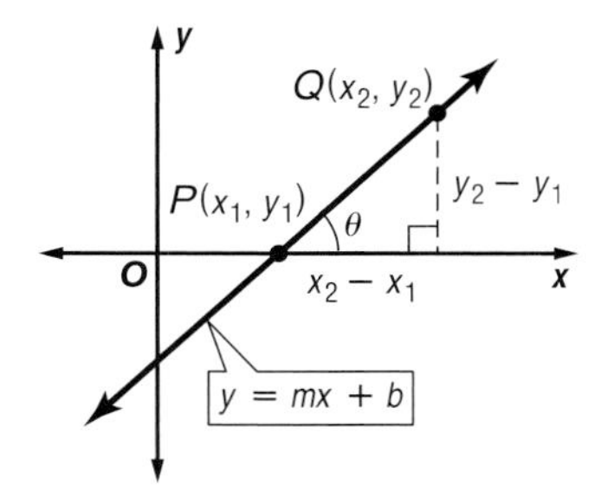

53. 37°
55.

x	0	$\frac{1}{2}$	$\frac{\sqrt{2}}{2}$	$\frac{\sqrt{3}}{2}$	1	$-\frac{1}{2}$	$-\frac{\sqrt{2}}{2}$	$-\frac{\sqrt{3}}{2}$	−1
y	$\frac{\pi}{2}$	$\frac{\pi}{2}$	$\frac{\pi}{2}$	$\frac{\pi}{2}$	$\frac{\pi}{2}$	$\frac{\pi}{2}$	$\frac{\pi}{2}$	$\frac{\pi}{2}$	$\frac{\pi}{2}$

57. From a right triangle perspective, if an acute angle θ has a given sine, say x, then the complementary angle $\frac{\pi}{2} - \theta$ has that same value as its cosine. This can be verified by looking at a right triangle. Therefore, the sum of the angle whose sine is x and the angle whose cosine is x should be $\frac{\pi}{2}$.
59. −1 **61.** sines; $B \approx 69°$, $C \approx 81°$, $c \approx 6.1$ or $B \approx 111°$, $C \approx 39°$, $c \approx 3.9$ **63.** 46, 39 **65.** 11, 109

Selected Answers

Pages 752–756 Chapter 13 Study Guide and Review

1. false, coterminal **3.** true **5.** true **7.** false, an angle that has its terminal side on an axis where x or y is equal to zero **9.** false, terminal **11.** $B = 65°, a \approx 2.5, b \approx 5.4$ **13.** $A = 7°, a \approx 0.7, c \approx 5.6$ **15.** $A \approx 76°, B \approx 14°, b \approx 1.0, c \approx 4.1$ **17.** $-\frac{7\pi}{6}$ **19.** $-720°$ **21.** $320°, -400°$ **23.** $\frac{\pi}{4}, -\frac{15\pi}{4}$ **25.** $\sin\theta = -\frac{8}{17}, \cos\theta = \frac{15}{17}, \tan\theta = -\frac{8}{15}, \csc\theta = -\frac{17}{8}, \sec\theta = \frac{17}{15}, \cot\theta = -\frac{15}{8}$ **27.** $-\sqrt{3}$ **29.** $\frac{2\sqrt{3}}{3}$ **31.** two; $B \approx 53°, C \approx 87°, c \approx 12.4$; $B \approx 127°, C \approx 13°, c \approx 3.0$ **33.** no **35.** one; $A = 51°, a = 70.2, c = 89.7$ **37.** sines; $C = 105°, a \approx 28.3, c \approx 38.6$ **39.** cosines; $A \approx 34°, B \approx 81°, c \approx 6.4$ **41.** cosines; $B \approx 26°, C \approx 125°, a \approx 8.3$ **43.** $\frac{1}{2}$ **45.** $-\frac{\sqrt{2}}{2}$ **47.** $-\sqrt{3}$ **49.** 1.05 **51.** 0

Chapter 14 Trigonometric Graphs and Identities

Page 761 Chapter 14 Getting Started

1. $\frac{\sqrt{2}}{2}$ **3.** 0 **5.** $-\frac{\sqrt{2}}{2}$ **7.** $-\frac{1}{2}$ **9.** $-\frac{\sqrt{3}}{2}$ **11.** 1 **13.** not defined **15.** $\frac{1}{2}$ **17.** $-5x(3x + 1)$ **19.** prime **21.** $(2x + 1)(x - 2)$ **23.** 8, −3 **25.** −8, 5 **27.** −4, $-\frac{3}{2}$

Pages 766–768 Lesson 14-1

1. Sample answer: Amplitude is half the difference between the maximum and minimum values of a graph; $y = \tan\theta$ has no maximum or minimum value. **3.** Jamile; The amplitude is 3 and the period is 3π. **5.** amplitude: 2; period: 360° or 2π

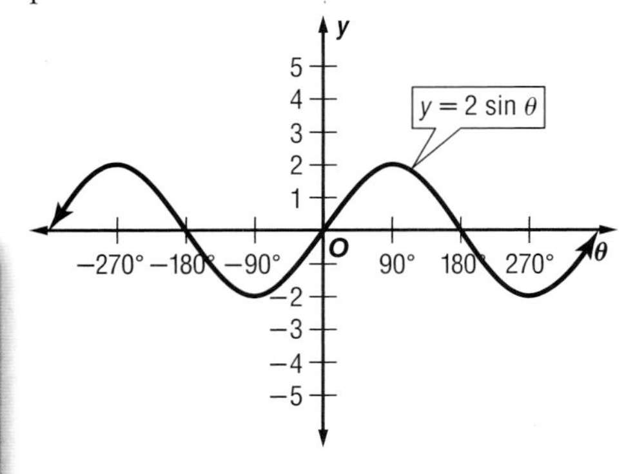

7. amplitude : does not exist; period: 180° or π

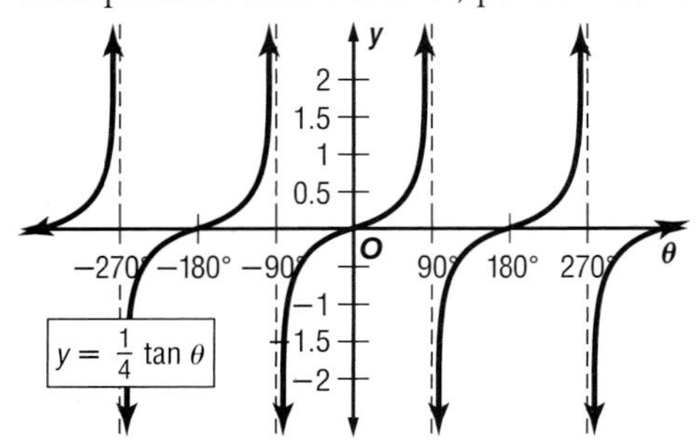

9. amplitude: 4; period: 180° or π

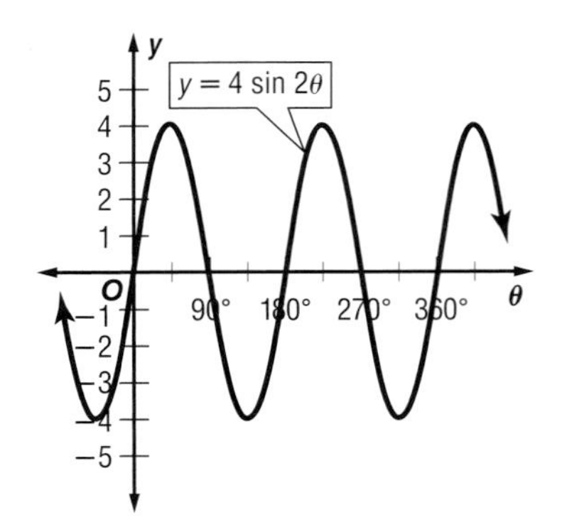

11. amplitude: does not exist; period: 120° or $\frac{2\pi}{3}$

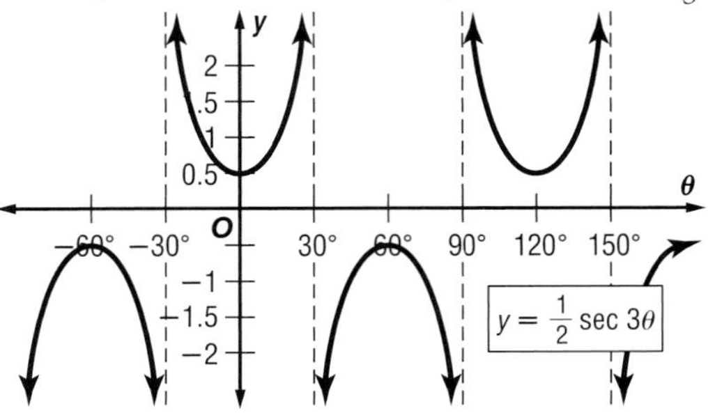

13. 12 months; Sample answer: The pattern in the population will repeat itself every 12 months.

15. amplitude: 3; period: 360° or 2π

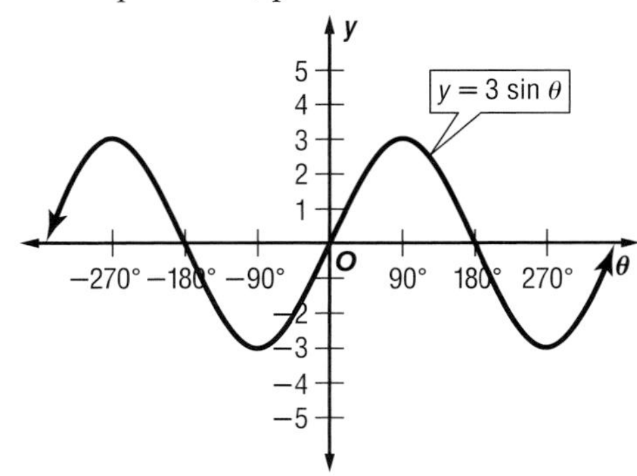

17. amplitude: does not exist; period: 360° or 2π

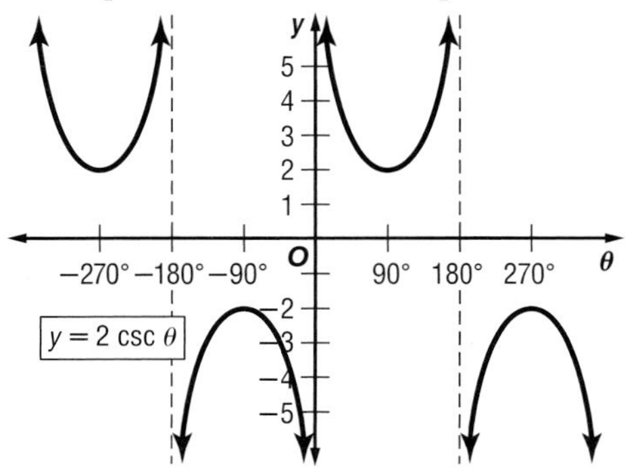

19. amplitude: $\frac{1}{5}$; period: 360° or 2π

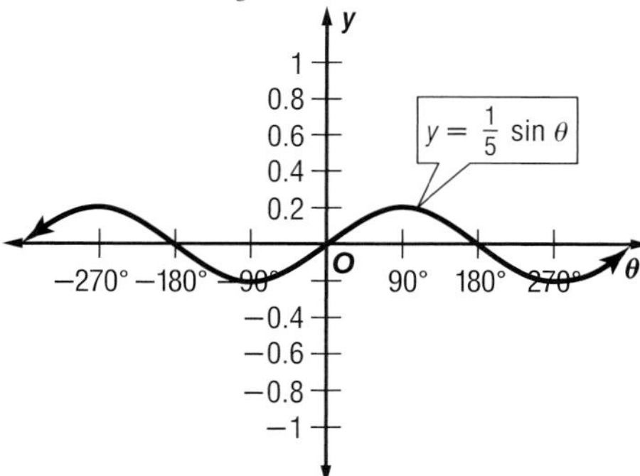

21. amplitude: 1; period 90° or $\frac{\pi}{2}$

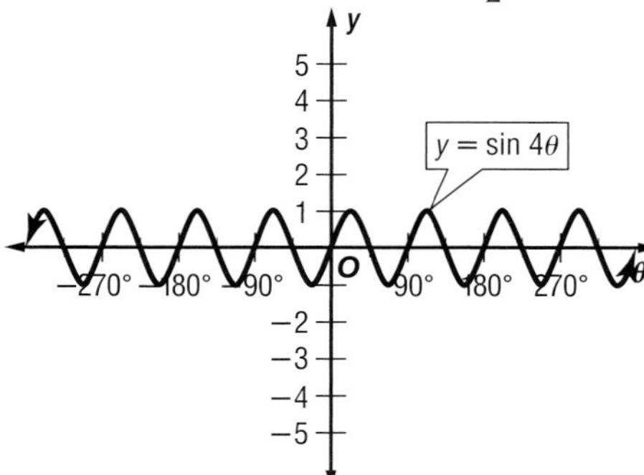

23. amplitude: does not exist; period: 120° or $\frac{2\pi}{3}$

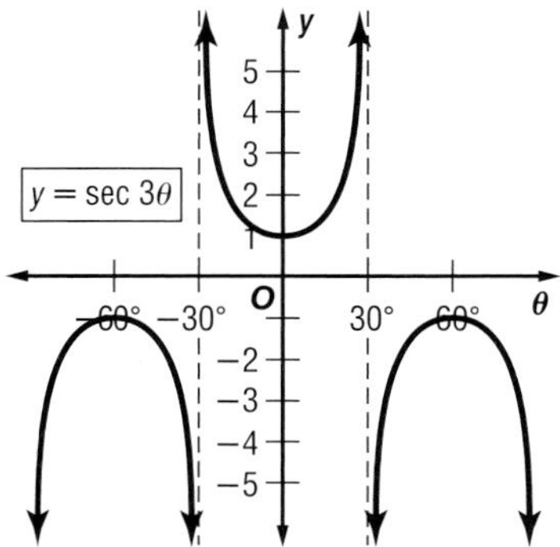

25. amplitude: does not exist; period: 540° or 3π

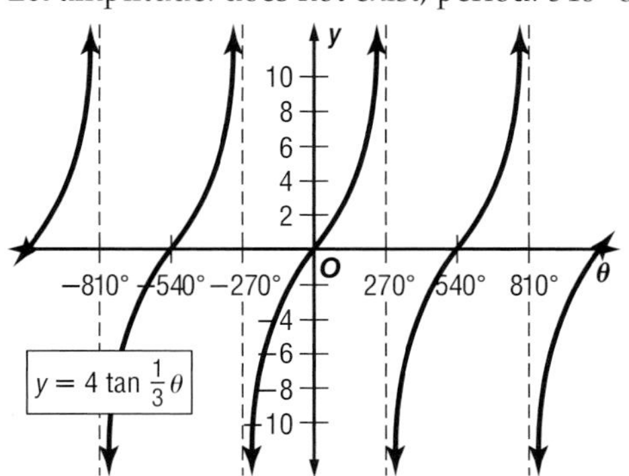

27. amplitude: 6; period: 540° or 3π

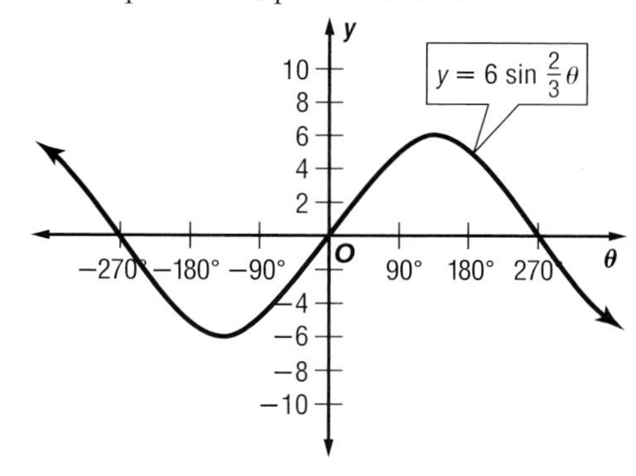

29. amplitude: does not exist; period: 720° or 4π

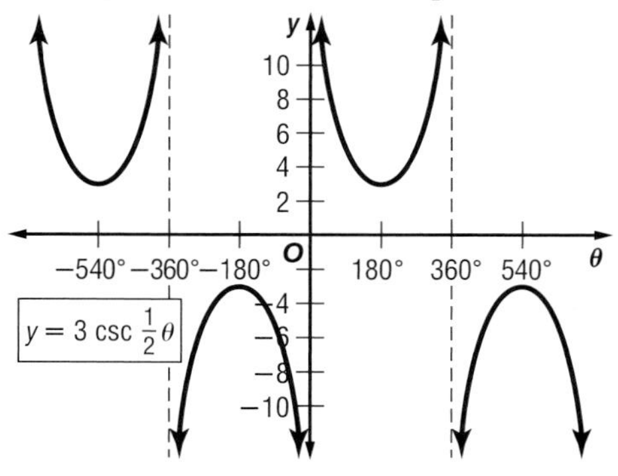

31. amplitude: does not exist; period: 180° or π

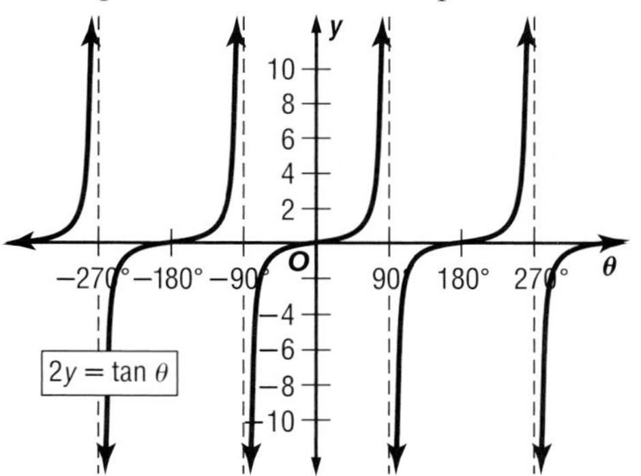

33.

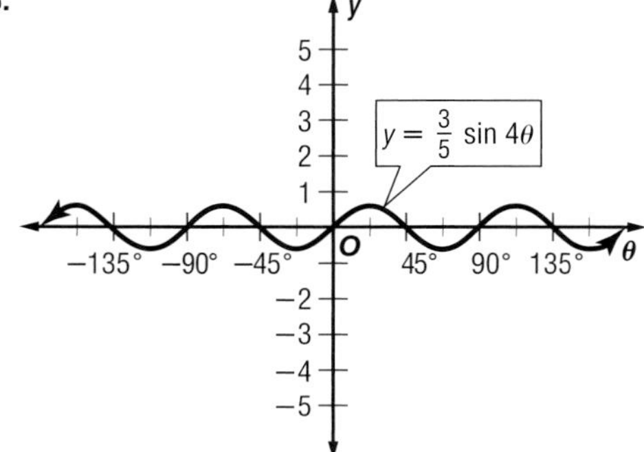

$y = \frac{3}{5}\sin 4\theta$ **35.** $\frac{1}{10^7}$ **37.** Sample answer: The amplitudes are the same. As the frequency increases, the period decreases.

39. $y = 2\sin\frac{\pi}{5}t$

41. about 1.9 ft **43.** A **45.** 90° **47.** 45° **49.** $\frac{\sqrt{2}}{2}$ **51.** $\frac{13}{16}$

53.

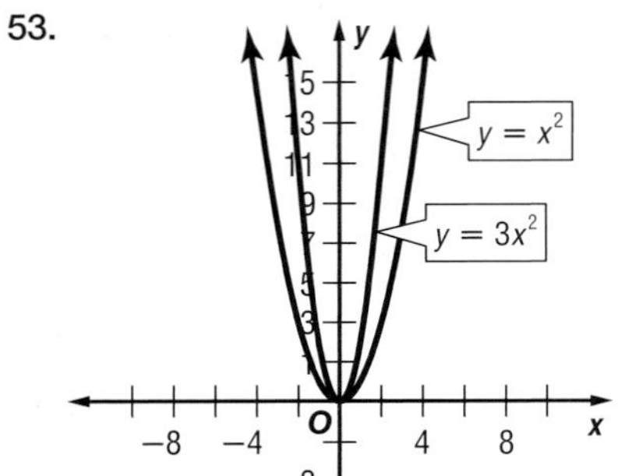

55.

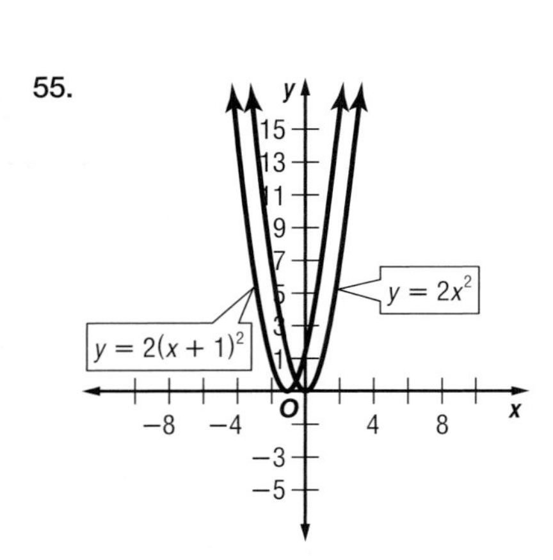

Pages 774–776 Lesson 14-2

1. vertical shift: 15; amplitude: 3; period: 180°; phase shift: 45° **3.** Sample answer: $y = \sin(\theta + 45°)$ **5.** no amplitude; 180°; −60°

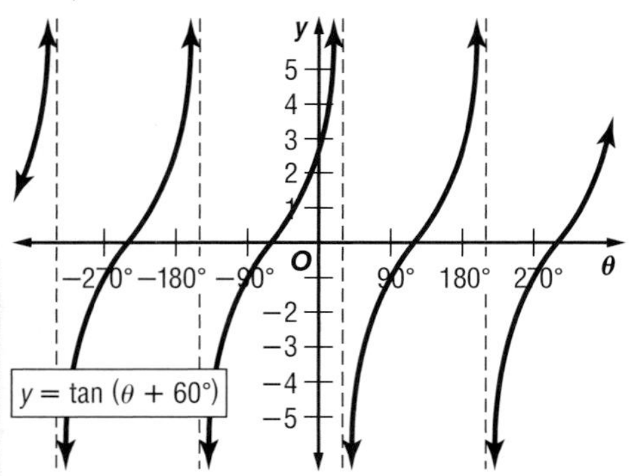

7. no amplitude; 2π; $-\frac{\pi}{3}$

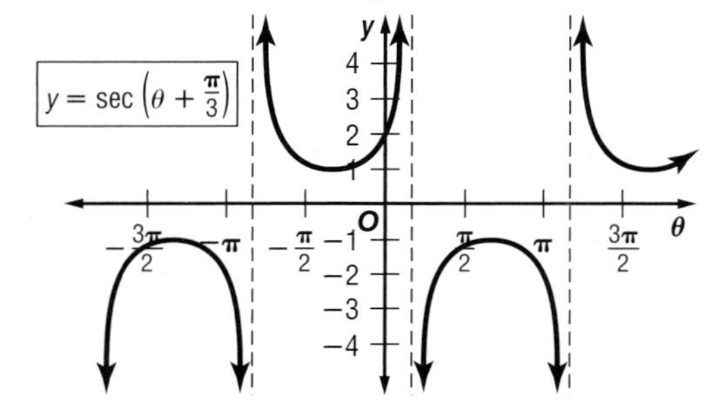

9. −5; $y = -5$; no amplitude; 360°

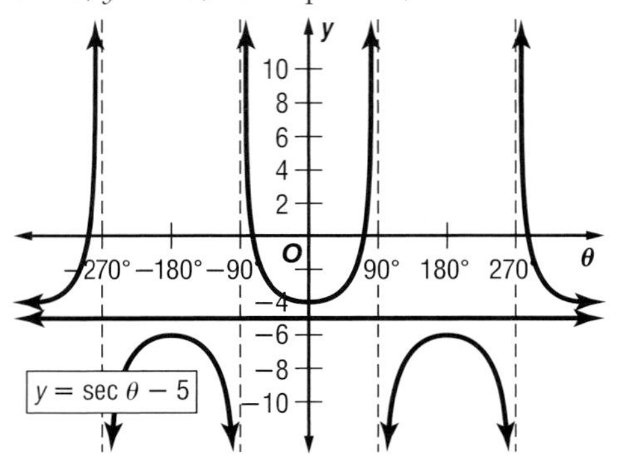

11. 0.25; $y = 0.25$; 1; 360°

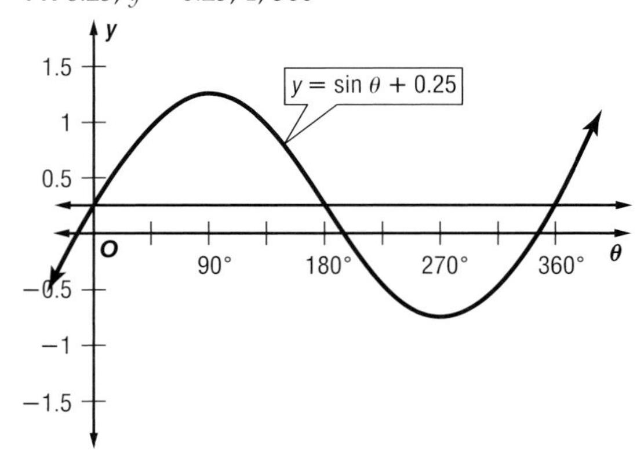

13. −6; no amplitude; 60°; −45°

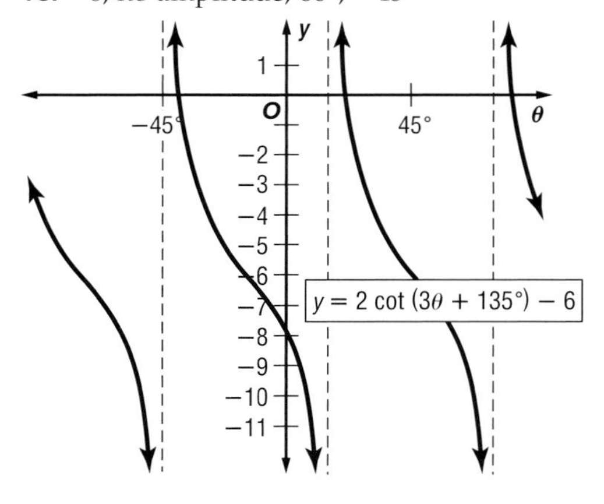

15. −2; $\frac{2}{3}$; 4π; $-\frac{\pi}{6}$

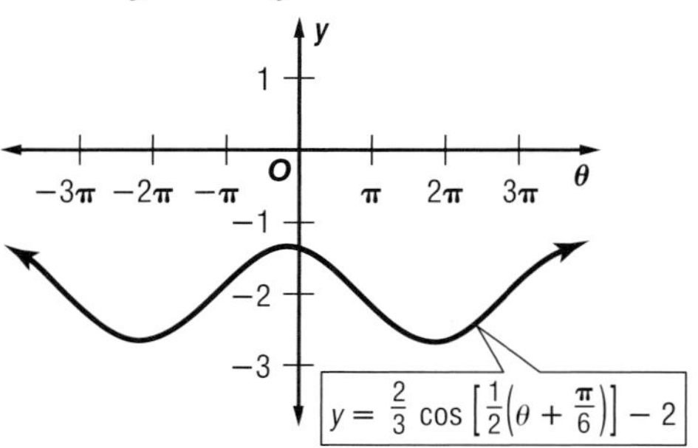

17. $h = 4 - \cos \frac{\pi}{2}t$ or $h = 4 - \cos 90°t$

19. 1; 360°; −90°

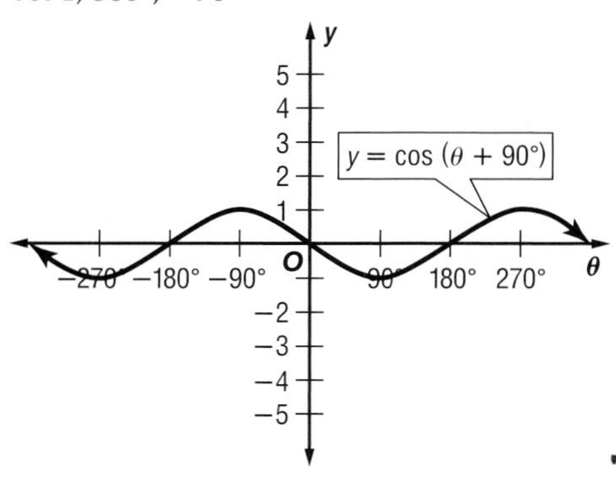

21. 1; 2π; $\frac{\pi}{4}$

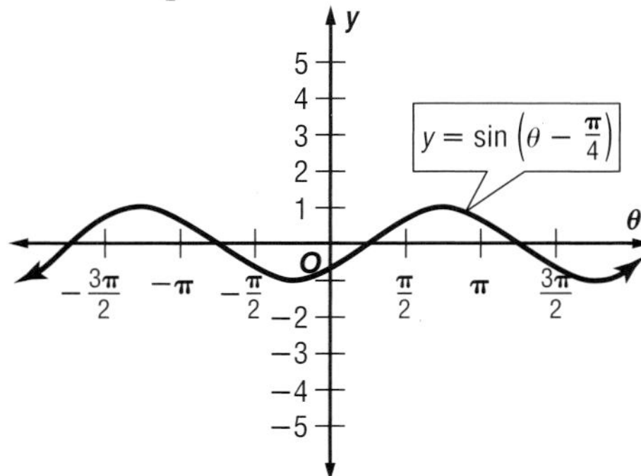

23. no amplitude; 180°; −22.5°

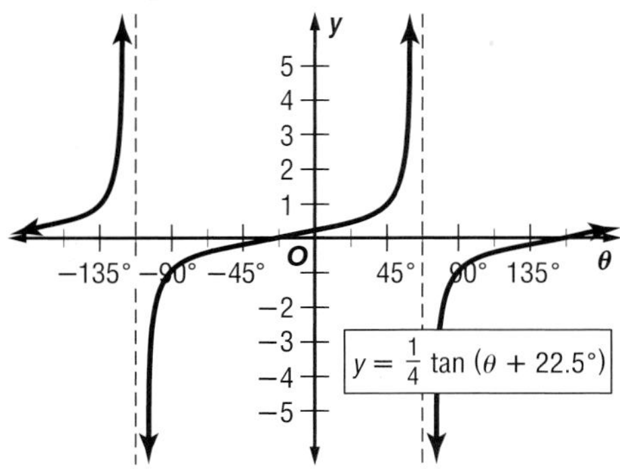

25. −1; $y = -1$; 1; 360°

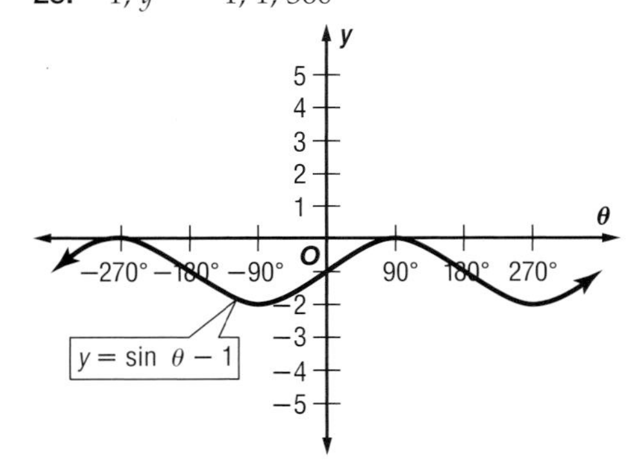

27. −5; $y = -5$; 1; 360°

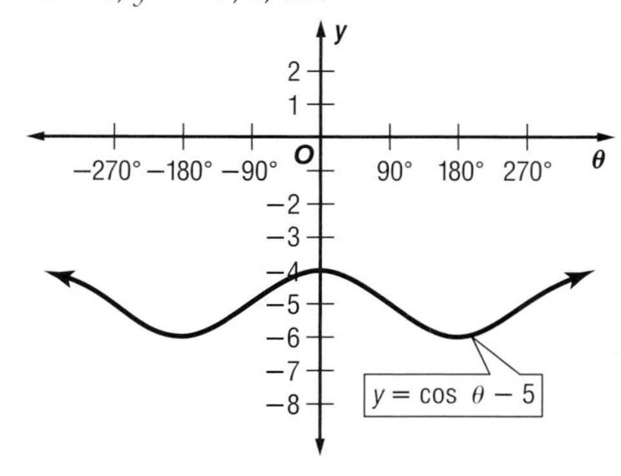

29. $\frac{1}{2}$; $y = \frac{1}{2}$; $\frac{1}{2}$; 360°

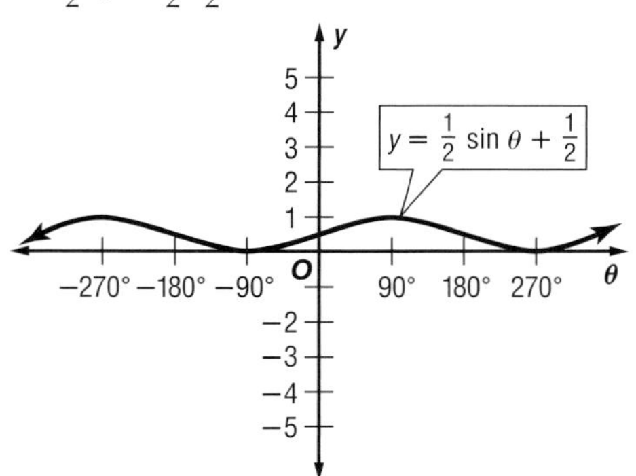

31.

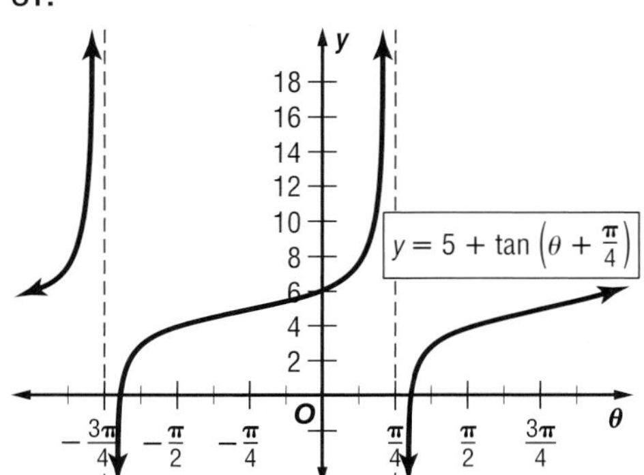

translation $\frac{\pi}{4}$ units left and 5 units up

33. 1; 2; 120°; 45°

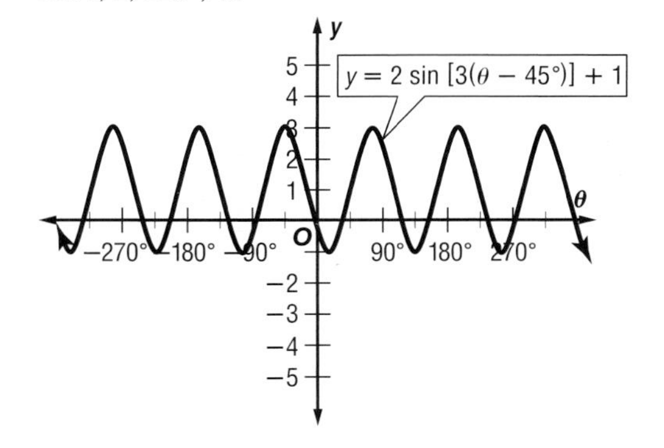

35. −3.5; does not exist; 720°; −60°

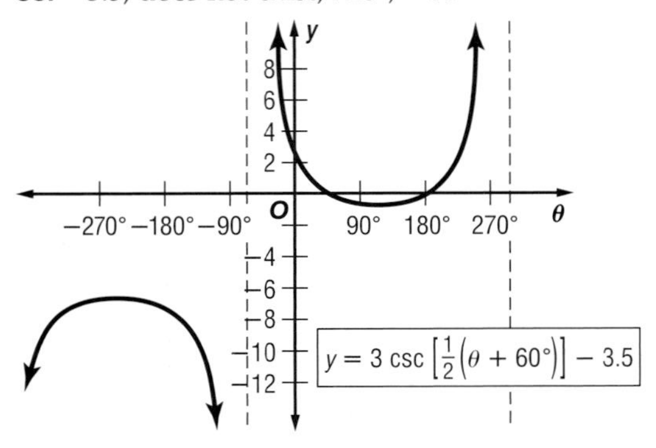

37. 1; $\frac{1}{4}$; 180°; 75°

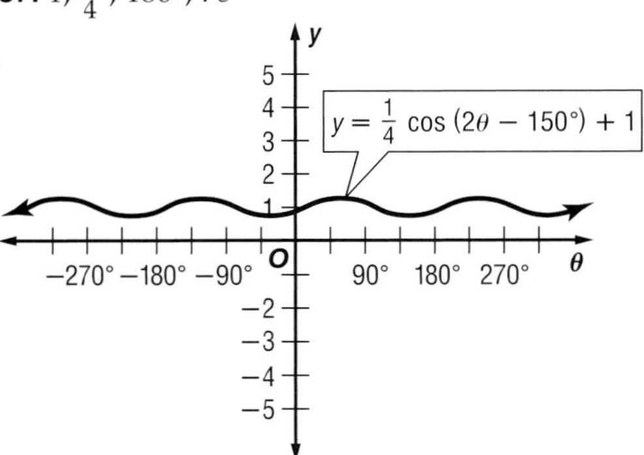

39. 3; 2; π; $-\frac{\pi}{4}$

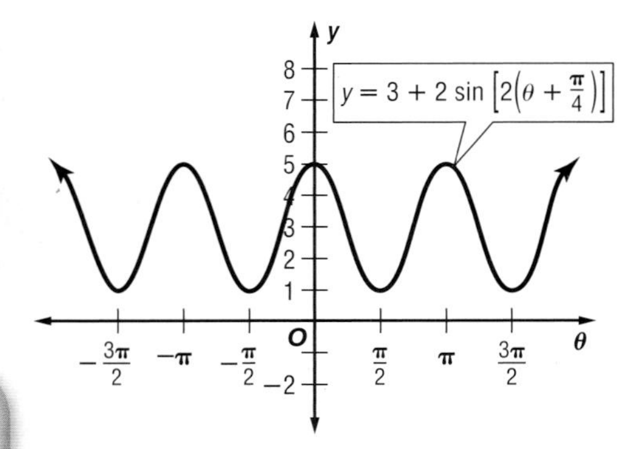

41.

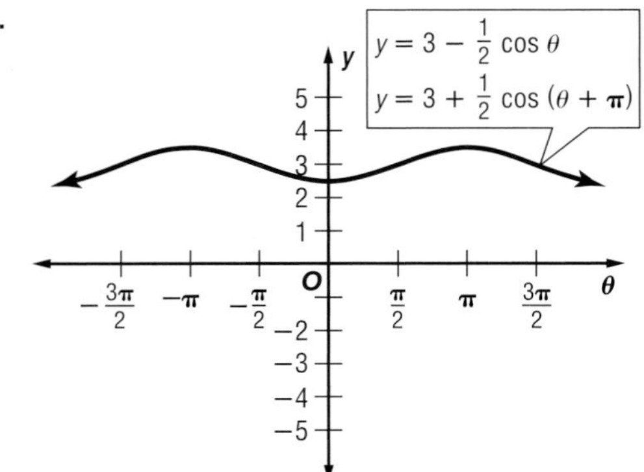

The graphs are identical. **43.** c **45.** 300; 14.5 yr

47. $h = 9 + 6 \sin \left[\frac{\pi}{9}(t - 1.5)\right]$

49. Sample answer: You can use changes in amplitude and period along with vertical and horizontal shifts to show an animal population's starting point and display changes to that population over a period of time. Answers should include the following information.

- The equation shows a rabbit population that begins at 1200, increases to a maximum of 1450 then decreases to a minimum of 950 over a period of 4 years.
- Relative to $y = a \cos bx$, $y = a \cos bx + k$ would have a vertical shift of k units, while $y = a \cos [b(x - h)]$ has a horizontal shift of h units.

51. D **53.** amplitude: 1; period: 720° or 4π

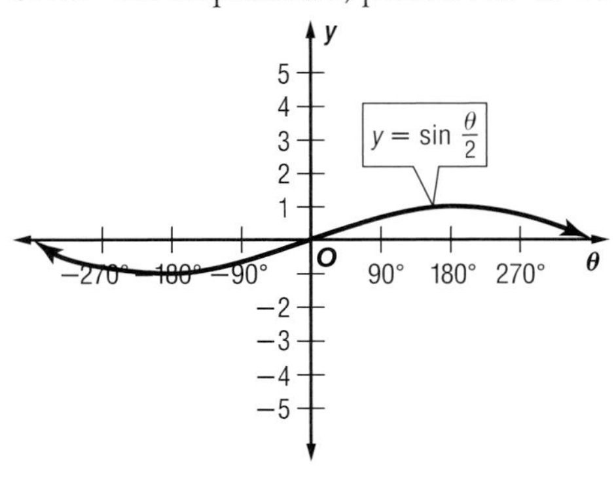

55. 0.75 **57.** 0.83 **59.** 35 **61.** 0.66 **63.** $\frac{5a - 13}{(a - 2)(a - 3)}$ **65.** $\frac{3y^2 + 10y + 5}{2(y - 5)(y + 3)}$ **67.** −1 **69.** $\frac{1}{2}$ **71.** $\frac{\sqrt{3}}{3}$ **73.** 1

Pages 779–781 Lesson 14-3

1. Sample answer: The sine function is negative in the third and fourth quadrants. Therefore, the terminal side of the angle must lie in one of those two quadrants.

3. Sample answer: Simplifying a trigonometric expression means writing the expression as a numerical value or in terms of a single trigonometric function, if possible.

5. $-\frac{5}{4}$ **7.** $\sqrt{2}$ **9.** $\tan^2 \theta$ **11.** $\csc \theta$ **13.** $\frac{1}{2}$ **15.** $-\sqrt{5}$ **17.** $\frac{5}{4}$ **19.** $\frac{\sqrt{3}}{2}$ **21.** $\frac{3}{4}$ **23.** $-\frac{4\sqrt{7}}{7}$ **25.** $\cot \theta$ **27.** $\cos \theta$ **29.** 2 **31.** $\cot^2 \theta$ **33.** 1 **35.** $\csc^2 \theta$ **37.** about 11.5° **39.** about 9.4° **41.** No; $R^2 = \frac{I \tan \theta \cos \theta}{E}$ simplifies to $E = \frac{I \sin \theta}{R^2}$. **43.** $P = I^2R - \frac{I^2R}{1 + \tan^2 2\pi ft}$.

45. Sample answer: You can use equations to find the height and the horizontal distance of a baseball after it has been hit. The equations involve using the initial angle the ball makes with the ground with the sine function. Answers should include the following information.

- Both equations are quadratic in nature with a leading negative coefficient. Thus, both are inverted parabolas which model the path of a baseball.
- model rockets, hitting a golf ball, kicking a rock

47. A **49.** 12; $y = 12$; no amplitude; 180°

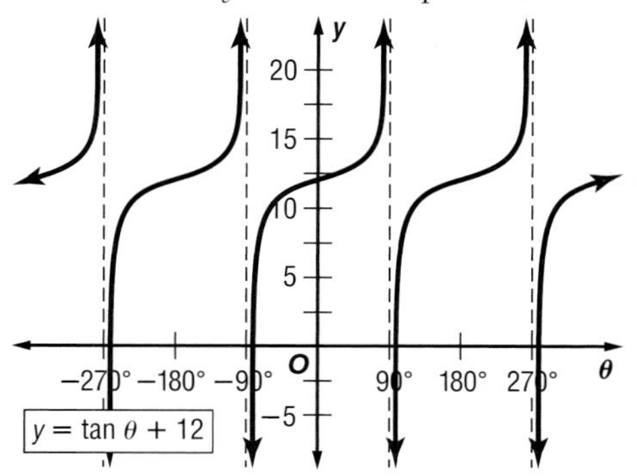

51. amplitude: 1; period: 120° or $\frac{2\pi}{3}$

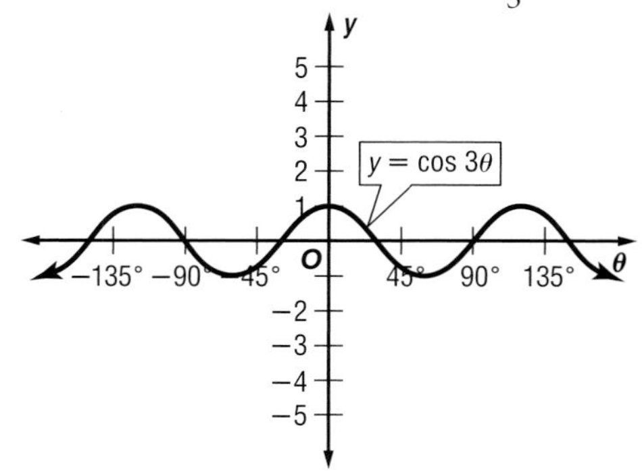

53. 93 **55.** Symmetric (=) **57.** Multiplication (=)

Page 781 Practice Quiz 1

1. $\frac{3}{4}$, 720° or 4π

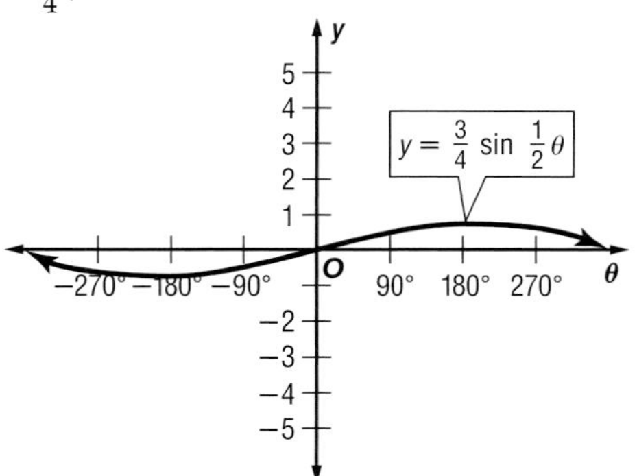

3. $-\frac{3}{5}$ **5.** $\frac{\sqrt{5}}{2}$

Pages 784–785 Lesson 14-4

1.

$\sin\theta\tan\theta \stackrel{?}{=} \sec\theta - \cos\theta$

$\sin\theta\tan\theta \stackrel{?}{=} \frac{1}{\cos\theta} - \cos\theta$ *$\sec\theta = \frac{1}{\cos\theta}$*

$\sin\theta\tan\theta \stackrel{?}{=} \frac{1}{\cos\theta} - \frac{\cos^2\theta}{\cos\theta}$ *Multiply by the LCD, $\cos\theta$.*

$\sin\theta\tan\theta \stackrel{?}{=} \frac{1-\cos^2\theta}{\cos\theta}$ *Subtract.*

$\sin\theta\tan\theta \stackrel{?}{=} \frac{\sin^2\theta}{\cos\theta}$ *$1 - \cos^2\theta = \sin^2\theta$*

$\sin\theta\tan\theta \stackrel{?}{=} \sin\theta \cdot \frac{\sin\theta}{\cos\theta}$ *Factor.*

$\sin\theta\tan\theta = \sin\theta\tan\theta$ *$\frac{\sin\theta}{\cos\theta} = \tan\theta$*

3. Sample answer: $\sin^2\theta = 1 + \cos^2\theta$; it is not an identity because $\sin^2\theta = 1 - \cos^2\theta$.

5.

$\tan^2\theta\cos^2\theta \stackrel{?}{=} 1 - \cos^2\theta$

$\frac{\sin^2\theta}{\cos^2\theta} \cdot \cos^2\theta \stackrel{?}{=} \sin^2\theta$

$\sin^2\theta = \sin^2\theta$

7.

$\frac{1+\tan^2\theta}{\csc\theta} \stackrel{?}{=} \tan^2\theta$

$\frac{\sec^2\theta}{\csc^2\theta} \stackrel{?}{=} \tan^2\theta$

$\frac{\frac{1}{\cos^2\theta}}{\frac{1}{\sin^2\theta}} \stackrel{?}{=} \tan^2\theta$

$\frac{1}{\cos^2\theta} \cdot \sin^2\theta \stackrel{?}{=} \tan^2\theta$

$\tan^2\theta = \tan^2\theta$

9.

$\frac{\sec\theta+1}{\tan\theta} \stackrel{?}{=} \frac{\tan\theta}{\sec\theta-1}$

$\frac{\sec\theta+1}{\tan\theta} \stackrel{?}{=} \frac{\tan\theta}{\sec\theta-1} \cdot \frac{\sec\theta+1}{\sec\theta+1}$

$\frac{\sec\theta+1}{\tan\theta} \stackrel{?}{=} \frac{\tan\theta\cdot(\sec\theta+1)}{\sec^2\theta-1}$

$\frac{\sec\theta+1}{\tan\theta} \stackrel{?}{=} \frac{\tan\theta\cdot(\sec\theta+1)}{\tan^2\theta}$

$\frac{\sec\theta+1}{\tan\theta} = \frac{\sec\theta+1}{\tan\theta}$

11.

$\cos^2\theta + \tan^2\theta\cos^2\theta \stackrel{?}{=} 1$

$\cos^2\theta + \frac{\sin^2\theta}{\cos^2\theta} \cdot \cos^2\theta \stackrel{?}{=} 1$

$\cos^2\theta + \sin^2\theta \stackrel{?}{=} 1$

$1 = 1$

13.

$1 + \sec^2\theta\sin^2\theta \stackrel{?}{=} \sec^2\theta$

$1 + \frac{1}{\cos^2\theta} \cdot \sin^2\theta \stackrel{?}{=} \sec^2\theta$

$1 + \tan^2\theta \stackrel{?}{=} \sec^2\theta$

$\sec^2\theta = \sec^2\theta$

15.

$\frac{1-\cos\theta}{1+\cos\theta} \stackrel{?}{=} (\csc\theta - \cot\theta)^2$

$\frac{1-\cos\theta}{1+\cos\theta} \stackrel{?}{=} \csc^2\theta - 2\cot\theta\csc\theta + \cot^2\theta$

$\frac{1-\cos\theta}{1+\cos\theta} \stackrel{?}{=} \frac{1}{\sin^2\theta} - 2\cdot\frac{\cos\theta}{\sin\theta}\cdot\frac{1}{\sin\theta} + \frac{\cos^2\theta}{\sin^2\theta}$

$\frac{1-\cos\theta}{1+\cos\theta} \stackrel{?}{=} \frac{1}{\sin^2\theta} - \frac{2\cos\theta}{\sin^2\theta} + \frac{\cos^2\theta}{\sin^2\theta}$

$\frac{1-\cos\theta}{1+\cos\theta} \stackrel{?}{=} \frac{1-2\cos\theta+\cos^2\theta}{\sin^2\theta}$

$\frac{1-\cos\theta}{1+\cos\theta} \stackrel{?}{=} \frac{(1-\cos\theta)(1-\cos\theta)}{1-\cos^2\theta}$

$\frac{1-\cos\theta}{1+\cos\theta} \stackrel{?}{=} \frac{(1-\cos\theta)(1-\cos\theta)}{(1-\cos\theta)(1+\cos\theta)}$

$\frac{1-\cos\theta}{1+\cos\theta} = \frac{1-\cos\theta}{1+\cos\theta}$

17.

$\cot\theta\csc\theta \stackrel{?}{=} \frac{\cot\theta+\csc\theta}{\sin\theta+\tan\theta}$

$\cot\theta\csc\theta \stackrel{?}{=} \frac{\frac{\cos\theta}{\sin\theta}+\frac{1}{\sin\theta}}{\sin\theta+\frac{\sin\theta}{\cos\theta}}$

$\cot\theta\csc\theta \stackrel{?}{=} \frac{\frac{\cos\theta+1}{\sin\theta}}{\frac{\sin\theta\cos\theta+\sin\theta}{\cos\theta}}$

$\cot\theta\csc\theta \stackrel{?}{=} \frac{\frac{\cos\theta+1}{\sin\theta}}{\frac{\sin\theta(\cos\theta+1)}{\cos\theta}}$

$\cot\theta\csc\theta \stackrel{?}{=} \frac{\cos\theta+1}{\sin\theta} \cdot \frac{\cos\theta}{\sin\theta(\cos\theta+1)}$

$\cot\theta\csc\theta \stackrel{?}{=} \frac{\cos\theta}{\sin\theta} \cdot \frac{1}{\sin\theta}$

$\cot\theta\csc\theta = \cot\theta\csc\theta$

Selected Answers

19.
$$\begin{aligned}
\frac{\sec\theta}{\sin\theta} - \frac{\sin\theta}{\cos\theta} &\stackrel{?}{=} \cot\theta\\
\frac{\frac{1}{\cos\theta}}{\sin\theta} - \frac{\sin\theta}{\cos\theta} &\stackrel{?}{=} \cot\theta\\
\frac{1}{\sin\theta\cos\theta} - \frac{\sin^2\theta}{\sin\theta\cos\theta} &\stackrel{?}{=} \cot\theta\\
\frac{1-\sin^2\theta}{\sin\theta\cos\theta} &\stackrel{?}{=} \cot\theta\\
\frac{\cos^2\theta}{\sin\theta\cos\theta} &\stackrel{?}{=} \cot\theta\\
\frac{\cos\theta}{\sin\theta} &\stackrel{?}{=} \cot\theta\\
\cot\theta &= \cot\theta
\end{aligned}$$

21.
$$\begin{aligned}
\frac{1+\sin\theta}{\sin\theta} &\stackrel{?}{=} \frac{\cot^2\theta}{\csc\theta - 1}\\
\frac{1+\sin\theta}{\sin\theta} &\stackrel{?}{=} \frac{\cot^2\theta}{\csc\theta - 1}\cdot\frac{\csc\theta+1}{\csc\theta+1}\\
\frac{1+\sin\theta}{\sin\theta} &\stackrel{?}{=} \frac{\cot^2\theta(\csc\theta+1)}{\csc^2\theta-1}\\
\frac{1+\sin\theta}{\sin\theta} &\stackrel{?}{=} \frac{\cot^2\theta(\csc\theta+1)}{\cot^2\theta}\\
\frac{1+\sin\theta}{\sin\theta} &\stackrel{?}{=} \csc\theta + 1\\
\frac{1+\sin\theta}{\sin\theta} &\stackrel{?}{=} \frac{1}{\sin\theta} + \frac{\sin\theta}{\sin\theta}\\
\frac{1+\sin\theta}{\sin\theta} &= \frac{1+\sin\theta}{\sin\theta}
\end{aligned}$$

23.
$$\begin{aligned}
\frac{1}{\sec^2\theta} + \frac{1}{\csc^2\theta} &\stackrel{?}{=} 1\\
\cos^2\theta + \sin^2\theta &\stackrel{?}{=} 1\\
1 &= 1
\end{aligned}$$

25.
$$\begin{aligned}
1 - \tan^4\theta &\stackrel{?}{=} 2\sec^2\theta - \sec^4\theta\\
(1-\tan^2\theta)(1+\tan^2\theta) &\stackrel{?}{=} \sec^2\theta\,(2-\sec^2\theta)\\
[1-(\sec^2\theta-1)](\sec^2\theta) &\stackrel{?}{=} (2-\sec^2\theta)(\sec^2\theta)\\
(2-\sec^2\theta)(\sec^2\theta) &= (2-\sec^2\theta)(\sec^2\theta)
\end{aligned}$$

27.
$$\begin{aligned}
\frac{1-\cos\theta}{\sin\theta} &\stackrel{?}{=} \frac{\sin\theta}{1+\cos\theta}\\
\frac{1-\cos\theta}{\sin\theta}\cdot\frac{1+\cos\theta}{1+\cos\theta} &\stackrel{?}{=} \frac{\sin\theta}{1+\cos\theta}\\
\frac{1-\cos^2\theta}{\sin\theta(1+\cos\theta)} &\stackrel{?}{=} \frac{\sin\theta}{1+\cos\theta}\\
\frac{\sin^2\theta}{\sin\theta(1+\cos\theta)} &\stackrel{?}{=} \frac{\sin\theta}{1+\cos\theta}\\
\frac{\sin\theta}{1+\cos\theta} &= \frac{\sin\theta}{1+\cos\theta}
\end{aligned}$$

29.
$$\begin{aligned}
\tan\theta\sin\theta\cos\theta\csc^2\theta &\stackrel{?}{=} 1\\
\frac{\sin\theta}{\cos\theta}\cdot\sin\theta\cdot\cos\theta\cdot\frac{1}{\sin^2\theta} &\stackrel{?}{=} 1\\
1 &= 1
\end{aligned}$$

31.
$$\begin{aligned}
\frac{v^2\tan^2\theta}{2\sec^2\theta} &= \frac{v^2\,\frac{\sin^2\theta}{\cos^2\theta}}{2g\,\frac{1}{\cos^2\theta}}\\
&= \frac{v^2}{2g}\cdot\frac{\sin^2\theta}{\cos^2\theta}\cdot\frac{\cos^2\theta}{1}\\
&= \frac{v^2\sin^2\theta}{2g}
\end{aligned}$$

33. Sample answer: Consider a right triangle *ABC* with right angle at *C*. If an angle, say *A*, has a sine of *x*, then angle *B* must have a cosine of *x*. Since *A* and *B* are both in a right triangle and neither is the right angle, their sum must be $\frac{\pi}{2}$. **35.** D

37.

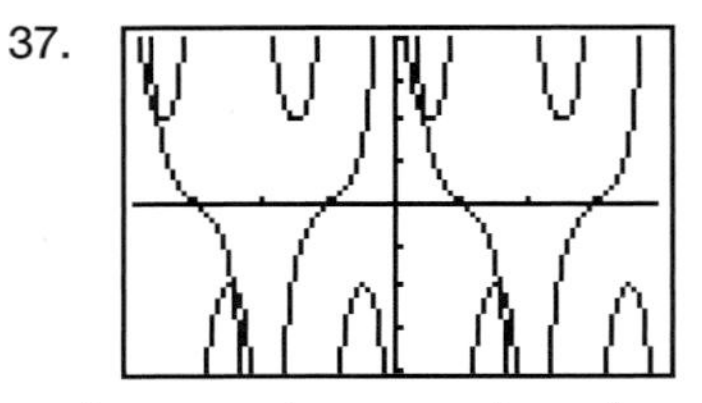

[−360, 360] scl: 90 by [−5, 5] scl: 1

is not

39.

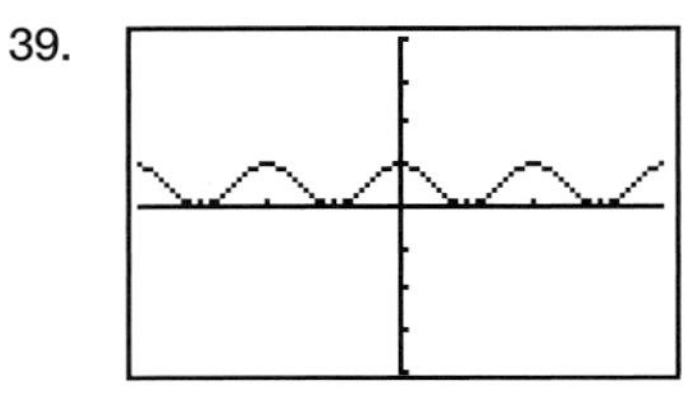

[−360, 360] scl: 90 by [−5, 5] scl: 1

may be

41.

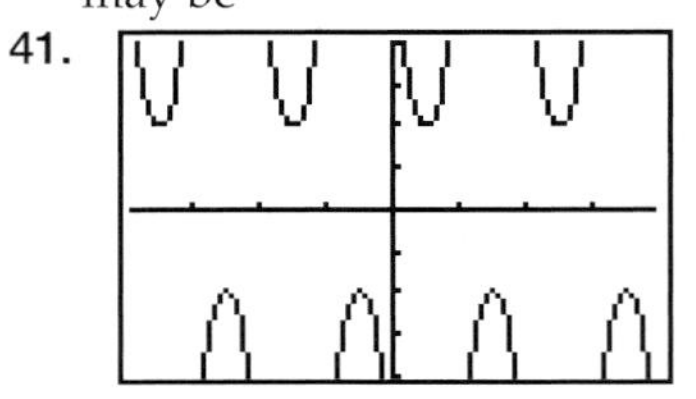

[−360, 360] scl: 90 by [−5, 5] scl: 1

may be

43. $\frac{\sqrt{5}}{2}$ **45.** $\frac{\sqrt{193}}{12}$ **47.** 1: 360°; 30°

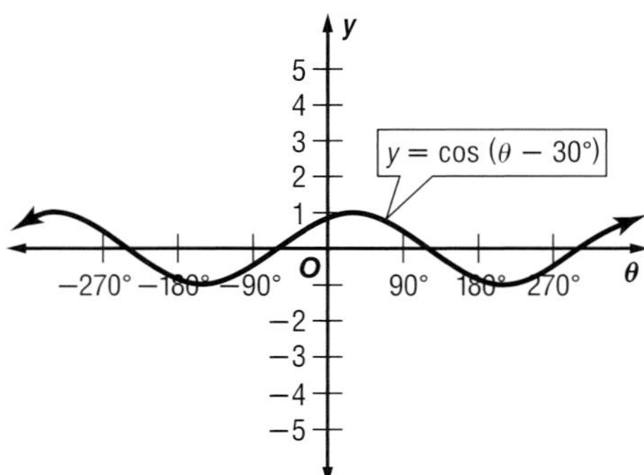

49. 3; 2π; $-\frac{\pi}{2}$

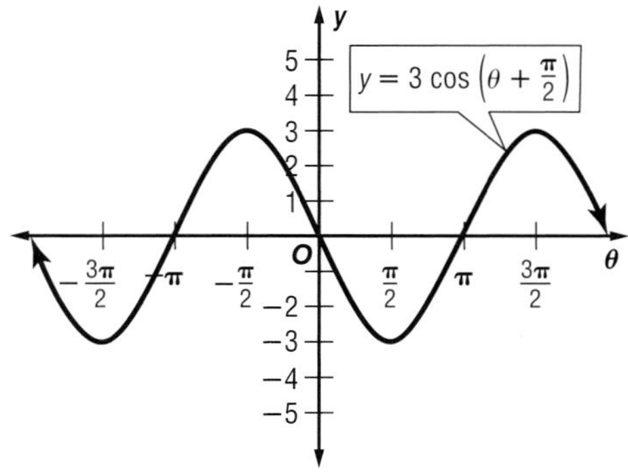

51. $\frac{\sqrt{6}}{4}$ **53.** $\frac{\sqrt{6}+2\sqrt{2}}{4}$

Pages 788–790 Lesson 14-5

1. $\sin(\alpha+\beta) - \sin\alpha + \sin\beta$

$\sin\alpha\cos\beta + \cos\alpha\sin\beta \neq \sin\alpha + \sin\beta$

3. Sometimes; sample answer: The cosine function can equal 1. **5.** $\frac{\sqrt{6}-\sqrt{2}}{4}$ **7.** $\frac{\sqrt{3}}{2}$

9. $-\frac{1}{2}$

11.
$$\sin\left(\theta+\frac{\pi}{2}\right) \stackrel{?}{=} \cos\theta$$
$$\sin\theta\cos\frac{\pi}{2}+\cos\theta\sin\frac{\pi}{2} \stackrel{?}{=} \cos\theta$$
$$\sin\theta\cdot 0+\cos\theta\cdot 1 \stackrel{?}{=} \cos\theta$$
$$\cos\theta = \cos\theta$$

13. $\frac{5-\sqrt{3}}{1+5\sqrt{3}}$ **15.** $\frac{\sqrt{2}-\sqrt{6}}{4}$ **17.** $\frac{-\sqrt{6}-\sqrt{2}}{4}$

19. $\frac{-\sqrt{6}-\sqrt{2}}{4}$ **21.** $-\frac{\sqrt{2}}{2}$ **23.** $\frac{\sqrt{2}}{2}$ **25.** $\frac{\sqrt{2}-\sqrt{6}}{4}$

27. $\frac{-\sqrt{6}-\sqrt{2}}{4}$

29.
$$\cos(90°+\theta) \stackrel{?}{=} \cos 90°\cos\theta-\sin 90°\sin\theta$$
$$\stackrel{?}{=} 0-1\sin\theta$$
$$= -\sin\theta$$

31.
$$\sin(90°-\theta) \stackrel{?}{=} \cos\theta$$
$$\sin 90°\cos\theta-\cos 90°\sin\theta \stackrel{?}{=} \cos\theta$$
$$1\cdot\cos\theta-0\cdot\sin\theta \stackrel{?}{=} \cos\theta$$
$$\cos\theta-0 \stackrel{?}{=} \cos\theta$$
$$\cos\theta = \cos\theta$$

33.
$$\cos(\pi-\theta) \stackrel{?}{=} -\cos\theta$$
$$\cos\pi\cos\theta+\sin\pi\sin\theta \stackrel{?}{=} -\cos\theta$$
$$-1\cdot\cos\theta+0\cdot\sin\theta \stackrel{?}{=} -\cos\theta$$
$$-\cos\theta = -\cos\theta$$

35.
$$\sin(\pi-\theta) \stackrel{?}{=} \sin\theta$$
$$\sin\pi\cos\theta-[\cos\pi\sin\theta] \stackrel{?}{=} \sin\theta$$
$$0\cdot\cos\theta-[-1\cdot\sin\theta] \stackrel{?}{=} \sin\theta$$
$$0-[-\sin\theta] \stackrel{?}{=} \sin\theta$$
$$\sin\theta = \sin\theta$$

37.
$$\sin\left(\theta+\frac{\pi}{3}\right)-\cos\left(\theta+\frac{\pi}{6}\right)$$
$$\stackrel{?}{=} \sin\theta\cos\frac{\pi}{3}+\cos\theta\sin\frac{\pi}{3}-\cos\theta\cos\frac{\pi}{6}+\sin\theta\sin\frac{\pi}{6}$$
$$\stackrel{?}{=} \frac{1}{2}\sin\theta+\frac{\sqrt{3}}{2}\cos\theta-\frac{\sqrt{3}}{2}\cos\theta+\frac{1}{2}\sin\theta$$
$$\stackrel{?}{=} \frac{1}{2}\sin\theta+\frac{1}{2}\sin\theta$$
$$= \sin\theta$$

39.
$$\cos(\alpha+\beta) \stackrel{?}{=} \frac{1-\tan\alpha\tan\beta}{\sec\alpha\sec\beta}$$
$$\cos(\alpha+\beta) \stackrel{?}{=} \frac{1-\frac{\sin\alpha}{\cos\alpha}\cdot\frac{\sin\beta}{\cos\beta}}{\frac{1}{\cos\alpha}\cdot\frac{1}{\cos\beta}}$$
$$\cos(\alpha+\beta) \stackrel{?}{=} \frac{1-\frac{\sin\alpha}{\cos\alpha}\cdot\frac{\sin\beta}{\cos\beta}}{\frac{1}{\cos\alpha}\cdot\frac{1}{\cos\beta}}\cdot\frac{\cos\alpha\cos\beta}{\cos\alpha\cos\beta}$$
$$\cos(\alpha+\beta) \stackrel{?}{=} \frac{\cos\alpha\cos\beta-\sin\alpha\sin\beta}{1}$$
$$\cos(\alpha+\beta) = \cos(\alpha+\beta)$$

41. Destructive; the resulting graph has a smaller amplitude than the two initial graphs. **43.** 0.4179 *E* **45.** 0.5564 *E*
47. Sample answer: To determine communication interference, you need to determine the sine or cosine of the sum or difference of two angles. Answers should include the following information.

- Interference occurs when waves pass through the same space at the same time. When the combined waves have a greater amplitude, constructive interference results and when the combined waves have a smaller amplitude, destructive interference results.

49. C

51.
$$\sin^2\theta+\tan^2\theta \stackrel{?}{=} (1-\cos^2\theta)+\frac{\sec^2\theta}{\csc^2\theta}$$
$$\sin^2\theta+\tan^2\theta \stackrel{?}{=} \sin^2\theta+\frac{\sec^2\theta}{\csc^2\theta}$$
$$\sin^2\theta+\tan^2\theta \stackrel{?}{=} \sin^2\theta+\frac{1}{\cos^2\theta}\div\frac{1}{\sin^2\theta}$$
$$\sin^2\theta+\tan^2\theta \stackrel{?}{=} \sin^2\theta+\frac{\sin^2\theta}{\cos^2\theta}$$
$$\sin^2\theta+\tan^2\theta = \sin^2\theta+\tan^2\theta$$

53.
$$\frac{\sec\theta}{\tan\theta} \stackrel{?}{=} \csc\theta$$
$$\frac{1}{\cos\theta}\div\frac{\sin\theta}{\cos\theta} \stackrel{?}{=} \csc\theta$$
$$\frac{1}{\cos\theta}\cdot\frac{\cos\theta}{\sin\theta} \stackrel{?}{=} \csc\theta$$
$$\frac{1}{\sin\theta} \stackrel{?}{=} \csc\theta$$
$$\csc\theta = \csc\theta$$

55. 4 **57.** $2\sec\theta$ **59.** $\sin\theta=-\frac{4}{5}$, $\cos\theta=-\frac{3}{5}$, $\tan\theta=\frac{4}{3}$, $\csc\theta=-\frac{5}{4}$, $\sec\theta=-\frac{5}{3}$, $\cot\theta=\frac{3}{4}$ **61.** 360 **63.** 56

65. about 228 mi **67.** $\pm\frac{\sqrt{5}}{2}$ **69.** $\pm\frac{\sqrt{5}}{5}$ **71.** $\pm\frac{\sqrt{6}}{2}$

73. $\frac{\sqrt{\sqrt{6}-\sqrt{2}}}{2}$

Pages 794–797 Lesson 14-6

1. Sample answer: If x is in the third quadrant, then $\frac{x}{2}$ is between 90° and 135°. Use the half-angle formula for cosine knowing that the value is negative. **3.** Sample answer: The identity used for $\cos 2\theta$ depends on whether you know the value of $\sin\theta$, $\cos\theta$ or both values.

5. $\frac{4\sqrt{5}}{9}, -\frac{1}{9}, \frac{\sqrt{30}}{6}, -\frac{\sqrt{6}}{6}$ **7.** $-\frac{3\sqrt{7}}{8}, -\frac{1}{8}, \frac{\sqrt{8-2\sqrt{7}}}{4}, -\frac{\sqrt{8+2\sqrt{7}}}{4}$ **9.** $\frac{\sqrt{2-\sqrt{3}}}{2}$

11.
$$\cos^2 2x+4\sin^2 x\cos^2 x \stackrel{?}{=} 1$$
$$\cos^2 2x+\sin^2 2x \stackrel{?}{=} 1$$
$$1 = 1$$

13. $-\frac{120}{169}, \frac{119}{169}, \frac{5\sqrt{26}}{26}, \frac{\sqrt{26}}{26}$ **15.** $\frac{4\sqrt{2}}{9}, -\frac{7}{9}, \frac{\sqrt{6}}{3}, -\frac{\sqrt{3}}{3}$

17. $-\frac{3\sqrt{55}}{32}, \frac{23}{32}, \frac{\sqrt{8-\sqrt{55}}}{4}, -\frac{\sqrt{8+\sqrt{55}}}{4}$

19. $\frac{\sqrt{35}}{18}, -\frac{17}{18}, \frac{\sqrt{15}}{6}, \frac{\sqrt{21}}{6}$ **21.** $-\frac{4\sqrt{2}}{9}, \frac{7}{9}, \frac{\sqrt{18-12\sqrt{2}}}{6}, -\frac{\sqrt{18-12\sqrt{2}}}{6}$ **23.** $\frac{4\sqrt{5}}{9}, -\frac{1}{9}, \frac{\sqrt{6}}{6}, \frac{\sqrt{30}}{6}$ **25.** $-\frac{\sqrt{2+\sqrt{3}}}{2}$

27. $-\frac{\sqrt{2+\sqrt{2}}}{2}$ **29.** $\frac{\sqrt{2-\sqrt{2}}}{2}$

31.
$$\sin 2x \stackrel{?}{=} 2\cot x\sin^2 x$$
$$2\sin x\cos x \stackrel{?}{=} 2\frac{\cos x}{\sin x}\cdot\sin^2 x$$
$$2\sin x\cos x = 2\sin x\cos x$$

33.
$$\sin^4 x-\cos^4 x \stackrel{?}{=} 2\sin^2 x-1$$
$$(\sin^2 x-\cos^2 x)(\sin^2 x+\cos^2 x) \stackrel{?}{=} 2\sin^2 x-1$$
$$(\sin^2 x-\cos^2 x)\cdot 1 \stackrel{?}{=} 2\sin^2 x-1$$
$$[\sin^2 x-(1-\sin^2 x)]\cdot 1 \stackrel{?}{=} 2\sin^2 x-1$$
$$\sin^2 x-1+\sin^2 x \stackrel{?}{=} 2\sin^2 x-1$$
$$2\sin^2 x-1 = 2\sin^2 x-1$$

35.
$$\tan^2\frac{x}{2} \stackrel{?}{=} \frac{1-\cos x}{1+\cos x}$$
$$\frac{\sin^2\frac{x}{2}}{\cos^2\frac{x}{2}} \stackrel{?}{=} \frac{1-\cos x}{1+\cos x}$$
$$\frac{\left(\pm\sqrt{\frac{1-\cos x}{2}}\right)^2}{\left(\pm\sqrt{\frac{1+\cos x}{2}}\right)^2} \stackrel{?}{=} \frac{1-\cos x}{1+\cos x}$$
$$\frac{1-\cos x}{1+\cos x} \stackrel{?}{=} \frac{1-\cos x}{1+\cos x}$$

Selected Answers

37. 46.3° **39.** $2 + \sqrt{3}$ **41.** $\frac{1}{4}\tan\theta$ **43.** The maxima occur at $x = \pm\frac{\pi}{2}$ and $\pm\frac{3\pi}{2}$. The minima occur at $x = 0, \pm\pi$ and $\pm 2\pi$. **45.** The graph of $f(x)$ crosses the x-axis at the points specified in Exercise 43. **47.** Sample answer: The sound waves associated with music can be modeled using trigonometric functions. Answers should include the following information.

- In moving from one harmonic to the next, the number of vibrations that appear as sine waves increase by 1.
- The period of the function as you move from the nth harmonic to the $(n + 1)$th harmonic decreases from $\frac{2\pi}{n}$ to $\frac{2\pi}{n+1}$.

49. B **51.** $\frac{\sqrt{6} - \sqrt{2}}{4}$ **53.** $-\frac{\sqrt{3}}{2}$ **55.** $\frac{1}{2}$

57. $\cos\theta(\cos\theta + \cot\theta) \stackrel{?}{=} \cot\theta\cos\theta(\sin\theta + 1)$

$$\cos\theta(\cos\theta + \cot\theta) \stackrel{?}{=} \frac{\cos\theta}{\sin\theta}\cos\theta\sin\theta + \cot\theta\cos\theta$$

$$\cos\theta(\cos\theta + \cot\theta) \stackrel{?}{=} \cos^2\theta + \cot\theta\cos\theta$$

$$\cos\theta(\cos\theta + \cot\theta) = \cos\theta(\cos\theta + \cot\theta)$$

59. $10^{2.5}$ or about 316 times greater **61.** 1, −1 **63.** $\frac{5}{2}$, −2 **65.** 0, $-\frac{1}{2}$

Page 797 Practice Quiz 2

1. $\sin\theta\sec\theta \stackrel{?}{=} \tan\theta$

$$\sin\theta \cdot \frac{1}{\cos\theta} \stackrel{?}{=} \tan\theta$$

$$\frac{\sin\theta}{\cos\theta} \stackrel{?}{=} \tan\theta$$

$$\tan\theta = \tan\theta$$

3. $\sin\theta + \tan\theta \stackrel{?}{=} \frac{\sin\theta(\cos\theta + 1)}{\cos\theta}$

$$\sin\theta + \tan\theta \stackrel{?}{=} \frac{\sin\theta\cos\theta + \sin\theta}{\cos\theta}$$

$$\sin\theta + \tan\theta \stackrel{?}{=} \frac{\sin\theta\cos\theta}{\cos\theta} + \frac{\sin\theta}{\cos\theta}$$

$$\sin\theta + \tan\theta = \sin\theta + \tan\theta$$

5. $\cos\left(\frac{3\pi}{2} - \theta\right) \stackrel{?}{=} -\sin\theta$

$$\cos\frac{3\pi}{2}\cos\theta + \sin\frac{3\pi}{2}\sin\theta \stackrel{?}{=} -\sin\theta$$

$$0 + (-1 \cdot \sin\theta) \stackrel{?}{=} -\sin\theta$$

$$-\sin\theta = -\sin\theta$$

7. $\frac{\sqrt{3}}{2}$ **9.** $\frac{\sqrt{2 - \sqrt{3}}}{2}$

Pages 802–804 Lesson 14-7

1. Sample answer: If $\sec\theta = 0$ then $\frac{1}{\cos\theta} = 0$. Since no value of θ makes $\frac{1}{\cos\theta} = 0$, there are no solutions. **3.** Sample answer: $\sin\theta = 2$ **5.** 135°, 225° **7.** $\frac{\pi}{6}$ **9.** $0 + k\pi$ **11.** $60° + k \cdot 360°, 300° + k \cdot 360°$ **13.** $\frac{\pi}{6} + 2k\pi, \frac{5\pi}{6} + 2k\pi, \frac{\pi}{2} + 2k\pi$ or $30° + k \cdot 360°$, $150° + k \cdot 360°, 90° + k \cdot 360°$ **15.** 60°, 300° **17.** 210°, 330° **19.** $\frac{\pi}{6}, \frac{5\pi}{6}, \frac{3\pi}{2}$ **21.** $\frac{7\pi}{6}, \frac{11\pi}{6}$ **23.** $\frac{\pi}{3} + 2k\pi, \frac{5\pi}{3} + 2k\pi$ **25.** $\frac{2\pi}{3} + 2k\pi, \frac{4\pi}{3} + 2k\pi$ **27.** $\frac{\pi}{3} + 2k\pi, \frac{5\pi}{3} + 2k\pi$ **29.** $45° + k \cdot 180°$ **31.** $270° + k \cdot 360°$ **33.** $0° + k \cdot 180°$, $60° + k \cdot 180°$ **35.** $0 + 2k\pi, \frac{\pi}{2} + 2k\pi, \frac{3\pi}{2} + 2k\pi$ or $0° + k \cdot 360°, 90° + k \cdot 360°, 270° + k \cdot 360°$ **37.** $0 + k\pi$ or $0° + k \cdot 180°$ **39.** $0 + 2k\pi, \frac{\pi}{3} + 2k\pi, \frac{5\pi}{3} + 2k\pi$, or $0° + k \cdot 360°$, $60° + k \cdot 360°, 300° + k \cdot 360°$ **41.** $S = \frac{352}{\tan\theta}$ or $S = 352\cot\theta$

43. $y = \frac{3}{2} + \frac{3}{2}\sin(\pi t)$

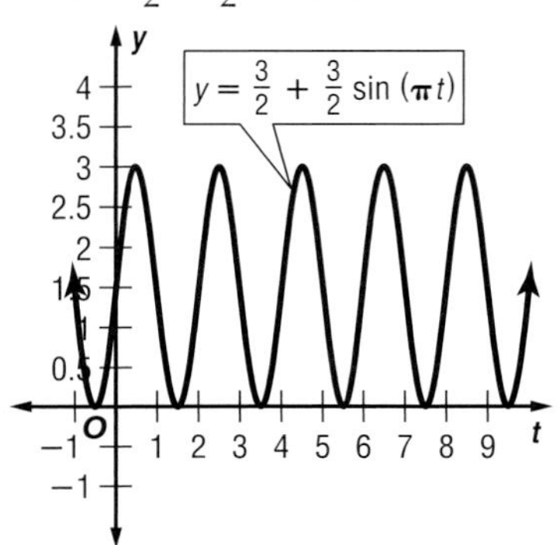

45. (4.964, −0.598) **47.** D **49.** $\frac{24}{25}, \frac{7}{25}, \frac{\sqrt{10}}{10}, \frac{3\sqrt{10}}{10}$ **51.** $\frac{5\sqrt{11}}{18}, \frac{7}{18}, \frac{\sqrt{3}}{6}, \frac{\sqrt{33}}{6}$ **53.** $-\frac{\sqrt{3}}{2}$ **55.** $b = 11.0, c = 12.2$, $m\angle C = 78$

Pages 805–808 Chapter 14 Study Guide and Review

1. h **3.** d **5.** e **7.** g **9.** amplitude: $\frac{1}{2}$; period: 360° or 2π

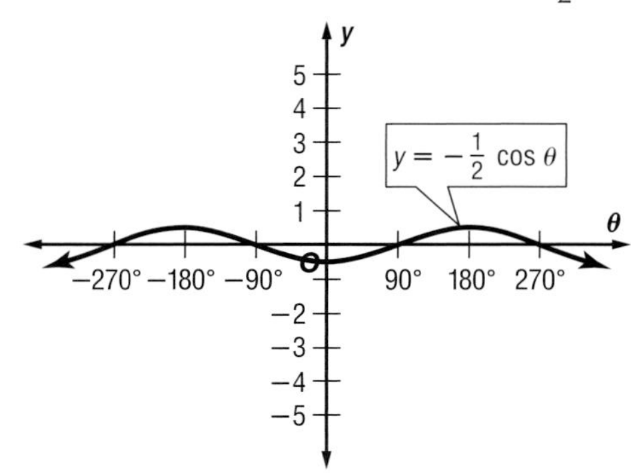

11. amplitude: 1; period: 720° or 4π

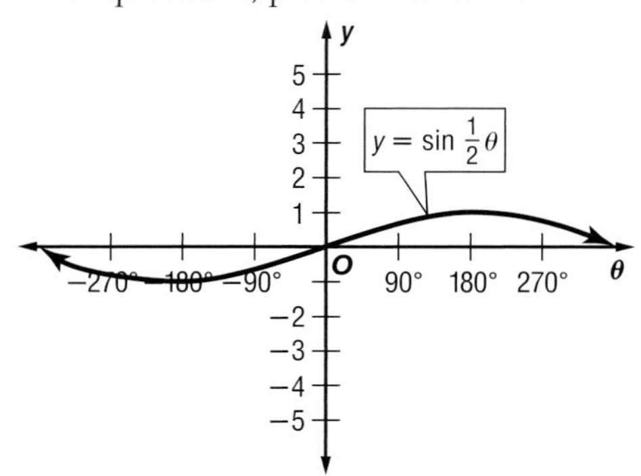

13. amplitude: does not exist; period: 540° or 3π

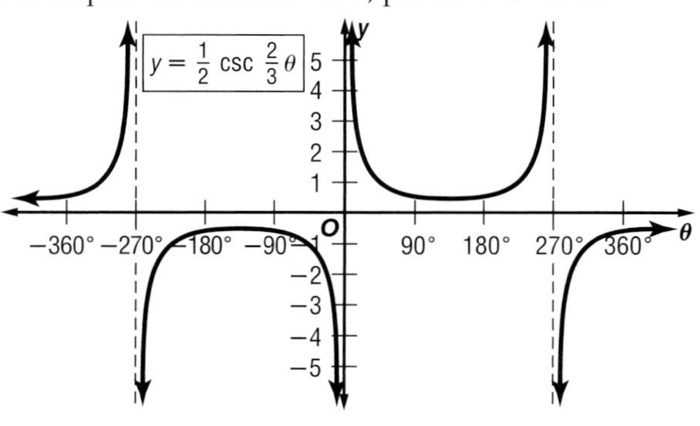

15. $-1, \frac{1}{2}, 180°, 60°$

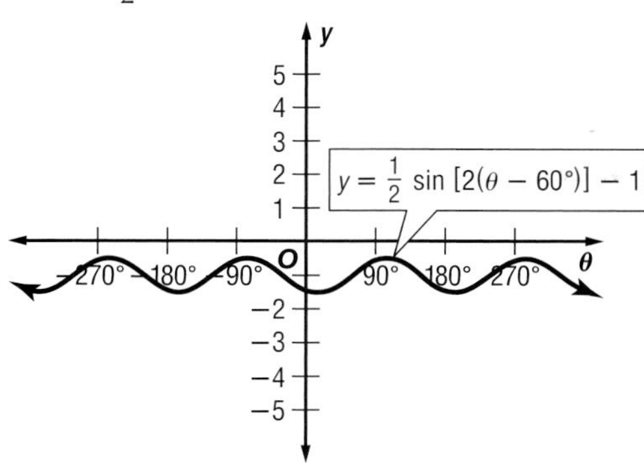

17. 1, does not exist, 4π, $-\frac{\pi}{4}$

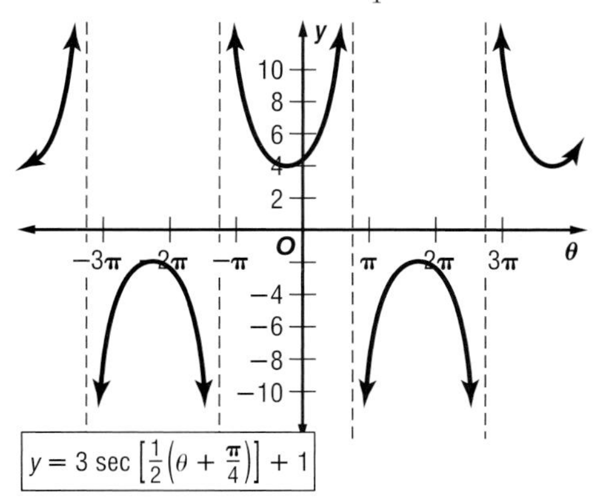

19. $-\frac{4}{3}$ **21.** $\sin^2 \theta$ **23.** $\sec \theta$

25.
$$\frac{\sin \theta}{1 - \cos \theta} \stackrel{?}{=} \csc \theta + \cot \theta$$
$$\frac{\sin \theta}{1 - \cos \theta} \stackrel{?}{=} \frac{1}{\sin \theta} + \frac{\cos \theta}{\sin \theta}$$
$$\frac{\sin \theta}{1 - \cos \theta} \stackrel{?}{=} \frac{1 + \csc \theta}{\sin \theta}$$
$$\frac{\sin \theta}{1 - \cos \theta} \stackrel{?}{=} \frac{1 + \cos \theta}{\sin \theta} \cdot \frac{1 - \cos \theta}{1 - \cos \theta}$$
$$\frac{\sin \theta}{1 - \cos \theta} \stackrel{?}{=} \frac{1 - \cos^2 \theta}{\sin \theta\,(1 - \cos \theta)}$$
$$\frac{\sin \theta}{1 - \cos \theta} \stackrel{?}{=} \frac{\sin^2 \theta}{\sin \theta\,(1 - \cos \theta)}$$
$$\frac{\sin \theta}{1 - \cos \theta} = \frac{\sin \theta}{1 - \cos \theta}$$

27.
$$\sec \theta(\sec \theta - \cos \theta) \stackrel{?}{=} \tan^2 \theta$$
$$\frac{1}{\cos \theta}\left(\frac{1}{\cos \theta} - \cos \theta\right) \stackrel{?}{=} \tan^2 \theta$$
$$\frac{1}{\cos^2 \theta} - 1 \stackrel{?}{=} \tan^2 \theta$$
$$\sec^2 \theta - 1 \stackrel{?}{=} \tan^2 \theta$$
$$\tan^2 \theta = \tan^2 \theta$$

29. $\frac{\sqrt{6} - \sqrt{2}}{4}$ **31.** $\frac{\sqrt{2} - \sqrt{6}}{4}$ **33.** $\frac{-\sqrt{6} - \sqrt{2}}{4}$

35.
$$\sin (30 - \theta) \stackrel{?}{=} \cos (60 + \theta)$$
$$\sin 30° \cos \theta - \cos 30° \sin \theta \stackrel{?}{=} \cos 60° \cos \theta - \sin 60° \sin \theta$$
$$\frac{1}{2} \cos \theta - \frac{\sqrt{3}}{2} \sin \theta = \frac{1}{2} \cos \theta - \frac{\sqrt{3}}{2} \sin \theta$$

37.
$$-\cos \theta \stackrel{?}{=} \cos (\pi + \theta)$$
$$-\cos \theta \stackrel{?}{=} \cos \pi \cos \theta - \sin \pi \sin \theta$$
$$-\cos \theta \stackrel{?}{=} -1 \cdot \cos \theta - 0 \cdot \sin \theta$$
$$-\cos \theta = -\cos \theta$$

39. $\frac{120}{169}, \frac{119}{169}, \frac{5\sqrt{26}}{26}, -\frac{\sqrt{26}}{26}$ **41.** $-\frac{120}{169}, \frac{119}{169}, \frac{\sqrt{26}}{26}, -\frac{5\sqrt{26}}{26}$

43. $0°$ **45.** $\frac{\pi}{6} + 2k\pi, \frac{5\pi}{6} + 2k\pi$

Selected Answers

Photo Credits

About the Cover: Alexander Calder (1898–1976) was one of America's most acclaimed sculptors. Renowned for his invention of the *mobile*, or movable sculpture, Calder also created sculptures called *stabiles*, or immovable sculptures. The cover photograph illustrates his *Grand Stabile Rouge*, located in Paris. One of Calder's last great public works, this sculpture is reminiscent of another of his stabiles, *Flamingo*, in Chicago. Both stabiles feature large red arches that resemble parabolas.

Cover Vanni Archive/CORBIS; **x** D & K Tapparel/Getty Images; **xi** Telegraph Colour Library/Getty Images; **xii** DEX Images Inc./CORBIS Stock Market; **xiii** AFP/CORBIS; **xiv** CORBIS; **xix** Jenny Hager/ImageState; **xv** Brownie Harris/CORBIS Stock Market; **xvi** Ray F. Hillstrom Jr.; **xvii** Kunio Owaki/CORBIS Stock Market; **xviii** Jane Burton/Bruce Coleman; **xx** Food & Drug Administration/SPL/Photo Researchers; **xxi** R. Ian Lloyd/Masterfile; **xxii** Getty Images; **2** David De Lossy/Getty Images; **2–3** Bryan Peterson/Getty Images; **4** Johnny Stockshooter/International Stock; **4–5** Orion/International Stock; **6** Mark Harmel/Getty Images; **14** Amy C. Etra/PhotoEdit; **16** Archivo Iconografico, S.A./CORBIS; **19** Aaron Haupt; **20** SuperStock; **23** Michael Newman/PhotoEdit; **26** Pictor; **28** Robert Yager/Getty Images; **31** E.L. Shay; **38** Lawrence Migdale; **40** Index Stock/Ewing Galloway; **43** PhotoDisc; **44** Rudi Von Briel/PhotoEdit; **54–55** Jack Dykinga/Getty Images; **56** William J. Weber; **61** Bettmann/CORBIS; **64** D & K Tapparel/Getty Images; **67** Lynn M. Stone; **72** (l)SuperStock, (r)Richard T. Nowitz/CORBIS; **80** VCG/Getty Images; **82** John Evans; **85** Matt Meadows; **94** David Ball/CORBIS Stock Market; **99** Getty Images; **108** PhotoDisc; **108–109** NASA/TSADO/Tom Stack & Assotes; **111** Dave Starrett/Masterfile; **114** Telegraph Colour Library/Getty Images; **121** Will Hart/PhotoEdit; **124** NASA; **126** Doug Martin; **129** AFP/CORBIS; **131** Caroline Penn/CORBIS; **138** S. Carmona/CORBIS; **140** M. Angelo/CORBIS; **143** Andy Lyons/Allsport; **152** CORBIS; **152–153** William Sallaz/DUOMO; **157** Bettman/CORBIS; **161** Tui De Roy/Bruce Coleman, Inc.; **165** PhotoDisc; **169** Andy Lyons STF/Allsport; **172** Mark Tomalty/Masterfile; **175** Mark Richards/PhotoEdit; **180** Michael Denora/Getty Images; **187** Jonathan Blair/CORBIS; **190** FDR Library; **193** DEX Images Inc./CORBIS Stock Market; **195** Jose Luis Pelaez Inc./CORBIS Stock Market; **197** Volker Steger/SPL/Photo Researchers; **203** Ken Eward/Science Source/Photo Researchers; **218** www.comstock.com; **218–219** Rafael Marcia/Photo Researchers; **220–221** Carl Purcell/Words and Pictures/PictureQuest; **225** AFP/CORBIS; **227** K.G. Murti/Visuals Unlimited; **229** David Umberger/Purdue University Photo; **243** Steve Rayer/CORBIS; **249** Roger Ressmeyer/CORBIS; **255** Roy Ooms/Masterfile; **259** AFP/CORBIS; **262** Victoria & Albert Museum, London/Art Resource, NY; **267** Lori Adamski Peek/Getty Images; **274** Kaluzny/Thatcher/Getty Images; **284** Allsport Concepts/Getty Images; **284–285** Ed Pritchard/Getty Images; **291** Aidan O'Rourke; **292** Getty Images; **298** SuperStock; **304** Matthew McVay/Stock Boston; **306** DUOMO/CORBIS; **311** CORBIS; **313** Jeff Kaufman/Getty Images; **318** Bruce Hands/Getty Images; **327** NASA; **329** Nick Wilson/Allsport; **331** Todd Rosenberg/Allsport; **334** Aaron Haupt; **344–345** Guy Grenier/Masterfile; **346** Brownie Harris/CORBIS Stock Market; **351** Martha Swope/Timepix; **355** VCG/Getty Images; **357** Michael Newman/PhotoEdit; **363** Gregg Mancuso/Stock Boston; **365** Boden/Ledingham/Masterfile; **372** National Library of Medicine/Mark Marten/Photo Researchers; **376** VCG/Getty Images; **381** Bob Krist/CORBIS; **383** Ed Bock/CORBIS Stock Market; **388** SuperStock; **392** Matt Meadows; **394** SuperStock; **395** Raymond Gehman/CORBIS; **396** Frank Rossotto/Stocktreck/CORBIS Stock Market; **398** Getty Images; **408** Michael S. Yamashita/CORBIS; **408–409** Jose Fuste Raga/eStock Photo; **410** Photographers Library LTD/eStock Photo; **410–411** James Hackett/eStock Photo; **424** James Rooney; **426** SuperStock; **432** Matt Meadows; **435** Ray F. Hillstrom Jr.; **439** James P. Blair/CORBIS; **443** CORBIS; **446** Bob Krist/Getty Images; **459** (l)Space Telescope Science Institute/NASA/SPL/Photo Researchers, (r)Michael Newman/PhotoEdit; **470–471** David Fleetham/Getty Images; **477** AFP/CORBIS; **483** Pascal Rondeau/Allsport; **487** Aaron Haupt; **489** Bettmann/CORBIS; **494** JPL/TSADO/Tom Stack & Associates; **496** Geoff Butler; **497** Lynn M. Stone/Bruce Coleman, Inc.; **499** Picture Press/CORBIS; **503** Kunio Owaki/CORBIS Stock Market; **505** (b)Phil Schermeister/CORBIS, (t)Bruce Ayres/Getty Images; **507** Reuters NewMedia Inc./CORBIS; **511** Keith Wood/Getty Images; **520–521** Michael S. Yamashita/CORBIS; **522** Aaron Haupt; **525** Ariel Skelley/CORBIS Stock Market; **529** Jeff Zaruba/CORBIS Stock Market; **536** (l)Mark Jones/Minden Pictures, (r)Jane Burton/Bruce Coleman, Inc.; **537** David Weintraub/Photo Researchers; **542** SuperStock; **545** Bettman/CORBIS; **558** Jim Craigmyle/Masterfile; **561** Richard T. Nowitz/Photo Researchers; **564** Karl Weatherly/CORBIS; **574** Amanda Kaye; **574–575** Bryan Barr; **576–577** Christine Osborne/CORBIS; **579** SuperStock; **583** Michele Wigginton; **584** Michelle Bridwell/PhotoEdit; **595** Hank Morgan/Photo Researchers; **599** IN THE BLEACHERS ©1997 Steve Moore. Reprinted with permission of Universal Press Syndicate. All rights reserved; **603** Jenny Hager/ImageState; **609** Jeff Greenberg/Visuals Unlimited; **612** SPL/Photo Researchers; **630** Steve Liss/Timepix; **630–631** Greg Mathieson/Timepix; **632** D.F. Harris; **635** ©1979 United Feature Syndicate, Inc.; **636** (l)Mitch Kezar/Getty Images, (r)Jim Erickson/CORBIS Stock Market; **638** Mark C. Burnett/Photo Researchers; **642** Matt Meadows; **644** CORBIS; **648** Food & Drug Administration/SPL/Photo Researchers; **651** Chris Trotman/DUOMO; **656** Charles Gupton/CORBIS Stock Market; **660** *The Born Loser* reprinted by permission of Newspaper Enterprise Association, Inc.; **662** Bob Daemmrich/Stock Boston; **667** Greg Fiume/New Sport/CORBIS; **668** SuperStock; **671** AFP/CORBIS; **675** Will & Deni McIntyre/Photo Researchers; **677** Gregg Forwerck/SportsChrome USA; **679** Steve Chenn/CORBIS; **683** Aaron Haupt; **685** HMS Images/Getty Images; **686** Aaron Haupt; **696** Photofest; **696–697** Ed and Chris Kumler; **698–699** Bill Ross/CORBIS; **705** John P. Kelly/Getty Images; **707** SuperStock; **709** L. Clarke/CORBIS; **713** Ray Juno/CORBIS Stock Market; **716** Aaron Haupt; **717** courtesy Skycoaster of Florida; **721** Reuters NewMedia Inc./CORBIS; **723** Otto Greule/Allsport; **729** Peter Miller/Photo Researchers; **731** SuperStock; **735** Roy Ooms/Masterfile; **737** John T. Carbone/Photonica; **744** R. Ian Lloyd/Masterfile; **746** Doug Plummer/Photonica; **748** SuperStock; **750** Steven E. Sutton/DUOMO; **760–761** Boden/Ledingham/Masterfile; **766** Larry Hamill; **773** Ben Edwards/Getty Images; **780** James Schot/Martha's Vineyard Preservation Trust; **789** Cosmo Condina/Getty Images; **795** SuperStock; **799** SuperStock; **803** (l)Getty Images, (r)Frank Wiewandt/Image Finders.

Index

Index

Index

G

Q

Index

Index

Formulas

Coordinate Geometry

Midpoint	$M = \left(\frac{x_1 + x_2}{2}, \frac{y_1 + y_2}{2}\right)$
Distance	$d = \sqrt{(x_2 - x_1)^2 + (y_2 - y_1)^2}$
Slope	$m = \frac{y_2 - y_1}{x_2 - x_1}$

Matrices

Matrix multiplication	$\begin{bmatrix} a_1 & b_1 \\ a_2 & b_2 \end{bmatrix} \cdot \begin{bmatrix} x_1 & y_1 \\ x_2 & y_2 \end{bmatrix} = \begin{bmatrix} a_1x_1 + b_1x_2 & a_1y_1 + b_1y_2 \\ a_2x_1 + b_2x_2 & a_2y_1 + b_2y_2 \end{bmatrix}$

Polynomials

Quadratic Formula	$x = \frac{-b \pm \sqrt{b^2 - 4ac}}{2a}$
Difference of Two Squares	$a^2 - b^2 = (a + b)(a - b)$
Perfect Square Trinomials	$a^2 + 2ab + b^2 = (a + b)^2$ $a^2 - 2ab + b^2 = (a - b)^2$

Logarithms

Change of Base Formula	$\log_a n = \frac{\log_b n}{\log_b a}$

Probability and Statistics

Permutations	$P(n, r) = \frac{n!}{(n - r)!}$
Combinations	$C(n, r) = \frac{n!}{(n - r)!r!}$
Standard deviation	$\sigma = \sqrt{\frac{(x_1 - \bar{x})^2 + (x_2 - \bar{x})^2 + \ldots + (x_n - \bar{x})^2}{n}}$

Trigonometry

Pythagorean Theorem	$a^2 + b^2 = c^2$		
Law of Sines	$\frac{\sin A}{a} = \frac{\sin B}{b} = \frac{\sin C}{c}$		
Law of Cosines	$a^2 = b^2 + c^2 - 2bc \cos A$		
Trigonometric functions	$\sin \theta = \frac{\text{opp}}{\text{hyp}}$ $\csc \theta = \frac{\text{hyp}}{\text{opp}}$	$\cos \theta = \frac{\text{adj}}{\text{hyp}}$ $\sec \theta = \frac{\text{hyp}}{\text{adj}}$	$\tan \theta = \frac{\text{opp}}{\text{adj}}$ $\cot \theta = \frac{\text{adj}}{\text{opp}}$
Quotient Identities	$\tan \theta = \frac{\sin \theta}{\cos \theta}$	$\cot \theta = \frac{\cos \theta}{\sin \theta}$	
Reciprocal Identities	$\csc \theta = \frac{1}{\sin \theta}$	$\sec \theta = \frac{1}{\cos \theta}$	$\cot \theta = \frac{1}{\tan \theta}$
Pythagorean Identities	$\cos^2 \theta + \sin^2 \theta = 1$	$\tan^2 \theta + 1 = \sec^2 \theta$	$\cot^2 \theta + 1 = \csc^2 \theta$